INTERMEDIATE ALGEBRA

K. Elayn Martin-Gay

University of New Orleans

PRENTICE I

Upper Saddle River, New Jersey

Library of Congress Cataloging-in-Publication Data

Martin-Gay, K. Elayn
Intermediate algebra / K. Elayn Martin-Gay.
p. cm.
Includes index.
ISBN 0-13-228800-1 (student ed.).—ISBN 0-13-862376-7 (annotated instructor's ed.)
1. Algebra. I. Title.
QA152.2.M368 1999 98-44925
512.9—dc21 CIP

To my husband, Clayton,
and our sons,
Eric and Bryan

Acquisitions Editor: Karin E. Wagner
Editor-in-Chief: Jerome Grant
Editorial Director: Tim Bozik
Editor-in-Chief, Development: Carol Trueheart
Senior Managing Editor: Linda Mihatov Behrens
Executive Managing Editor: Kathleen Schiaparelli
Assistant Vice President of Production and Manufacturing: David W. Riccardi
Marketing Manager: Jolene Howard
Manufacturing Buyer: Alan Fischer
Manufacturing Manager: Trudy Pisciotti
Editorial Assistant/Supplements Editor: Kate Marks
Associate Editor, Math/Statistics Media: Audra J. Walsh
Art Director/Cover Designer: Maureen Eide
Associate Creative Director: Amy Rosen
Director of Creative Services: Paula Maylahn
Assistant to Art Director: John Christiana
Art Manager: Gus Vibal
Art Editor: Grace Hazeldine
Cover image: Scarlet Macaw/Superstock, Inc.
Text Design and Project Management: Elm Street Publishing Services, Inc.
Photo Researcher: Beth Boyd
Photo Research Administrator: Melinda Reo
Art Studio: Academy Artworks

Printed in the United States of America
10 9 8 7 6 5 4 3 2 1

ISBN: 0-13-228800-1

Prentice-Hall International (UK) Limited, *London*
Prentice-Hall of Australia Pty. Limited, *Sydney*
Prentice-Hall Canada Inc., *Toronto*
Prentice-Hall Hispanoamericana, S.A., *Mexico*
Prentice-Hall of India Private Limited, *New Delhi*
Prentice-Hall of Japan, Inc., *Tokyo*
Simon & Schuster Asia Pte. Ltd., *Singapore*
Editora Prentice-Hall do Brasil, Ltda., *Rio de Janeiro*

Contents

CHAPTER 10 Exponential and Logarithmic Functions 747

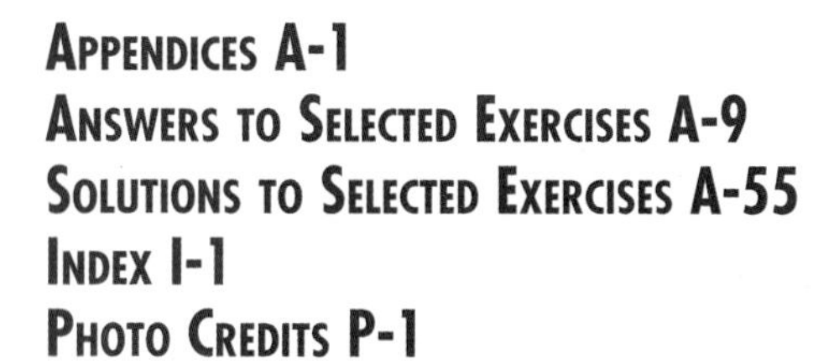

Preface

About This Book

This worktext was written to provide students with a solid foundation in algebra as well as to help develop their problem-solving skills. Specific care was taken to ensure that students have the most up-to-date relevant text preparation for their next mathematics course or for nonmathematical courses that require an understanding of algebraic fundamentals. I have tried to achieve this by writing a user-friendly text that is keyed to objectives and contains many worked-out examples. As suggested by the AMATYC Crossroads Document and the NCTM Standards (plus Addenda), real-life and real-data applications, data interpretation, conceptual understanding, problem solving, writing, cooperative learning, appropriate use of technology, mental mathematics, number sense, critical thinking, and geometric concepts are emphasized and integrated throughout the book.

Key Pedagogical Features

Readability and Connections I have tried to make the writing style as clear as possible while still retaining the mathematical integrity of the content. When a new topic is presented, an effort has been made to relate the new ideas to those that students may already know. Constant reinforcement and connections within problem-solving strategies, data interpretation, geometry, patterns, graphs, and situations from everyday life can help students gradually master both new and old information. In addition, each section begins with a list of objectives covered in the section. Clear organization of section material based on objectives further enhances readability.

Problem-Solving Process This is formally introduced in Chapter 2 with a four-step process that is integrated throughout the text. The four steps are Understand, Translate, Solve, and Interpret. The repeated use of these steps in a variety of examples shows their wide applicability. Reinforcing the steps can increase students' comfort level and confidence in tackling problems.

Applications and Connections Every effort was made to include as many interesting and relevant real-life applications as possible throughout the text in both worked-out examples and exercise sets. The applications help to motivate students and strengthen their understanding of mathematics in the real world. They show connections to a wide range of fields including agriculture, allied health, animal science, anthropology, astronomy, biology, business, chemistry, computer technology, construction, consumer affairs, criminal justice, demographics, earth science, economics, education, electronics, entertainment, environmental issues, finance, geography, government, health, highway safety, history, interior decorating, investments, manufacturing, medicine, merchandising, meteorology, nutrition, personal finance, physics, probability, psychology, real estate, space exploration, sports, telecommunications, transportation, and travel. Many of the applications are based on recent real data. Sources for data include newspapers, magazines, publicly held companies, government agencies, special-interest groups, research organizations, and reference books. Opportunities for obtaining your own real data are also included.

Exercise Sets Each text section ends with an Exercise Set. Each exercise in the set is keyed to one of the objectives of the section. Wherever possible, a specific example is also given.

Throughout the text exercises there is an emphasis on *Data and Graphical Interpretation.* There is an emphasis on data interpretation in exercises via tables and graphs. The ability to interpret data and read and create a variety of types of graphs is developed gradually so students become comfortable with it. In addition to the more than 4200 exercises in end-of-section exercise sets, exercises may also be found in the Pretests, Integrated Reviews, Chapter Reviews, Chapter Tests, and Cumulative Reviews. Each Exercise Set contains one or more of the following features:

Mental Math Found at the beginning of an exercise set, these mental warmups reinforce concepts found in the accompanying section and increase students' confidence before they tackle an exercise set. By relying on their own mental skills, students increase not only their confidence in themselves but also their number sense and estimation ability.

Review and Preview These exercises occur in each exercise set (except for those in Chapter 1) after the exercises keyed to the objectives of the section. Review and Preview problems are keyed to earlier sections and review concepts learned earlier in the text that are needed in the next section or in the next chapter. These exercises show the links between earlier topics and later material.

Combining Concepts These exercises are found at the end of each exercise set after the Review and Preview exercises. Combining Concepts exercises require students to combine several concepts from that section or to take the concepts of the section a step further by combining them with concepts learned in previous sections. For instance, sometimes students are required to combine the concepts of a section with the problem-solving process they learned in Chapter 2 to try their hand at solving an application problem.

Internet Excursions These exercises occur once per chapter. Internet Excursions require students to use the Internet as a data-collection tool to complete the exercises, allowing students first-hand experience with manipulating and working with real data.

Conceptual and Writing Exercises These exercises occur in almost every exercise set and are marked with the icon ◪. They require students to show an understanding of a concept learned in the corresponding section. This is accomplished by asking students questions that require them to use two or more concepts together. Some require students to stop, think, and explain in their own words the concept(s) used in the exercises they have just completed. Guidelines recommended by the American Mathematical Association of Two Year Colleges (AMATYC) and other professional groups recommend incorporating writing in mathematics courses to reinforce concepts.

Practice Problems Throughout the text, each worked-out example has a parallel Practice Problem placed next to the example in the margin. These invite students to be actively involved in the learning process before beginning the end-of-section exercise set. Practice Problems immediately reinforce a skill after it is developed.

Concept Checks These margin exercises are appropriately placed in many sections of the text. They allow students to gauge their grasp of an idea as it is being explained in the text. Concept Checks stress conceptual understanding at point of use and help suppress misconceived notions before they start.

Integrated Reviews These "mid-chapter reviews" are appropriately placed once per chapter. Integrated Reviews allow students to review and assimilate the many different skills learned separately over several sections before moving on to related material in the chapter.

Helpful Hints Helpful Hints contain practical advice on applying mathematical concepts. These are found throughout the text and strategically placed where students are most likely to need immediate reinforcement. Helpful Hints are highlighted for quick reference.

Focus On Appropriately placed throughout each chapter, these are divided into Focus on Study Skills, Focus on Mathematical Connections, Focus on Business and Career, Focus on the Real World, and Focus on History. They are written to help students develop effective habits for studying mathematics, engage in investigations of other branches of mathematics, understand the importance of mathematics in various careers and in the world of business, and see the relevance of mathematics in both the present and past through critical thinking exercises and group activities.

Calculator and Graphing Calculator Explorations These optional explorations offer point-of-use instruction, through examples and exercises, on the proper use of scientific and graphing calculators as tools in the mathematical problem-solving process. Placed appropriately throughout the text, Calculator and Graphing Calculator Explorations also reinforce concepts learned in the corresponding section and motivate discovery-based learning.

Additional exercises building on the skill developed in the Explorations may be found in exercise sets throughout the text. Exercises requiring a calculator are marked with the icon. Exercises requiring a graphing calculator are marked with the icon.

Chapter Activity These features occur once per chapter at the end of the chapter, often serving as a chapter wrap-up. For individual or group completion, the Chapter Activity, usually hands-on or data-based, complements and extends the concepts of the chapter, allowing students to make decisions and interpretations and to think and write about algebra.

Visual Reinforcement of Concepts The text contains a wealth of graphics, models, photographs, and illustrations to visually clarify and reinforce concepts. These include bar graphs, line graphs, calculator screens, application illustrations, and geometric figures.

Pretests Each chapter begins with a pretest that is designed to help students identify areas where they need to pay special attention in the upcoming chapter.

Chapter Highlights Found at the end of each chapter, these contain key definitions, concepts, and examples to help students understand and retain what they have learned.

Chapter Review and Test The end of each chapter contains a review of topics introduced in the chapter. The Chapter Review offers exercises that are keyed to sections of the chapter. The Chapter Test is a practice test and is not keyed to sections of the chapter.

Cumulative Review These are found at the end of each chapter (except Chapter 1). Each problem contained in the cumulative review is actually an earlier worked example in the text that is referenced in the back of the book along with the answer. Students who need to see a complete worked-out solution, with explanation, can do so by turning to the appropriate example in the text.

Student Resource Icons At the beginning of each section, videotape, software, and solutions manual icons are displayed. These icons help reinforce that these learning aids are available should students wish to use them to help them review concepts and skills at their own pace. These items have direct correlation to the text and emphasize the text's methods of solution.

Functional Use of Color and Design Elements of the text are highlighted with color or design to make it easier for students to read and study. Color is also used to clarify the problem-solving process in worked examples.

SUPPLEMENTS FOR THE INSTRUCTOR

Printed Supplements

Annotated Instructor's Edition (0-13-862376-7)

- ▲ Answers to all exercises printed on the same text page.
- ▲ Teaching Tips throughout the text placed at key points in the margin.

Instructor's Solution Manual (0-13-862392-9)

- ▲ Solutions to even-numbered section exercises.
- ▲ Solutions to every (even and odd) Mental Math exercise.
- ▲ Solutions to every (even and odd) Practice Problem (margin exercise).
- ▲ Solutions to every (even and odd) exercise found in the Chapter Pretests, Integrated Reviews (mid-chapter reviews), Chapter Reviews, Chapter Tests, Cumulative Reviews.

Instructor's Resource Manual with Tests (0-13-862384-8)

- ▲ Notes to the Instructor that includes an introduction to Interactive Learning, Interpreting Graphs and Data, Alternative Assessment, Using Technology and Helping Students Succeed.
- ▲ Two free-response Pretests per chapter.
- ▲ Eight Chapter Tests per chapter (3 multiple-choice, 5 free-response).
- ▲ Two Cumulative Review Tests (one multiple-choice, one free-response) every two chapters (after chapters 2, 4, 6, 8, 10).
- ▲ Eight Final Exams (3 multiple-choice, 5 free-response).
- ▲ Twenty additional exercises per section for added test exercises if needed.

Media Supplements

TestPro Computerized Testing

- ▲ Algorithmically driven, text-specific testing program.
- ▲ Networkable for administering tests and capturing grades on-line.
- ▲ Edit and add your own questions—create nearly unlimited number of tests and drill worksheets.

Companion Web site

- ▲ www.prenhall.com/martin-gay
- ▲ Links related to the Internet Excursions in each chapter allow you to collect data to solve specific internet exercises.
- ▲ Additional links to helpful, generic sites include Fun Math and For Additional Help.

SUPPLEMENTS FOR THE STUDENT

Printed Supplements

Student's Solution Manual (0-13-862434-8)

- ▲ Solutions to odd-numbered section exercises.
- ▲ Solutions to every (even and odd) Mental Math exercise.
- ▲ Solutions to every (even and odd) Practice Problem (margin exercise).
- ▲ Solutions to every (even and odd) exercise found in the Chapter Pretests, Integrated Reviews (mid-chapter reviews), Chapter Reviews, Chapter Tests, Cumulative Reviews.

New York Times *Themes of the Times*

- ▲ Have your instructor contact the local Prentice Hall sales representative.

How to Study Mathematics

- ▲ Have your instructor contact the local Prentice Hall sales representative.

Internet Guide

- ▲ Have your instructor contact the local Prentice Hall sales representative.

Media Supplements

MathPro4 computerized tutorial

- ▲ Keyed to each section of the text for text-specific tutorial exercises and instruction.
- ▲ Includes Warm-up exercises and graded Practice Problems.
- ▲ Algorithmically driven and fully networkable.
- ▲ Have your instructor contact the local Prentice Hall sales representative—also available for purchase for home use.

Videotape Series (0-13-862426-7)

- ▲ Written and presented by Elayn Martin-Gay.
- ▲ Keyed to each section of the text.
- ▲ Step-by-step solutions to exercises from each section of the text. Exercises that are worked in the videos are marked with a video icon ( ).

Companion Web site

- ▲ www.prenhall.com/martin-gay
- ▲ Links related to the Internet Excursions in each chapter allow you to collect data to solve specific internet exercises.
- ▲ Additional links to generic sites include Fun Math, For Additional Help . . .

ACKNOWLEDGMENTS

First, as usual, I would like to thank my husband, Clayton, for his constant encouragement. I would also like to thank my children, Eric and Bryan, for their sense of humor and especially for their suggestion of letting Dad cook the bacon that I always used to burn.

I would also like to thank my extended family for their invaluable help and also their sense of humor. Their contributions are too numerous to list. They are Rod and Karen Pasch; Michael, Christopher, Matthew, and Jessica Callac; Stuart, Earline, Melissa, Mandy and Bailey Martin; Mark, Sabrina, and Madison Martin; Leo and Barbara Miller; and Jewett Gay.

I would like to thank the following reviewers for their input and suggestions:

Janet Cater, *Bakersfield College*
Patrick Cross, *University of Oklahoma*
Dorothy French, *Community College of Philadelphia*

Mary Ellen Gallegos, *Santa Fe Community College*
James Harris, *John A. Logan College*
Brian Hayes, *Triton College*
Rosa Kavanaugh, *Ozarks Technical Community College*
Wendy McGuire, *Santa Fe Community College*
Michael Montaño, *Riverside Community College*
Ellen O'Connell, *Triton College*
Anthony Ponder, *Sinclair Community College*
Flauren Ricketts, *Normandale Community College*
Len Ruth, *Sinclair Community College*
Susan Santolucito, *Delgado Community College*
Susan Shulman, *Middlesex County College*

There were many people who helped me develop this text and I will attempt to thank some of them here. Penelope S. Arnold did an excellent job of providing answers. Cheryl Roberts Cantwell was invaluable for contributing to the overall accuracy of this text. Emily Keaton was also invaluable for her many suggestions and contributions during the development and writing of this first edition. Ingrid Mount at Elm Street Publishing Services provided guidance throughout the production process. I thank Terri Bittner, Cindy Trimble, Jeff Rector, and Teri Lovelace at Laurel Technical Services for all their work on the supplements, text and thorough accuracy check. Lastly, a special thank you to my editor, Karin Wagner, for her support and assistance throughout the development and production of this text and to all the staff at Prentice Hall: Linda Behrens, Alan Fischer, Maureen Eide, Karen Branson, Grace Hazeldine, Gus Vibal, Kate Marks, Jolene Howard, John Tweeddale, Jerome Grant, and Tim Bozik.

K. Elayn Martin-Gay

About the Author

K. Elayn Martin-Gay has taught mathematics at the University of New Orleans for 20 years. Her numerous teaching awards include the local University Alumni Association's Award for Excellence in Teaching, and Outstanding Developmental Educator at University of New Orleans, presented by the Louisiana Association of Developmental Educators.

Elayn is the author of an entire product line of highly successful textbooks. The Martin-Gay library includes the exciting new worktext series: *Basic College Mathematics*, *Introductory Algebra* and *Intermediate Algebra*, the successful hardbound series; *Beginning Algebra* 2e, *Intermediate Algebra* 2e, *Intermediate Algebra A Graphing Approach* (co-authored with Margaret Greene), *Prealgebra* 2e, and *Introductory and Intermediate Algebra,* a combined algebra text.

Prior to writing textbooks, Elayn developed an acclaimed series of lecture videos to support developmental mathematics students in their quest for success. These highly successful videos originally served as the foundation material for her texts. Today the tapes specifically support each book in the Martin-Gay series.

Index of Applications

How to Use This Worktext: A Guide for Students

Intermediate Algebra has been written and designed to help you succeed in this course. The goals that the author will help you achieve include:

- Exposure to real-world applications you will encounter in life and other courses
- An organized, integrated system of learning that combines the text with a comprehensive set of print and media tools
- Better preparation relevant to the *next* course you will take in mathematics

Take a few moments now to acquaint yourself with some of the features that have been built into *Intermediate Algebra* to help you excel.

Real Numbers and Algebraic Expressions

CHAPTER 1

In arithmetic, we add, subtract, multiply, divide, raise to powers, and take roots of numbers. In algebra, we add, subtract, multiply, divide, raise to powers, and take roots of variables. Letters, such as x, that represent numbers are called variables. Understanding algebraic expressions made up of combinations of variables and numbers depends on your understanding of arithmetic expressions. This chapter reviews the arithmetic operations on real numbers and the corresponding algebraic expressions. After this review, we will be prepared to explore how widely useful these algebraic expressions are for problem solving in diverse situations.

The world's first subway system was built in London, England, in 1863. The growing congestion on the streets of London led to relocating the existing steam-operated metropolitan railway underground. Thirty-four years later, America's first subway was built in Boston, Massachusetts. The Boston subway started out with single electrified streetcars running on rails underground. New York City soon followed suit and completed its own subway system in 1904. The New York subway's underground electric trains replaced surface electric train routes, elevated railways, and horse car tracks. Today, New York's subway system is one of the largest and most complex in the world, with 714 miles of subway track and 468 stations. In Exercise 87 on page 55, we will use scientific notation to write the very large number of passengers who ride on New York City subways each year.

1

The photo applications at the opening of every chapter, and applications throughout, introduce you to real world situations that are applicable to the mathematics you will learn in the upcoming chapter. These applications are often discussed later in the chapter as well.

Page 1

Connect the Concepts

Learning and succeeding in a math course require practice and a broader understanding of how everything works together. As you study, make connections using this text's organization to help you. All of these features have been included to enhance your understanding of mathematical concepts.

Concept Checks are special margin exercises found in most sections. Work these to gauge your grasp of the idea being explained in the text.

✓ **CONCEPT CHECK**

Without taking any solution steps, how do you know that the absolute value inequality $|3x - 2| > -9$ has a solution? What is its solution?

Page 144

Look for **Combining Concepts** exercises at the end of each exercise set. Solving these exercises will expose you to the way mathematical ideas build upon each other.

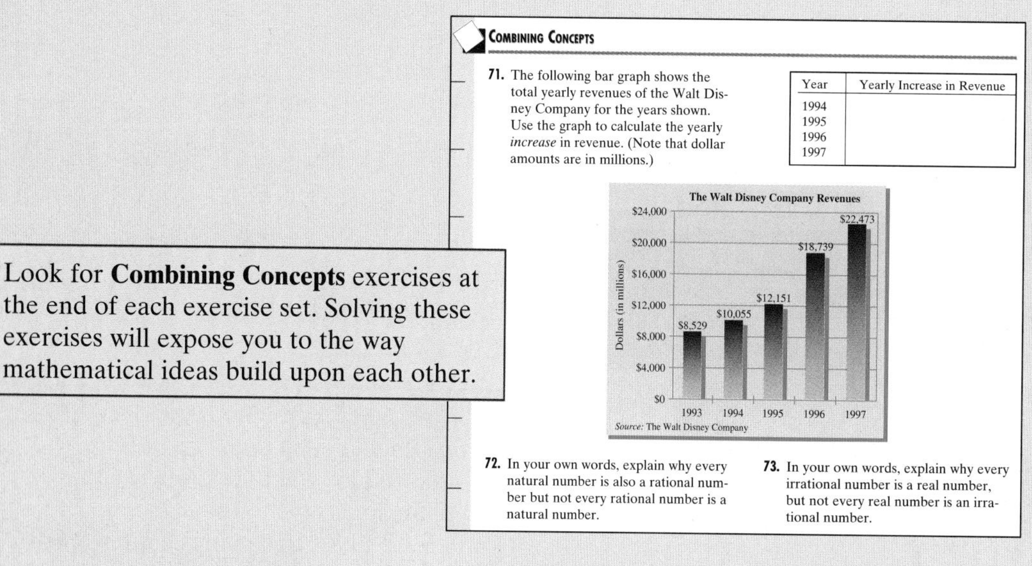

COMBINING CONCEPTS

71. The following bar graph shows the total yearly revenues of the Walt Disney Company for the years shown. Use the graph to calculate the yearly *increase* in revenue. (Note that dollar amounts are in millions.)

Year	Yearly Increase in Revenue
1994	
1995	
1996	
1997	

72. In your own words, explain why every natural number is also a rational number but not every rational number is a natural number.

73. In your own words, explain why every irrational number is a real number, but not every real number is an irrational number.

Page 12

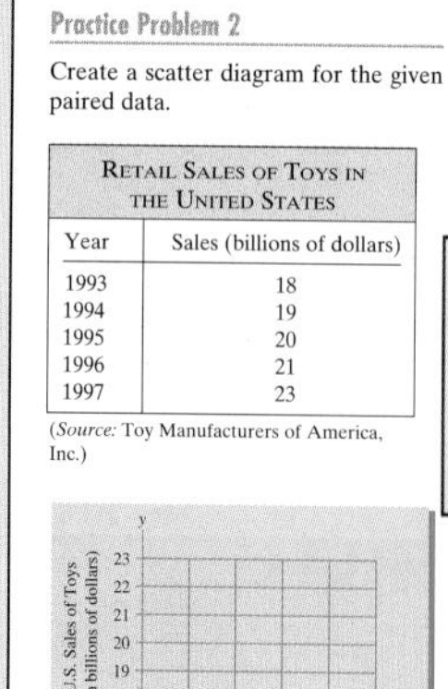

Practice Problem 2

Create a scatter diagram for the given paired data.

Retail Sales of Toys in the United States	
Year	Sales (billions of dollars)
1993	18
1994	19
1995	20
1996	21
1997	23

(*Source:* Toy Manufacturers of America, Inc.)

Practice Problems occur in the margins next to every Example. Work these problems after you read an Example to reinforce your understanding.

Page 172

Test Yourself and Check Your Understanding

Good exercise sets are an essential ingredient for a solid basic college mathematics textbook. The exercises you will find in this worktext are intended to help you build skills and understand concepts as well as motivate and challenge you. Note that the features like Chapter Highlights, Tests, and Cumulative Reviews are found at the end of each chapter to help you study and organize your notes.

Chapter 4 Pretest

Solve each system of equations by graphing.

1. $\begin{cases} 2x + y = -1 \\ x + y = -3 \end{cases}$ **2.** $\begin{cases} x - y = 2 \\ 3x - y = 0 \end{cases}$

Solve each system of equations by the substitution method.

3. $\begin{cases} 2x - 3y = -8 \\ y = x + 1 \end{cases}$ **4.** $\begin{cases} \frac{x}{6} + \frac{y}{2} = -2 \\ x - \frac{y}{2} = 2 \end{cases}$

Pretests open each chapter. Take a **Pretest** to evaluate where you need help the most before beginning a new chapter.

Page 260

Integrated Review—Linear Equations in Two Variables

Below is a review of equations of lines.

Forms of Linear Equations

$Ax + By = C$	**Standard form** of a linear equation A and B are not both 0.
$y = mx + b$	**Slope-intercept form** of a linear equation The slope is m, and the y-intercept is b.
$y - y_1 = m(x - x_1)$	**Point-slope form** of a linear equation The slope is m, and (x_1, y_1) is a point on the line.
$y = c$	**Horizontal line** The slope is 0, and the y-intercept is c.
$x = c$	**Vertical line** The slope is undefined and the x-intercept is c.

Integrated Reviews serve as a mid-chapter review and ask you to assimilate the new skills you have learned separately over several sections.

Page 217

Mental Math

Determine whether a line with the given slope slants upward, downward, horizontally, or vertically from left to right.

1. $m = \frac{7}{6}$ **2.** $m = -3$ **3.** $m = 0$ **4.** m is undefined.

Confidence-building **Mental Math** problems are in most sections.

Page 193

Review and Preview

Find each product by using the FOIL order of multiplying binomials. See Section 5.2.

43. $(x + 2)(x - 5)$ **44.** $(x - 7)(x - 1)$ **45.** $(x + 3)(x + 2)$

46. $(x - 4)(x + 2)$ **47.** $(y - 3)(y - 1)$ **48.** $(s + 8)(s + 10)$

Review and Preview

Page 377

Get Involved!

Discover how mathematics relates to and appears in the world around you. Evaluate and interpret real data in graphs, tables, and charts—make an educated guess on the answers and outcomes! Knowing how to use data and graphs to estimate and predict is a valuable skill in the workplace as well as in other courses.

Example 2 Create a scatter diagram for the given paired data.

National Average SAT Math Scores	
Year	Score
1994	504
1995	506
1996	508
1997	511

(*Source:* The College Board)

Real data is integrated throughout the worktext, drawn from current and familiar sources.

Page 172

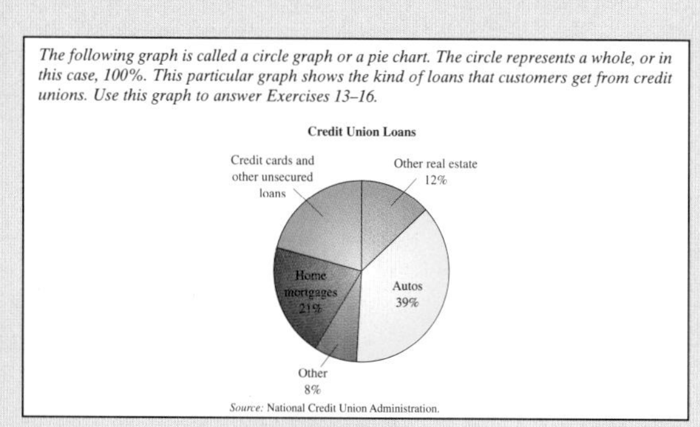
The following graph is called a circle graph or a pie chart. The circle represents a whole, or in this case, 100%. This particular graph shows the kind of loans that customers get from credit unions. Use this graph to answer Exercises 13–16.

Page 98

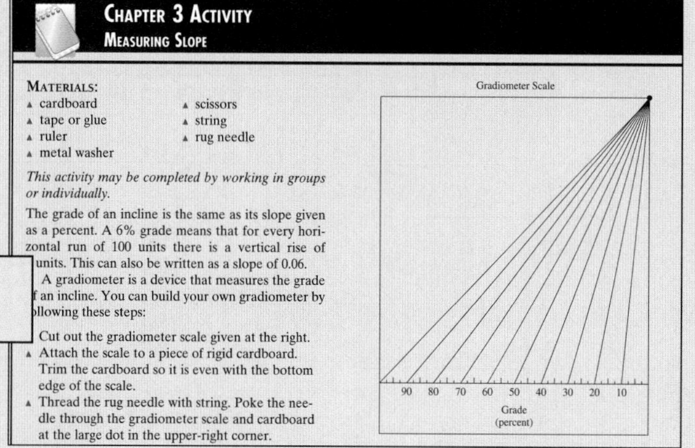
Chapter 3 Activity
Measuring Slope

Materials:
- cardboard
- tape or glue
- ruler
- metal washer
- scissors
- string
- rug needle

This activity may be completed by working in groups or individually.

The grade of an incline is the same as its slope given as a percent. A 6% grade means that for every horizontal run of 100 units there is a vertical rise of units. This can also be written as a slope of 0.06.

A gradiometer is a device that measures the grade an incline. You can build your own gradiometer by llowing these steps:

Cut out the gradiometer scale given at the right.
- Attach the scale to a piece of rigid cardboard. Trim the cardboard so it is even with the bottom edge of the scale.
- Thread the rug needle with string. Poke the needle through the gradiometer scale and cardboard at the large dot in the upper-right corner.

Graphics, models and illustrations provide visual reinforcement.

Page 241

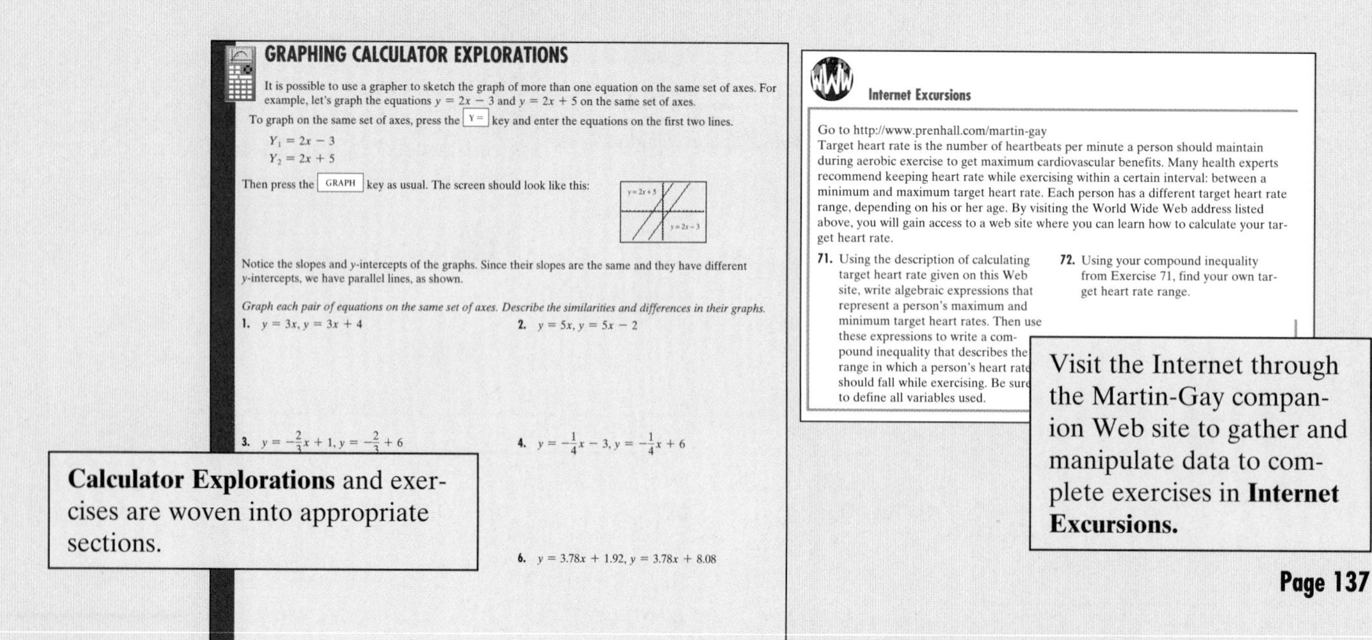
GRAPHING CALCULATOR EXPLORATIONS

It is possible to use a grapher to sketch the graph of more than one equation on the same set of axes. For example, let's graph the equations $y = 2x - 3$ and $y = 2x + 5$ on the same set of axes.

To graph on the same set of axes, press the Y= key and enter the equations on the first two lines.

$Y_1 = 2x - 3$
$Y_2 = 2x + 5$

Then press the GRAPH key as usual. The screen should look like this:

Notice the slopes and y-intercepts of the graphs. Since their slopes are the same and they have different y-intercepts, we have parallel lines, as shown.

Graph each pair of equations on the same set of axes. Describe the similarities and differences in their graphs.

1. $y = 3x, y = 3x + 4$
2. $y = 5x, y = 5x - 2$
3. $y = -\frac{2}{3}x + 1, y = -\frac{2}{3} + 6$
4. $y = -\frac{1}{4}x - 3, y = -\frac{1}{4}x + 6$
6. $y = 3.78x + 1.92, y = 3.78x + 8.08$

Internet Excursions

Go to http://www.prenhall.com/martin-gay

Target heart rate is the number of heartbeats per minute a person should maintain during aerobic exercise to get maximum cardiovascular benefits. Many health experts recommend keeping heart rate while exercising within a certain interval: between a minimum and maximum target heart rate. Each person has a different target heart rate range, depending on his or her age. By visiting the World Wide Web address listed above, you will gain access to a web site where you can learn how to calculate your target heart rate.

71. Using the description of calculating target heart rate given on this Web site, write algebraic expressions that represent a person's maximum and minimum target heart rates. Then use these expressions to write a compound inequality that describes the range in which a person's heart rate should fall while exercising. Be sure to define all variables used.

72. Using your compound inequality from Exercise 71, find your own target heart rate range.

Calculator Explorations and exercises are woven into appropriate sections.

Visit the Internet through the Martin-Gay companion Web site to gather and manipulate data to complete exercises in **Internet Excursions.**

Page 137

Page 192

Focus On boxes found throughout each chapter help you see the relevance of math through critical thinking exercises and group activities. Try these on your own or with another student.

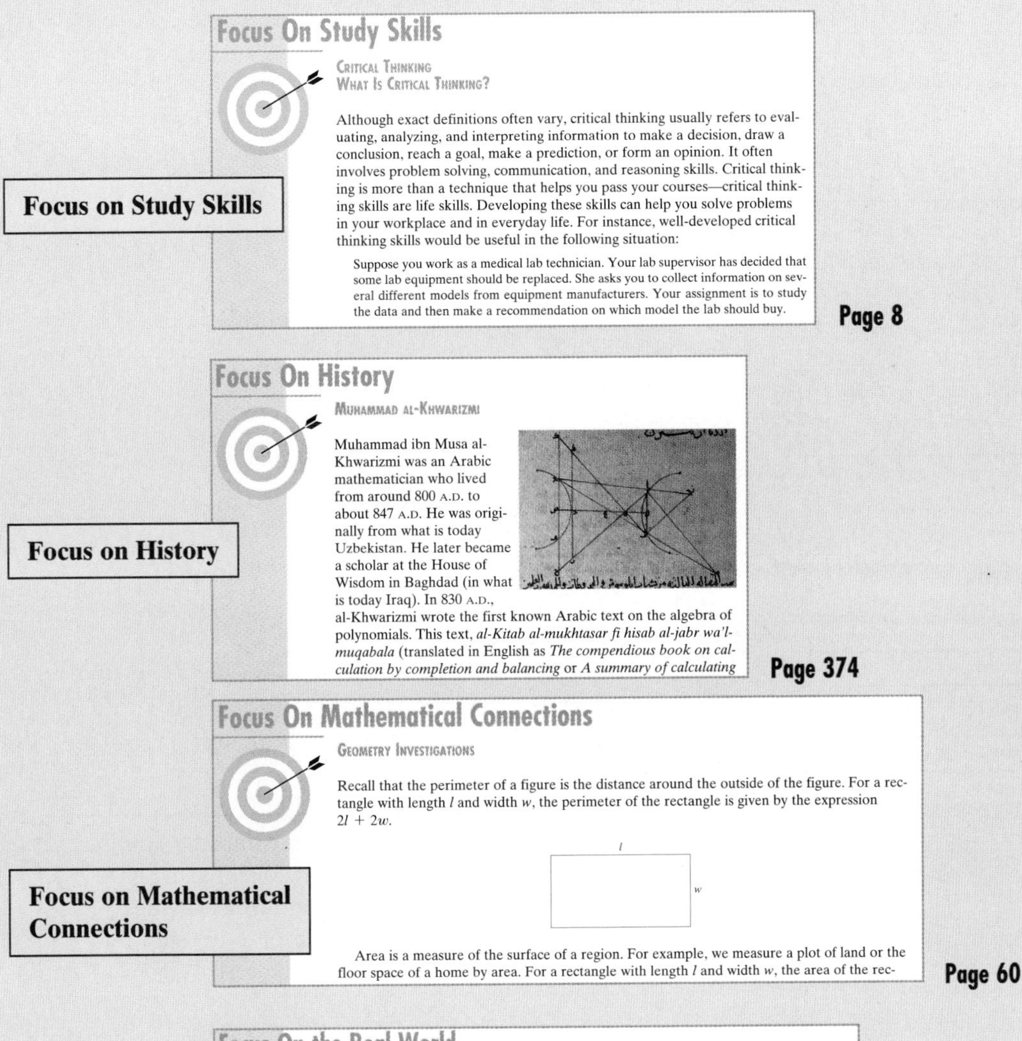

Focus on Study Skills

Focus On Study Skills

CRITICAL THINKING
WHAT IS CRITICAL THINKING?

Although exact definitions often vary, critical thinking usually refers to evaluating, analyzing, and interpreting information to make a decision, draw a conclusion, reach a goal, make a prediction, or form an opinion. It often involves problem solving, communication, and reasoning skills. Critical thinking is more than a technique that helps you pass your courses—critical thinking skills are life skills. Developing these skills can help you solve problems in your workplace and in everyday life. For instance, well-developed critical thinking skills would be useful in the following situation:

Suppose you work as a medical lab technician. Your lab supervisor has decided that some lab equipment should be replaced. She asks you to collect information on several different models from equipment manufacturers. Your assignment is to study the data and then make a recommendation on which model the lab should buy.

Page 8

Focus on History

Focus On History

MUHAMMAD AL-KHWARIZMI

Muhammad ibn Musa al-Khwarizmi was an Arabic mathematician who lived from around 800 A.D. to about 847 A.D. He was originally from what is today Uzbekistan. He later became a scholar at the House of Wisdom in Baghdad (in what is today Iraq). In 830 A.D., al-Khwarizmi wrote the first known Arabic text on the algebra of polynomials. This text, *al-Kitab al-mukhtasar fi hisab al-jabr wa'l-muqabala* (translated in English as *The compendious book on calculation by completion and balancing* or *A summary of calculating*

Page 374

Focus on Mathematical Connections

Focus On Mathematical Connections

GEOMETRY INVESTIGATIONS

Recall that the perimeter of a figure is the distance around the outside of the figure. For a rectangle with length l and width w, the perimeter of the rectangle is given by the expression $2l + 2w$.

Area is a measure of the surface of a region. For example, we measure a plot of land or the floor space of a home by area. For a rectangle with length l and width w, the area of the rec-

Page 60

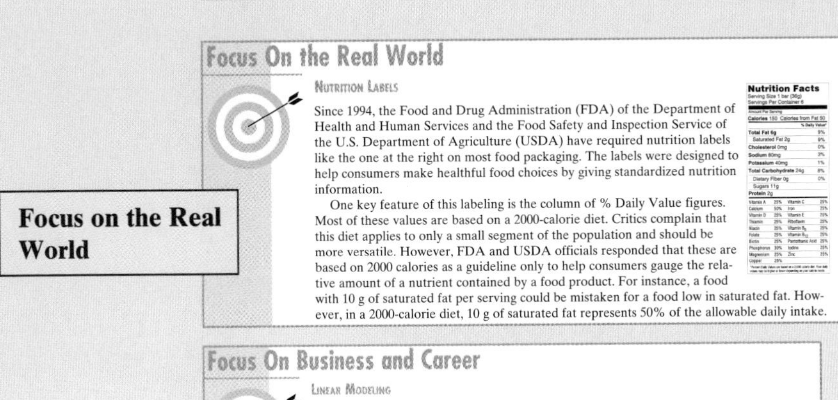

Focus on the Real World

Focus On the Real World

NUTRITION LABELS

Since 1994, the Food and Drug Administration (FDA) of the Department of Health and Human Services and the Food Safety and Inspection Service of the U.S. Department of Agriculture (USDA) have required nutrition labels like the one at the right on most food packaging. The labels were designed to help consumers make healthful food choices by giving standardized nutrition information.

One key feature of this labeling is the column of % Daily Value figures. Most of these values are based on a 2000-calorie diet. Critics complain that this diet applies to only a small segment of the population and should be more versatile. However, FDA and USDA officials responded that these are based on 2000 calories as a guideline only to help consumers gauge the relative amount of a nutrient contained by a food product. For instance, a food with 10 g of saturated fat per serving could be mistaken for a food low in saturated fat. However, in a 2000-calorie diet, 10 g of saturated fat represents 50% of the allowable daily intake.

Page 96

Focus on Business and Career

Focus On Business and Career

LINEAR MODELING

As we saw in Section 3.3, businesses often depend on equations that "closely fit" data. To *model* the data means to find an equation that describes the relationship between the paired data of two variables, such as time in years and profit. A model that accurately summarizes the relationship between two variables can be used to replace a potentially lengthy listing of the raw data. An accurate model might also be used to predict future trends by answering questions such as "If the trend seen in our company's performance in the last several years continues, what level of profit can we reasonably expect in 3 years?"

There are several ways to find a linear equation that models a set of data. If only two ordered pair data points are involved, an exact equation that contains both points can be found using the methods of Section 3.4. When more than two ordered pair data points are involved, it may be impossible to find a linear equation that contains all of the data points. In this case, the graph of the **best fit equation** should have a majority of the plotted ordered pair data points on the graph or close to it. In statistics, a technique called least squares regression is used to determine an equation that best fits a set of data. Various graphing utilities have built-in capabilities for finding an equation (called a regression equation) that best fits a set of ordered pair data points. Regression capabilities are often found with a graphing utility's statistics features.* A best fit equation can also be estimated using an algebraic method, which is outlined in the Group Activity below. In either case, a useful first step when finding a linear equation that models a set of data is creating a scatter diagram of the ordered pair data points to verify that a linear equation is an appropriate model.

Page 210

Enrich Your Learning

Seek out these additional Student Resources and tools to match your personal learning style.

Text-specific **videos** hosted by the award-winning teacher and author of *Intermediate Algebra* cover each objective in every chapter section as a supplementary review. Many of the examples and exercises worked out in the videos are selected directly from *Intermediate Algebra*.

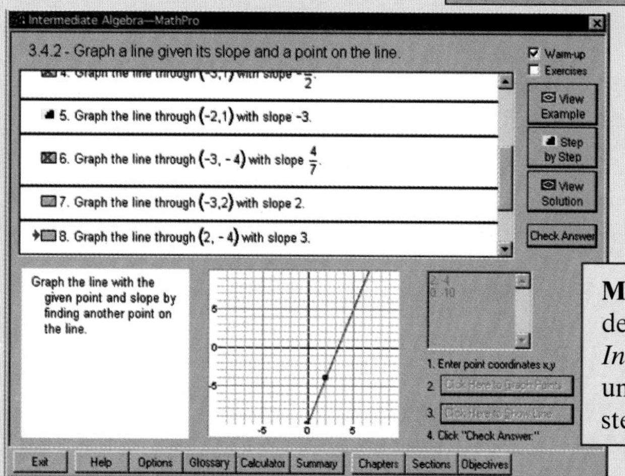

MathPro and **MathPro Explorer** tutorial software is developed around the content and concepts of *Intermediate Algebra*. **MathPro Explorer** provides unlimited practice problems and interactive step-by-step help as well as exploratory activities.

Also available:

Ask your instructor or bookstore about these additional study aids.

Real Numbers and Algebraic Expressions

CHAPTER 1

In arithmetic, we add, subtract, multiply, divide, raise to powers, and take roots of numbers. In algebra, we add, subtract, multiply, divide, raise to powers, and take roots of variables. Letters, such as x, that represent numbers are called variables. Understanding algebraic expressions made up of combinations of variables and numbers depends on your understanding of arithmetic expressions. This chapter reviews the arithmetic operations on real numbers and the corresponding algebraic expressions. After this review, we will be prepared to explore how widely useful these algebraic expressions are for problem solving in diverse situations.

The world's first subway system was built in London, England, in 1863. The growing congestion on the streets of London led to relocating the existing steam-operated metropolitan railway underground. Thirty-four years later, America's first subway was built in Boston, Massachusetts. The Boston subway started out with single electrified streetcars running on rails underground. New York City soon followed suit and completed its own subway system in 1904. The New York subway's underground electric trains replaced surface electric train routes, elevated railways, and horse car tracks. Today, New York's subway system is one of the largest and most complex in the world, with 714 miles of subway track and 468 stations. In Exercise 87 on page 55, we will use scientific notation to write the very large number of passengers who ride on New York City subways each year.

Answers

1. ______
2. ______
3. ______
4. ______
5. ______
6. ______
7. ______
8. ______
9. ______
10. ______
11. ______
12. ______
13. ______
14. ______
15. ______
16. ______
17. ______
18. ______
19. ______
20. ______

Name ______ Section ______ Date ______

Chapter 1 Pretest

1. Evaluate $2x - 3y^2$ when $x = 5$ and $y = 2$.

2. Translate the following phrase to an algebraic expression. Use x to represent the unknown number. Seven more than twice a number.

Insert $<$, $>$, or $=$ between each pair of numbers to form a true statement.

3. $\frac{2}{3} \quad \frac{10}{17}$

4. $-7 \quad -6$

Write the opposite (or additive inverse) of each number if one exists.

5. 9.25

6. $-\frac{7}{8}$

Simplify.

7. $-|-21|$

8. $-13 - (-22)$

9. $\frac{-5.1}{1.7}$

10. -5^2

11. $\sqrt{\frac{25}{121}}$

12. $9 - [(2 - 6) + (3 - 17)]$

13. $2(3x - 5) - 4(2 - x)$

Simplify. Write answers using positive exponents only.

14. $(-5b^6)(4b^8)$

15. $\frac{-45a^2b^7c}{-9ab^{12}c^2}$

16. $\frac{z^{-6}z^{12}}{z^{-9}}$

17. $5(y^4z)^{-3}$

18. $\left(\frac{2x^{-2}y^3}{6xz^5}\right)^{-4}$

19. Evaluate: $2x^0 + 3$

20. Write the following number in scientific notation: 65,400,000,000

1.1 ALGEBRAIC EXPRESSIONS AND SETS OF NUMBERS

Objectives

A Identify and evaluate algebraic expressions.

B Identify natural numbers, whole numbers, integers, rational, and irrational real numbers.

C Write phrases as algebraic expressions.

SSM CD-ROM Video 1.1

A EVALUATING ALGEBRAIC EXPRESSIONS

Recall that letters that represent numbers are called **variables**. An **algebraic expression** is formed by numbers and variables connected by the operations of addition, subtraction, multiplication, division, raising to powers, or taking roots. For example,

$$2x, \quad \frac{x+5}{6}, \quad \sqrt{y} - 1.6, \quad \text{and} \quad z^3$$

are algebraic expressions or, more simply, expressions. (Recall that the expression $2x$ means $2 \cdot x$.)

Algebraic expressions occur often during problem solving. For example, suppose that a television commercial for a watch is being filmed on the Golden Gate Bridge. A portion of this commercial consists of dropping a watch from the bridge. To determine the best camera angles and also whether the watch will survive the fall, it is important to know the speed of the watch at 1-second intervals. The algebraic expression

$$32t$$

gives the speed of the watch in feet per second for time t (in seconds).

To find the speed of the watch at, for example, 1 second, we replace the variable t with 1 and perform the indicated multiplication. This process is called **evaluating** an expression, and the result is called the **value** of the expression for the given replacement value.

When $t = 1$ second, $32t = 32 \cdot 1 = 32$ feet per second.
When $t = 2$ seconds, $32t = 32 \cdot 2 = 64$ feet per second.
When $t = 3$ seconds, $32t = 32 \cdot 3 = 96$ feet per second.

Example 1 Finding the Area of a Tile

The research department of a flooring company is considering a new flooring design that contains parallelograms. The area of a parallelogram with base b and height h is bh. Find the area of a parallelogram with base 10 centimeters and height 8.2 centimeters.

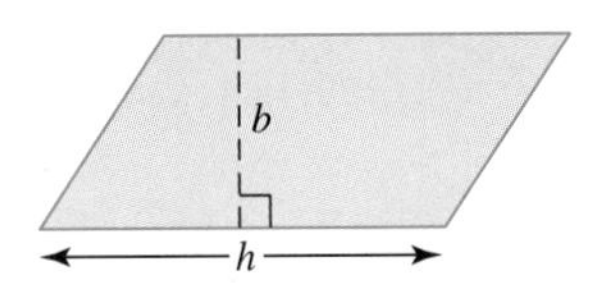

Solution: We replace b with 10 and h with 8.2 in the algebraic expression bh.

$$bh = 10 \cdot 8.2 = 82$$

The area is 82 square centimeters.

Practice Problem 1

The area of a triangle with base b and height h is $\frac{1}{2}bh$. Find the area of a triangle with base 12 inches and height 7 inches.

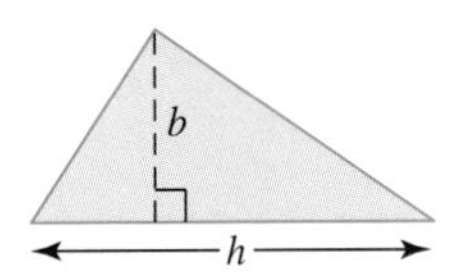

Example 2

Evaluate: $3x - y$ when $x = 15$ and $y = 4$

Solution: We replace x with 15 and y with 4 in the expression.

$$3x - y = 3 \cdot 15 - 4 = 45 - 4 = 41$$

Practice Problem 2

Evaluate: $2a - b$ when $a = 20$ and $b = 12$

Answers

1. 42 square inches, **2.** 28

Practice Problem 3

Evaluate: $\frac{m}{n}$ when $m = 70$ and $n = 10$

Example 3 Evaluate: $\frac{r}{s}$ when $r = 48$ and $s = 6$

Solution: We replace r with 48 and s with 6 in the expression.

$$\frac{r}{s} = \frac{48}{6} = 8$$

When evaluating an expression to solve a problem, we often need to think about the kind of number that is appropriate for the solution. For example, if we are asked to determine the maximum number of parking spaces for a parking lot to be constructed, an answer of $98\frac{1}{10}$ is not appropriate because $\frac{1}{10}$ of a parking space is not realistic.

B IDENTIFYING COMMON SETS OF NUMBERS

Let's review some common sets of numbers and their graphs on a number line. To construct a number line, we draw a line and label a point 0 with which we associate the number 0. This point is called the **origin**. If we choose a point to the right of 0 and label it 1, the distance from 0 to 1 is called the **unit distance** and can be used to locate more points. The **positive numbers** lie to the right of the origin, and the **negative numbers** lie to the left of the origin. The number 0 is neither positive nor negative.

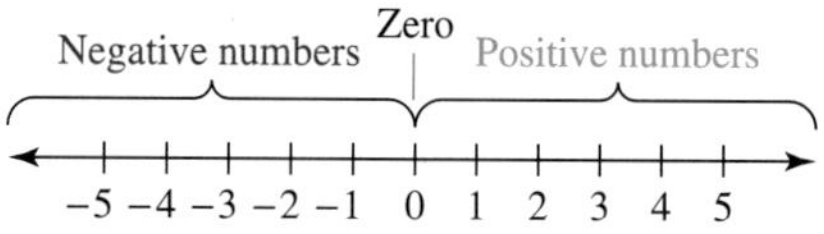

✓ CONCEPT CHECK

Use the definitions of positive numbers, negative numbers, and zero to describe the meaning of *nonnegative numbers*.

TRY THE CONCEPT CHECK IN THE MARGIN.

A number is **graphed** on a number line by shading the point on the number line that corresponds to the number. Some common sets of numbers and their graphs are shown next.

IDENTIFYING NUMBERS

Natural numbers: $\{1, 2, 3, \ldots\}$

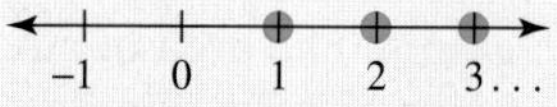

Whole numbers: $\{0, 1, 2, 3, \ldots\}$

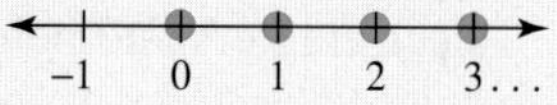

Integers: $\{\ldots, -3, -2, -1, 0, 1, 2, 3, \ldots\}$

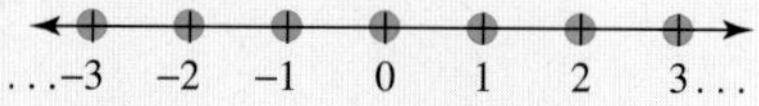

Each listing of three dots above, . . . , is called an **ellipsis** and means to continue in the same pattern.

A **set** is a collection of objects. The objects of a set are called its **elements**. When the elements of a set are listed, such as those displayed in the box, the set is written in **roster** form. A set can also be written in **set builder notation**, which describes the members of a set but does not list them. The set following is written in set builder notation.

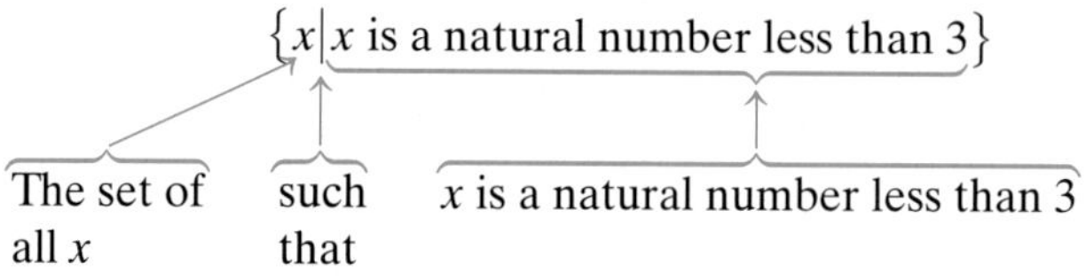

Answers

3. 7

✓ Concept Check: A number which is zero or positive.

This same set written in roster form is $\{1, 2\}$.

A set that contains *no* elements is called the **empty set**, symbolized by $\{\ \}$, or the **null set**, symbolized by $\emptyset$.

$$\{x|x \text{ is a month with 32 days}\} \text{ is } \emptyset \text{ or } \{\ \}$$

because no month has 32 days. The set has no elements.

HELPFUL HINT

Use $\{\ \}$ to write the empty set. $\{\emptyset\}$ is *not* the empty set because it has one element: $\emptyset$.

Examples Write each set in roster form.

4. $\{x|x \text{ is a whole number between 1 and 6}\}$
$\{2, 3, 4, 5\}$

5. $\{x|x \text{ is a natural number greater than 100}\}$
$\{101, 102, 103, \ldots\}$

The symbol $\in$ is used to denote that an element is in a particular set. The symbol $\in$ is read as "is an element of." For example, the true statement "3 is an element of $\{1, 2, 3, 4, 5\}$" can be written in symbols as

$$3 \in \{1, 2, 3, 4, 5\}$$

The symbol $\notin$ is read as "is not an element of." In symbols, we write the true statement "p is not an element of $\{a, 5, g, j, q\}$" as

$$p \notin \{a, 5, g, j, q\}$$

Examples Determine whether each statement is true or false.

6. $3 \in \{x|x \text{ is a natural number}\}$ True, since 3 is a natural number and therefore an element of the set.

7. $7 \notin \{1, 2, 3\}$ True, since 7 is not an element of the set $\{1, 2, 3\}$.

We can use set builder notation to describe three other common sets of numbers.

IDENTIFYING NUMBERS

Real numbers: $\{x|x \text{ corresponds to a point on the number line}\}$

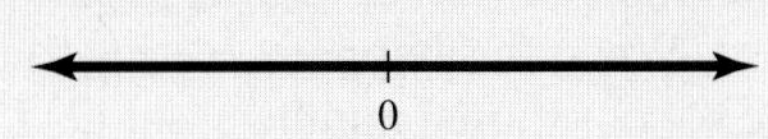

Rational numbers: $\left\{\frac{a}{b} \middle| a \text{ and } b \text{ are integers and } b \neq 0\right\}$

Irrational numbers: $\{x|x \text{ is a real number and } x \text{ is not a rational number}\}$

Notice that every integer is also a rational number since each integer can be written as the quotient of itself and 1:

$$3 = \frac{3}{1}, \quad 0 = \frac{0}{1}, \quad -8 = \frac{-8}{1}$$

Practice Problems 4–5

Write each set in roster form.

4. $\{x|x \text{ is a whole number between 0 and 4}\}$

5. $\{x|x \text{ is a natural number greater than 80}\}$

Practice Problems 6–7

Determine whether each statement is true or false.

6. $0 \in \{x|x \text{ is a natural number}\}$

7. $9 \notin \{4, 6, 8, 10\}$

Answers

4. $\{1, 2, 3\}$, **5.** $\{81, 82, 83, \ldots\}$, **6.** false, **7.** true

Not every rational number, however, is an integer. The rational number $\frac{2}{3}$, for example, is not an integer. Some square roots are rational numbers and some are irrational numbers. For example, $\sqrt{2}$, $\sqrt{3}$, and $\sqrt{7}$ are irrational numbers but $\sqrt{25}$ is a rational number because $\sqrt{25} = 5 = \frac{5}{1}$. The number π is an irrational number. To help you make the distinction between rational and irrational numbers, here are a few examples of each.

RATIONAL NUMBERS		IRRATIONAL NUMBERS
Number	Equivalent Quotient of Integers, $\frac{a}{b}$	
$-\frac{2}{3}$	$\frac{-2}{3}$ or $\frac{2}{-3}$	$\sqrt{5}$
$\sqrt{36}$	$\frac{6}{1}$	$\frac{\sqrt{6}}{7}$
5	$\frac{5}{1}$	$-\sqrt{3}$
0	$\frac{0}{1}$	π
1.2	$\frac{12}{10}$	$\frac{2}{\sqrt{3}}$
$3\frac{7}{8}$	$\frac{31}{8}$	

Every rational number can be written as a decimal that either repeats or terminates. For example,

$$\frac{1}{2} = 0.5 \qquad \frac{5}{4} = 1.25$$

$$\frac{2}{3} = 0.6666666\ldots = 0.\overline{6} \qquad \frac{1}{11} = 0.090909\ldots = 0.\overline{09}$$

An irrational number written as a decimal neither terminates nor repeats. When we perform calculations with irrational numbers, we often use decimal approximations that have been rounded. For example, consider the following irrational numbers along with a four-decimal-place approximation of each:

$$\pi \approx 3.1416 \qquad \sqrt{2} \approx 1.4142$$

Earlier we mentioned that every integer is also a rational number. In other words, all the elements of the set of integers are also elements of the set of rational numbers. When this happens, we say that the set of integers, set I, is a subset of the set of rational numbers, set Q. The natural numbers, whole numbers, integers, rational numbers, and irrational numbers are each a subset of the set of real numbers. The relationships among these sets of numbers are shown in the following diagram.

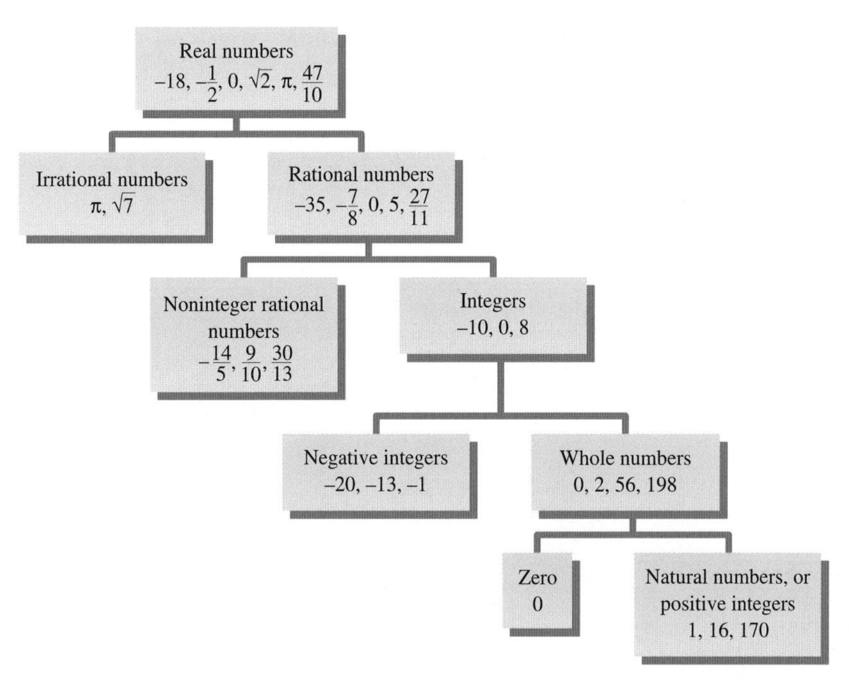

Examples Determine whether each statement is true or false.

8. 3 is a real number. True. Every whole number is a real number.

9. Every rational number is an integer. False. The number $\frac{2}{3}$, for example, is a rational number, but it is not an integer.

10. $\frac{1}{5}$ is an irrational number. False. The number $\frac{1}{5}$ is a rational number, since it is in the form $\frac{a}{b}$ with a and b integers and $b \neq 0$.

Practice Problems 8–10

Determine whether each statement is true or false.

8. 0 is a real number.
9. Every integer is a rational number.
10. $\sqrt{3}$ is a rational number.

C WRITING PHRASES AS ALGEBRAIC EXPRESSIONS

Often, solving problems involves translating a phrase into an algebraic expression. The following is a list of key words and phrases and their translations.

Addition	**Subtraction**	**Multiplication**	**Division**
sum	difference of	product	quotient
plus	minus	times	divide
added to	subtracted from	multiply	into
more than	less than	twice	ratio
increased by	decreased by	of	
total	less		

Answers

8. true, **9.** true, **10.** false

Practice Problems 11–15

Write each phrase as an algebraic expression. Use the variable x to represent each unknown number.

11. Twice a number
12. Five more than six times a number
13. The quotient of 6 and a number
14. One-fourth subtracted from three times a number
15. Eleven less than a number

Examples Write each phrase as an algebraic expression. Use the variable x to represent each unknown number.

11. Eight times a number — $8 \cdot x$ or $8x$
12. Three more than eight times a number — $8x + 3$
13. The quotient of a number and -7 — $x \div -7$ or $\frac{x}{-7}$
14. One and six-tenths subtracted from twice a number — $2x - 1.6$
15. Six less than a number — $x - 6$

Focus On Study Skills

CRITICAL THINKING
WHAT IS CRITICAL THINKING?

Although exact definitions often vary, critical thinking usually refers to evaluating, analyzing, and interpreting information to make a decision, draw a conclusion, reach a goal, make a prediction, or form an opinion. It often involves problem solving, communication, and reasoning skills. Critical thinking is more than a technique that helps you pass your courses—critical thinking skills are life skills. Developing these skills can help you solve problems in your workplace and in everyday life. For instance, well-developed critical thinking skills would be useful in the following situation:

> Suppose you work as a medical lab technician. Your lab supervisor has decided that some lab equipment should be replaced. She asks you to collect information on several different models from equipment manufacturers. Your assignment is to study the data and then make a recommendation on which model the lab should buy.

HOW CAN CRITICAL THINKING BE DEVELOPED?

Just as physical exercise can help to develop and strengthen certain muscles of the body, mental exercise can help to develop critical thinking skills. Mathematics is ideal for helping to develop such skills because it requires using logic and reasoning, recognizing patterns, making conjectures and educated guesses, and drawing conclusions. You will find many opportunities to build your critical thinking skills throughout *Intermediate Algebra*:

- In real-life application problems (see Exercise 89 in Section 1.5)
- In conceptual and writing exercises marked with the icon (see Exercise 55 in Section 1.2)
- In the Combining Concepts subsection of the exercise sets (see Exercise 117 in Section 1.3)
- In the Chapter Activities (see page 67)
- In the Critical Thinking and Group Activities questions found in Focus On features like this one throughout the book (see pages 96 and 138)

Answers

11. $2x$, **12.** $6x + 5$, **13.** $\frac{6}{x}$ or $6 \div x$, **14.** $3x - \frac{1}{4}$, **15.** $x - 11$

Name ______________________ Section ______ Date ________

EXERCISE SET 1.1

A *Evaluate each algebraic expression at the given replacement values. See Examples 1 through 3.*

1. $5x$ when $x = 7$

2. $3y$ when $y = 45$

3. $9.8z$ when $z = 3.1$

4. $7.1a$ when $a = 1.5$

5. $\frac{x}{y}$ when $x = 18$ and $y = 3$

6. $\frac{s}{t}$ when $s = 6$ and $t = 1$

7. ab when $a = \frac{1}{2}$ and $b = \frac{3}{4}$

8. yz when $y = \frac{2}{3}$ and $z = \frac{1}{5}$

9. $3x + y$ when $x = 6$ and $y = 4$

10. $2a - b$ when $a = 12$ and $b = 7$

11. $qr - s$ when $q = 1, r = 14$, and $s = 3$

12. $xy + z$ when $x = 9$, $y = 2$, and $z = 8$

13. The aircraft B737-400 flies an average speed of 414 miles per hour.

The algebraic expression $414t$ gives the distance traveled by the aircraft in t hours. Find the distance traveled by the B737-400 in 5 hours. (*Source:* Air Transport Association of America)

14. The algebraic expression $1.5x$ gives the total length of shelf space in inches needed for x encyclopedias. Find the length of shelf space needed for a set of 30 encyclopedias.

15. Employees of Wal-Mart constantly reorganize and reshelf merchandise. In doing so, they calculate floor space needed for displays. The algebraic expression $l \cdot w$ gives the floor space needed in square units for a display that measures length l units and width w units. Calculate the floor space needed for a display whose length is 5.1 feet and whose width is 4 feet.

16. The algebraic expression $\frac{x}{5}$ can be used to calculate the distance in miles that you are from a flash of lightning, where x is the number of seconds between the time you see a flash of lightning and the time you hear the thunder. Calculate the distance that you are from the flash of lightning if you hear the thunder 2 seconds after you see the lightning.

ANSWERS

1. ______
2. ______
3. ______
4. ______
5. ______
6. ______
7. ______
8. ______
9. ______
10. ______
11. ______
12. ______
13. ______
14. ______
15. ______
16. ______

17.

18.

19.

20.

21.

22.

23.

24.

25.

26.

27.

28.

29.

30.

31.

32.

Name

17. The B747-400 aircraft costs \$7075 dollars per hour to operate. The algebraic expression $7075t$ gives the total cost to operate the aircraft for t hours. Find the total cost to operate the B747-400 for 5.2 hours. (*Source:* Air Transport Association of America)

18. On October 15, 1997, Royal Air Force pilot Andy Green set a new one-mile land speed record in the rocket-powered car *Thrust SSC* in the Black Rock Desert, Nevada. His record-setting speed was 763.035 miles per hour. At this speed, the algebraic expression $763.035t$ gives the total distance covered in t hours. Find the distance covered by Green in the *Thrust SSC* in 0.75 hours. (*Source:* United States Auto Club)

B *Write each set in roster form. See Examples 4 and 5.*

19. $\{x|x$ is a natural number less than $6\}$

20. $\{x|x$ is a natural number greater than $6\}$

21. $\{x|x$ is a natural number between 10 and $17\}$

22. $\{x|x$ is an odd natural number$\}$

23. $\{x|x$ is a whole number that is not a natural number$\}$

24. $\{x|x$ is a natural number less than $1\}$

25. $\{x|x$ is an even whole number less than $9\}$

26. $\{x|x$ is an odd whole number less than $9\}$

List the elements of the set $\left\{3, 0, \sqrt{7}, \sqrt{36}, \frac{2}{5}, -134\right\}$ *that are also elements of the given set. See Examples 6 and 7.*

27. Whole numbers

28. Integers

29. Natural numbers

30. Rational numbers

31. Irrational numbers

32. Real numbers

Name ______________________________

Place $\in$ or $\notin$ in the space provided to make each statement true. See Examples 6 through 10.

33. -11 $\{x|x \text{ is an integer}\}$

34. -6 $\{2, 4, 6, \ldots\}$

35. 0 $\{x|x \text{ is a positive integer}\}$

36. 12 $\{1, 2, 3, \ldots\}$

37. 12 $\{1, 3, 5, \ldots\}$

38. $\frac{1}{2}$ $\{x|x \text{ is an irrational number}\}$

Determine whether each statement is true or false. See Examples 8 through 10.

39. Every whole number is a real number.

40. Every irrational number is a real number.

41. Some real numbers are irrational numbers.

42. Some real numbers are whole numbers.

43. Every whole number is a natural number.

44. Every irrational number is a rational number.

C *Write each phrase as an algebraic expression. Use the variable x to represent each unknown number. See Examples 11 through 15.*

45. Twice a number

46. Six times a number

47. Ten less than a number

48. A number minus seven

49. The sum of a number and two

50. The difference of twenty-five and a number

51. A number divided by eleven

52. The quotient of a number and thirteen

53. Four subtracted from a number

54. Seventeen subtracted from a number

55. A number plus twenty

56. Fifteen plus a number

57. A number less than ten

58. Twelve less than a number

59. Nine times a number

60. Nine minus a number

61. Nine added to a number

62. Nine divided by a number

33. ______
34. ______
35. ______
36. ______
37. ______
38. ______
39. ______
40. ______
41. ______
42. ______
43. ______
44. ______
45. ______
46. ______
47. ______
48. ______
49. ______
50. ______
51. ______
52. ______
53. ______
54. ______
55. ______
56. ______
57. ______
58. ______
59. ______
60. ______
61. ______
62. ______

63. ______
64. ______
65. ______
66. ______
67. ______
68. ______
69. ______
70. ______
71. ______
72. ______
73. ______
74. ______
75. ______
76. ______
77. ______

Name ______________________________

63. Five more than twice a number

64. One more than six times a number

65. Twelve minus three times a number

66. Four subtracted from three times a number

67. One plus twice a number

68. Three less than twice a number

69. Ten subtracted from five times a number

70. Four minus three times a number

Combining Concepts

71. The following bar graph shows the total yearly revenues of the Walt Disney Company for the years shown. Use the graph to calculate the yearly *increase* in revenue. (Note that dollar amounts are in millions.)

Year	Yearly Increase in Revenue
1994	
1995	
1996	
1997	

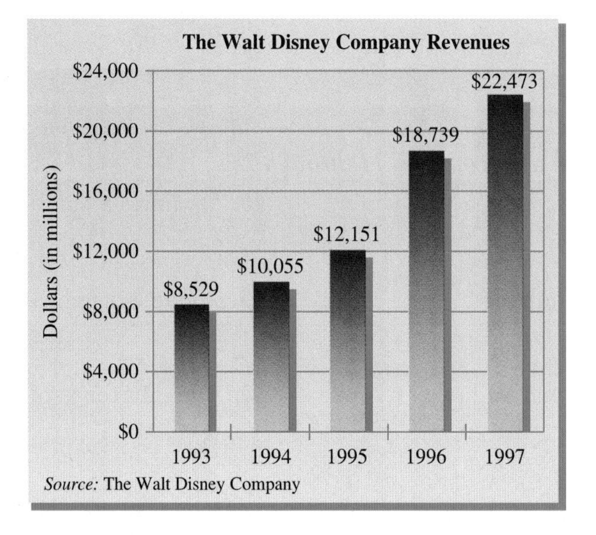

72. In your own words, explain why every natural number is also a rational number but not every rational number is a natural number.

73. In your own words, explain why every irrational number is a real number, but not every real number is an irrational number.

Write each phrase as an algebraic expression. Use the variable x to represent each unknown number.

74. Twice the sum of a number and three

75. The quotient of four and the sum of a number and one

76. The quotient of five and the difference of four and a number

77. Eight times the difference of a number and nine

1.2 PROPERTIES OF REAL NUMBERS

A WRITING SENTENCES AS EQUATIONS

When writing sentences as equations, we use the symbol $=$ to translate the phrase "**is equal to**." All of the following key words and phrases also mean equality.

> **EQUALITY**
>
> | equals | is/was | represents | is the same as |
> |---|---|---|---|
> | gives | yields | amounts to | is equal to |

Examples Write each sentence as an equation.

1. The sum of x and 5 is 20.

$$x + 5 = 20$$

2. The difference of 8 and x is the same as the product of 2 and x.

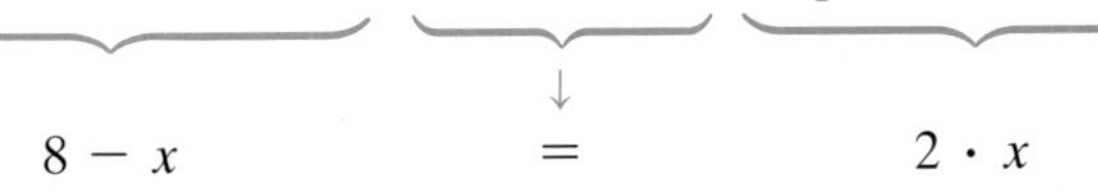

$$8 - x = 2 \cdot x$$

3. The quotient of z and 9 amounts to 9 plus z.

$$z \div 9 = 9 + z$$

or $$\frac{z}{9} = 9 + z$$

B USING INEQUALITY SYMBOLS

If we want to write in symbols that two numbers are not equal, we can use the symbol $\neq$, which means "**is not equal to.**" For example,

$$3 \neq 2$$

Graphing two numbers on a number line gives us a way to compare two numbers. For any two real numbers graphed on a number line, the number to the left is less than the number to the right. This means that the number to the right is greater than the number to the left.

The symbol $<$ means "**is less than.**" Since -4 is to the left of -1 on the number line, we write $-4 < -1$. The symbol $>$ means "**is greater than.**" Since -1 is to the right of -4 on the number line, we write $-1 > -4$.

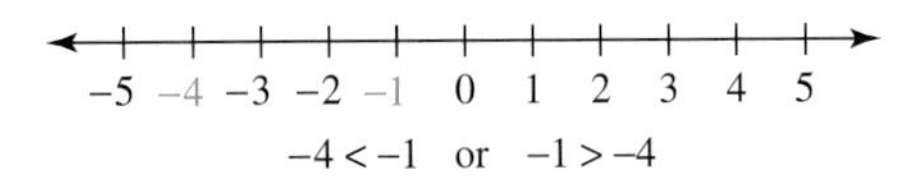

Notice that since $-4 < -1$, then we also know that $-1 > -4$. This is true for any two numbers, say, a and b.

> If $a < b$, then also $b > a$.

Objectives

A Write sentences as equations.

B Use inequality symbols.

C Find the opposite, or additive inverse, and the reciprocal, or multiplicative inverse, of a number.

D Identify and use the commutative, associative, and distributive properties.

SSM CD-ROM Video 1.2

Practice Problems 1–3

Write each sentence as an equation.

1. The difference of x and 7 is 45.
2. The product of 5 and x amounts to the sum of x and 14.
3. The quotient of y and 23 is the same as 20 subtracted from y.

Answers

1. $x - 7 = 45$, **2.** $5x = x + 14$, **3.** $\frac{y}{23} = y - 20$

Examples

Insert $<$, $>$, or $=$ between each pair of numbers to form a true statement.

4. -5 5 — -5 is to the left of 5 on a number line, so $-5 < 5$.
5. 3 -7 — 3 is to the right of -7, so $3 > -7$.
6. -16 -6 — -16 is to the left of -6, so $-16 < -6$.
7. -2.5 -2.1 — -2.5 is to the left of -2.1, so $-2.5 < -2.1$.
8. 6.36 6.360 — The true statement is $6.36 = 6.360$.
9. 4.3 -5.2 — 4.3 is to the right of -5.2, so $4.3 > -5.2$.
10. $\frac{5}{8}$ $\frac{3}{8}$ — The denominators are the same, so $\frac{5}{8} > \frac{3}{8}$ since $5 > 3$.
11. $\frac{2}{3}$ $\frac{3}{4}$ — By dividing, we see that $\frac{3}{4} = 0.75$ and $\frac{2}{3} = 0.666\ldots$. Thus $\frac{2}{3} < \frac{3}{4}$ since $0.666\ldots < 0.75$.

HELPFUL HINT

When inserting the $>$ or $<$ symbol, think of the symbols as arrowheads that "point" toward the smaller number when the statement is true.

In addition to $<$ and $>$, there are the inequality symbols $\leq$ and $\geq$. The symbol $\leq$ means **"is less than or equal to,"** and the symbol $\geq$ means **"is greater than or equal to."**

Examples

Determine whether each statement is true or false.

12. $-9 \leq 7$ — True, since $-9 < 7$ is true.
13. $-5 \leq -5$ — True, since $-5 = -5$ is true.
14. $-5 \geq -5$ — True, since $-5 = -5$ is true.
15. $-24 \geq -20$ — False, since neither $-24 > -20$ nor $-24 = -20$ is true.

C FINDING OPPOSITES AND RECIPROCALS

Of all the real numbers, two of them stand out as extraordinary: 0 and 1. Zero is the only real number that can be added to *any* real number and result in the same real number. Also, 1 is the only real number that can be multiplied by *any* real number and result in the same real number. This is why 0 is called the **additive identity** and 1 is called the **multiplicative identity**.

IDENTITY PROPERTIES

For every real number a,

Identity Property of 0: $a + 0 = 0 + a = a$

Also,

Identity Property of 1: $a \cdot 1 = 1 \cdot a = a$

Practice Problems 4–11

Insert $<$, $>$, or $=$ between each pair of numbers to form a true statement.

4. 7 -7
5. -1 11
6. -10 -12
7. -3.25 -3.025
8. 7.206 7.2060
9. 18.6 -14.2
10. $\frac{4}{7}$ $\frac{5}{7}$
11. $\frac{3}{8}$ $\frac{1}{3}$

Practice Problems 12–15

Determine whether each statement is true or false.

12. $-11 \leq 16$
13. $-7 \leq -7$
14. $-7 \geq -7$
15. $-25 \geq -30$

Answers

4. $>$, **5.** $<$, **6.** $>$, **7.** $<$, **8.** $=$, **9.** $>$, **10.** $<$, **11.** $>$, **12.** true, **13.** true, **14.** true, **15.** true

We use the identity property of 1 when we say that x, for example, means $1 \cdot x$ or $1x$. We also use this property when we write equivalent expressions. For example,

$$\underbrace{\frac{2}{3} = \frac{2}{3} \cdot 1}_{\text{identity property of 1}} = \frac{2}{3} \cdot \frac{5}{5} = \frac{10}{15} \quad \frac{5}{5} \text{ is another name for 1.}$$

Two numbers whose sum is the additive identity 0 are called **opposites** or **additive inverses** of each other. Each real number has a unique opposite.

OPPOSITES OR ADDITIVE INVERSES

If a is a real number, then the unique **opposite**, or **additive inverse**, of a is written as $-a$ and the following is true:

$$a + (-a) = 0$$

On the number line, we picture a real number and its opposite as being the same distance from 0 but on opposite sides of 0.

The opposite of 6 is -6.

6 units 6 units
−6 −5 −4 −3 −2 −1 0 1 2 3 4 5 6

The opposite of $\frac{2}{3}$ is $-\frac{2}{3}$.

$\frac{2}{3}$ unit $\frac{2}{3}$ unit
$-1 \quad -\frac{2}{3} \quad -\frac{1}{3} \quad 0 \quad \frac{1}{3} \quad \frac{2}{3} \quad 1$

The opposite of -4 is 4.

4 units 4 units
−5 −4 −3 −2 −1 0 1 2 3 4 5

We stated that the opposite or additive inverse of a number a is $-a$. This means that the opposite of -4 is $-(-4)$. But we stated above that the opposite of -4 is 4. This means that $-(-4) = 4$, and in general, we have the following property.

DOUBLE NEGATIVE PROPERTY

For every real number a, $-(-a) = a$.

Examples

Find the opposite, or additive inverse, or each number.

16. 8 The opposite of 8 is -8.

17. $-\frac{1}{5}$ The opposite of $-\frac{1}{5}$ is $-\left(-\frac{1}{5}\right)$ or $\frac{1}{5}$.

18. 0 The opposite of 0 is -0, or 0.

19. -3.5 The opposite of -3.5 is $-(-3.5)$ or 3.5.

Practice Problems 16–19

Find the opposite, or additive inverse, of each number.

16. 7

17. $\frac{2}{3}$

18. $-\frac{5}{7}$

19. -4.7

Two numbers whose product is 1 are called **reciprocals** or **multiplicative inverses** of each other. Just as each real number has a unique opposite, each nonzero real number also has a unique reciprocal.

Answers

16. -7, **17.** $-\frac{2}{3}$, **18.** $\frac{5}{7}$, **19.** 4.7

RECIPROCALS OR MULTIPLICATIVE INVERSES

If a is a nonzero real number, then its **reciprocal**, or **multiplicative inverse**, is $\frac{1}{a}$ and the following is true:

$$a \cdot \frac{1}{a} = 1$$

Practice Problems 20–22

Find the reciprocal, or multiplicative inverse, of each number.

20. 13

21. -5

22. $\frac{2}{3}$

Examples Find the reciprocal, or multiplicative inverse, of each number.

20. 11 The reciprocal of 11 is $\frac{1}{11}$.

21. -9 The reciprocal of -9 is $-\frac{1}{9}$.

22. $\frac{7}{4}$ The reciprocal of $\frac{7}{4}$ is $\frac{4}{7}$ $\left(\text{since } \frac{7}{4} \cdot \frac{4}{7} = 1\right)$.

HELPFUL HINT

The number 0 has no reciprocal. Why? There is no number that when multiplied by 0 gives a product of 1.

✓ CONCEPT CHECK

Can a number's additive inverse and multiplicative inverse ever be the same? Explain.

TRY THE CONCEPT CHECK IN THE MARGIN.

D USING THE COMMUTATIVE, ASSOCIATIVE, AND DISTRIBUTIVE PROPERTIES

In addition to these special real numbers, all real numbers have certain properties that allow us to write equivalent expressions—that is, expressions that have the same value. These properties will be especially useful in Chapter 2 when we solve equations.

The **commutative properties** state that the order in which two real numbers are added or multiplied does not affect their sum or product.

COMMUTATIVE PROPERTIES

For real numbers a and b,

Addition: $a + b = b + a$

Multiplication: $a \cdot b = b \cdot a$

For example,

$7 + 11 = 18$ and $11 + 7 = 18$ Addition

$7 \cdot 11 = 77$ and $11 \cdot 7 = 77$ Multiplication

The **associative properties** state that regrouping numbers that are added or multiplied does not affect their sum or product.

Answers

20. $\frac{1}{13}$, **21.** $-\frac{1}{5}$, **22.** $\frac{3}{2}$

✓ Concept Check: no; answers may vary

ASSOCIATIVE PROPERTIES

For real numbers a, b, and c,

Addition: $(a + b) + c = a + (b + c)$
Multiplication: $(a \cdot b) \cdot c = a \cdot (b \cdot c)$

For example,

$2 + (3 + 7) = 2 + 10 = 12$ Addition
$(2 + 3) + 7 = 5 + 7 = 12$
$2 \cdot (3 \cdot 7) = 2 \cdot 21 = 42$ Multiplication
$(2 \cdot 3) \cdot 7 = 6 \cdot 7 = 42$

Example 23 Use the commutative property of addition to write an expression equivalent to $7x + 5$.

Solution: $7x + 5 = 5 + 7x$

Example 24 Use the associative property of multiplication to write an expression equivalent to $4 \cdot (9y)$. Then simplify this equivalent expression.

Solution: $4 \cdot (9y) = (4 \cdot 9)y = 36y$

The **distributive property** states that multiplication distributes over addition.

DISTRIBUTIVE PROPERTIES

For real numbers a, b, and c,

$a(b + c) = ab + ac$

For example,

$3(6 + 2) = 3(8) = 24$
$3(6 + 2) = 3(6) + 3(2) = 18 + 6 = 24$

Examples Use the distributive property to multiply.

25. $3(2x - y) = 3 \cdot 2x - 3 \cdot y = 6x - 3y$

26. $-4(y + 5) = -4 \cdot y + (-4) \cdot 5 = -4y - 20$

27. $7x(y - 2) = 7x \cdot y - 7x \cdot 2 = 7xy - 14x$

TRY THE CONCEPT CHECK IN THE MARGIN.

Practice Problem 23

Use the commutative property of addition to write an expression equivalent to $9 + 4x$.

Practice Problem 24

Use the associative property of multiplication to write an expression equivalent to $5 \cdot (6x)$. Then simplify this equivalent expression.

Practice Problems 25–27

Use the distributive property to multiply.

25. $7(4x - y)$

26. $-8(3 + x)$

27. $5x(y - 4)$

✓ CONCEPT CHECK

Is the following true? Why or why not?

$6(2a)(3b) = 6(2a) \cdot 6(3b)$

Answers

23. $4x + 9$, **24.** $(5 \cdot 6)x = 30x$, **25.** $28x - 7y$, **26.** $-24 - 8x$, **27.** $5xy - 20x$

✓ Concept Check: no; $6(2a)(3b) = 6(6ab) = 36ab$

Focus On Study Skills

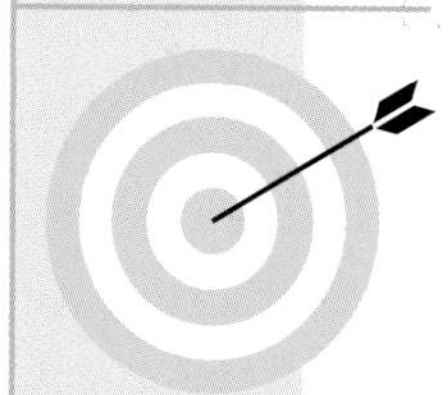

STUDY TIPS

Have you wondered what you can do to be successful in your algebra course? If so, that may well be your first step to success in algebra! Here are some tips on how to use this text and how to study mathematics in general.

USING THIS TEXT

1. Each example in the section has a parallel Practice Problem. As you read a section, try each Practice Problem after you've finished the corresponding example. This "learn-by-doing" approach will help you grasp ideas before you move on to other concepts.
2. The main section of each exercise set contains references to the objectives, (A, B, etc.) and examples in that section.
3. If you need extra help in a particular section, check at the beginning of the section to see what videotapes and software are available.
4. Integrated Reviews in each chapter offer you a chance to practice—in one place—the many concepts that you have learned separately over several sections.
5. There are many opportunities at the end of each chapter to help you understand the concepts of the chapter.
   - **Highlights** contain chapter summaries with examples.
   - **Chapter Review** contains review problems organized by section.
   - **Chapter Test** is a sample test to help you prepare for an exam.
   - **Cumulative Review** is a review consisting of material from the beginning of the book to the end of the particular chapter.

GENERAL TIPS

1. Choose to attend all class periods. If possible, sit near the front of the classroom. This way, you will see, hear, and focus on the presentation better. It may be easier for you to participate in classroom activities.
2. Do your homework. You've probably heard the phrase "practice makes perfect" in relation to music and sports. It also applies to mathematics. You will find that the more time you spend solving mathematics problems, the easier the process becomes. Be sure to block out enough time in your schedule to complete your assignments.
3. Check your work. Review the steps you made while working a problem. Learn to check your answers in the original problems. You can also compare your answers to the answers to selected exercises listed in the back of the book. If you have made a mistake, figure out what went wrong. Then correct your mistake.
4. Learn from your mistakes. Everyone, even your instructor, makes mistakes. You can use your mistakes to become a better math student. The key is finding and understanding your mistakes. Was your mistake a careless mistake? If so, you can try to work more slowly and make a conscious effort to carefully check your work. Did you make a mistake because you don't understand a concept? If so, take the time to review the concept or ask questions to better understand the concept.
5. Know how to get help if you need it. It's OK to ask for help. In fact, it's a good idea to ask for help whenever there is something that you don't understand. Make sure you know when your instructor has office hours and how to find his or her office. Find out if math tutoring services are available on your campus and check out the hours, location, and requirements of the tutoring service. You might also want to find another student in your class that you can call to discuss your assignment.

Name ______________________ Section ________ Date ________

ANSWERS

1. ____________
2. ____________
3. ____________
4. ____________
5. ____________
6. ____________
7. ____________
8. ____________
9. ____________
10. ____________
11. ____________
12. ____________
13. ____________
14. ____________
15. ____________
16. ____________
17. ____________
18. ____________
19. ____________
20. ____________
21. ____________
22. ____________
23. ____________
24. ____________

EXERCISE SET 1.2

A *Write each sentence as an equation. See Examples 1 through 3.*

1. The product of 4 and c is 7.

2. The sum of 10 and x is -12.

3. Twice x plus 5 is the same as -14.

4. The difference of y and 3 amounts to 12.

5. The quotient of n and 5 is 4 times n.

6. The quotient of 8 and y is 3 more than y.

7. The difference of z and 2 is the same as the product of z and 2.

8. Five added to twice q is the same as 4 more than q.

B *Insert $<$, $>$, or $=$ between each pair of numbers to form a true statement. See Examples 4 through 11.*

9. $0 \quad -2$

10. $-5 \quad 0$

11. $\frac{12}{3} \quad \frac{8}{2}$

12. $\frac{20}{5} \quad \frac{20}{4}$

13. $-7.9 \quad -7.09$

14. $-13.07 \quad -13.7$

15. $7.4 \quad 7.40$

16. $\frac{12}{4} \quad \frac{15}{5}$

17. $8.6 \quad -3.5$

18. $-4.7 \quad 3.8$

19. $\frac{7}{11} \quad \frac{9}{11}$

20. $\frac{9}{20} \quad \frac{3}{20}$

21. $\frac{1}{2} \quad \frac{5}{8}$

22. $\frac{3}{4} \quad \frac{7}{8}$

23. $-16 \quad -17$

24. $-14 \quad -24$

25. ____

26. ____

27. ____

28. ____

29. ____

30. ____

31. ____

32. ____

33. ____

34. ____

35. ____

36. ____

37. ____

38. ____

39. ____

40. ____

41. ____

42. ____

43. ____

44. ____

45. ____

46. ____

47. ____

48. ____

49. ____

50. ____

51. see table

52. see table

53. see table

54. see table

Name ____________________

Determine whether each statement is true or false. See Examples 12 through 15.

25. $-6 \leq 0$ **26.** $0 \leq -4$ **27.** $-3 \geq -3$ **28.** $-8 \leq -8$

29. $-14 \geq -1$ **30.** $-14 \leq -1$ **31.** $-3 \leq -3$ **32.** $-8 \geq -8$

C *Write the opposite (or additive inverse) of each number. See Examples 16 through 19.*

33. -6.2 **34.** -7.8 **35.** $\frac{4}{7}$ **36.** $\frac{9}{5}$

37. $-\frac{2}{3}$ **38.** $-\frac{14}{3}$ **39.** 0 **40.** 10.3

Write the reciprocal (or multiplicative inverse) of each number if one exists. See Examples 20 through 22.

41. 5 **42.** 9 **43.** -8 **44.** -4

45. $-\frac{1}{4}$ **46.** $\frac{1}{9}$ **47.** 0 **48.** $\frac{0}{6}$

49. $\frac{7}{8}$ **50.** $-\frac{23}{5}$

Fill in the chart. See Examples 16 through 22.

	Number	Opposite	Reciprocal
51.	5		
52.	-3		
53.	$\frac{2}{3}$		
54.	$-\frac{7}{11}$		

Name ______________________________

55. Name the only real number that has no reciprocal, and explain why this is so.

56. Name the only real number that is its own opposite, and explain why this is so.

D *Use a commutative property to write an equivalent expression. See Example 23.*

57. $7x + y$

58. $3a + 2b$

59. $z \cdot w$

60. $r \cdot s$

61. $\frac{1}{3} \cdot \frac{x}{5}$

62. $\frac{x}{2} \cdot \frac{9}{10}$

Use an associative property to write an equivalent expression. See Example 24.

63. $5 \cdot (7x)$

64. $3 \cdot (10z)$

65. $(x + 1.2) + y$

66. $5q + (2r + s)$

67. $(14z) \cdot y$

68. $(9.2x) \cdot y$

Use the distributive property to multiply. See Examples 25 through 27.

69. $3(x + 5)$

70. $7(y + 2)$

71. $8(2a + b)$

72. $9(c + 7d)$

73. $2(6x + 5y + 2z)$

74. $5(3a + b + 9c)$

75. $4(z - 6)$

76. $2(7 - y)$

77. $6x(y - 4)$

78. $11y(z - 2)$

Complete each statement to illustrate the given property.

79. $3x + 6 =$ ______ Commutative property of addition

80. $8 + 0 =$ ______ Additive identity property

81. $\frac{2}{3} + \left(-\frac{2}{3}\right) =$ ______ Additive inverse property

82. $4(x + 3) =$ ______ Distributive property

55. ______
56. ______
57. ______
58. ______
59. ______
60. ______
61. ______
62. ______
63. ______
64. ______
65. ______
66. ______
67. ______
68. ______
69. ______
70. ______
71. ______
72. ______
73. ______
74. ______
75. ______
76. ______
77. ______
78. ______
79. ______
80. ______
81. ______
82. ______

83. ______

84. ______

85. ______

86. ______

87. ______

88. ______

89. ______

90. ______

91. ______

92. ______

93. ______

94. ______

Name ______

83. $7 \cdot 1 =$ ______ Multiplicative identity property

84. $0 \cdot (-5.4) =$ ______ Multiplication property of zero

85. $10(2y) =$ ______ Associative property

86. $9y + (x + 3z) =$ ______ Associative property

Combining Concepts

In each statement, a property of real numbers has been incorrectly applied. Correct the right-hand side of each statement.

87. $3(x + 4) = 3x + 4$

88. $5(7y) = (5 \cdot 7)(5 \cdot y)$

89. $4 + 8y = 4y + 8$

90. Is subtraction commutative? Explain why or why not?

91. Is division commutative? Explain why or why not.

92. Evaluate $12 - (5 - 3)$ and $(12 - 5) - 3$. Use these two expressions and discuss whether subtraction is associative.

93. Evaluate $24 \div (6 \div 3)$ and $(24 \div 6) \div 3$. Use these two expressions and discuss whether division is associative.

94. To demonstrate the distributive property geometrically, represent the area of the larger rectangle in two ways: First as length a times width $b + c$, and second as the sum of the areas of the smaller rectangles.

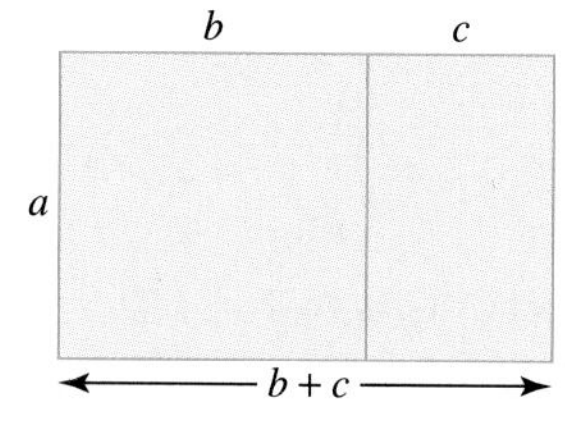

1.3 OPERATIONS ON REAL NUMBERS

A FINDING THE ABSOLUTE VALUE OF A NUMBER

In Section 1.2, we used the number line to compare two real numbers. The number line can also be used to visualize distance, which leads to the concept of absolute value. The **absolute value** of a number is the distance between the number and 0 on the number line. The symbol for absolute value is $| \ |$. For example, since -4 and 4 are both 4 units from 0 on the number line, each has an absolute value of 4.

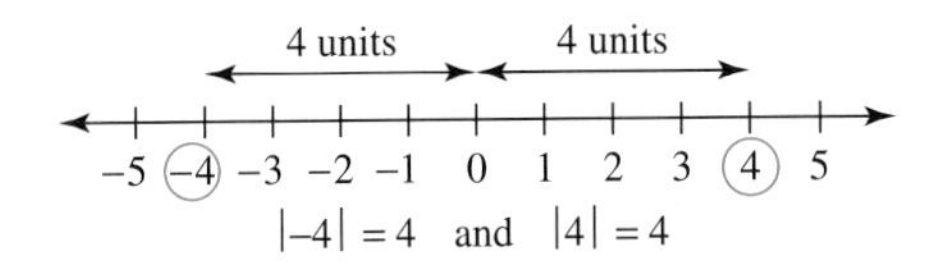

An equivalent definition of the absolute value of a real number a is given next.

ABSOLUTE VALUE

The absolute value of a, written as $|a|$, is

$$|a| = \begin{cases} a \text{ if } a \text{ is 0 or a positive number} \\ -a \text{ if } a \text{ is a negative number} \end{cases}$$

↑ the opposite of

Objectives

A Find the absolute value of a number.

B Add and subtract real numbers.

C Multiply and divide real numbers.

D Simplify expressions containing exponents.

E Find roots of numbers.

SSM

CD-ROM

Video 1.3

Examples Find each absolute value.

1. $|3| = 3$
2. $|0| = 0$
3. $|-4| = -(-4) = 4$ (↑ the opposite of)
4. $-|2| = -2$
5. $-|-8| = -8$ Since $|-8|$ is 8, $-|-8| = -8$.

Practice Problems 1–5

Find each absolute value.

1. $|7|$
2. $|-1|$
3. $|-9|$
4. $-|5|$
5. $-|-3|$

HELPFUL HINT

Since distance is always positive or zero, the absolute value of a number is always positive or zero.

TRY THE CONCEPT CHECK IN THE MARGIN.

✓ CONCEPT CHECK

Explain how you know that $|14| = -14$ is a false statement.

B ADDING AND SUBTRACTING REAL NUMBERS

When solving problems, we often need to add real numbers. For example, if the New Orleans Saints lose 5 yards in one play, then lose another 7 yards in the next play, their total loss may be described by $-5 + (-7)$.

The addition of two real numbers may be summarized by the following.

Answers

1. 7, **2.** 1, **3.** 9, **4.** -5, **5.** -3

✓ Concept Check: $|14| = 14$ since the absolute value of a number is the distance between the number and zero and distance cannot be negative.

ADDING REAL NUMBERS

1. To add two numbers with the *same* sign, add their absolute values and attach their common sign.
2. To add two numbers with *different* signs, subtract the smaller absolute value from the larger absolute value and attach the sign of the number with the larger absolute value.

For example, to add $-5 + (-7)$, we first add their absolute values.

$$|-5| = 5, \quad |-7| = 7, \quad \text{and} \quad 5 + 7 = 12$$

Next, we attach their common negative sign.

$$-5 + (-7) = -12$$

(This represents a total loss of 12 yards for the New Orleans Saints in the example above.)

To find $(-4) + 3$, we first subtract their absolute values.

$$|-4| = 4, \quad |3| = 3, \quad \text{and} \quad 4 - 3 = 1$$

Next, we attach the sign of the number with the larger absolute value.

$$(-4) + 3 = -1$$

Practice Problems 6–11

Add.

6. $-7 + (-10)$
7. $8 + (-12)$
8. $-14 + 20$
9. $-4.6 + (-1.9)$
10. $-\frac{2}{3} + \frac{1}{6}$
11. $-\frac{1}{7} + \frac{1}{2}$

Examples Add.

6. $-3 + (-11) = -(3 + 11) = -14$ — Add their absolute values, or $3 + 11 = 14$. Then attach the common negative sign.

7. $3 + (-7) = -4$ — Subtract their absolute values, or $7 - 3 = 4$. Since -7 has the larger absolute value, the answer is -4.

8. $-10 + 15 = 5$

9. $-8.3 + (-1.9) = -10.2$

10. $-\frac{1}{4} + \frac{1}{2} = -\frac{1}{4} + \frac{2}{4} = \frac{1}{4}$

11. $-\frac{2}{3} + \frac{3}{7} = -\frac{14}{21} + \frac{9}{21} = -\frac{5}{21}$

Subtraction of two real numbers may be defined in terms of addition.

SUBTRACTING REAL NUMBERS

If a and b are real numbers, then the difference of a and b, written $a - b$, is defined by

$$a - b = a + (-b)$$

In other words, to subtract a second real number from a first, we add the first number and the opposite of the second number.

Practice Problems 12–15

Subtract.

12. $7 - 14$
13. $-10 - (-2)$
14. $13.3 - (-8.9)$
15. $-\frac{1}{3} - \frac{1}{2}$

Examples Subtract.

12. $2 - 8 = 2 + (-8) = -6$ (add the opposite)

13. $-8 - (-1) = -8 + (1) = -7$ (add the opposite)

14. $10.7 - (-9.8) = 10.7 + 9.8 = 20.5$

Answers

6. -17, **7.** -4, **8.** 6, **9.** -6.5, **10.** $-\frac{1}{2}$, **11.** $\frac{5}{14}$, **12.** -7, **13.** -8, **14.** 22.2, **15.** $-\frac{5}{6}$

15. $-\frac{2}{3} - \frac{1}{4} = -\frac{2}{3} + \left(-\frac{1}{4}\right) =$

$$-\frac{2 \cdot 4}{3 \cdot 4} + \left(-\frac{1 \cdot 3}{4 \cdot 3}\right) = -\frac{8}{12} + \left(-\frac{3}{12}\right) = -\frac{11}{12}$$

To add or subtract three or more real numbers, we add or subtract from left to right.

Examples Simplify each expression.

16. $11 + 2 - 7 = 13 - 7 = 13 + (-7) = 6$

17. $-5 - 4 + 2 = -5 + (-4) + 2 = -9 + 2 = -7$

Practice Problems 16–17

Simplify each expression.

16. $18 + 3 - 4$

17. $-3 - 11 + 7$

C MULTIPLYING AND DIVIDING REAL NUMBERS

To discover sign patterns when you multiply real numbers, recall that multiplication by a positive integer is the same as repeated addition. For example,

$$3(2) = 2 + 2 + 2 = 6$$
$$3(-2) = (-2) + (-2) + (-2) = -6$$

Notice here that $3(-2) = -6$. This illustrates that the product of two numbers with different signs is negative. We summarize sign patterns for multiplying any two real numbers as follows.

MULTIPLYING TWO REAL NUMBERS

1. The product of two numbers with the *same* sign is positive.
2. The product of two numbers with *different* signs is negative.

Also recall that the product of zero and any real number is zero.

$$0 \cdot a = 0$$

Examples Multiply.

18. $-8(-1) = 8$

19. $2\left(-\frac{1}{6}\right) = \frac{2}{1} \cdot \left(-\frac{1}{6}\right) = -\frac{2}{6} \text{ or } -\frac{1}{3}$

20. $-1.2(0.3) = -0.36$

21. $7(-6) = -42$

22. $-\frac{1}{3}\left(-\frac{1}{2}\right) = \frac{1}{6}$

23. $(-4.6)(-2.5) = 11.5$

24. $0(-6) = 0$

Practice Problems 18–24

Multiply.

18. $-4(-2)$

19. $5\left(-\frac{1}{10}\right)$

20. $-3.2(0.1)$

21. $8(-6)$

22. $-\frac{2}{5}\left(-\frac{1}{3}\right)$

23. $(-1.3)(-1.5)$

24. $0(-10)$

Recall that $\frac{8}{4} = 2$ because $2 \cdot 4 = 8$. Likewise, $\frac{8}{-4} = -2$ because $(-2)(-4) = 8$. Also, $\frac{-8}{4} = -2$ because $(-2)4 = -8$, and $\frac{-8}{-4} = 2$ because $2(-4) = -8$. From these examples, we can see that the sign patterns for division are the same as for multiplication.

DIVIDING TWO REAL NUMBERS

1. The quotient of two numbers with the *same* sign is positive.
2. The quotient of two numbers with *different* signs is negative.

Answers

16. 17, **17.** -7, **18.** 8, **19.** $-\frac{1}{2}$, **20.** -0.32, **21.** -48, **22.** $\frac{2}{15}$, **23.** 1.95, **24.** 0

Notice from the above reasoning that we cannot divide by 0. Why? If $\frac{5}{0}$ did exist, it would equal a number such that the number times 0 would equal 5. There is no such number, so we cannot define division by 0. We say, for example, that $\frac{5}{0}$ is undefined.

Examples

Divide.

25. $\frac{20}{-4} = -5$

26. $\frac{-9}{-3} = 3$

27. $\frac{-40}{10} = -4$

28. $\frac{-8}{0}$ is undefined

29. $\frac{0}{-8} = 0$

30. $\frac{-10}{-80} = 0.125$

Practice Problems 25–30

Divide.

25. $\frac{45}{-9}$

26. $\frac{-16}{-4}$

27. $\frac{25}{-5}$

28. $\frac{-3}{0}$

29. $\frac{0}{-3}$

30. $\frac{-1}{-4}$

With sign rules for division, we can understand why the positioning of the negative sign in a fraction does not change the value of the fraction. For example,

$$\frac{-12}{3} = -4, \quad \frac{12}{-3} = -4, \quad \text{and} \quad -\frac{12}{3} = -4$$

Since all these fractions equal -4, we can say that

$$\frac{-12}{3} = \frac{12}{-3} = -\frac{12}{3}$$

In general, the following holds true:

> If a and b are real numbers and $b \neq 0$, then
>
> $$\frac{a}{-b} = \frac{-a}{b} = -\frac{a}{b}$$

Also recall that division by a nonzero real number b is the same as multiplication by its reciprocal $\frac{1}{b}$. In other words,

$$a \div b = a \cdot \frac{1}{b}$$

Examples

Divide.

31. $-\frac{1}{10} \div \left(-\frac{2}{5}\right) = -\frac{1}{10} \cdot \left(-\frac{5}{2}\right) = \frac{5}{20}$ or $\frac{1}{4}$

32. $-\frac{1}{4} \div \frac{3}{7} = -\frac{1}{4} \cdot \frac{7}{3} = -\frac{7}{12}$

Practice Problems 31–32

Divide.

31. $-\frac{3}{4} \div \left(-\frac{3}{8}\right)$ 32. $-\frac{1}{11} \div \frac{2}{7}$

Answers

25. -5, **26.** 4, **27.** -5, **28.** undefined, **29.** 0, **30.** 0.25, **31.** 2, **32.** $-\frac{7}{22}$

D SIMPLIFYING EXPRESSIONS CONTAINING EXPONENTS

Recall that when two numbers are multiplied, they are called **factors**. For example, in $3 \cdot 5 = 15$, the 3 and 5 are called factors.

A natural number *exponent* is a shorthand notation for repeated multiplication of the same factor. This repeated factor is called the **base**, and the number of times it is used as a factor is indicated by the **exponent**. For example,

$$4^3 = \underbrace{4 \cdot 4 \cdot 4}_{\text{4 is a factor 3 times}} = 64$$

(exponent: 3; base: 4)

EXPONENTS

If a is a real number and n is a natural number, then the ***n*th power of *a***, or ***a* raised to the *n*th power**, written as a^n, is the product of n factors, each of which is a.

$$a^n = \underbrace{a \cdot a \cdot a \cdot a \cdot \ldots \cdot a}_{a \text{ is a factor } n \text{ times}}$$

(exponent: n; base: a)

It is not necessary to write an exponent of 1. For example, 3 is assumed to be 3^1.

Examples

Find the value of each expression.

33. $3^2 = 3 \cdot 3 = 9$

34. $-5^2 = -(5 \cdot 5) = -25$

35. $-5^3 = -(5 \cdot 5 \cdot 5) = -125$

36. $\left(\frac{1}{2}\right)^4 = \left(\frac{1}{2}\right)\left(\frac{1}{2}\right)\left(\frac{1}{2}\right)\left(\frac{1}{2}\right) = \frac{1}{16}$

37. $(-5)^2 = (-5)(-5) = 25$

38. $(-5)^3 = (-5)(-5)(-5) = -125$

TRY THE CONCEPT CHECK IN THE MARGIN.

HELPFUL HINT

Be very careful when finding the value of expressions such as -5^2 and $(-5)^2$.

$$-5^2 = -(5 \cdot 5) = -25 \text{ and } (-5)^2 = (-5)(-5) = 25$$

Without parentheses, the base to square is 5, not -5.

Practice Problems 33–38

Simplify each expression.

33. 4^2
34. -2^2
35. -2^3
36. $\left(\frac{1}{3}\right)^4$
37. $(-2)^2$
38. $(-2)^3$

✓ CONCEPT CHECK

When $(-8.2)^7$ is evaluated, will the value be positive or negative? How can you tell without making any calculations?

Answers

33. 16, **34.** -4, **35.** -8, **36.** $\frac{1}{81}$, **37.** 4, **38.** -8

✓ Concept Check: negative; the exponent is an odd number

E FINDING THE ROOT OF A NUMBER

The opposite of squaring a number is taking the **square root** of a number. For example, since the square of 4, or 4^2, is 16, we say that a square root of 16 is 4. The notation $\sqrt{a}$ is used to denote the **positive**, or **principal**, **square root** of a nonnegative number a. We then have in symbols that

$$\sqrt{16} = 4$$

Practice Problems 39–41

Find each root.

39. $\sqrt{36}$ 40. $\sqrt{4}$ 41. $\sqrt{\frac{1}{9}}$

Examples

Find each root.

39. $\sqrt{9} = 3$, since 3 is positive and $3^2 = 9$.
40. $\sqrt{25} = 5$, since $5^2 = 25$.
41. $\sqrt{\frac{1}{4}} = \frac{1}{2}$, since $\left(\frac{1}{2}\right)^2 = \frac{1}{4}$.

We can find roots other than square roots. Since 2 cubed, written as 2^3, is 8, we say that the **cube root** of 8 is 2. This is written as

$$\sqrt[3]{8} = 2$$

Also, since $3^4 = 81$ and 3 is positive,

$$\sqrt[4]{81} = 3$$

Practice Problems 42–44

Find each root.

42. $\sqrt[3]{8}$

43. $\sqrt[4]{1}$

44. $\sqrt[5]{32}$

Examples

Find each root.

42. $\sqrt[3]{27} = 3$, since $3^3 = 27$.
43. $\sqrt[5]{1} = 1$, since $1^5 = 1$.
44. $\sqrt[4]{16} = 2$, since 2 is positive and $2^4 = 16$.

Of course, as mentioned in Section 1.1, not all roots simplify to rational numbers. We study radicals further in Chapter 7.

Answers

39. 6, **40.** 2, **41.** $\frac{1}{3}$, **42.** 2, **43.** 1, **44.** 2

Name ______________________ Section ________ Date ________

EXERCISE SET 1.3

A *Find each absolute value. See Examples 1 through 5.*

1. $|2|$ **2.** $|8|$ **3.** $|-4|$ **4.** $|-6|$

5. $|0|$ **6.** $|-1|$ **7.** $-|3|$ **8.** $-|11|$

9. $-|-2|$ **10.** $-|-14|$

B *Add or subtract as indicated. See Examples 6 through 17.*

11. $-3 + 8$ **12.** $-5 + (-9)$ **13.** $-14 + (-10)$

14. $12 + (-7)$ **15.** $-4.3 - 6.7$ **16.** $-8.2 - (-6.6)$

17. $13 - 17$ **18.** $15 - (-1)$ **19.** $\frac{11}{15} - \left(-\frac{3}{5}\right)$

20. $\frac{7}{10} - \frac{4}{5}$ **21.** $19 - 10 - 11$ **22.** $-13 - 4 + 9$

23. $-14 - 7$ **24.** $-6 - 31$ **25.** $-\frac{4}{5} - \left(-\frac{3}{10}\right)$

26. $-\frac{5}{2} - \left(-\frac{2}{3}\right)$ **27.** Subtract 14 from 8 **28.** Subtract 9 from -3

29. $-4 + 7$ **30.** $-9 + 15$ **31.** $-9 + (-3)$

32. $-17 + (-2)$ **33.** $-4 - (-19)$ **34.** $-5 - (-17)$

ANSWERS

1. ______
2. ______
3. ______
4. ______
5. ______
6. ______
7. ______
8. ______
9. ______
10. ______
11. ______
12. ______
13. ______
14. ______
15. ______
16. ______
17. ______
18. ______
19. ______
20. ______
21. ______
22. ______
23. ______
24. ______
25. ______
26. ______
27. ______
28. ______
29. ______
30. ______
31. ______
32. ______
33. ______
34. ______

35.
36.
37.
38.
39.
40.
41.
42.
43.
44.
45.
46.
47.
48.
49.
50.
51.
52.
53.
54.
55.
56.
57.
58.
59.
60.
61.
62.
63.
64.
65.
66.
67.
68.
69.
70.
71.
72.
73.

Name ______________________

35. $6.3 - 18.5$

36. $15.9 - 21.7$

37. $16 - 8 - 9$

38. $-14 - 3 + 6$

39. $-5 + (-7) - 10$

40. $-8 + (-10) - 6$

C *Multiply or divide as indicated. See Examples 18 through 32.*

41. $-5 \cdot 12$

42. $-3 \cdot 8$

43. $6(-3)$

44. $5(-4)$

45. $-8(-10)$

46. $-4(-11)$

47. $-9 \cdot 8$

48. $-6 \cdot 7$

49. $-17 \cdot 0$

50. $-5 \cdot 0$

51. $0(-1)$

52. $0(-34)$

53. $\frac{-9}{3}$

54. $\frac{-20}{5}$

55. $\frac{16}{-2}$

56. $\frac{35}{-7}$

57. $\frac{-12}{-4}$

58. $\frac{-36}{-6}$

59. $-18 \div 6$

60. $-42 \div 6$

61. $\frac{0}{-5}$

62. $\frac{0}{-11}$

63. $\frac{-18}{0}$

64. $\frac{-22}{0}$

65. $-4(-2)(-1)$

66. $-5(-3)(-2)$

67. $-7(-1)(5)$

68. $-6(2)(-3)$

69. $\frac{-6}{7} \div 2$

70. $\frac{-9}{13} \div (-3)$

71. $-\frac{2}{7} \cdot \left(-\frac{1}{6}\right)$

72. $\frac{5}{9} \cdot \left(-\frac{3}{5}\right)$

73. $-\frac{1}{6} \div \frac{9}{10}$

Name ______________________________

74. $\frac{4}{7} \div \left(-\frac{1}{8}\right)$

75. $-\frac{2}{3} \cdot \left(\frac{6}{4}\right)$

76. $\frac{5}{6} \cdot \left(\frac{-12}{15}\right)$

77. $-2(-3.6)$

78. $-5(-4.2)$

79. $-6(-5)(0)$

80. $4(-3)(0)$

81. $\frac{-5.2}{-1.3}$

82. $\frac{-7}{-1.4}$

83. $-25 \div (-5)$

84. $-88 \div (-11)$

85. $\frac{3}{5} \div \left(-\frac{2}{5}\right)$

86. $\frac{2}{7} \div \left(-\frac{1}{14}\right)$

87. $9.1 \div (-1.3)$

88. $22.5 \div (-2.5)$

D *Find the value of each expression. See Examples 33 through 38.*

89. -7^2

90. $(-7)^2$

91. $(-6)^2$

92. -6^2

93. $(-2)^3$

94. -2^3

95. $\left(-\frac{1}{3}\right)^3$

96. $\left(-\frac{1}{2}\right)^4$

97. Explain why -3^2 and $(-3)^2$ simplify to different numbers.

98. Explain why -3^3 and $(-3)^3$ simplify to the same number.

E *Find each root. See Examples 39 through 44.*

99. $\sqrt{49}$

100. $\sqrt{81}$

101. $\sqrt{64}$

102. $\sqrt{100}$

103. $\sqrt{\frac{1}{9}}$

104. $\sqrt{\frac{1}{25}}$

105. $\sqrt{\frac{1}{16}}$

106. $\sqrt{\frac{1}{49}}$

107. $\sqrt[3]{64}$

108. $\sqrt[5]{32}$

109. $\sqrt[4]{81}$

110. $\sqrt[3]{1}$

111. $\sqrt[3]{8}$

112. $\sqrt[3]{125}$

74. ______
75. ______
76. ______
77. ______
78. ______
79. ______
80. ______
81. ______
82. ______
83. ______
84. ______
85. ______
86. ______
87. ______
88. ______
89. ______
90. ______
91. ______
92. ______
93. ______
94. ______
95. ______
96. ______
97. ______
98. ______
99. ______
100. ______
101. ______
102. ______
103. ______
104. ______
105. ______
106. ______
107. ______
108. ______
109. ______
110. ______
111. ______
112. ______

113. ______

114. ______

115. ______

116. ______

117. ______

118. ______

119. ______

120. ______

Name ______

Combining Concepts

Each circle below represents a whole, or 1. Determine the unknown fractional part of each circle.

113.

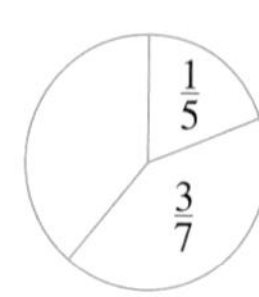

114.

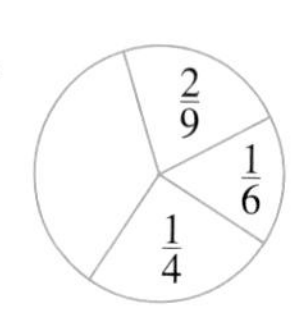

115. Most of Mount Kea, a volcano on Hawaii, lies below sea level. If this volcano begins at 5998 meters below sea level and then rises 10,203 meters, find the height of the volcano above sea level.

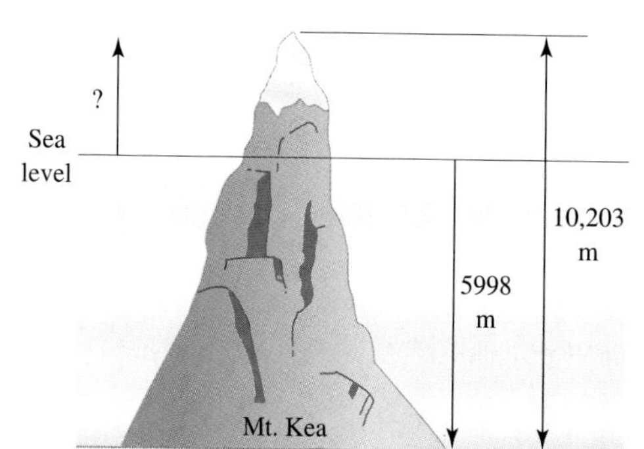

116. The highest point on land on Earth is the top of Mt. Everest, in the Himalayas, at an elevation of 29,028 feet above sea level. The lowest point on land is the Dead Sea, between Israel and Jordan, at 1312 feet below sea level. Find the difference in elevations. (*Source:* National Geographic Society)

A fair game is one in which each team or player has the same chance of winning. Suppose that a game consists of three players taking turns spinning a spinner. If the spinner lands on yellow, player 1 gets a point. If the spinner lands on red, player 2 gets a point, and if the spinner lands on blue, player 3 gets a point. After 12 spins, the player with the most points wins.

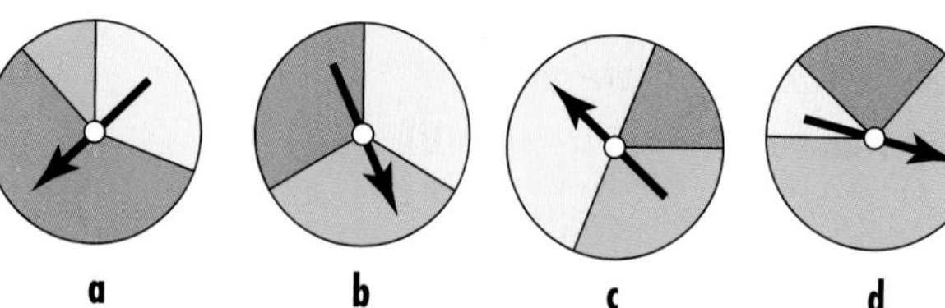

117. Which spinner would lead to a fair game?

118. If you are player 2 and want to win the game, which spinner would you choose?

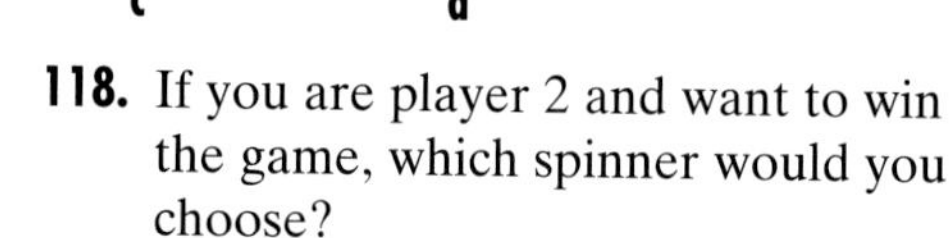

119. If you are player 1 and want to lose the game, which spinner would you choose?

120. Is it possible for the game to end in a three-way tie? If so, list the possible ending scores.

Name ____

121. Is it possible for the game to end in a two-way tie? If so, list the possible ending scores.

Use a calculator to approximate each square root. Round to four decimal places.

122. $\sqrt{10}$ **123.** $\sqrt{273}$ **124.** $\sqrt{7.9}$ **125.** $\sqrt{19.6}$

Investment firms often advertise their gains and losses in the form of bar graphs such as the one that follows. This graph shows investment risk over time for the S&P 500 Index by showing average annual compound returns for 1 year, 5 years, 15 years, and 25 years. For example, after 1 year, the annual compound return in percent for an investor is anywhere from a gain of 181.5% to a loss of 64%. Use this graph to answer Exercises 126–130.

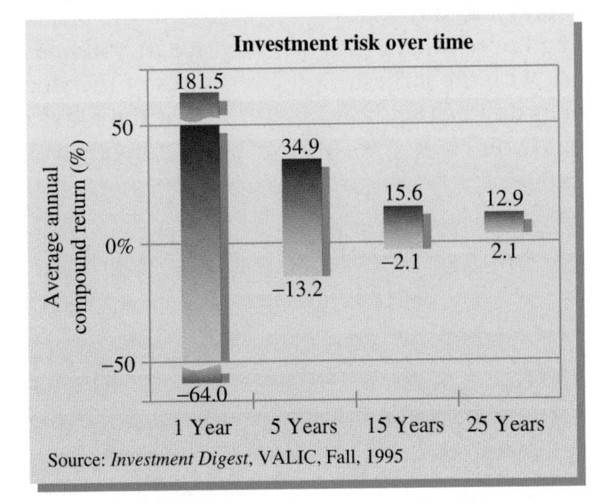

Source: *Investment Digest*, VALIC, Fall, 1995

126. A person investing in the S&P 500 Index may expect at most an average annual gain of what percent after 15 years?

127. A person investing in the S&P 500 Index may expect to lose at most an average per year of what percent after 5 years?

128. Find the difference in percent of the highest average annual return and the lowest average annual return after 15 years.

129. Find the difference in percent of the highest average annual return and the lowest average annual return after 25 years.

130. Do you think that the type of investment shown in the figure is recommended for short-term investments or long-term investments? Explain your answer.

121. ____________

122. ____________

123. ____________

124. ____________

125. ____________

126. ____________

127. ____________

128. ____________

129. ____________

130. ____________

131. ____________

Name ____

Internet Excursions

Go to http://www.prenhall.com/martin-gay
Publicly held corporations sell shares of their company's stock on a stock exchange such as the New York Stock Exchange (NYSE). Many sites on the World Wide Web allow you to track stock prices. By going to the World Wide Web site listed above, you will gain access to the CNN Financial Network Web site, or a related site. You will be able to find current information about the activity on the various stock markets in the United States.

131. Record the date and time of your visit to this web site. Submit a request for a listing of the NYSE gainers. Make a list of the five stocks with the largest gains. Which stock made the largest gain? The gain listed is from the previous day's closing price. If you had purchased 275 shares of the stock with the largest gain at the previous day's close, how much money would you have gained?

132. Record the date and time of your visit to this web site. Submit a request for a listing of the NYSE losers. Make a list of the five stocks with the biggest losses. Which stock had the largest loss? The loss listed is from the previous day's closing price. If you had purchased 425 shares of the stock with the largest loss at the previous day's close, how much money would you have lost?

132. ____________

Name ______________________ Section ________ Date ________

ANSWERS

1. ________
2. ________
3. ________
4. ________
5. ________
6. ________
7. ________
8. ________
9. ________
10. ________
11. ________
12. ________
13. ________
14. ________
15. ________
16. ________
17. ________
18. ________
19. ________
20. ________
21. ________
22. ________

INTEGRATED REVIEW—ALGEBRAIC EXPRESSIONS AND REAL NUMBERS

Evaluate each algebraic expression at the given replacement values.

1. $3x$ when $x = 1.7$

2. $6x - y$ when $x = 9$ and $y = 2$

Write each set by listing its elements.

3. $\{x|x$ is a natural number less than $4\}$

4. $\{x|x$ is an odd whole number less than $6\}$

5. $\{x|x$ is an even natural number greater than $7\}$

6. $\{x|x$ is a whole number between 10 and $15\}$

Insert $<$, $>$, or $=$ between each pair of numbers to form a true statement.

7. $-4 \quad -6$

8. $8.6 \quad 8.600$

9. $\frac{9}{10} \quad \frac{11}{10}$

10. $-6.1 \quad -6.01$

Write each sentence as an equation.

11. The product of 5 and x is the same as 20.

12. The sum of a and 12 amounts to 14.

13. The quotient of y and 10 is the same as the product of y and 10.

14. The sum of x and 1 equals the difference of x and 1.

Perform the indicated operations.

15. $-4 + 7$

16. $-11 + 20$

17. $-4(7)$

18. $-11(20)$

19. $-8 - (-13)$

20. $-12 - 16$

21. $\frac{-20}{-4}$

22. $\frac{-18}{6}$

23. ______

24. ______

25. ______

26. ______

27. ______

28. ______

29. ______

30. ______

31. ______

32. ______

33. ______

34. ______

35. ______

36. ______

37. ______

38. ______

Name ______

23. -5^2

24. $(-5)^2$

25. $-6 - 1 + 20$

26. $18 - 4 - 19$

27. $\frac{0}{-3}$

28. $\frac{5}{0}$

29. $-4(3)(2)$

30. $-5(-1)(6)$

31. $-\frac{1}{2} \cdot \frac{6}{7}$

32. $\frac{4}{5} \cdot \left(-\frac{1}{8}\right)$

33. $\frac{3}{10} - \frac{4}{5}$

34. $-\frac{2}{3} - \frac{1}{4}$

35. $\frac{1.6}{-0.2}$

36. $\frac{-4.8}{16}$

37. $6.7 - (-1.3)$

38. $-4.6 + 9$

1.4 ORDER OF OPERATIONS AND ALGEBRAIC EXPRESSIONS

Objectives

A Use the order of operations.
B Identify and evaluate algebraic expressions.
C Identify like terms and simplify algebraic expressions.

SSM CD-ROM Video 1.4

A USING ORDER OF OPERATIONS

The expression $3 + 2 \cdot 10$ represents the total number of disks shown.

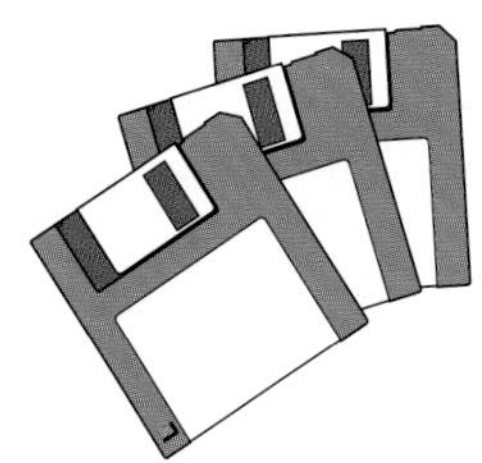

Expressions containing more than one operation are written to follow a particular agreed-upon **order of operations**. For example, when we write $3 + 2 \cdot 10$, we mean to multiply first, and then add.

> ORDER OF OPERATIONS
>
> Simplify expressions using the following order. If grouping symbols such as parentheses are present, simplify expressions within those first, starting with the innermost set. If fraction bars are present, simplify the numerator and the denominator separately.
>
> 1. Evaluate exponential expressions.
> 2. Perform multiplications or divisions in order from left to right.
> 3. Perform additions or subtractions in order from left to right.

Example 1 Simplify: $3 + 2 \cdot 10$

Solution: First we multiply; then we add.

$$3 + 2 \cdot 10 = 3 + 20 = 23$$

Practice Problem 1

Simplify: $15 - 2 \cdot 5$

Example 2 Simplify: $2(1 - 4)^2$

Solution:

$2(1 - 4)^2 = 2(-3)^2$ Simplify inside parentheses first.

$= 2(9)$ Write $(-3)^2$ as 9.

$= 18$ Multiply.

Practice Problem 2

Simplify: $5(2 - 6)^2$

Example 3 Simplify: $\dfrac{|-2|^3 + 1}{-7 - \sqrt{4}}$

Solution: We simplify the numerator and the denominator separately. Then we divide.

$\dfrac{|-2|^3 + 1}{-7 - \sqrt{4}} = \dfrac{2^3 + 1}{-7 - 2}$ Write $|-2|$ as 2 and $\sqrt{4}$ as 2.

$= \dfrac{8 + 1}{-9}$ Write 2^3 as 8.

$= \dfrac{9}{-9} = -1$ Simplify the numerator, then divide.

Practice Problem 3

Simplify: $\dfrac{|-3|^2 + 5}{\sqrt{9} - 10}$

Answers

1. 5, **2.** 80, **3.** −2

Example 4

Simplify: $\frac{(6+2)-(-4)}{2-(-3)}$

Solution:

$$\frac{(6+2)-(-4)}{2-(-3)} = \frac{8-(-4)}{2-(-3)}$$ Simplify inside parentheses first.

$$= \frac{8+4}{2+3}$$ Write subtraction as equivalent addition.

$$= \frac{12}{5}$$ Add in both the numerator and the denominator.

Practice Problem 4

Simplify: $\frac{(8-3)-(-6)}{4-(-1)}$

Besides parentheses, other symbols used for grouping expressions are brackets [ ] and braces { }. These other grouping symbols are commonly used when we group expressions that already contain parentheses.

Example 5

Simplify: $3-[(4-6)+2(5-9)]$

Solution:

$$3-[(4-6)+2(5-9)] = 3-[-2+2(-4)]$$ Simplify within the innermost sets of parentheses.

$$= 3-[-2+(-8)]$$

$$= 3-[-10]$$

$$= 13$$

Practice Problem 5

Simplify:
$7-[2(1-3)+5(10-12)]$

HELPFUL HINT

When grouping symbols occur within grouping symbols remember to perform operations on the innermost set first.

TRY THE CONCEPT CHECK IN THE MARGIN.

✓ CONCEPT CHECK

True or false? If two different people use the order of operations to simplify a numerical expression and neither makes a calculation error, it is not possible that they each obtain a different result. Explain.

B EVALUATING ALGEBRAIC EXPRESSIONS

Recall from Section 1.1 that an algebraic expression is formed by numbers and variables connected by the operations of addition, subtraction, multiplication, division, raising to powers, or taking roots. Also, if numbers are substituted for the variables in an algebraic expression and the operations performed, the result is called the value of the expression for the given replacement values. This entire process is called evaluating an expression.

Example 6 Converting Degrees Fahrenheit to Degrees Celsius

The algebraic expression $\frac{5(x-32)}{9}$ represents the equivalent temperature in degrees Celsius when x is degrees Fahrenheit. Complete the following table by evaluating this expression at given values of x.

Degrees Fahrenheit	x	−4	10	32
Degrees Celsius	$\frac{5(x-32)}{9}$			

Practice Problem 6

Use the algebraic expression given in Example 6 to complete the following table.

Degrees Fahrenheit	x	−13	0	41
Degrees Celsius	$\frac{5(x-32)}{9}$			

Answers

4. $\frac{11}{5}$, **5.** 21, **6.** $-25; -\frac{160}{9}$ or $-17\frac{7}{9}$; 5

✓ Concept Check: true; answers may vary

Solution: To complete the table, we evaluate $\frac{5(x-32)}{9}$ at each given replacement value.

When $x = -4$,

$$\frac{5(x-32)}{9} = \frac{5(-4-32)}{9} = \frac{5(-36)}{9} = -20$$

When $x = 10$,

$$\frac{5(x-32)}{9} = \frac{5(10-32)}{9} = \frac{5(-22)}{9} = \frac{-110}{9} \text{ or } -12\frac{2}{9}$$

When $x = 32$,

$$\frac{5(x-32)}{9} = \frac{5(32-32)}{9} = \frac{5 \cdot 0}{9} = 0$$

The completed table is:

Degrees Fahrenheit	x	-4	10	32
Degrees Celsius	$\frac{5(x-32)}{9}$	-20	$\frac{-110}{9}$ or $-12\frac{2}{9}$	0

Thus, $-4°F$ is equivalent to $-20°C$, $10°F$ is equivalent to $-\frac{110}{9}°$ C or $-12\frac{2}{9}°C$, and $32°F$ is equivalent to $0°C$.

C SIMPLIFYING ALGEBRAIC EXPRESSIONS BY COMBINING LIKE TERMS

Often, an expression may be **simplified** by removing grouping symbols and combining any like terms. The **terms** of an expression are the addends of the expression. For example, in the expression $3x^2 + 4x$, the terms are $3x^2$ and $4x$.

Expression	**Terms**
$-2x + y$	$-2x,\ y$
$3x^2 - \frac{y}{5} + 7$	$3x^2,\ -\frac{y}{5},\ 7$

Terms with the same variable(s) raised to the same power are called **like terms**. We can add or subtract like terms by using the distributive property. This process is called **combining like terms**.

Examples Simplify by combining like terms.

7. $3x - 5x + 4 = (3 - 5)x + 4$ Use the distributive property.
$= -2x + 4$

8. $y + 3y = 1y + 3y = (1 + 3)y$
$= 4y$

9. $7x + 9x + 6 - 10 = (7 + 9)x + (6 - 10)$
$= 16x - 4$

The associative and commutative properties may sometimes be needed to rearrange and group like terms when we simplify expressions.

Practice Problems 7–9

Simplify by combining like terms.

7. $9x - 15x + 7$

8. $8y + y$

9. $4x + 12x - 9 - 10$

Answers

7. $-6x + 7$, **8.** $9y$, **9.** $16x - 19$

Examples Simplify.

10. $-7x + 5 + 3x - 2 = -7x + 3x + 5 - 2$ Use the commutative property.
$= (-7 + 3)x + (5 - 2)$ Use the distributive property.
$= -4x + 3$ Simplify.

11. $3y - 2y + 5 - 7 + y = 3y - 2y + y + 5 - 7$ Use the commutative property.
$= (3 - 2 + 1)y + (5 - 7)$ Use the distributive property.
$= 2y - 2$ Simplify.

Examples Simplify by using the distributive property to multiply, and then combining like terms.

12. $-2(x + 3) = -2(x) + (-2)(3) = -2x - 6$

13. $7x + 3 - 5(x - 4) = 7x + 3 - 5x + 20$ Use the distributive property.
$= 2x + 23$ Combine like terms.

14. $(2.1x - 5.6) - (-x - 5.3) = (2.1x - 5.6) - 1(-x - 5.3)$
$= 2.1x - 5.6 + 1x + 5.3$ Use the distributive property.
$= 3.1x - 0.3$ Combine like terms.

TRY THE CONCEPT CHECK IN THE MARGIN.

Practice Problems 10–11

Simplify.

10. $-4x + 7 - 5x - 8$

11. $5y - 6y + 2 - 11 + y$

Practice Problems 12–14

Simplify by using the distributive property to multiply, and then combining like terms.

12. $-3(y + 1)$

13. $8x + 2 - 4(x - 9)$

14. $(3.2x - 4.1) - (-x + 7.6)$

✓ CONCEPT CHECK

Find and correct the error in the following:

$$x - 4(x - 5) = x - 4x - 20$$
$$= -3x - 20$$

Answers

10. $-9x - 1$, **11.** -9, **12.** $-3y - 3$, **13.** $4x + 38$, **14.** $4.2x - 11.7$

✓ Concept Check

$$x - 4(x - 5) = x - 4x + 20$$
$$= -3x + 20$$

Name ______________________ Section ______ Date ________

Exercise Set 1.4

A *Simplify each expression. See Examples 1 through 5.*

1. $3(5-7)^4$

2. $7(3-8)^2$

3. -3^2+2^3

4. -5^2-2^4

5. $\dfrac{3-(-12)}{-5}$

6. $\dfrac{-4-(-8)}{-4}$

7. $|3.6-7.2|+|3.6+7.2|$

8. $|8.6-1.9|-|2.1+5.3|$

9. $(-3)^2+2^3$

10. $(-15)^2-2^4$

11. $-3[6-(-2)]$

12. $-5[8-(-3)]$

13. $-9\cdot 8+5(-6)$

14. $-6\cdot 6+9(-5)$

15. $4[8-(2-4)]$

16. $3[11-(1-3)]$

17. $-8\left(-\frac{3}{4}\right)-8$

18. $-10\left(-\frac{2}{5}\right)-10$

19. $2-[(7-6)+(9-19)]$

20. $8-[(4-7)+(8-1)]$

21. 5^2-3^4

22. 6^2-5^3

23. $2\cdot 7-4\cdot 5$

24. $(2\cdot 7)-(4\cdot 5)$

25. $2\cdot(7-4\cdot 5)$

26. $(2\cdot 7-4)\cdot 5$

27. $18-3(-4)+7$

28. $25-2(-3)+10$

29. $\dfrac{(-9+6)(-1^2)}{-2-2}$

30. $\dfrac{(-1-2)(-3^2)}{-6-3}$

Answers

1. ______
2. ______
3. ______
4. ______
5. ______
6. ______
7. ______
8. ______
9. ______
10. ______
11. ______
12. ______
13. ______
14. ______
15. ______
16. ______
17. ______
18. ______
19. ______
20. ______
21. ______
22. ______
23. ______
24. ______
25. ______
26. ______
27. ______
28. ______
29. ______
30. ______

31. ____
32. ____
33. ____
34. ____
35. ____
36. ____
37. ____
38. ____
39. ____
40. ____
41. ____
42. ____
43. ____
44. ____
45. ____
46. ____
47. ____
48. ____
49. ____
50. ____

Name ____

31. $(\sqrt[3]{8})(-4) - (\sqrt{9})(-5)$

32. $(\sqrt[3]{27})(-5) - (\sqrt{25})(-3)$

33. $12 + \{6 - [5 - 2(-5)]\}$

34. $18 + \{9 - [1 - 6(-3)]\}$

35. $25 - [(3 - 5) + (14 - 18)]^2$

36. $10 - [(4 - 5)^2 + (12 - 14)]^4$

37. $\dfrac{(3 - \sqrt{9}) - (-5 - 1.3)}{-3}$

38. $\dfrac{-\sqrt{16} - (6 - 2.4)}{-2}$

39. $\dfrac{|3 - 9| - |-5|}{-3}$

40. $\dfrac{|-14| - |2 - 7|}{-15}$

41. $\dfrac{3(-2 + 1)}{5} - \dfrac{-7(2 - 4)}{1 - (-2)}$

42. $\dfrac{-1 - 2}{2(-3) + 10} - \dfrac{2(-5)}{-1(8) + 1}$

43. $\dfrac{\frac{1}{3} \cdot 9 - 7}{3 + \frac{1}{2} \cdot 4}$

44. $\dfrac{\frac{1}{5} \cdot 20 - 6}{10 + \frac{1}{4} \cdot 12}$

45. $3\{-2 + 5[1 - 2(-2 + 5)]\}$

46. $2\{-1 + 3[7 - 4(-10 + 12)]\}$

47. $-150(3.25 - 1.68)$

48. $-290(9.61 - 6.27)$

49. $\left(\dfrac{5.6 - 8.4}{1.9 - 2.7}\right)^2$

50. $\left(\dfrac{9.4 - 10.8}{8.7 - 7.9}\right)^2$

Name ______________________

51. a. see table

b. ______

52. a. see table

b. ______

53. a. see table

b. ______

54. a. see table

b. ______

55. ______

56. ______

57. ______

58. ______

B *Complete each table. See Example 6.*

51. The algebraic expression $8 + 2y$ represents the perimeter of a rectangle with width 4 and length y.

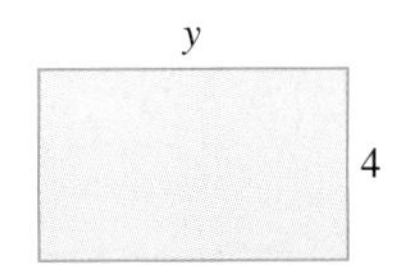

a. Complete the table by evaluating this expression at given values of y.

Length	y	5	7	10	100
Perimeter	$8 + 2y$				

b. Use the results of the table in (a) to answer the following question. As the width of a rectangle remains the same and the length increases, does the perimeter increase or decrease? Explain how you arrived at your answer.

52. The algebraic expression πr^2 represents the area of a circle with radius r.

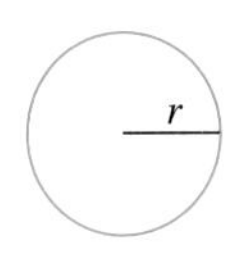

a. Complete the table by evaluating this expression at given values of r. (Use 3.14 for π.)

Radius	r	2	3	7	10
Area	πr^2				

b. As the radius of a circle increases, does its area increase or decrease? Explain your answer.

53. The algebraic expression $\frac{100x + 5000}{x}$ represents the cost per bookshelf (in dollars) of producing x bookshelves.

a. Complete the table.

Number of Bookshelves	x	10	100	1000
Cost per Bookshelf	$\frac{100x + 5000}{x}$			

b. As the number of bookshelves manufactured increases, does the cost per bookshelf increase or decrease? Why do you think that this is so?

54. If C is degrees Celsius, the algebraic expression $1.8C + 32$ represents the equivalent temperature in degrees Fahrenheit.

a. Complete the table.

Degrees Celsius	C	-10	0	50
Degrees Fahrenheit	$1.8C + 32$			

b. As degrees Celsius increase, do degrees Fahrenheit increase or decrease?

C *Simplify. See Examples 7 through 11.*

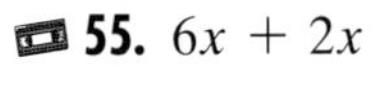

55. $6x + 2x$ **56.** $8x + 11x$ **57.** $9y - 11y$ **58.** $11y - 20y$

59. ______

60. ______

61. ______

62. ______

63. ______

64. ______

65. ______

66. ______

67. ______

68. ______

69. ______

70. ______

71. ______

72. ______

73. ______

74. ______

75. ______

76. ______

77. ______

78. ______

79. ______

80. ______

81. ______

82. ______

83. ______

84. ______

85. ______

86. ______

87. ______

88. ______

89. ______

90. ______

91. ______

92. ______

Name ______________________________

59. $7x + x$ **60.** $2y + y$ **61.** $19y - y$ **62.** $14x - x$

63. $6x - 4x + 10x$ **64.** $13y - 2y - 23y$ **65.** $9x - 8 - 10x$

66. $14x - 1 - 20x$ **67.** $10a + 7 + 4a + 8$ **68.** $9a + 6 + 2a + 9$

69. $-9 + 4x + 18 - 10x$ **70.** $5y - 14 + 7y - 20y$ **71.** $3a - 4b + a - 9b$

72. $11x - y + 11x - 6y$ **73.** $x - y + x - y$ **74.** $a - b + 3a - 3b$

75. $1.5x + 2.3 - 0.7x - 5.9$ **76.** $6.3y - 9.7 + 2.2y - 11.1$

77. $\frac{3}{4}b - \frac{1}{2} + \frac{1}{6}b - \frac{2}{3}$ **78.** $\frac{7}{8}a - \frac{11}{12} - \frac{1}{2}a + \frac{5}{6}$

Simplify. See Examples 12 through 14.

79. $2(3x + 7)$ **80.** $4(5y + 12)$ **81.** $-5(x - 1)$

82. $-6(b - 6)$ **83.** $3(2a - 3b + 4)$ **84.** $5(3x - 4y + 1)$

85. $5k - (3k - 10)$ **86.** $-11c - (4 - 2c)$ **87.** $(3x + 4) - (6x - 1)$

88. $(8 - 5y) - (4 + 3y)$ **89.** $3(x - 2) + x + 15$ **90.** $-4(y + 3) - 7y + 1$

91. $-(n + 5) + (5n - 3)$ **92.** $-(8 - t) + (2t - 6)$

Name ______________________

93. ______
94. ______
95. ______
96. ______
97. ______
98. ______
99. ______
100. ______
101. ______
102. ______
103. ______
104. ______

93. $4(6n - 3) - 3(8n + 4)$

94. $5(2z - 6) + 10(3 - z)$

95. $3x - 2(x - 5) + x$

96. $7n + 3(2n - 6) - 2$

97. $-1.2(5.7x - 3.6) + 8.75x$

98. $5.8(-9.6 - 31.2y) - 18.65$

99. $8.1z + 7.3(z + 5.2) - 6.85$

100. $6.5y - 4.4(1.8y - 3.3) + 10.95$

Combining Concepts

Insert parentheses so that each expression evaluates to the given number.

101. $2 + 7 \cdot 1 + 3$; evaluates to 36

102. $6 - 5 \cdot 2 + 2$; evaluates to -6

The following graph is called a broken-line graph, or simply a line graph. This particular graph shows the past, present, and future predicted population over 65. Just as with a bar graph, to find the population over 65 for a particular year, read the height of the corresponding point. To read the height, follow the point horizontally to the left until you reach the vertical axis. Use this graph to answer Exercises 103–108.

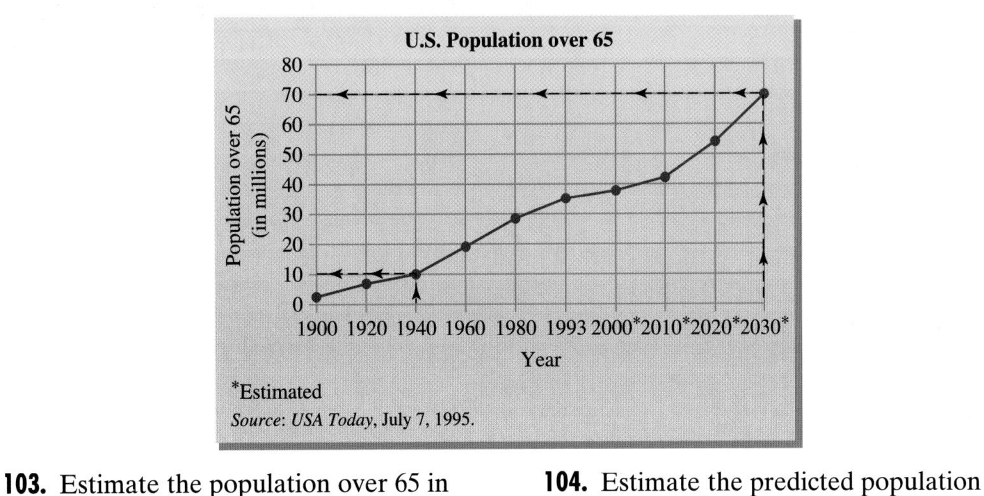

103. Estimate the population over 65 in the year 1940.

104. Estimate the predicted population over 65 in the year 2030.

105. ______

106. ______

107. ______

108. ______

109. ______

110. ______

Name ______

105. Estimate the predicted population over 65 in the year 2010.

106. Estimate the population over 65 in the year 1993.

107. Is the population over 65 increasing as time passes or decreasing? Explain how you arrived at your answer.

108. The percent of Americans over 65 approximately tripled from 1900 to 1993. If this percent in 1900 was 4.1%, estimate the percent of Americans over 65 in the year 1993.

Simplify. Round each result to the nearest ten thousandth.

109. $\dfrac{-1.682 - 17.895}{(-7.102)(-4.691)}$

110. $\dfrac{(-5.161)(3.222)}{7.955 - 19.676}$

1.5 EXPONENTS AND SCIENTIFIC NOTATION

A USING THE PRODUCT RULE

Recall that exponents may be used to write repeated factors in a more compact form. As we have seen in the previous sections, exponents can be used when the repeated factor is a number or a variable. For example,

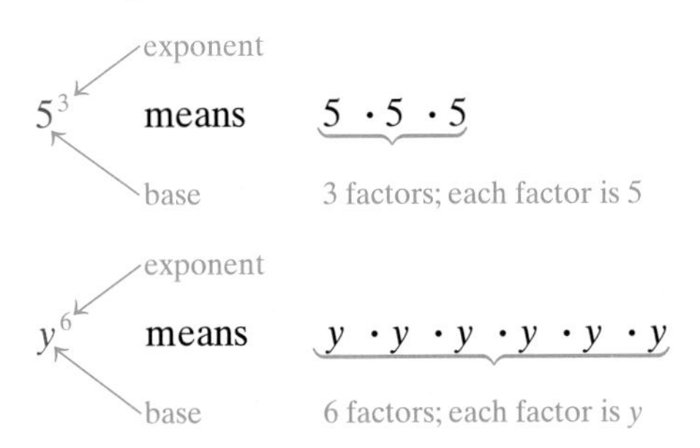

Expressions such as 5^3 and y^6 that contain exponents are called **exponential expressions**.

Exponential expressions can be multiplied, divided, added, subtracted, and themselves raised to powers. In this section, we review operations on exponential expressions.

We review multiplication first. To multiply x^2 by x^3, we use the definition of a^n:

$$x^2 \cdot x^3 = \underbrace{(x \cdot x)(x \cdot x \cdot x)}_{x \text{ is a factor 5 times}}$$
$$= x^5$$

Notice that the result is exactly the same if we add the exponents.

$$x^2 \cdot x^3 = x^{2+3} = x^5$$

This suggests the following rule.

> **PRODUCT RULE FOR EXPONENTS**
>
> If m and n are integers and a is a real number, then
>
> $$a^m \cdot a^n = a^{m+n}$$

In other words, the *product* of exponential expressions with a common base is the common base raised to a power equal to the *sum* of the exponents of the factors.

Examples Use the product rule to simplify.

1. $2^2 \cdot 2^5 = 2^{2+5} = 2^7$

2. $x^7 \cdot x^3 = x^{7+3} = x^{10}$

3. $y \cdot y^2 \cdot y^4 = (y^1 \cdot y^2) \cdot y^4$
$= y^3 \cdot y^4$
$= y^7$

Examples Use the product rule to simplify.

4. $(3x^6)(5x) = 3(5)x^6x^1 = 15x^7$

5. $(-2x^3p^2)(4xp^{10}) = -2(4)x^3x^1p^2p^{10} = -8x^4p^{12}$

Use properties of multiplication to group like bases.

Objectives

A Use the product rule for exponents.

B Evaluate expressions raised to the zero power.

C Use the quotient rule for exponents.

D Evaluate expressions raised to negative powers.

E Convert between scientific notation and standard notation.

SSM CD-ROM Video 1.5

Practice Problems 1–3

Use the product rule to simplify.

1. $5^2 \cdot 5^6$
2. $x^5 \cdot x^9$
3. $y \cdot y^4 \cdot y^3$

Practice Problems 4–5

Use the product rule to simplify.

4. $(7y^5)(6y)$
5. $(-3x^2y^7)(5xy^6)$

Answers

1. 5^8, **2.** x^{14}, **3.** y^8, **4.** $42y^6$, **5.** $-15x^3y^{13}$

B EVALUATING EXPRESSIONS RAISED TO THE ZERO POWER

The definition of a^n does not include the possibility that n might be 0. But if it did, then, by the product rule,

$$a^0 \cdot a^n = a^{0+n} = a^n = 1 \cdot a^n$$

From this, we reasonably define that $a^0 = 1$, as long as a does not equal 0.

> **ZERO EXPONENT**
>
> If a does not equal 0, then $a^0 = 1$.

Practice Problems 6–9

Evaluate each expression.

6. 8^0
7. -8^0
8. $(y - 3)^0$
9. $5x^0$

Examples

Evaluate each expression.

6. $7^0 = 1$

7. $-7^0 = -(7^0) = -(1) = -1$ Without parentheses, only 7 is raised to the 0 power.

8. $(2x + 5)^0 = 1$

9. $2x^0 = 2(1) = 2$

C USING THE QUOTIENT RULE

To find quotients of exponential expressions, we again begin with the definition of a^n to simplify $\frac{x^9}{x^2}$. For example,

$$\frac{x^9}{x^2} = \frac{x \cdot x \cdot x \cdot x \cdot x \cdot x \cdot x \cdot x \cdot x}{x \cdot x} = x^7$$

(Assume for the next two sections that denominators containing variables are not 0.) Notice that the result is exactly the same if we subtract the exponents.

$$\frac{x^9}{x^2} = x^{9-2} = x^7$$

This suggests the following rule.

> **QUOTIENT RULE FOR EXPONENTS**
>
> If a is a nonzero real number and m and n are integers, then
>
> $$\frac{a^m}{a^n} = a^{m-n}$$

In other words, the *quotient* of exponential expressions with a common base is the common base raised to a power equal to the *difference* of the exponents.

Practice Problems 10–13

Use the quotient rule to simplify.

10. $\frac{y^6}{y^2}$
11. $\frac{6^{10}}{6^2}$
12. $\frac{36x^5}{9x}$
13. $\frac{10a^7b^9}{15a^5b^9}$

Answers

6. 1, **7.** −1, **8.** 1, **9.** 5, **10.** y^4, **11.** 6^8, **12.** $4x^4$, **13.** $\frac{2}{3}a^2$

Examples

Use the quotient rule to simplify.

10. $\frac{x^7}{x^4} = x^{7-4} = x^3$

11. $\frac{5^8}{5^2} = 5^{8-2} = 5^6$

12. $\dfrac{20x^6}{4x^5} = 5x^{6-5} = 5x^1$, or $5x$

13. $\dfrac{12y^{10}z^7}{14y^8z^7} = \dfrac{6}{7}y^{10-8} \cdot z^{7-7} = \dfrac{6}{7}y^2z^0 = \dfrac{6}{7}y^2$, or $\dfrac{6y^2}{7}$

D EVALUATING EXPRESSIONS RAISED TO NEGATIVE POWERS

When the exponent of the denominator is larger than the exponent of the numerator, applying the quotient rule gives a negative exponent. For example,

$$\frac{x^3}{x^5} = x^{3-5} = x^{-2}$$

However, using the definition of a^n gives us

$$\frac{x^3}{x^5} = \frac{x \cdot x \cdot x}{x \cdot x \cdot x \cdot x \cdot x} = \frac{1}{x^2}$$

From this, we reasonably define $x^{-2} = \dfrac{1}{x^2}$ or, in general, $a^{-n} = \dfrac{1}{a^n}$.

NEGATIVE EXPONENTS

If a is a real number other than 0 and n is a positive integer, then

$$a^{-n} = \frac{1}{a^n}$$

Examples

Simplify and write with positive exponents only.

14. $5^{-2} = \dfrac{1}{5^2} = \dfrac{1}{25}$

15. $2x^{-3} = 2 \cdot \dfrac{1}{x^3} = \dfrac{2}{x^3}$ Without parentheses, only x is raised to the -3 power.

16. $(3x)^{-1} = \dfrac{1}{(3x)^1} = \dfrac{1}{3x}$ With parentheses, both 3 and x are raised to the -1 power.

17. $\dfrac{m^5}{m^{15}} = m^{5-15} = m^{-10} = \dfrac{1}{m^{10}}$

18. $\dfrac{3^3}{3^6} = 3^{3-6} = 3^{-3} = \dfrac{1}{3^3} = \dfrac{1}{27}$

19. $2^{-1} + 3^{-2} = \dfrac{1}{2^1} + \dfrac{1}{3^2} = \dfrac{1}{2} + \dfrac{1}{9} = \dfrac{9}{18} + \dfrac{2}{18} = \dfrac{11}{18}$

20. $\dfrac{1}{t^{-5}} = \dfrac{1}{\frac{1}{t^5}} = 1 \div \dfrac{1}{t^5} = 1 \cdot \dfrac{t^5}{1} = t^5$

Practice Problems 14–20

Simplify and write with positive exponents only.

14. 7^{-2}

15. $5x^{-4}$

16. $(2x)^{-1}$

17. $\dfrac{x^3}{x^{10}}$

18. $\dfrac{4^2}{4^5}$

19. $3^{-1} + 2^{-2}$

20. $\dfrac{1}{y^{-4}}$

HELPFUL HINTS

Notice that when a factor containing an exponent is moved from the numerator to the denominator or from the denominator to the numerator, the sign of its exponent changes.

$x^{-3} = \dfrac{1}{x^3}$ $\quad 5^{-2} = \dfrac{1}{5^2} = \dfrac{1}{25}$

$\dfrac{1}{y^{-4}} = y^4$ $\quad \dfrac{1}{2^{-3}} = 2^3 = 8$

Answers

14. $\frac{1}{49}$, **15.** $\frac{5}{x^4}$, **16.** $\frac{1}{2x}$, **17.** $\frac{1}{x^7}$, **18.** $\frac{1}{64}$, **19.** $\frac{7}{12}$, **20.** y^4

Practice Problems 21–25

Simplify and write using positive exponents only.

21. $\frac{y^{-10}}{y^3}$

22. $\frac{q^5}{q^{-4}}$

23. $\frac{5^{-4}}{5^{-2}}$

24. $\frac{10x^{-8}y^5}{20xy^{-5}}$

25. $\frac{(4x^{-1})(x^5)}{x^7}$

✓ Concept Check

Find and correct the error in the following:

$$\frac{y^{-6}}{y^{-2}} = y^{-6-2} = y^{-8} = \frac{1}{y^8}$$

Practice Problems 26–27

Simplify. Assume that n and m are nonzero integers and that x is not 0.

26. $x^{3m} \cdot x^n$

27. $\frac{x^{2m-2}}{x^{m-6}}$

Examples

Simplify and write using positive exponents only.

21. $\frac{x^{-9}}{x^2} = x^{-9-2} = x^{-11} = \frac{1}{x^{11}}$

22. $\frac{p^4}{p^{-3}} = p^{4-(-3)} = p^7$

23. $\frac{2^{-3}}{2^{-1}} = 2^{-3-(-1)} = 2^{-2} = \frac{1}{2^2} = \frac{1}{4}$

24. $\frac{2x^{-7}y^2}{10xy^{-5}} = \frac{x^{-7-1} \cdot y^{2-(-5)}}{5} = \frac{x^{-8}y^7}{5} = \frac{y^7}{5x^8}$

25. $\frac{(3x^{-3})(x^2)}{x^6} = \frac{3x^{-3+2}}{x^6} = \frac{3x^{-1}}{x^6} = 3x^{-1-6} = 3x^{-7} = \frac{3}{x^7}$

TRY THE CONCEPT CHECK IN THE MARGIN

Examples

Simplify. Assume that a and t are nonzero integers and that x is not 0.

26. $x^{2a} \cdot x^3 = x^{2a+3}$ Use the product rule.

27. $\frac{x^{2t-1}}{x^{t-5}} = x^{(2t-1)-(t-5)}$ Use the quotient rule.

$= x^{2t-1-t+5} = x^{t+4}$

E CONVERTING BETWEEN SCIENTIFIC NOTATION AND STANDARD NOTATION

Very large and very small numbers occur frequently in nature. For example, the distance between Earth and the sun is approximately 150,000,000 kilometers. A helium atom has a diameter of 0.000000022 centimeters. It can be tedious to write these very large and very small numbers in standard notation like this. **Scientific notation** is a convenient shorthand notation for writing very large and very small numbers.

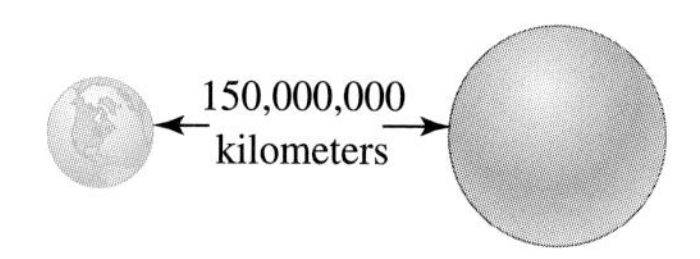

Helium Atom

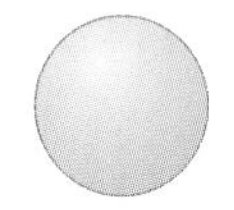

0.000 000 022

> **SCIENTIFIC NOTATION**
>
> A positive number is written in **scientific notation** if it is written as the product of a number a, where $1 \le a < 10$, and an integer power n of 10: $a \times 10^n$

For example,

2.03×10^2 7.362×10^7 8.1×10^{-5}

Answers

21. $\frac{1}{y^{13}}$, **22.** q^9, **23.** $\frac{1}{25}$, **24.** $\frac{y^{10}}{2x^9}$, **25.** $\frac{4}{x^3}$, **26.** x^{3m+n}, **27.** x^{m+4}

✓ **Concept Check:** $\frac{y^{-6}}{y^{-2}} = y^{-6-(-2)} = y^{-4} = \frac{1}{y^4}$

> **WRITING A NUMBER IN SCIENTIFIC NOTATION**
>
> **Step 1.** Move the decimal point in the original number until the new number has a value between 1 and 10.
>
> **Step 2.** Count the number of decimal places the decimal point was moved in Step 1. If the decimal point was moved to the left, the count is positive. If the decimal point was moved to the right, the count is negative.
>
> **Step 3.** Write the product of the new number in Step 1 and 10 raised to an exponent equal to the count found in Step 2.

Example 28 Write 730,000 in scientific notation.

Solution: **Step 1.** Move the decimal point until the number is between 1 and 10.

730,000.

Step 2. The decimal point is moved 5 places to the left, so the count is positive 5.

Step 3. $730{,}000 = 7.3 \times 10^5$

Example 29 Write 0.00000104 in scientific notation.

Solution: **Step 1.** Move the decimal point until the number is between 1 and 10.

0.000001 04

Step 2. The decimal point is moved 6 places to the right, so the count is -6.

Step 3. $0.00000104 = 1.04 \times 10^{-6}$

To write a scientific notation number in standard form, we reverse the preceding steps.

> **WRITING A SCIENTIFIC NOTATION NUMBER IN STANDARD NOTATION**
>
> Move the decimal point in the number the same number of places as the exponent on 10. If the exponent is positive, move the decimal point to the right. If the exponent is negative, move the decimal point to the left.

Examples Write each number in standard notation.

30. $7.7 \times 10^8 = 770{,}000{,}000$ Since the exponent is positive, move the decimal point 8 places to the right. Add zeros as needed.

31. $1.025 \times 10^{-3} = 0.001025$ Since the exponent is negative, move the decimal point 3 places to the left. Add zeros as needed.

TRY THE CONCEPT CHECK IN THE MARGIN.

Practice Problem 28

Write 1,760,000 in scientific notation.

Practice Problem 29

Write 0.00028 in scientific notation.

Practice Problems 30–31

Write each number in standard notation.

30. 8.6×10^7

31. 3.022×10^{-4}

✓ CONCEPT CHECK

Which of the following numbers have values that are less than 1?

a. 3.5×10^{-5}
b. 3.5×10^5
c. -3.5×10^5
d. -3.5×10^{-5}

Answers

28. 1.76×10^6, **29.** 2.8×10^{-4}, **30.** 86,000,000, **31.** 0.0003022

✓ Concept Check: a, c, d

CALCULATOR EXPLORATIONS

Multiply 5,000,000 by 700,000 on your calculator. The display should read [3.5 12] or [3.5 E 12], which is the product written in scientific notation. Both these notations mean 3.5×10^{12}.

To enter a number written in scientific notation on a calculator, find the key marked [EE]. (On some calculators, this key may be marked [EXP].)

To enter 7.26×10^{13}, press the keys

[7.26] [EE] [13]

The display will read [7.26 13] or [7.26 E 13].

Use your calculator to perform each indicated operation.

1. Multiply 3×10^{11} and 2×10^{32}.

2. Divide 6×10^{14} by 3×10^{9}.

3. Multiply 5.2×10^{23} and 7.3×10^{4}.

4. Divide 4.38×10^{41} by 3×10^{17}.

Name ________________ Section ______ Date ________

Mental Math

Use positive exponents to state each expression.

1. $5x^{-1}y^{-2}$ **2.** $7xy^{-4}$ **3.** $a^2b^{-1}c^{-5}$

4. $a^{-4}b^2c^{-6}$ **5.** $\frac{y^{-2}}{x^{-4}}$ **6.** $\frac{x^{-7}}{z^{-3}}$

Exercise Set 1.5

A *Use the product rule to simplify each expression. See Examples 1 through 5.*

1. $4^2 \cdot 4^3$ **2.** $3^3 \cdot 3^5$ **3.** $x^5 \cdot x^3$ **4.** $a^2 \cdot a^9$

5. $-7x^3 \cdot 20x^9$ **6.** $-3y \cdot -9y^4$ **7.** $(4xy)(-5x)$ **8.** $(7xy)(7aby)$

9. $(-4x^3p^2)(4y^3x^3)$ **10.** $(-6a^2b^3)(-3ab^3)$ **11.** $x^7 \cdot x^8$

12. $y^6 \cdot y$ **13.** $2x^3 \cdot 5x^7$ **14.** $-3z^4 \cdot 10z^7$

B *Evaluate each expression. See Examples 6 through 9.*

15. -8^0 **16.** $(-9)^0$ **17.** $(4x + 5)^0$ **18.** $8x^0 + 1$

19. $(5x)^0 + 5x^0$ **20.** $4y^0 - (4y)^0$ **21.** $4x^0 + 5$ **22.** $-5x^0$

23. $3^0 - 3t^0$ **24.** $4^0 + 4x^0$

25. Explain why $(-5)^0$ simplifies to 1 but -5^0 simplifies to -1.

26. Explain why both $4x^0 - 3y^0$ and $(4x - 3y)^0$ simplify to 1.

C *Use the quotient rule to simplify. See Examples 10 through 13.*

27. $\frac{a^5}{a^2}$ **28.** $\frac{x^9}{x^4}$ **29.** $\frac{x^9y^6}{x^8y^6}$ **30.** $\frac{a^{12}b^2}{a^9b}$

Mental Math Answers

1. ______
2. ______
3. ______
4. ______
5. ______
6. ______

Answers

1. ______
2. ______
3. ______
4. ______
5. ______
6. ______
7. ______
8. ______
9. ______
10. ______
11. ______
12. ______
13. ______
14. ______
15. ______
16. ______
17. ______
18. ______
19. ______
20. ______
21. ______
22. ______
23. ______
24. ______
25. ______
26. ______
27. ______
28. ______
29. ______
30. ______

31.
32.
33.
34.
35.
36.
37.
38.
39.
40.
41.
42.
43.
44.
45.
46.
47.
48.
49.
50.
51.
52.
53.
54.
55.
56.
57.
58.
59.
60.
61.
62.
63.
64.

Name ______________________________

31. $-\frac{26z^{11}}{2z^7}$

32. $\frac{16x^5}{8x}$

33. $\frac{-36a^5b^7c^{10}}{6ab^3c^4}$

34. $\frac{49a^3bc^{14}}{-7abc^8}$

35. $\frac{z^{12}}{z^{15}}$

36. $\frac{x^{11}}{x^{20}}$

D *Simplify and write using positive exponents only. See Examples 14 through 27.*

37. 4^{-2}

38. 2^{-3}

39. $\frac{x^7}{x^{15}}$

40. $\frac{z}{z^3}$

41. $5a^{-4}$

42. $10b^{-1}$

43. $\frac{x^{-2}}{x^5}$

44. $\frac{y^{-6}}{y^{-9}}$

45. $\frac{8r^4}{2r^{-4}}$

46. $\frac{3s^3}{15s^{-3}}$

47. $\frac{x^{-9}x^4}{x^{-5}}$

48. $\frac{y^{-7}y}{y^8}$

49. $4^{-1} + 3^{-2}$

50. $1^{-3} - 4^{-2}$

51. $\frac{y^{-3}}{y^{-7}}$

52. $\frac{z^{-12}}{z^{10}}$

53. $3x^{-1}$

54. $(4x)^{-1}$

55. $\frac{r^4}{r^{-4}}$

56. $\frac{x^{-5}}{x^3}$

57. $\frac{x^{-7}y^{-2}}{x^2y^2}$

58. $\frac{a^{-5}b^7}{a^{-2}b^{-3}}$

59. $\frac{2a^{-6}b^2}{18ab^{-5}}$

60. $\frac{18ab^{-6}}{3a^{-3}b^6}$

61. $\frac{(24x^8)(x)}{20x^{-7}}$

62. $\frac{(30z^2)(z^5)}{55z^{-4}}$

Simplify, Assume that variables in the exponents represent nonzero integers and the x, y, and z are not 0. See Examples 26 and 27.

63. $x^5 \cdot x^{7a}$

64. $y^{2p} \cdot y^{9p}$

Name ______________________________

65. $\frac{x^{3t-1}}{x^t}$

66. $\frac{y^{4p-2}}{y^{3p}}$

67. $x^{4a} \cdot x^7$

68. $x^{9y} \cdot x^{-7y}$

69. $\frac{z^{6x}}{z^7}$

70. $\frac{y^6}{y^{4z}}$

71. $\frac{x^{3t} \cdot x^{4t-1}}{x^t}$

72. $\frac{z^{5x} \cdot z^{x-7}}{z^x}$

E *Write each number in scientific notation. See Examples 28 and 29.*

73. 31,250,000

74. 678,000

75. 0.016

76. 0.007613

77. 67,413

78. 36,800,000

79. 0.0125

80. 0.00084

81. 0.000053

82. 98,700,000,000

Write each number in scientific notation.

83. The approximate distance between Jupiter and the sun is 778,300,000 kilometers. (*Source*: National Space Data Center)

84. Total revenues for Sears in fiscal year 1998 were $41,296,000,000. (*Source*: Sears Roebuck & Co.)

85. In 1997, the American toy industry had retail sales of $22,580,000,000. (*Source*: Toy Manufacturers of America, Inc.)

86. In April 1998, domestic airline flights carried a total of 42,602,000 passengers. (*Source*: Air Transport Association of America)

87. In 1997, the New York City subway system carried a total of 1,130,000,000 passengers. (*Source*: New York City Transit Authority)

88. The center of the sun is about 27,000,000°F.

65. ________
66. ________
67. ________
68. ________
69. ________
70. ________
71. ________
72. ________
73. ________
74. ________
75. ________
76. ________
77. ________
78. ________
79. ________
80. ________
81. ________
82. ________
83. ________
84. ________
85. ________
86. ________
87. ________
88. ________

89.

90.

91.

92.

93.

94.

95.

96.

97.

98.

99.

100.

101.

102.

Name

89. A pulsar is a rotating neutron star that gives off sharp, regular pulses of radio waves. For one particular pulsar, the rate of pulses is every 0.001 second.

90. To convert from cubic inches to cubic meters, multiply by 0.0000164.

Write each number in standard notation. See Examples 30 and 31.

91. 3.6×10^{-9}

92. 2.7×10^{-5}

93. 9.3×10^{7}

94. 6.378×10^{8}

95. 1.278×10^{6}

96. 7.6×10^{4}

97. 7.35×10^{12}

98. 1.66×10^{-5}

99. 4.03×10^{-7}

100. 8.007×10^{8}

Combining Concepts

101. Explain how to convert a number from standard notation to scientific notation.

102. Explain how to convert a number from scientific notation to standard notation.

1.6 MORE WORK WITH EXPONENTS AND SCIENTIFIC NOTATION

A USING THE POWER RULES

The volume of the cube shown whose side measures x^2 units is $(x^2)^3$ cubic units. To simplify an expression such as $(x^2)^3$, we use the definition of a^n:

$$(x^2)^3 = \underbrace{(x^2)(x^2)(x^2)}_{x^2 \text{ is a factor 3 times}} = x^{2+2+2} = x^6$$

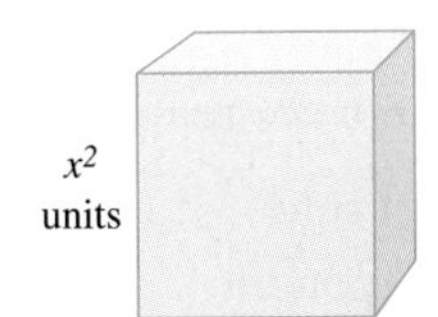

Notice that the result is exactly the same if the exponents are multiplied.

$$(x^2)^3 = x^{2 \cdot 3} = x^6$$

This suggests that the power of an exponential expression raised to a power is the product of the exponents. Two additional rules for exponents are given in the following box.

THE POWER RULE AND POWER OF A PRODUCT OR QUOTIENT RULES FOR EXPONENTS

If a and b are real numbers and m and n are integers, then

$(a^m)^n = a^{m \cdot n}$ Power rule

$(ab)^m = a^m b^m$ Power of a product

$\left(\frac{a}{b}\right)^n = \frac{a^n}{b^n} \quad (b \neq 0)$ Power of a quotient

Objectives

A Use the power rules for exponents.

B Use exponent rules and definitions to simplify exponential expressions.

C Use scientific notation to compute.

SSM CD-ROM Video 1.6

Examples

Use the power rule to simplify each expression. Write each answer using only positive exponents.

1. $(x^5)^7 = x^{5 \cdot 7} = x^{35}$

2. $(2^2)^3 = 2^{2 \cdot 3} = 2^6 = 64$

3. $(5^{-1})^2 = 5^{-1 \cdot 2} = 5^{-2} = \frac{1}{5^2} = \frac{1}{25}$

4. $(y^{-3})^{-4} = y^{-3(-4)} = y^{12}$

Practice Problems 1–4

Use the power rule to simplify each expression. Write each answer using positive exponents only.

1. $(y^2)^8$
2. $(3^3)^2$
3. $(6^2)^{-1}$
4. $(x^{-5})^{-7}$

Examples

Use the power rules to simplify each expression. Write each answer using positive exponents only.

5. $(5x^2)^3 = 5^3 \cdot (x^2)^3 = 5^3 \cdot x^{2 \cdot 3} = 125x^6$

6. $\left(\frac{2}{3}\right)^3 = \frac{2^3}{3^3} = \frac{8}{27}$

7. $\left(\frac{3p^4}{q^5}\right)^2 = \frac{(3p^4)^2}{(q^5)^2} = \frac{3^2 \cdot (p^4)^2}{(q^5)^2} = \frac{9p^8}{q^{10}}$

8. $\left(\frac{2^{-3}}{y}\right)^{-2} = \frac{(2^{-3})^{-2}}{y^{-2}}$

$= \frac{2^6}{y^{-2}} = 64y^2$ Use the negative exponent rule.

9. $(x^{-5}y^2z^{-1})^7 = (x^{-5})^7 \cdot (y^2)^7 \cdot (z^{-1})^7$

$= x^{-35}y^{14}z^{-7} = \frac{y^{14}}{x^{35}z^7}$

Practice Problems 5–9

Use the power rules to simplify each expression. Write each answer using positive exponents only.

5. $(3x^4)^3$
6. $\left(\frac{4}{5}\right)^2$
7. $\left(\frac{4m^5}{n^3}\right)^3$
8. $\left(\frac{2^{-1}}{y}\right)^{-3}$
9. $(a^{-4}b^3c^{-2})^6$

Answers

1. y^{16}, **2.** 729, **3.** $\frac{1}{36}$, **4.** x^{35}, **5.** $27x^{12}$, **6.** $\frac{16}{25}$, **7.** $\frac{64m^{15}}{n^9}$, **8.** $8y^3$, **9.** $\frac{b^{18}}{a^{24}c^{12}}$

B USING EXPONENT RULES TO SIMPLIFY EXPRESSIONS

In the next few examples, we practice the use of several of the rules and definitions for exponents. The following is a summary of these rules and definitions.

SUMMARY OF RULES FOR EXPONENTS

If a and b are real numbers and m and n are integers, then

Product rule	$a^m \cdot a^n = a^{m+n}$	
Zero exponent	$a^0 = 1$	$(a \neq 0)$
Negative exponent	$a^{-n} = \frac{1}{a^n}$	$(a \neq 0)$
Quotient rule	$\frac{a^m}{a^n} = a^{m-n}$	$(a \neq 0)$
Power rule	$(a^m)^n = a^{m \cdot n}$	
Power of a product	$(ab)^m = a^m \cdot b^m$	
Power of a quotient	$\left(\frac{a}{b}\right)^m = \frac{a^m}{b^m}$	$(b \neq 0)$

Practice Problems 10–13

Simplify each expression. Write each answer using positive exponents only.

10. $(7xy^{-2})^{-2}$

11. $\left(\frac{y^{-7}}{y^{-10}}\right)^{-3}$

12. $\left(\frac{3}{5}\right)^{-2}$

13. $\frac{6^{-2}x^{-4}y^{10}}{x^2y^{-6}}$

Examples

Simplify each expression. Write each answer using positive exponents only.

10. $(2x^0y^{-3})^{-2} = 2^{-2}(x^0)^{-2}(y^{-3})^{-2}$

$= 2^{-2}x^0y^6$

$= \frac{1(y^6)}{2^2}$ Write x^0 as 1.

$= \frac{y^6}{4}$

11. $\left(\frac{x^{-5}}{x^{-2}}\right)^{-3} = \frac{(x^{-5})^{-3}}{(x^{-2})^{-3}} = \frac{x^{15}}{x^6} = x^{15-6} = x^9$

12. $\left(\frac{2}{7}\right)^{-2} = \frac{2^{-2}}{7^{-2}} = \frac{7^2}{2^2} = \frac{49}{4}$

13. $\frac{5^{-2}x^{-3}y^{11}}{x^2y^{-5}} = (5^{-2})\left(\frac{x^{-3}}{x^2}\right)\left(\frac{y^{11}}{y^{-5}}\right) = 5^{-2}x^{-3-2}y^{11-(-5)} = 5^{-2}x^{-5}y^{16}$

$= \frac{y^{16}}{5^2x^5} = \frac{y^{16}}{25x^5}$

Practice Problems 14–15

Simplify each expression. Write each answer using positive exponents only.

14. $\left(\frac{4a^3b^2}{b^{-6}c}\right)^{-2}$

15. $\left(\frac{4x^3}{3y^{-1}}\right)^3\left(\frac{y^{-2}}{3x^{-1}}\right)^{-1}$

Examples

Simplify each expression. Write each answer using positive exponents only.

14. $\left(\frac{3x^2y}{y^{-9}z}\right)^{-2} = \left(\frac{3x^2y^{10}}{z}\right)^{-2} = \frac{3^{-2}x^{-4}y^{-20}}{z^{-2}} = \frac{z^2}{3^2x^4y^{20}} = \frac{z^2}{9x^4y^{20}}$

15. $\left(\frac{3a^2}{2x^{-1}}\right)^3\left(\frac{x^{-3}}{4a^{-2}}\right)^{-1} = \frac{27a^6}{8x^{-3}} \cdot \frac{x^3}{4^{-1}a^2}$

$= \frac{27 \cdot 4 \cdot a^6x^3x^3}{8 \cdot a^2} = \frac{27a^4x^6}{2}$

Answers

10. $\frac{y^4}{49x^2}$, **11.** $\frac{1}{y^9}$, **12.** $\frac{25}{9}$, **13.** $\frac{y^{16}}{36x^6}$, **14.** $\frac{c^2}{16a^6b^{16}}$, **15.** $\frac{64x^8y^5}{9}$

Examples Simplify. Assume that a and b are integers and that x and y are not 0.

16. $x^{-b}(2x^b)^2 = x^{-b}2^2x^{2b} = 4x^{-b+2b} = 4x^b$

17. $\frac{(y^{3a})^2}{y^{a-6}} = \frac{y^{2(3a)}}{y^{a-6}} = \frac{y^{6a}}{y^{a-6}} = y^{6a-(a-6)} = y^{6a-a+6} = y^{5a+6}$

Practice Problems 16–17

Simplify. Assume that m and n are integers and that x and y are not 0.

16. $x^{-n}(3x^n)^2$

17. $\frac{(y^{2m})^2}{y^{m-3}}$

C Using Scientific Notation to Compute

To perform operations on numbers written in scientific notation, we use properties of exponents.

Examples Perform each indicated operation. Write each answer in scientific notation.

18. $(8.1 \times 10^5)(5 \times 10^{-7}) = 8.1 \times 5 \times 10^5 \times 10^{-7}$

$= 40.5 \times 10^{-2}$

$= (4.05 \times 10^1) \times 10^{-2}$

$= 4.05 \times 10^{-1}$

19. $\frac{1.2 \times 10^4}{3 \times 10^{-2}} = \left(\frac{1.2}{3}\right)\left(\frac{10^4}{10^{-2}}\right) = 0.4 \times 10^{4-(-2)}$

$= 0.4 \times 10^6 = (4 \times 10^{-1}) \times 10^6 = 4 \times 10^5$

Practice Problems 18–19

Perform each indicated operation. Write each answer in scientific notation.

18. $(9.6 \times 10^6)(4 \times 10^{-8})$

19. $\frac{4.2 \times 10^7}{7 \times 10^{-3}}$

Example 20 Use scientific notation to simplify: $\frac{2000 \times 0.000021}{700}$

Solution:

$\frac{2000 \times 0.000021}{700} = \frac{(2 \times 10^3)(2.1 \times 10^{-5})}{7 \times 10^2} = \frac{2(2.1)}{7} \cdot \frac{10^3 \cdot 10^{-5}}{10^2}$

$= 0.6 \times 10^{-4}$

$= (6 \times 10^{-1}) \times 10^{-4}$

$= 6 \times 10^{-5}$

Practice Problem 20

Use scientific notation to simplify:

$\frac{3000 \times 0.000012}{400}$

Answers

16. $9x^n$, **17.** y^{3m+3}, **18.** 3.84×10^{-1}, **19.** 6×10^9, **20.** 9×10^{-5}

Focus On Mathematical Connections

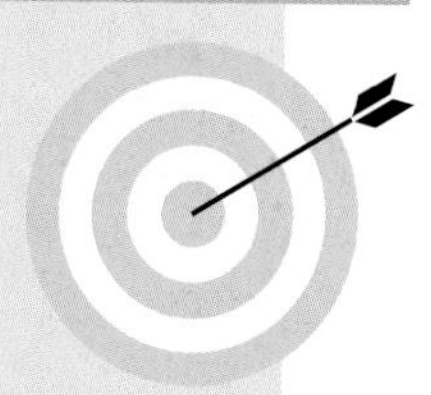

GEOMETRY INVESTIGATIONS

Recall that the perimeter of a figure is the distance around the outside of the figure. For a rectangle with length l and width w, the perimeter of the rectangle is given by the expression $2l + 2w$.

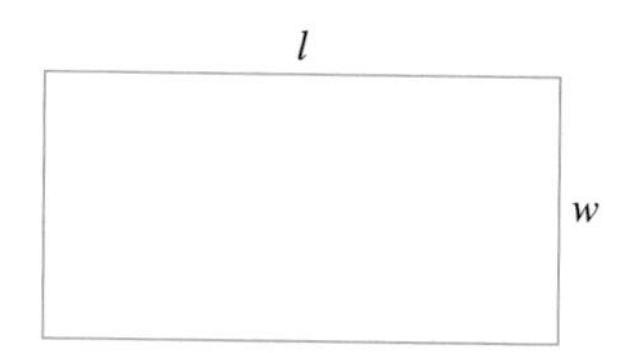

Area is a measure of the surface of a region. For example, we measure a plot of land or the floor space of a home by area. For a rectangle with length l and width w, the area of the rectangle is given by the expression lw.

A circular cylinder can be formed by rolling a rectangle into a tube. The surface area of the cylinder (excluding the two ends of the cylinder) is the same as the area of the rectangle used to form the cylinder. Recall that volume is a measure of the space inside a three-dimensional region. The volume of a circular cylinder with height h and radius r is given by the expression $\pi r^2 h$.

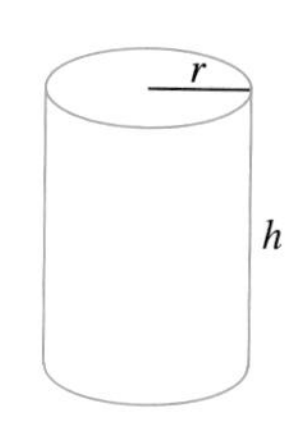

GROUP ACTIVITY

1. Work together to discover whether two rectangles with the same perimeter always have the same area. Explain your results. Give examples.
2. Do figures with the same surface area always have the same volume? To see, take two $8\frac{1}{2}$-by-11-inch sheets of paper and construct two cylinders using the following figures as a guide. Verify that both cylinders have the same surface area. Measure the height and radius of each resulting cylinder. Then find the volume of each cylinder to the nearest tenth of a cubic inch. Explain your results.

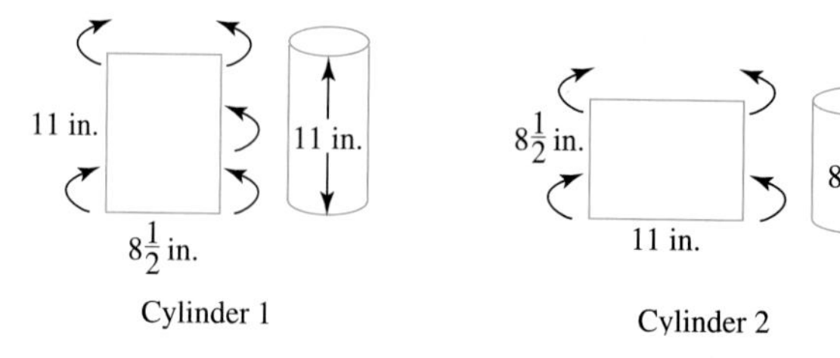

Name ______________________ Section ______ Date ________

MENTAL MATH ANSWERS

1. ________
2. ________
3. ________
4. ________
5. ________
6. ________
7. ________
8. ________
9. ________
10. ________

ANSWERS

1. ________
2. ________
3. ________
4. ________
5. ________
6. ________
7. ________
8. ________
9. ________
10. ________
11. ________
12. ________
13. ________
14. ________
15. ________
16. ________
17. ________
18. ________
19. ________
20. ________
21. ________
22. ________
23. ________
24. ________

MENTAL MATH

Simplify. See Examples 1 through 4.

1. $(x^4)^5$ **2.** $(5^6)^2$ **3.** $x^4 \cdot x^5$

4. $x^7 \cdot x^8$ **5.** $(y^6)^7$ **6.** $(x^3)^4$

7. $(z^4)^9$ **8.** $(z^3)^7$ **9.** $(z^{-6})^{-3}$

10. $(y^{-4})^{-2}$

EXERCISE SET 1.6

A *Simplify. Write each answer using positive exponents only. See Examples 1 through 9.*

1. $(3^{-1})^2$ **2.** $(2^{-2})^2$ **3.** $(x^4)^{-9}$

4. $(y^7)^{-3}$ **5.** $(y)^{-5}$ **6.** $(z^{-1})^{10}$

7. $(3x^2y^3)^2$ **8.** $(4x^3yz)^2$ **9.** $\left(\frac{2x^5}{y^{-3}}\right)^4$

10. $\left(\frac{3a^{-4}}{b^7}\right)^3$ **11.** $(a^2bc^{-3})^{-6}$ **12.** $(6x^{-6}y^7z^0)^{-2}$

13. $\left(\frac{x^7y^{-3}}{z^{-4}}\right)^{-5}$ **14.** $\left(\frac{a^{-2}b^{-5}}{c^{-11}}\right)^{-6}$ **15.** $(5^{-1})^3$

16. $(8^2)^{-1}$ **17.** $(x^7)^{-9}$ **18.** $(y^{-4})^5$

19. $\left(\frac{7}{8}\right)^3$ **20.** $\left(\frac{4}{3}\right)^2$ **21.** $(4x^2)^2$

22. $(-8x^3)^2$ **23.** $(-2^{-2}y)^3$ **24.** $(-4^{-6}y^{-6})^{-4}$

25. ______

26. ______

27. ______

28. ______

29. ______

30. ______

31. ______

32. ______

33. ______

34. ______

35. ______

36. ______

37. ______

38. ______

39. ______

40. ______

41. ______

42. ______

43. ______

44. ______

45. ______

46. ______

47. ______

48. ______

49. ______

50. ______

Name ______________________________

25. $\left(\frac{4^{-4}}{y^3x}\right)^{-2}$

26. $\left(\frac{7^{-3}}{ab^2}\right)^{-2}$

B *Simplify. Write each answer using positive exponents only. See Examples 10 through 15.*

27. $\left(\frac{a^{-4}}{a^{-5}}\right)^{-2}$

28. $\left(\frac{x^{-9}}{x^{-4}}\right)^{-3}$

29. $\left(\frac{2a^{-2}b^5}{4a^2b^7}\right)^{-2}$

30. $\left(\frac{5x^7y^4}{10x^3y^{-2}}\right)^{-3}$

31. $\frac{4^{-1}x^2yz}{x^{-2}yz^3}$

32. $\frac{8^{-2}x^{-3}y^{11}}{x^2y^{-5}}$

33. $\left(\frac{6p^6}{p^{12}}\right)^2$

34. $\left(\frac{4p^6}{p^9}\right)^3$

35. $(-8y^3xa^{-2})^{-3}$

36. $(-xy^0x^2a^3)^{-3}$

37. $\left(\frac{x^{-2}y^{-2}}{a^{-3}}\right)^{-7}$

38. $\left(\frac{x^{-1}y^{-2}}{5^{-3}}\right)^{-5}$

39. $\left(\frac{3x^5}{6x^4}\right)^4$

40. $\left(\frac{8^{-3}}{y^2}\right)^{-2}$

41. $\left(\frac{1}{4}\right)^{-3}$

42. $\left(\frac{1}{8}\right)^{-2}$

43. $\frac{(y^3)^{-4}}{y^3}$

44. $\frac{2(y^3)^{-3}}{y^{-3}}$

45. $\frac{8p^7}{4p^9}$

46. $\left(\frac{2x^4}{x^2}\right)^3$

47. $(4x^6y^5)^{-2}(6x^4y^3)$

48. $(5xy)^3(z^{-2})^{-3}$

49. $x^6(x^6bc)^{-6}$

50. $2(y^2b)^{-4}$

Name ____________

51. $\dfrac{2^{-3}x^2y^{-5}}{5^{-2}x^7y^{-1}}$

52. $\dfrac{7^{-1}a^{-3}b^5}{a^2b^{-2}}$

53. $\left(\dfrac{2x^2}{y^4}\right)^3 \cdot \left(\dfrac{2x^5}{y}\right)^{-2}$

54. $\left(\dfrac{3z^{-2}}{y}\right)^2 \cdot \left(\dfrac{9y^{-4}}{z^{-3}}\right)^{-1}$

Simplify. Assume that variables in the exponents represent nonzero integers and that all other variables are not 0. See Examples 16 and 17.

55. $(x^{3a+6})^3$

56. $(x^{2b+7})^2$

57. $\dfrac{x^{4a}(x^{4a})^3}{x^{4a-2}}$

58. $\dfrac{x^{-5y+2}x^{2y}}{x}$

59. $(b^{5x-2})^{2x}$

60. $(c^{2a+3})^3$

61. $\dfrac{(y^{2a})^8}{y^{a-3}}$

62. $\dfrac{(y^{4a})^7}{y^{2a-1}}$

63. $\left(\dfrac{2x^{3t}}{x^{2t-1}}\right)^4$

64. $\left(\dfrac{3y^{5a}}{y^{-a+1}}\right)^2$

C *Perform each indicated operation. Write each answer in scientific notation. See Examples 18 through 20.*

65. $(5 \times 10^{11})(2.9 \times 10^{-3})$

66. $(3.6 \times 10^{-12})(6 \times 10^{9})$

67. $(2 \times 10^5)^3$

68. $(3 \times 10^{-7})^3$

69. $\dfrac{3.6 \times 10^{-4}}{9 \times 10^2}$

70. $\dfrac{1.2 \times 10^9}{2 \times 10^{-5}}$

71. $\dfrac{0.0069}{0.023}$

72. $\dfrac{0.00048}{0.0016}$

51. ____________

52. ____________

53. ____________

54. ____________

55. ____________

56. ____________

57. ____________

58. ____________

59. ____________

60. ____________

61. ____________

62. ____________

63. ____________

64. ____________

65. ____________

66. ____________

67. ____________

68. ____________

69. ____________

70. ____________

71. ____________

72. ____________

73. ____________

Name ____________

74. ____________

75. ____________

76. ____________

77. ____________

78. ____________

79. ____________

80. ____________

81. ____________

82. ____________

83. ____________

84. ____________

85. ____________

73. $\dfrac{18{,}200 \times 100}{91{,}000}$

74. $\dfrac{0.0003 \times 0.0024}{0.0006 \times 20}$

75. $\dfrac{6000 \times 0.006}{0.009 \times 400}$

76. $\dfrac{0.00016 \times 300}{0.064 \times 100}$

77. $\dfrac{0.00064 \times 2000}{16{,}000}$

78. $\dfrac{0.00072 \times 0.003}{0.00024}$

79. $\dfrac{66{,}000 \times 0.001}{0.002 \times 0.003}$

80. $\dfrac{0.0007 \times 11{,}000}{0.001 \times 0.0001}$

81. $\dfrac{1.25 \times 10^{15}}{(2.2 \times 10^{-2})(6.4 \times 10^{-5})}$

82. $\dfrac{(2.6 \times 10^{-3})(4.8 \times 10^{-4})}{1.3 \times 10^{-12}}$

Solve.

83. A fast computer can add two numbers in about 10^{-8} second. Express in scientific notation how long it would take this computer to do this task 200,000 times.

84. To convert from square inches to square meters, multiply by 6.452×10^{-4}. The area of the following square is 4×10^{-2} square inches. Convert this area to square meters.

4×10^{-2}
square inches

85. To convert from cubic inches to cubic meters, multiply by 1.64×10^{-5}. A grain of salt is in the shape of a cube. If an average size of a grain of salt is 3.8×10^{-6} cubic inches, convert this volume to cubic meters.

Name ______________________________

COMBINING CONCEPTS

86. Each side of the cube shown is $\frac{2x^{-2}}{y}$ meters. Find its volume.

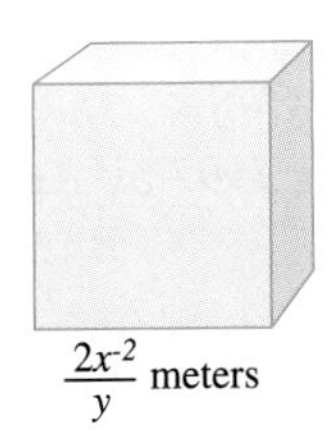

87. The lot shown is in the shape of a parallelogram with base $\frac{3x^{-1}}{y^{-3}}$ feet and height $5x^{-7}$ feet. Find its area.

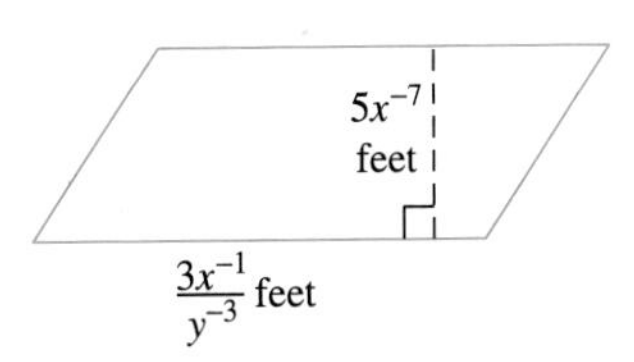

88. The density D of an object is equivalent to the quotient of its mass M and volume V. Thus $D = \frac{M}{V}$. Express in scientific notation the density of an object whose mass is 500,000 pounds and whose volume is 250 cubic feet.

89. The density of ordinary water is 3.12×10^{-2} tons per cubic foot. The volume of water in the largest of the Great Lakes, Lake Superior, is 4.269×10^{14} cubic feet. Use the formula $D = \frac{M}{V}$ (see Exercise 88) to find the mass (in tons) of the water in Lake Superior. Express your answer in scientific notation. (*Source:* National Ocean Service)

90. Is there a number a such that $a^{-1} = a^{1}$? If so, give the value of a.

91. Is there a number a such that a^{-2} is a negative number? If so, give the value of a.

92. Explain whether 0.4×10^{-5} is written in scientific notation.

93. The estimated population of the United States in 1997 was 2.680×10^{8}. The land area of the United States is 3.536×10^{6} square miles. Find the population density (number of people per square mile) for the United States in 1997. Round to the nearest whole number. (*Source:* U.S. Bureau of the Census)

86. ______

87. ______

88. ______

89. ______

90. ______

91. ______

92. ______

93. ______

94. ________________

Name ________________

94. The subway system with the largest passenger volume in the world in 1997 was the Moscow subway with 3.16×10^9 passengers. The tenth busiest subway system was São Paulo's with 7.01×10^8 passengers in 1997. How many times greater was the Moscow subway volume than the São Paulo volume? Round to the nearest tenth. (*Source:* New York City Transit Authority)

95. In 1997, China had the largest armed forces in the world. Its fighting force numbered 2.93×10^6 soldiers. Taiwan's fighting force numbered only 4.25×10^5. How many times greater was China's armed forces than Taiwan's? Round to the nearest whole number. (*Source: The Top 10 of Everything 1997* by Russell Ash)

95. ________________

Name ______________________ Section ________ Date ________

Chapter 1 Activity
Searching for Patterns

This activity may be completed by working in groups or individually.

Professor Jakow Trachtenberg was the founder of the Mathematical Institute in Zurich, Switzerland, in 1950. During World War II, he spent seven years in various Nazi concentration camps. To keep himself occupied and mentally sharp while a prisoner, he developed his own speed system of basic mathematics. Trachtenberg worked out his system of simplified mental arithmetic entirely without the use of pencil and paper. After escaping from a labor camp in Trieste in 1945, Trachtenberg fled to Switzerland and completed his speed system of mathematics. He began teaching his system of addition, subtraction, multiplication, and division to children who had trouble learning conventionally taught mathematics. His success lead him to found the Mathematical Institute in Zurich.

Trachtenberg's speed system is based on certain patterns that occur when basic operations are performed. For example, he developed a different rule for multiplying any number by each of the factors 2 through 12. The following steps describe how to quickly multiply any number by 11.

Trachtenberg's Rule for Multiplication by Eleven

a. The last digit of the product is the last digit of the number being multiplied by 11.

b. To find the middle digits in the product, add each digit of the number being multiplied by 11 to its immediate "neighbor" to the right. (It may be necessary to add a carried 1 to the result at any stage.)

c. The first digit of the product is the first digit of the number being multiplied by 11. (Adding a carried 1 to this digit may be necessary.)

Example 35,924 × 11

35,924	
× 11	
4	(last digit of 35,924)
6	(sum of 2 and its neighbor 4 in 35,924)
1	(sum of 9 and its neighbor 2; carry 1)
5	(sum of 5 and its neighbor 9 + 1; carry 1)
9	(sum of 3 and its neighbor 5 + 1)
3	(first digit of 35,924)
395,164	Product

1. Use the Trachtenberg rule to find the following products. Then check each result with a calculator.

a. 7234 × 11

b. 362,713 × 11

c. 4,386,275 × 11

d. 5845 × 11

2. Develop a Trachtenberg-like rule for finding the product of any number and 2. Begin by looking for patterns when numbers like 14, 19, 28, 374, 621, and 314,672 are multiplied by 2. What relationship do you observe between the digits of the product and the digits of the number being multiplied by 2?

3. Develop a Trachtenberg-like rule for finding the product of any number and 12. Begin by looking for patterns when numbers like 18, 28, 71, 89, 123, and 456 are multiplied by 12. Investigate additional products as necessary. What relationship do you observe between the digits of the product and the digits of the number being multiplied by 12? (*Hint:* To make your rule work out, you may find it helpful to add a place-holding 0 in front of each number being multiplied by 12. For instance, you may want to consider the numbers 018, 028, 071, 089, 0123, and 0456.)

CHAPTER 1 HIGHLIGHTS

DEFINITIONS AND CONCEPTS	EXAMPLES
SECTION 1.1 ALGEBRAIC EXPRESSIONS AND SETS OF NUMBERS	
Letters that represent numbers are called **variables**.	x, a, m, y
An **algebraic expression** is formed by numbers and variables connected by the operations of addition, subtraction, multiplication, division, raising to powers, or taking roots.	$7y, \quad -3, \quad \frac{x^2 - 9}{-2} + 14x, \quad \sqrt{3} + \sqrt{m}$
To **evaluate** an algebraic expression containing variables, substitute the given numbers for the variables and simplify. The result is called the **value** of the expression.	Evaluate: $2.7x$ when $x = 3$ $2.7x = 2.7(3)$ $= 8.1$
Natural numbers: $\{1, 2, 3, \ldots\}$ **Whole numbers:** $\{0, 1, 2, 3, \ldots\}$ **Integers:** $\{\ldots, -3, -2, -1, 0, 1, 2, 3, \ldots\}$ Each listing of three dots above is called an **ellipsis**. The members of a set are called its **elements**. **Set builder notation** describes the elements of a set but does not list them. **Real numbers:** $\{x \mid x$ corresponds to a point on the number line$\}$ **Rational numbers:** $\{\frac{a}{b} \mid a$ and b are integers and $b \neq 0\}$ **Irrational numbers:** $\{x \mid x$ is a real number and x is not a rational number$\}$	Given the set $\left\{-9.6, -5, -\sqrt{2}, 0, \frac{2}{5}, 101\right\}$, list the elements that belong to the set of Natural numbers 101 Whole numbers $0, 101$ Integers $-5, 0, 101$ Real numbers $-9.6, -5, -\sqrt{2}, 0, \frac{2}{5}, 101$ Rational numbers $-9.6, -5, 0, \frac{2}{5}, 101$ Irrational numbers $-\sqrt{2}$ Write the set $\{x \mid x$ is an integer between -2 and $5\}$ in roster form. $\{-1, 0, 1, 2, 3, 4\}$
If all the elements of set A are also in set B, we say that set A is a **subset** of set B.	The set of integers is a subset of the set of rational numbers.
SECTION 1.2 PROPERTIES OF REAL NUMBERS	
SYMBOLS	
$=$ is equal to	$-5 = -5$
$\neq$ is not equal to	$-5 \neq -3$
$>$ is greater than	$1.7 > 1.2$
$<$ is less than	$-1.7 < -1.2$
$\geq$ is greater than or equal to	$\frac{5}{3} \geq \frac{5}{3}$
$\leq$ is less than or equal to	$-\frac{1}{2} \leq \frac{1}{2}$

Section 1.2 (continued)

Identity	
$a + 0 = a$ $0 + a = a$	$3 + 0 = 3$ $0 + 3 = 3$
$a \cdot 1 = a$ $1 \cdot a = a$	$-1.8 \cdot 1 = -1.8$ $1 \cdot -1.8 = -1.8$
Inverse	
$a + (-a) = 0$ $-a + a = 0$	$7 + (-7) = 0$ $-7 + 7 = 0$
$a \cdot \frac{1}{a} = 1$ $\frac{1}{a} \cdot a = 1$	$5 \cdot \frac{1}{5} = 1$ $\frac{1}{5} \cdot 5 = 1$
Commutative	
$a + b = b + a$	$x + 7 = 7 + x$
$a \cdot b = b \cdot a$	$9 \cdot y = y \cdot 9$
Associative	
$(a + b) + c = a + (b + c)$	$(3 + 1) + 10 = 3 + (1 + 10)$
$(a \cdot b) \cdot c = a \cdot (b \cdot c)$	$(3 \cdot 1) \cdot 10 = 3 \cdot (1 \cdot 10)$
Distributive	
$a(b + c) = ab + ac$	$6(x + 5) = 6 \cdot x + 6 \cdot 5$ $= 6x + 30$

Section 1.3 Operations on Real Numbers

Absolute Value	
$\lvert a \rvert = \begin{cases} a \text{ if } a \text{ is 0 or a positive number} \\ -a \text{ if } a \text{ is a negative number} \end{cases}$	$\lvert 3 \rvert = 3, \lvert 0 \rvert = 0, \lvert -7.2 \rvert = 7.2$
Adding Real Numbers	
1. To add two numbers with the same sign, add their absolute values and attach their common sign. **2.** To add two numbers with different signs, subtract the smaller absolute value from the larger absolute value and attach the sign of the number with the larger absolute value.	$\frac{2}{7} + \frac{1}{7} = \frac{3}{7}$ $-5 + (-2.6) = -7.6$ $-18 + 6 = -12$ $20.8 + (-10.2) = 10.6$
Subtracting Real Numbers	
$a - b = a + (-b)$	$18 - 21 = 18 + (-21) = -3$
Multiplying and Dividing Real Numbers	
The product or quotient of two numbers with the same sign is positive.	$(-8)(-4) = 32$ $\frac{-8}{-4} = 2$ $8 \cdot 4 = 32$ $\frac{8}{4} = 2$

SECTION 1.3 (CONTINUED)

The product or quotient of two numbers with different signs is negative.	$-17 \cdot 2 = -34$ $\quad \frac{-14}{2} = -7$ $4(-1.6) = -6.4$ $\quad \frac{22}{-2} = -11$
A natural number **exponent** is a shorthand notation for repeated multiplication of the same factor.	$3^4 = 3 \cdot 3 \cdot 3 \cdot 3 = 81$
The notation $\sqrt{a}$ is used to denote the **positive**, or **principal, square root** of a nonnegative number a. $\sqrt{a} = b$ if $b^2 = a$ and b is positive	$\sqrt{49} = 7$
Also, $\sqrt[3]{a} = b$ if $b^3 = a$ $\sqrt[4]{a} = b$ if $b^4 = a$ and b is positive	$\sqrt[3]{64} = 4$ $\sqrt[4]{16} = 2$

SECTION 1.4 ORDER OF OPERATIONS AND ALGEBRAIC EXPRESSIONS

ORDER OF OPERATIONS Simplify expressions using the order that follows. If grouping symbols such as parentheses are present, simplify expressions within those first, starting with the innermost set. If fraction bars are present, simplify the numerator and denominator separately. **1.** Evaluate exponential expressions. **2.** Multiply or divide in order from left to right. **3.** Add or subtract in order from left to right.	Simplify: $\frac{42 - 2(3^2 - \sqrt{16})}{-8}$ $\frac{42 - 2(3^2 - \sqrt{16})}{-8} = \frac{42 - 2(9 - 4)}{-8}$ $= \frac{42 - 2(5)}{-8}$ $= \frac{42 - 10}{-8}$ $= \frac{32}{-8} = -4$

SECTION 1.5 EXPONENTS AND SCIENTIFIC NOTATION

PRODUCT RULE $a^m \cdot a^n = a^{m+n}$	$x^2 \cdot x^3 = x^5$
ZERO EXPONENT $a^0 = 1, a \neq 0$	$7^0 = 1, (-10)^0 = 1$
QUOTIENT RULE $\frac{a^m}{a^n} = a^{m-n}$	$\frac{y^{10}}{y^4} = y^{10-4} = y^6$
NEGATIVE EXPONENT $a^{-n} = \frac{1}{a^n}$	$3^{-2} = \frac{1}{3^2} = \frac{1}{9}, \frac{x^{-5}}{x^{-7}} = x^{-5-(-7)} = x^2$
A positive number is written in **scientific notation** if it is written as the product of a number a, where $1 \leq a < 10$, and an integer power of 10: $a \times 10^n$.	Numbers written in scientific notation: $568{,}000 = 5.68 \times 10^5$ $0.0002117 = 2.117 \times 10^{-4}$

SECTION 1.6 MORE WORK WITH EXPONENTS AND SCIENTIFIC NOTATION

POWER RULES $(a^m)^n = a^{m \cdot n}$ $(ab)^m = a^m b^m$ $\left(\frac{a}{b}\right)^n = \frac{a^n}{b^n}$	$(7^8)^2 = 7^{16}$ $(2y)^3 = 2^3 y^3 = 8y^3$ $\left(\frac{5x^{-3}}{x^2}\right)^{-2} = \frac{5^{-2}x^6}{x^{-4}}$ $= 5^{-2} \cdot x^{6-(-4)}$ $= \frac{x^{10}}{5^2}$, or $\frac{x^{10}}{25}$

Name ______________________ **Section** __________ **Date** __________

CHAPTER 1 REVIEW

(1.1) *Evaluate each algebraic expression at the given replacement values..*

1. $7x$ when $x = 3$

2. st when $s = 1.6$ and $t = 5$

3. The hummingbird has an average wing speed of 90 beats per second. The expression $90t$ gives the number of wing beats in t seconds. Calculate the number of wing beats in *1 hour* for the hummingbird.

Write each set in roster form.

4. $\{x | x$ is an odd integer between -2 and $4\}$

5. $\{x | x$ is an even integer between -3 and $7\}$

6. $\{x | x$ is a negative whole number$\}$

7. $\{x | x$ is a natural number that is not a rational number$\}$

8. $\{x | x$ is a whole number greater than $5\}$

9. $\{x | x$ is an integer less than $3\}$

Determine whether each statement is true or false if $A = \{6, 10, 12\}$, $B = \{5, 9, 11\}$, $C = \{\ldots, -3, -2, -1, 0, 1, 2, 3, \ldots\}$, $D = \{2, 4, 6, \ldots, 16\}$, $E = \{x | x$ *is a rational number*$\}$, $F = \{\ \}$, $G = \{x | x$ *is an irrational number*$\}$, *and* $H = \{x | x$ *is a real number*$\}$.

10. $10 \in D$

11. $59 \in B$

12. $\sqrt{169} \notin A$

13. $0 \notin F$

14. $\pi \in E$

15. $\pi \in H$

16. $\sqrt{4} \in G$

17. $-9 \in C$

List the elements of the set $\left\{5, -\frac{2}{3}, \frac{8}{2}, \sqrt{9}, 0.3, \sqrt{7}, 1\frac{5}{8}, -1, \pi\right\}$ *that are also elements of each given set.*

18. Whole numbers

19. Natural numbers

Name ______________________________

20. Rational numbers

21. Irrational numbers

22. Real numbers

23. Integers

(1.2) *Write each statement as an equation.*

24. Twelve is the product of x and negative 4.

25. The sum of n and twice n is negative fifteen.

26. Four times the sum of y and three is -1.

27. The difference of t and five, multiplied by six is four.

28. Seven subtracted from z is six.

29. Ten less than the product of x and nine is five.

30. The difference of x and 5 is the same as 12.

31. The opposite of four is equal to the product of y and seven.

32. Two-thirds is equal to twice the sum of n and one-fourth.

33. The sum of t and six amounts to negative twelve.

Find the opposite, or additive inverse, of each number.

34. $-\frac{3}{4}$

35. 0.6

36. 0

37. 1

Find the reciprocal, or multiplicative inverse, of each number.

38. $-\frac{3}{4}$

39. 0.6

40. 0

41. 1

Name each property illustrated.

42. $(M + 5) + P = M + (5 + P)$

43. $5(3x - 4) = 15x - 20$

Name ______________________________

44. $(-4) + 4 = 0$

45. $(3 + x) + 7 = 7 + (3 + x)$

46. $(XY)Z = (YZ)X$

47. $\left(-\frac{3}{5}\right) \cdot \left(-\frac{5}{3}\right) = 1$

48. $T \cdot 0 = 0$

49. $(ab)c = a(bc)$

50. $A + 0 = A$

51. $8 \cdot 1 = 8$

Complete each equation using the given property.

52. $5(x - 3z) =$ _____ Distributive property

53. $(7 + y) + (3 + x) =$ _____ Commutative property

54. $0 =$ _____ Additive inverse property

55. $1 =$ _____ Multiplicative inverse property

56. $[(3.4)(0.7)]5 =$ _____ Associative property

57. $7 =$ _____ Additive identity property

Insert $<$, $>$, or $=$ to make each statement true.

58. $-9 \quad -12$

59. $0 \quad -6$

60. $-3 \quad -1$

61. $7 \quad |-7|$

62. $-5 \quad -(-5)$

63. $-(-2) \quad -2$

(1.3) *Simplify.*

64. $-7 + 3$

65. $-10 + (-25)$

66. $5(-0.4)$

67. $(-3.1)(-0.1)$

68. $-7 - (-15)$

69. $9 - (-4.3)$

Name ______________________________

70. $\sqrt{16} - 2^3$

71. $\sqrt[3]{27} - 5^2$

72. $(-24) \div 0$

73. $0 \div (-45)$

74. $(-36) \div (-9)$

75. $(60) \div (-12)$

76. $\left(-\frac{4}{5}\right) - \left(-\frac{2}{3}\right)$

77. $\left(\frac{5}{4}\right) - \left(-2\frac{3}{4}\right)$

78. Determine the unknown fractional part.

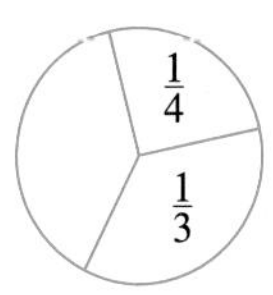

(1.4) *Simplify.*

79. $-5 + 7 - 3 - (-10)$

80. $8 - (-3) + (-4) + 6$

81. $3(4 - 5)^4$

82. $6(7 - 10)^2$

83. $\left(-\frac{8}{15}\right) \cdot \left(-\frac{2}{3}\right)^2$

84. $\left(-\frac{3}{4}\right)^2 \cdot \left(-\frac{10}{21}\right)$

85. $-\frac{6}{15} \div \frac{8}{25}$

86. $\frac{4}{9} \div -\frac{8}{45}$

87. $-\frac{3}{8} + 3(2) \div 6$

88. $5(-2) - (-3) - \frac{1}{6} + \frac{2}{3}$

89. $|2^3 - 3^2| - |5 - 7|$

90. $|5^2 - 2^2| + |9 \div (-3)|$

91. $(2^3 - 3^2) - (5 - 7)$

92. $(5^2 - 2^2) + [9 \div (-3)]$

93. $\dfrac{(8 - 10)^3 - (-4)^2}{2 + 8(2) \div 4}$

Name ______________________________

94. $\dfrac{(2+4)^2+(-1)^5}{12 \div 2 \cdot 3 - 3}$

95. $\dfrac{(4-9)+4-9}{10-12 \div 4 \cdot 8}$

96. $\dfrac{3-7-(7-3)}{15+30 \div 6 \cdot 2}$

97. $\dfrac{\sqrt{25}}{4+3 \cdot 7}$

98. $\dfrac{\sqrt{64}}{24-8 \cdot 2}$

99. The algebraic expression $2\pi r$ represents the circumference of (distance around) a circle of radius r.

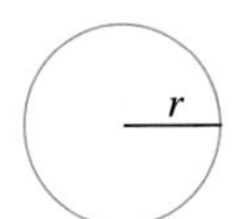

a. Complete the table by evaluating the expression at the given values of r. (Use 3.14 for π).

Radius	r	1	10	100
Circumference	$2\pi r$			

b. As the radius of a circle increases, does the circumference of the circle increase or decrease?

(1.5) *Evaluate.*

100. $(-2)^2$

101. $(-3)^4$

102. -2^2

103. -3^4

104. 8^0

105. -9^0

106. -4^{-2}

107. $(-4)^{-2}$

Simplify each expression. Write each answer with positive exponents only.

108. $-xy^2 \cdot y^3 \cdot xy^2z$

109. $(-4xy)(-3xy^2b)$

110. $a^{-14} \cdot a^5$

111. $\dfrac{a^{16}}{a^{17}}$

112. $\dfrac{x^{-7}}{x^4}$

113. $\dfrac{9a(a^{-3})}{18a^{15}}$

114. $\dfrac{y^{6p-3}}{y^{6p+2}}$

Write each number in scientific notation.

115. 36,890,000

116. -0.000362

Name ______________________________

Write each number without exponents.

117. 1.678×10^{-6}

118. 4.1×10^{5}

(1.6) *Simplify. Write each answer with positive exponents only.*

119. $(8^5)^3$

120. $\left(\frac{a}{4}\right)^2$

121. $(3x)^3$

122. $(-4x)^{-2}$

123. $\left(\frac{6x}{5}\right)^2$

124. $(8^6)^{-3}$

125. $\left(\frac{4}{3}\right)^{-2}$

126. $(-2x^3)^{-3}$

127. $\left(\frac{8p^6}{4p^4}\right)^{-2}$

128. $(-3x^{-2}y^2)^3$

129. $\left(\frac{x^{-5}y^{-3}}{z^3}\right)^{-5}$

130. $\frac{4^{-1}x^3yz}{x^{-2}yx^4}$

131. $(5xyz)^{-4}(x^{-2})^{-3}$

132. $\frac{2(3yz)^{-3}}{y^{-3}}$

Simplify each expression.

133. $x^{4a}(3x^{5a})^3$

134. $\frac{4y^{3x-3}}{2y^{2x+4}}$

Name ______________________ Section ________ Date ________

CHAPTER 1 TEST

Determine whether each statement is true or false.

1. $-2.3 > 2.33$

2. $-6^2 = (-6)^2$

3. $-5 - 8 = -(5 - 8)$

4. $(-2)(-3)(0) = \frac{(-4)}{0}$

5. All natural numbers are integers.

6. All rational numbers are integers.

Simplify.

7. $5 - 12 \div 3(2)$

8. $|4 - 6|^3 - (1 - 6^2)$

9. $(4 - 9)^3 - |-4 - 6|^2$

10. $[3|4 - 5|^5 - (-9)] \div (-6)$

11. $\frac{6(7 - 9)^3 + (-2)}{(-2)(-5)(-5)}$

Evaluate each expression when $q = 4$, $r = -2$, and $t = 1$.

12. $q^2 - r^2$

13. $\frac{5t - 3q}{3r - 1}$

14. The algebraic expression $5.75x$ represents the total cost for x adults to attend the theater.
   a. Complete the table that follows.
   b. As the number of adults increases does the total cost increase or decrease?

Adults	x	1	3	10	20
Total Cost	$5.75x$				

Write each statement as an equation.

15. Three times the quotient of n and five is the opposite of n.

16. Twenty is equal to six subtracted from twice x.

17. Negative two is equal to x divided by the sum of x and five.

ANSWERS

1. ________
2. ________
3. ________
4. ________
5. ________
6. ________
7. ________
8. ________
9. ________
10. ________
11. ________
12. ________
13. ________
14. **a.** see table
    b. ________
15. ________
16. ________
17. ________

18. ____________

19. ____________

20. ____________

21. ____________

22. ____________

23. ____________

24. ____________

25. ____________

26. ____________

27. ____________

28. ____________

29. ____________

30. ____________

31. ____________

Name ______________________________

Name each property illustrated.

18. $6(x - 4) = 6x - 24$

19. $(4 + x) + z = 4 + (x + z)$

20. $(-7) + 7 = 0$

21. $(-18)(0) = 0$

Simplify. Write answers using positive exponents only.

22. $(-9x)^{-2}$

23. $\frac{6^{-1}a^2b^{-3}}{3^{-2}a^{-5}b^2}$

24. $\left(\frac{-xy^{-5}z}{xy^3}\right)^{-5}$

Write each number in scientific notation.

25. 630,000,000

26. 0.01200

27. Write 5.0×10^{-6} without exponents.

28. Use scientific notation to find the quotient.

$$\frac{(0.0024)(0.00012)}{0.00032}$$

Use scientific notation to find the quotient. Express the quotient in scientific notation.

29. $\frac{(0.00012)(144{,}000)}{0.0003}$

Simplify. Write each answer using positive exponents only.

30. $\frac{27x^{-5}y^5}{18x^{-6}y^2} \cdot \frac{x^4y^{-2}}{x^{-2}y^3}$

31. $\frac{(x^w)^2}{(x^{w-4})^{-2}}$

Equations, Inequalities, and Problem Solving

CHAPTER 2

Mathematics is a tool for solving problems in such diverse fields as transportation, engineering, economics, medicine, business, and biology. We solve problems using mathematics by modeling real-world phenomena with mathematical equations or inequalities. Our ability to solve problems using mathematics, then, depends in part on our ability to solve equations and inequalities. In this chapter, we solve linear equations and inequalities in one variable and graph their solutions on number lines.

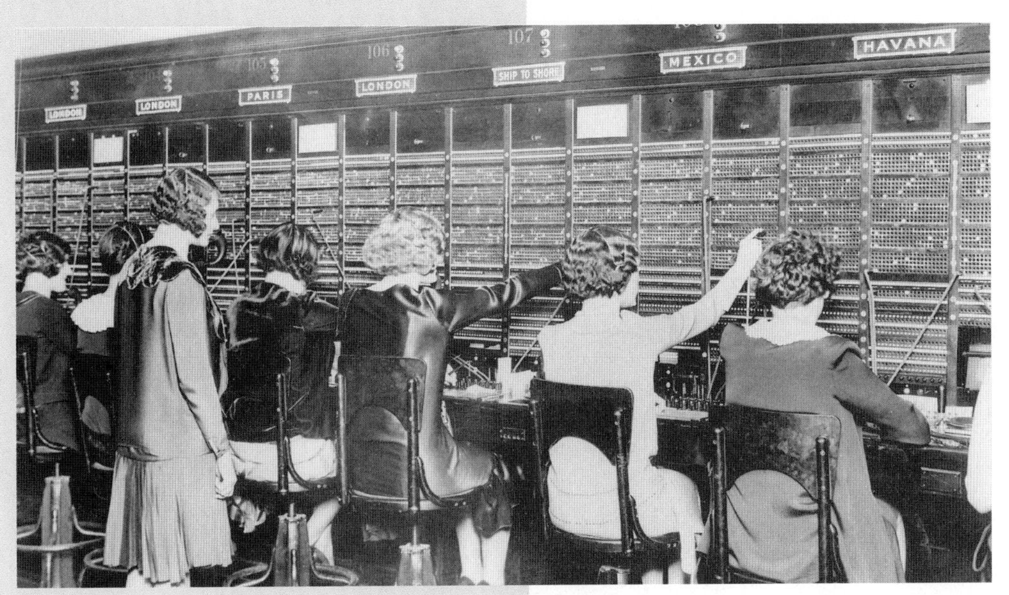

Alexander Graham Bell patented his telephone invention in 1876. He introduced it to the world later that year at the 1876 Centennial Exposition in Philadelphia. The following year he formed the Bell Telephone Company, and the modern telecommunications industry was born. At first, telephone service was available only on a subscription basis. Citizens paid to be connected to their neighbors within a local exchange. As exchanges were connected between cities, the telephone system grew. By 1885, there were 140,000 telephone subscribers in the United States. Well into the 20th century, telephone calls were connected manually at banks of switchboards by legions of telephone operators. The invention of automatic switching equipment allowed telephone calls to be connected first mechanically and later electronically. This development freed telephone operators to concentrate on providing services such as collect calls, third-party billing, and directory assistance to telephone customers. In Exercise 39 on page 101, we will analyze the change in employment of modern telephone operators.

ANSWERS

1. ____

2. ____

3. ____

4. ____

5. ____

6. ____

7. ____

8. ____

9. ____

10. ____

11. ____

12. ____

13. ____

14. ____

15. ____

16. ____

17. ____

18. ____

19. ____

20. ____

Name ____ Section ____ Date ____

CHAPTER 2 PRETEST

Solve each equation.

1. $2x - 17 = 21$

2. $3x - 2 + 14 = 8x + 4 - x$

3. $\frac{3y}{5} + 1 = \frac{4y}{3} - 2$

4. $7(t - 1) + 6 = 7t + 10$

5. $|8 - 3d| = 5$

6. $|2x - 1| = |-x + 4|$

Solve each equation for the specified variable.

7. $7y + 5x = 6$; for y

8. $S = 2LW + 2LH + 2WH$; for L

Solve each inequality.

9. $x + 12 \leq -8$

10. $\frac{5}{7}y > 20$

11. $2(x - 9) \leq 4x + 6$

12. $\frac{5x - 3}{6} - \frac{x + 4}{3} < -2$

13. $x \geq -1$ and $x \geq 2$

14. $x + 2 \geq -3$ and $x + 1 \leq 5$

15. $3x - 2 < 1$ or $2x < 10$

16. $|x + 9| \leq 6$

17. $|3 - x| \geq 1$

Solve.

18. Find two numbers such that the second number is 4 less than twice the first number and the sum of the two numbers is 50.

19. Find 18% of 900.

20. If the area of a triangular sign is 4 square feet and its base is 2 feet, find the height of the sign.

2.1 LINEAR EQUATIONS IN ONE VARIABLE

Objectives

A Decide whether a number is a solution of an equation.

B Solve linear equations using properties of equality.

C Solve linear equations that can be simplified by combining like terms.

D Solve linear equations containing fractions or decimals.

E Recognize identities and equations with no solution.

SSM

CD-ROM

Video 2.1

A DECIDING WHETHER A NUMBER IS A SOLUTION OF AN EQUATION

An **equation** is a statement that two expressions are equal. To solve problems, we need to be able to solve equations. In this section, we will solve a special type of equation called a **linear equation in one variable**.

> LINEAR EQUATION IN ONE VARIABLE
>
> A linear equation in one variable is an equation that can be written in the form
>
> $ax + b = c$
>
> where a, b, and c are real numbers and $a \neq 0$. For example,
>
> $3x = -15 \qquad 7 - y = 3y \qquad 4n - 9n + 6 = 0 \qquad z = -2$

When a variable in an equation is replaced by a number and the resulting equation is true, then that number is called a **solution** of the equation. For example, 1 is a solution of the equation $3x + 4 = 7$, since $3(1) + 4 = 7$ is a true statement. But 2 is not a solution of this equation, since $3(2) + 4 = 7$ is *not* a true statement. The **solution set** of an equation is the set of solutions of the equation. For example, the solution set of $3x + 4 = 7$ is $\{1\}$.

Example 1 Determine whether -15 is a solution of $x - 9 = -24$.

Solution: We replace x with -15 and see whether a true statement results.

$$x - 9 = -24$$
$$-15 - 9 \stackrel{?}{=} -24 \quad \text{Replace } x \text{ with } -15.$$
$$-24 = -24 \quad \text{True.}$$

Since a true statement results, -15 is a solution.

Practice Problem 1

Determine whether -7 is a solution of $14 - x = 21$.

Example 2 Determine whether 5 is a solution of $2x - 3 = x + 3$.

Solution:

$$2x - 3 = x + 3$$
$$2 \cdot 5 - 3 \stackrel{?}{=} 5 + 3 \quad \text{Replace } x \text{ with } 5.$$
$$7 = 8 \quad \text{False.}$$

Since a false statement results, 5 is not a solution.

Practice Problem 2

Determine whether 8 is a solution of $x - 10 = 2x - 14$.

B USING THE PROPERTIES OF EQUALITY

To **solve an equation** is to find the solution set of an equation. Equations with the same solution set are called **equivalent equations**. For example,

$$3x + 4 = 7 \qquad 3x = 3 \qquad x = 1$$

are equivalent equations because they all have the same solution set, namely, $\{1\}$. To solve an equation in x, we start with the given equation and write a series of simpler equivalent equations until we obtain an equation of the form

$$x = \textbf{number}$$

Answers

1. -7 is a solution, **2.** 8 is not a solution

To write equivalent equations, we use two important properties.

THE ADDITION PROPERTY OF EQUALITY

If a, b, and c, are real numbers, then

$$a = b \text{ and } a + c = b + c$$

are equivalent equations.

THE MULTIPLICATION PROPERTY OF EQUALITY

If $c \neq 0$, then

$$a = b \text{ and } ac = bc$$

are equivalent equations.

The **addition property of equality** guarantees that the same number may be added to (or subtracted from) both sides of an equation, and the result is an equivalent equation. The **multiplication property of equality** guarantees that both sides of an equation may be multiplied by (or divided by) the same nonzero number, and the result is an equivalent equation.

For example, to solve $2x + 5 = 9$, we use the addition and multiplication properties of equality to get x alone—that is, to write an equivalent equation of the form

$$x = \text{number}$$

We will do this in the next example.

Practice Problem 3

Solve: $3x + 6 = 21$

Example 3

Solve: $2x + 5 = 9$

Solution: First we use the addition property of equality and subtract 5 from both sides.

$$2x + 5 = 9$$
$$2x + 5 - 5 = 9 - 5 \quad \text{Subtract 5 from both sides.}$$
$$2x = 4 \quad \text{Simplify.}$$

Now we use the multiplication property of equality and divide both sides by 2.

$$\frac{2x}{2} = \frac{4}{2} \quad \text{Divide both sides by 2.}$$
$$x = 2 \quad \text{Simplify.}$$

Check: To check, we replace x in the original equation with 2.

$$2x + 5 = 9 \quad \text{Original equation}$$
$$2(2) + 5 \stackrel{?}{=} 9 \quad \text{Replace } x \text{ with 2.}$$
$$4 + 5 \stackrel{?}{=} 9$$
$$9 = 9 \quad \text{True.}$$

The solution set is $\{2\}$.

Practice Problem 4

Solve: $4.5 = 3 + 2.5x$

Answers

3. $\{5\}$, **4.** $\{0.6\}$

Example 4

Solve: $0.6 = 2 - 3.5c$

Solution: We use both the addition property and the multiplication property of equality.

$$0.6 = 2 - 3.5c$$
$$0.6 - 2 = 2 - 3.5c - 2 \quad \text{Subtract 2 from both sides.}$$
$$-1.4 = -3.5c \quad \text{Simplify.}$$
$$\frac{-1.4}{-3.5} = \frac{-3.5c}{-3.5} \quad \text{Divide both sides by } -3.5.$$
$$0.4 = c \quad \text{Simplify } \frac{-1.4}{-3.5}.$$

Check:

$$0.6 = 2 - 3.5c$$
$$0.6 \stackrel{?}{=} 2 - 3.5(0.4) \quad \text{Replace } c \text{ with } 0.4.$$
$$0.6 \stackrel{?}{=} 2 - 1.4 \quad \text{Multiply.}$$
$$0.6 = 0.6 \quad \text{True.}$$

The solution set is $\{0.4\}$.

HELPFUL HINT

Don't forget that

$0.4 = c$ and $c = 0.4$ are equivalent equations.

We may solve an equation so that the variable is alone on either side of the equation.

C SOLVING LINEAR EQUATIONS BY COMBINING LIKE TERMS

Often, an equation can be simplified by removing any grouping symbols and combining any like terms.

Example 5 Solve: $-6x - 1 + 5x = 3$

Solution: First we simplify the left side of this equation by combining the like terms $-6x$ and $5x$. Then we use the addition property of equality and add 1 to both sides of the equation.

$$-6x - 1 + 5x = 3$$
$$-x - 1 = 3 \quad \text{Combine like terms.}$$
$$-x - 1 + 1 = 3 + 1 \quad \text{Add 1 to both sides.}$$
$$-x = 4 \quad \text{Simplify.}$$

Notice that this equation is not solved for x since we have $-x$, or $-1x$, not x. To get x alone, we divide both sides by -1.

$$\frac{-x}{-1} = \frac{4}{-1} \quad \text{Divide both sides by } -1.$$
$$x = -4 \quad \text{Simplify.}$$

Check to see that the solution set is $\{-4\}$.

Practice Problem 5

Solve: $-2x + 2 - 4x = 20$

If an equation contains parentheses, we use the distributive property to remove them.

Example 6 Solve: $2(x - 3) = 5x - 9$

Solution: First we use the distributive property.

$$2(x - 3) = 5x - 9$$
$$2x - 6 = 5x - 9 \quad \text{Use the distributive property.}$$

Practice Problem 6

Solve: $4(x - 2) = 6x - 10$

Answers

5. $\{-3\}$, **6.** $\{1\}$

Next we get variable terms on the same side of the equation by using the addition property of equality.

$$2x - 6 - 5x = 5x - 9 - 5x \quad \text{Subtract } 5x \text{ from both sides.}$$
$$-3x - 6 = -9 \quad \text{Simplify.}$$
$$-3x - 6 + 6 = -9 + 6 \quad \text{Add 6 to both sides.}$$
$$-3x = -3 \quad \text{Simplify.}$$
$$\frac{-3x}{-3} = \frac{-3}{-3} \quad \text{Divide both sides by } -3.$$
$$x = 1$$

Check to see that $\{1\}$ is the solution set.

HELPFUL HINT

When we multiply both sides of an equation by a number, by the distributive property, each term of the equation is multiplied by that number.

D Solving Linear Equations Containing Fractions or Decimals

If an equation contains fractions, we first clear the equation of fractions by multiplying both sides of the equation by the *least common denominator* (LCD) of all fractions in the equation.

Example 7 Solve: $\frac{y}{3} - \frac{y}{4} = \frac{1}{6}$

Solution: First we clear the equation of fractions by multiplying both sides of the equation by 12, the LCD of the denominators 3, 4, and 6.

$$\frac{y}{3} - \frac{y}{4} = \frac{1}{6}$$
$$12\left(\frac{y}{3} - \frac{y}{4}\right) = 12\left(\frac{1}{6}\right) \quad \text{Multiply both sides by the LCD, 12.}$$
$$12\left(\frac{y}{3}\right) - 12\left(\frac{y}{4}\right) = 2 \quad \text{Use the distributive property.}$$
$$4y - 3y = 2 \quad \text{Simplify.}$$
$$y = 2 \quad \text{Simplify.}$$

Check: To check, we replace y with 2 in the original equation.

$$\frac{y}{3} - \frac{y}{4} = \frac{1}{6} \quad \text{Original equation}$$
$$\frac{2}{3} - \frac{2}{4} \stackrel{?}{=} \frac{1}{6} \quad \text{Replace } y \text{ with 2.}$$
$$\frac{8}{12} - \frac{6}{12} \stackrel{?}{=} \frac{1}{6} \quad \text{Write fractions with the LCD.}$$
$$\frac{2}{12} \stackrel{?}{=} \frac{1}{6} \quad \text{Subtract.}$$
$$\frac{1}{6} = \frac{1}{6} \quad \text{Simplify.}$$

Since a true statement results, the solution set is $\{2\}$.

Practice Problem 7

Solve: $\frac{x}{6} - \frac{x}{8} = \frac{1}{8}$

As a general guideline, the following steps may be used to solve a linear equation in one variable.

Answer

7. $\{3\}$

SOLVING A LINEAR EQUATION IN ONE VARIABLE

Step 1. Clear the equation of fractions by multiplying both sides of the equation by the least common denominator (LCD) of all denominators in the equation.

Step 2. Use the distributive property to remove grouping symbols such as parentheses.

Step 3. Combine like terms on each side of the equation.

Step 4. Use the addition property of equality to rewrite the equation as an equivalent equation, with variable terms on one side and numbers on the other side.

Step 5. Use the multiplication property of equality to get the variable alone.

Step 6. Check the proposed solution in the original equation.

Example 8

Solve: $\frac{x+5}{2} + \frac{1}{2} = 2x - \frac{x-3}{8}$

Solution: To begin, we multiply both sides of the equation by 8, the LCD of 2 and 8.

$$8\left(\frac{x+5}{2} + \frac{1}{2}\right) = 8\left(2x - \frac{x-3}{8}\right)$$ Multiply both sides by 8.

$$4(x+5) + 4 = 16x - (x-3)$$ Use the distributive property.

$$4x + 20 + 4 = 16x - x + 3$$ Use the distributive property to remove parentheses.

$$4x + 24 = 15x + 3$$ Combine like terms.

$$4x - 15x = 3 - 24$$ Subtract $15x$ and 24 from both sides.

$$-11x = -21$$ Simplify.

$$\frac{-11x}{-11} = \frac{-21}{-11}$$ Divide both sides by -11.

$$x = \frac{21}{11}$$ Simplify.

To check, verify that replacing x with $\frac{21}{11}$ makes the original equation true. The solution set is $\left\{\frac{21}{11}\right\}$.

If an equation contains decimals, you may want to first clear the equation of decimals.

Example 9

Solve: $0.3x + 0.1 = 0.27x - 0.02$

Solution: To clear this equation of decimals, we multiply both sides of the equation by 100. Recall that multiplying a number by 100 moves its decimal point two places to the right.

$$100(0.3x + 0.1) = 100(0.27x - 0.02)$$

$$100(0.3x) + 100(0.1) = 100(0.27x) - 100(0.02)$$ Use the distributive property.

$$30x + 10 = 27x - 2$$ Multiply.

$$30x - 27x = -2 - 10$$ Subtract $27x$ and 10 from both sides.

$$3x = -12$$ Simplify.

Practice Problem 8

Solve: $\frac{x-1}{3} + \frac{2}{3} = x - \frac{2x+3}{9}$

Practice Problem 9

Solve: $0.2x + 0.1 = 0.12x - 0.06$

Answers

8. $\left\{\frac{3}{2}\right\}$, **9.** $\{-2\}$

$$\frac{3x}{3} = \frac{-12}{3} \quad \text{Divide both sides by 3.}$$
$$x = -4 \quad \text{Simplify.}$$

Check to see that the solution set is $\{-4\}$.

TRY THE CONCEPT CHECK IN THE MARGIN.

✓ CONCEPT CHECK

Find and correct the error in the following solution.

$$3x - 5 = 16$$
$$3x = 11$$
$$\frac{3x}{3} = \frac{11}{3}$$
$$x = \frac{11}{3}$$

E RECOGNIZING IDENTITIES AND EQUATIONS WITH NO SOLUTION

So far, each linear equation that we have solved has had a single solution. We will now look at two other types of equations: contradictions and identities.

An equation in one variable that has no solution is called a **contradiction**, and an equation in one variable that has every number (for which the equation is defined) as a solution is called an **identity**. The next examples show how to recognize contradictions and identities.

Practice Problem 10

Solve: $5x - 1 = 5(x + 3)$

Example 10 Solve: $3x + 5 = 3(x + 2)$

Solution: First we use the distributive property and remove parentheses.

$$3x + 5 = 3(x + 2)$$
$$3x + 5 = 3x + 6 \quad \text{Use the distributive property.}$$
$$3x + 5 - 3x = 3x + 6 - 3x \quad \text{Subtract } 3x \text{ from both sides.}$$
$$5 = 6$$

The equation $5 = 6$ is a false statement no matter what value the variable x might have. Thus the original equation has no solution. Its solution set is written either as $\{\ \}$ or $\emptyset$. This equation is a contradiction.

Practice Problem 11

Solve: $-4(x - 1) = -4x - 9 + 13$

Example 11 Solve: $6x - 4 = 2 + 6(x - 1)$

Solution: First we use the distributive property and remove parentheses.

$$6x - 4 = 2 + 6(x - 1)$$
$$6x - 4 = 2 + 6x - 6 \quad \text{Use the distributive property.}$$
$$6x - 4 = 6x - 4 \quad \text{Combine like terms.}$$

At this point we might notice that both sides of the equation are the same, so replacing x by any real number gives a true statement. Thus the solution set of this equation is the set of real numbers, and the equation is an identity. Continuing to "solve" $6x - 4 = 6x - 4$, we eventually arrive at the same conclusion.

$$6x - 4 + 4 = 6x - 4 + 4 \quad \text{Add 4 to both sides.}$$
$$6x = 6x \quad \text{Simplify.}$$
$$6x - 6x = 6x - 6x \quad \text{Subtract } 6x \text{ from both sides.}$$
$$0 = 0 \quad \text{Simplify.}$$

Since $0 = 0$ is a true statement for every value of x, the solution set is the set of all real numbers, which can be written as $\{x|x \text{ is a real number}\}$. The equation is called an identity.

> **HELPFUL HINT**
>
> For linear equations, *any* false statement such as $5 = 6$, $0 = 1$, or $-2 = 2$ informs us that the original has no solution. Also, *any* true statement such as $0 = 0$, $2 = 2$, or $-5 = -5$ informs us that the original equation is an identity.

Answers

10. $\emptyset$, **11.** $\{x|x \text{ is a real number}\}$

✓ Concept Check:

$$3x - 5 = 16$$
$$3x = 21$$
$$x = 7$$

Therefore the correct solution set is $\{7\}$.

Name ______________________ Section ________ Date ________

Mental Math

Solve each equation mentally.

1. $3x = 18$
2. $2x = 60$
3. $x - 7 = 10$
4. $x - 2 = 15$
5. $\frac{x}{2} = 4$
6. $\frac{x}{3} = 5$
7. $x + 1 = 11$
8. $x + 4 = 20$

Exercise Set 2.1

A *Determine whether each number is a solution of the given equation. See Examples 1 and 2.*

1. $-24; \frac{x}{-6} = 4$
2. $15; \frac{x}{-3} = -5$
3. $-3; x - 17 = 20$
4. $-8; x - 10 = -2$
5. $-2; 5 + 3x = -1$
6. $-1; 6 - 2x = 4$
7. $5; x - 7 = x + 2$
8. $5; x - 1 = x - 1$
9. $5; 4(x - 3) = 12$
10. $12; 5(x - 6) = 30$
11. $-8; 4x - 2 = 5x + 6$
12. $2; 7x + 1 = 6x - 1$

B *Solve each equation and check. See Examples 3 and 4.*

13. $-5x = -30$
14. $-2x = 18$
15. $10 = x + 12$
16. $25 = y + 30$
17. $x + 2.8 = 1.9$
18. $y - 8.6 = -6.3$
19. $5x - 4 = 26$
20. $2y - 3 = 11$
21. $-4.1 - 7z = 3.6$
22. $10.3 - 6x = -2.3$
23. $5y + 12 = 2y - 3$
24. $4x + 14 = 6x + 8$

Mental Math Answers

1. ______
2. ______
3. ______
4. ______
5. ______
6. ______
7. ______
8. ______

Answers

1. ______
2. ______
3. ______
4. ______
5. ______
6. ______
7. ______
8. ______
9. ______
10. ______
11. ______
12. ______
13. ______
14. ______
15. ______
16. ______
17. ______
18. ______
19. ______
20. ______
21. ______
22. ______
23. ______
24. ______

25.

26.

27.

28.

29.

30.

31.

32.

33.

34.

35.

36.

37. a.

b.

c.

38.

39.

40.

41.

42.

43.

44.

Name

C *Solve each equation and check. See Examples 5 and 6.*

25. $8x - 5x + 3 = x - 7 + 10$

26. $6 + 3x + x = -x + 2 - 26$

27. $5x + 12 = 2(2x + 7)$

28. $2(x + 3) = x + 5$

29. $3(x - 6) = 5x$

30. $6x = 4(5 + x)$

31. $3x - 4 - 5x = x + 4 + x$

32. $13x - 15x + 8 = 4x + 2 - 24$

33. $-2(5y - 1) - y = -4(y - 3)$

34. $-3(2w - 7) - 10 = 9 - 2(5w + 4)$

35. $y + 0.2 = 0.6(y + 3)$

36. $-(w + 0.2) = 0.3(4 - w)$

37. a. Simplify the expression $4(x + 1) + 1$.

b. Solve the equation $4(x + 1) + 1 = -7$.

c. Explain the difference between solving an equation for a variable and simplifying an expression.

38. Explain why the multiplication property of equality does not include multiplying both sides of an equation by 0. (*Hint:* Write down a false statement and then multiply both sides by 0. Is the result true or false? What does this mean?

D *Solve each equation and check. See Examples 7 through 9.*

39. $\frac{x}{2} + \frac{2}{3} = \frac{3}{4}$

40. $\frac{x}{2} + \frac{x}{3} = \frac{5}{2}$

41. $\frac{3t}{4} - \frac{t}{2} = 1$

42. $\frac{4r}{5} - 7 = \frac{r}{10}$

43. $\frac{n - 3}{4} + \frac{n + 5}{7} = \frac{5}{14}$

44. $\frac{2 + h}{9} + \frac{h - 1}{3} = \frac{1}{3}$

Name ______________________

45. $0.6x - 10 = 1.4x - 14$

46. $0.3x + 2.4 = 0.1x + 4$

47. $\frac{3x - 1}{9} + x = \frac{3x + 1}{3} + 4$

48. $\frac{2z + 7}{8} - 2 = z + \frac{z - 1}{2}$

49. $1.5(4 - x) = 1.3(2 - x)$

50. $2.4(2x + 3) = -0.1(2x + 3)$

E *Solve each equation and check. Some of these equations are identities and some are contradictions. See Examples 10 and 11.*

51. $4(n + 3) = 2(6 + 2n)$

52. $6(4n + 4) = 8(3 + 3n)$

53. $3(x + 1) + 5 = 3x + 2$

54. $5x - (x + 4) = 4 + 4(x + 2)$

55. $2y + 5(y - 4) = 4y - 2(y - 10)$

56. $9c - 3(6 - 5c) = c - 2(3c + 9)$

57. $2(x - 8) + x = 3(x - 6) + 2$

58. $4(x + 5) = 3(x - 4) + x$

59. $\frac{3}{8} + \frac{b}{3} = \frac{5}{12}$

60. $\frac{a}{2} + \frac{7}{4} = 5$

61. $x - 10 = -6x + 4$

62. $4x - 7 = 2x - 7$

63. $5(x - 2) + 2x = 7(x + 4)$

64. $3x + 2(x + 4) = 5(x + 1) + 3$

65. $\frac{1}{4}(a + 2) = \frac{1}{6}(5 - a)$

66. $\frac{1}{3}(8 + 2c) = \frac{1}{5}(3c - 5)$

45. ______
46. ______
47. ______
48. ______
49. ______
50. ______
51. ______
52. ______
53. ______
54. ______
55. ______
56. ______
57. ______
58. ______
59. ______
60. ______
61. ______
62. ______
63. ______
64. ______
65. ______
66. ______

67. ______

68. ______

69. ______

70. ______

71. ______

72. ______

73. ______

74. ______

75. ______

76. ______

77. ______

78. ______

79. ______

80. ______

81. ______

82. ______

83. ______

84. ______

Name ______________________________

67. $6x - 2(x - 3) = 4(x + 1) + 4$

68. $10x - 2(x + 4) = 8(x - 2) + 6$

69. $\frac{m - 4}{3} - \frac{3m - 1}{5} = 1$

70. $\frac{n + 1}{8} - \frac{2 - n}{3} = \frac{5}{6}$

71. In your own words, explain why the equation $x + 7 = x + 6$ has no solution while the solution set of the equation $x + 7 = x + 7$ contains all real numbers.

72. In your own words, explain why the equation $x = -x$ has one solution—namely, 0—while the solution set of the equation $x = x$ is all real numbers.

Review and Preview

Translate each phrase into an expression. Use the variable x to represent each unknown number. See Section 1.1.

73. the quotient of 8 and a number

74. the sum of 8 and a number

75. the product of 8 and a number

76. the difference of 8 and a number

77. 2 more than three times a number

78. 5 subtracted from twice a number

Combining Concepts

Solve and check.

79. $-9.112y = -47.537304$

80. $2.86z - 8.1258 = -3.75$

81. $x(x - 6) + 7 = x(x + 1)$

82. $7x^2 + 2x - 3 = 6x(x + 4) + x^2$

Find the value of K such that the equations are equivalent.

83. $3.2x + 4 = 5.4x - 7$

$3.2x = 5.4x + K$

84. $\frac{x}{6} + 4 = \frac{x}{3}$

$x + K = 2x$

2.2 AN INTRODUCTION TO PROBLEM SOLVING

Objective

A Apply the steps for problem solving.

SSM

CD-ROM

Video 2.2

A SOLVING PROBLEMS

Our main purpose for studying algebra is to solve problems. The following problem-solving strategy will be used throughout this text and may also be used to solve real-life problems that occur outside the mathematics classroom.

HELPFUL HINT

You may want to begin this section by studying key words and phrases and their translations in Sections 1.1 Objective C and 1.2 Objective A.

GENERAL STRATEGY FOR PROBLEM SOLVING

1. UNDERSTAND the problem. During this step, become comfortable with the problem. Some ways of doing this are:

   Read and reread the problem.
   Choose a variable to represent the unknown.
   Construct a drawing.
   Propose a solution and check. Pay careful attention to how you check your proposed solution. This will help when writing an equation to model the problem.

2. TRANSLATE the problem into an equation.
3. SOLVE the equation.
4. INTERPRET the results: *Check* the proposed solution in the stated problem and *state* your conclusion.

Let's review this strategy by solving a problem involving unknown numbers.

Example 1 Finding Unknown Numbers

Find two numbers such that the second number is 3 more than twice the first number and the sum of the two numbers is 72.

Solution:

1. UNDERSTAND the problem. First let's read and reread the problem and then propose a solution. For example, if the first number is 25, then the second number is 3 more than twice 25, or 53. The sum of 25 and 53 is 78, not the required sum, but we have gained some valuable information about the problem. First, we know that the first number is less than 25 since our guess led to a sum greater than the required sum. Also, we have gained some information as to how to model the problem.

HELPFUL HINT

The purpose of guessing a solution is not to guess correctly but to gain confidence and to help understand the problem and how to model it.

Practice Problem 1

One number is three times another number. If their sum is 148, find the two numbers.

Answer

1. 37 and 111

Next let's assign a variable and use this variable to represent any other unknown quantities. If we let

the first number $= x$, then
the second number $= 2x + 3$

↑ 3 more than
twice the second number

2. TRANSLATE the problem into an equation. To do so, we use the fact that the sum of the numbers is 72. First let's write this relationship in words and then translate to an equation.

In words:	first number	added to	second number	is	72
	↓	↓	↓	↓	↓
Translate:	x	$+$	$(2x + 3)$	$=$	72

3. SOLVE the equation.

$$x + (2x + 3) = 72$$
$$x + 2x + 3 = 72 \quad \text{Remove parentheses.}$$
$$3x + 3 = 72 \quad \text{Combine like terms.}$$
$$3x = 69 \quad \text{Subtract 3 from both sides.}$$
$$x = 23 \quad \text{Divide both sides by 3.}$$

4. INTERPRET. Here, we *check* our work and *state* the solution. Recall that if the first number $x = 23$, then the second number $2x + 3 = 2 \cdot 23 + 3 = 49$.

Check: Is the second number 3 more than twice the first number? Yes, since 3 more than twice 23 is $46 + 3$, or 49. Also, their sum, $23 + 49 = 72$, is the required sum.

State: The two numbers are 23 and 49.

Many of today's rates and statistics are given as percents. Interest rates, tax rates, nutrition labeling, and percent of households in a given category are just a few examples. Before we practice solving problems containing percents, let's take a moment and review the meaning of percent and how to find a percent of a number.

The word *percent* means *per hundred*, and the symbol % is used to denote percent. This means that 23% is 23 per hundred, or $\frac{23}{100}$. Also,

$$41\% = \frac{41}{100} = 0.41$$

To find a percent of a number, we multiply.

Example 2

Find 16% of 25.

Solution: To find 16% of 25, we find the product of 16% (written as a decimal) and 25.

$$16\% \cdot 25 = 0.16 \cdot 25$$
$$= 4$$

Thus, 16% of 25 is 4.

Next, we solve a problem containing percent.

Practice Problem 2

Find 35% of 160.

Answer

2. 56

TRY THE CONCEPT CHECK IN THE MARGIN.

Example 3 Finding the Original Price of a Computer

Suppose that Service Merchandise just announced an 8% decrease in the price of their Compaq Presario computers. If one particular computer model sells for $2162 after the decrease, find the original price of this computer.

Solution:

1. UNDERSTAND. Read and reread the problem. Recall that a percent decrease means a percent of the original price. Let's guess that the original price of the computer is $2500. The amount of decrease is then 8% of $2500, or $(0.08)(\$2500) = \$200$. This means that the new price of the computer is the original price minus the decrease, or $\$2500 - \$200 = \$2300$. Our guess is incorrect, but we now have an idea of how to model this problem. In our model, we will let $x =$ the original price of the computer.

2. TRANSLATE.

In words:	original price of computer	minus	8% of original price	is	new price
	↓	↓	↓	↓	↓
Translate:	x	$-$	$0.08x$	$=$	2162

3. SOLVE the equation.

$$x - 0.08x = 2162$$

$$0.92x = 2162 \quad \text{Combine like terms.}$$

$$x = \frac{2162}{0.92} = 2350 \quad \text{Divide both sides by 0.92.}$$

4. INTERPRET.

Check: If the original price of the computer was $2350, the new price is

$$\$2350 - (0.08)(\$2350) = \$2350 - \$188$$

$$= \$2162 \quad \text{The given new price}$$

State: The original price of the computer was $2350.

Example 4 Finding the Lengths of a Triangle's Sides

A pennant in the shape of an isosceles triangle is to be constructed for the Slidell High School Athletic Club and

✓ CONCEPT CHECK

Suppose you are finding 112% of a number x. Which of the following is a correct description of the result? Explain.

a. The result is less than x.

b. The result is equal to x.

c. The result is greater than x.

Practice Problem 3

The price of a home was just decreased by 6%. If the decreased price is $83,660, find the original price of the home.

Practice Problem 4

A rectangle has a perimeter of 106 meters. Its length is 5 meters more than twice its width. Find the length and the width of the rectangle.

Answers

3. $89,000, **4.** length: 16 m, width: 37 m

✓ Concept Check: c; The result is greater than x.

sold at a fund-raiser. The company manufacturing the pennant charges according to perimeter, and the athletic club has determined that a perimeter of 149 centimeters should make a nice profit. If each equal side of the triangle is twice the length of the third side, increased by 12 centimeters, find the lengths of the sides of the triangular pennant.

Solution: **1.** UNDERSTAND. Read and reread the problem. Recall that the perimeter of a triangle is the distance around. Let's guess that the third side of the triangular pennant is 20 centimeters. This means that each equal side is twice 20 centimeters, increased by 12 centimeters, or $2(20) + 12 = 52$ centimeters.

This gives a perimeter of $20 + 52 + 52 = 124$ centimeters. Our guess is incorrect, but we now have a better understanding of how to model this problem.

52 centimeters
20 centimeters
Slidell High
52 centimeters

Now we let the third side of the triangle $= x$

the first side =	twice	the third side	increased by 12	
	↓	↓	↓	
=	2	x	+	12,

or $2x + 12$

the second side $= 2x + 12$

2. TRANSLATE.

In words:	first side	+	second side	+	third side	=	149
	↓		↓		↓		↓
Translate:	$(2x + 12)$	+	$(2x + 12)$	+	x	=	149

3. SOLVE the equation.

$$(2x + 12) + (2x + 12) + x = 149$$
$$2x + 12 + 2x + 12 + x = 149 \quad \text{Remove parentheses.}$$
$$5x + 24 = 149 \quad \text{Combine like terms.}$$
$$5x = 125 \quad \text{Subtract 24 from both sides.}$$
$$x = 25 \quad \text{Divide both sides by 5.}$$

4. INTERPRET. If the third side is 25 centimeters, then the first side is $2(25) + 12 = 62$ centimeters and the second side is 62 centimeters also.

Check: The first and second sides are each twice 25 centimeters increased by 12 centimeters or 62 centimeters. Also, the perimeter is $25 + 62 + 62 = 149$ centimeters, the required perimeter.

State: The dimensions of the triangle are 25 centimeters, 62 centimeters, and 62 centimeters.

Consecutive integers are integers that follow one another in order. Study the examples of consecutive, even, and odd integers and their representations.

Consecutive Integers:

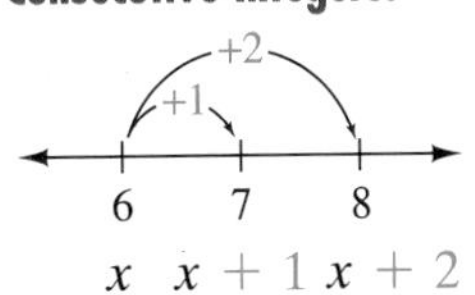

Consecutive Even Integers:

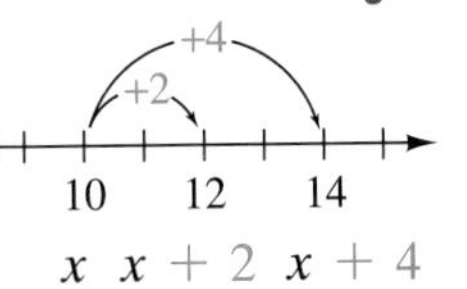

Consecutive Odd Integers:

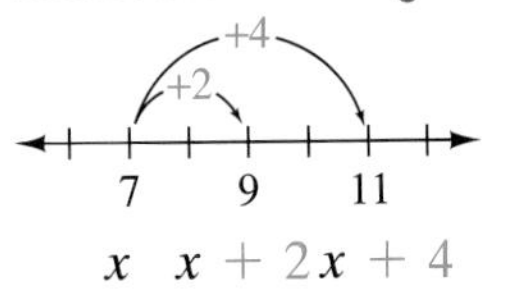

Example 5 Kelsey Ohleger was helping her friend Benji Burnstine study for an exam. Kelsey told Benji that her two latest art history quiz scores are two consecutive even integers whose sum is 174. Help Benji find the scores.

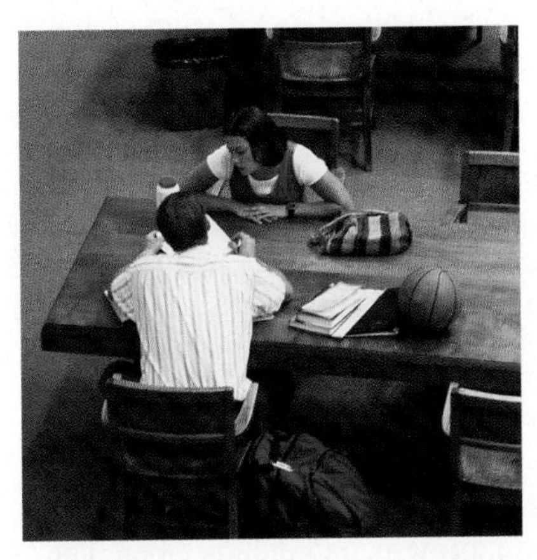

Solution: **1.** UNDERSTAND. Read and reread the problem. Since we are looking for consecutive even integers, let

$x =$ the first integer. Then

$x + 2 =$ the next consecutive even integer.

2. TRANSLATE.

In words:	first integer	+	next even integer	=	174
	↓		↓		↓
Translate:	x	+	$(x + 2)$	=	174

3. SOLVE.

$$x + (x + 2) = 174$$
$$2x + 2 = 174 \quad \text{Combine like terms.}$$
$$2x = 172 \quad \text{Subtract 2 from both sides.}$$
$$x = 86 \quad \text{Divide both sides by 2.}$$

4. INTERPRET. If $x = 86$, then $x + 2 = 86 + 2$ or 88.

Check: The numbers 86 and 88 are two consecutive even integer. Their sum is 174, the required sum.

State: Kelsey's art history quiz scores are 86 and 88.

Practice Problem 5

Find two consecutive integers whose sum is 251.

Answer

5. 125 and 126

Focus On the Real World

NUTRITION LABELS

Nutrition Facts
Serving Size 1 bar (36g)
Servings Per Container 6

Amount Per Serving
Calories 150 Calories from Fat 50

	% Daily Value*
Total Fat 6g	9%
Saturated Fat 2g	9%
Cholesterol 0mg	0%
Sodium 80mg	3%
Potassium 40mg	1%
Total Carbohydrate 24g	8%
Dietary Fiber 0g	0%
Sugars 11g	
Protein 2g	

Vitamin A	25%	Vitamin C	25%
Calcium	50%	Iron	25%
Vitamin D	25%	Vitamin E	25%
Thiamin	25%	Riboflavin	25%
Niacin	25%	Vitamin B_6	25%
Folate	25%	Vitamin B_{12}	25%
Biotin	25%	Pantothanic Acid	25%
Phosphorus	30%	Iodine	25%
Magnesium	25%	Zinc	25%
Copper	25%		

*Percent Daily Values are based on a 2,000 calorie diet. Your daily values may be higher or lower depending on your calorie needs.

Since 1994, the Food and Drug Administration (FDA) of the Department of Health and Human Services and the Food Safety and Inspection Service of the U.S. Department of Agriculture (USDA) have required nutrition labels like the one at the right on most food packaging. The labels were designed to help consumers make healthful food choices by giving standardized nutrition information.

One key feature of this labeling is the column of % Daily Value figures. Most of these values are based on a 2000-calorie diet. Critics complain that this diet applies to only a small segment of the population and should be more versatile. However, FDA and USDA officials responded that these are based on 2000 calories as a guideline only to help consumers gauge the relative amount of a nutrient contained by a food product. For instance, a food with 10 g of saturated fat per serving could be mistaken for a food low in saturated fat. However, in a 2000-calorie diet, 10 g of saturated fat represents 50% of the allowable daily intake. This percentage signals to consumers that this product is relatively high in saturated fat. Similarly, a food with 125 mg of sodium per serving could be mistaken for a high-sodium food. However, because a person should ingest no more than 2400 mg of sodium per day, a corresponding % Daily Value figure of about 5% conveys to a consumer that a serving of food with 125 mg of sodium is relatively low in sodium.

The % Daily Value figures that depend on the number of calories consumed per day include total fat, saturated fat, carbohydrate, protein, and dietary fiber. The % Daily Value figures for nutrients such as cholesterol, sodium, and potassium do not depend on calories.

For diets that include more or less than 2000 calories, the daily allowable amount of a nutrient that depends on calories can be figured with the following guidelines:

- The daily allowable amount of total fat is based on 30% of calories.
- The daily allowable amount of saturated fat is based on 10% of calories.
- The daily allowable amount of carbohydrate is based on 60% of calories.
- The daily allowable amount of protein is based on 10% of calories.
- The daily allowable amount of dietary fiber is based on 11.5 g of fiber per 1000 calories.

Additionally, each gram of protein and carbohydrate contains 4 calories. A gram of fat contains 9 calories.

The above information can be used to calculate % Daily Value figures for diets with other calorie levels. For instance, the daily allowable amount of total fat in a 2200-calorie diet is $0.30(2200) = 660$ calories from fat. For the nutrition label shown, one serving contains 6 g of fat or $6(9) = 54$ calories from fat. The % Daily Value for total fat that one serving of this food provides in a 2200-calorie diet is $54 \div 660 \approx 0.08$ or 8%.

GROUP ACTIVITY

1. Calculate the daily allowable amounts of total fat, saturated fat, carbohydrate, protein, and dietary fiber in 1500-calorie, 1800-calorie, 2500-calorie, and 2800-calorie diets. Summarize your results in a table.

2. Choose five different food products having Nutrition Facts labels. For each product, use the information given on the label to calculate the % Daily Value figures for total fat, saturated fat, carbohydrate, protein, and dietary fiber for (a) a 1500-calorie diet, (b) an 1800-calorie diet, (c) a 2500-calorie diet, and (d) a 2800-calorie diet. Create a chart showing your results for each food product.

3. Use the data given in the food product nutrition labels used in Question 2 to estimate the daily allowable amounts of cholesterol, sodium, and potassium. Recall that the allowable amounts for these nutrients do not depend on calorie intake.

Name ______________________ Section ______ Date ________

ANSWERS

EXERCISE SET 2.2

A *Solve. See Example 1.*

1. Four times the difference of a number and 2 is the same as 2 increased by 6 times the number. Find the number.

2. Twice the sum of a number and 3 is the same as 1 subtracted from the number. Find the number.

3. One number is 5 times another number. If the sum of the two numbers is 270, find the numbers.

4. One number is 6 less than another number. If the sum of the two numbers is 150, find the numbers.

Solve. See Examples 2 through 5.

5. Find 30% of 260.

6. Find 70% of 180.

7. Find 12% of 16.

8. Find 22% of 12.

9. The United States consists of 2271 million acres of land. Approximately 29% of this land is federally owned. Find the number of acres that are federally owned. (*Source:* U.S. General Services Administration)

10. The state of Nevada contains the most federally owned acres of land in the United States. If 90% of the state's 70 million acres of land is federally owned, find the number of federally owned acres. (*Source:* U.S. General Services Administration)

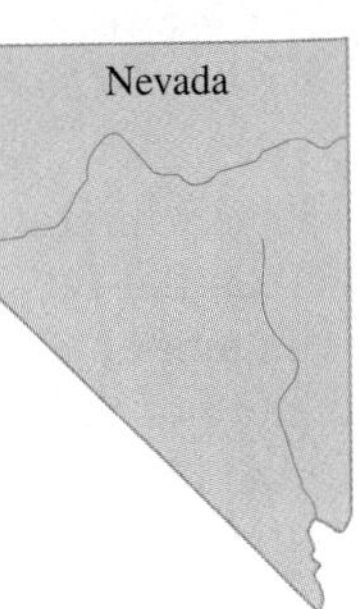

11. Recently, 47% of homes in the United States contained computers. If Charlotte, North Carolina, contains 110,000 homes, how many of these homes would you expect to have computers? (*Source: Telecommunication Research* survey)

12. Recently, 12% of homes in the United States contained on-line services. If Abilene, Texas, contains 40,000 homes, how many of these homes would you expect to have on-line services? (*Source: Telecommunication Research* survey)

1. ______
2. ______
3. ______
4. ______
5. ______
6. ______
7. ______
8. ______
9. ______
10. ______
11. ______
12. ______

13.

14.

15.

16.

17.

18.

19.

20.

Name

The following graph is called a circle graph or a pie chart. The circle represents a whole, or in this case, 100%. This particular graph shows the kind of loans that customers get from credit unions. Use this graph to answer Exercises 13–16.

Credit Union Loans

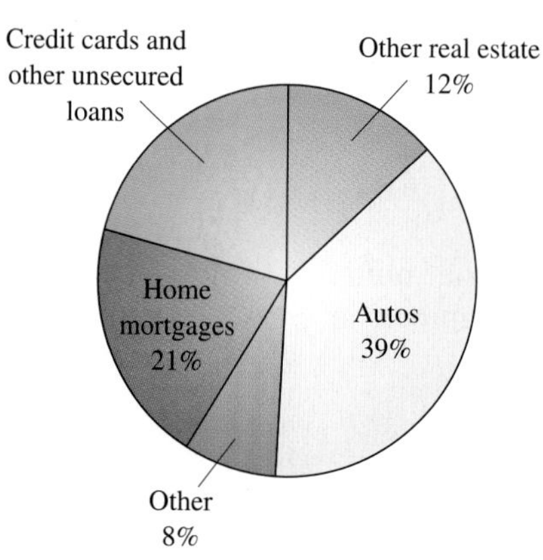

Source: National Credit Union Administration.

13. What percent of credit union loans are for credit cards and other unsecured loans?

14. What types of loans make up most credit union loans?

15. If the University of New Orleans Credit Union processed 300 loans last year, how many of these might we expect to be automobile loans?

16. If Homestead's Credit Union processed 537 loans last year, how many of these do you expect to be either home mortgages or other real estate? (Round to the nearest whole.)

17. The B767-300ER aircraft has 104 more seats than the B737-200 aircraft. If their total number of seats is 328, find the number of seats for each aircraft. (*Source:* Air Transport Association of America)

18. The governor of Connecticut makes $29,000 less per year than the governor of Delaware. If the total of these salaries is $185,000, find the salary of each governor. (*Source: 1998 World Almanac*)

19. A new FAX machine was recently purchased for an office in Hopedale for $464.40 including tax. If the tax rate in Hopedale is 8%, find the price of the FAX machine before taxes.

20. A premedical student at a local university was complaining that she had just paid $86.11 for her human anatomy book, including tax. Find the price of the book before taxes if the tax rate at this university is 9%.

Name ______________________________

21. Two frames are needed with the same outside perimeter: one frame in the shape of a square and one in the shape of an equilateral triangle. Each side of the triangle is 6 centimeters longer than each side of the square. Find the dimensions of each frame.

22. The length of a rectangular sign is 2 feet less than three times its width. Find the dimensions if the perimeter is 28 feet.

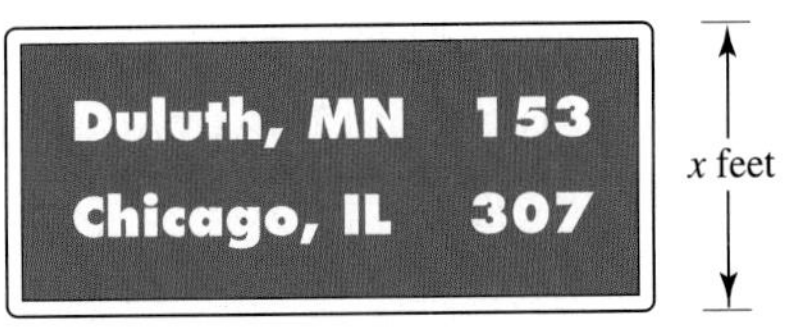

23. In a blueprint of a rectangular room, the length is to be 2 centimeters greater than twice its width. Find the dimensions if the perimeter is to be 40 centimeters.

24. A plant food solution contains 5 cups of water for every 1 cup of concentrate. If the solution contains 78 cups of these two ingredients, find the number of cups of concentrate in the solution.

25. Manufacturers claim that a CD-ROM disk will last 20 years. Recently, statements made by the U.S. National Archives and Records Administration suggest that 20 years decreased by 75% is a more realistic lifespan because the aluminum substratum on which the data is recorded can be affected by oxidation. Find the lifespan of a CD-ROM according to the U.S. National Archives and Records Administration.

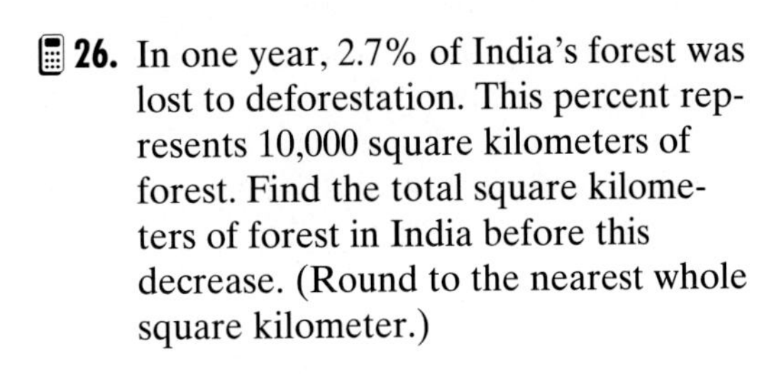

26. In one year, 2.7% of India's forest was lost to deforestation. This percent represents 10,000 square kilometers of forest. Find the total square kilometers of forest in India before this decrease. (Round to the nearest whole square kilometer.)

21. ______________

22. ______________

23. ______________

24. ______________

25. ______________

26. ______________

Name ______________________________

27. ______________

28. ______________

29. ______________

30. ______________

31. ______________

32. ______________

33. ______________

34. ______________

27. The external tank of a NASA Space Shuttle contains the propellants used for the first 8.5 minutes after launch. Its height is 5 times the sum of its width and 1. If the sum of the height and width is 55.4 meters, find the dimensions of this tank. (*Source:* NASA/Kennedy Space Center)

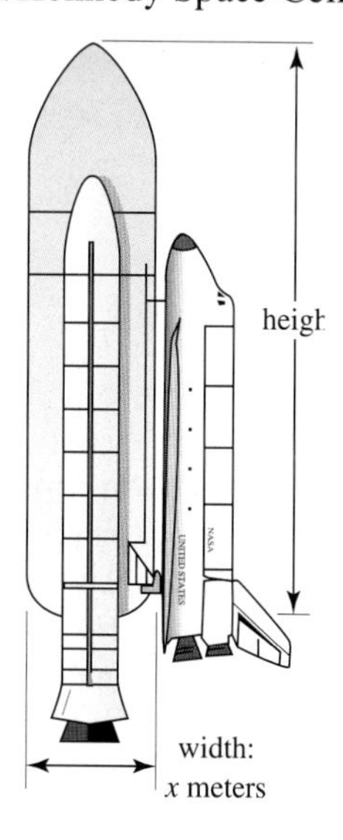

28. The blue whale is the largest of whales. Its average weight is 3 times the difference of the average weight of a humpback whale and 5 tons. If the total of the average weights is 117 tons, find the average weight of each type of whale.

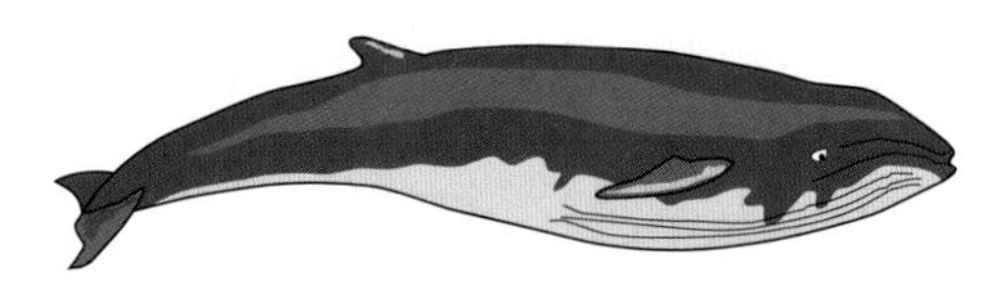

29. Recall that the sum of the angle measures of a triangle is 180°. Find the measures of the angles of a triangle if the measure of one angle is twice the measure of a second angle and the third angle measures 3 times the second angle decreased by 12.

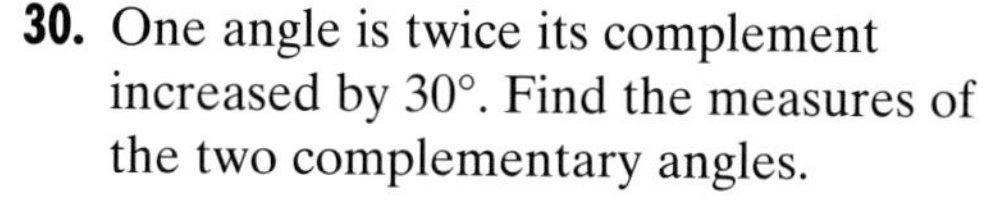

30. One angle is twice its complement increased by 30°. Find the measures of the two complementary angles.

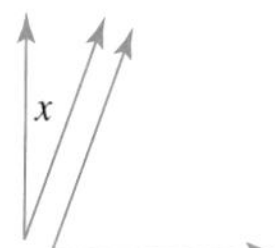

31. The sum of two consecutive integers is 151. Find the integers.

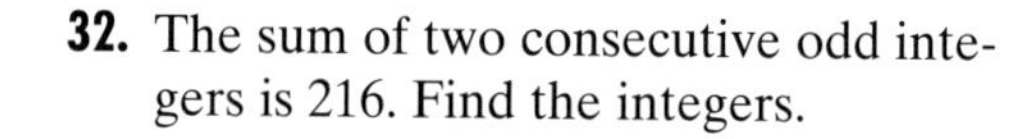

32. The sum of two consecutive odd integers is 216. Find the integers.

33. The zip codes of three Nevada locations—Fallon, Fernley, and Gardnerville Ranchos—are three consecutive even integers. If twice the first integer added to the third is 268,222, find each zip code. (*Source:* United States Postal Service)

34. During a recent year, the average SAT scores in math for the states of Alabama, Louisiana, and Michigan were three consecutive integers. If the sum of the first integer, second integer, and three times the third integer is 2637, find each score.

Name ______________________________

35. The sum of the angles in a triangle is 180°. Find the angles of a triangle whose two base angles are equal and whose third angle is 10° less than three times a base angle.

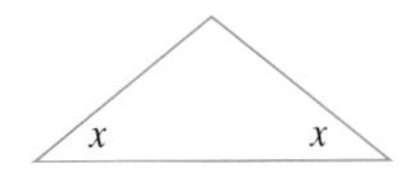

36. Find an angle such that its supplement is equal to twice its complement increased by 50°.

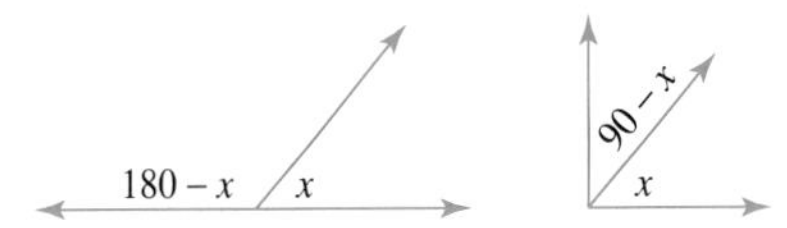

37. Coca-Cola Company is the largest producer of soft drinks in the world. In 1997, its net income was $4129 million. This represented an increase in net income of 18% from 1996. What was Coca-Cola's net income in 1996? (*Source:* Coca-Cola Company)

38. In 1997, there were a total of 8,157,000 aircraft departures from United States airports. This represents a decrease of 0.9% from the number of departures in 1996. Find the number of aircraft departures in 1996. Round to the nearest whole. (*Source:* Air Transport Association of America)

39. According to government statistics, the number of telephone company operators in the United States is expected to decrease to 26,000 by the year 2006. This represents a decrease of 47% from the number of telephone operators in 1996. (*Source:* U.S. Bureau of Labor Statistics)

a. Find the number of telephone company operators in 1996. Round to the nearest whole number.

b. In your own words, explain why you think that the need for telephone company operators is decreasing.

40. The number of deaths by tornadoes from the 1940s to the 1980s has decreased by 70.86%. There were 521 deaths from tornadoes in the 1980s. (*Source:* National Weather Service)

a. Find the number of deaths by tornadoes in the 1940s. Round to the nearest whole number.

b. In your own words, explain why you think that the number of deaths by tornadoes has decreased so much since the 1940s.

Review and Preview

Find the value of each expression for the given values. See Section 1.4.

41. $2a + b - c$; $a = 5$, $b = -1$, and $c = 3$

42. $-3a + 2c - b$; $a = -2$; $b = 6$, and $c = -7$

43. $4ab - 3bc$; $a = -5$, $b = -8$, and $c = 2$

44. $ab + 6bc$; $a = 0$, $b = -1$ and $c = 9$

35. ________
36. ________
37. ________
38. ________
39. a. ________
b. ________
40. a. ________
b. ________
41. ________
42. ________
43. ________
44. ________

45.

46.

47.

48.

49.

50.

51.

52.

53.

54.

55.

Name

45. $n^2 - m^2$; $n = -3$ and $m = -8$

46. $2n^2 + 3m^2$; $n = -2$ and $m = 7$

47. $P + PRT$; $P = 3000$, $R = 0.0325$, $T = 2$

48. $\frac{1}{3}lwh$; $l = 37.8$, $w = 5.6$, $h = 7.9$

Combining Concepts

49. Newsprint is either discarded or recycled. Americans recycle about 27% of all newsprint, but an amount of newsprint equivalent to 30 million trees is discarded every year. About how many trees' worth of newsprint is *recycled* in the United States each year? (*Source:* The EarthWorks Group)

To break even in a manufacturing business, income or revenue R must equal the cost of production C. Use this information to answer Exercises 50–55.

50. The cost C to produce x number of skateboards is $C = 100 + 20x$. The skateboards are sold wholesale for $24 each, so revenue R is given by $R = 24x$. Find how many skateboards the manufacturer needs to produce and sell to break even. (*Hint:* Set the cost expression equal to the revenue expression and solve for x.)

51. The revenue R from selling x number of computer boards is given by $R = 60x$, and the cost C of producing them is given by $C = 50x + 5000$. Find how many boards must be sold to break even. Find how much money is needed to produce the break-even number of boards.

52. In your own words, explain what happens if a company makes and sells fewer products than the break-even number.

53. In your own words, explain what happens if more products than the break-even number are made and sold.

54. Determine whether there are three consecutive integers such that their sum is three times the second integer.

55. Determine whether there are two consecutive odd integers such that 7 times the first exceeds 5 times the second by 54.

2.3 FORMULAS AND PROBLEM SOLVING

Objectives

A Solve a formula for a specified variable.

B Use formulas to solve problems.

A SOLVING FORMULAS FOR SPECIFIED VARIABLES

Solving problems that we encounter in the real world sometimes requires us to express relationships among measured quantities. A **formula** is an equation that describes a known relationship among measured phenomena, such as time, area, and gravity. Some examples of formulas are

Formula	Meaning
$I = PRT$	Interest = principal · rate · time
$A = lw$	Area of a rectangle = length · width
$d = rt$	Distance = rate · time
$C = 2\pi r$	Circumference of a circle = 2 · π · radius
$V = lwh$	Volume of a rectangular solid = length · width · height

Other formulas are listed in the front cover of this text. Notice that the formula for the volume of a rectangular solid $V = lwh$ is solved for V since V is by itself on one side of the equation with no V's on the other side of the equation. Suppose that the volume of a rectangular solid is known as well as its width and its length, and we wish to find its height. One way to find its height is to begin by solving the formula $V = lwh$ for h.

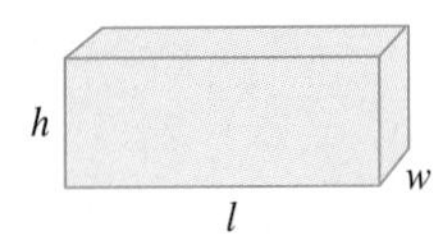

Example 1 Solve $V = lwh$ for h.

Solution: To solve $V = lwh$ for h, we want to get h alone on one side of the equation. To do so, we divide both sides of the equation by lw.

$$V = lwh$$

$$\frac{V}{lw} = \frac{lwh}{lw} \quad \text{Divide both sides by } lw.$$

$$\frac{V}{lw} = h \quad \text{Simplify.}$$

Thus we see that to find the height of a rectangular solid, we divide its volume by the product of its length and its width.

Practice Problem 1

Solve $d = rt$ for r.

The following steps may be used to solve formulas and equations for a specified variable.

Answer

1. $r = \frac{d}{t}$

SOLVING EQUATIONS FOR A SPECIFIED VARIABLE

Step 1. Clear the equation of fractions by multiplying each side of the equation by the least common denominator.

Step 2. Use the distributive property to remove grouping symbols such as parentheses.

Step 3. Combine like terms on each side of the equation.

Step 4. Use the addition property of equality to rewrite the equation as an equivalent equation with terms containing the specified variable on one side and all other terms on the other side.

Step 5. Use the distributive property and the multiplication property of equality to get the specified variable alone.

Practice Problem 2

Solve $2y + 5x = 10$ for y.

Example 2

Solve $3y - 2x = 7$ for y.

Solution: This is a linear equation in two variables. Often an equation such as this is solved for y in order to reveal some properties about the graph of this equation, which we will learn more about in Chapter 3. Since there are no fractions or grouping symbols, we begin with Step 4 and get the term containing the specified variable y on one side by adding $2x$ to both sides of the equation.

$$3y - 2x = 7$$

$$3y - 2x + 2x = 7 + 2x \quad \text{Add } 2x \text{ to both sides.}$$

$$3y = 7 + 2x$$

To solve for y, we divide both sides by 3.

$$\frac{3y}{3} = \frac{7 + 2x}{3} \quad \text{Divide both sides by 3.}$$

$$y = \frac{7 + 2x}{3} \quad \text{or} \quad y = \frac{7}{3} + \frac{2x}{3}$$

Practice Problem 3

Solve $A = \frac{1}{2}(B + b)h$ for B.

Example 3

Solve $A = \frac{1}{2}(B + b)h$ for b.

Solution: Since this formula for finding the area of a trapezoid contains fractions, we begin by multiplying both sides of the equation by the LCD, 2.

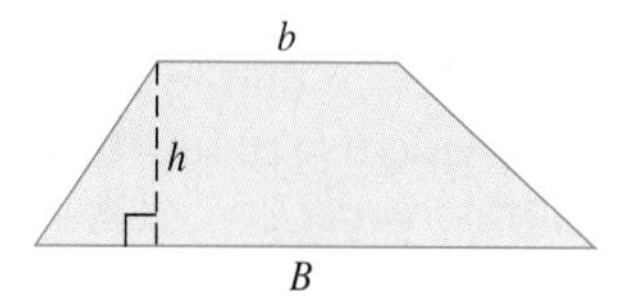

$$A = \frac{1}{2}(B + b)h$$

$$2 \cdot A = 2 \cdot \frac{1}{2}(B + b)h \quad \text{Multiply both sides by 2.}$$

$$2A = (B + b)h \quad \text{Simplify.}$$

$$2A = Bh + bh \quad \text{Use the distributive property.}$$

$$2A - Bh = bh \quad \text{Get the term containing } b \text{ alone by subtracting } Bh \text{ from both sides.}$$

Answers

2. $y = \frac{10 - 5x}{2}$, **3.** $B = \frac{2A - bh}{h}$

$$\frac{2A - Bh}{h} = \frac{bh}{h}$$ Divide both sides by h.

$$\frac{2A - Bh}{h} = b \quad \text{or} \quad b = \frac{2A - Bh}{h}$$

HELPFUL HINT

Remember that we may get the specified variable alone on either side of the equation.

B USING FORMULAS TO SOLVE PROBLEMS

In this section, we also solve problems that can be modeled by known formulas. We use the same problem-solving steps that were introduced in the previous section.

Formulas are very useful in problem solving. For example, the compound interest formula

$$A = P\left(1 + \frac{r}{n}\right)^{nt}$$

is used by banks to compute the amount A in an account that pays compound interest. The variable P represents the principal or amount invested in the account, r is the annual rate of interest, t is the time in years, and n is the number of times compounded per year.

Example 4 **Find the Amount in a Savings Account**

Marial Callier just received an inheritance of $10,000 and plans to place all the money in a savings account that pays 5% compounded quarterly to help her son go to college in 3 years. How much money will be in the account in 3 years?

Solution:

1. UNDERSTAND. Read and reread the problem. The appropriate formula needed to solve this problem is the compound interest formula

$$A = P\left(1 + \frac{r}{n}\right)^{nt}$$

Make sure that you understand the meaning of all the variables in this formula:

A = amount in the account after t years
P = principal or amount invested
t = time in years
r = annual rate of interest
n = number of times compounded per year

2. TRANSLATE. Use the compound interest formula and let $P = \$10{,}000$, $r = 5\% = 0.05$, $t = 3$ years, and $n = 4$ since the account is compounded quarterly, or 4 times a year.

Practice Problem 4

If $5000 is invested in an account paying 4% compounded monthly, determine how much money will be in the account in 2 years. Use the formula from Example 4.

Answer

4. $5415.71

Formula: $A = P\left(1 + \frac{r}{n}\right)^{nt}$

Substitute: $A = 10{,}000\left(1 + \frac{0.05}{4}\right)^{4 \cdot 3}$

3. SOLVE. We simplify the right side of the equation.

$A = 10{,}000\left(1 + \frac{0.05}{4}\right)^{4 \cdot 3}$

$A = 10{,}000(1.0125)^{12}$ Simplify $1 + \frac{0.05}{4}$ and write $4 \cdot 3$ as 12.

$A \approx 10{,}000(1.160754518)$ Approximate $(1.0125)^{12}$.

$A \approx 11{,}607.55$ Multiply and round to two decimal places.

4. INTERPRET.

Check: Repeat your calculations to make sure that no error was made. Notice that \$11,607.55 is a reasonable amount to have in the account after 3 years.

State: In 3 years, the account will contain \$11,607.55.

Name ______________________ Section ________ Date ________

MENTAL MATH

Solve each equation for the specified variable. See Examples 1 through 3.

1. $2x + y = 5$; for y

2. $7x - y = 3$; for y

3. $a - 5b = 8$; for a

4. $7r + s = 10$; for s

5. $5j + k - h = 6$; for k

6. $w - 4y + z = 0$; for z

EXERCISE SET 2.3

A *Solve each equation for the specified variable. See Examples 1 through 3.*

1. $d = rt$ for t

2. $W = gh$ for g

3. $I = PRT$ for R

4. $C = 2\pi r$ for r

5. $P = a + b + c$ for c

6. $a^2 + b^2 = c^2$ for b^2

7. $9x - 4y = 16$ for y

8. $2x + 3y = 17$ for y

9. $P = 2l + 2w$ for l

10. $P = 2l + 2w$ for w

11. $E = I(r + R)$ for r

12. $A = P(1 + rt)$ for t

13. $S = 2LW + 2LH + 2WH$ for H

14. $S = 2\pi r^2 + 2\pi rh$ for h

B *Solve. Round all dollar amounts to two decimal places. See Example 4.*

15. Complete the table and find the balance A if \$3500 is invested at an annual percent rate of 3% for 10 years and compounded n times a year.

n	1	2	4	12	365
A					

MENTAL MATH ANSWERS

1. ________
2. ________
3. ________
4. ________
5. ________
6. ________

ANSWERS

1. ________
2. ________
3. ________
4. ________
5. ________
6. ________
7. ________
8. ________
9. ________
10. ________
11. ________
12. ________
13. ________
14. ________
15. see table

16. see table

17. a.

b.

c.

18. a.

b.

c.

19.

20.

21.

22.

23.

24.

Name

16. Complete the table and find the balance A if \$5000 is invested at an annual percent rate of 6% for 15 years and compounded n times a year.

n	1	2	4	12	365
A					

17. A principal of \$6000 is invested in an account paying an annual percent rate of 4%. Find the amount in the account after 5 years if the account is compounded

a. semiannually.

b. quarterly.

c. monthly.

18. A principal of \$25,000 is invested in an account paying an annual percent rate of 5%. Find the amount in the account after 2 years if the account is compounded

a. semiannually.

b. quarterly

c. monthly.

19. One day's high temperature in Phoenix, Arizona, was recorded as 104°F. Write 104°F as degrees Celsius. (Use the formula $C = \frac{5}{9}(F - 32)$).

20. One year's low temperature in Nome, Alaska, was recorded as −15°C. Write −15°C as degrees Fahrenheit. (Use the formula $F = \frac{9}{5}C + 32$.)

21. Omaha, Nebraska, is about 90 miles from Lincoln, Nebraska. Irania Schmidt must go to the law library in Lincoln to get a document for the law firm she works for. Find how long it takes her to drive *round-trip* if she averages 50 mph.

22. It took the Selby family $5\frac{1}{2}$ hours round-trip to drive from their house to their beach house 154 miles away. Find their average speed.

23. A package of floor tiles contains 24 one-foot-square tiles. Find how many packages should be bought to cover a square ballroom floor whose side measures 64 feet.

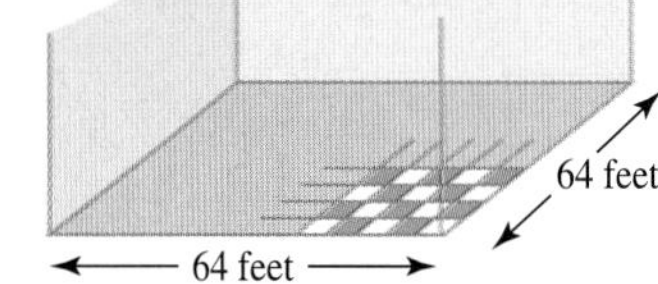

24. One-foot-square ceiling tiles are sold in packages of 50. Find how many packages must be bought for a rectangular ceiling 18 feet by 12 feet.

Name ______________________________

25. If the area of a triangular kite is 18 square feet and its base is 4 feet, find the height of the kite.

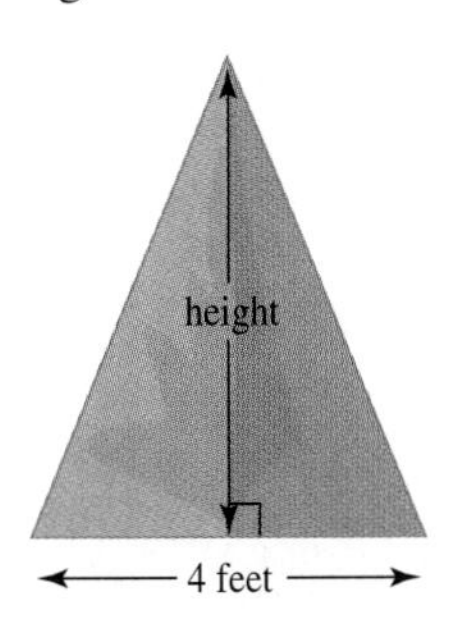

26. Bryan, Eric, Mandy, and Melissa would like to go to Disneyland in 3 years. The total cost should be $4500. If each invests $1000 in a savings account paying 5.5% interest, compounded semiannually, will they have enough in 3 years?

27. A gallon of latex paint can cover 500 square feet. Find how many gallon containers of paint should be bought to paint two coats on each wall of a rectangular room whose dimensions are 14 feet by 16 feet (assume 8-foot ceilings).

28. A gallon of enamel paint can cover 300 square feet. Find how many gallon containers of paint should be bought to paint three coats on a wall measuring 21 feet by 8 feet.

29. A portion of the external tank of the Space Shuttle *Endeavour* is a liquid hydrogen tank. If the ends of the tank are hemispheres, find the volume of the tank. To do so, answer parts (a) through (c). (*Source:* NASA/Kennedy Space Center)

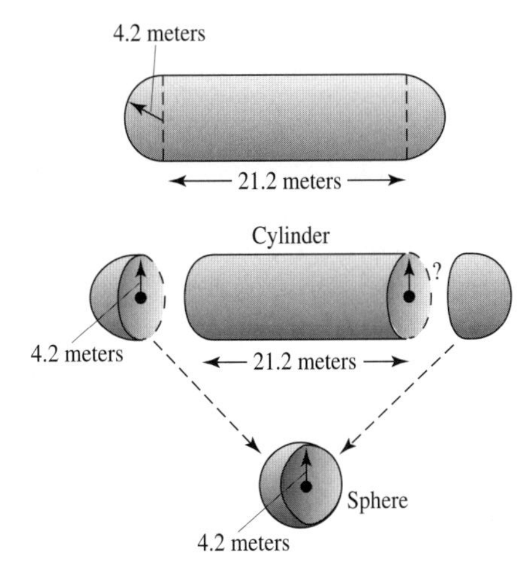

a. Find the volume of the cylinder shown. Round to two decimal places.

b. Find the volume of the sphere shown. Round to two decimal places.

c. Add the results of parts (a) and (b). This sum is the approximate volume of the tank.

25. ____________

26. ____________

27. ____________

28. ____________

29 a. ____________

b. ____________

c. ____________

Name ______________________________

30. ______________

30. The Cassini spacecraft mission to Saturn was launched October 15, 1997. It will take more than six and a half years to reach Saturn, arriving in July 2004. During its mission, Cassini will travel a total distance of 2 billion miles in 80.5 months. Find the average speed of the spacecraft in miles per hour. (*Hint:* Convert 80.5 months to hours using 1 month = 30 days and then use the formula $d = rt$) (*Source:* NASA Jet Propulsion Laboratory)

31. ______________

32. ______________

31. In 1945, Arthur C. Clarke, a scientist and science-fiction writer, predicted that an artificial satellite placed at a height of 22,248 miles directly above the equator would orbit the globe at the same speed with which the Earth was rotating. This belt along the equator is known as the Clarke belt. Use the formula for circumference of a circle and approximate the "length" of the Clarke belt. (*Hint:* Recall that the radius of the Earth is approximately 4000 miles. Round to the nearest whole mile.)

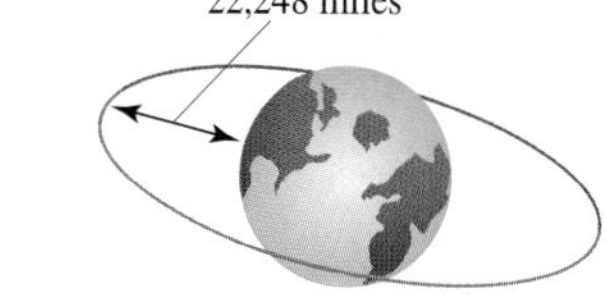

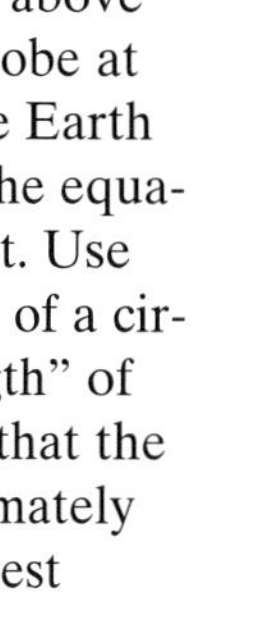

32. The *Endeavour* Space Shuttle has a cargo bay that is in the shape of a cylinder whose length is 18.3 meters and whose diameter is 4.6 meters. Find its volume.

33. ______________

34. ______________

33. The deepest hole in the ocean floor is beneath the Pacific Ocean and is called Hole 504B. It is located off the coast of Ecuador. Scientists are drilling it to learn more about the Earth's history. Currently, the hole is in the shape of a cylinder whose volume is approximately 3800 cubic feet and whose length is 1.3 miles. Find the radius of the hole to the nearest hundredth of a foot. (*Hint:* Make sure the same units of measurement are used.)

34. The deepest man-made hole is called the Kola Superdeep Borehole. It is approximately 8 miles deep and is located near a small Russian town in the Arctic Circle. If it takes 7.5 hours to remove the drill from the bottom of the hole, find the rate that the drill can be retrieved in feet per second. Round to the nearest tenth. (*Hint:* Write 8 miles as feet, 7.5 hours as seconds, and then use the formula $d = rt$.)

Name ______________________________

35. ______________

Review and Preview

Determine which numbers in the set $\{-3, -2, -1, 0, 1, 2, 3\}$ *are solutions of each inequality. See Sections 1.2 and 2.1.*

35. $x < 0$

36. $x > 1$

36. ______________

37. $x + 5 \leq 6$

38. $x - 3 \geq -7$

37. ______________

39. In your own words, explain what real numbers are solutions of $x < 0$.

40. In your own words, explain what real numbers are solutions of $x > 1$.

38. ______________

Combining Concepts

41. Solar System distances are so great that units other than miles or kilometers are often used. For example, the astronomical unit (AU) is the average distance between the Earth and the sun, or 92,900,000 miles. Use this information to convert each planet's distance in miles from the sun to astronomical units. Round to three decimal places. (*Source:* National Space Science Data Center)

39. ______________

Planet	Miles from the Sun	AU from the Sun	Planet	Miles from the Sun	AU from the Sun
Mercury	36 million		Saturn	886.1 million	
Venus	67.2 million		Uranus	1783 million	
Earth	92.9 million		Neptune	2793 million	
Mars	141.5 million		Pluto	3670 million	
Jupiter	483.3 million				

40. ______________

42. An orbit such as Clarke's belt in Exercise 31 is called a geostationary orbit. In your own words, why do you think that communications satellites are placed in geostationary orbits?

41. see table

42. ______________

43. ____________

44. ____________

45. ____________

46. ____________

Name ______________________________

43. How much do you think it costs each American to build a space shuttle? Write down your estimate. The Space Shuttle *Endeavour* was completed in 1992 and cost approximately \$1.7 billion. If the population of the United States in 1992 was 250 million, find the cost per person to build the *Endeavour*. How close was your estimate?

44. Find *how much interest* \$10,000 earns in 2 years in a certificate of deposit account paying 8.5% interest compounded quarterly.

45. If you are investing money in a savings account paying a rate of r, which account should you choose—an account compounded 4 times a year or 12 times a year? Explain your choice.

46. To borrow money at a rate of r, which bank should you choose—one compounding 4 times a year or 12 times a year? Explain your choice.

2.4 LINEAR INEQUALITIES AND PROBLEM SOLVING

Objective

A Use interval notation.

B Solve linear inequalities using the addition property of inequality.

C Solve linear inequalities using the multiplication property of inequality.

D Solve linear inequalities using both properties of inequality.

E Solve problems that can be modeled by linear inequalities.

SSM CD-ROM Video 2.4

Relationships among measurable quantities are not always described by equations. For example, suppose that a salesperson earns a base of \$600 per month plus a commission of 20% of sales. Find the minimum amount of sales needed to receive a total income of *at least* \$1500 per month. Here, the phrase "at least" implies that an income of \$1500 *or more* is acceptable. In symbols, we can write

$$\text{income} \geq 1500$$

This is an example of an inequality, which we will solve in Example 11.

A **linear inequality** is similar to a linear equation except that the equality symbol is replaced with an inequality symbol, such as $<$, $>$, $\leq$, or $\geq$.

LINEAR INEQUALITY IN ONE VARIABLE

A linear inequality in one variable is an inequality that can be written in the form

$$ax + b < c$$

where a, b, and c are real numbers and $a \neq 0$. For example,

$$3x + 5 \geq 4 \qquad 2y < 0 \qquad 4n \geq n - 3$$

$$3(x - 4) < 5x \qquad \frac{x}{3} \leq 5$$

In this section, when we make definitions, state properties, or list steps about an inequality containing the symbol $<$, we mean that the definition, property, or steps apply to an inequality containing the symbols $>$, $\leq$, and $\geq$, also.

A USING INTERVAL NOTATION

A **solution** of an inequality is a value of the variable that makes the inequality a true statement. The **solution set** of an inequality is the set of all solutions. Notice that the solution set of the inequality $x > 2$, for example, contains all numbers greater than 2. Its graph is an interval on the number line since an infinite number of values satisfy the variable. If we use open/closed-circle notation, the graph of $\{x|x > 2\}$ looks like the following:

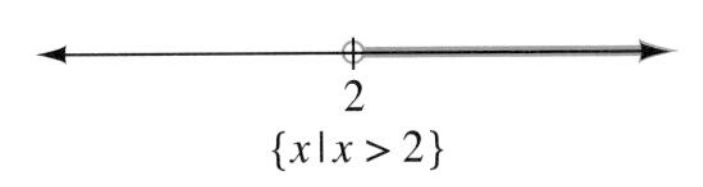

In this text, a different graphing notation will be used to help us understand **interval notation**. Instead of an open circle, we use a parenthesis; instead of a closed circle, we use a bracket. With this new notation, the graph of $\{x|x > 2\}$ now looks like

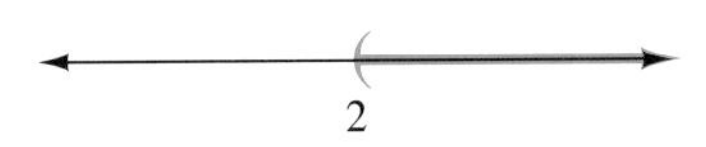

and can be represented in interval notation as $(2, \infty)$. The symbol ∞ is read "infinity" and indicates that the interval includes *all* numbers greater than 2. The left parenthesis indicates that 2 *is not* included in the interval. Using a left bracket, [, would indicate that 2 *is* included in the interval. The following table shows three equivalent ways to describe an interval: in set notation, as a graph, and in interval notation.

Set Notation	Graph	Interval Notation
$\{x \mid x < a\}$	a	$(-\infty, a)$
$\{x \mid x > a\}$	a	(a, ∞)
$\{x \mid x \leq a\}$	a	$(-\infty, a]$
$\{x \mid x \geq a\}$	a	$[a, \infty)$
$\{x \mid a < x < b\}$	a b	(a, b)
$\{x \mid a \leq x \leq b\}$	a b	$[a, b]$
$\{x \mid a < x \leq b\}$	a b	$(a, b]$
$\{x \mid a \leq x < b\}$	a b	$[a, b)$

Notice that a parenthesis is always used to enclose ∞ and $-\infty$.

Examples Graph each set on a number line and then write it in interval notation.

1. $\{x \mid x \geq 2\}$ −2 −1 0 1 2 3 4 $[2, \infty)$

2. $\{x \mid x < -1\}$ −3 −2 −1 0 1 2 3 $(-\infty, -1)$

3. $\{x \mid 0.5 < x \leq 3\}$ 0.5 −1 0 1 2 3 $(0.5, 3]$

TRY THE CONCEPT CHECK IN THE MARGIN.

USING THE ADDITION PROPERTY OF INEQUALITY

Interval notation can be used to write solutions of linear inequalities. To solve a linear inequality, we use a process similar to the one used to solve a linear equation. We use properties of inequalities to write equivalent inequalities until the variable is alone on one side of the inequality.

Practice Problems 1–3

Graph each set on a number line and then write it in interval notation.

1. $\{x \mid x > -3\}$

−5 −4 −3 −2 −1 0 1 2 3 4 5

2. $\{x \mid x \leq 0\}$

−5 −4 −3 −2 −1 0 1 2 3 4 5

3. $\{x \mid -0.5 \leq x < 2\}$

−5 −4 −3 −2 −1 0 1 2 3 4 5

✓ CONCEPT CHECK

Explain what is wrong with writing the interval $(5, \infty]$.

Answers

1. $(-3, \infty)$,

−5 −4 −3 −2 −1 0 1 2 3 4 5

2. $(-\infty, 0]$,

−5 −4 −3 −2 −1 0 1 2 3 4 5

3. $[-0.5, 2)$

−0.5

−5 −4 −3 −2 −1 0 1 2 3 4 5

✓ Concept Check: Should be $(5, \infty)$ since a parenthesis is always used to enclose ∞.

ADDITION PROPERTY OF INEQUALITY

If a, b, and c are real numbers, then

$$a < b \quad \text{and} \quad a + c < b + c$$

are equivalent inequalities.

In other words, we may add the same real number to both sides of an inequality, and the resulting inequality will have the same solution set. This property also allows us to subtract the same real number from both sides.

Example 4

Solve: $x - 2 < 5$. Graph the solution set.

Solution:

$x - 2 < 5$

$x - 2 + 2 < 5 + 2$ Add 2 to both sides.

$x < 7$ Simplify.

The solution set is $\{x | x < 7\}$, which in interval notation is $(-\infty, 7)$. The graph of the solution set is

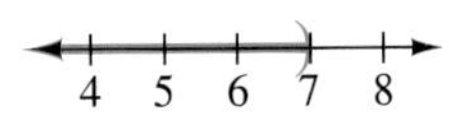

Practice Problem 4

Solve: $x + 3 < 1$. Graph the solution set.

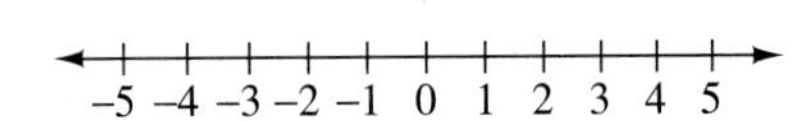

HELPFUL HINT

In Example 4, the solution set is $\{x | x < 7\}$. This means that *all* numbers less than 7 are solutions. For example, 6.9, 0, $-\pi$, 1, and -56.7 are solutions, just to name a few. To see this, replace x in $x - 2 < 5$ with each of these numbers and see that the result is a true inequality.

Example 5

Solve: $4x - 2 < 5x$. Graph the solution set.

Solution: To get x alone on one side of the inequality, we subtract $4x$ from both sides.

$4x - 2 < 5x$

$4x - 2 - 4x < 5x - 4x$ Subtract $4x$ from both sides.

$-2 < x$ or $x > -2$ Simplify.

HELPFUL HINT

Don't forget that $-2 < x$ means the same as $x > -2$.

The solution set is $\{x | x > -2\}$, which in interval notation is $(-2, \infty)$. The graph is

–3 –2 –1 0 1 2

Practice Problem 5

Solve: $3x - 4 < 4x$. Graph the solution set.

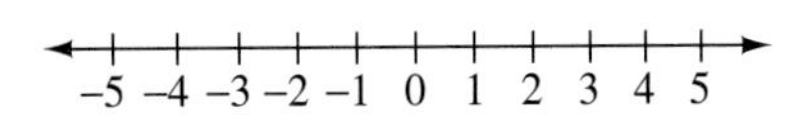

Answers

4. $\{x | x < -2\}$, $(-\infty, -2)$,

–5 –4 –3 –2 –1 0 1 2 3 4 5

5. $\{x | x > -4\}$, $(-4, \infty)$,

–5 –4 –3 –2 –1 0 1 2 3 4 5

Practice Problem 6

Solve: $5x - 1 \geq 4x + 4$. Graph the solution set.

−5 −4 −3 −2 −1 0 1 2 3 4 5

Example 6 Solve: $3x + 4 \geq 2x - 6$. Graph the solution set.

Solution:

$$3x + 4 \geq 2x - 6$$

$$3x + 4 - 2x \geq 2x - 6 - 2x \quad \text{Subtract } 2x \text{ from both sides.}$$

$$x + 4 \geq -6 \quad \text{Combine like terms.}$$

$$x + 4 - 4 \geq -6 - 4 \quad \text{Subtract 4 from both sides.}$$

$$x \geq -10 \quad \text{Simplify.}$$

The solution set is $\{x|x \geq -10\}$, which in interval notation is $[-10, \infty)$. The graph of the solution set is

−11 −10 −9 −8 −7 −6

C USING THE MULTIPLICATION PROPERTY OF INEQUALITY

Next, we introduce and use the multiplication property of inequality to solve linear inequalities. To understand this property, let's start with the true statement $-3 < 7$ and multiply both sides by 2.

$$-3 < 7$$

$$-3(2) < 7(2) \quad \text{Multiply both sides by 2.}$$

$$-6 < 14 \quad \text{True.}$$

The statement remains true.

Notice what happens if both sides of $-3 < 7$ are multiplied by -2.

$$-3 < 7$$

$$-3(-2) < 7(-2)$$

$$6 < -14 \quad \text{False.}$$

The inequality $6 < -14$ is a false statement. However, *if the direction of the inequality sign is reversed,* the result is

$$6 > -14 \quad \text{True.}$$

These examples suggest the following property.

MULTIPLICATION PROPERTY OF INEQUALITY

If a, b, and c are real numbers and c is **positive**, then $a < b$ and $ac < bc$ are equivalent inequalities.
If a, b, and c are real numbers and c is **negative**, then $a < b$ and $ac > bc$ are equivalent inequalities.

In other words, we may multiply both sides of an inequality by the same positive real number, and the result is an equivalent inequality. We may also multiply both sides of an inequality by the same *negative number* and *reverse the direction of the inequality symbol,* and the result is an equivalent inequality. The multiplication property holds for division also, since division is defined in terms of multiplication.

HELPFUL HINT

Whenever both sides of an inequality are multiplied or divided by a negative number, the direction of the inequality symbol *must be* reversed to form an equivalent inequality.

Answer

6. $\{x|x \geq 5\}, [5, \infty)$

−5 −4 −3 −2 −1 0 1 2 3 4 5

Example 7 Solve: $\frac{1}{4}x \leq \frac{3}{2}$. Graph the solution set.

Solution:

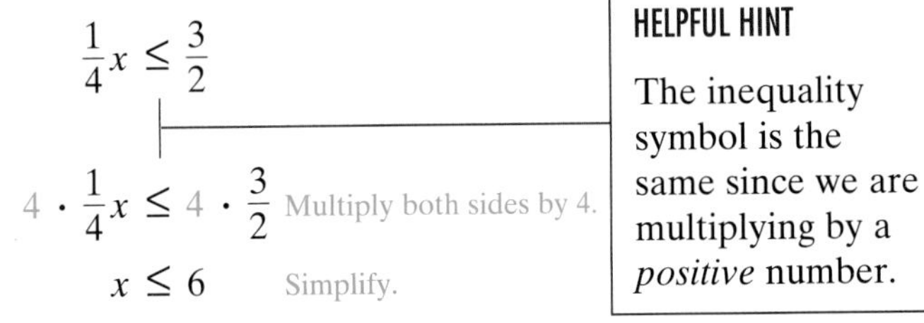

$$\frac{1}{4}x \leq \frac{3}{2}$$

$$4 \cdot \frac{1}{4}x \leq 4 \cdot \frac{3}{2} \quad \text{Multiply both sides by 4.}$$

$$x \leq 6 \quad \text{Simplify.}$$

> **HELPFUL HINT**
> The inequality symbol is the same since we are multiplying by a *positive* number.

The solution set is $\{x|x \leq 6\}$, which in interval notation is $(-\infty, 6]$. The graph of this solution set is

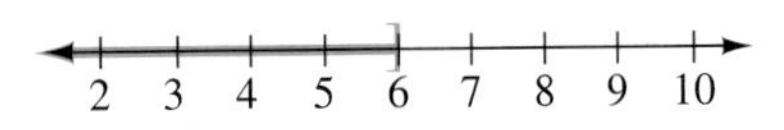

Practice Problem 7

Solve: $\frac{1}{6}x \leq \frac{2}{3}$. Graph the solution set.

−5 −4 −3 −2 −1 0 1 2 3 4 5

Example 8 Solve: $-2.3x < 6.9$. Graph the solution set.

Solution: $-2.3x < 6.9$

$$\frac{-2.3x}{-2.3} > \frac{6.9}{-2.3} \quad \text{Divide both sides by } -2.3 \text{ and reverse the inequality symbol.}$$

$$x > -3 \quad \text{Simplify.}$$

> **HELPFUL HINT**
> The inequality symbol is *reversed* since we divided by a *negative* number.

The solution set is $\{x|x > -3\}$, which is $(-3, \infty)$ in interval notation. The graph of the solution set is

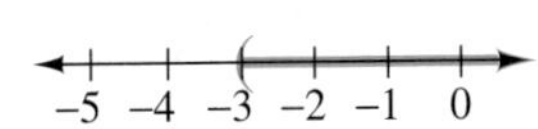

Practice Problem 8

Solve: $-1.1x < 5.5$. Graph the solution set.

−5 −4 −3 −2 −1 0 1 2 3 4 5

TRY THE CONCEPT CHECK IN THE MARGIN.

To solve linear inequalities in general, we follow steps similar to those for solving linear equations.

SOLVING A LINEAR INEQUALITY IN ONE VARIABLE

Step 1. Clear the equation of fractions by multiplying both sides of the inequality by the least common denominator (LCD) of all fractions in the inequality.

Step 2. Use the distributive property to remove grouping symbols such as parentheses.

Step 3. Combine like terms on each side of the inequality.

Step 4. Use the addition property of inequality to write the inequality as an equivalent inequality with variable terms on one side and numbers on the other side.

Step 5. Use the multiplication property of inequality to get the variable alone on one side of the inequality.

✓ CONCEPT CHECK

In which of the following inequalities must the inequality symbol be reversed during the solution process?

a. $-2x > 7$

b. $2x - 3 > 10$

c. $-x + 4 + 3x < 5$

d. $-x + 4 < 5$

D USING BOTH PROPERTIES OF INEQUALITY

Many problems require us to use both properties of inequality.

Example 9 Solve: $5 - x \leq 4x - 15$. Write the solution set in interval notation.

Practice Problem 9

Solve: $6 - 2x \leq 8x - 14$. Write the solution set in interval notation.

Answers

7. $\{x|x \leq 4\}, (-\infty, 4]$,

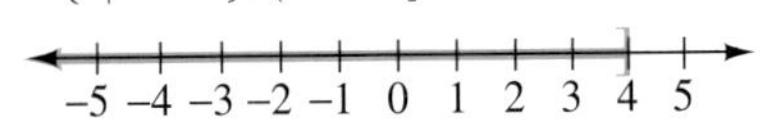

8. $\{x|x > -5\}, (-5, \infty)$,

−5 −4 −3 −2 −1 0 1 2 3 4 5

9. $[2, \infty)$

✓ Concept Check: a, d

Solution:

$$5 - x \le 4x - 15$$
$$5 - x + x \le 4x - 15 + x \quad \text{Add } x \text{ to both sides.}$$
$$5 \le 5x - 15 \quad \text{Combine like terms.}$$
$$5 + 15 \le 5x - 15 + 15 \quad \text{Add 15 to both sides.}$$
$$20 \le 5x \quad \text{Combine like terms.}$$
$$\frac{20}{5} \le \frac{5x}{5} \quad \text{Divide both sides by 5.}$$
$$4 \le x \quad \text{or} \quad x \ge 4 \quad \text{Simplify.}$$

The solution set is $[4, \infty)$.

Practice Problem 10

Solve: $\frac{3}{4}(x + 2) \ge x - 6$. Write the solution set in interval notation.

Example 10

Solve: $\frac{2}{5}(x - 6) \ge x - 1$. Write the solution set in interval notation.

Solution:

$$\frac{2}{5}(x - 6) \ge x - 1$$
$$5\left[\frac{2}{5}(x - 6)\right] \ge 5(x - 1) \quad \text{Multiply both sides by 5 to eliminate fractions.}$$
$$2x - 12 \ge 5x - 5 \quad \text{Use the distributive property.}$$
$$-3x - 12 \ge -5 \quad \text{Subtract } 5x \text{ from both sides.}$$
$$-3x \ge 7 \quad \text{Add 12 to both sides.}$$
$$\frac{-3x}{-3} \le \frac{7}{-3} \quad \text{Divide both sides by } -3 \text{ and reverse the inequality symbol.}$$
$$x \le -\frac{7}{3} \quad \text{Simplify.}$$

The solution set is $\left(-\infty, -\frac{7}{3}\right]$.

E LINEAR INEQUALITIES AND PROBLEM SOLVING

Problems containing words such as "at least," "at most," "between," "no more than," and "no less than" usually indicate that an inequality is to be solved instead of an equation. In solving applications involving linear inequalities, we use the same four-step strategy as when we solved applications involving linear equations.

Practice Problem 11

A salesperson earns \$1000 a month plus a commission of 15% of sales. Find the minimum amount of sales needed to receive a total income of at least \$4000 per month.

Example 11 Calculating Income with Commission

A salesperson earns \$600 per month plus a commission of 20% of sales. Find the minimum amount of sales needed to receive a total income of at least \$1500 per month.

Solution:

1. UNDERSTAND. Read and reread the problem. Let

   $x =$ amount of sales

2. TRANSLATE. As stated in the beginning of this section, we want the income to be greater than or equal to \$1500. To write an inequality, notice that the salesperson's income consists of \$600 plus a commission (20% of sales).

Answers

10. $(-\infty, 30]$, **11.** \$20,000

In words:	600	+	commission (20% of sales)	≥	1500
	↓		↓		↓
Translate:	600	+	$0.20x$	≥	1500

3. SOLVE the inequality for x.

$$600 + 0.20x \geq 1500$$
$$600 + 0.20x - 600 \geq 1500 - 600$$
$$0.20x \geq 900$$
$$x \geq 4500$$

4. INTERPRET.

Check: The income for sales of \$4500 is

$600 + 0.20(4500)$, or 1500

Thus, if sales are greater than or equal to \$4500, income is greater than or equal to \$1500.

State: The minimum amount of sales needed for the salesperson to earn at least \$1500 per month is \$4500.

Example 12 Finding the Annual Consumption

In the United States, the annual consumption of cigarettes is declining. The consumption c in billions of cigarettes per year since the year 1985 can be approximated by the formula

$$c = -14.25t + 598.69$$

where t is the number of years after 1985. Use this formula to predict the years that the consumption of cigarettes will be less than 200 billion per year.

Solution:

1. UNDERSTAND. Read and reread the problem. To become familiar with the given formula, let's find the cigarette consumption after 20 years, which would be the year $1985 + 20$, or 2005. To do so, we substitute 20 for t in the given formula.

$$c = -14.25(20) + 598.69 = 313.69$$

Thus, in 2005, we predict cigarette consumption to be about 313.69 billion.

Variables have already been assigned in the given formula. For review, they are

$c =$ the annual consumption of cigarettes in the United States in billions of cigarettes

$t =$ the number of years after 1985

2. TRANSLATE. We are looking for the years that the consumption of cigarettes c is less than 200. Since we are finding years t, we substitute the expression in the formula given for c, or

$$-14.25t + 598.69 < 200$$

Practice Problem 12

Use the formula given in Example 12 to predict when the consumption of cigarettes will be less than 100 billion per year.

Answer

12. after the year 2020

3. SOLVE the inequality.

$$-14.25t + 598.69 < 200$$ Subtract 598.69 from both sides.

$$-14.25t < -398.69$$ Divide both sides by −14.25 and round the result.

$$t > 27.98$$

4. INTERPRET.

Check: We substitute a number greater than 27.98 and see that c is less than 200.

State: The annual consumption of cigarettes will be less than 200 billion for the years more than 27.98 years after 1985, or in approximately $28 + 1985 = 2013$.

Focus On Study Skills

STUDYING FOR AND TAKING A MATH EXAM

Remember that one of the best ways to start preparing for an exam is to keep current with your assignments as they are made. Make an effort to clear up any confusion on topics as you cover them. Begin reviewing for your exam a few days in advance. If you find a topic during your review that you still don't understand, you'll have plenty of time to ask your instructor, another student in your class, or a math tutor for help. Don't wait until the last minute to "cram" for the test.

- Reread your notes and carefully review the Chapter Highlights at the end of each chapter to be covered.
- Try solving a few exercises from each section.
- Pay special attention to any new terminology or definitions in the chapter. Be sure you can state the meanings of definitions in your own words.
- Find a quiet place to take the Chapter Test found at the end of the chapter to be covered. This gives you a chance to practice taking the real exam, so try the Chapter Test without referring to your notes or looking up anything in your book. Give yourself the same amount of time to take the Chapter Test as you will have to take the exam for which you are preparing. If your exam covers more than one chapter, you should try taking the Chapter Tests for each chapter covered. You may also find working through the Cumulative Reviews helpful when preparing for a multi-chapter test.
- When you have finished taking the Chapter Test, check your answers against those in the back of the book. Redo any of the problems you missed. Then spend extra time solving similar problems.
- If you tend to get anxious while taking an exam, try to visualize yourself taking the exam in advance. Picture yourself being calm, clearheaded, and successful. Picture yourself remembering concepts and definitions with no trouble. When you are well prepared for an exam, a lot of nervousness can be avoided through positive thinking.
- Get lots of rest the night before the exam. It's hard to show how well you know the material if your brain is foggy from lack of sleep.

When it's time to take your exam, remember these hints:

- Make sure you have all the tools you will need to take the exam, including an extra pencil and eraser, paper (if needed), and calculator (if allowed).
- Try to relax. Taking a few deep breaths, inhaling and then exhaling slowly before you begin, might help.
- Are there any special definitions or solution steps that you'll need to remember during the exam? As soon as you get your exam, write these down at the top, bottom, or on the back of your paper.
- Scan the entire test to get an idea of what questions are being asked.
- Start with the questions that are easiest for you. This will help build your confidence. Then return to the harder ones.
- Read all directions carefully. Make sure that your final result answers the question being asked.
- Show all of your work. Try to work neatly.
- Don't spend too much time on a single problem. If you get stuck, try moving on to other problems so you can increase your chances of finishing the test. If you have time, you can return to the problem giving you trouble.
- Before turning in your exam, check your work carefully if time allows. Be on the lookout for careless mistakes.

Name ______________________ Section ______ Date ______

MENTAL MATH ANSWERS

1. ______
2. ______
3. ______
4. ______
5. ______
6. ______
7. ______
8. ______

MENTAL MATH

Solve each inequality mentally.

1. $x - 2 < 4$ **2.** $x - 1 > 6$ **3.** $x + 5 \geq 15$ **4.** $x + 1 \leq 8$

5. $3x > 12$ **6.** $5x < 20$ **7.** $\frac{x}{2} \leq 1$ **8.** $\frac{x}{4} \geq 2$

EXERCISE SET 2.4

A *Graph the solution set of each inequality on a number line and then write it in interval notation. See Examples 1 through 3.*

1. $\{x|x < -3\}$

−5 −4 −3 −2 −1 0 1 2 3 4 5

2. $\{x|x \geq -7\}$

−10 −8 −6 −4 −2 0 2 4 6 8 10

3. $\{x|x \geq 0.3\}$

−5 −4 −3 −2 −1 0 1 2 3 4 5

4. $\{x|x < -0.2\}$

−5 −4 −3 −2 −1 0 1 2 3 4 5

5. $\{x|5 < x\}$

−5 −4 −3 −2 −1 0 1 2 3 4 5

6. $\{x|-7 \geq x\}$

−10 −8 −6 −4 −2 0 2 4 6 8 10

7. $\{x|-2 < x < 5\}$

−5 −4 −3 −2 −1 0 1 2 3 4 5

8. $\{x|-5 \leq x \leq -1\}$

−5 −4 −3 −2 −1 0 1 2 3 4 5

9. $\{x|5 > x > -1\}$

−5 −4 −3 −2 −1 0 1 2 3 4 5

10. $\{x|-3 \geq x \geq -7\}$

−10 −8 −6 −4 −2 0 2 4 6 8 10

11. When graphing the solution set of an inequality, explain how you know whether to use a parenthesis or a bracket.

12. Explain what is wrong with the interval notation $(-6, -\infty)$.

ANSWERS

1. ______
2. ______
3. ______
4. ______
5. ______
6. ______
7. ______
8. ______
9. ______
10. ______
11. ______
12. ______

13. ______

14. ______

15. ______

16. ______

17. ______

18. ______

19. ______

20. ______

21. ______

22. ______

23. ______

24. ______

25. ______

26. ______

Name ______________________________

B *Solve. Graph the solution set and write it in interval notation. See Examples 4 through 6.*

13. $x - 7 \geq -9$

−5 −4 −3 −2 −1 0 1 2 3 4 5

14. $x + 2 \leq -1$

−5 −4 −3 −2 −1 0 1 2 3 4 5

15. $7x < 6x + 1$

−5 −4 −3 −2 −1 0 1 2 3 4 5

16. $11x < 10x + 5$

−5 −4 −3 −2 −1 0 1 2 3 4 5

17. $8x - 7 \leq 7x - 5$

−5 −4 −3 −2 −1 0 1 2 3 4 5

18. $7x - 1 \geq 6x - 1$

−5 −4 −3 −2 −1 0 1 2 3 4 5

C *Solve. Write the solution set in interval notation and then graph it. See Examples 7 and 8.*

19. $\frac{3}{4}x \geq 2$

−5 −4 −3 −2 −1 0 1 2 3 4 5

20. $\frac{5}{6}x \geq -8$

−10 −8 −6 −4 −2 0 2 4 6 8 10

21. $5x < -23.5$

−5 −4 −3 −2 −1 0 1 2 3 4 5

22. $4x > -11.2$

−5 −4 −3 −2 −1 0 1 2 3 4 5

23. $-3x \geq 9$

−5 −4 −3 −2 −1 0 1 2 3 4 5

24. $-4x \geq 15$

−5 −4 −3 −2 −1 0 1 2 3 4 5

25. $-x < -4$

−5 −4 −3 −2 −1 0 1 2 3 4 5

26. $-x > -2$

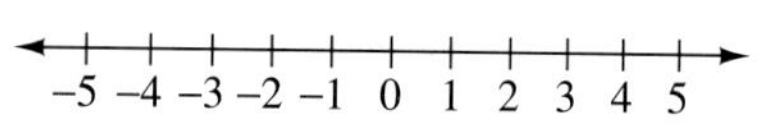

Name ______________________________

D *Solve. Write the solution set using interval notation. See Examples 9 and 10.*

27. $-2x + 7 \geq 9$

28. $8 - 5x \leq 23$

29. $15 + 2x \geq 4x - 7$

30. $20 + x < 6x$

31. $3(x - 5) < 2(2x - 1)$

32. $5(x + 4) \leq 4(2x + 3)$

33. $\frac{1}{2} + \frac{2}{3} \geq \frac{x}{6}$

34. $\frac{3}{4} - \frac{2}{3} > \frac{x}{6}$

35. $-5x + 4 \leq -4(x - 1)$

36. $-6x + 2 < -3(x + 4)$

37. $\frac{1}{4}(x - 7) \geq x + 2$

38. $\frac{3}{5}(x + 1) \leq x + 1$

39. $0.8x + 0.6x \geq 4.2$

40. $0.7x - x > 0.45$

41. $4(2x + 1) > 4$

42. $6(2 - x) \geq 12$

43. $\frac{5x + 1}{7} - \frac{2x - 6}{4} \geq -4$

44. $\frac{1 - 2x}{3} + \frac{3x + 7}{7} > 1$

45. $4(x - 6) + 2x - 4 \geq 3(x - 7) + 10x$

46. $7(2x + 3) + 4x \leq 7 + 5(3x - 4)$

E *Solve. See Examples 11 and 12.*

47. Shureka Washburn has scores of 72, 67, 82, and 79 on her algebra tests. Use an inequality to find the minimum score she can make on the final exam to pass the course with an average of 60 or higher, given that the final exam counts as two tests.

48. In a Winter Olympics 5000-meter speed-skating event, Hans Holden scored times of 6.85, 7.04, and 6.92 minutes on his first three trials. Use an inequality to find the maximum time he can score on his last trial so that his average time is under 7.0 minutes.

27. ____________

28. ____________

29. ____________

30. ____________

31. ____________

32. ____________

33. ____________

34. ____________

35. ____________

36. ____________

37. ____________

38. ____________

39. ____________

40. ____________

41. ____________

42. ____________

43. ____________

44. ____________

45. ____________

46. ____________

47. ____________

48. ____________

49. ______

50. ______

51. ______

52. ______

53. ______

54. ______

55. ______

56. ______

57. a. ______

b. ______

Name ______

49. A small plane's maximum takeoff weight is 2000 pounds. Six passengers weigh an average of 160 pounds each. Use an inequality to find the maximum weight of luggage and cargo the plane can carry.

50. A clerk must use the elevator to move boxes of paper. The elevator's weight limit is 1500 pounds. If each box of paper weighs 66 pounds and the clerk weighs 147 pounds, use an inequality to find the maximum number of boxes she can move on the elevator at one time.

51. To mail an envelope first class, the U.S. Post Office charges 32 cents for the first ounce and 23 cents per ounce for each additional ounce. Use an inequality to find the maximum weight that can be mailed for $4.00.

52. A shopping mall parking garage charges $1 for the first half-hour and 60 cents for each additional half-hour or a portion of a half-hour. Use an inequality to find how long you can park if you have only $4.00 in cash.

53. Northeast Telephone Company offers two billing plans for local calls. Plan 1 charges $25 per month for unlimited calls, and plan 2 charges $13 per month plus 6 cents per call. Use an inequality to find the number of monthly calls for which plan 1 is more economical than plan 2.

54. A car rental company offers two subcompact rental plans. Plan A charges $32 per day for unlimited mileage, and plan B charges $24 per day plus 15 cents per mile. Use an inequality to find the number of daily miles for which plan A is more economical than plan B.

55. At room temperature, glass used in windows actually has some properties of a liquid. It has a very slow, viscous flow. (Viscosity is the property of a fluid that resists internal flow. For example, lemonade flows more easily than fudge syrup. Fudge syrup has a higher viscosity than lemonade.) Glass does not become a true liquid until temperatures are greater than or equal to 500°C. Find the Fahrenheit temperatures for which glass is a liquid. (Use the formula $F = \frac{9}{5}C + 32$.)

56. Stibnite is a silvery white mineral with a metalic luster. It is one of the few minerals that melts easily in a match flame or at temperatures of approximately 977°F or greater. Find the Celsius temperatures for which stibnite melts. (Use the formula $C = \frac{5}{9}(F - 32)$.)

57. Although beginning salaries vary greatly according to your field of study, the equation $s = 651.2t + 27{,}821$ can be used to approximate and to predict average beginning salaries for candidates with bachelor's degrees. The variable s is the starting salary and t is the number of years after 1989.

a. Approximate when beginning salaries for candidates will be greater than $35,000.

b. Determine the year you plan to graduate from college. Use this year to find the corresponding value of t and approximate your beginning salary.

Name ______________________

58. ______

59. ______

60. ______

61. ______

62. ______

63. ______

64. ______

65. ______

66. ______

58. Use the formula in Example 12 to estimate the years that the consumption of cigarettes will be less than 50 billion per year.

The average consumption per person per year of whole milk w can be approximated by the equation

$$w = -2.71t + 89.67$$

where t is the number of years after 1990. The average consumption of skim milk s per person per year can be approximated by the equation

$$s = 1.86t + 21.96$$

where t is the number of years after 1990. The consumption of whole milk is shown on the graph in blue and the consumption of skim milk is shown on the graph in red. Use this information to answer Exercises 59–67.

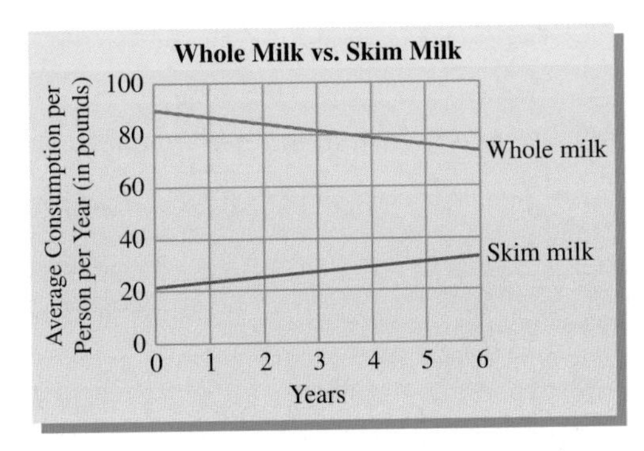

(*Source:* Based on data from Economic Research Service, U.S. Department of Agriculture, *Agricultural Outlook*, June/July 1998)

59. Is the consumption of whole milk increasing or decreasing over time? Explain how you arrived at your answer.

60. Is the consumption of skim milk increasing or decreasing over time? Explain how you arrived at your answer.

61. Predict the consumption of whole milk in the year 2000. (*Hint:* Find the value of t that corresponds to the year 2000.)

62. Predict the consumption of skim milk in the year 2000. (*Hint:* Find the value of t that corresponds to the year 2000.)

63. Determine when the consumption of whole milk will be less than 55 pounds per person per year.

64. Determine when the consumption of skim milk will be greater than 45 pounds per person per year.

65. For 1990 through 1996, the consumption of whole milk was greater than the consumption of skim milk. Explain how this can be determined from the graph.

66. How will the two lines in the graph appear when the consumption of whole milk is the same as the consumption of skim milk?

67. ______

68. ______

69. ______

70. ______

71. ______

72. ______

73. ______

74. ______

75. ______

76. ______

77. ______

78. ______

79. ______

80. ______

81. ______

Name ______

67. The consumption of whole milk will be the same as the consumption of skim milk when $w = s$. Find when this will occur by substituting the given equivalent expression for w and the given equivalent expression for s and solving for t. Round the value of t to the nearest whole and estimate the year when this will occur.

REVIEW AND PREVIEW

List or describe the integers that make both inequalities true.

68. $x < 5$ and $x > 1$

69. $x \geq 0$ and $x \leq 7$

70. $x \geq -2$ and $x \geq 2$

71. $x < 6$ and $x < -5$

Graph each set on a number line and write it in interval notation. See Section 2.4.

72. $\{x | 0 \leq x \leq 5\}$

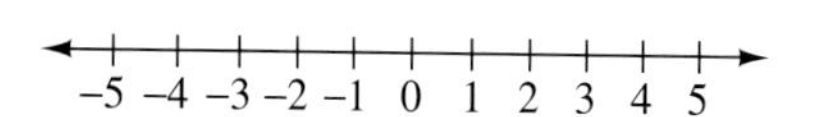

73. $\{x | -7 < x \leq 1\}$

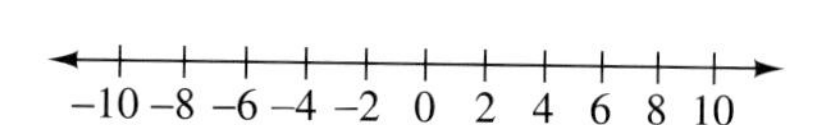

74. $\left\{x \middle| -\frac{1}{2} < x < \frac{3}{2}\right\}$

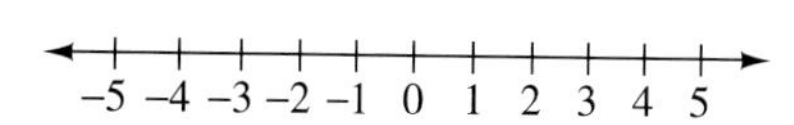

75. $\{x | -2.5 \leq x < 5.3\}$

−5 −4 −3 −2 −1 0 1 2 3 4 5

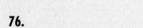

COMBINING CONCEPTS

Solve each inequality.

76. $4(x - 1) \geq 4x - 8$

77. $3x + 1 < 3(x - 2)$

78. $7x < 7(x - 2)$

79. $8(x + 3) \leq 7(x + 5) + x$

80. Explain how solving a linear inequality is similar to solving a linear equation.

81. Explain how solving a linear inequality is different from solving a linear equation.

Name ______________________ Section ________ Date ________

Integrated Review—Linear Equations and Inequalities

Solve each equation or inequality.

1. $-4x = 20$

2. $-4x < 20$

3. $\frac{3x}{4} \geq 2$

4. $5x + 3 \geq 2 + 4x$

5. $6(y - 4) = 3(y - 8)$

6. $-4x \leq \frac{2}{5}$

7. $-3x \geq \frac{1}{2}$

8. $5(y + 4) = 4(y + 5)$

9. $7x < 7(x - 2)$

10. $\frac{-5x + 11}{2} \leq 7$

11. $-5x + 1.5 = -19.5$

12. $-5x + 4 = -26$

13. $5 + 2x - x = -x + 3 - 14$

14. $12x + 14 < 11x - 2$

15. $\frac{x}{5} - \frac{x}{4} = \frac{x - 2}{2}$

16. $12x - 12 = 8(x - 1)$

17. $2(x - 3) > 70$

18. $-3x - 4.7 = 11.8$

Answers

1. ________
2. ________
3. ________
4. ________
5. ________
6. ________
7. ________
8. ________
9. ________
10. ________
11. ________
12. ________
13. ________
14. ________
15. ________
16. ________
17. ________
18. ________

19. ____________

20. ____________

21. ____________

22. ____________

23. ____________

24. ____________

Name ______________________________

19. $-2(b - 4) - (3b - 1) = 5b + 3$

20. $8(x + 3) < 7(x + 5) + x$

21. $\frac{3t + 1}{8} = \frac{5 + 2t}{7} + 2$

22. $4(x - 6) - x = 8(x - 3) - 5x$

23. $\frac{x + 3}{12} + \frac{x - 5}{15} < \frac{2}{3}$

24. $\frac{y}{3} + \frac{y}{5} = \frac{y + 3}{10}$

2.5 Sets and Compound Inequalities

Objectives

A Find the intersection of two sets.

B Solve compound inequalities containing "**and.**"

C Find the union of two sets.

D Solve compound inequalities containing "**or.**"

SSM CD-ROM Video 2.5

Two inequalities joined by the words **and** or **or** are called **compound inequalities.**

Compound Inequalities

$x + 3 < 8$ and $x > 2$

$\frac{2x}{3} \geq 5$ or $-x + 10 < 7$

A Finding the Intersection of Two Sets

The solution set of a compound inequality formed by the word **and** is the **intersection** of the solution sets of the two inequalities.

> **Intersection of Two Sets**
>
> The intersection of two sets, A and B, is the set of all elements common to both sets. A intersect B is denoted by
>
> $A \cap B$

Example 1 Find the intersection: $\{2, 4, 6, 8\} \cap \{3, 4, 5, 6\}$

Solution: The numbers 4 and 6 are in both sets. The intersection is $\{4, 6\}$.

Practice Problem 1

Find the intersection:
$\{1, 2, 3, 4, 5\} \cap \{3, 4, 5, 6\}$

B Solving Compound Inequalities Containing "and"

A value of x is a solution of a compound inequality formed by the word **and** if it is a solution of *both* inequalities. For example, the solution set of the compound inequality $x \leq 5$ and $x \geq 3$ contains all values of x that make the inequality $x \leq 5$ a true statement **and** the inequality $x \geq 3$ a true statement. The first graph shown below is the graph of $x \leq 5$, the second graph is the graph of $x \geq 3$, and the third graph shows the intersection of the two graphs. The third graph is the graph of $x \leq 5$ **and** $x \geq 3$.

$\{x|x \leq 5\}$ (number line −1 0 1 2 3 4 5 6) $(-\infty, 5]$

$\{x|x \geq 3\}$ (number line −1 0 1 2 3 4 5 6) $[3, \infty)$

$\{x|x \leq 5 \text{ and } x \geq 3\}$ (number line −1 0 1 2 3 4 5 6) $[3, 5]$

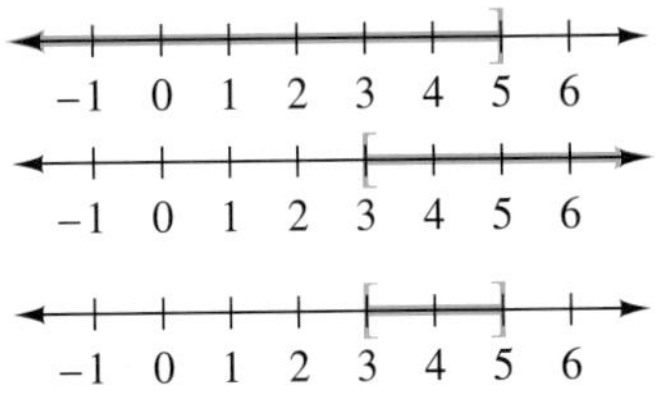

In interval notation, the set $\{x|x \leq 5 \text{ and } x \geq 3\}$ is written as $[3, 5]$.

Example 2 Solve: $x - 7 < 2$ and $2x + 1 < 9$

Solution: First we solve each inequality separately.

$$x - 7 < 2 \text{ and } 2x + 1 < 9$$
$$x < 9 \text{ and } \quad 2x < 8$$
$$x < 9 \text{ and } \quad x < 4$$

Now we can graph the two intervals on two number lines and find their intersection.

Practice Problem 2

Solve: $x + 5 < 9$ and $3x - 1 < 2$

Answers

1. $\{3, 4, 5\}$, **2.** $(-\infty, 1)$

$\{x|x < 9\}$ $(-\infty, 9)$

$\{x|x < 4\}$ $(-\infty, 4)$

$\{x|x < 9 \text{ and } x < 4\}$ $(-\infty, 4)$
$= \{x|x < 4\}$

The solution set is $(-\infty, 4)$.

Practice Problem 3

Solve: $4x \geq 0$ and $2x + 4 \geq 2$

Example 3

Solve: $2x \geq 0$ and $4x - 1 \leq -9$

Solution: First we solve each inequality separately.

$$2x \geq 0 \quad \text{and} \quad 4x - 1 \leq -9$$
$$x \geq 0 \quad \text{and} \quad 4x \leq -8$$
$$x \geq 0 \quad \text{and} \quad x \leq -2$$

Now we can graph the two intervals and find their intersection.

$\{x|x \geq 0\}$ $[0, \infty)$

$\{x|x \leq -2\}$ $(-\infty, -2]$

$\{x|x \geq 0 \text{ and } x \leq -2\}$
$= \emptyset$

There is no number that is greater than or equal to 0 *and* less than or equal to -2. The solution set is $\emptyset$.

Some compound inequalities containing the word **and** can be written in a more compact form. The compound inequality $2 \leq x$ and $x \leq 6$ can be written as

$$2 \leq x \leq 6$$

Recall from Section 2.4 that the graph of $2 \leq x \leq 6$ is all numbers between 2 and 6, including 2 and 6.

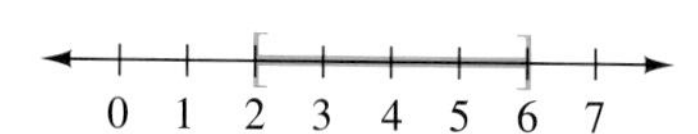

The set $\{x|2 \leq x \leq 6\}$ written in interval notation is $[2, 6]$.

To solve a compound inequality like $2 < 4 - x < 7$, we get x alone on the "middle side." Since a compound inequality is really two inequalities in one statement, we must perform the same operation to all three "sides" of the inequality.

Practice Problem 4

Solve: $5 < 1 - x < 9$

Example 4

Solve: $2 < 4 - x < 7$

Solution: To get x alone, we first subtract 4 from all three sides.

$$2 < 4 - x < 7$$
$$2 - 4 < 4 - x - 4 < 7 - 4 \quad \text{Subtract 4 from all three sides.}$$
$$-2 < -x < 3 \quad \text{Simplify.}$$
$$\frac{-2}{-1} > \frac{-x}{-1} > \frac{3}{-1} \quad \text{Divide all three sides by } -1 \text{ and reverse the inequality symbols.}$$
$$2 > x > -3$$

HELPFUL HINT

Don't forget to reverse both inequality symbols.

Answers

3. $[0, \infty)$, 4. $(-8, -4)$

This is equivalent to $-3 < x < 2$, and its graph is shown. The solution set in interval notation is $(-3, 2)$.

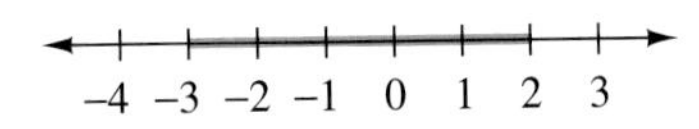

Example 5 Solve: $-1 \le \frac{2x}{3} + 5 \le 2$

Solution: First, clear the inequality of fractions by multiplying all three sides by the LCD of 3.

$$-1 \le \frac{2x}{3} + 5 \le 2$$

$$3(-1) \le 3\left(\frac{2x}{3} + 5\right) \le 3(2)$$ Multiply all three sides by the LCD of 3.

$$-3 \le 2x + 15 \le 6$$ Use the distributive property and multiply.

$$-3 - 15 \le 2x + 15 - 15 \le 6 - 15$$ Subtract 15 from all three sides.

$$-18 \le 2x \le -9$$ Simplify.

$$\frac{-18}{2} \le \frac{2x}{2} \le \frac{-9}{2}$$ Divide all three sides by 2.

$$-9 \le x \le -\frac{9}{2}$$ Simplify.

The graph of the solution is shown.

$-\frac{9}{2}$

−10 −9 −8 −7 −6 −5 −4 −3

The solution set in interval notation is $\left[-9, -\frac{9}{2}\right]$.

Practice Problem 5

Solve: $-3 \le \frac{x}{2} + 1 \le 5$

C FINDING THE UNION OF TWO SETS

The solution set of a compound inequality formed by the word **or** is the **union** of the solution sets of the two inequalities.

> **UNION OF TWO SETS**
>
> The union of two sets, A and B, is the set of elements that belong to *either* of the sets. A union B is denoted by
>
> $A \cup B$

Example 6 Find the union: $\{2, 4, 6, 8\} \cup \{3, 4, 5, 6\}$

Solution: The numbers that are in either set or both sets are $\{2, 3, 4, 5, 6, 8\}$. This set is the union.

Practice Problem 6

Find the union:
$\{1, 2, 3, 4, 5\} \cup \{3, 4, 5, 6\}$

D SOLVING COMPOUND INEQUALITIES CONTAINING "OR"

A value of x is a solution of a compound inequality formed by the word **or** if it is a solution of **either** inequality. For example, the solution set of the compound inequality $x \le 1$ **or** $x \ge 3$ contains all numbers that make the inequality $x \le 1$ a true statement **or** the inequality $x \ge 3$ a true statement.

Answers

5. $[-8, 8]$, **6.** $\{1, 2, 3, 4, 5, 6\}$

-1 0 1 2 3 4 5 6 $(-\infty, 1]$

-1 0 1 2 3 4 5 6 $[3, \infty)$

-1 0 1 2 3 4 5 6 $(-\infty, 1] \cup [3, \infty)$

In interval notation, the set $\{x | x \leq 1 \text{ or } x \geq 3\}$ is written as $(-\infty, 1] \cup [3, \infty)$.

Practice Problem 7

Solve: $3x - 2 \geq 10$ or $x - 6 \leq -4$

Example 7

Solve: $5x - 3 \leq 10$ or $x + 1 \geq 5$

Solution: First we solve each inequality separately.

$$5x - 3 \leq 10 \text{ or } x + 1 \geq 5$$
$$5x \leq 13 \text{ or } x \geq 4$$
$$x \leq \frac{13}{5} \text{ or } x \geq 4$$

Now we can graph each interval and find their union.

$\left\{x | x \leq \frac{13}{5}\right\}$ -1 0 1 2 3 4 5 6 $\left(-\infty, \frac{13}{5}\right]$

$\{x | x \geq 4\}$ -1 0 1 2 3 4 5 6 $[4, \infty)$

$\left\{x \middle| x \leq \frac{13}{5} \text{ or } x \geq 4\right\}$ -1 0 1 2 3 4 5 6 $\left(-\infty, \frac{13}{5}\right] \cup [4, \infty)$

The solution set is $\left(-\infty, \frac{13}{5}\right] \cup [4, \infty)$.

Practice Problem 8

Solve: $x - 7 \leq -1$ or $2x - 6 \geq 2$

Example 8

Solve: $-2x - 5 < -3$ or $6x < 0$

Solution: First we solve each inequality separately.

$$-2x - 5 < -3 \text{ or } 6x < 0$$
$$-2x < 2 \text{ or } x < 0$$
$$x > -1 \text{ or } x < 0$$

Now we can graph each interval and find their union.

$\{x | x > -1\}$ -4 -3 -2 -1 0 1 2 3 $(-1, \infty)$

$\{x | x < 0\}$ -4 -3 -2 -1 0 1 2 3 $(-\infty, 0)$

$\{x | x > -1 \text{ or } x < 0\}$ $(-\infty, \infty)$

= all real numbers

The solution set is $(-\infty, \infty)$.

✓ Concept Check

Which of the following is *not* a correct way to represent the set of all numbers between -3 and 5?

a. $\{x | -3 < x < 5\}$

b. $-3 < x$ or $x < 5$

c. $(-3, 5)$

d. $x > -3$ and $x < 5$

TRY THE CONCEPT CHECK IN THE MARGIN.

Answers

7. $(-\infty, 2] \cup [4, \infty)$, **8.** $(-\infty, \infty)$

✓ **Concept Check:** b is not correct

Name ______________________ Section ________ Date ________

Exercise Set 2.5

A *If $A = \{x \mid x \text{ is an even integer}\}$, $B = \{x \mid x \text{ is an odd integer}\}$, $C = \{2, 3, 4, 5\}$, and $D = \{4, 5, 6, 7\}$, list the elements of each set. See Example 1.*

1. $A \cap C$

2. $B \cap D$

3. $A \cap B$

4. $C \cap D$

5. $B \cap C$

6. $A \cap D$

B *Solve each compound inequality. Graph the two inequalities on the first two number lines and the solution set on the third number line. See Examples 2 and 3.*

7. $x < 1$ and $x > -3$

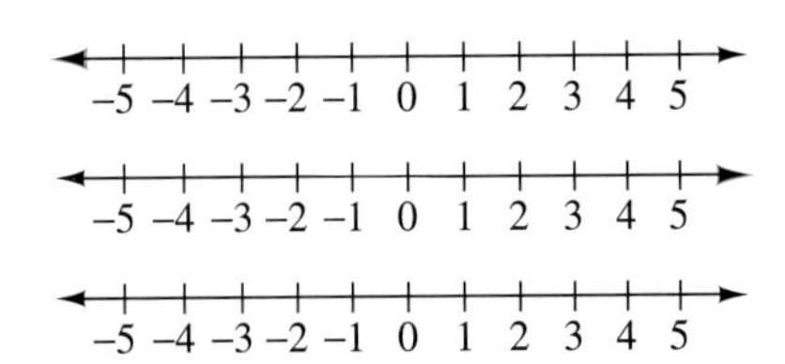

8. $x \leq 0$ and $x \geq -2$

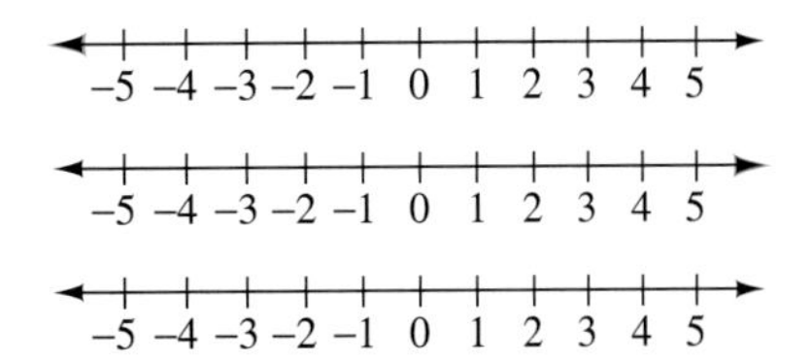

9. $x \leq -3$ and $x \geq -2$

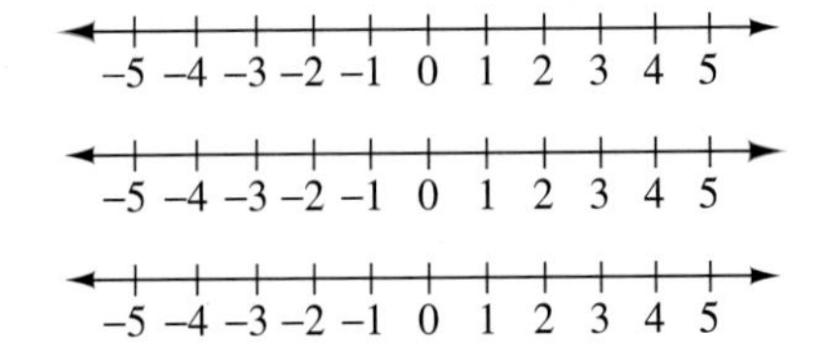

10. $x < 2$ and $x > 4$

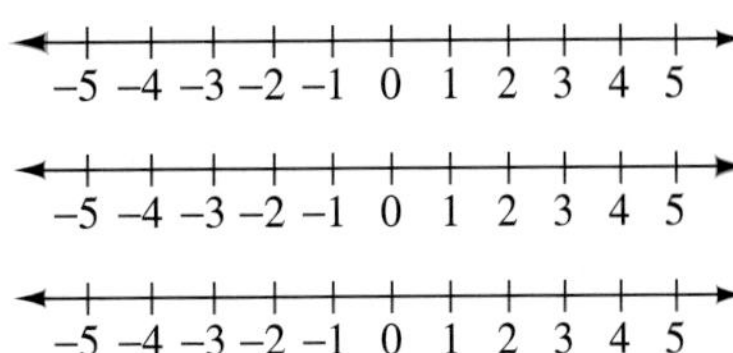
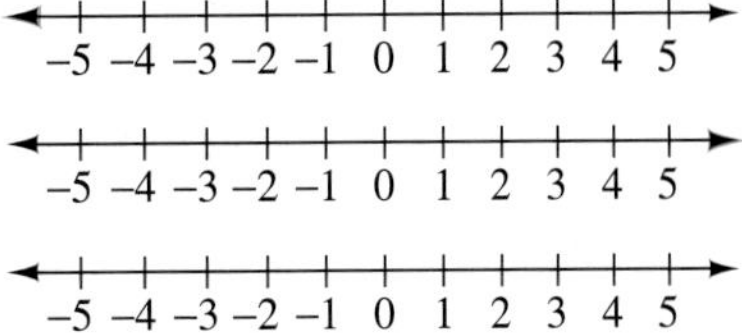

11. $x < -1$ and $x < 1$

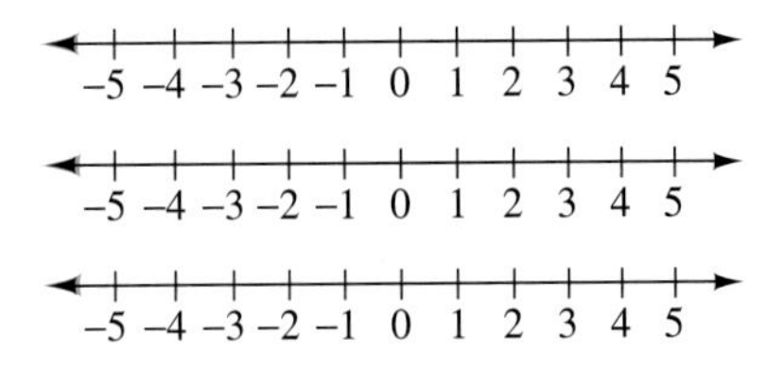

12. $x \geq -4$ and $x > 1$

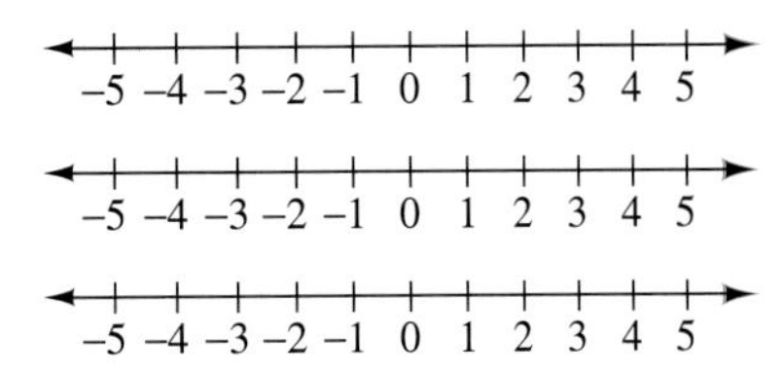

Solve each compound inequality. See Examples 2 and 3.

13. $x < 5$ and $x > -2$

14. $x \leq 7$ and $x \leq 1$

15. $x + 1 \geq 7$ and $3x - 1 \geq 5$

16. $-2x < -8$ and $x - 5 < 5$

Answers

1. ________
2. ________
3. ________
4. ________
5. ________
6. ________
7. see number lines
8. see number lines
9. see number lines
10. see number lines
11. see number lines
12. see number lines
13. ________
14. ________
15. ________
16. ________

17.

18.

19.

20.

21.

22.

23.

24.

25.

26.

27.

28.

29.

30.

31.

32.

33.

34.

35.

36.

37.

38.

Name

17. $4x + 2 \leq -10$ and $2x \leq 0$

18. $x + 4 > 0$ and $4x > 0$

19. $x + 3 \geq 3$ and $x + 3 \leq 2$

20. $2x - 1 \geq 3$ and $-x > 2$

Solve each compound inequality. See Examples 4 and 5.

21. $5 < x - 6 < 11$

22. $-2 \leq x + 3 \leq 0$

23. $-2 \leq 3x - 5 \leq 7$

24. $1 < 4 + 2x < 7$

25. $1 \leq \frac{2}{3}x + 3 \leq 4$

26. $-2 < \frac{1}{2}x - 5 < 1$

27. $-5 \leq \frac{x + 1}{4} \leq -2$

28. $-4 \leq \frac{2x + 5}{3} \leq 1$

29. $0 \leq 2x - 3 \leq 9$

30. $3 < 5x + 1 < 11$

31. $-6 < 3(x - 2) \leq 8$

32. $-5 < 2(x + 4) < 8$

C *If $A = \{1, 2, 3, 4, 5, 6, 7, 8\}$, $B = \{1, 5\}$, $C = \{2, 4, 6, 8\}$ and $D = \{6\}$, list the elements of each set. See Example 6.*

33. $A \cup B$

34. $A \cup C$

35. $B \cup D$

36. $B \cup C$

37. $C \cup D$

38. $D \cup B$

Name ______________________

D *Solve each compound inequality. Graph the two given inequalities on the first two number lines and the solution set on the third number line. See Examples 7 and 8.*

39. $x < 4$ or $x < 5$

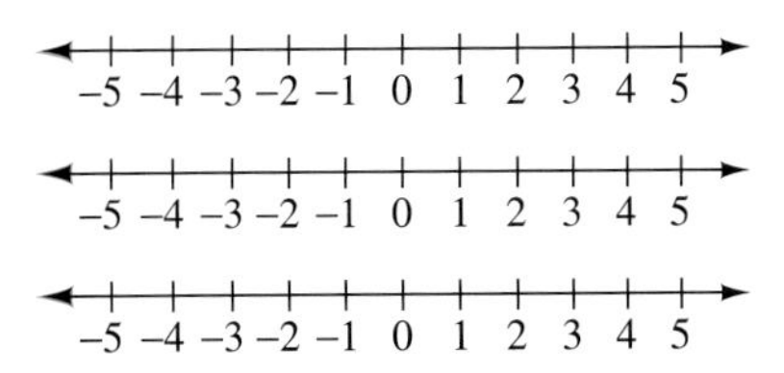

40. $x \geq -2$ or $x \leq 2$

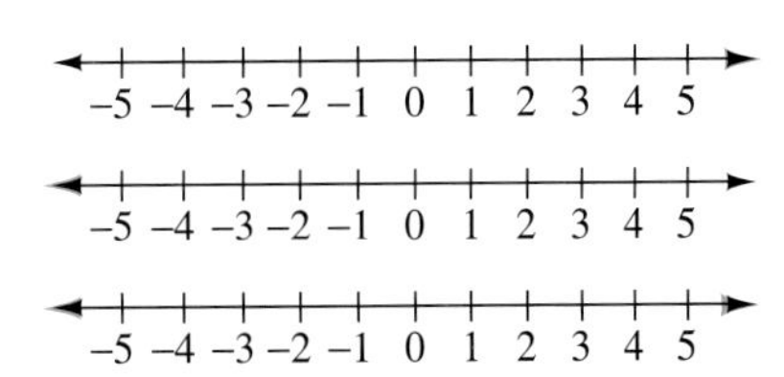

41. $x \leq -4$ or $x \geq 1$

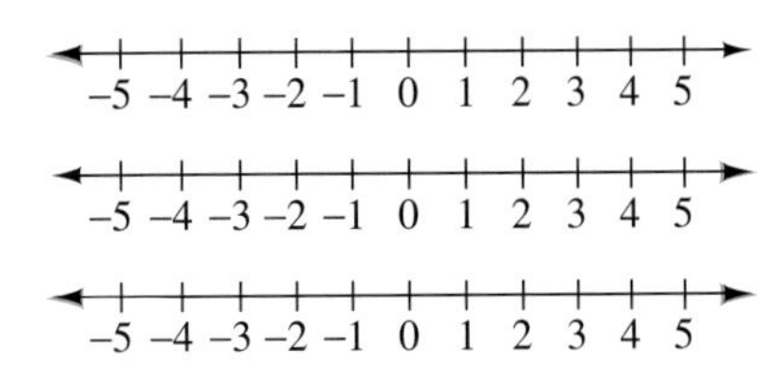

42. $x < 0$ or $x < 1$

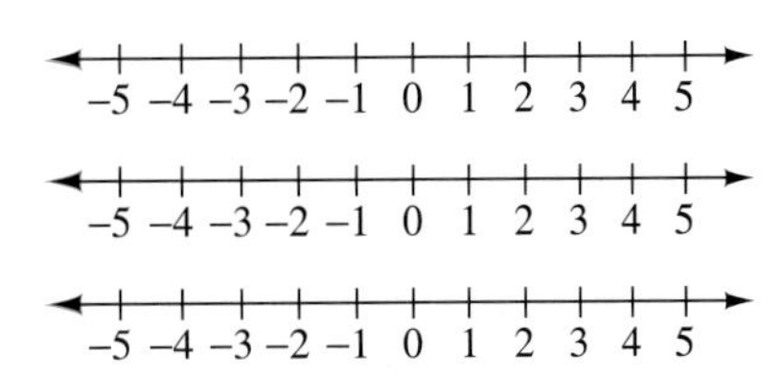

43. $x > 0$ or $x < 3$

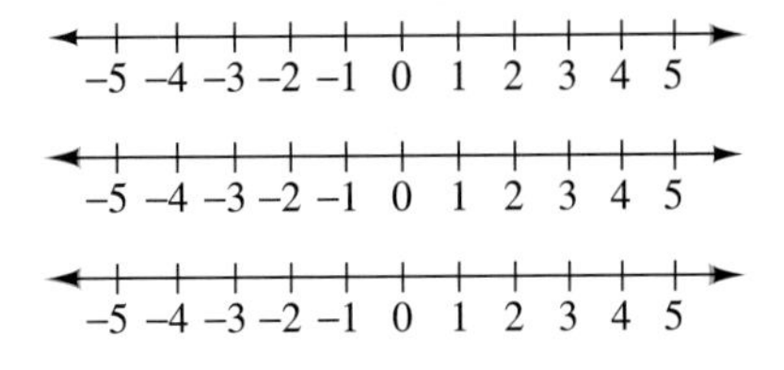

44. $x \geq -3$ or $x \leq -4$

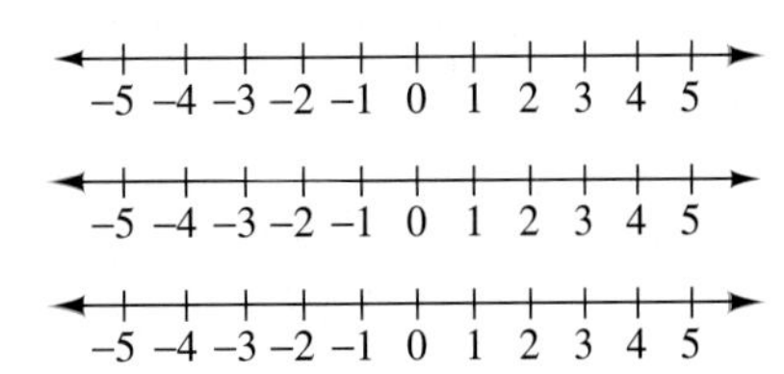

Solve each compound inequality. See Examples 7 and 8.

45. $x < -1$ or $x > 0$

46. $x \leq 1$ or $x \leq -3$

47. $-2x \leq -4$ or $5x - 20 \geq 5$

48. $x + 4 < 0$ or $6x > -12$

49. $3(x - 1) < 12$ or $x + 7 > 10$

50. $5(x - 1) \geq -5$ or $5 - x \leq 11$

51. $3x + 2 \leq 5$ or $7x > 29$

52. $-x < 7$ or $3x + 1 < -20$

39. see number lines

40. see number lines

41. see number lines

42. see number lines

43. see number lines

44. see number lines

45. ______

46. ______

47. ______

48. ______

49. ______

50. ______

51. ______

52. ______

53.

54.

55.

56.

57.

58.

59.

60.

61.

62.

63.

64.

65.

66.

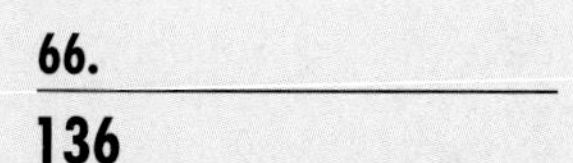

Name ____________________

53. $3x \geq 5$ or $-x - 6 < 1$

54. $\frac{3}{8}x + 1 \leq 0$ or $-2x < -4$

55. $6x - 4 > 2x$ or $4x - 1 < x + 5$

56. $6x - 2 > 5x + 3$ or $4x - 3 < x$

Review and Preview

Evaluate. See Section 1.4.

57. $|-7| - |19|$

58. $|-7 - 19|$

59. $-(-6) - |-10|$

60. $|-4| - (-4) + |-20|$

Find by inspection all values for x that make each equation true.

61. $|x| = 7$

62. $|x| = 5$

63. $|x| = 0$

64. $|x| = -2$

Combining Concepts

Use the graph to answer Exercises 65–66.

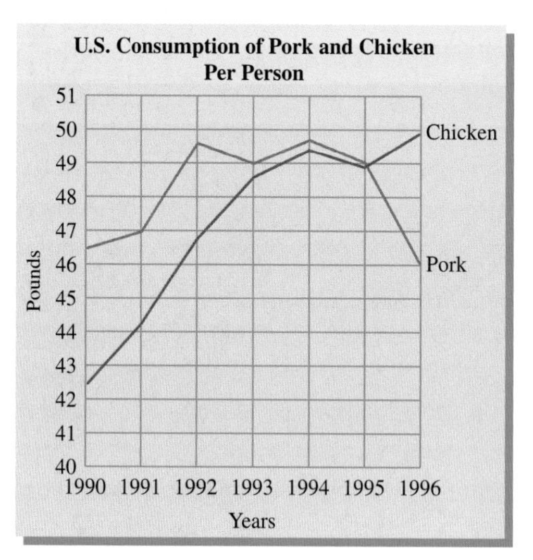

(*Source:* Based on data from Economic Research Service, U.S. Department of Agriculture, *Agricultural Outlook*, June/July 1998)

65. For what years was the consumption of pork greater than 48 pounds per person *and* the consumption of chicken greater than 48 pounds per person?

66. For what years was the consumption of pork less than 48 pounds per person *or* the consumption of chicken greater than 49 pounds per person?

Name ______________________

To solve a compound inequality such as $x - 6 < 3x < 2x + 5$, *we solve*

$$x - 6 < 3x \quad \text{and} \quad 3x < 2x + 5$$

Use this information to solve the inequalities in Exercises 67–70.

67. $x - 6 < 3x < 2x + 5$

68. $2x - 3 < 3x + 1 < 4x - 5$

69. $x + 3 < 2x + 1 < 4x + 6$

70. $-3(x - 2) \leq 3 - 2x \leq 10 - 3x$

Internet Excursions

Go to http://www.prenhall.com/martin-gay

Target heart rate is the number of heartbeats per minute a person should maintain during aerobic exercise to get maximum cardiovascular benefits. Many health experts recommend keeping heart rate while exercising within a certain interval: between a minimum and maximum target heart rate. Each person has a different target heart rate range, depending on his or her age. By visiting the World Wide Web address listed above, you will gain access to a web site where you can learn how to calculate your target heart rate.

71. Using the description of calculating target heart rate given on this Web site, write algebraic expressions that represent a person's maximum and minimum target heart rates. Then use these expressions to write a compound inequality that describes the range in which a person's heart rate should fall while exercising. Be sure to define all variables used.

72. Using your compound inequality from Exercise 71, find your own target heart rate range.

67. ______

68. ______

69. ______

70. ______

71. ______

72. ______

Focus On the Real World

Energy Audit

Have you ever been surprised by high electric bills? Has it made you wonder where all of your electricity expenditure is going or how to lower your bill? If so, one approach to learning more about your electricity consumption is performing an energy audit of your home or apartment. Once you understand your patterns of electricity usage, you can make informed decisions on where to cut back or whether or not to replace an older appliance with a newer, more energy efficient one.

To perform your own energy audit, fill out the table below. First, make a list of all the electrical appliances in your home. Be sure to include components of your heating, hot water, and/or air-conditioning systems, major appliances, indoor and outdoor lights, computer and audio-visual components, and small kitchen or personal care appliances. Don't forget to include easily overlooked items such as room space heaters, ceiling fans, and water bed heaters.

Next, estimate how many hours each item is run per 30-day month. For items used nearly every day, estimate daily usage in hours and multiply by 30. For items used less often, estimate how many hours they are used per week. Then divide by 7 and multiply by 30 to get an estimate for a 30-day month.

For each item on the list, record its wattage. This information can usually be found on its serial number plate. Wattage is abbreviated W, so look for a number like 13W or 200W. If wattage is not listed on the plate, look for information on voltage (abbreviated V for volts) and amperage (abbreviated A for amps). Wattage can be estimated by multiplying volts times amps. (*Note:* Sometimes a range of values is listed for voltage. If the range includes 120 V, use 120 in the wattage calculation. Otherwise, use the maximum value of the voltage range for the wattage calculation.)

Fill in the fourth column of the table by multiplying the number of hours each appliance is run during a month by its wattage to find watt-hours. Then fill in the fifth column of the table by dividing watt-hours by 1000 to find kilowatt-hours. For the last column, consult your electricity bill to find the price charged by your electric company per kilowatt-hour (often abbreviated KWH). Alternatively, contact the local electric company to ask its standard charge per kilowatt-hour for residential customers. Fill in the last column of the table by figuring the cost to run an appliance for one month: Multiply the number of kilowatt-hours per month by what the electric company charges per kilowatt-hour.

Appliance	Hours run per 30-day Month	Wattage (or use volts $\times$ amps)	Watt-Hours (hours $\times$ wattage)	Kilowatt-Hours (watt-hours $\div$ 1000)	Cost to Run Appliance for One Month (kilowatt-hours $\times$ cost per KWH)

Critical Thinking

Do an energy audit of your home, apartment, or dormitory room.

1. Which item is the most expensive to operate over the course of the month? Does this surprise you? Why or why not?
2. Which item is the least expensive to operate? Does this surprise you? Why or why not?
3. Are there any items on the list whose usage could be cut back to save energy costs? Which ones would be the most viable choices? How much could usage be cut and how much money would that save? Explain.
4. Are there any appliances that could be replaced with more energy-efficient models? Conduct research to find a more recent model that would be a better choice. If the usage of the new model is the same as the old model, how much money could be saved each month by switching? How long will it take for the monthly energy cost savings to "pay off" the price of buying the new appliance? Explain.
5. In what other ways could you lower your electric bill?

2.6 ABSOLUTE VALUE EQUATIONS AND INEQUALITIES

Objectives

A Solve absolute value equations.

B Solve absolute value inequalities.

SSM CD-ROM Video 2.6

In Chapter 1, we defined the absolute value of a number as its distance from 0 on a number line.

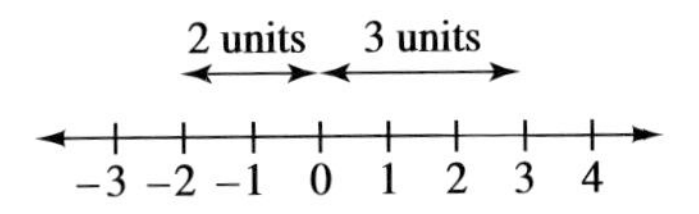

$|-2| = 2$ and $|3| = 3$

In this section, we concentrate on solving equations and inequalities containing the absolute value of a variable or a variable expression. Examples of absolute value equations and inequalities are

$$|x| = 3 \qquad -5 \geq |2y + 7| \qquad |z - 6.7| = |3z + 1.2| \qquad |x - 3| > 7$$

Absolute value equations and inequalities are extremely useful in data analysis, especially for calculating acceptable measurement error and errors that result from the way numbers are sometimes represented in computers.

A SOLVING ABSOLUTE VALUE EQUATIONS

To begin let's solve a few absolute value equations by inspection.

Example 1 Solve: $|x| = 3$

Solution: The solution set of this equation will contain all numbers whose distance from 0 is 3 units. Two numbers are 3 units away from 0 on the number line: 3 and −3.

3 units 3 units

−4 −3 −2 −1 0 1 2 3 4

Check: To check, let $x = 3$ and $x = -3$ in the original equation.

$\lvert x\rvert = 3$		$\lvert x\rvert = 3$	
$\lvert 3\rvert \stackrel{?}{=} 3$	Let $x = 3$.	$\lvert -3\rvert \stackrel{?}{=} 3$	Let $x = -3$.
$3 = 3$	True.	$3 = 3$	True.

Both solutions check. Thus the solution set of the equation $|x| = 3$ is $\{3, -3\}$.

Practice Problem 1

Solve: $|y| = 5$

Example 2 Solve: $|x| = -2$

Solution: The absolute value of a number is never negative, so this equation has no solution. The solution set is $\{\ \}$ or $\varnothing$.

Practice Problem 2

Solve: $|p| = -4$

Example 3 Solve: $|y| = 0$

Solution: We are looking for all numbers whose distance from 0 is zero units. The only number is 0. The solution set is $\{0\}$.

Practice Problem 3

Solve: $|x| = 0$

From the above examples, we have the following.

Answers

1. $\{-5, 5\}$, **2.** $\varnothing$, **3.** $\{0\}$

THE ABSOLUTE VALUE PROPERTY

Solve:

$$|X| = a \quad \begin{cases} \text{If } a \text{ is positive, then solve } X = a \text{ or } X = -a. \\ \text{If } a \text{ is 0, then } X = 0. \\ \text{If } a \text{ is negative, the equation } |X| = a \text{ has no solution.} \end{cases}$$

HELPFUL HINT

For the equation $|X| = a$ in the box above, X can be a single variable, or a variable expression.

When we are solving absolute value equations, if $|X|$ is not alone on one side of the equation we first use properties of equality to get $|X|$ alone.

Practice Problem 4

Solve: $3|y| - 4 = 17$

Example 4 Solve: $2|x| + 25 = 37$

Solution: First we get $|x|$ alone.

$2|x| + 25 = 37$

$2|x| = 12$ Subtract 25 from both sides.

$|x| = 6$ Divide both sides by 2.

$x = 6$ or $x = -6$ Use the absolute value property.

The solution set is $\{-6, 6\}$.

If the expression inside the absolute value bars is more complicated than a single variable x, we can still use the absolute value property.

Practice Problem 5

Solve: $|x - 4| = 11$

Example 5 Solve: $|w + 3| = 7$

Solution: If we think of the expression $w + 3$ as X in the absolute value property, we have that

$|w + 3| = 7$

$w + 3 = 7$ or $w + 3 = -7$ Use the absolute value property.

$w = 4$ or $w = -10$

The solution set is $\{4, -10\}$.

Don't forget that to use the absolute value property you must first make sure that the absolute value expression is alone on one side of the equation.

HELPFUL HINT

If the equation has a single absolute value expression containing variables, get the absolute value expression alone. Then use the absolute value property.

Practice Problem 6

Solve: $|4x + 2| + 1 = 7$

Example 6 Solve: $|2x - 1| + 5 = 6$

Solution: We want the absolute value expression alone on one side of the equation, so we begin by subtracting 5 from both sides. Then we use the absolute value property.

Answers

4. $\{-7, 7\}$, **5.** $\{15, -7\}$, **6.** $\{1, -2\}$

$$|2x - 1| + 5 = 6$$
$$|2x - 1| = 1 \quad \text{Subtract 5 from both sides.}$$
$$2x - 1 = 1 \quad \text{or} \quad 2x - 1 = -1 \quad \text{Use the absolute value property.}$$
$$2x = 2 \quad \text{or} \quad 2x = 0$$
$$x = 1 \quad \text{or} \quad x = 0 \quad \text{Solve.}$$

The solution set is $\{0, 1\}$.

Given two absolute value expressions, we might ask, when are the absolute values of two expressions equal? To see the answer, notice that

$$|2| = |2| \qquad |-2| = |-2| \qquad |-2| = |2| \qquad |2| = |-2|$$

same   same   opposites   opposites

Two absolute value expressions are equal when the expressions inside the absolute value bars are equal to or are opposites of each other.

Example 7 Solve: $|3x + 2| = |5x - 8|$

Solution: This equation is true if the expressions inside the absolute value bars are equal to or are opposites of each other.

$$3x + 2 = 5x - 8 \quad \text{or} \quad 3x + 2 = -(5x - 8)$$

Next we solve each equation.

$$3x + 2 = 5x - 8 \quad \text{or} \quad 3x + 2 = -5x + 8$$
$$-2x + 2 = -8 \quad \text{or} \quad 8x + 2 = 8$$
$$-2x = -10 \quad \text{or} \quad 8x = 6$$
$$x = 5 \quad \text{or} \quad x = \frac{3}{4}$$

Check to see that replacing x with 5 or with $\frac{3}{4}$ results in a true statement. The solution set is $\left\{\frac{3}{4}, 5\right\}$.

Practice Problem 7

Solve: $|4x - 5| = |3x + 5|$

Example 8 Solve: $|x - 3| = |5 - x|$

Solution:

$$x - 3 = 5 - x \quad \text{or} \quad x - 3 = -(5 - x)$$
$$2x - 3 = 5 \quad \text{or} \quad x - 3 = -5 + x$$
$$2x = 8 \quad \text{or} \quad x - 3 - x = -5 + x - x$$
$$x = 4 \quad \text{or} \quad -3 = -5 \quad \text{False.}$$

Recall from Section 2.1 that when an equation simplifies to a false statement, the equation has no solution. Thus the only solution for the original absolute value equation is 4, and the solution set is $\{4\}$.

Practice Problem 8

Solve: $|x + 2| = |4 - x|$

✓ CONCEPT CHECK

True or false? Absolute value equations always have two solutions. Explain your answer.

TRY THE CONCEPT CHECK IN THE MARGIN.

B SOLVING ABSOLUTE VALUE INEQUALITIES

To begin, let's solve a few absolute value inequalities by inspection.

Example 9 Solve: $|x| < 2$ using a number line.

Practice Problem 9

Solve $|x| < 4$ using a number line.

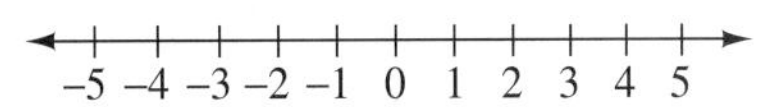

Answers

7. $\{0, 10\}$, **8.** $\{1\}$

9.
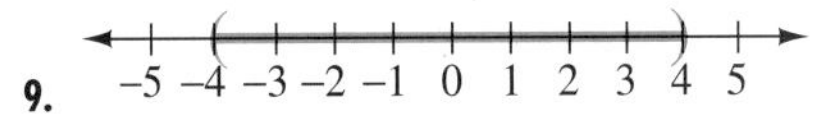

✓ Concept Check: false; answers may vary

Solution: The solution set contains all numbers whose distance from 0 is less than 2 units on the number line.

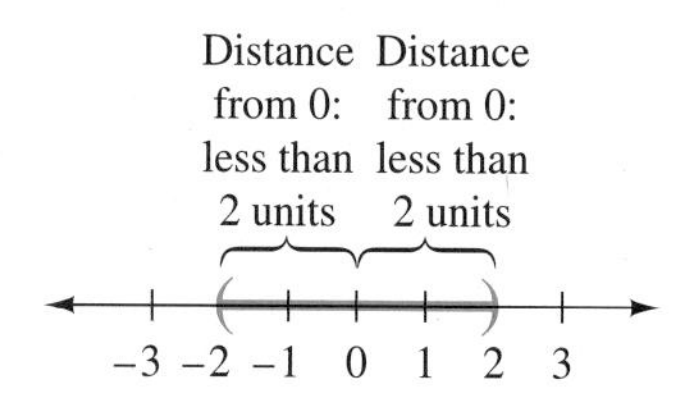

The solution set is $\{x|-2 < x < 2\}$, or $(-2, 2)$ in interval notation.

Practice Problem 10

Solve $|x| \geq 5$ using a number line.

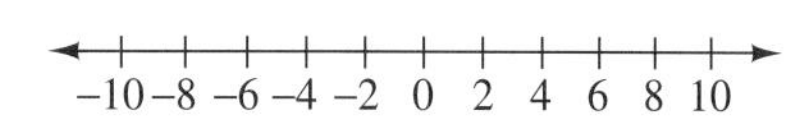

Example 10

Solve $|x| \geq 3$ using a number line.

Solution: The solution set contains all numbers whose distance from 0 is 3 or more units. Thus the graph of the solution set contains 3 and all points to the right of 3 on the number line or −3 and all points to the left of −3 on the number line.

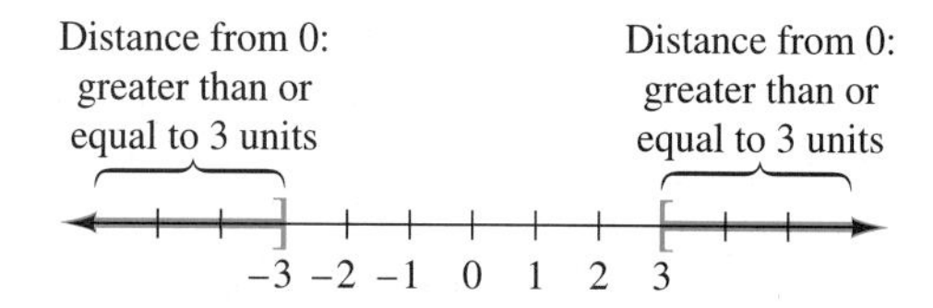

This solution set is $\{x|x \leq -3 \text{ or } x \geq 3\}$. In interval notation, the solution set is $(-\infty, -3] \cup [3, \infty)$, since *or* means union.

The following box summarizes solving absolute value equations and inequalities.

> **SOLVING ABSOLUTE VALUE EQUATIONS AND INEQUALITIES**
>
> If a is a positive number,
>
> To solve $|X| = a$, solve $X = a$ or $X = -a$. ($-a$, a)
>
> To solve $|X| < a$, solve $-a < X < a$. ($-a$, a)
>
> To solve $|X| > a$, solve $X < -a$ or $X > a$.

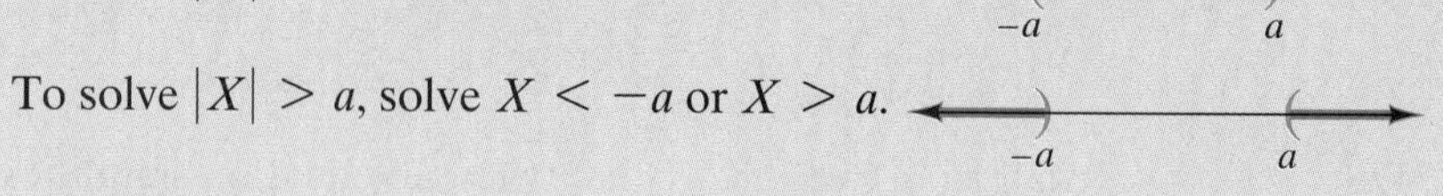

Practice Problem 11

Solve: $|x + 2| > 4$. Graph the solution set.

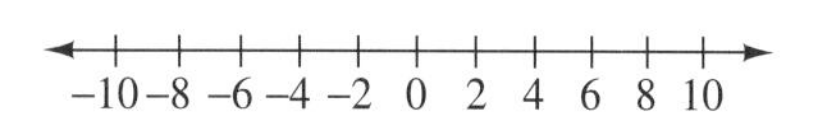

Example 11

Solve: $|x - 3| > 7$

Solution: Since 7 is positive, to solve $|x - 3| > 7$, we solve the compound inequality $x - 3 < -7$ or $x - 3 > 7$.

$$x - 3 < -7 \quad \text{or} \quad x - 3 > 7$$
$$x < -4 \quad \text{or} \quad x > 10 \qquad \text{Add 3 to both sides.}$$

The solution set is $\{x|x < -4 \text{ or } x > 10\}$ or $(-\infty, -4) \cup (10, \infty)$ in interval notation. Its graph is shown.

Answers

10.

11. $(-\infty, -6) \cup (2, \infty)$

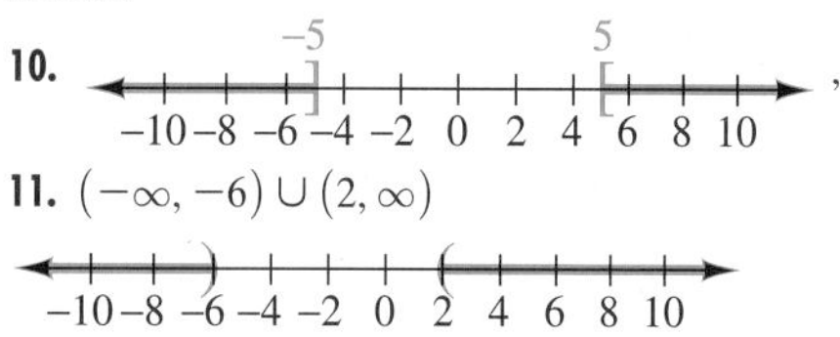

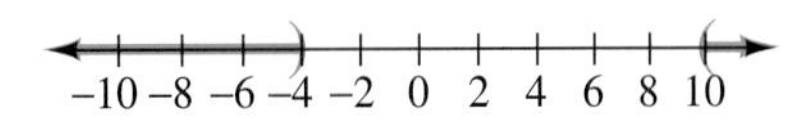

Let's review the differences in solving absolute value equations and inequalities by solving an absolute value equation.

Example 12 Solve: $|x + 1| = 6$

Solution: This is an equation, so we solve

$$x + 1 = 6 \quad \text{or} \quad x + 1 = -6$$
$$x = 5 \quad \text{or} \quad x = -7$$

The solution set is $\{-7, 5\}$. Its graph is shown.

−7 5
−10 −8 −6 −4 −2 0 2 4 6 8 10

Practice Problem 12

Solve: $|x - 3| = 5$. Graph the solution set.

−10 −8 −6 −4 −2 0 2 4 6 8 10

Example 13 Solve: $|x - 6| \le 2$

Solution: To solve $|x - 6| \le 2$, we solve

$$-2 \le x - 6 \le 2$$
$$-2 + 6 \le x - 6 + 6 \le 2 + 6 \quad \text{Add 6 to all three sides.}$$
$$4 \le x \le 8 \quad \text{Simplify.}$$

The solution set is $\{x|4 \le x \le 8\}$ or $[4, 8]$ in interval notation. Its graph is shown below.

3 4 5 6 7 8 9

Practice Problem 13

Solve: $|x - 2| \le 1$. Graph the solution set.

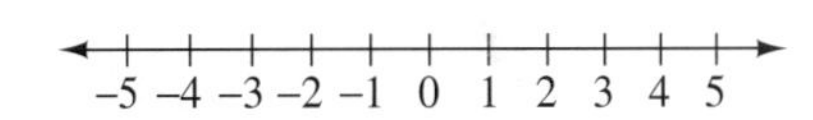

HELPFUL HINT

Before using an absolute value inequality property, get an absolute value expression alone on one side of the inequality.

Example 14 Solve: $|5x + 1| + 1 \le 10$

Solution: First we get the absolute value expression alone by subtracting 1 from both sides.

$$|5x + 1| + 1 \le 10$$
$$|5x + 1| \le 10 - 1 \quad \text{Subtract 1 from both sides.}$$
$$|5x + 1| \le 9 \quad \text{Simplify.}$$

Since 9 is positive, to solve $|5x + 1| \le 9$, we solve

$$-9 \le 5x + 1 \le 9$$
$$-9 - 1 \le 5x + 1 - 1 \le 9 - 1 \quad \text{Subtract 1 from all three sides.}$$
$$-10 \le 5x \le 8 \quad \text{Simplify.}$$
$$-2 \le x \le \frac{8}{5} \quad \text{Divide all three sides by 5.}$$

The solution set is $\left[-2, \frac{8}{5}\right]$.

Practice Problem 14

Solve: $|2x - 5| + 2 \le 9$

The next few examples are special cases of absolute value inequalities.

Answers

12. $\{-2, 8\}$

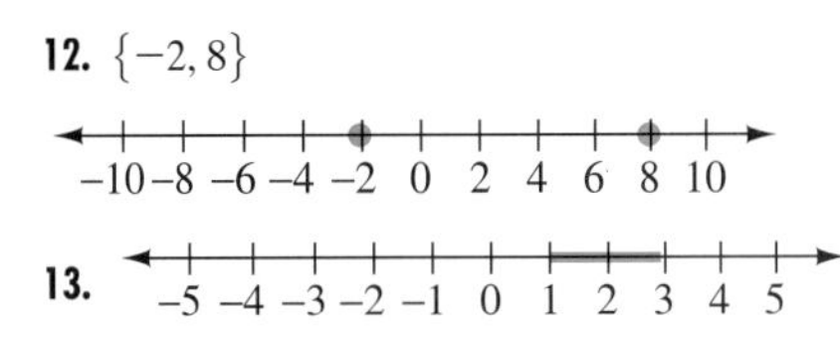

13.

14. $[-1, 6]$

Example 15 Solve: $|x| \leq -3$

Solution: The absolute value of a number is never negative. Thus it will then never be less than or equal to -3. The solution set is $\{ \}$ or $\emptyset$.

Example 16 Solve: $|x - 1| > -2$

Solution: The absolute value of a number is always nonnegative. Thus it will always be greater than -2. The solution set contains all real numbers, or $(-\infty, \infty)$.

TRY THE CONCEPT CHECK IN THE MARGIN.

Practice Problem 15

Solve: $|x| < -1$

Practice Problem 16

Solve: $|x + 1| \geq -3$

✓ CONCEPT CHECK

Without taking any solution steps, how do you know that the absolute value inequality $|3x - 2| > -9$ has a solution? What is its solution?

Answers

15. $\emptyset$, **16.** $(-\infty, \infty)$

✓ **Concept Check:** $(-\infty, \infty)$ since the absolute value is always nonnegative

Name ______________________ Section ________ Date __________

MENTAL MATH

Simplify each expression.

1. $|-7|$ **2.** $|-8|$ **3.** $-|5|$

4. $-|10|$ **5.** $-|-6|$ **6.** $-|-3|$

7. $|-3| + |-2| + |-7|$ **8.** $|-1| + |-6| + |-8|$

EXERCISE SET 2.6

A *Solve. See Examples 1 through 6.*

1. $|x| = 7$ **2.** $|y| = 15$ **3.** $|x| = -4$

4. $|x| = -20$ **5.** $|3x| = 12.6$ **6.** $|6n| = 12.6$

7. $|x - 9| = 14$ **8.** $|x + 2| = 8$ **9.** $|2x - 5| = 9$

10. $|6 + 2n| = 4$ **11.** $\left|\frac{x}{2} - 3\right| = 1$ **12.** $\left|\frac{n}{3} + 2\right| = 4$

13. $|z| + 4 = 9$ **14.** $|x| + 1 = 3$ **15.** $|3x| + 5 = 14$

16. $|2x| - 6 = 4$ **17.** $|2x| = 0$ **18.** $|7z| = 0$

19. $|4n + 1| + 10 = 4$ **20.** $|3z - 2| + 8 = 1$

Solve. See Examples 7 and 8.

21. $|5x - 7| = |3x + 11|$ **22.** $|9y + 1| = |6y + 4|$

23. $|z + 8| = |z - 3|$ **24.** $|2x - 5| = |2x + 5|$

MENTAL MATH ANSWERS

1. ______
2. ______
3. ______
4. ______
5. ______
6. ______
7. ______
8. ______

ANSWERS

1. ______
2. ______
3. ______
4. ______
5. ______
6. ______
7. ______
8. ______
9. ______
10. ______
11. ______
12. ______
13. ______
14. ______
15. ______
16. ______
17. ______
18. ______
19. ______
20. ______
21. ______
22. ______
23. ______
24. ______

25. ______
26. ______
27. ______
28. ______
29. ______
30. ______
31. ______
32. ______
33. ______
34. ______
35. ______
36. ______
37. ______
38. ______
39. ______
40. ______
41. ______
42. ______

Name ______________________________

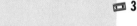

25. $|2y - 3| = |9 - 4y|$

26. $|5z - 1| = |7 - z|$

27. $|2x - 6| = |10 - 2x|$

28. $|4n + 5| = |4n + 3|$

29. $|x + 4| = |7 - x|$

30. $|8 - y| = |y + 2|$

31. $|5x + 1| = |4x - 7|$

32. $|3 + 6n| = |4n + 11|$

B *Solve. Graph the solution set. See Examples 9 through 16.*

33. $|x| \leq 4$

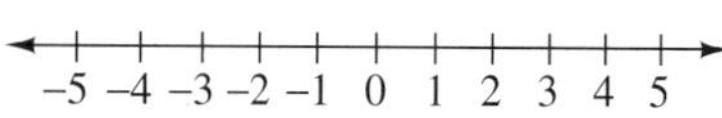

34. $|x| < 6$

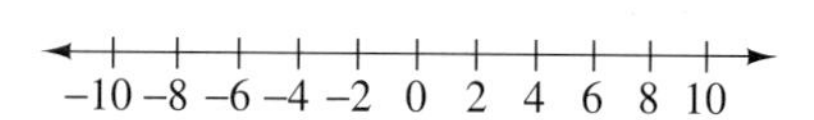

35. $|x| > 3$

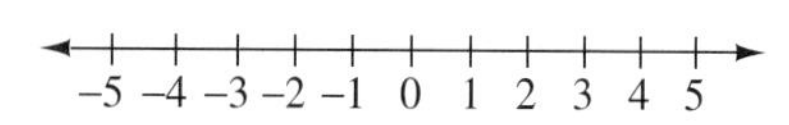

36. $|y| \geq 4$

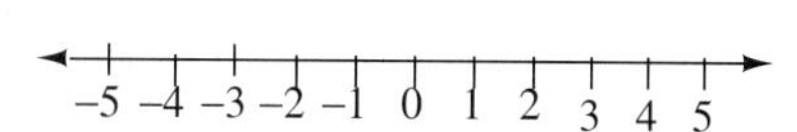

37. $|x + 3| < 2$

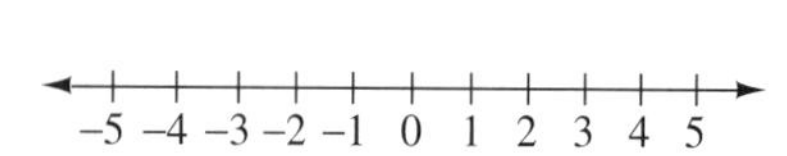

38. $|x + 4| < 6$

–10 –8 –6 –4 –2 0 2 4 6 8 10

39. $|y - 6| \geq 7$

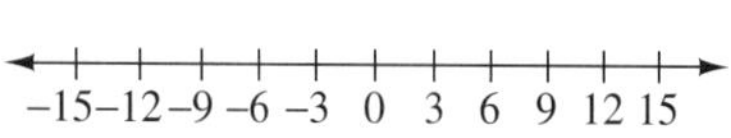

40. $|x - 3| \geq 10$

–15 –12 –9 –6 –3 0 3 6 9 12 15

41. $|x| + 7 \leq 12$

–5 –4 –3 –2 –1 0 1 2 3 4 5

42. $|x| + 6 \leq 7$

–5 –4 –3 –2 –1 0 1 2 3 4 5

Name ______________________________

43. $|x| + 2 > 6$

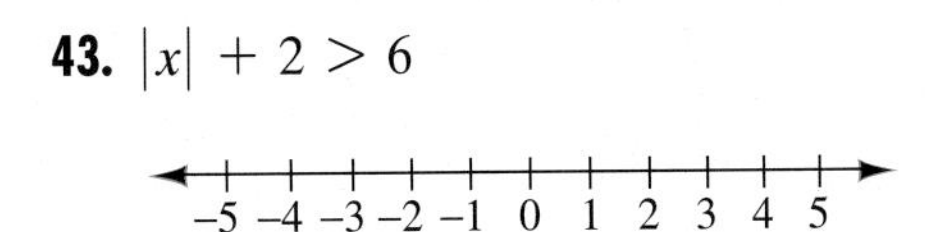

44. $|x| - 1 > 3$

45. $|2x + 7| \leq 13$

46. $|5x - 3| \leq 18$

47. $|x + 10| \geq 14$

48. $|x - 9| \geq 2$

49. $|2x - 7| \leq 11$

50. $|5x + 2| < 8$

51. $|x| > -4$

52. $|x| \leq -7$

53. $6 + |4x - 1| \leq 9$

54. $-3 + |5x - 2| \leq 4$

55. $|6x - 8| + 3 > 7$

56. $|10 + 3x| + 1 > 2$

43. ______

44. ______

45. ______

46. ______

47. ______

48. ______

49. ______

50. ______

51. ______

52. ______

53. ______

54. ______

55. ______

56. ______

57. ______
58. ______
59. ______
60. ______
61. ______
62. ______
63. ______
64. ______
65. ______
66. ______
67. ______
68. ______
69. ______
70. ______
71. ______
72. ______
73. ______
74. ______
75. ______
76. ______
77. ______
78. ______
79. ______
80. ______

Name ______

57. $|5x + 3| < -6$

−5 −4 −3 −2 −1 0 1 2 3 4 5

58. $|4 + 9x| \geq -6$

−5 −4 −3 −2 −1 0 1 2 3 4 5

59. $\left|\frac{x + 6}{3}\right| > 2$

−15 −12 −9 −6 −3 0 3 6 9 12 15

60. $\left|\frac{7 + x}{2}\right| \geq 4$

−15 −12 −9 −6 −3 0 3 6 9 12 15

Solve each equation or inequality for x. See Examples 1 through 16.

61. $|2x - 3| < 7$

62. $|2x - 3| > 7$

63. $|2x - 3| = 7$

64. $|5 - 6x| = 29$

65. $|x - 5| \geq 12$

66. $|x + 4| \geq 20$

67. $|9 + 4x| = 0$

68. $|9 + 4x| \geq 0$

69. $|2x + 1| + 4 < 7$

70. $8 + |5x - 3| \geq 11$

71. $|3x - 5| + 4 = 5$

72. $|x - 1| + 7 = 11$

73. $|x + 11| = -1$

74. $|4x - 4| = -3$

75. $\left|\frac{2x - 1}{3}\right| = 6$

76. $\left|\frac{6 - x}{4}\right| = 5$

77. $\left|\frac{3x - 5}{6}\right| > 5$

78. $\left|\frac{4x - 7}{5}\right| < 2$

79. $|6x - 3| = |4x + 5|$

80. $|3x + 1| = |4x + 10|$

Name ______________________

Review and Preview

The circle graph shows the sources of Walt Disney Company's operating income for the first half of fiscal year 1998. Use this graph to answer Exercises 81–83. See Section 2.2.

Walt Disney Company
Operating Income 1998

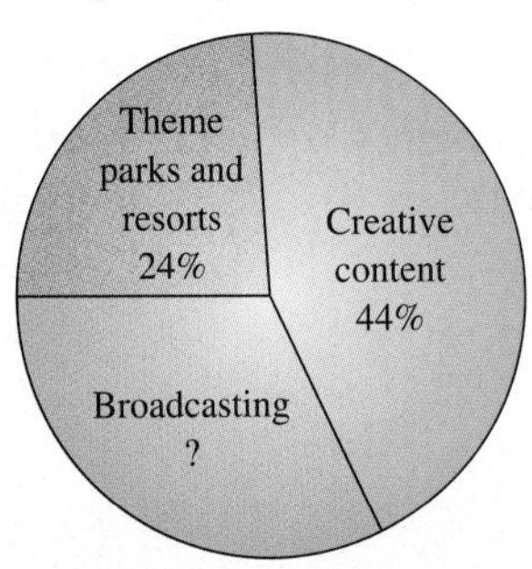

Source: Walt Disney Company

81. What percent of Disney's operating income came from the broadcasting segment?

82. A circle contains 360°. Find the number of degrees found in the 24% sector for theme parks and resorts.

83. If Disney's operating income for all of 1998 was \$4.5 billion, find the amount of income expected from the creative content segment.

Consider the equation $3x - 4y = 12$. For each value of x or y given, find the corresponding value of the other variable that makes the statement true. See Section 2.3.

84. If $x = 2$, find y.

85. If $y = -1$, find x.

86. If $y = -3$, find x.

87. If $x = 4$, find y.

Combining Concepts

88. Write an absolute value equation representing all numbers x whose distance from 0 is 5 units.

89. Write an absolute value equation representing all numbers x whose distance from 0 is 2 units.

81. ______
82. ______
83. ______
84. ______
85. ______
86. ______
87. ______
88. ______
89. ______

90. ______

91. ______

92. ______

93. ______

94. ______

95. ______

Name ______

90. Write an absolute value inequality representing all numbers x whose distance from 0 is less than 7 units.

91. Write an absolute value inequality representing all numbers x whose distance from 0 is greater than 4 units.

92. Write $-5 \leq x \leq 5$ as an equivalent inequality containing an absolute value.

93. Write $x > 1$ or $x < -1$ as an equivalent inequality containing an absolute value.

The expression $|x_T - x|$ is defined to be the absolute error in x, where x_T is the true value of a quantity and x is the measured value or value as stored in a computer.

94. If the true value of a quantity is 3.5 and the absolute error must be less than 0.05, find the acceptable measured values.

95. If the true value of a quantity is 0.2 and the approximate value stored in a computer is $\frac{51}{256}$, find the absolute error.

Name ________________________________ Section ________ Date _________

Chapter 2 Activity

Room Redecorating

This activity may be completed by working in groups or individually.

Have you ever stopped to think about how much math and geometry is involved in redecorating a room? In this project, you will plan for a redecorating project by estimating the necessary amount of materials and their costs. Be sure to show and explain all of your work.

1. Choose an actual room that you can use as the model for your redecorating project. You could use your math classroom, your living room, your dormitory room, or any room that can be easily measured. Measure each dimension of the room with a tape measure. Make a sketch of the room and label each dimension with its measurement. Be sure to include the sizes and locations of items such as doors and windows.

2. **Paint** The first task in your redecorating project will be painting the walls of the room.
   a. To estimate the amount of paint you will need, first find the total wall area to be painted. Be sure to subtract the areas of any regions, such as doors or windows, that will not be painted. (*Hint:* Use the geometric formulas on the inside front cover of this text to help you find the necessary areas.)
   b. Using newspaper flyers or a trip to a local paint store, choose a paint that would be suitable for your redecorating project. List its price per gallon and its normal surface coverage. Using the information on coverage, estimate how many gallons of paint will be needed for the project. (*Note:* If surface coverage data, normally found on the paint can label, is unavailable, use the guideline that a gallon of paint generally covers 400 square feet of wall area.)
   c. How much will all of the paint that is needed cost?

3. **Wallpaper Border** The second task in your redecorating project will be installing a wallpaper border all the way around the room just below the ceiling.
   a. Calculate the length of the border needed.
   b. Using newspaper flyers or a trip to a local paint/decorating store, choose a border that would be suitable for your redecorating project. List its price per roll and the length of border on each roll. How many rolls of wallpaper border will be needed for the project?
   c. Find the cost of the border necessary for the project.

4. **Wall-to-Wall Carpeting** The third task in your redecorating project will be installing wall-to-wall carpeting.
   a. How much carpeting will be needed?
   b. Using newspaper flyers or a trip to a local carpeting store, choose a wall-to-wall carpet that would be suitable for your redecorating project. List its price per square unit.
   c. How much will carpeting the room cost? Be sure to include any fixed fees for installation or delivery, and so on.

5. Redecorating projects often involve buying additional utensils and supplies such as paint brushes, wallpaper adhesive, and so on. A reasonable cost estimate for these extras is 20% of the cost of the basic project supplies (paint, border, and carpeting in this case). Taking these extras into consideration, what is your estimate of the total cost of the redecorating project?

CHAPTER 2 HIGHLIGHTS

DEFINITIONS AND CONCEPTS	EXAMPLES
SECTION 2.1 LINEAR EQUATIONS IN ONE VARIABLE	
An **equation** is a statement that two expressions are equal.	$5 = 5 \quad 7x + 2 = -14 \quad 3(x-1)^2 = 9x^2 - 6$
A **linear equation in one variable** is an equation that can be written in the form $ax + b = c$, where a, b, and c are real numbers and a is not 0.	$7x + 2 = -14 \quad x = -3$ $5(2y - 7) = -2(8y - 1)$
A **solution** of an equation is a value for the variable that makes the equation a true statement.	Determine whether -1 is a solution of $3(x - 1) = 4x - 2$. $3(-1 - 1) \stackrel{?}{=} 4(-1) - 2$ $3(-2) \stackrel{?}{=} -4 - 2$ $-6 = -6$ True. Thus, -1 is a solution.
Equivalent equations have the same solution.	$x - 12 = 14$ and $x = 26$ are equivalent equations.
The **addition property of equality** guarantees that the same number may be added to (or subtracted from) both sides of an equation, and the result is an equivalent equation. The **multiplication property of equality** guarantees that both sides of an equation may be multiplied by (or divided by) the same nonzero number, and the result is an equivalent equation.	Solve: $-3x - 2 = 10$ $-3x - 2 + 2 = 10 + 2$ Add 2 to both sides. $-3x = 12$ $\frac{-3x}{-3} = \frac{12}{-3}$ Divide both sides by -3. $x = -4$
SOLVING A LINEAR EQUATION IN ONE VARIABLE	Solve: $x - \frac{x-2}{6} = \frac{x-7}{3} + \frac{2}{3}$
1. Clear the equation of fractions.	**1.** $6\left(x - \frac{x-2}{6}\right) = 6\left(\frac{x-7}{3} + \frac{2}{3}\right)$ Multiply both sides by 6. $6x - (x - 2) = 2(x - 7) + 2(2)$
2. Remove grouping symbols such as parentheses.	**2.** $6x - x + 2 = 2x - 14 + 4$ Use the distributive property.
3. Simplify by combining like terms.	**3.** $5x + 2 = 2x - 10$
4. Write variable terms on one side and numbers on the other side using the addition property of equality.	**4.** $5x + 2 - 2 = 2x - 10 - 2$ Subtract 2 from both sides. $5x = 2x - 12$ $5x - 2x = 2x - 12 - 2x$ Subtract $2x$ from both sides. $3x = -12$
5. Get the variable alone by using the multiplication property of equality.	**5.** $\frac{3x}{3} = \frac{-12}{3}$ Divide both sides by 3. $x = -4$

Section 2.1 (continued)

6. Check the proposed solution in the original equation.	**6.** $-4 - \frac{-4-2}{6} \stackrel{?}{=} \frac{-4-7}{3} + \frac{2}{3}$ Replace x with -4 in the original equation. $-4 - \frac{-6}{6} \stackrel{?}{=} \frac{-11}{3} + \frac{2}{3}$ $-4 - (-1) \stackrel{?}{=} \frac{-9}{3}$ $-3 = -3$ True.

Section 2.2 An Introduction to Problem Solving

Problem-Solving Strategy	Colorado is shaped like a rectangle whose length is about 1.3 times its width. If the perimeter of Colorado is 2070 kilometers, find its dimensions.
1. UNDERSTAND the problem.	**1.** Read and reread the problem. Guess a solution and check your guess. Let x = width of Colorado in kilometers. Then $1.3x$ = length of Colorado in kilometers 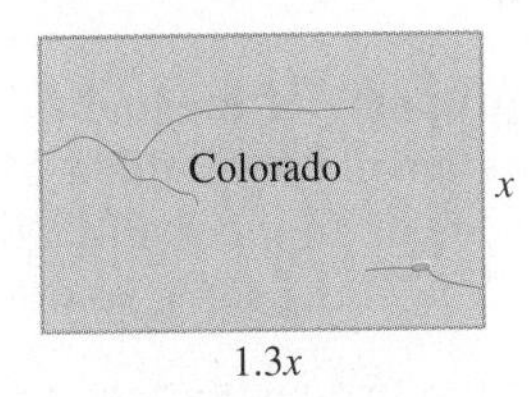
2. TRANSLATE the problem.	**2.** In words: twice the length + twice the width = perimeter Translate: $2(1.3x) + 2x = 2070$
3. SOLVE the equation.	**3.** $2.6x + 2x = 2070$ $4.6x = 2070$ $x = 450$
4. INTERPRET the results.	**4.** If $x = 450$ kilometers, then $1.3x = 1.3(450) = 585$ kilometers. *Check:* The perimeter of a rectangle whose width is 450 kilometers and length is 585 kilometers is $2(450) + 2(585) = 2070$ kilometers, the required perimeter. *State:* The dimensions of Colorado are 450 kilometers by 585 kilometers.

SECTION 2.3 FORMULAS AND PROBLEM SOLVING

An equation that describes a known relationship among quantities is called a **formula**.

$A = \pi r^2$ (area of a circle)
$I = PRT$ (interest = principal · rate · time)

To solve a formula for a specified variable, use the steps for solving an equation. Treat the specified variable as the only variable of the equation.

Solve $A = 2HW + 2LW + 2LH$ for H.

$A - 2LW = 2HW + 2LH$ Subtract $2LW$ from both sides.

$A - 2LW = H(2W + 2L)$ Use the distributive property.

$\frac{A - 2LW}{2W + 2L} = \frac{H(2W + 2L)}{2W + 2L}$ Divide both sides by $2W + 2L$.

$\frac{A - 2LW}{2W + 2L} = H$ Simplify.

SECTION 2.4 LINEAR INEQUALITIES AND PROBLEM SOLVING

A **linear inequality in one variable** is an inequality that can be written in the form $ax + b < c$, where a, b, and c are real numbers and $a \neq 0$. (The inequality symbols $\leq$, $>$, and $\geq$ also apply here.)

$5x - 2 \leq -7 \qquad 3y > 1 \qquad \frac{z}{7} < -9(z - 3)$

The **addition property of inequality** guarantees that the same number may be added to (or subtracted from) both sides of an inequality, and the resulting inequality will have the same solution set.

$x - 9 \leq -16$

$x - 9 + 9 \leq -16 + 9$ Add 9 to both sides.

$x \leq -7$

The **multiplication property of inequality** guarantees that both sides of an inequality may be multiplied by (or divided by) the same **positive** number, and the resulting inequality will have the same solution set.

$6x < -66$ Divide both sides by 6. Do not reverse the direction of the inequality symbol.

$\frac{6x}{6} < \frac{-66}{6}$

$x < -11$

We may also multiply (or divide) both sides of an inequality by the same **negative** number and **reverse the direction of the inequality symbol**, and the result is an inequality with the same solution set.

$-6x < -66$ Divide both sides by -6. Reverse the direction of the inequality symbol.

$\frac{-6x}{-6} > \frac{-66}{-6}$

$x > 11$

SOLVING A LINEAR INEQUALITY IN ONE VARIABLE

Solve: $\frac{3}{7}(x - 4) \geq x + 2$

1. Clear the equation of fractions.
2. Remove grouping symbols such as parentheses.
3. Simplify by combining like terms.
4. Write variable terms on one side and numbers on the other side using the addition property of inequality.
5. Get the variable alone using the multiplication property of inequality.

1. $7\left[\frac{3}{7}(x - 4)\right] \geq 7(x + 2)$ Multiply both sides by 7.

$3(x - 4) \geq 7(x + 2)$

2. $3x - 12 \geq 7x + 14$ Use the distributive property.

4. $-4x - 12 \geq 14$ Subtract $7x$ from both sides.

$-4x \geq 26$ Add 12 to both sides.

$\frac{-4x}{-4} \leq \frac{26}{-4}$ Divide both sides by -4. Reverse the direction of the inequality symbol.

$x \leq -\frac{13}{2}$

Section 2.5 Sets and Compound Inequalities

Two inequalities joined by the words **and** or **or** are called **compound inequalities**.

$x - 7 \leq 4$ and $x \geq -21$

$2x + 7 > x - 3$ or $5x + 2 > -3$

The solution set of a compound inequality formed by the word **and** is the **intersection** $\cap$ of the solution sets of the two inequalities.

Solve:

$x < 5$ and $x < 3$

$\{x|x < 5\}$ –1 0 1 2 3 4 5 6 $(-\infty, 5)$

$\{x|x < 3\}$ –1 0 1 2 3 4 5 6 $(-\infty, 3)$

$\{x|x < 3$ and $x < 5\}$ –1 0 1 2 3 4 5 6 $(-\infty, 3)$

The solution set of a compound inequality formed by the word **or** is the **union** $\cup$ of the solution sets of the two inequalities.

Solve:

$x - 2 \geq -3$ or $2x \leq -4$

$x \geq -1$ or $x \leq -2$

$\{x|x \geq -1\}$ –3 –2 –1 0 1 $[-1, \infty)$

$\{x|x \leq -2\}$ –3 –2 –1 0 1 $(-\infty, -2]$

$\{x|x \leq -2$ or $x \geq -1\}$ –3 –2 –1 0 1 $(-\infty, -2] \cup [-1, \infty)$

Section 2.6 Absolute Value Equations and Inequalities

If a is a positive number, then $|X| = a$ is equivalent to $X = a$ or $X = -a$.

Solve: $|5y - 1| - 7 = 4$

$|5y - 1| = 11$ Add 7 to both sides.

$5y - 1 = 11$ or $5y - 1 = -11$ Add 1 to both sides.

$5y = 12$ $5y = -10$

$y = \frac{12}{5}$ $y = -2$ Divide both sides by 5.

The solution set is $\left\{-2, \frac{12}{5}\right\}$

If a is negative, then $|X| = a$ has no solution.

Solve: $\left|\frac{x}{2} - 7\right| = -1$

The solution set is $\{\ \}$, or $\emptyset$.

SECTION 2.6 (CONTINUED)	
If an absolute value equation is of the form $\lvert X \rvert = \lvert Y \rvert$, solve $X = Y$ or $X = -Y$.	Solve: $\lvert x - 7 \rvert = \lvert 2x + 1 \rvert$ $x - 7 = 2x + 1$ or $x - 7 = -(2x + 1)$ $x = 2x + 8$ $\quad$ $x - 7 = -2x - 1$ $-x = 8$ $\quad$ $x = -2x + 6$ $x = -8$ $\quad$ $3x = 6$ $x = 2$ The solution set is $\{-8, 2\}$.
If a is a positive number, then $\lvert X \rvert < a$ is equivalent to $-a < X < a$.	Solve: $\lvert y - 5 \rvert \le 3$ $-3 \le y - 5 \le 3$ $-3 + 5 \le y - 5 + 5 \le 3 + 5$ Add 5 to all three sides. $2 \le y \le 8$ The solution set is $[2, 8]$.
If a is a positive number, then $\lvert X \rvert > a$ is equivalent to $X < -a$ or $X > a$.	Solve: $\left\lvert \frac{x}{2} - 3 \right\rvert > 7$ $\frac{x}{2} - 3 < -7$ or $\frac{x}{2} - 3 > 7$ Multiply both sides by 2. $x - 6 < -14$ $\quad$ $x - 6 > 14$ $x < -8$ $\quad$ $x > 20$ Add 6 to both sides. The solution set is $(-\infty, -8) \cup (20, \infty)$.

Name ______________________ Section __________ Date __________

Chapter 2 Review

(2.1) *Solve each linear equation.*

1. $4(x - 5) = 2x - 14$

2. $x + 7 = -2(x + 8)$

3. $3(2y - 1) = -8(6 + y)$

4. $-(z + 12) = 5(2z - 1)$

5. $n - (8 + 4n) = 2(3n - 4)$

6. $4(9v + 2) = 6(1 + 6v) - 10$

7. $0.3(x - 2) = 1.2$

8. $1.5 = 0.2(c - 0.3)$

9. $-4(2 - 3x) = 2(3x - 4) + 6x$

10. $6(m - 1) + 3(2 - m) = 0$

11. $6 - 3(2g + 4) - 4g = 5(1 - 2g)$

12. $20 - 5(p + 1) + 3p = -(2p - 15)$

13. $\frac{x}{3} - 4 = x - 2$

14. $\frac{9}{4}y = \frac{2}{3}y$

15. $\frac{3n}{8} - 1 = 3 + \frac{n}{6}$

16. $\frac{z}{6} + 1 = \frac{z}{2} + 2$

17. $\frac{y}{4} - \frac{y}{2} = -8$

18. $\frac{2x}{3} - \frac{8}{3} = x$

19. $\frac{b - 2}{3} = \frac{b + 2}{5}$

20. $\frac{2t - 1}{3} = \frac{3t + 2}{15}$

Name ____________________

21. $\frac{2(t+1)}{3} = \frac{2(t-1)}{3}$

22. $\frac{3a-3}{6} = \frac{4a+1}{15} + 2$

23. $\frac{x-2}{5} + \frac{x+2}{2} = \frac{x+4}{3}$

24. $\frac{2z-3}{4} - \frac{4-z}{2} = \frac{z+1}{3}$

(2.2) *Solve.*

25. Twice the difference of a number and 3 is the same as 1 added to three times the number. Find the number.

26. One number is 5 more than another number. If the sum of the numbers is 285, find the numbers.

27. Find 40% of 130.

28. Find 1.5% of 8.

29. In 1996, the average annual earnings for a worker with a bachelor's degree was $39,136. This represents a 77.25% increase over the average annual earnings for a high school graduate in 1996. Find the average annual earnings for a high school graduate in 1996. Round to the nearest whole dollar. (*Source:* U.S. Bureau of the Census)

30. Find four consecutive integers such that twice the first subtracted from the sum of the other three integers is 16.

31. Determine whether there are two consecutive odd integers such that 5 times the first exceeds 3 times the second by 54.

32. The length of a rectangular playing field is 5 meters less than twice its width. If 230 meters of fencing goes around the field, find the dimensions of the field.

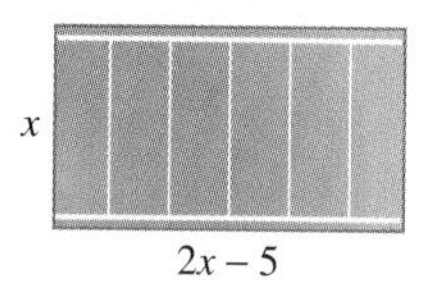

33. A car rental company charges $19.95 per day for a compact car plus 12 cents per mile for every mile over 100 miles driven per day. If Mr. Woo's bill for 2 days use is $46.86, find how many miles he drove.

34. The cost C of producing x number of scientific calculators is given by $C = 4.50x + 3000$ and the revenue R from selling them is given by $R = 16.50x$. Find the number of calculators that must be sold to break even. (Recall that to break even, revenue = cost.)

Name ______________________________

35. An entrepreneur can sell her plants that vibrate to music for \$40 each, while her cost C to produce x number of plants is given by $C = 20x + 100$. Find her break-even number. Find her revenue if she sells exactly that number of plants.

(2.3) *Solve each equation for the specified variable.*

36. $V = LWH$ for W

37. $C = 2\pi r$ for r

38. $5x - 4y = -12$ for y

39. $5x - 4y = -12$ for x

40. $y - y_1 = m(x - x_1)$ for m

41. $y - y_1 = m(x - x_1)$ for x

42. $E = I(R + r)$ for r

43. $S = vt + gt^2$ for g

44. $T = gr + gvt$ for g

45. $I = Prt + P$ for P

46. $A = \frac{h}{2}(B + b)$ for B

47. $V = \frac{1}{3}\pi r^2 h$ for h

Solve.

48. A principal of \$3000 is invested in an account paying an annual percentage rate of 3%. Find the amount (to the nearest cent) in the account after 7 years if the amount is compounded

a. semiannually.

b. weekly.

49. The high temperature in Slidell, Louisiana, one day was 90° Fahrenheit. Convert this temperature to degrees Celsius.

Name ______________________________

50. Angie Applegate has a photograph in which the length is 2 inches longer than the width. If she increases each dimension by 4 inches, the area is increased by 88 square inches. Find the original dimensions.

51. One-square-foot floor tiles come 24 to a package. Find how many packages are needed to cover a rectangular floor 18 feet by 21 feet.

52. Determine which container holds more ice cream, an 8 inch × 5 inch × 3 inch box or a cylinder with radius of 3 inches and height of 6 inches.

53. Erasmos Gonzalez left Los Angeles at 11 A.M. and drove nonstop to San Diego, 130 miles away. If he arrived at 1:15 P.M., find his average speed, rounded to the nearest mile per hour.

(2.4) *Solve each linear inequality. Write your answer in interval notation.*

54. $3(x-5) > -(x+3)$

55. $-2(x+7) \geq 3(x+2)$

56. $4x - (5+2x) < 3x - 1$

57. $3(x-8) < 7x + 2(5-x)$

58. $24 \geq 6x - 2(3x-5) + 2x$

59. $48 + x \geq 5(2x+4) - 2x$

60. $\frac{x}{3} + \frac{1}{2} > \frac{2}{3}$

61. $x + \frac{3}{4} < \frac{-x}{2} + \frac{9}{4}$

62. $\frac{x-5}{2} \leq \frac{3}{8}(2x+6)$

63. $\frac{3(x-2)}{5} > \frac{-5(x-2)}{3}$

Solve.

64. George Boros can pay his housekeeper $15 per week to do his laundry, or he can have the laundromat do it at a cost of 50 cents per pound for the first 10 pounds and 40 cents for each additional pound. Use an inequality to find the weight at which it is more economical to use the housekeeper than the laundromat.

65. Ceramic firing temperatures usually range from 500° to 1000° Fahrenheit. Use a compound inequality to convert this range to the Celsius scale. Round to the nearest degree.

Name ______________________________

66. In the Olympic gymnastics competition, Nana must average a score of 9.65 to win the silver medal. Seven of the eight judges have reported scores of 9.5, 9.7, 9.9. 9.7, 9.7, 9.6, and 9.5. Use an inequality to find the minimum score that the last judge can give so that Nana wins the silver medal.

67. Carol would like to pay cash for a car when she graduates from college and estimates that she can afford a car that costs between $4000 and $8000. She has saved $500 so far and plans to earn the rest of the money by working the next two summers. If Carol plans to save the same amount each summer, use a compound inequality to find the range of money she must save each summer to buy the car.

(2.5) *Solve each inequality. Write your answers in interval notation.*

68. $1 \leq 4x - 7 \leq 3$

69. $-2 \leq 8 + 5x < -1$

70. $-3 < 4(2x - 1) < 12$

71. $-6 < x - (3 - 4x) < -3$

72. $\frac{1}{6} < \frac{4x - 3}{3} \leq \frac{4}{5}$

73. $0 \leq \frac{2(3x + 4)}{5} \leq 3$

74. $x \leq 2$ and $x > -5$

75. $x \leq 2$ or $x > -5$

76. $3x - 5 > 6$ or $-x < -5$

77. $-2x \leq 6$ and $-2x + 3 < -7$

(2.6) *Solve each absolute value equation.*

78. $|x - 7| = 9$

79. $|8 - x| = 3$

80. $|2x + 9| = 9$

81. $|-3x + 4| = 7$

82. $|3x - 2| + 6 = 10$

83. $5 + |6x + 1| = 5$

84. $-5 = |4x - 3|$

85. $|5 - 6x| + 8 = 3$

Name ______________________________

86. $|7x| - 26 = -5$

87. $-8 = |x - 3| - 10$

88. $\left|\frac{3x - 7}{4}\right| = 2$

89. $\left|\frac{9 - 2x}{5}\right| = -3$

90. $|6x + 1| = |15 + 4x|$

91. $|x - 3| = |7 + 2x|$

Solve each absolute value inequality. Graph the solution set and write it in interval notation.

92. $|5x - 1| < 9$

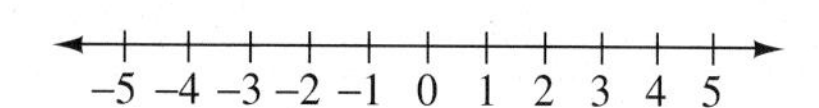

93. $|6 + 4x| \geq 10$

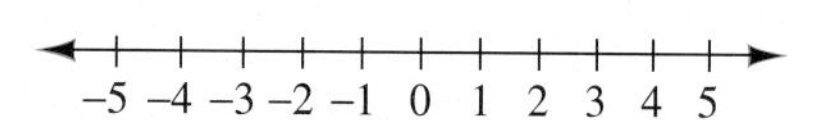

94. $|3x| - 8 > 1$

–5 –4 –3 –2 –1 0 1 2 3 4 5

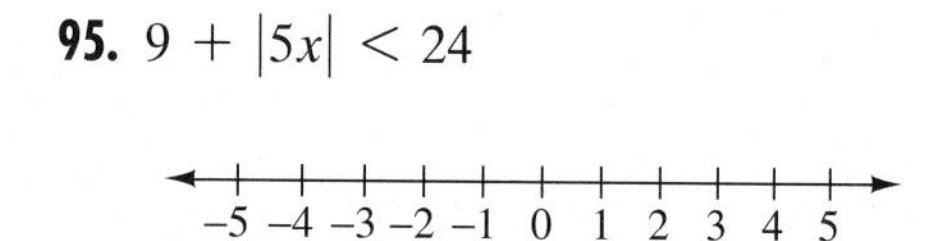

95. $9 + |5x| < 24$

–5 –4 –3 –2 –1 0 1 2 3 4 5

96. $|6x - 5| \leq -1$

–5 –4 –3 –2 –1 0 1 2 3 4 5

97. $|6x - 5| \geq -1$

–5 –4 –3 –2 –1 0 1 2 3 4 5

98. $\left|3x + \frac{2}{5}\right| \geq 4$

–5 –4 –3 –2 –1 0 1 2 3 4 5

99. $\left|\frac{4x - 3}{5}\right| < 1$

–5 –4 –3 –2 –1 0 1 2 3 4 5

100. $\left|\frac{x}{3} + 6\right| - 8 > -5$

–40 –30 –20 –10 0 10 20 30 40

101. $\left|\frac{4(x - 1)}{7}\right| + 10 < 2$

–5 –4 –3 –2 –1 0 1 2 3 4 5

Name ______________________ Section ________ Date ________

Chapter 2 Test

Solve each equation.

1. $8x + 14 = 5x + 44$

2. $3(x + 2) = 11 - 2(2 - x)$

3. $3(y - 4) + y = 2(6 + 2y)$

4. $7n - 6 + n = 2(4n - 3)$

5. $\frac{z}{2} + \frac{z}{3} = 10$

6. $\frac{7w}{4} + 5 = \frac{3w}{10} + 1$

7. $|6x - 5| = 1$

8. $|8 - 2t| = -6$

9. $|2x - 3| = |4x + 5|$

Solve each equation for the specified variable.

10. $3x - 4y = 8$ for y

11. $4(2n - 3m) - 3(5n - 7m) = 0$ for n

12. $S = gt^2 + gvt$ for g

13. $F = \frac{9}{5}C + 32$ for C

Solve each inequality. Write your answer in interval notation.

14. $3(2x - 7) - 4x > -(x + 6)$

15. $8 - \frac{x}{2} \leq 7$

16. $-3 < 2(x - 3) \leq 4$

17. $|3x + 1| > 5$

18. $|x - 6| + 4 \leq 9$

19. $x \geq 5$ and $x \geq 4$

20. $x \geq 5$ or $x \geq 4$

21. $-x > 1$ and $3x + 3 \geq x - 3$

22. $6x + 1 > 5x + 4$ or $1 - x > -4$

23. Find 12% of 80.

Answers

1. ________
2. ________
3. ________
4. ________
5. ________
6. ________
7. ________
8. ________
9. ________
10. ________
11. ________
12. ________
13. ________
14. ________
15. ________
16. ________
17. ________
18. ________
19. ________
20. ________
21. ________
22. ________
23. ________

24. ______

25. ______

26. ______

27. ______

Name ______

Solve.

24. In 2006, the number of people employed as database administrators, computer support specialists, and all other computer scientists is expected to be 461,000 in the United States. This represents a 118% increase over the number of people employed in these occupations in 1996. Find the number of database administrators, computer support specialists, and all other computer scientists employed in 1996. (*Source:* U.S. Bureau of Labor Statistics)

25. A circular dog pen has a circumference of 78.5 feet. Approximate π by 3.14 and estimate how many hunting dogs could be safely kept in the pen if each dog needs at least 60 square feet of room.

26. The company that makes Photoray sunglasses figures that the cost C to make x number of sunglasses weekly is given by $C = 3910 + 2.8x$, and the weekly revenue R is given by $R = 7.4x$. Use an inequality to find the number of sunglasses that must be made and sold to make a profit. (Recall that revenue must exceed cost in order to make a profit.)

27. Find the amount of money in an account after 10 years if a principal of $2500 is invested at 3.5% interest compounded quarterly. (Round to the nearest cent.)

Name ____________ Section ______ Date ______

ANSWERS

1. ______

2. ______

3. ______

4. ______

5. ______

6. ______

7. ______

8. ______

9. ______

10. ______

11. ______

12. ______

13. ______

14. ______

15. ______

16. ______

Cumulative Review

1. Evaluate $3x - y$ when $x = 15$ and $y = 4$.

Write each sentence as an equation.

2. The sum of x and 5 is 20.

3. The quotient of z and 9 amounts to 9 plus z.

Find the opposite or additive inverse of each number.

4. 8

5. $-\frac{1}{5}$

Add.

6. $-3 + (-11)$

7. $-10 + 15$

8. $-\frac{2}{3} + \frac{3}{7}$

Divide.

9. $\frac{20}{-4}$

10. $\frac{0}{-8}$

11. $\frac{-10}{-80}$

12. Simplify $3 + 2 \cdot 10$

Simplify by combining like terms.

13. $3x - 5x + 4$

14. $y + 3y$

Use the quotient rule to simplify.

15. $\frac{x^7}{x^4}$

16. $\frac{20x^6}{4x^5}$

Name ______________________________

17. ______________

18. ______________

19. ______________

20. ______________

21. ______________

22. ______________

23. ______________

24. ______________

25. ______________

Simplify each expression. Write each answer using positive exponents only.

17. $\left(\dfrac{3x^2y}{y^{-9}z}\right)^{-2}$

18. $\left(\dfrac{3a^2}{2x^{-1}}\right)^3\left(\dfrac{x^{-3}}{4a^{-2}}\right)^{-1}$

19. Solve: $2x + 5 = 9$

20. Solve: $6x - 4 = 2 + 6(x - 1)$

21. Suppose that Service Merchandise just announced an 8% decrease in the price of their Compaq Presario computers. If one particular computer model sells for $2162 after the decrease, find the original price of this computer.

22. Solve $V = lwh$ for h.

23. Solve: $\frac{1}{4}x \leq \frac{3}{2}$. Graph the solution set.

−10 −8 −6 −4 −2 0 2 4 6 8 10

24. Solve: $-2x - 5 < -3$ or $6x < 0$

25. Solve: $|x| = 3$

Graphs and Functions

CHAPTER 3

The linear equations we explored in Chapter 2 are statements about a single variable. This chapter examines statements about two variables: linear equations and inequalities in two variables. We focus particularly on graphs of these equations and inequalities, which lead to the notion of relation and to the notion of function, perhaps the single most important and useful concept in all of mathematics.

At the beginning of the 20th century, there were approximately 237,600 students enrolled in the 977 institutions of higher education in the United States. At that time, only 19% of bachelor's degree recipients were women. By the year 2000, the projected 3800 colleges and universities in the United States will have an estimated 14,800,000 students. Roughly 56% of bachelor's degree recipients are expected to be women. The phenomenal growth of colleges and universities can also be seen in the average tuition costs at these institutions of higher learning. For instance, the average annual tuition at a private four-year college or university has increased from $1809 in 1970 to $13,664 in 1998, an increase of about 655%! In Exercises 33 and 34 on page 204, we will use linear equations to predict the future cost of annual tuition at both two-year and four-year public colleges and universities.

Answers

1. ____
2. a. ____
   b. ____
   c. ____
   d. ____
3. see graph
4. see graph
5. see graph
6. see graph
7. ____
8. ____
9. ____
10. ____
11. ____
12. ____
13. ____
14. ____
15. ____
16. ____
17. ____
18. ____
19. see graph
20. see graph

Name ______________________ Section ______ Date ________

Chapter 3 Pretest

1. Determine whether each ordered pair is a solution of the given equation.
$2x - 5y = -13;\ (-1, -3), (-4, 1)$

2. Name the quadrant (or axis) in which each point lies.
a. $(6, -5)$, **b.** $(0, -4)$, **c.** $\left(-\frac{1}{2}, -\frac{2}{3}\right)$,
d. $(7, 1.8)$

Graph each line.

3. $2x + y = 4$

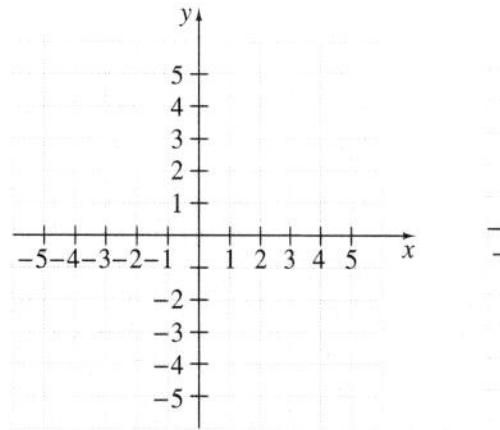

4. $y = -x + 3$

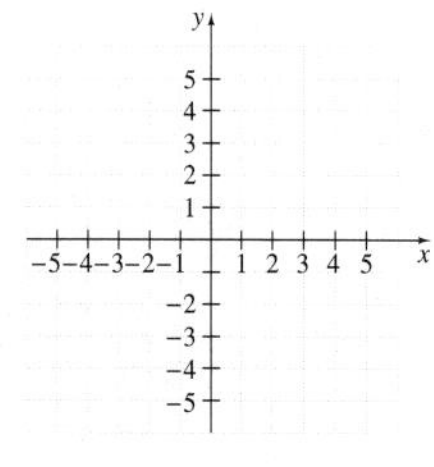

5. $y = -2$

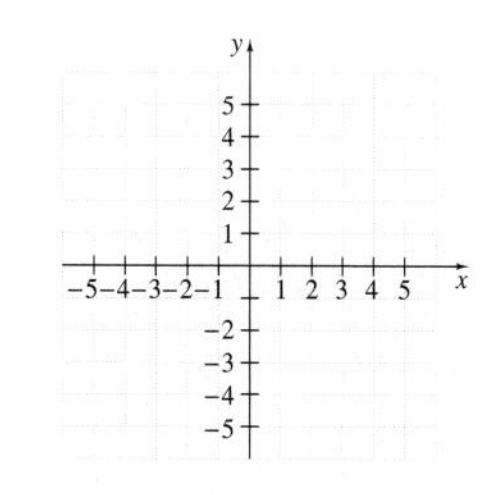

6. $x - 5 = 0$

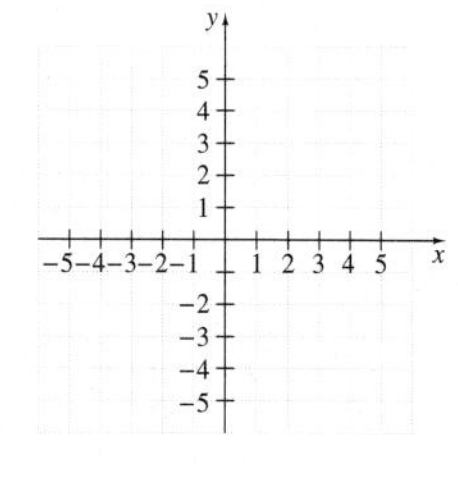

7. Find the slope of the line that passes through $(5, 4)$ and $(-6, 2)$.

8. Find the slope and the y-intercept of the line $4x - 5y = 2$.

9. Find the slope of the line $y = 5$.

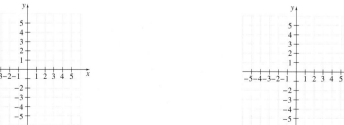

Use the slope-intercept form of the linear equation to write the equation of each line with the given slope and y-intercept.

10. Slope $\frac{1}{3}$; y-intercept 6

11. Slope -7; y-intercept 0

Find an equation of each line satisfying the conditions given. Write the equations in standard form.

12. Slope 2; through $(-3, -7)$

13. Through $(5, 4)$ and $(-1, 6)$

14. Horizontal; through $(9, 10)$

15. Perpendicular to $3y - x = 6$; through $(8, 0)$

16. Find the domain and the range of the given relation. Also determine whether the relation is a function. $\{(-2, 5), (3, -7), (2, 5)\}$

If $f(x) = 7x - 1$ and $g(x) = 2x^2 + x - 5$, find the following.

17. $f(-3)$

18. $g(0)$

Graph each inequality.

19. $x - y < 4$

20. The intersection of $x \leq 3$ and $y > -1$

3.1 GRAPHING LINEAR EQUATIONS

Graphs are widely used today in newspapers, magazines, and all forms of newsletters. A few examples of graphs are shown here.

Objectives

A Plot ordered pairs on a rectangular coordinate system.

B Determine whether an ordered pair of numbers is a solution of an equation in two variables.

C Graph linear equations.

D Graph vertical and horizontal lines.

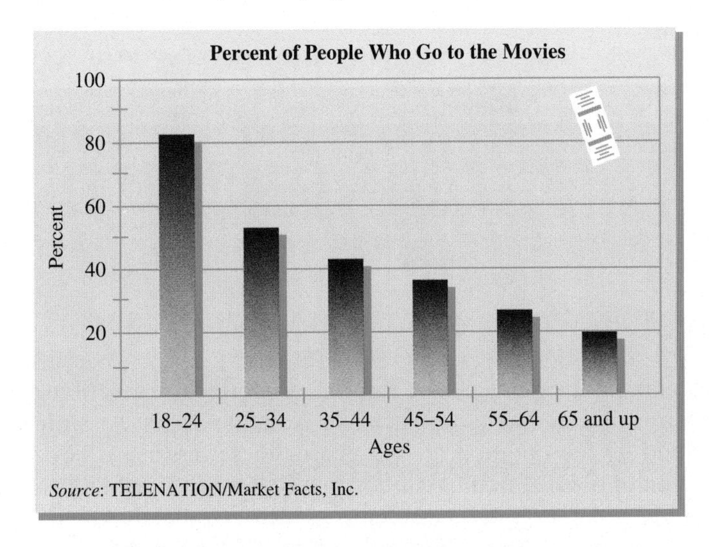

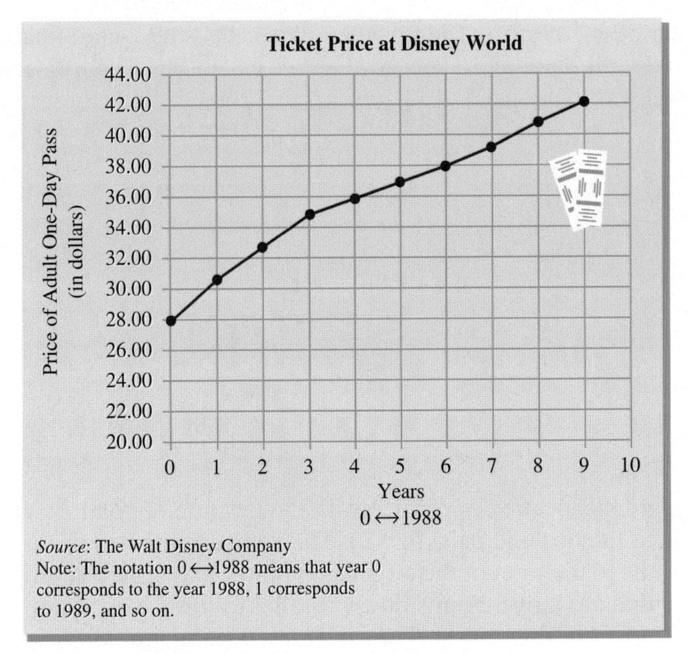

To help us understand how to read these graphs, we will review their origin—the rectangular coordinate system.

A PLOTTING ORDERED PAIRS ON A RECTANGULAR COORDINATE SYSTEM

One way to locate points on a plane is by using a **rectangular coordinate system**, which is also called a **Cartesian coordinate system** after its inventor, René Descartes (1596–1650). A rectangular coordinate system consists of two number lines that intersect at right angles at their 0 coordinates. We position these axes on paper such that one number line is horizontal and the other number line is then vertical. The horizontal number line is called the ***x*-axis** (or the axis of the **abscissa**), and the vertical number line is called the ***y*-axis** (or the axis of the **ordinate**). The point of intersection of these axes is named the **origin**.

Notice that the axes divide the plane into four regions. These regions are called **quadrants**. The top-right region is quadrant I. Quadrants II, III, and IV are numbered counterclockwise from the first quadrant as shown. The x-axis and the y-axis are not in any quadrant.

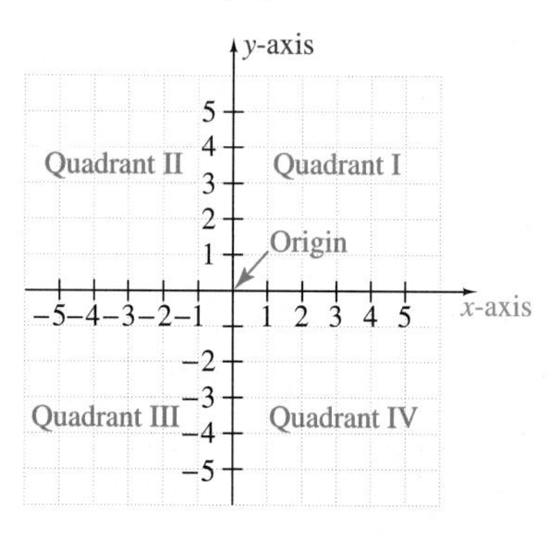

Each point in the plane can be located, or **plotted**, by describing its position in terms of distances along each axis from the origin. An **ordered pair**, represented by the notation (x, y), records these distances. For example, the location of point A in the figure below is described as 2 units to the left of the origin along the x-axis and 5 units upward parallel to the y-axis. Thus, we identify point A with the ordered pair $(-2, 5)$. Notice that the *order* of these numbers is *critical*. The x-value -2 is called the **x-coordinate** and is associated with the x-axis. The y-value 5 is called the **y-coordinate** and is associated with the y-axis.

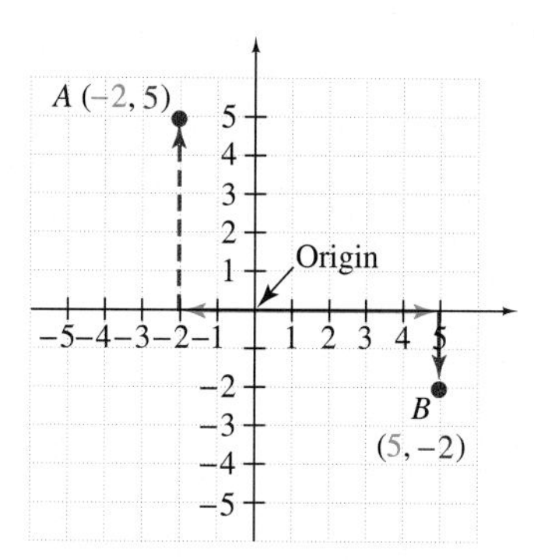

Compare the location of point A with the location of point B, which corresponds to the ordered pair $(5, -2)$. The x-coordinate 5 indicates that we move 5 units to the right of the origin along the x-axis. The y-coordinate -2 indicates that we move 2 units down parallel to the y-axis. Point A is in a different position than point B. Two ordered pairs are considered equal and correspond to the same point if and only if their x-coordinates are equal and their y-coordinates are equal.

Keep in mind that *each ordered pair corresponds to exactly one point in the real plane and that each point in the plane corresponds to exactly one ordered pair*. Thus, we may refer to the ordered pair (x, y) as the **point** (x, y).

Example 1

Plot each ordered pair on a rectangular coordinate system and name the quadrant in which the point is located.

a. $(2, -1)$ **b.** $(0, 5)$ **c.** $(-3, 5)$

d. $(-2, 0)$ **e.** $\left(-\frac{1}{2}, -4\right)$ **f.** $(1.5, 1.5)$

Practice Problem 1

Plot each ordered pair on a rectangular coordinate system and name the quadrant in which the point is located.

a. $(3, -2)$ b. $(0, 3)$

c. $(-4, 1)$ d. $(-1, 0)$

e. $\left(-2\frac{1}{2}, -3\right)$ f. $(3.5, 4.5)$

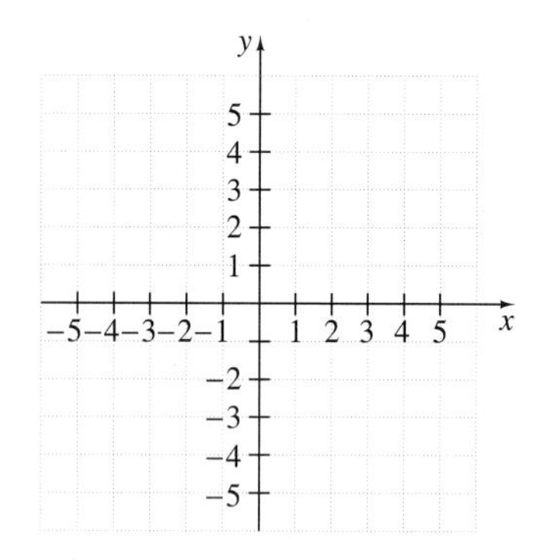

Answers

1.

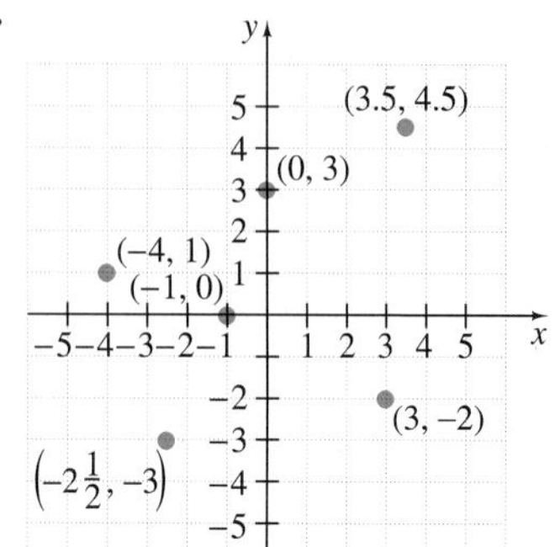

a. quadrant IV, **b.** not in any quadrant, **c.** quadrant II, **d.** not in any quadrant, **e.** quadrant III, **f.** quadrant I

Solution: The six points are graphed as shown:

a. $(2, -1)$ lies in quadrant IV.

b. $(0, 5)$ is not in any quadrant.

c. $(-3, 5)$ lies in quadrant II.

d. $(-2, 0)$ is not in any quadrant.

e. $\left(-\frac{1}{2}, -4\right)$ is in quadrant III.

f. $(1.5, 1.5)$ is in quadrant I.

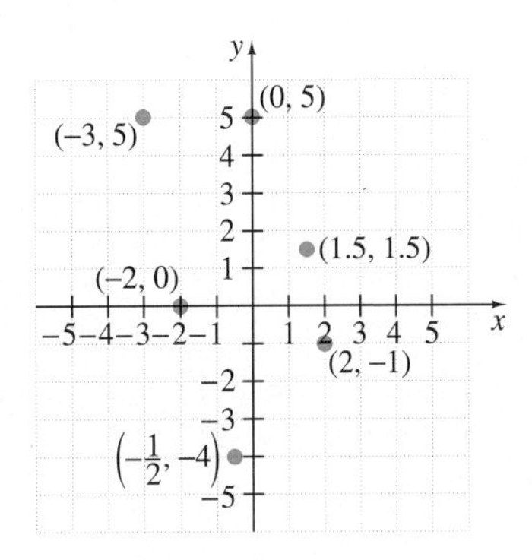

Notice that the y-coordinate of any point on the x-axis is 0. For example, the point with coordinates $(-2, 0)$ lies on the x-axis. Also, the x-coordinate of any point on the y-axis is 0. For example, the point with coordinates $(0, 5)$ lies on the y-axis. A point on an axis is called a **quadrantel** point.

TRY THE CONCEPT CHECK IN THE MARGIN.

Many types of real-world data occur in pairs. For example, the data pairs below were used for the Disney World ticket graph at the beginning of this section. The graph of paired data, such as the one below, is called a **scatter diagram**. Such diagrams are used to look for patterns and relationships in paired data.

Paired Data

Year, x	Price (in dollars), y
0	28.00
1	30.65
2	32.75
3	34.85
4	35.90
5	36.95
6	38.00
7	39.22
8	40.81
9	42.14

Note: The notation 0⟷1988 beneath the graph means that year 0 corresponds to the year 1988, 1 corresponds to 1989, and so on.

Scatter Diagram

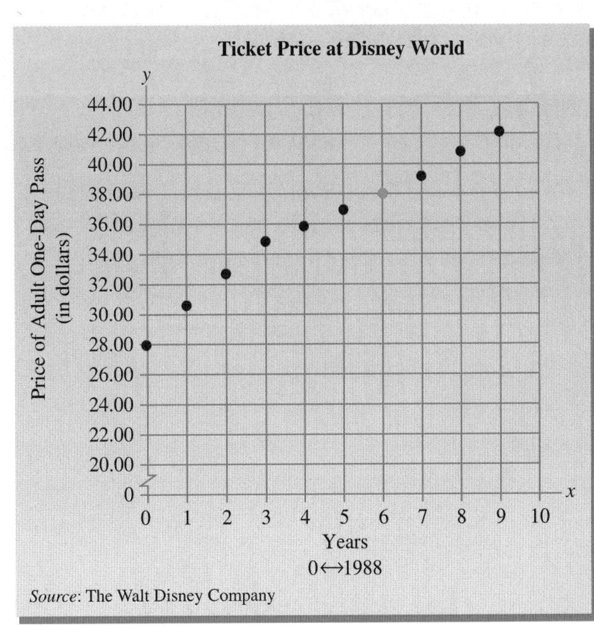

✓ CONCEPT CHECK

Which of the following correctly describes the location of the point $(3, -6)$ in a rectangular coordinate system?

a. 3 units to the left of the y-axis and 6 units above the x-axis.

b. 3 units above the x-axis and 6 units to the left of the y-axis.

c. 3 units to the right of the y-axis and 6 units below the x-axis.

d. 3 units below the x-axis and 6 units to the right of the y-axis.

HELPFUL HINT

Notice, for example, the paired data (6, 38.00) and its corresponding plotted point, both in blue.

Answer

✓ **Concept Check:** c

Practice Problem 2

Create a scatter diagram for the given paired data.

RETAIL SALES OF TOYS IN THE UNITED STATES	
Year	Sales (billions of dollars)
1993	18
1994	19
1995	20
1996	21
1997	23

(*Source:* Toy Manufacturers of America, Inc.)

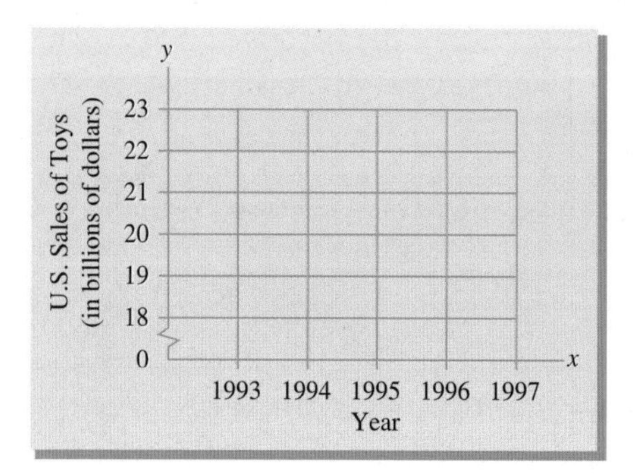

Example 2

Create a scatter diagram for the given paired data.

NATIONAL AVERAGE SAT MATH SCORES	
Year	Score
1994	504
1995	506
1996	508
1997	511

(*Source:* The College Board)

Solution: To graph the paired data in the table, we use the first column for the x- (or horizontal) axis and the second column for the y- (or vertical) axis.

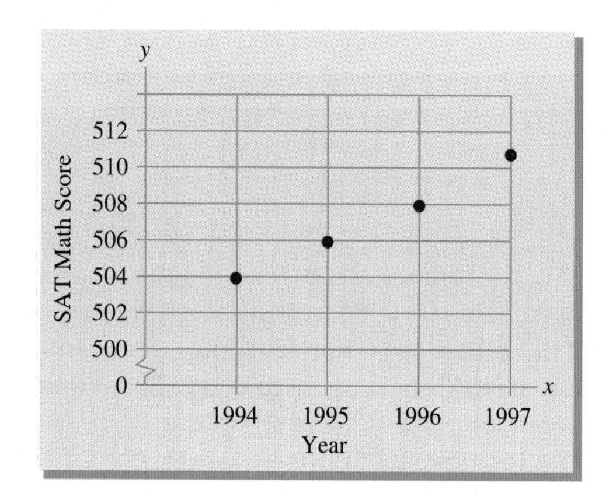

B DETERMINING WHETHER AN ORDERED PAIR OF NUMBERS IS A SOLUTION OF AN EQUATION

Solutions of equations in two variables consist of two numbers that can be written as ordered pairs of numbers. Unless we are told otherwise, we will assume that variable values are written as ordered pairs in alphabetical order (that is, x first and then y).

Practice Problem 3

Determine whether $(0, -6)$, $(1, 4)$, and $(-1, -4)$ are solutions of the equation $2x + y = -6$.

Example 3

Determine whether $(0, -12)$, $(1, 9)$, and $(2, -6)$ are solutions of the equation $3x - y = 12$.

Solution: To check each ordered pair, we replace x with each x-coordinate and y with each y-coordinate and see whether a true statement results.

Let $x = 0$ and $y = -12$.

$$3x - y = 12$$
$$3(0) - (-12) \stackrel{?}{=} 12$$
$$0 + 12 \stackrel{?}{=} 12$$
$$12 = 12 \quad \text{True.}$$

Let $x = 1$ and $y = 9$.

$$3x - y = 12$$
$$3(1) - 9 \stackrel{?}{=} 12$$
$$3 - 9 \stackrel{?}{=} 12$$
$$-6 = 12 \quad \text{False.}$$

Let $x = 2$ and $y = -6$.

$$3x - y = 12$$
$$3(2) - (-6) \stackrel{?}{=} 12$$
$$6 + 6 \stackrel{?}{=} 12$$
$$12 = 12 \quad \text{True.}$$

Answers

2.

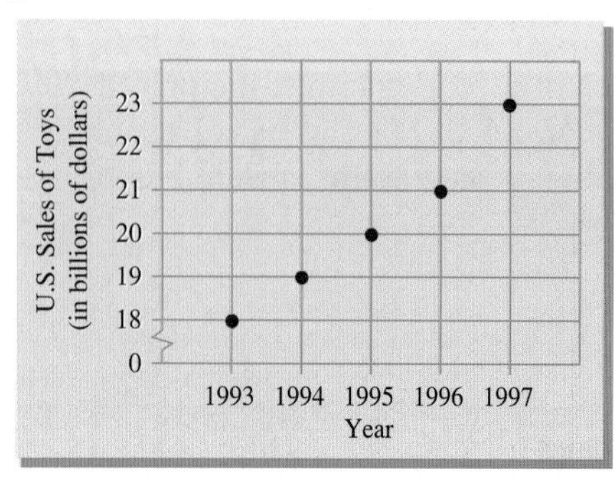

3. yes, no, yes

We see that $(1, 9)$ is not a solution but both $(2, -6)$ and $(0, -12)$ are solutions.

C Graphing Linear Equations

As we saw in Example 3, some linear equations have more than one ordered pair solution. In fact, the equation $3x - y = 12$ has an infinite number of ordered pair solutions. Since it is impossible to list all solutions, we visualize them by graphing them.

A few more ordered pairs that satisfy $3x - y = 12$ are $(4, 0)$, $(3, -3)$, $(5, 3)$, and $(1, -9)$. These ordered pair solutions, along with the ordered pair solutions from Example 3 are plotted on the following graph. The graph of $3x - y = 12$ is the single line containing these points. Every ordered pair solution of the equation corresponds to a point on this line, and every point on this line corresponds to an ordered pair solution.

x	y
4	0
1	−9
0	−12
2	−6
5	3
3	−3

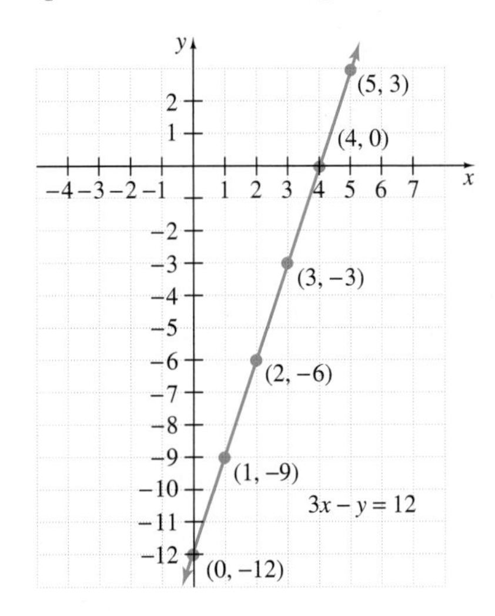

The equation $3x - y = 12$ is called a **linear equation in two variables**, and *the graph of every linear equation in two variables is a line.*

Linear Equation in Two Variables

A linear equation in two variables is an equation that can be written in the form

$$Ax + By = C$$

where A, B, and C are real numbers, and A and B are not both 0. The graph of a linear equation in two variables is a line.

A linear equation written in the form $Ax + By = C$ is said to be written in **standard form**. Some examples are

$$3x - y = 12$$
$$-2.1x + 5.6y = 0$$

HELPFUL HINT

Remind students that a linear equation is in standard form when variable terms are on one side of the equation and the constant is on the other side.

Recall from geometry that a line is determined by two points. This means that to graph a linear equation in two variables, just two solutions are needed. We will find a third solution, just to check our work. To find ordered-pair solutions of linear equations in two variables, we can choose an x-value and find its corresponding y-value, or we can choose a y-value and find its corresponding x-value. The number 0 is often a convenient value to choose for x and also for y.

Practice Problem 4

Graph: $3x - 2y = 6$

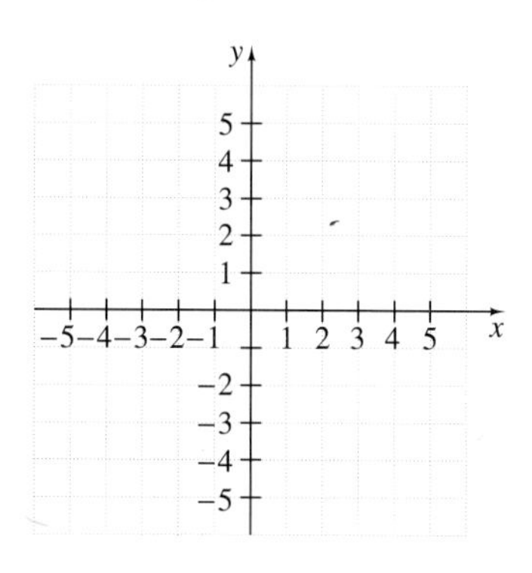

Example 4

Graph: $5x - 2y = 10$

Solution: First we find three ordered pair solutions, and then we plot the ordered pairs. The line through the plotted points is the graph. Let's let x be 0, let y be 0, and then let x be 1 to find our three ordered pair solutions.

Let $x = 0$.

$$5x - 2y = 10$$
$$5 \cdot 0 - 2y = 10$$
$$-2y = 10 \quad \text{Simplify.}$$
$$y = -5 \quad \text{Divide by } -2.$$

Let $y = 0$.

$$5x - 2y = 10$$
$$5x - 2 \cdot 0 = 10$$
$$5x = 10 \quad \text{Simplify.}$$
$$x = 2 \quad \text{Divide by 5.}$$

Let $x = 1$.

$$5x - 2y = 10$$
$$5 \cdot 1 - 2y = 10$$
$$5 - 2y = 10 \quad \text{Multiply.}$$
$$-2y = 5 \quad \text{Subtract 5.}$$
$$y = -\frac{5}{2}, \text{ or } -2\frac{1}{2}$$

The three ordered pair solutions—$(0, -5)$, $(2, 0)$, and $\left(1, -2\frac{1}{2}\right)$—are listed in the table, and the graph of $5x - 2y = 10$ is shown.

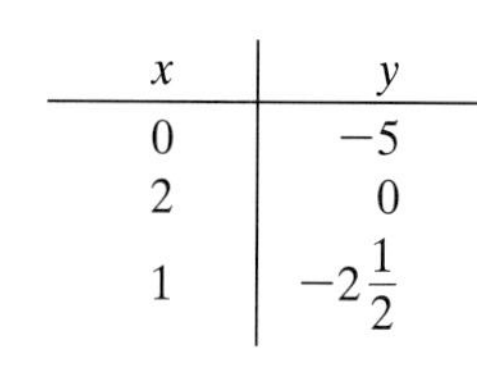

x	y
0	-5
2	0
1	$-2\frac{1}{2}$

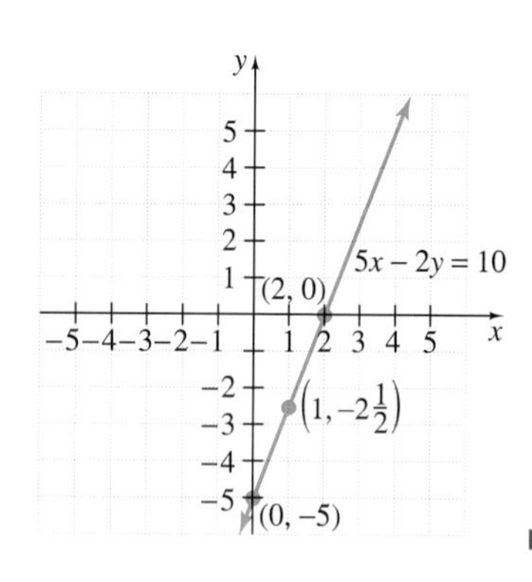

Answer

4.

Notice that the graph in Example 4 crosses the x-axis at the point $(2, 0)$. This point is called the ***x*-intercept point**, and 2 is called the ***x*-intercept**. This graph also crosses the y-axis at the point $(0, -5)$. This point is called the ***y*-intercept point**, and -5 is called the ***y*-intercept**.

In general, to find the y-intercept of the graph of an equation, let $x = 0$ since any point on the y-axis has an x-coordinate of 0. To find the x-intercept of a line, let $y = 0$ since any point on the x-axis has a y-coordinate of 0.

> **Finding x- and y-Intercepts**
>
> To find an x-intercept, let $y = 0$ and solve for x.
> To find a y-intercept, let $x = 0$ and solve for y.

Example 5 Find the intercepts and graph: $x + 4y = -4$

Solution: To find the y-intercept, we let $x = 0$ and solve for y. To find the x-intercept, we let $y = 0$ and solve for x. Let's let $x = 0$, $y = 0$, and then let $x = 2$ to find our third check point.

Let $x = 0$.	Let $y = 0$.	Let $x = 2$.
$x + 4y = -4$	$x + 4y = -4$	$x + 4y = -4$
$0 + 4y = -4$	$x + 4 \cdot 0 = -4$	$2 + 4y = -4$
$4y = -4$	$x = -4$	$4y = -6$
$y = -1$		$y = -\frac{6}{4} = -1\frac{1}{2}$
$(0, -1)$	$(-4, 0)$	$(2, -1\frac{1}{2})$

The ordered pairs are $(0, -1)$, $(-4, 0)$, and $(2, -1\frac{1}{2})$. We plot these points to obtain the graph shown.

	x	y	
	0	-1	← y-intercept
x-intercept →	-4	0	
	2	$-1\frac{1}{2}$	

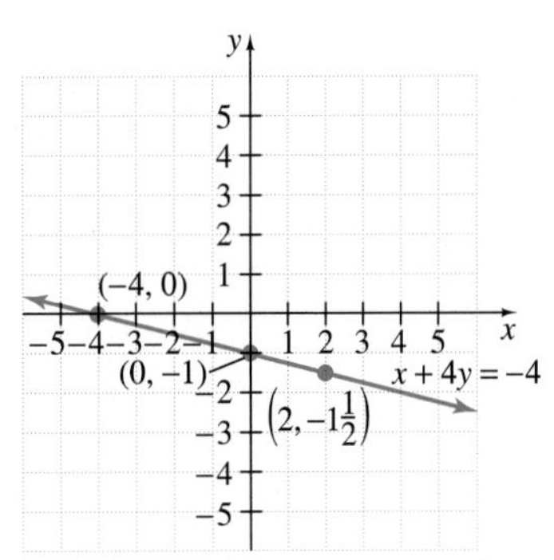

Example 6 Find the intercepts and graph: $x = -2y$

Solution: We let $y = 0$ to find the x-intercept and $x = 0$ to find the y-intercept.

Let $y = 0$.	Let $x = 0$.
$x = -2y$	$x = -2y$
$x = -2(0)$	$0 = -2y$
$x = 0$	$0 = y$

Both the x-intercept and y-intercept are 0. In other words, when $x = 0$, then $y = 0$, which gives the ordered pair $(0, 0)$. Also, when $y = 0$, then $x = 0$, which gives the same ordered pair $(0, 0)$. This happens when the graph passes through the origin. Since two points are needed to determine a line, we must find at least one more ordered pair that satisfies $x = -2y$. Let's let $y = -1$ to find a second ordered pair solution and let $y = 1$ as a checkpoint.

Practice Problem 5

Find the intercepts and graph $x + 3y = -6$

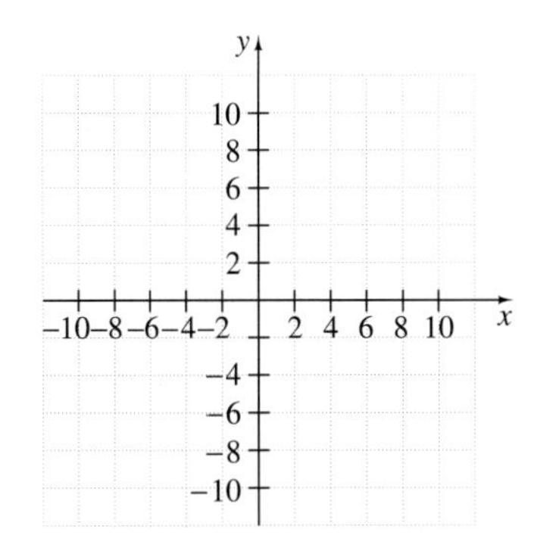

Practice Problem 6

Find the intercepts and graph: $y = 2x$

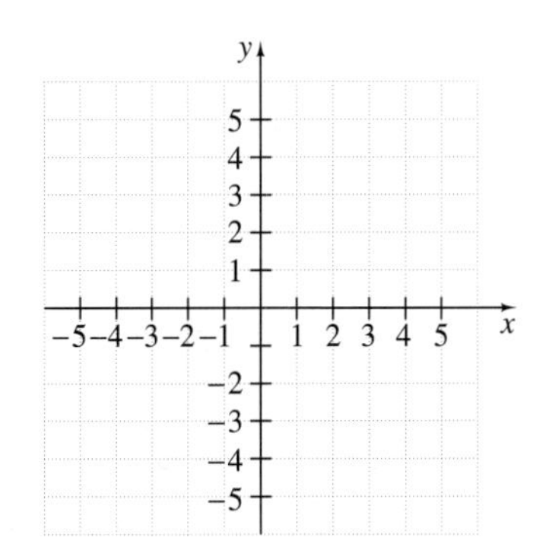

Answers

5.

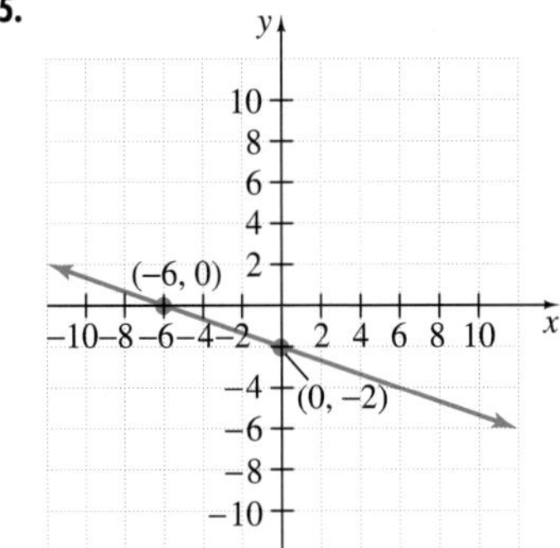

6.

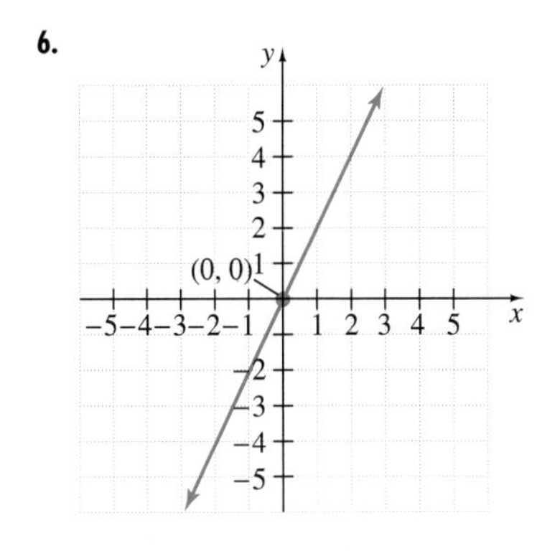

Let $y = -1$. $\quad x = -2(-1)$ $\quad x = 2$

Let $y = 1$. $\quad x = -2(1)$ $\quad x = -2$

HELPFUL HINT

Since the equation, $x = -2y$ is solved for x, we choose y-values for finding a second and third point. This way, we simply need to evaluate an expression to find the y-value, as shown.

The ordered pairs are $(0, 0)$, $(2, -1)$, and $(-2, 1)$. We plot these points to obtain the graph shown.

	x	y	
x-intercept →	0	0	← y-intercept
	2	−1	
	−2	1	

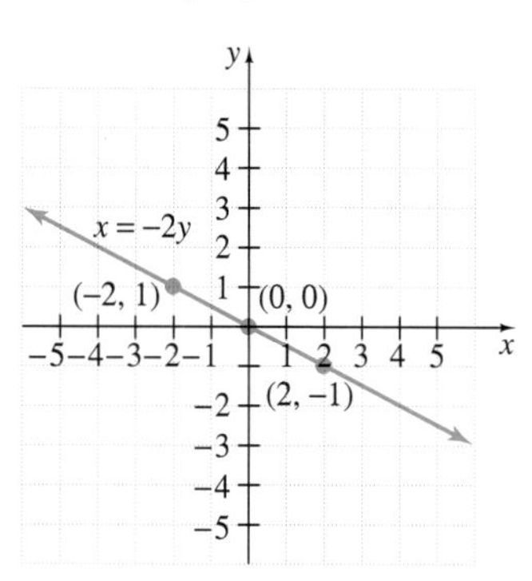

D GRAPHING VERTICAL AND HORIZONTAL LINES

The equation $x = c$, where c is a real number constant, is a linear equation in two variables because it can be written in the form $x + 0y = c$. The graph of this equation is a vertical line, as shown in the next example.

Practice Problem 7

Graph: $x = -1$

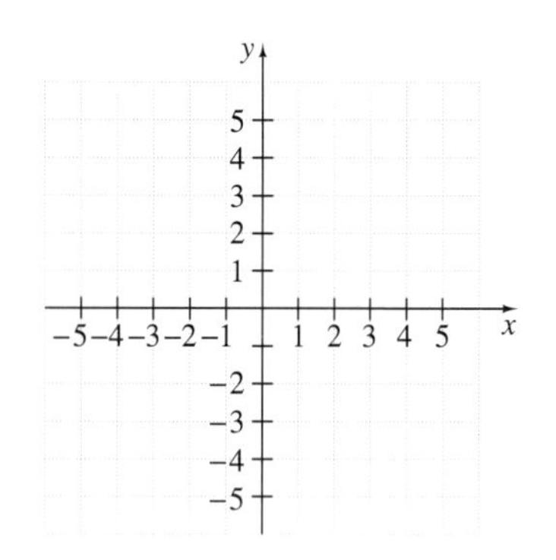

Example 7 Graph: $x = 2$

Solution: The equation $x = 2$ can be written as $x + 0y = 2$. For any y-value chosen, notice that x is 2. No other value for x satisfies $x + 0y = 2$. Any ordered pair whose x-coordinate is 2 is a solution to $x + 0y = 2$ because 2 added to 0 times any value of y is $2 + 0$, or 2. We will use the ordered pairs $(2, 3)$, $(2, 0)$ and $(2, -3)$ to graph $x = 2$.

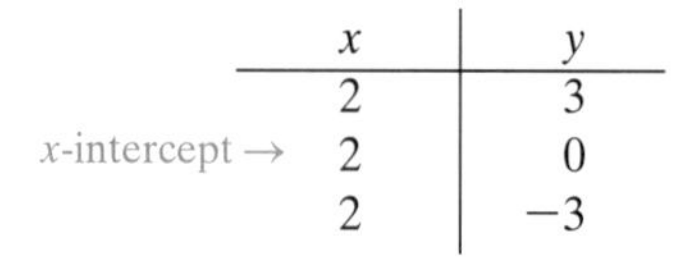

	x	y
	2	3
x-intercept →	2	0
	2	−3

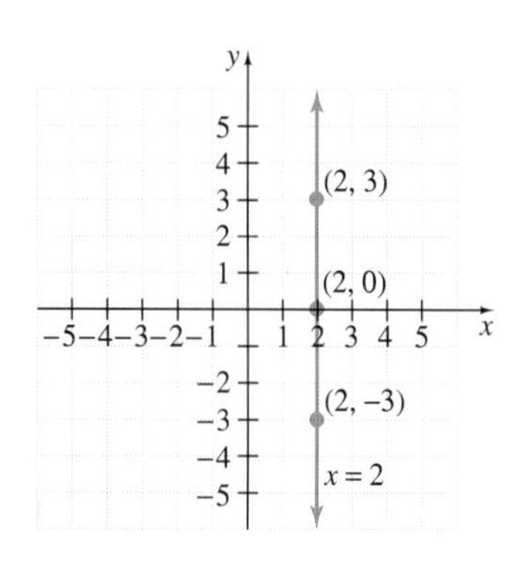

The graph is a vertical line with x-intercept 2. It has no y-intercept because x is never 0.

Answer

7.

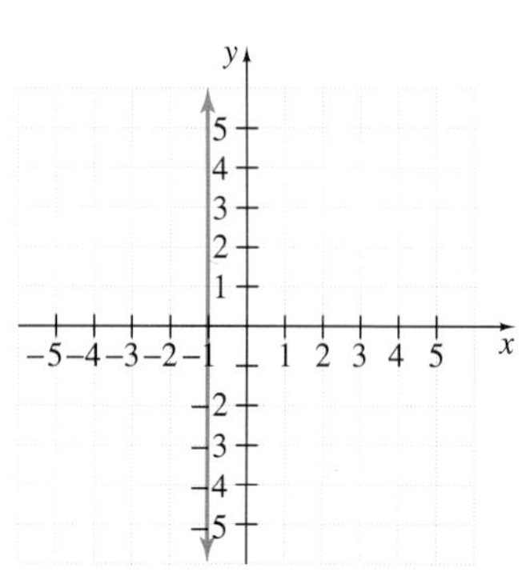

Example 8 Graph: $y = -3$

Solution: The equation $y = -3$ can be written as $0x + y = -3$. For any x-value chosen, y is -3. If we choose 4, 0, and -2 as x-values, the ordered pair solutions are $(4, -3)$, $(0, -3)$, and $(-2, -3)$. We will use these ordered pairs to graph $y = -3$.

x	y
4	-3
0	-3 ← y-intercept
-2	-3

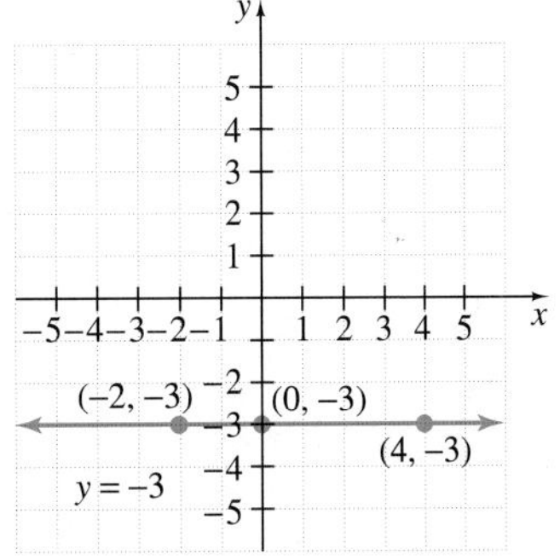

The graph is a horizontal line with y-intercept -3 and no x-intercept.

From Examples 7 and 8, we have the following generalization.

GRAPHING VERTICAL AND HORIZONTAL LINES

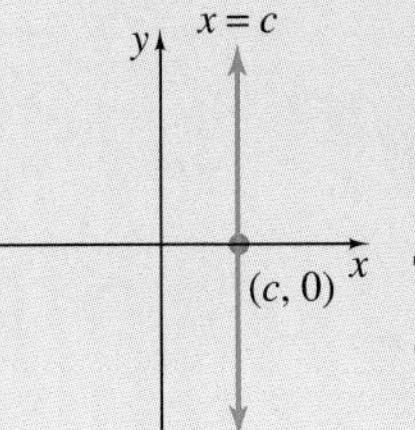

The graph of $x = c$, where c is a real number, is a vertical line with x-intercept c.

The graph of $y = c$, where c is a real number, is a horizontal line with y-intercept c.

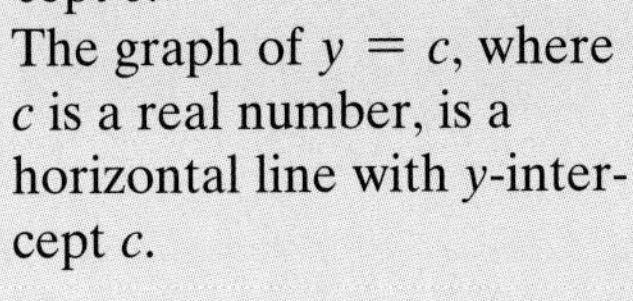

Practice Problem 8

Graph: $y = 2$

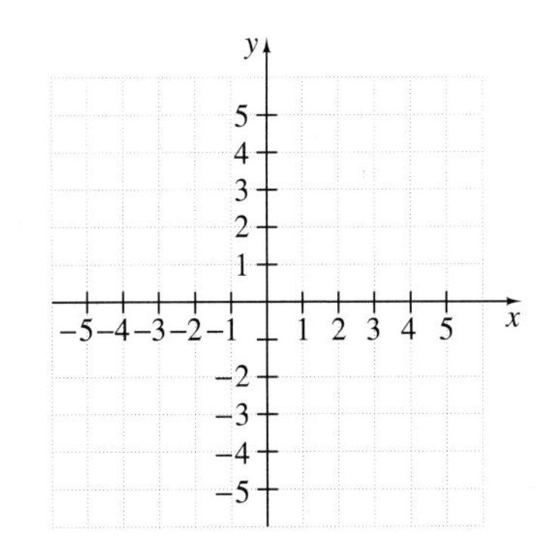

Answer

8.

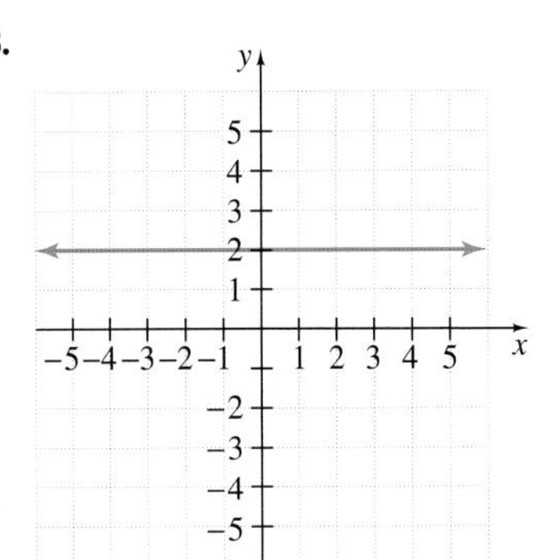

GRAPHING CALCULATOR EXPLORATIONS

In this section, we begin a study of graphing calculators and graphing software packages for computers. These graphers use the same point-plotting technique that we introduced in this section. The advantage of this graphing technology is, of course, that graphing calculators and computers can find and plot ordered-pair solutions much faster than we can. Note, however, that the features described in these boxes may not be available on all graphing calculators.

The rectangular screen where a portion of the rectangular coordinate system is displayed is called a **window**. We call it a **standard window** for graphing when both the x- and y-axes display coordinates between -10 and 10. This information is often displayed in the window menu on a graphing calculator as

Xmin = −10
Xmax = 10
Xscl = 1 The scale on the x-axis is one unit per tick mark.
Ymin = −10
Ymax = 10
Yscl = 1 The scale on the y-axis is one unit per tick mark.

To use a graphing calculator to graph the equation $y = -5x + 4$, press the Y= key and enter the keystrokes

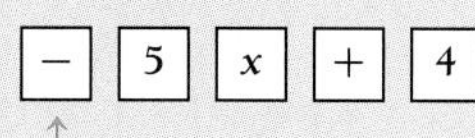

(Check your owner's manual to make sure the "negative" key is pressed here and not the "subtraction" key.)

The top row should now read $Y_1 = -5x + 4$. Next press the GRAPH key, and the display should look like this:

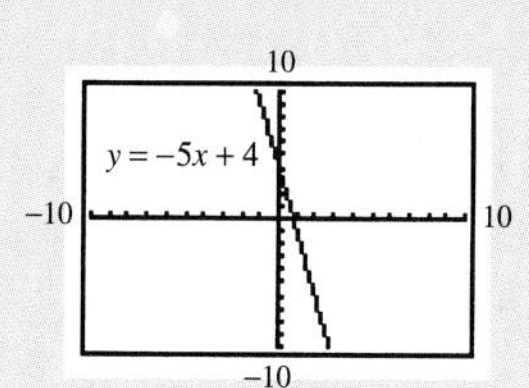

Use a standard window and graph the following linear equations. (Unless otherwise stated, we will use a standard window when graphing.)

1. $y = 6x - 1$

2. $y = 3x - 2$

3. $y = -3.2x + 7.9$

4. $y = -x + 5.85$

5. $y = \frac{1}{4}x - \frac{2}{3}$ (Parentheses may need to be inserted around $\frac{1}{4}$.)

6. $y = \frac{2}{3}x - \frac{1}{5}$ (Parentheses may need to be inserted around $\frac{2}{3}$.)

Name ______________________ Section ______ Date ________

MENTAL MATH

Determine the coordinates of each point on the graph.

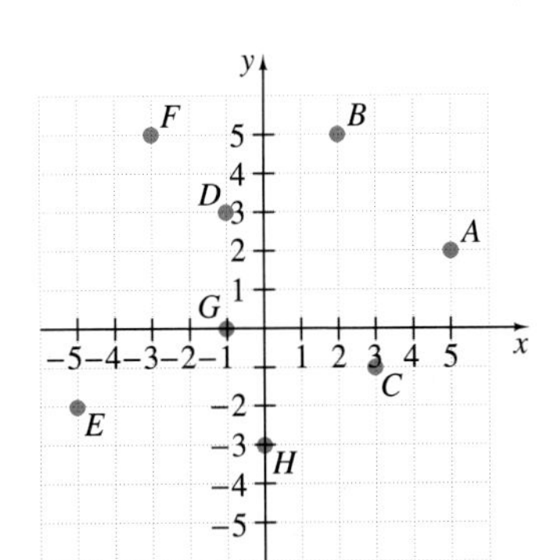

1. Point A 2. Point B 3. Point C

4. Point D 5. Point E 6. Point F

7. Point G 8. Point H

EXERCISE SET 3.1

A *Plot each ordered pair on a rectangular coordinate system and name the quadrant (or axis) in which the point lies. See Example 1.*

1. $(3, 2)$

$(-5, 3)$

$\left(5\frac{1}{2}, -4\right)$

$(0, 3.5)$

$(-2, -4)$

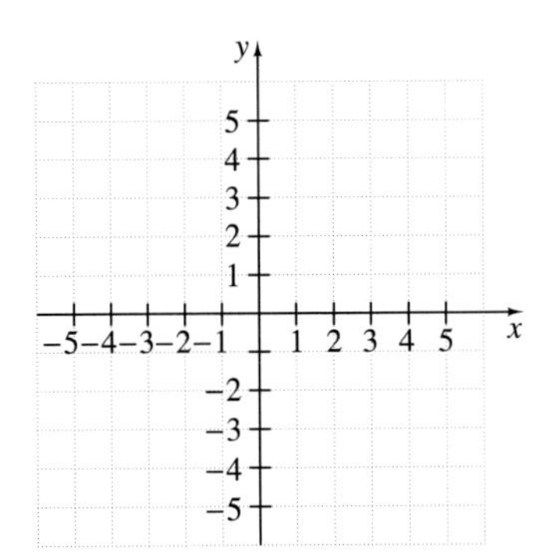

2. $(2, -1)$

$(-3, -1)$

$\left(-2, 6\frac{1}{3}\right)$

$(-2, 4)$

$(-4.2, 0)$

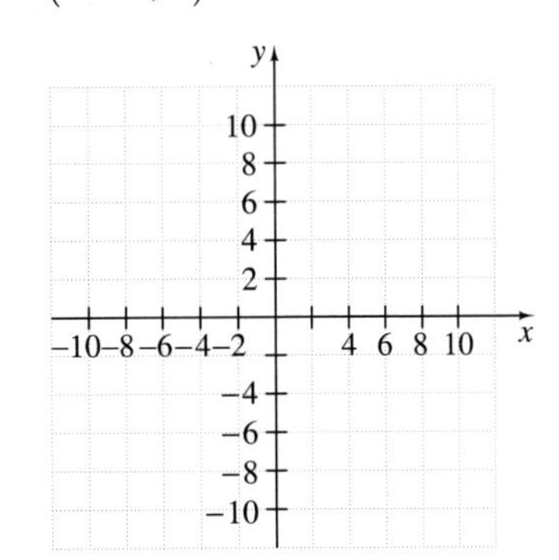

Given that x is a positive number and y is a positive number, determine the quadrant (or axis) in which each point lies. See Example 1.

3. $(x, -y)$ **4.** $(-x, y)$ **5.** $(x, 0)$

6. $(0, -y)$ **7.** $(-x, -y)$ **8.** $(0, 0)$

MENTAL MATH ANSWERS

1. ________
2. ________
3. ________
4. ________
5. ________
6. ________
7. ________
8. ________

ANSWERS

1. ________
2. ________
3. ________
4. ________
5. ________
6. ________
7. ________
8. ________

9. see graph

10. see graph

11. ______

12. ______

13. ______

14. ______

15. ______

16. ______

Name ______________________

Create a scatter diagram for the given paired data. See Example 2.

9.

AIRLINE REVENUES FROM PASSENGERS IN THE UNITED STATES	
Year	Revenue (billions of dollars)
1993	64
1994	65
1995	70
1996	75
1997	79

(*Source:* Air Transport Association of America)

U.S. Airline Revenues from Passengers

Revenue (in billions of dollars)

y

80

75

70

65

60

0

x

1993 1994 1995 1996 1997

Year

10.

U.S. DEATH RATE FROM TUBERCULOSIS	
Year, x	Death Rate, y (per million population)
1950	225
1960	61
1970	26
1980	9
1990	7

(*Source: 1998 World Almanac*)

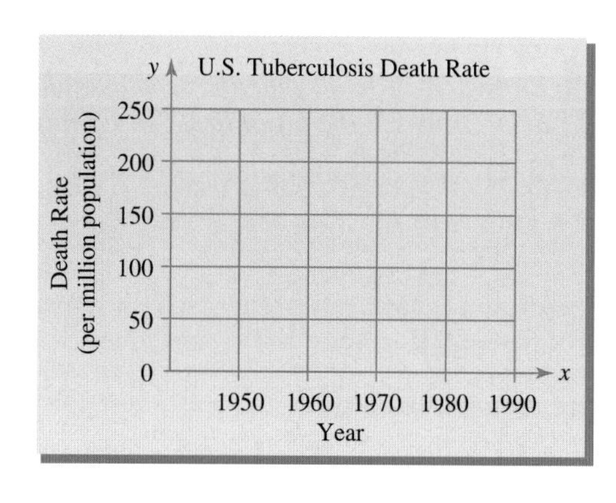

B *Determine whether each ordered pair is a solution of the given equation. See Example 3.*

11. $y = 3x - 5; (0, 5), (-1, -8)$

12. $y = -2x + 7; (1, 5), (-2, 3)$

13. $-6x + 5y = -6; (1, 0), \left(2, \frac{6}{5}\right)$

14. $5x - 3y = 9; (0, 3), \left(\frac{12}{5}, -1\right)$

15. $y = -3; (1, -3), (-3, 6)$

16. $y = 2; (2, 5), (0, 2)$

Name ______________________________

C D *Graph each linear equation. See Examples 4 through 8.*

17. $x - 2y = 4$

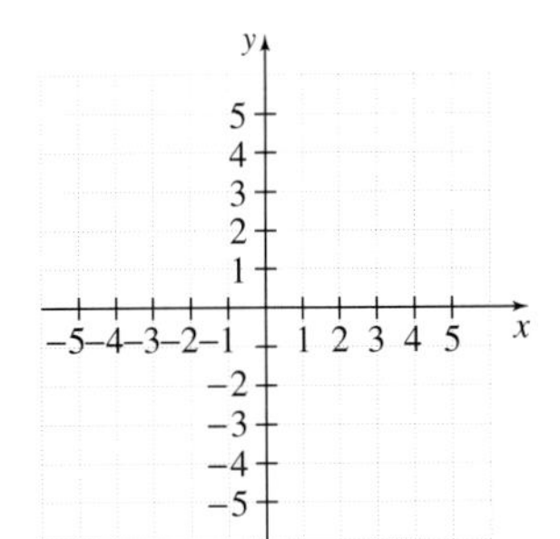

18. $y - 2x = 4$

19. $3x + 2y = 6$

20. $2x + 4y = 8$

21. $x = 4$

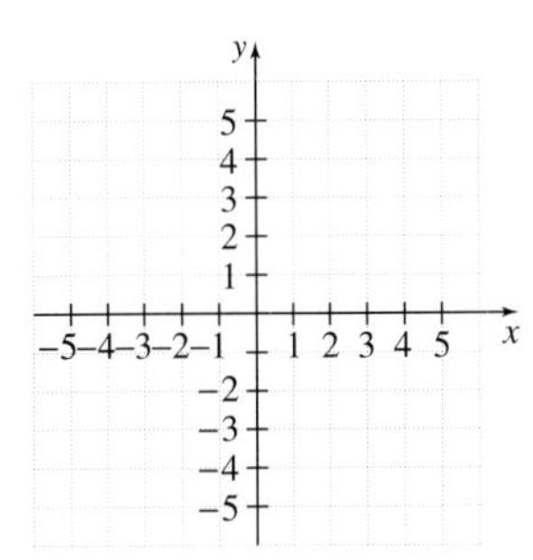

22. $y = 5$

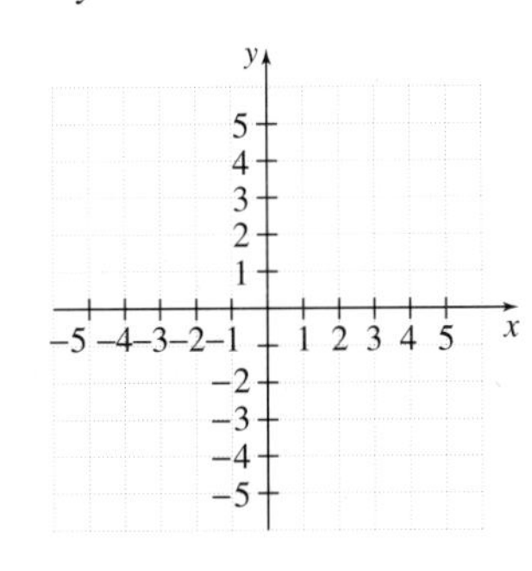

23. $x - 3y = 6$

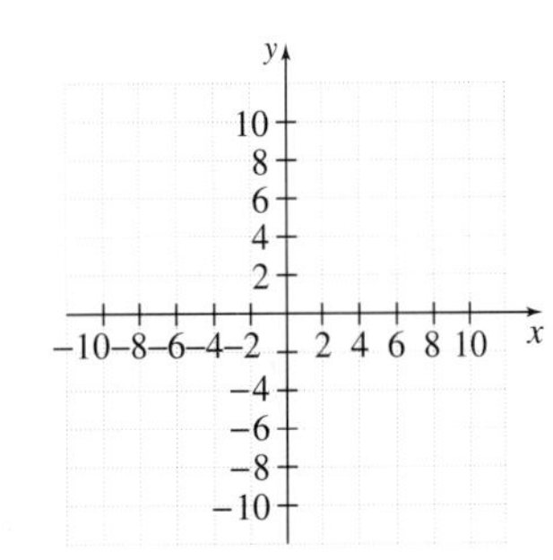

24. $x - 2y = 4$

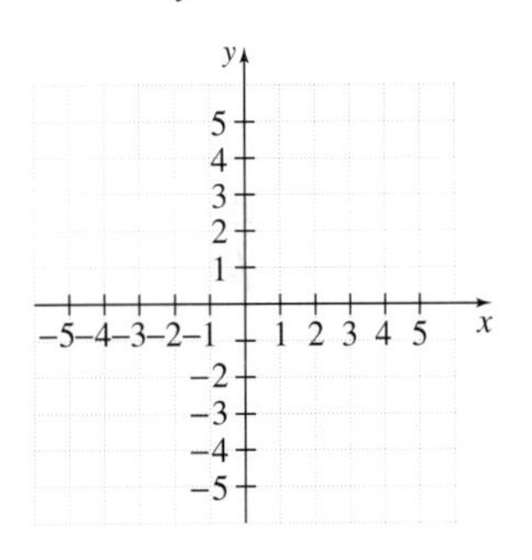

25. $y = 3x$

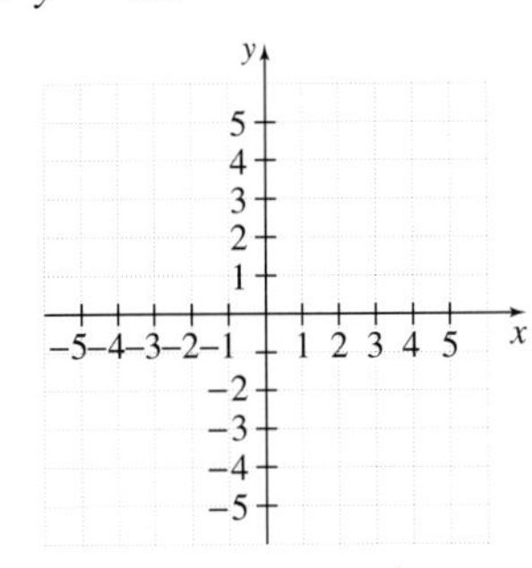

26. $y = -4x$

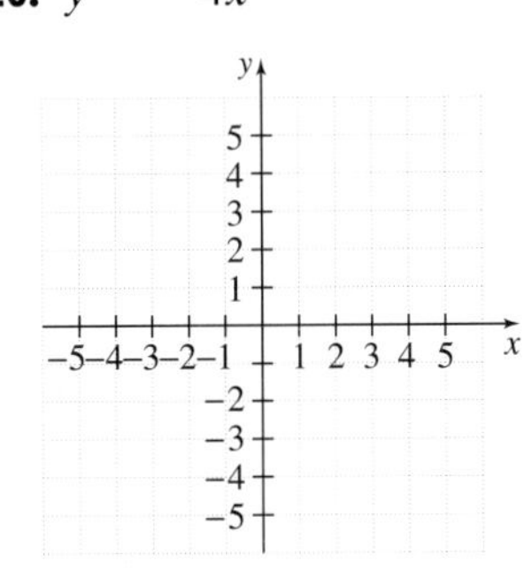

27. $y = -2$

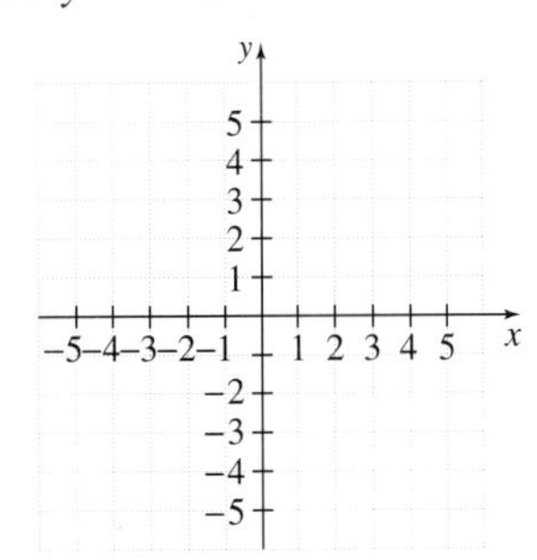

28. $x = -3$

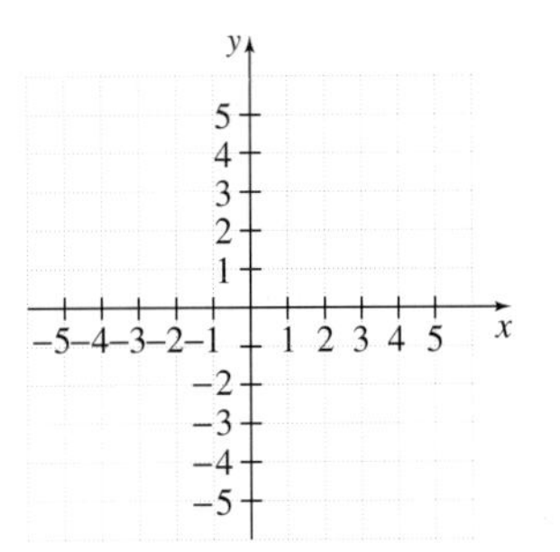

29. $4x + 5y = 15$

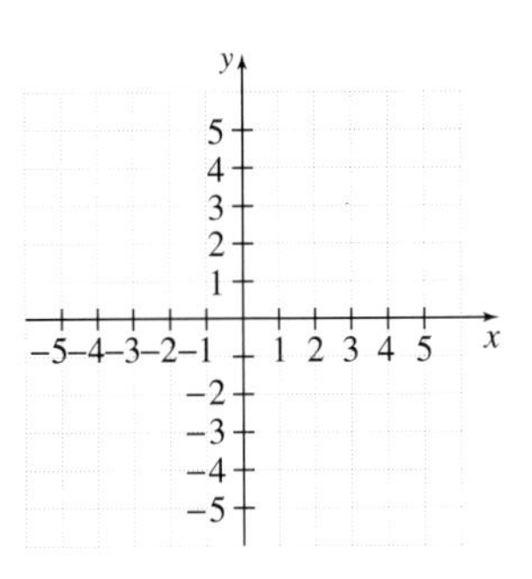

30. $2x + 3y = 9$

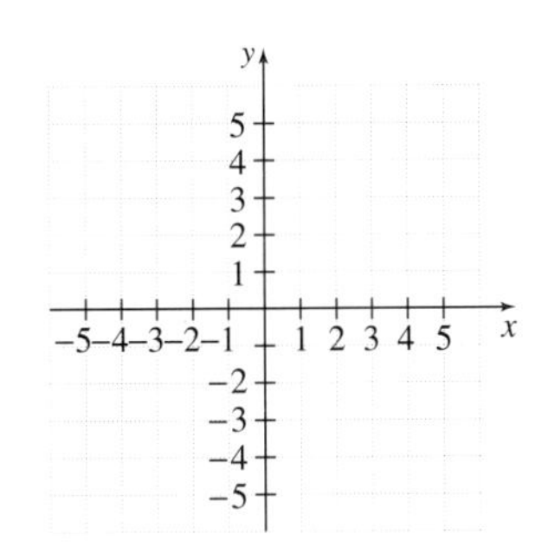

31. $5y = x - 10$

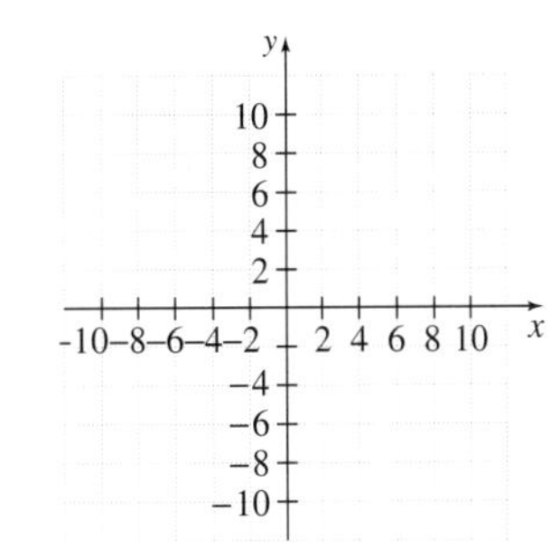

32. $3y = x - 3$

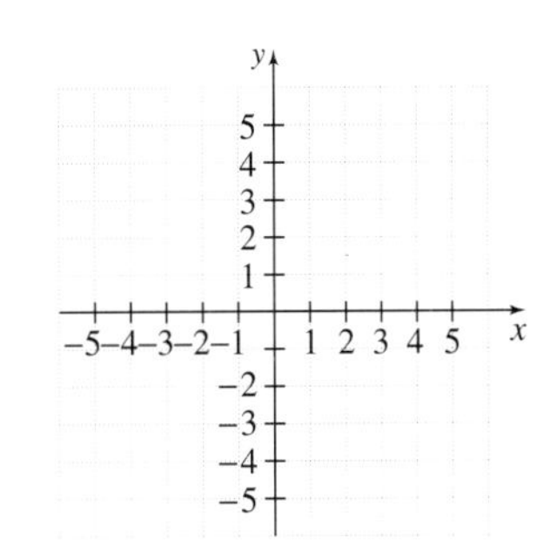

Name ____________________

33. $x = \frac{1}{2}$

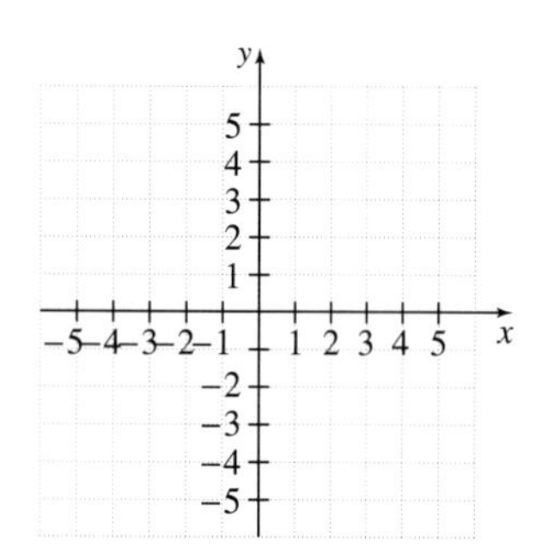

34. $y = -\frac{5}{2}$

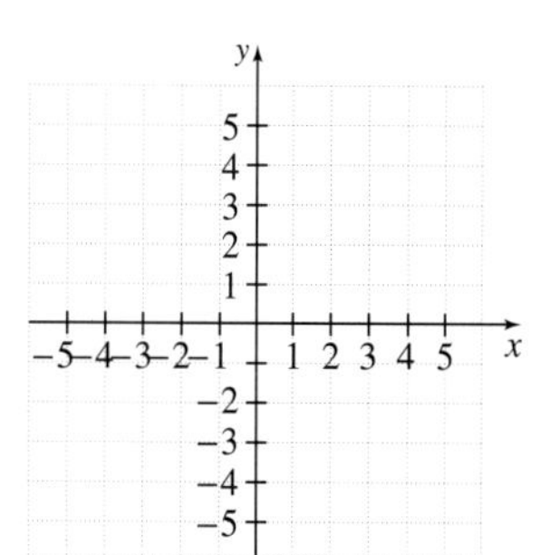

35. $y = \frac{1}{2}x$

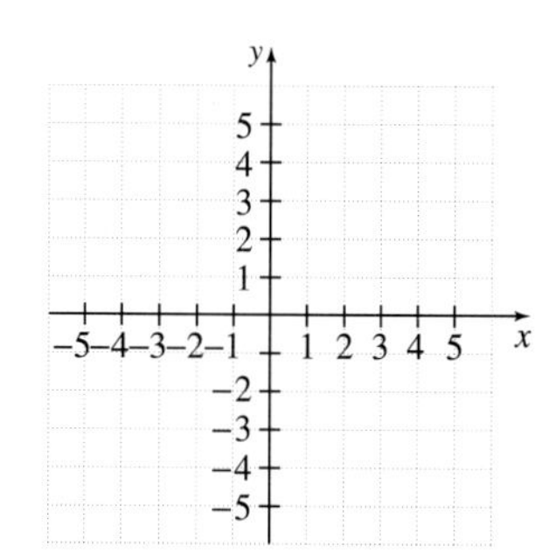

36. $x = \frac{1}{2}y$

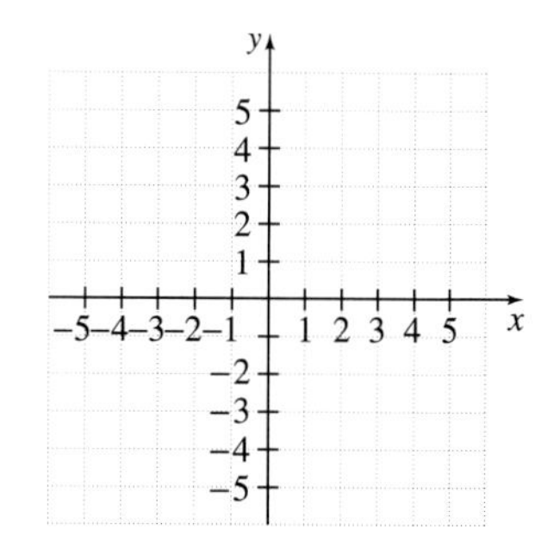

37. $y = -4x + 1$

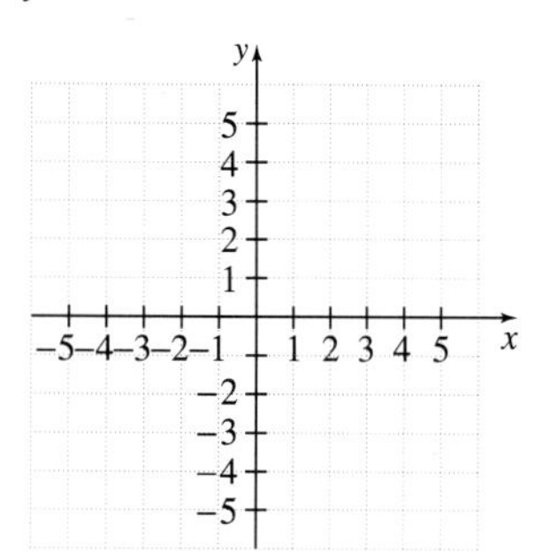

38. $y = -3x + 1$

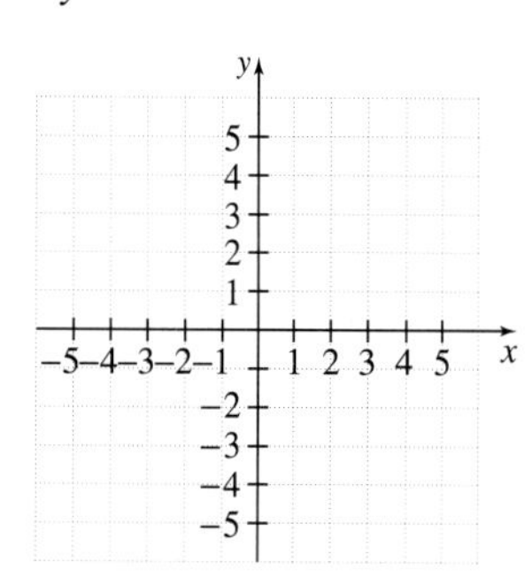

39. $2y - 6 = 0$

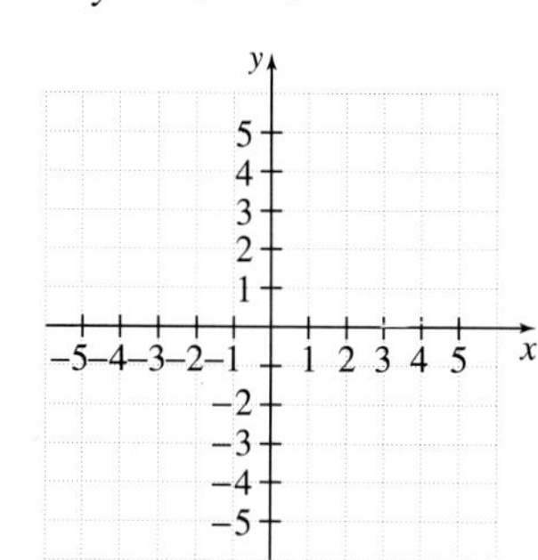

40. $3x + 6 = 0$

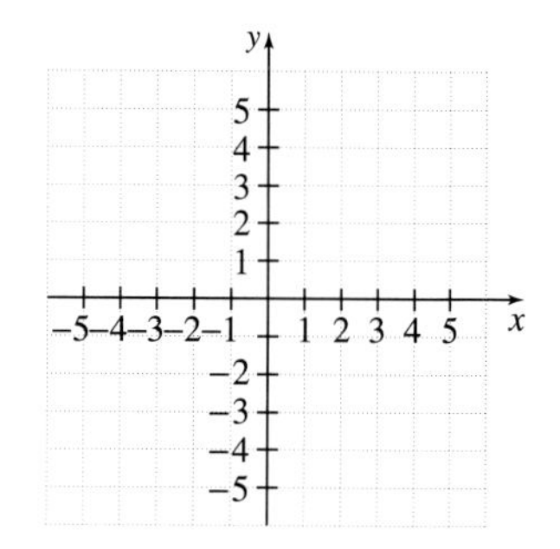

Review and Preview

Match each equation with its graph.

A

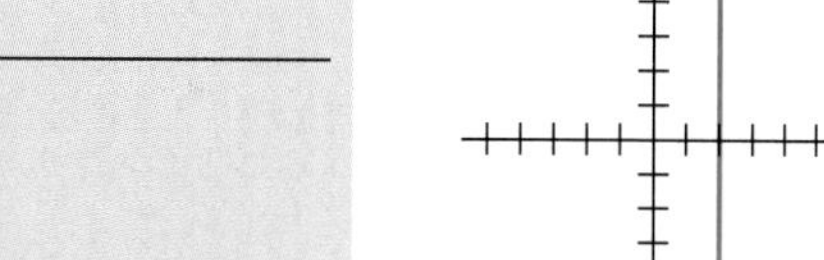

B

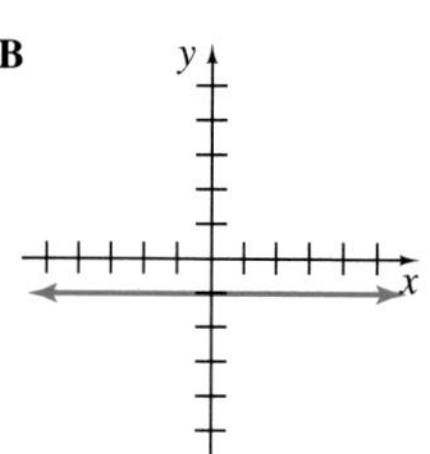

C

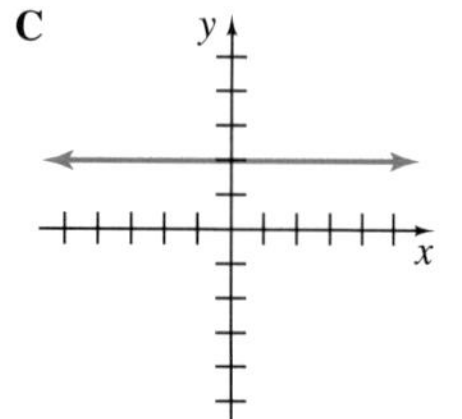

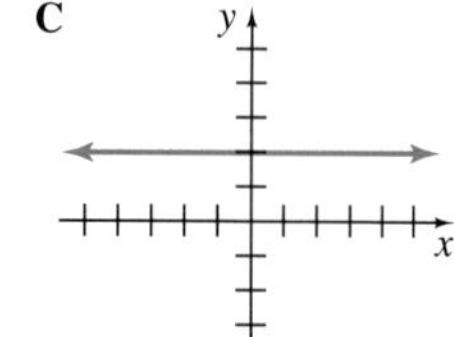

D

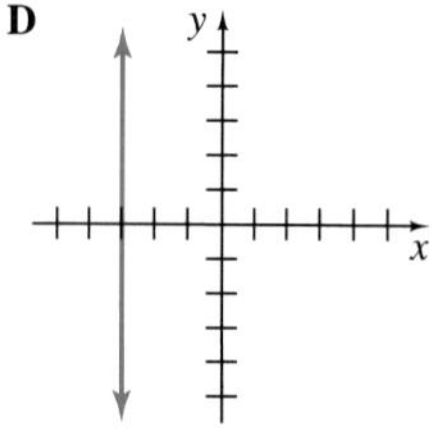

41. $y = 2$

42. $x = -3$

43. $x - 2 = 0$

44. $y + 1 = 0$

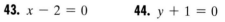

Simplify. See Section 1.4.

45. $\frac{-6 - 3}{2 - 8}$

46. $\frac{4 - 5}{-1 - 0}$

47. $\frac{-8 - (-2)}{-3 - (-2)}$

41. ____________________

42. ____________________

43. ____________________

44. ____________________

45. ____________________

46. ____________________

47. ____________________

Name ______________________

48. $\dfrac{12 - 3}{10 - 9}$

49. $\dfrac{0 - 6}{5 - 0}$

50. $\dfrac{2 - 2}{3 - 5}$

Combining Concepts

51. Broyhill Furniture found that it takes 2 hours to manufacture each table for one of its special dining room sets. Each chair takes 3 hours to manufacture. A total of 1500 hours is available to produce tables and chairs of this style. The linear equation that models this situation is $2x + 3y = 1500$, where x represents the number of tables produced and y the number of chairs produced.

a. Complete the ordered pair solution $(0, \ )$ of this equation. Describe the manufacturing situation to which this solution corresponds.

b. Complete the ordered pair solution $(\ , 0)$ for this equation. Describe the manufacturing situation to which this solution corresponds.

c. If 50 tables are produced, find the greatest number of chairs the company can make.

52. While manufacturing two different camera models, Kodak found that the basic model costs \$55 to produce, whereas the deluxe model costs \$75. The weekly budget for those two models is limited to \$33,000 in production costs. The linear equation that models this situation is $55x + 75y = 33{,}000$, where x represents the number of basic models and y the number of deluxe models.

a. Complete the ordered pair solution $(0, \ )$ of this equation. Describe the manufacturing situation to which this solution corresponds.

b. Complete the ordered pair solution $(\ , 0)$ of this equation. Describe the manufacturing situation to which this solution corresponds.

c. If 350 deluxe models are produced, find the greatest number of basic models that can be made in one week.

53. On the same set of axes, graph $y = 2x$, $y = 2x - 5$, and $y = 2x + 5$. What patterns do you see in these graphs?

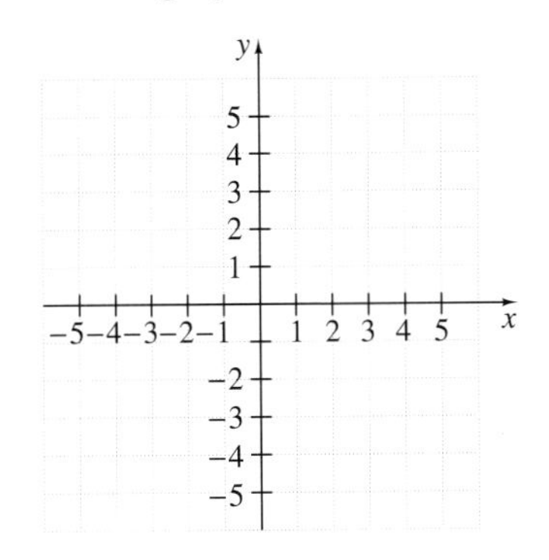

54. Explain why we generally use three points to graph a line, when only two points are needed.

48. ______

49. ______

50. ______

51. a. ______

b. ______

c. ______

52. a. ______

b. ______

c. ______

53. see graph

54. ______

55. see calculator screen

56. see calculator screen

57. see calculator screen

58. see calculator screen

59.

60.

Name ____

Use a grapher to verify the graphs of each exercise.

55. Exercise 25

56. Exercise 26

57. Exercise 37

58. Exercise 38

59. Discuss whether a vertical line ever has a y-intercept.

60. Discuss whether a horizontal line ever has an x-intercept.

3.2 THE SLOPE OF A LINE

You may have noticed by now that different lines often tilt differently. It is very important in many fields to be able to measure and compare the tilt, or **slope**, of lines. For example, a wheelchair ramp with a slope of $\frac{1}{12}$ means that the ramp rises 1 foot for every 12 horizontal feet. A road with a slope or grade of 11% $\left(\text{or } \frac{11}{100}\right)$ means that the road rises 11 feet for every 100 horizontal feet.

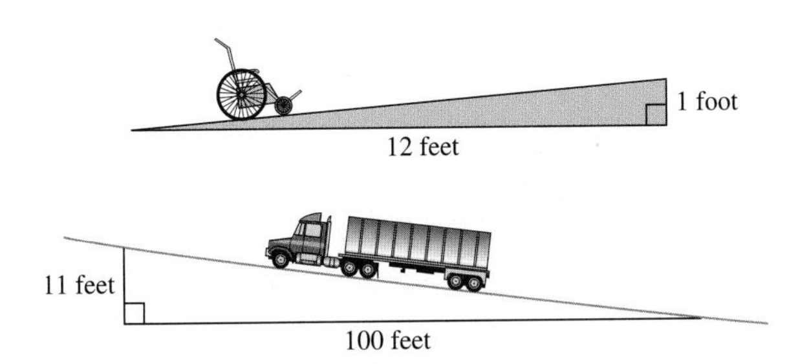

We measure the slope of a line as a ratio of **vertical change** to **horizontal change**. Slope is usually designated by the letter m.

Objective

A Find the slope of a line given two points on the line.

B Find the slope of a line given the equation of the line.

C Find the slopes of horizontal and vertical lines.

D Compare the slopes of parallel and perpendicular lines.

A FINDING SLOPE FROM TWO POINTS

Suppose that we want to measure the slope of the following line.

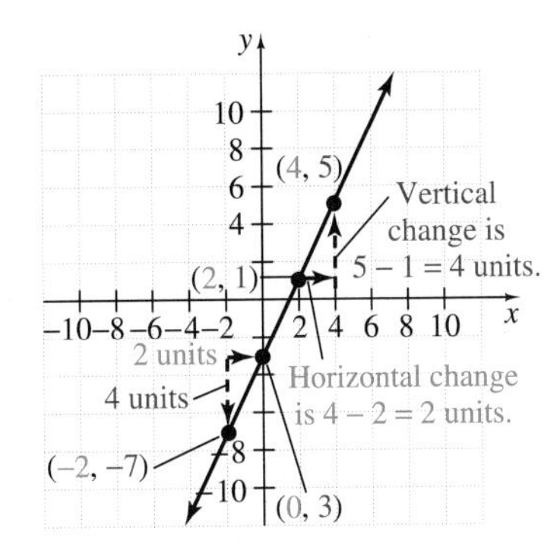

The vertical change between both pairs of points on the line is 4 units per horizontal change of 2 units. Then

$$\text{slope } m = \frac{\text{change in } y \text{ (vertical change)}}{\text{change in } x \text{ (horizontal change)}} = \frac{4}{2} = 2$$

Notice that slope is a rate of change between points. A slope of 2 or $\frac{2}{1}$ means that between pairs of points on the line, the rate of change is a vertical change of 2 units per horizontal change of 1 unit.

In general, consider the line which passes through the points (x_1, y_1) and (x_2, y_2). (The notation x_1 is read "x-sub-one.") The vertical change, or *rise*, between these points is the difference in the y-coordinates: $y_2 - y_1$. The horizontal change, or *run*, between the points is the difference of the x-coordinates: $x_2 - x_1$.

SLOPE OF A LINE

Given a line passing through points (x_1, y_1) and (x_2, y_2) the **slope** m of the line is

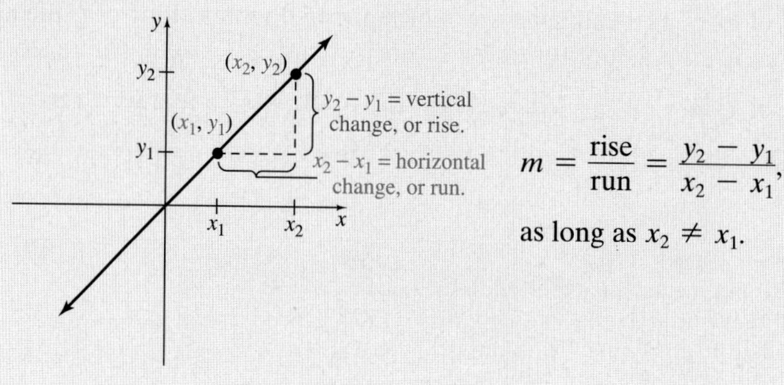

$$m = \frac{\text{rise}}{\text{run}} = \frac{y_2 - y_1}{x_2 - x_1},$$

as long as $x_2 \neq x_1$.

✓ CONCEPT CHECK

In the definition of slope, we state that $x_2 \neq x_1$. Explain why.

Practice Problem 1

Find the slope of the line containing the points $(-1, -2)$ and $(2, 5)$. Graph the line.

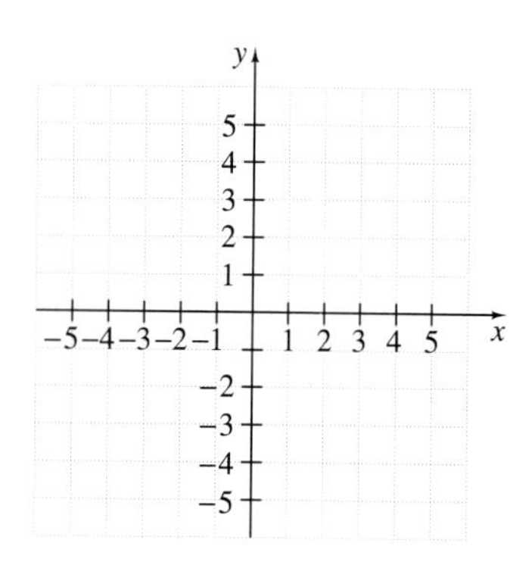

TRY THE CONCEPT CHECK IN THE MARGIN.

Example 1 Find the slope of the line containing the points $(0, 3)$ and $(2, 5)$. Graph the line.

Solution: We use the slope formula. It does not matter which point we call (x_1, y_1) and which point we call (x_2, y_2). We'll let $(x_1, y_1) = (0, 3)$ and $(x_2, y_2) = (2, 5)$.

$$m = \frac{y_2 - y_1}{x_2 - x_1}$$

$$= \frac{5 - 3}{2 - 0} = \frac{2}{2} = 1$$

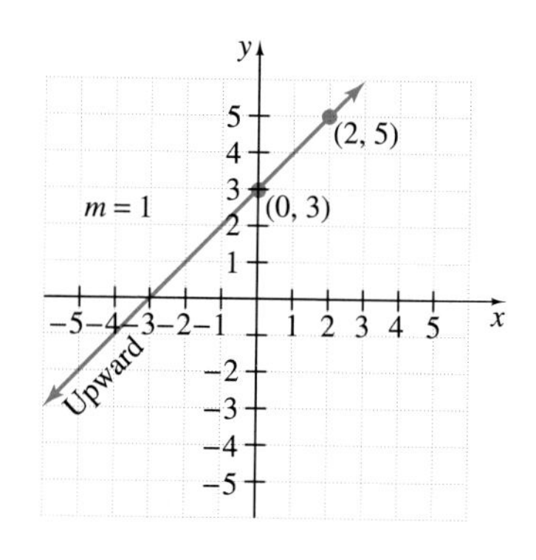

Notice in this example that the slope is positive and that the graph of the line containing $(0, 3)$ and $(2, 5)$ moves upward, or increases, as we go from left to right.

HELPFUL HINT

When we are trying to find the slope of a line through two given points, it makes no difference which given point is called (x_1, y_1) and which is called (x_2, y_2). Once an x-coordinate is called x_1, however, make sure its corresponding y-coordinate is called y_1.

Answers

1. $m = \frac{7}{3}$,

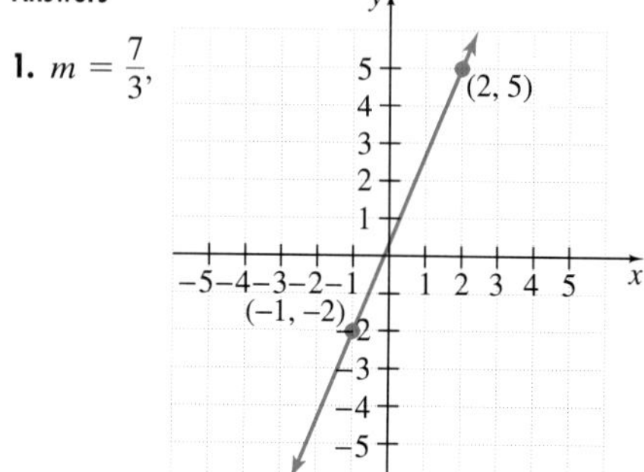

✓ Concept Check: $x_2 \neq x_1$, so that the denominator is never 0.

Example 2 Find the slope of the line containing the points $(5, -4)$ and $(-3, 3)$. Graph the line.

Solution: We use the slope formula, and let $(x_1, y_1) = (5, -4)$ and $(x_2, y_2) = (-3, 3)$.

$$m = \frac{y_2 - y_1}{x_2 - x_1}$$

$$= \frac{3 - (-4)}{-3 - 5} = \frac{7}{-8} = -\frac{7}{8}$$

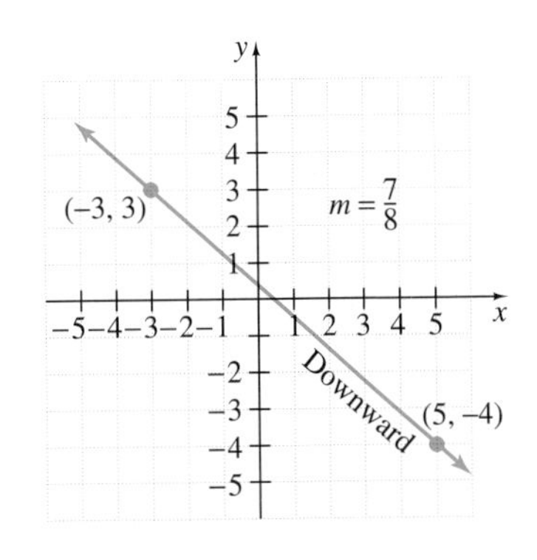

Notice in this example that the slope is negative and that the graph of the line through $(5, -4)$ and $(-3, 3)$ moves downward, or decreases, as we go from left to right.

TRY THE CONCEPT CHECK IN THE MARGIN.

B FINDING SLOPE FROM AN EQUATION

As we have seen, the slope of a line is defined by two points on the line. Thus, if we know the equation of a line, we can find its slope.

Example 3 Find the slope of the line $y = 3x + 2$.

Solution: Two points are needed on the line defined by $y = 3x + 2$ to find its slope. We will let $x = 0$, then $x = 1$ to find the required points.

If $x = 0$, then	If $x = 1$, then
$y = 3x + 2$	$y = 3x + 2$
$y = 3 \cdot 0 + 2$	$y = 3 \cdot 1 + 2$
$y = 2$	$y = 5$

Now we use the points $(0, 2)$ and $(1, 5)$ to find the slope. We'll let (x_1, y_1) be $(0, 2)$ and (x_2, y_2) be $(1, 5)$. Then

$$m = \frac{y_2 - y_1}{x_2 - x_1} = \frac{5 - 2}{1 - 0} = \frac{3}{1} = 3$$

Analyzing the results of Example 3, you may notice a striking pattern:

The slope of $y = 3x + 2$ is 3, the same as the coefficient of x.
The y-intercept is 2, the same as the constant term.

When a linear equation is written in the form $y = mx + b$, m is the slope of the line and b is its y-intercept. The form $y = mx + b$ is appropriately called the **slope-intercept form**.

Practice Problem 2

Find the slope of the line containing the points $(1, -1)$ and $(-2, 4)$. Graph the line.

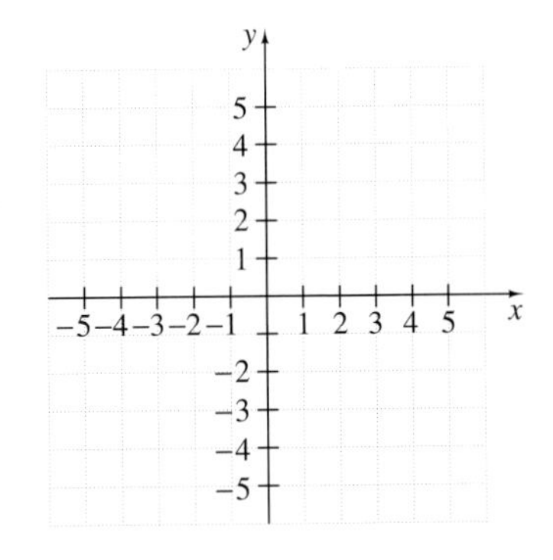

✓ CONCEPT CHECK

Find and correct the error in the following calculation of slope of the line containing the points $(12, 2)$ and $(4, 7)$.

$$m = \frac{12 - 4}{2 - 7} = \frac{8}{-5} = -\frac{8}{5}$$

Practice Problem 3

Find the slope of the line $y = 2x + 4$.

Answers

2. $m = -\frac{5}{3}$,

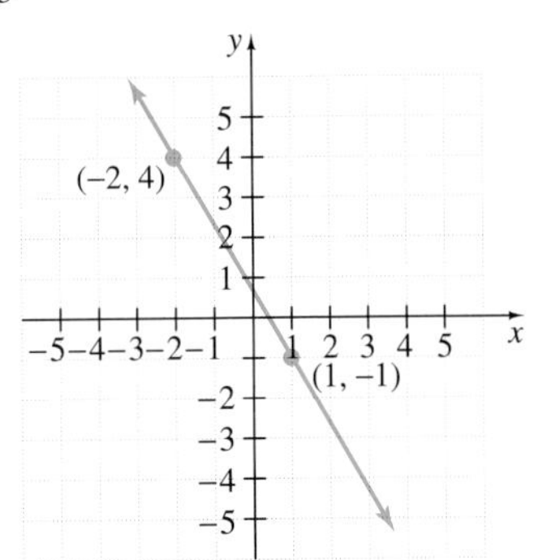

3. $m = 2$

✓ **Concept Check:** $m = \frac{2 - 7}{12 - 4} = \frac{-5}{8} = -\frac{5}{8}$

SLOPE-INTERCEPT FORM

When a linear equation in two variables is written in slope-intercept form,

$$\underset{\text{slope}}{y = \overset{\downarrow}{m}x + \overset{\downarrow}{b}}$$

(slope ↓ m, y-intercept ↓ b)

then m is the slope of the line and b is the y-intercept of the line.

Practice Problem 4

Find the slope and the y-intercept of the line $2x - 4y = 8$.

Example 4

Find the slope and the y-intercept of the line $3x - 4y = 4$.

Solution: We write the equation in slope-intercept form by solving for y.

$$3x - 4y = 4$$

$$-4y = -3x + 4 \quad \text{Subtract } 3x \text{ from both sides.}$$

$$\frac{-4y}{-4} = \frac{-3x}{-4} + \frac{4}{-4} \quad \text{Divide both sides by } -4.$$

$$y = \frac{3}{4}x - 1 \quad \text{Simplify.}$$

The coefficient of x, $\frac{3}{4}$, is the slope, and the constant term, -1, is the y-intercept.

The graphs of $y = \frac{1}{2}x + 1$ and $y = 5x + 1$ are shown below. Recall that the graph of $y = \frac{1}{2}x + 1$ has a slope of $\frac{1}{2}$ and that the graph of $y = 5x + 1$ has a slope of 5.

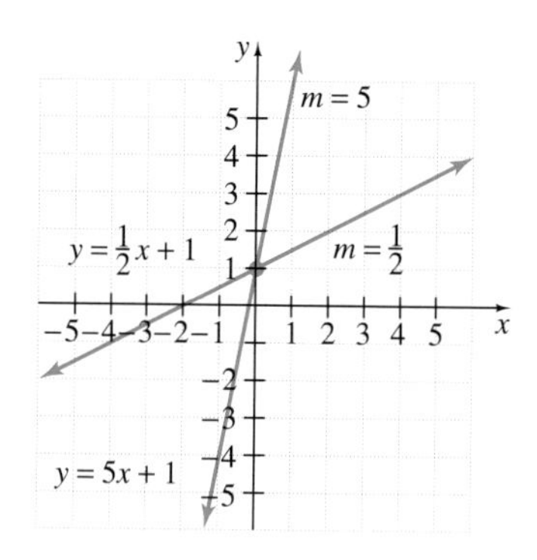

Notice that the line with the slope of 5 is steeper than the line with the slope of $\frac{1}{2}$. This is true in general for positive slopes.

For a line with positive slope m, as m increases, the line becomes steeper.

C FINDING SLOPES OF HORIZONTAL AND VERTICAL LINES

Next we find the slopes of two special types of lines: vertical lines and horizontal lines.

Answer

4. slope: $\frac{1}{2}$; y-intercept: -2

Example 5 Find the slope of the line $x = -5$.

Solution: Recall that the graph of $x = -5$ is a vertical line with x-intercept -5. To find the slope, we find two ordered pair solutions of $x = -5$. Of course, solutions of $x = -5$ must have an x-value of -5. We will let $(x_1, y_1) = (-5, 0)$ and $(x_2, y_2) = (-5, 4)$. Then

$$m = \frac{y_2 - y_1}{x_2 - x_1} = \frac{4 - 0}{-5 - (-5)} = \frac{4}{0}$$

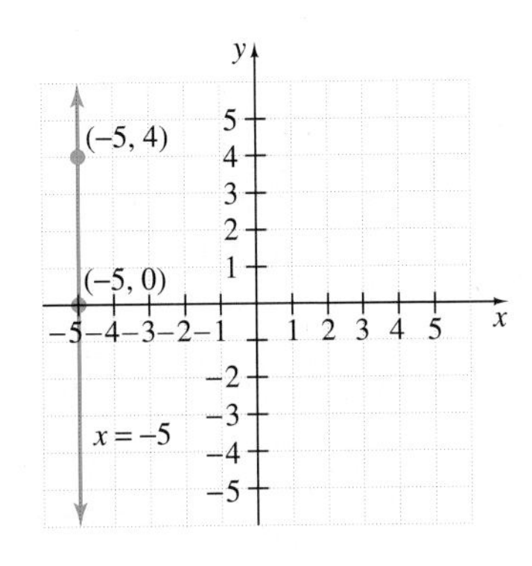

Since $\frac{4}{0}$ is undefined, we say that the slope of the vertical line $x = -5$ is undefined.

Practice Problem 5

Find the slope of the line $x = 3$.

Example 6 Find the slope of the line $y = 2$.

Solution: Recall that the graph of $y = 2$ is a horizontal line with y-intercept 2. To find the slope, we find two points on the line, such as $(0, 2)$ and $(1, 2)$, and use these points to find the slope.

$$m = \frac{2 - 2}{1 - 0} = \frac{0}{1} = 0$$

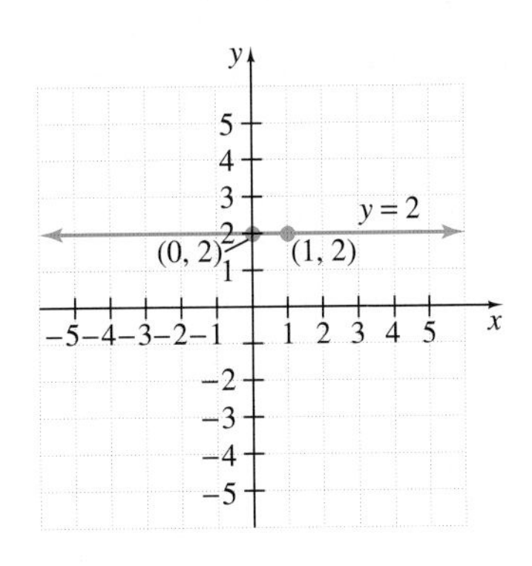

The slope of the horizontal line $y = 2$ is 0.

Practice Problem 6

Find the slope of the line $y = -3$.

From the above examples, we have the following generalization.

> The slope of any vertical line is undefined.
> The slope of any horizontal line is 0.

HELPFUL HINT

Slope of 0 and undefined slope are not the same. Vertical lines have undefined slope, whereas horizontal lines have slope of 0.

Answers

5. undefined, **6.** 0

The following four graphs summarize the overall appearance of lines with positive, negative, zero, or undefined slopes.

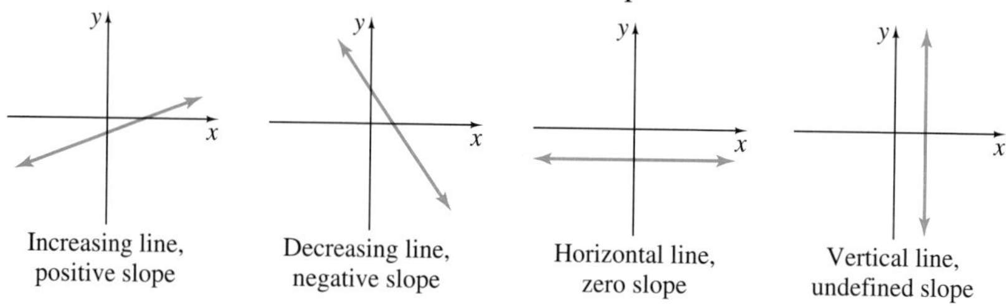

D COMPARING SLOPES OF PARALLEL AND PERPENDICULAR LINES

Slopes of lines can help us determine whether lines are parallel. Parallel lines are distinct lines with the same steepness, so it follows that they have the same slope.

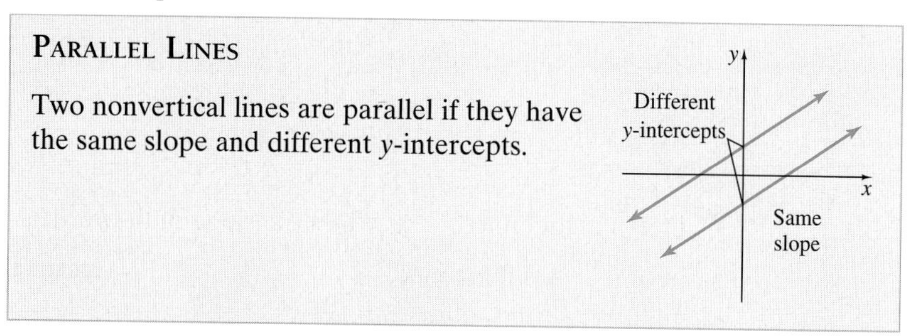

PARALLEL LINES

Two nonvertical lines are parallel if they have the same slope and different y-intercepts.

How do the slopes of perpendicular lines compare? (Two lines intersecting at right angles are called **perpendicular lines**.) Suppose that a line has a slope of $\frac{a}{b}$. If the line is rotated 90°, the rise and run are now switched, except that the run is now negative. This means that the new slope is $-\frac{b}{a}$. Notice that

$$\left(\frac{a}{b}\right) \cdot \left(-\frac{b}{a}\right) = -1$$

This is how we tell whether two lines are perpendicular.

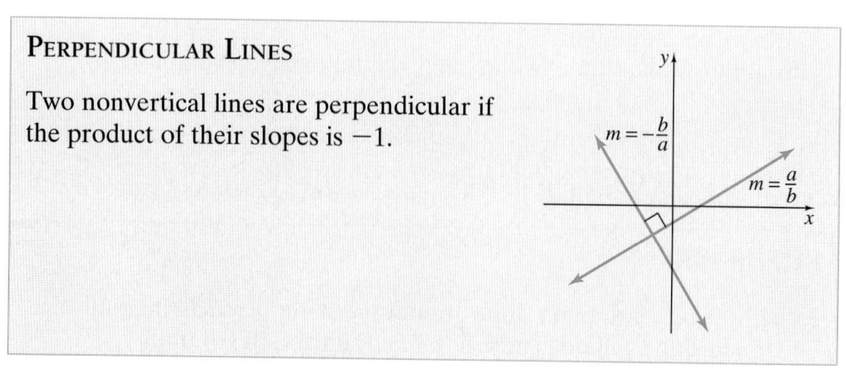

PERPENDICULAR LINES

Two nonvertical lines are perpendicular if the product of their slopes is -1.

In other words, two nonvertical lines are perpendicular if the slope of one is the negative reciprocal of the slope of the other.

Example 7

Determine whether the two lines are parallel, perpendicular, or neither.

$$3x + 7y = 4$$
$$6x + 14y = 7$$

Solution: We find the slope of each line by solving each equation for y.

$$3x + 7y = 4 \qquad\qquad 6x + 14y = 7$$
$$7y = -3x + 4 \qquad\qquad 14y = -6x + 7$$
$$\frac{7y}{7} = \frac{-3x}{7} + \frac{4}{7} \qquad\qquad \frac{14y}{14} = \frac{-6x}{14} + \frac{7}{14}$$
$$y = -\frac{3}{7}x + \frac{4}{7} \qquad\qquad y = -\frac{3}{7}x + \frac{1}{2}$$

↑ slope ↖ y-intercept ↑ slope ↑ y-intercept

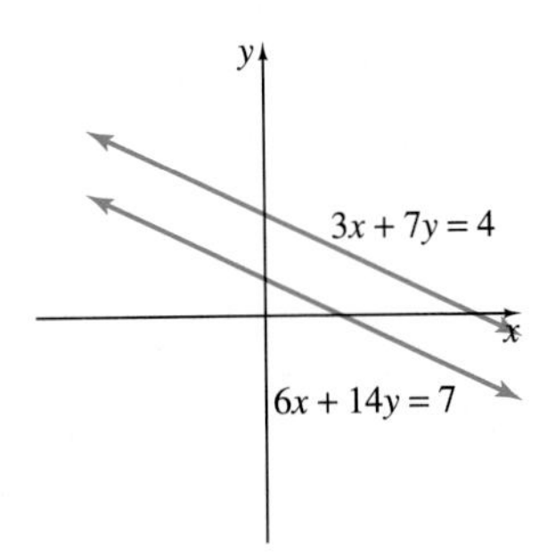

The slopes of both lines are $-\frac{3}{7}$. The y-intercepts are different. Therefore, the lines are parallel.

Practice Problem 7

Determine whether the two lines are parallel, perpendicular, or neither.

$$2x + 5y = 1$$
$$4x + 10y = 3$$

Example 8

Determine whether the two lines are parallel, perpendicular, or neither.

$$-x + 3y = 2$$
$$2x + 6y = 5$$

Solution: When we solve each equation for y, we have:

$$-x + 3y = 2 \qquad\qquad 2x + 6y = 5$$
$$3y = x + 2 \qquad\qquad 6y = -2x + 5$$
$$\frac{3y}{3} = \frac{x}{3} + \frac{2}{3} \qquad\qquad \frac{6y}{6} = \frac{-2x}{6} + \frac{5}{6}$$
$$y = \frac{1}{3}x + \frac{2}{3} \qquad\qquad y = -\frac{1}{3}x + \frac{5}{6}$$

↑ slope ↖ y-intercept ↑ slope ↖ y-intercept

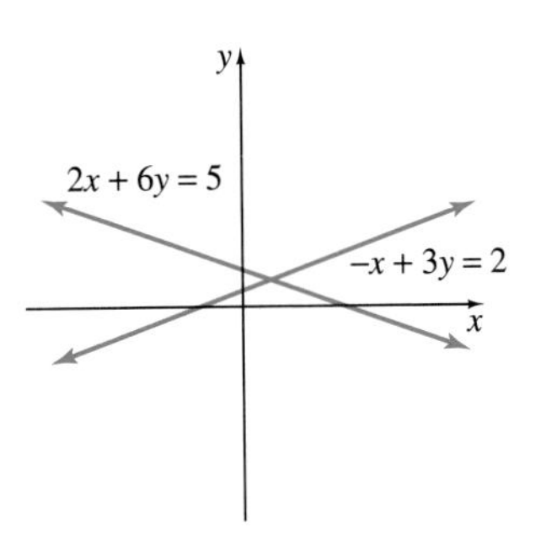

Practice Problem 8

Determine whether the two lines are parallel, perpendicular, or neither.

$$x - 4y = 3$$
$$3x + 12y = 7$$

Answers

7. The lines are parallel., **8.** The lines are neither parallel nor perpendicular.

✓ CONCEPT CHECK

What is *different* about the equations of two parallel lines?

Answer

✓ Concept Check: The y-intercepts are different.

The slopes are not the same and their product is not -1. $\left[\left(\frac{1}{3}\right)\cdot\left(-\frac{1}{3}\right) = -\frac{1}{9}\right]$ Therefore, the lines are neither parallel nor perpendicular.

TRY THE CONCEPT CHECK IN THE MARGIN.

GRAPHING CALCULATOR EXPLORATIONS

It is possible to use a grapher to sketch the graph of more than one equation on the same set of axes. For example, let's graph the equations $y = 2x - 3$ and $y = 2x + 5$ on the same set of axes.

To graph on the same set of axes, press the [Y=] key and enter the equations on the first two lines.

$$Y_1 = 2x - 3$$
$$Y_2 = 2x + 5$$

Then press the [GRAPH] key as usual. The screen should look like this:

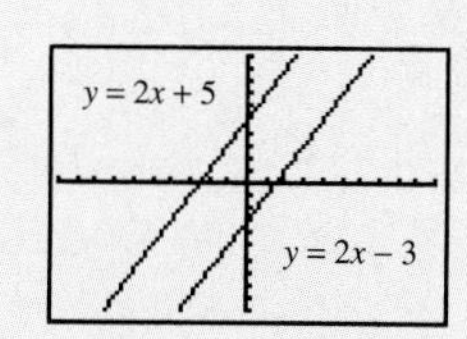

Notice the slopes and y-intercepts of the graphs. Since their slopes are the same and they have different y-intercepts, we have parallel lines, as shown.

Graph each pair of equations on the same set of axes. Describe the similarities and differences in their graphs.

1. $y = 3x,\ y = 3x + 4$

2. $y = 5x,\ y = 5x - 2$

3. $y = -\frac{2}{3}x + 1,\ y = -\frac{2}{3} + 6$

4. $y = -\frac{1}{4}x - 3,\ y = -\frac{1}{4}x + 6$

5. $y = 4.61x - 1.86,\ y = 4.61x + 2.11$

6. $y = 3.78x + 1.92,\ y = 3.78x + 8.08$

Name ______________________ Section ______ Date ________

MENTAL MATH ANSWERS

1. ______
2. ______
3. ______
4. ______

ANSWERS

1. ______
2. ______
3. ______
4. ______
5. ______
6. ______
7. ______
8. ______
9. ______
10. ______
11. ______
12. ______
13. ______
14. ______
15. ______
16. ______
17. ______
18. ______
19. ______
20. ______
21. ______
22. ______

MENTAL MATH

Determine whether a line with the given slope slants upward, downward, horizontally, or vertically from left to right.

1. $m = \frac{7}{6}$ **2.** $m = -3$ **3.** $m = 0$ **4.** m is undefined.

EXERCISE SET 3.2

A *Find the slope of the line containing each pair of points. See Examples 1 and 2.*

1. $(3, 2), (8, 11)$ **2.** $(1, 6), (7, 11)$ **3.** $(3, 1), (1, 8)$

4. $(2, 9), (6, 4)$ **5.** $(-2, 8), (4, 3)$ **6.** $(3, 7), (-2, 11)$

7. $(-2, -6), (4, -4)$ **8.** $(-3, -4), (-1, 6)$ **9.** $(-3, -1), (-12, 11)$

10. $(3, -1), (-6, 5)$ **11.** $(-2, 5), (3, 5)$ **12.** $(4, 2), (4, 0)$

Find each slope. See Examples 1 and 2.

13. Find the pitch, or slope, of the roof shown.

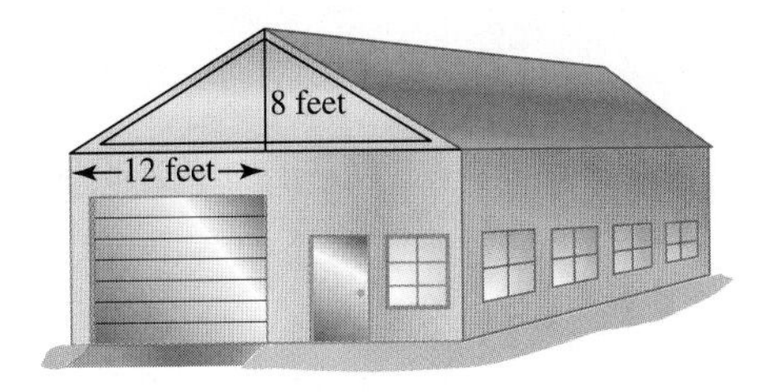

14. Upon takeoff, a Delta Airlines jet climbs to 3 miles as it passes over 25 miles of land below it. Find the slope of its climb.

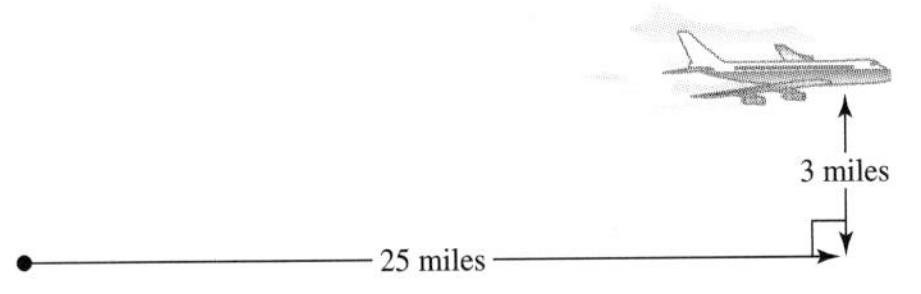

15. Find the grade, or slope, of the road shown.

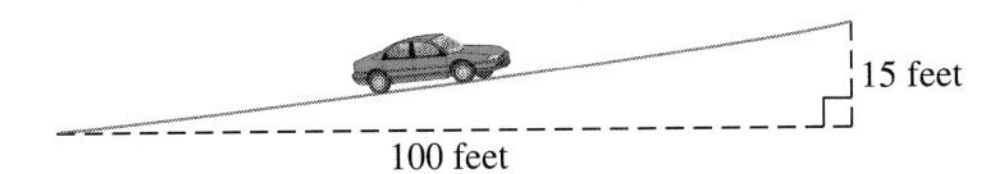

16. Driving down Bald Mountain in Wyoming, Bob Dean finds that he descends 1600 feet in elevation by the time he is 2.5 miles (horizontally) away from the high point on the mountain road. Find the slope of his descent. (*Hint:* 1 mile = 5280 feet.)

B *Find the slope and the y-intercept of each line. See Examples 3 and 4.*

17. $y = 5x - 2$ **18.** $y = -2x + 6$ **19.** $2x + y = 7$

20. $-5x + y = 10$ **21.** $2x - 3y = 10$ **22.** $-3x - 4y = 6$

23.
24.
25.
26.
27.
28.
29.
30.
31.
32.
33.
34.
35.
36.
37.
38.
39.
40.
41.
42.
43.
44.
45.
46.
47.
48.

Name ______________________

23. $y = \frac{1}{2}x$ **24.** $y = -\frac{1}{4}x$ **25.** $3x + 9 = y$ **26.** $2y - 7 = x$

Match each graph with its equation.

A

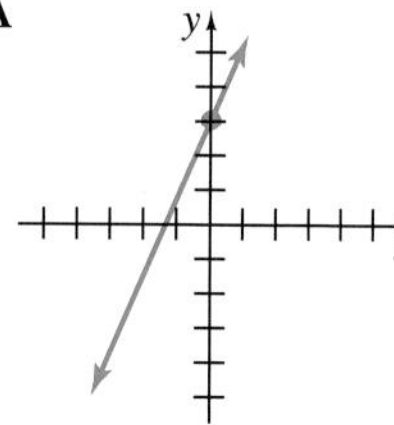

B

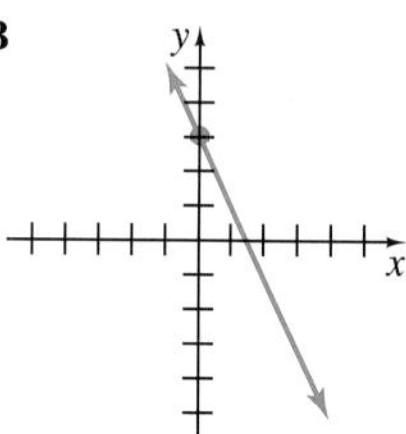

C

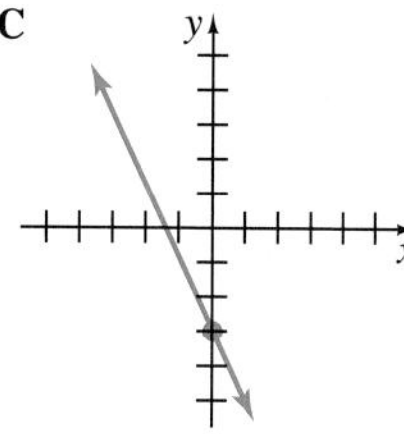

D

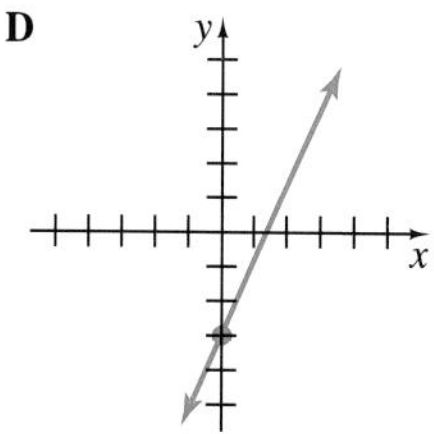

27. $y = 2x + 3$ **28.** $y = 2x - 3$

29. $y = -2x + 3$ **30.** $y = -2x - 3$

B C *Find the slope of each line. See Examples 3 through 6.*

31. $x = 1$ **32.** $y = -2$ **33.** $y = -x + 5$

34. $y = x + 2$ **35.** $-6x + 5y = 30$ **36.** $4x - 7y = 28$

37. $x = 4$ **38.** $y = -3$ **39.** $y = 7x$

40. $y = \frac{1}{7}x$ **41.** $x + 2 = 0$ **42.** $y - 7 = 0$

Two lines are graphed on each set of axes. Determine whether l_1 or l_2 has the greater slope on each graph.

43.

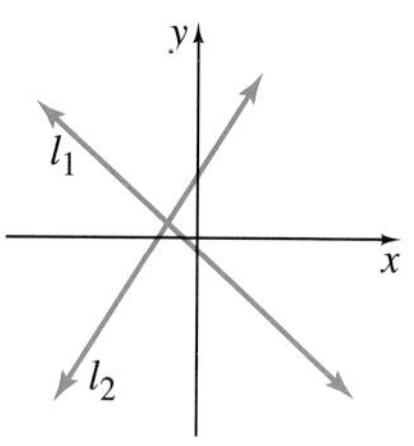

44.

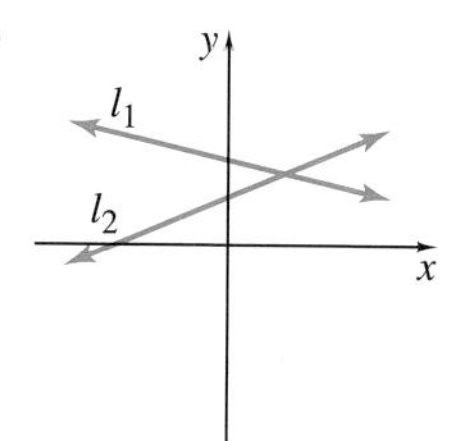

45.

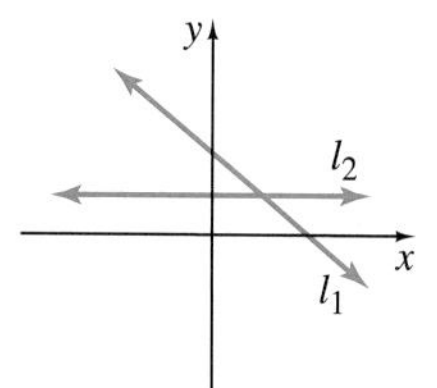

46.

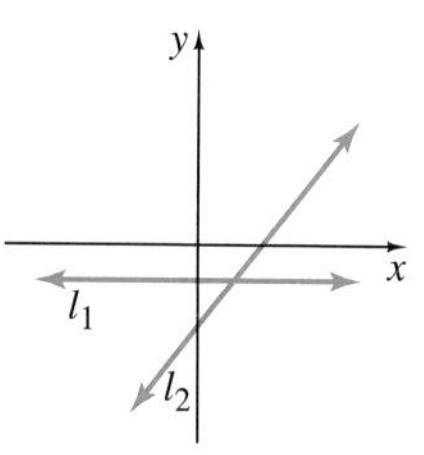

47.

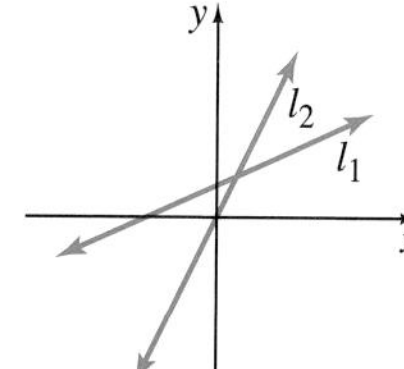

48.

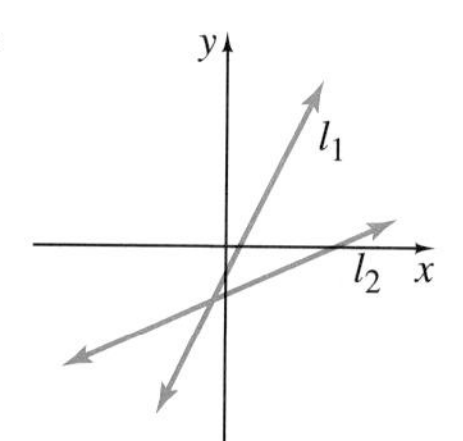

Name ______________________________

D *Determine whether each pair of lines is parallel, perpendicular, or neither. See Examples 7 and 8.*

49. $y = -3x + 6$
$y = 3x + 5$

50. $y = 5x - 6$
$y = 5x + 2$

51. $-4x + 2y = 5$
$2x - y = 7$

52. $2x - y = -10$
$2x + 4y = 2$

53. $-2x + 3y = 1$
$3x + 2y = 12$

54. $x + 4y = 7$
$2x - 5y = 0$

55. $y = -9x + 3$
$y = \frac{3}{2}x - 7$

56. $y = 2x - 12$
$y = \frac{1}{2}x - 6$

57. $y = 12x + 6$
$y = 12x - 2$

58. $y = -5x + 8$
$y = -5x - 8$

59. Find the slope of a line parallel to the line $y = -\frac{7}{2}x - 6$.

60. Find the slope of a line perpendicular to the line $y = -\frac{7}{2}x - 6$.

61. Find the slope of a line parallel to the line $5x - 2y = 6$.

62. Find the slope of a line parallel to the line $-3x + 4y = 10$.

Review and Preview

Solve. See Section 2.6.

63. $|x - 3| = 6$

64. $|x + 2| < 4$

65. $|2x + 5| > 3$

66. $|5x| = 10$

67. $|3x - 4| \leq 2$

68. $|7x - 2| \geq 5$

Combining Concepts

69. Each line below has negative slope.

a. Find the slope of each line.

b. Use the result of part (a) to fill in the blank: For lines with negative slopes, the steeper line has the ______ (greater/lesser) slope.

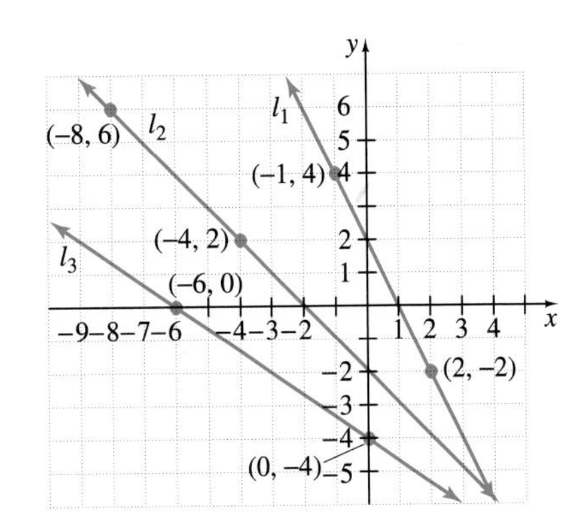

49. ______
50. ______
51. ______
52. ______
53. ______
54. ______
55. ______
56. ______
57. ______
58. ______
59. ______
60. ______
61. ______
62. ______
63. ______
64. ______
65. ______
66. ______
67. ______
68. ______
69. a. ______
b. ______

70.

71.

72.

73.

74.

75.

76.

see calculator
77. a. screen

see calculator
b. screen

c.

Name ______________________

The following graph shows the altitude of a seagull in flight over a time period of 30 seconds. Use this graph to answer Exercises 70–73.

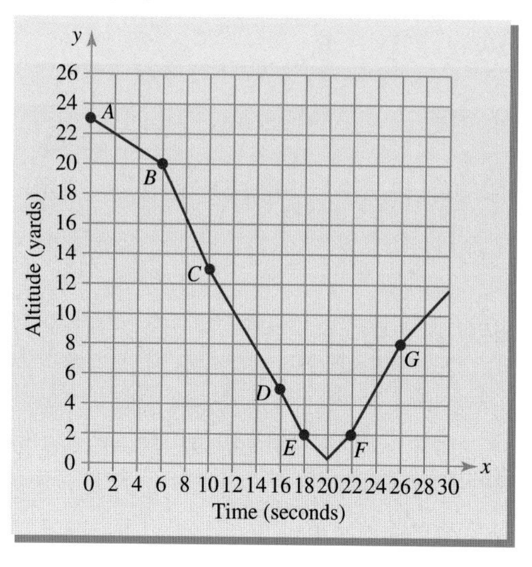

70. Find the coordinates of point B.

71. Find the coordinates of point C.

72. Find the rate of change of altitude between points B and C. (Recall that the rate of change between points is the slope between points. This rate of change will be in yards per second.)

73. Find the rate of change of altitude (in yards per second) between points F and G.

74. Professional plumbers suggest that a sewer pipe should be sloped 0.25 inch for every foot. Find the recommended slope for a sewer pipe. (*Source: Rules of Thumb* by Tom Parker, 1983, Houghton Mifflin Company)

75. Explain whether two lines, both with positive slopes, can be perpendicular.

76. Explain how merely looking at a line can tell us whether its slope is negative, positive, undefined, or zero.

77. a. On a single screen of a graphing calculator, graph $y = \frac{1}{2}x + 1$, $y = x + 1$, and $y = 2x + 1$. Notice the change in slope for each graph.

b. On a single screen of a graphing calculator, graph $y = -\frac{1}{2}x + 1$, $y = -x + 1$, and $y = -2x + 1$. Notice the change in slope for each graph.

c. Determine whether the following statement is true or false for slope m of a given line. As $|m|$ becomes greater, the line becomes steeper.

3.3 THE SLOPE-INTERCEPT FORM

A GRAPHING A LINE USING SLOPE AND y-INTERCEPT

In the last section, we learned that the slope-intercept form of a linear equation is $y = mx + b$. When an equation is written in this form, the slope of the line is the same as the coefficient m of x. Also, the y-intercept of the line is the same as the constant term b. For example, the slope of the line defined by $y = 2x + 3$ is 2 and its y-intercept is 3.

We may also use the slope-intercept form to graph a linear equation.

Example 1 Graph: $y = \frac{1}{4}x - 3$

Solution: Recall that the slope of the graph of $y = \frac{1}{4}x - 3$ is $\frac{1}{4}$ and the y-intercept is -3. To graph the line, we first plot the y-intercept point $(0, -3)$. To find another point on the line, we recall that slope is $\frac{\text{rise}}{\text{run}} = \frac{1}{4}$. Another point may then be plotted by starting at $(0, -3)$, rising 1 unit up, and then running 4 units to the right. We are now at the point $(4, -2)$. The graph of $y = \frac{1}{4}x - 3$ is the line through points $(0, -3)$ and $(4, -2)$, as shown.

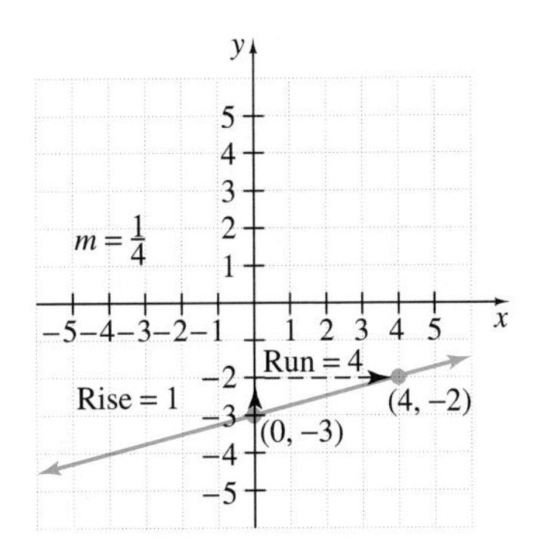

Example 2 Graph: $2x + y = 3$

Solution: First, we solve the equation for y to write it in slope-intercept form. In slope-intercept form, the equation is $y = -2x + 3$. Next we plot the y-intercept point, $(0, 3)$. To find another point on the line, we use the slope -2, which can be written as $\frac{\text{rise}}{\text{run}} = \frac{-2}{1}$. We start at $(0, 3)$ and move vertically 2 units down, since the numerator of the slope is -2; then we move horizontally 1 unit to the right since the denominator of the slope is 1. We arrive at the point $(1, 1)$. The line through $(1, 1)$ and $(0, 3)$ will have the required slope of -2.

Objectives

A Graph a line using its slope and y-intercept.

B Use the slope-intercept form to write an equation of the line.

C Interpret the slope-intercept form in an application.

Practice Problem 1

Graph: $y = \frac{2}{3}x + 1$

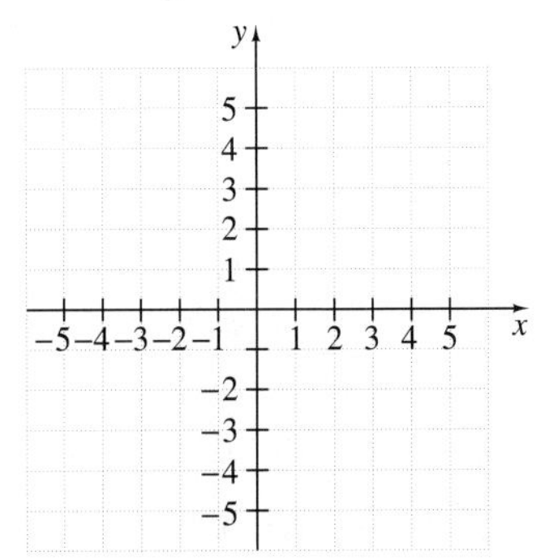

Practice Problem 2

Graph: $3x + y = -2$

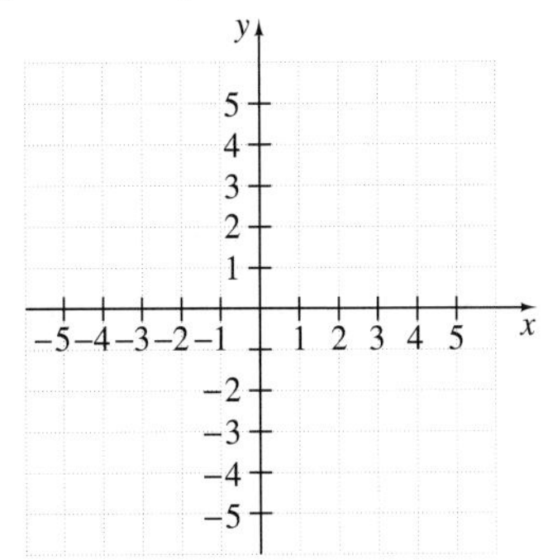

Answers

1.

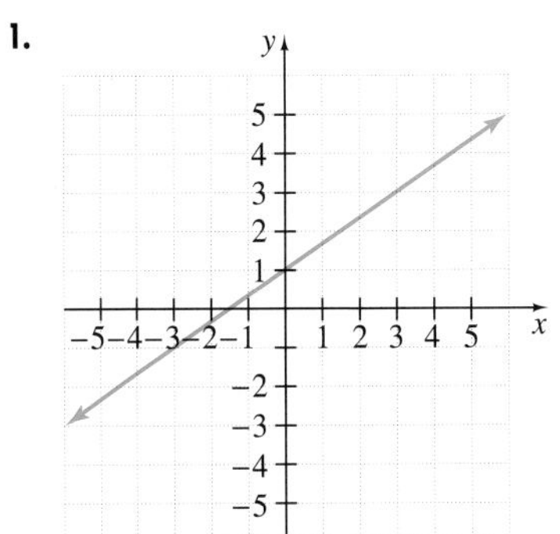

2. (answer on next page)

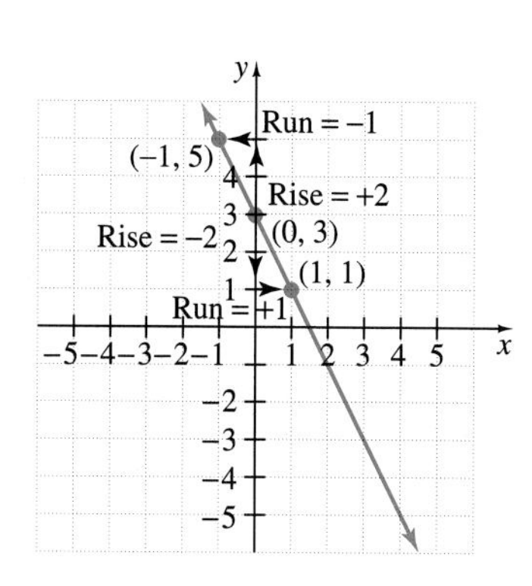

The slope -2 can also be written as $\frac{2}{-1}$, so to find another point for Example 2 we could start at $(0, 3)$ and move 2 units up and then 1 unit left. We would stop at the point $(-1, 5)$. The line through $(-1, 5)$ and $(0, 3)$ will have the required slope and will be the same line as shown previously through $(1, 1)$ and $(0, 3)$.

B USING THE SLOPE-INTERCEPT FORM TO WRITE AN EQUATION

Given the slope and y-intercept of a line, we may write its equation as well as graph the line.

Example 3 Write an equation of the line with y-intercept -3 and slope of $\frac{1}{4}$.

Solution: We are given the slope and the y-intercept. We let $m = \frac{1}{4}$ and $b = -3$, and write the equation in slope-intercept form, $y = mx + b$.

$y = mx + b$

$y = \frac{1}{4}x + (-3)$ Let $m = \frac{1}{4}$ and $b = -3$.

$y = \frac{1}{4}x - 3$ Simplify.

Notice that the graph of this equation has slope $\frac{1}{4}$ and y-intercept -3, as desired.

TRY THE CONCEPT CHECK IN THE MARGIN.

C INTERPRETING THE SLOPE-INTERCEPT FORM

Recall from Section 3.1 the graph of an adult one-day pass price for Disney World. Notice that the graph resembles the graph of a line. Often, businesses depend on equations that "closely fit" lines like this one to model the data and predict future trends. For example, by a method called least squares regression, the linear equation $y = 1.462x + 29.35$ approximates the data shown, where x is the number of years since 1988 and y is the ticket price for that year.

Practice Problem 3

Write an equation of the line with slope $\frac{2}{3}$ and y-intercept 1.

✓ CONCEPT CHECK

What is wrong with the following equation of a line with y-intercept 4 and slope 2?
$y = 4x + 2$

Answers

2.

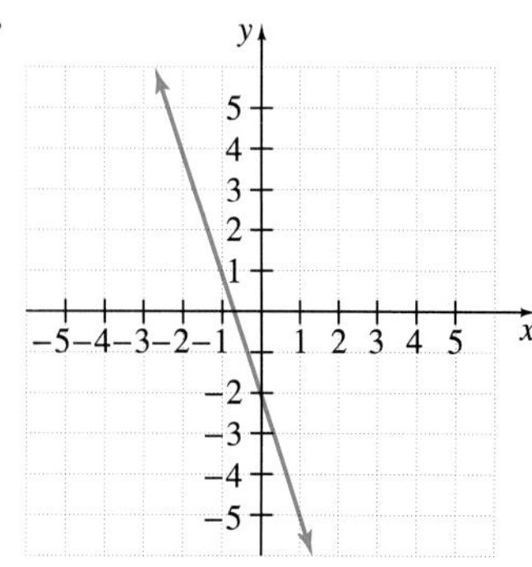

3. $y = \frac{2}{3}x + 1$

✓ Concept Check: The y-intercept and slope were switched. It should be $y = 2x + 4$.

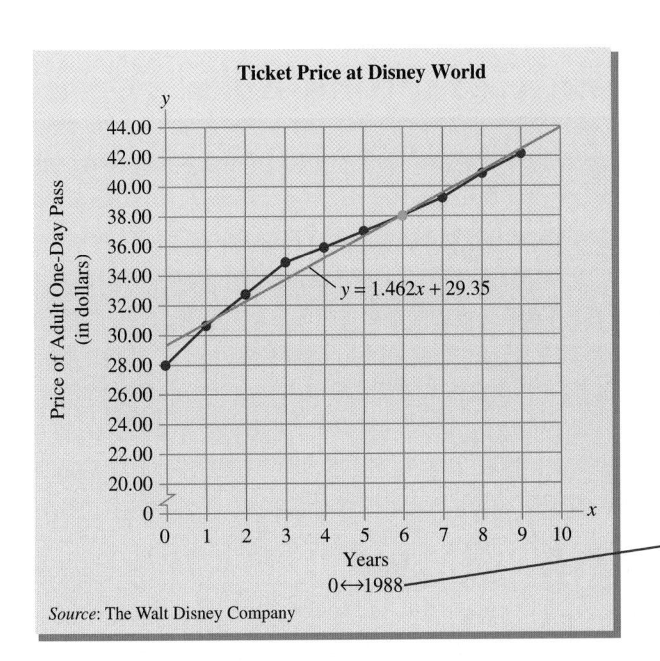

HELPFUL HINT

The notation $0 \leftrightarrow 1988$ means that the number 0 corresponds to the year 1988, 1 corresponds to the year 1989, and so on.

Example 4 Predicting Future Prices

The adult one-day pass price y for Disney World is given by

$y = 1.462x + 29.35$

where x is the number of years since 1988.

a. Use this equation to predict the ticket price for the year 2002.
b. What does the slope of this equation mean?
c. What does the y-intercept of this equation mean?

Solution:

a. To predict the price of a pass in 2002, we need to find y when x is 14. (Since year 1988 corresponds to $x = 0$, year 2002 corresponds to $x = 2002 - 1988 = 14$.

$$y = 1.462x + 29.35$$
$$= 1.462(14) + 29.35 \quad \text{Let } x = 14.$$
$$= 49.818$$

We predict that in the year 2002, the price of an adult one-day pass to Disney World will be about \$49.82.

b. The slope of $y = 1.462x + 29.35$ is 1.462. We can think of this number as $\frac{\text{rise}}{\text{run}}$ or $\frac{1.462}{1}$. This means that the ticket price increases on the average by \$1.462 every 1 year.

c. The y-intercept of $y = 1.462x + 29.35$ is 29.35. Notice that it corresponds to the point of the graph $(0, 29.35)$
↑ year ↖ price

This means that at year $x = 0$ or 1988, the ticket price was \$29.35.

Practice Problem 4

The yearly average income y of an American woman with some high school education but no diploma is given by the equation

$$y = 356.5x + 8912.2,$$

where x is the number of years since 1991. (*Source*: Based on data from the U.S. Bureau of the Census, 1991–1996)

a. Predict the income for the year 2001.
b. What does the slope of this equation mean?
c. What does the y-intercept of this equation mean?

Answers

4. a. \$12,477.20, **b.** The yearly average income increases by \$356.50 every year, **c.** At year $x = 0$, or 1991, the yearly average income was \$8912.20.

GRAPHING CALCULATOR EXPLORATIONS

You may have noticed by now that to use the [Y=] key on a grapher to graph an equation, the equation must be solved for y.

Graph each equation by first solving the equation for y.

1. $x = 3.5y$

2. $-2.7y = x$

3. $5.78x + 2.31y = 10.98$

4. $-7.22x + 3.89y = 12.57$

5. $y - x = 3.78$

6. $3y - 5x = 6x - 4$

7. $y - 5.6x = 7.7x + 1.5$

8. $y + 2.6x = -3.2$

Name ______________________ **Section** ________ **Date** ________

Mental Math

Find the slope and the y-intercept of each line.

1. $y = -4x + 12$ **2.** $y = \frac{2}{3}x - \frac{7}{2}$ **3.** $y = 5x$

4. $y = -x$ **5.** $y = \frac{1}{2}x + 6$ **6.** $y = -\frac{2}{3}x + 5$

Exercise Set 3.3

A *Graph each line passing through the given point with the given slope. See Examples 1 and 2.*

1. Through $(1, 3)$ with slope $\frac{3}{2}$

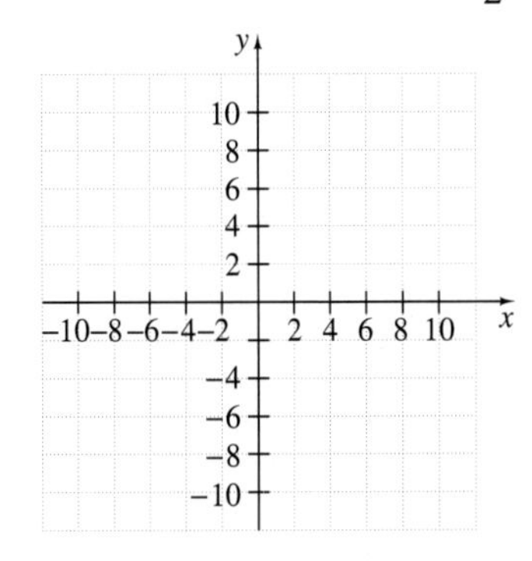

2. Through $(-2, -4)$ with slope $\frac{2}{5}$

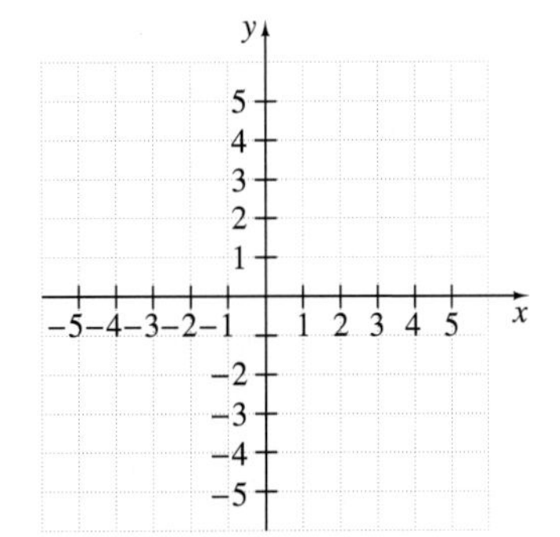

3. Through $(0, 0)$ with slope 5

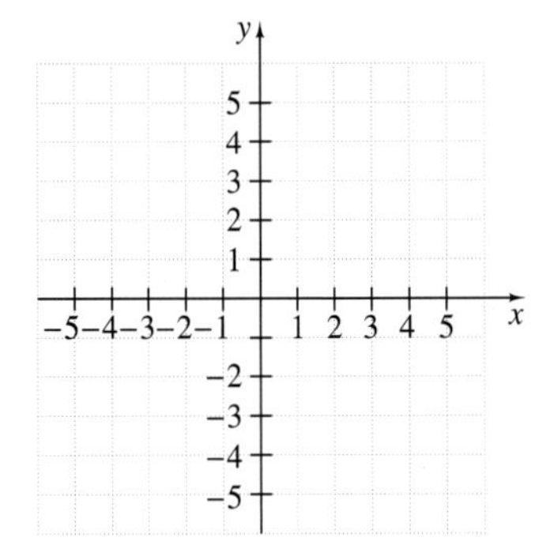

4. Through $(-5, 2)$ with slope 2

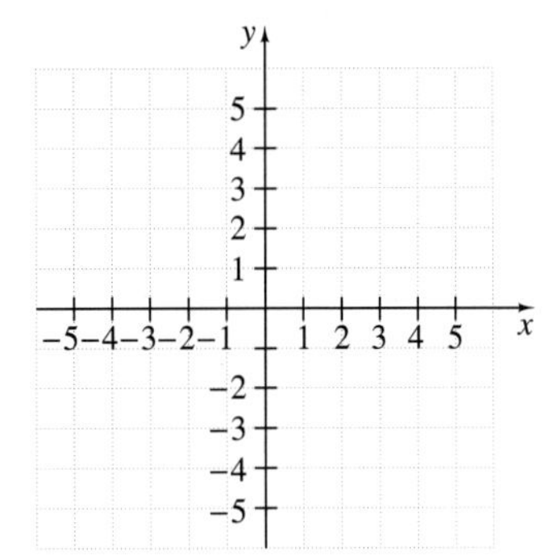

5. Through $(0, 7)$ with slope -1

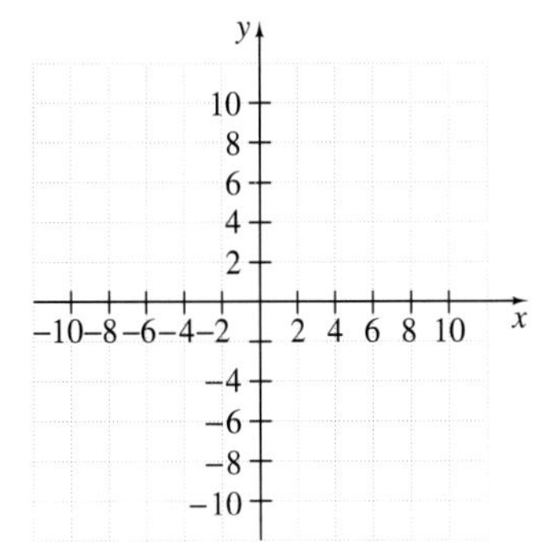

6. Through $(3, 0)$ with slope -3

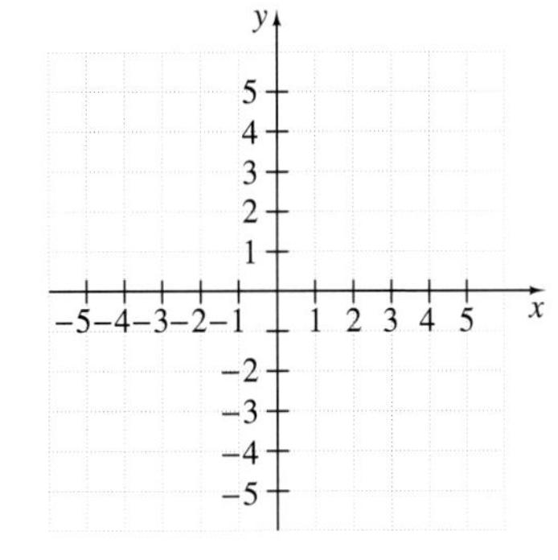

Mental Math Answers

1. ________
2. ________
3. ________
4. ________
5. ________
6. ________

Answers

1. see graph
2. see graph
3. see graph
4. see graph
5. see graph
6. see graph

Name ______________________________

Graph each linear equation using the slope and y-intercept. See Examples 1 and 2.

7. $y = -2x$

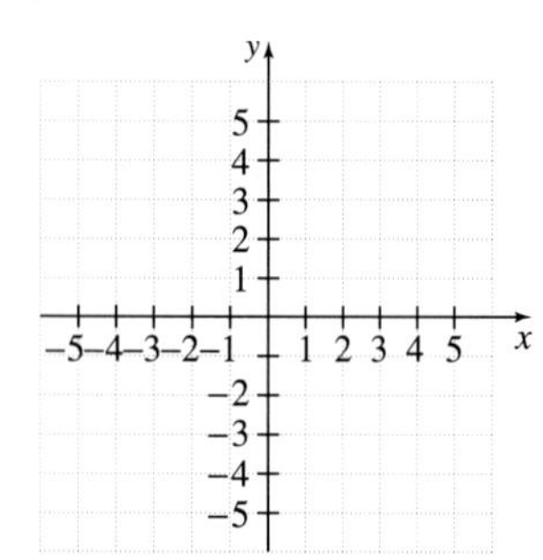

8. $y = 2x$

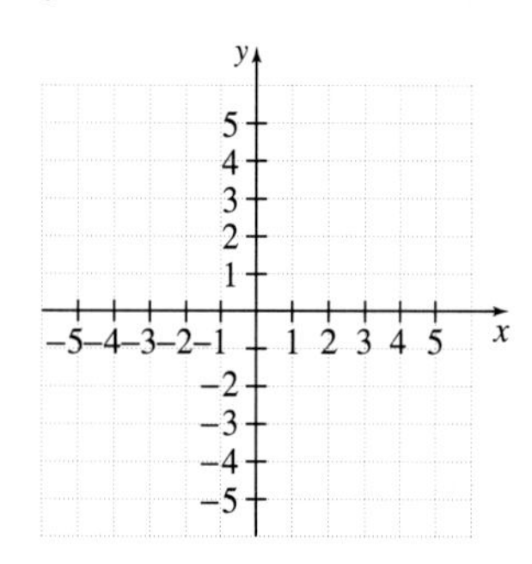

9. $y = -2x + 3$

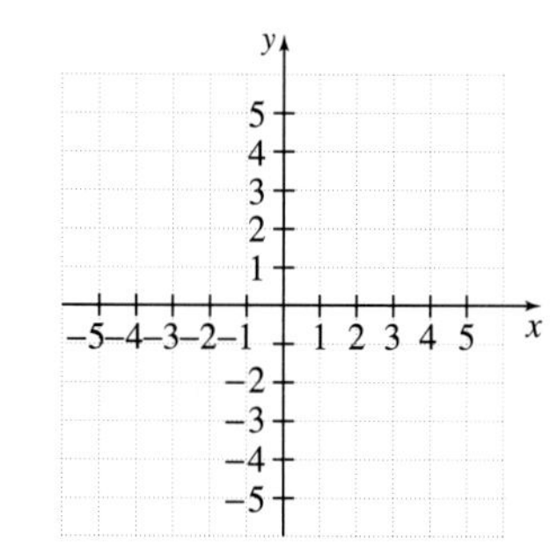

10. $y = 2x + 6$

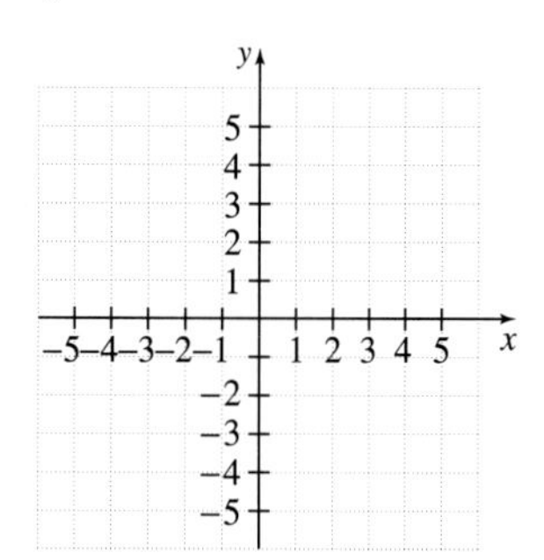

11. $y = \frac{1}{2}x$

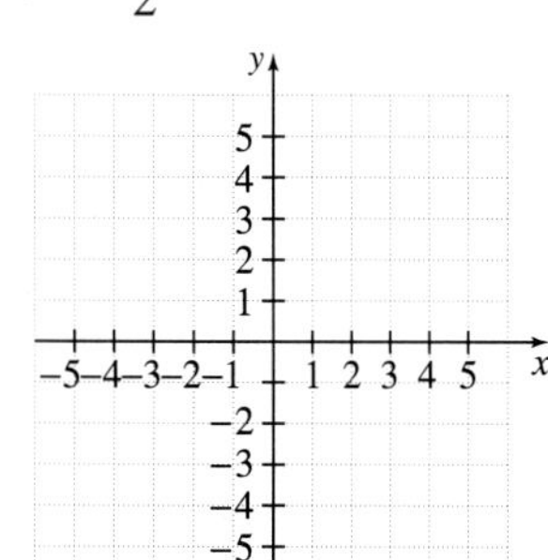

12. $y = \frac{1}{3}x$

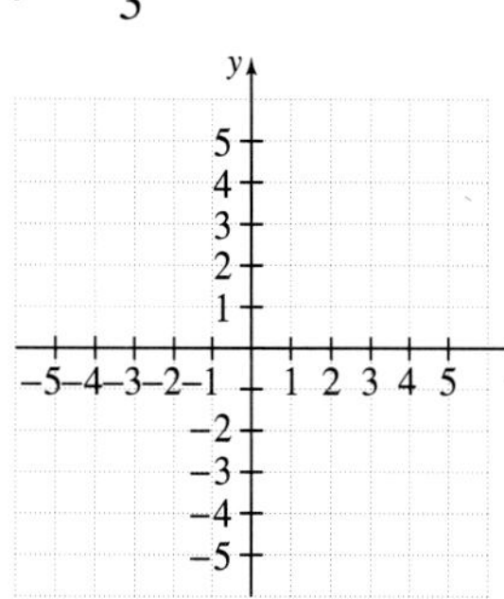

13. $y = \frac{1}{2}x - 4$

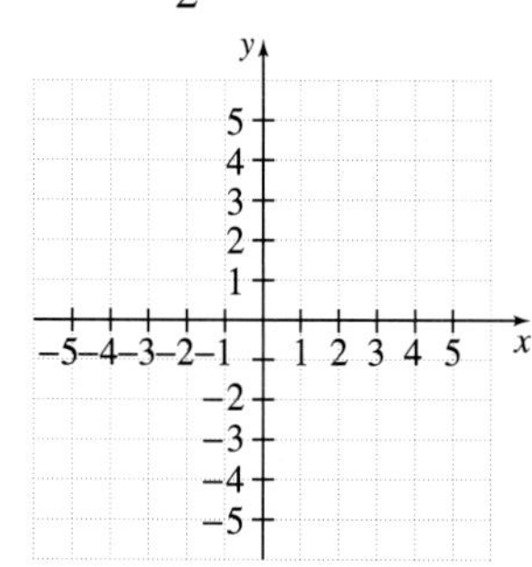

14. $y = \frac{1}{3}x - 2$

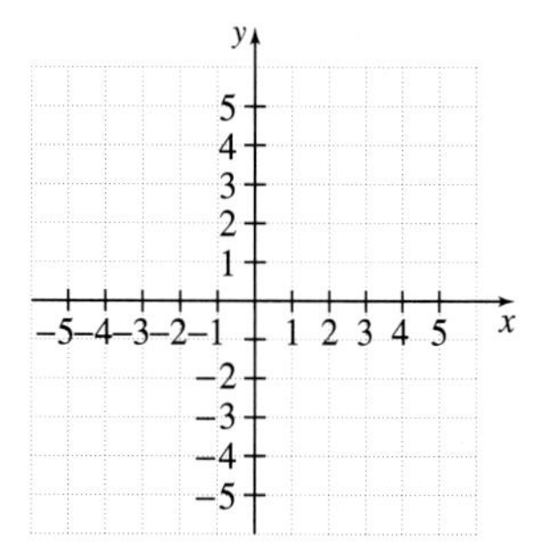

15. $x - y = 3$

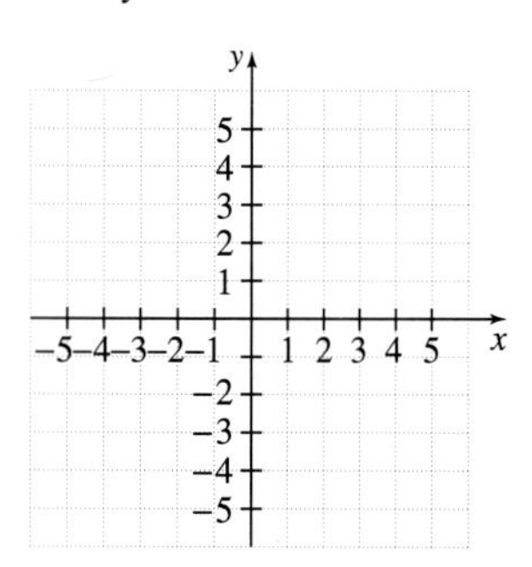

16. $x - y = -4$

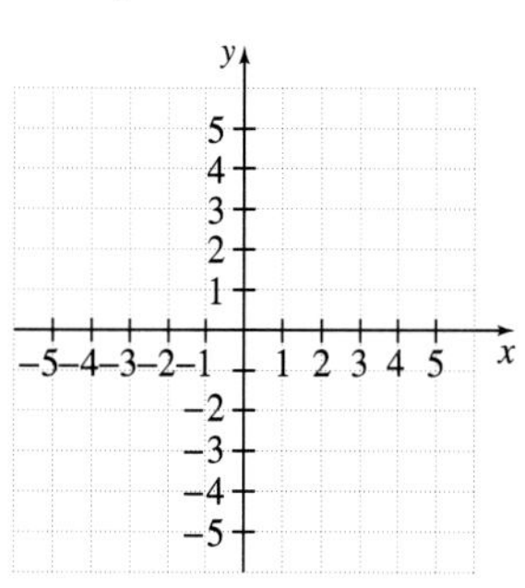

17. $x + 2y = 8$

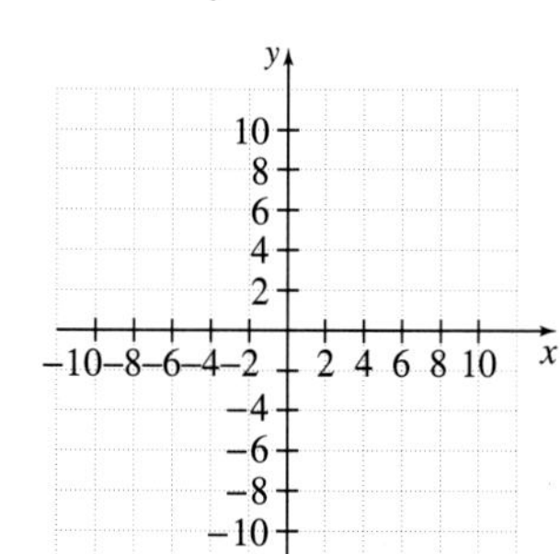

18. $x - 3y = 3$

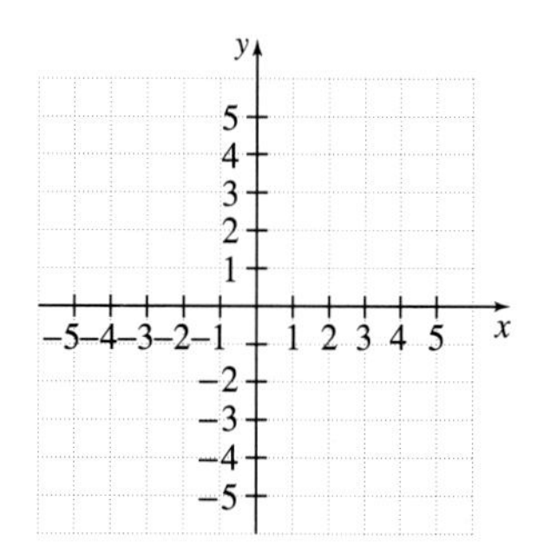

The graph of $y = 5x$ follows. Use this graph to match each linear equation with its graph. See Examples 1 and 2.

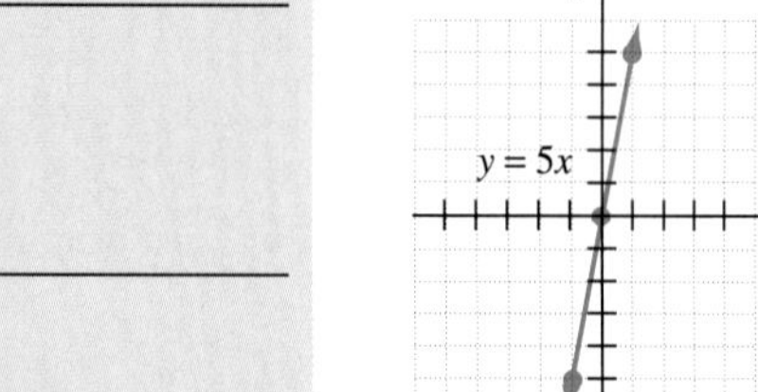

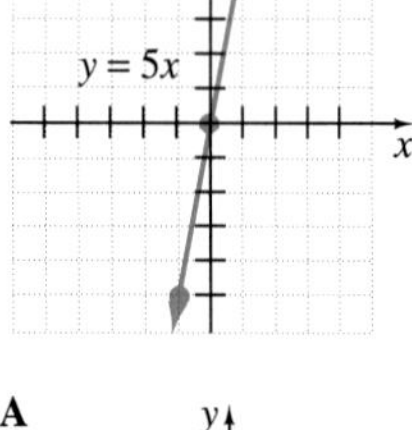

19. $y = 5x - 3$

20. $y = 5x - 2$

21. $y = 5x + 1$

22. $y = 5x + 3$

A

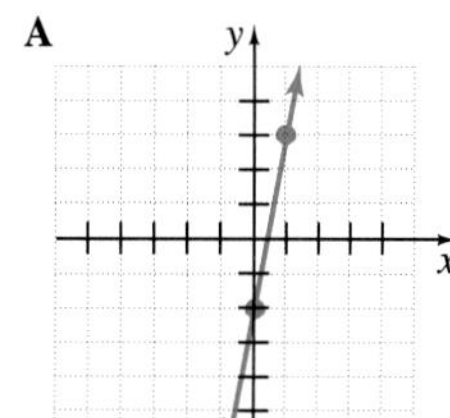

B **C**

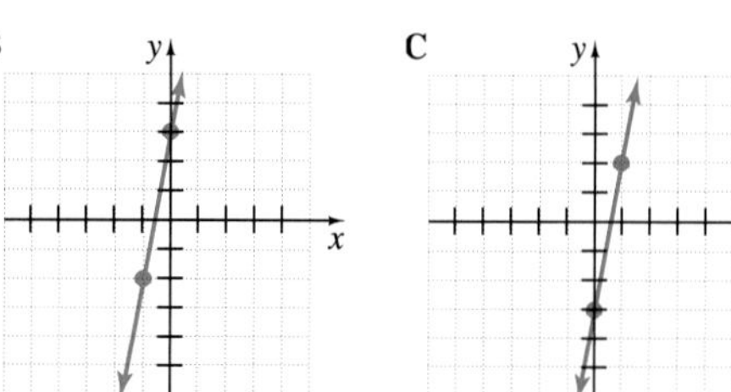

D

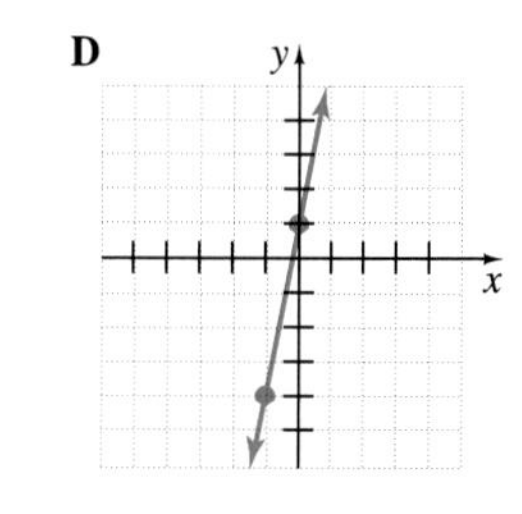

19. ______________

20. ______________

21. ______________

22. ______________

Name ______________________________

B *Use the slope-intercept form of a linear equation to write the equation of each line with the given slope and y-intercept. See Example 3.*

23. Slope -1; y-intercept 1

24. Slope $\frac{1}{2}$; y-intercept -6

25. Slope 2; y-intercept $\frac{3}{4}$

26. Slope -3; y-intercept $-\frac{1}{5}$

27. Slope $\frac{2}{7}$; y-intercept 0

28. Slope $-\frac{4}{5}$; y-intercept 0

C *Solve. See Example 4.*

29. The annual average income y of an American man with an associate's degree is given by the linear equation $y = 1431.5x + 31{,}775.2$ where x is the number of years after 1992. (*Source:* Based on data from the U.S. Bureau of the Census, 1992–1996)

a. Find the average income of an American man with an associate's degree in 1996.

b. Find and interpret the slope of the equation.

c. Find and interpret the y-intercept of the equation.

30. The annual income of an American woman with a bachelor's degree is given by the linear equation $y = 1054.7x + 23{,}285.9$ where x is the number of years after 1991. (*Source:* Based on data from the U.S. Bureau of the Census, 1991–1996)

a. Find the average income of an American woman with a bachelor's degree in 1996.

b. Find and interpret the slope of the equation.

c. Find and interpret the y-intercept of the equation.

31. One of the top ten occupations in terms of job growth in the next few years is expected to be home health aide. The number of people y in thousands employed as home health aides in the United States can be estimated by the linear equation

$378x - 10y = -4950,$

where x is the number of years after 1996. (*Source:* Based on projections from the U.S. Bureau of Labor Statistics, 1996–2006)

a. Find the slope and y-intercept of the linear equation.

b. What does the slope mean in this context?

c. What does the y-intercept mean in this context?

32. One of the faster growing occupations over the next few years is expected to be paralegal. The number of people y in thousands employed as paralegals in the United States can be estimated by the linear equation $-76x + 10y = 1130$, where x is the number of years after 1996. (*Source:* Based on projections from the U.S. Bureau of Labor Statistics, 1996–2006)

a. Find the slope and y-intercept of the linear equation.

b. What does the slope mean in this context?

c. What does the y-intercept mean in this context?

23. ________
24. ________
25. ________
26. ________
27. ________
28. ________
29. a. ________
b. ________
c. ________
30. a. ________
b. ________
c. ________
31. a. ________
b. ________
c. ________
32. a. ________
b. ________
c. ________

33. a. ____
b. ____
c. ____
34. a. ____
b. ____
c. ____
35. ____
36. ____
37. ____
38. ____
39. ____
40. ____
41. ____

Name ____

33. The yearly cost of tuition and required fees for attending a public four-year college full-time can be estimated by the linear equation

$y = 131.4x + 2037.4,$

where x is the number of years after 1990 and y is the total cost in dollars. (*Source:* Based on data from the National Center for Education Statistics, U.S. Department of Education, 1990–1998)

a. Use this equation to approximate the yearly cost of attending a four-year public college in the year 2010.

b. Use this equation to predict in what year the yearly cost of tuition and required fees will exceed \$4000. (*Hint:* Let $y = 4000$ and solve for x.)

c. Use this equation to approximate the yearly cost of attending a four-year college in the present year. If you attend a four-year college, is this amount greater than or less than the amount that is currently charged by the college that you attend?

34. The yearly cost of tuition and required fees for attending a public two-year college full-time can be estimated by the linear equation

$y = 97.6x + 737.9,$

where x is the number of years after 1990 and y is the total cost in dollars. (*Source:* Based on data from the National Center for Education Statistics, U.S. Department of Education, 1990–1998)

a. Use this equation to approximate the yearly cost of attending a two-year public college in the year 2010.

b. Use this equation to predict in what year the yearly cost of tuition and required fees will exceed \$2000. (*Hint:* Let $y = 2000$ and solve for x.)

c. Use this equation to approximate the yearly cost of attending a two-year college in the present year. If you attend a two-year college, is this amount greater than or less than the amount that is currently charged by the college that you attend?

Review and Preview

Simplify and solve for y. See Section 2.3.

35. $y - 2 = 5(x + 6)$

36. $y - 0 = -3[x - (-10)]$

37. $y - (-1) = 2(x - 0)$

38. $y - 9 = -8[x - (-4)]$

Combining Concepts

39. In your own words explain how to graph an equation using its slope and y-intercept.

40. Suppose that the revenue of a company has increased at a steady rate of \$42,000 per year since 1990. Also, the company's revenue in 1990 was \$2,900,000. Write an equation that describes the company's revenue since 1990.

41. Suppose that a bird dives off a 500-foot cliff and descends at a rate of 7 feet per second. Write an equation that describes the bird's height at any time x.

3.4 MORE EQUATIONS OF LINES

Objectives

A Use the point-slope form to write the equation of a line.

B Write equations of vertical and horizontal lines.

C Write equations of parallel and perpendicular lines.

D Use the point-slope form in real-world applications.

SSM

CD-ROM

Video 3.4

A USING THE POINT-SLOPE FORM TO WRITE AN EQUATION

When the slope of a line and a point on the line are known, the equation of the line can also be found. To do this, we use the slope formula to write the slope of a line that passes through points (x, y), and (x_1, y_1). We have

$$m = \frac{y - y_1}{x - x_1}$$

We multiply both sides of this equation by $x - x_1$ to obtain

$$y - y_1 = m(x - x_1)$$

This form is called the **point-slope form** of the equation of a line.

> POINT-SLOPE FORM OF THE EQUATION OF A LINE
>
> The **point-slope form** of the equation of a line is
>
> $$y - y_1 = \overset{\text{slope}}{\overset{\downarrow}{m}}(x - x_1)$$
>
> (point: x_1, y_1)
>
> where m is the slope of the line and (x_1, y_1) is a point on the line.

Example 1 Write an equation of the line with slope -3 containing the point $(1, -5)$.

Solution: Because we know the slope and a point on the line, we use the point-slope form with $m = -3$ and $(x_1, y_1) = (1, -5)$.

$$\begin{aligned} y - y_1 &= m(x - x_1) && \text{Point-slope form} \\ y - (-5) &= -3(x - 1) && \text{Let } m = -3 \text{ and } (x_1, y_1) = (1, -5). \\ y + 5 &= -3x + 3 && \text{Use the distributive property.} \\ y &= -3x - 2 \end{aligned}$$

The equation is $y = -3x - 2$.

Practice Problem 1

Write an equation of the line with slope -2 containing the point $(2, -4)$. Write the equation in slope-intercept form, $y = mx + b$.

Example 2 Write an equation of the line through points $(4, 0)$ and $(-4, -5)$.

Solution: First we find the slope of the line.

$$m = \frac{-5 - 0}{-4 - 4} = \frac{-5}{-8} = \frac{5}{8}$$

Next we make use of the point-slope form. We replace (x_1, y_1) by either $(4, 0)$ or $(-4, -5)$ in the point-slope equation. We will choose the point $(4, 0)$. The line through $(4, 0)$ with slope $\frac{5}{8}$ is

Practice Problem 2

Write an equation of the line through points $(3, 0)$ and $(-2, 4)$. Write the equation in slope-intercept form, $y = mx + b$.

Answers

1. $y = -2x$, **2.** $y = -\frac{4}{5}x + \frac{12}{5}$

$$y - y_1 = m(x - x_1) \quad \text{Point-slope form}$$

$$y - 0 = \frac{5}{8}(x - 4) \quad \text{Let } m = \frac{5}{8} \text{ and } (x_1, y_1) = (4, 0).$$

$$y = \frac{5}{8}x - \frac{5}{8} \cdot 4 \quad \text{Use the distributive property.}$$

$$y = \frac{5}{8}x - \frac{5}{2} \quad \text{Simplify.}$$

The equation is $y = \frac{5}{8}x - \frac{5}{2}$. If we choose to use the point $(-4, -5)$, we have $y - (-5) = \frac{5}{8}[x - (-4)]$, which also simplifies to $y = \frac{5}{8}x - \frac{5}{2}$.

HELPFUL HINT

If two points of a line are given, either one may be used with the slope-intercept form to write an equation of the line.

B WRITING EQUATIONS OF VERTICAL AND HORIZONTAL LINES

A few special types of linear equations are those whose graphs are vertical and horizontal lines.

Practice Problem 3

Write an equation of the horizontal line containing the point $(-1, 6)$.

Example 3 Write an equation of the horizontal line containing the point $(2, 3)$.

Solution: Recall from Section 3.1, that a horizontal line has an equation of the form $y = b$. Since the line contains the point $(2, 3)$, the equation is $y = 3$.

Practice Problem 4

Write an equation of the line containing the point $(4, 7)$ with undefined slope.

Example 4 Write an equation of the line containing the point $(2, 3)$ with undefined slope.

Solution: Since the line has undefined slope, the line must be vertical. A vertical line has an equation of the form $x = c$, and since the line contains the point $(2, 3)$, the equation is $x = 2$.

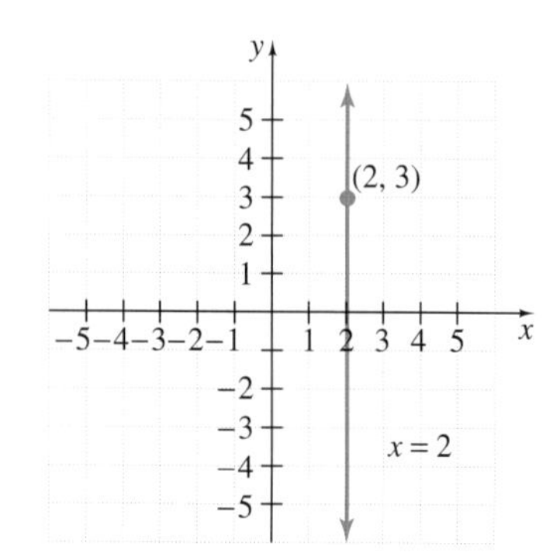

C WRITING EQUATIONS OF PARALLEL AND PERPENDICULAR LINES

Next, we write equations of parallel and perpendicular lines.

Answers

3. $y = 6$, **4.** $x = 4$

Example 5 Write an equation of the line containing the point $(4, 4)$ and parallel to the line $2x + y = -6$.

Solution: Because the line we want to find is *parallel* to the line $2x + y = -6$, the two lines must have equal slopes. So we first find the slope of $2x + y = -6$ by solving it for y to write it in the form $y = mx + b$. Here $y = -2x - 6$ so the slope is -2.

Now we use the point-slope form to write the equation of a line through $(4, 4)$ with slope -2.

$y - y_1 = m(x - x_1)$

$y - 4 = -2(x - 4)$ Let $m = -2$, $x_1 = 4$, and $y_1 = 4$.

$y - 4 = -2x + 8$ Use the distributive property.

$y = -2x + 12$

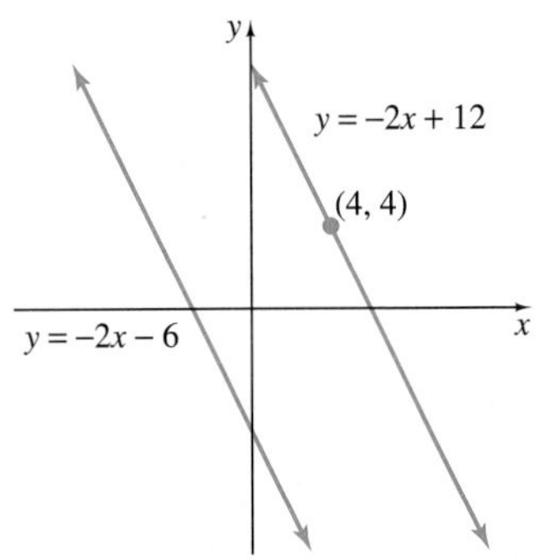

The equation, $y = -2x - 6$, and the new equation, $y = -2x + 12$, have the same slope but different y-intercepts so their graphs are parallel. Also, the graph of $y = -2x + 12$ contains the point $(4, 4)$, as desired.

Practice Problem 5

Write an equation of the line containing the point $(-1, 2)$ and parallel to the line $3x + y = 5$. Write the equation in the form $y = mx + b$.

Example 6 Write an equation of the line containing the point $(-2, 1)$ and perpendicular to the line $3x + 5y = 4$.

Solution: First we find the slope of $3x + 5y = 4$ by solving it for y.

$5y = -3x + 4$

$y = -\frac{3}{5}x + \frac{4}{5}$

The slope of the given line is $-\frac{3}{5}$. A line perpendicular to this line will have a slope that is the negative reciprocal of $-\frac{3}{5}$, or $\frac{5}{3}$. We use the point-slope form to write an equation of a new line through $(-2, 1)$ with slope $\frac{5}{3}$.

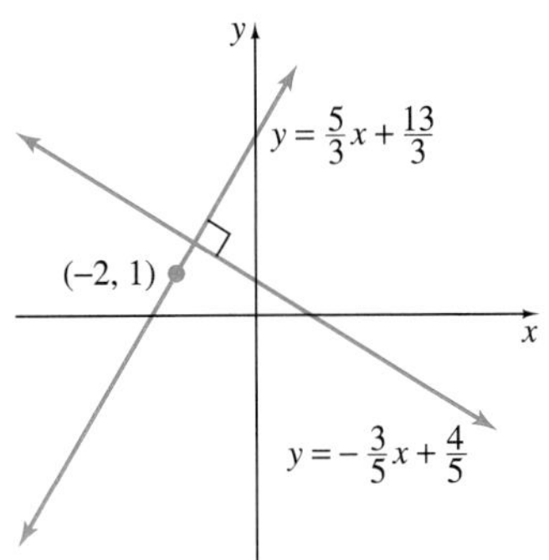

Practice Problem 6

Write an equation of the line containing the point $(3, 4)$ and perpendicular to the line $2x + 4y = 5$. Write the equation in standard form.

Answers

5. $y = -3x - 1$, **6.** $2x - y = 2$

$$y - 1 = \frac{5}{3}[x - (-2)]$$

$$y - 1 = \frac{5}{3}(x + 2) \quad \text{Simplify.}$$

$$y - 1 = \frac{5}{3}x + \frac{10}{3} \quad \text{Use the distributive property.}$$

$$y = \frac{5}{3}x + \frac{13}{3} \quad \text{Add 1 to both sides.}$$

The equation $y = -\frac{3}{5}x + \frac{4}{5}$ and the new equation $y = \frac{5}{3}x + \frac{13}{3}$ have negative reciprocal slopes so their graphs are perpendicular. Also, the graph of $y = \frac{5}{3}x + \frac{13}{3}$ contains the point $(-2, 1)$, as desired.

D USING THE POINT-SLOPE FORM IN APPLICATIONS

The point-slope form of an equation is very useful for solving real-world problems.

Practice Problem 7

Southwest Regional is an established office product maintenance company that has enjoyed constant growth in new maintenance contracts since 1985. In 1990, the company obtained 15 new contracts and in 1997, the company obtained 36 new contracts. Use these figures to predict the number of new contracts this company can expect in 2004.

Example 7 Predicting Sales

Southern Star Realty is an established real estate company that has enjoyed constant growth in sales since 1990. In 1992 the company sold 200 houses, and in 1997 the company sold 275 houses. Use these figures to predict the number of houses this company will sell in the year 2002.

Solution:

1. UNDERSTAND. Read and reread the problem. Then let

$x =$ the number of years after 1990 and

$y =$ the number of houses sold in the year corresponding to x.

The information provided then gives the ordered pairs $(2, 200)$ and $(7, 275)$. To better visualize the sales of Southern Star Realty, we graph the linear equation that passes through the points $(2, 200)$ and $(7, 275)$.

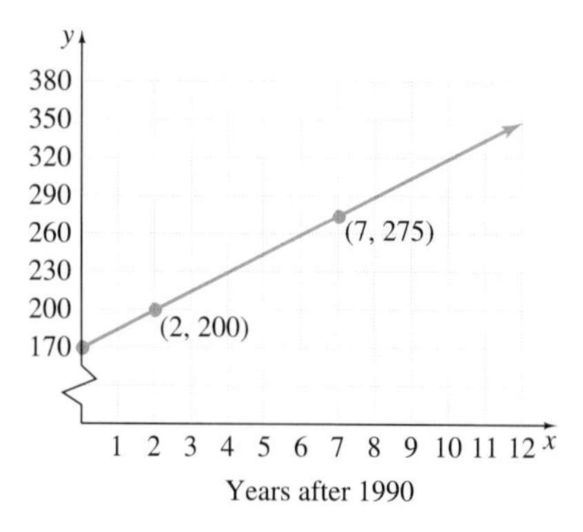

2. TRANSLATE. We write a linear equation that passes through the points $(2, 200)$ and $(7, 275)$. To do so, we first find the slope of the line.

Answer

7. 57 new contracts

$$m = \frac{275 - 200}{7 - 2} = \frac{75}{5} = 15$$

Then, using the point-slope form to write the equation, we have

$y - y_1 = m(x - x_1)$

$y - 200 = 15(x - 2)$ Let $m = 15$ and $(x_1, y_1) = (2, 200)$.

$y - 200 = 15x - 30$ Multiply.

$y = 15x + 170$ Add 200 to both sides.

3. SOLVE. To predict the number of houses sold in the year 2002, we use $y = 15x + 170$ and complete the ordered pair $(12, \ )$, since $2002 - 1990 = 12$.

$y = 15(12) + 170$ Let $x = 12$.

$y = 350$

4. INTERPRET.

Check: Verify that the point $(12, 350)$ is a point on the line graphed in step 1.

State: Southern Star Realty should expect to sell 350 houses in the year 2002.

GRAPHING CALCULATOR EXPLORATIONS

Many graphing calculators have a TRACE feature. This feature allows you to trace along a graph and see the corresponding x- and y-coordinates appear on the screen. Use this feature for the following exercises.

Graph each equation and then use the TRACE feature to complete each ordered pair solution. (Many times the tracer will not show an exact x- or y-value asked for. In each case, trace as closely as you can to the given x- or y-coordinate and approximate the other, unknown, coordinate to one decimal place.)

1. $y = 2.3x + 6.7$
$x = 5.1, y = ?$

2. $y = -4.8x + 2.9$
$x = -1.8, y = ?$

3. $y = -5.9x - 1.6$
$x = ?, y = 7.2$

4. $y = 0.4x - 8.6$
$x = ?, y = -4.4$

5. $y = 5.2x - 3.3$
$x = 2.3, y = ?$
$x = ?, y = 36$

6. $y = -6.2x - 8.3$
$x = 3.2, y = ?$
$x = ?, y = 12$

Focus On Business and Career

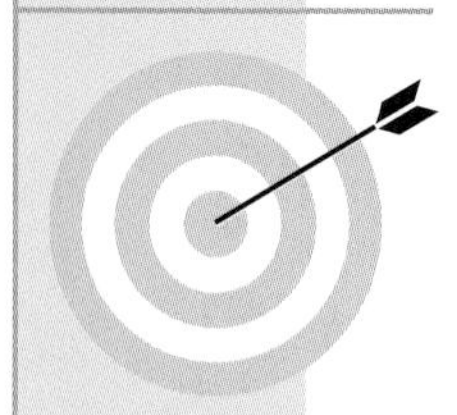

LINEAR MODELING

As we saw in Section 3.3, businesses often depend on equations that "closely fit" data. To *model* the data means to find an equation that describes the relationship between the paired data of two variables, such as time in years and profit. A model that accurately summarizes the relationship between two variables can be used to replace a potentially lengthy listing of the raw data. An accurate model might also be used to predict future trends by answering questions such as "If the trend seen in our company's performance in the last several years continues, what level of profit can we reasonably expect in 3 years?"

There are several ways to find a linear equation that models a set of data. If only two ordered pair data points are involved, an exact equation that contains both points can be found using the methods of Section 3.4. When more than two ordered pair data points are involved, it may be impossible to find a linear equation that contains all of the data points. In this case, the graph of the **best fit equation** should have a majority of the plotted ordered pair data points on the graph or close to it. In statistics, a technique called least squares regression is used to determine an equation that best fits a set of data. Various graphing utilities have built-in capabilities for finding an equation (called a regression equation) that best fits a set of ordered pair data points. Regression capabilities are often found with a graphing utility's statistics features.* A best fit equation can also be estimated using an algebraic method, which is outlined in the Group Activity below. In either case, a useful first step when finding a linear equation that models a set of data is creating a scatter diagram of the ordered pair data points to verify that a linear equation is an appropriate model.

GROUP ACTIVITY

Year	1993	1994	1995	1996	1997
Net Profit (in billions of dollars)	2.2	2.6	3.0	3.5	4.1

(*Source:* The Coca-Cola Company)

Coca-Cola Company is the world's largest producer of soft drinks and juices. The table shows Coca-Cola's net profit (in billions of dollars) for the years 1993–1997. Use the table along with your answers to the questions below to find a linear function $f(x)$ that represents net profit (in billions of dollars) as a function of the number of years after 1993.

1. Create a scatter diagram of the paired data given in the table. Does a linear model seem appropriate for the data?
2. Use a straight edge to draw on your graph what appears to be the line that "best fits" the data you plotted.
3. Estimate the coordinates of two points that fall on your best fit line. Use these points to find a linear function $f(x)$ for the line.
4. Use your linear function to find $f(8)$, and interpret its meaning in context.
5. Compare your group's linear function with other groups' functions. Are they the same or different? Explain why.
6. (Optional) Enter the data from the table into a graphing utility and use the linear regression feature to find a linear function that models the data. Compare this function with the one you found in Question 3. How are they alike or different?
7. (Optional) Using corporation annual reports or articles from magazines or newspapers, search for a set of business-related data that could be modeled with a linear function. Explain how modeling this data could be useful to a business. Then find the best fit equation for the data.

*To find out more about using a graphing utility to find a regression equation, consult the user's manual for your graphing utility.

Name ______________ Section ______ Date ______

MENTAL MATH

Find the slope and a point of the graph of each equation.

1. $y - 4 = -2(x - 1)$

2. $y - 6 = -3(x - 4)$

3. $y - 0 = \frac{1}{4}(x - 2)$

4. $y - 1 = -\frac{2}{3}(x - 0)$

5. $y + 2 = 5(x - 3)$

6. $y - 7 = 4(x + 6)$

EXERCISE SET 3.4

A *Write an equation of each line with the given slope and containing the given point. Write the equation in the form $y = mx + b$. See Example 1.*

1. Slope 3; through $(1, 2)$

2. Slope 4; through $(5, 1)$

3. Slope -2; through $(1, -3)$

4. Slope -4; through $(2, -4)$

5. Slope $\frac{1}{2}$; through $(-6, 2)$

6. Slope $\frac{2}{3}$; through $(-9, 4)$

7. Slope $-\frac{9}{10}$; through $(-3, 0)$

8. Slope $-\frac{1}{5}$; through $(4, -6)$

9. Slope 2; through $(-2, 3)$

10. Slope 3; through $(-4, 2)$

11. Slope $-\frac{4}{3}$; through $(-5, 0)$

12. Slope $-\frac{3}{5}$; through $(4, -1)$

Write an equation of the line passing through the given points. Write the equation in the form $y = mx + b$. See Example 2.

13. $(2, 0)$ and $(4, 6)$

14. $(3, 0)$ and $(7, 8)$

15. $(-2, 5)$ and $(-6, 13)$

16. $(7, -4)$ and $(2, 6)$

17. $(-2, -4)$ and $(-4, -3)$

18. $(-9, -2)$ and $(-3, 10)$

MENTAL MATH ANSWERS

1. ______
2. ______
3. ______
4. ______
5. ______
6. ______

ANSWERS

1. ______
2. ______
3. ______
4. ______
5. ______
6. ______
7. ______
8. ______
9. ______
10. ______
11. ______
12. ______
13. ______
14. ______
15. ______
16. ______
17. ______
18. ______

19. ______

20. ______

21. ______

22. ______

23. ______

24. ______

25. ______

26. ______

27. ______

28. ______

29. ______

30. ______

31. ______

32. ______

33. ______

34. ______

Name ______

19. $(-3, -8)$ and $(-6, -9)$

20. $(8, -3)$ and $(4, -8)$

21. $(-7, -4)$ and $(0, -6)$

22. $(2, -8)$ and $(-4, -3)$

23. $\left(\frac{3}{5}, \frac{4}{10}\right)$ and $\left(-\frac{1}{5}, \frac{7}{10}\right)$

24. $\left(\frac{1}{2}, -\frac{1}{4}\right)$ and $\left(\frac{3}{2}, \frac{3}{4}\right)$

B *Write an equation of each line. See Examples 3 and 4.*

25. Vertical; through $(2, 6)$

26. Slope 0; through $(-2, -4)$

27. Horizontal; through $(-3, 1)$

28. Vertical; through $(4, 7)$

29. Undefined slope; through $(0, 5)$

30. Horizontal; through $(0, 5)$

C *Write an equation of each line. Write the equation in the form $y = mx + b$. See Examples 5 and 6.*

31. Through $(3, 8)$; parallel to $y = 4x - 2$

32. Through $(1, 5)$; parallel to $y = 3x - 4$

33. Through $(2, -5)$; perpendicular to $y = -2x - 6$

34. Through $(-4, 8)$; perpendicular to $y = -4x - 1$

Name ______________________________

35. Through $(-2, -3)$; parallel to $3x + 2y = 5$

36. Through $(-2, -3)$; perpendicular to $3x + 2y = 5$

37. Through $(3, 5)$; perpendicular to $2x - y = 8$

38. Through $(6, 1)$; parallel to $8x - y = 9$

39. Through $(6, -2)$; parallel to $2x + 4y = 9$

40. Through $(8, -3)$; parallel to $6x + 2y = 5$

41. Through $(-1, 5)$; perpendicular to $x - 4y = 4$

42. Through $(2, -3)$; perpendicular to $x - 5y = 10$

D *Solve. See Example 7*

43. A rock is dropped from the top of a 400-foot building. After 1 second, the rock is traveling 32 feet per second. After 3 seconds, the rock is traveling 96 feet per second. Let y be the rate of descent and x be the number of seconds since the rock was dropped.

a. Write a linear equation that relates time x to rate y (*Hint:* Use the ordered pairs $(1, 32)$ and $(3, 96)$.)

b. Use this equation to determine the rate of the rock 4 seconds after it was dropped.

44. The Whammo Company has learned that, by pricing a newly released Frisbee at \$6, sales will reach 2000 per day. Raising the price to \$8 will cause the sales to fall to 1500 per day. Assume that the ratio of change in price to change in daily sales is constant, and let x be the price of the Frisbee and y be number of sales.

a. Find the linear equation that models the price–sales relationship for this Frisbee. (*Hint:* The line must pass through $(6, 2000)$ and $(8, 1500)$.)

b. Use this equation to predict the daily sales of Frisbees if the price is set at \$7.50.

35. ______________

36. ______________

37. ______________

38. ______________

39. ______________

40. ______________

41. ______________

42. ______________

43. a. ______________

b. ______________

44. a. ______________

b. ______________

45. a. ______

b. ______

c. ______

46. a. ______

b. ______

47. a. ______

b. ______

48. a. ______

b. ______

Name ______

45. Del Monte Fruit Company recently released a new applesauce. By the end of its first year, profits on this product amounted to \$30,000. The anticipated profit for the end of the fourth year is \$66,000. The ratio of change in time to change in profit is constant. Let x be years and y be profit.

a. Write a linear equation that relates profit and time.

b. Use this equation to predict the company's profit at the end of the seventh year.

c. Predict when the profit should reach \$126,000.

46. The Pool Fun Company has learned that, by pricing a newly released Fun Noodle at \$3, sales will reach 10,000 Fun Noodles per day during the summer. Raising the price to \$5 will cause the sales to fall to 8000 Fun Noodles per day. Let x be price and y be the number sold.

a. Assume that the relationship between sales price and number of Fun Noodles sold is linear and write an equation describing this relationship.

b. Use this equation to predict the daily sales of Fun Noodles if the price is \$3.50.

47. In 1994, the median price of an existing home in the United States was \$109,900. In 1997, the median price of an existing home was \$123,200. Let y be the median price of an existing home in the year x, where $x = 0$ represents 1994. (*Source*: National Association of REALTORS®)

a. Write a linear equation that models the median existing home price in terms of the year x. (*Hint:* The line must pass through the points $(0, 109{,}900)$ and $(3, 123{,}200)$.)

b. Use this equation to predict the median existing home price for the year 2002.

48. The number of births (in thousands) in the United States in 1996 was 3915. The number of births (in thousands) in the United States in 1991 was 4111. Let y be the number of births (in thousands) in the year x, where $x = 0$ represents 1991. (*Source:* National Center for Health Statistics)

a. Write a linear equation that models the number of births (in thousands) in terms of the year x. (See hint for Exercise 47a.)

b. Use this equation to predict the number of births in the United States for the year 2001.

Name ______________________

49. The number of people employed in the United States as medical assistants was 225 thousand in 1996. By the year 2006, this number is expected to rise to 391 thousand. Let y be the number of medical assistants (in thousands) employed in the United States in the year x, where $x = 0$ represents 1996. (*Source:* Bureau of Labor Statistics)

a. Write a linear equation that models the number of people (in thousands) employed as medical assistants in the year x. (See hint for Exercise 47a.)

b. Use this equation to estimate the number of people who will be employed as medical assistants in the year 2004.

50. The number of people employed in the United States as systems analysts was 506 thousand in 1996. By the year 2006, this number is expected to rise to 1025 thousand. Let y be the number of systems analysts (in thousands) employed in the United States in the year x, where $x = 0$ represents 1996. (*Source:* Bureau of Labor Statistics)

a. Write a linear equation that models the number of people (in thousands) employed as systems analysts in the year x. (See hint for Exercise 47a.)

b. Use this equation to estimate the number of people who will be employed as systems analysts in the year 2002.

Review and Preview

Complete each ordered pair for the given equation. See Section 3.1.

51. $y = 7x + 3; (4, \ \ )$

52. $y = 2x - 6; (2, \ \ )$

53. $y = 4.2x; (-2, \ \ )$

54. $y = -1.3x; (6, \ \ )$

55. $y = x^2 + 2x + 1; (1, \ \ )$

56. $y = x^2 - 6x + 4; (0, \ \ )$

Combining Concepts

Answer true or false.

57. A vertical line is always perpendicular to a horizontal line.

58. A vertical line is always parallel to a vertical line.

49. a. ______
b. ______
50. a. ______
b. ______
51. ______
52. ______
53. ______
54. ______
55. ______
56. ______
57. ______
58. ______

59. ____________

Name ____________

Write an equation of each line.

59. Through $(5, -6)$; perpendicular to $y = 9$

60. Through $(-3, -5)$; parallel to $y = 9$

60. ____________

Use a grapher with a TRACE feature to see the results of each exercise.

61. Exercise 31; graph the equation and verify that it passes through $(3, 8)$ and is parallel to $y = 4x - 2$.

62. Exercise 32; graph the equation and verify that it passes through $(1, 5)$ and is parallel to $y = 3x - 4$.

61. see calculator screen

62. see calculator screen

Internet Excursions

Go to http://www.prenhall.com/martin-gay

The U.S. Bureau of Labor Statistics (BLS) is the principal fact-finding agency for the federal government in the broad field of labor economics and statistics. The BLS regularly makes data such as unemployment figures, average earnings, and job growth data available to the American public, government, and businesses. The World Wide Web address listed above will provide you with access to the homepage of the BLS National Current Employment Statistics, or a related site. You will be able to research information needed to answer the questions below.

63. ____________

63. Choose an industry listed in the table summarizing average hourly earnings. Write down two ordered pairs for this industry. Describe what each ordered pair represents. Then use the ordered pairs to find an equation for the line between these two points.

64. Choose an industry listed in the table summarizing average weekly earnings. Write down two ordered pairs for this industry. Describe what each ordered pair represents. Use the ordered pairs to find an equation for the line between these two points. Then make a prediction using your equation and explain its significance.

64. ____________

Name ____________ Section ______ Date ______

ANSWERS

INTEGRATED REVIEW—LINEAR EQUATIONS IN TWO VARIABLES

Below is a review of equations of lines.

FORMS OF LINEAR EQUATIONS

$Ax + By = C$	**Standard form** of a linear equation A and B are not both 0.
$y = mx + b$	**Slope-intercept form** of a linear equation The slope is m, and the y-intercept is b.
$y - y_1 = m(x - x_1)$	**Point-slope form** of a linear equation The slope is m, and (x_1, y_1) is a point on the line.
$y = c$	**Horizontal line** The slope is 0, and the y-intercept is c.
$x = c$	**Vertical line** The slope is undefined and the x-intercept is c.

PARALLEL AND PERPENDICULAR LINES

Nonvertical parallel lines have the same slope.
The product of the slopes of two nonvertical perpendicular lines is -1.

Graph each linear equation.

1. $y = -2x$

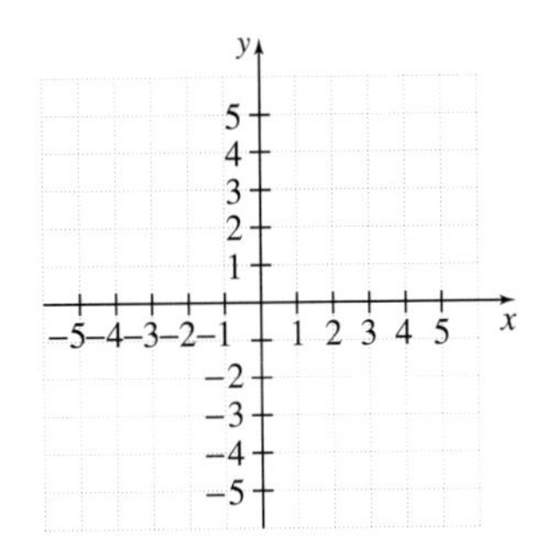

2. $3x - 2y = 6$

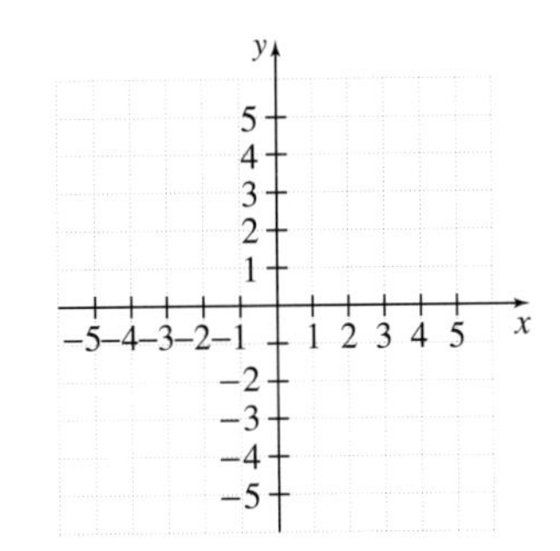

3. $x = -3$

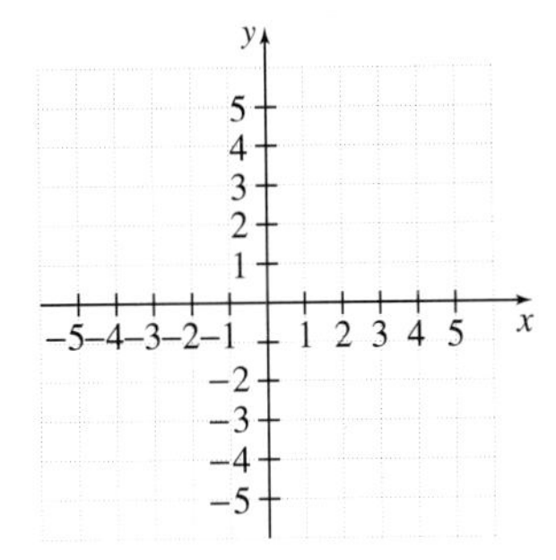

4. $y = 1.5$

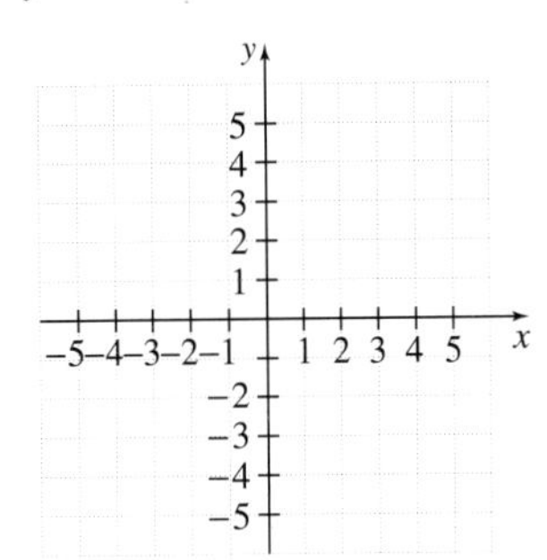

Find the slope of the line containing each pair of points.

5. $(-2, -5), (3, -5)$

6. $(5, 2), (0, 5)$

1. see graph
2. see graph
3. see graph
4. see graph
5. ______
6. ______

7. ______

8. ______

9. ______

10. ______

11. ______

12. ______

13. ______

14. ______

15. ______

16. ______

17. ______

18. ______

19. ______

20. ______

21. ______

22. ______

23. ______

24. ______

Name ______________________________

Find the slope and y-intercept of each line.

7. $y = 3x - 5$

8. $5x - 2y = 7$

Determine whether each pair of lines is parallel, perpendicular, or neither.

9. $y = 8x - 6$
$y = 8x + 6$

10. $y = \frac{2}{3}x + 1$
$2y + 3x = 1$

Find the equation of each line. Write the equation in the form $x = a$, $y = b$, or $y = mx + b$ form.

11. Through $(1, 6)$ and $(5, 2)$

12. Vertical line; through $(-2, -10)$

13. Horizontal line; through $(1, 0)$

14. Through $(2, -8)$ and $(-6, -5)$

15. Through $(-2, 4)$ with slope -5

16. Slope -4; y-intercept $\frac{1}{3}$

17. Slope $\frac{1}{2}$; y-intercept -1

18. Through $\left(\frac{1}{2}, 0\right)$ with slope 3

19. Through $(-1, -5)$; parallel to $3x - y = 5$

20. Through $(0, 4)$; perpendicular to $4x - 5y = 10$

21. Through $(2, -3)$; perpendicular to $4x + y = \frac{2}{3}$

22. Through $(-1, 0)$; parallel to $5x + 2y = 2$

23. undefined slope; through $(-1, 3)$

24. $m = 0$; through $(-1, 3)$

3.5 INTRODUCTION TO FUNCTIONS

Objectives

- **A** Define relation, domain, and range.
- **B** Identify functions.
- **C** Use the vertical line test for functions.
- **D** Find the domain and range of a relation from its graph.
- **E** Use function notation.
- **F** Graph a linear function.

SSM

CD-ROM

Video 3.5

A DEFINING RELATION, DOMAIN, AND RANGE

Equations in two variables, such as $y = 2x + 1$, describe **relations** between x-values and y-values. For example, if $x = 1$, then this equation describes how to find the y-value related to $x = 1$. In words, the equation $y = 2x + 1$ says that twice the x-value increased by 1 gives the corresponding y-value. The x-value of 1 corresponds to the y-value of $2(1) + 1 = 3$ for this equation, and we have the ordered pair $(1, 3)$.

There are other ways of describing relations or correspondences between two numbers or, in general, a first set (sometimes called the set of *inputs*) and a second set (sometimes called the set of *outputs*). For example,

First Set: Input	Correspondence	Second Set: Output
People in a certain city	Each person's age	The set of nonnegative integers

A few examples of ordered pairs from this relation might be (Ana, 4); (Bob, 36); (Trey, 21); and so on.

Below are just a few other ways of describing relations between two sets and the ordered pairs that they generate.

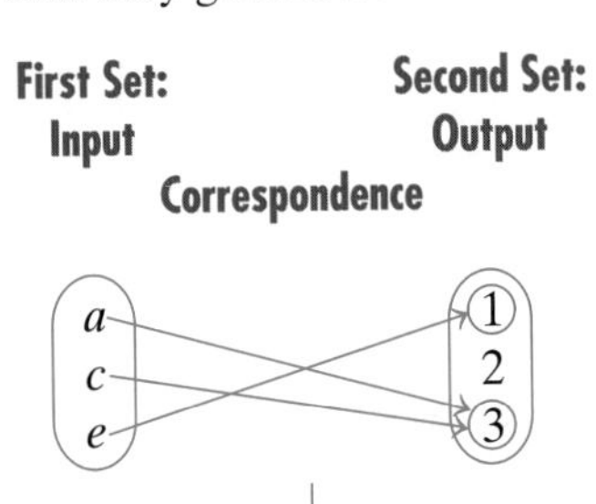

Ordered Pairs
$(a, 3), (c, 3), (e, 1)$

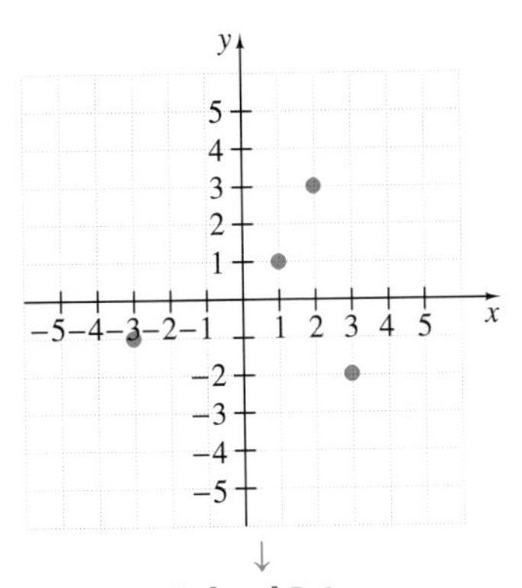

Ordered Pairs
$(-3, -1), (1, 1), (2, 3), (3, -2)$

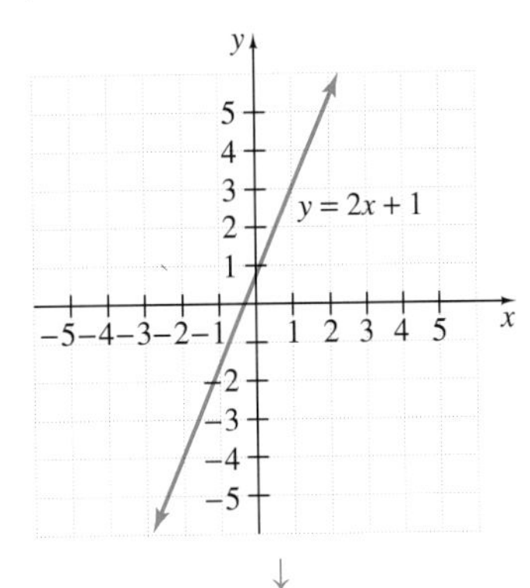

Some Ordered Pairs
$(1, 3), (0, 1)$, and so on

> **RELATION, DOMAIN, AND RANGE**
>
> A **relation** is a set of ordered pairs.
> The **domain** of the relation is the set of all first components of the ordered pairs.
> The **range** of the relation is the set of all second components of the ordered pairs.

For example, the domain for our middle relation above is $\{a, c, e\}$ and the range is $\{1, 3\}$. Notice that the range does not include the element 2 of the second set. This is because no element of the first set is assigned to this element. If a relation is defined in terms of x- and y-values, we will agree that the domain corresponds to x-values and that the range corresponds to y-values.

Examples

Determine the domain and range of each relation.

1. $\{(2, 3), (2, 4), (0, -1), (3, -1)\}$
The domain is the set of all first coordinates of the ordered pairs, $\{2, 0, 3\}$.
The range is the set of all second coordinates, $\{3, 4, -1\}$.

2.

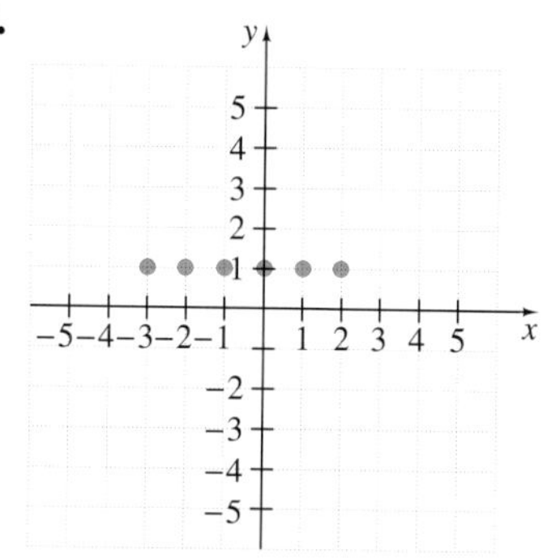

The relation is $\{(-3, 1), (-2, 1), (-1, 1), (0, 1), (1, 1), (2, 1)\}$.
The domain is $\{-3, -2, -1, 0, 1, 2\}$.
The range is $\{1\}$.

3.

Input: Cities **Output: Population (in thousands)**

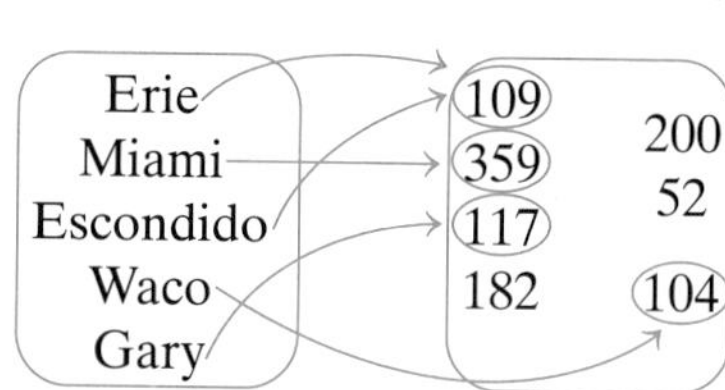

The domain is the first set, $\{$ Erie, Escondido, Gary, Miami, Waco$\}$.
The range is the numbers in the second set that correspond to elements in the first set, $\{104, 109, 117, 359\}$.

Practice Problems 1–3

Determine the domain and range of each relation.

1. $\{(1, 6), (2, 8), (0, 3), (0, -2)\}$

2.

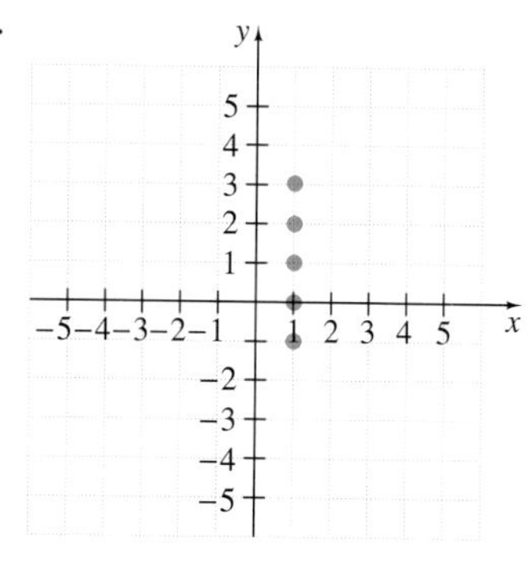

3.

Input: States **Output: Number of Representatives**

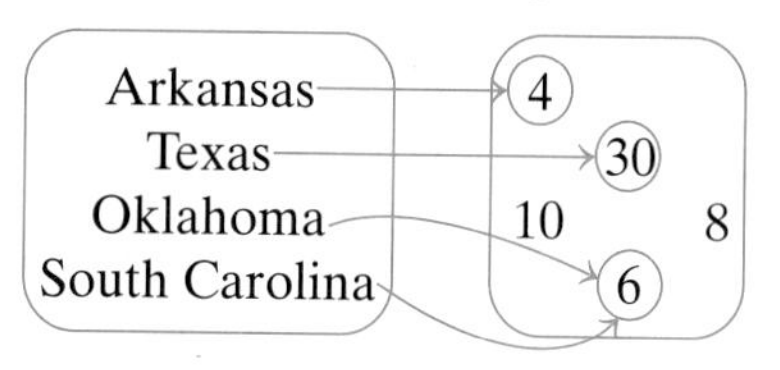

Answers

1. domain: $\{1, 2, 0\}$, range: $\{6, 8, 3, -2\}$, **2.** domain: $\{1\}$, range: $\{-1, 0, 1, 2, 3\}$, **3.** domain: $\{$Arkansas, Texas, Oklahoma, South Carolina$\}$, range: $\{4, 30, 6\}$

B IDENTIFYING FUNCTIONS

Now we consider a special kind of relation called a function.

FUNCTION

A **function** is a relation in which each first component in the ordered pairs corresponds to *exactly one* second component.

HELPFUL HINT

A function is a special type of relation, so all functions are relations, but not all relations are functions.

Examples

Determine whether each relation is also a function.

4. $\{(-2, 5), (2, 7), (-3, 5), (9, 9)\}$

Although the ordered pairs $(-2, 5)$ and $(-3, 5)$ have the same y-value, each x-value is assigned to only one y-value, so this set of ordered pairs is a function.

5.

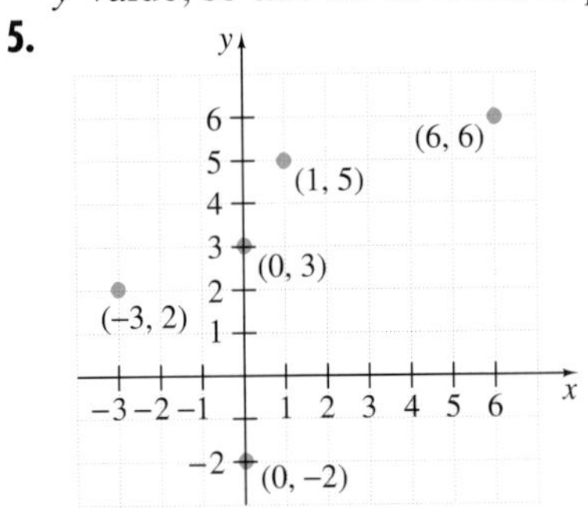

The x-value 0 is assigned to two y-values, -2 and 3, in this graph so this relation is not a function.

6.

Input	Correspondence	Output
People in a certain city	Each person's age	The set of nonnegative integers

This relation is a function because although two different people may have the same age, each person has only one age. This means that each element in the first set is assigned to only one element in the second set.

TRY THE CONCEPT CHECK IN THE MARGIN.

We will call an equation such as $y = 2x + 1$ a relation since this equation defines a set of ordered pair solutions.

Example 7

Determine whether the relation $y = 2x + 1$ is also a function.

Solution: The relation $y = 2x + 1$ is a function if each x-value corresponds to just one y-value. For each x-value substituted in the equation $y = 2x + 1$, the multiplication and addition performed gives a single result, so only one y-value will be associated with each x-value. Thus, $y = 2x + 1$ is a function.

Example 8

Determine whether the relation $x = y^2$ is also a function.

Solution: In $x = y^2$, if $y = 3$, then $x = 9$. Also, if $y = -3$, then $x = 9$. In other words, we have the ordered pairs $(9, 3)$ and $(9, -3)$. Since the x-value 9 corresponds to two y-values, 3 and -3, $x = y^2$ is not a function.

Practice Problems 4–6

Determine whether each relation is also a function.

4. $\{(-3, 7), (1, 7), (2, 2)\}$

5.

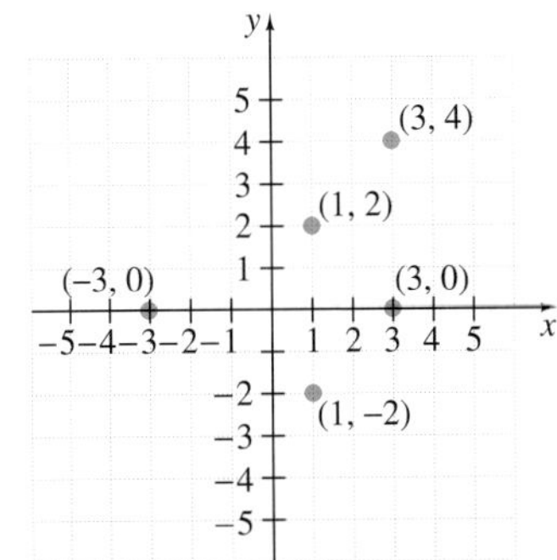

6.

Input	Correspondence	Output
People in a certain state	County/ Parish that a person lives in	Counties of that state

✓ CONCEPT CHECK

Explain why a function can contain both the ordered pairs $(1, 3)$ and $(2, 3)$ but not both $(3, 1)$ and $(3, 2)$.

Practice Problem 7

Determine whether the relation $y = 3x + 2$ is also a function.

Practice Problem 8

Determine whether the relation $x = y^2 + 1$ is also a function.

Answers

4. function, **5.** not a function, **6.** function **7.** yes, **8.** no

✓ Concept Check: Two different ordered pairs can have the same y-value, but not the same x-value in a function.

Practice Problems 9–13

Use the vertical line test to determine which are graphs of functions.

9.

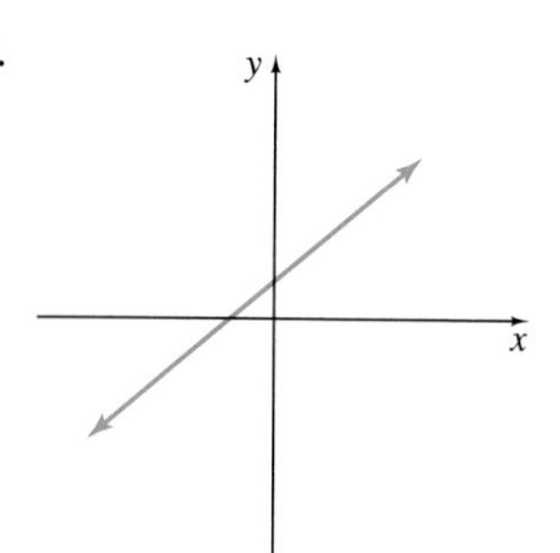

10.

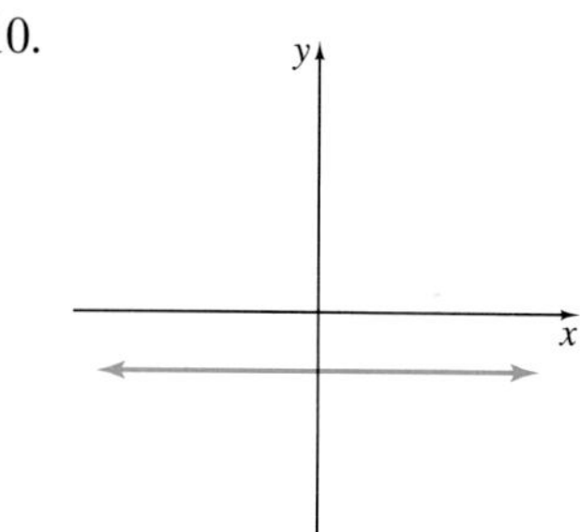

11.

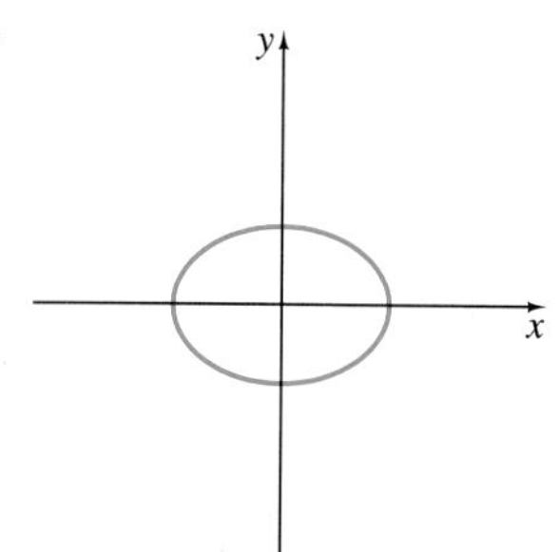

12.

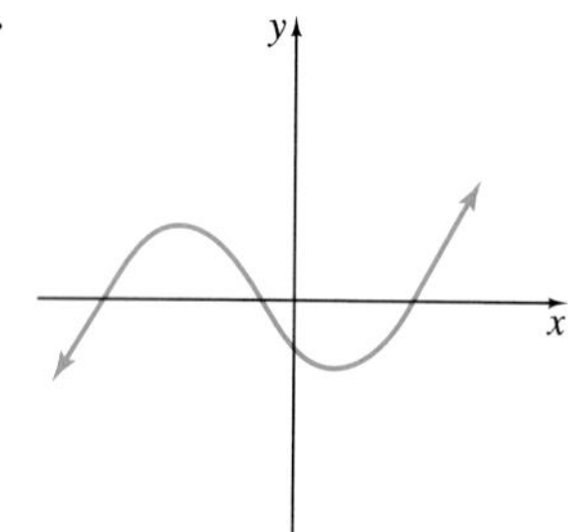

13.

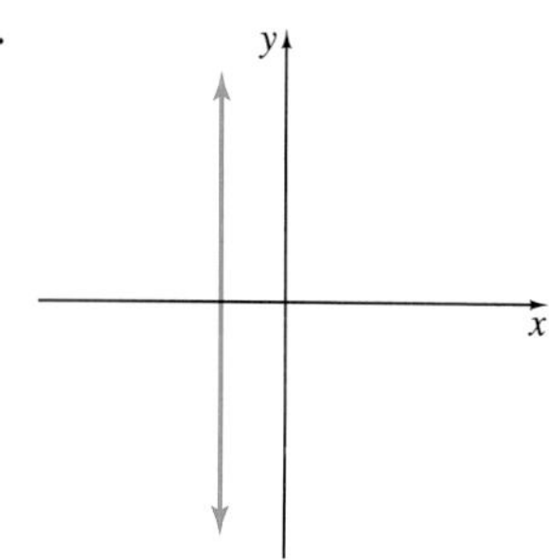

Answers

9. function, **10.** function, **11.** not a function, **12.** function, **13.** not a function

C Using the Vertical Line Test

As we have seen, not all relations are functions. Consider the graphs of $y = 2x + 1$ and $x = y^2$ shown next. On the graph of $y = 2x + 1$, notice that each x-value corresponds to only one y-value. Recall from Example 7 that $y = 2x + 1$ is a function.

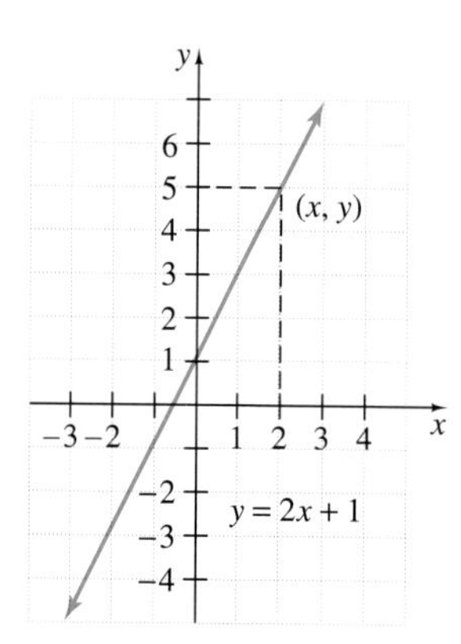

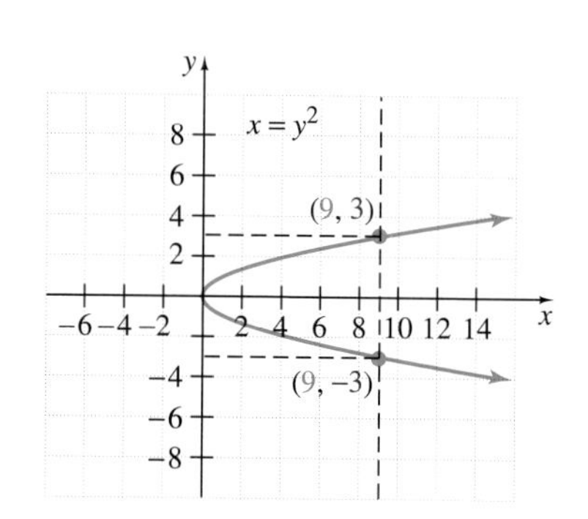

On the graph of $x = y^2$, the x-value 9, for example, corresponds to two y-values, 3 and -3, as shown by the vertical line. Recall from Example 8 that $x = y^2$ is not a function.

Graphs can be used to help determine whether a relation is also a function by the following vertical line test.

Vertical Line Test

If no vertical line can be drawn so that it intersects a graph more than once, the graph is the graph of a function.

Examples

Use the vertical line test to determine which are graphs of functions.

9.

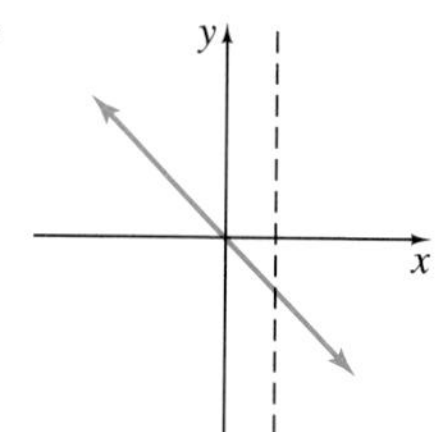

This is the graph of a function since no vertical line will intersect this graph more than once.

10.

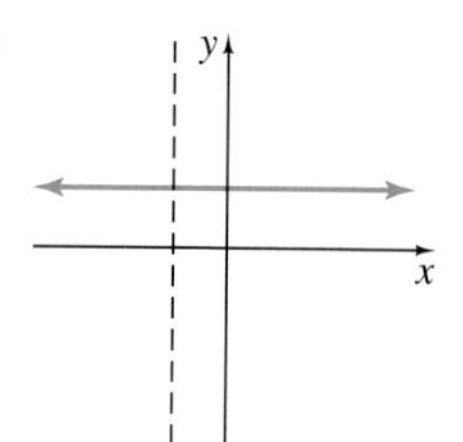

This is the graph of a function.

11.

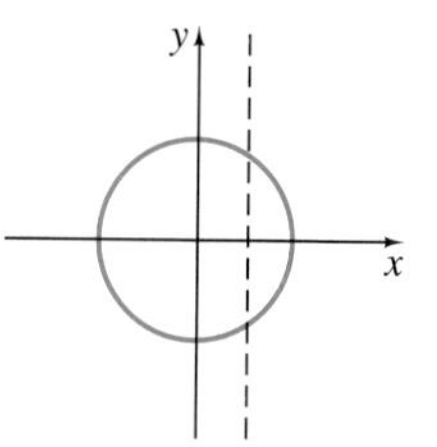

This is not the graph of a function. Note that vertical lines can be drawn that intersect the graph in two points.

12.

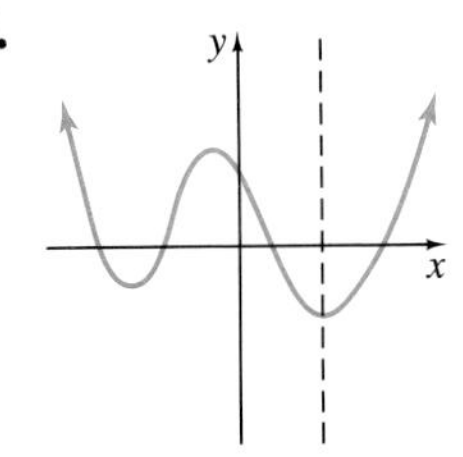

This is the graph of a function.

13.

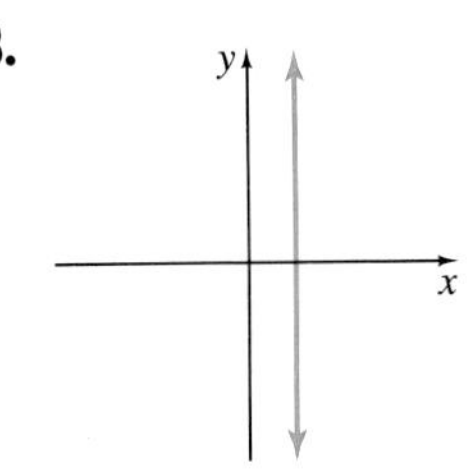

This is not the graph of a function. A vertical line can be drawn that intersects this line at every point.

TRY THE CONCEPT CHECK IN THE MARGIN.

D FINDING DOMAIN AND RANGE FROM A GRAPH

Next we practice finding the domain and range of a relation from its graph.

Examples Find the domain and range of each relation.

Solutions:

14.

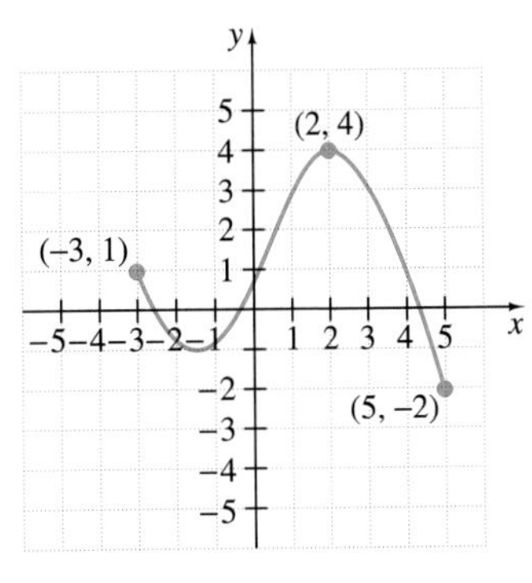

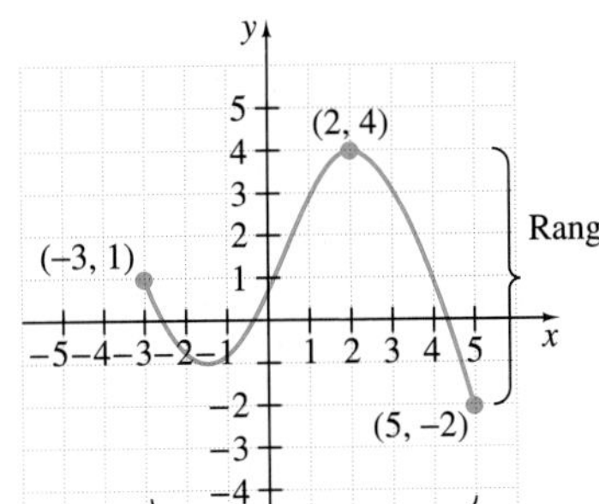

Range: The y-values graphed are from −2 to 4, or [−2, 4]

Domain: The x-values graphed are from −3 to 5, or [−3, 5]

15.

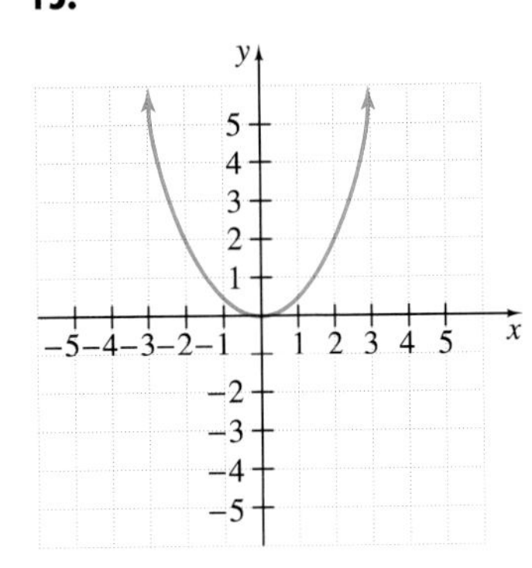

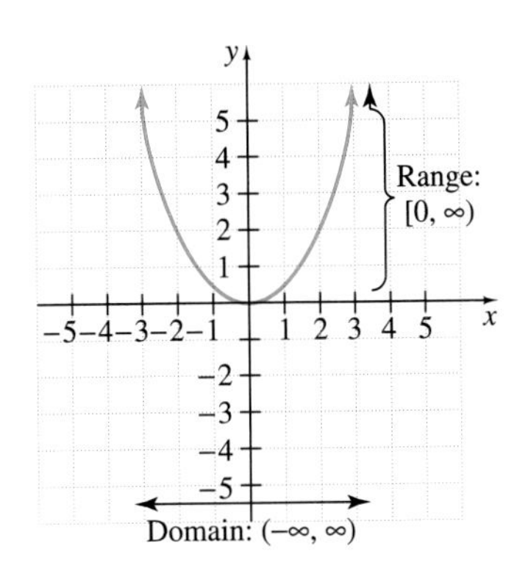

Range: [0, ∞)

Domain: (−∞, ∞)

✓ CONCEPT CHECK

Determine which equations represent functions. Explain your answer.

a. $y = 14$
b. $x = -5$
c. $x + y = 6$

Practice Problems 14–17

Find the domain and range of each relation.

14.

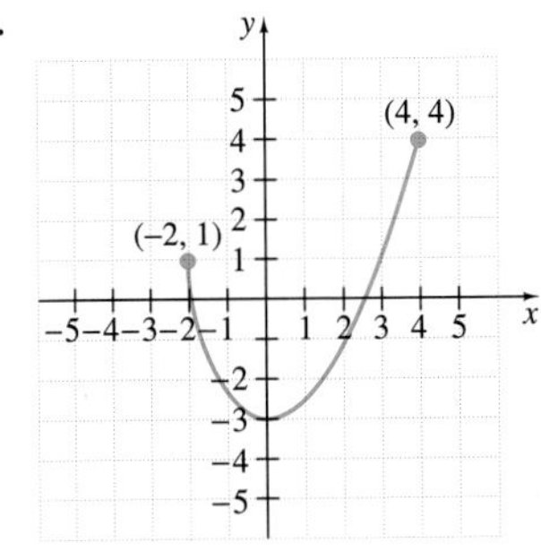

15.

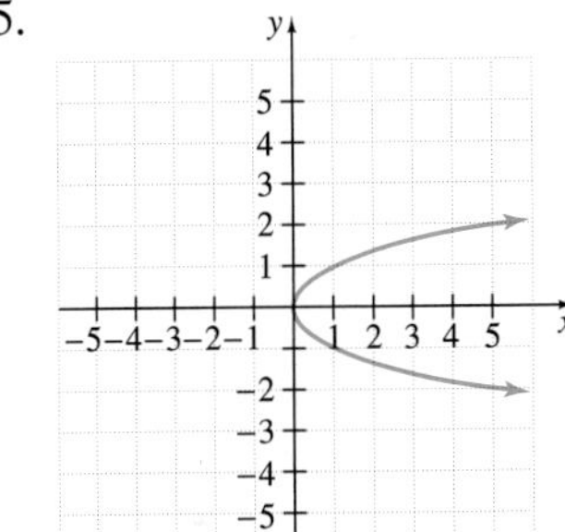

16.

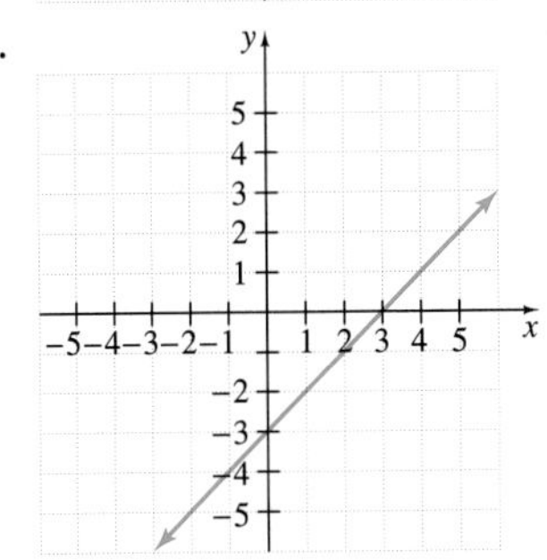

17.

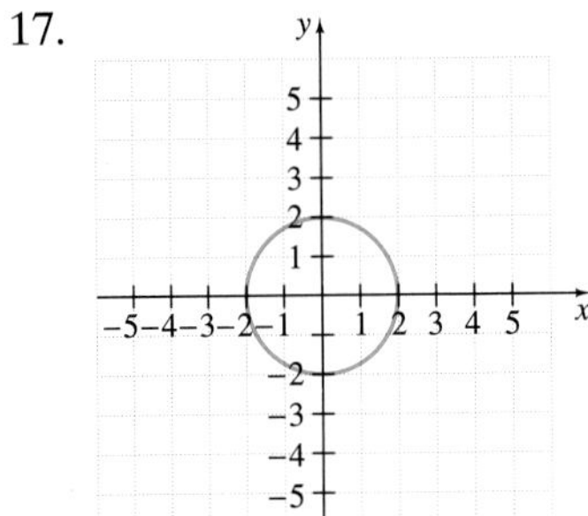

Answers

14. domain: $[-2, 4]$, range: $[-3, 4]$, **15.** domain: $[0, \infty)$, range: $(-\infty, \infty)$, **16.** domain: $(-\infty, \infty)$, range: $(-\infty, \infty)$, **17.** domain: $[-2, 2]$, range: $[-2, 2]$

✓ Concept Check: a, c

16.

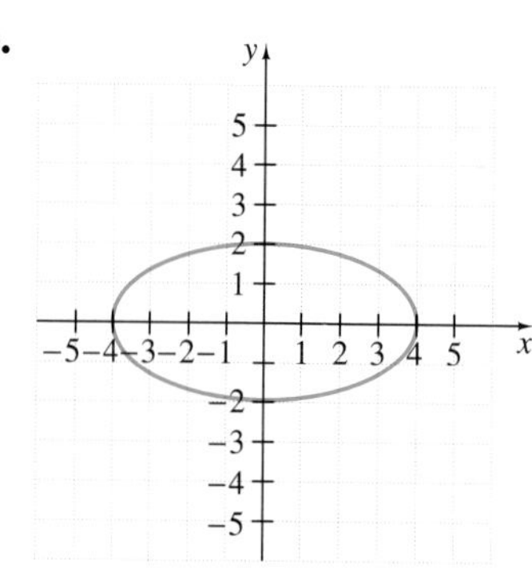

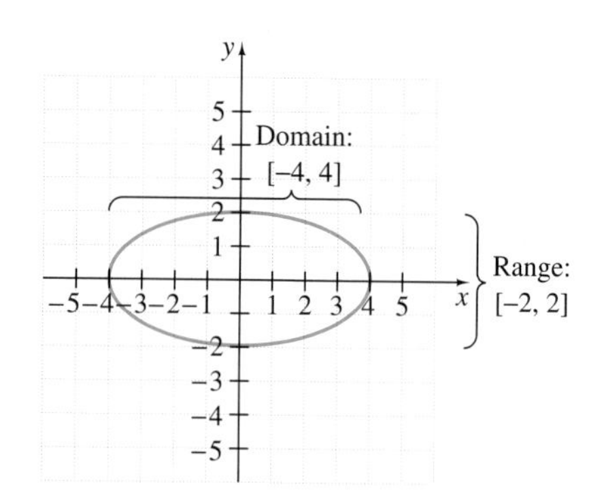

17.

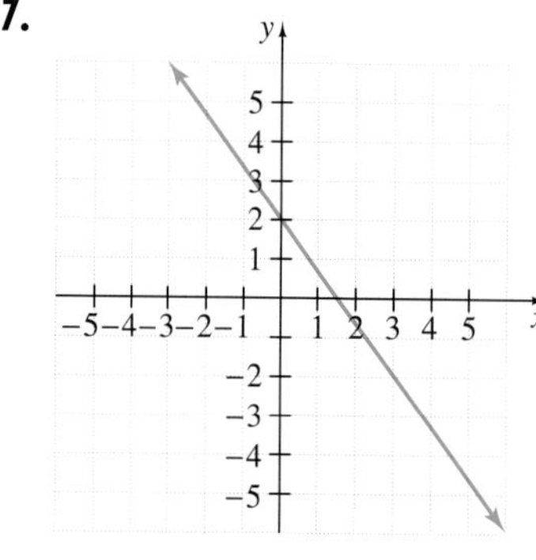

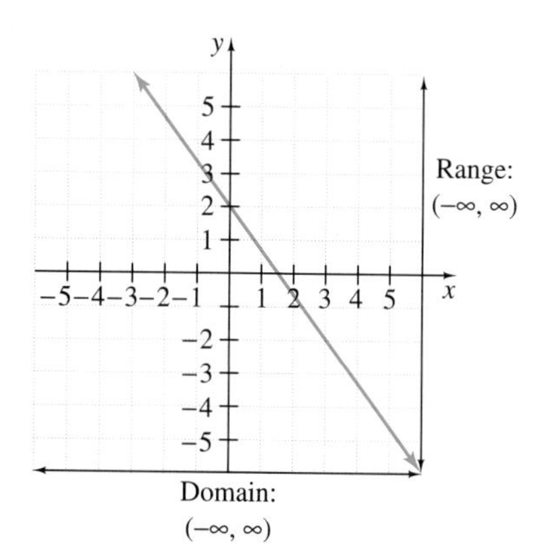

E USING FUNCTION NOTATION

Many times letters such as f, g, and h are used to name functions. To denote that y is a function of x, we can write

$$y = f(x)$$

This means that **y is a function of x** or that *y depends on x*. For this reason, y is called the **dependent variable** and x the **independent variable**. The notation $f(x)$ is read "f of x" and is called **function notation**.

For example, to use function notation with the function $y = 4x + 3$, we write $f(x) = 4x + 3$. The notation $f(1)$ means to replace x with 1 and find the resulting y or function value. Since

$$f(x) = 4x + 3$$

then

$$f(1) = 4(1) + 3 = 7$$

This means that when $x = 1$, y or $f(x) = 7$. The corresponding ordered pair is $(1, 7)$. Here, the input is 1 and the output is $f(1)$ or 7. Now let's find $f(2)$, $f(0)$, and $f(-1)$.

$$\begin{aligned} f(x) &= 4x + 3 \\ f(2) &= 4(2) + 3 \\ &= 8 + 3 \\ &= 11 \end{aligned} \qquad \begin{aligned} f(x) &= 4x + 3 \\ f(0) &= 4(0) + 3 \\ &= 0 + 3 \\ &= 3 \end{aligned} \qquad \begin{aligned} f(x) &= 4x + 3 \\ f(-1) &= 4(-1) + 3 \\ &= -4 + 3 \\ &= -1 \end{aligned}$$

Ordered Pairs:

$(2, 11)$ $\qquad$ $(0, 3)$ $\qquad$ $(-1, -1)$

HELPFUL HINT

Note that $f(x)$ is a special symbol in mathematics used to denote a function. The symbol $f(x)$ is read "f of x." It does *not* mean $f \cdot x$ (f times x).

Examples Find each function value.

18. If $g(x) = 3x - 2$, find $g(1)$.

$g(1) = 3(1) - 2 = 1$

19. If $g(x) = 3x - 2$, find $g(0)$.

$g(0) = 3(0) - 2 = -2$

20. If $f(x) = 7x^2 - 3x + 1$, find $f(1)$.

$f(1) = 7(1)^2 - 3(1) + 1 = 5$

21. If $f(x) = 7x^2 - 3x + 1$, find $f(-2)$.

$f(-2) = 7(-2)^2 - 3(-2) + 1 = 35$

TRY THE CONCEPT CHECK IN THE MARGIN.

F GRAPHING LINEAR FUNCTIONS

Recall that the graph of a linear equation in two variables is a line, and a line that is not vertical will always pass the vertical line test. Thus, *all linear equations are functions except those whose graph is a vertical line.* We call such functions **linear functions**.

> LINEAR FUNCTION
>
> A **linear function** is a function that can be written in the form
>
> $f(x) = mx + b$

Example 22 Graph the function $f(x) = 2x + 1$.

Solution: Since $y = f(x)$, we could replace $f(x)$ with y and graph as usual. The graph of $y = 2x + 1$ has slope 2 and y-intercept 1. Its graph is shown.

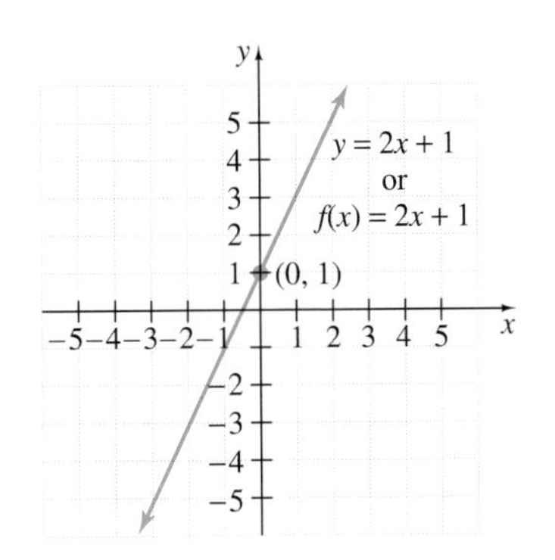

Practice Problems 18–21

Find each function value.

18. If $g(x) = 4x + 5$, find $g(0)$.
19. If $g(x) = 4x + 5$, find $g(-5)$.
20. If $f(x) = 3x^2 - x + 2$, find $f(2)$.
21. If $f(x) = 3x^2 - x + 2$, find $f(-1)$.

✓ CONCEPT CHECK

Suppose $y = f(x)$ and we are told that $f(3) = 9$. Which is not true?

a. When $x = 3$, $y = 9$.
b. A possible function is $f(x) = x^2$.
c. A point on the graph of the function is $(3, 9)$.
d. A possible function is $f(x) = 2x + 4$.

Practice Problem 22

Graph the function $f(x) = 3x - 2$.

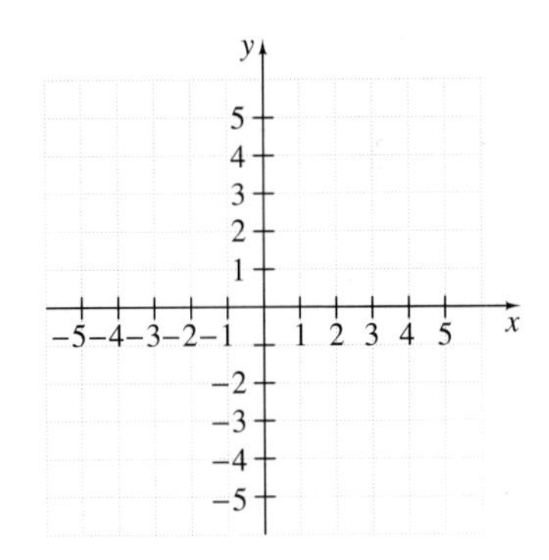

Answers

18. $g(0) = 5$, **19.** $g(-5) = -15$, **20.** $f(2) = 12$, **21.** $f(-1) = 6$

22.

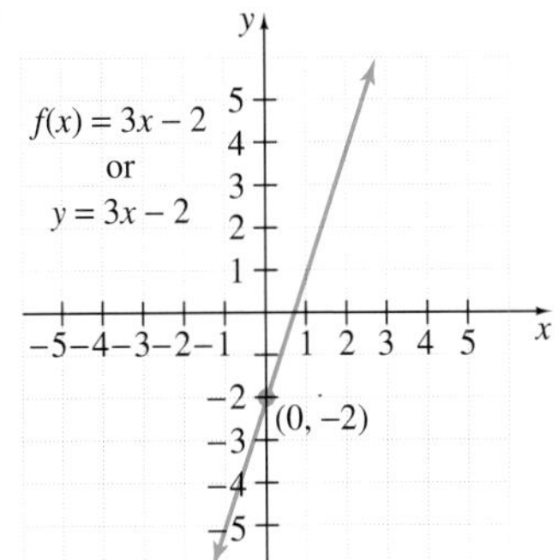

✓ **Concept Check:** d

Focus On Mathematical Connections

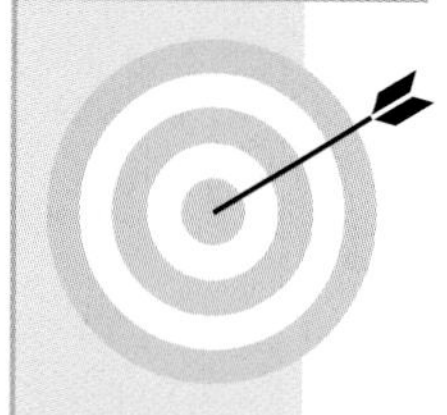

PERPENDICULAR BISECTORS

A **perpendicular bisector** is a line that is perpendicular to a given line segment and divides the segment into two equal lengths. A perpendicular bisector crosses the line segment at the point that is located exactly halfway between the two endpoints of the line segment. That point is called the **midpoint** of the line segment. If a line segment has the endpoints (x_1, y_1) and (x_2, y_2), then the midpoint of this line segment is the point with coordinates $\left(\frac{x_1 + x_2}{2}, \frac{y_1 + y_2}{2}\right)$. An example of a line segment and its perpendicular bisector is shown in the figure.

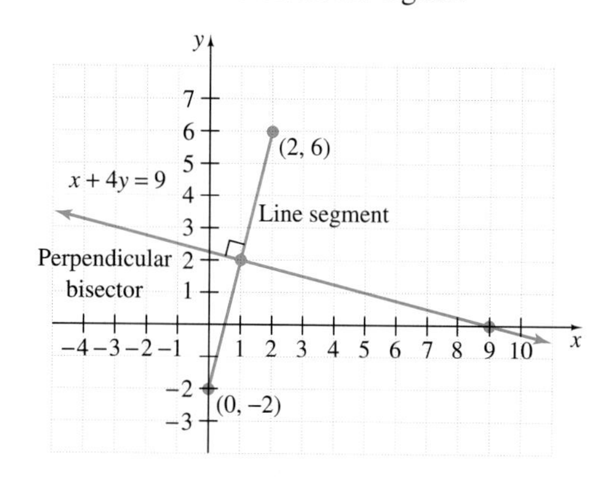

To find the equation of a line segment's perpendicular bisector, follow these steps:

Step 1. Find the midpoint of the line segment.
Step 2. Find the slope of the line segment.
Step 3. Find the slope of a line that is perpendicular to the line segment.
Step 4. Use the midpoint and the slope of the perpendicular line to find the equation of the perpendicular bisector.

CRITICAL THINKING

Use the steps given above and what you have learned in this chapter to find the equation of the perpendicular bisector of each line segment whose endpoints are given.

1. $(3, -1); (-5, 1)$

2. $(-6, -3); (-8, -1)$

3. $(-2, 6); (-22, -4)$

4. $(5, 8); (7, 2)$

5. $(2, 3); (-4, 7)$

6. $(-6, 8); (-4, -2)$

Name ______________________ Section ______ Date ______

EXERCISE SET 3.5

A B *Find the domain and the range of each relation. Also determine whether the relation is a function. See Examples 1 through 6.*

1. $\{(-1, 7), (0, 6), (-2, 2), (5, 6)\}$

2. $\{(4, 9), (-4, 9), (2, 3), (10, -5)\}$

3. $\{(-2, 4), (6, 4), (-2, -3), (-7, -8)\}$

4. $\{(6, 6), (5, 6), (5, -2), (7, 6)\}$

5. $\{(1, 1), (1, 2), (1, 3), (1, 4)\}$

6. $\{(1, 1), (2, 1), (3, 1), (4, 1)\}$

7. $\left\{\left(\frac{3}{2}, \frac{1}{2}\right), \left(1\frac{1}{2}, -7\right), \left(0, \frac{4}{5}\right)\right\}$

8. $\{(\pi, 0), (0, \pi), (-2, 4), (4, -2)\}$

9. $\{(-3, -3), (0, 0), (3, 3)\}$

10. $\left\{\left(\frac{1}{2}, \frac{1}{4}\right), \left(0, \frac{7}{8}\right), (0.5, \pi)\right\}$

11.

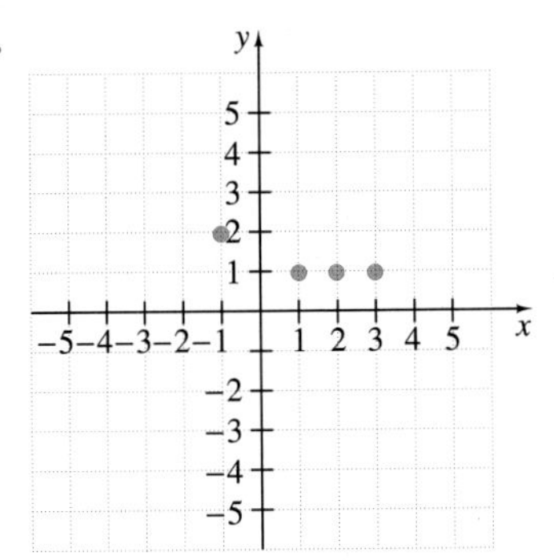

12.

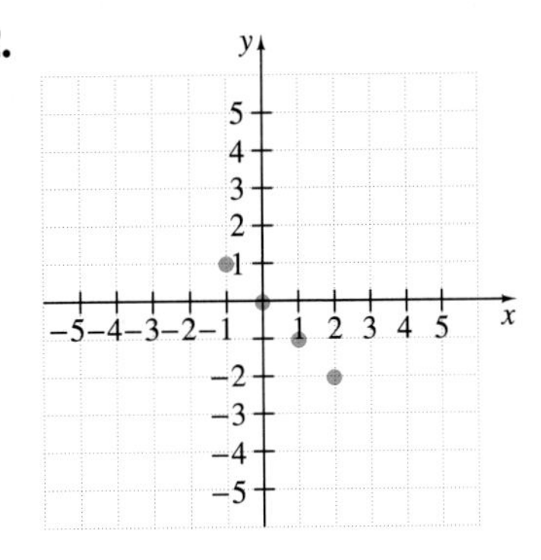

13.

Input: **State**

Output: **Number of Congressional Representatives**

Colorado
Alaska
Delaware
Illinois
Connecticut
Texas

6
1
20
30

14.

Input: **Animal**

Output: **Average Life Span (in years)**

Polar Bear
Cow
Chimpanzee
Giraffe
Gorilla
Kangaroo
Red Fox

20
15
10
7

ANSWERS

1. ______
2. ______
3. ______
4. ______
5. ______
6. ______
7. ______
8. ______
9. ______
10. ______
11. ______
12. ______
13. ______
14. ______

Name ______________________________

15. ______
16. ______
17. ______
18. ______
19. ______
20. ______
21. ______
22. ______
23. ______
24. ______
25. ______
26. ______
27. ______
28. ______
29. ______
30. ______

15. **Input: Degrees Fahrenheit** **Output: Degrees Celsius**

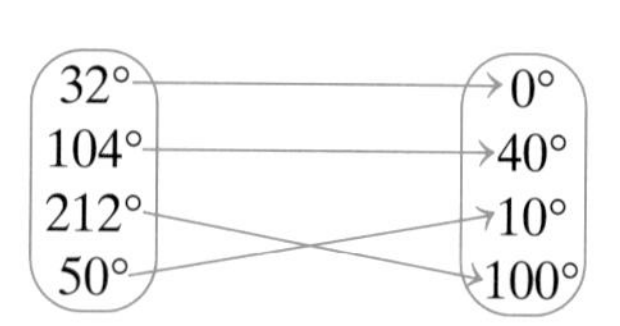

16. **Input: Words** **Output: Number of Letters**

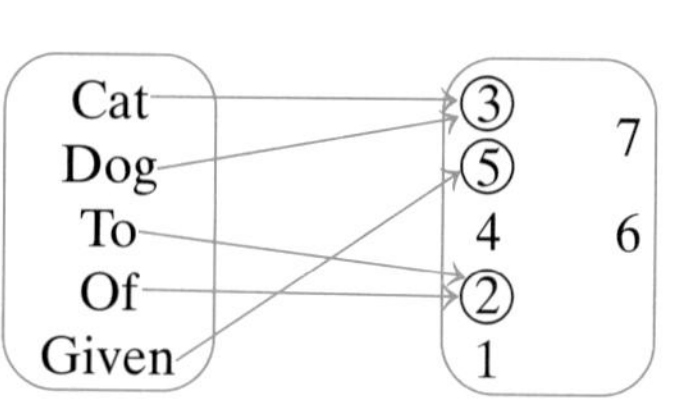

17. **Input:** **Output:**

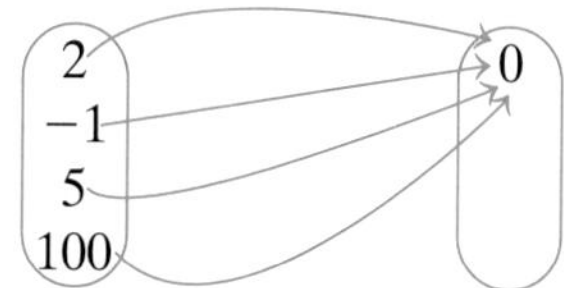

18. **Input:** **Output:**

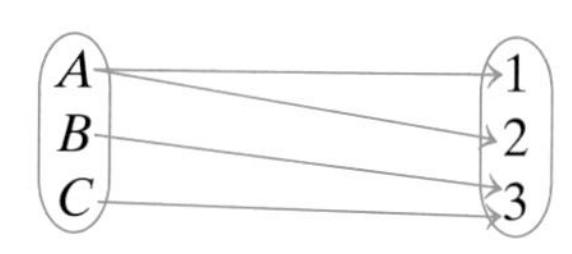

Determine whether each relation is a function. See Examples 4 through 6.

19.

First Set: Input	Correspondence	Second Set: Output
Class of algebra students	Grade average	Set of nonnegative numbers

20.

First Set: Input	Correspondence	Second Set: Output
People in New Orleans (population 500,000)	Birthdate	Days of the year

Determine whether each relation is also a function. See Examples 7 and 8.

21. $y = x + 1$ **22.** $y = x - 1$ **23.** $x = 2y^2$

24. $y = x^2$ **25.** $y - x = 7$ **26.** $2x - 3y = 9$

C *Use the vertical line test to determine whether each graph is the graph of a function. See Examples 9 through 13.*

27.

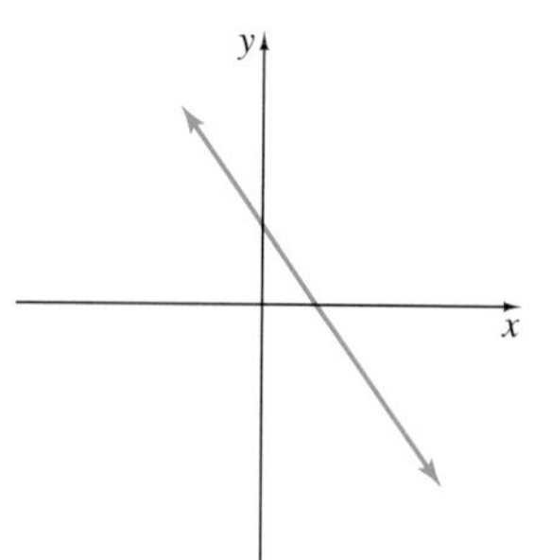

28.

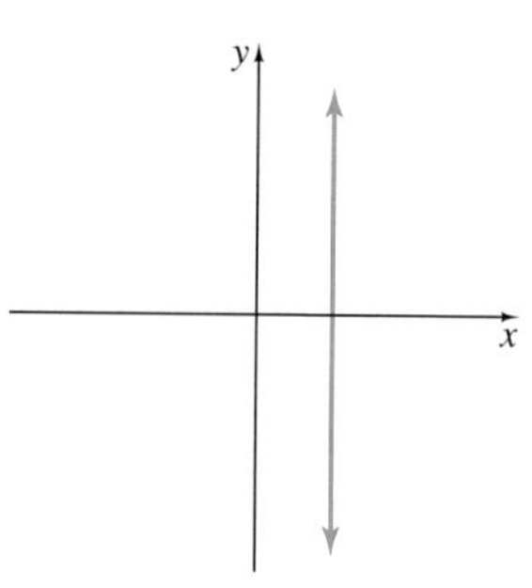

29.

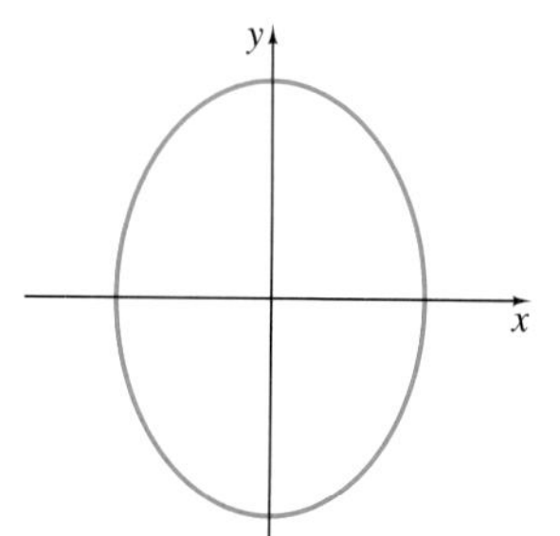

30.

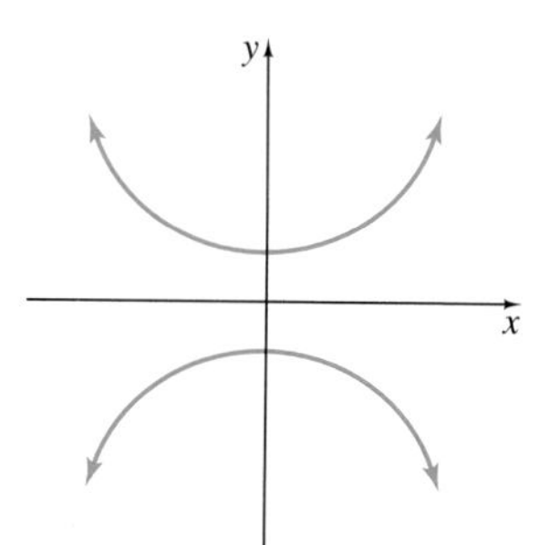

Name ______________________________

31. ______
32. ______
33. ______
34. ______
35. ______
36. ______
37. ______
38. ______
39. ______
40. ______

31.

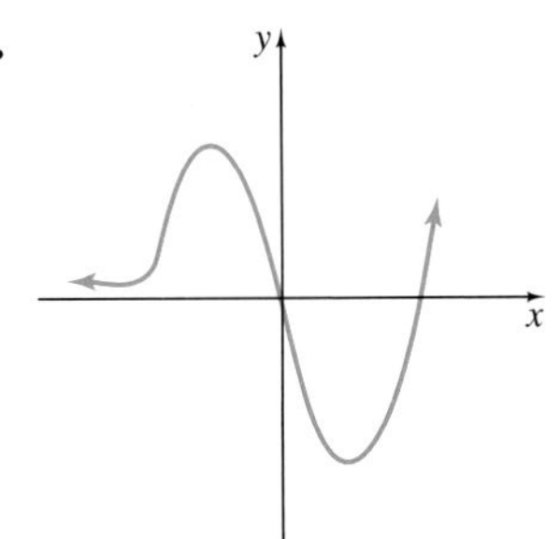

32.

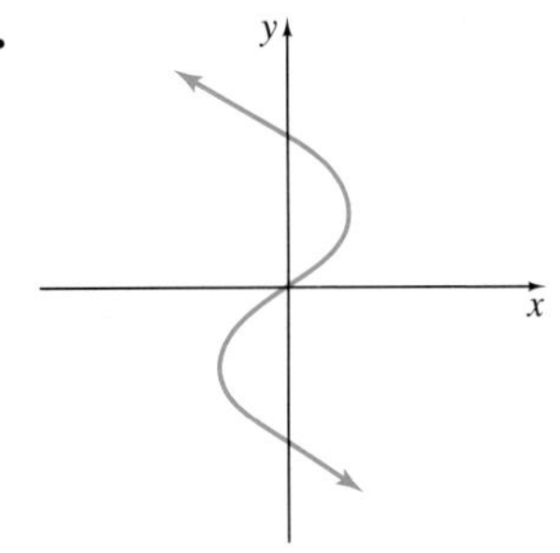

D *Find the domain and the range of each relation. Use the vertical line test to determine whether each graph is the graph of a function. See Examples 14 through 17.*

33.

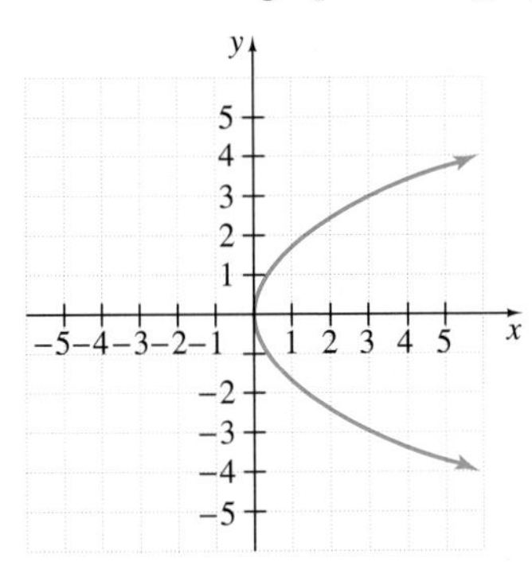

34.

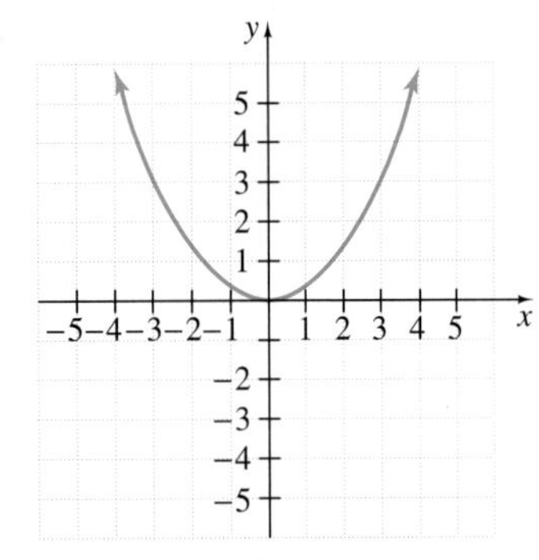

35.

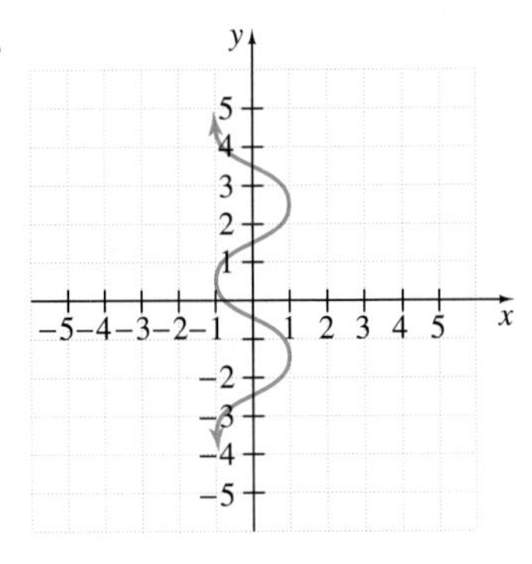

36.

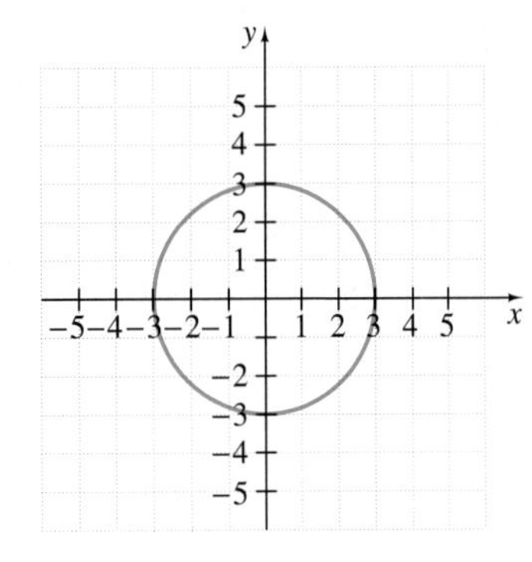

37.

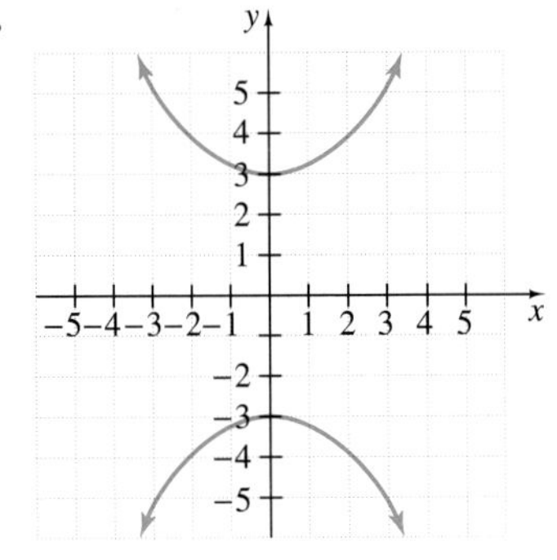

38.

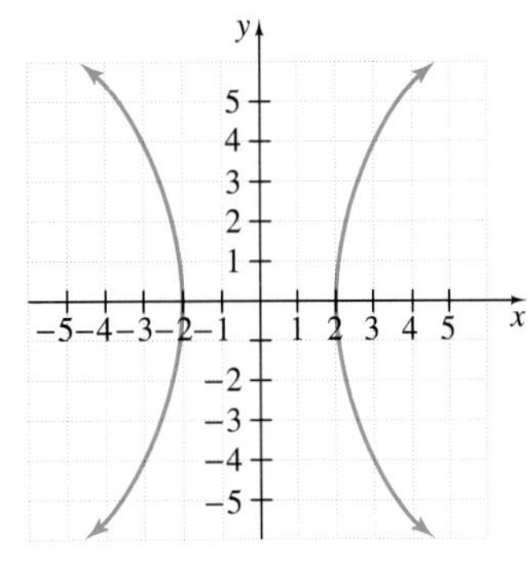

39.

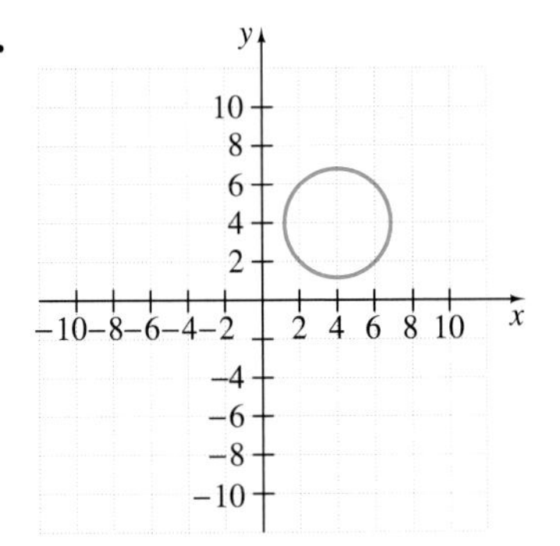

40.

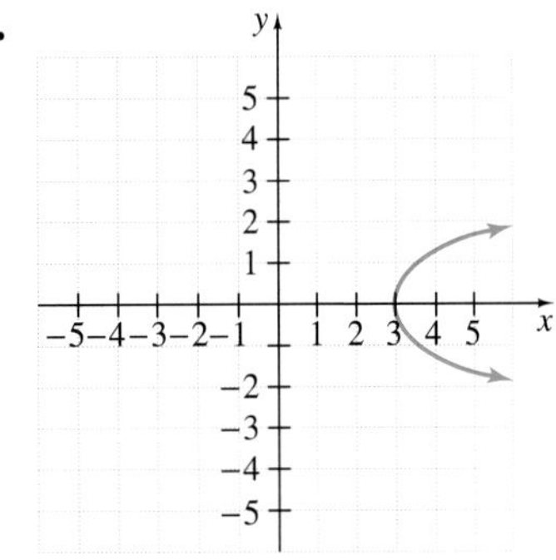

Name ______________________________

41. ______
42. ______
43. ______
44. ______
45. ______
46. ______
47. ______
48. ______
49. ______
50. ______
51. ______
52. ______
53. a. ______
b. ______
c. ______
54. a. ______
b. ______
c. ______
55. a. ______
b. ______
c. ______
56. a. ______
b. ______
c. ______
57. ______
58. ______

41.

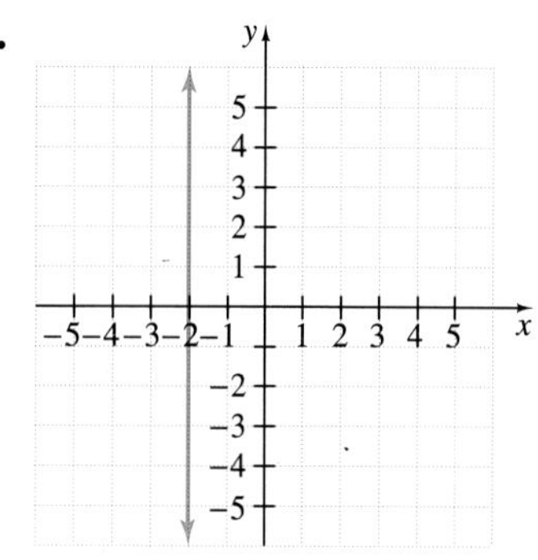

42.

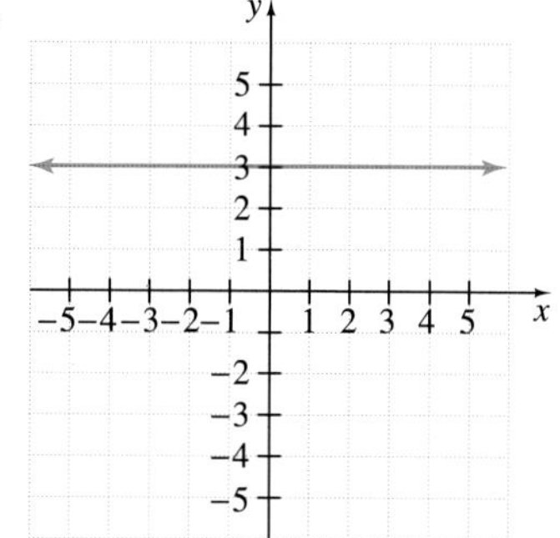

43.

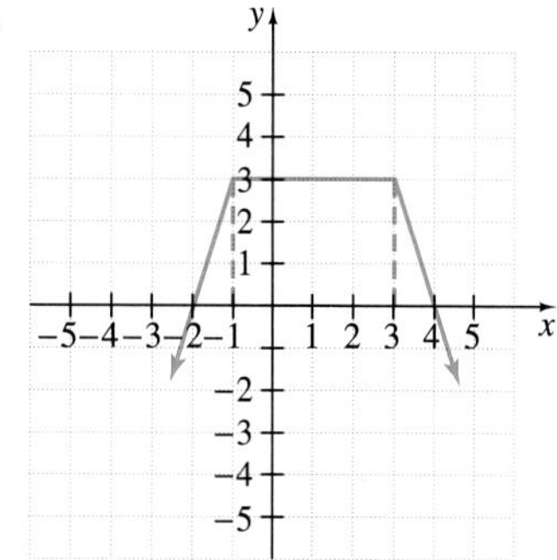

44.

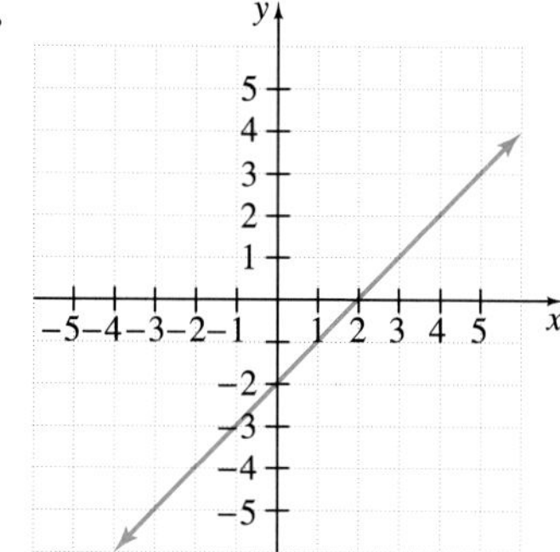

E *If $f(x) = 3x + 3$, $g(x) = 4x^2 - 6x + 3$, and $h(x) = 5x^2 - 7$, find each function value. See Examples 18 through 21.*

45. $f(4)$ **46.** $f(-1)$ **47.** $h(-3)$ **48.** $h(0)$

49. $g(2)$ **50.** $g(1)$ **51.** $g(0)$ **52.** $h(-2)$

For each function, find the indicated values. See Examples 18 through 21.

53. $f(x) = \frac{1}{2}$;
a. $f(0)$
b. $f(2)$
c. $f(-2)$

54. $g(x) = -\frac{1}{3}x$;
a. $g(0)$
b. $g(-1)$
c. $g(3)$

55. $f(x) = -5$;
a. $f(2)$
b. $f(0)$
c. $f(606)$

56. $h(x) = 7$;
a. $h(7)$
b. $h(542)$
c. $h\left(-\frac{3}{4}\right)$

The function $A(r) = \pi r^2$ may be used to find the area of a circle if we are given its radius. Use this function to answer Exercises 57–58.

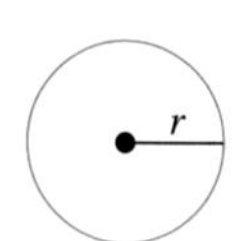

57. Find the area of a circle whose radius is 5 centimeters. (Do not approximate π.)

58. Find the area of a circular garden whose radius is 8 feet. (Do not approximate π.)

Name ______________________________

59. ____________

The function $V(x) = x^3$ may be used to find the volume of a cube if we are given the length x of a side. Use this function to answer Exercises 59–60.

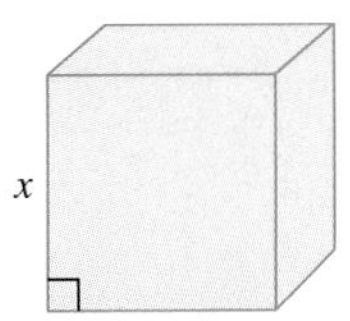

60. ____________

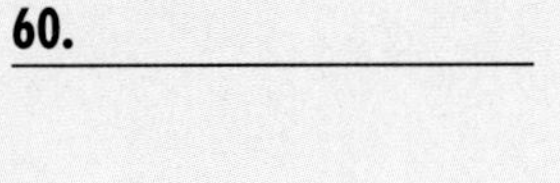

59. Find the volume of a cube whose side is 14 inches.

60. Find the volume of a die whose side is 1.7 centimeters.

61. ____________

Forensic scientists use the following functions to find the height of a woman if they are given the height of her femur bone (f) or her tibia bone (t) in centimeters.

$$H(f) = 2.59f + 47.24$$
$$H(t) = 2.72t + 61.28$$

62. ____________

Use these functions to answer Exercises 61–62.

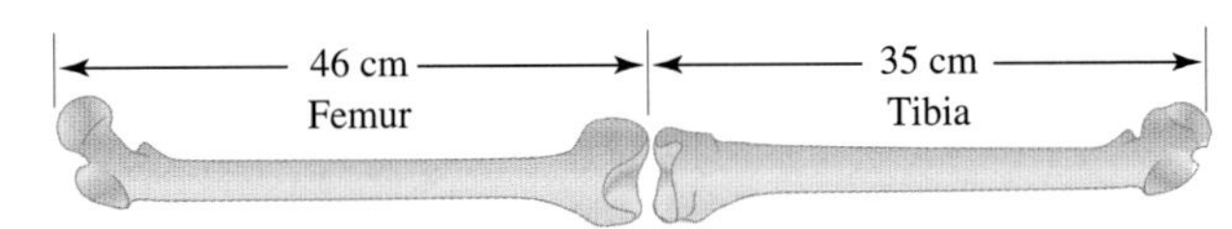

63. ____________

61. Find the height of a woman whose femur measures 46 centimeters.

62. Find the height of a woman whose tibia measures 35 centimeters.

64. ____________

The dosage in milligrams D of Ivermectin, a heartworm preventive, for a dog who weighs x pounds is given by

$$D(x) = \frac{136}{25}x$$

Use this function to answer Exercises 63–64.

63. Find the proper dosage for a dog that weighs 30 pounds.

64. Find the proper dosage for a dog that weighs 50 pounds.

65. a. ____________

b. ____________

65. The per capita consumption (in pounds) of all poultry in the United States is given by the function $C(x) = 1.7x + 88$, where x is the number of years since 1995. (*Source:* Based on actual and estimated data from the Economic Research Service, U.S. Department of Agriculture, 1995–1999)

a. Find and interpret $C(2)$.

b. Estimate the per capita consumption of all poultry in the United States in 1999.

66. The number of passengers (in millions) aboard airline flights in the United States is given by the function $P(x) = 25.4x + 448.4$, where x is the number of years since 1991. (*Source:* Based on data from the Air Transport Association of America, 1991–1997)

a. Find and interpret $P(4)$.

b. Estimate the number of airline passengers in 1997.

66. a. ____________

b. ____________

Name ______________________________

F *Graph each linear function. See Example 22.*

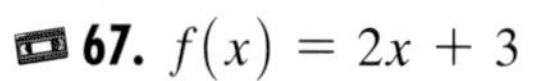

67. $f(x) = 2x + 3$

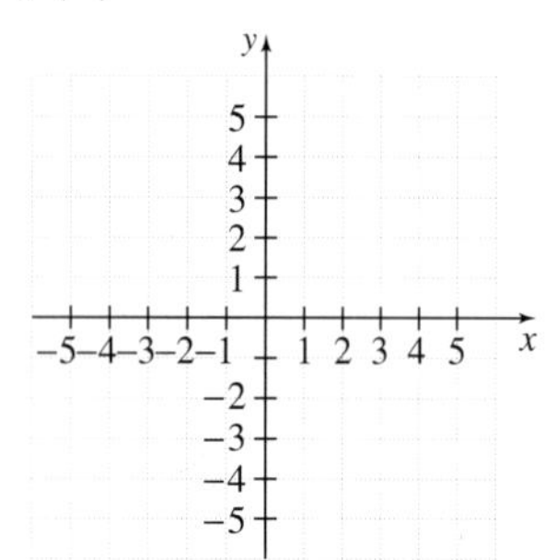

68. $f(x) = 5x - 1$

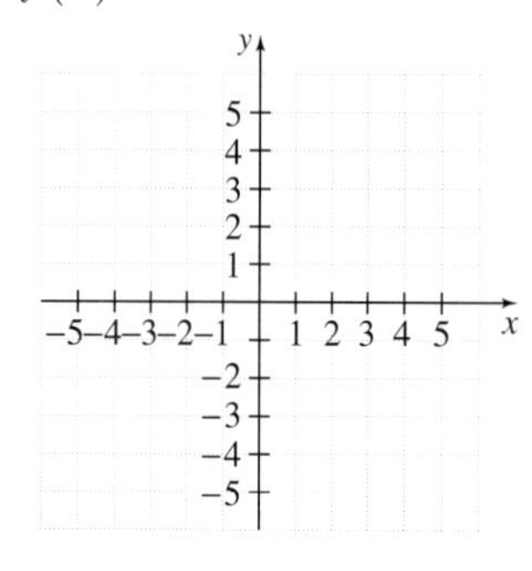

69. $f(x) = -3x$

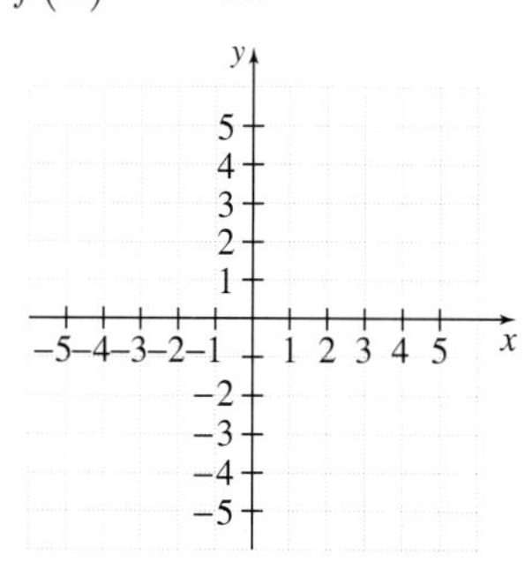

70. $f(x) = -4x$

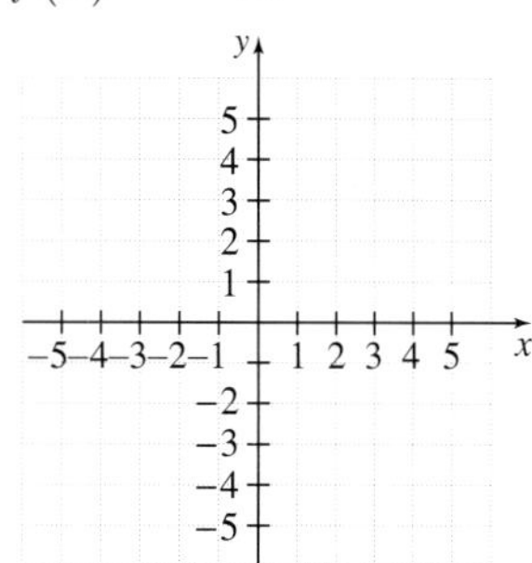

71. $f(x) = -x + 2$

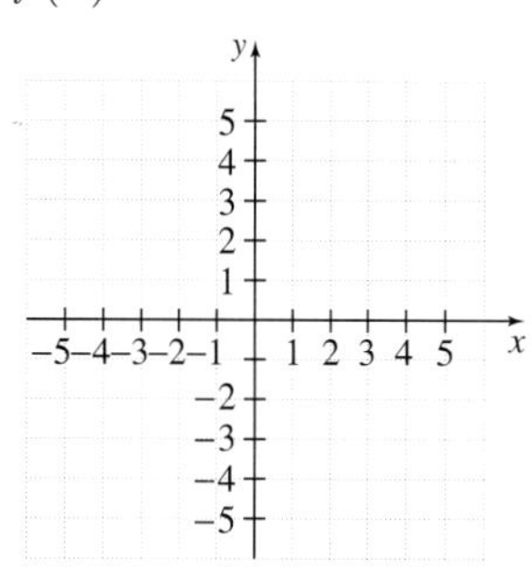

72. $f(x) = -x + 1$

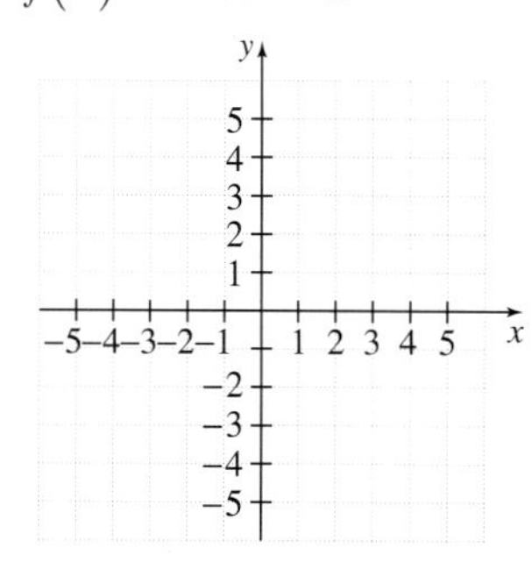

Review and Preview

Solve. See Section 2.4.

73. $2x - 7 \leq 21$

74. $-3x + 1 > 0$

75. $5(x - 2) \geq 3(x - 1)$

76. $-2(x + 1) \leq -x + 10$

77. $\frac{x}{2} + \frac{1}{4} < \frac{1}{8}$

78. $\frac{x}{5} - \frac{3}{10} \geq \frac{x}{2} - 1$

Combining Concepts

79. If $f(x) = 1.3x^2 - 2.6x + 5.1$, find

a. $f(2)$ **b.** $f(-2)$ **c.** $f(3.1)$

For each function, find the indicated values.

80. $f(x) = 2x + 7$;

a. $f(2)$ **b.** $f(a)$

81. $f(x) = x^2 - 12$;

a. $f(12)$ **b.** $f(a)$

82. Describe a function whose domain is the set of people in your home town.

83. Describe a function whose domain is the set of people in your algebra class.

84. Since $y = x + 7$ describes a function, rewrite the equation using function notation.

85. In your own words, explain how to find the domain of a function given its graph.

86. Explain the vertical line test and how it is used.

73. ____________
74. ____________
75. ____________
76. ____________
77. ____________
78. ____________
79. a. ____________
b. ____________
c. ____________
80. a. ____________
b. ____________
81. a. ____________
b. ____________
82. ____________
83. ____________
84. ____________
85. ____________
86. ____________

3.6 GRAPHING LINEAR INEQUALITIES

Objectives

A Graph linear inequalities.

B Graph the intersection or union of two linear inequalities.

A GRAPHING LINEAR INEQUALITIES

Recall that the graph of a linear equation in two variables is the graph of all ordered pairs that satisfy the equation, and we determined that the graph is a line. Here we graph **linear inequalities** in two variables; that is, we graph all the ordered pairs that satisfy the inequality.

If the equal sign in a linear equation in two variables is replaced with an inequality symbol, the result is a linear inequality in two variables. Some examples are

$3x + 5y \geq 6$ $\quad 2x - 4y < -3$

$4x > 2$ $\quad y \leq 5$

To graph the linear inequality $x + y < 3$, for example, we first graph the related equation $x + y = 3$. The resulting **boundary line** contains all ordered pairs the sum of whose coordinates is 3. The line separates the plane into two regions called **half-planes**. All points "above" the boundary line $x + y = 3$ have coordinates that satisfy the inequality $x + y > 3$, and all points "below" the line have coordinates that satisfy the inequality $x + y < 3$.

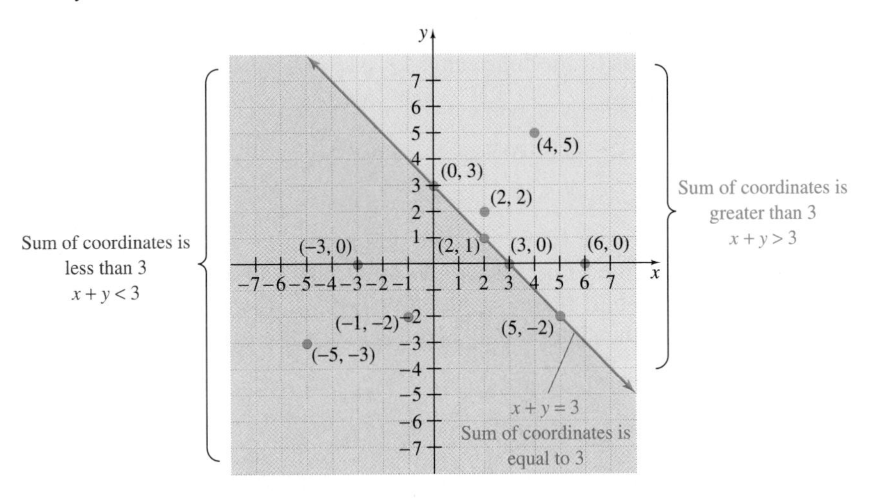

The graph, or **solution region**, for $x + y < 3$, then, is the half-plane below the boundary line and is shown shaded below. The boundary line is shown dashed since it is not a part of the solution region. These ordered pairs on this line satisfy $x + y = 3$, but not $x + y < 3$.

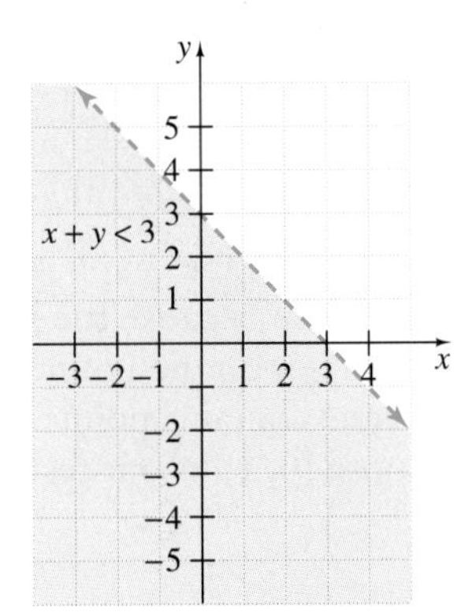

The following steps may be used to graph linear inequalities in two variables.

Graphing a Linear Inequality in Two Variables

Step 1. Graph the boundary line found by replacing the inequality sign with an equal sign. If the inequality sign is $<$ or $>$, graph a dashed line indicating that points on the line are not solutions of the inequality. If the inequality sign is $\leq$ or $\geq$, graph a solid line indicating that points on the line are solutions of the inequality.

Step 2. Choose a **test point** *not on the boundary line* and substitute the coordinates of this test point into the *original inequality.*

Step 3. If a true statement is obtained in *Step 2*, shade the half-plane that contains the test point. If a false statement is obtained, shade the half-plane that does not contain the test point.

Practice Problem 1

Graph: $x + 3y > 4$

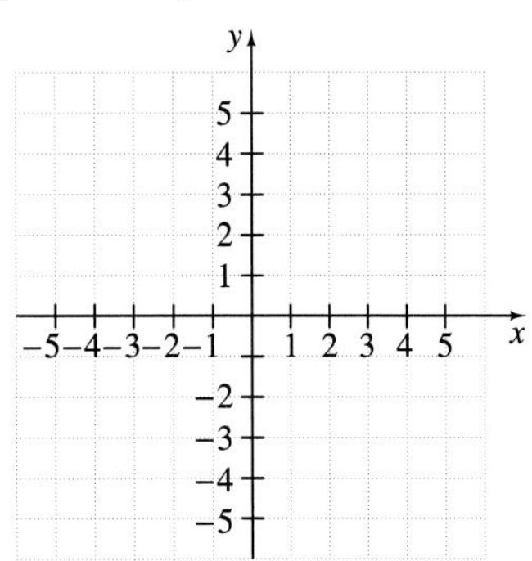

Practice Problem 2

Graph: $x \leq 2y$

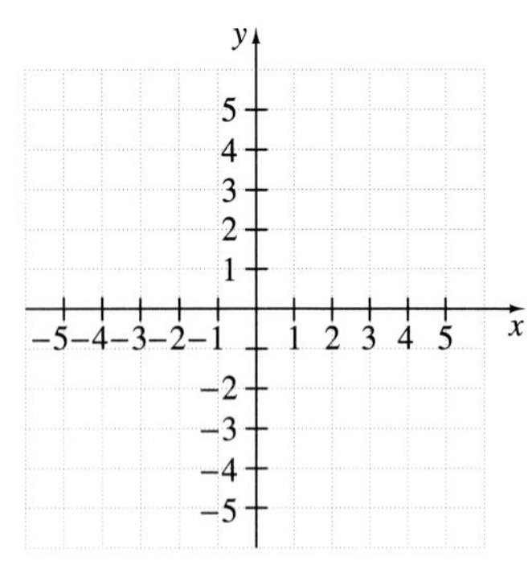

Answer

1.

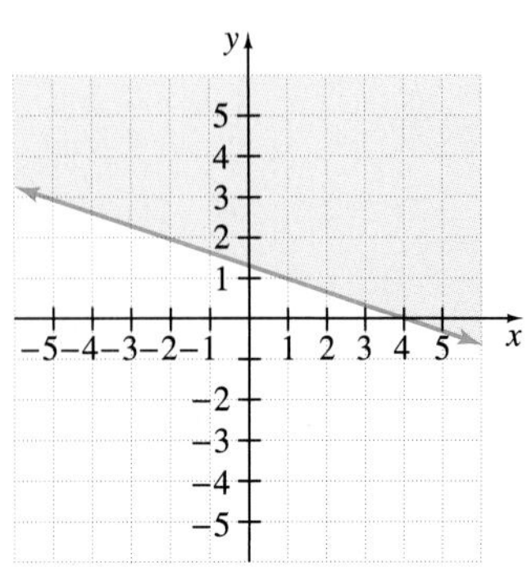

2. (answer on next page)

Example 1

Graph: $2x - y < 6$

Solution: The boundary line for this inequality is the graph of $2x - y = 6$. We graph a dashed boundary line because the inequality symbol is $<$. Next we choose a test point on either side of the boundary line. The point (0, 0) is not on the boundary line, so we use this point. Replacing x with 0 and y with 0 in the *original inequality* $2x - y < 6$ leads to the following:

$$2x - y < 6$$
$$2(0) - 0 < 6 \quad \text{Let } x = 0 \text{ and } y = 0.$$
$$0 < 6 \quad \text{True.}$$

Because (0, 0) satisfies the inequality, so does every point on the same side of the boundary line as (0, 0). We shade the half-plane that contains (0, 0), as shown. Every point in the shaded half-plane satisfies the original inequality.

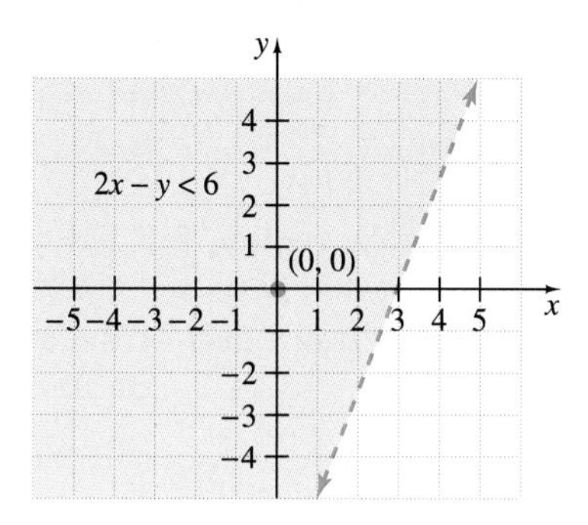

Example 2

Graph: $3x \geq y$

Solution: The boundary line is the graph of $3x = y$. We graph a solid boundary line because the inequality symbol is $\geq$. We test a point not on the boundary line to determine which half-plane contains points that satisfy the inequality. Let's choose (0, 1) as our test point.

$$3x \geq y$$
$$3(0) \geq 1 \quad \text{Let } x = 0 \text{ and } y = 1.$$
$$0 \geq 1 \quad \text{False.}$$

This point does not satisfy the inequality, so the correct half-plane is on the opposite side of the boundary line from (0, 1). The graph of $3x \geq y$ is the boundary line together with the shaded region, as shown.

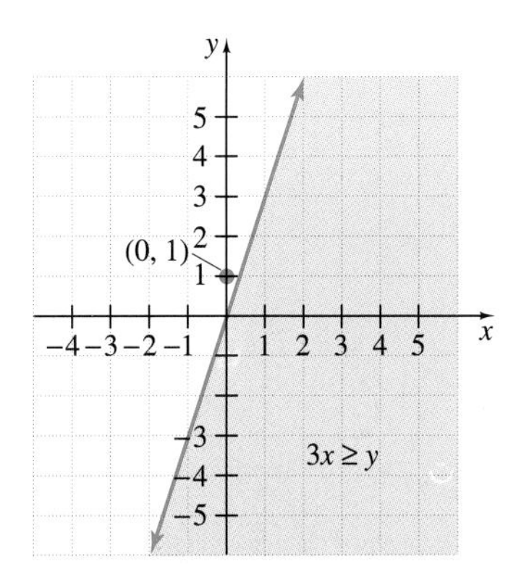

TRY THE CONCEPT CHECK IN THE MARGIN.

B GRAPHING INTERSECTIONS AND UNIONS

The intersections and the unions of linear inequalities can also be graphed, as shown in the next two examples.

Example 3 Graph the intersection of $x \geq 1$ and $y \geq 2x - 1$.

Solution: First we graph each inequality. The intersection of the two graphs is all points common to both regions, as shown by the *heaviest* shading in the third graph.

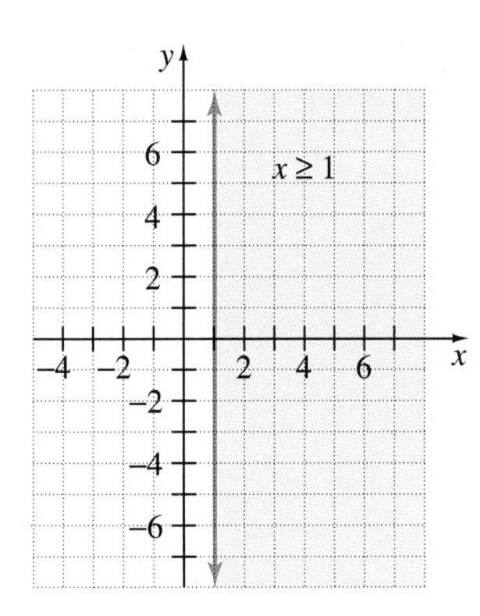

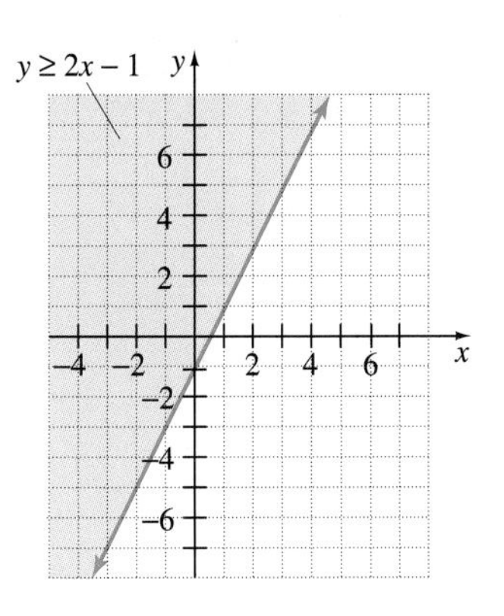

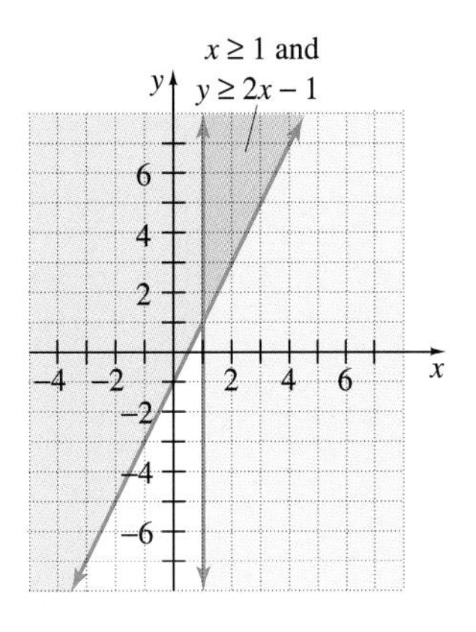

✓ CONCEPT CHECK

If a point on the boundary line is included in the solution of an inequality in two variables, should the graph of the boundary line be solid or dashed?

Practice Problem 3

Graph the intersection of $x \leq 2$ and $y \geq x + 1$.

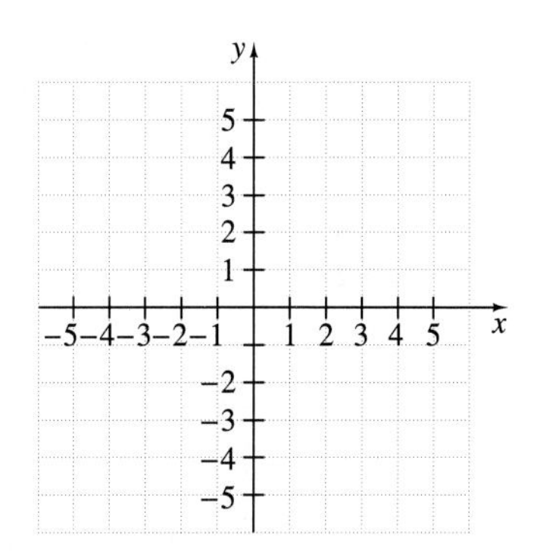

Answers

2.

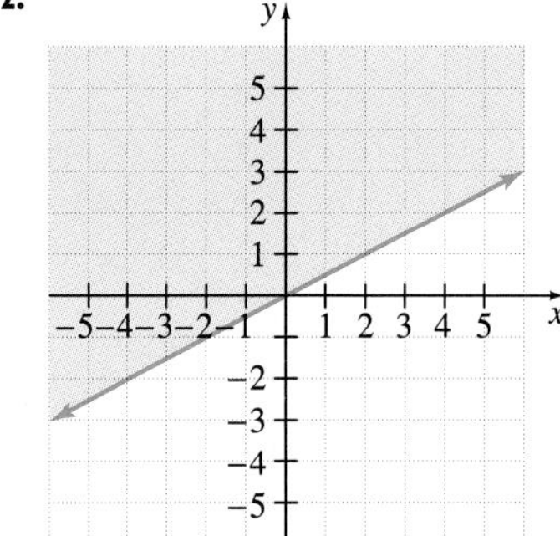

3.

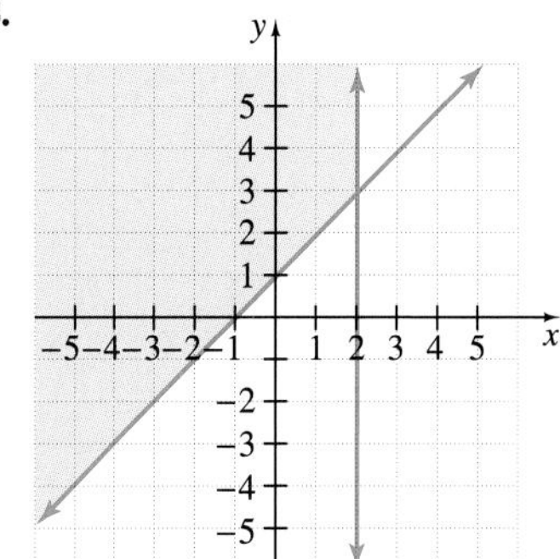

✓ Concept Check: solid

Practice Problem 4

Graph the union of $x + 2y \leq 4$ or $y \geq -1$

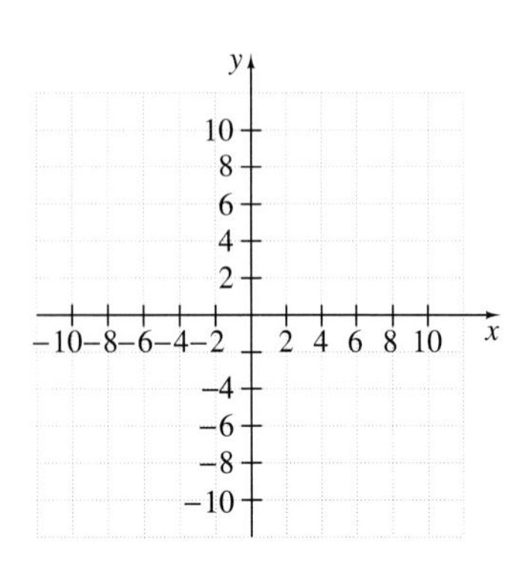

Example 4 Graph the union of $2x + y \geq -8$ or $y \leq -2$.

Solution: First we graph each inequality. The union of the two inequalities is both shaded regions, including the solid boundary lines, as shown in the third graph.

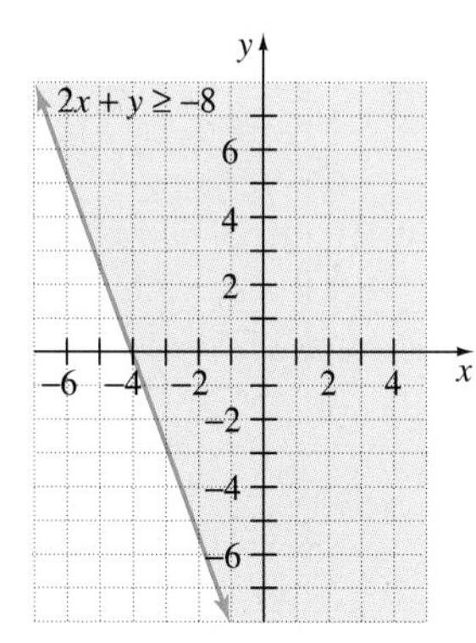

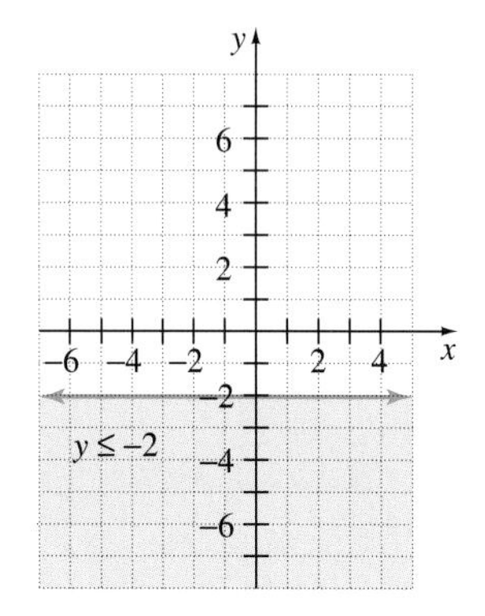

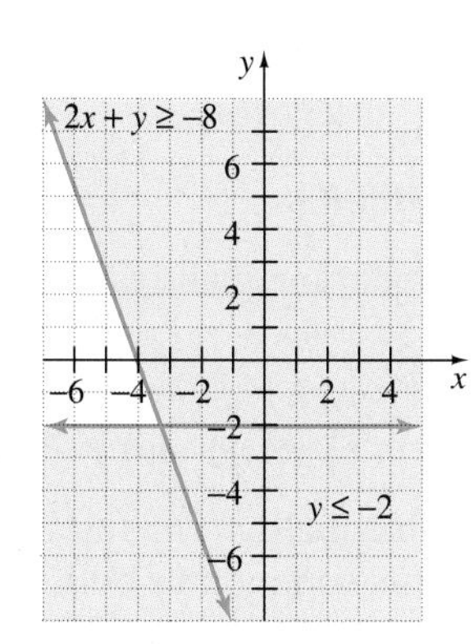

Answer

4.

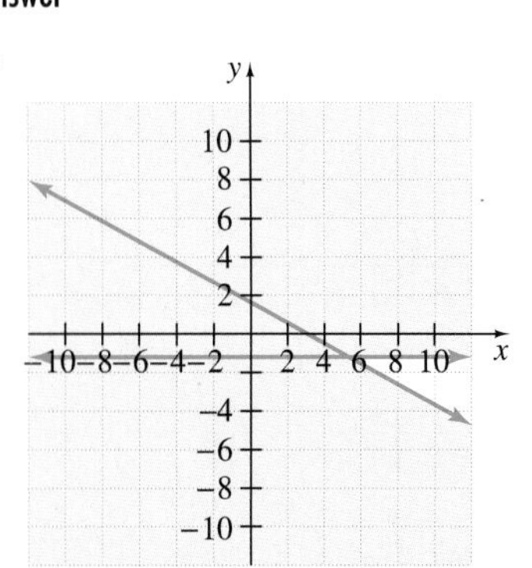

Name ______________________ Section ________ Date __________

Exercise Set 3.6

A *Graph each inequality. See Examples 1 and 2.*

1. $x < 2$

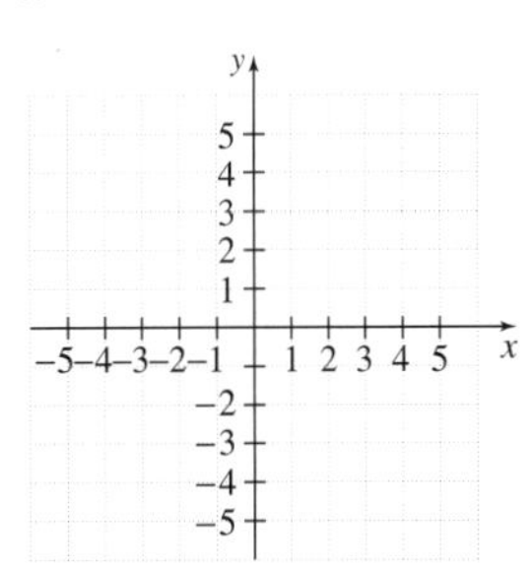

2. $x > -3$

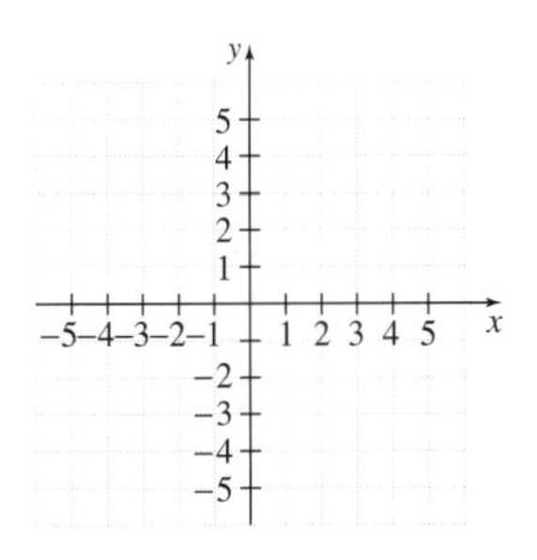

3. $x - y \geq 7$

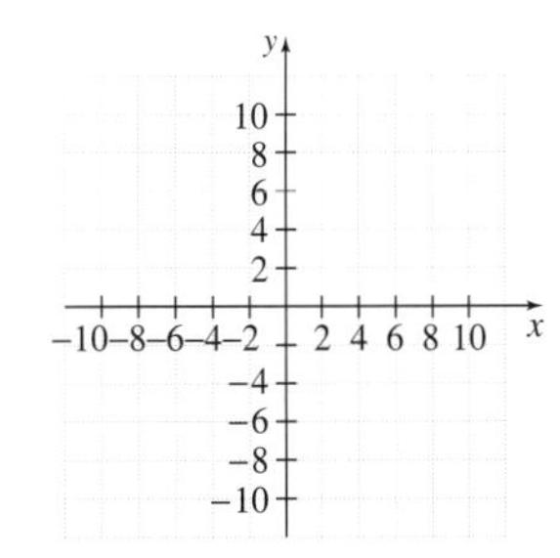

4. $3x + y \leq 1$

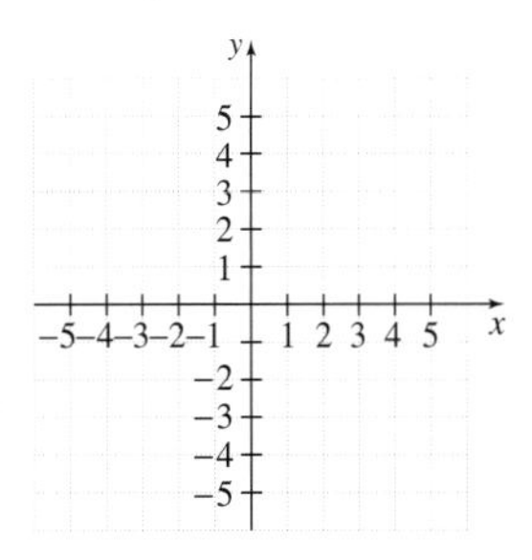

5. $3x + y > 6$

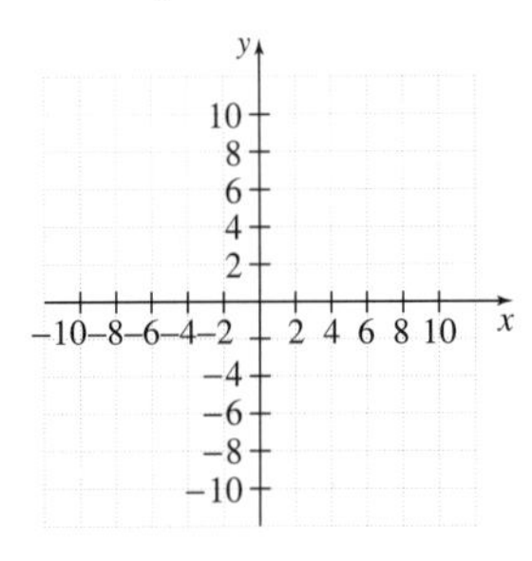

6. $2x + y > 2$

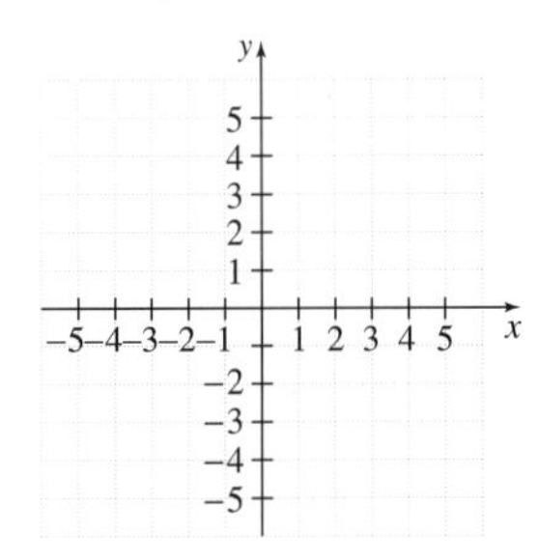

7. $y \leq -2x$

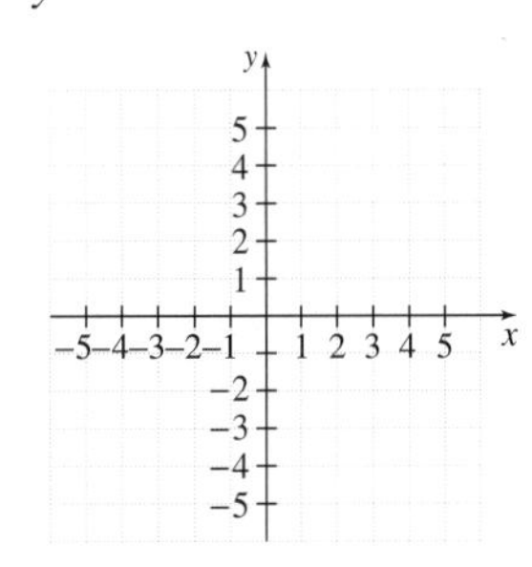

8. $y \leq 3x$

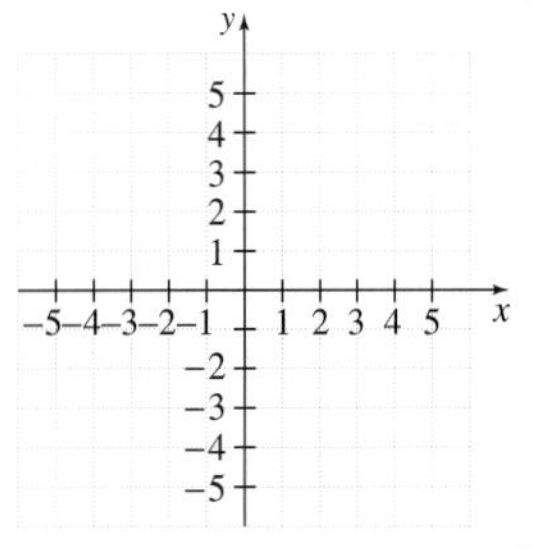

9. $2x + 4y \geq 8$

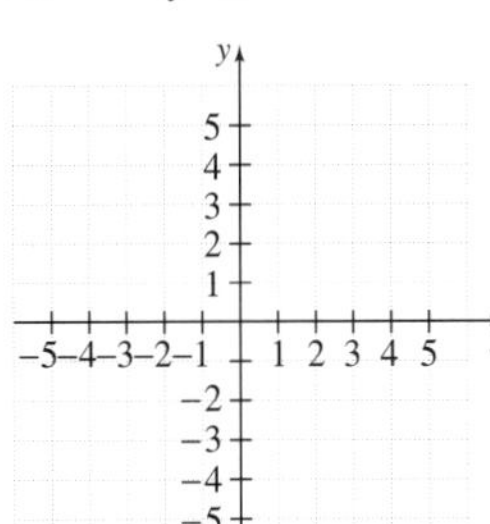

10. $2x + 6y \leq 12$

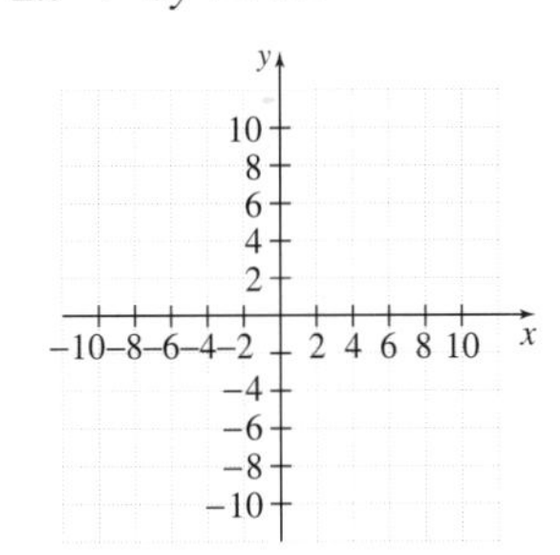

11. $5x + 3y > -15$

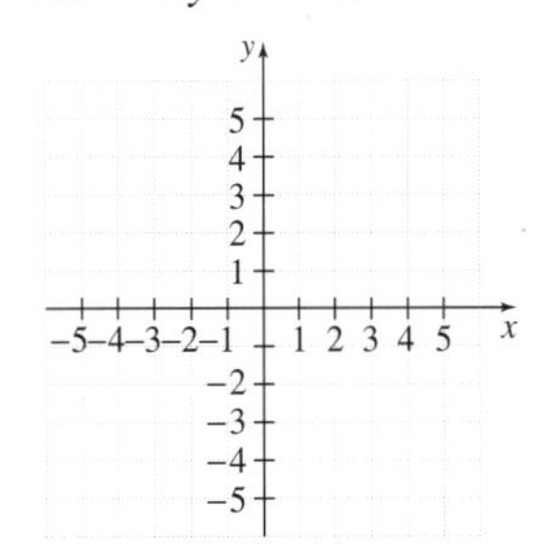

12. $2x + 5y < -20$

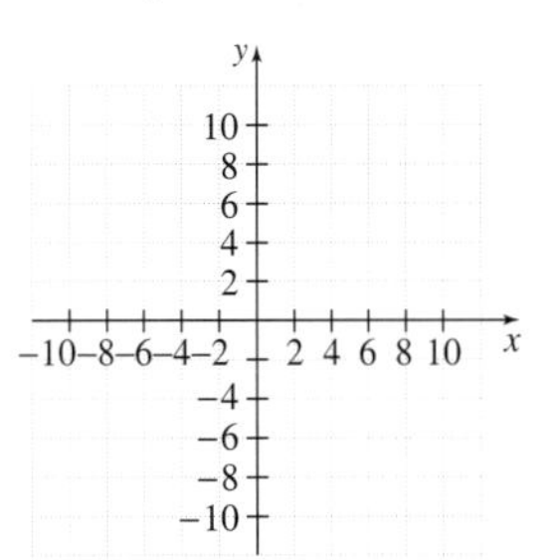

13. Explain when a dashed boundary line should be used in the graph of an inequality.

14. Explain why, after the boundary line is sketched, we test a point on either side of this boundary in the original inequality.

13. ______________

14. ______________

Name ____________________

B *Graph each union or intersection. See Examples 3 and 4.*

15. The intersection of $x \geq 3$ and $y \leq -2$

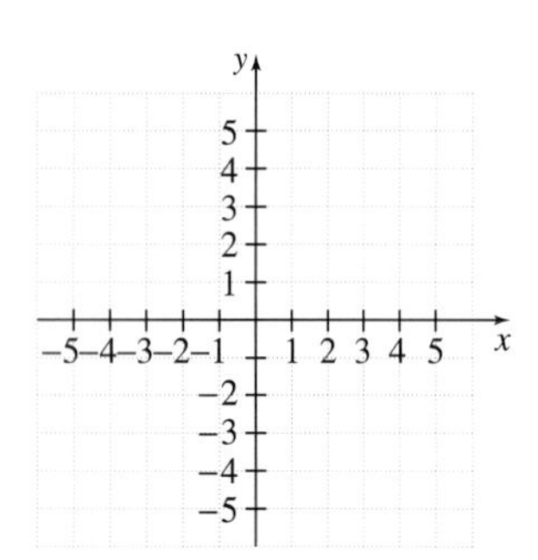

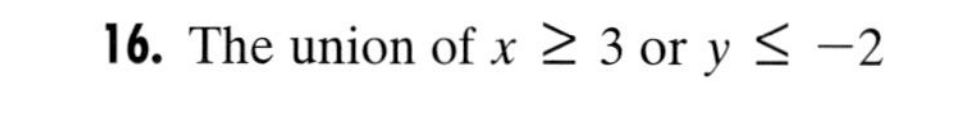

16. The union of $x \geq 3$ or $y \leq -2$

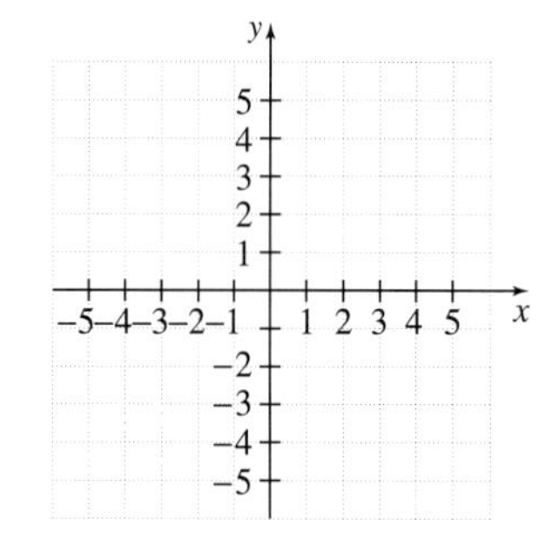

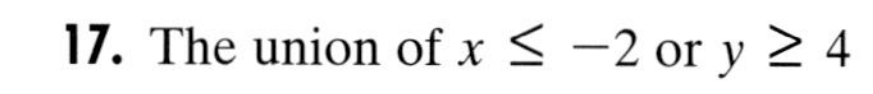

17. The union of $x \leq -2$ or $y \geq 4$

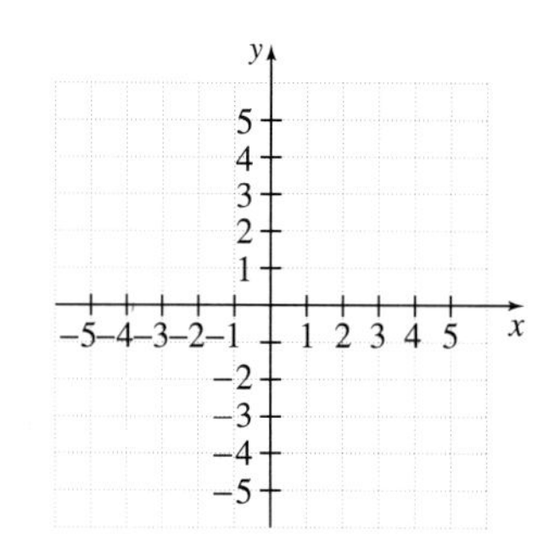

18. The intersection of $x \leq -2$ and $y \geq 4$

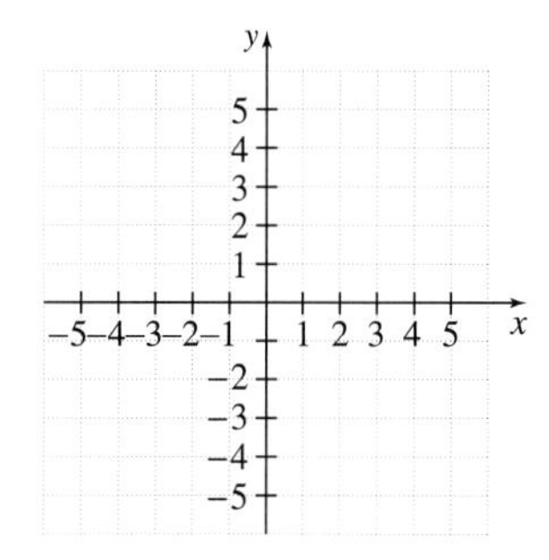

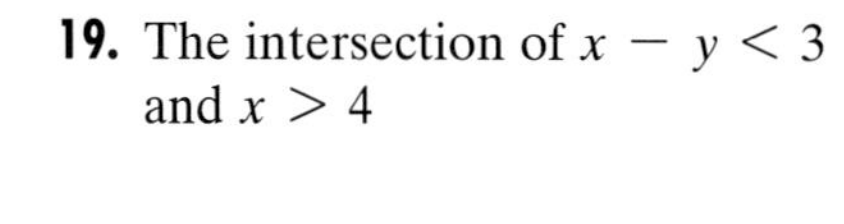

19. The intersection of $x - y < 3$ and $x > 4$

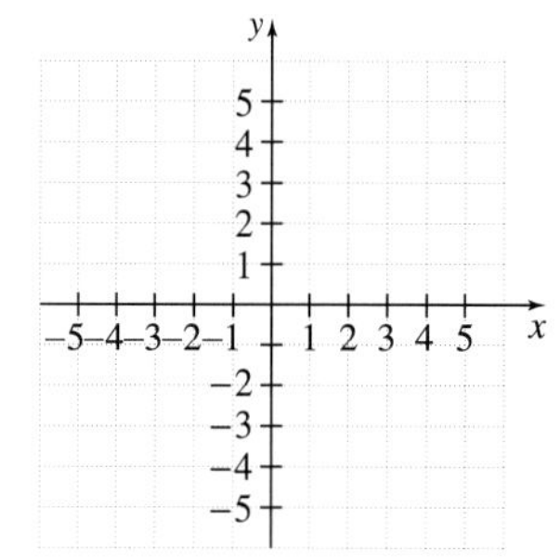

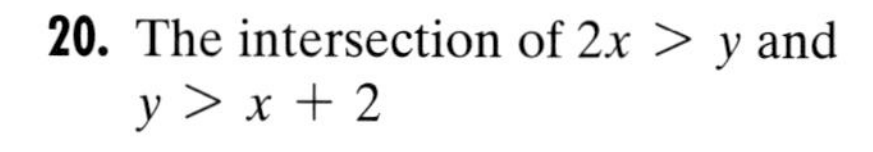

20. The intersection of $2x > y$ and $y > x + 2$

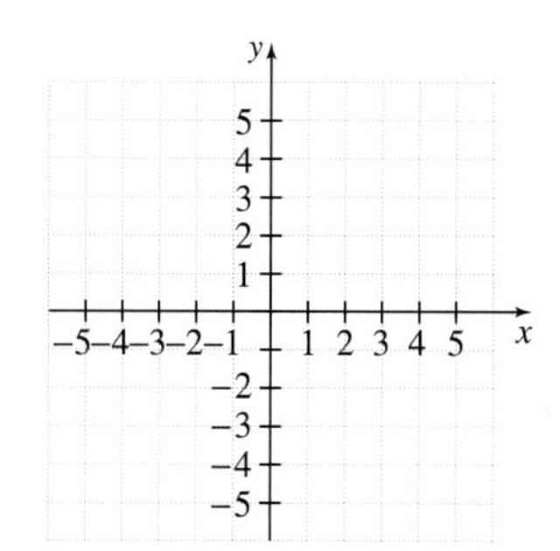

21. The union of $x + y \leq 3$ or $x - y \geq 5$

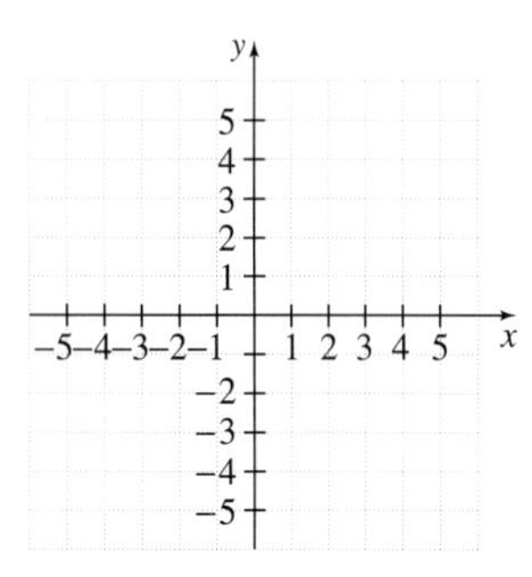

22. The union of $x - y \leq 3$ or $x + y > -1$

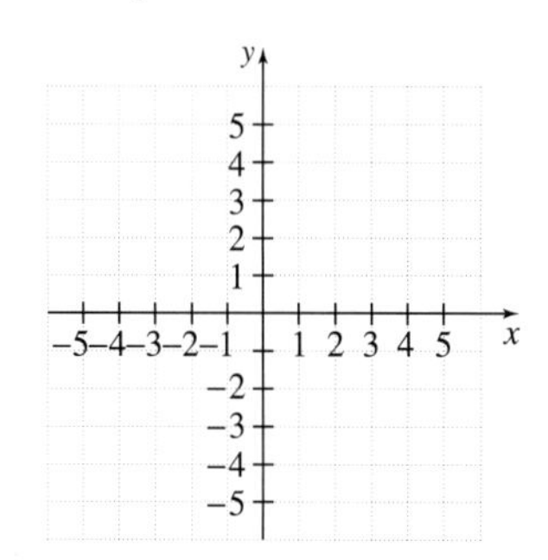

23. The union of $x - y \geq 2$ or $y < 5$

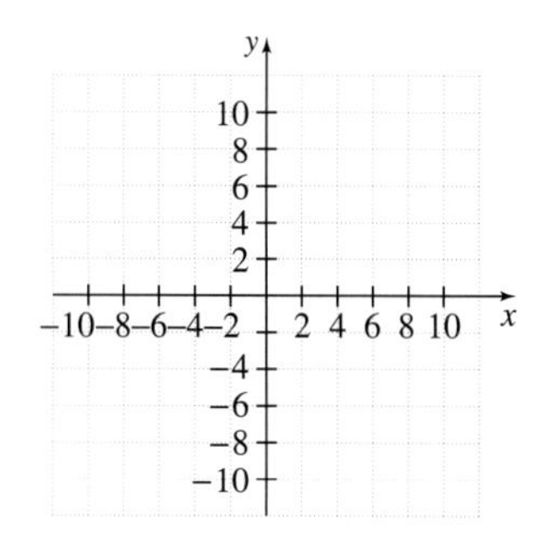

Name ______________________________

24. The union of $x - y < 3$ or $x > 4$

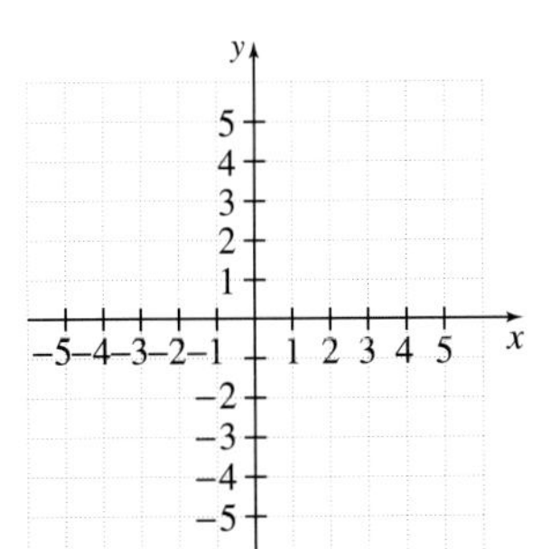

25. The intersection of $x + y \leq 1$ and $y \leq -1$

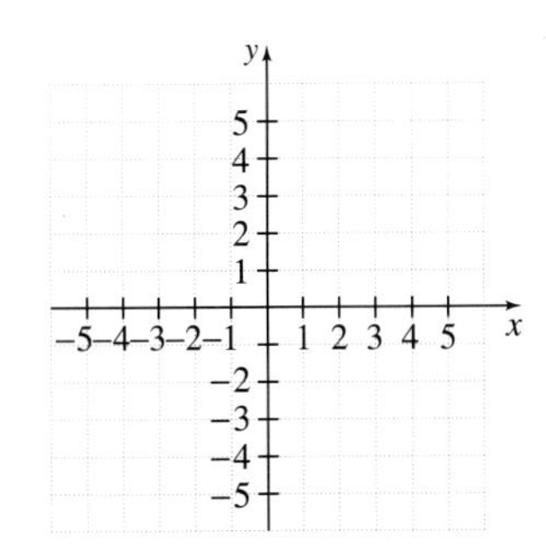

26. The intersection of $y \geq x$ and $2x - 4y \geq 6$

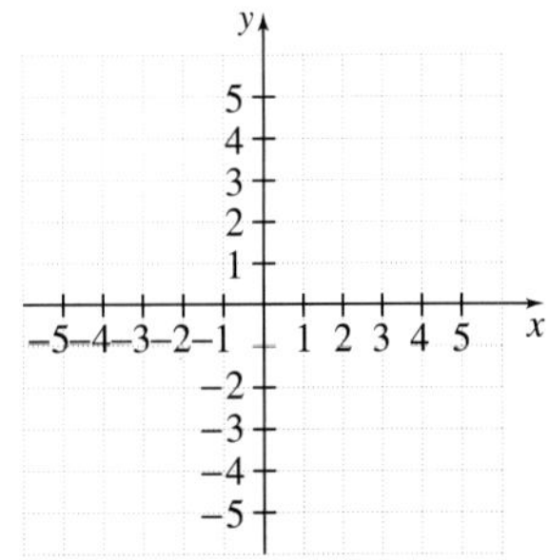

Match each inequality with its graph.

A

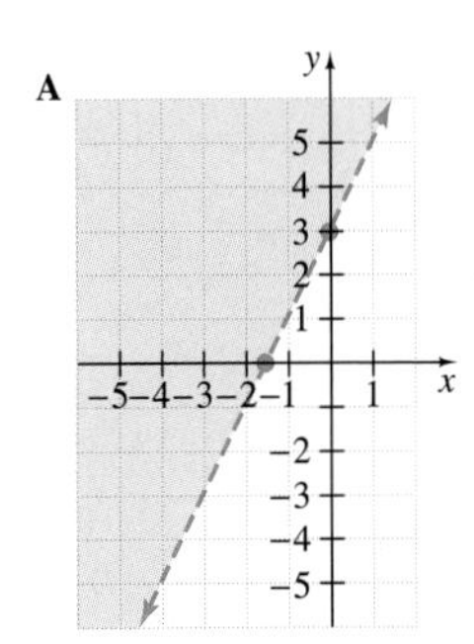

B

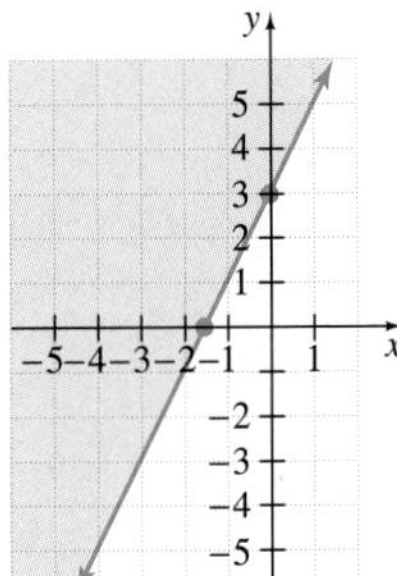

C

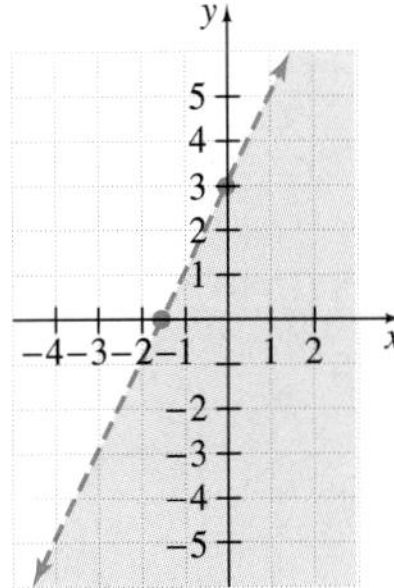

D

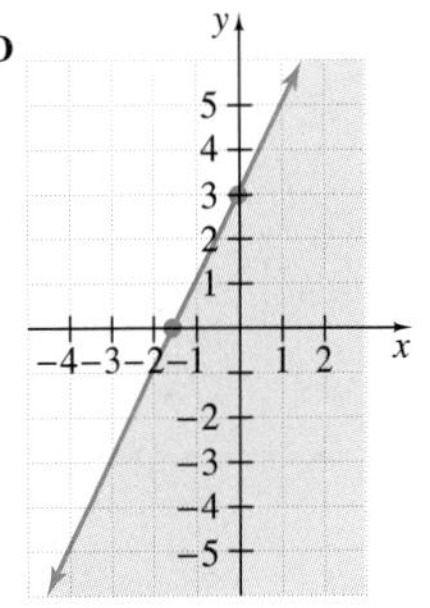

27. $y \leq 2x + 3$ **28.** $y < 2x + 3$ **29.** $y > 2x + 3$ **30.** $y \geq 2x + 3$

Review and Preview

Determine whether the ordered pair is a solution of both equations. See Section 3.1.

31. $(3, -1)$; $x - y = 4$, $x + 2y = 1$

32. $(0, 2)$; $x + 3y = 6$, $4x - y = -2$

33. $(-4, 0)$; $3x + 2y = -12$, $x = 4y$

34. $(-5, 2)$; $x + y = -3$, $2x - y = -8$

27. ____________

28. ____________

29. ____________

30. ____________

31. ____________

32. ____________

33. ____________

34. ____________

35. see graph

36. see graph

Name ______________________________

Combining Concepts

Solve.

35. Chris-Craft manufactures boats out of Fiberglas and wood. Fiberglas hulls require 2 hours of work, whereas wood hulls require 4 hours of work. Employees work at most 40 hours a week. The following inequalities model these restrictions, where x represents the number of Fiberglas hulls produced and y represents the number of wood hulls produced.

$$\begin{cases} x \geq 0 \\ y \geq 0 \\ 2x + 4y \leq 40 \end{cases}$$

Graph the intersection of these inequalities.

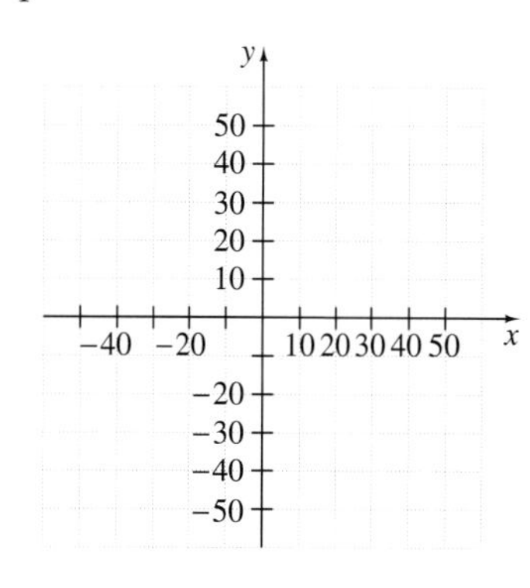

36. Rheem Abo-Zahrah decides that she will study at most 20 hours every week and that she must work at least 10 hours every week. Let x represent the hours studying and y represent the hours working. Write two inequalities that model this situation and graph their intersection.

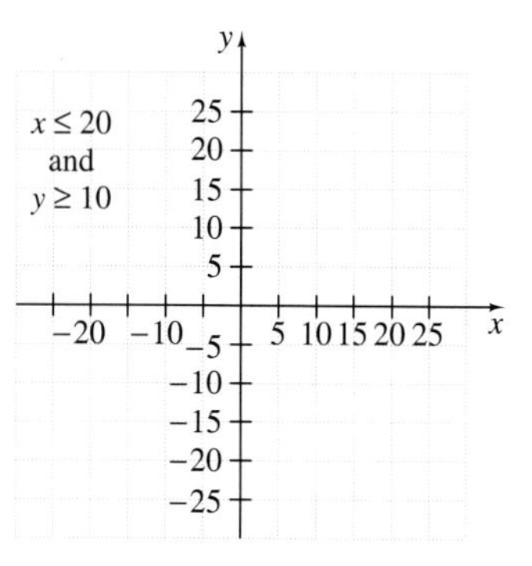

Name ______________________ Section __________ Date __________

CHAPTER 3 ACTIVITY
MEASURING SLOPE

MATERIALS:

- cardboard
- tape or glue
- ruler
- metal washer
- scissors
- string
- rug needle

This activity may be completed by working in groups or individually.

The grade of an incline is the same as its slope given as a percent. A 6% grade means that for every horizontal run of 100 units there is a vertical rise of 6 units. This can also be written as a slope of 0.06.

A gradiometer is a device that measures the grade of an incline. You can build your own gradiometer by following these steps:

- Cut out the gradiometer scale given at the right.
- Attach the scale to a piece of rigid cardboard. Trim the cardboard so it is even with the bottom edge of the scale.
- Thread the rug needle with string. Poke the needle through the gradiometer scale and cardboard at the large dot in the upper-right corner.
- Tie a large knot in the portion of the string hanging out the back. Pull the string from the front until the knot blocks the hole.
- Tie the washer to the portion of the string hanging across the front of the scale so that, when the gradiometer is held upright, the washer hangs in the portion marked "Grade (percent)" at the bottom.
- Attach the gradiometer to a ruler (roughly in the middle) so the bottom edge of the gradiometer is aligned with the bottom edge of the ruler.
- To use your gradiometer to measure the grade of an incline, place the bottom edge of the ruler along the incline. The point at which the string attached to the washer crosses the scale at the bottom of the gradiometer corresponds to the grade of the incline. (See figure.)

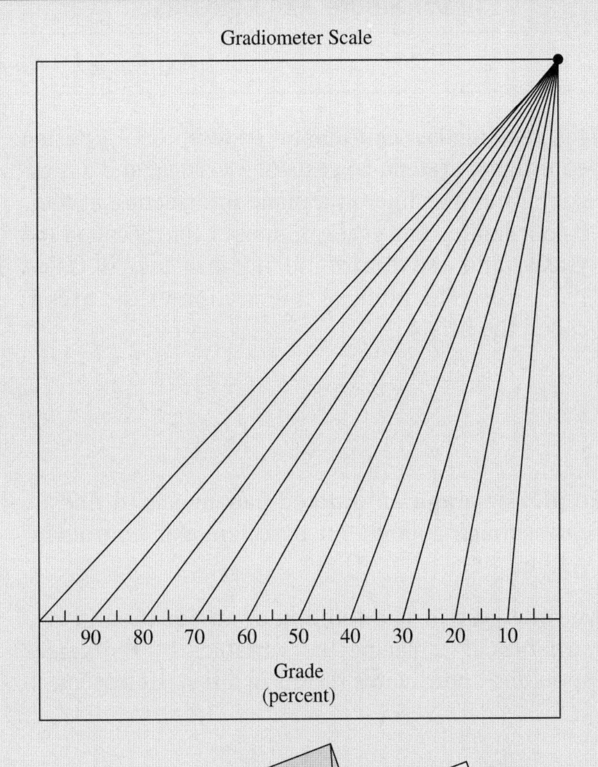

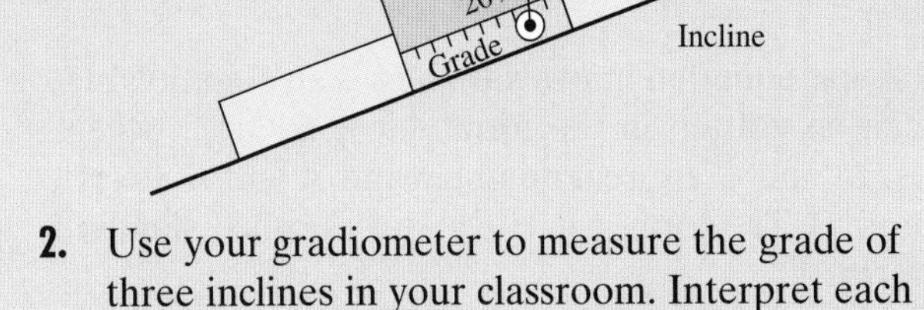

1. Refer to the graph on page 197. Express the slope of the line as a percent (grade). Use your gradiometer to measure the grade of the line. How close is your gradiometer reading to the actual grade? (*Hint:* You will need to hold the text book upright on a flat surface to measure the grade of the line.)

2. Use your gradiometer to measure the grade of three inclines in your classroom. Interpret each measurement. (*Hint:* Consider measuring the inclines of desks, lecterns, ramps, or steps.)

3. Notice that your gradiometer directly measures only positive slopes—inclines that rise from left to right. Explain how you could use your gradiometer to measure a negative slope (an incline that falls from left to right).

4. (Optional) According to the Americans with Disabilities Act (1990), handicapped-accessible ramps should have a grade of no more than 8.3%. Ramps with vertical rises greater than 6 inches and grades greater than 5% must provide handrails. Use your gradiometer to measure the grades of several wheelchair ramps on campus. Do they comply with the Americans with Disabilities Act guidelines?

CHAPTER 3 HIGHLIGHTS

DEFINITIONS AND CONCEPTS	EXAMPLES
SECTION 3.1 GRAPHING LINEAR EQUATIONS	
The **rectangular coordinate system**, or **Cartesian coordinate system**, consists of a vertical and a horizontal number line on a plane intersecting at their 0 coordinate. The vertical number line is called the **y-axis**, and the horizontal number line is called the **x-axis**. The point of intersection of the axes is called the **origin**.	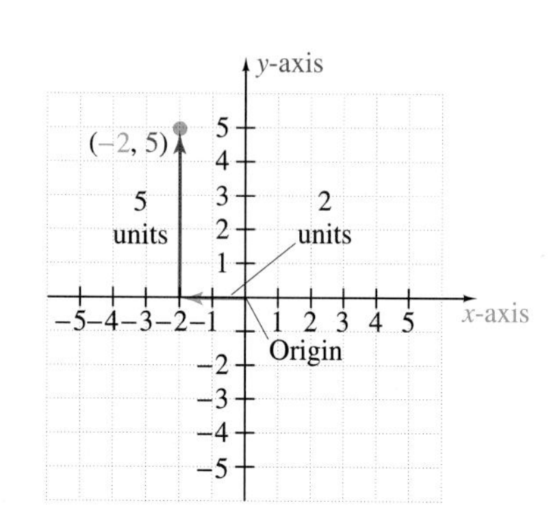
To **plot** or **graph** an ordered pair means to find its corresponding point on a rectangular coordinate system.	Plot or graph the ordered pair $(-2, 5)$. Start at the origin. Move 2 units to the left along the x-axis, then 5 units upward parallel to the y-axis.
An ordered pair is a **solution** of an equation in two variables if replacing the variables by the corresponding coordinates results in a true statement.	Determine whether $(-2, 3)$ is a solution of $3x + 2y = 0$. $3(-2) + 2(3) = 0$ $-6 + 6 = 0$ $0 = 0$ True. $(-2, 3)$ is a solution.
A **linear equation in two variables** is an equation that can be written in the form $Ax + By = C$, where A, B, and C are real numbers and A and B are not both 0. The form $Ax + By = C$ is called **standard form**.	$y = -2x + 5, x = 7,$ $y - 3 = 0, 6x - 4y = 10$ $6x - 4y = 10$ is in standard form.
The **graph of a linear equation** in two variables is a line. To graph a linear equation in two variables, find three ordered pair solutions. Plot the solution points, and draw the line connecting the points. Finding x- and y-intercepts To find an x-intercept, let $y = 0$ and solve for x. To find a y-intercept, let $x = 0$ and solve for y.	Graph: $3x + y = -6$ 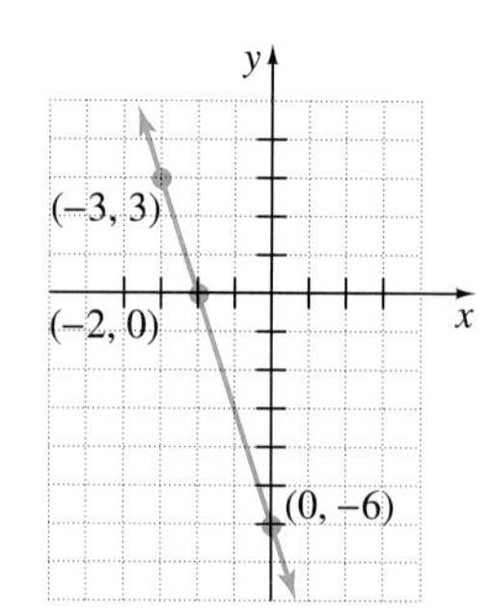 x: 0, −2, −3; y: −6, 0, 3
The graph of $x = c$ is a vertical line with x-intercept c. The graph of $y = c$ is a horizontal line with y-intercept c.	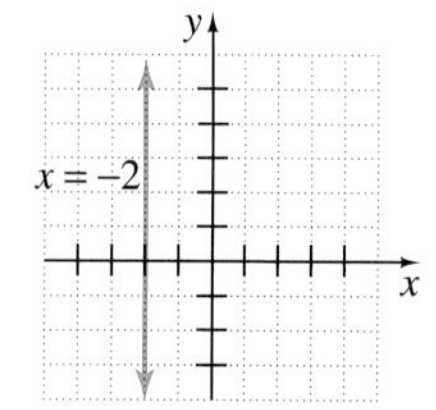 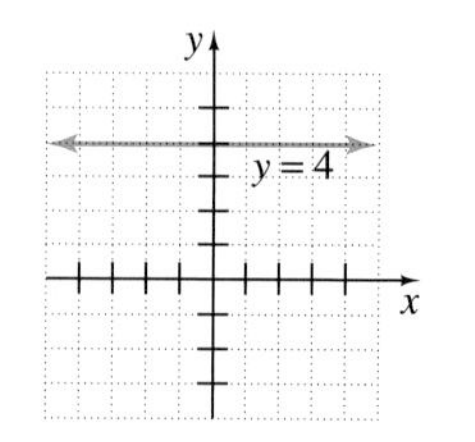

x	y
0	−6
−2	0
−3	3

Section 3.2 The Slope of a Line

The **slope** m of the line through (x_1, y_1) and (x_2, y_2) is given by

$$m = \frac{y_2 - y_1}{x_2 - x_1}$$

as long as $x_2 \neq x_1$

Find the slope of the line through $(-1, 7)$ and $(-2, -3)$.

$$m = \frac{y_2 - y_1}{x_2 - x_1} = \frac{-3 - 7}{-2 - (-1)} = \frac{-10}{-1} = 10$$

The **slope-intercept form** of a linear equation is

$$y = mx + b$$

where m is the slope of the line and b is the y-intercept.

Find the slope and y-intercept of $-3x + 2y = -8$.

$$2y = 3x - 8$$

$$\frac{2y}{2} = \frac{3x}{2} - \frac{8}{2}$$

$$y = \frac{3}{2}x - 4$$

The slope of the line is $\frac{3}{2}$, and the y-intercept is -4.

The slope of a horizontal line is 0.
The slope of a vertical line is undefined.
Nonvertical parallel lines have the same slope.

The slope of $y = -2$ is 0.
The slope of $x = 5$ is undefined.

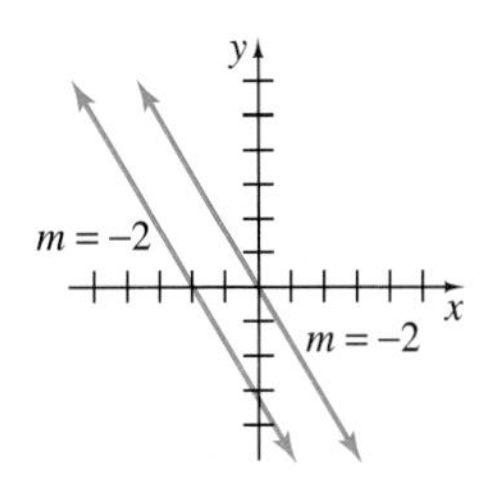

If the product of the slopes of two lines is -1, then the lines are perpendicular.

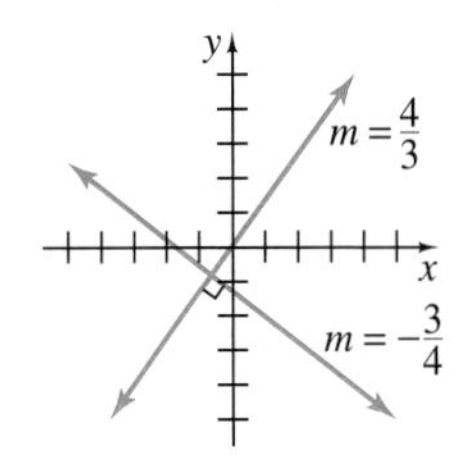

Section 3.3 The Slope-Intercept Form

We can use the slope-intercept form to write an equation of a line given its slope and y-intercept.

Write an equation of the line with y-intercept -1 and slope $\frac{2}{3}$.

$$y = mx + b$$

$$y = \frac{2}{3}x - 1$$

Section 3.4 More Equations of Lines

The **point-slope form** of the equation of a line is

$$y - y_1 = m(x - x_1)$$

where m is the slope of the line and (x_1, y_1) is a point on the line.

Find an equation of the line with slope 2 containing the point $(1, -4)$. Write the equation in standard form: $Ax + By = C$.

$$y - y_1 = m(x - x_1)$$

$$y - (-4) = 2(x - 1)$$

$$y + 4 = 2x - 2$$

$$-2x + y = -6 \quad \text{Standard form}$$

SECTION 3.5 INTRODUCTION TO FUNCTIONS

A **relation** is a set of ordered pairs. The **domain** of the relation is the set of all first components of the ordered pairs. The **range** of the relation is the set of all second components of the ordered pairs.

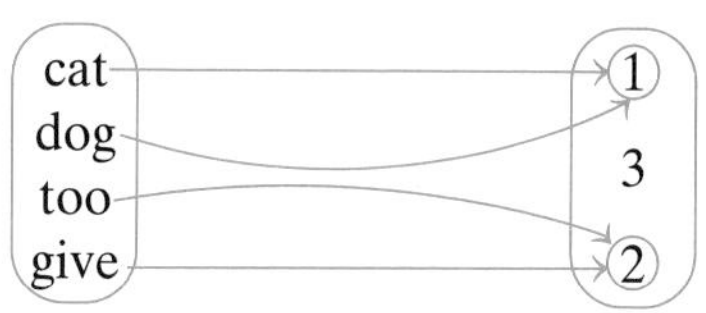

Domain: {cat, dog, too, give}
Range: {1, 2}

A **function** is a relation in which each element of the first set corresponds to exactly one element of the second set.

The previous relation is a function. Each word contains one exact number of vowels.

VERTICAL LINE TEST

If no vertical line can be drawn so that it intersects a graph more than once, the graph is the graph of a function.

Find the domain and the range of the relation. Also determine whether the relation is a function.

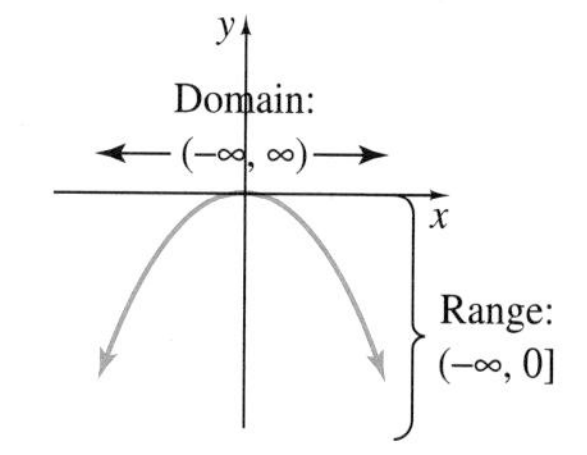

By the vertical line test, this is the graph of a function.

The symbol $f(x)$ means **function of x** and is called **function notation**.

If $f(x) = 2x^2 - 5$, find $f(-3)$.

$$f(-3) = 2(-3)^2 - 5 = 2(9) - 5 = 13$$

A **linear function** is a function that can be written in the form

$$f(x) = mx + b$$

$$f(x) = -3, g(x) = 5x, h(x) = -\frac{1}{3}x - 7$$

To graph a linear function, use the slope and y-intercept.

Graph: $f(x) = -2x$
(or $y = -2x + 0$)

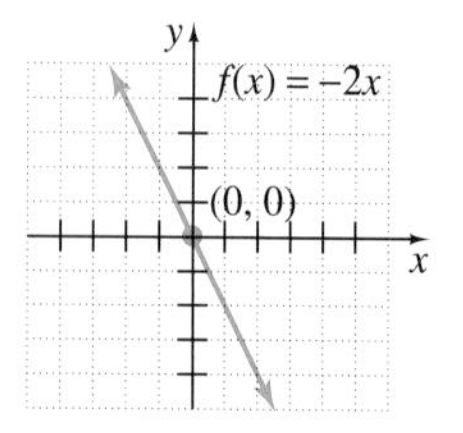

The slope is $\frac{2}{-1}$.
y-intercept is 0, or the point (0, 0).

Section 3.6 Graphing Linear Inequalities

If the equal sign in a linear equation in two variables is replaced with an inequality symbol, the result is a **linear inequality in two variables**.

$x \leq -5y$, $y \geq 2$,
$3x - 2y > 7$, $x < -5$

Graphing a Linear Inequality

Graph: $2x - 4y > 4$

1. Graph the **boundary line** by graphing the related equation. Draw a solid line if the inequality symbol is $\leq$ or $\geq$. Draw a dashed line if the inequality symbol is $<$ or $>$.

1. Graph $2x - 4y = 4$. Draw a dashed line because the inequality symbol is $>$.

2. Choose a **test point** not on the line. Substitute its coordinates into the original inequality.

2. Check the test point (0, 0) in the inequality $2x - 4y > 4$.

$2 \cdot 0 - 4 \cdot 0 > 4$ Let $x = 0$ and $y = 0$.

$0 > 4$ False.

3. If the resulting inequality is true, shade the **half-plane** that contains the test point. If the inequality is not true, shade the half-plane that does not contain the test point.

3. The inequality is false, so shade the half-plane that does not contain (0, 0).

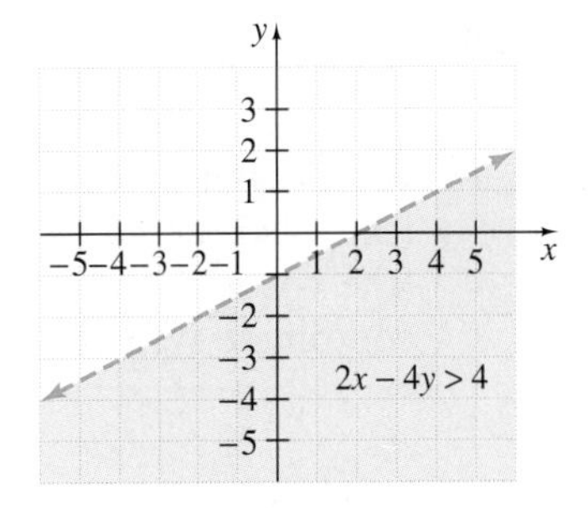

Focus On History

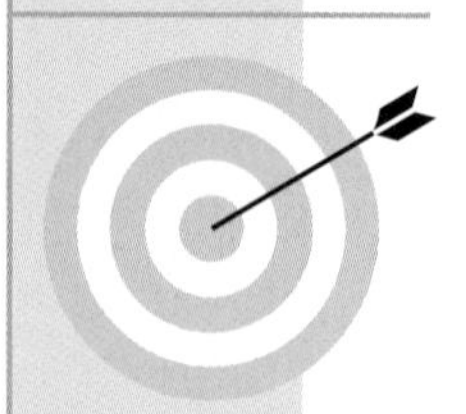

CARTESIAN COORDINATE SYSTEM

The French mathematician and philosopher René Descartes (1596–1650) is generally credited with devising the rectangular coordinate system that we use in mathematics today. It is said that Descartes thought of describing the location of a point in a plane using a fixed frame of reference while watching a fly crawl on his ceiling as he laid in bed one morning meditating. He incorporated this idea of defining a point's position in the plane by giving its distances, x and y, to two fixed axes in his text *La Géométrie*.

Although Descartes is credited with the concept of the rectangular coordinate system, nowhere in his written works does an example of the modern gridlike coordinate system appear. He also never referred to a point's location as we do today with (x, y)-notation giving an ordered pair of coordinates. In fact, Descartes never even used the term *coordinate*! Instead, his basic ideas were expanded upon by later mathematicians. The Dutch mathematician Frans van Schooten (1615–1660) is credited with making Descartes' concept of a coordinate system widely accepted with his text, *Geometria a Renato Des Cartes* (*Geometry by René Descartes*). The German mathematician Gottfried Wilhelm Leibniz (1646–1716) later contributed the terms *abscissa* (the x-axis in the modern rectangular coordinate system), *ordinate* (the y-axis), and *coordinate* to the development of the rectangular coordinate system.

Name ______________________ Section __________ Date __________

CHAPTER 3 REVIEW

(3.1) *Plot the points and name the quadrant in which each point lies.*

1. $A(2, -1), B(-2, 1), C(0, 3), D(-3, -5)$

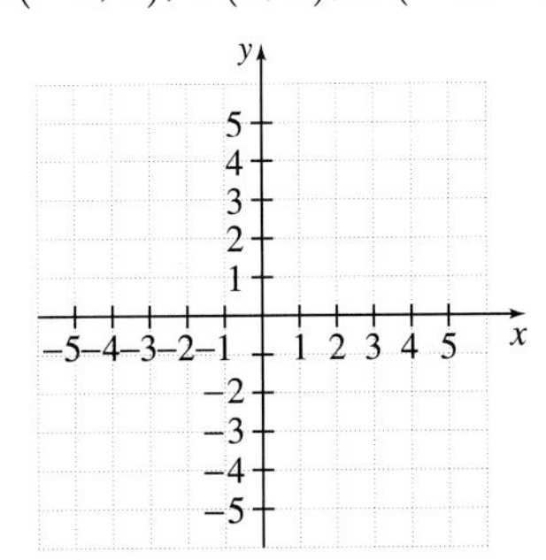

A, quadrant IV; B, quadrant II; C, no quadrant; D, quadrant III

2. $A(-3, 4), B(4, -3), C(-2, 0), D(-4, 1)$

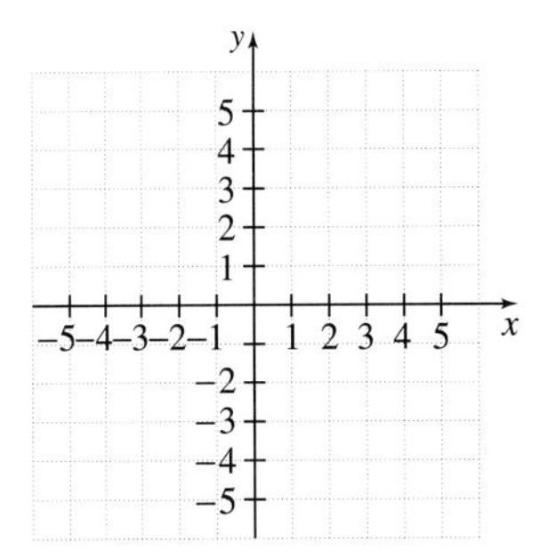

A and D, quadrant II; C, no quadrant; B, quadrant IV

Create a scatter diagram for the given paired data.

3.

PER PERSON CONSUMPTION OF CITRUS FRUIT IN THE UNITED STATES	
Year	Per Person Consumption (pounds per person)
1993	26
1994	25
1995	24
1996	25
1997	28
1998	29

(*Source:* Economic Research Service, U.S. Department of Agriculture)

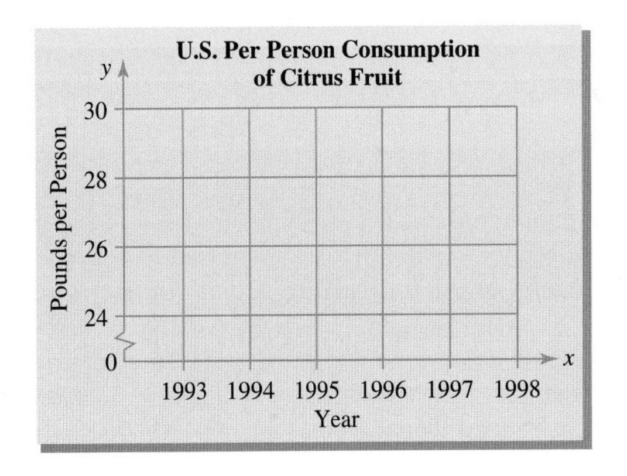

4.

U.S. ARMED FORCES ACTIVE DUTY PERSONNEL	
Year	Armed Forces (in millions)
1950	1.5
1960	2.5
1970	3
1980	2
1990	2
1996	1.5

(*Source: 1998 World Almanac*)

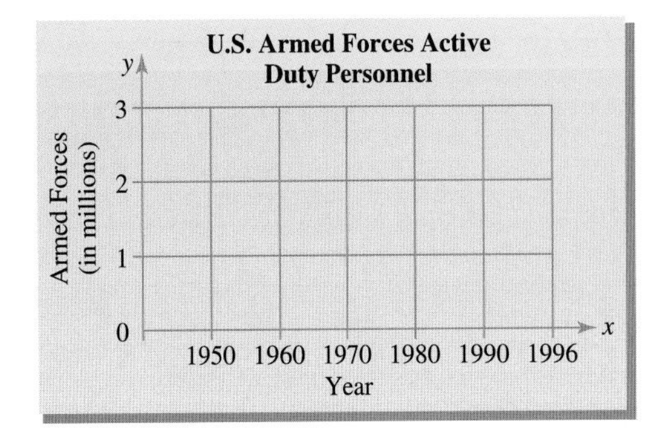

Determine whether each ordered pair is a solution to the given equation.

5. $7x - 8y = 56; (0, 56), (8, 0)$

6. $-2x + 5y = 10; (-5, 0), (1, 1)$

7. $x = 13; (13, 5), (13, 13)$

8. $y = 2; (7, 2), (2, 7)$

Name ______________________________________

Graph each linear equation.

9. $3x - y = 3$

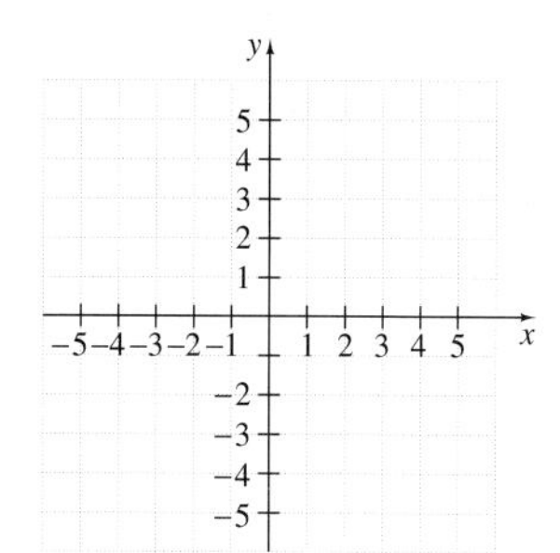

10. $2x - y = 4$

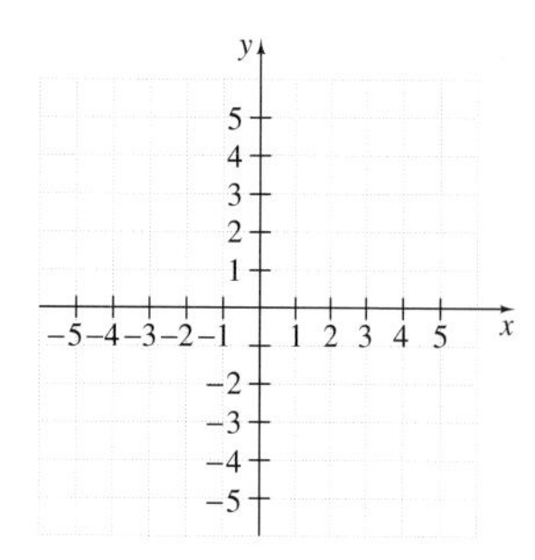

11. $4x + 5y = 20$

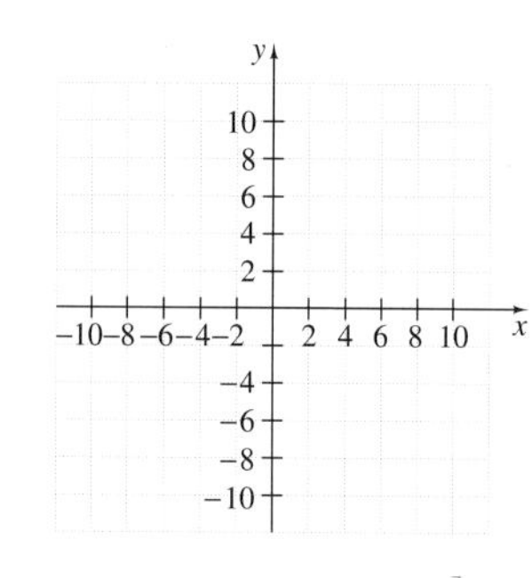

12. $3x - 2y = -9$

13. $y = 5$

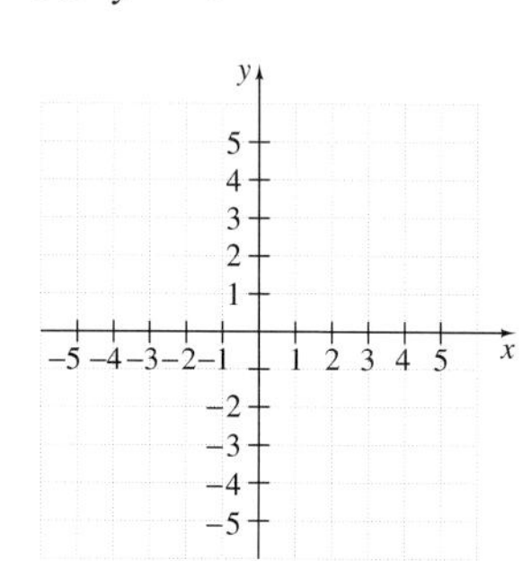

14. $x = -2$

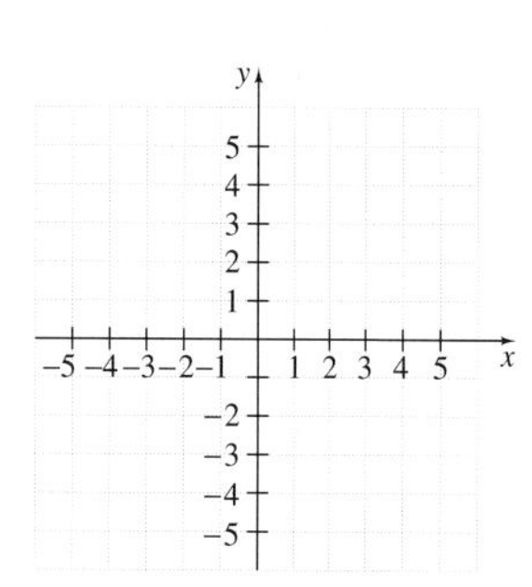

15. $y = \frac{1}{3}x$

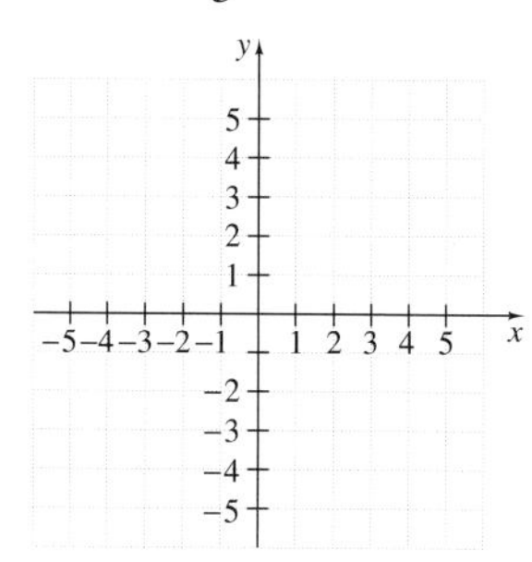

16. $x = -4y$

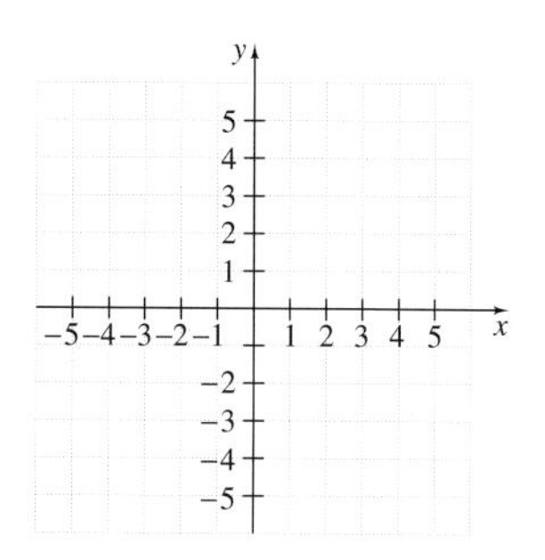

(3.2) *Find the slope of the line through each pair of points.*

17. $(2, 8)$ and $(6, -4)$

18. $(-3, 9)$, and $(5, 13)$

19. $(-7, -4)$ and $(-3, 6)$

20. $(7, -2)$ and $(-5, 7)$

Determine the slope of each line.

21.

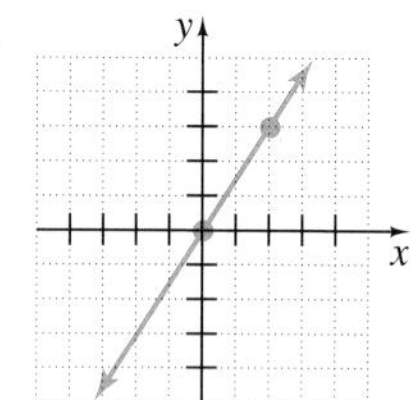

22.

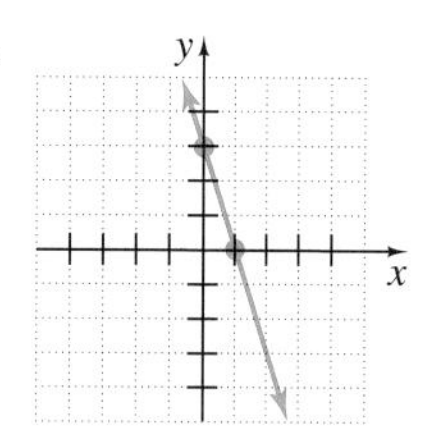

23.

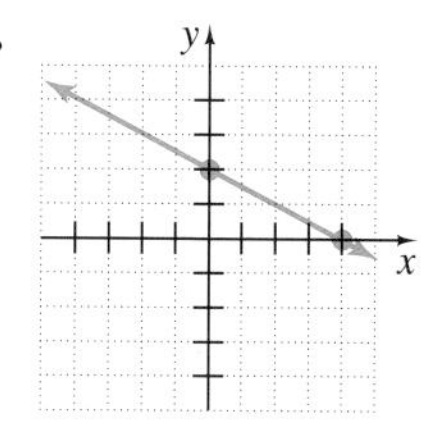

24.

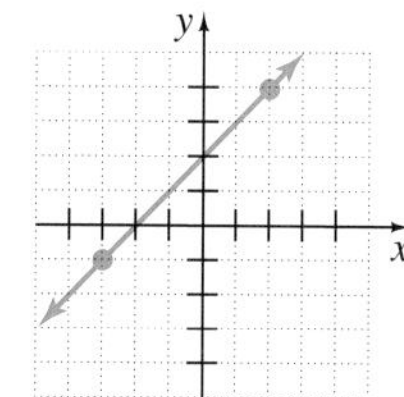

Two lines are graphed on each set of axes. Determine whether l_1 or l_2 has the greater slope.

25.

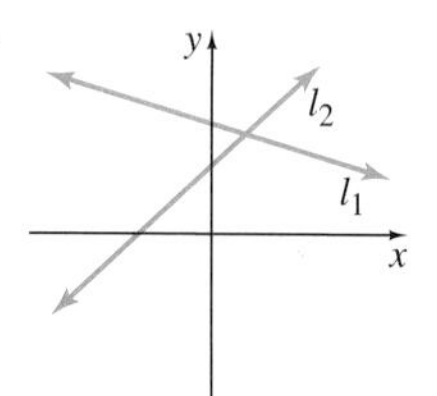

26.

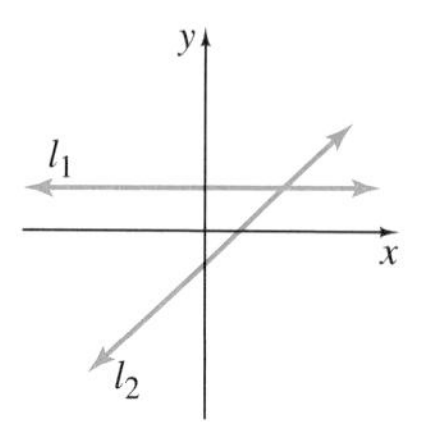

27.

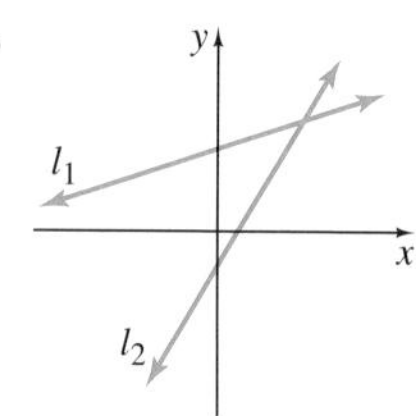

28.

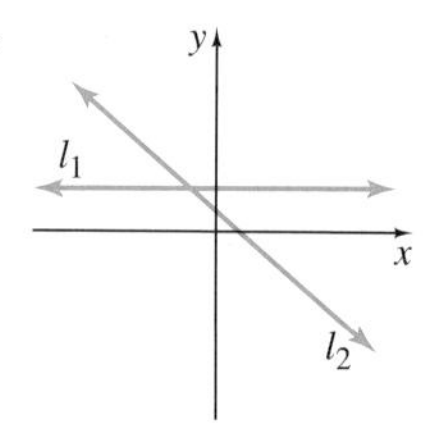

Find the slope and y-intercept of each line.

29. $y = -3x + \frac{1}{2}$

30. $y = 2x + 4$

31. $6x - 15y = 20$

32. $4x + 14y = 21$

Name ______________________________

Find the slope of each line.

33. $y - 3 = 0$

34. $x = -5$

Determine whether each pair of lines are parallel, perpendicular, or neither.

35. $y = -2x + 6$
$y = 2x - 1$

36. $-x + 3y = 2$
$6x - 18y = 3$

37. $y = \frac{3}{4}x + 1$
$y = -\frac{4}{3}x + 1$

38. $x - 2y = 6$
$4x + y = 8$

(3.3) *Graph each line passing through the given point with the given slope.*

39. Through $(2, -3)$ with slope $\frac{2}{3}$

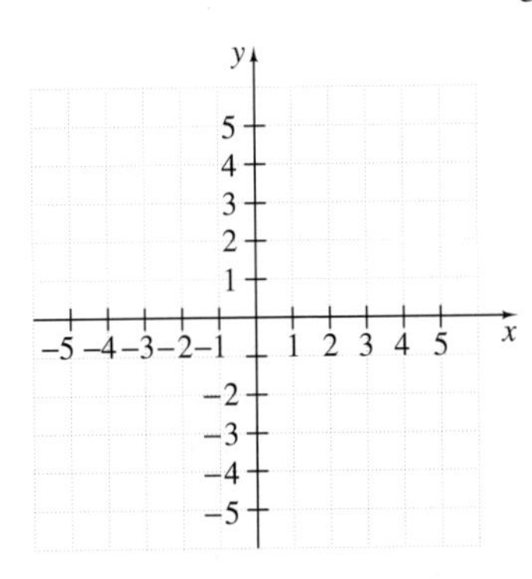

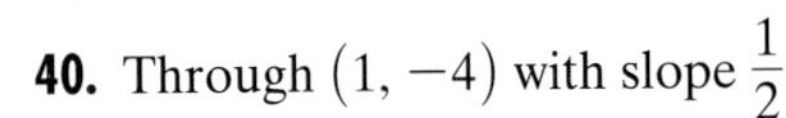

40. Through $(1, -4)$ with slope $\frac{1}{2}$

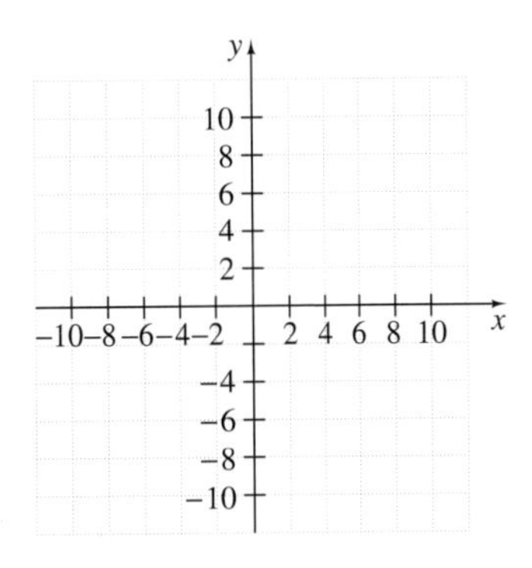

41. Through $(0, 1)$ with slope 2

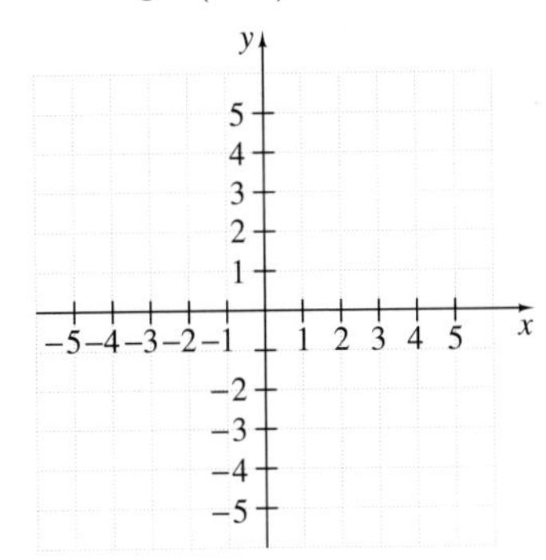

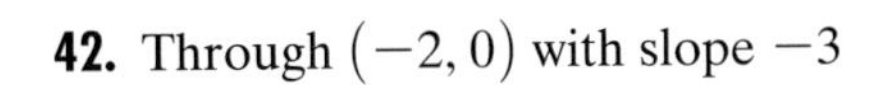

42. Through $(-2, 0)$ with slope -3

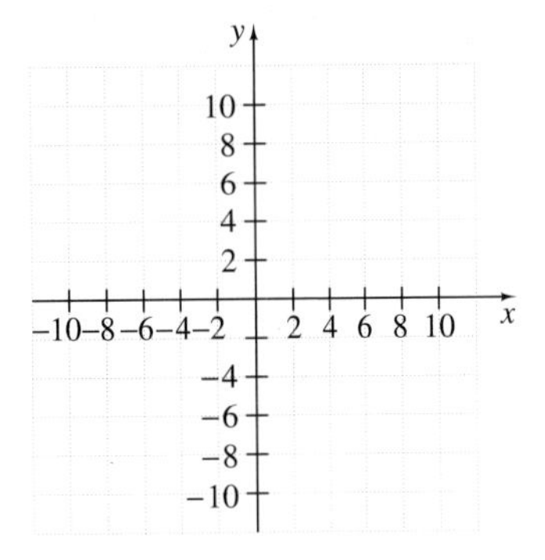

Graph each linear equation using the slope and y-intercept.

43. $y = -x + 1$

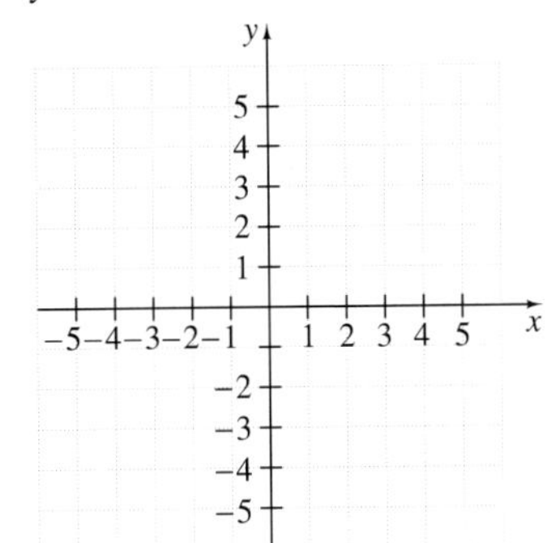

44. $y = 4x - 3$

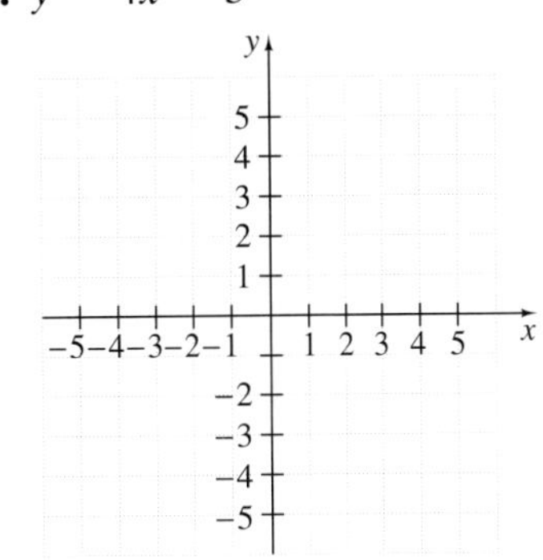

Name ______________________________

45. $3x - y = 6$

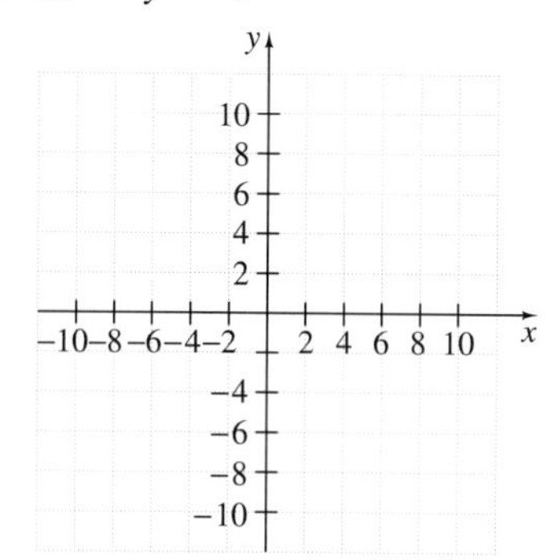

46. $y = -5x$

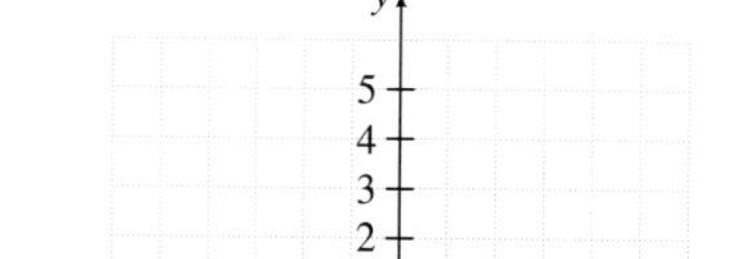
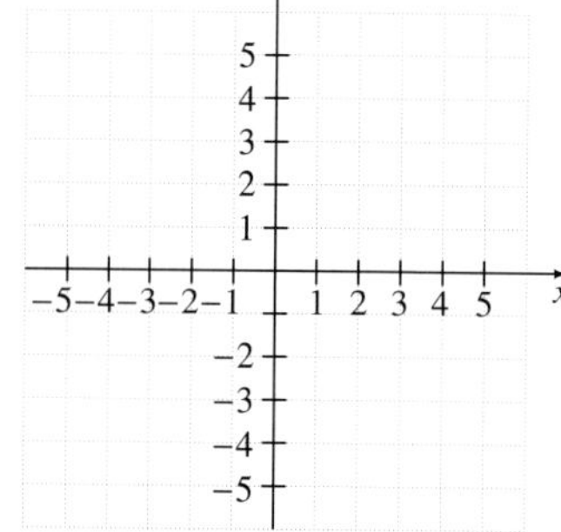

47. The cost C, in dollars, of renting a minivan for a day is given by the linear equation $y = 0.3x + 42$, where x is number of miles driven.

a. Find the cost of renting the minivan for a day and driving it 150 miles.

b. Find and interpret the slope of this equation.

c. Find and interpret the y-intercept of this equation.

(3.4) *Write an equation of the line satisfying each set of conditions.*

48. Horizontal; through $(3, -1)$

49. Vertical; through $(-2, -4)$

50. Slope undefined; through $(-4, -3)$

51. Slope 0; through $(2, 5)$

Write the equation of the line satisfying each set of the conditions. Write the equation in the form $y = mx + b$.

52. Through $(-3, 5)$; slope 3

53. Slope 2; through $(5, -2)$

54. Through $(-6, -1)$ and $(-4, -2)$

55. Through $(-5, 3)$ and $(-4, -8)$

56. Through $(2, -6)$; parallel to $y = -2x + 3$

57. Through $(-4, -2)$; parallel to $y = -\frac{3}{2}x + 1$

58. Through $(-6, -1)$; perpendicular to $4x + 3y = 5$

59. Through $(-4, 5)$; perpendicular to $2x - 3y = 6$

60. The value of a computer bought in 1996 depreciates, or decreases, as time passes. Two years after the computer was bought, it was worth \$2600; four years after it was bought, it was worth \$1000.

a. Assuming that this relationship between number of years past 1996 and value of computer is linear, write an equation describing this relationship. (*Hint:* Use ordered pairs of the form (years past 1996, value of computer).)

b. Use this equation to estimate the value of the computer in the year 2001.

61. The value of a building bought in 1980 appreciates, or increases, as time passes. Seven years after the building was bought, it was worth \$165,000; 12 years after it was bought, it was worth \$180,000.

a. Assuming that this relationship between number of years past 1980 and value of building is linear, write an equation describing this relationship. (*Hint:* Use ordered pairs of the form (years past 1980, value of building).)

b. Use this equation to estimate the value of the building in the year 2005.

Name ____

62. $\left\{\left(-\frac{1}{2}, \frac{3}{4}\right), (6, 0.65), (0, -12), (25, 25)\right\}$

63. $\left\{\left(\frac{3}{4}, -\frac{1}{2}\right), (0.65, 6), (-12, 0), (25, 25)\right\}$

64.

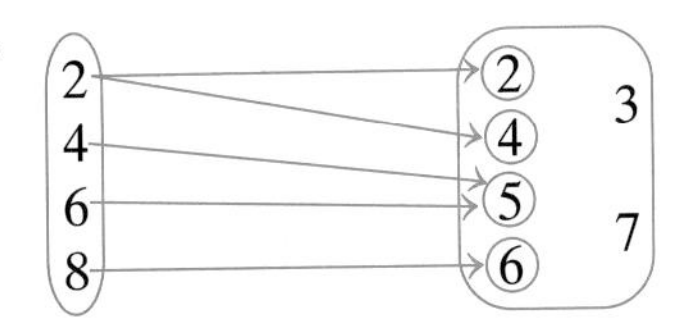

65.

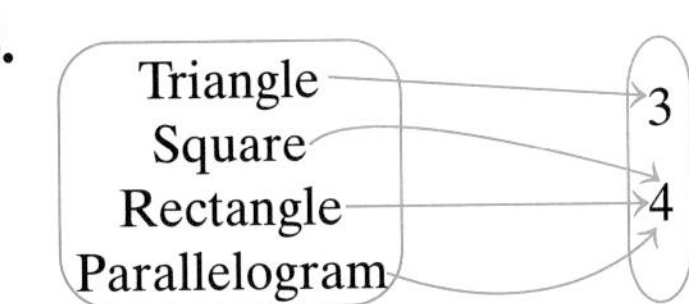

66.

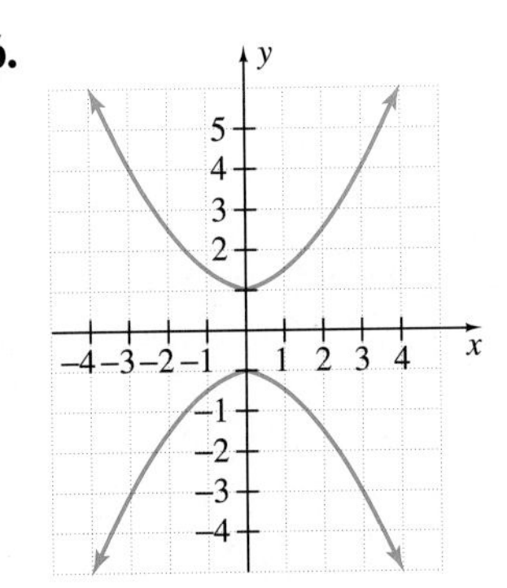

67.

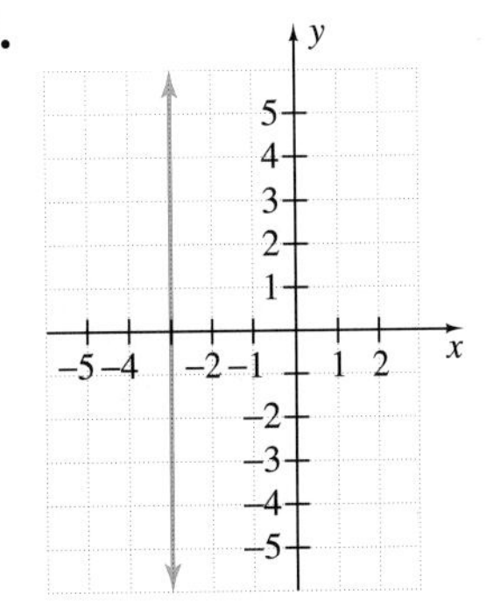

68.

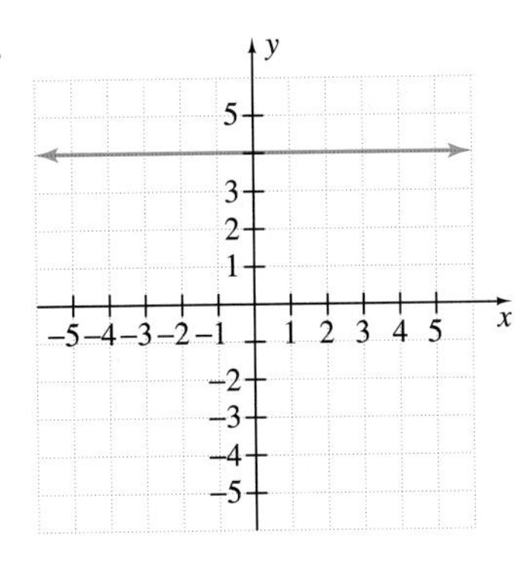

69.

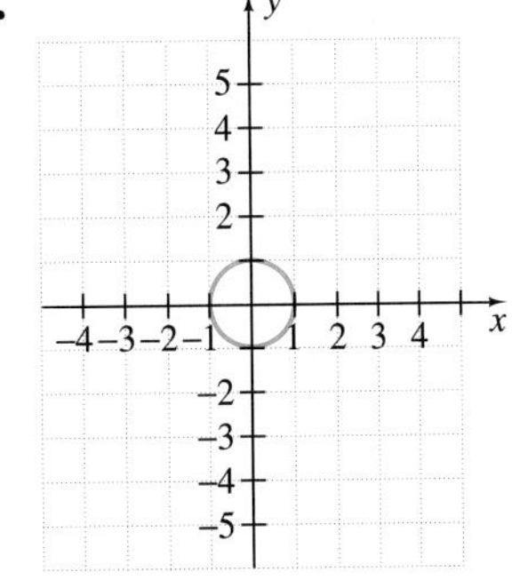

If $f(x) = x - 5$, $g(x) = -3x$, and $h(x) = 2x^2 - 6x + 1$, find each function value.

70. $f(2)$

71. $g(0)$

72. $g(-6)$

73. $h(-1)$

74. $h(1)$

75. $f(5)$

The function $J(x) = 2.54x$ may be used to calculate the weight of an object on Jupiter (J) given its weight on Earth (x).

76. If a person weighs 150 pounds on Earth, find the equivalent weight on Jupiter.

77. A 2000-pound probe on Earth weighs how many pounds on Jupiter?

Name ______________________________

Graph each linear function.

78. $f(x) = x + 2$

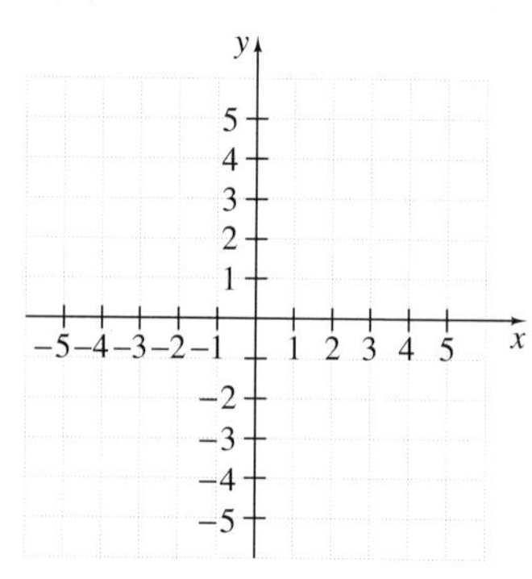

79. $f(x) = -\frac{1}{2}x + 3$

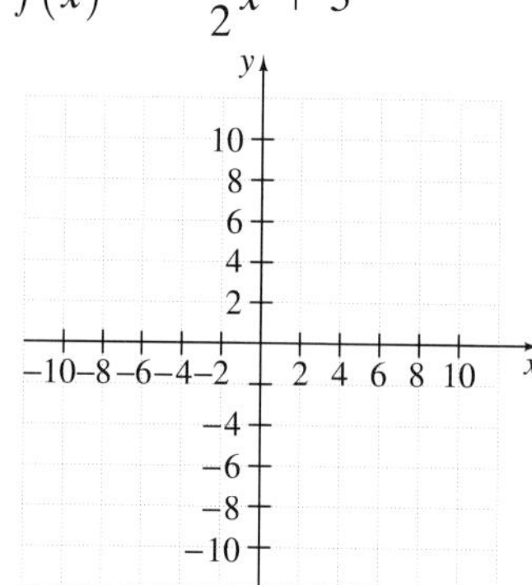

(3.6) *Graph each linear inequality.*

80. $3x + y > 4$

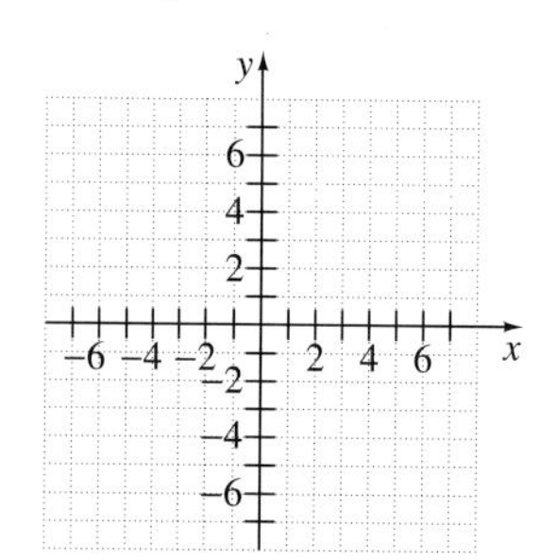

81. $\frac{1}{2}x - y < 2$

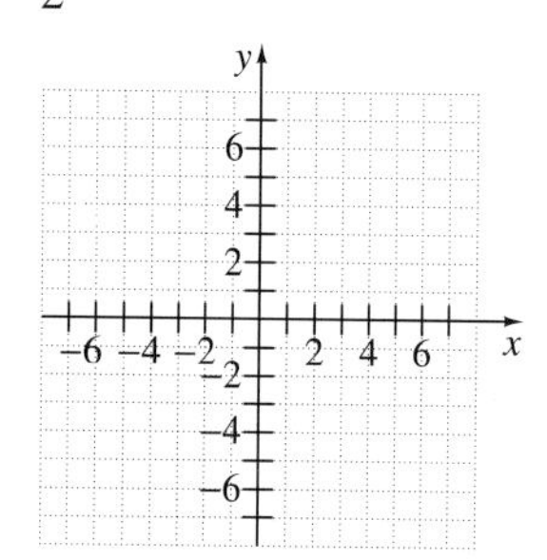

82. $5x - 2y \leq 9$

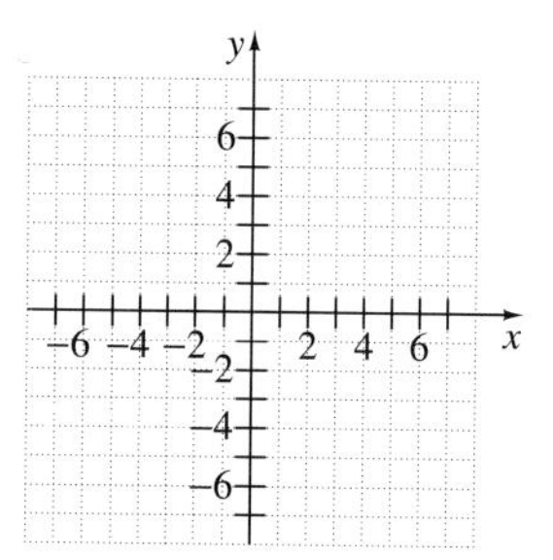

83. $3y \geq x$

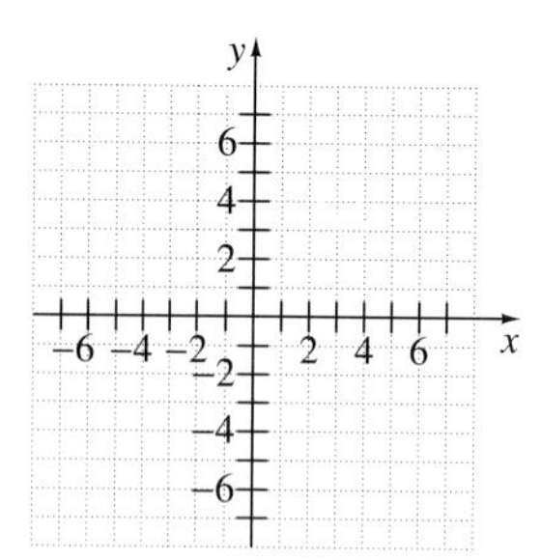

84. $y < 1$

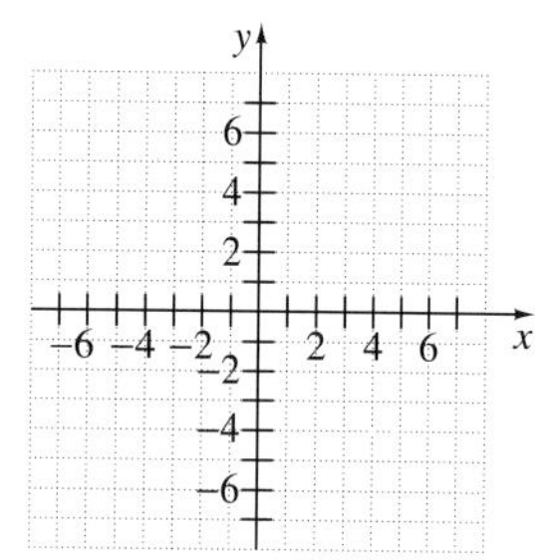

85. $x > -2$

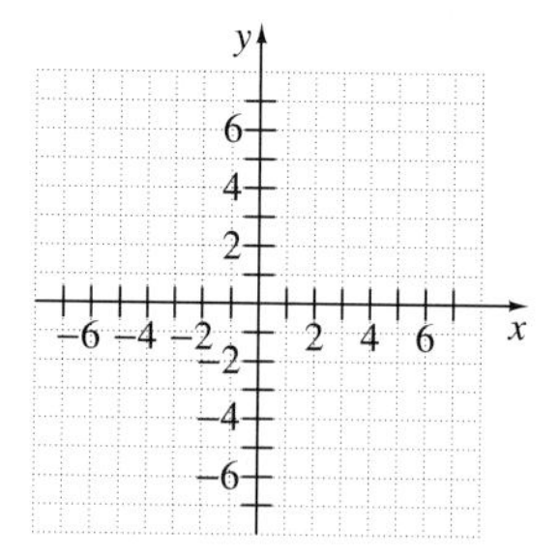

86. Graph the union of $y > 2x + 3$ or $x \leq -3$.

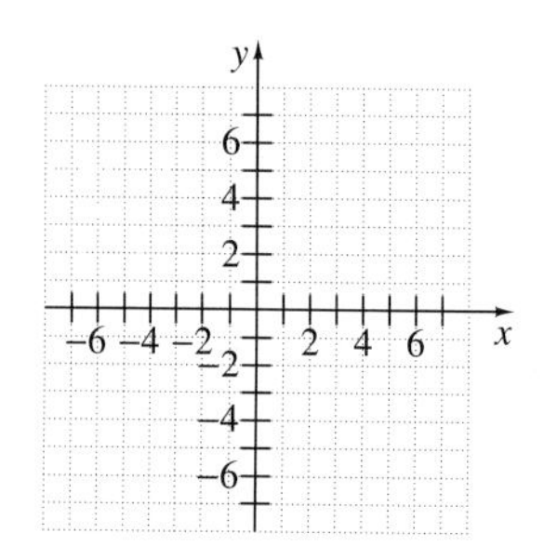

87. Graph the intersection of $2x < 3y + 8$ and $y \geq -2$.

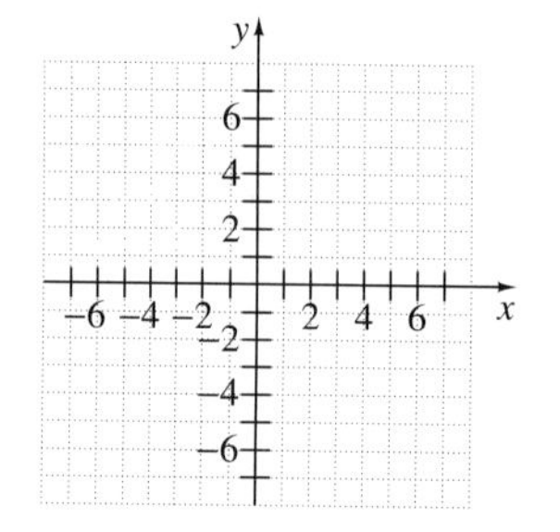

Name ______________________ Section ________ Date ________

ANSWERS

CHAPTER 3 TEST

1. Plot the points, and name the quadrant in which each is located: $A(6, -2), B(4, 0), C(-1, 6)$.

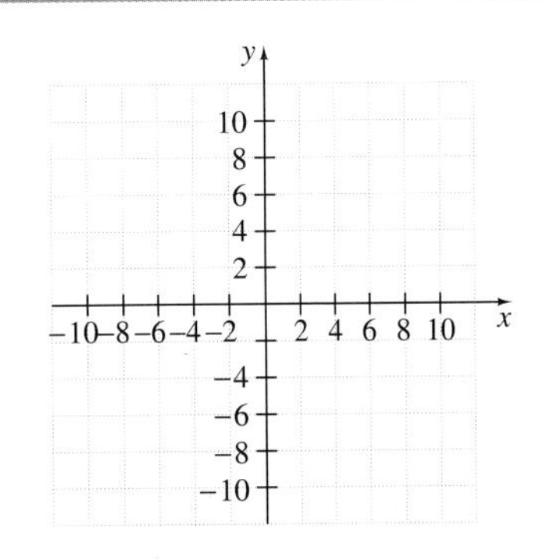

1. see graph

2. Create a scatter diagram for the paired data.

2. see graph

U.S. CELLULAR PHONE SUBSCRIBERS	
Year	Number (in millions)
1990	5
1991	8
1992	11
1993	16
1994	24
1995	34
1996	44

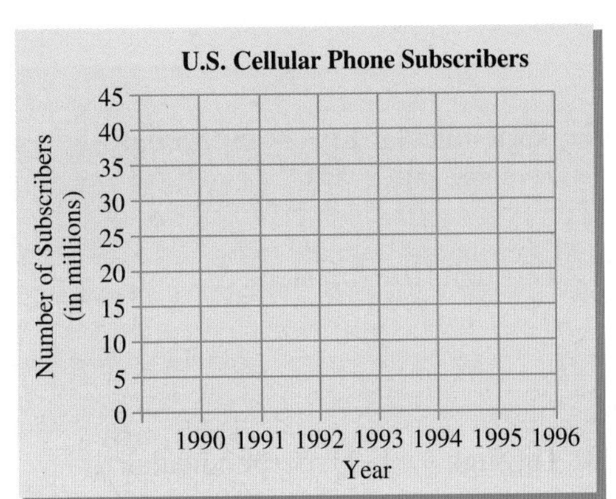

Graph each linear equation.

3. $-3x + y = -3$

4. $2x - 3y = -6$

5. $4x + 6y = 8$

6. $y = -3$

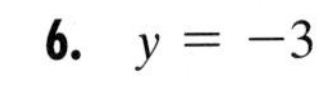

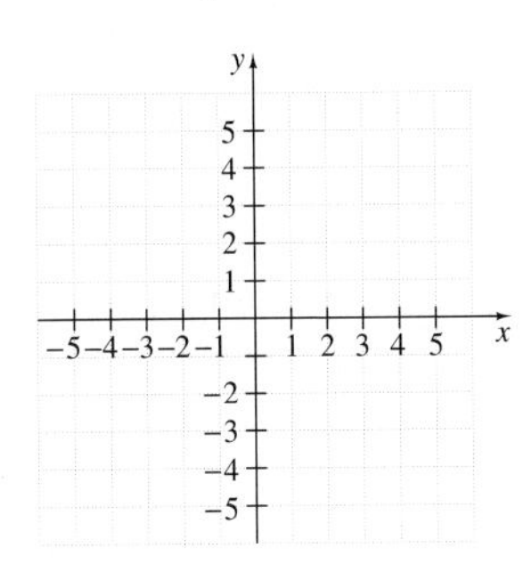

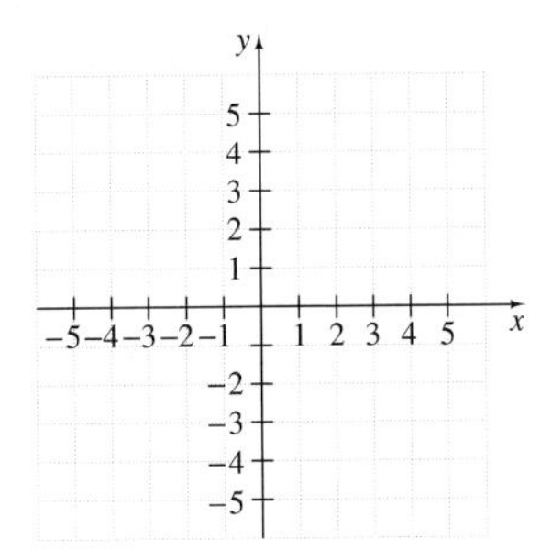

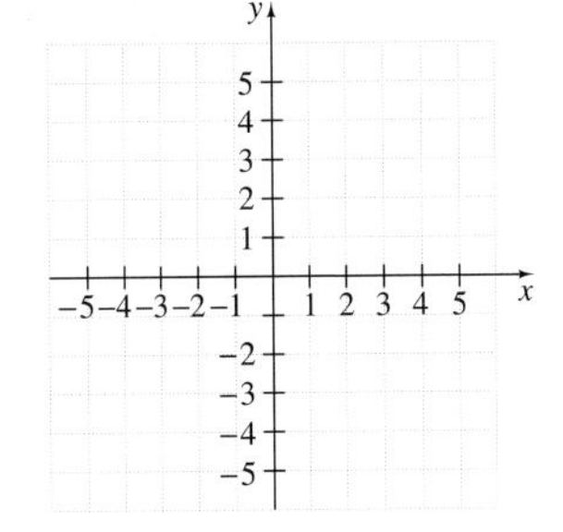

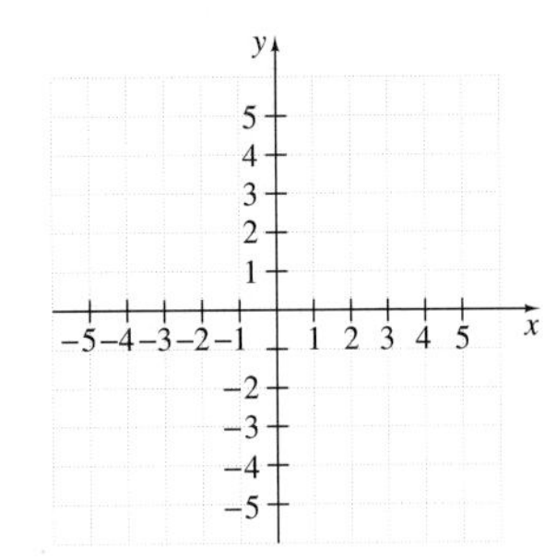

7. Find the slope of the line that passes through $(5, -8)$ and $(-7, 10)$.

8. Find the slope and the y-intercept of the line $3x + 12y = 8$.

7. ____________

8. ____________

Match each equation with its graph.

9. $f(x) = 3x + 1$

10. $f(x) = 3x - 2$

9. ____________

10. ____________

11. ______

12. ______

13. ______

14. ______

15. ______

16. ______

17. ______

18. ______

19. ______

20. ______

21. ______

22. a. ______

b. ______

c. ______

d. ______

e. ______

Name ______________________________

11. $f(x) = 3x + 2$ 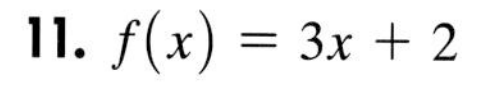

12. $f(x) = 3x - 5$ 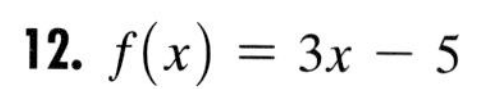

A

B 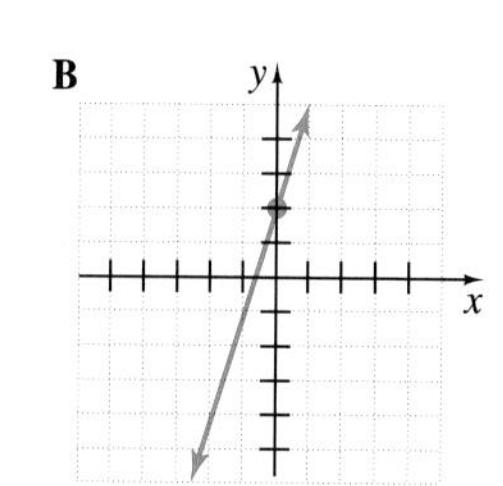

C 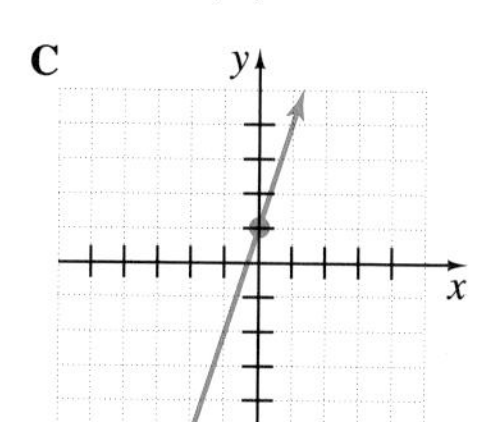

D 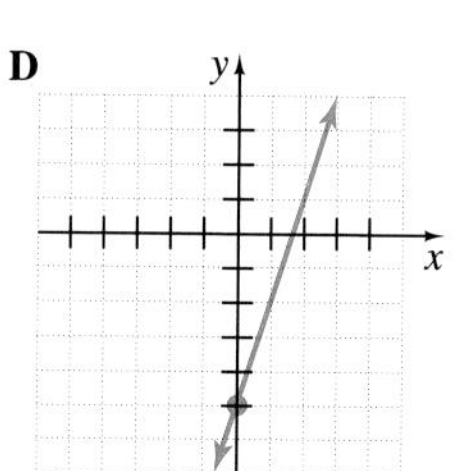

Find an equation of the line satisfying each set of conditions. Write the equations in the form $x = a$, $y = b$, or $y = mx + b$.

13. Horizontal; through $(2, -8)$

14. Vertical; through $(-4, -3)$

15. Perpendicular to $x = 5$; through $(3, -2)$

16. Through $(4, -1)$; slope -3

17. Through $(0, -2)$; slope 5

18. Through $(4, -2)$ and $(6, -3)$

19. Through $(-1, 2)$; perpendicular to $3x - y = 4$

20. Parallel to $2y + x = 3$; through $(3, -2)$

21. Line L_1 has the equation $2x - 5y = 8$. Line L_2 passes through the points $(1, 4)$ and $(-1, -1)$. Determine whether these lines are parallel, perpendicular, or neither.

22. The average yearly earnings for high school graduates age 18 and older is given by the linear equation

$$y = 708x + 13{,}570$$

where x is the number of years since 1985 that a person graduated.

a. Find the average earnings in 1990 for high school graduates.

b. Predict the average earnings for high school graduates in the year 2000.

c. Predict the first year that the average earnings for high school graduates will be greater than \$30,000.

d. Find and interpret the slope of the equation.

e. Find and interpret the y-intercept of the equation.

Name ______________________________

Graph each inequality.

23. $x \leq -4$

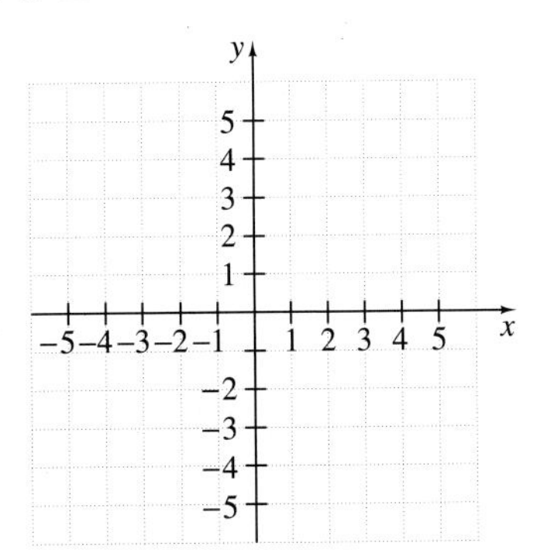

24. $2x - y > 5$

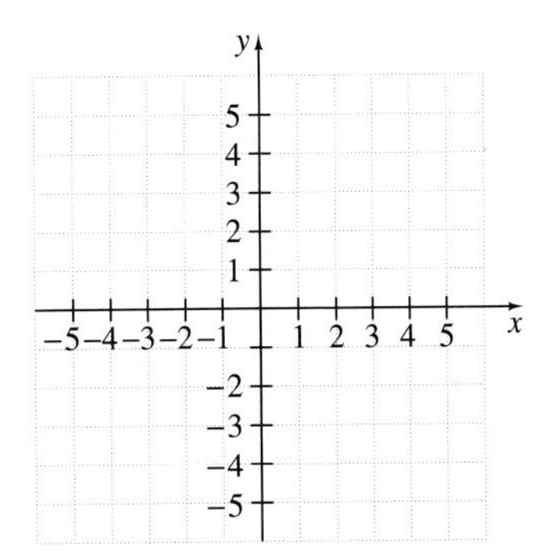

25. The intersection of $2x + 4y < 6$ and $y \leq -4$

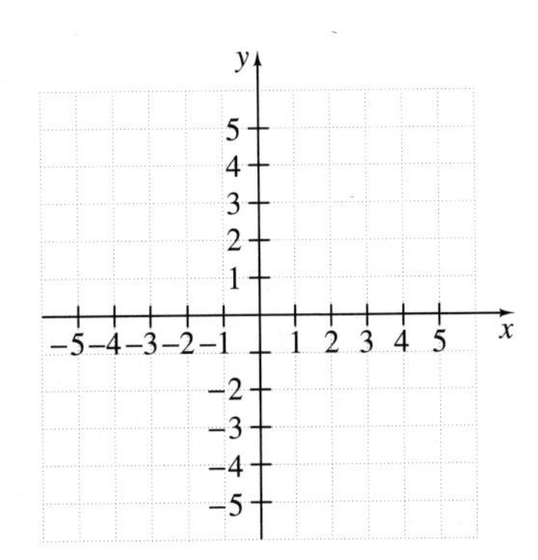

Find the domain and range of each relation. Also determine whether the relation is a function.

26.

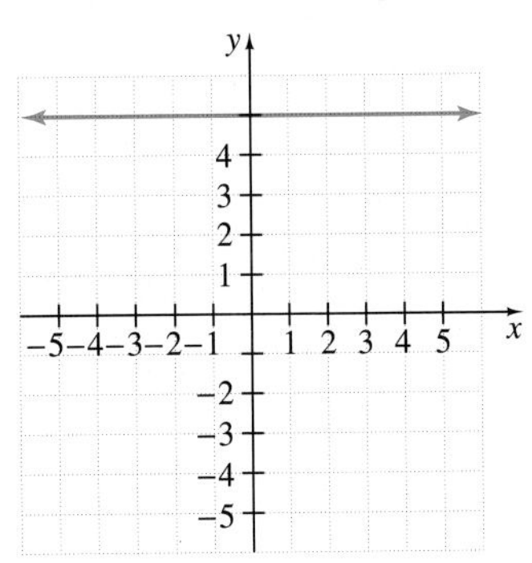

27.

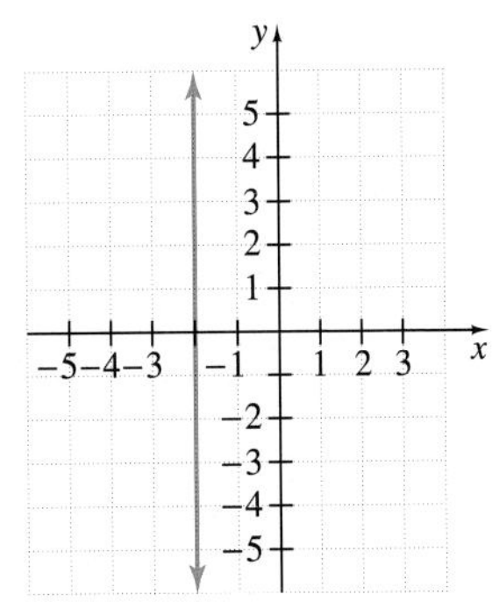

28.

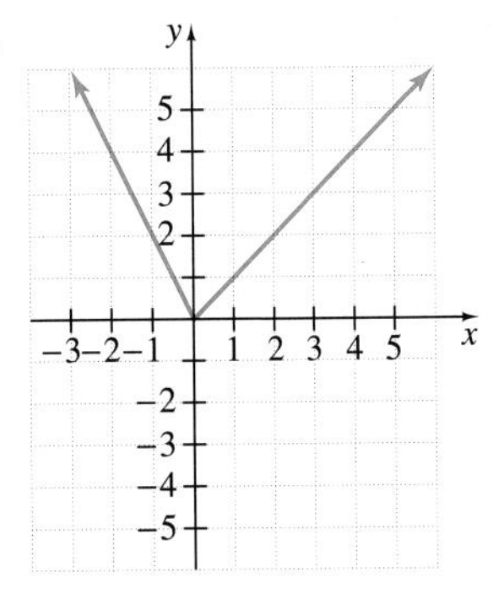

29.

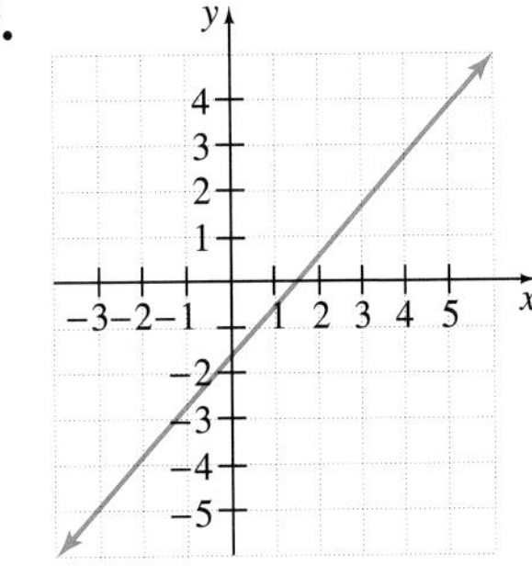

23. see graph

24. see graph

25. see graph

26. ______________

27. ______________

28. ______________

29. ______________

Focus On the Real World

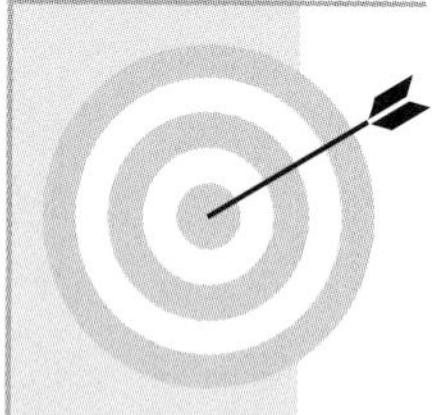

Centers of Mass

The **center of mass**, also known as **center of gravity**, of an object is the point at which the mass of the object may be considered to be concentrated. For a two-dimensional object or surface, such as a flat board, the center of mass can be described as the point on which the surface would balance.

The idea of center of mass is an important one in many disciplines, especially physics and its applications. The following list describes situations in which an object's center of mass is important.

- Geographers are sometimes concerned with pinpointing the *geographic center* of a county, state, country, or continent. A geographic center is actually the center of mass of a geographic region if it is considered as a two-dimensional surface. The geographic center of the 48 contiguous United States is near Lebanon, Kansas. The geographic center of the North American continent is 6 miles west of Balta, North Dakota.
- A top-loading washing machine is designed so its center of mass is located within its agitator post. During the spin cycle, the washer tub spins around its center of mass. If clothes aren't carefully distributed within the tub, they can bunch up and throw off the center of mass. This causes the machine to vibrate, sometimes jumping or shaking wildly. Some washing machine models will stop operating when the loads become "unbalanced" in this way.
- Single-hulled boats are normally designed so that their centers of mass are below the water line. This provides a boat with stability. Otherwise, with a center of mass above the water line, the boat would have a tendency to tip over in the water.
- Most small airplanes must be carefully loaded so as not to affect the location of the airplane's center of mass. The center of mass of an airplane must be near the center of the wings. Pilots of small aircraft usually try to balance the weight of their cargo around the airplane's center of mass.

Group Activity

Attach a piece of graph paper to a piece of cardboard. Cut out a triangle and label the vertices with their coordinates. Lay the triangle on a horizontal table top with the graph paper face down. Slide the triangle toward the edge of the table until it is balanced on the edge, just about to tip over the side of the table. Firmly hold the triangle in place while another group member uses the straight edge of the table to draw a line on the graph paper side of the triangle marking the position of the table edge. Rotate the triangle a quarter turn and rebalance the triangle on the edge of the table. Draw a second line on the graph paper side of the triangle marking the position of the table edge.

1. The point where the two lines drawn on the triangle intersect is the center of mass of the triangle. Find the coordinates of this point.
2. Verify that the point you have located is roughly the center of mass of the triangle by balancing it at this point on the tip of a pencil or pen.
3. List the coordinates of the vertices of your triangle. What is the relationship between the coordinates of the center of mass and the coordinates of the triangle's vertices? (*Hint:* You may find it helpful to examine the sum of the x-coordinates and the sum of the y-coordinates of the vertices of the triangle.)
4. Test your observation in Question 3. Cut out another triangle. Label its vertices and, using your observation from Question 3, predict the location of the center of mass of the triangle. Use the balancing procedure to find the center of mass. How close was your prediction?

Name ______________________ Section ________ Date ________

ANSWERS

CUMULATIVE REVIEW

Write each set in roster form.

1. $\{x|x$ is a whole number between 1 and 6$\}$

2. $\{x|x$ is a natural number greater than 100$\}$

Find the reciprocal, or multiplicative inverse, of each number.

3. -9

4. $\frac{7}{4}$

Find each absolute value.

5. $|3|$

6. $-|2|$

7. Simplify: $3 - [(4 - 6) + 2(5 - 9)]$

Use the product rule to simplify.

8. $2^2 \cdot 2^5$

9. $y \cdot y^2 \cdot y^4$

Use the power rules to simplify. Write each result using positive exponents.

10. $(5x^2)^3$

11. $\left(\frac{2^{-3}}{y}\right)^{-2}$

12. Solve: $0.6 = 2 - 3.5c$

13. Find two numbers such that the second number is 3 more than twice the first number and the sum of the two numbers is 72.

14. Marial Callier just received an inheritance of $10,000 and plans to place all the money in a savings account that pays 5% compounded quarterly to help her son go to college in 3 years. How much money will be in the account in 3 years?

1. ________
2. ________
3. ________
4. ________
5. ________
6. ________
7. ________
8. ________
9. ________
10. ________
11. ________
12. ________
13. ________
14. ________

Name ______________________________

Graph each set on a number line and then write it in interval notation.

15. $\{x | x \geq 2\}$

−5 −4 −3 −2 −1 0 1 2 3 4 5

16. $\{x | 0.5 < x \leq 3\}$

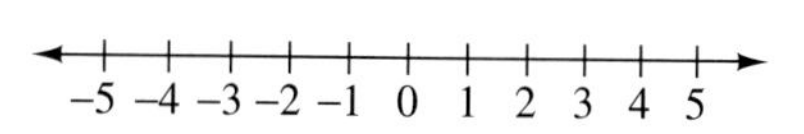

17. Solve: $x - 7 < 2$ and $2x + 1 < 9$

18. Solve: $|x - 3| = |5 - x|$

19. Solve: $|x - 3| > 7$

20. Find the intercepts and graph:
$x + 4y = -4$

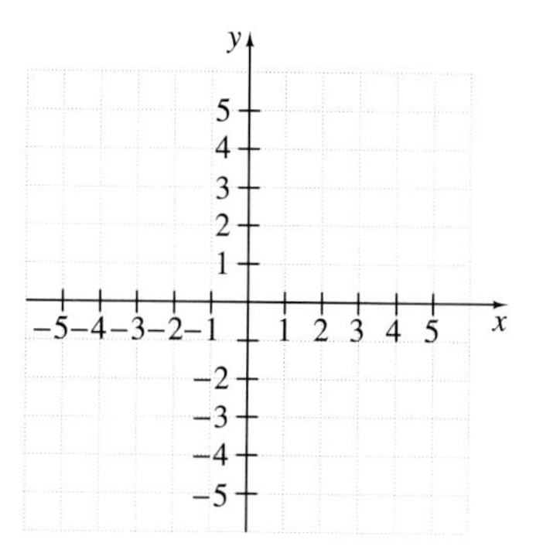

21. Find the slope of the line $y = 3x + 2$.

22. Write an equation of the line with y-intercept -3 and slope of $\frac{1}{4}$.

23. Find an equation of the horizontal line containing the point $(2, 3)$.

Find the following.

24. If $f(x) = 7x^2 - 3x + 1$, find $f(1)$.

25. If $g(x) = 3x - 2$, find $g(0)$.

15. ____________

16. ____________

17. ____________

18. ____________

19. ____________

20. see graph

21. ____________

22. ____________

23. ____________

24. ____________

25. ____________

Systems of Equations and Inequalities

CHAPTER 4

In this chapter, two or more equations in two or more variables are solved simultaneously. Such a collection of equations is called a **system of equations**. Systems of equations are good mathematical models for many real-world problems because these problems may involve several related patterns. We will study various methods for solving systems of equations and will conclude with a look at systems of inequalities.

Lightning, most often produced during thunderstorms, is a rapid discharge of high-current electricity into the atmosphere. At any given moment around the world, there are about 2000 thunderstorms in progress producing approximately 100 lightning flashes per second. In the United States, lightning causes an average of 75 fatalities per year. An estimated 5% of all residential insurance claims in the United States are due to lightning damage, totaling more than $1 billion per year. In addition, roughly 30% of all power outages in the United States are lightning related. Because of lightning's potentially destructive nature, meteorologists track lightning activity by recording and plotting the positions of lightning strikes. In the Chapter Activity on page 321, we will see how systems of equations can be used to pinpoint the location of a lightning strike.

Answers

1. ______
2. ______
3. ______
4. ______
5. ______
6. ______
7. ______
8. ______
9. ______
10. ______
11. ______
12. ______
13. ______
14. ______
15. ______
16. ______
17. see graph
18. see graph
19. ______
20. ______

Name ______________________ Section ________ Date ________

Chapter 4 Pretest

Solve each system of equations by graphing.

1. $\begin{cases} 2x + y = -1 \\ x + y = -3 \end{cases}$

2. $\begin{cases} x - y = 2 \\ 3x - y = 0 \end{cases}$

Solve each system of equations by the substitution method.

3. $\begin{cases} 2x - 3y = -8 \\ y = x + 1 \end{cases}$

4. $\begin{cases} \dfrac{x}{6} + \dfrac{y}{2} = -2 \\ x - \dfrac{y}{2} = 2 \end{cases}$

Solve each system of equations by the elimination method.

5. $\begin{cases} x - 4y = -29 \\ 5x + y = 44 \end{cases}$

6. $\begin{cases} 2x + 3y = -34 \\ 3x - 5y = 25 \end{cases}$

Solve each system.

7. $\begin{cases} x - y = 3 \\ 7y = -7 \\ 2x + y + 3z = -6 \end{cases}$

8. $\begin{cases} 3x - y + 2z = -1 \\ 2x + 5y - z = 4 \\ 4x + 6y + z = 2 \end{cases}$

Use matrices to solve each system.

9. $\begin{cases} 3x + 4y = -10 \\ x - y = 20 \end{cases}$

10. $\begin{cases} -2x + y = 6 \\ 4x - 2y = 12 \end{cases}$

11. $\begin{cases} 3x + 2z = -5 \\ x + y + z = -2 \\ -x + 2y - 3z = -5 \end{cases}$

12. $\begin{cases} x - 6y + z = -48 \\ 5x + y + 3z = 8 \\ 2x - y - z = -8 \end{cases}$

Evaluate.

13. $\begin{vmatrix} 5 & -1 \\ 2 & 6 \end{vmatrix}$

14. $\begin{vmatrix} 2 & 0 & -1 \\ 3 & 1 & -2 \\ 0 & 4 & 3 \end{vmatrix}$

Use Cramer's rule to solve each system.

15. $\begin{cases} 8x - y = -5 \\ -x + 5y = -53 \end{cases}$

16. $\begin{cases} x + y - 2z = 8 \\ 8x + 3y + z = 15 \\ -5x + 4y - 3z = 13 \end{cases}$

Graph the solution of each system of linear inequalities.

17. $\begin{cases} y \leq x + 1 \\ y > 3x - 2 \end{cases}$

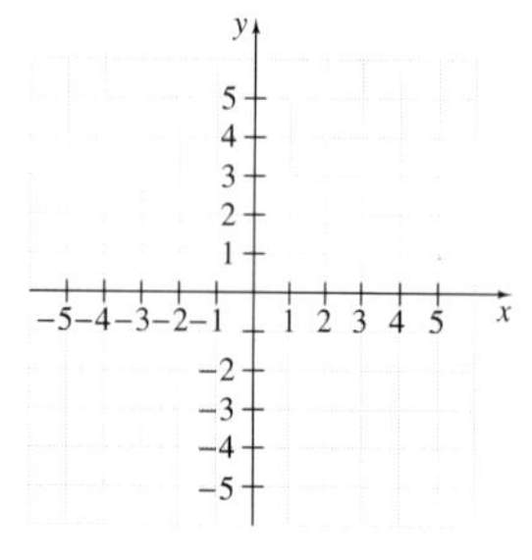

18. 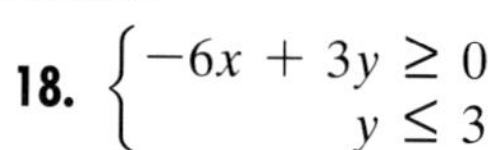
$\begin{cases} -6x + 3y \geq 0 \\ y \leq 3 \end{cases}$

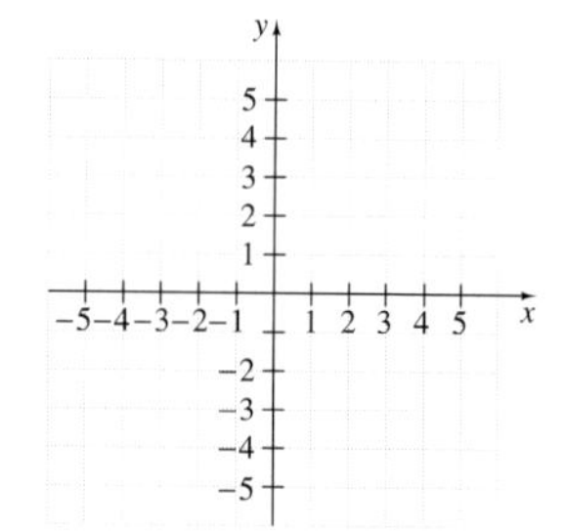

Solve.

19. Six times one number minus a second is 12, and the sum of the numbers is 16. Find the numbers.

20. The measure of the largest angle of a triangle is five times the measure of the smallest angle, and the measure of the remaining angle is 40° more than the measure of the smallest angle. Find the measure of each angle.

4.1 SOLVING SYSTEMS OF LINEAR EQUATIONS IN TWO VARIABLES

Recall from Chapter 3 that the graph of a linear equation in two variables is a line. Two or more linear equations form a **system of linear equations**. Some examples of systems of linear equations in two variables are

$$\begin{cases} x - 2y = -7 \\ 3x + y = 0 \end{cases} \qquad \begin{cases} x = 5 \\ x + \dfrac{y}{2} = 9 \end{cases} \qquad \begin{cases} x - 3 = 2y + 6 \\ y = 1 \end{cases}$$

Objectives

A Determine whether an ordered pair is a solution of a system of two linear equations.

B Solve a system of two equations by graphing.

C Solve a system using substitution.

D Solve a system using elimination.

SSM CD-ROM Video 4.1

A DETERMINING WHETHER AN ORDERED PAIR IS A SOLUTION

Recall that a solution of an equation in two variables is an ordered pair (x, y) that makes the equation true. A **solution of a system** of two equations in two variables is an ordered pair (x, y) that makes both equations true.

Example 1 Determine whether the ordered pair $(-1, 1)$ is a solution of the system.

$$\begin{cases} -x + y = 2 \\ 2x - y = -3 \end{cases}$$

Solution: We replace x with -1 and y with 1 in each equation.

$$\begin{aligned} -x + y &= 2 && \text{First equation} \\ -(-1) + (1) &= 2 && \text{Let } x = -1 \text{ and } y = 1. \\ 1 + 1 &= 2 \\ 2 &= 2 && \text{True.} \end{aligned}$$

$$\begin{aligned} 2x - y &= -3 && \text{Second equation} \\ 2(-1) - (1) &= -3 && \text{Let } x = -1 \text{ and } y = 1. \\ -2 - 1 &= -3 \\ -3 &= -3 && \text{True.} \end{aligned}$$

Since $(-1, 1)$ makes both equations true, it is a solution.

Practice Problem 1

Determine whether the ordered pair $(4, 1)$ is a solution of the system.

$$\begin{cases} x - y = 3 \\ 2x - 3y = 5 \end{cases}$$

Example 2 Determine whether the ordered pair $(-2, 3)$ is a solution of the system.

$$\begin{cases} 5x + 3y = -1 \\ x - y = 1 \end{cases}$$

Solution: We replace x with -2 and y with 3 in each equation.

$$\begin{aligned} 5x + 3y &= -1 && \text{First equation} \\ 5(-2) + 3(3) &= -1 && \text{Let } x = -2 \text{ and } y = 3. \\ -10 + 9 &= -1 \\ -1 &= -1 && \text{True.} \end{aligned}$$

$$\begin{aligned} x - y &= 1 && \text{Second equation} \\ (-2) - (3) &= 1 && \text{Let } x = -2 \text{ and } y = 3. \\ -5 &= 1 && \text{False.} \end{aligned}$$

Since the ordered pair $(-2, 3)$ does not make both equations true, it is not a solution of the system.

Practice Problem 2

Determine whether the ordered pair $(-3, 3)$ is a solution of the system.

$$\begin{cases} 3x - y = -12 \\ x - y = 0 \end{cases}$$

Answers

1. Yes, it is a solution., **2.** No, it is not a solution.

Practice Problems 3–4

Solve each system by graphing. If the system has just one solution, estimate the solution.

3. $\begin{cases} x - y = 2 \\ x + 3y = 6 \end{cases}$

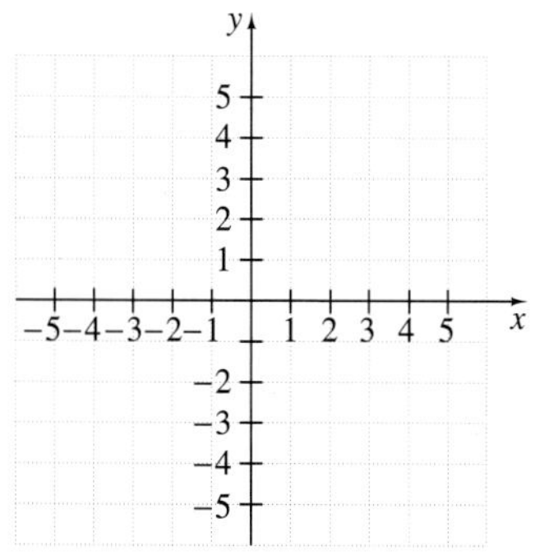

4. $\begin{cases} y = -3x \\ 6x + 2y = 4 \end{cases}$

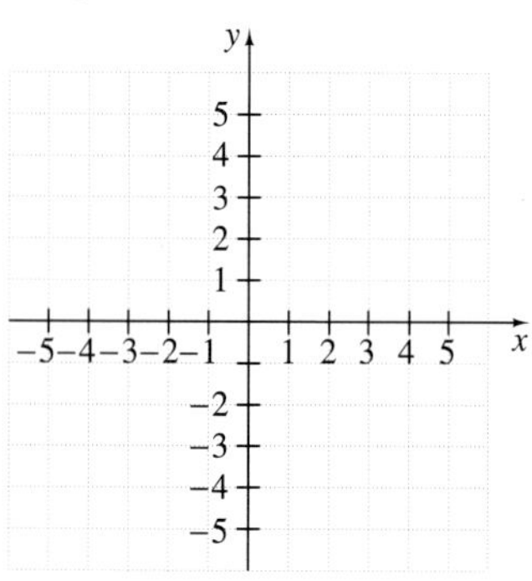

B Solving a System by Graphing

The graph of each linear equation in a system is a line. Each point on each line corresponds to an ordered pair solution of its equation. If the lines intersect, the point of intersection lies on both lines and corresponds to an ordered pair solution of both equations. In other words, the point of intersection corresponds to an ordered pair solution of the system. Therefore, we can estimate the solutions of a system by graphing each equation on the same rectangular coordinate system and estimating the coordinates of any point of intersection.

Example 3 Solve the system by graphing. If the system has just one solution, estimate the solution.

$$\begin{cases} x + y = 2 \\ 3x - y = -2 \end{cases}$$

Solution: First we graph each linear equation on the same rectangular coordinate system.

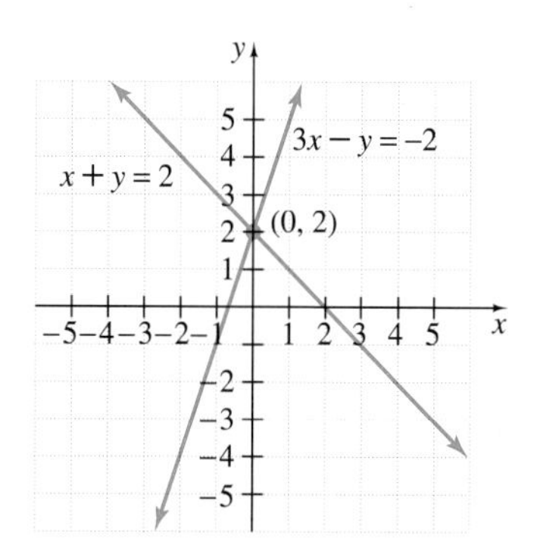

These lines intersect at one point as shown. The coordinates of the point of intersection appear to be $(0, 2)$. Check this estimated solution by replacing x with 0 and y with 2 in *both* equations.

$x + y = 2$	First equation	$3x - y = -2$	Second equation
$0 + 2 = 2$	Let $x = 0$ and $y = 2$.	$3(0) - 2 = -2$	Let $x = 0$ and $y = 2$.
$2 = 2$	True.	$-2 = -2$	True.

The ordered pair $(0, 2)$ is the solution of the system. A system that has at least one solution, such as this one, is said to be **consistent**.

Example 4 Solve the system by graphing. If the system has just one solution, estimate the solution.

$$\begin{cases} x - 2y = 4 \\ x = 2y \end{cases}$$

Solution: We graph each linear equation.

Answers

3. (3, 1)

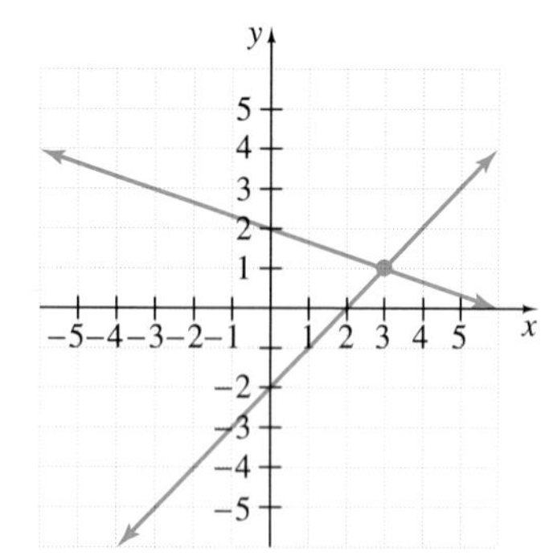

4.

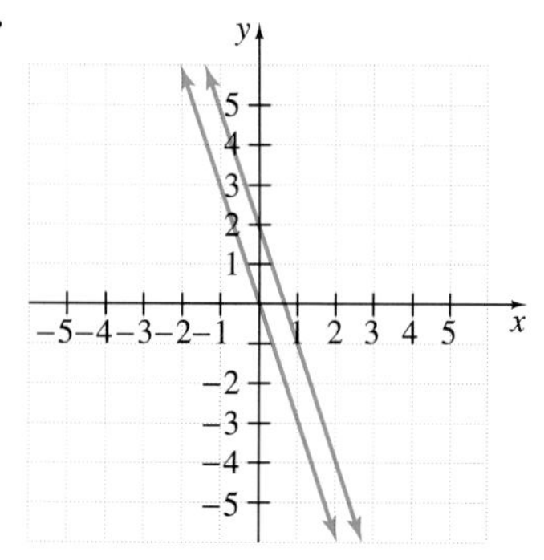

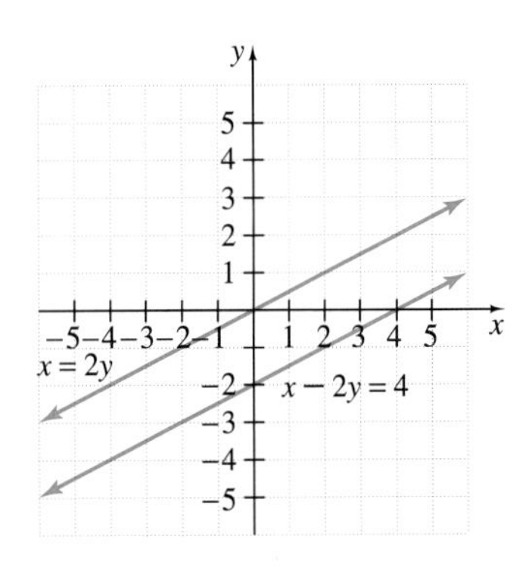

The lines appear to be parallel. To be sure, let's write each equation in point-slope form, $y = mx + b$, and solve for y.

$x - 2y = 4$	First equation	$x = 2y$	Second equation
$-2y = -x + 4$	Subtract x from both sides.	$\frac{1}{2}x = y$	Divide both sides by 2.
$y = \frac{1}{2}x - 2$	Divide both sides by -2.	$y = \frac{1}{2}x$	

The graphs of these equations have the same slope $\frac{1}{2}$, but different y-intercepts, so these lines are parallel. Therefore, the system has no solution since the equations have no common solution (there are no intersection points). A system that has no solution is said to be **inconsistent**.

TRY THE CONCEPT CHECK IN THE MARGIN.

Example 5

Solve the system by graphing. If the system has just one solution, estimate the solution.

$$\begin{cases} 2x + 4y = 10 \\ x + 2y = 5 \end{cases}$$

Solution: We graph each linear equation.

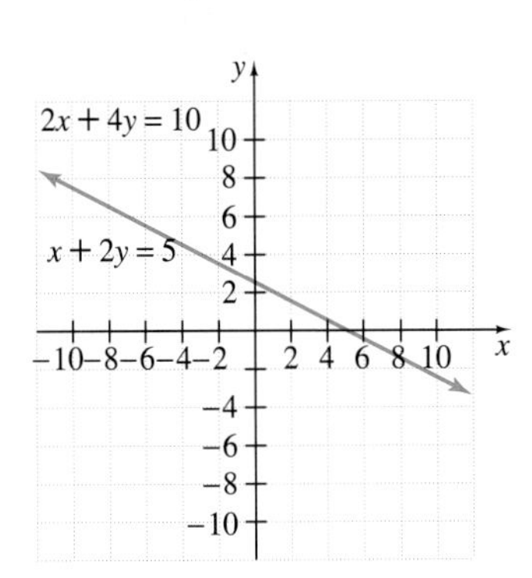

We see that the graphs of both equations are the same line. To confirm this, notice that if both sides of the second equation are multiplied by 2, the result is the first

✓ CONCEPT CHECK

How can you tell just by looking at the following system that it has no solution?

$$\begin{cases} y = 3x + 5 \\ y = 3x - 7 \end{cases}$$

Practice Problem 5

Solve the system by graphing.

5. $\begin{cases} -2x + y = 1 \\ 4x - 2y = -2 \end{cases}$

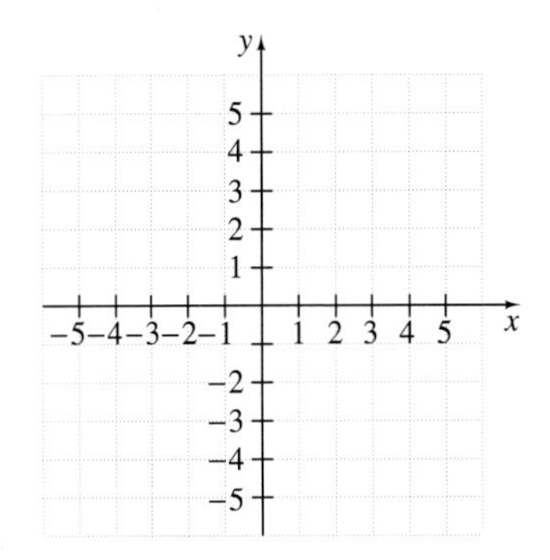

Answers

5.

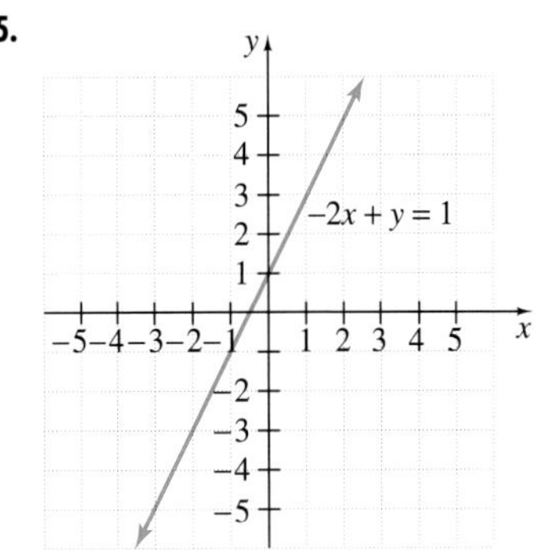

✓ Concept Check: answers may vary

equation. This means that the equations have identical solutions. Any ordered pair solution of one equation satisfies the other equation also. These equations are said to be **dependent equations**. The solution set of the system is $\{(x, y)|x + 2y = 5\}$ or, equivalently, $\{(x, y)|2x + 4y = 10\}$ since the lines describe identical ordered pairs. Written the second way, the solution set is read "the set of all ordered pairs (x, y), such that $2x + 4y = 10$." There are an infinite number of solutions to this system.

✓ CONCEPT CHECK

How can you tell just by looking at the following system that it has infinitely many solutions?

$$\begin{cases} x + y = 5 \\ 2x + 2y = 10 \end{cases}$$

TRY THE CONCEPT CHECK IN THE MARGIN.

We can summarize the information discovered in Examples 3 through 5 as follows.

POSSIBLE SOLUTIONS TO SYSTEMS OF TWO EQUATIONS

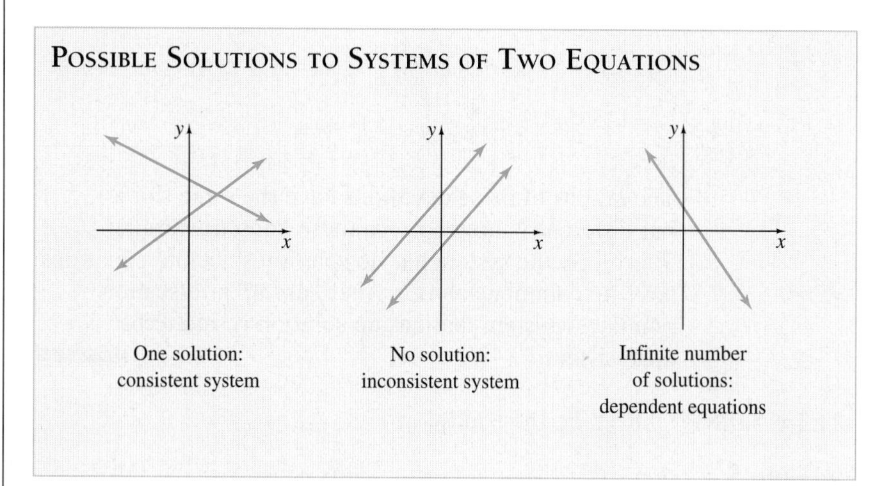

C SOLVING A SYSTEM USING SUBSTITUTION

Graphing the equations of a system by hand is often a good method for finding approximate solutions of a system, but it is not a reliable method for finding exact solutions. To find an exact solution, we need to use *algebra*. One such *algebraic* method is called the **substitution method**.

SOLVING A SYSTEM OF TWO EQUATIONS USING THE SUBSTITUTION METHOD

Step 1. Solve one of the equations for one of its variables.

Step 2. Substitute the expression for the variable found in Step 1 into the other equation.

Step 3. Find the value of one variable by solving the equation from Step 2.

Step 4. Find the value of the other variable by substituting the value found in Step 3 into the equation from Step 1.

Step 5. Check the ordered pair solution in *both* original equations.

Practice Problem 6

Use the substitution method to solve the system:

$$\begin{cases} 6x - 4y = 10 \\ y = 3x - 3 \end{cases}$$

Example 6 Use the substitution method to solve the system:

$$\begin{cases} 2x + 4y = -6 & \text{First equation} \\ x = 2y - 5 & \text{Second equation} \end{cases}$$

Answers

6. $\left(\frac{1}{3}, -2\right)$

✓ Concept Check: answers may vary

Solution: In the second equation, we are told that x is equal to $2y - 5$. Since they are equal, we can *substitute* $2y - 5$ for x in the first equation. This will give us an equation in one variable, which we can solve for y.

$$2x + 4y = -6 \quad \text{First equation}$$

$$2(2y - 5) + 4y = -6 \quad \text{Substitute } 2y - 5 \text{ for } x.$$

$$4y - 10 + 4y = -6$$

$$8y = 4$$

$$y = \frac{4}{8} = \frac{1}{2} \quad \text{Solve for } y.$$

The y-coordinate of the solution is $\frac{1}{2}$. To find the x-coordinate, we replace y with $\frac{1}{2}$ in the second equation, $x = 2y - 5$.

$$x = 2y - 5$$

$$x = 2\left(\frac{1}{2}\right) - 5 = 1 - 5 = -4$$

The ordered pair solution is $\left(-4, \frac{1}{2}\right)$. Check to see that $\left(-4, \frac{1}{2}\right)$ satisfies both equations of the system.

Example 7 Use the substitution method to solve the system:

$$\begin{cases} -\frac{x}{6} + \frac{y}{2} = \frac{1}{2} \\ \frac{x}{3} - \frac{y}{6} = -\frac{3}{4} \end{cases}$$

Solution: First we multiply each equation by its least common denominator to clear the system of fractions. We multiply the first equation by 6 and the second equation by 12.

$$\begin{cases} 6\left(-\frac{x}{6} + \frac{y}{2}\right) = 6\left(\frac{1}{2}\right) \\ 12\left(\frac{x}{3} - \frac{y}{6}\right) = 12\left(-\frac{3}{4}\right) \end{cases} \text{ simplifies to } \begin{cases} -x + 3y = 3 & \text{First equation} \\ 4x - 2y = -9 & \text{Second equation} \end{cases}$$

We now solve the first equation for x.

$$-x + 3y = 3 \quad \text{First equation}$$

$$3y - 3 = x \quad \text{Solve for } x.$$

Next we replace x with $3y - 3$ in the second equation.

$$4x - 2y = -9 \quad \text{Second equation}$$

$$4(3y - 3) - 2y = -9$$

$$12y - 12 - 2y = -9$$

$$10y = 3$$

$$y = \frac{3}{10} \quad \text{Solve for } y.$$

Practice Problem 7

Use the substitution method to solve the system: $\begin{cases} -\frac{x}{2} + \frac{y}{4} = \frac{1}{2} \\ \frac{x}{2} + \frac{y}{2} = -\frac{1}{8} \end{cases}$

Answer

7. $\left(-\frac{3}{4}, \frac{1}{2}\right)$

The y-coordinate is $\frac{3}{10}$. To find the x-coordinate, we replace y with $\frac{3}{10}$ in the equation $x = 3y - 3$. Then

$$x = 3\left(\frac{3}{10}\right) - 3 = \frac{9}{10} - 3 = \frac{9}{10} - \frac{30}{10} = -\frac{21}{10}$$

The ordered pair solution is $\left(-\frac{21}{10}, \frac{3}{10}\right)$. Check to see that this solution satisfies both original equations.

HELPFUL HINT

If a system of equations contains equations with fractions, first take a step and clear the equations of fractions.

D Solving a System Using Elimination

The **elimination method**, or **addition method**, is a second algebraic technique for solving systems of equations. For this method, we rely on a version of the addition property of equality, which states that "equals added to equals are equal."

If $A = B$ and $C = D$ then $A + C = B + D$

Solving a System of Two Linear Equations Using the Elimination Method

Step 1. Rewrite each equation in standard form, $Ax + By = C$.

Step 2. If necessary, multiply one or both equations by some nonzero number so that the coefficient of one variable in one equation is the opposite of its coefficient in the other equation.

Step 3. Add the equations.

Step 4. Find the value of one variable by solving the equation from Step 3.

Step 5. Find the value of the second variable by substituting the value found in Step 4 into either original equation.

Step 6. Check the proposed ordered pair solution in *both* original equations.

Practice Problem 8

Use the elimination method to solve the system: $\begin{cases} 3x - y = 1 \\ 4x + y = 6 \end{cases}$

Answer

8. $(1, 2)$

Example 8 Use the elimination method to solve the system:

$$\begin{cases} x - 5y = -12 & \text{First equation} \\ -x + y = 4 & \text{Second equation} \end{cases}$$

Solution: Since the left side of each equation is equal to the right side, we add equal quantities by adding the left sides of the equations and the right sides of the equations. This sum gives us an equation in one variable, y, which we can solve for y.

$$\begin{array}{rll} x - 5y = & -12 & \text{First equation} \\ -x + y = & 4 & \text{Second equation} \\ \hline -4y = & -8 & \text{Add.} \\ y = & 2 & \text{Solve for } y. \end{array}$$

The y-coordinate of the solution is 2. To find the corresponding x-coordinate, we replace y with 2 in either original equation of the system. Let's use the second equation.

$$\begin{array}{rll} -x + y = & 4 & \text{Second equation} \\ -x + 2 = & 4 & \text{Let } y = 2. \\ -x = & 2 & \\ x = & -2 & \end{array}$$

The ordered pair solution is $(-2, 2)$. Check to see that $(-2, 2)$ satisfies both equations of the system.

Example 9 Use the elimination method to solve the system:

$$\begin{cases} 3x + \frac{y}{2} = 2 \\ 6x + y = 5 \end{cases}$$

Solution: If we add the two equations, the sum will still be an equation in two variables. Notice, however, that if we multiply both sides of the first equation by -2, the coefficients of x in the two equations will be opposites. Then

$$\begin{cases} -2\left(3x + \frac{y}{2}\right) = -2(2) \\ 6x + y = 5 \end{cases} \quad \text{simplifies to} \quad \begin{cases} -6x - y = -4 \\ 6x + y = 5 \end{cases}$$

Now we can add the left sides and add the right sides.

$$\begin{array}{rll} -6x - y = & -4 & \\ 6x + y = & 5 & \\ \hline 0 = & 1 & \text{False.} \end{array}$$

The resulting equation, $0 = 1$, is false for all values of y or x. Thus, the system has no solution. The solution set is $\{\ \}$ or $\varnothing$. This system is inconsistent, and the graphs of the equations are parallel lines.

Practice Problem 9

Use the elimination method to solve the system: $\begin{cases} \frac{x}{3} + 2y = -1 \\ x + 6y = 2 \end{cases}$

Example 10 Use the elimination method to solve the system:

$$\begin{cases} 3x - 2y = 10 \\ 4x - 3y = 15 \end{cases}$$

Solution: To eliminate y when the equations are added, we multiply both sides of the first equation by 3 and both sides of the second equation by -2. Then

$$\begin{cases} 3(3x - 2y) = 3(10) \\ -2(4x - 3y) = -2(15) \end{cases} \quad \text{simplifies to} \quad \begin{cases} 9x - 6y = 30 \\ -8x + 6y = -30 \end{cases}$$

Next we add the left sides and add the right sides.

$$\begin{array}{rl} 9x - 6y = & 30 \\ -8x + 6y = & -30 \\ \hline x \quad = & 0 \end{array}$$

Practice Problem 10

Use the elimination method to solve the system: $\begin{cases} 2x - 5y = 6 \\ 3x - 4y = 9 \end{cases}$

Answers

9. no solution or $\varnothing$, **10.** $(3, 0)$

To find y, we let $x = 0$ in either equation of the system.

$$3x - 2y = 10 \quad \text{First equation}$$
$$3(0) - 2y = 10 \quad \text{Let } x = 0.$$
$$-2y = 10$$
$$y = -5$$

The ordered pair solution is $(0, -5)$. Check to see that $(0, -5)$ satisfies both equations.

Practice Problem 11

Use the elimination method to solve the system: $\begin{cases} 4x - 7y = 10 \\ -8x + 14y = -20 \end{cases}$

Example 11 Use the elimination method to solve the system:

$$\begin{cases} -5x - 3y = 9 \\ 10x + 6y = -18 \end{cases}$$

Solution: To eliminate x when the equations are added, we multiply both sides of the first equation by 2. Then

$$\begin{cases} 2(-5x - 3y) = 2(9) \\ 10x + 6y \quad = -18 \end{cases} \text{ simplifies to } \begin{cases} -10x - 6y = 18 \\ 10x + 6y = -18 \end{cases}$$

Next we add the equations.

$$\begin{array}{r} -10x - 6y = 18 \\ 10x + 6y = -18 \\ \hline 0 = 0 \end{array}$$

The resulting equation, $0 = 0$, is true for all possible values of y or x. Notice in the original system that if both sides of the first equation are multiplied by -2, the result is the second equation. This means that the two equations are equivalent. They have the same solution set and there are an infinite number of solutions. Thus, the equations of this system are dependent, and the solution set of the system is

$$\{(x, y)|-5x - 3y = 9\} \text{ or, equivalently, } \{(x, y)|10x + 6y = -18\}.$$

Answer

11. $\{(x, y)|4x - 7y = 10\}$

GRAPHING CALCULATOR EXPLORATIONS

A grapher may be used to approximate solutions of systems of equations by graphing both equations on the same set of axes and approximating any points of intersection. For example, let's approximate the solution of the system.

$$\begin{cases} y = -2.6x + 5.6 \\ y = 4.3x - 4.9 \end{cases}$$

We use a standard window and graph both equations on a single screen.

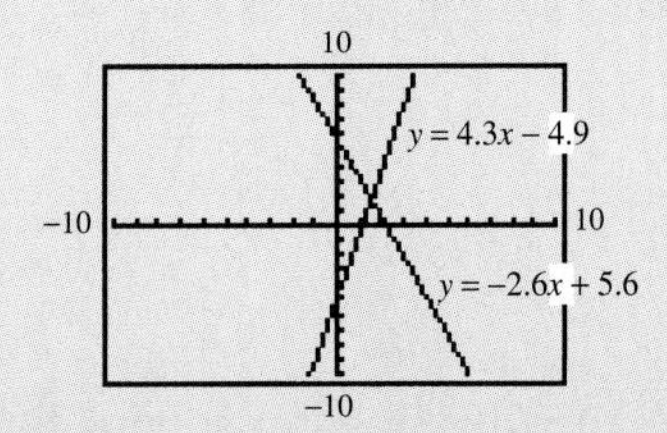

The two lines intersect. To approximate the point of intersection, we trace to the point of intersection and use an Intersect feature of the grapher, a Zoom In feature of the grapher, or redefine the window to [0,3] by [0,3]. If we redefine the window to [0, 3] by [0, 3], the screen should look like the following:

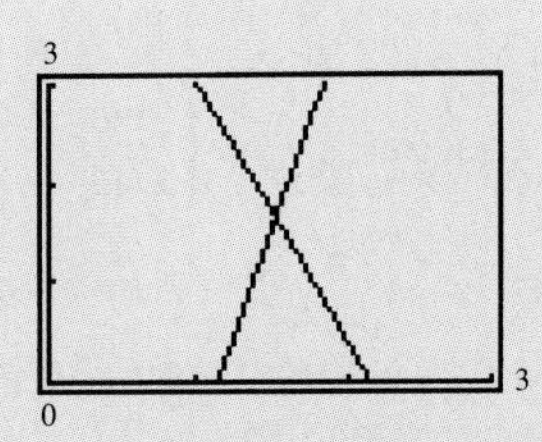

By tracing along the curves, we can see that the point of intersection has an x-value between 1.5 and 1.532. We can continue to zoom and trace or redefine the window until the coordinates of the point of intersection can be determined to the nearest hundredth. The approximate point of intersection is (1.52, 1.64).

Solve each system of equations. Approximate each solution to two decimal places.

1. $y = -1.65x + 3.65$
$y = 4.56x - 9.44$

2. $y = 7.61x + 3.48$
$y = -1.26x - 6.43$

3. $2.33x - 4.72y = 10.61$
$5.86x + 6.22y = -8.89$

4. $-7.89x - 5.68y = 3.26$
$-3.65x + 4.98y = 11.77$

Focus On Mathematical Connections

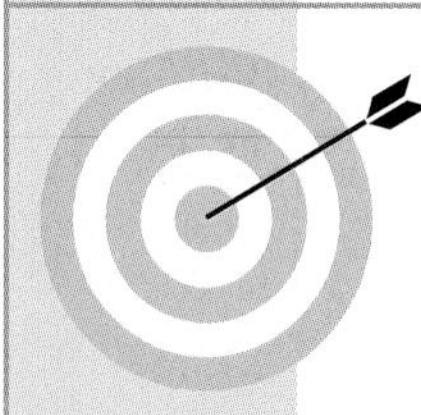

SOLVING NONLINEAR SYSTEMS

Recall that a linear equation in two variables is an equation that can be written in the form

$Ax + By = C$. By this definition, we can see that an equation of the form $\frac{A}{x} + \frac{B}{y} = C$ is clearly not linear. However, with a slight adjustment, we can solve a nonlinear system such as

$$\begin{cases} \frac{A}{x} + \frac{B}{y} = C \\ \frac{D}{x} + \frac{E}{y} = F \end{cases}$$

using the methods we already know for solving linear systems.

To solve such a system, first make the following substitutions. Let $w = \frac{1}{x}$ and $z = \frac{1}{y}$ in both equations. Then

$$\begin{cases} \frac{A}{x} + \frac{B}{y} = C \\ \frac{D}{x} + \frac{E}{y} = F \end{cases} \quad \text{becomes} \quad \begin{cases} Aw + Bz = C \\ Dw + Ez = F \end{cases}$$

This new system of equations is linear and can be solved with any of the techniques we already know. Once the values of w and z have been found, simply substitute into the equations $w = \frac{1}{x}$ and $z = \frac{1}{y}$. Then solve each equation to find the value of x and the value of y.

CRITICAL THINKING

Apply the method described above to solve each nonlinear system.

1. $\begin{cases} \frac{2}{x} + \frac{3}{y} = 5 \\ \frac{5}{x} - \frac{3}{y} = 2 \end{cases}$

2. $\begin{cases} x + \frac{2}{y} = 7 \\ 3x + \frac{3}{y} = 6 \end{cases}$

3. $\begin{cases} \frac{3}{x} - \frac{2}{y} = -18 \\ \frac{2}{x} + \frac{3}{y} = 1 \end{cases}$

4. $\begin{cases} \frac{2}{x} - \frac{4}{y} = 5 \\ \frac{1}{x} - \frac{2}{y} = \frac{3}{2} \end{cases}$

Name ______________________ **Section** ______ **Date** ________

MENTAL MATH ANSWERS

1. ________
2. ________
3. ________
4. ________

MENTAL MATH

Match each graph with the solution of the corresponding system.

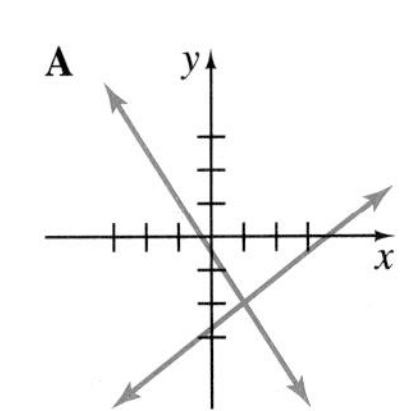

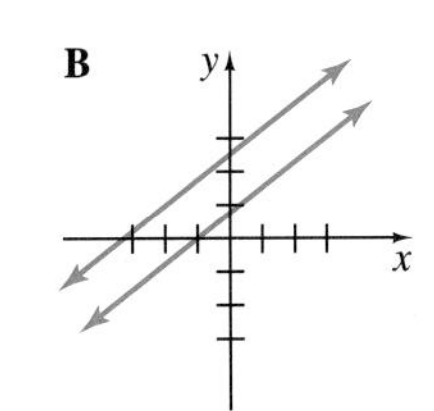

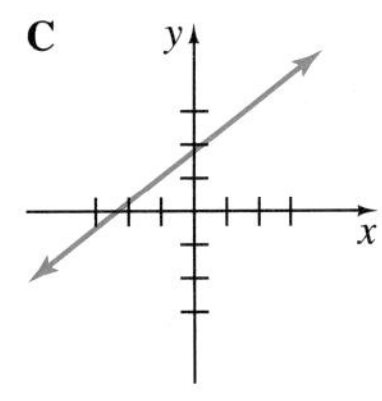

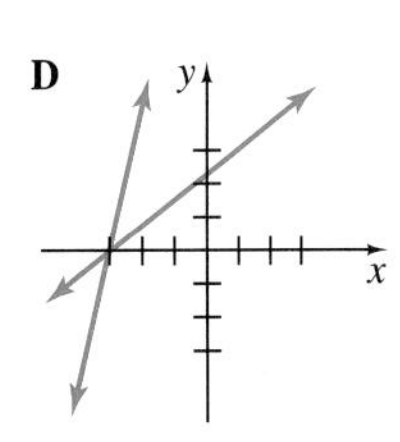

1. no solution

2. Infinite number of solutions

3. $(1, -2)$

4. $(-3, 0)$

EXERCISE SET 4.1

A *Determine whether each given ordered pair is a solution of each system. See Examples 1 and 2.*

1. $\begin{cases} x - y = 3 \\ 2x - 4y = 8 \end{cases}$ $(2, -1)$

2. $\begin{cases} x - y = -4 \\ 2x + 10y = 4 \end{cases}$ $(-3, 1)$

3. $\begin{cases} 2x - 3y = -9 \\ 4x + 2y = -2 \end{cases}$ $(3, 5)$

4. $\begin{cases} 2x - 5y = -2 \\ 3x + 4y = 4 \end{cases}$ $(4, 2)$

5. $\begin{cases} y = -5x \\ x = -2 \end{cases}$ $(-2, 10)$

6. $\begin{cases} y = 6 \\ x = -2y \end{cases}$ $(-12, 6)$

B *Solve each system by graphing. See Examples 3 through 5.*

7. $\begin{cases} x + y = 1 \\ x - 2y = 4 \end{cases}$

8. $\begin{cases} 2x - y = 8 \\ x + 3y = 11 \end{cases}$

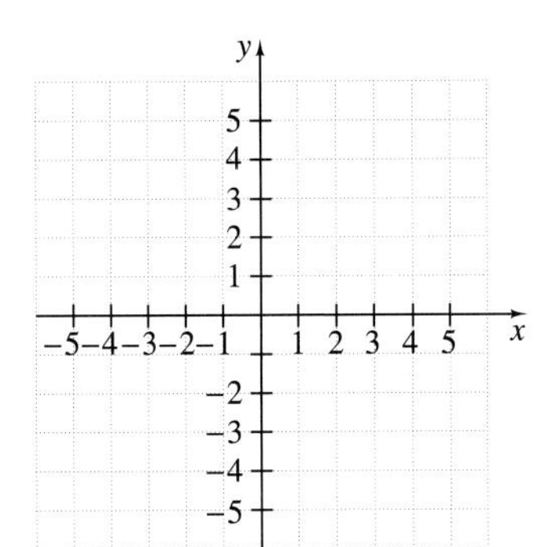

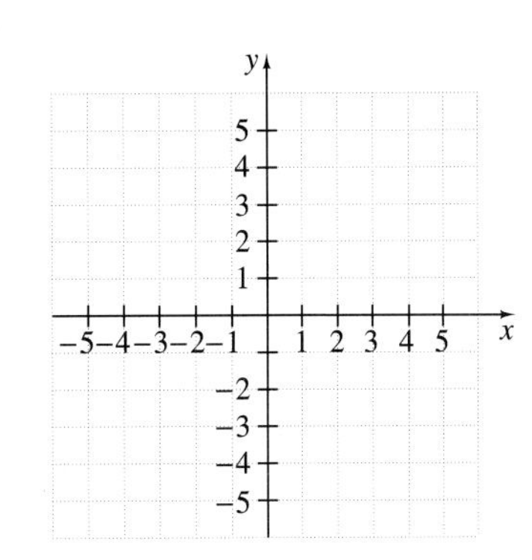

ANSWERS

1. ________
2. ________
3. ________
4. ________
5. ________
6. ________
7. ________
8. ________

9. ____________

10. ____________

11. ____________

12. ____________

13. ____________

14. ____________

15. ____________

16. ____________

17. ____________

Name ______________________________

9. $\begin{cases} 2y - 4 = 0 \\ x + 2y = 5 \end{cases}$

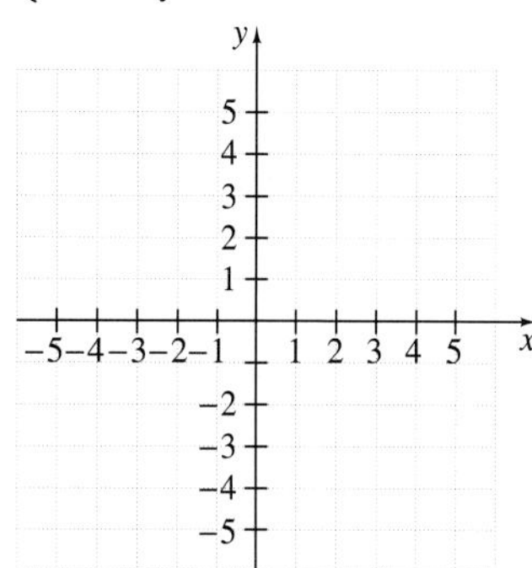

10. $\begin{cases} 4x - y = 6 \\ x - y = 0 \end{cases}$

11. $\begin{cases} 3x - y = 4 \\ 6x - 2y = 4 \end{cases}$

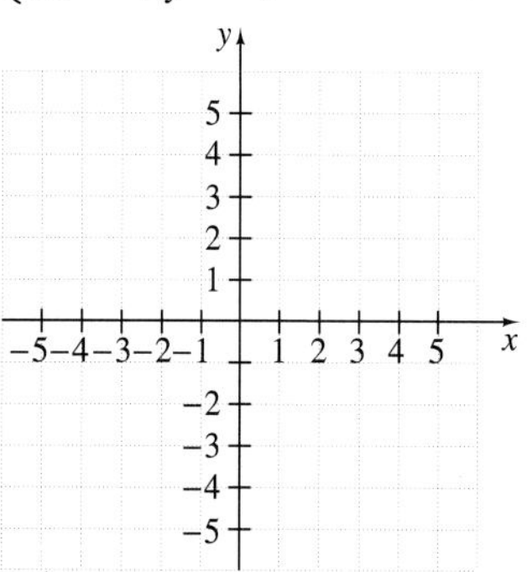

12. $\begin{cases} -x + 3y = 6 \\ 3x - 9y = 9 \end{cases}$

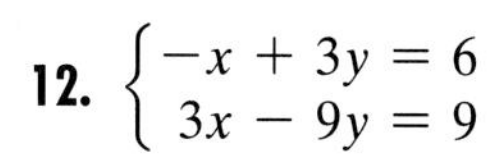

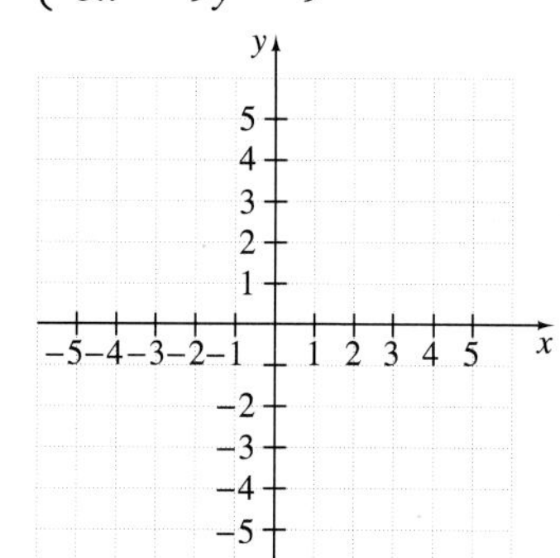

13. $\begin{cases} y = -3x \\ 2x - y = -5 \end{cases}$

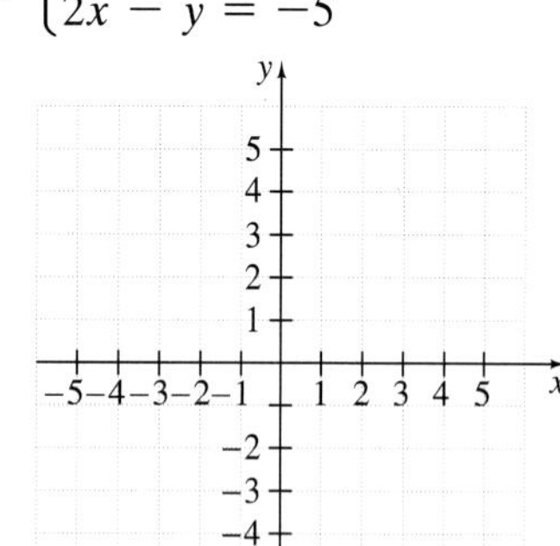

14. $\begin{cases} y = -2x \\ -3x + y = 10 \end{cases}$

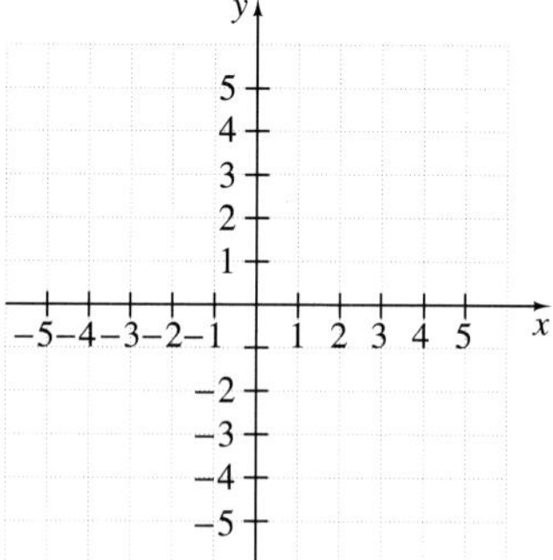

C *Use the substitution method to solve each system of equations. See Examples 6 and 7.*

15. $\begin{cases} x + y = 10 \\ y = 4x \end{cases}$

16. $\begin{cases} 5x + 2y = -17 \\ x = 3y \end{cases}$

17. $\begin{cases} 4x - y = 9 \\ 2x + 3y = -27 \end{cases}$

Name ______________________

18. $\begin{cases} 3x - y = 6 \\ -4x + 2y = -8 \end{cases}$

19. $\begin{cases} \frac{1}{2}x + \frac{3}{4}y = -\frac{1}{4} \\ \frac{3}{4}x - \frac{1}{4}y = 1 \end{cases}$

20. $\begin{cases} \frac{2}{5}x + \frac{1}{5}y = -1 \\ x + \frac{2}{5}y = -\frac{8}{5} \end{cases}$

21. $\begin{cases} \frac{x}{3} + y = \frac{4}{3} \\ -x + 2y = 11 \end{cases}$

22. $\begin{cases} \frac{x}{8} - \frac{y}{2} = 1 \\ \frac{x}{3} - y = 2 \end{cases}$

23. $\begin{cases} 2x - y = -1 \\ y = -2x \end{cases}$

24. $\begin{cases} x = \frac{1}{5}y \\ x - y = -4 \end{cases}$

25. $\begin{cases} 2x = 6 \\ y = 5 - x \end{cases}$

26. $\begin{cases} x = 3y + 4 \\ -y = 5 \end{cases}$

D *Use the elimination method to solve each system of equations. See Examples 8 through 11.*

27. $\begin{cases} 2x - 4y = 0 \\ x + 2y = 5 \end{cases}$

28. $\begin{cases} 2x - 3y = 0 \\ 2x + 6y = 3 \end{cases}$

29. $\begin{cases} 5x + 2y = 1 \\ x - 3y = 7 \end{cases}$

30. $\begin{cases} 6x - y = -5 \\ 4x - 2y = 6 \end{cases}$

31. $\begin{cases} 5x - 2y = 27 \\ -3x + 5y = 18 \end{cases}$

32. $\begin{cases} 3x + 4y = 2 \\ 2x + 5y = -1 \end{cases}$

33. $\begin{cases} 3x - 5y = 11 \\ 2x - 6y = 2 \end{cases}$

34. $\begin{cases} 6x - 3y = -3 \\ 4x + 5y = -9 \end{cases}$

35. $\begin{cases} x - 2y = 4 \\ 2x - 4y = 4 \end{cases}$

36. $\begin{cases} -x + 3y = 6 \\ 3x - 9y = 9 \end{cases}$

37. $\begin{cases} 3x + y = 1 \\ 2y = 2 - 6x \end{cases}$

38. $\begin{cases} y = 2x - 5 \\ 8x - 4y = 20 \end{cases}$

18. ______
19. ______
20. ______
21. ______
22. ______
23. ______
24. ______
25. ______
26. ______
27. ______
28. ______
29. ______
30. ______
31. ______
32. ______
33. ______
34. ______
35. ______
36. ______
37. ______
38. ______

39.	
40.	
41.	
42.	
43.	
44.	
45.	
46.	
47.	
48.	
49.	
50.	
51.	
52.	
53.	
54.	
55.	
56.	

Name ______________________

39. $\begin{cases} x = 3y - 2 \\ 5x - 15y = 0 \end{cases}$

40. $\begin{cases} x = 3y - 1 \\ 2x - 6y = -2 \end{cases}$

41. $\begin{cases} 4x + 2y = 5 \\ 2x + y = -1 \end{cases}$

42. $\begin{cases} 3x + 6y = 15 \\ 2x + 4y = 3 \end{cases}$

43. $\begin{cases} \frac{3}{4}x + \frac{5}{2}y = 11 \\ \frac{1}{16}x - \frac{3}{4}y = -1 \end{cases}$

44. $\begin{cases} \frac{2}{3}x + \frac{1}{4}y = -\frac{3}{2} \\ \frac{1}{2}x - \frac{1}{4}y = -2 \end{cases}$

45. $\begin{cases} \frac{2}{3}x - \frac{3}{4}y = -1 \\ -\frac{1}{6}x + \frac{3}{8}y = 1 \end{cases}$

46. $\begin{cases} \frac{1}{2}x - \frac{1}{3}y = -3 \\ \frac{1}{8}x + \frac{1}{6}y = 0 \end{cases}$

47. $\begin{cases} 0.7x - 0.2y = -1.6 \\ 0.2x - y = -1.4 \end{cases}$

48. $\begin{cases} -0.7x + 0.6y = 1.3 \\ 0.5x - 0.3y = -0.8 \end{cases}$

49. $\begin{cases} 10y - 2x = 1 \\ 5y = 4 - 6x \end{cases}$

50. $\begin{cases} 3x + 4y = 0 \\ 7x = 3y \end{cases}$

51. $\begin{cases} x = 3y + 2 \\ 5x - 15y = 10 \end{cases}$

52. $\begin{cases} y = \frac{1}{7}x + 3 \\ x - 7y = -21 \end{cases}$

REVIEW AND PREVIEW

Determine whether the given replacement values make each equation true or false. See Section 2.1.

53. $3x - 4y + 2z = 5$; $x = 1$, $y = 2$, and $z = 5$

54. $x + 2y - z = 7$; $x = 2$, $y = -3$, and $z = 3$

55. $-x - 5y + 3z = 15$; $x = 0$, $y = -1$, and $z = 5$

56. $-4x + y - 8z = 4$; $x = 1$, $y = 0$, and $z = -1$

Name ______________________

Add the equations. See Section 4.1.

57. $\begin{cases} 3x + 2y - 5z = 10 \\ -3x + 4y + z = 15 \end{cases}$

58. $\begin{cases} x + 4y - 5z = 20 \\ 2x - 4y - 2z = -17 \end{cases}$

59. $\begin{cases} 10x + 5y + 6z = 14 \\ -9x + 5y - 6z = -12 \end{cases}$

60. $\begin{cases} -9x - 8y - z = 31 \\ 9x + 4y - z = 12 \end{cases}$

Combining Concepts

61. Can a system consisting of two linear equations have exactly two solutions? Explain why or why not.

62. Suppose the graph of the equations in a system of two equations in two variables consists of a circle and a line. Discuss the possible number of solutions for this system.

The concept of supply and demand is used often in business. In general, as the unit price of a commodity increases, the demand for that commodity decreases. Also, as a commodity's unit price increases, the manufacturer normally increases the supply. The point where supply is equal to demand is called the equilibrium point. The following shows the graph of a demand equation and the graph of a supply equation for ties. The x-axis represents number of ties in thousands, and the y-axis represents the cost of a tie. Use this graph to answer Exercises 63–66.

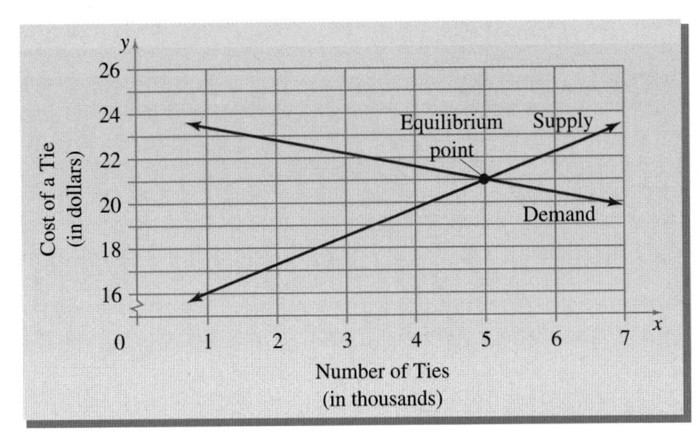

63. Find the number of ties and the price per tie when supply equals demand.

64. When x is between 3 and 4, is supply greater than demand or is demand greater than supply?

57. ______

58. ______

59. ______

60. ______

61. ______

62. ______

63. ______

64. ______

Name ______________________

65. ______

66. ______

67. a. ______

b. ______

c. ______

65. When x is greater than 7, is supply greater than demand or is demand greater than supply?

66. For what x-values are the y-values corresponding to the supply equation greater than the y-values corresponding to the demand equation?

67. The amount y of red meat consumed per person in the United States (in pounds) in the year x can be modeled by the linear equation $y = -0.6x + 121.2$. The amount y of all poultry consumed per person in the United States (in pounds) in the year x can be modeled by the linear equation $y = 1.7x + 88$. In both models, $x = 0$ represents the year 1995. (*Source:* Based on data and forecasts from the Economic Research Service, U.S. Department of Agriculture, 1995–1999)

a. What does the slope of each equation tell you about the patterns of red meat and poultry consumption in the United States?

b. Solve this system of equations. (Round your final results to the nearest whole numbers.)

c. Explain the meaning of your answer to part (b).

4.2 Solving Systems of Linear Equations in Three Variables

Objective

A Solve a system of three linear equations in three variables.

SSM CD-ROM Video 4.2

In this section, we solve systems of three linear equations in three variables. We call the equation $3x - y + z = -15$, for example, a **linear equation in three variables** since there are three variables and each variable is raised only to the power 1. A solution of this equation is an **ordered triple (x, y, z)** that makes the equation a true statement.

For example, the ordered triple $(2, 0, -21)$ is a solution of $3x - y + z = -15$ since replacing x with 2, y with 0, and z with -21 yields the true statement

$$3(2) - 0 + (-21) = -15$$

The graph of this equation is a plane in three-dimensional space, just as the graph of a linear equation in two variables is a line in two-dimensional space.

Although we will not discuss the techniques for graphing equations in three variables, visualizing the possible patterns of intersecting planes gives us insight into the possible patterns of solutions of a system of three three-variable linear equations. There are four possible patterns.

1. Three planes have a single point in common. This point represents the single solution of the system. This system is **consistent**.

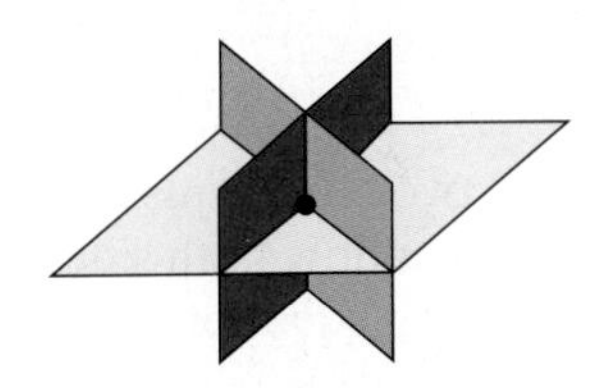

2. Three planes intersect at no point common to all three. This system has no solution. A few ways that this can occur are shown. This system is **inconsistent**.

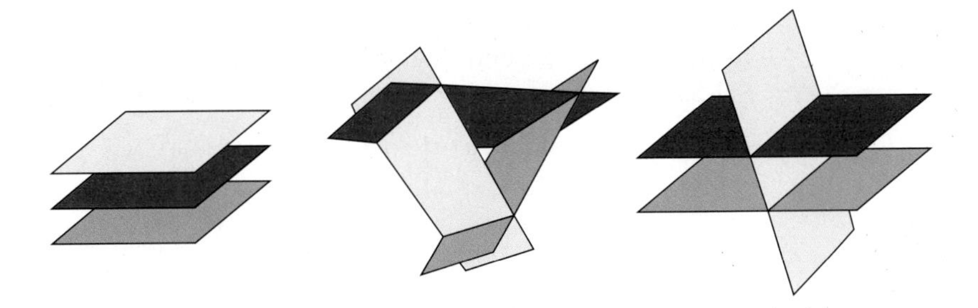

3. Three planes intersect at all the points of a single line. The system has infinitely many solutions. This system is **consistent**.

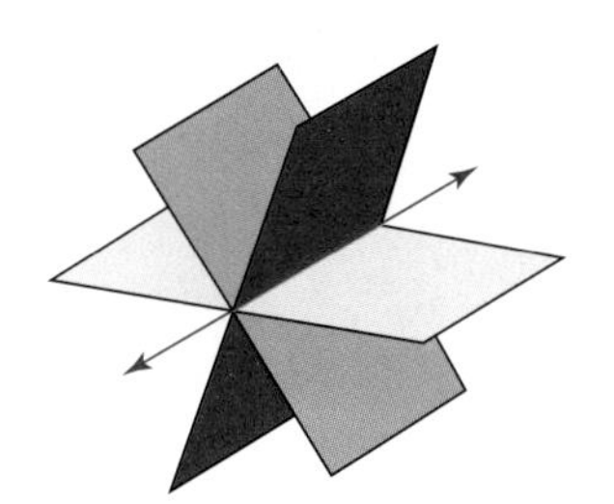

SOLVING A SYSTEM OF THREE LINEAR EQUATIONS BY THE ELIMINATION METHOD

Step 1. Write each equation in standard form, $Ax + By + Cz = D$.
Step 2. Choose a pair of equations and use the equations to eliminate a variable.
Step 3. Choose any other pair of equations and eliminate the *same variable* as in Step 2.
Step 4. Two equations in two variables should be obtained from Step 2 and Step 3. Use methods from Section 4.1 to solve this system for both variables.
Step 5. To solve for the third variable, substitute the values of the variables found in Step 4 into any of the original equations containing the third variable.
Step 6. Check the ordered triple solution in *all three* original equations.

Practice Problem 1

Solve the system:

$$\begin{cases} 2x - y + 3z = 13 \\ x + y - z = -2 \\ 3x + 2y + 2z = 13 \end{cases}$$

Answer
1. (1, 1, 4)

4. Three planes coincide at all points on the plane. The system is consistent, and the equations are **dependent**.

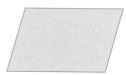

A SOLVING A SYSTEM OF THREE LINEAR EQUATIONS IN THREE VARIABLES

Just as with systems of two equations in two variables, we can use the elimination method to solve a system of three equations in three variables. To do so, we eliminate a variable and obtain a system of two equations in two variables. Then we use the methods we learned in the previous section to solve the system of two equations. See the box in the margin for steps.

Example 1 Solve the system:

$$\begin{cases} 3x - y + z = -15 & \text{Equation (1)} \\ x + 2y - z = 1 & \text{Equation (2)} \\ 2x + 3y - 2z = 0 & \text{Equation (3)} \end{cases}$$

Solution: We add equations (1) and (2) to eliminate z.

$$\begin{array}{r} 3x - y + z = -15 \\ x + 2y - z = 1 \\ \hline 4x + y = -14 \end{array} \quad \text{Equation (4)}$$

Next we add two *other* equations and *eliminate z again.* To do so, we multiply both sides of equation (1) by 2 and add this resulting equation to equation (3). Then

$$\begin{cases} 2(3x - y + z) = 2(-15) \\ 2x + 3y - 2z = 0 \end{cases} \text{ simplifies to } \begin{cases} 6x - 2y + 2z = -30 \\ 2x + 3y - 2z = 0 \end{cases}$$
$$8x + y = -30 \quad \text{Equation (5)}$$

Now we solve equations (4) and (5) for x and y. To solve by elimination, we multiply both sides of equation (4) by -1 and add this resulting equation to equation (5). Then

$$\begin{cases} -1(4x + y) = -1(-14) \\ 8x + y = -30 \end{cases} \text{ simplifies to } \begin{cases} -4x - y = 14 \\ 8x + y = -30 \end{cases}$$
$$4x = -16 \quad \text{Add the equations.}$$
$$x = -4 \quad \text{Solve for } x.$$

We now replace x with -4 in equation (4) or (5).

$$4x + y = -14 \quad \text{Equation (4)}$$
$$4(-4) + y = -14 \quad \text{Let } x = -4.$$
$$y = 2 \quad \text{Solve for } y.$$

Finally, we replace x with -4 and y with 2 in equation (1), (2), or (3).

$$x + 2y - z = 1 \quad \text{Equation (2)}$$
$$-4 + 2(2) - z = 1 \quad \text{Let } x = -4 \text{ and } y = 2.$$
$$-4 + 4 - z = 1$$
$$-z = 1$$
$$z = -1$$

The ordered triple solution is $(-4, 2, -1)$. To check, let $x = -4$, $y = 2$, and $z = -1$ in *all three* original equations of the system.

Equation (1)

$$\begin{aligned} 3x - y + z &= -15 \\ 3(-4) - 2 + (-1) &= -15 \\ -12 - 2 - 1 &= -15 \\ -15 &= -15 \end{aligned}$$

True.

Equation (2)

$$\begin{aligned} x + 2y - z &= 1 \\ -4 + 2(2) - (-1) &= 1 \\ -4 + 4 + 1 &= 1 \\ 1 &= 1 \end{aligned}$$

True.

Equation (3)

$$\begin{aligned} 2x + 3y - 2z &= 0 \\ 2(-4) + 3(2) - 2(-1) &= 0 \\ -8 + 6 + 2 &= 0 \\ 0 &= 0 \end{aligned}$$

True.

All three statements are true, so the ordered triple solution is $(-4, 2, -1)$.

Example 2 Solve the system:

$$\begin{cases} 2x - 4y + 8z = 2 & (1) \\ -x - 3y + z = 11 & (2) \\ x - 2y + 4z = 0 & (3) \end{cases}$$

Solution: When we add equations (2) and (3) to eliminate x, the new equation is

$$-5y + 5z = 11 \quad (4)$$

To eliminate x again, we multiply both sides of equation (2) by 2 and add the resulting equation to equation (1). Then

$$\begin{cases} 2x - 4y + 8z = 2 \\ 2(-x - 3y + z) = 2(11) \end{cases} \text{ simplifies to } \begin{cases} 2x - 4y + 8z = 2 \\ -2x - 6y + 2z = 22 \end{cases}$$
$$-10y + 10z = 24 \quad (5)$$

Next we solve for y and z using equations (4) and (5). To do so, we multiply both sides of equation (4) by -2 and add the resulting equation to equation (5).

$$\begin{cases} -2(-5y + 5z) = -2(11) \\ -10y + 10z = 24 \end{cases} \text{ simplifies to } \begin{cases} 10y - 10z = -22 \\ -10y + 10z = 24 \end{cases}$$
$$0 = 2 \quad \text{False.}$$

Since the statement is false, this system is inconsistent and has no solution. The solution set is the empty set $\{\ \}$ or $\emptyset$.

TRY THE CONCEPT CHECK IN THE MARGIN.

Example 3 Solve the system:

$$\begin{cases} 2x + 4y = 1 & (1) \\ 4x - 4z = -1 & (2) \\ y - 4z = -3 & (3) \end{cases}$$

Practice Problem 2

Solve the system:

$$\begin{cases} 2x + 4y - 2z = 3 \\ -x + y - z = 6 \\ x + 2y - z = 1 \end{cases}$$

✓ CONCEPT CHECK

In the system

$$\begin{cases} x + y + z = 6 & \text{Equation (1)} \\ 2x - y + z = 3 & \text{Equation (2)} \\ x + 2y + 3z = 14 & \text{Equation (3)} \end{cases}$$

Equations (1) and (2) are used to eliminate y. Which action could be used to finish solving? Why?

(a) Use (1) and (2) to eliminate z
(b) Use (2) and (3) to eliminate y
(c) Use (1) and (3) to eliminate x

Practice Problem 3

Solve the system:

$$\begin{cases} 3x + 2y = -1 \\ 6x - 2z = 4 \\ y - 3z = 2 \end{cases}$$

Answers

2. $\emptyset$, **3.** $\left(\frac{1}{3}, -1, -1\right)$

✓ Concept Check: b

Solution: Notice that equation (2) has no term containing the variable y. Let us eliminate y using equations (1) and (3). We multiply both sides of equation (3) by -4 and add the resulting equation to equation (1). Then

$$\begin{cases} 2x + 4y = 1 \\ -4(y - 4z) = -4(-3) \end{cases} \text{ simplifies to } \begin{cases} 2x + 4y = 1 \\ -4y + 16z = 12 \\ \hline 2x + 16z = 13 \quad (4) \end{cases}$$

Next we solve for z using equations (4) and (2). We multiply both sides of equation (4) by -2 and add the resulting equation to equation (2).

$$\begin{cases} -2(2x + 16z) = -2(13) \\ 4x - 4z = -1 \end{cases} \text{ simplifies to } \begin{cases} -4x - 32z = -26 \\ 4x - 4z = -1 \\ \hline -36z = -27 \\ z = \frac{3}{4} \end{cases}$$

Now we replace z with $\frac{3}{4}$ in equation (3) and solve for y.

$$y - 4\left(\frac{3}{4}\right) = -3 \quad \text{Let } z = \frac{3}{4} \text{ in equation (3).}$$
$$y - 3 = -3$$
$$y = 0$$

Finally, we replace y with 0 in equation (1) and solve for x.

$$2x + 4(0) = 1 \quad \text{Let } y = 0 \text{ in equation (1).}$$
$$2x = 1$$
$$x = \frac{1}{2}$$

The ordered triple solution is $\left(\frac{1}{2}, 0, \frac{3}{4}\right)$. Check to see that this solution satisfies *all three* equations of the system.

Practice Problem 4

Solve the system:

$$\begin{cases} x - 3y + 4z = 2 \\ -2x + 6y - 8z = -4 \\ \frac{1}{2}x - \frac{3}{2}y + 2z = 1 \end{cases}$$

Example 4 Solve the system:

$$\begin{cases} x - 5y - 2z = 6 & (1) \\ -2x + 10y + 4z = -12 & (2) \\ \frac{1}{2}x - \frac{5}{2}y - z = 3 & (3) \end{cases}$$

Solution: We multiply both sides of equation (3) by 2 to eliminate fractions, and we multiply both sides of equation (2) by $-\frac{1}{2}$ so that the coefficient of x is 1. The resulting system is then

$$\begin{cases} x - 5y - 2z = 6 & (1) \\ x - 5y - 2z = 6 & \text{Multiply (2) by } -\frac{1}{2}. \\ x - 5y - 2z = 6 & \text{Multiply (3) by 2.} \end{cases}$$

All three resulting equations are identical, and therefore equations (1), (2), and (3) are all equivalent. There are infinitely many solutions of this system. The equations are dependent. The solution set can be written as $\{(x, y, z) | x - 5y - 2z = 6\}$.

Answer

4. $\{(x, y, z) | x - 3y + 4z = 2\}$

Name ______________________ Section ______ Date ________

Exercise Set 4.2

A *Solve each system. See Examples 1 through 4.*

1. $\begin{cases} x + y = 3 \\ 2y = 10 \\ 3x + 2y - 3z = 1 \end{cases}$

2. $\begin{cases} 5x = 5 \\ 2x + y = 4 \\ 3x + y - 4z = -15 \end{cases}$

3. $\begin{cases} 2x + 2y + z = 1 \\ -x + y + 2z = 3 \\ x + 2y + 4z = 0 \end{cases}$

4. $\begin{cases} 2x - 3y + z = 5 \\ x + y + z = 0 \\ 4x + 2y + 4z = 4 \end{cases}$

5. $\begin{cases} x - 2y + z = -5 \\ -3x + 6y - 3z = 15 \\ 2x - 4y + 2z = -10 \end{cases}$

6. $\begin{cases} 3x + y - 2z = 2 \\ -6x - 2y + 4z = -2 \\ 9x + 3y - 6z = 6 \end{cases}$

7. $\begin{cases} 4x - y + 2z = 5 \\ 2y + z = 4 \\ 4x + y + 3z = 10 \end{cases}$

8. $\begin{cases} 5y - 7z = 14 \\ 2x + y + 4z = 10 \\ 2x + 6y - 3z = 30 \end{cases}$

9. $\begin{cases} x + 5z = 0 \\ 5x + y = 0 \\ y - 3z = 0 \end{cases}$

10. $\begin{cases} x - 5y = 0 \\ x - z = 0 \\ -x + 5z = 0 \end{cases}$

11. $\begin{cases} 6x - 5z = 17 \\ 5x - y + 3z = -1 \\ 2x + y = -41 \end{cases}$

12. $\begin{cases} x + 2y = 6 \\ 7x + 3y + z = -33 \\ x - z = 16 \end{cases}$

13. $\begin{cases} x + y + z = 8 \\ 2x - y - z = 10 \\ x - 2y - 3z = 22 \end{cases}$

14. $\begin{cases} 5x + y + 3z = 1 \\ x - y + 3z = -7 \\ -x + y = 1 \end{cases}$

15. $\begin{cases} x + 2y - z = 5 \\ 6x + y + z = 7 \\ 2x + 4y - 2z = 5 \end{cases}$

16. $\begin{cases} 4x - y + 3z = 10 \\ x + y - z = 5 \\ 8x - 2y + 6z = 10 \end{cases}$

17. $\begin{cases} 2x - 3y + z = 2 \\ x - 5y + 5z = 3 \\ 3x + y - 3z = 5 \end{cases}$

18. $\begin{cases} 4x + y - z = 8 \\ x - y + 2z = 3 \\ 3x - y + z = 6 \end{cases}$

19. $\begin{cases} -2x - 4y + 6z = -8 \\ x + 2y - 3z = 4 \\ 4x + 8y - 12z = 16 \end{cases}$

20. $\begin{cases} -6x + 12y + 3z = -6 \\ 2x - 4y - z = 2 \\ -x + 2y + \frac{z}{2} = -1 \end{cases}$

21. $\begin{cases} 2x + 2y - 3z = 1 \\ y + 2z = -14 \\ 3x - 2y = -1 \end{cases}$

22. $\begin{cases} 7x + 4y = 10 \\ x - 4y + 2z = 6 \\ y - 2z = -1 \end{cases}$

23. $\begin{cases} \frac{3}{4}x - \frac{1}{3}y + \frac{1}{2}z = 9 \\ \frac{1}{6}x + \frac{1}{3}y - \frac{1}{2}z = 2 \\ \frac{1}{2}x - y + \frac{1}{2}z = 2 \end{cases}$

24. $\begin{cases} \frac{1}{3}x - \frac{1}{4}y + z = -9 \\ \frac{1}{2}x - \frac{1}{3}y - \frac{1}{4}z = -6 \\ x - \frac{1}{2}y - z = -8 \end{cases}$

Answers

1. ______
2. ______
3. ______
4. ______
5. ______
6. ______
7. ______
8. ______
9. ______
10. ______
11. ______
12. ______
13. ______
14. ______
15. ______
16. ______
17. ______
18. ______
19. ______
20. ______
21. ______
22. ______
23. ______
24. ______

25. ____________

26. ____________

27. ____________

28. ____________

29. ____________

30. ____________

31. ____________

32. ____________

33. ____________

34. ____________

35. ____________

36. ____________

37. ____________

38. ____________

Name ____

25. The fraction $\frac{1}{24}$ can be written as the following sum:

$$\frac{1}{24} = \frac{x}{8} + \frac{y}{4} + \frac{z}{3}$$

where the numbers x, y, and z are solutions of

$$\begin{cases} x + y + z = 1 \\ 2x - y + z = 0 \\ -x + 2y + 2z = -1 \end{cases}$$

Solve the system and see that the sum of the fractions is $\frac{1}{24}$.

26. The fraction $\frac{1}{18}$ can be written as the following sum:

$$\frac{1}{18} = \frac{x}{2} + \frac{y}{3} + \frac{z}{9}$$

where the numbers x, y, and z are solutions of

$$\begin{cases} x + 3y + z = -3 \\ -x + y + 2z = -14 \\ 3x + 2y - z = 12 \end{cases}$$

Solve the system and see that the sum of the fractions is $\frac{1}{18}$.

Review and Preview

Solve. See Section 2.2.

27. The sum of two numbers is 45 and one number is twice the other. Find the numbers.

28. The difference between two numbers is 5. Twice the smaller number added to five times the larger number is 53. Find the numbers.

Solve. See Section 2.1.

29. $2(x - 1) - 3x = x - 12$

30. $7(2x - 1) + 4 = 11(3x - 2)$

31. $-y - 5(y + 5) = 3y - 10$

32. $z - 3(z + 7) = 6(2z + 1)$

Combining Concepts

33. Write a linear equation in three variables that has $(-1, 2, -4)$ as a solution. (There are many possibilities.)

34. Write a system of three linear equations in three variables that has $(2, 1, 5)$ as a solution. (There are many possibilities.)

Solving systems involving more than three variables can be approached using methods similar to those encountered in this section. Apply what you already know to solve each system of equations in four variables.

35. $$\begin{cases} x + y - w = 0 \\ y + 2z + w = 3 \\ x - z = 1 \\ 2x - y - w = -1 \end{cases}$$

36. $$\begin{cases} 5x + 4y = 29 \\ y + z - w = -2 \\ 5x + z = 23 \\ y - z + w = 4 \end{cases}$$

37. $$\begin{cases} x + y + z + w = 5 \\ 2x + y + z + w = 6 \\ x + y + z = 2 \\ x + y = 0 \end{cases}$$

38. $$\begin{cases} 2x - z = -1 \\ y + z + w = 9 \\ y - 2w = -6 \\ x + y = 3 \end{cases}$$

4.3 Systems of Linear Equations and Problem Solving

A Solving Problems Modeled by Systems of Two Equations

Thus far, we have solved problems by writing one-variable equations and solving for the variable. Some of these problems can be solved, perhaps more easily, by writing a system of equations, as illustrated in this section. We begin with a problem about numbers.

Example 1 Finding Unknown Numbers

A first number is 4 less than a second number. Four times the first number is 6 more than twice the second. Find the numbers.

Solution:

1. UNDERSTAND. Read and reread the problem and guess a solution. If one number is 10 and this is 4 less than a second number, the second number is 14. Four times the first number is $4(10)$, or 40. This is not equal to 6 more than twice the second number, which is $2(14) + 6$ or 34. Although we guessed incorrectly, we now have a better understanding of the problem.

Since we are looking for two numbers, we will let

$x =$ first number

$y =$ second number

2. TRANSLATE. Since we have assigned two variables to this problem, we will translate the given facts into two equations. For the first statement we have

In words:

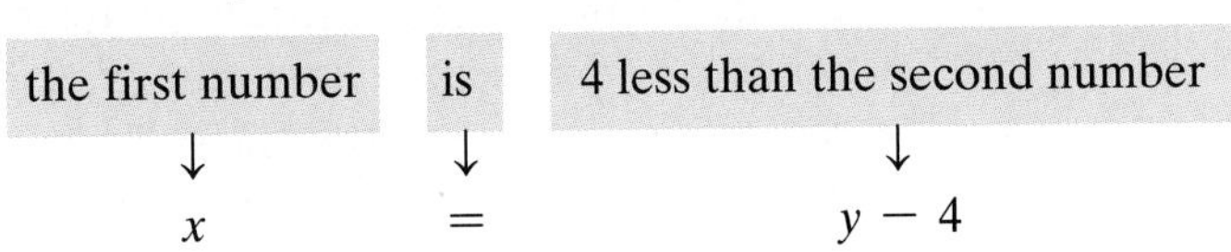
the first number
is
4 less than the second number

Translate: $x \quad = \quad y - 4$

Next we translate the second statement into an equation.

In words:

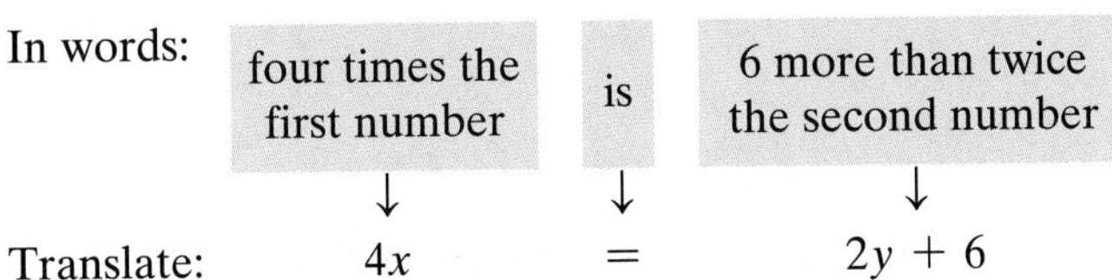
four times the first number
is
6 more than twice the second number

Translate: $4x \quad = \quad 2y + 6$

3. SOLVE. Here we solve the system

$$\begin{cases} x = y - 4 \\ 4x = 2y + 6 \end{cases}$$

Since the first equation expresses x in terms of y, we will use substitution. We substitute $y - 4$ for x in the second equation and solve for y.

$4x = 2y + 6$ Second equation

$4(y - 4) = 2y + 6$ Let $x = y - 4$.

$$4y - 16 = 2y + 6$$

$$2y = 22$$

$$y = 11$$

Objectives

A Solve problems that can be modeled by a system of two linear equations.

B Solve problems that can be modeled by a system of three linear equations.

C Solve problems with cost and revenue functions.

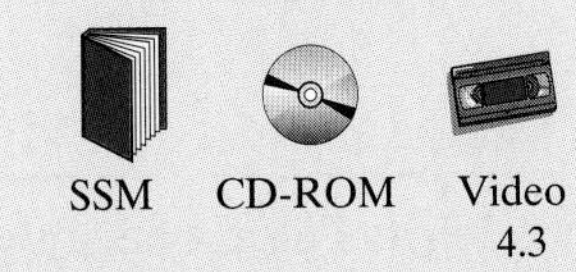

SSM CD-ROM Video 4.3

Practice Problem 1

A first number is 7 greater than a second number. Twice the first number is 4 more than three times the second. Find the numbers.

Answer

1. 17 and 10

Now we replace y with 11 in the equation $x = y - 4$ and solve for x. Then $x = y - 4$ becomes $x = 11 - 4 = 7$. The ordered pair solution of the system is $(7, 11)$.

4. INTERPRET. Since the solution of the system is $(7, 11)$, then the first number we are looking for is 7 and the second number is 11.

Check: Notice that 7 *is* 4 less than 11, and 4 times 7 *is* 6 more than twice 11. The proposed numbers, 7 and 11, are correct.

State: The numbers are 7 and 11.

Practice Problem 2

Two trains leave Tulsa, one traveling north and the other south. After 4 hours, they are 376 miles apart. If one train is traveling 10 mph faster than the other, what is the speed of each?

Example 2 Finding the Rate of Speed

Two cars leave Indianapolis, one traveling east and the other west. After 3 hours they are 297 miles apart. If one car is traveling 5 mph faster than the other, what is the speed of each?

Solution: 1. UNDERSTAND. Read and reread the problem. Let's guess a solution and use the formula $d = r \cdot t$ to check. Suppose that one car is traveling at a rate of 55 miles per hour. This means that the other car is traveling at a rate of 50 miles per hour since we are told that one car is traveling 5 mph faster than the other. To find the distance apart after 3 hours, we will first find the distance traveled by each car. One car's distance is rate $\cdot$ time $= 55(3) = 165$ miles. The other car's distance is rate $\cdot$ time $= 50(3) = 150$ miles. Since one car is traveling east and the other west, their distance apart is the sum of their distances, or 165 miles $+$ 150 miles $=$ 315 miles. Although this distance apart is not the required distance of 297 miles, we now have a better understanding of the problem.

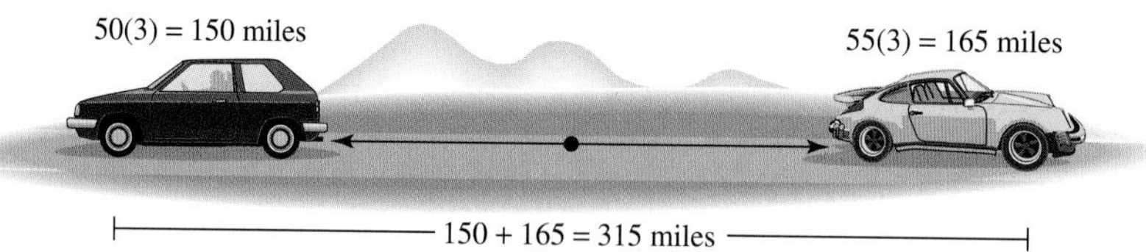

Let's model the problem with a system of equations. We will let

$x =$ speed of one car

$y =$ speed of the other car

We summarize the information on the following chart. Both cars have traveled 3 hours. Since distance $=$ rate $\cdot$ time, their distances are $3x$ and $3y$ miles, respectively.

Answer

2. 42 mph and 52 mph

	Rate	· Time	= Distance
One Car	x	3	$3x$
Other Car	y	3	$3y$

2. TRANSLATE. We can now translate the stated conditions into two equations.

In words:

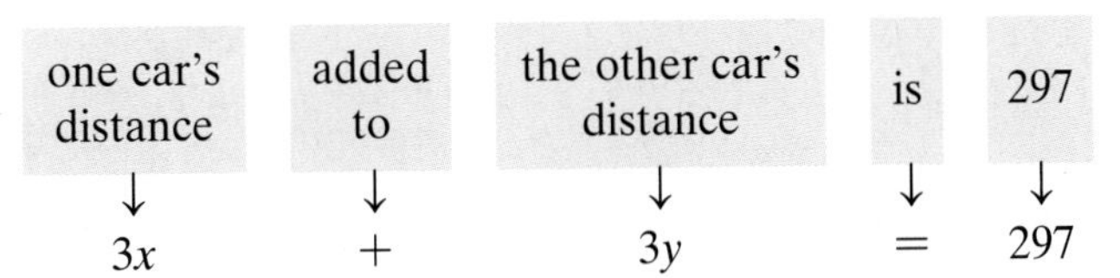
one car's distance
added to
the other car's distance
is
297

Translate: $3x \quad + \quad 3y \quad = \quad 297$

In words: one car's speed — is — 5 mph faster than the other

Translate: $x \quad = \quad y + 5$

3. SOLVE. Here we solve the system

$$\begin{cases} 3x + 3y = 297 \\ x = y + 5 \end{cases}$$

Again, the substitution method is appropriate. We replace x with $y + 5$ in the first equation and solve for y.

$$3x + 3y = 297 \quad \text{First equation}$$

$$3(y + 5) + 3y = 297 \quad \text{Let } x = y + 5.$$

$$3y + 15 + 3y = 297$$

$$6y = 282$$

$$y = 47$$

To find x, we replace y with 47 in the equation $x = y + 5$. Then $x = 47 + 5 = 52$. The ordered pair solution of the system is $(52, 47)$.

4. INTERPRET. The solution $(52, 47)$ means that the cars are traveling at 52 mph and 47 mph, respectively.

Check: Notice that one car is traveling 5 mph faster than the other. Also, if one car travels 52 mph for 3 hours, the distance is $3(52) = 156$ miles. The other car traveling for 3 hours at 47 mph travels a distance of $3(47) = 141$ miles. The sum of the distances $156 + 141$ is 297 miles, the required distance.

State: The cars are traveling at 52 mph and 47 mph.

Example 3 Mixing Solutions

Lynn Pike, a pharmacist, needs 70 liters of a 50% alcohol solution. She has available a 30% alcohol solution and an 80% alcohol solution. How many liters of each solution

Practice Problem 3

One solution contains 20% acid and a second solution contains 50% acid. How many ounces of each solution should be mixed in order to have 60 ounces of a 30% acid solution?

Answer

3. 40 ounces of the 20% solution and 20 ounces of the 50% solution

should she mix to obtain 70 liters of a 50% alcohol solution?

Solution:

1. UNDERSTAND. Read and reread the problem. Next, guess the solution. Suppose that we need 20 liters of the 30% solution. Then we need $70 - 20 = 50$ liters of the 80% solution. To see if this gives us 70 liters of a 50% alcohol solution, let's find the amount of pure alcohol in each solution.

number of liters	×	alcohol strength	=	amount of pure alcohol
↓		↓		↓
20 liters	×	0.30	=	6 liters
50 liters	×	0.80	=	40 liters
70 liters	×	0.50	=	35 liters

Since 6 liters + 40 liters = 46 liters and not 35 liters, our guess is incorrect, but we have gained some insight as to how to model and check this problem.

We will let

$x =$ amount of 30% solution, in liters

$y =$ amount of 80% solution, in liters

and use a table to organize the given data.

	Number of Liters	Alcohol Strength	Amount of Pure Alcohol
30% Solution	x	30%	$0.30x$
80% Solution	y	80%	$0.80y$
50% Solution Needed	70	50%	$(0.50)(70)$

2. TRANSLATE. We translate the stated conditions into two equations.

In words:	amount of 30% solution	+	amount of 80% solution	=	70
	↓		↓		↓
Translate:	x	+	y	=	70

In words:	amount of pure alcohol in 30% solution	+	amount of pure alcohol in 80% solution	=	amount of pure alcohol in 50% solution
	↓		↓		↓
Translate:	$0.30x$	+	$0.80y$	=	$(0.50)(70)$

3. SOLVE. Here we solve the system

$$\begin{cases} x + y = 70 \\ 0.30x + 0.80y = (0.50)(70) \end{cases}$$

To solve this system, we use the elimination method. We multiply both sides of the first equation by -3 and both sides of the second equation by 10. Then

$$\begin{cases} -3(x + y) = -3(70) \\ 10(0.30x + 0.80y) = 10(0.50)(70) \end{cases} \text{ simplifies to } \begin{cases} -3x - 3y = -210 \\ 3x + 8y = 350 \end{cases}$$

$$\begin{aligned} 5y &= 140 \\ y &= 28 \end{aligned}$$

Now we replace y with 28 in the equation $x + y = 70$ and find that $x + 28 = 70$, or $x = 42$.

The ordered pair solution of the system is $(42, 28)$.

4. INTERPRET.

Check: Check the solution in the same way that we checked our guess.

State: The pharmacist needs to mix 42 liters of 30% solution and 28 liters of 80% solution to obtain 70 liters of 50% solution.

TRY THE CONCEPT CHECK IN THE MARGIN.

B SOLVING PROBLEMS MODELED BY SYSTEMS OF THREE EQUATIONS

To introduce problem solving by writing a system of three linear equations in three variables, we solve a problem about triangles.

Example 4 Finding Angle Measures

The measure of the largest angle of a triangle is 80° more than the measure of the smallest angle, and the measure of the remaining angle is 10° more than the measure of the smallest angle. Find the measure of each angle.

Solution:

1. UNDERSTAND. Read and reread the problem. Recall that the sum of the measures of the angles of a triangle is 180°. Then guess a solution. If the smallest angle measures 20°, the measure of the largest angle is 80° more, or 20° + 80° = 100°. The measure of the remaining angle is 10° more than the measure of the smallest angle, or 20° + 10° = 30°. The sum of these three angles is 20° + 100° + 30° = 150°, not the required 180°. We now know that the measure of the smallest angle is greater than 20°.

To model this problem we will let

x = degree measure of the smallest angle
y = degree measure of the largest angle
z = degree measure of the remaining angle

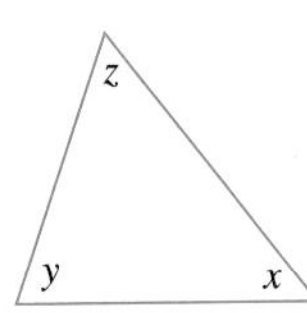

2. TRANSLATE. We translate the given information into three equations.

In words:	the sum of the measures	=	180
	↓		↓
Translate:	$x + y + z$	=	180

✓ CONCEPT CHECK

Suppose you mix an amount of 25% acid solution with an amount of 60% acid solution. You then calculate the acid strength of the resulting acid mixture. For which of the following results should you suspect an error in your calculation? Why?

a. 14%
b. 32%
c. 55%

Practice Problem 4

The measure of the largest angle of a triangle is 90° more than the measure of the smallest angle, and the measure of the remaining angle is 30° more than the measure of the smallest angle. Find the measure of each angle.

Answers

4. 20°, 50°, 110°

✓ Concept Check: a; answers may vary

In words:	the largest angle	is	80 more than the smallest angle
	↓	↓	↓
Translate:	y	$=$	$x + 80$
In words:	the remaining angle	is	10 more than the smallest angle
	↓	↓	↓
Translate:	z	$=$	$x + 10$

3. SOLVE. We solve the system

$$\begin{cases} x + y + z = 180 \\ y = x + 80 \\ z = x + 10 \end{cases}$$

Since y and z are both expressed in terms of x, we will solve using the subsitution method. We substitute $y = x + 80$ and $z = x + 10$ in the first equation. Then

$$x + y + z = 180 \quad \text{First equation}$$

$$x + (x + 80) + (x + 10) = 180 \quad \text{Let } y = x + 80 \text{ and } z = x + 10.$$

$$3x + 90 = 180$$

$$3x = 90$$

$$x = 30$$

Then $y = x + 80 = 30 + 80 = 110$, and $z = x + 10 = 30 + 10 = 40$. The ordered triple solution is $(30, 110, 40)$.

4. INTERPRET.

Check: Notice that $30° + 40° + 110° = 180°$. Also, the measure of the largest angle, 110°, is 80° more than the measure of the smallest angle, 30°. The measure of the remaining angle, 40°, is 10° more than the measure of the smallest angle, 30°.

State: The angles measure 30°, 110°, and 40°.

C Solving Problems with Cost and Revenue Functions

Recall that businesses are often computing cost and revenue functions or equations to predict sales, to determine whether prices need to be adjusted, and to see whether the company is making or losing money. Recall also that the value at which revenue equals cost is called the break-even point. When revenue is less than cost, the company is losing money; when revenue is greater than cost, the company is making money.

Example 5 Finding a Break-Even Point

A manufacturing company recently purchased $3000 worth of new equipment to offer new personalized

Practice Problem 5

A company that manufactures boxes recently purchased $2000 worth of new equipment to offer gift boxes to its customers. The cost of producing a package of gift boxes is $1.50 and it is sold for $4.00. Find the number of packages that must be sold for the company to break even.

Answer

5. 800 packages

stationery to its customers. The cost of producing a package of personalized stationery is $3.00, and it is sold for $5.50. Find the number of packages that must be sold for the company to break even.

Solution:

1. UNDERSTAND. Read and reread the problem. Notice that the cost to the company will include a one-time cost of $3000 for the equipment and then $3.00 per package produced. The revenue will be $5.50 per package sold.

To model this problem, we will let

x = number of packages of personalized stationery
$C(x)$ = total cost for producing x packages of stationery
$R(x)$ = total revenue for selling x packages of stationery

2. TRANSLATE. The revenue equation is

In words:	revenue for selling x packages of stationery	=	price per package	·	number of packages
	↓		↓		↓
Translate:	$R(x)$	=	5.5	·	x

The cost equation is

In words:	cost for producing x packages of stationery	=	cost per package	·	number of packages	+	cost for equipment
	↓		↓		↓		↓
Translate:	$C(x)$	=	3	·	x	+	3000

Since the break-even point is when $R(x) = C(x)$, we solve the equation

$$5.5x = 3x + 3000$$

3. SOLVE.

$5.5x = 3x + 3000$
$2.5x = 3000$ Subtract $3x$ from both sides.
$x = 1200$ Divide both sides by 2.5.

4. INTERPRET.

Check: To see whether the break-even point occurs when 1200 packages are produced and sold, see if revenue equals cost when $x = 1200$. When $x = 1200$,

$R(x) = 5.5x = 5.5(1200) = 6600$ and
$C(x) = 3x + 3000 = 3(1200) + 3000 = 6600.$
Since $R(1200) = C(1200) = 6600$, the break-even point is 1200.

State: The company must sell 1200 packages of stationery to break even. The graph of this system is shown.

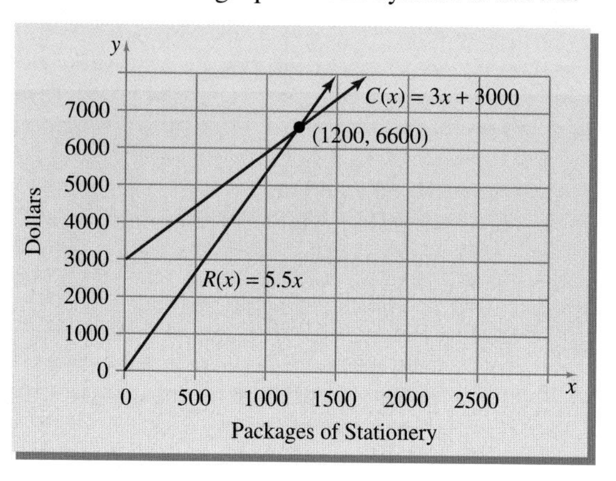

Name ______________________ Section ________ Date ________

ANSWERS

1. ____________

2. ____________

3. ____________

4. ____________

5. ____________

6. ____________

7. ____________

8. ____________

9. ____________

10. ____________

EXERCISE SET 4.3

A *Solve. See Examples 1 through 3.*

1. One number is two more than a second number. Twice the first is 4 less than 3 times the second. Find the numbers.

2. Three times one number minus a second is 8, and the sum of the numbers is 12. Find the numbers.

3. A Delta 727 traveled 560 mph with the wind and 480 mph against the wind. Find the speed of the plane in still air and the speed of the wind.

4. Terry Watkins can row about 10.6 kilometers in 1 hour downstream and 6.8 kilometers upstream in 1 hour. Find how fast he can row in still water, and find the speed of the current.

5. Find how many quarts of 4% butterfat milk and 1% butterfat milk should be mixed to yield 60 quarts of 2% butterfat milk.

6. A pharmacist needs 500 milliliters of a 20% phenobarbital solution but has only 5% and 25% phenobarbital solutions available. Find how many milliliters of each she should mix to get the desired solution.

7. Karen Karlin bought some large frames for $15 each and some small frames for $8 each at a closeout sale. If she bought 22 frames for $239, find how many of each type she bought.

8. Hilton University Drama Club sold 311 tickets for a play. Student tickets cost 50 cents each; nonstudent tickets cost $1.50. If total receipts were $385.50, find how many tickets of each type were sold.

9. One number is two less than a second number. Twice the first is 4 more than 3 times the second. Find the numbers.

10. Twice one number plus a second number is 42, and the first number minus the second number is -6. Find the numbers.

11. ____________

12. ____________

13. ____________

14. ____________

15. ____________

16. ____________

17. ____________

18. ____________

19. ____________

20. ____________

Name ____

11. An office supply store in San Diego sells 7 writing tablets and 4 pens for $6.40. Also, 2 tablets and 19 pens cost $5.40. Find the price of each.

12. A Candy Barrel shop manager mixes M&M's worth $2.00 per pound with trail mix worth $1.50 per pound. Find how many pounds of each she should use to get 50 pounds of a party mix worth $1.80 per pound.

13. An airplane takes 3 hours to travel a distance of 2160 miles with the wind. The return trip takes 4 hours against the wind. Find the speed of the plane in still air and the speed of the wind.

14. Two cyclists start at the same point and travel in opposite directions. One travels 4 mph faster than the other. In 4 hours they are 112 miles apart. Find how fast each is traveling.

15. The perimeter of a quadrilateral (four-sided polygon) is 29 inches. The longest side is twice as long as the shortest side. The other two sides are equally long and are 2 inches longer than the shortest side. Find the length of all four sides.

16. The perimeter of a triangle is 93 centimeters. If two sides are equally long and the third side is 9 centimeters longer than the others, find the lengths of the three sides.

17. The sum of three numbers is 40. One number is five more than a second and twice the third. Find the numbers.

18. The sum of the digits of a three-digit number is 15. The tens-place digit is twice the hundreds-place digit, and the ones-place digit is 1 less than the hundreds-place digit. Find the three-digit number.

19. Jack Reinholt, a car salesman, has a choice of two pay arrangements; a weekly salary of $200 plus 5% commission on sales, or a straight 15% commission. Find the amount of weekly sales for which Jack's earnings are the same regardless of the pay arrangement.

20. Hertz car rental agency charges $25 daily plus 10 cents per mile. Budget charges $20 daily plus 25 cents per mile. Find the daily mileage for which the Budget charge for the day is twice that of the Hertz charge for the day.

Name ______________________________

21. Carroll Blakemore, a drafting student, bought three templates and a pencil one day for \$6.45. Another day he bought two pads of paper and four pencils for \$7.50. If the price of a pad of paper is three times the price of a pencil, find the price of each type of item.

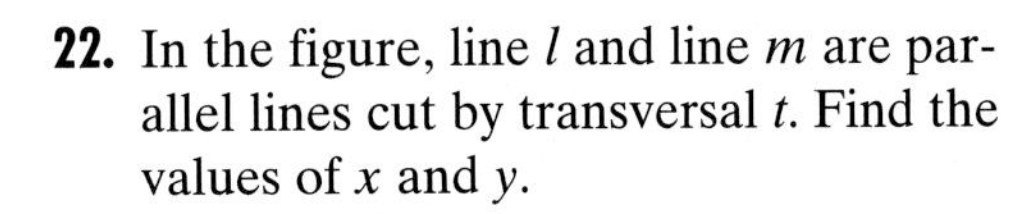

22. In the figure, line l and line m are parallel lines cut by transversal t. Find the values of x and y.

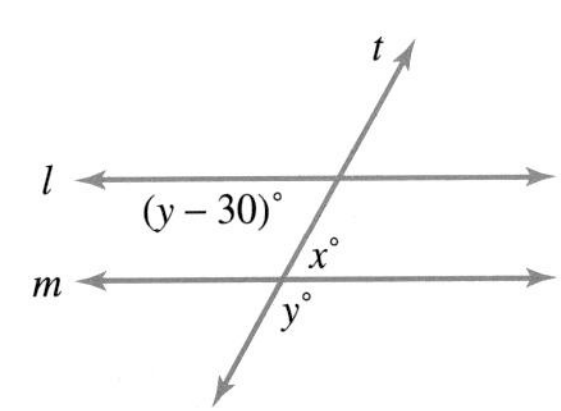

23. Find the values of x and y in the following isosceles triangle.

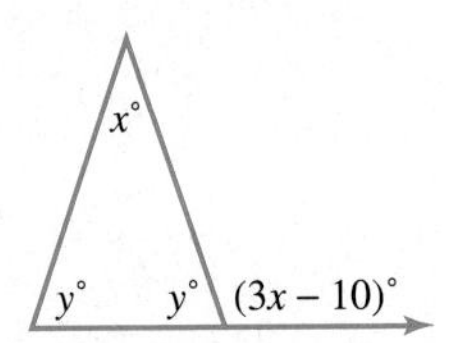

B *Solve. See Example 4.*

24. Rabbits in a lab are to be kept on a strict daily diet to include 30 grams of protein, 16 grams of fat, and 24 grams of carbohydrates. The scientist has only three food mixes available with the following grams of nutrients per unit.

	Protein	Fat	Carbohydrate
Mix A	4	6	3
Mix B	6	1	2
Mix C	4	1	12

Find how many units of each mix are needed daily to meet each rabbit's dietary need.

25. Gerry Gundersen mixes different solutions with concentrations of 25%, 40%, and 50% to get 200 liters of a 32% solution. If he uses twice as much of the 25% solution as of the 40% solution, find how many liters of each kind he uses.

21. ______________

22. ______________

23. ______________

24. ______________

25. ______________

Name ______________________________

26. ____________

27. ____________

28. ____________

29. ____________

30. ____________

31. ____________

26. In 1997 the WNBA's top scorer was Cynthia Cooper of the Houston Comets. She scored a total of 621 points during the regular season. The number of two-point field goals Cooper made was 10 less than twice the number of three-point field goals she made. She also made 48 more free throws (each worth one point) than two-point field goals. Find how many free throws, two-point field goals, and three-point field goals Cynthia Cooper made during the 1997 season. (*Source:* Women's National Basketball Association)

27. During the 1997–1998 regular NBA season, the top-scoring player was the Chicago Bulls' Michael Jordan. Jordan scored a total of 2357 points during the regular season. The number of free throws (each worth 1 point) he made was 25 more than 18 times the number of three-point field goals he made. Jordan also made 286 more two-point field goals than free throws. How many free throws, two-point field goals, and three-point field goals did Michael Jordan make during the 1997–1998 season? (*Source:* National Basketball Association)

28. Find the values of x, y, and z in the following triangle.

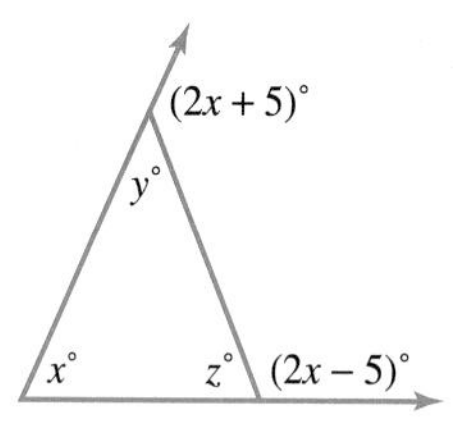

29. The sum of the measures of the angles of a quadrilateral is 360°. Find the value of x, y, and z in the following quadrilateral.

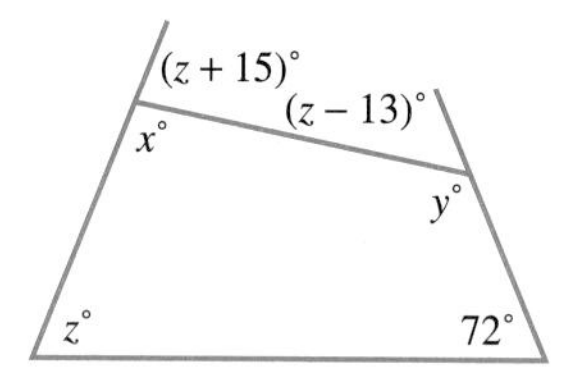

C *Given the cost function $C(x)$ and the revenue function $R(x)$, find the number of units x that must be sold to break even. See Example 5.*

30. $C(x) = 30x + 10{,}000$
$R(x) = 46x$

31. $C(x) = 12x + 15{,}000$
$R(x) = 32x$

Name ______

32. $C(x) = 1.2x + 1500$
$R(x) = 1.7x$

33. $C(x) = 0.8x + 900$
$R(x) = 2x$

34. $C(x) = 75x + 160{,}000$
$R(x) = 200x$

35. $C(x) = 105x + 70{,}000$
$R(x) = 245x$

36. The planning department of Abstract Office Supplies has been asked to determine whether the company should introduce a new computer desk next year. The department estimates that \$6000 of new equipment will need to be purchased and that the cost of manufacturing each desk will be \$200. The department also estimates that the revenue from each desk will be \$450.
   a. Determine the revenue function $R(x)$ from the sale of x desks.
   b. Determine the cost function $C(x)$ for manufacturing x desks.
   c. Find the break-even point.

37. Baskets, Inc., is planning to introduce a new woven basket. The company estimates that \$500 worth of new equipment will be needed to manufacture this new type of basket and that it will cost \$15 per basket to manufacture. The company also estimates that the revenue from each basket will be \$31.
   a. Determine the revenue function $R(x)$ from the sale of x baskets.
   b. Determine the cost function $C(x)$ for manufacturing x baskets.
   c. Find the break-even point.

32. ______
33. ______
34. ______
35. ______
36. a. ______
b. ______
c. ______
37. a. ______
b. ______
c. ______

Review and Preview

Multiply both sides of equation (1) by 2, and add the resulting equation to equation (2). See Section 4.2.

38. $3x - y + z = 2$ (1)
$-x + 2y + 3z = 6$ (2)

39. $2x + y + 3z = 7$ (1)
$-4x + y + 2z = 4$ (2)

Multiply both sides of equation (1) by −3, and add the resulting equation to equation (2). See Section 4.2.

40. $x + 2y - z = 0$ (1)
$3x + y - z = 2$ (2)

41. $2x - 3y + 2z = 5$ (1)
$x - 9y + z = -1$ (2)

38. ______
39. ______
40. ______
41. ______

42.

43. a.

b.

c.

Name ______

Combining Concepts

42. The number of personal bankruptcy petitions filed in the United States has been on the rise since the early 1980s. In 1997, the number of petitions filed was 4.5 times the number of petitions filed in 1980. This is equivalent to an increase of 1,050,000 petitions filed from 1980 to 1997. Find how many personal bankruptcy petitions were filed in each year. (*Source:* Based on data from the Administrative Office of the United States Courts)

43. Chlorofluorocarbons (CFCs) and hydrochlorofluorocarbons (HCFCs) are chemical compounds that destroy the ozone layer. Production of CFCs has stopped in the United States, but CFC emissions into the atmosphere still continue as existing CFC applications, such as refrigeration and fire extinguishers, are used. HCFC production in the United States will be phased out by the year 2030. In 1996, the combined CFC/HCFC emissions in the United States were 194,857 metric tons. There were 58,657 more metric tons of HCFCs released into the atmosphere than CFCs in 1996. (*Source:* Based on data from the U.S. Environmental Protection Agency)

a. How many metric tons of CFCs were released into the atmosphere in 1996?

b. How many metric tons of HCFCs were released into the atmosphere in 1996?

c. In 1990, there were 394,750 metric tons of CFC emissions and 79,789 metric tons of HCFC emissions. Did CFC emissions, HCFC emissions, and total CFC/HCFC emissions increase or decrease between 1990 and 1996? Give a possible explanation for these trends.

Name ______________________ Section ______ Date ________

INTEGRATED REVIEW—SYSTEMS OF LINEAR EQUATIONS

The graphs of the equations of systems of equations are shown. Match each graph with the solution of its corresponding system.

A

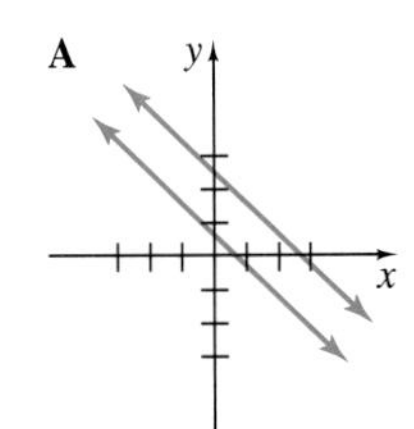

B

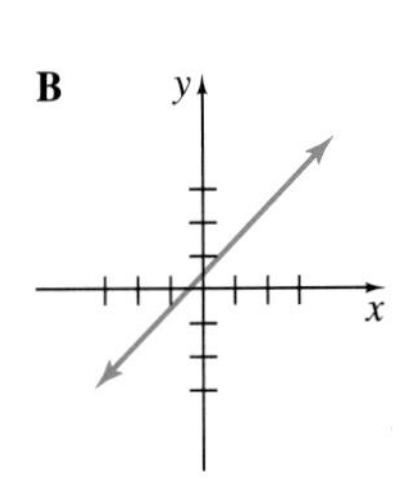

C

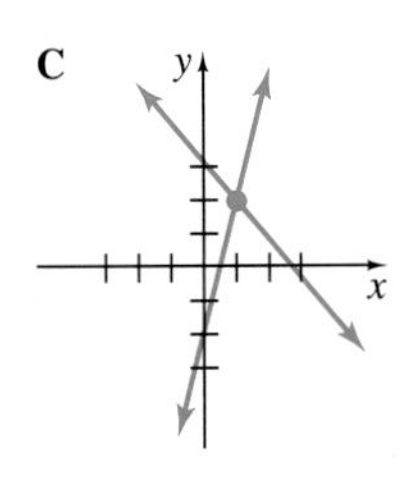

D

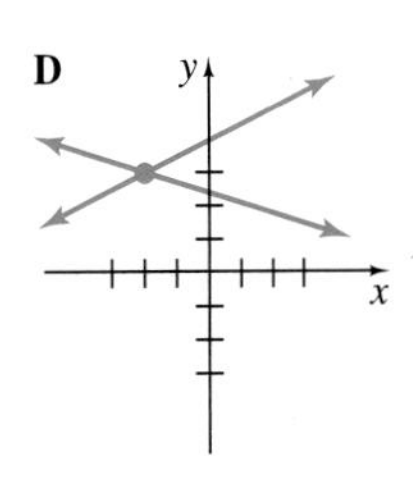

1. solution: $(1, 2)$

2. solution: $(-2, 3)$

3. no solution

4. infinite number of solutions

Solve each system by elimination or substitution.

5. $\begin{cases} x + y = 4 \\ y = 3x \end{cases}$

6. $\begin{cases} x - y = -4 \\ y = 4x \end{cases}$

7. $\begin{cases} x + y = 1 \\ x - 2y = 4 \end{cases}$

8. $\begin{cases} 2x - y = 8 \\ x + 3y = 11 \end{cases}$

9. $\begin{cases} 2x + 5y = 8 \\ 6x + y = 10 \end{cases}$

10. $\begin{cases} x - 4y = -5 \\ -3x - 8y = 0 \end{cases}$

11. $\begin{cases} 4x - 7y = 7 \\ 12x - 21y = 24 \end{cases}$

12. $\begin{cases} 2x - 5y = 3 \\ -4x + 10y = -6 \end{cases}$

13. $\begin{cases} x + y = 2 \\ -3y + z = -7 \\ 2x + y - z = -1 \end{cases}$

14. $\begin{cases} y + 2z = -3 \\ x - 2y = 7 \\ 2x - y + z = 5 \end{cases}$

ANSWERS

1. ________

2. ________

3. ________

4. ________

5. ________

6. ________

7. ________

8. ________

9. ________

10. ________

11. ________

12. ________

13. ________

14. ________

15. ______

16. ______

17. ______

18. ______

19. ______

20. ______

Name ______________________________

15. $\begin{cases} 2x + 4y - 6z = 3 \\ -x + y - z = 6 \\ x + 2y - 3z = 1 \end{cases}$

16. $\begin{cases} x - y + 3z = 2 \\ -2x + 2y - 6z = -4 \\ 3x - 3y + 9z = 6 \end{cases}$

17. $\begin{cases} x + y - 4z = 5 \\ x - y + 2z = -2 \\ 3x + 2y + 4z = 18 \end{cases}$

18. $\begin{cases} 2x - y + 3z = 2 \\ x + y - 6z = 0 \\ 3x + 4y - 3z = 6 \end{cases}$

19. A first number is 8 less than a second number. Twice the first number is 11 more than the second number. Find the numbers.

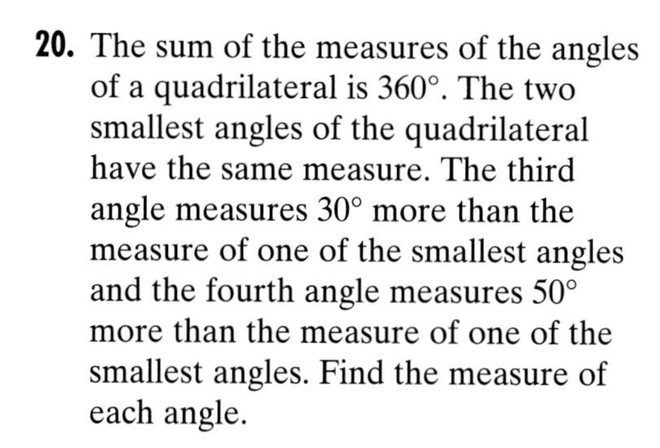

20. The sum of the measures of the angles of a quadrilateral is 360°. The two smallest angles of the quadrilateral have the same measure. The third angle measures 30° more than the measure of one of the smallest angles and the fourth angle measures 50° more than the measure of one of the smallest angles. Find the measure of each angle.

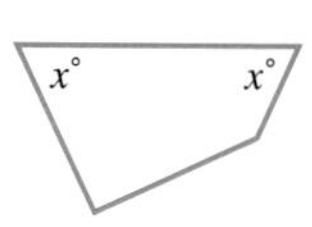

4.4 SOLVING SYSTEMS OF EQUATIONS USING MATRICES

Objectives

A Use matrices to solve a system of two equations.

B Use matrices to solve a system of three equations.

SSM CD-ROM Video 4.4

By now, you may have noticed that the solution of a system of equations depends on the coefficients of the equations in the system and not on the variables. In this section, we introduce how to solve a system of equations using a **matrix**.

A USING MATRICES TO SOLVE A SYSTEM OF TWO EQUATIONS

A **matrix** (plural: **matrices**) is a rectangular array of numbers. The following are examples of matrices.

$$\begin{bmatrix} 1 & 0 \\ 0 & 1 \end{bmatrix} \quad \begin{bmatrix} 2 & 1 & 3 & -1 \\ 0 & -1 & 4 & 5 \\ -6 & 2 & 1 & 0 \end{bmatrix} \quad \begin{bmatrix} a & b & c \\ d & e & f \end{bmatrix}$$

The numbers aligned horizontally in a matrix are in the same **row**. The numbers aligned vertically are in the same **column**.

$$\begin{matrix} \text{row 1} \rightarrow \\ \text{row 2} \rightarrow \end{matrix} \begin{bmatrix} 2 & 1 & 0 \\ -1 & 6 & 2 \end{bmatrix}$$

column 1, column 2, column 3

This matrix has 2 rows and 3 columns. It is called a 2×3 (read "two by three") matrix.

To see the relationship between systems of equations and matrices, study the example below.

System of Equations

$$\begin{cases} 2x - 3y = 6 & \text{Equation 1} \\ x + y = 0 & \text{Equation 2} \end{cases}$$

Corresponding Matrix

$$\left[\begin{array}{cc|c} 2 & -3 & 6 \\ 1 & 1 & 0 \end{array}\right] \begin{matrix} \text{Row 1} \\ \text{Row 2} \end{matrix}$$

Notice that the rows of the matrix correspond to the equations in the system. The coefficients of each variable are placed to the left of a vertical dashed line. The constants are placed to the right. Each of these numbers in the matrix is called an **element**.

The method of solving systems by matrices is to write this matrix as an equivalent matrix from which we easily identify the solution. Two matrices are equivalent if they represent systems that have the same solution set. The following **row operations** can be performed on matrices, and the result is an equivalent matrix.

ELEMENTARY ROW OPERATIONS

1. Any two rows in a matrix may be interchanged.
2. The elements of any row may be multiplied (or divided) by the same nonzero number.
3. The elements of any row may be multiplied (or divided) by a nonzero number and added to their corresponding elements in any other row.

HELPFUL HINT

Notice that these *row* operations are the same operations that we can perform on *equations* in a system.

Practice Problem 1

Use matrices to solve the system:

$$\begin{cases} x + 2y = -4 \\ 2x - 3y = 13 \end{cases}$$

Practice Problem 2

Use matrices to solve the system:

$$\begin{cases} -3x + y = 0 \\ -6x + 2y = 2 \end{cases}$$

Answers

1. $(2, -3)$, **2.** no solution

Example 1

Use matrices to solve the system:

$$\begin{cases} x + 3y = 5 \\ 2x - y = -4 \end{cases}$$

Solution: The corresponding matrix is $\left[\begin{array}{rr|r} 1 & 3 & 5 \\ 2 & -1 & -4 \end{array}\right]$. We use elementary row operations to write an equivalent matrix that looks like $\left[\begin{array}{cc|c} 1 & a & b \\ 0 & 1 & c \end{array}\right]$.

For the matrix given, the element in the first row, first column is already 1, as desired. Next we write an equivalent matrix with a 0 below the 1. To do this, we multiply row 1 by -2 and add to row 2. *We will change only row 2.*

$$\left[\begin{array}{c|c} 1 & 3 & 5 \\ \underset{\substack{\uparrow \\ \text{row 1} \\ \text{element}}}{-2(1)} + \underset{\substack{\uparrow \\ \text{row 2} \\ \text{element}}}{2} & \underset{\substack{\uparrow \\ \text{row 1} \\ \text{element}}}{-2(3)} + \underset{\substack{\uparrow \\ \text{row 2} \\ \text{element}}}{(-1)} & \underset{\substack{\uparrow \\ \text{row 1} \\ \text{element}}}{-2(5)} + \underset{\substack{\uparrow \\ \text{row 2} \\ \text{element}}}{(-4)} \end{array}\right] \text{ simplifies to } \left[\begin{array}{rr|r} 1 & 3 & 5 \\ 0 & -7 & -14 \end{array}\right]$$

Now we change the -7 to a 1 by use of an elementary row operation. We divide row 2 by -7, then

$$\left[\begin{array}{cc|c} 1 & 3 & 5 \\ \frac{0}{-7} & \frac{-7}{-7} & \frac{-14}{-7} \end{array}\right] \text{ simplifies to } \left[\begin{array}{cc|c} 1 & 3 & 5 \\ 0 & 1 & 2 \end{array}\right]$$

This last matrix corresponds to the system

$$\begin{cases} x + 3y = 5 \\ y = 2 \end{cases}$$

To find x, we let $y = 2$ in the first equation, $x + 3y = 5$.

$$\begin{aligned} x + 3y &= 5 && \text{First equation} \\ x + 3(2) &= 5 && \text{Let } y = 2. \\ x &= -1 \end{aligned}$$

The ordered pair solution is $(-1, 2)$. Check to see that this ordered pair satisfies both equations.

Example 2

Use matrices to solve the system:

$$\begin{cases} 2x - y = 3 \\ 4x - 2y = 5 \end{cases}$$

Solution: The corresponding matrix is $\left[\begin{array}{rr|r} 2 & -1 & 3 \\ 4 & -2 & 5 \end{array}\right]$. To get 1 in the row 1, column 1 position, we divide the elements of row 1 by 2.

$$\left[\begin{array}{cc|c} \frac{2}{2} & -\frac{1}{2} & \frac{3}{2} \\ 4 & -2 & 5 \end{array}\right] \text{ simplifies to } \left[\begin{array}{cc|c} 1 & -\frac{1}{2} & \frac{3}{2} \\ 4 & -2 & 5 \end{array}\right]$$

To get 0 under the 1, we multiply the elements of row 1 by -4 and add the new elements to the elements of row 2.

$$\left[\begin{array}{cc|c} 1 & -\frac{1}{2} & \frac{3}{2} \\ -4(1)+4 & -4\left(-\frac{1}{2}\right)-2 & -4\left(\frac{3}{2}\right)+5 \end{array}\right] \text{ simplifies to } \left[\begin{array}{cc|c} 1 & -\frac{1}{2} & \frac{3}{2} \\ 0 & 0 & -1 \end{array}\right]$$

The corresponding system is $\begin{cases} x - \frac{1}{2}y = \frac{3}{2} \\ 0 = -1 \end{cases}$. The equation $0 = -1$ is false for all y or x values; hence the system is inconsistent and has no solution.

TRY THE CONCEPT CHECK IN THE MARGIN.

B USING MATRICES TO SOLVE A SYSTEM OF THREE EQUATIONS

To solve a system of three equations in three variables using matrices, we will write the corresponding matrix in the form

$$\left[\begin{array}{ccc|c} 1 & a & b & d \\ 0 & 1 & c & e \\ 0 & 0 & 1 & f \end{array}\right]$$

Example 3 Use matrices to solve the system:

$$\begin{cases} x + 2y + z = 2 \\ -2x - y + 2z = 5 \\ x + 3y - 2z = -8 \end{cases}$$

Solution: The corresponding matrix is $\left[\begin{array}{ccc|c} 1 & 2 & 1 & 2 \\ -2 & -1 & 2 & 5 \\ 1 & 3 & -2 & -8 \end{array}\right]$.

Our goal is to write an equivalent matrix with 1s along the diagonal (see the numbers in red) and 0s below the 1s. The element in row 1, column 1 is already 1. Next we get 0s for each element in the rest of column 1. To do this, first we multiply the elements of row 1 by 2 and add the new elements to row 2. Also, we multiply the elements of row 1 by -1 and add the new elements to the elements of row 3. We *do not change row 1*. Then

$$\left[\begin{array}{ccc|c} 1 & 2 & 1 & 2 \\ 2(1)-2 & 2(2)-1 & 2(1)+2 & 2(2)+5 \\ -1(1)+1 & -1(2)+3 & -1(1)-2 & -1(2)-8 \end{array}\right] \text{ simplifies to } \left[\begin{array}{ccc|c} 1 & 2 & 1 & 2 \\ 0 & 3 & 4 & 9 \\ 0 & 1 & -3 & -10 \end{array}\right]$$

✓ CONCEPT CHECK

Consider the system
$\begin{cases} 2x - 3y = 8 \\ x + 5y = -3 \end{cases}$
What is wrong with its corresponding matrix shown below?
$\left[\begin{array}{cc|c} 2 & 3 & 8 \\ 0 & 5 & 3 \end{array}\right]$

Practice Problem 3

Use matrices to solve the system:

$$\begin{cases} x + 3y + z = 5 \\ -3x + y - 3z = 5 \\ x + 2y - 2z = 9 \end{cases}$$

Answers

3. $(1, 2, -2)$

✓ Concept Check: Its matrix is $\left[\begin{array}{cc|c} 2 & -3 & 8 \\ 1 & 5 & -3 \end{array}\right]$.

We continue down the diagonal and use elementary row operations to get 1 where the element 3 is now. To do this, we interchange rows 2 and 3.

$$\left[\begin{array}{ccc|c} 1 & 2 & 1 & 2 \\ 0 & 3 & 4 & 9 \\ 0 & 1 & -3 & -10 \end{array}\right] \text{ is equivalent to } \left[\begin{array}{ccc|c} 1 & 2 & 1 & 2 \\ 0 & 1 & -3 & -10 \\ 0 & 3 & 4 & 9 \end{array}\right]$$

Next we want the new row 3, column 2 element to be 0. We multiply the elements of row 2 by -3 and add the result to the elements of row 3.

$$\left[\begin{array}{ccc|c} 1 & 2 & 1 & 2 \\ 0 & 1 & -3 & -10 \\ -3(0)+0 & -3(1)+3 & -3(-3)+4 & -3(-10)+9 \end{array}\right] \text{ simplifies to } \left[\begin{array}{ccc|c} 1 & 2 & 1 & 2 \\ 0 & 1 & -3 & -10 \\ 0 & 0 & 13 & 39 \end{array}\right]$$

Finally, we divide the elements of row 3 by 13 so that the final diagonal element is 1.

$$\left[\begin{array}{ccc|c} 1 & 2 & 1 & 2 \\ 0 & 1 & -3 & -10 \\ \frac{0}{13} & \frac{0}{13} & \frac{13}{13} & \frac{39}{13} \end{array}\right] \text{ simplifies to } \left[\begin{array}{ccc|c} 1 & 2 & 1 & 2 \\ 0 & 1 & -3 & -10 \\ 0 & 0 & 1 & 3 \end{array}\right]$$

This matrix corresponds to the system

$$\begin{cases} x + 2y + z = 2 \\ y - 3z = -10 \\ z = 3 \end{cases}$$

We identify the z-coordinate of the solution as 3. Next we replace z with 3 in the second equation and solve for y.

$$\begin{aligned} y - 3z &= -10 && \text{Second equation} \\ y - 3(3) &= -10 && \text{Let } z = 3. \\ y &= -1 \end{aligned}$$

To find x, we let $z = 3$ and $y = -1$ in the first equation.

$$\begin{aligned} x + 2y + z &= 2 && \text{First equation} \\ x + 2(-1) + 3 &= 2 && \text{Let } z = 3 \text{ and } y = -1. \\ x &= 1 \end{aligned}$$

The ordered triple solution is $(1, -1, 3)$. Check to see that it satisfies all three equations in the original system.

Name ______________ **Section** ______ **Date** ______

Exercise Set 4.4

A *Use matrices to solve each system of linear equations. See Examples 1 and 2.*

1. $\begin{cases} x + y = 1 \\ x - 2y = 4 \end{cases}$

2. $\begin{cases} 2x - y = 8 \\ x + 3y = 11 \end{cases}$

3. $\begin{cases} x + 3y = 2 \\ x + 2y = 0 \end{cases}$

4. $\begin{cases} 4x - y = 5 \\ 3x - 3 = 0 \end{cases}$

5. $\begin{cases} x - 2y = 4 \\ 2x - 4y = 4 \end{cases}$

6. $\begin{cases} -x + 3y = 6 \\ 3x - 9y = 9 \end{cases}$

7. $\begin{cases} 3x - 3y = 9 \\ 2x - 2y = 6 \end{cases}$

8. $\begin{cases} 9x - 3y = 6 \\ -18x + 6y = -12 \end{cases}$

9. $\begin{cases} x - 4 = 0 \\ x + y = 1 \end{cases}$

10. $\begin{cases} 3y = 6 \\ x + y = 7 \end{cases}$

B *Use matrices to solve each system of linear equations. See Example 3.*

11. $\begin{cases} x + y = 3 \\ 2y = 10 \\ 3x + 2y - 4z = 12 \end{cases}$

12. $\begin{cases} 5x = 5 \\ 2x + y = 4 \\ 3x + y - 5z = -15 \end{cases}$

13. $\begin{cases} 2y - z = -7 \\ x + 4y + z = -4 \\ 5x - y + 2z = 13 \end{cases}$

14. $\begin{cases} 4y + 3z = -2 \\ 5x - 4y = 1 \\ -5x + 4y + z = -3 \end{cases}$

15. $\begin{cases} x + y + z = 2 \\ 2x - z = 5 \\ 3y + z = 2 \end{cases}$

16. $\begin{cases} x + 2y + z = 5 \\ x - y - z = 3 \\ y + z = 2 \end{cases}$

17. $\begin{cases} 4x + y + z = 3 \\ -x + y - 2z = -11 \\ x + 2y + 2z = -1 \end{cases}$

18. $\begin{cases} x + y + z = 9 \\ 3x - y + z = -1 \\ -2x + 2y - 3z = -2 \end{cases}$

Answers

1. ______
2. ______
3. ______
4. ______
5. ______
6. ______
7. ______
8. ______
9. ______
10. ______
11. ______
12. ______
13. ______
14. ______
15. ______
16. ______
17. ______
18. ______

19. ______

Name ______

Review and Preview

Evaluate. See Section 1.3.

20. ______

19. $(-1)(-5) - (6)(3)$

20. $(2)(-8) - (-4)(1)$

21. ______

21. $(4)(-10) - (2)(-2)$

22. $(-7)(3) - (-2)(-6)$

22. ______

23. $(-3)(-3) - (-1)(-9)$

24. $(5)(6) - (10)(10)$

23. ______

Combining Concepts

24. ______

25. a. ______

b. ______

c. ______

25. The percent y of U.S. households that owned a black-and-white television set between the years 1980 and 1993 can be modeled by the linear equation $2.3x + y = 52$, where x represents the number of years after 1980. Similarly, the percent y of U.S. households that owned a microwave oven during this same period can be modeled by the linear equation $-5.4x + y = 14$. (*Source:* Based on data from the Energy Information Administration, U.S. Department of Energy)

a. The data used to form these two models was incomplete. It is impossible to tell from the data the year in which the percent of households owning black-and-white television sets was the same as the percent of households owning microwave ovens. Use matrix methods to estimate the year in which this occurred.

b. Did more households own black-and-white television sets or microwave ovens in 1980? In 1993? What trends do these models show? Does this seem to make sense? Why or why not?

c. According to the models, when will the percent of households owning black-and-white television sets reach 0%?

4.5 SOLVING SYSTEMS OF EQUATIONS USING DETERMINANTS

We have solved systems of two linear equations in two variables in four different ways: graphically, by substitution, by elimination, and by matrices. Now we analyze another method called **Cramer's rule**.

Objectives

A Define and evaluate a 2 × 2 determinant.

B Use Cramer's rule to solve a system of two linear equations in two variables.

C Define and evaluate a 3 × 3 determinant.

D Use Cramer's rule to solve a linear system of three equations in three variables.

SSM CD-ROM Video 4.5

A EVALUATING 2 × 2 DETERMINANTS

Recall that a matrix is a rectangular array of numbers. If a matrix has the same number of rows and columns, it is called a **square matrix**. Examples of square matrices are

$$\begin{bmatrix} 1 & 6 \\ 5 & 2 \end{bmatrix} \qquad \begin{bmatrix} 2 & 4 & 1 \\ 0 & 5 & 2 \\ 3 & 6 & 9 \end{bmatrix}$$

A **determinant** is a real number associated with a square matrix. The determinant of a square matrix is denoted by placing vertical bars about the array of numbers. Thus,

The determinant of the square matrix $\begin{bmatrix} 1 & 6 \\ 5 & 2 \end{bmatrix}$ is $\begin{vmatrix} 1 & 6 \\ 5 & 2 \end{vmatrix}$.

The determinant of the square matrix $\begin{bmatrix} 2 & 4 & 1 \\ 0 & 5 & 2 \\ 3 & 6 & 9 \end{bmatrix}$ is $\begin{vmatrix} 2 & 4 & 1 \\ 0 & 5 & 2 \\ 3 & 6 & 9 \end{vmatrix}$.

We define the determinant of a 2 × 2 matrix first. (Recall that 2 × 2 is read "two by two." It means that the matrix has 2 rows and 2 columns.)

DETERMINANT OF A 2 × 2 MATRIX

$$\begin{vmatrix} a & b \\ c & d \end{vmatrix} = ad - bc$$

Example 1 Evaluate each determinant.

a. $\begin{vmatrix} -1 & 2 \\ 3 & -4 \end{vmatrix}$ **b.** $\begin{vmatrix} 2 & 0 \\ 7 & -5 \end{vmatrix}$

Solution: First we identify the values of a, b, c, and d. Then we perform the evaluation.

a. Here $a = -1, b = 2, c = 3$, and $d = -4$.

$$\begin{vmatrix} -1 & 2 \\ 3 & -4 \end{vmatrix} = ad - bc = (-1)(-4) - (2)(3) = -2$$

b. In this example, $a = 2, b = 0, c = 7$, and $d = -5$.

$$\begin{vmatrix} 2 & 0 \\ 7 & -5 \end{vmatrix} = ad - bc = 2(-5) - (0)(7) = -10$$

Practice Problem 1

Evaluate each determinant.

a. $\begin{vmatrix} -3 & 6 \\ 2 & 1 \end{vmatrix}$

b. $\begin{vmatrix} 4 & 5 \\ 0 & -5 \end{vmatrix}$

Answers

1. a. -15, **b.** -20

B USING CRAMER'S RULE TO SOLVE A SYSTEM OF TWO LINEAR EQUATIONS

To develop Cramer's rule, we solve the system $\begin{cases} ax + by = h \\ cx + dy = k \end{cases}$ using elimination. First, we eliminate y by multiplying both sides of the first equation by d and both sides of the second equation by $-b$ so that the coefficients of y are opposites. The result is that

$$\begin{cases} d(ax + by) = d \cdot h \\ -b(cx + dy) = -b \cdot k \end{cases} \quad \text{simplifies to} \quad \begin{cases} adx + bdy = hd \\ -bcx - bdy = -kb \end{cases}$$

We now add the two equations and solve for x.

$$\begin{aligned} adx + bdy &= hd \\ -bcx - bdy &= -kb \\ \hline adx - bcx \qquad &= hd - kb && \text{Add the equations.} \\ (ad - bc)x &= hd - kb \\ x &= \frac{hd - kb}{ad - bc} && \text{Solve for } x. \end{aligned}$$

When we replace x with $\frac{hd - kb}{ad - bc}$ in the equation $ax + by = h$ and solve for y, we find that $y = \frac{ak - ch}{ad - bc}$.

Notice that the numerator of the value of x is the determinant of

$$\begin{vmatrix} h & b \\ k & d \end{vmatrix} = hd - kb$$

Also, the numerator of the value of y is the determinant of

$$\begin{vmatrix} a & h \\ c & k \end{vmatrix} = ak - hc$$

Finally, the denominators of the values of x and y are the same and are the determinant of

$$\begin{vmatrix} a & b \\ c & d \end{vmatrix} = ad - bc$$

This means that the values of x and y can be written in determinant notation:

$$x = \frac{\begin{vmatrix} h & b \\ k & d \end{vmatrix}}{\begin{vmatrix} a & b \\ c & d \end{vmatrix}} \quad \text{and} \quad y = \frac{\begin{vmatrix} a & h \\ c & k \end{vmatrix}}{\begin{vmatrix} a & b \\ c & d \end{vmatrix}}$$

For convenience, we label the determinants D, D_x, and D_y.

x-coefficients

y-coefficients

$$\begin{vmatrix} a & b \\ c & d \end{vmatrix} = D \qquad \begin{vmatrix} h & b \\ k & d \end{vmatrix} = D_x \qquad \begin{vmatrix} a & h \\ c & k \end{vmatrix} = D_y$$

x-column replaced by constants

y-column replaced by constants

These determinant formulas for the coordinates of the solution of a system are known as **Cramer's rule**.

CRAMER'S RULE FOR TWO LINEAR EQUATIONS IN TWO VARIABLES

The solution of the system $\begin{cases} ax + by = h \\ cx + dy = k \end{cases}$ is given by

$$x = \frac{\begin{vmatrix} h & b \\ k & d \end{vmatrix}}{\begin{vmatrix} a & b \\ c & d \end{vmatrix}} = \frac{D_x}{D} \qquad y = \frac{\begin{vmatrix} a & h \\ c & k \end{vmatrix}}{\begin{vmatrix} a & b \\ c & d \end{vmatrix}} = \frac{D_y}{D}$$

as long as $D = ad - bc$ is not 0.

When $D = 0$, the system is either inconsistent or the equations are dependent. When this happens, we need to use another method to see which is the case.

Example 2 Use Cramer's rule to solve the system:

$$\begin{cases} 3x + 4y = -7 \\ x - 2y = -9 \end{cases}$$

Solution: First we find D, D_x, and D_y.

$$\begin{cases} \overset{\overset{a}{\downarrow}}{3}x + \overset{\overset{b}{\downarrow}}{4}y = \overset{\overset{h}{\downarrow}}{-7} \\ \underset{\underset{c}{\uparrow}}{x} - \underset{\underset{d}{\uparrow}}{2}y = \underset{\underset{k}{\uparrow}}{-9} \end{cases}$$

$$D = \begin{vmatrix} a & b \\ c & d \end{vmatrix} = \begin{vmatrix} 3 & 4 \\ 1 & -2 \end{vmatrix} = 3(-2) - 4(1) = -10$$

$$D_x = \begin{vmatrix} h & b \\ k & d \end{vmatrix} = \begin{vmatrix} -7 & 4 \\ -9 & -2 \end{vmatrix} = (-7)(-2) - 4(-9) = 50$$

$$D_y = \begin{vmatrix} a & h \\ c & k \end{vmatrix} = \begin{vmatrix} 3 & -7 \\ 1 & -9 \end{vmatrix} = 3(-9) - (-7)(1) = -20$$

Then $x = \frac{D_x}{D} = \frac{50}{-10} = -5$ and $y = \frac{D_y}{D} = \frac{-20}{-10} = 2$.

The ordered pair solution is $(-5, 2)$.

As always, check the solution in both original equations.

Practice Problem 2

Use Cramer's rule to solve the system.

$$\begin{cases} x - y = -4 \\ 2x + 3y = 2 \end{cases}$$

Example 3 Use Cramer's rule to solve the system:

$$\begin{cases} 5x + y = 5 \\ -7x - 2y = -7 \end{cases}$$

Solution: First we find D, D_x, and D_y.

$$D = \begin{vmatrix} 5 & 1 \\ -7 & -2 \end{vmatrix} = 5(-2) - (-7)(1) = -3$$

$$D_x = \begin{vmatrix} 5 & 1 \\ -7 & -2 \end{vmatrix} = 5(-2) - (-7)(1) = -3$$

$$D_y = \begin{vmatrix} 5 & 5 \\ -7 & -7 \end{vmatrix} = 5(-7) - 5(-7) = 0$$

Practice Problem 3

Use Cramer's rule to solve the system.

$$\begin{cases} 4x + y = 3 \\ 2x - 3y = -9 \end{cases}$$

Answers

2. $(-2, 2)$, **3.** $(0, 3)$

Then

$$x = \frac{D_x}{D} = \frac{-3}{-3} = 1 \qquad y = \frac{D_y}{D} = \frac{0}{-3} = 0$$

The ordered pair solution is $(1, 0)$.

C EVALUATING 3 × 3 DETERMINANTS

A 3 × 3 determinant can be used to solve a system of three equations in three variables. The determinant of a 3 × 3 matrix, however, is considerably more complex than a 2 × 2 one.

DETERMINANT OF A 3 × 3 MATRIX

$$\begin{vmatrix} a_1 & b_1 & c_1 \\ a_2 & b_2 & c_2 \\ a_3 & b_3 & c_3 \end{vmatrix} = a_1 \cdot \begin{vmatrix} b_2 & c_2 \\ b_3 & c_3 \end{vmatrix} - a_2 \cdot \begin{vmatrix} b_1 & c_1 \\ b_3 & c_3 \end{vmatrix} + a_3 \cdot \begin{vmatrix} b_1 & c_1 \\ b_2 & c_2 \end{vmatrix}$$

Notice that the determinant of a 3 × 3 matrix is related to the determinants of three 2 × 2 matrices. Each determinant of these 2 × 2 matrices is called a **minor**, and every element of a 3 × 3 matrix has a minor associated with it. For example, the minor of c_2 is the determinant of the 2 × 2 matrix found by deleting the row and column containing c_2.

$$\begin{matrix} a_1 & b_1 & c_1 \\ a_2 & b_2 & c_2 \\ a_3 & b_3 & c_3 \end{matrix} \quad \text{The minor of } c_2 \text{ is} \quad \begin{vmatrix} a_1 & b_1 \\ a_3 & b_3 \end{vmatrix}$$

Also, the minor of element a_1 is the determinant of the 2 × 2 matrix that has no row or column containing a_1.

$$\begin{matrix} a_1 & b_1 & c_1 \\ a_2 & b_2 & c_2 \\ a_3 & b_3 & c_3 \end{matrix} \quad \text{The minor of } a_1 \text{ is} \quad \begin{vmatrix} b_2 & c_2 \\ b_3 & c_3 \end{vmatrix}$$

So the determinant of a 3 × 3 matrix can be written as

$$a_1 \cdot (\text{minor of } a_1) - a_2 \cdot (\text{minor of } a_2) + a_3 \cdot (\text{minor of } a_3)$$

Finding the determinant by using minors of elements in the first column is called **expanding** by the minors of the first column. *The value of a determinant can be found by expanding by the minors of any row or column.* The following **array of signs** is helpful in determining whether to add or subtract the product of an element and its minor.

$$\begin{matrix} + & - & + \\ - & + & - \\ + & - & + \end{matrix}$$

If an element is in a position marked +, we add. If marked −, we subtract.

TRY THE CONCEPT CHECK IN THE MARGIN.

Example 4 Evaluate by expanding by the minors of the given row or column.

$$\begin{vmatrix} 0 & 5 & 1 \\ 1 & 3 & -1 \\ -2 & 2 & 4 \end{vmatrix}$$

a. First column **b.** Second row

✓ CONCEPT CHECK

Suppose you are interested in finding the determinant of a 4 × 4 matrix. Study the pattern shown in the array of signs for a 3 × 3 matrix. Use the pattern to expand the array of signs for use with a 4 × 4 matrix.

Practice Problem 4

Evaluate by expanding by the minors of the given row or column.

a. First column b. Third row

$$\begin{vmatrix} 2 & 0 & 1 \\ -1 & 3 & 2 \\ 5 & 1 & 4 \end{vmatrix}$$

Answers

4. a. 4, **b.** 4

✓ Concept Check: $\begin{matrix} + & - & + & - \\ - & + & - & + \\ + & - & + & - \\ - & + & - & + \end{matrix}$

Solution: **a.** The elements of the first column are 0, 1, and −2. The first column of the array of signs is +, −, +.

$$\begin{vmatrix} 0 & 5 & 1 \\ 1 & 3 & -1 \\ -2 & 2 & 4 \end{vmatrix} = 0 \cdot \begin{vmatrix} 3 & -1 \\ 2 & 4 \end{vmatrix} - 1 \cdot \begin{vmatrix} 5 & 1 \\ 2 & 4 \end{vmatrix} + (-2) \cdot \begin{vmatrix} 5 & 1 \\ 3 & -1 \end{vmatrix}$$

$$= 0(12 - (-2)) - 1(20 - 2) + (-2)(-5 - 3)$$

$$= 0 - 18 + 16 = -2$$

b. The elements of the second row are 1, 3, and −1. This time, the signs begin with − and again alternate.

$$\begin{vmatrix} 0 & 5 & 1 \\ 1 & 3 & -1 \\ -2 & 2 & 4 \end{vmatrix} = -1 \cdot \begin{vmatrix} 5 & 1 \\ 2 & 4 \end{vmatrix} + 3 \cdot \begin{vmatrix} 0 & 1 \\ -2 & 4 \end{vmatrix} - (-1) \cdot \begin{vmatrix} 0 & 5 \\ -2 & 2 \end{vmatrix}$$

$$= -1(20 - 2) + 3(0 - (-2)) - (-1)(0 - (-10))$$

$$= -18 + 6 + 10 = -2$$

Notice that the determinant of the 3 × 3 matrix is the same regardless of the row or column you select to expand by.

TRY THE CONCEPT CHECK IN THE MARGIN.

D USING CRAMER'S RULE TO SOLVE A SYSTEM OF THREE LINEAR EQUATIONS

A system of three equations in three variables may be solved with Cramer's rule also. Using the elimination process to solve a system with unknown constants as coefficients leads to the following.

CRAMER'S RULE FOR THREE EQUATIONS IN THREE VARIABLES

The solution of the system $\begin{cases} a_1x + b_1y + c_1z = k_1 \\ a_2x + b_2y + c_2z = k_2 \\ a_3x + b_3y + c_3z = k_3 \end{cases}$ is given by

$$x = \frac{D_x}{D} \qquad y = \frac{D_y}{D} \qquad \text{and} \qquad z = \frac{D_z}{D}$$

where

$$D = \begin{vmatrix} a_1 & b_1 & c_1 \\ a_2 & b_2 & c_2 \\ a_3 & b_3 & c_3 \end{vmatrix} \quad D_x = \begin{vmatrix} k_1 & b_1 & c_1 \\ k_2 & b_2 & c_2 \\ k_3 & b_3 & c_3 \end{vmatrix}$$

$$D_y = \begin{vmatrix} a_1 & k_1 & c_1 \\ a_2 & k_2 & c_2 \\ a_3 & k_3 & c_3 \end{vmatrix} \quad D_z = \begin{vmatrix} a_1 & b_1 & k_1 \\ a_2 & b_2 & k_2 \\ a_3 & b_3 & k_3 \end{vmatrix}$$

as long as D is not 0.

Example 5 Use Cramer's rule to solve the system:

$$\begin{cases} x - 2y + z = 4 \\ 3x + y - 2z = 3 \\ 5x + 5y + 3z = -8 \end{cases}$$

✓ CONCEPT CHECK

Why would expanding by minors of the second row be a good choice for the determinant $\begin{vmatrix} 3 & 4 & -2 \\ 5 & 0 & 0 \\ 6 & -3 & 7 \end{vmatrix}$?

Practice Problem 5

Use Cramer's rule to solve the system:

$$\begin{cases} x + 2y - z = 3 \\ 2x - 3y + z = -9 \\ -x + y - 2z = 0 \end{cases}$$

Answers

5. $(-1, 3, 2)$

✓ Concept Check: Two elements of the second row are 0, which makes calculations easier.

Solution: First we find D, D_x, D_y, and D_z. Beginning with D, we expand by the minors of the first column.

$$D = \begin{vmatrix} 1 & -2 & 1 \\ 3 & 1 & -2 \\ 5 & 5 & 3 \end{vmatrix} = 1 \cdot \begin{vmatrix} 1 & -2 \\ 5 & 3 \end{vmatrix} - 3 \cdot \begin{vmatrix} -2 & 1 \\ 5 & 3 \end{vmatrix} + 5 \cdot \begin{vmatrix} -2 & 1 \\ 1 & -2 \end{vmatrix}$$
$$= 1(3 - (-10)) - 3(-6 - 5) + 5(4 - 1)$$
$$= 13 + 33 + 15 = 61$$

$$D_x = \begin{vmatrix} 4 & -2 & 1 \\ 3 & 1 & -2 \\ -8 & 5 & 3 \end{vmatrix} = 4 \cdot \begin{vmatrix} 1 & -2 \\ 5 & 3 \end{vmatrix} - 3 \cdot \begin{vmatrix} -2 & 1 \\ 5 & 3 \end{vmatrix} + (-8) \cdot \begin{vmatrix} -2 & 1 \\ 1 & -2 \end{vmatrix}$$
$$= 4(3 - (-10)) - 3(-6 - 5) + (-8)(4 - 1)$$
$$= 52 + 33 - 24 = 61$$

$$D_y = \begin{vmatrix} 1 & 4 & 1 \\ 3 & 3 & -2 \\ 5 & -8 & 3 \end{vmatrix} = 1 \cdot \begin{vmatrix} 3 & -2 \\ -8 & 3 \end{vmatrix} - 3 \cdot \begin{vmatrix} 4 & 1 \\ -8 & 3 \end{vmatrix} + 5 \cdot \begin{vmatrix} 4 & 1 \\ 3 & -2 \end{vmatrix}$$
$$= 1(9 - 16) - 3(12 - (-8)) + 5(-8 - 3)$$
$$= -7 - 60 - 55 = -122$$

$$D_z = \begin{vmatrix} 1 & -2 & 4 \\ 3 & 1 & 3 \\ 5 & 5 & -8 \end{vmatrix} = 1 \cdot \begin{vmatrix} 1 & 3 \\ 5 & -8 \end{vmatrix} - 3 \cdot \begin{vmatrix} -2 & 4 \\ 5 & -8 \end{vmatrix} + 5 \cdot \begin{vmatrix} -2 & 4 \\ 1 & 3 \end{vmatrix}$$
$$= 1(-8 - 15) - 3(16 - 20) + 5(-6 - 4)$$
$$= -23 + 12 - 50 = -61$$

From these determinants, we calculate the solution:

$$x = \frac{D_x}{D} = \frac{61}{61} = 1 \qquad y = \frac{D_y}{D} = \frac{-122}{61} = -2 \qquad z = \frac{D_z}{D} = \frac{-61}{61} = -1$$

The ordered triple solution is $(1, -2, -1)$. Check this solution by verifying that it satisfies each equation of the system.

Name ______________________ Section ________ Date ________

MENTAL MATH ANSWERS

1. ______
2. ______
3. ______
4. ______
5. ______
6. ______

ANSWERS

1. ______
2. ______
3. ______
4. ______
5. ______
6. ______
7. ______
8. ______
9. ______
10. ______
11. ______
12. ______
13. ______
14. ______
15. ______
16. ______
17. ______
18. ______
19. ______
20. ______
21. ______

MENTAL MATH

Evaluate each determinant mentally.

1. $\begin{vmatrix} 7 & 2 \\ 0 & 8 \end{vmatrix}$ **2.** $\begin{vmatrix} 6 & 0 \\ 1 & 2 \end{vmatrix}$ **3.** $\begin{vmatrix} -4 & 2 \\ 0 & 8 \end{vmatrix}$

4. $\begin{vmatrix} 5 & 0 \\ 3 & -5 \end{vmatrix}$ **5.** $\begin{vmatrix} -2 & 0 \\ 3 & -10 \end{vmatrix}$ **6.** $\begin{vmatrix} -1 & 4 \\ 0 & -18 \end{vmatrix}$

EXERCISE SET 4.5

A *Evaluate each determinant. See Example 1.*

1. $\begin{vmatrix} 3 & 5 \\ -1 & 7 \end{vmatrix}$ **2.** $\begin{vmatrix} -5 & 1 \\ 1 & -4 \end{vmatrix}$ **3.** $\begin{vmatrix} 9 & -2 \\ 4 & -3 \end{vmatrix}$

4. $\begin{vmatrix} 4 & -1 \\ 9 & 8 \end{vmatrix}$ **5.** $\begin{vmatrix} -2 & 9 \\ 4 & -18 \end{vmatrix}$ **6.** $\begin{vmatrix} -40 & 8 \\ 70 & -14 \end{vmatrix}$

7. $\begin{vmatrix} \frac{3}{4} & \frac{5}{2} \\ -\frac{1}{6} & \frac{7}{3} \end{vmatrix}$ **8.** $\begin{vmatrix} \frac{5}{7} & \frac{1}{3} \\ \frac{6}{7} & \frac{2}{3} \end{vmatrix}$

B *Use Cramer's rule, if possible, to solve each system of linear equations. See Examples 2 and 3.*

9. $\begin{cases} 2y - 4 = 0 \\ x + 2y = 5 \end{cases}$ **10.** $\begin{cases} 4x - y = 5 \\ 3x - 3 = 0 \end{cases}$ **11.** $\begin{cases} 3x + y = 1 \\ 2y = 2 - 6x \end{cases}$

12. $\begin{cases} y = 2x - 5 \\ 8x - 4y = 20 \end{cases}$ **13.** $\begin{cases} 5x - 2y = 27 \\ -3x + 5y = 18 \end{cases}$ **14.** $\begin{cases} 4x - y = 9 \\ 2x + 3y = -27 \end{cases}$

15. $\begin{cases} 2x - 5y = 4 \\ x + 2y = -7 \end{cases}$ **16.** $\begin{cases} 3x - y = 2 \\ -5x + 2y = 0 \end{cases}$

17. $\begin{cases} \frac{2}{3}x - \frac{3}{4}y = -1 \\ -\frac{1}{6}x + \frac{3}{4}y = \frac{5}{2} \end{cases}$ **18.** $\begin{cases} \frac{1}{2}x - \frac{1}{3}y = -3 \\ \frac{1}{8}x + \frac{1}{6}y = 0 \end{cases}$

C *Evaluate. See Example 4.*

19. $\begin{vmatrix} 2 & 1 & 0 \\ 0 & 5 & -3 \\ 4 & 0 & 2 \end{vmatrix}$ **20.** $\begin{vmatrix} -6 & 4 & 2 \\ 1 & 0 & 5 \\ 0 & 3 & 1 \end{vmatrix}$ **21.** $\begin{vmatrix} 4 & -6 & 0 \\ -2 & 3 & 0 \\ 4 & -6 & 1 \end{vmatrix}$

22. ______

23. ______

24. ______

25. ______

26. ______

27. ______

28. ______

29. ______

30. ______

31. ______

32. ______

33. ______

34. ______

35. ______

36. ______

37. ______

38. ______

39. ______

40. ______

41. ______

42. ______

Name ______

22. $\begin{vmatrix} 5 & 2 & 1 \\ 3 & -6 & 0 \\ -2 & 8 & 0 \end{vmatrix}$

23. $\begin{vmatrix} 1 & 0 & 4 \\ 1 & -1 & 2 \\ 3 & 2 & 1 \end{vmatrix}$

24. $\begin{vmatrix} 0 & 1 & 2 \\ 3 & -1 & 2 \\ 3 & 2 & -2 \end{vmatrix}$

25. $\begin{vmatrix} 3 & 6 & -3 \\ -1 & -2 & 3 \\ 4 & -1 & 6 \end{vmatrix}$

26. $\begin{vmatrix} 2 & -2 & 1 \\ 4 & 1 & 3 \\ 3 & 1 & 2 \end{vmatrix}$

D *Use Cramer's rule, if possible, to solve each system of linear equations. See Example 5.*

27. $\begin{cases} 3x + z = -1 \\ -x - 3y + z = 7 \\ 3y + z = 5 \end{cases}$

28. $\begin{cases} 4y - 3z = -2 \\ 8x - 4y = 4 \\ -8x + 4y + z = -2 \end{cases}$

29. $\begin{cases} x + y + z = 8 \\ 2x - y - z = 10 \\ x - 2y + 3z = 22 \end{cases}$

30. $\begin{cases} 5x + y + 3z = 1 \\ x - y - 3z = -7 \\ -x + y = 1 \end{cases}$

31. $\begin{cases} 2x + 2y + z = 1 \\ -x + y + 2z = 3 \\ x + 2y + 4z = 0 \end{cases}$

32. $\begin{cases} 2x - 3y + z = 5 \\ x + y + z = 0 \\ 4x + 2y + 4z = 4 \end{cases}$

33. $\begin{cases} x - 2y + z = -5 \\ 3y + 2z = 4 \\ 3x - y = -2 \end{cases}$

34. $\begin{cases} 4x + 5y = 10 \\ 3y + 2z = -6 \\ x + y + z = 3 \end{cases}$

Review and Preview

Solve each linear inequality. See Section 2.4.

35. $-2x - 5 < 6$

36. $-2x \geq x - 9$

37. $5(x - 6) \leq 2(x - 7)$

38. $\frac{x}{3} > \frac{1}{2} - \frac{x}{2}$

Combining Concepts

Find the value of x that will make each a true statement.

39. $\begin{vmatrix} 1 & x \\ 2 & 7 \end{vmatrix} = -3$

40. $\begin{vmatrix} 6 & 1 \\ -2 & x \end{vmatrix} = 26$

41. If all the elements in a single row of a determinant are zero, what is the value of the determinant? Explain your answer.

42. If all the elements in a single column of a determinant are 0, what is the value of the determinant? Explain your answer.

4.6 SYSTEMS OF LINEAR INEQUALITIES

A GRAPHING SYSTEMS OF LINEAR INEQUALITIES

Objective

A Graph a system of linear inequalities.

In Section 3.6 we solved linear inequalities in two variables. Just as two linear equations make a system of linear equations, two linear inequalities make a **system of linear inequalities**. Systems of inequalities are very important in a process called linear programming. Many businesses use linear programming to find the most profitable way to use limited resources such as employees, machines, or buildings.

A **solution of a system of linear inequalities** is an ordered pair that satisfies each inequality in the system. The set of all such ordered pairs is the solution set of the system. Graphing this set gives us a picture of the solution set. We can graph a system of inequalities by graphing each inequality in the system and identifying the region of overlap.

GRAPHING THE SOLUTIONS OF A SYSTEM OF LINEAR INEQUALITIES

Step 1. Graph each inequality in the system on the same set of axes.

Step 2. The solutions of the system are the points common to the graphs of all the inequalities in the system.

Example 1 Graph the solutions of the system $\begin{cases} 3x \geq y \\ x + 2y \leq 8 \end{cases}$

Solution: We begin by graphing each inequality on the *same* set of axes. The graph of the solutions of the system is the region contained in the graphs of both inequalities. In other words, it is their intersection.

First let's graph $3x \geq y$. The boundary line is the graph of $3x = y$. We sketch a solid boundary line since the inequality $3x \geq y$ means $3x > y$ or $3x = y$. The test point $(1, 0)$ satisfies the inequality, so we shade the half-plane that includes $(1, 0)$.

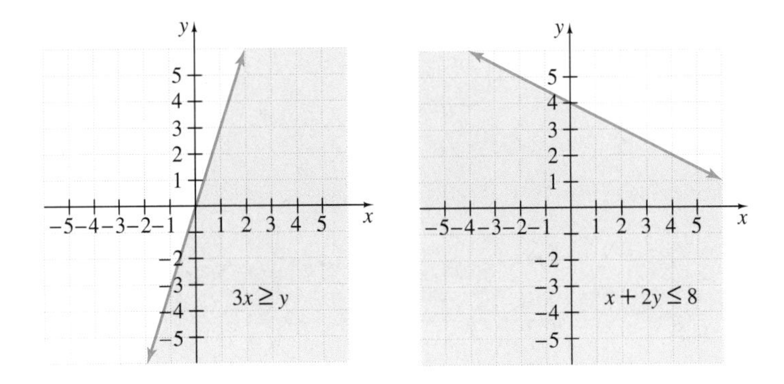

Next we sketch a solid boundary line $x + 2y = 8$ on the same set of axes. The test point $(0, 0)$ satisfies the inequality $x + 2y \leq 8$, so we shade the half-plane that includes $(0, 0)$. (For clarity, the graph of $x + 2y \leq 8$ is shown here on a separate set of axes.)

An ordered pair solution of the system must satisfy both inequalities. These solutions are points that lie in

Practice Problem 1

Graph the solutions of the system:

$\begin{cases} 2x \leq y \\ x + 4y \geq 4 \end{cases}$

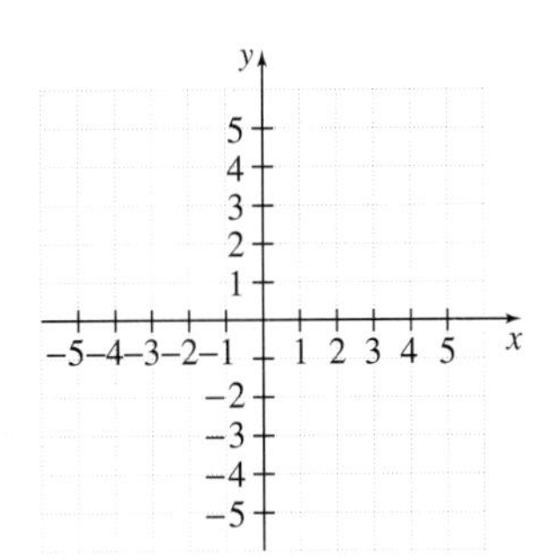

Answer

1.

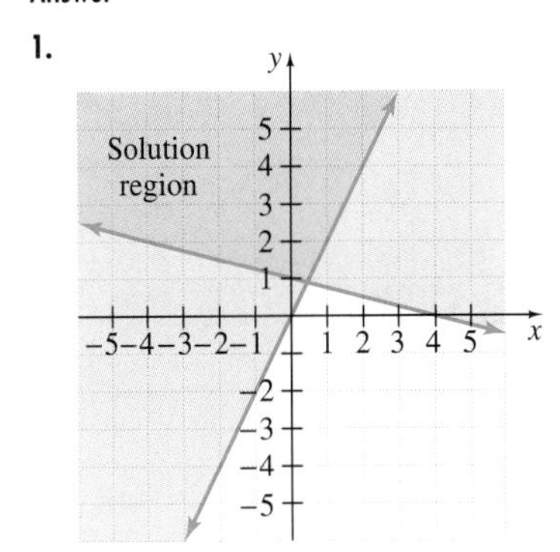

both shaded regions. The solution of the system is the darkest shaded region. This solution includes parts of both boundary lines.

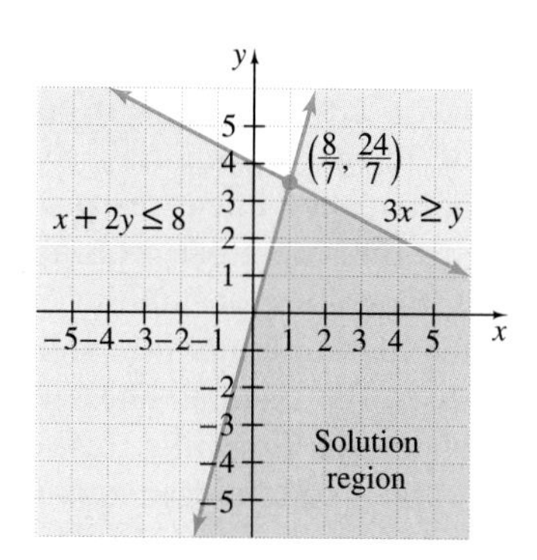

In linear programming, it is sometimes necessary to find the coordinates of the **corner point**: the point at which the two boundary lines intersect. To find the point of intersection for the system of Example 1, we solve the related linear system

$$\begin{cases} 3x = y \\ x + 2y = 8 \end{cases}$$

using either the subsitution or the elimination method. The lines intersect at $\left(\frac{8}{7}, \frac{24}{7}\right)$, the corner point of the graph.

Example 2

Graph the solutions of the system: $\begin{cases} x - y < 2 \\ x + 2y > -1 \\ y < 2 \end{cases}$

Solution: First we graph all three inequalities on the same set of axes. All boundary lines are dashed lines since the inequality symbols are $<$ and $>$. The solution of the system is the region shown by the darkest shading. In this example, the boundary lines are *not* a part of the solution.

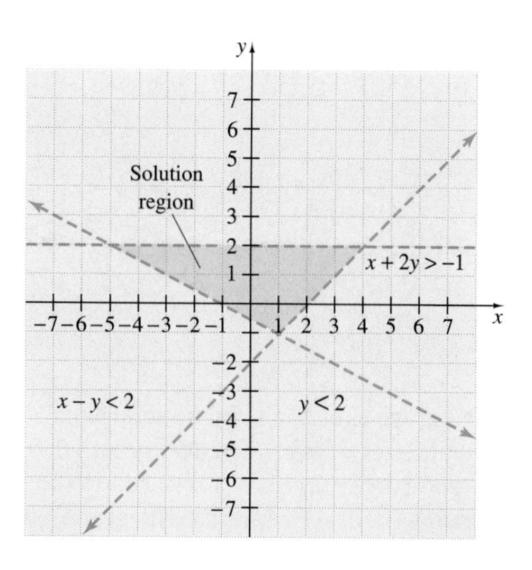

Practice Problem 2

Graph the solutions of the system:

$$\begin{cases} -x + y < 3 \\ y < 1 \\ 2x + y > -2 \end{cases}$$

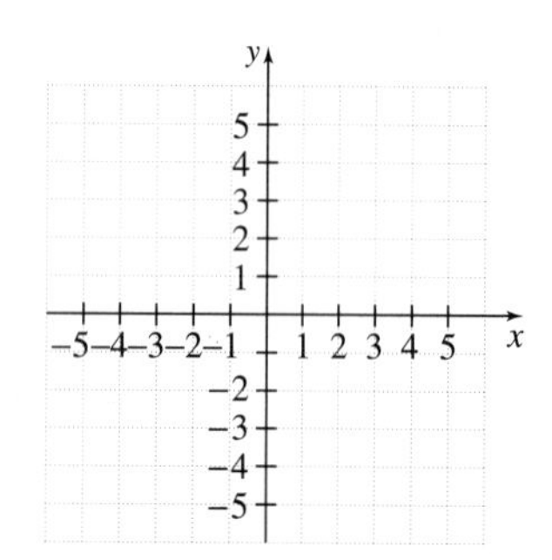

Answer

2.

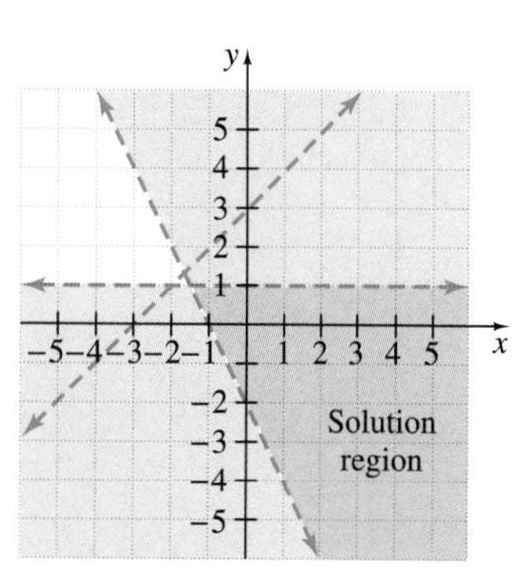

TRY THE CONCEPT CHECK IN THE MARGIN.

Example 3

Graph the solutions of the system $\begin{cases} -3x + 4y \leq 12 \\ x \leq 3 \\ x \geq 0 \\ y \geq 0 \end{cases}$

Solution: We graph the inequalities on the same set of axes. The intersection of the inequalities is the solution region. It is the only region shaded in this graph and includes portions of all four boundary lines.

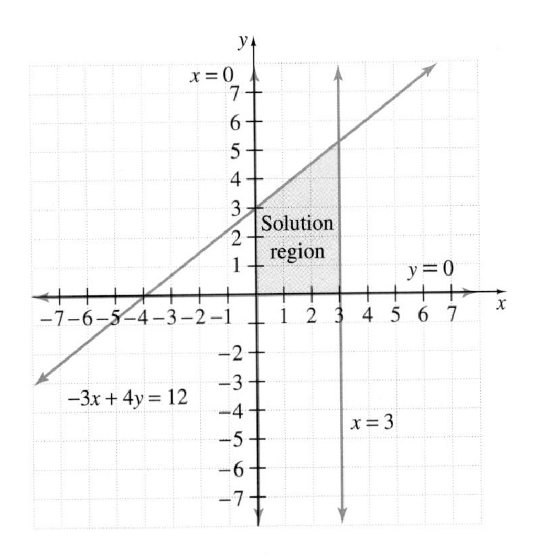

✓ CONCEPT CHECK

Describe the solution of the system of inequalities:

$\begin{cases} x \leq 2 \\ x \geq 2 \end{cases}$

Practice Problem 3

Graph the solutions of the system:

$\begin{cases} 2x - 3y \leq 6 \\ y \geq 0 \\ y \leq 4 \\ x \geq 0 \end{cases}$

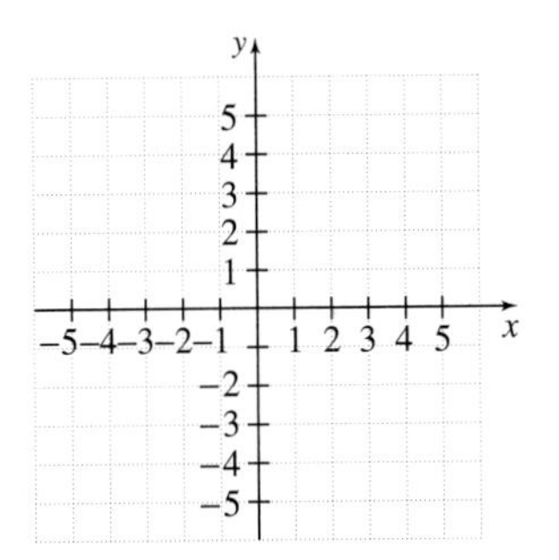

Answers

3.

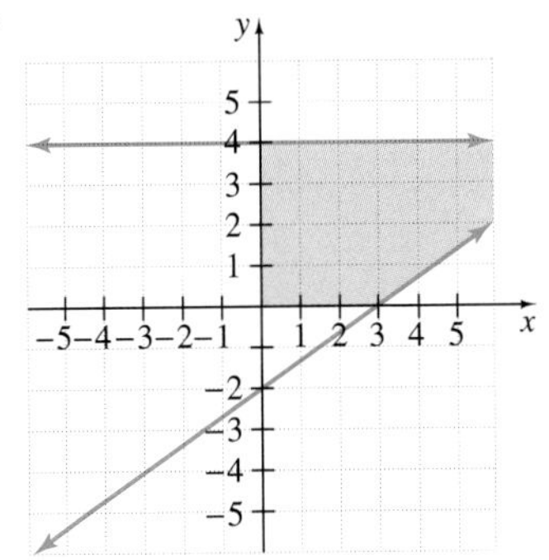

✓ Concept Check: It is the line $x = 2$.

Focus On the Real World

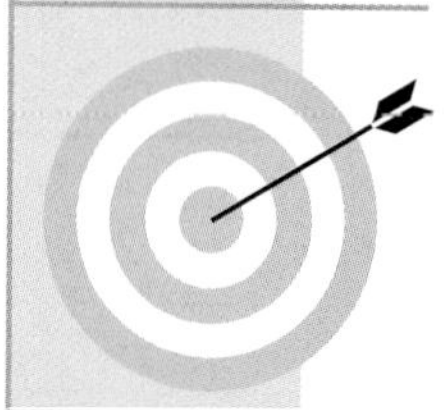

ANOTHER MATHEMATICAL MODEL

Sometimes mathematical models other than linear models are appropriate for data. Suppose that an equation of the form $y = ax^2 + bx + c$ is an appropriate model for the ordered pairs (x_1, y_1), (x_2, y_2), and (x_3, y_3). Then it is necessary to find the values of a, b, and c such that the given ordered pairs are solutions of the equation $y = ax^2 + bx + c$. To do so, substitute each ordered pair into the equation. Each time, the result is an equation in three unknowns: a, b, and c. Solving the resulting system of three linear equation in three unknowns will give the required values of a, b, and c.

GROUP ACTIVITY

1. The table gives the average annual fatalities from lightning in each of the years listed.
   a. Write the data as ordered pairs of the form (x, y) where y is the number of lightning fatalities in the year x ($x = 0$ represents 1900).
   b. Find the values of a, b, and c such that the equation $y = ax^2 + bx + c$ models this data.
   c. Verify that the model you found in part (b) gives each of the ordered pair solutions from part (a).
   d. According to the model, what was the average annual number of fatalities from lightning in 1955?

AVERAGE ANNUAL LIGHTNING FATALITIES

Year	Fatalities
1940	337
1950	184
1960	133

(*Source:* National Weather Service)

2. The table gives the world production of red meat (in millions of metric tons) for each of the years listed.
   a. Write the data as ordered pairs of the form (x, y) where y is the red meat production (in millions of metric tons) in the year x ($x = 0$ represents 1990).
   b. Find the values of a, b, and c such that the equation $y = ax^2 + bx + c$ models this data.
   c. According to the model, what is the world production of red meat in 1999?

WORLD PRODUCTION OF RED MEAT

Year	Millions of metric tons
1993	119.3
1996	135.5
1998	140.1

(*Source:* Economic Research Service, U.S. Department of Agriculture)

3. a. Make up an equation of the form $y = ax^2 + bx + c$.
   b. Find three ordered pair solutions of the equation.
   c. Without revealing your equation from part (a), exchange lists of ordered pair solutions with another group.
   d. Use the method described above to find the values of a, b, and c such that the equation $y = ax^2 + bx + c$ has the ordered pair solutions you received from the other group.
   e. Check with the other group to see if your equation from part (d) is the correct one.

Name ____________________ Section ______ Date ________

Exercise Set 4.6

A *Graph the solutions of each system of linear inequalities. See Examples 1 through 3.*

1. $\begin{cases} y \geq x + 1 \\ y \geq 3 - x \end{cases}$

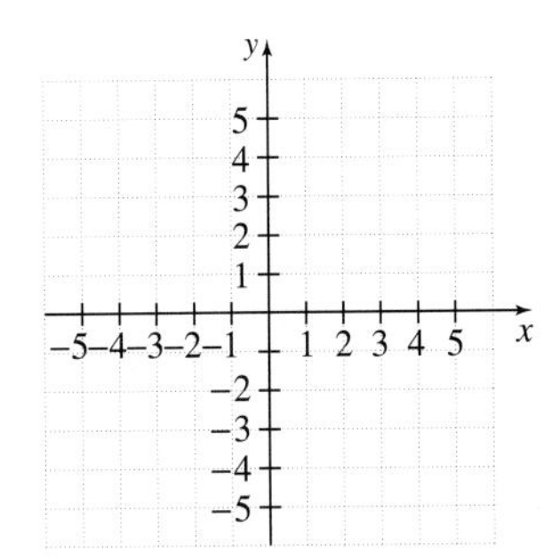

2. $\begin{cases} y \geq x - 3 \\ y \geq -1 - x \end{cases}$

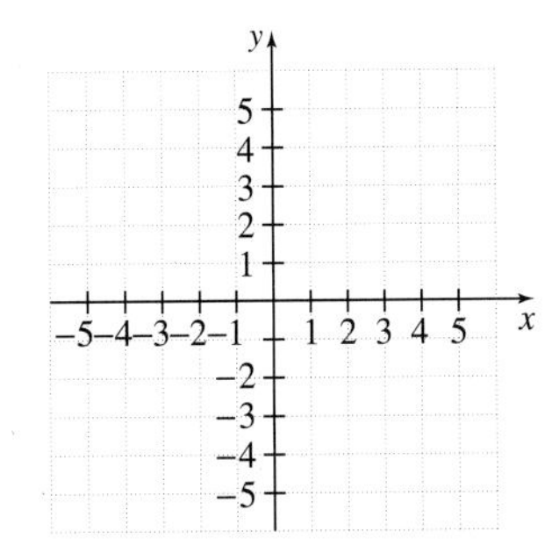

3. $\begin{cases} y < 3x - 4 \\ y \leq x + 2 \end{cases}$

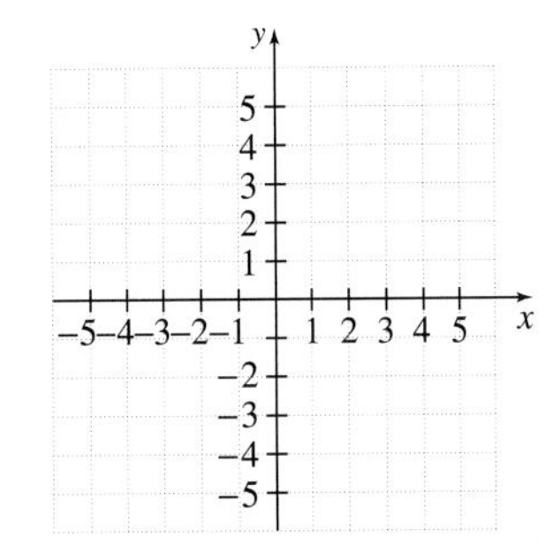

4. $\begin{cases} y \leq 2x + 1 \\ y > x + 2 \end{cases}$

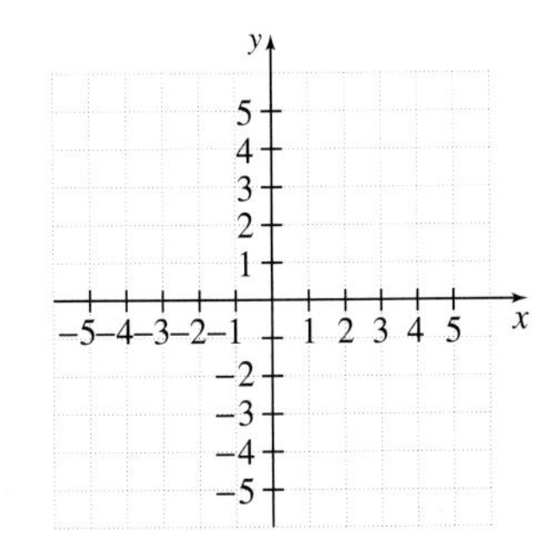

5. $\begin{cases} y \leq -2x - 2 \\ y \geq x + 4 \end{cases}$

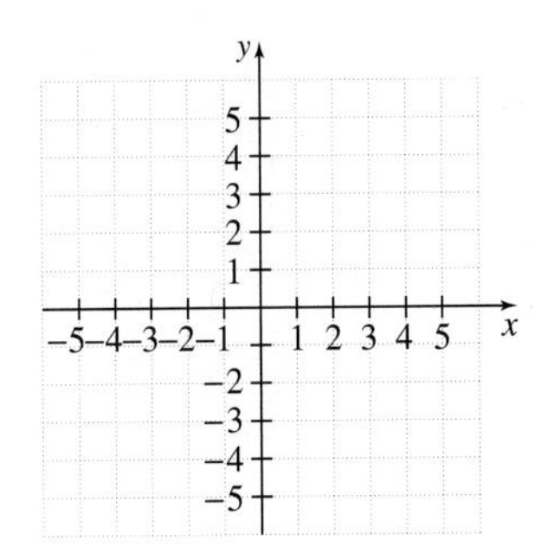

6. $\begin{cases} y \leq 2x + 4 \\ y \geq -x - 5 \end{cases}$

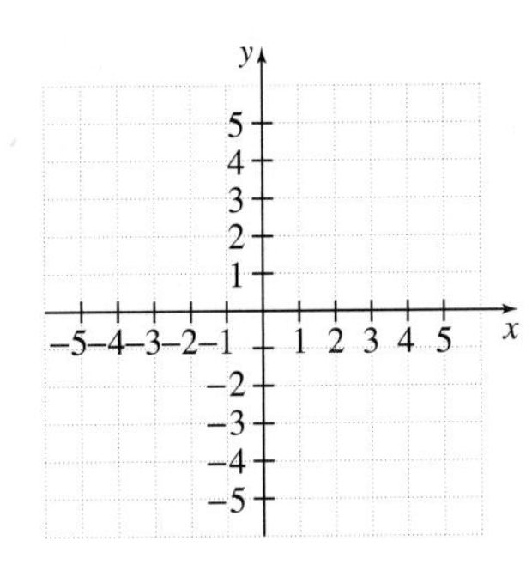

7. $\begin{cases} y \geq -x + 2 \\ y \leq 2x + 5 \end{cases}$

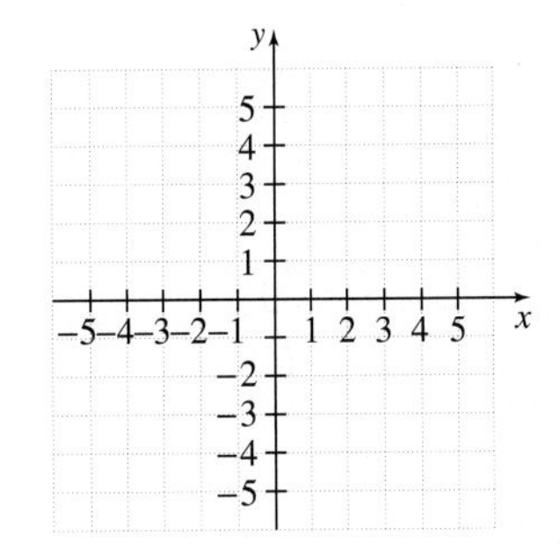

8. $\begin{cases} y \geq x - 5 \\ y \leq -3x + 3 \end{cases}$

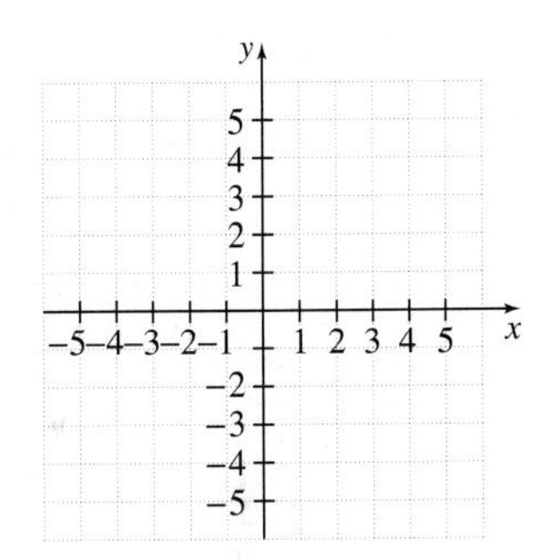

9. $\begin{cases} x \geq 3y \\ x + 3y \leq 6 \end{cases}$

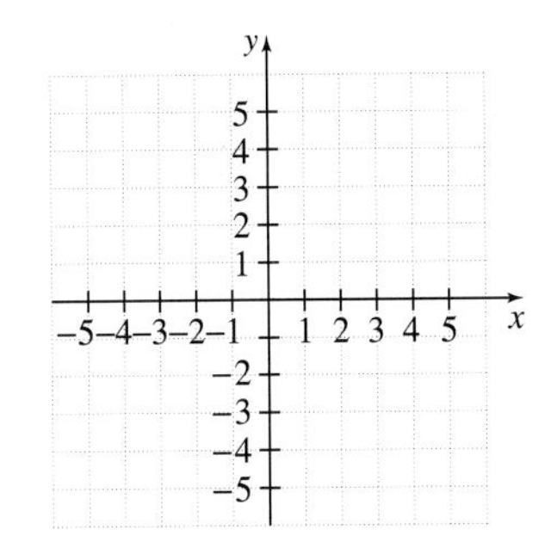

10. $\begin{cases} -2x < y \\ x + 2y < 3 \end{cases}$

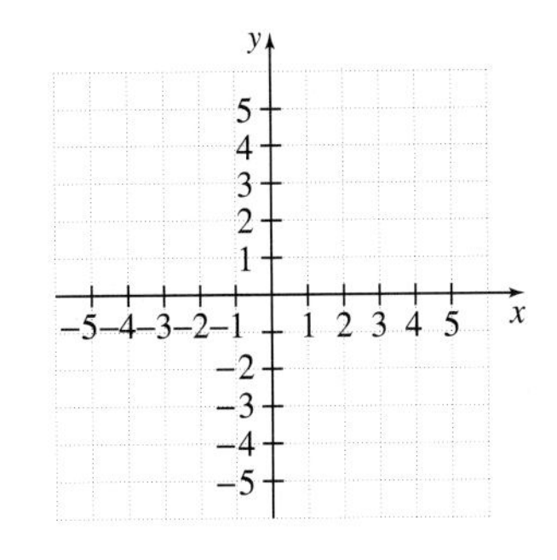

11. $\begin{cases} x \leq 2 \\ y \geq -3 \end{cases}$

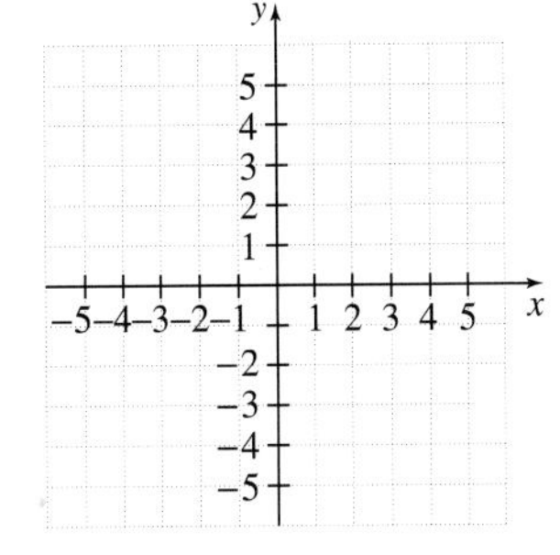

12. $\begin{cases} x \geq -3 \\ y \geq -2 \end{cases}$

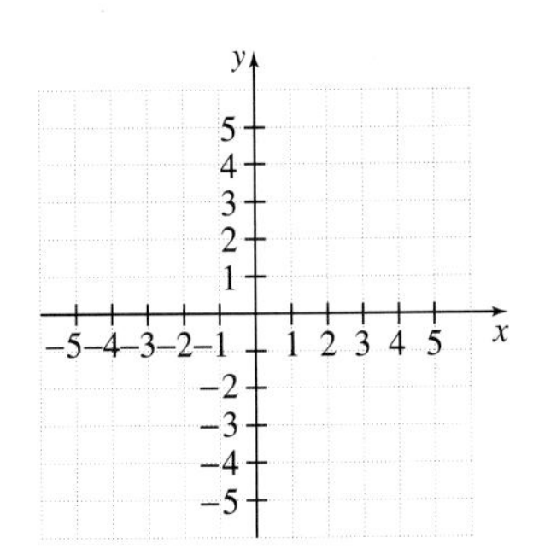

Name ______________________________

13. $\begin{cases} y \geq 1 \\ x < -3 \end{cases}$

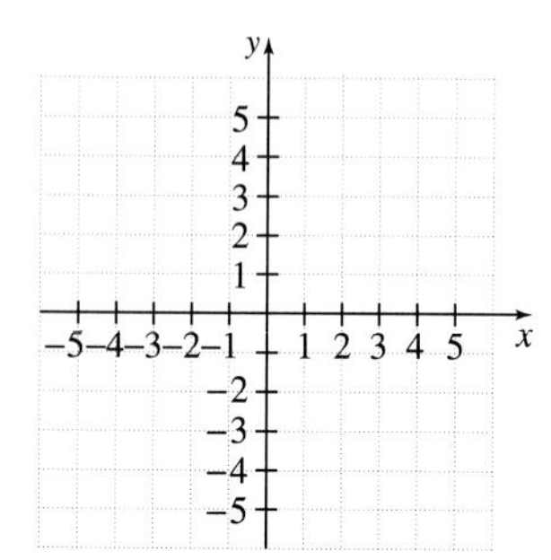

14. $\begin{cases} y > 2 \\ x \geq -1 \end{cases}$

15. $\begin{cases} y + 2x \geq 0 \\ 5x - 3y \leq 12 \\ y \leq 2 \end{cases}$

16. $\begin{cases} y + 2x \leq 0 \\ 5x + 3y \geq -2 \\ y \leq 4 \end{cases}$

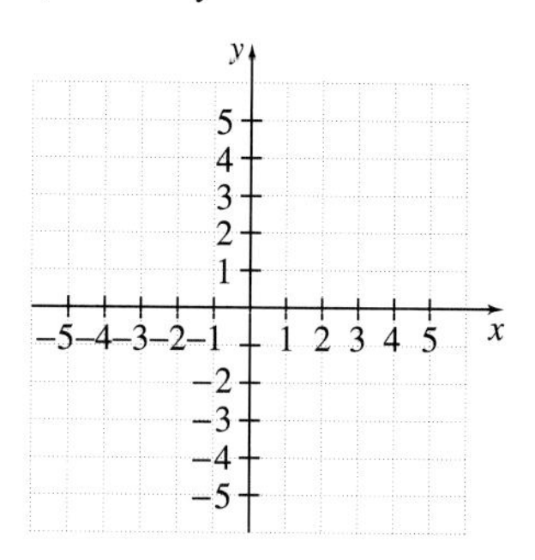

17. $\begin{cases} 3x - 4y \geq -6 \\ 2x + y \leq 7 \\ y \geq -3 \end{cases}$

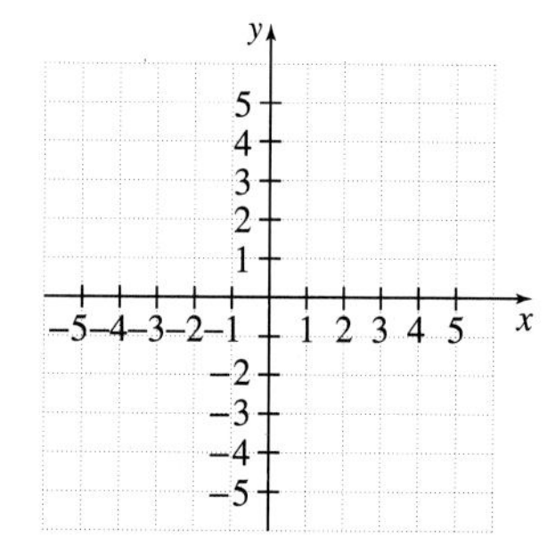

18. $\begin{cases} 4x - y \geq -2 \\ 2x + 3y \leq -8 \\ y \geq -5 \end{cases}$

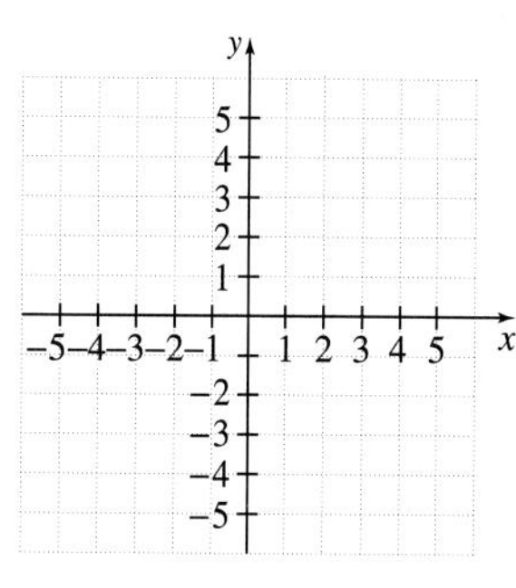

19. $\begin{cases} 2x + y \leq 5 \\ x \leq 3 \\ x \geq 0 \\ y \geq 0 \end{cases}$

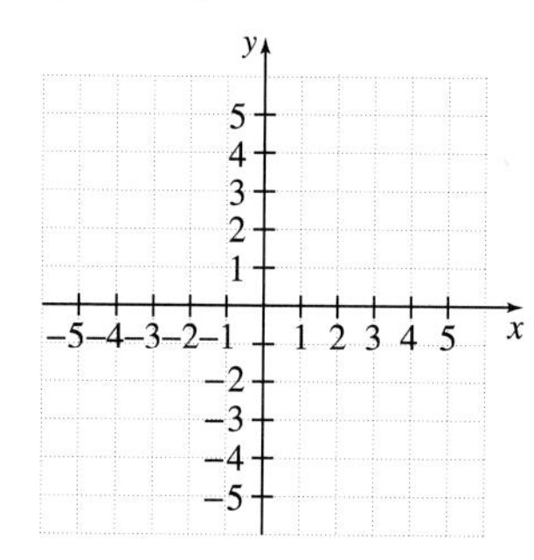

20. $\begin{cases} 3x + y \leq 4 \\ x \leq 4 \\ x \geq 0 \\ y \geq 0 \end{cases}$

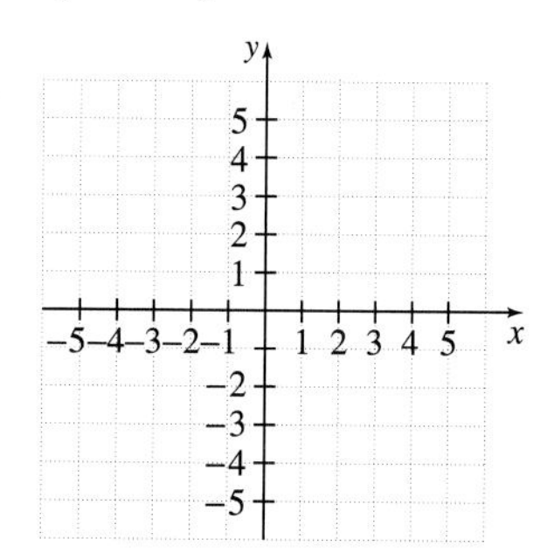

Match each system of inequalities to the corresponding graph.

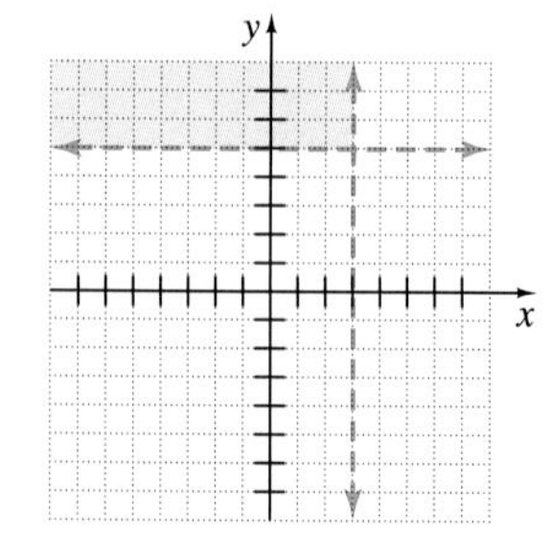

A

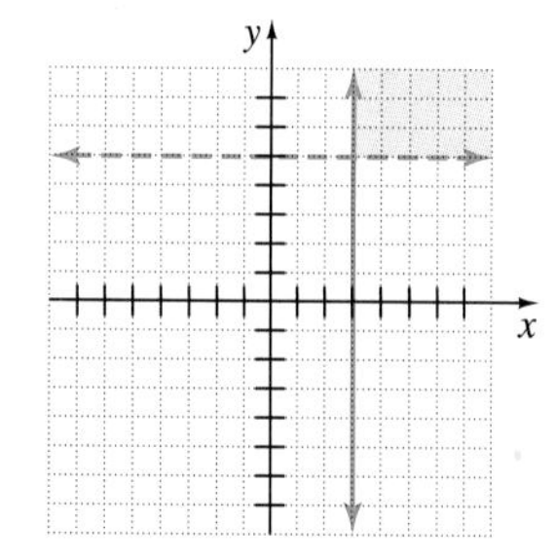

B

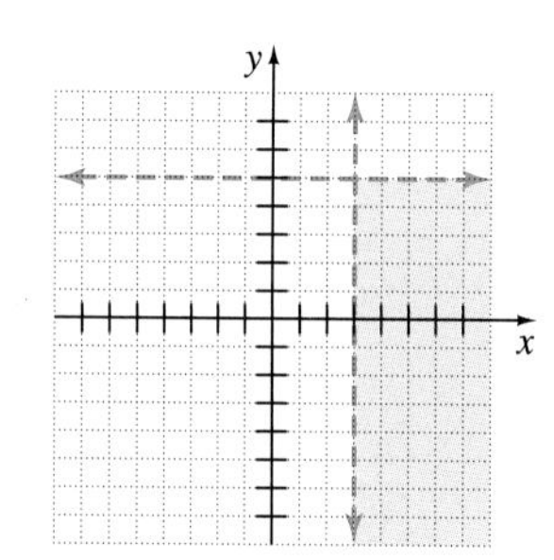

C

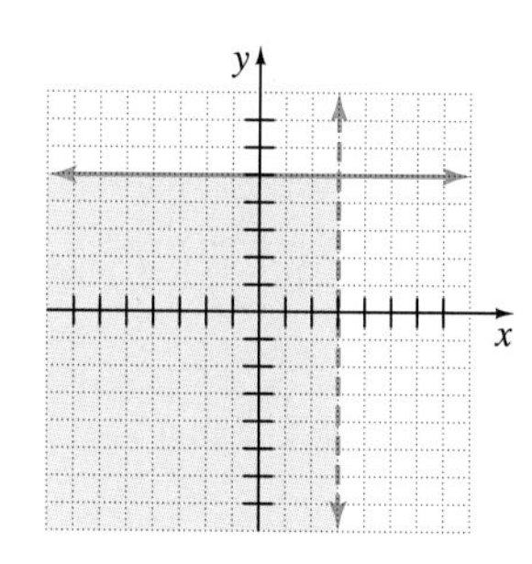

D

Name ______________________________

21. $\begin{cases} y < 5 \\ x > 3 \end{cases}$

22. $\begin{cases} y > 5 \\ x < 3 \end{cases}$

23. $\begin{cases} y \leq 5 \\ x < 3 \end{cases}$

24. $\begin{cases} y > 5 \\ x \geq 3 \end{cases}$

REVIEW AND PREVIEW

Evaluate each expression. See Section 1.6.

25. $(-3)^2$

26. $(-5)^3$

27. $\left(\frac{2}{3}\right)^2$

28. $\left(\frac{3}{4}\right)^3$

Perform each indicated operation.

29. $(-2)^2 - (-3) + 2(-1)$

30. $5^2 - 11 + 3(-5)$

31. $8^2 + (-13) - 4(-2)$

32. $(-12)^2 + (-1)(2) - 6$

COMBINING CONCEPTS

33. Tony Noellert budgets his time at work today. Part of the day he can write bills; the rest of the day he can use to write purchase orders. The total time available is at most 8 hours. Less than 3 hours is to be spent writing bills.

a. Write a system of inequalities to describe the situation. (Let $x =$ hours available for writing bills and $y =$ hours available for writing purchase orders.)

b. Graph the solutions of the system.

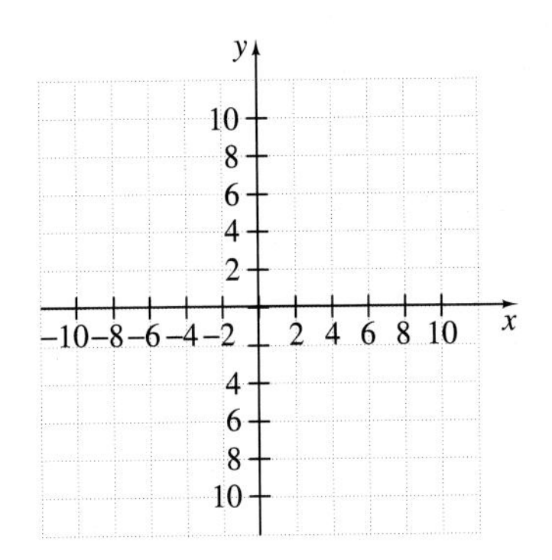

34. Explain how to decide which region to shade to show the solution region of the following system.

$\begin{cases} x \geq 3 \\ y \geq -2 \end{cases}$

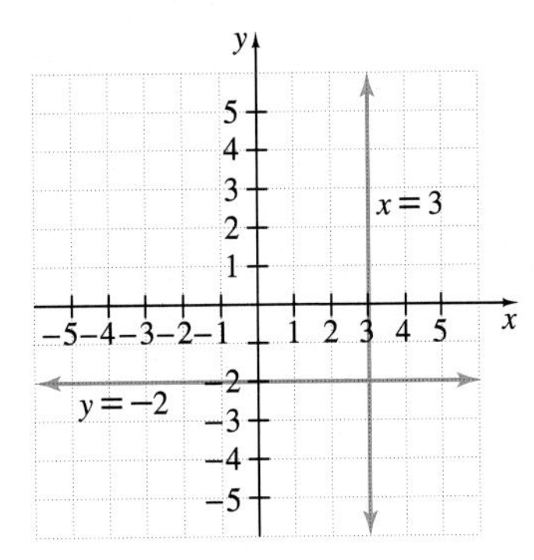

ANSWERS

21. ______

22. ______

23. ______

24. ______

25. ______

26. ______

27. ______

28. ______

29. ______

30. ______

31. ______

32. ______

33. a. ______

b. see graph

34. ______

Focus On the Real World

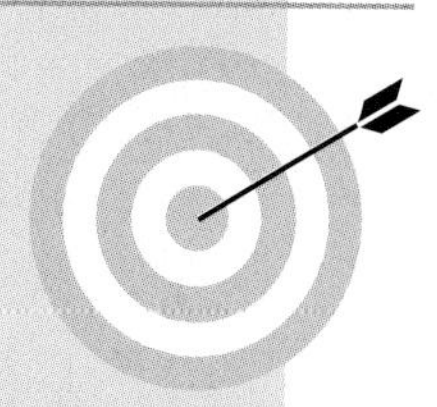

LINEAR MODELING

In Chapter 3, we learned several ways to find a linear model when given either two ordered pairs or an ordered pair and slope. Another way to find a linear model of the form $y = mx + b$ for two ordered pairs (x_1, y_1) and (x_2, y_2) is to solve the following system of linear equations for m and b:

$$\begin{cases} y_1 = mx_1 + b \\ y_2 = mx_2 + b \end{cases}$$

For example, suppose a researcher wishes to find a linear model for the number of traffic accidents involving teenagers. The researcher locates statistics stating that there were 5215 teenage deaths from motor vehicle accidents in the United States in 1992. By 1996, this number had increased to 5805 teenage deaths in motor vehicle accidents. (*Source:* Insurance Institute for Highway Safety)

This data gives two ordered pairs: (1992, 5215) and (1996, 5805). Alternatively, the ordered pairs could be written as (2, 5215) and (6, 5805), where the x-coordinate represents the number of years after 1990. (Adjusting data given as years in this way often simplifies calculations.) By substituting the coordinates of the second set of ordered pairs into the general linear system, we obtain the system

$$\begin{cases} 5215 = 2m + b \\ 5805 = 6m + b \end{cases}$$

The solution of this system is $m = 147.5$ and $b = 4920$. We can use these values to write the model the researcher wished to find: $y = 147.5x + 4920$, where y is the number of teenage deaths in motor vehicle accidents x years after 1990.

Internet Excursions

Go to http://www.prenhall.com/martin-gay

The Insurance Institute for Highway Safety (IIHS) is a nonprofit research organization supported by automobile insurers that collects, studies, and distributes motor vehicle safety statistics. This page offers access to a wide variety of traffic fatality statistics organized by categories such as alcohol-related fatalities, fatalities of children, and fatalities involving bicycles. (Alternatively, you can visit the IIHS homepage at http://www.hwysafety.org and look for the Fatality Facts option.)

1. Browse the list to find a set of data that interests you. Make a list of the ordered pairs that make up the set of data.
2. Create a scatter diagram of the data. Does the data appear approximately linear? If not, is there a portion of the data that appears approximately linear? If so, indicate which portion is approximately linear. If not, start over with Question #1.
3. Pick two ordered pairs from the linear portion of the data. Use these ordered pairs to form a system of linear equations.
4. Solve the system from Question #3. Find the linear equation that models your data.
5. Add the graph of your linear model to the scatter diagram from Question #2. How well does your model "fit" the data?
6. What trend does your model describe over the linear portion of your data?

Name ______________________ Section ________ Date ________

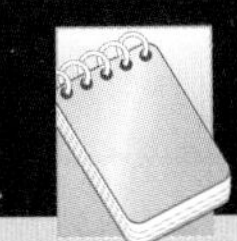

Chapter 4 Activity
Locating Lightning Strikes

MATERIALS:
- ▲ calculator
- ▲ graphing utility (optional)

This activity may be completed by working in groups or individually.

Weather-recording stations use a directional antenna to detect and measure the electromagnetic field emitted by a lightning bolt. The antenna can determine the angle between a fixed point and the position of the lightning strike but cannot determine the distance to the lightning strike. However, the angle measured by the antenna can be used to find the slope of the line connecting the positions of the weather station and the lightning strike. From there, the equation of the line connecting the positions of the weather station and the lightning strike may be found. If two such lines may be found—that is, if another weather station's antenna detects the same lightning flash—the coordinates of the lightning strike's position may be pinpointed.

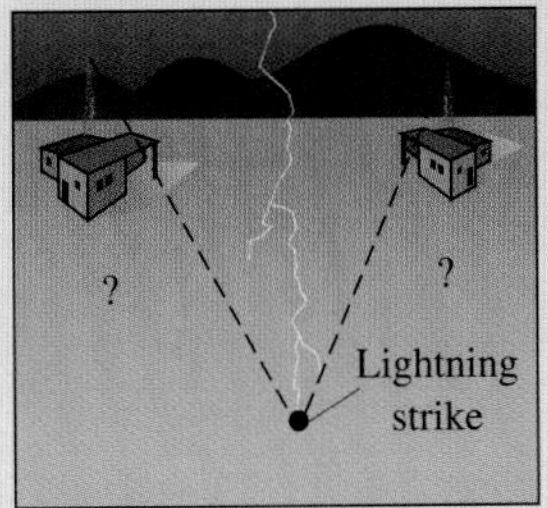

1. A weather-recording station A is located at the coordinates $(35, 28)$. A second weather-recording station B is located at the coordinates $(52, 12)$. Plot the position of the two weather-recording stations.

2. A lightning strike is detected by both stations. Station A uses a measured angle to find the slope of the line from the station to the lightning strike as $m = -1.732$. Station B computes a slope of $m = 0.577$ from the angle it measured. Use this information to find the equations of the lines connecting each station to the position of the lightning strike.

3. Solve the resulting system of equations in each of the following ways (or work with other students in your class so that each student solves the system in one of the following ways):
   - **a.** Using a graph. Graph the two equations on your plot of the positions of the two weather-recording stations. Estimate the coordinates of their point of intersection.
   - **b.** Using either the method of substitution or of elimination (whichever you prefer)
   - **c.** Using matrices
   - **d.** Using Cramer's rule
   - **e.** (Optional) Using a graphing utility to graph the lines and use an intersect feature to estimate the coordinates of their point of intersection.

4. Compare the results from each method. What are the coordinates of the lightning strike? Which method do you prefer? Why?

CHAPTER 4 HIGHLIGHTS

DEFINITIONS AND CONCEPTS	EXAMPLES
SECTION 4.1 SOLVING SYSTEMS OF LINEAR EQUATIONS IN TWO VARIABLES	
A **system of linear equations** consists of two or more linear equations.	$\begin{cases} x - 3y = 6 \\ y = \frac{1}{2}x \end{cases}$ $\begin{cases} x + 2y - z = 1 \\ 3x - y + 4z = 0 \\ 5y + z = 6 \end{cases}$
A **solution** of a system of two equations in two variables is an ordered pair (x, y) that makes both equations true.	Determine whether $(2, -5)$ is a solution of the system. $\begin{cases} x + y = -3 \\ 2x - 3y = 19 \end{cases}$ Replace x with 2 and y with -5 in both equations. $x + y = -3$; $2 + (-5) = -3$; $-3 = -3$ True. $2x - 3y = 19$; $2(2) - 3(-5) = 19$; $4 + 15 = 19$; $19 = 19$ True. $(2, -5)$ is a solution of the system.
Geometrically, a solution of a system in two variables is a point common to the graphs of the equations.	Solve by graphing: $\begin{cases} y = 2x - 1 \\ x + 2y = 13 \end{cases}$ 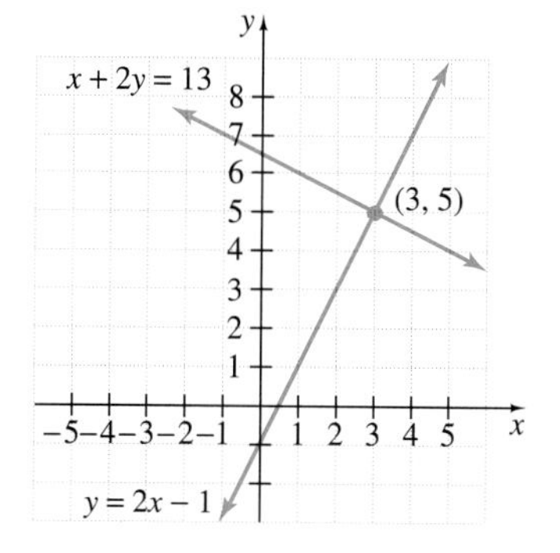
A system of equations with at least one solution is a **consistent system**. A system that has no solution is an **inconsistent system**. If the graphs of two linear equations are identical, the equations are **dependent**. If their graphs are different, the equations are **independent**.	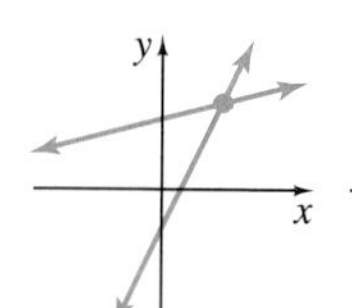 One solution; consistent 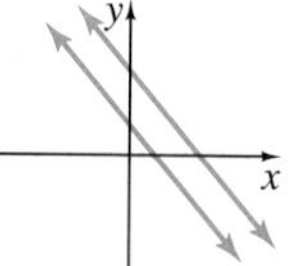 No solution; inconsistent 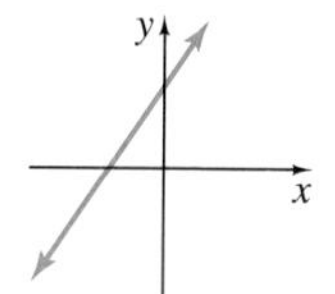 Infinite number of solutions; dependent

Section 4.1 (continued)

Solving a System of Linear Equations by the Substitution Method

Step 1. Solve one equation for a variable.
Step 2. Substitute the expression for the variable into the other equation.
Step 3. Solve the equation from Step 2 to find the value of one variable.
Step 4. Substitute the value from Step 3 in either original equation to find the value of the other variable.
Step 5. Check the solution in both equations.

Solve by substitution:

$$\begin{cases} y = x + 2 \\ 3x - 2y = -5 \end{cases}$$

Substitute $x + 2$ for y in the second equation.

$$\begin{aligned} 3x - 2y &= -5 && \text{Second equation} \\ 3x - 2(x + 2) &= -5 && \text{Let } y = x + 2. \\ 3x - 2x - 4 &= -5 \\ x - 4 &= -5 && \text{Simplify.} \\ x &= -1 && \text{Add 4.} \end{aligned}$$

To find y, let $x = -1$ in $y = x + 2$, so $y = -1 + 2 = 1$. The solution $(-1, 1)$ checks.

Solving a System of Linear Equations by the Elimination Method

Step 1. Rewrite each equation in standard form, $Ax + By = C$.
Step 2. Multiply one or both equations by a nonzero number so that the coefficients of a variable are opposites.
Step 3. Add the equations.
Step 4. Find the value of one variable by solving the resulting equation.
Step 5. Substitute the value from Step 4 into either original equation to find the value of the other variable.
Step 6. Check the solution in both equations.

Solve by elimination:

$$\begin{cases} x - 3y = -3 \\ -2x + y = 6 \end{cases}$$

Multiply both sides of the first equation by 2.

$$\begin{aligned} 2x - 6y &= -6 \\ -2x + y &= 6 \\ \hline -5y &= 0 && \text{Add.} \\ y &= 0 && \text{Divide by } -5. \end{aligned}$$

To find x, let $y = 0$ in an original equation.

$$\begin{aligned} x - 3y &= -3 \\ x - 3 \cdot 0 &= -3 \\ x &= -3 \end{aligned}$$

The solution $(-3, 0)$ checks.

Section 4.2 Solving Systems of Linear Equations in Three Variables

A **solution** of an equation in three variables x, y and z is an **ordered triple** (x, y, z) that makes the equation a true statement.

Verify that $(-2, 1, 3)$ is a solution of $2x + 3y - 2z = -7$. Replace x with -2, y with 1, and z with 3.

$$\begin{aligned} 2(-2) + 3(1) - 2(3) &= -7 \\ -4 + 3 - 6 &= -7 \\ -7 &= -7 && \text{True.} \end{aligned}$$

$(-2, 1, 3)$ is a solution.

SECTION 4.2 (CONTINUED)

SOLVING A SYSTEM OF THREE LINEAR EQUATIONS BY THE ELIMINATION METHOD

Step 1. Write each equation in standard form, $Ax + By + Cz = D$.
Step 2. Choose a pair of equations and use the equations to eliminate a variable.
Step 3. Choose any other pair of equations and eliminate the same variable.
Step 4. Solve the system of two equations in two variables from Steps 2 and 3.
Step 5. Solve for the third variable by substituting the values of the variables from Step 4 into any of the original equations.
Step 6. Check the solution in all three original equations.

Solve:

$$\begin{cases} 2x + y - z = 0 & (1) \\ x - y - 2z = -6 & (2) \\ -3x - 2y + 3z = -22 & (3) \end{cases}$$

1. Each equation is written in standard form.

2.
$$\begin{array}{rl} 2x + y - z = 0 & (1) \\ x - y - 2z = -6 & (2) \\ \hline 3x \quad - 3z = -6 & (4) \text{ Add.} \end{array}$$

3. Eliminate y from equations (1) and (3) also.

$$\begin{array}{rll} 4x + 2y - 2z = 0 & & \text{Multiply equation} \\ -3x - 2y + 3z = -22 & (3) & \text{(1) by 2.} \\ \hline x \quad + z = -22 & (5) & \text{Add.} \end{array}$$

4. Solve.

$$\begin{cases} 3x - 3z = -6 & (4) \\ x + z = -22 & (5) \end{cases}$$

$$\begin{array}{rll} x - z = -2 & & \text{Divide equation (4) by 3.} \\ x + z = -22 & (5) & \\ \hline 2x \quad = -24 & & \\ x \quad = -12 & & \end{array}$$

To find z, use equation (5).

$$\begin{aligned} x + z &= -22 \\ -12 + z &= -22 \\ z &= -10 \end{aligned}$$

5. To find y, use equation (1).

$$\begin{aligned} 2x + y - z &= 0 \\ 2(-12) + y - (-10) &= 0 \\ -24 + y + 10 &= 0 \\ y &= 14 \end{aligned}$$

The solution $(-12, 14, -10)$ checks.

SECTION 4.3 SYSTEMS OF EQUATIONS AND PROBLEM SOLVING

1. UNDERSTAND the problem.

Two numbers have a sum of 11. Twice one number is 3 less than 3 times the other. Find the numbers.

1. Read and reread.

x = one number

y = other number

Section 4.3 (continued)

2. TRANSLATE.	**2.** In words: sum of numbers ↓ is ↓ 11 ↓ Translate: $x + y = 11$ In words: twice one number ↓ is ↓ 3 less than 3 times the other number ↓ Translate: $2x = 3y - 3$
3. SOLVE.	**3.** Solve the system: $\begin{cases} x + y = 11 \\ 2x = 3y - 3 \end{cases}$ In the first equation $x = 11 - y$. Substitute into the other equation. $2x = 3y - 3$ $2(11 - y) = 3y - 3$ $22 - 2y = 3y - 3$ $-5y = -25$ $y = 5$ Replace y with 5 in the equation $x = 11 - y$. Then $x = 11 - 5 = 6$. The solution is $(6, 5)$.
4. INTERPRET.	**4.** *Check*: See that $6 + 5 = 11$ is the required sum and that twice 6 is 3 times 5 less 3. *State*: The numbers are 6 and 5.

Section 4.4 Solving Systems of Equations Using Matrices

A **matrix** is a rectangular array of numbers.	$\begin{bmatrix} -7 & 0 & 3 \\ 1 & 2 & 4 \end{bmatrix}$ $\begin{bmatrix} a & b & c \\ d & e & f \\ g & h & i \end{bmatrix}$
The **corresponding matrix of the system** is obtained by writing a matrix composed of the coefficients of the variables and the constants of the system.	The corresponding matrix of the system $\begin{cases} x - y = 1 \\ 2x + y = 11 \end{cases}$ is $\left[\begin{array}{cc\|c} 1 & -1 & 1 \\ 2 & 1 & 11 \end{array}\right]$

SECTION 4.4 (CONTINUED)

The following **row operations** can be performed on matrices, and the result is an equivalent matrix.

Elementary row operations:

1. Interchange any two rows.
2. Multiply (or divide) the elements of one row by the same nonzero number.
3. Multiply (or divide) the elements of one row by the same nonzero number and add to its corresponding elements in any other row.

Use matrices to solve: $\begin{cases} x - y = 1 \\ 2x + y = 11 \end{cases}$

The corresponding matrix is

$$\left[\begin{array}{cc|c} 1 & -1 & 1 \\ 2 & 1 & 11 \end{array}\right]$$

Use row operations to write an equivalent matrix with 1s along the diagonal and 0s below each 1 in the diagonal. Multiply row 1 by -2 and add to row 2. Change row 2 only.

$$\left[\begin{array}{cc|c} 1 & -1 & 1 \\ -2(1)+2 & -2(-1)+1 & -2(1)+11 \end{array}\right]$$

simplifies to $\left[\begin{array}{cc|c} 1 & -1 & 1 \\ 0 & 3 & 9 \end{array}\right]$

Divide row 2 by 3.

$$\left[\begin{array}{cc|c} 1 & -1 & 1 \\ \frac{0}{3} & \frac{3}{3} & \frac{9}{3} \end{array}\right] \text{ simplifies to } \left[\begin{array}{cc|c} 1 & -1 & 1 \\ 0 & 1 & 3 \end{array}\right]$$

This matrix corresponds to the system

$$\begin{cases} x - y = 1 \\ \quad\; y = 3 \end{cases}$$

Let $y = 3$ in the first equation.

$$x - 3 = 1$$
$$x = 4$$

The ordered pair solution is $(4, 3)$.

SECTION 4.5 SOLVING SYSTEMS OF EQUATIONS USING DETERMINANTS

A **square matrix** is a matrix with the same number of rows and columns.

$$\begin{bmatrix} -2 & 1 \\ 6 & 8 \end{bmatrix} \quad \begin{bmatrix} 4 & -1 & 6 \\ 0 & 2 & 5 \\ 1 & 1 & 2 \end{bmatrix}$$

A **determinant** is a real number associated with a square matrix. To denote the determinant, place vertical bars about the array of numbers.

The determinant of $\begin{bmatrix} -2 & 1 \\ 6 & 8 \end{bmatrix}$ is $\begin{vmatrix} -2 & 1 \\ 6 & 8 \end{vmatrix}$.

The determinant of a 2×2 matrix is

$$\begin{vmatrix} a & b \\ c & d \end{vmatrix} = ad - bc$$

$$\begin{vmatrix} -2 & 1 \\ 6 & 8 \end{vmatrix} = -2 \cdot 8 - 1 \cdot 6 = -22$$

Section 4.5 (continued)

Cramer's Rule for Two Linear Equations in Two Variables

The solution of the system $\begin{cases} ax + by = h \\ cx + dy = k \end{cases}$ is given by

$$x = \frac{\begin{vmatrix} h & b \\ k & d \end{vmatrix}}{\begin{vmatrix} a & b \\ c & d \end{vmatrix}} = \frac{D_x}{D} \qquad y = \frac{\begin{vmatrix} a & h \\ c & k \end{vmatrix}}{\begin{vmatrix} a & b \\ c & d \end{vmatrix}} = \frac{D_y}{D}$$

as long as $D = ad - bc$ is not 0.

Use Cramer's rule to solve

$$\begin{cases} 3x + 2y = 8 \\ 2x - y = -11 \end{cases}$$

$$D = \begin{vmatrix} 3 & 2 \\ 2 & -1 \end{vmatrix} = 3(-1) - 2(2) = -7$$

$$D_x = \begin{vmatrix} 8 & 2 \\ -11 & -1 \end{vmatrix} = 8(-1) - 2(-11) = 14$$

$$D_y = \begin{vmatrix} 3 & 8 \\ 2 & -11 \end{vmatrix} = 3(-11) - 8(2) = -49$$

$$x = \frac{D_x}{D} = \frac{14}{-7} = -2 \qquad y = \frac{D_y}{D} = \frac{-49}{-7} = 7$$

The ordered pair solution is $(-2, 7)$.

Determinant of a 3 × 3 Matrix

$$\begin{vmatrix} a_1 & b_1 & c_1 \\ a_2 & b_2 & c_2 \\ a_3 & b_3 & c_3 \end{vmatrix} = a_1 \cdot \begin{vmatrix} b_2 & c_2 \\ b_3 & c_3 \end{vmatrix} - a_2 \cdot \begin{vmatrix} b_1 & c_1 \\ b_3 & c_3 \end{vmatrix} + a_3 \cdot \begin{vmatrix} b_1 & c_1 \\ b_2 & c_2 \end{vmatrix}$$

Each 2 × 2 matrix above is called a **minor**.

$$\begin{vmatrix} 0 & 2 & -1 \\ 5 & 3 & 0 \\ 2 & -2 & 4 \end{vmatrix} = 0\begin{vmatrix} 3 & 0 \\ -2 & 4 \end{vmatrix} - 2\begin{vmatrix} 5 & 0 \\ 2 & 4 \end{vmatrix} + (-1)\begin{vmatrix} 5 & 3 \\ 2 & -2 \end{vmatrix}$$

$$= 0(12 - 0) - 2(20 - 0) - 1(-10 - 6)$$

$$= 0 - 40 + 16 = -24$$

Cramer's Rule for Three Equations in Three Variables

The solution of the system $\begin{cases} a_1x + b_1y + c_1z = k_1 \\ a_2x + b_2y + c_2z = k_2 \\ a_3x + b_3y + c_3z = k_3 \end{cases}$

is given by

$$x = \frac{D_x}{D} \quad y = \frac{D_y}{D} \quad \text{and} \quad z = \frac{D_z}{D}$$

where

$$D = \begin{vmatrix} a_1 & b_1 & c_1 \\ a_2 & b_2 & c_2 \\ a_3 & b_3 & c_3 \end{vmatrix} \qquad D_x = \begin{vmatrix} k_1 & b_1 & c_1 \\ k_2 & b_2 & c_2 \\ k_3 & b_3 & c_3 \end{vmatrix}$$

$$D_y = \begin{vmatrix} a_1 & k_1 & c_1 \\ a_2 & k_2 & c_2 \\ a_3 & k_3 & c_3 \end{vmatrix} \qquad D_z = \begin{vmatrix} a_1 & b_1 & k_1 \\ a_2 & b_2 & k_2 \\ a_3 & b_3 & k_3 \end{vmatrix}$$

as long as D is not 0.

Use Cramer's rule to solve

$$\begin{cases} 3y + 2z = 8 \\ x + y + z = 3 \\ 2x - y + z = 2 \end{cases}$$

$$D = \begin{vmatrix} 0 & 3 & 2 \\ 1 & 1 & 1 \\ 2 & -1 & 1 \end{vmatrix} = -3$$

$$D_x = \begin{vmatrix} 8 & 3 & 2 \\ 3 & 1 & 1 \\ 2 & -1 & 1 \end{vmatrix} = 3$$

$$D_y = \begin{vmatrix} 0 & 8 & 2 \\ 1 & 3 & 1 \\ 2 & 2 & 1 \end{vmatrix} = 0$$

$$D_z = \begin{vmatrix} 0 & 3 & 8 \\ 1 & 1 & 3 \\ 2 & -1 & 2 \end{vmatrix} = -12$$

$$x = \frac{D_x}{D} = \frac{3}{-3} = -1 \qquad y = \frac{D_y}{D} = \frac{0}{-3} = 0$$

$$z = \frac{D_z}{D} = \frac{-12}{-3} = 4$$

The ordered triple solution is $(-1, 0, 4)$.

SECTION 4.6 SYSTEMS OF LINEAR INEQUALITIES

A system of linear inequalities consists of two or more linear inequalities.	$\begin{cases} x - y \geq 3 \\ y \leq -2x \end{cases}$
To graph a system of inequalities, graph each inequality in the system. The overlapping region is the solution of the system.	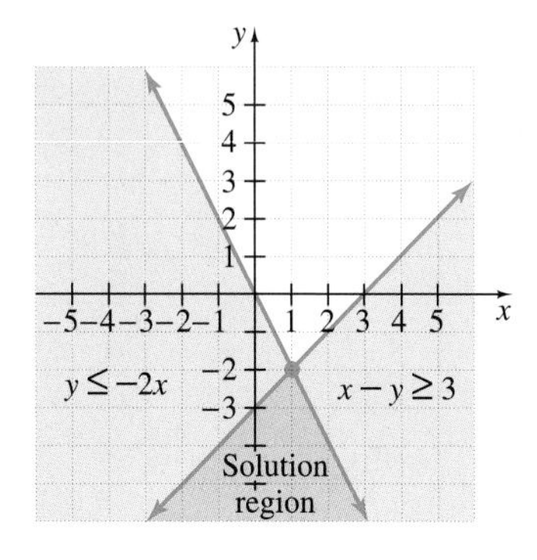

Name ______________________ Section __________ Date __________

Chapter 4 Review

(4.1) *Solve each system of equations in two variables by each method: (1) graphing, (2) substitution, and (3) elimination.*

1. $\begin{cases} 3x + 10y = 1 \\ x + 2y = -1 \end{cases}$

2. $\begin{cases} y = \frac{1}{2}x + \frac{2}{3} \\ 4x + 6y = 4 \end{cases}$

3. $\begin{cases} 2x - 4y = 22 \\ 5x - 10y = 15 \end{cases}$

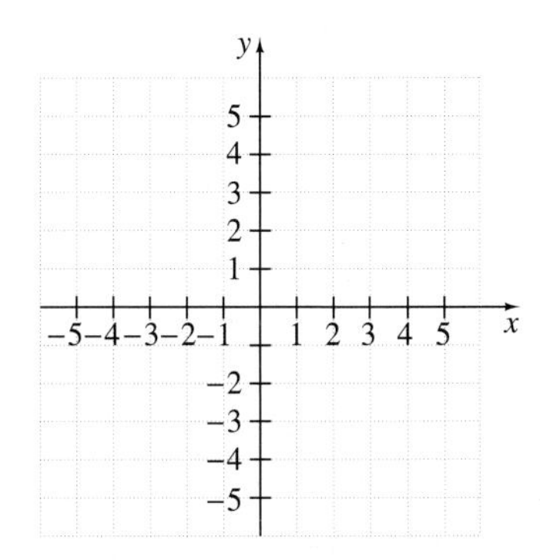

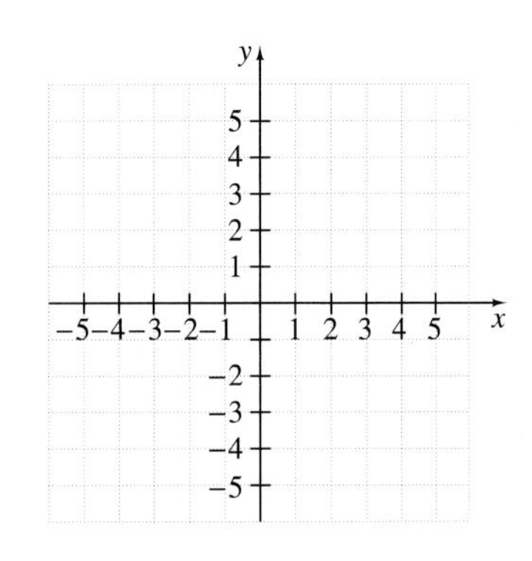

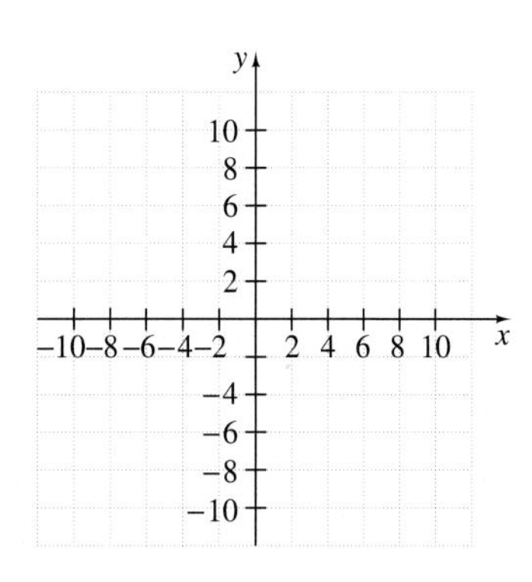

4. $\begin{cases} 3x - 6y = 12 \\ 2y = x - 4 \end{cases}$

5. 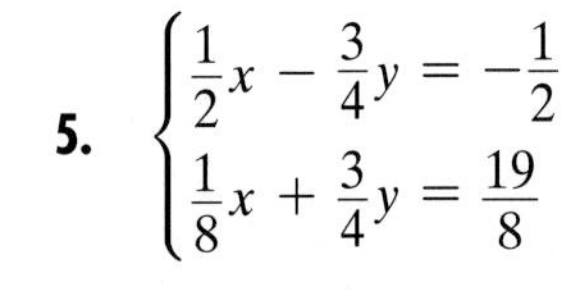

$\begin{cases} \frac{1}{2}x - \frac{3}{4}y = -\frac{1}{2} \\ \frac{1}{8}x + \frac{3}{4}y = \frac{19}{8} \end{cases}$

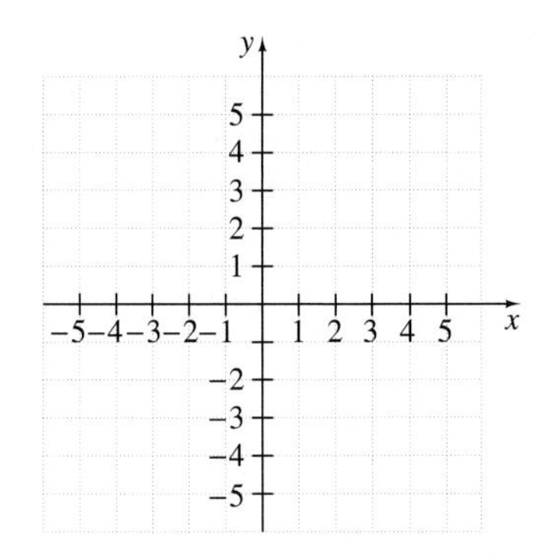

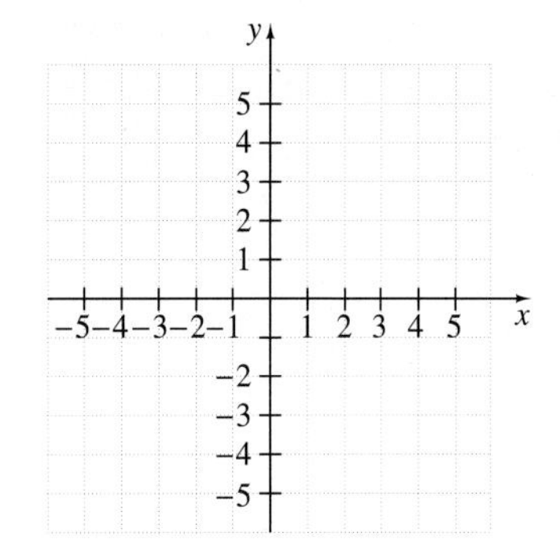

6. The revenue equation for a certain style of backpack is $y = 32x$, where x is the number of backpacks sold and y is the income in dollars for selling x backpacks. The cost equation for these units is $y = 15x + 25{,}500$, where x is the number of backpacks manufactured and y is the cost in dollars for manufacturing x backpacks. Find the number of units to be sold for the company to break even. (*Hint:* Solve the system of equations formed by the two given equations.)

(4.2) *Solve each system of equations in three variables.*

7. $\begin{cases} x + z = 4 \\ 2x - y = 4 \\ x + y - z = 0 \end{cases}$

8. $\begin{cases} 2x + 5y = 4 \\ x - 5y + z = -1 \\ 4x - z = 11 \end{cases}$

9. $\begin{cases} 4y + 2z = 5 \\ 2x + 8y = 5 \\ 6x + 4z = 1 \end{cases}$

10. $\begin{cases} 5x + 7y = 9 \\ 14y - z = 28 \\ 4x + 2z = -4 \end{cases}$

Name ______________________________

11. $\begin{cases} 3x - 2y + 2z = 5 \\ -x + 6y + z = 4 \\ 3x + 14y + 7z = 20 \end{cases}$

12. $\begin{cases} x + 2y + 3z = 11 \\ y + 2z = 3 \\ 2x + 2z = 10 \end{cases}$

13. $\begin{cases} 7x - 3y + 2z = 0 \\ 4x - 4y - z = 2 \\ 5x + 2y + 3z = 1 \end{cases}$

14. $\begin{cases} x - 3y - 5z = -5 \\ 4x - 2y + 3z = 13 \\ 5x + 3y + 4z = 22 \end{cases}$

(4.3) *Use systems of equations to solve.*

15. The sum of three numbers is 98. The sum of the first and second is two more than the third number, and the second is four times the first. Find the numbers.

16. One number is 3 times a second number, and twice the sum of the numbers is 168. Find the numbers.

17. Two cars leave Chicago, one traveling east and the other west. After 4 hours they are 492 miles apart. If one car is traveling 7 mph faster than the other, find the speed of each.

18. The foundation for a rectangular Hardware Warehouse has a length three times the width and is 296 feet around. Find the dimensions of the building.

19. James Callahan has available a 10% alcohol solution and a 60% alcohol solution. Find how many liters of each solution he should mix to make 50 liters of a 40% alcohol solution.

20. An employee at See's Candy Store needs a special mixture of candy. She has creme-filled chocolates that sell for \$3.00 per pound, chocolate-covered nuts that sell for \$2.70 per pound, and chocolate-covered raisins that sell for \$2.25 per pound. She wants to have twice as many raisins as nuts in the mixture. Find how many pounds of each she should use to make 45 pounds worth \$2.80 per pound.

21. Chris Kringler has \$2.77 in her coin jar—all in pennies, nickels, and dimes. If she has 53 coins in all and four more nickels than dimes, find how many of each type of coin she has.

22. If \$10,000 and \$4000 are invested such that \$1250 in interest is earned in one year, and if the rate of interest on the larger investment is 2% more than that of the smaller investment, find the rates of interest.

23. The perimeter of an isosceles (two sides equal) triangle is 73 centimeters. If two sides are of equal length and the third side is 7 centimeters longer than the others, find the lengths of the three sides.

24. The sum of three numbers is 295. One number is five more than a second and twice the third. Find the numbers.

Name ______________________________

(4.4) *Use matrices to solve each system.*

25. $\begin{cases} 3x + 10y = 1 \\ x + 2y = -1 \end{cases}$

26. $\begin{cases} 3x - 6y = 12 \\ 2y = x - 4 \end{cases}$

27. $\begin{cases} 3x - 2y = -8 \\ 6x + 5y = 11 \end{cases}$

28. $\begin{cases} 6x - 6y = -5 \\ 10x - 2y = 1 \end{cases}$

29. $\begin{cases} 3x - 6y = 0 \\ 2x + 4y = 5 \end{cases}$

30. $\begin{cases} 5x - 3y = 10 \\ -2x + y = -1 \end{cases}$

31. $\begin{cases} 0.2x - 0.3y = -0.7 \\ 0.5x + 0.3y = 1.4 \end{cases}$

32. $\begin{cases} 3x + 2y = 8 \\ 3x - y = 5 \end{cases}$

33. $\begin{cases} x + z = 4 \\ 2x - y = 0 \\ x + y - z = 0 \end{cases}$

34. $\begin{cases} 2x + 5y = 4 \\ x - 5y + z = -1 \\ 4x - z = 11 \end{cases}$

35. $\begin{cases} 3x - y = 11 \\ x + 2z = 13 \\ y - z = -7 \end{cases}$

36. $\begin{cases} 5x + 7y + 3z = 9 \\ 14y - z = 28 \\ 4x + 2z = -4 \end{cases}$

37. $\begin{cases} 7x - 3y + 2z = 0 \\ 4x - 4y - z = 2 \\ 5x + 2y + 3z = 1 \end{cases}$

38. $\begin{cases} x + 2y + 3z = 14 \\ y + 2z = 3 \\ 2x - 2z = 10 \end{cases}$

(4.5) *Evaluate.*

39. $\begin{vmatrix} -1 & 3 \\ 5 & 2 \end{vmatrix}$

40. $\begin{vmatrix} 3 & -1 \\ 2 & 5 \end{vmatrix}$

41. $\begin{vmatrix} 2 & -1 & -3 \\ 1 & 2 & 0 \\ 3 & -2 & 2 \end{vmatrix}$

42. $\begin{vmatrix} -2 & 3 & 1 \\ 4 & 4 & 0 \\ 1 & -2 & 3 \end{vmatrix}$

Use Cramer's rule to solve each system of equations.

43. $\begin{cases} 3x - 2y = -8 \\ 6x + 5y = 11 \end{cases}$

44. $\begin{cases} 6x - 6y = -5 \\ 10x - 2y = 1 \end{cases}$

45. $\begin{cases} 3x + 10y = 1 \\ x + 2y = -1 \end{cases}$

46. $\begin{cases} y = \frac{1}{2}x + \frac{2}{3} \\ 4x + 6y = 4 \end{cases}$

47. $\begin{cases} 2x - 4y = 22 \\ 5x - 10y = 16 \end{cases}$

48. $\begin{cases} 3x - 6y = 12 \\ 2y = x - 4 \end{cases}$

Name ______________________________

49. $\begin{cases} x + z = 4 \\ 2x - y = 0 \\ x + y - z = 0 \end{cases}$

50. $\begin{cases} 2x + 5y = 4 \\ x - 5y + z = -1 \\ 4x - z = 11 \end{cases}$

51. $\begin{cases} x + 3y - z = 5 \\ 2x - y - 2z = 3 \\ x + 2y + 3z = 4 \end{cases}$

52. $\begin{cases} 2x - z = 1 \\ 3x - y + 2z = 3 \\ x + y + 3z = -2 \end{cases}$

53. $\begin{cases} x + 2y + 3z = 14 \\ y + 2z = 3 \\ 2x - 2z = 10 \end{cases}$

54. $\begin{cases} 5x + 7y = 9 \\ 14y - z = 28 \\ 4x + 2z = -4 \end{cases}$

(4.6) *Graph the solution of the following systems of linear inequalities.*

55. $\begin{cases} y \geq 2x - 3 \\ y \leq -2x + 1 \end{cases}$

56. $\begin{cases} y \leq -3x - 3 \\ y \leq 2x + 7 \end{cases}$

57. $\begin{cases} x + 2y > 0 \\ x - y \leq 6 \end{cases}$

58. $\begin{cases} x - 2y \geq 7 \\ x + y \leq -5 \end{cases}$

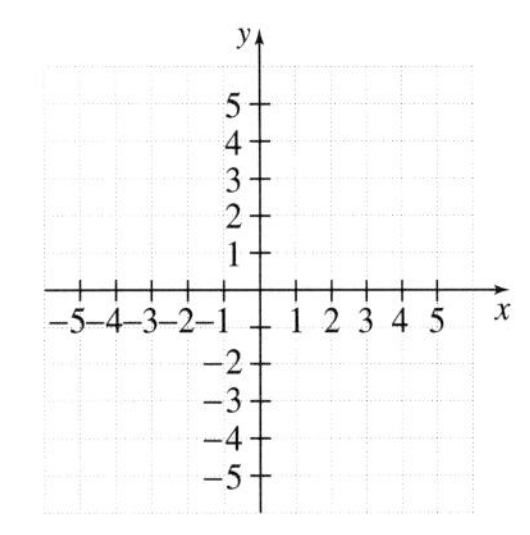

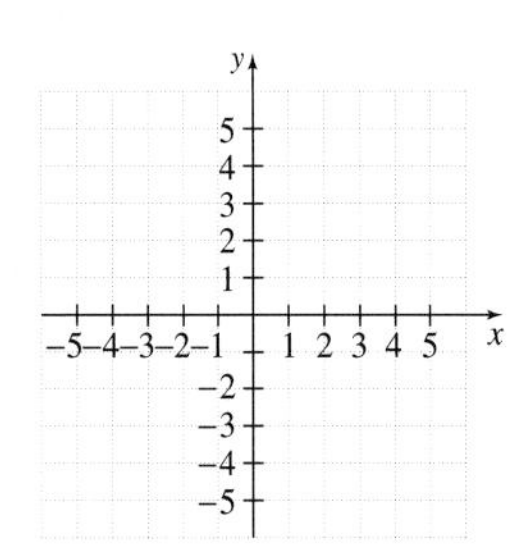

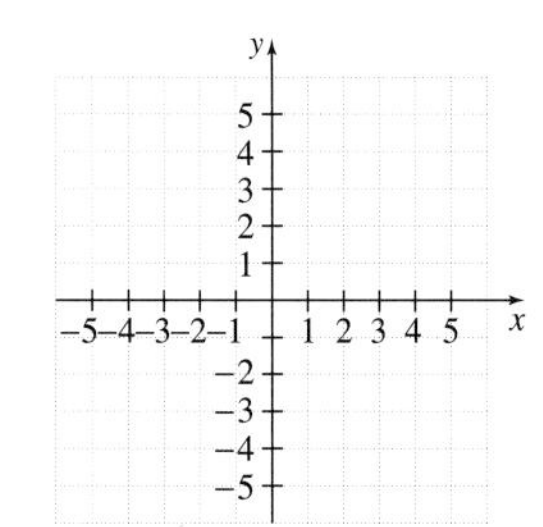

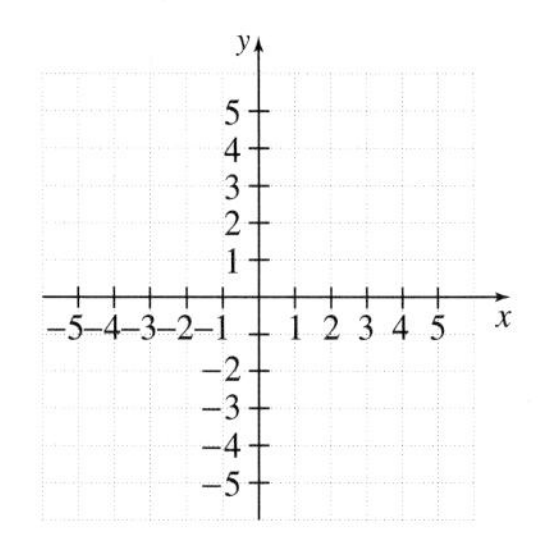

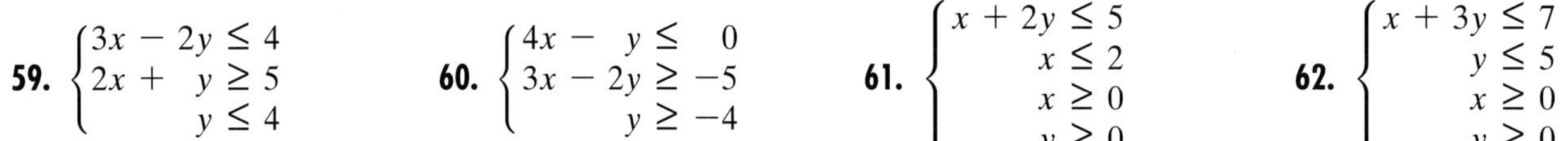

59. $\begin{cases} 3x - 2y \leq 4 \\ 2x + y \geq 5 \\ y \leq 4 \end{cases}$

60. $\begin{cases} 4x - y \leq 0 \\ 3x - 2y \geq -5 \\ y \geq -4 \end{cases}$

61. $\begin{cases} x + 2y \leq 5 \\ x \leq 2 \\ x \geq 0 \\ y \geq 0 \end{cases}$

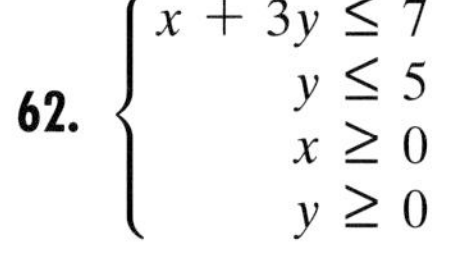

62. $\begin{cases} x + 3y \leq 7 \\ y \leq 5 \\ x \geq 0 \\ y \geq 0 \end{cases}$

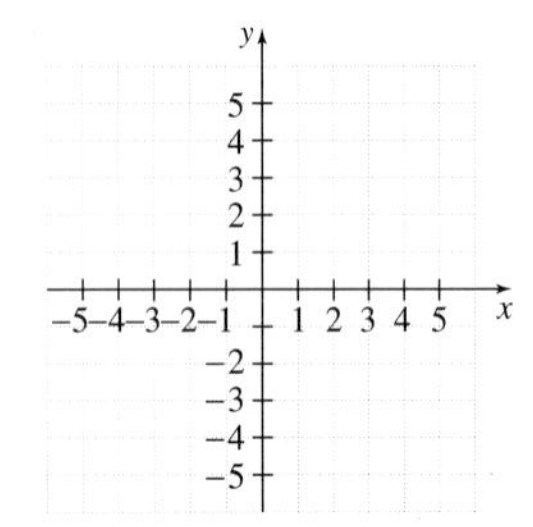

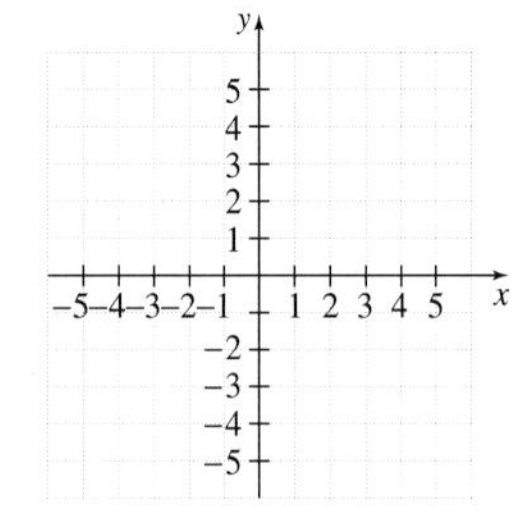

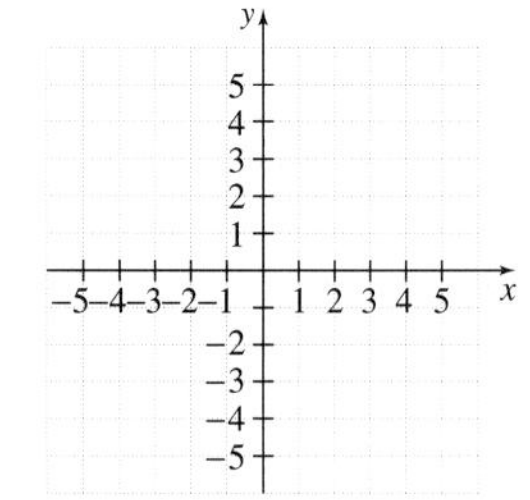

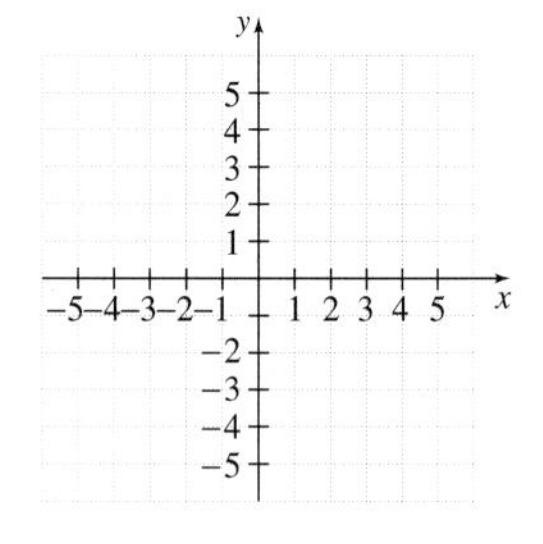

Name ____________________ Section ______ Date ________

Chapter 4 Test

Evaluate each determinant.

1. $\begin{vmatrix} 4 & -7 \\ 2 & 5 \end{vmatrix}$

2. $\begin{vmatrix} 4 & 0 & 2 \\ 1 & -3 & 5 \\ 0 & -1 & 2 \end{vmatrix}$

Solve each system of equations graphically and then solve by the elimination method or the substitution method.

3. $\begin{cases} 2x - y = -1 \\ 5x + 4y = 17 \end{cases}$

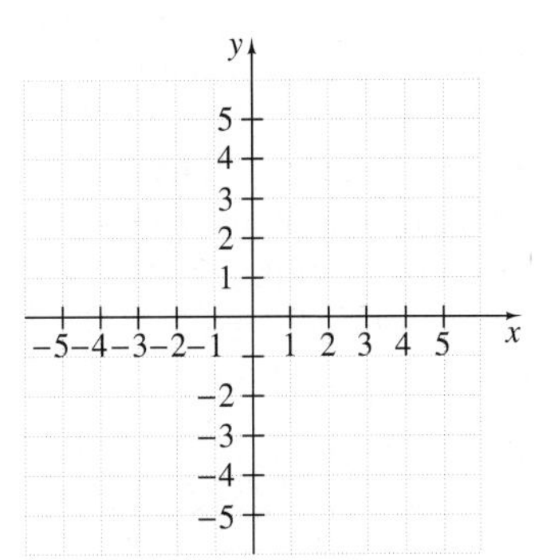

4. $\begin{cases} 7x - 14y = 5 \\ x = 2y \end{cases}$

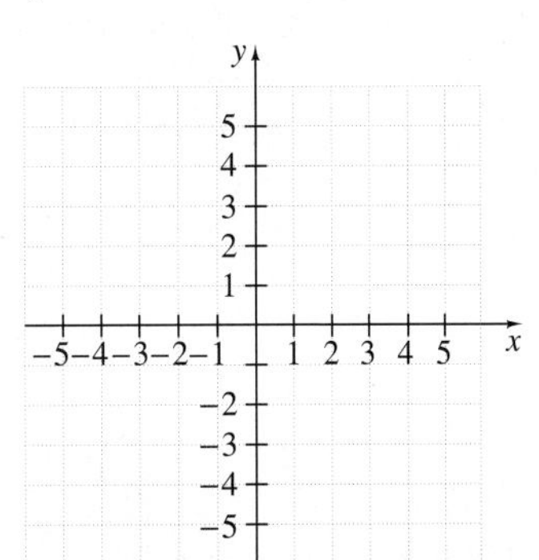

Solve each system.

5. $\begin{cases} 4x - 7y = 29 \\ 2x + 5y = -11 \end{cases}$

6. $\begin{cases} 15x + 6y = 15 \\ 10x + 4y = 10 \end{cases}$

7. $\begin{cases} 2x - 3y = 4 \\ 3y + 2z = 2 \\ x - z = -5 \end{cases}$

8. $\begin{cases} 3x - 2y - z = -1 \\ 2x - 2y = 4 \\ 2x - 2z = -12 \end{cases}$

9. $\begin{cases} \dfrac{x}{2} + \dfrac{y}{4} = -\dfrac{3}{4} \\ x + \dfrac{3}{4}y = -4 \end{cases}$

Use Cramer's rule to solve each system.

10. $\begin{cases} 3x - y = 7 \\ 2x + 5y = -1 \end{cases}$

11. $\begin{cases} 4x - 3y = -6 \\ -2x + y = 0 \end{cases}$

12. $\begin{cases} x + y + z = 4 \\ 2x + 5y = 1 \\ x - y - 2z = 0 \end{cases}$

13. $\begin{cases} 3x + 2y + 3z = 3 \\ x - z = 9 \\ 4y + z = -4 \end{cases}$

Answers

1. ________
2. ________
3. ________
4. ________
5. ________
6. ________
7. ________
8. ________
9. ________
10. ________
11. ________
12. ________
13. ________

14. ______

15. ______

16. ______

17. ______

18. ______

19. ______

20. ______

21. see graph

22. see graph

Name ______________________________

Use matrices to solve each system.

14. $\begin{cases} x - y = -2 \\ 3x - 3y = -6 \end{cases}$

15. $\begin{cases} x + 2y = -1 \\ 2x + 5y = -5 \end{cases}$

16. $\begin{cases} x - y - z = 0 \\ 3x - y - 5z = -2 \\ 2x + 3y = -5 \end{cases}$

17. $\begin{cases} 2x - y + 3z = 4 \\ 3x - 3z = -2 \\ -5x + y = 0 \end{cases}$

18. A motel in New Orleans charges $90 per day for double occupancy and $80 per day for single occupancy. If 80 rooms are occupied for a total of $6930, how many rooms of each kind are occupied.

19. The research department of a company that manufactures children's fruit drinks is experimenting with a new flavor. A 17.5% fructose solution is needed, but only 10% and 20% solutions are available. How many gallons of a 10% fructose solution should be mixed with a 20% fructose solution to obtain 20 gallons of a 17.5% fructose solution?

20. Frame Masters, Inc., recently purchased $5500 worth of new equipment to offer a new style of eyeglass frame. The marketing department of Frame Masters estimates that the cost of producing this new frame is $18 and that the frame will be sold to stores for $38. Find the number of frames that must be sold in order to break even.

Graph the solutions of each system of linear inequalities.

21. $\begin{cases} 2y - x \geq 1 \\ x + y \geq -4 \\ y \leq 2 \end{cases}$

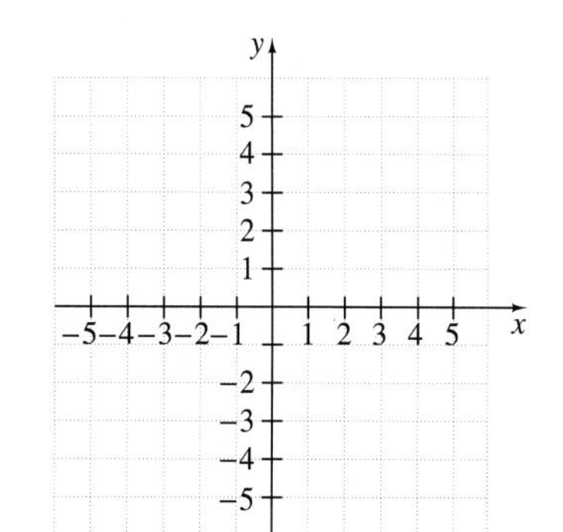

22. 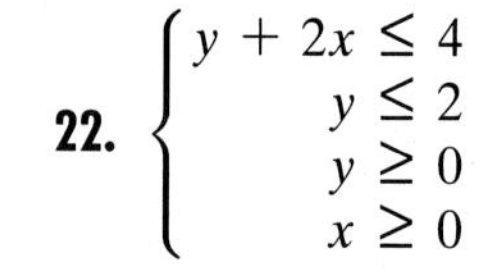

$\begin{cases} y + 2x \leq 4 \\ y \leq 2 \\ y \geq 0 \\ x \geq 0 \end{cases}$

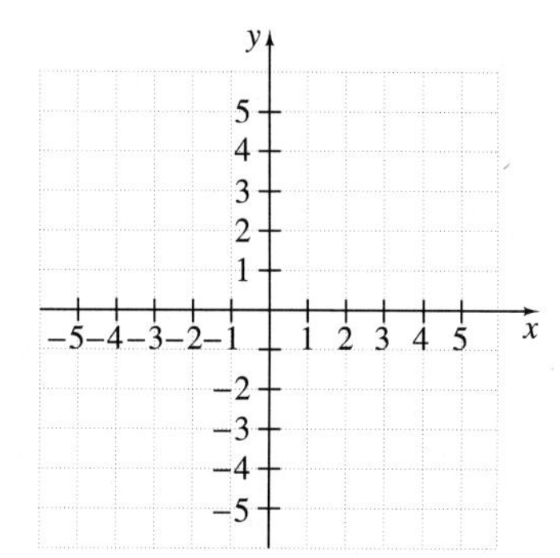

Name ______________________ Section ________ Date __________

Cumulative Review

Determine whether each statement is true or false.

1. $7 \notin \{1, 2, 3\}$

2. $\frac{1}{5}$ is an irrational number.

Insert $<$, $>$, or $=$ between each pair of numbers to form a true statement.

3. $-5 \quad 5$

4. $-2.5 \quad -2.1$

5. $\frac{2}{3} \quad \frac{3}{4}$

Find each absolute value.

6. $|-4|$

7. $-|-8|$

8. Simplify: $\frac{|-2|^3 + 1}{-7 - \sqrt{4}}$

Simplify and write with positive exponents only.

9. $(3x)^{-1}$

10. $2^{-1} + 3^{-2}$

11. Determine whether -15 is a solution of $x - 9 = -24$.

12. Find 16% of 25.

13. Solve $A = \frac{1}{2}(B + b)h$ for b.

14. Solve: $4x - 2 < 5x$

15. Find the intersection: $\{2, 4, 6, 8\} \cap \{3, 4, 5, 6\}$

16. Solve: $|y| = 0$

17. Graph: $y = -3$

18. Find the slope and the y-intercept of the line $3x - 4y = 4$.

Answers

1. ________
2. ________
3. ________
4. ________
5. ________
6. ________
7. ________
8. ________
9. ________
10. ________
11. ________
12. ________
13. ________
14. ________
15. ________
16. ________
17. see graph ________
18. ________

Name ______________________

19. ______

20. ______

21. see graph

22. ______

23. ______

24. ______

25. a. ______

b. ______

19. Write an equation of the line with slope -3 containing the point $(1, -5)$.

20. Determine whether the relation $y = 2x + 1$ is also a function.

21. Graph the intersection of $x \geq 1$ and $y \geq 2x - 1$.

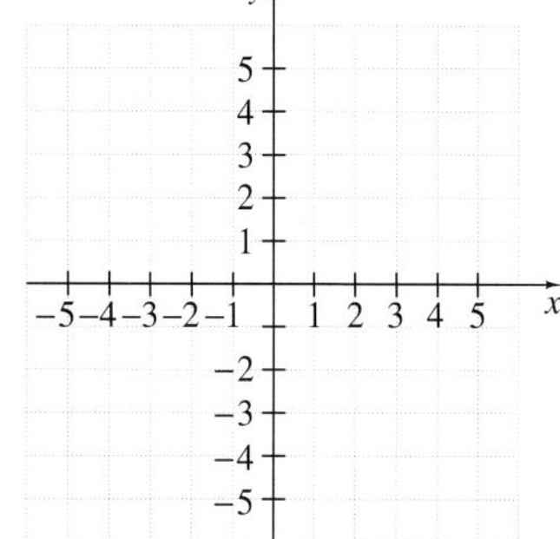

22. Use the elimination method to solve the system: $\begin{cases} 3x - 2y = 10 \\ 4x - 3y = 15 \end{cases}$

23. Solve the system:
$$\begin{cases} 2x - 4y + 8z = 2 \\ -x - 3y + z = 11 \\ x - 2y + 4z = 0 \end{cases}$$

24. A first number is 4 less than a second number. Four times the first number is 6 more than twice the second. Find the numbers.

25. Evaluate each determinant.

a. $\begin{vmatrix} -1 & 2 \\ 3 & -4 \end{vmatrix}$

b. $\begin{vmatrix} 2 & 0 \\ 7 & -5 \end{vmatrix}$

Polynomials and Polynomial Functions

CHAPTER 5

Linear equations are important for solving problems. They are not sufficient, however, to solve all problems. Many real-world phenomena are modeled by polynomials. In the first portion of this chapter we will study operations on polynomials. We then look at how polynomials can be used in problem solving. We conclude with a study of graphs of polynomial functions.

The Eiffel Tower in Paris, France, is recognized throughout the world. It was built in 1889 for that year's World's Fair (held in Paris) as well as to commemorate the 100th anniversary of the French Revolution. Alexandre Gustave Eiffel built the 984-foot-tall tower with 7000 tons of wrought iron. Workers on the mammoth structure assembled over 18,000 individual parts with more than 2.5 million rivets. At the time of its construction, the Eiffel Tower was the tallest man-made structure in the world and remained so until the Empire State Building was completed in 1931 in New York City. In Exercise 74 on page 349, we will use a polynomial function to find the height of an object thrown from the top of the Eiffel Tower.

ANSWERS

1. ____
2. ____
3. ____
4. ____
5. ____
6. ____
7. ____
8. ____
9. ____
10. ____
11. ____
12. ____
13. ____
14. ____
15. ____
16. ____
17. ____
18. ____
19. ____
20. see graph

Name ________________ Section ______ Date ________

CHAPTER 5 PRETEST

1. Find the degree of the polynomial $2x^4 - 3xy^5 + y^3$

2. If $P(x) = -x^2 + 2x + 6$, find $P(-1)$.

Perform the indicated operations.

3. $(-2x + 7) + (3x^2 + 6x - 6)$

4. $(-8y^2 - 3y + 5) - (4y^2 - 6y - 1)$

5. $(2x - 1)(3x + 5)$

6. $(8y + 3)^2$

7. $(2m + 5)(2m - 5)$

8. $\dfrac{6t^3 - 4t^2 + 5t}{2t}$

9. $(2x^3 - 5x^2 - 10x - 4) \div (2x + 3)$

Factor each polynomial completely.

10. $6x^4 - 12x^3 + 10x^2$

11. $3ac - 4ad + 6bc - 8bd$

12. $x^2 + 2x - 63$

13. $6x^2 + 19x + 3$

14. $(x - 4)^2 - 7(x - 4) + 10$

15. $8t^3 - 1$

16. $12x^3 - 27xy^2$

Solve each equation.

17. $2n^2 + 13n = -15$

18. $x^3 + x^2 = 25x + 25$

19. One number exceeds another by seven, and their product is 120. Find the numbers.

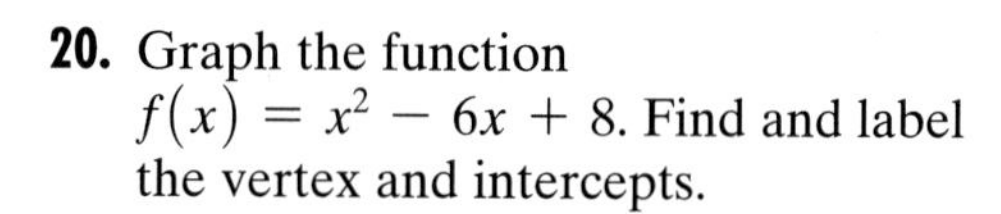

20. Graph the function $f(x) = x^2 - 6x + 8$. Find and label the vertex and intercepts.

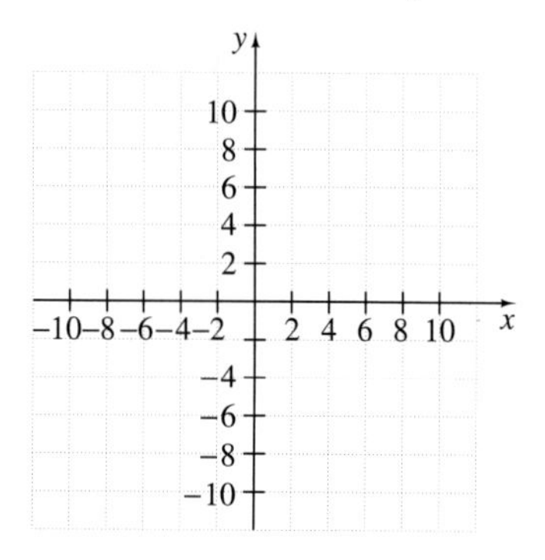

5.1 ADDING AND SUBTRACTING POLYNOMIALS

Objectives

A Define term, constant, polynomial, monomial, binomial, trinomial, degree of a term, and degree of a polynomial.
B Combine like terms.
C Add polynomials.
D Subtract polynomials.
E Evaluate polynomial functions.

SSM CD-ROM Video 5.1

A DEFINING A POLYNOMIAL AND RELATED TERMS

A **term** is a number or the product of a number and one or more variables raised to powers. The **numerical coefficient**, or simply the **coefficient**, is the numerical factor of a term.

Term	Numerical Coefficient
$-12x^5$	-12
x^3y	1
$-z$	-1
2	2

If a term contains only a number, it is called a **constant term**, or simply a **constant**.

A **polynomial** is a finite sum of terms in which all variables are raised to nonnegative integer powers and no variables appear in any denominator.

Polynomials	Not Polynomials	
$4x^5y + 7xz$	$5x^{-3} + 2x$	Negative integer exponent
$-5x^3 + 2x + \frac{2}{3}$	$\frac{6}{x^2} - 5x + 1$	Variable in denominator

A polynomial that contains only one variable is called a **polynomial in one variable**. For example, $3x^2 - 2x + 7$ is a **polynomial in *x***. This polynomial in x is written in *descending order* since the terms are listed in descending order of the variable's exponents. (The term 7 can be thought of as $7x^0$.) The following examples are polynomials in one variable written in **descending order**:

$$4x^3 - 7x^2 + 5 \qquad y^2 - 4 \qquad 8a^4 - 7a^2 + 4a$$

A **monomial** is a polynomial consisting of one term. A **binomial** is a polynomial consisting of two terms. A **trinomial** is a polynomial consisting of three terms.

Monomials	Binomials	Trinomials
ax^2	$x + y$	$x^2 + 4xy + y^2$
$-3x$	$6y^2 - 2$	$-x^4 + 3x^3 + 1$
4	$\frac{5}{7}z^3 - 2z$	$8y^2 - 2y - 10$

By definition, all monomials, binomials, and trinomials are also polynomials.

Each term of a polynomial has a **degree**.

DEGREE OF A TERM

The **degree of a term** is the sum of the exponents on the *variables* contained in the term.

Practice Problems 1–5

Find the degree of each term.

1. $2x^3$
2. 7^2x^4
3. x
4. $15xy^2z^4$
5. 9

Examples

Find the degree of each term.

1. $3x^2$ The exponent on x is 2, so the degree of the term is 2.

2. -2^3x^5 The exponent on x is 5, so the degree of the term is 5. (Recall that the degree is the sum of the exponents on *only* the *variables*.)

3. y The degree of y, or y^1, is 1.

4. $12x^2yz^3$ The degree is the sum of the exponents on the variables, or $2 + 1 + 3 = 6$.

5. 5 The degree of 5, which can be written as $5x^0$, is 0.

From the preceding examples, we can say that the degree of a constant is 0. Also, the term 0 has no degree.

Each polynomial also has a degree.

> DEGREE OF A POLYNOMIAL
>
> The **degree of a polynomial** is the largest degree of any of its terms.

Practice Problems 6–8

Find the degree of each polynomial and indicate whether the polynomial is also a monomial, binomial, or trinomial.

6. $4x^5 + 7x^3 - 1$
7. $-2xy^2z$
8. $y^3 + 6y$

Examples

Find the degree of each polynomial and also indicate whether the polynomial is a monomial, binomial, or trinomial.

	Polynomial	Degree	Classification
6.	$7x^3 - 3x + 2$	3	Trinomial
7.	$-xyz$	$1 + 1 + 1 = 3$	Monomial
8.	$x^4 - 16$	4	Binomial

Practice Problem 9

Find the degree of the polynomial $7x^2y - 6x^2yz + 2 - 4y^3$.

Example 9

Find the degree of the polynomial $3xy + x^2y^2 - 5x^2 - 6$.

Solution: The degree of each term is

$$\begin{array}{rcccc} & 3xy & +\ x^2y^2 & -\ 5x^2 & -\ 6 \\ & \downarrow & \downarrow & \downarrow & \downarrow \\ \text{Degree:} & 2 & 4 & 2 & 0 \end{array}$$

The largest degree of any term is 4, so the degree of this polynomial is 4.

B COMBINING LIKE TERMS

Before we add polynomials, recall from Section 1.4 that terms are considered to be **like terms** if they contain exactly the same variables raised to exactly the same powers.

Like Terms	Unlike Terms
$-5x^2, -x^2$	$4x^2, 3x$
$7xy^3z, -2xzy^3$	$12x^2y^3, -2xy^3$

To simplify, a polynomial, we **combine like terms** by using the distributive property. For example, by the distributive property,

$$5x + 7x = (5 + 7)x = 12x$$

Answers

1. 3, **2.** 4, **3.** 1, **4.** 7, **5.** 0, **6.** 5, trinomial, **7.** 4, monomial, **8.** 3, binomial, **9.** 4

Examples Simplify each polynomial by combining like terms.

10. $-12x^2 + 7x^2 - 6x = (-12 + 7)x^2 - 6x = -5x^2 - 6x$

11. $3xy - 2x + 5xy - x = 3xy + 5xy - 2x - x$
$= (3 + 5)xy + (-2 - 1)x$
$= 8xy - 3x$

Practice Problems 10–11

Simplify each polynomial by combining like terms.

10. $10x^3 - 12x^3 - 3x$

11. $-6ab + 2a + 12ab - a$

C ADDING POLYNOMIALS

Now we have reviewed the skills we need to add polynomials.

> **ADDING POLYNOMIALS**
>
> To add polynomials, combine all like terms.

Example 12 Add $11x^3 - 12x^2 + x - 3$ and $x^3 - 10x + 5$.

Solution:

$$(11x^3 - 12x^2 + x - 3) + (x^3 - 10x + 5)$$
$$= 11x^3 + x^3 - 12x^2 + x - 10x - 3 + 5 \quad \text{Group like terms.}$$
$$= 12x^3 - 12x^2 - 9x + 2 \quad \text{Combine like terms.}$$

Practice Problem 12

Add: $14x^4 - 6x^3 + x^2 - 6$ and $x^3 - 5x^2 + 1$.

Sometimes it is more convenient to add polynomials vertically. To do this, we line up like terms beneath one another and then add like terms.

Example 13 Add $11x^3 - 12x^2 + x - 3$ and $x^3 - 10x + 5$ vertically.

Solution:

$$\begin{array}{rrrr} 11x^3 & - 12x^2 & + x & - 3 \\ x^3 & & - 10x & + 5 \\ \hline 12x^3 & - 12x^2 & - 9x & + 2 \end{array}$$

Line up like terms.
Combine like terms.

This example is the same as Example 12, only here we added vertically.

Practice Problem 13

Add $10y^3 - y^2 + 4y - 11$ and $y^3 - 4y^2 + 3y$ vertically.

Example 14 Add: $(7x^3y - xy^3 + 11) + (6x^3y - 4)$

Solution: To add these polynomials, we remove the parentheses and group like terms.

$$(7x^3y - xy^3 + 11) + (6x^3y - 4)$$
$$= 7x^3y - xy^3 + 11 + 6x^3y - 4 \quad \text{Remove parentheses.}$$
$$= 7x^3y + 6x^3y - xy^3 + 11 - 4 \quad \text{Group like terms.}$$
$$= 13x^3y - xy^3 + 7 \quad \text{Combine like terms.}$$

Practice Problem 14

Add: $(4x^2y - xy^2 + 5) + (-6x^2y - 1)$

D SUBTRACTING POLYNOMIALS

The definition of subtraction of real numbers can be extended to apply to polynomials. To subtract a number, we add its opposite:

$$a - b = a + (-b)$$

Likewise, to subtract a polynomial we add its opposite. In other words, if P and Q are polynomials, then

$$P - Q = P + (-Q)$$

Answers

10. $-2x^3 - 3x$, **11.** $6ab + a$, **12.** $14x^4 - 5x^3 - 4x^2 - 5$, **13.** $11y^3 - 5y^2 + 7y - 11$, **14.** $-2x^2y - xy^2 + 4$

The polynomial $-Q$ is the **opposite**, or **additive inverse**, of the polynomial Q. We can find $-Q$ by changing the sign of each term of Q.

✓ CONCEPT CHECK

Which polynomial is the opposite of $16x^3 - 5x + 7$?
a. $-16x^3 - 5x + 7$
b. $-16x^3 + 5x - 7$
c. $16x^3 + 5x + 7$
d. $-16x^3 + 5x + 7$

TRY THE CONCEPT CHECK IN THE MARGIN.

> **SUBTRACTING POLYNOMIALS**
>
> To subtract polynomials, change the signs of the terms of the polynomial being subtracted and then add.

For example,

To subract, change the signs; then

Add

$$(3x^2 + 4x - 7) - (3x^2 - 2x - 5) = (3x^2 + 4x - 7) + (-3x^2 + 2x + 5)$$
$$= 3x^2 + 4x - 7 - 3x^2 + 2x + 5$$
$$= 6x - 2 \quad \text{Combine like terms.}$$

Practice Problem 15

Subtract:
$(7x^4 - 8x^2 + x) - (9x^4 + x^2 - 18)$

Example 15

Subtract: $(12z^5 - 12z^3 + z) - (-3z^4 + z^3 + 12z)$

Solution: First we change the sign of each term of the second polynomial, and then we add the result to the first polynomial.

$$(12z^5 - 12z^3 + z) - (-3z^4 + z^3 + 12z)$$
$$= 12z^5 - 12z^3 + z + 3z^4 - z^3 - 12z \quad \text{Change signs and add.}$$
$$= 12z^5 + 3z^4 - 12z^3 - z^3 + z - 12z \quad \text{Group like terms.}$$
$$= 12z^5 + 3z^4 - 13z^3 - 11z \quad \text{Combine like terms.}$$

Practice Problem 16

Subtract $(2y^4 + 4y) - (6y^4 + 7y^3 - 3y)$ vertically.

Example 16

Subtract $(10x^3 - 7x^2) - (4x^3 - 3x^2 + 2)$ vertically.

Solution: To subtract these polynomials, we add the opposite of the second polynomial to the first one.

$$\begin{array}{r} 10x^3 - 7x^2 \\ -(4x^3 - 3x^2 + 2) \\ \hline \end{array} \quad \text{is equivalent to} \quad \begin{array}{r} 10x^3 - 7x^2 \\ -4x^3 + 3x^2 - 2 \\ \hline 6x^3 - 4x^2 - 2 \end{array} \quad \text{Add.}$$

✓ CONCEPT CHECK

Why is the following subtraction incorrect?
$(7z - 5) - (3z - 4)$
$= 7z - 5 - 3z - 4$
$= 4z - 9$

TRY THE CONCEPT CHECK IN THE MARGIN.

Practice Problem 17

Subtract $3a^2b^3 - 4ab^2 + 6a$ from $7a^2b^3 - ab^2$.

Example 17

Subtract $4x^3y^2 - 3x^2y^2 + 2y^2$ from $10x^3y^2 - 7x^2y^2$.

Solution: Notice the order of the numbers, and then write "Subtract $4x^3y^2 - 3x^2y^2 + 2y^2$ from $10x^3y^2 - 7x^2y^2$" as a mathematical expression. (For example, if we subtract 2 from 8, we would write $8 - 2 = 6$.)

$$(10x^3y^2 - 7x^2y^2) - (4x^3y^2 - 3x^2y^2 + 2y^2)$$
$$= 10x^3y^2 - 7x^2y^2 - 4x^3y^2 + 3x^2y^2 - 2y^2 \quad \text{Remove parentheses.}$$
$$= 6x^3y^2 - 4x^2y^2 - 2y^2 \quad \text{Combine like terms.}$$

Answers

15. $-2x^4 - 9x^2 + x + 18$, **16.** $-4y^4 - 7y^3 + 7y$, **17.** $4a^2b^3 + 3ab^2 - 6a$

✓ Concept Check: b

✓ Concept Check: With parentheses removed, the expression should be $7z - 5 - 3z + 4 = 4z - 1$

E EVALUATING POLYNOMIAL FUNCTIONS

Recall function notation first introduced in Section 3.5. At times it is convenient to use function notation to represent polynomials. For example, we may write $P(x)$ to represent the polynomial $3x^2 - 2x - 5$. In symbols, we would write

$$P(x) = 3x^2 - 2x - 5$$

This function is called a **polynomial function** because the expression $3x^2 - 2x - 5$ is a polynomial.

HELPFUL HINT

Recall that the symbol $P(x)$ **does not mean** P times x. It is a special symbol used to denote a function.

Examples

If $P(x) = 3x^2 - 2x - 5$, find each function value.

18. $P(1) = 3(1)^2 - 2(1) - 5 = -4$ Let $x = 1$ in the function $P(x)$.

19. $P(-2) = 3(-2)^2 - 2(-2) - 5 = 11$ Let $x = -2$ in the function $P(x)$.

Practice Problems 18–19

If $P(x) = 5x^2 - 3x + 7$, find each function value.

18. $P(2)$

19. $P(-1)$

Many real-world phenomena are modeled by polynomial functions. If the polynomial function model is given, we can often find the solution of a problem by evaluating the function at a certain value.

Example 20 Finding the Height of an Object

The world's highest bridge, Royal Gorge suspension bridge in Colorado, is 1053 feet above the Arkansas River. An object is dropped from the top of this bridge. Neglecting air resistance, the height of the object at time t seconds is given by the polynomial function $P(t) = -16t^2 + 1053$. Find the height of the object when $t = 1$ second and when $t = 8$ seconds.

Solution: To find the height of the object at 1 second, we find $P(1)$.

$$P(t) = -16t^2 + 1053$$
$$P(1) = -16(1)^2 + 1053$$
$$P(1) = 1037$$

When $t = 1$ second, the height of the object is 1037 feet.

Practice Problem 20

Use the polynomial function in Example 20 to find the height of the object when $t = 3$ seconds and $t = 7$ seconds.

Answers

18. $P(2) = 21$, **19.** $P(-1) = 15$, **20.** At 3 seconds, height is 909 feet; at 7 seconds, height is 269 feet

To find the height of the object at 8 seconds, we find $P(8)$.

$$P(t) = -16t^2 + 1037$$
$$P(8) = -16(8)^2 + 1037$$
$$P(8) = -1024 + 1037$$
$$P(8) = 13$$

When $t = 8$ seconds, the height of the object is 13 feet. Notice that as time t increases, the height of the object decreases.

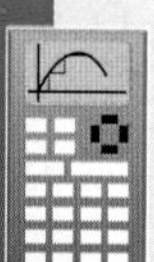

GRAPHING CALCULATOR EXPLORATIONS

A grapher may be used to visualize addition and subtraction of polynomials in one variable. For example, to visualize the following polynomial subtraction statement:

$$(3x^2 - 6x + 9) - (x^2 - 5x + 6) = 2x^2 - x + 3$$

graph both

$$Y_1 = (3x^2 - 6x + 9) - (x^2 - 5x + 6) \quad \text{Left side of equation}$$

and

$$Y_2 = 2x^2 - x + 3 \quad \text{Right side of equation}$$

on the same screen and see that their graphs coincide. (*Note:* If the graphs do not coincide, we can be sure that a mistake has been made either in combining polynomials or in calculator keystrokes. However, if the graphs appear to coincide, we cannot be sure that our work is correct. This is because it is possible for the graphs to differ so slightly that we do not notice it.)

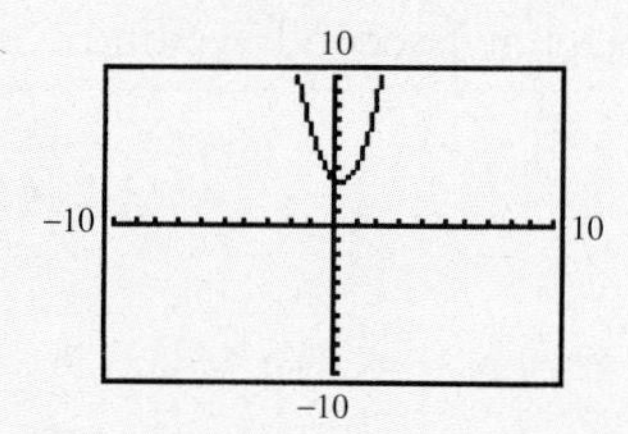

The graphs of Y_1 and Y_2 are shown. The graphs appear to coincide so the subtraction statement

$$(3x^2 - 6x + 9) - (x^2 - 5x + 6) = 2x^2 - x + 3$$

appears to be correct.

Perform each indicated operation. Then use the procedure described above to visualize each statement.

1. $(2x^2 + 7x + 6) + (x^3 - 6x^2 - 14)$

2. $(-14x^3 - x + 2) + (-x^3 + 3x^2 + 4x)$

3. $(1.8x^2 - 6.8x - 1.7) - (3.9x^2 - 3.6x)$

4. $(-4.8x^2 + 12.5x - 7.8) - (3.1x^2 - 7.8x)$

5. $(1.29x - 5.68) + (7.69x^2 - 2.55x + 10.98)$

6. $(-0.98x^2 - 1.56x + 5.57) + (4.36x - 3.71)$

Name ______________________ Section ________ Date ________

Mental Math

Add or subtract as indicated.

1. $7x + 3x$ **2.** $8x - 2x$ **3.** $14y - 9y$

4. $14y + 9y$ **5.** $3z - 12z$ **6.** $2z - 6z$

Exercise Set 5.1

A *Find the degree of each term. See Examples 1 through 5.*

1. 4 **2.** 7 **3.** $5x^2$

4. $-z^3$ **5.** $-3xy^2$ **6.** $12x^3z$

Find the degree of each polynomial and indicate whether the polynomial is a monomial, binomial, trinomial, or none of these. See Examples 6 through 9.

7. $6x + 3$ **8.** $7x - 8$ **9.** $3x^2 - 2x + 5$

10. $5x^2 - 3x^2y - 2x^3$ **11.** $-xyz$ **12.** -9

13. $x^2y - 4xy^2 + 5x + y$ **14.** $-2x^2y - 3y^2 + 4x + y^5$

15. In your own words, describe how to find the degree of a term.

16. In your own words, describe how to find the degree of a polynomial.

B *Simplify each polynomial by combining like terms. See Examples 10 and 11.*

17. $5y + y$ **18.** $-x + 3x$

19. $4x + 7x - 3$ **20.** $-8y + 9y + 4y^2$

21. $4xy + 2x - 3xy - 1$ **22.** $-8xy^2 + 4x - x + 2xy^2$

Mental Math Answers

1. ____
2. ____
3. ____
4. ____
5. ____
6. ____

Answers

1. ____
2. ____
3. ____
4. ____
5. ____
6. ____
7. ____
8. ____
9. ____
10. ____
11. ____
12. ____
13. ____
14. ____
15. ____
16. ____
17. ____
18. ____
19. ____
20. ____
21. ____
22. ____

23. ____
24. ____
25. ____
26. ____
27. ____
28. ____
29. ____
30. ____
31. ____
32. ____
33. ____
34. ____
35. ____
36. ____
37. ____
38. ____
39. ____
40. ____
41. ____
42. ____

Name ____

C *Add. See Examples 12 through 14.*

23. $(9y^2 - 8) + (9y^2 - 9)$

24. $(x^2 + 4x - 7) + (8x^2 + 9x - 7)$

25. $(x^2 + xy - y^2)$ and $(2x^2 - 4xy + 7y^2)$

26. $(4x^3 - 6x^2 + 5x + 7)$ and $(2x^2 + 6x - 3)$

27.
$$\begin{array}{r} x^2 - 6x + 3 \\ + \quad (2x + 5) \\ \hline \end{array}$$

28.
$$\begin{array}{r} -2x^2 + 3x - 9 \\ + \quad (2x - 3) \\ \hline \end{array}$$

29.
$$\begin{array}{r} 3x^2 + 15x + 8 \\ + \ (2x^2 + 7x + 8) \\ \hline \end{array}$$

30.
$$\begin{array}{r} 9x^2 + 9x - 4 \\ + (7x^2 - 3x - 4) \\ \hline \end{array}$$

31. $(-3x + 8) + (-3x^2 + 3x - 5)$

32. $(5y^2 - 2y + 4) + (3y + 7)$

33. $(5y^4 - 7y^2 + x^2 - 3) + (-3y^4 + 2y^2 + 4)$

34. $(8x^4 - 14x^2 + 6) + (-12x^6 - 21x^4 - 9x^2)$

35. $(5x - 11) + (-x - 2)$

36. $(3x^2 - 2x) + (5x^2 - 9x)$

37. $(3x^3 - b + 2a - 6) + (-4x^3 + b + 6a - 6)$

38. $(5x^2 - 6) + (2x^2 - 4x + 8)$

39. $(-3 + 4x^2 + 7xy^2) + (2x^3 - x^2 + xy^2)$

40. $(-3x^2y + 4) + (-7x^2y - 8y)$

41. $(7x^3y - 4xy + 8) + (5x^3y + 4xy + 8x)$

42. $(9xyz + 4x - y) + (-9xyz - 3x + y + 2)$

Name ______________________

43. $(0.6x^3 + 1.2x^2 - 4.5x + 9.1) + (3.9x^3 - x^2 + 0.7x)$

44. $(9.3y^2 - y + 12.8) + (2.6y^2 + 4.4y - 8.9)$

D *Subtract. See Examples 15 through 17.*

45. $(9y^2 - 7y + 5) - (8y^2 - 7y + 2)$

46. $(2x^2 + 3x + 12) - (5x - 7)$

47. Subtract $(6x^2 - 3x)$ from $(4x^2 + 2x)$

48. Subtract $(y^2 + x - y)$ from $(y^2 + x - 3)$

49. $\begin{array}{r} 3x^2 - 4x + 8 \\ - \quad (5x^2 - 7) \\ \hline \end{array}$

50. $\begin{array}{r} -3x^2 - 4x + 8 \\ - \quad (5x + 12) \\ \hline \end{array}$

51. $\begin{array}{r} 6y^2 - 6y + 4 \\ -(-y^2 - 6y + 7) \\ \hline \end{array}$

52. $\begin{array}{r} -4x^3 + 4x^2 - 4x \\ -(2x^3 - 2x^2 + 3x) \\ \hline \end{array}$

53. $(4x^2 - 6x + 2) - (-x^2 + 3x + 5)$

54. $(5x^2 + x + 9) - (2x^2 - 9)$

55. $(7x^2 + x + 1) - (6x^2 + x - 1)$

56. $(4x - 4) - (-x - 4)$

57. $(9x^3 - 2x^2 + 4x - 7) - (2x^3 - 6x^2 - 4x + 3)$

58. $(3x^2 + 6xy + 3y^2) - (8x^2 - 6xy - y^2)$

59. Subtract $(y^2 + 4yx + 7)$ from $(-19y^2 + 7yx + 7)$

60. Subtract $(x^2y - 4)$ from $(3x^2 - 4x^2y + 5)$

61. Subtract $(3x + 7)$ from the sum of $(7x^2 + 4x + 9)$ and $(8x^2 + 7x - 8)$

62. Subtract $(9x + 8)$ from the sum of $(3x^2 - 2x - x^3 + 2)$ and $(5x^2 - 8x - x^3 + 4)$

43. ______
44. ______
45. ______
46. ______
47. ______
48. ______
49. ______
50. ______
51. ______
52. ______
53. ______
54. ______
55. ______
56. ______
57. ______
58. ______
59. ______
60. ______
61. ______
62. ______

63.
64.
65.
66.
67.
68.
69.
70.
71.
72.
73. a.
b.
c.
d.
e.
f.

Name

63. $(14ab - 10a^2b + 6b^2) - (18a^2 - 20a^2b - 6b^2)$

64. $(13x^2 - 26x^2y^2 + 4) - (19x^2 + x^2y^2 - 11)$

65. $\left(\frac{2}{3}x^2 - \frac{1}{6}x + \frac{5}{6}\right) - \left(\frac{1}{3}x^2 + \frac{5}{6}x - \frac{1}{6}\right)$

66. $\left(\frac{3}{16}x^2 + \frac{5}{8}x - \frac{1}{4}\right) - \left(\frac{5}{16}x^2 - \frac{3}{8}x + \frac{3}{4}\right)$

E *If $P(x) = x^2 + x + 1$ and $Q(x) = 5x^2 - 1$, find each function value. See Examples 18 and 19.*

67. $P(7)$

68. $Q(4)$

69. $Q(-10)$

70. $P(-4)$

71. $P(0)$

72. $Q(0)$

Solve. See Example 20.

73. A projectile is fired upward from the ground with an initial velocity of 300 feet per second. Neglecting air resistance, the height of the projectile at any time t can be described by the polynomial function $P(t) = -16t^2 + 300t$. Find the height of the projectile at each given time.

a. $t = 1$ second

b. $t = 2$ seconds

c. $t = 3$ seconds

d. $t = 4$ seconds

e. Explain why the height increases and then decreases as time passes.

f. Approximate (to the nearest second) how long before the object hits the ground.

Name ________________

74. ________

75. ________

76. ________

77. ________

78. ________

79. ________

80. ________

81. ________

82. a. ________

b. ________

c. ________

d. ________

74. An object is thrown upward with an initial velocity of 25 feet per second from the top of the 984-foot-high Eiffel Tower in Paris, France. The height of the object at any time t can be described by the polynomial function $P(t) = -16t^2 + 25t + 984$. Find the height of the projectile when $t = 1$ second, $t = 3$ seconds, and $t = 5$ seconds. (*Source:* Council on Tall Buildings and Urban Habitat, Lehigh University)

75. The polynomial function $P(x) = 45x - 100{,}000$ models the relationship between the number of lamps x that Sherry's Lamp Shop sells and the profit the shop makes, $P(x)$. Find $P(4000)$, the profit from selling 4000 lamps.

76. The total cost (in dollars) for MCD, Inc., Manufacturing Company to produce x blank audiocassette tapes per week is given by the polynomial function $C(x) = 0.8x + 10{,}000$. Find the total cost of producing 20,000 tapes per week.

77. The total revenues (in dollars) for MCD, Inc., Manufacturing Company to sell x blank audiocasette tapes per week is given by the polynomial function $R(x) = 2x$. Find the total revenue from selling 20,000 tapes per week.

Review and Preview

Multiply. See Section 1.2.

78. $5(3x - 2)$

79. $-7(2z - 6y)$

80. $-2(x^2 - 5x + 6)$

81. $5(-3y^2 - 2y + 7)$

Combining Concepts

82. The function $f(x) = 0.46x^2 + 0.39x + 33.13$ can be used to approximate the number of Americans enrolled in health maintenance organizations (HMOs), where x is the number of years after 1990 and $f(x)$ is the number of millions of Americans. (*Source:* Based on data from the National Center for Health Statistics, 1990–1996)

a. Approximate the number of Americans enrolled in HMOs in the year 1991.

b. Approximate the number of Americans enrolled in HMOs in the year 1996.

c. Use the function to predict the number of Americans enrolled in HMOs in the year 2001.

d. From parts (a), (b), and (c), determine whether the number of Americans enrolled in HMOs is rising at a steady rate. Explain why or why not.

83. ______

84. ______

85. ______

86. ______

87. ______

88. ______

Name ______

If $P(x) = 3x + 3$, $Q(x) = 4x^2 - 6x + 3$, *and* $R(x) = 5x^2 - 7$, *find each function.*

83. $P(x) + Q(x)$

84. $Q(x) - R(x)$

85. If $P(x) = 2x - 3$, find $P(a)$, $P(-x)$, and $P(x + h)$.

Perform each indicated operation.

86. $(8x^{2y} - 7x^y + 3) + (-4x^{2y} + 9x^y - 14)$

87. $(14z^{5x} + 3z^{2x} + z) - (2z^{5x} - 10z^{2x} + 3z)$

Find the perimeter.

88.

$(x + 5y)$ units

$(3x^2 - x + 2y)$ units

5.2 MULTIPLYING POLYNOMIALS

A MULTIPLYING ANY TWO POLYNOMIALS

Properties of real numbers and exponents are used continually in the process of multiplying polynomials. To multiply monomials, for example, we apply the commutative and associative properties of real numbers and the product rule for exponents.

Examples Multiply.

1. $(2x^3)(5x^6) = 2(5)(x^3)(x^6) = 10x^{3+6} = 10x^9$ — Group like bases and apply the product rule for exponents.
2. $(7y^4z^4)(-xy^{11}z^5) = 7(-1)x(y^4y^{11})(z^4z^5) = -7xy^{4+11}z^{4+5} = -7xy^{15}z^9$

HELPFUL HINT

See Sections 1.5 and 1.6 to review exponential expressions further.

To multiply a monomial by a polynomial other than a monomial, we use an expanded form of the distributive property:

$$a(b + c + d + \cdots + z) = ab + ac + ad + \cdots + az$$

Notice that the monomial a is multiplied by each term of the polynomial.

Examples Multiply.

3. $2x(5x - 4) = 2x(5x) + 2x(-4)$ — Use the distributive property.
$= 10x^2 - 8x$ — Multiply.

4. $-3x^2(4x^2 - 6x + 1) = -3x^2(4x^2) + (-3x^2)(-6x) + (-3x^2)(1)$
$= -12x^4 + 18x^3 - 3x^2$

5. $-xy(7x^2y + 3xy - 11) = -xy(7x^2y) + (-xy)(3xy) + (-xy)(-11)$
$= -7x^3y^2 - 3x^2y^2 + 11xy$

TRY THE CONCEPT CHECK IN THE MARGIN.

To multiply any two polynomials, we can use the following.

MULTIPLYING ANY TWO POLYNOMIALS

To multiply any two polynomials, use the distributive property and multiply each term of one polynomial by each term of the other polynomial. Then combine any like terms.

Example 6 Multiply: $(x + 3)(2x + 5)$

Solution: We multiply each term of $(x + 3)$ by $(2x + 5)$.

$(x + 3)(2x + 5) = x(2x + 5) + 3(2x + 5)$ — Use the distributive property.
$= 2x^2 + 5x + 6x + 15$ — Use the distributive property again.
$= 2x^2 + 11x + 15$ — Combine like terms.

Objectives

- **A** Multiply any two polynomials.
- **B** Multiply binomials.
- **C** Square binomials.
- **D** Multiply the sum and difference of two terms.
- **E** Multiply three or more polynomials.

SSM

CD-ROM

Video 5.2

Practice Problems 1–2

Multiply.

1. $(7y^2)(4y^5)$
2. $(-a^2b^3c)(10ab^2c^{12})$

Practice Problems 3–5

Multiply.

3. $4x(3x - 2)$
4. $-2y^3(5y^2 - 2y + 6)$
5. $-a^2b(4a^3 - 2ab + b^2)$

✓ CONCEPT CHECK

Find the error:

$4x(x - 5) + 2x$
$= 4x(x) + 4x(-5) + 4x(2x)$
$= 4x^2 - 20x + 8x^2$
$= 12x^2 - 20x$

Practice Problem 6

Multiply: $(x + 2)(3x + 1)$

Answers

1. $28y^7$, **2.** $-10a^3b^5c^{13}$, **3.** $12x^2 - 8x$, **4.** $-10y^5 + 4y^4 - 12y^3$, **5.** $-4a^5b + 2a^3b^2 - a^2b^3$, **6.** $3x^2 + 7x + 2$

✓ Concept Check: $4x(x - 5) + 2x$
$= 4x(x) + 4x(-5) + 2x$
$= 4x^2 - 20x + 2x$
$= 4x^2 - 18x$

Practice Problem 7

Multiply: $(5x - 1)(2x^2 - x + 4)$

Example 7

Multiply: $(2x - 3)(5x^2 - 6x + 7)$

Solution: We multiply each term of $(2x - 3)$ by each term of $(5x^2 - 6x + 7)$.

$$\begin{aligned}(2x - 3)(5x^2 - 6x + 7) &= 2x(5x^2 - 6x + 7) + (-3)(5x^2 - 6x + 7)\\ &= 10x^3 - 12x^2 + 14x - 15x^2 + 18x - 21\\ &= 10x^3 - 27x^2 + 32x - 21 \quad \text{Combine like terms.}\end{aligned}$$

Sometimes polynomials are easier to multiply vertically, in the same way we multiply real numbers. When multiplying vertically, we line up like terms in the **partial products** vertically. This makes combining like terms easier.

Practice Problem 8

Multiply vertically: $(3x^2 + 5)(x^2 - 6x + 1)$

Example 8

Multiply vertically: $(4x^2 + 7)(x^2 + 2x + 8)$

Solution:

$$\begin{array}{rrrrrl} & & x^2 & +\ 2x & +\ 8 & \\ & & 4x^2 & & +\ 7 & \\ \hline & & 7x^2 & +\ 14x & +\ 56 & 7(x^2 + 2x + 8)\\ 4x^4 & +\ 8x^3 & +\ 32x^2 & & & 4x^2(x^2 + 2x + 8)\\ \hline 4x^4 & +\ 8x^3 & +\ 39x^2 & +\ 14x & +\ 56 & \text{Combine like terms.}\end{array}$$

B MULTIPLYING BINOMIALS

When multiplying a binomial by a binomial, we can use a special order of multiplying terms, called the **FOIL** order. The letters of FOIL stand for "**F**irst–**O**uter–**I**nner–**L**ast." To illustrate this method, let's multiply $(2x - 3)$ by $(3x + 1)$.

Multiply the **F**irst terms of each binomial. $(2x - 3)(3x + 1)$ **F** $2x(3x) = 6x^2$

Multiply the **O**uter terms of each binomial. $(2x - 3)(3x + 1)$ **O** $2x(1) = 2x$

Multiply the **I**nner terms of each binomial. $(2x - 3)(3x + 1)$ **I** $-3(3x) = -9x$

Multiply the **L**ast terms of each binomial. $(2x - 3)(3x + 1)$ **L** $-3(1) = -3$

Combine like terms.

$$6x^2 + 2x - 9x - 3 = 6x^2 - 7x - 3$$

Practice Problem 9

Use the FOIL order to multiply $(x - 7)(x + 5)$.

Example 9

Use the FOIL order to multiply $(x - 1)(x + 2)$.

Solution:

First, Outer, Inner, Last

$$\begin{aligned}(x - 1)(x + 2) &= x \cdot x + 2 \cdot x + (-1)x + (-1)(2)\\ &= x^2 + 2x - x - 2\\ &= x^2 + x - 2 \quad \text{Combine like terms.}\end{aligned}$$

Answers

7. $10x^3 - 7x^2 + 21x - 4$, **8.** $3x^4 - 18x^3 + 8x^2 - 30x + 5$, **9.** $x^2 - 2x - 35$

Examples Use the FOIL order to multiply.

$$\begin{array}{l} \text{10. } (2x-7)(3x-4) = \overset{\text{First}}{2x(3x)} + \overset{\text{Outer}}{2x(-4)} + \overset{\text{Inner}}{(-7)(3x)} + \overset{\text{Last}}{(-7)(-4)} \\ \qquad = 6x^2 - 8x - 21x + 28 \\ \qquad = 6x^2 - 29x + 28 \end{array}$$

$$\begin{array}{l} \text{11. } (3x+y)(5x-2y) = \overset{F}{15x^2} - \overset{O}{6xy} + \overset{I}{5xy} - \overset{L}{2y^2} \\ \qquad = 15x^2 - xy - 2y^2 \end{array}$$

Practice Problems 10–11

Use the FOIL order to multiply.

10. $(4x - 3)(x - 6)$
11. $(6x + 5y)(2x - y)$

C SQUARING BINOMIALS

The **square of a binomial** is a special case of the product of two binomials. By the FOIL order for multiplying two binomials, we have

$$(a + b)^2 = (a + b)(a + b)$$

$$= \overset{F}{a^2} + \overset{O}{ab} + \overset{I}{ba} + \overset{L}{b^2}$$

$$= a^2 + 2ab + b^2$$

This product can be visualized geometrically by analyzing areas.

Area of square in the margin: $(a + b)^2$

Sum of areas of smaller rectangles: $a^2 + 2ab + b^2$

Thus, $(a + b)^2 = a^2 + 2ab + b^2$

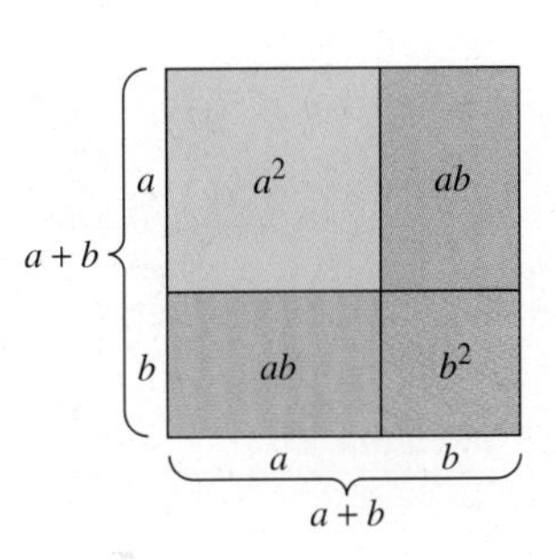

The same pattern occurs for the square of a difference. In general,

> **SQUARE OF A BINOMIAL**
>
> $(a + b)^2 = a^2 + 2ab + b^2$ $\qquad$ $(a - b)^2 = a^2 - 2ab + b^2$

In other words, a binomial squared is the sum of the first term squared, twice the product of both terms, and the second term squared.

Examples Multiply.

$$(a+b)^2 = a^2 + 2 \cdot a \cdot b + b^2$$

12. $(x + 5)^2 = x^2 + 2 \cdot x \cdot 5 + 5^2 = x^2 + 10x + 25$
13. $(x - 9)^2 = x^2 - 2 \cdot x \cdot 9 + 9^2 = x^2 - 18x + 81$
14. $(3x + 2z)^2 = (3x)^2 + 2(3x)(2z) + (2z)^2 = 9x^2 + 12xz + 4z^2$
15. $(4m^2 - 3n)^2 = (4m^2)^2 - 2(4m^2)(3n) + (3n)^2 = 16m^4 - 24m^2n + 9n^2$

Practice Problems 12–15

Multiply.

12. $(x + 3)^2$
13. $(y - 6)^2$
14. $(2x + 5y)^2$
15. $(6a^2 - 2b)^2$

> **HELPFUL HINT**
>
> Note that $(a + b)^2 = a^2 + 2ab + b^2$, not $a^2 + b^2$. Also, $(a - b)^2 = a^2 - 2ab + b^2$ *not* $a^2 - b^2$.

Answers

10. $4x^2 - 27x + 18$, **11.** $12x^2 + 4xy - 5y^2$, **12.** $x^2 + 6x + 9$, **13.** $y^2 - 12y + 36$, **14.** $4x^2 + 20xy + 25y^2$, **15.** $36a^4 - 24a^2b + 4b^2$

D MULTIPLYING THE SUM AND DIFFERENCE OF TWO TERMS

Another special product applies to the sum and difference of the same two terms. Multiply $(a + b)(a - b)$ to see a pattern.

$$(a + b)(a - b) = a^2 - ab + ba - b^2$$
$$= a^2 - b^2$$

> **PRODUCT OF THE SUM AND DIFFERENCE OF TWO TERMS**
>
> $$(a + b)(a - b) = a^2 - b^2$$

In other words, the product of the sum and difference of the same two terms is the difference of the first term squared and the second term squared.

Practice Problems 16–19

Multiply.

16. $(x + 4)(x - 4)$
17. $(3m - 6)(3m + 6)$
18. $(a^2 + 5y)(a^2 - 5y)$
19. $\left(4y^2 - \frac{1}{3}\right)\left(4y^2 + \frac{1}{3}\right)$

Examples

Multiply.

$(a + b)\ (a - b) = a^2 - b^2$

16. $(x + 3)(x - 3) = x^2 - 3^2 = x^2 - 9$

17. $(4y - 1)(4y + 1) = (4y)^2 - 1^2 = 16y^2 - 1$

18. $(x^2 + 2y)(x^2 - 2y) = (x^2)^2 - (2y)^2 = x^4 - 4y^2$

19. $\left(3m^2 - \frac{1}{2}\right)\left(3m^2 + \frac{1}{2}\right) = (3m^2)^2 - \left(\frac{1}{2}\right)^2 = 9m^4 - \frac{1}{4}$

Practice Problem 20

Multiply: $[(2x + 3y) - 2][(2x + 3y) + 2]$

Example 20

Multiply: $[(5x - 2y) - 1][(5x - 2y) + 1]$

Solution: We can think of $(5x - 2y)$ as the first term and 1 as the second term. Then we can apply the method for the product of the sum and difference of two terms.

$$\underbrace{[(5x - 2y)}_{a} \underbrace{- 1]}_{-b}\underbrace{[(5x - 2y)}_{a} \underbrace{+ 1]}_{+b} = \underbrace{(5x - 2y)^2}_{a^2} \underbrace{- 1^2}_{-b^2}$$
$$= (5x)^2 - 2(5x)(2y) + (2y)^2 - 1 \quad \text{Square } (5x - 2y).$$
$$= 25x^2 - 20xy + 4y^2 - 1$$

E MULTIPLYING THREE OR MORE POLYNOMIALS

To multiply three or more polynomials, more than one method may be needed.

Practice Problem 21

Multiply: $(y - 2)(y + 2)(y^2 - 4)$

Example 21

Multiply: $(x - 3)(x + 3)(x^2 - 9)$

Solution: We multiply the first two binomials, the sum and difference of two terms. Then we multiply the resulting two binomials, the square of a binomial.

$$(x - 3)(x + 3)(x^2 - 9) = (x^2 - 9)(x^2 - 9) \quad \text{Multiply } (x - 3)(x + 3).$$
$$= (x^2 - 9)^2$$
$$= x^4 - 18x^2 + 81 \quad \text{Square } (x^2 - 9).$$

Answers

16. $x^2 - 4$, **17.** $9m^2 - 36$, **18.** $a^4 - 25y^2$, **19.** $16y^4 - \frac{1}{9}$, **20.** $4x^2 + 12xy + 9y^2 - 4$, **21.** $y^4 - 8y^2 + 16$

GRAPHING CALCULATOR EXPLORATIONS

In the previous section, we used a grapher to visualize addition and subtraction of polynomials in one variable. In this section, the same method is used to visualize multiplication of polynomials in one variable. For example, to see that

$$(x - 2)(x + 1) = x^2 - x - 2$$

graph both $Y_1 = (x - 2)(x + 1)$ and $Y_2 = x^2 - x - 2$ on the same screen and see whether their graphs coincide.

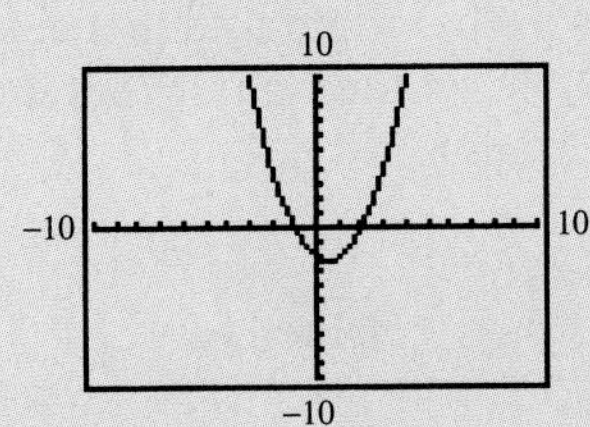

By tracing along both graphs, we see that the graphs of Y_1 and Y_2 appear to coincide, and thus $(x - 2)(x + 1) = x^2 - x - 2$ appears to be correct.

Multiply. Then use a grapher to visualize the results.

1. $(x + 4)(x - 4)$

2. $(x + 3)(x + 3)$

3. $(3x - 7)^2$

4. $(5x - 2)^2$

5. $(5x + 1)(x^2 - 3x - 2)$

6. $(7x + 4)(2x^2 + 3x - 5)$

Focus On Mathematical Connections

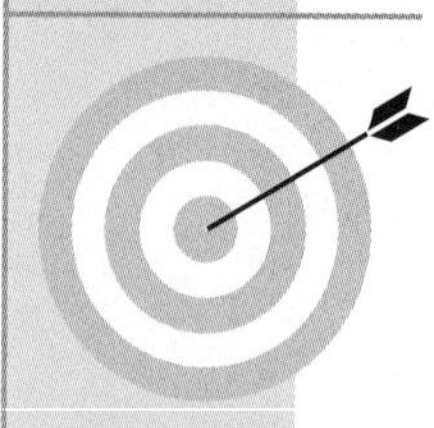

FINITE DIFFERENCES

When polynomial functions are evaluated at successive integer values, a list of values called a **sequence** is generated. The differences between successive pairs of numbers in such a sequence have special properties. Let's investigate these properties, beginning with a first-degree polynomial function, the linear function.

Notice in the table below on the left that *first differences* are the differences between the successive pairs of numbers in the original sequence. Find the first differences for any other linear function and fill in the table on the right. What do you notice? (*Note:* You may wish to try several different linear functions.)

x	Original Sequence $f(x) = 3x + 4$	First Differences
8	28	
7	25	3
6	22	3
5	19	3
4	16	3
3	13	3
2	10	3
1	7	3

x	Original Sequence $f(x) =$	First Differences

Now let's look at differences for a second-degree polynomial. Notice in the table below on the left that *second differences* are the differences between successive pairs of first differences. Find first and second differences for any other second-degree polynomial function and fill in the table on the right. What do you notice? (*Note:* You may wish to try several different second-degree polynomial functions.)

x	Original Sequence $f(x) = 2x^2 - 3x + 4$	First Differences	Second Differences
8	108		
7	81	27	
6	58	23	4
5	39	19	4
4	24	15	4
3	13	11	4
2	6	7	4
1	3	3	4

x	Original Sequence $f(x) =$	First Differences	Second Differences

CRITICAL THINKING

1. As you might guess, third differences are the differences between successive pairs of second differences. Find the first, second, and third differences for any two third-degree polynomial functions. What do you notice?
2. What would you expect to be true about the differences for a fourth-degree polynomial function?
3. What would you expect to be true about the differences for an nth-degree polynomial function?

Name ______________________ Section ________ Date ________

Exercise Set 5.2

A *Multiply. See Examples 1 through 8.*

1. $(-4x^3)(3x^2)$

2. $(-6a)(4a)$

3. $3x(4x + 7)$

4. $5x(6x - 4)$

5. $-6xy(4x + y)$

6. $-8y(6xy + 4x)$

7. $-4ab(xa^2 + ya^2 - 3)$

8. $-6b^2z(z^2a + baz - 3b)$

9. $(x - 3)(2x + 4)$

10. $(y + 5)(3y - 2)$

11. $(2x + 3)(x^3 - x + 2)$

12. $(a + 2)(3a^2 - a + 5)$

13. $\begin{array}{r} 3x - 2 \\ \underline{5x + 1} \end{array}$

14. $\begin{array}{r} 2z - 4 \\ \underline{6z - 2} \end{array}$

15. $\begin{array}{r} 3m^2 + 2m - 1 \\ \underline{5m + 2} \end{array}$

16. $\begin{array}{r} 2x^2 - 3x - 4 \\ \underline{x + 5} \end{array}$

17. $\begin{array}{r} 3x^2 + 4x - 4 \\ \underline{3x + 6} \end{array}$

18. $\begin{array}{r} 6x^2 + 2x - 1 \\ \underline{3x - 6} \end{array}$

19. $-6a^2b^2(5a^2b^2 - 6a - 6b)$

20. $7x^2y^3(-3ax - 4xy + z)$

21. $(2x^3 + 5)(5x^2 + 4x + 1)$

22. $(3y^3 - 1)(3y^3 - 6y + 1)$

23. $(3x^2 + 2x - 1)^2$

24. $(4x^2 + 4x - 4)^2$

25. $(3x + 1)(4x^2 - 2x + 5)$

26. $(2x - 1)(5x^2 - x - 2)$

27. Explain how to multiply a polynomial by a polynomial.

28. Explain why $(3x + 2)^2$ does not equal $9x^2 + 4$.

Answers

1. ________
2. ________
3. ________
4. ________
5. ________
6. ________
7. ________
8. ________
9. ________
10. ________
11. ________
12. ________
13. ________
14. ________
15. ________
16. ________
17. ________
18. ________
19. ________
20. ________
21. ________
22. ________
23. ________
24. ________
25. ________
26. ________
27. ________
28. ________

29. ______
30. ______
31. ______
32. ______
33. ______
34. ______
35. ______
36. ______
37. ______
38. ______
39. ______
40. ______
41. ______
42. ______
43. ______
44. ______
45. ______
46. ______
47. ______
48. ______
49. ______
50. ______
51. ______
52. ______
53. ______
54. ______
55. ______
56. ______

Name ______________________________

B *Use the FOIL order to multiply. See Examples 9 through 11.*

29. $(x - 3)(x + 4)$

30. $(c - 3)(c + 1)$

31. $(5x + 8y)(2x - y)$

32. $(2n - 9m)(n - 7m)$

33. $(3x - 1)(x + 3)$

34. $(5d - 3)(d + 6)$

35. $(a - 4)(2a - 4)$

36. $(2x - 3)(x + 1)$

37. $(y - 4)(y - 3)$

38. $(c - 8)(c + 2)$

39. $(3x + 1)(3x + 5)$

40. $(4x - 5)(5x + 6)$

41. $\left(4x + \frac{1}{3}\right)\left(4x - \frac{1}{2}\right)$

42. $\left(4y - \frac{1}{3}\right)\left(3y - \frac{1}{8}\right)$

43. $(5x^2 - 2y^2)(x^2 - 3y^2)$

44. $(4x^2 - 5y^2)(x^2 - 2y^2)$

C D *Use special products to multiply. See Examples 12 through 20.*

45. $(x + 4)^2$

46. $(x - 5)^2$

47. $(6y - 1)(6y + 1)$

48. $(x - 9)(x + 9)$

49. $(3x - y)^2$

50. $(4x - z)^2$

51. $(7ab + 3c)(7ab - 3c)$

52. $(3xy - 2b)(3xy + 2b)$

53. $(m - 4)^2$

54. $(x + 2)^2$

55. $(3x + 1)^2$

56. $(4x + 6)^2$

Name ______________________

57. $(3b - 6y)(3b + 6y)$

58. $(2x - 4y)(2x + 4y)$

59. $(7x - 3)(7x + 3)$

60. $(4x + 1)(4x - 1)$

61. $\left(3x + \frac{1}{2}\right)\left(3x - \frac{1}{2}\right)$

62. $\left(2x - \frac{1}{3}\right)\left(2x + \frac{1}{3}\right)$

63. $(6x + 1)^2$

64. $(4x + 7)^2$

65. $(x^2 + 2y)(x^2 - 2y)$

66. $(3x + 2y)(3x - 2y)$

67. $[3 + (4b + 1)]^2$

68. $[5 - (3b - 3)]^2$

69. $[(2s - 3) - 1][(2s - 3) + 1]$

70. $[(2y + 5) + 6][(2y + 5) - 6]$

71. $[(xy + 4) - 6]^2$

72. $[(2a^2 + 4a) + 1]^2$

E *Multiply. See Example 21.*

73. $(x + y)(2x - 1)(x + 1)$

74. $(z + 2)(z - 3)(2z + 1)$

75. $(x - 2)^4$

76. $(x - 1)^4$

77. $(x - 5)(x + 5)(x^2 + 25)$

78. $(x + 3)(x - 3)(x^2 + 9)$

Review and Preview

Simplify. See Section 1.5.

79. $\frac{6x^3}{3x}$

80. $\frac{4x^7}{x^2}$

81. $\frac{20a^3b^5}{18ab^2}$

82. $\frac{15x^7y^2}{6xy^2}$

83. $\frac{8m^4n}{12mn}$

84. $\frac{6n^6p}{8np}$

57. ______
58. ______
59. ______
60. ______
61. ______
62. ______
63. ______
64. ______
65. ______
66. ______
67. ______
68. ______
69. ______
70. ______
71. ______
72. ______
73. ______
74. ______
75. ______
76. ______
77. ______
78. ______
79. ______
80. ______
81. ______
82. ______
83. ______
84. ______

85. ______

86. ______

87. ______

88. ______

89. ______

90. ______

91. ______

92. ______

93. ______

94. ______

95. ______

96. ______

Name ______

Combining Concepts

85. Find the area of the circle. Do not approximate π.

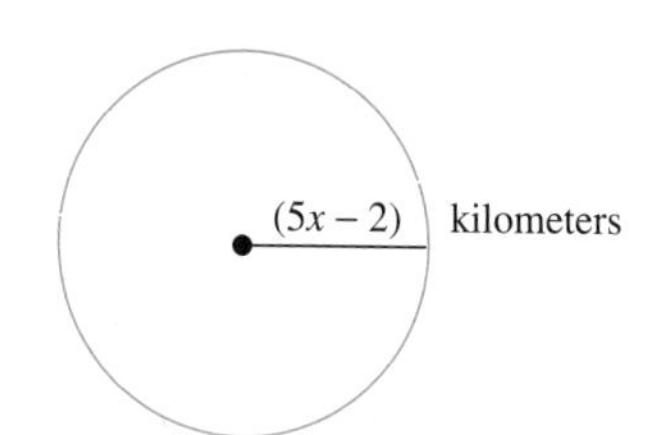

86. Find the volume of the cylinder. Do not approximate π.

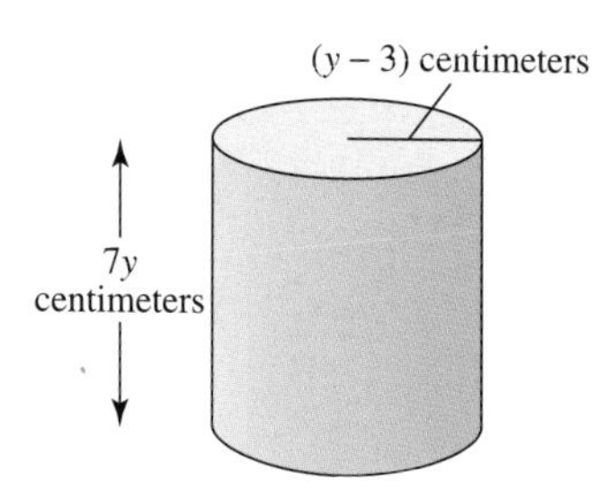

Multiply. Assume that variables represent positive integers.

87. $5x^2y^n(6y^{n+1} - 2)$

88. $-3yz^n(2y^3z^{2n} - 1)$

89. $(x^a + 5)(x^{2a} - 3)$

90. $(x^a + y^{2b})(x^a - y^{2b})$

91. Perform each indicated operation. Explain the difference between the two problems.

a. $(3x + 5) + (3x + 7)$

b. $(3x + 5)(3x + 7)$

92. Explain when the FOIL method can be used to multiply polynomials.

If $R(x) = x + 5$, $Q(x) = x^2 - 2$, and $P(x) = 5x$, find each function.

93. $P(x) \cdot R(x)$

94. $P(x) \cdot Q(x)$

If $f(x) = x^2 - 3x$, find each function value.

95. $f(a)$

96. $f(a + h)$

5.3 DIVIDING POLYNOMIALS

Objectives

A Divide a polynomial by a monomial.

B Divide by a polynomial.

C Use synthetic division.

SSM CD-ROM Video 5.3

Now that we have added, subtracted, and multiplied polynomials, we will learn how to divide them.

A DIVIDING A POLYNOMIAL BY A MONOMIAL

Recall the following addition fact for fractions with a common denominator:

$$\frac{a}{c} + \frac{b}{c} = \frac{a+b}{c}$$

If a, b, and c are monomials, we can read this equation from right to left and gain insight into how to divide a polynomial by a monomial.

> **DIVIDING A POLYNOMIAL BY A MONOMIAL**
>
> To divide a polynomial by a monomial, divide each term in the polynomial by the monomial.
>
> $$\frac{a+b}{c} = \frac{a}{c} + \frac{b}{c}, \qquad c \neq 0$$

Example 1 Divide $10x^3 - 5x^2 + 20x$ by $5x$.

Solution: We divide each term of $10x^3 - 5x^2 + 20x$ by $5x$ and simplify.

$$\frac{10x^3 - 5x^2 + 20x}{5x} = \frac{10x^3}{5x} - \frac{5x^2}{5x} + \frac{20x}{5x} = 2x^2 - x + 4$$

To check, see that (quotient) (divisor) = dividend, or

$$(2x^2 - x + 4)(5x) = 10x^3 - 5x^2 + 20x$$

Practice Problem 1

Divide $16y^3 - 8y^2 + 6y$ by $2y$.

Example 2 Divide: $\dfrac{3x^5y^2 - 15x^3y - x^2y - 6x}{x^2y}$

Solution: We divide each term in the numerator by x^2y.

$$\frac{3x^5y^2 - 15x^3y - x^2y - 6x}{x^2y} = \frac{3x^5y^2}{x^2y} - \frac{15x^3y}{x^2y} - \frac{x^2y}{x^2y} - \frac{6x}{x^2y}$$

$$= 3x^3y - 15x - 1 - \frac{6}{xy}$$

Practice Problem 2

Divide: $\dfrac{9a^3b^3 - 6a^2b^2 + a^2b - 4a}{a^2b}$

B DIVIDING BY A POLYNOMIAL

To divide a polynomial by a polynomial other than a monomial, we use **long division**. Polynomial long division is similar to long division of real numbers. We review long division of real numbers by dividing 7 into 296.

```
             42
Divisor:  7)296
           -28     4(7) = 28
            16     Subtract and bring down the next digit in the dividend.
           -14     2(7) = 14
             2     Subtract. The remainder is 2.
```

Answers

1. $8y^2 - 4y + 3$, **2.** $9ab^2 - 6b + 1 - \dfrac{4}{ab}$

The quotient is $42\frac{2}{7}\frac{\text{(remainder)}}{\text{(divisor)}}$. To check, notice that

$$42(7) + 2 = 296 \quad \text{The dividend}$$

This same division process can be applied to polynomials, as shown next.

Practice Problem 3

Divide $6x^2 + 11x - 2$ by $x + 2$.

Example 3 Divide $2x^2 - x - 10$ by $x + 2$.

Solution: $2x^2 - x - 10$ is the dividend, and $x + 2$ is the divisor.

Step 1. Divide $2x^2$ by x.

$$x + 2\overline{)2x^2 - x - 10} \quad \text{quotient: } 2x$$

$\frac{2x^2}{x} = 2x$, so $2x$ is the first term of the quotient.

Step 2. Multiply $2x(x + 2)$.

$$\begin{array}{r} 2x \\ x + 2\overline{)2x^2 - x - 10} \\ 2x^2 + 4x \end{array}$$

$2x(x + 2)$
Like terms are lined up vertically.

Step 3. Subtract $(2x^2 + 4x)$ from $(2x^2 - x - 10)$ by changing the signs of $(2x^2 + 4x)$ and adding.

$$\begin{array}{r} 2x \\ x + 2\overline{)\ 2x^2 - x - 10} \\ \underline{-2x^2 - 4x} \\ -5x \end{array}$$

Step 4. Bring down the next term, -10, and start the process over.

$$\begin{array}{r} 2x \\ x + 2\overline{)\ 2x^2 - x - 10} \\ \underline{-2x^2 - 4x} \\ -5x - 10 \end{array}$$

Step 5. Divide $-5x$ by x.

$$\begin{array}{r} 2x - 5 \\ x + 2\overline{)\ 2x^2 - x - 10} \\ \underline{-2x^2 - 4x} \\ -5x - 10 \end{array}$$

$\frac{-5x}{x} = -5$ so -5 is the second term of the quotient.

Step 6. Multiply $-5(x + 2)$.

$$\begin{array}{r} 2x - 5 \\ x + 2\overline{)\ 2x^2 - x - 10} \\ \underline{-2x^2 - 4x} \\ -5x - 10 \\ \underline{-5x - 10} \end{array}$$

$-5(x + 2)$
Like terms are lined up vertically.

Step 7. Subtract $(-5x - 10)$ from $(-5x - 10)$.

$$\begin{array}{r} 2x - 5 \\ x + 2\overline{)\ 2x^2 - x - 10} \\ \underline{-2x^2 - 4x} \\ -5x - 10 \\ \underline{+5x + 10} \\ 0 \end{array}$$

Answer

3. $6x - 1$

Then $\dfrac{2x^2 - x - 10}{x + 2} = 2x - 5$. There is no remainder.

Check this result by multiplying $2x - 5$ by $x + 2$. Their product is

$(2x - 5)(x + 2) = 2x^2 - x - 10$ The dividend

Example 4 Divide: $(6x^2 - 19x + 12) \div (3x - 5)$

Solution:

$$\begin{array}{r|l} 2x & \\ 3x - 5\,\overline{)\,6x^2 - 19x + 12} & \text{Divide: } \frac{6x^2}{3x} = 2x. \\ \underline{6x^2 - 10x} & \text{Multiply: } 2x(3x - 5). \\ -9x + 12 & \text{Subtract: } 6x^2 - 19x - (6x^2 - 10x) = -9x. \\ & \text{Bring down the next term, } + 12. \end{array}$$

$$\begin{array}{r|l} 2x - 3 & \\ 3x - 5\,\overline{)\,6x^2 - 19x + 12} & \text{Divide: } \frac{-9x}{3x} = -3. \\ \underline{6x^2 - 10x} & \\ -9x + 12 & \\ \underline{-9x + 15} & \text{Multiply } -3(3x - 5). \\ -3 & \text{Subtract: } -9x + 12 - (-9x + 15) = -3. \end{array}$$

Check: divisor · quotient + remainder

$$(3x - 5)(2x - 3) + (-3) = 6x^2 - 19x + 15 - 3$$
$$= 6x^2 - 19x + 12$$

The dividend

The division checks, so

$$\frac{6x^2 - 19x + 12}{3x - 5} = 2x - 3 - \frac{3}{3x - 5}$$

HELPFUL HINT

This fraction is the remainder over the divisor.

Example 5 Divide: $3x^4 + 2x^3 - 8x + 6$ by $x^2 - 1$.

Solution: Before dividing, we represent any "missing powers" by the product of 0 and the variable raised to the missing power. There is no x^2 term in the dividend, so we include $0x^2$ to represent the missing term. Also, there is no x term in the divisor, so we include $0x$ in the divisor.

$$\begin{array}{r|l} 3x^2 + 2x + 3 & \frac{3x^4}{x^2} = 3x^2 \\ x^2 + 0x - 1\,\overline{)\,3x^4 + 2x^3 + 0x^2 - 8x + 6} & \\ \underline{3x^4 + 0x^3 - 3x^2} & 3x^2(x^2 + 0x - 1) \\ 2x^3 + 3x^2 - 8x & \text{Subtract. Bring down } -8x. \\ & 2x^3/x^2 = 2x\text{, a term of the quotient.} \\ \underline{2x^3 + 0x^2 - 2x} & 2x(x^2 + 0x - 1) \\ 3x^2 - 6x + 6 & \text{Subtract. Bring down 6.} \\ & 3x^2/x^2 = 3\text{, a term of the quotient.} \\ \underline{3x^2 + 0x - 3} & 3(x^2 + 0x - 1) \\ -6x + 9 & \text{Subtract.} \end{array}$$

Practice Problem 4

Divide: $(10x^2 - 17x + 5) \div (5x - 1)$

Practice Problem 5

Divide: $3x^4 + 4x^2 - 6x + 1$ by $x^2 + 1$.

Answers

4. $2x - 3 + \dfrac{2}{5x - 1}$,

5. $3x^2 + 1 - \dfrac{6x}{x^2 + 1}$

The division process is finished when the degree of the remainder polynomial is less than the degree of the divisor. Thus,

$$\frac{3x^4 + 2x^3 - 8x + 6}{x^2 - 1} = 3x^2 + 2x + 3 + \frac{-6x + 9}{x^2 - 1}$$

Practice Problem 6

Divide $64x^3 - 27$ by $4x - 3$.

Example 6 Divide: $27x^3 + 8$ by $3x + 2$.

Solution: We replace the missing terms in the dividend with $0x^2$ and $0x$.

$$\begin{array}{r l}
 & 9x^2 - 6x + 4 \\
3x + 2 \overline{)} & 27x^3 + 0x^2 + 0x + 8 \\
 & 27x^3 + 18x^2 \qquad 9x^2(3x + 2) \\
 & -18x^2 + 0x \qquad \text{Subtract. Bring down } 0x. \\
 & -18x^2 - 12x \qquad -6x(3x + 2) \\
 & 12x + 8 \qquad \text{Subtract. Bring down 8.} \\
 & 12x + 8 \qquad 4(3x + 2)
\end{array}$$

Thus $\frac{27x^3 + 8}{3x + 2} = 9x^2 - 6x + 4$.

✓ CONCEPT CHECK

In a division problem, the divisor is $4x^3 - 5$. The division process can be stopped when which of these possible remainder polynomials is reached?

a. $2x^4 + x^2 - 3$

b. $x^3 - 5x^2$

c. $4x^2 + 25$

TRY THE CONCEPT CHECK IN THE MARGIN.

C USING SYNTHETIC DIVISION

When a polynomial is to be divided by a binomial of the form $x - c$, a shortcut process called **synthetic division** may be used. On the left is an example of long division, and on the right is the same example showing the coefficients of the variables only.

$$\begin{array}{r l}
 & 2x^2 + 5x + 2 \\
x - 3 \overline{)} & 2x^3 - x^2 - 13x + 1 \\
 & 2x^3 - 6x^2 \\
 & 5x^2 - 13x \\
 & 5x^2 - 15x \\
 & 2x + 1 \\
 & 2x - 6 \\
 & 7
\end{array}
\qquad
\begin{array}{r l}
 & 2 \quad 5 \quad 2 \\
1 - 3 \overline{)} & 2 - 1 - 13 + 1 \\
 & 2 - 6 \\
 & 5 - 13 \\
 & 5 - 15 \\
 & 2 + 1 \\
 & 2 - 6 \\
 & 7
\end{array}$$

Notice that as long as we keep coefficients of powers of x in the same column, we can perform division of polynomials by performing algebraic operations on the coefficients only. This shortest process of dividing with coefficients only in a special format is called synthetic division. To find $(2x^3 - x^2 - 13x + 1) \div (x - 3)$ by synthetic division, follow the next example.

Practice Problem 7

Use synthetic division to divide $3x^3 - 2x^2 + 5x + 4$ by $x - 2$.

Example 7 Use synthetic division to divide $2x^3 - x^2 - 13x + 1$ by $x - 3$.

Solution: To use synthetic division, the divisor must be in the form $x - c$. Since we are dividing by $x - 3$, c is 3. We write down 3 and the coefficients of the dividend.

Answers

6. $16x^2 + 12x + 9$,

7. $3x^2 + 4x + 13 + \frac{30}{x - 2}$,

✓ Concept Check: c

$$\begin{array}{r|rrrr} 3 & 2 & -1 & -13 & 1 \\ \hline & 2 & & & \end{array}$$

Next, draw a line and bring down the first coefficient of the dividend.

$$\begin{array}{r|rrrr} 3 & 2 & -1 & -13 & 1 \\ & & 6 & & \\ \hline & 2 & & & \end{array}$$

Multiply $3 \cdot 2$ and write down the product, 6.

$$\begin{array}{r|rrrr} 3 & 2 & -1 & -13 & 1 \\ & & 6 & & \\ \hline & 2 & 5 & & \end{array}$$

Add $-1 + 6$. Write down the sum, 5.

$$\begin{array}{r|rrrr} 3 & 2 & -1 & -13 & 1 \\ & & 6 & 15 & \\ \hline & 2 & 5 & 2 & \end{array}$$

$3 \cdot 5 = 15$.
$-13 + 15 = 2$.

$$\begin{array}{r|rrrr} 3 & 2 & -1 & -13 & 1 \\ & & 6 & 15 & 6 \\ \hline & 2 & 5 & 2 & 7 \end{array}$$

$3 \cdot 2 = 6$.
$1 + 6 = 7$.

The quotient is found in the bottom row. The numbers 2, 5, and 2 are the coefficients of the quotient polynomial, and the number 7 is the remainder. The degree of the quotient polynomial is one less than the degree of the dividend. In our example, the degree of the dividend is 3, so the degree of the quotient polynomial is 2. As we found when we performed the long division, the quotient is

$2x^2 + 5x + 2, \qquad$ remainder 7

or

$$2x^2 + 5x + 2 + \frac{7}{x - 3}$$

When using synthetic division, if there are missing powers of the variable, insert 0s as coefficients.

Example 8 Use synthetic division to divide $x^4 - 2x^3 - 11x^2 + 34$ by $x + 2$.

Solution: The divisor is $x + 2$, which in the form $x - c$ is $x - (-2)$. Thus, c is -2. There is no x-term in the dividend, so we insert a coefficient of 0. The dividend coefficients are 1, -2, -11, 0, and 34.

$$\begin{array}{r|rrrrr} -2 & 1 & -2 & -11 & 0 & 34 \\ & & -2 & 8 & 6 & -12 \\ \hline & 1 & -4 & -3 & 6 & 22 \end{array}$$

Practice Problem 8

Use synthetic division to divide $x^4 + 3x^3 - 5x + 4$ by $x + 1$.

Answer

8. $x^3 + 2x^2 - 2x - 3 + \dfrac{7}{x + 1}$

The dividend is a fourth-degree polynomial, so the quotient polynomial is a third-degree polynomial. The quotient is $x^3 - 4x^2 - 3x + 6$ with a remainder of 22. Thus,

$$\frac{x^4 - 2x^3 - 11x^2 + 34}{x + 2} = x^3 - 4x^2 - 3x + 6 + \frac{22}{x + 2}$$

HELPFUL HINT

Before dividing by synthetic division, write the dividend in descending order of variable exponents. Any "missing powers" of the variable must be represented by 0 times the variable raised to the missing power.

TRY THE CONCEPT CHECK IN THE MARGIN.

✓ CONCEPT CHECK

Which division problems are candidates for the synthetic division process?

a. $(3x^2 + 5) \div (x + 4)$

b. $(x^3 - x^2 + 2) \div (3x^3 - 2)$

c. $(y^4 + y - 3) \div (x^2 + 1)$

d. $x^5 \div (x - 5)$

Focus On Business and Career

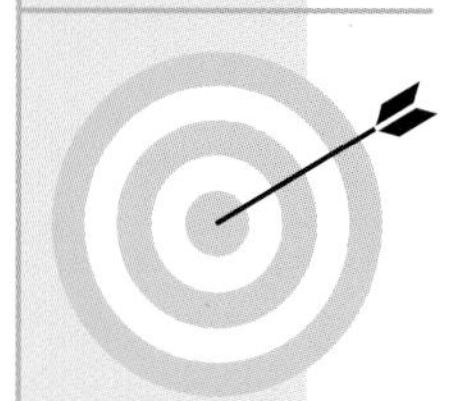

BUSINESS TERMS

For most businesses, a financial goal is to "make money." But what does that mean from a mathematical point of view? To find out, we must first discuss some common business terms.

- **Revenue** is the amount of money a business takes in. A company's annual revenue is the amount of money it collects during its fiscal, or business, year. For most companies, the largest source of revenue is from the sales of their products or services. For instance, a computer manufacturer's annual revenue is the amount of money it collects during the year from selling computers to customers. Large companies may also have revenues from investment interest or leases. When revenue can be expressed as a function of another variable, it is often denoted as $R(x)$.
- **Expenses** are the costs of doing business. For instance, a large part of a computer manufacturer's expenses include the cost of the computer components it buys from wholesalers to use in the manufacturing or assembling process. Other expenses include salaries, mortgage payments, equipment, taxes, advertising, and so on. Some businesses refer to their expenses simply as cost. When cost can be expressed as a function of another variable, it is often denoted as $C(x)$.
- **Net income/loss** is the difference between a company's annual revenues and expenses. This difference may also be referred to as net earnings. Positive net earnings—that is, a positive difference—result in a net income or net profit. Posting a net income can be interpreted as "making money." Negative net earnings—that is, a negative difference—result in a net loss. Posting a net loss can be interpreted as "losing money." A profit function can be expressed as $P(x) = R(x) - C(x)$. In this case, a negative profit is interpreted as a net loss.

GROUP ACTIVITY

Locate several corporate annual reports. Using the data in the reports, verify that the net income or net earnings given in a report was calculated as the difference between revenue and expenses. If this was not the case, can you tell what caused the variation? If so, explain.

Answer

✓ Concept Check: a and d

Name ______________________ Section ______ Date ________

EXERCISE SET 5.3

A *Divide. See Examples 1 and 2.*

1. $4a^2 + 8a$ by $2a$

2. $6x^4 - 3x^3$ by $3x^2$

3. $\dfrac{12a^5b^2 + 16a^4b}{4a^4b}$

4. $\dfrac{4x^3y + 12x^2y^2 - 4xy^3}{4xy}$

5. $\dfrac{4x^2y^2 + 6xy^2 - 4y^2}{2x^2y}$

6. $\dfrac{6x^5 + 74x^4 + 24x^3}{2x^3}$

B *Divide. See Examples 3 through 6.*

7. $(x^2 + 3x + 2) \div (x + 2)$

8. $(y^2 + 7y + 10) \div (y + 5)$

9. $(2x^2 - 6x - 8) \div (x + 1)$

10. $(3x^2 + 19x + 20) \div (x + 5)$

11. $2x^2 + 3x - 2$ by $2x + 4$

12. $6x^2 - 17x - 3$ by $3x - 9$

13. $(4x^3 + 7x^2 + 8x + 20) \div (2x + 4)$

14. $(18x^3 + x^2 - 90x - 5) \div (9x^2 - 45)$

15. $(6x^3 + 2x^2 - 18x - 6) \div (3x + 1)$

16. $(10x^3 - 15x^2 + 4x - 6) \div (2x - 3)$

17. $(2x^3 - 6x^2 - 4) \div (x - 4)$

18. $(3x^3 + 4x - 10) \div (x + 2)$

19. $(10x^3 - 5x^2 - 12x + 1) \div (2x - 1)$

20. $(20x^3 - 8x^2 + 5x - 5) \div (5x - 2)$

21. $(3x^5 - x^3 + 4x^2 - 12x - 8) \div (x^2 - 2)$

22. $(2x^5 - 6x^4 + x^3 - 4x + 3) \div (x^2 - 3)$

ANSWERS

1. ______
2. ______
3. ______
4. ______
5. ______
6. ______
7. ______
8. ______
9. ______
10. ______
11. ______
12. ______
13. ______
14. ______
15. ______
16. ______
17. ______
18. ______
19. ______
20. ______
21. ______
22. ______

Name ______________________

23. ______
24. ______
25. ______
26. ______
27. ______
28. ______
29. ______
30. ______
31. ______
32. ______
33. ______
34. ______
35. ______
36. ______
37. ______
38. ______
39. ______
40. ______

23. $\left(2x^4 + \frac{1}{2}x^3 + x^2 + x\right) \div (x - 2)$

24. $\left(x^4 - \frac{2}{3}x^3 + x\right) \div (x - 3)$

C *Use synthetic division to divide. See Examples 7 and 8.*

25. $\frac{x^2 + 3x - 40}{x - 5}$

26. $\frac{x^2 - 14x + 24}{x - 2}$

27. $\frac{x^2 + 5x - 6}{x + 6}$

28. $\frac{x^2 + 12x + 32}{x + 4}$

29. $\frac{x^3 - 7x^2 - 13x + 5}{x - 2}$

30. $\frac{x^3 + 6x^2 + 4x - 7}{x + 5}$

31. $\frac{4x^2 - 9}{x - 2}$

32. $\frac{3x^2 - 4}{x - 1}$

33. $\frac{2x^4 - 13x^3 + 16x^2 - 9x + 20}{x - 5}$

34. $\frac{3x^4 + 5x^3 - x^2 + x - 2}{x + 2}$

35. $\frac{7x^2 - 4x + 12 + 3x^3}{x + 1}$

36. $\frac{x^4 + 4x^3 - x^2 - 16x - 4}{x - 2}$

37. $\frac{3x^3 + 2x^2 - 4x + 1}{x - \frac{1}{3}}$

38. $\frac{9y^3 + 9y^2 - y + 2}{y + \frac{2}{3}}$

39. $\frac{x^3 - 1}{x - 1}$

40. $\frac{y^3 - 8}{y - 2}$

Name ______________________________

Review and Preview

Multiply. See Section 5.2.

41. $6x(x + 3) + 5(x + 3)$

42. $7y(y - 1) + 2(y - 1)$

Solve each inequality. See Section 2.7.

43. $|x + 5| < 4$

44. $|x - 1| \leq 8$

45. $|2x + 7| \geq 9$

46. $|4x + 2| > 10$

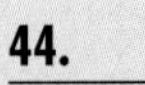

Combining Concepts

47. A board of length $(3x^4 + 6x^2 - 18)$ meters is to be cut into three pieces of the same length. Find the length of each piece.

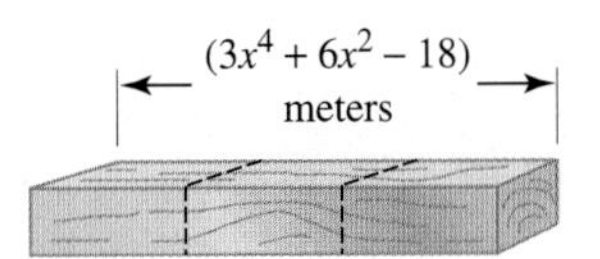

48. The perimeter of a regular hexagon is given to be $(12x^5 - 48x^3 + 3)$ miles. Find the length of each side.

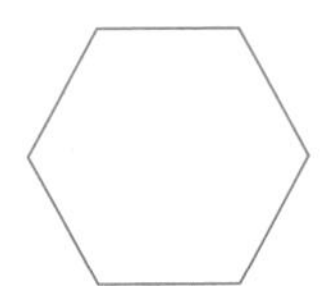

49. If the area of the rectangle is $(15x^2 - 29x - 14)$ square inches, and its length is $(5x + 2)$ inches, find its width.

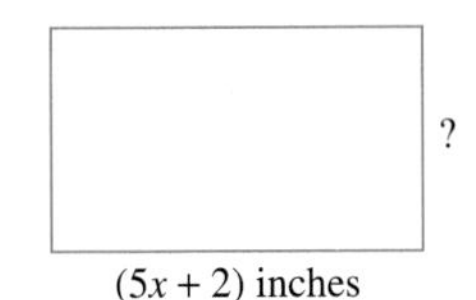

50. If the area of a parallelogram is $(2x^2 - 17x + 35)$ square centimeters and its base is $(2x - 7)$ centimeters, find its height.

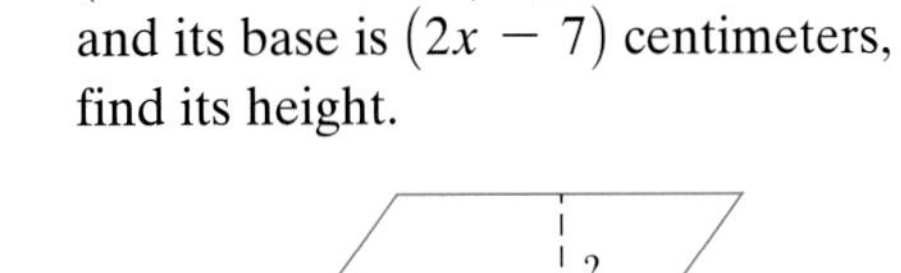

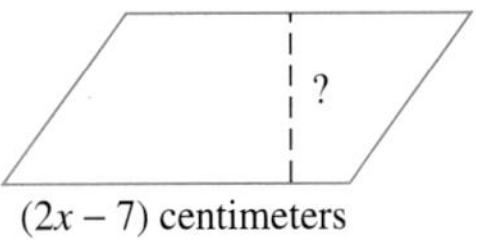

51. Find $P(1)$ for the polynomial function $P(x) = 3x^3 + 2x^2 - 4x + 3$. Next divide $3x^3 + 2x^2 - 4x + 3$ by $x - 1$. Compare the remainder with $P(1)$.

52. Find $P(-2)$ for the polynomial function $P(x) = x^3 - 4x^2 - 3x + 5$. Next divide $x^3 - 4x^2 - 3x + 5$ by $x + 2$. Compare the remainder with $P(-2)$.

41. ______
42. ______
43. ______
44. ______
45. ______
46. ______
47. ______
48. ______
49. ______
50. ______
51. ______
52. ______

53.

54.

55. a.

b.

Name

53. If a polynomial is divided by $x + 3$, the quotient is $x^2 - x + 10$ and the remainder is -2. Find the original polynomial.

54. Explain an advantage of using synthetic division instead of long division.

55. Gateway 2000, Inc., is a leading direct marketer of personal computers. Gateway's annual net profit can be modeled by the polynomial function $P(x) = -24.98x^3 + 130.36x^2 - 119.40x + 102.08$, where $P(x)$ is net profit in millions of dollars in the year x. Gateway's annual revenue can be modeled by the function $R(x) = 1145.80x + 1596.02$, where $R(x)$ is revenue in millions of dollars in the year x. In both models, $x = 0$ represents the year 1993. (*Source*: Gateway 2000, Inc., 1993–1997)

a. Suppose that a market analyst has found the model $P(x)$, and another analyst at the same firm has found the model $R(x)$. The analysts have been asked by their manager to work together to find a model for Gateway's net profit margin. The analysts know that a company's net profit margin is the ratio of its net profit to its revenue. Describe how these two analysts could collaborate to find a function $m(x)$ that models Gateway's net profit margin based on the work they have done independently.

b. Without actually finding $m(x)$, give a general description of what you would expect the form of the result to be.

5.4 THE GREATEST COMMON FACTOR AND FACTORING BY GROUPING

Objectives

A Factor out the greatest common factor of a polynomial's terms.

B Factor polynomials by grouping.

SSM CD-ROM Video 5.4

Factoring is the reverse process of multiplying. It is the process of writing a polynomial as a product.

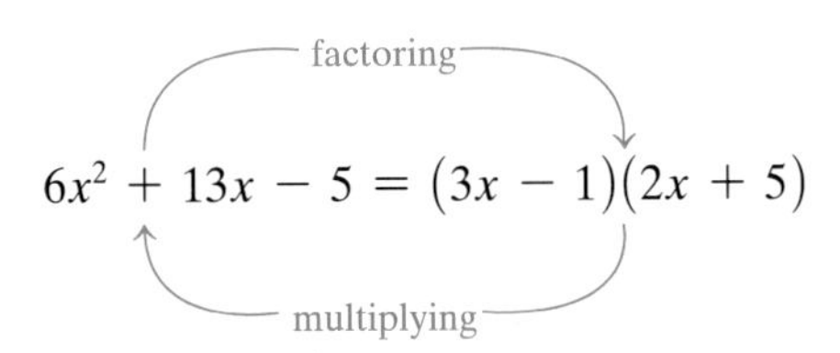

In the next few sections, we review techniques for factoring polynomials.

A FACTORING OUT THE GREATEST COMMON FACTOR

To factor a polynomial, we first **factor out** the greatest common factor of its terms, using the distributive property. The **greatest common factor** of a polynomial is the common factor with the largest exponents and the largest numerical coefficient.

For example, some common factors of

$6x^3 - 18x^2$ are 6 and x^2

but the product $6x^2$ is the greatest common factor. By the distributive property

$$6x^3 + 18x^2 = 6x^2 \cdot x - 6x^2 \cdot 3 = 6x^2(x - 3)$$

Example 1 Factor: $8x + 4$

Solution: The greatest common factor of the terms $8x$ and 4 is 4.

$8x + 4 = 4(2x) + 4(1)$ Factor out 4 from each term.

$= 4(2x + 1)$ Use the distributive property.

The factored form of $8x + 4$ is $4(2x + 1)$. To check, multiply $4(2x + 1)$ to see that the product is $8x + 4$.

Practice Problem 1

Factor: $9x + 3$

Examples Factor.

2. $6x^2 - 3x^3 = 3x^2(2) - 3x^2(x)$ Factor out the greatest common factor, $3x^2$.

$= 3x^2(2 - x)$ Use the distributive property.

3. $3y + 1$ There is no common factor other than 1.

4. $17x^3y^2 - 34x^4y^2 = 17x^3y^2(1) - 17x^3y^2(2x)$ Factor out the greatest common factor, $17x^3y^2$.

$= 17x^3y^2(1 - 2x)$ Use the distributive property.

Practice Problems 2–4

Factor each polynomial.

2. $20y^2 - 4y^3$

3. $6a - 7$

4. $6a^4b^2 - 3a^2b^2$

HELPFUL HINT

If the greatest common factor happens to be one of the terms in the polynomial, a factor of 1 will remain for this term when the greatest common factor is factored out. For example, in the polynomial $21x^2 + 7x$, the greatest common factor of $21x^2$ and $7x$ is $7x$, so

$$21x^2 + 7x = 7x(3x) + 7x(1) = 7x(3x + 1)$$

Answers

1. $3(3x + 1)$, **2.** $4y^2(5 - y)$, **3.** $6a - 7$, **4.** $3a^2b^2(2a^2 - 1)$

✓ Concept Check

Which factorization of $12x^2 + 9x - 3$ is correct?

a. $3(4x^2 + 3x + 1)$
b. $3(4x^2 + 3x - 1)$
c. $3(4x^2 + 3x - 3)$
d. $3(4x^2 + 3x)$

Try the Concept Check in the margin.

> **HELPFUL HINT**
>
> To check that the greatest common factor has been factored out correctly, multiply the factors together and see that their product is the original polynomial.

Practice Problem 5

Factor: $-2x^2y - 4xy + 10y$

Example 5 Factor: $-3x^3y + 2x^2y - 5xy$

Solution: Two possibilities are shown for factoring this polynomial. First, the common factor xy is factored out.

$$-3x^3y + 2x^2y - 5xy = xy(-3x^2 + 2x - 5)$$

Also, the common factor $-xy$ can be factored out as shown.

$$\begin{aligned}-3x^3y + 2x^2y - 5xy &= -xy(3x^2) + (-xy)(-2x) + (-xy)(5) \\ &= -xy(3x^2 - 2x + 5)\end{aligned}$$

Both of these are correct.

Practice Problem 6

Factor: $3(x + 7) + 5y(x + 7)$

Example 6 Factor: $2(x - 5) + 3a(x - 5)$

Solution: The greatest common factor is the binomial factor $(x - 5)$.

$$2(x - 5) + 3a(x - 5) = (x - 5)(2 + 3a)$$

Practice Problem 7

Factor: $6a(2a + 3b) - (2a + 3b)$

Example 7 Factor: $7x(x^2 + 5y) - (x^2 + 5y)$

Solution:

> **HELPFUL HINT**
>
> Notice that we wrote $-(x^2 + 5y)$ as $-1(x^2 + 5y)$ to aid in factoring.

$$\begin{aligned}7x(x^2 + 5y) - (x^2 + 5y) &= 7x(x^2 + 5y) - 1(x^2 + 5y) \\ &= (x^2 + 5y)(7x - 1)\end{aligned}$$

B Factoring by Grouping

Sometimes it is possible to factor a polynomial by grouping the terms of the polynomial and looking for common factors in each group. This method of factoring is called **factoring by grouping**.

Practice Problem 8

Factor: $xy - 5y + 3x - 15$

Example 8 Factor: $ab - 6a + 2b - 12$

Solution:

$$\begin{aligned}ab - 6a + 2b - 12 &= (ab - 6a) + (2b - 12) && \text{Group pairs of terms.} \\ &= a(b - 6) + 2(b - 6) && \text{Factor each binomial.} \\ &= (b - 6)(a + 2) && \text{Factor out the greatest common factor, } (b - 6).\end{aligned}$$

To check, multiply $(b - 6)$ and $(a + 2)$ to see that the product is $ab - 6a + 2b - 12$.

Answers

5. $-2y(x^2 + 2x - 5)$, **6.** $(x + 7)(3 + 5y)$, **7.** $(2a + 3b)(6a - 1)$, **8.** $(x - 5)(y + 3)$

✓ **Concept Check:** b

HELPFUL HINT

Notice that the polynomial $a(b-6)+2(b-6)$ is *not* in factored form. It is a *sum,* not a *product*. The factored form is $(b-6)(a+2)$.

Example 9 Factor: $x^3+5x^2+3x+15$

Solution:

$$\begin{aligned} x^3+5x^2+3x+15 &= (x^3+5x^2)+(3x+15) && \text{Group pairs of terms.} \\ &= x^2(x+5)+3(x+5) && \text{Factor each binomial.} \\ &= (x+5)(x^2+3) && \text{Factor out the common factor, } (x+5). \end{aligned}$$

Practice Problem 9

Factor: $y^3+6y^2+4y+24$

Example 10 Factor: $m^2n^2+m^2-2n^2-2$

Solution:

$$\begin{aligned} m^2n^2+m^2-2n^2-2 &= (m^2n^2+m^2)+(-2n^2-2) && \text{Group pairs of terms.} \\ &= m^2(n^2+1)-2(n^2+1) && \text{Factor each binomial.} \\ &= (n^2+1)(m^2-2) && \text{Factor out the common factor, } (n^2+1). \end{aligned}$$

Practice Problem 10

Factor: $a^2b^2+a^2-3b^2-3$

Example 11 Factor: $xy+2x-y-2$

Solution:

$$\begin{aligned} xy+2x-y-2 &= (xy+2x)+(-y-2) && \text{Group pairs of terms.} \\ &= x(y+2)-1(y+2) && \text{Factor each binomial.} \\ &= (y+2)(x-1) && \text{Factor out the common factor, } y+2. \end{aligned}$$

Practice Problem 11

Factor: $ab+5a-b-5$

Answers

9. $(y+6)(y^2+4)$, **10.** $(b^2+1)(a^2-3)$, **11.** $(b+5)(a-1)$

Focus On History

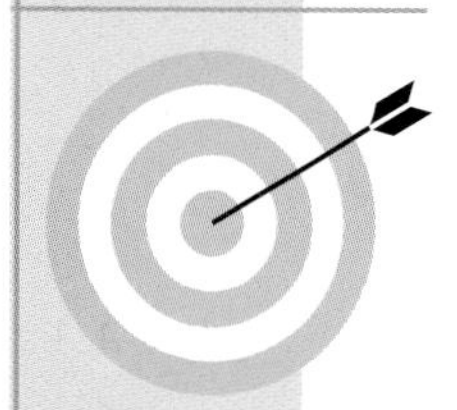

MUHAMMAD AL-KHWARIZMI

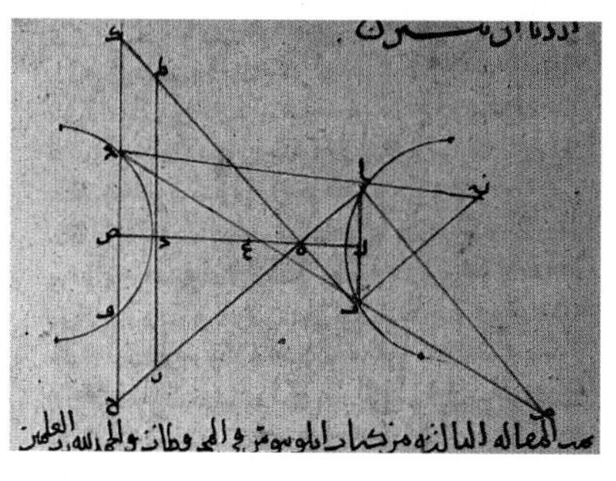

Muhammad ibn Musa al-Khwarizmi was an Arabic mathematician who lived from around 800 A.D. to about 847 A.D. He was originally from what is today Uzbekistan. He later became a scholar at the House of Wisdom in Baghdad (in what is today Iraq). In 830 A.D., al-Khwarizmi wrote the first known Arabic text on the algebra of polynomials. This text, *al-Kitab al-mukhtasar fi hisab al-jabr wa'l-muqabala* (translated in English as *The compendious book on calculation by completion and balancing* or *A summary of calculating through the reunion of broken parts*), discusses first-degree and second-degree polynomial equations and their application to practical matters. Al-Khwarizmi's *Al-jabr wa'l-muqabala* is believed to be a summary of Arab mathematics and the mathematical techniques and algebraic theories known to the Arabs through Hindu, Greek, Hebrew, and Babylonian influences.

Al-Khwarizmi and his *Al-jabr wa'l-muqabala* are important to the history of mathematics for several reasons. First, it is through this textbook that many diverse lines of mathematical thought were merged into what was to become a more unified theory of algebra. Second, a later translation of *Al-jabr wa'l-muqabala* into Latin was responsible for spreading algebraic concepts to Europe during the Middle Ages. Third, al-Khwarizmi's textbook is actually responsible for the naming of algebra. The word *algebra* is a corruption of the Arabic word *al-jabr* in the text's title. Finally, the Latin form of al-Khwarizmi's name is the source of the word *algorithm* in the English language.

Al-Khwarizmi was the first to encourage writing out mathematical calculations rather than relying solely on calculating with an abacus. He is also thought to be responsible for spreading the use of the modern decimal system for writing numbers and may have introduced the use of zero as a placeholder in the decimal system.

Muhammad ibn Musa al-Khwarizmi's many contributions to mathematics in general and algebra in particular were rewarded when a crater of the moon, Crater Al-Khwarizmi, was named in his honor by the International Astronomical Union in 1973.

Name ______________________ Section ________ Date ________

MENTAL MATH

Find the greatest common factor of each list of monomials.

1. 6, 12 **2.** 9, 27 **3.** $15x, 10$ **4.** $9x, 12$

5. $13x, 2x$ **6.** $4y, 5y$ **7.** $7x, 14x$ **8.** $8z, 4z$

EXERCISE SET 5.4

A *Factor out the greatest common factor. See Examples 1 through 7.*

1. $18x - 12$

2. $21x + 14$

3. $4y^2 - 16xy^3$

4. $3z - 21xz^4$

5. $6x^5 - 8x^4 + 2x^3$

6. $9x + 3x^2 - 6x^3$

7. $8a^3b^3 - 4a^2b^2 + 4ab + 16ab^2$

8. $12a^3b - 6ab + 18ab^2 - 18a^2b$

9. $6(x + 3) + 5a(x + 3)$

10. $2(x - 4) + 3y(x - 4)$

11. $2x(z + 7) + (z + 7)$

12. $x(y - 2) + (y - 2)$

13. $3x(x^2 + 5) - 2(x^2 + 5)$

14. $4x(2y + 3) - 5(2y + 3)$

15. The material needed to manufacture a tin can is given by the polynomial $2\pi r^2 + 2\pi rh$,

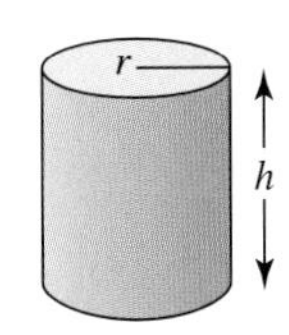

where the radius is r and height is h. Factor this expression.

16. The amount E of current in an electrical circuit is given by the formula $IR_1 + IR_2 = E$. Write an equivalent equation by factoring the expression $IR_1 + IR_2$.

MENTAL MATH ANSWERS

1. ______
2. ______
3. ______
4. ______
5. ______
6. ______
7. ______
8. ______

ANSWERS

1. ______
2. ______
3. ______
4. ______
5. ______
6. ______
7. ______
8. ______
9. ______
10. ______
11. ______
12. ______
13. ______
14. ______
15. ______
16. ______

17. ______

18. ______

19. ______

20. ______

21. ______

22. ______

23. ______

24. ______

25. ______

26. ______

27. ______

28. ______

29. ______

30. ______

31. ______

32. ______

Name ______

17. At the end of T years, the amount of money A in a savings account earning simple interest from an initial investment of P dollars at rate R is given by the formula $A = P + PRT$. Write an equivalent equation by factoring the expression $P + PRT$.

18. An open-topped box has a square base and a height of 10 inches. If each of the bottom edges of the box has length x inches, find the amount of material needed to construct the box. Write the answer in factored form.

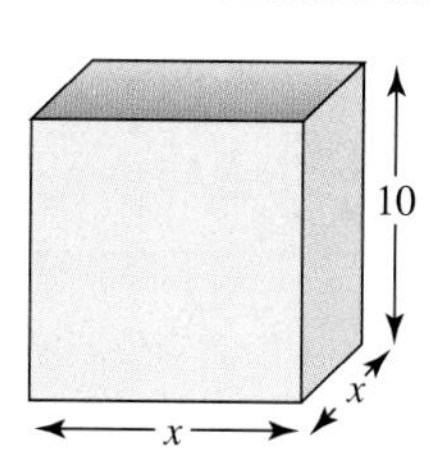

19. When $3x^2 - 9x + 3$ is factored, the result is $3(x^2 - 3x + 1)$. Explain why it is necessary to include the term 1 in this factored form.

20. Construct a trinomial whose greatest common factor is $5x^2y^3$.

B *Factor each polynomial by grouping. See Examples 8 through 11.*

21. $ab + 3a + 2b + 6$

22. $ab + 2a + 5b + 10$

23. $ac + 4a - 2c - 8$

24. $bc + 8b - 3c - 24$

25. $2xy - 3x - 4y + 6$

26. $12xy - 18x - 10y + 15$

27. $12xy - 8x - 3y + 2$

28. $20xy - 15x - 4y + 3$

29. $x^3 + 3x^2 + 4x + 12$

30. $x^3 + 4x^2 + 3x + 12$

31. $x^3 - x^2 - 2x + 2$

32. $x^3 - 2x^2 - 3x + 6$

Name ______________________________

33. $2x^2 + 3xy + 4x + 6y$

34. $3x^2 + 12x + 4xy + 16y$

35. $5x^2 + 5xy - 3x - 3y$

36. $4x^2 + 2xy - 10x - 5y$

37. $6xy + 10x + 9y + 15$

38. $15xy + 20x + 6y + 8$

39. $xy + 3y - 5x - 15$

40. $xy + 4y - 3x - 12$

41. $9abc^2 + 6a^2bc - 6ab + 3bc$

42. $4a^2b^2c - 6ab^2c - 4ac + 8a$

Review and Preview

Find each product by using the FOIL order of multiplying binomials. See Section 5.2.

43. $(x + 2)(x - 5)$

44. $(x - 7)(x - 1)$

45. $(x + 3)(x + 2)$

46. $(x - 4)(x + 2)$

47. $(y - 3)(y - 1)$

48. $(s + 8)(s + 10)$

Combining Concepts

49. A factored polynomial can be in many forms. For example, a factored form of $xy - 3x - 2y + 6$ is $(x - 2)(y - 3)$. Which of the following (if any) is not a factored form of $xy - 3x - 2y + 6$?

a. $(2 - x)(3 - y)$

b. $(-2 + x)(-3 + y)$

c. $(y - 3)(x - 2)$

d. $(-x + 2)(-y + 3)$

33. ______
34. ______
35. ______
36. ______
37. ______
38. ______
39. ______
40. ______
41. ______
42. ______
43. ______
44. ______
45. ______
46. ______
47. ______
48. ______
49. ______

Name ______________________

Factor out the greatest common factor. Assume that variables used as exponents represent positive integers.

50. $x^{3n} - 2x^{2n} + 5x^{n}$

51. $3y^{n} + 3y^{2n} + 5y^{8n}$

52. $6x^{8a} - 2x^{5a} - 4x^{3a}$

53. $3x^{5a} - 6x^{3a} + 9x^{2a}$

54. An object is thrown upward from the ground with an initial velocity of 64 feet per second. The height $h(t)$ of the object after t seconds is given by the polynomial function $h(t) = -16t^2 + 64t$.

a. Write an equivalent factored expression for the function $h(t)$ by factoring $-16t^2 + 64t$.

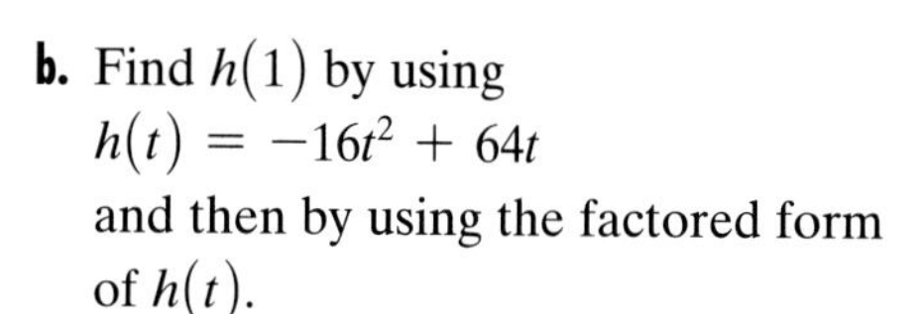

b. Find $h(1)$ by using $h(t) = -16t^2 + 64t$ and then by using the factored form of $h(t)$.

c. Explain why the values found in part (b) are the same.

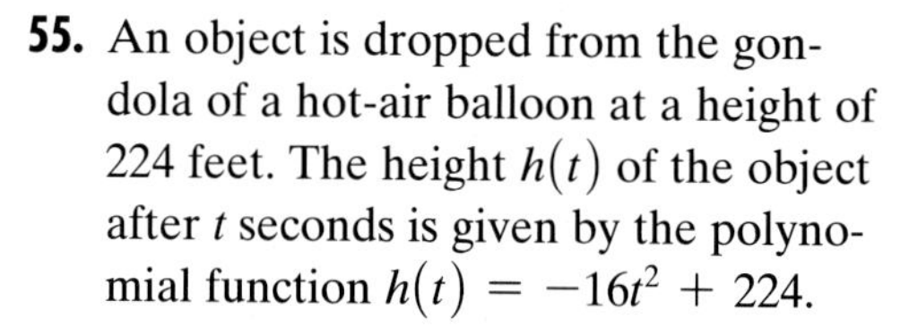

55. An object is dropped from the gondola of a hot-air balloon at a height of 224 feet. The height $h(t)$ of the object after t seconds is given by the polynomial function $h(t) = -16t^2 + 224$.

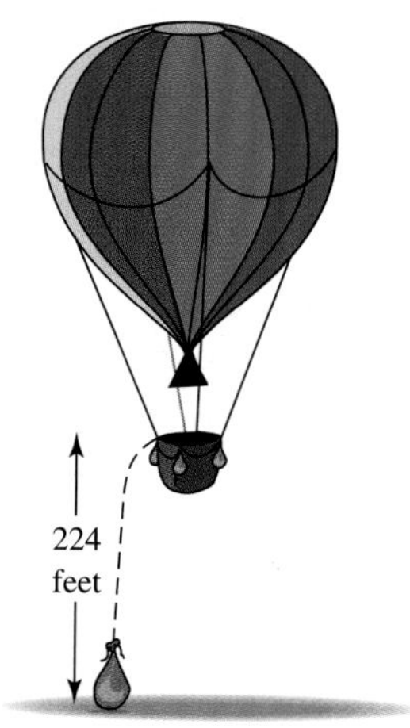

a. Write an equivalent factored expression for the function $h(t)$ by factoring $-16t^2 + 224$.

b. Find $h(2)$ by using $h(t) = -16t^2 + 224$ and then by using the factored form of the function.

c. Explain why the values found in part (b) are the same.

50. ______

51. ______

52. ______

53. ______

54. a. ______

b. ______

c. ______

55. a. ______

b. ______

c. ______

5.5 FACTORING TRINOMIALS

Objectives

A Factor trinomials of the form $x^2 + bx + c$.

B Factor trinomials of the form $ax^2 + bx + c$.

C Factor by substitution.

SSM CD-ROM Video 5.5

A FACTORING TRINOMIALS OF THE FORM $x^2 + bx + c$

In the previous section, we used factoring by grouping to factor four-term polynomials. In this section, we present techniques for factoring trinomials. Since $(x - 2)(x + 5) = x^2 + 3x - 10$, we say that $(x - 2)(x + 5)$ is a factored form of $x^2 + 3x - 10$. Taking a close look at how $(x - 2)$ and $(x + 5)$ are multiplied suggests a pattern for factoring trinomials of the form $x^2 + bx + c$.

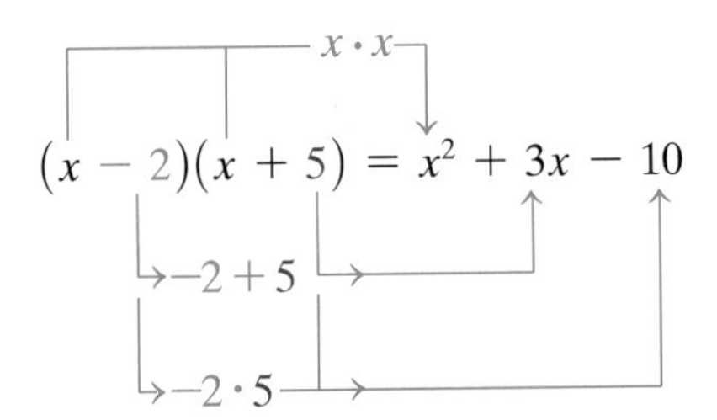

The pattern for factoring is summarized next.

FACTORING A TRINOMIAL OF THE FORM $x^2 + bx + c$

Find two numbers whose product is c and whose sum is b. The factored form of $x^2 + bx + c$ is

$(x + \text{one number})(x + \text{other number})$

Example 1 Factor: $x^2 + 10x + 16$

Solution: We look for two integers whose product is 16 and whose sum is 10. Since our integers must have a positive product and a positive sum, we look at only positive factors of 16.

Positive Factors of 16	Sum of Factors	
1, 16	$1 + 16 = 17$	
4, 4	$4 + 4 = 8$	
2, 8	$2 + 8 = 10$	Correct pair

The correct pair of numbers is 2 and 8 because their product is 16 and their sum is 10. Thus,

$$x^2 + 10x + 16 = (x + 2)(x + 8)$$

To check, see that $(x + 2)(x + 8) = x^2 + 10x + 16$.

Practice Problem 1

Factor: $x^2 + 8x + 15$

Example 2 Factor: $x^2 - 12x + 35$

Solution: We need to find two integers whose product is 35 and whose sum is -12. Since our integers must have a positive product and a negative sum, we consider only negative factors of 35. The numbers are -5 and -7.

$$\begin{aligned} x^2 - 12x + 35 &= [x + (-5)][x + (-7)] \\ &= (x - 5)(x - 7) \end{aligned}$$

To check, see that $(x - 5)(x - 7) = x^2 - 12x + 35$.

Practice Problem 2

Factor: $x^2 - 10x + 24$

Answers

1. $(x + 5)(x + 3)$, **2.** $(x - 4)(x - 6)$

Practice Problem 3

Factor: $6x^3 + 24x^2 - 30x$

Example 3 Factor: $5x^3 - 30x^2 - 35x$

Solution: First we factor out the greatest common factor, $5x$.

$$5x^3 - 30x^2 - 35x = 5x(x^2 - 6x - 7)$$

Next we factor $x^2 - 6x - 7$ by finding two numbers whose product is -7 and whose sum is -6. The numbers are 1 and -7.

$$\begin{aligned} 5x^3 - 30x^2 - 35x &= 5x(x^2 - 6x - 7) \\ &= 5x(x + 1)(x - 7) \end{aligned}$$

HELPFUL HINT

If the polynomial to be factored contains a common factor that is factored out, don't forget to include that common factor in the final factored form of the original polynomial.

Practice Problem 4

Factor: $3y^2 + 6y + 6$

Example 4 Factor: $2n^2 - 38n + 80$

Solution: The terms of this polynomial have a greatest common factor of 2, which we factor out first.

$$2n^2 - 38n + 80 = 2(n^2 - 19n + 40)$$

Next we factor $n^2 - 19n + 40$ by finding two numbers whose product is 40 and whose sum is -19. Both numbers must be negative since their sum is -19. Possibilities are

-1 and -40, -2 and -20, -4 and -10, -5 and -8

None of the pairs has a sum of -19, so no further factoring with integers is possible. The factored form of $2n^2 - 38n + 80$ is

$$2n^2 - 38n + 80 = 2(n^2 - 19n + 40)$$

We call a polynomial such as $n^2 - 19n + 40$ that cannot be factored further, a **prime polynomial**.

B FACTORING TRINOMIALS OF THE FORM $ax^2 + bx + c$

Next, we factor trinomials of the form $ax^2 + bx + c$, where the coefficient a of x^2 is not 1. Don't forget that the first step in factoring any polynomial is to factor out the greatest common factor of its terms.

FACTORING $ax^2 + bx + c$ BY TRIAL AND CHECK

Practice Problem 5

Factor: $3x^2 + 13x + 4$

Example 5 Factor: $2x^2 + 11x + 15$

Solution: Factors of $2x^2$ are $2x$ and x. Let's try these factors as first terms of the binomials.

Answers

3. $6x(x - 1)(x + 5)$, **4.** $3(y^2 + 2y + 2)$, **5.** $(3x + 1)(x + 4)$

$2x^2 + 11x + 15 = (2x + \quad)(x + \quad)$

Next we try combinations of factors of 15 until the correct middle term, $11x$, is obtained. We will try only positive factors of 15 since the coefficient of the middle term, 11, is positive. Positive factors of 15 are 1 and 15 and 3 and 5.

$(2x + 1)(x + 15)$: $1x + 30x = 31x$ Incorrect middle term

$(2x + 15)(x + 1)$: $15x + 2x = 17x$ Incorrect middle term

$(2x + 3)(x + 5)$: $3x + 10x = 13x$ Incorrect middle term

$(2x + 5)(x + 3)$: $5x + 6x = 11x$ Correct middle term

Thus, the factored form of $2x^2 + 11x + 15$ is $(2x + 5)(x + 3)$.

FACTORING A TRINOMIAL OF THE FORM $ax^2 + bx + c$

Step 1. Write all pairs of factors of ax^2.

Step 2. Write all pairs of factors of c, the constant term.

Step 3. Try various combinations of these factors until the correct middle term bx is found.

Step 4. If no combination exists, the polynomial is **prime**.

Example 6 Factor: $3x^2 - x - 4$

Solution: Factors of $3x^2$: $3x \cdot x$

Factors of -4: $-1 \cdot 4$, $1 \cdot -4$, $-2 \cdot 2$, $2 \cdot -2$

Let's try possible combinations of these factors.

$(3x - 1)(x + 4)$: $-1x + 12x = 11x$ Incorrect middle term

$(3x + 4)(x - 1)$: $4x + (-3x) = 1x$ Incorrect middle term

$(3x - 4)(x + 1)$: $-4x + 3x = -1x$ Correct middle term

Thus, $3x^2 - x - 4 = (3x - 4)(x + 1)$.

Practice Problem 6

Factor: $5x^2 + 13x - 6$

Answer

6. $(x + 3)(5x - 2)$

HELPFUL HINT—SIGN PATTERNS

A positive constant in a trinomial tells us to look for two numbers with the same sign. The sign of the coefficient of the middle term tells us whether the signs are both positive or both negative.

both positive ↓ same sign ↓

$2x^2 + 7x + 3 = (2x + 1)(x + 3)$

both negative ↓ same sign ↓

$2x^2 - 7x + 3 = (2x - 1)(x - 3)$

A negative constant in a trinomial tells us to look for two numbers with opposite signs.

opposite signs ↓

$2x^2 - 5x - 3 = (2x + 1)(x - 3)$

opposite signs ↓

$2x^2 + 5x - 3 = (2x - 1)(x + 3)$

Practice Problem 7

Factor: $24x^2y^2 - 42xy^2 + 9y^2$

Example 7 Factor: $12x^3y - 22x^2y + 8xy$

Solution: First we factor out the greatest common factor of the terms of this trinomial, $2xy$.

$$12x^3y - 22x^2y + 8xy = 2xy(6x^2 - 11x + 4)$$

Now we try to factor the trinomial $6x^2 - 11x + 4$.

Factors of $6x^2$: $2x \cdot 3x$, $6x \cdot x$

Let's try $2x$ and $3x$.

$$2xy(6x^2 - 11x + 4) = 2xy(2x + \quad)(3x + \quad)$$

The constant term, 4, is positive and the coefficient of the middle term, -11, is negative, so we factor 4 into negative factors only.

Negative factors of 4: $-4(-1)$, $-2(-2)$

Let's try -4 and -1.

$$2xy(2x - 4)(3x - 1)$$

$-12x$
$-2x$
$-14x$ Incorrect middle term

This combination cannot be correct, because one of the factors $(2x - 4)$ has a common factor of 2. This cannot happen if the polynomial $6x^2 - 11x + 4$ has no common factors.

Now let's try -1 and -4.

$$2xy(2x - 1)(3x - 4)$$

$-3x$
$-8x$
$-11x$ Correct middle term

Thus,

$$12x^3y - 22x^2y + 8xy = 2xy(2x - 1)(3x - 4)$$

Answer

7. $3y^2(2x - 3)(4x - 1)$

If this combination had not worked, we would try -2 and -2 as factors of 4 and then $6x$ and x as factors of $6x^2$.

HELPFUL HINT

If a trinomial has no common factor (other than 1), then none of its binomial factors will contain a common factor (other than 1).

Example 8 Factor: $16x^2 + 24xy + 9y^2$

Solution: No greatest common factor can be factored out of this trinomial.

Factors of $16x^2$: $16x \cdot x$, $8x \cdot 2x$, $4x \cdot 4x$
Factors of $9y^2$: $y \cdot 9y$, $3y \cdot 3y$

We try possible combinations until the correct factorization is found.

$$16x^2 + 24xy + 9y^2 = (4x + 3y)(4x + 3y) \quad \text{or} \quad (4x + 3y)^2$$

Practice Problem 8

Factor: $4x^2 + 28xy + 49y^2$

The trinomial $16x^2 + 24xy + 9y^2$ in Example 8 is an example of a **perfect square trinomial** since its factors are two identical binomials. In the next section, we examine a special method for factoring perfect square trinomials.

FACTORING $ax^2 + bx + c$ BY GROUPING

There is another method we can use when factoring trinomials of the form $ax^2 + bx + c$: Write the trinomial as a four-term polynomial, and then factor by grouping.

FACTORING A TRINOMIAL OF THE FORM $ax^2 + bx + c$ BY GROUPING

Step 1. Find two numbers whose product is $a \cdot c$ and whose sum is b.
Step 2. Write the term bx as a sum by using the factors found in Step 1.
Step 3. Factor by grouping.

Example 9 Factor: $6x^2 + 13x + 6$

Solution: In this trinomial, $a = 6$, $b = 13$, and $c = 6$.

Step 1. Find two numbers whose product is $a \cdot c$, or $6 \cdot 6 = 36$, and whose sum is b, 13. The two numbers are 4 and 9.

Step 2. Write the middle term $13x$ as the sum $4x + 9x$.

$$6x^2 + 13x + 6 = 6x^2 + 4x + 9x + 6$$

Step 3. Factor $6x^2 + 4x + 9x + 6$ by grouping.

$$(6x^2 + 4x) + (9x + 6) = 2x(3x + 2) + 3(3x + 2)$$
$$= (3x + 2)(2x + 3)$$

Practice Problem 9

Factor: $12x^2 + 11x + 2$

✓ CONCEPT CHECK

Name at least two ways that a factorization can be checked.

TRY THE CONCEPT CHECK IN THE MARGIN.

Answers

8. $(2x + 7y)^2$, **9.** $(3x + 2)(4x + 1)$

✓ Concept Check: Answers may vary. Samples are:
1. By multiplying the factors to see that the product is the original polynomial.
2. Checking to see that all common factors are factored out.

C FACTORING BY SUBSTITUTION

A complicated-looking polynomial may be a simpler trinomial "in disguise." Revealing the simpler trinomial is possible by substitution.

Practice Problem 10

Factor: $3(z + 2)^2 - 19(z + 2) + 6$

Example 10 Factor: $2(a + 3)^2 - 5(a + 3) - 7$

Solution: The quantity $(a + 3)$ is in two of the terms of this polynomial. If we *substitute* x for $(a + 3)$, the result is the following simpler trinomial.

$$2(a + 3)^2 - 5(a + 3) - 7 \quad \text{Original trinomial}$$
$$= 2(x)^2 - 5(x) - 7 \quad \text{Substitute } x \text{ for } (a + 3).$$

Now we can factor $2x^2 - 5x - 7$.

$2x^2 - 5x - 7 = (2x - 7)(x + 1)$

But the quantity in the original polynomial was $(a + 3)$, not x. Thus we need to reverse the substitution and replace x with $(a + 3)$.

$$(2x - 7)(x + 1) \quad \text{Factored expression}$$
$$= [2(a + 3) - 7][(a + 3) + 1] \quad \text{Substitute } (a + 3) \text{ for } x.$$
$$= (2a + 6 - 7)(a + 3 + 1) \quad \text{Remove inside parentheses.}$$
$$= (2a - 1)(a + 4) \quad \text{Simplify.}$$

Thus, $2(a + 3)^2 - 5(a + 3) - 7 = (2a - 1)(a + 4)$.

Practice Problem 11

Factor: $14x^4 + 23x^2 + 3$

Example 11 Factor: $5x^4 + 29x^2 - 42$

Solution: Again, substitution may help us factor this polynomial more easily. We will let $y = x^2$, so $y^2 = (x^2)^2$, or x^4. Then

$$5x^4 + 29x^2 - 42$$

becomes

$$5y^2 + 29y - 42$$

which factors as

$$5y^2 + 29y - 42 = (5y - 6)(y + 7)$$

Now we replace y with x^2 to get

$$(5x^2 - 6)(x^2 + 7)$$

Answers

10. $(3z + 5)(z - 4)$,
11. $(2x^2 + 3)(7x^2 + 1)$

Name ______________ Section ______ Date ______

Mental Math

1. Find two numbers whose product is 10 and whose sum is 7.
2. Find two numbers whose product is 12 and whose sum is 8.
3. Find two numbers whose product is 24 and whose sum is 11.
4. Find two numbers whose product is 30 and whose sum is 13.

Exercise Set 5.5

A *Factor each trinomial. See Examples 1 through 4.*

1. $x^2 + 9x + 18$
2. $x^2 + 9x + 20$
3. $x^2 - 12x + 32$
4. $x^2 - 12x + 27$
5. $x^2 + 10x - 24$
6. $x^2 + 3x - 54$
7. $x^2 - 2x - 24$
8. $x^2 - 9x - 36$
9. $3x^2 - 18x + 24$
10. $x^2y^2 + 4xy^2 + 3y^2$
11. $4x^2z + 28xz + 40z$
12. $5x^2 - 45x + 70$
13. $2x^2 + 30x - 108$
14. $3x^2 + 12x - 96$
15. $x^2 - 24x - 81$
16. $x^2 - 48x - 100$
17. $x^2 - 15x - 54$
18. $x^2 - 15x + 54$
19. $3x^2 - 6x + 3$
20. $2x^2 + 4x + 2$
21. $2x^2 + 2x - 12$
22. $3x^2 + 6x - 45$
23. $x^2 + 6xy + 5y^2$
24. $x^2 + 6xy + 8y^2$

Mental Math Answers

1. ______
2. ______
3. ______
4. ______

Answers

1. ______
2. ______
3. ______
4. ______
5. ______
6. ______
7. ______
8. ______
9. ______
10. ______
11. ______
12. ______
13. ______
14. ______
15. ______
16. ______
17. ______
18. ______
19. ______
20. ______
21. ______
22. ______
23. ______
24. ______

25. ______

26. ______

27. ______

28. ______

29. ______

30. ______

31. ______

32. ______

33. ______

34. ______

35. ______

36. ______

37. ______

38. ______

39. ______

40. ______

41. ______

42. ______

43. ______

44. ______

45. ______

46. ______

47. ______

48. ______

49. ______

50. ______

51. ______

52. ______

53. ______

54. ______

55. ______

Name ______________________

25. The volume $V(x)$ of a box in terms of its height x is given by the function $V(x) = x^3 + 2x^2 - 8x$. Factor this expression for $V(x)$.

26. Based on your results from Exercise 25, find the length and width of the box if the height is 5 inches and the dimensions of the box are whole numbers.

27. Find all positive and negative integers b such that $x^2 + bx + 6$ factors.

28. Find all positive and negative integers b such that $x^2 + bx - 10$ factors.

B *Factor each trinomial. See Examples 5 through 9.*

29. $5x^2 + 16x + 3$

30. $3x^2 + 8x + 4$

31. $2x^2 - 11x + 12$

32. $3x^2 - 19x + 20$

33. $2x^2 + 25x - 20$

34. $6x^2 - 13x - 8$

35. $4x^2 - 12x + 9$

36. $25x^2 - 30x + 9$

37. $12x^2 + 10x - 50$

38. $12y^2 - 48y + 45$

39. $3y^4 - y^3 - 10y^2$

40. $2x^2z + 5xz - 12z$

41. $6x^3 + 8x^2 + 24x$

42. $18y^3 + 12y^2 + 2y$

43. $x^2 + 8xz + 7z^2$

44. $a^2 - 2ab - 15b^2$

45. $2x^2 - 5xy - 3y^2$

46. $6x^2 + 11xy + 4y^2$

47. $x^2 - x - 12$

48. $x^2 + 4x - 5$

49. $28y^2 + 22y + 4$

50. $24y^3 - 2y^2 - y$

51. $2x^2 + 15x - 27$

52. $3x^2 + 14x + 15$

53. $3x^2 - 5x - 2$

54. $5x^2 - 14x - 3$

55. $8x^2 - 26x + 15$

Name ______________________________

56. $12x^2 - 17x + 6$

57. $18x^4 + 21x^3 + 6x^2$

58. $20x^5 + 54x^4 + 10x^3$

59. $3a^2 + 12ab + 12b^2$

60. $2x^2 + 16xy + 32y^2$

61. $6x^3 - x^2 - x$

62. $12x^3 + x^2 - x$

63. $12a^2 - 29ab + 15b^2$

64. $16y^2 + 6yx - 27x^2$

65. $9x^2 + 30x + 25$

66. $4x^2 + 6x + 9$

67. $3x^2y - 11xy + 8y$

68. $5xy^2 - 9xy + 4x$

C *Use substitution to factor each polynomial completely. See Examples 10 and 11.*

69. $x^4 + x^2 - 6$

70. $x^4 - x^2 - 20$

71. $(5x + 1)^2 + 8(5x + 1) + 7$

72. $(3x - 1)^2 + 5(3x - 1) + 6$

73. $x^6 - 7x^3 + 12$

74. $x^6 - 4x^3 - 12$

75. $(a + 5)^2 - 5(a + 5) - 24$

76. $(3c + 6)^2 + 12(3c + 6) - 28$

77. $(x - 4)^2 + 3(x - 4) - 18$

78. $(x - 3)^2 - 2(x - 3) - 8$

79. $2x^6 + 3x^3 - 9$

80. $3x^6 - 14x^3 + 8$

81. $2(x + 4)^2 + 3(x + 4) - 5$

82. $3(x + 3)^2 + 2(x + 3) - 5$

83. $x^4 - 5x^2 - 6$

84. $x^4 - 5x^2 + 6$

56. ______
57. ______
58. ______
59. ______
60. ______
61. ______
62. ______
63. ______
64. ______
65. ______
66. ______
67. ______
68. ______
69. ______
70. ______
71. ______
72. ______
73. ______
74. ______
75. ______
76. ______
77. ______
78. ______
79. ______
80. ______
81. ______
82. ______
83. ______
84. ______

85. ______

86. ______

87. ______

88. ______

89. ______

90. ______

91. ______

92. ______

93. ______

94. ______

95. ______

96. ______

97. ______

98. ______

99. ______

100. ______

101. ______

102. ______

Name ______

Review and Preview

Multiply. See Section 5.2.

85. $(x - 3)(x + 3)$

86. $(x - 4)(x + 4)$

87. $(2x + 1)^2$

88. $(3x + 5)^2$

89. $(x - 2)(x^2 + 2x + 4)$

90. $(y + 1)(y^2 - y + 1)$

Combining Concepts

Factor. Assume that variables used as exponents represent positive integers.

91. $x^{2n} + 10x^n + 16$

92. $x^{2n} - 7x^n + 12$

93. $x^{2n} - 3x^n - 18$

94. $x^{2n} + 7x^n - 18$

95. $2x^{2n} + 11x^n + 5$

96. $3x^{2n} - 8x^n + 4$

97. $4x^{2n} - 12x^n + 9$

98. $9x^{2n} + 24x^n + 16$

Recall that a grapher may be used to visualize addition, subtraction, and multiplication of polynomials. In the same manner, a grapher may be used to visualize factoring of polynomials in one variable. For example, to see that

$$2x^3 - 9x^2 - 5x = x(2x + 1)(x - 5)$$

graph $Y_1 = 2x^3 - 9x^2 - 5x$ *and* $Y_2 = x(2x + 1)(x - 5)$. *Then trace along both graphs to see that they coincide. Factor the following and use this method to check your results.*

99. $x^4 + 6x^3 + 5x^2$

100. $x^3 + 6x^2 + 8x$

101. $30x^3 + 9x^2 - 3x$

102. $-6x^4 + 10x^3 - 4x^2$

5.6 FACTORING BY SPECIAL PRODUCTS

Objectives

A Factor a perfect square trinomial.

B Factor the difference of two squares.

C Factor the sum or difference of two cubes.

SSM CD-ROM Video 5.6

A FACTORING PERFECT SQUARE TRINOMIALS

In the previous section, we considered a variety of ways to factor trinomials of the form $ax^2 + bx + c$. In Example 8, we factored $16x^2 + 24xy + 9y^2$ as

$$16x^2 + 24xy + 9y^2 = (4x + 3y)^2$$

Recall that we called $16x^2 + 24xy + 9y^2$ a perfect square trinomial because its factors are two identical binomials. A trinomial is a perfect square trinomial if it can be written so that its first term is the square of some quantity a, its last term is the square of some quantity b, and its middle term is twice the product of the quantities a and b.

The following special formulas can be used to factor perfect square trinomials.

PERFECT SQUARE TRINOMIALS

$$a^2 + 2ab + b^2 = (a + b)^2$$
$$a^2 - 2ab + b^2 = (a - b)^2$$

Notice that these equations are the same special products from Section 5.2 for the square of a binomial.

From

$$a^2 + 2ab + b^2 = (a + b)^2$$

we see that

$$16x^2 + 24xy + 9y^2 = (4x)^2 + 2(4x)(3y) + (3y)^2 = (4x + 3y)^2$$

Example 1 Factor: $m^2 + 10m + 25$

Solution: Notice that the first term is a square: $m^2 = (m)^2$,
the last term is a square: $25 = 5^2$,
and $10m = 2 \cdot 5 \cdot m$

This is a perfect square trinomial. Thus,

$$m^2 + 10m + 25 = m^2 + 2(m)(5) + 5^2 = (m + 5)^2$$

Practice Problem 1

Factor: $x^2 + 8x + 16$

Examples Factor each trinomial.

2. $4x^2 + 2x + 1 = (2x)^2 + 2 \cdot 2x \cdot 1 + 1^2$ See whether it is a perfect square trinomial.
$= (2x + 1)^2$ Factor.

3. $9x^2 - 12x + 4 = (3x)^2 - 2(3x)(2) + 2^2$ See whether it is a perfect square trinomial.
$= (3x - 2)^2$ Factor.

Practice Problems 2–3

Factor.

2. $9x^2 + 6x + 1$

3. $25x^2 - 20x + 4$

Example 4 Factor: $3a^2x - 12abx + 12b^2x$

Solution: The terms of this trinomial have a greatest common factor of $3x$, which we factor out first.

$$3a^2x - 12abx + 12b^2x = 3x(a^2 - 4ab + 4b^2)$$

Practice Problem 4

Factor: $4x^3 - 32x^2y + 64xy^2$

Answers

1. $(x + 4)^2$, **2.** $(3x + 1)^2$, **3.** $(5x - 2)^2$, **4.** $4x(x - 4y)^2$

The polynomial $a^2 - 4ab + 4b^2$ is a perfect square trinomial. Notice that the first term is a square: $a^2 = (a)^2$, the last term is a square: $4b^2 = (2b)^2$, and $4ab = 2(a)(2b)$. The factoring can now be completed as

$$3x(a^2 - 4ab + 4b^2) = 3x(a - 2b)^2$$

HELPFUL HINT

If you recognize a trinomial as a perfect square trinomial, use the special formulas to factor. However, methods for factoring trinomials in general from Section 5.5 will also result in the correct factored form.

B FACTORING THE DIFFERENCE OF TWO SQUARES

We now factor special types of binomials, beginning with the **difference of two squares**. The special product pattern presented in Section 5.2 for the product of a sum and a difference of two terms is used again here. However, the emphasis is now on factoring rather than on multiplying.

DIFFERENCE OF TWO SQUARES

$$a^2 - b^2 = (a + b)(a - b)$$

Notice that a binomial is a difference of two squares when it is the difference of the square of some quantity a and the square of some quantity b.

Practice Problems 5–8

Factor.

5. $x^2 - 49$

6. $4y^2 - 81$

7. $12 - 3a^2$

8. $y^2 - \frac{1}{25}$

Examples Factor.

5. $x^2 - 9 = x^2 - 3^2$
$= (x + 3)(x - 3)$

6. $16y^2 - 9 = (4y)^2 - 3^2$
$= (4y + 3)(4y - 3)$

7. $50 - 8y^2 = 2(25 - 4y^2)$ Factor out the common factor of 2.
$= 2[5^2 - (2y)^2]$
$= 2(5 + 2y)(5 - 2y)$

8. $x^2 - \frac{1}{4} = x^2 - \left(\frac{1}{2}\right)^2$
$= \left(x + \frac{1}{2}\right)\left(x - \frac{1}{2}\right)$

The binomial $x^2 + 9$ is a **sum of two squares** and cannot be factored by using real numbers. *In general, except for factoring out a greatest common factor, the sum of two squares usually cannot be factored by using real numbers.*

HELPFUL HINT

The sum of two squares whose greatest common factor is 1 usually cannot be factored by using real numbers.

Answers

5. $(x + 7)(x - 7)$, **6.** $(2y + 9)(2y - 9)$, **7.** $3(2 + a)(2 - a)$, **8.** $\left(y + \frac{1}{5}\right)\left(y - \frac{1}{5}\right)$

Example 9 Factor: $p^4 - 16$

Solution:

$$p^4 - 16 = (p^2)^2 - 4^2$$
$$= (p^2 + 4)(p^2 - 4)$$

The binomial factor $p^2 + 4$ cannot be factored by using real numbers, but the binomial factor $p^2 - 4$ is a difference of squares.

$$(p^2 + 4)(p^2 - 4) = (p^2 + 4)(p + 2)(p - 2)$$

Practice Problem 9

Factor: $a^4 - 81$

TRY THE CONCEPT CHECK IN THE MARGIN.

✓ CONCEPT CHECK

Is $(x - 4)(y^2 - 9)$ completely factored? Why or why not?

Example 10 Factor: $(x + 3)^2 - 36$

Solution:

$$(x + 3)^2 - 36 = (x + 3)^2 - 6^2$$
$$= [(x + 3) + 6][(x + 3) - 6] \quad \text{Factor as the difference of two squares.}$$
$$= [x + 3 + 6][x + 3 - 6] \quad \text{Remove parentheses.}$$
$$= (x + 9)(x - 3) \quad \text{Simplify.}$$

Practice Problem 10

$(x + 1)^2 - 9$

Example 11 Factor: $x^2 + 4x + 4 - y^2$

Solution: Factoring by grouping comes to mind since the sum of the first three terms of this polynomial is a perfect square trinomial.

$$x^2 + 4x + 4 - y^2 = (x^2 + 4x + 4) - y^2 \quad \text{Group the first three terms.}$$
$$= (x + 2)^2 - y^2 \quad \text{Factor the perfect square trinomial.}$$

This is not completely factored yet since we have a *difference*, not a *product*. Since $(x + 2)^2 - y^2$ is a difference of squares, we have

$$(x + 2)^2 - y^2 = [(x + 2) + y][(x + 2) - y]$$
$$= (x + 2 + y)(x + 2 - y)$$

Practice Problem 11

Factor: $a^2 + 2a + 1 - b^2$

C FACTORING THE SUM OR DIFFERENCE OF TWO CUBES

Although the sum of two squares usually cannot be factored, the sum of two cubes, as well as the difference of two cubes, can be factored as follows.

SUM AND DIFFERENCE OF TWO CUBES

$$a^3 + b^3 = (a + b)(a^2 - ab + b^2)$$
$$a^3 - b^3 = (a - b)(a^2 + ab + b^2)$$

To check the first pattern, let's find the product of $(a + b)$ and $(a^2 - ab + b^2)$.

$$(a + b)(a^2 - ab + b^2) = a(a^2 - ab + b^2) + b(a^2 - ab + b^2)$$
$$= a^3 - a^2b + ab^2 + a^2b - ab^2 + b^3$$
$$= a^3 + b^3$$

Answers

9. $(a^2 + 9)(a + 3)(a - 3)$, **10.** $(x - 2)(x + 4)$, **11.** $(a + 1 + b)(a + 1 - b)$

✓ Concept Check: no; $(y^2 - 9)$ can be factored

Practice Problem 12

Factor: $x^3 + 27$

Practice Problem 13

Factor: $x^3 + 64y^3$

Practice Problem 14

Factor: $y^3 - 8$

Practice Problem 15

Factor: $27a^2 - b^3a^2$

Answers

12. $(x + 3)(x^2 - 3x + 9)$,
13. $(x + 4y)(x^2 - 4xy + 16y^2)$,
14. $(y - 2)(y^2 + 2y + 4)$,
15. $a^2(3 - b)(9 + 3b + b^2)$

Example 12 Factor: $x^3 + 8$

Solution: First we write the binomial in the form $a^3 + b^3$. Then we use the formula

$$a^3 + b^3 = (a + b)(a^2 - a \cdot b + b^2), \text{ where } a \text{ is } x \text{ and } b \text{ is } 2$$

$$x^3 + 8 = x^3 + 2^3 = (x + 2)(x^2 - x \cdot 2 + 2^2)$$

Thus, $x^3 + 8 = (x + 2)(x^2 - 2x + 4)$

Example 13 Factor: $p^3 + 27q^3$

Solution:

$$\begin{aligned} p^3 + 27q^3 &= p^3 + (3q)^3 \\ &= (p + 3q)[p^2 - (p)(3q) + (3q)^2] \\ &= (p + 3q)(p^2 - 3pq + 9q^2) \end{aligned}$$

Example 14 Factor: $y^3 - 64$

Solution: This is a difference of cubes since $y^3 - 64 = y^3 - 4^3$.

From $a^3 - b^3 = (a - b)(a^2 + a \cdot b + b^2)$ we have that

$$\begin{aligned} y^3 - 4^3 &= (y - 4)(y^2 + y \cdot 4 + 4^2) \\ &= (y - 4)(y^2 + 4y + 16) \end{aligned}$$

HELPFUL HINT

When factoring sums or differences of cubes, be sure to notice the sign patterns.

$$x^3 + y^3 = (x + y)(x^2 - xy + y^2)$$

same sign; opposite sign; always positive

$$x^3 - y^3 = (x - y)(x^2 + xy + y^2)$$

same sign; opposite sign; always positive

Example 15 Factor: $125q^2 - n^3q^2$

Solution: First we factor out a common factor of q^2.

$$\begin{aligned} 125q^2 - n^3q^2 &= q^2(125 - n^3) \\ &= q^2(5^3 - n^3) \\ &= q^2(5 - n)[5^2 + (5)(n) + (n^2)] \\ &= q^2(5 - n)(25 + 5n + n^2) \end{aligned}$$

opposite sign; positive

Thus $125q^2 - n^3q^2 = q^2(5 - n)(25 + 5n + n^2)$. The trinomial $25 + 5n + n^2$ cannot be factored further.

Name ______________________ Section ________ Date __________

EXERCISE SET 5.6

A *Factor. See Examples 1 through 4.*

1. $x^2 + 6x + 9$

2. $x^2 - 10x + 25$

3. $4x^2 - 12x + 9$

4. $25x^2 + 10x + 1$

5. $3x^2 - 24x + 48$

6. $x^3 + 14x^2 + 49x$

7. $9y^2x^2 + 12yx^2 + 4x^2$

8. $32x^2 - 16xy + 2y^2$

9. $4a^2 + 12a + 9$

10. $9a^2 - 30a + 25$

B *Factor. See Examples 5 through 11.*

11. $x^2 - 25$

12. $y^2 - 100$

13. $9 - 4z^2$

14. $16x^2 - y^2$

15. $(y + 2)^2 - 49$

16. $(x - 1)^2 - z^2$

17. $64x^2 - 100$

18. $4x^2 - 36$

19. $18x^2y - 2y$

20. $12xy^2 - 108x$

21. $9x^2 - 49$

22. $25x^2 - 4$

23. $x^4 - 81$

24. $x^4 - 256$

25. $(x + 2y)^2 - 9$

ANSWERS

1. ________
2. ________
3. ________
4. ________
5. ________
6. ________
7. ________
8. ________
9. ________
10. ________
11. ________
12. ________
13. ________
14. ________
15. ________
16. ________
17. ________
18. ________
19. ________
20. ________
21. ________
22. ________
23. ________
24. ________
25. ________

26. ______
27. ______
28. ______
29. ______
30. ______
31. ______
32. ______
33. ______
34. ______
35. ______
36. ______
37. ______
38. ______
39. ______
40. ______
41. ______
42. ______
43. ______
44. ______
45. ______
46. ______
47. ______

Name ______________________

26. $(3x + y)^2 - 25$

27. $x^2 + 16x + 64 - x^4$

28. $x^2 + 20x + 100 - x^4$

29. $x^2 - 10x + 25 - y^2$

30. $x^2 - 18x + 81 - y^2$

31. $4x^2 + 4x + 1 - z^2$

32. $9y^2 + 12y + 4 - x^2$

C *Factor. See Examples 12 through 15.*

33. $x^3 + 27$

34. $y^3 + 1$

35. $z^3 - 1$

36. $x^3 - 8$

37. $m^3 + n^3$

38. $r^3 + 125$

39. $x^3y^2 - 27y^2$

40. $64 - p^3$

41. $a^3b + 8b^4$

42. $8ab^3 + 27a^4$

43. $125y^3 - 8x^3$

44. $54y^3 - 128$

45. $x^6 - y^3$

46. $x^3 - y^6$

47. $8x^3 + 27y^3$

Name ______

48. $125x^3 + 8y^3$ **49.** $x^3 - 1$ **50.** $x^3 - 8$

51. $x^3 + 125$ **52.** $x^3 + 216$ **53.** $3x^6y^2 + 81y^2$ **54.** $x^2y^9 + x^2y^3$

Review and Preview

Solve each equation. See Section 2.1.

55. $x - 5 = 0$ **56.** $x + 7 = 0$ **57.** $3x + 1 = 0$

58. $5x - 15 = 0$ **59.** $-2x = 0$ **60.** $3x = 0$

61. $-5x + 25 = 0$ **62.** $-4x - 16 = 0$

Combining Concepts

63. The manufacturer of Antonio's Metal Washers needs to determine the cross-sectional area of each washer. If the outer radius of the washer is R and the radius of the hole is r, express the area of the washer as a polynomial. Factor this polynomial completely.

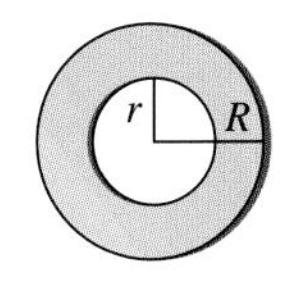

64. Express the area of the shaded region as a polynomial. Factor the polynomial completely.

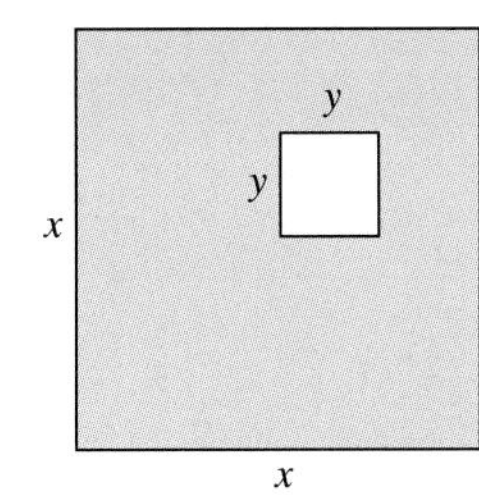

48. ______
49. ______
50. ______
51. ______
52. ______
53. ______
54. ______
55. ______
56. ______
57. ______
58. ______
59. ______
60. ______
61. ______
62. ______
63. ______
64. ______

65. ____

66. ____

67. ____

68. ____

69. a. ____

b. ____

c. ____

70. ____

71. ____

72. ____

73. ____

74. ____

75. ____

Name ____

Find a value of c that makes each trinomial a perfect square trinomial.

65. $x^2 + 6x + c$

66. $y^2 + 10y + c$

67. $m^2 - 14m + c$

68. $n^2 - 2n + c$

69. Factor $x^6 - 1$ completely, using the following methods from this chapter.

a. Factor the expression by treating it as the difference of two squares $(x^3)^2 - 1^2$.

b. Factor the expression treating it as the difference of two cubes $(x^2)^3 - 1^3$.

c. Are the answers to parts (a) and (b) the same? Why or why not?

Factor. Assume that variables used as exponents represent positive integers.

70. $x^{2n} - 25$

71. $x^{2n} - 36$

72. $36x^{2n} - 49$

73. $25x^{2n} - 81$

74. $x^{4n} - 16$

75. $x^{4n} - 625$

Name ______________________ Section ________ Date ________

Integrated Review—Operations on Polynomials and Factoring Strategies

Operations on Polynomials

Perform the indicated operation.

1. $(-y^2 + 6y - 1) + (3y^2 - 4y - 10)$

2. $(5z^4 - 6z^2 + z + 1) - (7z^4 - 2z + 1)$

3. Subtract $(x - 5)$ from $(x^2 - 6x + 2)$

4. $(2x^2 + 6x - 5) + (5x^2 - 10x)$

5. $(5x - 3)^2$

6. $(5x^2 - 14x - 3) \div (5x + 1)$

7. $(2x^4 - 3x^2 + 5x - 2) \div (x + 2)$

8. $(4x - 1)(x^2 - 3x - 2)$

Factoring Strategies

The key to proficiency in factoring polynomials is to practice until you are comfortable with each technique. A strategy for factoring polynomials completely is given next.

Factoring a Polynomial

Step 1. Are there any common factors? If so, factor out the greatest common factor.

Step 2. How many terms are in the polynomial?

a. If there are *two* terms, decide if one of the following formulas may be applied:

- **i.** Difference of two squares: $a^2 - b^2 = (a - b)(a + b)$
- **ii.** Difference of two cubes: $a^3 - b^3 = (a - b)(a^2 + ab + b^2)$
- **iii.** Sum of two cubes: $a^3 + b^3 = (a + b)(a^2 - ab + b^2)$

b. If there are *three* terms, try one of the following:

- **i.** Perfect square trinomial: $a^2 + 2ab + b^2 = (a + b)^2$
  $a^2 - 2ab + b^2 = (a - b)^2$
- **ii.** If not a perfect square trinomial factor by using the methods presented in Section 5.5.

c. If there are *four* or more terms, try factoring by grouping.

Step 3. See if any factors in the factored polynomial can be factored further.

Factor completely.

9. $x^2 - 8x + 16 - y^2$

10. $12x^2 - 22x - 20$

11. $x^4 - x$

12. $(2x + 1)^2 - 3(2x + 1) + 2$

13. $14x^2y - 2xy$

14. $24ab^2 - 6ab$

Answers

1. ____
2. ____
3. ____
4. ____
5. ____
6. ____
7. ____
8. ____
9. ____
10. ____
11. ____
12. ____
13. ____
14. ____

15. ______
16. ______
17. ______
18. ______
19. ______
20. ______
21. ______
22. ______
23. ______
24. ______
25. ______
26. ______
27. ______
28. ______
29. ______
30. ______
31. ______

Name ______

15. $4x^2 - 16$

16. $9x^2 - 81$

17. $3x^2 - 8x - 11$

18. $5x^2 - 2x - 3$

19. $4x^2 + 8x - 12$

20. $6x^2 - 6x - 12$

21. $4x^2 + 36x + 81$

22. $25x^2 + 40x + 16$

23. $8x^3 + 125y^3$

24. $27x^3 - 64y^3$

25. $64x^2y^3 - 8x^2$

26. $27x^5y^4 - 216x^2y$

27. $(x + 5)^3 + y^3$

28. $(y - 1)^3 + 27x^3$

29. $(5a - 3)^2 - 6(5a - 3) + 9$

30. $(4r + 1)^2 + 8(4r + 1) + 16$

31. Express the area of the shaded region as a polynomial. Factor the polynomial completely.

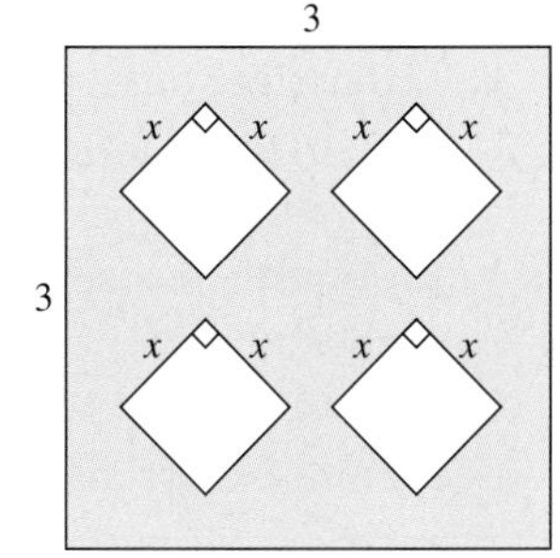

5.7 SOLVING EQUATIONS BY FACTORING AND SOLVING PROBLEMS

Objectives

A Solve polynomial equations by factoring.

B Solve problems that can be modeled by polynomial equations.

SSM CD-ROM Video 5.7

A SOLVING POLYNOMIAL EQUATIONS BY FACTORING

In this section, your efforts to learn factoring will start to pay off. We use factoring to solve polynomial equations.

A **polynomial equation** is the result of setting two polynomials equal to each other. Examples are

$$3x^3 - 2x^2 = x^2 + 2x - 1 \qquad 2.6x + 7 = -1.3$$
$$-5x^2 - 5 = -9x^2 - 2x + 1$$

A polynomial equation is in **standard form** if one side of the equation is 0. For example,

$$3x^3 - 3x^2 - 2x + 1 = 0 \qquad 2.6x + 8.3 = 0$$
$$4x^2 + 2x - 6 = 0$$

The degree of a simplified polynomial equation in standard form is the same as the highest degree of any of its terms. A polynomial equation of degree 2 is also called a **quadratic equation**.

A solution of a polynomial equation in one variable is a value of the variable that makes the equation true. The method presented in this section for solving polynomial equations is called the **factoring method**. This method is based on the **zero-factor property**.

> **ZERO-FACTOR PROPERTY**
>
> If a and b are real numbers and $a \cdot b = 0$, then $a = 0$ or $b = 0$. This property is true for three or more factors also.

In other words, if the product of two or more real numbers is zero, then at least one of the numbers must be zero.

Example 1 Solve: $(x + 2)(x - 6) = 0$

Solution: By the zero-factor property, $(x + 2)(x - 6) = 0$ only if $x + 2 = 0$ or $x - 6 = 0$.

$x + 2 = 0$ or $x - 6 = 0$ Use the zero-factor property.

$x = -2$ $\qquad x = 6$ Solve each linear equation.

To check, let $x = -2$ and then let $x = 6$ in the original equation.

Let $x = -2$.	Let $x = 6$.
$(x + 2)(x - 6) = 0$	$(x + 2)(x - 6) = 0$
$(-2 + 2)(-2 - 6) = 0$	$(6 + 2)(6 - 6) = 0$
$(0)(-8) = 0$	$(8)(0) = 0$
$0 = 0$ True.	$0 = 0$ True.

Both -2 and 6 check, and the solution set is $\{-2, 6\}$.

Practice Problem 1

Solve: $(x - 3)(x + 5) = 0$

Answer

1. $\{-5, 3\}$,

Practice Problem 2

Solve: $3x^2 + 5x - 2 = 0$

Example 2 Solve: $2x^2 + 9x - 5 = 0$

Solution: To use the zero-factor property, one side of the equation must be 0, and the other side must be in factored form.

$$2x^2 + 9x - 5 = 0$$

$$(2x - 1)(x + 5) = 0 \quad \text{Factor.}$$

$$2x - 1 = 0 \quad \text{or} \quad x + 5 = 0 \quad \text{Set each factor equal to zero.}$$

$$2x = 1 \qquad x = -5 \quad \text{Solve each linear equation.}$$

$$x = \frac{1}{2}$$

To check, let $x = \frac{1}{2}$ in the original equation; then let $x = -5$ in the original equation. The solution set is $\left\{-5, \frac{1}{2}\right\}$.

SOLVING POLYNOMIAL EQUATIONS BY FACTORING

Step 1. Write the equation in standard form so that one side of the equation is 0.
Step 2. Factor the polynomial completely.
Step 3. Set each factor containing a variable equal to 0.
Step 4. Solve the resulting equations.
Step 5. Check each solution in the original equation.

Since it is not always possible to factor a polynomial, not all polynomial equations can be solved by factoring. Other methods of solving polynomial equations are presented in Chapter 8.

Practice Problem 3

Solve: $x(5x - 7) = -2$

Example 3 Solve: $x(2x - 7) = 4$

Solution: We first write the equation in standard form; then we factor.

$$x(2x - 7) = 4$$

$$2x^2 - 7x = 4 \quad \text{Multiply.}$$

$$2x^2 - 7x - 4 = 0 \quad \text{Write in standard form.}$$

$$(2x + 1)(x - 4) = 0 \quad \text{Factor.}$$

$$2x + 1 = 0 \quad \text{or} \quad x - 4 = 0 \quad \text{Set each factor equal to zero.}$$

$$2x = -1 \qquad x = 4 \quad \text{Solve.}$$

$$x = -\frac{1}{2}$$

Check both solutions in the original equation. The solution set is $\left\{-\frac{1}{2}, 4\right\}$.

Answers

2. $\left\{-2, \frac{1}{3}\right\}$ **3.** $\left\{\frac{2}{5}, 1\right\}$

HELPFUL HINT

To apply the zero-factor property, one side of the equation must be 0, and the other side of the equation must be factored. To solve the equation $x(2x - 7) = 4$, for example, you may *not* set each factor equal to 4.

Example 4

Solve: $3(x^2 + 4) + 5 = -6(x^2 + 2x) + 13$

Solution: We rewrite the equation so that one side is 0.

$$3(x^2 + 4) + 5 = -6(x^2 + 2x) + 13$$

$$3x^2 + 12 + 5 = -6x^2 - 12x + 13 \quad \text{Use the distributive property.}$$

$$9x^2 + 12x + 4 = 0 \quad \text{Rewrite the equation so that one side is 0.}$$

$$(3x + 2)(3x + 2) = 0 \quad \text{Factor.}$$

$$3x + 2 = 0 \quad \text{or} \quad 3x + 2 = 0 \quad \text{Set each factor equal to 0.}$$

$$3x = -2 \qquad 3x = -2 \quad \text{Solve each equation.}$$

$$x = -\frac{2}{3} \qquad x = -\frac{2}{3}$$

Check by substituting $-\frac{2}{3}$ into the original equation. The solution set is $\left\{-\frac{2}{3}\right\}$.

Practice Problem 4

Solve: $2(x^2 + 5) + 10 = -2(x^2 + 10x) - 5$

If the equation contains fractions, we clear the equation of fractions as a first step.

Example 5

Solve: $2x^2 = \frac{17}{3}x + 1$

Solution:

$$2x^2 = \frac{17}{3}x + 1$$

$$3(2x^2) = 3\left(\frac{17}{3}x + 1\right) \quad \text{Clear the equation of fractions.}$$

$$6x^2 = 17x + 3 \quad \text{Use the distributive property.}$$

$$6x^2 - 17x - 3 = 0 \quad \text{Rewrite the equation in standard form.}$$

$$(6x + 1)(x - 3) = 0 \quad \text{Factor.}$$

$$6x + 1 = 0 \quad \text{or} \quad x - 3 = 0 \quad \text{Set each factor equal to zero.}$$

$$6x = -1 \qquad x = 3 \quad \text{Solve each equation.}$$

$$x = -\frac{1}{6}$$

Check by substituting into the original equation. The solution set is $\left\{-\frac{1}{6}, 3\right\}$.

Practice Problem 5

Solve: $2x^2 + \frac{5}{2}x = 3$

Answers

4. $\left\{-\frac{5}{2}\right\}$, **5.** $\left\{-2, \frac{3}{4}\right\}$

Practice Problem 6

Solve: $x^3 = x^2 + 6x$

✓ Concept Check

Which solution strategies are incorrect? Why?

a. Solve $(y - 2)(y + 2) = 4$ by setting each factor equal to 4.
b. Solve $(x + 1)(x + 3) = 0$ by setting each factor equal to 0.
c. Solve $z^2 + 5z + 6 = 0$ by factoring $z^2 + 5z + 6$ and setting each factor equal to 0.
d. Solve $x^2 + 6x + 8 = 10$ by factoring $x^2 + 6x + 8$ and setting each factor equal to 0.

Practice Problem 7

A model rocket is launched from the ground. Its height h at time t is approximated by the equation

$h = -16t^2 + 112t$

Find how long it takes the rocket to return to the ground.

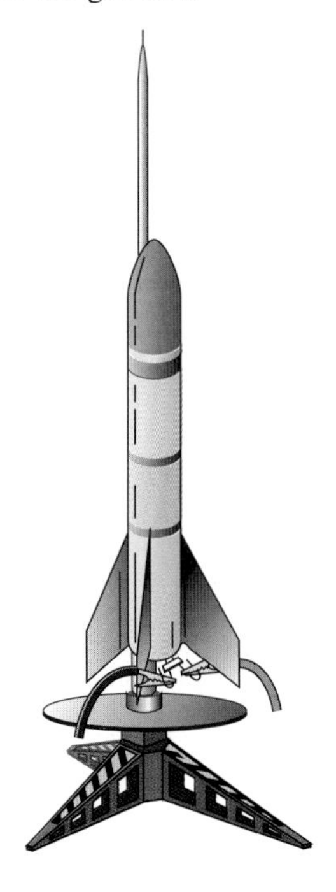

Answers

6. $\{-2, 0, 3\}$ **7.** 7 seconds

✓ Concept Check: a and d; the zero factor property works only if one side of the equation is 0

Example 6 Solve: $x^3 = 4x$

Solution:

$$x^3 = 4x$$

$$x^3 - 4x = 0 \quad \text{Rewrite the equation so that one side is 0.}$$

$$x(x^2 - 4) = 0 \quad \text{Factor out the greatest common factor.}$$

$$x(x + 2)(x - 2) = 0 \quad \text{Factor the difference of squares.}$$

$$x = 0 \quad \text{or} \quad x + 2 = 0 \quad \text{or} \quad x - 2 = 0 \quad \text{Set each factor equal to 0.}$$

$$x = -2 \qquad x = 2 \quad \text{Solve each equation.}$$

Check by substituting into the original equation. The solution set is $\{-2, 0, 2\}$.

Notice that the *third*-degree equation of Example 6 yielded *three* solutions.

TRY THE CONCEPT CHECK IN THE MARGIN.

B Solving Problems Modeled by Polynomial Equations

Some problems may be modeled by polynomial equations. To solve these problems, we use the same problem-solving steps that were introduced in Section 2.2. When solving these problems, keep in mind that a solution of an equation that models a problem is not always a solution to the problem. For example, a person's weight or the length of a side of a geometric figure is always a positive number. Discard solutions that do not make sense as solutions of the problem.

Example 7 Finding the Return Time of a Rocket

An Alpha III model rocket is launched from the ground with an A8–3 engine. Without a parachute the height of the rocket h at time t seconds is approximated by the equation.

$h = -16t^2 + 144t$

Find how long it takes the rocket to return to the ground.

Solution:

1. UNDERSTAND. Read and reread the problem. The equation $h = -16t^2 + 144t$ models the height of the rocket. Familiarize yourself with this equation by finding a few values.
When $t = 1$ second, the height of the rocket is

$h = -16(1)^2 + 144(1) = 128$ feet

When $t = 2$ seconds, the height of the rocket is

$h = -16(2)^2 + 144(2) = 224$ feet

2. TRANSLATE. To find how long it takes the rocket to return to the ground, we want to know what value of t makes the height h equal to 0. That is, we want to solve $h = 0$.

$-16t^2 + 144t = 0$

3. SOLVE the quadratic equation by factoring.

$$-16t^2 + 144t = 0$$
$$-16t(t - 9) = 0$$
$$-16t = 0 \text{ or } t - 9 = 0$$
$$t = 0 \qquad t = 9$$

4. INTERPRET. The height h is 0 feet at time 0 seconds (when the rocket is launched) and at time 9 seconds.

Check: See that the height of the rocket at 9 seconds equals 0.

$$h = -16(9)^2 + 144(9) = -1296 + 1296 = 0$$

State: The rocket returns to the ground 9 seconds after it is launched.

Some of the exercises at the end of this section make use of the **Pythagorean theorem**. Before we review this theorem, recall that a **right triangle** is a triangle that contains a 90° angle, or right angle. The **hypotenuse** of a right triangle is the side opposite the right angle and is the longest side of the triangle. The **legs** of a right triangle are the other sides of the triangle.

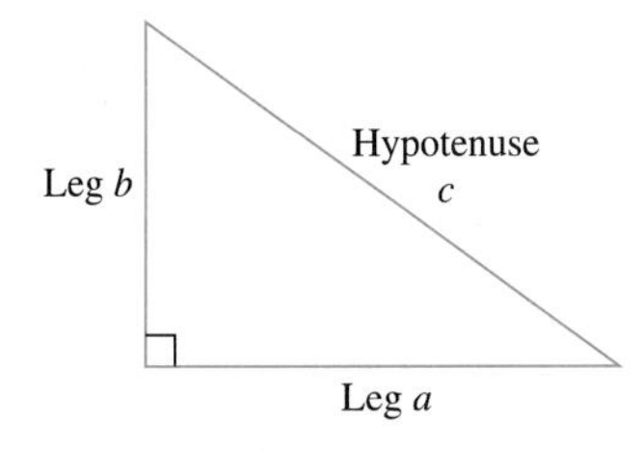

Pythagorean Theorem

In a right triangle, the sum of the squares of the lengths of the two legs is equal to the square of the length of the hypotenuse.

$$(\text{leg})^2 + (\text{leg})^2 = (\text{hypotenuse})^2 \quad \text{or} \quad a^2 + b^2 = c^2$$

Example 8 Using the Pythagorean Theorem

While framing an addition to an existing home, Kim Menzies, a carpenter, used the Pythagorean theorem to determine whether a wall was "square"—that is, whether the wall formed a right angle with the floor. He used a triangle whose sides are three consecutive integers. Find a right triangle whose sides are three consecutive integers.

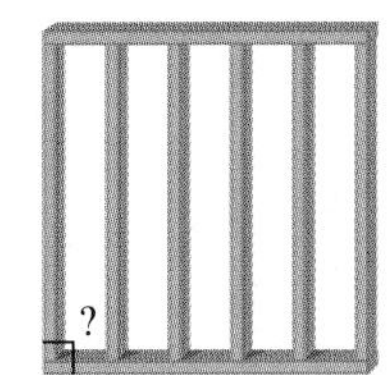

Practice Problem 8

Find a right triangle whose sides are three consecutive even integers.

Answer

8. 6, 8, and 10 units

Solution:

1. UNDERSTAND. Read and reread the problem.
Let x, $x + 1$, and $x + 2$ be three consecutive integers. Since these integers represent lengths of the sides of a right triangle, we have

$$x = \text{one leg}$$
$$x + 1 = \text{other leg}$$
$$x + 2 = \text{hypotenuse (longest side)}$$

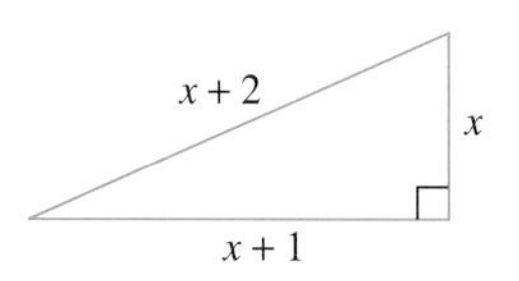

2. TRANSLATE. By the Pythagorean theorem, we have

In words: $(\text{leg})^2 + (\text{leg})^2 = (\text{hypotenuse})^2$

Translate: $(x)^2 + (x + 1)^2 = (x + 2)^2$

3. SOLVE the equation.

$$x^2 + (x + 1)^2 = (x + 2)^2$$
$$x^2 + x^2 + 2x + 1 = x^2 + 4x + 4 \quad \text{Multiply.}$$
$$2x^2 + 2x + 1 = x^2 + 4x + 4$$
$$x^2 - 2x - 3 = 0 \quad \text{Write in standard form.}$$
$$(x - 3)(x + 1) = 0$$
$$x - 3 = 0 \quad \text{or} \quad x + 1 = 0$$
$$x = 3 \qquad\qquad x = -1$$

4. INTERPRET. Discard $x = -1$ since length cannot be negative. If $x = 3$, then $x + 1 = 4$ and $x + 2 = 5$.

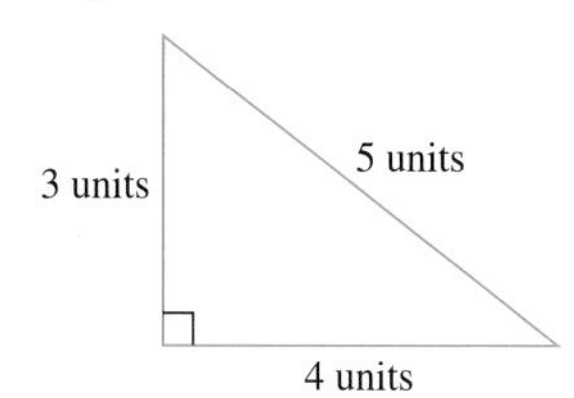

Check: To check, see that $(\text{leg})^2 + (\text{leg})^2 = (\text{hypotenuse})^2$

$$3^2 + 4^2 = 5^2$$
$$9 + 16 = 25 \quad \text{True.}$$

State: The lengths of the sides of the right triangle are 3, 4, and 5 units. Kim used this information, for example, by marking off lengths of 3 and 4 feet on the floor and framing respectively. If the diagonal length between these marks is 5 feet, the wall was "square." If not, adjustments were made.

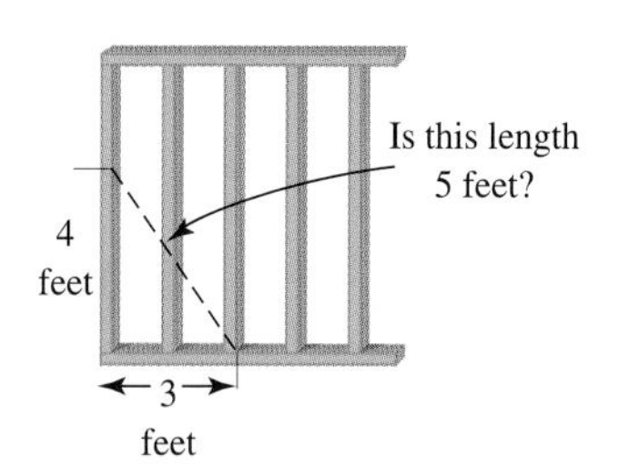

Name ____________________ Section ______ Date ________

MENTAL MATH ANSWERS
1. ______
2. ______
3. ______
4. ______
5. ______
6. ______

ANSWERS
1. ______ 2. ______ 3. ______ 4. ______ 5. ______ 6. ______ 7. ______ 8. ______ 9. ______ 10. ______ 11. ______ 12. ______ 13. ______ 14. ______ 15. ______ 16. ______ 17. ______ 18. ______ 19. ______ 20. ______ 21. ______ 22. ______ 23. ______ 24. ______ 25. ______ 26. ______

MENTAL MATH

Solve each equation for the variable. See Example 1.

1. $(x - 3)(x + 5) = 0$ **2.** $(y + 5)(y + 3) = 0$ **3.** $(z - 3)(z + 7) = 0$

4. $(c - 2)(c - 4) = 0$ **5.** $x(x - 9) = 0$ **6.** $w(w + 7) = 0$

EXERCISE SET 5.7

A *Solve each equation. See Example 1.*

1. $(x + 3)(3x - 4) = 0$ **2.** $(5x + 1)(x - 2) = 0$

3. $3(2x - 5)(4x + 3) = 0$ **4.** $8(3x - 4)(2x - 7) = 0$

Solve each equation. See Examples 2 through 5.

5. $x^2 + 11x + 24 = 0$ **6.** $y^2 - 10y + 24 = 0$ **7.** $12x^2 + 5x - 2 = 0$

8. $3y^2 - y - 14 = 0$ **9.** $z^2 + 9 = 10z$ **10.** $n^2 + n = 72$

11. $x(5x + 2) = 3$ **12.** $n(2n - 3) = 2$ **13.** $x^2 - 6x = x(8 + x)$

14. $n(3 + n) = n^2 + 4n$ **15.** $\frac{z^2}{6} - \frac{z}{2} - 3 = 0$ **16.** $\frac{c^2}{20} - \frac{c}{4} + \frac{1}{5} = 0$

17. $\frac{x^2}{2} + \frac{x}{20} = \frac{1}{10}$ **18.** $\frac{y^2}{30} = \frac{y}{15} + \frac{1}{2}$ **19.** $\frac{4t^2}{5} = \frac{t}{5} + \frac{3}{10}$

20. $\frac{5x^2}{6} - \frac{7x}{2} + \frac{2}{3} = 0$

Solve each equation. See Example 6.

21. $(x + 2)(x - 7)(3x - 8) = 0$ **22.** $(4x + 9)(x - 4)(x + 1) = 0$

23. $y^3 = 9y$ **24.** $n^3 = 16n$

25. $x^3 - x = 2x^2 - 2$ **26.** $m^3 = m^2 + 12m$

27. ______

28. ______

29. ______

30. ______

31. ______

32. ______

33. ______

34. ______

35. ______

36. ______

37. ______

38. ______

39. ______

40. ______

41. ______

42. ______

43. ______

44. ______

45. ______

46. ______

47. ______

48. ______

49. ______

50. ______

51. ______

52. ______

Name ______

27. Explain how solving $2(x-3)(x-1)=0$ differs from solving $2x(x-3)(x-1)=0$.

28. Explain why the zero-factor property works for more than two numbers whose product is 0.

Solve each equation. See Examples 1 through 6.

29. $(2x+7)(x-10)=0$

30. $(x+4)(5x-1)=0$

31. $3x(x-5)=0$

32. $4x(2x+3)=0$

33. $x^2-2x-15=0$

34. $x^2+6x-7=0$

35. $12x^2+2x-2=0$

36. $8x^2+13x+5=0$

37. $w^2-5w=36$

38. $x^2+32=12x$

39. $25x^2-40x+16=0$

40. $9n^2+30n+25=0$

41. $2r^3+6r^2=20r$

42. $-2t^3=108t-30t^2$

43. $z(5z-4)(z+3)=0$

44. $2r(r-3)(5r+4)=0$

45. $2z(z+6)=2z^2+12z-8$

46. $3c^2-8c+2=c(3c-8)$

47. $-3(x-4)+x=5(3-x)$

48. $-4(a+1)-3a=-7(2a-3)$

B *Solve. See Examples 7 and 8.*

49. One number exceeds another by five, and their product is 66. Find the numbers.

50. If the sum of two numbers is 4 and their product is $\frac{15}{4}$, find the numbers.

51. An electrician needs to run a cable from the top of a 60-foot tower to a transmitter box located 45 feet away from the base of the tower. Find how long he should cut the cable.

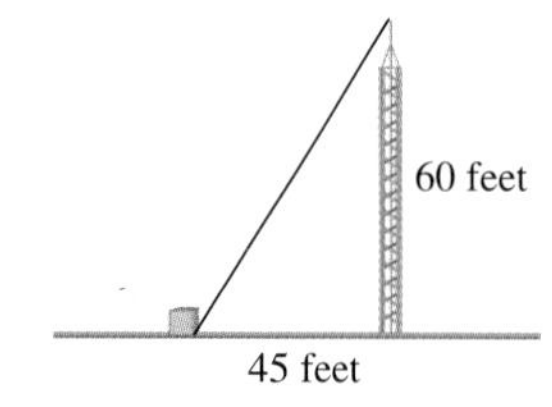

52. A stereo-system installer needs to run speaker wire along the two diagonals of a rectangular room whose dimensions are 40 feet by 75 feet. Find how much speaker wire she needs.

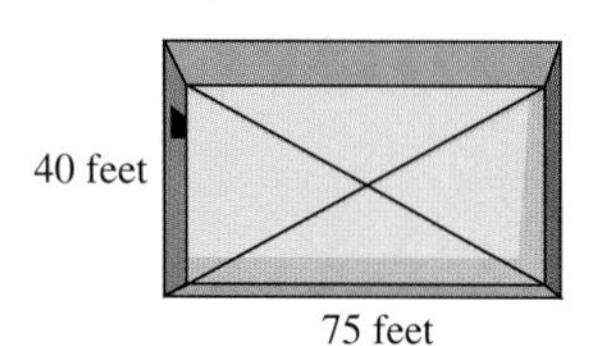

Name ______________________

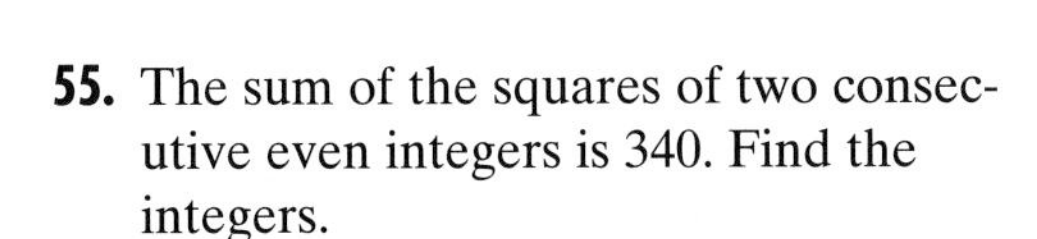

53. The shorter leg of a right triangle is two feet less than the other leg. Find the length of the two legs if the hypotenuse is 10 feet.

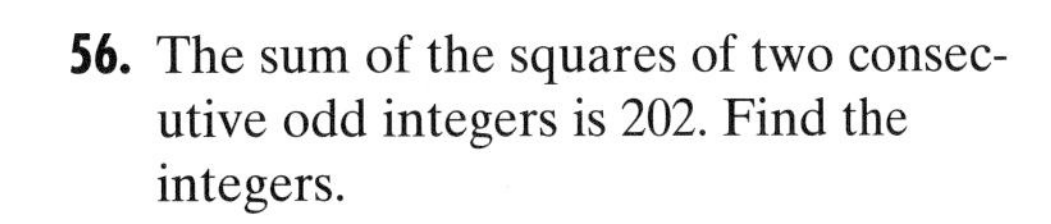

54. The shorter leg of a right triangle is 3 centimeters less than the other leg. Find the length of the two legs if the hypotenuse is 15 centimeters.

55. The sum of the squares of two consecutive even integers is 340. Find the integers.

56. The sum of the squares of two consecutive odd integers is 202. Find the integers.

57. While hovering near the top of Ribbon Falls in Yosemite National Park at 1600 feet, a helicopter pilot accidentally drops his sunglasses. The height h of the sunglasses after t seconds is given by the polynomial equation

$$h = -16t^2 + 1600$$

When will the sunglasses hit the ground?

58. After t seconds, the height h of a model rocket launched from the ground into the air is given by the equation

$$h = -16t^2 + 80t$$

Find how long it takes the rocket to reach a height of 96 feet.

59. The floor of a shed has an area of 91 square feet. The floor is in the shape of a rectangle whose length is 6 feet more than the width. Find the length and the width of the floor of the shed.

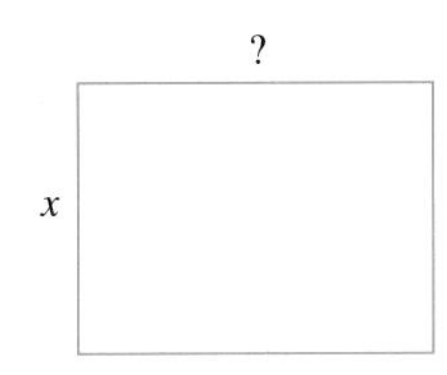

60. A vegetable garden with an area of 143 square feet is to be fertilized. If the width of the garden is 2 feet less than the length, find the dimensions of the garden.

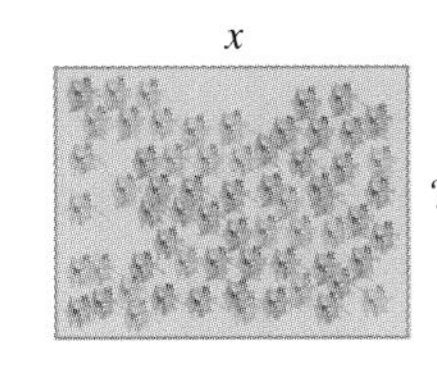

53. ______________

54. ______________

55. ______________

56. ______________

57. ______________

58. ______________

59. ______________

60. ______________

61. ______

62. ______

63. ______

64. ______

65. ______

66. ______

67. ______

68. ______

69. ______

70. ______

71. ______

Name ______

61. Marie Mulroney has a rectangular board 12 inches by 16 inches around which she wants to put a uniform border of shells. If she has enough shells for a border whose area is 128 square inches, determine the width of the border.

62. A gardener has a rose garden that measures 30 feet by 20 feet. He wants to put a uniform border of pine bark around the outside of the garden. Find how wide the border should be if he has enough pine bark to cover 336 square feet.

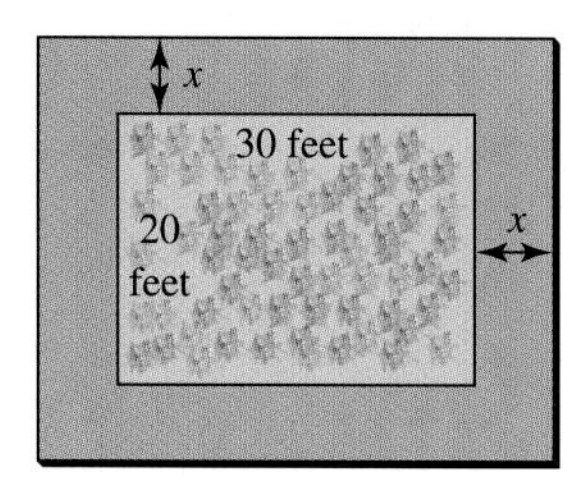

Review and Preview

Write the x- and y-intercept points for each graph. See Section 3.1.

63.

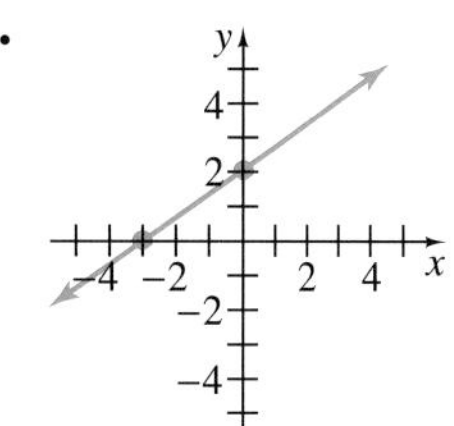

64.

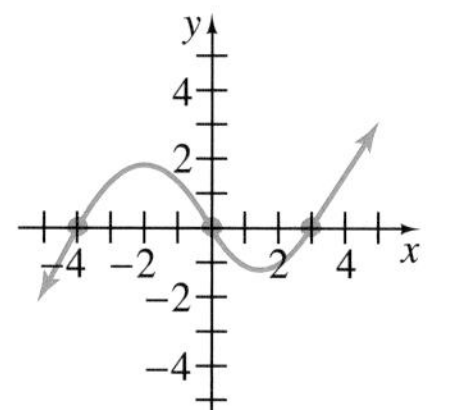

65.

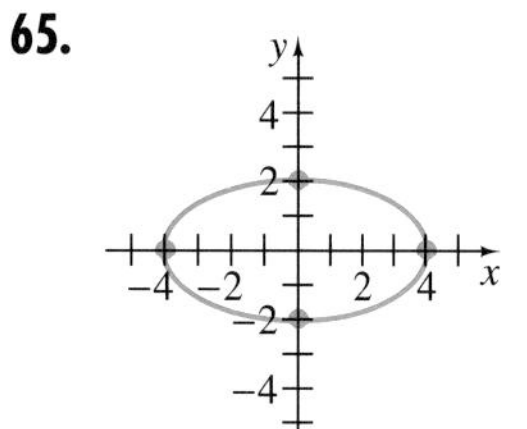

66.

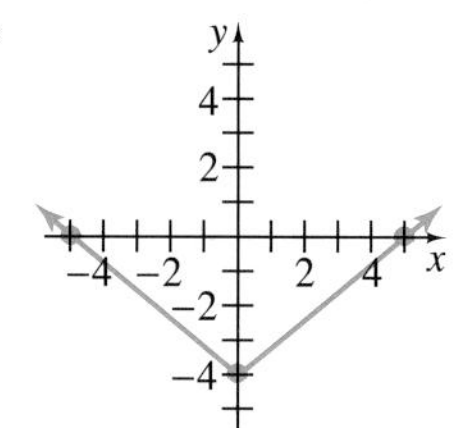

Combining Concepts

Solve.

67. $(x^2 + x - 6)(3x^2 - 14x - 5) = 0$

68. $(x^2 - 9)(x^2 + 8x + 16) = 0$

69. Is the following step correct? Why or why not?

$$x(x - 3) = 5$$
$$x = 5 \text{ or } x - 3 = 5$$

Write a quadratic equation that has the given numbers as solutions.

70. 5, 3

71. 6, 7

5.8 An Introduction to Graphing Polynomial Functions

A Analyzing Graphs of Polynomial Functions

In Section 5.1, we introduced polynomial functions. Some polynomial functions are given special names according to their degree. For example,

$f(x) = 2x - 6$ is called a **linear function**; its **degree is one**.
$f(x) = 5x^2 - x + 3$ is called a **quadratic function**; its **degree is two**.
$f(x) = 7x^3 + 3x^2 - 1$ is called a **cubic function**; its **degree is three**.
$f(x) = -8x^4 - 3x^3 + 2x^2 + 20$ is called a **quartic function**; its **degree is four**.

All the above functions are also polynomial functions.

Objectives

A Analyze the graph of a polynomial function.

B Graph quadratic functions.

C Find the vertex of a parabola by using the vertex formula.

D Graph cubic functions.

SSM CD-ROM Video 5.8

Example 1 Given the graph of the function $g(x)$ below to the left:

a. Find the domain and the range of the function.
b. List the x- and y-intercept points.
c. Find the coordinates of the point with the greatest y-value.
d. Find the coordinates of the point with the least y-value.

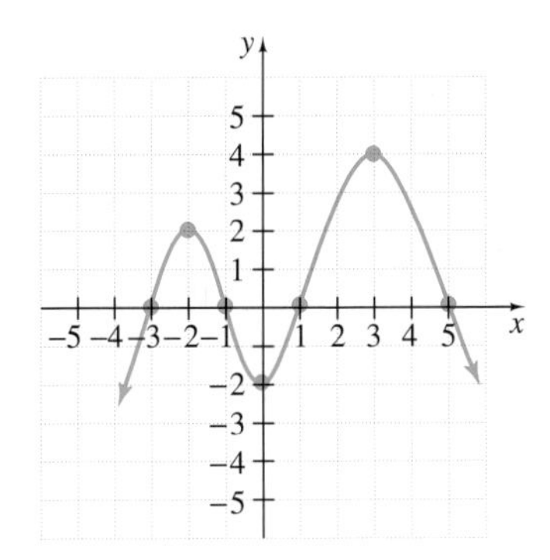

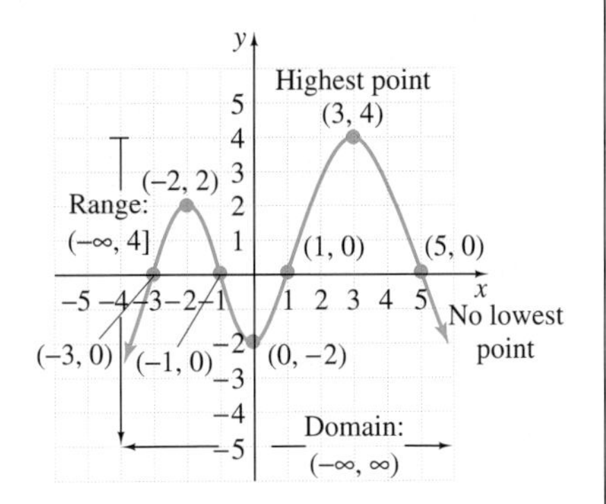

Solution:

a. The domain is the set of all real numbers, or in interval notation, $(-\infty, \infty)$. The range is $(-\infty, 4]$.
b. The x-intercept points are $(-3, 0)$, $(-1, 0)$, $(1, 0)$, and $(5, 0)$. The y-intercept point is $(0, -2)$.
c. The point with the greatest y-value corresponds to the "highest" point. This is the point with coordinates $(3, 4)$. (This means that for all real number values for x, the greatest y-value, or $g(x)$ value, is 4.)
d. The point with the least y-value corresponds to the "lowest" point. This graph contains no "lowest" point, so there is no point with the least y-value.

The graph of any polynomial function (linear, quadratic, cubic, and so on) can be sketched by plotting a sufficient number of ordered pairs that satisfy the function and connecting them to form a smooth curve. The graphs of all polynomial functions will pass the vertical line test since they are graphs of functions.

Practice Problem 1

Given the graph of the function $f(x)$:

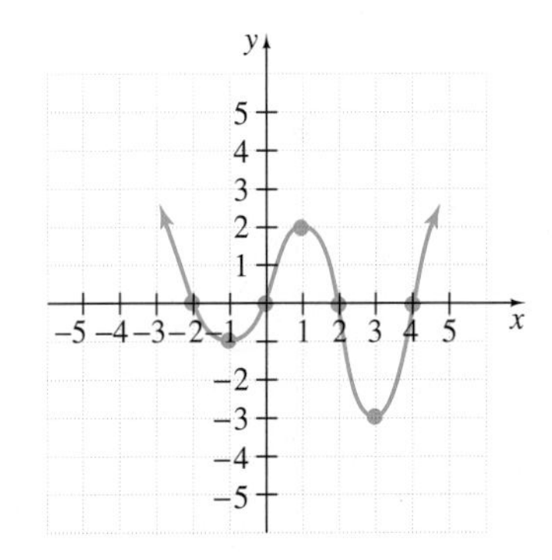

a. Find the domain and the range of the function.
b. List the x- and y-intercept points.
c. Find the coordinates of the point with the greatest y-value.
d. Find the coordinates of the point with the least y-value.

Answers

1. a. domain: $(-\infty, \infty)$; range: $[-3, \infty)$, **b.** $(-2, 0)$, $(0, 0)$, $(2, 0)$, $(4, 0)$, **c.** no greatest y-value point, **d.** $(3, -3)$

B GRAPHING QUADRATIC FUNCTIONS

Since we know how to graph linear functions (see Section 3.1), we will now graph quadratic functions and discuss special characteristics of their graphs.

> **QUADRATIC FUNCTION**
>
> A quadratic function is a function that can be written in the form
>
> $$f(x) = ax^2 + bx + c$$
>
> where a, b, and c are real numbers and $a \neq 0$.

We know that an equation of the form $f(x) = ax^2 + bx + c$ may be written as $y = ax^2 + bx + c$. Thus, both $f(x) = ax^2 + bx + c$ and $y = ax^2 + bx + c$ define quadratic functions as long as a is not 0.

Practice Problem 2

Graph the function $f(x) = -x^2$.

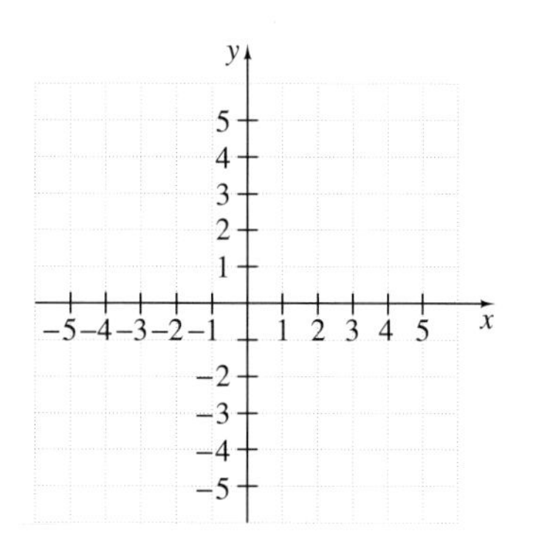

Example 2 Graph the function $f(x) = x^2$ by plotting points.

Solution: This function is not linear, and its graph is not a line. We begin by finding ordered pair solutions. Then we plot the points and draw a smooth curve through them.

If $x = -3$, then $f(-3) = (-3)^2$, or 9.
If $x = -2$, then $f(-2) = (-2)^2$, or 4.
If $x = -1$, then $f(-1) = (-1)^2$, or 1.
If $x = 0$, then $f(0) = 0^2$, or 0.
If $x = 1$, then $f(1) = 1^2$, or 1.
If $x = 2$, then $f(2) = 2^2$, or 4.
If $x = 3$, then $f(3) = 3^2$, or 9.

x	$y = f(x)$
−3	9
−2	4
−1	1
0	0
1	1
2	4
3	9

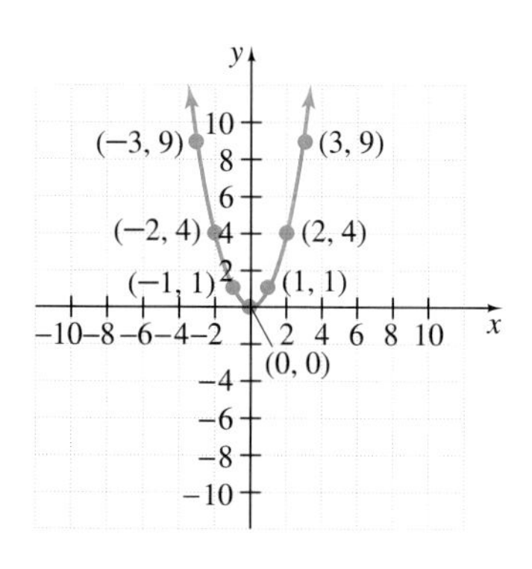

Notice that the graph of Example 2 passes the vertical line test, as it should since it is a function. This curve is called a **parabola**. The highest point on a parabola that opens downward or the lowest point on a parabola that opens upward is called the **vertex** of the parabola. The vertex of this parabola is $(0, 0)$, the lowest point on the graph. If we fold the graph along the y-axis, we can see that the two sides of the graph coincide. This means that this curve is symmetric about the y-axis, and the y-axis, or the line $x = 0$, is called the **axis of symmetry**. The graph of every quadratic function is a parabola and has an axis of symmetry: the vertical line that passes through the vertex of the parabola.

Answer

2.

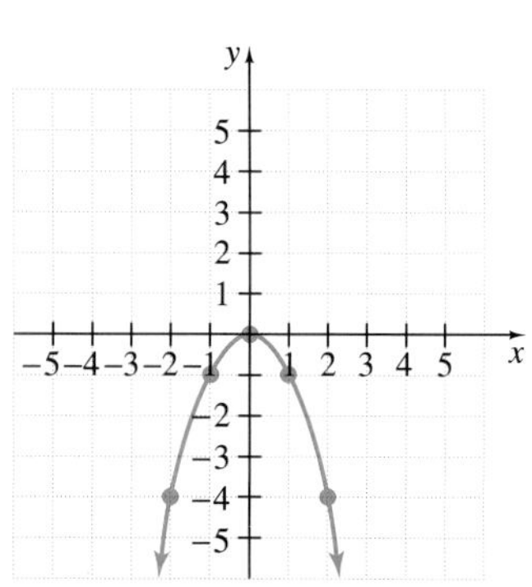

Example 3 Graph the quadratic function $f(x) = -x^2 + 2x - 3$ by plotting points.

Solution: To graph, we choose values for x and find corresponding $f(x)$ or y-values. Then we plot the points and draw a smooth curve through them.

x	$y = f(x)$
-2	-11
-1	-6
0	-3
1	-2
2	-3
3	-6

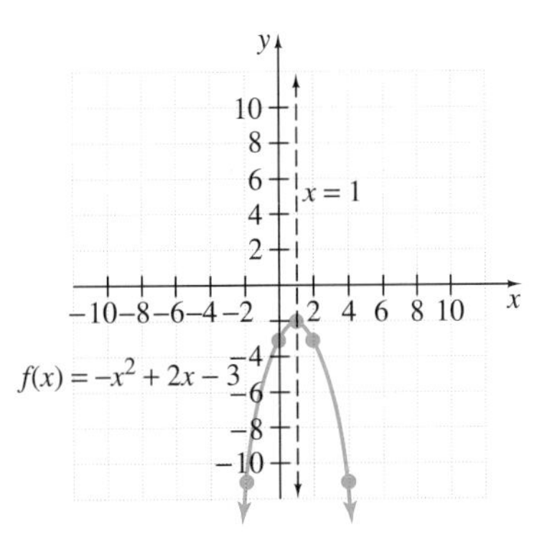

The vertex of this parabola is $(1, -2)$, the highest point on the graph. The vertical line $x = 1$ is the axis of symmetry. Recall that to find the x-intercepts of a graph, we let $y = 0$. Using function notation, this is the same as letting $f(x) = 0$. Since this graph has no x-intercepts, it means that $0 = -x^2 + 2x - 3$ has no real number solutions.

Practice Problem 3

Graph the quadratic function $f(x) = -x^2 - 2x - 3$.

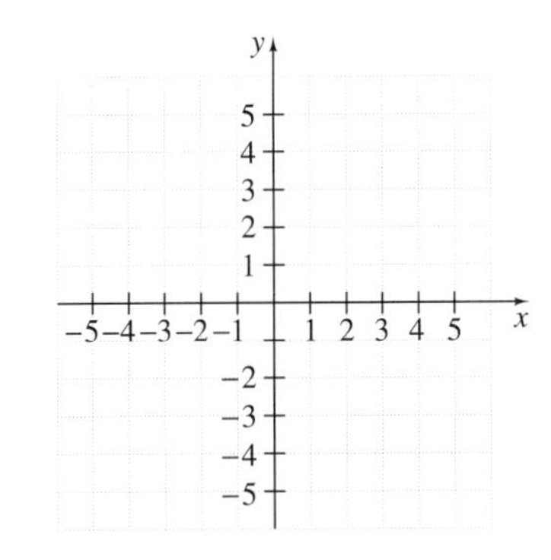

Notice that the parabola $f(x) = -x^2 + 2x - 3$ opens downward, whereas $f(x) = x^2$ opens upward. When the equation of a quadratic function is written in the form $f(x) = ax^2 + bx + c$, the coefficient of the squared variable, a, determines whether the parabola opens downward or upward. If $a > 0$, the parabola opens upward, and if $a < 0$, the parabola opens downward.

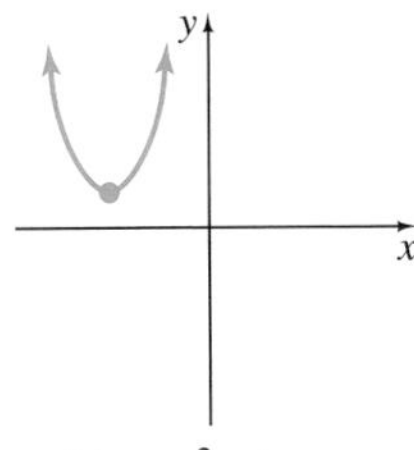

$f(x) = ax^2 + bx + c$,
$a > 0$, opens upward

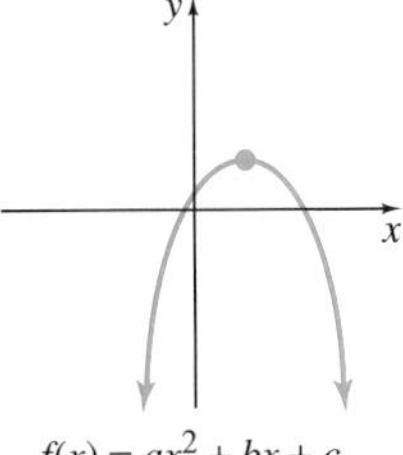

$f(x) = ax^2 + bx + c$,
$a < 0$, opens downward

C FINDING THE VERTEX OF A PARABOLA

In both $f(x) = x^2$ and $f(x) = -x^2 + 2x - 3$, the vertex happens to be one of the points we chose to plot. Since this is not always the case, and since plotting the vertex allows us to draw the graph quickly, we need a consistent method for finding the vertex. One method is to use the following formula, which we shall derive in Chapter 8.

Answer

3.

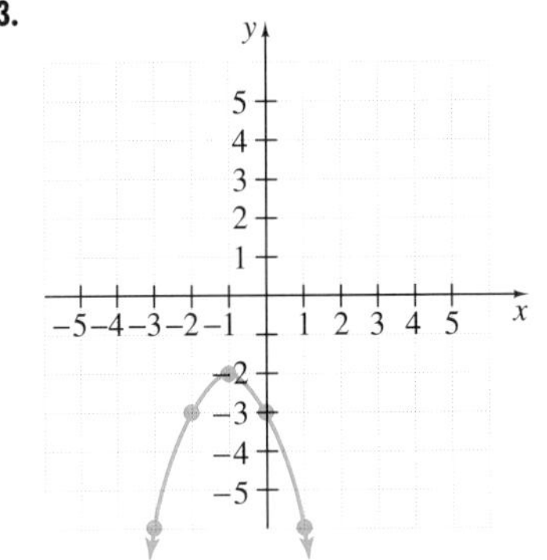

VERTEX FORMULA

The graph of $f(x) = ax^2 + bx + c$, $a \neq 0$, is a parabola with vertex

$$\left(\frac{-b}{2a}, f\left(\frac{-b}{2a}\right)\right)$$

We can also find the x- and y-intercepts of a parabola to aid in graphing. Recall that x-intercepts of the graph of any equation may be found by letting $y = 0$ or $f(x) = 0$ in the equation and solving for x. Also, y-intercepts may be found by letting $x = 0$ in the equation and solving for y or $f(x)$.

Practice Problem 4

Graph $f(x) = x^2 - 2x - 3$. Find the vertex and any intercepts.

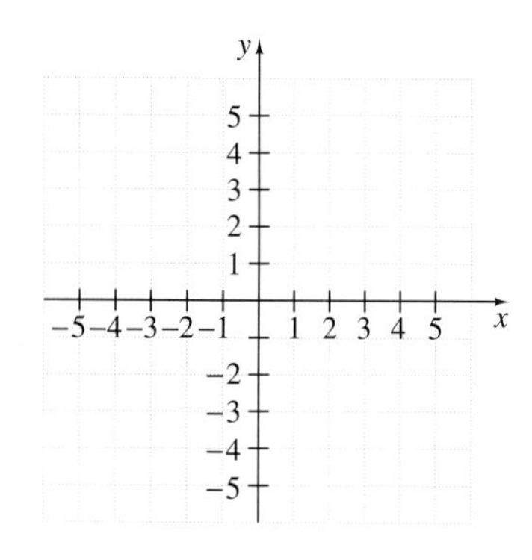

Example 4 Graph $f(x) = x^2 + 2x - 3$. Find the vertex and any intercepts.

Solution: To find the vertex, we use the vertex formula. For the function $f(x) = x^2 + 2x - 3$, $a = 1$ and $b = 2$. Thus,

$$x = \frac{-b}{2a} = \frac{-2}{2(1)} = -1$$

$$f(-1) = (-1)^2 + 2(-1) - 3 \quad \text{Find } f(-1).$$
$$= 1 - 2 - 3$$
$$= -4$$

The vertex is $(-1, -4)$, and since $a = 1$ is greater than 0, this parabola opens upward. Graph the vertex and notice that this parabola will have two x-intercepts because its vertex lies below the x-axis and it opens upward. To find the x-intercepts, we let y or $f(x) = 0$ and solve for x.

To find the y-intercept, let $x = 0$.

$$f(x) = x^2 + 2x - 3$$
$$0 = x^2 + 2x - 3$$
$$0 = (x + 3)(x - 1)$$
$$x + 3 = 0 \quad \text{or} \quad x - 1 = 0$$
$$x = -3 \qquad\qquad x = 1$$

$$f(x) = x^2 + 2x - 3$$
$$f(0) = 0^2 + 2(0) - 3$$
$$f(0) = -3$$

The x-intercepts are -3 and 1 and the corresponding points are $(-3, 0)$ and $(1, 0)$. The y-intercept is -3 and the corresponding point is $(0, -3)$.

Now we plot these points and connect them with a smooth curve.

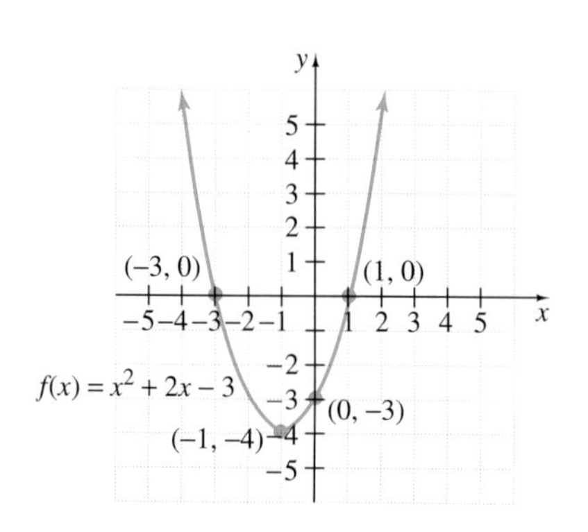

Answer

4.

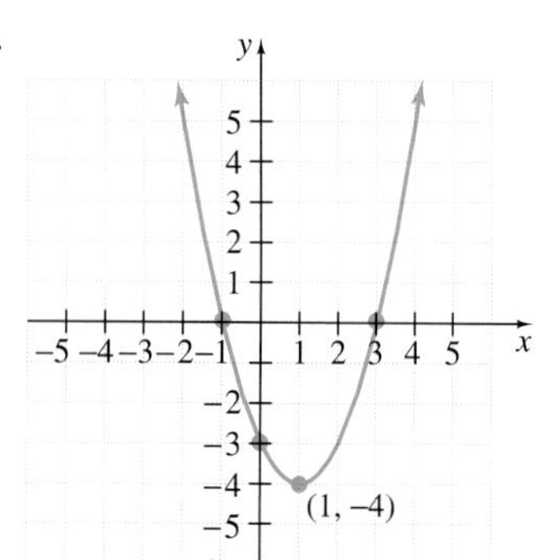

HELPFUL HINT

Not all graphs of parabolas have x-intercepts. To see this, first plot the vertex of the parabola and decide whether the parabola opens upward or downward. Then use this information to decide whether the graph of the parabola has x-intercepts.

Example 5 Graph $f(x) = 3x^2 - 12x + 13$. Find the vertex and any intercepts.

Solution: To find the vertex, we use the vertex formula. For the function $y = 3x^2 - 12x + 13$, $a = 3$ and $b = -12$. Thus

$$x = \frac{-b}{2a} = \frac{-(-12)}{2(3)} = \frac{12}{6} = 2 \quad f(2) = 3(2)^2 - 12(2) + 13 \quad \text{Find } f(2).$$
$$= 3(4) - 24 + 13$$
$$= 1$$

The vertex is $(2, 1)$. Also, this parabola opens upward since $a = 3$ is greater than 0. Graph the vertex and notice that this parabola has no x-intercepts: Its vertex lies above the x-axis, and it opens upward.

To find the y-intercept, let $x = 0$.

$$f(0) = 3(0)^2 - 12(0) + 13$$
$$= 0 - 0 + 13$$
$$= 13$$

The y-intercept is 13. Use this information along with symmetry of a parabola to sketch the graph of $f(x) = 3x^2 - 12x + 13$.

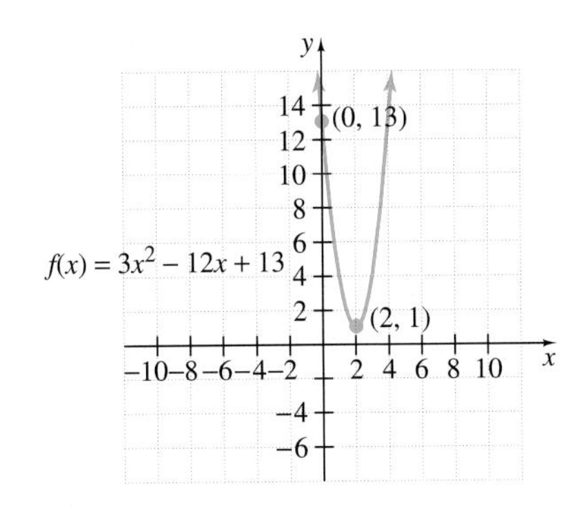

In Section 8.5 we study the graphing of quadratic functions further.

D GRAPHING CUBIC FUNCTIONS

To sketch the graph of a cubic function, we again plot points and then connect the points with a smooth curve. The general shapes of cubic graphs are given below.

Practice Problem 5

Graph $f(x) = 2x^2 + 4x + 4$. Find the vertex and any intercepts.

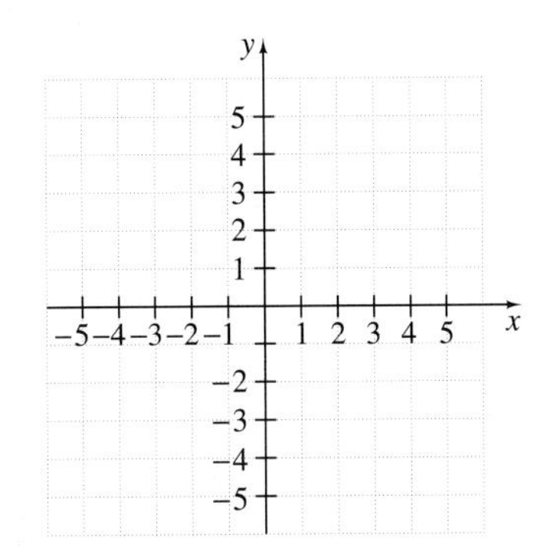

Answer

5.

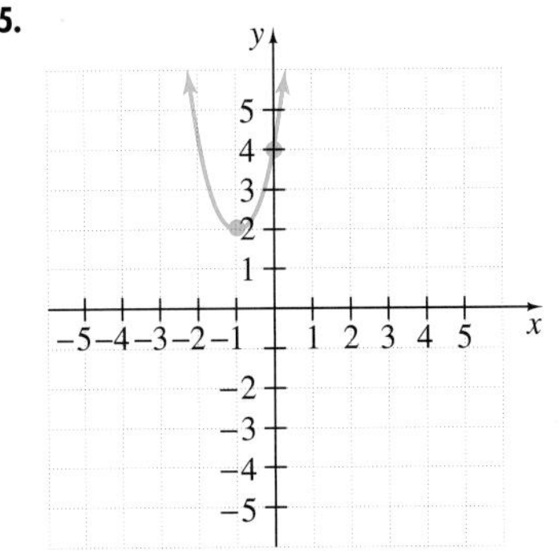

Graph of a Cubic Function
(Degree 3)

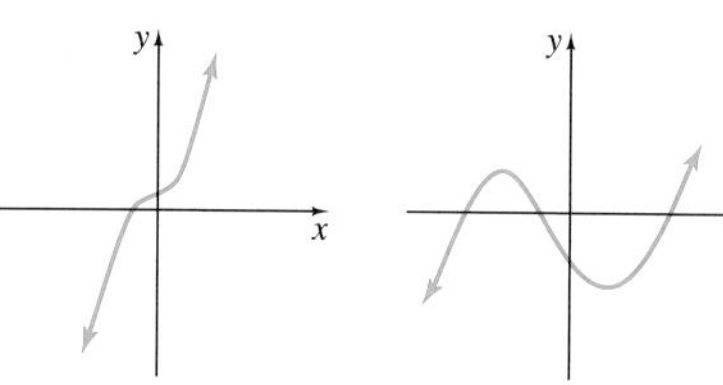
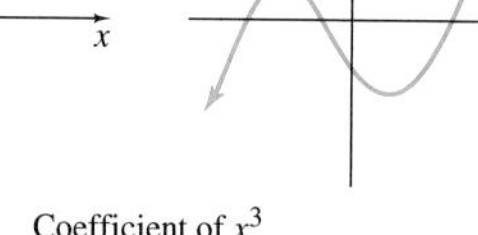

Coefficient of x^3 is a positive number.

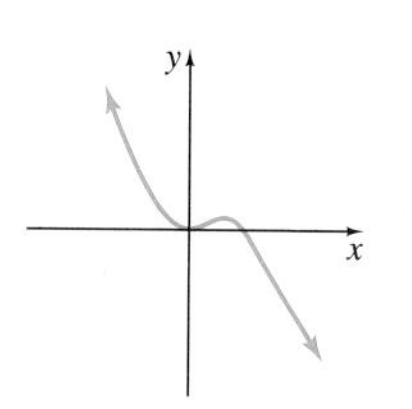
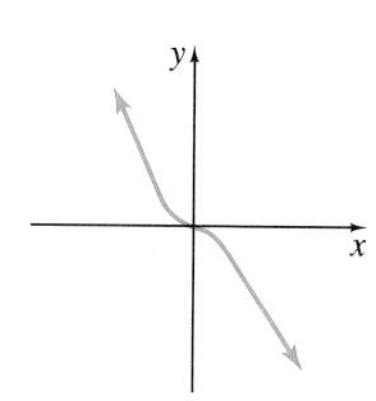

Coefficient of x^3 is a negative number.

Practice Problem 6

Graph $f(x) = x^3 - 9x$. Find any intercepts.

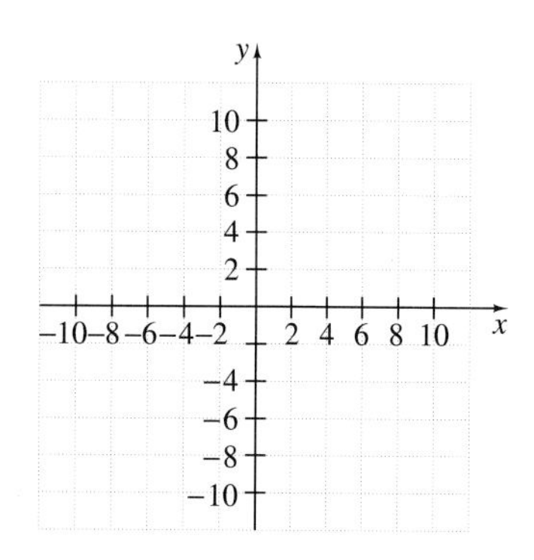

Example 6

Graph $f(x) = x^3 - 4x$. Find any intercepts.

Solution: To find x-intercepts, we let y or $f(x) = 0$ and solve for x.

$$f(x) = x^3 - 4x$$

$$0 = x^3 - 4x \quad \text{Let } f(x) = 0.$$

$$0 = x(x^2 - 4)$$

$$0 = x(x + 2)(x - 2) \quad \text{Factor.}$$

$$x = 0 \quad \text{or} \quad x + 2 = 0 \quad \text{or} \quad x - 2 = 0 \quad \text{Set each factor equal to 0.}$$

$$x = 0 \qquad x = -2 \qquad x = 2 \quad \text{Solve.}$$

This graph has three x-intercepts. They are 0, -2, and 2.

To find the y-intercept, we let $x = 0$.

$$f(0) = 0^3 - 4(0) = 0$$

Next let's select some x-values and find their corresponding $f(x)$ or y-values.

$$f(x) = x^3 - 4x$$

$$f(-3) = (-3)^3 - 4(-3) = -27 + 12 = -15$$

$$f(-1) = (-1)^3 - 4(-1) = -1 + 4 = 3$$

$$f(1) = 1^3 - 4(1) = 1 - 4 = -3$$

$$f(3) = 3^3 - 4(3) = 27 - 12 = 15$$

x	$f(x)$
-3	-15
-1	3
1	-3
3	15

Finally, we plot the intercepts and points and connect them with a smooth curve.

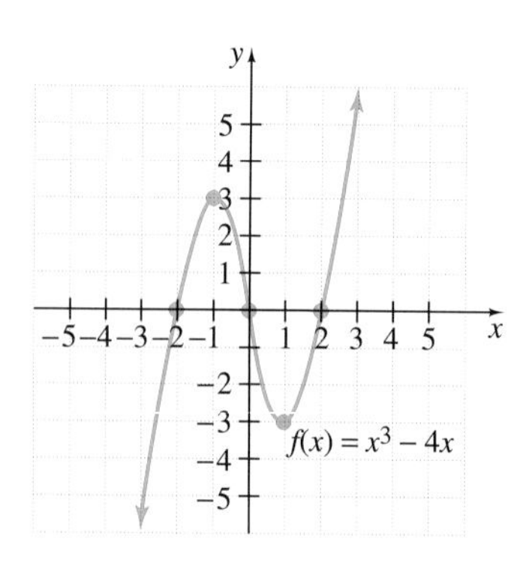

HELPFUL HINT

When a graph has an x-intercept of 0, notice that the y-intercept will also be 0.

Answer

6.

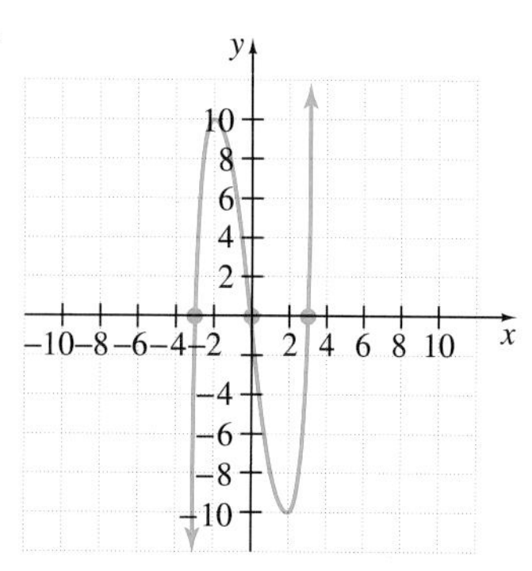

HELPFUL HINT

If you are unsure about the graph of a function, plot more points.

Example 7

Graph $f(x) = -x^3$. Find any intercepts.

Solution: To find x-intercepts, we let y or $f(x) = 0$ and solve for x.

$$f(x) = -x^3$$
$$0 = -x^3$$
$$0 = x$$

The only x-intercept is 0. This means that the y-intercept is 0 also.

Next we choose some x-values and find their corresponding y-values.

$$f(x) = -x^3$$
$$f(-2) = -(-2)^3 = 8$$
$$f(-1) = -(-1)^3 = 1$$
$$f(1) = -(1)^3 = -1$$
$$f(2) = -2^3 = -8$$

x	$f(x)$
−2	8
−1	1
1	−1
2	−8

Now we plot the points and sketch the graph of $f(x) = -x^3$.

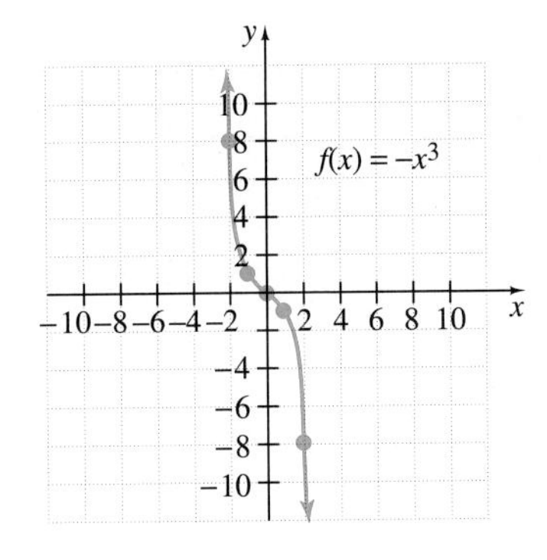

Practice Problem 7

Graph $f(x) = 2x^3$. Find any intercepts.

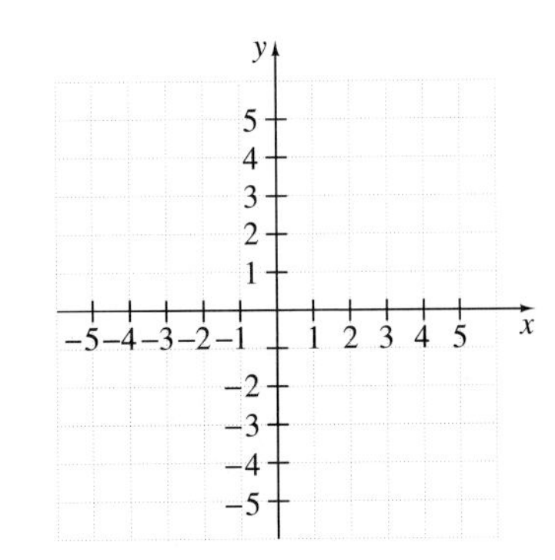

Answer

7.

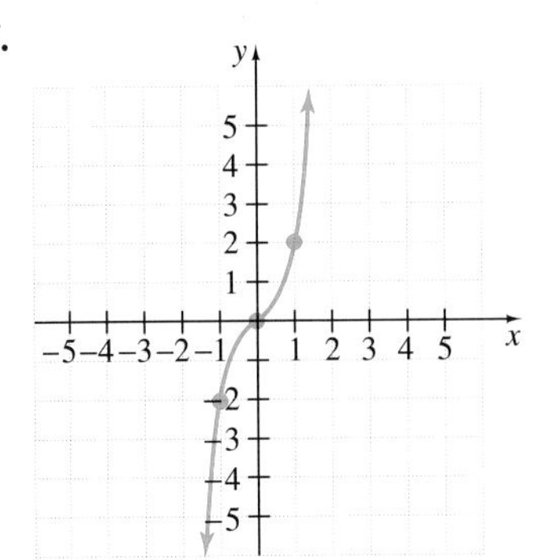

GRAPHING CALCULATOR EXPLORATIONS

We can use a grapher to approximate real number solutions of any quadratic equation in standard form, whether the associated polynomial is factorable or not. For example, let's solve the quadratic equation $x^2 - 2x - 4 = 0$. The solutions of this equation will be the x-intercepts of the graph of the function $f(x) = x^2 - 2x - 4$. (Recall that to find x-intercepts, we let $f(x) = 0$, or $y = 0$.) When we use a standard window, the graph of this function looks like this:

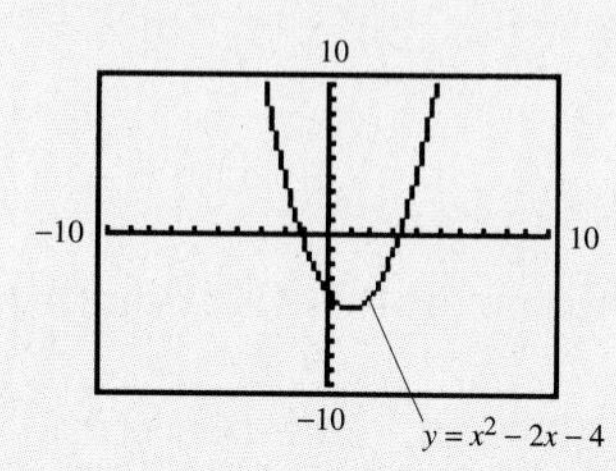

The graph appears to have one x-intercept between -2 and -1 and one between 3 and 4. To find the x-intercept between 3 and 4 to the nearest hundredth, we can use a Root feature, a Zoom feature, which magnifies a portion of the graph around the cursor, or we can redefine our window. If we redefine our window to

Xmin = 2 Ymin = −1
Xmax = 5 Ymax = 1
Xscl = 1 Yscl = 1

the resulting screen is

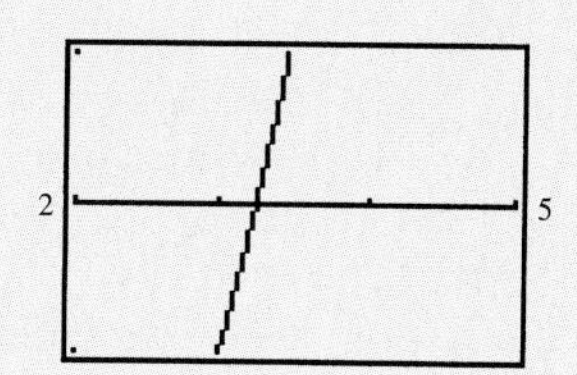

By using the Trace feature, we can now see that one of the intercepts is between 3.21 and 3.25. To approximate to the nearest hundredth, Zoom again or redefine the window to

Xmin = 3.2 Ymin = −0.1
Xmax = 3.3 Ymax = 0.1
Xscl = 1 Yscl = 1

If we use the Trace feature again, we see that, to the nearest thousandth, the x-intercept is 3.236. By repeating this process, we can approximate the other x-intercept to be -1.236.

To check, find $f(3.236)$ and $f(-1.236)$. Both of these values should be close to 0. (They will not be exactly 0 since we approximated these solutions.)

$$f(3.236) = -0.000304 \quad \text{and} \quad f(-1.236) = -0.000304$$

Solve each of these quadratic equations by graphing a related function and approximating the x-intercepts to the nearest thousandth.

1. $x^2 + 3x - 2 = 0$

2. $5x^2 - 7x + 1 = 0$

3. $2.3x^2 - 4.4x - 5.6 = 0$

4. $0.2x^2 + 6.2x + 2.1 = 0$

5. $0.09x^2 - 0.13x - 0.08 = 0$

6. $x^2 + 0.08x - 0.01 = 0$

Name ______________________ Section ________ Date ________

Mental Math

State whether the graph of each quadratic function, a parabola, opens upward or downward.

1. $f(x) = 2x^2 + 7x + 10$

2. $f(x) = -3x^2 - 5x$

3. $f(x) = -x^2 + 5$

4. $f(x) = x^2 + 3x + 7$

Exercise Set 5.8

A *For the graph of each function $f(x)$ answer the following. See Example 1.*

a. Find the domain and the range of the function.

b. List the x- and y-intercept points.

c. Find the coordinates of the point with the greatest y-value.

d. Find the coordinates of the point with the least y-value.

1.

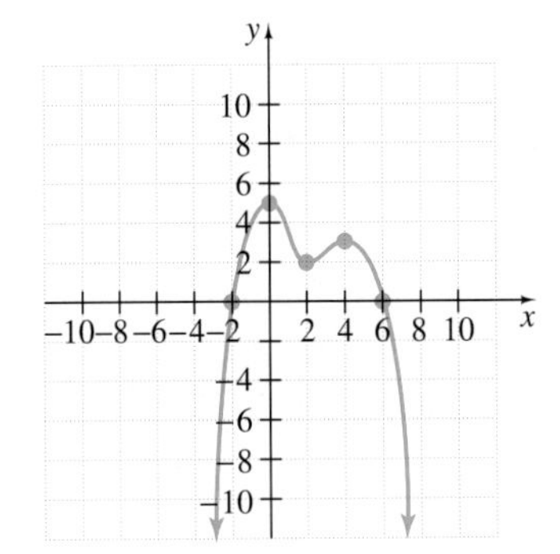

2.

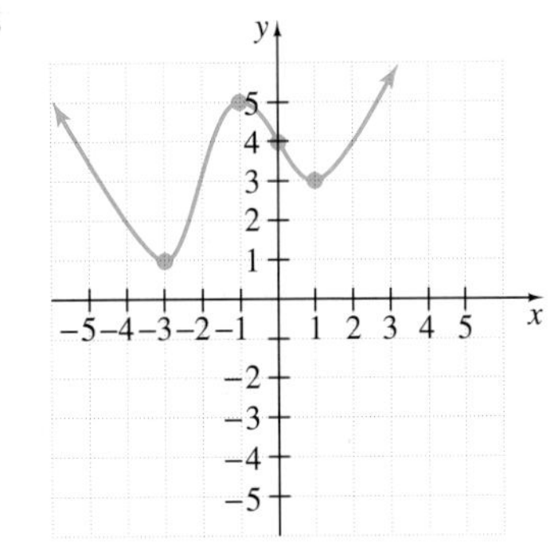

3. The graph in Example 4 of this section

4. The graph in Example 5 of this section

5. The graph in Example 6 of this section

6. The graph in Example 7 of this section

Mental Math Answers

1. ________
2. ________
3. ________
4. ________

Answers

1. a. ________
   b. ________
   c. ________
   d. ________
2. a. ________
   b. ________
   c. ________
   d. ________
3. a. ________
   b. ________
   c. ________
   d. ________
4. a. ________
   b. ________
   c. ________
   d. ________
5. a. ________
   b. ________
   c. ________
   d. ________
6. a. ________
   b. ________
   c. ________
   d. ________

Name ____

B *Graph each quadratic function by plotting points. See Examples 2 and 3.*

7. $f(x) = 2x^2$

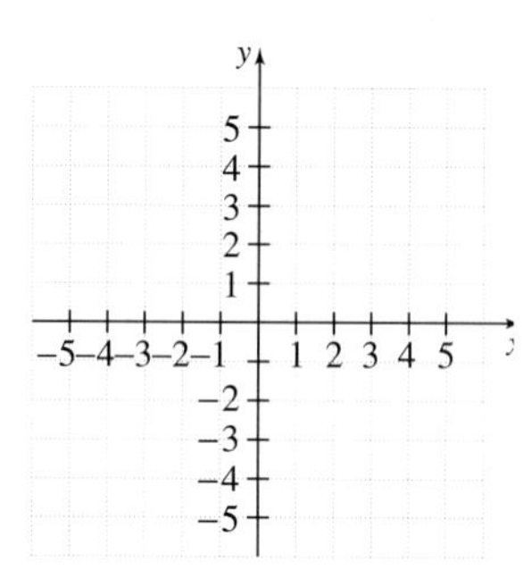

8. $f(x) = -3x^2$

9. $f(x) = x^2 + 1$

10. $f(x) = x^2 - 2$

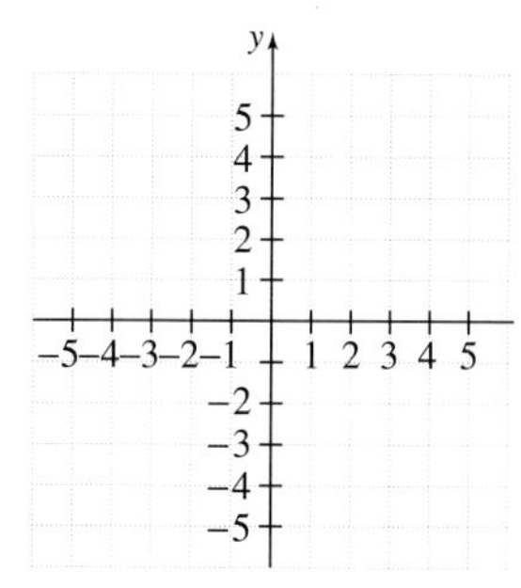

11. $f(x) = -x^2$

12. $f(x) = \frac{1}{2}x^2$

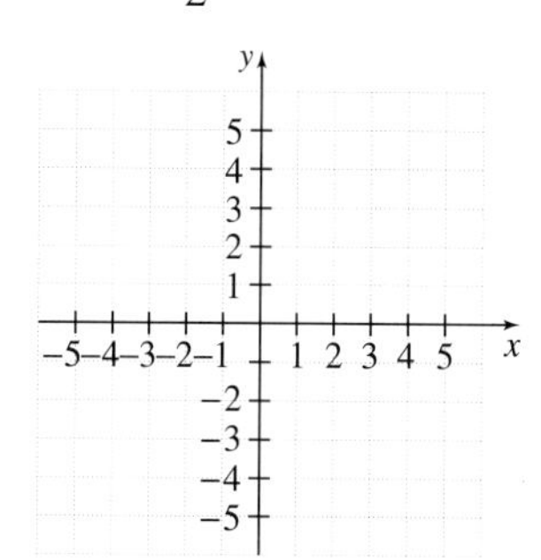

C *Graph each quadratic function. Find and label the vertex and intercepts. See Examples 4 and 5.*

13. $f(x) = x^2 + 8x + 7$

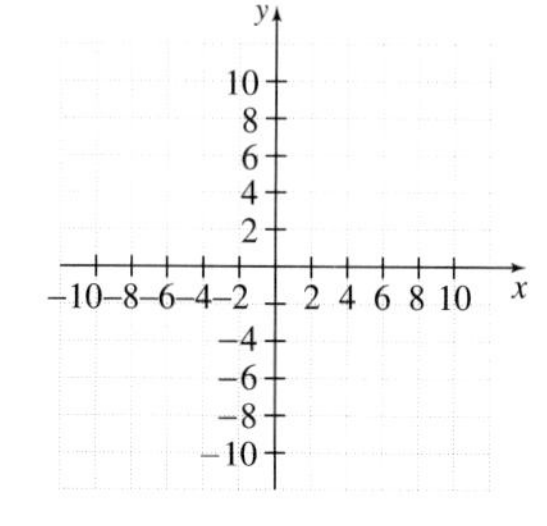

14. $f(x) = x^2 + 6x + 5$

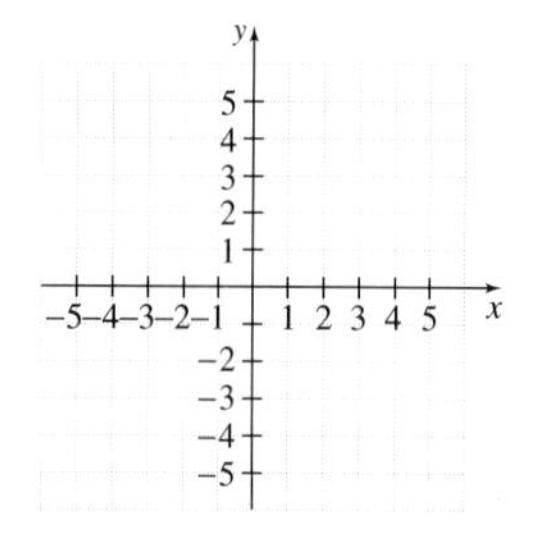

15. $f(x) = x^2 - 2x - 24$

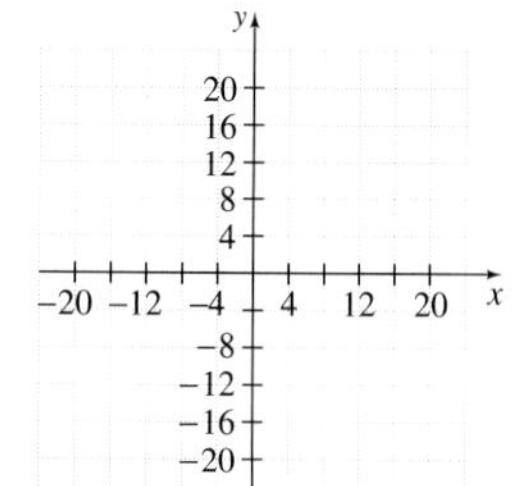

16. $f(x) = x^2 - 12x + 35$

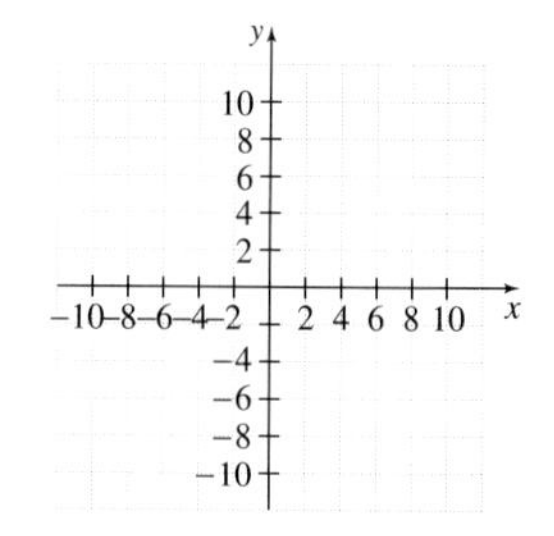

Name ________________________________

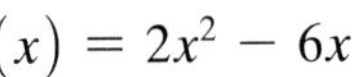

17. $f(x) = 2x^2 - 6x$

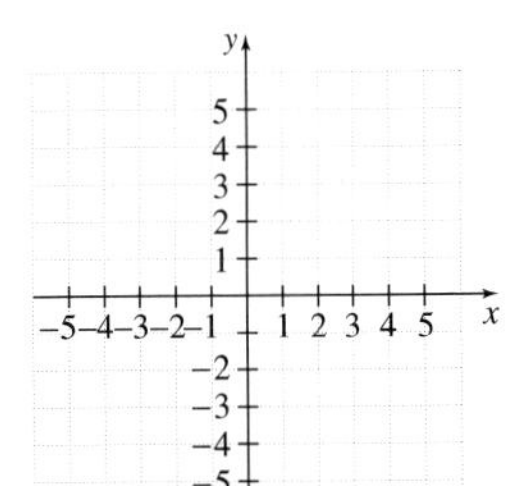

18. $f(x) = -3x^2 + 6x$

19. $f(x) = x^2 + 1$

20. $f(x) = x^2 + 4$

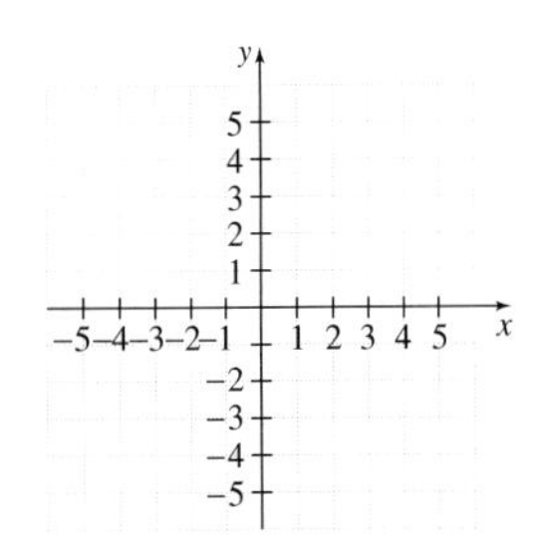

21. If the vertex of a parabola lies below the x-axis and the parabola opens upward, how many x-intercepts will the graph have?

22. If the vertex of a parabola lies below the x-axis and the parabola opens downward, how many x-intercepts will the graph have?

23. If the vertex of a parabola lies above the x-axis and the parabola opens upward, how many x-intercepts will the graph have?

24. If the vertex of a parabola lies above the x-axis and the parabola opens downward, how many x-intercepts will the graph have?

21. ____________________

22. ____________________

23. ____________________

24. ____________________

Name ______________________

Graph each cubic function. Find any intercepts. See Examples 6 and 7.

25. $f(x) = 4x^3 - 9x$

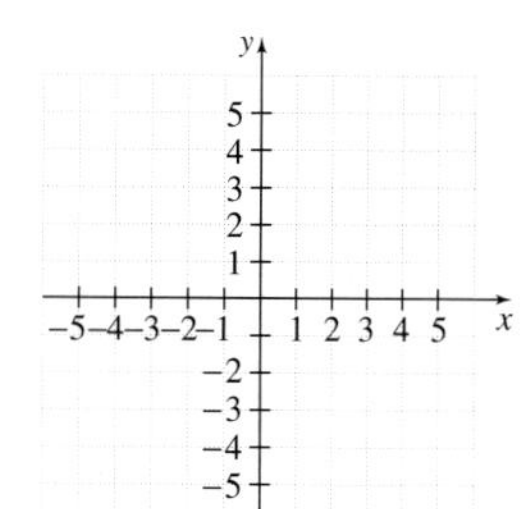

26. $f(x) = 2x^3 - 5x^2 - 3x$

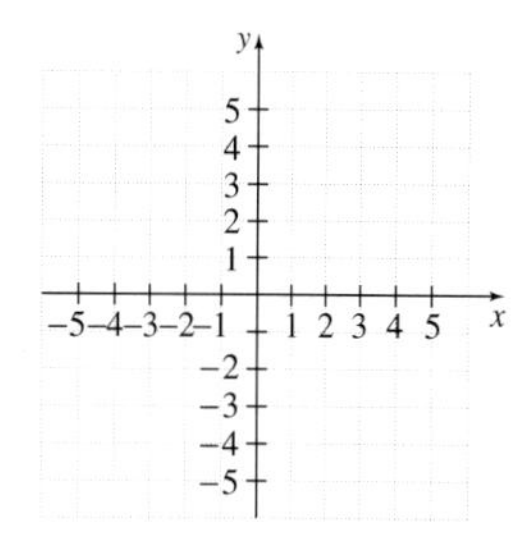

27. $f(x) = x^3 + 3x^2 - x - 3$

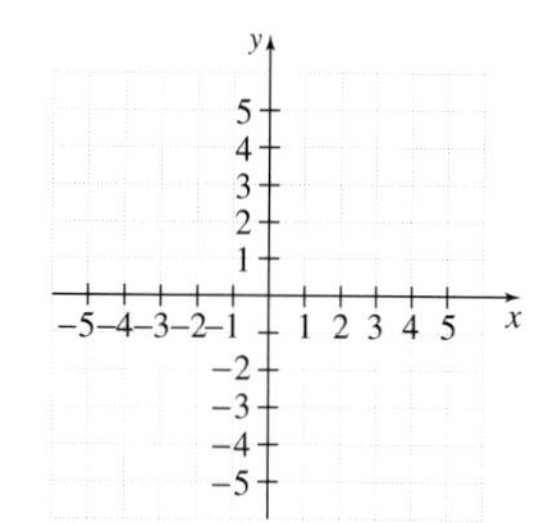

28. $f(x) = x^3 + x^2 - 4x - 4$

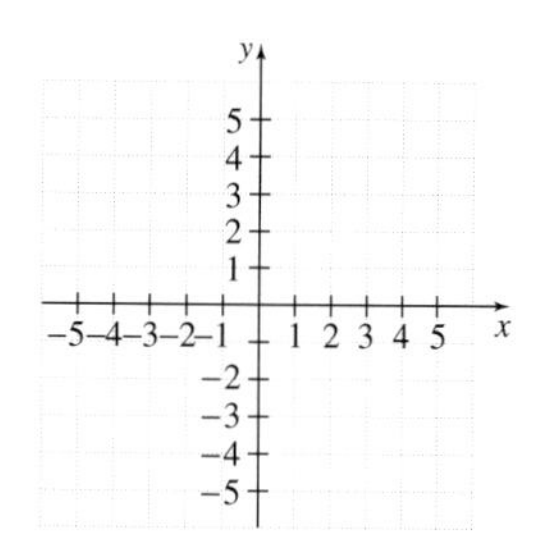

29. $f(x) = x(x - 4)(x + 2)$

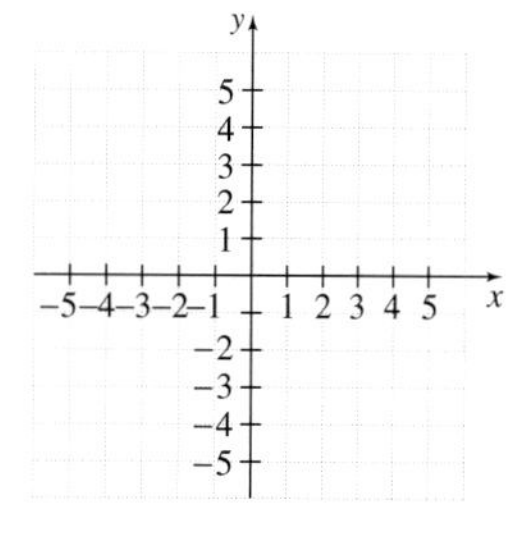

30. $f(x) = 3x(x - 3)(x + 5)$

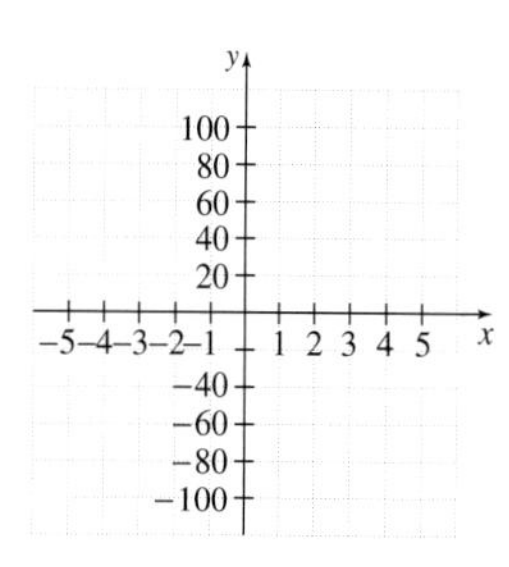

31. $g(x) = x(x - 2)(x + 3)(x + 5)$

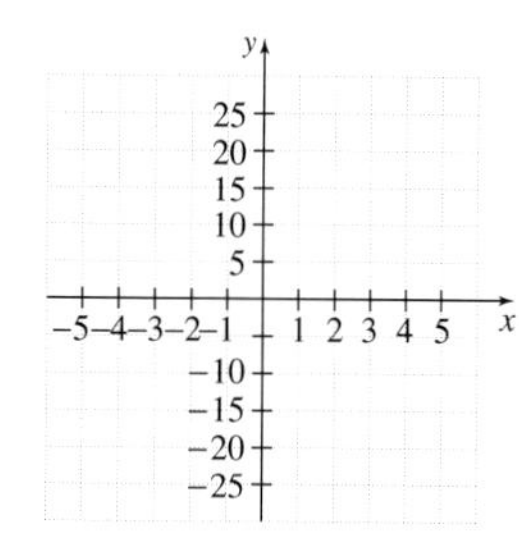

32. $h(x) = (x - 4)(x - 2)(2x + 1)(x + 3)$

y
50
25
-5-4-3-2-1 1 2 3 4 5 x
-25
-50

33. Can the graph of a function ever have more than one y-intercept point? Why or why not?

34. In general, is there a limit to the number of x-intercepts for the graph of a function?

33. ______________________

34. ______________________

Name ______________________________

Review and Preview

Simplify each fraction. See Sections 1.5 and 1.6.

35. $-\dfrac{8}{10}$

36. $-\dfrac{45}{100}$

37. $\dfrac{x^7y^{10}}{x^3y^{15}}$

38. $\dfrac{a^{14}b^2}{ab^4}$

39. $\dfrac{7n^{-9}m^{-2}}{14nm^{-5}}$

40. $\dfrac{20x^{-3}y^5}{25y^{-2}x}$

Match each polynomial function (A–F) with its graph.

41. $f(x) = (x - 2)(x + 5)$

42. $f(x) = (x + 1)(x - 6)$

43. $f(x) = x(x + 3)(x - 3)$

44. $f(x) = (x + 1)(x - 2)(x + 5)$

45. $f(x) = 2x^2 + 9x + 4$

46. $f(x) = 2x^2 - 7x - 4$

A

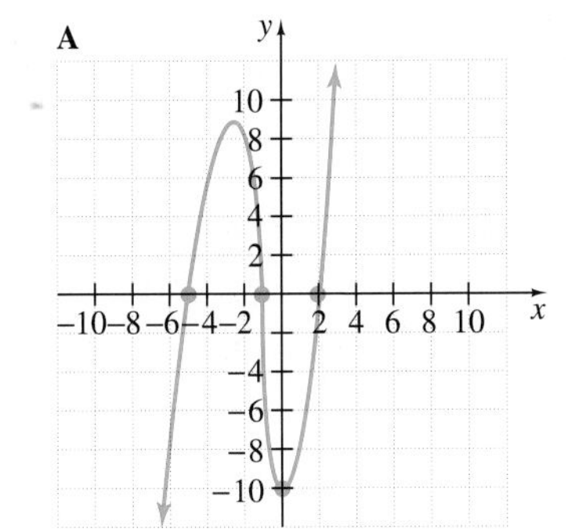

B

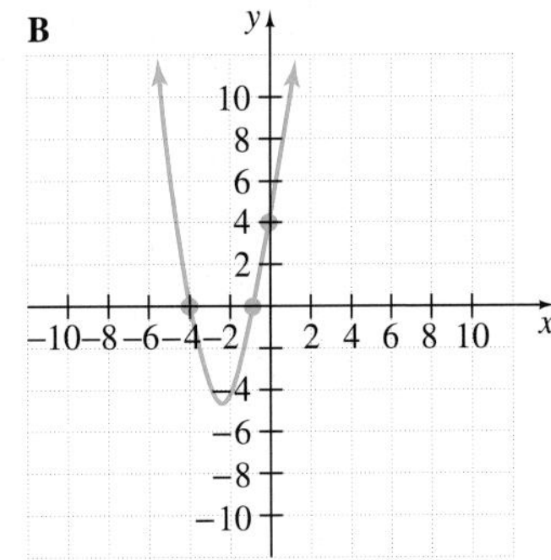

C

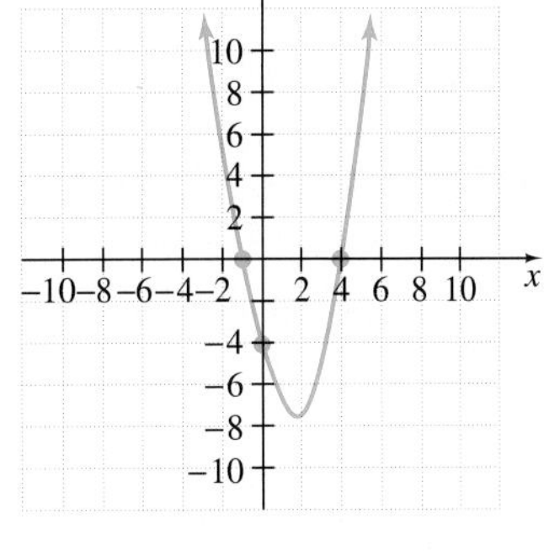
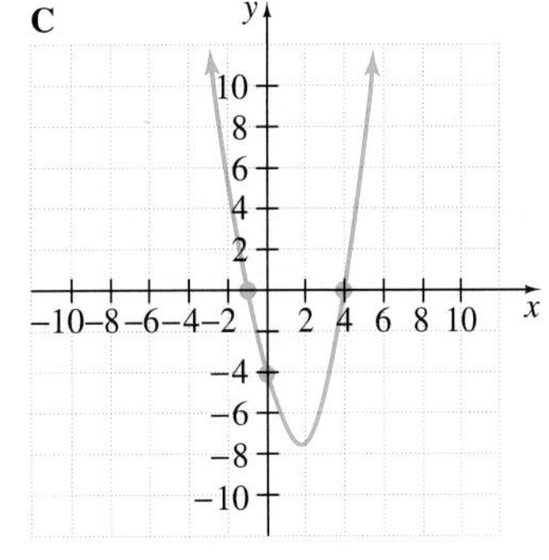

D

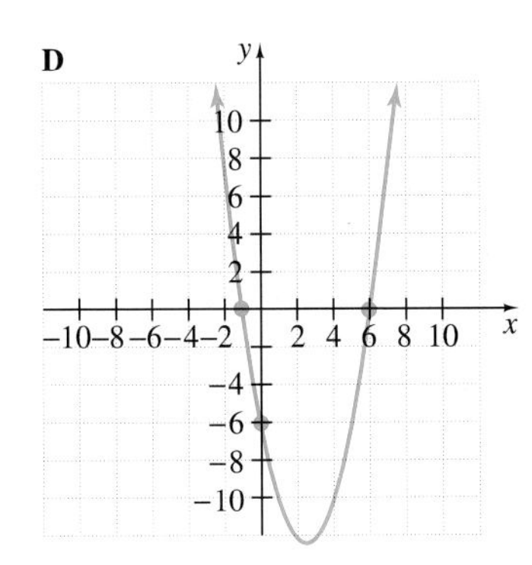

E

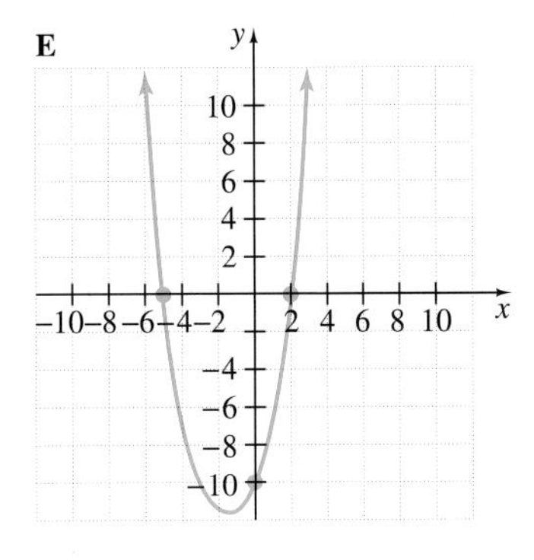

F

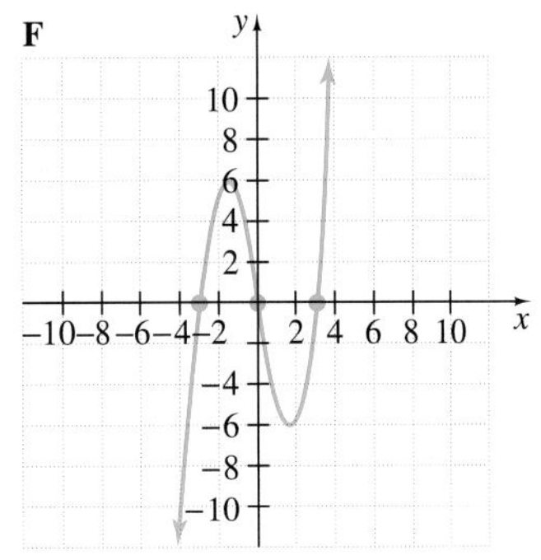

35. ____________

36. ____________

37. ____________

38. ____________

39. ____________

40. ____________

41. ____________

42. ____________

43. ____________

44. ____________

45. ____________

46. ____________

47. see calculator screen

48. see calculator screen

49. see calculator screen

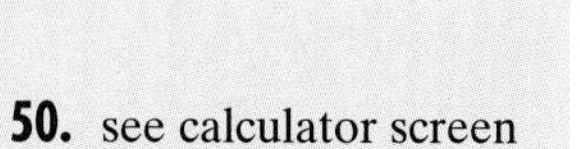

50. see calculator screen

51. ____________

52. ____________

Name ____

Combining Concepts

Use a grapher to verify the graph in each exercise.

47. Exercise 13

48. Exercise 14

49. Exercise 25

50. Exercise 26

Internet Excursions

Go to http://www.prenhall.com/martin-gay
This World Wide Web address will give you access to a web site where you can graph a second-degree polynomial or quadratic function of the form $f(x) = (a_2)x^2 + (a_1)x + (a_0)$. Specifiy the coefficients a_2, a_1, and a_0 in order to answer the questions below.

51. Leaving the values of a_1 and a_0 set equal to 0, investigate the effect of the value of a_2, the coefficient of the x^2-term of the quadratic function, on the graph of the function. Describe the shape of the graph for larger versus smaller positive values of a_2. Similarly describe the shape of the graph for larger versus smaller negative values of a_2. What patterns do you notice?

52. Leaving the value of a_1 set equal to 0 and the value of a_2 set equal to any value except 0, investigate the effect of the value of a_0, the constant term of the quadratic function, on the graph of the function. As you change the value of a_0, describe how the graph changes. What relationship do you notice between the value of a_0 and the position of the graph?

Name ______________________ Section ________ Date ________

Chapter 5 Activity
Finding the Largest Area

This activity may be completed by working in groups or individually.

A picture framer has a piece of wood that measures 1 inch wide by 50 inches long. She would like to make a picture frame with the largest possible interior area. Complete the following activity to help her determine the dimensions of the frame that she should use to achieve her goal.

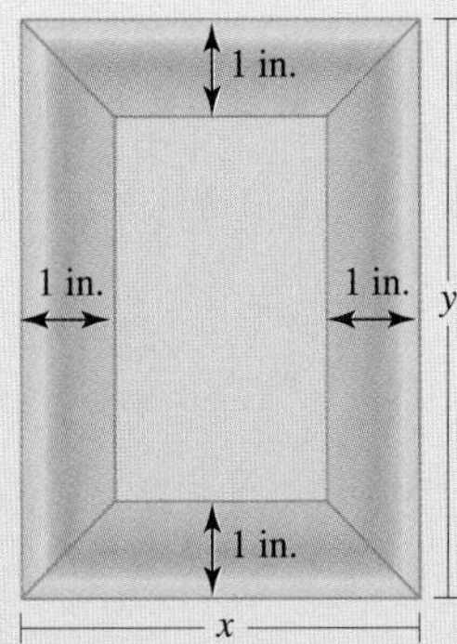

1. Use the situation shown in the figure to write an equation in x and y for the *outer* perimeter of the frame. (Remember that the outer perimeter will equal 50 inches.)
2. Use your equation from Question 1 to help you find the value of y for each value of x given in the table. Complete the y column of the table. (*Note:* The first two columns of the table give possible combinations for the outer dimensions of the frame.)
3. How is the interior width of the frame related to the exterior width of the frame? How is the interior height of the frame related to the exterior height of the frame? Use these relationships to complete the two columns of the table labeled "Frame's Interior Dimensions."
4. Complete the last column of the table labeled "Interior Area" by using the columns of dimensions for the interior width and height.
5. From the table, what appears to be the largest interior area of the frame? Which exterior dimensions of the frame provide this area?
6. Use the patterns in the table to write an algebraic expression in terms of x for the interior width of the frame.
7. Use the patterns in the table to write an algebraic expression in terms of y for the interior height of the frame.
8. Use the perimeter equation from Question 1 to rewrite the algebraic expression for the interior height of the frame in terms of x.

Frame's Interior Dimensions				
x	y	Interior Width	Interior Height	Interior Area
2.0				
2.5				
3.0				
3.5				
4.0				
4.5				
5.0				
5.5				
6.0				
6.5				
7.0				
7.5				
8.0				
8.5				
9.0				
9.5				
10.0				
10.5				
11.0				
11.5				
12.0				
12.5				
13.0				
13.5				
14.0				
14.5				
15.0				

9. Find a function $A(x)$ that gives the interior area of the frame in terms of its exterior width x. (*Hint:* Study the patterns in the table. How could the expressions from Questions 6 and 8 be used to write this function?)
10. Graph the function $A(x)$. Locate and label the point from the table that represents the maximum interior area. Describe the location of the point in relation to the rest of the graph.

CHAPTER 5 HIGHLIGHTS

DEFINITIONS AND CONCEPTS	EXAMPLES
SECTION 5.1 ADDING AND SUBTRACTING POLYNOMIALS	
A **polynomial** is a finite sum of terms in which all variables have exponents raised to nonnegative integer powers and no variables appear in any denominator.	$1.3x^2$ Monomial $-\frac{1}{3}y + 5$ Binomial $6z^2 - 5z + 7$ Trinomial
To add polynomials, combine all like terms.	Add: $(3y^2x - 2yx + 11) + (-5y^2x - 7)$ $= -2y^2x - 2yx + 4$
To subtract polynomials, change the signs of the terms of the polynomial being subtracted; then add.	Subtract: $(-2z^3 - z + 1) - (3z^3 + z - 6)$ $= -2z^3 - z + 1 - 3z^3 - z + 6$ $= -5z^3 - 2z + 7$
A function P is a **polynomial function** if $P(x)$ is a polynomial.	For the polynomial function $P(x) = -x^2 + 6x - 12$ find $P(-2)$. $P(-2) = -(-2)^2 + 6(-2) - 12 = -28$
SECTION 5.2 MULTIPLYING POLYNOMIALS	
TO MULTIPLY TWO POLYNOMIALS Use the distributive property and multiply each term of one polynomial by each term of the other polynomial; then combine like terms.	Multiply. $(x^2 - 2x)(3x^2 - 5x + 1)$ $= 3x^4 - 5x^3 + x^2 - 6x^3 + 10x^2 - 2x$ $= 3x^4 - 11x^3 + 11x^2 - 2x$
SPECIAL PRODUCTS $(a + b)^2 = a^2 + 2ab + b^2$ $(a - b)^2 = a^2 - 2ab + b^2$ $(a + b)(a - b) = a^2 - b^2$	$(3m + 2n)^2 = 9m^2 + 12mn + 4n^2$ $(z^2 - 5)^2 = z^4 - 10z^2 + 25$ $(7y + 1)(7y - 1) = 49y^2 - 1$
The **FOIL order** may be used when multiplying two binomials.	Multiply. $(x^2 + 5)(2x^2 - 9)$ F O I L $= x^2(2x^2) + x^2(-9) + 5(2x^2) + 5(-9)$ $= 2x^4 - 9x^2 + 10x^2 - 45$
SECTION 5.3 DIVIDING POLYNOMIALS	
TO DIVIDE A POLYNOMIAL BY A MONOMIAL Divide each term in the polynomial by the monomial.	$\frac{12a^5b^3 - 6a^2b^2 + ab}{6a^2b^2}$ $= \frac{12a^5b^3}{6a^2b^2} - \frac{6a^2b^2}{6a^2b^2} + \frac{ab}{6a^2b^2}$ $= 2a^3b - 1 + \frac{1}{6ab}$

Section 5.3 (continued)

To Divide a Polynomial by a Polynomial Other Than a Monomial Use **long division**.	Divide $2x^3 - x^2 - 8x - 1$ by $x - 2$ $$\begin{array}{r} 2x^2 + 3x - 2 \\ x - 2 \overline{)2x^3 - x^2 - 8x - 1} \\ \underline{2x^3 - 4x^2} \qquad\quad \\ 3x^2 - 8x - 1 \\ \underline{3x^2 - 6x} \quad\;\; \\ -2x - 1 \\ \underline{-2x + 4} \\ -5 \end{array}$$ The quotient is $2x^2 + 3x - 2 - \frac{5}{x-2}$.
A shortcut method called **synthetic division** may be used to divide a polynomial by a binomial of the form $x - c$.	Use synthetic division to divide $2x^3 - x^2 - 8x - 1$ by $x - 2$. $$\begin{array}{r\|rrrr} 2 & 2 & -1 & -8 & -1 \\ & & 4 & 6 & -4 \\ \hline & 2 & 3 & -2 & -5 \end{array}$$ The quotient is $2x^2 + 3x - 2 - \frac{5}{x-2}$.

Section 5.4 The Greatest Common Factor and Factoring by Grouping

The greatest common factor of the terms of a polynomial is the product of the greatest common factor of the numerical coefficients and the greatest common factor of the variable factors.	Factor: $14xy^3 - 2xy^2 = 2 \cdot 7 \cdot x \cdot y^3 - 2 \cdot x \cdot y^2$ The greatest common factor is $2 \cdot x \cdot y^2$, or $2xy^2$. $14xy^3 - 2xy^2 = 2xy^2(7y - 1)$
To Factor a Polynomial by Grouping Group the terms so that each group has a common factor. Factor out these common factors. Then see if the new groups have a common factor.	Factor: $x^4y - 5x^3 + 2xy - 10$ $= x^3(xy - 5) + 2(xy - 5)$ $= (xy - 5)(x^3 + 2)$

Section 5.5 Factoring Trinomials

To Factor $ax^2 + bx + c$ Step 1. Write all pairs of factors of ax^2. Step 2. Write all pairs of factors of c. Step 3. Try combinations of these factors until the middle term bx is found.	Factor: $28x^2 - 27x - 10$ Factors of $28x^2$: $28x$ and x, $2x$ and $14x$, $4x$ and $7x$. Factors of -10: -2 and 5, 2 and -5, -10 and 1, 10 and -1. $28x^2 - 27x - 10 = (7x + 2)(4x - 5)$

Section 5.6 Factoring by Special Products

Perfect Square Trinomial $a^2 + 2ab + b^2 = (a + b)^2$ $a^2 - 2ab + b^2 = (a - b)^2$	Factor. $25x^2 + 30x + 9 = (5x + 3)^2$ $49z^2 - 28z + 4 = (7z - 2)^2$

SECTION 5.6 (CONTINUED)

DIFFERENCE OF TWO SQUARES

$$a^2 - b^2 = (a + b)(a - b)$$

Factor.

$$36x^2 - y^2 = (6x + y)(6x - y)$$

SUM AND DIFFERENCE OF TWO CUBES

$$a^3 + b^3 = (a + b)(a^2 - ab + b^2)$$
$$a^3 - b^3 = (a - b)(a^2 + ab + b^2)$$

Factor.

$$8y^3 + 1 = (2y + 1)(4y^2 - 2y + 1)$$
$$27p^3 - 64q^3 = (3p - 4q)(9p^2 + 12pq + 16q^2)$$

SECTION 5.7 SOLVING EQUATIONS BY FACTORING AND SOLVING PROBLEMS

TO SOLVE POLYNOMIAL EQUATIONS BY FACTORING

Step 1. Write the equation so that one side is 0.
Step 2. Factor the polynomial completely.
Step 3. Set each factor equal to 0.
Step 4. Solve the resulting equations.
Step 5. Check each solution.

Solve: $2x^3 - 5x^2 = 3x$

$$2x^3 - 5x^2 - 3x = 0$$
$$x(2x + 1)(x - 3) = 0$$
$$x = 0 \quad \text{or} \quad 2x + 1 = 0 \quad \text{or} \quad x - 3 = 0$$
$$x = 0 \qquad x = -\frac{1}{2} \qquad x = 3$$

SECTION 5.8 AN INTRODUCTION TO GRAPHING POLYNOMIAL FUNCTIONS

TO GRAPH A POLYNOMIAL FUNCTION

Find and plot x- and y-intercepts and a sufficient number of ordered pair solutions. Then connect the plotted points with a smooth curve.

Graph: $f(x) = x^3 + 2x^2 - 3x$

$$0 = x^3 + 2x^2 - 3x$$
$$0 = x(x - 1)(x + 3)$$
$$x = 0 \quad \text{or} \quad x = 1 \quad \text{or} \quad x = -3$$

The x-intercept points are $(0, 0)$, $(1, 0)$, and $(-3, 0)$.

$$f(0) = 0^3 + 2 \cdot 0^2 - 3 \cdot 0 = 0$$

The y-intercept point is $(0, 0)$.

x	$f(x)$
-4	-20
-2	6
-1	4
$\frac{1}{2}$	$-\frac{7}{8}$
2	10

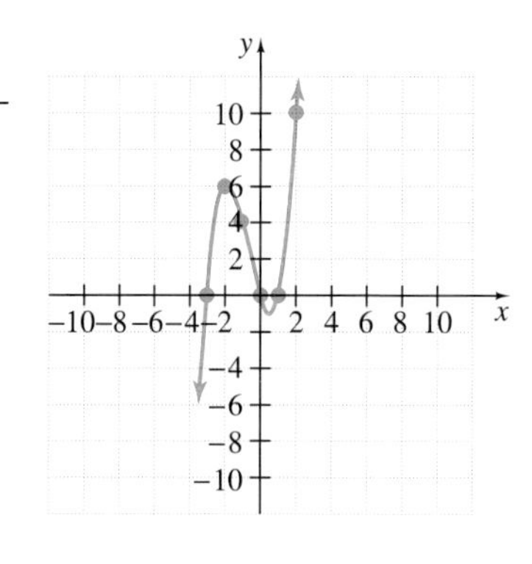

A **quadratic function** is a function that can be written in the form

$$f(x) = ax^2 + bx + c, \quad a \neq 0$$

The graph of this quadratic function is a **parabola** with **vertex** $\left(\frac{-b}{2a}, f\left(\frac{-b}{2a}\right)\right)$.

Find the vertex of the graph of the quadratic function

$$f(x) = 2x^2 - 8x + 1$$

Here $a = 2$ and $b = -8$.

$$\frac{-b}{2a} = \frac{-(-8)}{2 \cdot 2} = 2$$
$$f(2) = 2 \cdot 2^2 - 8 \cdot (2) + 1 = -7$$

The vertex has coordinates $(2, -7)$.

Name ______________________ Section __________ Date __________

Chapter 5 Review

(5.1) *Find the degree of each polynomial.*

1. $x^2y - 3xy^3z + 5x + 7y$

2. $3x + 2$

Simplify by combining like terms.

3. $4x + 8x - 6x^2 - 6x^2y$

4. $-8xy^3 + 4xy^3 - 3x^3y$

Add or subtract as indicated.

5. $(3x + 7y) + (4x^2 - 3x + 7) + (y - 1)$

6. $(4x^2 - 6xy + 9y^2) - (8x^2 - 6xy - y^2)$

7. $(3x^2 - 4b + 28) + (9x^2 - 30) - (4x^2 - 6b + 20)$

8. Add $(9xy + 4x^2 + 18)$ and $(7xy - 4x^3 - 9x)$.

9. Subtract $(x - 7)$ from the sum of $(3x^2y - 7xy - 4)$ and $(9x^2y + x)$.

10.
$$\begin{array}{r} x^2 - 5x + 7 \\ \underline{-\quad (x + 4)} \end{array}$$

11.
$$\begin{array}{r} x^3 + 2xy^2 - y \\ \underline{+\ (x - 4xy^2 \quad - 7)} \end{array}$$

If $P(x) = 9x^2 - 7x + 8$, find each function value.

12. $P(6)$

13. $P(-2)$

14. $P(-3)$

If $P(x) = 2x - 1$ and $Q(x) = x^2 + 2x - 5$, find each function.

15. $P(x) + Q(x)$

16. $2P(x) - Q(x)$

17. Find the perimeter of the rectangle.

$x^2y + 5$ cm

$2x^2y - 6x + 1$ cm

Name ______________________________

(5.2) *Multiply.*

18. $-6x(4x^2 - 6x + 1)$

19. $-4ab^2(3ab^3 + 7ab + 1)$

20. $(x - 4)(2x + 9)$

21. $(-3xa + 4b)^2$

22. $(9x^2 + 4x + 1)(4x - 3)$

23. $(5x - 9y)(3x + 9y)$

24. $\left(x - \frac{1}{3}\right)\left(x + \frac{2}{3}\right)$

25. $(x^2 + 9x + 1)^2$

26. $(2x - 1)(x^2 + 2x - 5)$

Use special products to multiply.

27. $(3x - y)^2$

28. $(4x + 9)^2$

29. $(x + 3y)(x - 3y)$

30. $[4 + (3a - b)][4 - (3a - b)]$

31. Find the area of the rectangle.

3y − 7z units

3y + 7z units

(5.3) *Divide.*

32. $(4xy + 2x^2 - 9) \div (4xy)$

33. $12xb^2 + 16xb^4$ by $4xb^3$

Name ______________________________

34. $(3x^4 - 25x^2 - 20) \div (x - 3)$

35. $(-x^2 + 2x^4 + 5x - 12) \div (x - 3)$

36. $(2x^4 - x^3 + 2x^2 - 3x + 1) \div \left(x - \frac{1}{2}\right)$

37. $(x^3 + 3x^2 - 2x + 2) \div \left(x - \frac{1}{2}\right)$

38. $(3x^4 + 5x^3 + 7x^2 + 3x - 2) \div (x^2 + x + 2)$

39. $(9x^4 - 6x^3 + 3x^2 - 12x - 30) \div (3x^2 - 2x - 5)$

Use synthetic division to find each quotient.

40. $(3x^3 + 12x - 4) \div (x - 2)$

41. $(4x^3 + 2x^2 - 4x - 2) \div \left(x + \frac{3}{2}\right)$

42. $(x^5 - 1) \div (x + 1)$

43. $(x^3 - 81) \div (x - 3)$

44. $(x^3 - x^2 + 3x^4 - 2) \div (x - 4)$

45. $(3x^4 - 2x^2 + 10) \div (x + 2)$

Name ______________________________

(5.4) *Factor out the greatest common factor.*

46. $16x^3 - 24x^2$

47. $36y - 24y^2$

48. $6ab^2 + 8ab - 4a^2b^2$

49. $14a^2b^2 - 21ab^2 + 7ab$

50. $6a(a + 3b) - 5(a + 3b)$

51. $4x(x - 2y) - 5(x - 2y)$

Factor by grouping.

52. $xy - 6y + 3x - 18$

53. $ab - 8b + 4a - 32$

54. $pq - 3p - 5q + 15$

55. $x^3 - x^2 - 2x + 2$

56. A smaller square is cut from a larger rectangle. Write the area of the shaded region as a factored polynomial.

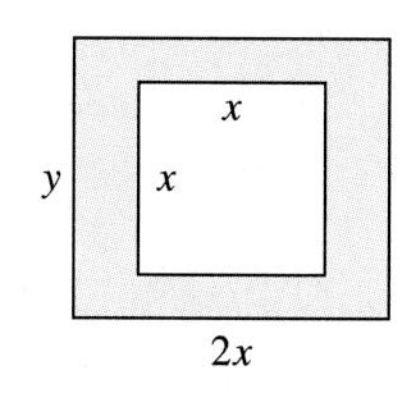

(5.5) *Factor each polynomial completely.*

57. $x^2 - 14x - 72$

58. $x^2 + 16x - 80$

59. $2x^2 - 18x + 28$

60. $3x^2 + 33x + 54$

61. $2x^3 - 7x^2 - 9x$

62. $3x^2 + 2x - 16$

63. $6x^2 + 17x + 10$

64. $15x^2 - 91x + 6$

65. $4x^2 + 2x - 12$

Name ____

66. $9x^2 - 12x - 12$

67. $y^2(x + 6)^2 - 2y(x + 6)^2 - 3(x + 6)^2$

68. $(x + 5)^2 + 6(x + 5) + 8$

69. $x^4 - 6x^2 - 16$

70. $x^4 + 8x^2 - 20$

(5.6) *Factor each polynomial completely.*

71. $x^2 - 100$

72. $x^2 - 81$

73. $2x^2 - 32$

74. $6x^2 - 54$

75. $81 - x^4$

76. $16 - y^4$

77. $(y + 2)^2 - 25$

78. $(x - 3)^2 - 16$

79. $x^3 + 216$

80. $y^3 + 512$

81. $8 - 27y^3$

82. $1 - 64y^3$

83. $6x^4y + 48xy$

84. $2x^5 + 16x^2y^3$

85. $x^2 - 2x + 1 - y^2$

86. $x^2 - 6x + 9 - 4y^2$

87. $4x^2 + 12x + 9$

88. $16a^2 - 40ab + 25b^2$

Name ______________________________

89. The volume of the cylindrical shell is $\pi R^2h - \pi r^2h$ cubic units. Write this volume as a factored expression.

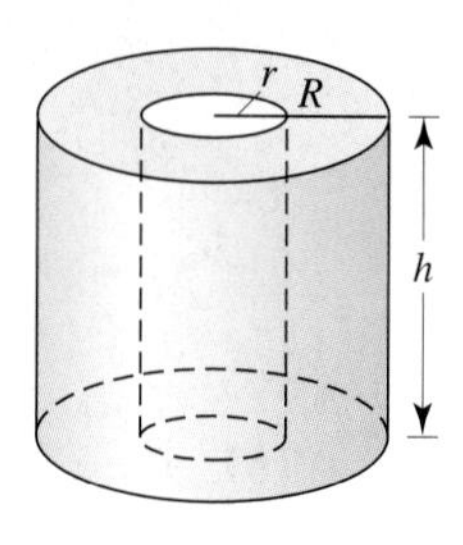

(5.7) *Solve each polynomial equation for the variable.*

90. $(3x - 1)(x + 7) = 0$

91. $3(x + 5)(8x - 3) = 0$

92. $5x(x - 4)(2x - 9) = 0$

93. $6(x + 3)(x - 4)(5x + 1) = 0$

94. $2x^2 = 12x$

95. $4x^3 - 36x = 0$

96. $(1 - x)(3x + 2) = -4x$

97. $2x(x - 12) = -40$

98. $3x^2 + 2x = 12 - 7x$

99. $2x^2 + 3x = 35$

100. $x^3 - 18x = 3x^2$

101. $19x^2 - 42x = -x^3$

102. $12x = 6x^3 + 6x^2$

103. $8x^3 + 10x^2 = 3x$

104. The sum of a number and twice its square is 105. Find the number.

105. The length of a rectangular piece of carpet is 2 meters less than 5 times its width. Find the dimensions of the carpet if its area is 16 square meters.

Name ____

106. A scene from an adventure film calls for a stunt dummy to be dropped from above the second-story platform of the Eiffel Tower, a distance of 400 feet. Its height h at the time t seconds is given by $h = -16t^2 + 400$. Determine how long before the stunt dummy reaches the ground.

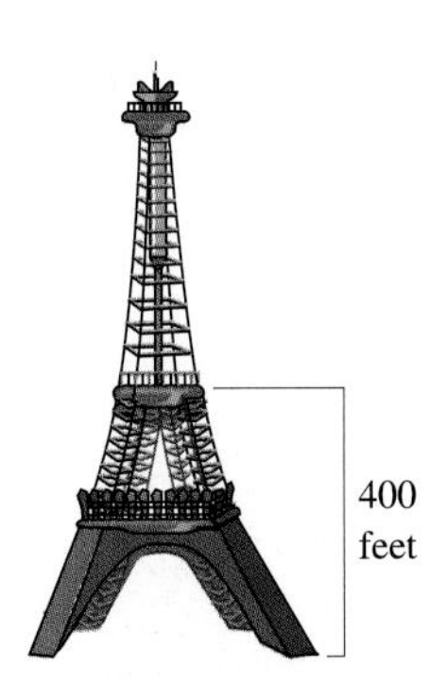

(5.8) *For Exercises 107–110, refer to the following graph.*

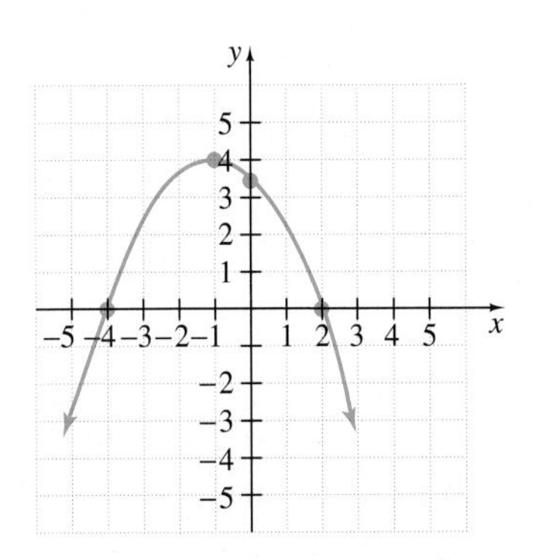

107. Find the domain and the range of the function.

108. List the x- and y-intercept points.

109. Find the coordinates of the point with the greatest y-value.

110. List the x-values for which the y-values are greater than 0.

Graph each polynomial function defined by the equation. Find all intercepts. If the function is a quadratic function, find the vertex.

111. $f(x) = x^2 + 6x + 9$

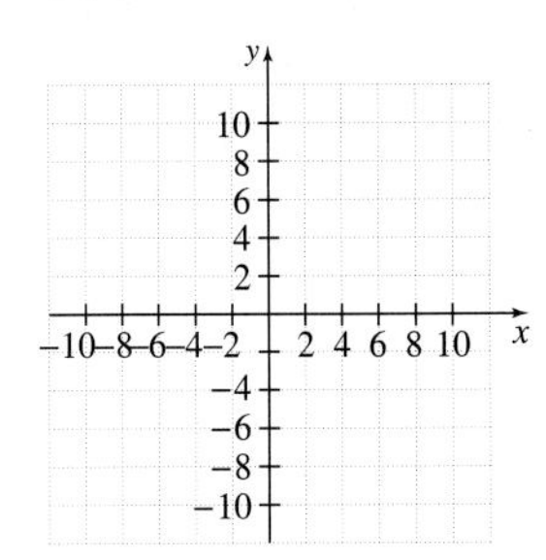

112. $f(x) = x^2 - 5x + 4$

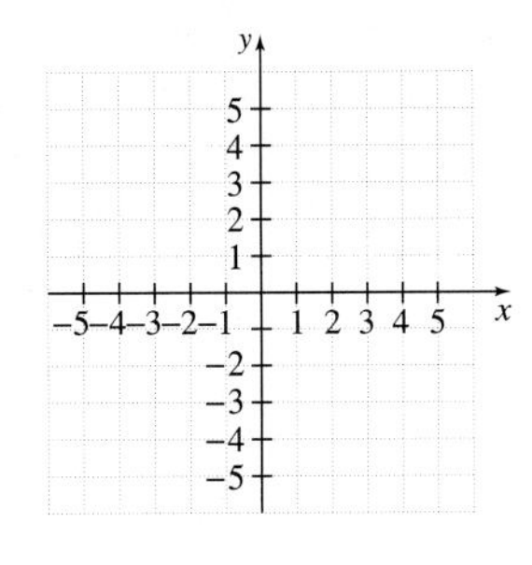

113. $f(x) = (x - 1)(x^2 - 2x - 3)$

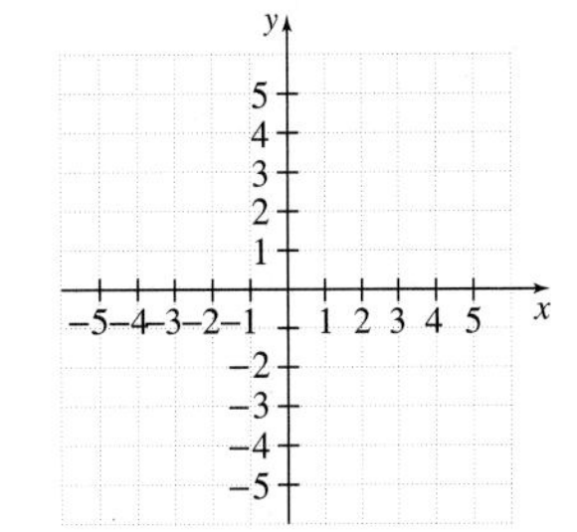

114. $f(x) = (x + 3)(x^2 - 4x + 3)$

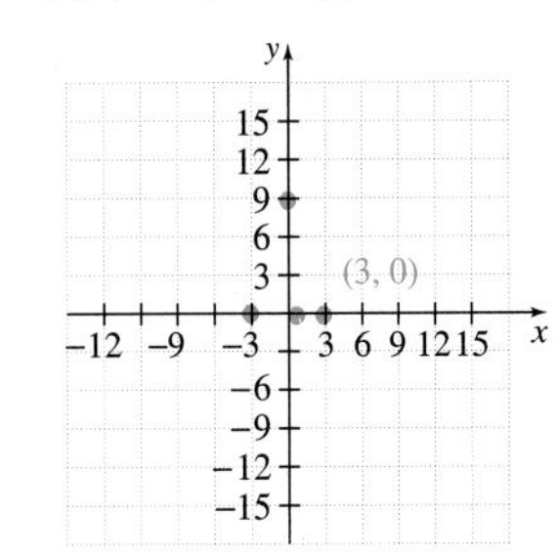

Name ______________________________

115. $f(x) = 2x^2 - 4x + 5$

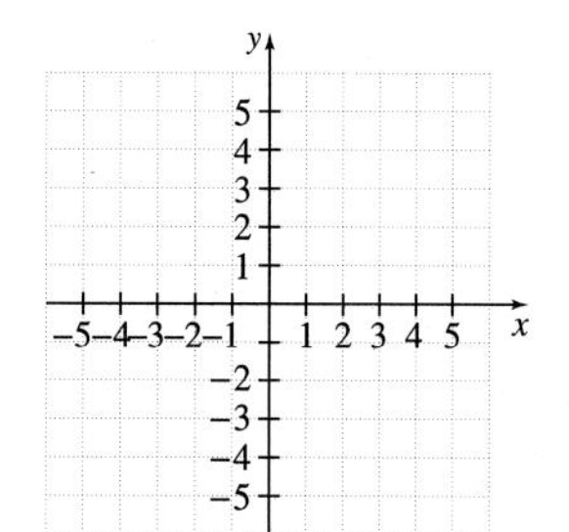

116. $f(x) = x^2 - 2x + 3$

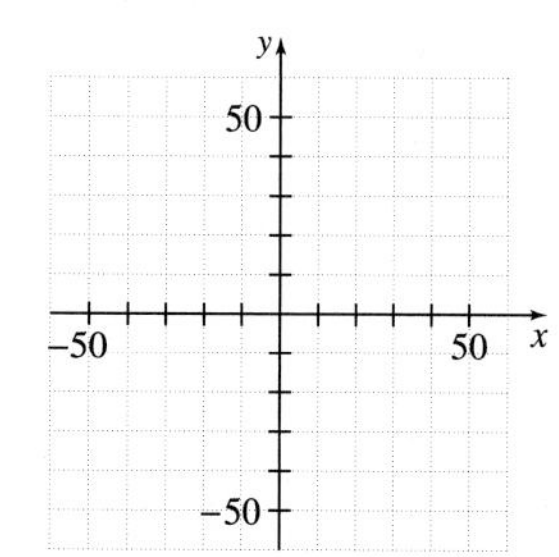

117. $f(x) = x^3 - 16x$

118. $f(x) = x^3 + 5x^2 + 6x$

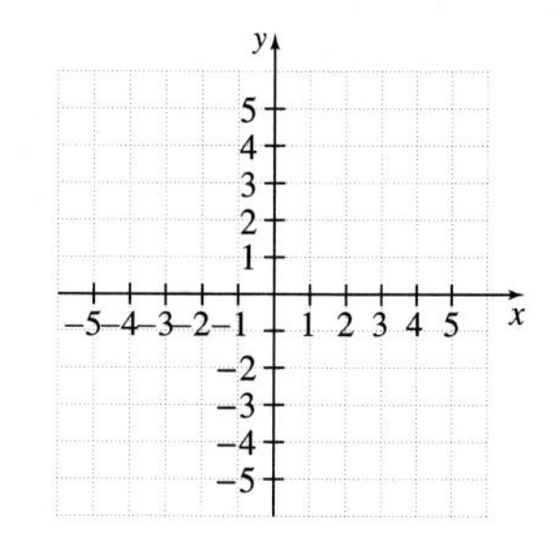

Name ______________________ Section ________ Date ________

Chapter 5 Test

Perform each indicated operation.

1. $(4x^3 - 3x - 4) - (9x^3 + 8x + 5)$

2. $-3xy(4x + y)$

3. $(3x + 4)(4x - 7)$

4. $(5a - 2b)(5a + 2b)$

5. $(6m + n)^2$

6. $(2x - 1)(x^2 - 6x + 4)$

7. $(4x^2y + 9x + z) \div (3xz)$

8. $(4x^5 - 2x^4 + 4x^2 - 6x + 3) \div (2x - 1)$

9. Use synthetic division to divide $4x^4 - 3x^3 + 2x^2 - x - 1$ by $x + 3$.

Factor each polynomial completely.

10. $16x^3y - 12x^2y^4$

11. $x^2 - 13x - 30$

12. $4y^2 + 20y + 25$

13. $6x^2 - 15x - 9$

14. $4x^2 - 25$

15. $x^3 + 64$

16. $3x^2y - 27y^3$

17. $6x^2 + 24$

18. $x^2y - 9y - 3x^2 + 27$

Solve each equation for the variable.

19. $3(n - 4)(7n + 8) = 0$

20. $(x + 2)(x - 2) = 5(x + 4)$

Answers

1. ________
2. ________
3. ________
4. ________
5. ________
6. ________
7. ________
8. ________
9. ________
10. ________
11. ________
12. ________
13. ________
14. ________
15. ________
16. ________
17. ________
18. ________
19. ________
20. ________

Name ______________________________

21. ______________

22. ______________

23. a. ______________

b. ______________

c. ______________

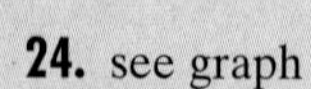

24. see graph ______________

25. see graph ______________

21. $2x^3 + 5x^2 - 8x - 20 = 0$

22. Write the area of the shaded region as a factored polynomial.

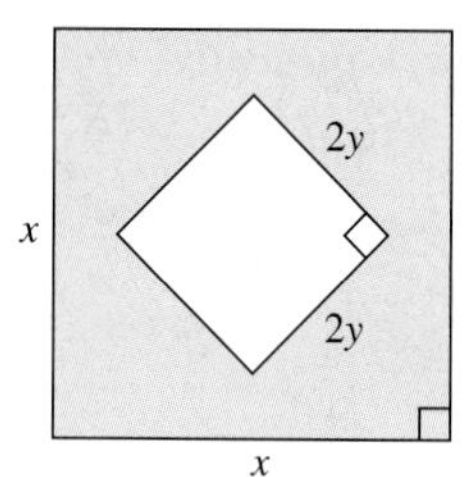

23. A pebble is hurled upward from the top of the 880-foot-tall Canada Trust Tower with an initial velocity of 96 feet per second. Neglecting air resistance, the height h of the pebble after t seconds is given by the polynomial function.

$$h = -16t^2 + 96t + 880$$

a. Find the height of the pebble when $t = 1$.

b. Find the height of the pebble when $t = 5.1$.

c. When will the pebble hit the ground?

Graph. Find and label x- and y-intercepts. If the graph is a parabola, find its vertex.

24. $f(x) = x^2 - 4x - 5$

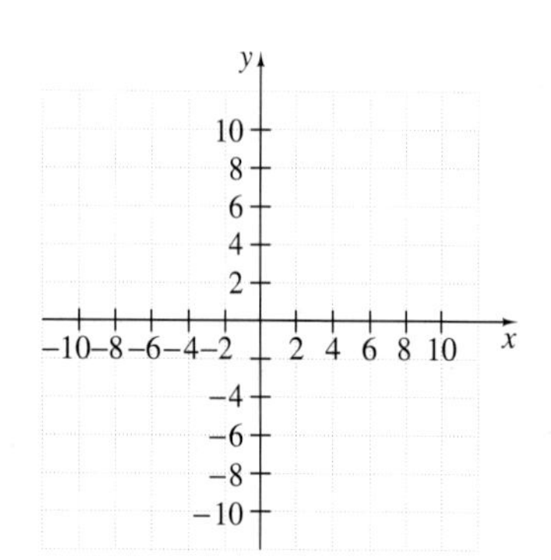

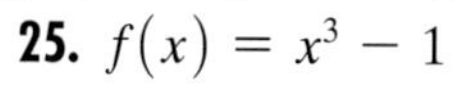

25. $f(x) = x^3 - 1$

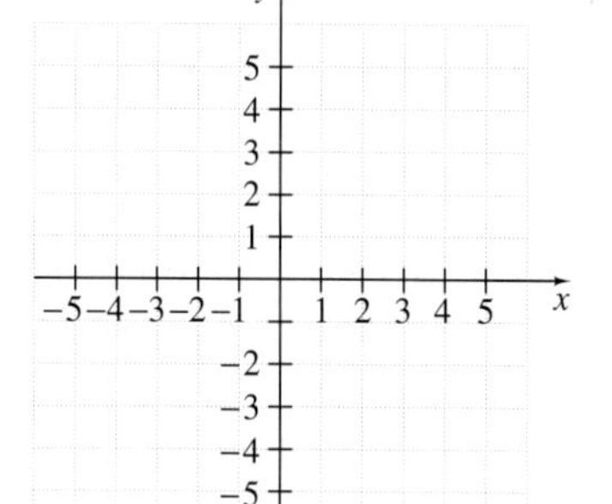

Name ______________________ Section ________ Date ________

ANSWERS

1. ______

2. ______

3. ______

4. ______

5. ______

6. ______

7. ______

8. ______

9. ______

10. see graph ______

11. ______

12. ______

13. ______

14. ______

CUMULATIVE REVIEW

Simplify.

1. $-2(x + 3)$

2. $7x + 3 - 5(x - 4)$

3. Use scientific notation to simplify: $\dfrac{2000 \times 0.000021}{700}$

4. Solve: $2(x - 3) = 5x - 9$

5. Solve $3y - 2x = 7$ for y.

6. Solve: $5 - x \leq 4x - 15$. Write the solution set in interval notation.

7. Solve: $-1 \leq \dfrac{2x}{3} + 5 \leq 2$

8. Solve: $|3x + 2| = |5x - 8|$

9. Find the slope of the line $x = -5$.

10. Graph: $y = \dfrac{1}{4}x - 3$

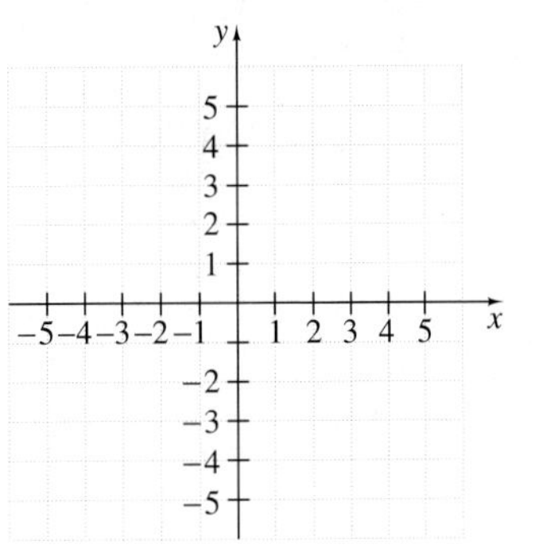

11. Write an equation of the line containing the point $(-2, 1)$ and perpendicular to the line $3x + 5y = 4$.

12. Determine the domain and range of the relation: $\{(2, 3), (2, 4), (0, -1), (3, -1)\}$

Find each function value.

13. If $g(x) = 3x - 2$, find $g(0)$.

14. If $f(x) = 7x^2 - 3x + 1$, find $f(1)$.

15. see graph

16.

17.

18.

19.

20.

21. see graph

22.

23.

24.

25.

Name ____________________

15. Graph: $2x - y < 6$

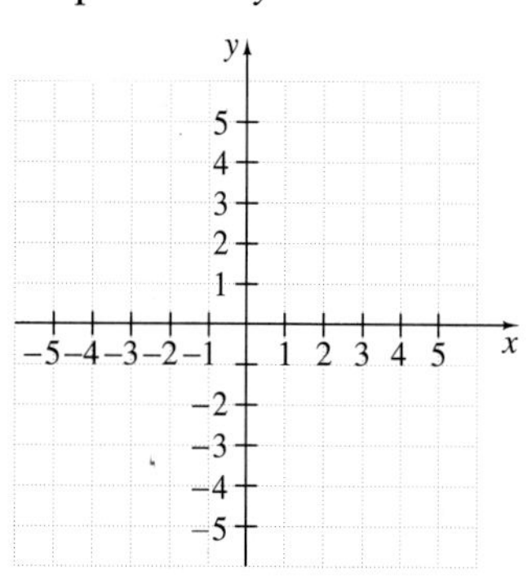

16. Use the substitution method to solve the system: $\begin{cases} 2x + 4y = -6 \\ x = 2y - 5 \end{cases}$

17. Solve the system:
$$\begin{cases} 3x - y + z = -15 \\ x + 2y - z = 1 \\ 2x + 3y - 2z = 0 \end{cases}$$

18. Use matrices to solve the system:
$$\begin{cases} x + 3y = 5 \\ 2x - y = -4 \end{cases}$$

Use Cramer's rule to solve each system.

19. $\begin{cases} 3x + 4y = -7 \\ x - 2y = -9 \end{cases}$

20. $\begin{cases} x - 2y + z = 4 \\ 3x + y - 2z = 3 \\ 5x + 5y + 3z = -8 \end{cases}$

21. Graph the solutions of the system
$$\begin{cases} x - y < 2 \\ x + 2y > -1 \\ y < 2 \end{cases}$$

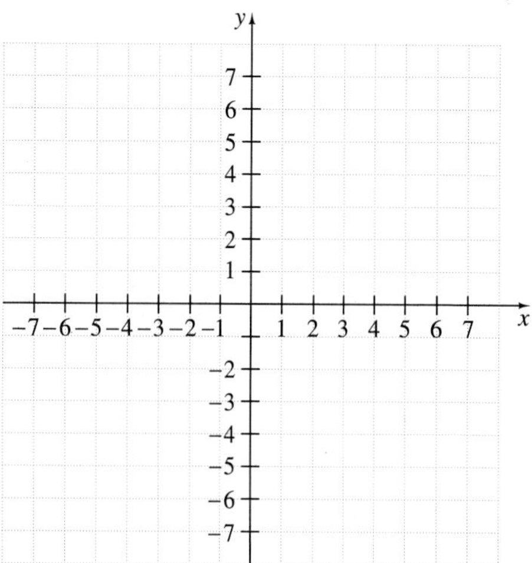

22. Find the degree of the polynomial $3xy + x^2y^2 - 5x^2 - 6$.

Multiply.

23. $2x(5x - 4)$

24. $-xy(7x^2y + 3xy - 11)$

25. Divide $10x^3 - 5x^2 + 20x$ by $5x$.

Rational Expressions

CHAPTER 6

Dividing one polynomial by another in Section 5.3, we found quotients of polynomials. When the remainder part of the quotient was not 0, the remainder was a fraction, such as $\frac{x}{x^2 + 1}$. This fraction is not a polynomial, since it cannot be written as the sum of whole number powers. Instead, it is called a *rational expression.* In this chapter, we study these algebraic forms, the operations that can be performed on them, and the *rational functions* they generate.

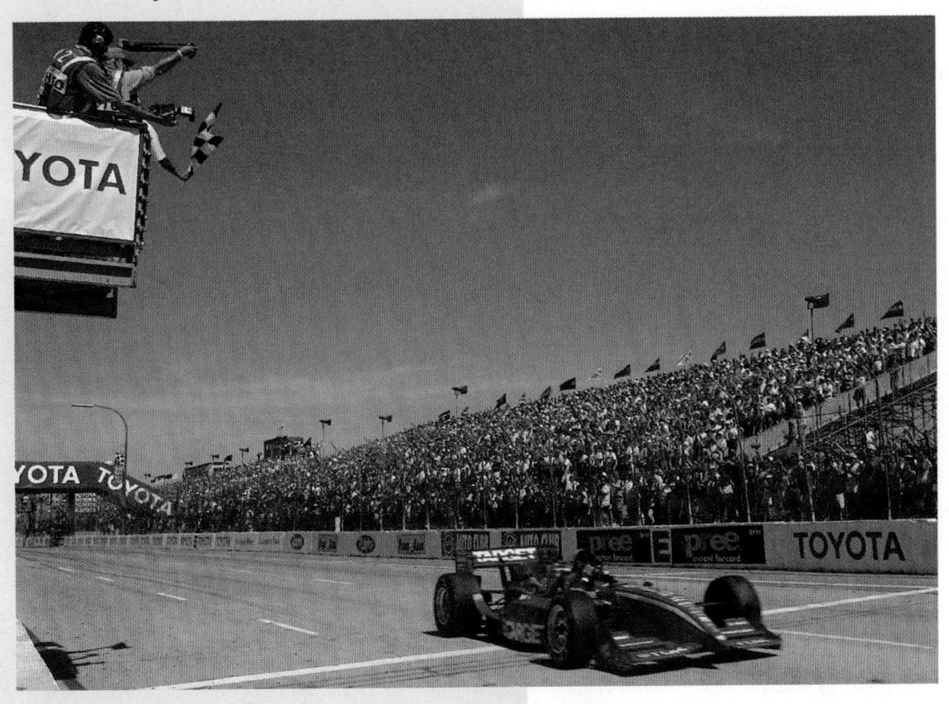

Have you ever watched an auto race, either in person or on television, and noticed that the pitch of the racecar's engine sounds higher as it approaches on the track and drops lower once it passes and moves away? If so, then you have experienced the Doppler effect. This apparent change in pitch of a moving sound can also be heard when a vehicle sounding its horn or siren, or playing its stereo passes you on the street. This effect was discovered by Austrian physicist Christian Doppler in 1842. The Doppler effect has applications in many diverse fields: from meteorology to medicine to astronomy. In Exercise 63 on page 490, we will analyze situations involving the Doppler effect with a rational equation.

For more on the Doppler effect, including an audio demonstration, visit http://www.haystack.mit.edu/midas/doppler.html on the Internet.

ANSWERS

1. ______
2. ______
3. ______
4. ______
5. ______
6. ______
7. ______
8. ______
9. ______
10. ______
11. ______
12. ______
13. ______
14. ______
15. ______
16. ______
17. ______
18. ______
19. ______
20. ______

Name ______ Section ______ Date ______

CHAPTER 6 PRETEST

1. Find all numbers for which the rational expression $\frac{x-1}{x^2-5x+6}$ is undefined.

Simplify each rational expression.

2. $\frac{2x-4x^2}{2x}$

3. $\frac{3x^2+7x+2}{3x^2-11x-4}$

Perform the indicated operation.

4. $\frac{3x-6}{8}\cdot\frac{4}{2-x}$

5. $\frac{x^2+3x+2}{x^2+4x+3}\cdot\frac{2x^2-x-21}{2x^2-x-10}$

6. $\frac{8m^2n^3}{m^2-25}\div\frac{4mn}{m+5}$

7. $\frac{2y-y^2}{y^3-8}\div\frac{y}{y^2+2y+4}$

8. $\frac{7}{x-7}+\frac{x}{x-7}$

9. $\frac{4}{3a^2}-\frac{2}{5a}$

10. $\frac{y+4}{y-3}-\frac{y+1}{y+3}$

11. $\frac{5}{x^2+7x+6}+\frac{x}{2x^2+13x+6}$

Simplify each complex fraction.

12. $\frac{\frac{9}{2x}}{\frac{3}{8x}}$

13. $\frac{\frac{4}{y}+\frac{5}{y^2}}{\frac{16}{y^2}-\frac{25}{y}}$

14. $\frac{x^{-2}}{x^{-1}+y^{-1}}$

Solve each equation for x.

15. $\frac{x}{6}-\frac{x}{5}=-2$

16. $\frac{x+3}{x^2+9x+18}=\frac{1}{2x+6}-\frac{1}{x+6}$

17. Solve for L: $S=\frac{n(a+L)}{2}$

Solve.

18. The sum of a number and 6 times its reciprocal is 5. Find the number(s).

19. Suppose y varies directly as x. If $y=6$ when $x=24$, find y when $x=30$.

20. Suppose that W is inversely proportional to V. If $W=100$ when $V=8$, find W when $V=25$.

6.1 MULTIPLYING AND DIVIDING RATIONAL EXPRESSIONS

Objectives

A Find values for which a rational expression is undefined.

B Simplify rational expressions.

C Multiply rational expressions.

D Divide rational expressions.

E Use rational functions in real-world applications.

SSM CD-ROM Video 6.1

Recall that a *rational number*, or *fraction*, is a number that can be written as the quotient $\frac{p}{q}$ of two integers p and q as long as q is not 0. A **rational expression** is an expression that can be written as the quotient $\frac{P}{Q}$ of two polynomials P and Q as long as Q is not 0. Examples are

$$\frac{3x+7}{2} \qquad \frac{5x^2-3}{x-1} \qquad \frac{7x-2}{2x^2+7x+6}$$

A FINDING VALUES FOR WHICH A RATIONAL EXPRESSION IS UNDEFINED

As with numerical fractions, a rational expression is **undefined** if the denominator is 0. If a variable in a rational expression is replaced with a number that makes the denominator 0, we say that the rational expression is **undefined** for this value of the variable. For example, the rational expression

$\frac{x^2+2}{x-3}$ is undefined when x is 3

because replacing x with 3 results in a denominator of 0.

$$\frac{x^2+2}{x-3} = \frac{3^2+2}{3-3} = \frac{11}{0}$$ This rational expression is undefined when $x = 3$.

Examples Find all numbers for which each rational expression is undefined.

1. $\frac{5x^2-3}{x-1}$ is undefined when the denominator $x-1$ is 0.

$x - 1 = 0$ or $x = 1$ Set the denominator equal to 0 and solve.

If x is replaced with 1, the rational expression is undefined.

2. $\frac{7x-2}{x^2-2x-15}$ is undefined when the denominator is 0.

$x^2 - 2x - 15 = 0$ Set the denominator equal to 0 and solve.

$(x-5)(x+3) = 0$

$x - 5 = 0$ or $x + 3 = 0$

$x = 5$ or $x = -3$

If x is replaced with 5 or with -3, the rational expression is undefined.

3. $\frac{5x-1}{3}$ is undefined when the denominator is 0. No matter what value x is replaced with, the denominator—3—is never 0. No real number makes this rational expression undefined. That is, this expression is defined for all real numbers.

TRY THE CONCEPT CHECK IN THE MARGIN.

Practice Problems 1–3

Find all numbers for which each rational expression is undefined.

1. $\frac{x^2+1}{x-6}$
2. $\frac{5x+4}{x^2-3x-10}$
3. $\frac{x^2-9}{4}$

✓ CONCEPT CHECK

For which of these values (if any) is the rational expression $\frac{x-3}{x^2+2}$ undefined? Explain.

a. 2
b. 3
c. -2
d. 0
e. None of these

Answers

1. 6, **2.** 5, -2, **3.** no real number

✓ **Concept Check:** e

B SIMPLIFYING RATIONAL EXPRESSIONS

Recall that a fraction is in lowest terms or simplest form if the numerator and denominator have no common factors other than 1 (or −1). For example, $\frac{3}{13}$ is in lowest terms since 3 and 13 have no common factors other than 1 (or −1).

To **simplify** a rational expression, or to write it in lowest terms, we use the fundamental principle of rational expressions.

> **FUNDAMENTAL PRINCIPLE OF RATIONAL EXPRESSIONS**
>
> For any rational expression $\frac{P}{Q}$ and any polynomial R, $R \neq 0$,
>
> $$\frac{PR}{QR} = \frac{P}{Q}$$

Thus, the fundamental principle says that multiplying or dividing the numerator and denominator of a rational expression by the same nonzero polynomial yields an equivalent rational expression.

To simplify a rational expression such as $\frac{(x+2)^2}{x^2-4}$, factor the numerator and the denominator and then use the fundamental principle of rational expressions to divide out common factors.

$$\frac{(x+2)^2}{x^2-4} = \frac{(x+2)(x+2)}{(x+2)(x-2)}$$

$$= \frac{x+2}{x-2}$$

This means that the rational expression $\frac{(x+2)^2}{x^2-4}$ has the same value as the rational expression $\frac{x+2}{x-2}$ for all values of x except 2 and −2. (Remember that when x is 2, the denominators of both rational expressions are 0 and that when x is −2, the original rational expression has a denominator of 0.)

As we simplify rational expressions, we will assume that the simplified rational expression is equivalent to the original rational expression for all real numbers except those for which either denominator is 0.

In general, the following steps may be used to simplify rational expressions or to write a rational expression in lowest terms.

> **SIMPLIFYING OR WRITING A RATIONAL EXPRESSION IN LOWEST TERMS**
>
> **Step 1.** Completely factor the numerator and denominator of the rational expression.
>
> **Step 2.** Apply the fundamental principle of rational expressions to divide out factors common to both the numerator and denominator.

For now, we assume that variables in a rational expression do not represent values that make the denominator 0.

Examples

Simplify each rational expression.

4. $\frac{24x^6y^5}{8x^7y} = \frac{(8x^6y)3y^4}{(8x^6y)x}$ Factor the numerator and denominator.

$= \frac{3y^4}{x}$ Apply the fundamental principle and divide out common factors.

5. $\frac{2x^2}{10x^3 - 2x^2} = \frac{2x^2 \cdot 1}{2x^2(5x - 1)}$ Factor the numerator and denominator.

$= \frac{1}{5x - 1}$ Apply the fundamental principle and divide out common factors.

Practice Problems 4–5

Simplify each rational expression.

4. $\frac{20a^7b^4}{5a^3b^5}$

5. $\frac{3y^3}{6y^4 - 3y^3}$

Examples

Simplify each rational expression.

6. $\frac{2 + x}{x + 2} = \frac{x + 2}{x + 2} = 1$ By the commutative property of addition, $2 + x = x + 2$.

7. $\frac{2 - x}{x - 2}$

The terms in the numerator of $\frac{2 - x}{x - 2}$ differ by sign from the terms of the denominator, so the polynomials are opposites of each other and the expression simplifies to -1. To see this, we factor out -1 from the numerator or the denominator. If -1 is factored from the numerator, then

$$\frac{2 - x}{x - 2} = \frac{-1(-2 + x)}{x - 2} = \frac{-1(x - 2)}{x - 2} = \frac{-1}{1} = -1$$

If -1 is factored from the denominator, the result is the same.

$$\frac{2 - x}{x - 2} = \frac{2 - x}{-1(-x + 2)} = \frac{2 - x}{-1(2 - x)} = \frac{1}{-1} = -1$$

HELPFUL HINT

When the numerator and the denominator of a rational expression are opposites of each other, the expression simplifies to -1.

8. $\frac{18 - 2x^2}{x^2 - 2x - 3} = \frac{2(9 - x^2)}{(x + 1)(x - 3)}$ Factor.

$= \frac{2(3 + x)(3 - x)}{(x + 1)(x - 3)}$ Factor completely.

Notice the opposites $3 - x$ and $x - 3$. We write $3 - x$ as $-1(x - 3)$ and simplify.

$$\frac{2(3 + x)(3 - x)}{(x + 1)(x - 3)} = \frac{2(3 + x) \cdot -1(x - 3)}{(x + 1)(x - 3)} = -\frac{2(3 + x)}{x + 1}$$

Practice Problems 6–8

Simplify each rational expression.

6. $\frac{5 + x}{x + 5}$

7. $\frac{5 - x}{x - 5}$

8. $\frac{3 - 3x^2}{x^2 + x - 2}$

Answers

4. $\frac{4a^4}{b}$, **5.** $\frac{1}{2y - 1}$, **6.** 1, **7.** -1, **8.** $-\frac{3(x + 1)}{x + 2}$

HELPFUL HINT

Recall that for a fraction $\frac{a}{b}$,

$$\frac{a}{-b} = \frac{-a}{b} = -\frac{a}{b}$$

For example,

$$\frac{-(x + 1)}{(x + 2)} = \frac{(x + 1)}{-(x + 2)} = -\frac{x + 1}{x + 2}$$

✓ CONCEPT CHECK

Which of the following expressions are equivalent to $\frac{x}{8 - x}$?

a. $\frac{-x}{x - 8}$ b. $\frac{-x}{8 - x}$

c. $\frac{x}{x - 8}$ d. $\frac{-x}{-8 + x}$

TRY THE CONCEPT CHECK IN THE MARGIN.

Practice Problems 9–10

Simplify each rational expression.

9. $\frac{x^3 + 27}{x + 3}$

10. $\frac{3x^2 + 6}{x^3 - 3x^2 + 2x - 6}$

Examples

Simplify each rational expression.

9. $\frac{x^3 + 8}{x + 2} = \frac{(x + 2)(x^2 - 2x + 4)}{x + 2}$ Factor the sum of the two cubes.

$= x^2 - 2x + 4$ Divide out common factors.

10. $\frac{2y^2 + 2}{y^3 - 5y^2 + y - 5} = \frac{2(y^2 + 1)}{(y^3 - 5y^2) + (y - 5)}$ Factor the numerator.

$= \frac{2(y^2 + 1)}{y^2(y - 5) + 1(y - 5)}$ Factor the denominator by grouping.

$= \frac{2(y^2 + 1)}{(y - 5)(y^2 + 1)}$

$= \frac{2}{y - 5}$ Divide out common factors.

✓ CONCEPT CHECK

Does $\frac{n}{n + 2}$ simplify to $\frac{1}{2}$? Why or why not?

TRY THE CONCEPT CHECK IN THE MARGIN.

C MULTIPLYING RATIONAL EXPRESSIONS

Arithmetic operations on rational expressions are performed in the same way as they are on rational numbers. To multiply rational expressions, we multiply numerators and multiply denominators.

When multiplying rational expressions, we will factor each numerator and denominator first. This will help when we apply the fundamental principle to simplify the product.

The following steps may be used to multiply rational expressions.

MULTIPLYING RATIONAL EXPRESSIONS

Step 1. Completely factor each numerator and denominator.

Step 2. Multiply the numerators and multiply the denominators.

Step 3. Simplify the product by applying the fundamental property and dividing the numerator and denominator by their common factors.

Answers

9. $x^2 - 3x + 9$, **10.** $\frac{3}{x - 3}$

✓ Concept Check: a and d

✓ Concept Check: no; answers may vary

Examples

Multiply.

11. $\frac{3n+1}{2n} \cdot \frac{2n-4}{3n^2-2n-1} = \frac{3n+1}{2n} \cdot \frac{2(n-2)}{(3n+1)(n-1)}$ Factor.

$= \frac{(3n+1) \cdot 2(n-2)}{2n(3n+1)(n-1)}$ Multiply.

$= \frac{n-2}{n(n-1)}$ Divide out common factors.

12. $\frac{x^3-1}{-3x+3} \cdot \frac{15x^2}{x^2+x+1} = \frac{(x-1)(x^2+x+1)}{-3(x-1)} \cdot \frac{15x^2}{x^2+x+1}$ Factor.

$= \frac{(x-1)(x^2+x+1) \cdot 3 \cdot 5x^2}{-1 \cdot 3(x-1)(x^2+x+1)}$ Factor.

$= \frac{5x^2}{-1}$ Divide out common factors.

$= -5x^2$

D DIVIDING RATIONAL EXPRESSIONS

Recall that two numbers are reciprocals of each other if their product is 1. Similarly, if $\frac{P}{Q}$ is a rational expression, then $\frac{Q}{P}$ is its **reciprocal**, since

$$\frac{P}{Q} \cdot \frac{Q}{P} = \frac{P \cdot Q}{Q \cdot P} = 1$$

The following are examples of expressions and their reciprocals.

Expression	Reciprocal
$\frac{3}{x}$	$\frac{x}{3}$
$\frac{2+x^2}{4x-3}$	$\frac{4x-3}{2+x^2}$
x^3	$\frac{1}{x^3}$
0	no reciprocal

DIVIDING RATIONAL EXPRESSIONS

To divide by a rational expression, multiply by its reciprocal. Then simplify if possible.

Examples

Divide.

13. $\frac{3x}{5y} \div \frac{9y}{x^5} = \frac{3x}{5y} \cdot \frac{x^5}{9y}$ Multiply by the reciprocal of the divisor.

$= \frac{x^6}{15y^2}$ Simplify.

Practice Problems 11–12

Multiply.

11. $\frac{2x-3}{5x} \cdot \frac{5x+5}{2x^2-x-3}$

12. $\frac{x^3+27}{-2x-6} \cdot \frac{4x^3}{x^2-3x+9}$

Practice Problems 13–14

Divide.

13. $\frac{7x}{2y^2} \div \frac{8y}{3x^4}$

14. $\frac{12y^3}{5y^2-5} \div \frac{6}{1-y}$

Answers

11. $\frac{1}{x}$, **12.** $-2x^3$, **13.** $\frac{21x^5}{16y^3}$, **14.** $-\frac{2y^3}{5(y+1)}$

14. $$\frac{8m^2}{3m^2 - 12} \div \frac{40}{2 - m} = \frac{8m^2}{3m^2 - 12} \cdot \frac{2 - m}{40}$$ Multiply by the reciprocal of the divisor.

$$= \frac{8m^2(2 - m)}{3(m + 2)(m - 2) \cdot 40}$$ Factor and multiply.

$$= \frac{8m^2 \cdot -1(m - 2)}{3(m + 2)(m - 2) \cdot 8 \cdot 5}$$ Write $(2 - m)$ as $-1(m - 2)$.

$$= -\frac{m^2}{15(m + 2)}$$ Simplify.

HELPFUL HINT

When dividing rational expressions, do not divide out common factors until the division problem is rewritten as a multiplication problem.

Practice Problem 15

Perform each indicated operation.

$$\frac{(x + 3)^2}{x^2 - 9} \cdot \frac{2x - 6}{5x} \div \frac{x^2 + 7x + 12}{x}$$

Example 15 Perform each indicated operation.

$$\frac{x^2 - 25}{(x + 5)^2} \cdot \frac{3x + 15}{4x} \div \frac{x^2 - 3x - 10}{x}$$

Solution:

$$\frac{x^2 - 25}{(x + 5)^2} \cdot \frac{3x + 15}{4x} \div \frac{x^2 - 3x - 10}{x}$$

$$= \frac{x^2 - 25}{(x + 5)^2} \cdot \frac{3x + 15}{4x} \cdot \frac{x}{x^2 - 3x - 10}$$ To divide, multiply by the reciprocal.

$$= \frac{(x + 5)(x - 5)}{(x + 5)(x + 5)} \cdot \frac{3(x + 5)}{4x} \cdot \frac{x}{(x - 5)(x + 2)}$$

$$= \frac{3}{4(x + 2)}$$

E APPLICATIONS WITH RATIONAL FUNCTIONS

Rational expressions are sometimes used to describe functions. For example, we call the function $f(x) = \frac{x^2 + 2}{x - 3}$ a **rational function** since $\frac{x^2 + 2}{x - 3}$ is a rational expression in one variable.

The domain of a rational function such as $f(x) = \frac{x^2 + 2}{x - 3}$ is the set of all possible replacement values for x. In other words, since the rational expression $\frac{x^2 + 2}{x - 3}$ is not defined when $x = 3$, we say that the domain of $f(x) = \frac{x^2 + 2}{x - 3}$ is all real numbers except 3. We can write the domain as:

$$\{x | x \text{ is a real number and } x \neq 3\}$$

Answer

15. $\frac{2}{5(x + 4)}$

Example 16 Finding Unit Cost

For the ICL Production Company, the rational function $C(x) = \frac{2.6x + 10{,}000}{x}$ describes the company's cost per disc of pressing x compact discs. Find the cost per disc for pressing:

a. 100 compact discs
b. 1000 compact discs

Solution:

a. $C(100) = \frac{2.6(100) + 10{,}000}{100} = \frac{10{,}260}{100} = 102.6$

The cost per disc for pressing 100 compact discs is \$102.60.

b. $C(1000) = \frac{2.6(1000) + 10{,}000}{1000} = \frac{12{,}600}{1000} = 12.6$

The cost per disc for pressing 1000 compact discs is \$12.60. Notice that as more compact discs are produced, the cost per disc decreases.

Practice Problem 16

A company's cost per book for printing x particular books is given by the rational function $C(x) = \frac{0.8x + 5000}{x}$.

Find the cost per book for printing:

a. 100 books

b. 1000 books

Answers

16. a. \$50.80, **b.** \$5.80

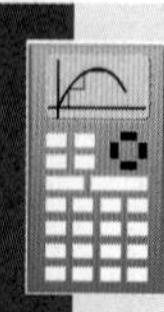

GRAPHING CALCULATOR EXPLORATIONS

Recall that since the rational expression $\frac{7x - 2}{(x - 2)(x + 5)}$ is not defined when $x = 2$ or when $x = -5$, we say that the domain of the rational function $f(x) = \frac{7x - 2}{(x - 2)(x + 5)}$ is all real numbers except 2 and -5. This domain can be written as $\{x|x \text{ is a real number and } x \neq 2, x \neq -5\}$. This means that the graph of $f(x)$ should not cross the vertical lines $x = 2$ and $x = -5$. The graph of $f(x)$ in *connected* mode follows. In connected mode the grapher tries to connect all dots of the graph so that the result is a smooth curve. This is what has happened in the graph. Notice that the graph appears to contain vertical lines at $x = 2$ and at $x = -5$. We know that this cannot happen because the function is not defined at $x = 2$ and at $x = -5$. We also know that this cannot happen because the graph of this function would not pass the vertical line test.

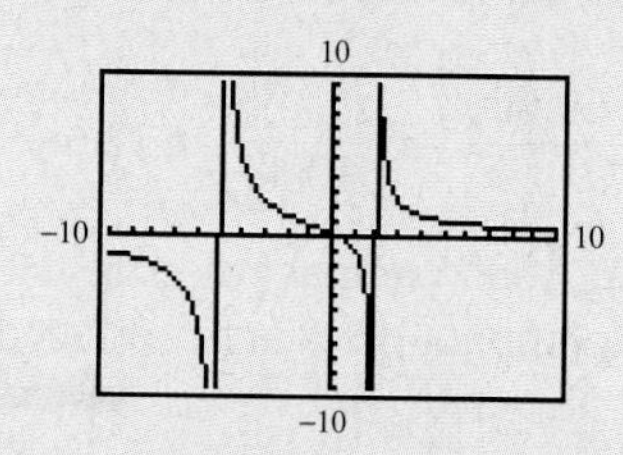

If we graph $f(x)$ in *dot* mode, the graph appears as follows. In dot mode the grapher will not connect dots with a smooth curve. Notice that the vertical lines have disappeared, and we have a better picture of the graph. The graph, however, actually appears more like the hand-drawn graph to its right. By using a Table feature, a Calculate Value feature, or by tracing, we can see that the function is not defined at $x = 2$ and at $x = -5$.

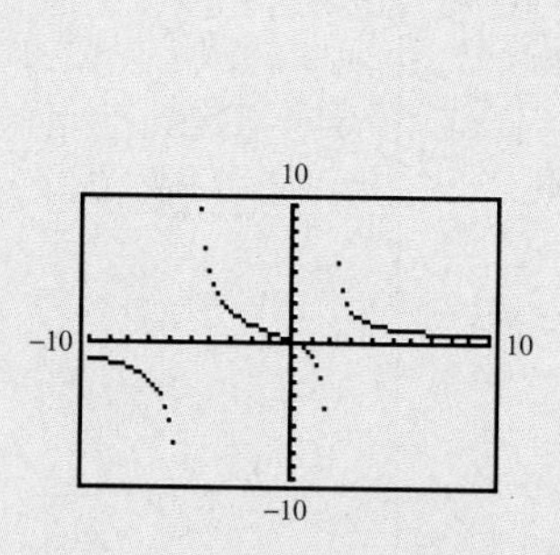

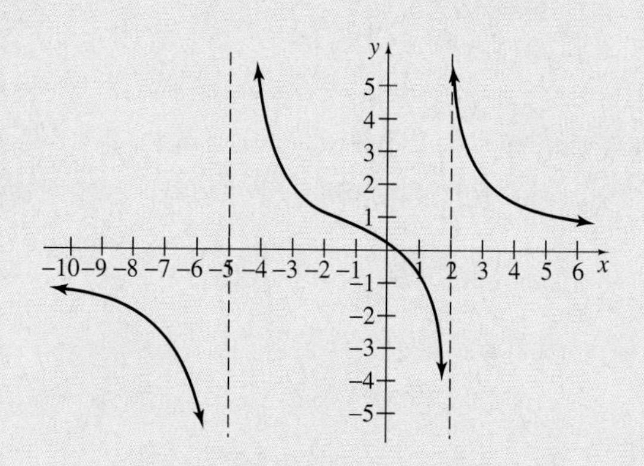

Find the domain of each rational function. Then graph each rational function and use the graph to confirm the domain.

1. $f(x) = \frac{x + 1}{x^2 - 4}$

2. $g(x) = \frac{5x}{x^2 - 9}$

3. $h(x) = \frac{x^2}{2x^2 + 7x - 4}$

4. $f(x) = \frac{3x + 2}{4x^2 - 19x - 5}$

Name ______________________ Section ________ Date ________

MENTAL MATH ANSWERS

1. ________
2. ________
3. ________
4. ________
5. ________
6. ________

MENTAL MATH

Multiply.

1. $\frac{x}{5} \cdot \frac{y}{2}$
2. $\frac{y}{6} \cdot \frac{z}{5}$
3. $\frac{2}{x} \cdot \frac{y}{3}$
4. $\frac{a}{5} \cdot \frac{7}{b}$
5. $\frac{m}{6} \cdot \frac{m}{6}$
6. $\frac{9}{x} \cdot \frac{8}{x}$

EXERCISE SET 6.1

A *Find all numbers for which each rational expression is undefined. See Examples 1 through 3.*

1. $\frac{x+3}{x-2}$
2. $\frac{x+5}{x-1}$
3. $\frac{2x}{5x+1}$
4. $\frac{5x}{7x+2}$
5. $\frac{x^2+1}{3x}$
6. $\frac{x^2+7}{5x}$
7. $\frac{x-7}{4}$
8. $\frac{4-3x}{2}$
9. $\frac{3+2x}{x^3+x^2-2x}$
10. $\frac{5-3x}{2x^3-14x^2+20x}$
11. $\frac{x+3}{x^2-4}$
12. $\frac{5}{x^2-7x}$

B *Simplify each rational expression. See Examples 4 through 8.*

13. $\frac{10x^3}{18x}$
14. $-\frac{48a^7}{16a^{10}}$
15. $\frac{9x^6y^3}{18x^2y^5}$
16. $\frac{10ab^5}{15a^3b^5}$
17. $\frac{8x-16x^2}{8x}$
18. $\frac{3x-6x^2}{3x}$
19. $\frac{x^2-9}{x-3}$
20. $\frac{x^2-25}{x+5}$
21. $\frac{9y-18}{7y-14}$
22. $\frac{6y-18}{2y-6}$
23. $\frac{x^2+6x-40}{x+10}$
24. $\frac{x^2-8x+16}{x-4}$
25. $\frac{x-9}{9-x}$
26. $\frac{x-4}{4-x}$
27. $\frac{x^2-49}{7-x}$

ANSWERS

1. ________
2. ________
3. ________
4. ________
5. ________
6. ________
7. ________
8. ________
9. ________
10. ________
11. ________
12. ________
13. ________
14. ________
15. ________
16. ________
17. ________
18. ________
19. ________
20. ________
21. ________
22. ________
23. ________
24. ________
25. ________
26. ________
27. ________

28. ______

29. ______

30. ______

31. ______

32. ______

33. ______

34. ______

35. ______

36. ______

37. ______

38. ______

39. ______

40. ______

41. ______

42. ______

43. ______

44. ______

45. ______

46. ______

47. ______

48. ______

49. ______

50. ______

51. ______

52. ______

Name ______________________________

28. $\frac{x^2 - y^2}{y - x}$ **29.** $\frac{2x^2 - 7x - 4}{x^2 - 5x + 4}$ **30.** $\frac{3x^2 - 11x + 10}{x^2 - 7x + 10}$

Simplify each rational expression. See Examples 9 and 10.

31. $\frac{x^3 - 125}{2x - 10}$ **32.** $\frac{4x + 4}{x^3 + 1}$ **33.** $\frac{3x^2 - 5x - 2}{6x^3 + 2x^2 + 3x + 1}$

34. $\frac{2x^2 - x - 3}{2x^3 - 3x^2 + 2x - 3}$ **35.** $\frac{9x^2 - 15x + 25}{27x^3 + 125}$ **36.** $\frac{8x^3 - 27}{4x^2 + 6x + 9}$

C *Multiply and simplify. See Examples 11 and 12.*

37. $\frac{4}{x} \cdot \frac{x^2}{8}$ **38.** $\frac{x}{3} \cdot \frac{9}{x^3}$

39. $\frac{2x - 4}{15} \cdot \frac{6}{2 - x}$ **40.** $\frac{10 - 2x}{7} \cdot \frac{14}{5x - 25}$

41. $\frac{18a - 12a^2}{4a^2 + 4a + 1} \cdot \frac{4a^2 + 8a + 3}{4a^2 - 9}$ **42.** $\frac{a - 5b}{a^2 + ab} \cdot \frac{b^2 - a^2}{10b - 2a}$

43. $\frac{9x + 9}{4x + 8} \cdot \frac{2x + 4}{3x^2 - 3}$ **44.** $\frac{2x^2 - 2}{10x + 30} \cdot \frac{12x + 36}{3x - 3}$

45. $\frac{2x^3 - 16}{6x^2 + 6x - 36} \cdot \frac{9x + 18}{3x^2 + 6x + 12}$ **46.** $\frac{x^2 - 3x + 9}{5x^2 - 20x - 105} \cdot \frac{x^2 - 49}{x^3 + 27}$

47. $\frac{a^3 + a^2b + a + b}{5a^3 + 5a} \cdot \frac{6a^2}{2a^2 - 2b^2}$ **48.** $\frac{4a^2 - 8a}{ab - 2b + 3a - 6} \cdot \frac{8b + 24}{3a + 6}$

49 $\frac{x^2 - 6x - 16}{2x^2 - 128} \cdot \frac{x^2 + 16x + 64}{3x^2 + 30x + 48}$ **50.** $\frac{2x^2 + 12x - 32}{x^2 + 16x + 64} \cdot \frac{x^2 + 10x + 16}{x^2 - 3x - 10}$

51. Find the area of the rectangle.

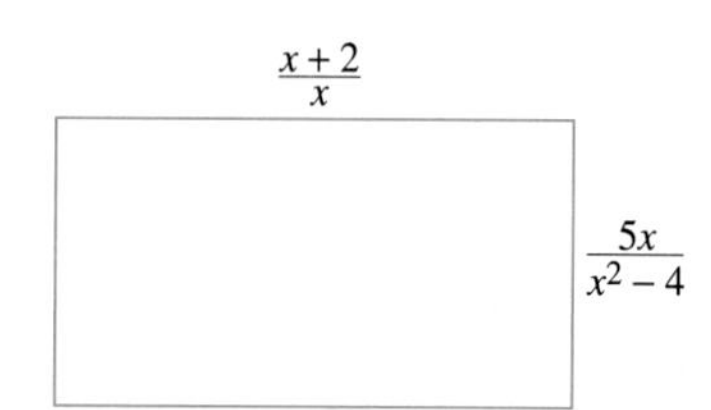

52. Find the area of the triangle.

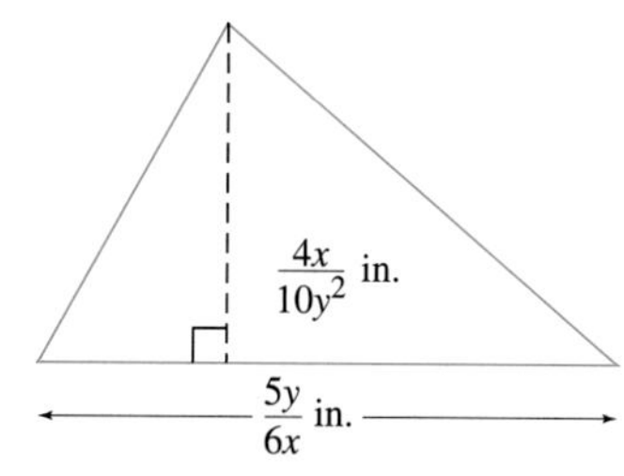

Name ______________________

D *Divide and simplify. See Examples 13 and 14.*

53. $\frac{4}{x} \div \frac{8}{x^2}$

54. $\frac{x}{3} \div \frac{x^3}{9}$

55. $\frac{2x}{5} \div \frac{6x + 12}{5x + 10}$

56. $\frac{7}{3x} \div \frac{14 - 7x}{18 - 9x}$

57. $\frac{a + b}{ab} \div \frac{a^2 - b^2}{4a^3b}$

58. $\frac{6a^2b^2}{a^2 - 4} \div \frac{3ab^2}{a - 2}$

59. $\frac{x^2 - 6x + 9}{x^2 - x - 6} \div \frac{x^2 - 9}{4}$

60. $\frac{x^2 - 4}{3x + 6} \div \frac{2x^2 - 8x + 8}{x^2 + 4x + 4}$

61. $\frac{x^2 - 6x - 16}{2x^2 - 128} \div \frac{x^2 + 10x + 16}{x^2 + 16x + 64}$

62. $\frac{a^2 - a - 6}{a^2 - 81} \div \frac{a^2 - 7a - 18}{4a + 36}$

63. $\frac{3x - x^2}{x^3 - 27} \div \frac{x}{x^2 + 3x + 9}$

64. $\frac{x^2 - 3x}{x^3 - 27} \div \frac{2x}{2x^2 + 6x + 18}$

65. $\frac{8b + 24}{3a + 6} \div \frac{ab - 2b + 3a - 6}{a^2 - 4a + 4}$

66. $\frac{2a^2 - 2b^2}{a^3 + a^2b + a + b} \div \frac{6a^2}{a^3 + a}$

Perform each indicated operation. See Example 15.

67. $\frac{4}{x} \div \frac{3xy}{x^2} \cdot \frac{6x^2}{x^4}$

68. $\frac{4}{x} \cdot \frac{3xy}{x^2} \div \frac{6x^2}{x^4}$

69. $\frac{3x^2 - 5x - 2}{y^2 + y - 2} \cdot \frac{y^2 + 4y - 5}{12x^2 + 7x + 1} \div \frac{5x^2 - 9x - 2}{8x^2 - 2x - 1}$

70. $\frac{x^2 + x - 2}{3y^2 - 5y - 2} \cdot \frac{12y^2 + y - 1}{x^2 + 4x - 5} \div \frac{8y^2 - 6y + 1}{5y^2 - 9y - 2}$

E *Find each function value. See Example 16.*

71. If $f(x) = \frac{x + 8}{2x - 1}$, find $f(2)$, $f(0)$, and $f(-1)$.

72. If $f(x) = \frac{x - 2}{-5 + x}$, find $f(-5)$, $f(0)$, and $f(10)$.

53. ______
54. ______
55. ______
56. ______
57. ______
58. ______
59. ______
60. ______
61. ______
62. ______
63. ______
64. ______
65. ______
66. ______
67. ______
68. ______
69. ______
70. ______
71. ______
72. ______

73. a. ______

b. ______

c. ______

74. a. ______

b. ______

c. ______

75. ______

76. ______

77. ______

78. ______

79. ______

80. ______

81. ______

82. ______

83. ______

84. ______

85. ______

86. ______

Name ______________________

73. The total revenue from the sale of a popular book is approximated by the rational function $R(x) = \frac{1000x^2}{x^2 + 4}$, where x is the number of years since publication and $R(x)$ is the total revenue in millions of dollars.

a. Find the total revenue at the end of the first year.

b. Find the total revenue at the end of the second year.

c. Find the revenue during the second year only.

74. The function $f(x) = \frac{100{,}000x}{100 - x}$ models the cost in dollars for removing x percent of the pollutants from a bayou in which a nearby company dumped creosol.

a. Find the cost of removing 20% of the pollutants from the bayou. (*Hint:* Find $f(20)$.)

b. Find the cost of removing 60% of the pollutants and then 80% of the pollutants.

c. Find $f(90)$, then $f(95)$, and then $f(99)$. What happens to the cost as x approaches 100%?

Review and Preview

Perform each indicated operation. See Section 1.3.

75. $\frac{4}{5} + \frac{3}{5}$

76. $\frac{4}{10} - \frac{7}{10}$

77. $\frac{5}{28} - \frac{2}{21}$

78. $\frac{5}{13} + \frac{2}{7}$

79. $\frac{3}{8} + \frac{1}{2} - \frac{3}{16}$

80. $\frac{2}{9} - \frac{1}{6} + \frac{2}{3}$

Combining Concepts

81. A parallelogram has an area of $\frac{x^2 + x - 2}{x^3}$ square feet and a height of $\frac{x^2}{x - 1}$ feet. Express the length of its base as a rational expression in x. (*Hint:* Since $A = b \cdot h$, then $b = \frac{A}{h}$ or $b = A \div h$.)

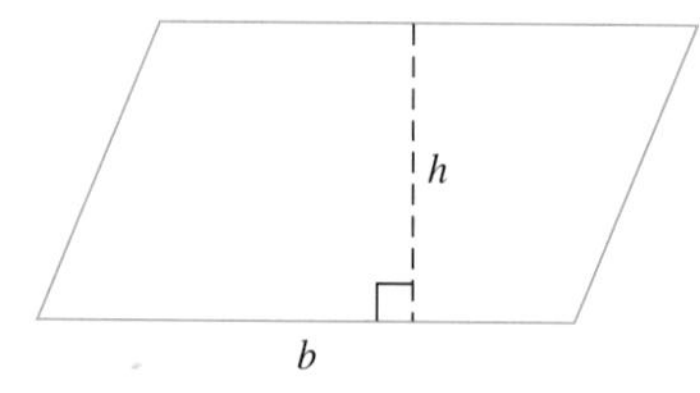

82. A lottery prize of $\frac{15x^3}{y^2}$ dollars is to be divided among $5x$ people. Express the amount of money each person is to receive as a rational expression in x and y.

Simplify. Assume that no denominator is 0.

83. $\frac{p^x - 4}{4 - p^x}$

84. $\frac{3 + q^n}{q^n + 3}$

85. $\frac{x^n + 4}{x^{2n} - 16}$

86. $\frac{x^{2k} - 9}{3 + x^k}$

6.2 ADDING AND SUBTRACTING RATIONAL EXPRESSIONS

A ADDING OR SUBTRACTING RATIONAL EXPRESSIONS WITH THE SAME DENOMINATOR

We add or subtract rational expressions just as we add or subtract fractions.

> **ADDING OR SUBTRACTING RATIONAL EXPRESSIONS WITH THE SAME DENOMINATOR**
>
> To add or subtract rational expressions with the same denominator, add or subtract the numerators. Write the result over the common denominator.

Examples Add.

1. $\frac{5}{7}+\frac{x}{7}=\frac{5+x}{7}$ Add the numerators and write the result over the common denominator.

2. $\frac{x}{4}+\frac{5x}{4}=\frac{x+5x}{4}=\frac{6x}{4}=\frac{3x}{2}$

Examples Subtract.

3. $\frac{x^2}{x+7}-\frac{49}{x+7}=\frac{x^2-49}{x+7}$ Subtract the numerators and write the result over the common denominator.

$=\frac{(x+7)(x-7)}{x+7}$ Factor the numerator.

Simplify.

$=x-7$

HELPFUL HINT

Be sure to insert parentheses here so that the entire numerator is subtracted.

4. $\frac{x}{3y^2}-\frac{x+1}{3y^2}=\frac{x-(x+1)}{3y^2}$ Subtract the numerators.

$=\frac{x-x-1}{3y^2}$ Use the distributive property.

$=-\frac{1}{3y^2}$ Simplify.

TRY THE CONCEPT CHECK IN THE MARGIN.

B FINDING THE LCD OF RATIONAL EXPRESSIONS

To add or subtract rational expressions with unlike, or different, denominators, we first write the rational expressions as equivalent rational expressions with common denominators.

The **least common denominator (LCD)** is usually the easiest common denominator to work with.

Objectives

A Add or subtract rational expressions with the same denominator.

B Find the least common denominator (LCD) of two or more rational expressions.

C Add and subtract rational expressions with different denominators.

SSM CD-ROM Video 6.2

Practice Problems 1–2

Add.

1. $\frac{9}{11}+\frac{y}{11}$
2. $\frac{x}{6}+\frac{7x}{6}$

Practice Problems 3–4

Subtract.

3. $\frac{x^2}{x+3}-\frac{9}{x+3}$
4. $\frac{a}{5b^3}-\frac{a+2}{5b^3}$

✓ CONCEPT CHECK

Find and correct the error.

$\frac{3+2y}{y^2-1}-\frac{y+3}{y^2-1}$

$=\frac{3+2y-y+3}{y^2-1}$

$=\frac{y+6}{y^2-1}$

Answers

1. $\frac{9+y}{11}$, **2.** $\frac{4x}{3}$, **3.** $x-3$, **4.** $-\frac{2}{5b^3}$

✓ Concept Check

$\frac{3+2y}{y^2-1}-\frac{y+3}{y^2-1}=\frac{3+2y-y-3}{y^2-1}=\frac{y}{y^2-1}$

The following steps can be used to find the LCD.

FINDING THE LEAST COMMON DENOMINATOR (LCD)

Step 1. Factor each denominator completely.

Step 2. The LCD is the product of all unique factors each raised to the greatest power that appears in any factored denominator.

Practice Problem 5

Find the LCD of the rational expressions in each list.

a. $\frac{7}{20a^2b^3}, \frac{9}{15ab^4}$

b. $\frac{6x}{x-2}, \frac{5}{x+2}$

c. $\frac{x+4}{x^2-36}, \frac{x}{x^2+12x+36}, \frac{x^3}{3x^2+19x+6}$

d. $\frac{6}{x^2-1}, \frac{7}{2-2x}$

Example 5

Find the LCD of the rational expressions in each list.

a. $\frac{2}{3x^5y^2}, \frac{3z}{5xy^3}$

b. $\frac{7}{z+1}, \frac{z}{z-1}$

c. $\frac{m-1}{m^2-25}, \frac{2m}{2m^2-9m-5}, \frac{7}{m^2-10m+25}$

d. $\frac{x}{x^2-4}, \frac{11}{6-3x}$

Solution: **a.** First we factor each denominator.

$$3x^5y^2 = 3 \cdot x^5 \cdot y^2$$
$$5xy^3 = 5 \cdot x \cdot y^3$$
$$\text{LCD} = 3 \cdot 5 \cdot x^5 \cdot y^3 = 15x^5y^3$$

HELPFUL HINT

The greatest power of x is 5, so we have a factor of x^5. The greatest power of y is 3, so we have a factor of y^3.

b. The denominators $z + 1$ and $z - 1$ do not factor further.

$$(z+1) = (z+1)$$
$$(z-1) = (z-1)$$
$$\text{LCD} = (z+1)(z-1)$$

c. We first factor each denominator.

$$m^2 - 25 = (m+5)(m-5)$$
$$2m^2 - 9m - 5 = (2m+1)(m-5)$$
$$m^2 - 10m + 25 = (m-5)(m-5)$$
$$\text{LCD} = (m+5)(2m+1)(m-5)^2$$

d. Factor each denominator.

$$x^2 - 4 = (x+2)(x-2)$$
$$6 - 3x = 3(2-x) = 3(-1)(x-2)$$
$$\text{LCD} = 3(-1)(x+2)(x-2)$$
$$= -3(x+2)(x-2)$$

HELPFUL HINT

$(x - 2)$ and $(2 - x)$ are opposite factors. Notice that a -1 was factored from $(2 - x)$ so that the factors are identical.

Answers

5. a. $60a^2b^4$, **b.** $(x-2)(x+2)$, **c.** $(x-6)(3x+1)(x+6)^2$, **d.** $-2(x+1)(x-1)$

HELPFUL HINT

If opposite factors occur, do not use both in the LCD. Instead, factor -1 from one of the opposite factors so that the factors are then identical.

C Adding or Subtracting Rational Expressions with Different Denominators

To add or subtract rational expressions with different denominators, we write each rational expression as an equivalent rational expression with the LCD as denominator. To do this, we use the multiplication property of 1 and multiply the numerator and the denominator by the same factor so that the denominator becomes the LCD.

ADDING OR SUBTACTING RATIONAL EXPRESSIONS WITH DIFFERENT DENOMINATORS

Step 1. Find the LCD of the rational expressions.

Step 2. Write each rational expression as an equivalent rational expression whose denominator is the LCD found in Step 1.

Step 3. Add or subtract numerators, and write the result over the common denominator.

Step 4. Simplify the resulting rational expression.

Example 6 Add: $\frac{2}{x^2} + \frac{5}{3x^3}$

Solution: The LCD is $3x^3$, so we write each rational expression as an equivalent rational expression with denominator $3x^3$.

$$\frac{2}{x^2} + \frac{5}{3x^3} = \frac{2 \cdot 3x}{x^2 \cdot 3x} + \frac{5}{3x^3}$$ The second expression already has a denominator of $3x^3$.

$$= \frac{6x}{3x^3} + \frac{5}{3x^3}$$

$$= \frac{6x + 5}{3x^3}$$ Add the numerators.

Practice Problem 6

Add: $\frac{7}{a^3} + \frac{9}{2a^4}$

Example 7 Add: $\frac{3}{x + 2} + \frac{2x}{x - 2}$

Solution: The LCD is the product of the two denominators: $(x + 2)(x - 2)$.

$$\frac{3}{x + 2} + \frac{2x}{x - 2} = \frac{3 \cdot (x - 2)}{(x + 2) \cdot (x - 2)} + \frac{2x \cdot (x + 2)}{(x - 2) \cdot (x + 2)}$$ Write equivalent rational expressions.

$$= \frac{3(x - 2) + 2x(x + 2)}{(x + 2)(x - 2)}$$ Add the numerators.

$$= \frac{3x - 6 + 2x^2 + 4x}{(x + 2)(x - 2)}$$ Multiply in the numerator.

$$= \frac{2x^2 + 7x - 6}{(x + 2)(x - 2)}$$ Simplify.

Practice Problem 7

Add: $\frac{1}{x + 5} + \frac{6x}{x - 5}$

Answers

6. $\frac{14a + 9}{2a^4}$, **7.** $\frac{6x^2 + 31x - 5}{(x + 5)(x - 5)}$

Practice Problem 8

Subtract: $\dfrac{m}{m-6} - \dfrac{8}{6-m}$

Example 8

Subtract: $\dfrac{x}{x-1} - \dfrac{4}{1-x}$

Solution: The LCD is either $x - 1$ or $1 - x$. To get a common denominator of $x - 1$, we factor -1 from the denominator of the second rational expression.

$$\frac{x}{x-1} - \frac{4}{1-x} = \frac{x}{x-1} - \frac{4}{-1(x-1)} \quad \text{Write } 1-x \text{ as } -1(x-1).$$

$$= \frac{x}{x-1} - \frac{-1 \cdot 4}{x-1} \quad \text{Write } \frac{4}{-1(x-1)} \text{ as } \underline{\qquad}.$$

$$= \frac{x-(-4)}{x-1}$$

$$= \frac{x+4}{x-1} \quad \text{Simplify.}$$

Practice Problem 9

Subtract: $\dfrac{2x}{x^2-9} - \dfrac{3}{x^2-4x+3}$

Example 9

Subtract: $\dfrac{5k}{k^2-4} - \dfrac{2}{k^2+k-2}$

Solution: Factor each denominator to find the LCD.

$$\frac{5k}{k^2-4} - \frac{2}{k^2+k-2} = \frac{5k}{(k+2)(k-2)} - \frac{2}{(k+2)(k-1)}$$

The LCD is $(k+2)(k-2)(k-1)$. We write equivalent rational expressions with the LCD as denominators.

$$\frac{5k}{(k+2)(k-2)} - \frac{2}{(k+2)(k-1)} =$$

$$\frac{5k \cdot (k-1)}{(k+2)(k-2) \cdot (k-1)} - \frac{2 \cdot (k-2)}{(k+2)(k-1) \cdot (k-2)}$$

$$= \frac{5k(k-1) - 2(k-2)}{(k+2)(k-2)(k-1)} \quad \text{Subtract the numerators.}$$

$$= \frac{5k^2 - 5k - 2k + 4}{(k+2)(k-2)(k-1)} \quad \text{Multiply in the numerator.}$$

$$= \frac{5k^2 - 7k + 4}{(k+2)(k-2)(k-1)} \quad \text{Simplify.}$$

Practice Problem 10

Add: $\dfrac{x+1}{x^2+x-12} + \dfrac{2x-1}{x^2+6x+8}$

Example 10

Add: $\dfrac{2x-1}{2x^2-9x-5} + \dfrac{x+3}{6x^2-x-2}$

Solution: Factor the denominators.

$$\frac{2x-1}{2x^2-9x-5} + \frac{x+3}{6x^2-x-2} = \frac{2x-1}{(2x+1)(x-5)} + \frac{x+3}{(2x+1)(3x-2)}$$

The LCD is $(2x+1)(x-5)(3x-2)$.

$$= \frac{(2x-1) \cdot (3x-2)}{(2x+1)(x-5) \cdot (3x-2)} + \frac{(x+3) \cdot (x-5)}{(2x+1)(3x-2) \cdot (x-5)}$$

$$= \frac{(2x-1)(3x-2) + (x+3)(x-5)}{(2x+1)(x-5)(3x-2)} \quad \text{Add the numerators.}$$

$$= \frac{6x^2 - 7x + 2 + x^2 - 2x - 15}{(2x+1)(x-5)(3x-2)} \quad \text{Multiply in the numerator.}$$

$$= \frac{7x^2 - 9x - 13}{(2x+1)(x-5)(3x-2)} \quad \text{Simplify.}$$

Answers

8. $\dfrac{m+8}{m-6}$, **9.** $\dfrac{2x^2-5x-9}{(x+3)(x-3)(x-1)}$, **10.** $\dfrac{3x^2-4x+5}{(x+2)(x-3)(x+4)}$

Example 11 Perform each indicated operation:

$$\frac{7}{x-1}+\frac{10x}{x^2-1}-\frac{5}{x+1}$$

Solution:

$$\frac{7}{x-1}+\frac{10x}{x^2-1}-\frac{5}{x+1}=\frac{7}{x-1}+\frac{10x}{(x-1)(x+1)}-\frac{5}{x+1}$$

The LCD is $(x-1)(x+1)$.

$$=\frac{7\cdot(x+1)}{(x-1)\cdot(x+1)}+\frac{10x}{(x-1)\cdot(x+1)}-\frac{5\cdot(x-1)}{(x+1)\cdot(x-1)}$$

$$=\frac{7(x+1)+10x-5(x-1)}{(x-1)(x+1)}$$ Add and subtract the numerators.

$$=\frac{7x+7+10x-5x+5}{(x-1)(x+1)}$$ Multiply in the numerator.

$$=\frac{12x+12}{(x-1)(x+1)}$$ Simplify.

$$=\frac{12(x+1)}{(x-1)(x+1)}$$ Factor the numerator.

$$=\frac{12}{x-1}$$ Divide out common factors.

Practice Problem 11

Perform each indicated operation.

$$\frac{6}{x-5}+\frac{x-35}{x^2-5x}-\frac{2}{x}$$

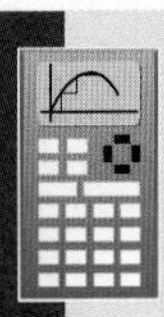

GRAPHING CALCULATOR EXPLORATIONS

A grapher can be used to support the results of operations on rational expressions. For example, to verify the result of Example 7, graph

$$Y_1=\frac{3}{x+2}+\frac{2x}{x-2} \text{ and } Y_2=\frac{2x^2+7x-6}{(x+2)(x-2)}$$

on the same set of axes. The graphs should be the same. Use a Table feature or a Trace feature to see that this is true.

Answer

11. $\frac{5}{x}$

Focus On Study Skills

PROBLEM SOLVING

When you are faced with solving a problem in real life that involves math, it may not always be immediately obvious how to approach the problem or what type of math is needed. Here are some problem-solving strategies that may be helpful:

- ▲ Break a complicated problem up into simpler problems or steps. For instance, this strategy might be useful if you need to find the area or volume of a complicated geometric shape.
- ▲ In problems that involve considering many different scenarios, try listing the possibilities. This strategy would help in a situation such as finding the number of four-person committees that could be formed from a pool of 12 club members.
- ▲ Look at several specific examples of a general problem. Then see if a pattern emerges from these results that can be used to solve the problem.
- ▲ Make a table of values related to the problem. If done methodically, this may reveal patterns that can be used to solve the problem.
- ▲ Work backward from the desired results until suitable initial conditions for the problem are reached.
- ▲ When not all of the necessary initial conditions are given in problem, make some assumptions. Try to use reasonable estimates or educated guesses for the missing information so that your solution is meaningful for the problem.
- ▲ Try a simpler or similar problem. Starting with smaller numbers or "round" numbers may help you see how to handle the problem.
- ▲ Make a guess and check to see if it satisfies the problem. If not, revise your guess and check again. Continue guessing and checking until a guess checks or until this process gives enough insight into the problem that another method can be used.
- ▲ Draw a diagram. This strategy can be helpful in many different problem-solving situations. This can help you visualize a physical relationship that might make the problem easier or clearer.

CRITICAL THINKING

Try solving the following problems. Explain how you found each solution. You may find some of the strategies described above useful.

1. Jamie Webb can slice through a loaf of cinnamon bread in 15 seconds. How long will it take her to cut a 12-inch loaf into half-inch slices?
2. How many times does the digit "5" appear in the numbers from 500 (inclusive) through 600 (inclusive)?
3. Two cars are racing on an oval racetrack. One car can complete a lap in 60 seconds. The other car can complete a lap in 65 seconds. If both cars cross the start/finish line at the same time, how long will it be before the faster car passes the slower car on the track?
4. At a carnival, there are three gambling games. Game A costs $10 to play; Game B costs $50 to play; and Game C costs $100 to play. For each game, the player must pay the game operator the required fee. If the player loses the game, he or she receives nothing. If the player wins the game, he or she receives twice what was paid to play the game. A gambler plays Game A four times, Game B three times, and Game C twice. She wins each type of game once and ends up with $180. How much money did she start out with?
5. A couple has three children. If each of their children has three children, and each of their children has three children, and so on, how many great-great-great-grandchildren will the original couple have?
6. Numbering the pages in a book requires a total of 1350 digits. If the book numbering starts on page 1, how many pages are in the book?

Name ______________________ Section ______ Date ________

Answers

Exercise Set 6.2

A *Add or subtract as indicated. Simplify each answer. See Examples 1 through 4.*

1. $\frac{2}{x} - \frac{5}{x}$

2. $\frac{4}{x^2} + \frac{2}{x^2}$

3. $\frac{2}{x-2} + \frac{x}{x-2}$

4. $\frac{x}{5-x} + \frac{7}{5-x}$

5. $\frac{x^2}{x+2} - \frac{4}{x+2}$

6. $\frac{x^2}{x+6} - \frac{36}{x+6}$

7. $\frac{2x-6}{x^2+x-6} + \frac{3-3x}{x^2+x-6}$

8. $\frac{5x+2}{x^2+2x-8} + \frac{2-4x}{x^2+2x-8}$

9. $\frac{x-5}{2x} - \frac{x+5}{2x}$

10. $\frac{x+4}{4x} - \frac{x-4}{4x}$

11. Find the perimeter of the square.

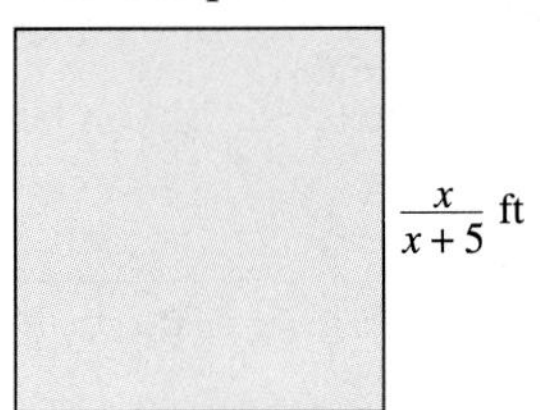

12. Find the perimeter of the quadrilateral.

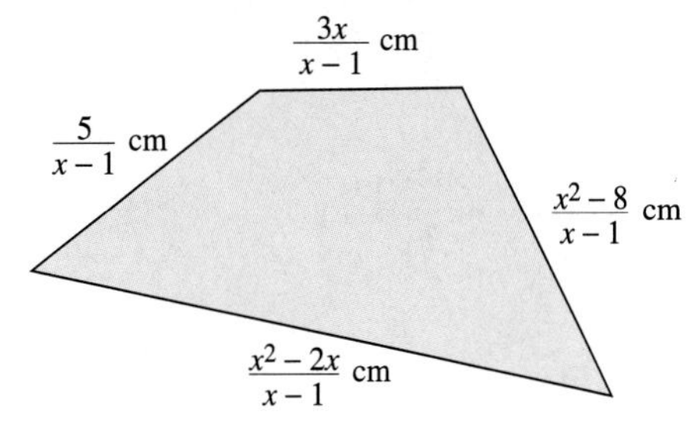

B *Find the LCD of the rational expressions in each list. See Example 5.*

13. $\frac{2}{7}, \frac{3}{5x}$

14. $\frac{4}{5y}, \frac{3}{4y^2}$

15. $\frac{3}{x}, \frac{2}{x+1}$

1. ________
2. ________
3. ________
4. ________
5. ________
6. ________
7. ________
8. ________
9. ________
10. ________
11. ________
12. ________
13. ________
14. ________
15. ________

16. ______
17. ______
18. ______
19. ______
20. ______
21. ______
22. ______
23. ______
24. ______
25. ______
26. ______
27. ______
28. ______
29. ______
30. ______
31. ______
32. ______
33. ______
34. ______
35. ______
36. ______
37. ______
38. ______
39. ______
40. ______
41. ______

Name ______________________

16. $\frac{5}{2x}, \frac{7}{2 + x}$

17. $\frac{12}{x + 7}, \frac{8}{x - 7}$

18. $\frac{1}{2x - 1}, \frac{x}{2x + 1}$

19. $\frac{5}{3x + 6}, \frac{2x}{2x - 4}$

20. $\frac{2}{3a + 9}, \frac{5}{5a - 15}$

21. $\frac{2a}{a^2 - b^2}, \frac{1}{a^2 - 2ab + b^2}$

22. $\frac{2a}{a^2 + 8a + 16}, \frac{7a}{a^2 + a - 12}$

23. $\frac{x}{x^2 - 9}, \frac{5}{x}, \frac{7}{12 - 4x}$

24. $\frac{9}{x^2 - 25}, \frac{1}{50 - 10x}, \frac{6}{x}$

C *Add or subtract as indicated. Simplify each answer. See Examples 6 through 10.*

25. $\frac{4}{3x} + \frac{3}{2x}$

26. $\frac{10}{7x} - \frac{5}{2x}$

27. $\frac{5}{2y^2} - \frac{2}{7y}$

28. $\frac{4}{11x^4} - \frac{1}{4x^2}$

29. $\frac{x - 3}{x + 4} - \frac{x + 2}{x - 4}$

30. $\frac{x - 1}{x - 5} - \frac{x + 2}{x + 5}$

31. $\frac{1}{x - 5} + \frac{2x - 19}{x^2 - x - 20}$

32. $\frac{4x - 2}{x^2 - x - 20} - \frac{2}{x + 4}$

33. $\frac{3}{2x + 10} + \frac{8}{3x + 15}$

34. $\frac{10}{3x - 3} + \frac{1}{7x - 7}$

35. $\frac{-2}{x^2 - 3x} - \frac{1}{x^3 - 3x^2}$

36. $\frac{-3}{2a + 8} - \frac{8}{a^2 + 4a}$

37. $\frac{1}{a - b} + \frac{1}{b - a}$

38. $\frac{1}{a - 3} - \frac{1}{3 - a}$

39. $\frac{5}{x - 2} + \frac{x + 4}{2 - x}$

40. $\frac{3}{5 - x} + \frac{x + 2}{x - 5}$

41. $\frac{y + 1}{y^2 - 6y + 8} - \frac{3}{y^2 - 16}$

Name ______________________________

42. $\dfrac{x+2}{x^2-36} - \dfrac{x}{x^2+9x+18}$

43. $\dfrac{7}{x^2-x-2} + \dfrac{x}{x^2+4x+3}$

44. $\dfrac{a}{a^2+10a+25} + \dfrac{4}{a^2+6a+5}$

45. $\dfrac{x+4}{3x^2+11x+6} + \dfrac{x}{2x^2+x-15}$

46. $\dfrac{x+3}{5x^2+12x+4} + \dfrac{6}{x^2-x-6}$

47. $\dfrac{2}{a^2+2a+1} + \dfrac{3}{a^2-1}$

48. $\dfrac{9x+2}{3x^2-2x-8} + \dfrac{7}{3x^2+x-4}$

49. $\dfrac{ab}{a^2-b^2} + \dfrac{b}{2a+2b}$

50. $\dfrac{2}{3x-15} + \dfrac{x}{25-x^2}$

51. $\dfrac{5}{x^2-4} - \dfrac{4}{x^2+4x+4}$

52. $\dfrac{3z}{z^2-9} - \dfrac{2}{3-z}$

Perform each indicated operation. Simplify each answer. See Example 11.

53. $\dfrac{2}{x+1} - \dfrac{3x}{3x+3} + \dfrac{1}{2x+2}$

54. $\dfrac{5}{3x-6} - \dfrac{x}{x-2} + \dfrac{3+2x}{5x-10}$

55. $\dfrac{3}{x+3} + \dfrac{5}{x^2+6x+9} - \dfrac{x}{x^2-9}$

56. $\dfrac{x+2}{x^2-2x-3} + \dfrac{x}{x-3} - \dfrac{4}{x+1}$

57. $\dfrac{x}{x^2-9} + \dfrac{3}{x^2-6x+9} - \dfrac{1}{x+3}$

58. $\dfrac{3}{x^2-9} - \dfrac{x}{x^2-6x+9} + \dfrac{1}{x+3}$

59. $\left(\dfrac{1}{x} + \dfrac{2}{3}\right) - \left(\dfrac{1}{x} - \dfrac{2}{3}\right)$

60. $\left(\dfrac{1}{2} + \dfrac{2}{x}\right) - \left(\dfrac{1}{2} - \dfrac{1}{x}\right)$

42. ______________

43. ______________

44. ______________

45. ______________

46. ______________

47. ______________

48. ______________

49. ______________

50. ______________

51. ______________

52. ______________

53. ______________

54. ______________

55. ______________

56. ______________

57. ______________

58. ______________

59. ______________

60. ______________

61.

62.

63.

64.

65.

66.

67.

68.

69.

Name

Review and Preview

Use the distributive property to multiply each expression. See Section 1.2.

61. $12\left(\frac{2}{3}+\frac{1}{6}\right)$

62. $14\left(\frac{1}{7}+\frac{3}{14}\right)$

63. $x^2\left(\frac{4}{x^2}+1\right)$

64. $5y^2\left(\frac{1}{y^2}-\frac{1}{5}\right)$

Combining Concepts

65. In your own words, explain how to add rational expressions with different denominators.

Perform each indicated operation. (Hint: First write each expression with positive exponents.)

66. $x^{-1}+(2x)^{-1}$

67. $y^{-1}+(4y)^{-1}$

68. $4x^{-2}-3x^{-1}$

69. $(4x)^{-2}-(3x)^{-1}$

Use a grapher to support the results of each exercise.

70. Exercise 3

71. Exercise 4

6.3 SIMPLIFYING COMPLEX FRACTIONS

Objectives

A Simplify complex fractions by simplifying the numerator and denominator and then dividing.

B Simplify complex fractions by multiplying by the least common denominator (LCD).

C Simplify expressions with negative exponents.

SSM CD-ROM Video 6.3

A rational expression whose numerator, denominator, or both contain one or more rational expressions is called a **complex rational expression** or a **complex fraction**. Examples are

$$\frac{\frac{1}{a}}{\frac{b}{2}} \qquad \frac{\frac{x}{2y^2}}{\frac{6x-2}{9y}} \qquad \frac{x+\frac{1}{y}}{y+1}$$

The parts of a complex fraction are

$$\frac{\frac{x}{y+2}}{7+\frac{1}{y}}$$

← Numerator of complex fraction
← Main fraction bar
← Denominator of complex fraction

Our goal in this section is to simplify complex fractions. A complex fraction is simplified when it is in the form $\frac{P}{Q}$, where P and Q are polynomials that have no common factors. Two methods of simplifying complex fractions are introduced.

A METHOD 1: SIMPLIFYING A COMPLEX FRACTION BY SIMPLIFYING THE NUMERATOR AND DENOMINATOR AND THEN DIVIDING

In the first method we study, we simplify complex fractions by simplifying and dividing.

> SIMPLIFYING A COMPLEX FRACTION: METHOD 1
>
> **Step 1.** Simplify the numerator and the denominator of the complex fraction so that each is a single fraction.
>
> **Step 2.** Perform the indicated division by multiplying the numerator of the complex fraction by the reciprocal of the denominator of the complex fraction.
>
> **Step 3.** Simplify if possible.

Example 1 Simplify: $\frac{\frac{2x}{27y^2}}{\frac{6x^2}{9}}$

Solution: The numerator of the complex fraction is already a single fraction, and so is the denominator. Thus we perform the indicated division by multiplying the numerator, $\frac{2x}{27y^2}$, by the reciprocal of the denominator, $\frac{6x^2}{9}$. Then we simplify.

Practice Problem 1

Use Method 1 to simplify: $\frac{\frac{3a}{25b^2}}{\frac{9a^2}{5b}}$

Answer

1. $\frac{1}{15ab}$

$$\frac{\frac{2x}{27y^2}}{\frac{6x^2}{9}} = \frac{2x}{27y^2} \div \frac{6x^2}{9}$$

$$= \frac{2x}{27y^2} \cdot \frac{9}{6x^2} \quad \text{Multiply by the reciprocal of } \frac{6x^2}{9}.$$

$$= \frac{2x \cdot 9}{27y^2 \cdot 6x^2}$$

$$= \frac{1}{9xy^2} \quad \text{Simplify.}$$

Example 2

Simplify: $\dfrac{\frac{5x}{x+2}}{\frac{10}{x-2}}$

Solution:

$$\frac{\frac{5x}{x+2}}{\frac{10}{x-2}} = \frac{5x}{x+2} \cdot \frac{x-2}{10} \quad \text{Multiply by the reciprocal of } \frac{10}{x-2}.$$

$$= \frac{5x(x-2)}{2 \cdot 5(x+2)}$$

$$= \frac{x(x-2)}{2(x+2)} \quad \text{Simplify.}$$

TRY THE CONCEPT CHECK IN THE MARGIN.

Example 3

Simplify: $\dfrac{\frac{x}{y^2}+\frac{1}{y}}{\frac{y}{x^2}+\frac{1}{x}}$

Solution: First we simplify the numerator and the denominator of the complex fraction separately so that each is a single fraction.

$$\frac{\frac{x}{y^2}+\frac{1}{y}}{\frac{y}{x^2}+\frac{1}{x}} = \frac{\frac{x}{y^2}+\frac{1 \cdot y}{y \cdot y}}{\frac{y}{x^2}+\frac{1 \cdot x}{x \cdot x}} \quad \begin{array}{l}\text{The LCD is } y^2.\\ \text{The LCD is } x^2.\end{array}$$

$$= \frac{\frac{x+y}{y^2}}{\frac{y+x}{x^2}} \quad \begin{array}{l}\text{Add.}\\ \text{Add.}\end{array}$$

$$= \frac{x+y}{y^2} \cdot \frac{x^2}{y+x} \quad \text{Multiply by the reciprocal of } \frac{y+x}{x^2}.$$

$$= \frac{x^2(x+y)}{y^2(y+x)}$$

$$= \frac{x^2}{y^2} \quad \text{Simplify.}$$

Practice Problem 2

Use Method 1 to simplify: $\dfrac{\frac{6x}{x-5}}{\frac{12}{x+5}}$

✓ CONCEPT CHECK

Which of the following are equivalent to $\dfrac{\frac{1}{x}}{\frac{3}{y}}$?

a. $\frac{1}{x} \div \frac{3}{y}$ b. $\frac{1}{x} \cdot \frac{y}{3}$ c. $\frac{1}{x} \div \frac{y}{3}$

Practice Problem 3

Use Method 1 to simplify: $\dfrac{\frac{x}{y^2}-\frac{1}{y}}{\frac{y}{x^2}-\frac{1}{x}}$

Answers

2. $\frac{x(x+5)}{2(x-5)}$, **3.** $-\frac{x^2}{y^2}$

✓ Concept Check: a, b

B METHOD 2: SIMPLIFYING A COMPLEX FRACTION BY MULTIPLYING THE NUMERATOR AND DENOMINATOR BY THE LCD

With this method, we multiply the numerator and the denominator of the complex fraction by the least common denominator (LCD) of all fractions in the complex fraction.

> **SIMPLIFYING A COMPLEX FRACTION: METHOD 2**
>
> **Step 1.** Multiply the numerator and the denominator of the complex fraction by the LCD of the fractions in both the numerator and the denominator.
>
> **Step 2.** Simplify.

Example 4 Simplify: $\dfrac{\dfrac{5x}{x+2}}{\dfrac{10}{x-2}}$

Solution: The least common denominator of $\dfrac{5x}{x+2}$ and $\dfrac{10}{x-2}$ is $(x+2)(x-2)$. We multiply both the numerator, $\dfrac{5x}{x+2}$, and the denominator, $\dfrac{10}{x-2}$, by this LCD.

$$\frac{\dfrac{5x}{x+2}}{\dfrac{10}{x-2}} = \frac{\left(\dfrac{5x}{x+2}\right)\cdot(x+2)(x-2)}{\left(\dfrac{10}{x-2}\right)\cdot(x+2)(x-2)}$$ Multiply the numerator and denominator by the LCD.

$$= \frac{5x\cdot(x-2)}{2\cdot 5\cdot(x+2)}$$ Simplify.

$$= \frac{x(x-2)}{2(x+2)}$$ Simplify.

Practice Problem 4

Use Method 2 to simplify: $\dfrac{\dfrac{6x}{x-5}}{\dfrac{12}{x+5}}$

Example 5 Simplify: $\dfrac{\dfrac{x}{y^2}+\dfrac{1}{y}}{\dfrac{y}{x^2}+\dfrac{1}{x}}$

Solution: The least common denominator of $\dfrac{x}{y^2}, \dfrac{1}{y}, \dfrac{y}{x^2}$, and $\dfrac{1}{x}$ is x^2y^2.

$$\frac{\dfrac{x}{y^2}+\dfrac{1}{y}}{\dfrac{y}{x^2}+\dfrac{1}{x}} = \frac{\left(\dfrac{x}{y^2}+\dfrac{1}{y}\right)\cdot x^2y^2}{\left(\dfrac{y}{x^2}+\dfrac{1}{x}\right)\cdot x^2y^2}$$ Multiply the numerator and denominator by the LCD.

$$= \frac{\dfrac{x}{y^2}\cdot x^2y^2+\dfrac{1}{y}\cdot x^2y^2}{\dfrac{y}{x^2}\cdot x^2y^2+\dfrac{1}{x}\cdot x^2y^2}$$ Use the distributive property.

Practice Problem 5

Use Method 2 to simplify: $\dfrac{\dfrac{x}{y^2}-\dfrac{1}{y}}{\dfrac{y}{x^2}-\dfrac{1}{x}}$

Answers

4. $\dfrac{x(x+5)}{2(x-5)}$, **5.** $-\dfrac{x^2}{y^2}$

$$= \frac{x^3 + x^2y}{y^3 + xy^2} \quad \text{Simplify.}$$

$$= \frac{x^2(x + y)}{y^2(y + x)} \quad \text{Factor.}$$

$$= \frac{x^2}{y^2} \quad \text{Simplify.}$$

C SIMPLIFYING EXPRESSIONS WITH NEGATIVE EXPONENTS

Some expressions containing negative exponents can be written as complex fractions. To simplify these expressions, we first write them as equivalent expressions with positive exponents.

Practice Problem 6

Simplify: $\frac{2x^{-1} + 3y^{-1}}{x^{-1} - 2y^{-1}}$

Example 6 Simplify: $\frac{x^{-1} + 2xy^{-1}}{x^{-2} - x^{-2}y^{-1}}$

Solution: This fraction does not appear to be a complex fraction. However, if we write it by using only positive exponents we see that it is a complex fraction.

$$\frac{x^{-1} + 2xy^{-1}}{x^{-2} - x^{-2}y^{-1}} = \frac{\frac{1}{x} + \frac{2x}{y}}{\frac{1}{x^2} - \frac{1}{x^2y}}$$

The LCD of $\frac{1}{x}$, $\frac{2x}{y}$, $\frac{1}{x^2}$, and $\frac{1}{x^2y}$ is x^2y. We multiply both the numerator and denominator by x^2y.

$$= \frac{\left(\frac{1}{x} + \frac{2x}{y}\right) \cdot x^2y}{\left(\frac{1}{x^2} - \frac{1}{x^2y}\right) \cdot x^2y}$$

$$= \frac{\frac{1}{x} \cdot x^2y + \frac{2x}{y} \cdot x^2y}{\frac{1}{x^2} \cdot x^2y - \frac{1}{x^2y} \cdot x^2y} \quad \text{Use the distributive property.}$$

$$= \frac{xy + 2x^3}{y - 1} \quad \text{Simplify.}$$

Answer

6. $\frac{2y + 3x}{y - 2x}$

Name ______________ Section ______ Date ______

Exercise Set 6.3

A B *Simplify each complex fraction. See Examples 1 through 5.*

1. $\dfrac{\frac{10}{3x}}{\frac{5}{6x}}$

2. $\dfrac{\frac{15}{2x}}{\frac{5}{6x}}$

3. $\dfrac{1 + \frac{2}{5}}{2 + \frac{3}{5}}$

4. $\dfrac{2 + \frac{1}{7}}{3 - \frac{4}{7}}$

5. $\dfrac{\frac{4}{x - 1}}{\frac{x}{x - 1}}$

6. $\dfrac{\frac{x}{x + 2}}{\frac{2}{x + 2}}$

7. $\dfrac{1 - \frac{2}{x}}{x + \frac{4}{9x}}$

8. $\dfrac{5 - \frac{3}{x}}{x + \frac{2}{3x}}$

9. $\dfrac{\frac{4x^2 - y^2}{xy}}{\frac{2}{y} - \frac{1}{x}}$

10. $\dfrac{\frac{x^2 - 9y^2}{xy}}{\frac{1}{y} - \frac{3}{x}}$

11. $\dfrac{\frac{x + 1}{3}}{\frac{2x - 1}{6}}$

12. $\dfrac{\frac{x + 3}{12}}{\frac{4x - 5}{15}}$

13. $\dfrac{\frac{2}{x} + \frac{3}{x^2}}{\frac{4}{x^2} - \frac{9}{x}}$

14. $\dfrac{\frac{2}{x^2} + \frac{1}{x}}{\frac{4}{x^2} - \frac{1}{x}}$

15. $\dfrac{\frac{1}{x} + \frac{2}{x^2}}{x + \frac{8}{x^2}}$

16. $\dfrac{\frac{1}{y} + \frac{3}{y^2}}{y + \frac{27}{y^2}}$

17. $\dfrac{\frac{4}{5 - x} + \frac{5}{x - 5}}{\frac{2}{x} + \frac{3}{x - 5}}$

18. $\dfrac{\frac{3}{x - 4} - \frac{2}{4 - x}}{\frac{2}{x - 4} - \frac{2}{x}}$

19. $\dfrac{\frac{x + 2}{x} - \frac{2}{x - 1}}{\frac{x + 1}{x} + \frac{x + 1}{x - 1}}$

20. $\dfrac{\frac{5}{a + 2} - \frac{1}{a - 2}}{\frac{3}{2 + a} + \frac{6}{2 - a}}$

C *Simplify. See Example 6.*

21. $\dfrac{x^{-1}}{x^{-2} + y^{-2}}$

22. $\dfrac{a^{-3} + b^{-1}}{a^{-2}}$

23. $\dfrac{2a^{-1} + 3b^{-2}}{a^{-1} - b^{-1}}$

Answers

1. ______
2. ______
3. ______
4. ______
5. ______
6. ______
7. ______
8. ______
9. ______
10. ______
11. ______
12. ______
13. ______
14. ______
15. ______
16. ______
17. ______
18. ______
19. ______
20. ______
21. ______
22. ______
23. ______

Name ______________________________

24. ______
25. ______
26. ______
27. ______
28. ______
29. ______
30. ______
31. ______
32. ______
33. ______
34. ______
35. ______
36. ______
37. ______

24. $\dfrac{x^{-1} + y^{-1}}{3x^{-2} + 5y^{-2}}$

25. $\dfrac{1}{x - x^{-1}}$

26. $\dfrac{x^{-2}}{x + 3x^{-1}}$

Review and Preview

Solve each equation for x. See Sections 2.1 and 5.7.

27. $7x + 2 = x - 3$

28. $4 - 2x = 17 - 5x$

29. $x^2 = 4x - 4$

30. $5x^2 + 10x = 15$

31. $\frac{x}{3} - 5 = 13$

32. $\frac{2x}{9} + 1 = \frac{7}{9}$

Combining Concepts

33. When the source of a sound is traveling toward a listener, the pitch that the listener hears due to the Doppler effect is given by the complex rational compression $\dfrac{a}{1 - \dfrac{s}{770}}$, where a is the actual pitch of the sound and s is the speed of the sound source. Simplify this expression.

Simplify.

34. $\dfrac{1}{1 + (1 + x)^{-1}}$

35. $\dfrac{(x + 2)^{-1} + (x - 2)^{-1}}{(x^2 - 4)^{-1}}$

36. $\dfrac{x}{1 - \dfrac{1}{1 + \dfrac{1}{x}}}$

37. $\dfrac{x}{1 - \dfrac{1}{1 - \dfrac{1}{x}}}$

6.4 SOLVING EQUATIONS CONTAINING RATIONAL EXPRESSIONS

A SOLVING EQUATIONS CONTAINING RATIONAL EXPRESSIONS

In this section, we solve equations containing rational expressions. Before beginning this section, make sure that you understand the difference between an *equation* and an *expression*. An **equation** contains an equal sign and an **expression** does not.

Equation	Expression
$\frac{x}{2} + \frac{x}{6} = \frac{2}{3}$	$\frac{x}{2} + \frac{x}{6}$

SOLVING EQUATIONS CONTAINING RATIONAL EXPRESSIONS

To solve *equations* containing rational expressions, first clear the equation of fractions by multiplying both sides of the equation by the LCD of all rational expressions. Then solve as usual.

HELPFUL HINT

The method described above is for equations only. It may *not* be used for performing operations on expressions.

TRY THE CONCEPT CHECK IN THE MARGIN.

Example 1 Solve: $\frac{4x}{5} + \frac{3}{2} = \frac{3x}{10}$

Solution: The LCD of $\frac{4x}{5}, \frac{3}{2}$, and $\frac{3x}{10}$ is 10. We multiply both sides of the equation by 10.

$$\frac{4x}{5} + \frac{3}{2} = \frac{3x}{10}$$

$$10\left(\frac{4x}{5} + \frac{3}{2}\right) = 10\left(\frac{3x}{10}\right) \quad \text{Multiply both sides by the LCD.}$$

$$10 \cdot \frac{4x}{5} + 10 \cdot \frac{3}{2} = 10 \cdot \frac{3x}{10} \quad \text{Use the distributive property.}$$

$$8x + 15 = 3x \quad \text{Simplify.}$$

$$15 = -5x \quad \text{Subtract } 8x \text{ from both sides.}$$

$$-3 = x \quad \text{Solve.}$$

Verify this solution by replacing x with -3 in the original equation.

Check:

$$\frac{4x}{5} + \frac{3}{2} = \frac{3x}{10}$$

$$\frac{4(-3)}{5} + \frac{3}{2} \stackrel{?}{=} \frac{3(-3)}{10}$$

$$\frac{-12}{5} + \frac{3}{2} \stackrel{?}{=} \frac{-9}{10}$$

$$-\frac{24}{10} + \frac{15}{10} \stackrel{?}{=} -\frac{9}{10}$$

$$-\frac{9}{10} = -\frac{9}{10} \quad \text{True.}$$

The solution set is $\{-3\}$.

Objectives

A Solve equations containing rational expressions.

SSM CD-ROM Video 6.4

✓ CONCEPT CHECK

True or false? Clearing fractions is valid when solving an equation and when simplifying rational expressions. Explain.

Practice Problem 1

Solve: $\frac{5x}{6} + \frac{1}{2} = \frac{x}{3}$

Answers

1. $\{-1\}$

✓ Concept Check: false; answers may vary

The important difference about the equations in this section is that the denominator of a rational expression may contain a variable. Recall that a rational expression is undefined for values of the variable that make the denominator 0. If a proposed solution makes the denominator 0, then it must be rejected as a solution of the original equation. Such proposed solutions are called **extraneous solutions**.

Practice Problem 2

Solve: $\frac{5}{x} - \frac{3x+6}{2x} = \frac{7}{2}$

Example 2 Solve: $\frac{3}{x} - \frac{x+21}{3x} = \frac{5}{3}$

Solution: The LCD of the denominators x, $3x$, and 3 is $3x$. We multiply both sides by $3x$.

$$\frac{3}{x} - \frac{x+21}{3x} = \frac{5}{3}$$

$$3x\left(\frac{3}{x} - \frac{x+21}{3x}\right) = 3x\left(\frac{5}{3}\right)$$ Multiply both sides by the LCD.

$$3x \cdot \frac{3}{x} - 3x \cdot \frac{x+21}{3x} = 3x \cdot \frac{5}{3}$$ Use the distributive property.

$$9 - (x+21) = 5x$$ Simplify.

$$9 - x - 21 = 5x$$

$$-12 = 6x$$

$$-2 = x$$ Solve.

The proposed solution is -2.

Check: Check the proposed solution in the original equation.

$$\frac{3}{x} - \frac{x+21}{3x} = \frac{5}{3}$$

$$\frac{3}{-2} - \frac{-2+21}{3(-2)} \stackrel{?}{=} \frac{5}{3}$$

$$-\frac{9}{6} + \frac{19}{6} \stackrel{?}{=} \frac{5}{3}$$

$$\frac{10}{6} = \frac{5}{3}$$ True.

The solution set is $\{-2\}$.

Practice Problem 3

Solve: $\frac{x+5}{x-3} = \frac{2(x+1)}{x-3}$

Example 3 Solve: $\frac{x+6}{x-2} = \frac{2(x+2)}{x-2}$

Solution: First we multiply both sides of the equation by the LCD, $x - 2$.

$$\frac{x+6}{x-2} = \frac{2(x+2)}{x-2}$$

$$(x-2) \cdot \frac{x+6}{x-2} = (x-2) \cdot \frac{2(x+2)}{x-2}$$ Multiply both sides by $x - 2$.

$$x + 6 = 2(x+2)$$ Simplify.

$$x + 6 = 2x + 4$$ Use the distributive property.

$$2 = x$$ Solve.

Answers

2. $\left\{\frac{2}{5}\right\}$, **3.** $\emptyset$

Check: The proposed solution is 2. Notice that 2 makes the denominator 0 in the original equation. This can also be seen in a check. Check the proposed solution 2 in the original equation.

$$\frac{x+6}{x-2} = \frac{2(x+2)}{x-2}$$
$$\frac{2+6}{2-2} = \frac{2(2+2)}{2-2}$$
$$\frac{8}{0} = \frac{2(4)}{0}$$

The denominators are 0, so 2 is not a solution of the original equation. The solution set is $\varnothing$ or $\{\ \}$.

Example 4 Solve: $\frac{2x}{2x-1} + \frac{1}{x} = \frac{1}{2x-1}$

Solution: The LCD is $x(2x-1)$. Multiply both sides by $x(2x-1)$. By the distributive property, this is the same as multiplying each term by $x(2x-1)$.

$$x(2x-1)\cdot\frac{2x}{2x-1} + x(2x-1)\cdot\frac{1}{x} = x(2x-1)\cdot\frac{1}{2x-1}$$
$$x(2x) + (2x-1) = x$$
$$2x^2 + 2x - 1 - x = 0$$
$$2x^2 + x - 1 = 0$$
$$(x+1)(2x-1) = 0$$
$$x+1=0 \quad \text{or} \quad 2x-1=0$$
$$x=-1 \qquad\qquad x=\frac{1}{2}$$

The number $\frac{1}{2}$ makes the denominator $2x-1$ equal 0, so it is not a solution. The solution set is $\{-1\}$.

Practice Problem 4

Solve: $\frac{3x}{3x-1} + \frac{1}{x} = \frac{1}{3x-1}$

Example 5 Solve: $\frac{2x}{x-3} + \frac{6-2x}{x^2-9} = \frac{x}{x+3}$

Solution: We factor the second denominator to find that the LCD is $(x+3)(x-3)$. We multiply both sides of the equation by $(x+3)(x-3)$. By the distributive property, this is the same as multiplying each term by $(x+3)(x-3)$.

$$\frac{2x}{x-3} + \frac{6-2x}{x^2-9} = \frac{x}{x+3}$$
$$(x+3)(x-3)\cdot\frac{2x}{x-3} + (x+3)(x-3)\cdot\frac{6-2x}{(x+3)(x-3)}$$
$$= (x+3)(x-3)\left(\frac{x}{x+3}\right)$$
$$2x(x+3) + (6-2x) = x(x-3) \quad \text{Simplify.}$$
$$2x^2 + 6x + 6 - 2x = x^2 - 3x \quad \text{Use the distributive property.}$$

Practice Problem 5

Solve: $\frac{2x}{x-4} + \frac{10-5x}{x^2-16} = \frac{x}{x+4}$

Answers

4. $\{-1\}$, **5.** $\{-5, -2\}$

Next we solve this quadratic equation by the factoring method. To do so, we first write the equation so that one side is 0.

$$x^2 + 7x + 6 = 0$$
$$(x + 6)(x + 1) = 0 \quad \text{Factor.}$$
$$x = -6 \quad \text{or} \quad x = -1 \quad \text{Set each factor equal to 0.}$$

Neither -6 nor -1 makes any denominator 0. The solution set is $\{-6, -1\}$.

Focus On Business and Career

FASTEST-GROWING OCCUPATIONS

According to U.S. Bureau of Labor Statistics projections, the careers listed below will be the top ten fastest-growing jobs into the next century, according to percent increase in the number of jobs.

Occupation	Employment (in thousands) 1996	2006	% Change
Computer engineers216	451	109	
Database administrators, computer support specialists, and all other computer scientists	212	461	
Desktop publishing specialists	30	53	
Home health aides	495	873	
Medical assistants	225	391	
Occupational therapy assistants and aides	16	26	
Personal and home care aides	202	374	
Physical and corrective therapy assistants and aides	84	151	
Physical therapists	115	196	
Systems analysts	506	1025	

(*Source:* Bureau of Labor Statistics, *Monthly Labor Review*, November 1997)

What do all of these fast-growing occupations have in common? They all require a knowledge of math! For some careers, such as desktop publishing specialists, medical assistants, and computer engineers, the ways math is used on the job may be obvious. For other occupations, the use of math may not be quite as apparent. However, tasks common to many jobs like filling in a time sheet, writing up an expense or mileage report, planning a budget, figuring a bill, ordering supplies, and even making a work schedule all require math.

GROUP ACTIVITY

1. Find the percent change in the number of jobs available from 1996 to 2006 for each occupation in the list.

2. Rank these top-ten occupations according to percent growth, from greatest to least.

3. Which occupation will be the fastest growing during this period?

4. How many occupations will have more than double the number of positions in 2006 than in 1996?

5. Which of these occupations will be the slowest growing during this period?

Name ____________________ Section ________ Date ________

MENTAL MATH

Determine whether each is an equation or an expression. Do not solve or simplify.

1. $\frac{x}{2} = \frac{3x}{5} + \frac{x}{6}$
2. $\frac{3x}{5} + \frac{x}{6}$
3. $\frac{x}{x-1} + \frac{2x}{x+1}$
4. $\frac{x}{x-1} + \frac{2x}{x+1} = 5$
5. $\frac{y+7}{2} = \frac{y+1}{6} + \frac{1}{y}$
6. $\frac{y+1}{6} + \frac{1}{y}$

EXERCISE SET 6.4

A *Solve each equation. See Examples 1 through 5.*

1. $\frac{x}{2} - \frac{x}{3} = 12$
2. $x = \frac{x}{2} - 4$
3. $\frac{5}{x} = \frac{20}{12}$
4. $\frac{2}{x} = \frac{10}{5}$
5. $1 - \frac{4}{a} = 5$
6. $7 + \frac{6}{a} = 5$
7. $\frac{x}{3} = \frac{1}{6} + \frac{x}{4}$
8. $\frac{x}{2} = \frac{21}{10} - \frac{x}{5}$
9. $\frac{2}{x} + \frac{1}{2} = \frac{5}{x}$
10. $\frac{5}{3x} + 1 = \frac{7}{6}$
11. $\frac{x+3}{x} = \frac{5}{x}$
12. $\frac{4-3x}{2x} = -\frac{8}{2x}$
13. $\frac{5}{x-2} - \frac{2}{x+4} = -\frac{4}{x^2+2x-8}$
14. $\frac{1}{x-1} + \frac{1}{x+1} = \frac{2}{x^2-1}$
15. $\frac{1}{x-1} = \frac{2}{x+1}$
16. $\frac{6}{x+3} = \frac{4}{x-3}$
17. $\frac{1}{x-4} - \frac{3x}{x^2-16} = \frac{2}{x+4}$
18. $\frac{3}{2x+3} - \frac{1}{2x-3} = \frac{4}{4x^2-9}$
19. $\frac{1}{x-4} = \frac{8}{x^2-16}$
20. $\frac{2}{x^2-4} = \frac{1}{2x-4}$
21. $\frac{1}{x-2} - \frac{2}{x^2-2x} = 1$
22. $\frac{12}{3x^2+12x} = 1 - \frac{1}{x+4}$

MENTAL MATH ANSWERS

1. ________
2. ________
3. ________
4. ________
5. ________
6. ________

ANSWERS

1. ________
2. ________
3. ________
4. ________
5. ________
6. ________
7. ________
8. ________
9. ________
10. ________
11. ________
12. ________
13. ________
14. ________
15. ________
16. ________
17. ________
18. ________
19. ________
20. ________
21. ________
22. ________

23. ____
24. ____
25. ____
26. ____
27. ____
28. ____
29. ____
30. ____
31. ____
32. ____
33. ____
34. ____
35. ____
36. ____
37. ____
38. ____
39. ____
40. ____
41. ____
42. ____

Name ____

23. $\frac{1}{2x} - \frac{1}{x+1} = \frac{1}{3x^2 + 3x}$

24. $\frac{2}{x-5} + \frac{1}{2x} = \frac{5}{3x^2 - 15x}$

25. $\frac{1}{x} - \frac{x}{25} = 0$

26. $\frac{x}{4} + \frac{5}{x} = 3$

27. $5 - \frac{2}{2y-5} = \frac{3}{2y-5}$

28. $1 - \frac{5}{y+7} = \frac{4}{y+7}$

29. $\frac{x+3}{x+2} = \frac{1}{x+2}$

30. $\frac{2x+1}{4-x} = \frac{9}{4-x}$

31. $\frac{1}{a-3} + \frac{2}{a+3} = \frac{1}{a^2 - 9}$

32. $\frac{12}{9-a^2} + \frac{3}{3+a} = \frac{2}{3-a}$

33. $\frac{64}{x^2 - 16} + 1 = \frac{2x}{x-4}$

34. $2 + \frac{3}{x} = \frac{2x}{x+3}$

35. $\frac{-15}{4y+1} + 4 = y$

36. $\frac{36}{x^2 - 9} + 1 = \frac{2x}{x+3}$

37. $\frac{28}{x^2 - 9} + \frac{2x}{x-3} + \frac{6}{x+3} = 0$

38. $\frac{x^2 - 20}{x^2 - 7x + 12} = \frac{3}{x-3} + \frac{5}{x-4}$

39. $\frac{x+2}{x^2 + 7x + 10} = \frac{1}{3x+6} - \frac{1}{x+5}$

40. $\frac{3}{2x-5} + \frac{2}{2x+3} = 0$

Review and Preview

Write each sentence as an equation and solve. See Section 2.2.

41. Four more than 3 times a number is 19.

42. The sum of two consecutive integers is 147.

Name ______________________________

43. ______________

43. The length of a rectangle is 5 inches more than the width. Its perimeter is 50 inches. Find the length and width.

44. The sum of a number and its reciprocal is $\frac{5}{2}$.

44. ______________

The following graph is from a survey of state and federal prisons. Use this histogram to answer Exercises 45–49.

45. ______________

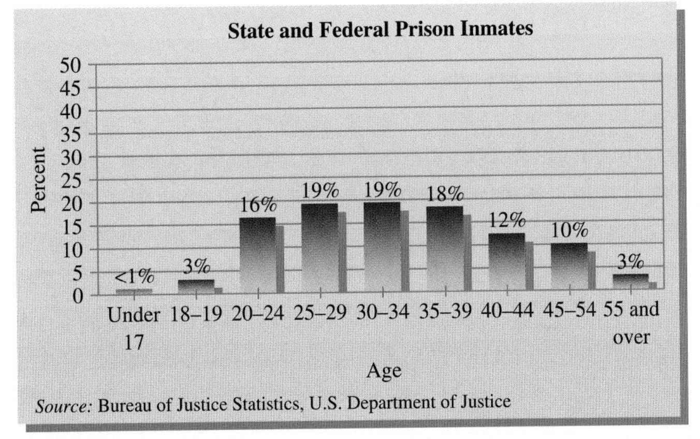

46. ______________

45. What percent of state and federal prison inmates are aged 45 to 54?

46. What percent of state and federal prison inmates are 55 years old or older?

47. ______________

47. What age category shows the highest percent of prison inmates?

48. What percent of state and federal prison inmates are 20 to 34 years old?

48. ______________

49. At the end of 1997, there were 29,265 inmates under the jurisdiction of state and federal correctional authorities in the state of Louisiana. Approximately how many 25- to 29-year-old inmates would you expect to have been held in Louisiana at the end of 1997? Round to the nearest whole. (*Source:* Bureau of Justice Statistics)

49. ______________

Combining Concepts

50. ______________

50. In your own words, explain the differences between equations and expressions.

51. In your own words, explain why it is necessary to check solutions to equations containing rational expressions.

51. ______________

52.

53.

54.

55.

56.

57.

Name ______________________

Solve each equation. Begin by writing each equation with positive exponents only.

52. $x^{-2} - 19x^{-1} + 48 = 0$

53. $x^{-2} - 5x^{-1} - 36 = 0$

Solve each equation. Round solutions to two decimal places.

54. $\frac{1.4}{x - 2.6} = \frac{-3.5}{x + 7.1}$

55. $\frac{10.6}{y} - 14.7 = \frac{9.92}{3.2} + 7.6$

56. The average cost of producing x game disks for a computer is given by the function $f(x) = 3.3 + \frac{5400}{x}$. Find the number of game disks that must be produced for the average cost to be \$5.10.

57. The average cost of producing x electric pencil sharpeners is given by the function $f(x) = 20 + \frac{4000}{x}$. Find the number of electric pencil sharpeners that must be produced for the average cost to be \$25.

Use a grapher to verify the solution of each given exercise.

58. Exercise 20

59. Exercise 21

Name ______________________ Section ________ Date ________

ANSWERS

1. ________

2. ________

3. ________

4. ________

5. ________

6. ________

7. ________

8. ________

9. ________

10. ________

Integrated Review—Expressions and Equations Containing Rational Expressions

It is very important that you understand the difference between an expression and an equation containing rational expressions. An equation contains an equal sign; an expression does not.

Expression to be Simplified

$$\frac{x}{2} + \frac{x}{6}$$

Write both rational expressions with the LCD, 6, as the denominator.

$$\frac{x}{2} + \frac{x}{6} = \frac{x \cdot 3}{2 \cdot 3} + \frac{x}{6}$$

$$= \frac{3x}{6} + \frac{x}{6}$$

$$= \frac{4x}{6} = \frac{2x}{3}$$

Equation to be Solved

$$\frac{x}{2} + \frac{x}{6} = \frac{2}{3}$$

Multiply both sides by the LCD, 6.

$$6\left(\frac{x}{2} + \frac{x}{6}\right) = 6\left(\frac{2}{3}\right)$$

$$3x + x = 4$$

$$4x = 4$$

$$x = 1$$

Check to see that the solution is 1.

HELPFUL HINT

Remember: Equations can be cleared of fractions, expressions cannot.

Perform each indicated operation and simplify, or solve the equation for the variable.

1. $\frac{x}{2} = \frac{1}{8} + \frac{x}{4}$

2. $\frac{x}{4} = \frac{3}{2} + \frac{x}{10}$

3. $\frac{1}{8} + \frac{x}{4}$

4. $\frac{3}{2} + \frac{x}{10}$

5. $\frac{4}{x + 2} - \frac{2}{x - 1}$

6. $\frac{5}{x - 2} - \frac{10}{x + 4}$

7. $\frac{4}{x + 2} = \frac{2}{x - 1}$

8. $\frac{5}{x - 2} = \frac{10}{x + 4}$

9. $\frac{2}{x^2 - 4} = \frac{1}{x + 2} - \frac{3}{x - 2}$

10. $\frac{3}{x^2 - 25} = \frac{1}{x + 5} + \frac{2}{x - 5}$

11. ______

12. ______

13. ______

14. ______

15. ______

16. ______

17. ______

18. ______

19. ______

20. ______

21. ______

22. ______

23. ______

24. ______

25. ______

26. ______

Name ______

11. $\dfrac{5}{x^2 - 3x} + \dfrac{4}{2x - 6}$

12. $\dfrac{5}{x^2 - 3x} \div \dfrac{4}{2x - 6}$

13. $\dfrac{x - 1}{x + 1} + \dfrac{x + 7}{x - 1} = \dfrac{4}{x^2 - 1}$

14. $\left(1 - \frac{y}{x}\right) \div \left(1 - \frac{x}{y}\right)$

15. $\dfrac{a^2 - 9}{a - 6} \cdot \dfrac{a^2 - 5a - 6}{a^2 - a - 6}$

16. $\dfrac{2}{a - 6} + \dfrac{3a}{a^2 - 5a - 6} - \dfrac{a}{5a + 5}$

17. $\dfrac{2x + 3}{3x - 2} = \dfrac{4x + 1}{6x + 1}$

18. $\dfrac{5x - 3}{2x} = \dfrac{10x + 3}{4x + 1}$

19. $\dfrac{a}{9a^2 - 1} + \dfrac{2}{6a - 2}$

20. $\dfrac{3}{4a - 8} - \dfrac{a + 2}{a^2 - 2a}$

21. $-\dfrac{3}{x^2} - \dfrac{1}{x} + 2 = 0$

22. $\dfrac{x}{2x + 6} + \dfrac{5}{x^2 - 9}$

23. $\dfrac{x - 8}{x^2 - x - 2} + \dfrac{2}{x - 2}$

24. $\dfrac{x - 8}{x^2 - x - 2} + \dfrac{2}{x - 2} = \dfrac{3}{x + 1}$

25. $\dfrac{3}{a} - 5 = \dfrac{7}{a} - 1$

26. $\dfrac{7}{3z - 9} + \dfrac{5}{z}$

6.5 RATIONAL EQUATIONS AND PROBLEM SOLVING

Objectives

A Solve an equation containing rational expressions for a specified variable.

B Solve problems by writing equations containing rational expressions.

SSM CD-ROM Video 6.5

A SOLVING RATIONAL EQUATIONS FOR A SPECIFIED VARIABLE

In Section 2.3 we solved equations for a specified variable. In this section, we continue practicing this skill by solving equations containing rational expressions for a specified variable. The steps given in Section 2.3 for solving equations for a specified variable are repeated here.

SOLVING EQUATIONS FOR A SPECIFIED VARIABLE

Step 1. Clear the equation of fractions or rational expressions by multiplying each side of the equation by the least common denominator (LCD) of all denominators in the equation.

Step 2. Use the distributive property to remove grouping symbols such as parentheses.

Step 3. Combine like terms on each side of the equation.

Step 4. Use the addition property of equality to rewrite the equation as an equivalent equation with terms containing the specified variable on one side and all other terms on the other side.

Step 5. Use the distributive property and the multiplication property of equality to get the specified variable alone.

Example 1 Solve: $\frac{1}{x}+\frac{1}{y}=\frac{1}{z}$ for x.

Solution: To clear this equation of fractions, we multiply both sides of the equation by xyz, the LCD of $\frac{1}{x}$, $\frac{1}{y}$, and $\frac{1}{z}$.

$$\frac{1}{x}+\frac{1}{y}=\frac{1}{z}$$

$$xyz\left(\frac{1}{x}+\frac{1}{y}\right)=xyz\left(\frac{1}{z}\right) \quad \text{Multiply both sides by } xyz.$$

$$xyz\left(\frac{1}{x}\right)+xyz\left(\frac{1}{y}\right)=xyz\left(\frac{1}{z}\right) \quad \text{Use the distributive property.}$$

$$yz+xz=xy \quad \text{Simplify.}$$

Notice the two terms that contain the specified variable x.

Next, we subtract xz from both sides so that all terms containing the specified variable x are on one side of the equation and all other terms are on the other side.

$$yz=xy-xz$$

Now we use the distributive property to factor x from $xy-xz$ and then the multiplication property of equality to solve for x.

$$yz=x(y-z)$$

$$\frac{yz}{y-z}=x \quad \text{or} \quad x=\frac{yz}{y-z} \quad \text{Divide both sides by } y-z.$$

Practice Problem 1

Solve: $\frac{1}{x}+\frac{1}{y}=\frac{1}{z}$ for y.

Answer

1. $y=\frac{xz}{x-z}$

B SOLVING PROBLEMS MODELED BY RATIONAL EQUATIONS

Problem solving sometimes involves modeling a described situation with an equation containing rational expressions. In Examples 2 through 5, we practice solving such problems and use the problem-solving steps first introduced in Section 2.2.

Practice Problem 2

Find the number that, when added to the numerator and subtracted from the denominator of $\frac{1}{20}$, results in a fraction equivalent to $\frac{2}{5}$.

Example 2 Finding an Unknown Number

If a certain number is subtracted from the numerator and added to the denominator of $\frac{9}{19}$, the new fraction is equivalent to $\frac{1}{3}$. Find the number.

Solution:

1. UNDERSTAND the problem. Read and reread the problem and try guessing the solution. For example, if the unknown number is 3, we have

$$\frac{9-3}{19+3} = \frac{1}{3}$$

To see if this is a true statement, we simplify the fraction on the left side.

$$\frac{6}{22} = \frac{1}{3} \quad \text{or} \quad \frac{3}{11} = \frac{1}{3} \qquad \text{False.}$$

Since this is not a true statement, 3 is not the correct number. Remember that the purpose of this step is not to guess the correct solution but to gain an understanding of the problem posed.

We will let $n =$ the number to be subtracted from the numerator and added to the denominator.

2. TRANSLATE the problem.

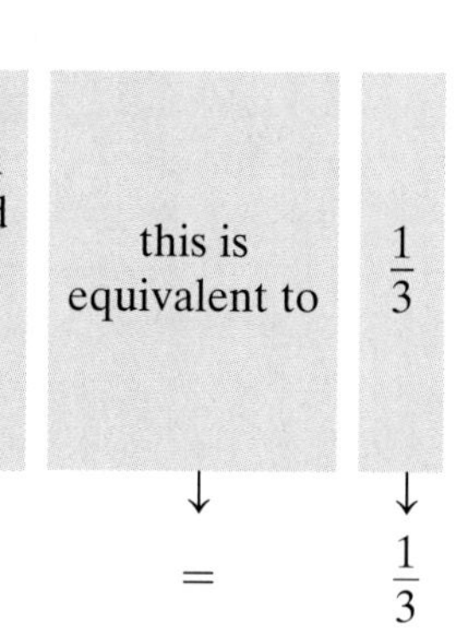

In words:	when the number is subtracted from the numerator and added to the denominator of the fraction $\frac{9}{19}$	this is equivalent to	$\frac{1}{3}$
	↓	↓	↓
Translate:	$\frac{9-n}{19+n}$	$=$	$\frac{1}{3}$

3. SOLVE the equation for n.

$$\frac{9-n}{19+n} = \frac{1}{3}$$

To solve for n, we begin by multiplying both sides by the LCD, $3(19+n)$.

$$3(19+n) \cdot \frac{9-n}{19+n} = 3(19+n) \cdot \frac{1}{3} \qquad \text{Multiply both sides by the LCD.}$$

$$3(9-n) = 19+n \qquad \text{Simplify.}$$

$$27 - 3n = 19 + n$$

$$8 = 4n$$

$$2 = n \qquad \text{Solve.}$$

Answer

2. 5

4. INTERPRET the results.

Check: If we subtract 2 from the numerator and add 2 to the denominator of $\frac{9}{19}$, we have $\frac{9-2}{19+2} = \frac{7}{21} = \frac{1}{3}$, and the problem checks.

State: The unknown number is 2.

Example 3 Finding the Distance of a Light Source

The intensity I of light, as measured in foot-candles, x feet from its source is given by the rational equation

$$I = \frac{320}{x^2}$$

How far away is the source if the intensity of light is 5 foot-candles?

Solution:

1. UNDERSTAND. Read and reread the problem, and guess a solution. Since an equation has been given that describes the relationship between I and x, we replace x with a few values to help us become familiar with the equation.

To find the intensity I of light 1 foot from the source, we let $x = 1$.

$$I = \frac{320}{1^2} = \frac{320}{1} = 320 \text{ foot-candles}$$

To find the intensity I of light 3 feet from the source, we let $x = 3$.

$$I = \frac{320}{3^2} = \frac{320}{9} = 35\frac{5}{9} \text{ foot-candles}$$

Notice that as x increases, I decreases. That is, as the number of feet from the light source increases, the intensity decreases, as expected.

2. TRANSLATE. We are given that the intensity I is 5 foot-candles, and we are asked to find how far away is the light source, x. To do so, we let $I = 5$.

$$I = \frac{320}{x^2}$$

$$5 = \frac{320}{x^2} \quad \text{Let } I = 5.$$

3. SOLVE the equation for x.

Practice Problem 3

Use the formula given in Example 3 to find how far away the light source is if the intensity of light is 20-foot candles.

Answer

3. 4 ft

$$5 = \frac{320}{x^2}$$

$$x^2 \cdot 5 = x^2 \cdot \frac{320}{x^2}$$ Multiply both sides by x^2.

$$5x^2 = 320$$ Simplify.

$$5x^2 - 320 = 0$$ Subtract 320.

$$5(x^2 - 64) = 0$$ Factor.

$$5(x + 8)(x - 8) = 0$$ Factor.

$$x = -8 \quad \text{or} \quad x = 8$$

4. INTERPRET. Since x represents distance and distance cannot be negative, the proposed solution -8 must be rejected. *Check* the solution 8 feet in the given formula. Then *state* the conclusion: The source of light is 8 feet away when the intensity is 5-foot candles.

The following work example leads to an equation containing rational expressions.

Practice Problem 4

Greg Guillot can paint a room alone in 3 hours. His brother Phillip can do the same job alone in 5 hours. How long would it take them to paint the room if they work together?

Example 4 Calculating Work Hours

Melissa Scarlatti can clean the house in 4 hours, whereas her husband, Zack, can do the same job in 5 hours. They have agreed to clean together so that they can finish in time to watch a movie on TV that starts in 2 hours. How long will it take them to clean the house together? Can they finish before the movie starts?

Solution:

1. UNDERSTAND. Read and reread the problem. The key idea here is the relationship between the *time* (in hours) it takes to complete the job and the *part of the job* completed in 1 unit of time (1 hour). For example, if the *time* it takes Melissa to complete the job is 4 hours, the *part of the job* she can complete in 1 hour is $\frac{1}{4}$. Similarly, Zack can complete $\frac{1}{5}$ of the job in 1 hour.

We will let $t =$ the *time* in hours it takes Melissa and Zack to clean the house together. Then $\frac{1}{t}$ represents the *part of the job* they complete in 1 hour. We summarize the given information on a chart.

	Hours to Complete the Job	Part of Job Completed in 1 Hour
Melissa Alone	4	$\frac{1}{4}$
Zack Alone	5	$\frac{1}{5}$
Together	t	$\frac{1}{t}$

2. TRANSLATE.

Answer

4. $1\frac{7}{8}$ hours

In words:

part of job Melissa can complete in 1 hour	added to	part of job Zack can complete in 1 hour	is equal to	part of job they can complete together in 1 hour
↓	↓	↓	↓	↓
$\frac{1}{4}$	$+$	$\frac{1}{5}$	$=$	$\frac{1}{t}$

Translate: $\frac{1}{4} + \frac{1}{5} = \frac{1}{t}$

3. SOLVE.

$$\frac{1}{4} + \frac{1}{5} = \frac{1}{t}$$

$$20t\left(\frac{1}{4} + \frac{1}{5}\right) = 20t\left(\frac{1}{t}\right)$$ Multiply both sides by the LCD, $20t$.

$$5t + 4t = 20$$

$$9t = 20$$

$$t = \frac{20}{9} \text{ or } 2\frac{2}{9}$$ Solve.

4. INTERPRET.

Check: The proposed solution is $2\frac{2}{9}$. That is, Melissa and Zack would take $2\frac{2}{9}$ hours to clean the house together. This proposed solution is reasonable since $2\frac{2}{9}$ hours is more than half of Melissa's time and less than half of Zack's time. Check this solution in the originally started problem.

State: Melissa and Zack can clean the house together in $2\frac{2}{9}$ hours. They cannot complete the job before the movie starts.

Example 5 Finding the Speed of a Current

Steve Deitmer takes $1\frac{1}{2}$ times as long to go 72 miles upstream in his boat as he does to return. If the boat cruises at 30 mph in still water, what is the speed of the current?

Solution:

1. UNDERSTAND. Read and reread the problem. Guess a solution. Suppose that the current is 4 mph. The speed of the boat upstream is slowed down by the current: $30 - 4$, or 26 mph, and the speed of the boat downstream is speeded up by the current: $30 + 4$, or 34 mph. Next let's find out how long it takes to travel 72 miles upstream and 72 miles downstream. To do so, we use the formula $d = r \cdot t$, or $\frac{d}{r} = t$.

Practice Problem 5

A fisherman traveling on the Pearl River takes $\frac{3}{2}$ times longer to travel 60 miles upstream in his boat than to return. If the boat's speed is 25 miles per hour in still water, find the speed of the current.

Answer

5. 5 mph

Upstream	**Downstream**
$\frac{d}{r} = t$	$\frac{d}{r} = t$
$\frac{72}{26} = t$	$\frac{72}{34} = t$
$2\frac{10}{13} = t$	$2\frac{2}{17} = t$

Since the time upstream $\left(2\frac{10}{13}\text{ hours}\right)$ is not $1\frac{1}{2}$ times the time downstream $\left(2\frac{2}{17}\text{ hours}\right)$, our guess is not correct. We do, however, have a better understanding of the problem.

We will let

$x =$ the speed of the current
$30 + x =$ the speed of the boat downstream
$30 - x =$ the speed of the boat upstream

This information is summarized in the following chart, where we use the formula $\frac{d}{r} = t$.

	Distance	Rate	Time $\left(\frac{d}{r}\right)$
Upstream	72	$30 - x$	$\frac{72}{30 - x}$
Downstream	72	$30 + x$	$\frac{72}{30 + x}$

2. TRANSLATE. Since the time spent traveling upstream is $1\frac{1}{2}$ times the time spent traveling downstream, we have

In words:	time upstream	is	$1\frac{1}{2}$	times	time downstream
	↓	↓	↓	↓	↓
Translate:	$\frac{72}{30 - x}$	$=$	$\frac{3}{2}$	$\cdot$	$\frac{72}{30 + x}$

3. SOLVE. $\frac{72}{30 - x} = \frac{3}{2} \cdot \frac{72}{30 + x}$

First we multiply both sides by the LCD, $2(30 + x)(30 - x)$.

$$2(30 + x)(30 - x) \cdot \frac{72}{30 - x} = 2(30 + x)(30 - x)\left(\frac{3}{2} \cdot \frac{72}{30 + x}\right)$$

$$72 \cdot 2(30 + x) = 3 \cdot 72 \cdot (30 - x) \quad \text{Simplify.}$$

$$2(30 + x) = 3(30 - x) \quad \text{Divide both sides by 72.}$$

$$60 + 2x = 90 - 3x \quad \text{Use the distributive property.}$$

$$5x = 30$$

$$x = 6 \quad \text{Solve.}$$

4. INTERPRET.

Check: Check the proposed solution of 6 mph in the originally stated problem.

State: The current's speed is 6 mph.

Name ______________ Section ______ Date ________

EXERCISE SET 6.5

A *Solve each equation for the specified variable. See Example 1.*

1. $F = \frac{9}{5}C + 32$ for C (Meteorology)

2. $V = \frac{1}{3}\pi r^2 h$ for h (Volume)

3. $Q = \frac{A - I}{L}$ for I (Finance)

4. $P = 1 - \frac{C}{S}$ for S (Finance)

5. $\frac{1}{R} = \frac{1}{R_1} + \frac{1}{R_2}$ for R (Electronics)

6. $\frac{1}{R} = \frac{1}{R_1} + \frac{1}{R_2}$ for R_1 (Electronics)

7. $S = \frac{n(a + L)}{2}$ for n (Sequences)

8. $S = \frac{n(a + L)}{2}$ for a (Sequences)

9. $A = \frac{h(a + b)}{2}$ for b (Geometry)

10. $A = \frac{h(a + b)}{2}$ for h (Geometry)

11. $\frac{P_1 V_1}{T_1} = \frac{P_2 V_2}{T_2}$ for T_2 (Chemistry)

12. $H = \frac{kA(T_1 - T_2)}{L}$ for T_2 (Physics)

13. $f = \frac{f_1 f_2}{f_1 + f_2}$ for f_2

14. $I = \frac{E}{R + r}$ for r (Electronics)

15. $\lambda = \frac{2L}{n}$ for L

16. $S = \frac{a_1 - a_n r}{1 - r}$ for a_1 (Sequences)

17. $\frac{\theta}{\omega} = \frac{2L}{c}$ for C

18. $F = \frac{-GMm}{r^2}$ for M (Physics)

B *Solve. See Example 2.*

19. The sum of a number and 5 times its reciprocal is 6. Find the number(s).

20. The quotient of a number and 9 times its reciprocal is 1. Find the number(s).

ANSWERS

1. ______
2. ______
3. ______
4. ______
5. ______
6. ______
7. ______
8. ______
9. ______
10. ______
11. ______
12. ______
13. ______
14. ______
15. ______
16. ______
17. ______
18. ______
19. ______
20. ______

21. ______

22. ______

23. ______

24. ______

25. ______

26. ______

27. ______

28. ______

29. ______

Name ______

21. If a number is added to the numerator of $\frac{12}{41}$ and twice the number is added to the denominator of $\frac{12}{41}$, the resulting fraction is equivalent to $\frac{1}{3}$. Find the number.

22. If a number is subtracted from the numerator of $\frac{13}{8}$ and added to the denominator of $\frac{13}{8}$, the resulting fraction is equivalent to $\frac{2}{5}$. Find the number.

In electronics, the relationship among the resistances R_1 and R_2 of two resistors wired in a parallel circuit and their combined resistance R is described by the formula $\frac{1}{R} = \frac{1}{R_1} + \frac{1}{R_2}$. Use this formula to solve Exercises 23 through 25. See Example 3.

23. If the combined resistance is 2 ohms and one of the two resistances is 3 ohms, find the other resistance.

24. Find the combined resistance of two resistors of 12 ohms each when they are wired in a parallel circuit.

25. The relationship among resistance of two resistors wired in a parallel circuit and their combined resistance may be extended to three resistors of resistances R_1, R_2 and R_3. Write an equation you believe may describe the relationship, and use it to find the combined resistance if R_1 is 5, R_2 is 6 and R_3 is 2.

Solve. See Example 4.

26. Alan Cantrell can word process a research paper in 6 hours. With Steve Isaac's help, the paper can be processed in 4 hours. Find how long it takes Steve to word process the paper alone.

27. An experienced roofer can roof a house in 26 hours. A beginning roofer needs 39 hours to complete the same job. Find how long it takes for the two to do the job together.

28. A new printing press can print newspapers twice as fast as the old one can. The old one can print the afternoon edition in 4 hours. Find how long it takes to print the afternoon edition if both printers are operating.

29. Three postal workers can sort a stack of mail in 20 minutes, 30 minutes, and 60 minutes, respectively. Find how long it takes to sort the mail if all three work together.

Name ______________________________

Solve. See Example 5

30. An F-100 plane and a Toyota Truck leave the same town at sunrise and head for a town 450 miles away. The speed of the plane is three times the speed of the truck, and the plane arrives 6 hours ahead of the truck. Find the speed of the truck.

31. Mattie Evans drove 150 miles in the same amount of time that it took a turbopropeller plane to travel 600 miles. The speed of the plane was 150 mph faster than the speed of the car. Find the speed of the plane.

32. The speed of a boat in still water is 24 mph. If the boat travels 54 miles upstream in the same time that it takes to travel 90 miles downstream, find the speed of the current.

33. The speed of Lazy River's current is 5 mph. If a boat travels 20 miles downstream in the same time that it takes to travel 10 miles upstream, find the speed of the boat in still water.

Solve.

34. The sum of the reciprocals of two consecutive odd integers is $\frac{20}{99}$. Find the two integers.

35. The sum of the reciprocals of two consecutive integers is $-\frac{15}{56}$. Find the two integers.

36. If Sarah Clark can do a job in 5 hours and Dick Belli and Sarah working together can do the same job in 2 hours, find how long it takes Dick to do the job alone.

37. One hose can fill a goldfish pond in 45 minutes, and two hoses can fill the same pond in 20 minutes. Find how long it takes the second hose alone to fill the pond.

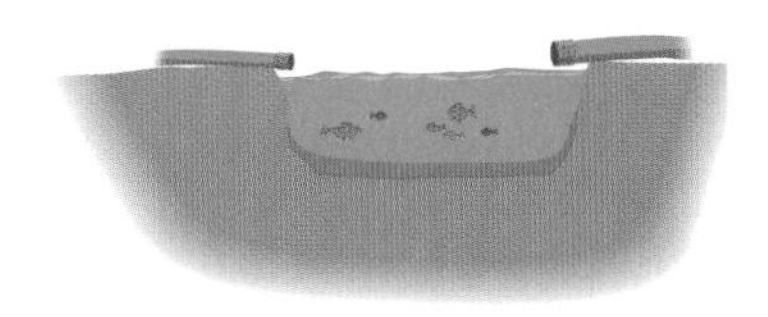

38. The speed of a bicyclist is 10 mph faster than the speed of a walker. If the bicyclist travels 26 miles in the same amount of time that the walker travels 6 miles, find the speed of the bicyclist.

39. Two trains going in opposite directions leave at the same time. One train travels 15 mph faster than the other. In 6 hours the trains are 630 miles apart. Find the speed of each.

30. ____________

31. ____________

32. ____________

33. ____________

34. ____________

35. ____________

36. ____________

37. ____________

38. ____________

39. ____________

40. ______

41. ______

42. ______

43. ______

44. ______

45. ______

46. ______

47. ______

48. ______

49. ______

Name ______

40. The numerator of a fraction is 4 less than the denominator. If both the numerator and the denominator are increased by 2, the resulting fraction is equivalent to $\frac{2}{3}$. Find the fraction.

41. Fabio Casartelli of Italy won the individual road race in cycling during the 1992 Summer Olympics. An amateur cyclist training for a road race rode the first 20-mile portion of his workout at a constant rate. For the 16-mile cooldown portion of his workout, he reduced his speed by 2 miles per hour. Each portion of the workout took equal time. Find the cyclist's rate during the first portion and his rate during the cooldown portion.

42. The denominator of a fraction is 1 more than the numerator. If both the numerator and the denominator are decreased by 3, the resulting fraction is equivalent to $\frac{4}{5}$. Find the fraction.

43. Moo Dairy has three machines to fill half-gallon milk cartons. The machines can fill the daily quota in 5 hours, 6 hours, and 7.5 hours, respectively. Find how long it takes to fill the daily quota if all three machines are running.

44. The inlet pipe of an oil tank can fill the tank in 1 hour 30 minutes. The outlet pipe can empty the tank in 1 hour. Find how long it takes to empty a full tank if both pipes are open.

45. A plane flies 465 miles with the wind and 345 miles against the wind in the same length of time. If the speed of the wind is 20 mph, find the speed of the plane in still air.

46. Two rockets are launched. The first travels at 9000 mph. Fifteen minutes later the second is launched at 10,000 mph. Find the distance at which both rockets are an equal distance from Earth.

47. Two joggers, one averaging 8 mph and one averaging 6 mph, start from a designated initial point. The slower jogger arrives at the end of the run a half-hour after the other jogger. Find the distance of the run.

48. A semi truck travels 300 miles through the flatland in the same amount of time that it travels 180 miles through the Great Smoky Mountains. The rate of the truck is 20 miles per hour slower in the mountains than in the flatland. Find both the flatland rate and mountain rate.

49. Smith Engineering is in the process of reviewing the salaries of their surveyors. During this review, the company found that an experienced surveyor surveys a roadbed in 4 hours. An apprentice surveyor needs 5 hours to survey the same stretch of road. If the two work together, find how long it takes them to complete the job.

Name ______________________________

50. ______

51. ______

52. ______

53. ______

54. ______

55. ______

56. ______

57. ______

58. ______

59. ______

60. ______

50. An experienced bricklayer constructs a small wall in 3 hours. An apprentice completes the job in 6 hours. Find how long it takes if they work together.

51. A marketing manager travels 1080 miles in a corporate jet and then an additional 240 miles by car. If the car ride takes 1 hour longer, and if the rate of the jet is 6 times the rate of the car, find the time the manager travels by jet and find the time she travels by car.

52. Gary Marcus and Tony Alva work at Lombardo's Pipe and Concrete. Mr. Lombardo is preparing an estimate for a customer. He knows that Gary lays a slab of concrete in 6 hours. Tony lays the same size slab in 4 hours. If both work on the job and the cost of labor is $45.00 per hour, determine what the labor estimate should be.

53. In 2 minutes, a conveyor belt moves 300 pounds of recyclable aluminum from the delivery truck to a storage area. A smaller belt moves the same quantity of cans the same distance in 6 minutes. If both belts are used, find how long it takes to move the cans to the storage area.

54. Mr. Dodson can paint his house by himself in four days. His son needs an additional day to complete the job if he works by himself. If they work together, find how long it takes to paint the house.

55. While road testing a new make of car, the editor of a consumer magazine finds that she can go 10 miles into a 3-mile-per-hour wind in the same amount of time that she can go 11 miles with a 3-mile-per-hour wind behind her. Find the speed of the car in still air.

56. The world record for the largest white bass caught is held by Ronald Sprouse of Virginia. The bass weighed 6 pounds 13 ounces. If Ronald rows to his favorite fishing spot 9 miles downstream in the same amount of time that he rows 3 miles upstream and if the current is 6 miles per hour, find how long it takes him to cover the 12 miles.

Review and Preview

Solve each equation for x. See Section 2.1.

57. $\frac{x}{5} = \frac{x + 2}{3}$

58. $\frac{x}{4} = \frac{x + 3}{6}$

59. $\frac{x - 3}{2} = \frac{x - 5}{6}$

60. $\frac{x - 6}{4} = \frac{x - 2}{5}$

61. ______

62. ______

63. ______

64. ______

65. ______

Name ______

Combining Concepts

Calculating body-mass index is a way to gauge whether a person should lose weight. Doctors recommend that body-mass index values fall between 19 and 25. The formula for body-mass index B is $B = \frac{705w}{h^2}$*, where w is weight in pounds and h is height in inches. Use this formula to answer Exercises 61 and 62.*

61. A patient is 5 ft 8 in. tall. What should his or her weight be to have a body-mass index of 25? Round to the nearest whole pound.

62. A doctor recorded a body-mass index of 47 on a patient's chart. Later, a nurse notices that the doctor recorded the patient's weight as 240 pounds but neglected to record the patient's height. Explain how the nurse can use the information from the chart to find the patient's height. Then find the height.

In physics, when the source of a sound is traveling toward an observer, the relationship between the actual pitch a of the sound and the pitch h that the observer hears due to the Doppler effect is described by the formula $h = \frac{a}{1 - \frac{s}{770}}$*, where s is the speed of the sound source in miles per hour. Use this formula to answer Exercise 63.*

63. An emergency vehicle has a single-tone siren with the pitch of the musical note E. As it approaches an observer standing by the road, the vehicle is traveling 50 miles per hour. Is the pitch that the observer hears due to the Doppler effect lower or higher than the actual pitch? To which musical note is the pitch that the observer hears closest?

Pitch of an Octave of Musical Notes in Hertz (Hz)

Note	Pitch
Middle C	261.63
D	293.66
E	329.63
F	349.23
G	392.00
A	440.00
B	493.88

Note: Greater numbers indicate higher pitches (acoustically).
(*Source:* American Standards Association)

Internet Excursions

http://www.prenhall.com/martin-gay
Body-mass index (BMI) determines a person's risk for weight-related health problems. One way to calculate BMI is to use the formula given with Exercises 61 and 62. Another way is to use one of the many sites on the World Wide Web that offer an interactive BMI calculator. The World Wide Web address listed above will direct you to a website with an Interactive BMI calculator. You can calculate BMI by entering height in feet and inches and weight in pounds.

64. Use the interactive BMI calculator to find the BMI for a person who is 5 ft 5 in. and weighs 145 pounds. Then verify your result using the formula given with Exercises 61 and 62.

65. Use the interactive BMI calculator to find your own BMI.

6.6 VARIATION AND PROBLEM SOLVING

Objectives

A Solve problems involving direct variation.

B Solve problems involving inverse variation.

C Solve problems involving joint variation.

D Solve problems involving combined variation.

SSM CD-ROM Video 6.6

A SOLVING PROBLEMS INVOLVING DIRECT VARIATION

A very familiar example of direct variation is the relationship of the circumference C of a circle to its radius r. The formula $C = 2\pi r$ expresses that the circumference is always 2π times the radius. In other words, C is always a constant multiple (2π) of r. Because it is, we say that ***C* varies directly as *r***, that ***C* varies directly with *r***, or that ***C* is directly proportional to *r***.

DIRECT VARIATION

y* varies directly as *x, or ***y* is directly proportional to *x***, if there is a nonzero constant k such that

$$y = kx$$

The number k is called the **constant of variation** or the **constant of proportionality**.

In the above definition, the relationship described between x and y is a linear one. In other words, the graph of $y = kx$ is a line. The slope of the line is k, and the line passes through the origin.

For example, the graph of the direct variation equation $C = 2\pi r$ is shown. The horizontal axis represents the radius r, and the vertical axis is the circumference C. From the graph we can read that when the radius is 6 units, the circumference is approximately 38 units. Also, when the circumference is 45 units, the radius is between 7 and 8 units. Notice that as the radius increases, the circumference increases.

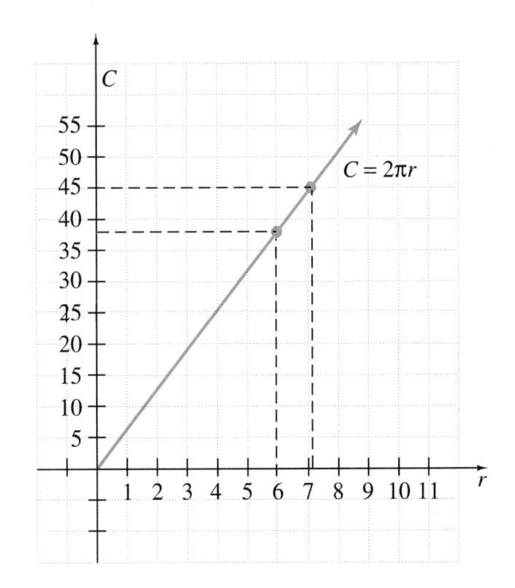

Example 1 Suppose that y varies directly as x. If y is 5 when x is 30, find the constant of variation and the direct variation equation.

Solution: Since y varies directly as x, we write $y = kx$. If $y = 5$ when $x = 30$, we have that

$y = kx$

$5 = k(30)$ Replace y with 5 and x with 30.

$\frac{1}{6} = k$ Solve for k.

Practice Problem 1

Suppose that y varies directly as x. If y is 24 when x is 8, find the constant of variation and the direct variation equation.

Answer

1. $k = 3$; $y = 3x$

The constant of variation is $\frac{1}{6}$.

After finding the constant of variation k, the direct variation equation can be written as $y = \frac{1}{6}x$.

Practice Problem 2

Use Hooke's law as stated in Example 2. If a 56-pound weight attached to a spring stretches the spring 8 inches, find the distance that an 85-pound weight attached to the spring stretches the spring.

Example 2 Using Direct Variation and Hooke's Law

Hooke's law states that the distance a spring stretches is directly proportional to the weight attached to the spring. If a 40-pound weight attached to the spring stretches the spring 5 inches, find the distance that a 65-pound weight attached to the spring stretches the spring.

Solution:

1. UNDERSTAND. Read and reread the problem. Notice that we are given that the distance a spring stretches is **directly proportional** to the weight attached. We let

$d =$ the distance stretched

$w =$ the weight attached

The constant of variation is represented by k.

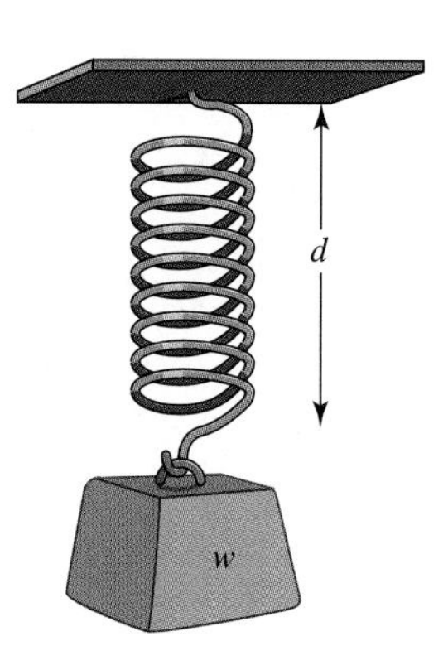

2. TRANSLATE. Because d is directly proportional to w, we write

$d = kw$

3. SOLVE. When a weight of 40 pounds is attached, the spring stretches 5 inches. That is, when $w = 40$, $d = 5$.

$5 = k(40)$ Replace d with 5 and w with 40.

$\frac{1}{8} = k$ Solve for k.

Now when we replace k with $\frac{1}{8}$ in the equation $d = kw$, we have

$d = \frac{1}{8}w$

To find the stretch when a weight of 65 pounds is attached, we replace w with 65 to find d.

Answer

2. $12\frac{1}{7}$ inches

$$d = \frac{1}{8}(65)$$

$$= \frac{65}{8} = 8\frac{1}{8} \quad \text{or} \quad 8.125$$

4. INTERPRET.

Check: Check the proposed solution of 8.125 inches in the original problem.

State: The spring stetches 8.125 inches when a 65-pound weight is attached.

B SOLVING PROBLEMS INVOLVING INVERSE VARIATION

When y is proportional to the **reciprocal** of another variable x, we say that **y varies inversely as x**, or that **y is inversely proportional to x**. An example of the inverse variation relationship is the relationship between the pressure that a gas exerts and the volume of its container. As the volume of a container decreases, the pressure of the gas it contains increases.

> INVERSE VARIATION
>
> **y varies inversely as x,** or **y is inversely proportional to x,** if there is a nonzero constant k such that
>
> $$y = \frac{k}{x}$$
>
> The number k is called the **constant of variation** or the **constant of proportionality**.

Notice that $y = \frac{k}{x}$ is a rational equation. Its graph for $k > 0$ and $x > 0$ is shown. From the graph, we can see that as x increases, y decreases.

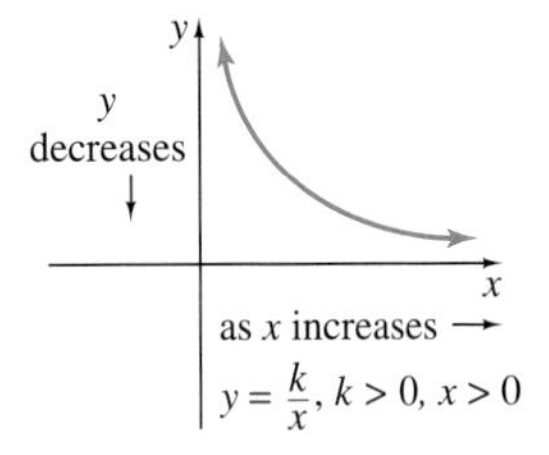

Example 3 Suppose that u varies inversely as w. If u is 3 when w is 5, find the constant of variation and the inverse variation equation.

Solution: Since u varies inversely as w, we have $u = \frac{k}{w}$. We let $u = 3$ and $w = 5$, and we solve for k.

$$u = \frac{k}{w}$$

$$3 = \frac{k}{5} \quad \text{Let } u = 3 \text{ and } w = 5.$$

$$15 = k \quad \text{Multiply both sides by 5.}$$

Practice Problem 3

Suppose that y varies inversely as x. If y is 6 when x is 3, find the constant of variation and the inverse variation equation.

Answer

3. $k = 18$; $y = \frac{18}{x}$

The constant of variation k is 15. This gives the inverse variation equation

$$u = \frac{15}{w}$$

Practice Problem 4

The speed r to drive a constant distance is inversely proportional to the time t. A fixed distance can be driven in 5 hours at a rate of 24 miles per hour. Find the rate needed to drive the same distance in 4 hours.

Example 4 Using Inverse Variation and Boyle's Law

Boyle's law says that if the temperature stays the same, the pressure P of a gas is inversely proportional to the volume V. If a cylinder in a steam engine has a pressure of 960 kilopascals when the volume is 1.4 cubic meters, find the pressure when the volume increases to 2.5 cubic meters.

Solution:

1. UNDERSTAND. Read and reread the problem. Notice that we are given that the pressure of a gas is *inversely proportional* to the volume. We will let $P =$ the pressure and $V =$ the volume. The constant of variation is represented by k.
2. TRANSLATE. Because P is inversely proportional to V, we write

$$P = \frac{k}{V}$$

When $P = 960$ kilopascals, the volume $V = 1.4$ cubic meters. We use this information to find k.

$$960 = \frac{k}{1.4} \quad \text{Let } P = 960 \text{ and } V = 1.4.$$

$$1344 = k \quad \text{Multiply both sides by 1.4.}$$

Thus, the value of k is 1344. Replacing k with 1344 in the variation equation, we have

$$P = \frac{1344}{V}$$

Next we find P when V is 2.5 cubic meters.

3. SOLVE:

$$P = \frac{1344}{2.5} \quad \text{Let } V = 2.5.$$

$$= 537.6$$

4. INTERPRET. *Check* the proposed solution in the original problem.

State: When the volume is 2.5 cubic meters, the pressure is 537.6 kilopascals.

Answer

4. 30 mph

C SOLVING PROBLEMS INVOLVING JOINT VARIATION

Sometimes the ratio of a variable to the product of many other variables is constant. For example, the ratio of distance traveled to the product of speed and time traveled is constantly 1:

$$\frac{d}{rt} = 1 \quad \text{or} \quad d = rt$$

Such a relationship is called **joint variation.**

> **JOINT VARIATION**
>
> If the ratio of a variable y to the product of two or more variables is constant, then **y varies jointly as**, or **is jointly proportional to**, the other variables. If
>
> $$y = kxz$$
>
> then the number k is the **constant of variation** or the **constant of proportionality**.

TRY THE CONCEPT CHECK IN THE MARGIN.

✓ CONCEPT CHECK

Which type of variation is represented by the equation $xy = 8$? Why?

a. direct variation
b. inverse variation
c. joint variation

Example 5 The surface area of a cylinder varies jointly as its radius and height. Express surface area S in terms of radius r and height h.

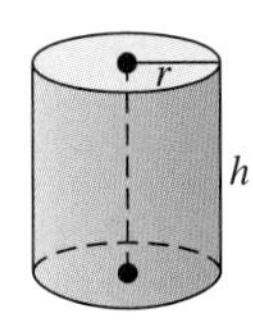

Solution: Because the surface area varies jointly as the radius r and the height h, we equate S to a constant multiple of r and h:

$$S = krh$$

In the equation, $S = krh$, it can be determined that the constant k is 2π, and we then have the formula $S = 2\pi rh$.

Practice Problem 5

The area of a triangle varies jointly as its base and height. Express the area in terms of base b and height h.

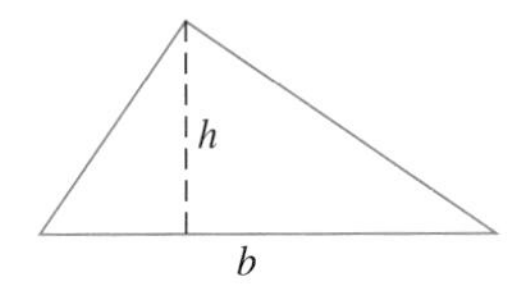

D SOLVING PROBLEMS INVOLVING COMBINED VARIATION

Some examples of variation involve combinations of direct, inverse, and joint variation. We will call these variations **combined variation.**

Example 6 The maximum weight that a circular column can support is directly proportional to the fourth power of its diameter and inversely proportional to the square of its height. A 2-meter column that is 8 meters in height can support 1 ton. Find the weight that a 1-meter column that is 4 meters in height can support.

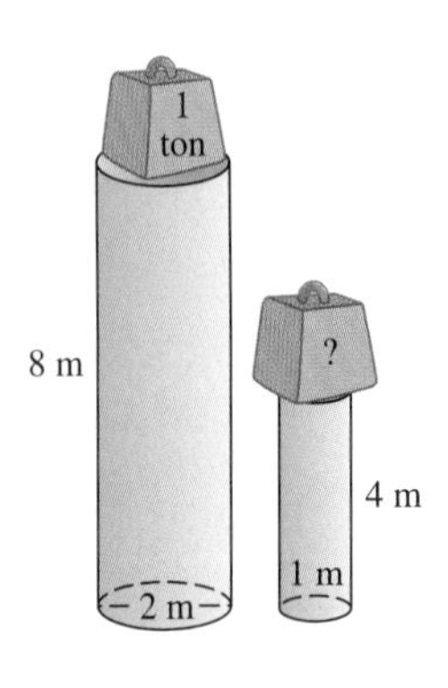

Practice Problem 6

The maximum weight that a rectangular beam can support varies jointly as its width and the square of its height and inversely as its length. If a beam $\frac{1}{3}$ foot wide, 1 foot high, and 10 feet long can support 3 tons, find how much weight a similar beam can support if it is 1 foot wide, $\frac{1}{3}$ foot high, and 9 feet long.

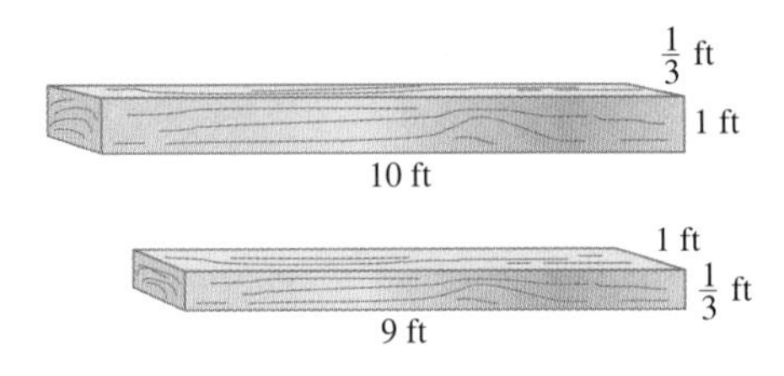

Answers

5. $A = kbh$, **6.** $1\frac{1}{9}$ tons

✓ **Concept Check:** b

Solution:

1. UNDERSTAND. Read and reread the problem. Let $w =$ weight, $d =$ diameter, $h =$ height, and $k =$ the constant of variation.
2. TRANSLATE. Since w is directly proportional to d^4 and inversely proportional to h^2, we have:

$$w = \frac{kd^4}{h^2}$$

3. SOLVE. To find k, we are given that a 2-meter column that is 8 meters in height can support 1 ton. That is, $w = 1$ when $d = 2$ and $h = 8$, or

$$1 = \frac{k \cdot 2^4}{8^2} \quad \text{Let } w = 1, d = 2, \text{ and } h = 8.$$

$$1 = \frac{k \cdot 16}{64}$$

$$4 = k \quad \text{Solve for } k.$$

Now replace k with 4 in the equation $w = \frac{kd^4}{h^2}$ and we have

$$w = \frac{4d^4}{h^2}$$

To find weight, w, for a 1-meter column that is 4 meters in height, let $d = 1$ and $h = 4$.

$$w = \frac{4 \cdot 1^4}{4^2}$$

$$w = \frac{4}{16} = \frac{1}{4}$$

4. INTERPRET. *Check* the proposed solution in the original problem. *State:* The 1-meter column that is 4 meters in height can hold $\frac{1}{4}$ ton of weight.

Name ______________________ Section ______ Date ________

ANSWERS

1. ______
2. ______
3. ______
4. ______
5. ______
6. ______
7. ______
8. ______
9. ______
10. ______
11. ______
12. ______
13. ______
14. ______

EXERCISE SET 6.6

A *If y varies directly as x, find the constant of variation and the direct variation equation for each situation. See Example 1.*

1. $y = 4$ when $x = 20$

2. $y = 5$ when $x = 30$

3. $y = 6$ when $x = 4$

4. $y = 12$ when $x = 8$

5. $y = 7$ when $x = \frac{1}{2}$

6. $y = 11$ when $x = \frac{1}{3}$

7. $y = 0.2$ when $x = 0.8$

8. $y = 0.4$ when $x = 2.5$

Solve. See Example 2.

9. The weight of a synthetic ball varies directly with the cube of its radius. A ball with a radius of 2 inches weighs 1.20 pounds. Find the weight of a ball of the same material with a 3-inch radius.

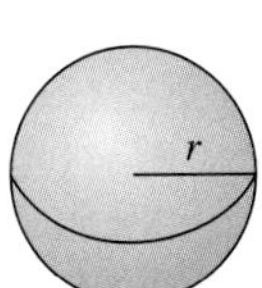

10. At sea, the distance to the horizon is directly proportional to the square root of the elevation of the observer. If a person who is 36 feet above the water can see 7.4 miles, find how far a person 64 feet above the water can see.

11. The amount P of pollution varies directly with the population N of people. Kansas City has a population of 450,000 and produces 260,000 tons of pollutants. Find how many tons of pollution we should expect St. Louis to produce, if we know that its population is 980,000.

12. Charles's law states that if the pressure P stays the same, the volume V of a gas is directly proportional to its temperature T. If a balloon is filled with 20 cubic meters of a gas at a temperature of 300 K, find the new volume if the temperature rises to 360 K while the pressure stays the same.

B *If y varies inversely as x, find the constant of variation and the inverse variation equation for each situation. See Example 3.*

13. $y = 6$ when $x = 5$

14. $y = 20$ when $x = 9$

Name ______________________________

15. ______

16. ______

17. ______

18. ______

19. ______

20. ______

21. ______

22. ______

23. ______

24. ______

15. $y = 100$ when $x = 7$

16. $y = 63$ when $x = 3$

17. $y = \frac{1}{8}$ when $x = 16$

18. $y = \frac{1}{10}$ when $x = 40$

19. $y = 0.2$ when $x = 0.7$

20. $y = 0.6$ when $x = 0.3$

Solve. See Example 4.

21. Pairs of markings a set distance apart are made on highways so that police can detect drivers exceeding the speed limit. Over a fixed distance, the speed R varies inversely with the time T. In one particular pair of markings, R is 45 mph when T is 6 seconds. Find the speed of a car that travels the given distance in 5 seconds.

22. The weight of an object on or above the surface of Earth varies inversely as the square of the distance between the object and Earth's center. If a person weighs 160 pounds on Earth's surface, find the individual's weight if he moves 200 miles above Earth. (Assume that Earth's radius is 4000 miles.)

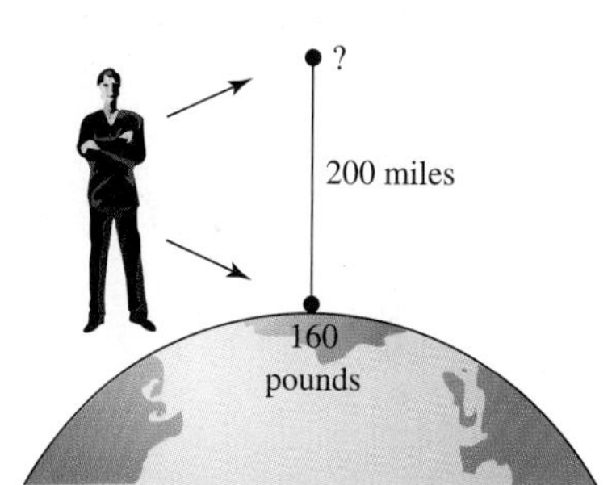

23. If the voltage V in an electric circuit is held constant, the current I is inversely proportional to the resistance R. If the current is 40 amperes when the resistance is 270 ohms, find the current when the resistance is 150 ohms.

24. Because it is more efficient to produce larger numbers of items, the cost of producing Dysan computer disks is inversely proportional to the number produced. If 4000 can be produced at a cost of \$1.20 each, find the cost per disk when 6000 are produced.

Name ______________________________

25. The intensity I of light varies inversely as the square of the distance d from the light source. If the distance from the light source is doubled (see the figure), determine what happens to the intensity of light at the new location.

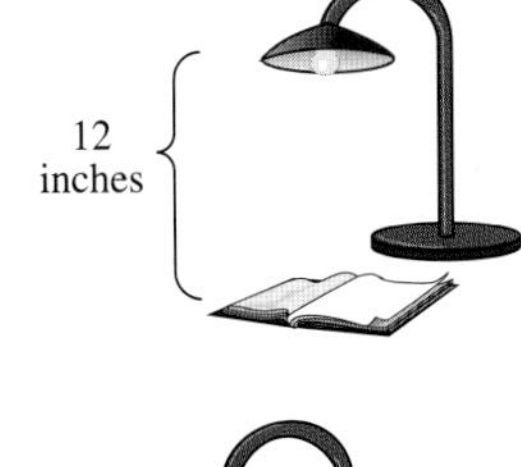

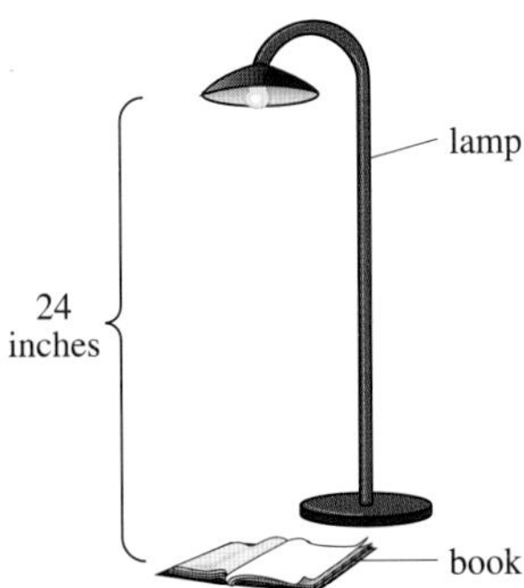

26. The maximum weight that a circular column can hold is inversely proportional to the square of its height. If an 8-foot column can hold 2 tons, find how much weight a 10-foot column can hold.

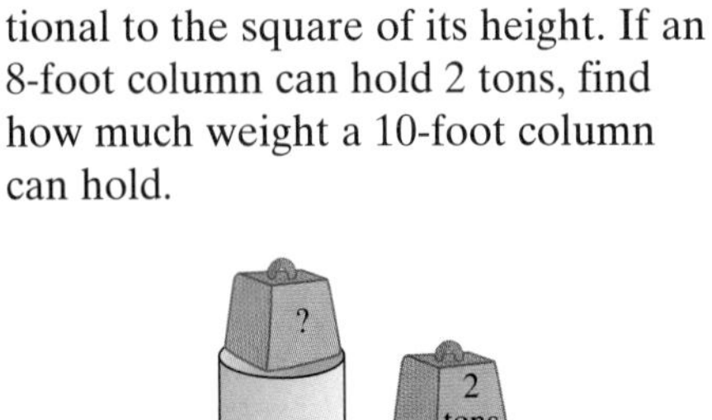

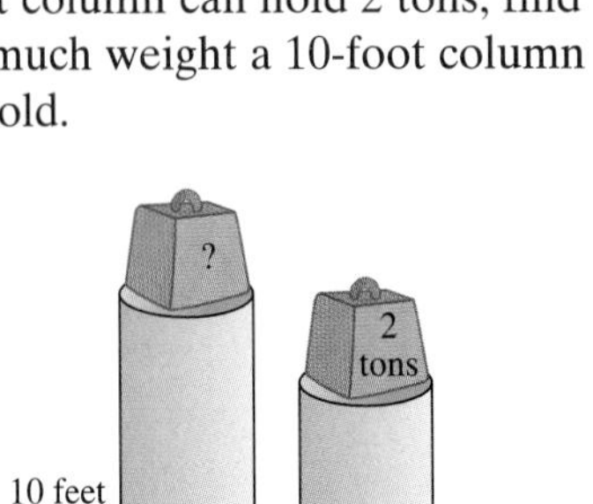

C D *Write each statement as an equation. See Examples 5 and 6.*

27. x varies jointly as y and z.

28. P varies jointly as R and the square of S.

29. r varies jointly as s and the cube of t.

30. a varies jointly as b and c.

Solve. See Examples 5 and 6.

31. The maximum weight that a rectangular beam can support varies jointly as its width and the square of its height and inversely as its length. If a beam $\frac{1}{2}$ foot wide, $\frac{1}{3}$ foot high, and 10 feet long can support 12 tons, find how much a similar beam can support if the beam is $\frac{2}{3}$ foot wide, $\frac{1}{2}$ foot high, and 16 feet long.

32. The number of cars manufactured on an assembly line at a General Motors plant varies jointly as the number of workers and the time they work. If 200 workers can produce 60 cars in 2 hours, find how many cars 240 workers should be able to make in 3 hours.

25. ______

26. ______

27. ______

28. ______

29. ______

30. ______

31. ______

32. ______

33. ______

34. ______

35. ______

36. ______

37. ______

38. ______

39. ______

Name ______

33. The volume of a cone varies jointly as the square of its radius and its height. If the volume of a cone is 32π cubic inches when the radius is 4 inches and the height is 6 inches, find the volume of a cone when the radius is 3 inches and the height is 5 inches.

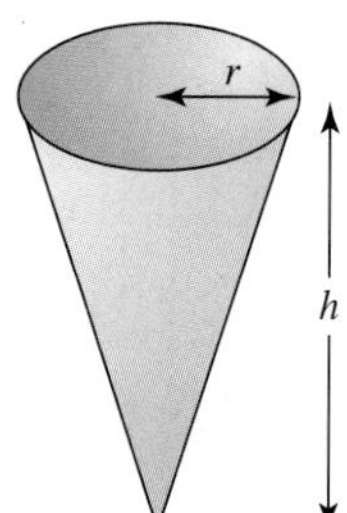

34. When a wind blows perpendicularly against a flat surface, its force is jointly proportional to the surface area and the speed of the wind. A sail whose surface area is 12 square feet experiences a 20-pound force when the wind speed is 10 miles per hour. Find the force on an 8-square-foot sail if the wind speed is 12 miles per hour.

35. The horsepower that can be safely transmitted to a shaft varies jointly as the shaft's angular speed of rotation (in revolutions per minute) and the cube of its diameter. A 2-inch shaft making 120 revolutions per minute safely transmits 40 horsepower. Find how much horsepower can be safely transmitted by a 3-inch shaft making 80 revolutions per minute.

Review and Preview

Find the exact circumference and area of each circle. See the inside cover for a list of geometric formulas.

36.

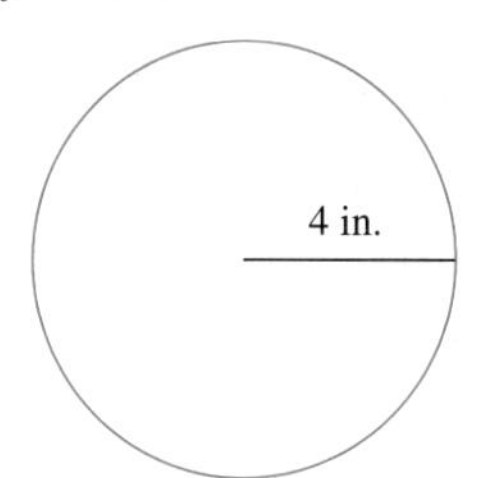

37.

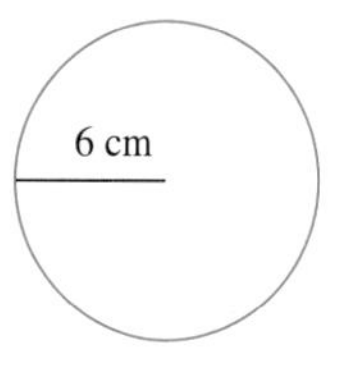

38.

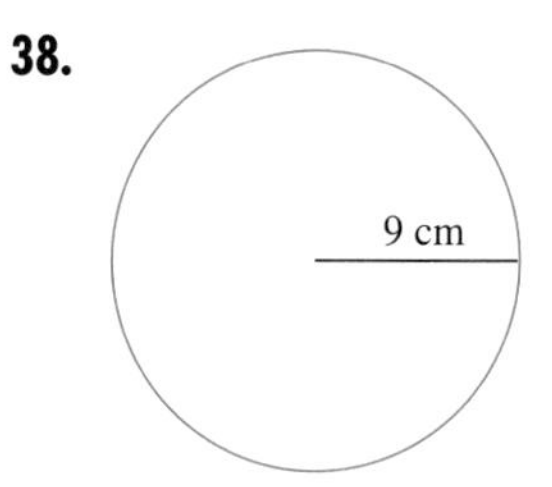

39.

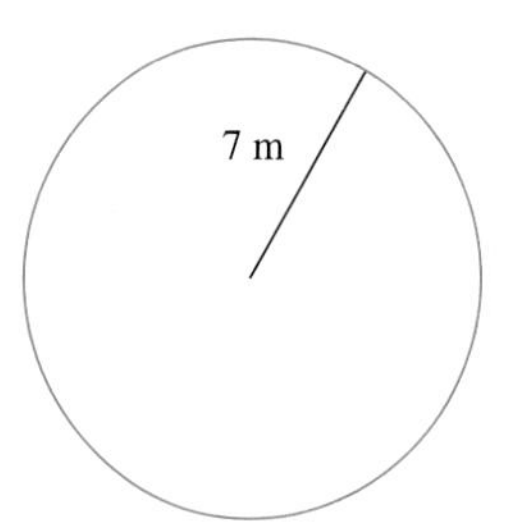

Name

Find the slope of the line containing each pair of points. See Section 3.2.

40. $(-5, -2), (0, 7)$

41. $(3, 6), (-2, 6)$

42. $(2, 1), (2, -3)$

43. $(4, -1), (5, -2)$

Combining Concepts

44. The horsepower to drive a boat varies directly as the cube of the speed of the boat. If the speed of the boat is to double, determine the corresponding increase in horsepower required.

45. The volume of a cylinder varies jointly as the height and the square of the radius. If the height is halved and the radius is doubled, determine what happens to the volume.

46. Suppose that y varies directly as x. If x is doubled, what is the effect on y?

47. Suppose that y varies directly as x^2. If x is doubled, what is the effect on y?

40. ____________

41. ____________

42. ____________

43. ____________

44. ____________

45. ____________

46. ____________

47. ____________

Name ________________________ Section ________ Date ________

Chapter 6 Activity

Estimating Population Sizes

Materials:

- ▲ dried beans
- ▲ felt-tip marker
- ▲ bowl or bag
- ▲ measuring cup or small paper cup

This activity may be completed by working in groups or individually.

In wildlife management, conservationists sometimes need to know the size of a certain wildlife population—the number of a certain species of fish in a lake or the number of certain birds in a geographic region. In manufacturing, it might be necessary to estimate the number of parts in a large storage bin. In either case, actually counting the number of animals or parts would be either too difficult or too time consuming.

When it is necessary to know the approximate size of a population, it is sometimes useful to use sampling techniques to estimate the population size. There are several different ways to take samples, but no matter which way is used, a rational equation is solved to estimate the size of the population. One sampling method is the **capture–recapture method**. In this method, an initial sample is taken (or captured) from the population and then tagged or marked. The tagged sample is counted and then returned to the population. After the marked portion of the population has been allowed to thoroughly mix with the unmarked portion, a second, or recapture, sample is taken. If the population has been well mixed and the second sample is taken randomly, the fraction of marked units in the recapture sample should approximate the fraction of marked units in the entire population. This information leads to the following equation that can be solved for the size of the population x:

$$\frac{\text{number of marked units in population}}{x} = \frac{\text{number of marked units in recapture sample}}{\text{total size of recapture sample}}$$

Another sampling method is the **addition method**. In this method, rather than capturing, marking, and returning an initial sample, a known number of nearly identical marked units are *added* to the original population. It is important to understand that these marked units were not part of the original population. Once the marked units have been thoroughly mixed with the original unmarked population, a sample is taken. This method leads to the following equation that can be solved for the size of the *original* population x:

$$\frac{\text{number of marked units added to original population}}{x + \text{number of marked units added to original population}} = \frac{\text{number of marked units in sample}}{\text{total size of sample}}$$

The addition method is very useful in situations where it would be (1) unwise to mark or damage existing members of the population, (2) difficult to take an initial sample from the population to be reintroduced after tagging, and/or (3) feasible to introduce nearly identical but marked units into the original population. In some situations, such as in wildlife management, where it would be impossible or unwise to introduce nearly identical units into an existing population, the capture–recapture method is the better choice.

Method 1

1. Fill a bowl or bag with an unknown number of dried beans. Your goal is to estimate the size of this population of beans.

2. Using a measuring cup or small paper cup, take a sample of beans from the bowl. Count and record the number of beans in the sample. Mark each bean with several Xs using a felt-tip marker. Then return the marked beans to the bowl.

3. Thoroughly mix the beans. Take a sample from the bowl. Count and record the total number of beans in the sample and the number of marked beans in the sample. Then return the bean sample to the bowl. Repeat this sampling process a total of five times to complete the table for Method 1.

4. Which sampling method does this procedure represent?

5. For each sample in the table, calculate an estimate of the number of beans.

6. If you had to give a single estimate of the number of beans, what would it be? Explain.

Name ____________________

7. Count the total bean population. How close are your individual estimates? Single estimate?

Method 1			
Sample	Total Number of Beans in Sample	Number of Marked Beans in Sample	Population Estimate
1			
2			
3			
4			
5			

Method 2

8. Gather or make a small supply of marked beans. Count and record this number of marked beans.

9. Start with a new and different population of dried beans in a bowl or bag. (*Note:* If reusing beans from Method 1, make sure that any and all marked beans have been removed.) Your goal is to estimate the size of this population of beans.

10. Add the marked beans you gathered to the bowl.

11. Thoroughly mix the beans. Take a sample from the bowl. Count and record the total number of beans in the sample and the number of marked beans in the sample. Then return the bean sample to the bowl. Repeat this sampling process a total of five times to complete the table for Method 2.

12. Which sampling method does this procedure represent?

13. For each sample in the table, calculate an estimate of the number of beans.

14. If you had to give a single estimate of the number of beans, what would it be? Explain.

15. Count the original bean population. How close are your individual estimates? Single estimate?

Method 2			
Sample	Total Number of Beans in Sample	Number of Marked Beans in Sample	Population Estimate
1			
2			
3			
4			
5			

CHAPTER 6 HIGHLIGHTS

DEFINITIONS AND CONCEPTS	EXAMPLES
SECTION 6.1 MULTIPLYING AND DIVIDING RATIONAL EXPRESSIONS	
A **rational expression** is the quotient $\frac{P}{Q}$ of two polynomials P and Q, as long as Q is not 0.	$\frac{2x-6}{7}$, $\frac{t^2-3t+5}{t-1}$
TO SIMPLIFY A RATIONAL EXPRESSION Step 1. Completely factor the numerator and the denominator. Step 2. Apply the fundamental principle of rational expressions.	Simplify. $\frac{2x^2+9x-5}{x^2-25} = \frac{(2x-1)(x+5)}{(x-5)(x+5)} = \frac{2x-1}{x-5}$
TO MULTIPLY RATIONAL EXPRESSIONS Step 1. Completely factor numerators and denominators. Step 2. Multiply the numerators and multiply the denominators. Step 3. Apply the fundamental principle of rational expressions.	Multiply: $\frac{x^3+8}{12x-18} \cdot \frac{14x^2-21x}{x^2+2x}$. $= \frac{(x+2)(x^2-2x+4)}{6(2x-3)} \cdot \frac{7x(2x-3)}{x(x+2)}$ $= \frac{7(x^2-2x+4)}{6}$
TO DIVIDE RATIONAL EXPRESSIONS Multiply the first rational expression by the reciprocal of the second rational expression.	Divide: $\frac{x^2+6x+9}{5xy-5y} \div \frac{x+3}{10y}$ $= \frac{(x+3)(x+3)}{5y(x-1)} \cdot \frac{2 \cdot 5y}{x+3}$ $= \frac{2(x+3)}{x-1}$
A **rational function** is a function described by a rational expression.	$f(x) = \frac{2x-6}{7}$, $h(t) = \frac{t^2-3t+5}{t-1}$
SECTION 6.2 ADDING AND SUBTRACTING RATIONAL EXPRESSIONS	
TO ADD OR SUBTRACT RATIONAL EXPRESSIONS Step 1. Find the LCD. Step 2. Write each rational expression as an equivalent rational expression whose denominator is the LCD. Step 3. Add or subtract numerators and write the sum or difference over the common denominator. Step 4. Simplify the result.	Subtract: $\frac{3}{x+2} - \frac{x+1}{x-3}$ $= \frac{3 \cdot (x-3)}{(x+2) \cdot (x-3)} - \frac{(x+1) \cdot (x+2)}{(x-3) \cdot (x+2)}$ $= \frac{3(x-3)-(x+1)(x+2)}{(x+2)(x-3)}$ $= \frac{3x-9-(x^2+3x+2)}{(x+2)(x-3)}$ $= \frac{3x-9-x^2-3x-2}{(x+2)(x-3)}$ $= \frac{-x^2-11}{(x+2)(x-3)}$

SECTION 6.3 SIMPLIFYING COMPLEX FRACTIONS

Method 1: Simplify the numerator and the denominator so that each is a single fraction. Then perform the indicated division and simplify if possible.

Simplify: $\dfrac{\dfrac{x+2}{x}}{x-\dfrac{4}{x}}$

Method 1: $\dfrac{\dfrac{x+2}{x}}{\dfrac{x \cdot x}{1 \cdot x}-\dfrac{4}{x}} = \dfrac{\dfrac{x+2}{x}}{\dfrac{x^2-4}{x}}$

$= \dfrac{x+2}{x} \cdot \dfrac{x}{(x+2)(x-2)} = \dfrac{1}{x-2}$

Method 2: Multiply the numerator and the denominator of the complex fraction by the LCD of the fractions in both the numerator and the denominator. Then simplify if possible.

Method 2: $\dfrac{\left(\dfrac{x+2}{x}\right) \cdot x}{\left(x-\dfrac{4}{x}\right) \cdot x} = \dfrac{x+2}{x \cdot x - \dfrac{4}{x} \cdot x}$

$= \dfrac{x+2}{x^2-4} = \dfrac{x+2}{(x+2)(x-2)} = \dfrac{1}{x-2}$

SECTION 6.4 SOLVING EQUATIONS CONTAINING RATIONAL EXPRESSIONS

TO SOLVE AN EQUATION CONTAINING RATIONAL EXPRESSIONS

Multiply both sides of the equation by the LCD of all rational expressions. Then use the distributive property and simplify. Solve the resulting equation and then check each proposed solution to see whether it makes any denominator 0. Discard any solutions that do.

Solve: $x - \dfrac{3}{x} = \dfrac{1}{2}$

$2x\left(x - \dfrac{3}{x}\right) = 2x\left(\dfrac{1}{2}\right)$ The LCD is $2x$.

$2x \cdot x - 2x\left(\dfrac{3}{x}\right) = 2x\left(\dfrac{1}{2}\right)$ Distribute.

$2x^2 - 6 = x$

$2x^2 - x - 6 = 0$ Subtract x from both sides.

$(2x+3)(x-2) = 0$ Factor.

$x = -\dfrac{3}{2}$ or $x = 2$

Both $-\dfrac{3}{2}$ and 2 check. The solution set is $\left\{2, -\dfrac{3}{2}\right\}$.

SECTION 6.5 RATIONAL EQUATIONS AND PROBLEM SOLVING

TO SOLVE AN EQUATION FOR A SPECIFIED VARIABLE

Treat the specified variable as the only variable of the equation and solve as usual.

Solve for x.

$A = \dfrac{2x+3y}{5}$

$5A = 2x + 3y$ Multiply both sides by 5.

$5A - 3y = 2x$ Subtract $3y$ from both sides.

$\dfrac{5A-3y}{2} = x$ Divide both sides by 2.

PROBLEM-SOLVING STEPS FOLLOW

Jeanee and David Dillon volunteer every year to clean a strip of Lake Ponchartrain beach. Jeanee can clean all the trash in this area of beach in 6 hours; David takes 5 hours. Find how long it will take them to clean the area of beach together.

SECTION 6.5 (CONTINUED)

1. UNDERSTAND

1. Read and reread the problem.
Let $x =$ time in hours that it takes Jeanee and David to clean the beach together.

	Hours to Complete	Part Completed in 1 Hour
Jeanee Alone	6	$\frac{1}{6}$
David Alone	5	$\frac{1}{5}$
Together	x	$\frac{1}{x}$

2. TRANSLATE.

2. In words:

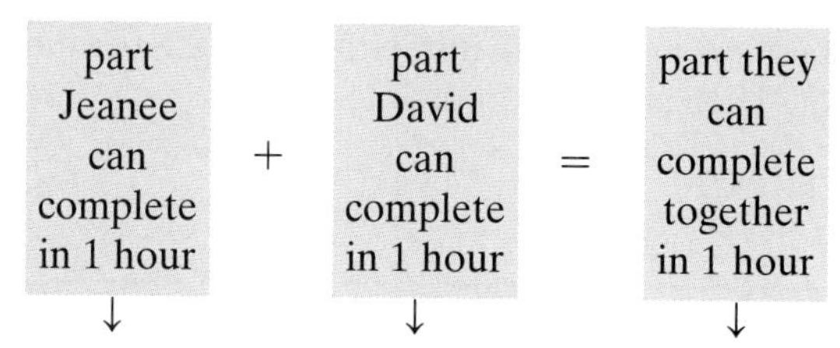

part Jeanee can complete in 1 hour
+
part David can complete in 1 hour
=
part they can complete together in 1 hour

Translate:

$$\frac{1}{6} \quad + \quad \frac{1}{5} \quad = \quad \frac{1}{x}$$

3. SOLVE.

3.

$$\frac{1}{6} + \frac{1}{5} = \frac{1}{x} \qquad \text{Multiply both sides by } 30x.$$

$$5x + 6x = 30$$

$$11x = 30$$

$$x = \frac{30}{11} \quad \text{or} \quad 2\frac{8}{11}$$

4. INTERPRET.

4. *Check* and then *state*. Together, they can clean the beach in $2\frac{8}{11}$ hours.

SECTION 6.6 VARIATION AND PROBLEM SOLVING

y **varies directly** as x, or y is **directly proportional** to x, if there is a nonzero constant k such that

$$y = kx$$

The circumference of a circle C varies directly as its radius r.

$$C = \underbrace{2\pi}_{k} r$$

y **varies inversely** as x, or y is **inversely proportional** to x, if there is a nonzero constant k such that

$$y = \frac{k}{x}$$

Pressure P varies inversely with volume V.

$$P = \frac{k}{V}$$

y **varies jointly** as x and z or y is **jointly proportional** to x and z if there is a nonzero constant k such that

$$y = kxz$$

The surface area S of a cylinder varies jointly as its radius r and height h.

$$S = \underbrace{2\pi}_{k} rh$$

Name ______________________ Section __________ Date __________

Chapter 6 Review

(6.1) *Find all numbers for which each rational expression is undefined.*

1. $\frac{3-5x}{7}$

2. $\frac{2x+4}{11}$

3. $\frac{-3x^2}{x-5}$

4. $\frac{4x}{3x-12}$

5. $\frac{x^3+2}{x^2+8x}$

6. $\frac{20}{3x^2-48}$

Write each rational expression in lowest terms.

7. $\frac{15x^4}{45x^2}$

8. $\frac{x+2}{2+x}$

9. $\frac{18m^6p^2}{10m^4p}$

10. $\frac{x-12}{12-x}$

11. $\frac{5x-15}{25x-75}$

12. $\frac{22x+8}{11x+4}$

13. $\frac{2x}{2x^2-2x}$

14. $\frac{x+7}{x^2-49}$

15. $\frac{2x^2+4x-30}{x^2+x-20}$

16. $\frac{xy-3x+2y-6}{x^2+4x+4}$

17. The average cost of manufacturing x bookcases is given by the rational function.

$$C(x) = \frac{35x+4200}{x}$$

a. Find the average cost per bookcase of manufacturing 50 bookcases.

b. Find the average cost per bookcase of manufacturing 100 bookcases.

c. As the number of bookcases increases, does the average cost per bookcase increase or decrease? (See parts (a) and (b).)

Perform each indicated operation. Write your answers in lowest terms.

18. $\frac{5}{x^3} \cdot \frac{x^2}{15}$

19. $\frac{3x^4yz^3}{15x^2y^2} \cdot \frac{10xy}{z^6}$

20. $\frac{4-x}{5} \cdot \frac{15}{2x-8}$

21. $\frac{x^2-6x+9}{2x^2-18} \cdot \frac{4x+12}{5x-15}$

22. $\frac{a-4b}{a^2+ab} \cdot \frac{b^2-a^2}{8b-2a}$

Name ______________________________

23. $\dfrac{x^2 - x - 12}{2x^2 - 32} \cdot \dfrac{x^2 + 8x + 16}{3x^2 + 21x + 36}$

24. $\dfrac{2x^3 + 54}{5x^2 + 5x - 30} \cdot \dfrac{6x + 12}{3x^2 - 9x + 27}$

25. $\dfrac{3}{4x} \div \dfrac{8}{2x^2}$

26. $\dfrac{4x + 8y}{3} \div \dfrac{5x + 10y}{9}$

27. $\dfrac{5ab}{14c^3} \div \dfrac{10a^4b^2}{6ac^5}$

28. $\dfrac{2}{5x} \div \dfrac{4 - 18x}{6 - 27x}$

29. $\dfrac{x^2 - 25}{3} \div \dfrac{x^2 - 10x + 25}{x^2 - x - 20}$

30. $\dfrac{a - 4b}{a^2 + ab} \div \dfrac{20b - 5a}{b^2 - a^2}$

31. $\dfrac{7x + 28}{2x + 4} \div \dfrac{x^2 + 2x - 8}{x^2 - 2x - 8}$

32. $\dfrac{3x + 3}{x - 1} \div \dfrac{x^2 - 6x - 7}{x^2 - 1}$

33. $\dfrac{2x - x^2}{x^3 - 8} \div \dfrac{x^2}{x^2 + 2x + 4}$

34. $\dfrac{5a^2 - 20}{a^3 + 2a^2 + a + 2} \div \dfrac{7a}{a^3 + a}$

35. $\dfrac{2a}{21} \div \dfrac{3a^2}{7} \cdot \dfrac{4}{a}$

36. $\dfrac{5x - 15}{3 - x} \cdot \dfrac{x + 2}{10x + 20} \cdot \dfrac{x^2 - 9}{x^2 - x - 6}$

37. $\dfrac{4a + 8}{5a^2 - 20} \cdot \dfrac{3a^2 - 6a}{a + 3} \div \dfrac{2a^2}{5a + 15}$

(6.2) *Find the LCD of the rational expressions in each list.*

38. $\dfrac{4}{9}, \dfrac{5}{2}$

39. $\dfrac{5}{4x^2y^5}, \dfrac{3}{10x^2y^4}, \dfrac{x}{6y^4}$

40. $\dfrac{5}{2x}, \dfrac{7}{x - 2}$

41. $\dfrac{3}{5x}, \dfrac{2}{x - 5}$

42. $\dfrac{1}{5x^3}, \dfrac{4}{x^2 + 3x - 28}, \dfrac{11}{10x^2 - 30x}$

Perform each indicated operation. Write your answers in lowest terms.

43. $\dfrac{2}{15} + \dfrac{4}{15}$

44. $\dfrac{4}{x - 4} + \dfrac{x}{x - 4}$

45. $\dfrac{4}{3x^2} + \dfrac{2}{3x^2}$

Name ______________________________

46. $\dfrac{1}{x-2} - \dfrac{1}{4-2x}$

47. $\dfrac{2x+1}{x^2+x-6} + \dfrac{2-x}{x^2+x-6}$

48. $\dfrac{7}{2x} + \dfrac{5}{6x}$

49. $\dfrac{1}{3x^2y^3} - \dfrac{1}{5x^4y}$

50. $\dfrac{1}{10-x} + \dfrac{x-1}{x-10}$

51. $\dfrac{x-2}{x+1} - \dfrac{x-3}{x-1}$

52. $\dfrac{x}{9-x^2} - \dfrac{2}{5x-15}$

53. $2x + 1 - \dfrac{1}{x-3}$

54. $\dfrac{2}{a^2-2a+1} + \dfrac{3}{a^2-1}$

55. $\dfrac{x}{9x^2+12x+16} - \dfrac{3x+4}{27x^3-64}$

Perform each indicated operation. Write your answers in lowest terms.

56. $\dfrac{2}{x-1} - \dfrac{3x}{3x-3} + \dfrac{1}{2x-2}$

57. $\dfrac{2}{x^2-16} - \dfrac{3x}{x^2+8x+16} + \dfrac{3}{x+4}$

58. Find the perimeter of the heptagon (a polygon with seven sides.)

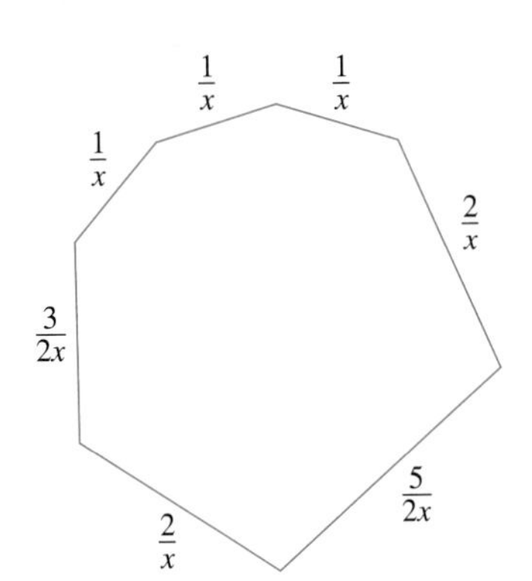

Name ____________________

(6.3) *Simplify each complex fraction.*

59. $\dfrac{\frac{2x}{5}}{\frac{3x}{5}}$

60. $\dfrac{1-\frac{3x}{4}}{2+\frac{x}{4}}$

61. $\dfrac{\frac{1}{x}-\frac{2}{3x}}{\frac{5}{2x}-\frac{1}{3}}$

62. $\dfrac{\frac{x^2}{15}}{\frac{x+1}{5x}}$

63. $\dfrac{\frac{3}{y^2}}{\frac{6}{y^3}}$

64. $\dfrac{\frac{x+2}{3}}{\frac{5}{x-2}}$

65. $\dfrac{2-\frac{3}{2x}}{x-\frac{2}{5x}}$

66. $\dfrac{1+\frac{x}{y}}{\frac{x^2}{y^2}-1}$

67. $\dfrac{\frac{5}{x}+\frac{1}{xy}}{\frac{3}{x^2}}$

68. $\dfrac{\frac{x}{3}-\frac{3}{x}}{1+\frac{3}{x}}$

69. $\dfrac{\frac{1}{x-1}+1}{\frac{1}{x+1}-1}$

70. $\dfrac{2}{1-\frac{2}{x}}$

71. $\dfrac{1}{1+\frac{2}{1-\frac{1}{x}}}$

72. $\dfrac{\frac{x^2+5x-6}{4x+3}}{\frac{(x+6)^2}{8x+6}}$

73. $\dfrac{\frac{x-3}{x+3}+\frac{x+3}{x-3}}{\frac{x-3}{x+3}-\frac{x+3}{x-3}}$

74. $\dfrac{\frac{3}{x-1}-\frac{2}{1-x}}{\frac{2}{x-1}-\frac{2}{x}}$

(6.4) *Solve each equation.*

75. $\frac{2}{5}=\frac{x}{15}$

76. $\frac{3}{x}+\frac{1}{3}=\frac{5}{x}$

77. $4+\frac{8}{x}=8$

78. $\frac{2x+3}{5x-9}=\frac{3}{2}$

79. $\frac{1}{x-2}-\frac{3x}{x^2-4}=\frac{2}{x+2}$

Name ____________________

80. $\frac{7}{x} - \frac{x}{7} = 0$

81. $\frac{x-2}{x^2 - 7x + 10} = \frac{1}{5x - 10} - \frac{1}{x - 5}$

Solve each equation or perform each indicated operation. Simplify.

82. $\frac{5}{x^2 - 7x} + \frac{4}{2x - 14}$

83. $3 - \frac{5}{x} - \frac{2}{x^2} = 0$

84. $\frac{4}{3 - x} - \frac{7}{2x - 6} + \frac{5}{x}$

(6.5) *Solve each equation for the specified variable.*

85. $A = \frac{h(a + b)}{2}$ for a

86. $\frac{1}{R} = \frac{1}{R_1} + \frac{1}{R_2}$ for R_2

87. $I = \frac{E}{R + r}$ for R

88. $A = P + Prt$ for r

89. $H = \frac{kA(T_1 - T_2)}{L}$ for A

Solve.

90. The sum of a number and twice its reciprocal is 3. Find the number(s).

91. If a number is added to the numerator of $\frac{3}{7}$, and twice that number is added to the denominator of $\frac{3}{7}$, the result is equivalent to $\frac{10}{21}$. Find the number.

92. The denominator of a fraction is 2 more than the numerator. If the numerator is decreased by 3 and the denominator is increased by 5, the resulting fraction is equivalent to $\frac{2}{3}$. Find the fraction.

93. The sum of the reciprocals of two consecutive even integers is $-\frac{9}{40}$. Find the two integers.

94. Three boys can paint a fence in 4 hours, 5 hours, and 6 hours, respectively. Find how long it will take all three boys to paint the fence.

95. If Sue Katz can type a certain number of mailing labels in 6 hours and Tom Neilson and Sue working together can type the same number of mailing labels in 4 hours, find how long it takes Tom alone to type the mailing labels.

Name ______________________________

96. The inlet pipe of a water tank can fill the tank in 2 hours and 30 minutes. The outlet pipe can empty the tank in 2 hours. Find how long it takes to empty a full tank if both pipes are open.

97. Timmy Garnica drove 210 miles in the same amount of time that it took a DC-10 jet to travel 1715 miles. The speed of the jet was 430 mph faster than the speed of the car. Find the speed of the jet.

98. The combined resistance R of two resistors in parallel with resistances R_1 and R_2 is given by the formula $\frac{1}{R} = \frac{1}{R_1} + \frac{1}{R_2}$. If the combined resistance is $\frac{30}{11}$ ohms and the resistance of one of the two resistors is 5 ohms, find the resistance of the other resistor.

99. The speed of a Ranger boat in still water is 32 mph. If the boat travels 72 miles upstream in the same time that it takes to travel 120 miles downstream, find the current of the stream.

100. A B737 jet flies 445 miles with the wind and 355 miles against the wind in the same length of time. If the speed of the jet in still air is 400 mph, find the speed of the wind.

101. The speed of a jogger is 3 mph faster than the speed of a walker. If the jogger travels 14 miles in the same amount of time that the walker travels 8 miles, find the speed of the walker.

102. Two Amtrak trains traveling on parallel tracks leave Tucson at the same time. In 6 hours the faster train is 382 miles from Tucson and the trains are 112 miles apart. Find how fast each train is traveling.

(6.6) *Solve each variation problem.*

103. A is directly proportional to B. If $A = 6$ when $B = 14$, find A when $B = 21$.

104. C is inversely proportional to D. If $C = 12$ when $D = 8$, find C when $D = 24$.

105. According to Boyle's law, the pressure exerted by a gas is inversely proportional to the volume, as long as the temperature stays the same. If a gas exerts a pressure of 1250 pounds per square inch when the volume is 2 cubic feet, find the volume when the pressure is 800 pounds per square inch.

106. The surface area of a sphere varies directly as the square of its radius. If the surface area is 36π square inches when the radius is 3 inches, find the surface area when the radius is 4 inches.

Name ______________________ Section ________ Date __________

Chapter 6 Test

Find all numbers for which each rational expression is undefined.

1. $\frac{5x^2}{1 - x}$

2. $\frac{9x^2 - 9}{x^2 + 4x + 3}$

Write each rational expression in lowest terms.

3. $\frac{5x^7}{3x^4}$

4. $\frac{7x - 21}{24 - 8x}$

5. $\frac{x^2 - 4x}{x^2 + 5x - 36}$

Perform each indicated operation. Write your answers in lowest terms.

6. $\frac{x}{x - 2} \cdot \frac{x^2 - 4}{5x}$

7. $\frac{2x^3 + 16}{6x^2 + 12x} \cdot \frac{5}{x^2 - 2x + 4}$

8. $\frac{26ab}{7c} \div \frac{13a^2c^5}{14a^4b^3}$

9. $\frac{3x^2 - 12}{x^2 + 2x - 8} \div \frac{6x + 18}{x + 4}$

10. $\frac{4x - 12}{2x - 9} \div \frac{3 - x}{4x^2 - 81} \cdot \frac{x + 3}{5x + 15}$

11. $\frac{5}{4x^3} + \frac{7}{4x^3}$

12. $\frac{3 + 2x}{10 - x} + \frac{13 + x}{x - 10}$

13. $\frac{3}{x^2 - x - 6} + \frac{2}{x^2 - 5x + 6}$

14. $\frac{5}{x - 7} - \frac{2x}{3x - 21} + \frac{x}{2x - 14}$

15. $\frac{3x}{5} \cdot \left(\frac{5}{x} - \frac{5}{2x}\right)$

Simplify each complex fraction.

16. $\frac{\frac{4x}{13}}{\frac{20x}{13}}$

17. $\frac{\frac{5}{x} - \frac{7}{3x}}{\frac{9}{8x} - \frac{1}{x}}$

18. $\frac{\frac{x^2 - 5x + 6}{x + 3}}{\frac{x^2 - 4x + 4}{x^2 - 9}}$

Answers

1. ______________
2. ______________
3. ______________
4. ______________
5. ______________
6. ______________
7. ______________
8. ______________
9. ______________
10. ______________
11. ______________
12. ______________
13. ______________
14. ______________
15. ______________
16. ______________
17. ______________
18. ______________

19. ______

20. ______

21. ______

22. ______

23. ______

24. ______

25. ______

26. ______

27. ______

Name ______

Solve each equation.

19. $\dfrac{5x + 3}{3x - 7} = \dfrac{19}{7}$

20. $\dfrac{5}{x - 5} + \dfrac{x}{x + 5} = -\dfrac{29}{21}$

21. $\dfrac{x}{x - 4} = 3 - \dfrac{4}{x - 4}$

22. Solve $\dfrac{x + b}{a} = \dfrac{4x - 7a}{b}$ for x.

23. The product of one more than a num ber and twice the reciprocal of the number is $\frac{12}{5}$. Find the number.

24. If Jan Ewing can weed the garden in 2 hours and her husband can weed it in 1 hour and 30 minutes, find how long it takes them to weed the garden together.

25. Suppose that W is inversely proportional to V. If $W = 20$ when $V = 12$, find W when $V = 15$.

26. Suppose that Q is jointly proportional to R and the square of S. If $Q = 24$ when $R = 3$ and $S = 4$, find Q when $R = 2$ and $S = 3$.

27. When an anvil is dropped into a gorge, the speed at which it strikes the ground is directly proportional to the square root of the distance it falls. An anvil that falls 400 feet hits the ground at a speed of 160 feet per second. Find the height of a cliff over the gorge if a dropped anvil hits the ground at a speed of 128 feet per second.

Name ______________________ Section ________ Date ________

ANSWERS

1. ____________

2. ____________

3. ____________

4. ____________

5. ____________

6. ____________

7. ____________

8. ____________

9. ____________

10. ____________

11. ____________

12. ____________

CUMULATIVE REVIEW

1. Evaluate: $\frac{r}{s}$ when $r = 48$ and $s = 6$

Evaluate each expression.

2. -7^0

3. $(2x + 5)^0$

4. Solve: $2x \geq 0$ and $4x - 1 \leq -9$

5. Solve: $|w + 3| = 7$

6. Determine whether $(0, -12)$, $(1, 9)$, and $(2, -6)$ are solutions of the equation $3x - y = 12$.

7. Determine whether the two lines are parallel, perpendicular, or neither.

$$3x + 7y = 4$$
$$6x + 14y = 7$$

8. Write an equation of the line through points $(4, 0)$ and $(-4, -5)$.

Find the domain and range of each relation.

9.

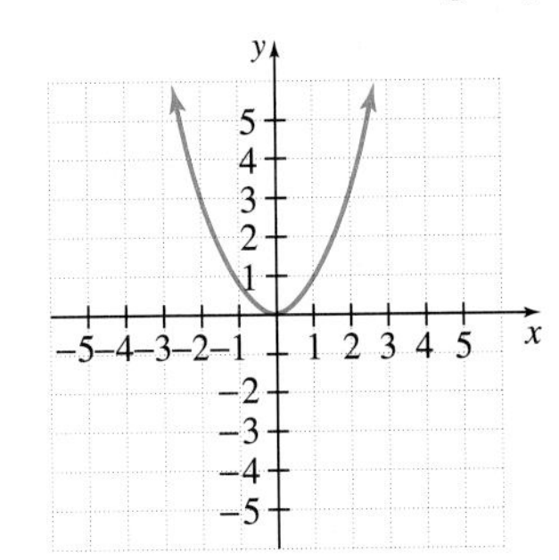

10.

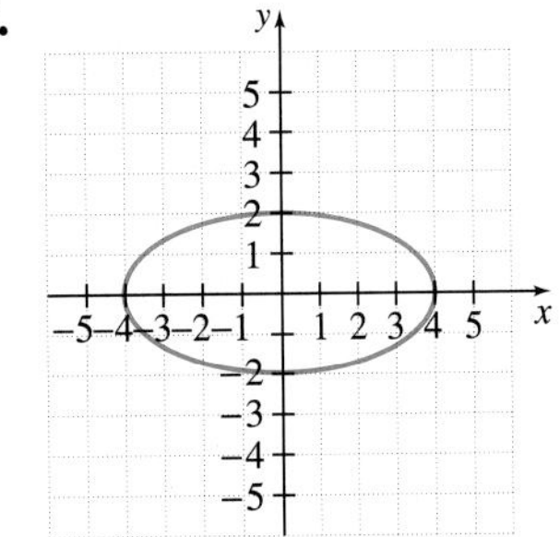

11. Use the substitution method to solve the system:

$$\begin{cases} -\frac{x}{6} + \frac{y}{2} = \frac{1}{2} \\ \frac{x}{3} - \frac{y}{6} = -\frac{3}{4} \end{cases}$$

12. Solve the system:

$$\begin{cases} 2x + 4y = 1 \\ 4x - 4z = -1 \\ y - 4z = -3 \end{cases}$$

Name ______________________

13. ______

14. ______

15. ______

16. see graph. ______

17. ______

18. ______

19. ______

20. ______

21. ______

22. ______

23. ______

24. ______

25. ______

13. Two cars leave Indianapolis, one traveling east and the other west. After 3 hours they are 297 miles apart. If one car is traveling 5 mph faster than the other, what is the speed of each?

14. Use matrices to solve the system:

$$\begin{cases} x + 2y + z = 2 \\ -2x - y + 2z = 5 \\ x + 3y - 2z = -8 \end{cases}$$

15. Use Cramer's rule to solve the system,

$$\begin{cases} 5x + y = 5 \\ -7x - 2y = -7 \end{cases}$$

16. Graph the solutions of the system

$$\begin{cases} 3x \geq y \\ x + 2y \leq 8 \end{cases}$$

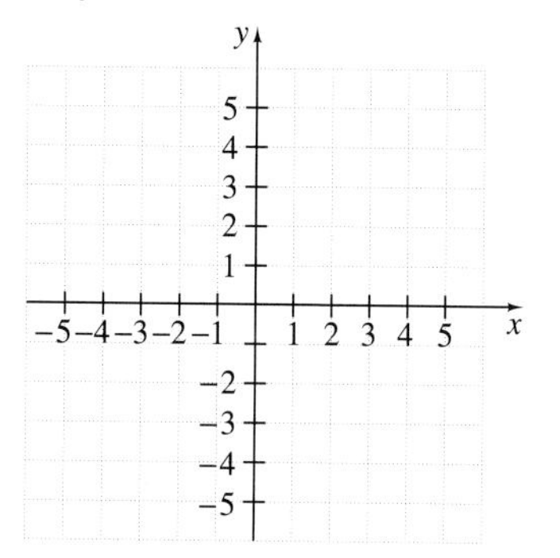

Simplify by combining like terms.

17. $-12x^2 + 7x^2 - 6x$

18. $3xy - 2x + 5xy - x$

19. Multiply: $(4x^2 + 7)(x^2 + 2x + 8)$

20. Divide $2x^2 - x - 10$ by $x + 2$.

21. Factor: $-3x^3y + 2x^2y - 5xy$

22. Factor: $x^2 + 10x + 16$

23. Factor: $p^4 - 16$

24. Solve: $(x + 2)(x - 6) = 0$

25. Simplify: $\dfrac{\frac{2x}{27y^2}}{\frac{6x^2}{9}}$

Rational Exponents, Radicals, and Complex Numbers

CHAPTER 7

In this chapter, radical notation is reviewed, and then rational exponents are introduced. As the name implies, rational exponents are exponents that are rational numbers. We present an interpretation of rational exponents that is consistent with the meaning and rules already established for integer exponents, and we present two forms of notation for roots: radical and exponent. We conclude this chapter with complex numbers, a natural extension of the real number system.

Mount Vesuvius is the only active volcano on the European continent. Although the Romans thought the volcano was extinct, Vesuvius erupted violently in 79 A.D. The eruption buried the cities of Pompeii, Herculaneum, and Stabiae under up to 60 feet of ash and mud, killing approximately 16,000 people. Although Pompeii was completely engulfed, the city was far from destroyed. The blanket of mud and ash perfectly preserved much of the city in a snapshot of daily life of the ancient Romans. Pompeii lay undisturbed for over 1500 years until the first excavations were made and its archaeological significance was proven. The 79 A.D. eruption of Vesuvius seemed to be the volcano's renewal. It has erupted with varying degrees of violence more than 30 times since Pompeii's burial, most recently in 1944. In Exercise 84 on page 542, we will use a radical expression to find the surface area of Mount Vesuvius.

ANSWERS

1. ______
2. ______
3. ______
4. ______
5. ______
6. ______
7. ______
8. ______
9. ______
10. ______
11. ______
12. ______
13. ______
14. ______
15. ______
16. ______
17. ______
18. ______
19. ______
20. ______

Name ______ Section ______ Date ______

CHAPTER 7 PRETEST

Find each root. Assume that variables represent positive numbers.

1. $\sqrt{81x^2}$

2. $\sqrt[3]{-27y^9}$

3. $\sqrt[5]{x^{30}}$

4. If $f(x) = \sqrt{7x + 2}$, find $f(2)$.

Simplify. Write with positive exponents.

5. $144^{1/2}$

6. $81^{3/4}$

7. $\dfrac{3}{2m^{-5/6}}$

8. $(9x^4y^{-6})^{3/2}$

Perform the indicated operations. Assume that all variables represent positive numbers.

9. $\sqrt{13} \cdot \sqrt{3x}$

10. $\sqrt{\dfrac{6}{49}}$

11. $3\sqrt{18} - 6\sqrt{50}$

12. $(4\sqrt{x} + 1)(2\sqrt{x} - 3)$

13. $(\sqrt{5} - y)^2$

Simplify.

14. $\sqrt[3]{135a^2b^{12}}$

Rationalize the denominator. Assume that all variables represent positive numbers.

15. $\dfrac{2}{\sqrt{12x}}$

16. $\dfrac{14}{2 + \sqrt{11}}$

17. Solve: $\sqrt{2x + 1} - 5 = 0$

Perform the indicated operation and simplify. Write the result in the form $a + bi$.

18. $(10 - 3i) - (6 + 5i)$

19. $(2 - 5i)^2$

20. $\dfrac{3 + i}{5 + i}$

7.1 Radical Expressions

A Finding Square Roots

Recall from Section 1.3 that to find a **square root** of a number a, we find a number that was squared to get a.

> **Square Root**
>
> The number b is a **square root** of a if $b^2 = a$.

Examples Find the square roots of each number.

1. 25 Since $5^2 = 25$ and $(-5)^2 = 25$, the square roots of 25 are 5 and -5.

2. 49 Since $7^2 = 49$ and $(-7)^2 = 49$, the square roots of 49 are 7 and -7.

3. -4 There is no real number whose square is -4. The number -4 has no real number square root.

Recall that we denote the **nonnegative**, or **principal**, **square root** with the **radical sign**:

$$\sqrt{25} = 5$$

We denote the **negative square root** with the **negative radical sign**:

$$-\sqrt{25} = -5$$

An expression containing a radical sign is called a **radical expression**. An expression within, or "under," a radical sign is called a **radicand**.

radical expression: $\sqrt{a}$ (radical sign: $\sqrt{\ }$; radicand: a)

> **Principal and Negative Square Roots**
>
> The **principal square root** of a nonnegative number a is its nonnegative square root. The principal square root is written as $\sqrt{a}$. The **negative square root** of a is written as $-\sqrt{a}$.

Examples Find each square root. Assume that all variables represent nonnegative real numbers.

4. $\sqrt{36} = 6$ because $6^2 = 36$.

5. $\sqrt{0} = 0$ because $0^2 = 0$.

6. $\sqrt{\frac{4}{49}} = \frac{2}{7}$ because $\left(\frac{2}{7}\right)^2 = \frac{4}{49}$.

7. $\sqrt{0.25} = 0.5$ because $(0.5)^2 = 0.25$.

8. $\sqrt{x^6} = x^3$ because $(x^3)^2 = x^6$.

9. $\sqrt{9x^{10}} = 3x^5$ because $(3x^5)^2 = 9x^{10}$.

10. $-\sqrt{81} = -9$. The negative in front of the radical indicates the negative square root of 81.

11. $\sqrt{-81}$ is not a real number.

Objectives

A Find square roots.

B Approximate roots using a calculator.

C Find cube roots.

D Find nth roots.

E Find $\sqrt[n]{a^n}$ when a is any real number.

F Find function values of radical functions.

Practice Problems 1–3

Find the square roots of each number.

1. 36
2. 81
3. -16

Practice Problems 4–11

Find each square root. Assume that all variables represent nonnegative real numbers.

4. $\sqrt{25}$ 5. $\sqrt{0}$
6. $\sqrt{\frac{9}{25}}$ 7. $\sqrt{0.36}$
8. $\sqrt{x^{10}}$ 9. $\sqrt{36x^6}$
10. $-\sqrt{25}$ 11. $\sqrt{-25}$

Answers

1. $6, -6$, **2.** $9, -9$, **3.** no real number square root, **4.** 5, **5.** 0, **6.** $\frac{3}{5}$, **7.** 0.6, **8.** x^5, **9.** $6x^3$, **10.** -5, **11.** not a real number

HELPFUL HINT

Don't forget that the square root of a negative number is not a real number. For example,

$\sqrt{-9}$ it not a real number

because there is no real number that when multiplied by itself would give a product of -9. In Section 7.7, we will see what kind of a number $\sqrt{-9}$ is.

B APPROXIMATING ROOTS

Recall that numbers such as 1, 4, 9, and 25 are called **perfect squares**, since $1 = 1^2$, $4 = 2^2$, $9 = 3^2$, and $25 = 5^2$. Square roots of perfect square radicands simplify to rational numbers. What happens when we try to simplify a root such as $\sqrt{3}$? Since 3 is not a perfect square, $\sqrt{3}$ is not a rational number. It is called an **irrational number**, and we can find a decimal **approximation** of it. To find decimal approximations, we can use the table in the appendix or a calculator. For example, an approximation for $\sqrt{3}$ is

$$\sqrt{3} \approx 1.732$$

↑ approximation symbol

To see if the approximation is reasonable, notice that since

$1 < 3 < 4$, then
$\sqrt{1} < \sqrt{3} < \sqrt{4}$, or
$1 < \sqrt{3} < 2$.

We found $\sqrt{3} \approx 1.732$, a number between 1 and 2, so our result is reasonable.

Practice Problem 12

Use a calculator or the appendix to approximate $\sqrt{30}$. Round the approximation to 3 decimal places and check to see that your approximation is reasonable.

Example 12 Use a calculator or the appendix to approximate $\sqrt{20}$. Round the approximation to 3 decimal places and check to see that your approximation is reasonable.

$\sqrt{20} \approx 4.472$

Is this reasonable? Since $16 < 20 < 25$, then $\sqrt{16} < \sqrt{20} < \sqrt{25}$, or $4 < \sqrt{20} < 5$. The approximation is between 4 and 5 and is thus reasonable.

C FINDING CUBE ROOTS

Finding roots can be extended to other roots such as cube roots. For example, since $2^3 = 8$, we call 2 the **cube root** of 8. In symbols, we write

$$\sqrt[3]{8} = 2$$

CUBE ROOT

The **cube root** of a real number a is written as $\sqrt[3]{a}$, and

$$\sqrt[3]{a} = b \text{ only if } b^3 = a$$

From this definition, we have

$\sqrt[3]{64} = 4$ since $4^3 = 64$
$\sqrt[3]{-27} = -3$ since $(-3)^3 = -27$
$\sqrt[3]{x^3} = x$ since $x^3 = x^3$

Notice that, unlike with square roots, *it is possible to have a negative radicand when finding a cube root.* This is so because the *cube* of a negative number is a negative number. Therefore, the *cube root* of a negative number is a negative number.

Practice Problems 13–17

Find each cubic root.

13. $\sqrt[3]{0}$
14. $\sqrt[3]{-8}$
15. $\sqrt[3]{\frac{1}{64}}$
16. $\sqrt[3]{x^9}$
17. $\sqrt[3]{-64x^6}$

Examples Find each cube root.

13. $\sqrt[3]{1} = 1$ because $1^3 = 1$.
14. $\sqrt[3]{-64} = -4$ because $(-4)^3 = -64$.

Answers

12. 5.477, **13.** 0, **14.** -2, **15.** $\frac{1}{4}$, **16.** x^3, **17.** $-4x^2$

15. $\sqrt[3]{\frac{8}{125}} = \frac{2}{5}$ because $\left(\frac{2}{5}\right)^3 = \frac{8}{125}$.
16. $\sqrt[3]{x^6} = x^2$ because $(x^2)^3 = x^6$.
17. $\sqrt[3]{-8x^9} = -2x^3$ because $(-2x^3)^3 = -8x^9$.

D FINDING nTH ROOTS

Just as we can raise a real number to powers other than 2 or 3, we can find roots other than square roots and cube roots. In fact, we can find the **nth root** of a number, where n is any natural number. In symbols, the nth root of a is written as $\sqrt[n]{a}$, where n is called the **index**. The index 2 is usually omitted for square roots.

HELPFUL HINT

If the index is even, such as $\sqrt{}$, $\sqrt[4]{}$, $\sqrt[6]{}$, and so on, the radicand must be nonnegative for the root to be a real number. For example,

$\sqrt[4]{16} = 2$, but $\sqrt[4]{-16}$ is not a real number,
$\sqrt[6]{64} = 2$, but $\sqrt[6]{-64}$ is not a real number.

If the index is odd, such as $\sqrt[3]{}$, $\sqrt[5]{}$, and so on, the radicand may be any real number. For example,

$\sqrt[3]{64} = 4$ and $\sqrt[3]{-64} = -4$,
$\sqrt[5]{32} = 2$ and $\sqrt[5]{-32} = -2$.

TRY THE CONCEPT CHECK IN THE MARGIN.

✓ CONCEPT CHECK

Which one is not a real number?
a. $\sqrt[3]{-15}$
b. $\sqrt[4]{-15}$
c. $\sqrt[5]{-15}$
d. $\sqrt{(-15)^2}$

Examples

Find each root.

18. $\sqrt[4]{81} = 3$ because $3^4 = 81$ and 3 is positive.
19. $\sqrt[5]{-243} = -3$ because $(-3)^5 = -243$.
20. $-\sqrt{25} = -5$ because -5 is the opposite of $\sqrt{25}$.
21. $\sqrt[4]{-81}$ is not a real number. There is no real number that, when raised to the fourth power, is -81.
22. $\sqrt[3]{64x^3} = 4x$ because $(4x)^3 = 64x^3$.

Practice Problems 18–22

Find each root.

18. $\sqrt[4]{16}$
19. $\sqrt[5]{-32}$
20. $-\sqrt{36}$
21. $\sqrt[4]{-16}$
22. $\sqrt[3]{8x^6}$

E FINDING $\sqrt[n]{a^n}$ WHEN a IS ANY REAL NUMBER

Recall that the notation $\sqrt{a^2}$ indicates the positive square root of a^2 only. For example,

$$\sqrt{(-5)^2} = \sqrt{25} = 5$$

When variables are present in the radicand and it is *unclear whether the variable represents a positive number or a negative number*, absolute value bars are sometimes needed to ensure that the result is a positive number. For example,

$$\sqrt{x^2} = |x|$$

This ensures that the result is positive. This same situation may occur when the index is any *even* positive integer. When the index is any *odd* positive integer, absolute value bars are not necessary.

Answers
18. 2, **19.** -2, **20.** -6, **21.** not a real number, **22.** $2x^2$

✓ Concept Check: b

> FINDING $\sqrt[n]{a^n}$
>
> If n is an *even* positive integer, then $\sqrt[n]{a^n} = |a|$.
> If n is an *odd* positive integer, then $\sqrt[n]{a^n} = a$.

Examples

Simplify. Assume that the variables represent any real number.

23. $\sqrt{(-3)^2} = |-3| = 3$ — When the index is even, the absolute value bars ensure that the result is not negative.

24. $\sqrt{x^2} = |x|$

25. $\sqrt[4]{(x-2)^4} = |x-2|$

26. $\sqrt[3]{(-5)^3} = -5$ — Absolute value bars are not needed when the index is odd.

27. $\sqrt[5]{(2x-7)^5} = 2x - 7$

28. $\sqrt{25x^2} = 5|x|$

29. $\sqrt{x^2 + 2x + 1} = \sqrt{(x+1)^2} = |x+1|$

Practice Problems 23–29

Simplify. Assume that the variables represent any real number.

23. $\sqrt{(-5)^2}$
24. $\sqrt{x^6}$
25. $\sqrt[4]{(x+6)^4}$
26. $\sqrt[3]{(-3)^3}$
27. $\sqrt[5]{(7x-1)^5}$
28. $\sqrt{36x^2}$
29. $\sqrt{x^2 + 6x + 9}$

F FINDING FUNCTION VALUES

Functions of the form

$$f(x) = \sqrt[n]{x}$$

are called **radical functions**. Recall that the domain of a function in x is the set of all possible replacement values of x. This means that if n is even, the domain is the set of all nonnegative numbers, or $\{x|x \geq 0\}$. If n is odd, the domain is the set of all real numbers. Keep this in mind as we find function values. In Chapter 9, we will graph these functions.

Examples

If $f(x) = \sqrt{x-4}$ and $g(x) = \sqrt[3]{x+2}$, find each function value.

30. $f(8) = \sqrt{8-4} = \sqrt{4} = 2$

31. $f(6) = \sqrt{6-4} = \sqrt{2}$

32. $g(-1) = \sqrt[3]{-1+2} = \sqrt[3]{1} = 1$

33. $g(1) = \sqrt[3]{1+2} = \sqrt[3]{3}$

Practice Problems 30–33

If $f(x) = \sqrt{x+2}$ and $g(x) = \sqrt[3]{x-1}$, find each function value.

30. $f(7)$
31. $g(9)$
32. $f(0)$
33. $g(10)$

HELPFUL HINT

Notice that for the function $f(x) = \sqrt{x-4}$, the domain includes all real numbers that make the radicand ≥ 0. To see what numbers these are, solve $x - 4 \geq 0$ and find that $x \geq 4$. The domain is $\{x|x \geq 4\}$.
The domain of the cube root function $g(x) = \sqrt[3]{x+2}$ is the set of real numbers.

Answers

23. 5, **24.** $|x^3|$, **25.** $|x+6|$, **26.** -3, **27.** $7x-1$, **28.** $6|x|$, **29.** $|x+3|$, **30.** 3, **31.** 2, **32.** $\sqrt{2}$, **33.** $\sqrt[3]{9}$

Name ______________________ Section ________ Date __________

EXERCISE SET 7.1

A *Find the square roots of each number. See Examples 1 through 3.*

1. 4 **2.** 9 **3.** -25

4. -49 **5.** 100 **6.** 64

Find each square root. Assume that all variables represent nonnegative real numbers. See Examples 4 through 11.

7. $\sqrt{100}$ **8.** $\sqrt{400}$ **9.** $\sqrt{\frac{1}{4}}$ **10.** $\sqrt{\frac{9}{25}}$

11. $\sqrt{0.0001}$ **12.** $\sqrt{0.04}$ **13.** $-\sqrt{36}$ **14.** $-\sqrt{9}$

15. $\sqrt{x^{10}}$ **16.** $\sqrt{x^{16}}$ **17.** $\sqrt{16y^6}$ **18.** $\sqrt{64y^{20}}$

B *Use a calculator or the appendix to approximate each square root to 3 decimal places. Check to see that each approximation is reasonable. See Example 12.*

19. $\sqrt{7}$ **20.** $\sqrt{11}$ **21.** $\sqrt{38}$

22. $\sqrt{56}$ **23.** $\sqrt{200}$ **24.** $\sqrt{300}$

C *Find each cube root. See Examples 13 through 17.*

25. $\sqrt[3]{64}$ **26.** $\sqrt[3]{27}$ **27.** $\sqrt[3]{\frac{1}{8}}$

28. $\sqrt[3]{\frac{27}{64}}$ **29.** $\sqrt[3]{-1}$ **30.** $\sqrt[3]{-125}$

31. $\sqrt[3]{x^{12}}$ **32.** $\sqrt[3]{x^{15}}$ **33.** $\sqrt[3]{-27x^9}$ **34.** $\sqrt[3]{-64x^6}$

ANSWERS

1. ____
2. ____
3. ____
4. ____
5. ____
6. ____
7. ____
8. ____
9. ____
10. ____
11. ____
12. ____
13. ____
14. ____
15. ____
16. ____
17. ____
18. ____
19. ____
20. ____
21. ____
22. ____
23. ____
24. ____
25. ____
26. ____
27. ____
28. ____
29. ____
30. ____
31. ____
32. ____
33. ____
34. ____

35. ____
36. ____
37. ____
38. ____
39. ____
40. ____
41. ____
42. ____
43. ____
44. ____
45. ____
46. ____
47. ____
48. ____
49. ____
50. ____
51. ____
52. ____
53. ____
54. ____
55. ____
56. ____
57. ____
58. ____
59. ____
60. ____
61. ____
62. ____
63. ____
64. ____
65. ____
66. ____
67. ____
68. ____

Name ____________________

D *Find each root. Assume that all variables represent nonnegative real numbers. See Examples 18 through 22.*

35. $-\sqrt[4]{16}$ **36.** $\sqrt[5]{-243}$ **37.** $\sqrt[4]{-16}$ **38.** $\sqrt{-16}$

39. $\sqrt[5]{-32}$ **40.** $\sqrt[5]{-1}$ **41.** $\sqrt[5]{x^{20}}$ **42.** $\sqrt[4]{x^{20}}$

43. $\sqrt[6]{64x^{12}}$ **44.** $\sqrt[5]{-32x^{15}}$ **45.** $\sqrt{81x^4}$ **46.** $\sqrt[4]{81x^4}$

47. $\sqrt[4]{256x^8}$ **48.** $\sqrt{256x^8}$

E *Simplify. Assume that the variables represent any real number. See Examples 23 through 29.*

49. $\sqrt{(-8)^2}$ **50.** $\sqrt{(-7)^2}$ **51.** $\sqrt[3]{(-8)^3}$ **52.** $\sqrt[5]{(-7)^5}$

53. $\sqrt{4x^2}$ **54.** $\sqrt[4]{16x^4}$ **55.** $\sqrt[3]{x^3}$ **56.** $\sqrt[5]{x^5}$

57. $\sqrt[4]{(x-2)^4}$ **58.** $\sqrt[6]{(2x-1)^6}$

59. $\sqrt{x^2+4x+4}$
(*Hint:* Factor the polynomial first.)

60. $\sqrt{x^2-8x+16}$
(*Hint:* Factor the polynomial first.)

F *If $f(x) = \sqrt{2x+3}$ and $g(x) = \sqrt[3]{x-8}$, find each function value. See Examples 30 through 33.*

61. $f(0)$ **62.** $g(0)$ **63.** $g(7)$ **64.** $f(-1)$

65. $g(-19)$ **66.** $f(3)$ **67.** $f(2)$ **68.** $g(1)$

Name ______________________________

Review and Preview

Simplify each exponential expression. See Sections 1.5 and 1.6.

69. $(-2x^3y^2)^5$

70. $(4y^6z^7)^3$

71. $(-3x^2y^3z^5)(20x^5y^7)$

72. $(-14a^5bc^2)(2abc^4)$

73. $\dfrac{7x^{-1}y}{14(x^5y^2)^{-2}}$

74. $\dfrac{(2a^{-1}b^2)^3}{(8a^2b)^{-2}}$

Combining Concepts

75. Suppose that a friend tells you that $\sqrt{13} \approx 5.7$. Without a calculator, how can you convince your friend that he must have made an error?

76. Escape velocity is the minimum speed that an object must reach to escape a planet's pull of gravity. Escape velocity v is given by the equation $v = \sqrt{\dfrac{2Gm}{r}}$, where m is the mass of the planet, r is its radius, and G is the universal gravitational constant which has a value of $G = 6.67 \times 10^{-11}$ $m^3 / kg \cdot s^2$. The mass of Earth is 5.97×10^{24} kg and its radius is 6.37×10^6 m. Use this information to find the escape velocity for Earth. Round to the nearest whole number. (*Source:* National Space Science Data Center)

69. ______________

70. ______________

71. ______________

72. ______________

73. ______________

74. ______________

75. ______________

76. ______________

77. ____________

78. ____________

Name ____________

Internet Excursions

Go to http://www.prenhall.com/martin-gay
The National Space Science Data Center (NSSDC) is a division of NASA that provides access to information collected from NASA experiments and space flight missions. One of the many areas in which the NSSDC maintains a library of information is the planetary sciences. The given World Wide Web site gives a listing of planetary fact sheets from which the user may choose. (Alternatively, you can visit the NSSDC homepage at http://nssdc.gsfc.nasa.gov and navigate to Planetary Sciences. Then look for the Planetary Fact Sheets option.)

77. Choose any one of the fact sheets for a planet, asteroid, or the sun. Use the information given in the fact sheet and the escape velocity formula given in Exercise 76 to compute the escape velocity for that body. Then compare your calculation to the escape velocity given in the fact sheet. How close is your calculation? (*Note:* You should use the "volumetric mean radius" for the planet's radius in your calculation.)

78. Repeat Exercise 77 for one of the other listed bodies of the solar system.

7.2 RATIONAL EXPONENTS

Objectives

A Understand the meaning of $a^{1/n}$.
B Understand the meaning of $a^{m/n}$.
C Understand the meaning of $a^{-m/n}$.
D Use rules for exponents to simplify expressions that contain rational exponents.
E Use rational exponents to simplify radical expressions.

SSM

CD-ROM

Video 7.2

A UNDERSTANDING $a^{1/n}$

So far in this text, we have not defined expressions with rational exponents such as $3^{1/2}$, $x^{2/3}$, and $-9^{-1/4}$. We will define these expressions so that the rules for exponents shall apply to these rational exponents as well.

Suppose that $x = 5^{1/3}$. Then

$$x^3 = (5^{1/3})^3 = 5^{1/3 \cdot 3} = 5^1 \text{ or } 5$$

using rules for exponents

Since $x^3 = 5$, then x is the number whose cube is 5, or $x = \sqrt[3]{5}$. Notice that we also know that $x = 5^{1/3}$. This means that

$$5^{1/3} = \sqrt[3]{5}$$

DEFINITION OF $a^{1/n}$

If n is a positive integer greater than 1 and $\sqrt[n]{a}$ is a real number, then

$$a^{1/n} = \sqrt[n]{a}$$

Notice that the denominator of the rational exponent corresponds to the index of the radical.

Examples Use radical notation to write each expression. Simplify if possible.

1. $4^{1/2} = \sqrt{4} = 2$
2. $64^{1/3} = \sqrt[3]{64} = 4$
3. $x^{1/4} = \sqrt[4]{x}$
4. $-9^{1/2} = -\sqrt{9} = -3$
5. $(81x^8)^{1/4} = \sqrt[4]{81x^8} = 3x^2$
6. $5y^{1/3} = 5\sqrt[3]{y}$

Practice Problems 1–6

Use radical notation to write each expression. Simplify if possible.

1. $25^{1/2}$
2. $27^{1/3}$
3. $x^{1/5}$
4. $-25^{1/2}$
5. $(-27y^6)^{1/3}$
6. $7x^{1/5}$

B UNDERSTANDING $a^{m/n}$

As we expand our use of exponents to include $\frac{m}{n}$, we define their meaning so that rules for exponents still hold true. For example, by properties of exponents,

$$8^{2/3} = (8^{1/3})^2 = (\sqrt[3]{8})^2 \quad \text{or}$$
$$8^{2/3} = (8^2)^{1/3} = \sqrt[3]{8^2}$$

DEFINITION OF $a^{m/n}$

If m and n are positive integers greater than 1 with $\frac{m}{n}$ in lowest terms, then

$$a^{m/n} = \sqrt[n]{a^m} = (\sqrt[n]{a})^m$$

as long as $\sqrt[n]{a}$ is a real number.

Answers

1. 5, **2.** 3, **3.** $\sqrt[5]{x}$, **4.** -5, **5.** $-3y^2$, **6.** $7\sqrt[5]{x}$

Notice that the denominator n of the rational exponent corresponds to the index of the radical. The numerator m of the rational exponent indicates that the base is to be raised to the mth power. This means that

$$8^{2/3} = \sqrt[3]{8^2} = \sqrt[3]{64} = 4 \qquad \text{or}$$
$$8^{2/3} = \left(\sqrt[3]{8}\right)^2 = 2^2 = 4$$

HELPFUL HINT

Most of the time, $(\sqrt[n]{a})^m$ will be easier to calculate than $\sqrt[n]{a^m}$.

Practice Problems 7–11

Use radical notation to write each expression. Simplify if possible.

7. $9^{3/2}$
8. $-256^{3/4}$
9. $(-32)^{2/5}$
10. $\left(\frac{1}{4}\right)^{3/2}$
11. $(2x + 1)^{2/7}$

Examples

Use radical notation to write each expression. Simplify if possible.

7. $4^{3/2} = \left(\sqrt{4}\right)^3 = 2^3 = 8$

8. $-16^{3/4} = -\left(\sqrt[4]{16}\right)^3 = -(2)^3 = -8$

9. $(-27)^{2/3} = \left(\sqrt[3]{-27}\right)^2 = (-3)^2 = 9$

10. $\left(\frac{1}{9}\right)^{3/2} = \left(\sqrt{\frac{1}{9}}\right)^3 = \left(\frac{1}{3}\right)^3 = \frac{1}{27}$

11. $(4x - 1)^{3/5} = \sqrt[5]{(4x - 1)^3}$

HELPFUL HINT

The *denominator* of a rational exponent is the index of the corresponding radical. For example, $x^{1/5} = \sqrt[5]{x}$ and $z^{2/3} = \sqrt[3]{z^2}$, or $z^{2/3} = \left(\sqrt[3]{z}\right)^2$.

C UNDERSTANDING $a^{-m/n}$

The rational exponents we have given meaning to exclude negative rational numbers. To complete the set of definitions, we define $a^{-m/n}$.

DEFINITION OF $a^{-m/n}$

$$a^{-m/n} = \frac{1}{a^{m/n}}$$

as long as $a^{m/n}$ is a nonzero real number.

Practice Problems 12–13

Write each expression with a positive exponent. Then simplify.

12. $27^{-2/3}$
13. $-256^{-3/4}$

Examples

Write each expression with a positive exponent. Then simplify.

12. $16^{-3/4} = \frac{1}{16^{3/4}} = \frac{1}{\left(\sqrt[4]{16}\right)^3} = \frac{1}{2^3} = \frac{1}{8}$

13. $(-27)^{-2/3} = \frac{1}{(-27)^{2/3}} = \frac{1}{\left(\sqrt[3]{-27}\right)^2} = \frac{1}{(-3)^2} = \frac{1}{9}$

Answers

7. 27, **8.** −64, **9.** 4, **10.** $\frac{1}{8}$, **11.** $\sqrt[7]{(2x + 1)^2}$, **12.** $\frac{1}{9}$, **13.** $-\frac{1}{64}$

HELPFUL HINT

If an expression contains a negative rational exponent, you may want to first write the expression with a positive exponent, then interpret the rational exponent. Notice that the sign of the base is not affected by the sign of its exponent. For example,

$$9^{-3/2} = \frac{1}{9^{3/2}} = \frac{1}{(\sqrt{9})^3} = \frac{1}{27}$$

Also,

$$(-27)^{-1/3} = \frac{1}{(-27)^{1/3}} = -\frac{1}{3}$$

TRY THE CONCEPT CHECK IN THE MARGIN.

D USING RULES FOR EXPONENTS

It can be shown that the properties of integer exponents hold for rational exponents. By using these properties and definitions, we can now simplify expressions that contain rational exponents. These rules are repeated here for review.

SUMMARY OF EXPONENT RULES

If m and n are rational numbers, and a, b, and c are numbers for which the expressions below exist, then

Product rule for exponents:	$a^m \cdot a^n = a^{m+n}$
Power rule for exponents:	$(a^m)^n = a^{m \cdot n}$
Power rules for products and quotients:	$(ab)^n = a^n b^n$ and
	$\left(\frac{a}{c}\right)^n = \frac{a^n}{c^n}, c \neq 0$
Quotient rule for exponents:	$\frac{a^m}{a^n} = a^{m-n}, a \neq 0$
Zero exponent:	$a^0 = 1, a \neq 0$
Negative exponent:	$a^{-n} = \frac{1}{a^n}, a \neq 0$

Examples Use the properties of exponents to simplify.

14. $x^{1/2}x^{1/3} = x^{1/2+1/3} = x^{3/6+2/6} = x^{5/6}$ Use the product rule.

15. $\frac{7^{1/3}}{7^{4/3}} = 7^{1/3-4/3} = 7^{-3/3} = 7^{-1} = \frac{1}{7}$ Use the quotient rule.

16. $\frac{(2x^{2/5})^5}{x^2} = \frac{2^5(x^{2/5})^5}{x^2}$ Use the power rule.

$= \frac{32x^2}{x^2}$ Simplify.

$= 32x^{2-2}$ Use the quotient rule.

$= 32x^0$ Simplify.

$= 32 \cdot 1$ or 32 Substitute 1 for x^0.

✓ CONCEPT CHECK

Which one is correct?

a. $-8^{2/3} = \frac{1}{4}$

b. $8^{-2/3} = -\frac{1}{4}$

c. $8^{-2/3} = -4$

d. $-8^{-2/3} = -\frac{1}{4}$

Practice Problems 14–16

Use the properties of exponents to simplify.

14. $x^{1/3}x^{1/4}$

15. $\frac{9^{2/5}}{9^{12/5}}$

16. $\frac{(3x^{2/3})^3}{x^2}$

Answers

14. $x^{7/12}$, **15.** $\frac{1}{81}$, **16.** 27

✓ **Concept Check:** d

E USING RATIONAL EXPONENTS TO SIMPLIFY RADICAL EXPRESSIONS

Some radical expressions are easier to simplify when we first write them with rational exponents. We can simplify some radical expressions by first writing the expression with rational exponents. Use properties of exponents to simplify, and then convert back to radical notation.

Examples Use rational exponents to simplify. Assume that all variables represent positive numbers.

17. $\sqrt[8]{x^4} = x^{4/8}$ Write with rational exponents.
$= x^{1/2}$ Simplify the exponent.
$= \sqrt{x}$ Write with radical notation.

18. $\sqrt[6]{25} = 25^{1/6}$ Write with rational exponents.
$= (5^2)^{1/6}$ Write 25 as 5^2.
$= 5^{2/6}$ Use the power rule.
$= 5^{1/3}$ Simplify the exponent.
$= \sqrt[3]{5}$ Write with radical notation.

19. $\sqrt[6]{r^2s^4} = (r^2s^4)^{1/6}$ Write with rational exponents.
$= r^{2/6}s^{4/6}$ Use the power rule.
$= r^{1/3}s^{2/3}$ Simplify the exponents.
$= (rs^2)^{1/3}$ Use $a^nb^n = (ab)^n$.
$= \sqrt[3]{rs^2}$ Write with radical notation.

Practice Problems 17–19

Use rational exponents to simplify. Assume that all variables represent positive numbers.

17. $\sqrt[10]{y^5}$

18. $\sqrt[4]{9}$

19. $\sqrt[9]{a^6b^3}$

Examples Use rational exponents to write as a single radical.

20. $\sqrt{x} \cdot \sqrt[4]{x} = x^{1/2} \cdot x^{1/4} = x^{1/2+1/4}$
$= x^{3/4} = \sqrt[4]{x^3}$

21. $\dfrac{\sqrt{x}}{\sqrt[3]{x}} = \dfrac{x^{1/2}}{x^{1/3}} = x^{1/2-1/3} = x^{3/6-2/6}$
$= x^{1/6} = \sqrt[6]{x}$

22. $\sqrt[3]{3} \cdot \sqrt{2} = 3^{1/3} \cdot 2^{1/2}$ Write with rational exponents.
$= 3^{2/6} \cdot 2^{3/6}$ Write the exponents so that they have the same denominator.
$= (3^2 \cdot 2^3)^{1/6}$ Use $a^nb^n = (ab)^n$.
$= \sqrt[6]{3^2 \cdot 2^3}$ Write with radical notation.
$= \sqrt[6]{72}$ Multiply $3^2 \cdot 2^3$.

Practice Problems 20–22

Use rational exponents to write as a single radical.

20. $\sqrt{y} \cdot \sqrt[3]{y}$

21. $\dfrac{\sqrt[3]{x}}{\sqrt[4]{x}}$

22. $\sqrt{5} \cdot \sqrt[3]{2}$

Answers

17. $\sqrt{y}$, **18.** $\sqrt{3}$, **19.** $\sqrt[3]{a^2b}$, **20.** $\sqrt[6]{y^5}$, **21.** $\sqrt[12]{x}$, **22.** $\sqrt[6]{500}$

Name ____________ Section ______ Date ______

EXERCISE SET 7.2

A *Use radical notation to write each expression. Simplify if possible. See Examples 1 through 6.*

1. $49^{1/2}$ **2.** $64^{1/3}$ **3.** $27^{1/3}$

4. $8^{1/3}$ **5.** $\left(\frac{1}{16}\right)^{1/4}$ **6.** $\left(\frac{1}{64}\right)^{1/2}$

7. $169^{1/2}$ **8.** $81^{1/4}$ **9.** $2m^{1/3}$

10. $(2m)^{1/3}$ **11.** $(9x^4)^{1/2}$ **12.** $(16x^8)^{1/2}$

13. $(-27)^{1/3}$ **14.** $-64^{1/2}$ **15.** $-16^{1/4}$ **16.** $(-32)^{1/5}$

B *Use radical notation to write each expression. Simplify if possible. See Examples 7 through 11.*

17. $16^{3/4}$ **18.** $4^{5/2}$ **19.** $(-64)^{2/3}$ **20.** $(-8)^{4/3}$

21. $(-16)^{3/4}$ **22.** $(-9)^{3/2}$ **23.** $(2x)^{3/5}$ **24.** $2x^{3/5}$

25. $(7x + 2)^{2/3}$ **26.** $(x - 4)^{3/4}$ **27.** $\left(\frac{16}{9}\right)^{3/2}$ **28.** $\left(\frac{49}{25}\right)^{3/2}$

C *Write with positive exponents. Simplify if possible. See Examples 12 and 13.*

29. $8^{-4/3}$ **30.** $64^{-2/3}$ **31.** $(-64)^{-2/3}$ **32.** $(-8)^{-4/3}$

ANSWERS

1. ______
2. ______
3. ______
4. ______
5. ______
6. ______
7. ______
8. ______
9. ______
10. ______
11. ______
12. ______
13. ______
14. ______
15. ______
16. ______
17. ______
18. ______
19. ______
20. ______
21. ______
22. ______
23. ______
24. ______
25. ______
26. ______
27. ______
28. ______
29. ______
30. ______
31. ______
32. ______

33.
34.
35.
36.
37.
38.
39.
40.
41.
42.
43.
44.
45.
46.
47.
48.
49.
50.
51.
52.
53.
54.
55.
56.

Name ______________________________

33. $(-4)^{-3/2}$

34. $(-16)^{-5/4}$

35. $x^{-1/4}$

36. $y^{-1/6}$

37. $\frac{1}{a^{-2/3}}$

38. $\frac{1}{n^{-8/9}}$

39. $\frac{5}{7x^{-3/4}}$

40. $\frac{2}{3y^{-5/7}}$

41. Explain how writing x^{-7} with positive exponents is similar to writing $x^{-1/4}$ with positive exponents.

42. Explain how writing $2x^{-5}$ with positive exponents is similar to writing $2x^{-3/4}$ with positive exponents.

D *Use the properties of exponents to simplify each expression. Write with positive exponents. See Examples 14 through 16.*

43. $a^{2/3}a^{5/3}$

44. $b^{9/5}b^{8/5}$

45. $x^{-2/5} \cdot x^{7/5}$

46. $y^{4/3} \cdot y^{-1/3}$

47. $3^{1/4} \cdot 3^{3/8}$

48. $5^{1/2} \cdot 5^{1/6}$

49. $\frac{y^{1/3}}{y^{1/6}}$

50. $\frac{x^{3/4}}{x^{1/8}}$

51. $(4u^2)^{3/2}$

52. $(32^{1/5}x^{2/3})^3$

53. $\frac{b^{1/2}b^{3/4}}{-b^{1/4}}$

54. $\frac{a^{1/4}a^{-1/2}}{a^{2/3}}$

55. $\frac{(3x^{1/4})^3}{x^{1/12}}$

56. $\frac{(2x^{1/5})^4}{x^{3/10}}$

Name ______________________

E *Use rational exponents to simplify each radical. Assume that all variables represent positive numbers. See Examples 17 through 19.*

57. $\sqrt[6]{x^3}$
58. $\sqrt[9]{a^3}$
59. $\sqrt[6]{4}$
60. $\sqrt[4]{36}$

61. $\sqrt[4]{16x^2}$
62. $\sqrt[8]{4y^2}$
63. $\sqrt[4]{(x+3)^2}$
64. $\sqrt[8]{(y+1)^4}$

65. $\sqrt[8]{x^4y^4}$
66. $\sqrt[9]{y^6z^3}$
67. $\sqrt[12]{a^8b^4}$
68. $\sqrt[10]{a^5b^5}$

Use rational expressions to write as a single radical expression. See Examples 20 through 22.

69. $\sqrt[3]{y} \cdot \sqrt[5]{y^2}$
70. $\sqrt[3]{y^2} \cdot \sqrt[6]{y}$
71. $\dfrac{\sqrt[3]{b^2}}{\sqrt[4]{b}}$

72. $\dfrac{\sqrt[4]{a}}{\sqrt[5]{a}}$
73. $\sqrt[3]{x} \cdot \sqrt[4]{x} \cdot \sqrt[8]{x^3}$
74. $\sqrt[6]{y} \cdot \sqrt[3]{y} \cdot \sqrt[5]{y^2}$

75. $\dfrac{\sqrt[3]{a^2}}{\sqrt[6]{a}}$
76. $\dfrac{\sqrt[5]{b^2}}{\sqrt[10]{b^3}}$
77. $\sqrt{3} \cdot \sqrt[3]{4}$

78. $\sqrt[3]{5} \cdot \sqrt{2}$
79. $\sqrt[5]{7} \cdot \sqrt[3]{y}$
80. $\sqrt[4]{5} \cdot \sqrt[3]{x}$

81. $\sqrt{5r} \cdot \sqrt[3]{s}$
82. $\sqrt[3]{b} \cdot \sqrt[5]{4a}$

Review and Preview

Write each integer as a product of two integers such that one of the factors is a perfect square. For example, write 18 as $9 \cdot 2$, because 9 is a perfect square.

83. 75
84. 20
85. 48
86. 45

57. ______
58. ______
59. ______
60. ______
61. ______
62. ______
63. ______
64. ______
65. ______
66. ______
67. ______
68. ______
69. ______
70. ______
71. ______
72. ______
73. ______
74. ______
75. ______
76. ______
77. ______
78. ______
79. ______
80. ______
81. ______
82. ______
83. ______
84. ______
85. ______
86. ______

87. ______

88. ______

89. ______

90. ______

91. ______

92. ______

93. ______

94. ______

95. ______

96. ______

97. ______

98. ______

Name ______

Write each integer as a product of two integers such that one of the factors is a perfect cube. For example, write 24 as 8 · 3, because 8 is a perfect cube.

87. 16 **88.** 56 **89.** 54 **90.** 80

Combining Concepts

Fill in each box with the correct expression.

91. $\square \cdot a^{2/3} = a^{3/3}$, or a

92. $\square \cdot x^{1/8} = x^{4/8}$, or $x^{1/2}$

93. $\dfrac{\square}{x^{-2/5}} = x^{3/5}$

94. $\dfrac{\square}{y^{-3/4}} = y^{4/4}$, or y

Use a calculator to write a four-decimal-place approximation of each number.

95. $8^{1/4}$

96. $18^{3/5}$

97. In physics, the speed of a wave traveling over a stretched string with tension t and density u is given by the expression $\dfrac{\sqrt{t}}{\sqrt{u}}$. Write this expression with rational exponents.

98. In electronics, the angular frequency of oscillations in a certain type of circuit is given by the expression $(LC)^{-1/2}$. Use radical notation to write this expression.

7.3 Simplifying Radical Expressions

Objectives

A Use the product rule for radicals.
B Use the quotient rule for radicals.
C Simplify radicals.

SSM CD-ROM Video 7.3

A Using the Product Rule

It is possible to simplify some radicals that do not evaluate to rational numbers. To do so, we use a product rule and a quotient rule for radicals. To discover the product rule, notice the following pattern:

$$\sqrt{9} \cdot \sqrt{4} = 3 \cdot 2 = 6$$
$$\sqrt{9 \cdot 4} = \sqrt{36} = 6$$

Since both expressions simplify to 6, it is true that

$$\sqrt{9} \cdot \sqrt{4} = \sqrt{9 \cdot 4}$$

This pattern suggests the following product rule for exponents.

Product Rule for Radicals

If $\sqrt[n]{a}$ and $\sqrt[n]{b}$ are real numbers, then

$$\sqrt[n]{a} \cdot \sqrt[n]{b} = \sqrt[n]{ab}$$

Notice that the product rule is the relationship $a^{1/n} \cdot b^{1/n} = (ab)^{1/n}$ stated in radical notation.

Examples Use the product rule to multiply.

1. $\sqrt{3} \cdot \sqrt{5} = \sqrt{3 \cdot 5} = \sqrt{15}$
2. $\sqrt{21} \cdot \sqrt{x} = \sqrt{21x}$
3. $\sqrt[3]{4} \cdot \sqrt[3]{2} = \sqrt[3]{4 \cdot 2} = \sqrt[3]{8} = 2$
4. $\sqrt[4]{5} \cdot \sqrt[4]{2x^3} = \sqrt[4]{5 \cdot 2x^3} = \sqrt[4]{10x^3}$
5. $\sqrt{\frac{2}{a}} \cdot \sqrt{\frac{b}{3}} = \sqrt{\frac{2}{a} \cdot \frac{b}{3}} = \sqrt{\frac{2b}{3a}}$

Practice Problems 1–5

Use the product rule to multiply.

1. $\sqrt{2} \cdot \sqrt{7}$
2. $\sqrt{17} \cdot \sqrt{y}$
3. $\sqrt[3]{2} \cdot \sqrt[3]{32}$
4. $\sqrt[4]{6} \cdot \sqrt[4]{3x^2}$
5. $\sqrt{\frac{3}{x}} \cdot \sqrt{\frac{y}{2}}$

B Using the Quotient Rule

To discover the quotient rule for radicals, notice the following pattern:

$$\sqrt{\frac{4}{9}} = \frac{2}{3}$$
$$\frac{\sqrt{4}}{\sqrt{9}} = \frac{2}{3}$$

Since both expressions simplify to $\frac{2}{3}$, it is true that

$$\sqrt{\frac{4}{9}} = \frac{\sqrt{4}}{\sqrt{9}}$$

This pattern suggests the following quotient rule for radicals.

Quotient Rule for Radicals

If $\sqrt[n]{a}$ and $\sqrt[n]{b}$ are real numbers and $\sqrt[n]{b}$ is not zero, then

$$\sqrt[n]{\frac{a}{b}} = \frac{\sqrt[n]{a}}{\sqrt[n]{b}}$$

Answers

1. $\sqrt{14}$, **2.** $\sqrt{17y}$, **3.** 4, **4.** $\sqrt[4]{18x^2}$, **5.** $\sqrt{\frac{3y}{2x}}$

Notice that the quotient rule is the relationship $\left(\frac{a}{b}\right)^{1/n} = \frac{a^{1/n}}{b^{1/n}}$ stated in radical notation. We can use the quotient rule to simplify radical expressions by reading the rule from left to right or to divide radicals by reading the rule from right to left.

For example,

$$\sqrt{\frac{x}{16}} = \frac{\sqrt{x}}{\sqrt{16}} = \frac{\sqrt{x}}{4} \qquad \text{Using } \sqrt[n]{\frac{a}{b}} = \frac{\sqrt[n]{a}}{\sqrt[n]{b}}$$

$$\frac{\sqrt{75}}{\sqrt{3}} = \sqrt{\frac{75}{3}} = \sqrt{25} = 5 \qquad \text{Using } \frac{\sqrt[n]{a}}{\sqrt[n]{b}} = \sqrt[n]{\frac{a}{b}}$$

Note: *For the remainder of this chapter, we will assume that variables represent positive real numbers. If this is so, we need not insert absolute value bars when we simplify even roots.*

Practice Problems 6–9

Use the quotient rule to simplify.

6. $\sqrt{\frac{9}{25}}$

7. $\sqrt{\frac{y}{36}}$

8. $\sqrt[3]{\frac{27}{64}}$

9. $\sqrt[5]{\frac{7}{32x^5}}$

Examples

Use the quotient rule to simplify.

6. $\sqrt{\frac{25}{49}} = \frac{\sqrt{25}}{\sqrt{49}} = \frac{5}{7}$

7. $\sqrt{\frac{x}{9}} = \frac{\sqrt{x}}{\sqrt{9}} = \frac{\sqrt{x}}{3}$

8. $\sqrt[3]{\frac{8}{27}} = \frac{\sqrt[3]{8}}{\sqrt[3]{27}} = \frac{2}{3}$

9. $\sqrt[4]{\frac{3}{16y^4}} = \frac{\sqrt[4]{3}}{\sqrt[4]{16y^4}} = \frac{\sqrt[4]{3}}{2y}$

C SIMPLIFYING RADICALS

Both the product and quotient rules can be used to simplify a radical. If the product rule is read from right to left, we have that $\sqrt[n]{ab} = \sqrt[n]{a} \cdot \sqrt[n]{b}$. We use this to simplify the following radicals.

Practice Problem 10

Simplify: $\sqrt{18}$

Example 10

Simplify: $\sqrt{50}$

Solution: We factor 50 such that one factor is the largest perfect square that divides 50. The largest perfect square factor of 50 is 25, so we write 50 as 25 · 2 and use the product rule for radicals to simplify.

$$\sqrt{50} = \sqrt{25 \cdot 2} = \sqrt{25} \cdot \sqrt{2} = 5\sqrt{2}$$

The largest perfect square factor of 50.

HELPFUL HINT

Don't forget that, for example, $5\sqrt{2}$ means $5 \cdot \sqrt{2}$.

Practice Problems 11–13

Simplify:

11. $\sqrt[3]{40}$

12. $\sqrt{14}$

13. $\sqrt[4]{162}$

Examples

Simplify.

11. $\sqrt[3]{24} = \sqrt[3]{8 \cdot 3} = \sqrt[3]{8} \cdot \sqrt[3]{3} = 2\sqrt[3]{3}$

The largest perfect cube factor of 24.

12. $\sqrt{26}$ The largest perfect square factor of 26 is 1, so $\sqrt{26}$ cannot be simplified further.

13. $\sqrt[4]{32} = \sqrt[4]{16 \cdot 2} = \sqrt[4]{16} \cdot \sqrt[4]{2} = 2\sqrt[4]{2}$

The largest 4th power factor of 32.

Answers

6. $\frac{3}{5}$, **7.** $\frac{\sqrt{y}}{6}$, **8.** $\frac{3}{4}$, **9.** $\frac{\sqrt[5]{7}}{2x}$, **10.** $3\sqrt{2}$, **11.** $2\sqrt[3]{5}$, **12.** $\sqrt{14}$, **13.** $3\sqrt[4]{2}$

After simplifying a radical such as a square root, always check the radicand to see that it contains no other perfect square factors. It may, if the largest perfect square factor of the radicand was not originally recognized. For example,

$$\sqrt{200} = \sqrt{4 \cdot 50} = \sqrt{4} \cdot \sqrt{50} = 2\sqrt{50}$$

Notice that the radicand 50 still contains the perfect square factor 25. This is because 4 is not the largest perfect square factor of 200. We continue as follows:

$$2\sqrt{50} = 2\sqrt{25 \cdot 2} = 2 \cdot \sqrt{25} \cdot \sqrt{2} = 2 \cdot 5 \cdot \sqrt{2} = 10\sqrt{2}$$

The radical is now simplified since 2 contains no perfect square factors (other than 1).

HELPFUL HINT

To help you recognize largest perfect power factors of a radicand, it will help if you are familiar with some perfect powers. A few are listed below

Perfect Squares	1, 4, 9, 16, 25, 36, 49, 64, 81, 100, 121, 144
Perfect Cubes	1, 8, 27, 64, 125
Perfect 4th powers	1, 16, 81, 256

HELPFUL HINT

We say that a radical of the form $\sqrt[n]{a}$ is simplified when the radicand a contains no factors that are perfect nth powers (other than 1 or -1).

Examples Simplify.

14. $\sqrt{25x^3} = \sqrt{25 \cdot x^2 \cdot x}$ Find the largest perfect square factor.
$= \sqrt{25 \cdot x^2} \cdot \sqrt{x}$ Use the product rule.
$= 5x\sqrt{x}$ Simplify.

15. $\sqrt[3]{54x^6y^8} = \sqrt[3]{27 \cdot 2 \cdot x^6 \cdot y^6 \cdot y^2}$ Factor the radicand and identify perfect cube factors.
$= \sqrt[3]{27 \cdot x^6 \cdot y^6 \cdot 2y^2}$
$= \sqrt[3]{27 \cdot x^6 \cdot y^6} \cdot \sqrt[3]{2y^2}$ Use the product rule.
$= 3x^2y^2\sqrt[3]{2y^2}$ Simplify.

16. $\sqrt[4]{81z^{11}} = \sqrt[4]{81 \cdot z^8 \cdot z^3}$ Factor the radicand and identify perfect fourth power factors.
$= \sqrt[4]{81 \cdot z^8} \cdot \sqrt[4]{z^3}$ Use the product rule.
$= 3z^2\sqrt[4]{z^3}$ Simplify.

Examples Use the quotient rule to divide. Then simplify if possible.

17. $\dfrac{\sqrt{20}}{\sqrt{5}} = \sqrt{\dfrac{20}{5}}$ Use the quotient rule.
$= \sqrt{4}$ Simplify.
$= 2$ Simplify.

18. $\dfrac{\sqrt{50x}}{2\sqrt{2}} = \dfrac{1}{2} \cdot \sqrt{\dfrac{50x}{2}}$ Use the quotient rule.

Practice Problems 14–16

Simplify.

14. $\sqrt{49a^5}$
15. $\sqrt[3]{24x^9y^7}$
16. $\sqrt[4]{16z^9}$

Practice Problems 17–20

Use the quotient rule to divide. Then simplify if possible.

17. $\dfrac{\sqrt{75}}{\sqrt{3}}$
18. $\dfrac{\sqrt{80y}}{3\sqrt{5}}$
19. $\dfrac{5\sqrt[3]{162x^8}}{\sqrt[3]{3x^2}}$
20. $\dfrac{3\sqrt[4]{243x^9y^6}}{\sqrt[4]{x^{-3}y}}$

Answers

14. $7a^2\sqrt{a}$, **15.** $2x^3y^2\sqrt[3]{3y}$, **16.** $2z^2\sqrt[4]{z}$, **17.** 5, **18.** $\frac{4}{3}\sqrt{y}$, **19.** $15x^2\sqrt[3]{2}$, **20.** $9x^3y\sqrt[4]{3y}$

$= \frac{1}{2} \cdot \sqrt{25x}$ Simplify.

$= \frac{1}{2} \cdot \sqrt{25} \cdot \sqrt{x}$ Factor $25x$.

$= \frac{1}{2} \cdot 5 \cdot \sqrt{x}$ Simplify.

$= \frac{5}{2}\sqrt{x}$

19. $\frac{7\sqrt[3]{48y^4}}{\sqrt[3]{2y}} = 7\sqrt[3]{\frac{48y^4}{2y}} = 7\sqrt[3]{24y^3} = 7\sqrt[3]{8 \cdot y^3 \cdot 3}$

$= 7\sqrt[3]{8 \cdot y^3} \cdot \sqrt[3]{3} = 7 \cdot 2y\sqrt[3]{3} = 14y\sqrt[3]{3}$

20. $\frac{2\sqrt[4]{32a^8b^6}}{\sqrt[4]{a^{-1}b^2}} = 2\sqrt[4]{\frac{32a^8b^6}{a^{-1}b^2}} = 2\sqrt[4]{32a^9b^4} = 2\sqrt[4]{16 \cdot a^8 \cdot b^4 \cdot 2 \cdot a}$

$= 2\sqrt[4]{16 \cdot a^8 \cdot b^4} \cdot \sqrt[4]{2 \cdot a} = 2 \cdot 2a^2b \cdot \sqrt[4]{2a} = 4a^2b\sqrt[4]{2a}$

TRY THE CONCEPT CHECK IN THE MARGIN.

✓ CONCEPT CHECK

Find and correct the error:

$\frac{\sqrt[3]{27}}{\sqrt{9}} = \sqrt[3]{\frac{27}{9}} = \sqrt[3]{3}$

Focus On History

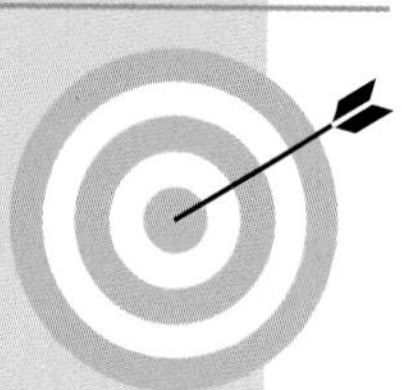

DEVELOPMENT OF THE RADICAL SYMBOL

The first mathematician to use the symbol we use today to denote a square root was Christoff Rudolff (1499–1545). In 1525, Rudolff wrote and published the first German algebra text, *Die Coss*. In it, he used √ to represent a square root, the symbol ʌʌ/ to represent a cube root, and the symbol ʌʌʌ/ to represent a fourth root, It was another 100 years before the square root symbol was extended with an overbar called a *vinculum*, $\sqrt{}$, to indicate the inclusion of several terms under the radical symbol. This innovation was introduced by René Descartes (1596–1650) in 1637 in his text *La Géométrie*. The modern use of a numeral as part of a radical sign to indicate the index of the radical for higher roots did not appear until 1690 when this notation was used by French mathematician Michel Rolle (1652–1719) in his text *Traité d'Algébre*.

Answer

✓ Concept Check: $\frac{\sqrt[3]{27}}{\sqrt{9}} = \frac{3}{3} = 1$

Name ______________________ Section ______ Date ______

Exercise Set 7.3

A *Use the product rule to multiply. See Examples 1 through 5.*

1. $\sqrt{7} \cdot \sqrt{2}$

2. $\sqrt{11} \cdot \sqrt{10}$

3. $\sqrt[4]{8} \cdot \sqrt[4]{2}$

4. $\sqrt[4]{27} \cdot \sqrt[4]{3}$

5. $\sqrt[3]{4} \cdot \sqrt[3]{9}$

6. $\sqrt[3]{10} \cdot \sqrt[3]{5}$

7. $\sqrt{2} \cdot \sqrt{3x}$

8. $\sqrt{3y} \cdot \sqrt{5x}$

9. $\sqrt{\frac{7}{x}} \cdot \sqrt{\frac{2}{y}}$

10. $\sqrt{\frac{6}{m}} \cdot \sqrt{\frac{n}{5}}$

11. $\sqrt[4]{4x^3} \cdot \sqrt[4]{5}$

12. $\sqrt[4]{ab^2} \cdot \sqrt[4]{27ab}$

B *Use the quotient rule to simplify. See Examples 6 through 9.*

13. $\sqrt{\frac{6}{49}}$

14. $\sqrt{\frac{8}{81}}$

15. $\sqrt{\frac{2}{49}}$

16. $\sqrt{\frac{5}{121}}$

17. $\sqrt[4]{\frac{x^3}{16}}$

18. $\sqrt[4]{\frac{y}{81x^4}}$

19. $\sqrt[3]{\frac{4}{27}}$

20. $\sqrt[3]{\frac{3}{64}}$

21. $\sqrt[4]{\frac{8}{x^8}}$

22. $\sqrt[4]{\frac{a^3}{81}}$

23. $\sqrt[3]{\frac{2x}{81y^{12}}}$

24. $\sqrt[3]{\frac{3}{8x^6}}$

25. $\sqrt{\frac{x^2y}{100}}$

26. $\sqrt{\frac{y^2z}{36}}$

27. $\sqrt{\frac{5x^2}{4y^2}}$

Answers

1. ______
2. ______
3. ______
4. ______
5. ______
6. ______
7. ______
8. ______
9. ______
10. ______
11. ______
12. ______
13. ______
14. ______
15. ______
16. ______
17. ______
18. ______
19. ______
20. ______
21. ______
22. ______
23. ______
24. ______
25. ______
26. ______
27. ______

28. ______
29. ______
30. ______
31. ______
32. ______
33. ______
34. ______
35. ______
36. ______
37. ______
38. ______
39. ______
40. ______
41. ______
42. ______
43. ______
44. ______
45. ______
46. ______
47. ______
48. ______
49. ______
50. ______
51. ______
52. ______
53. ______
54. ______
55. ______
56. ______
57. ______
58. ______

Name ______________________________

28. $\sqrt{\dfrac{y^{10}}{9x^6}}$ **29.** $-\sqrt[3]{\dfrac{z^7}{27x^3}}$ **30.** $-\sqrt[3]{\dfrac{64a}{b^9}}$

C *Simplify. See Examples 10 through 16.*

31. $\sqrt{32}$ **32.** $\sqrt{27}$ **33.** $\sqrt[3]{192}$

34. $\sqrt[3]{108}$ **35.** $5\sqrt{75}$ **36.** $3\sqrt{8}$

37. $\sqrt{24}$ **38.** $\sqrt{20}$ **39.** $\sqrt{100x^5}$

40. $\sqrt{64y^9}$ **41.** $\sqrt[3]{16y^7}$ **42.** $\sqrt[3]{64y^9}$

43. $\sqrt[4]{a^8b^7}$ **44.** $\sqrt[5]{32z^{12}}$ **45.** $\sqrt{y^5}$

46. $\sqrt[3]{y^5}$ **47.** $\sqrt{25a^2b^3}$ **48.** $\sqrt{9x^5y^7}$

49. $\sqrt[5]{-32x^{10}y}$ **50.** $\sqrt[5]{-243z^9}$ **51.** $\sqrt[3]{50x^{14}}$

52. $\sqrt[3]{40y^{10}}$ **53.** $-\sqrt{32a^8b^7}$ **54.** $-\sqrt{20ab^6}$

55. $\sqrt{9x^7y^9}$ **56.** $\sqrt{12r^9s^{12}}$ **57.** $\sqrt[3]{125r^9s^{12}}$ **58.** $\sqrt[3]{8a^6b^9}$

Name ______________________________

Use the quotient rule to divide. Then simplify if possible. See Examples 17 through 20.

59. $\dfrac{\sqrt{14}}{\sqrt{7}}$

60. $\dfrac{\sqrt{45}}{\sqrt{9}}$

61. $\dfrac{\sqrt[3]{24}}{\sqrt[3]{3}}$

62. $\dfrac{\sqrt[3]{10}}{\sqrt[3]{2}}$

63. $\dfrac{5\sqrt[4]{48}}{\sqrt[4]{3}}$

64. $\dfrac{7\sqrt[4]{162}}{\sqrt[4]{2}}$

65. $\dfrac{\sqrt{x^5y^3}}{\sqrt{xy}}$

66. $\dfrac{\sqrt{a^7b^6}}{\sqrt{a^3b^2}}$

67. $\dfrac{8\sqrt[3]{54m^7}}{\sqrt[3]{2m}}$

68. $\dfrac{\sqrt[3]{128x^3}}{-3\sqrt[3]{2x}}$

69. $\dfrac{3\sqrt{100x^2}}{2\sqrt{2x^{-1}}}$

70. $\dfrac{\sqrt{270y^2}}{5\sqrt{3y^{-4}}}$

71. $\dfrac{\sqrt[4]{96a^{10}b^3}}{\sqrt[4]{3a^2b^3}}$

72. $\dfrac{\sqrt[5]{64x^{10}y^3}}{\sqrt[5]{2x^3y^{-7}}}$

Review and Preview

Perform each indicated operation. See Sections 1.4 and 5.2.

73. $6x + 8x$

74. $(6x)(8x)$

75. $(2x + 3)(x - 5)$

76. $(2x + 3) + (x - 5)$

77. $9y^2 - 8y^2$

78. $(9y^2)(-8y^2)$

79. $-3(x + 5)$

80. $-3 + x + 5$

81. $(x - 4)^2$

82. $(2x + 1)^2$

59. ____________
60. ____________
61. ____________
62. ____________
63. ____________
64. ____________
65. ____________
66. ____________
67. ____________
68. ____________
69. ____________
70. ____________
71. ____________
72. ____________
73. ____________
74. ____________
75. ____________
76. ____________
77. ____________
78. ____________
79. ____________
80. ____________
81. ____________
82. ____________

83. a. ______________

b. ______________

84. ______________

85. a. ______________

b. ______________

c. ______________

Name ____

COMBINING CONCEPTS

83. The formula for the surface area A of a cone with height h and radius r is given by

$$A = \pi r\sqrt{r^2 + h^2}$$

a. Find the surface area of a cone whose height is 3 centimeters and whose radius is 4 centimeters.

b. Approximate to two decimal places the surface area of a cone whose height is 7.2 feet and whose radius is 6.8 feet.

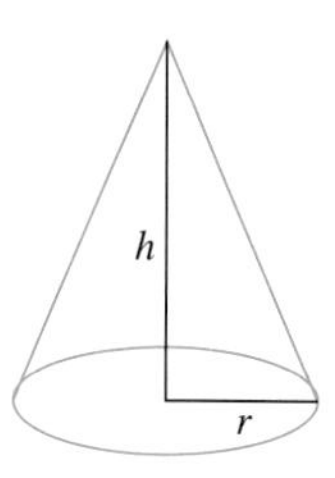

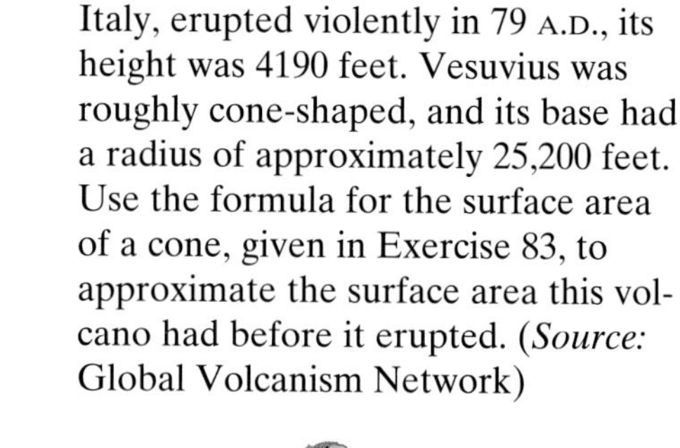

84. Before Mount Vesuvius, a volcano in Italy, erupted violently in 79 A.D., its height was 4190 feet. Vesuvius was roughly cone-shaped, and its base had a radius of approximately 25,200 feet. Use the formula for the surface area of a cone, given in Exercise 83, to approximate the surface area this volcano had before it erupted. (*Source:* Global Volcanism Network)

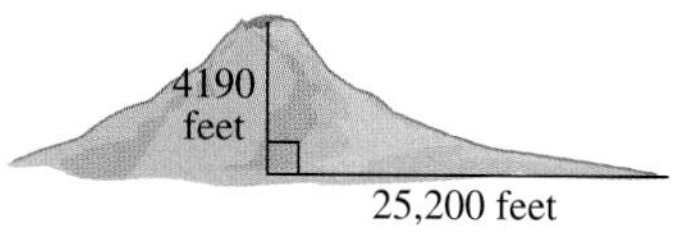

85. The owner of Knightime Video has determined that the demand equation for renting older released tapes is given by the equation $F(x) = 0.6\sqrt{49 - x^2}$, where x is the price in dollars per two-day rental and $F(x)$ is the number of times the video is demanded per week.

a. Approximate to one decimal place the demand per week of an older released video if the rental price is \$3 per two-day rental.

b. Approximate to one decimal place the demand per week of an older released video if the rental price is \$5 per two-day rental.

c. Explain how the owner of the video store can use this equation to predict the number of copies of each tape that should be in stock.

7.4 ADDING, SUBTRACTING, AND MULTIPLYING RADICAL EXPRESSIONS

Objectives

A Add or subtract radical expressions.

B Multiply radical expressions.

SSM CD-ROM Video 7.4

A ADDING OR SUBTRACTING RADICAL EXPRESSIONS

We have learned that the sum or difference of like terms can be simplified. To simplify these sums or differences, we use the distributive property. For example,

$$2x + 3x = (2 + 3)x = 5x$$

The distributive property can also be used to add **like radicals**.

LIKE RADICALS

Radicals with the same index and the same radicand are like radicals. For example,

$$2\sqrt{7} + 3\sqrt{7} = (2 + 3)\sqrt{7} = 5\sqrt{7}$$

HELPFUL HINT

The expression

$$5\sqrt{7} - 3\sqrt{6}$$

does not contain like radicals and cannot be simplified further.

Examples Add or subtract as indicated.

1. $4\sqrt{11} + 8\sqrt{11} = (4 + 8)\sqrt{11} = 12\sqrt{11}$

2. $5\sqrt[3]{3x} - 7\sqrt[3]{3x} = (5 - 7)\sqrt[3]{3x} = -2\sqrt[3]{3x}$

3. $2\sqrt{7} + 2\sqrt[3]{7}$ This expression cannot be simplified since $2\sqrt{7}$ and $2\sqrt[3]{7}$ do not contain like radicals.

Practice Problems 1–3

Add or subtract as indicated.

1. $5\sqrt{15} + 2\sqrt{15}$
2. $9\sqrt[3]{2y} - 15\sqrt[3]{2y}$
3. $6\sqrt{10} - 3\sqrt[3]{10}$

✓ **CONCEPT CHECK**

True or false?
$\sqrt{a} + \sqrt{b} = \sqrt{a + b}$
Explain.

TRY THE CONCEPT CHECK IN THE MARGIN.

When adding or subtracting radicals, always check first to see whether any radicals can be simplified.

Examples Add or subtract as indicated.

4.
$$\begin{aligned}\sqrt{20} + 2\sqrt{45} &= \sqrt{4 \cdot 5} + 2\sqrt{9 \cdot 5} && \text{Factor 20 and 45.}\\ &= \sqrt{4} \cdot \sqrt{5} + 2 \cdot \sqrt{9} \cdot \sqrt{5} && \text{Use the product rule.}\\ &= 2 \cdot \sqrt{5} + 2 \cdot 3 \cdot \sqrt{5} && \text{Simplify } \sqrt{4} \text{ and } \sqrt{9}.\\ &= 2\sqrt{5} + 6\sqrt{5} && \text{Add like radicals.}\\ &= 8\sqrt{5}\end{aligned}$$

5.
$$\begin{aligned}&\sqrt[3]{54} - 5\sqrt[3]{16} + \sqrt[3]{2}\\ &= \sqrt[3]{27} \cdot \sqrt[3]{2} - 5 \cdot \sqrt[3]{8} \cdot \sqrt[3]{2} + \sqrt[3]{2} && \text{Factor and use the product rule.}\\ &= 3 \cdot \sqrt[3]{2} - 5 \cdot 2 \cdot \sqrt[3]{2} + \sqrt[3]{2} && \text{Simplify } \sqrt[3]{27} \text{ and } \sqrt[3]{8}.\\ &= 3\sqrt[3]{2} - 10\sqrt[3]{2} + \sqrt[3]{2} && \text{Write } 5 \cdot 2 \text{ as 10.}\\ &= -6\sqrt[3]{2} && \text{Combine like radicals.}\end{aligned}$$

Practice Problems 4–8

Add or subtract as indicated.

4. $\sqrt{50} + 5\sqrt{18}$
5. $\sqrt[3]{24} - 4\sqrt[3]{192} + \sqrt[3]{3}$
6. $\sqrt{20x} - 6\sqrt{16x} + \sqrt{45x}$
7. $\sqrt[4]{32} + \sqrt{32}$
8. $\sqrt[3]{8y^5} + \sqrt[3]{27y^5}$

Answers

1. $7\sqrt{15}$, **2.** $-6\sqrt[3]{2y}$, **3.** $6\sqrt{10} - 3\sqrt[3]{10}$, **4.** $20\sqrt{2}$, **5.** $-13\sqrt[3]{3}$, **6.** $5\sqrt{5x} - 24\sqrt{x}$, **7.** $2\sqrt[4]{2} + 4\sqrt{2}$, **8.** $5y\sqrt[3]{y^2}$

✓ **Concept Check:** false; answers may vary

6. $\sqrt{27x} - 2\sqrt{9x} + \sqrt{72x}$

$= \sqrt{9} \cdot \sqrt{3x} - 2 \cdot \sqrt{9} \cdot \sqrt{x} + \sqrt{36} \cdot \sqrt{2x}$ — Factor and use the product rule.

$= 3 \cdot \sqrt{3x} - 2 \cdot 3 \cdot \sqrt{x} + 6 \cdot \sqrt{2x}$ — Simplify $\sqrt{9}$ and $\sqrt{36}$.

$= 3\sqrt{3x} - 6\sqrt{x} + 6\sqrt{2x}$ — Write $2 \cdot 3$ as 6.

HELPFUL HINT

None of these terms contain like radicals. We can simplify no further.

7. $\sqrt[3]{98} + \sqrt{98} = \sqrt[3]{98} + \sqrt{49} \cdot \sqrt{2}$ — Factor and use the product rule.

$= \sqrt[3]{98} + 7\sqrt{2}$ — No further simplification is possible.

8. $\sqrt[3]{48y^4} + \sqrt[3]{6y^4} = \sqrt[3]{8y^3} \cdot \sqrt[3]{6y} + \sqrt[3]{y^3} \cdot \sqrt[3]{6y}$ — Factor and use the product rule.

$= 2y\sqrt[3]{6y} + y\sqrt[3]{6y}$ — Simplify $\sqrt[3]{8y^3}$ and $\sqrt[3]{y^3}$.

$= 3y\sqrt[3]{6y}$ — Combine like radicals.

Practice Problems 9–10

Add or subtract as indicated.

9. $\dfrac{\sqrt{75}}{9} - \dfrac{\sqrt{3}}{2}$

10. $\sqrt[3]{\dfrac{5x}{27}} + 4\sqrt[3]{5x}$

Examples

Add or subtract as indicated.

9. $\dfrac{\sqrt{45}}{4} - \dfrac{\sqrt{5}}{3} = \dfrac{3\sqrt{5}}{4} - \dfrac{\sqrt{5}}{3}$ — To subtract, notice that the LCD is 12.

$= \dfrac{3\sqrt{5} \cdot 3}{4 \cdot 3} - \dfrac{\sqrt{5} \cdot 4}{3 \cdot 4}$ — Write each expression as an equivalent expression with a denominator of 12.

$= \dfrac{9\sqrt{5}}{12} - \dfrac{4\sqrt{5}}{12}$ — Multiply factors in the numerators and the denominators.

$= \dfrac{5\sqrt{5}}{12}$ — Subtract.

10. $\sqrt[3]{\dfrac{7x}{8}} + 2\sqrt[3]{7x} = \dfrac{\sqrt[3]{7x}}{\sqrt[3]{8}} + 2\sqrt[3]{7x}$ — Use the quotient rule for radicals.

$= \dfrac{\sqrt[3]{7x}}{2} + 2\sqrt[3]{7x}$ — Simplify.

$= \dfrac{\sqrt[3]{7x}}{2} + \dfrac{2\sqrt[3]{7x} \cdot 2}{2}$ — Write each expression as an equivalent expression with a denominator of 2.

$= \dfrac{\sqrt[3]{7x}}{2} + \dfrac{4\sqrt[3]{7x}}{2}$

$= \dfrac{5\sqrt[3]{7x}}{2}$ — Add.

B MULTIPLYING RADICAL EXPRESSIONS

We can multiply radical expressions by using many of the same properties used to multiply polynomial expressions. For instance, to multiply $\sqrt{2}(\sqrt{6} - 3\sqrt{2})$, we use the distributive property and multiply $\sqrt{2}$ by each term inside the parentheses.

$\sqrt{2}(\sqrt{6} - 3\sqrt{2}) = \sqrt{2}(\sqrt{6}) - \sqrt{2}(3\sqrt{2})$ — Use the distributive property.

$= \sqrt{2 \cdot 6} - 3\sqrt{2 \cdot 2}$ — Use the product rule for radicals.

$= \sqrt{2 \cdot 2 \cdot 3} - 3 \cdot 2$

$= 2\sqrt{3} - 6$

Answers

9. $\dfrac{19\sqrt{3}}{18}$, **10.** $\dfrac{13\sqrt[3]{5x}}{3}$

Example 11 Multiply: $\sqrt{3}(5 + \sqrt{30})$

Solution:

$$\begin{aligned}\sqrt{3}(5 + \sqrt{30}) &= \sqrt{3}(5) + \sqrt{3}(\sqrt{30})\\ &= 5\sqrt{3} + \sqrt{3 \cdot 30}\\ &= 5\sqrt{3} + \sqrt{3 \cdot 3 \cdot 10}\\ &= 5\sqrt{3} + 3\sqrt{10}\end{aligned}$$

Examples Multiply.

12. $(\sqrt{5} - \sqrt{6})(\sqrt{7} + 1) = \underset{\textbf{First}}{\sqrt{5} \cdot \sqrt{7}} + \underset{\textbf{Outer}}{\sqrt{5} \cdot 1} - \underset{\textbf{Inner}}{\sqrt{6} \cdot \sqrt{7}} - \underset{\textbf{Last}}{\sqrt{6} \cdot 1}$ Using the FOIL order.

$= \sqrt{35} + \sqrt{5} - \sqrt{42} - \sqrt{6}$ Simplify.

13. $(\sqrt{2x} + 5)(\sqrt{2x} - 5) = (\sqrt{2x})^2 - 5^2$ Multiply the sum and difference of two terms: $(a + b)(a - b) = a^2 - b^2$.

$= 2x - 25$

14. $(\sqrt{3} - 1)^2 = (\sqrt{3})^2 - 2 \cdot \sqrt{3} \cdot 1 + 1^2$ Square the binomial: $(a - b)^2 = a^2 - 2ab + b^2$.

$= 3 - 2\sqrt{3} + 1$

$= 4 - 2\sqrt{3}$

Square the binomial: $(a + b)^2 = a^2 + 2ab + b^2$

15. $(\underbrace{\sqrt{x - 3}}_{a} + \underbrace{5}_{b})^2 = \underbrace{(\sqrt{x - 3})^2}_{a^2} \underbrace{+ 2 \cdot}_{+2\cdot} \underbrace{\sqrt{x - 3}}_{a} \underbrace{\cdot 5}_{\cdot b} \underbrace{+ 5^2}_{+b^2}$

$= x - 3 + 10\sqrt{x - 3} + 25$ Simplify.

$= x + 22 + 10\sqrt{x - 3}$ Combine like terms.

Practice Problem 11

Multiply: $\sqrt{2}(6 + \sqrt{10})$

Practice Problems 12–15

Multiply.

12. $(\sqrt{3} - \sqrt{5})(\sqrt{2} + 7)$

13. $(\sqrt{5y} + 2)(\sqrt{5y} - 2)$

14. $(\sqrt{3} - 7)^2$

15. $(\sqrt{x + 1} + 2)^2$

Answers

11. $6\sqrt{2} + 2\sqrt{5}$, **12.** $\sqrt{6} + 7\sqrt{3} - \sqrt{10} - 7\sqrt{5}$, **13.** $5y - 4$, **14.** $52 - 14\sqrt{3}$, **15.** $x + 5 + 4\sqrt{x + 1}$

Focus On the Real World

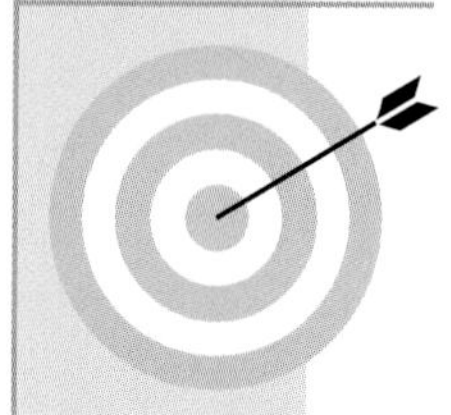

DIFFUSION

Diffusion is the spontaneous movement of the molecules of a substance from a region of higher concentration to a region of lower concentration until a uniform concentration throughout the region is reached. For example, if a drop of food coloring is added to a glass of water, the molecules of the coloring are diffused so that the entire glass of water is colored evenly without any kind of stirring. Diffusion is also mostly responsible for the spread of the smell of baking brownies throughout a house.

Diffusion is used or seen in important aspects of many disciplines. The following list describes situations in which diffusion plays a role.

- In the commercial production of sugar, sugar can be extracted from sugar cane through a diffusion process.
- Solid-state diffusion plays a role in the manufacturing process of silicon computer chips.
- In biology, the diffusion phenomenon allows water molecules, nutrient molecules, and dissolved gas molecules (such as oxygen and carbon dioxide) to pass through the semipermeable membranes of cell walls.
- During a human pregnancy, the fetus is nourished from the mother's blood supply via diffusion through the placenta. Waste materials from the fetus are also diffused through the placenta to be carried away by the mother's circulatory system.
- The medical treatment known as kidney dialysis, in which waste materials are removed from the blood of a patient without kidney function, is made possible by diffusion.
- A diffusion process is widely used to separate the uranium isotope U-235, which can be used as a fuel in nuclear power plants, from the uranium isotope U-238, which cannot be used to create nuclear energy.

In chemistry, Graham's law states that the diffusion rate of a substance in its gaseous state is inversely proportional to the square root of its molecular weight. Another useful property of diffusion is that the distance a material diffuses over time is directly proportional to the square root of the time.

CRITICAL THINKING

1. Write an equation for the relationship described by Graham's law. Be sure to define the variables and constants that you use.

2. According to Graham's law, which molecule will diffuse more rapidly: a molecule with a molecular weight of 58.4 or a molecule with a molecular weight of 180.2? Explain your reasoning.

3. Write an equation for the relationship between the distance that a material diffuses and time. Again, be sure to define the variables and constants that you use.

4. Suppose it takes sugar 1 week to diffuse a distance of 1 cm from its starting point in a particular liquid. How long will it take the sugar to diffuse a total of 3 cm from its starting point in the liquid?

Name ______________________ Section ______ Date ________

MENTAL MATH

Simplify. Assume that all variables represent positive real numbers.

1. $2\sqrt{3} + 4\sqrt{3}$ **2.** $5\sqrt{7} + 3\sqrt{7}$ **3.** $8\sqrt{x} - 5\sqrt{x}$ **4.** $3\sqrt{y} + 10\sqrt{y}$

5. $7\sqrt[3]{x} + 5\sqrt[3]{x}$ **6.** $8\sqrt[3]{z} - 2\sqrt[3]{z}$ **7.** $(\sqrt{3})^2$ **8.** $(\sqrt{4x} + 1)^2$

EXERCISE SET 7.4

A *Add or subtract as indicated. See Examples 1 through 10.*

1. $\sqrt{8} - \sqrt{32}$

2. $\sqrt{27} - \sqrt{75}$

3. $2\sqrt{2x^3} + 4x\sqrt{8x}$

4. $3\sqrt{45x^3} + x\sqrt{5x}$

5. $2\sqrt{50} - 3\sqrt{125} + \sqrt{98}$

6. $4\sqrt{32} - \sqrt{18} + 2\sqrt{128}$

7. $\sqrt[3]{16x} - \sqrt[3]{54x}$

8. $2\sqrt[3]{3a^4} - 3a\sqrt[3]{81a}$

9. $\sqrt{9b^3} - \sqrt{25b^3} + \sqrt{49b^3}$

10. $\sqrt{4x^7} + 9x^2\sqrt{x^3} - 5x\sqrt{x^5}$

11. $\dfrac{5\sqrt{2}}{3} + \dfrac{2\sqrt{2}}{5}$

12. $\dfrac{\sqrt{3}}{2} + \dfrac{4\sqrt{3}}{3}$

13. $\sqrt[3]{\dfrac{11}{8}} - \dfrac{\sqrt[3]{11}}{6}$

14. $\dfrac{2\sqrt[3]{4}}{7} - \dfrac{\sqrt[3]{4}}{14}$

15. $\dfrac{\sqrt{20x}}{9} + \sqrt{\dfrac{5x}{9}}$

16. $\dfrac{3x\sqrt{7}}{5} + \sqrt{\dfrac{7x^2}{100}}$

17. $7\sqrt{9} - 7 + \sqrt{3}$

18. $\sqrt{16} - 5\sqrt{10} + 7$

19. $2 + 3\sqrt{y^2} - 6\sqrt{y^2} + 5$

20. $3\sqrt{7} - \sqrt[3]{x} + 4\sqrt{7} - 3\sqrt[3]{x}$

MENTAL MATH ANSWERS

1. ________
2. ________
3. ________
4. ________
5. ________
6. ________
7. ________
8. ________

ANSWERS

1. ________
2. ________
3. ________
4. ________
5. ________
6. ________
7. ________
8. ________
9. ________
10. ________
11. ________
12. ________
13. ________
14. ________
15. ________
16. ________
17. ________
18. ________
19. ________
20. ________

21. ________
22. ________
23. ________
24. ________
25. ________
26. ________
27. ________
28. ________
29. ________
30. ________
31. ________
32. ________
33. ________
34. ________
35. ________
36. ________
37. ________
38. ________
39. ________
40. ________
41. ________
42. ________
43. ________
44. ________

Name ____________________

21. $3\sqrt{108} - 2\sqrt{18} - 3\sqrt{48}$

22. $-\sqrt{75} + \sqrt{12} - 3\sqrt{3}$

23. $-5\sqrt[3]{625} + \sqrt[3]{40}$

24. $-2\sqrt[3]{108} - \sqrt[3]{32}$

25. $\sqrt{9b^3} - \sqrt{25b^3} + \sqrt{16b^3}$

26. $\sqrt{4x^7y^5} + 9x^2\sqrt{x^3y^5} - 5xy\sqrt{x^5y^3}$

27. $5y\sqrt{8y} + 2\sqrt{50y^3}$

28. $3\sqrt{8x^2y^3} - 2x\sqrt{32y^3}$

29. $\sqrt[3]{54xy^3} - 5\sqrt[3]{2xy^3} + y\sqrt[3]{128x}$

30. $2\sqrt[3]{24x^3y^4} + 4x\sqrt[3]{81y^4}$

31. $6\sqrt[3]{11} + 8\sqrt{11} - 12\sqrt{11}$

32. $3\sqrt[3]{5} + 4\sqrt{5}$

33. $-2\sqrt[4]{x^7} + 3\sqrt[4]{16x^7}$

34. $6\sqrt[3]{24x^3} - 2\sqrt[3]{81x^3} - x\sqrt[3]{3}$

35. $\frac{4\sqrt{3}}{3} - \frac{\sqrt{12}}{3}$

36. $\frac{\sqrt{45}}{10} + \frac{7\sqrt{5}}{10}$

37. $\frac{\sqrt[3]{8x^4}}{7} + \frac{3x\sqrt[3]{x}}{7}$

38. $\frac{\sqrt[4]{48}}{5x} - \frac{2\sqrt[4]{3}}{10x}$

39. $\sqrt{\frac{28}{x^2}} + \sqrt{\frac{7}{4x^2}}$

40. $\frac{\sqrt{99}}{5x} - \sqrt{\frac{44}{x^2}}$

41. $\sqrt[3]{\frac{16}{27}} - \frac{\sqrt[3]{54}}{6}$

42. $\frac{\sqrt[3]{3}}{10} + \sqrt[3]{\frac{24}{125}}$

43. $-\frac{\sqrt[3]{2x^4}}{9} + \sqrt[3]{\frac{250x^4}{27}}$

44. $\frac{\sqrt[3]{y^5}}{8} + \frac{5y\sqrt[3]{y^2}}{4}$

Name ______________________________

45. Find the perimeter of the trapezoid.

$2\sqrt{12}$ in., $3\sqrt{3}$ in., $\sqrt{12}$ in., $2\sqrt{27}$ in.

46. Find the perimeter of the triangle.

$\sqrt{8}$ m, $\sqrt{32}$ m, $\sqrt{45}$ m

B *Multiply. Then simplify if possible. See Examples 11 through 15.*

47. $\sqrt{7}(\sqrt{5} + \sqrt{3})$

48. $\sqrt{5}(\sqrt{15} - \sqrt{35})$

49. $(\sqrt{5} - \sqrt{2})^2$

50. $(3x - \sqrt{2})(3x - \sqrt{2})$

51. $\sqrt{3x}(\sqrt{3} - \sqrt{x})$

52. $\sqrt{5y}(\sqrt{y} + \sqrt{5})$

53. $(2\sqrt{x} - 5)(3\sqrt{x} + 1)$

54. $(8\sqrt{y} + z)(4\sqrt{y} - 1)$

55. $(\sqrt[3]{a} - 4)(\sqrt[3]{a} + 5)$

56. $(\sqrt[3]{a} + 2)(\sqrt[3]{a} + 7)$

57. $6(\sqrt{2} - 2)$

58. $\sqrt{5}(6 - \sqrt{5})$

59. $\sqrt{2}(\sqrt{2} + x\sqrt{6})$

60. $\sqrt{3}(\sqrt{3} - 2\sqrt{5x})$

61. $(2\sqrt{7} + 3\sqrt{5})(\sqrt{7} - 2\sqrt{5})$

62. $(\sqrt{6} - 4\sqrt{2})(3\sqrt{6} + 1)$

63. $(\sqrt{x} - y)(\sqrt{x} + y)$

64. $(3\sqrt{x} + 2)(\sqrt{3x} - 2)$

65. $(\sqrt{3} + x)^2$

66. $(\sqrt{y} - 3x)^2$

67. $(\sqrt{5x} - 3\sqrt{2})(\sqrt{5x} - 3\sqrt{3})$

68. $(5\sqrt{3x} - \sqrt{y})(4\sqrt{x} + 1)$

45. ______
46. ______
47. ______
48. ______
49. ______
50. ______
51. ______
52. ______
53. ______
54. ______
55. ______
56. ______
57. ______
58. ______
59. ______
60. ______
61. ______
62. ______
63. ______
64. ______
65. ______
66. ______
67. ______
68. ______

69.

70.

71.

72.

73.

74.

75.

76.

77.

78.

79.

80.

81.

82.

83.

84.

85.

86.

87.

Name

69. $(\sqrt[3]{4}+2)(\sqrt[3]{2}-1)$

70. $(\sqrt[3]{3}+\sqrt[3]{2})(\sqrt[3]{9}-\sqrt[3]{4})$

71. $(\sqrt[3]{x}+1)(\sqrt[3]{x}-4\sqrt{x}+7)$

72. $(\sqrt[3]{3x}+3)(\sqrt[3]{2x}-3x-1)$

73. $(\sqrt{x-1}+5)^2$

74. $(\sqrt{3x+1}+2)^2$

75. $(\sqrt{2x+5}-1)^2$

76. $(\sqrt{x-6}-7)^2$

Review and Preview

Factor each numerator and denominator. Then simplify if possible. See Section 6.1.

77. $\dfrac{2x-14}{2}$

78. $\dfrac{8x-24y}{4}$

79. $\dfrac{7x-7y}{x^2-y^2}$

80. $\dfrac{x^3-8}{4x-8}$

81. $\dfrac{6a^2b-9ab}{3ab}$

82. $\dfrac{14r-28r^2s^2}{7rs}$

83. $\dfrac{-4+2\sqrt{3}}{6}$

84. $\dfrac{-5+10\sqrt{7}}{5}$

Combining Concepts

85. Find the perimeter and area of the rectangle.

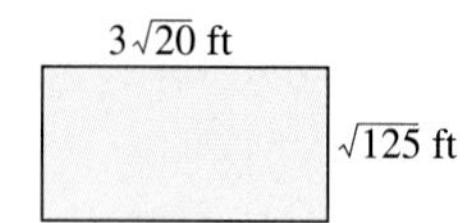

86. Find the area and perimeter of the trapezoid. (*Hint:* The area of a trapezoid is the product of half the height $6\sqrt{3}$ meters and the sum of the bases $2\sqrt{63}$ and $7\sqrt{7}$ meters.)

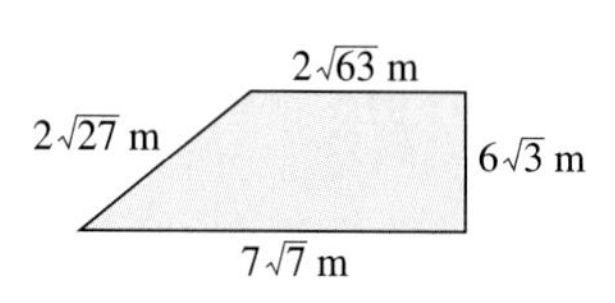

87. Multiply: $(\sqrt{2}+\sqrt{3}-1)^2$

7.5 RATIONALIZING NUMERATORS AND DENOMINATORS OF RADICAL EXPRESSIONS

Objectives

A Rationalize denominators.

B Rationalize numerators.

C Rationalize denominators or numerators having two terms.

SSM CD-ROM Video 7.5

A RATIONALIZING DENOMINATORS

Often in mathematics, it is helpful to write a radical expression such as $\frac{\sqrt{3}}{\sqrt{2}}$ either without a radical in the denominator or without a radical in the numerator. The process of writing this expression as an equivalent expression but without a radical in the denominator is called **rationalizing the denominator**. To rationalize the denominator of $\frac{\sqrt{3}}{\sqrt{2}}$, we use the fundamental principle of fractions and multiply the numerator and the denominator by $\sqrt{2}$. Recall that this is the same as multiplying by $\frac{\sqrt{2}}{\sqrt{2}}$, which simplifies to 1.

$$\frac{\sqrt{3}}{\sqrt{2}} = \frac{\sqrt{3} \cdot \sqrt{2}}{\sqrt{2} \cdot \sqrt{2}} = \frac{\sqrt{6}}{\sqrt{4}} = \frac{\sqrt{6}}{2}$$

Example 1 Rationalize the denominator of $\frac{2}{\sqrt{5}}$.

Solution: To rationalize the denominator, we multiply the numerator and denominator by a factor that makes the radicand in the denominator a perfect square.

$$\frac{2}{\sqrt{5}} = \frac{2 \cdot \sqrt{5}}{\sqrt{5} \cdot \sqrt{5}} = \frac{2\sqrt{5}}{5}$$ The denominator is now rationalized.

Practice Problem 1

Rationalize the denominator of $\frac{7}{\sqrt{2}}$.

Example 2 Rationalize the denominator of $\frac{2\sqrt{16}}{\sqrt{9x}}$.

Solution: First we simplify the radicals; then we rationalize the denominator.

$$\frac{2\sqrt{16}}{\sqrt{9x}} = \frac{2(4)}{\sqrt{9} \cdot \sqrt{x}} = \frac{8}{3\sqrt{x}}$$

To rationalize the denominator, we multiply the numerator and the denominator by $\sqrt{x}$.

$$\frac{8}{3\sqrt{x}} = \frac{8 \cdot \sqrt{x}}{3\sqrt{x} \cdot \sqrt{x}} = \frac{8\sqrt{x}}{3x}$$

Practice Problem 2

Rationalize the denominator of $\frac{2\sqrt{9}}{\sqrt{16y}}$.

Example 3 Rationalize the denominator of $\sqrt[3]{\frac{1}{2}}$.

Solution:

$$\sqrt[3]{\frac{1}{2}} = \frac{\sqrt[3]{1}}{\sqrt[3]{2}} = \frac{1}{\sqrt[3]{2}}$$

Now we rationalize the denominator. Since $\sqrt[3]{2}$ is a cube root, we want to multiply by a value that will make the radicand 2 a perfect cube. If we multiply by $\sqrt[3]{2^2}$, we get $\sqrt[3]{2^3} = 2$. Thus,

$$\frac{1 \cdot \sqrt[3]{2^2}}{\sqrt[3]{2} \cdot \sqrt[3]{2^2}} = \frac{\sqrt[3]{4}}{\sqrt[3]{2^3}} = \frac{\sqrt[3]{4}}{2}$$ Multiply numerator and denominator by $\sqrt[3]{2^2}$ and then simplify.

Practice Problem 3

Rationalize the denominator of $\sqrt[3]{\frac{2}{25}}$.

TRY THE CONCEPT CHECK IN THE MARGIN.

✓ CONCEPT CHECK

In each case, determine by which number both the numerator and denominator should be multiplied to rationalize the denominator of the radical expression.

a. $\frac{1}{\sqrt[3]{7}}$ b. $\frac{1}{\sqrt[4]{8}}$

Answers

1. $\frac{7\sqrt{2}}{2}$, **2.** $\frac{3\sqrt{y}}{2y}$, **3.** $\frac{\sqrt[3]{10}}{5}$

✓ Concept Check: a. $\sqrt[3]{7^2}$ or $\sqrt[3]{49}$, b. $\sqrt[4]{2}$

Practice Problem 4

Rationalize the denominator of $\sqrt{\frac{5m}{11n}}$.

Practice Problem 5

Rationalize the denominator of

$\frac{\sqrt[5]{a^2}}{\sqrt[5]{32b^{12}}}$.

Practice Problem 6

Rationalize the numerator of $\frac{\sqrt{18}}{\sqrt{75}}$.

Practice Problem 7

Rationalize the numerator of $\frac{\sqrt[3]{3a}}{\sqrt[3]{7b}}$

Answers

4. $\frac{\sqrt{55mn}}{11n}$, **5.** $\frac{\sqrt[5]{a^2b^3}}{2b^3}$, **6.** $\frac{6}{5\sqrt{6}}$, **7.** $\frac{3a}{\sqrt[3]{63a^2b}}$

Example 4 Rationalize the denominator of $\sqrt{\frac{7x}{3y}}$.

Solution:

$$\sqrt{\frac{7x}{3y}} = \frac{\sqrt{7x}}{\sqrt{3y}}$$ Use the quotient rule. No radical may be simplified further.

$$= \frac{\sqrt{7x}\cdot\sqrt{3y}}{\sqrt{3y}\cdot\sqrt{3y}}$$ Multiply numerator and denominator by $\sqrt{3y}$ so that the radicand in the denominator is a perfect square.

$$= \frac{\sqrt{21xy}}{3y}$$ Use the product rule in the numerator and denominator. Remember that $\sqrt{3y}\cdot\sqrt{3y} = 3y$.

Example 5 Rationalize the denominator of $\frac{\sqrt[4]{x}}{\sqrt[4]{81y^5}}$.

Solution: First, simplify each radical if possible.

$$\frac{\sqrt[4]{x}}{\sqrt[4]{81y^5}} = \frac{\sqrt[4]{x}}{\sqrt[4]{81y^4}\cdot\sqrt[4]{y}}$$ Use the product rule in the denominator.

$$= \frac{\sqrt[4]{x}}{3y\sqrt[4]{y}}$$ Write $\sqrt[4]{81y^4}$ as $3y$.

$$= \frac{\sqrt[4]{x}\cdot\sqrt[4]{y^3}}{3y\sqrt[4]{y}\cdot\sqrt[4]{y^3}}$$ Multiply numerator and denominator by $\sqrt[4]{y^3}$ so that the radicand in the denominator is a perfect 4th power.

$$= \frac{\sqrt[4]{xy^3}}{3y\sqrt[4]{y^4}}$$ Use the product rule in the numerator and denominator.

$$= \frac{\sqrt[4]{xy^3}}{3y^2}$$ In the denominator, $\sqrt[4]{y^4} = y$ and $3y\cdot y = 3y^2$.

B RATIONALIZING NUMERATORS

As mentioned earlier, it is also often helpful to write an expression such as $\frac{\sqrt{3}}{\sqrt{2}}$ as an equivalent expression without a radical in the numerator. This process is called **rationalizing the numerator**. To rationalize the numerator of $\frac{\sqrt{3}}{\sqrt{2}}$, we multiply the numerator and the denominator by $\sqrt{3}$.

$$\frac{\sqrt{3}}{\sqrt{2}} = \frac{\sqrt{3}\cdot\sqrt{3}}{\sqrt{2}\cdot\sqrt{3}} = \frac{\sqrt{9}}{\sqrt{6}} = \frac{3}{\sqrt{6}}$$

Example 6 Rationalize the numerator of $\frac{\sqrt{7}}{\sqrt{45}}$.

Solution: First we simplify $\sqrt{45}$.

$$\frac{\sqrt{7}}{\sqrt{45}} = \frac{\sqrt{7}}{\sqrt{9\cdot 5}} = \frac{\sqrt{7}}{3\sqrt{5}}$$

Next we rationalize the numerator by multiplying the numerator and the denominator by $\sqrt{7}$.

$$\frac{\sqrt{7}}{3\sqrt{5}} = \frac{\sqrt{7}\cdot\sqrt{7}}{3\sqrt{5}\cdot\sqrt{7}} = \frac{7}{3\sqrt{5\cdot 7}} = \frac{7}{3\sqrt{35}}$$

Example 7 Rationalize the numerator of $\frac{\sqrt[3]{2x^2}}{\sqrt[3]{5y}}$.

Solution:

$$\frac{\sqrt[3]{2x^2}}{\sqrt[3]{5y}} = \frac{\sqrt[3]{2x^2}\cdot\sqrt[3]{2^2x}}{\sqrt[3]{5y}\cdot\sqrt[3]{2^2x}}$$ Multiply the numerator and denominator by $\sqrt[3]{2^2x}$ so that the radicand in the numerator is a perfect cube.

$$= \frac{\sqrt[3]{2^3x^3}}{\sqrt[3]{5y\cdot 2^2x}}$$ Use the product rule in the numerator and denominator.

$$= \frac{2x}{\sqrt[3]{20xy}}$$ Simplify.

C RATIONALIZING DENOMINATORS OR NUMERATORS HAVING TWO TERMS

Remember the product of the sum and difference of two terms?

$$(a+b)(a-b) = a^2 - b^2$$

These two expressions are called conjugates of each other.

To rationalize a numerator or denominator that is a sum or difference of two terms, we use conjugates. To see how and why this works, let's rationalize the denominator of the expression $\frac{5}{\sqrt{3}-2}$. To do so, we multiply both the numerator and the denominator by $\sqrt{3}+2$, the **conjugate** of the denominator $\sqrt{3}-2$ and see what happens.

$$\frac{5}{\sqrt{3}-2} = \frac{5(\sqrt{3}+2)}{(\sqrt{3}-2)(\sqrt{3}+2)}$$

$$= \frac{5(\sqrt{3}+2)}{(\sqrt{3})^2 - 2^2}$$ Multiply the sum and difference of two terms: $(a+b)(a-b) = a^2 - b^2$.

$$= \frac{5(\sqrt{3}+2)}{3-4}$$

$$= \frac{5(\sqrt{3}+2)}{-1}$$

$$= -5(\sqrt{3}+2) \quad \text{or} \quad -5\sqrt{3}-10$$

Notice in the denominator that the product of $(\sqrt{3}-2)$ and its conjugate, $(\sqrt{3}+2)$, is -1. In general, the product of an expression and its conjugate will contain no radical terms. This is why, when rationalizing a denominator or a numerator containing two terms, we multiply by its conjugate. Examples of conjugates are

$\sqrt{a}-\sqrt{b}$ and $\sqrt{a}+\sqrt{b}$

$x+\sqrt{y}$ and $x-\sqrt{y}$

Example 8 Rationalize the denominator of $\frac{2}{3\sqrt{2}+4}$.

Solution: We multiply the numerator and the denominator by the conjugate of $3\sqrt{2}+4$.

$$\frac{2}{3\sqrt{2}+4} = \frac{2(3\sqrt{2}-4)}{(3\sqrt{2}+4)(3\sqrt{2}-4)}$$

$$= \frac{2(3\sqrt{2}-4)}{(3\sqrt{2})^2 - 4^2}$$ Multiply the sum and difference of two terms: $(a+b)(a-b) = a^2 - b^2$.

Practice Problem 8

Rationalize the denominator of $\frac{3}{2\sqrt{5}+1}$.

Answer

8. $\frac{3(2\sqrt{5}-1)}{19}$

$$= \frac{2(3\sqrt{2} - 4)}{18 - 16} \quad \text{Write } (3\sqrt{2})^2 \text{ as } 9 \cdot 2 \text{ or } 18 \text{ and } 4^2 \text{ as } 16.$$

$$= \frac{2(3\sqrt{2} - 4)}{2} \quad \text{or} \quad 3\sqrt{2} - 4$$

As we saw in Example 8, it is often helpful to leave a numerator in factored form to help determine whether the expression can be simplified.

Practice Problem 9

Rationalize the denominator of $\frac{\sqrt{5} + 3}{\sqrt{3} - \sqrt{2}}$.

Example 9 Rationalize the denominator of $\frac{\sqrt{6} + 2}{\sqrt{5} - \sqrt{3}}$.

Solution: We multiply the numerator and the denominator by the conjugate of $\sqrt{5} - \sqrt{3}$.

$$\frac{\sqrt{6} + 2}{\sqrt{5} - \sqrt{3}} = \frac{(\sqrt{6} + 2)(\sqrt{5} + \sqrt{3})}{(\sqrt{5} - \sqrt{3})(\sqrt{5} + \sqrt{3})}$$

$$= \frac{\sqrt{6}\sqrt{5} + \sqrt{6}\sqrt{3} + 2\sqrt{5} + 2\sqrt{3}}{(\sqrt{5})^2 - (\sqrt{3})^2}$$

$$= \frac{\sqrt{30} + \sqrt{18} + 2\sqrt{5} + 2\sqrt{3}}{5 - 3}$$

$$= \frac{\sqrt{30} + 3\sqrt{2} + 2\sqrt{5} + 2\sqrt{3}}{2}$$

Practice Problem 10

Rationalize the denominator of $\frac{3}{2 - \sqrt{x}}$.

Example 10 Rationalize the denominator of $\frac{2\sqrt{m}}{3\sqrt{x} + \sqrt{m}}$.

Solution: We multiply by the conjugate of $3\sqrt{x} + \sqrt{m}$ to eliminate the radicals from the denominator.

$$\frac{2\sqrt{m}}{3\sqrt{x} + \sqrt{m}} = \frac{2\sqrt{m}(3\sqrt{x} - \sqrt{m})}{(3\sqrt{x} + \sqrt{m})(3\sqrt{x} - \sqrt{m})} = \frac{6\sqrt{mx} - 2m}{(3\sqrt{x})^2 - (\sqrt{m})^2}$$

$$= \frac{6\sqrt{mx} - 2m}{9x - m}$$

Practice Problem 11

Rationalize the numerator of $\frac{\sqrt{x} + 5}{3}$.

Example 11 Rationalize the numerator of $\frac{\sqrt{x} + 2}{5}$.

Solution: We multiply the numerator and the denominator by the conjugate of $\sqrt{x} + 2$, the numerator.

$$\frac{\sqrt{x} + 2}{5} = \frac{(\sqrt{x} + 2)(\sqrt{x} - 2)}{5(\sqrt{x} - 2)} \quad \text{Multiply by } \sqrt{x} - 2 \text{, the conjugate of } \sqrt{x} + 2.$$

$$= \frac{(\sqrt{x})^2 - 2^2}{5(\sqrt{x} - 2)} \quad (a + b)(a - b) = a^2 - b^2.$$

$$= \frac{x - 4}{5(\sqrt{x} - 2)}$$

Answers

9. $\sqrt{15} + \sqrt{10} + 3\sqrt{3} + 3\sqrt{2}$, **10.** $\frac{6 + 3\sqrt{x}}{4 - x}$, **11.** $\frac{x - 25}{3(\sqrt{x} - 5)}$

Name ______________________ Section ______ Date ________

Mental Math Answers

1. ______
2. ______
3. ______
4. ______
5. ______
6. ______

Answers

1. ______
2. ______
3. ______
4. ______
5. ______
6. ______
7. ______
8. ______
9. ______
10. ______
11. ______
12. ______
13. ______
14. ______
15. ______
16. ______
17. ______
18. ______
19. ______
20. ______
21. ______
22. ______
23. ______
24. ______
25. ______

Mental Math

Find the conjugate of each expression.

1. $\sqrt{2} + x$ **2.** $\sqrt{3} + y$ **3.** $5 - \sqrt{a}$

4. $6 - \sqrt{b}$ **5.** $7\sqrt{4} + 8\sqrt{x}$ **6.** $9\sqrt{2} - 6\sqrt{y}$

Exercise Set 7.5

A *Rationalize each denominator. See Examples 1 through 5.*

1. $\frac{\sqrt{2}}{\sqrt{7}}$ **2.** $\frac{\sqrt{3}}{\sqrt{2}}$ **3.** $\sqrt{\frac{1}{5}}$

4. $\sqrt{\frac{1}{2}}$ **5.** $\frac{4}{\sqrt[3]{3}}$ **6.** $\frac{6}{\sqrt[3]{9}}$

7. $\frac{3}{\sqrt{8x}}$ **8.** $\frac{5}{\sqrt{27a}}$ **9.** $\frac{3}{\sqrt[3]{4x^2}}$

10. $\frac{5}{\sqrt[3]{3y}}$ **11.** $\frac{9}{\sqrt{3a}}$ **12.** $\frac{x}{\sqrt{5}}$

13. $\frac{3}{\sqrt[3]{2}}$ **14.** $\frac{5}{\sqrt[3]{9}}$ **15.** $\frac{2\sqrt{3}}{\sqrt{7}}$

16. $\frac{-5\sqrt{2}}{\sqrt{11}}$ **17.** $\sqrt{\frac{2x}{5y}}$ **18.** $\sqrt{\frac{13a}{2b}}$

19. $\sqrt[4]{\frac{16}{9x^7}}$ **20.** $\sqrt[5]{\frac{32}{m^6n^{13}}}$ **21.** $\frac{5a}{\sqrt[5]{8a^9b^{11}}}$ **22.** $\frac{9y}{\sqrt[4]{4y^9}}$

B *Rationalize each numerator. See Examples 6 and 7.*

23. $\sqrt{\frac{5}{3}}$ **24.** $\sqrt{\frac{3}{2}}$ **25.** $\sqrt{\frac{18}{5}}$

26.

27.

28.

29.

30.

31.

32.

33.

34.

35.

36.

37.

38.

39.

40.

41.

42.

43.

44.

45.

46.

47.

48.

49.

50.

51.

Name ______________________

26. $\sqrt{\frac{12}{7}}$ **27.** $\frac{\sqrt{4x}}{7}$ **28.** $\frac{\sqrt{3x^5}}{6}$

29. $\frac{\sqrt[3]{5y^2}}{\sqrt[3]{4x}}$ **30.** $\frac{\sqrt[3]{4x}}{\sqrt[3]{z^4}}$ **31.** $\sqrt{\frac{2}{5}}$

32. $\sqrt{\frac{3}{7}}$ **33.** $\frac{\sqrt{2x}}{11}$ **34.** $\frac{\sqrt{y}}{7}$

35. $\sqrt[3]{\frac{7}{8}}$ **36.** $\sqrt[3]{\frac{25}{2}}$ **37.** $\frac{\sqrt[3]{3x^5}}{10}$

38. $\sqrt[3]{\frac{9y}{7}}$ **39.** $\sqrt{\frac{18x^4y^6}{3z}}$ **40.** $\sqrt{\frac{8x^5y}{2z}}$

41. When rationalizing the denominator of $\frac{\sqrt{5}}{\sqrt{7}}$, explain why both the numerator and the denominator must be multiplied by $\sqrt{7}$.

42. When rationalizing the numerator of $\frac{\sqrt{5}}{\sqrt{7}}$, explain why both the numerator and the denominator must be multiplied by $\sqrt{5}$.

C *Rationalize each denominator. See Examples 8 through 10.*

43. $\frac{6}{2 - \sqrt{7}}$ **44.** $\frac{3}{\sqrt{7} - 4}$ **45.** $\frac{-7}{\sqrt{x} - 3}$

46. $\frac{-8}{\sqrt{y} + 4}$ **47.** $\frac{\sqrt{2} - \sqrt{3}}{\sqrt{2} + \sqrt{3}}$ **48.** $\frac{\sqrt{3} + \sqrt{4}}{\sqrt{2} + \sqrt{3}}$

49. $\frac{\sqrt{a} + 1}{2\sqrt{a} - \sqrt{b}}$ **50.** $\frac{2\sqrt{a} - 3}{2\sqrt{a} - \sqrt{b}}$ **51.** $\frac{8}{1 + \sqrt{10}}$

Name ______________________________

52. $\dfrac{-3}{\sqrt{6} - 2}$

53. $\dfrac{\sqrt{x}}{\sqrt{x} + \sqrt{y}}$

54. $\dfrac{2\sqrt{a}}{2\sqrt{x} - \sqrt{y}}$

55. $\dfrac{2\sqrt{3} + \sqrt{6}}{4\sqrt{3} - \sqrt{6}}$

56. $\dfrac{4\sqrt{5} + \sqrt{2}}{2\sqrt{5} - \sqrt{2}}$

Rationalize each numerator. See Example 11.

57. $\dfrac{2 - \sqrt{11}}{6}$

58. $\dfrac{\sqrt{15} + 1}{2}$

59. $\dfrac{2 - \sqrt{7}}{-5}$

60. $\dfrac{\sqrt{5} + 2}{\sqrt{2}}$

61. $\dfrac{\sqrt{x} + 3}{\sqrt{x}}$

62. $\dfrac{5 + \sqrt{2}}{\sqrt{2x}}$

63. $\dfrac{\sqrt{2} - 1}{\sqrt{2} + 1}$

64. $\dfrac{\sqrt{8} - \sqrt{3}}{\sqrt{2} + \sqrt{3}}$

65. $\dfrac{\sqrt{x} + 1}{\sqrt{x} - 1}$

66. $\dfrac{\sqrt{x} + \sqrt{y}}{\sqrt{x} - \sqrt{y}}$

Review and Preview

Solve each equation. See Sections 2.1 and 5.7.

67. $2x - 7 = 3(x - 4)$

68. $9x - 4 = 7(x - 2)$

69. $(x - 6)(2x + 1) = 0$

70. $(y + 2)(5y + 4) = 0$

71. $x^2 - 8x = -12$

72. $x^3 = x$

52. ______
53. ______
54. ______
55. ______
56. ______
57. ______
58. ______
59. ______
60. ______
61. ______
62. ______
63. ______
64. ______
65. ______
66. ______
67. ______
68. ______
69. ______
70. ______
71. ______
72. ______

73. ______

Name ______

Combining Concepts

73. The formula of the radius of a sphere r with surface area A is given by the formula

$$r = \sqrt{\frac{A}{4\pi}}$$

Rationalize the denominator of the radical expression in this formula.

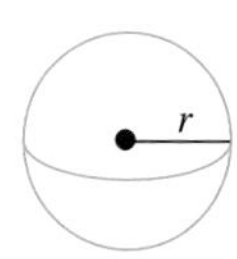

74. The formula for the radius of a cone r with height 7 centimeters and volume V is given by the formula

$$r = \sqrt{\frac{3V}{7\pi}}$$

Rationalize the numerator of the radical expression in this formula.

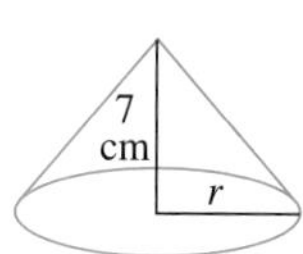

74. ______

75. Explain why rationalizing the denominator does not change the value of the original expression.

76. Explain why rationalizing the numerator does not change the value of the original expression.

75. ______

76. ______

Name ______________________ Section ______ Date ________

INTEGRATED REVIEW—RADICALS AND RATIONAL EXPONENTS

Throughout this review, assume that variables represent positive real numbers. Find each root.

1. $\sqrt{81}$ **2.** $\sqrt[3]{-8}$ **3.** $\sqrt[4]{\frac{1}{16}}$ **4.** $\sqrt{x^6}$

5. $\sqrt[3]{y^9}$ **6.** $\sqrt{4y^{10}}$ **7.** $\sqrt[5]{-32y^5}$ **8.** $\sqrt[4]{81b^{12}}$

Use radical notation to rewrite each expression. Simplify if possible.

9. $36^{1/2}$ **10.** $(3y)^{1/4}$ **11.** $64^{-2/3}$ **12.** $(x + 1)^{3/5}$

Use the properties of exponents to simplify each expression. Write with positive exponents.

13. $y^{-1/6} \cdot y^{7/6}$ **14.** $\frac{(2x^{1/3})^4}{x^{5/6}}$ **15.** $\frac{x^{1/4}x^{3/4}}{x^{-1/4}}$ **16.** $4^{1/3} \cdot 4^{2/5}$

Use rational exponents to simplify each radical.

17. $\sqrt[3]{8x^6}$ **18.** $\sqrt[12]{a^9b^6}$

Use rational exponents to write each as a single radical expression.

19. $\sqrt[4]{x} \cdot \sqrt{x}$ **20.** $\sqrt{5} \cdot \sqrt[3]{2}$

Simplify.

21. $\sqrt{40}$ **22.** $\sqrt[4]{16x^7y^{10}}$

23. $\sqrt[3]{54x^4}$ **24.** $\sqrt[5]{-64b^{10}}$

ANSWERS

1. ____
2. ____
3. ____
4. ____
5. ____
6. ____
7. ____
8. ____
9. ____
10. ____
11. ____
12. ____
13. ____
14. ____
15. ____
16. ____
17. ____
18. ____
19. ____
20. ____
21. ____
22. ____
23. ____
24. ____

25. ______
26. ______
27. ______
28. ______
29. ______
30. ______
31. ______
32. ______
33. ______
34. ______
35. ______
36. ______
37. ______
38. ______
39. ______
40. ______

Name ______________________________

Multiply or divide. Then simplify if possible.

25. $\sqrt{5} \cdot \sqrt{x}$

26. $\sqrt[3]{8x} \cdot \sqrt[3]{8x^2}$

27. $\dfrac{\sqrt{98y^6}}{\sqrt{2y}}$

28. $\dfrac{\sqrt[4]{48a^9b^3}}{\sqrt[4]{ab^3}}$

Perform indicated operations.

29. $\sqrt{20} - \sqrt{75} + 5\sqrt{7}$

30. $\sqrt[3]{54y^4} - y\sqrt[3]{16y}$

31. $\sqrt{3}(\sqrt{5} - \sqrt{2})$

32. $(\sqrt{7} + \sqrt{3})^2$

33. $(2x - \sqrt{5})(2x + \sqrt{5})$

34. $(\sqrt{x+1} - 1)^2$

Rationalize each denominator.

35. $\sqrt{\dfrac{7}{3}}$

36. $\dfrac{5}{\sqrt[3]{2x^2}}$

37. $\dfrac{\sqrt{3} - \sqrt{7}}{2\sqrt{3} + \sqrt{7}}$

Rationalize each numerator.

38. $\sqrt{\dfrac{7}{3}}$

39. $\sqrt[3]{\dfrac{9y}{11}}$

40. $\dfrac{\sqrt{x} - 2}{\sqrt{x}}$

7.6 RADICAL EQUATIONS AND PROBLEM SOLVING

Objectives

A Solve equations that contain radical expressions.

B Use the Pythagorean theorem to model problems.

SSM CD-ROM Video 7.6

A SOLVING EQUATIONS THAT CONTAIN RADICAL EXPRESSIONS

In this section, we present techniques to solve equations containing radical expressions such as

$$\sqrt{2x-3} = 9$$

We use the power rule to help us solve these radical equations.

> **POWER RULE**
>
> If both sides of an equation are raised to the same power, *all* solutions of the original equation are *among* the solutions of the new equation.

This property *does not* say that raising both sides of an equation to a power yields an equivalent equation. A solution of the new equation *may or may not* be a solution of the original equation. Thus, *each solution of the new equation must be checked* to make sure it is a solution of the original equation. Recall that a proposed solution that is not a solution of the original equation is called an extraneous solution.

Example 1 Solve: $\sqrt{2x-3} = 9$

Solution: We use the power rule to square both sides of the equation to eliminate the radical.

$$\begin{aligned} \sqrt{2x-3} &= 9 \\ (\sqrt{2x-3})^2 &= 9^2 \\ 2x-3 &= 81 \\ 2x &= 84 \\ x &= 42 \end{aligned}$$

Now we check the solution in the original equation.

Check:

$$\begin{aligned} \sqrt{2x-3} &= 9 \\ \sqrt{2(42)-3} &\stackrel{?}{=} 9 \quad \text{Let } x = 42. \\ \sqrt{84-3} &\stackrel{?}{=} 9 \\ \sqrt{81} &\stackrel{?}{=} 9 \\ 9 &= 9 \quad \text{True.} \end{aligned}$$

The solution checks, so we conclude that the solution set is $\{42\}$.

Practice Problem 1

Solve: $\sqrt{3x-2} = 5$

To solve a radical equation, first isolate a radical on one side of the equation.

Example 2 Solve: $\sqrt{-10x-1} + 3x = 0$

Solution: First, isolate the radical on one side of the equation. To do this, we subtract $3x$ from both sides.

$$\begin{aligned} \sqrt{-10x-1} + 3x &= 0 \\ \sqrt{-10x-1} + 3x - 3x &= 0 - 3x \\ \sqrt{-10x-1} &= -3x \end{aligned}$$

Practice Problem 2

Solve: $\sqrt{9x-2} - 2x = 0$

Answers

1. $\{9\}$, **2.** $\left\{\frac{1}{4}, 2\right\}$

Next we use the power rule to eliminate the radical.

$$(\sqrt{-10x - 1})^2 = (-3x)^2$$
$$-10x - 1 = 9x^2$$

Since this is a quadratic equation, we can set the equation equal to 0 and try to solve by factoring.

$$9x^2 + 10x + 1 = 0$$
$$(9x + 1)(x + 1) = 0 \quad \text{Factor.}$$
$$9x + 1 = 0 \quad \text{or} \quad x + 1 = 0 \quad \text{Set each factor equal to 0.}$$
$$x = -\frac{1}{9} \qquad x = -1$$

Check: Let $x = -\frac{1}{9}$.

$$\sqrt{-10x - 1} + 3x = 0$$
$$\sqrt{-10\left(-\frac{1}{9}\right) - 1} + 3\left(-\frac{1}{9}\right) \stackrel{?}{=} 0$$
$$\sqrt{\frac{10}{9} - \frac{9}{9}} - \frac{3}{9} \stackrel{?}{=} 0$$
$$\sqrt{\frac{1}{9}} - \frac{1}{3} \stackrel{?}{=} 0$$
$$\frac{1}{3} - \frac{1}{3} = 0 \quad \text{True.}$$

Let $x = -1$.

$$\sqrt{-10x - 1} + 3x = 0$$
$$\sqrt{-10(-1) - 1} + 3(-1) \stackrel{?}{=} 0$$
$$\sqrt{10 - 1} - 3 \stackrel{?}{=} 0$$
$$\sqrt{9} - 3 \stackrel{?}{=} 0$$
$$3 - 3 = 0 \quad \text{True.}$$

Both solutions check. The solution set is $\left\{-\frac{1}{9}, -1\right\}$.

The following steps may be used to solve a radical equation.

SOLVING A RADICAL EQUATION

Step 1. Isolate one radical on one side of the equation.

Step 2. Raise each side of the equation to a power equal to the index of the radical and simplify.

Step 3. If the equation still contains a radical term, repeat Steps 1 and 2. If not, solve the equation.

Step 4. Check all proposed solutions in the original equation.

Practice Problem 3

Solve: $\sqrt[3]{x - 5} + 2 = 1$

Example 3 Solve: $\sqrt[3]{x + 1} + 5 = 3$

Solution: First we isolate the radical by subtracting 5 from both sides of the equation.

$$\sqrt[3]{x + 1} + 5 = 3$$
$$\sqrt[3]{x + 1} = -2$$

Next we raise both sides of the equation to the third power to eliminate the radical.

Answer

3. $\{4\}$

$$(\sqrt[3]{x+1})^3 = (-2)^3$$
$$x + 1 = -8$$
$$x = -9$$

The solution checks in the original equation, so the solution set is $\{-9\}$.

Example 4 Solve: $\sqrt{4-x} = x - 2$

Solution:

$$\sqrt{4-x} = x - 2$$
$$(\sqrt{4-x})^2 = (x-2)^2$$
$$4 - x = x^2 - 4x + 4$$
$$x^2 - 3x = 0 \quad \text{Write the quadratic equation in standard form.}$$
$$x(x-3) = 0 \quad \text{Factor.}$$
$$x = 0 \text{ or } x - 3 = 0 \quad \text{Set each factor equal to 0.}$$
$$x = 3$$

Check:

$\sqrt{4-x} = x - 2$

$\sqrt{4-0} \stackrel{?}{=} 0 - 2$ Let $x = 0$.

$2 = -2$ False.

$\sqrt{4-x} = x - 2$

$\sqrt{4-3} \stackrel{?}{=} 3 - 2$ Let $x = 3$.

$1 = 1$ True.

The proposed solution 3 checks, but 0 does not. Since 0 is an extraneous solution, the solution set is $\{3\}$.

Practice Problem 4

Solve: $\sqrt{9+x} = x + 3$

HELPFUL HINT

In Example 4, notice that $(x-2)^2 = x^2 - 4x + 4$. Make sure binomials are squared correctly.

✓ CONCEPT CHECK

How can you immediately tell that the equation $\sqrt{2y+3} = -4$ has no real solution?

TRY THE CONCEPT CHECK IN THE MARGIN.

Example 5 Solve: $\sqrt{2x+5} + \sqrt{2x} = 3$

Solution: We get one radical alone by subtracting $\sqrt{2x}$ from both sides.

$$\sqrt{2x+5} + \sqrt{2x} = 3$$
$$\sqrt{2x+5} = 3 - \sqrt{2x}$$

Now we use the power rule to begin eliminating the radicals. First we square both sides.

$$(\sqrt{2x+5})^2 = (3 - \sqrt{2x})^2$$
$$2x + 5 = 9 - 6\sqrt{2x} + 2x \quad \text{Multiply: } (3-\sqrt{2x})(3-\sqrt{2x}).$$

There is still a radical in the equation, so we get the radical alone again. Then we square both sides.

Practice Problem 5

Solve: $\sqrt{3x+1} + \sqrt{3x} = 2$

Answers

4. $\{0\}$, **5.** $\left\{\frac{3}{16}\right\}$

✓ Concept Check: answers may vary

$$2x + 5 = 9 - 6\sqrt{2x} + 2x$$
$$6\sqrt{2x} = 4 \quad \text{Get the radical alone.}$$
$$(6\sqrt{2x})^2 = 4^2 \quad \text{Square both sides of the equation to eliminate the radical.}$$
$$36(2x) = 16$$
$$72x = 16 \quad \text{Multiply.}$$
$$x = \frac{16}{72} \quad \text{Solve.}$$
$$x = \frac{2}{9} \quad \text{Simplify.}$$

The proposed solution, $\frac{2}{9}$ checks in the original equation.

The solution set is $\left\{\frac{2}{9}\right\}$.

HELPFUL HINT

Make sure expressions are squared correctly. In Example 5, we squared $(3 - \sqrt{2x})$ as

$$(3 - \sqrt{2x})^2 = (3 - \sqrt{2x})(3 - \sqrt{2x})$$
$$= 3 \cdot 3 - 3\sqrt{2x} - 3\sqrt{2x} + \sqrt{2x} \cdot \sqrt{2x}$$
$$= 9 - 6\sqrt{2x} + 2x$$

✓ CONCEPT CHECK

What is wrong with the following solution?

$$\sqrt{2x+5} + \sqrt{4-x} = 8$$
$$(\sqrt{2x+5} + \sqrt{4-x})^2 = 8^2$$
$$(2x + 5) + (4 - x) = 64$$
$$x + 9 = 64$$
$$x = 55$$

TRY THE CONCEPT CHECK IN THE MARGIN.

B USING THE PYTHAGOREAN THEOREM

Recall that the Pythagorean theorem states that in a right triangle, the length of the hypotenuse squared equals the sum of the lengths of each of the legs squared.

PYTHAGOREAN THEOREM

If a and b are the lengths of the legs of a right triangle and c is the length of the hypotenuse, then $a^2 + b^2 = c^2$.

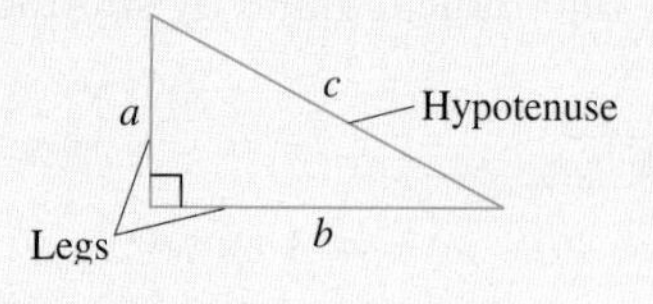

Practice Problem 6

Find the length of the unknown leg of the right triangle.

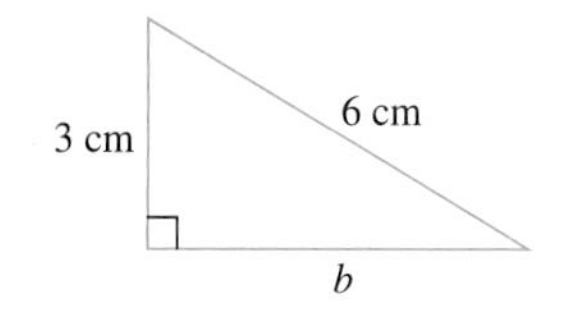

Example 6 Find the length of the unknown leg of the right triangle.

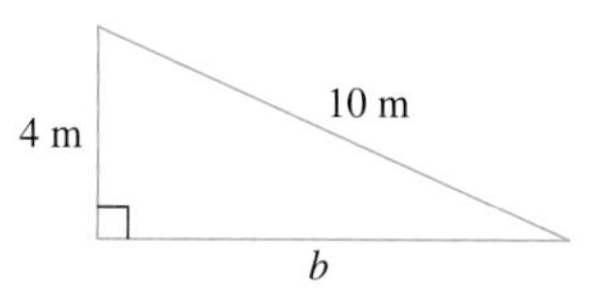

Solution: In the formula $a^2 + b^2 = c^2$, c is the hypotenuse. Here, $c = 10$, the length of the hypotenuse, and $a = 4$. We solve for b. Then $a^2 + b^2 = c^2$ becomes

Answers

6. $3\sqrt{3}$ cm

✓ **Concept Check:** answers may vary

$$4^2 + b^2 = 10^2$$
$$16 + b^2 = 100$$
$$b^2 = 84 \quad \text{Subtract 16 from both sides.}$$

Since b is a length and thus is positive, we have that

$$b = \sqrt{84} = \sqrt{4 \cdot 21} = 2\sqrt{21}$$

The unknown leg of the triangle is $2\sqrt{21}$ meters long.

Example 7 Calculating Placement of a Wire

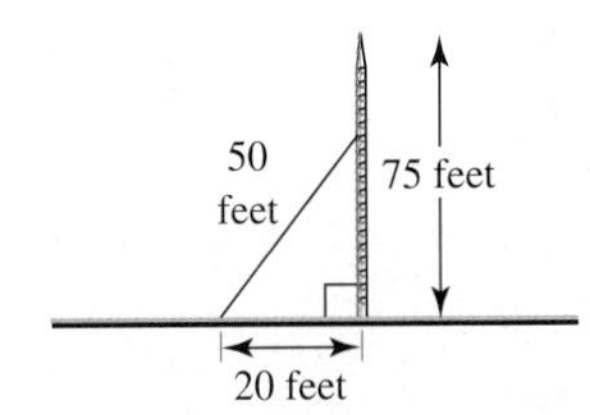

A 50-foot supporting wire is to be attached to a 75-foot antenna. Because of surrounding buildings, sidewalks, and roadways, the wire must be anchored exactly 20 feet from the base of the antenna.

a. How high from the base of the antenna is the wire attached?

b. Local regulations require that a supporting wire be attached at a height no less than $\frac{3}{5}$ of the total height of the antenna. From part (a), have local regulations been met?

Solution:

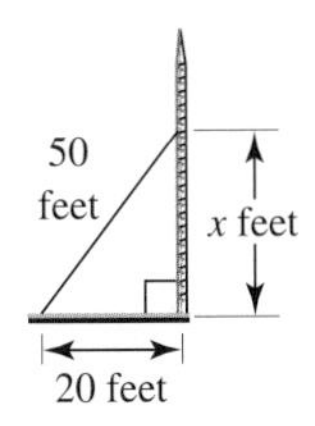

1. UNDERSTAND. Read and reread the problem. From the diagram we notice that a right triangle is formed with hypotenuse 50 feet and one leg 20 feet. Let $x =$ the height from the base of the antenna to the attached wire.

2. TRANSLATE. Use the Pythagorean theorem.

$$(a)^2 + (b)^2 = (c)^2$$
$$(20)^2 + x^2 = (50)^2 \quad a = 20, c = 50$$

3. SOLVE.
$$(20)^2 + x^2 = (50)^2$$
$$400 + x^2 = 2500$$
$$x^2 = 2100 \quad \text{Subtract 400 from both sides.}$$
$$x = \sqrt{2100}$$
$$= 10\sqrt{21}$$

4. INTERPRET. *Check* the work and *state* the solution.

a. The wire is attached exactly $10\sqrt{21}$ feet from the base of the pole, or approximately 45.8 feet.

b. The supporting wire must be attached at a height no less than $\frac{3}{5}$ of the total height of the antenna. This height is $\frac{3}{5}$(75 feet), or 45 feet. Since we know from part (a) that the wire is to be attached at a height of approximately 45.8 feet, local regulations have been met.

Practice Problem 7

A furniture upholsterer wishes to cut a strip from a piece of fabric that is 45 inches by 45 inches. The strip must be cut on the bias of the fabric. What is the longest strip that can be cut? Give an exact answer and a two decimal place approximation.

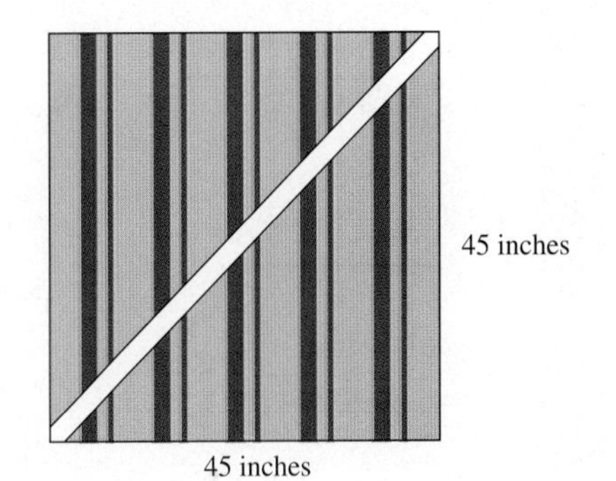

Answer

7. $45\sqrt{2}$ in. ≈ 63.64 in.

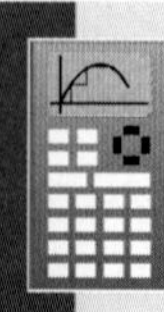

GRAPHING CALCULATOR EXPLORATIONS

We can use a grapher to solve radical equations. For example, to use a grapher to approximate the solutions of the equation solved in Example 4, we graph the following:

$Y_1 = \sqrt{4 - x}$ and $Y_2 = x - 2$

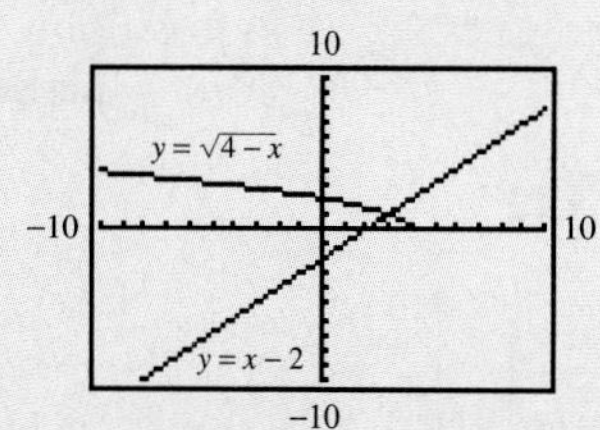

The x-value of the point of intersection is the solution. Use the Intersect feature or the Zoom and Trace features of your grapher to see that the solution is 3.

Use a grapher to solve each radical equation. Round all solutions to the nearest hundredth.

1. $\sqrt{x + 7} = x$

2. $\sqrt{3x + 5} = 2x$

3. $\sqrt{2x + 1} = \sqrt{2x + 2}$

4. $\sqrt{10x - 1} = \sqrt{-10x + 10} - 1$

5. $1.2x = \sqrt{3.1x + 5}$

6. $\sqrt{1.9x^2 - 2.2} = -0.8x + 3$

Name ______________________ Section ________ Date __________

Exercise Set 7.6

A *Solve. See Examples 1 and 2.*

1. $\sqrt{2x} = 4$

2. $\sqrt{3x} = 3$

3. $\sqrt{x - 3} = 2$

4. $\sqrt{x + 1} = 5$

5. $\sqrt{2x} = -4$

6. $\sqrt{5x} = -5$

7. $\sqrt{4x - 3} - 5 = 0$

8. $\sqrt{x - 3} - 1 = 0$

9. $\sqrt{2x - 3} - 2 = 1$

10. $\sqrt{3x + 3} - 4 = 8$

Solve. See Example 3.

11. $\sqrt[3]{6x} = -3$

12. $\sqrt[3]{4x} = -2$

13. $\sqrt[3]{x - 2} - 3 = 0$

14. $\sqrt[3]{2x - 6} - 4 = 0$

Solve. See Examples 4 and 5.

15. $\sqrt{13 - x} = x - 1$

16. $\sqrt{2x - 3} = 3 - x$

17. $x - \sqrt{4 - 3x} = -8$

18. $2x + \sqrt{x + 1} = 8$

19. $\sqrt{y + 5} = 2 - \sqrt{y - 4}$

20. $\sqrt{x + 3} + \sqrt{x - 5} = 3$

21. $\sqrt{x - 3} + \sqrt{x + 2} = 5$

22. $\sqrt{2x - 4} - \sqrt{3x + 4} = -2$

Answers

1. ______

2. ______

3. ______

4. ______

5. ______

6. ______

7. ______

8. ______

9. ______

10. ______

11. ______

12. ______

13. ______

14. ______

15. ______

16. ______

17. ______

18. ______

19. ______

20. ______

21. ______

22. ______

23. ______

24. ______

25. ______

26. ______

27. ______

28. ______

29. ______

30. ______

31. ______

32. ______

33. ______

34. ______

35. ______

36. ______

37. ______

38. ______

39. ______

40. ______

41. ______

42. ______

43. ______

44. ______

45. ______

46. ______

47. ______

48. ______

Name ______

Solve. See Examples 1 through 5.

23. $\sqrt{3x - 2} = 5$

24. $\sqrt{5x - 4} = 9$

25. $-\sqrt{2x} + 4 = -6$

26. $-\sqrt{3x + 9} = -12$

27. $\sqrt{3x + 1} + 2 = 0$

28. $\sqrt{3x + 1} - 2 = 0$

29. $\sqrt[4]{4x + 1} - 2 = 0$

30. $\sqrt[4]{2x - 9} - 3 = 0$

31. $\sqrt{4x - 3} = 5$

32. $\sqrt{3x + 9} = 12$

33. $\sqrt[3]{6x - 3} - 3 = 0$

34. $\sqrt[3]{3x} + 4 = 7$

35. $\sqrt[3]{2x - 3} - 2 = -5$

36. $\sqrt[3]{x - 4} - 5 = -7$

37. $\sqrt{x + 4} = \sqrt{2x - 5}$

38. $\sqrt{3y + 6} = \sqrt{7y - 6}$

39. $x - \sqrt{1 - x} = -5$

40. $x - \sqrt{x - 2} = 4$

41. $\sqrt[3]{-6x - 1} = \sqrt[3]{-2x - 5}$

42. $x + \sqrt{x + 5} = 7$

43. $\sqrt{5x - 1} - \sqrt{x} + 2 = 3$

44. $\sqrt{2x - 1} - 4 = -\sqrt{x - 4}$

45. $\sqrt{2x - 1} = \sqrt{1 - 2x}$

46. $\sqrt{7x - 4} = \sqrt{4 - 7x}$

47. $\sqrt{3x + 4} - 1 = \sqrt{2x + 1}$

48. $\sqrt{x - 2} + 3 = \sqrt{4x + 1}$

Name ________________________________

49. $\sqrt{y+3} - \sqrt{y-3} = 1$

50. $\sqrt{x+1} - \sqrt{x-1} = 2$

B *Find the length of the unknown side in each triangle. See Example 6.*

51.

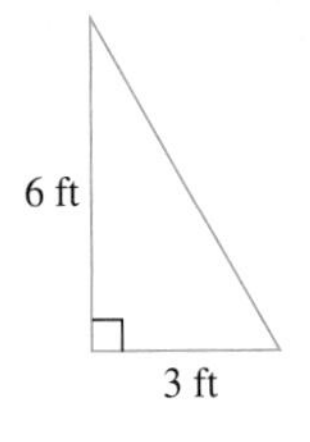

52.

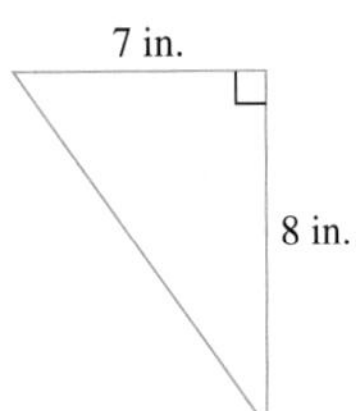

53.

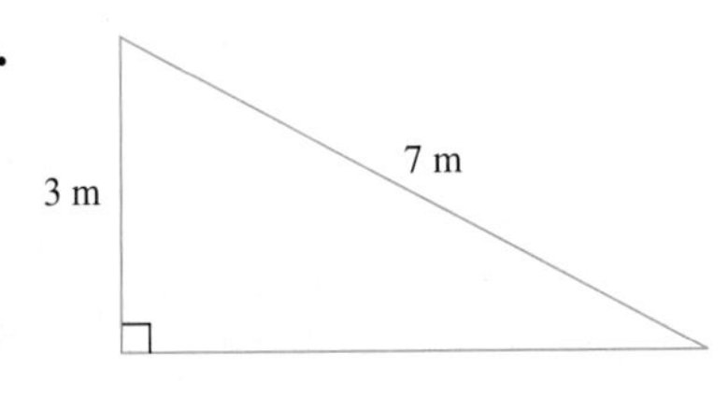

54.

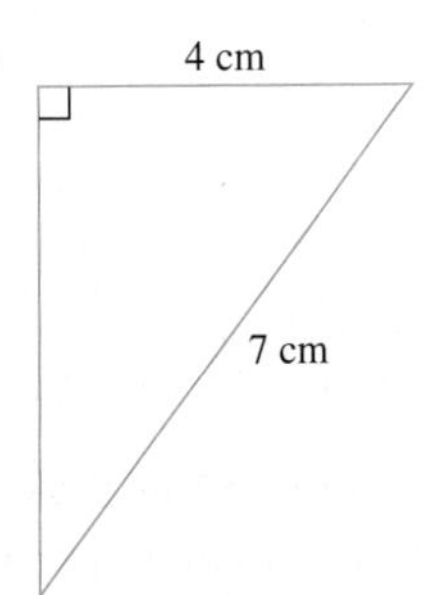

Find the length of the unknown side of each triangle. Give the exact length and a one-decimal-place approximation. See Example 6.

55.

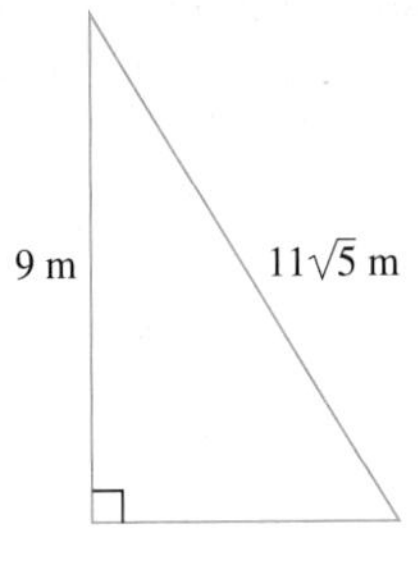

56.

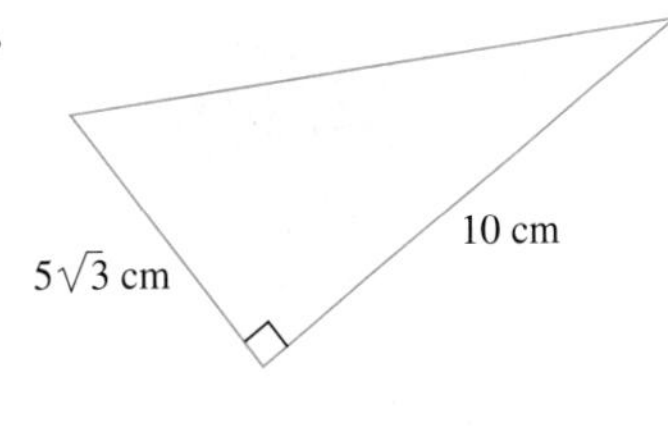

57.

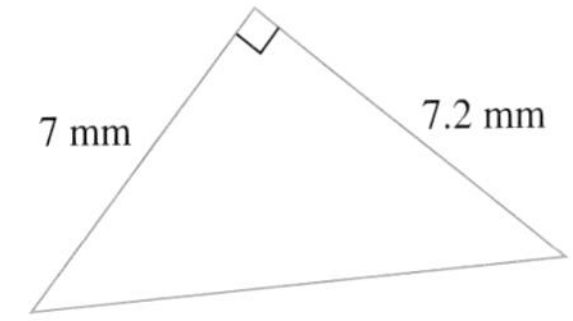

58.

2.7 in.

2.3 in.

49. ____________

50. ____________

51. ____________

52. ____________

53. ____________

54. ____________

55. ____________

56. ____________

57. ____________

58. ____________

59. ______

Name ______

Solve. See Example 7. Give exact answers and two-decimal-place approximations where appropriate.

59. A wire is needed to support a vertical pole 15 feet high. The cable will be anchored to a stake 8 feet from the base of the pole. How much cable is needed?

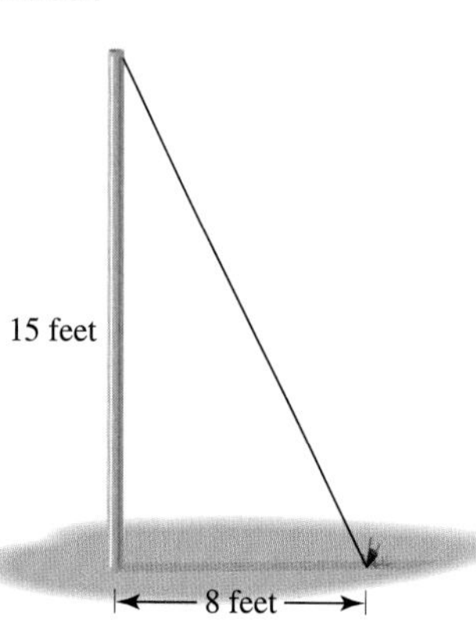

60. The tallest structure in the United States is a TV tower in Blanchard, North Dakota. Its height is 2063 feet. A 2382-foot length of wire is to be used as a guy wire attached to the top of the tower. Approximate to the nearest foot how far from the base of the tower the guy wire must be anchored. (*Source:* U.S. Geological Survey)

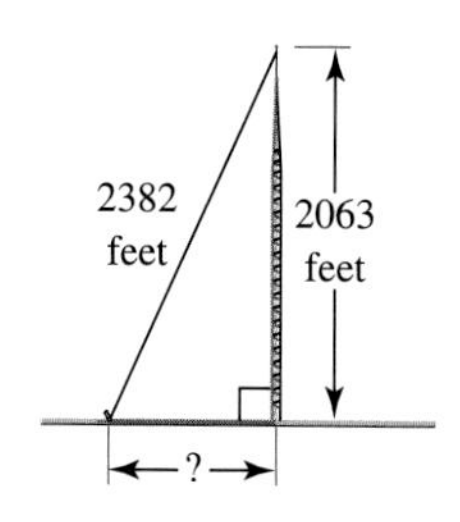

60. ______

61. A spotlight is mounted on the eaves of a house 12 feet above the ground. A flower bed runs between the house and the sidewalk, so the closest the ladder can be placed to the house is 5 feet. How long a ladder is needed so that an electrician can reach the place where the light is mounted?

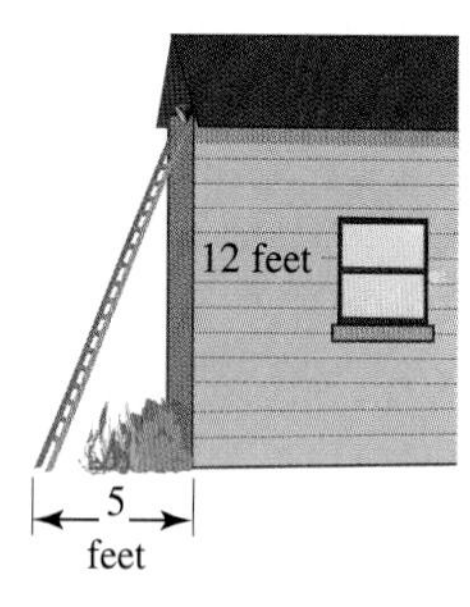

61. ______

62. A wire is to be attached to support a telephone pole. Because of surrounding buildings, sidewalks, and roadway, the wire must be anchored exactly 15 feet from the base of the pole. Telephone company workers have only 30 feet of cable, and 2 feet of that must be used to attach the cable to the pole and to the stake on the ground. How high from the base of the pole can the wire be attached?

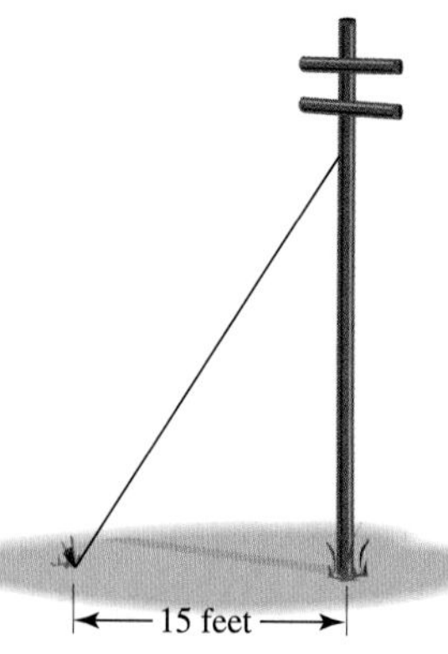

62. ______

Name ______________________________

63. ______________

64. ______________

65. ______________

66. ______________

67. ______________

68. ______________

69. ______________

70. ______________

63. The radius of the Moon is 1080 miles. Use the formula for the radius r of a sphere given its surface area A,

$$r = \sqrt{\frac{A}{4\pi}}$$

to find the surface area of the Moon. Round to the nearest square mile. (*Source:* National Space Science Data Center)

64. Police departments find it very useful to be able to approximate the speed of a car when they are given the distance that the car skidded before it came to a stop. If the road surface is wet concrete, the function $S(x) = \sqrt{10.5x}$ is used, where $S(x)$ is the speed of the car in miles per hour and x is the distance skidded in feet. Find how fast a car was moving if it skidded 280 feet on wet concrete.

65. The formula $v = \sqrt{2gh}$ relates the velocity v, in feet per second, of an object after it falls h feet accelerated by gravity g, in feet per second squared. If g is approximately 32 feet per second squared, find how far an object has fallen if its velocity is 80 feet per second.

66. Two tractors are pulling a tree stump from a field. If two forces A and B pull at right angles (90°) to each other, the size of the resulting force R is given by the formula $R = \sqrt{A^2 + B^2}$. If tractor A is exerting 600 pounds of force and the resulting force is 850 pounds, find how much force tractor B is exerting.

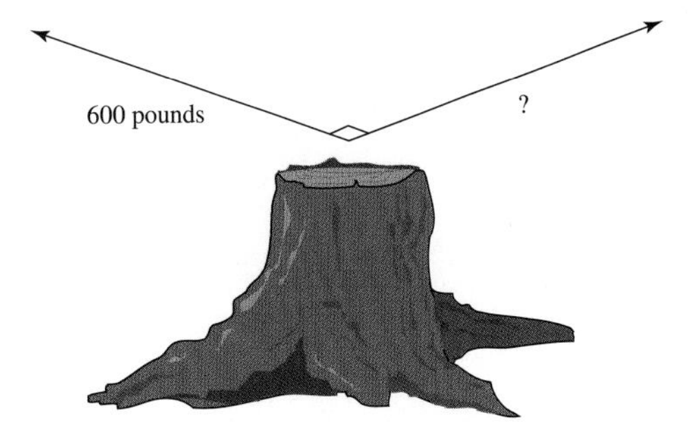

Review and Preview

Simplify. See Section 6.3.

67. $\dfrac{\frac{x}{6}}{\frac{2x}{3} + \frac{1}{2}}$

68. $\dfrac{\frac{1}{y} + \frac{4}{5}}{\frac{-3}{20}}$

69. $\dfrac{\frac{z}{5} + \frac{1}{10}}{\frac{z}{20} - \frac{z}{5}}$

70. $\dfrac{\frac{1}{y} + \frac{1}{x}}{\frac{1}{y} - \frac{1}{x}}$

71. ____________

72. ____________

73. ____________

74. ____________

75. a. ____________

b. ____________

Name ____________________

Combining Concepts

71. Solve: $\sqrt{\sqrt{x+3}+\sqrt{x}}=\sqrt{3}$

72. The maximum distance $D(h)$ that a person can see from a height h kilometers above the ground is given by the function $D(h) = 111.7\sqrt{h}$. Find the height that would allow a person to see 80 kilometers.

73. The cost $C(x)$ in dollars per day to operate a small delivery service is given by $C(x) = 80\sqrt[3]{x} + 500$, where x is the number of deliveries per day. In July, the manager decides that it is necessary to keep delivery costs below \$1620.00. Find the greatest number of deliveries this company can make per day and still keep overhead below \$1620.00.

74. Explain why proposed solutions of radical equations must be checked.

75. Consider the equations $\sqrt{2x} = 4$ and $\sqrt[3]{2x} = 4$.

a. Explain the difference in solving these equations.

b. Explain the similarity in solving these equations.

7.7 COMPLEX NUMBERS

Objectives

A Write square roots of negative numbers in the form bi.
B Add or subtract complex numbers.
C Multiply complex numbers.
D Divide complex numbers.
E Raise i to powers.

SSM CD-ROM Video 7.7

A WRITING NUMBERS IN THE FORM *bi*

Our work with radical expressions has excluded expressions such as $\sqrt{-16}$ because $\sqrt{-16}$ is not a real number; there is no real number whose square is -16. In this section, we discuss a number system that includes roots of negative numbers. This number system is the **complex number system**, and it includes the set of real numbers as a subset. The complex number system allows us to solve equations such as $x^2 + 1 = 0$ that have no real number solutions. The set of complex numbers includes the **imaginary unit**.

> IMAGINARY UNIT
>
> The **imaginary unit**, written i, is the number whose square is -1. That is,
>
> $$i^2 = -1 \quad \text{and} \quad i = \sqrt{-1}$$

To write the square root of a negative number in terms of i, we use the property that if a is a positive number, then

$$\sqrt{-a} = \sqrt{-1} \cdot \sqrt{a}$$
$$= i \cdot \sqrt{a}$$

Using i, we can write $\sqrt{-16}$ as

$$\sqrt{-16} = \sqrt{-1 \cdot 16} = \sqrt{-1} \cdot \sqrt{16} = i \cdot 4 \text{ or } 4i$$

Examples Write using i notation.

1. $\sqrt{-36} = \sqrt{-1 \cdot 36} = \sqrt{-1} \cdot \sqrt{36} = i \cdot 6$ or $6i$

2. $\sqrt{-5} = \sqrt{-1(5)} = \sqrt{-1} \cdot \sqrt{5} = i\sqrt{5}$. Since $\sqrt{5}i$ can easily be confused with $\sqrt{5i}$, we write $\sqrt{5}i$ as $i\sqrt{5}$.

3. $-\sqrt{-20} = -\sqrt{-1 \cdot 20} = -\sqrt{-1} \cdot \sqrt{4 \cdot 5} = -i \cdot 2\sqrt{5} = -2i\sqrt{5}$

Practice Problems 1–3

Write using i notation.

1. $\sqrt{-25}$
2. $\sqrt{-3}$
3. $-\sqrt{-50}$

The product rule for radicals does not necessarily hold true for imaginary numbers. *To multiply square roots of negative numbers, first we write each number in terms of the imaginary unit i.* For example, to multiply $\sqrt{-4}$ and $\sqrt{-9}$, we first write each number in the form bi:

$$\sqrt{-4}\sqrt{-9} = 2i(3i) = 6i^2 = 6(-1) = -6$$

We will also use this method to simplify quotients of square roots of negative numbers.

Examples Multiply or divide as indicated.

4. $\sqrt{-3} \cdot \sqrt{-5} = i\sqrt{3}(i\sqrt{5}) = i^2\sqrt{15} = -1\sqrt{15} = -\sqrt{15}$

5. $\sqrt{-36} \cdot \sqrt{-1} = 6i(i) = 6i^2 = 6(-1) = -6$

6. $\sqrt{8} \cdot \sqrt{-2} = 2\sqrt{2}(i\sqrt{2}) = 2i(\sqrt{2}\sqrt{2}) = 2i(2) = 4i$

7. $\dfrac{\sqrt{-125}}{\sqrt{5}} = \dfrac{i\sqrt{125}}{\sqrt{5}} = i\sqrt{25} = 5i$

Practice Problems 4–7

Multiply or divide as indicated.

4. $\sqrt{-2} \cdot \sqrt{-7}$
5. $\sqrt{-25} \cdot \sqrt{-1}$
6. $\sqrt{27} \cdot \sqrt{-3}$
7. $\dfrac{\sqrt{-8}}{\sqrt{2}}$

Now that we have practiced working with the imaginary unit, we define complex numbers.

Answers
1. $5i$, **2.** $i\sqrt{3}$, **3.** $-5i\sqrt{2}$, **4.** $-\sqrt{14}$, **5.** -5, **6.** $9i$, **7.** $2i$

> **COMPLEX NUMBERS**
>
> A **complex number** is a number that can be written in the form $a + bi$, where a and b are real numbers.

Notice that the set of real numbers is a subset of the complex numbers since any real number can be written in the form of a complex number. For example,

$$16 = 16 + 0i$$

In general, a complex number $a + bi$ is a real number if $b = 0$. Also, a complex number is called an **imaginary number** if $a = 0$. For example,

$$3i = 0 + 3i \quad \text{and} \quad i\sqrt{7} = 0 + i\sqrt{7}$$

are imaginary numbers.

The following diagram shows the relationship between complex numbers and their subsets.

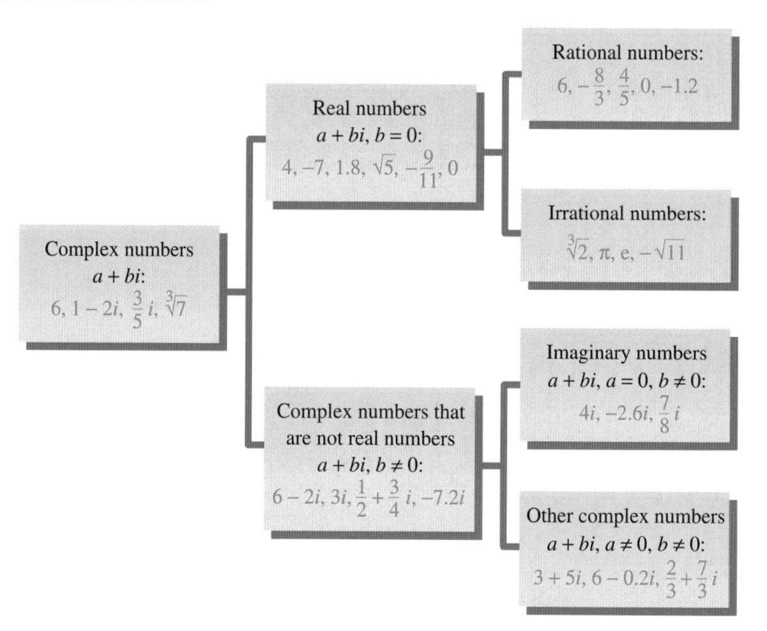

✓ **CONCEPT CHECK**

True or false? Every complex number is also a real number.

TRY THE CONCEPT CHECK IN THE MARGIN.

B ADDING OR SUBTRACTING COMPLEX NUMBERS

Two complex numbers $a + bi$ and $c + di$ are equal if and only if $a = c$ and $b = d$. Complex numbers can be added or subtracted by adding or subtracting their real parts and then adding or subtracting their imaginary parts.

> **SUM OR DIFFERENCE OF COMPLEX NUMBERS**
>
> If $a + bi$ and $c + di$ are complex numbers, then their sum is
>
> $$(a + bi) + (c + di) = (a + c) + (b + d)i$$
>
> Their difference is
>
> $$(a + bi) - (c + di) = a + bi - c - di = (a - c) + (b - d)i$$

Answer

✓ **Concept Check:** false

Examples

Add or subtract as indicated.

8. $(2 + 3i) + (-3 + 2i) = (2 - 3) + (3 + 2)i = -1 + 5i$

9. $5i - (1 - i) = 5i - 1 + i$
$= -1 + (5 + 1)i$
$= -1 + 6i$

10. $(-3 - 7i) - (-6) = -3 - 7i + 6$
$= (-3 + 6) - 7i$
$= 3 - 7i$

Practice Problems 8–10

Add or subtract as indicated.

8. $(5 + 2i) + (4 - 3i)$
9. $6i - (2 - i)$
10. $(-2 - 4i) - (-3)$

C Multiplying Complex Numbers

To multiply two complex numbers of the form $a + bi$, we multiply as though they are binomials. Then use the relationship $i^2 = -1$ to simplify.

Examples

Multiply.

11. $-7i \cdot 3i = -21i^2$
$= -21(-1)$ Replace i^2 with -1.
$= 21$

12. $3i(2 - i) = 3i \cdot 2 - 3i \cdot i$ Use the distributive property.
$= 6i - 3i^2$ Multiply.
$= 6i - 3(-1)$ Replace i^2 with -1.
$= 6i + 3$
$= 3 + 6i$

Use the FOIL method. (First, Outer, Inner, Last)

13. $(2 - 5i)(4 + i) = \underset{F}{2(4)} + \underset{O}{2(i)} - \underset{I}{5i(4)} - \underset{L}{5i(i)}$
$= 8 + 2i - 20i - 5i^2$
$= 8 - 18i - 5(-1)$ $i^2 = -1$.
$= 8 - 18i + 5$
$= 13 - 18i$

14. $(2 - i)^2 = (2 - i)(2 - i)$
$= 2(2) - 2(i) - 2(i) + i^2$
$= 4 - 4i + (-1)$ $i^2 = -1$.
$= 3 - 4i$

15. $(7 + 3i)(7 - 3i) = 7(7) - 7(3i) + 3i(7) - 3i(3i)$
$= 49 - 21i + 21i - 9i^2$
$= 49 - 9(-1)$ $i^2 = -1$.
$= 49 + 9$
$= 58$

Practice Problems 11–15

Multiply.

11. $-5i \cdot 3i$
12. $-2i(6 - 2i)$
13. $(3 - 4i)(6 + i)$
14. $(1 - 2i)^2$
15. $(6 + 5i)(6 - 5i)$

Notice that if you add, subtract, or multiply two complex numbers, the result is a complex number.

D Dividing Complex Numbers

From Example 15, notice that the product of $7 + 3i$ and $7 - 3i$ is a real number. These two complex numbers are called **complex conjugates** of one another. In general, we have the following definition.

Answers

8. $9 - i$, **9.** $-2 + 7i$, **10.** $1 - 4i$, **11.** 15, **12.** $-4 - 12i$, **13.** $22 - 21i$, **14.** $-3 - 4i$, **15.** 61

COMPLEX CONJUGATES

The complex numbers $(a + bi)$ and $(a - bi)$ are called **complex conjugates** of each other, and

$$(a + bi)(a - bi) = a^2 + b^2$$

To see that the product of a complex number $a + bi$ and its conjugate $a - bi$ is the real number $a^2 + b^2$, we multiply:

$$\begin{aligned}(a + bi)(a - bi) &= a^2 - abi + abi - b^2i^2 \\ &= a^2 - b^2(-1) \\ &= a^2 + b^2\end{aligned}$$

We will use complex conjugates to divide by a complex number.

Practice Problem 16

Divide and write in the form $a + bi$: $\dfrac{3 + i}{2 - 3i}$

Example 16 Divide and write in the form $a + bi$: $\dfrac{2 + i}{1 - i}$

Solution: We multiply the numerator and the denominator by the complex conjugate of $1 - i$ to eliminate the imaginary number in the denominator.

$$\begin{aligned}\frac{2 + i}{1 - i} &= \frac{(2 + i)(1 + i)}{(1 - i)(1 + i)} \\ &= \frac{2(1) + 2(i) + 1(i) + i^2}{1^2 - i^2} \\ &= \frac{2 + 3i - 1}{1 + 1} \\ &= \frac{1 + 3i}{2} = \frac{1}{2} + \frac{3}{2}i\end{aligned}$$

Practice Problem 17

Divide and write in the form $a + bi$: $\dfrac{6}{5i}$

Example 17 Divide and write in the form $a + bi$: $\dfrac{7}{3i}$

Solution: We multiply the numerator and the denominator by the conjugate of $3i$. Note that $3i = 0 + 3i$, so its conjugate is $0 - 3i$ or $-3i$.

$$\frac{7}{3i} = \frac{7(-3i)}{(3i)(-3i)} = \frac{-21i}{-9i^2} = \frac{-21i}{-9(-1)} = \frac{-21i}{9} = \frac{-7i}{3} = -\frac{7}{3}i$$

E FINDING POWERS OF i

We can use the fact that $i^2 = -1$ to simplify i^3 and i^4.

$$i^3 = i^2 \cdot i = (-1)i = -i$$
$$i^4 = i^2 \cdot i^2 = (-1) \cdot (-1) = 1$$

We continue this process and use the fact that $i^4 = 1$ and $i^2 = -1$ to simplify i^5 and i^6.

$$i^5 = i^4 \cdot i = 1 \cdot i = i$$
$$i^6 = i^4 \cdot i^2 = 1 \cdot (-1) = -1$$

Answers

16. $\frac{3}{13} + \frac{11}{13}i$, 17. $-\frac{6}{5}i$

If we continue finding powers of i, we generate the following pattern. Notice that the values i, -1, $-i$, and 1 repeat as i is raised to higher and higher powers.

$$\begin{array}{lll} i^1 = i & i^5 = i & i^9 = i \\ i^2 = -1 & i^6 = -1 & i^{10} = -1 \\ i^3 = -i & i^7 = -i & i^{11} = -i \\ i^4 = 1 & i^8 = 1 & i^{12} = 1 \end{array}$$

This pattern allows us to find other powers of i. To do so, we will use the fact that $i^4 = 1$ and rewrite a power of i in terms of i^4.
For example, $i^{22} = i^{20} \cdot i^2 = (i^4)^5 \cdot i^2 = 1^5 \cdot (-1) = 1 \cdot (-1) = -1$.

Examples

Find each power of i.

18. $i^7 = i^4 \cdot i^3 = 1(-i) = -i$

19. $i^{20} = (i^4)^5 = 1^5 = 1$

20. $i^{46} = i^{44} \cdot i^2 = (i^4)^{11} \cdot i^2 = 1^{11}(-1) = -1$

21. $i^{-12} = \dfrac{1}{i^{12}} = \dfrac{1}{(i^4)^3} = \dfrac{1}{(1)^3} = \dfrac{1}{1} = 1$

Practice Problems 18–21

Find the powers of i.

18. i^{11}

19. i^{40}

20. i^{50}

21. i^{-10}

Answers

18. $-i$, **19.** 1, **20.** -1, **21.** -1

Focus On History

HERON OF ALEXANDRIA

Heron (also Hero) was a Greek mathematician and engineer. He lived and worked in Alexandria, Egypt, around 75 A.D. During his prolific work life, Heron developed a rotary steam engine called an aeolipile, a surveying tool called a dioptra, as well as a wind organ and a fire engine. As an engineer, he must have had the need to approximate square roots because he described an iterative method for doing so in his work *Metrica*. Heron's method for approximating a square root can be summarized as follows:

> Suppose that x is not a perfect square and a^2 is the nearest perfect square to x. For a rough estimate of the value of $\sqrt{x}$, find the value of $y_1 = \frac{1}{2}\left(a + \frac{x}{a}\right)$. This estimate can be improved by calculating a second estimate using the first estimate y_1 in place of a: $y_2 = \frac{1}{2}\left(y_1 + \frac{x}{y_1}\right)$. Repeating this process several times will give more and more accurate estimates of $\sqrt{x}$.

CRITICAL THINKING

1. **a.** Which perfect square is closest to 80?
   b. Use Heron's method for approximating square roots to calculate the first estimate of the square root of 80.
   c. Use the first estimate of the square root of 80 to find a more refined second estimate.
   d. Use a calculator to find the actual value of the square root of 80. List all digits shown on your calculator's display.
   e. Compare the actual value from part (d) to the values of the first and second estimates. What do you notice?
   f. How many iterations of this process are necessary to get an estimate that differs no more than one digit from the actual value recorded in part (d)?

2. Repeat Question 1 for finding an estimate of the square root of 30.

3. Repeat Question 1 for finding an estimate of the square root of 4572.

4. Why would this iterative method have been important to people of Heron's era? Would you say that this method is as important today? Why or why not?

Name ______________________ Section ________ Date ________

Mental Math

Simplify. See Example 1.

1. $\sqrt{-81}$
2. $\sqrt{-49}$
3. $\sqrt{-7}$
4. $\sqrt{-3}$
5. $-\sqrt{16}$
6. $-\sqrt{4}$
7. $\sqrt{-64}$
8. $\sqrt{-100}$

Exercise Set 7.7

A *Write using i notation. See Examples 1 through 3.*

1. $\sqrt{-24}$
2. $\sqrt{-32}$
3. $-\sqrt{-36}$
4. $-\sqrt{-121}$
5. $8\sqrt{-63}$
6. $4\sqrt{-20}$
7. $-\sqrt{54}$
8. $\sqrt{-63}$

Multiply or divide as indicated. See Examples 4 through 7.

9. $\sqrt{-2} \cdot \sqrt{-7}$
10. $\sqrt{-11} \cdot \sqrt{-3}$
11. $\sqrt{-5} \cdot \sqrt{-10}$
12. $\sqrt{-2} \cdot \sqrt{-6}$
13. $\sqrt{16} \cdot \sqrt{-1}$
14. $\sqrt{3} \cdot \sqrt{-27}$
15. $\dfrac{\sqrt{-9}}{\sqrt{3}}$
16. $\dfrac{\sqrt{49}}{\sqrt{-10}}$
17. $\dfrac{\sqrt{-80}}{\sqrt{-10}}$
18. $\dfrac{\sqrt{-40}}{\sqrt{-8}}$

B *Add or subtract as indicated. Write your answers in the form $a + bi$. See Examples 8 through 10.*

19. $(4 - 7i) + (2 + 3i)$
20. $(2 - 4i) - (2 - i)$
21. $(6 + 5i) - (8 - i)$
22. $(8 - 3i) + (-8 + 3i)$
23. $6 - (8 + 4i)$
24. $(9 - 4i) - 9$
25. $(6 - 3i) - (4 - 2i)$
26. $(-2 - 4i) - (6 - 8i)$
27. $(5 - 6i) - 4i$
28. $(6 - 2i) + 7i$

Mental Math Answers

1. ________
2. ________
3. ________
4. ________
5. ________
6. ________
7. ________
8. ________

Answers

1. ________
2. ________
3. ________
4. ________
5. ________
6. ________
7. ________
8. ________
9. ________
10. ________
11. ________
12. ________
13. ________
14. ________
15. ________
16. ________
17. ________
18. ________
19. ________
20. ________
21. ________
22. ________
23. ________
24. ________
25. ________
26. ________
27. ________
28. ________

29. ______
30. ______
31. ______
32. ______
33. ______
34. ______
35. ______
36. ______
37. ______
38. ______
39. ______
40. ______
41. ______
42. ______
43. ______
44. ______
45. ______
46. ______
47. ______
48. ______
49. ______
50. ______
51. ______
52. ______
53. ______
54. ______
55. ______
56. ______
57. ______
58. ______

Name ______________________

29. $(2 + 4i) + (6 - 5i)$

30. $(5 - 3i) + (7 - 8i)$

C *Multiply. Write your answers in the form a + bi. See Examples 11 through 15.*

31. $6i \cdot 2i$

32. $5i \cdot 7i$

33. $-9i \cdot 7i$

34. $-6i \cdot 4i$

35. $-10i \cdot -4i$

36. $-2i \cdot -11i$

37. $6i(2 - 3i)$

38. $5i(4 - 7i)$

39. $-3i(-1 + 9i)$

40. $-5i(-2 + i)$

41. $(4 + i)(5 + 2i)$

42. $(3 + i)(2 + 4i)$

43. $(\sqrt{3} + 2i)(\sqrt{3} - 2i)$

44. $(\sqrt{5} - 5i)(\sqrt{5} + 5i)$

45. $(4 - 2i)^2$

46. $(6 - 3i)^2$

47. $(6 - 2i)(3 + i)$

48. $(2 - 4i)(2 - i)$

49. $(1 - i)(1 + i)$

50. $(6 + 2i)(6 - 2i)$

51. $(9 + 8i)^2$

52. $(4 + 7i)^2$

53. $(1 - i)^2$

54. $(2 - 2i)^2$

D *Divide. Write your answers in the form a + bi. See Examples 16 and 17.*

55. $\frac{4}{i}$

56. $\frac{5}{6i}$

57. $\frac{7}{4 + 3i}$

58. $\frac{9}{1 - 2i}$

Name ______________________________

59. $\dfrac{6i}{1 - 2i}$

60. $\dfrac{3i}{5 + i}$

61. $\dfrac{3 + 5i}{1 + i}$

62. $\dfrac{6 + 2i}{4 - 3i}$

63. $\dfrac{4 - 5i}{2i}$

64. $\dfrac{6 + 8i}{3i}$

65. $\dfrac{16 + 15i}{-3i}$

66. $\dfrac{2 - 3i}{-7i}$

67. $\dfrac{2}{3 + i}$

68. $\dfrac{5}{3 - 2i}$

69. $\dfrac{2 - 3i}{2 + i}$

70. $\dfrac{6 + 5i}{6 - 5i}$

E *Find each power of i. See Examples 18 through 21.*

71. i^8

72. i^{10}

73. i^{21}

74. i^{15}

75. i^{11}

76. i^{40}

77. i^{-6}

78. i^{-9}

79. $(2i)^6$

80. $(5i)^4$

81. $(-3i)^5$

82. $(-2i)^7$

Review and Preview

Thirty people were recently polled about their average monthly balance in their checking account. The results of this poll are shown in the bar graph. Use this graph to answer Exercises 83–88. See Section 1.1.

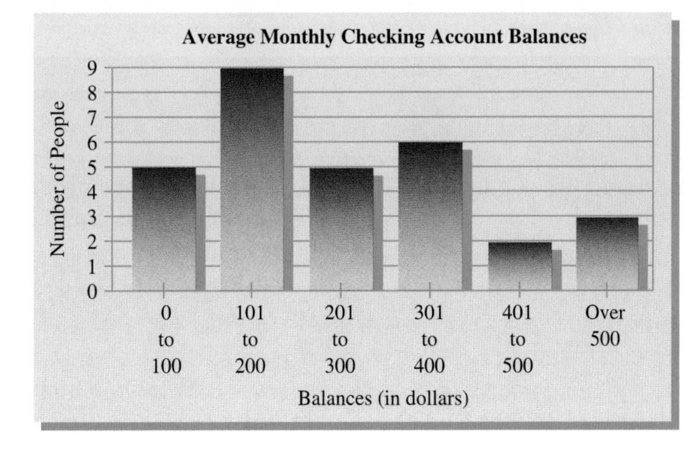

83. How many people polled reported an average checking balance of \$201 to \$300?

84. How many people polled reported an average checking balance of \$0 to \$100?

59. ______
60. ______
61. ______
62. ______
63. ______
64. ______
65. ______
66. ______
67. ______
68. ______
69. ______
70. ______
71. ______
72. ______
73. ______
74. ______
75. ______
76. ______
77. ______
78. ______
79. ______
80. ______
81. ______
82. ______
83. ______
84. ______

85.

86.

87.

88.

89.

90.

91.

92.

93.

94.

95.

96.

97.

98.

99.

100.

101.

102.

103.

Name ______________________________

85. How many people polled reported an average checking balance of $200 or less?

86. How many people polled reported an average checking balance of $301 or more?

87. What percent of people polled reported an average checking balance of $201 to $300?

88. What percent of people polled reported an average checking balance of 0 to $100?

Combining Concepts

Write in the form a + bi.

89. $i^3 + i^4$

90. $i^8 - i^7$

91. $i^6 + i^8$

92. $i^4 + i^{12}$

93. $2 + \sqrt{-9}$

94. $5 - \sqrt{-16}$

95. $\frac{6 + \sqrt{-18}}{3}$

96. $\frac{4 - \sqrt{-8}}{2}$

97. $\frac{5 - \sqrt{-75}}{10}$

98. Describe how to find the conjugate of a complex number.

99. Explain why the product of a complex number and its complex conjugate is a real number.

Simplify.

100. $(8 - \sqrt{-3}) - (2 + \sqrt{-12})$

101. $(8 - \sqrt{-4}) - (2 + \sqrt{-16})$

102. Determine whether $2i$ is a solution of $x^2 + 4 = 0$.

103. Determine whether $-1 + i$ is a solution of $x^2 + 2x = -2$.

Name ______________________ Section ________ Date ________

Chapter 7 Activity
Calculating the Length and Period of a Pendulum

Materials

- string (at least 1 meter long)
- meter stick
- calculator
- weight
- stopwatch

This activity may be completed by working in groups or individually.

Make a simple pendulum by securely tying the string to a weight.

The formula relating a pendulum's period T (in seconds) to its length l (in centimeters) is

$$T = 2\pi\sqrt{\frac{l}{980}}$$

The **period** of a pendulum is defined as the time it takes the pendulum to complete one full back-and-forth swing. In this activity, you will be measuring your simple pendulum's period with a stopwatch. Because the periods will be only a few seconds long, it will be more accurate for you to time a total of five complete swings and then find the average time of one complete swing.

1. For each of the pendulum (string) lengths given in Table 1, measure the time required for 5 complete swings and record it in the appropriate column. Next, divide this value by 5 to find the measured period of the pendulum for the given length and record it in the Measured Period T_m column in the table. Use the given formula to calculate the theoretical period T for the same pendulum length and record it in the appropriate column. (Round to two decimal places.) Find and record in the last column the difference between the measured period and the theoretical period.
2. For each of the periods T given in Table 2, use the given formula and calculate the theoretical pendulum length l required to yield the given period. Record l in the appropriate column; round to one decimal place. Next, use this length l and measure and record the time for 5 complete swings. Divide this value by 5 to find the measured period T_m, and record it. Then find and record in the last column the difference between the theoretical period and the measured period.
3. Use the general trends you find in the tables to describe the relationship between a pendulum's period and its length.
4. Discuss the differences you found between the values of the theoretical period and the measured period. What factors contributed to these differences?

Table 1

Length l (centimeters)	Time for 5 Swings (seconds)	Measured Period T_m (seconds)	Theoretical Period T (seconds)	Difference $\lvert T - T_m \rvert$
30				
55				
70				

Table 2

Period T (seconds)	Theoretical Length l (centimeters)	Time for 5 Swings (seconds)	Measured Period T_m (seconds)	Difference $\lvert T - T_m \rvert$
1				
1.25				
2				

CHAPTER 7 HIGHLIGHTS

DEFINITIONS AND CONCEPTS	EXAMPLES
SECTION 7.1 RADICAL EXPRESSIONS	
The **positive**, or **principal**, **square root** of a nonnegative number a is written as $\sqrt{a}$. $\sqrt{a} = b$ only if $b^2 = a$ and $b \geq 0$	$\sqrt{36} = 6$ $\sqrt{\frac{9}{100}} = \frac{3}{10}$
The **negative square root** of a is written as $-\sqrt{a}$.	$-\sqrt{36} = -6$ $\sqrt{0.04} = 0.2$
The **cube root** of a real number a is written as $\sqrt[3]{a}$. $\sqrt[3]{a} = b$ only if $b^3 = a$	$\sqrt[3]{27} = 3$ $\sqrt[3]{-\frac{1}{8}} = -\frac{1}{2}$ $\sqrt[3]{y^6} = y^2$ $\sqrt[3]{64x^9} = 4x^3$
If n is an even positive integer, then $\sqrt[n]{a^n} = \lvert a \rvert$. If n is an odd positive integer, then $\sqrt[n]{a^n} = a$.	$\sqrt{(-3)^2} = \lvert -3 \rvert = 3$ $\sqrt[3]{(-7)^3} = -7$
A **radical function** in x is a function defined by an expression containing a root of x.	If $f(x) = \sqrt{x} + 2$, $f(1) = \sqrt{1} + 2 = 1 + 2 = 3$ $f(3) = \sqrt{3} + 2 \approx 3.73$
SECTION 7.2 RATIONAL EXPONENTS	
$a^{1/n} = \sqrt[n]{a}$ if $\sqrt[n]{a}$ is a real number.	$81^{1/2} = \sqrt{81} = 9$ $(-8x^3)^{1/3} = \sqrt[3]{-8x^3} = -2x$
If m and n are positive integers greater than 1 with $\frac{m}{n}$ in lowest terms and $\sqrt[n]{a}$ is a real number, then $a^{m/n} = (a^{1/n})^m = (\sqrt[n]{a})^m$	$4^{5/2} = (\sqrt{4})^5 = 2^5 = 32$ $27^{2/3} = (\sqrt[3]{27})^2 = 3^2 = 9$
$a^{-m/n} = \frac{1}{a^{m/n}}$ as long as $a^{m/n}$ is a nonzero number.	$16^{-3/4} = \frac{1}{16^{3/4}} = \frac{1}{(\sqrt[4]{16})^3} = \frac{1}{2^3} = \frac{1}{8}$
Exponent rules are true for rational exponents.	$x^{2/3} \cdot x^{-5/6} = x^{2/3-5/6} = x^{-1/6} = \frac{1}{x^{1/6}}$ $(8^{14})^{1/7} = 8^2 = 64$ $\frac{a^{4/5}}{a^{-2/5}} = a^{4/5-(-2/5)} = a^{6/5}$
SECTION 7.3 SIMPLIFYING RADICAL EXPRESSIONS	
PRODUCT AND QUOTIENT RULES If $\sqrt[n]{a}$ and $\sqrt[n]{b}$ are real numbers, $\sqrt[n]{a} \cdot \sqrt[n]{b} = \sqrt[n]{a \cdot b}$ $\frac{\sqrt[n]{a}}{\sqrt[n]{b}} = \sqrt[n]{\frac{a}{b}}$, provided $\sqrt[n]{b} \neq 0$	Multiply or divide as indicated: $\sqrt{11} \cdot \sqrt{3} = \sqrt{33}$ $\frac{\sqrt[3]{40x}}{\sqrt[3]{5x}} = \sqrt[3]{8} = 2$
A radical of the form $\sqrt[n]{a}$ is **simplified** when a contains no factors that are perfect nth powers.	$\sqrt{40} = \sqrt{4 \cdot 10} = 2\sqrt{10}$ $\sqrt{36x^5} = \sqrt{36x^4 \cdot x} = 6x^2\sqrt{x}$ $\sqrt[3]{24x^7y^3} = \sqrt[3]{8x^6y^3 \cdot 3x} = 2x^2y\sqrt[3]{3x}$

Section 7.4 Adding, Subtracting, and Multiplying Radical Expressions

Radicals with the same index and the same radicand are **like radicals**.	$5\sqrt{6} + 2\sqrt{6} = (5 + 2)\sqrt{6} = 7\sqrt{6}$
The distributive property can be used to add like radicals.	$-\sqrt[3]{3x} - 10\sqrt[3]{3x} + 3\sqrt[3]{10x}$ $= (-1 - 10)\sqrt[3]{3x} + 3\sqrt[3]{10x}$ $= -11\sqrt[3]{3x} + 3\sqrt[3]{10x}$
Radical expressions are multiplied by using many of the same properties used to multiply polynomials.	Multiply: $(\sqrt{5} - \sqrt{2x})(\sqrt{2} + \sqrt{2x})$ $= \sqrt{10} + \sqrt{10x} - \sqrt{4x} - 2x$ $= \sqrt{10} + \sqrt{10x} - 2\sqrt{x} - 2x$ $(2\sqrt{3} - \sqrt{8x})(2\sqrt{3} + \sqrt{8x})$ $= 4(3) - 8x = 12 - 8x$

Section 7.5 Rationalizing Numerators and Denominators of Radical Expressions

The **conjugate** of $a + b$ is $a - b$.	The conjugate of $\sqrt{7} + \sqrt{3}$ is $\sqrt{7} - \sqrt{3}$.
The process of writing the denominator of a radical expression without a radical is called **rationalizing the denominator**.	Rationalize each denominator: $\frac{\sqrt{5}}{\sqrt{3}} = \frac{\sqrt{5} \cdot \sqrt{3}}{\sqrt{3} \cdot \sqrt{3}} = \frac{\sqrt{15}}{3}$ $\frac{6}{\sqrt{7} + \sqrt{3}} = \frac{6(\sqrt{7} - \sqrt{3})}{(\sqrt{7} + \sqrt{3})(\sqrt{7} - \sqrt{3})}$ $= \frac{6(\sqrt{7} - \sqrt{3})}{7 - 3}$ $= \frac{6(\sqrt{7} - \sqrt{3})}{4} = \frac{3(\sqrt{7} - \sqrt{3})}{2}$
The process of writing the numerator of a radical expression without a radical is called **rationalizing the numerator**.	Rationalize each numerator: $\frac{\sqrt[3]{9}}{\sqrt[3]{5}} = \frac{\sqrt[3]{9} \cdot \sqrt[3]{3}}{\sqrt[3]{5} \cdot \sqrt[3]{3}} = \frac{\sqrt[3]{27}}{\sqrt[3]{15}} = \frac{3}{\sqrt[3]{15}}$ $\frac{\sqrt{9} + \sqrt{3x}}{12} = \frac{(\sqrt{9} + \sqrt{3x})(\sqrt{9} - \sqrt{3x})}{12(\sqrt{9} - \sqrt{3x})}$ $= \frac{9 - 3x}{12(\sqrt{9} - \sqrt{3x})}$ $= \frac{3(3 - x)}{3 \cdot 4(3 - \sqrt{3x})} = \frac{3 - x}{4(3 - \sqrt{3x})}$

Section 7.6 Radical Equations and Problem Solving

To Solve a Radical Equation	Solve: $x = \sqrt{4x + 9} + 3$
Step 1. Write the equation so that one radical is by itself on one side of the equation.	1. $x - 3 = \sqrt{4x + 9}$
Step 2. Raise each side of the equation to a power equal to the index of the radical and simplify.	2. $(x - 3)^2 = (\sqrt{4x + 9})^2$ $x^2 - 6x + 9 = 4x + 9$
Step 3. If the equation still contains a radical, repeat Steps 1 and 2. If not, solve the equation.	3. $x^2 - 10x = 0$ $x(x - 10) = 0$ $x = 0$ or $x = 10$
Step 4. Check all proposed solutions in the original equation.	4. The proposed solution 10 checks, but 0 does not. The solution is $\{10\}$.

SECTION 7.7 COMPLEX NUMBERS

A **complex number** is a number that can be written in the form $a + bi$, where a and b are real numbers.

$i^2 = -1$ and $i = \sqrt{-1}$

Simplify: $\sqrt{-9}$

$\sqrt{-9} = \sqrt{-1 \cdot 9} = \sqrt{-1} \cdot \sqrt{9} = i \cdot 3$, or $3i$

Complex Numbers	Written in Form $a + bi$
12	$12 + 0i$
$-5i$	$0 + (-5)i$
$-2 - 3i$	$-2 + (-3)i$

Multiply.

$$\begin{aligned}\sqrt{-3} \cdot \sqrt{-7} &= i\sqrt{3} \cdot i\sqrt{7} \\ &= i^2\sqrt{21} \\ &= -\sqrt{21}\end{aligned}$$

To add or subtract complex numbers, add or subtract their real parts and then add or subtract their imaginary parts.

Perform each indicated operation.

$$\begin{aligned}(-3 + 2i) - (7 - 4i) &= -3 + 2i - 7 + 4i \\ &= -10 + 6i\end{aligned}$$

To multiply complex numbers, multiply as though they are binomials.

$$\begin{aligned}(-7 - 2i)(6 + i) &= -42 - 7i - 12i - 2i^2 \\ &= -42 - 19i - 2(-1) \\ &= -42 - 19i + 2 \\ &= -40 - 19i\end{aligned}$$

The complex numbers $(a + bi)$ and $(a - bi)$ are called **complex conjugates**.

The complex conjugate of

$(3 + 6i)$ is $(3 - 6i)$.

Their product is a real number:

$$\begin{aligned}(3 - 6i)(3 + 6i) &= 9 - 36i^2 \\ &= 9 - 36(-1) = 9 + 36 = 45\end{aligned}$$

To divide complex numbers, multiply the numerator and the denominator by the conjugate of the denominator.

Divide:

$$\begin{aligned}\frac{4}{2 - i} &= \frac{4(2 + i)}{(2 - i)(2 + i)} \\ &= \frac{4(2 + i)}{4 - i^2} \\ &= \frac{4(2 + i)}{5} \\ &= \frac{8 + 4i}{5} = \frac{8}{5} + \frac{4}{5}i\end{aligned}$$

Name ______________________ **Section** __________ **Date** __________

CHAPTER 7 REVIEW

(7.1) *Find each root. Assume that all variables represent positive numbers.*

1. $\sqrt{81}$

2. $\sqrt[4]{81}$

3. $\sqrt[3]{-8}$

4. $\sqrt[4]{-16}$

5. $-\sqrt{\frac{1}{49}}$

6. $\sqrt{x^{64}}$

7. $-\sqrt{36}$

8. $\sqrt[3]{64}$

9. $\sqrt[3]{-a^6b^9}$

10. $\sqrt{16a^4b^{12}}$

11. $\sqrt[5]{32a^5b^{10}}$

12. $\sqrt[5]{-32x^{15}y^{20}}$

13. $\sqrt{\frac{x^{12}}{36y^2}}$

14. $\sqrt[3]{\frac{27y^3}{z^{12}}}$

Simplify. Use absolute value bars when necessary.

15. $\sqrt{x^2}$

16. $\sqrt[4]{(x^2 - 4)^4}$

17. $\sqrt[3]{(-27)^3}$

18. $\sqrt[5]{(-5)^5}$

19. $-\sqrt[5]{x^5}$

20. $\sqrt[4]{16(2y + z)^4}$

21. $\sqrt{25(x - y)^2}$

22. $\sqrt[5]{y^5}$

23. $\sqrt[6]{x^6}$

24. If $f(x) = \sqrt{x} + 3$, find $f(0)$ and $f(9)$.

25. If $g(x) = \sqrt[3]{x - 3}$, find $g(11)$ and $g(20)$.

(7.2) *Evaluate.*

26. $\left(\frac{1}{81}\right)^{1/4}$

27. $\left(-\frac{1}{27}\right)^{1/3}$

28. $(-27)^{-1/3}$

Name ______________________________

29. $(-64)^{-1/3}$

30. $-9^{3/2}$

31. $64^{-1/3}$

32. $(-25)^{5/2}$

33. $\left(\frac{25}{49}\right)^{-3/2}$

34. $\left(\frac{8}{27}\right)^{-2/3}$

35. $\left(-\frac{1}{36}\right)^{-1/4}$

Write with rational exponents.

36. $\sqrt[3]{x^2}$

37. $\sqrt[5]{5x^2y^3}$

Write using radical notation.

38. $y^{4/5}$

39. $5(xy^2z^5)^{1/3}$

40. $(x + 2y)^{-1/2}$

Simplify each expression. Assume that all variables represent positive numbers. Write with only positive exponents.

41. $a^{1/3}a^{4/3}a^{1/2}$

42. $\frac{b^{1/3}}{b^{4/3}}$

43. $(a^{1/2}a^{-2})^3$

44. $(x^{-3}y^6)^{1/3}$

45. $\left(\frac{b^{3/4}}{a^{-1/2}}\right)^8$

46. $\frac{x^{1/4}x^{-1/2}}{x^{2/3}}$

47. $\left(\frac{49c^{5/3}}{a^{-1/4}b^{5/6}}\right)^{-1}$

48. $a^{-1/4}(a^{5/4} - a^{9/4})$

Use a calculator and write a three-decimal-place approximation of each number.

49. $\sqrt{20}$

50. $\sqrt[3]{-39}$

51. $\sqrt[4]{726}$

52. $56^{1/3}$

53. $-78^{3/4}$

54. $105^{-2/3}$

Name ____

Use rational exponents to write each as a single radical.

55. $\sqrt[3]{2} \cdot \sqrt{7}$

56. $\sqrt[3]{3} \cdot \sqrt[4]{x}$

(7.3) *Perform each indicated operation and then simplify if possible. For the remainder of this review, assume that variables represent positive numbers only.*

57. $\sqrt{3} \cdot \sqrt{8}$

58. $\sqrt[3]{7y} \cdot \sqrt[3]{x^2z}$

59. $\dfrac{\sqrt{44x^3}}{\sqrt{11x}}$

60. $\dfrac{\sqrt[4]{a^6b^{13}}}{\sqrt[4]{a^2b}}$

Simplify.

61. $\sqrt{60}$

62. $-\sqrt{75}$

63. $\sqrt[3]{162}$

64. $\sqrt[3]{-32}$

65. $\sqrt{36x^7}$

66. $\sqrt[3]{24a^5b^7}$

67. $\sqrt{\dfrac{p^{17}}{121}}$

68. $\sqrt[3]{\dfrac{y^5}{27x^6}}$

69. $\sqrt[4]{\dfrac{xy^6}{81}}$

70. $\sqrt{\dfrac{2x^3}{49y^4}}$

71. The formula for the radius r of a circle of area A is

$$r = \sqrt{\frac{A}{\pi}}$$

a. Find the exact radius of a circle whose area is 25 square meters.

b. Approximate to two decimal places the radius of a circle whose area is 104 square inches.

Name ______________________________

(7.4) *Perform each indicated operation.*

72. $\sqrt{20} + \sqrt{45} - 7\sqrt{5}$

73. $x\sqrt{75x} - \sqrt{27x^3}$

74. $\sqrt[3]{128} + \sqrt[3]{250}$

75. $3\sqrt[4]{32a^5} - a\sqrt[4]{162a}$

76. $\frac{5}{\sqrt{4}} + \frac{\sqrt{3}}{3}$

77. $\sqrt{\frac{8}{x^2}} - \sqrt{\frac{50}{16x^2}}$

78. $2\sqrt{50} - 3\sqrt{125} + \sqrt{98}$

79. $2a\sqrt[4]{32b^5} - 3b\sqrt[4]{162a^4b} + \sqrt[4]{2a^4b^5}$

Multiply and then simplify if possible.

80. $\sqrt{3}(\sqrt{27} - \sqrt{3})$

81. $(\sqrt{x} - 3)^2$

82. $(\sqrt{5} - 5)(2\sqrt{5} + 2)$

83. $(2\sqrt{x} - 3\sqrt{y})(2\sqrt{x} + 3\sqrt{y})$

84. $(\sqrt{a} + 3)(\sqrt{a} - 3)$

85. $(\sqrt[3]{a} + 2)^2$

86. $(\sqrt[3]{5x} + 9)(\sqrt[3]{5x} - 9)$

87. $(\sqrt[3]{a} + 4)(\sqrt[3]{a^2} - 4\sqrt[3]{a} + 16)$

(7.5) *Rationalize each denominator.*

88. $\frac{3}{\sqrt{7}}$

89. $\sqrt{\frac{x}{12}}$

90. $\frac{5}{\sqrt[3]{4}}$

91. $\sqrt{\frac{24x^5}{3y^2}}$

92. $\sqrt[3]{\frac{15x^6y^7}{z^2}}$

93. $\sqrt[4]{\frac{81}{8x^{10}}}$

94. $\frac{3}{\sqrt{y} - 2}$

95. $\frac{\sqrt{2} - \sqrt{3}}{\sqrt{2} + \sqrt{3}}$

Name ______________________________

Rationalize each numerator.

96. $\frac{\sqrt{11}}{3}$

97. $\sqrt{\frac{18}{y}}$

98. $\frac{\sqrt[3]{9}}{7}$

99. $\sqrt{\frac{24x^5}{3y^2}}$

100. $\sqrt[3]{\frac{xy^2}{10z}}$

101. $\frac{\sqrt{x}+5}{-3}$

(7.6) *Solve each equation.*

102. $\sqrt{y-7} = 5$

103. $\sqrt{2x} + 10 = 4$

104. $\sqrt[3]{2x-6} = 4$

105. $\sqrt{x+6} = \sqrt{x+2}$

106. $2x - 5\sqrt{x} = 3$

107. $\sqrt{x+9} = 2 + \sqrt{x-7}$

Find each unknown length.

108.

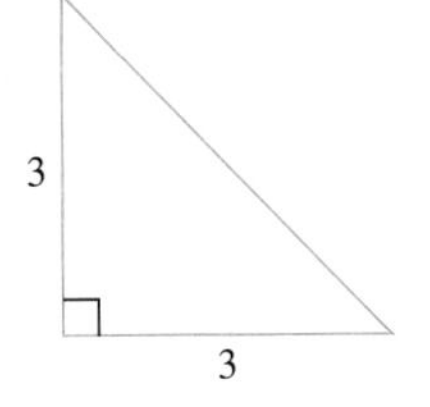

109.

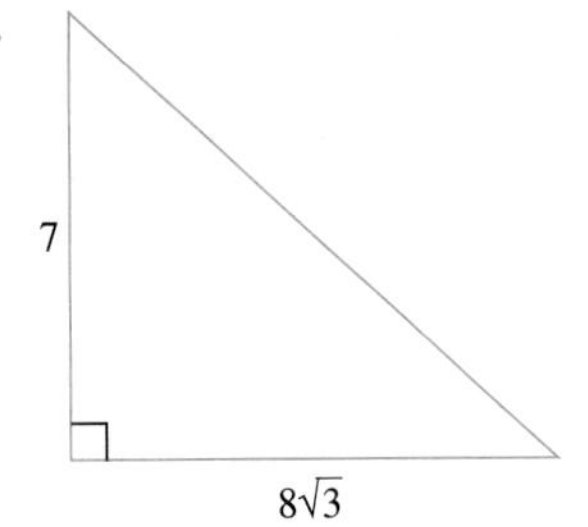

110. Craig and Daniel Cantwell want to determine the distance x across a pond on their property. They are able to measure the distances shown on the following diagram. Find how wide the lake is at the crossing point indicated by the triangle to the nearest tenth of a foot.

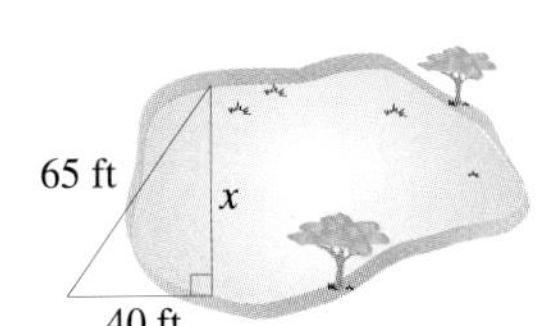

111. Andrea Roberts, a pipefitter, needs to connect two underground pipelines that are offset by 3 feet, as pictured in the diagram. Neglecting the joints needed to join the pipes, find the length of the shortest possible connecting pipe rounded to the nearest hundredth of a foot.

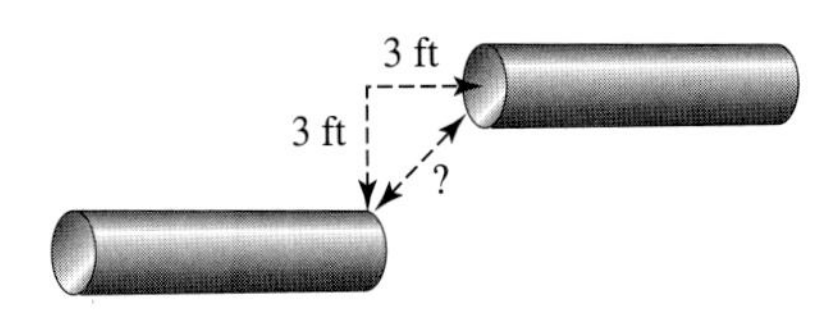

Name ______________________________

(7.7) *Perform each indicated operation and simplify. Write the results in the form $a + bi$.*

112. $\sqrt{-8}$

113. $-\sqrt{-6}$

114. $\sqrt{-4} + \sqrt{-16}$

115. $\sqrt{-2} \cdot \sqrt{-5}$

116. $(12 - 6i) + (3 + 2i)$

117. $(-8 - 7i) - (5 - 4i)$

118. $(2i)^6$

119. $2i(2 - 5i)$

120. $-3i(6 - 4i)$

121. $(3 + 2i)(1 + i)$

122. $(2 - 3i)^2$

123. $(\sqrt{6} - 9i)(\sqrt{6} + 9i)$

124. $\dfrac{2 + 3i}{2i}$

125. $\dfrac{1 + i}{-3i}$

Name ________ Section ______ Date ________

Chapter 7 Test

Raise to the power or find the root. Assume that all variables represent positive numbers. Write with only positive exponents.

1. $\sqrt{216}$ **2.** $-\sqrt[4]{x^{64}}$ **3.** $\left(\frac{1}{125}\right)^{1/3}$ **4.** $\left(\frac{1}{125}\right)^{-1/3}$

5. $\left(\frac{8x^3}{27}\right)^{2/3}$ **6.** $\sqrt[3]{-a^{18}b^9}$

7. $\left(\frac{64c^{4/3}}{a^{-2/3}b^{5/6}}\right)^{1/2}$ **8.** $a^{-2/3}(a^{5/4} - a^3)$

Find each root. Use absolute value bars when necessary.

9. $\sqrt[4]{(4xy)^4}$ **10.** $\sqrt[3]{(-27)^3}$

Rationalize each denominator. Assume that all variables represent positive numbers.

11. $\sqrt{\frac{9}{y}}$ **12.** $\frac{4 - \sqrt{x}}{4 + 2\sqrt{x}}$ **13.** $\sqrt[3]{\frac{8}{9x}}$

14. Rationalize the numerator of $\frac{\sqrt{6} + x}{8}$ and simplify.

Perform each indicated operation. Assume that all variables represent positive numbers.

15. $\sqrt{125x^3} - 3\sqrt{20x^3}$ **16.** $\sqrt{3}(\sqrt{16} - \sqrt{2})$

17. $(\sqrt{x} + 1)^2$ **18.** $(\sqrt{2} - 4)(\sqrt{3} + 1)$ **19.** $(\sqrt{5} + 5)(\sqrt{5} - 5)$

Answers

1. ________
2. ________
3. ________
4. ________
5. ________
6. ________
7. ________
8. ________
9. ________
10. ________
11. ________
12. ________
13. ________
14. ________
15. ________
16. ________
17. ________
18. ________
19. ________

20. ______

21. ______

22. ______

23. ______

24. ______

25. ______

26. ______

27. ______

28. ______

29. ______

30. ______

31. ______

32. ______

33. ______

34. ______

Name ______________________________

Use a calculator to approximate each number to three decimal places.

20. $\sqrt{561}$

21. $386^{-2/3}$

Solve.

22. $x = \sqrt{x - 2} + 2$

23. $\sqrt{x^2 - 7} + 3 = 0$

24. $\sqrt{x + 5} = \sqrt{2x - 1}$

Perform each indicated operation and simplify. Write the results in the form $a + bi$.

25. $\sqrt{-2}$

26. $-\sqrt{-8}$

27. $(12 - 6i) - (12 - 3i)$

28. $(6 - 2i)(6 + 2i)$

29. $(4 + 3i)^2$

30. $\frac{1 + 4i}{1 - i}$

31. Find x.

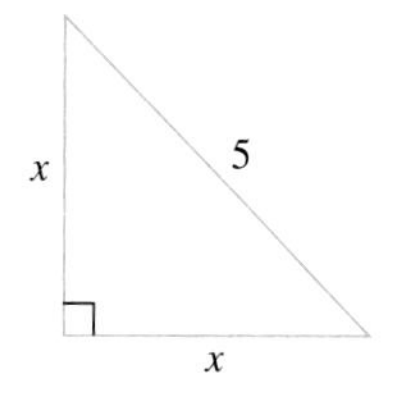

32. If $g(x) = \sqrt{x + 2}$, find $g(0)$ and $g(23)$.

Solve.

33. The function $V = \sqrt{2.5r}$ can be used to estimate the maximum safe velocity, V, in miles per hour, at which a car can travel if it is driven along a curved road with a *radius of curvature*, r, in feet. To the nearest whole number, find the maximum safe speed if a cloverleaf exit on an expressway has a radius of curvature of 300 feet.

34. Use the formula from Exercise 33 to find the radius of curvature if the safe velocity is 30 mph.

Name ______________ Section ______ Date ________

Cumulative Review

Simplify. Assume that a and b are integers and that x and y are not 0.

1. $x^{-b}(2x^b)^2$

2. $\dfrac{(y^{3a})^2}{y^{a-6}}$

3. Solve: $|x + 1| = 6$

4. Write an equation of the line containing the point $(4, 4)$ and parallel to the line $2x + y = -6$.

5. Use the elimination method to solve the system: $\begin{cases} x - 5y = -12 \\ -x + y = 4 \end{cases}$

6. Solve the system:

$$\begin{cases} x - 5y - 2z = 6 \\ -2x + 10y + 4z = -12 \\ \frac{1}{2}x - \frac{5}{2}y - z = 3 \end{cases}$$

7. Use matrices to solve the system:

$$\begin{cases} 2x - y = 3 \\ 4x - 2y = 5 \end{cases}$$

8. Add:
$(7x^3y - xy^3 + 11) + (6x^3y - 4)$

If $P(x) = 3x^2 - 2x - 5$, *find the following:*

9. $P(1)$

10. $P(-2)$

Multiply.

11. $(x + 5)^2$

12. $(4m^2 - 3n)^2$

13. Use synthetic division to divide $2x^3 - x^2 - 13x + 1$ by $x - 3$.

Factor.

14. $ab - 6a + 2b - 12$

15. $2n^2 - 38n + 80$

16. $16x^2 + 24xy + 9y^2$

17. $50 - 8y^2$

Answers

1. ______

2. ______

3. ______

4. ______

5. ______

6. ______

7. ______

8. ______

9. ______

10. ______

11. ______

12. ______

13. ______

14. ______

15. ______

16. ______

17. ______

Name

18. ______

19. see graph ______

20. ______

21. ______

22. ______

23. ______

24. ______

25. ______

18. Solve: $x(2x - 7) = 4$

19. Graph $f(x) = x^2 + 2x - 3$. Find the vertex and any intercepts.

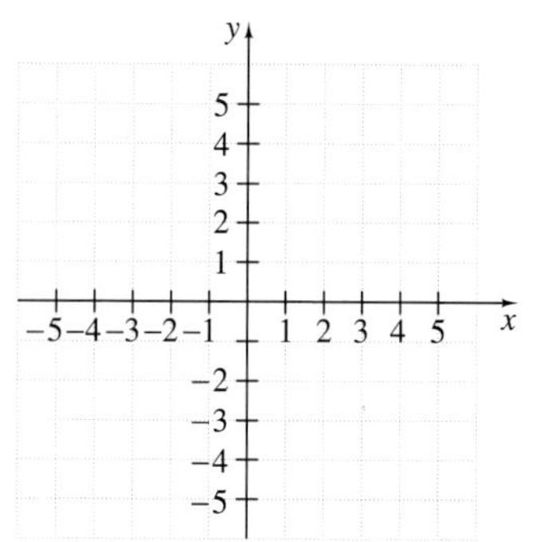

Simplify each rational expression.

20. $\dfrac{24x^6y^5}{8x^7y}$

21. $\dfrac{2x^2}{10x^3 - 2x^2}$

22. Add: $\dfrac{5}{7} + \dfrac{x}{7}$

23. Solve: $\dfrac{3}{x} - \dfrac{x + 21}{3x} = \dfrac{5}{3}$

Write each expression with a positive exponent, and then simplify.

24. $16^{-3/4}$

25. $(-27)^{-2/3}$

Quadratic Equations and Functions

Chapter 8

An important part of algebra is learning to model and solve problems. Often, the model of a problem is a quadratic equation or a function containing a second-degree polynomial. In this chapter, we continue the work from Chapter 5, solving polynomial equations in one variable by factoring. Two other methods of solving quadratic equations are analyzed in this chapter, with methods of solving nonlinear inequalities in one variable and the graphs of quadratic functions.

The surface of the Earth is heated by the sun and then slowly radiated into outer space. Sometimes, certain gases in the atmosphere reflect the heat radiation back to Earth, preventing it from escaping. The gradual warming of the atmosphere is known as the greenhouse effect. This effect is compounded by the increase of certain gases (greenhouse gases) such as carbon dioxide, methane, and nitrous oxide. Although these gases occur naturally and are needed to keep the surface of the Earth at a temperature that is hospitable to life, the recent buildup of these gases is due primarily to human activities. According to the Natural Resources Defense Council, carbon dioxide concentrations have increased by 30% globally over the past century. In Exercise 54 on page 670, we will use a quadratic function to analyze the U.S. level of emissions of methane gas.

ANSWERS

1. ______

2. ______

3. ______

4. ______

5. ______

6. ______

7. ______

8. ______

9. ______

10. ______

11. ______

12. see graph

13. see graph

14. see graph

15. see graph

16. ______

Name ______ Section ______ Date ______

CHAPTER 8 PRETEST

Solve each equation for the variable.

1. $x^2 = 54$

2. $(3y + 2)^2 = 12$

Solve the equation for the variable by completing the square.

3. $x^2 + 4x = 10$

4. $3y^2 + 18y - 1 = 0$

Use the quadratic formula to solve each equation.

5. $2x^2 - x + 4 = 0$

6. $3y^2 + \frac{1}{2}y - \frac{1}{3} = 0$

Solve.

7. $x^3 = 64$

8. $x^{2/3} - 5x^{1/3} + 4 = 0$

Solve each inequality for x. Write the solution set in interval notation.

9. $(x - 6)(x + 5) \geq 0$

10. $x^2 < -x$

11. $\frac{x + 2}{x + 3} > 1$

Graph each quadratic function. Label the vertex, y-intercept, and x-intercepts (if any).

12. $f(x) = x^2 - 5$

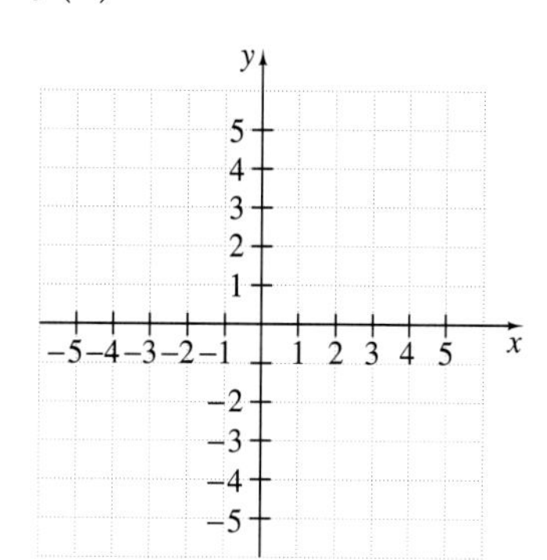

13. $g(x) = (x - 2)^2$

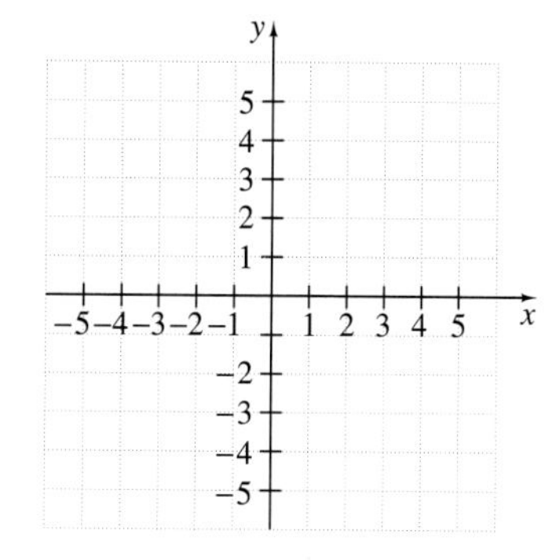

14. $h(x) = (x + 3)^2 - 1$

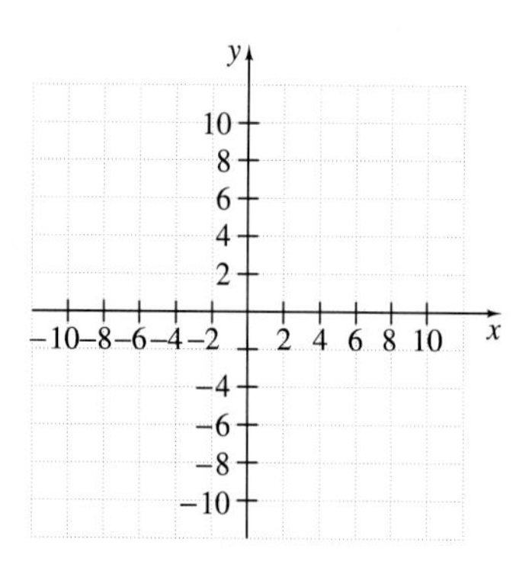

15. $f(x) = -x^2 - 4x + 2$

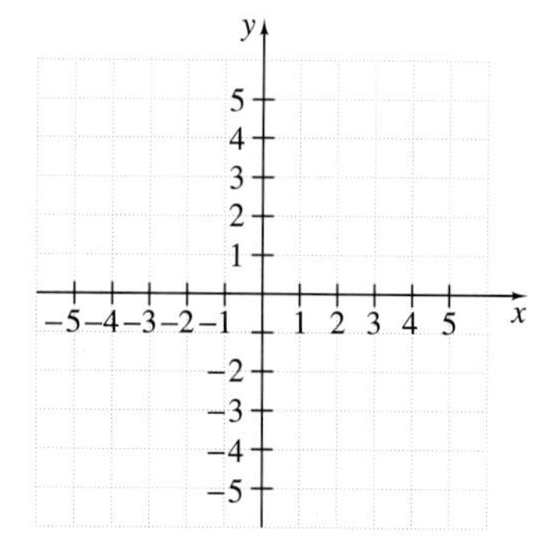

Solve.

16. The width of a rectangle is $\frac{1}{4}$ its length. If its area is 100 square centimeters, find its length and width.

8.1 SOLVING QUADRATIC EQUATIONS BY COMPLETING THE SQUARE

Objectives

A Use the square root property to solve quadratic equations.

B Write perfect square trinomials.

C Solve quadratic equations by completing the square.

D Use quadratic equations to solve problems.

SSM CD-ROM Video 8.1

A USING THE SQUARE ROOT PROPERTY

In Chapter 5, we solved quadratic equations by factoring. Recall that a **quadratic**, or **second-degree**, **equation** is an equation that can be written in the form $ax^2 + bx + c = 0$, where a, b, and c are real numbers and a is not 0. To solve a quadratic equation such as $x^2 = 9$ by factoring, we use the zero-factor theorem. To use the zero-factor theorem, the equation must first be written in standard form, $ax^2 + bx + c = 0$.

$x^2 = 9$

$x^2 - 9 = 0$ Subtract 9 from both sides to write in standard form.

$(x + 3)(x - 3) = 0$ Factor.

$x + 3 = 0$ or $x - 3 = 0$ Set each factor equal to 0.

$x = -3$ $\quad x = 3$ Solve.

The solution set is $\{-3, 3\}$, the positive and negative square roots of 9.

Not all quadratic equations can be solved by factoring, so we need to explore other methods. Notice that the solutions of the equation $x^2 = 9$ are two numbers whose square is 9:

$3^2 = 9$ and $(-3)^2 = 9$

Thus, we can solve the equation $x^2 = 9$ by taking the square root of both sides. Be sure to include both $\sqrt{9}$ and $-\sqrt{9}$ as solutions since both $\sqrt{9}$ and $-\sqrt{9}$ are numbers whose square is 9.

$x^2 = 9$

$x = \pm\sqrt{9}$ The notation $\pm\sqrt{9}$ (read as plus or minus $\sqrt{9}$) indicates the pair of numbers $+\sqrt{9}$ and $-\sqrt{9}$.

$x = \pm 3$

This illustrates the square root property.

HELPFUL HINT

The notation ± 3, for example, is read as "plus or minus 3." It is a shorthand notation for the pair of numbers $+3$ and -3.

SQUARE ROOT PROPERTY

If b is a real number and if $a^2 = b$, then $a = \pm\sqrt{b}$.

Example 1 Use the square root property to solve $x^2 = 50$.

Solution:

$x^2 = 50$

$x = \pm\sqrt{50}$ Use the square root property.

$x = \pm 5\sqrt{2}$ Simplify the radical.

Practice Problem 1

Use the square root property to solve $x^2 = 20$.

Answer

1. $\{2\sqrt{5}, -2\sqrt{5}\}$

Check:

Let $x = 5\sqrt{2}$.

$$x^2 = 50$$
$$(5\sqrt{2})^2 \stackrel{?}{=} 50$$
$$25 \cdot 2 \stackrel{?}{=} 50$$
$$50 = 50 \quad \text{True.}$$

Let $x = -5\sqrt{2}$.

$$x^2 = 50$$
$$(-5\sqrt{2})^2 \stackrel{?}{=} 50$$
$$25 \cdot 2 \stackrel{?}{=} 50$$
$$50 = 50 \quad \text{True.}$$

The solution set is $\{5\sqrt{2}, -5\sqrt{2}\}$.

Example 2

Use the square root property to solve $2x^2 = 14$.

Solution: First we get the squared variable alone on one side of the equation.

$$2x^2 = 14$$
$$x^2 = 7 \quad \text{Divide both sides by 2.}$$
$$x = \pm\sqrt{7} \quad \text{Use the square root property.}$$

Check:

Let $x = \sqrt{7}$.

$$2x^2 = 14$$
$$2(\sqrt{7})^2 \stackrel{?}{=} 14$$
$$2 \cdot 7 \stackrel{?}{=} 14$$
$$14 = 14 \quad \text{True.}$$

Let $x = -\sqrt{7}$.

$$2x^2 = 14$$
$$2(-\sqrt{7})^2 \stackrel{?}{=} 14$$
$$2 \cdot 7 \stackrel{?}{=} 14$$
$$14 = 14 \quad \text{True.}$$

The solution set is $\{\sqrt{7}, -\sqrt{7}\}$.

Practice Problem 2

Use the square root property to solve $5x^2 = 55$.

Example 3

Use the square root property to solve $(x + 1)^2 = 12$.

Solution:

$$(x + 1)^2 = 12$$
$$x + 1 = \pm\sqrt{12} \quad \text{Use the square root property.}$$
$$x + 1 = \pm 2\sqrt{3} \quad \text{Simplify the radical.}$$
$$x = -1 \pm 2\sqrt{3} \quad \text{Subtract 1 from both sides.}$$

Check: Below is a check for $-1 + 2\sqrt{3}$. The check for $-1 - 2\sqrt{3}$ is almost the same and is left for you to do on your own.

$$(x + 1)^2 = 12$$
$$(-1 + 2\sqrt{3} + 1)^2 \stackrel{?}{=} 12$$
$$(2\sqrt{3})^2 \stackrel{?}{=} 12$$
$$4 \cdot 3 \stackrel{?}{=} 12$$
$$12 = 12 \quad \text{True.}$$

The solution set is $\{-1 + 2\sqrt{3}, -1 - 2\sqrt{3}\}$.

Practice Problem 3

Use the square root property to solve $(x + 2)^2 = 18$.

Example 4

Use the square root property to solve $(2x - 5)^2 = -16$.

Solution:

$$(2x - 5)^2 = -16$$
$$2x - 5 = \pm\sqrt{-16} \quad \text{Use the square root property.}$$
$$2x - 5 = \pm 4i \quad \text{Simplify the radical.}$$
$$2x = 5 \pm 4i \quad \text{Add 5 to both sides.}$$
$$x = \frac{5 \pm 4i}{2} \quad \text{Divide both sides by 2.}$$

Check each proposed solution in the original equation to see that the solution set is $\left\{\frac{5 + 4i}{2}, \frac{5 - 4i}{2}\right\}$.

TRY THE CONCEPT CHECK IN THE MARGIN.

Practice Problem 4

Use the square root property to solve $(3x - 1)^2 = -4$.

✓ Concept Check

How do you know just by looking that $(x - 2)^2 = -4$ has complex solutions?

Answers

2. $\{\sqrt{11}, -\sqrt{11}\}$,
3. $\{-2 + 3\sqrt{2}, -2 - 3\sqrt{2}\}$,
4. $\left\{\frac{1 - 2i}{3}, \frac{1 + 2i}{3}\right\}$

✓ **Concept Check:** answers may vary

B WRITING PERFECT SQUARE TRINOMIALS

Notice from Examples 3 and 4 that, if we write a quadratic equation so that one side is the square of a binomial, we can solve by using the square root property. To write the square of a binomial, we write perfect square trinomials. Recall that a perfect square trinomial is a trinomial that can be factored into two identical binomial factors, that is, as a binomial squared.

Perfect Square Trinomials	Factored Form
$x^2 + 8x + 16$	$(x + 4)^2$
$x^2 - 6x + 9$	$(x - 3)^2$
$x^2 + 3x + \frac{9}{4}$	$\left(x + \frac{3}{2}\right)^2$

Notice that for each perfect square trinomial, *the constant term of the trinomial is the square of half the coefficient of the x-term*. For example,

$$x^2 + 8x + 16 \qquad\qquad x^2 - 6x + 9$$

$$\frac{1}{2}(8) = 4 \text{ and } 4^2 = 16 \qquad \frac{1}{2}(-6) = -3 \text{ and } (-3)^2 = 9$$

Example 5 Add the proper constant to $x^2 + 6x$ so that the result is a perfect square trinomial. Then factor.

Solution: We add the square of half the coefficient of x.

$$x^2 + 6x + 9 \qquad = (x + 3)^2 \quad \text{In factored form}$$

$$\frac{1}{2}(6) = 3 \text{ and } 3^2 = 9$$

Example 6 Add the proper constant to $x^2 - 3x$ so that the result is a perfect square trinomial. Then factor.

Solution: We add the square of half the coefficient of x.

$$x^2 - 3x + \frac{9}{4} \qquad = \left(x - \frac{3}{2}\right)^2 \quad \text{In factored form}$$

$$\frac{1}{2}(-3) = -\frac{3}{2} \text{ and } \left(-\frac{3}{2}\right)^2 = \frac{9}{4}$$

C SOLVING BY COMPLETING THE SQUARE

The process of writing a quadratic equation so that one side is a perfect square trinomial is called **completing the square**. We will use this process in the next examples.

Example 7 Solve $p^2 + 2p = 4$ by completing the square.

Solution: First we add the square of half the coefficient of p to both sides so that the resulting trinomial will be a perfect square trinomial. The coefficient of p is 2.

$$\frac{1}{2}(2) = 1 \qquad \text{and} \qquad 1^2 = 1$$

Practice Problem 5

Add the proper constant to $x^2 + 12x$ so that the result is a perfect square trinomial. Then factor.

Practice Problem 6

Add the proper constant to $y^2 - 5y$ so that the result is a perfect square trinomial. Then factor.

Practice Problem 7

Solve $x^2 + 8x = 1$ by completing the square.

Answers

5. $x^2 + 12x + 36 = (x + 6)^2$,
6. $y^2 - 5y + \frac{25}{4} = \left(x - \frac{5}{2}\right)^2$,
7. $\{-4 - \sqrt{17}, -4 + \sqrt{17}\}$

Now we add 1 to both sides of the original equation.

$$p^2 + 2p = 4$$

$$p^2 + 2p + 1 = 4 + 1 \quad \text{Add 1 to both sides.}$$

$$(p + 1)^2 = 5 \quad \text{Factor the trinomial; simplify the right side.}$$

We may now use the square root property and solve for p.

$$p + 1 = \pm\sqrt{5} \quad \text{Use the square root property.}$$

$$p = -1 \pm \sqrt{5} \quad \text{Subtract 1 from both sides.}$$

Notice that there are two solutions: $-1 + \sqrt{5}$ and $-1 - \sqrt{5}$. The solution set is $\{-1 + \sqrt{5}, -1 - \sqrt{5}\}$.

Practice Problem 8

Solve $y^2 - 5y + 2 = 0$ by completing the square.

Example 8 Solve $m^2 - 7m - 1 = 0$ by completing the square.

Solution: First we add 1 to both sides of the equation so that the left side has no constant term. We can then add the constant term on both sides that will make the left side a perfect square trinomial.

$$m^2 - 7m - 1 = 0$$

$$m^2 - 7m = 1$$

Now we find the constant term that makes the left side a perfect square trinomial by squaring half the coefficient of m. We add this constant to both sides of the equation.

$$\frac{1}{2}(-7) = -\frac{7}{2} \quad \text{and} \quad \left(-\frac{7}{2}\right)^2 = \frac{49}{4}$$

$$m^2 - 7m + \frac{49}{4} = 1 + \frac{49}{4} \quad \text{Add } \frac{49}{4} \text{ to both sides of the equation.}$$

$$\left(m - \frac{7}{2}\right)^2 = \frac{53}{4} \quad \text{Factor the perfect square trinomial and simplify the right side.}$$

$$m - \frac{7}{2} = \pm\sqrt{\frac{53}{4}} \quad \text{Use the square root property.}$$

$$m = \frac{7}{2} \pm \frac{\sqrt{53}}{2} \quad \text{Add } \frac{7}{2} \text{ to both sides and simplify } \sqrt{\frac{53}{4}}.$$

$$m = \frac{7 \pm \sqrt{53}}{2} \quad \text{Simplify.}$$

The solution set is $\left\{\frac{7 + \sqrt{53}}{2}, \frac{7 - \sqrt{53}}{2}\right\}$.

The following steps may be used to solve a quadratic equation such as $ax^2 + bx + c = 0$ by completing the square. This method may be used whether or not the polynomial $ax^2 + bx + c$ is factorable.

Answer

8. $\left\{\frac{5 - \sqrt{17}}{2}, \frac{5 + \sqrt{17}}{2}\right\}$

SOLVING A QUADRATIC EQUATION IN x BY COMPLETING THE SQUARE

Step 1. If the coefficient of x^2 is 1, go to Step 2. Otherwide, divide both sides of the equation by the coefficient of x^2.

Step 2. Get all variable terms alone on one side of the equation.

Step 3. Complete the square for the resulting binomial by adding the square of half of the coefficient of x to both sides of the equation.

Step 4. Factor the resulting perfect square trinomial and write it as the square of a binomial.

Step 5. Use the square root property to solve for x.

Example 9 Solve $4x^2 - 24x + 41 = 0$ by completing the square.

Solution: First we divide both sides of the equation by 4 so that the coefficient of x^2 is 1.

$$4x^2 - 24x + 41 = 0$$

Step 1. $x^2 - 6x + \frac{41}{4} = 0$ Divide both sides of the equation by 4.

Step 2. $x^2 - 6x = -\frac{41}{4}$ Subtract — from both sides.

Since $\frac{1}{2}(-6) = -3$ and $(-3)^2 = 9$, we add 9 to both sides of the equation.

Step 3. $x^2 - 6x + 9 = -\frac{41}{4} + 9$ Add 9 to both sides.

Step 4. $(x - 3)^2 = -\frac{41}{4} + \frac{36}{4}$ Factor the perfect square trinomial.

$$(x - 3)^2 = -\frac{5}{4}$$

Step 5. $x - 3 = \pm\sqrt{-\frac{5}{4}}$ Use the square root property.

$x - 3 = \pm\frac{i\sqrt{5}}{2}$ Simplify the radical.

$x = 3 \pm \frac{i\sqrt{5}}{2}$ Add 3 to both sides.

$= \frac{6}{2} \pm \frac{i\sqrt{5}}{2}$ Find a common denominator.

$= \frac{6 \pm i\sqrt{5}}{2}$ Simplify.

The solution set is $\left\{\frac{6 + i\sqrt{5}}{2}, \frac{6 - i\sqrt{5}}{2}\right\}$.

Practice Problem 9

Solve $2x^2 - 2x + 7 = 0$ by completing the square.

Answer

9. $\left\{\frac{1 + i\sqrt{13}}{2}, \frac{1 - i\sqrt{13}}{2}\right\}$

D SOLVING PROBLEMS MODELED BY QUADRATIC EQUATIONS

Recall the **simple interest** formula $I = Prt$, where I is the interest earned, P is the principal, r is the rate of interest, and t is time. If \$100 is invested at a simple interest rate of 5% annually, at the end of 3 years the total interest I earned is

$$I = P \cdot r \cdot t$$

or

$$I = 100 \cdot 0.05 \cdot 3 = \$15$$

and the new principal is

$$\$100 + \$15 = \$115$$

Most of the time, the interest computed on money borrowed or money deposited is **compound interest**. Compound interest, unlike simple interest, is computed on original principal *and* on interest already earned. To see the difference between simple interest and compound interest, suppose that \$100 is invested at a rate of 5% compounded annually. To find the total amount of money at the end of 3 years, we calculate as follows:

$$I \quad = \quad P \cdot r \cdot t$$

First year:	Interest $= \$100 \cdot 0.05 \cdot 1 = \$5.00$
	New principal $= \$100.00 + \$5.00 = \$105.00$
Second year:	Interest $= \$105.00 \cdot 0.05 \cdot 1 = \$5.25$
	New principal $= \$105.00 + \$5.25 = \$110.25$
Third year:	Interest $= \$110.25 \cdot 0.05 \cdot 1 \approx \$5.51$
	New principal $= \$110.25 + \$5.51 = \$115.76$

At the end of the third year, the total compound interest earned is \$15.76, whereas the total simple interest earned is \$15.

It is tedious to calculate compound interest as we did above, so we use a compound interest formula. The formula for calculating the total amount of money when interest is compounded annually is

$$A = P(1 + r)^t$$

where P is the original investment, r is the interest rate per compounding period, and t is the number of periods. For example, the amount of money A at the end of 3 years if \$100 is invested at 5% compounded annually is

$$A = \$100(1 + 0.05)^3 \approx \$100(1.1576) = \$115.76$$

as we previously calculated.

Practice Problem 10

Use the formula from Example 10 to find the interest rate r if \$1600 compounded annually grows to \$1764 in 2 years.

Example 10 Finding Interest Rates

Find the interest rate r if \$2000 compounded annually grows to \$2420 in 2 years.

Solution:

1. UNDERSTAND the problem. For this example, make sure that you understand the formula for compounding interest annually.
2. TRANSLATE. We substitute the given values into the formula:

$$A = P(1 + r)^t$$

$$2420 = 2000(1 + r)^2 \quad \text{Let } A = 2420,\ P = 2000, \text{ and } t = 2.$$

Answer

10. 5%

3. SOLVE. Solve the equation for r.

$$2420 = 2000(1 + r)^2$$

$$\frac{2420}{2000} = (1 + r)^2 \qquad \text{Divide both sides by 2000.}$$

$$\frac{121}{100} = (1 + r)^2 \qquad \text{Simplify the fraction.}$$

$$\pm\sqrt{\frac{121}{100}} = 1 + r \qquad \text{Use the square root property.}$$

$$\pm\frac{11}{10} = 1 + r \qquad \text{Simplify.}$$

$$-1 \pm \frac{11}{10} = r$$

$$-\frac{10}{10} \pm \frac{11}{10} = r$$

$$\frac{1}{10} = r \text{ or } -\frac{21}{10} = r$$

4. INTERPRET. The rate cannot be negative, so we reject $-\frac{21}{10}$.

Check: $\frac{1}{10} = 0.10 = 10\%$ per year. If we invest \$2000 at 10% compounded annually, in 2 years the amount in the account would be $2000(1 + 0.10)^2 = 2420$ dollars, the desired amount.

State: The interest rate is 10% compounded annually. ■

GRAPHING CALCULATOR EXPLORATIONS

In Section 5.8, we showed how we can use a grapher to approximate real number solutions of a quadratic equation written in standard form. We can also use a grapher to solve a quadratic equation when it is not written in standard form. For example, to solve $(x + 1)^2 = 12$, the quadratic equation in Example 3, we graph the following on the same set of axes. We use Xmin $= -10$, Xmax $= 10$, Ymin $= -13$, and Ymax $= 13$.

$$Y_1 = (x + 1)^2 \text{ and } Y_2 = 12$$

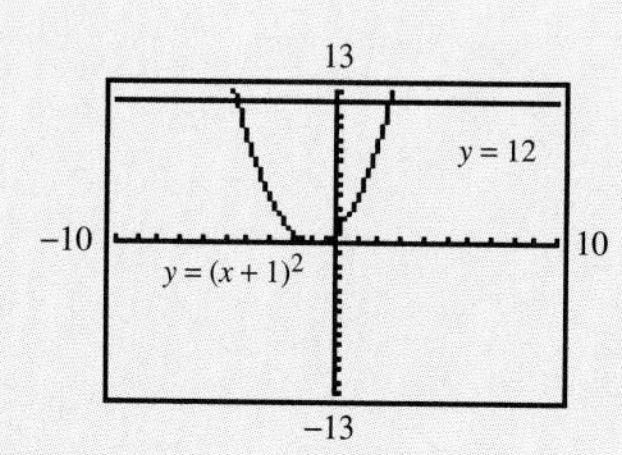

Use the Intersect feature or the Zoom and Trace features to locate the points of intersection of the graphs. The x-values of these points are the solutions of $(x + 1)^2 = 12$. The solutions, rounded to two decimal points, are 2.46 and -4.46.

Check to see that these numbers are approximations of the exact solutions $-1 \pm 2\sqrt{3}$.

Use a grapher to solve each quadratic equation. Round all solutions to the nearest hundredth.

1. $x(x - 5) = 8$

2. $x(x + 2) = 5$

3. $x^2 + 0.5x = 0.3x + 1$

4. $x^2 - 2.6x = -2.2x + 3$

5. Use a grapher and solve $(2x - 5)^2 = -16$, Example 4 in this section, using the window

Xmin $= -20$
Xmax $= 20$
Xscl $= 1$
Ymin $= -20$
Ymax $= 20$
Yscl $= 1$

Explain the results. Compare your results with the solution found in Example 4.

6. What are the advantages and disadvantages of using a grapher to solve quadratic equations?

Name ______________________ Section ________ Date ________

Exercise Set 8.1

A *Use the square root property to solve each equation. See Examples 1 through 4.*

1. $x^2 = 16$

2. $x^2 = 49$

3. $x^2 - 7 = 0$

4. $x^2 - 11 = 0$

5. $x^2 = 18$

6. $y^2 = 20$

7. $3z^2 - 30 = 0$

8. $2x^2 = 4$

9. $(x + 5)^2 = 9$

10. $(y - 3)^2 = 4$

11. $(z - 6)^2 = 18$

12. $(y + 4)^2 = 27$

13. $(2x - 3)^2 = 8$

14. $(4x + 9)^2 = 6$

15. $x^2 + 9 = 0$

16. $x^2 + 4 = 0$

17. $x^2 - 6 = 0$

18. $y^2 - 10 = 0$

19. $2z^2 + 16 = 0$

20. $3p^2 + 36 = 0$

21. $(x - 1)^2 = -16$

22. $(y + 2)^2 = -25$

23. $(z + 7)^2 = 5$

24. $(x + 10)^2 = 11$

25. $(x + 3)^2 = -8$

26. $(y - 4)^2 = -18$

Answers

1. ________
2. ________
3. ________
4. ________
5. ________
6. ________
7. ________
8. ________
9. ________
10. ________
11. ________
12. ________
13. ________
14. ________
15. ________
16. ________
17. ________
18. ________
19. ________
20. ________
21. ________
22. ________
23. ________
24. ________
25. ________
26. ________

27. ______

28. ______

29. ______

30. ______

31. ______

32. ______

33. ______

34. ______

35. ______

36. ______

37. ______

38. ______

39. ______

40. ______

41. ______

42. ______

43. ______

44. ______

45. ______

46. ______

47. ______

48. ______

49. ______

Name ______________________________

B *Add the proper constant to each binomial so that the resulting trinomial is a perfect square trinomial. Then factor the trinomial. See Examples 5 and 6.*

27. $x^2 + 16x$ **28.** $y^2 + 2y$ **29.** $z^2 - 12z$ **30.** $x^2 - 8x$

31. $p^2 + 9p$ **32.** $n^2 + 5n$ **33.** $r^2 - 3r$ **34.** $p^2 - 7p$

C *Solve each equation by completing the square. See Examples 7 through 9.*

35. $x^2 + 8x = -15$ **36.** $y^2 + 6y = -8$ **37.** $x^2 + 6x + 2 = 0$

38. $x^2 - 2x - 2 = 0$ **39.** $x^2 + x - 1 = 0$ **40.** $x^2 + 3x - 2 = 0$

41. $x^2 + 2x - 5 = 0$ **42.** $y^2 + y - 7 = 0$ **43.** $3p^2 - 12p + 2 = 0$

44. $2x^2 + 14x - 1 = 0$ **45.** $2x^2 + 7x = 4$ **46.** $3x^2 - 4x = 4$

47. $x^2 + 8x + 1 = 0$ **48.** $x^2 - 10x + 2 = 0$ **49.** $3y^2 + 6y - 4 = 0$

Name ______________________________

50. $2y^2 + 12y + 3 = 0$

51. $y^2 + 2y + 2 = 0$

52. $x^2 + 4x + 6 = 0$

53. $x^2 - 6x + 3 = 0$

54. $x^2 - 7x - 1 = 0$

55. $2a^2 + 8a = -12$

56. $3x^2 + 12x = -14$

57. $2x^2 - x + 6 = 0$

58. $4x^2 - 2x + 5 = 0$

59. $x^2 + 10x + 28 = 0$

60. $y^2 + 8y + 18 = 0$

61. $z^2 + 3z - 4 = 0$

62. $y^2 + y - 2 = 0$

63. $2x^2 - 4x + 3 = 0$

64. $9x^2 - 36x = -40$

65. $3x^2 + 3x = 5$

66. $5y^2 - 15y = 1$

D *Use the formula $A = P(1 + r)^t$ to solve Exercises 67–70. See Example 10.*

67. Find the rate r at which \$3000 grows to \$4320 in 2 years.

68. Find the rate r at which \$800 grows to \$882 in 2 years.

69. Find the rate at which \$810 grows to \$1000 in 2 years.

70. Find the rate at which \$2000 grows to \$2880 in 2 years.

71. In your own words, what is the difference between simple interest and compound interest?

72. If you are depositing money in an account that pays 4%, would you prefer the interest to be simple or compound?

73. If you are borrowing money at a rate of 10%, would you prefer the interest to be simple or compound?

50. ______
51. ______
52. ______
53. ______
54. ______
55. ______
56. ______
57. ______
58. ______
59. ______
60. ______
61. ______
62. ______
63. ______
64. ______
65. ______
66. ______
67. ______
68. ______
69. ______
70. ______
71. ______
72. ______
73. ______

74. ____

75. ____

76. ____

77. ____

78. ____

79. ____

80. ____

81. ____

82. ____

83. ____

84. ____

85. ____

Name ________________

Review and Preview

Simplify each expression. See Section 7.5.

74. $\dfrac{6 + 4\sqrt{5}}{2}$

75. $\dfrac{10 - 20\sqrt{3}}{2}$

76. $\dfrac{3 - 9\sqrt{2}}{6}$

77. $\dfrac{12 - 8\sqrt{7}}{16}$

Evaluate $\sqrt{b^2 - 4ac}$ *for each set of values. See Section 7.3.*

78. $a = 2, b = 4, c = -1$

79. $a = 1, b = 6, c = 2$

80. $a = 3, b = -1, c = -2$

81. $a = 1, b = -3, c = -1$

Combining Concepts

Find two possible missing terms so that each is a perfect square trinomial.

82. $x^2 + \square + 16$

83. $y^2 + \square + 9$

Neglecting air resistance, the distance $s(t)$ in feet traveled by a freely falling object is given by the function $s(t) = 16t^2$, where t is time in seconds. Use this formula to solve Exercises 84–87. Round answers to two decimal places.

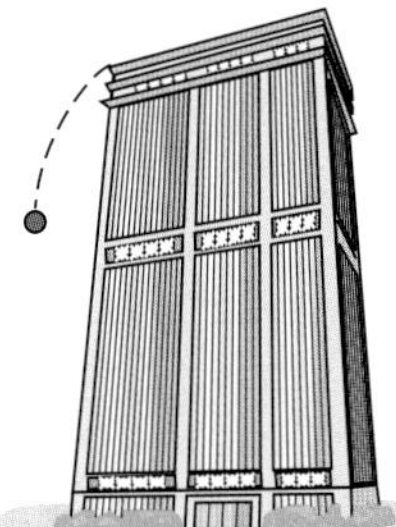

84. The Petronas Towers in Kuala Lumpur, built in 1997, are the tallest buildings in Malaysia. Each tower is 1483 feet tall. How long would it take an object to fall to the ground from the top of one of the towers? (*Source:* Council on Tall Buildings and Urban Habitat, Lehigh University)

85. The height of the Chicago Beach Tower Hotel, built in 1998 in Dubai, United Arab Emirates, is 1053 feet. How long would it take an object to fall to the ground from the top of the building? (*Source:* Council on Tall Buildings and Urban Habitat, Lehigh University)

Name ______________________________

86. The height of the Nurek Dam in Tajikistan (part of the former USSR that borders Afghanistan) is 984 feet. How long would it take an object to fall from the top to the base of the dam? (*Source:* U.S. Committee on Large Dams of the International Commission on Large Dams)

87. The Hoover Dam, located on the Colorado River on the border of Nevada and Arizona near Las Vegas, is 725 feet tall. How long would it take an object to fall from the top to the base of the dam? (*Source:* U.S. Committee on Large Dams of the International Commission on Large Dams)

Internet Excursions

http://www.prenhall.com/martin-gay

This World Wide Web address listed above will direct you to a website that contains current interest rates—both composite rates and those offered by individual savings institutions—on a wide variety of financial products such as savings deposits and auto loans.

88. Choose a financial product. Using actual data for a current interest rate on that type of product, write a problem similar to Exercises 67–70. When you have finished writing your problem, trade with another student in your class to solve. Then check each other's work.

89. Choose a different financial product from this web site. Write another interest rate problem using current interest rates. Then exchange problems with another student in your class to solve. Check each other's work.

86. ______

87. ______

88. ______

89. ______

Focus On Business and Career

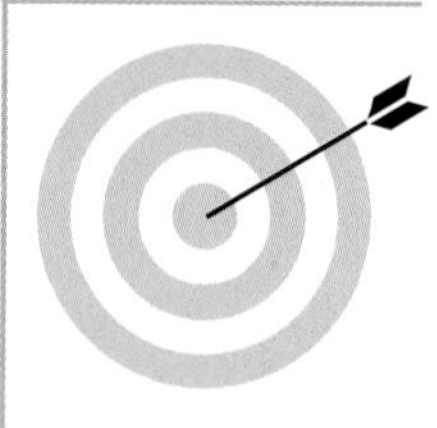

Flowcharts

We saw in the Focus On Business and Career feature in Chapter 6 that three of the top 10 fastest-growing jobs into the 21st century are computer related. A useful skill in computer-related careers is *flowcharting*. A **flowchart** is a diagram showing a sequence of procedures used to complete a task. Flowcharts are commonly used in computer programming to help a programmer plan the steps and commands needed to write a program. Flowcharts are also used in other types of careers, such as manufacturing or finance, to describe the sequence of events needed in a certain process.

A flowchart usually uses the following symbols to represent certain types of actions.

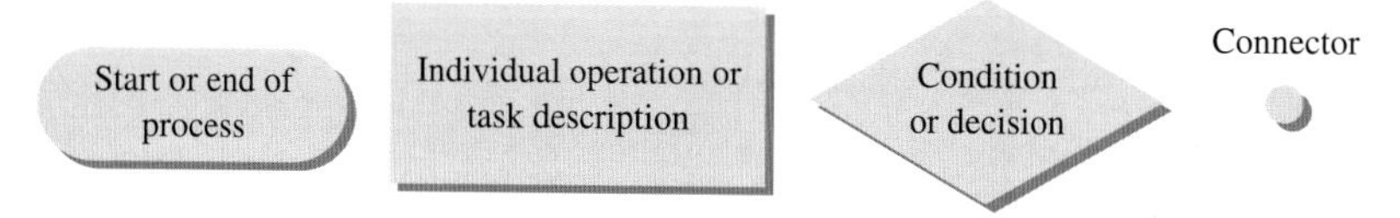

For instance, suppose we want to write a flowchart for the process of computing a household's monthly electric bill. The flowchart might look something like the one here.

Critical Thinking

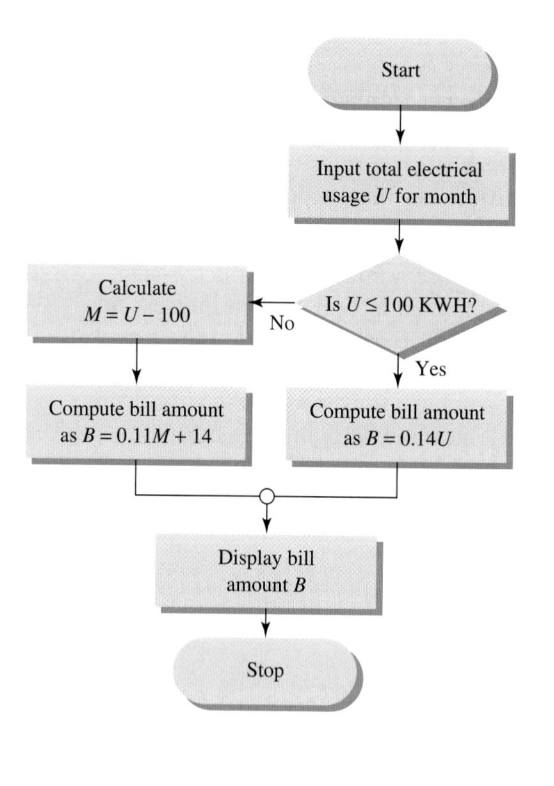

1. Using the given flowchart as a guide, describe in words this utility company's pricing structure for household electricity usage.

2. Make a flowchart for the process of computing the human-equivalent age for the age of a dog if 1 dog year is equivalent to 7 human years.

3. Make a flowchart for the process of determining the number and type of solutions of a quadratic equation of the form $ax^2 + bx + c = 0$ using the discriminant.

4. (Optional) Write a programmable calculator program for your graphing calculator that determines the number and type of solutions of a quadratic equation of the form $ax^2 + bx + c = 0$.

8.2 SOLVING QUADRATIC EQUATIONS BY THE QUADRATIC FORMULA

A SOLVING EQUATIONS BY USING THE QUADRATIC FORMULA

Any quadratic equation can be solved by completing the square. Since the same sequence of steps is repeated each time we complete the square, let's complete the square for a general quadratic equation, $ax^2 + bx + c = 0$. By doing so, we find a pattern for the solutions of a quadratic equation known as the **quadratic formula**.

Recall that to complete the square for an equation such as $ax^2 + bx + c = 0, a \neq 0$, we first divide both sides by the coefficient of x^2.

$$ax^2 + bx + c = 0$$

$$x^2 + \frac{b}{a}x + \frac{c}{a} = 0 \quad \text{Divide both sides by } a \text{, the coefficient of } x^2.$$

$$x^2 + \frac{b}{a}x = -\frac{c}{a} \quad \text{Subtract the constant } \frac{c}{a} \text{ from both sides.}$$

Next we find the square of half $\frac{b}{a}$, the coefficient of x.

$$\frac{1}{2}\left(\frac{b}{a}\right) = \frac{b}{2a} \quad \text{and} \quad \left(\frac{b}{2a}\right)^2 = \frac{b^2}{4a^2}$$

Now we add this result to both sides of the equation.

$$x^2 + \frac{b}{a}x + \frac{b^2}{4a^2} = -\frac{c}{a} + \frac{b^2}{4a^2} \quad \text{Add } \frac{b^2}{4a^2} \text{ to both sides.}$$

$$x^2 + \frac{b}{a}x + \frac{b^2}{4a^2} = \frac{-c \cdot 4a}{a \cdot 4a} + \frac{b^2}{4a^2} \quad \text{Find a common denominator on the right side.}$$

$$x^2 + \frac{b}{a}x + \frac{b^2}{4a^2} = \frac{b^2 - 4ac}{4a^2} \quad \text{Simplify the right side.}$$

$$\left(x + \frac{b}{2a}\right)^2 = \frac{b^2 - 4ac}{4a^2} \quad \text{Factor the perfect square trinomial on the left side.}$$

$$x + \frac{b}{2a} = \pm\sqrt{\frac{b^2 - 4ac}{4a^2}} \quad \text{Use the square root property.}$$

$$x + \frac{b}{2a} = \pm\frac{\sqrt{b^2 - 4ac}}{2a} \quad \text{Simplify the radical.}$$

$$x = -\frac{b}{2a} \pm \frac{\sqrt{b^2 - 4ac}}{2a} \quad \text{Subtract } \frac{b}{2a} \text{ from both sides.}$$

$$x = \frac{-b \pm \sqrt{b^2 - 4ac}}{2a} \quad \text{Simplify.}$$

The resulting equation identifies the solutions of the general quadratic equation in standard form and is called the quadratic formula. It can be used to solve any equation written in standard form $ax^2 + bx + c = 0$ as long as a is not 0.

QUADRATIC FORMULA

A quadratic equation written in the form $ax^2 + bx + c = 0$ has the solutions

$$x = \frac{-b \pm \sqrt{b^2 - 4ac}}{2a}$$

Objectives

A Solve quadratic equations by using the quadratic formula.

B Determine the number and type of solutions of a quadratic equation by using the discriminant.

C Solve geometric problems modeled by quadratic equations.

SSM CD-ROM Video 8.2

Practice Problem 1

Solve: $2x^2 + 9x + 10 = 0$

Example 1

Solve: $3x^2 + 16x + 5 = 0$

Solution: This equation is in standard form with $a = 3, b = 16$, and $c = 5$. We substitute these values into the quadratic formula.

$$x = \frac{-b \pm \sqrt{b^2 - 4ac}}{2a} \quad \text{Quadratic formula}$$

$$= \frac{-16 \pm \sqrt{16^2 - 4(3)(5)}}{2(3)} \quad \text{Let } a = 3, b = 16, \text{ and } c = 5.$$

$$= \frac{-16 \pm \sqrt{256 - 60}}{6}$$

$$= \frac{-16 \pm \sqrt{196}}{6} = \frac{-16 \pm 14}{6}$$

$$x = \frac{-16 + 14}{6} = -\frac{1}{3} \quad \text{or} \quad x = \frac{-16 - 14}{6} = -\frac{30}{6} = -5$$

The solution set is $\left\{-\frac{1}{3}, -5\right\}$.

HELPFUL HINT

To replace a, b, and c correctly in the quadratic formula, write the quadratic equation in standard form $ax^2 + bx + c = 0$.

Practice Problem 2

Solve: $2x^2 - 6x - 1 = 0$

Example 2

Solve: $2x^2 - 4x = 3$

Solution: First we write the equation in standard form by subtracting 3 from both sides.

$2x^2 - 4x - 3 = 0$

Now $a = 2, b = -4$, and $c = -3$. We substitute these values into the quadratic formula.

$$x = \frac{-b \pm \sqrt{b^2 - 4ac}}{2a}$$

$$= \frac{-(-4) \pm \sqrt{(-4)^2 - 4(2)(-3)}}{2(2)}$$

$$= \frac{4 \pm \sqrt{16 + 24}}{4}$$

$$= \frac{4 \pm \sqrt{40}}{4} = \frac{4 \pm 2\sqrt{10}}{4}$$

$$= \frac{2(2 \pm \sqrt{10})}{2 \cdot 2} = \frac{2 \pm \sqrt{10}}{2}$$

The solution set is $\left\{\frac{2 + \sqrt{10}}{2}, \frac{2 - \sqrt{10}}{2}\right\}$.

TRY THE CONCEPT CHECK IN THE MARGIN.

✓ CONCEPT CHECK

For the quadratic equation $x^2 = 7$, which substitution is correct?

a. $a = 1, b = 0$, and $c = -7$
b. $a = 1, b = 0$, and $c = 7$
c. $a = 0, b = 0$, and $c = 7$
d. $a = 1, b = 1$, and $c = -7$

Answers

1. $\left\{-\frac{5}{2}, -2\right\}$, **2.** $\left\{\frac{3 + \sqrt{11}}{2}, \frac{3 - \sqrt{11}}{2}\right\}$

✓ Concept Check: a

HELPFUL HINT

To simplify the expression $\frac{4 \pm 2\sqrt{10}}{4}$ in the preceding example, note that 2 is factored out of both terms of the numerator *before* simplifying.

$$\frac{4 \pm 2\sqrt{10}}{4} = \frac{2(2 \pm \sqrt{10})}{2 \cdot 2} = \frac{2 \pm \sqrt{10}}{2}$$

Example 3 Solve: $\frac{1}{4}m^2 - m + \frac{1}{2} = 0$

Solution: We could use the quadratic formula with $a = \frac{1}{4}$, $b = -1$, and $c = \frac{1}{2}$. Instead, let's find a simpler, equivalent standard form equation whose coefficients are not fractions.

First we multiply both sides of the equation by 4 to clear the fractions.

$$4\left(\frac{1}{4}m^2 - m + \frac{1}{2}\right) = 4 \cdot 0$$

$$m^2 - 4m + 2 = 0 \quad \text{Simplify.}$$

Now we can substitute $a = 1$, $b = -4$, and $c = 2$ into the quadratic formula and simplify.

$$m = \frac{-(-4) \pm \sqrt{(-4)^2 - 4(1)(2)}}{2(1)}$$

$$= \frac{4 \pm \sqrt{16 - 8}}{2}$$

$$= \frac{4 \pm \sqrt{8}}{2} = \frac{4 \pm 2\sqrt{2}}{2} = \frac{2(2 \pm \sqrt{2})}{2} = 2 \pm \sqrt{2}$$

The solution set is $\{2 + \sqrt{2}, 2 - \sqrt{2}\}$.

Practice Problem 3

Solve: $\frac{1}{6}x^2 - \frac{1}{2}x - 1 = 0$

Example 4 Solve: $p = -3p^2 - 3$

Solution: The equation in standard form is $3p^2 + p + 3 = 0$. Thus, $a = 3$, $b = 1$, and $c = 3$ in the quadratic formula.

$$p = \frac{-1 \pm \sqrt{1^2 - 4(3)(3)}}{2(3)} = \frac{-1 \pm \sqrt{1 - 36}}{6}$$

$$= \frac{-1 \pm \sqrt{-35}}{6} = \frac{-1 \pm i\sqrt{35}}{6}$$

The solution set is $\left\{\frac{-1 + i\sqrt{35}}{6}, \frac{-1 - i\sqrt{35}}{6}\right\}$.

Practice Problem 4

Solve: $x = -4x^2 - 4$

TRY THE CONCEPT CHECK IN THE MARGIN.

✓ CONCEPT CHECK

What is the first step in solving $-3x^2 = 5x - 4$ using the quadratic formula?

Answers

3. $\left\{\frac{3 + \sqrt{33}}{2}, \frac{3 - \sqrt{33}}{2}\right\}$,
4. $\left\{\frac{-1 - 3i\sqrt{7}}{8}, \frac{-1 + 3i\sqrt{7}}{8}\right\}$

✓ Concept Check answers may vary

B USING THE DISCRIMINANT

In the quadratic formula $x = \dfrac{-b \pm \sqrt{b^2 - 4ac}}{2a}$, the radicand $b^2 - 4ac$ is called the **discriminant** because when we know its value, we can **discriminate** among the possible number and type of solutions of a quadratic equation. Possible values of the discriminant and their meanings are summarized next.

DISCRIMINANT

The following table corresponds the discriminant $b^2 - 4ac$ of a quadratic equation of the form $ax^2 + bx + c = 0$ with the number and type of solutions of the equation.

$b^2 - 4ac$	Number and Type of Solutions
Positive	Two real solutions
Zero	One real solution
Negative	Two complex but not real solutions

Practice Problem 5

Use the discriminant to determine the number and type of solutions of $x^2 + 4x + 4 = 0$.

Example 5

Use the discriminant to determine the number and type of solutions of $x^2 + 2x + 1$.

Solution: In $x^2 + 2x + 1 = 0$, $a = 1$, $b = 2$, and $c = 1$. Thus,

$$b^2 - 4ac = 2^2 - 4(1)(1) = 0$$

Since $b^2 - 4ac = 0$, this quadratic equation has one real solution.

Practice Problem 6

Use the discriminant to determine the number and type of solutions of $5x^2 + 7 = 0$.

Example 6

Use the discriminant to determine the number and type of solutions of $3x^2 + 2 = 0$.

Solution: In this equation, $a = 3$, $b = 0$, and $c = 2$. Then $b^2 - 4ac = 0^2 - 4(3)(2) = -24$. Since $b^2 - 4ac$ is negative, this quadratic equation has two complex but not real solutions.

Practice Problem 7

Use the discriminant to determine the number and type of solutions of $3x^2 - 2x - 2 = 0$.

Example 7

Use the discriminant to determine the number and type of solutions of $2x^2 - 7x - 4 = 0$.

Solution: In this equation, $a = 2$, $b = -7$, and $c = -4$. Then

$$b^2 - 4ac = (-7)^2 - 4(2)(-4) = 81$$

Since $b^2 - 4ac$ is positive, this quadratic equation has two real solutions.

C SOLVING PROBLEMS MODELED BY QUADRATIC EQUATIONS

The quadratic formula is useful in solving problems that are modeled by quadratic equations.

Answers

5. 1 real solution, **6.** 2 complex but not real solutions, **7.** 2 real solutions

Example 8 Calculating Distance Saved

At a local university, students often leave the sidewalk and cut across the lawn to save walking distance. Given the diagram below of a favorite place to cut across the lawn, approximate how many feet of walking distance a student saves by cutting across the lawn instead of walking on the sidewalk.

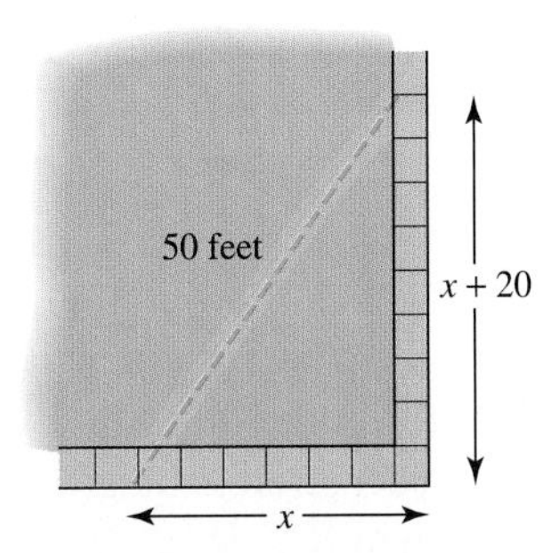

Solution:

1. UNDERSTAND. Read and reread the problem. You may want to review the Pythagorean theorem.

2. TRANSLATE. By the Pythagorean theorem, we have

In words: $(\text{leg})^2 + (\text{leg})^2 = (\text{hypotenuse})^2$

Translate: $x^2 + (x + 20)^2 = 50^2$

3. SOLVE. Use the quadratic formula to solve.

$$x^2 + x^2 + 40x + 400 = 2500 \quad \text{Square } (x + 20) \text{ and } 50.$$
$$2x^2 + 40x - 2100 = 0 \quad \text{Set the equation equal to 0.}$$
$$x^2 + 20x - 1050 = 0 \quad \text{Divide by 2.}$$

Here, $a = 1, b = 20, c = -1050$. By the quadratic formula,

$$x = \frac{-20 \pm \sqrt{20^2 - 4(1)(-1050)}}{2 \cdot 1}$$
$$= \frac{-20 \pm \sqrt{400 + 4200}}{2} = \frac{-20 \pm \sqrt{4600}}{2}$$
$$= \frac{-20 \pm \sqrt{100 \cdot 46}}{2} = \frac{-20 \pm 10\sqrt{46}}{2}$$
$$= -10 \pm 5\sqrt{46} \quad \text{Simplify.}$$

Check:

4. INTERPRET. Check your calculations in the quadratic formula. The length of a side of a triangle can't be negative, so we reject $-10 - 5\sqrt{46}$. Since $-10 + 5\sqrt{46} \approx 24$ feet, the walking distance along the sidewalk is

$$x + (x + 20) \approx 24 + (24 + 20) = 68 \text{ feet.}$$

State: A person saves 68 − 50 or 18 feet of walking distance by cutting across the lawn.

Practice Problem 8

Given the diagram below, approximate to the nearest foot how many feet of walking distance a person saves by cutting across the lawn instead of walking on the sidewalk.

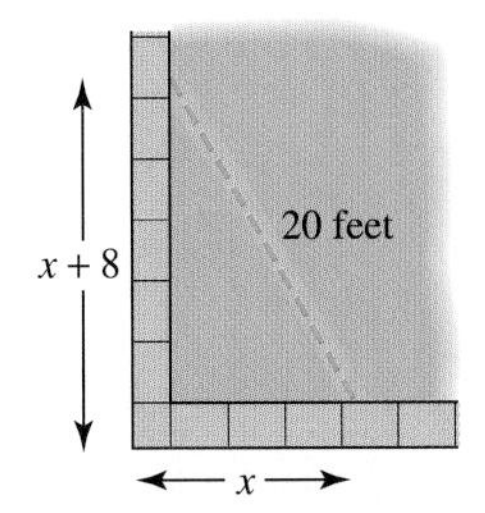

Answer

8. 9 feet

Practice Problem 9

Use the equation given in Example 9 to find how long after the object is thrown it will be 100 feet from the ground. Round to the nearest tenth of a second.

Example 9

An object is thrown upward from the top of a 200-foot cliff with a velocity of 12 feet per second. The height h of the object after t seconds is

$$h = -16t^2 + 12t + 200$$

How long after the object is thrown will it strike the ground? Round to the nearest tenth of a second.

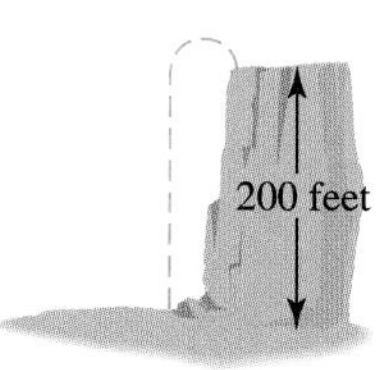

Solution:

1. UNDERSTAND. Read and reread the problem.
2. TRANSLATE. Since we want to know when the object strikes the ground, we want to know when the height $h = 0$, or

   $$0 = -16t^2 + 12t + 200$$

3. SOLVE. First, divide both sides of the equation by -4.

   $$0 = 4t^2 - 3t - 50 \quad \text{Divide both sides by } -4.$$

   Here, $a = 4$, $b = -3$, and $c = -50$. By the quadratic formula,

   $$t = \frac{-(-3) \pm \sqrt{(-3)^2 - 4(4)(-50)}}{2 \cdot 4}$$

   $$= \frac{3 \pm \sqrt{9 + 800}}{8}$$

   $$= \frac{3 \pm \sqrt{809}}{8}$$

Check:

4. INTERPRET. Check your calculations in the quadratic formula. Since the time won't be negative, we reject the proposed solution $\frac{3 - \sqrt{809}}{8}$.

State: The time it takes for the object to strike the ground is exactly $\frac{3 + \sqrt{809}}{8}$ seconds $\approx$ 3.9 seconds.

Answer

9. 1.7 sec

Name ______________________ Section ______ Date ________

Mental Math

Identify the values of a, b, and c in each quadratic equation.

1. $x^2 + 3x + 1 = 0$ **2.** $2x^2 - 5x - 7 = 0$ **3.** $7x^2 - 4 = 0$

4. $x^2 + 9 = 0$ **5.** $6x^2 - x = 0$ **6.** $5x^2 + 3x = 0$

Exercise Set 8.2

A *Use the quadratic formula to solve each equation. See Examples 1 through 4.*

1. $m^2 + 5m - 6 = 0$ **2.** $p^2 + 11p - 12 = 0$ **3.** $2y = 5y^2 - 3$

4. $5x^2 - 3 = 14x$ **5.** $x^2 - 6x + 9 = 0$ **6.** $y^2 + 10y + 25 = 0$

7. $x^2 + 7x + 4 = 0$ **8.** $y^2 + 5y + 3 = 0$ **9.** $8m^2 - 2m = 7$

10. $11n^2 - 9n = 1$ **11.** $3m^2 - 7m = 3$ **12.** $x^2 - 13 = 5x$

13. $\frac{1}{2}x^2 - x - 1 = 0$ **14.** $\frac{1}{6}x^2 + x + \frac{1}{3} = 0$ **15.** $\frac{2}{5}y^2 + \frac{1}{5}y = \frac{3}{5}$

16. $\frac{1}{8}x^2 + x = \frac{5}{2}$ **17.** $\frac{1}{3}y^2 - y - \frac{1}{6} = 0$ **18.** $\frac{1}{2}y^2 = y + \frac{1}{2}$

19. $10y^2 + 10y + 3 = 0$ **20.** $3y^2 + 6y + 5 = 0$ **21.** $x^2 + 5x = -2$

22. $y^2 - 8 = 4y$ **23.** $(m + 2)(2m - 6) = 5(m - 1) - 12$

Mental Math Answers

1. ______
2. ______
3. ______
4. ______
5. ______
6. ______

Answers

1. ______
2. ______
3. ______
4. ______
5. ______
6. ______
7. ______
8. ______
9. ______
10. ______
11. ______
12. ______
13. ______
14. ______
15. ______
16. ______
17. ______
18. ______
19. ______
20. ______
21. ______
22. ______
23. ______

24. ______
25. ______
26. ______
27. ______
28. ______
29. ______
30. ______
31. ______
32. ______
33. ______
34. ______
35. ______
36. ______
37. ______
38. ______
39. ______
40. ______
41. ______
42. ______
43. ______
44. ______
45. ______
46. ______
47. ______
48. ______
49. ______
50. ______

Name ______________________________

24. $7p(p - 2) + 2(p + 4) = 3$

25. $\frac{x^2}{3} - x = \frac{5}{3}$

26. $\frac{x^2}{2} - 3 = -\frac{9}{2}x$

27. $x(6x + 2) - 3 = 0$

28. $x(7x + 1) = 2$

29. Solve Exercise 1 by factoring. Explain the result.

30. Solve Exercise 2 by factoring. Explain the result.

Use the quadratic formula to solve each equation. See Examples 1 through 4.

31. $6 = -4x^2 + 3x$

32. $9x^2 + x + 2 = 0$

33. $(x + 5)(x - 1) = 2$

34. $x(x + 6) = 2$

35. $x^2 + 6x + 13 = 0$

36. $x^2 + 2x + 2 = 0$

37. $\frac{2}{5}y^2 + \frac{1}{5}y + \frac{3}{5} = 0$

38. $\frac{1}{8}x^2 + x + \frac{5}{2} = 0$

39. $\frac{1}{2}y^2 = y - \frac{1}{2}$

40. $\frac{2}{3}x^2 - \frac{20}{3}x = -\frac{100}{6}$

41. $(n - 2)^2 = 2n$

42. $\left(p - \frac{1}{2}\right)^2 = \frac{p}{2}$

B *Use the discriminant to determine the number and types of solutions of each equation. See Examples 5 through 7.*

43. $9x - 2x^2 + 5 = 0$

44. $5 - 4x + 12x^2 = 0$

45. $4x^2 + 12x = -9$

46. $9x^2 + 1 = 6x$

47. $3x = -2x^2 + 7$

48. $3x^2 = 5 - 7x$

49. $6 = 4x - 5x^2$

50. $8x = 3 - 9x^2$

Name ____________________

C *Solve. See Examples 8 and 9.*

51. Nancy, Thelma, and John Varner live on a corner lot. Often, neighborhood children cut across their lot to save walking distance. Given the diagram below, approximate to the nearest foot how many feet of walking distance is saved by cutting across their property instead of walking around the lot.

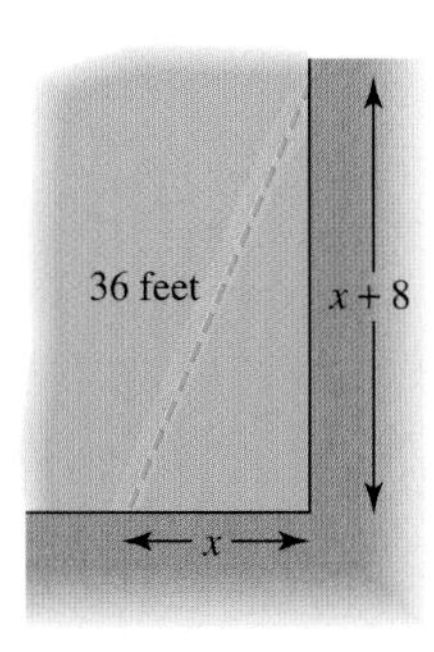

52. Given the diagram below, approximate to the nearest foot how many feet of walking distance a person saves by cutting across the lawn instead of walking on the sidewalk.

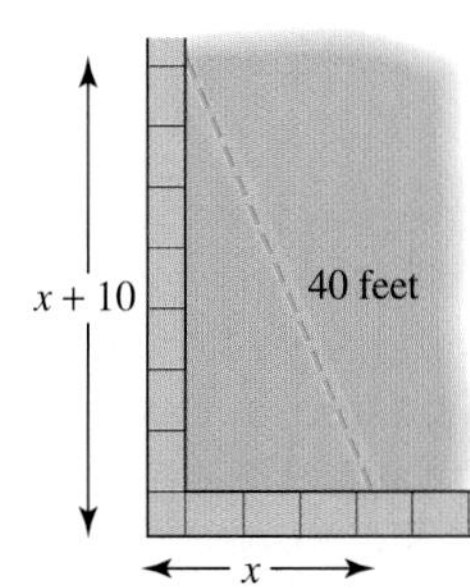

53. The hypotenuse of an isosceles right triangle is 2 centimeters longer than either of its legs. Find the exact length of each side. (*Hint:* An isosceles right triangle is a right triangle whose legs are the same length.)

54. The hypotenuse of an isosceles right triangle is one meter longer than either of its legs. Find the length of each side.

55. Uri Chechov's rectangular dog pen for his Irish setter must have an area of 400 square feet. Also, the length must be 10 feet longer than the width. Find the dimensions of the pen.

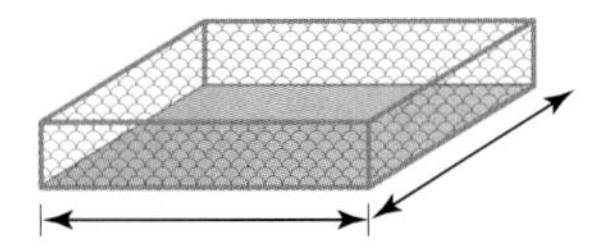

56. An entry in the Peach Festival Poster Contest must be rectangular and have an area of 1200 square inches. Furthermore, its length must be 20 inches longer than its width. Find the dimensions each entry must have.

57. A holding pen for cattle must be square and have a diagonal length of 100 meters.

a. Find the length of a side of the pen.

b. Find the area of the pen.

58. A rectangle is three times longer than it is wide. It has a diagonal of length 50 centimeters.

a. Find the dimensions of the rectangle.

b. Find the perimeter of the rectangle.

50 cm

51. ____________

52. ____________

53. ____________

54. ____________

55. ____________

56. ____________

57. a. ____________

b. ____________

58. a. ____________

b. ____________

59.

60.

61.

62.

63.

64.

65.

66.

67.

68.

Name

59. If a point B divides a line segment such that the smaller portion is to the larger portion as the larger is to the whole, the whole is the length of the *golden ratio*.

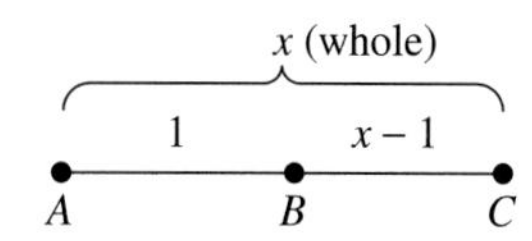

The golden ratio was thought by the Greeks as the most pleasing to the eye, and many of their buildings contained numerous examples of the golden ratio. The value of the golden ratio is the positive solution of

$$\frac{x-1}{1} = \frac{1}{x}$$

(smaller)/(larger) = (larger)/(whole)

Find this value.

60. The base of a triangle is four more than twice its height. If the area of the triangle is 42 square centimeters, find its base and height.

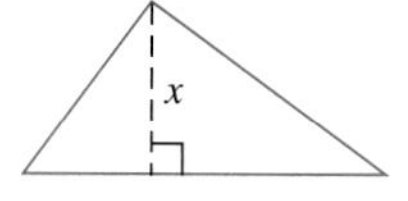

The Wollomombi Falls in Australia have a height of 1100 feet. A pebble is thrown upward from the top of the falls with an initial velocity of 20 feet per second. The height of the pebble h after t seconds is given by the equation $h = -16t^2 + 20t + 1100$. Use this equation for Exercises 61 and 62.

61. How long after the pebble is thrown will it hit the ground. Round to the nearest tenth of a second.

62. How long after the pebble is thrown will it be 550 feet from the ground. Round to the nearest tenth of a second.

A ball is thrown downward from the top of a 180-foot building with an initial velocity of 20 feet per second. The height of the ball h after t seconds is given by the equation $h = -16t^2 - 20t + 180$. Use this equation to answer Exercises 63 and 64.

63. How long after the ball is thrown will it strike the ground? Round the result to the nearest tenth of a second.

64. How long after the ball is thrown will it be 50 feet from the ground? Round the result to the nearest tenth of a second.

Review and Preview

Solve each equation. See Sections 6.4 and 7.6.

65. $\sqrt{5x - 2} = 3$

66. $\sqrt{y + 2} + 7 = 12$

67. $\frac{1}{x} + \frac{2}{5} = \frac{7}{x}$

68. $\frac{10}{z} = \frac{5}{z} - \frac{1}{3}$

Name ______________________________

Factor. See Section 5.6.

69. $x^4 + x^2 - 20$

70. $2y^4 + 11y^2 - 6$

71. $z^4 - 13z^2 + 36$

72. $x^4 - 1$

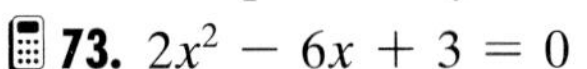

Combining Concepts

Use the quadratic formula and a calculator to approximate each solution to the nearest tenth.

73. $2x^2 - 6x + 3 = 0$

74. $3.6x^2 + 1.8x - 4.3 = 0$

The graph shows the daily low temperatures for one week in New Orleans, Louisiana. Use this graph to answer Exercises 75–78.

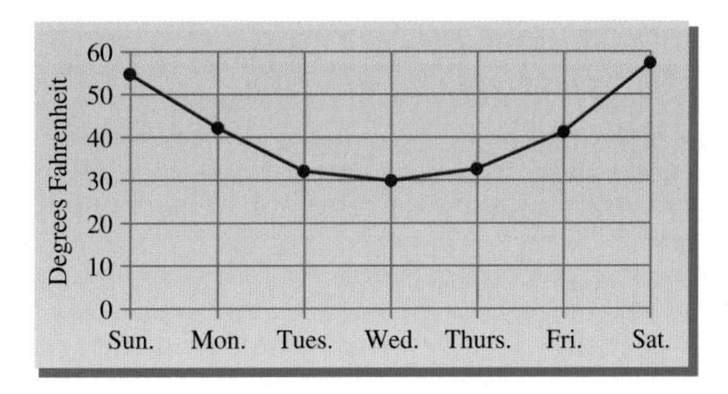

75. Which day of the week shows the greatest decrease in temperature low?

76. Which day of the week shows the greatest increase in temperature low?

77. Which day of the week had the lowest temperature?

78. Use the graph to estimate the low temperature on Thursday.

Notice that the shape of the temperature graph is similar to a parabola (see Section 5.9). In fact, this graph can be approximated by the quadratic function $f(x) = 3x^2 - 18x + 57$, where $f(x)$ is the temperature in degrees Fahrenheit and x is the number of days from Sunday. Use this function to answer Exercises 79 and 80.

79. Use the quadratic function given to approximate the temperature on Thursday. Does your answer agree with the graph above?

80. Use the function given and the quadratic formula to find when the temperature was 35°F. (*Hint:* Let $f(x) = 35$ and solve for x.) Round your answer to one decimal place and interpret your result. Does your answer agree with the graph above?

69. ______________

70. ______________

71. ______________

72. ______________

73. ______________

74. ______________

75. ______________

76. ______________

77. ______________

78. ______________

79. ______________

80. ______________

83. a. ______

b. ______

84. a. ______

b. ______

c. ______

Name ______

81. Use a grapher to solve Exercise 73.

82. Use a grapher to solve Exercise 74.

83. Wal-Mart Stores' net income can be modeled by the quadratic function $f(x) = 128.5x^2 - 69.5x + 2681$, where $f(x)$ is net income in millions of dollars and x is the number of years after 1995. (*Source:* Based on data from Wal-Mart Stores, Inc., 1995–1997)

a. Find Wal-Mart's net income in 1997.

b. If the trend described by the model continues, predict the year after 1995 in which Wal-Mart's net income will be $15,000 million. Round to the nearest whole year.

84. The number of inmates in custody in U.S. prisons and jails can be modeled by the quadratic function $p(x) = -2464.4x^2 + 99{,}699.5x + 1{,}148{,}702$, where $p(x)$ is the number of inmates and x is the number of years after 1990. (*Source:* Based on data from the Bureau of Justice Statistics, U.S. Department of Justice, 1990–1997) Round **a** and **b** to the nearest ten-thousand.

a. Find the number of prison inmates in the United States in 1992.

b. Find the number of prison inmates in the United States in 1997.

c. If the trend described by the model continues, predict the years in which the number of prisoners will be 2,050,000. Round to the nearest whole year.

8.3 SOLVING EQUATIONS BY USING QUADRATIC METHODS

Objectives

A Solve various equations that are quadratic in form.

B Solve problems that lead to quadratic equations.

SSM CD-ROM Video 8.3

A SOLVING EQUATIONS THAT ARE QUADRATIC IN FORM

In this section, we discuss various types of equations that can be solved in part by using the methods for solving quadratic equations.

Once each equation is simplified, you may want to use these steps when deciding what method to use to solve the quadratic equation.

> SOLVING A QUADRATIC EQUATION
>
> **Step 1.** If the equation is in the form $(ax + b)^2 = c$, use the square root property and solve. If not, go to Step 2.
>
> **Step 2.** Write the equation in standard form: $ax^2 + bx + c = 0$.
>
> **Step 3.** Try to solve the equation by the factoring method. If not possible, go to Step 4.
>
> **Step 4.** Solve the equation by the quadratic formula.

The first example is a radical equation that becomes a quadratic equation once we square both sides.

Example 1 Solve: $x - \sqrt{x} - 6 = 0$

Solution: Recall that to solve a radical equation, first get the radical alone on one side of the equation. Then square both sides.

$$x - 6 = \sqrt{x} \quad \text{Add } \sqrt{x} \text{ to both sides.}$$
$$x^2 - 12x + 36 = x \quad \text{Square both sides.}$$
$$x^2 - 13x + 36 = 0 \quad \text{Set the equation equal to 0.}$$
$$(x - 9)(x - 4) = 0$$
$$x - 9 = 0 \quad \text{or} \quad x - 4 = 0$$
$$x = 9 \qquad\qquad x = 4$$

Check:

Let $x = 9$

$$x - \sqrt{x} - 6 = 0$$
$$9 - \sqrt{9} - 6 \stackrel{?}{=} 0$$
$$9 - 3 - 6 \stackrel{?}{=} 0$$
$$0 = 0 \quad \text{True}$$

Let $x = 4$

$$x - \sqrt{x} - 6 = 0$$
$$4 - \sqrt{4} - 6 \stackrel{?}{=} 0$$
$$4 - 2 - 6 \stackrel{?}{=} 0$$
$$-4 = 0 \quad \text{False}$$

The solution set is $\{9\}$.

Practice Problem 1

Solve: $x - \sqrt{x - 1} - 3 = 0$

Example 2 Solve: $\dfrac{3x}{x - 2} - \dfrac{x + 1}{x} = \dfrac{6}{x(x - 2)}$

Solution: In this equation, x cannot be either 2 or 0, because these values cause denominators to equal zero. To solve for x, we first multiply both sides of the equation by $x(x - 2)$ to clear the fractions. By the distributive property, this means that we multiply each term by $x(x - 2)$.

Practice Problem 2

Solve: $\dfrac{2x}{x - 1} - \dfrac{x + 2}{x} = \dfrac{5}{x(x - 1)}$

Answers

1. $\{5\}$, **2.** $\left\{\dfrac{1 + \sqrt{13}}{2}, \dfrac{1 - \sqrt{13}}{2}\right\}$

$$x(x-2)\left(\frac{3x}{x-2}\right) - x(x-2)\left(\frac{x+1}{x}\right) = x(x-2)\left[\frac{6}{x(x-2)}\right]$$

$$3x^2 - (x-2)(x+1) = 6 \quad \text{Simplify.}$$
$$3x^2 - (x^2 - x - 2) = 6 \quad \text{Multiply.}$$
$$3x^2 - x^2 + x + 2 = 6$$
$$2x^2 + x - 4 = 0 \quad \text{Simplify.}$$

This equation cannot be factored using integers, so we solve by the quadratic formula.

$$x = \frac{-1 \pm \sqrt{1^2 - 4(2)(-4)}}{2 \cdot 2} \quad \text{Let } a = 2, b = 1, \text{ and } c = -4 \text{ in the quadratic formula.}$$
$$= \frac{-1 \pm \sqrt{1 + 32}}{4} \quad \text{Simplify.}$$
$$= \frac{-1 \pm \sqrt{33}}{4}$$

Neither proposed solution will make the denominators 0.

The solution set is $\left\{\frac{-1 + \sqrt{33}}{4}, \frac{-1 - \sqrt{33}}{4}\right\}$.

Practice Problem 3

Solve: $x^4 - 5x^2 - 36 = 0$

Example 3 Solve: $p^4 - 3p^2 - 4 = 0$

Solution: First we factor the trinomial.

$$p^4 - 3p^2 - 4 = 0$$
$$(p^2 - 4)(p^2 + 1) = 0 \quad \text{Factor.}$$
$$(p-2)(p+2)(p^2+1) = 0 \quad \text{Factor further.}$$
$$p - 2 = 0 \quad \text{or} \quad p + 2 = 0 \quad \text{or} \quad p^2 + 1 = 0 \quad \text{Set each factor equal to 0 and solve.}$$
$$p = 2 \qquad p = -2 \qquad p^2 = -1$$
$$p = \pm\sqrt{-1} = \pm i$$

The solution set is $\{2, -2, i, -i\}$.

✓ CONCEPT CHECK

a. *True or False?* The maximum number of solutions that a quadratic equation can have is 2.
b. *True or False?* The maximum number of solutions that an equation in quadratic form can have is 2.

TRY THE CONCEPT CHECK IN THE MARGIN.

Practice Problem 4

Solve: $(x + 4)^2 - (x + 4) - 6 = 0$

Example 4 Solve: $(x-3)^2 - 3(x-3) - 4 = 0$

Solution: Notice that the quantity $(x - 3)$ is repeated in this equation. Sometimes it is helpful to substitute a variable (in this case other than x) for the repeated quantity. We will let $y = x - 3$. Then

$$(x-3)^2 - 3(x-3) - 4 = 0$$

becomes

$$y^2 - 3y - 4 = 0 \quad \text{Let } x - 3 = y.$$
$$(y-4)(y+1) = 0 \quad \text{Factor.}$$

To solve, we use the zero factor property.

$$y - 4 = 0 \quad \text{or} \quad y + 1 = 0 \quad \text{Set each factor equal to 0.}$$
$$y = 4 \qquad y = -1 \quad \text{Solve.}$$

Answers

3. $\{3, -3, 2i, -2i\}$ **4.** $\{-1, -6\}$

✓ Concept Check: **a.** true, **b.** false

To find values of x, we substitute back. That is, we substitute $x - 3$ for y.

$$x - 3 = 4 \quad \text{or} \quad x - 3 = -1$$
$$x = 7 \qquad\qquad x = 2$$

HELPFUL HINT

When using substitution, don't forget to substitute back to the original variable.

Both 2 and 7 check. The solution is $\{2, 7\}$.

Example 5 Solve: $x^{2/3} - 5x^{1/3} + 6 = 0$

Solution: The key to solving this equation is recognizing that $x^{2/3} = (x^{1/3})^2$. We replace $x^{1/3}$ with m so that

$$(x^{1/3})^2 - 5x^{1/3} + 6 = 0$$

becomes

$$m^2 - 5m + 6 = 0$$

Now we solve by factoring.

$$m^2 - 5m + 6 = 0$$
$$(m - 3)(m - 2) = 0 \qquad \text{Factor.}$$
$$m - 3 = 0 \quad \text{or} \quad m - 2 = 0 \qquad \text{Set each factor equal to 0.}$$
$$m = 3 \qquad\qquad m = 2$$

Since $m = x^{1/3}$, we have

$$x^{1/3} = 3 \quad \text{or} \quad x^{1/3} = 2$$
$$x = 3^3 = 27 \quad \text{or} \quad x = 2^3 = 8$$

Both 8 and 27 check. The solution set is $\{8, 27\}$.

HELPFUL HINT

Example 3 can be solved using substitution also. Think of $p^4 - 3p^2 - 4 = 0$ as

$$(p^2)^2 - 3p^2 - 4 = 0$$
$$\downarrow \qquad \swarrow$$
$$x^2 - 3x - 4 = 0$$

Then let $x = p^2$, and solve and substitute back. The solution set will be the same.

B SOLVING PROBLEMS THAT LEAD TO QUADRATIC EQUATIONS

The next example is a work problem. This problem is modeled by a rational equation that simplifies to a quadratic equation.

Example 6 **Finding Work Time**

Together, an experienced typist and an apprentice typist can process a document in 6 hours. Alone, the experienced typist can process the document 2 hours faster than the apprentice typist can. Find the time in which each person can process the document alone.

Practice Problem 5

Solve: $x^{2/3} - 7x^{1/3} + 10 = 0$

Practice Problem 6

Together, Karen and Doug Lewis can clean a strip of beach in 5 hours. Alone, Karen can clean the strip of beach one hour faster than Doug. Find the time that each person can clean the strip of beach alone. Give an exact answer and a one decimal place approximation.

Answers

5. $\{8, 125\}$, **6.** Doug, $\frac{11 + \sqrt{101}}{2} \approx 10.5$ hours; Karen, $\frac{9 + \sqrt{101}}{2} \approx 9.5$ hours

Solution: **1.** UNDERSTAND. Read and reread the problem. The key idea here is the relationship between the *time* (hours) it takes to complete the job and the *part of the job* completed in one unit of time (hour). For example, because they can complete the job together in 6 hours, the *part of the job* they can complete in 1 hour is $\frac{1}{6}$. Let

$x =$ the *time* in hours it takes the apprentice typist to complete the job alone

$x - 2 =$ the time in hours it takes the experienced typist to complete the job alone

We can summarize in a chart the information discussed.

	Total Hours to Complete Job	Part of Job Completed in 1 Hour
Apprentice Typist	x	$\frac{1}{x}$
Experienced Typist	$x - 2$	$\frac{1}{x-2}$
Together	6	$\frac{1}{6}$

2. TRANSLATE.

In words:	part of job completed by apprentice typist in 1 hour	added to	part of job completed by experienced typist in 1 hour	is equal to	part of job completed together in 1 hour
	↓	↓	↓	↓	↓
Translate:	$\frac{1}{x}$	$+$	$\frac{1}{x-2}$	$=$	$\frac{1}{6}$

3. SOLVE.

$$\frac{1}{x} + \frac{1}{x-2} = \frac{1}{6}$$

$$6x(x-2)\left(\frac{1}{x} + \frac{1}{x-2}\right) = 6x(x-2) \cdot \frac{1}{6}$$ Multiply both sides by the LCD, $6x(x - 2)$.

$$6x(x-2) \cdot \frac{1}{x} + 6x(x-2) \cdot \frac{1}{x-2} = 6x(x-2) \cdot \frac{1}{6}$$ Use the distributive property.

$$6(x-2) + 6x = x(x-2)$$

$$6x - 12 + 6x = x^2 - 2x$$

$$0 = x^2 - 14x + 12$$

Now we can substitute $a = 1$, $b = -14$, and $c = 12$ into the quadratic formula and simplify.

$$x = \frac{-(-14) \pm \sqrt{(-14)^2 - 4(1)(12)}}{2(1)} = \frac{14 \pm \sqrt{148}}{2}$$

Using a calculator or a square root table, we see that $\sqrt{148} \approx 12.2$ rounded to one decimal place. Thus,

$$x \approx \frac{14 \pm 12.2}{2}$$

$$x \approx \frac{14 + 12.2}{2} = 13.1 \quad \text{or} \quad x \approx \frac{14 - 12.2}{2} = 0.9$$

4. INTERPRET.

Check: If the apprentice typist completes the job alone in 0.9 hours, the experienced typist completes the job alone in $x - 2 = 0.9 - 2 = -1.1$ hours. Since this is not possible, we reject the solution of 0.9. The approximate solution is thus 13.1 hours.

State: The apprentice typist can complete the job alone in approximately 13.1 hours, and the experienced typist can complete the job alone in approximately $x - 2 = 13.1 - 2 = 11.1$ hours.

Example 7

Beach and Fargo are about 400 miles apart. A salesperson travels from Fargo to Beach one day at a certain speed. She returns to Fargo the next day and drives 10 mph faster. Her total travel time was $14\frac{2}{3}$ hours. Find her speed to Beach and the return speed to Fargo.

Beach x mph Fargo
400 miles
$x + 10$ mph
400 miles

Solution: **1.** UNDERSTAND. Read and reread the problem. Let

$x =$ the speed to Beach, so

$x + 10 =$ the return speed to Fargo.

Then organize the given information in a table.

	distance	= rate	· time
To Beach	400	x	$\frac{400}{x}$
Return to Fargo	400	$x + 10$	$\frac{400}{x + 10}$

2. TRANSLATE.

In words: time to Beach + return time to Fargo = $14\frac{2}{3}$ hours

Translate: $$\frac{400}{x} + \frac{400}{x + 10} = \frac{44}{3}$$

Practice Problem 7

A family drives 500 miles to the beach for a vacation. The return trip was made at a speed that was 10 mph faster. The total traveling time was $18\frac{1}{3}$ hours. Find the speed to the beach and the return speed.

Answer

7. 50 mph to the beach; 60 mph returning

3. SOLVE.

$$\frac{400}{x} + \frac{400}{x+10} = \frac{44}{3}$$

$$\frac{100}{x} + \frac{100}{x+10} = \frac{11}{3}$$ Divide both sides by 4.

$$3x(x+10)\left(\frac{100}{x} + \frac{100}{x+10}\right) = 3x(x+10) \cdot \frac{11}{3}$$ Multiply both sides by the LCD, $3x(x+10)$.

$$3x(x+10) \cdot \frac{100}{x} + 3x(x+10) \cdot \frac{100}{x+10} = 3x(x+10) \cdot \frac{11}{3}$$ Use the distributive property.

$$3(x+10) \cdot 100 + 3x \cdot 100 = x(x+10) \cdot 11$$

$$300x + 3000 + 300x = 11x^2 + 110x$$

$$0 = 11x^2 - 490x - 3000$$ Set equation equal to 0.

$$0 = (11x+60)(x-50)$$ Factor.

$$11x + 60 = 0 \quad \text{or} \quad x - 50 = 0$$ Set each factor equal to 0.

$$x = -\frac{60}{11} \text{ or } -5\frac{5}{11} \qquad x = 50$$

4. INTERPRET.

Check: The speed is not negative, so it's not $-5\frac{5}{11}$. The number 50 does check.

State: The speed to Beach was 50 mph and her return speed to Fargo was 60 mph.

Name ______ Section ______ Date ______

Exercise Set 8.3

A *Solve. See Example 1.*

1. $2x = \sqrt{10 + 3x}$

2. $3x = \sqrt{8x + 1}$

3. $x - 2\sqrt{x} = 8$

4. $x - \sqrt{2x} = 4$

5. $\sqrt{9x} = x + 2$

6. $\sqrt{16x} = x + 3$

Solve. See Example 2.

7. $\frac{2}{x} + \frac{3}{x - 1} = 1$

8. $\frac{6}{x^2} = \frac{3}{x + 1}$

9. $\frac{3}{x} + \frac{4}{x + 2} = 2$

10. $\frac{5}{x - 2} + \frac{4}{x + 2} = 1$

11. $\frac{7}{x^2 - 5x + 6} = \frac{2x}{x - 3} - \frac{x}{x - 2}$

12. $\frac{11}{2x^2 + x - 15} = \frac{5}{2x - 5} - \frac{x}{x + 3}$

Solve. See Example 3.

13. $p^4 - 16 = 0$

14. $x^4 + 2x^2 - 3 = 0$

15. $4x^4 + 11x^2 = 3$

16. $z^4 = 81$

17. $z^4 - 13z^2 + 36 = 0$

18. $9x^4 + 5x^2 - 4 = 0$

Solve. See Examples 4 and 5.

19. $x^{2/3} - 3x^{1/3} - 10 = 0$

20. $x^{2/3} + 2x^{1/3} + 1 = 0$

21. $(5n + 1)^2 + 2(5n + 1) - 3 = 0$

22. $(m - 6)^2 + 5(m - 6) + 4 = 0$

23. $2x^{2/3} - 5x^{1/3} = 3$

24. $3x^{2/3} + 11x^{1/3} = 4$

Answers

1. ______
2. ______
3. ______
4. ______
5. ______
6. ______
7. ______
8. ______
9. ______
10. ______
11. ______
12. ______
13. ______
14. ______
15. ______
16. ______
17. ______
18. ______
19. ______
20. ______
21. ______
22. ______
23. ______
24. ______

25.

26.

27.

28.

29.

30.

31.

32.

33.

34.

35.

36.

37.

38.

39.

40.

41.

42.

43.

44.

45.

46.

47.

48.

Name

25. $1 + \frac{2}{3t - 2} = \frac{8}{(3t - 2)^2}$

26. $2 - \frac{7}{x + 6} = \frac{15}{(x + 6)^2}$

27. $20x^{2/3} - 6x^{1/3} - 2 = 0$

28. $4x^{2/3} + 16x^{1/3} = -15$

Solve. See Examples 1 through 5.

29. $a^4 - 5a^2 + 6 = 0$

30. $x^4 - 12x^2 + 11 = 0$

31. $\frac{2x}{x - 2} + \frac{x}{x + 3} = \frac{-5}{x + 3}$

32. $\frac{5}{x - 3} + \frac{x}{x + 3} = \frac{19}{x^2 - 9}$

33. $(p + 2)^2 = 9(p + 2) - 20$

34. $2(4m - 3)^2 - 9(4m - 3) = 5$

35. $2x = \sqrt{11x + 3}$

36. $4x = \sqrt{2x + 3}$

37. $x^{2/3} - 8x^{1/3} + 15 = 0$

38. $x^{2/3} - 2x^{1/3} - 8 = 0$

39. $y^3 + 9y - y^2 - 9 = 0$

40. $x^3 + x - 3x^2 - 3 = 0$

41. $2x^{2/3} + 3x^{1/3} - 2 = 0$

42. $6x^{2/3} - 25x^{1/3} - 25 = 0$

43. $x^{-2} - x^{-1} - 6 = 0$

44. $y^{-2} - 8y^{-1} + 7 = 0$

45. $x - \sqrt{x} = 2$

46. $x - \sqrt{3x} = 6$

47. $\frac{x}{x - 1} + \frac{1}{x + 1} = \frac{2}{x^2 - 1}$

48. $\frac{x}{x - 5} + \frac{5}{x + 5} = \frac{-1}{x^2 - 25}$

Name ____________________

49. $p^4 - p^2 - 20 = 0$

50. $x^4 - 10x^2 + 9 = 0$

51. $2x^3 = -54$

52. $y^3 - 216 = 0$

53. $1 = \dfrac{4}{x - 7} + \dfrac{5}{(x - 7)^2}$

54. $3 + \dfrac{1}{(2p + 4)} = \dfrac{10}{(2p + 4)^2}$

55. $27y^4 + 15y^2 = 2$

56. $8z^4 + 14z^2 = -5$

B *Solve. See Examples 6 and 7.*

57. A jogger ran 3 miles, decreased her speed by 1 mile per hour and then ran another 4 miles. If her total time jogging was $1\frac{3}{5}$ hours, find her speed for each part of her run.

58. Mark Keaton's workout consists of jogging for 3 miles, and then riding his bike for 5 miles at a speed 4 miles per hour faster than he jogs. If his total workout time is 1 hour, find his jogging speed and his biking speed.

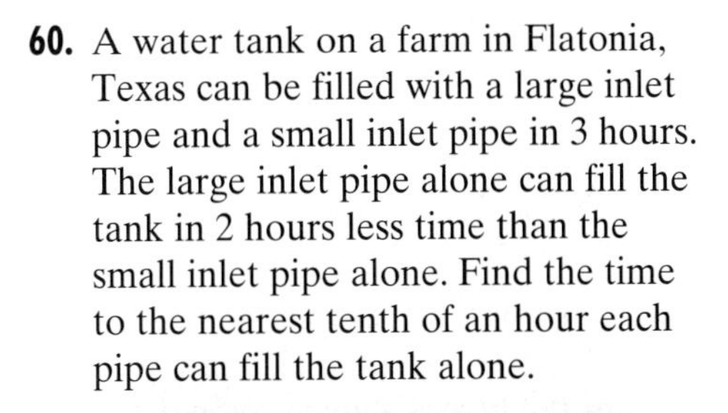

59. A chinese restaurant in Mandeville, Louisiana has a large gold fish pond around the restaurant. Suppose that an inlet pipe and a hose together can fill the pond in 8 hours. The inlet pipe alone can complete the job in one hour less time than the hose alone. Find the time that the hose can complete the job alone and the time that the inlet pipe can complete the job alone. Round each to the nearest tenth of an hour.

60. A water tank on a farm in Flatonia, Texas can be filled with a large inlet pipe and a small inlet pipe in 3 hours. The large inlet pipe alone can fill the tank in 2 hours less time than the small inlet pipe alone. Find the time to the nearest tenth of an hour each pipe can fill the tank alone.

49. ____________

50. ____________

51. ____________

52. ____________

53. ____________

54. ____________

55. ____________

56. ____________

57. ____________

58. ____________

59. ____________

60. ____________

61.

62.

63.

64.

65.

66.

67. a.

b.

c.

68.

Name

61. Roma Sherry drove 330 miles from her home town to Tucson. During her return trip, she was able to increase her speed by 11 mph. If her return trip took 1 hour less time, find her original speed and her speed returning home.

62. A salesperson drove to Portland, a distance of 300 miles. During the last 80 miles of his trip, heavy rainfall forced him to decrease his speed by 15 mph. If his total driving time was 6 hours, find his original speed and his speed during the rainfall.

63. Bill Shaughnessy and his son Billy can clean the house together in 4 hours. When the son works alone, it takes him an hour longer to clean than it takes his dad alone. Find how long to the nearest tenth of an hour it takes the son to clean alone.

64. Together, Noodles and Freckles eat a 50-pound bag of dog food in 30 days. Noodles by himself eats a 50-pound bag in 2 weeks less time than Freckles does by himself. How many days to the nearest whole day would a 50-pound bag of dog food last Freckles?

65. The product of a number and 4 less than the number is 96. Find the number.

66. A whole number increased by its square is two more than twice itself. Find the number.

67. Suppose that an open box is to be made from a square sheet of cardboard by cutting out squares from each corner as shown and then folding along the dotted lines. If the box is to have a volume of 300 cubic centimeters, find the original dimensions of the sheet of cardboard.

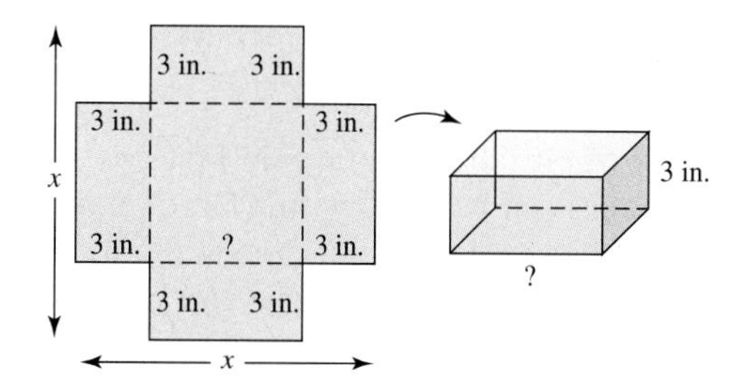

a. The ? in the drawing to the left will be the length (and also the width) of the box as shown in the drawing to the right. Represent this length in terms of x.

b. Use the formula for volume of a box, $V = l \cdot w \cdot h$, to write an equation in x.

c. Solve the equation for x and give the dimensions of the sheet of cardboard. Check your solution.

68. Suppose that an open box is to be made from a square sheet of cardboard by cutting out squares from each corner as shown and then folding along the dotted lines. If the box is to have a volume of 128 cubic inches, find the original dimensions of the sheet of cardboard. (*Hint:* Use Exercise 67 parts a, b, and c to help you.)

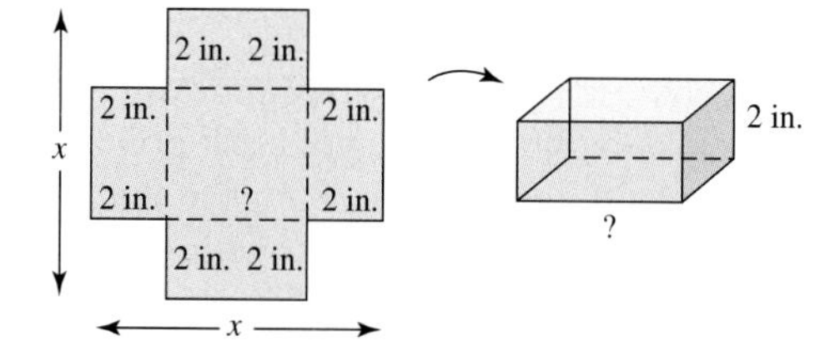

Name ____________________

Review and Preview

Solve each inequality. See Section 2.4

69. $\frac{5x}{3} + 2 \leq 7$

70. $\frac{2x}{3} + \frac{1}{6} \geq 2$

71. $\frac{y-1}{15} > \frac{-2}{5}$

72. $\frac{z-2}{12} < \frac{1}{4}$

Combining Concepts

73. Write a polynomial equation that has three solutions: 2, 5, and -7.

74. Write a polynomial equation that has three solutions: 0, $2i$, and $-2i$.

75. During the 1998 U.S. 500 auto race held in Brooklyn, Michigan, Richie Hearn posted the fastest speed but Greg Moore won the race. The track is 10,560 feet (2 miles) long. Hearn's fastest speed was 6.64 feet per second faster than Moore's fastest speed. Traveling at these fastest speeds, Moore would have taken 0.535 seconds longer than Hearn to complete a lap. (*Source:* Championship Auto Racing Teams, Inc. Round answers to parts a, b, and c to 2 decimal places.)

a. Find Richie Hearn's fastest speed during the race.

b. Find Greg Moore's fastest speed during the race.

c. Convert each speed to miles per hour.

76. Use a grapher to solve Exercise 29. Compare the solution with the solution from Exercise 29. Explain any differences.

69. ____________________

70. ____________________

71. ____________________

72. ____________________

73. ____________________

74. ____________________

75. a. ____________________

b. ____________________

c. ____________________

76. ____________________

Focus On History

THE EVOLUTION OF SOLVING QUADRATIC EQUATIONS

The ancient Babylonians (circa 2000 B.C.) are sometimes credited with being the first to solve quadratic equations. This is only partially true because the Babylonians had no concept of an equation. However, what they did develop was a method for completing the square to apply to problems that today would be solved with a quadratic equation. The Babylonians only recognized positive solutions to such problems and did not acknowledge the existence of negative solutions at all.

Babylonian mathematical knowledge influenced much of the ancient world, most notably Hindu Indians. The Hindus were the first culture to denote debts in everyday business affairs with negative numbers. With this level of comfort with negative numbers, the Indian mathematician Brahmagupta (598–665 A.D.) extended the Babylonian methods and was the first to recognize negative solutions to quadratic equations. Later Hindu mathematicians noted that every positive number has two square roots: a positive square root and a negative square root. Hindus allowed irrational solutions (quite an innovation in the ancient world!) to quadratic equations and were the first to realize that quadratic equations could have 0, 1, or 2 real number solutions. They did not, however, acknowledge complex numbers and, therefore, could not solve equations with solutions requiring the square root of a negative number.

Complex numbers were finally developed by European mathematicians during the 17th and 18th centuries. Up until that time, what we would today consider complex solutions to quadratic equations were routinely ignored by mathematicians.

Name ______________________ Section ________ Date ________

Integrated Review—Summary on Solving Quadratic Equations

Use the square root property to solve each equation.

1. $x^2 - 10 = 0$

2. $x^2 - 14 = 0$

3. $(x - 1)^2 = 8$

4. $(x + 5)^2 = 12$

Solve each equation by completing the square.

5. $x^2 + 2x - 12 = 0$

6. $x^2 - 12x + 11 = 0$

7. $3x^2 + 3x = 5$

8. $16y^2 + 16y = 1$

Use the quadratic formula to solve each equation.

9. $2x^2 - 4x + 1 = 0$

10. $\frac{1}{2}x^2 + 3x + 2 = 0$

11. $x^2 + 4x = -7$

12. $x^2 + x = -3$

Solve each equation. Use a method of your choice.

13. $x^2 + 3x + 6 = 0$

14. $2x^2 + 18 = 0$

15. $x^2 + 17x = 0$

16. $4x^2 - 2x - 3 = 0$

17. $(x - 2)^2 = 27$

18. $\frac{1}{2}x^2 - 2x + \frac{1}{2} = 0$

Answers

1. ________
2. ________
3. ________
4. ________
5. ________
6. ________
7. ________
8. ________
9. ________
10. ________
11. ________
12. ________
13. ________
14. ________
15. ________
16. ________
17. ________
18. ________

19.

20.

21.

22.

23.

24.

Name

19. $3x^2 + 2x = 8$

20. $2x^2 = -5x - 1$

21. $x(x - 2) = 5$

22. The diagonal of a square room measures 20 feet. Find the exact length of a side of the room. Then approximate the length to the nearest tenth of a foot.

20 feet

x

23. Diane Gray and Lucy Hoag together can prepare a crawfish boil for a large party in 4 hours. Lucy alone can complete the job in 2 hours less time than Diane alone. Find the time that each person can prepare the crawfish boil alone. Round each time to the nearest tenth of an hour.

24. Kraig Blackwelder exercises at Total Body Gym. On the treadmill, he runs 5 miles, then increases his speed 1 mph and runs an additional 2 miles. If his total time on the treadmill is $1\frac{1}{3}$ hours, find his speed during each part of his run.

8.4 NONLINEAR INEQUALITIES IN ONE VARIABLE

Objectives

A Solve polynomial inequalities of degree 2 or greater.

B Solve inequalities that contain rational expressions with variables in the denominator.

SSM CD-ROM Video 8.4

A SOLVING POLYNOMIAL INEQUALITIES

Just as we can solve linear inequalities in one variable, so we can also solve quadratic inequalities in one variable. A **quadratic inequality** is an inequality that can be written so that one side is a quadratic expression and the other side is 0. Here are examples of quadratic inequalities in one variable. Each is written in **standard form.**

$x^2 - 10x + 7 \le 0$ $\quad 3x^2 + 2x - 6 > 0$

$2x^2 + 9x - 2 < 0$ $\quad x^2 - 3x + 11 \ge 0$

A solution of a quadratic inequality in one variable is a value of the variable that makes the inequality a true statement.

The value of an expression such as $x^2 - 3x - 10$ will sometimes be positive, sometimes negative, and sometimes 0, depending on the value substituted for x. To solve the inequality $x^2 - 3x - 10 < 0$, we are looking for all values of x that make the expression $x^2 - 3x - 10$ **less than 0**, or **negative**. To understand how we find these values, we'll study the graph of the quadratic function $y = x^2 - 3x - 10$.

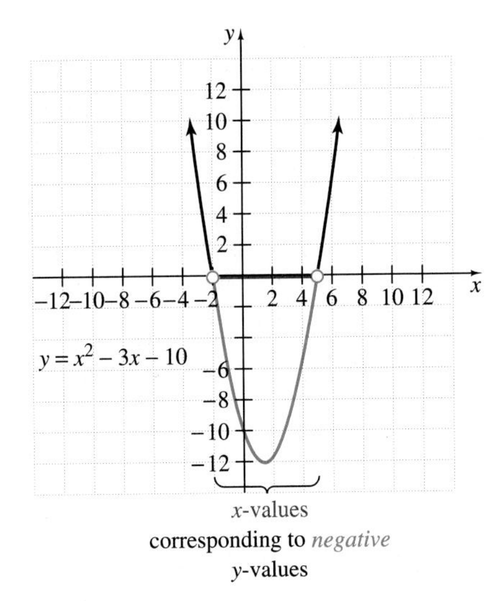

Notice that the x-values for which y or $x^2 - 3x - 10$ is positive are separated from the x values for which y or $x^2 - 3x - 10$ is negative by the values for which y or $x^2 - 3x - 10$ is 0, the x-intercepts. Thus, the solution set of $x^2 - 3x - 10 < 0$ consists of all real numbers from -2 to 5, or in interval notation, $(-2, 5)$.

It is not necessary to graph $y = x^2 - 3x - 10$ to solve the related inequality $x^2 - 3x - 10 < 0$. Instead, we can draw a number line representing the x-axis and keep the following in mind: *A region on the number line for which the value of $x^2 - 3x - 10$ is positive is separated from a region on the number line for which the value of $x^2 - 3x - 10$ is negative by a value for which the expression is 0.*

Let's find these values for which the expression is 0 by solving the related equation:

$$x^2 - 3x - 10 = 0$$

$$(x - 5)(x + 2) = 0 \quad \text{Factor.}$$

$$x - 5 = 0 \quad \text{or} \quad x + 2 = 0 \quad \text{Set each factor equal to 0.}$$

$$x = 5 \qquad\qquad x = -2 \quad \text{Solve.}$$

These two numbers -2 and 5, divide the number line into three regions. We will call the regions A, B, and C. These regions are important because, if the value of $x^2 - 3x - 10$ is negative when a number from a region is substituted for x, then $x^2 - 3x - 10$ is negative when any number in that region is substituted for x. The same is true if the value of $x^2 - 3x - 10$ is positive for a particular value of x in a region.

To see whether the inequality $x^2 - 3x - 10 < 0$ is true or false in each region, we choose a test point from each region and substitute its value for x in the inequality $x^2 - 3x - 10 < 0$. If the resulting inequality is true, the region containing the test point is a solution region.

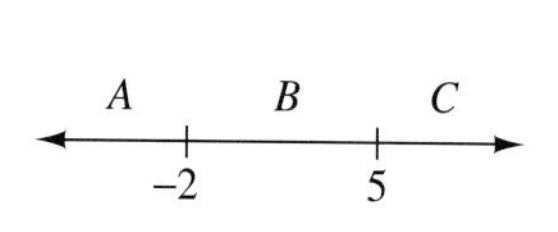

Region	Test Point Value	$(x - 5)(x + 2) < 0$	Result
A	-3	$(-8)(-1) < 0$	False.
B	0	$(-5)(2) < 0$	True.
C	6	$(1)(8) < 0$	False.

The values in region B satisfy the inequality. The numbers -2 and 5 are not included in the solution set since the inequality symbol is $<$. The solution set is $(-2, 5)$, and its graph is shown.

Practice Problem 1

Solve: $(x - 2)(x + 4) > 0$

SOLVING A POLYNOMIAL INEQUALITY

Step 1. Write the inequality in standard form and then solve the related equation.

Step 2. Separate the number line into regions with the solutions from Step 1.

Step 3. For each region, choose a test point and determine whether its value satisfies *the original inequality.*

Step 4. The solution set includes the regions whose test point value is a solution. If the inequality symbol is $\leq$ or $\geq$, the values from Step 1 are solutions; if $<$ or $>$, they are not.

Example 1

Solve: $(x + 3)(x - 3) > 0$

Solution: First we solve the related equation $(x + 3)(x - 3) = 0$.

$$(x + 3)(x - 3) = 0$$

$$x + 3 = 0 \quad \text{or} \quad x - 3 = 0$$

$$x = -3 \qquad\qquad x = 3$$

The two numbers -3 and 3 separate the number line into three regions, A, B, and C.

Now we substitute the value of a test point from each region. If the test value satisfies the inequality, every value in the region containing the test value is a solution.

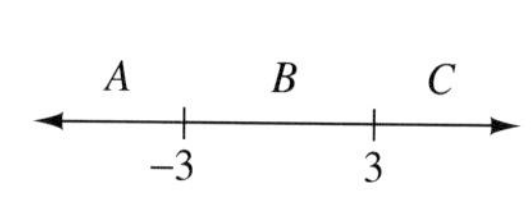

Region	Test Point Value	$(x + 3)(x - 3) > 0$	Result
A	-4	$(-1)(-7) > 0$	True.
B	0	$(3)(-3) > 0$	False.
C	4	$(7)(1) > 0$	True.

The points in regions A and C satisfy the inequality. The numbers -3 and 3 are not included in the solution since the inequality symbol is $>$. The solution set is $(-\infty, -3) \cup (3, \infty)$, and its graph is shown.

✓ CONCEPT CHECK

When choosing a test point in Step 4, why would the solutions from Step 2 not make good choices for test points?

The steps in the margin may be used to solve a polynomial inequality.

TRY THE CONCEPT CHECK IN THE MARGIN.

Answer

1. $(-\infty, -4) \cup (2, \infty)$

✓ **Concept Check:** The solutions found in Step 2 have a value of zero in the original inequality.

Example 2 Solve: $x^2 - 4x \le 0$

Solution: First we solve the related equation $x^2 - 4x = 0$.

$$x^2 - 4x = 0$$
$$x(x - 4) = 0$$
$$x = 0 \text{ or } x = 4$$

The numbers 0 and 4 separate the number line into three regions, A, B and C.

Check a test value in each region in the original inequality. Values in region B satisfy the inequality. The numbers 0 and 4 are included in the solution since the inequality symbol is $\le$. The solution set is $[0, 4]$, and its graph is shown.

Practice Problem 2

Solve: $x^2 - 6x \le 0$

Example 3 Solve: $(x + 2)(x - 1)(x - 5) \le 0$

Solution: First we solve $(x + 2)(x - 1)(x - 5) = 0$. By inspection, we see that the solutions are -2, 1, and 5. They separate the number line into four regions, A, B, C, and D. Next we check test points from each region.

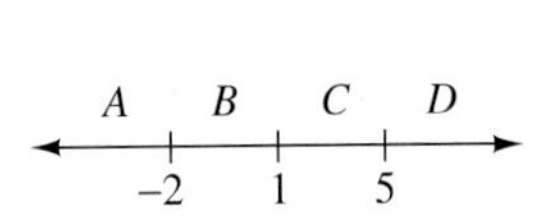

Region	Test Point Value	$(x + 2)(x - 1)(x - 5) \le 0$	Result
A	−3	$(-1)(-4)(-8) \le 0$	True.
B	0	$(2)(-1)(-5) \le 0$	False.
C	2	$(4)(1)(-3) \le 0$	True.
D	6	$(8)(5)(1) \le 0$	False.

The solution set is $(-\infty, -2] \cup [1, 5]$, and its graph is shown. We include the numbers -2, 1 and 5 because the inequality symbol is $\le$.

Practice Problem 3

Solve: $(x - 2)(x + 1)(x + 5) \le 0$

B Solving Rational Inequalities

Inequalities containing rational expressions with variables in the denominator are solved by using a similar procedure.

Example 4 Solve: $\dfrac{x + 2}{x - 3} \le 0$

Solution: First we find all values that make the denominator equal to 0. To do this, we solve $x - 3 = 0$, or $x = 3$.

Next, we solve the related equation $\dfrac{x + 2}{x - 3} = 0$.

Practice Problem 4

Solve: $\dfrac{x - 3}{x + 5} \le 0$

Answers

2. $[0, 6]$, **3.** $(-\infty, -5] \cup [-1, 2]$, **4.** $(-5, 3]$

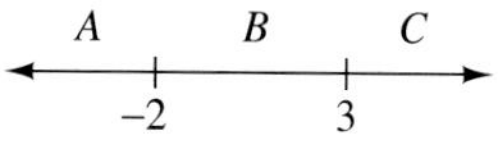

$$\frac{x+2}{x-3} = 0 \quad \text{Multiply both sides by the LCD, } x-3.$$
$$x + 2 = 0$$
$$x = -2$$

Now we place these numbers on a number line and proceed as before, checking test point values in the original inequality.

Choose −3 from region A.

$$\frac{x+2}{x-3} \le 0$$
$$\frac{-3+2}{-3-3} \le 0$$
$$\frac{-1}{-6} \le 0$$
$$\frac{1}{6} \le 0 \quad \text{False.}$$

Choose 0 from region B.

$$\frac{x+2}{x-3} \le 0$$
$$\frac{0+2}{0-3} \le 0$$
$$-\frac{2}{3} \le 0 \quad \text{True.}$$

Choose 4 from region C.

$$\frac{x+2}{x-3} \le 0$$
$$\frac{4+2}{4-3} \le 0$$
$$6 \le 0 \quad \text{False.}$$

The solution set is $[-2, 3)$. This interval includes -2 because -2 satisfies the original inequality. This interval does not include 3, because 3 would make the denominator 0.

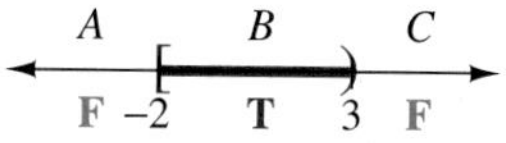

SOLVING A RATIONAL INEQUALITY

Step 1. Solve for values that make all denominators 0.

Step 2. Solve the related equation.

Step 3. Separate the number line into regions with the solutions from Steps 1 and 2.

Step 4. For each region, choose a test point and determine whether its value satisfies the *original inequality*.

Step 5. The solution set includes the regions whose test point value is a solution. Check whether to include values from Step 2. Be sure *not* to include values that make any denominator 0.

The steps in the margin may be used to solve a rational inequality with variables in the denominator.

Practice Problem 5

Solve: $\frac{3}{x-2} < 2$

Example 5 Solve: $\frac{5}{x+1} < -2$

Solution: First we find values for x that make the denominator equal to 0.

$$x + 1 = 0$$
$$x = -1$$

Next we solve $\frac{5}{x+1} = -2$.

$$(x+1) \cdot \frac{5}{x+1} = (x+1) \cdot -2 \quad \text{Multiply both sides by the LCD, } x+1.$$
$$5 = -2x - 2 \quad \text{Simplify.}$$
$$7 = -2x$$
$$-\frac{7}{2} = x$$

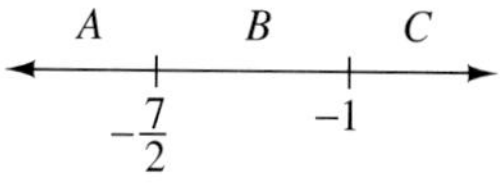

We use these two solutions to divide a number line into three regions and choose test points. Only a test point value from region B satisfies the *original inequality*. The solution set is $\left(-\frac{7}{2}, -1\right)$, and its graph is shown.

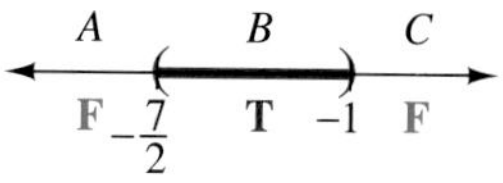

Answer

5. $(-\infty, 2) \cup \left(\frac{7}{2}, \infty\right)$

Name ______________________ Section ________ Date ________

Exercise Set 8.4

A *Solve. See Examples 1 through 3.*

1. $(x + 1)(x + 5) > 0$

2. $(x + 1)(x + 5) \le 0$

3. $(x - 3)(x + 4) \le 0$

4. $(x + 4)(x - 1) > 0$

5. $x^2 - 7x + 10 \le 0$

6. $x^2 + 8x + 15 \ge 0$

7. $3x^2 + 16x < -5$

8. $2x^2 - 5x < 7$

9. $(x - 6)(x - 4)(x - 2) > 0$

10. $(x - 6)(x - 4)(x - 2) \le 0$

11. $x(x - 1)(x + 4) \le 0$

12. $x(x - 6)(x + 2) > 0$

13. $(x^2 - 9)(x^2 - 4) > 0$

14. $(x^2 - 16)(x^2 - 1) \le 0$

15. $x^2 - x - 56 > 0$

16. $x^2 - 4x - 5 < 0$

17. $6x^2 + 5x \le 4$

18. $12x^2 - 5x \ge 3$

19. $x^2 > x$

20. $x^2 < 25$

21. $(2x - 8)(x + 4)(x - 6) \le 0$

22. $(3x - 12)(x + 5)(2x - 3) \ge 0$

B *Solve. See Examples 4 and 5.*

23. $\frac{x + 7}{x - 2} < 0$

24. $\frac{x - 5}{x - 6} > 0$

25. $\frac{5}{x + 1} > 0$

26. $\frac{3}{y - 5} < 0$

27. $\frac{x + 1}{x - 4} \ge 0$

28. $\frac{x + 1}{x - 4} \le 0$

29. $\frac{x + 2}{x - 3} < 1$

30. $\frac{x - 1}{x + 4} > 2$

31. $\frac{x}{x - 10} < 0$

32. $\frac{x + 10}{x - 10} > 0$

33. $\frac{x - 5}{x + 4} \ge 0$

34. $\frac{x - 3}{x + 2} \le 0$

Answers

1. ______
2. ______
3. ______
4. ______
5. ______
6. ______
7. ______
8. ______
9. ______
10. ______
11. ______
12. ______
13. ______
14. ______
15. ______
16. ______
17. ______
18. ______
19. ______
20. ______
21. ______
22. ______
23. ______
24. ______
25. ______
26. ______
27. ______
28. ______
29. ______
30. ______
31. ______
32. ______
33. ______
34. ______

35.

36.

37.

38.

39.

40.

41. see table

42. see table

43. see table

44. see table

45.

46.

47.

48.

49.

50.

51. see calculator screen

52. see calculator screen

Name ____________________

35. $\dfrac{x(x+6)}{(x-7)(x+1)} \geq 0$

36. $\dfrac{(x-2)(x+2)}{(x+1)(x-4)} \leq 0$

37. $\dfrac{-1}{x-1} > -1$

38. $\dfrac{4}{y+2} < -2$

39. $\dfrac{x}{x+4} \leq 2$

40. $\dfrac{4x}{x-3} \geq 5$

Review and Preview

Fill in the table so that each ordered pair is a solution of the function. See Section 3.5.

41. $f(x) = x^2$

x	y
0	
1	
−1	
2	
−2	

42. $f(x) = 2x^2$

x	y
0	
1	
−1	
2	
−2	

43. $f(x) = -x^2$

x	y
0	
1	
−1	
2	
−2	

44. $f(x) = -3x^2$

x	y
0	
1	
−1	
2	
−2	

Combining Concepts

45. Explain why $\dfrac{x+2}{x-3} > 0$ and $(x+2)(x-3) > 0$ have the same solutions.

46. Explain why $\dfrac{x+2}{x-3} \geq 0$ and $(x+2)(x-3) \geq 0$ do not have the same solutions.

Find all numbers that satisfy each statement.

47. A number minus its reciprocal is less than zero. Find the numbers.

48. Twice a number added to its reciprocal is nonnegative. Find the numbers.

49. The total profit function $P(x)$ for a company producing x thousand units is given by $P(x) = -2x^2 + 26x - 44$. Find the values of x for which the company makes a profit. [*Hint:* The company makes a profit when $P(x) > 0$.]

50. A projectile is fired straight up from the ground with an initial velocity of 80 feet per second. Its height $s(t)$ in feet at any time t is given by the function $s(t) = -16t^2 + 80t$. Find the interval of time for which the height of the projectile is greater than 96 feet.

Use a graphing calculator to check each exercise.

51. Exercise 15

52. Exercise 16

8.5 QUADRATIC FUNCTIONS AND THEIR GRAPHS

Objectives

A Graph quadratic functions of the form $f(x) = x^2 + k$.

B Graph quadratic functions of the form $f(x) = (x - h)^2$.

C Graph quadratic functions of the form $f(x) = (x - h)^2 + k$.

D Graph quadratic functions of the form $f(x) = ax^2$.

E Graph quadratic functions of the form $f(x) = a(x - h)^2 + k$.

SSM CD-ROM Video 8.5

A GRAPHING $f(x) = x^2 + k$

We first graphed the quadratic function $f(x) = x^2$ in Section 5.8. In that section, we discovered that the graph of a quadratic function is a parabola opening upward or downward. Now, as we continue our study, we will discover more details about quadratic functions and their graphs.

First, let's recall the definition of a quadratic function.

> **QUADRATIC FUNCTION**
>
> A quadratic function is a function that can be written in the form $f(x) = ax^2 + bx + c$, where a, b, and c are real numbers and $a \neq 0$.

Notice that equations of the form $y = ax^2 + bx + c$, where $a \neq 0$, also define quadratic functions since y is a function of x or $y = f(x)$.

Recall that if $a > 0$, the parabola opens upward and if $a < 0$, the parabola opens downward. Also, the vertex of a parabola is the lowest point if the parabola opens upward and the highest point if the parabola opens downward. The axis of symmetry is the vertical line that passes through the vertex.

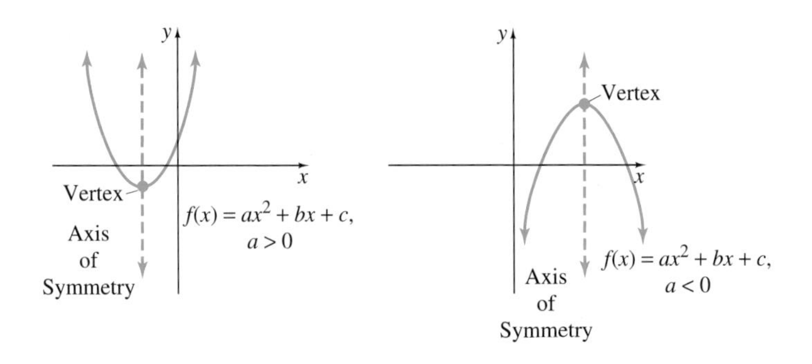

Example 1 Graph $f(x) = x^2$ and $g(x) = x^2 + 3$ on the same set of axes.

Solution: First we construct a table of values for $f(x)$ amd plot the points. Notice that for each x-value, the corresponding value of $g(x)$ must be 3 more than the corresponding value of $f(x)$ since $f(x) = x^2$ and $g(x) = x^2 + 3$. In other words, the graph of $g(x) = x^2 + 3$ is the same as the graph of $f(x) = x^2$ shifted upward 3 units. The axis of symmetry for both graphs is the y-axis.

Practice Problem 1

Graph $f(x) = x^2$ and $g(x) = x^2 + 4$ on the same set of axes.

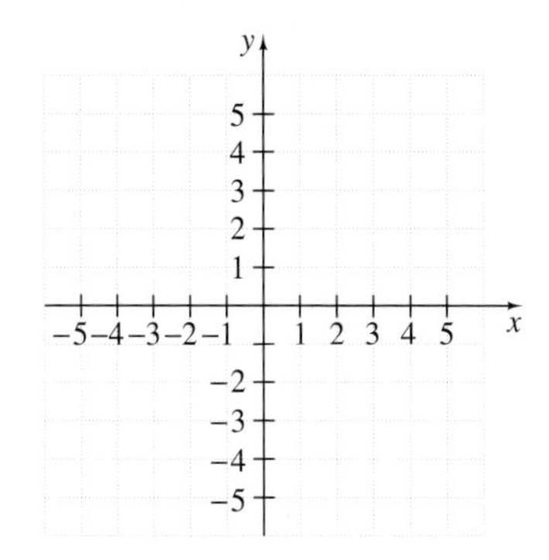

Answer

1.

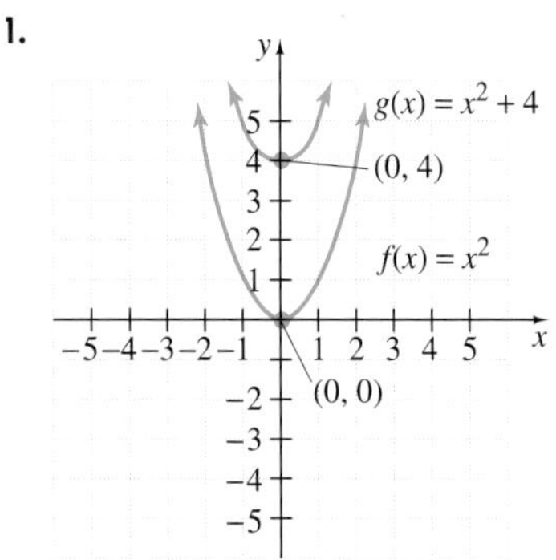

x	$f(x) = x^2$	$g(x) = x^2 + 3$
−2	4	7
−1	1	4
0	0	3
1	1	4
2	4	7

Each y-value is increased by 3.

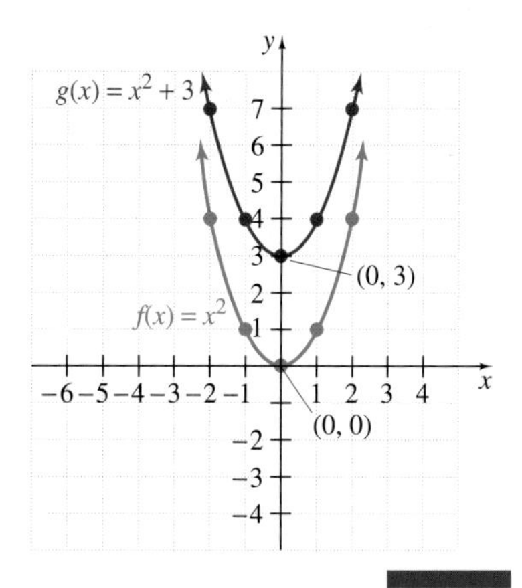

In general, we have the following properties.

GRAPHING THE PARABOLA DEFINED BY $f(x) = x^2 + k$

If k is positive, the graph of $f(x) = x^2 + k$ is the graph of $y = x^2$ shifted upward k units.

If k is negative, the graph of $f(x) = x^2 + k$ is the graph of $y = x^2$ shifted downward $|k|$ units.

The vertex is $(0, k)$, and the axis of symmetry is the y-axis.

Examples

Graph each function.

2. $F(x) = x^2 + 2$

The graph of $F(x) = x^2 + 2$ is obtained by shifting the graph of $y = x^2$ upward 2 units.

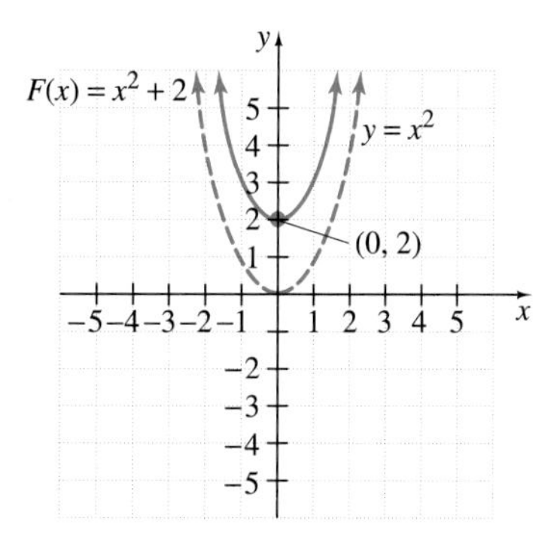

3. $g(x) = x^2 - 3$

The graph of $g(x) = x^2 - 3$ is obtained by shifting the graph of $y = x^2$ downward 3 units.

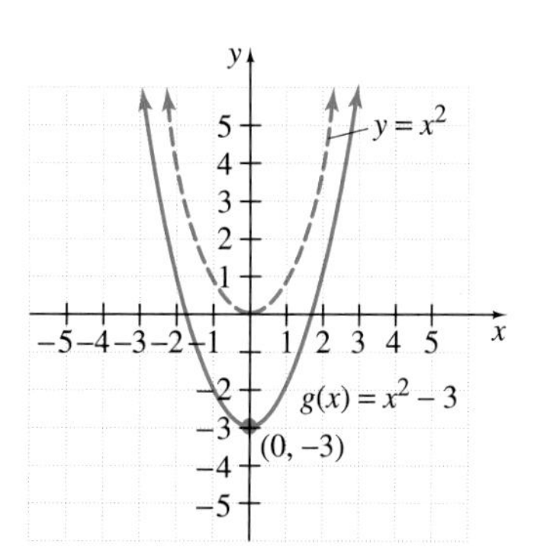

Practice Problems 2–3

Graph each function.

2. $F(x) = x^2 + 1$

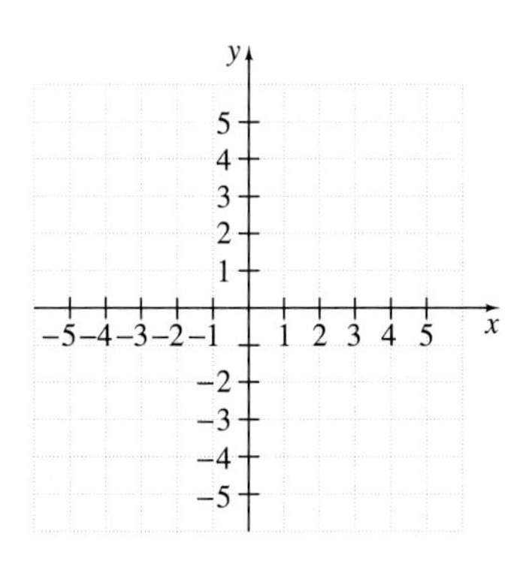

3. $g(x) = x^2 - 2$

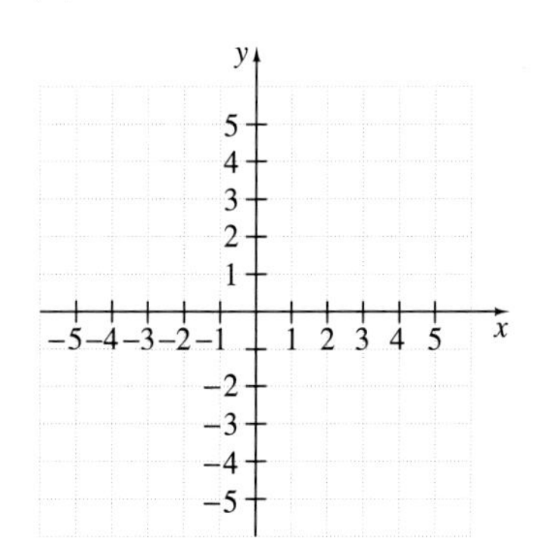

Answers

2.

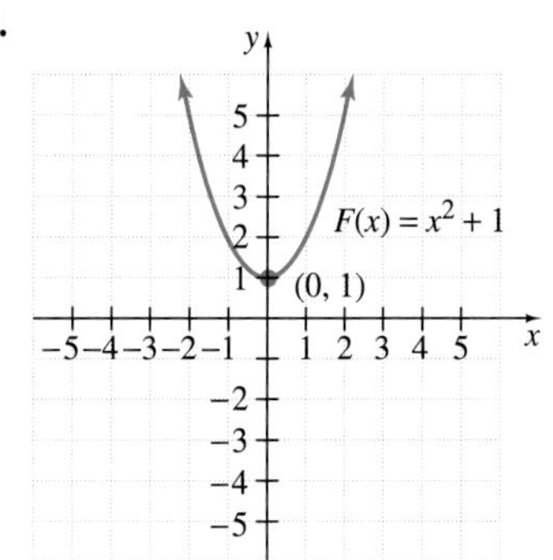

3.

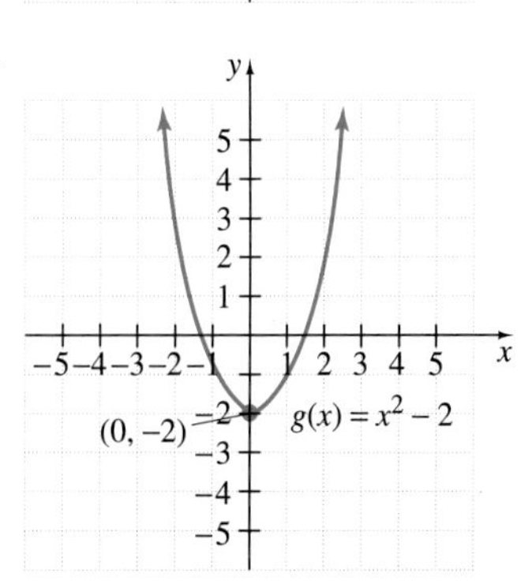

B GRAPHING $f(x) = (x - h)^2$

Now we will graph functions of the form $f(x) = (x - h)^2$.

Example 4 Graph $f(x) = x^2$ and $g(x) = (x - 2)^2$ on the same set of axes.

Solution: By plotting points, we see that for each x-value, the corresponding value of $g(x)$ is the same as the value of $f(x)$ when the x-value is increased by 2. Thus, the graph of $g(x) = (x - 2)^2$ is the graph of $f(x) = x^2$ shifted to the right 2 units. The axis of symmetry for the graph of $g(x) = (x - 2)^2$ is also shifted 2 units to the right and is the line $x = 2$.

x	$f(x) = x^2$	x	$g(x) = (x - 2)^2$
−2	4	0	4
−1	1	1	1
0	0	2	0
1	1	3	1
2	4	4	4

Each x-value increased by 2 corresponds to same y-value.

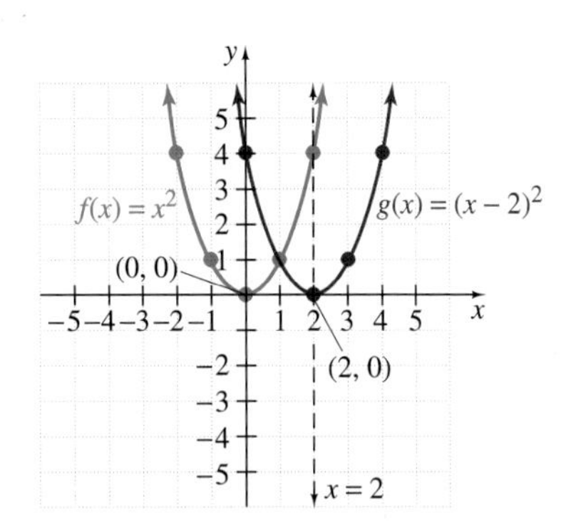

In general, we have the following properties.

> **GRAPHING THE PARABOLA DEFINED BY $f(x) = (x - h)^2$**
>
> If h is positive, the graph of $f(x) = (x - h)^2$ is the graph of $y = x^2$ shifted to the right h units.
>
> If h is negative, the graph of $f(x) = (x - h)^2$ is the graph of $y = x^2$ shifted to the left $|h|$ units.
>
> The vertex is $(h, 0)$, and the axis of symmetry is the vertical line $x = h$.

Examples Graph each function.

5. $G(x) = (x - 3)^2$

The graph of $G(x) = (x - 3)^2$ is obtained by shifting the graph of $y = x^2$ to the right 3 units.

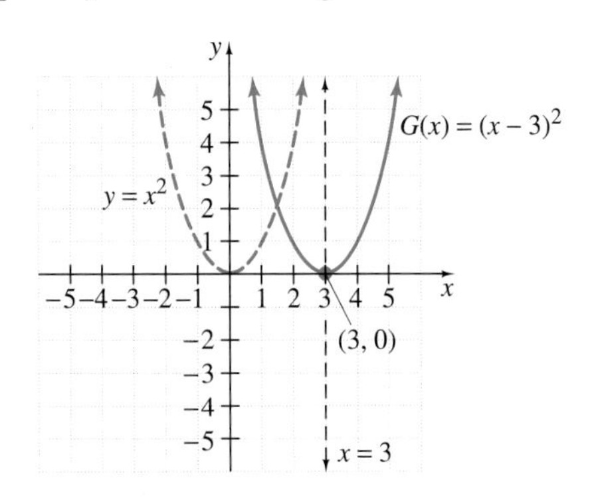

Practice Problem 4

Graph $f(x) = x^2$ and $g(x) = (x - 1)^2$ on the same set of axes.

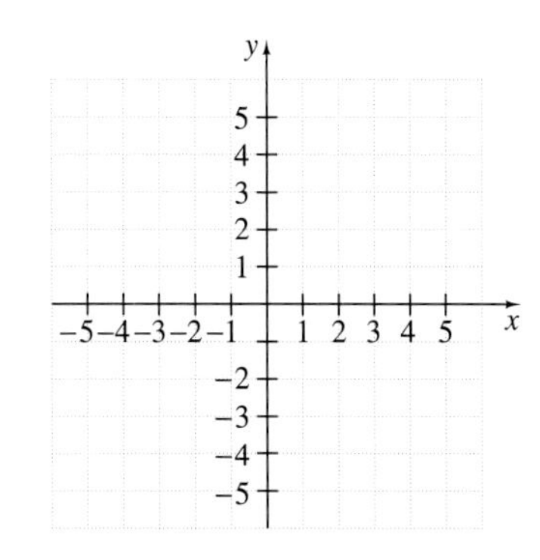

Answer

4.

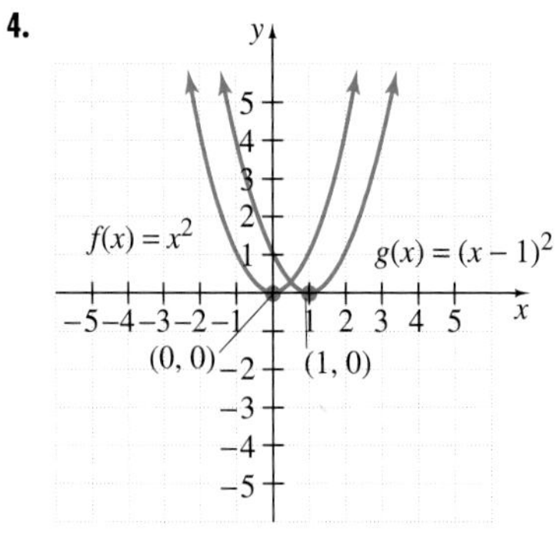

Practice Problems 5–6

Graph each function.

5. $G(x) = (x - 4)^2$

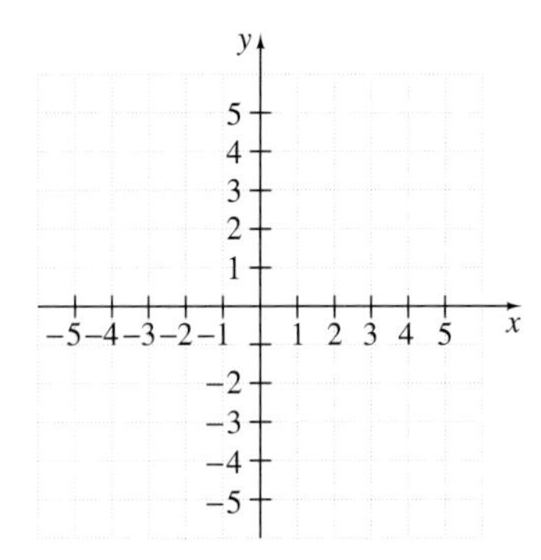

6. $F(x) = (x + 2)^2$

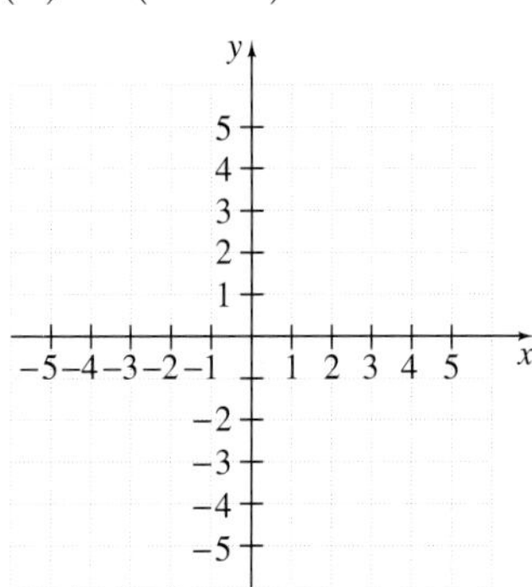

Answers

5.

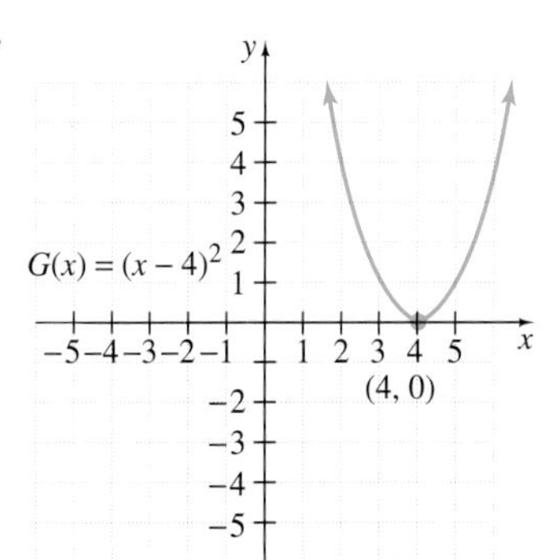

6.

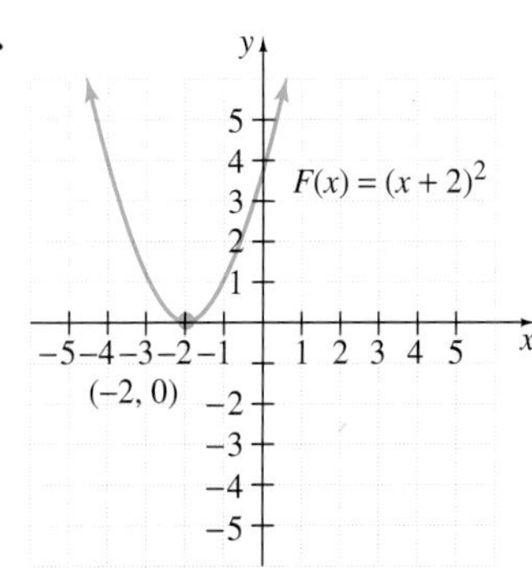

6. $F(x) = (x + 1)^2$

The equation $F(x) = (x + 1)^2$ can be written as $F(x) = [x - (-1)]^2$. The graph of $F(x) = [x - (-1)]^2$ is obtained by shifting the graph of $y = x^2$ to the left 1 unit.

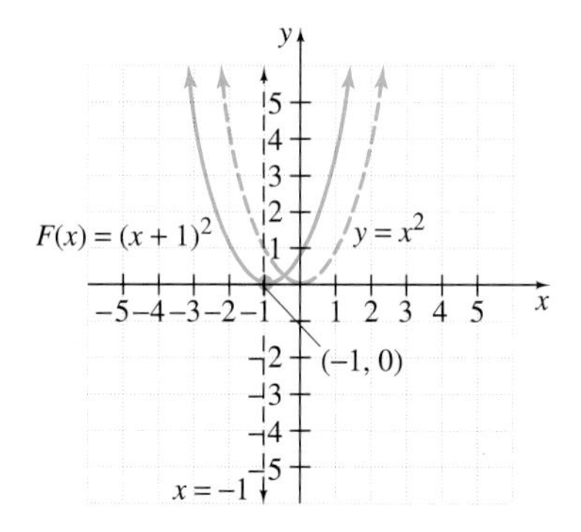

C GRAPHING $f(x) = (x - h)^2 + k$

As we will see in graphing functions of the form $f(x) = (x - h)^2 + k$, it is possible to combine vertical and horizontal shifts.

> **GRAPHING THE PARABOLA DEFINED BY $f(x) = (x - h)^2 + k$**
>
> The parabola has the same shape as $y = x^2$.
> The vertex is (h, k), and the axis of symmetry is the vertical line $x = h$.

Example 7 Graph: $F(x) = (x - 3)^2 + 1$

Solution: The graph of $F(x) = (x - 3)^2 + 1$ is the graph of $y = x^2$ shifted 3 units to the right and 1 unit up. The vertex is then $(3, 1)$, and the axis of symmetry is $x = 3$. A few ordered pair solutions are plotted to aid in graphing.

x	$F(x) = (x - 3)^2 + 1$
1	5
2	2
4	2
5	5

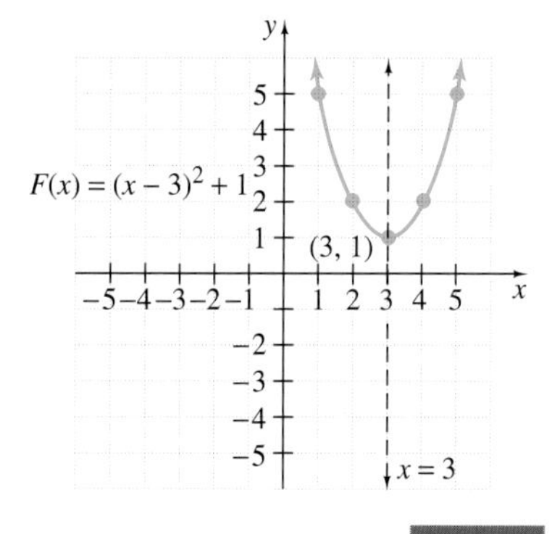

D GRAPHING $f(x) = ax^2$

Next, we discover the change in the shape of the graph when the coefficient of x^2 is not 1.

Example 8 Graph $f(x) = x^2$, $g(x) = 3x^2$, and $h(x) = \frac{1}{2}x^2$ on the same set of axes.

Solution: Comparing the table of values, we see that for each x-value, the corresponding value of $g(x)$ is triple the corresponding value of $f(x)$. Similarly, the value of $h(x)$ is half the value of $f(x)$.

x	$f(x) = x^2$	x	$g(x) = 3x^2$
−2	4	−2	12
−1	1	−1	3
0	0	0	0
1	1	1	3
2	4	2	12

x	$h(x) = \frac{1}{2}x^2$
−2	2
−1	$\frac{1}{2}$
0	0
1	$\frac{1}{2}$
2	2

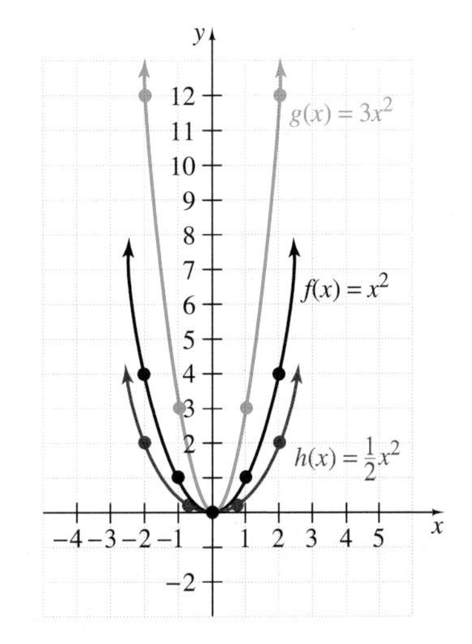

The result is that the graph of $g(x) = 3x^2$ is narrower than the graph of $f(x) = x^2$ and the graph of $h(x) = \frac{1}{2}x^2$ is wider. The vertex for each graph is $(0, 0)$, and the axis of symmetry is the y-axis.

GRAPHING THE PARABOLA DEFINED BY $f(x) = ax^2$

If a is positive, the parabola opens upward, and if a is negative, the parabola opens downard.

If $|a| > 1$, the graph of the parabola is narrower than the graph of $y = x^2$.

If $|a| < 1$, the graph of the parabola is wider than the graph of $y = x^2$.

Example 9 Graph: $f(x) = -2x^2$

Solution: Because $a = -2$, a negative value, this parabola opens downward. Since $|-2| = 2$ and $2 > 1$, the parabola is narrower than the graph of $y = x^2$. The vertex is $(0, 0)$, and the axis of symmetry is the y-axis. We verify this by plotting a few points.

Practice Problem 7

Graph: $F(x) = (x - 2)^2 + 3$

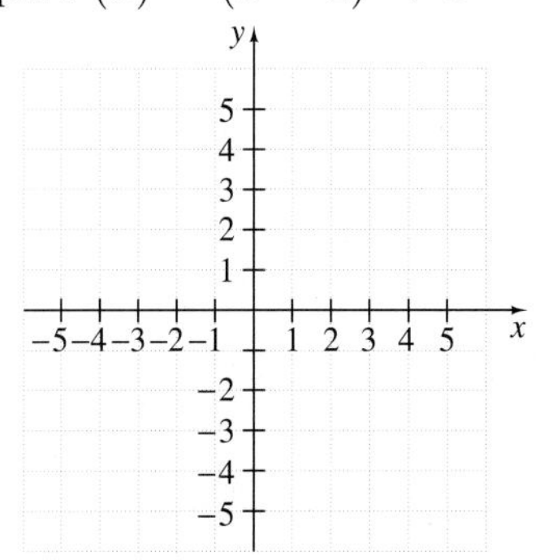

Practice Problem 8

Graph $f(x) = x^2$, $g(x) = 2x^2$, and $h(x) = \frac{1}{3}x^2$ on the same set of axes.

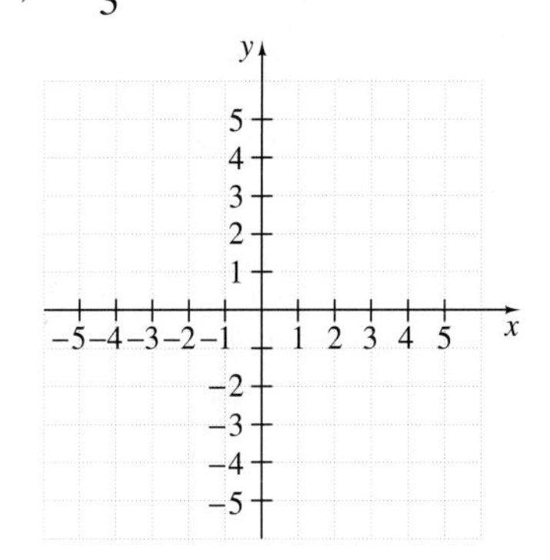

Answers

7.

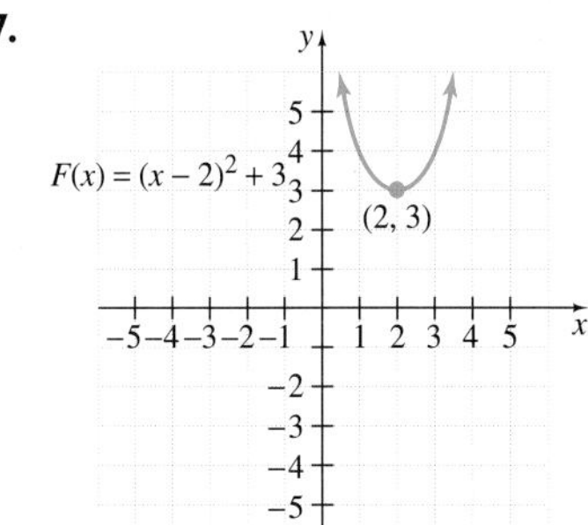

8.

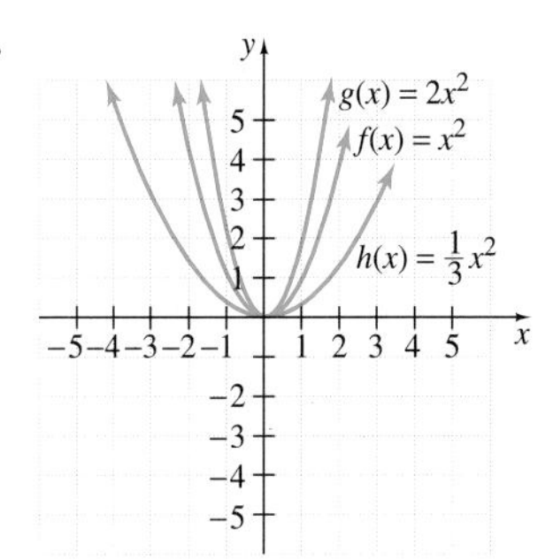

Practice Problem 9

Graph: $f(x) = -3x^2$

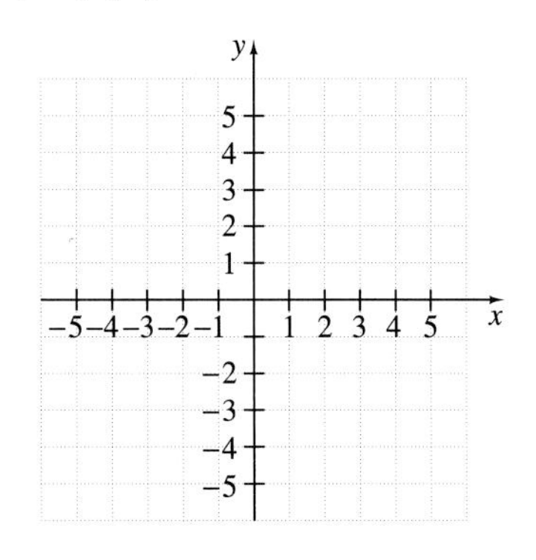

Practice Problem 10

Graph: $f(x) = 2(x + 3)^2 - 4$. Find the vertex and axis of symmetry.

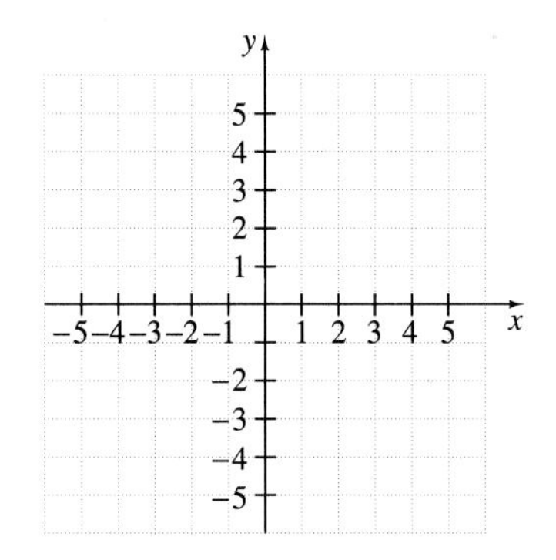

Answers

9.

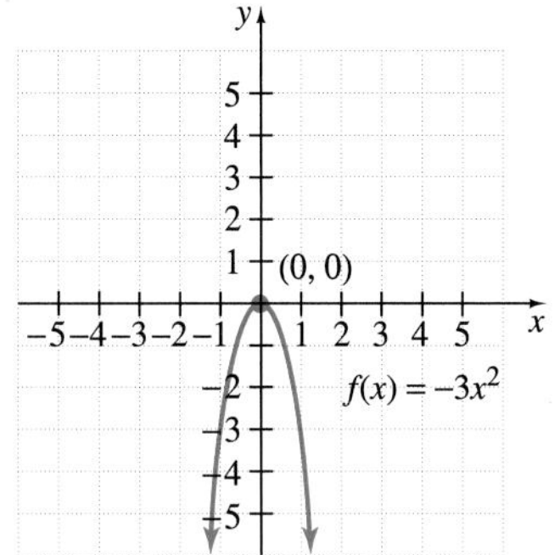

10.

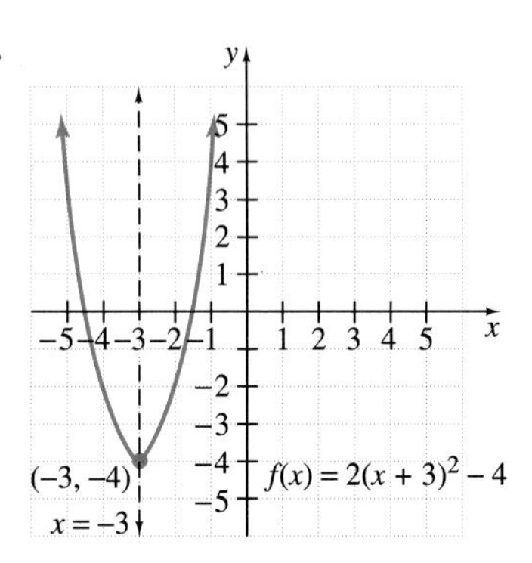

x	$f(x) = -2x^2$
-2	-8
-1	-2
0	0
1	-2
2	-8

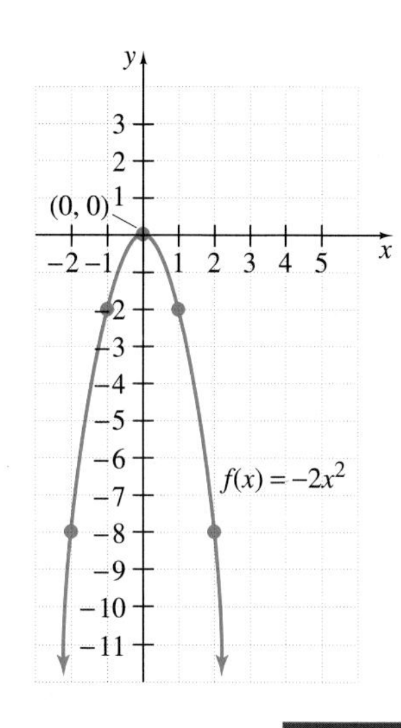

E GRAPHING $f(x) = a(x - h)^2 + k$

Now we will see the shape of the graph of a quadratic function of the form $f(x) = a(x - h)^2 + k$.

Example 10 Graph: $g(x) = \frac{1}{2}(x + 2)^2 + 5$. Find the vertex and the axis of symmetry.

Solution: The function $g(x) = \frac{1}{2}(x + 2)^2 + 5$ may be written as $g(x) = \frac{1}{2}[x - (-2)]^2 + 5$. Thus, this graph is the same as the graph of $y = x^2$ shifted 2 units to the left and 5 units up, and it is wider because a is $\frac{1}{2}$. The vertex is $(-2, 5)$, and the axis of symmetry is $x = -2$. We plot a few points to verify.

x	$g(x) = \frac{1}{2}(x + 2)^2 + 5$
-4	7
-3	$5\frac{1}{2}$
-2	5
-1	$5\frac{1}{2}$
0	7

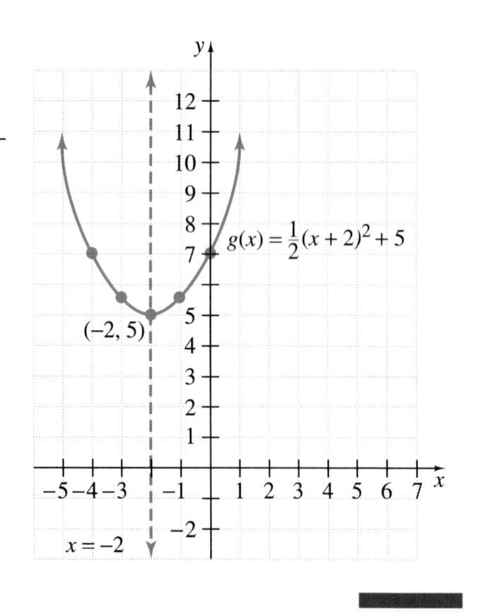

In general, the following holds.

GRAPH OF A QUADRATIC FUNCTION

The graph of a quadratic function written in the form $f(x) = a(x - h)^2 + k$ is a parabola with vertex (h, k). If $a > 0$, the parabola opens upward, and if $a < 0$, the parabola opens downward. The axis of symmetry is the line whose equation is $x = h$.

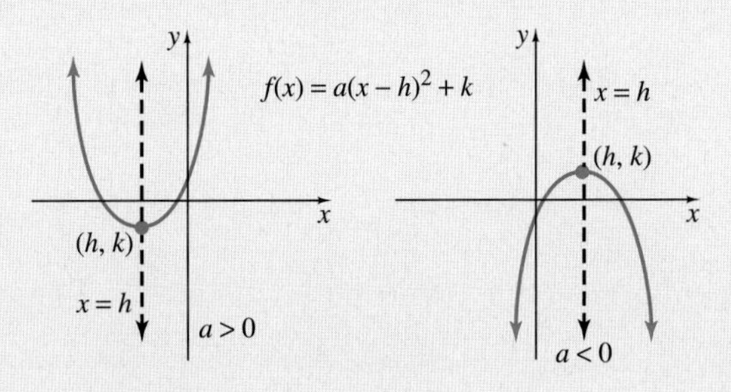

TRY THE CONCEPT CHECK IN THE MARGIN.

✓ CONCEPT CHECK

Which description of the graph of $f(x) = -0.35(x + 3)^2 - 4$ is correct?

a. The graph opens downward and has its vertex at $(-3, 4)$.
b. The graph opens upward and has its vertex at $(-3, 4)$.
c. The graph opens downward and has its vertex at $(-3, -4)$.
d. The graph is narrower than the graph of $y = x^2$.

Answer
✓ **Concept Check:** c

GRAPHING CALCULATOR EXPLORATIONS

Use a grapher to graph the first function of each pair that follows. Then use its graph to predict the graph of the second function. Check your prediction by graphing both on the same set of axes.

1. $F(x) = \sqrt{x}; G(x) = \sqrt{x} + 1$

2. $g(x) = x^3; H(x) = x^3 - 2$

3. $H(x) = |x|; f(x) = |x - 5|$

4. $h(x) = x^3 + 2; g(x) = (x - 3)^3 + 2$

5. $f(x) = |x + 4|; F(x) = |x + 4| + 3$

6. $G(x) = \sqrt{x} - 2; g(x) = \sqrt{x - 4} - 2$

Name ______________________ Section ______ Date ________

Mental Math Answers

1. ______
2. ______
3. ______
4. ______
5. ______
6. ______
7. ______
8. ______

Mental Math

State the vertex of the graph of each quadratic function.

1. $f(x) = x^2$ **2.** $f(x) = -5x^2$ **3.** $g(x) = (x - 2)^2$

4. $g(x) = (x + 5)^2$ **5.** $f(x) = 2x^2 + 3$ **6.** $h(x) = x^2 - 1$

7. $g(x) = (x + 1)^2 + 5$ **8.** $h(x) = (x - 10)^2 - 7$

Exercise Set 8.5

A B *Graph each quadratic function. Label the vertex, and sketch and label the axis of symmetry. See Examples 1 through 6.*

1. $f(x) = x^2 - 1$

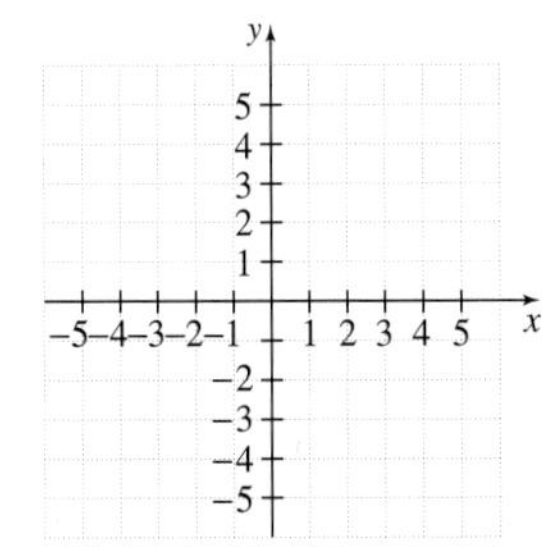

2. $g(x) = x^2 + 3$

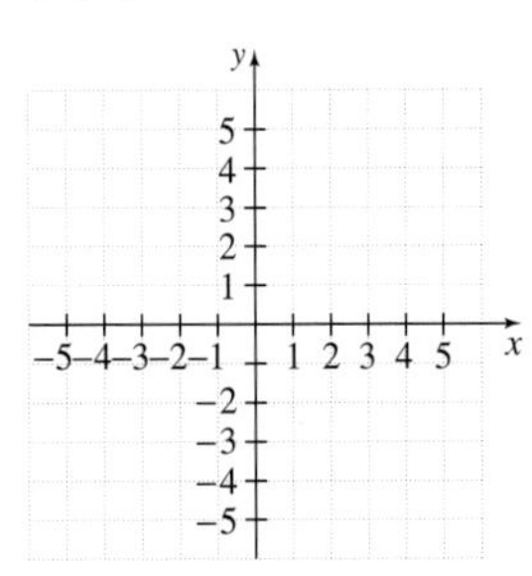

3. $f(x) = (x - 5)^2$

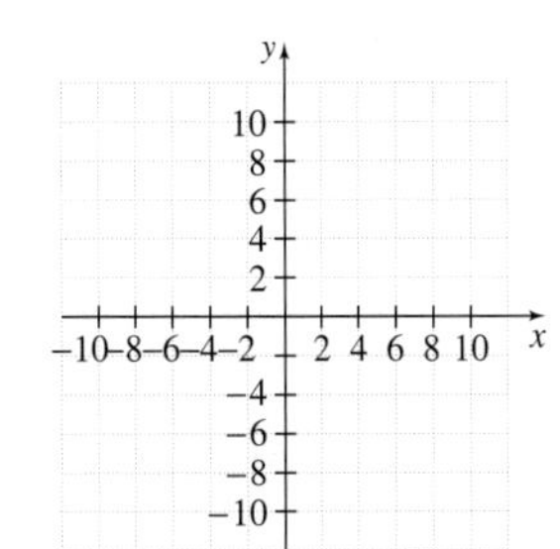

4. $g(x) = (x + 5)^2$

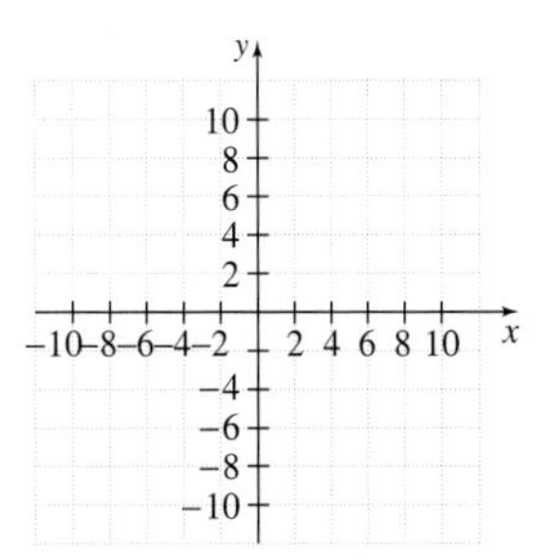

5. $h(x) = x^2 + 5$

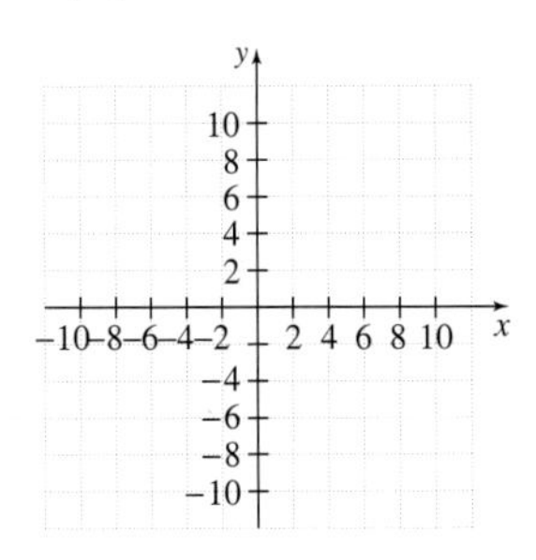

6. $h(x) = x^2 - 4$

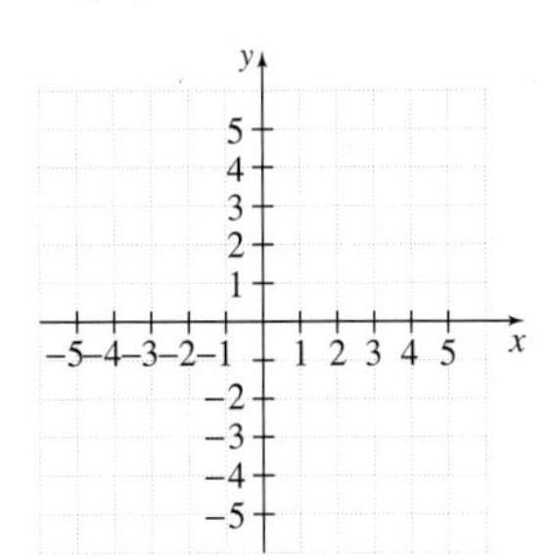

7. $h(x) = (x + 2)^2$

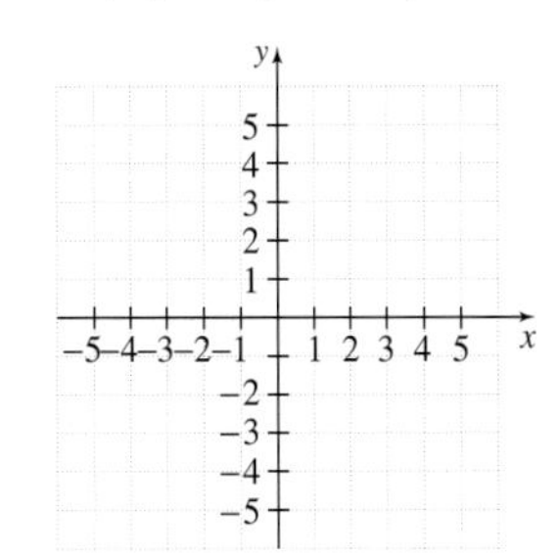

8. $H(x) = (x - 1)^2$

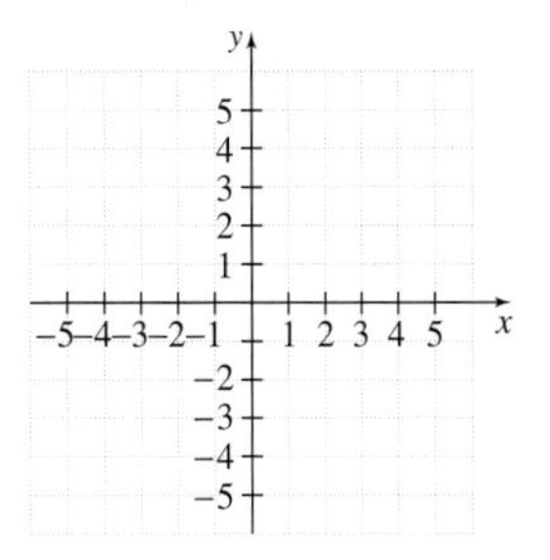

9. $g(x) = x^2 + 7$

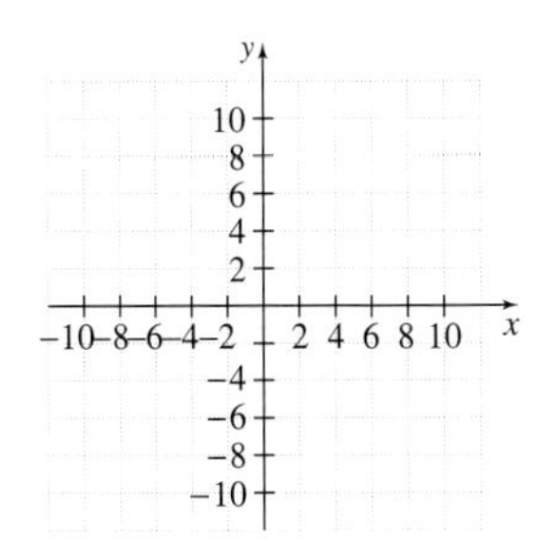

10. $f(x) = x^2 - 2$

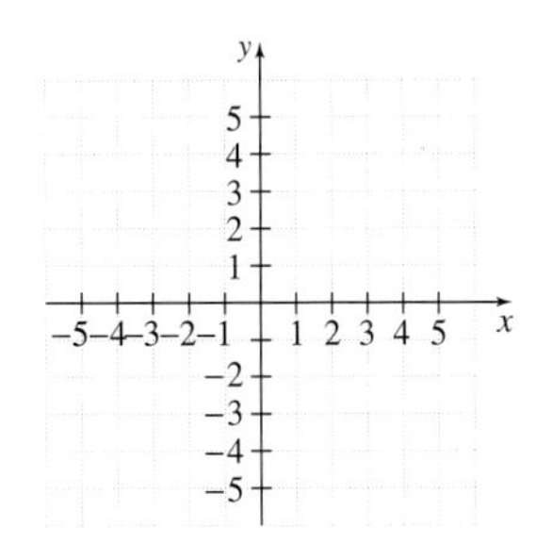

11. $G(x) = (x + 3)^2$

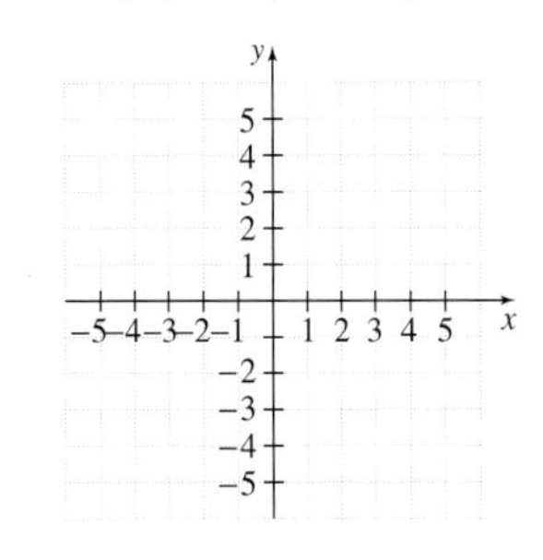

12. $f(x) = (x - 6)^2$

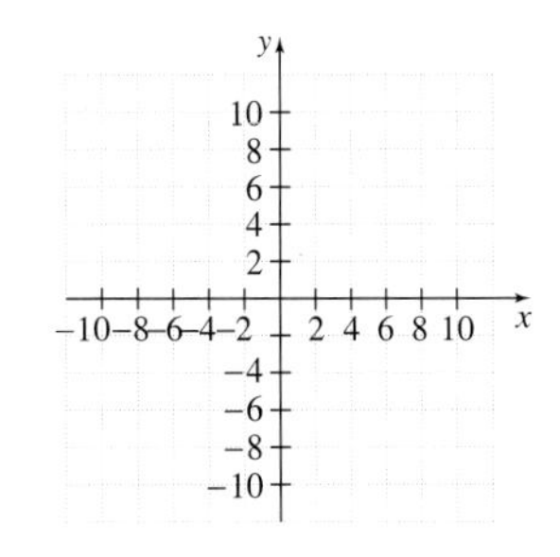

Name ______________________________

C *Graph each quadratic function. Label the vertex, and sketch and label the axis of symmetry. See Example 7.*

13. $f(x) = (x - 2)^2 + 5$

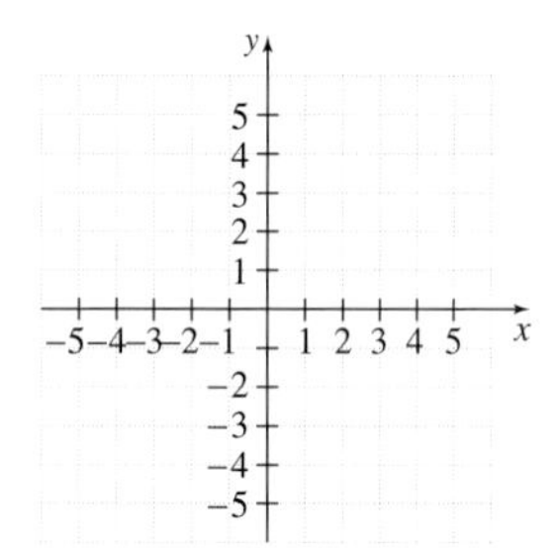

14. $g(x) = (x - 6)^2 + 1$

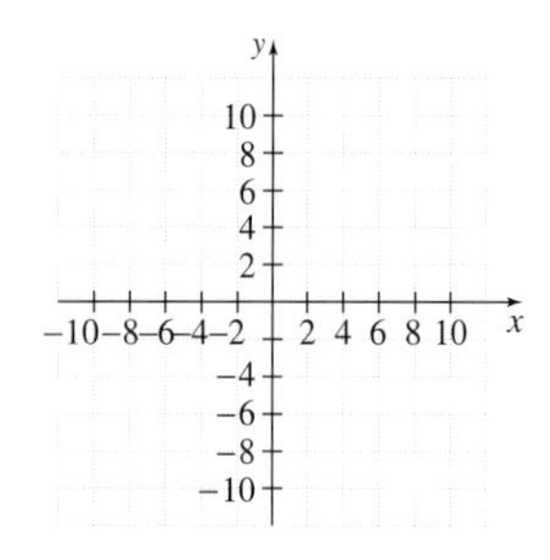

15. $h(x) = (x + 1)^2 + 4$

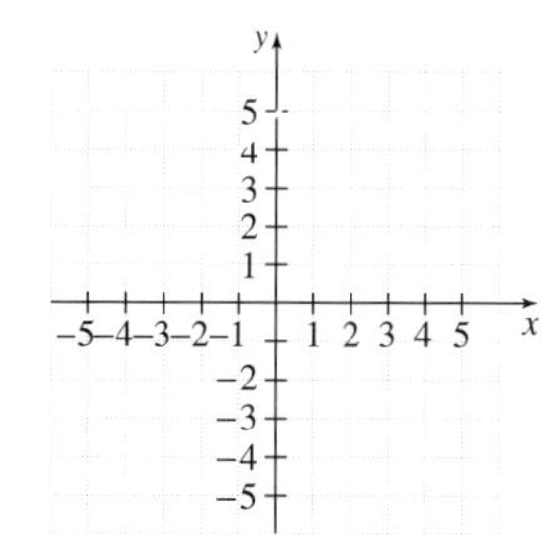

16. $G(x) = (x + 3)^2 + 3$

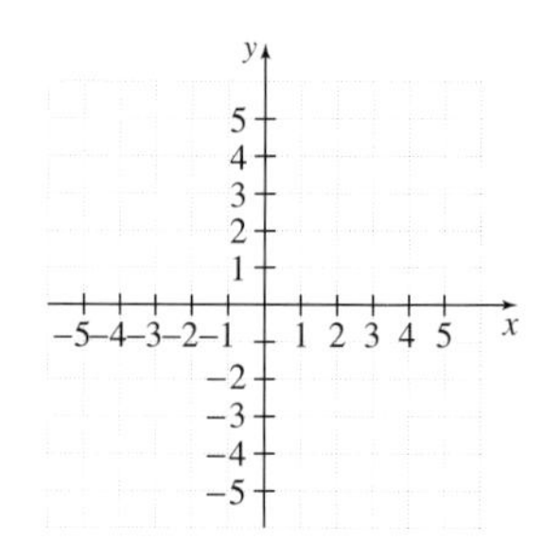

17. $g(x) = (x + 2)^2 - 5$

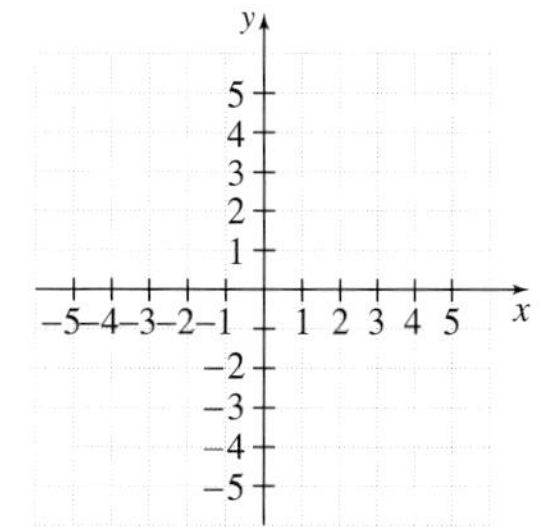

18. $h(x) = (x + 4)^2 - 6$

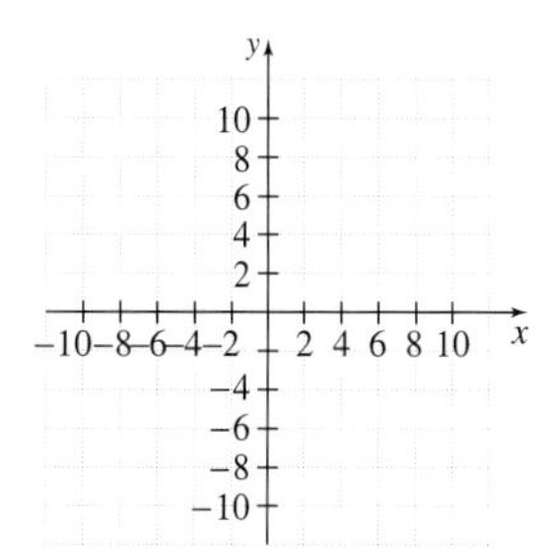

19. $h(x) = (x - 3)^2 + 2$

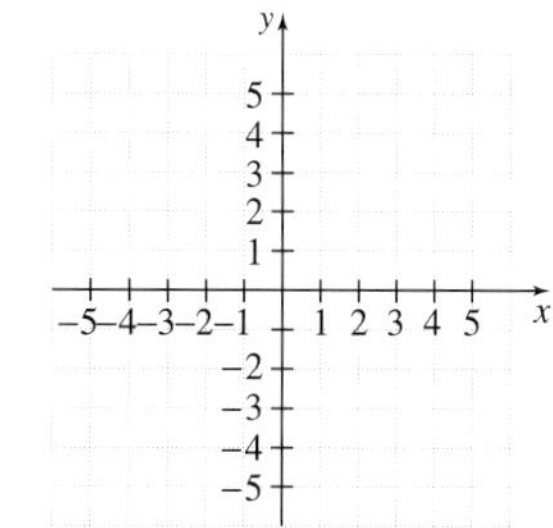

20. $F(x) = (x - 2)^2 - 3$

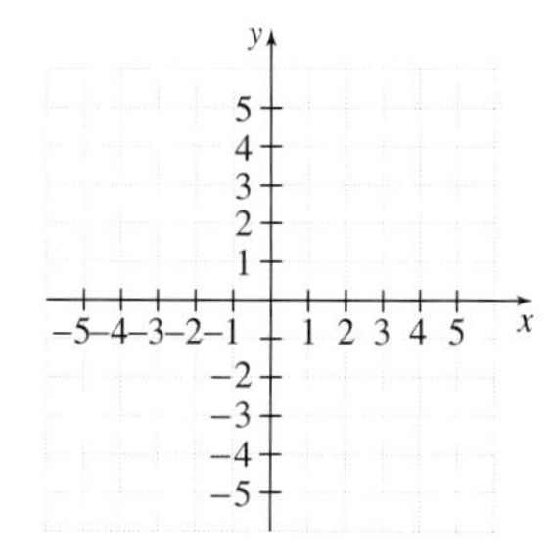

D *Graph each quadratic function. Label the vertex, and sketch and label the axis of symmetry. See Examples 8 and 9.*

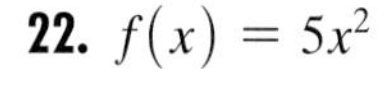

21. $g(x) = -x^2$

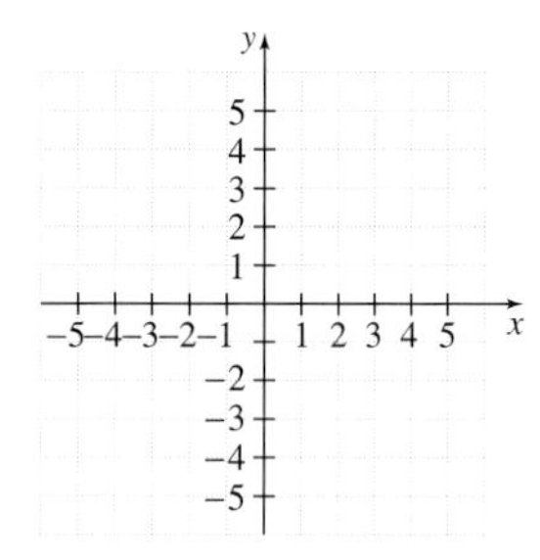

22. $f(x) = 5x^2$

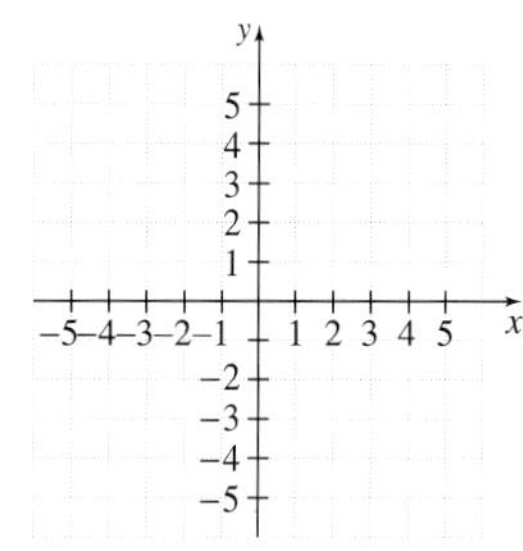

23. $h(x) = \frac{1}{3}x^2$

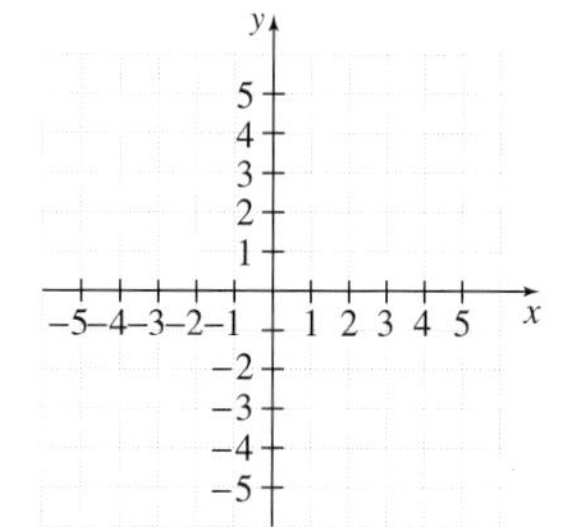

24. $g(x) = -3x^2$

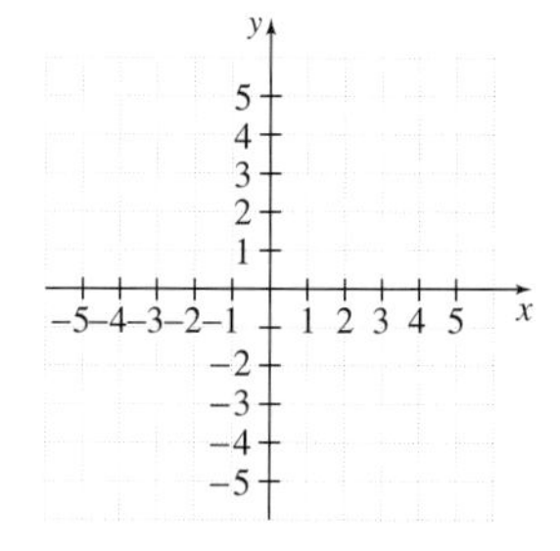

25. $H(x) = 2x^2$

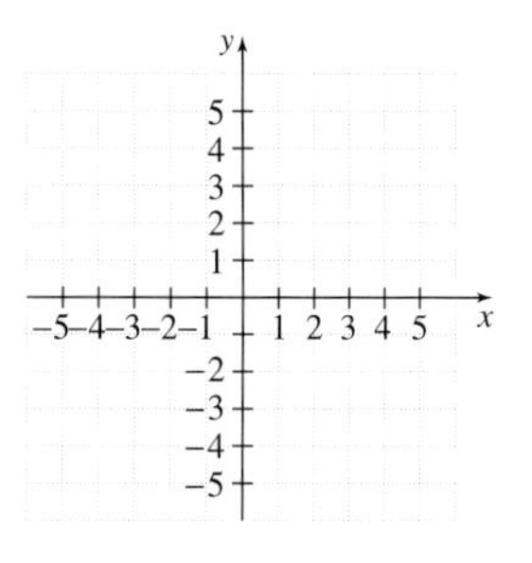

26. $f(x) = -\frac{1}{4}x^2$

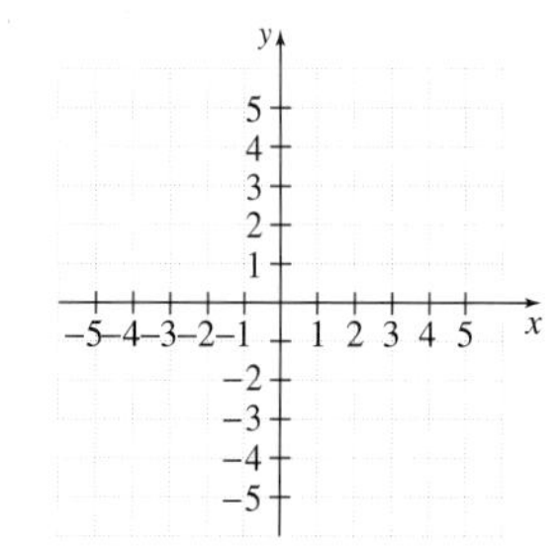

27. $F(x) = -4x^2$

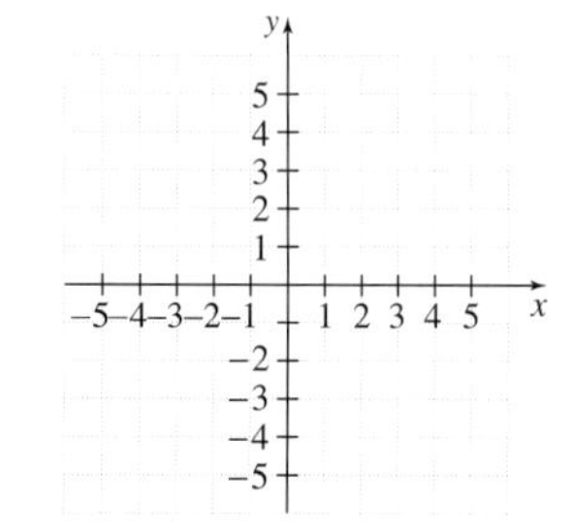

28. $G(x) = \frac{1}{5}x^2$

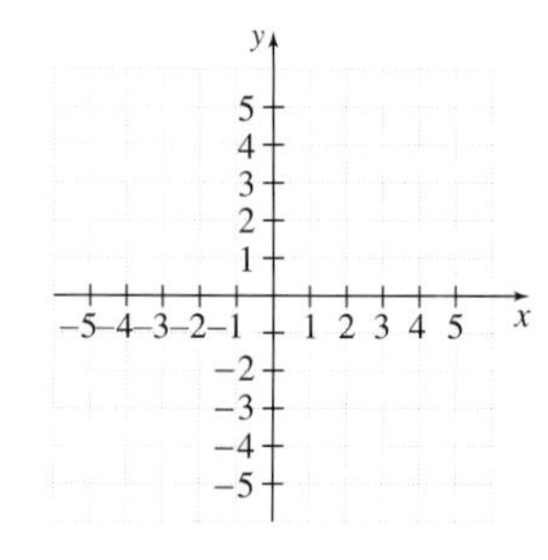

Name ______________________________

E *Graph each quadratic function. Label the vertex, and sketch and label the axis of symmetry. See Example 10.*

29. $f(x) = 10(x + 4)^2 - 6$

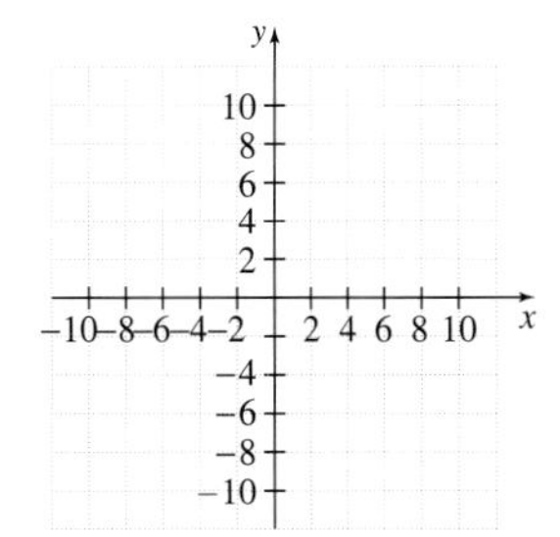

30. $g(x) = 4(x - 4)^2 + 2$

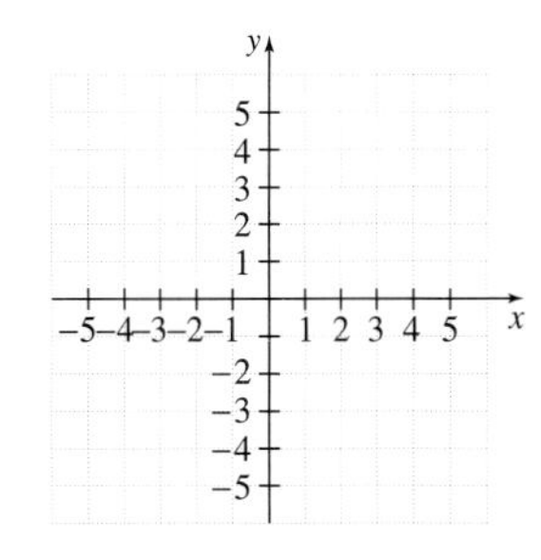

31. $h(x) = -3(x + 3)^2 + 1$

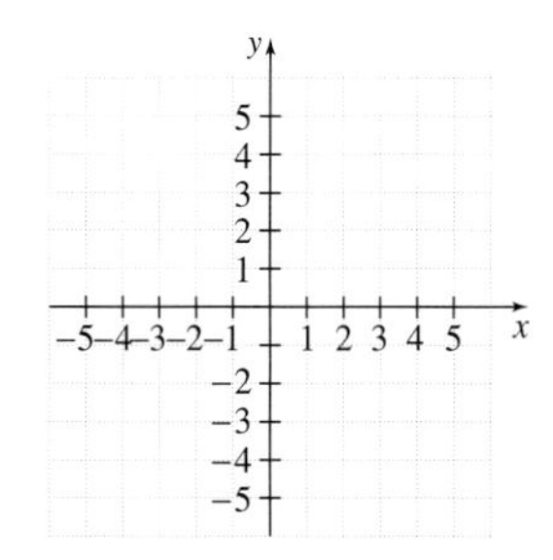

32. $f(x) = -(x - 2)^2 - 6$

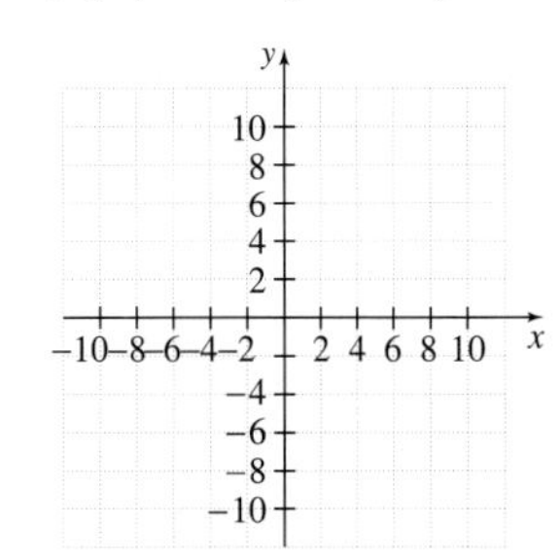

33. $H(x) = \frac{1}{2}(x - 6)^2 - 3$

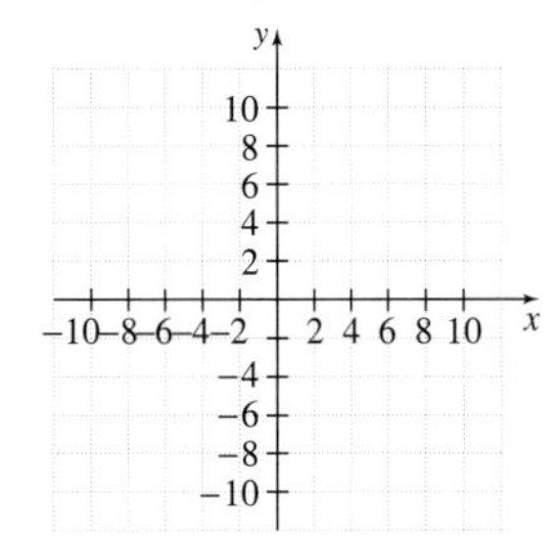

34. $G(x) = \frac{1}{5}(x + 4)^2 + 3$

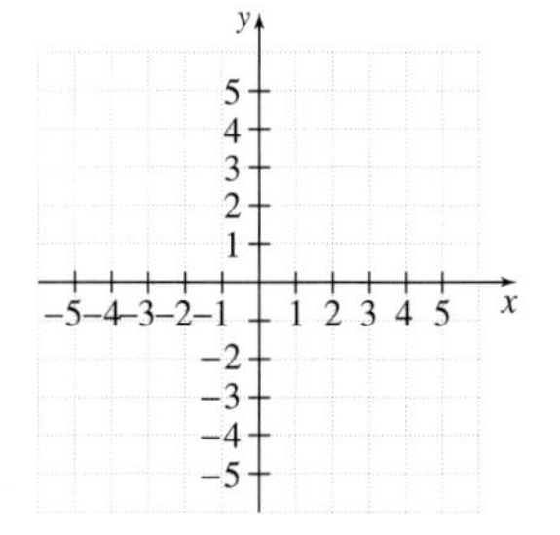

35. $f(x) = -(x - 1)^2$

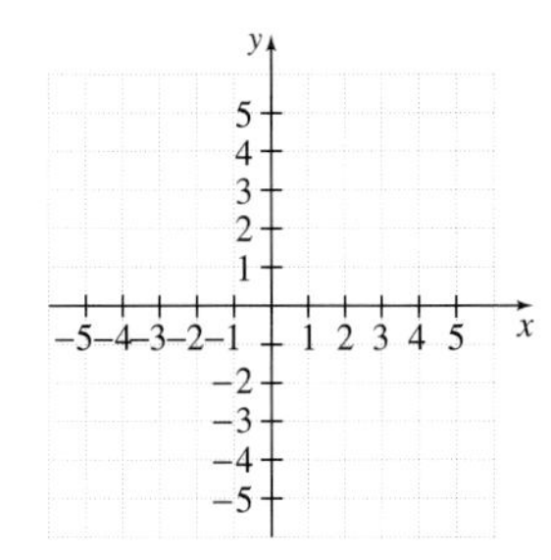

36. $f(x) = 2(x + 3)^2$

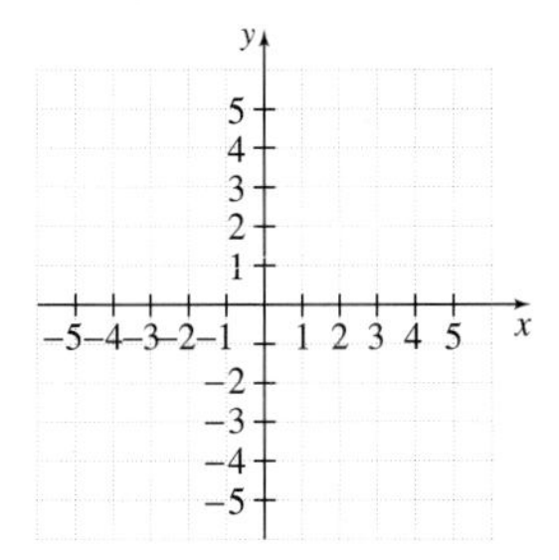

37. $F(x) = \left(x + \frac{1}{2}\right)^2 - 2$

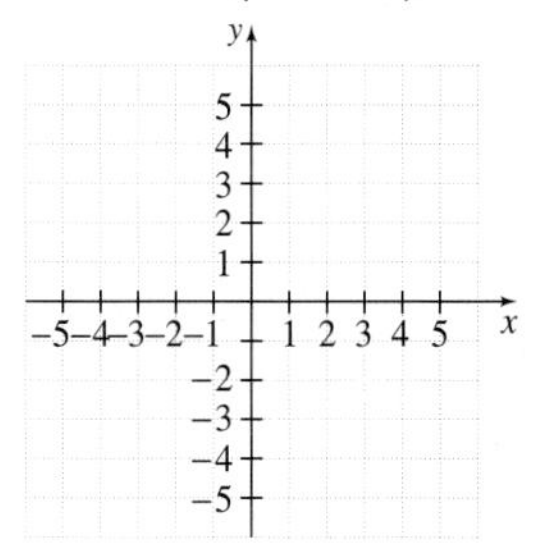

38. $H(x) = \left(x + \frac{1}{4}\right)^2 - 3$

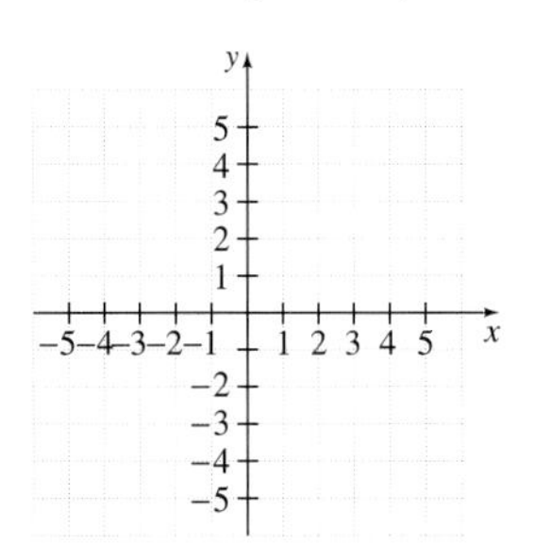

39. $F(x) = -x^2 + 2$

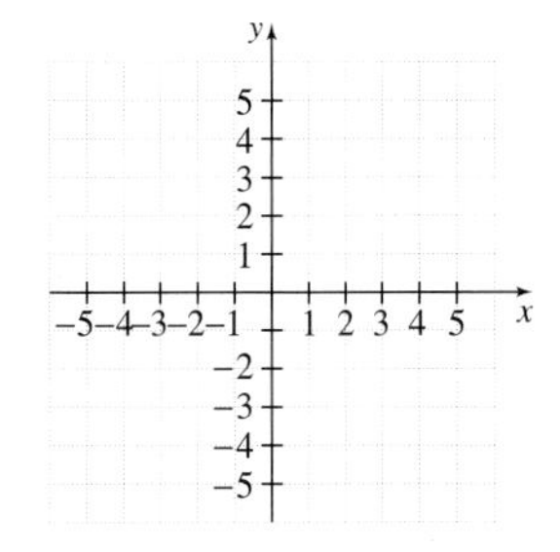

40. $G(x) = 3x^2 + 1$

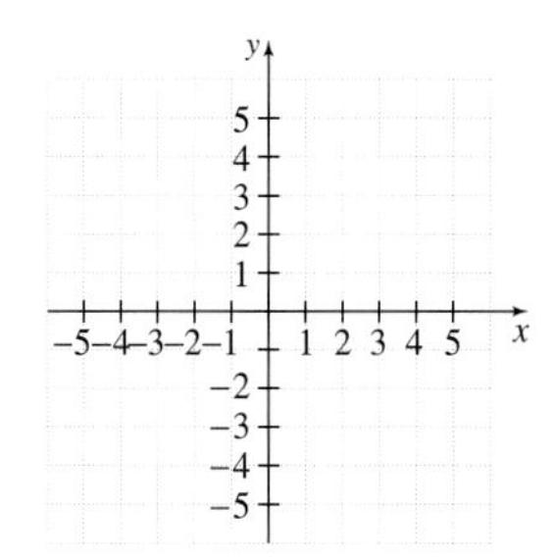

41. ______

42. ______

43. ______

44. ______

45. ______

46. ______

47. ______

48. ______

49. ______

50. ______

51. see graph

52. see graph

Name ______________________

Review and Preview

Add the proper constant to each binomial so that the resulting trinomial is a perfect square trinomial. See Section 8.1

41. $x^2 + 8x$ **42.** $y^2 + 4y$ **43.** $z^2 - 16z$

44. $x^2 - 10x$ **45.** $y^2 + y$ **46.** $z^2 - 3z$

Combining Concepts

Write the equation of the parabola that has the same shape as $f(x) = 5x^2$ but with each given vertex.

47. $(2, 3)$ **48.** $(1, 6)$ **49.** $(-3, 6)$ **50.** $(4, -1)$

The shifting properties covered in this section apply to the graphs of all functions. Given the accompanying graph of $y = f(x)$, graph each function.

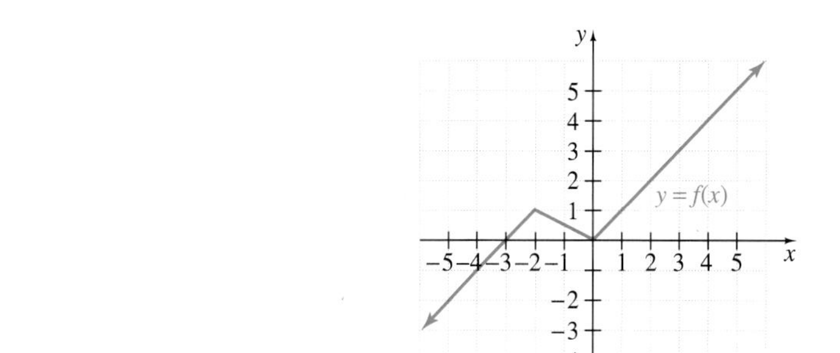

51. $y = f(x) + 1$

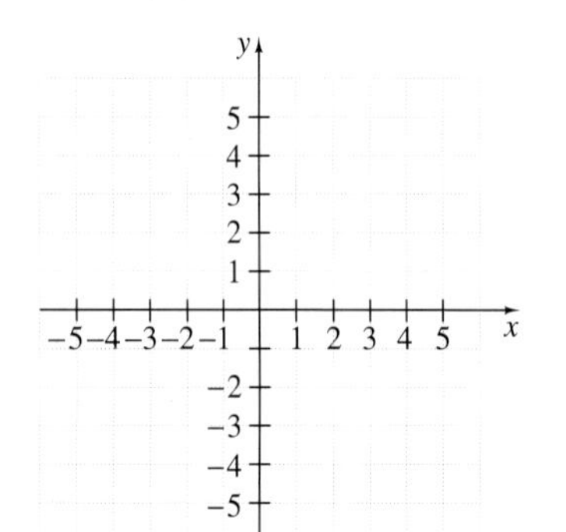

52. $y = f(x) - 2$

Name ____________________

53. $y = f(x - 3)$

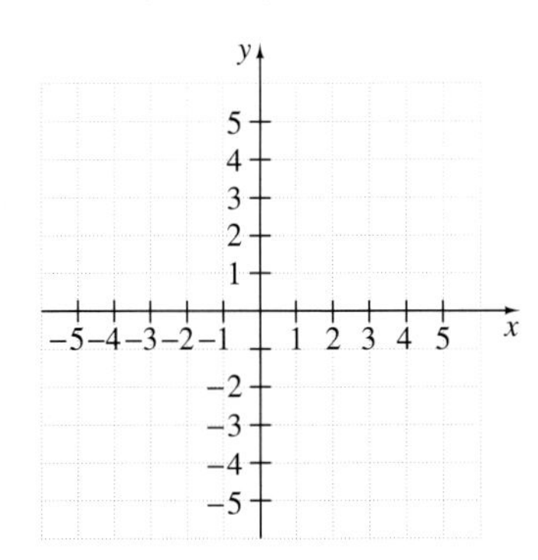

54. $y = f(x + 3)$

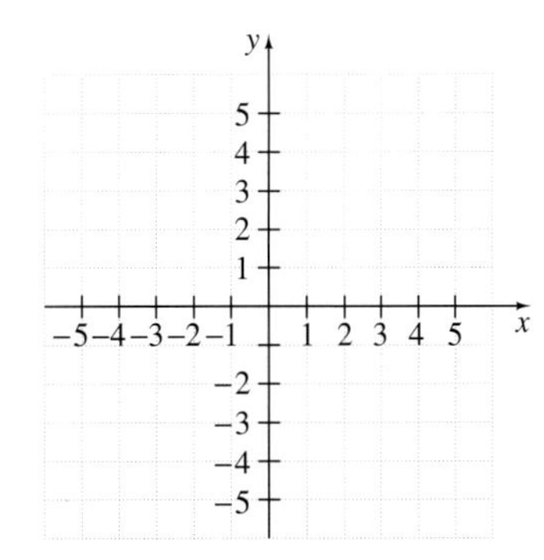

55. $y = f(x + 2) + 2$

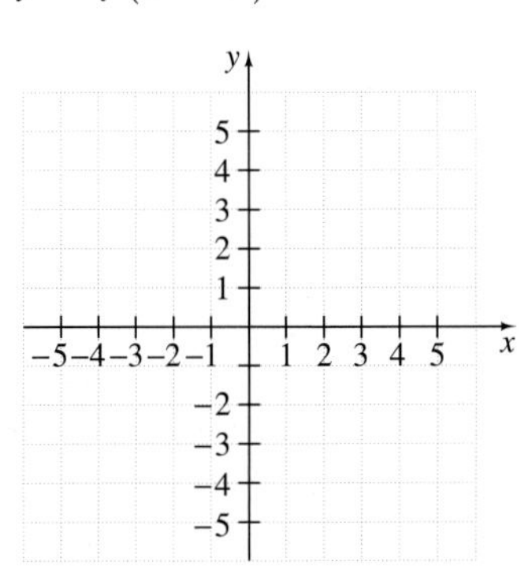

56. $y = f(x - 1) + 1$

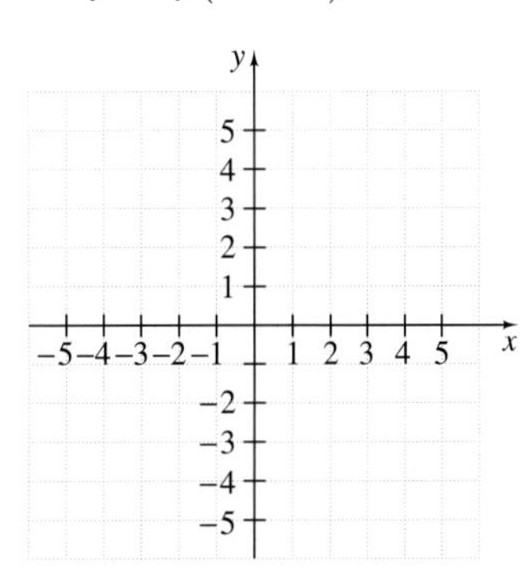

53. see graph

54. see graph

55. see graph

56. see graph

Focus On Business and Career

Financial Ratios

A financial ratio is a number found with a rational expression that tells something about a company's activities. Such ratios allow a comparison between the financial positions of two companies, even if the values of the companies' financial data are very different. Here are some common financial ratios:

- The **current ratio** gauges a company's ability to pay its short-term debts. It is given by the formula

  $$\text{current ratio} = \frac{\text{total current assets}}{\text{total current liabilities}}$$

  The higher the value of this ratio, the better able the company is to pay off its short-term debts.
- The **total asset turnover ratio** gauges how effectively a company is using all of its resources to generate sales of its products and services. It is given by the formula

  $$\text{total asset turnover ratio} = \frac{\text{sales}}{\text{total assets}}$$

  The higher the value of this ratio, the more effective the company is at utilizing its resources for sales generation.
- The **gross profit margin ratio** gauges how effectively the company is making pricing decisions as well as controlling production costs. It is given by the formula

  $$\text{gross profit margin ratio} = \frac{\text{sales} - \text{cost of sales}}{\text{sales}}$$

  The higher the value of this ratio, the better the company is doing with regards to controlling costs and pricing products.
- The **price-to-earnings (P/E) ratio** gauges the stock market's view of a company with respect to risk. It is given by the formula

  $$\text{P/E ratio} = \frac{\text{stock market price per share}}{\text{earnings per share}}$$

  A company with low risk will generally have a high P/E ratio. A high P/E ratio also translates into better growth potential for the company's earnings.

For all of these ratios, a higher-than-industry-average ratio is generally considered to be a sign of good financial health.

Additional Definitions

- **Assets**—things of value that are owned by a company. *Current assets* include cash and assets that can be converted into cash quickly. *Total assets* are all things of value, including property and equipment, owned by the company. Current assets and total assets may be found on a company's consolidated balance sheet or statement of financial position in an annual report.
- **Liabilities**—what a company owes to creditors. *Current liabilities* include any debts expected to come due within the next year. Current liabilities may be found on a company's consolidated balance sheet or statement of financial position in an annual report.
- **Sales**—the total amount of money collected by a company from the sales of its goods or services. Sales (or sometimes noted as "net sales") may be found on a company's consolidated statement of income/earnings/operations in an annual report.
- **Cost of sales**—a company's cost of inventory actually sold to customers. This is also sometimes referred to as "cost of goods/merchandise sold." Cost of sales may be found on a company's consolidated statement of income/earnings/operations in an annual report.
- **Earnings per share**—the value of a company's earnings available for each share of common stock held by stockholders. Earnings per share (or sometimes noted as net income per share) may be found on a company's consolidated statement of income/earnings/operations in an annual report.
- **Stock market price per share**—the current price of a company's share of stock as given on one of the major stock markets. Current share prices may be found in newspapers or on the World-Wide Web.

Group Activity

Locate annual reports for two companies involved in similar industries. Using the information and definitions given above, compute these four financial ratios for each company. Then compare the companies' ratios and discuss what the ratios indicate about the two companies. Which company do you think is in better overall financial health? Why?

8.6 FURTHER GRAPHING OF QUADRATIC FUNCTIONS

A WRITING QUADRATIC FUNCTIONS IN THE FORM $y = a(x - h)^2 + k$

We know that the graph of a quadratic function is a parabola. If a quadratic function is written in the form

$$f(x) = a(x - h)^2 + k$$

we can easily find the vertex (h, k) and graph the parabola. To write a quadratic function in this form, we need to complete the square. (See Section 8.1 for a review of completing the square.)

Example 1 Graph: $f(x) = x^2 - 4x - 12$. Find the vertex and any intercepts.

Solution: The graph of this quadratic function is a parabola. To find the vertex of the parabola, we complete the square on the binomial $x^2 - 4x$. To simplify our work, we let $f(x) = y$.

$y = x^2 - 4x - 12$ Let $f(x) = y$.

$y + 12 = x^2 - 4x$ Add 12 to both sides to get the x-variable terms alone.

Now we add the square of half of -4 to both sides.

$$\frac{1}{2}(-4) = -2 \quad \text{and} \quad (-2)^2 = 4$$

$y + 12 + 4 = x^2 - 4x + 4$ Add 4 to both sides.

$y + 16 = (x - 2)^2$ Factor the trinomial.

$y = (x - 2)^2 - 16$ Subtract 16 from both sides.

$f(x) = (x - 2)^2 - 16$ Replace y with $f(x)$.

From this equation, we can see that the vertex of the parabola is $(2, -16)$, a point in quadrant IV, and the axis of symmetry is the line $x = 2$.

Notice that $a = 1$. Since $a > 0$, the parabola opens upward. This parabola opening upward with vertex $(2, -16)$ will have two x-intercepts.

To find them, we let $f(x)$ or $y = 0$.

$$0 = x^2 - 4x - 12$$
$$0 = (x - 6)(x + 2)$$
$$0 = x - 6 \quad \text{or} \quad 0 = x + 2$$
$$6 = x \qquad\qquad -2 = x$$

The two x-intercepts are 6 and -2. To find the y-intercept, we let $x = 0$.

$$f(0) = 0^2 - 4 \cdot 0 - 12 = -12$$

The y-intercept is -12. The sketch of $f(x) = x^2 - 4x - 12$ is shown.

Objectives

A Write quadratic functions in the form $y = a(x - h)^2 + k$.

B Derive a formula for finding the vertex of a parabola.

C Find the minimum or maximum value of a quadratic function.

SSM

CD-ROM

Video 8.6

Practice Problem 1

Graph: $f(x) = x^2 - 4x - 5$. Find the vertex and any intercepts.

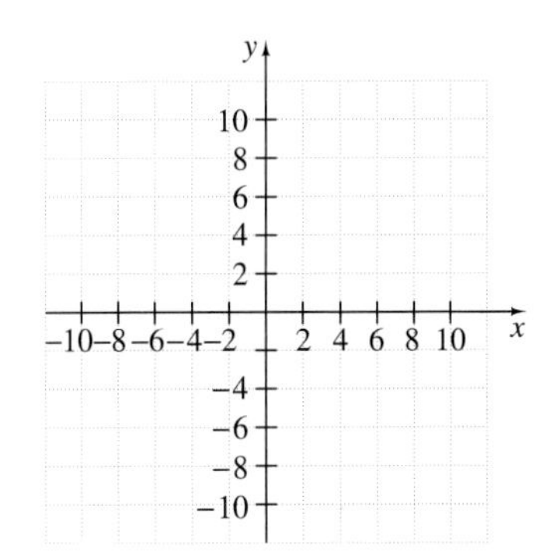

Answer

1. vertex: $(2, -9)$; x-intercepts: $-1, 5$; y-intercept: -5

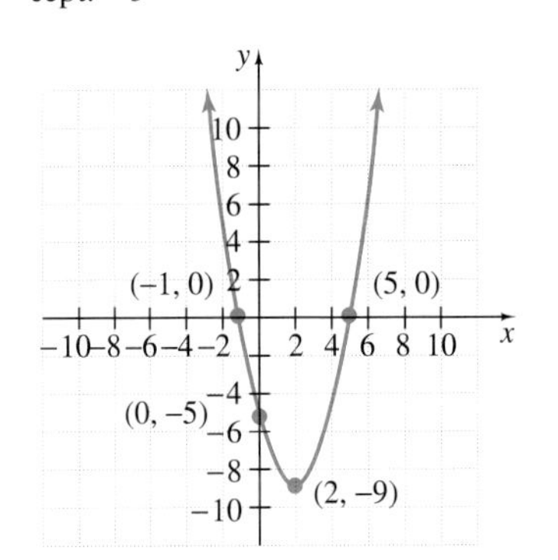

HELPFUL HINT

Parabola Opens Upward
Vertex in I or II: no x-intercepts
Vertex in III or IV: 2 x-intercepts

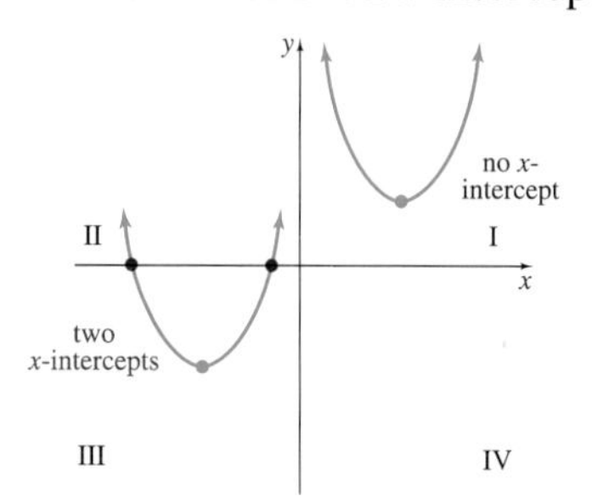

Parabola Opens Downward
Vertex in I or II: 2 x-intercepts
Vertex in III or IV: no x-intercepts.

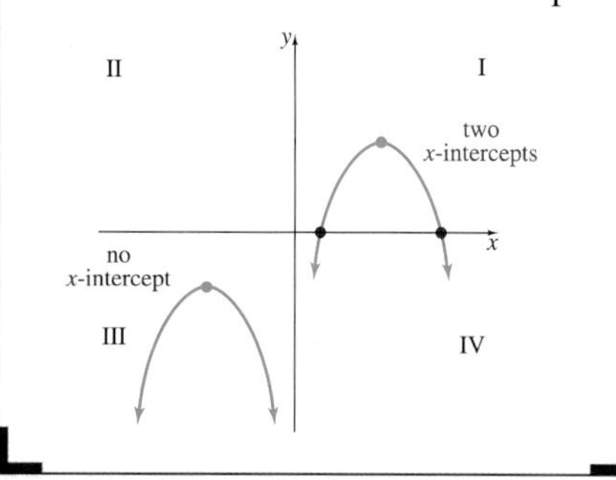

Practice Problem 2

Graph: $f(x) = 2x^2 + 2x + 5$. Find the vertex and any intercepts.

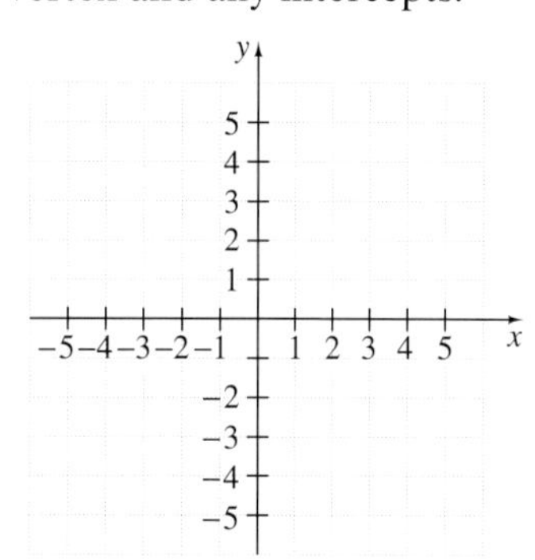

Answer

2. vertex: $\left(-\frac{1}{2}, \frac{9}{2}\right)$; y-intercept: 5

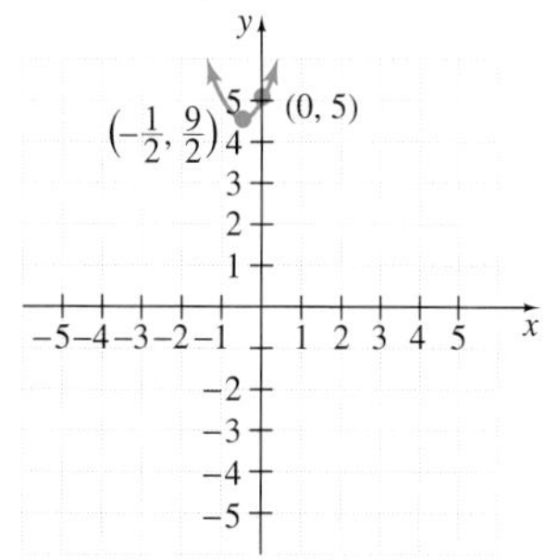

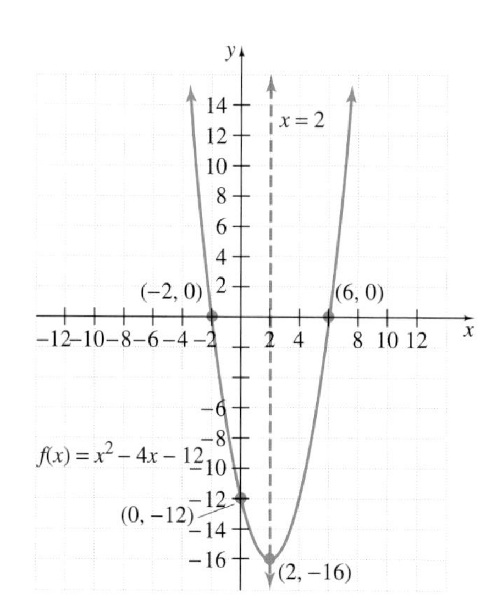

Example 2

Graph: $f(x) = 3x^2 + 3x + 1$. Find the vertex and any intercepts.

Solution: We replace $f(x)$ with y and complete the square on x to write the equation in the form $y = a(x - h)^2 + k$.

$y = 3x^2 + 3x + 1$ Replace $f(x)$ with y.

$y - 1 = 3x^2 + 3x$ Get the x-variable terms alone.

Next we factor 3 from the terms $3x^2 + 3x$ so that the coefficient of x^2 is 1.

$y - 1 = 3(x^2 + x)$ Factor out 3.

The coefficient of x is 1. Then $\frac{1}{2}(1) = \frac{1}{2}$ and $\left(\frac{1}{2}\right)^2 = \frac{1}{4}$.

Since we are adding $\frac{1}{4}$ inside the parentheses, we are really adding $3\left(\frac{1}{4}\right)$, so we *must* add $3\left(\frac{1}{4}\right)$ to the left side.

$$y - 1 + 3\left(\frac{1}{4}\right) = 3\left(x^2 + x + \frac{1}{4}\right)$$

$$y - \frac{1}{4} = 3\left(x + \frac{1}{2}\right)^2 \quad \text{Simplify the left side and factor the right side.}$$

$$y = 3\left(x + \frac{1}{2}\right)^2 + \frac{1}{4} \quad \text{Add } \frac{1}{4} \text{ to both sides.}$$

$$f(x) = 3\left(x + \frac{1}{2}\right)^2 + \frac{1}{4} \quad \text{Replace } y \text{ with } f(x).$$

Then $a = 3$, $h = -\frac{1}{2}$, and $k = \frac{1}{4}$. This means that the parabola opens upward with vertex $\left(-\frac{1}{2}, \frac{1}{4}\right)$ and that the axis of symmetry is the line $x = -\frac{1}{2}$.

To find the y-intercept, we let $x = 0$. Then

$f(0) = 3(0)^2 + 3(0) + 1 = 1$

This parabola has no x-intercepts since the vertex is in the second quadrant and it opens upward. We use the vertex, axis of symmetry, and y-intercept to graph the parabola.

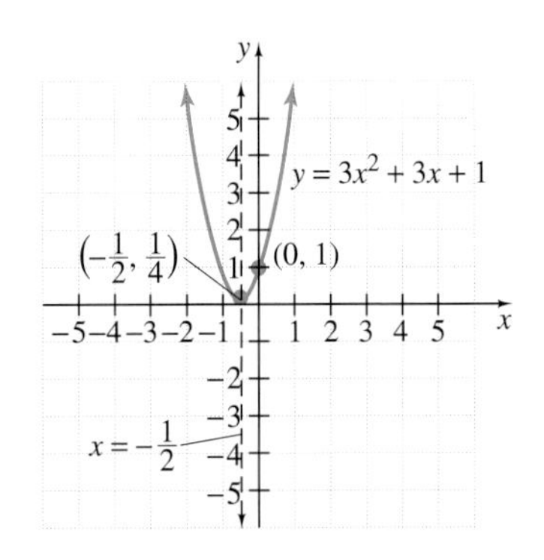

Example 3 Graph: $f(x) = -x^2 - 2x + 3$. Find the vertex and any intercepts.

Solution: We write $f(x)$ in the form $a(x - h)^2 + k$ by completing the square. First we replace $f(x)$ with y.

$$f(x) = -x^2 - 2x + 3$$

$$y = -x^2 - 2x + 3$$

$$y - 3 = -x^2 - 2x \quad \text{Subtract 3 from both sides to get the } x\text{-variable terms alone.}$$

$$y - 3 = -1(x^2 + 2x) \quad \text{Factor } -1 \text{ from the terms } -x^2 - 2x.$$

The coefficient of x is 2. Then $\frac{1}{2}(2) = 1$ and $1^2 = 1$. We add 1 to the right side inside the parentheses and add $-1(1)$ to the left side.

$$y - 3 - 1(1) = -1(x^2 + 2x + 1)$$

$$y - 4 = -1(x + 1)^2 \quad \text{Simplify the left side and factor the right side.}$$

$$y = -1(x + 1)^2 + 4 \quad \text{Add 4 to both sides.}$$

$$f(x) = -1(x + 1)^2 + 4 \quad \text{Replace } y \text{ with } f(x).$$

Since $a = -1$, the parabola opens downward with vertex $(-1, 4)$ and axis of symmetry $x = -1$.

To find the y-intercept, we let $x = 0$ and solve for y. Then

$$f(0) = -0^2 - 2(0) + 3 = 3$$

Thus, 3 is the y-intercept.

To find the x-intercepts, we let y or $f(x) = 0$ and solve for x.

$$f(x) = -x^2 - 2x + 3$$

$$0 = -x^2 - 2x + 3 \quad \text{Let } f(x) = 0.$$

Now we divide both sides by -1 so that the coefficient of x^2 is 1:

Practice Problem 3

Graph: $f(x) = -x^2 - 2x + 8$. Find the vertex and any intercepts.

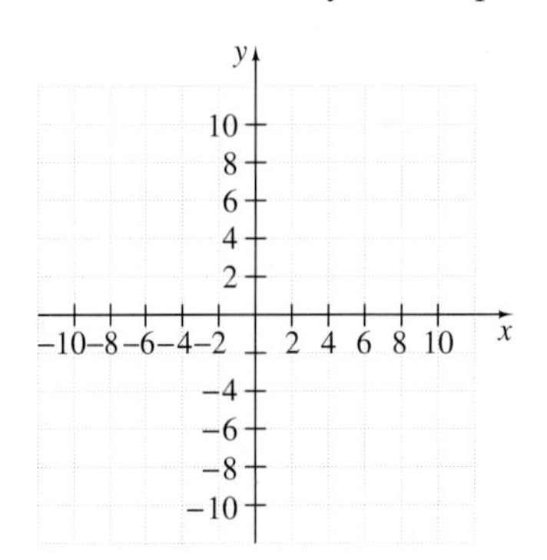

HELPFUL HINT

This can be written as $f(x) = -1[x - (-1)]^2 + 4$. Notice that the vertex is $(-1, 4)$.

Answer

3. vertex: $(-1, 9)$; x-intercepts: $-4, 2$; y-intercept: 8

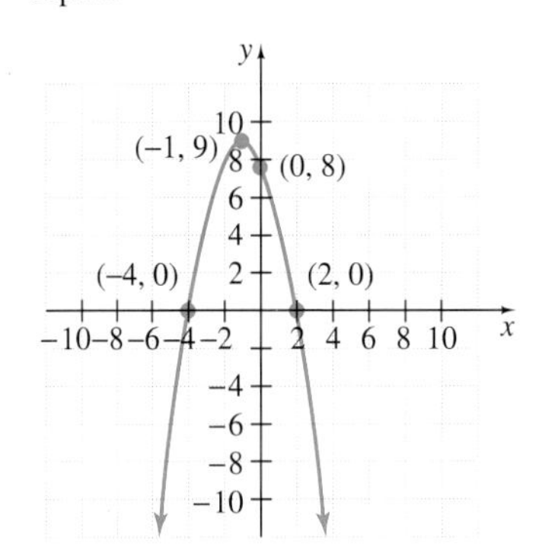

$$\frac{0}{-1} = \frac{-x^2}{-1} - \frac{2x}{-1} + \frac{3}{-1} \quad \text{Divide both sides by } -1.$$

$$0 = x^2 + 2x - 3 \quad \text{Simplify.}$$

$$0 = (x + 3)(x - 1) \quad \text{Factor.}$$

$$x + 3 = 0 \quad \text{or} \quad x - 1 = 0 \quad \text{Set each factor equal to 0.}$$

$$x = -3 \qquad x = 1 \quad \text{Solve.}$$

The x-intercepts are -3 and 1. We use these points to graph the parabola.

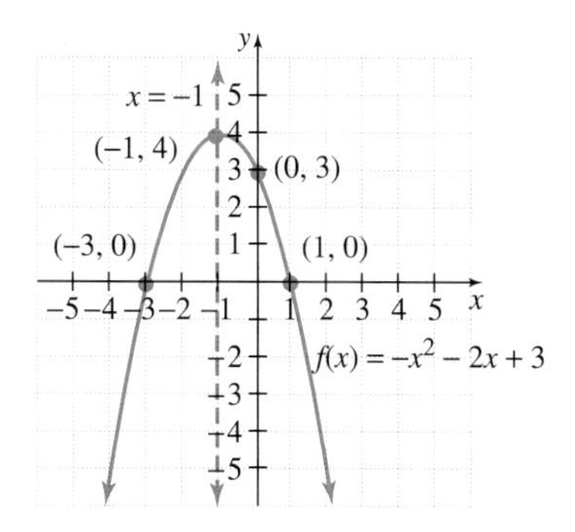

B DERIVING A FORMULA FOR FINDING THE VERTEX

Recall from Section 5.8 that we introduced a formula for finding the vertex of a parabola. Now that we have practiced completing the square, we will show that the x-coordinate of the vertex of the graph of $f(x)$ or $y = ax^2 + bx + c$ can be found by the formula $x = \frac{-b}{2a}$. To do so, we complete the square on x and write the equation in the form $y = (x - h)^2 + k$.

First we get the x-variable terms alone by subtracting c from both sides.

$$y = ax^2 + bx + c$$

$$y - c = ax^2 + bx$$

$$y - c = a\left(x^2 + \frac{b}{a}x\right) \quad \text{Factor } a \text{ from the terms } ax^2 + bx.$$

Now we add the square of half of $\frac{b}{a}$, or $\left(\frac{b}{2a}\right)^2 = \frac{b^2}{4a^2}$, to the right side inside the parentheses. Because of the factor a, what we really added is $a\left(\frac{b^2}{4a^2}\right)$ and this must be added to the left side.

$$y - c + a\left(\frac{b^2}{4a^2}\right) = a\left(x^2 + \frac{b}{a}x + \frac{b^2}{4a^2}\right)$$

$$y - c + \frac{b^2}{4a} = a\left(x + \frac{b}{2a}\right)^2 \quad \text{Simplify the left side and factor the right side.}$$

$$y = a\left(x + \frac{b}{2a}\right)^2 + c - \frac{b^2}{4a} \quad \text{Add } c \text{ to both sides and subtract } \frac{b^2}{4a} \text{ from both sides.}$$

Compare this form with $f(x)$ or $y = a(x - h)^2 + k$ and see that h is $\frac{-b}{2a}$, which means that the x-coordinate of the vertex of the graph of $f(x) = ax^2 + bx + c$ is $\frac{-b}{2a}$.

Let's use this vertex formula in the margin to find the vertex of the parabola we graphed in Example 1.

Example 4 Find the vertex of the graph of $f(x) = x^2 - 4x - 12$.

Solution: In the quadratic function $f(x) = x^2 - 4x - 12$, notice that $a = 1$, $b = -4$, and $c = -12$. Then

$$\frac{-b}{2a} = \frac{-(-4)}{2(1)} = 2$$

The x-value of the vertex is 2. To find the corresponding $f(x)$ or y-value, find $f(2)$. Then

$$f(2) = 2^2 - 4(2) - 12 = 4 - 8 - 12 = -16$$

The vertex is $(2, -16)$. These results agree with our findings in Example 1.

VERTEX FORMULA

The graph of $f(x) = ax^2 + bx + c$, when $a \neq 0$, is a parabola with vertex

$$\left(\frac{-b}{2a}, f\left(\frac{-b}{2a}\right)\right)$$

Practice Problem 4

Find the vertex of the graph of $f(x) = x^2 - 4x - 5$. Compare your result with the result of Practice Problem 1.

C FINDING MINIMUM AND MAXIMUM VALUES

The quadratic function whose graph is a parabola that opens upward has a minimum value, and the quadratic function whose graph is a parabola that opens downward has a maximum value. The $f(x)$ or y-value of the vertex is the minimum or maximum value of the function.

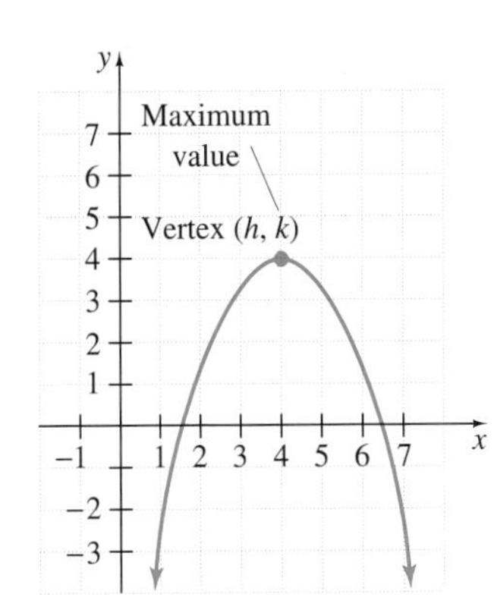

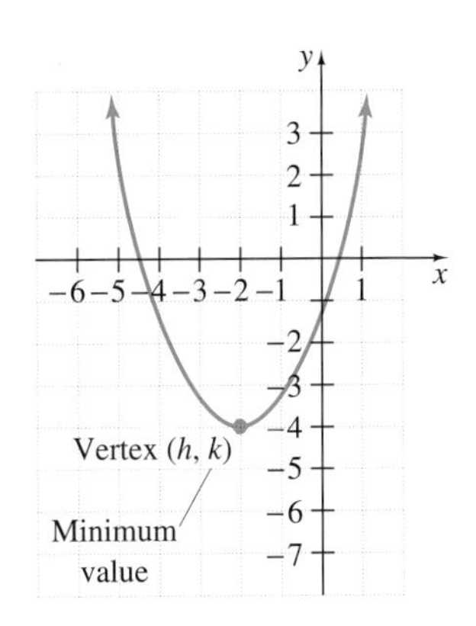

Recall from Section 8.2 that the discriminant, $b^2 - 4ac$, tells us how many solutions the quadratic equation $0 = ax^2 + bx + c$ has. It also tells us how many x-intercepts the graph of a quadratic equation $y = ax^2 + bx + c$ has.

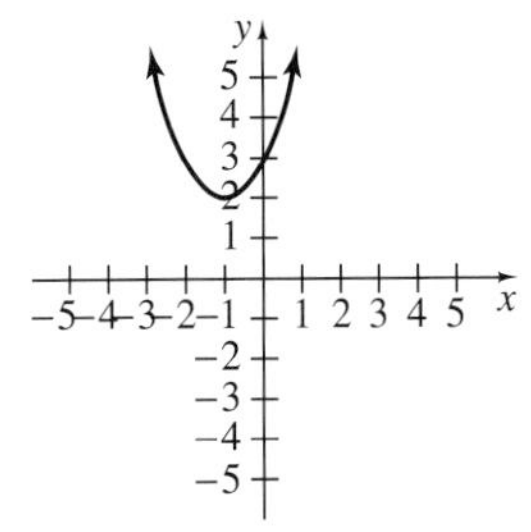

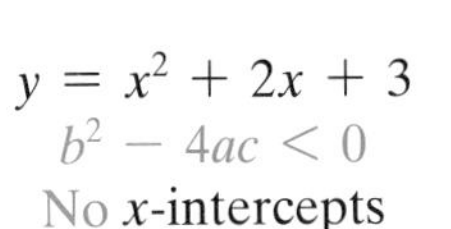

$y = x^2 + 2x + 3$
$b^2 - 4ac < 0$
No x-intercepts

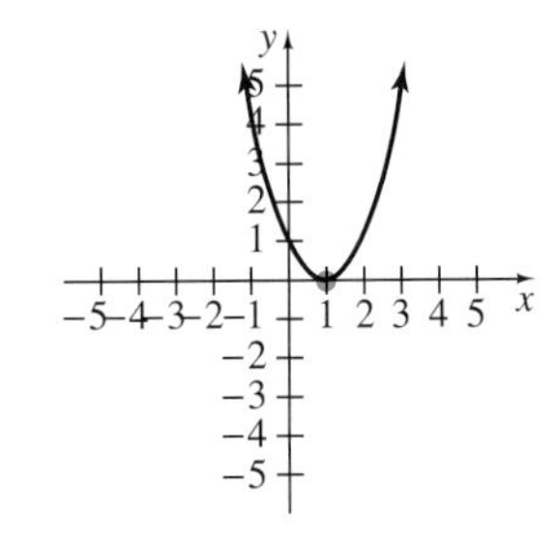

$y = x^2 - 2x + 1$
$b^2 - 4ac = 0$
One x-intercept

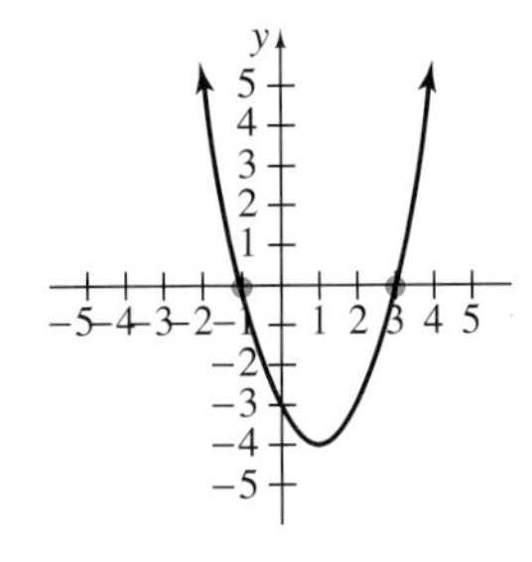

$y = x^2 - 2x - 3$
$b^2 - 4ac > 0$
Two x-intercepts

Answer

4. $(2, -9)$

✓ Concept Check

Without making any calculations, tell whether the graph of $f(x) = 7 - x - 0.3x^2$ has a maximum value or a minimum value. Explain your reasoning.

Practice Problem 5

An object is thrown upward from the top of a 100-foot cliff. Its height above ground after t seconds is given by the function $f(t) = -16t^2 + 10t + 100$. Find the maximum height of the object and the number of seconds it took for the object to reach its maximum height.

Try the Concept Check in the margin.

Example 5 Finding Maximum Height

A rock is thrown upward from the ground. Its height in feet above ground after t seconds is given by the function $f(t) = -16t^2 + 20t$. Find the maximum height of the rock and the number of seconds it took for the rock to reach its maximum height.

Solution:

1. UNDERSTAND. The maximum height of the rock is the largest value of $f(t)$. Since the function $f(t) = -16t^2 + 20t$ is a quadratic function, its graph is a parabola. It opens downward since $-16 < 0$. Thus, the maximum value of $f(t)$ is the $f(t)$ or y-value of the vertex of its graph.

2. TRANSLATE. To find the vertex (h, k), notice that for $f(t) = -16t^2 + 20t$, $a = -16$, $b = 20$, and $c = 0$. We will use these values and the vertex formula

$$\left(\frac{-b}{2a}, f\left(\frac{-b}{2a}\right)\right)$$

3. SOLVE. $h = \dfrac{-b}{2a} = \dfrac{-20}{-32} = \dfrac{5}{8}$

$$f\left(\frac{5}{8}\right) = -16\left(\frac{5}{8}\right)^2 + 20\left(\frac{5}{8}\right) = -16\left(\frac{25}{64}\right) + \frac{25}{2} = -\frac{25}{4} + \frac{50}{4} = \frac{25}{4}$$

4. INTERPRET. The graph of $f(t)$ is a parabola opening downward with vertex $\left(\frac{5}{8}, \frac{25}{4}\right)$. This means that the rock's maximum height is $\frac{25}{4}$ feet, or $6\frac{1}{4}$ feet, which was reached in $\frac{5}{8}$ second.

Answers

5. Maximum height: $101\frac{9}{16}$ feet in $\frac{5}{16}$ seconds

✓ Concept Check: $f(x)$ has a maximum value since it opens downward.

Name ______________________ Section ______ Date ________

Exercise Set 8.6

A B *Find the vertex of the graph of each quadratic function by completing the square or using the vertex formula. See Examples 1 through 4.*

1. $f(x) = x^2 + 8x + 7$

2. $f(x) = x^2 + 6x + 5$

3. $f(x) = -x^2 + 10x + 5$

4. $f(x) = -x^2 - 8x + 2$

5. $f(x) = 5x^2 - 10x + 3$

6. $f(x) = -3x^2 + 6x + 4$

7. $f(x) = -x^2 + x + 1$

8. $f(x) = x^2 - 9x + 8$

Match each function with its graph. See Examples 1 through 4.

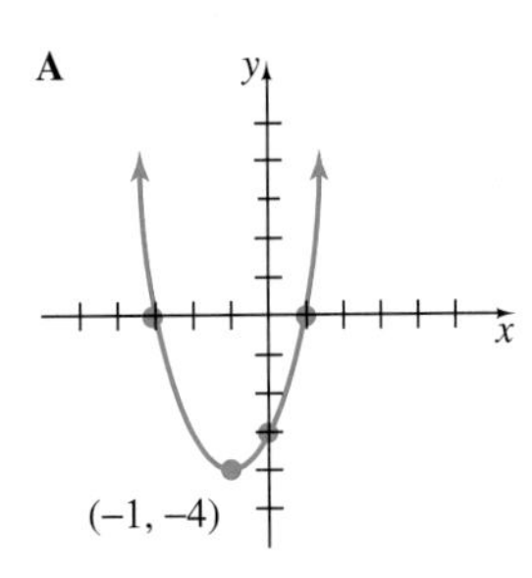

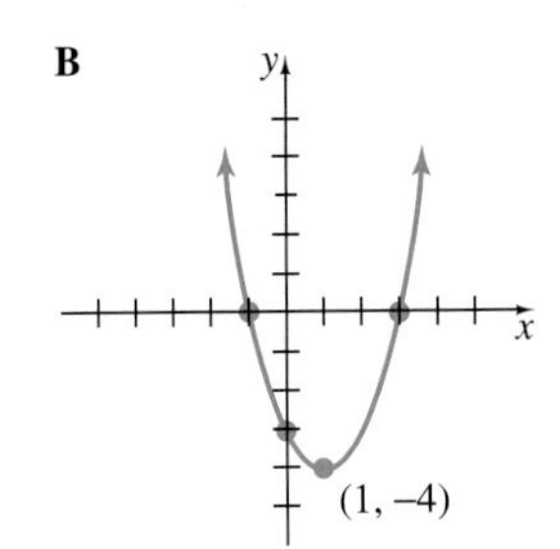

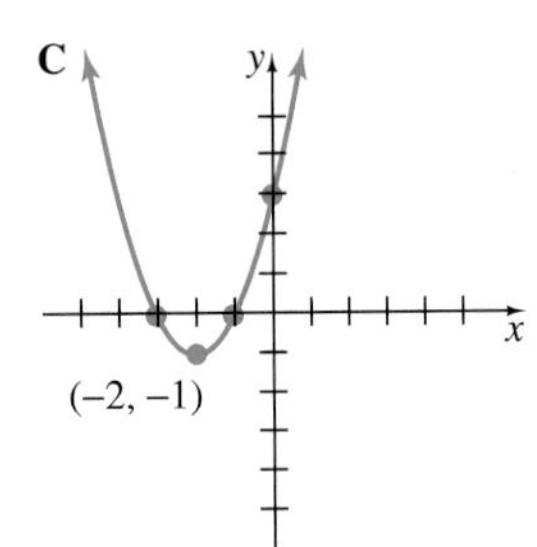

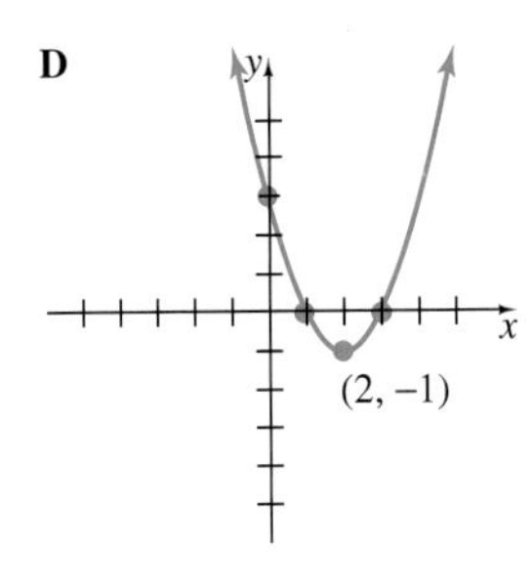

9. $f(x) = x^2 - 4x + 3$

10. $f(x) = x^2 + 2x - 3$

11. $f(x) = x^2 - 2x - 3$

12. $f(x) = x^2 + 4x + 3$

Answers

1. ________

2. ________

3. ________

4. ________

5. ________

6. ________

7. ________

8. ________

9. ________

10. ________

11. ________

12. ________

Name ______________________________

13. ______________

14. ______________

15. ______________

16. ______________

17. ______________

18. ______________

19. ______________

20. ______________

Find the vertex of the graph of each quadratic function. Determine whether the graph opens upward or downward, find any intercepts, and graph the function. See Examples 1 through 4.

13. $f(x) = x^2 + 4x - 5$

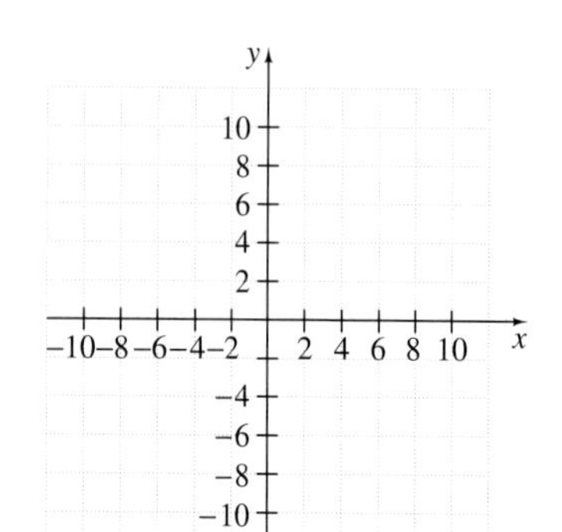

14. $f(x) = x^2 + 2x - 3$

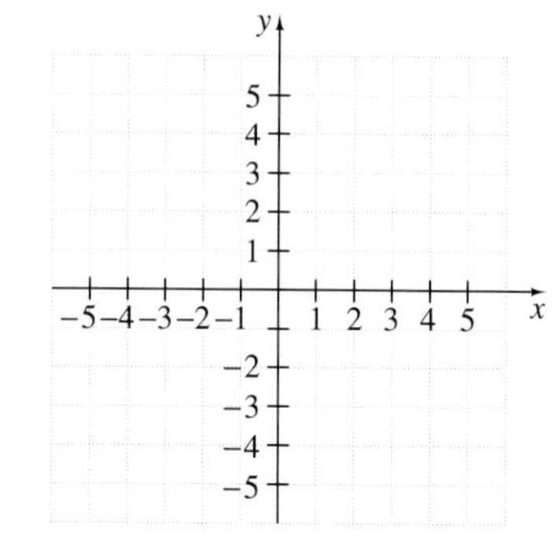

15. $f(x) = -x^2 + 2x - 1$

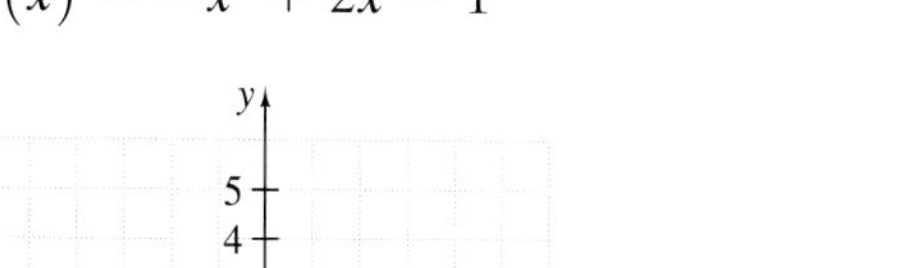

16. $f(x) = -x^2 + 4x - 4$

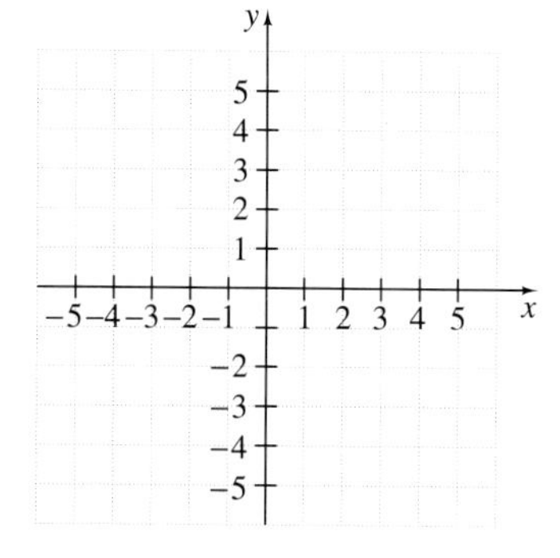

17. $f(x) = x^2 - 4$

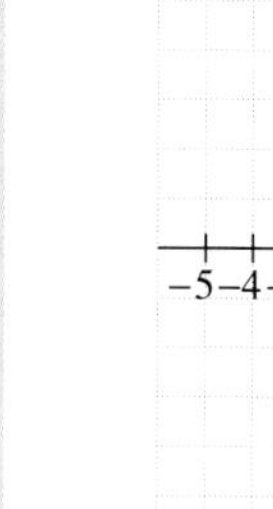

18. $f(x) = x^2 - 1$

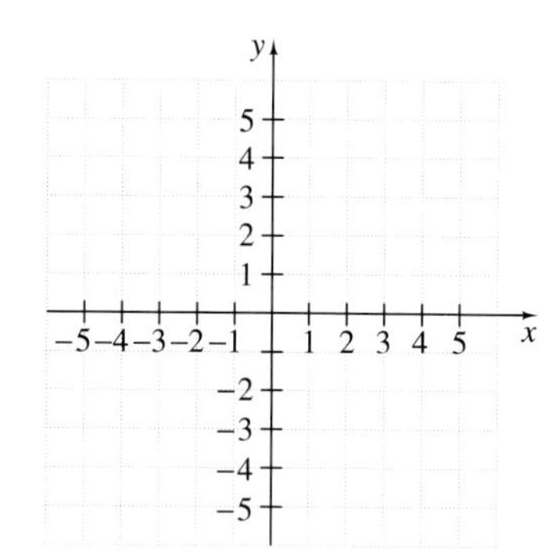

19. $f(x) = 4x^2 + 4x - 3$

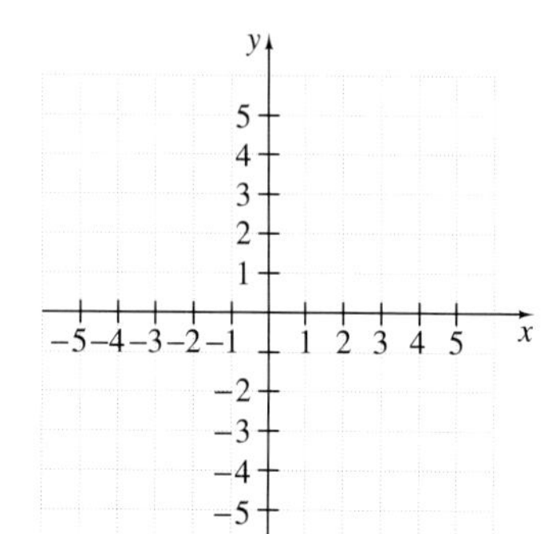

20. $f(x) = 2x^2 - x - 3$

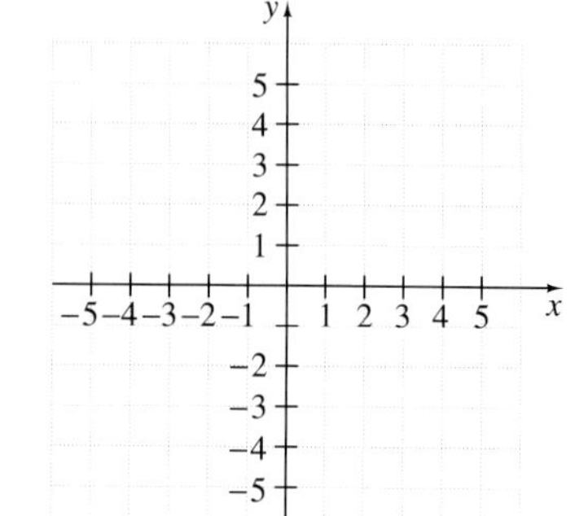

Name ______________________________

21. $f(x) = \frac{1}{2}x^2 + 4x + \frac{15}{2}$

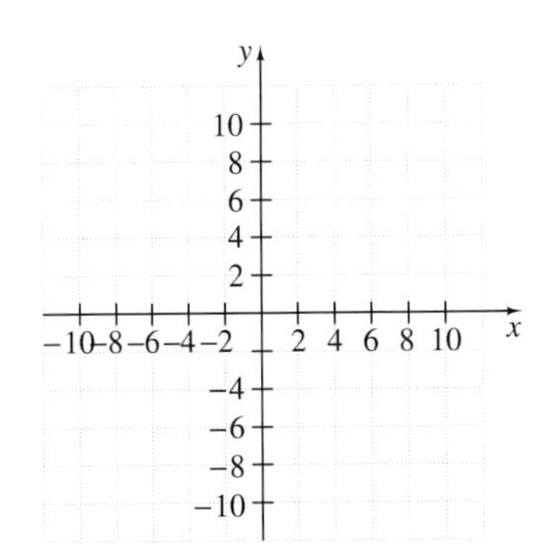

22. $f(x) = \frac{1}{5}x^2 + 2x + \frac{9}{5}$

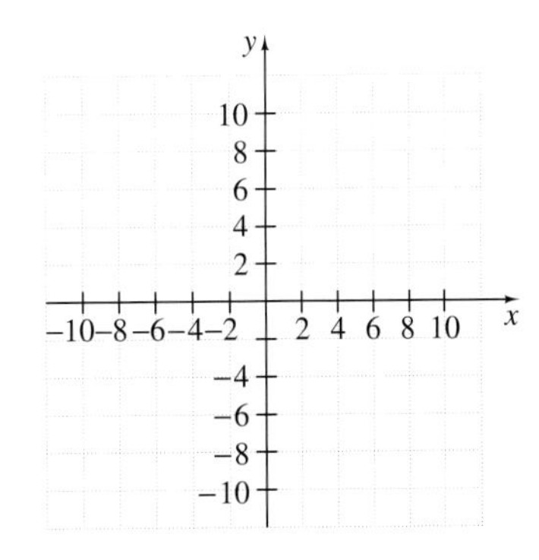

23. $f(x) = x^2 - 4x + 5$

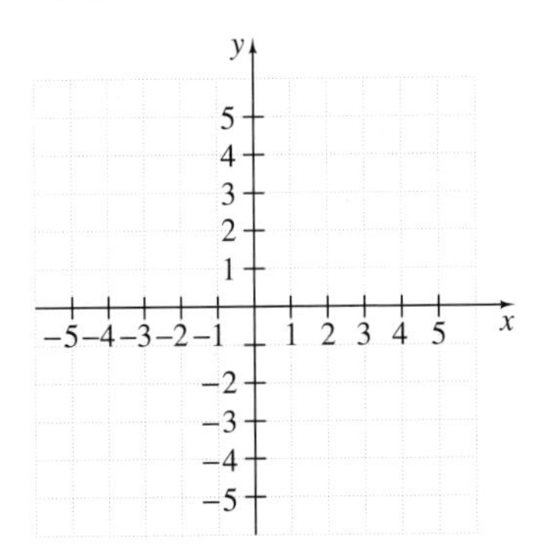

24. $f(x) = x^2 - 6x + 11$

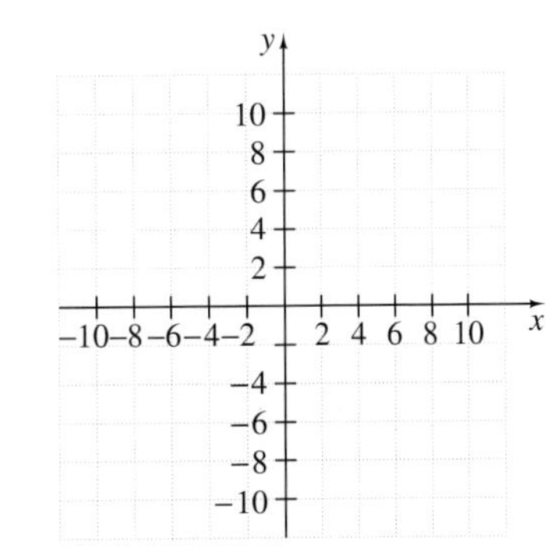

25. $f(x) = 2x^2 + 4x + 5$

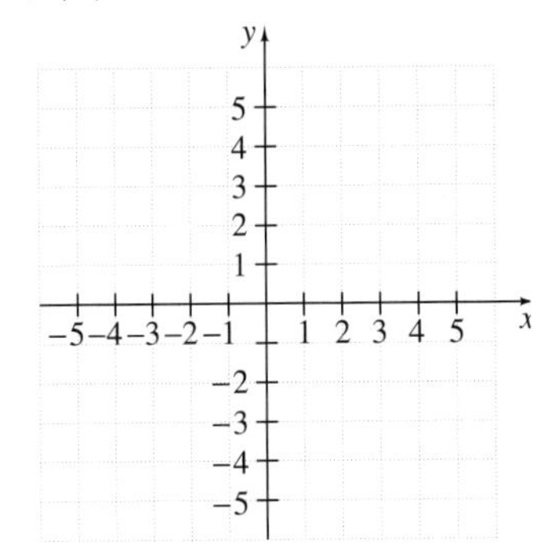

26. $f(x) = 3x^2 + 12x + 16$

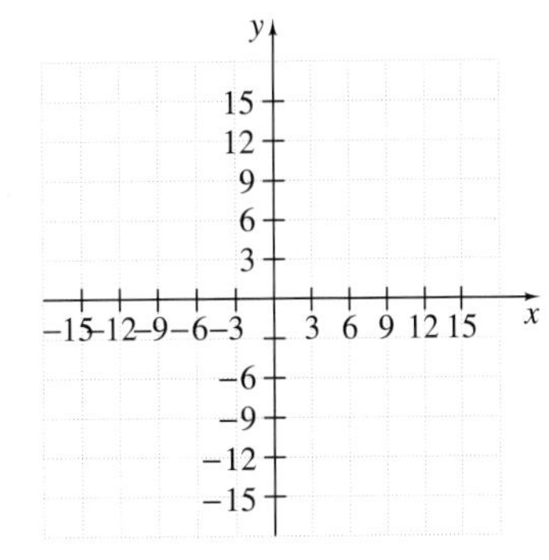

27. $f(x) = -2x^2 + 12x$

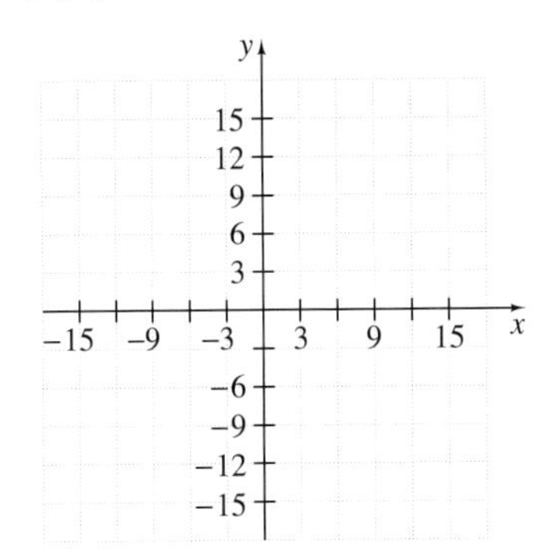

28. $f(x) = -4x^2 + 8x$

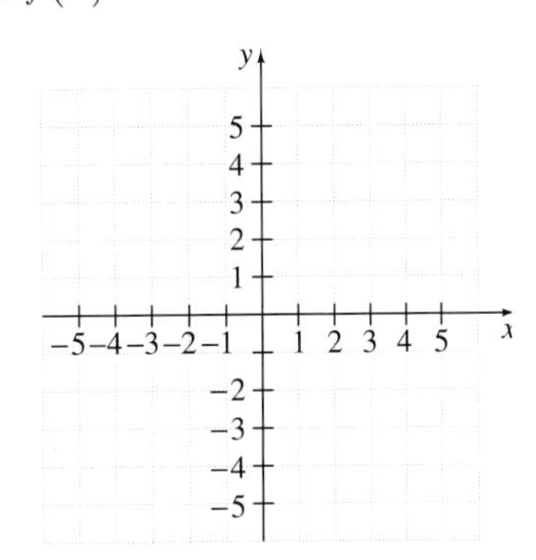

21. ______________

22. ______________

23. ______________

24. ______________

25. ______________

26. ______________

27. ______________

28. ______________

29. ______

Name ______

30. a. ______

b. ______

31. ______

32. a. ______

b. ______

33. ______

34. ______

35. ______

36. ______

37. ______

38. ______

C *Solve. See Example 5.*

29. If a projectile is fired straight upward from the ground with an initial speed of 96 feet per second, then its height h in feet after t seconds is given by the function $h(t) = -16t^2 + 96t$. Find the maximum height of the projectile.

30. The cost C in dollars of manufacturing x bicycles at Holladay's Production Plant is given by the function $C(x) = 2x^2 - 800x + 92{,}000$.

a. Find the number of bicycles that must be manufactured to minimize the cost.

b. Find the minimum cost.

31. If Rheam Gaspar throws a ball upward with an initial speed of 32 feet per second, then its height h in feet after t seconds is given by the function $h(t) = -16t^2 + 32t$. Find the maximum height of the ball.

32. The Utah Ski Club sells calendars to raise money. The profit P, in cents, from selling x calendars is given by the function $P(x) = 360x - x^2$.

a. Find how many calendars must be sold to maximize profit.

b. Find the maximum profit.

33. Find two numbers whose sum is 60 and whose product is as large as possible. [*Hint:* Let x and $60 - x$ be the two positive numbers. Their product can be described by the function $f(x) = x(60 - x)$.]

34. Find two numbers whose sum is 11 and whose product is as large as possible. (Use the hint for Exercise 33.)

35. Find two numbers whose difference is 10 and whose product is as small as possible. (Use the hint for Exercise 33.)

36. Find two numbers whose difference is 8 and whose product is as small as possible.

37. The length and width of a rectangle must have a sum of 40. Find the dimensions of the rectangle that will have the maximum area. (Use the hint for Exercise 33.)

38. The length and width of a rectangle must have a sum of 50. Find the dimensions of the rectangle that will have maximum area.

Name ______________________________

Review and Preview

Find the vertex of the graph of each function. See Section 8.5.

39. $f(x) = x^2 + 2$ **40.** $f(x) = (x - 3)^2$ **41.** $g(x) = x + 2$

42. $h(x) = x - 3$ **43.** $f(x) = (x + 5)^2 + 2$ **44.** $f(x) = 2(x - 3)^2 + 2$

45. $f(x) = 3(x - 4)^2 + 1$ **46.** $f(x) = (x + 1)^2 + 4$

Combining Concepts

Find the vertex of the graph of each quadratic function. Determine whether the graph opens upward or downward, find the y-intercept, approximate the x-intercepts to one decimal place, and graph the function.

47. $f(x) = x^2 + 10x + 15$

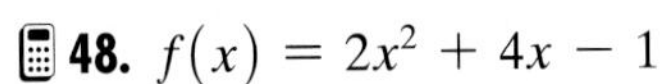

48. $f(x) = 2x^2 + 4x - 1$

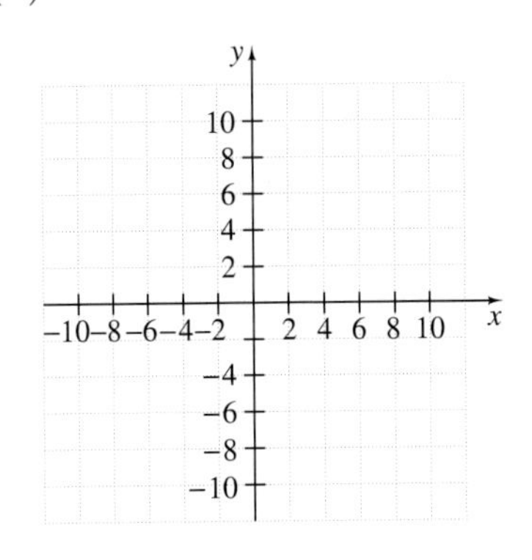

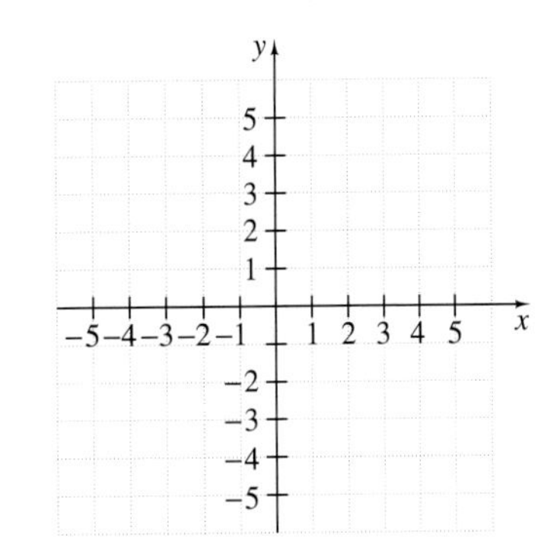

Use a grapher to verify the graph of each exercise.

49. Exercise 21. **50.** Exercise 22.

Find the maximum or minimum value of each function. Approximate to two decimal places.

51. $f(x) = 2.3x^2 - 6.1x + 3.2$ **52.** $f(x) = 7.6x^2 + 9.8x - 2.1$

39. ______

40. ______

41. ______

42. ______

43. ______

44. ______

45. ______

46. ______

47. ______

48. ______

49. see calculator screen

50. see calculator screen

51. ______

52. ______

53. a. ______

b. ______

c. ______

54. a. ______

b. ______

c. ______

Name ______

53. The number of inmates in custody in U.S. prisons and jails can be modeled by the quadratic function $p(x) = -2464.4x^2 + 99{,}699.5x + 1{,}148{,}702$, where $p(x)$ is the number of inmates and x is the number of years after 1990. (*Source:* based on data from the Bureau of Justice Statistics, U.S. Department of Justice, 1990–1997)

a. Will this function have a maximum or a minimum? How can you tell?

b. According to this model, when will the number of prison inmates in custody in the United States be at its maximum/minimum?

c. What is the maximum number of inmates predicted for that year?

54. Methane is a gas produced by landfills, natural gas systems, and coal mining that contributes to the greenhouse effect and global warming. Methane emissions in the United States can be modeled by the quadratic function $f(x) = -1.25x^2 + 5.89x + 170.34$, where $f(x)$ is the amount of methane produced in million metric tons and x is the number of years after 1993. (*Source:* based on data from the U.S. Environmental Protection Agency, 1993–1996)

a. If this trend continues, what will U.S. emissions of methane be in 2000?

b. In what year were methane emissions in the United States at their maximum? Round to the nearest whole year.

c. What was the maximum methane emissions level?

Name ______________________ Section ________ Date ________

Chapter 8 Activity

Recognizing Linear and Quadratic Models

This activity may be completed by working in groups or individually.

We have seen in this chapter and in previous chapters that data can be modeled by both linear models and quadratic models. However, when we are given a set of data to model, how can we tell which type of model—linear or quadratic—is appropriate? The best answer depends on looking at a scatter diagram of the data. If the plotted data points fall roughly on a line, a linear model is usually the better choice. If the plotted data points seem to fall on a definite curve or if a maximum or minimum point is apparent, a quadratic model is usually the better choice.

One of the sets of data shown in the tables is best modeled by a linear function and one is best modeled by a quadratic function. In each case, the variable x represents the number of years after 1987.

Annual Motor Vehicle Accident Injury Rate per 100,000 U.S. Population						
Year	1989	1991	1992	1993	1994	1996
x	2	4	5	6	7	9
Injury Rate, y	1330	1228	1204	1221	1254	1323

(*Source:* National Highway Traffic Safety Administration)

Number of Domestic Wal-Mart Stores and Supercenters						
Year	1987	1989	1990	1992	1994	1997
x	0	2	3	5	7	10
Number of Stores, y	980	1262	1405	1724	2022	2304

(*Source:* Wal-Mart Stores, Inc.)

1. Make a scatter diagram for each set of data. Which type of model should be used for each set of data?

2. For the set of data that you have determined to be linear, find a linear function that fits the data points. Explain the method that you used. (*Hint:* See the Focus on Business and Career in Chapter 3 or the Focus on the Real World in Chapter 4 for more information.)

3. For the set of data that you have determined to be quadratic, identify the point on your scatter diagram that appears to be the vertex of the parabola. Use the coordinates of this vertex in the quadratic model $f(x) = a(x - h)^2 + k$.

4. Solve for the remaining unknown constant in the quadratic model by substituting the coordinates for another data point into the function. Write the final form of the quadratic model for this data set.

5. Use your models to find the motor vehicle accident injury rate and number of domestic Wal-Mart stores in 1995.

6. (Optional) For each set of data, enter the data from the table into a graphing utility and use either the linear regression feature or the quadratic regression feature to find an appropriate function that models the data.* Compare these functions with the ones you found by hand. How are they alike or different?

*To find out more about using your graphing utility to find a regression equation, consult the user's manual for your graphing utility.

CHAPTER 8 HIGHLIGHTS

DEFINITIONS AND CONCEPTS	EXAMPLES
SECTION 8.1 SOLVING QUADRATIC EQUATIONS BY COMPLETING THE SQUARE	
SQUARE ROOT PROPERTY If b is a real number and if $a^2 = b$, then $a = \pm\sqrt{b}$.	Solve: $(x + 3)^2 = 14$ $x + 3 = \pm\sqrt{14}$ $x = -3 \pm \sqrt{14}$
TO SOLVE A QUADRATIC EQUATION IN x BY COMPLETING THE SQUARE	Solve: $3x^2 - 12x - 18 = 0$
Step 1. If the coefficient of x^2 is not 1, divide both sides of the equation by the coefficient of x^2.	**1.** $x^2 - 4x - 6 = 0$
Step 2. Get the variable terms alone.	**2.** $x^2 - 4x = 6$
Step 3. Complete the square by adding the square of half of the coefficient of x to both sides.	**3.** $\frac{1}{2}(-4) = -2$ and $(-2)^2 = 4$ $x^2 - 4x + 4 = 6 + 4$
Step 4. Write the resulting trinomial as the square of a binomial.	**4.** $(x - 2)^2 = 10$
Step 5. Use the square root property.	**5.** $x - 2 = \pm\sqrt{10}$ $x = 2 \pm \sqrt{10}$
SECTION 8.2 SOLVING QUADRATIC EQUATIONS BY THE QUADRATIC FORMULA	
QUADRATIC FORMULA A quadratic equation written in the form $ax^2 + bx + c = 0$ has solutions $x = \dfrac{-b \pm \sqrt{b^2 - 4ac}}{2a}$	Solve: $x^2 - x - 3 = 0$ $a = 1, b = -1, c = -3$ $x = \dfrac{-(-1) \pm \sqrt{(-1)^2 - 4(1)(-3)}}{2 \cdot 1}$ $x = \dfrac{1 \pm \sqrt{13}}{2}$
SECTION 8.3 SOLVING EQUATIONS BY USING QUADRATIC METHODS	
Substitution is often helpful in solving an equation that contains a repeated variable expression.	Solve: $(2x + 1)^2 - 5(2x + 1) + 6 = 0$ Let $m = 2x + 1$. Then $m^2 - 5m + 6 = 0$ Let $m = 2x + 1$. $(m - 3)(m - 2) = 0$ $m = 3$ or $m = 2$ $2x + 1 = 3$ $2x + 1 = 2$ Substitute back. $x = 1$ $x = \frac{1}{2}$

Section 8.4 Nonlinear Inequalities in One Variable

To Solve a Polynomial Inequality

Step 1. Write the inequality in standard form.
Step 2. Solve the related equation.
Step 3. Use solutions from Step 2 to separate the number line into regions.
Step 4. Use a test point to determine whether values in each region satisfy the original inequality.
Step 5. Write the solution set as the union of regions whose test point values are solutions.

Solve: $x^2 \geq 6x$

1. $x^2 - 6x \geq 0$

2. $x^2 - 6x = 0$

$x(x - 6) = 0$

$x = 0$ or $x = 6$

3. Number line with regions A, B, C separated at 0 and 6.

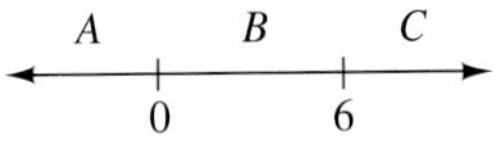

4.

Region	Test Point Value	$x^2 \geq 6x$	Result
A	-2	$(-2)^2 \geq 6(-2)$	True.
B	1	$1^2 \geq 6(1)$	False.
C	7	$7^2 \geq 6(7)$	True.

5. Number line shaded to the left of 0 (bracket at 0) and to the right of 6 (bracket at 6).

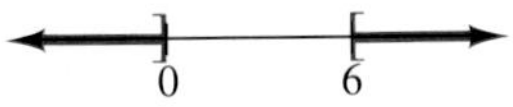

The solution set is $(-\infty, 0] \cup [6, \infty)$.

To Solve a Rational Inequality

Step 1. Solve for values that make all denominators 0.
Step 2. Solve the related equation.
Step 3. Use solutions from Steps 1 and 2 to separate the number line into regions.
Step 4. Use a test point to determine whether values in each region satisfy the original inequality.
Step 5. Write the solution set as the union of regions whose test point value is a solution.

Solve: $\dfrac{6}{x - 1} < -2$

1. $x - 1 = 0$ Set denominator equal to 0.

$x = 1$

2. $\dfrac{6}{x - 1} = -2$

$6 = -2(x - 1)$ Multiply by $(x - 1)$.

$6 = -2x + 2$

$4 = -2x$

$-2 = x$

3. Number line with regions A, B, C separated at -2 and 1.

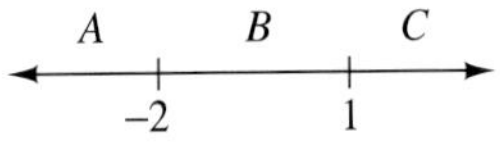

4. Only a test value from region B satisfies the original inequality.

5. Number line shaded between -2 and 1 (parentheses at both ends).

The solution set is $(-2, 1)$.

SECTION 8.5 QUADRATIC FUNCTIONS AND THEIR GRAPHS

GRAPH OF A QUADRATIC FUNCTION

The graph of a quadratic function written in the form $f(x) = a(x - h)^2 + k$ is a parabola with vertex (h, k). If $a > 0$, the parabola opens upward; if $a < 0$, the parabola opens downward. The axis of symmetry is the line whose equation is $x = h$.

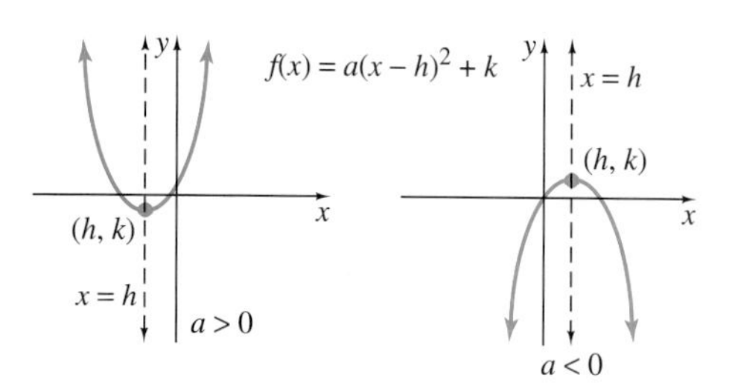

Graph: $g(x) = 3(x - 1)^2 + 4$

The graph is a parabola with vertex $(1, 4)$ and axis of symmetry $x = 1$. Since $a = 3$ is positive, the graph opens upward.

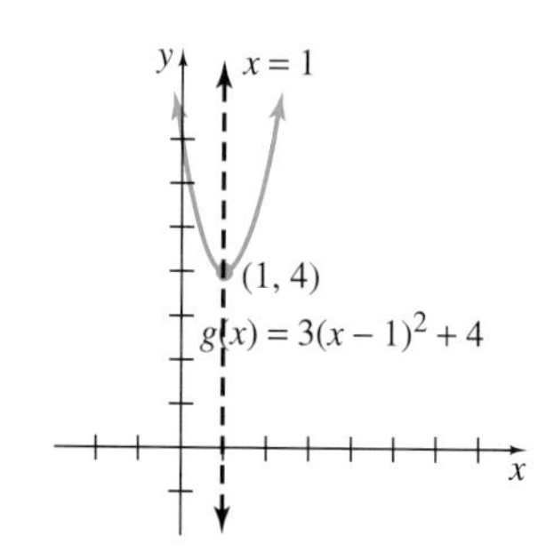

SECTION 8.6 FURTHER GRAPHING OF QUADRATIC FUNCTIONS

The graph of $f(x) = ax^2 + bx + c$, $a \neq 0$, is a parabola with vertex

$$\left(\frac{-b}{2a}, f\left(\frac{-b}{2a}\right)\right).$$

Graph: $f(x) = x^2 - 2x - 8$. Find the vertex and x- and y-intercepts.

$$\frac{-b}{2a} = \frac{-(-2)}{2 \cdot 1} = 1$$

$$f(1) = 1^2 - 2(1) - 8 = -9$$

The vertex is $(1, -9)$.

$$0 = x^2 - 2x - 8$$
$$0 = (x - 4)(x + 2)$$
$$x = 4 \text{ or } x = -2$$

The x-intercept points are $(4, 0)$ and $(-2, 0)$.

$$f(0) = 0^2 - 2 \cdot 0 - 8 = -8$$

The y-intercept point is $(0, -8)$.

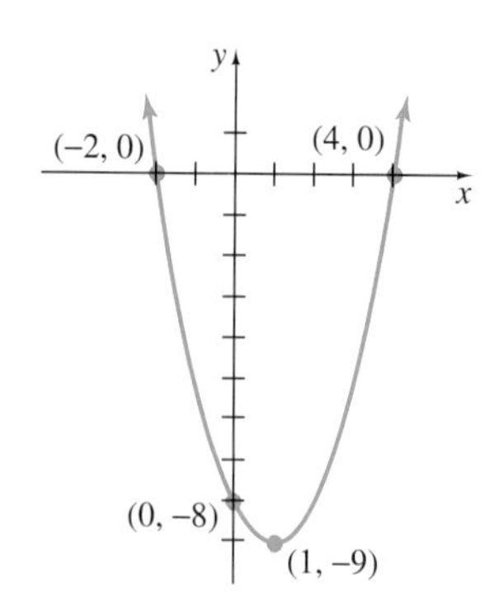

Name ______________________ Section __________ Date __________

Chapter 8 Review

(8.1) *Solve by factoring.*

1. $x^2 - 15x + 14 = 0$

2. $x^2 - x - 30 = 0$

3. $10x^2 = 3x + 4$

4. $7a^2 = 29a + 30$

Use the square root property to solve each equation.

5. $4m^2 = 196$

6. $9y^2 = 36$

7. $(9n + 1)^2 = 9$

8. $(5x - 2)^2 = 2$

Solve by completing the square.

9. $z^2 + 3z + 1 = 0$

10. $x^2 + x + 7 = 0$

11. $(2x + 1)^2 = x$

12. $(3x - 4)^2 = 10x$

13. If P dollars are originally invested, the formula $A = P(1 + r)^2$ gives the amount A in an account paying interest rate r compounded annually after 2 years. Find the interest rate r such that \$2500 increases to \$2717 in 2 years. Round the result to the nearest hundredth of a percent.

14. Two ships leave a port at the same time and travel at the same speed. One ship is traveling due north and the other due east. In a few hours, the ships are 150 miles apart. How many miles has each ship traveled? Give an exact answer and a one-decimal-place approximation.

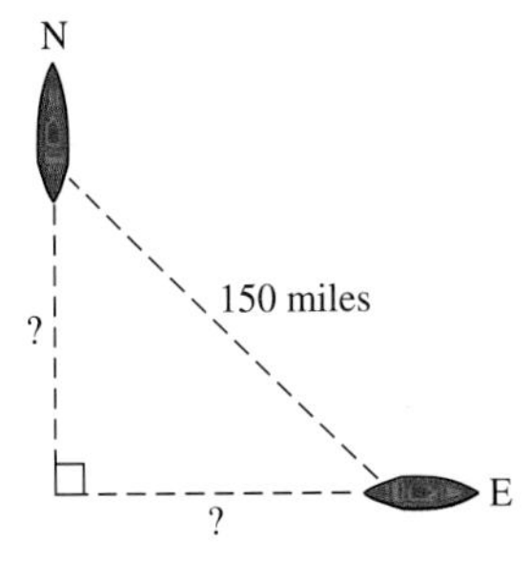

(8.2) *If the discriminant of a quadratic equation has the given value, determine the number and type of solutions of the equation.*

15. -8

16. 48

17. 100

18. 0

Use the quadratic formula to solve each equation.

19. $x^2 - 16x + 64 = 0$

20. $x^2 + 5x = 0$

21. $x^2 + 11 = 0$

22. $2x^2 + 3x = 5$

Name ______________________________

23. $6x^2 + 7 = 5x$

24. $9a^2 + 4 = 2a$

25. $(5a - 2)^2 - a = 0$

26. $(2x - 3)^2 = x$

27. Cadets graduating from military school usually toss their hats high into the air at the end of the ceremony. One cadet threw his hat so that its distance $d(t)$ in feet above the ground t seconds after it was thrown was $d(t) = -16t^2 + 30t + 6$.

a. Find the distance above the ground of the hat 1 second after it was thrown.

b. Find the time it takes the hat to hit the ground. Give an exact time and a one-decimal-place approximation.

28. The hypotenuse of an isosceles right triangle is 6 centimeters longer than either of the legs. Find the length of the legs.

x

x

(8.3) *Solve each equation.*

29. $x^3 = 27$

30. $y^3 = -64$

31. $\frac{5}{x} + \frac{6}{x - 2} = 3$

32. $\frac{7}{8} = \frac{8}{x^2}$

33. $x^4 - 21x^2 - 100 = 0$

34. $5(x + 3)^2 - 19(x + 3) = 4$

35. $x^{2/3} - 6x^{1/3} + 5 = 0$

36. $x^{2/3} - 6x^{1/3} = -8$

37. $a^6 - a^2 = a^4 - 1$

38. $y^{-2} + y^{-1} = 20$

39. Two postal workers, Jerome Grant and Tim Bozik, can sort a stack of mail in 5 hours. Working alone, Tim can sort the mail in 1 hour less time than Jerome can. Find the time that each postal worker can sort the mail alone. Round the result to one decimal place.

40. A negative number decreased by its reciprocal is $-\frac{24}{5}$. Find the number.

(8.4) *Solve each inequality for x. Write each solution set in interval notation.*

41. $2x^2 - 50 \leq 0$

42. $\frac{1}{4}x^2 < \frac{1}{16}$

43. $(2x - 3)(4x + 5) \geq 0$

44. $(x^2 - 16)(x^2 - 1) > 0$

45. $\frac{x - 5}{x - 6} < 0$

46. $\frac{x(x + 5)}{4x - 3} \geq 0$

Name ______________________________

47. $\dfrac{(4x + 3)(x - 5)}{x(x + 6)} > 0$

48. $(x + 5)(x - 6)(x + 2) \leq 0$

49. $x^3 + 3x^2 - 25x - 75 > 0$

50. $\dfrac{x^2 + 4}{3x} \leq 1$

51. $\dfrac{(5x + 6)(x - 3)}{x(6x - 5)} < 0$

52. $\dfrac{3}{x - 2} > 2$

(8.5) *Graph each function. Label the vertex and the axis of symmetry of each graph.*

53. $f(x) = x^2 - 4$

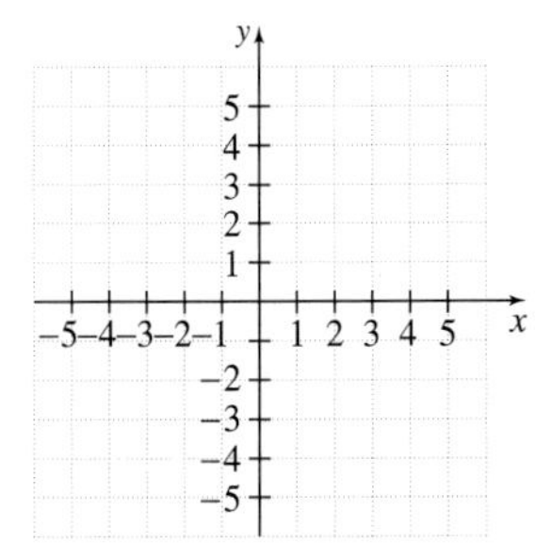

54. $g(x) = x^2 + 7$

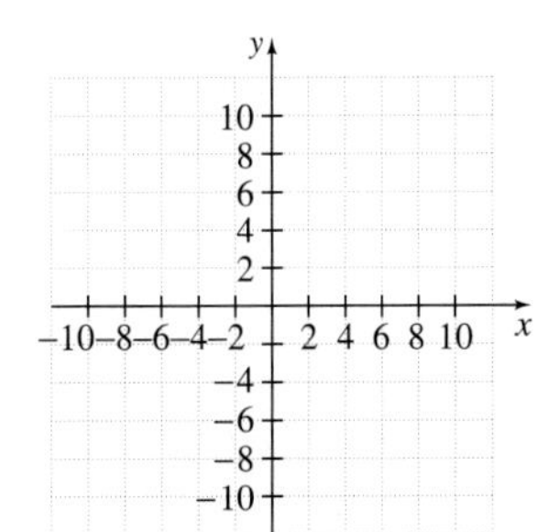

55. $H(x) = 2x^2$

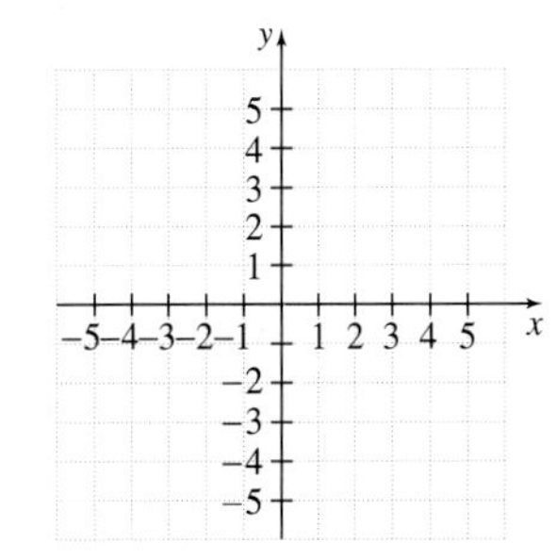

56. $h(x) = -\frac{1}{3}x^2$

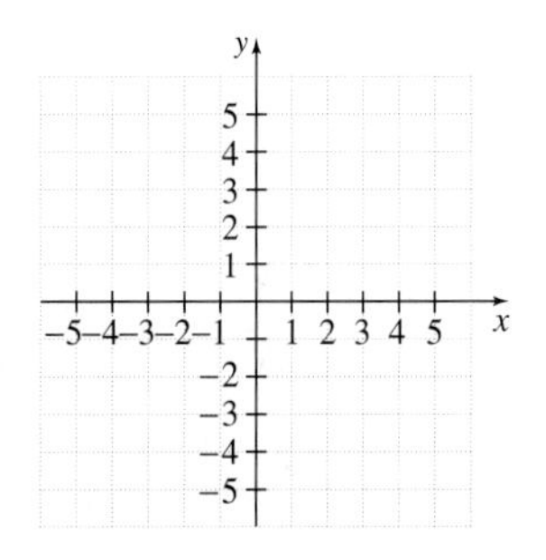

57. $F(x) = (x - 1)^2$

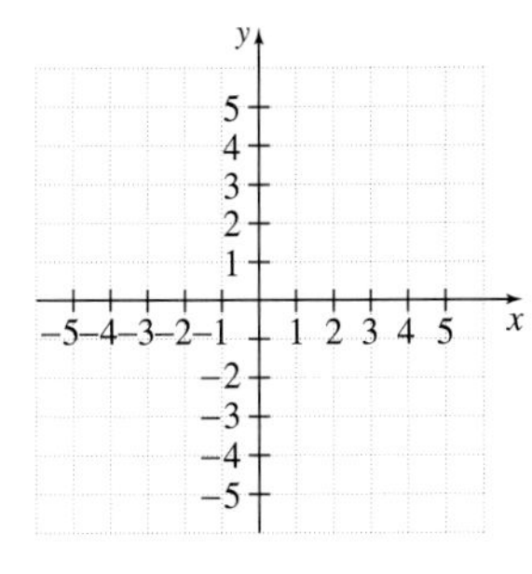

58. $G(x) = (x + 5)^2$

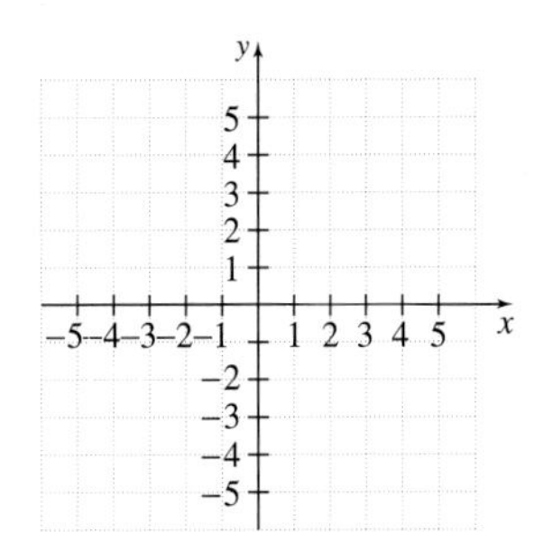

59. $f(x) = (x - 4)^2 - 2$

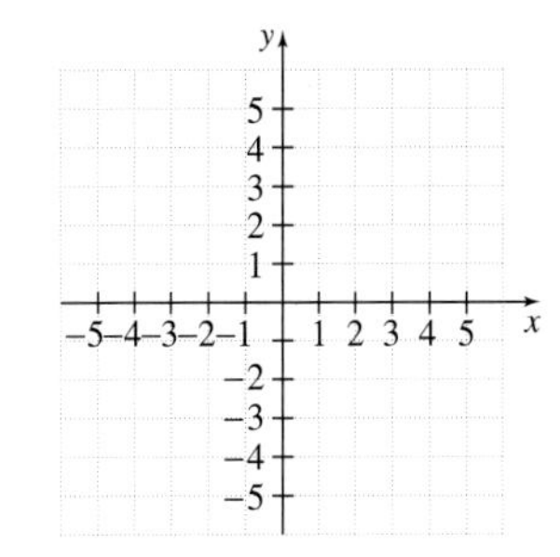

60. $f(x) = -3(x - 1)^2 + 1$

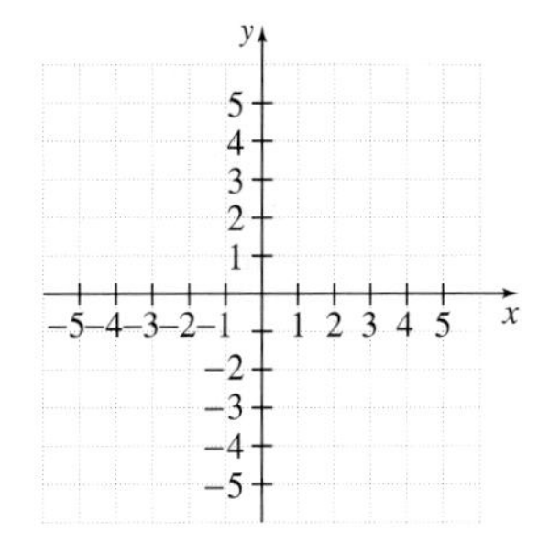

(8.6) *Graph each function. Find the vertex and any intercepts of each graph.*

61. $f(x) = x^2 + 10x + 25$

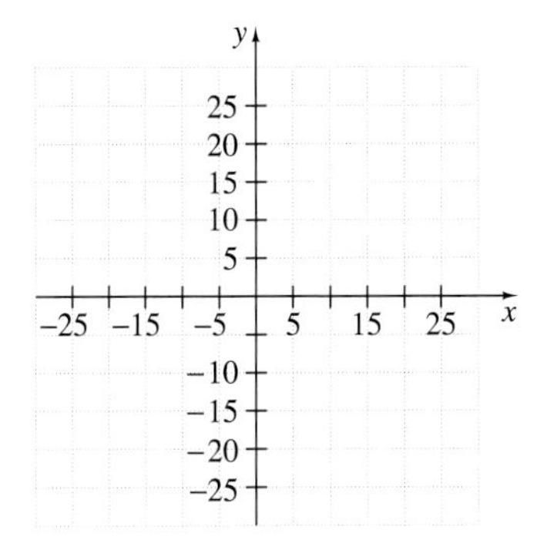

62. $f(x) = -x^2 + 6x - 9$

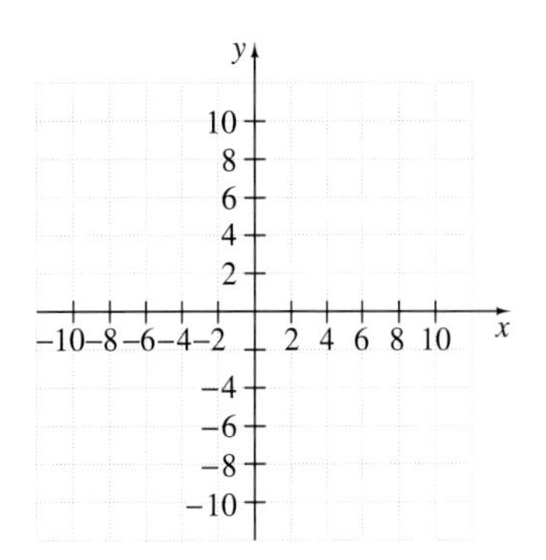

Name ______________________________

63. $f(x) = 4x^2 - 1$

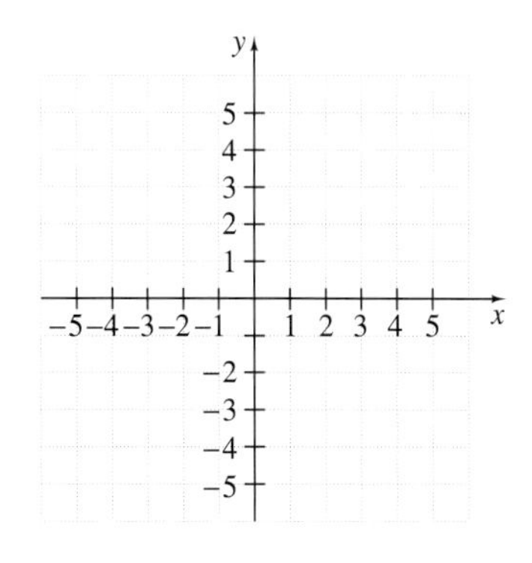

64. $f(x) = -5x^2 + 5$

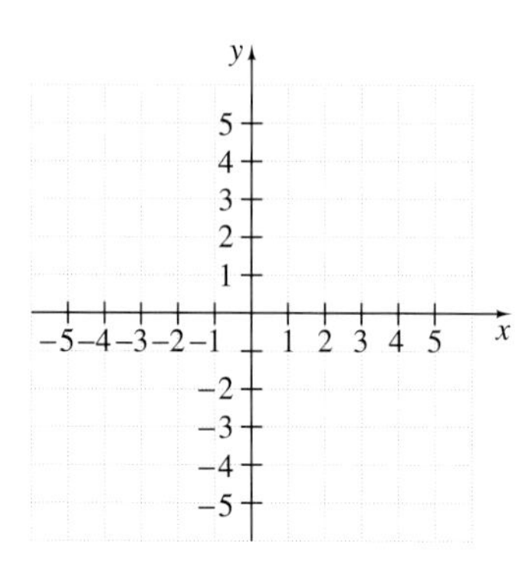

65. Find the vertex of the graph of $f(x) = -3x^2 - 5x + 4$. Determine whether the graph opens upward or downward, find the y-intercept, approximate the x-intercepts to one decimal place, and graph the function.

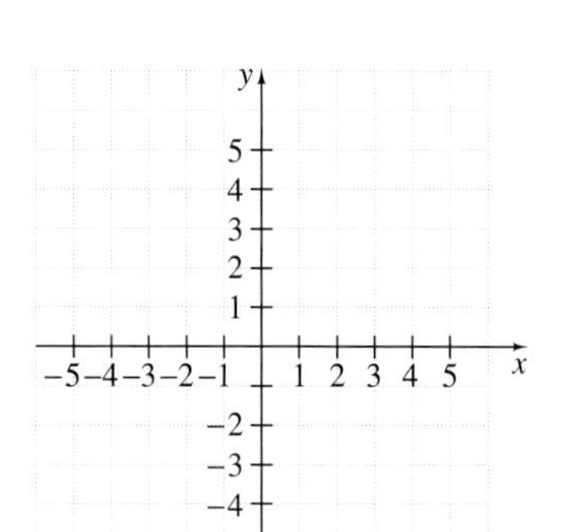

66. The function $h(t) = -16t^2 + 120t + 300$ gives the height in feet of a projectile fired from the top of a building in t seconds.

a. When will the object reach a height of 350 feet? Round your answer to one decimal place.

b. Explain why part (a) has two answers.

67. Find two numbers whose sum is 420 and whose product is as large as possible.

Name ____________________ Section ______ Date ________

CHAPTER 8 TEST

Solve each equation.

1. $5x^2 - 2x = 7$

2. $(x + 1)^2 = 10$

3. $m^2 - m + 8 = 0$

4. $u^2 - 6u + 2 = 0$

5. $7x^2 + 8x + 1 = 0$

6. $a^2 - 3a = 5$

7. $\dfrac{4}{x + 2} + \dfrac{2x}{x - 2} = \dfrac{6}{x^2 - 4}$

8. $x^4 - 8x^2 - 9 = 0$

9. $x^6 + 1 = x^4 + x^2$

10. $(x + 1)^2 - 15(x + 1) + 56 = 0$

Solve by completing the square.

11. $x^2 - 6x = -2$

12. $2a^2 + 5 = 4a$

Solve each inequality. Write the solution set in interval notation.

13. $2x^2 - 7x > 15$

14. $(x^2 - 16)(x^2 - 25) > 0$

15. $\dfrac{5}{x + 3} < 1$

16. $\dfrac{7x - 14}{x^2 - 9} \leq 0$

Graph each function. Label the vertex for each graph.

17. $f(x) = 3x^2$

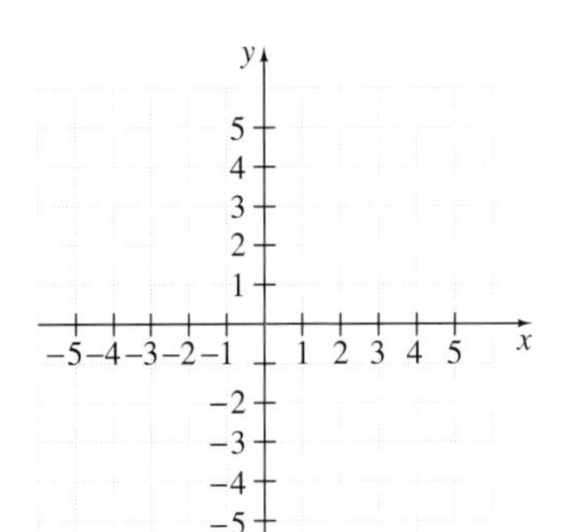

18. $G(x) = -2(x - 1)^2 + 5$

ANSWERS

1. ____________

2. ____________

3. ____________

4. ____________

5. ____________

6. ____________

7. ____________

8. ____________

9. ____________

10. ____________

11. ____________

12. ____________

13. ____________

14. ____________

15. ____________

16. ____________

17. see graph

18. see graph

19. see graph

20. see graph

21.

22.

23. a.

b.

Name ____________________

Graph each function. Find and label the vertex, y-intercept, and x-intercepts (if any) for each graph.

19. $h(x) = x^2 - 4x + 4$

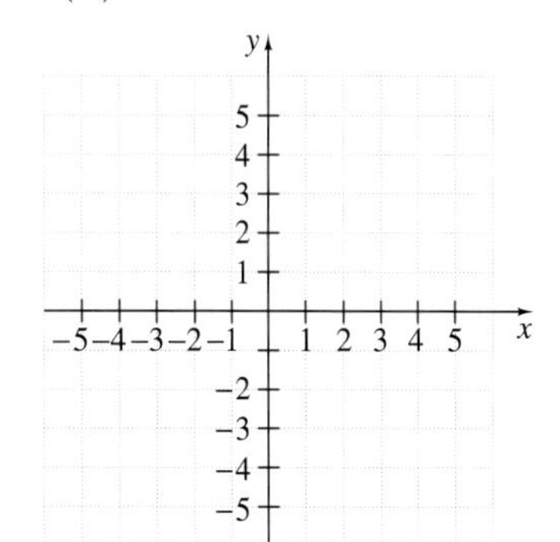

20. $F(x) = 2x^2 - 8x + 9$

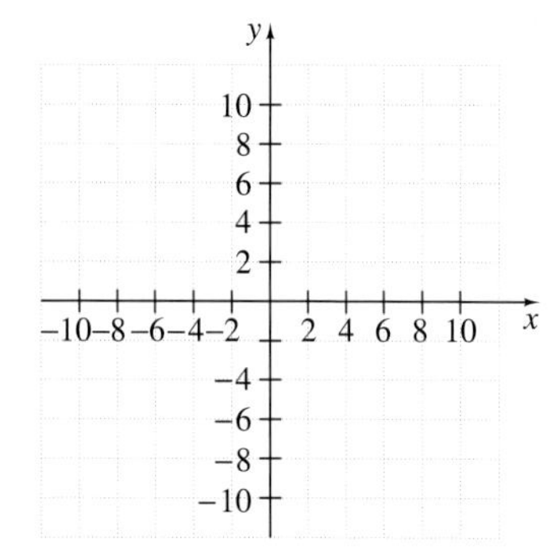

21. A 10-foot ladder is leaning against a house. The distance from the bottom of the ladder to the house is 4 feet less than the distance from the top of the ladder to the ground. Find how far the top of the ladder is from the ground. Give an exact answer and a one-decimal-place approximation.

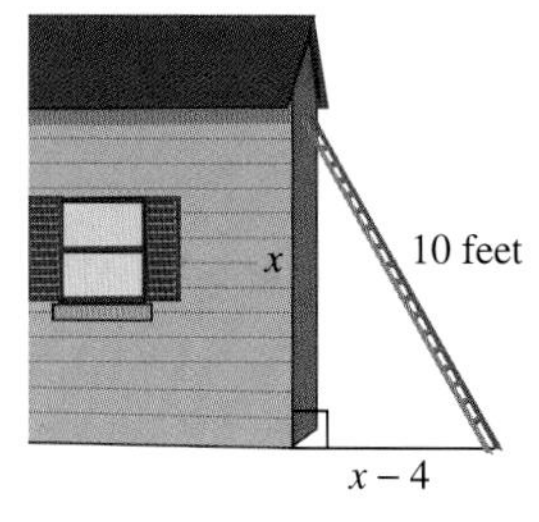

22. Dave and Sandy Hartranft can paint a room together in 4 hours. Working alone, Dave can paint the room in 2 hours less time than Sandy can. Find how long it takes Sandy to paint the room alone.

23. A stone is thrown upward from a bridge. The stone's height in feet, $s(t)$, above the water t seconds after the stone is thrown is given by the function $s(t) = -16t^2 + 32t + 256$.

a. Find the maximum height of the stone.

b. Find the time it takes the stone to hit the water.

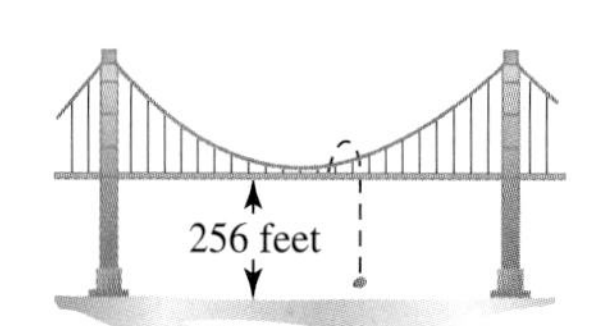

Name ______________________ Section ________ Date ________

Cumulative Review

1. Write 730,000 in scientific notation.

2. Solve: $|2x - 1| + 5 = 6$

3. Determine whether the relation $x = y^2$ is also a function.

4. Use the elimination method to solve the system: $\begin{cases} 3x + \dfrac{y}{2} = 2 \\ 6x + y = 5 \end{cases}$

5. Subtract $(10x^3 - 7x^2) - (4x^3 - 3x^2 + 2)$ vertically.

Use the FOIL order to multiply.

6. $(2x - 7)(3x - 4)$

7. $(3x + y)(5x - 2y)$

8. Divide $3x^4 + 2x^3 - 8x + 6$ by $x^2 - 1$.

9. Factor: $6x^2 - 3x^3$

10. Factor: $2(a + 3)^2 - 5(a + 3) - 7$

11. Factor: $3a^2x - 12abx + 12b^2x$

12. Solve: $2x^2 = \dfrac{17}{3}x + 1$

Multiply.

13. $\dfrac{3n + 1}{2n} \cdot \dfrac{2n - 4}{3n^2 - 2n - 1}$

14. $\dfrac{x^3 - 1}{-3x + 3} \cdot \dfrac{15x^2}{x^2 + x + 1}$

15. Add: $\dfrac{2}{x^2} + \dfrac{5}{3x^3}$

16. Simplify $\dfrac{\dfrac{x}{y^2} + \dfrac{1}{y}}{\dfrac{y}{x^2} + \dfrac{1}{x}}$

17. Solve: $\dfrac{2x}{x - 3} + \dfrac{6 - 2x}{x^2 - 9} = \dfrac{x}{x + 3}$

18. Solve $\dfrac{1}{x} + \dfrac{1}{y} = \dfrac{1}{z}$ for x.

Answers

1. ____________
2. ____________
3. ____________
4. ____________
5. ____________
6. ____________
7. ____________
8. ____________
9. ____________
10. ____________
11. ____________
12. ____________
13. ____________
14. ____________
15. ____________
16. ____________
17. ____________
18. ____________

19.

20.

21.

22.

23.

24.

25.

Name

19. The surface area of a cylinder varies jointly as its radius and height. Express surface area S in terms of radius r and height h.

Find the cube roots.

20. $\sqrt[3]{-64}$

21. $\sqrt[3]{\frac{8}{125}}$

Multiply.

22. $\sqrt[3]{4} \cdot \sqrt[3]{2}$

23. $\sqrt{\frac{2}{a}} \cdot \sqrt{\frac{b}{3}}$

Add or subtract.

24. $\sqrt[3]{54} - 5\sqrt[3]{16} + \sqrt[3]{2}$

25. $\sqrt[3]{\frac{7x}{8}} + 2\sqrt[3]{7x}$

Conic Sections

CHAPTER 9

In Chapter 8, we analyzed some of the important connections between a parabola and its equation. Parabolas are interesting in their own right but are more interesting still because they are part of a collection of curves known as conic sections. This chapter is devoted to quadratic equations in two variables and their conic section graphs: the parabola, circle, ellipse, and hyperbola.

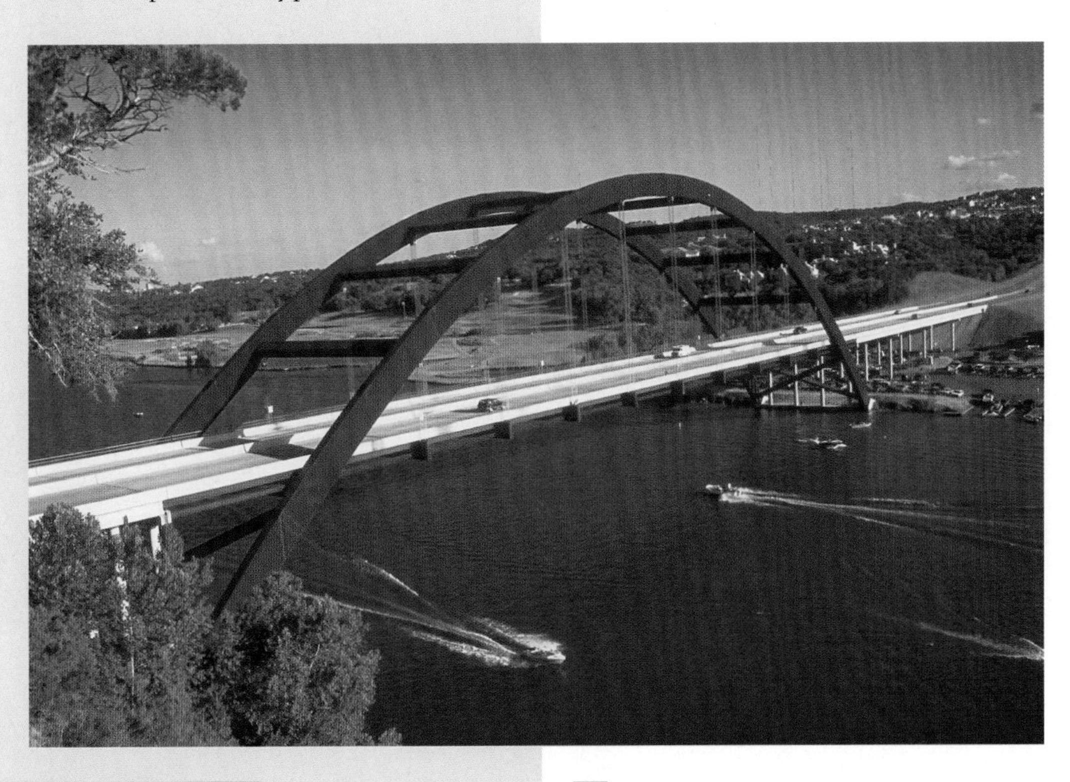

The shapes of conic sections are used in a variety of applications. They are used in architecture in the design of bridges, arches, and vaults. They are also used in astronomy to model the orbits of planets, comets, and satellites. Conic sections are used also in engineering in the design of certain gears and reflectors. In blueprints or diagrams of any of these situations, the exact shape of the conic section involved must be depicted. In Exercises 66 and 67 on page 698, we will see how architects and engineers use conic sections in their work.

ANSWERS

1. ______

2. ______

3. see graph; ______

4. see graph; ______

5. see graph; ______

6. see graph; ______

7. see graph; ______

8. see graph; ______

9. see graph; ______

10. see graph; ______

11. ______

12. see graph; ______

13. ______

14. ______

Name ______ Section ______ Date ______

CHAPTER 9 PRETEST

1. Find the distance between the points $(-4, 6)$ and $(1, 9)$.

2. Find the midpoint of the line segment whose end points are $(9, -15)$ and $(10, 22)$.

Sketch the graph of each equation.

3. $x = (y - 2)^2 + 1$

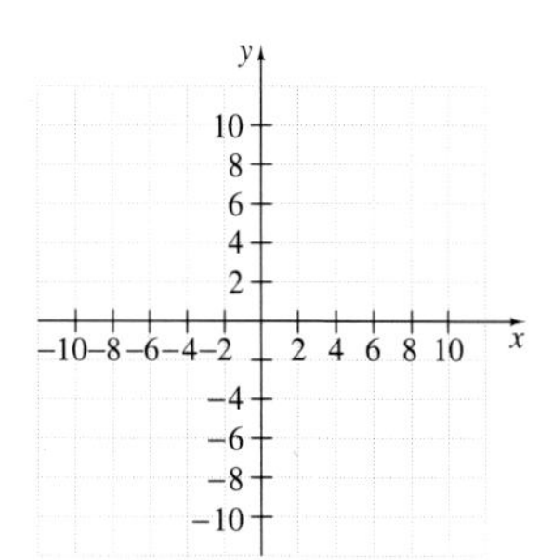

4. $y = x^2 + 4x - 5$

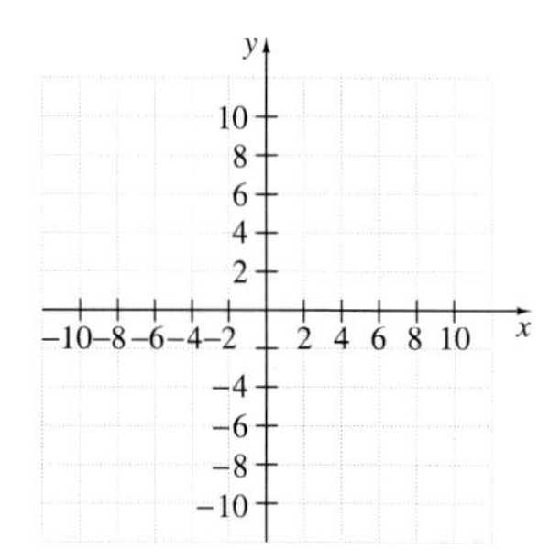

5. $x^2 + y^2 = 4$

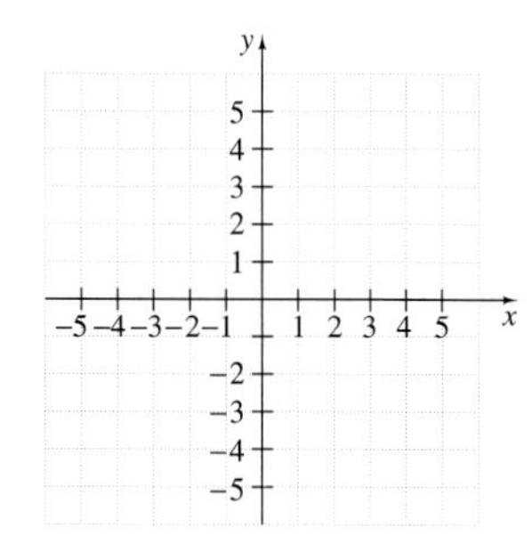

6. $(x + 2)^2 + (y - 3)^2 = 9$

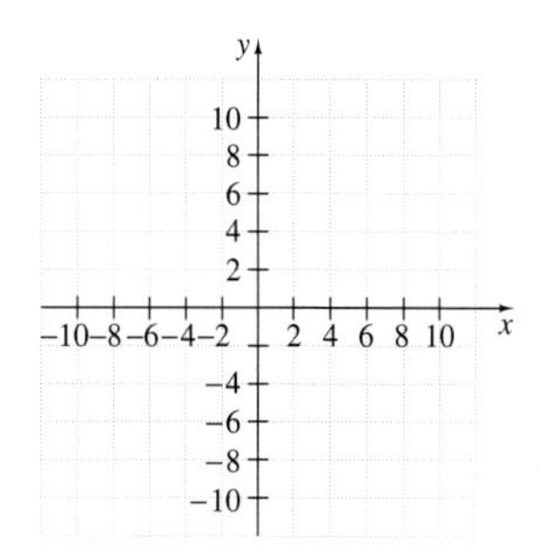

7. $4x^2 + 25y^2 = 100$

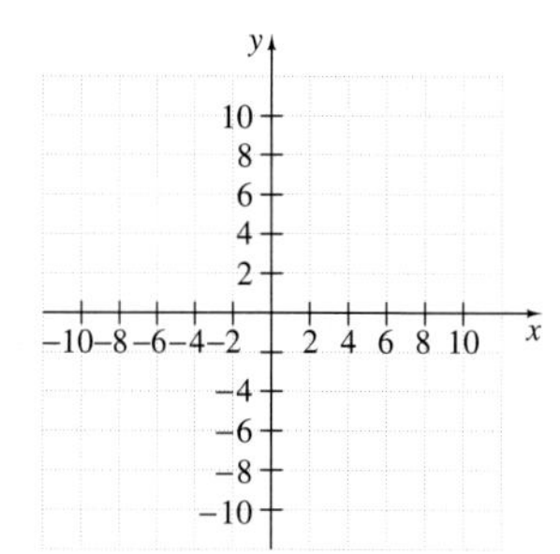

8. $\frac{x^2}{25} - \frac{y^2}{25} = 1$

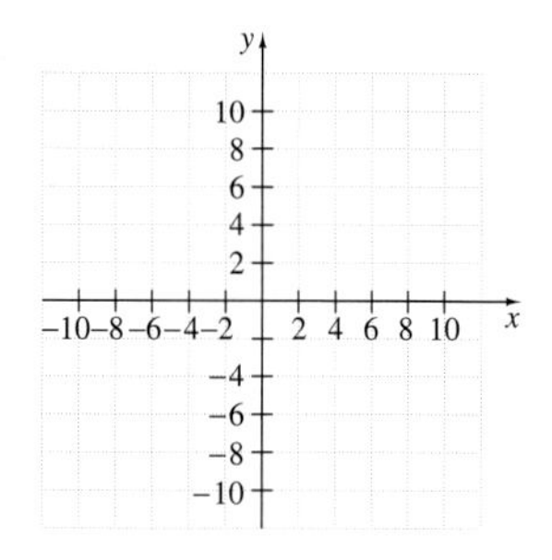

Graph each function.

9. $f(x) = |x| - 5$

10. $f(x) = \sqrt{x + 1} + 3$

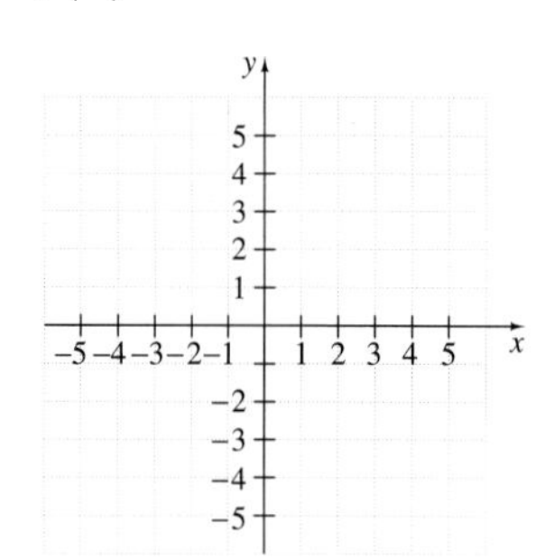

11. Solve the system:

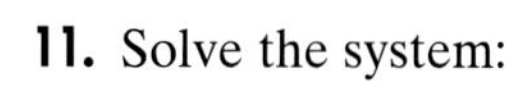

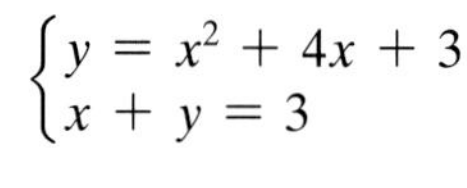

$$\begin{cases} y = x^2 + 4x + 3 \\ x + y = 3 \end{cases}$$

12. Graph the solution set of the system.

$$\begin{cases} y \geq \frac{1}{4}x^2 + 2 \\ x^2 + (y - 3)^2 \leq 4 \end{cases}$$

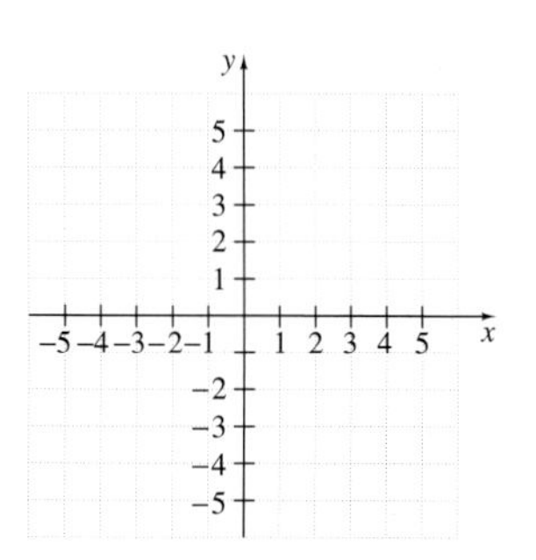

13. Write an equation of the circle with center $(2, 5)$ and radius 8.

14. Find the center and radius of the circle defined by $x^2 + y^2 + 6x - 8y = -16$.

9.1 THE PARABOLA AND THE CIRCLE

Conic sections derive their name because each conic section is the intersection of a right circular cone and a plane. The circle, parabola, ellipse, and hyperbola are the conic sections.

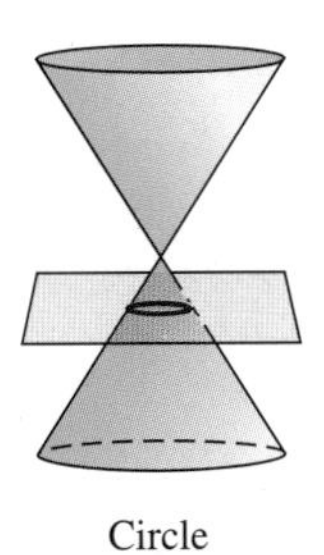

Circle

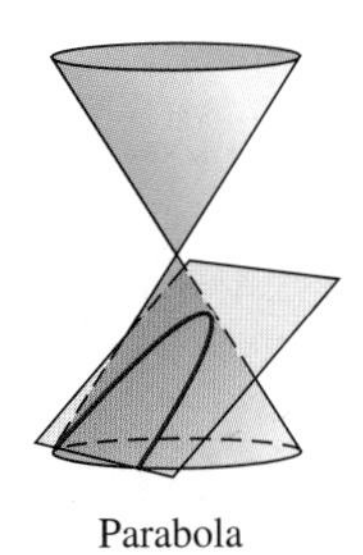

Parabola

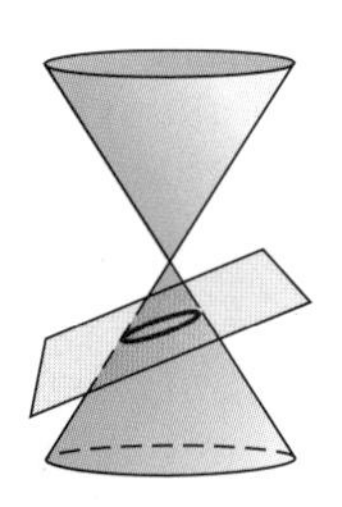

Ellipse

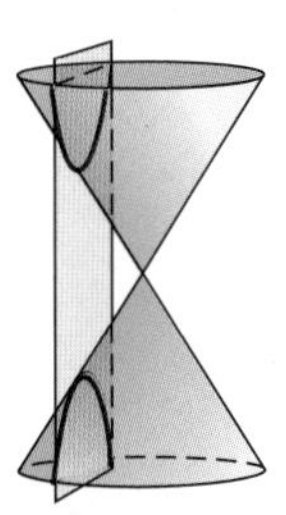

Hyperbola

A GRAPHING PARABOLAS

Thus far, we have seen that $f(x)$ or $y = a(x - h)^2 + k$ is the equation of a parabola that opens upward if $a > 0$ or downward if $a < 0$. Parabolas can also open left or right, or even on a slant. Equations of these parabolas are not functions of x, of course, since a parabola opening any way other than upward or downward fails the vertical line test. In this section, we introduce parabolas that open to the left and to the right. Parabolas opening on a slant will not be developed in this book.

Just as $y = a(x - h)^2 + k$ is the equation of a parabola that opens upward or downward, $x = a(y - k)^2 + h$ is the equation of a parabola that opens to the right or to the left. The parabola opens to the right if $a > 0$ and to the left if $a < 0$. The parabola has vertex (h, k), and its axis of symmetry is the line $y = k$.

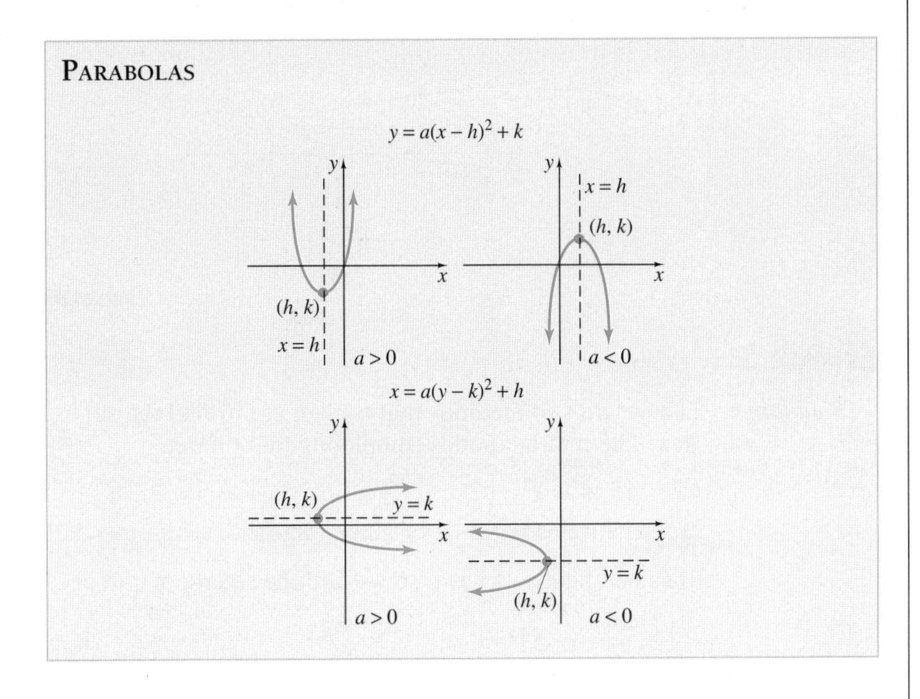

The forms $y = a(x - h)^2 + k$ and $x = a(y - k)^2 + h$ are called **standard forms**.

TRY THE CONCEPT CHECK IN THE MARGIN.

Objectives

A Graph parabolas of the forms $x = a(y - k)^2 + h$ and $y = a(x - h)^2 + k$.

B Use the distance formula and the midpoint formula.

C Graph circles of the form $(x - h)^2 + (y - k)^2 = r^2$.

D Write the equation of a circle, given its center and radius.

E Find the center and the radius of a circle, given its equation.

SSM

CD-ROM

Video 9.1

✓ CONCEPT CHECK

Does the graph of the parabola given by the equation $x = -3y^2$ open to the left, to the right, upward, or downward?

Answer

✓ Concept Check: opens to the left

Practice Problem 1

Graph: $x = 3y^2$

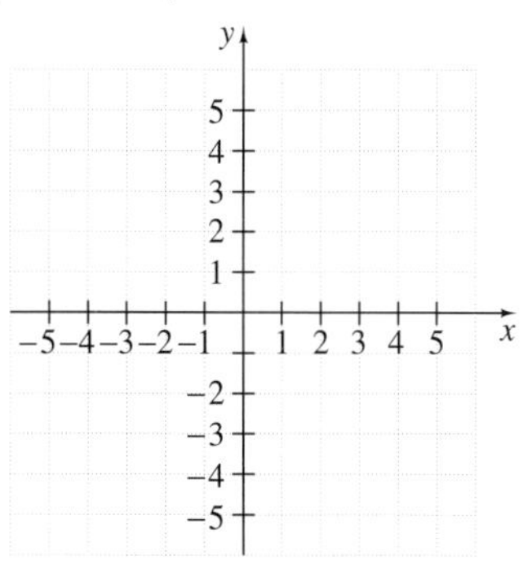

Example 1

Graph: $x = 2y^2$

Solution: Written in standard form, the equation $x = 2y^2$ is $x = 2(y - 0)^2 + 0$ with $a = 2$, $h = 0$, and $k = 0$. Its graph is a parabola with vertex $(0, 0)$, and its axis of symmetry is the line $y = 0$. Since $a > 0$, this parabola opens to the right. We use a table to obtain a few more ordered pair solutions to help us graph $x = 2y^2$.

x	y
8	−2
2	−1
0	0
2	1
8	2

Practice Problem 2

Graph: $x = -2(y - 3)^2 + 1$

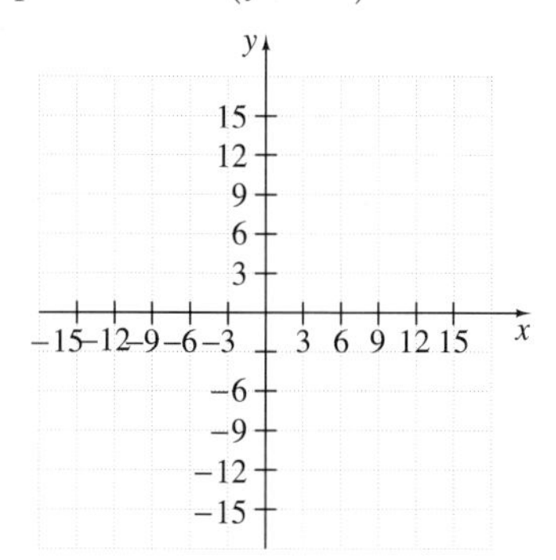

Example 2

Graph: $x = -3(y - 1)^2 + 2$

Solution: The equation $x = -3(y - 1)^2 + 2$ is in the form $x = a(y - k)^2 + h$ with $a = -3$, $k = 1$, and $h = 2$. Since $a < 0$, the parabola opens to the left. The vertex (h, k) is $(2, 1)$, and the axis of symmetry is the horizontal line $y = 1$. When $y = 0$, the x-intercept is $x = -1$.

Again, we use a table to obtain a few ordered pair solutions and then graph the parabola.

x	y
2	1
−1	0
−1	2

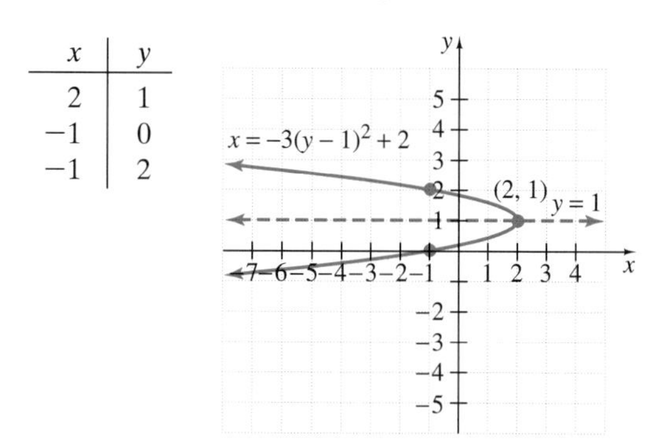

Example 3

Graph: $y = -x^2 - 2x + 15$

Solution: There are two methods that we can use to find the vertex. The first method is completing the square.

$y - 15 = -x^2 - 2x$ Subtract 15 from both sides.

$y - 15 = -1(x^2 + 2x)$ Factor -1 from the terms $-x^2 - 2x$.

The coefficient of x is 2, so we find the square of half of 2.

$$\frac{1}{2}(2) = 1 \text{ and } 1^2 = 1$$

$y - 15 - 1(1) = -1(x^2 + 2x + 1)$ Add $-1(1)$ to both sides.

$y - 16 = -1(x + 1)^2$ Simplify the left side and factor the right side.

$y = -(x + 1)^2 + 16$ Add 16 to both sides.

Answers

1.

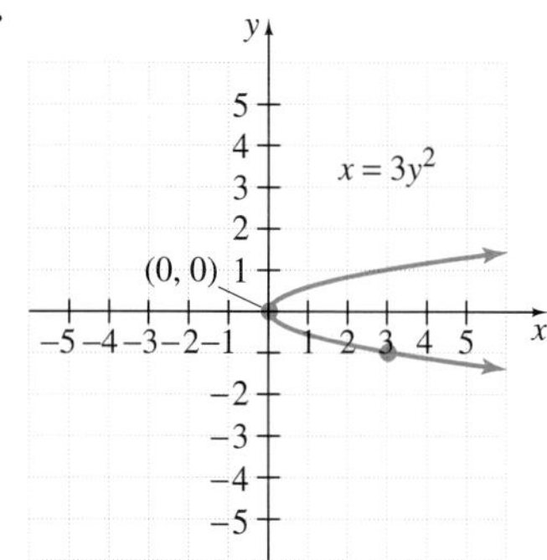

2.

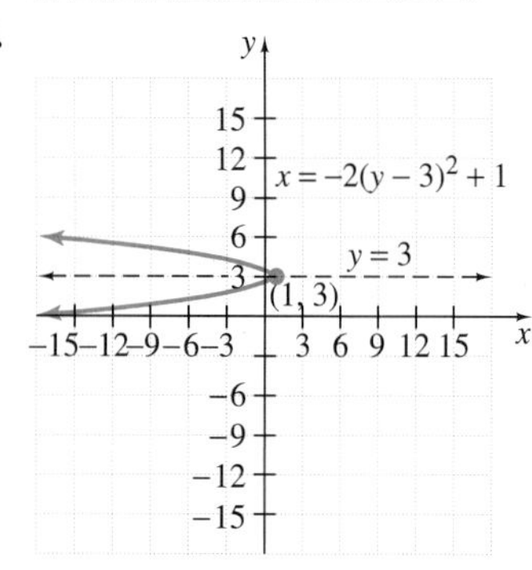

The vertex is $(-1, 16)$.

The second method for finding the vertex is by using the formula $\dfrac{-b}{2a}$.

$$x = \frac{-(-2)}{2(-1)} = \frac{2}{-2} = -1$$
$$y = -(-1)^2 - 2(-1) + 15 = -1 + 2 + 15 = 16$$

Again, we see that the vertex is $(-1, 16)$, and the axis of symmetry is the vertical line $x = -1$. The y-intercept is 15. Now we can use a few more ordered pair solutions to graph the parabola.

x	y
-1	16
0	15
-2	15
1	12
-3	12
3	0
-5	0

Example 4

Graph: $x = 2y^2 + 4y + 5$

Solution: Notice that this equation is quadratic in y, so its graph is a parabola that opens to the left or the right. We can complete the square on y or we can use the formula $\dfrac{-b}{2a}$ to find the vertex.

Since the equation is quadratic in y, the formula gives us the y-value of the vertex.

$$y = \frac{-4}{2 \cdot 2} = \frac{-4}{4} = -1$$
$$x = 2(-1)^2 + 4(-1) + 5 = 2 \cdot 1 - 4 + 5 = 3$$

The vertex is $(3, -1)$, and the axis of symmetry is the line $y = -1$. The parabola opens to the right since $a > 0$. The x-intercept is 5.

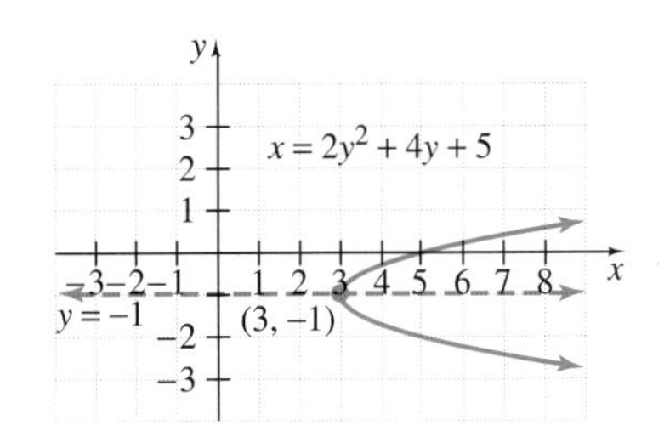

Practice Problem 3

Graph: $y = -x^2 - 4x + 12$

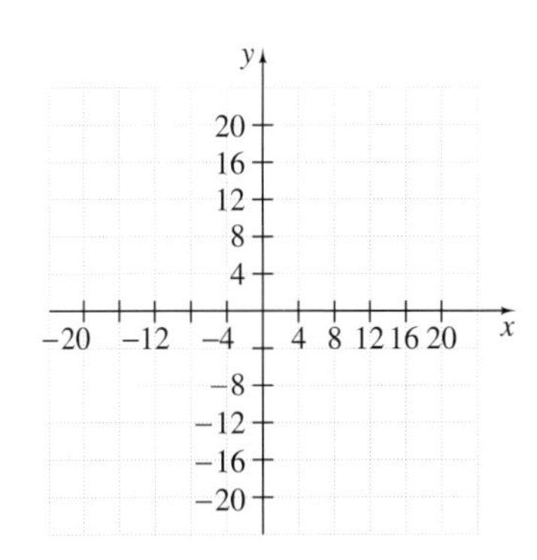

Practice Problem 4

Graph: $x = 3y^2 + 12y + 13$

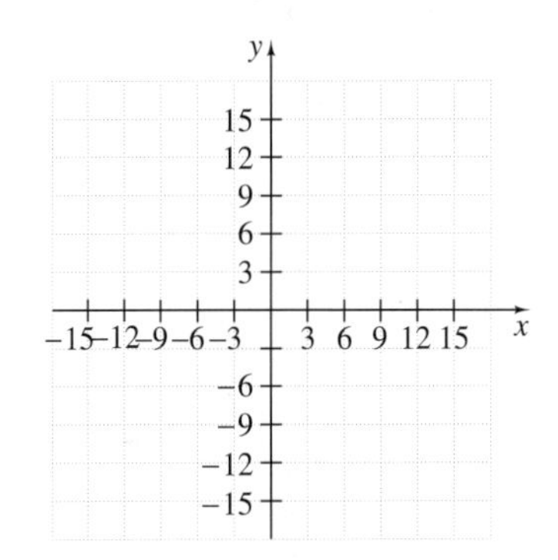

Answers

3.

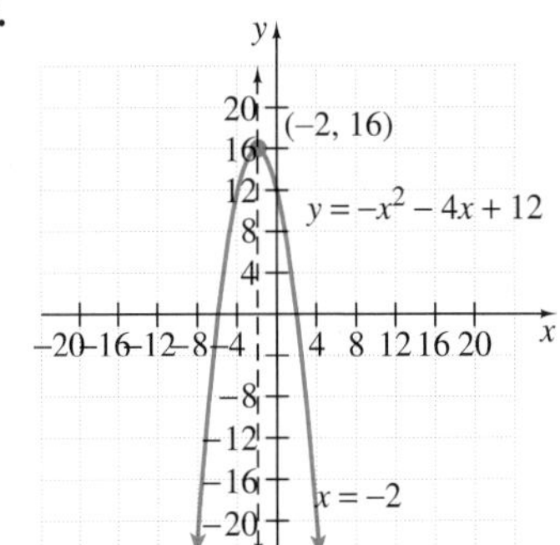

4.

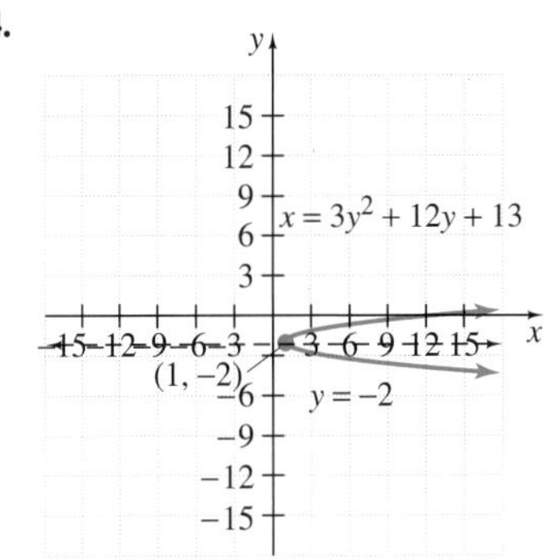

B USING THE DISTANCE AND MIDPOINT FORMULAS

The Cartesian coordinate system helps us visualize a distance between points. To find the distance between two points, we use the distance formula, which is derived from the Pythagorean theorem.

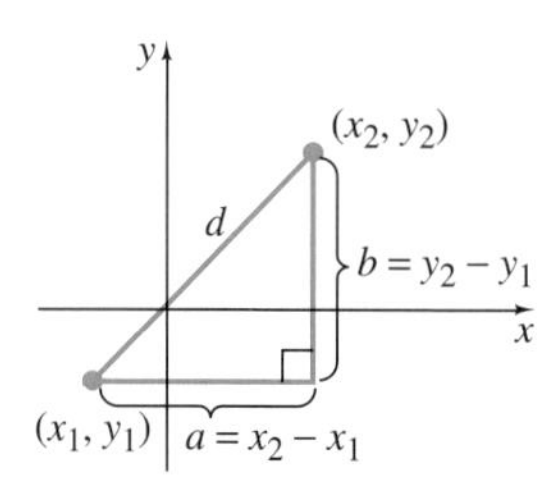

To find the distance d between two points (x_1, y_1) and (x_2, y_2) as shown next, draw vertical and horizontal lines so that a right triangle is formed, as shown. Notice that the length of leg a is $x_2 - x_1$ and that the length of leg b is $y_2 - y_1$. Thus, the Pythagorean theorem tell us that

$$d^2 = a^2 + b^2$$

or

$$d^2 = (x_2 - x_1)^2 + (y_2 - y_1)^2$$

or

$$d = \sqrt{(x_2 - x_1)^2 + (y_2 - y_1)^2}$$

This formula gives us the distance between any two points on the real plane.

DISTANCE FORMULA

The distance d between two points (x_1, y_1) and (x_2, y_2) is given by

$$d = \sqrt{(x_2 - x_1)^2 + (y_2 - y_1)^2}$$

Practice Problem 5

Find the distance between $(-1, 3)$ and $(-2, 6)$. Give an exact distance and a three-decimal-place approximation.

Example 5 Find the distance between $(2, -5)$ and $(1, -4)$. Give an exact distance and a three-decimal-place approximation.

Solution: To use the distance formula, it makes no difference which point we call (x_1, y_1) and which point we call (x_2, y_2). We will let $(x_1, y_1) = (2, -5)$ and $(x_2, y_2) = (1, -4)$.

$$\begin{aligned} d &= \sqrt{(x_2 - x_1)^2 + (y_2 - y_1)^2} \\ &= \sqrt{(1 - 2)^2 + [-4 - (-5)]^2} \\ &= \sqrt{(-1)^2 + (1)^2} \\ &= \sqrt{1 + 1} \\ &= \sqrt{2} \approx 1.414 \end{aligned}$$

The distance between the two points is exactly $\sqrt{2}$ units, or approximately 1.414 units.

The **midpoint** of a line segment is the **point** located exactly halfway between the two endpoints of the line segment. On the graph below, the

Answer

5. $\sqrt{10} \approx 3.162$

point M is the midpoint of line segment PQ. Thus, the distance between M and P equals the distance between M and Q.

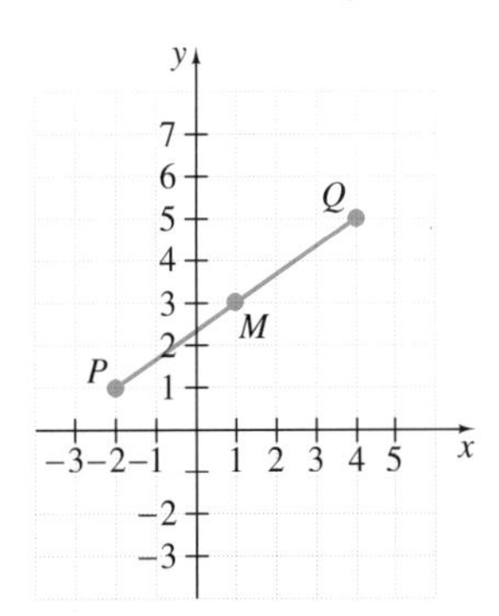

The x-coordinate of M is at half the distance between the x-coordinates of P and Q, and the y-coordinate of M is at half the distance between the y-coordinates of P and Q. That is, the x-coordinate of M is the average of the x-coordinates of P and Q; the y-coordinate of M is the average of the y-coordinates of P and Q.

MIDPOINT FORMULA

The midpoint of the line segment whose endpoints are (x_1, y_1) and (x_2, y_2) is the point with coordinates

$$\left(\frac{x_1 + x_2}{2}, \frac{y_1 + y_2}{2}\right)$$

Example 6 Find the midpoint of the line segment that joins points $P(-3, 3)$ and $Q(1, 0)$.

Solution: To use the midpoint formula, it makes no difference which point we call (x_1, y_1) or which point we call (x_2, y_2). We will let $(x_1, y_1) = (-3, 3)$ and $(x_2, y_2) = (1, 0)$.

$$\begin{aligned} \text{midpoint} &= \left(\frac{x_1 + x_2}{2}, \frac{y_1 + y_2}{2}\right) \\ &= \left(\frac{-3 + 1}{2}, \frac{3 + 0}{2}\right) \\ &= \left(\frac{-2}{2}, \frac{3}{2}\right) \\ &= \left(-1, \frac{3}{2}\right) \end{aligned}$$

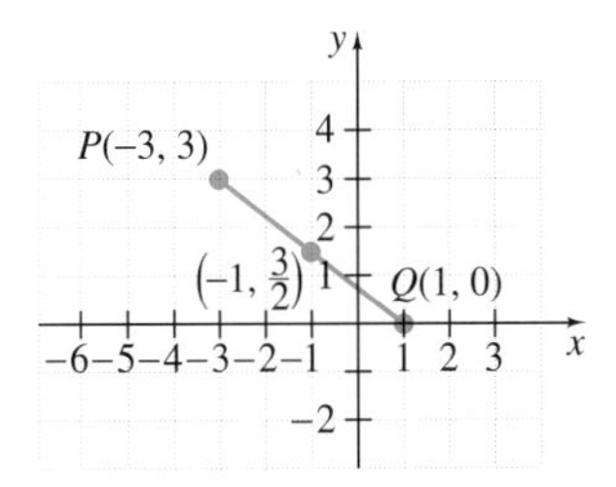

The midpoint of the segment is $\left(-1, \frac{3}{2}\right)$.

Practice Problem 6

Find the midpoint of the line segment that joins points $P(-2, 5)$ and $Q(4, -6)$.

Answer

6. $\left(1, -\frac{1}{2}\right)$

C GRAPHING CIRCLES

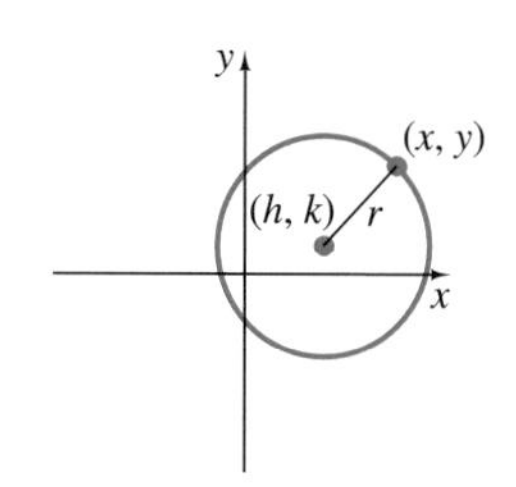

Another conic section is the **circle**. A circle is the set of all points in a plane that are the same distance from a fixed point called the **center**. The distance is called the **radius** of the circle. To find a standard equation for a circle, let (h, k) represent the center of the circle, and let (x, y) represent any point on the circle. The distance between (h, k) and (x, y) is defined to be the radius, r units. We can find this distance r by using the distance formula.

$$r = \sqrt{(x - h)^2 + (y - k)^2}$$

$$r^2 = (x - h)^2 + (y - k)^2 \quad \text{Square both sides.}$$

CIRCLE

The graph of $(x - h)^2 + (y - k)^2 = r^2$ is a circle with center (h, k) and radius r.

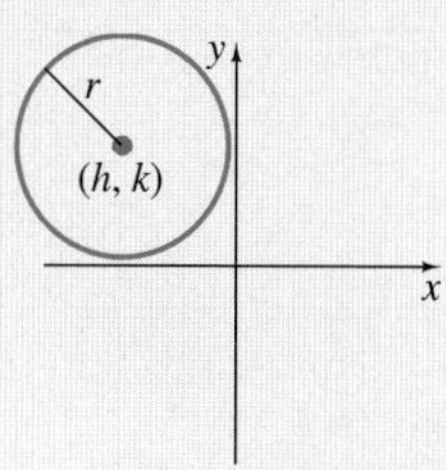

The form $(x - h)^2 + (y - k)^2 = r^2$ is called **standard form**.

If an equation can be written in the standard form

$$(x - h)^2 + (y - k)^2 = r^2$$

then its graph is a circle, which we can draw by graphing the center (h, k) and using the radius r.

HELPFUL HINT

Notice that the radius is the *distance* from the center of the circle to any point of the circle. Also notice that the *midpoint* of a diameter of a circle is the center of the circle.

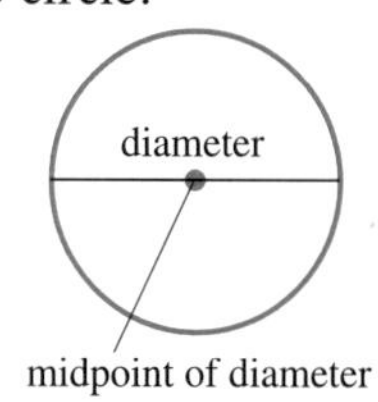

Example 7 Graph: $x^2 + y^2 = 4$

Solution: The equation can be written in standard form as

Practice Problem 7

Graph: $x^2 + y^2 = 9$

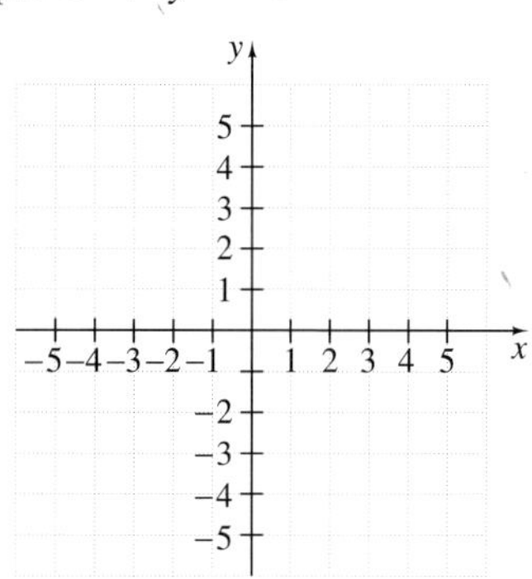

Answer

7.

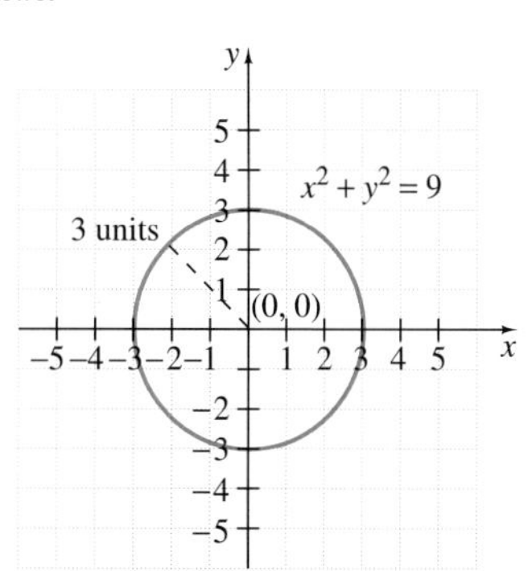

$$(x - 0)^2 + (y - 0)^2 = 2^2$$

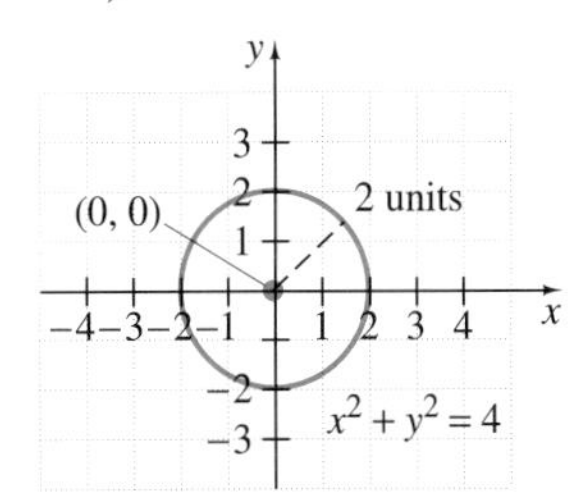

The center of the circle is $(0, 0)$, and the radius is 2, Its graph is shown above.

HELPFUL HINT

Notice the difference between the equation of a circle and the equation of a parabola. The equation of a circle contains both x^2 and y^2 terms on the same side of the equation with equal coefficients. The equation of a parabola has either an x^2 term or a y^2 term but not both.

Example 8 Graph: $(x + 1)^2 + y^2 = 8$

Solution: The equation can be written as $(x + 1)^2 + (y - 0)^2 = 8$ with $h = -1$, $k = 0$, and $r = \sqrt{8}$. The center is $(-1, 0)$, and the radius is $\sqrt{8} = 2\sqrt{2} \approx 2.8$.

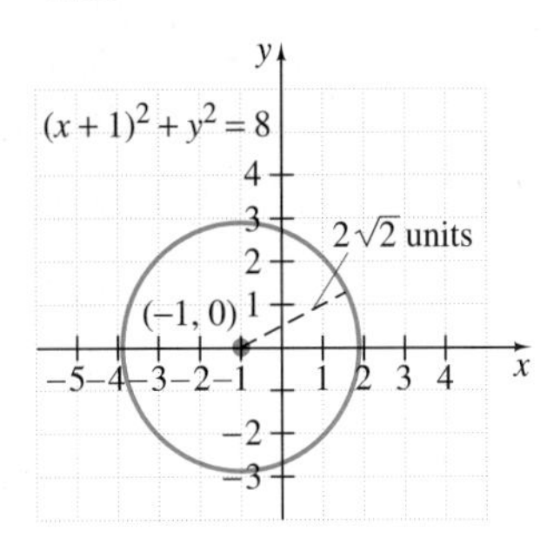

TRY THE CONCEPT CHECK IN THE MARGIN.

D WRITING EQUATIONS OF CIRCLES

Since a circle is determined entirely by its center and radius, this information is all we need to write the equation of a circle.

Example 9 Write an equation of the circle with center $(-7, 3)$ and radius 10.

Solution: Using the given values $h = -7$, $k = 3$, and $r = 10$, we write the equation

$$(x - h)^2 + (y - k)^2 = r^2$$

or

$$[x - (-7)]^2 + (y - 3)^2 = 10^2 \quad \text{Substitute the given values.}$$

or

$$(x + 7)^2 + (y - 3)^2 = 100$$

✓ CONCEPT CHECK

In the graph of the equation $(x - 3)^2 + (y - 2)^2 = 5$, what is the distance between the center of the circle and any point on the circle?

Practice Problem 8

Graph: $x^2 + (y + 2)^2 = 6$

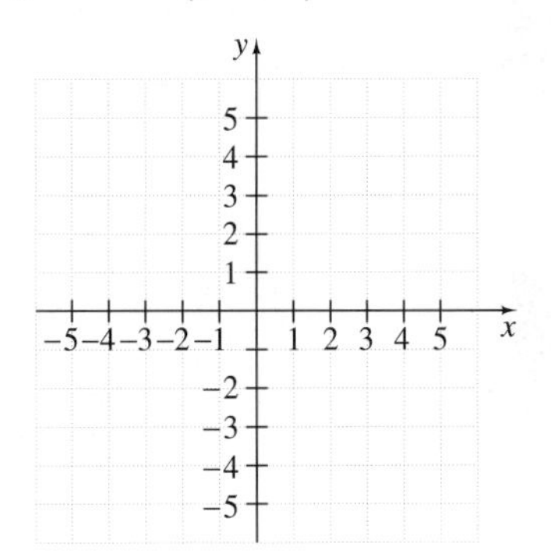

Practice Problem 9

Write an equation of the circle with the center $(2, -5)$ and radius 7.

Answers

8.

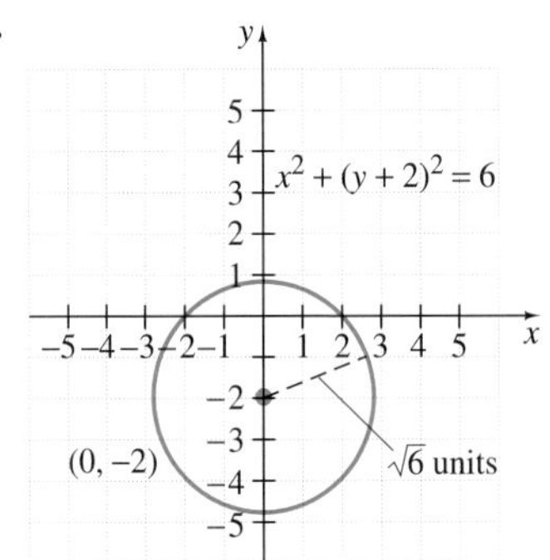

9. $(x - 2)^2 + (y + 5)^2 = 49$

✓ Concept Check: the distance is $\sqrt{5}$ units

E FINDING THE CENTER AND THE RADIUS OF A CIRCLE

To find the center and the radius of a circle from its equation, we write the equation in standard form. To write the equation of a circle in standard form, we complete the square on both x and y.

Example 10 Graph: $x^2 + y^2 + 4x - 8y = 16$

Solution: Since this equation contains x^2 and y^2 terms on the same side of the equation with equal coefficients, its graph is a circle. To write the equation in standard form, we group the terms involving x and the terms involving y, and then complete the square on each variable.

$$(x^2 + 4x) + (y^2 - 8y) = 16$$

Thus, $\frac{1}{2}(4) = 2$ and $2^2 = 4$. Also, $\frac{1}{2}(-8) = -4$ and $(-4)^2 = 16$. We add 4 and then 16 to both sides.

$$(x^2 + 4x + 4) + (y^2 - 8y + 16) = 16 + 4 + 16$$

$$(x + 2)^2 + (y - 4)^2 = 36 \quad \text{Factor.}$$

This circle has the center $(-2, 4)$ and radius 6, as shown.

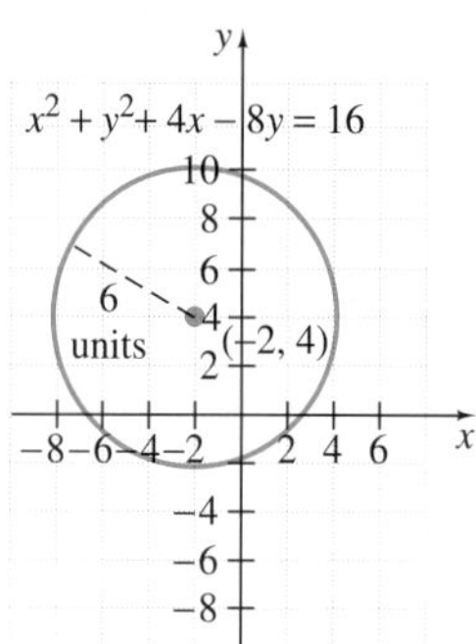

Practice Problem 10

Graph: $x^2 + y^2 - 2x + 6y = 6$

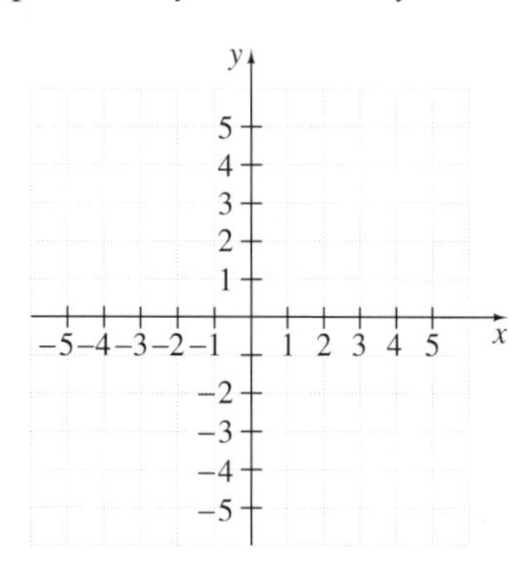

Answer

10.

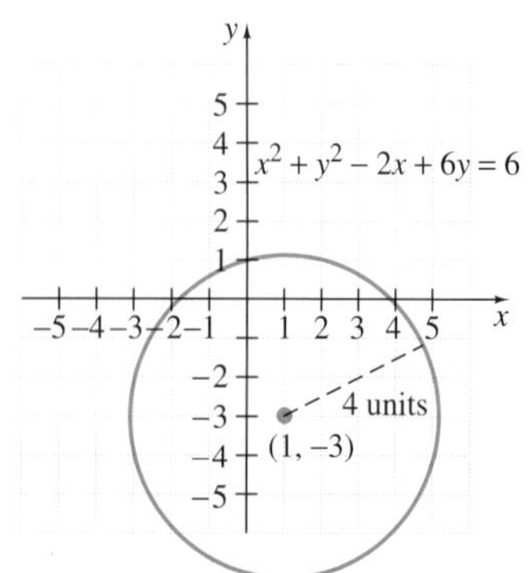

GRAPHING CALCULATOR EXPLORATIONS

To graph an equation such as $x^2 + y^2 = 25$ with a grapher, we first solve the equation for y.

$$x^2 + y^2 = 25$$
$$y^2 = 25 - x^2$$
$$y = \pm\sqrt{25 - x^2}$$

The graph of $y = \sqrt{25 - x^2}$ will be the top half of the circle, and the graph of $y = -\sqrt{25 - x^2}$ will be the bottom half of the circle.

To graph, press [Y=] and enter $Y_1 = \sqrt{25 - x^2}$ and $Y_2 = -\sqrt{25 - x^2}$. Insert parentheses about $25 - x^2$ so that $\sqrt{25 - x^2}$ and not $\sqrt{25} - x^2$ is graphed.

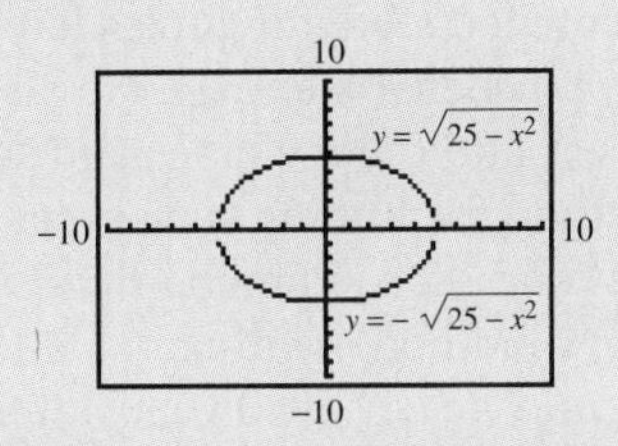

The graph does not appear to be a circle because we are currently using a standard window and the screen is rectangular. This causes the tick marks on the x-axis to be farther apart than the tick marks on the y-axis and thus creates the distorted circle. If we want the graph to appear circular, we define a square window by using a feature of the grapher or redefine the window to show the x-axis from -15 to 15 and the y-axis from -10 to 10. Using a square window, the graph appears as follows:

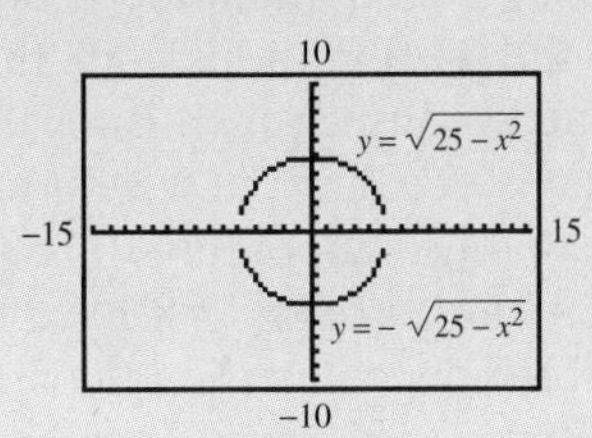

Use a grapher to graph each circle.

1. $x^2 + y^2 = 55$

2. $x^2 + y^2 = 20$

3. $7x^2 + 7y^2 - 89 = 0$

4. $3x^2 + 3y^2 - 35 = 0$

Focus On History

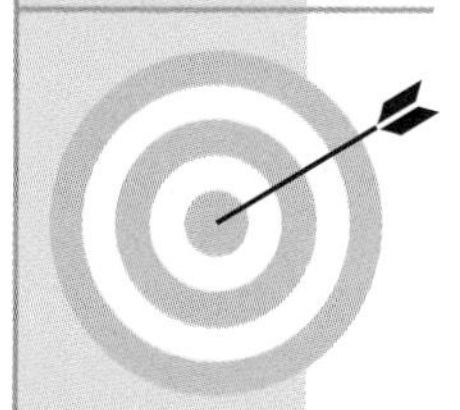

Conic Sections

It is believed that the conic sections were discovered by the Greek mathematician Menaechmus (380 B.C.–320 B.C.). He was the first to realize that the shapes of the parabola, ellipse, and hyperbola are formed by cutting a right circular cone with a plane in various ways. However, these conic sections were not given their names until later. Another Greek mathematician, Apollonius of Perga (262 B.C.–190 B.C.), was responsible for coining the terms *parabola*, *ellipse*, and *hyperbola* in his set of eight texts titled *Treatise on Conic Sections*. In these texts, Apollonius discussed the basic properties of conic sections as well as how they are drawn.

A contemporary of Apollonius, Archimedes of Syracuse (287 B.C.–212 B.C.), is probably the most famous of the Greek mathematicians who made contributions to the base of knowledge on conic sections. His detailed study of circles led to an important contribution: a calculation of the value of π as being between $3\frac{10}{71}$ and $3\frac{1}{7}$. He also studied the areas of conic sections, including parabolas and ellipses, and other shapes and solids that arise from the conic sections. He developed a special method for finding such areas by dividing the area of a figure up into infinitely narrow rectangles and then summing these individual areas to find the area of the entire figure. This revolutionary method eventually led to the discovery of the branch of advanced mathematics called *calculus* nearly 2000 years later.

As important as was his work with conics, Archimedes is probably best remembered for his work on practical matters for the king of Syracuse. For instance, Archimedes is credited with inventing the catapult, at the king's request, as a defense measure against a Roman invasion. Another time, the king asked Archimedes to help prove that a gold crown that he had commissioned was made partially from silver as well. The story goes that while Archimedes pondered this question, he was taking a bath and noticed that the amount of water that overflowed the tub was proportional to the portion of his body that was under water. He had discovered what is now known as Archimedes' Principle of Buoyancy: that an object immersed in water is buoyed up by a force that is equal to the weight of the water it displaces. Archimedes was so excited by his discovery that he supposedly ran naked straight from his bath through the streets of Syracuse shouting "Eureka, eureka!" ("I have found it!"). He immediately applied this discovery to the crown problem by comparing the amount of water displaced by a crown made from the same weight of pure gold to the amount of water displaced by the suspect crown. Because these amounts of water were not the same, Archimedes proved that the maker of the crown had cheated the king by using silver, a cheaper metal, in place of some of the gold.

Name ______________________ Section ________ Date ________

MENTAL MATH ANSWERS

1. ______
2. ______
3. ______
4. ______
5. ______
6. ______

MENTAL MATH

The graph of each equation is a parabola. Determine whether the parabola opens upward, downward, to the left, or to the right.

1. $y = x^2 - 7x + 5$
2. $y = -x^2 + 16$
3. $x = -y^2 - y + 2$
4. $x = 3y^2 + 2y - 5$
5. $y = -x^2 + 2x + 1$
6. $x = -y^2 + 2y - 6$

EXERCISE SET 9.1

A *The graph of each equation is a parabola. Find the vertex of the parabola and then graph it. See Examples 1 through 4.*

1. $x = 3y^2$

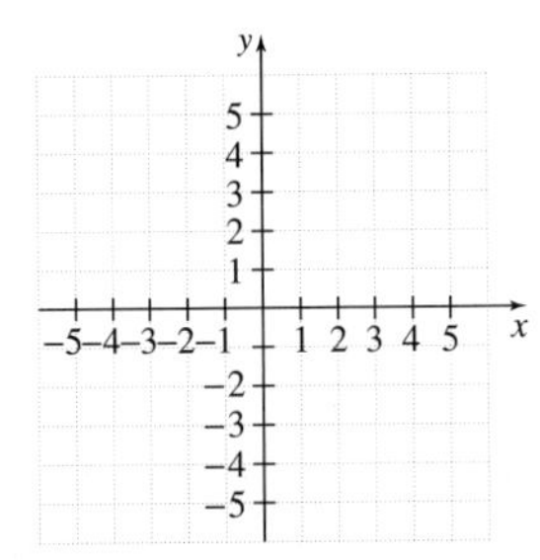

2. $x = -2y^2$

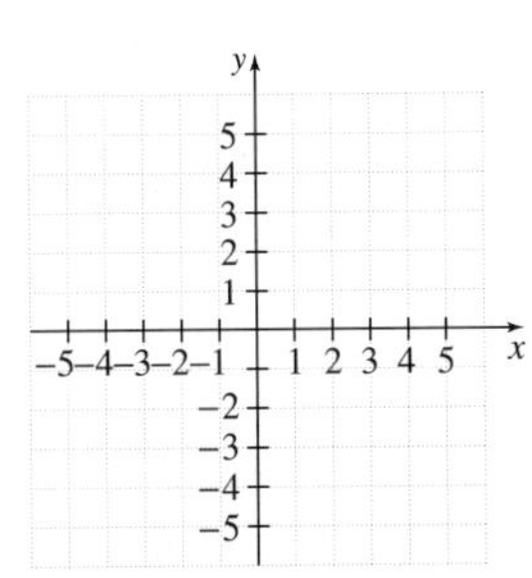

3. $x = (y - 2)^2 + 3$

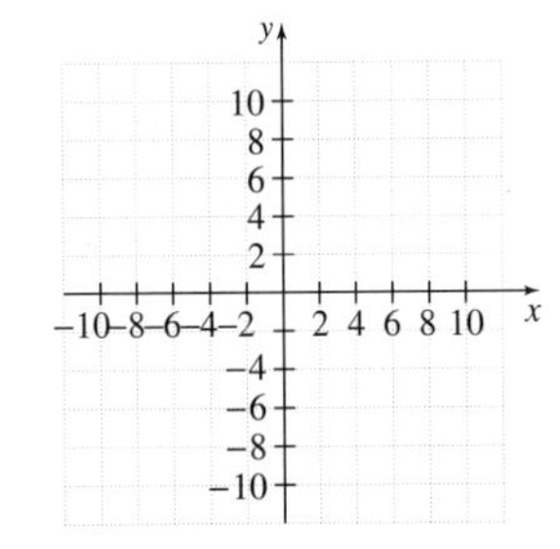

4. $x = (y - 4)^2 - 1$

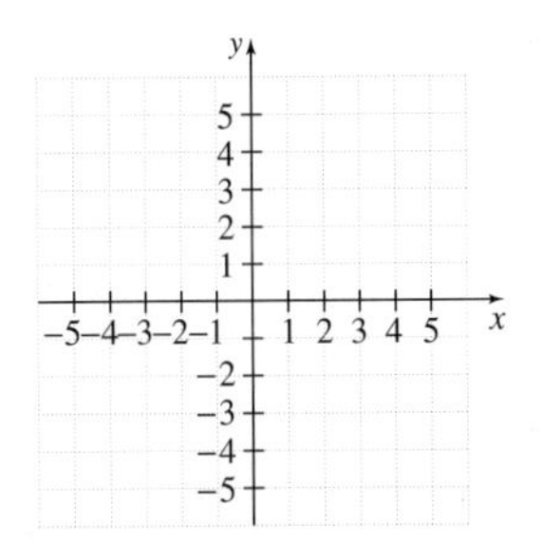

5. $y = 3(x - 1)^2 + 5$

6. $x = -4(y - 2)^2 + 2$

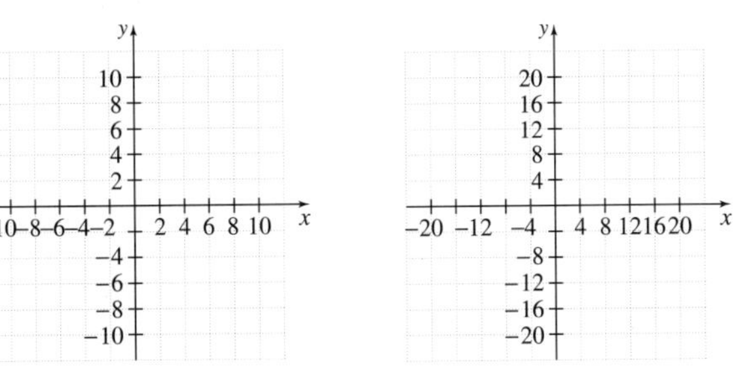

7. $x = y^2 + 6y + 8$

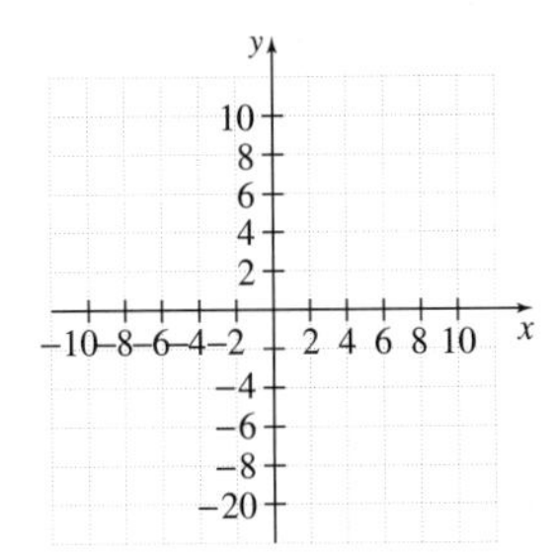

8. $x = y^2 - 6y + 6$

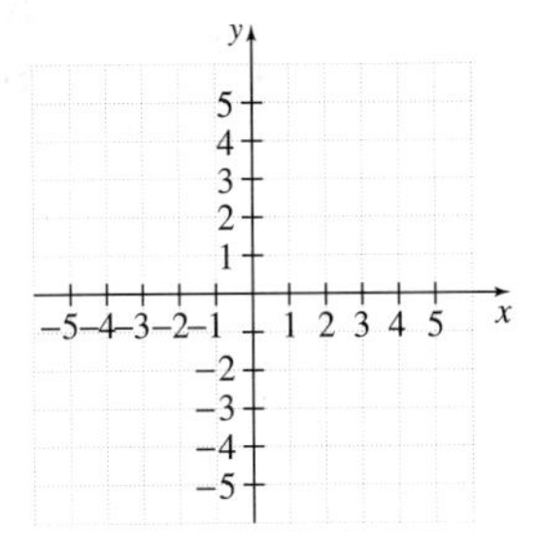

9. $y = x^2 + 10x + 20$

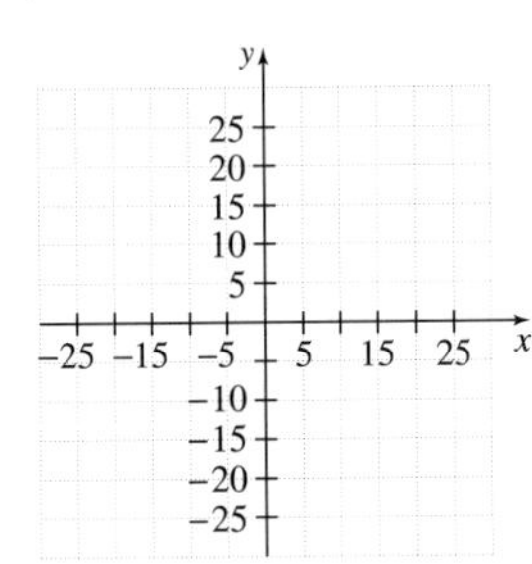

10. $y = x^2 + 4x - 5$

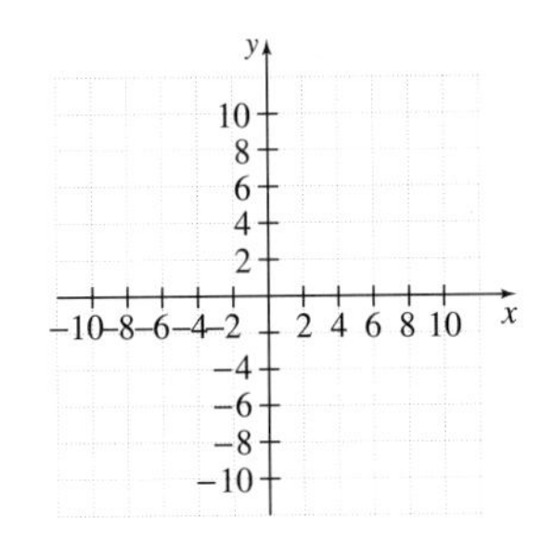

11. $x = -2y^2 + 4y + 6$

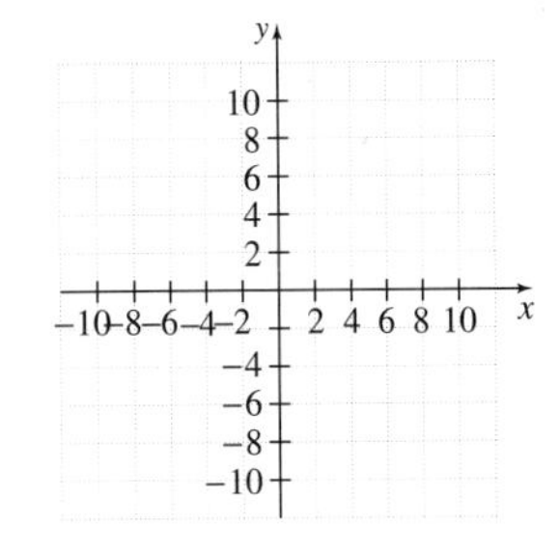

12. $x = 3y^2 + 6y + 7$

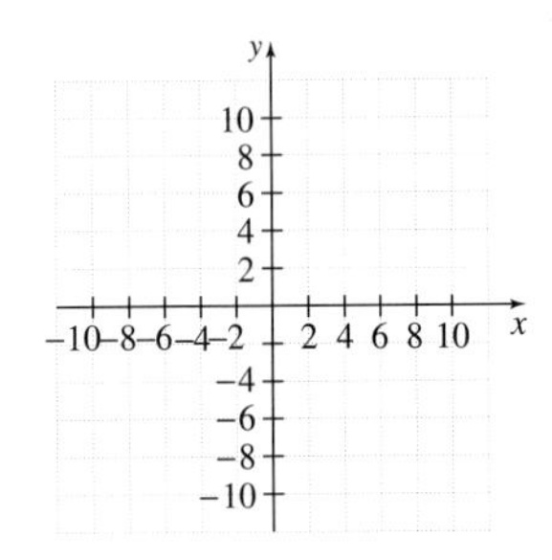

ANSWERS

13. ______
14. ______
15. ______
16. ______
17. ______
18. ______
19. ______
20. ______
21. ______
22. ______
23. ______
24. ______
25. ______
26. ______
27. ______
28. ______
29. ______
30. ______
31. ______
32. ______
33. ______
34. ______
35. ______
36. ______

Name ______________________________

B *Find the distance between each pair of points. Give an exact distance and a three-decimal-place approximation. See Example 5.*

13. $(5, 1)$ and $(8, 5)$

14. $(2, 3)$ and $(14, 8)$

15. $(-3, 2)$ and $(1, -3)$

16. $(3, -2)$ and $(-4, 1)$

17. $(-9, 4)$ and $(-8, 1)$

18. $(-5, -2)$ and $(-6, -6)$

19. $(0, -\sqrt{2})$ and $(\sqrt{3}, 0)$

20. $(-\sqrt{5}, 0)$ and $(0, \sqrt{7})$

21. $(1.7, -3.6)$ and $(-8.6, 5.7)$

22. $(9.6, 2.5)$ and $(-1.9, -3.7)$

23. $(2\sqrt{3}, \sqrt{6})$ and $(-\sqrt{3}, 4\sqrt{6})$

24. $(5\sqrt{2}, -4)$ and $(-3\sqrt{2}, -8)$

Find the midpoint of each line segment whose endpoints are given. See Example 6.

25. $(6, -8), (2, 4)$

26. $(3, 9), (7, 11)$

27. $(-2, -1), (-8, 6)$

28. $(-3, -4), (6, -8)$

29. $(7, 3), (-1, -3)$

30. $(-2, 5), (-1, 6)$

31. $\left(\frac{1}{2}, \frac{3}{8}\right), \left(-\frac{3}{2}, \frac{5}{8}\right)$

32. $\left(-\frac{2}{5}, \frac{7}{15}\right), \left(-\frac{2}{5}, -\frac{4}{15}\right)$

33. $(\sqrt{2}, 3\sqrt{5}), (\sqrt{2}, -2\sqrt{5})$

34. $(\sqrt{8}, -\sqrt{12}), (3\sqrt{2}, 7\sqrt{3})$

35. $(4.6, -3.5), (7.8, -9.8)$

36. $(-4.6, 2.1), (-6.7, 1.9)$

C E *The graph of each equation is a circle. Find the center and the radius, and then graph each circle. See Examples 7, 8, and 10.*

37. $x^2 + y^2 = 9$

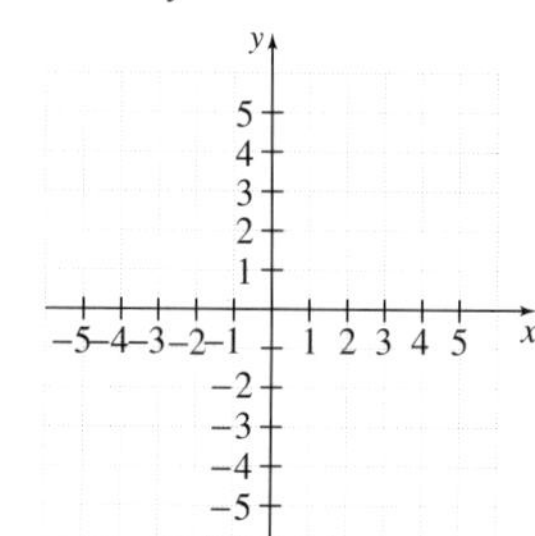

38. $x^2 + y^2 = 25$

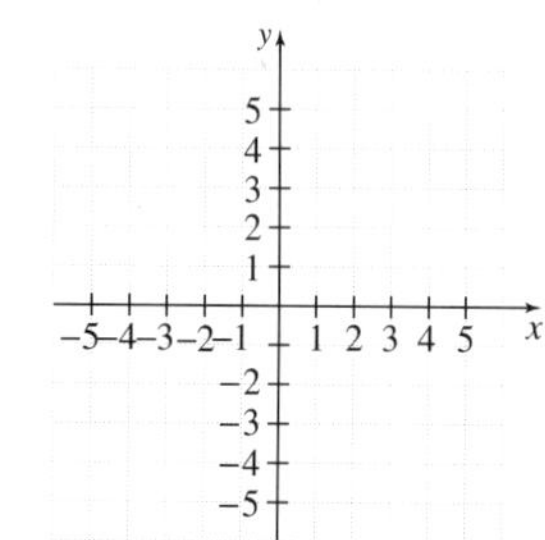

39. $x^2 + (y - 2)^2 = 1$

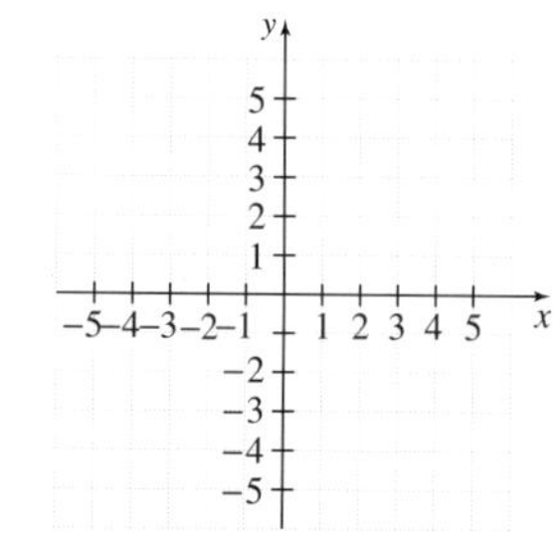

40. $(x - 3)^2 + y^2 = 9$

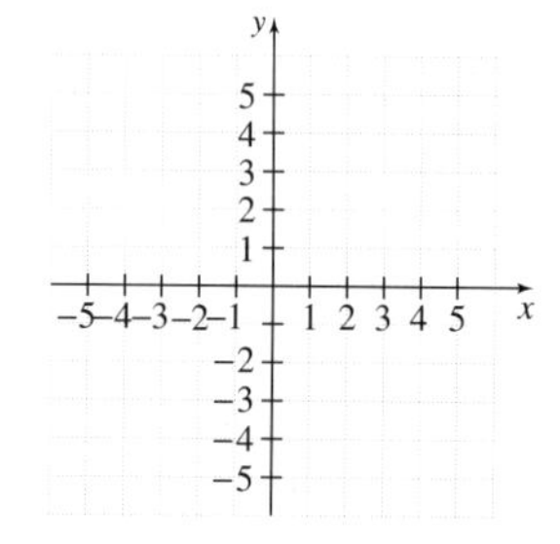

Name ____________________

41. $(x - 5)^2 + (y + 2)^2 = 1$

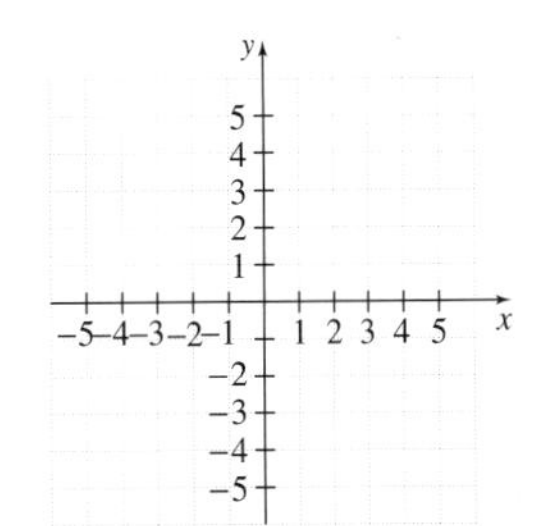

42. $(x + 3)^2 + (y + 3)^2 = 4$

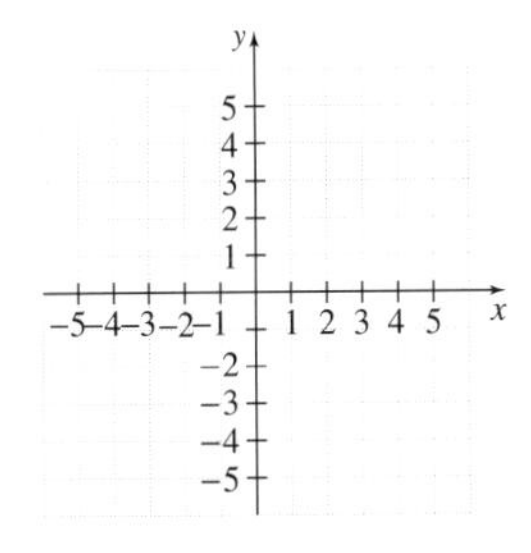

43. $x^2 + y^2 + 6y = 0$

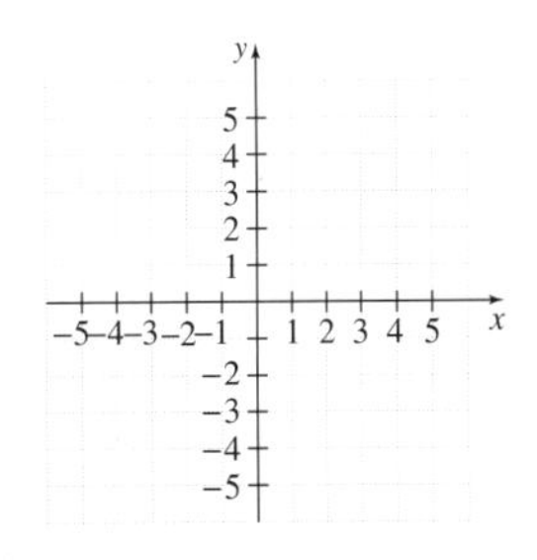

44. $x^2 + 10x + y^2 = 0$

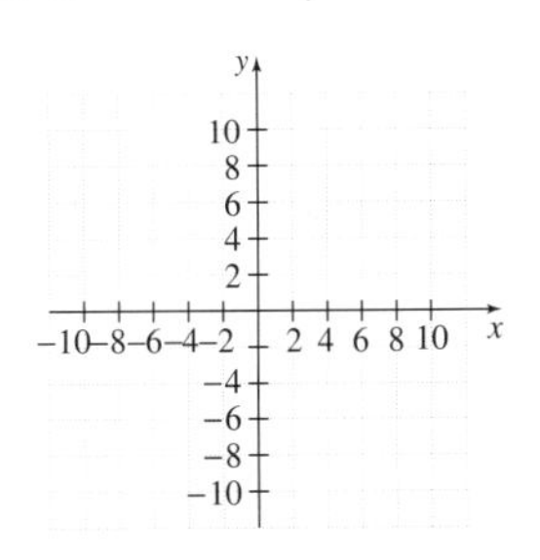

45. $x^2 + y^2 + 2x - 4y = 4$

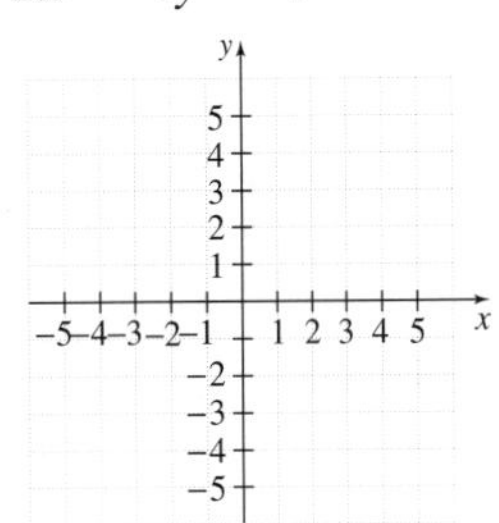

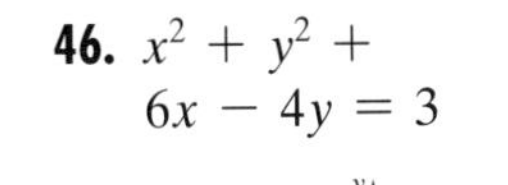

46. $x^2 + y^2 + 6x - 4y = 3$

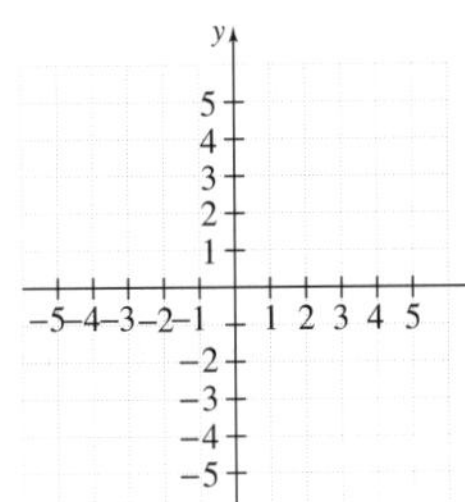

47. $x^2 + y^2 - 4x - 8y - 2 = 0$

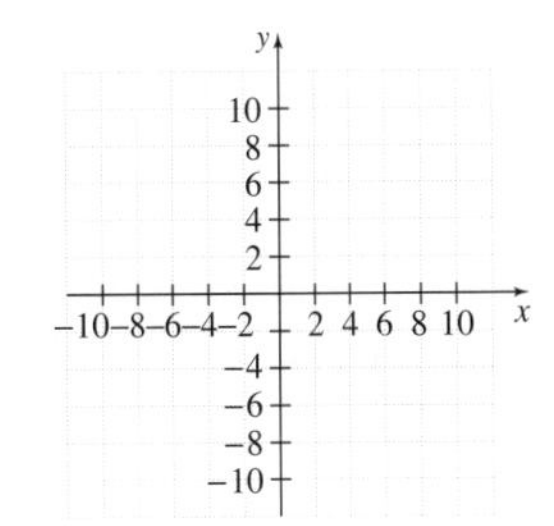

48. $x^2 + y^2 - 2x - 6y - 5 = 0$

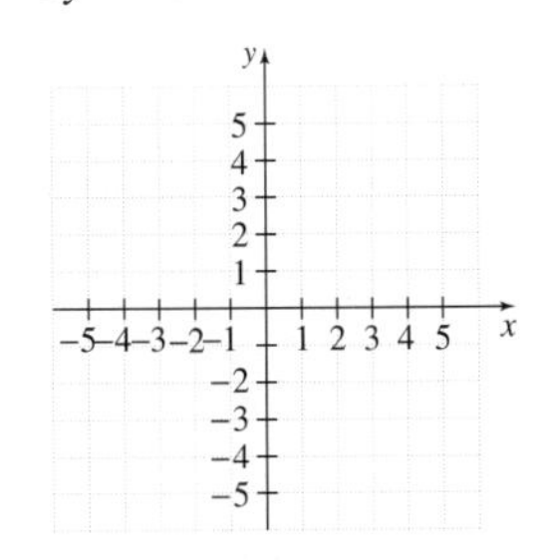

D *Write an equation of the circle with the given center and radius. See Example 9.*

49. $(2, 3); 6$

50. $(-7, 6); 2$

51. $(0, 0); \sqrt{3}$

52. $(0, -6); \sqrt{2}$

53. $(-5, 4); 3\sqrt{5}$

54. the origin; $4\sqrt{7}$

49. ____________________

50. ____________________

51. ____________________

52. ____________________

53. ____________________

54. ____________________

Review and Preview

Graph each equation. See Section 3.3.

55. $y = 2x + 5$

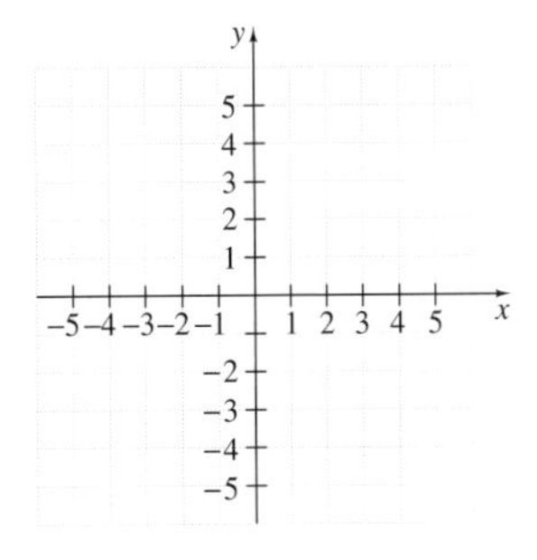

56. $y = -3x + 3$

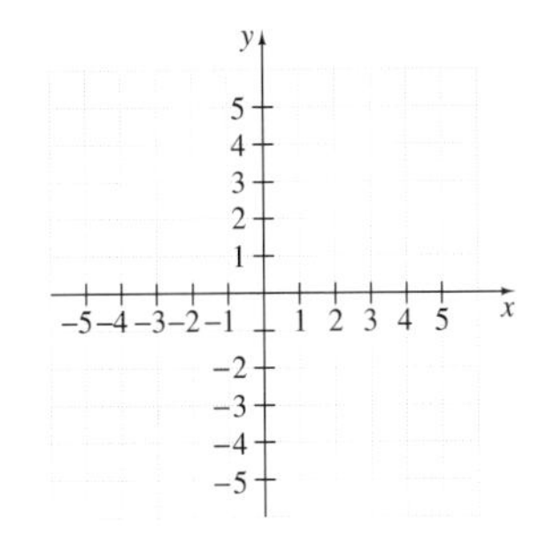

57. $y = 3$

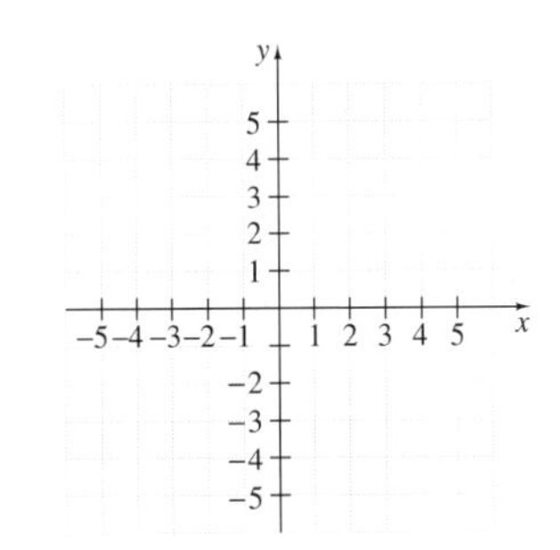

58. $x = -2$

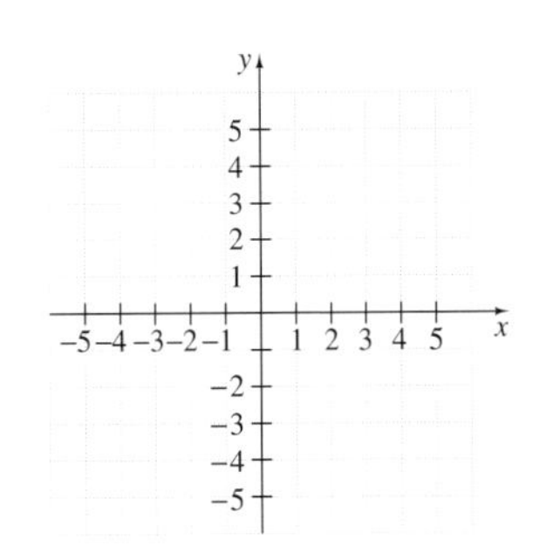

59. ______

60. ______

61. ______

62. ______

63. ______

64. ______

65. ______

66. a. see graph

b. ______

c. ______

67. ______

Name ______

Rationalize each denominator and simplify if possible. See Section 7.5

59. $\frac{1}{\sqrt{3}}$

60. $\frac{\sqrt{5}}{\sqrt{8}}$

61. $\frac{4\sqrt{7}}{\sqrt{6}}$

62. $\frac{10}{\sqrt{5}}$

Combining Concepts

63. If you are given a list of equations of circles and parabolas and none are in standard form, explain how you would determine which is an equation of a circle and which is an equation of a parabola. Explain also how you would distinguish the upward or downward parabolas from the left-opening or right-opening parabolas.

64. Determine whether the triangle with vertices $(2, 6)$, $(0, -2)$, and $(5, 1)$ is an isosceles triangle.

Solve.

65. Two surveyors need to find the distance across a lake. They place a reference pole at point A in the diagram. Point B is 3 meters east and 1 meter north of the reference point A. Point C is 19 meters east and 13 meters north of point A. Find the distance across the lake, from B to C.

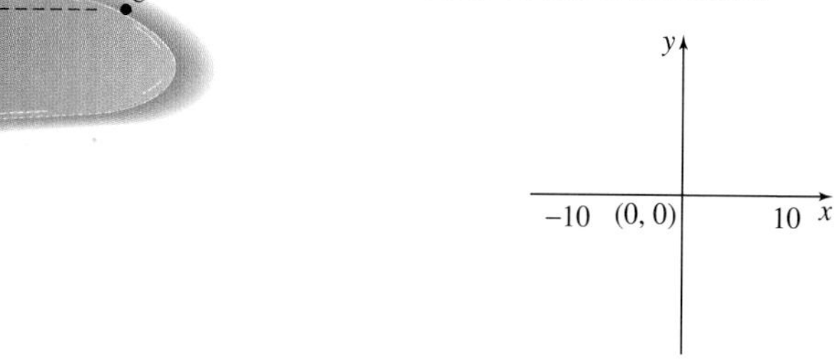

66. Cindy Brown, an architect, is drawing plans on grid paper for a circular pool with a fountain in the middle. The paper is marked off in centimeters, and each centimeter represents 1 foot. On the paper, the diameter of the "pool" is 20 centimeters, and "fountain" is the point $(0, 0)$.

a. Sketch the architect's drawing. Be sure to label the axes.

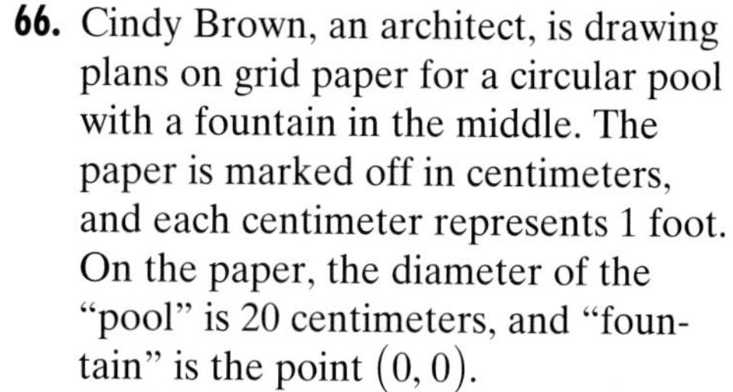

b. Write an equation that describes the circular pool.

c. Cindy plans to place a circle of lights around the fountain such that each light is 5 feet from the fountain. Write an equation for the circle of lights and sketch the circle on your drawing.

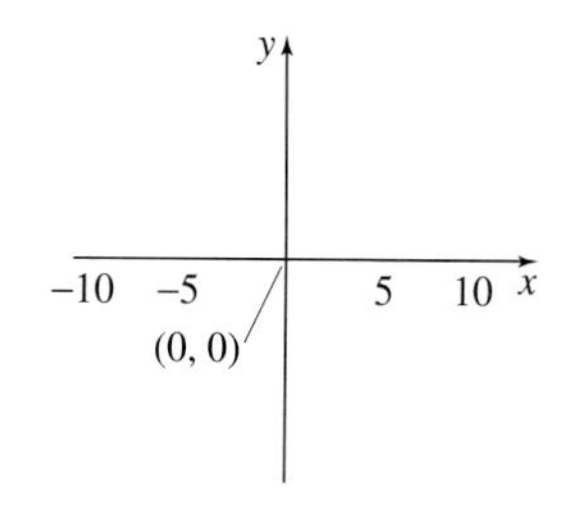

67. A bridge constructed over a bayou has a supporting arch in the shape of a parabola. Find an equation of the parabolic arch if the length of the road over the arch is 100 meters and the maximum height of the arch is 40 meters.

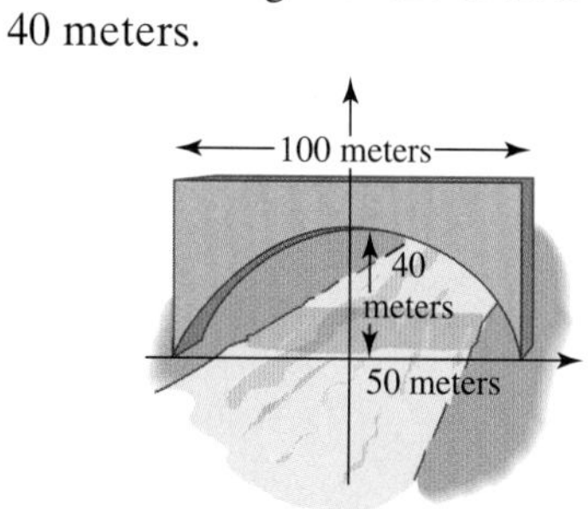

9.2 THE ELLIPSE AND THE HYPERBOLA

Objectives

A Define and graph an ellipse.
B Define and graph a hyperbola.

A GRAPHING ELLIPSES

An **ellipse** can be thought of as the set of points in a plane such that the sum of the distances of those points from two fixed points is constant. Each of the two fixed points is called a **focus**. The plural of focus is **foci**. The point midway between the foci is called the **center**.

An ellipse may be drawn by hand by using two tacks, a piece of string, and a pencil. Secure the two tacks into a piece of cardboard, for example, and tie each end of the string to a tack. Use your pencil to pull the string tight and draw the ellipse. The two tacks are the foci of the drawn ellipse.

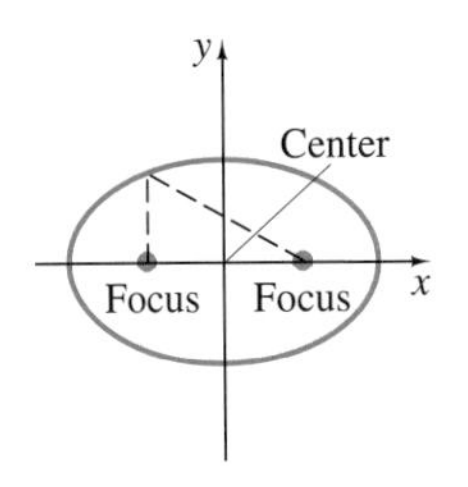

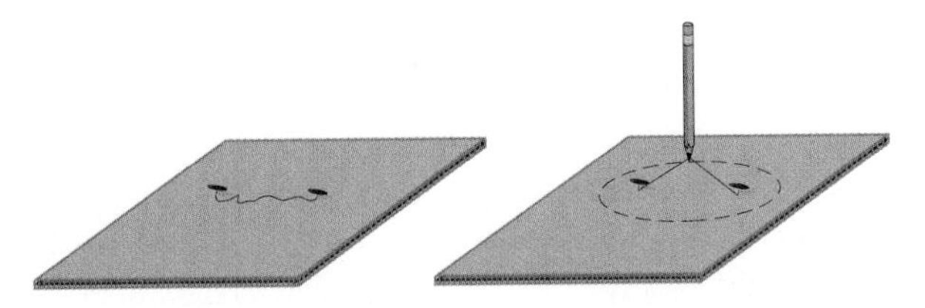

ELLIPSE WITH CENTER (0, 0)

The graph of an equation of the form $\frac{x^2}{a^2} + \frac{y^2}{b^2} = 1$ is an ellipse with center $(0, 0)$. The x-intercepts are a and $-a$, and the y-intercepts are b and $-b$.

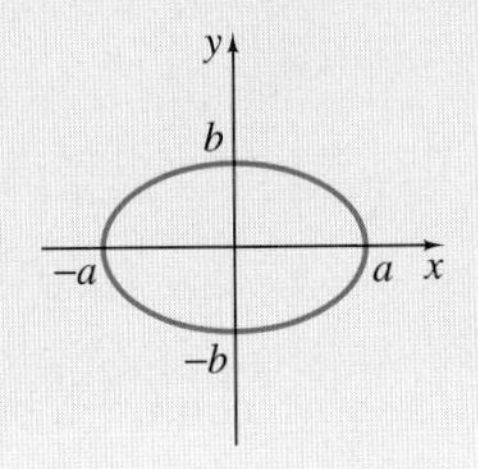

The **standard form** of an ellipse with center $(0, 0)$ is $\frac{x^2}{a^2} + \frac{y^2}{b^2} = 1$.

Example 1 Graph: $\frac{x^2}{9} + \frac{y^2}{16} = 1$

Solution: The equation is of the form $\frac{x^2}{a^2} + \frac{y^2}{b^2} = 1$, with $a = 3$ and $b = 4$, so its graph is an ellipse with center $(0, 0)$, x-intercepts 3 and -3, and y-intercepts 4 and -4.

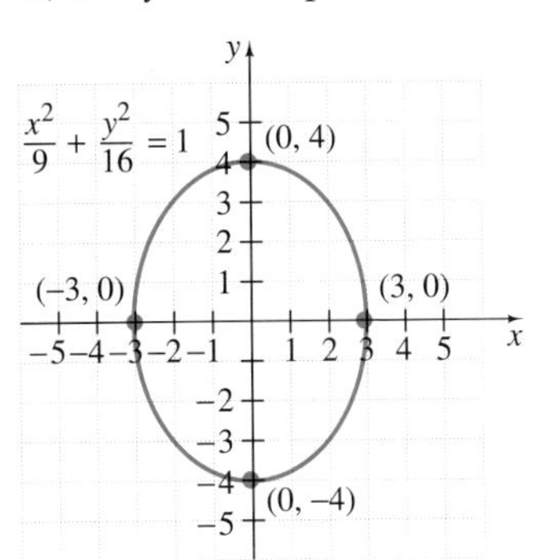

Practice Problem 1

Graph: $\frac{x^2}{4} + \frac{y^2}{9} = 1$

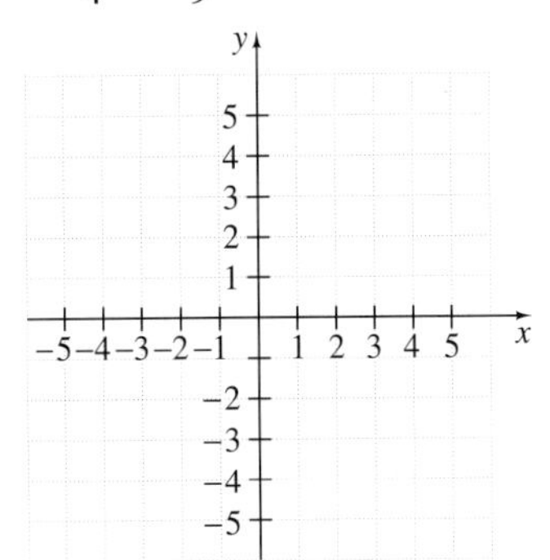

Answer

1.

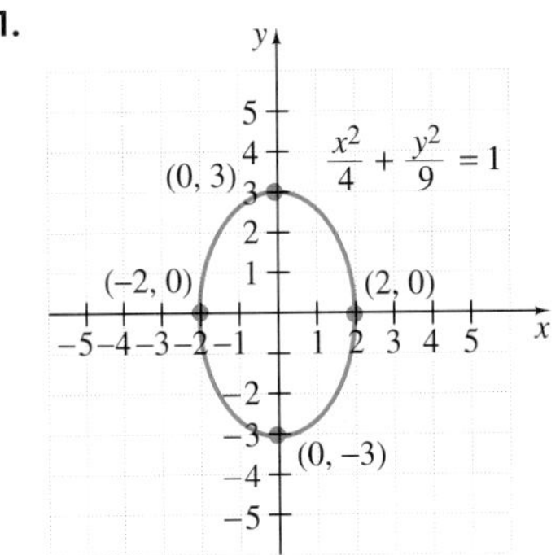

Practice Problem 2

Graph: $4x^2 + 25y^2 = 100$

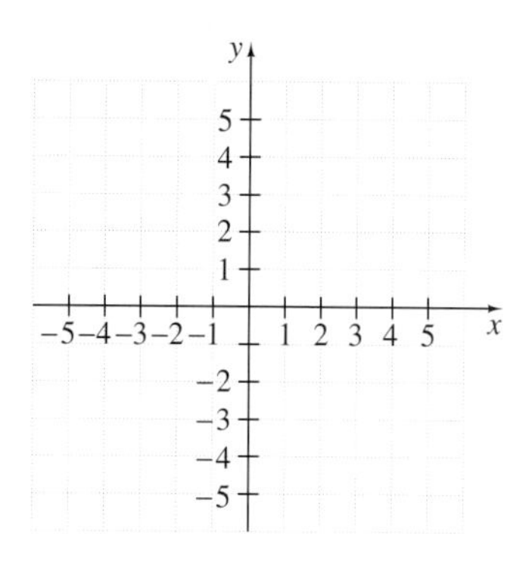

Practice Problem 3

Graph: $\dfrac{(x-1)^2}{9} + \dfrac{(y-3)^2}{16} = 1$

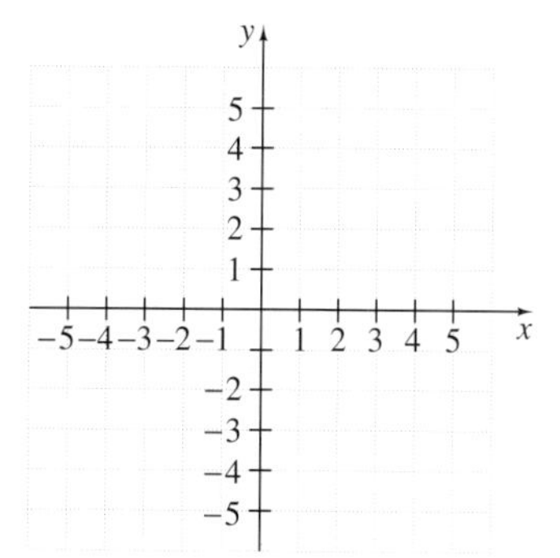

Answers

2.

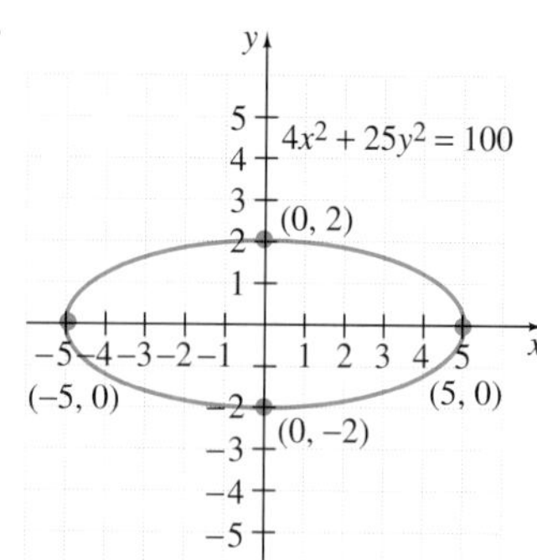

3.

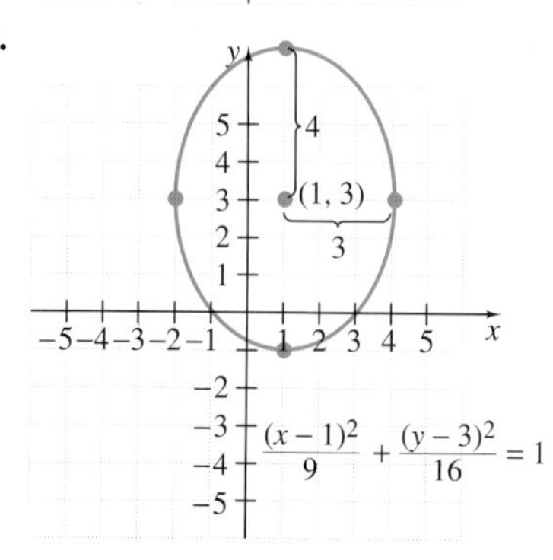

Example 2 Graph: $4x^2 + 16y^2 = 64$

Solution: Although this equation contains a sum of squared terms in x and y on the same side of an equation, this is not the equation of a circle since the coefficients of x^2 and y^2 are not the same. When this happens, the graph is an ellipse. Since the standard form of the equation of an ellipse has 1 on one side, we divide both sides of this equation by 64 to get it in standard form.

$$4x^2 + 16y^2 = 64$$

$$\frac{4x^2}{64} + \frac{16y^2}{64} = \frac{64}{64} \quad \text{Divide both sides by 64.}$$

$$\frac{x^2}{16} + \frac{y^2}{4} = 1 \quad \text{Simplify.}$$

We now recognize the equation of an ellipse with center $(0, 0)$, x-intercepts 4 and -4, and y-intercepts 2 and -2.

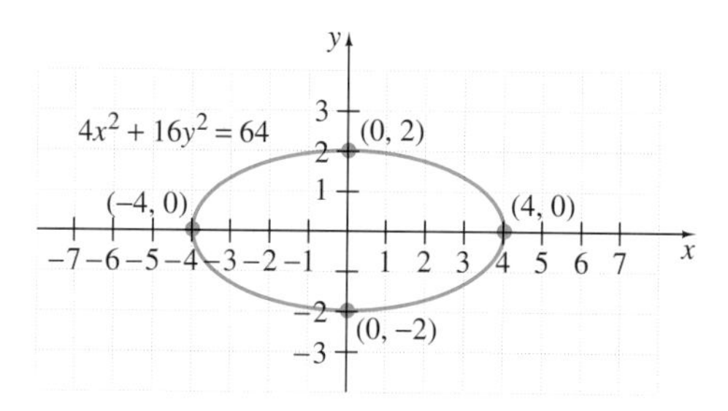

The center of an ellipse is not always $(0, 0)$, as shown in the next example.

The standard form of an ellipse with center (h, k) is

$$\frac{(x-h)^2}{a^2} + \frac{(y-k)^2}{b^2} = 1$$

Example 3 Graph: $\dfrac{(x+3)^2}{25} + \dfrac{(y-2)^2}{36} = 1$

Solution: This ellipse has center $(-3, 2)$. Notice that $a = 5$ and $b = 6$. To find four points on the graph of the ellipse, we first graph the center, $(-3, 2)$. Since $a = 5$, we count 5 units right and then 5 units left of the point with coordinates $(-3, 2)$. Next, since $b = 6$, we start at $(-3, 2)$ and count 6 units up and then 6 units down to find two more points on the ellipse.

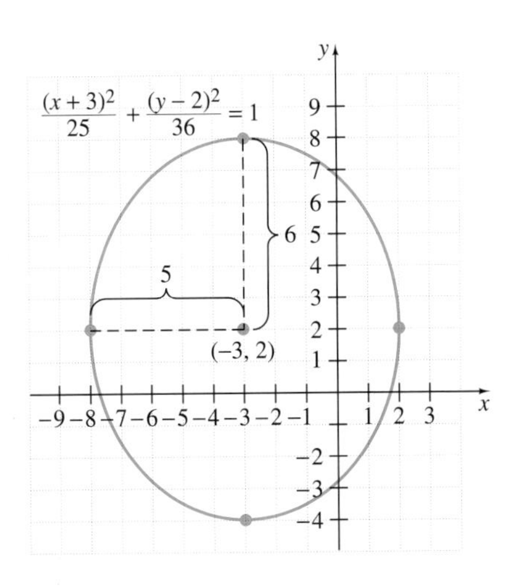

TRY THE CONCEPT CHECK IN THE MARGIN.

B GRAPHING HYPERBOLAS

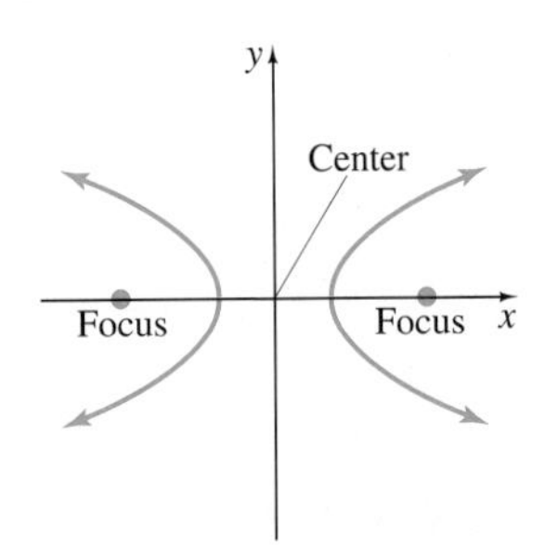

The final conic section is the **hyperbola**. A hyperbola is the set of points in a plane such that the absolute value of the difference of the distances from two fixed points is constant. Each of the two fixed points is called a **focus**. The point midway between the foci is called the **center**.

Using the distance formula, we can show that the graph of $\frac{x^2}{a^2} - \frac{y^2}{b^2} = 1$ is a hyperbola with center $(0, 0)$ and x-intercepts a and $-a$. Also, the graph of $\frac{y^2}{b^2} - \frac{x^2}{a^2} = 1$ is a hyperbola with center $(0, 0)$ and y-intercepts b and $-b$.

HYPERBOLA WITH CENTER (0, 0)

The graph of an equation of the form $\frac{x^2}{a^2} - \frac{y^2}{b^2} = 1$ is a hyperbola with center $(0, 0)$ and x-intercepts a and $-a$.

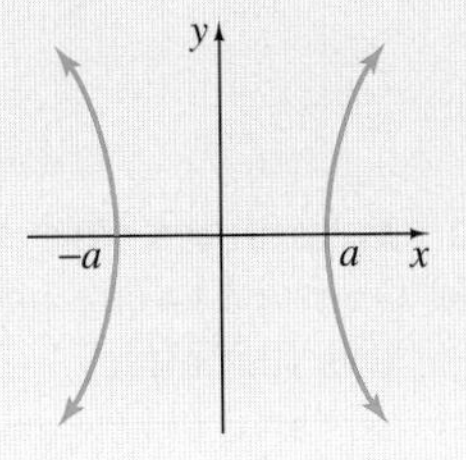

✓ CONCEPT CHECK

In the graph of the equation $\frac{x^2}{64} + \frac{y^2}{36} = 1$, which distance is longer: the distance between the x-intercepts or the distance between the y-intercepts? How much longer? Explain.

Answer

✓ **Concept Check:** x-intercepts by 4 units

The graph of an equation of the form $\frac{y^2}{b^2} - \frac{x^2}{a^2} = 1$ is a hyperbola with center $(0, 0)$ and y-intercepts b and $-b$.

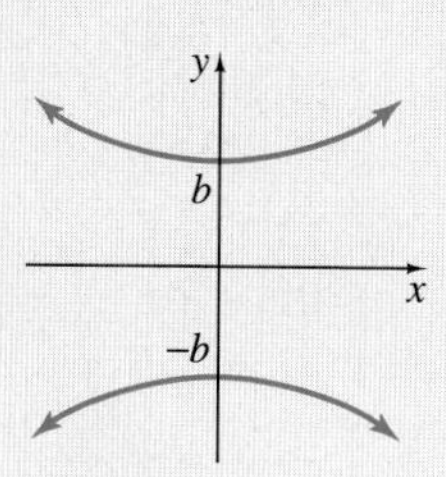

The equations $\frac{x^2}{a^2} - \frac{y^2}{b^2} = 1$ and $\frac{y^2}{b^2} - \frac{x^2}{a^2} = 1$ are the **standard forms** for the equation of a hyperbola.

HELPFUL HINT

Notice the difference between the equation of an ellipse and a hyperbola. The equation of the ellipse contains x^2 and y^2 terms on the same side of the equation with same sign coefficients. For a hyperbola, the coefficients on the same side of the equation have different signs.

Graphing a hyperbola such as $\frac{y^2}{b^2} - \frac{x^2}{a^2} = 1$ is made easier by recognizing one of its important characteristics. Examining the figure below, notice how the sides of the branches of the hyperbola extend indefinitely and seem to approach, but not intersect, the dashed lines in the figure. These dashed lines are called the **asymptotes** of the hyperbola.

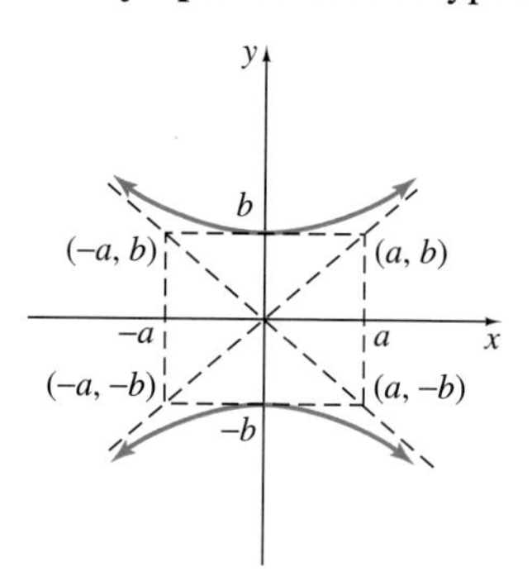

To sketch these lines, or asymptotes, draw a rectangle with vertices (a, b), $(-a, b)$, $(a, -b)$, and $(-a, -b)$. The asymptotes of the hyperbola are the extended diagonals of this rectangle.

Practice Problem 4

Graph: $\frac{x^2}{4} - \frac{y^2}{9} = 1$

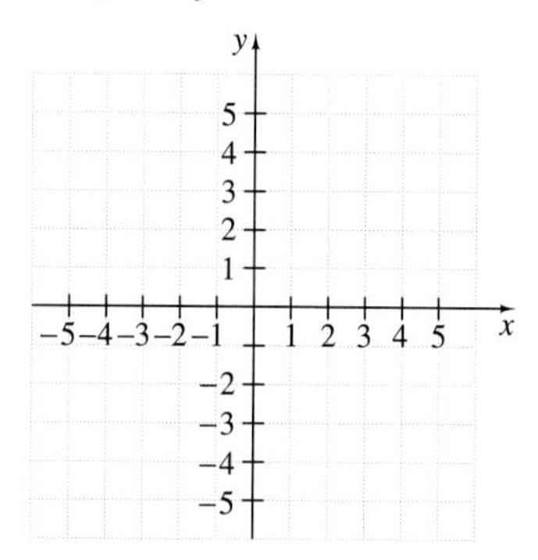

Answer

4.

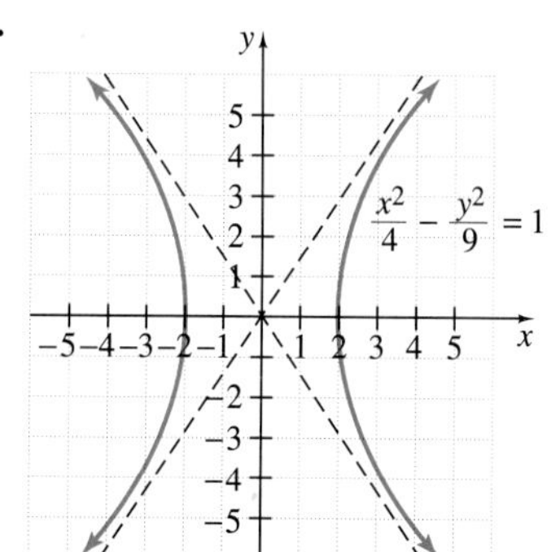

Example 4

Graph: $\frac{x^2}{16} - \frac{y^2}{25} = 1$

Solution: This equation has the form $\frac{x^2}{a^2} - \frac{y^2}{b^2} = 1$, with $a = 4$ and $b = 5$. Thus, its graph is a hyperbola with center $(0, 0)$ and x-intercepts of 4 and -4. To aid in graphing the hyperbola, we first sketch its asymptotes. The extended diagonals of the rectangle with coordinates $(4, 5)$, $(4, -5)$, $(-4, 5)$, and $(-4, -5)$ are the asymptotes of the hyperbola. Then we use the asymptotes to aid in graphing the hyperbola.

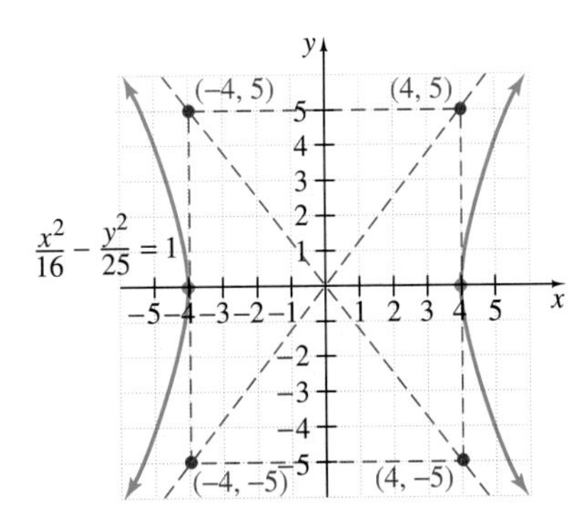

Example 5 Graph: $4y^2 - 9x^2 = 36$

Solution: Since this is a difference of squared terms in x and y on the same side of the equation, its graph is a hyperbola, as opposed to an ellipse or a circle. The standard form of the equation of a hyperbola has a 1 on one side, so we divide both sides of the equation by 36 to get it in standard form.

$$4y^2 - 9x^2 = 36$$

$$\frac{4y^2}{36} - \frac{9x^2}{36} = \frac{36}{36} \quad \text{Divide both sides by 36.}$$

$$\frac{y^2}{9} - \frac{x^2}{4} = 1 \quad \text{Simplify.}$$

The equation is of the form $\frac{y^2}{b^2} - \frac{x^2}{a^2} = 1$, with $a = 2$ and $b = 3$, so the hyperbola is centered at $(0, 0)$ with y-intercepts 3 and -3.

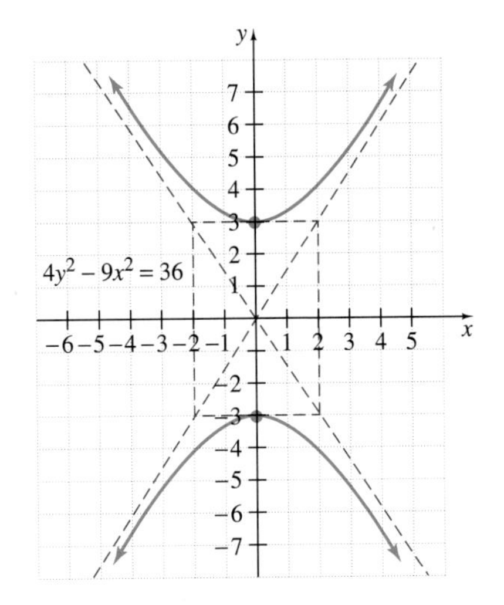

Practice Problem 5

Graph: $9y^2 - 16x^2 = 144$

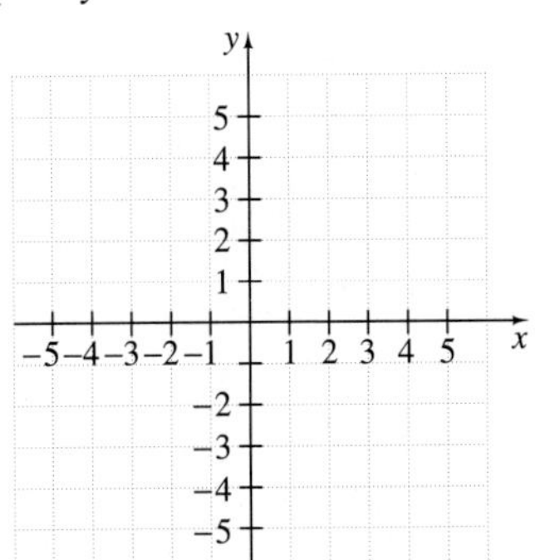

Answer

5.

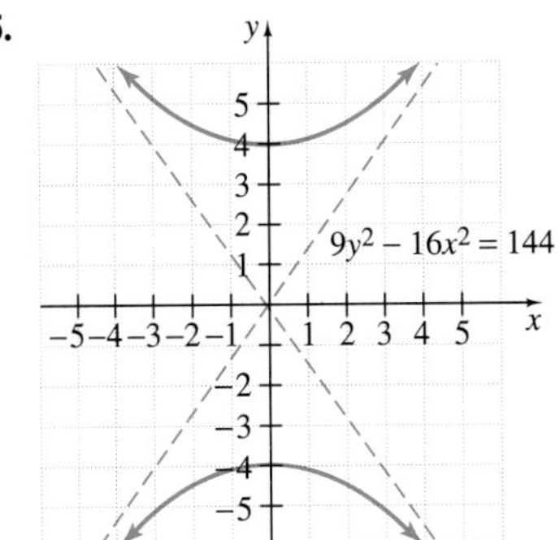

GRAPHING CALCULATOR EXPLORATIONS

To find the graph of an ellipse by using a grapher, use the same procedure as for graphing a circle. For example, to graph $x^2 + 3y^2 = 22$, first solve for y.

$$3y^2 = 22 - x^2$$

$$y^2 = \frac{22 - x^2}{3}$$

$$y = \pm\sqrt{\frac{22 - x^2}{3}}$$

Next press the $\boxed{Y=}$ key and enter $Y_1 = \sqrt{\frac{22 - x^2}{3}}$ and $Y_2 = -\sqrt{\frac{22 - x^2}{3}}$. (Insert two sets of parentheses in the radicand as $\sqrt{((22 - x^2)/3)}$ so that the desired graph is obtained.) The graph appears as follows:

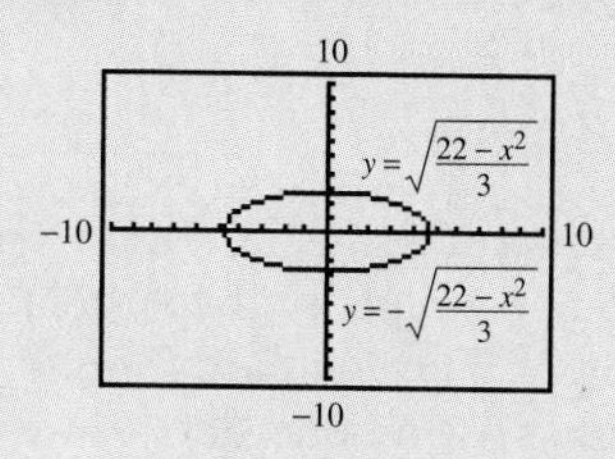

Use a grapher to graph each ellipse.

1. $10x^2 + y^2 = 32$

2. $20x^2 + 5y^2 = 100$

3. $7.3x^2 + 15.5y^2 = 95.2$

4. $18.8x^2 + 36.1y^2 = 205.8$

Name ______________________ Section ________ Date ________

MENTAL MATH ANSWERS

1. ______
2. ______
3. ______
4. ______
5. ______
6. ______

MENTAL MATH

Identify the graph of each as an ellipse or a hyperbola.

1. $\dfrac{x^2}{16} + \dfrac{y^2}{4} = 1$
2. $\dfrac{x^2}{16} - \dfrac{y^2}{4} = 1$
3. $x^2 - 5y^2 = 3$
4. $-x^2 + 5y^2 = 3$
5. $-\dfrac{y^2}{25} + \dfrac{x^2}{36} = 1$
6. $\dfrac{y^2}{25} + \dfrac{x^2}{36} = 1$

EXERCISE SET 9.2

A *Graph each equation. See Examples 1 and 2.*

1. $\dfrac{x^2}{4} + \dfrac{y^2}{25} = 1$
2. $\dfrac{x^2}{9} + y^2 = 1$
3. $\dfrac{x^2}{16} + \dfrac{y^2}{9} = 1$
4. $x^2 + \dfrac{y^2}{4} = 1$

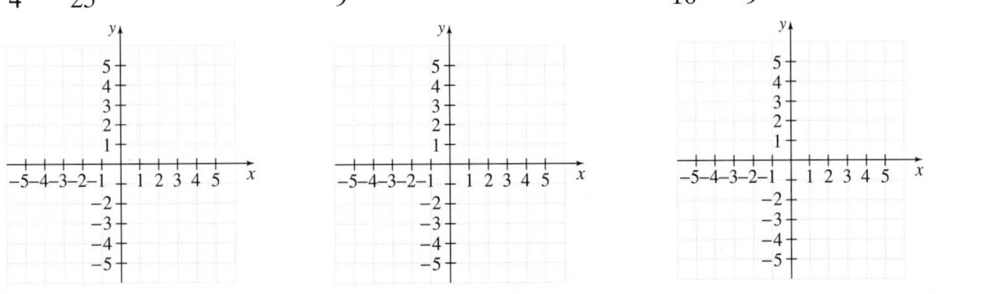

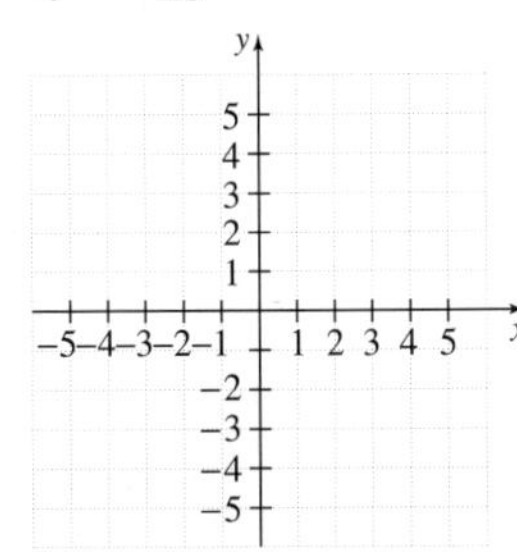

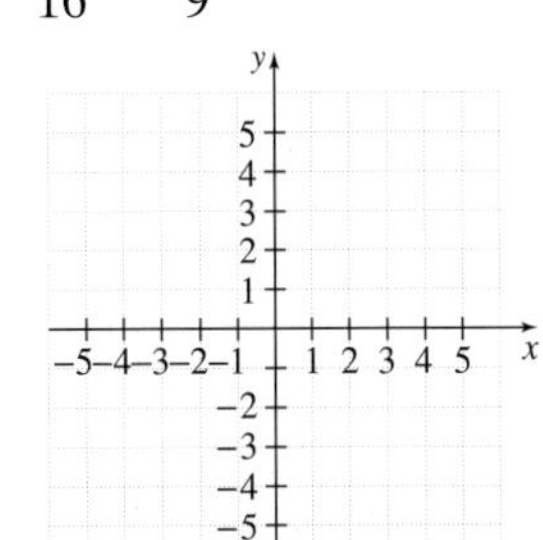

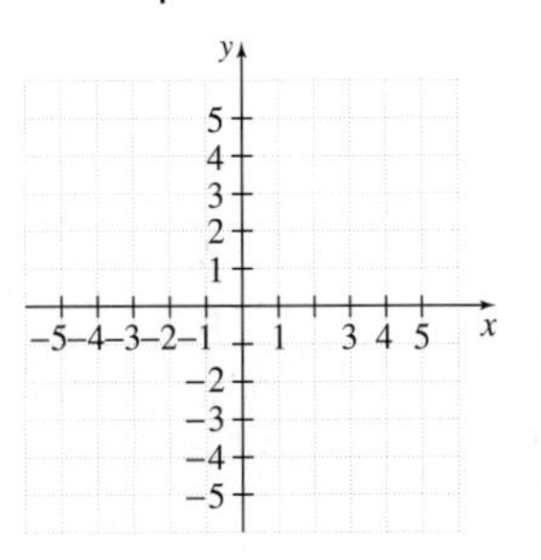

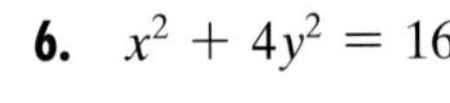

5. $9x^2 + 4y^2 = 36$
6. $x^2 + 4y^2 = 16$
7. $4x^2 + 25y^2 = 100$
8. $36x^2 + y^2 = 36$

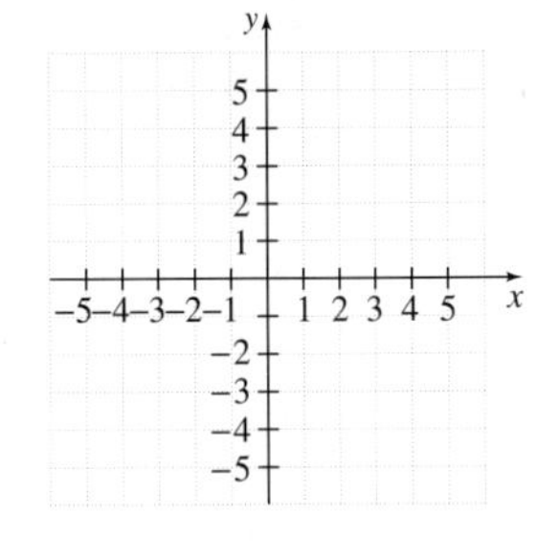

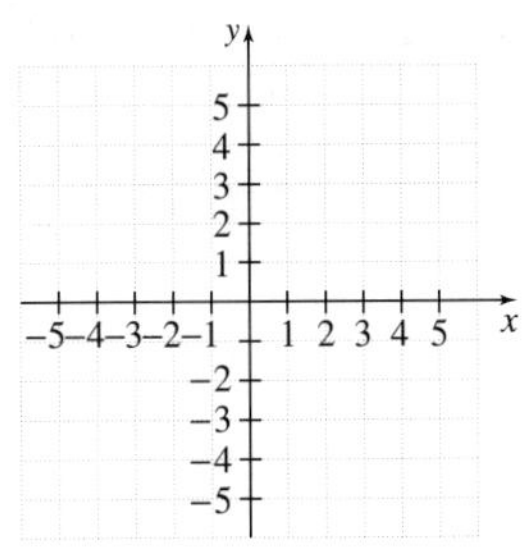

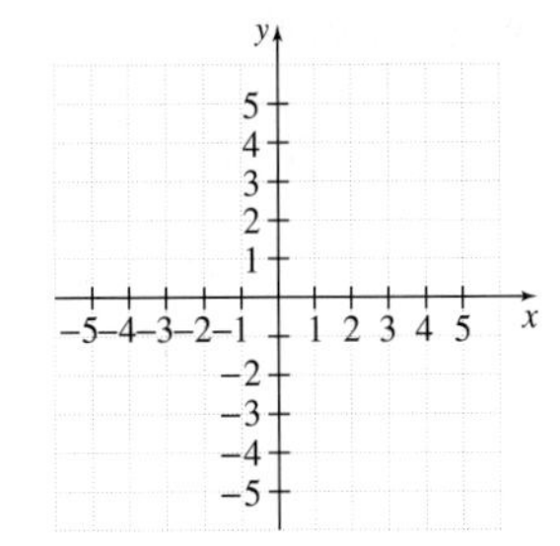

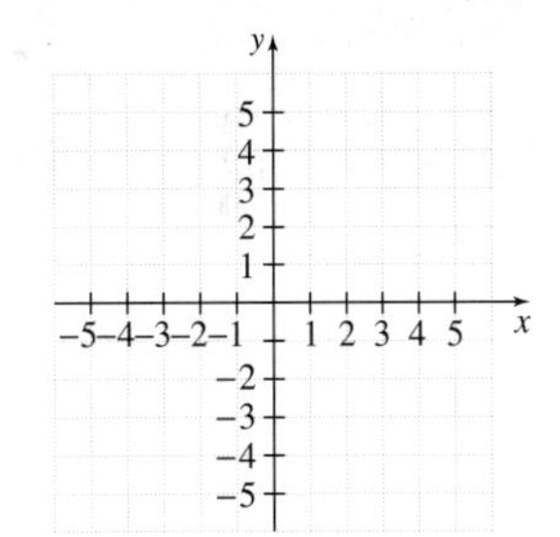

Graph each equation. See Example 3.

9. $\dfrac{(x+1)^2}{36} + \dfrac{(y-2)^2}{49} = 1$
10. $\dfrac{(x-3)^2}{9} + \dfrac{(y+3)^2}{16} = 1$
11. $\dfrac{(x-1)^2}{4} + \dfrac{(y-1)^2}{25} = 1$
12. $\dfrac{(x+3)^2}{16} + \dfrac{(y+2)^2}{4} = 1$

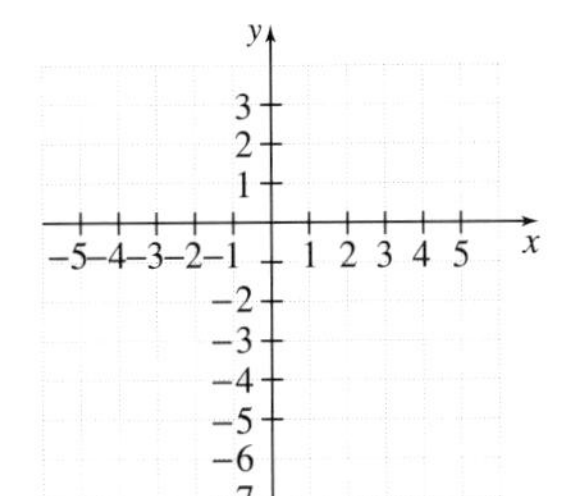

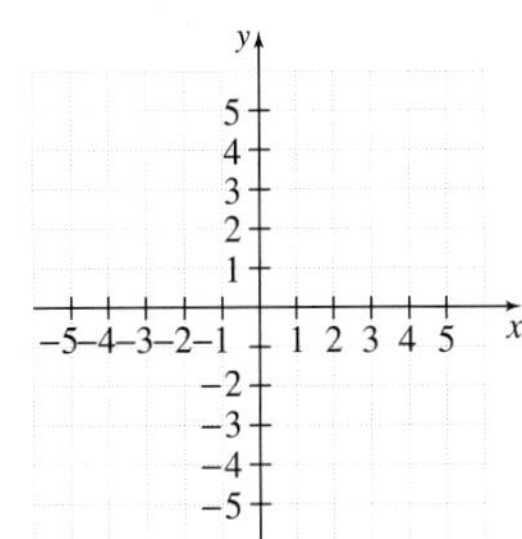

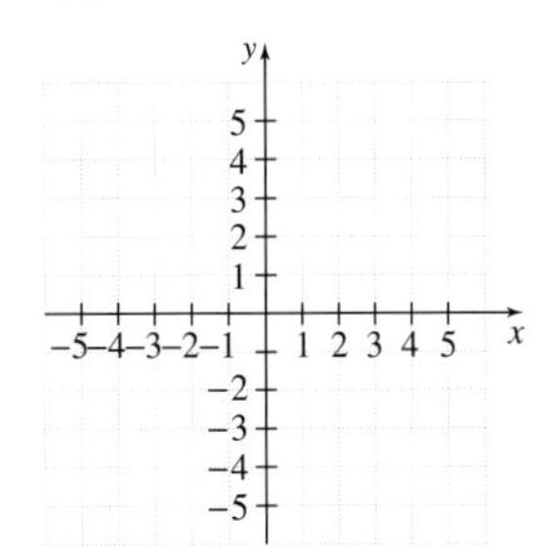

Name ______________________________

B *Graph each equation. See Examples 4 and 5.*

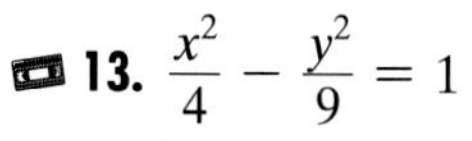

13. $\dfrac{x^2}{4} - \dfrac{y^2}{9} = 1$

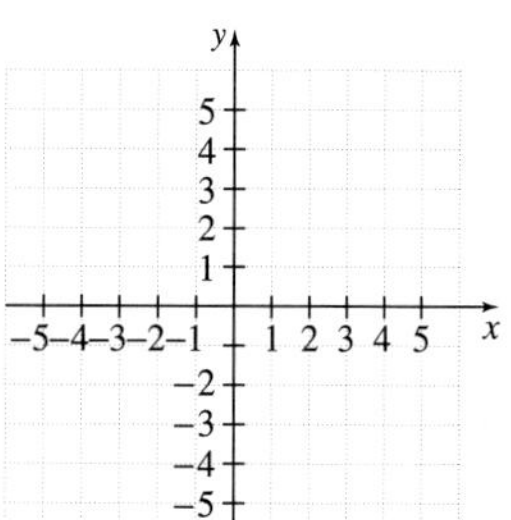

14. $\dfrac{x^2}{36} - \dfrac{y^2}{36} = 1$

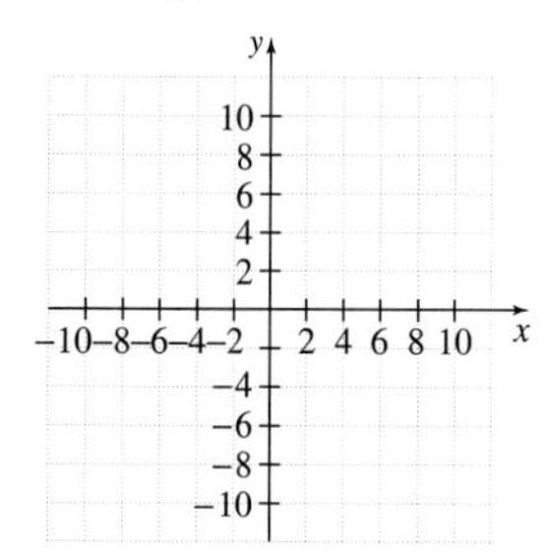

15. $\dfrac{y^2}{25} - \dfrac{x^2}{16} = 1$

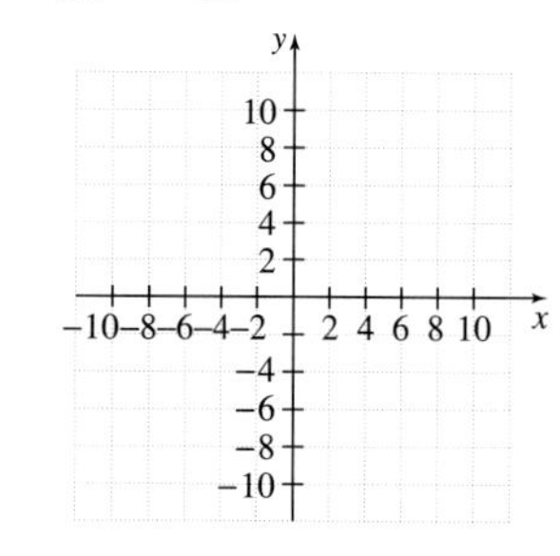

16. $\dfrac{y^2}{25} - \dfrac{x^2}{49} = 1$

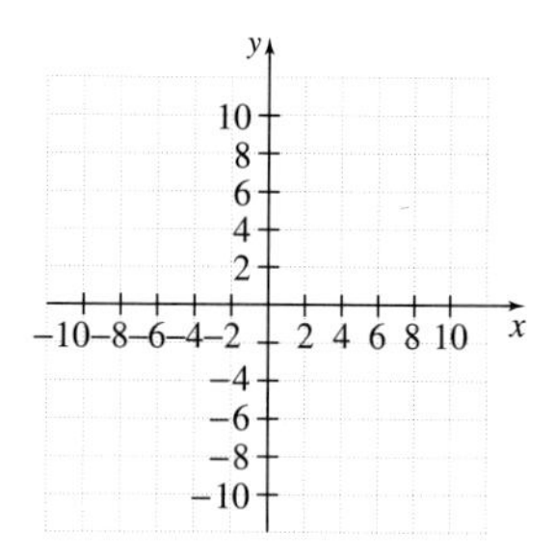

17. $x^2 - 4y^2 = 16$

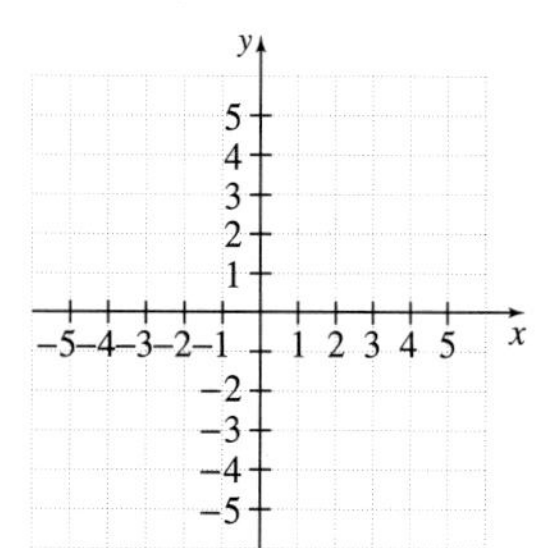

18. $4x^2 - y^2 = 36$

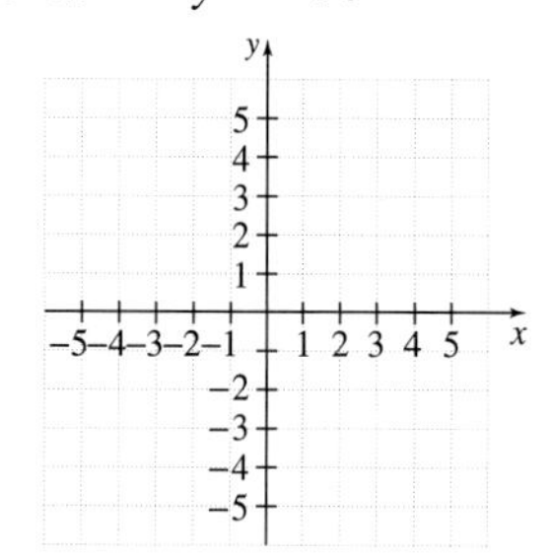

19. $16y^2 - x^2 = 16$

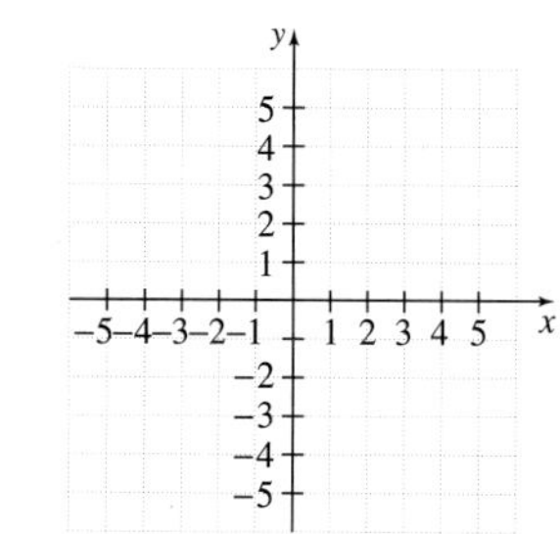

20. $4y^2 - 25x^2 = 100$

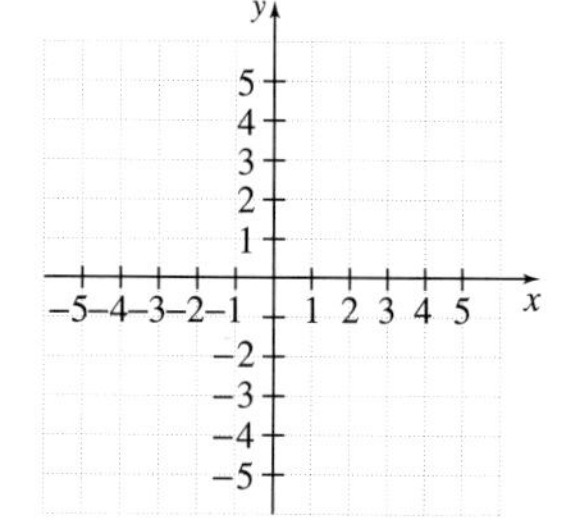

Review and Preview

Perform each indicated operation. See Sections 5.1 and 5.2.

21. $(2x^3)(-4x^2)$

22. $2x^3 - 4x^3$

23. $-5x^2 + x^2$

24. $(-5x^2)(x^2)$

Combining Concepts

25. We know that $x^2 + y^2 = 25$ is the equation of a circle. Rewrite the equation so that the right side is equal to 1. Which type of conic section does this equation form resemble? In fact, the circle is a special case of this type of conic section. Describe the conditions under which this type of conic section is a circle.

Answers

21. ____________

22. ____________

23. ____________

24. ____________

25. ____________

Name ______________________________

The orbits of stars, planets, comets, asteroids, and satellites all have the shape of one of the conic sections. Astronomers use a measure called eccentricity *to describe the shape and elongation of an orbital path. For the circle and ellipse, eccentricity e is calculated with the formula* $e = \frac{c}{d}$, *where* $c^2 = |a^2 - b^2|$ *and d is the larger value of a or b. For a hyperbola, eccentricity e is calculated with the formula* $e = \frac{c}{d}$, *where* $c^2 = a^2 + b^2$ *and the value of d is equal to a if the hyperbola has x-intercepts or equal to b if the hyperbola has y-intercepts.*

A $\frac{x^2}{36} - \frac{y^2}{13} = 1$ **B** $\frac{x^2}{4} + \frac{y^2}{4} = 1$ **C** $\frac{x^2}{25} + \frac{y^2}{16} = 1$ **D** $\frac{y^2}{25} - \frac{x^2}{39} = 1$

E $\frac{x^2}{17} + \frac{y^2}{81} = 1$ **F** $\frac{x^2}{36} + \frac{y^2}{36} = 1$ **G** $\frac{x^2}{16} - \frac{y^2}{65} = 1$ **H** $\frac{x^2}{144} + \frac{y^2}{140} = 1$

26. Identify the type of conic section represented by each of the equations A–H.

27. For each of the equations A–H, identify the values of a^2 and b^2.

28. For each of the equations A–H, calculate the value of c^2 and c.

29. For each of the equations A–H, find the value of d.

30. For each of the equations A–H, calculate the eccentricity e.

31. What do you notice about the values of e for the equations you identified as ellipses?

32. What do you notice about the values of e for the equations you identified as circles?

33. What do you notice about the values of e for the equations you identified as hyperbolas?

34. The eccentricity of a parabola is exactly 1. Use this information and the observations you made in Exercises 31, 32 and 33 to describe a way that could be used to identify the type of conic section based on its eccentricity value.

26. ______________

27. ______________

28. ______________

29. ______________

30. ______________

31. ______________

32. ______________

33. ______________

34. ______________

Name ____________________

35. Graph each of the ellipses given in equations A–H. What do you notice about the shape of these ellipses for increasing values of eccentricity? Which is the most elliptical? Which is the least elliptical, that is, the most circular?

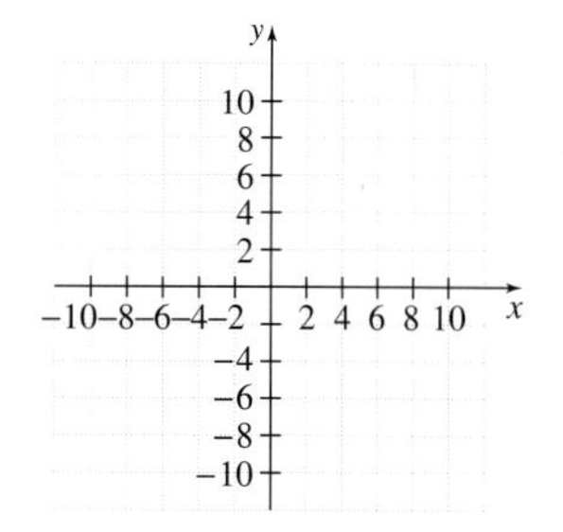

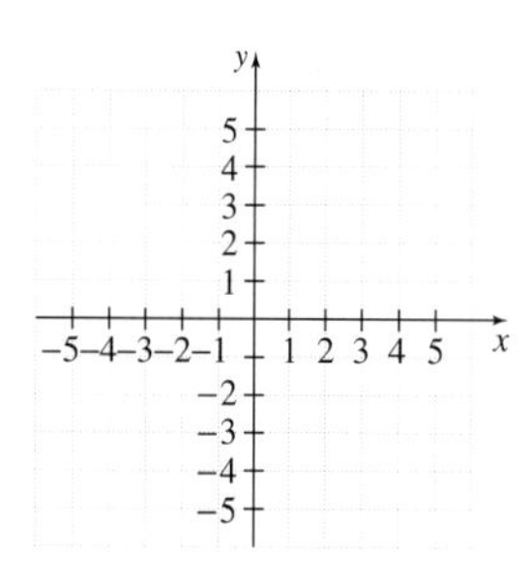

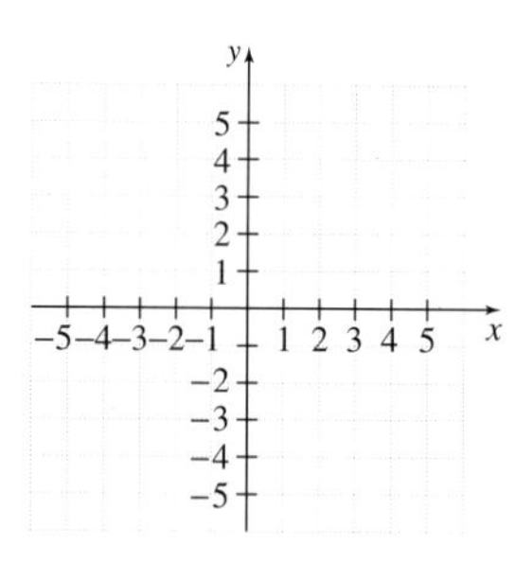

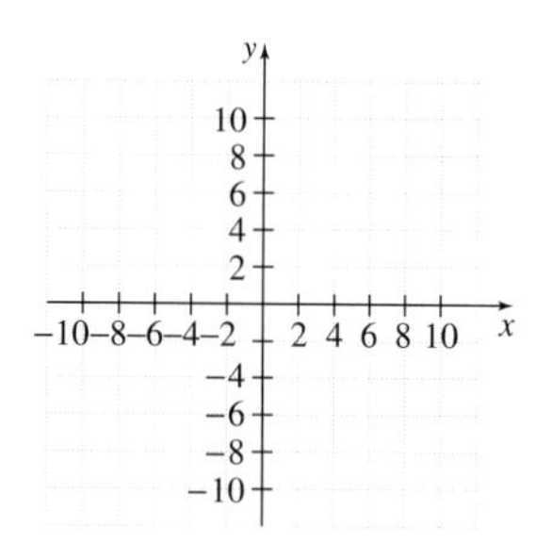

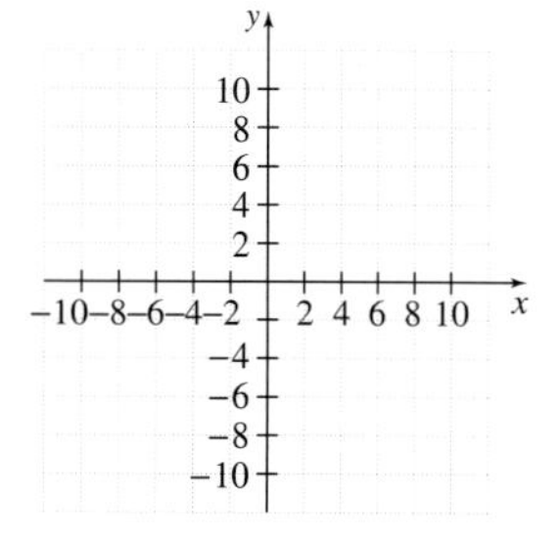

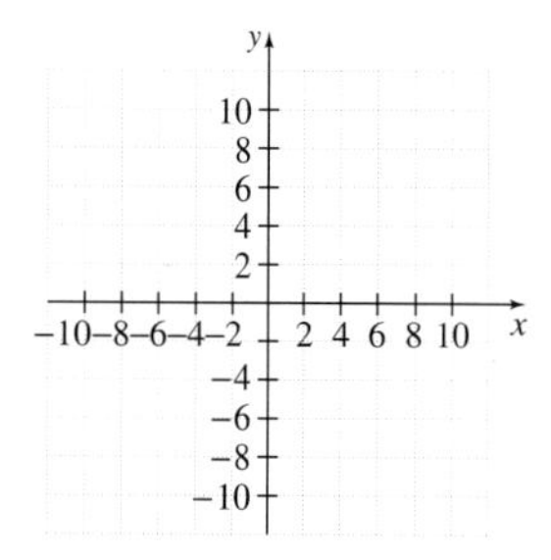

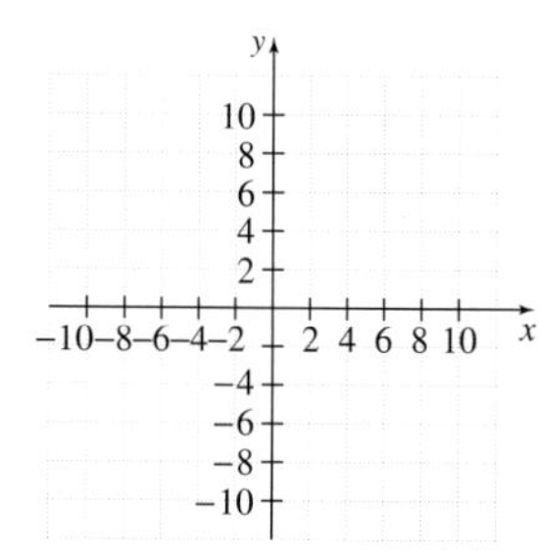

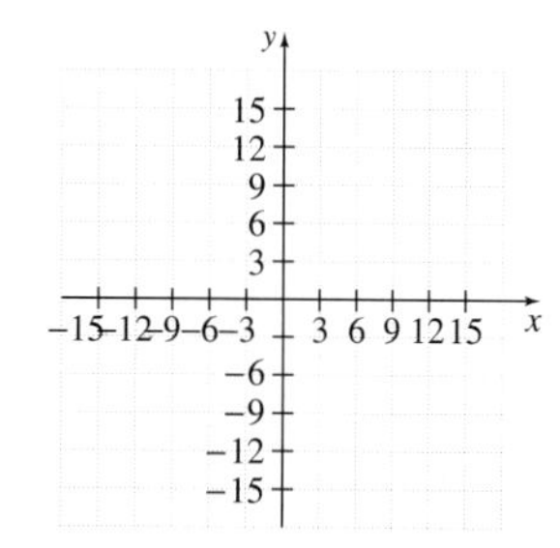

35. ____________________

36. ____________________

37. ____________________

Internet Excursions

[http://ssd.jpl.nasa.gov]
Go to http://www.prenhall.com/martin-gay
By going to the World Wide Web address listed above, you will be directed to a web site where you can look up information to help you answer the questions below.

36. Under Planets, select the Mean Orbital Elements option on this home page. This gives data about the nine planets of our solar system, including the eccentricities of their orbits. Using the data for the eccentricities of the planets' orbits and your conclusions from Exercise 34, decide which type of conic describes the orbital paths of all the planets. Which planet has the most circular path? Which planet has the most elliptical path?

37. Return to the JPL Solar System Dynamics page. Under Comets and Asteroids, select the Orbital Elements option. Then choose the Comets option. This gives data about the orbits of comets known to pass through our solar system, including their eccentricities. Using the data for the eccentricities of the comets' orbits and your conclusions from Exercise 34, decide which type of conic section describes the orbital paths of the majority of the comets. Which comets are the exceptions? What type of orbital paths do these comets have?

Name ______________________ Section ________ Date ________

Answers

Integrated Review—Graphing Conic Sections

Following is a summary of conic sections.

Conic Sections

	Standard Form	Graph
Parabola	$y = a(x - h)^2 + k$	
Parabola	$x = a(y - k)^2 + h$	
Circle	$(x - h)^2 + (y - k)^2 = r^2$	
Ellipse	$\frac{x^2}{a^2} + \frac{y^2}{b^2} = 1$	
Hyperbola	$\frac{x^2}{a^2} - \frac{y^2}{b^2} = 1$	
Hyperbola	$\frac{y^2}{b^2} - \frac{x^2}{a^2} = 1$	

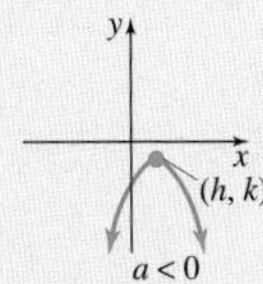

y
x
(h, k)
a > 0

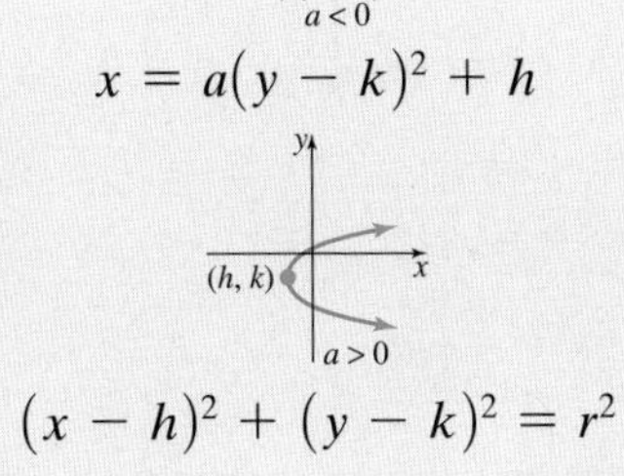

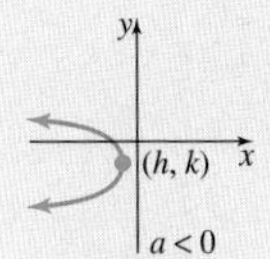

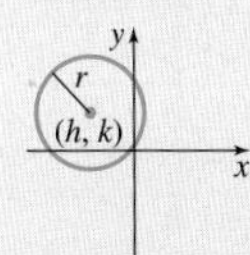

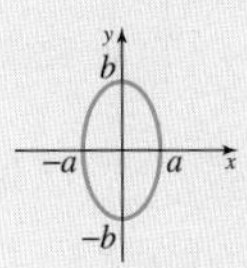

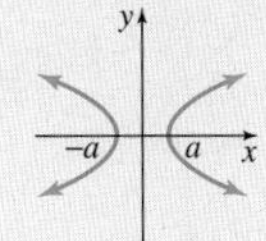

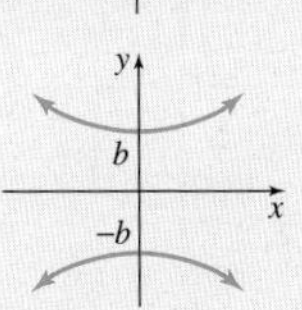

Identify whether each equation, when graphed, will be a parabola, circle, ellipse, or hyperbola. Then graph each equation.

1. $(x - 7)^2 + (y - 2)^2 = 4$

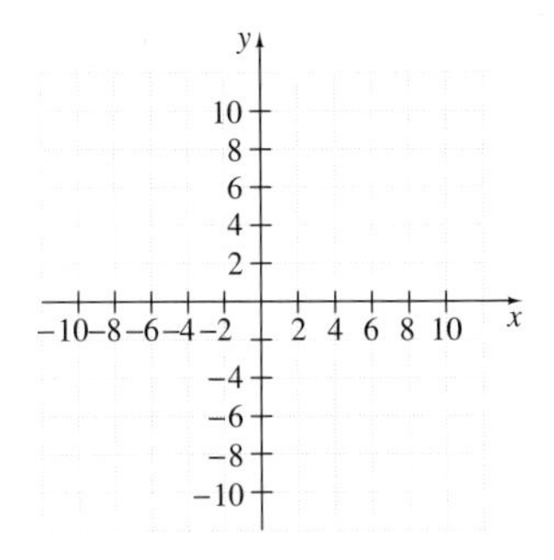

2. $y = x^2 + 4$

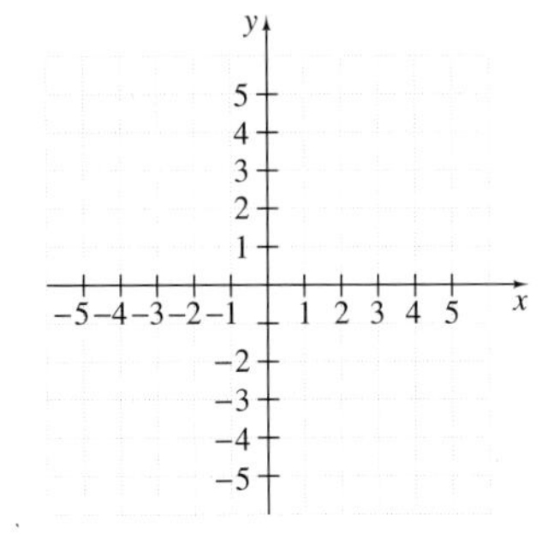

3. $y = x^2 + 12x + 36$

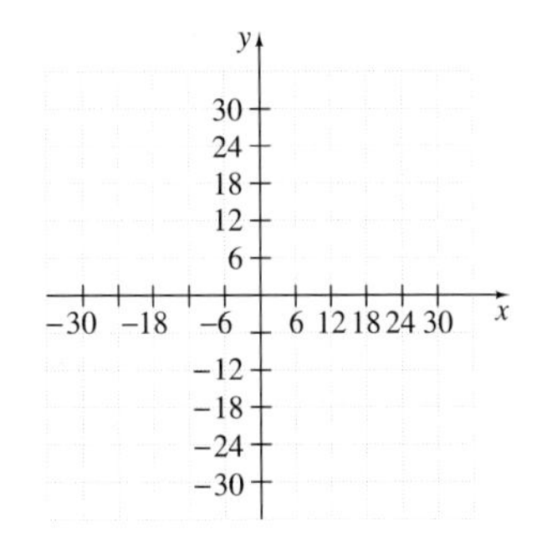

1. ____________

2. ____________

3. ____________

4. ______

5. ______

6. ______

7. ______

8. ______

9. ______

10. ______

11. ______

12. ______

13. ______

14. ______

15. ______

Name ______

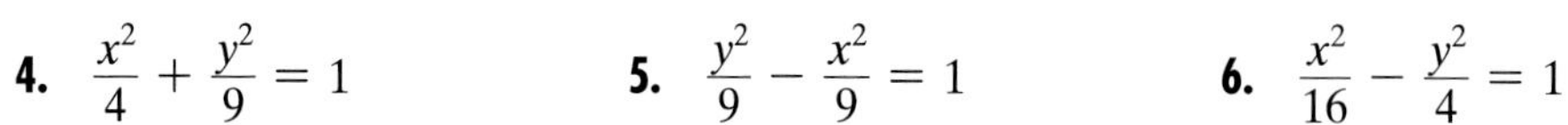

4. $\frac{x^2}{4} + \frac{y^2}{9} = 1$

5. $\frac{y^2}{9} - \frac{x^2}{9} = 1$

6. $\frac{x^2}{16} - \frac{y^2}{4} = 1$

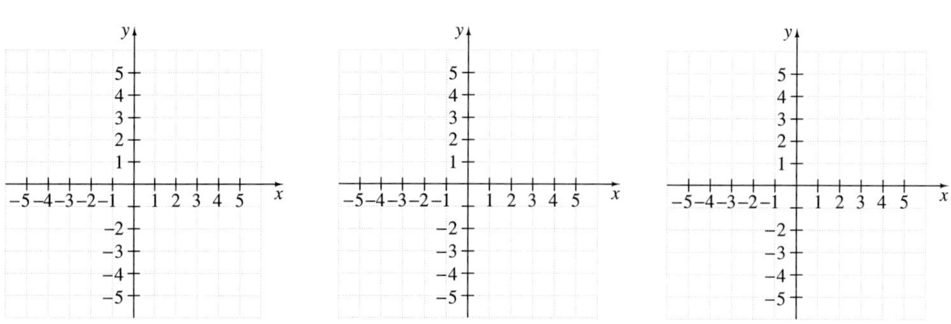

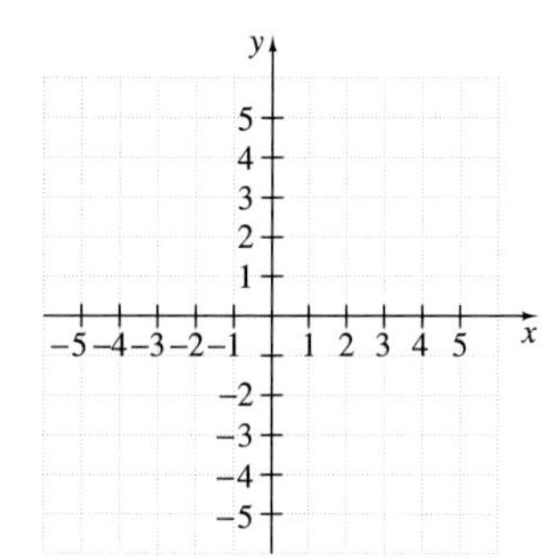

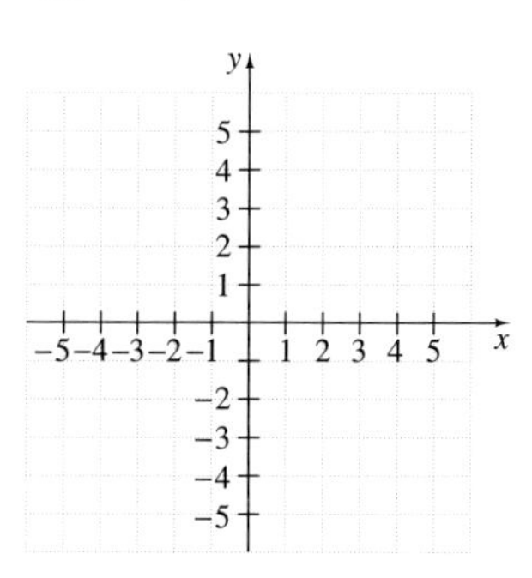

7. $\frac{x^2}{16} + \frac{y^2}{4} = 1$

8. $x^2 + y^2 = 16$

9. $x = y^2 + 4y - 1$

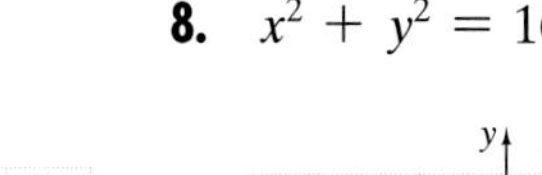

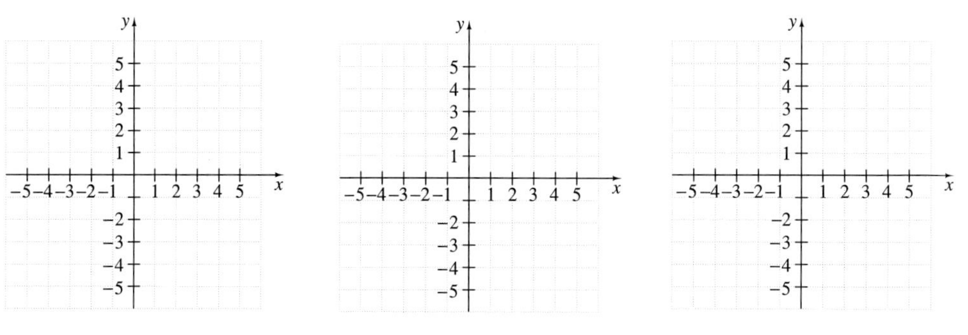

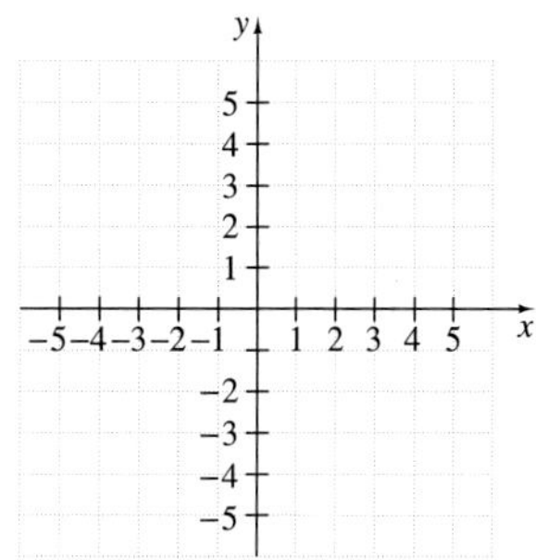

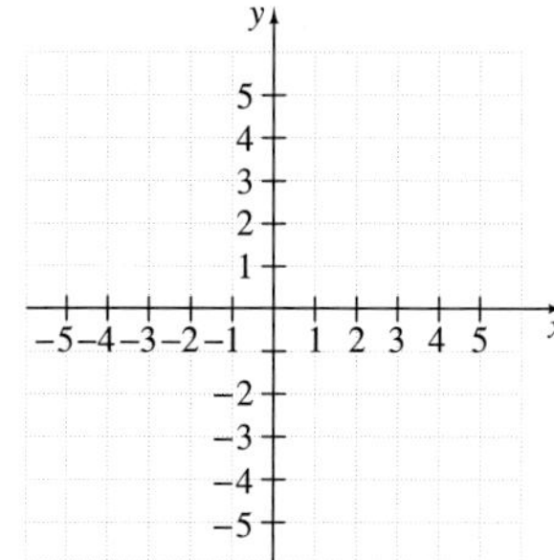

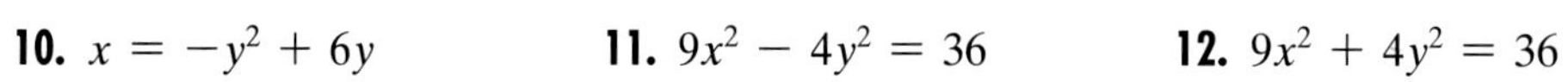

10. $x = -y^2 + 6y$

11. $9x^2 - 4y^2 = 36$

12. $9x^2 + 4y^2 = 36$

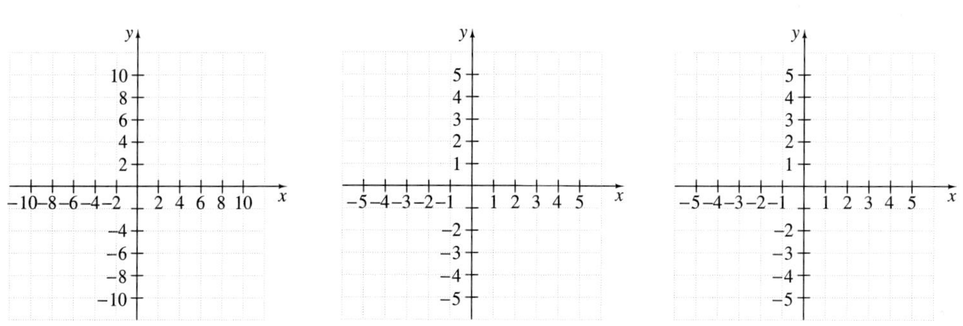

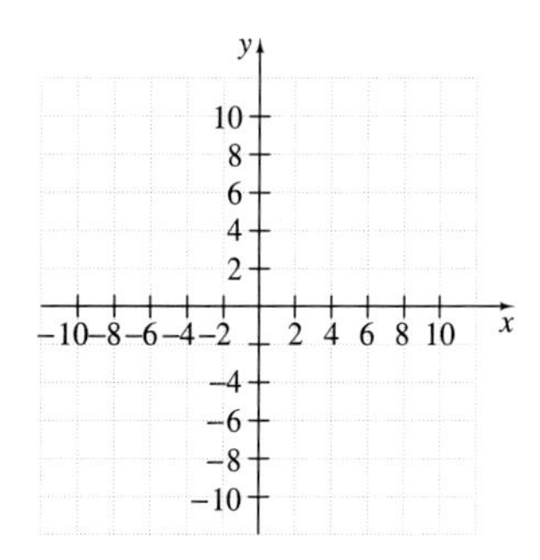

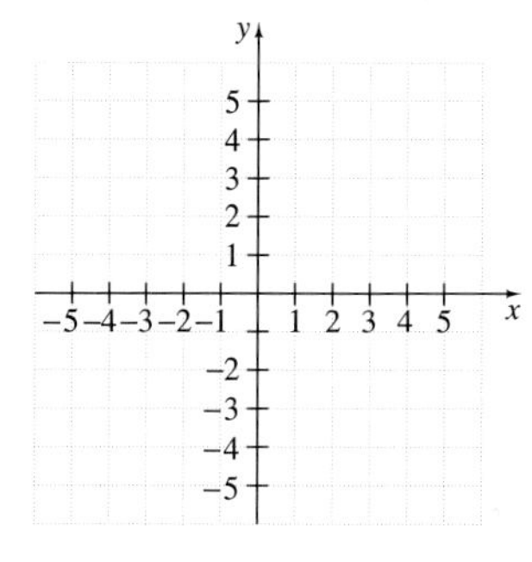

13. $\frac{(x-1)^2}{49} + \frac{(y+2)^2}{25} = 1$

14. $y^2 = x^2 + 16$

15. $\left(x + \frac{1}{2}\right)^2 + \left(y - \frac{1}{2}\right)^2 = 1$

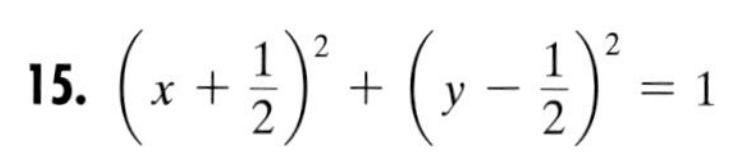

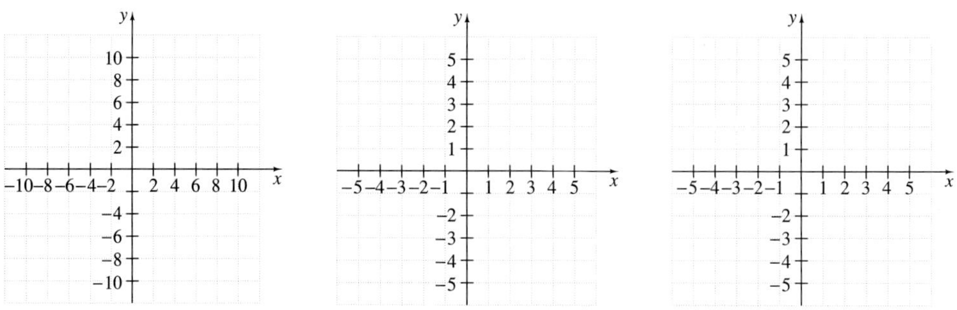

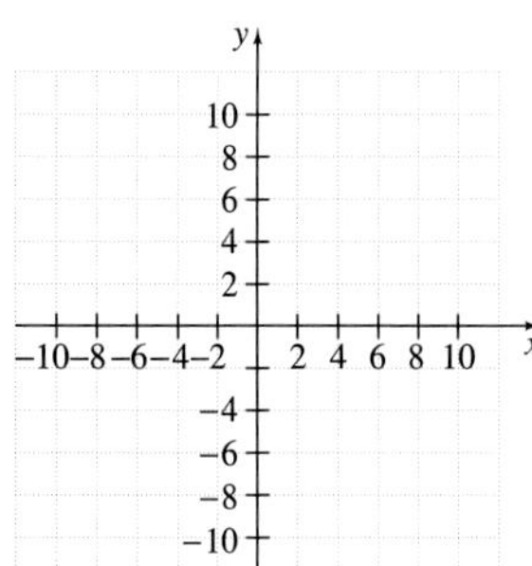

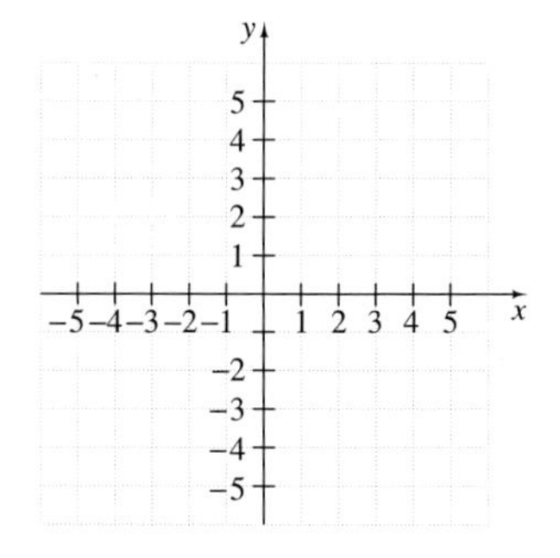

9.3 GRAPHING NONLINEAR FUNCTIONS

A GRAPHING NONLINEAR FUNCTIONS

Recall that the graph of $f(x) = x^2$ is a parabola with vertex $(0, 0)$. How does the graph of $g(x) = (x - 3)^2 + 2$ compare? Its graph is also a parabola, but with vertex $(3, 2)$. In other words, the graph of $g(x)$ is the same as the graph of $f(x)$, except that it has been shifted 3 units to the right and 2 units up. Keep this in mind as we graph other elementary functions. Remember, we are graphing functions, so all graphs should pass the vertical line test.

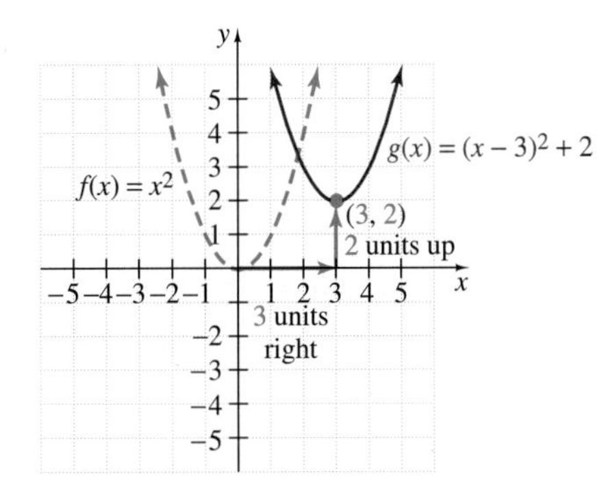

Objectives

A Graph the nonlinear functions $f(x) = |x|$ and $f(x) = \sqrt{x}$.

SSM CD-ROM Video 9.3

Example 1 Graph: $f(x) = |x|$

Solution: This is not a linear function, and its graph is not a line. Because we do not know the shape of this graph, we find many ordered pair solutions. We will choose x-values and substitute to find corresponding y-values.

If $x = -3$, then $y = |-3|$, or 3.
If $x = -2$, then $y = |-2|$, or 2.
If $x = -1$, then $y = |-1|$, or 1.
If $x = 0$, then $y = |0|$, or 0.
If $x = 1$, then $y = |1|$, or 1.
If $x = 2$, then $y = |2|$, or 2.
If $x = 3$, then $y = |3|$, or 3.

x	y
−3	3
−2	2
−1	1
0	0
1	1
2	2
3	3

Study the table of values for a moment and notice any patterns.

From the plotted ordered pairs, we see that the graph of this absolute value function is V-shaped.

Practice Problem 1

Graph: $f(x) = |x| + 1$

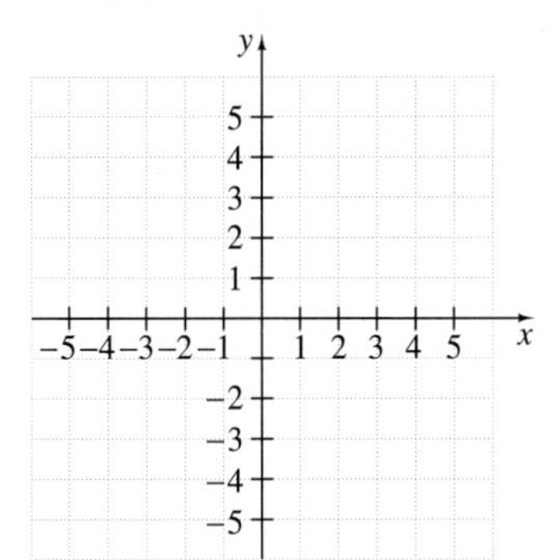

Answer

1.

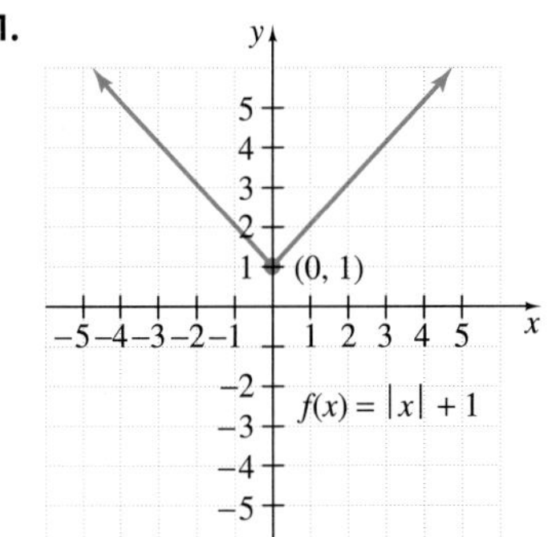

Practice Problem 2

Graph $f(x) = |x| - 1$

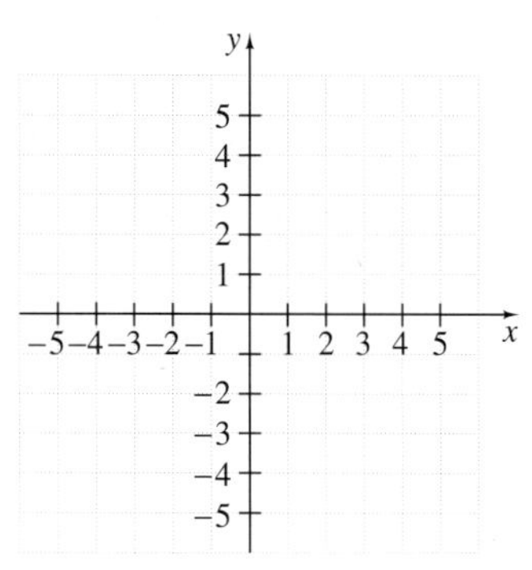

Practice Problem 3

Graph: $f(x) = |x - 1|$

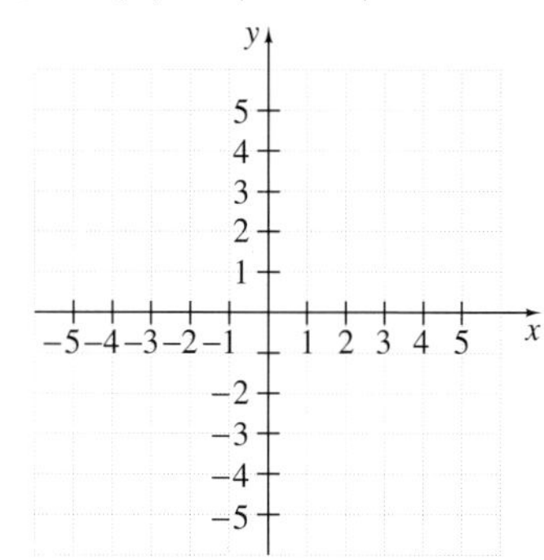

2.

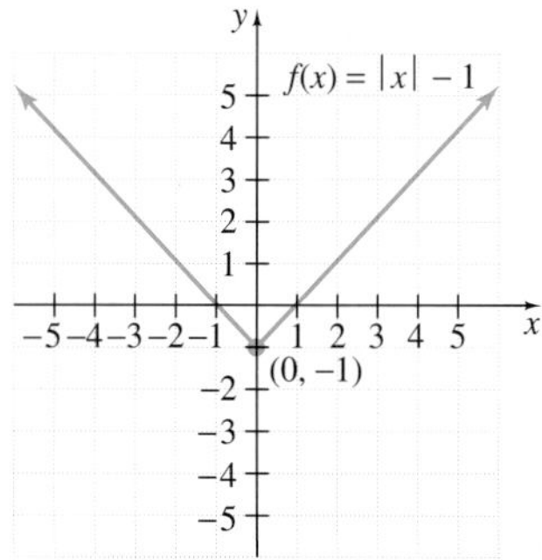

3.

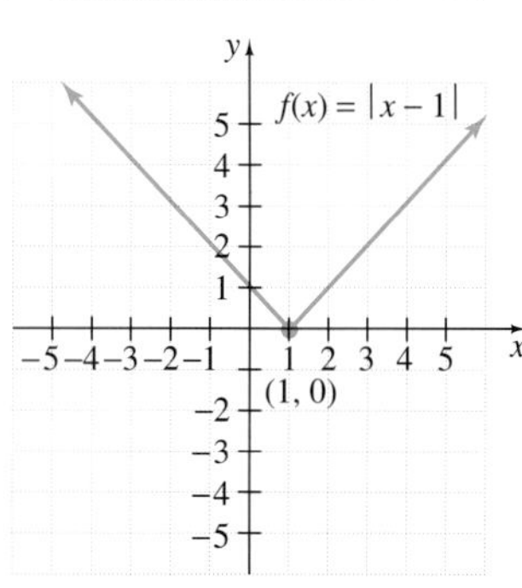

Example 2 Graph: $f(x) = |x| - 3$

Solution: To graph $f(x)$ or $y = |x| - 3$, we choose x-values and substitute to find corresponding y-values.

x	y
−3	0
−2	−1
−1	−2
0	−3
1	−2
2	−1
3	0

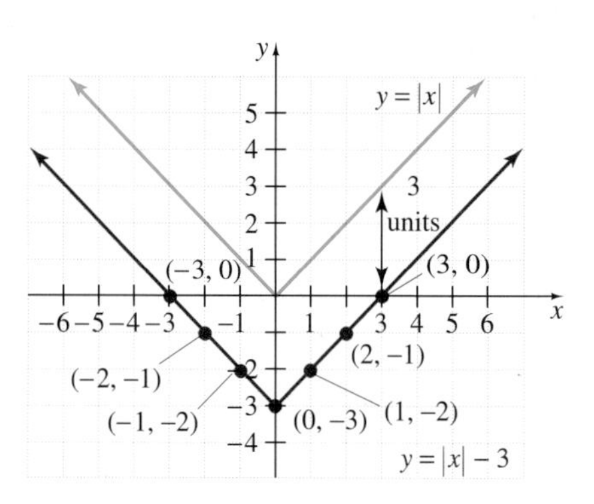

Recall that the graph of $y = x^2 - 3$ is the same as the graph of $y = x^2$ lowered 3 units. Now compare the graph of $y = |x|$ with the graph of $y = |x| - 3$. The graph of $y = |x| - 3$ is the same as the graph of $y = |x|$ lowered 3 units.

Example 3 Graph: $f(x) = |x - 2|$

Solution: First let's think about the graph of $y = (x - 2)^2$. The vertex of this graph is $(2, 0)$. In other words, the graph of $y = (x - 2)^2$ is the same as the graph of $y = x^2$ shifted to the right 2 units.

In the same manner, the graph of $y = |x - 2|$ is the same as the graph of $y = |x|$ shifted to the right 2 units. We use this knowledge along with a table of ordered-pair solutions to graph the function.

x	y
0	2
1	1
2	0
3	1
4	2

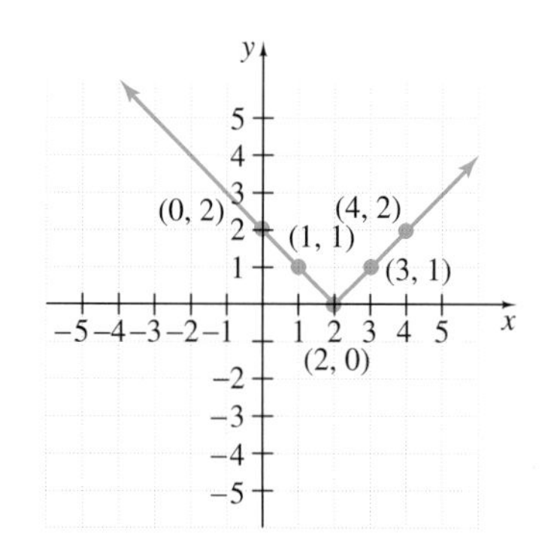

Example 4 Graph: $f(x) = |x - 1| + 2$

Solution: The graph of $y = (x - 1)^2 + 2$ has vertex $(1, 2)$. In other words, it is the graph of $y = x^2$ shifted 1 unit to the right and 2 units up. Similarly, the graph of $y = |x - 1| + 2$ is the graph of $y = |x|$ shifted 1 unit to the right and 2 units up. We use this knowledge along with a table of ordered-pair solutions to graph the function.

x	y
−1	4
0	3
1	2
2	3
3	4

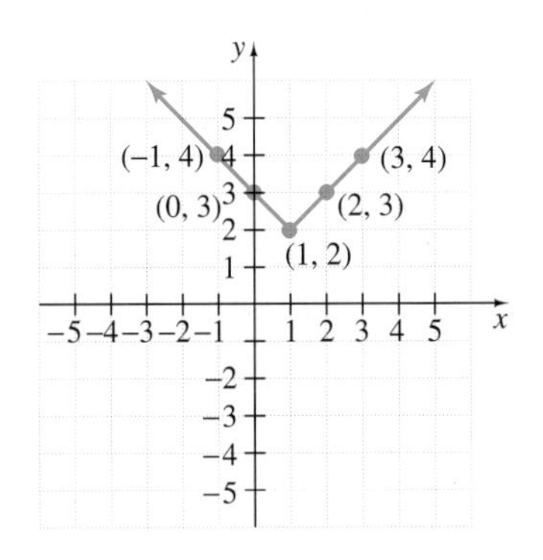

Practice Problem 4

Graph: $f(x) = |x - 3| + 2$

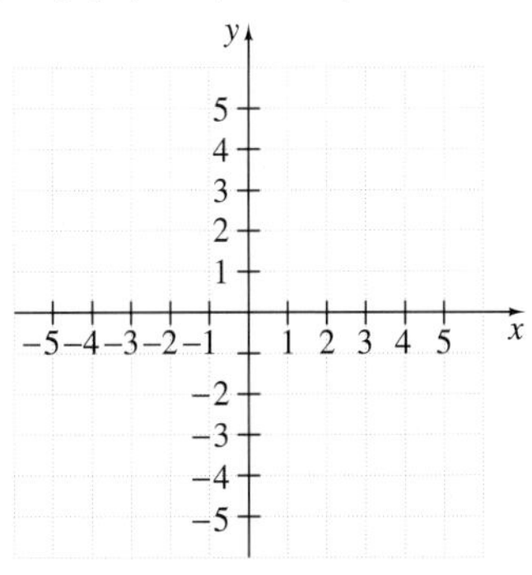

Recall that the domain of a function is basically the set of all possible x-values for that function. The domains of the functions thus far in this section have been the set of all real numbers. This is not the case for our next function, the square root function.

Example 5 Graph the square root function $f(x) = \sqrt{x}$.

Solution: Recall that the square root of a negative number is not a real number. To graph this function, evaluate the function for several values of x, plot the resulting points, and connect the points with a smooth curve. This means that the domain of this function is the set of all nonnegative numbers or $\{x|x \geq 0\}$.

x	y
0	0
1	1
3	$\sqrt{3} \approx 1.7$
4	2
9	3

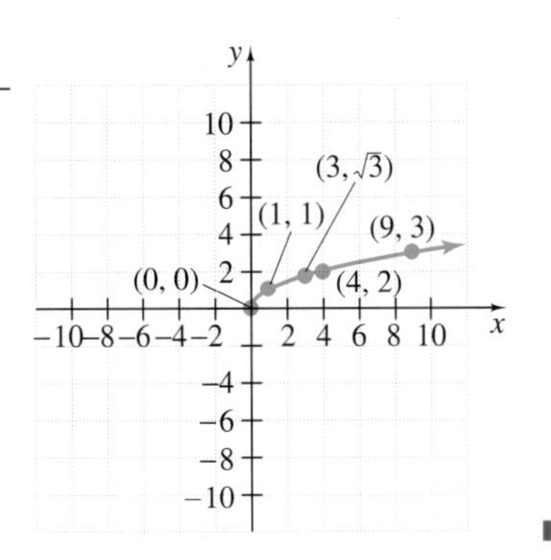

Practice Problem 5

Graph: $f(x) = \sqrt{x} + 2$

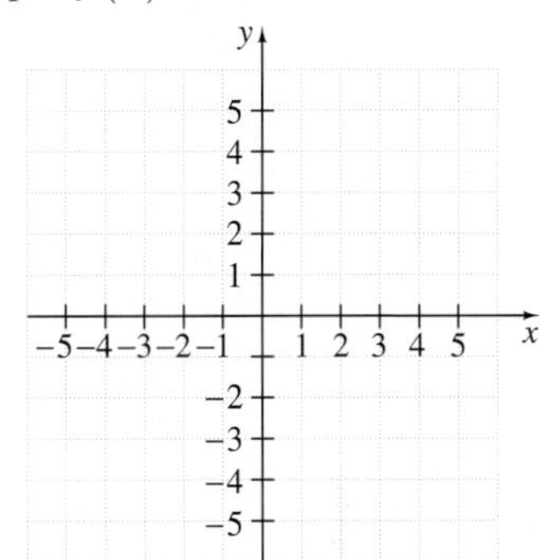

Answers

4.

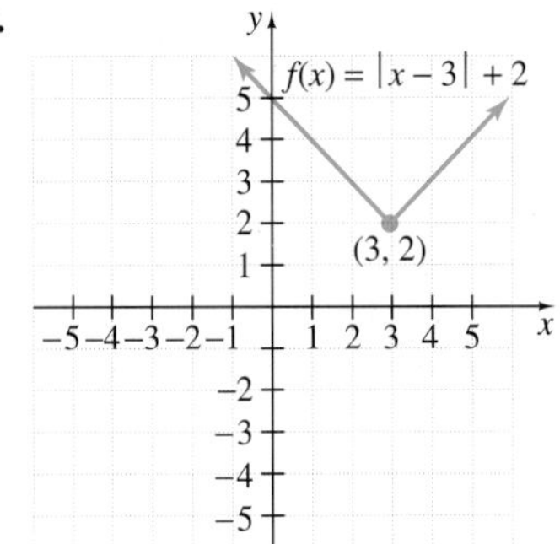

5.

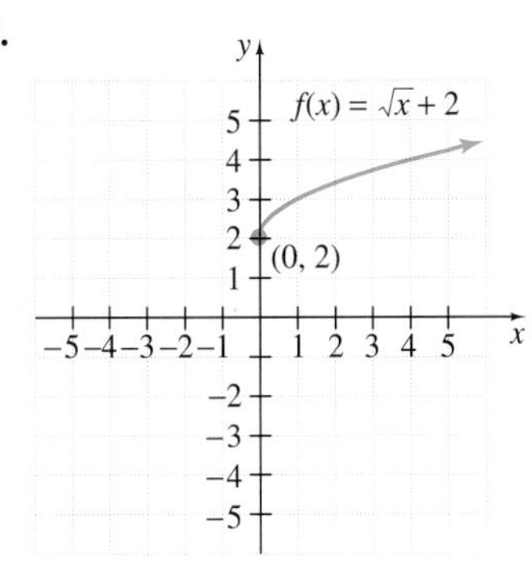

Practice Problem 6

Graph: $f(x) = \sqrt{x+1} + 2$

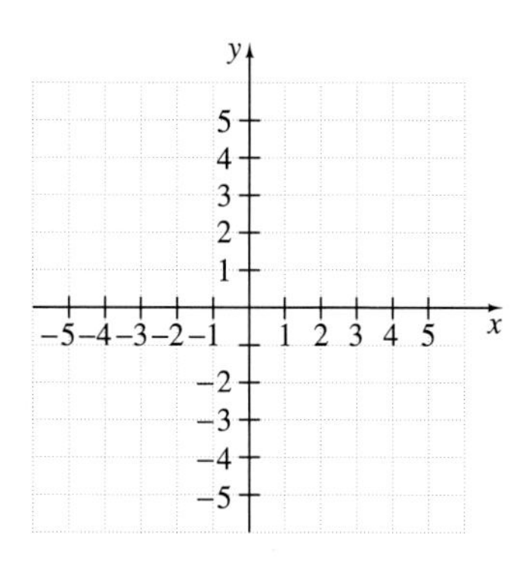

Example 6 Graph: $f(x) = \sqrt{x+2} + 3$

Solution: Recall the graph of $y = (x+2)^2 + 3$. The vertex is at $(-2, 3)$ and it is the graph of $y = x^2$ shifted 2 units left and 3 units up. Similarly, the graph of $y = \sqrt{x+2} + 3$ is the graph of $y = \sqrt{x}$ shifted 2 units left and 3 units up. We use this knowledge along with a table of values to graph the function.

x	y
-2	3
-1	4
2	5

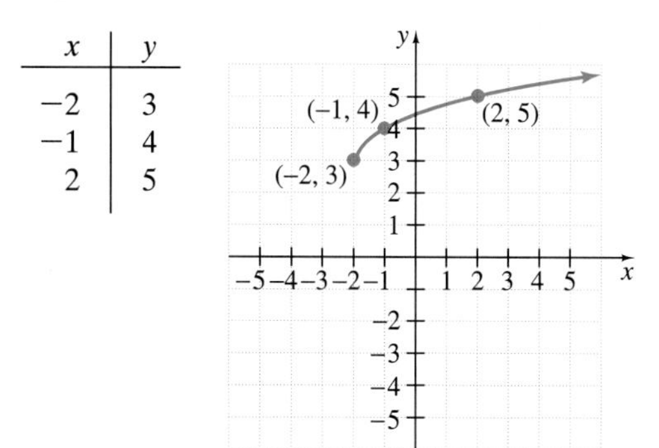

This vertical and horizontal shifting works for any function.

THE GRAPH OF THE FUNCTION

$$F(x) = f(x-h) + k$$

is the same as the graph of the function $f(x)$ except that it has been shifted left or right h units and up or down k units. It is shifted to the right if $h > 0$ and left if $h < 0$. It is shifted upward if $k > 0$ and downward if $k < 0$.

Answer

6.

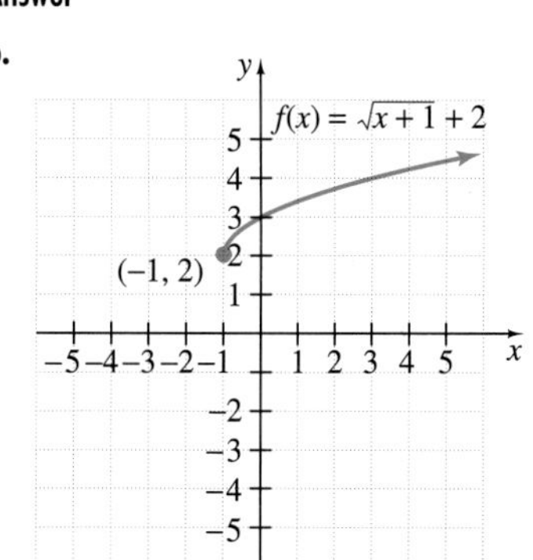

Name ________________________ Section ________ Date ________

Exercise Set 9.3

A *Graph each function. See Examples 1 through 6.*

1. $f(x) = |x| + 3$

2. $f(x) = |x| - 2$

3. $f(x) = \sqrt{x} - 2$

4. $f(x) = \sqrt{x} + 3$

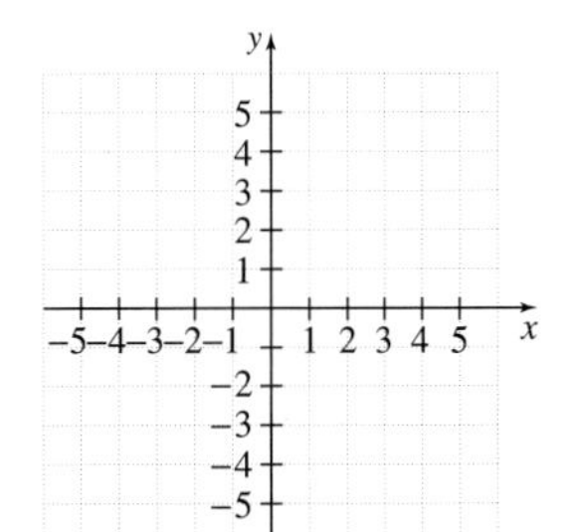
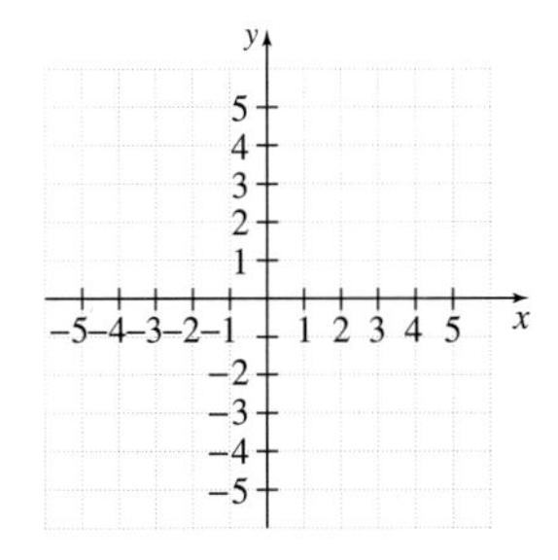
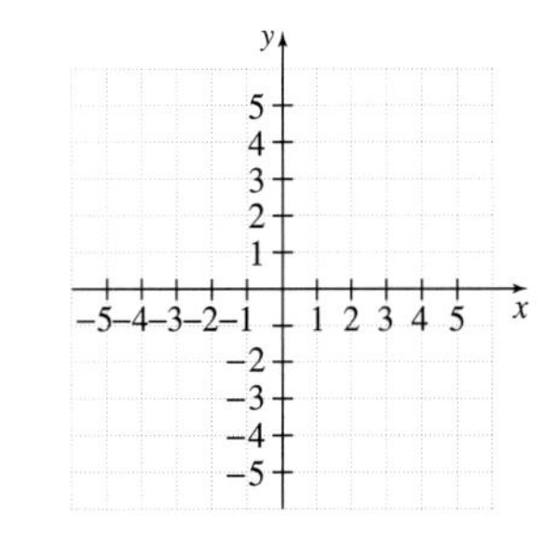

5. $f(x) = |x - 4|$

6. $f(x) = |x + 3|$

7. $f(x) = \sqrt{x + 2}$

8. $f(x) = \sqrt{x - 2}$

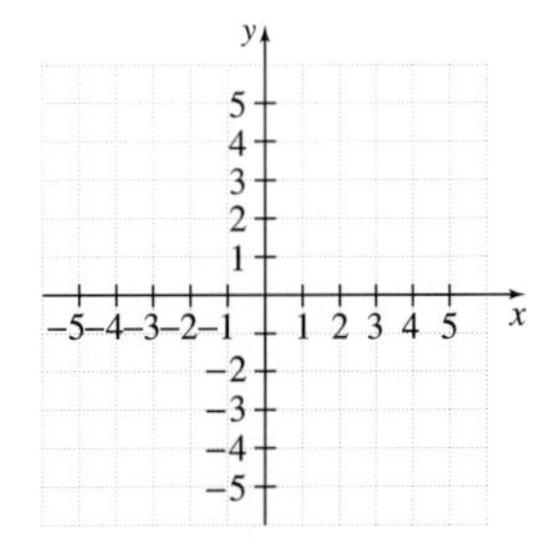
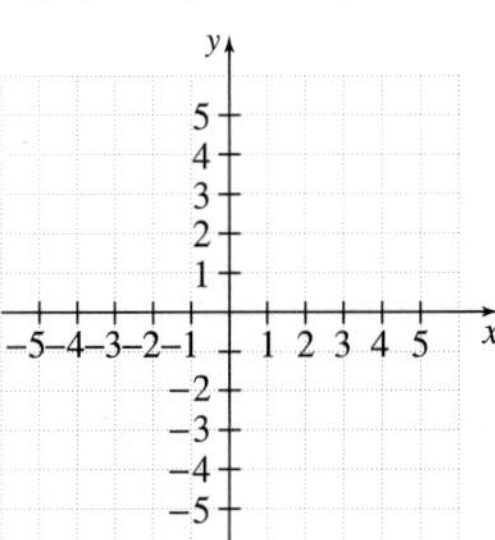
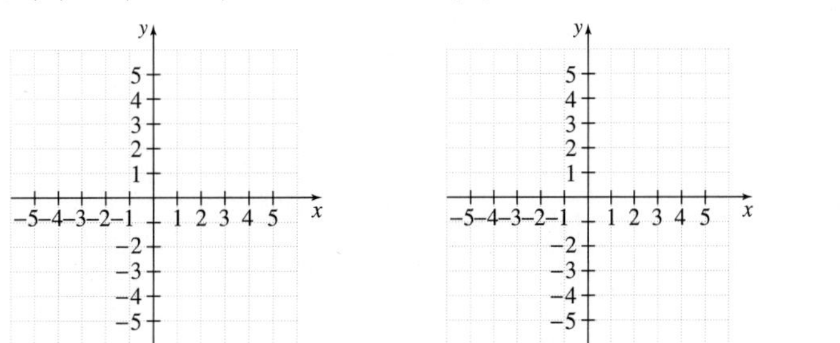
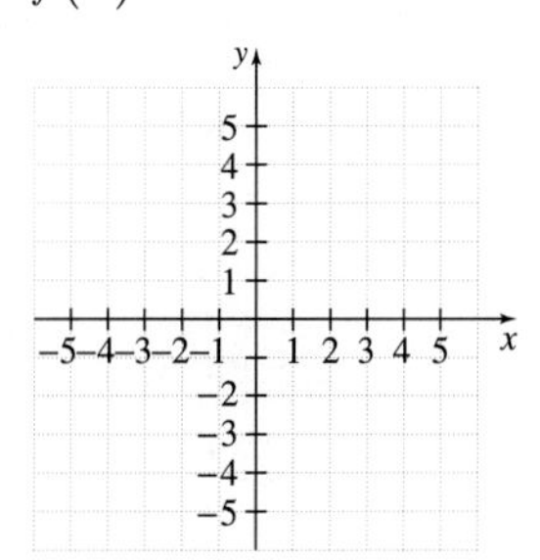

9. $f(x) = \sqrt{x - 2} + 3$

10. $f(x) = \sqrt{x - 1} + 3$

11. $f(x) = |x - 1| + 5$

12. $f(x) = |x - 3| + 2$

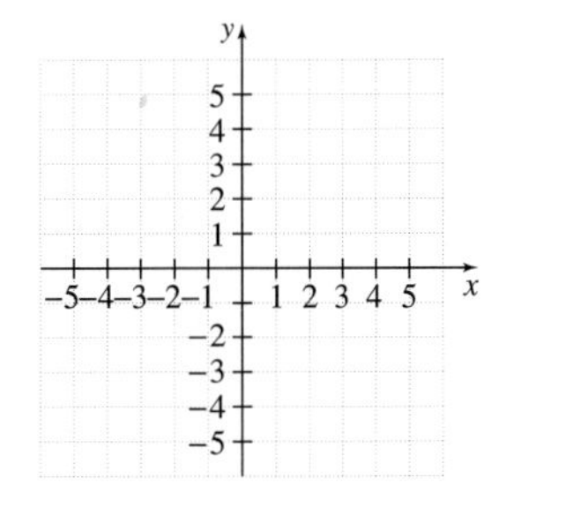
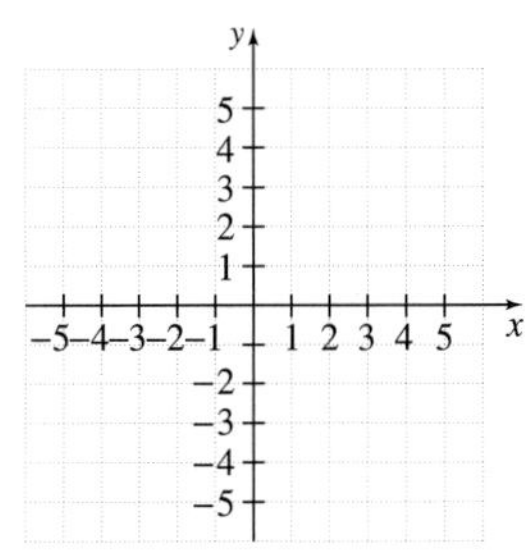
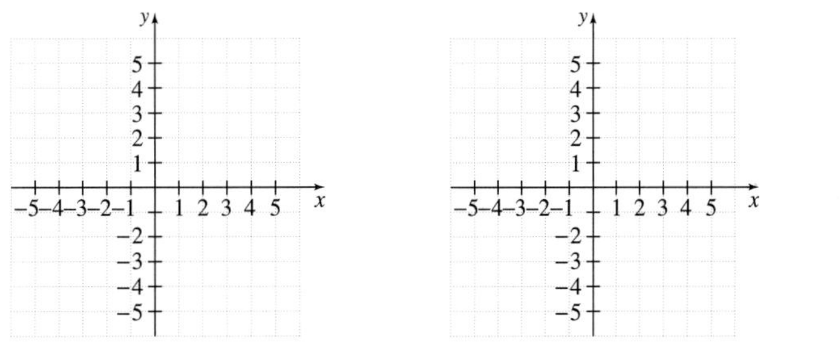
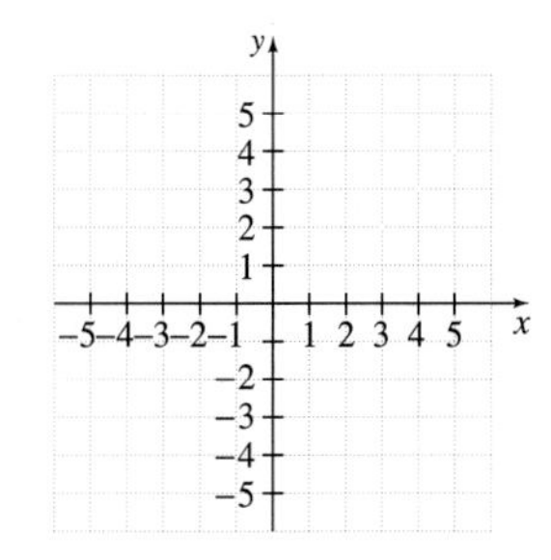

13. $f(x) = \sqrt{x + 1} + 1$

14. $f(x) = \sqrt{x + 3} + 2$

15. $f(x) = |x + 3| - 1$

16. $f(x) = |x + 1| - 4$

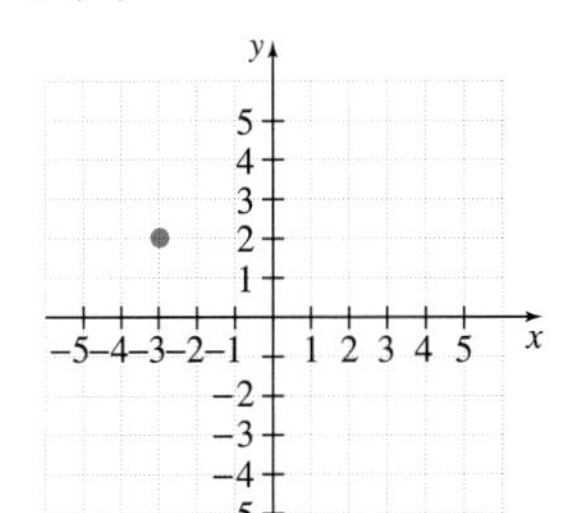
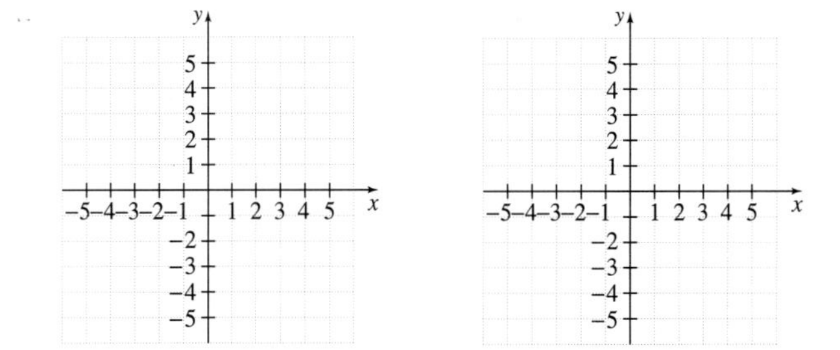

ANSWERS

17. ______

18. ______

19. ______

20. ______

Name ______________________

REVIEW AND PREVIEW

Solve each system of equations. See Section 4.1.

17. $\begin{aligned} x + y &= 6 \\ x - y &= 10 \end{aligned}$

18. $\begin{aligned} x + y &= -2 \\ -x + y &= -8 \end{aligned}$

19. $\begin{aligned} 2x + 3y &= 7 \\ -x + 4y &= 13 \end{aligned}$

20. $\begin{aligned} 4x - 3y &= -4 \\ 3x - y &= 10 \end{aligned}$

COMBINING CONCEPTS

Graph each function. Recall that the domain of the cube function and the cube root function is the set of all real numbers. For Exercises 22 and 24, predict the location and appearance of the graph and then use a graphing calculator to verify.

21. $f(x) = x^3$

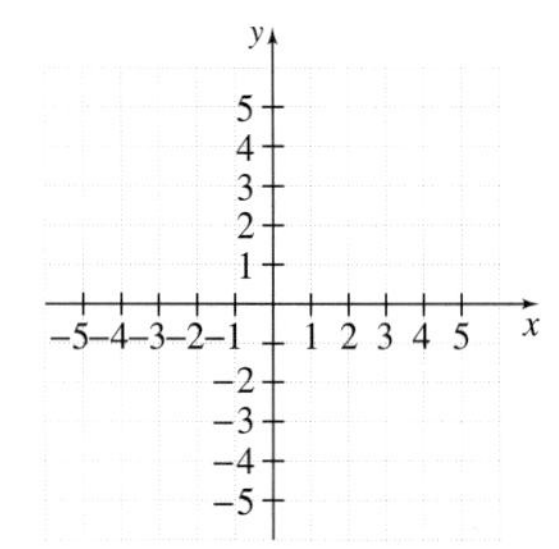

22. $f(x) = (x - 1)^3 + 2$

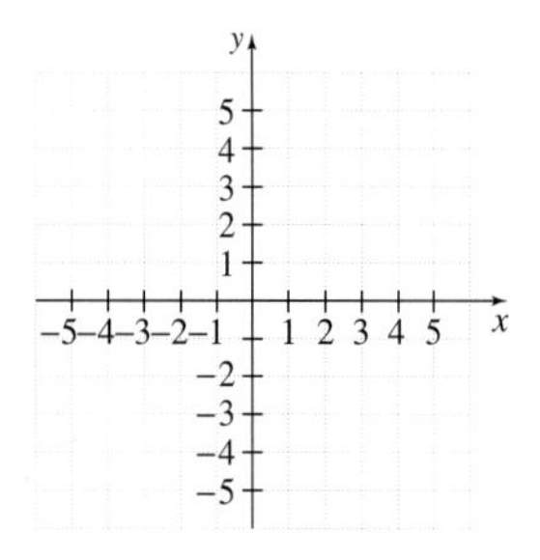

23. $f(x) = \sqrt[3]{x}$

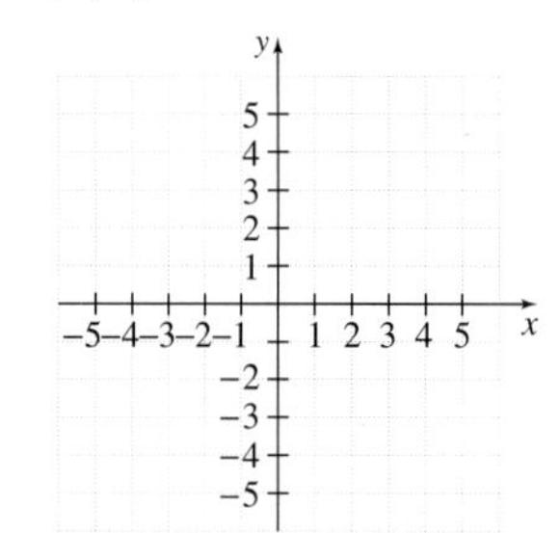

24. $f(x) = \sqrt[3]{x - 3} + 1$

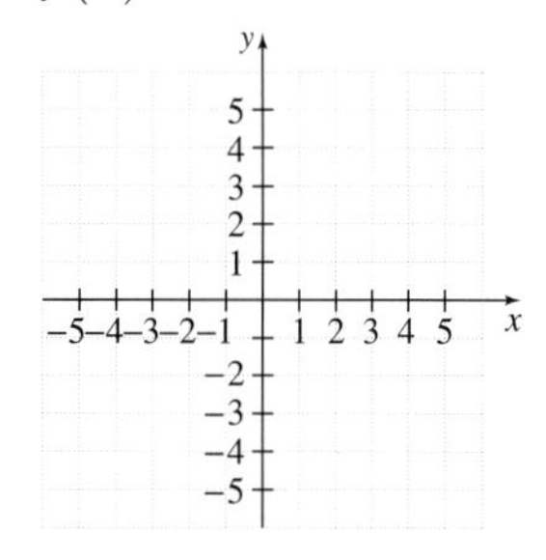

9.4 SOLVING NONLINEAR SYSTEMS OF EQUATIONS

In Section 4.1, we used graphing, substitution, and elimination methods to find solutions of systems of linear equations in two variables. We now apply these same methods to nonlinear systems of equations in two variables. A **nonlinear system of equations** is a system of equations at least one of which is not linear. Since we will be graphing the equations in each system, we are interested in real number solutions only.

Objectives

A Solve a nonlinear system by substitution.

B Solve a nonlinear system by elimination.

A SOLVING NONLINEAR SYSTEMS BY SUBSTITUTION

First we solve nonlinear systems by the substitution method.

Example 1

Solve the system:

$$\begin{cases} x^2 - 3y = 1 \\ x - y = 1 \end{cases}$$

Solution: We can solve this system by substitution if we solve one equation for one of the variables. Solving the first equation for x is not the best choice since doing so introduces a radical. Also, solving for y in the first equation introduces a fraction. Thus, we solve the second equation for y.

$x - y = 1$ Second equation

$x - 1 = y$ Solve for y.

Now we replace y with $x - 1$ in the first equation, and then solve for x.

$$x^2 - 3y = 1 \quad \text{First equation}$$

$$x^2 - 3(x - 1) = 1 \quad \text{Replace } y \text{ with } x - 1.$$

$$x^2 - 3x + 3 = 1$$

$$x^2 - 3x + 2 = 0$$

$$(x - 2)(x - 1) = 0$$

$$x = 2 \quad \text{or} \quad x = 1$$

Now we let $x = 2$ and then $x = 1$ in the equation $y = x - 1$ to find corresponding y-values.

Let $x = 2$.	Let $x = 1$.
$y = x - 1$	$y = x - 1$
$y = 2 - 1 = 1$	$y = 1 - 1 = 0$

The solution set is $\{(2, 1), (1, 0)\}$. When we check both solutions in both equations, we find that both solutions satisfy both equations. So both are solutions of the system. The graph of each equation in the system is shown.

Practice Problem 1

Solve the system: $\begin{cases} x^2 - 2y = 5 \\ x + y = -1 \end{cases}$

Answer

1. $\{(-3, 2), (1, -2)\}$

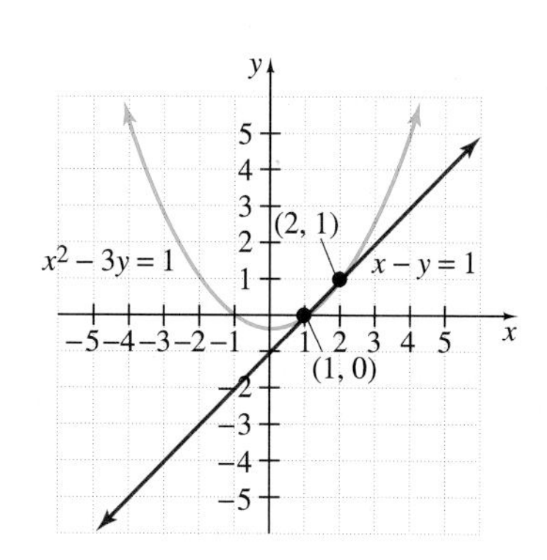

Practice Problem 2

Solve the system: $\begin{cases} y = \sqrt{x} \\ x^2 + y^2 = 12 \end{cases}$

Example 2 Solve the system:

$$\begin{cases} y = \sqrt{x} \\ x^2 + y^2 = 6 \end{cases}$$

Solution: This system is ideal for the substitution method since y is expressed in terms of x in the first equation. Notice that if $y = \sqrt{x}$, then both x and y must be nonnegative if they are real numbers. Let's substitute $\sqrt{x}$ for y in the second equation, and solve for x.

$$x^2 + y^2 = 6$$
$$x^2 + (\sqrt{x})^2 = 6 \quad \text{Let } y = \sqrt{x}.$$
$$x^2 + x = 6$$
$$x^2 + x - 6 = 0$$
$$(x + 3)(x - 2) = 0$$
$$x = -3 \quad \text{or} \quad x = 2$$

The solution -3 is discarded because we have noted that x must be nonnegative. To see this, we let $x = -3$ and $x = 2$ in the first equation to find the corresponding y-values.

Let $x = -3$.
$y = \sqrt{x}$
$y = \sqrt{-3}$ Not a real number

Let $x = 2$.
$y = \sqrt{x}$
$y = \sqrt{2}$

Since we are interested only in real number solutions, the only solution is $(2, \sqrt{2})$. The solution set is $\{(2, \sqrt{2})\}$. Check to see that this solution satisfies both equations. The graph of each equation in this system is shown.

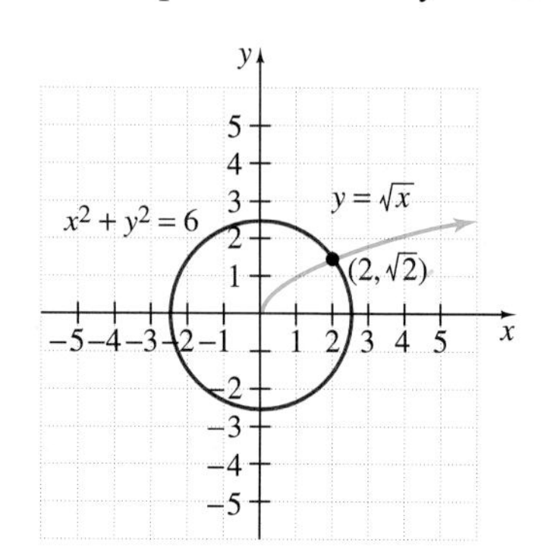

Answer

2. $\{(3, \sqrt{3})\}$

Example 3 Solve the system:

$$\begin{cases} x^2 + y^2 = 4 \\ x + y = 3 \end{cases}$$

Solution: We use the substitution method and solve the second equation for x.

$x + y = 3$ Second equation

$x = 3 - y$

Now we let $x = 3 - y$ in the first equation.

$x^2 + y^2 = 4$ First equation

$(3 - y)^2 + y^2 = 4$ Let $x = 3 - y$.

$9 - 6y + y^2 + y^2 = 4$

$2y^2 - 6y + 5 = 0$

By the quadratic formula, where $a = 2, b = -6$, and $c = 5$, we have

$$y = \frac{6 \pm \sqrt{(-6)^2 - 4 \cdot 2 \cdot 5}}{2 \cdot 2} = \frac{6 \pm \sqrt{-4}}{4}$$

Since $\sqrt{-4}$ is not a real number, there is no solution. Graphically, the circle and the line do not intersect, as shown.

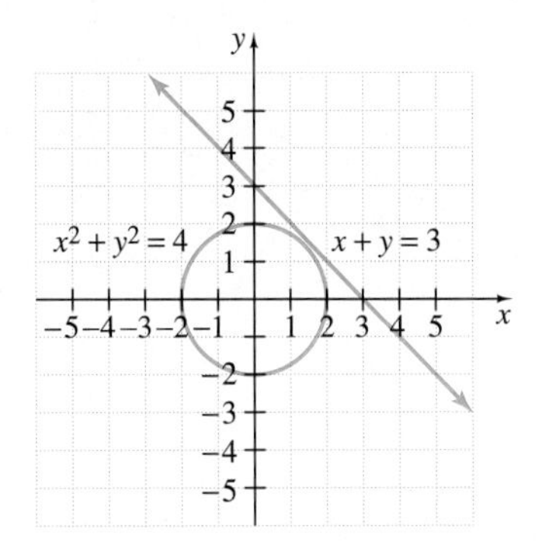

Practice Problem 3

Solve the system: $\begin{cases} x^2 + y^2 = 1 \\ x + y = 4 \end{cases}$

TRY THE CONCEPT CHECK IN THE MARGIN.

B SOLVING NONLINEAR SYSTEMS BY ELIMINATION

Some nonlinear systems may be solved by the elimination method.

Example 4 Solve the system:

$$\begin{cases} x^2 + 2y^2 = 10 \\ x^2 - y^2 = 1 \end{cases}$$

Solution: We will use the elimination, or addition, method to solve this system. To eliminate x^2 when we add the two equations, multiply both sides of the second equation by -1. Then

✓ CONCEPT CHECK

Without solving, how can you tell that $x^2 + y^2 = 9$ and $x^2 + y^2 = 16$ do not have any points of intersection?

Practice Problem 4

Solve the equation: $\begin{cases} x^2 + 3y^2 = 21 \\ x^2 - y^2 = 1 \end{cases}$

Answers

3. No solution **4.** $\{(\sqrt{6}, \sqrt{5}), (\sqrt{6}, -\sqrt{5}), (-\sqrt{6}, \sqrt{5}), (-\sqrt{6}, -\sqrt{5})\}$

✓ Concept Check: $x^2 + y^2 = 9$ is a circle inside the circle $x^2 + y^2 = 16$, therefore they do not have any points of intersection.

$$\begin{cases} x^2 + 2y^2 = 10 \\ (-1)(x^2 - y^2) = -1 \cdot 1 \end{cases} \text{ is equivalent to } \begin{cases} x^2 + 2y^2 = 10 \\ -x^2 + y^2 = -1 \end{cases}$$

$$3y^2 = 9 \quad \text{Add.}$$

$$y^2 = 3 \quad \text{Divide both sides by 3.}$$

$$y = \pm\sqrt{3}$$

To find the corresponding x-values, we let $y = \sqrt{3}$ and $y = -\sqrt{3}$ in either original equation. We choose the second equation.

Let $y = \sqrt{3}$.

$$x^2 - y^2 = 1$$
$$x^2 - (\sqrt{3})^2 = 1$$
$$x^2 - 3 = 1$$
$$x^2 = 4$$
$$x = \pm\sqrt{4} = \pm 2$$

Let $y = -\sqrt{3}$.

$$x^2 - y^2 = 1$$
$$x^2 - (-\sqrt{3})^2 = 1$$
$$x^2 - 3 = 1$$
$$x^2 = 4$$
$$x = \pm\sqrt{4} = \pm 2$$

The solution set is $\{(2, \sqrt{3}), (-2, \sqrt{3}), (2, -\sqrt{3}), (-2, -\sqrt{3})\}$. Check all four ordered pairs in both equations of the system. The graph of each equation in this system is shown.

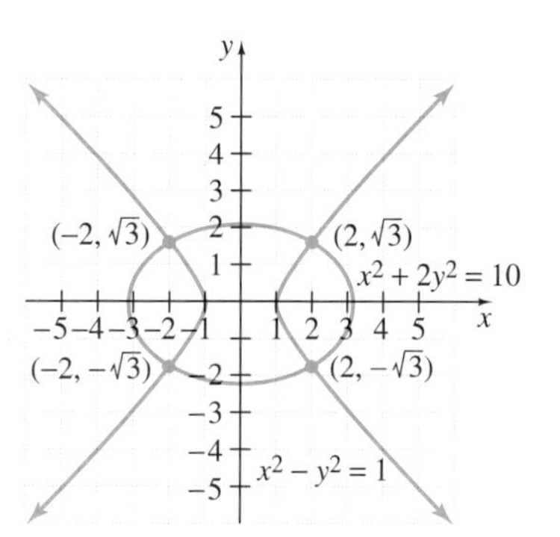

Name ______________________ Section ________ Date __________

Exercise Set 9.4

A B *Solve each nonlinear system of equations. See Examples 1 through 4.*

1. $\begin{cases} x^2 + y^2 = 25 \\ 4x + 3y = 0 \end{cases}$

2. $\begin{cases} x^2 + y^2 = 25 \\ 3x + 4y = 0 \end{cases}$

3. $\begin{cases} x^2 + 4y^2 = 10 \\ y = x \end{cases}$

4. $\begin{cases} 4x^2 + y^2 = 10 \\ y = x \end{cases}$

5. $\begin{cases} y^2 = 4 - x \\ x - 2y = 4 \end{cases}$

6. $\begin{cases} x^2 + y^2 = 4 \\ x + y = -2 \end{cases}$

7. $\begin{cases} x^2 + y^2 = 9 \\ 16x^2 - 4y^2 = 64 \end{cases}$

8. $\begin{cases} 4x^2 + 3y^2 = 35 \\ 5x^2 + 2y^2 = 42 \end{cases}$

9. $\begin{cases} x^2 + 2y^2 = 2 \\ x - y = 2 \end{cases}$

10. $\begin{cases} x^2 + 2y^2 = 2 \\ x^2 - 2y^2 = 6 \end{cases}$

11. $\begin{cases} y = x^2 - 3 \\ 4x - y = 6 \end{cases}$

12. $\begin{cases} y = x + 1 \\ x^2 - y^2 = 1 \end{cases}$

13. $\begin{cases} y = x^2 \\ 3x + y = 10 \end{cases}$

14. $\begin{cases} 6x - y = 5 \\ xy = 1 \end{cases}$

15. $\begin{cases} y = 2x^2 + 1 \\ x + y = -1 \end{cases}$

16. $\begin{cases} x^2 + y^2 = 9 \\ x + y = 5 \end{cases}$

17. $\begin{cases} y = x^2 - 4 \\ y = x^2 - 4x \end{cases}$

18. $\begin{cases} x = y^2 - 3 \\ x = y^2 - 3y \end{cases}$

Answers

1. ______________
2. ______________
3. ______________
4. ______________
5. ______________
6. ______________
7. ______________
8. ______________
9. ______________
10. ______________
11. ______________
12. ______________
13. ______________
14. ______________
15. ______________
16. ______________
17. ______________
18. ______________

19. ____

20. ____

21. ____

22. ____

23. ____

24. ____

25. ____

26. ____

27. ____

28. ____

29. ____

30. ____

31. ____

32. ____

Name ____

19. $\begin{cases} 2x^2 + 3y^2 = 14 \\ -x^2 + y^2 = 3 \end{cases}$

20. $\begin{cases} 4x^2 - 2y^2 = 2 \\ -x^2 + y^2 = 2 \end{cases}$

21. $\begin{cases} x^2 + y^2 = 1 \\ x^2 + (y + 3)^2 = 4 \end{cases}$

22. $\begin{cases} x^2 + 2y^2 = 4 \\ x^2 - y^2 = 4 \end{cases}$

23. $\begin{cases} y = x^2 + 2 \\ y = -x^2 + 4 \end{cases}$

24. $\begin{cases} x = -y^2 - 3 \\ x = y^2 - 5 \end{cases}$

25. $\begin{cases} 3x^2 + y^2 = 9 \\ 3x^2 - y^2 = 9 \end{cases}$

26. $\begin{cases} x^2 + y^2 = 25 \\ x = y^2 - 5 \end{cases}$

27. $\begin{cases} x^2 + 3y^2 = 6 \\ x^2 - 3y^2 = 10 \end{cases}$

28. $\begin{cases} x^2 + y^2 = 1 \\ y = x^2 - 9 \end{cases}$

29. $\begin{cases} x^2 + y^2 = 36 \\ y = \frac{1}{6}x^2 - 6 \end{cases}$

30. $\begin{cases} x^2 + y^2 = 16 \\ y = -\frac{1}{4}x^2 + 4 \end{cases}$

31. How many real solutions are possible for a system of equations whose graphs are a circle and a parabola?

32. How many real solutions are possible for a system of equations whose graphs are an ellipse and a line?

Name ______________________________

Review and Preview

Graph each inequality in two variables. See Section 3.6.

33. $x > -3$

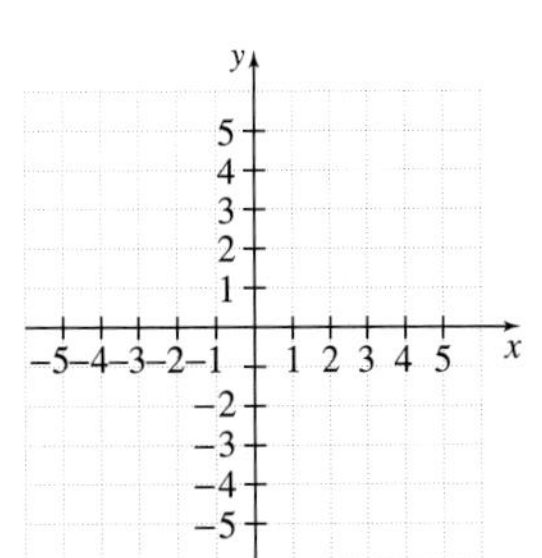

34. $y \leq 1$

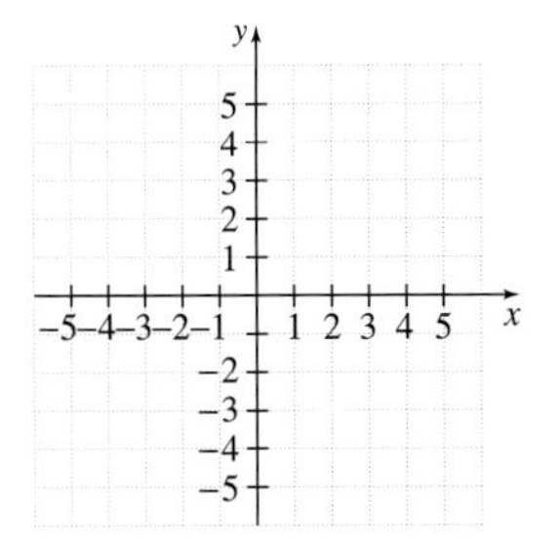

35. $y < 2x - 1$

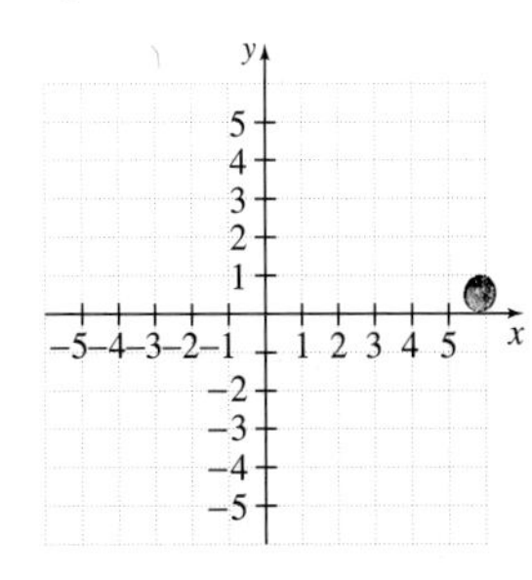

36. $3x - y \leq 4$

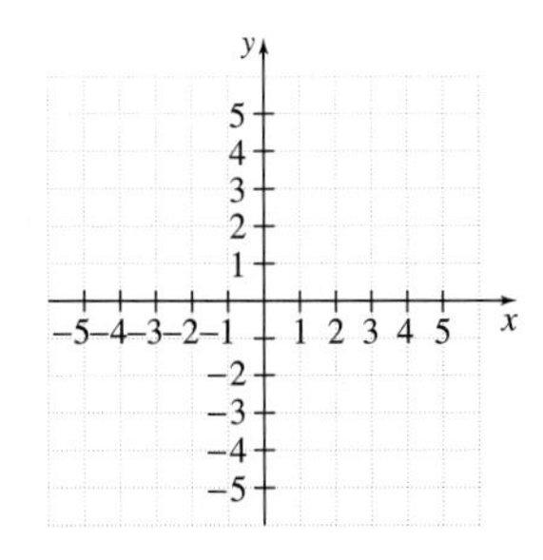

Combining Concepts

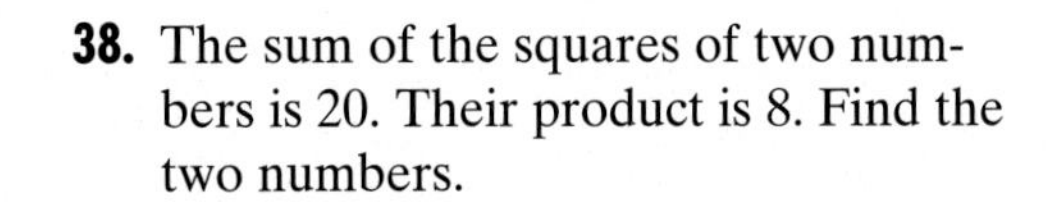

Solve.

37. The sum of the squares of two numbers is 130. The difference of the squares of the two numbers is 32. Find the two numbers.

38. The sum of the squares of two numbers is 20. Their product is 8. Find the two numbers.

39. During the development stage of a new rectangular keypad for a security system, it was decided that the area of the rectangle should be 285 square centimeters and the perimeter should be 68 centimeters. Find the dimensions of the keypad.

40. A rectangular holding pen for cattle is to be designed so that its perimeter is 92 feet and its area is 525 feet. Find the dimensions of the holding pen.

37. ______________

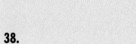

38. ______________

39. ______________

40. ______________

41. ______

42. ______

Name ______

Recall that in business, a demand function expresses the quantity of a commodity demanded as a function of the commodity's unit price. A supply function expresses the quantity of a commodity supplied as a function of the commodity's unit price. When the quantity produced and supplied is equal to the quantity demanded, then we have what is called ***market equilibrium****. Use this information for Exercises 41–42.*

41. The demand function for a certain compact disc is given by the function $p = -0.01x^2 - 0.2x + 9$ and the corresponding supply function is given by $p = 0.01x^2 - 0.1x + 3$, where p is in dollars and x is in thousands of units. Find the equilibrium quantity and the corresponding price by solving the system consisting of the two given equations.

42. The demand function for a certain style of picture frame is given by the function $p = -2x^2 + 90$ and the corresponding supply function is given by $p = 9x + 34$, where p is in dollars and x is in thousands of units. Find the equilibrium quantity and the corresponding price by solving the system consisting of the two given equations.

Use a grapher to verify the results of each exercise.

43. Exercise 3

44. Exercise 4

45. Exercise 23

46. Exercise 24

9.5 NONLINEAR INEQUALITIES AND SYSTEMS OF INEQUALITIES

Objectives

A Graph a nonlinear inequality.

B Graph a system of nonlinear inequalities.

SSM CD-ROM Video 9.5

A GRAPHING NONLINEAR INEQUALITIES

We can graph a nonlinear inequality in two variables such as $\frac{x^2}{9} + \frac{y^2}{16} \leq 1$ in a way similar to the way we graphed a linear inequality in two variables in Section 3.6. First, we graph the related equation $\frac{x^2}{9} + \frac{y^2}{16} = 1$. The graph of the equation is our boundary. Then, using test points, we determine and shade the region whose points satisfy the inequality.

Example 1 Graph: $\frac{x^2}{9} + \frac{y^2}{16} \leq 1$

Solution: First we graph the equation $\frac{x^2}{9} + \frac{y^2}{16} = 1$. We sketch a solid curve because of the inequality symbol $\leq$. It means that the graph of $\frac{x^2}{9} + \frac{y^2}{16} \leq 1$ includes the graph of $\frac{x^2}{9} + \frac{y^2}{16} = 1$. The graph is an ellipse, and it divides the plane into two regions, the "inside" and the "outside" of the ellipse. Recall from Section 3.6 that to determine which region contains the solutions, we select a test point in either region and determine whether the coordinates of the point satisfy the inequality. We choose $(0, 0)$ as the test point.

$$\frac{x^2}{9} + \frac{y^2}{16} \leq 1$$

$$\frac{0^2}{9} + \frac{0^2}{16} \leq 1 \quad \text{Let } x = 0 \text{ and } y = 0.$$

$$0 \leq 1 \quad \text{True.}$$

Since this statement is true, the solution set is the region containing $(0, 0)$. The graph of the solution set includes the points on and inside the ellipse, as shaded in the figure.

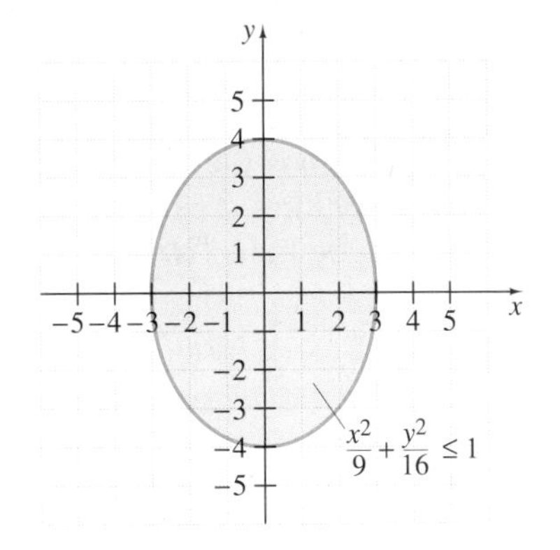

Practice Problem 1

Graph: $\frac{x^2}{25} + \frac{y^2}{4} \leq 1$

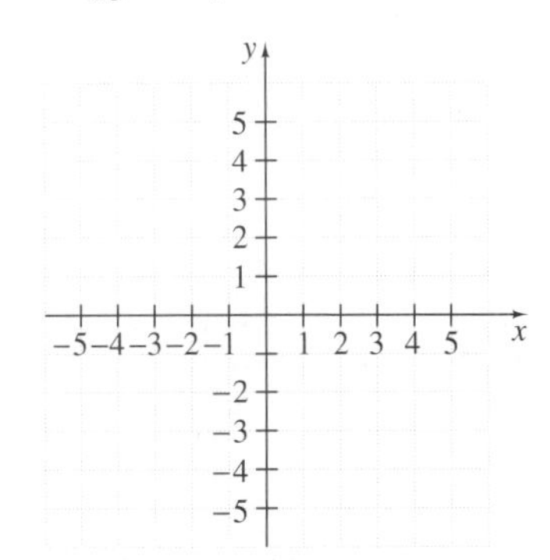

Answer

1.

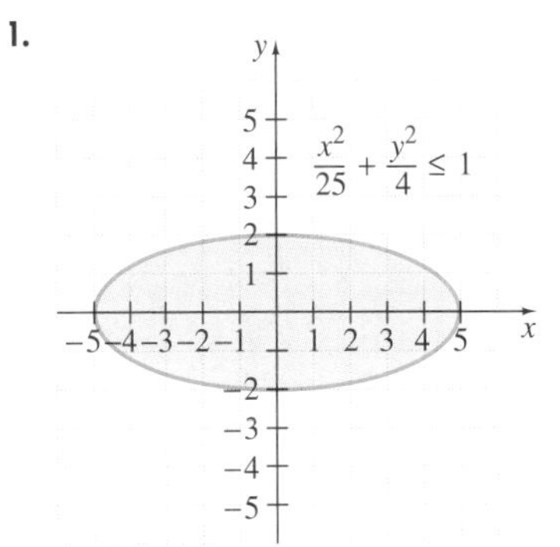

Practice Problem 2

Graph: $9x^2 > 4y^2 + 144$

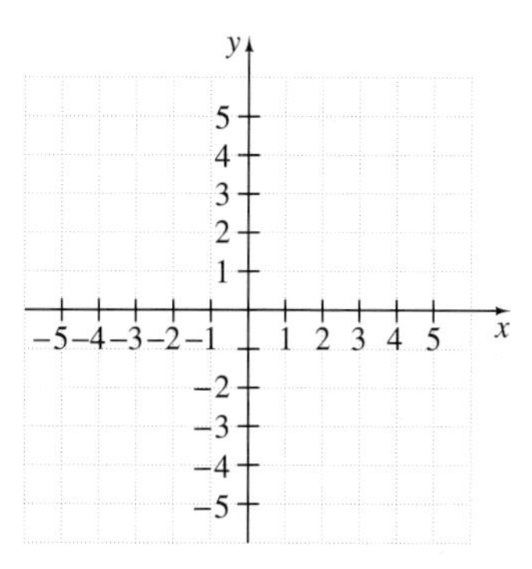

Practice Problem 3

Graph the system:

$$\begin{cases} y \ge x^2 \\ y \le -4x + 2 \end{cases}$$

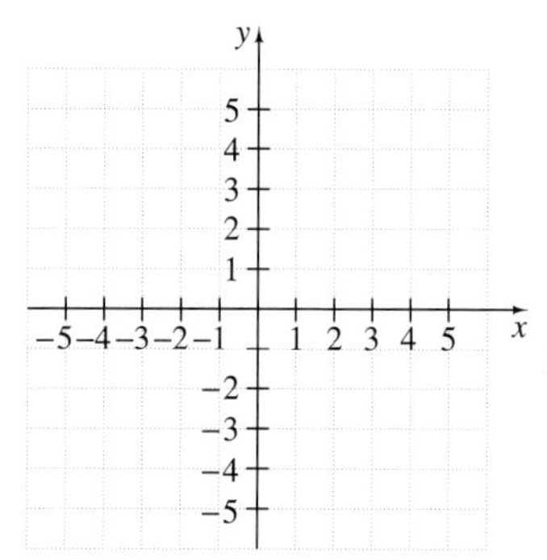

Answers

2.

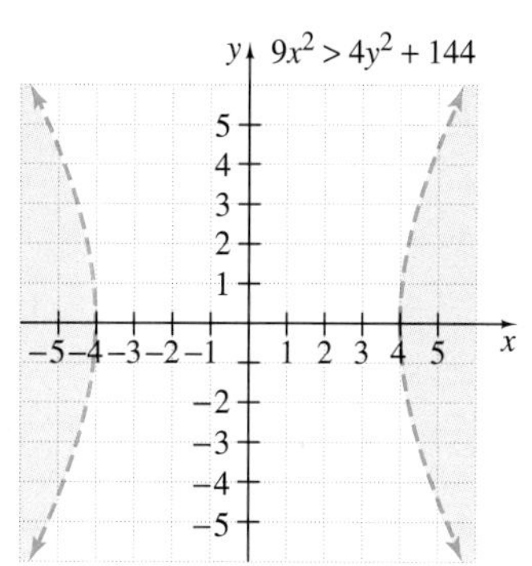

3.

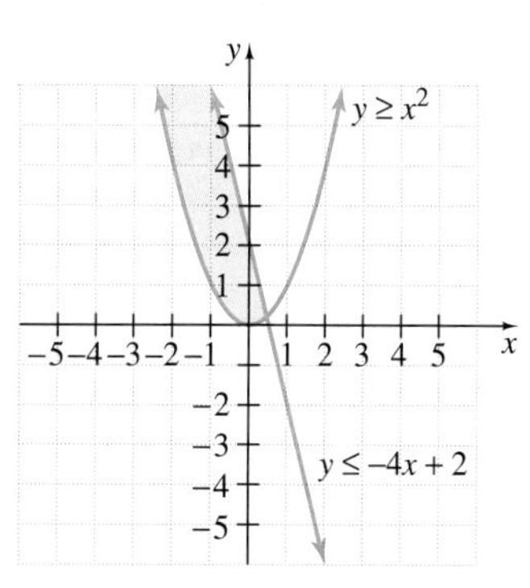

Example 2 Graph: $4y^2 > x^2 + 16$

Solution: The related equation is $4y^2 = x^2 + 16$, or $\frac{y^2}{4} - \frac{x^2}{16} = 1$, which is a hyperbola. We graph the hyperbola as a dashed curve because of the inequality symbol $>$. It means that the graph of $4y^2 > x^2 + 16$ does *not* include the graph of $4y^2 = x^2 + 16$. The hyperbola divides the plane into three regions. We select a test point in each region—not on a boundary line—to determine whether that region contains solutions of the inequality.

Test region *A* with (0, 4)	**Test region *B* with (0, 0)**	**Test region *C* with (0, −4)**
$4y^2 > x^2 + 16$	$4y^2 > x^2 + 16$	$4y^2 > x^2 + 16$
$4(4)^2 > 0^2 + 16$	$4(0)^2 > 0^2 + 16$	$4(-4)^2 > 0^2 + 16$
$64 > 16$ True.	$0 > 16$ False.	$64 > 16$ True.

The graph of the solution set includes the shaded regions *A* and *C* only, not the boundary.

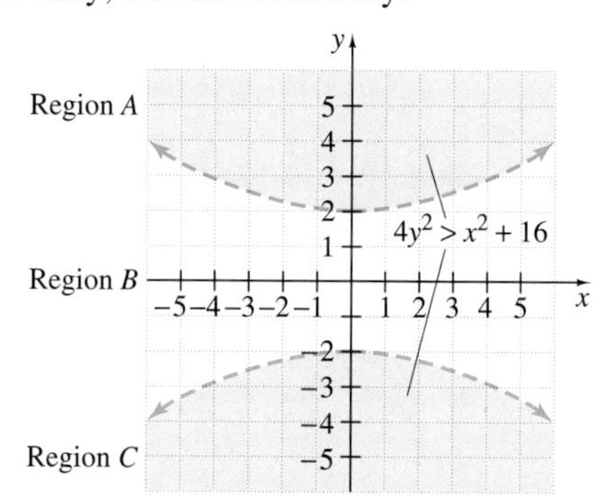

B GRAPHING A SYSTEM OF NONLINEAR INEQUALITIES

In Section 4.6 we graphed systems of linear inequalities. Recall that the graph of a system of inequalities is the intersection of the graphs of the inequalities.

Example 3 Graph the system:

$$\begin{cases} x \le 1 - 2y \\ y \le x^2 \end{cases}$$

Solution: We graph each inequality on the same set of axes. The intersection is the darkest shaded region along with its boundary lines. The coordinates of the points of intersection can be found by solving the related system.

$$\begin{cases} x = 1 - 2y \\ y = x^2 \end{cases}$$

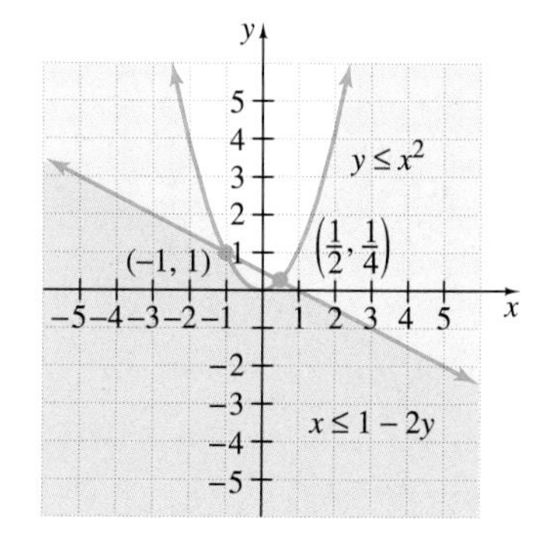

Example 4 Graph the system:

$$\begin{cases} x^2 + y^2 < 25 \\ \dfrac{x^2}{9} - \dfrac{y^2}{25} < 1 \\ y < x + 3 \end{cases}$$

Solution: We graph each inequality. The graph of $x^2 + y^2 < 25$ contains points "inside" the circle that has center $(0, 0)$ and radius 5. The graph of $\frac{x^2}{9} - \frac{y^2}{25} < 1$ is the region between the two branches of the hyperbola with x-intercepts -3 and 3 and center $(0, 0)$. The graph of $y < x + 3$ is the region "below" the line with the slope 1 and y-intercept 3. The graph of the solution set of the system is the intersection of all the graphs, the darkest shaded region shown. The boundary of this region is not part of the solution.

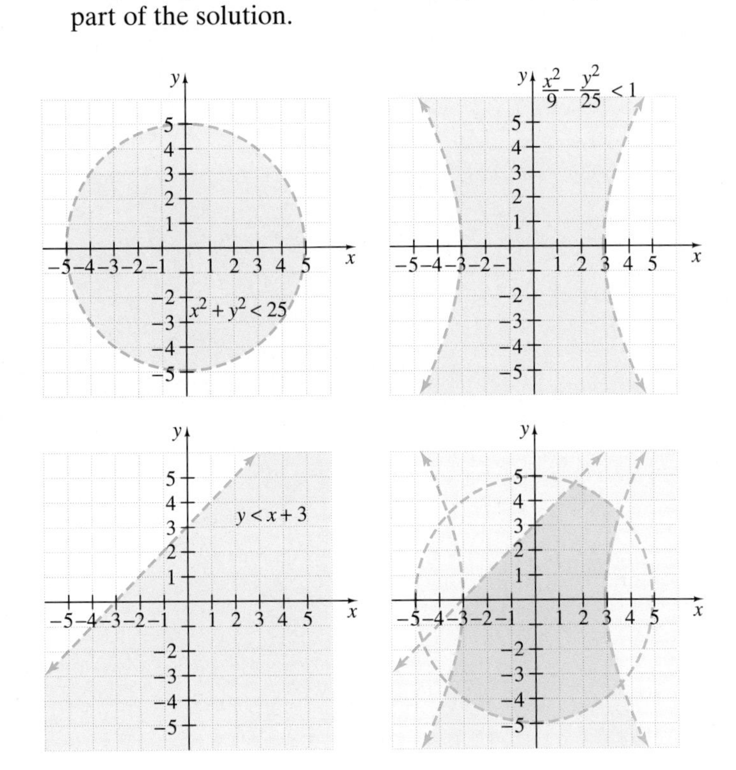

Practice Problem 4

Graph the system:

$$\begin{cases} x^2 + y^2 < 9 \\ \dfrac{x^2}{9} - \dfrac{y^2}{4} < 1 \\ y > x - 2 \end{cases}$$

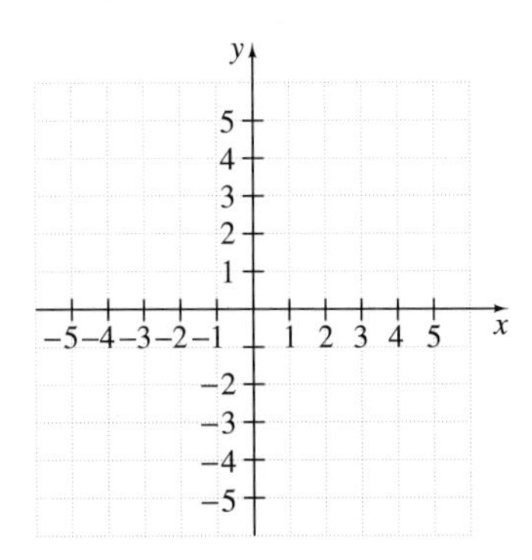

Answer

4.

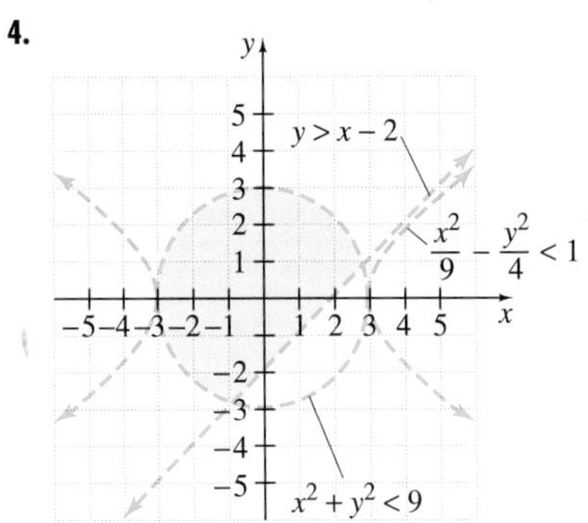

Name ______________________ **Section** ________ **Date** ________

EXERCISE SET 9.5

A *Graph each inequality. See Examples 1 and 2.*

1. $y < x^2$

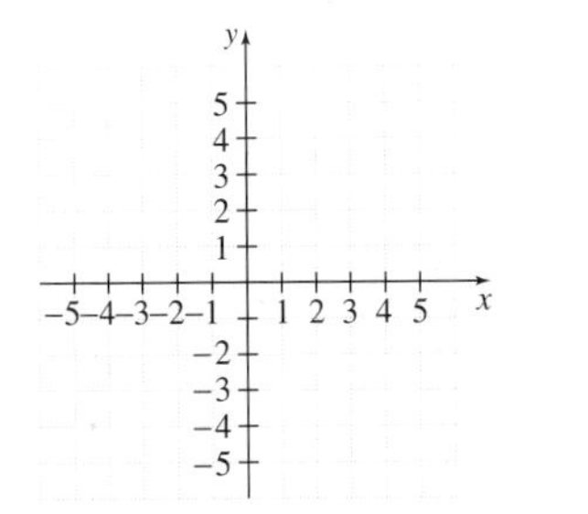

2. $y < -x^2$

3. $x^2 + y^2 \geq 16$

4. $x^2 + y^2 < 36$

5. $\frac{x^2}{4} - y^2 < 1$

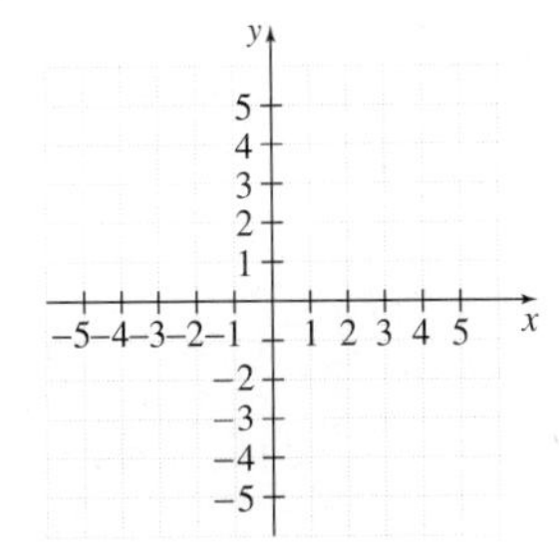

6. $x^2 - \frac{y^2}{9} \geq 1$

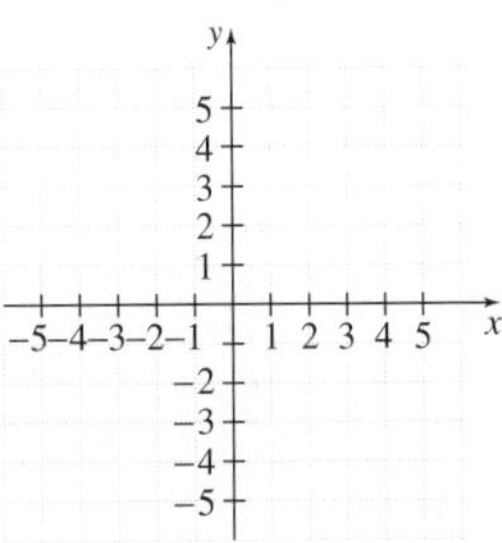

7. $y > (x - 1)^2 - 3$

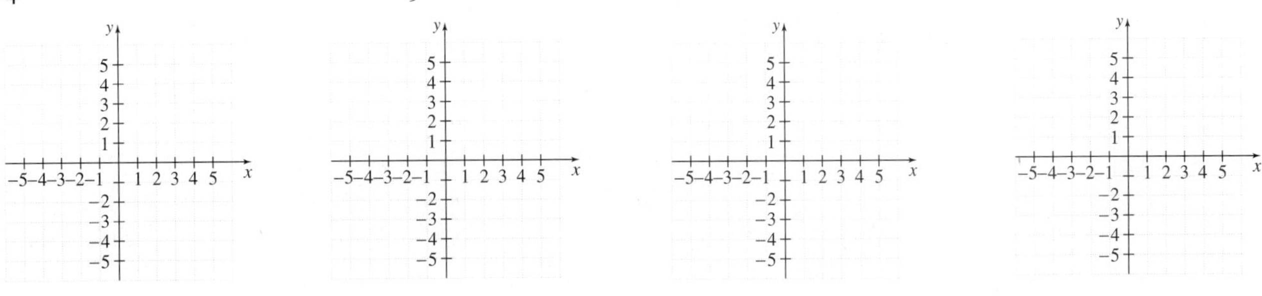

8. $y > (x + 3)^2 + 2$

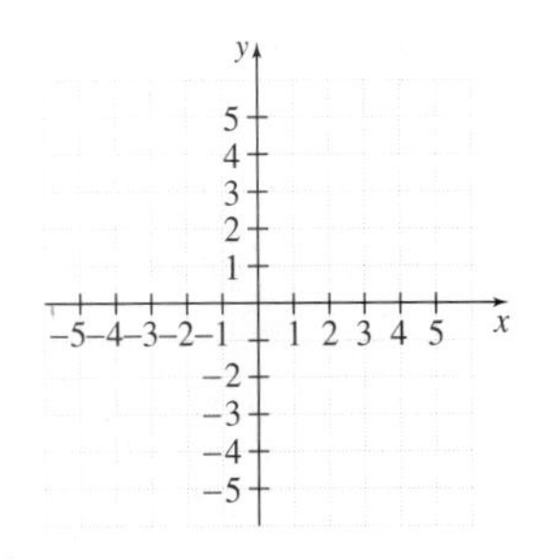

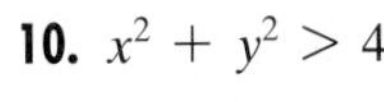

9. $x^2 + y^2 \leq 9$

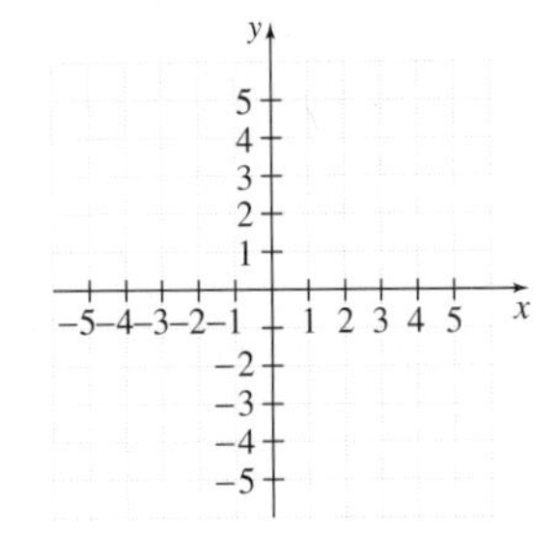

10. $x^2 + y^2 > 4$

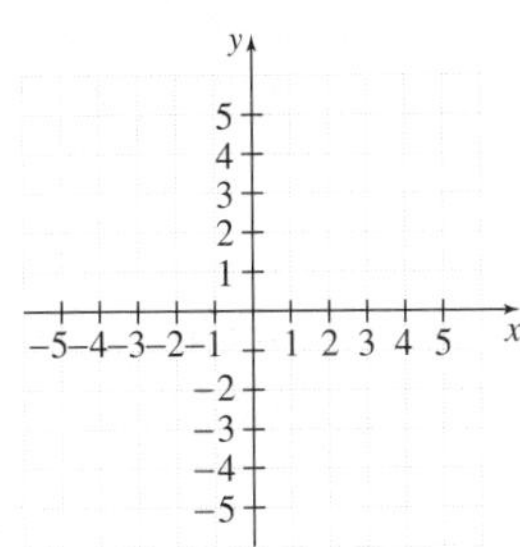

11. $y > -x^2 + 5$

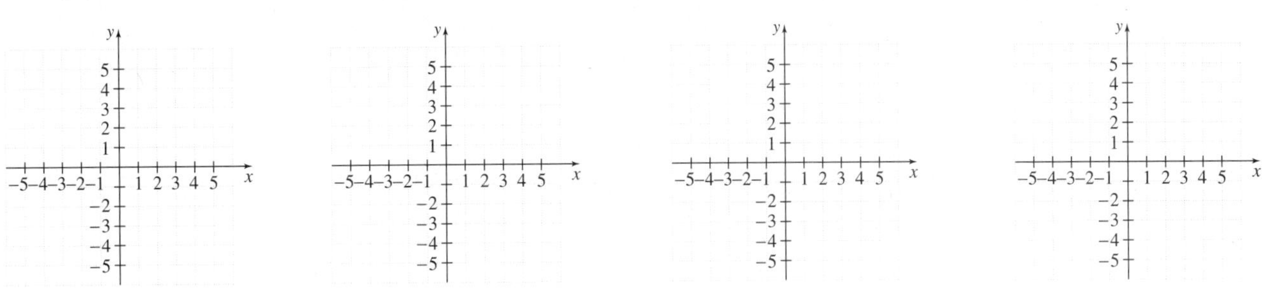

12. $y < -x^2 + 5$

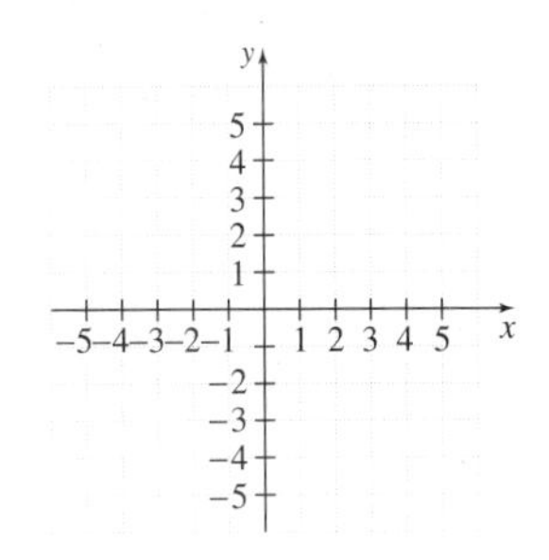

13. $\frac{x^2}{4} + \frac{y^2}{9} \leq 1$

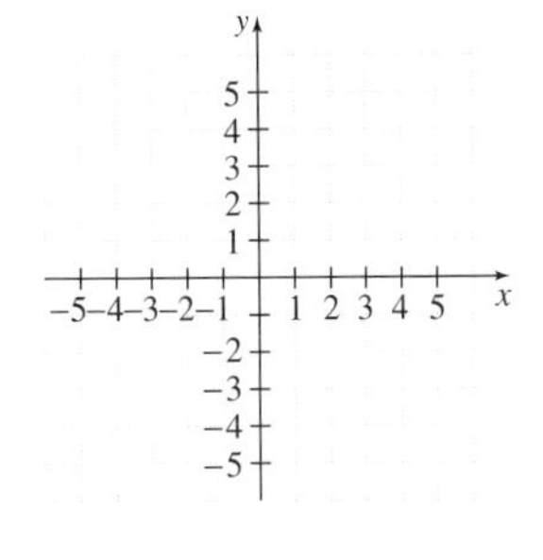

14. $\frac{x^2}{25} + \frac{y^2}{4} \geq 1$

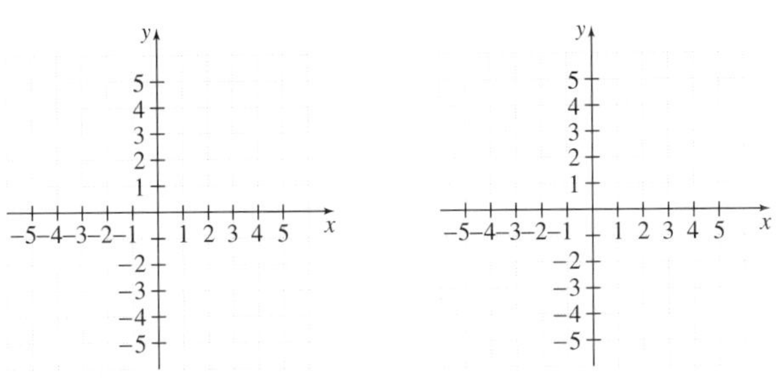

15. $\frac{y^2}{4} - x^2 \leq 1$

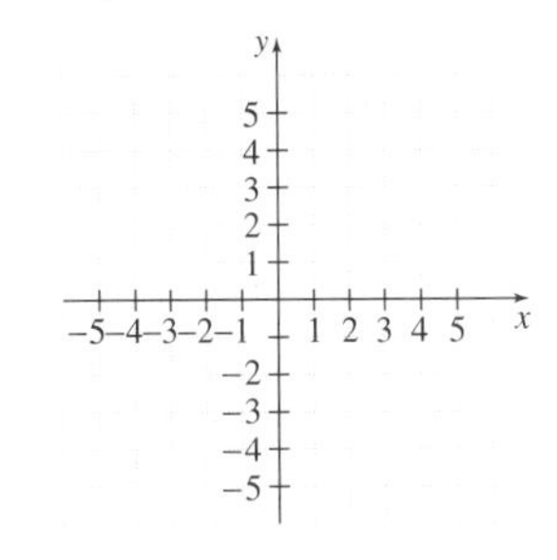

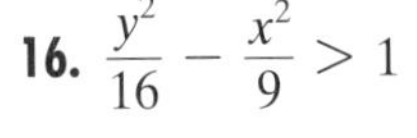

16. $\frac{y^2}{16} - \frac{x^2}{9} > 1$

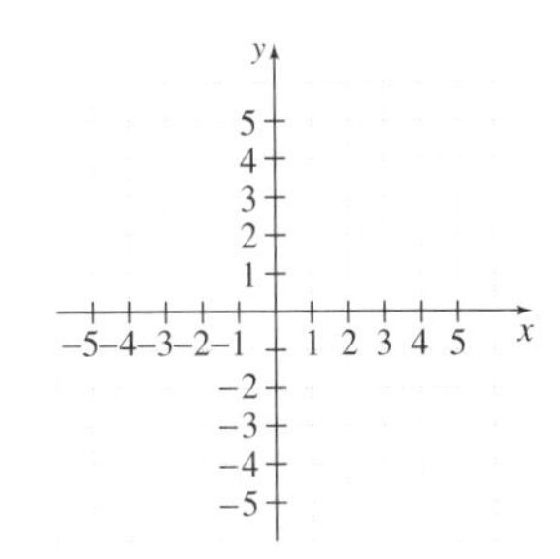

Name ______________________________

17. $y < (x - 2)^2 + 1$

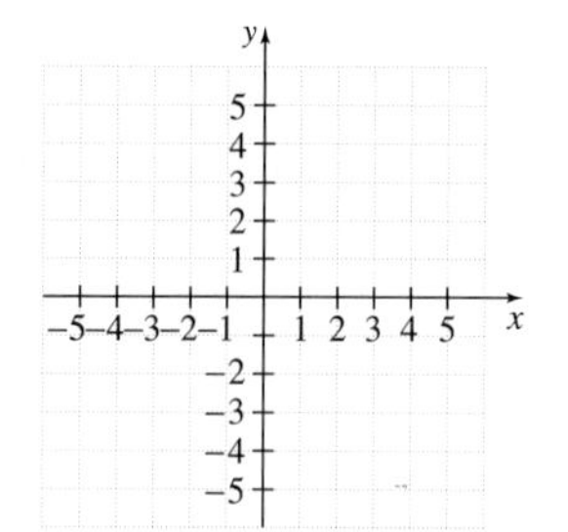

18. $y > (x - 2)^2 + 1$

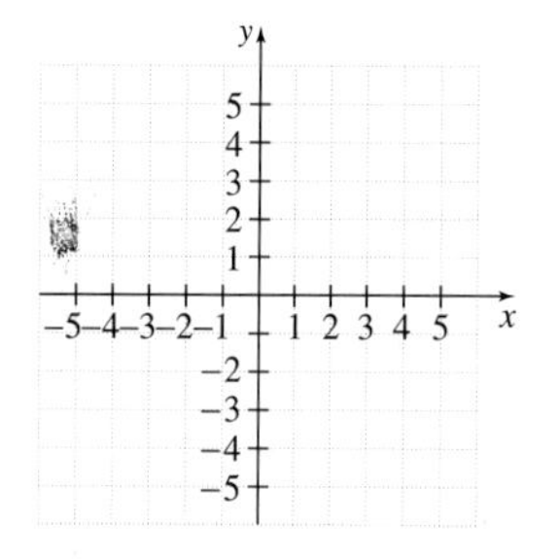

19. $y \le x^2 + x - 2$

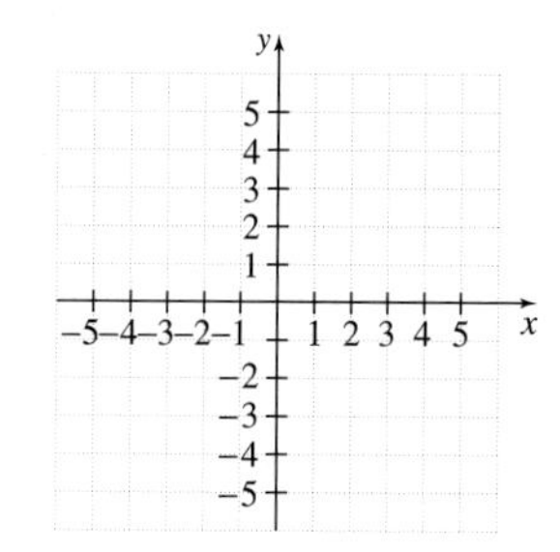

20. $y > x^2 + x - 2$

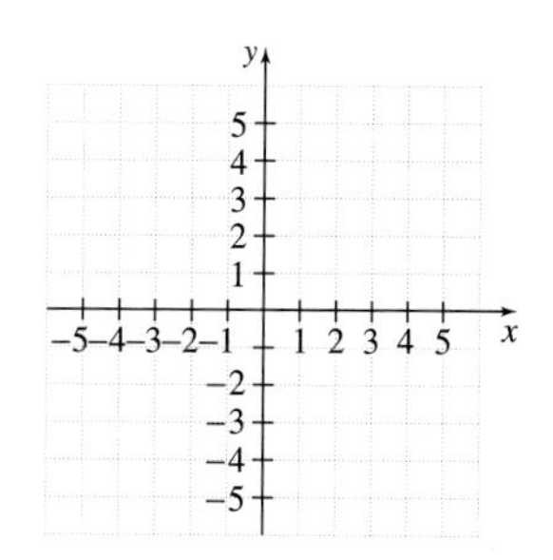

B *Graph each system. See Examples 3 and 4.*

21. $\begin{cases} 4x + 3y \ge 12 \\ x^2 + y^2 < 16 \end{cases}$

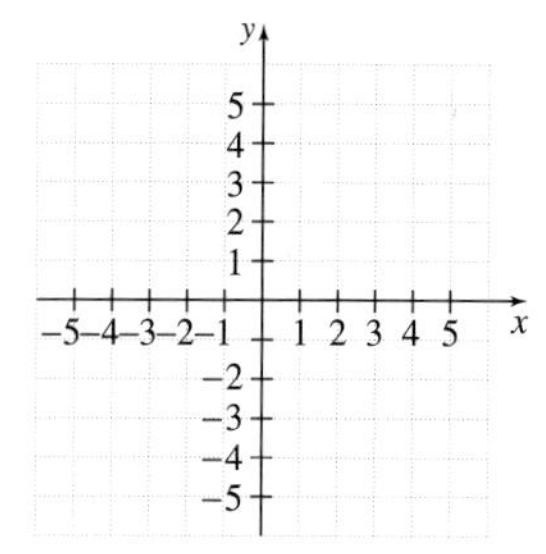

22. $\begin{cases} 3x - 4y \le 12 \\ x^2 + y^2 < 16 \end{cases}$

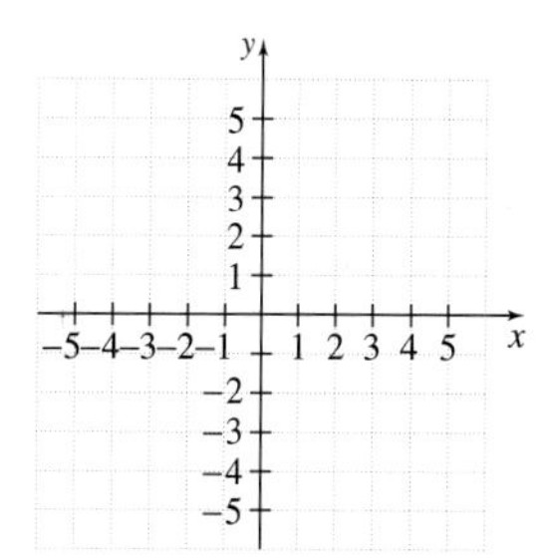

23. $\begin{cases} x^2 + y^2 \le 9 \\ x^2 + y^2 \ge 1 \end{cases}$

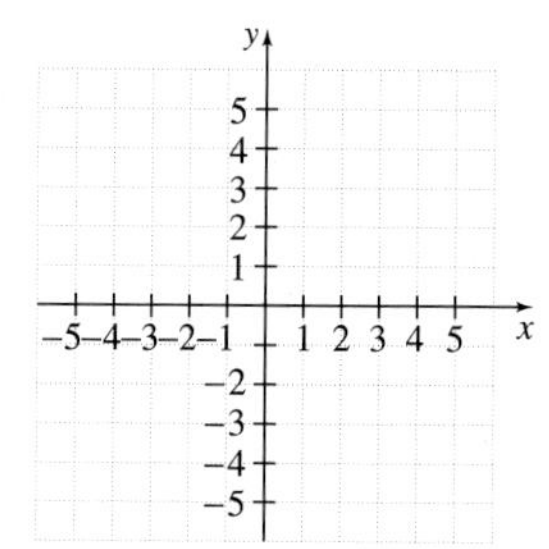

24. $\begin{cases} x^2 + y^2 \ge 9 \\ x^2 + y^2 \ge 16 \end{cases}$

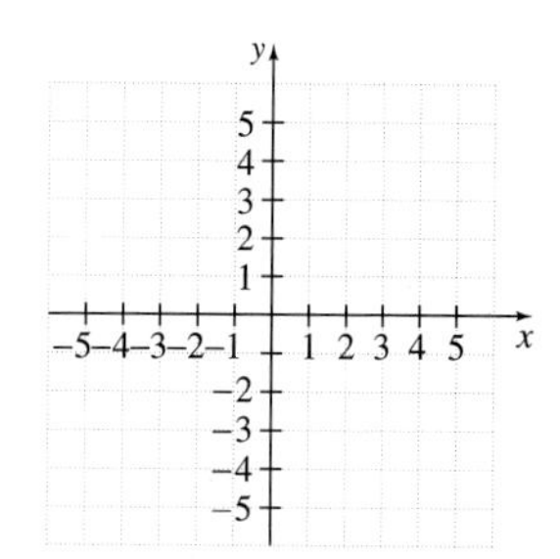

25. $\begin{cases} y > x^2 \\ y \ge 2x + 1 \end{cases}$

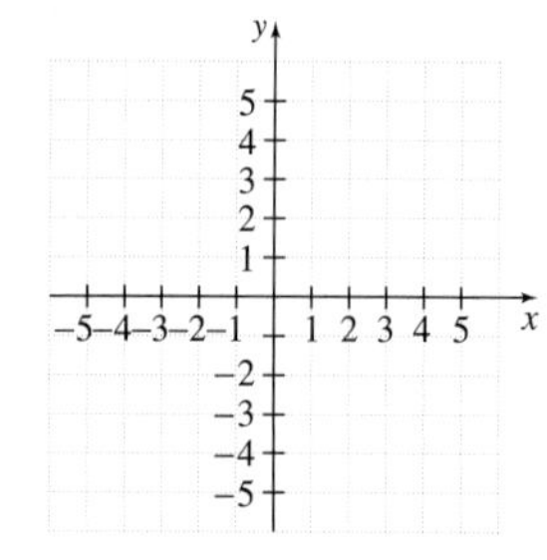

26. $\begin{cases} y \le -x^2 + 3 \\ y \le 2x - 1 \end{cases}$

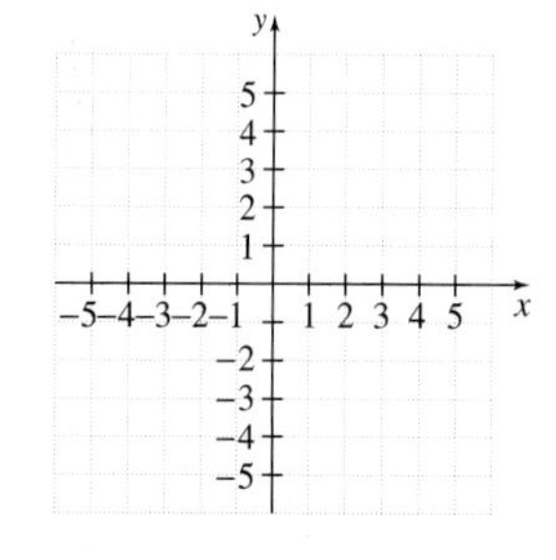

27. $\begin{cases} x^2 + y^2 > 9 \\ y > x^2 \end{cases}$

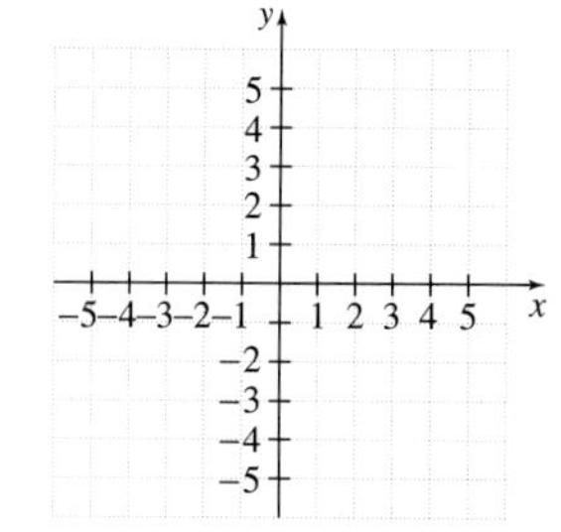

28. $\begin{cases} x^2 + y^2 \le 9 \\ y < x^2 \end{cases}$

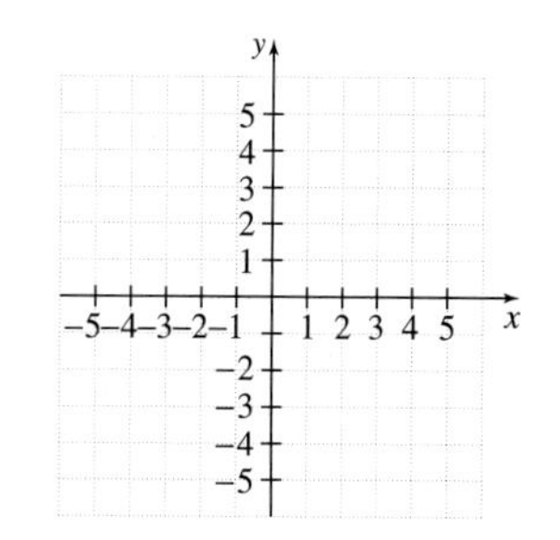

Name ______________________________

29. $\begin{cases} \dfrac{x^2}{4} + \dfrac{y^2}{9} \geq 1 \\ x^2 + y^2 \geq 4 \end{cases}$

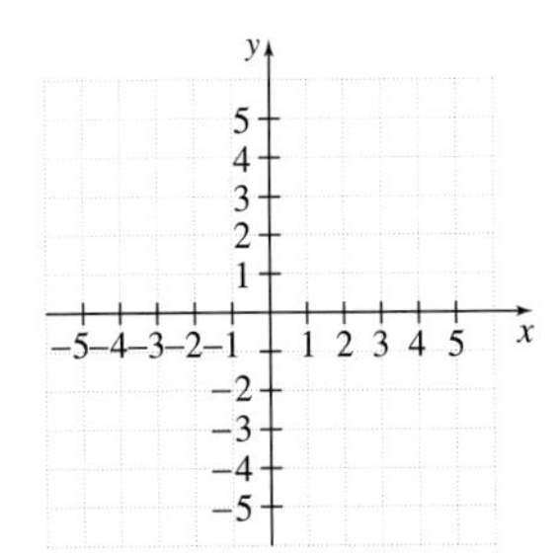

30. $\begin{cases} x^2 + (y - 2)^2 \geq 9 \\ \dfrac{x^2}{4} + \dfrac{y^2}{25} < 1 \end{cases}$

31. $\begin{cases} x^2 - y^2 \geq 1 \\ y \geq 0 \end{cases}$

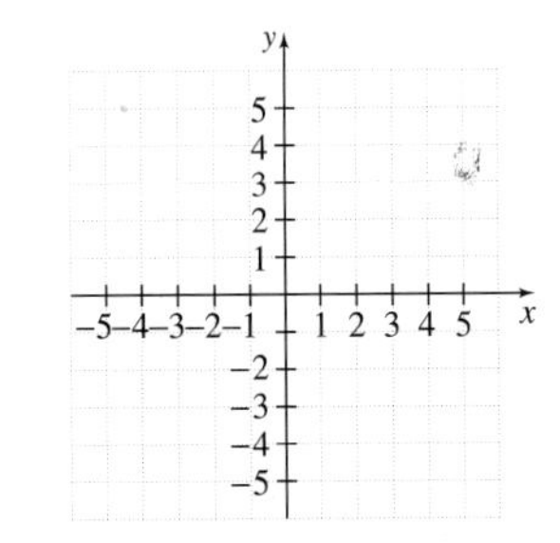

32. $\begin{cases} x^2 - y^2 \geq 1 \\ x \geq 0 \end{cases}$

33. $\begin{cases} x + y \geq 1 \\ 2x + 3y < 1 \\ x > -3 \end{cases}$

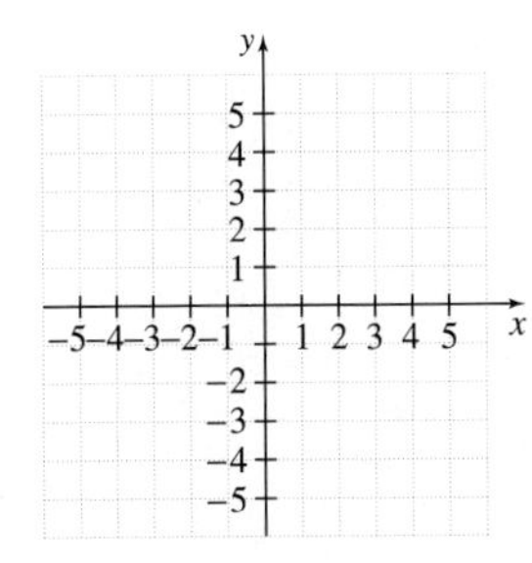

34. $\begin{cases} x - y < -1 \\ 4x - 3y > 0 \\ y > 0 \end{cases}$

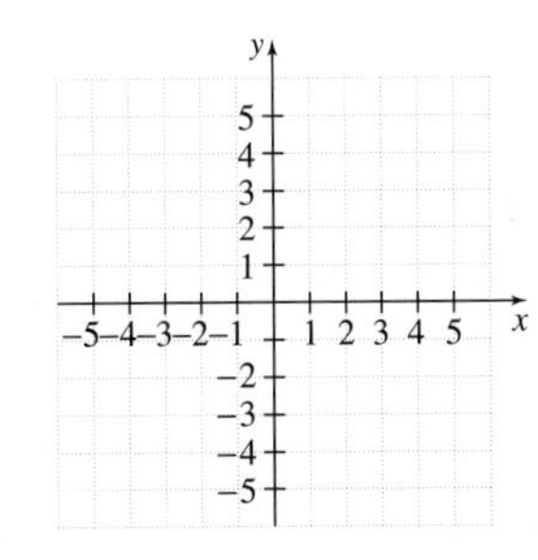

35. $\begin{cases} x^2 - y^2 < 1 \\ \dfrac{x^2}{16} + y^2 \leq 1 \\ x \geq -2 \end{cases}$

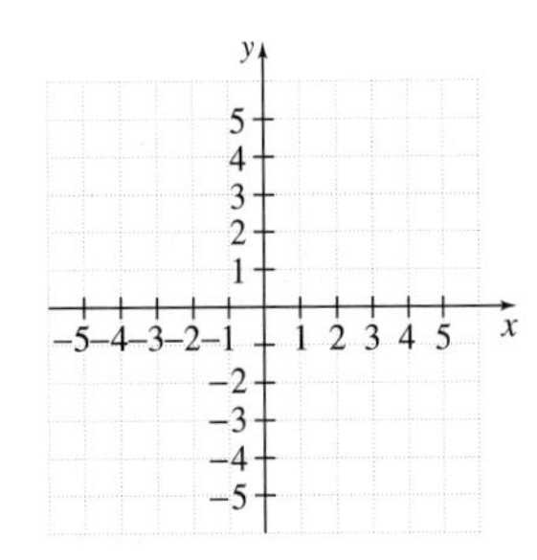

36. $\begin{cases} x^2 - y^2 \geq 1 \\ \dfrac{x^2}{16} + \dfrac{y^2}{4} \leq 1 \\ y \geq 1 \end{cases}$

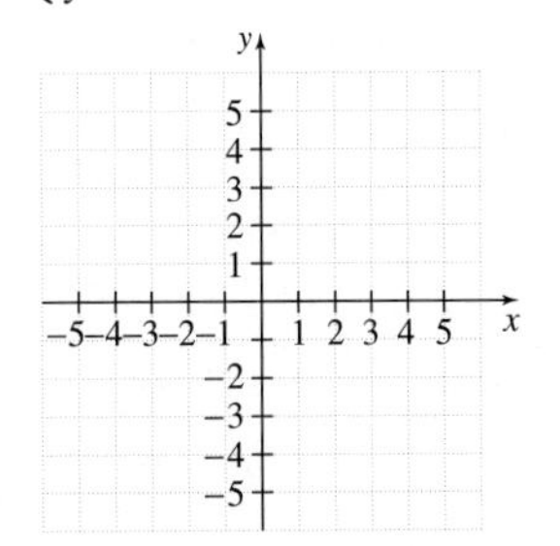

Review and Preview

Determine which graph is the graph of a function. See Section 3.5.

37.

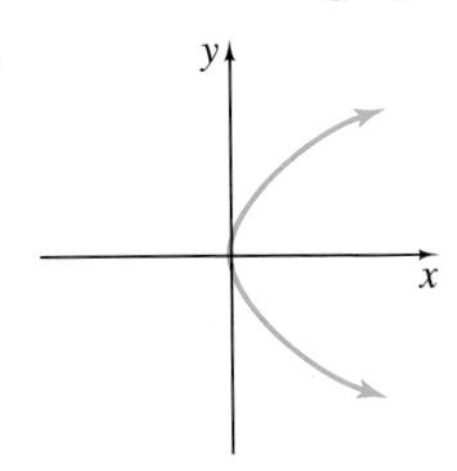

38.

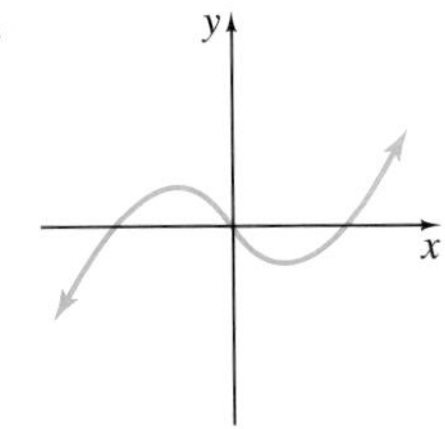

39.

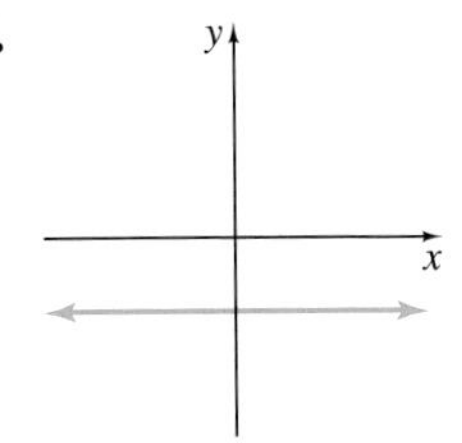

40.

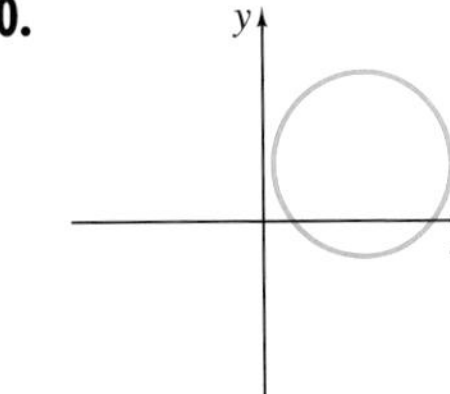

Answers

37. ______

38. ______

39. ______

40. ______

41. ______________

Name ______________________________

Combining Concepts

41. Discuss how graphing a linear inequality such as $x + y < 9$ is similar to graphing a nonlinear inequality such as $x^2 + y^2 < 9$.

42. Discuss how graphing a linear inequality such as $x + y < 9$ is different from graphing a nonlinear inequality such as $x^2 + y^2 < 9$.

43. Graph the system.

$$\begin{cases} y \leq x^2 \\ y \geq x + 2 \\ x \geq 0 \\ y \geq 0 \end{cases}$$

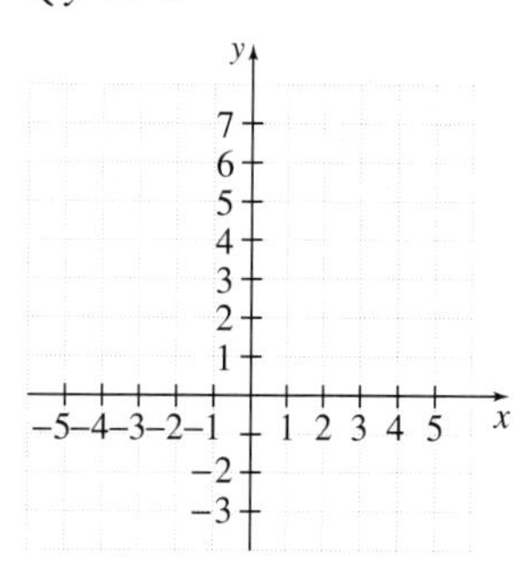

42. ______________

43. see graph

Name ______________________ Section ________ Date ________

Chapter 9 Activity
Modeling Conic Sections

MATERIALS

- Two thumbtacks (or nails)
- graph paper
- cardboard
- tape
- string
- pencil
- ruler

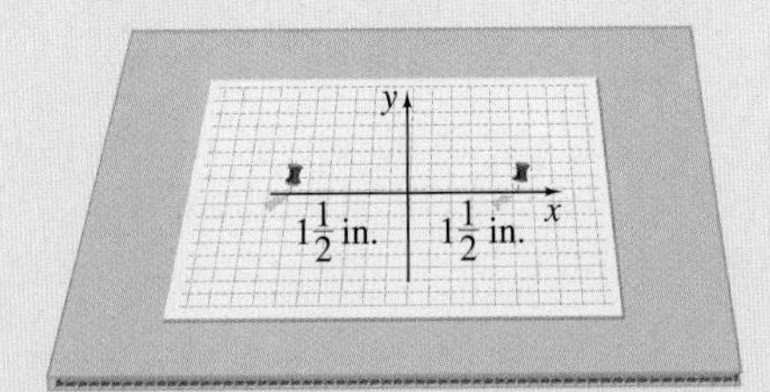

Figure 1

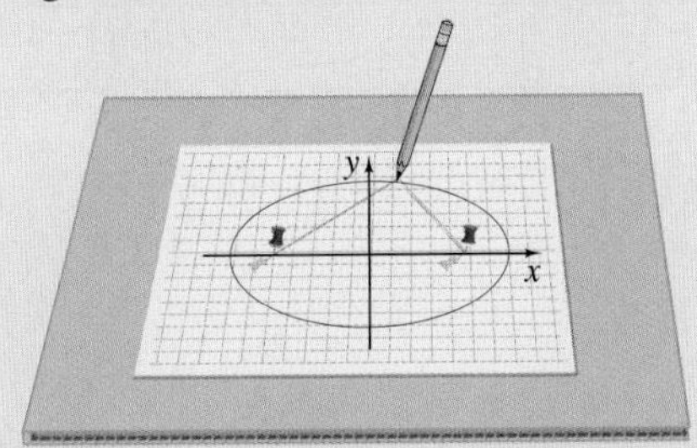

Figure 2

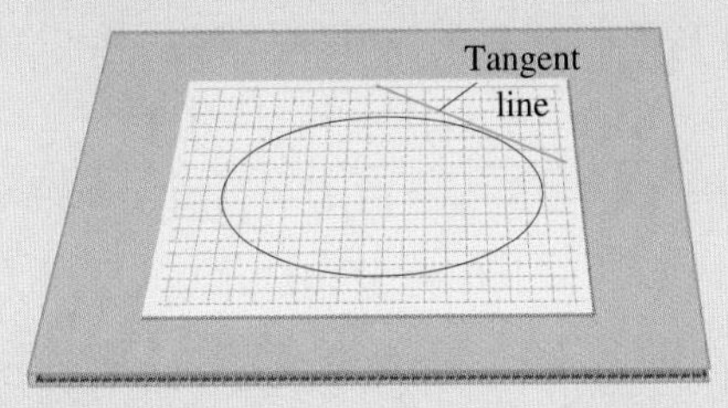

Figure 3

This activity may be completed by working in groups or individually.

1. Draw an x-axis and a y-axis on the graph paper as shown in Figure 1.

2. Place the graph paper on the cardboard and use tape to attach.

3. Locate two points on the x-axis each about $1\frac{1}{2}$ inches from the origin and on opposite sides of the origin (see Figure 1). Insert thumbtacks (or nails) at each of these locations.

4. Fasten a 9-inch piece of string to the thumbtacks as shown in Figure 2. Use your pencil to draw and keep the string taut while you carefully move the pencil in a path all around the thumbtacks.

5. Using the grid of the graph paper as a guide, find an approximate equation of the ellipse you drew.

6. Experiment by moving the tacks closer together or farther apart and drawing new ellipses. What do you observe?

7. Write a paragraph explaining why the figure drawn by the pencil is an ellipse. How might you use the same materials to draw a circle?

8. (Optional) Choose one of the ellipses you drew with the string and pencil. Use a ruler to draw any six tangent lines to the ellipse. (A line is tangent to the ellipse if it intersects, or just touches, the ellipse at only one point. See Figure 3.) Extend the tangent lines to yield six points of intersection among the tangents. Use a straight edge to draw a line connecting each pair of opposite points of intersection. What do you observe? Repeat with a different ellipse. Can you make a conjecture about the relationship among the lines that connect opposite points of intersection?

CHAPTER 9 HIGHLIGHTS

DEFINITIONS AND CONCEPTS | **EXAMPLES**

SECTION 9.1 THE PARABOLA AND THE CIRCLE

PARABOLAS

$$y = a(x - h)^2 + k$$

$$x = a(y - k)^2 + h$$

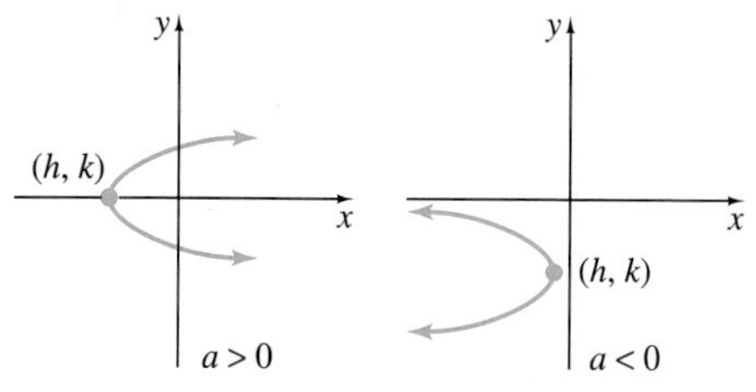

Graph: $x = 3y^2 - 12y + 13$

$$x - 13 = 3(y^2 - 4y)$$
$$x - 13 + 3(4) = 3(y^2 - 4y + 4)$$
$$x = 3(y - 2)^2 + 1$$

Since $a = 3$, this parabola opens to the right with vertex $(1, 2)$. Its axis of symmetry is $y = 2$. The x-intercept is 13.

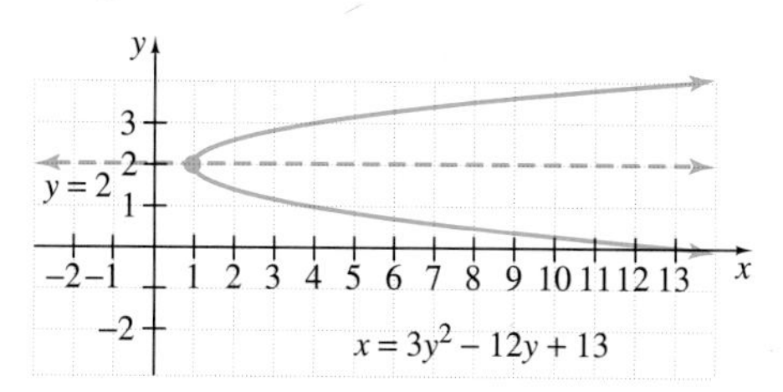

DISTANCE FORMULA

The distance d between two points (x_1, y_1) and (x_2, y_2) is given by

$$d = \sqrt{(x_2 - x_1)^2 + (y_2 - y_1)^2}$$

Find the distance between points $(-1, 6)$ and $(-2, -4)$.

Let $(x_1, y_1) = (-1, 6)$ and $(x_2, y_2) = (-2, -4)$.

$$d = \sqrt{(x_2 - x_1)^2 + (y_2 - y_1)^2}$$
$$= \sqrt{(-2 - (-1))^2 + (-4 - 6)^2}$$
$$= \sqrt{1 + 100} = \sqrt{101}$$

MIDPOINT FORMULA

The midpoint of the line segment whose endpoints are (x_1, y_1) and (x_2, y_2) is the point with coordinates

$$\left(\frac{x_1 + x_2}{2}, \frac{y_1 + y_2}{2}\right)$$

Find the midpoint of the line segment whose endpoints are $(-1, 6)$ and $(-2, -4)$.

$$\left(\frac{-1 + (-2)}{2}, \frac{6 + (-4)}{2}\right)$$

The midpoint is $\left(-\frac{3}{2}, 1\right)$

CIRCLE

The graph $(x - h)^2 + (y - k)^2 = r^2$ is a circle with center (h, k) and radius r.

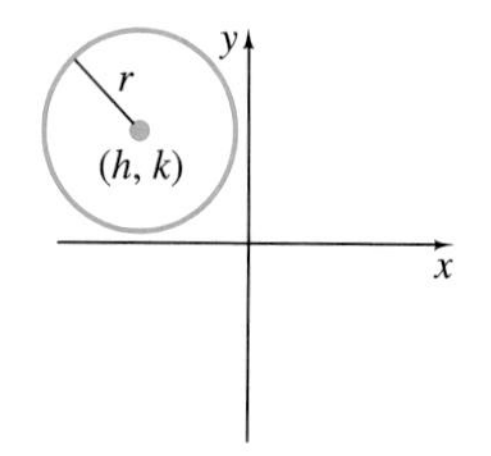

Graph: $x^2 + (y + 3)^2 = 5$

This equation can be written as

$$(x - 0)^2 + (y + 3)^2 = 5$$

with $h = 0$, $k = -3$, and $r = \sqrt{5}$. The center of this circle is $(0, -3)$, and the radius is $\sqrt{5}$.

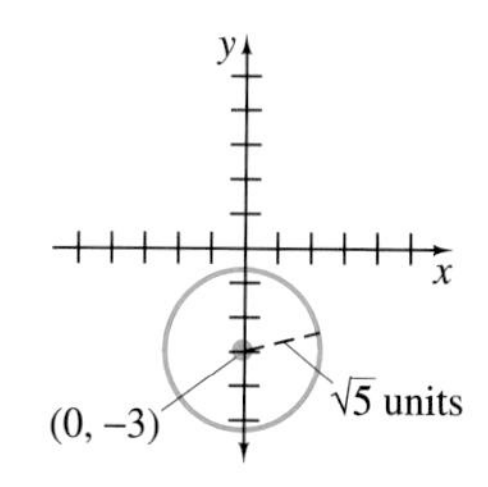

Section 9.2 The Ellipse and the Hyperbola

Ellipse with Center (0, 0)

The graph of an equation of the form $\frac{x^2}{a^2} + \frac{y^2}{b^2} = 1$ is an ellipse with center $(0, 0)$. The x-intercepts are a and $-a$, and the y-intercepts are b and $-b$.

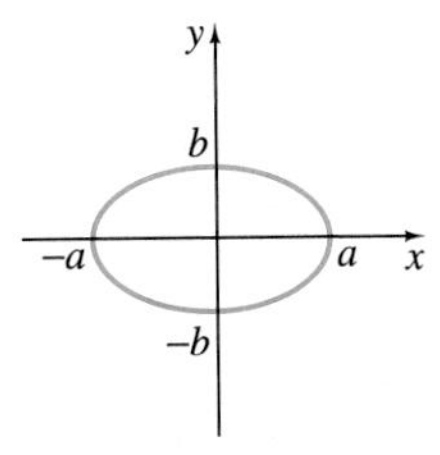

Graph: $4x^2 + 9y^2 = 36$

$$\frac{x^2}{9} + \frac{y^2}{4} = 1 \quad \text{Divide both sides by 36.}$$

$$\frac{x^2}{3^2} + \frac{y^2}{2^2} = 1$$

The ellipse has center $(0, 0)$, x-intercepts 3 and -3, and y-intercepts 2 and -2.

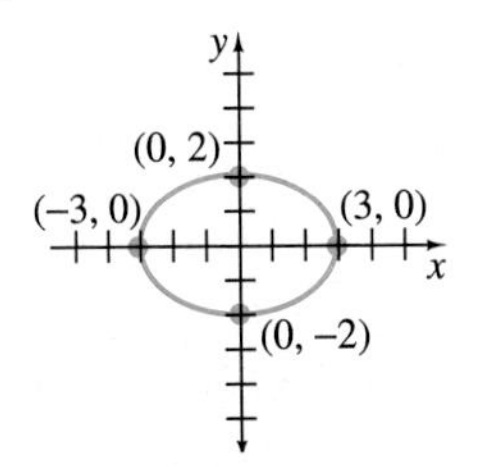

Hyperbola with Center (0, 0)

The graph of an equation of the form $\frac{x^2}{a^2} - \frac{y^2}{b^2} = 1$ is a hyperbola with center $(0, 0)$ and x-intercepts a and $-a$.

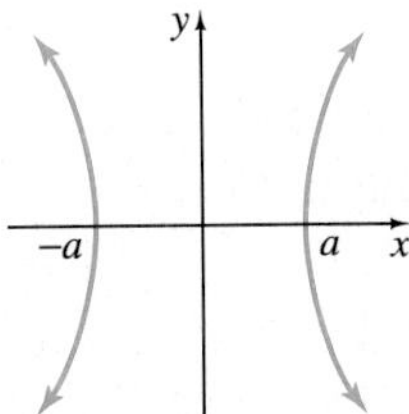

The graph of an equation of the form $\frac{y^2}{b^2} - \frac{x^2}{a^2} = 1$ is a hyperbola with center $(0, 0)$ and y-intercepts b and $-b$.

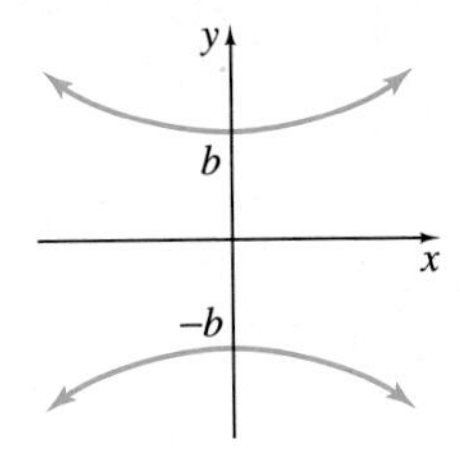

Graph: $\frac{x^2}{9} - \frac{y^2}{4} = 1$. Here $a = 3$ and $b = 2$.

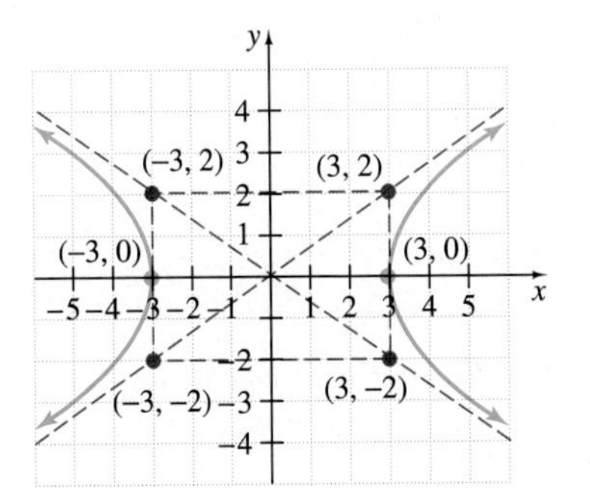

Section 9.3 Graphing Nonlinear Functions

$f(x) = |x|$

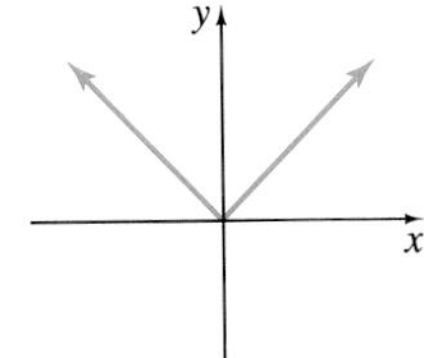

$f(x) = \sqrt{x}$

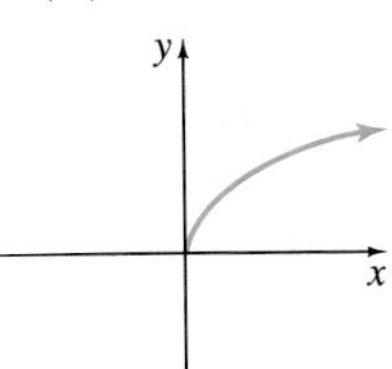

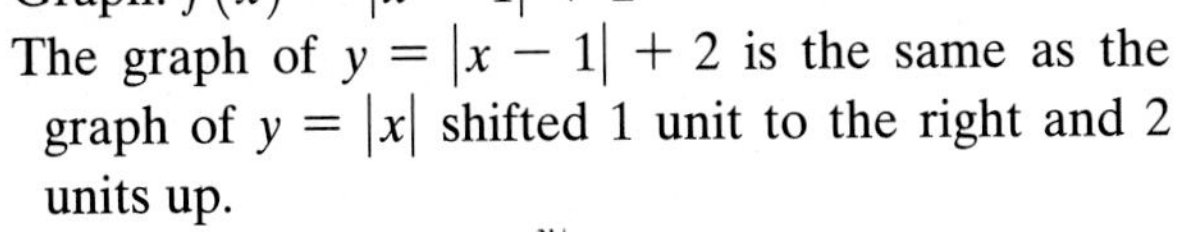

Graph: $f(x) = |x - 1| + 2$

The graph of $y = |x - 1| + 2$ is the same as the graph of $y = |x|$ shifted 1 unit to the right and 2 units up.

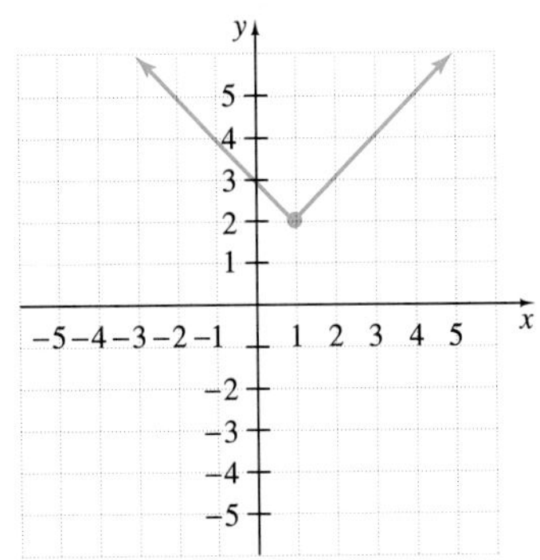

SECTION 9.4 SOLVING NONLINEAR SYSTEMS OF EQUATIONS

A **nonlinear system of equations** is a system of equations at least one of which is not linear. Both the substitution method and the elimination method may be used to solve a nonlinear system of equations.

Solve the nonlinear system $\begin{cases} y = x + 2 \\ 2x^2 + y^2 = 3 \end{cases}$

Substitute $x + 2$ for y in the second equation:

$$2x^2 + y^2 = 3$$
$$2x^2 + (x + 2)^2 = 3$$
$$2x^2 + x^2 + 4x + 4 = 3$$
$$3x^2 + 4x + 1 = 0$$
$$(3x + 1)(x + 1) = 0$$
$$x = -\frac{1}{3} \quad \text{or} \quad x = -1$$

If $x = -\frac{1}{3}$, $y = x + 2 = -\frac{1}{3} + 2 = \frac{5}{3}$.

If $x = -1$, $y = x + 2 = -1 + 2 = 1$.

The solution set is $\left\{\left(-\frac{1}{3}, \frac{5}{3}\right), (-1, 1)\right\}$

SECTION 9.5 NONLINEAR INEQUALITIES AND SYSTEMS OF INEQUALITIES

The **graph of a system of inequalities** is the intersection of the graphs of the inequalities.

Graph the system: $\begin{cases} x \geq y^2 \\ x + y \leq 4 \end{cases}$

The graph of the system is the darkest shaded region along with its boundary lines.

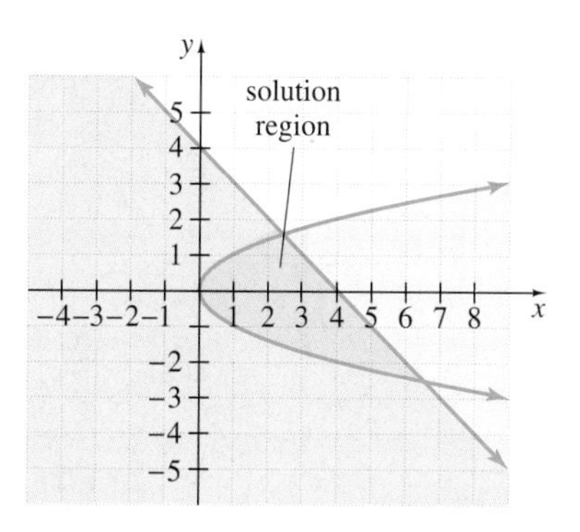

Name ______________________ Section __________ Date __________

Chapter 9 Review

(9.1) *Find the distance between each pair of points. Give an exact value and a three-decimal-place approximation.*

1. $(-6, 3)$ and $(8, 4)$

2. $(3, 5)$ and $(8, 9)$

3. $(-4, -6)$ and $(-1, 5)$

4. $(-1, 5)$ and $(2, -3)$

5. $(-\sqrt{2}, 0)$ and $(0, -4\sqrt{6})$

6. $(-\sqrt{5}, -\sqrt{11})$ and $(-\sqrt{5}, -3\sqrt{11})$

7. $(7.4, -8.6)$ and $(-1.2, 5.6)$

8. $(2.3, 1.8)$ and $(10.7, -9.2)$

Find the midpoint of each line segment whose endpoints are given.

9. $(2, 6)$ and $(-12, 4)$

10. $(-3, 8)$ and $(11, 24)$

11. $(-6, -5)$ and $(-9, 7)$

12. $(4, -6)$ and $(-15, 2)$

13. $\left(0, -\frac{3}{8}\right)$ and $\left(\frac{1}{10}, 0\right)$

14. $\left(\frac{3}{4}, -\frac{1}{7}\right)$ and $\left(-\frac{1}{4}, -\frac{3}{7}\right)$

15. $(\sqrt{3}, -2\sqrt{6})$ and $(\sqrt{3}, -4\sqrt{6})$

16. $(-5\sqrt{3}, 2\sqrt{7})$ and $(-3\sqrt{3}, 10\sqrt{7})$

Write an equation of each circle with the given center and radius or diameter.

17. center $(-4, 4)$, radius 3

18. center $(5, 0)$, diameter 10

19. center $(-7, -9)$, radius $\sqrt{11}$

20. center $(0, 0)$, diameter 7

Name ____________________

Graph each equation. If the graph is a circle, find its center and radius. If the graph is a parabola, find its vertex.

21. $x^2 + y^2 = 7$

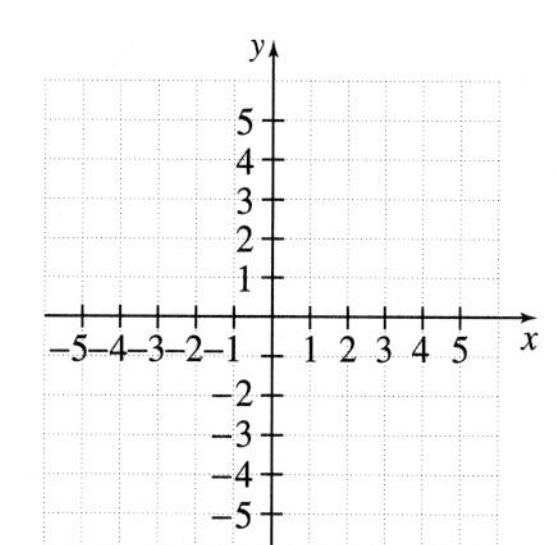

22. $x = 2(y - 5)^2 + 4$

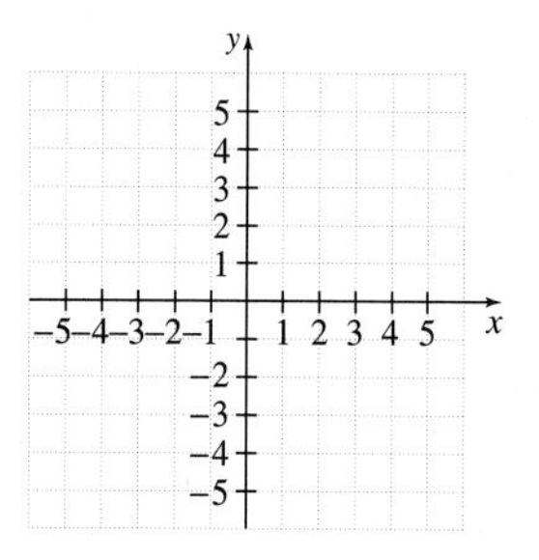

23. $x = -(y + 2)^2 + 3$

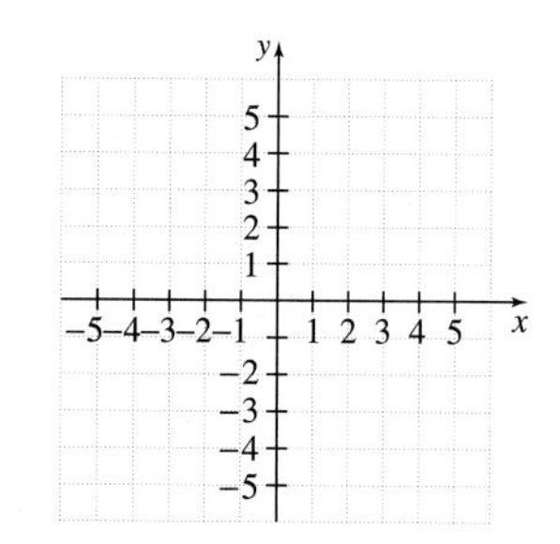

24. $(x - 1)^2 + (y - 2)^2 = 4$

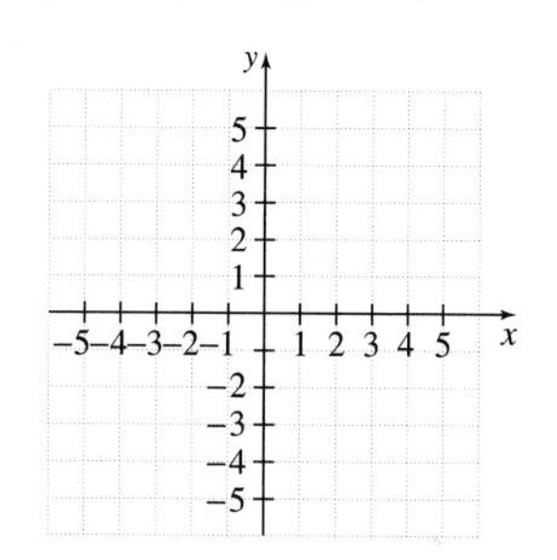

25. $y = -x^2 + 4x + 10$

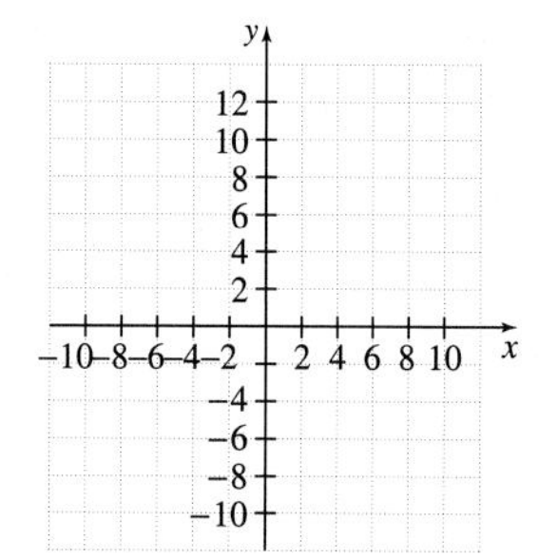

26. $x = -y^2 - 4y + 6$

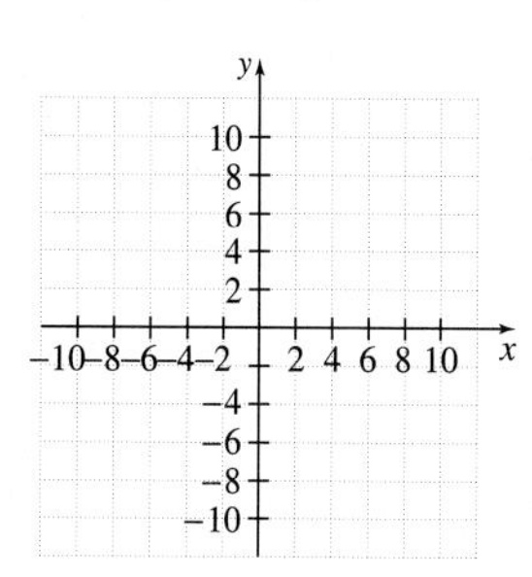

27. $x = \frac{1}{2}y^2 + 2y + 1$

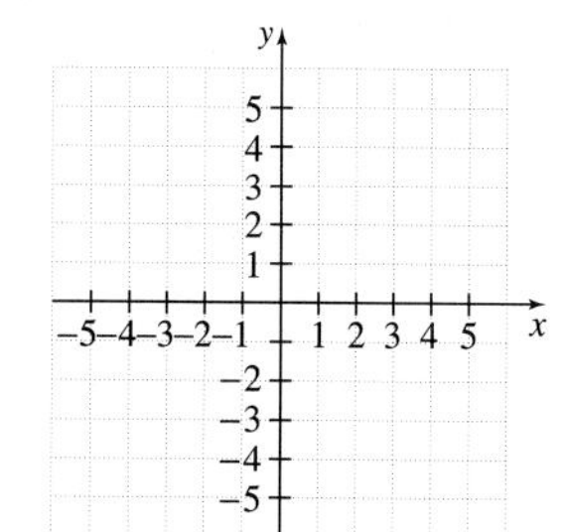

28. $y = -3x^2 + \frac{1}{2}x + 4$

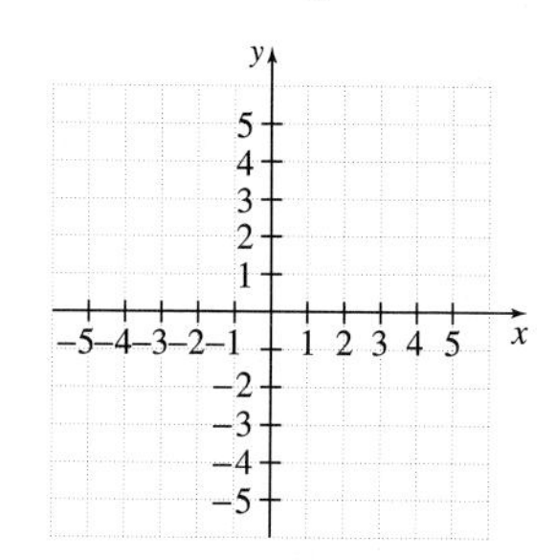

29. $x^2 + y^2 + 2x + y = \frac{3}{4}$

30. $x^2 + y^2 - 3y = \frac{7}{4}$

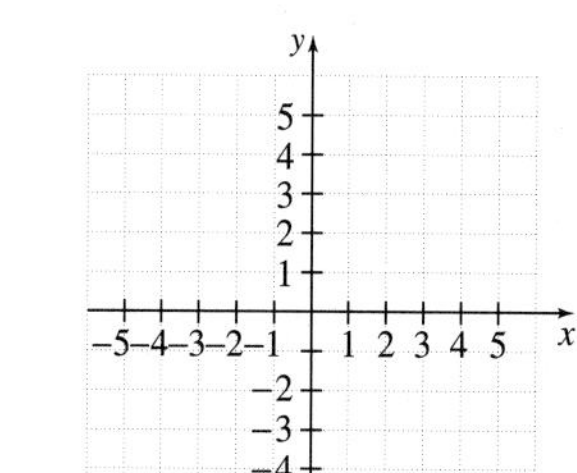

31. $4x^2 + 4y^2 + 16x + 8y = 1$

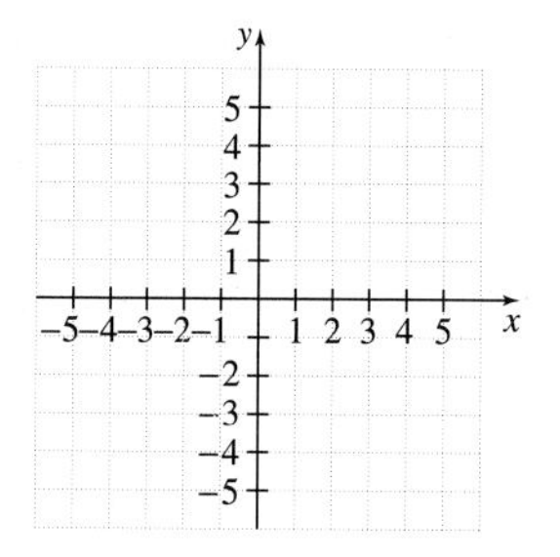

32. $3x^2 + 6x + 3y^2 = 9$

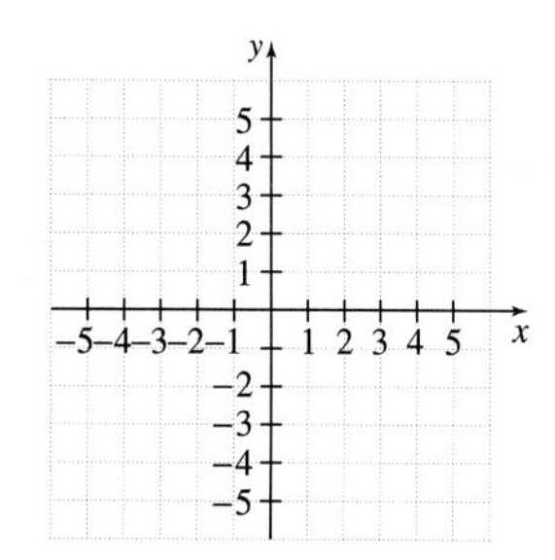

33. $y = x^2 + 6x + 9$

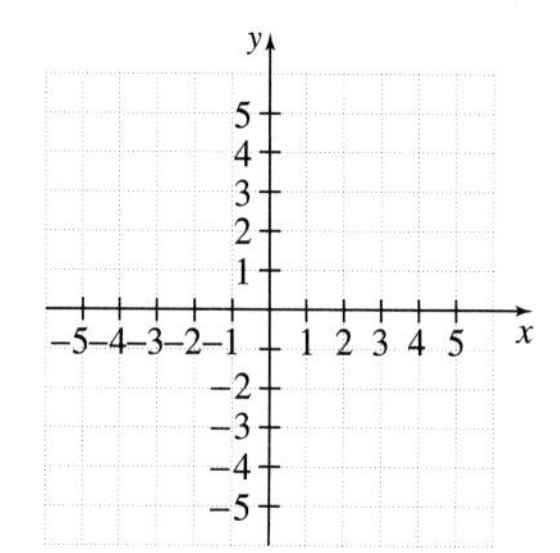

34. $x = y^2 + 6y + 9$

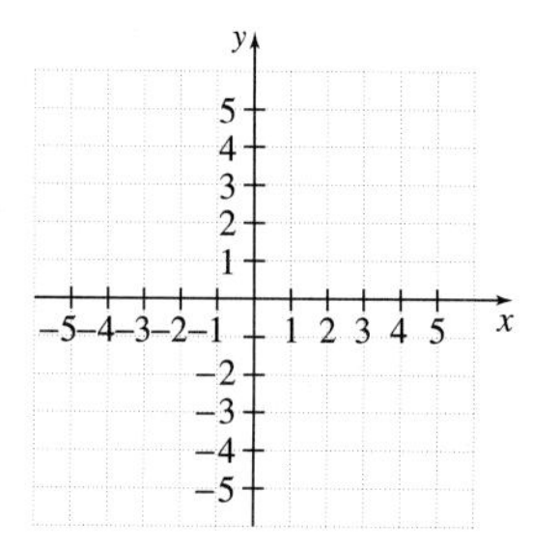

Name ______________________________________

(9.2) *Graph each equation.*

35. $x^2 + \dfrac{y^2}{4} = 1$

36. $x^2 - \dfrac{y^2}{4} = 1$

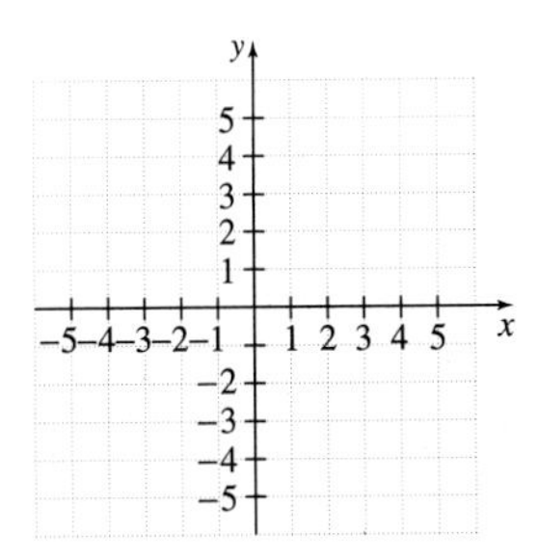

37. $\dfrac{y^2}{4} - \dfrac{x^2}{16} = 1$

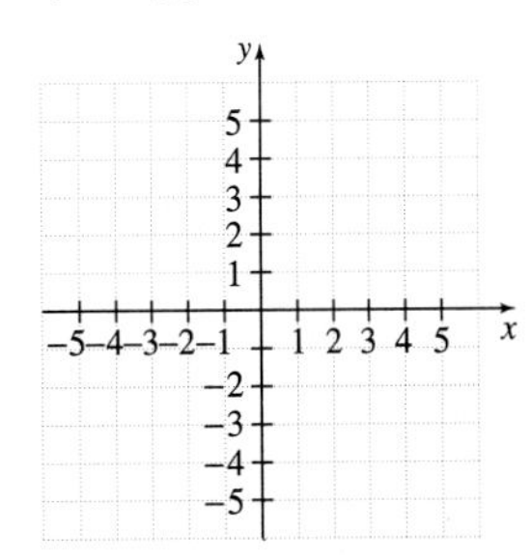

38. $\dfrac{y^2}{4} + \dfrac{x^2}{16} = 1$

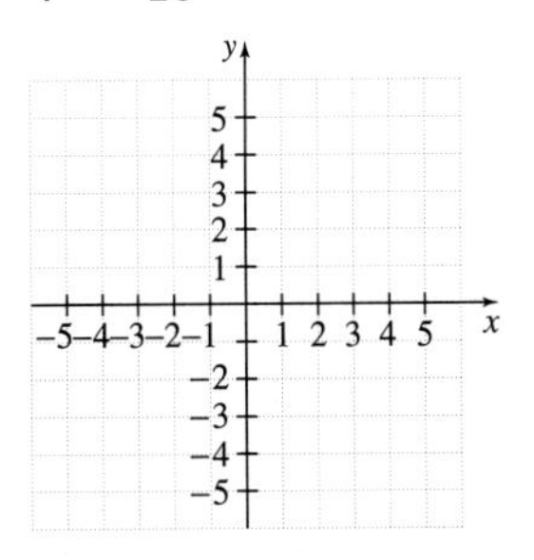

39. $-5x^2 + 25y^2 = 125$

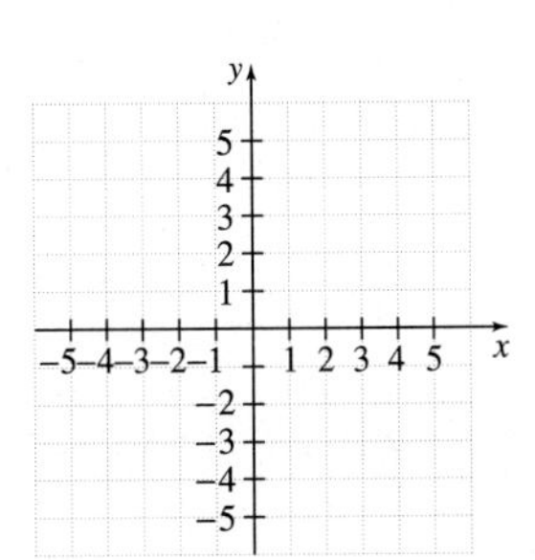

40. $4y^2 + 9x^2 = 36$

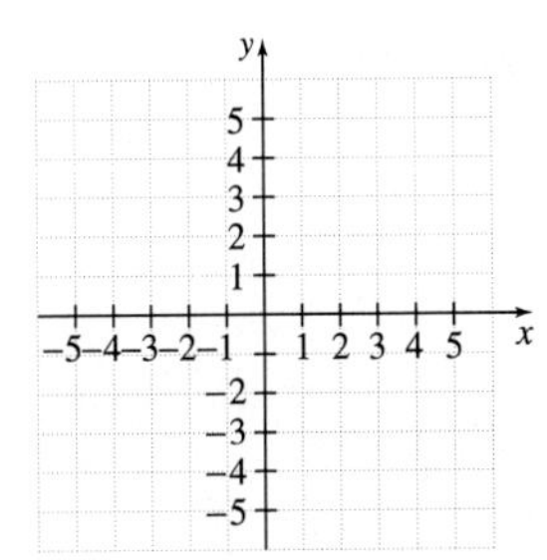

41. $\dfrac{(x-2)^2}{4} + (y-1)^2 = 1$

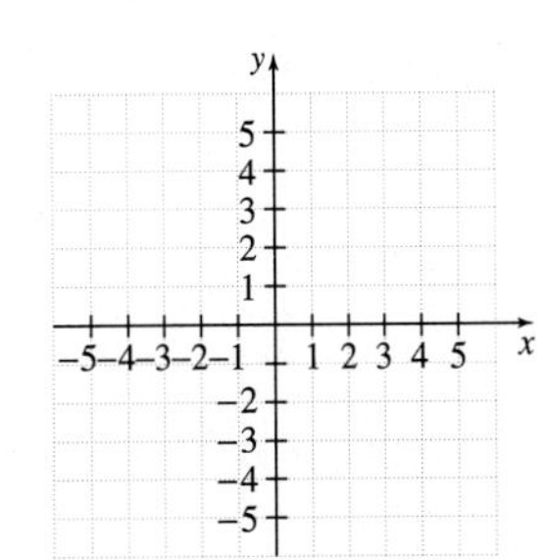

42. $\dfrac{(x+3)^2}{9} + \dfrac{(y-4)^2}{25} = 1$

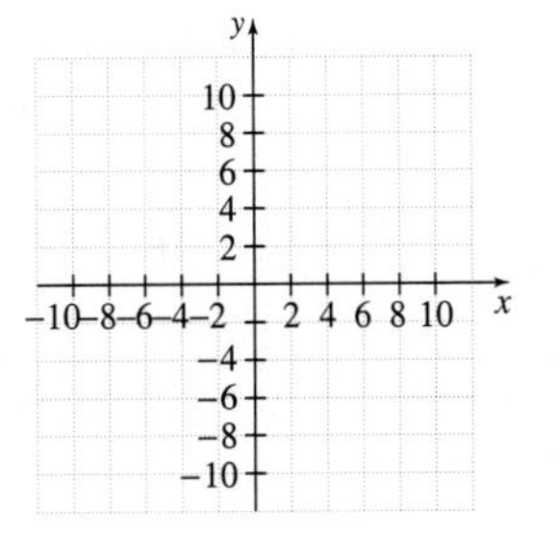

43. $x^2 - y^2 = 1$

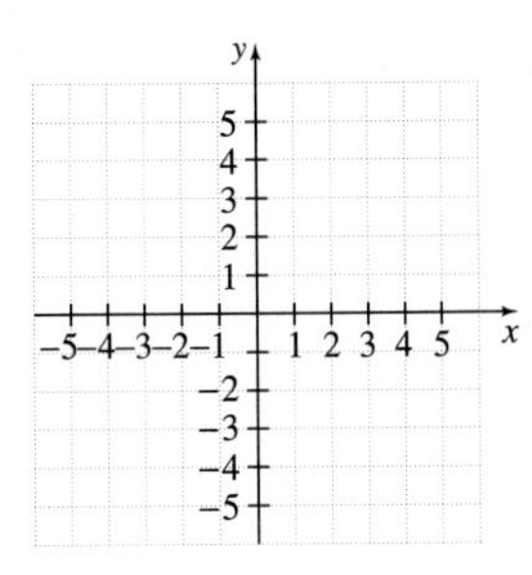

44. $36y^2 - 49x^2 = 1764$

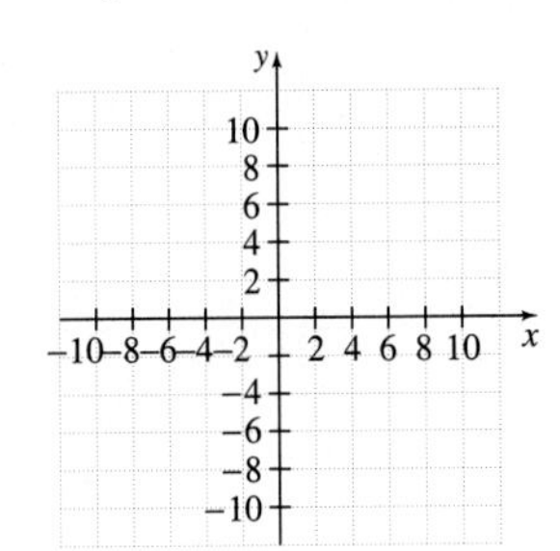

45. $y = x^2 + 9$

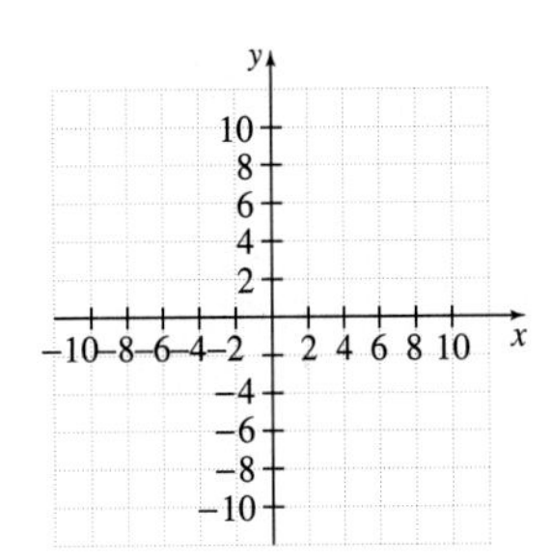

46. $x = 4y^2 - 16$

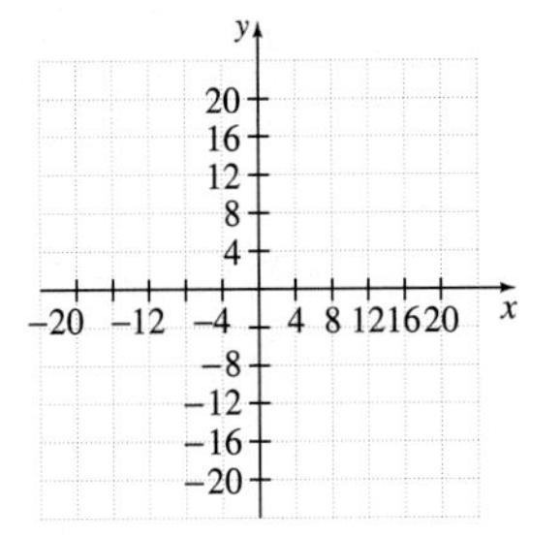

Graph each equation.

47. $y = x^2 + 4x + 6$

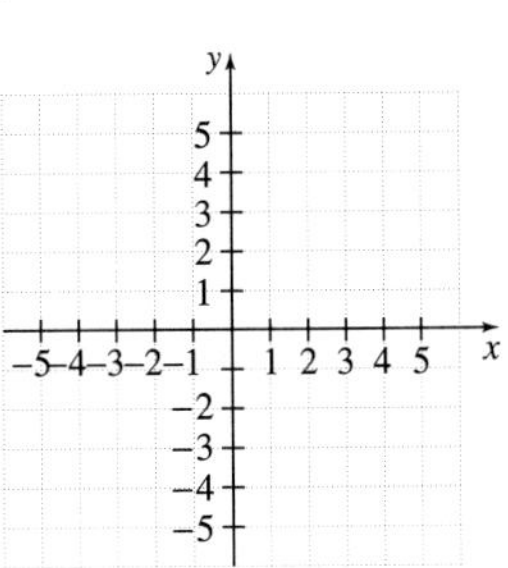

48. $y^2 = x^2 + 6$

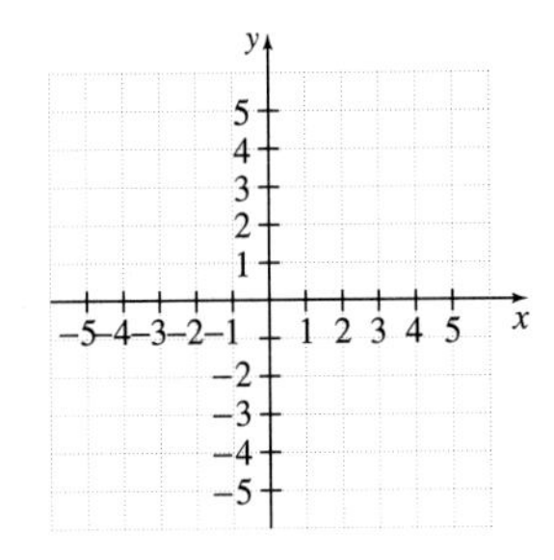

49. $y^2 + x^2 = 4x + 6$

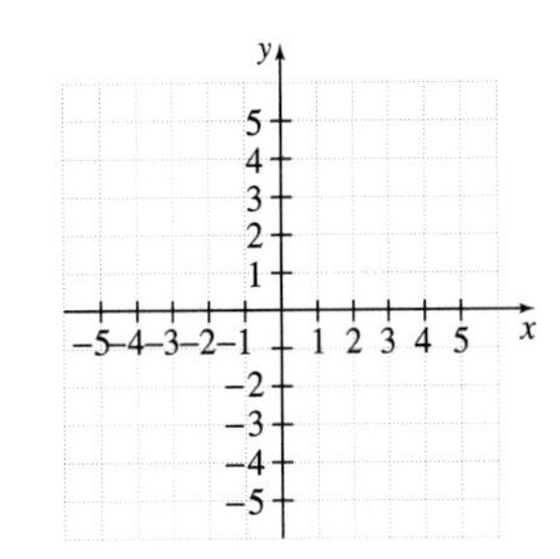

50. $y^2 + 2x^2 = 4x + 6$

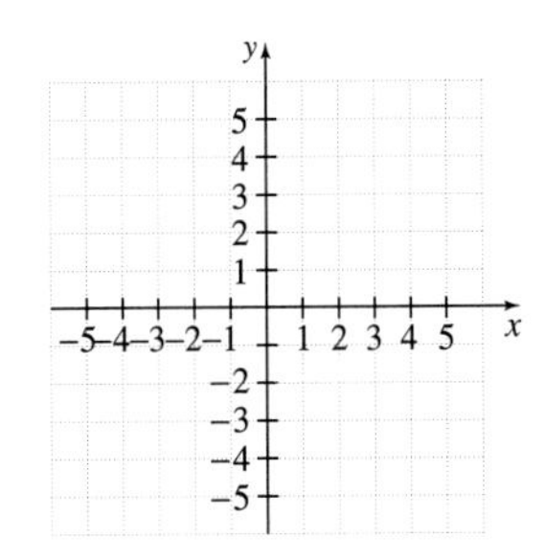

Name ______________________________

51. $x^2 + y^2 - 8y = 0$

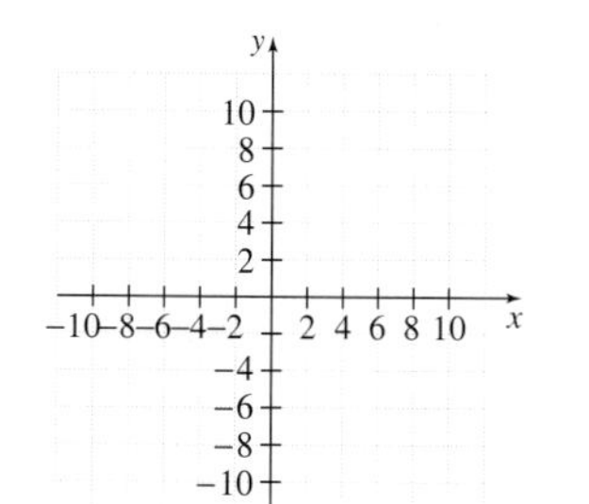

52. $x - 4y = y^2$

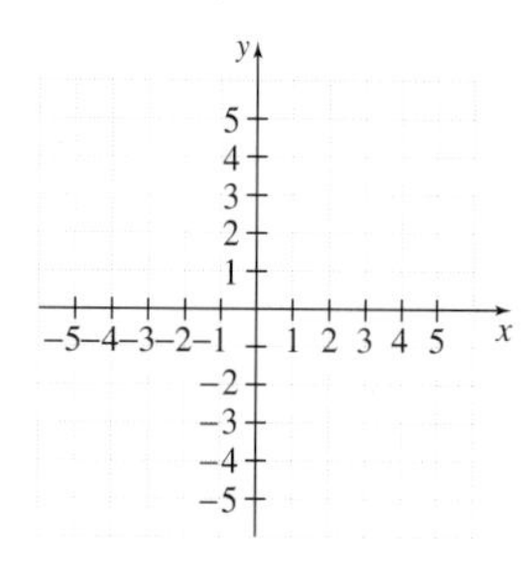

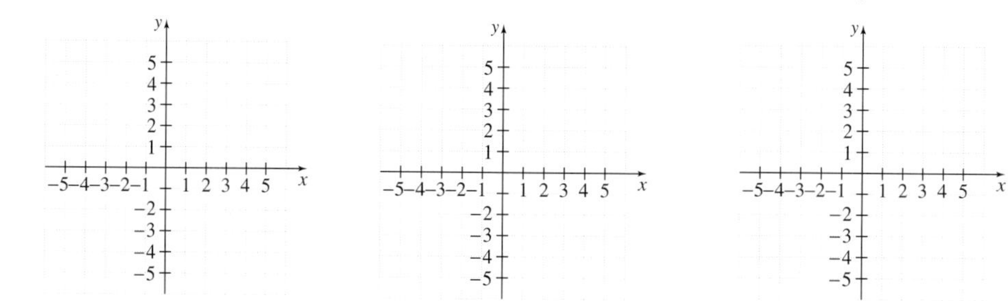

53. $x^2 - 4 = y^2$

54. $x^2 = 4 - y^2$

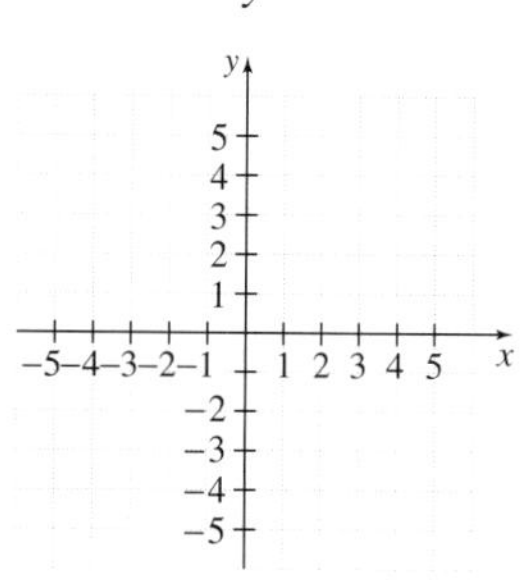

55. $6(x - 2)^2 + 9(y + 5)^2 = 36$

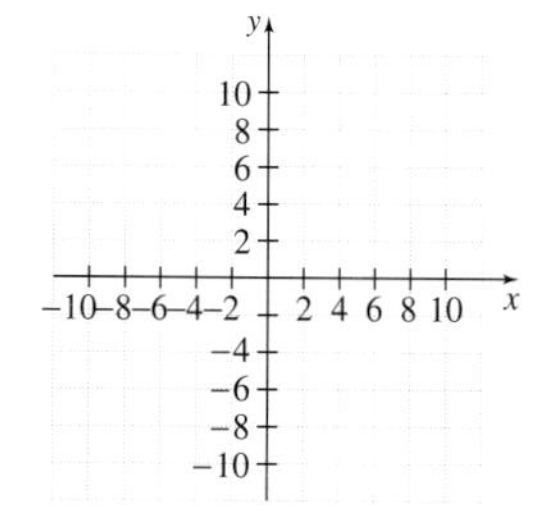

56. $36y^2 = 576 + 16x^2$

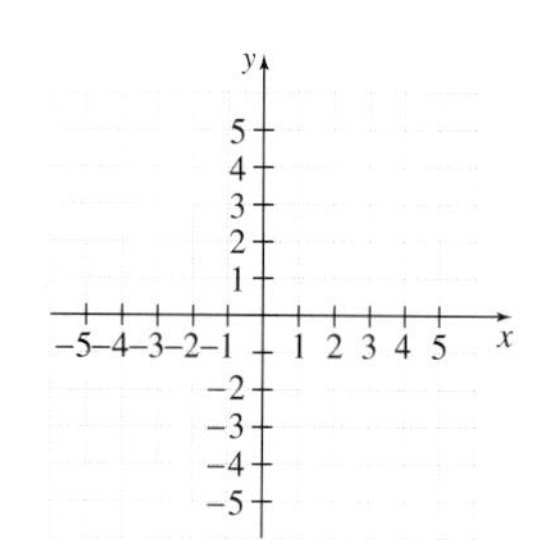

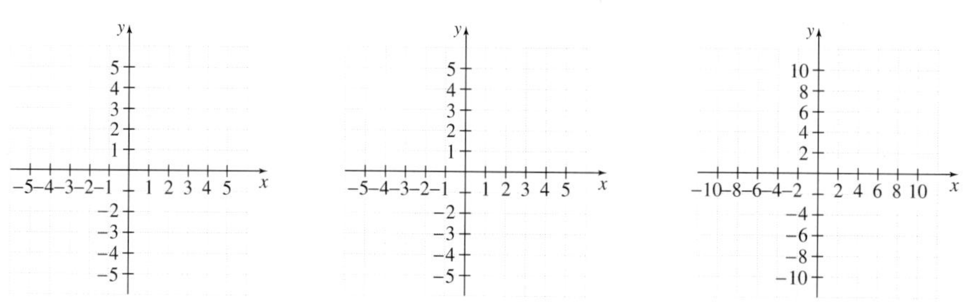

57. $\dfrac{x^2}{16} - \dfrac{y^2}{25} = 1$

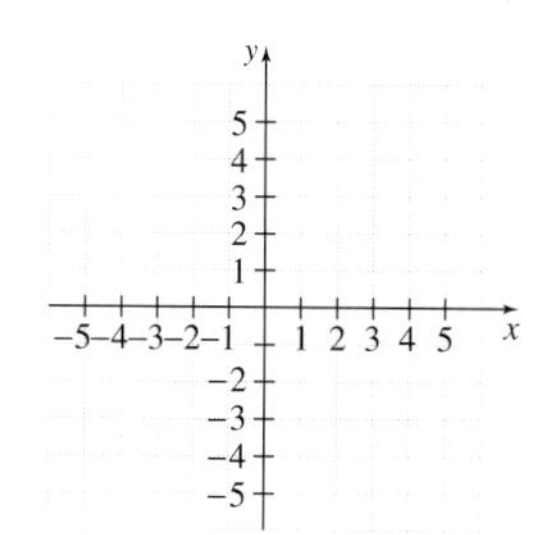

58. $3(x - 7)^2 + 3(y + 4)^2 = 1$

(9.3) *Graph each function.*

59. $f(x) = |x| + 2$

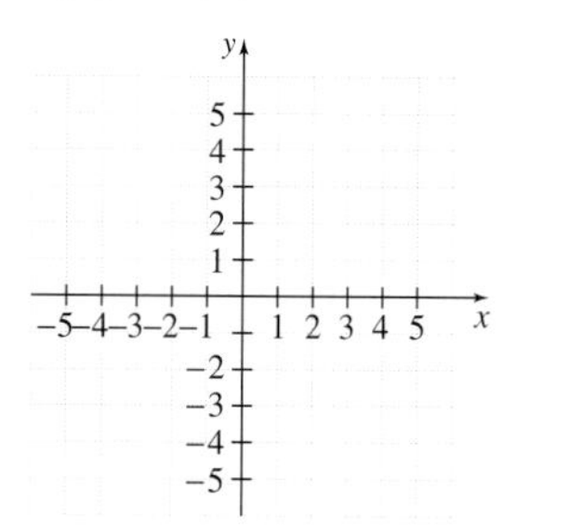

60. $f(x) = \sqrt{x} - 1$

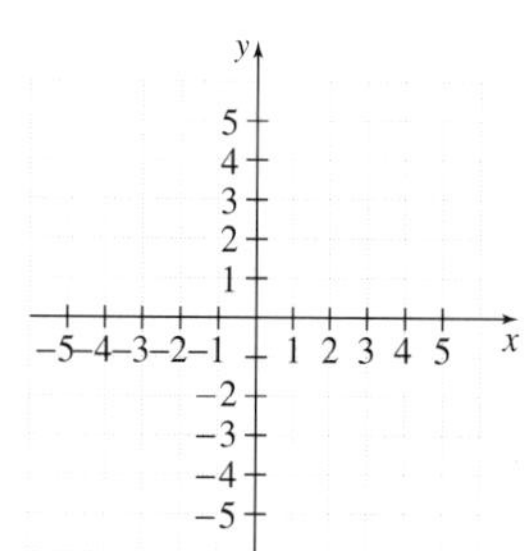

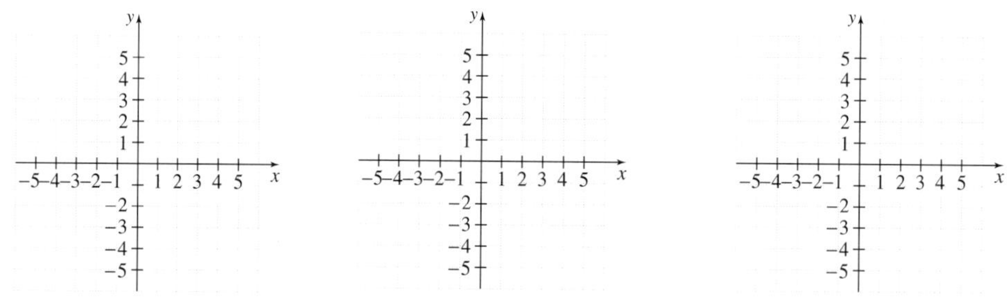

61. $f(x) = \sqrt{x - 4}$

62. $f(x) = |x + 1|$

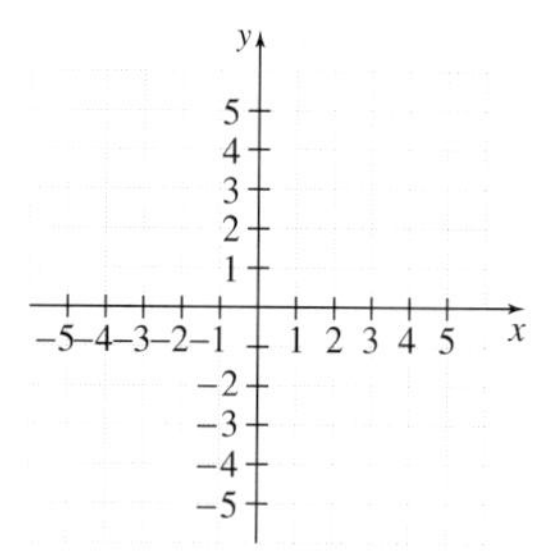

63. $f(x) = \sqrt{x - 3} + 1$

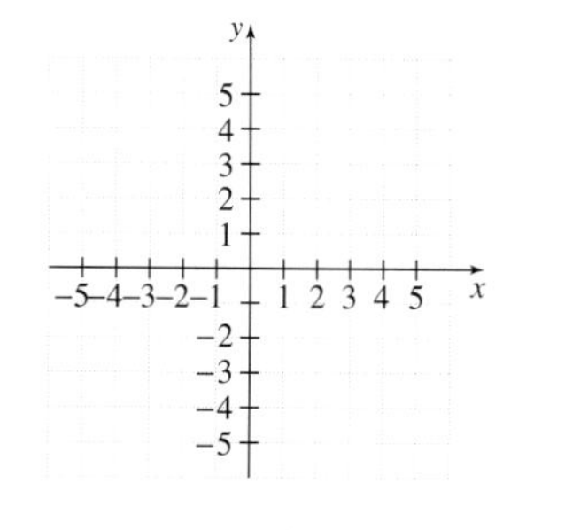

64. $f(x) = |x - 1| + 3$

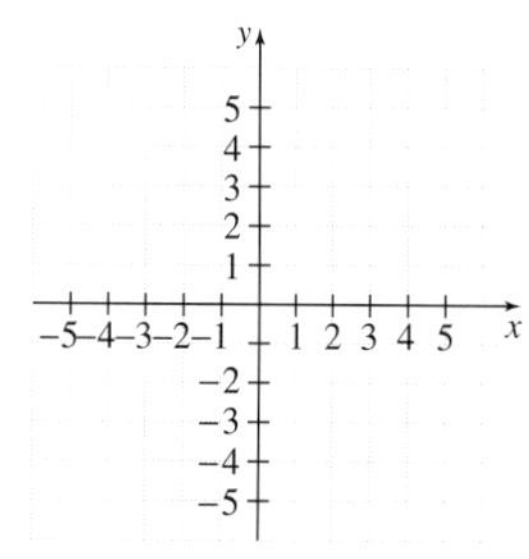

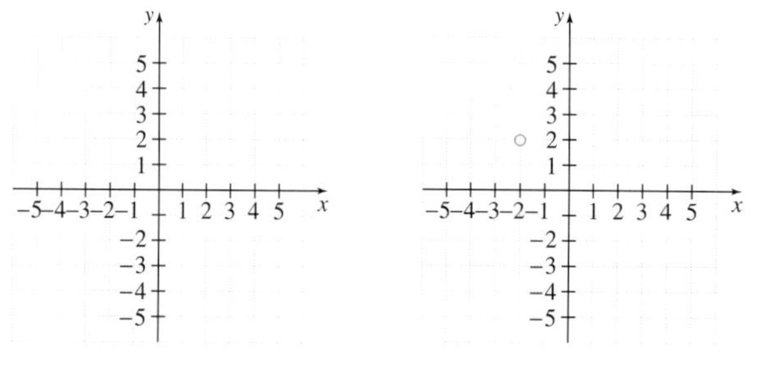

65. $f(x) = |x + 2| + 2$

66. $f(x) = \sqrt{x + 2} + 2$

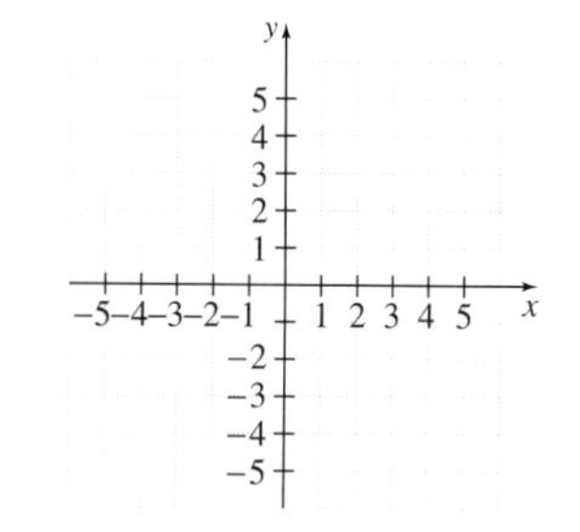

Name ______________________________

(9.4) *Solve each system of equations.*

67. $\begin{cases} y = 2x - 4 \\ y^2 = 4x \end{cases}$

68. $\begin{cases} x^2 + y^2 = 4 \\ x - y = 4 \end{cases}$

69. $\begin{cases} y = x + 2 \\ y = x^2 \end{cases}$

70. $\begin{cases} y = x^2 - 5x + 1 \\ y = -x + 6 \end{cases}$

71. $\begin{cases} 4x - y^2 = 0 \\ 2x^2 + y^2 = 16 \end{cases}$

72. $\begin{cases} x^2 + 4y^2 = 16 \\ x^2 + y^2 = 4 \end{cases}$

73. $\begin{cases} x^2 + y^2 = 10 \\ 9x^2 + y^2 = 18 \end{cases}$

74. $\begin{cases} x^2 + 2y = 9 \\ 5x - 2y = 5 \end{cases}$

75. $\begin{cases} y = 3x^2 + 5x - 4 \\ y = 3x^2 - x + 2 \end{cases}$

76. $\begin{cases} x^2 - 3y^2 = 1 \\ 4x^2 + 5y^2 = 21 \end{cases}$

77. Find the length and the width of a room whose area is 150 square feet and whose perimeter is 50 feet.

78. What is the greatest number of real solutions possible for a system of two equations whose graphs are an ellipse and a hyperbola?

(9.5) *Graph each inequality or system of inequalities.*

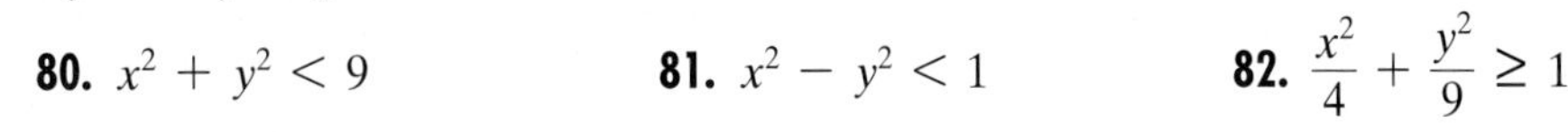

79. $y \le -x^2 + 3$

80. $x^2 + y^2 < 9$

81. $x^2 - y^2 < 1$

82. $\frac{x^2}{4} + \frac{y^2}{9} \ge 1$

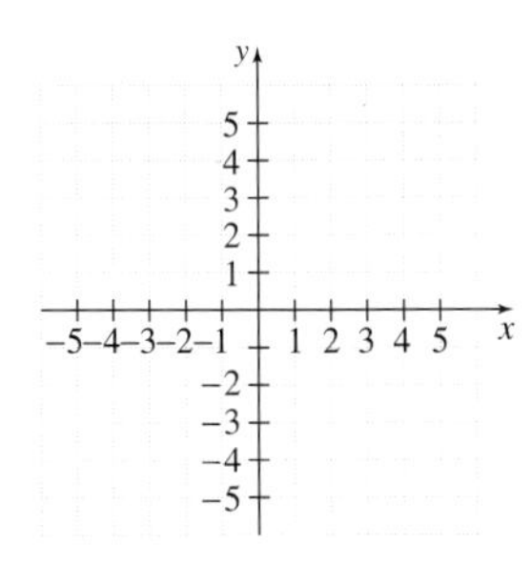

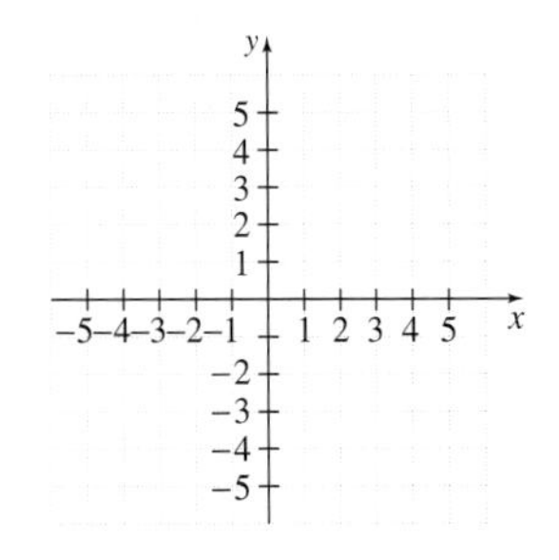

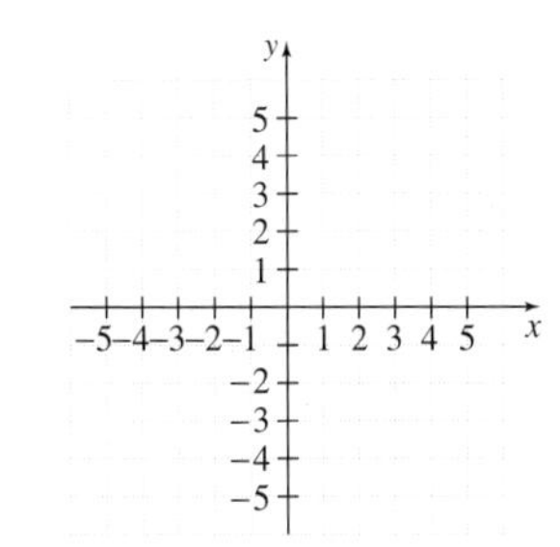

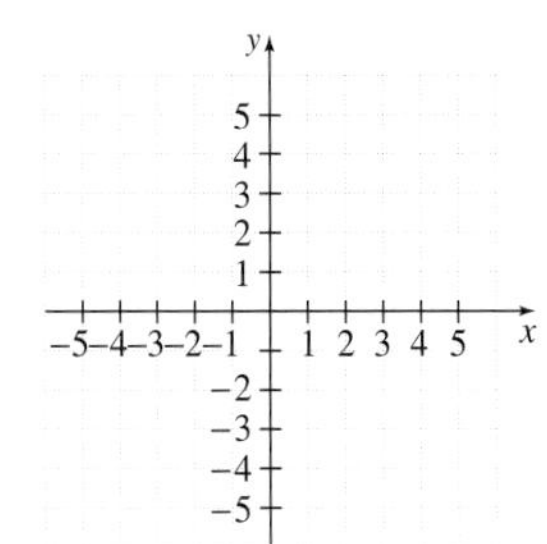

Name ______________________________

83. $\begin{cases} 2x \leq 4 \\ x + y \geq 1 \end{cases}$

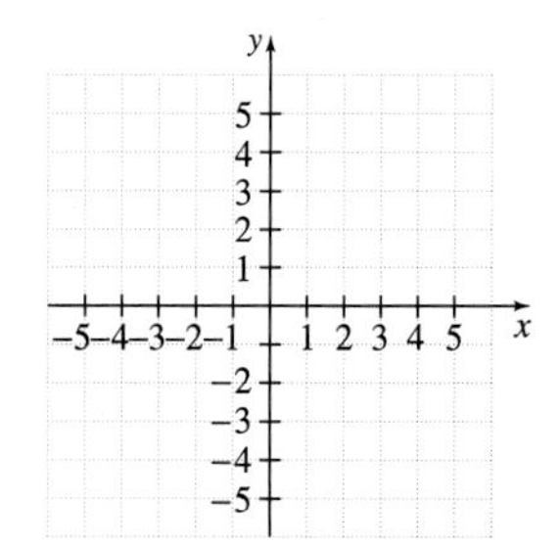

84. $\begin{cases} 3x + 4y \leq 12 \\ x - 2y > 6 \end{cases}$

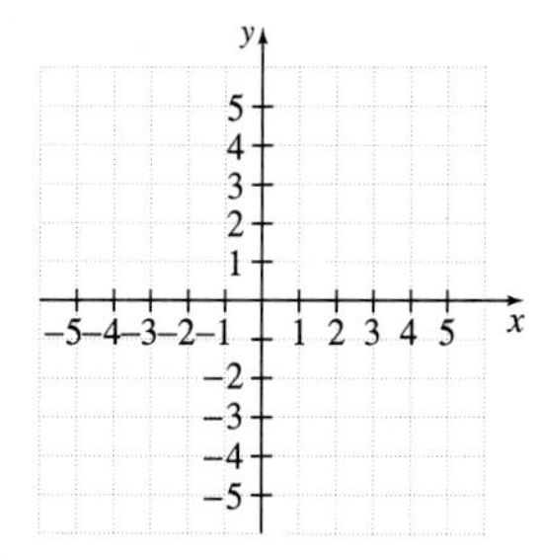

85. $\begin{cases} y > x^2 \\ x + y \geq 3 \end{cases}$

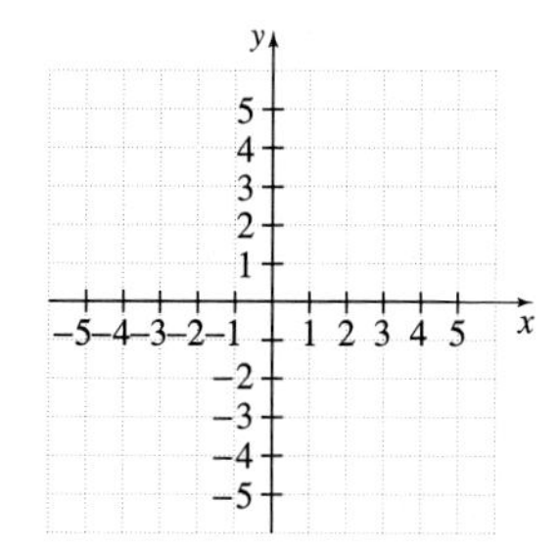

86. $\begin{cases} x^2 + y^2 \leq 16 \\ x^2 + y^2 \geq 4 \end{cases}$

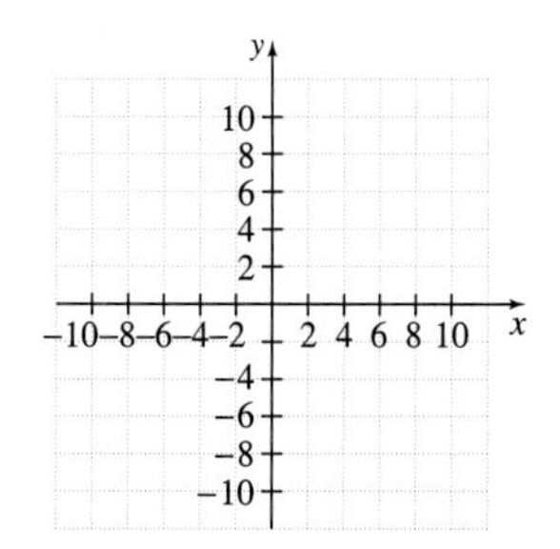

87. $\begin{cases} x^2 + y^2 < 4 \\ x^2 - y^2 \leq 1 \end{cases}$

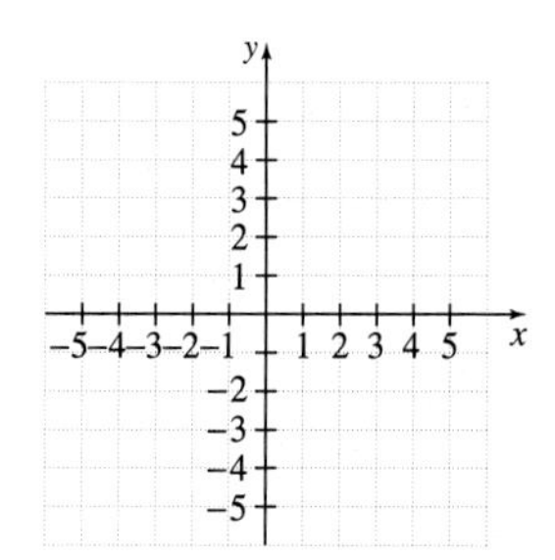

88. $\begin{cases} x^2 + y^2 < 4 \\ y \geq x^2 - 1 \\ x \geq 0 \end{cases}$

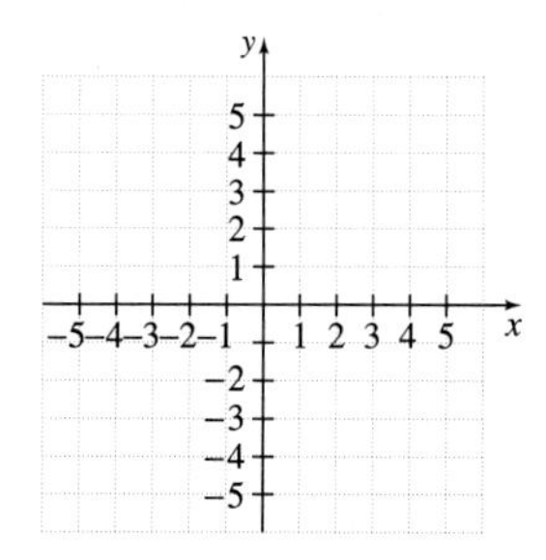

Name ______________________ Section ________ Date ________

ANSWERS

1. ____________

2. ____________

3. ____________

4. ____________

CHAPTER 9 TEST

1. Find the distance between the points $(-6, 3)$ and $(-8, -7)$.

2. Find the distance between the points $(-2\sqrt{5}, \sqrt{10})$ and $(-\sqrt{5}, 4\sqrt{10})$.

3. Find the midpoint of the line segment whose endpoints are $(-2, -5)$ and $(-6, 12)$.

4. Find the midpoint of the line segment whose endpoints are $\left(-\frac{2}{3}, -\frac{1}{5}\right)$ and $\left(-\frac{1}{3}, \frac{4}{5}\right)$.

Graph each equation.

5. $x^2 + y^2 = 36$

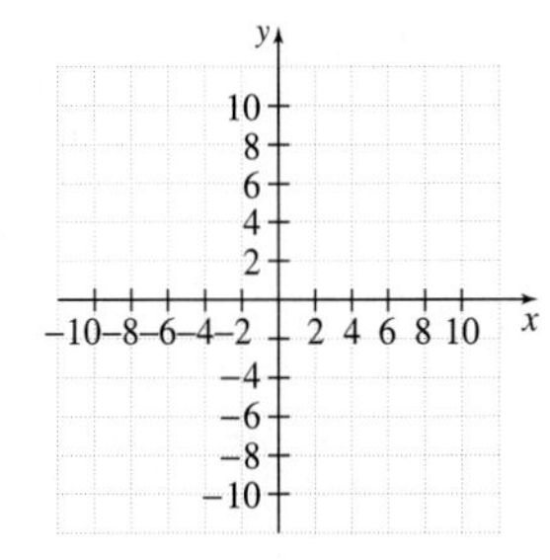

6. $x^2 - y^2 = 36$

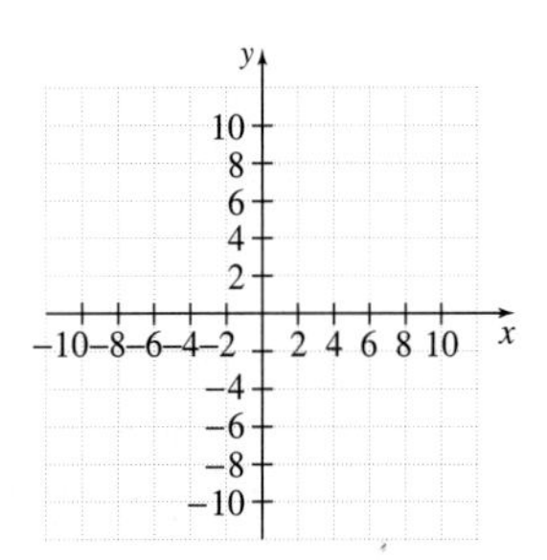

7. $16x^2 + 9y^2 = 144$

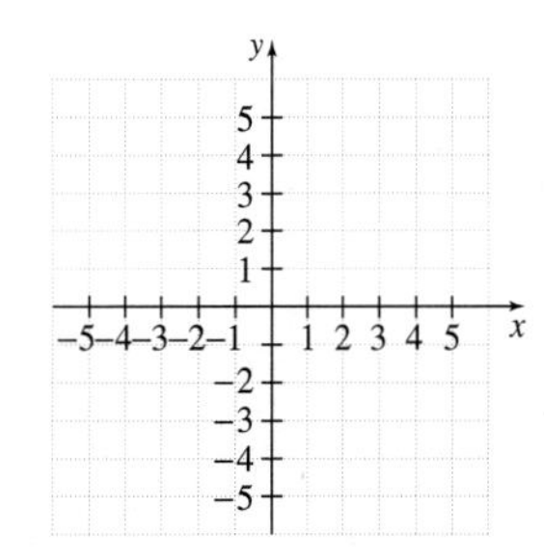

8. $y = x^2 - 8x + 16$

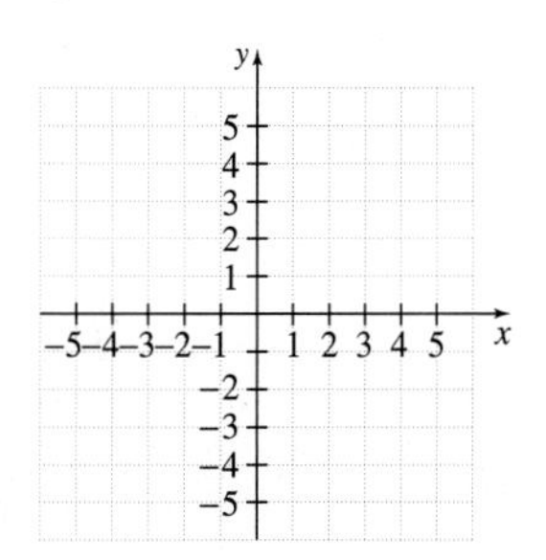

9. $x^2 + y^2 + 6x = 16$

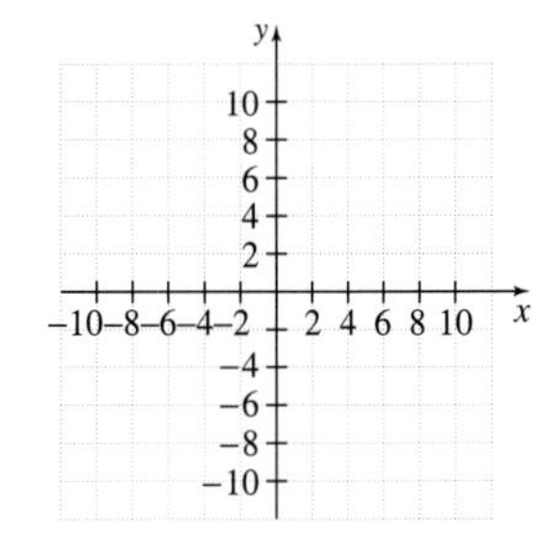

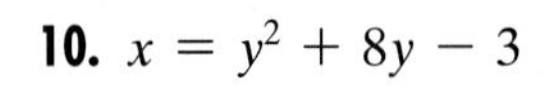

10. $x = y^2 + 8y - 3$

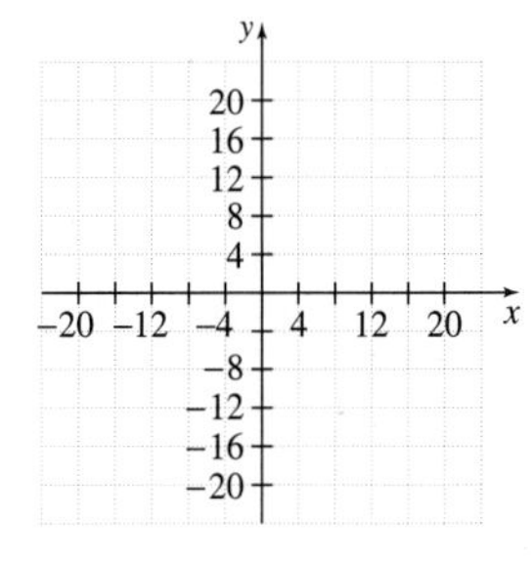

11. $\dfrac{(x-4)^2}{16} + \dfrac{(y-3)^2}{9} = 1$

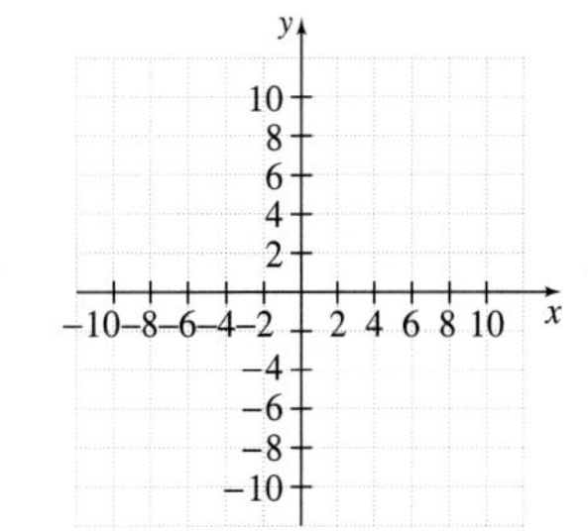

12. $y^2 - x^2 = 1$

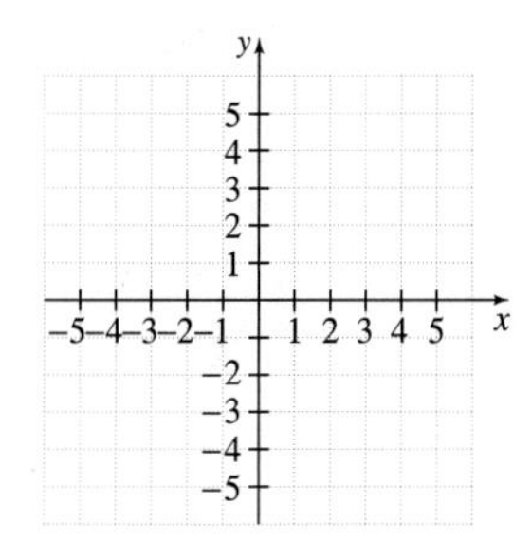

Solve each system.

13. $\begin{cases} x^2 + y^2 = 169 \\ 5x + 12y = 0 \end{cases}$

14. $\begin{cases} x^2 + y^2 = 26 \\ x^2 - y^2 = 24 \end{cases}$

15. $\begin{cases} y = x^2 - 5x + 6 \\ y = 2x \end{cases}$

16. $\begin{cases} x^2 + 4y^2 = 5 \\ y = x \end{cases}$

13. ____________

14. ____________

15. ____________

16. ____________

Name ______________________________

Graph each system.

17. $\begin{cases} 2x + 5y \geq 10 \\ y \geq x^2 + 1 \end{cases}$

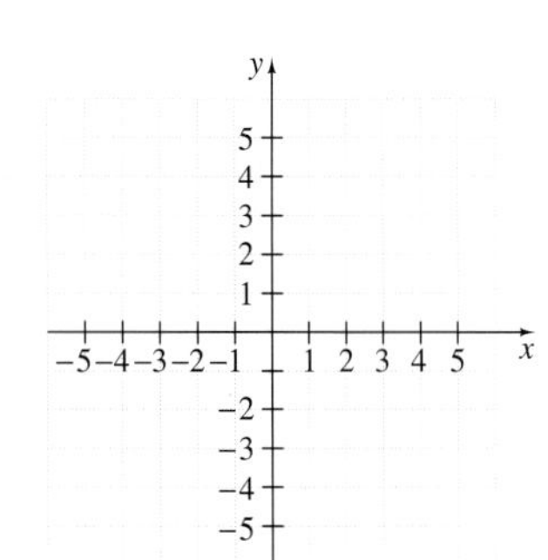

18. $\begin{cases} \frac{x^2}{4} + y^2 \leq 1 \\ x + y > 1 \end{cases}$

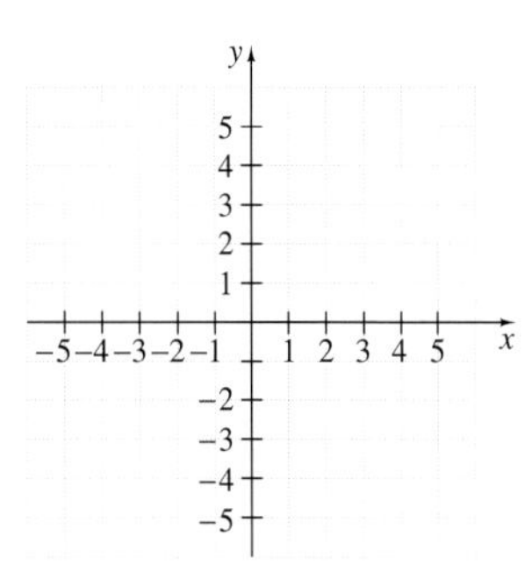

19. $\begin{cases} x^2 + y^2 > 1 \\ \frac{x^2}{4} - y^2 \geq 1 \end{cases}$

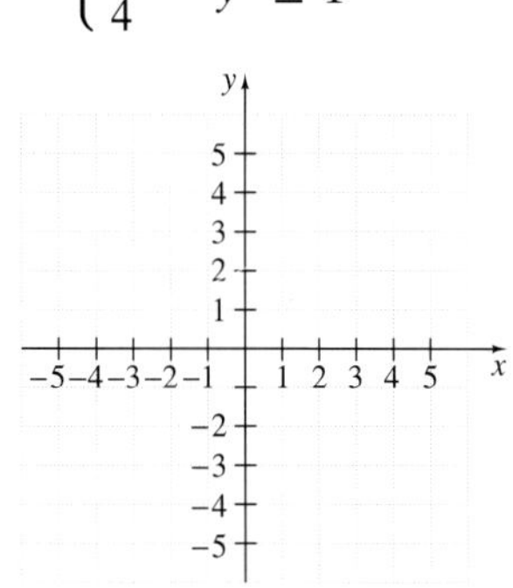

20. $\begin{cases} x^2 + y^2 \geq 4 \\ x^2 + y^2 < 16 \\ y \geq 0 \end{cases}$

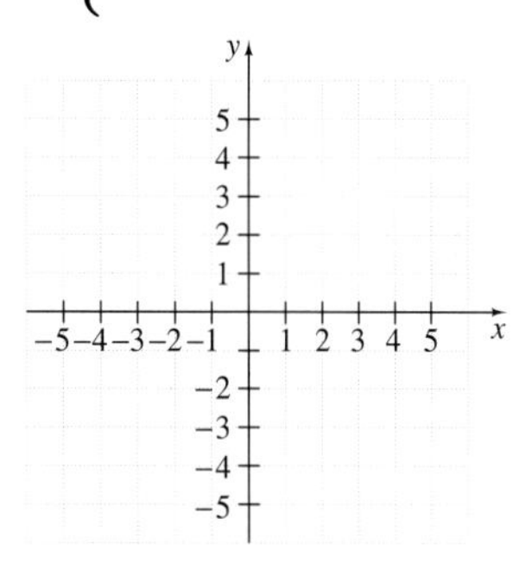

21. ______________________________

22. ______________________________

21. Which graph best resembles the graph of $x = a(y - k)^2 + h$ if $a > 0$, $h < 0$, and $k > 0$?

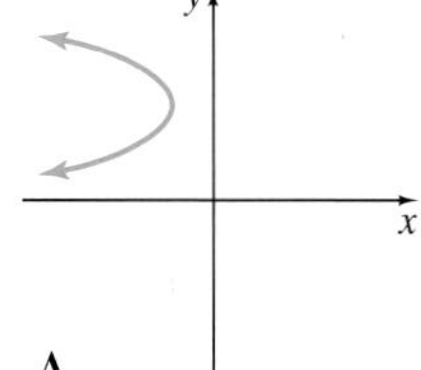

A

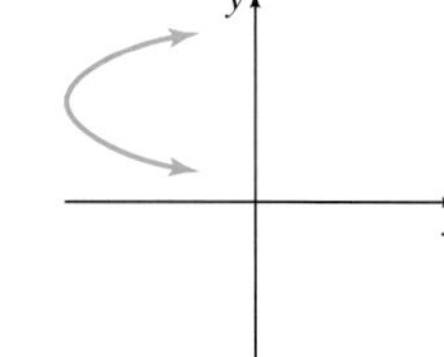

B

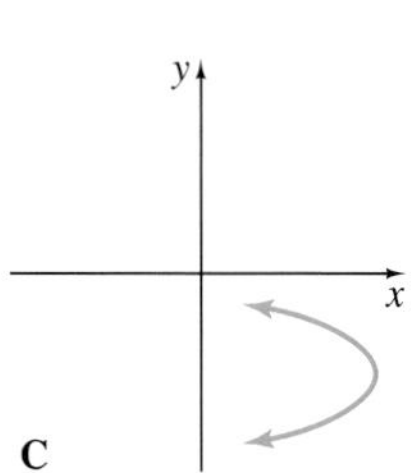

C

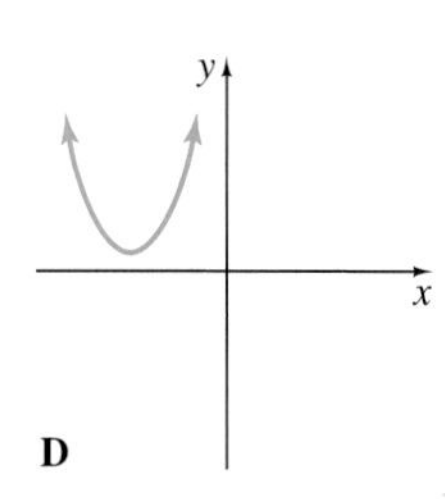

D

22. A bridge has an arch in the shape of half an ellipse. If the equation of the ellipse, measured in feet, is $100x^2 + 225y^2 = 22{,}500$, find the height of the arch from the road and the width of the arch.

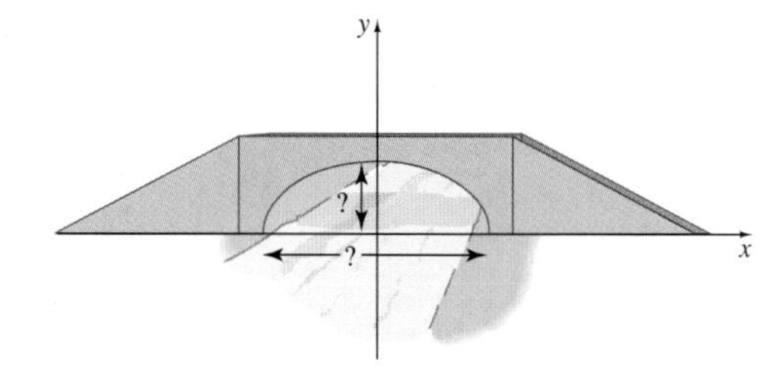

Graph each function.

23. $f(x) = \sqrt{x - 2}$

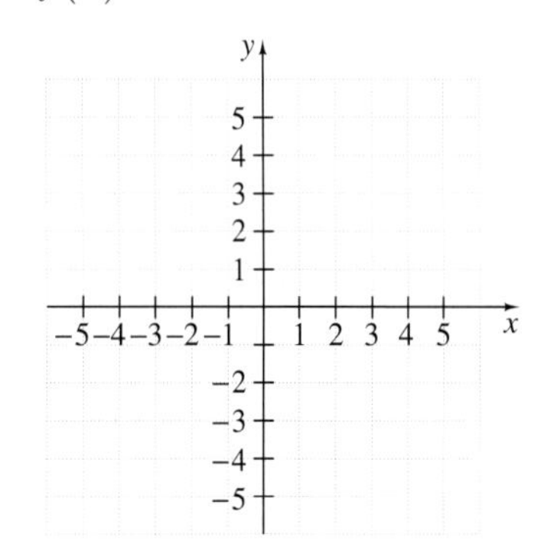

24. $f(x) = |x + 1| - 3$

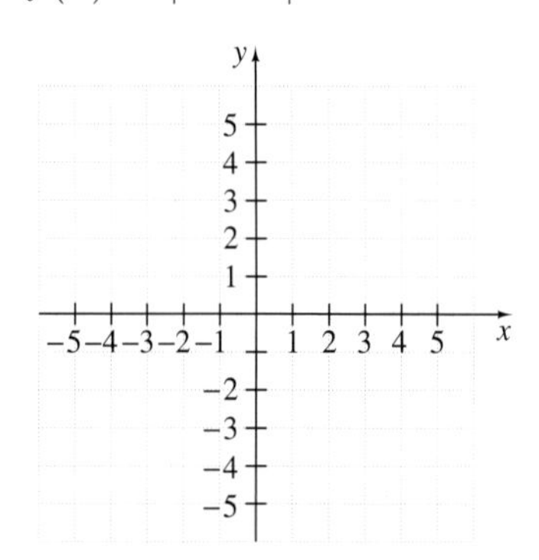

Name ______________________ Section ________ Date ________

ANSWERS

1. ____________
2. ____________
3. ____________
4. ____________
5. ____________
6. ____________
7. ____________
8. ____________
9. ____________
10. ____________
11. ____________
12. ____________
13. ____________
14. ____________
15. ____________
16. ____________

CUMULATIVE REVIEW

1. Add $11x^3 - 12x^2 + x - 3$ and $x^3 - 10x + 5$.

2. Multiply: $(x + 3)(2x + 5)$

3. Use synthetic division to divide $x^4 - 2x^3 - 11x^2 + 34$ by $x + 2$.

4. Factor: $2(x - 5) + 3a(x - 5)$

5. Factor: $x^2 - 12x + 35$

6. Factor: $(x + 3)^2 - 36$

7. Solve: $2x^2 + 9x - 5 = 0$

Simplify each rational expression.

8. $\dfrac{2 + x}{x + 2}$

9. $\dfrac{2 - x}{x - 2}$

Subtract.

10. $\dfrac{x^2}{x + 7} - \dfrac{49}{x + 7}$

11. $\dfrac{x}{3y^2} - \dfrac{x + 1}{3y^2}$

12. Simplify: $\dfrac{\dfrac{5x}{x + 2}}{\dfrac{10}{x - 2}}$

13. Solve: $\dfrac{2x}{x - 3} + \dfrac{6 - 2x}{x^2 - 9} = \dfrac{x}{x + 3}$

14. If a certain number is subtracted from the numerator and added to the denominator of $\frac{9}{19}$, the new fraction is equivalent to $\frac{1}{3}$. Find the number.

Find each square root.

15. $\sqrt{0}$

16. $\sqrt{0.25}$

17. ______

18. ______

19. ______

20. ______

21. ______

22. ______

23. ______

24. ______

25. ______

Name ____________________

Use rational exponents to simplify. Assume that all variables represent positive numbers.

17. $\sqrt[8]{x^4}$

18. $\sqrt[6]{r^2s^4}$

Simplify.

19. $\sqrt[3]{24}$

20. $\sqrt[4]{32}$

21. Rationalize the denominator of $\frac{2}{\sqrt{5}}$.

22. Solve: $\sqrt{-10x - 1} + 3x = 0$

23. Multiply: $(2 - 5i)(4 + i)$

24. Solve $p^2 + 2p = 4$ by completing the square.

25. Solve: $2x^2 - 4x = 3$

Exponential and Logarithmic Functions

CHAPTER 10

In this chapter, we discuss two closely related functions: exponential and logarithmic functions. These functions are vital in applications in economics, finance, engineering, the sciences, education, and other fields. Models of tumor growth and learning curves are two examples of the uses of exponential and logarithmic functions.

An earthquake is a series of vibrations in the crust of the earth. The size, or magnitude, of an earthquake is measured on the Richter scale. The magnitude of an earthquake can range broadly, from barely detectable (2.5 or less on the Richter scale) to massively destructive (7.0 or greater on the Richter scale). According to the United States Geological Survey, earthquakes are an everyday occurrence. In 1997, there were a total of 20,824 earthquakes around the world, or an average of 57 earthquakes per day. However, most of these were minor tremors with magnitudes of 3.9 or less, and many could only be detected by seismographs. Only 0.09% of the earthquakes occurring during 1997 could be classified as major earthquakes (7.0 or greater on the Richter scale). Even so, earthquakes were responsible for 2907 deaths that year. In Exercises 75 through 78 on page 798 and Exercises 61 and 62 on page 806, we will investigate the role of logarithms in finding the magnitude of an earthquake.

ANSWERS

1. ______

2. ______

3. ______

4. ______

5. ______

6. see graph

7. ______

8. ______

9. ______

10. ______

11. ______

12. ______

13. see graph

14. ______

15. ______

16. ______

17. ______

18. ______

19. ______

20. ______

Name ______ Section ______ Date ______

CHAPTER 10 PRETEST

If $f(x) = x^2 - 2x$ *and* $g(x) = 5x + 3$, *find the following.*

1. $(f + g)(x)$

2. $(f \circ g)(x)$

Determine whether the given functions are one-to-one.

3. $\{(-2, 7), (7, -3), (2, 1), (5, -8)\}$

4.

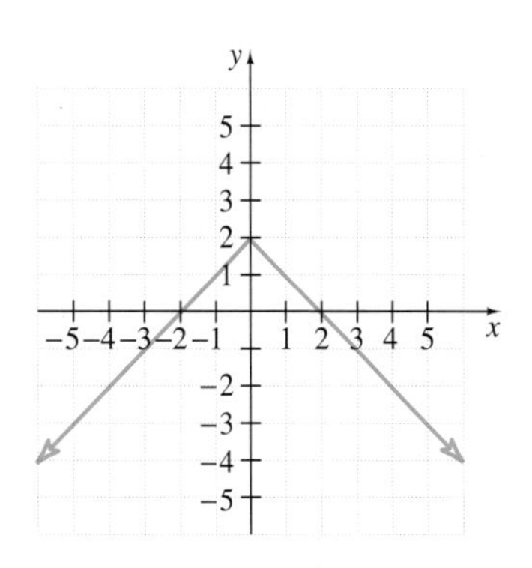

5. Given $f(x) = 7x - 12$, find $f^{-1}(x)$.

6. Graph $y = 2^x - 1$.

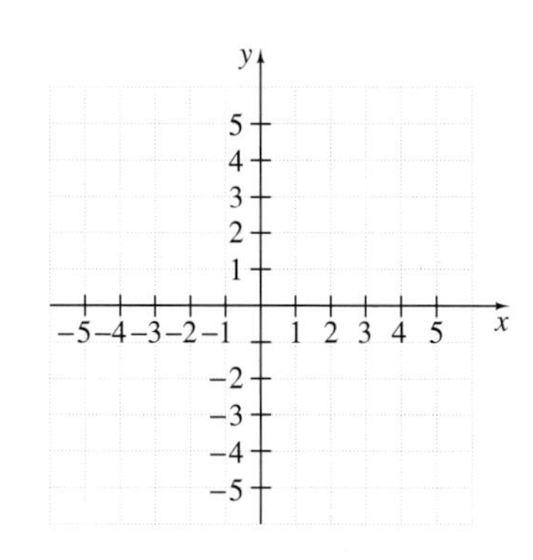

Solve each equation for x.

7. $6^x = 216$

8. $27^{4x+1} = 3$

9. $\log_5 x = 4$

10. $\log_2 \dfrac{1}{64} = x$

11. $\log_x 1000 = 3$

12. Simplify: $\log_7 7^5$

13. Graph $y = \log_4 x$.

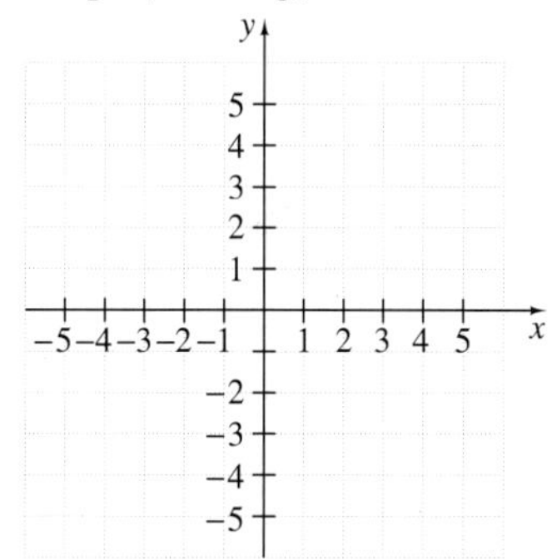

Use the properties of logarithms to write each expression as a single logarithm.

14. $\log_5 3 + \log_5 2$

15. $2 \log_6 a - 7 \log_6 (a + 1)$

16. Write the expression $\log_7 \dfrac{3y}{5x^2}$ as the sum or difference of multiples of logarithms.

Find the exact value.

17. $\log \dfrac{1}{100}$

18. $\ln e^8$

19. Approximate $\log_3 15$ to four decimal places.

20. Solve: $\log_2 10 + \log_2 (x + 5) = 3$

10.1 THE ALGEBRA OF FUNCTIONS

Objectives

A Add, subtract, multiply, and divide functions.

B Compose functions.

SSM

CD-ROM

Video 10.1

A ADDING, SUBTRACTING, MULTIPLYING, AND DIVIDING FUNCTIONS

As we have seen in earlier chapters, it is possible to add, subtract, multiply, and divide functions. Although we have not stated it as such, the sums, differences, products, and quotients of functions are themselves functions. For example, if $f(x) = 3x$ and $g(x) = x + 1$, their product, $f(x) \cdot g(x) = 3x(x + 1) = 3x^2 + 3x$, is a new function. We can use the notation $(f \cdot g)(x)$ to denote this new function. Finding the sum, difference, product, and quotient of functions to generate new functions is called the **algebra of functions**.

ALGEBRA OF FUNCTIONS

Let f and g be functions. New functions from f and g are defined as follows:

Sum	$(f + g)(x) = f(x) + g(x)$
Difference	$(f - g)(x) = f(x) - g(x)$
Product	$(f \cdot g)(x) = f(x) \cdot g(x)$
Quotient	$\left(\frac{f}{g}\right)(x) = \frac{f(x)}{g(x)}$

Example 1 If $f(x) = x - 1$ and $g(x) = 2x - 3$, find:

a. $(f + g)(x)$
b. $(f - g)(x)$
c. $(f \cdot g)(x)$
d. $\left(\frac{f}{g}\right)(x)$

Solution: Use the algebra of functions and replace $f(x)$ by $x - 1$ and $g(x)$ by $2x - 3$. Then we simplify.

a. $$(f + g)(x) = f(x) + g(x) = (x - 1) + (2x - 3) = 3x - 4$$

b. $$(f - g)(x) = f(x) - g(x) = (x - 1) - (2x - 3) = x - 1 - 2x + 3 = -x + 2$$

c. $$(f \cdot g)(x) = f(x) \cdot g(x) = (x - 1)(2x - 3) = 2x^2 - 5x + 3$$

d. $$\left(\frac{f}{g}\right)(x) = \frac{f(x)}{g(x)} = \frac{x - 1}{2x - 3}, \text{ where } x \neq \frac{3}{2}$$

Practice Problem 1

If $f(x) = x + 3$ and $g(x) = 3x - 1$, find:

a. $(f + g)(x)$
b. $(f - g)(x)$
c. $(f \cdot g)(x)$
d. $\left(\frac{f}{g}\right)(x)$

There is an interesting but not surprising relationship between the graphs of functions and the graph of their sum, difference, product, and quotient. For example, the graph of $(f + g)(x)$ can be found by adding the graph of $f(x)$ to the graph of $g(x)$. We add two graphs by adding corresponding y-values.

Answers

1. a. $4x + 2$, **b.** $-2x + 4$, **c.** $3x^2 + 8x - 3$, **d.** $\frac{x + 3}{3x - 1}$ where $x \neq \frac{1}{3}$

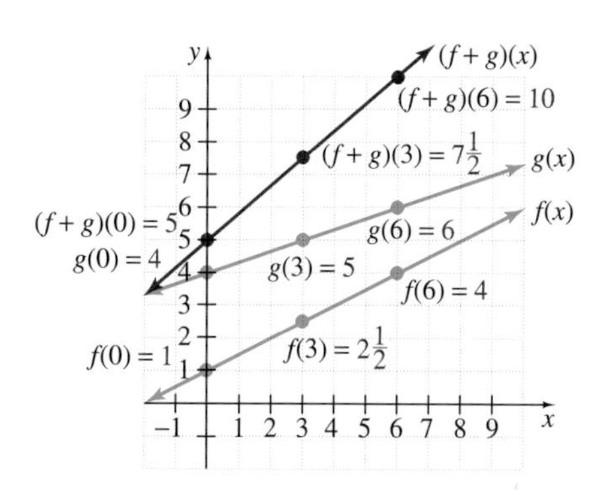

B COMPOSITION OF FUNCTIONS

Another way to combine functions is called **function composition**. To understand this new way of combining functions, study the tables below. They show degrees Fahrenheit converted to equivalent degrees Celsius, and then degrees Celsius converted to equivalent degrees Kelvin. (The Kelvin scale is a temperature scale devised by Lord Kelvin in 1848.)

x = Degrees Fahrenheit (Input)	-31	-13	32	68	149	212
$C(x)$ = Degrees Celsius (Output)	-35	-25	0	20	65	100

C = Degrees Celsius (Input)	-35	-25	0	20	65	100
$K(C)$ = Kelvins (Output)	238.15	248.15	273.15	293.15	338.15	373.15

Suppose that we want a table that shows a direct conversion from degrees Fahrenheit to kelvins. In other words, suppose that a table is needed that shows kelvins as a function of degrees Fahrenheit. This can easily be done because in the tables, the output of the first table is the same as the input of the second table. The new table is as follows.

x = Degrees Fahrenheit (Input)	-31	-13	32	68	149	212
$K(C(x))$ = Kelvins (Output)	238.15	248.15	273.15	293.15	338.15	373.15

Since the output of the first table is used as the input of the second table, we write the new function as $K(C(x))$. The new function is formed from the composition of the other two functions. The mathematical symbol for this composition is $(K \circ C)(x)$. Thus, $(K \circ C)(x) = K(C(x))$.

It is possible to find an equation for the composition of the two functions $C(x)$ and $K(x)$. In other words, we can find a function that converts degrees Fahrenheit directly to kelvins. The function $C(x) = \frac{5}{9}(x - 32)$ converts degrees Fahrenheit to degrees Celsius, and the function $K(C) = C + 273.15$ converts degrees Celsius to kelvins. Thus,

$$(K \circ C)(x) = K(C(x)) = K\left(\frac{5}{9}(x - 32)\right) = \frac{5}{9}(x - 32) + 273.15$$

In general, the notation $\boldsymbol{f(g(x))}$ means "f composed with g" and can be written as $\boldsymbol{(f \circ g)(x)}$. Also $g(f(x))$, or $(g \circ f)(x)$, means "g composed with f."

COMPOSITE FUNCTIONS

The composition of functions f and g is

$$(f \circ g)(x) = f(g(x))$$

HELPFUL HINT

$(f \circ g)(x)$ does not mean the same as $(f \cdot g)(x)$.

$$(f \circ g)(x) = f(g(x)) \text{ while } (f \cdot g)(x) = f(x) \cdot g(x)$$

Example 2 If $f(x) = x^2$ and $g(x) = x + 3$, find each composition.

a. $(f \circ g)(2)$ and $(g \circ f)(2)$
b. $(f \circ g)(x)$ and $(g \circ f)(x)$

Solution:

a. $(f \circ g)(2) = f(g(2))$
$= f(5)$ Since $g(x) = x + 3$, then $g(2) = 2 + 3 = 5$.
$= 5^2 = 25$

$(g \circ f)(2) = g(f(2))$
$= g(4)$ Since $f(x) = x^2$, then $f(2) = 2^2 = 4$.
$= 4 + 3 = 7$

b. $(f \circ g)(x) = f(g(x))$
$= f(x + 3)$ Replace $g(x)$ with $x + 3$.
$= (x + 3)^2$ $f(x + 3) = (x + 3)^2$
$= x^2 + 6x + 9$ Square $(x + 3)$.

$(g \circ f)(x) = g(f(x))$
$= g(x^2)$ Replace $f(x)$ with x^2.
$= x^2 + 3$ $g(x^2) = x^2 + 3$

Example 3 If $f(x) = |x|$ and $g(x) = x - 2$, find each composition.

a. $(f \circ g)(x)$
b. $(g \circ f)(x)$

Solution: **a.** $(f \circ g)(x) = f(g(x)) = f(x - 2) = |x - 2|$
b. $(g \circ f)(x) = g(f(x)) = g(|x|) = |x| - 2$

HELPFUL HINT

In Examples 2 and 3, notice that $(g \circ f)(x) \neq (f \circ g)(x)$. In general, $(g \circ f)(x)$ *may* or *may not* equal $(f \circ g)(x)$.

Example 4 If $f(x) = 5x$, $g(x) = x - 2$, and $h(x) = \sqrt{x}$, write each function as a composition with f, g, or h.

a. $F(x) = \sqrt{x - 2}$
b. $G(x) = 5x - 2$

Practice Problem 2

If $f(x) = x^2$ and $g(x) = 2x + 1$, find each composition.

a. $(f \circ g)(3)$ and $(g \circ f)(3)$

b. $(f \circ g)(x)$ and $(g \circ f)(x)$

Practice Problem 3

If $f(x) = \sqrt{x}$ and $g(x) = x + 1$, find each composition.

a. $(f \circ g)(x)$

b. $(g \circ f)(x)$

Practice Problem 4

If $f(x) = 2x$, $g(x) = x + 5$, and $h(x) = |x|$, write each function as a composition of f, g, or h.

a. $F(x) = |x + 5|$

b. $G(x) = 2x + 5$

Answers

2. a. 49; 19, **b.** $4x^2 + 4x + 1$; $2x^2 + 1$, **3. a.** $\sqrt{x + 1}$, **b.** $\sqrt{x} + 1$, **4. a.** $(h \circ g)(x)$, **b.** $(g \circ f)(x)$

Solution:

a. Notice the order in which the function F operates on an input value x. First, 2 is subtracted from x, and then the square root of that result is taken. This means that $F = h \circ g$. To check, we find $h \circ g$:

$$(h \circ g)(x) = h(g(x)) = h(x - 2) = \sqrt{x - 2}$$

b. Notice the order in which the function G operates on an input value x. First, x is multiplied by 5, and then 2 is subtracted from the result. This means that $G = g \circ f$. To check, we find $g \circ f$:

$$(g \circ f)(x) = g(f(x)) = g(5x) = 5x - 2$$

GRAPHING CALCULATOR EXPLORATIONS

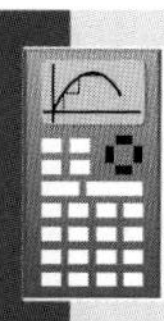

If $f(x) = \frac{1}{2}x + 2$ and $g(x) = \frac{1}{3}x^2 + 4$, then

$$\begin{aligned}(f + g)(x) &= f(x) + g(x) \\ &= \left(\frac{1}{2}x + 2\right) + \left(\frac{1}{3}x^2 + 4\right) \\ &= \frac{1}{3}x^2 + \frac{1}{2}x + 6.\end{aligned}$$

To visualize this addition of functions with a grapher, graph

$$Y_1 = \frac{1}{2}x + 2, \quad Y_2 = \frac{1}{3}x^2 + 4, \quad Y_3 = \frac{1}{3}x^2 + \frac{1}{2}x + 6$$

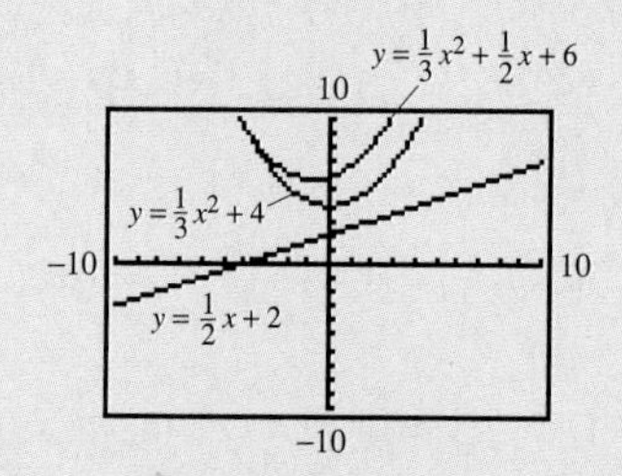

Use a TABLE feature to verify that for a given x value, $Y_1 + Y_2 = Y_3$. For example, verify that when $x = 0$, $Y_1 = 2$, $Y_2 = 4$ and $Y_3 = 2 + 4 = 6$.

Name ______________________ Section ________ Date __________

EXERCISE SET 10.1

A *For the functions f and g, find a.* $(f + g)(x)$, *b.* $(f - g)(x)$, *c.* $(f \cdot g)(x)$, *and d.* $\left(\frac{f}{g}\right)(x)$. *See Example 1.*

1. $f(x) = x - 7, g(x) = 2x + 1$

2. $f(x) = x + 4, g(x) = 5x - 2$

3. $f(x) = x^2 + 1, g(x) = 5x$

4. $f(x) = x^2 - 2, g(x) = 3x$

5. $f(x) = \sqrt{x}, g(x) = x + 5$

6. $f(x) = \sqrt[3]{x}, g(x) = x - 3$

7. $f(x) = -3x, g(x) = 5x^2$

8. $f(x) = 4x^3, g(x) = -6x$

B *If* $f(x) = x^2 - 6x + 2$, $g(x) = -2x$, *and* $h(x) = \sqrt{x}$, *find each composition. See Example 2.*

9. $(f \circ g)(2)$

10. $(h \circ f)(-2)$

11. $(g \circ f)(-1)$

12. $(f \circ h)(1)$

13. $(g \circ h)(0)$

14. $(h \circ g)(0)$

Find $(f \circ g)(x)$ *and* $(g \circ f)(x)$. *See Examples 2 and 3.*

15. $f(x) = x^2 + 1, g(x) = 5x$

16. $f(x) = x - 3, g(x) = x^2$

17. $f(x) = 2x - 3, g(x) = x + 7$

18. $f(x) = x + 10, g(x) = 3x + 1$

ANSWERS

1. a. ______
   b. ______
   c. ______
   d. ______
2. a. ______
   b. ______
   c. ______
   d. ______
3. a. ______
   b. ______
   c. ______
   d. ______
4. a. ______
   b. ______
   c. ______
   d. ______
5. a. ______
   b. ______
   c. ______
   d. ______
6. a. ______
   b. ______
   c. ______
   d. ______
7. a. ______
   b. ______
   c. ______
   d. ______
8. a. ______
   b. ______
   c. ______
   d. ______
9. ______
10. ______
11. ______
12. ______
13. ______
14. ______
15. ______
16. ______
17. ______
18. ______

Name ______

19. ______
20. ______
21. ______
22. ______
23. ______
24. ______
25. ______
26. ______
27. ______
28. ______
29. ______
30. ______
31. ______
32. ______
33. ______
34. ______
35. ______
36. ______
37. ______
38. ______
39. ______
40. ______
41. ______
42. ______

19. $f(x) = x^3 + x - 2, g(x) = -2x$

20. $f(x) = -4x, g(x) = x^3 + x^2 - 6$

21. $f(x) = \sqrt{x}, g(x) = -5x + 2$

22. $f(x) = 7x - 1, g(x) = \sqrt[3]{x}$

If $f(x) = 3x$, $g(x) = \sqrt{x}$, and $h(x) = x^2 + 2$, write each function as a composition with f, g, or h. See Example 4.

23. $H(x) = \sqrt{x^2 + 2}$

24. $G(x) = \sqrt{3x}$

25. $F(x) = 9x^2 + 2$

26. $H(x) = 3x^2 + 6$

27. $G(x) = 3\sqrt{x}$

28. $F(x) = x + 2$

Find $f(x)$ and $g(x)$ so that the given function $h(x) = (f \circ g)(x)$.

29. $h(x) = (x + 2)^2$

30. $h(x) = |x - 1|$

31. $h(x) = \sqrt{x + 5} + 2$

32. $h(x) = (3x + 4)^2 + 3$

33. $h(x) = \dfrac{1}{2x - 3}$

34. $h(x) = \dfrac{1}{x + 10}$

Review and Preview

Solve each equation for y. See Section 2.3.

35. $x = y + 2$

36. $x = y - 5$

37. $x = 3y$

38. $x = -6y$

39. $x = -2y - 7$

40. $x = 4y + 7$

Combining Concepts

41. Business people are concerned with cost functions, revenue functions, and profit functions. Recall that the profit $P(x)$ obtained from x units of a product is equal to the revenue $R(x)$ from selling the x units minus the cost $C(x)$ of manufacturing the x units. Write an equation expressing this relationship among $C(x)$, $R(x)$, and $P(x)$.

42. Suppose the revenue $R(x)$ for x units of a product can be described by $R(x) = 25x$, and the cost $C(x)$ can be described by $C(x) = 50 + x^2 + 4x$. Find the profit $P(x)$ for x units.

10.2 INVERSE FUNCTIONS

In the next section, we begin a study of two new functions: exponential and logarithmic functions. As we learn more about these functions, we will discover that they share a special relation to each other; they are inverses of each other.

Before we study these functions, we need to learn about inverses. We begin by defining one-to-one functions.

Objectives

A Determine whether a function is a one-to-one function.

B Use the horizontal line test to decide whether a function is a one-to-one function.

C Find the inverse of a function.

D Find the equation of the inverse of a function.

E Graph functions and their inverses.

SSM CD-ROM Video 10.2

A DETERMINING WHETHER A FUNCTION IS ONE-TO-ONE

Study the following table.

Degrees Fahrenheit (Input)	−31	−13	32	68	149	212
Degrees Celsius (Output)	−35	−25	0	20	65	100

Recall that since each degrees Fahrenheit (input) corresponds to exactly one degrees Celsius (output), this table of inputs and outputs does describe a function. Also notice that each output corresponds to a different input. This type of function is given a special name—a one-to-one function.

Does the set $f = \{(0, 1), (2, 2), (-3, 5), (7, 6)\}$ describe a one-to-one function? It is a function since each x-value corresponds to a unique y-value. For this particular function f, each y-value corresponds to a unique x-value. Thus, this function is also a **one-to-one function**.

ONE-TO-ONE FUNCTION

For a **one-to-one function**, each x-value (input) corresponds to only one y-value (output) and each y-value (output) corresponds to only one x-value (input).

Examples

Determine whether each function described is one-to-one.

1. $f\{(6, 2), (5, 4), (-1, 0), (7, 3)\}$

The function f is one-to-one since each y-value corresponds to only one x-value.

2. $g = \{(3, 9), (-4, 2), (-3, 9), (0, 0)\}$

The function g is not one-to-one because the y-value 9 in $(3, 9)$ and $(-3, 9)$ corresponds to two different x-values.

3. $h = \{(1, 1), (2, 2), (10, 10), (-5, -5)\}$

The function h is one-to-one since each y-value corresponds to only one x-value.

4.

Mineral (Input)	Talc	Gypsum	Diamond	Topaz	Stibnite
Hardness on the Mohs Scale (Output)	1	2	10	8	2

Practice Problems 1–5

Determine whether each function described is one-to-one.

1. $f = \{(7, 3), (-1, 1), (5, 0), (4, -2)\}$
2. $g = \{(-3, 2), (6, 3), (2, 14), (-6, 2)\}$
3. $h = \{(0, 0), (1, 2), (3, 4), (5, 6)\}$
4.

State (Input)	Colorado	Mississippi	Nevada	New Mexico	Utah
Number of Colleges and Universities (Output)	9	44	13	44	21

Source: The Chronicle of Higher Education, Vol. XLV, No. 1, August 28, 1998.

5.

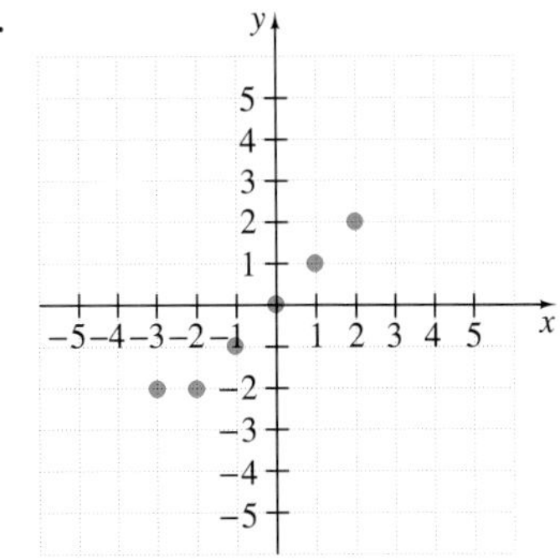

Answers

1. one-to-one, **2.** not one-to-one, **3.** one-to-one, **4.** not one-to-one, **5.** not one-to-one

This table does not describe a one-to-one function since the output 2 corresponds to two different inputs, gypsum and stibnite.

5.

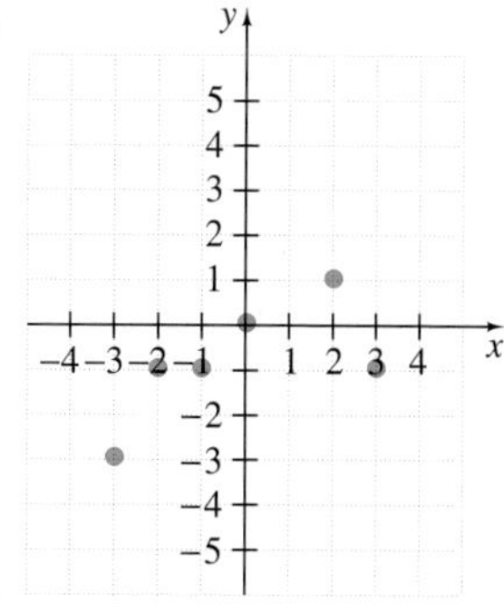

This graph does not describe a one-to-one function since the y-value -1 corresponds to three different x-values, -2, -1, and 3, as shown to the right.

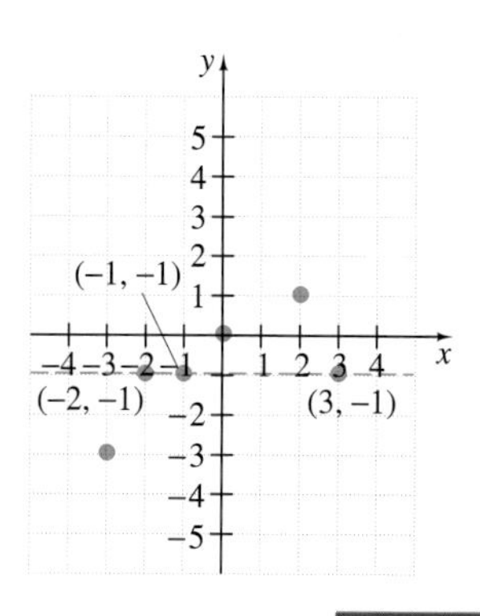

B USING THE HORIZONTAL LINE TEST

Recall that we recognize the graph of a function when it passes the vertical line test. Since every x-value of the function corresponds to exactly one y-value, each vertical line intersects the function's graph at most once. The graph shown next, for instance, is the graph of a function.

Is this function a *one-to-one* function? The answer is no. To see why not, notice that the y-value of the ordered pair $(-3, 3)$, for example, is the same as the y-value of the ordered pair $(3, 3)$. This function is therefore not one-to-one.

To test whether a graph is the graph of a one-to-one function, we can apply the vertical line test to see if it is a function, and then apply a similar **horizontal line test** to see if it is a one-to-one function.

HORIZONTAL LINE TEST

If every horizontal line intersects the graph of a function at most once, then the function is a one-to-one function.

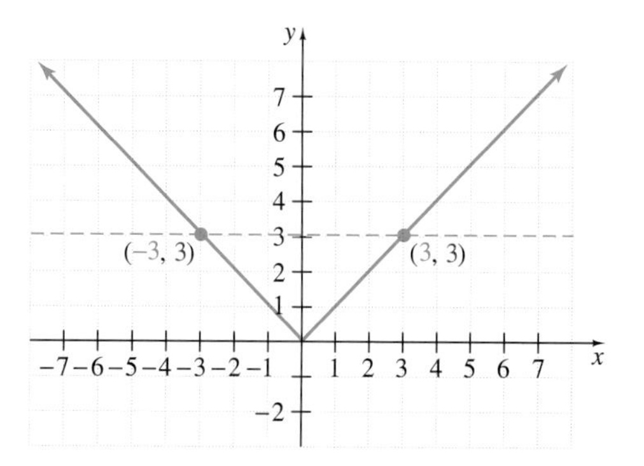

Example 6 Use the vertical and horizontal line tests to determine whether each graph is the graph of a one-to-one function.

a.

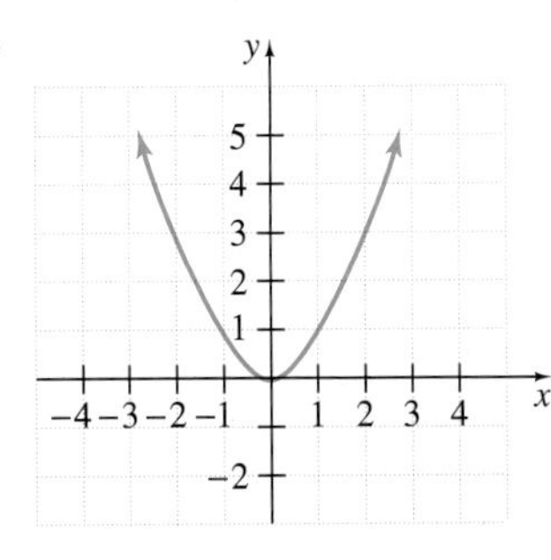

b.

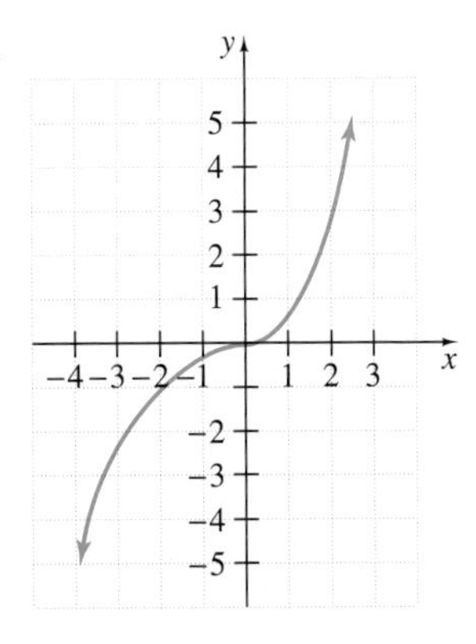

c.

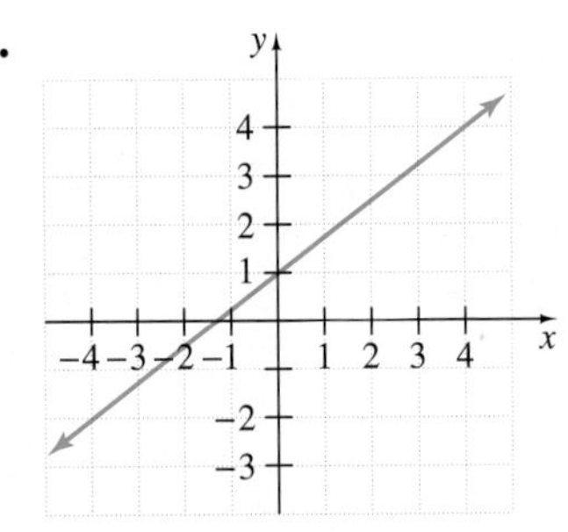

d.

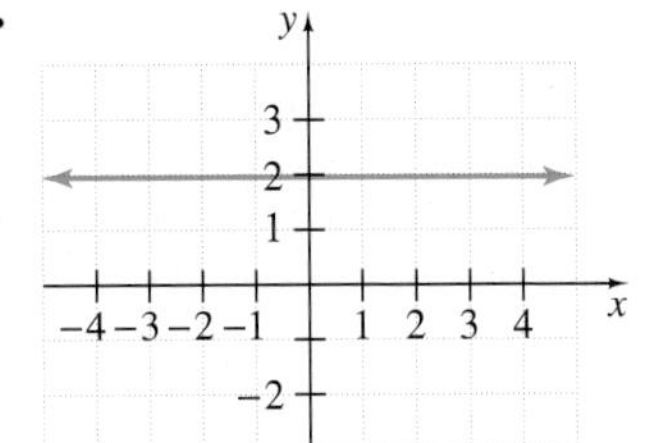

e.

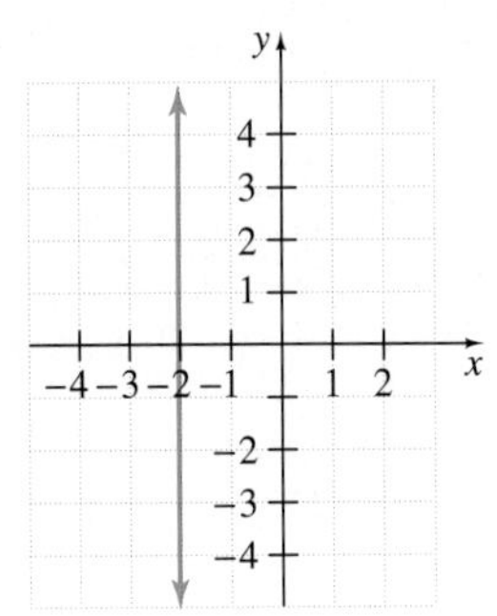

Solution: Graphs **a**, **b**, **c**, and **d** all pass the vertical line test, so only these graphs are graphs of functions. But, of these, only **b** and **c** pass the horizontal line test, so only **b** and **c** are graphs of one-to-one functions.

HELPFUL HINT

All linear equations are one-to-one functions except those whose graphs are horizontal or vertical lines. A vertical line does not pass the vertical line test and hence is not the graph of a function. A horizontal line is the graph of a function but does not pass the horizontal line test and hence is not the graph of a one-to-one function.

C FINDING THE INVERSE OF A FUNCTION

One-to-one functions are special in that their graphs pass the vertical and horizontal line tests. They are special, too, in another sense: We can find its

Practice Problem 6

Use the vertical and horizontal line tests to determine whether each graph is the graph of a one-to-one function.

a.

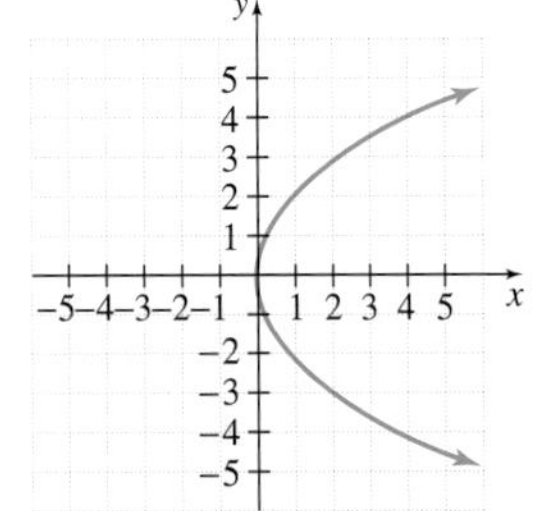

b.

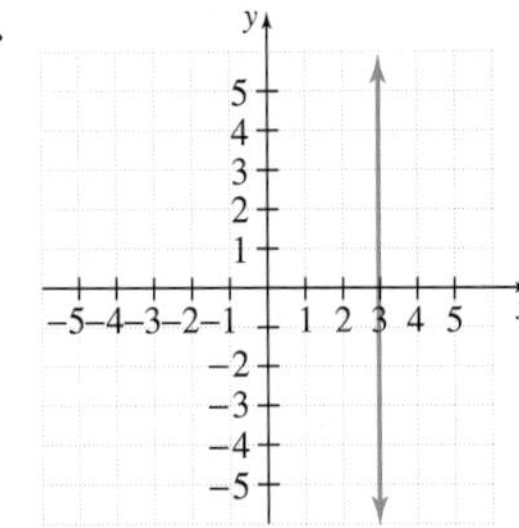

c.

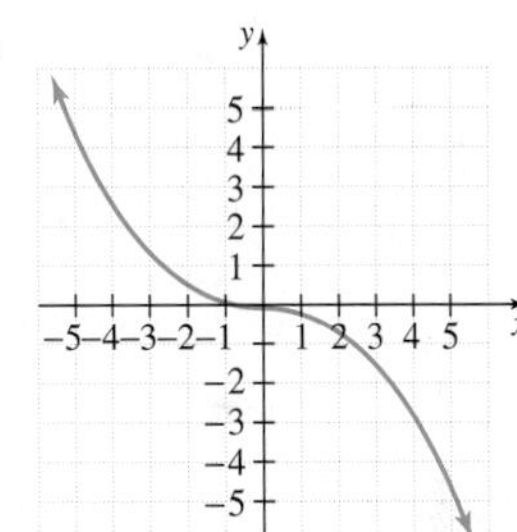

d.

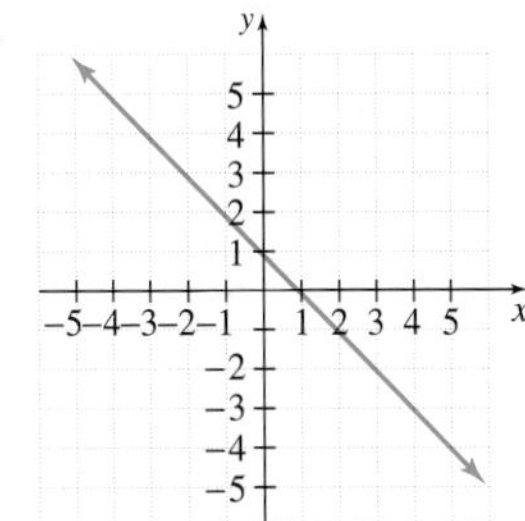

e.

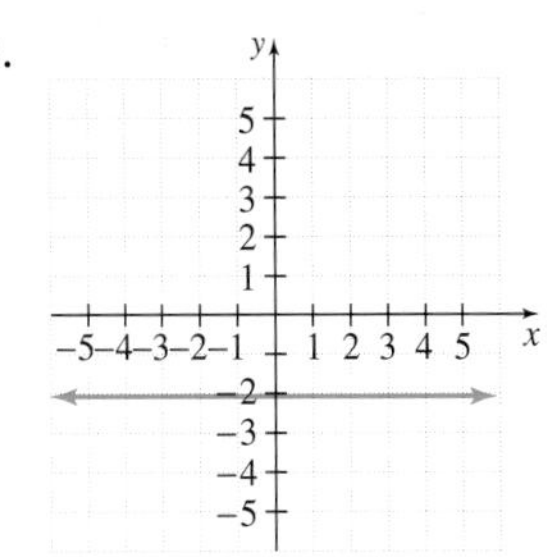

Answers

6. a. not one-to-one, **b.** not one-to-one, **c.** one-to-one, **d.** one-to-one, **e.** not one-to-one

inverse function for any one-to-one function by switching the coordinates of the ordered pairs of the function, or the inputs and the outputs. For example, the inverse of the one-to-one function

Degrees Fahrenheit (Input)	−31	−13	32	68	149	212
Degrees Celsius (Output)	−35	−25	0	20	65	100

is the function

Degrees Celsius (Input)	−35	−25	0	20	65	100
Degrees Fahrenheit (Output)	−31	−13	32	68	149	212

Notice that the ordered pair $(-31, -35)$ of the function, for example, becomes the ordered pair $(-35, -31)$ of its inverse.

Also, the inverse of the one-to-one function $f = \{(2, -3), (5, 10), (9, 1)\}$ is $\{(-3, 2), (10, 5), (1, 9)\}$. For a function f, we use the notation f^{-1}, read "f inverse," to denote its inverse function. Notice that since the coordinates of each ordered pair have been switched, the domain (set of inputs) of f is the range (set of outputs) of f^{-1}, and the range of f is the domain of f^{-1}. See the definition of inverse function in the margin.

INVERSE FUNCTION

The inverse of a one-to-one function f is the one-to-one function f^{-1} that consists of the set of all ordered pairs (y, x) where (x, y) belongs to f.

Example 7 Find the inverse of the one-to-one function:

$$f = \{(0, 1), (-2, 7), (3, -6), (4, 4)\}$$

Solution: $f^{-1} = \{(1, 0), (7, -2), (-6, 3), (4, 4)\}$ Switch coordinates of each ordered pair.

Practice Problem 7

Find the inverse of the one-to-one function: $f = \{(2, -4), (-1, 13), (0, 0), (-7, -8)\}$

TRY THE CONCEPT CHECK IN THE MARGIN.

✓ CONCEPT CHECK

Suppose that $f(x)$ is a one-to-one function. If the ordered pair $(1, 5)$ belongs to f, name one point that we know must belong to the inverse function f^{-1}.

D FINDING THE EQUATION OF THE INVERSE OF A FUNCTION

If a one-to-one function f is defined as a set of ordered pairs, we can find f^{-1} by interchanging the x- and y-coordinates of the ordered pairs. If a one-to-one function f is given in the form of an equation, we can find the equation of f^{-1} by using a similar procedure.

HELPFUL HINT

The symbol f^{-1} is the single symbol used to denote the inverse of the function f. It is read as "f inverse." This symbol *does not mean* $\frac{1}{f}$.

FINDING AN EQUATION OF THE INVERSE OF A ONE-TO-ONE FUNCTION $f(x)$

Step 1. Replace $f(x)$ with y.
Step 2. Interchange x with y.
Step 3. Solve the equation for y.
Step 4. Replace y with the notation $f^{-1}(x)$.

Practice Problem 8

Find the equation of the inverse of $f(x) = x - 6$.

Example 8 Find an equation of the inverse of $f(x) = x + 3$.

Solution: $f(x) = x + 3$

Step 1. $y = x + 3$ Replace $f(x)$ with y.
Step 2. $x = y + 3$ Interchange x and y.

Answers

7. $f^{-1} = \{(-4, 2), (13, -1), (0, 0), (-8, -7)\}$, **8.** $f^{-1}(x) = x + 6$

✓ Concept Check: $(5, 1)$ belongs to f^{-1}

Step 3. $x - 3 = y$ Solve for y.

Step 4. $f^{-1}(x) = x - 3$ Replace y with $f^{-1}(x)$.

The inverse of $f(x) = x + 3$ is $f^{-1}(x) = x - 3$. Notice that, for example,

$f(1) = 1 + 3 = 4$ and $f^{-1}(4) = 4 - 3 = 1$

Ordered pair: (1, 4) Ordered pair: (4, 1)

The coordinates are switched, as expected.

Example 9 Find the equation of the inverse of $f(x) = 3x - 5$. Graph f and f^{-1} on the same set of axes.

Solution: $f(x) = 3x - 5$

Step 1. $y = 3x - 5$ Replace $f(x)$ with y.

Step 2. $x = 3y - 5$ Interchange x and y.

Step 3. $3y = x + 5$ Solve for y.

$$y = \frac{x+5}{3}$$

Step 4. $f^{-1}(x) = \frac{x+5}{3}$ Replace y with $f^{-1}(x)$.

Now we graph $f(x)$ and $f^{-1}(x)$ on the same set of axes. Both $f(x) = 3x - 5$ and $f^{-1}(x) = \frac{x+5}{3}$ are linear functions, so each graph is a line.

$f(x) = 3x - 5$

x	$y = f(x)$
1	−2
0	−5
$\frac{5}{3}$	0

$f^{-1}(x) = \frac{x+5}{3}$

x	$y = f^{-1}(x)$
−2	1
−5	0
0	$\frac{5}{3}$

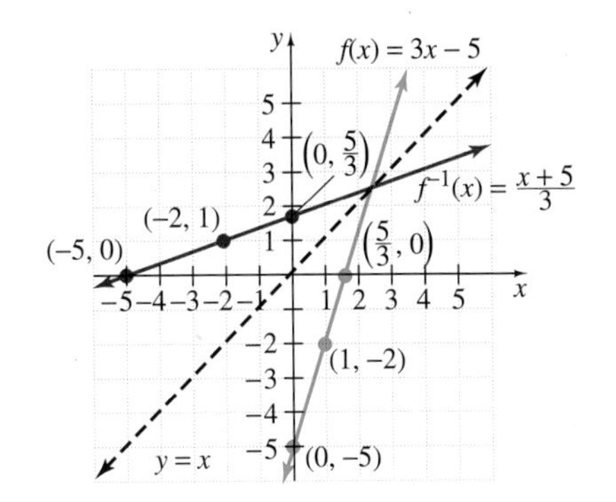

E GRAPHING INVERSE FUNCTIONS

Notice that the graphs of f and f^{-1} in Example 9 are mirror images of each other, and the "mirror" is the dashed line $y = x$. This is true for every function and its inverse. For this reason, we say that *the graphs of f and f^{-1} are symmetric about the line $y = x$.*

Practice Problem 9

Find an equation of the inverse of $f(x) = 2x + 3$. Graph f and f^{-1} on the same set of axes.

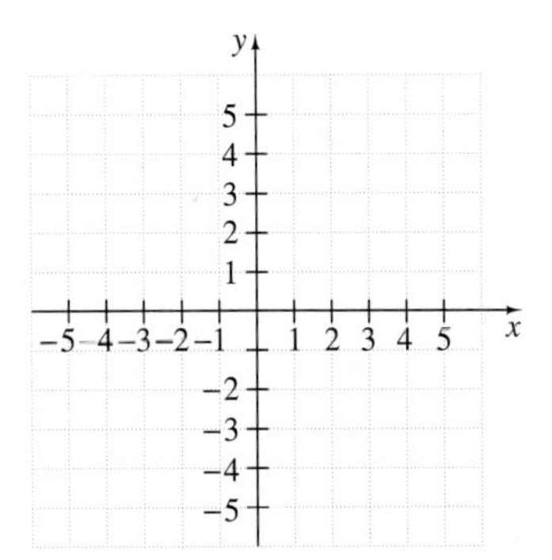

Answers

9. $f^{-1}(x) = \frac{x-3}{2}$

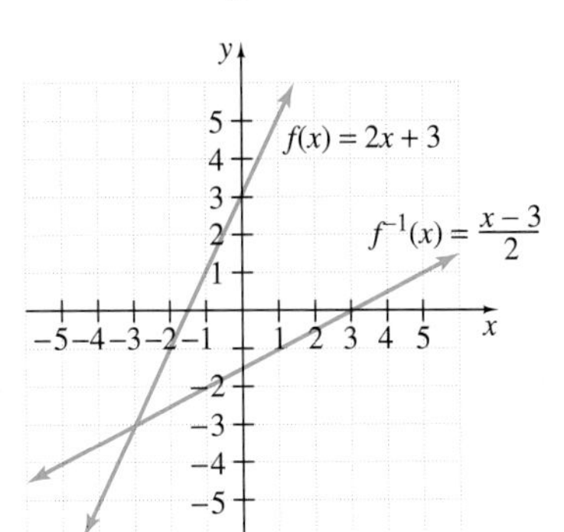

To see why this happens, study the graph of a few ordered pairs and their switched coordinates.

Example 10 Graph the inverse of each function.

Solution: The function is graphed in blue and the inverse is graphed in red.

a.

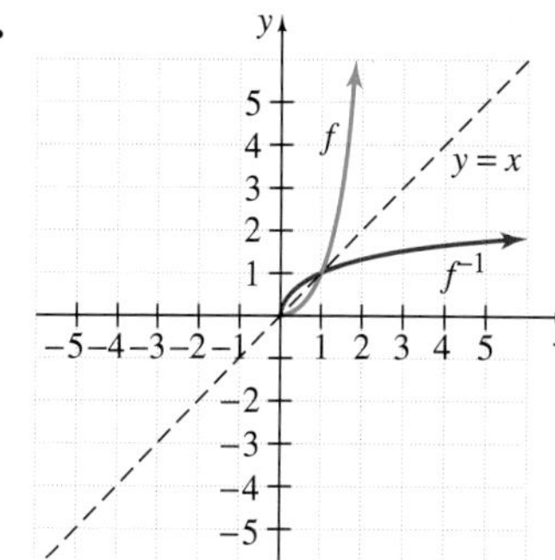

b.

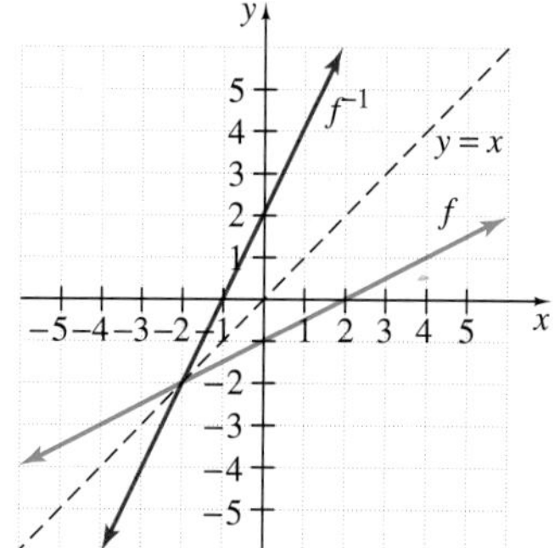

GRAPHING CALCULATOR EXPLORATIONS

A grapher can be used to visualize functions and their inverses. Recall that the graph of a function f and its inverse f^{-1} are mirror images of each other across the line $y = x$. To see this for the function $f(x) = 3x + 2$, use a square window and graph

the given function: $Y_1 = 3x + 2$

its inverse: $Y_2 = \dfrac{x - 2}{3}$

and the line: $Y_3 = x$

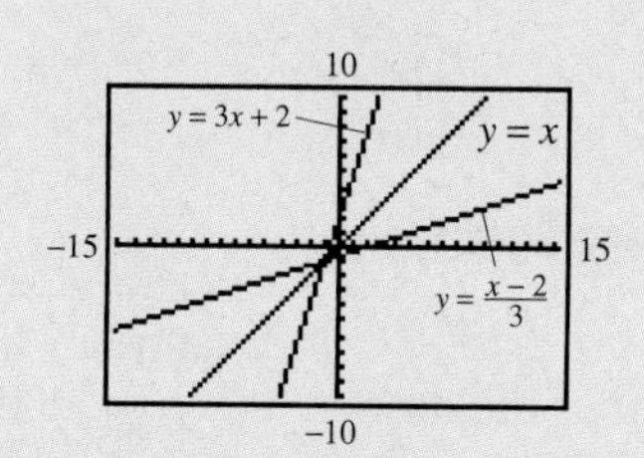

Exercises will follow in Exercise Set 10.2.

Practice Problem 10

Graph the inverse of each function.

a.

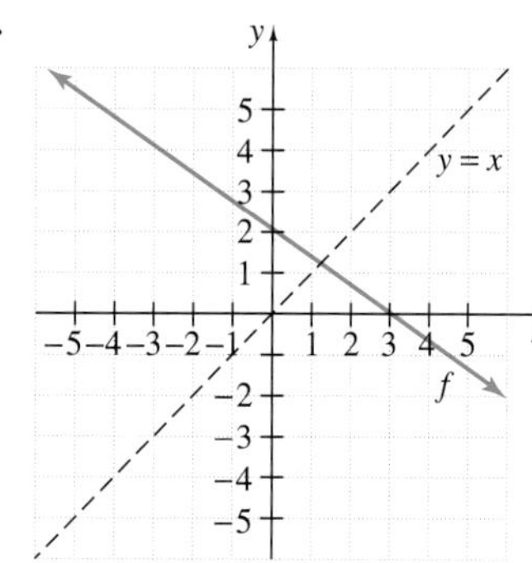

b.

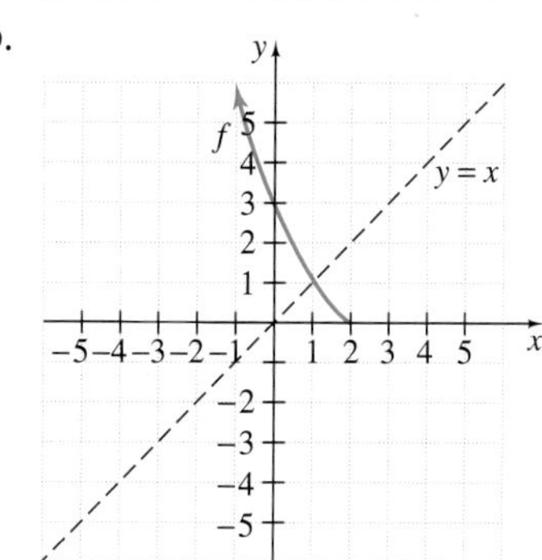

Answers

10. a.

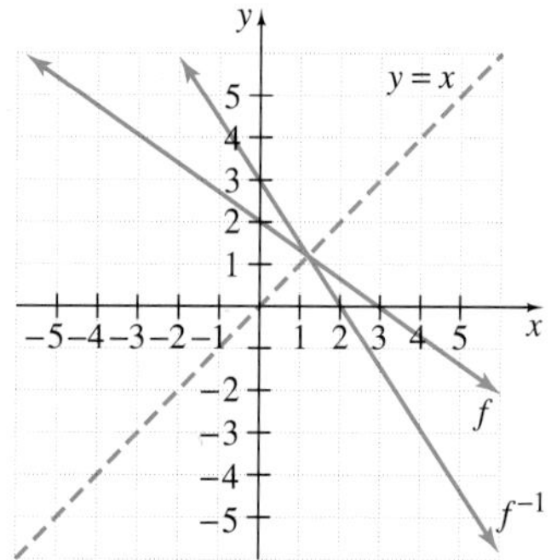

b.

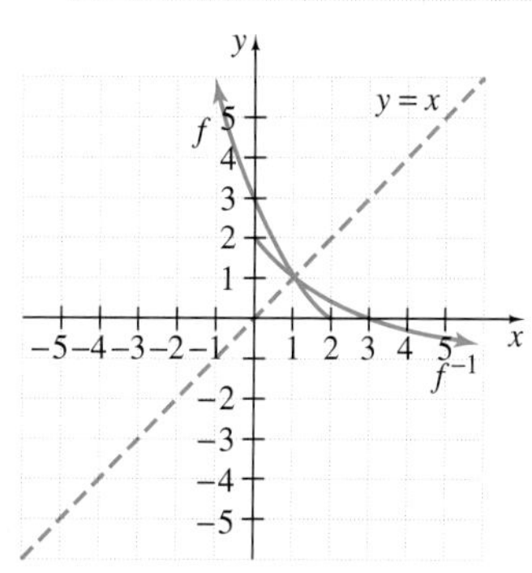

Name ______________________ Section ______ Date ________

ANSWERS

EXERCISE SET 10.2

A C *Determine whether each function is a one-to-one function. If it is one-to-one, list the inverse function by switching coordinates, or inputs and outputs. See Examples 1 through 5, and 7.*

1. $f = \{(-1, -1), (1, 1), (0, 2), (2, 0)\}$

2. $g = \{(8, 6), (9, 6), (3, 4), (-4, 4)\}$

3. $h = \{(10, 10)\}$

4. $r = \{(1, 2), (3, 4), (5, 6), (6, 7)\}$

5. $f = \{(11, 12), (4, 3), (3, 4), (6, 6)\}$

6. $g = \{(0, 3), (3, 7), (6, 7), (-2, -2)\}$

7.

Month of 1998 (Input)	January	February	March	April	May	June
Thousands of Houses on Sale at Month-End (Output)	282	277	281	285	282	287

(*Source:* U.S. Department of Housing and Urban Development)

8.

State (Input)	Washington	Ohio	Georgia	Colorado	California	Arizona
Electoral Votes (Output)	11	21	13	8	54	8

(*Source:* U.S. Bureau of the Census)

9.

State (Input)	California	Vermont	Virginia	Texas	South Dakota
Rank in Population (Output)	1	49	12	2	45

(*Source:* U.S. Bureau of the Census)

10.

Shape (Input)	Triangle	Pentagon	Quadrilateral	Hexagon	Decagon
Number of Sides (Output)	3	5	4	6	10

Given the one-to-one function $f(x) = x^3 + 2$, find the following. (Hint: you do not need to find the equation for $f^{-1}(x)$.)

11. a. $f(1)$
b. $f^{-1}(3)$

12. a. $f(0)$
b. $f^{-1}(2)$

13. a. $f(-1)$
b. $f^{-1}(1)$

14. a. $f(-2)$
b. $f^{-1}(-6)$

1. ______
2. ______
3. ______
4. ______
5. ______
6. ______
7. ______
8. ______
9. ______
10. ______
11. a. ______
b. ______
12. a. ______
b. ______
13. a. ______
b. ______
14. a. ______
b. ______

15. ______

16. ______

17. ______

18. ______

19. ______

20. ______

21. ______

22. ______

Name ______________________________

B *Determine whether the graph of each function is the graph of a one-to-one function. See Example 6.*

15.

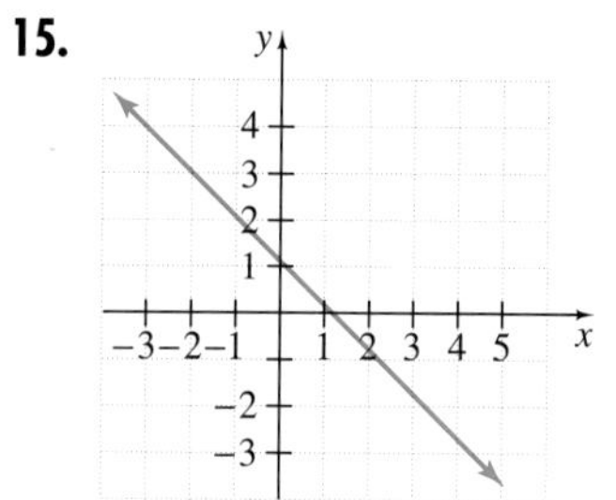

16.

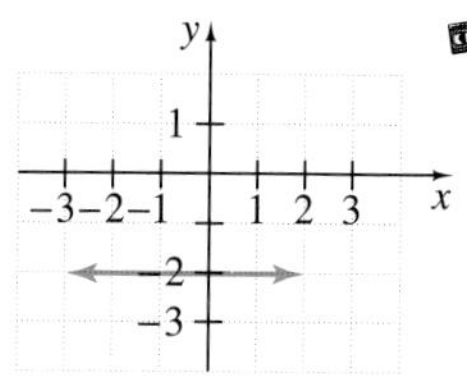

17.

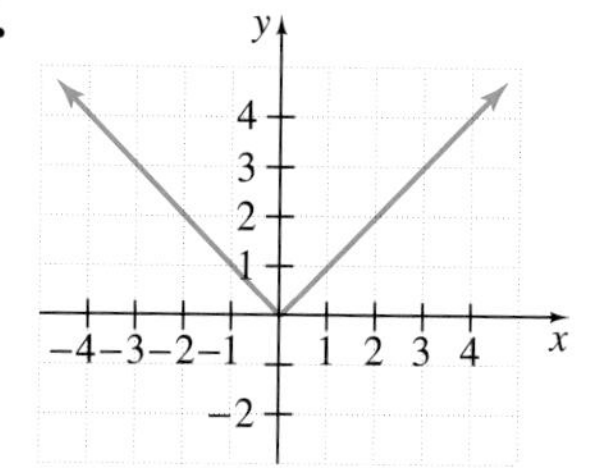

18.

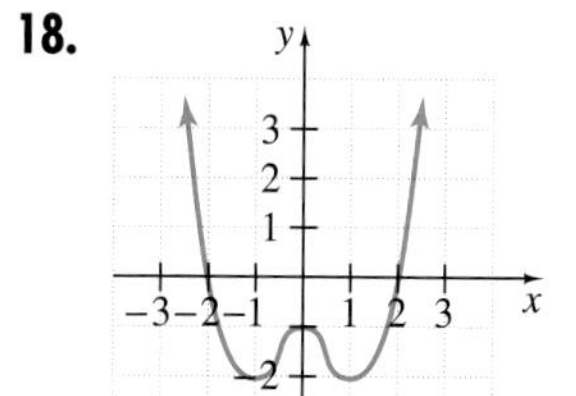

19.

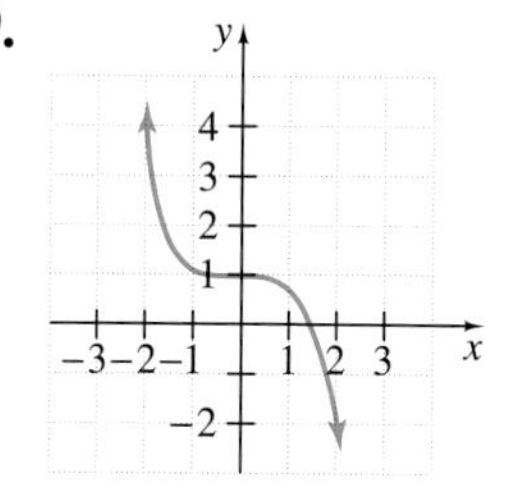

20.

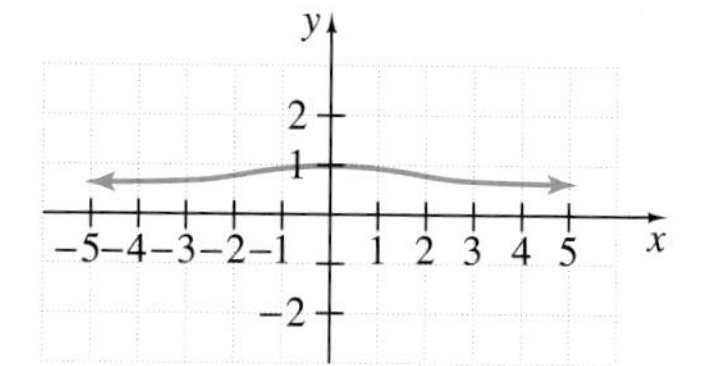

21.

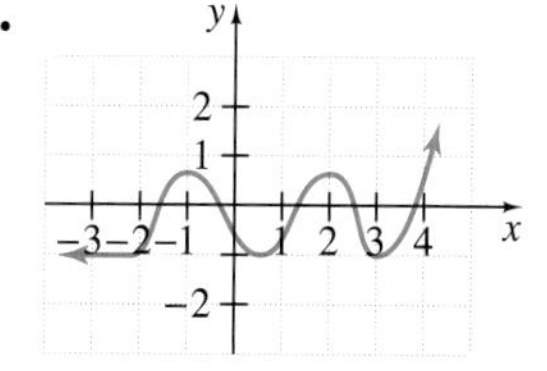

22.

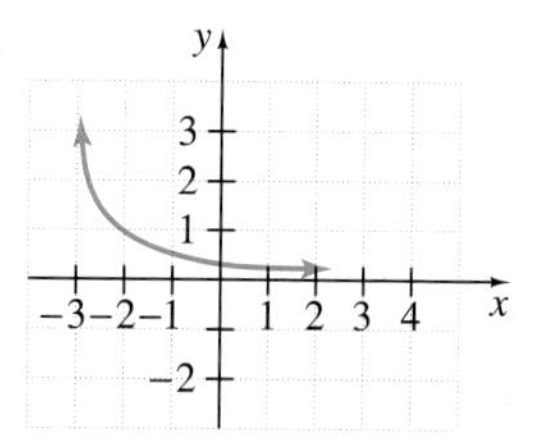

D E *Each of the following functions is one-to-one. Find the inverse of each function and graph the function and its inverse on the same set of axes. See Examples 8 and 9.*

23. $f(x) = x + 4$

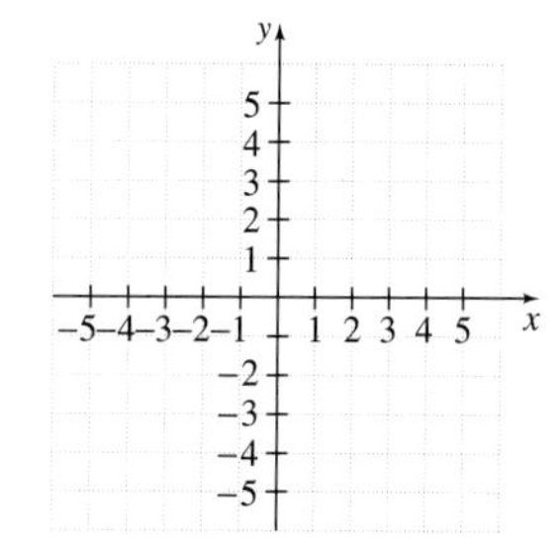

24. $f(x) = x - 5$

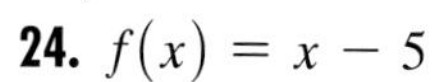

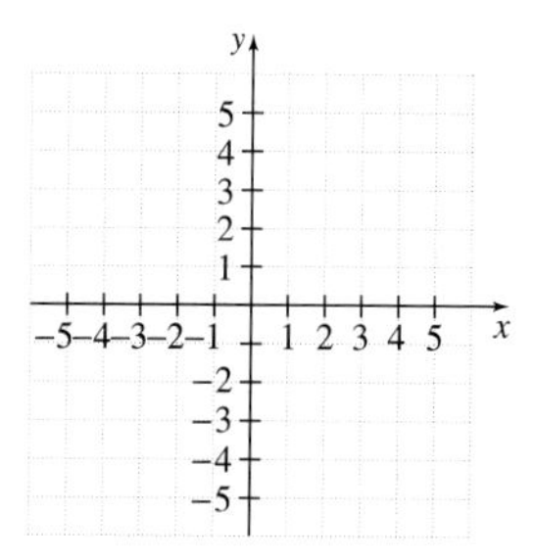

25. $f(x) = 2x - 3$

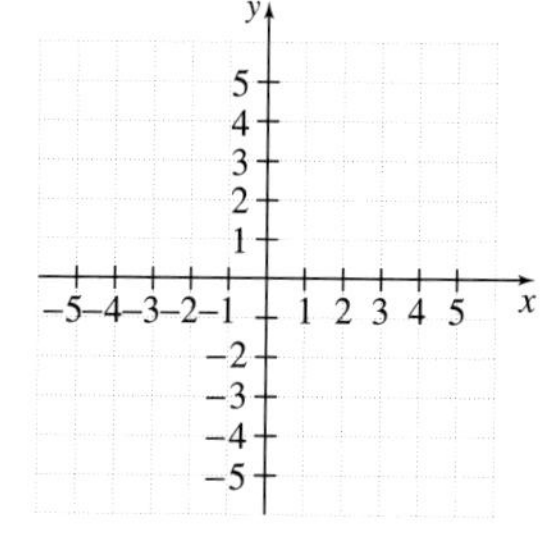

26. $f(x) = 4x + 9$

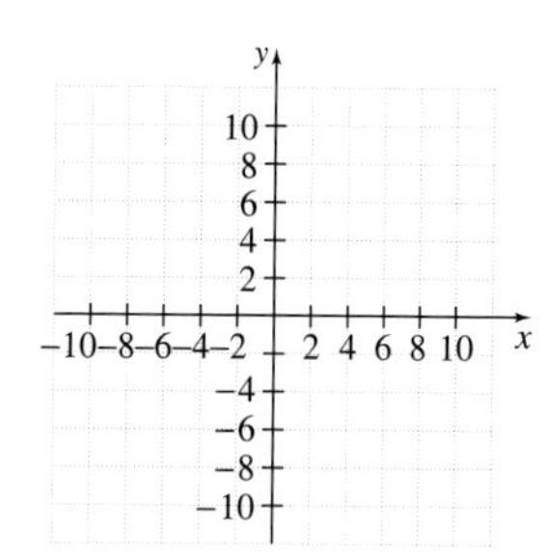

27. $f(x) = \frac{1}{2}x - 1$

28. $f(x) = -\frac{1}{2}x + 2$

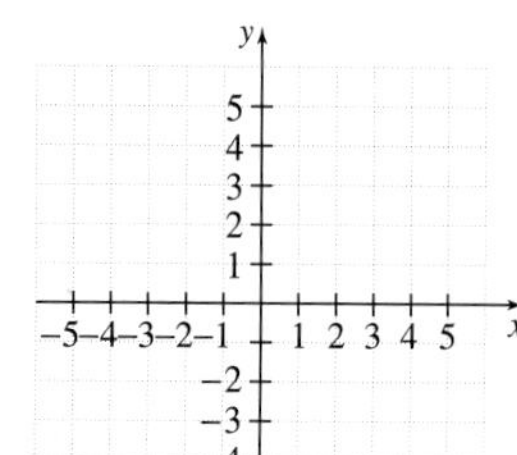

29. $f(x) = x^3$

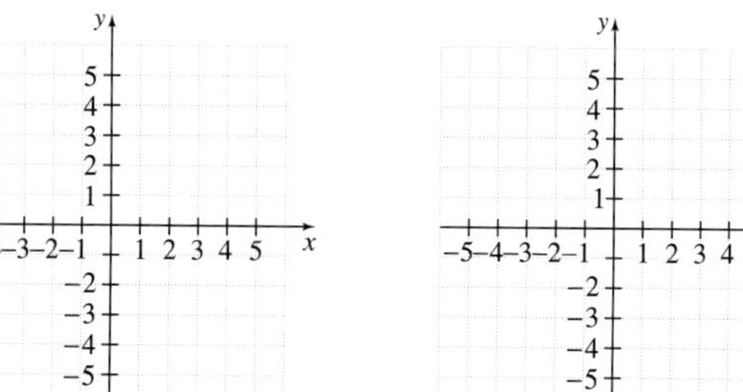

30. $f(x) = x^3 - 1$

Name ______________________________

Find the inverse of each one-to-one function. See Examples 8 and 9.

31. $f(x) = \dfrac{x-2}{5}$

32. $f(x) = \dfrac{4x-3}{2}$

33. $f(x) = \sqrt[3]{x}$

34. $f(x) = \sqrt[3]{x+1}$

35. $f(x) = \dfrac{5}{3x+1}$

36. $f(x) = \dfrac{7}{2x+4}$

37. $f(x) = (x+2)^3$

38. $f(x) = (x-5)^3$

31. ______
32. ______
33. ______
34. ______
35. ______
36. ______
37. ______
38. ______

Graph the inverse of each function on the same set of axes. See Example 10.

39.

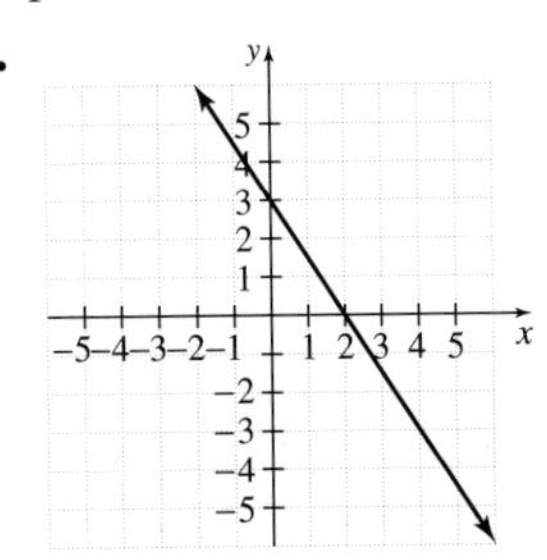

40.

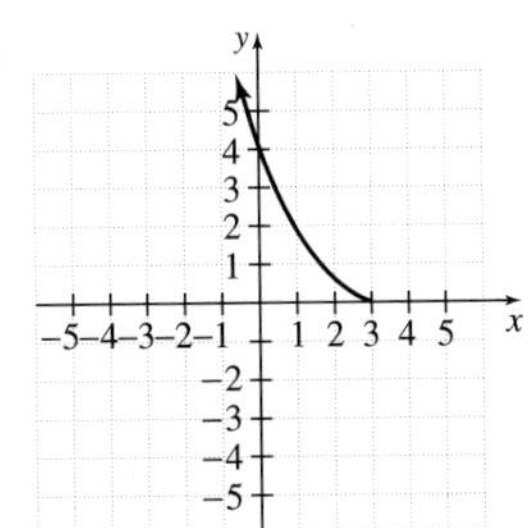

41.

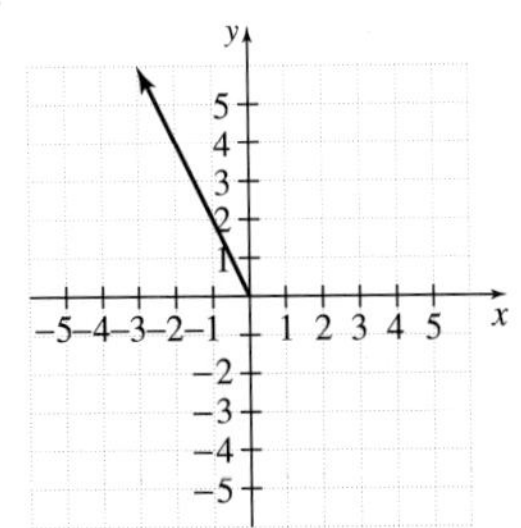

42.

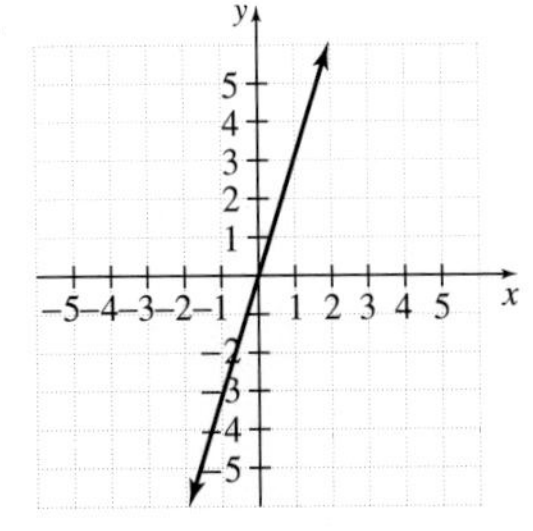

Review and Preview

Evaluate each exponential expression. See Section 7.2.

43. $25^{1/2}$

44. $49^{1/2}$

45. $16^{3/4}$

46. $27^{2/3}$

47. $9^{-3/2}$

48. $81^{-3/4}$

If $f(x) = 3^x$, find each value. In Exercises 51 and 52 give an exact answer and a two-decimal-place approximation. See Section 3.5.

49. $f(2)$

50. $f(0)$

51. $f\left(\dfrac{1}{2}\right)$

52. $f\left(\dfrac{2}{3}\right)$

43. ______
44. ______
45. ______
46. ______
47. ______
48. ______
49. ______
50. ______
51. ______
52. ______

Combining Concepts

For Exercises 53 and 54,

a. *Write the ordered pairs for $f(x)$ whose points are highlighted. (Include the points whose coordinates are given.)*

b. *Write the corresponding ordered pairs for the inverse of f, f^{-1}.*

c. *Graph the ordered pairs for f^{-1} found in part (b).*

d. *Graph $f^{-1}(x)$ by drawing a smooth curve through the plotted points.*

Name ______________________________

53. a. ______

b. ______

c. see graph ______

d. see graph ______

54. a. ______

b. ______

c. see graph ______

d. see graph ______

55. ______

56. ______

57. ______

58. ______

53. a.

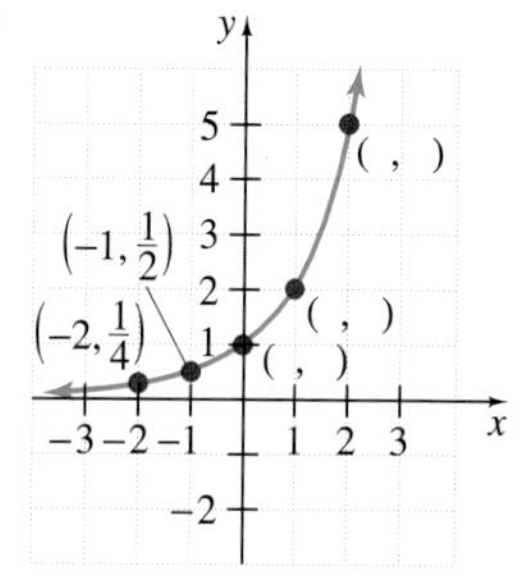

54. a.

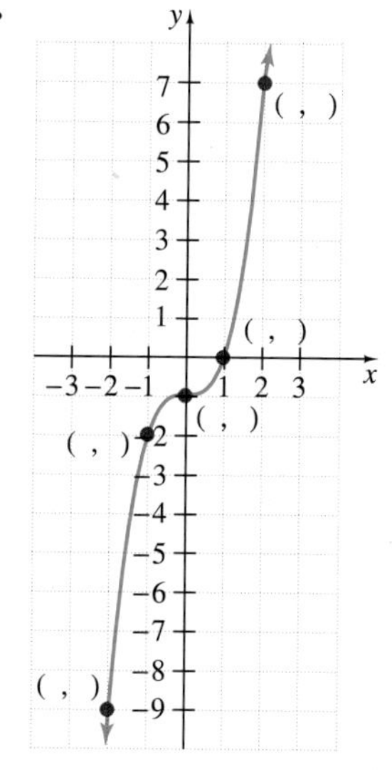

b. c. d.

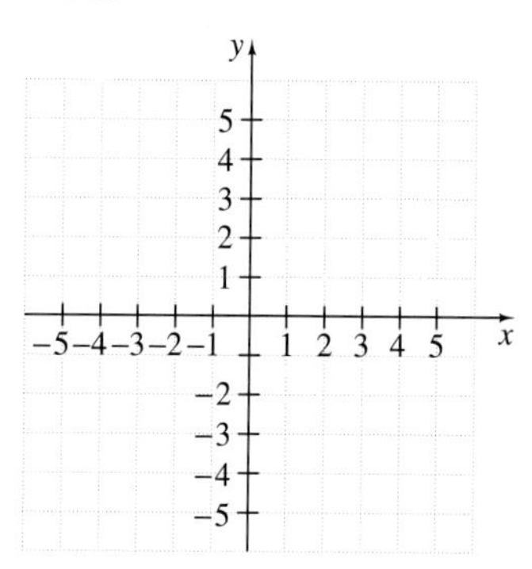

b. c. d.

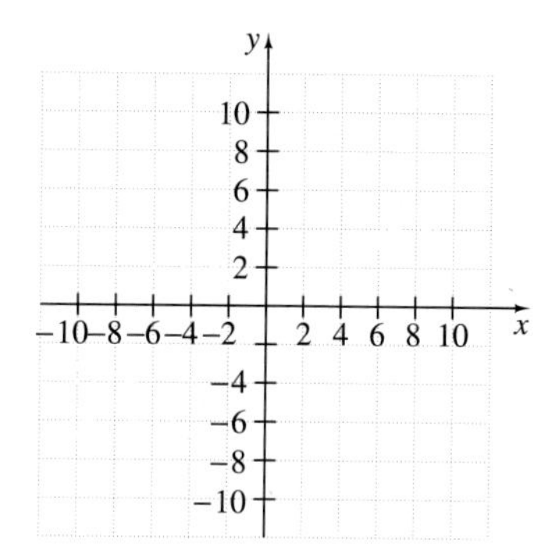

Find the inverse of each one-to-one function. Then graph the function and its inverse on a square window.

55. $f(x) = 3x + 1$

56. $f(x) = -2x - 6$

57. $f(x) = \sqrt[3]{x + 3}$

58. $f(x) = x^3 - 3$

10.3 EXPONENTIAL FUNCTIONS

Objectives

A Graph exponential functions.

B Solve equations of the form $b^x = b^y$.

C Solve problems modeled by exponential equations.

SSM

CD-ROM

Video 10.3

In earlier chapters, we gave meaning to exponential expressions such as 2^x, where x is a rational number. For example,

$$2^3 = 2 \cdot 2 \cdot 2 \quad \text{Three factors; each factor is 2}$$
$$2^{3/2} = (2^{1/2})^3 = \sqrt{2} \cdot \sqrt{2} \cdot \sqrt{2} \quad \text{Three factors; each factor is } \sqrt{2}$$

When x is an irrational number (for example, $\sqrt{3}$), what meaning can we give to $2^{\sqrt{3}}$?

It is beyond the scope of this book to give precise meaning to 2^x if x is irrational. We can confirm your intuition and say that $2^{\sqrt{3}}$ is a real number, and since $1 < \sqrt{3} < 2$, then $2^1 < 2^{\sqrt{3}} < 2^2$. We can also use a calculator and approximate $2^{\sqrt{3}}$: $2^{\sqrt{3}} \approx 3.321997$. In fact, as long as the base b is positive, b^x is a real number for all real numbers x. Finally, the rules of exponents apply whether x is rational or irrational, as long as b is positive.

In this section, we are interested in functions of the form $f(x) = b^x$, where $b > 0$. A function of this form is called an **exponential function.**

EXPONENTIAL FUNCTION

A function of the form

$$f(x) = b^x$$

is called an **exponential function** if $b > 0$, b is not 1, and x is a real number.

A GRAPHING EXPONENTIAL FUNCTIONS

Now let's practice graphing exponential functions.

Example 1 Graph the exponential functions defined by $f(x) = 2^x$ and $g(x) = 3^x$ on the same set of axes.

Solution: To graph these functions, we find some ordered pair solutions, plot the points, and connect them with a smooth curve.

$f(x) = 2^x$

x	0	1	2	3	−1	−2
$f(x)$	1	2	4	8	$\frac{1}{2}$	$\frac{1}{4}$

$g(x) = 3^x$

x	0	1	2	3	−1	−2
$g(x)$	1	3	9	27	$\frac{1}{3}$	$\frac{1}{9}$

Practice Problem 1

Graph the exponential function $f(x) = 4^x$.

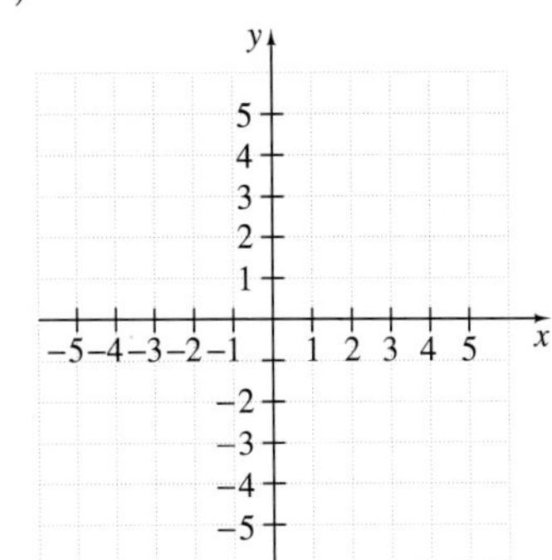

Answer

1.

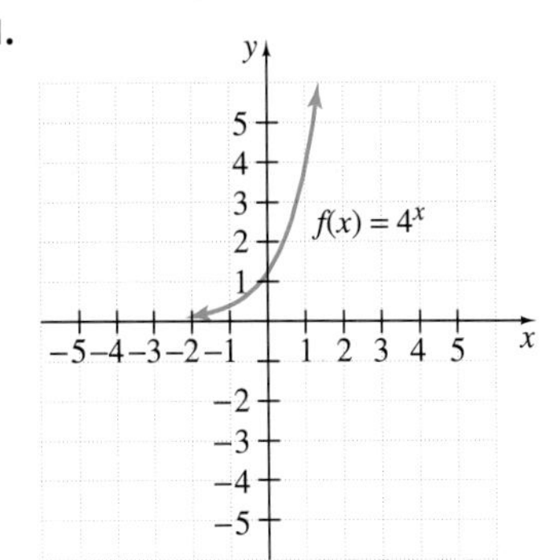

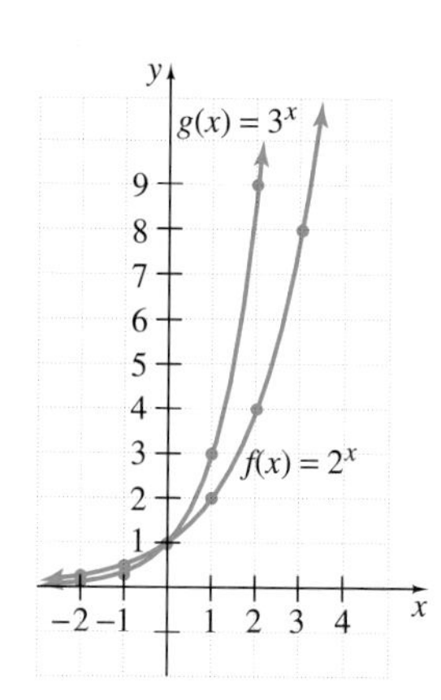

A number of things should be noted about the two graphs of exponential functions in Example 1. First, the graphs show that $f(x) = 2^x$ and $g(x) = 3^x$ are one-to-one functions since each graph passes the vertical and horizontal line tests. The y-intercept of each graph is 1, but neither graph has an x-intercept. From the graph, we can also see that the domain of each function is all real numbers and that the range is $(0, \infty)$. We can also see that as x-values are increasing, y-values are increasing also.

Practice Problem 2

Graph the exponential function $f(x) = \left(\frac{1}{5}\right)^x$.

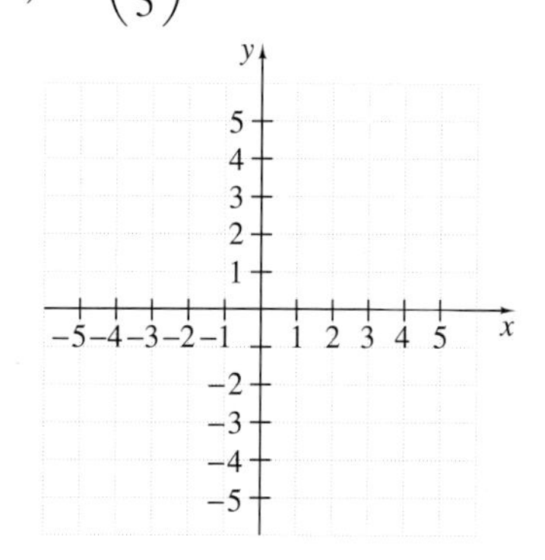

Example 2 Graph the exponential functions $y = \left(\frac{1}{2}\right)^x$ and $y = \left(\frac{1}{3}\right)^x$ on the same set of axes.

Solution: As before, we find some ordered pair solutions, plot the points, and connect them with a smooth curve.

$y = \left(\frac{1}{2}\right)^x$

x	0	1	2	3	−1	−2
y	1	$\frac{1}{2}$	$\frac{1}{4}$	$\frac{1}{8}$	2	4

$y = \left(\frac{1}{3}\right)^x$

x	0	1	2	3	−1	−2
y	1	$\frac{1}{3}$	$\frac{1}{9}$	$\frac{1}{27}$	3	9

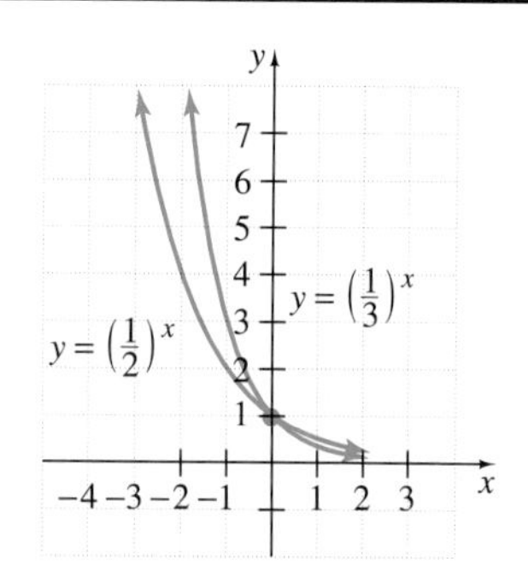

Each function in Example 2 again is a one-to-one function. The y-intercept of both is 1. The domain is the set of all real numbers, and the range is $(0, \infty)$.

Notice the difference between the graphs of Example 1 and the graphs of Example 2. An exponential function is always increasing if the base is

Answer

2.

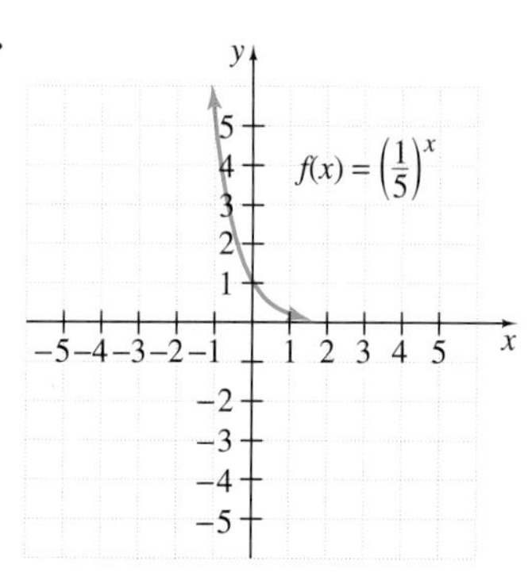

greater than 1. When the base is between 0 and 1, the graph is always decreasing. The following figures summarize these characteristics of exponential functions.

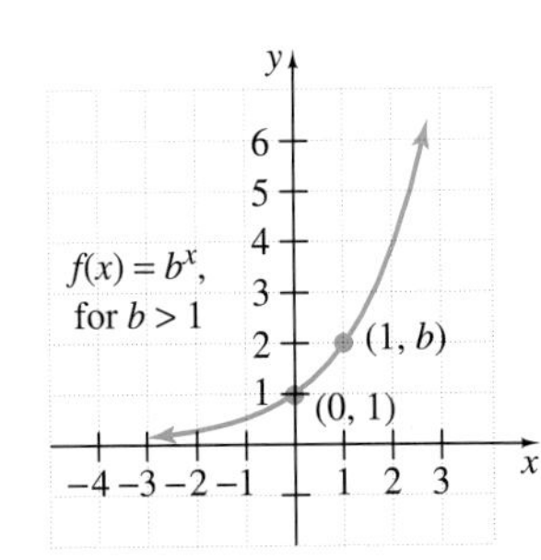

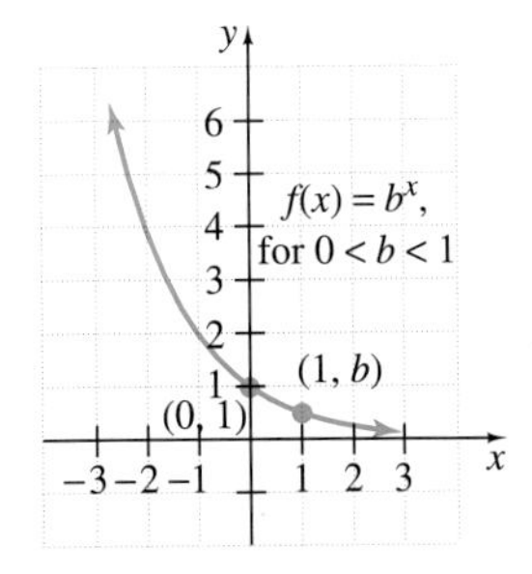

Example 3 Graph the exponential function $f(x) = 3^{x+2}$.

Solution: As before, we find and plot a few ordered pair solutions. Then we connect the points with a smooth curve.

$y = 3^{x+2}$

x	0	-1	-2	-3	-4
y	9	3	1	$\frac{1}{3}$	$\frac{1}{9}$

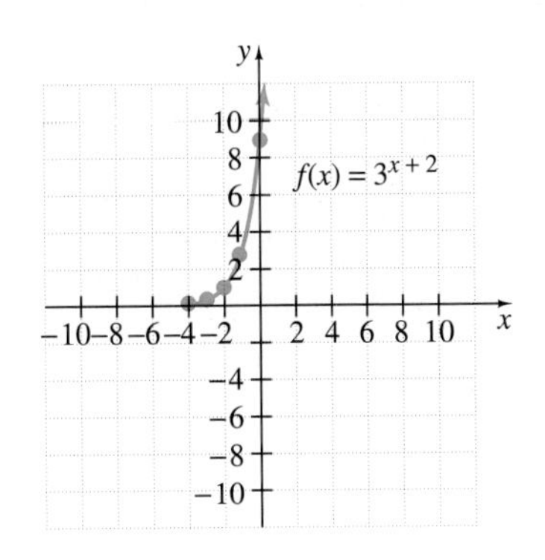

TRY THE CONCEPT CHECK IN THE MARGIN.

B SOLVING EQUATIONS OF THE FORM $b^x = b^y$

We have seen that an exponential function $y = b^x$ is a one-to-one function. Another way of stating this fact is a property that we can use to solve exponential equations.

> **UNIQUENESS OF b^x**
>
> Let $b > 0$ and $b \neq 1$. Then $b^x = b^y$ is equivalent to $x = y$.

Example 4 Solve: $2^x = 16$

Solution: We write 16 as a power of 2 and then use the uniqueness of b^x to solve.

$2^x = 16$

$2^x = 2^4$

Practice Problem 3

Graph the exponential function $f(x) = 2^{x-1}$.

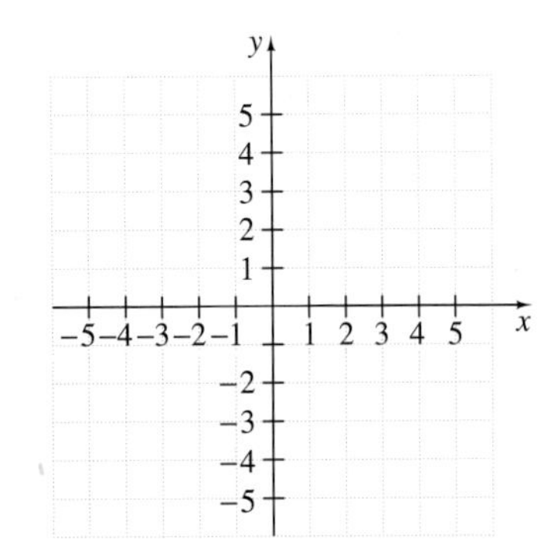

✓ CONCEPT CHECK

Which functions are exponential functions?

a. $f(x) = x^3$ b. $g(x) = \left(\frac{2}{3}\right)^x$

c. $h(x) = 5^{x-2}$ d. $w(x) = (2x)^2$

Answers

3.

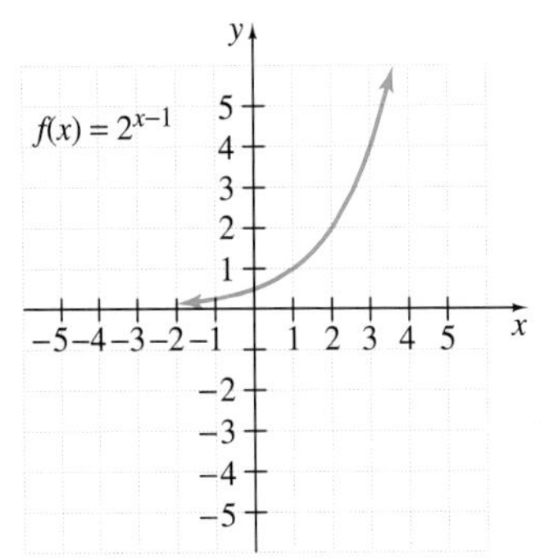

✓ **Concept Check:** b and c are exponential

Practice Problem 4

Solve: $5^x = 125$

Since the bases are the same and are nonnegative, by the uniqueness of b^x we then have that the exponents are equal. Thus,

$x = 4$

To check, we replace x with 4 in the original equation. The solution set is $\{4\}$.

Practice Problem 5

Solve: $4^x = 8$

Example 5

Solve: $9^x = 27$

Solution: Since both 9 and 27 are powers of 3, we can use the uniqueness of b^x.

$$9^x = 27$$
$$(3^2)^x = 3^3 \quad \text{Write 9 and 27 as powers of 3.}$$
$$3^{2x} = 3^3$$
$$2x = 3 \quad \text{Use the uniqueness of } b^x.$$
$$x = \frac{3}{2} \quad \text{Divide both sides by 2.}$$

To check, we replace x with $\frac{3}{2}$ in the original equation. The solution set is $\left\{\frac{3}{2}\right\}$.

Practice Problem 6

Solve: $9^{x-1} = 27^x$

Example 6

Solve: $4^{x+3} = 8^x$

Solution: We write both 4 and 8 as powers of 2, and then use the uniqueness of b^x.

$$4^{x+3} = 8^x$$
$$(2^2)^{x+3} = (2^3)^x$$
$$2^{2x+6} = 2^{3x}$$
$$2x + 6 = 3x \quad \text{Use the uniqueness of } b^x.$$
$$6 = x \quad \text{Subtract } 2x \text{ from both sides.}$$

Check to see that the solution set is $\{6\}$.

There is one major problem with the preceding technique. Often the two sides of an equation, $4 = 3^x$ for example, cannot easily be written as powers of a common base. We explore how to solve such an equation with the help of **logarithms** later.

C SOLVING PROBLEMS MODELED BY EXPONENTIAL EQUATIONS

The bar graph on the next page shows the increase in the number of cellular phone users. Notice that the graph of the exponential function $y = 3.638\,(1.443)^x$ approximates the heights of the bars. This is just one example of how the world abounds with patterns that can be modeled by exponential functions. To make these applications realistic, we use numbers that warrant a calculator. Another application of an exponential function has to do with interest rates on loans.

Answers

4. $\{3\}$, **5.** $\left\{\frac{3}{2}\right\}$, **6.** $\{-2\}$

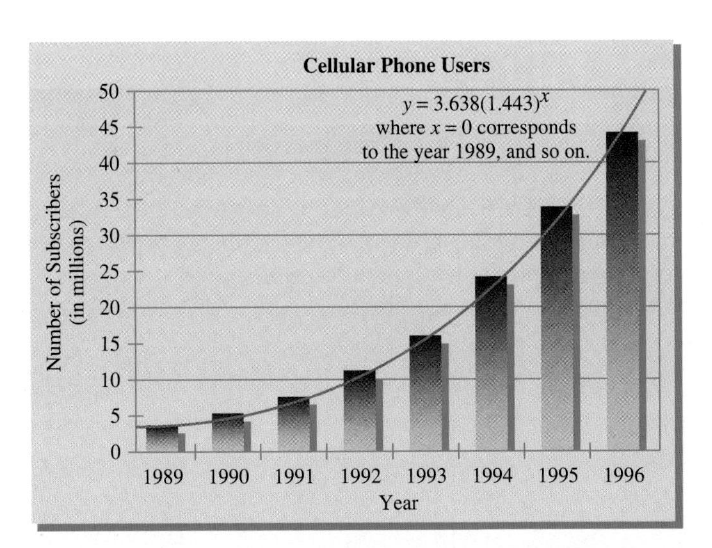

The exponential function defined by $A = P\left(1 + \frac{r}{n}\right)^{nt}$ models the pattern relating the dollars A accrued (or owed) after P dollars are invested (or loaned) at an annual rate of interest r compounded n times each year for t years. This function is known as the compound interest formula.

Example 7 Using the Compound Interest Formula

Find the amount owed at the end of 5 years if \$1600 is loaned at a rate of 9% compounded monthly.

Solution: We use the formula $A = P\left(1 + \frac{r}{n}\right)^{nt}$, with the following values:

$P = \$1600$ (the amount of the loan)

$r = 9\% = 0.09$ (the annual rate of interest)

$n = 12$ (the number of times interest is compounded each year)

$t = 5$ (the duration of the loan, in years)

$$A = P\left(1 + \frac{r}{n}\right)^{nt} \quad \text{Compound interest formula}$$

$$= 1600\left(1 + \frac{0.09}{12}\right)^{12(5)} \quad \text{Substitute known values.}$$

$$= 1600(1.0075)^{60}$$

To approximate A, use the $\boxed{y^x}$ or $\boxed{\wedge}$ key on your calculator.

2505.0896

Thus, the amount A owed is approximately \$2505.09.

Practice Problem 7

As a result of the Chernobyl nuclear accident, radioactive debris was carried through the atmosphere. One immediate concern was the impact that debris had on the milk supply. The percent y of radioactive material in raw milk after t days is estimated by $y = 100\,(2.7)^{-0.1t}$. Estimate the expected percent of radioactive material in the milk after 30 days.

Answer

7. approximately 5.08%

GRAPHING CALCULATOR EXPLORATIONS

We can use a graphing calculator and its TRACE feature to solve Practice Problem 7 graphically.

To estimate the expected percent of radioactive material in the milk after 30 days, enter $Y_1 = 100(2.7)^{-0.1x}$. The graph does not appear on a standard viewing window, so we need to determine an appropriate viewing window. Because it doesn't make sense to look at radioactivity *before* the Chernobyl nuclear accident, we use Xmin $= 0$. We are interested in finding the percent of radioactive material in the milk when $x = 30$, so we choose Xmax $= 35$ to leave enough space to see the graph at $x = 30$. Because the values of y are percents, it seems appropriate that $0 \leq y \leq 100$. (We also use Xscl $= 1$ and Yscl $= 10$.) Now we graph the function.

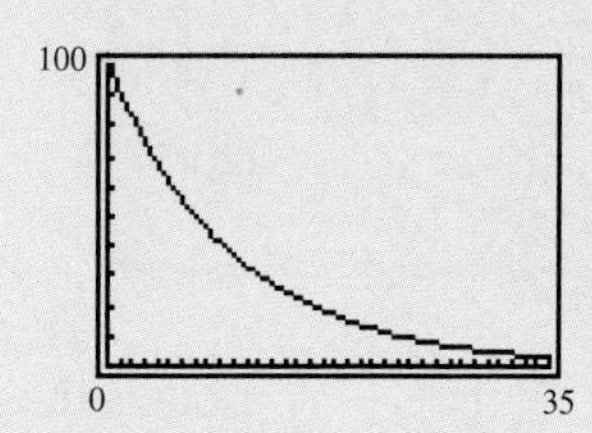

We can use the TRACE feature to obtain an approximation of the expected percent of radioactive material in the milk when $x = 30$. (A TABLE feature may also be used to approximate the percent.) To obtain a better approximation, let's use the ZOOM feature several times to zoom in near $x = 30$.

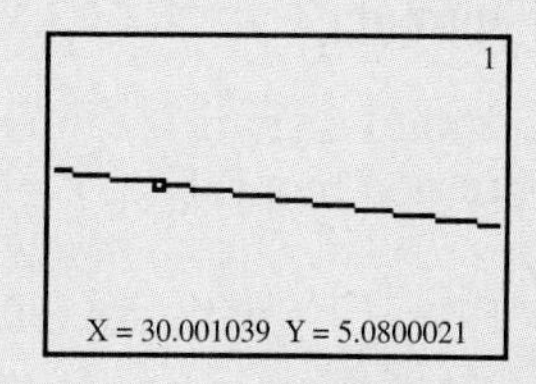

The percent of radioactive material in the milk 30 days after the Chernobyl accident was 5.08%, accurate to two decimal places.

Use a grapher to find each percent. Approximate your solutions so that they are accurate to two decimal places.

1. Estimate the expected percent of radioactive material in the milk 2 days after the Chernobyl nuclear accident.

2. Estimate the expected percent of radioactive material in the milk 10 days after the Chernobyl nuclear accident.

3. Estimate the expected percent of radioactive material in the milk 15 days after the Chernobyl nuclear accident.

4. Estimate the expected percent of radioactive material in the milk 25 days after the Chernobyl nuclear accident.

Name ______________________ Section ______ Date ________

Exercise Set 10.3

A *Graph each exponential function. See Examples 1 through 3.*

1. $y = 4^x$

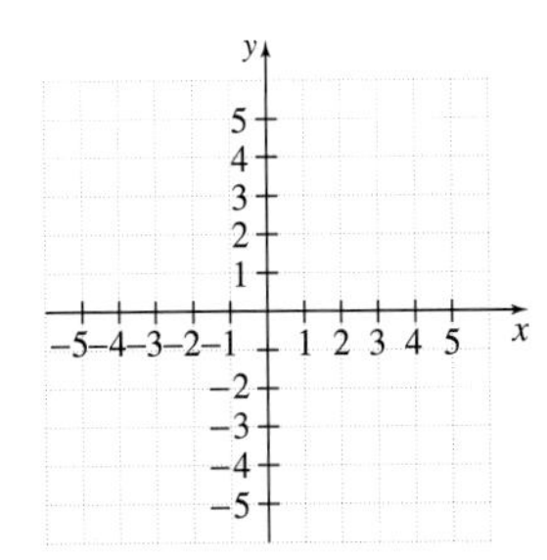

2. 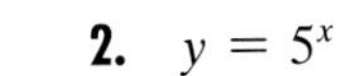 $y = 5^x$

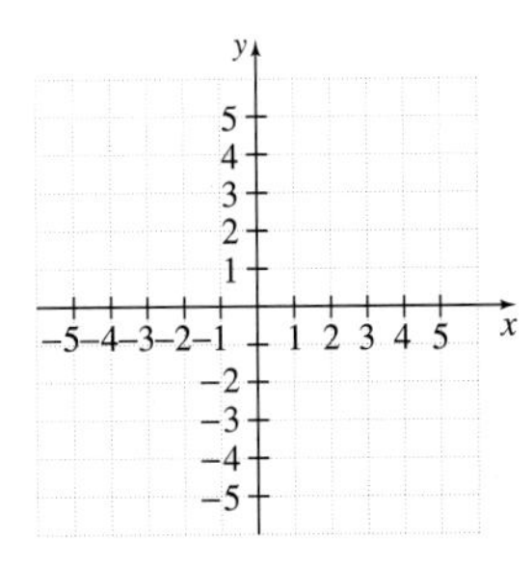

3. $y = 1 + 2^x$

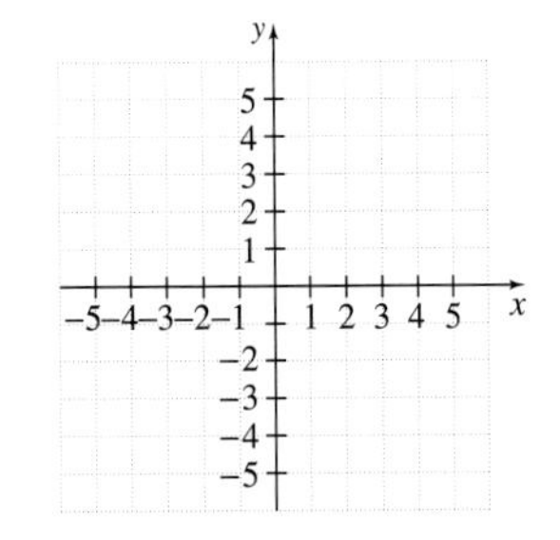

4. $y = 3^x - 1$

5. $y = \left(\frac{1}{4}\right)^x$

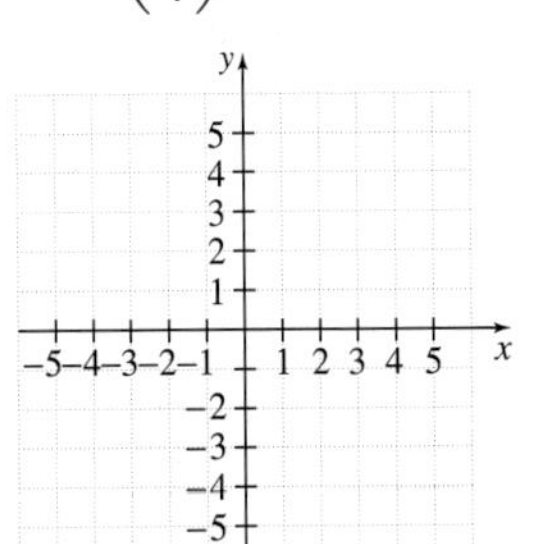

6. $y = \left(\frac{1}{5}\right)^x$

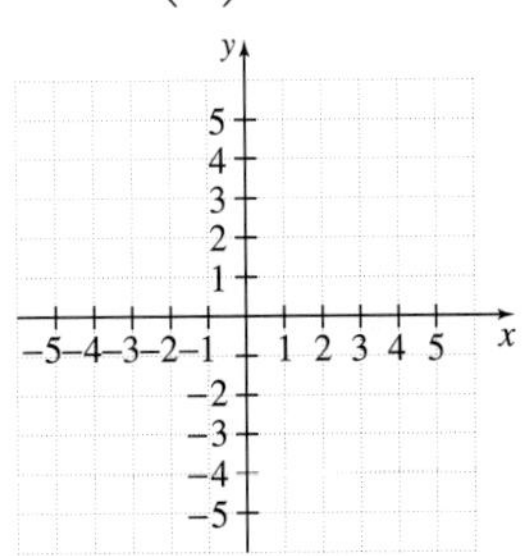

7. $y = \left(\frac{1}{2}\right)^x - 2$

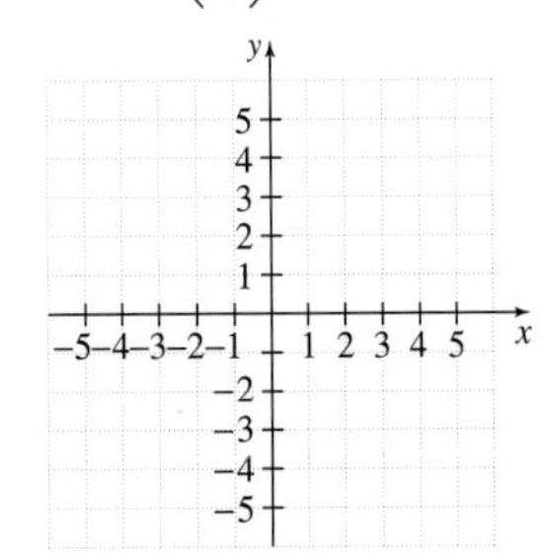

8. $y = \left(\frac{1}{3}\right)^x + 2$

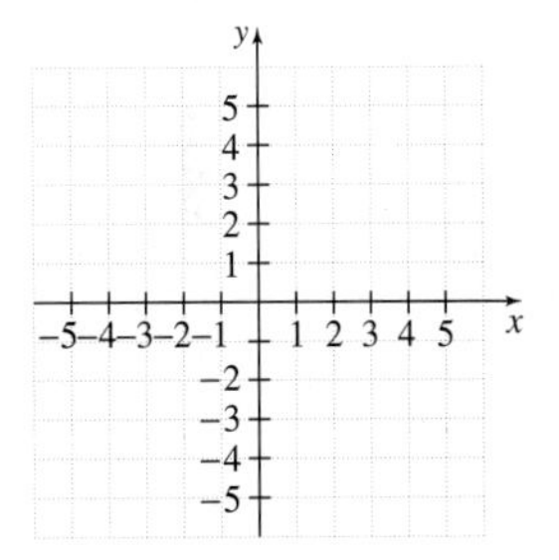

9. $y = -2^x$

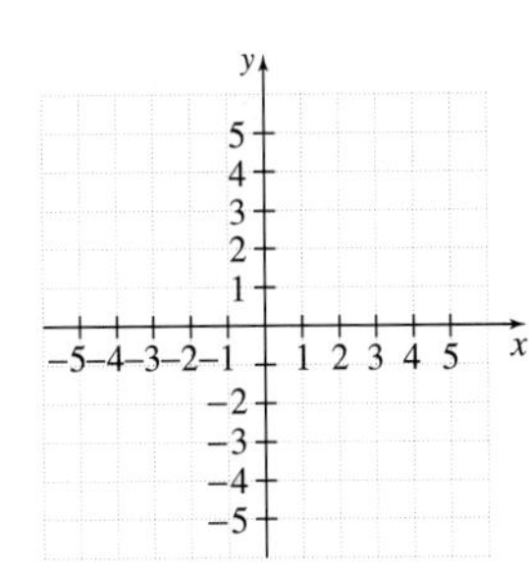

10. $y = -3^x$

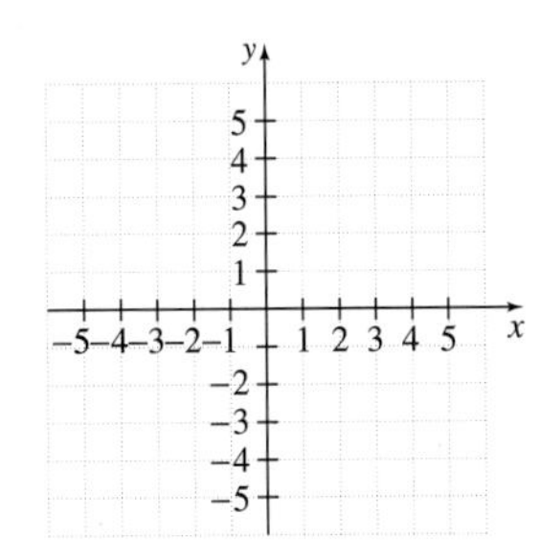

11. $y = 3^x - 2$

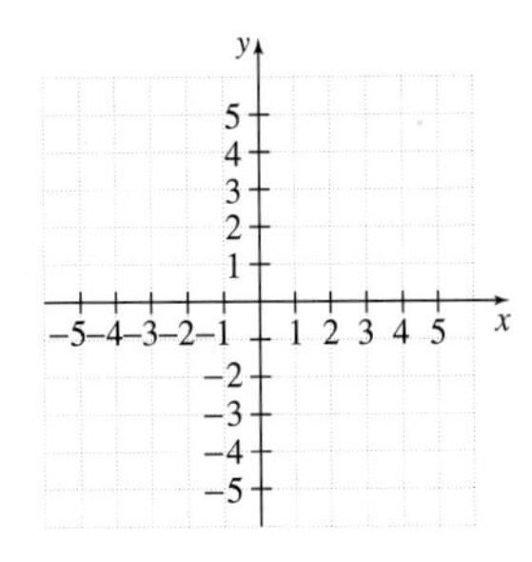

12. $y = 2^x - 3$

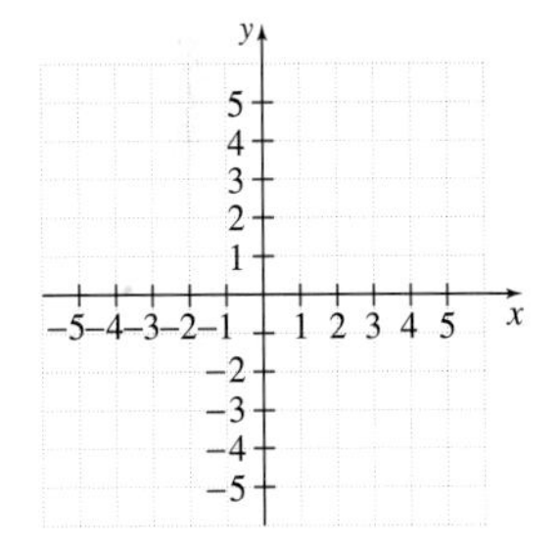

13. $y = -\left(\frac{1}{4}\right)^x$

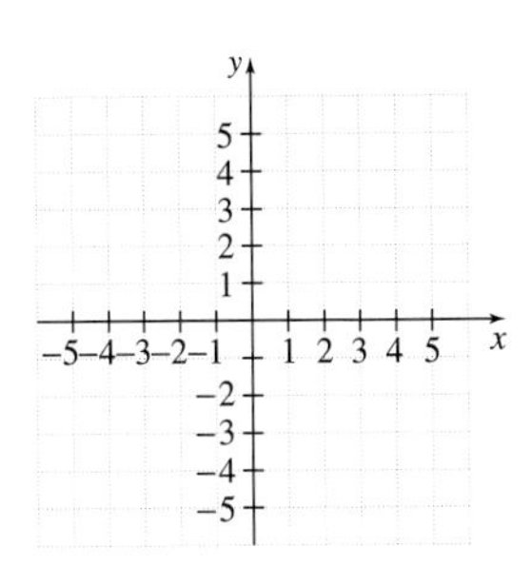

14. $y = -\left(\frac{1}{5}\right)^x$

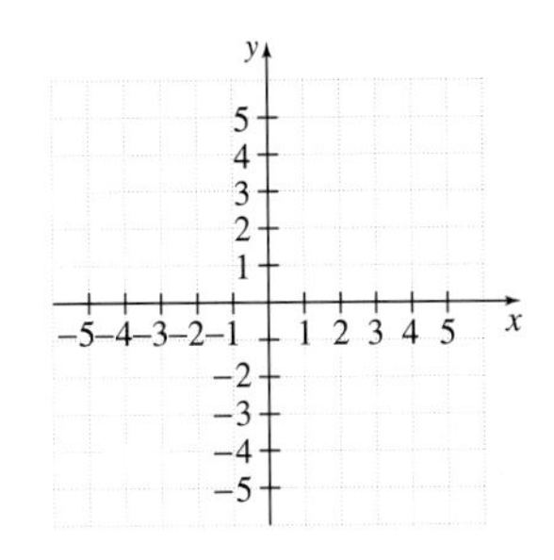

15. $y = \left(\frac{1}{3}\right)^x + 1$

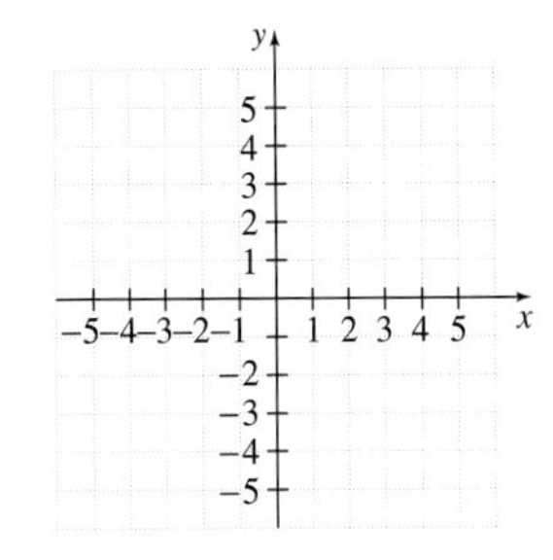

16. $y = \left(\frac{1}{2}\right)^x - 2$

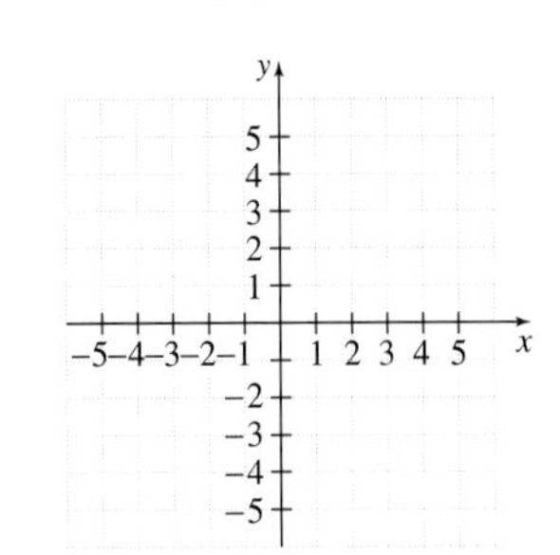

ANSWERS

17. ______

18. ______

19. ______

20. ______

21. ______

22. ______

23. ______

24. ______

25. ______

26. ______

27. ______

28. ______

29. ______

30. ______

31. ______

32. ______

33. ______

34. ______

35. a. ______

b. ______

Name ______

B *Solve. See Examples 4 through 6.*

17. $3^x = 27$

18. $6^x = 36$

19. $16^x = 8$

20. $64^x = 16$

21. $32^{2x-3} = 2$

22. $9^{2x+1} = 81$

23. $\frac{1}{4} = 2^{3x}$

24. $\frac{1}{27} = 3^{2x}$

25. $4^x = 8$

26. $32^x = 4$

27. $27^{x+1} = 9$

28. $125^{x-2} = 25$

29. $81^{x-1} = 27^{2x}$

30. $4^{3x-7} = 32^{2x}$

C *Solve. Unless otherwise indicated, round results to one decimal place. See Example 7.*

31. One type of uranium has a daily radioactive decay rate of 0.4%. If 30 pounds of this uranium is available today, how much will still remain after 50 days? Use $y = 30(2.7)^{-0.004t}$, and let t be 50.

32. The nuclear waste from an atomic energy plant decays at a rate of 3% each century. If 150 pounds of nuclear waste is disposed of, how much of it will still remain after 10 centuries? Use $y = 150(2.7)^{-0.03t}$, and let t be 10.

33. National Park Service personnel are trying to increase the size of the bison population of Theodore Roosevelt National Park. If 260 bison currently live in the park, and if the population's rate of growth is 2.5% annually, how many bison (rounded to the nearest whole) should there be in 10 years? Use $y = 260(2.7)^{0.025t}$.

34. The world population is currently growing at a rate of 1.32% annually. In 1998, the midyear population of the world was 5,926,466,814 people. Predict the midyear world population (to the nearest million) in 2005. Use $y = 5{,}926{,}466{,}814(2.7)^{0.0132t}$, where t is the number of years after 1998. (*Source:* Based on data from the U.S. Bureau of the Census, International Data Base)

35. Retail revenue from shopping on the Internet is expected to grow at a rate of 64% per year. In 1997, a total of \$2.4 billion in revenue was collected through Internet retail sales. To make the following predictions, use $y = 2.4(1.64)^t$, where t is the number of years after 1997. (*Source:* Based on data from Forrester Research Inc.)

a. What level of retail revenues from Internet shopping is expected in 2001?

b. Predict the level of Internet shopping revenues in 2010.

Name ____________________

36. a. ______

b. ______

37. ______

38. ______

39. ______

40. ______

41. ______

42. ______

43. ______

44. ______

36. Carbon dioxide (CO_2) is a greenhouse gas that contributes to global warming. Due to the combustion of fossil fuels, the amount of CO_2 in Earth's atmosphere has been increasing by 0.4% annually over the past century. In 1994, the concentration of CO_2 in the atmosphere was 358 parts per million by volume. To make the following predictions, use $y = 358(1.004)^t$, where t is the number of years after 1994. (*Source:* Based on data from the United Nations Environment Programme's Information Unit for Conventions)

a. Predict the concentration of CO_2 in the atmosphere in the year 2004.

b. Predict the concentration of CO_2 in the atmosphere in the year 2025.

The formula $y = 3.638(1.443)^x$ gives the number of cellular phone users y (in millions) in the United States for the years 1989 through 1996. In this formula, $x = 0$ corresponds to 1989, $x = 1$ corresponds to 1990, and so on. Use this formula to solve exercises 37 and 38.

37. Use this model to predict the number of cellular phone users in the year 2005.

38. Use this model to predict the number of cellular phone users in the year 2008.

Solve: Use $A = P\left(1 + \frac{r}{n}\right)^{nt}$. Round answers to two decimal places. See Example 7.

39. Find the amount Erica Entada owes at the end of 3 years if \$6000 is loaned to her at a rate of 8% compounded monthly.

40. Find the amount owed at the end of 5 years if \$3000 is loaned at a rate of 10% compounded quarterly.

Review and Preview

Solve each equation. See Sections 2.1 and 5.7.

41. $5x - 2 = 18$

42. $3x - 7 = 11$

43. $3x - 4 = 3(x + 1)$

44. $2 - 6x = 6(1 - x)$

45. ____________

46. ____________

47. ____________

48. ____________

49. ____________

50. ____________

51. see calculator screen

52. ____________

53. ____________

Name ____________________________

Combining Concepts

Match each exponential function with its graph.

45. $f(x) = \left(\frac{1}{2}\right)^x$

46. $f(x) = 2^x$

47. $f(x) = \left(\frac{1}{4}\right)^x$

48. $f(x) = 3^x$

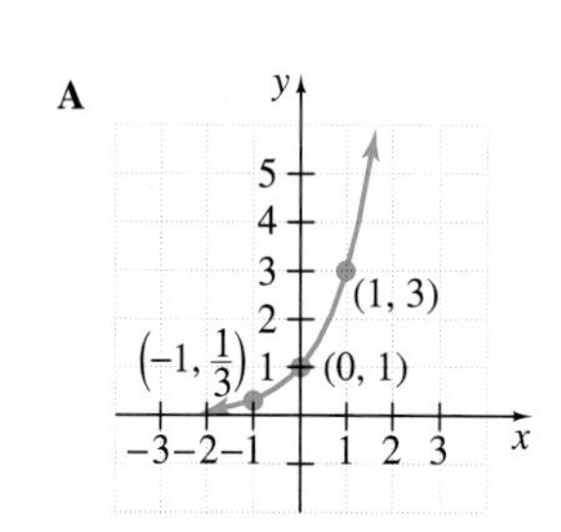

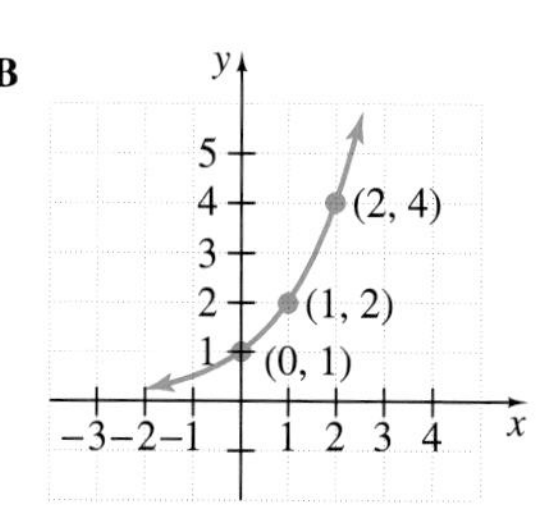

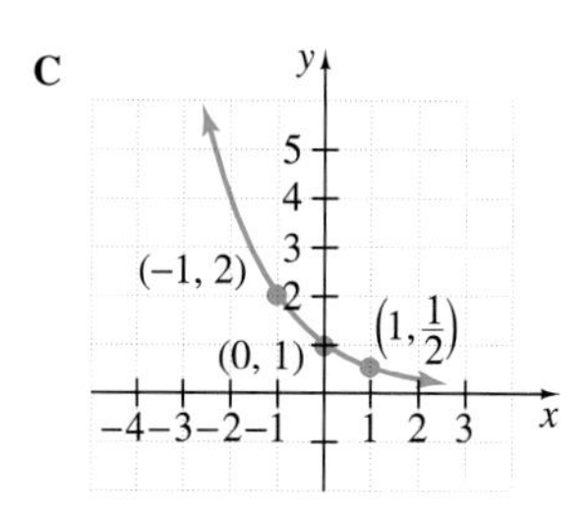

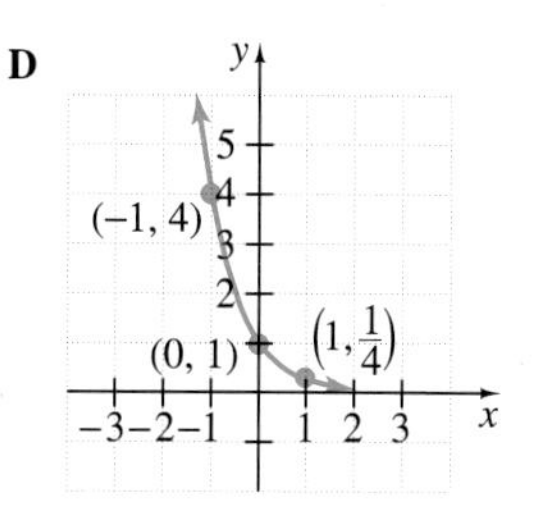

49. Explain why the graph of an exponential function $y = b^x$ contains the point $(1, b)$.

50. Explain why an exponential function $y = b^x$ has a y-intercept of 1.

Use a grapher to solve. Estimate your results to two decimal places.

51. Verify the results of Exercise 31.

52. From Exercise 31, estimate the number of pounds of uranium that will be available after 100 days.

53. From Exercise 31, estimate the number of pounds of uranium that will be available after 120 days.

10.4 LOGARITHMIC FUNCTIONS

Objectives

A Write exponential equations with logarithmic notation and write logarithmic equations with exponential notation.

B Solve logarithmic equations by using exponential notation.

C Identify and graph logarithmic functions.

A USING LOGARITHMIC NOTATION

Since the exponential function $f(x) = 2^x$ is a one-to-one function, it has an inverse. We can create a table of values for f^{-1} by switching the coordinates in the accompanying table of values for $f(x) = 2^x$.

x	$y = f(x)$
-3	$\frac{1}{8}$
-2	$\frac{1}{4}$
-1	$\frac{1}{2}$
0	1
1	2
2	4
3	8

x	$y = f^{-1}(x)$
$\frac{1}{8}$	-3
$\frac{1}{4}$	-2
$\frac{1}{2}$	-1
1	0
2	1
4	2
8	3

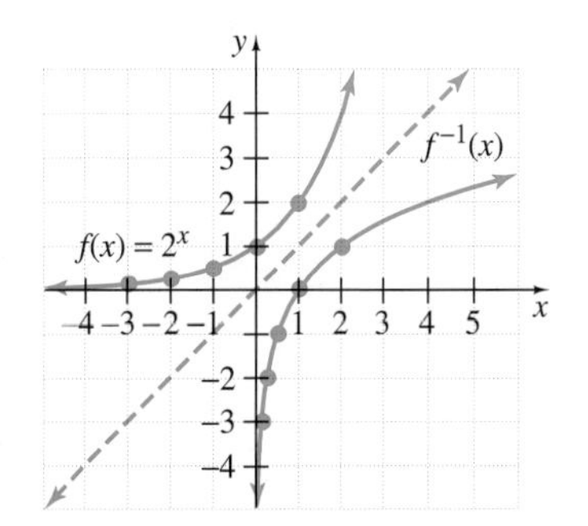

The graphs of $f(x)$ and its inverse are shown in the margin. Notice that the graphs of f and f^{-1} are symmetric about the line $y = x$, as expected.

Now we would like to be able to write an equation for f^{-1}. To do so, we follow the steps for finding the equation of an inverse:

$$f(x) = 2^x$$

Step 1. Replace $f(x)$ by y. $\quad y = 2^x$

Step 2. Interchange x and y. $\quad x = 2^y$

Step 3. Solve for y.

At this point, we are stuck. To solve this equation for y, a new notation, the **logarithmic notation**, is needed.

The symbol $\log_b x$ means "the power to which b is raised to produce a result of x." In other words,

$$\log_b x = y \text{ means } b^y = x$$

We say that $\log_b x$ is "the logarithm of x to the base b" or "the log of x to the base b." See the logarithmic definition in the margin.

LOGARITHMIC DEFINITION

If $b > 0$, and $b \neq 1$, then

$$y = \log_b x \text{ means } x = b^y$$

for every $x > 0$ and every real number y.

Before returning to the function $x = 2^y$ and solving it for y in terms of x, let's practice using the new notation $\log_b x$.

It is important to be able to write exponential equations with logarithmic notation, and vice versa. The following table shows examples of both forms.

Logarithmic Equation	Corresponding Exponential Equation
$\log_3 9 = 2$	$3^2 = 9$
$\log_6 1 = 0$	$6^0 = 1$
$\log_2 8 = 3$	$2^3 = 8$
$\log_4 \frac{1}{16} = -2$	$4^{-2} = \frac{1}{16}$
$\log_8 2 = \frac{1}{3}$	$8^{1/3} = 2$

HELPFUL HINT

Notice that a *logarithm* is an *exponent*. In other words, $\log_3 9$ is the *power* that we raise 3 to in order to get 9.

Practice Problems 1–3

Write as an exponential equation.

1. $\log_7 49 = 2$
2. $\log_8 \frac{1}{8} = -1$
3. $\log_3 \sqrt{3} = \frac{1}{2}$

Examples

Write as an exponential equation.

1. $\log_5 25 = 2$ means $5^2 = 25$

2. $\log_6 \frac{1}{6} = -1$ means $6^{-1} = \frac{1}{6}$

3. $\log_2 \sqrt{2} = \frac{1}{2}$ means $2^{1/2} = \sqrt{2}$

Practice Problems 4–6

Write as a logarithmic equation.

4. $3^4 = 81$
5. $2^{-3} = \frac{1}{8}$
6. $7^{1/3} = \sqrt[3]{7}$

Examples

Write as a logarithmic equation.

4. $9^3 = 729$ means $\log_9 729 = 3$

5. $6^{-2} = \frac{1}{36}$ means $\log_6 \frac{1}{36} = -2$

6. $5^{1/3} = \sqrt[3]{5}$ means $\log_5 \sqrt[3]{5} = \frac{1}{3}$

Practice Problem 7

Find the value of each logarithmic expression.

a. $\log_2 8$
b. $\log_3 \frac{1}{9}$
c. $\log_{25} 5$

Example 7

Find the value of each logarithmic expression.

a. $\log_4 16$

b. $\log_{10} \frac{1}{10}$

c. $\log_9 3$

Solution:

a. $\log_4 16 = 2$ because $4^2 = 16$.

b. $\log_{10} \frac{1}{10} = -1$ because $10^{-1} = \frac{1}{10}$.

c. $\log_9 3 = \frac{1}{2}$ because $9^{1/2} = \sqrt{9} = 3$.

B SOLVING LOGARITHMIC EQUATIONS

The ability to interchange the logarithmic and exponential forms of a statement is often the key to solving logarithmic equations.

Practice Problem 8

Solve: $\log_2 x = 4$

Practice Problem 9

Solve: $\log_x 9 = 2$

Practice Problem 10

Solve: $\log_2 1 = x$

Example 8

Solve: $\log_5 x = 3$

Solution:

$\log_5 x = 3$

$5^3 = x$ Write as an exponential equation.

$125 = x$

The solution set is $\{125\}$.

Example 9

Solve: $\log_x 25 = 2$

Solution:

$\log_x 25 = 2$

$x^2 = 25$ Write as an exponential equation.

$x = 5$

Even though $(-5)^2 = 25$, the base b of a logarithm must be positive. The solution set is $\{5\}$.

PROPERTIES OF LOGARITHMS

If b is a real number, $b > 0$ and $b \neq 1$, then

1. $\log_b 1 = 0$

2. $\log_b b^x = x$

3. $b^{\log_b x} = x$

Example 10

Solve: $\log_3 1 = x$

Solution:

$\log_3 1 = x$

$3^x = 1$ Write as an exponential equation.

$3^x = 3^0$ Write 1 as 3^0.

$x = 0$ Use the uniqueness of b^x.

The solution set is $\{0\}$.

Answers

1. $7^2 = 49$, **2.** $8^{-1} = \frac{1}{8}$, **3.** $3^{1/2} = \sqrt{3}$, **4.** $\log_3 81 = 4$, **5.** $\log_2 \frac{1}{8} = -3$, **6.** $\log_7 \sqrt[3]{7} = \frac{1}{3}$, **7. a.** 3, **b.** -2, **c.** $\frac{1}{2}$ **8.** $\{16\}$, **9.** $\{3\}$, **10.** $\{0\}$

In Example 10, we illustrated an important property of logarithms. That is, $\log_b 1$ is always 0. This property as well as two important others are given in the margin on page 776.

To see that $\log_b b^x = x$, change the logarithmic form to exponential form. Then, $\log_b b^x = x$ means $b^x = b^x$. In exponential form, the statement is true, so in logarithmic form, the statement is also true.

Example 11 Simplify.

a. $\log_3 3^2$ **b.** $\log_7 7^{-1}$
c. $5^{\log_5 3}$ **d.** $2^{\log_2 6}$

Solution:

a. From property 2, $\log_3 3^2 = 2$.
b. From property 2, $\log_7 7^{-1} = -1$.
c. From property 3, $5^{\log_5 3} = 3$.
d. From property 3, $2^{\log_2 6} = 6$.

C GRAPHING LOGARITHMIC FUNCTIONS

Let us now return to the function $f(x) = 2^x$ and write an equation for its inverse, $f^{-1}(x)$. Recall our earlier work.

$$f(x) = 2^x$$

Step 1. Replace $f(x)$ by y. $\quad y = 2^x$
Step 2. Interchange x and y. $\quad x = 2^y$

Having gained proficiency with the notation $\log_b x$, we can now complete the steps for writing the inverse equation.

Step 3. Solve for y. $\quad y = \log_2 x$
Step 4. Replace y with $f^{-1}(x)$. $\quad f^{-1}(x) = \log_2 x$

Thus, $f^{-1}(x) = \log_2 x$ defines a function that is the inverse function of the function $f(x) = 2^x$. The function $f^{-1}(x)$ or $y = \log_2 x$ is called a **logarithmic function**.

TRY THE CONCEPT CHECK IN THE MARGIN.

We can explore logarithmic functions by graphing them.

Example 12 Graph the logarithmic function $y = \log_2 x$.

Solution: First we write the equation with exponential notation as $2^y = x$. Then we find some ordered pair solutions that satisfy this equation. Finally, we plot the points and connect them with a smooth curve. The domain of this function is $(0, \infty)$, and the range is all real numbers.

Since $x = 2^y$ is solved for x, we choose y-values and compute corresponding x-values.

If $y = 0, x = 2^0 = 1$

If $y = 1, x = 2^1 = 2$

If $y = 2, x = 2^2 = 4$

If $y = -1, x = 2^{-1} = \frac{1}{2}$

$x = 2^y$	y
1	0
2	1
4	2
$\frac{1}{2}$	−1

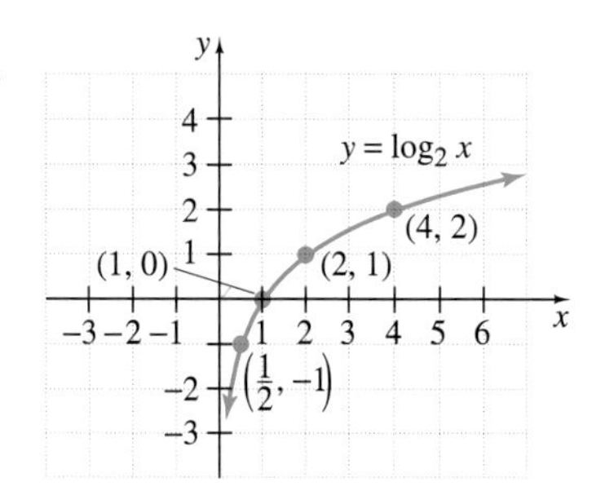

Practice Problem 11

Simplify.

a. $\log_6 6^3$ b. $\log_{11} 11^{-4}$
c. $7^{\log_7 5}$ d. $3^{\log_3 10}$

LOGARITHMIC FUNCTION

If x is a positive real number, b is a constant positive real number, and b is not 1, then a **logarithmic function** is a function that can be defined by

$$f(x) = \log_b x$$

The domain of f is the set of positive real numbers, and the range of f is the set of real numbers.

✓ CONCEPT CHECK

Let $f(x) = \log_3 x$ and $g(x) = 3^x$. These two functions are inverses of each other. Since $(2, 9)$ is an ordered pair solution of $g(x)$, what ordered pair do we know to be a solution of $f(x)$? Explain why.

Answers

11. a. 3, **b.** −4, **c.** 5, **d.** 10

✓ Concept Check: $(9, 2)$; answers may vary

Practice Problem 12

Graph the logarithmic function $y = \log_4 x$.

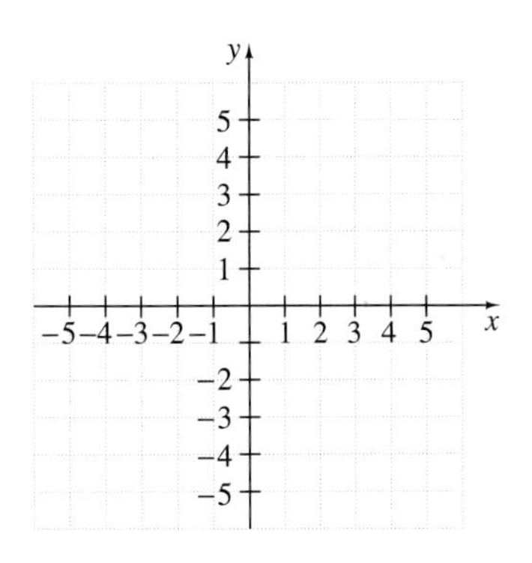

Practice Problem 13

Graph the logarithmic function $f(x) = \log_{1/2} x$.

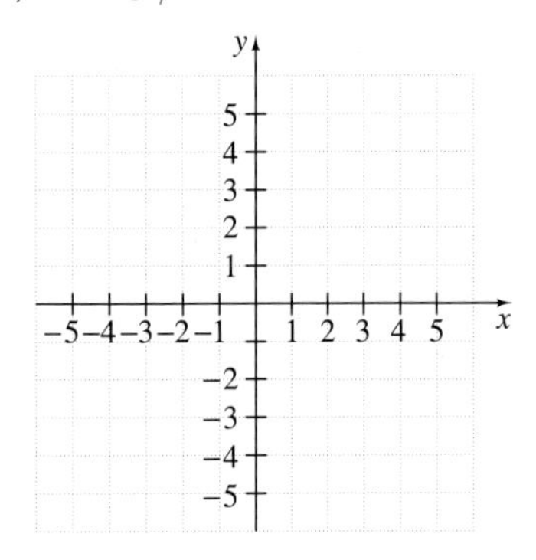

Answers

12.

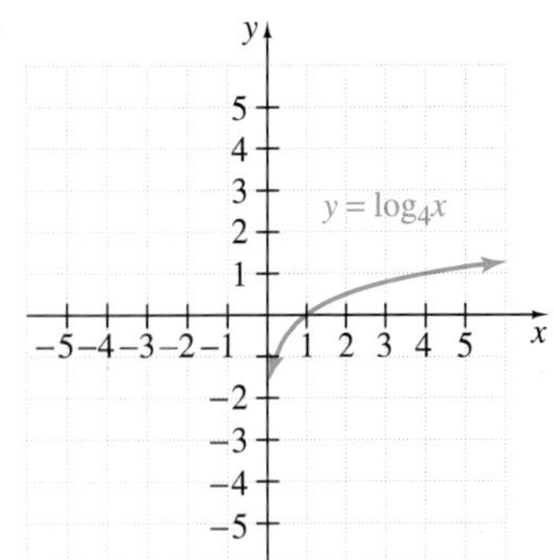

13.

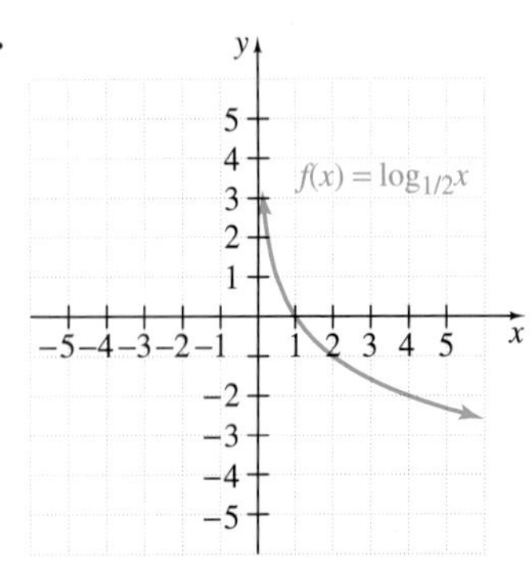

Example 13 Graph the logarithmic function $f(x) = \log_{1/3} x$.

Solution: We can replace $f(x)$ with y, and write the result with exponential notation.

$$f(x) = \log_{1/3} x$$

$$y = \log_{1/3} x \quad \text{Replace } f(x) \text{ with } y.$$

$$\left(\frac{1}{3}\right)^y = x \quad \text{Write in exponential form.}$$

Now we can find ordered pair solutions that satisfy $\left(\frac{1}{3}\right)^y = x$, plot these points, and connect them with a smooth curve.

If $y = 0$, $x = \left(\frac{1}{3}\right)^0 = 1$

If $y = 1$, $x = \left(\frac{1}{3}\right)^1 = \frac{1}{3}$

If $y = -1$, $x = \left(\frac{1}{3}\right)^{-1} = 3$

If $y = -2$, $x = \left(\frac{1}{3}\right)^{-2} = 9$

$x = \left(\frac{1}{3}\right)^y$	y
1	0
$\frac{1}{3}$	1
3	−1
9	−2

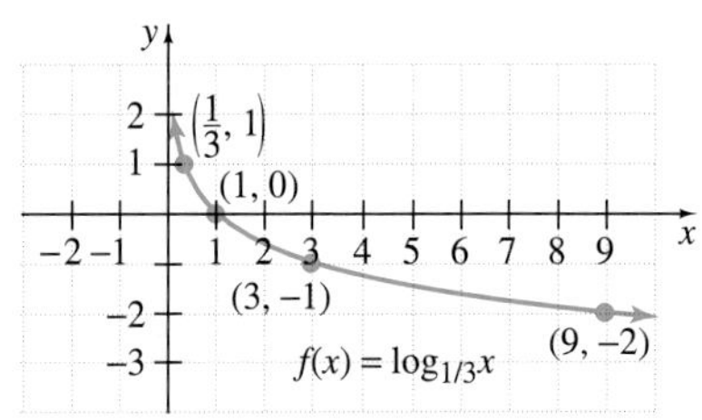

The domain of this function is $(0, \infty)$, and the range is the set of all real numbers.

The following figures summarize characteristics of logarithmic functions.

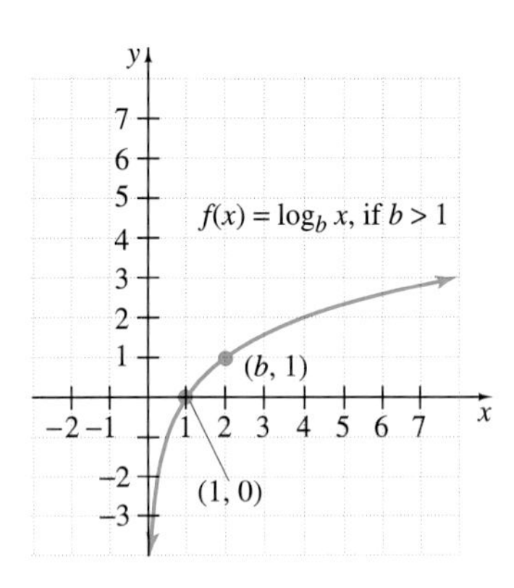

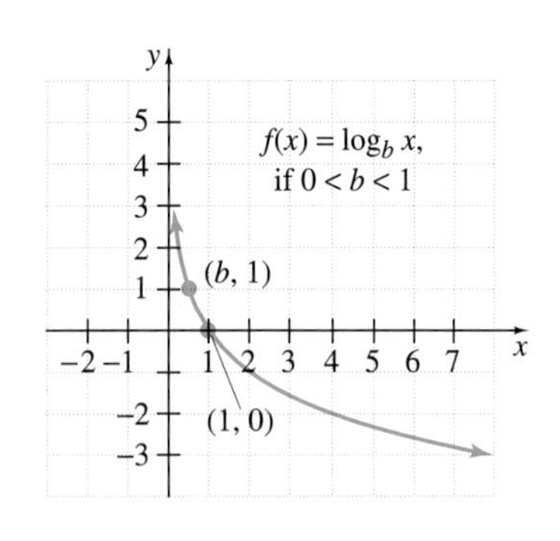

Name ______________________ Section ______ Date ________

Exercise Set 10.4

A *Write each as an exponential equation. See Examples 1 through 3.*

1. $\log_6 36 = 2$

2. $\log_2 32 = 5$

3. $\log_3 \frac{1}{27} = -3$

4. $\log_5 \frac{1}{25} = -2$

5. $\log_{10} 1000 = 3$

6. $\log_{10} 10 = 1$

7. $\log_e x = 4$

8. $\log_e \frac{1}{e} = -1$

9. $\log_e \frac{1}{e^2} = -2$

10. $\log_e y = 7$

11. $\log_7 \sqrt{7} = \frac{1}{2}$

12. $\log_{11} \sqrt[4]{11} = \frac{1}{4}$

Write each as a logarithmic equation. See Examples 4 through 6.

13. $2^4 = 16$

14. $5^3 = 125$

15. $10^2 = 100$

16. $10^4 = 1000$

17. $e^3 = x$

18. $e^5 = y$

19. $10^{-1} = \frac{1}{10}$

20. $10^{-2} = \frac{1}{100}$

21. $4^{-2} = \frac{1}{16}$

22. $3^{-4} = \frac{1}{81}$

23. $5^{1/2} = \sqrt{5}$

24. $4^{1/3} = \sqrt[3]{4}$

Find the value of each logarithmic expression. See Example 7.

25. $\log_2 8$

26. $\log_3 9$

27. $\log_3 \frac{1}{9}$

28. $\log_2 \frac{1}{32}$

29. $\log_{25} 5$

30. $\log_8 \frac{1}{2}$

Answers

1. ______
2. ______
3. ______
4. ______
5. ______
6. ______
7. ______
8. ______
9. ______
10. ______
11. ______
12. ______
13. ______
14. ______
15. ______
16. ______
17. ______
18. ______
19. ______
20. ______
21. ______
22. ______
23. ______
24. ______
25. ______
26. ______
27. ______
28. ______
29. ______
30. ______

31.
32.
33.
34.
35.
36.
37.
38.
39.
40.
41.
42.
43.
44.
45.
46.
47.
48.
49.
50.
51.
52.
53.
54.
55.
56.
57.
58.
59.
60.
61.

Name ______________________

31. $\log_{1/2} 2$

32. $\log_{2/3} \frac{4}{9}$

33. $\log_6 1$

34. $\log_9 9$

35. $\log_2 2^4$

36. $\log_6 6^{-2}$

37. $\log_{10} 100$

38. $\log_{10} \frac{1}{10}$

39. $3^{\log_3 5}$

40. $5^{\log_5 7}$

41. $\log_3 81$

42. $\log_2 16$

43. $\log_4 \frac{1}{64}$

44. $\log_3 \frac{1}{9}$

45. Explain why negative numbers are not included as logarithmic bases.

46. Explain why 1 is not included as a logarithmic base.

B *Solve. See Examples 8 through 10.*

47. $\log_3 9 = x$

48. $\log_2 8 = x$

49. $\log_3 x = 4$

50. $\log_2 x = 3$

51. $\log_x 49 = 2$

52. $\log_x 8 = 3$

53. $\log_2 \frac{1}{8} = x$

54. $\log_3 \frac{1}{81} = x$

55. $\log_3 \frac{1}{27} = x$

56. $\log_5 \frac{1}{125} = x$

57. $\log_8 x = \frac{1}{3}$

58. $\log_9 x = \frac{1}{2}$

59. $\log_4 16 = x$

60. $\log_2 16 = x$

61. $\log_{3/4} x = 3$

Name ______________________________

62. ______
63. ______
64. ______
65. ______
66. ______
67. ______
68. ______
69. ______
70. ______

62. $\log_{2/3} x = 2$

63. $\log_x 100 = 2$

64. $\log_x 27 = 3$

Simplify. See Example 11.

65. $\log_5 5^3$

66. $\log_6 6^2$

67. $2^{\log_2 3}$

68. $7^{\log_7 4}$

69. $\log_9 9$

70. $\log_8 (8)^{-1}$

C *Graph each logarithmic function. See Examples 12 and 13.*

71. $y = \log_3 x$

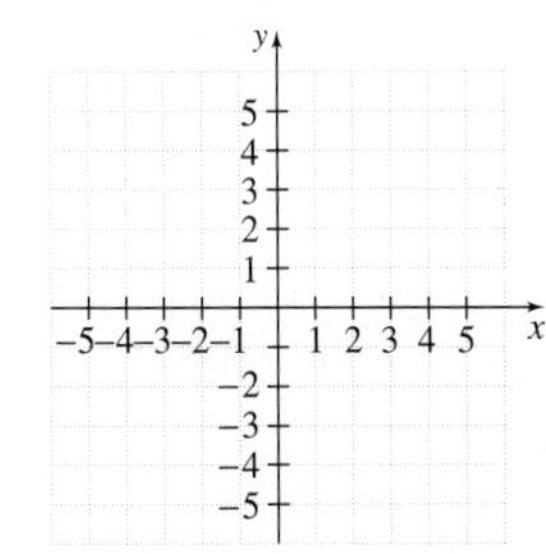

72. $y = \log_2 x$

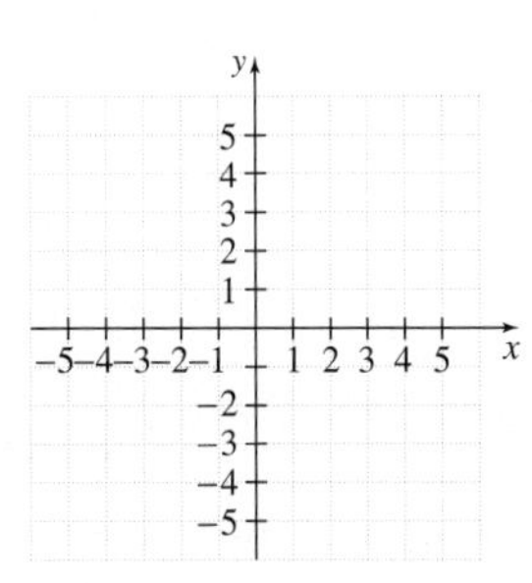

73. $f(x) = \log_{1/4} x$

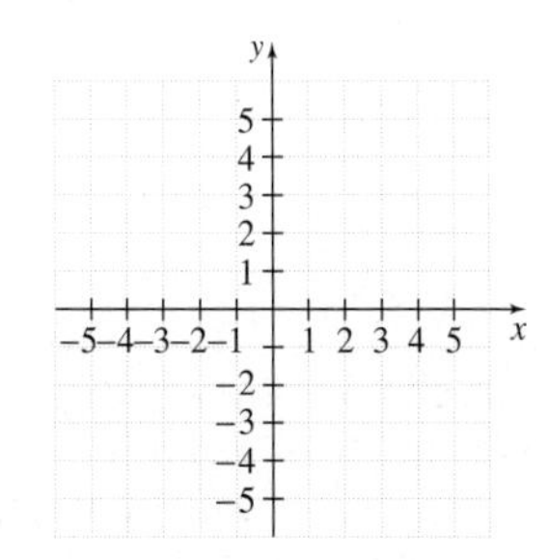

74. $f(x) = \log_{1/2} x$

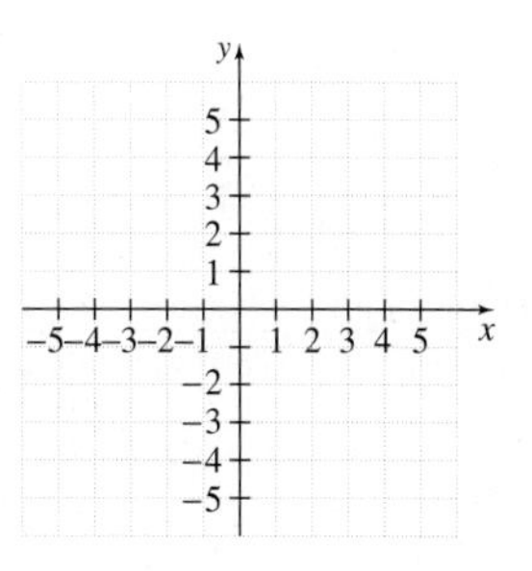

75. $f(x) = \log_5 x$

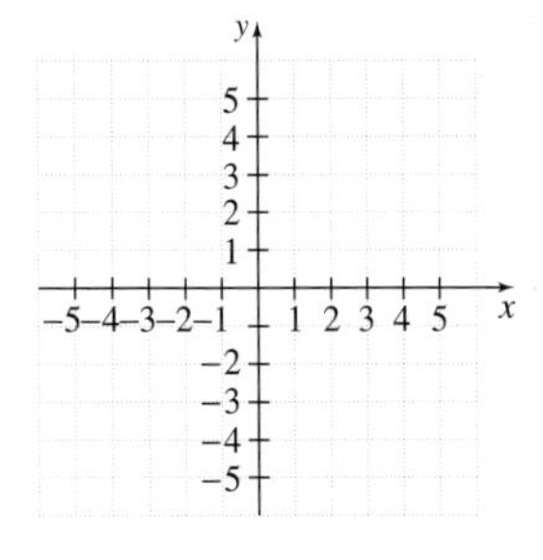

76. $f(x) = \log_6 x$

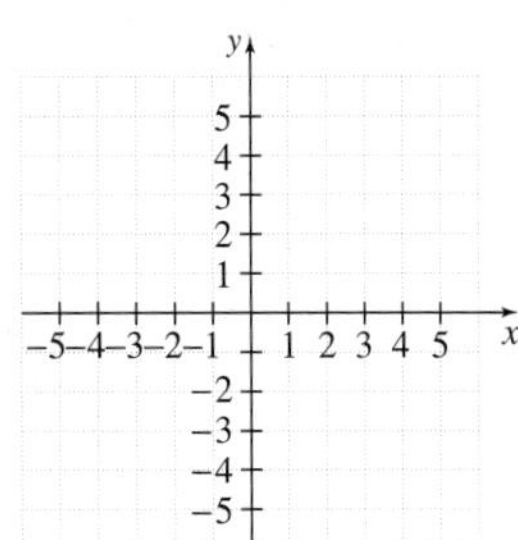

77. $f(x) = \log_{1/6} x$

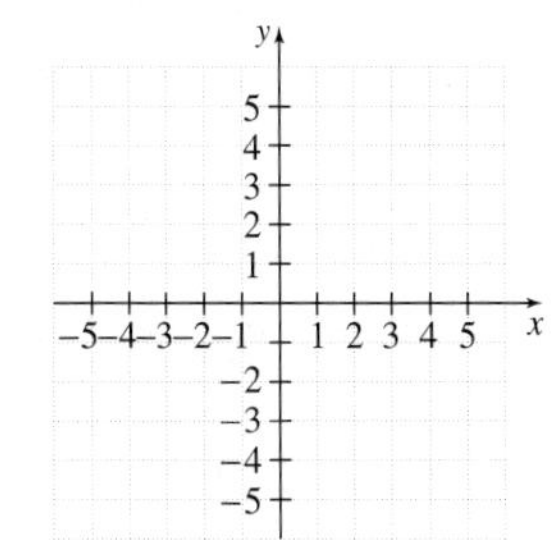

78. $f(x) = \log_{1/5} x$

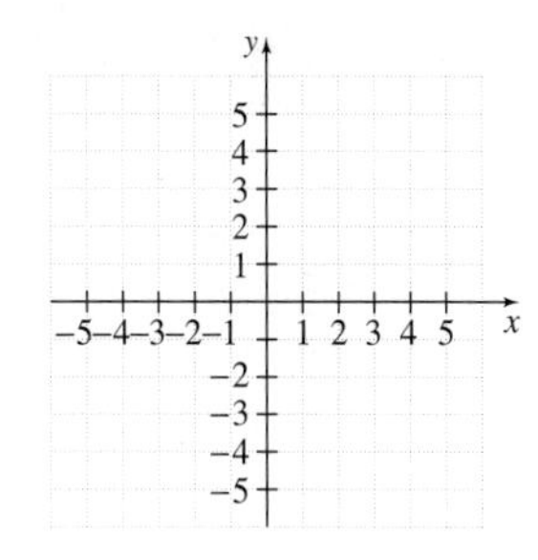

Review and Preview

Simplify each rational expression. See Section 6.1.

79. $\dfrac{x + 3}{3 + x}$

80. $\dfrac{x - 5}{5 - x}$

81. $\dfrac{x^2 - 8x + 16}{2x - 8}$

82. $\dfrac{x^2 - 3x - 10}{2 + x}$

79. ______
80. ______
81. ______
82. ______

Name ____________________

Combining Concepts

Graph each function and its inverse on the same set of axes.

83. $y = 4^x$; $y = \log_4 x$

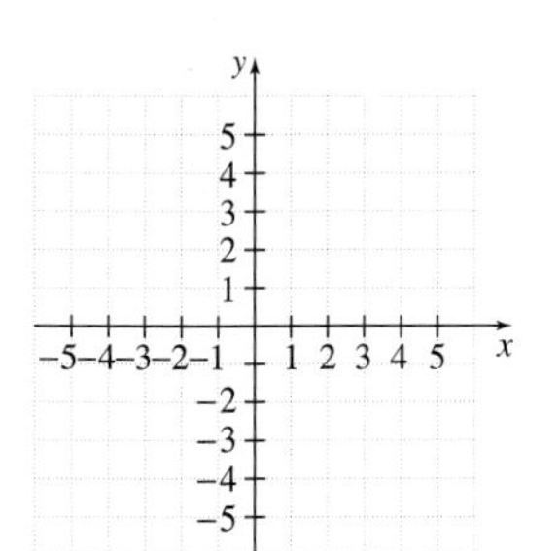

84. $y = 3^x$, $y = \log_3 x$

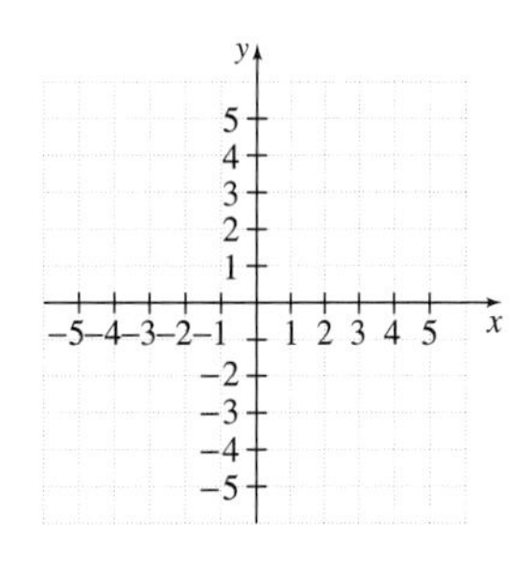

85. $y = \left(\frac{1}{3}\right)^x$; $y = \log_{1/3} x$

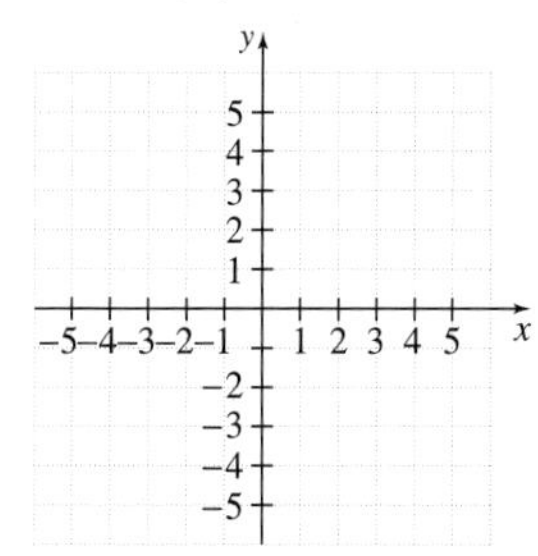

86. $y = \left(\frac{1}{2}\right)^x$; $y = \log_{1/2} x$

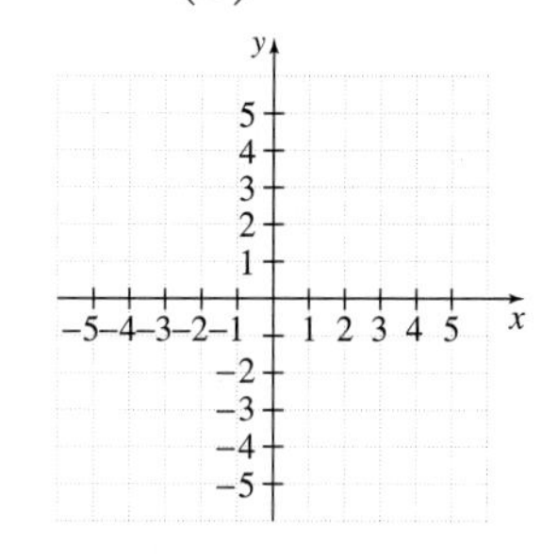

87. ____________________

88. ____________________

89. ____________________

87. The formula $\log_{10}(1 - k) = \dfrac{-0.3}{H}$ models the relationship between the half-life H of a radioactive material and its rate of decay k. Find the rate of decay of the iodine isotope I-131 if its half-life is 8 days. Round to 4 decimal places.

88. Explain why the graph of the function $y = \log_b x$ contains the point $(1, 0)$ no matter what b is.

89. $\log_3 10$ is between which two integers? Explain your answer.

10.5 PROPERTIES OF LOGARITHMS

Objectives

A Use the product property of logarithms.

B Use the quotient property of logarithms.

C Use the power property of logarithms

D Use the properties of logarithms together.

SSM CD-ROM Video 10.5

In the previous section we explored some basic properties of logarithms. We now introduce and explore additional properties. Because a logarithm is an exponent, logarithmic properties are just restatements of exponential properties.

A USING THE PRODUCT PROPERTY

The first of these properties is called the **product property of logarithms** because it deals with the logarithm of a product.

PRODUCT PROPERTY OF LOGARITHMS

If x, y, and b are positive real numbers and $b \neq 1$, then

$$\log_b xy = \log_b x + \log_b y$$

To prove this, we let $\log_b x = M$ and $\log_b y = N$. Now we write each logarithm with exponential notation.

$$\log_b x = M \quad \text{is equivalent to} \quad b^M = x$$
$$\log_b y = N \quad \text{is equivalent to} \quad b^N = y$$

When we multiply the left sides and the right sides of the exponential equations, we have that

$$xy = (b^M)(b^N) = b^{M+N}$$

If we write the equation $xy = b^{M+N}$ in equivalent logarithmic form, we have

$$\log_b xy = M + N$$

But since $M = \log_b x$ and $N = \log_b y$, we can write

$$\log_b xy = \log_b x + \log_b y \quad \text{Let } M = \log_b x \text{ and } N = \log_b y.$$

In other words, the logarithm of a product is the sum of the logarithms of the factors. This property is sometimes used to simplify logarithmic expressions.

Example 1 Write as a single logarithm: $\log_{11} 10 + \log_{11} 3$

Solution:
$$\log_{11} 10 + \log_{11} 3 = \log_{11}(10 \cdot 3) \quad \text{Use the product property.}$$
$$= \log_{11} 30$$

Practice Problem 1

Write as a single logarithm: $\log_2 7 + \log_2 5$

Example 2 Write as a single logarithm: $\log_2(x + 2) + \log_2 x$

Solution:

$$\log_2(x + 2) + \log_2 x = \log_2[(x + 2) \cdot x] = \log_2(x^2 + 2x)$$

Practice Problem 2

Write as a single logarithm: $\log_3 x + \log_3(x - 9)$

B USING THE QUOTIENT PROPERTY

The second property is the **quotient property of logarithms**.

Answers

1. $\log_2 35$, **2.** $\log_3(x^2 - 9x)$

✓ CONCEPT CHECK

Which of the following is the correct way to rewrite $\log_5 \frac{7}{2}$?

a. $\log_5 7 - \log_5 2$
b. $\log_5 (7 - 2)$
c. $\frac{\log_5 7}{\log_5 2}$
d. $\log_5 14$

Practice Problem 3

Write as a single logarithm: $\log_7 40 - \log_7 8$

Practice Problem 4

Write as a single logarithm: $\log_3 (x^3 + 4) - \log_3 (x^2 + 2)$

Practice Problems 5–6

Use the power property to rewrite each expression.

5. $\log_3 x^5$ 6. $\log_7 \sqrt[3]{4}$

Practice Problems 7–8

Write as a single logarithm.

7. $3 \log_4 2 + 2 \log_4 5$

8. $5 \log_2 (2x - 1) - \log_2 x$

Answers

3. $\log_7 5$, **4.** $\log_3 \frac{x^3 + 4}{x^2 + 2}$, **5.** $5 \log_3 x$, **6.** $\frac{1}{3} \log_7 4$, **7.** $\log_4 200$, 8, **8.** $\log_2 \frac{(2x - 1)^5}{x}$

✓ **Concept Check:** a

QUOTIENT PROPERTY OF LOGARITHMS

If x, y, and b are positive real numbers and $b \neq 1$, then

$$\log_b \frac{x}{y} = \log_b x - \log_b y$$

The proof of the quotient property of logarithms is similar to the proof of the product property. Notice that the quotient property says that the logarithm of a quotient is the difference of the logarithms of the dividend and divisor.

TRY THE CONCEPT CHECK IN THE MARGIN.

Example 3 Write as a single logarithm: $\log_{10} 27 - \log_{10} 3$

Solution:

$$\log_{10} 27 - \log_{10} 3 = \log_{10} \frac{27}{3} \quad \text{Use the quotient property.}$$
$$= \log_{10} 9$$

Example 4 Write as a single logarithm: $\log_3 (x^2 + 5) - \log_3 (x^2 + 1)$

Solution:

$$\log_3 (x^2 + 5) - \log_3 (x^2 + 1) = \log_3 \frac{x^2 + 5}{x^2 + 1} \quad \text{Use the quotient property.}$$

C USING THE POWER PROPERTY

The third and final property we introduce is the **power property of logarithms**.

POWER PROPERTY OF LOGARITHMS

If x and b are positive real numbers, $b \neq 1$, and r is a real number, then

$$\log_b x^r = r \log_b x$$

Examples Use the power property to rewrite each expression.

5. $\log_5 x^3 = 3 \log_5 x$

6. $\log_4 \sqrt{2} = \log_4 2^{1/2} = \frac{1}{2} \log_4 2$

D USING MORE THAN ONE PROPERTY

Many times we must use more than one property of logarithms to simplify a logarithmic expression.

Examples Write as a single logarithm.

7.
$$2 \log_5 3 + 3 \log_5 2 = \log_5 3^2 + \log_5 2^3 \quad \text{Use the power property.}$$
$$= \log_5 9 + \log_5 8$$
$$= \log_5 (9 \cdot 8) \quad \text{Use the product property.}$$
$$= \log_5 72$$

8. $3 \log_9 x - \log_9 (x + 1) = \log_9 x^3 - \log_9 (x + 1)$ Use the power property.

$= \log_9 \frac{x^3}{x + 1}$ Use the quotient property.

Examples

Write each expression as sums or differences of logarithms.

9. $\log_3 \frac{5 \cdot 7}{4} = \log_3 (5 \cdot 7) - \log_3 4$ Use the quotient property.

$= \log_3 5 + \log_3 7 - \log_3 4$ Use the product property.

10. $\log_2 \frac{x^5}{y^2} = \log_2 (x^5) - \log_2 (y^2)$ Use the quotient property.

$= 5 \log_2 x - 2 \log_2 y$ Use the power property.

HELPFUL HINT

Notice that we are not able to simplify further a logarithmic expression such as $\log_5 (2x - 1)$. None of the basic properties gives a way to write the logarithm of a difference in some equivalent form.

TRY THE CONCEPT CHECK IN THE MARGIN.

Examples

If $\log_b 2 = 0.43$ and $\log_b 3 = 0.68$, use the properties of logarithms to evaluate each expression.

11. $\log_b 6 = \log_b (2 \cdot 3)$ Write 6 as $2 \cdot 3$.

$= \log_b 2 + \log_b 3$ Use the product property.

$= 0.43 + 0.68$ Substitute given values.

$= 1.11$ Simplify.

12. $\log_b 9 = \log_b 3^2$ Write 9 as 3^2.

$= 2 \log_b 3$ Use the power property.

$= 2(0.68)$ Substitute the given value.

$= 1.36$ Simplify.

13. $\log_b \sqrt{2} = \log_b 2^{1/2}$ Write $\sqrt{2}$ as $2^{1/2}$.

$= \frac{1}{2} \log_b 2$ Use the power property.

$= \frac{1}{2}(0.43)$ Substitute the given value.

$= 0.215$ Simplify.

Practice Problems 9–10

Write each expression as sums or differences of logarithms.

9. $\log_7 \frac{6 \cdot 2}{5}$

10. $\log_3 \frac{x^4}{y^3}$

✓ Concept Check

What is wrong with the following?

$\log_{10} (x^2 + 5) = \log_{10} x^2 + \log_{10} 5$
$= 2 \log_{10} x + \log_{10} 5$

Use a numerical example to demonstrate that the result is incorrect.

Practice Problems 11–13

If $\log_b 4 = 0.86$ and $\log_b 7 = 1.21$, use the properties of logarithms to evaluate each expression.

11. $\log_b 28$

12. $\log_b 49$

13. $\log_b \sqrt[3]{4}$

Answers

9. $\log_7 6 + \log_7 2 - \log_7 5$, **10.** $4 \log_3 x - 3 \log_3 y$, **11.** 2.07, **12.** 2.42, **13.** $0.28\overline{6}$

✓ Concept Check: The properties do not give any way to simplify the logarithm of a sum; answers may vary

Focus On the Real World

SOUND INTENSITY

The decibel (dB) measures sound intensity, or the relative loudness or strength of a sound. One decibel is the smallest difference in sound levels that is detectable by humans. The decibel is a logarithmic unit. This means that for approximately every 3-decibel increase in sound intensity, the relative loudness of the sound is doubled. For example, a 35 dB sound is twice as loud as a 32 dB sound.

In the modern world, noise pollution has increasingly become a concern. Sustained exposure to high sound intensities can lead to hearing loss. Regular exposure to 90 dB sounds can eventually lead to loss of hearing. Sounds of 130 dB and more can cause permanent loss of hearing instantaneously.

The relative loudness of a sound D in decibels is given by the equation

$$D = 10 \log_{10} \frac{I}{10^{-16}}$$

where I is the intensity of a sound given in watts per square centimeter. Some sound intensities of common noises are listed in the table in order of increasing sound intensity.

SOME SOUND INTENSITIES OF COMMON NOISES

Noise	Intensity (watts/cm^2)
Whispering	10^{-15}
Rustling leaves	$10^{-14.2}$
Normal conversation	10^{-13}
Background noise in a quiet residence	$10^{-12.2}$
Typewriter	10^{-11}
Air conditioning	10^{-10}
Freight train at 50 feet	$10^{-8.5}$
Vacuum cleaner	10^{-8}
Nearby thunder	10^{-7}
Air hammer	$10^{-6.5}$
Jet plane at takeoff	10^{-6}
Threshold of pain	10^{-4}

GROUP ACTIVITY

1. Work together to create a table of the relative loudness (in decibels) of the sounds listed in the table.
2. Research the loudness of other common noises. Add these sounds and their decibel levels to your table. Be sure to list the sounds in order of increasing sound intensity.

Name ______________________ Section ______ Date ________

Exercise Set 10.5

A *Write each sum as a single logarithm. Assume that variables represent positive numbers. See Examples 1 and 2.*

1. $\log_5 2 + \log_5 7$

2. $\log_3 8 + \log_3 4$

3. $\log_4 9 + \log_4 x$

4. $\log_2 x + \log_2 y$

5. $\log_{10} 5 + \log_{10} 2 + \log_{10}(x^2 + 2)$

6. $\log_6 3 + \log_6(x + 4) + \log_6 5$

B *Write each difference as a single logarithm. Assume that variables represent positive numbers. See Examples 3 and 4.*

7. $\log_5 12 - \log_5 4$

8. $\log_7 20 - \log_7 4$

9. $\log_2 x - \log_2 y$

10. $\log_3 12 - \log_3 z$

11. $\log_3 8 - \log_3 2$

12. $\log_5 12 - \log_5 3$

C *Use the power property to rewrite each expression. See Examples 5 and 6.*

13. $\log_3 x^2$

14. $\log_2 x^5$

15. $\log_4 5^{-1}$

16. $\log_6 7^{-2}$

17. $\log_5 \sqrt{y}$

18. $\log_5 \sqrt[3]{x}$

D *Write each as a single logarithm. Assume that variables represent positive numbers. See Examples 7 and 8.*

19. $2 \log_2 5$

20. $3 \log_5 2$

21. $3 \log_5 x + 6 \log_5 z$

22. $2 \log_7 y + 6 \log_7 z$

23. $\log_{10} x - \log_{10}(x + 1) + \log_{10}(x^2 - 2)$

24. $\log_9(4x) - \log_9(x - 3) + \log_9(x^3 + 1)$

25. $\log_4 2 + \log_4 10 - \log_4 5$

26. $\log_6 18 + \log_6 2 - \log_6 9$

27. $\log_7 6 + \log_7 3 - \log_7 4$

28. $\log_8 5 + \log_8 15 - \log_8 20$

29. $3 \log_4 2 + \log_4 6$

30. $2 \log_3 5 + \log_3 2$

31. $3 \log_2 x + \frac{1}{2} \log_2 x - 2 \log_2(x + 1)$

32. $2 \log_5 x + \frac{1}{3} \log_5 x - 3 \log_5(x + 5)$

33. $2 \log_8 x - \frac{2}{3} \log_8 x + 4 \log_8 x$

34. $5 \log_6 x - \frac{3}{4} \log_6 x + 3 \log_6 x$

Write each expression as a sum or difference of logarithms. Assume that variables represent positive numbers. See Examples 9 and 10.

35. $\log_3 \frac{4y}{5}$

36. $\log_4 \frac{2}{9z}$

37. $\log_2 \frac{x^3}{y}$

38. $\log_5 \frac{x}{y^4}$

Answers

1. ______
2. ______
3. ______
4. ______
5. ______
6. ______
7. ______
8. ______
9. ______
10. ______
11. ______
12. ______
13. ______
14. ______
15. ______
16. ______
17. ______
18. ______
19. ______
20. ______
21. ______
22. ______
23. ______
24. ______
25. ______
26. ______
27. ______
28. ______
29. ______
30. ______
31. ______
32. ______
33. ______
34. ______
35. ______
36. ______
37. ______
38. ______

39. ______
40. ______
41. ______
42. ______
43. ______
44. ______
45. ______
46. ______
47. ______
48. ______
49. ______
50. ______
51. ______
52. ______
53. ______
54. ______
55. ______
56. ______
57. ______
58. ______
59. see graph
60. ______
61. ______
62. ______
63. ______
64. ______
65. ______
66. ______
67. ______
68. ______
69. ______

Name ______

39. $\log_b \sqrt{7x}$ **40.** $\log_b \sqrt{\frac{3}{y}}$ **41.** $\log_7 \frac{5x}{4}$ **42.** $\log_9 \frac{7}{y}$

43. $\log_5 x^3 (x + 1)$ **44.** $\log_2 y^3 z$ **45.** $\log_6 \frac{x^2}{x + 3}$ **46.** $\log_3 \frac{(x + 5)^2}{x}$

If $\log_b 3 = 0.5$ *and* $\log_b 5 = 0.7$*, evaluate each expression. See Examples 11 through 13.*

47. $\log_b \frac{5}{3}$ **48.** $\log_b 25$ **49.** $\log_b 15$

50. $\log_b \frac{3}{5}$ **51.** $\log_b \sqrt{5}$ **52.** $\log_b \sqrt[4]{3}$

If $\log_b 2 = 0.43$ *and* $\log_b 3 = 0.68$*, evaluate each expression. See Examples 11 through 13.*

53. $\log_b 8$ **54.** $\log_b 81$ **55.** $\log_b \frac{3}{9}$

56. $\log_b \frac{4}{32}$ **57.** $\log_b \sqrt{\frac{2}{3}}$ **58.** $\log_b \sqrt{\frac{3}{2}}$

Review and Preview

59. Graph the functions $y = 10^x$ and $y = \log_{10} x$ on the same set of axes. See Section 10.4.

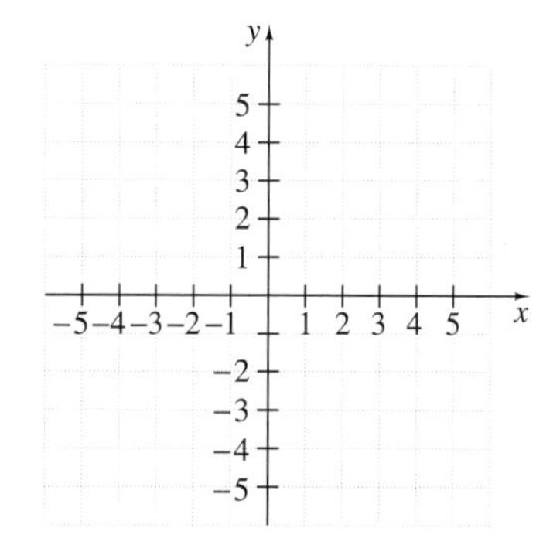

Evaluate each expression. See Section 10.4.

60. $\log_{10} 100$ **61.** $\log_{10} \frac{1}{10}$ **62.** $\log_7 7^2$ **63.** $\log_7 \sqrt{7}$

Combining Concepts

Determine whether each statement is true or false.

64. $\log_2 x^3 = 3 \log_2 x$

65. $\log_3 (x + y) = \log_3 x + \log_3 y$

66. $\frac{\log_7 10}{\log_7 5} = \log_7 2$

67. $\log_7 \frac{14}{8} = \log_7 14 - \log_7 8$

68. $\frac{\log_7 x}{\log_7 y} = (\log_7 x) - (\log_7 y)$

69. $(\log_3 6) \cdot (\log_3 4) = \log_3 24$

Name ______________________ Section ________ Date ________

ANSWERS

1. ________

2. ________

3. ________

4. ________

5. ________

6. ________

7. ________

8. ________

9. ________

10. ________

11. ________

12. ________

13. ________

14. ________

INTEGRATED REVIEW—FUNCTIONS AND PROPERTIES OF LOGARITHMS

If $f(x) = x - 6$ *and* $g(x) = x^2 + 1$, *find each value.*

1. $(f + g)(x)$ **2.** $(f - g)(x)$ **3.** $(f \cdot g)(x)$ **4.** $\left(\frac{f}{g}\right)(x)$

If $f(x) = \sqrt{x}$ *and* $g(x) = 3x - 1$, *find each value.*

5. $(f \circ g)(x)$ **6.** $(g \circ f)(x)$

Determine whether each is a one-to-one function. If it is, find its inverse.

7. $f = \{(-2, 6), (4, 8), (2, -6), (3, 3)\}$ **8.** $g = \{(4, 2), (-1, 3), (5, 3), (7, 1)\}$

Determine whether the graph of each function is one-to-one.

9.

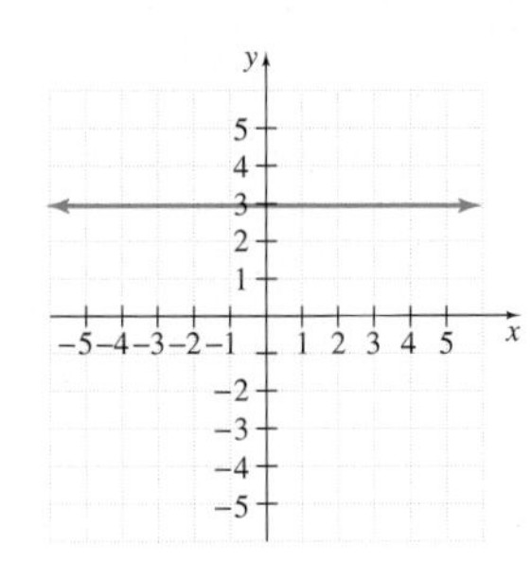

10.

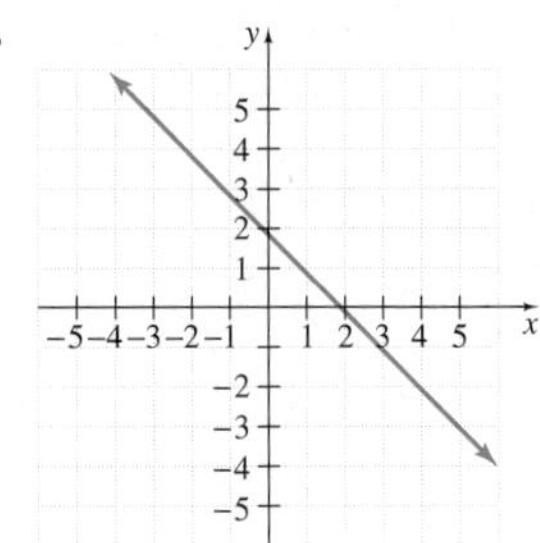

11.

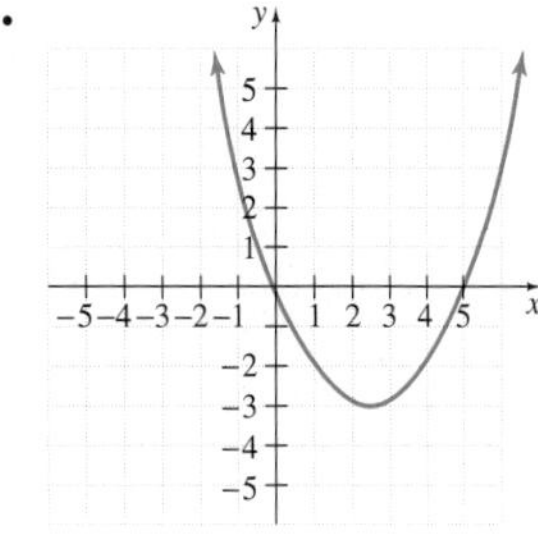

Each function listed is one-to-one. Find the inverse of each function.

12. $f(x) = 3x$ **13.** $f(x) = x + 4$ **14.** $f(x) = 5x - 1$

Graph each function.

15. $y = \left(\frac{1}{2}\right)^x$

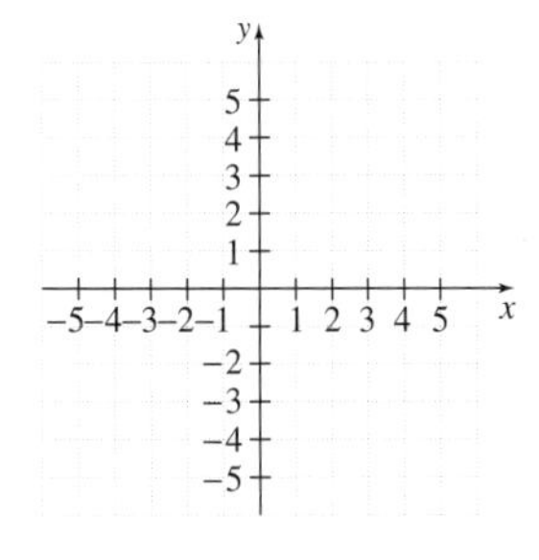

16. $y = 2^x + 1$

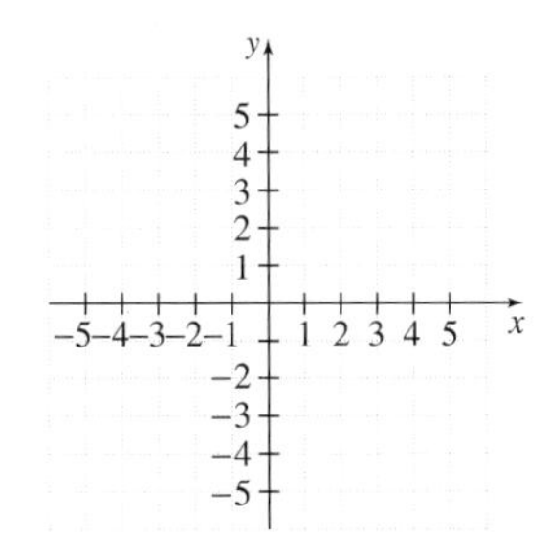

17. $y = \log_3 x$

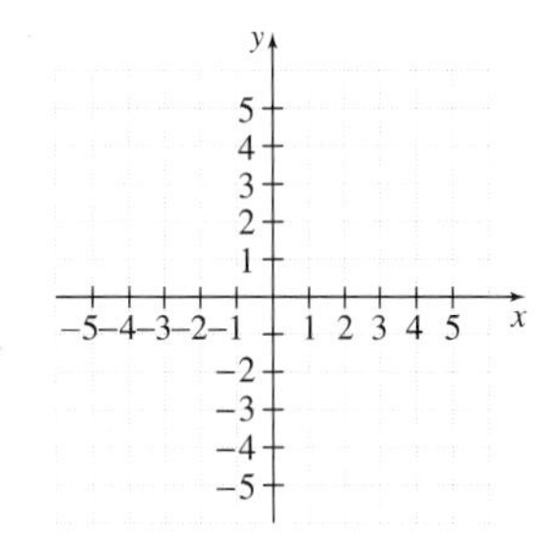

18. $y = \log_{1/3} x$

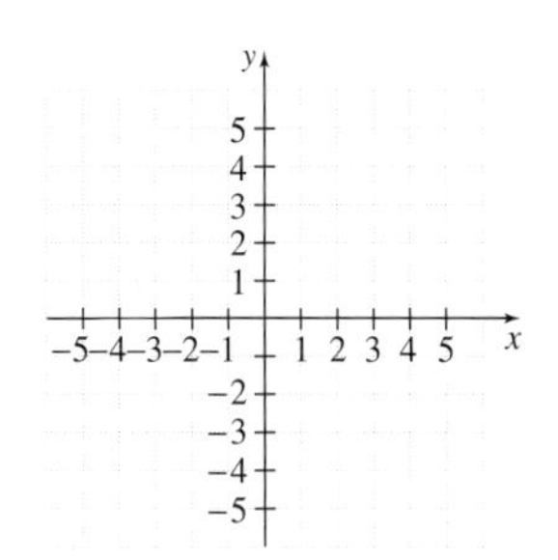

19. ____________

20. ____________

21. ____________

22. ____________

23. ____________

24. ____________

25. ____________

26. ____________

27. ____________

28. ____________

Name ____

Solve.

19. $2^x = 8$

20. $4^{x-1} = 8^{x+2}$

21. $\log_4 16 = x$

22. $\log_{49} 7 = x$

23. $\log_2 x = 5$

24. $\log_x 64 = 3$

25. $\log_x \frac{1}{125} = -3$

26. $\log_3 x = -2$

Write each as a single logarithm.

27. $3 \log_5 x - 5 \log_5 y$

28. $\log_2 x + \log_2 (x - 3) - \log_2 (x^2 + 4)$

10.6 COMMON LOGARITHMS, NATURAL LOGARITHMS, AND CHANGE OF BASE

Objectives

A Identify common logarithms and approximate them with a calculator.
B Evaluate common logarithms of powers of 10.
C Identify natural logarithms and approximate them with a calculator.
D Evaluate natural logarithms of powers of e.
E Use the change of base formula.

SSM CD-ROM Video 10.6

In this section we look closely at two particular logarithmic bases. These two logarithmic bases are used so frequently that logarithms to their bases are given special names. **Common logarithms** are logarithms to base 10. **Natural logarithms** are logarithms to base e, which we introduce in this section. The work in this section is based on the use of the calculator which has both the common "log" [LOG] and the natural "log" [LN] keys.

A APPROXIMATING COMMON LOGARITHMS

Logarithms to base 10—common logarithms—are used frequently because our number system is a base 10 decimal system. The notation $\log x$ means the same as $\log_{10} x$.

COMMON LOGARITHMS

$\log x$ means $\log_{10} x$

Example 1 Use a calculator to approximate log 7 to four decimal places.

Solution: Press the following sequence of keys:

[7] [LOG] or [LOG] [7] [ENTER]

To four decimal places,

$\log 7 \approx 0.8451$

Practice Problem 1

Use a calculator to aproximate log 21 to four decimal places.

B EVALUATING COMMON LOGARITHMS OF POWERS OF 10

To evaluate the common log of a power of 10, a calculator is not needed. According to the property of logarithms,

$$\log_b b^x = x$$

It follows that if b is replaced with 10, we have

$$\log 10^x = x$$

HELPFUL HINT

The base of this logarithm is understood to be 10.

Examples Find the exact value of each logarithm.

2. $\log 10 = \log 10^1 = 1$

3. $\log \frac{1}{10} = \log 10^{-1} = -1$

4. $\log 1000 = \log 10^3 = 3$

5. $\log \sqrt{10} = \log 10^{1/2} = \frac{1}{2}$

As we will soon see, equations containing common logs are useful models of many natural phenomena.

Practice Problems 2–5

Find the exact value of each logarithm.

2. $\log 100$
3. $\log \frac{1}{100}$
4. $\log 10{,}000$
5. $\log \sqrt[3]{10}$

Answers

1. 1.3222, **2.** 2, **3.** −2, **4.** 4, **5.** $\frac{1}{3}$

Practice Problem 6

Solve: $\log x = 2.9$. Give an exact solution, and then approximate the solution to four decimal places.

Example 6

Solve: $\log x = 1.2$. Give an exact solution, and then approximate the solution to four decimal places.

Solution: Remember that the base of a common log is understood to be 10.

HELPFUL HINT

The understood base is 10.

$$\log x = 1.2$$

$$10^{1.2} = x \quad \text{Write with exponential notation.}$$

The exact solution is $10^{1.2}$. To four decimal places, $x \approx 15.8489$.

C APPROXIMATING NATURAL LOGARITHMS

Natural logarithms are also frequently used, especially to describe natural events; hence the label "natural logarithm." Natural logarithms are logarithms to the base e, which is a constant approximately equal to 2.7183. The number e is an irrational number, as is π. The notation $\log_e x$ is usually abbreviated to $\ln x$. (The abbreviation ln is read "el en.")

NATURAL LOGARITHMS

$\ln x$ means $\log_e x$

Practice Problem 7

Use a calculator to approximate ln 11 to four decimal places.

Example 7

Use a calculator to approximate ln 8 to four decimal places.

Solution: Press the following sequence of keys:

[8] [ln] or [ln] [8] [ENTER]

To four decimal places,

$\ln 8 \approx 2.0794$

D EVALUATING NATURAL LOGARITHMS OF POWERS OF *e*

As a result of the property $\log_b b^x = x$, we know that $\log_e e^x = x$, or $\ln e^x = x$.

Practice Problems 8–9

Find the exact value of each natural logarithm.

8. $\ln e^5$

9. $\ln \sqrt{e}$

Examples

Find the exact value of each natural logarithm.

8. $\ln e^3 = 3$

9. $\ln \sqrt[5]{e} = \ln e^{1/5} = \frac{1}{5}$

Practice Problem 10

Solve: $\ln 7x = 10$. Give an exact solution, and then approximate the solution to four decimal places.

Example 10

Solve: $\ln 3x = 5$. Give an exact solution, and then approximate the solution to four decimal places.

Solution: Remember that the base of a natural logarithm is understood to be e.

HELPFUL HINT

The understood base is e.

$$\ln 3x = 5$$

$$e^5 = 3x \quad \text{Write with exponential notation.}$$

$$\frac{e^5}{3} = x \quad \text{Solve for } x.$$

The exact solution is $\frac{e^5}{3}$. To four decimal places, $x \approx 49.4711$.

Answers

6. $x = 10^{2.9}$; $x \approx 794.3282$, **7.** 2.3979, **8.** 5, **9.** $\frac{1}{2}$, **10.** $x = \frac{e^{10}}{7}$; $x \approx 3146.6380$

Recall from Section 10.3 the formula $A = P\left(1 + \frac{r}{n}\right)^{nt}$ for compound interest, where n represents the number of compoundings per year. When interest is compounded continuously, the formula $A = Pe^{rt}$ is used, where r is the annual interest rate and interest is compounded continuously for t years.

Example 11 Finding the Amount Owed on a Loan

Find the amount owed at the end of 5 years if $1600 is loaned at a rate of 9% compounded continuously.

Solution: We use the formula $A = Pe^{rt}$, where

$P = \$1600$ (the amount of the loan)

$r = 9\% = 0.09$ (the rate of interest)

$t = 5$ (the 5-year duration of the loan)

$$A = Pe^{rt}$$
$$= 1600e^{0.09(5)} \quad \text{Substitute known values.}$$
$$= 1600e^{0.45}$$

Now we can use a calculator to approximate the solution.

$A \approx 2509.30$

The total amount of money owed is approximately $2509.30.

Practice Problem 11

Find the amount owed at the end of 3 years if $1200 is loaned at a rate of 8% compounded continuously.

E Using the Change of Base Formula

Calculators are handy tools for approximating natural and common logarithms. Unfortunately, some calculators cannot be used to approximate logarithms to bases other than e or 10—at least not directly. In such cases, we use the change of base formula.

CHANGE OF BASE

If a, b, and c are positive real numbers and neither b nor c is 1, then

$$\log_b a = \frac{\log_c a}{\log_c b}$$

Example 12

Approximate $\log_5 3$ to four decimal places.

Solution: We use the change of base property to write $\log_5 3$ as a quotient of logarithms to base 10.

$$\log_5 3 = \frac{\log 3}{\log 5} \quad \text{Use the change of base property.}$$
$$\approx \frac{0.4771213}{0.69897} \quad \text{Approximate the logarithms by calculator.}$$
$$\approx 0.6826063 \quad \text{Simplify by calculator.}$$

To four decimal places, $\log_5 3 \approx 0.6826$.

TRY THE CONCEPT CHECK IN THE MARGIN.

Practice Problem 12

Approximate $\log_7 5$ to four decimal places.

✓ Concept Check

If a graphing calculator cannot directly evaluate logarithms to base 5, describe how you could use the graphing calculator to graph the function $f(x) = \log_5 x$.

Answers

11. $1525.50, **12.** 0.8271

✓ Concept Check: $f(x) = \frac{\log x}{\log 5}$

Focus On History

THE INVENTION OF LOGARITHMS

Logarithms were the invention of John Napier (1550–1617), a Scottish land owner and theologian. Napier was also fascinated by mathematics and made it his hobby. Over a period of 20 years in his spare time, he developed his theory of logarithms, which were explained in his Latin text *Mirifici Logarithmorum Canonis Descriptio* (A Description of an Admirable Table of Logarithms), published in 1614. He hoped that his discovery would help to simplify the many time-consuming calculations required in astronomy. In fact, Napier's logarithms revolutionized astronomy and many other advanced mathematical fields by replacing "the multiplications, divisions, square and cubical extractions of great numbers, which besides the tedious expense of time are for the most part subject to many slippery errors" with related numbers that can be easily added and subtracted instead. His discovery was a great time-saving device. Some historians suggest that the use of logarithms to simplify calculations enabled German astronomer Johannes Kepler to develop his three laws of planetary motion, which in turn helped English physicist Sir Isaac Newton develop his theory of gravitation. Two hundred years after Napier's discovery, the French mathematician Pierre de Laplace wrote that logarithms, "by shortening the labors, doubled the life of the astronomer."

Napier's original logarithm tables had several flaws: They did not actually use a particular logarithmic base per se and log 1 was not defined to be equal to 0. An English mathematician, Henry Briggs, read Napier's Latin text soon after it was published and was very impressed by his ideas. Briggs wrote to Napier, asking to meet in person to discuss his wonderful discovery and to offer several improvements. The two mathematicians met in the summer of 1615. Briggs suggested redefining logarithms to base 10 and defining $\log 1 = 0$. Napier had also thought of using base 10 but hadn't been well enough to start a new set of tables. He asked Briggs to undertake the construction of a new set of base 10 tables. And so it was that the first table of common logarithms was constructed by Briggs over the next two years. Napier died in 1617 before Briggs was able to complete his new tables.

CRITICAL THINKING

Locate a table of common logarithms and describe how to use it. Give several examples. Explain why a table of common logarithms would have been invaluable to many calculations before the invention of the hand-held calculator.

Name ______________________ Section ________ Date ________

Exercise Set 10.6

A C *Use a calculator to approximate each logarithm to four decimal places. See Examples 1 and 7.*

1. $\log 8$ **2.** $\log 6$ **3.** $\log 2.31$ **4.** $\log 4.86$

5. $\ln 2$ **6.** $\ln 3$ **7.** $\ln 0.0716$ **8.** $\ln 0.0032$

9. $\log 12.6$ **10.** $\log 25.9$ **11.** $\ln 5$ **12.** $\ln 7$

13. $\log 41.5$ **14.** $\ln 41.5$

15. Use a calculator to try to approximate $\log 0$. Describe what happens and explain why.

16. Use a calculator to try to approximate $\ln 0$. Describe what happens and explain why.

B D *Find the exact value of each logarithm. See Examples 2 through 5, 8, and 9.*

17. $\log 100$ **18.** $\log 10{,}000$ **19.** $\log \frac{1}{1000}$ **20.** $\log \frac{1}{10}$

21. $\ln e^2$ **22.** $\ln e^4$ **23.** $\ln \sqrt[4]{e}$ **24.** $\ln \sqrt[5]{e}$

25. $\log 10^3$ **26.** $\ln e^5$ **27.** $\ln e^2$ **28.** $\log 10^7$

29. $\log 0.0001$ **30.** $\log 0.001$ **31.** $\ln \sqrt{e}$ **32.** $\log \sqrt{10}$

ANSWERS

1. ____
2. ____
3. ____
4. ____
5. ____
6. ____
7. ____
8. ____
9. ____
10. ____
11. ____
12. ____
13. ____
14. ____
15. ____
16. ____
17. ____
18. ____
19. ____
20. ____
21. ____
22. ____
23. ____
24. ____
25. ____
26. ____
27. ____
28. ____
29. ____
30. ____
31. ____
32. ____

33. ______

34. ______

35. ______

36. ______

37. ______

38. ______

39. ______

40. ______

41. ______

42. ______

43. ______

44. ______

45. ______

46. ______

47. ______

48. ______

49. ______

50. ______

51. ______

52. ______

Name ______________________

Solve each equation. Give an exact solution and a four-decimal-place approximation. See Examples 6 and 10.

33. $\log x = 1.3$

34. $\log x = 2.1$

35. $\log 2x = 1.1$

36. $\log 3x = 1.3$

37. $\ln x = 1.4$

38. $\ln x = 2.1$

39. $\ln(3x - 4) = 2.3$

40. $\ln(2x + 5) = 3.4$

41. $\log x = 2.3$

42. $\log x = 3.1$

43. $\ln x = -2.3$

44. $\ln x = -3.7$

45. $\log(2x + 1) = -0.5$

46. $\log(3x - 2) = -0.8$

47. $\ln 4x = 0.18$

48. $\ln 3x = 0.76$

Use the formula $A = Pe^{rt}$ to solve. See Example 11.

49. How much money does Dana Jones have after 12 years if $1400 is invested at 8% interest compounded continuously?

50. Determine the size of an account in which $3500 earns 6% interest compounded continuously for 1 year.

51. How much money does Barbara Mack owe at the end of 4 years if 6% interest is compounded continuously on her $2000 debt?

52. Find the amount of money for which a $2500 certificate of deposit is redeemable if it has been paying 10% interest compounded continuously for 3 years.

Name ______________________________

E *Approximate each logarithm to four decimal places. See Example 12.*

53. $\log_2 3$

54. $\log_3 2$

55. $\log_{1/2} 5$

56. $\log_{1/3} 2$

57. $\log_4 9$

58. $\log_9 4$

59. $\log_3 \frac{1}{6}$

60. $\log_6 \frac{2}{3}$

61. $\log_8 6$

62. $\log_6 8$

Review and Preview

Solve for x. See Sections 2.1 and 5.7.

63. $6x - 3(2 - 5x) = 6$

64. $2x + 3 = 5 - 2(3x - 1)$

65. $2x + 3y = 6x$

66. $4x - 8y = 10x$

67. $x^2 + 7x = -6$

68. $x^2 + 4x = 12$

53. ______

54. ______

55. ______

56. ______

57. ______

58. ______

59. ______

60. ______

61. ______

62. ______

63. ______

64. ______

65. ______

66. ______

67. ______

68. ______

Combining Concepts

Graph each function by finding ordered pair solutions, plotting the solutions, and then drawing a smooth curve through the plotted points.

69. $f(x) = e^x$

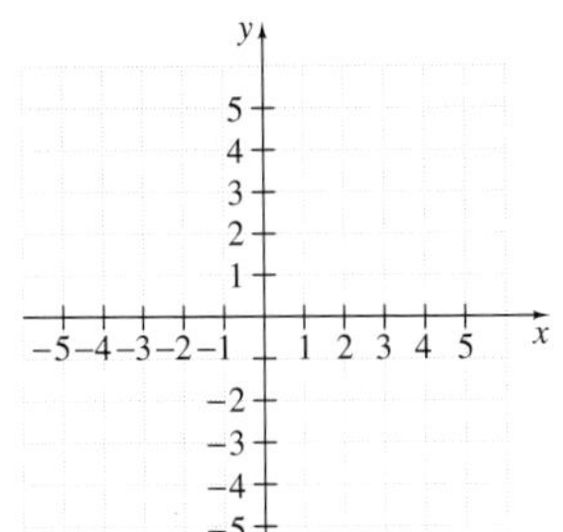

70. $f(x) = e^{2x}$

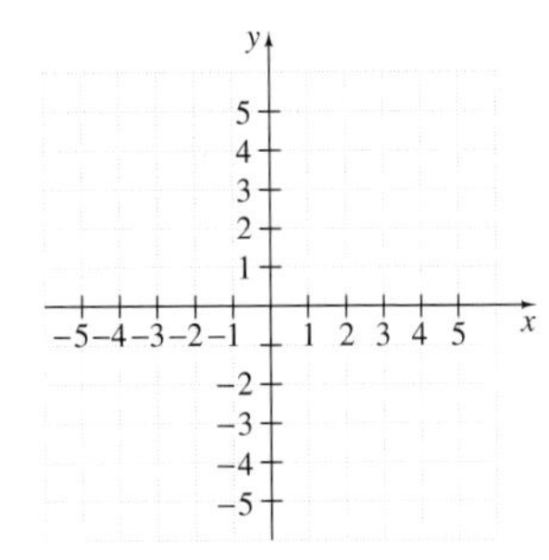

71. $f(x) = \ln x$

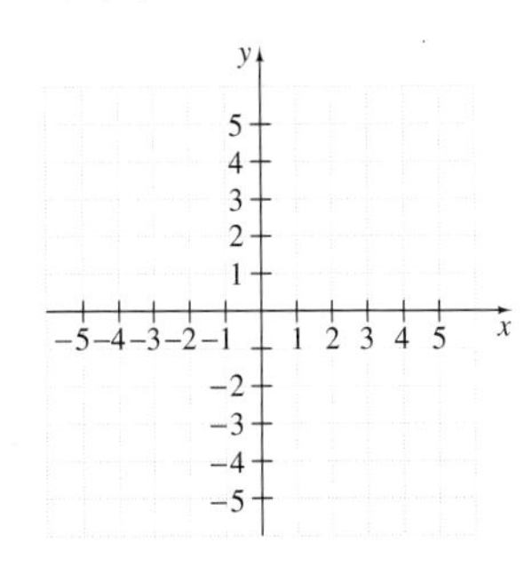

72. $f(x) = \log x$

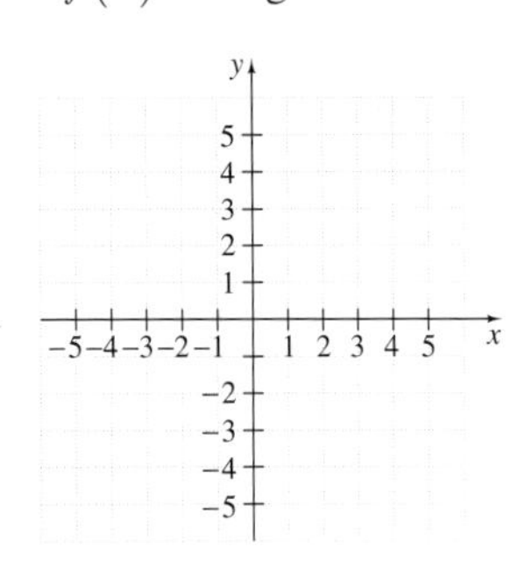

73. ______

74. ______

75. ______

76. ______

77. ______

78. ______

Name ______

73. Without using a calculator, explain which of log 50 or ln 50 must be larger.

74. Without using a calculator, explain which of log 50^{-1} or ln 50^{-1} must be larger.

The Richter scale measures the intensity, or magnitude, of an earthquake. The formula for the magnitude R of an earthquake is $R = \log\left(\frac{a}{T}\right) + B$, *where a is the amplitude in micrometers of the vertical motion of the ground at the recording station, T is the number of seconds between successive seismic waves, and B is an adjustment factor that takes into account the weakening of the seismic wave as the distance increases from the epicenter of the earthquake.*

Use the Richter scale formula to find the magnitude R of the earthquakes that fit the descriptions given. Round answers to one decimal place.

75. Amplitude a is 200 micrometers, time T between waves is 1.6 seconds, and B is 2.1.

76. Amplitude a is 150 micrometers, time T between waves is 3.6 seconds, and B is 1.9.

77. Amplitude a is 400 micrometers, time T between waves is 2.6 seconds, and B is 3.1.

78. Amplitude a is 450 micrometers, time T between waves is 4.2 seconds, and B is 2.7.

10.7 EXPONENTIAL AND LOGARITHMIC EQUATIONS AND PROBLEM SOLVING

Objectives

A Solve exponential equations.

B Solve logarithmic equations.

C Solve problems that can be modeled by exponential and logarithmic equations.

SSM CD-ROM Video 10.7

A SOLVING EXPONENTIAL EQUATIONS

In Section 10.3 we solved exponential equations such as $2^x = 16$ by writing 16 as a power of 2 and using the uniqueness of b^x:

$$2^x = 16$$
$$2^x = 2^4 \quad \text{Write 16 as } 2^4.$$
$$x = 4 \quad \text{Use the uniqueness of } b^x.$$

To solve an equation such as $3^x = 7$, we use the fact that $f(x) = \log_b x$ is a one-to-one function. Another way of stating this fact is as a property of equality.

LOGARITHM PROPERTY OF EQUALITY

Let a, b, and c be real numbers such that $\log_b a$ and $\log_b c$ are real numbers and b is not 1. Then

$$\log_b a = \log_b c \quad \text{is equivalent to} \quad a = c$$

Example 1 Solve: $3^x = 7$. Give an exact answer and a four-decimal-place approximation.

Solution: We use the logarithm property of equality and take the logarithm of both sides. For this example, we use the common logarithm.

$$3^x = 7$$
$$\log 3^x = \log 7 \quad \text{Take the common log of both sides.}$$
$$x \log 3 = \log 7 \quad \text{Use the power property of logarithms.}$$
$$x = \frac{\log 7}{\log 3} \quad \text{Divide both sides by log 3.}$$

The exact solution is $\frac{\log 7}{\log 3}$. To approximate to four decimal places, we have

$$\frac{\log 7}{\log 3} \approx \frac{0.845098}{0.4771213} \approx 1.7712$$

The solution set is $\left\{\frac{\log 7}{\log 3}\right\}$, or approximately $\{1.7712\}$.

Practice Problem 1

Solve: $2^x = 5$. Give an exact answer and a four-decimal-place aproximation.

B SOLVING LOGARITHMIC EQUATIONS

By applying the appropriate properties of logarithms, we can solve a broad variety of logarithmic equations.

Example 2 Solve: $\log_4 (x - 2) = 2$

Solution: Notice that $x - 2$ must be positive, so x must be greater than 2. With this in mind, we first write the equation with exponential notation.

Practice Problem 2

Solve: $\log_3 (x + 5) = 2$

Answers

1. $\left\{\frac{\log 5}{\log 2}\right\}$; $\{2.3219\}$, **2.** $\{4\}$

$$\log_4 (x - 2) = 2$$
$$4^2 = x - 2$$
$$16 = x - 2$$
$$18 = x \quad \text{Add 2 to both sides.}$$

To check, we replace x with 18 in the original equation.

$$\log_4 (x - 2) = 2$$
$$\log_4 (18 - 2) \stackrel{?}{=} 2 \quad \text{Let } x = 18.$$
$$\log_4 16 \stackrel{?}{=} 2$$
$$4^2 = 16 \quad \text{True.}$$

The solution set is $\{18\}$.

Practice Problem 3

Solve: $\log_6 x + \log_6 (x + 1) = 1$

Example 3

Solve: $\log_2 x + \log_2 (x - 1) = 1$

Solution: Notice that $x - 1$ must be positive, so x must be greater than 1. We use the product property on the left side of the equation.

$$\log_2 x + \log_2 (x - 1) = 1$$
$$\log_2 [x(x - 1)] = 1 \quad \text{Use the product property.}$$
$$\log_2 (x^2 - x) = 1$$

Next we write the equation with exponential notation and solve for x.

$$2^1 = x^2 - x$$
$$0 = x^2 - x - 2 \quad \text{Subtract 2 from both sides.}$$
$$0 = (x - 2)(x + 1) \quad \text{Factor.}$$
$$0 = x - 2 \text{ or } 0 = x + 1 \quad \text{Set each factor equal to 0.}$$
$$2 = x \qquad -1 = x$$

Recall that -1 cannot be a solution because x must be greater than 1. If we forgot this, we would still reject -1 after checking. To see this, we replace x with -1 in the original equation.

$$\log_2 x + \log_2 (x - 1) = 1$$
$$\log_2 (-1) + \log_2 (-1 - 1) \stackrel{?}{=} 1 \quad \text{Let } x = -1.$$

Because the logarithm of a negative number is undefined, -1 is rejected. Check to see that the solution set is $\{2\}$.

Practice Problem 4

Solve: $\log (x - 1) - \log x = 1$

Example 4

Solve: $\log (x + 2) - \log x = 2$

Solution: We use the quotient property of logarithms on the left side of the equation.

$$\log (x + 2) - \log x = 2$$
$$\log \frac{x + 2}{x} = 2 \quad \text{Use the quotient property.}$$
$$10^2 = \frac{x + 2}{x} \quad \text{Write using exponential notation.}$$
$$100 = \frac{x + 2}{x}$$

Answers

3. $\{2\}$, **4.** $\varnothing$

$$100x = x + 2 \quad \text{Multiply both sides by } x.$$
$$99x = 2 \quad \text{Subtract } x \text{ from both sides.}$$
$$x = \frac{2}{99} \quad \text{Divide both sides by 99.}$$

Check to see that the solution set is $\left\{\frac{2}{99}\right\}$.

C SOLVING PROBLEMS MODELED BY EXPONENTIAL AND LOGARITHMIC EQUATIONS

Logarithmic and exponential functions are used in a variety of scientific, technical, and business settings. A few examples follow.

Example 5 Estimating Population Size

The population size y of a community of lemmings varies according to the relationship $y = y_0e^{0.15t}$. In this formula, t is time in months, and y_0 is the initial population at time 0. Estimate the population after 6 months if there were originally 5000 lemmings.

Solution: We substitute 5000 for y_0 and 6 for t.

$$y = y_0e^{0.15t}$$
$$= 5000e^{0.15(6)} \quad \text{Let } t = 6 \text{ and } y_0 = 5000.$$
$$= 5000e^{0.9} \quad \text{Multiply.}$$

Using a calculator, we find that $y \approx 12{,}298.016$. In 6 months the population will be approximately 12,300 lemmings.

Practice Problem 5

Use the equation in Example 5 to estimate the lemming population in 8 months.

Example 6 Doubling an Investment

How long does it take an investment of $2000 to double if it is invested at 5% interest compounded quarterly?

The necessary formula is $A = P\left(1 + \frac{r}{n}\right)^{nt}$, where A is the accrued amount, P is the principal invested, r is the annual rate of interest, n is the number of compounding periods per year, and t is the number of years.

Solution: We are given that $P = \$2000$ and $r = 5\% = 0.05$. Compounding quarterly means 4 times a year, so $n = 4$. The investment is to double, so A must be $4000. We substitute these values and solve for t.

$$A = P\left(1 + \frac{r}{n}\right)^{nt}$$
$$4000 = 2000\left(1 + \frac{0.05}{4}\right)^{4t} \quad \text{Substitute known values.}$$
$$4000 = 2000(1.0125)^{4t} \quad \text{Simplify } 1 + \frac{0.05}{4}.$$
$$2 = (1.0125)^{4t} \quad \text{Divide both sides by 2000.}$$
$$\log 2 = \log 1.0125^{4t} \quad \text{Take the logarithm of both sides.}$$
$$\log 2 = 4t(\log 1.0125) \quad \text{Use the power property.}$$

Practice Problem 6

How long does it take an investment of $1000 to double if it is invested at 6% compounded quarterly?

Answers

5. approximately 16,601 lemmings, **6.** $11\frac{3}{4}$ years

$$\frac{\log 2}{4 \log 1.0125} = t \qquad \text{Divide both sides by 4 log 1.0125.}$$

$$13.949408 \approx t \qquad \text{Approximate by calculator.}$$

It takes 14 years for the money to double in value.

GRAPHING CALCULATOR EXPLORATIONS

Use a grapher to find how long it takes an investment of \$1500 to triple if it is invested at 8% interest compounded monthly.

First, let $P = \$1500$, $r = 0.08$, and $n = 12$ (for 12 months) in the formula

$$A = P\left(1 + \frac{r}{n}\right)^{nt}$$

Notice that when the investment has tripled, the accrued amount A is \$4500. Thus,

$$4500 = 1500\left(1 + \frac{0.08}{12}\right)^{12t}$$

Determine an appropriate viewing window and enter and graph the equations

$$Y_1 = 1500\left(1 + \frac{0.08}{12}\right)^{12x}$$

and

$$Y_2 = 4500$$

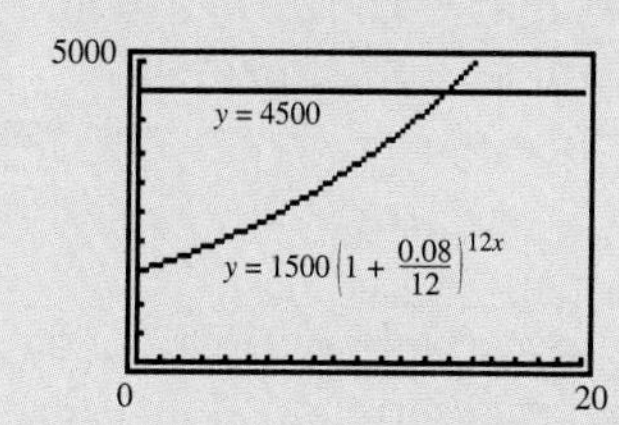

The point of intersection of the two curves is the solution. The x-coordinate tells how long it takes for the investment to triple.

Use a TRACE feature or an INTERSECT feature to approximate the coordinates of the point of intersection of the two curves. It takes approximately 13.78 years, or 13 years and 10 months, for the investment to triple in value to \$4500.

Use this graphical solution method to solve each problem. Round each answer to the nearest hundredth.

1. Find how long it takes an investment of \$5000 to grow to \$6000 if it is invested at 5% interest compounded quarterly.

2. Find how long it takes an investment of \$1000 to double if it is invested at 4.5% interest compounded daily. (Use 365 days in a year.)

3. Find how long it takes an investment of \$10,000 to quadruple if it is invested at 6% interest compounded monthly.

4. Find how long it takes \$500 to grow to \$800 if it is invested at 4% interest compounded semiannually.

Name ______________ Section ______ Date ______

Exercise Set 10.7

A *Solve each equation. Give an exact solution and a four-decimal-place approximation. See Example 1.*

1. $3^x = 6$ **2.** $4^x = 7$ **3.** $3^{2x} = 3.8$ **4.** $5^{3x} = 5.6$

5. $2^{x-3} = 5$ **6.** $8^{x-2} = 12$ **7.** $9^x = 5$ **8.** $3^x = 11$

9. $4^{x+7} = 3$ **10.** $6^{x+3} = 2$ **11.** $7^{3x-4} = 11$ **12.** $5^{2x-6} = 12$

13. $e^{6x} = 5$ **14.** $e^{2x} = 8$

B *Solve each equation. See Examples 2 through 4*

15. $\log_2(x+5) = 4$ **16.** $\log_6(x^2 - x) = 1$ **17.** $\log_4 2 + \log_4 x = 0$

18. $\log_3 5 + \log_3 x = 1$ **19.** $\log_2 6 - \log_2 x = 3$ **20.** $\log_4 10 - \log_4 x = 2$

21. $\log_4 x + \log_4(x+6) = 2$ **22.** $\log_3 x + \log_3(x+6) = 3$

23. $\log_5(x+3) - \log_5 x = 2$ **24.** $\log_6(x+2) - \log_6 x = 2$

25. $\log_3(x-2) = 2$ **26.** $\log_2(x-5) = 3$

27. $\log_4(x^2 - 3x) = 1$ **28.** $\log_8(x^2 - 2x) = 1$

29. $\log_2 x + \log_2(3x+1) = 1$ **30.** $\log_3 x + \log_3(x-8) = 2$

Answers

1. ______
2. ______
3. ______
4. ______
5. ______
6. ______
7. ______
8. ______
9. ______
10. ______
11. ______
12. ______
13. ______
14. ______
15. ______
16. ______
17. ______
18. ______
19. ______
20. ______
21. ______
22. ______
23. ______
24. ______
25. ______
26. ______
27. ______
28. ______
29. ______
30. ______

31. ______

32. ______

33. ______

34. ______

35. ______

36. ______

37. ______

38. ______

39. ______

40. ______

41. ______

42. ______

Name ______

C *Solve. See Example 5.*

31. The size of the wolf population at Isle Royale National Park increases at a rate of 4.3% per year. If the size of the current population is 83 wolves, find how many there should be in 5 years. Use $y = y_0e^{0.043t}$ and round to the nearest whole number.

32. The number of victims of a flu epidemic is increasing at a rate of 7.5% per week. If 20,000 people are currently infected, in how many days can we expect 45,000 people to have the flu? Use $y = y_0e^{0.075t}$ and round to the nearest whole number.

33. The size of the population of Senegal is increasing at a rate of 3.3% per year. If 9,723,000 people lived in Senegal in 1998, find how many inhabitants there will be by 2002. Use $y = y_0e^{0.033t}$ and round to the nearest ten thousandth. (*Source:* U.S. Bureau of the Census, International Data Base)

34. In 1998, 984 million people were citizens of India. Find how long it will take India's population to reach a size of 1500 million (that is, 1.5 billion) if the population size is growing at a rate of 1.7% per year. Use $y = y_0e^{0.017t}$ and round to the nearest tenth. (*Source:* U.S. Bureau of the Census, International Data Base)

35. In 1998, Russia had a population of 146,861,000. At that time, Russia's population was declining at a rate of 0.5% per year. How long will it take for Russia's population to reach 120,000,000? Use $y = y_0e^{-0.005t}$ and round to the nearest tenth. (*Source:* U.S. Bureau of the Census, International Data Base)

36. The population of Italy has been decreasing at a rate of 0.1% per year. If there were 56,783,000 people living in Italy in 1998, how many inhabitants will there be by 2020? Use $y = y_0e^{-0.001t}$ and round to the nearest whole number. (*Source:* U.S. Bureau of the Census, International Data Base)

Use the formula $A = P\left(1 + \frac{r}{n}\right)^{nt}$ *to solve these compound interest problems. Round to the nearest tenth. See Example 6.*

37. How long does it take for $600 to double if it is invested at 7% interest compounded monthly?

38. How long does it take for $600 to double if it is invested at 12% interest compounded monthly?

39. How long does it take for a $1200 investment to earn $200 interest if it is invested at 9% interest compounded quarterly?

40. How long does it take for a $1500 investment to earn $200 interest if it is invested at 10% compounded semiannually?

41. How long does it take for $1000 to double if it is invested at 8% interest compounded semiannually?

42. How long does it take for $1000 to double if it is invested at 8% interest compounded monthly?

Name ______________________________

The formula $w = 0.00185h^{2.67}$ is used to estimate the normal weight w of a boy h inches tall. Use this formula to solve Exercises 43 and 44. Round to the nearest tenth.

43. Find the expected height of a boy who weighs 85 pounds.

44. Find the expected height of a boy who weighs 140 pounds.

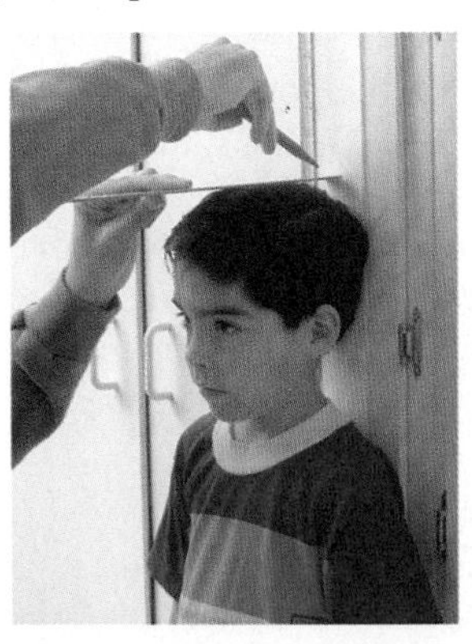

The formula $P = 14.7e^{-0.21x}$ gives the average atmospheric pressure P, in pounds per square inch, at an altitude x, in miles above sea level. Use this formula to solve Exercises 45–48. Round to the nearest tenth.

45. Find the average atmospheric pressure of Denver, which is 1 mile above sea level.

46. Find the average atmospheric pressure of Pikes Peak, which is 2.7 miles above sea level.

47. Find the elevation of a Delta jet if the atmospheric pressure outside the jet is 7.5 lb/in.^{2}.

48. Find the elevation of a remote Himalayan peak if the atmospheric pressure atop the peak is 6.5 lb/in.^{2}.

Psychologists call the graph of the formula $t = \frac{1}{c}\ln\left(\frac{A}{A - N}\right)$ the learning curve since the formula relates time t passed, in weeks, to a measure N of learning achieved, to a measure A of maximum learning possible, and to a measure c of an individual's learning style. Use this formula to answer Exercises 49–52. Round to the nearest week.

49. Norman Weidner is learning to type. If he wants to type at a rate of 50 words per minute (N is 50) and his expected maximum rate is 75 words per minute (A is 75), how many weeks should it take him to achieve his goal? Assume that c is 0.09.

50. An experiment with teaching chimpanzees sign language shows that a typical chimp can master a maximum of 65 signs. How many weeks should it take a chimpanzee to master 30 signs if c is 0.03?

51. Janine Jenkins is working on her dictation skills. She wants to take dictation at a rate of 150 words per minute and believes that the maximum rate she can hope for is 210 words per minute. How many weeks should it take her to achieve the 150-word level if c is 0.07?

52. A psychologist is measuring human capability to memorize nonsense syllables. How many weeks should it take a subject to learn 15 nonsense syllables if the maximum possible to learn is 24 syllables and c is 0.17?

Review and Preview

If $x = -2$, $y = 0$, and $z = 3$, find the value of each expression. See Section 1.4.

53. $\frac{x^2 - y + 2z}{3x}$

54. $\frac{x^3 - 2y + z}{2z}$

55. $\frac{3z - 4x + y}{x + 2z}$

56. $\frac{4y - 3x + z}{2x + y}$

43. ______
44. ______
45. ______
46. ______
47. ______
48. ______
49. ______
50. ______
51. ______
52. ______
53. ______
54. ______
55. ______
56. ______

57. ______________

58. ______________

59. ______________

60. ______________

61. ______________

62. ______________

Name ______________________________

Combining Concepts

The formula $y = y_0e^{kt}$ gives the population size y of a population that experiences an annual rate of population growth k (given as a decimal). In this formula, t is time in years and y_0 is the initial population at time 0. Use this formula to solve Exercises 57 and 58.

57. In 1990, the population of Nevada was 1,202,000. By 1997, the population had grown to 1,677,000. Find the annual rate of population growth over this period. Round your answer to the nearest tenth of a percent. (*Source:* U.S. Bureau of the Census)

58. In 1998, the population of Honduras was 5,862,000. By 2010, the population is expected to grow to 7,280,000. Find the expected annual rate of population growth over this period. Round your answer to the nearest tenth of a percent. (*Source:* U.S. Bureau of the Census, International Data Base)

Use a graphing calculator to solve each equation. For example, to solve Exercise 59, let $Y_1 = e^{0.3x}$, $Y_2 = 8$, and graph the equations. The x-value of the point of intersection is the solution. Round all solutions to two decimal places.

59. $e^{0.3x} = 8$

60. $10^{0.5x} = 7$

Internet Excursions

Go to http://www.prenhall.com/martin-gay

In Section 10.6 Combining Concepts, we learned that the Richter scale measures the magnitude of an earthquake. The relationship between a Richter scale reading R and an intensity I of the earthquake's shock wave is given by the equation $R = \log I$. Given the Richter scale magnitudes of two earthquakes, we can compare their intensities. First we use this relationship to find the intensity of each earthquake. Then we can use a ratio of the resulting intensities to conclude that one earthquake was so many times more intense than the other.

By going to the World Wide Web address listed above, you will gain access to a website where you can look up current earthquake information to help you answer Exercises 61 and 62.

61. Scan the list of the recent earthquakes to find the earthquake events with the highest and lowest Richter scale magnitudes. Report the date, time, location, and magnitude of each. How many times more intense was the earthquake with the higher Richter scale reading than the one with the lowest reading?

62. Scan the list of recent earthquakes to find the most recent and least recent earthquake events that are listed. Report the date, time, location, and magnitude of each. How many times more intense was the earthquake with the higher Richter scale reading than the other earthquake?

Name ______________________________ Section ________ Date _________

Chapter 10 Activity
Modeling Temperature

Method 1 Materials:
- A container of either cold or hot liquid
- thermometer
- stopwatch
- grapher with curve-fitting capabilities (optional)

Method 2 Materials:
- A container of either cold or hot liquid
- TI-82, TI-83, or TI-85 graphing calculator with unit-to-unit link cable
- TI-CBL (Calculator-Based Laboratory) unit with temperature probe

This activity may be completed by working in groups or individually.

Newton's law of cooling relates the temperature of an object to the time elapsed since its warming or cooling began. In this activity you will investigate experimental data to find a mathematical model for this relationship. You may collect the temperature data by using either Method 1 (stopwatch and thermometer) or Method 2 (CBL).

Method 1:

a. Insert the thermometer into the liquid and allow a thermometer reading to register. Take a temperature reading T as you start the stopwatch (at $t = 0$) and record it in the accompanying data table.

t	T
0	

b. Continue taking temperature readings at uniform intervals anywhere between 5 and 10 minutes long. At each reading, use the stopwatch to measure the length of time that has elapsed *since the temperature readings started*. Record your time t and liquid temperature T in the data table. Gather data for six to twelve readings.

c. Plot the data from the data table. Plot t on the horizontal axis and T on the vertical axis.

Method 2

a. Prepare the CBL unit and TI-82, TI-83, or TI-85 graphing calculator. Insert the temperature probe into the liquid.

b. Start the HEAT program on the TI graphing calculator and follow its instructions to begin collecting data. The program will collect 36 temperature readings in degrees Celsius and plot them in real time with t on the horizontal axis and T on the vertical axis.

1. Which of the following mathematical models best fits the data you collected? Explain your reasoning. (Assume $a > 0$.)

a. $T = ab^t + c$

b. $T = ab^{-t} + c$

c. $T = -ab^{-t} + c$

d. $T = \ln(-ax + b) + c$

e. $T = -\ln(-ax + b) + c$

2. What does the constant c represent in the model you chose? What is the value of c in this activity?

3. (Optional) Subtract the value of c from each of your observations of T. Enter the new ordered pairs $(t, T - c)$ into a grapher. Use the exponential or logarithmic curve-fitting feature to find a model for your experimental data. Graph the ordered pairs $(t, T - c)$ with the model you found. How well does the model fit the data? How does the model compare with your selection from Question 1?

CHAPTER 10 HIGHLIGHTS

DEFINITIONS AND CONCEPTS	EXAMPLES

SECTION 10.1 THE ALGEBRA OF FUNCTIONS

ALGEBRA OF FUNCTIONS

Sum	$(f+g)(x) = f(x) + g(x)$
Difference	$(f-g)(x) = f(x) - g(x)$
Product	$(f \cdot g)(x) = f(x) \cdot g(x)$
Quotient	$\left(\frac{f}{g}\right)(x) = \frac{f(x)}{g(x)}$

If $f(x) = 7x$ and $g(x) = x^2 + 1$,

$$(f+g)(x) = f(x) + g(x) = 7x + x^2 + 1$$
$$(f-g)(x) = f(x) - g(x) = 7x - (x^2+1) = 7x - x^2 - 1$$
$$(f \cdot g)(x) = f(x) \cdot g(x) = 7x(x^2+1) = 7x^3 + 7x^2$$
$$\left(\frac{f}{g}\right)(x) = \frac{f(x)}{g(x)} = \frac{7x}{x^2+1}$$

COMPOSITE FUNCTIONS

The notation $(f \circ g)(x)$ means "f composed with g."

$$(f \circ g)(x) = f(g(x))$$
$$(g \circ f)(x) = g(f(x))$$

If $f(x) = x^2 + 1$ and $g(x) = x - 5$, find $(f \circ g)(x)$.

$$(f \circ g)(x) = f(g(x)) = f(x-5) = (x-5)^2 + 1 = x^2 - 10x + 26$$

SECTION 10.2 INVERSE FUNCTIONS

If f is a function, then f is a **one-to-one function** only if each y-value (output) corresponds to only one x-value (input).

HORIZONTAL LINE TEST

If every horizontal line intersects the graph of a function at most once, then the function is a one-to-one function.

Determine whether each graph is a one-to-one function.

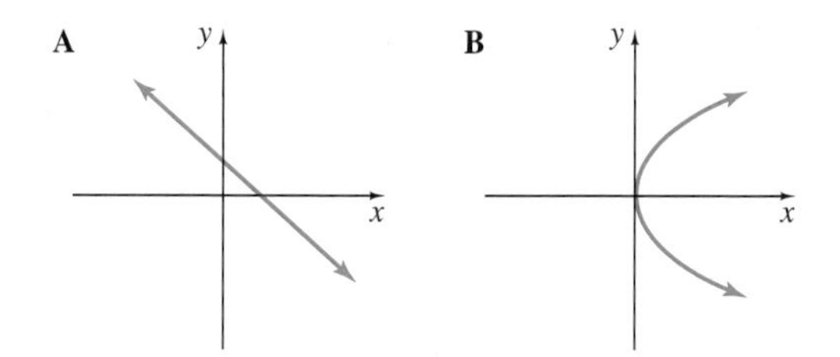

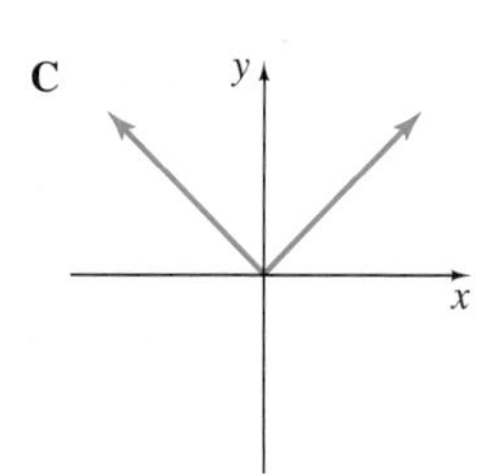

Graphs A and C pass the vertical line test, so only these are graphs of functions. Of graphs A and C, only graph A passes the horizontal line test, so only graph A is the graph of a one-to-one function.

Section 10.2 (continued)

The **inverse** of a one-to-one function f is the one-to-one function f^{-1} that is the set of all ordered pairs (b, a) such that (a, b) belongs to f.

To Find the Inverse of a One-to-One Function $f(x)$

Step 1. Replace $f(x)$ with y.
Step 2. Interchange x and y.
Step 3. Solve for y.
Step 4. Replace y with $f^{-1}(x)$.

Find the inverse of $f(x) = 2x + 7$.

$y = 2x + 7$ Replace $f(x)$ with y.

$x = 2y + 7$ Interchange x and y.

$2y = x - 7$ Solve for y.

$y = \frac{x - 7}{2}$

$f^{-1}(x) = \frac{x - 7}{2}$ Replace y with $f^{-1}(x)$.

The inverse of $f(x) = 2x + 7$ is $f^{-1}(x) = \frac{x - 7}{2}$.

Section 10.3 Exponential Functions

A function of the form $f(x) = b^x$ is an **exponential function**, where $b > 0$, $b \neq 1$, and x is a real number.

Graph the exponential function $y = 4^x$.

x	y
-2	$\frac{1}{16}$
-1	$\frac{1}{4}$
0	1
1	4
2	16

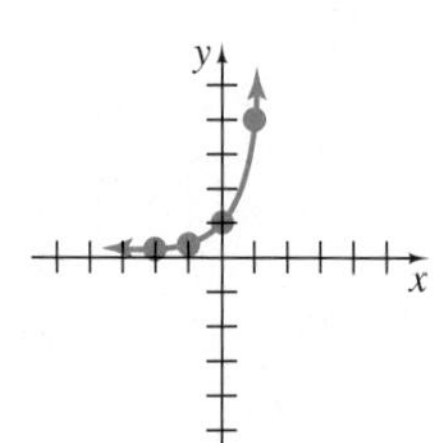

Uniqueness of b^x

If $b > 0$ and $b \neq 1$, then $b^x = b^y$ is equivalent to $x = y$.

Solve: $2^{x+5} = 8$

$2^{x+5} = 2^3$ Write 8 as 2^3.

$x + 5 = 3$ Use the uniqueness of b^x.

$x = -2$ Subtract 5 from both sides.

Section 10.4 Logarithmic Functions

Logarithmic Definition

If $b > 0$ and $b \neq 1$, then

$$y = \log_b x \quad \text{means} \quad x = b^y$$

for any positive number x and real number y.

Logarithmic Form	Corresponding Exponential Statement
$\log_5 25 = 2$	$5^2 = 25$
$\log_9 3 = \frac{1}{2}$	$9^{1/2} = 3$

Properties of Logarithms

If b is a real number, $b > 0$ and $b \neq 1$, then

$$\log_b 1 = 0, \quad \log_b b^x = x, \quad b^{\log_b x} = x$$

$$\log_5 1 = 0, \quad \log_7 7^2 = 2, \quad 3^{\log_3 6} = 6$$

SECTION 10.4 (CONTINUED)

LOGARITHMIC FUNCTION

If $b > 0$ and $b \neq 1$, then a **logarithmic function** is a function that can be defined as

$$f(x) = \log_b x$$

The domain of f is the set of positive real numbers, and the range of f is the set of real numbers.

Graph: $y = \log_3 x$

Write $y = \log_3 x$ as $3^y = x$. Plot the ordered pair solutions listed in the table, and connect them with a smooth curve.

x	y
3	1
1	0
$\frac{1}{3}$	-1
$\frac{1}{9}$	-2

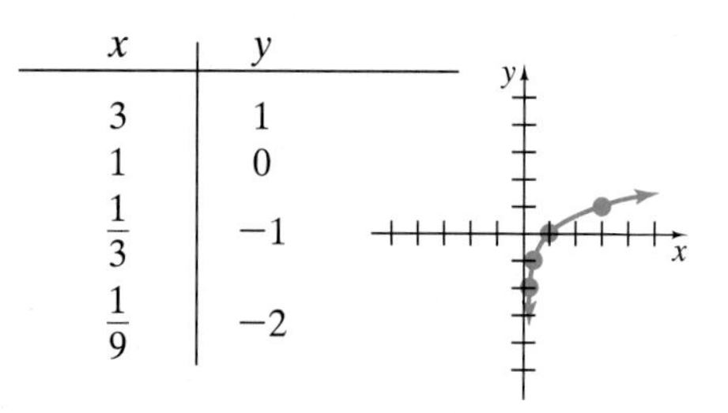

SECTION 10.5 PROPERTIES OF LOGARITHMS

Let x, y, and b be positive numbers and $b \neq 1$.

PRODUCT PROPERTY

$$\log_b xy = \log_b x + \log_b y$$

QUOTIENT PROPERTY

$$\log_b \frac{x}{y} = \log_b x - \log_b y$$

POWER PROPERTY

$$\log_b x^r = r \log_b x$$

Write as a single logarithm:

$2 \log_5 6 + \log_5 x - \log_5 (y + 2)$

$= \log_5 6^2 + \log_5 x - \log_5 (y + 2)$ Power property

$= \log_5 36 \cdot x - \log_5 (y + 2)$ Product property

$= \log_5 \dfrac{36x}{y + 2}$ Quotient property

SECTION 10.6 COMMON LOGARITHMS, NATURAL LOGARITHMS, AND CHANGE OF BASE

COMMON LOGARITHMS

$\log x$ means $\log_{10} x$

NATURAL LOGARITHMS

$\ln x$ means $\log_e x$

CONTINUOUSLY COMPOUNDED INTEREST FORMULA

$$A = Pe^{rt}$$

where r is the annual interest rate for P dollars invested for t years.

$\log 5 = \log_{10} 5 \approx 0.69897$

$\ln 7 = \log_e 7 \approx 1.94591$

Find the amount in an account at the end of 3 years if \$1000 is invested at an interest rate of 4% compounded continuously.

Here, $t = 3$ years, $P = \$1000$, and $r = 0.04$.

$A = Pe^{rt}$

$= 1000e^{0.04(3)}$

$\approx \$1127.50$

SECTION 10.7 EXPONENTIAL AND LOGARITHMIC EQUATIONS AND PROBLEM SOLVING

LOGARITHM PROPERTY OF EQUALITY

Let $\log_b a$ and $\log_b c$ be real numbers and $b \neq 1$. Then

$\log_b a = \log_b c$ is equivalent to $a = c$

Solve: $2^x = 5$

$\log 2^x = \log 5$ Log property of equality

$x \log 2 = \log 5$ Power property

$x = \dfrac{\log 5}{\log 2}$ Divide both sides by log 2.

$x \approx 2.3219$ Use a calculator.

Name ______________________ Section __________ Date __________

Chapter 10 Review

(10.1) *If* $f(x) = x - 5$ *and* $g(x) = 2x + 1$, *find:*

1. $(f + g)(x)$

2. $(f - g)(x)$

3. $(f \cdot g)(x)$

4. $\left(\frac{g}{f}\right)(x)$

If $f(x) = x^2 - 2$, $g(x) = x + 1$, *and* $h(x) = x^3 - x^2$, *find each composition.*

5. $(f \circ g)(x)$

6. $(g \circ f)(x)$

7. $(h \circ g)(2)$

8. $(f \circ f)(x)$

9. $(f \circ g)(-1)$

10. $(h \circ h)(2)$

(10.2) *Determine whether each function is a one-to-one function. If it is one-to-one, list the elements of its inverse.*

11. $h = \{(-9, 14), (6, 8), (-11, 12), (15, 15)\}$

12. $f = \{(-5, 5), (0, 4), (13, 5), (11, -6)\}$

13.

U.S. Region (Input)	West	Midwest	South	Northeast
Rank in Automobile Thefts (Output)	2	4	1	3

14.

Shape (Input)	Square	Triangle	Parallelogram	Rectangle
Number of Sides (Output)	4	3	4	4

Determine whether each function is a one-to-one function.

15.

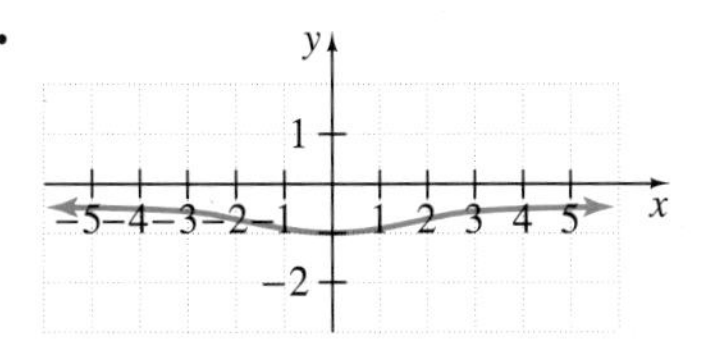

16.

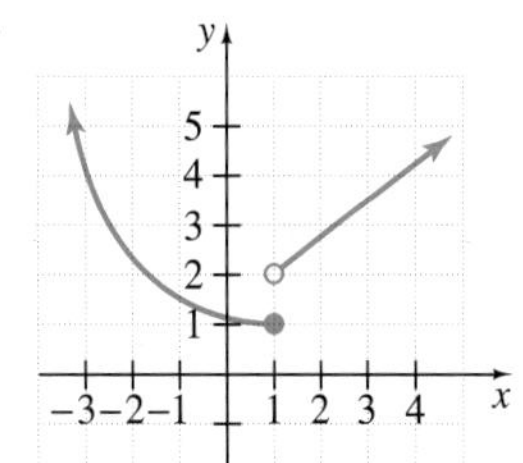

Name ______________________________

17.

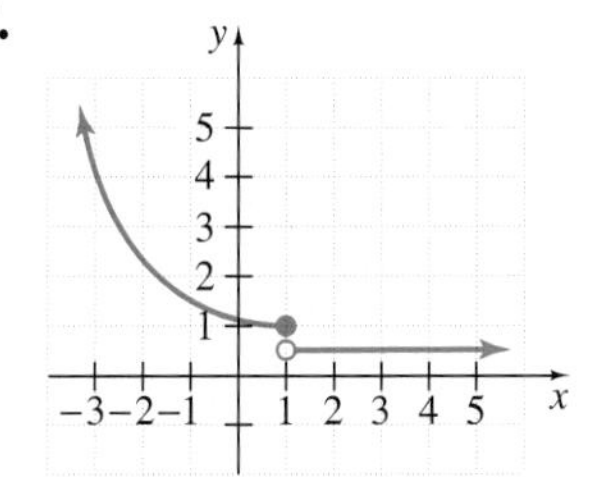

18.

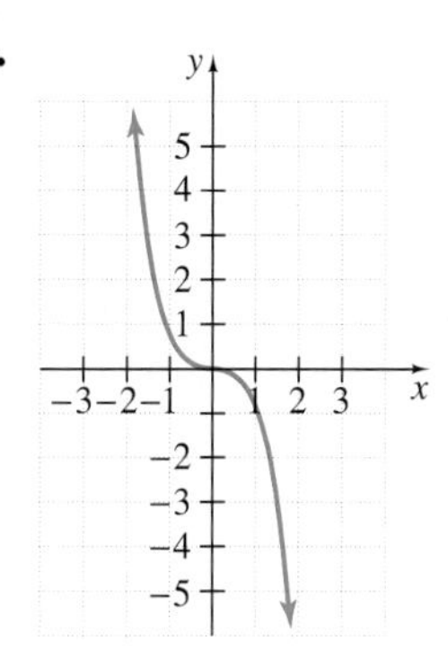

Find an equation defining the inverse function of each given one-to-one function.

19. $f(x) = 6x + 11$ **20.** $f(x) = 12x$ **21.** $f(x) = 3x - 5$ **22.** $f(x) = 2x + 1$

Graph each one-to-one function and its inverse on the same set of axes.

23. $f(x) = -2x + 3$

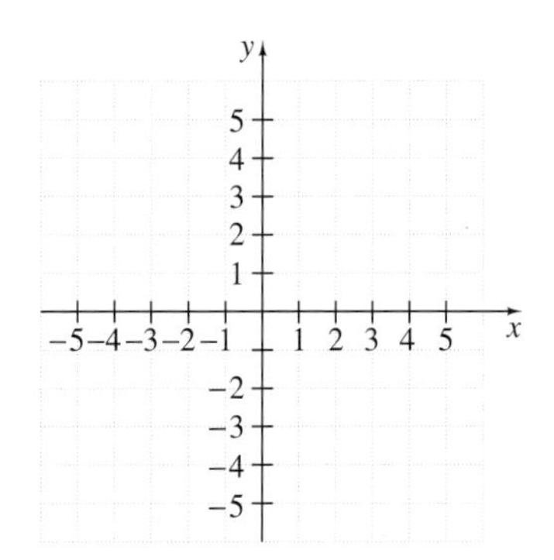

24. $f(x) = 5x - 5$

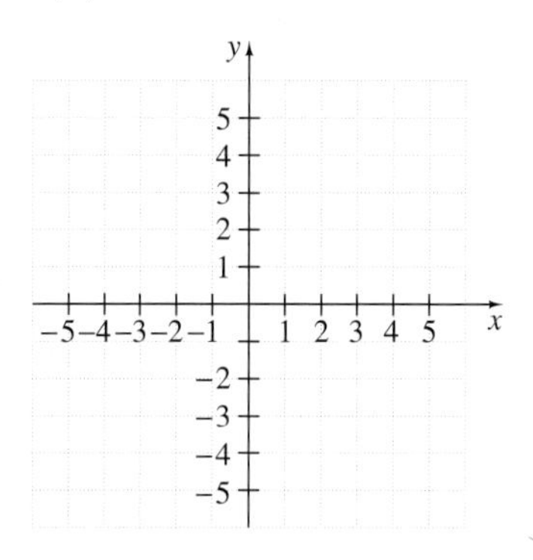

(10.3) *Solve each equation.*

25. $4^x = 64$ **26.** $3^x = \frac{1}{9}$ **27.** $2^{3x} = \frac{1}{16}$

28. $5^{2x} = 125$ **29.** $9^{x+1} = 243$ **30.** $8^{3x-2} = 4$

Graph each exponential function.

31. $y = 3^x$

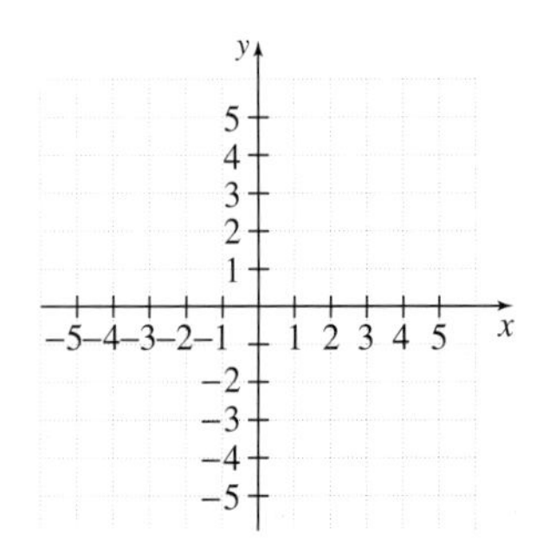

32. $y = \left(\frac{1}{3}\right)^x$

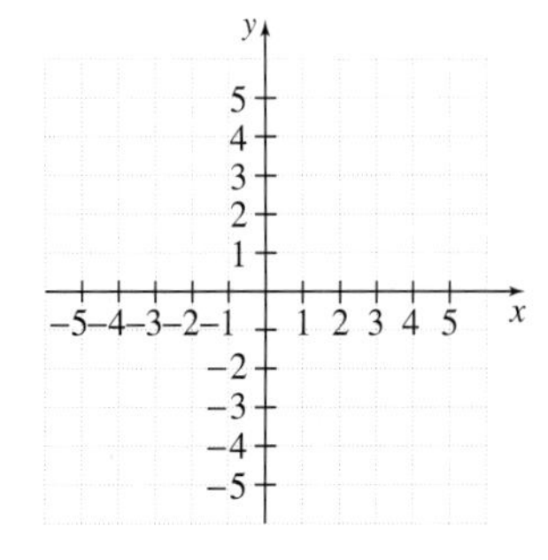

33. $y = 4 \cdot 2^x$

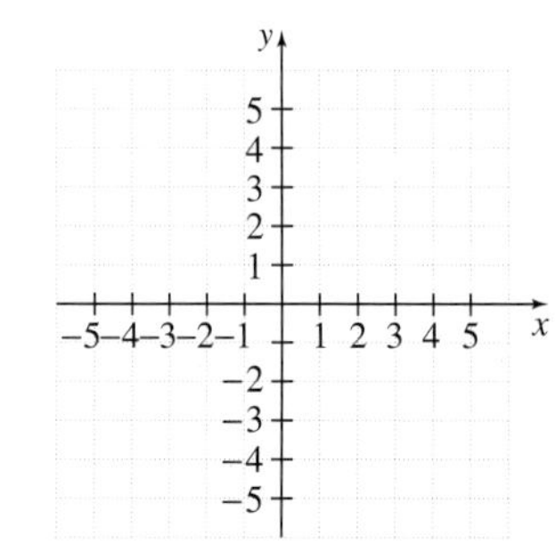

34. $y = 2^x + 4$

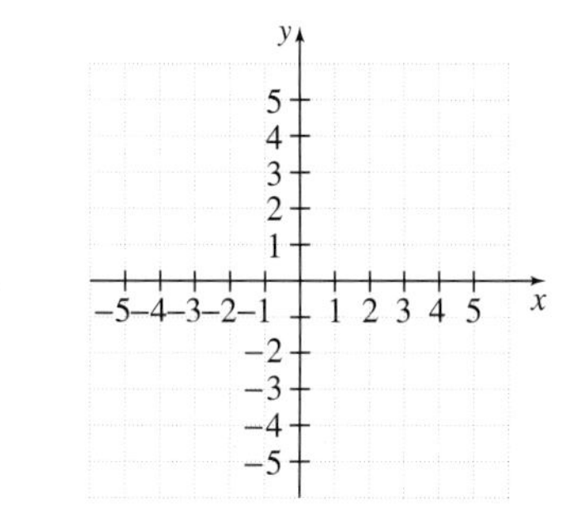

Name ______________________________

Use the formula $A = P\left(1 + \frac{r}{n}\right)^{nt}$ *to solve Exercises 35–36. In this formula,*

A = amount accrued (or owed)
P = principal invested (or loaned)
r = rate of interest
n = number of compounding periods per year
t = time in years

35. Find the amount accrued if \$1600 is invested at 9% interest compounded semiannually for 7 years.

36. A total of \$800 is invested in a 7% certificate of deposit for which interest is compounded quarterly. Find the value that this certificate will have at the end of 5 years.

(10.4) *Write each equation with logarithmic notation.*

37. $49 = 7^2$

38. $2^{-4} = \frac{1}{16}$

Write each logarithmic equation with exponential notation.

39. $\log_{1/2} 16 = -4$

40. $\log_{0.4} 0.064 = 3$

Solve.

41. $\log_4 x = -3$

42. $\log_3 x = 2$

43. $\log_3 1 = x$

44. $\log_4 64 = x$

45. $\log_x 64 = 2$

46. $\log_x 81 = 4$

47. $\log_4 4^5 = x$

48. $\log_7 7^{-2} = x$

49. $5^{\log_5 4} = x$

50. $2^{\log_2 9} = x$

51. $\log_2 (3x - 1) = 4$

52. $\log_3 (2x + 5) = 2$

53. $\log_4 (x^2 - 3x) = 1$

54. $\log_8 (x^2 + 7x) = 1$

Name ______________________________

Graph each pair of equations on the same set of axes.

55. $y = 2^x$ and $y = \log_2 x$

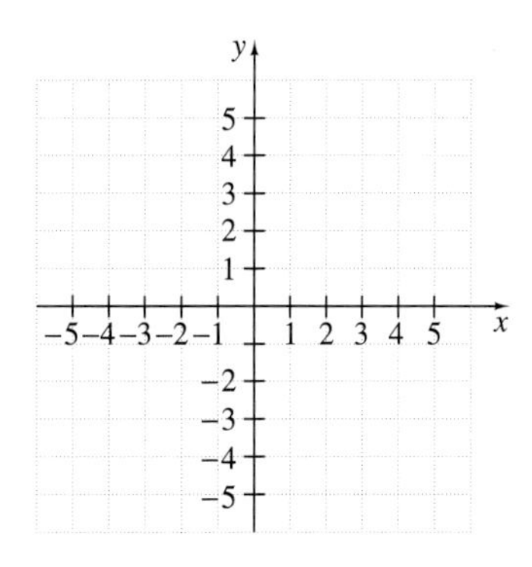

56. $y = \left(\frac{1}{2}\right)^x$ and $y = \log_{1/2} x$

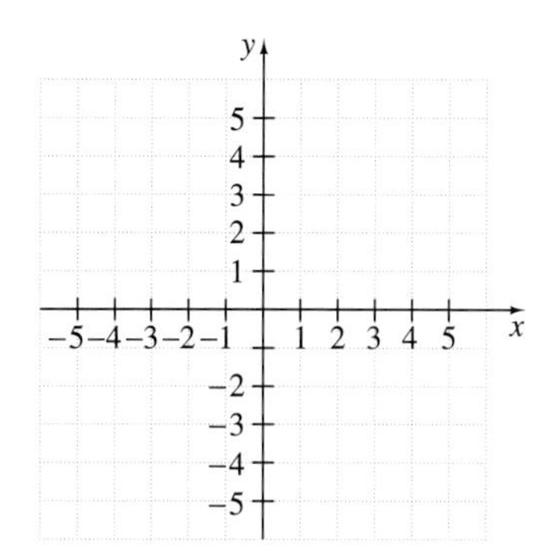

(10.5) *Write each expression as a single logarithm.*

57. $\log_3 8 + \log_3 4$

58. $\log_2 6 + \log_2 3$

59. $\log_7 15 - \log_7 20$

60. $\log 18 - \log 12$

61. $\log_{11} 8 + \log_{11} 3 - \log_{11} 6$

62. $\log_5 14 + \log_5 3 - \log_5 21$

63. $2 \log_5 x - 2 \log_5 (x + 1) + \log_5 x$

64. $4 \log_3 x - \log_3 x + \log_3 (x + 2)$

Use properties of logarithms to write each expression as a sum or difference of logarithms.

65. $\log_3 \frac{x^3}{x + 2}$

66. $\log_4 \frac{x + 5}{x^2}$

67. $\log_2 \frac{3x^2y}{z}$

68. $\log_7 \frac{yz^3}{x}$

If $\log_b 2 = 0.36$ *and* $\log_b 5 = 0.83$, *evaluate each expression.*

69. $\log_b 50$

70. $\log_b \frac{4}{5}$

Name ______________________________

(10.6) *Use a calculator to approximate each logarithm to four decimal places.*

71. $\log 3.6$ **72.** $\log 0.15$ **73.** $\ln 1.25$ **74.** $\ln 4.63$

Find the exact value of each logarithm.

75. $\log 1000$ **76.** $\log \frac{1}{10}$ **77.** $\ln\left(\frac{1}{e}\right)$ **78.** $\ln\left(e^4\right)$

Solve each equation.

79. $\ln(2x) = 2$ **80.** $\ln(3x) = 1.6$ **81.** $\ln(2x - 3) = -1$ **82.** $\ln(3x + 1) = 2$

Approximate each logarithm to four decimal places.

83. $\log_5 1.6$ **84.** $\log_3 4$

Use the formula $A = Pe^{rt}$ to solve Exercises 85–86, in which interest is compounded continuously. In this formula,

A = amount accrued (or owed)
P = principal invested (or loaned)
r = rate of interest
t = time in years

85. Bank of New York offers a 5-year 6% continuously compounded investment option. Find the amount accrued if \$1450 is invested.

86. Find the amount to which a \$940 investment grows if it is invested at 11% compounded continuously for 3 years.

(10.7) *Solve each exponential equation. Given an exact solution and a four-decimal-place approximation.*

87. $3^{2x} = 7$ **88.** $6^{3x} = 5$ **89.** $3^{2x+1} = 6$

90. $4^{3x+2} = 9$ **91.** $5^{3x-5} = 4$ **92.** $8^{4x-2} = 3$

Name ______________________________

93. $2 \cdot 5^{x-1} = 1$

94. $3 \cdot 4^{x+5} = 2$

Solve each equation.

95. $\log_5 2 + \log_5 x = 2$

96. $\log_3 x + \log_3 10 = 2$

97. $\log(5x) - \log(x + 1) = 4$

98. $\ln(3x) - \ln(x - 3) = 2$

99. $\log_2 x + \log_2 2x - 3 = 1$

100. $-\log_6(4x + 7) + \log_6 x = 1$

Use the formula $y = y_0 e^{kt}$ to solve Exercises 101–105. In this formula,

y = size of population
y_0 = initial count of population
k = rate of growth
t = time

Round each answer to the nearest whole number.

101. The population of mallard ducks in Nova Scotia is expected to grow at a rate of 6% per week during the spring migration. If 155,000 ducks are already in Nova Scotia, how many are expected by the end of 4 weeks?

102. The population of Indonesia is growing at a rate of 1.5% per year. If the population in 1998 was 212,942,000, find the expected population by the year 2006. (*Source:* U.S. Bureau of the Census, International Data Base)

103. Japan is experiencing an annual growth rate of 0.2%. In 1998, the population of Japan was 125,932,000. How long will it take for the population to be 140,000,000? (*Source:* U.S. Bureau of the Census, International Data Base)

104. In 1998, Canada had a population of 30,675,000. How long will it take the city to double in population if its growth rate is 1.1% annually? (*Source:* U.S. Bureau of the Census, International Data Base)

105. Egypt's population is increasing at a rate of 1.9% per year. How long will it take for its 1998 population of 66,050,000 to double in size? (*Source:* U.S. Bureau of the Census, International Data Base)

Use the compound interest equation $A = P\left(1 + \frac{r}{n}\right)^{nt}$ to solve Exercises 106–107. (See the directions for Exercises 35 and 36 for an explanation of this formula.) Round answers to the nearest tenth.

106. How long does it take for a $5000 investment to grow to $10,000 if it is invested at 8% interest compounded quarterly?

107. An investment of $6000 has grown to $10,000 while the money was invested at 6% interest compounded monthly. How long was it invested?

Name ______________________ Section ______ Date ________

CHAPTER 10 TEST

If $f(x) = x$ and $g(x) = 2x - 3$, find each function.

1. $(f \cdot g)(x)$

2. $(f - g)(x)$

If $f(x) = x$, $g(x) = x - 7$, and $h(x) = x^2 - 6x + 5$, find each composition.

3. $(f \circ h)(0)$

4. $(g \circ f)(x)$

5. $(g \circ h)(x)$

Graph the given one-to-one function and its inverse on the same set of axes.

6. $f(x) = 7x - 14$

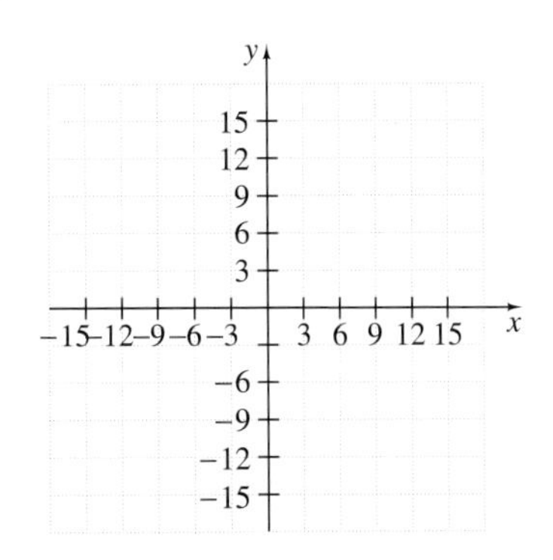

Determine whether each graph is the graph of a one-to-one function.

7.

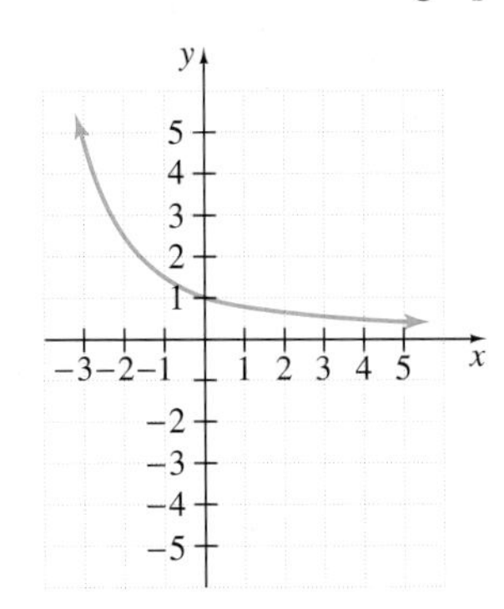

8.

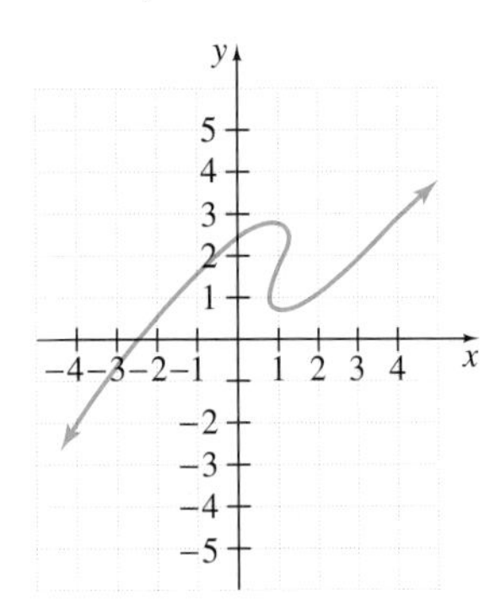

Determine whether each function is one-to-one. If it is one-to-one, find an equation or a set of ordered pairs that defines the inverse function of the given function.

9. $y = 6 - 2x$

10. $f = \{(0, 0), (2, 3), (-1, 5)\}$

11.

Word (Input)	Dog	Cat	House	Desk	Circle
First Letter of Word (Output)	d	c	h	d	c

Use the properties of logarithms to write each expression as a single logarithm.

12. $\log_3 6 + \log_3 4$

13. $\log_5 x + 3 \log_5 x - \log_5 (x + 1)$

14. Write the expression $\log_6 \frac{2x}{y^3}$ as the sum or difference of logarithms.

15. If $\log_b 3 = 0.79$ and $\log_b 5 = 1.16$, find the value of $\log_b \left(\frac{3}{25}\right)$.

ANSWERS

1. ________

2. ________

3. ________

4. ________

5. ________

6. see graph

7. ________

8. ________

9. ________

10. ________

11. ________

12. ________

13. ________

14. ________

15. ________

16. ______

17. ______

18. ______

19. ______

20. ______

21. ______

22. ______

23. ______

24. ______

25. see graph

26. see graph

27. ______

28. ______

29. ______

30. ______

Name ______

16. Approximate $\log_7 8$ to four decimal places.

17. Solve $8^{x-1} = \frac{1}{64}$ for x. Give an exact solution.

18. Solve $3^{2x+5} = 4$ for x. Give an exact solution and a four-decimal-place approximation.

Solve each logarithmic equation. Give an exact solution.

19. $\log_3 x = -2$

20. $\ln \sqrt{e} = x$

21. $\log_8 (3x - 2) = 2$

22. $\log_5 x + \log_5 3 = 2$

23. $\log_4 (x + 1) - \log_4 (x - 2) = 3$

24. Solve $\ln (3x + 7) = 1.31$ accurate to four decimal places.

25. Graph $y = \left(\frac{1}{2}\right)^x + 1$.

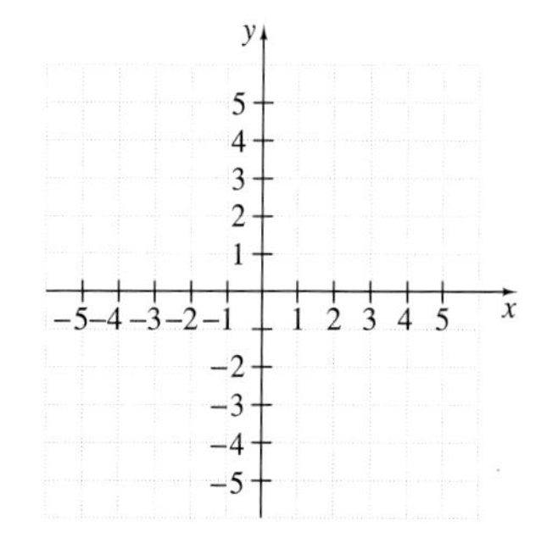

26. Graph the functions $y = 3^x$ and $y = \log_3 x$ on the same set of axes.

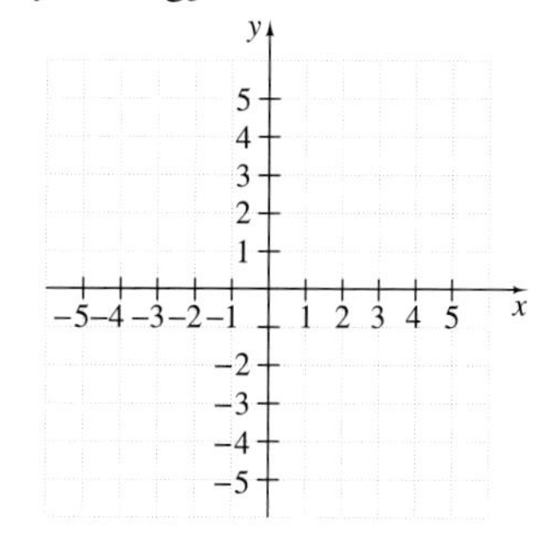

Use the formula $A = P\left(1 + \frac{r}{n}\right)^{nt}$ to solve Exercises 27 and 28.

27. Find the amount in the account if \$4000 is invested for 3 years at 9% interest compounded monthly.

28. How long will it take \$2000 to grow to \$3000 if the money is invested at 7% interest compounded semiannually? Round to the nearest whole.

Use the population growth formula $y = y_0 e^{kt}$ to solve Exercises 29 and 30.

29. The prairie dog population of the Grand Forks area now stands at 57,000 animals. If the population is growing at a rate of 2.6% annually, how many prairie dogs will there be in that area 5 years from now?

30. In an attempt to save an endangered species of wood duck, naturalists would like to increase the wood duck population from 400 to 1000 ducks. If the annual population growth rate is 6.2%, how long will it take the naturalists to reach their goal? Round to the nearest whole year.

Name ______________ Section ______ Date ________

Cumulative Review

Find each root.

1. $\sqrt[3]{27}$

2. $\sqrt[4]{16}$

3. The measure of the largest angle of a triangle is 80° more than the measure of the smallest angle, and the measure of the remaining angle is 10° more than the measure of the smallest angle. Find the measure of each angle.

4. Factor: $7x(x^2 + 5y) - (x^2 + 5y)$

5. Subtract: $\dfrac{5k}{k^2 - 4} - \dfrac{2}{k^2 + k - 2}$

If $f(x) = \sqrt{x - 4}$ and $g(x) = \sqrt[3]{x + 2}$, find each function value.

6. $f(8)$

7. $g(-1)$

Use the properties of exponents to simplify.

8. $x^{1/2}x^{1/3}$

9. $\dfrac{(2x^{2/5})^5}{x^2}$

Use the quotient rule to simplify.

10. $\sqrt{\dfrac{x}{9}}$

11. $\sqrt[4]{\dfrac{3}{16y^4}}$

Multiply.

12. $(\sqrt{2x} + 5)(\sqrt{2x} - 5)$

13. $(\sqrt{3} - 1)^2$

14. Rationalize the numerator of $\dfrac{\sqrt{7}}{\sqrt{45}}$.

15. Solve: $\sqrt{4 - x} = x - 2$

ANSWERS

1. ______
2. ______
3. ______
4. ______
5. ______
6. ______
7. ______
8. ______
9. ______
10. ______
11. ______
12. ______
13. ______
14. ______
15. ______

16.

17.

18.

19.

20.

21. see graph

22.

23. see graph

24.

25.

Name ______________________________

16. Add: $(2 + 3i) + (-3 + 2i)$

17. Solve $4x^2 - 24x + 41 = 0$ by completing the square.

18. Solve: $\frac{1}{4}m^2 - m + \frac{1}{2} = 0$

19. Solve: $x^{2/3} - 5x^{1/3} + 6 = 0$

20. Solve: $\frac{5}{x + 1} < -2$

21. Graph $f(x) = 3x^2 + 3x + 1$. Find the vertex and any intercepts.

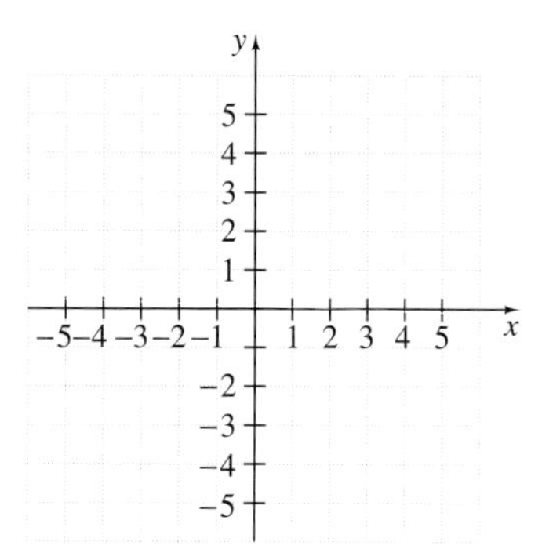

22. Find the midpoint of the line segment that joins points $P(-3, 3)$ and $Q(1, 0)$.

23. Graph: $4y^2 - 9x^2 = 36$

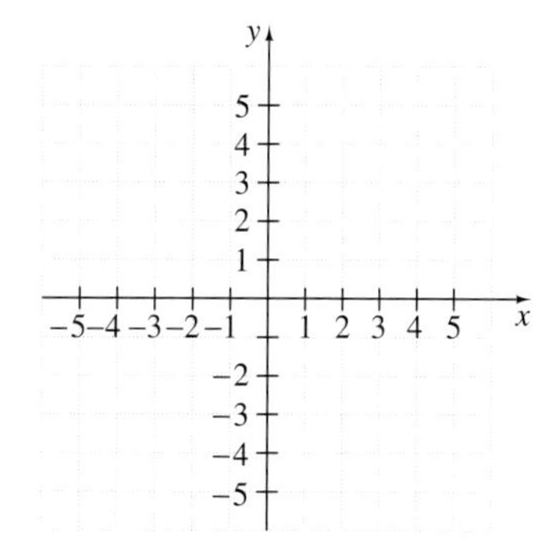

24. Solve the system $\begin{cases} y = \sqrt{x} \\ x^2 + y^2 = 6 \end{cases}$

25. Solve: $\log(x + 2) - \log x = 2$

APPENDIX

An Introduction to Using a Graphing Utility

VIEWING WINDOW AND INTERPRETING WINDOW SETTINGS

In this appendix, we will use the term **graphing utility** to mean a graphing calculator or a computer software graphing package. All graphing utilities graph equations by plotting points on a screen. While plotting several points can be slow and sometimes tedious for us, a graphing utility can quickly and accurately plot hundreds of points. How does a graphing utility show plotted points? A computer or calculator screen is made up of a grid of small rectangular areas called **pixels**. If a pixel contains a point to be plotted, the pixel is turned "on"; otherwise, the pixel remains "off." The graph of an equation is then a collection of pixels turned "on." The graph of $y = 3x + 1$ from a graphing calculator is shown in Figure A-1. Notice the irregular shape of the line caused by the rectangular pixels.

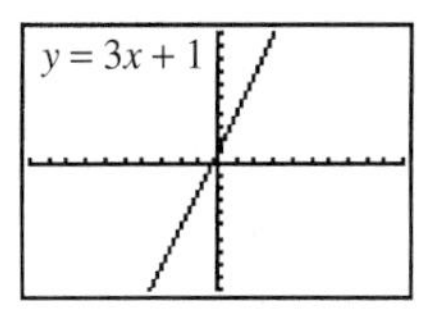

Figure A-1

The portion of the coordinate plane shown on the screen in Figure A-1 is called the **viewing window** or the **viewing rectangle**. Notice the x-axis and the y-axis on the graph. While tick marks are shown on the axes, they are not labeled. This means that from this screen alone, we do not know how many units each tick mark represents. To see what each tick mark represents and the minimum and maximum values on the axes, check the *window setting* of the graphing utility. It defines the viewing window. The window of the graph of $y = 3x + 1$ shown in Figure A-1 has the following setting (Figure A-2):

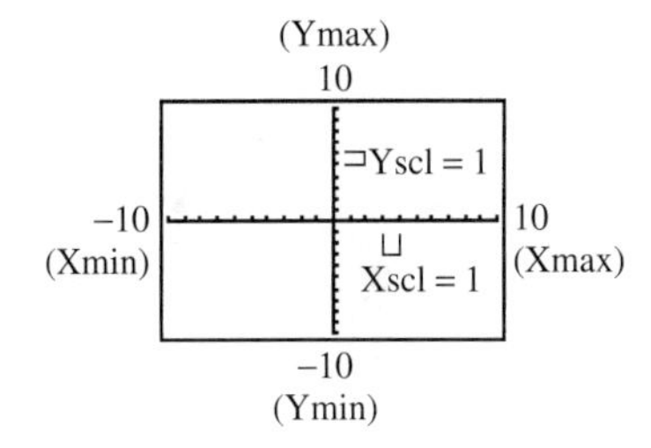

Figure A-2

Xmin $= -10$	The minimum x-value is -10.
Xmax $= 10$	The maximum x-value is 10.
Xscl $= 1$	The x-axis scale is 1 unit per tick mark.
Ymin $= -10$	The minimum y-value is -10.
Ymax $= 10$	The maximum y-value is 10.
Yscl $= 1$	The y-axis scale is 1 unit per tick mark.

By knowing the scale, we can find the minimum and the maximum values on the axes simply by counting tick marks. For example, if both the Xscl (x-axis scale) and the Yscl are 1 unit per tick mark on the graph in Figure A-3, we can count the tick marks and find that the minimum x-value is -10 and the maximum x-value is 10. Also, the minimum y-value is -10 and the maximum y-value is 10. If the Xscl (x-axis scale) changes to 2 units per tick mark (shown in Figure A-4), by counting tick marks, we see that the minimum x-value is now -20 and the maximum x-value is now 20.

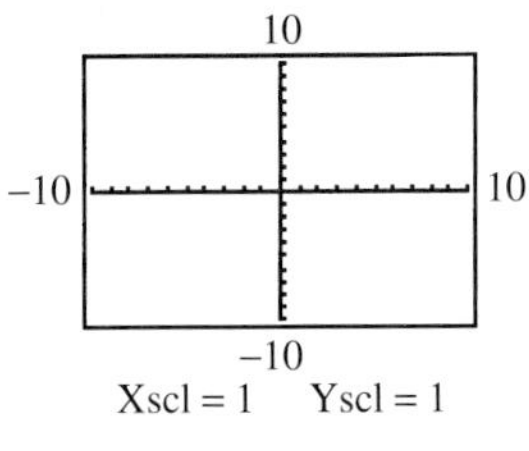

Figure A-3

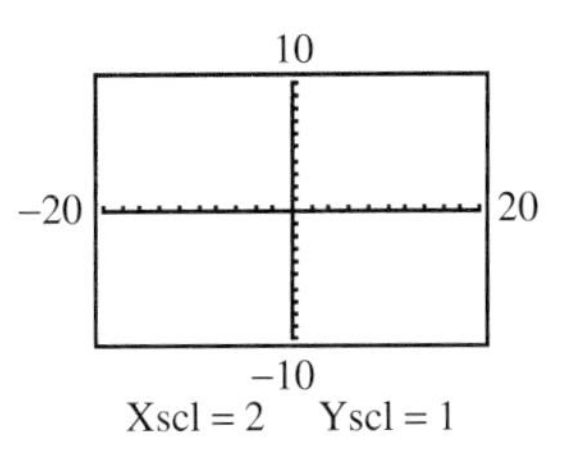

Figure A-4

It is also true that if we know the Xmin and the Xmax values, we can calculate the Xscl by the displayed axes. For example, the Xscl of the graph in

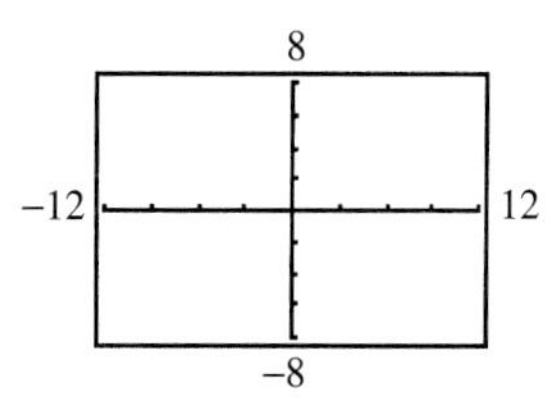

Figure A-5

Figure A-5 must be 3 units per tick mark for the maximum and minimum x-values to be as shown. Also, the Yscl of that graph must be 2 units per tick mark for the maximum and minimum y-values to be as shown.

We will call the viewing window in Figure A-3 a *standard* viewing window or rectangle. Although a standard viewing window is sufficient for much of this text, special care must be taken to ensure that all key features of a graph are shown. Figures A-6, A-7, and A-8 show the graph of $y = x^2 + 11x - 1$ on three different viewing windows. Note that certain viewing windows for this equation are misleading.

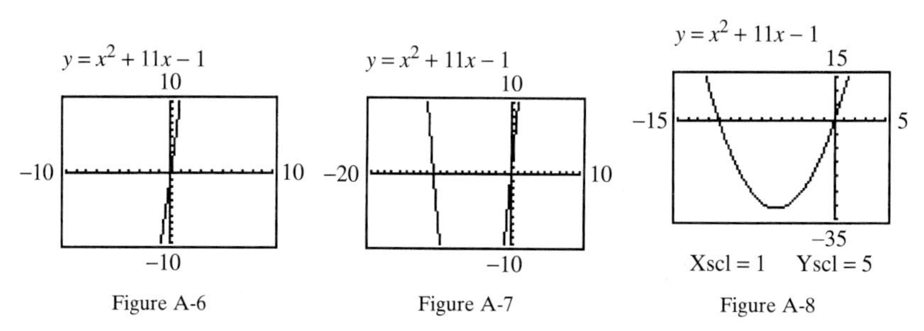

Figure A-6 Figure A-7 Figure A-8

How do we ensure that all distinguishing features of the graph of an equation are shown? It helps to know about the equation that is being graphed. For example, the equation $y = x^2 + 11x - 1$ is not a linear equation, and its graph is not a line. This equation is a quadratic equation, and therefore its graph is a parabola. By knowing this information, we know that the graph shown in Figure A-6, although correct, is misleading. Of the three viewing rectangles shown, the graph in Figure A-8 is best because it shows more of the distinguishing features of the parabola. Properties of equations needed for graphing will be studied in this text.

VIEWING WINDOW AND INTERPRETING WINDOW SETTINGS EXERCISE SET

In Exercises 1–4, determine whether all ordered pairs listed will lie within a standard viewing rectangle.

1. $(-9, 0), (5, 8), (1, -8)$

2. $(4, 7), (0, 0), (-8, 9)$

3. $(-11, 0), (2, 2), (7, -5)$

4. $(3, 5), (-3, -5), (15, 0)$

In Exercises 5–10, choose an Xmin, Xmax, Ymin, and Ymax so that all ordered pairs listed will lie within the viewing rectangle.

5. $(-90, 0), (55, 80), (0, -80)$

6. $(4, 70), (20, 20), (-18, 90)$

7. $(-11, 0), (2, 2), (7, -5)$

8. $(3, 5), (-3, -5), (15, 0)$

9. $(200, 200), (50, -50), (70, -50)$

10. $(40, 800), (-30, 500), (15, 0)$

ANSWERS

1. ____
2. ____
3. ____
4. ____
5. ____
6. ____
7. ____
8. ____
9. ____
10. ____

Name ______________________________

Write the window setting for each viewing window shown. Use the following format:

Xmin =	Ymin =
Xmax =	Ymax =
Xscl =	Yscl =

11.

12

−12 12

−12

12.

20

−20 20

−20

13.

12

−9 9

−12

14.

6

−27 27

−6

15.

25

−10 10

−25

16.

20

−50 50

−20

17.

Xscl = 1, Yscl = 3

18.

Xscl = 10, Yscl = 2

19.

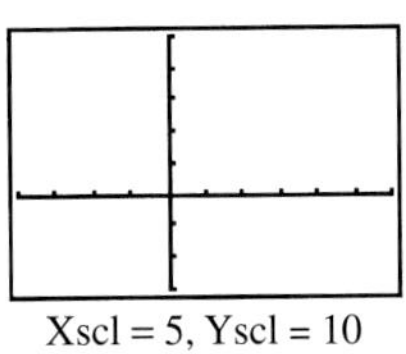

Xscl = 5, Yscl = 10

20.

Xscl = 100, Yscl = 200

11. ____________

12. ____________

13. ____________

14. ____________

15. ____________

16. ____________

17. ____________

18. ____________

19. ____________

20. ____________

GRAPHING EQUATIONS AND SQUARE VIEWING WINDOW

In general, the following steps may be used to graph an equation on a standard viewing window.

> **TO GRAPH AN EQUATION IN X AND Y WITH A GRAPHIING UTILITY ON A STANDARD VIEWING WINDOW**
>
> **Step 1.** Solve the equation for y.
>
> **Step 2.** Use your graphing utility and enter the equation in the form Y = *expression involving x*
>
> **Step 3.** Activate the graphing utility.

Special care must be taken when entering the *expression involving x* in *Step 2*. You must be sure that the graphing utility you are using interprets the expression as you want it to. For example, let's graph $3y = 4x$. To do so,

Step 1. Solve the equation for y.

$$3y = 4x$$

$$\frac{3y}{3} = \frac{4x}{3}$$

$$y = \frac{4}{3}x$$

Step 2. Using your graphing utility, enter the expression $\frac{4}{3}x$ after the Y = prompt. In order for your graphing utility to correctly interpret the expression, you may need to enter $(4/3)x$ or $(4 \div 3)x$.

Step 3. Activate the graphing utility. The graph should appear as in Figure A-9.

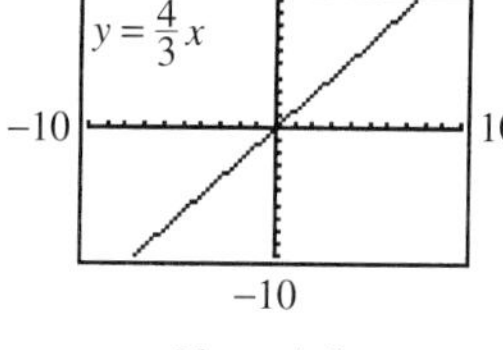

Figure A-9

Distinguishing features of the graph of a line include showing all the intercepts of the line. For example, the window of the graph of the line in Figure A-10 does not show both intercepts of the line, but the window of the graph of the same line in Figure A-11 does show both intercepts. Notice the notation below each graph. This is a shorthand notation of the range setting of the graph. This notation means [Xmin, Xmax] by [Ymin, Ymax].

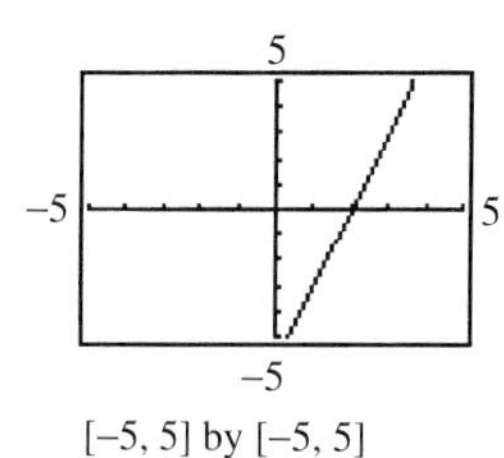

[−5, 5] by [−5, 5]

Figure A-10

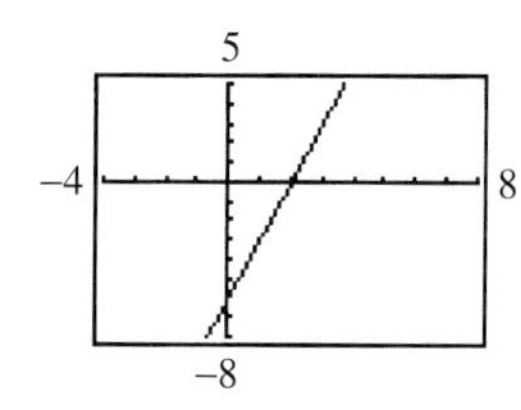

[−4, 8] by [−8, 5]

Figure A-11

On a standard viewing window, the tick marks on the y-axis are closer than the tick marks on the x-axis. This happens because the viewing window is a rectangle, and so 10 equally spaced tick marks on the positive y-axis will be closer together than 10 equally spaced tick marks on the positive x-axis. This causes the appearance of graphs to be distorted.

For example, notice the different appearances of the same line graphed using different viewing windows. The line in Figure A-12 is distorted because the tick marks along the x-axis are farther apart than the tick marks along the y-axis. The graph of the same line in Figure A-13 is not distorted because the viewing rectangle has been selected so that there is equal spacing between tick marks on both axes.

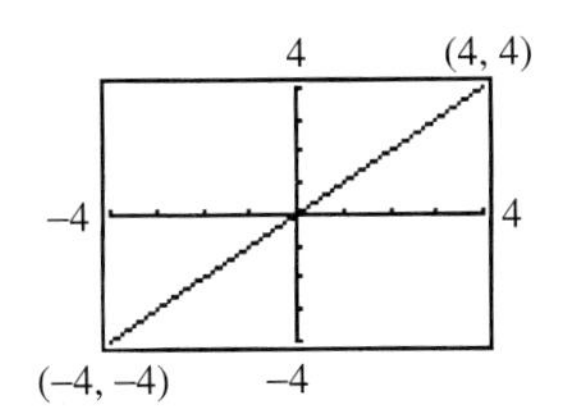

Figure A-12

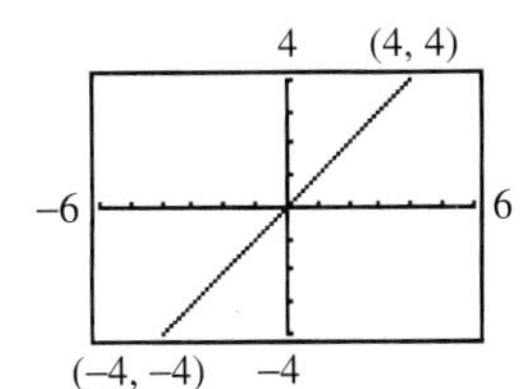

Figure A-13

We say that the line in Figure A-13 is graphed on a *square* setting. Some graphing utilities have a built-in program that, if activated, will automatically provide a square setting. A square setting is especially helpful when we are graphing perpendicular lines, circles, or when a true geometric perspective is desired. Some examples of square screens are shown in Figures A-14 and A-15.

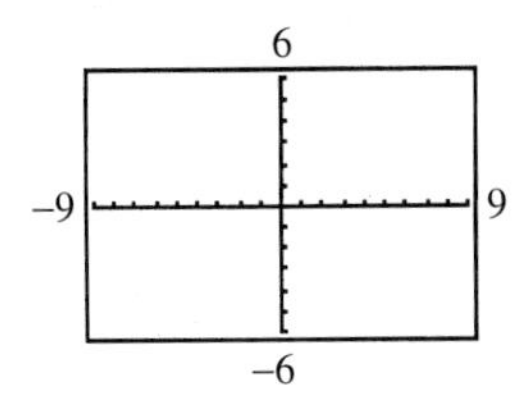

Figure A-14

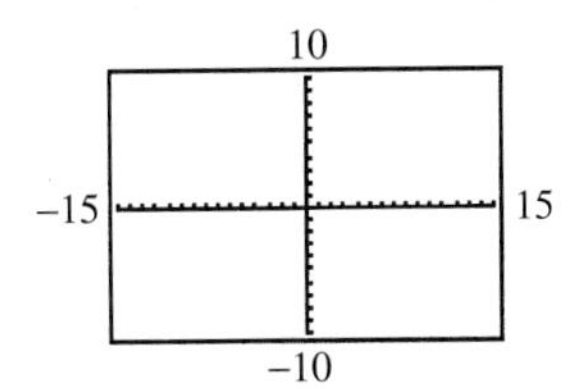

Figure A-15

Other features of a graphing utility such as Trace, Zoom, Intersect, and Table are discussed in appropriate Graphing Calculator Explorations in this text.

Graphing Equations and Square Viewing Window Exercise Set

Graph each linear equation in two variables, using the two different range settings given. Determine which setting shows all intercepts of a line.

1. $y = 2x + 12$
Setting A: $[-10, 10]$ by $[-10, 10]$
Setting B: $[-10, 10]$ by $[-10, 15]$

2. $y = -3x + 25$
Setting A: $[-5, 5]$ by $[-30, 10]$
Setting B: $[-10, 10]$ by $[-10, 30]$

3. $y = -x - 41$
Setting A: $[-50, 10]$ by $[-10, 10]$
Setting B: $[-50, 10]$ by $[-50, 15]$

4. $y = 6x - 18$
Setting A: $[-10, 10]$ by $[-20, 10]$
Setting B: $[-10, 10]$ by $[-10, 10]$

5. $y = \frac{1}{2}x - 15$
Setting A: $[-10, 10]$ by $[-20, 10]$
Setting B: $[-10, 35]$ by $[-20, 15]$

6. $y = -\frac{2}{3}x - \frac{29}{3}$
Setting A: $[-10, 10]$ by $[-10, 10]$
Setting B: $[-15, 5]$ by $[-15, 5]$

Answers

1. ____________
2. ____________
3. ____________
4. ____________
5. ____________
6. ____________

The graph of each equation is a line. Use a graphing utility and a standard viewing window to graph each equation.

7. $3x = 5y$

8. $7y = -3x$

9. $9x - 5y = 30$

10. $4x + 6y = 20$

11. $y = -7$

12. $y = 2$

13. $x + 10y = -5$

14. $x - 5y = 9$

Graph the following equations using the square setting given. Some keystrokes that may be helpful are given.

15. $y = \sqrt{x}$ $[-12, 12]$ by $[-8, 8]$
Suggested keystrokes: $\sqrt{\ }\ x$

16. $y = \sqrt{2x}$ $[-12, 12]$ by $[-8, 8]$
Suggested keystrokes: $\sqrt{\ }\ (2x)$

17. $y = x^2 + 2x + 1$ $[-15, 15]$ by $[-10, 10]$
Suggested keystrokes: $x \wedge 2 + 2\,x + 1$

18. $y = x^2 - 5$ $[-15, 15]$ by $[-10, 10]$
Suggested keystrokes: $x \wedge 2 - 5$

19. $y = |x|$ $[-9, 9]$ by $[-6, 6]$
Suggested keystrokes: ABS x

20. $y = |x - 2|$ $[-9, 9]$ by $[-6, 6]$
Suggested keystrokes: ABS $(x - 2)$

Graph the line on a single set of axes. Use a standard viewing window; then, if necessary, change the viewing window so that all intercepts of the line show.

21. $x + 2y = 30$

22. $1.5x - 3.7y = 40.3$

Answers to Selected Exercises

Chapter 1

Chapter 1 Pretest

1. -2; 1.1A **2.** $7 + 2x$; 1.1C **3.** $>$; 1.2B **4.** $<$; 1.2B **5.** -9.25; 1.2C **6.** $\frac{7}{8}$; 1.2C **7.** -21; 1.3A **8.** 9; 1.3B **9.** -3; 1.3C **10.** -25; 1.3D **11.** $\frac{5}{11}$; 1.3E **12.** 27; 1.4A **13.** $10x - 18$; 1.4C **14.** $-20b^{14}$; 1.5A **15.** $\frac{5a}{b^5c}$; 1.5C **16.** z^{15}; 1.5D **17.** $\frac{5}{y^{12}z^3}$; 1.6A **18.** $\frac{81x^{12}z^{20}}{y^{12}}$; 1.6B **19.** 5; 1.5B **20.** 6.54×10^{10}; 1.5E

Exercise Set 1.1

1. 35 **3.** 30.38 **5.** 6 **7.** $\frac{3}{8}$ **9.** 22 **11.** 11 **13.** 2070 mi **15.** 20.4 sq. ft **17.** $36,790.00 **19.** $\{1, 2, 3, 4, 5\}$ **21.** $\{11, 12, 13, 14, 15, 16\}$ **23.** $\{0\}$ **25.** $\{0, 2, 4, 6, 8\}$ **27.** $\{3, 0, \sqrt{36}\}$ **29.** $\{3, \sqrt{36}\}$ **31.** $\{\sqrt{7}\}$ **33.** $\in$ **35.** $\notin$ **37.** $\notin$ **39.** true **41.** true **43.** false **45.** $2x$ **47.** $x - 10$ **49.** $x + 2$ **51.** $\frac{x}{11}$ or $x \div 11$ **53.** $x - 4$ **55.** $x + 20$ **57.** $10 - x$ **59.** $9x$ **61.** $x + 9$ **63.** $2x + 5$ **65.** $12 - 3x$ **67.** $1 + 2x$ **69.** $5x - 10$ **71.** $1,526,000,000; $2,096,000,000; $6,588,000,000; $3,734,000,000 **73.** answers may vary **75.** $\frac{4}{x+1}$ **77.** $8(x - 9)$

Exercise Set 1.2

1. $4c = 7$ **3.** $2x + 5 = -14$ **5.** $\frac{n}{5} = 4n$ **7.** $z - 2 = 2z$ **9.** $>$ **11.** $=$ **13.** $<$ **15.** $=$ **17.** $>$ **19.** $<$ **21.** $<$ **23.** $>$ **25.** true **27.** true **29.** false **31.** true **33.** 6.2 **35.** $-\frac{4}{7}$ **37.** $\frac{2}{3}$ **39.** 0 **41.** $\frac{1}{5}$ **43.** $-\frac{1}{8}$ **45.** -4 **47.** undefined **49.** $\frac{8}{7}$ **51.** $-5; \frac{1}{5}$ **53.** $-\frac{2}{3}; \frac{3}{2}$ **55.** zero **57.** $y + 7x$ **59.** $w \cdot z$ **61.** $\frac{x}{5} \cdot \frac{1}{3}$ **63.** $(5 \cdot 7)x$ **65.** $(x + 1.2) + y$ **67.** $14(z \cdot y)$ **69.** $3x + 15$ **71.** $16a + 8b$ **73.** $12x + 10y + 4z$ **75.** $4z - 24$ **77.** $6xy - 24x$ **79.** $6 + 3x$ **81.** 0 **83.** 7 **85.** $(10 \cdot 2)y$ **87.** $3x + 12$ **89.** $8y + 4$ **91.** no **93.** answers may vary

Exercise Set 1.3

1. 2 **3.** 4 **5.** 0 **7.** -3 **9.** -2 **11.** 5 **13.** -24 **15.** -11 **17.** -4 **19.** $\frac{4}{3}$ **21.** -2 **23.** -21 **25.** $-\frac{1}{2}$ **27.** -6 **29.** 3 **31.** -12 **33.** 15 **35.** -12.2 **37.** -1 **39.** -22 **41.** -60 **43.** -18 **45.** 80 **47.** -72 **49.** 0 **51.** 0 **53.** -3 **55.** -8 **57.** 3 **59.** -3 **61.** 0 **63.** undefined **65.** -8 **67.** 35 **69.** $-\frac{3}{7}$ **71.** $\frac{1}{21}$ **73.** $-\frac{5}{27}$ **75.** -1 **77.** 7.2 **79.** 0 **81.** 4 **83.** 5 **85.** $-\frac{3}{2}$ **87.** -7 **89.** -49 **91.** 36 **93.** -8 **95.** $-\frac{1}{27}$ **97.** answers may vary **99.** 7 **101.** 8 **103.** $\frac{1}{3}$ **105.** $\frac{1}{4}$ **107.** 4 **109.** 3 **111.** 2 **113.** $\frac{13}{35}$ **115.** 4205 m **117.** b **119.** d **121.** yes; two players have 6 points each (the third player has 0 points), or two players have 5 points each (the third player has 2 points) **123.** 16.5227 **125.** 4.4272 **127.** 13.2% **129.** 10.8%

Integrated Review

1. 5.1 **2.** 52 **3.** $\{1, 2, 3\}$ **4.** $\{1, 3, 5\}$ **5.** $\{8, 10, 12, \ldots\}$ **6.** $\{11, 12, 13, 14\}$ **7.** $>$ **8.** $=$ **9.** $<$ **10.** $<$ **11.** $5x = 20$ **12.** $a + 12 = 14$ **13.** $\frac{y}{10} = y \cdot 10$ **14.** $x + 1 = x - 1$ **15.** 3 **16.** 9 **17.** -28 **18.** -220 **19.** 5 **20.** -28 **21.** 5 **22.** -3 **23.** -25 **24.** 25 **25.** 13 **26.** -5 **27.** 0 **28.** undefined **29.** -24 **30.** 30 **31.** $-\frac{3}{7}$ **32.** $-\frac{1}{10}$ **33.** $-\frac{1}{2}$ **34.** $-\frac{11}{12}$ **35.** -8 **36.** -0.3 **37.** 8 **38.** 4.4

Exercise Set 1.4

1. 48 **3.** -1 **5.** -3 **7.** 14.4 **9.** 17 **11.** -24 **13.** -102 **15.** 40 **17.** -2 **19.** 11 **21.** -56 **23.** -6

25. -26 27. 37 29. $-\frac{3}{4}$ 31. 7 33. 3 35. -11 37. -2.1 39. $-\frac{1}{3}$ 41. $-\frac{79}{15}$ 43. $-\frac{4}{5}$ 45. -81
47. -235.5 49. 12.25 51. a. 18; 22; 28; 208 b. increase 53. a. 600; 150; 105 b. decrease 55. $8x$ 57. $-2y$ 59. $8x$
61. $18y$ 63. $12x$ 65. $-x-8$ 67. $14a+15$ 69. $-6x+9$ 71. $4a-13b$ 73. $2x-2y$ 75. $0.8x-3.6$
77. $\frac{11}{12}b-\frac{7}{6}$ 79. $6x+14$ 81. $-5x+5$ 83. $6a-9b+12$ 85. $2k+10$ 87. $-3x+5$ 89. $4x+9$ 91. $4n-8$
93. -24 95. $2x+10$ 97. $1.91x+4.32$ 99. $15.4z+31.11$ 101. $(2+7)\cdot(1+3)$ 103. 10 million 105. 42 million
107. increasing 109. -0.5876

Calculator Explorations

1. 6×10^{43} 3. 3.796×10^{28}

Mental Math

1. $\frac{5}{xy^2}$ 3. $\frac{a^2}{bc^5}$ 5. $\frac{x^4}{y^2}$

Exercise Set 1.5

1. 4^5 3. x^8 5. $-140x^{12}$ 7. $-20x^2y$ 9. $-16x^6y^3p^2$ 11. x^{15} 13. $10x^{10}$ 15. -1 17. 1 19. 6 21. 9 23. -2
25. answers may vary 27. a^3 29. x 31. $-13z^4$ 33. $-6a^4b^4c^6$ 35. $\frac{1}{z^3}$ 37. $\frac{1}{16}$ 39. $\frac{1}{x^8}$ 41. $\frac{5}{a^4}$ 43. $\frac{1}{x^7}$ 45. $4r^8$
47. 1 49. $\frac{13}{36}$ 51. y^4 53. $\frac{3}{x}$ 55. r^8 57. $\frac{1}{x^9y^4}$ 59. $\frac{b^7}{9a^7}$ 61. $\frac{6x^{16}}{5}$ 63. x^{7a+5} 65. x^{2t-1} 67. x^{4a+7}
69. z^{6x-7} 71. x^{6t-1} 73. 3.125×10^7 75. 1.6×10^{-2} 77. 6.7413×10^4 79. 1.25×10^{-2} 81. 5.3×10^{-5}
83. 7.78×10^8 85. 2.258×10^{10} 87. 1.13×10^9 89. 1.0×10^{-3} 91. 0.0000000036 93. 93,000,000 95. 1,278,000
97. 7,350,000,000,000 99. 0.000000403 101. answers may vary

Mental Math

1. x^{20} 3. x^9 5. y^{42} 7. z^{36} 9. z^{18}

Exercise Set 1.6

1. $\frac{1}{9}$ 3. $\frac{1}{x^{36}}$ 5. $\frac{1}{y^5}$ 7. $9x^4y^6$ 9. $16x^{20}y^{12}$ 11. $\frac{c^{18}}{a^{12}b^6}$ 13. $\frac{y^{15}}{x^{35}z^{20}}$ 15. $\frac{1}{125}$ 17. $\frac{1}{x^{63}}$ 19. $\frac{343}{512}$ 21. $16x^4$
23. $-\frac{y^3}{64}$ 25. $4^8x^2y^6$ 27. $\frac{1}{a^2}$ 29. $4a^8b^4$ 31. $\frac{x^4}{4z^2}$ 33. $\frac{36}{p^{12}}$ 35. $-\frac{a^6}{512x^3y^9}$ 37. $\frac{x^{14}y^{14}}{a^{21}}$ 39. $\frac{x^4}{16}$ 41. 64 43. $\frac{1}{y^{15}}$
45. $\frac{2}{p^2}$ 47. $\frac{3}{8x^8y^7}$ 49. $\frac{1}{x^{30}b^6c^6}$ 51. $\frac{25}{8x^5y^4}$ 53. $\frac{2}{x^4y^{10}}$ 55. x^{9a+18} 57. x^{12a+2} 59. b^{10x^2-4x} 61. y^{15a+3}
63. $16x^{4t+4}$ 65. 1.45×10^9 67. 8×10^{15} 69. 4×10^{-7} 71. 3×10^{-1} 73. 2×10^1 75. 1×10^1 77. 8×10^{-5}
79. 1.1×10^7 81. $8.877840909 \times 10^{20}$ 83. $0.002 = 2 \times 10^{-3}$ sec 85. 6.232×10^{-11} cu. m 87. $\frac{15y^3}{x^8}$ sq. ft
89. 1.331928×10^{13} 91. no 93. 76 people per sq. mi 95. 7 times

Chapter 1 Review

1. 21 2. 8 3. $324{,}000$ 4. $\{-1, 1, 3\}$ 5. $\{-2, 0, 2, 4, 6\}$ 6. $\varnothing$ 7. $\varnothing$ 8. $\{6, 7, 8, \ldots\}$ 9. $\{\ldots, -1, 0, 1, 2\}$
10. true 11. false 12. true 13. true 14. false 15. true 16. false 17. true 18. $\left\{5, \frac{8}{2}, \sqrt{9}\right\}$ 19. $\left\{5, \frac{8}{2}, \sqrt{9}\right\}$
20. $\left\{5, -\frac{2}{3}, \frac{8}{2}, \sqrt{9}, 0.3, 1\frac{5}{8}, -1\right\}$ 21. $\{\sqrt{7}, \pi\}$ 22. $\left\{5, -\frac{2}{3}, \frac{8}{2}, \sqrt{9}, 0.3, \sqrt{7}, 1\frac{5}{8}, -1, \pi\right\}$ 23. $\left\{5, \frac{8}{2}, \sqrt{9}, -1\right\}$ 24. $12 = -4x$
25. $n + 2n = -15$ 26. $4(y+3) = -1$ 27. $6(t-5) = 4$ 28. $z - 7 = 6$ 29. $9x - 10 = 5$ 30. $x - 5 = 12$
31. $-4 = 7y$ 32. $\frac{2}{3} = 2\left(n + \frac{1}{4}\right)$ 33. $t + 6 = -12$ 34. $\frac{3}{4}$ 35. -0.6 36. 0 37. -1 38. $-\frac{4}{3}$ 39. $\frac{1}{0.6}$
40. undefined 41. 1 42. associative property of addition 43. distributive property 44. additive inverse property
45. commutative property of addition 46. associative and commutative properties of multiplication 47. multiplicative inverse property
48. multiplication property of zero 49. associative property of multiplication 50. additive identity property
51. multiplicative identity property 52. $5x - 15z$ 53. $(3+x)+(7+y)$ 54. $2+(-2)$, for example 55. $2 \cdot \frac{1}{2}$, for example
56. $(3.4)[(0.7)5]$ 57. $7+0$ 58. $>$ 59. $>$ 60. $<$ 61. $=$ 62. $<$ 63. $>$ 64. -4 65. -35 66. -2
67. 0.31 68. 8 69. 13.3 70. -4 71. -22 72. undefined 73. 0 74. 4 75. -5 76. $-\frac{2}{15}$ 77. 4 78. $\frac{5}{12}$
79. 9 80. 13 81. 3 82. 54 83. $-\frac{32}{135}$ 84. $-\frac{15}{56}$ 85. $-\frac{5}{4}$ 86. $-\frac{5}{2}$ 87. $\frac{5}{8}$ 88. $-6\frac{1}{2}$ 89. -1 90. 24
91. 1 92. 18 93. -4 94. $\frac{7}{3}$ 95. $\frac{5}{7}$ 96. $-\frac{8}{25}$ 97. $\frac{1}{5}$ 98. 1 99. a. 6.28; 62.8; 628 b. increase 100. 4
101. 81 102. -4 103. -81 104. 1 105. -1 106. $-\frac{1}{16}$ 107. $\frac{1}{16}$ 108. $-x^2y^7z$ 109. $12x^2y^3b$ 110. $\frac{1}{a^9}$

111. $\frac{1}{a}$ **112.** $\frac{1}{x^{11}}$ **113.** $\frac{1}{2a^{17}}$ **114.** $\frac{1}{y^5}$ **115.** 3.689×10^7 **116.** -3.62×10^{-4} **117.** 0.000001678 **118.** 410,000

119. 8^{15} **120.** $\frac{a^2}{16}$ **121.** $27x^3$ **122.** $\frac{1}{16x^2}$ **123.** $\frac{36x^2}{25}$ **124.** $\frac{1}{8^{18}}$ **125.** $\frac{9}{16}$ **126.** $-\frac{1}{8x^9}$ **127.** $\frac{1}{4p^4}$ **128.** $-\frac{27y^6}{x^6}$

129. $x^{25}y^{15}z^{15}$ **130.** $\frac{xz}{4}$ **131.** $\frac{x^2}{625y^4z^4}$ **132.** $\frac{2}{27z^3}$ **133.** $27x^{19a}$ **134.** $2y^{x-7}$

Chapter 1 Test

1. false **2.** false **3.** false **4.** false **5.** true **6.** false **7.** -3 **8.** 43 **9.** -225 **10.** -2 **11.** 1 **12.** 12 **13.** 1 **14. a.** 5.75; 17.25; 57.50; 115.00 **b.** increase **15.** $3\left(\frac{n}{5}\right) = -n$ **16.** $20 = 2x - 6$ **17.** $-2 = \frac{x}{x + 5}$ **18.** distributive property **19.** associative property of addition **20.** additive inverse property **21.** multiplication property of zero **22.** $\frac{1}{81x^2}$ **23.** $\frac{3a^7}{2b^5}$ **24.** $-\frac{y^{40}}{z^5}$ **25.** 6.3×10^8 **26.** 1.2×10^{-2} **27.** 0.000005 **28.** 9×10^{-4} or 0.0009 **29.** 5.76×10^4 **30.** $\frac{3x^7}{2y^2}$ **31.** x^{4w-8}

Chapter 2

Chapter 2 Pretest

1. $\{19\}$; 2.1B **2.** $\{2\}$; 2.1C **3.** $\left\{\frac{45}{11}\right\}$; 2.1D **4.** $\{\ \}$; 2.1E **5.** $\left\{1, \frac{13}{3}\right\}$; 2.6A **6.** $\left\{-3, \frac{5}{3}\right\}$; 2.6A **7.** $y = \frac{-5x + 6}{7}$; 2.3A **8.** $L = \frac{S - 2WH}{2W + 2H}$; 2.3A **9.** $(-\infty, -20]$; 2.4B **10.** $(28, \infty)$; 2.4C **11.** $[-12, \infty)$; 2.4D **12.** $\left(-\infty, -\frac{1}{3}\right)$; 2.4D **13.** $[2, \infty)$; 2.5B **14.** $[-5, 4]$; 2.5B **15.** $(-\infty, 5)$; 2.5D **16.** $[-15, -3]$; 2.6B **17.** $(-\infty, 2] \cup [4, \infty)$; 2.6B **18.** 18, 32; 2.2A **19.** 162; 2.2A **20.** 4 ft; 2.3B

Mental Math

1. $\{6\}$ **3.** $\{17\}$ **5.** $\{8\}$ **7.** $\{10\}$

Exercise Set 2.1

1. -24 is a solution **3.** -3 is not a solution **5.** -2 is a solution **7.** 5 is not a solution **9.** 5 is not a solution **11.** -8 is a solution **13.** $\{6\}$ **15.** $\{-2\}$ **17.** $\{-0.9\}$ **19.** $\{6\}$ **21.** -1.1 **23.** $\{-5\}$ **25.** $\{0\}$ **27.** $\{2\}$ **29.** $\{-9\}$ **31.** $\{-2\}$ **33.** $\left\{-\frac{10}{7}\right\}$ **35.** $\{4\}$ **37. a.** $4x + 5$ **b.** $\{-3\}$ **c.** answers may vary **39.** $\left\{\frac{1}{6}\right\}$ **41.** $\{4\}$ **43.** $\{1\}$ **45.** $\{5\}$ **47.** $\left\{\frac{40}{3}\right\}$ **49.** $\{17\}$ **51.** $\{n|n$ is a real number$\}$ **53.** $\varnothing$ **55.** $\{8\}$ **57.** $\{x|x$ is a real number$\}$ **59.** $\left\{\frac{1}{8}\right\}$ **61.** $\{2\}$ **63.** $\varnothing$ **65.** $\left\{\frac{4}{5}\right\}$ **67.** $\varnothing$ **69.** $\{-8\}$ **71.** answers may vary **73.** $\frac{8}{x}$ **75.** $8x$ **77.** $3x + 2$ **79.** $\{5.217\}$ **81.** $\{1\}$ **83.** $K = -11$

Exercise Set 2.2

1. -5 **3.** 45, 225 **5.** 78 **7.** 1.92 **9.** approximately 658.59 million acres **11.** 51,700 homes **13.** 20% **15.** 117 automobile loans **17.** B767-300ER, 216 seats; B737-200, 112 seats **19.** $430.00 **21.** 18 cm; 24 cm **23.** length, 14 cm; width, 6 cm **25.** 5 yrs **27.** width 8.4m; height 47 m **29.** 64°, 32°, 84° **31.** 75, 76 **33.** Fallon's zip code, 89406; Fernley's zip code, 89408; Gardnerville's zip code, 89410 **35.** 38°, 38°, 104° **37.** $3499 million **39. a.** 49,057 telephone company operators **b.** answers may vary **41.** 6 **43.** 208 **45.** -55 **47.** 3195 **49.** 11 million trees **51.** 500 boards; $30,000 **53.** The company makes a profit. **55.** no solution

Mental Math

1. $y = 5 - 2x$ **3.** $a = 5b + 8$ **5.** $k = h - 5j + 6$

Exercise Set 2.3

1. $t = \frac{d}{r}$ **3.** $R = \frac{I}{PT}$ **5.** $c = P - a - b$ **7.** $y = \frac{9x - 16}{4}$ **9.** $l = \frac{P - 2w}{2}$ **11.** $r = \frac{E}{I} - R$ **13.** $H = \frac{S - 2LW}{2L + 2W}$ **15.** $4703.71; $4713.99; $4719.22; $4722.74; $4724.45 **17. a.** $7313.97 **b.** $7321.14 **c.** $7325.98 **19.** 40°C **21.** 3 hr, 36 min **23.** 171 packages **25.** 9 ft **27.** 2 gal **29. a.** 1174.86 cu. m **b.** 310.34 cu. m **c.** 1485.20 cu. m **31.** 164,921 mi **33.** 0.42 ft **35.** $\{-3, -2, -1\}$ **37.** $\{-3, -2, -1, 0, 1\}$ **39.** answers may vary **41.** Mercury, 0.388; Venus, 0.723; Earth, 1.00; Mars, 1.523; Jupiter, 5.202; Saturn, 9.538; Uranus, 19.193; Neptune, 30.065; Pluto, 39.505 **43.** $6.80 per person **45.** answers may vary

Mental Math

1. $\{x|x < 6\}$ **3.** $\{x|x \geq 10\}$ **5.** $\{x|x > 4\}$ **7.** $\{x|x \leq 2\}$

Exercise Set 2.4

1. 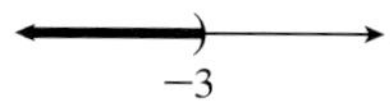 -3 ; $(-\infty, -3)$ 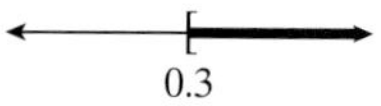 3. 0.3 ; $(0.3, \infty)$ 5. 5 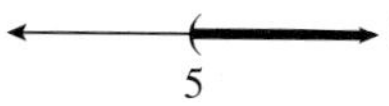 ; $(5, \infty)$ 7.  -2 5 ; $(-2, 5)$ 9.  -1 5 ; $(-1, 5)$ 11. answers may vary 13. 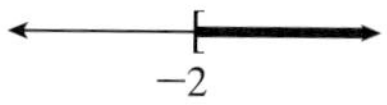 -2 ; $[-2, \infty)$ 15. 1 ; $(-\infty, 1)$ 17. 2 ; $(-\infty, 2]$ 19. 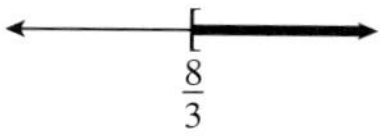 $\frac{8}{3}$; $\left[\frac{8}{3}, \infty\right)$ 21. -4.7 ; $(-\infty, -4.7)$ 23. -3 ; $(-\infty, -3]$ 25. 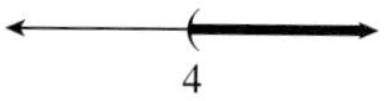 4 ; $(4, \infty)$ 27. $(-\infty, -1]$ 29. $(-\infty, 11]$ 31. $(-13, \infty)$ 33. $(-\infty, 7]$ 35. $[0, \infty)$ 37. $(-\infty, -5]$ 39. $[3, \infty)$ 41. $(0, \infty)$ 43. $\left[-\frac{79}{3}, \infty\right)$ 45. $(-\infty, -1]$ 47. minimum score, 30 49. 1040 lb 51. 17 oz 53. more than 200 calls 55. $F \geq 932°$ 57. a. the year 2000 b. answers may vary 59. decreasing 61. 62.57 lb 63. 2003 65. answers may vary 67. 2005 69. $\{0, 1, 2, 3, 4, 5, 6, 7\}$ 71. $\{..., -9, -8, -7, -6\}$ 73.  -7 1 ; $(-7, 1]$ 75.  -2.5 5.3 ; $[-2.5, 5.3)$ 77. $\varnothing$ 79. $(-\infty, \infty)$ 81. answers may vary

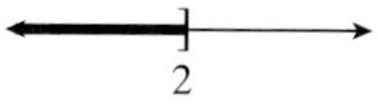

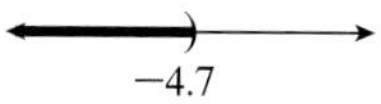

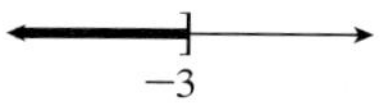

Integrated Review

1. $\{-5\}$ 2. $(-5, \infty)$ 3. $\left[\frac{8}{3}, \infty\right)$ 4. $[-1, \infty)$ 5. $\{0\}$ 6. $\left[-\frac{1}{10}, \infty\right)$ 7. $\left(-\infty, -\frac{1}{6}\right]$ 8. $\{0\}$ 9. $\varnothing$ 10. $\left[-\frac{3}{5}, \infty\right)$ 11. $\{4.2\}$ 12. $\{6\}$ 13. $\{-8\}$ 14. $(-\infty, -15)$ 15. $\left\{\frac{20}{11}\right\}$ 16. $\{1\}$ 17. $(38, \infty)$ 18. $\{-5, -5\}$ 19. $\left\{\frac{3}{5}\right\}$ 20. $(-\infty, \infty)$ 21. $\{29\}$ 22. $\{x|x \text{ is a real number}\}$ 23. $(-\infty, 5)$ 24. $\left\{\frac{9}{13}\right\}$

Exercise Set 2.5

1. $\{2, 4\}$ 3. $\varnothing$ 5. $\{3, 5\}$ 7.  -3 1 9.  11. 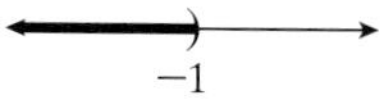 -1 13. $(-2, 5)$ 15. $[6, \infty)$ 17. $(-\infty, -3]$ 19. $\varnothing$ 21. $(11, 17)$ 23. $[1, 4]$ 25. $\left[-3, \frac{3}{2}\right]$ 27. $[-21, -9]$ 29. $\left[\frac{3}{2}, 6\right]$ 31. $\left(0, \frac{14}{3}\right]$ 33. $\{1, 2, 3, 4, 5, 6, 7, 8\}$ 35. $\{1, 5, 6\}$ 37. $\{2, 4, 6, 8\}$ 39.  5 41.  -4 1 43.  45. $(-\infty, -1) \cup (0, \infty)$ 47. $[2, \infty)$ 49. $(-\infty, \infty)$ 51. $(-\infty, 1] \cup \left(\frac{29}{7}, \infty\right)$ 53. $(-7, \infty)$ 55. $(-\infty, \infty)$ 57. -12 59. -4 61. $-7, 7$ 63. 0 65. 1993, 1994, 1995 67. $(-3, 5)$ 69. $(2, \infty)$

Mental Math

1. 7 3. -5 5. -6 7. 12

Exercise Set 2.6

1. $\{7, -7\}$ 3. $\varnothing$ 5. $\{4.2, -4.2\}$ 7. $\{-5, 23\}$ 9. $\{7, -2\}$ 11. $\{8, 4\}$ 13. $\{5, -5\}$ 15. $\{3, -3\}$ 17. $\{0\}$ 19. $\varnothing$ 21. $\left\{-\frac{1}{2}, 9\right\}$ 23. $\left\{-\frac{5}{2}\right\}$ 25. $\{3, 2\}$ 27. $\{4\}$ 29. $\left\{\frac{3}{2}\right\}$ 31. $\left\{-8, \frac{2}{3}\right\}$ 33.  -4 4 ; $[-4, 4]$ 35. 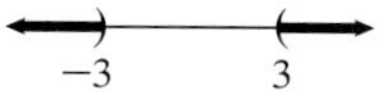 -3 3 ; $(-\infty, -3) \cup (3, \infty)$ 37.  -5 -1 ; $(-5, -1)$ 39.  -1 13 ; $(-\infty, -1] \cup [13, \infty)$ 41. -5 5 ; $[-5, 5]$ 43.  -4 4 ; $(-\infty, -4) \cup (4, \infty)$ 45.  -10 3 ; $[-10, 3]$ 47.  -24 4 ; $(-\infty, -24] \cup [4, \infty)$ 49. -2 9 ; $[-2, 9]$ 51.  ; $(-\infty, \infty)$ 53. 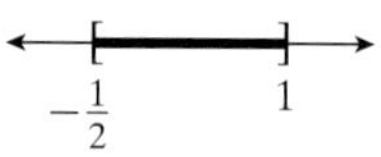 $-\frac{1}{2}$ 1 ; $\left[-\frac{1}{2}, 1\right]$ 55. 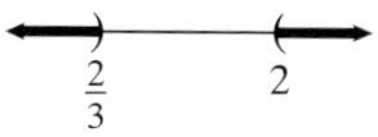 $\frac{2}{3}$ 2 ; $\left(-\infty, \frac{2}{3}\right) \cup (2, \infty)$ 57.  ; $\varnothing$ 59.  -12 0 ; $(-\infty, -12) \cup (0, \infty)$ 61. $(-2, 5)$ 63. $\{5, -2\}$ 65. $(-\infty, -7] \cup [17, \infty)$ 67. $\left\{-\frac{9}{4}\right\}$ 69. $(-2, 1)$

ANSWERS

71. $\left\{2, \frac{4}{3}\right\}$ **73.** ∅ **75.** $\left\{-\frac{17}{2}, \frac{19}{2}\right\}$ **77.** $\left(-\infty, -\frac{25}{3}\right) \cup \left(\frac{35}{3}, \infty\right)$ **79.** $\left\{4, -\frac{1}{5}\right\}$ **81.** 32% **83.** $1.98 billion **85.** $\frac{8}{3}$ **87.** 0 **89.** $|x| = 2$ **91.** $|x| > 4$ **93.** $|x| > 1$ **95.** $\frac{1}{1280}$ or 0.00078125

Chapter 2 Review

1. {3} **2.** $\left\{-\frac{23}{3}\right\}$ **3.** $\left\{-\frac{45}{14}\right\}$ **4.** $\left\{-\frac{7}{11}\right\}$ **5.** {0} **6.** ∅ **7.** {6} **8.** {7.8} **9.** $\{x|x$ is a real number$\}$ **10.** {0} **11.** ∅ **12.** $\{p|p$ is a real number$\}$ **13.** {−3} **14.** {0} **15.** $\left\{\frac{96}{5}\right\}$ **16.** {−3} **17.** {32} **18.** {−8} **19.** {8} **20.** {1} **21.** ∅ **22.** {11} **23.** {2} **24.** $\left\{\frac{37}{8}\right\}$ **25.** −7 **26.** 140, 145 **27.** 52 **28.** 0.12 **29.** $22,080 **30.** 10, 11, 12, 13 **31.** no such integers exist **32.** width, 40 m; length, 75 m **33.** 258 mi **34.** 250 calculators **35.** 5 plants, $200 **36.** $W = \frac{V}{LH}$ **37.** $r = \frac{C}{2\pi}$ **38.** $y = \frac{5x + 12}{4}$ **39.** $x = \frac{4y - 12}{5}$ **40.** $m = \frac{y - y_1}{x - x_1}$ **41.** $x = \frac{y - y_1 + mx_1}{m}$ **42.** $r = \frac{E - IR}{I}$ **43.** $g = \frac{S - vt}{t^2}$ **44.** $g = \frac{T}{r + vt}$ **45.** $P = \frac{I}{1 + rt}$ **46.** $B = \frac{2A - hb}{h}$ **47.** $h = \frac{3V}{\pi r^2}$ **48. a.** $3695.27 **b.** $3700.81 **49.** $\left(\frac{290}{9}\right)^\circ$C ≈ 32.2°C **50.** length, 10 in.; width, 8 in. **51.** 16 packages **52.** The cylinder holds more ice cream. **53.** 58 mph **54.** $(3, \infty)$ **55.** $(-\infty, -4]$ **56.** $(-4, \infty)$ **57.** $(-17, \infty)$ **58.** $(-\infty, 7]$ **59.** $(-\infty, 4]$ **60.** $\left(\frac{1}{2}, \infty\right)$ **61.** $(-\infty, 1)$ **62.** $[-19, \infty)$ **63.** $(2, \infty)$ **64.** It is more economical to use the housekeeper for more than 35 lb per week. **65.** $260° \le C \le 538°$ **66.** minimum score, 9.6 **67.** $1750 to $3750 **68.** $\left[2, \frac{5}{2}\right]$ **69.** $\left[-2, -\frac{9}{5}\right)$ **70.** $\left(\frac{1}{8}, 2\right)$ **71.** $\left(-\frac{3}{5}, 0\right)$ **72.** $\left(\frac{7}{8}, \frac{27}{20}\right]$ **73.** $\left[-\frac{4}{3}, \frac{7}{6}\right]$ **74.** $(-5, 2]$ **75.** $(-\infty, \infty)$ **76.** $\left(\frac{11}{3}, \infty\right)$ **77.** $(5, \infty)$ **78.** {16, −2} **79.** {5, 11} **80.** {0, −9} **81.** $\left\{-1, \frac{11}{3}\right\}$ **82.** $\left\{2, -\frac{2}{3}\right\}$ **83.** $\left\{-\frac{1}{6}\right\}$ **84.** ∅ **85.** ∅ **86.** {3, −3} **87.** {1, 5} **88.** $\left\{5, -\frac{1}{3}\right\}$ **89.** ∅ **90.** $\left\{7, -\frac{8}{5}\right\}$ **91.** $\left\{-10, -\frac{4}{3}\right\}$ **92.** [number line: $-\frac{8}{5}$, 2]; $\left(-\frac{8}{5}, 2\right)$ **93.**  [number line: −4, −1]; $(-\infty, -4] \cup [1, \infty)$ **94.** [number line: −3, 3]; $(-\infty, -3) \cup (3, \infty)$ **95.** [number line: −3, 3]; $(-3, 3)$ **96.** [number line]; ∅ **97.** [number line]; $(-\infty, \infty)$ **98.** 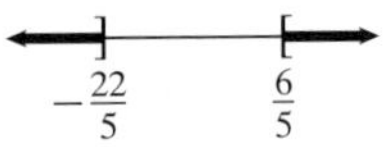 [number line: $-\frac{22}{5}$, $\frac{6}{5}$]; $\left(-\infty, -\frac{22}{15}\right] \cup \left[\frac{6}{5}, \infty\right)$ **99.** [number line: $-\frac{1}{2}$, 2]; $\left(-\frac{1}{2}, 2\right)$ **100.**  [number line: −27, −9]; $(-\infty, -27) \cup (-9, \infty)$ **101.** [number line]; ∅

Chapter 2 Test

1. {10} **2.** {1} **3.** ∅ **4.** $\{n|n$ is a real number$\}$ **5.** {12} **6.** $\left\{-\frac{80}{29}\right\}$ **7.** $\left\{1, \frac{2}{3}\right\}$ **8.** ∅ **9.** $\left\{-4, -\frac{1}{3}\right\}$ **10.** $y = \frac{3x - 8}{4}$ **11.** $n = \frac{9}{7}m$ **12.** $g = \frac{S}{t^2 + vt}$ **13.** $C = \frac{5}{9}(F - 32)$ **14.** $(5, \infty)$ **15.** $[2, \infty)$ **16.** $\left(\frac{3}{2}, 5\right]$ **17.** $(-\infty, -2) \cup \left(\frac{4}{3}, \infty\right)$ **18.** $[1, 11]$ **19.** $[5, \infty)$ **20.** $[4, \infty)$ **21.** $[-3, -1)$ **22.** $(-\infty, \infty)$ **23.** 9.6 **24.** 211,468 people **25.** approximately 8 hunting dogs **26.** more than 850 sunglasses **27.** $3542.27

Cumulative Review

1. 41; Sec. 1.1, Ex. 2 **2.** $x + 5 = 20$; Sec. 1.2, Ex. 1 **3.** $\frac{z}{9} = 9 + z$; Sec. 1.2, Ex. 3 **4.** −8; Sec. 1.2, Ex. 16 **5.** $\frac{1}{5}$; Sec. 1.2, Ex. 17 **6.** −14; Sec. 1.3, Ex. 6 **7.** 5; Sec. 1.3, Ex. 8 **8.** $-\frac{5}{21}$; Sec. 1.3, Ex. 11 **9.** −5; Sec. 1.3, Ex. 25 **10.** 0; Sec. 1.3, Ex. 29 **11.** 0.125; Sec. 1.3, Ex. 30 **12.** 23; Sec. 1.4, Ex. 1 **13.** $-2x + 4$; Sec. 1.4, Ex. 7 **14.** $4y$; Sec. 1.4, Ex. 8 **15.** x^3; Sec. 1.5, Ex. 10 **16.** $5x$; Sec. 1.5, Ex. 12 **17.** $\frac{z^2}{9x^4y^{20}}$; Sec. 1.6, Ex. 14 **18.** $\frac{27a^4x^6}{2}$; Sec. 1.6, Ex. 15 **19.** {2}; Sec. 2.1, Ex. 3 **20.** $\{x|x$ is a real number$\}$; Sec. 2.1, Ex. 11 **21.** $2350; Sec. 2.2, Ex. 3 **22.** $\frac{V}{lw} = h$; Sec. 2.3, Ex. 1 **23.** $(-\infty, 6]$, [number line: 6]; Sec. 2.4, Ex. 7 **24.** $(-\infty, \infty)$; Sec. 2.5, Ex. 8 **25.** {3, −3}; Sec. 2.6, Ex. 1

ANSWERS

Chapter 3

Chapter 3 Pretest

1. no, yes; 3.1B **2. a.** quadrant IV **b.** y-axis **c.** quadrant III **d.** quadrant I; 3.1A

3. 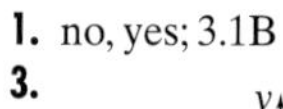 3.1C, 3.3A **4.** 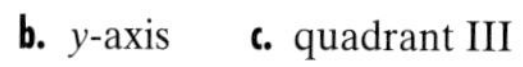 3.1C, 3.3A **5.** 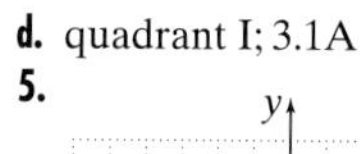 3.1D **6.** 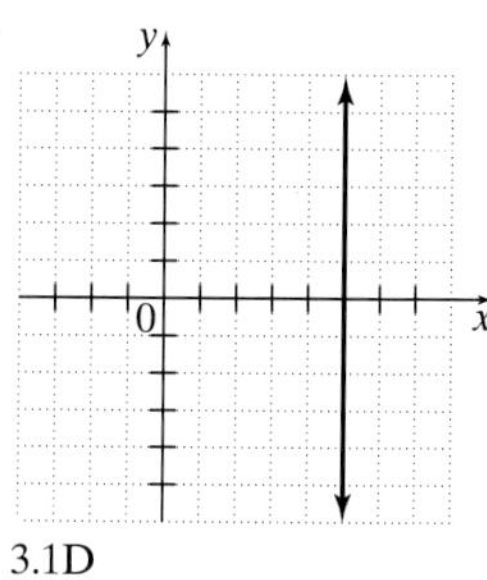 3.1D

7. $\frac{2}{11}$; 3.2A **8.** $m = \frac{4}{5}, b = -\frac{2}{5}$; 3.2B **9.** 0; 3.2C **10.** $y = \frac{1}{3}x + 6$; 3.3B **11.** $y = -7x$; 3.3B **12.** $2x - y = 1$; 3.4A **13.** $x + 3y = 17$; 3.4A **14.** $y = 10$; 3.4B **15.** $3x + y = 24$; 3.4C **16.** domain: $\{-2, 3, 2\}$; range: $\{5, -7\}$; function; 3.5A, B **17.** -22; 3.5E **18.** -5; 3.5E

19. 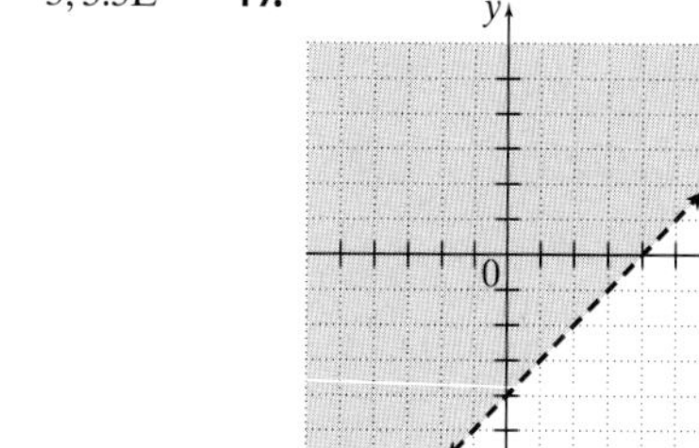 3.6A **20.** 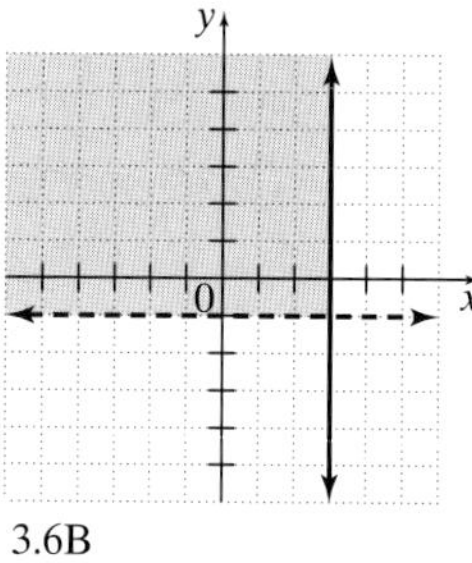 3.6B

Calculator Explorations

1. 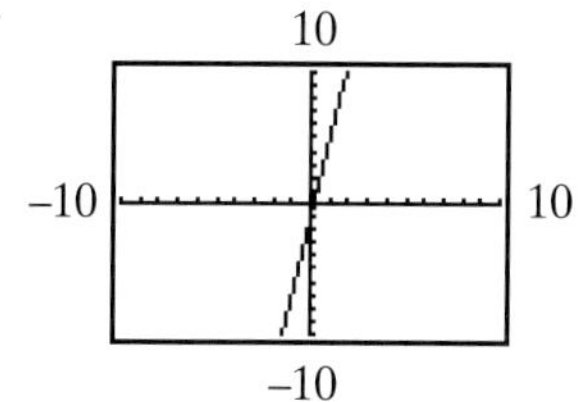

3. 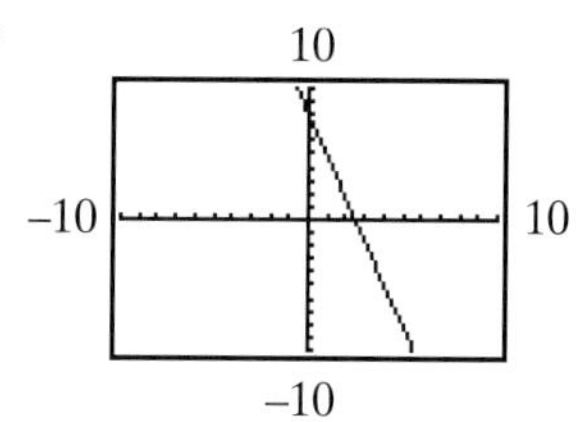

5. 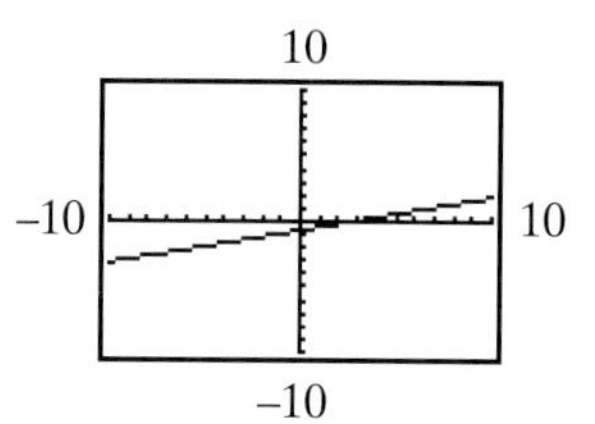

Mental Math

1. $(5, 2)$ **3.** $(3, -1)$ **5.** $(-5, -2)$ **7.** $(-1, 0)$

Exercise Set 3.1

1. quadrant I; quadrant II; quadrant IV; y-axis; quadrant III

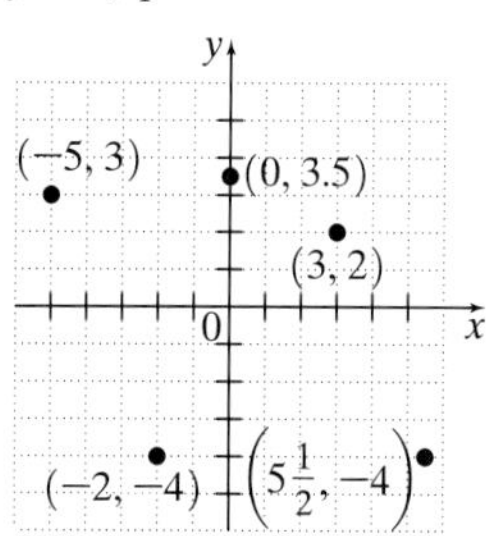

3. quadrant IV **5.** quadrant IV **7.** quadrant III

9.

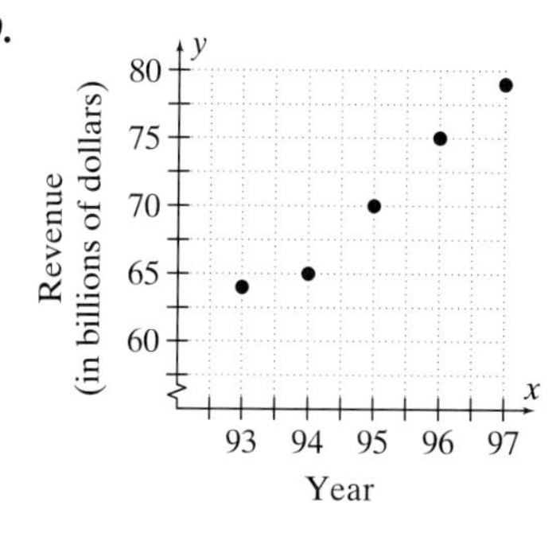

11. no; yes **13.** yes; yes **15.** yes; no

17.

19. 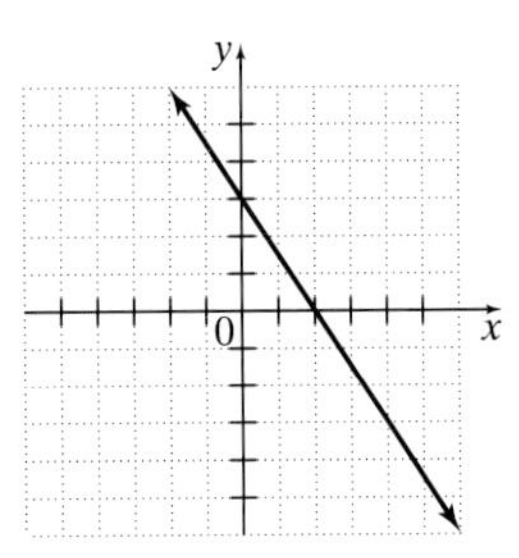

21. 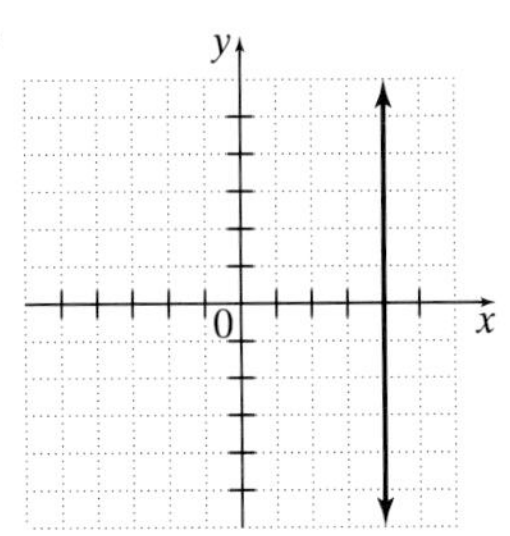

23. 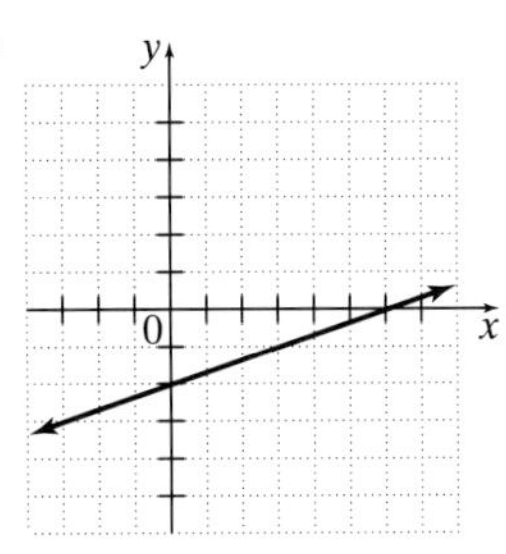

25. 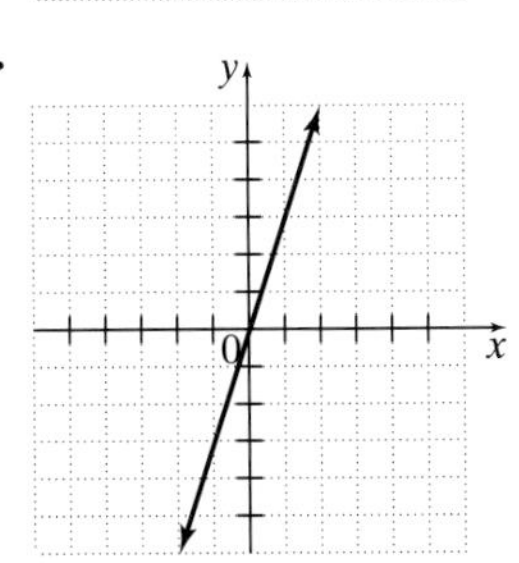

27. 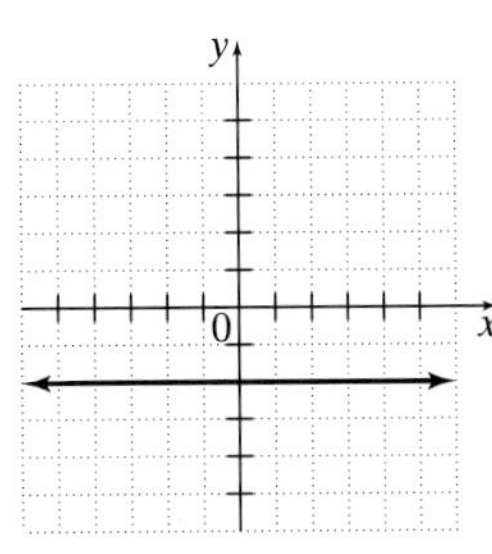

29. 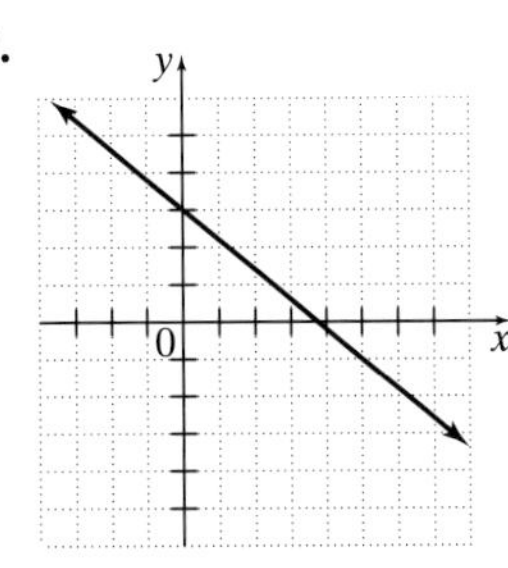

31. 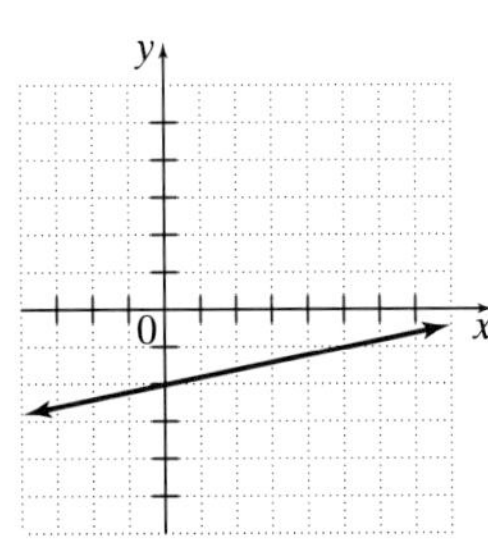

33. 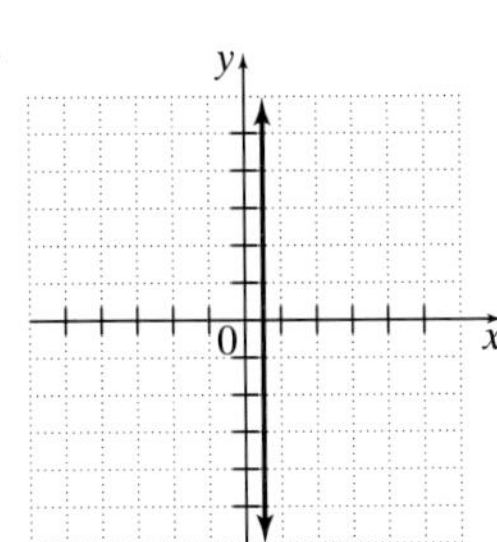

35. 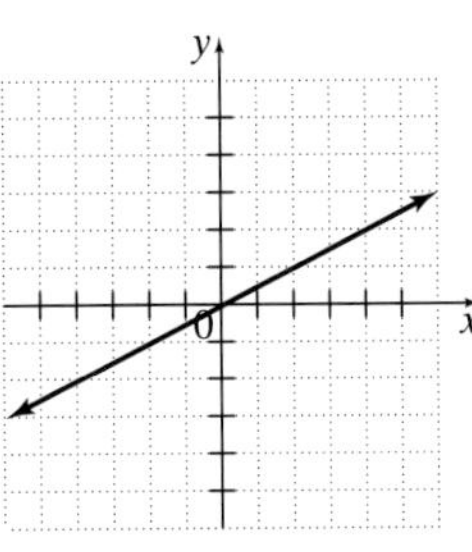

37. 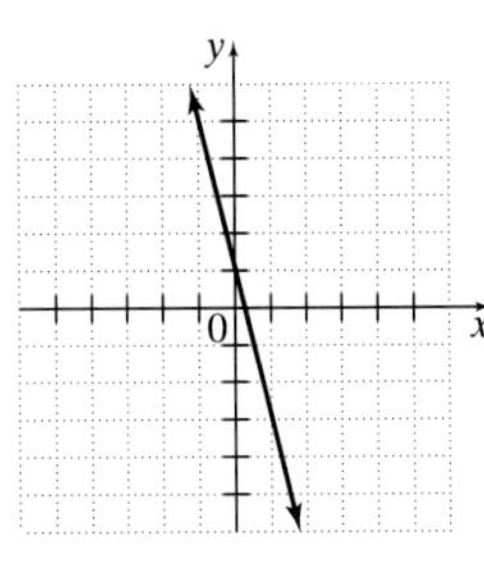

39. 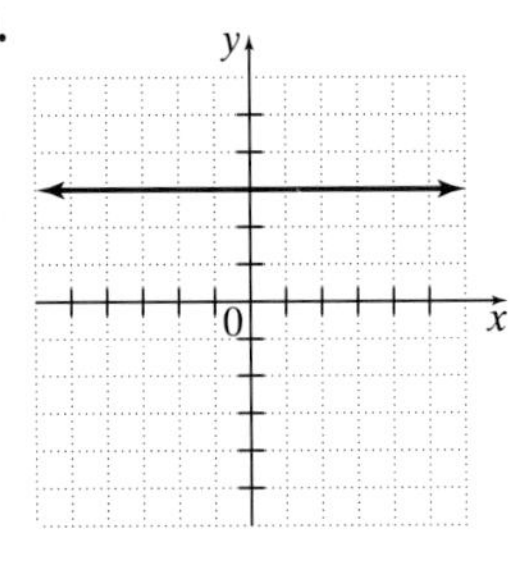

41. C **43.** A **45.** $\frac{3}{2}$ **47.** 6 **49.** $-\frac{6}{5}$

51. a. $(0, 500)$; 0 tables and 500 chairs are produced **b.** $(750, 0)$; 750 tables and 0 chairs are produced **c.** 466 chairs

53. $y = 2x + 5$, $y = 2x$, $y = 2x - 5$

55. 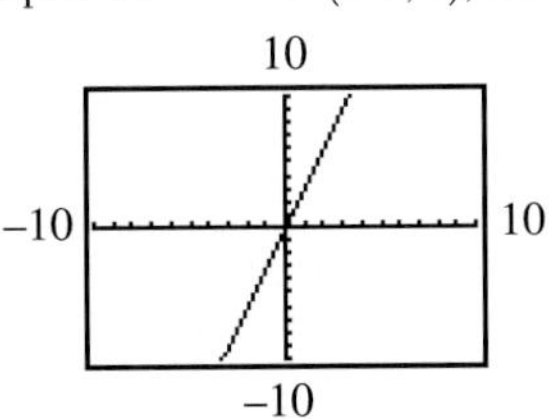

57. 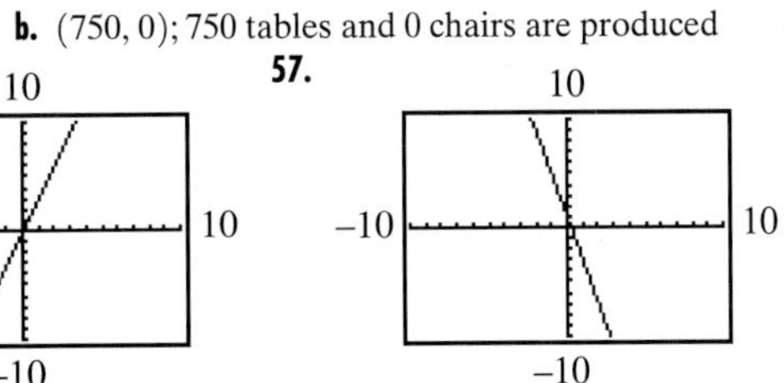

59. The vertical line $x = 0$ has y-intercepts.

Calculator Explorations

1. 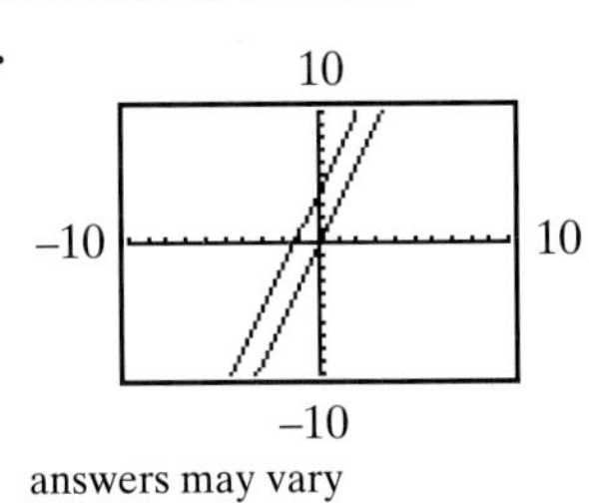

answers may vary

3. 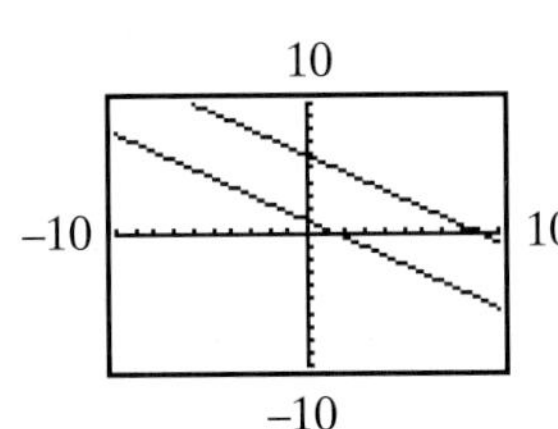

answers may vary

5. 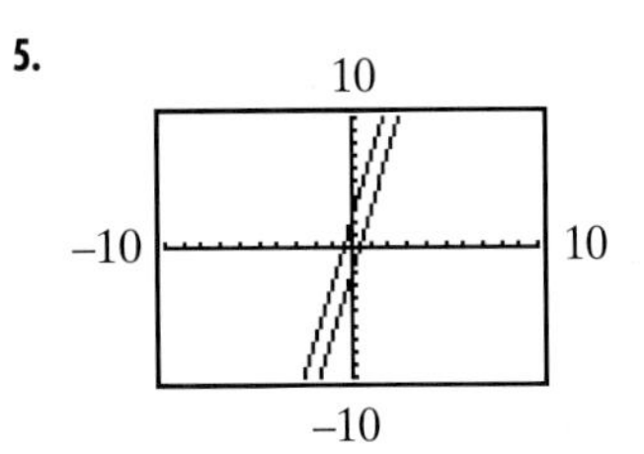

answers may vary

Mental Math

1. upward **3.** horizontal

Exercise Set 3.2

1. $\frac{9}{5}$ **3.** $-\frac{7}{2}$ **5.** $-\frac{5}{6}$ **7.** $\frac{1}{3}$ **9.** $-\frac{4}{3}$ **11.** 0 **13.** $\frac{2}{3}$ **15.** $\frac{3}{20}$ **17.** $m = 5, b = -2$ **19.** $m = -2, b = 7$ **21.** $m = \frac{2}{3}, b = -\frac{10}{3}$ **23.** $m = \frac{1}{2}, b = 0$ **25.** $m = 3, b = 9$ **27.** A **29.** B **31.** undefined **33.** -1 **35.** $\frac{6}{5}$ **37.** undefined **39.** 7 **41.** undefined **43.** l_2 **45.** l_2 **47.** l_2 **49.** neither **51.** parallel **53.** perpendicular **55.** neither **57.** parallel **59.** $-\frac{7}{2}$ **61.** $\frac{5}{2}$ **63.** $\{9, -3\}$ **65.** $(-\infty, -4) \cup (-1, \infty)$ **67.** $\left[\frac{2}{3}, 2\right]$ **69. a.** l_1: -2, l_2: -1, l_3: $-\frac{2}{3}$

ANSWERS

b. lesser **71.** (10, 13) **73.** $\frac{3}{2}$ yd per sec **75.** answers may vary

77. a.

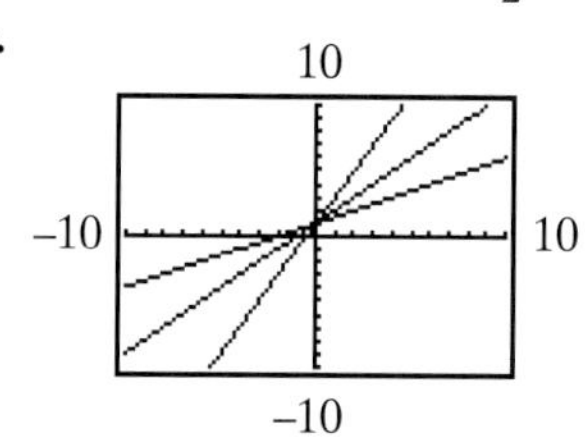

b.

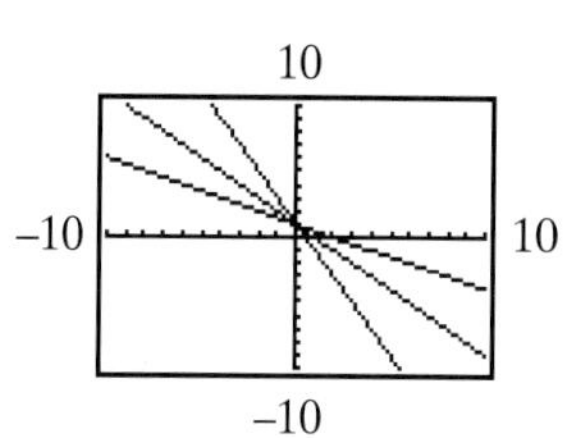

c. true

Calculator Explorations

1. $y = \frac{1}{3.5}x$

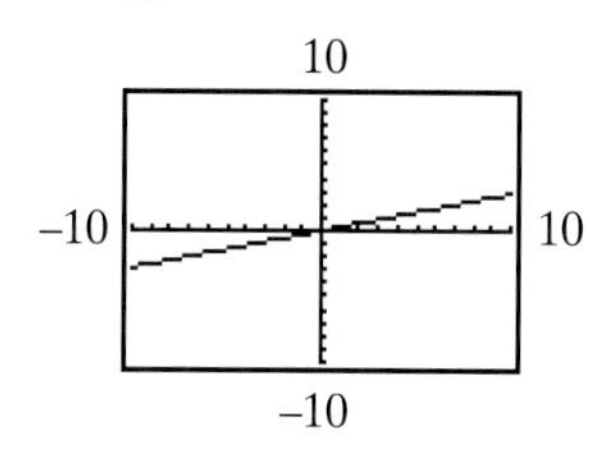

3. $y = -\frac{5.78}{2.31}x + \frac{10.98}{2.31}$

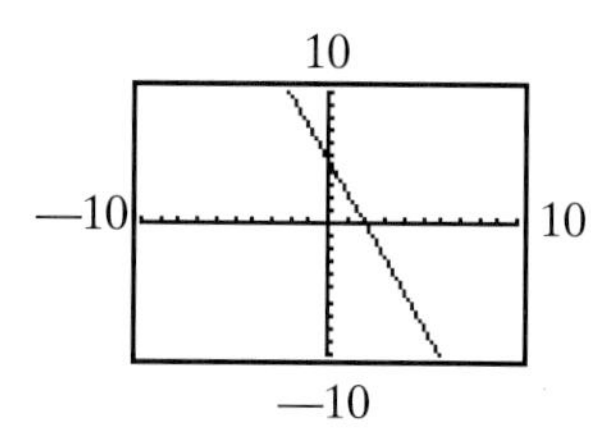

5. $y = x + 3.78$

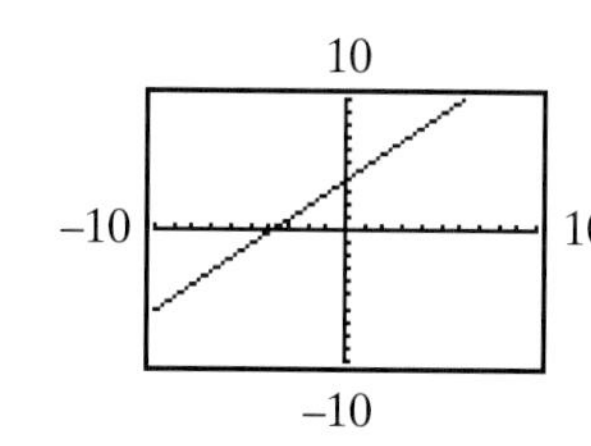

7. $y = 13.3x + 1.5$

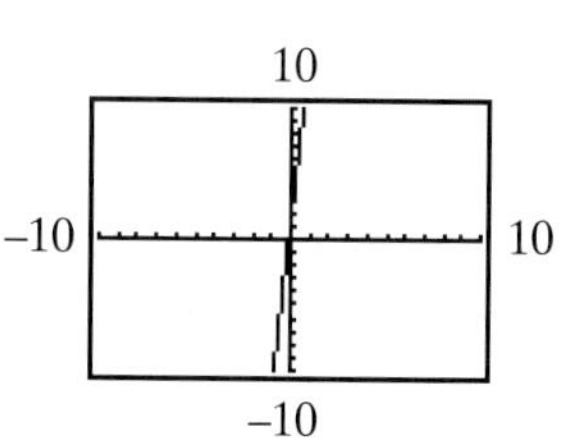

Mental Math

1. $m = -4, b = 12$ **3.** $m = 5, b = 0$ **5.** $m = \frac{1}{2}, b = 6$

Exercise Set 3.3

1.

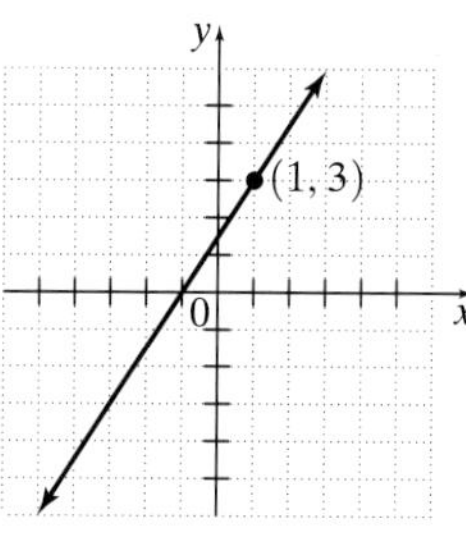

3.

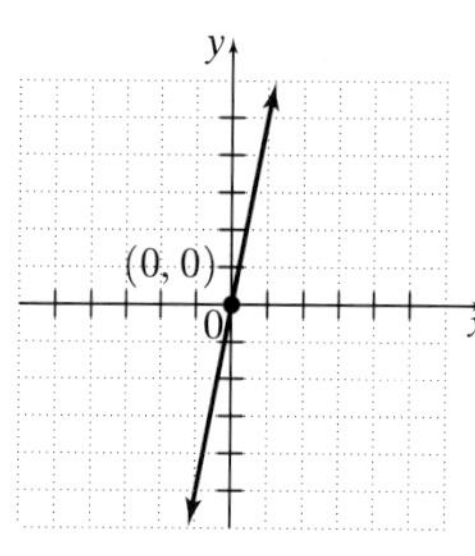

5.

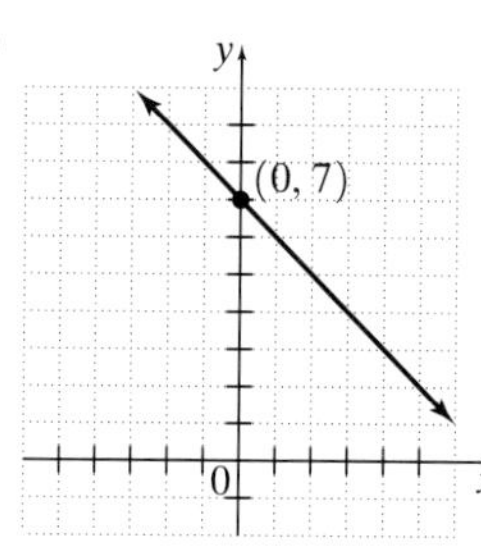

7.

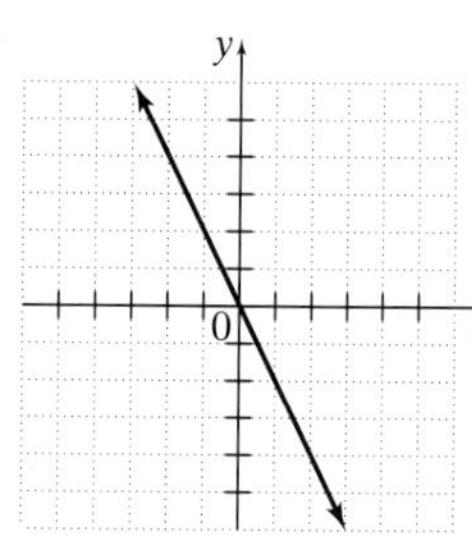

9.

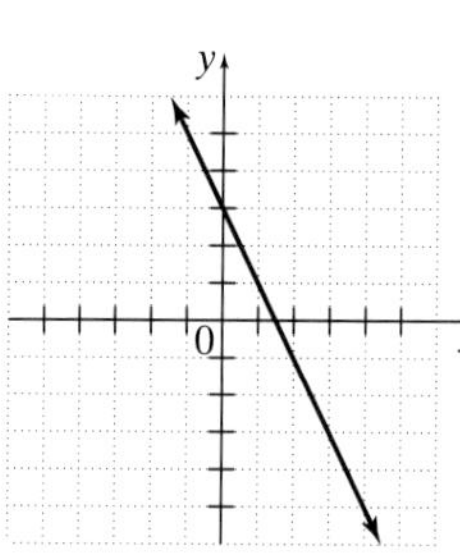

11.

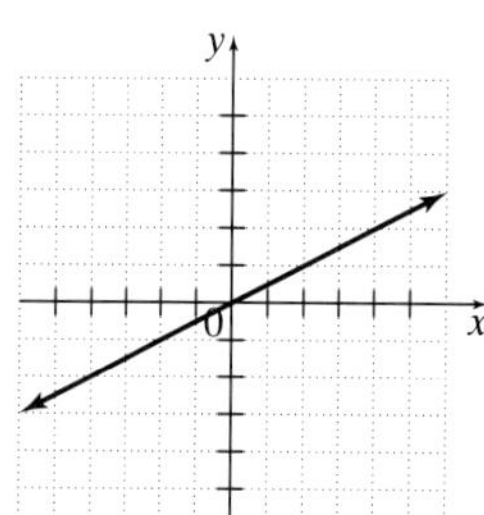

13.

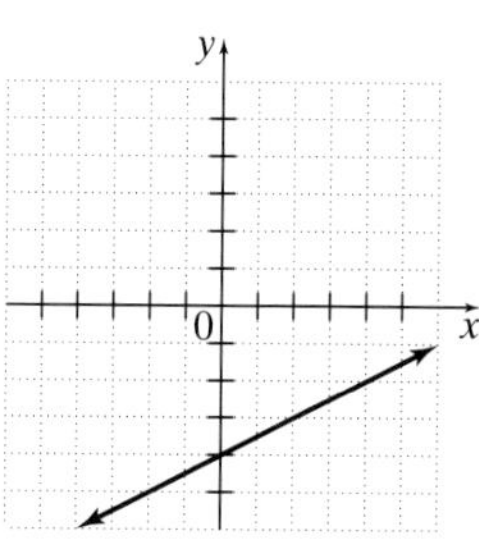

15.

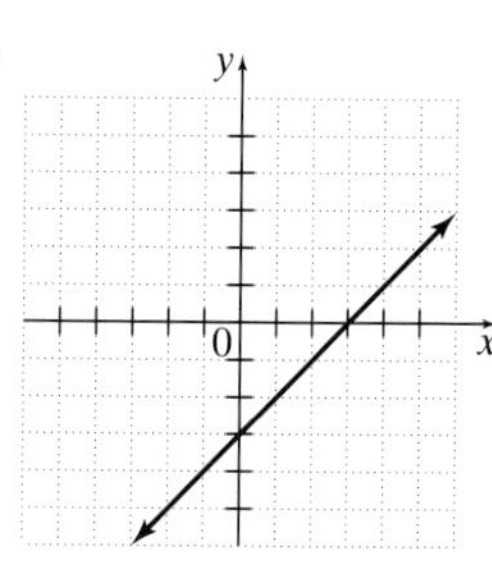

17.

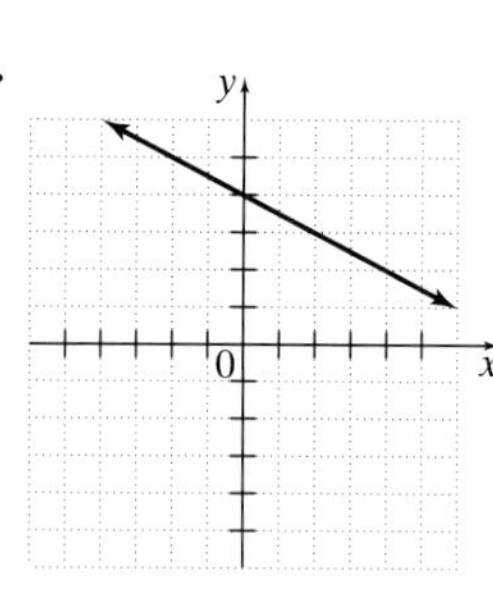

19. C **21.** D **23.** $y = -x + 1$ **25.** $y = 2x + \frac{3}{4}$ **27.** $y = \frac{2}{7}x$
29. a. $37,501.20 **b.** $m = 1431.5$; The annual income increases $1431.50 every year.
c. $b = 31{,}775.2$; At year $x = 0$, or 1992, the annual average income was $31,775.20.
31. a. $m = 37.8, b = 495$ **b.** The number of people employed as home health aids increases 37.8 thousand for every 1 year. **c.** There were 495 thousand home health aids employed in 1996.
33. a. $4665.40 **b.** 2005 **c.** answers may vary **35.** $y = 5x + 32$ **37.** $y = 2x - 1$
39. answers may vary **41.** $y = -7x + 500$ where y is the height at time x.

Calculator Explorations

1. 18.4 **3.** −1.5 **5.** 8.7, 7.6

Mental Math

1. $m = -2; (1, 4)$ **3.** $m = \frac{1}{4}; (2, 0)$ **5.** $m = 5; (3, -2)$

Exercise Set 3.4

1. $y = 3x - 1$ **3.** $y = -2x - 1$ **5.** $y = \frac{1}{2}x + 5$ **7.** $y = -\frac{9}{10}x - \frac{27}{10}$ **9.** $y = 2x + 7$ **11.** $y = -\frac{4}{3}x - \frac{20}{3}$ **13.** $y = 3x - 6$ **15.** $y = -2x + 1$ **17.** $y = -\frac{1}{2}x - 5$ **19.** $y = \frac{1}{3}x - 7$ **21.** $y = -\frac{2}{7}x - 6$ **23.** $y = -\frac{3}{8}x + \frac{5}{8}$ **25.** $x = 2$ **27.** $y = 1$ **29.** $x = 0$ **31.** $y = 4x - 4$ **33.** $y = \frac{1}{2}x - 6$ **35.** $y = -\frac{3}{2}x - 6$ **37.** $y = -\frac{1}{2}x + \frac{13}{2}$ **39.** $y = -\frac{1}{2}x + 1$ **41.** $y = -4x + 1$ **43. a.** $y = 32x$ **b.** 128 ft per sec **45. a.** $y = 12{,}000x + 18{,}000$ **b.** \$102,000 **c.** 9 yrs **47. a.** $y = \frac{13{,}300}{3}x + 109{,}900$ **b.** \$145,366.67 **49. a.** $y = 16.6x + 225$ **b.** 357.8 thousand people **51.** 31 **53.** −8.4 **55.** 4 **57.** true **59.** $x = 5$ **61.**

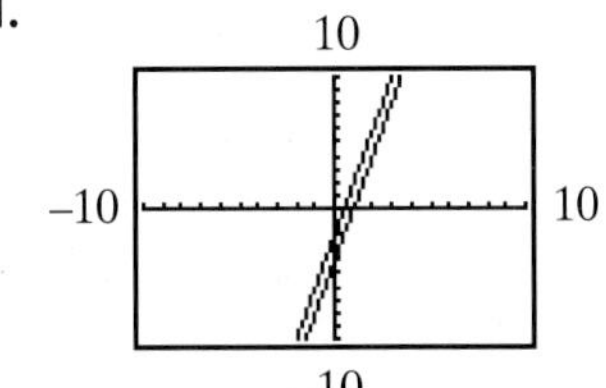

Integrated Review

1.

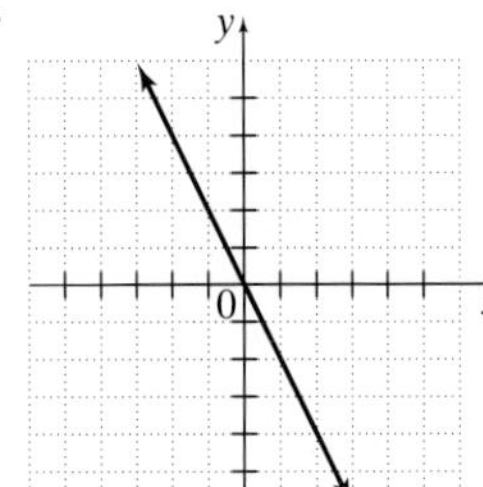

2.

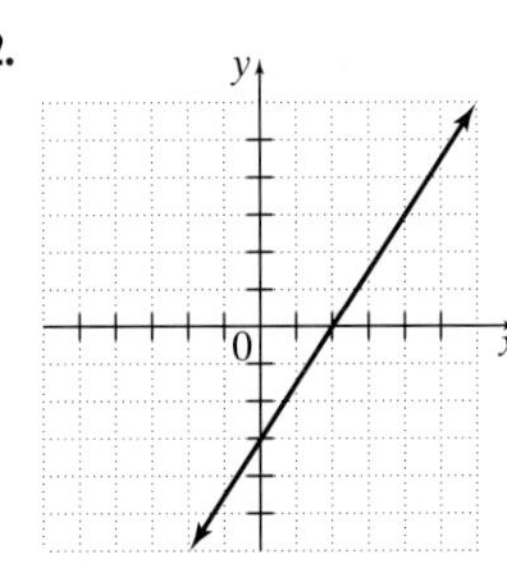

3.

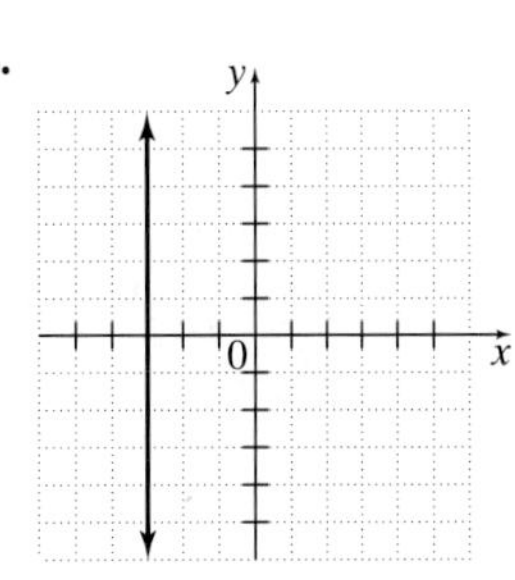

4.

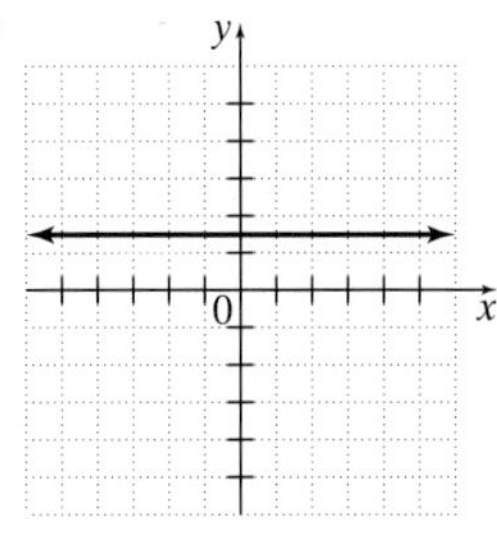

5. 0 **6.** $-\frac{3}{5}$ **7.** $m = 3; b = -5$ **8.** $m = \frac{5}{2}; b = -\frac{7}{2}$ **9.** parallel **10.** perpendicular **11.** $y = -x + 7$ **12.** $x = -2$ **13.** $y = 0$ **14.** $y = -\frac{3}{8}x - \frac{29}{4}$ **15.** $y = -5x - 6$ **16.** $y = -4x + \frac{1}{3}$ **17.** $y = \frac{1}{2}x - 1$ **18.** $y = 3x - \frac{3}{2}$ **19.** $y = 3x - 2$ **20.** $y = -\frac{5}{4}x + 4$ **21.** $y = \frac{1}{4}x - \frac{7}{2}$ **22.** $y = -\frac{5}{2}x - \frac{5}{2}$ **23.** $x = -1$ **24.** $y = 3$

Exercise Set 3.5

1. domain: $\{-1, 0, -2, 5\}$; range: $\{7, 6, 2\}$; function **3.** domain: $\{-2, 6, -7\}$; range: $\{4, -3, -8\}$; not a function **5.** domain: $\{1\}$; range: $\{1, 2, 3, 4\}$; not a function **7.** domain: $\left\{\frac{3}{2}, 0\right\}$; range: $\left\{\frac{1}{2}, -7, \frac{4}{5}\right\}$; not a function **9.** domain: $\{-3, 0, 3\}$; range: $\{-3, 0, 3\}$; function **11.** domain: $\{-1, 1, 2, 3\}$; range: $\{2, 1\}$; function **13.** domain: {Colorado, Alaska, Delaware, Illinois, Connecticut, Texas}; range: $\{6, 1, 20, 30\}$; function **15.** domain: $\{32°, 104°, 212°, 50°\}$; range: $\{0°, 40°, 10°, 100°\}$; function **17.** domain: $\{2, -1, 5, 100\}$; range: $\{0\}$; function **19.** function **21.** yes **23.** no **25.** yes **27.** function **29.** not a function **31.** function **33.** domain: $[0, \infty)$; range: $(-\infty, \infty)$; not a function **35.** domain: $[-1, 1]$; range: $(-\infty, \infty)$; not a function **37.** domain: $(-\infty, \infty)$; range: $(-\infty, -3] \cup [3, \infty)$; not a function **39.** domain: $[1, 7]$; range: $[1, 7]$; not a function **41.** domain: $\{-2\}$; range: $(-\infty, \infty)$; not a function **43.** domain: $(-\infty, \infty)$; range: $(-\infty, 3]$; function **45.** 15 **47.** 38 **49.** 7 **51.** 3 **53. a.** 0 **b.** 1 **c.** −1 **55. a.** −5 **b.** −5 **c.** −5 **57.** 25π sq. cm **59.** 2744 cu. in. **61.** 166.38 cm **63.** 163.2 mg **65. a.** 91.4; The per capita consumption of poultry was 91.4 lb in 1997. **b.** 94.8 lb

67.

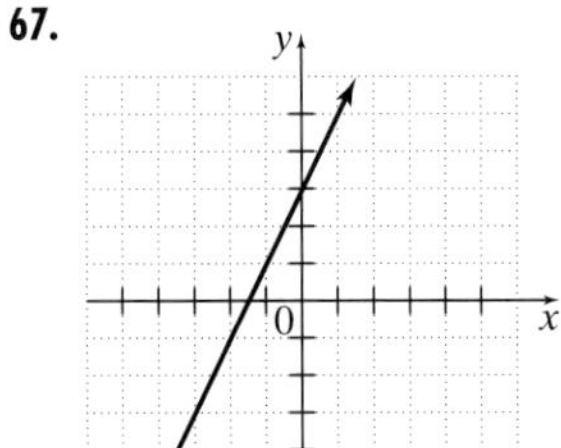

69.

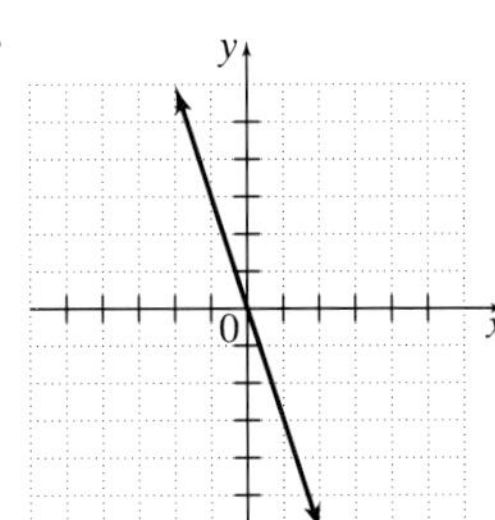

71.

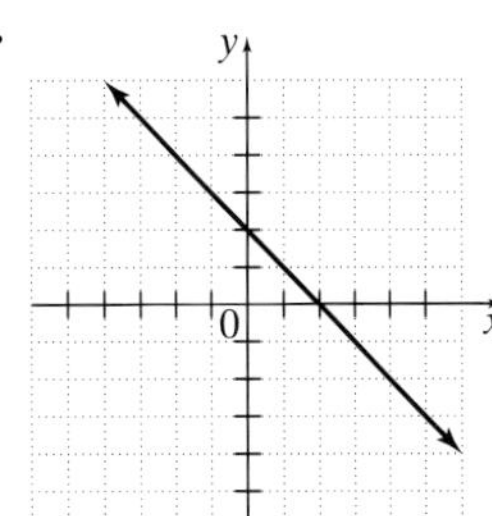

73. $(-\infty, 14]$ **75.** $\left[\frac{7}{2}, \infty\right)$ **77.** $\left(-\infty, -\frac{1}{4}\right)$ **79. a.** 5.1 **b.** 15.5 **c.** 9.533 **81. a.** 132 **b.** $a^2 - 12$ **83.** answers may vary **85.** answers may vary

ANSWERS

Exercise Set 3.6

1.
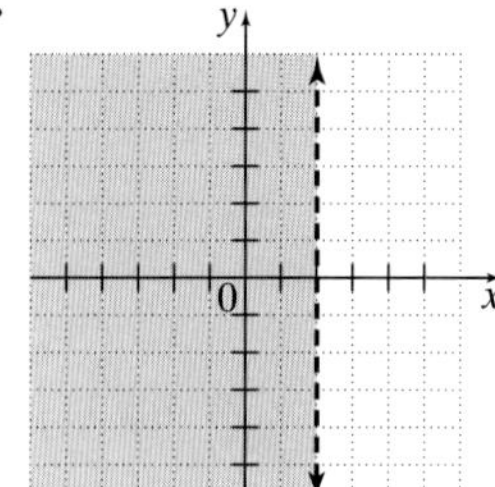

3.
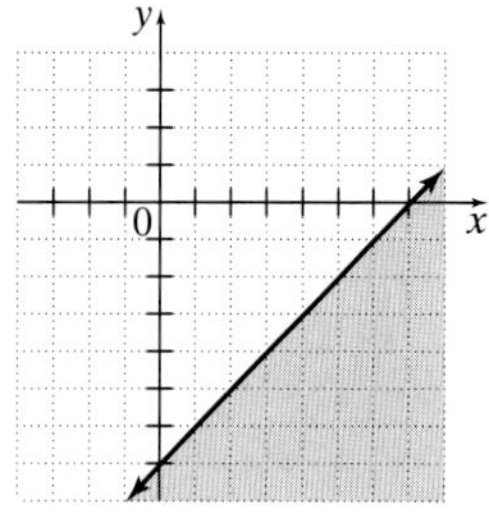

5.
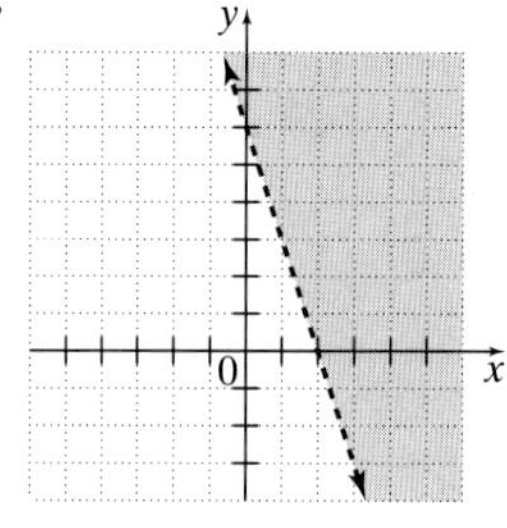

7.
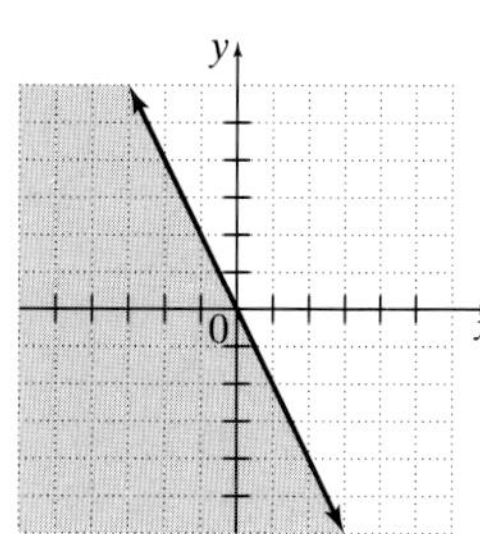

9.
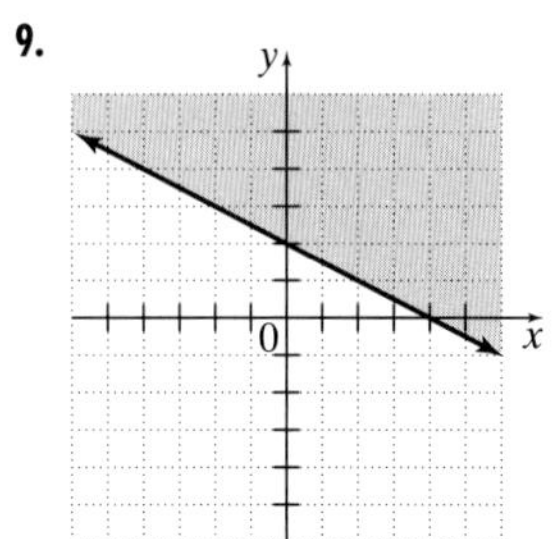

11.
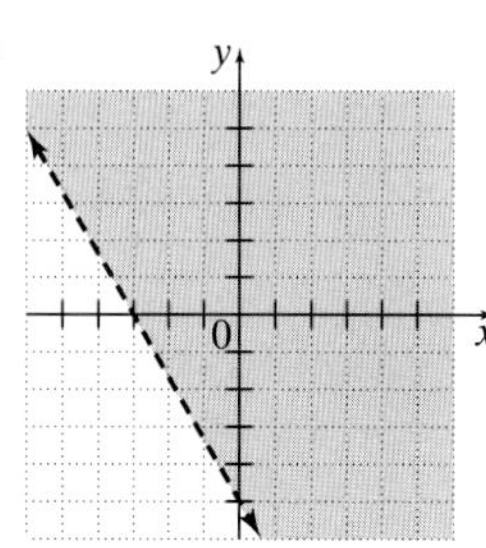

13. with $<$ or $>$

15.
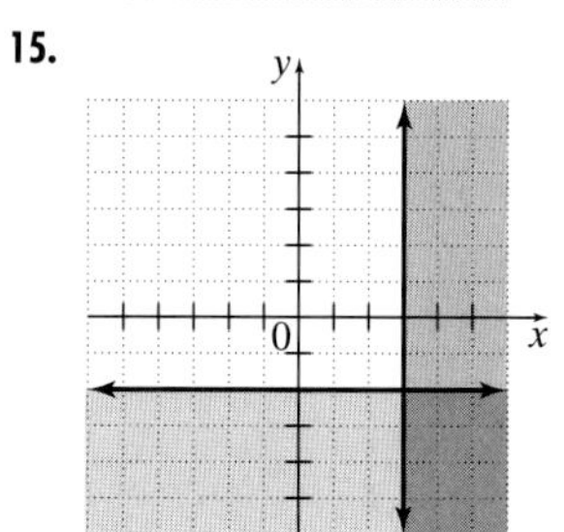

17.
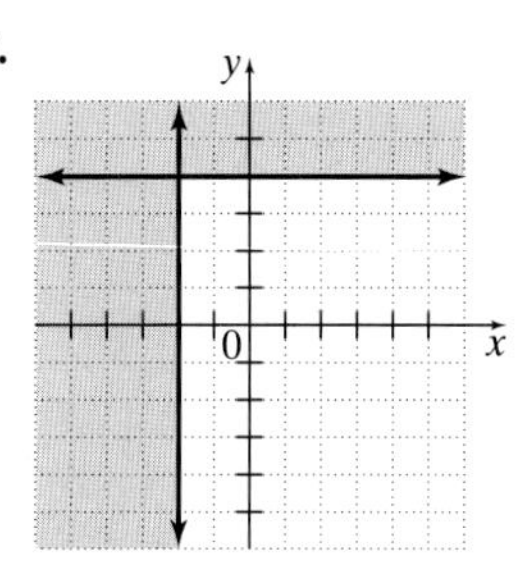

19.
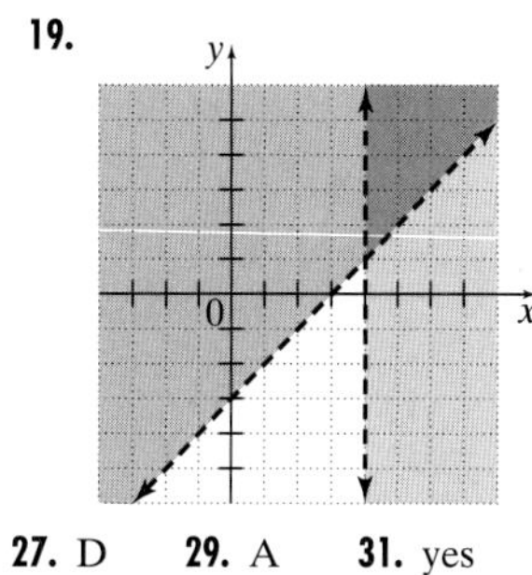

21.
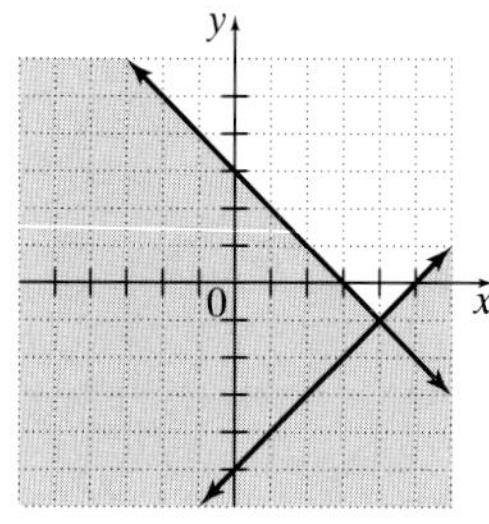

23.
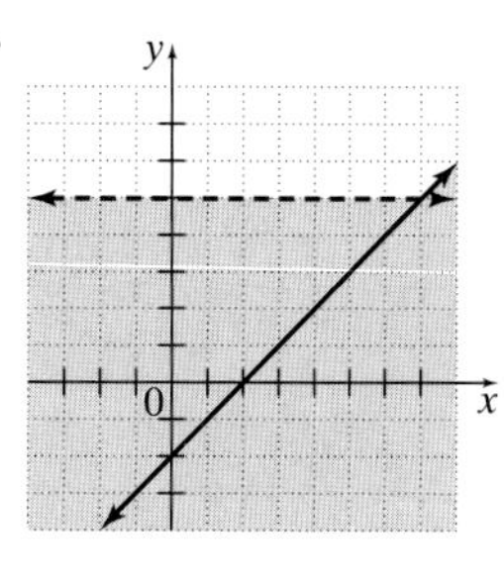

25.
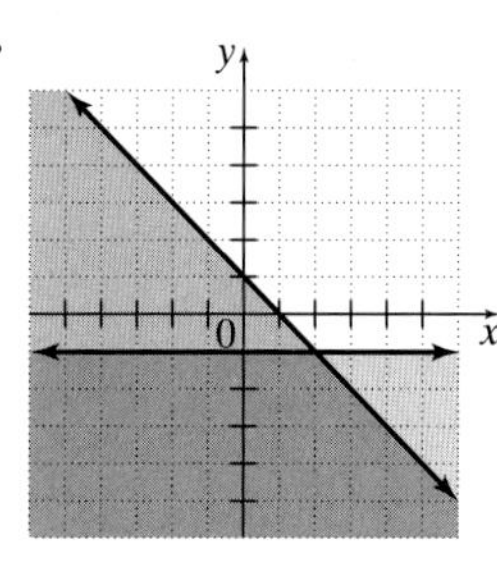

27. D 29. A 31. yes 33. no

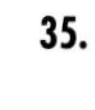

35.
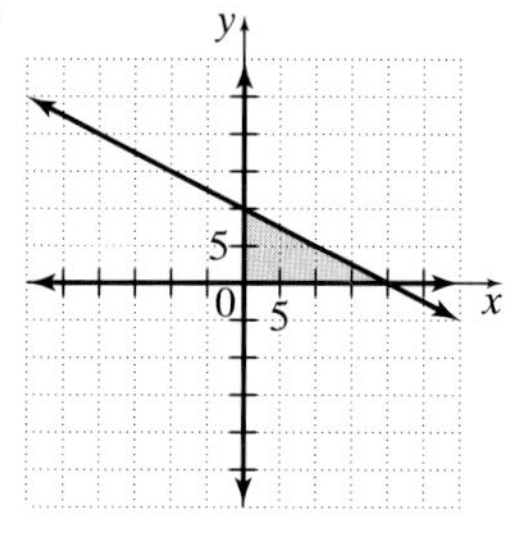

Chapter 3 Review

1.
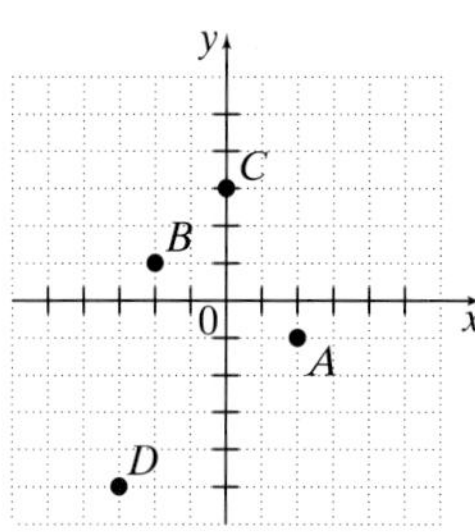

A lies in quadrant IV.
B lies in quadrant II.
C lies on the *y*-axis, no quadrant.
D lies in quadrant III.

2.
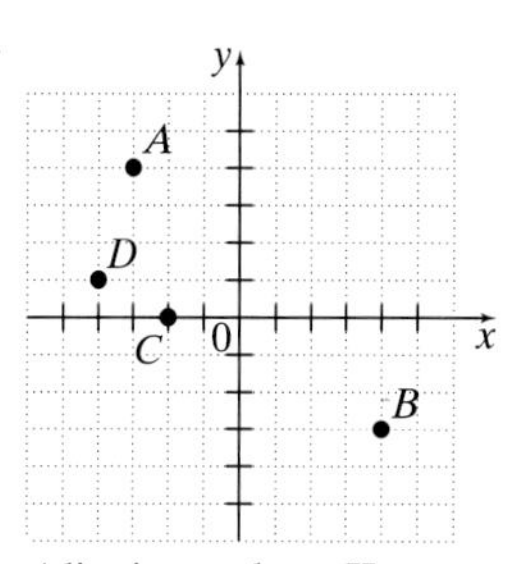

A lies in quadrant II.
B lies in quadrant IV.
C lies on the *x*-axis, no quadrant.
D lies in quadrant II.

3.
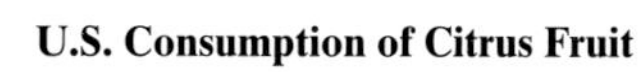
U.S. Consumption of Citrus Fruit

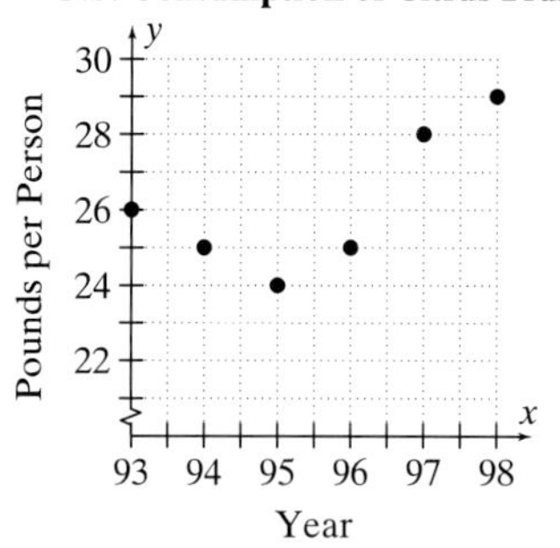

4. U.S. Armed Forces Active Duty Personnel

Armed Forces (in millions)

Year

5. no, yes **6.** yes, no **7.** yes, yes **8.** yes, no

9.

10.

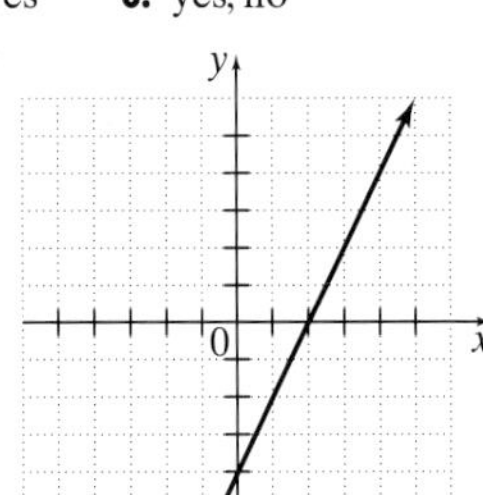

11.

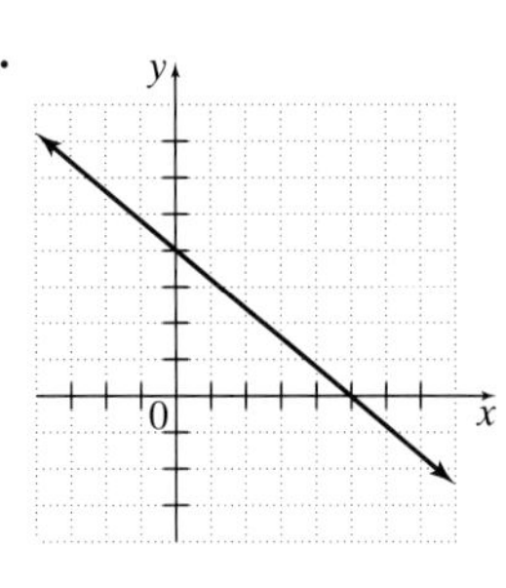

12.

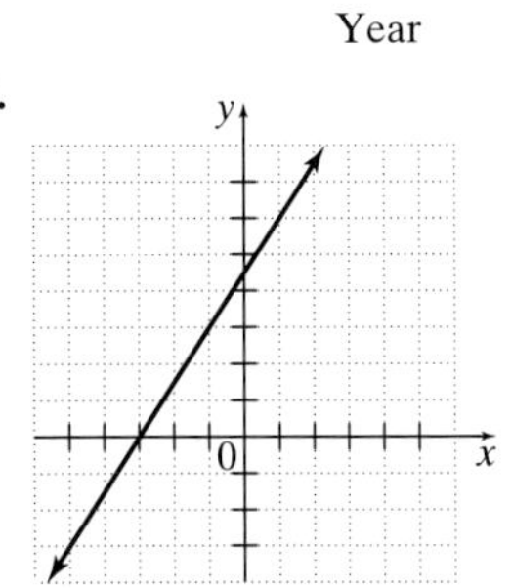

13.

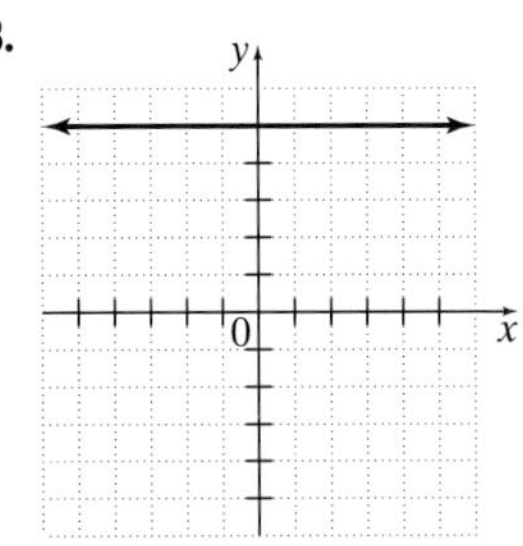

14.

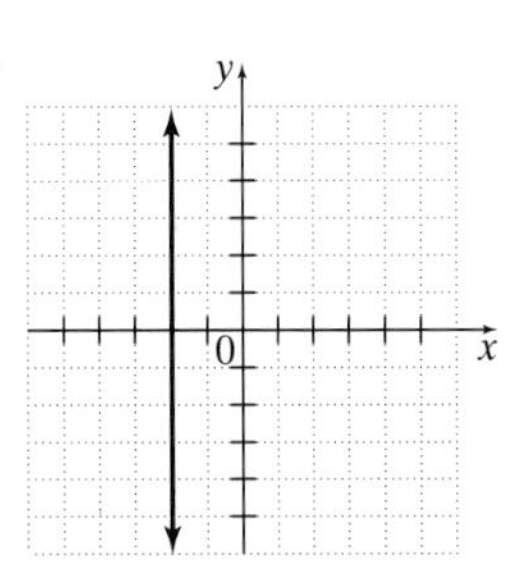

15.

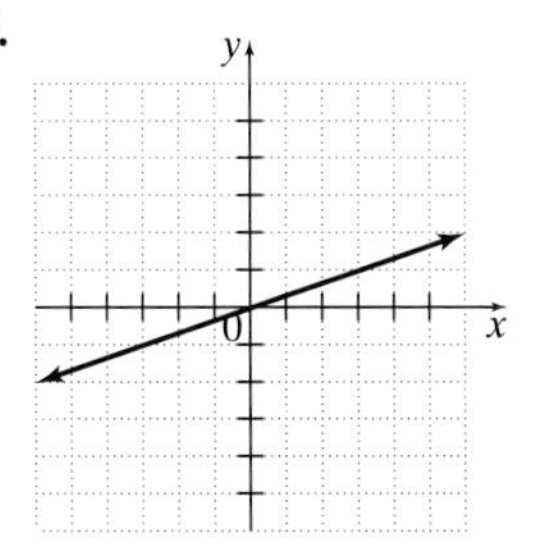

16.

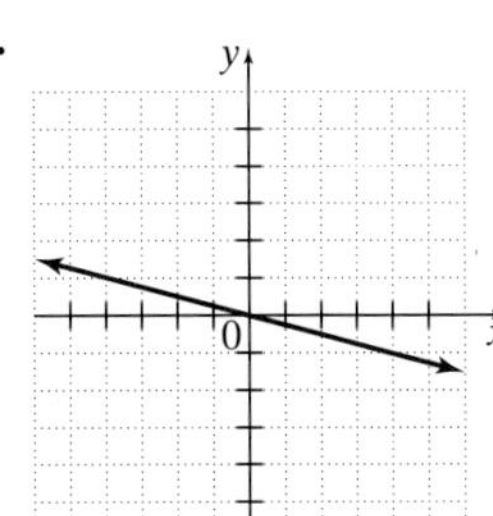

17. -3 **18.** $\frac{1}{2}$ **19.** $\frac{5}{2}$ **20.** $-\frac{3}{4}$ **21.** $\frac{3}{2}$ **22.** -3 **23.** $-\frac{1}{2}$ **24.** 1 **25.** l_2 **26.** l_2 **27.** l_2 **28.** l_1 **29.** $m = -3, b = \frac{1}{2}$ **30.** $m = 2, b = 4$ **31.** $m = \frac{2}{5}, b = -\frac{4}{3}$ **32.** $m = -\frac{2}{7}, b = \frac{3}{2}$ **33.** 0 **34.** undefined **35.** neither **36.** parallel **37.** perpendicular **38.** neither

39.

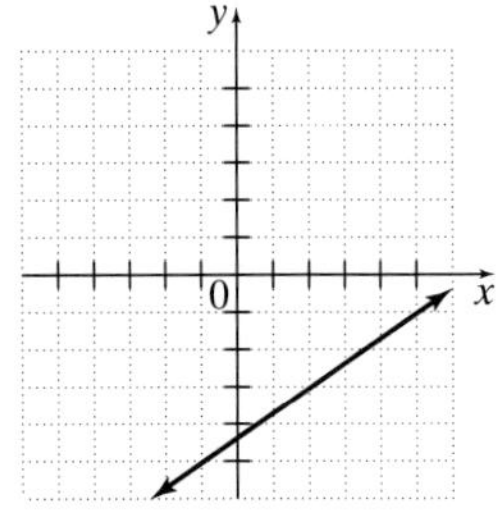

40.

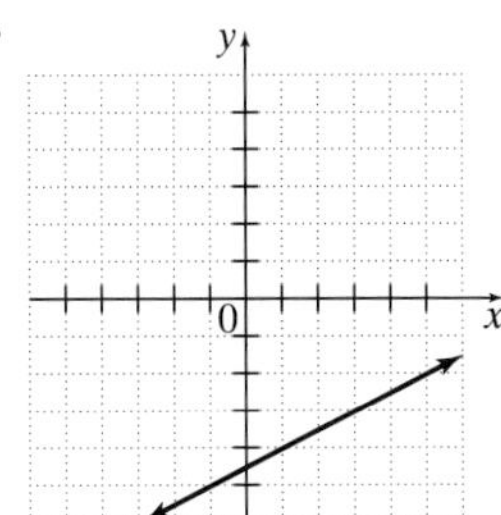

41.

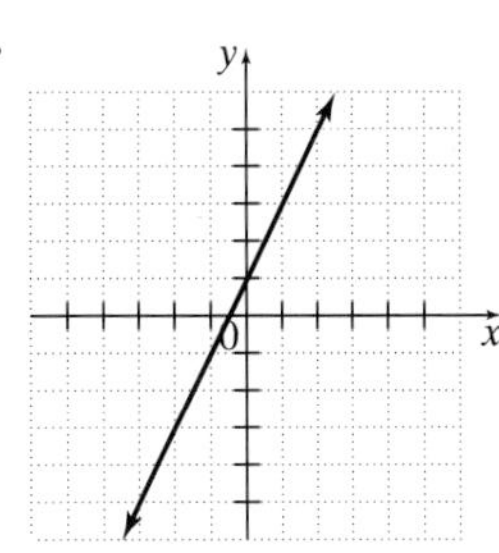

42.

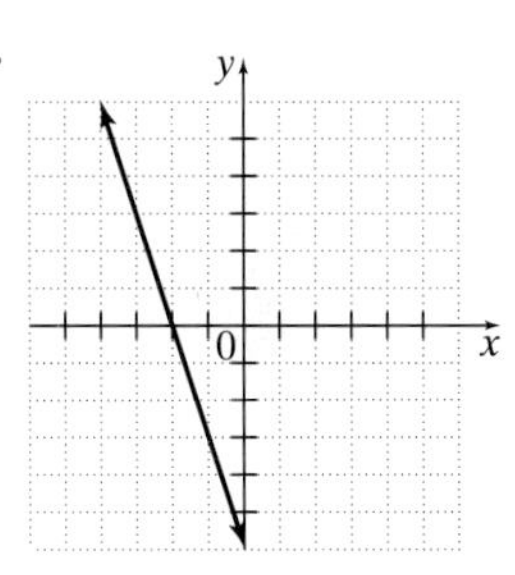

43.

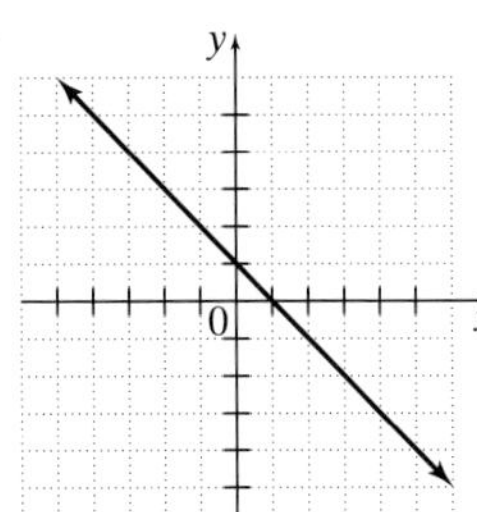

44.

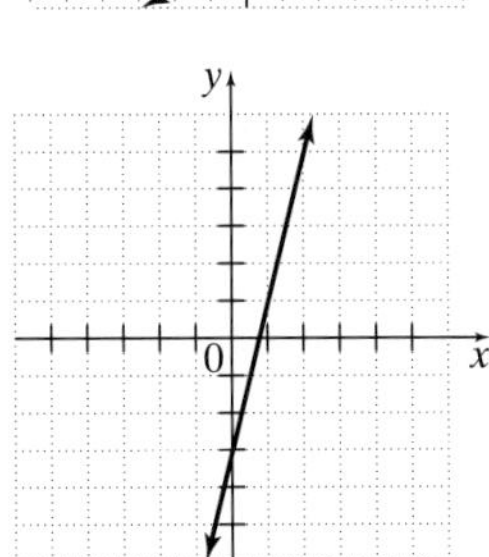

45.

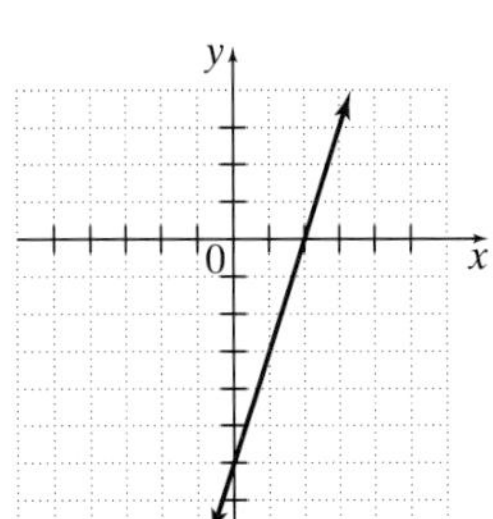

46.

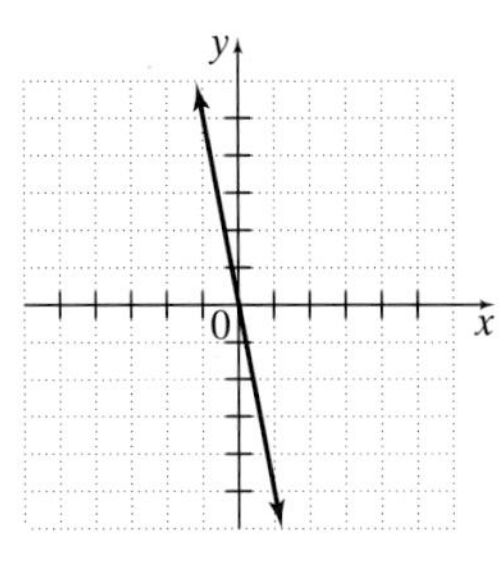

47. a. $C(150) = 87$ **b.** $m = 0.3$; The cost increases by \$0.30 for each additional mile driven. **c.** $b = 42$; The cost for 0 miles driven is \$42. **48.** $y = -1$ **49.** $x = -2$ **50.** $x = -4$ **51.** $y = 5$ **52.** $y = 3x + 14$ **53.** $y = 2x - 12$ **54.** $y = -\frac{1}{2}x - 4$ **55.** $y = -11x - 52$ **56.** $y = -2x - 2$ **57.** $y = -\frac{3}{2}x - 8$ **58.** $y = \frac{3}{4}x + \frac{7}{2}$ **59.** $y = -\frac{3}{2}x - 1$ **60. a.** $y = -800x + 4200$ **b.** \$200 **61. a.** $y = 3000x + 144{,}000$ **b.** \$219,000 **62.** domain: $\left\{-\frac{1}{2}, 6, 0, 25\right\}$; range: $\left\{\frac{3}{4}, 0.65, -12, 25\right\}$; function **63.** domain: $\left\{\frac{3}{4}, 0.65, -12, 25\right\}$; range: $\left\{-\frac{1}{2}, 6, 0, 25\right\}$; function **64.** domain: $\{2, 4, 6, 8\}$; range: $\{2, 4, 5, 6\}$; not a function **65.** domain: {triangle, square, rectangle, parallelogram}; range: $\{3, 4\}$; function **66.** domain: $(-\infty, \infty)$; range: $(-\infty, -1] \cup [1, \infty)$; not a function **67.** domain: $\{-3\}$; range: $(-\infty, \infty)$; not a function **68.** domain: $(-\infty, \infty)$; range: $\{4\}$; function

69. domain: $[-1, 1]$; range: $[-1, 1]$; not a function **70.** -3 **71.** 0 **72.** 18 **73.** 9 **74.** -3 **75.** 0 **76.** 381 lb **77.** 5080 lb

78.

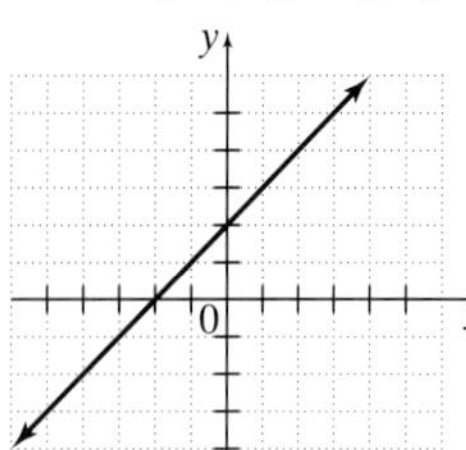

79.

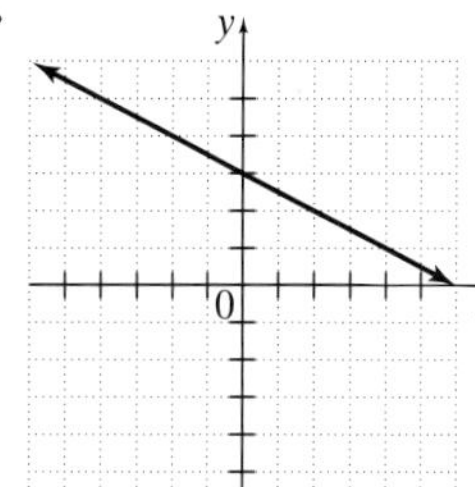

80.

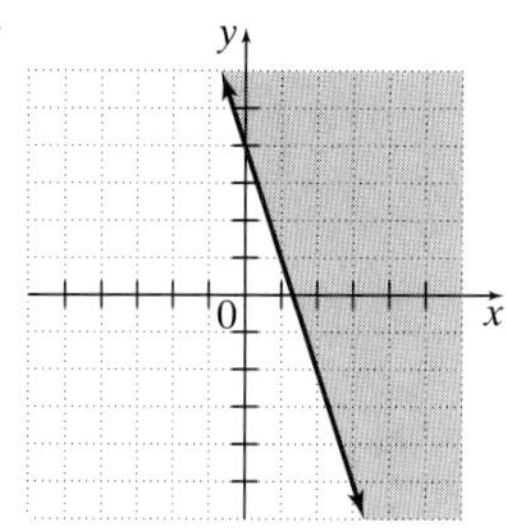

81.

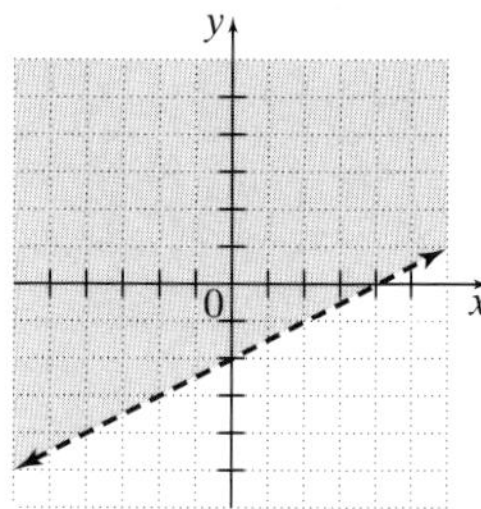

82.

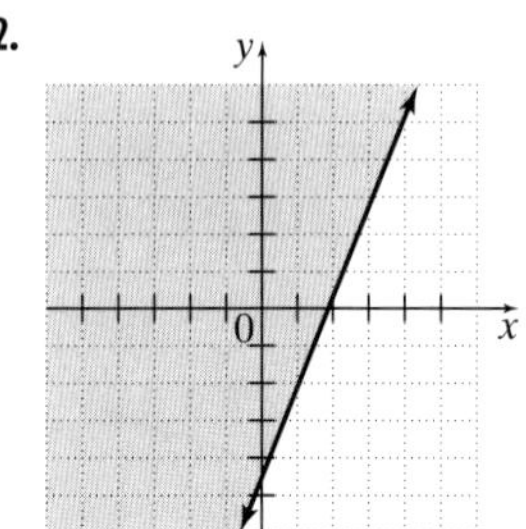

83.

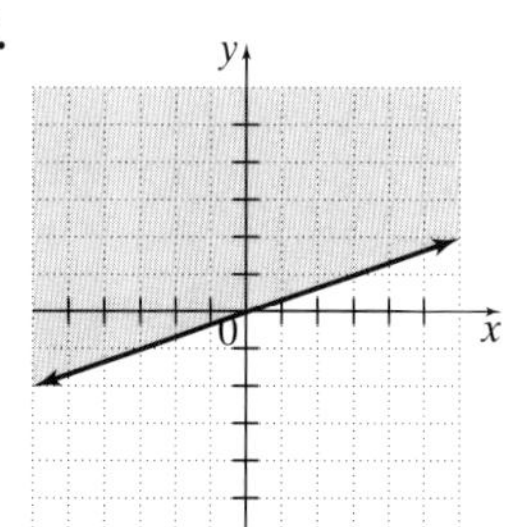

84.

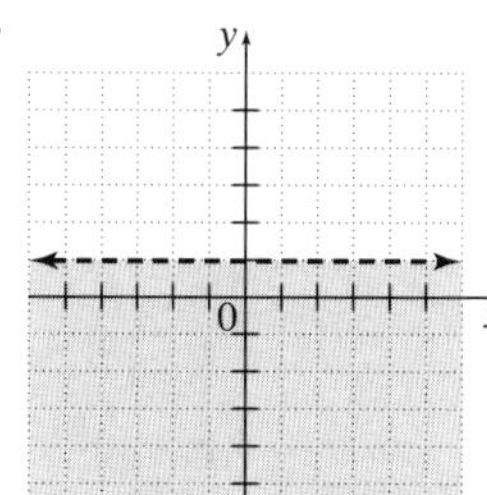

85.

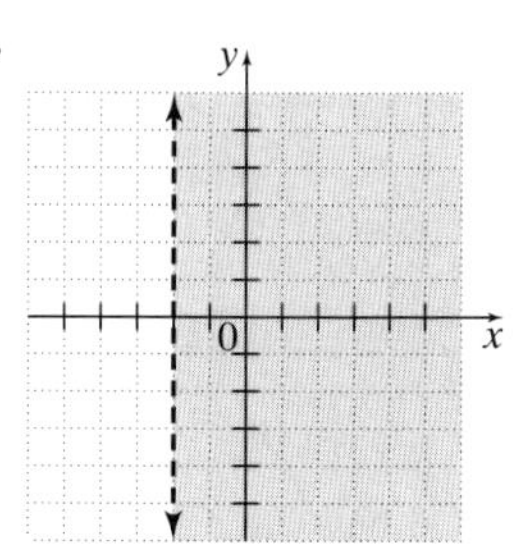

86.

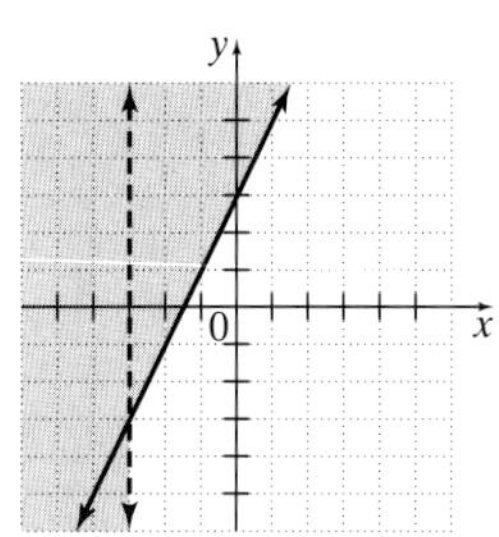

87.

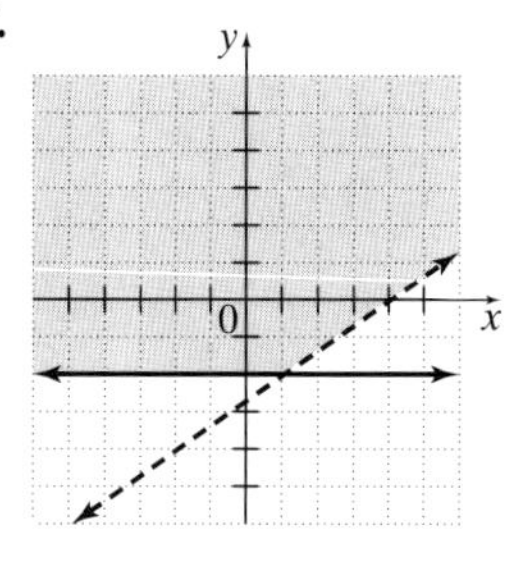

Chapter 3 Test

1.

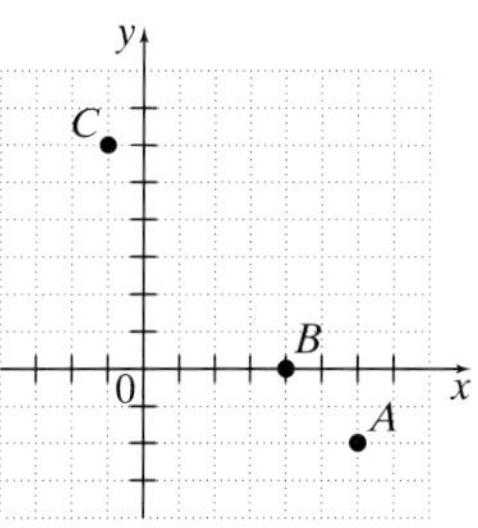

A is in quadrant IV.
B is on the x-axis, no quadrant.
C is in quadrant II.

2.

U.S. Cellular Phone Subscribers

Number of Subscribers (in millions)

Year

3.

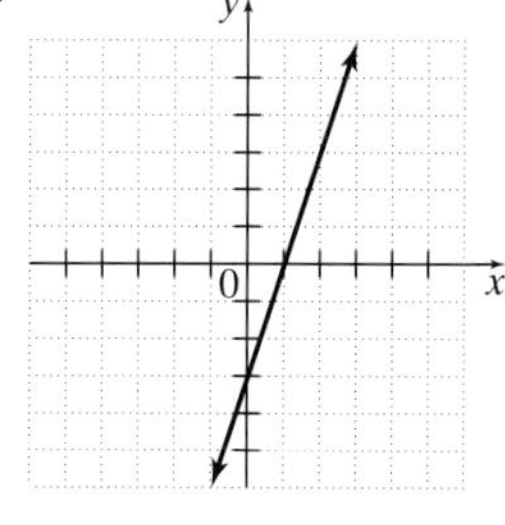

4.

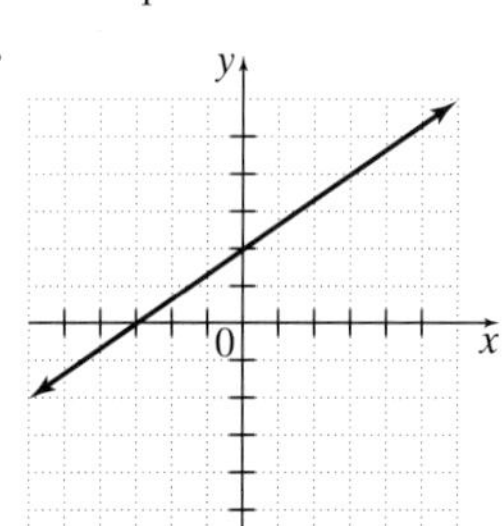

5.

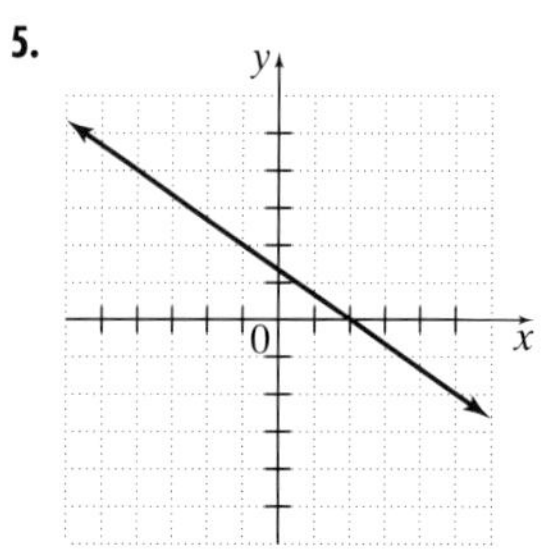

6.

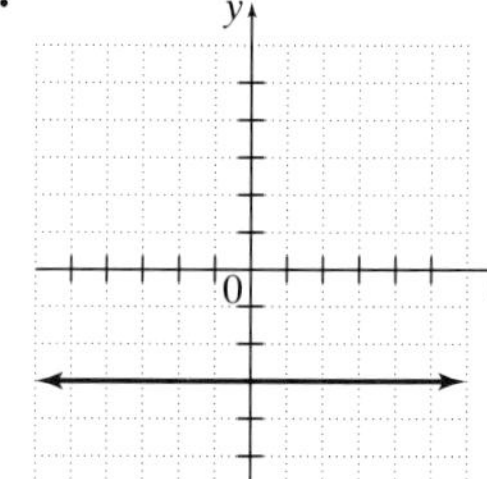

7. $m = -\frac{3}{2}$

8. $m = -\frac{1}{4}, b = \frac{2}{3}$ **9.** C

10. A **11.** B **12.** D

13. $y = -8$ **14.** $x = -4$

15. $y = -2$

16. $y = -3x + 11$

17. $y = 5x - 2$ **18.** $y = -\frac{1}{2}x$ **19.** $y = -\frac{1}{3}x + \frac{5}{3}$ **20.** $y = -\frac{1}{2}x - \frac{1}{2}$ **21.** neither **22. a.** \$17,110 **b.** \$24,190 **c.** 2009 **d.** 708; The earnings for high school graduates increase \$708 per year. **e.** 13,570; When $x = 0$, or in 1985, the average yearly earnings for a high school graduate was \$13,570.

23.

24.

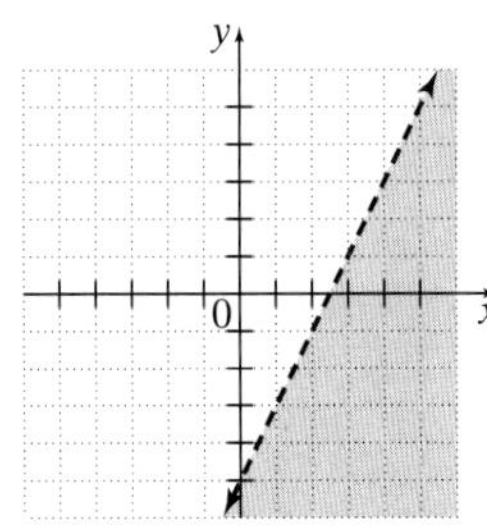

25.

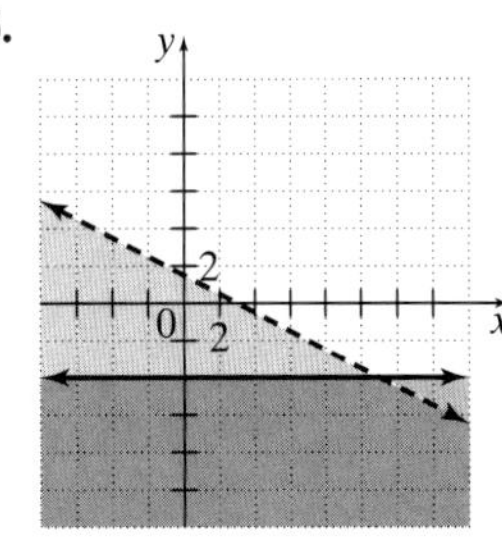

26. domain: $(-\infty, \infty)$; range: $\{5\}$; function
27. domain: $\{-2\}$; range: $(-\infty, \infty)$; not a function
28. domain: $(-\infty, \infty)$, range: $[0, \infty)$; function
29. domain: $(-\infty, \infty)$; range: $(-\infty, \infty)$; function

Cumulative Review

1. $\{2, 3, 4, 5\}$; Sec. 1.1, Ex. 4 **2.** $\{101, 102, 103, ...\}$; Sec. 1.1, Ex. 5 **3.** $-\frac{1}{9}$; Sec. 1.2, Ex. 21 **4.** $\frac{4}{7}$; Sec. 1.2, Ex. 22 **5.** 3; Sec. 1.3, Ex. 1
6. -2; Sec. 1.3, Ex. 4 **7.** 13; Sec. 1.4, Ex. 5 **8.** 2^7; Sec. 1.5, Ex. 1 **9.** y^7; Sec. 1.5, Ex. 3 **10.** $125x^6$; Sec. 1.6, Ex. 5
11. $64y^2$; Sec. 1.6, Ex. 8 **12.** $c = 0.4$; Sec. 2.1, Ex. 4 **13.** 23, 49; Sec. 2.2, Ex. 1 **14.** \$11,607.55; Sec. 2.3, Ex. 4
15. −4 −3 −2 −1 0 1 2 3 4 ; $[2, \infty)$; Sec. 2.4, Ex. 1 **16.** −4 −3 −2 −1 0 1 2 3 4 ; $(0.5, 3]$; Sec. 2.4, Ex. 3
17. $(-\infty, 4)$; Sec. 2.5, Ex. 2 **18.** $\{4\}$; Sec. 2.6, Ex. 8 **19.** $(-\infty, -4) \cup (10, \infty)$; Sec. 2.6, Ex. 11
20.

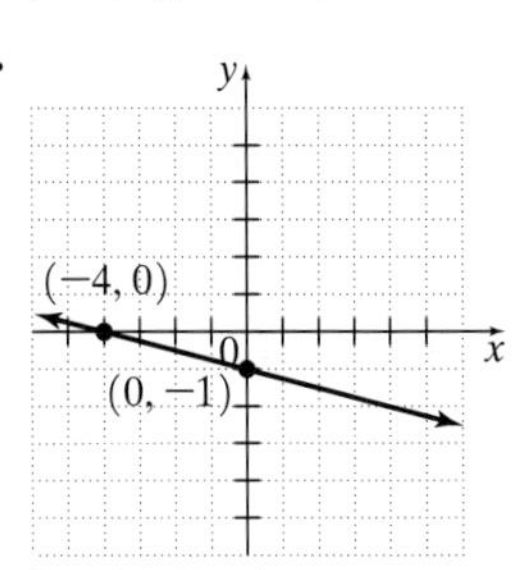

Sec. 3.1, Ex. 5

21. $m = 3$; Sec. 3.2, Ex. 3 **22.** $y = \frac{1}{4}x - 3$; Sec. 3.3, Ex. 3 **23.** $y = 3$; Sec. 3.4, Ex. 3
24. 5; Sec. 3.5, Ex. 20 **25.** -2; Sec. 3.5, Ex. 19

Chapter 4

Chapter 4 Pretest

1. $(2, -5)$; 4.1B **2.** $(-1, -3)$; 4.1B **3.** $(5, 6)$; 4.1C **4.** $(0, -4)$; 4.1C **5.** $(7, 9)$; 4.1D **6.** $(-5, -8)$; 4.1D **7.** $(2, -1, -3)$; 4.2A
8. $(1, 0, -2)$; 4.2A **9.** $(10, -10)$; 4.4A **10.** $\varnothing$; 4.4A **11.** $(-3, -1, 2)$; 4.4B **12.** $(0, 8, 0)$; 4.4B **13.** 32; 4.5A **14.** 10; 4.5C
15. $(-2, -11)$; 4.5B
16. $(1, 3, -2)$; 4.5D
17.

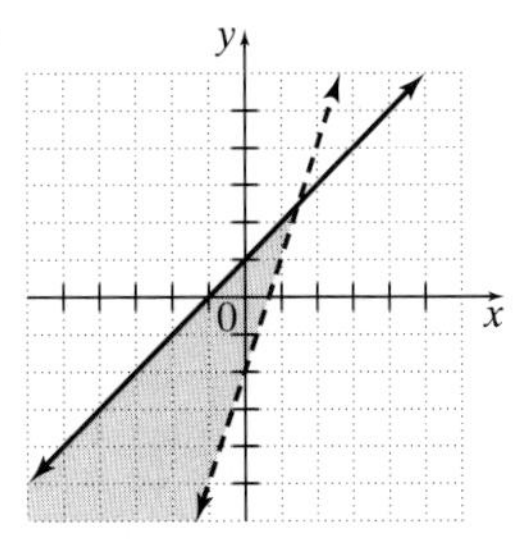

18.

19. 4 and 12; 4.3A **20.** 20°, 60°, 100°; 4.3B

Calculator Explorations

1. $(2.11, 0.17)$ **3.** $(0.57, -1.97)$

Mental Math

1. B **3.** A

Exercise Set 4.1

1. yes **3.** no **5.** yes **7.**

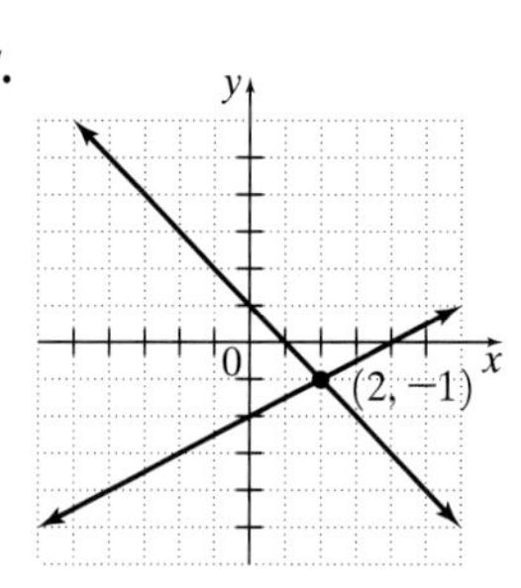

9.

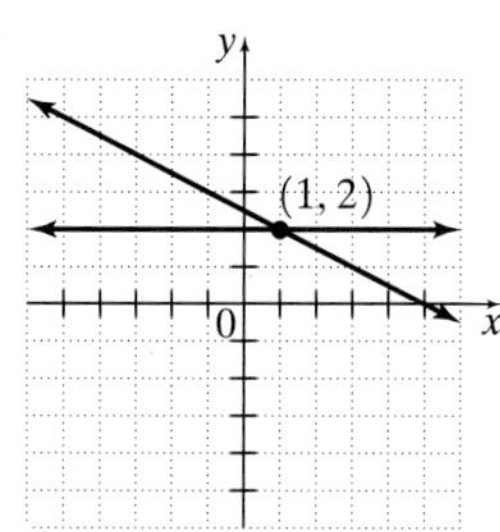

11. ∅

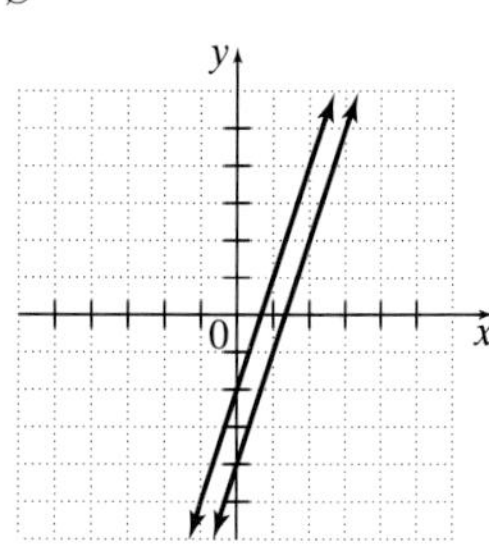

13.

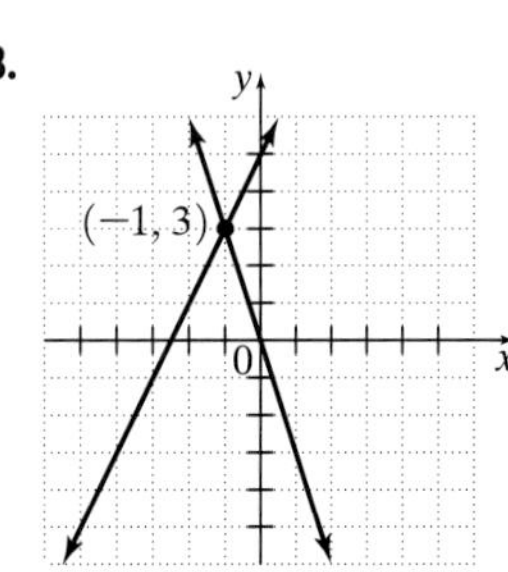

15. $(2, 8)$ **17.** $(0, -9)$ **19.** $(1, -1)$ **21.** $(-5, 3)$ **23.** $\left(-\frac{1}{4}, \frac{1}{2}\right)$ **25.** $(3, 2)$ **27.** $\left(\frac{5}{2}, \frac{5}{4}\right)$ **29.** $(1, -2)$ **31.** $(9, 9)$ **33.** $(7, 2)$ **35.** ∅ **37.** $\{(x, y)|3x + y = 1\}$ **39.** ∅ **41.** ∅ **43.** $(8, 2)$ **45.** $(3, 4)$ **47.** $(-2, 1)$ **49.** $\left(\frac{1}{2}, \frac{1}{5}\right)$ **51.** $\{(x, y)|x = 3y + 2\}$ **53.** true **55.** false **57.** $6y - 4z = 25$ **59.** $x + 10y = 2$ **61.** no **63.** 5000 ties; \$21 **65.** Supply is greater than demand. **67. a.** Consumption of red meat is decreasing while consumption of poultry is increasing. **b.** $(14, 113)$ **c.** In the year 2009, red meat and poultry consumption will each be about 113 lb per person.

Exercise Set 4.2

1. $(-2, 5, 1)$ **3.** $(-2, 3, -1)$ **5.** $\{(x, y, z)|x - 2y + z = -5\}$ **7.** ∅ **9.** $(0, 0, 0)$ **11.** $(-3, -35, -7)$ **13.** $(6, 22, -20)$ **15.** ∅ **17.** $(3, 2, 2)$ **19.** $\{(x, y, z)|x + 2y - 3z = 4\}$ **21.** $(-3, -4, -5)$ **23.** $(12, 6, 4)$ **25.** $(1, 1, -1)$ **27.** 15 and 30 **29.** $\{5\}$ **31.** $\left\{-\frac{15}{9}\right\}$ **33.** answers may vary **35.** $(1, 1, 0, 2)$ **37.** $(1, -1, 2, 3)$

Exercise Set 4.3

1. 10 and 8 **3.** plane, 520 mph; wind, 40 mph **5.** 20 qts of 4%; 40 qts of 1% **7.** 9 large frames; 13 small frames **9.** −10 and −8 **11.** tablets: \$0.80; pens: \$0.20 **13.** speed of plane: 630 mph; speed of wind: 90 mph **15.** 5 in., 7 in., 7 in., 10 in. **17.** 18, 13, and 9 **19.** \$2000 **21.** template: \$1.90; pencil: \$0.75; pad: \$2.25 **23.** $x = 40$; $y = 70$ **25.** 120 L of 25%, 60 L of 40% and 20 L of 50% **27.** 565 free throws, 851 two-point field goals, 30 three-point field goals **29.** $x = 95$, $y = 123$, $z = 70$ **31.** 750 units **33.** 750 units **35.** 500 units **37. a.** $R(x) = 31x$ **b.** $C(x) = 15x + 500$ **c.** 31.25, or 32 baskets **39.** $3y + 8z = 18$ **41.** $-5x - 5z = -16$ **43. a.** 68,100 metric tons **b.** 126,757 metric tons **c.** CFC emissions decreased, HCFC emissions increased and total CFC/HCFC emissions decreased; answers may vary

Integrated Review

1. C **2.** D **3.** A **4.** B **5.** $(1, 3)$ **6.** $\left(\frac{4}{3}, \frac{16}{3}\right)$ **7.** $(2, -1)$ **8.** $(5, 2)$ **9.** $\left(\frac{3}{2}, 1\right)$ **10.** $\left(-2, \frac{3}{4}\right)$ **11.** ∅ **12.** $\{(x, y)|2x - 5y = 3\}$ **13.** $(-1, 3, 2)$ **14.** $(1, -3, 0)$ **15.** ∅ **16.** $\{(x, y, z)|x - y + 3z = 2\}$ **17.** $\left(2, 5, \frac{1}{2}\right)$ **18.** $\left(1, 1, \frac{1}{3}\right)$ **19.** 19 and 27 **20.** 70°, 70°, 100°, 120°

Exercise Set 4.4

1. $(2, -1)$ **3.** $(-4, 2)$ **5.** ∅ **7.** $\{(x, y)|x - y = 3\}$ **9.** $(4, -3)$ **11.** $(-2, 5, -2)$ **13.** $(1, -2, 3)$ **15.** $(2, 1, -1)$ **17.** $(1, -4, 3)$ **19.** −13 **21.** −36 **23.** 0 **25. a.** end of 1984 **b.** black-and-white sets; microwave ovens; The percent of households owning black-and-white television sets is decreasing and the percent of households owning microwave ovens is increasing.; answers may vary **c.** in 2002

Mental Math

1. 56 **3.** −32 **5.** 20

Exercise Set 4.5

1. 26 **3.** −19 **5.** 0 **7.** $\frac{13}{6}$ **9.** $(1, 2)$ **11.** $\{(x, y)|3x + y = 1\}$ **13.** $(9, 9)$ **15.** $(-3, -2)$ **17.** $(3, 4)$ **19.** 8 **21.** 0 **23.** 15 **25.** 54 **27.** $(-2, 0, 5)$ **29.** $(6, -2, 4)$ **31.** $(-2, 3, -1)$ **33.** $(0, 2, -1)$ **35.** $x > -\frac{11}{2}$ **37.** $x \leq \frac{16}{3}$ **39.** 5 **41.** 0

Exercise Set 4.6

1.

3.

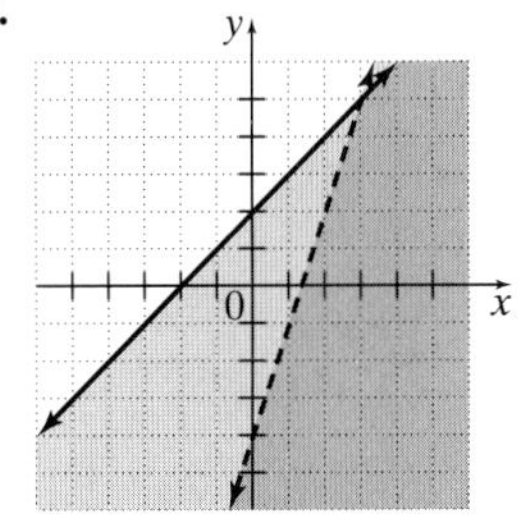

5.

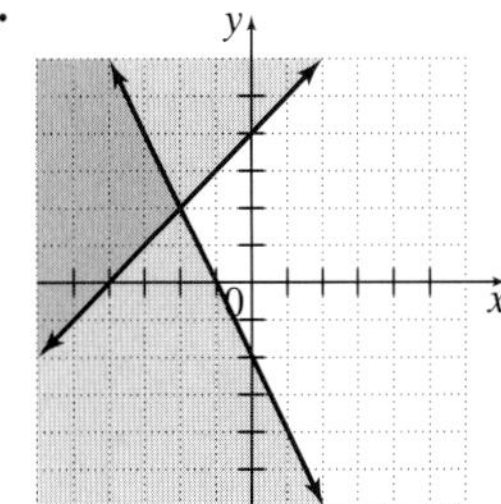

7.

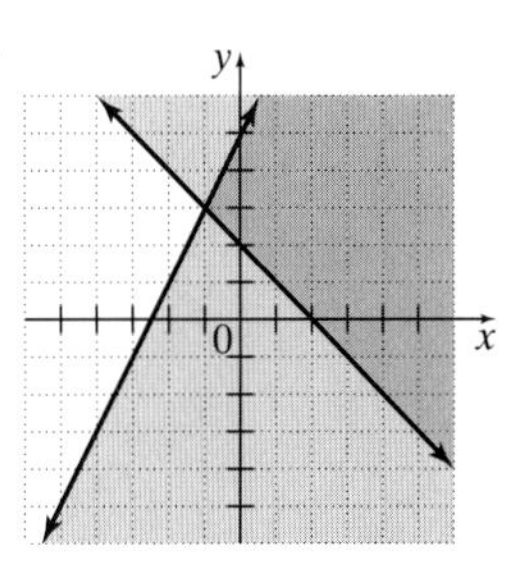

9.

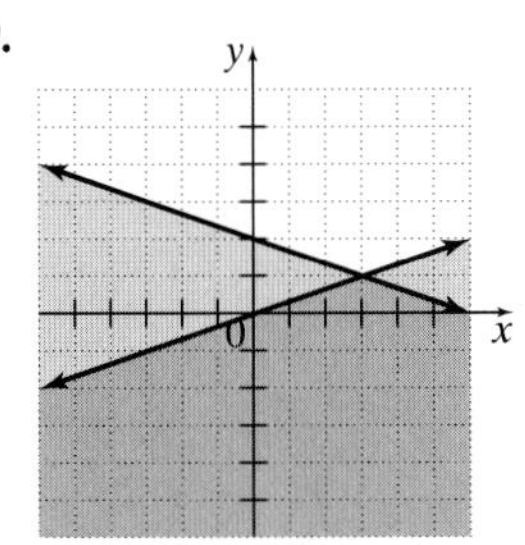

11.

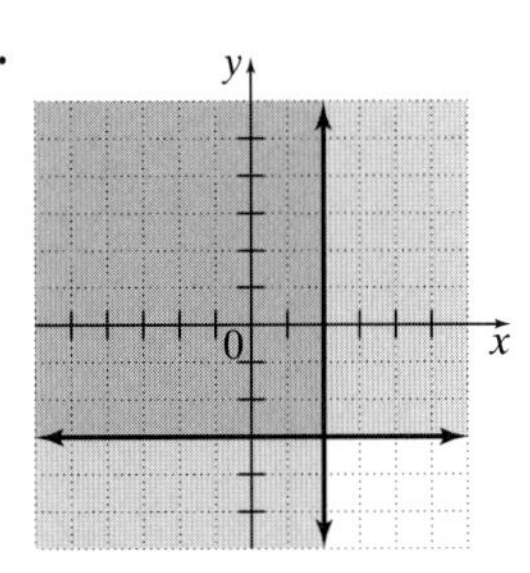

13.

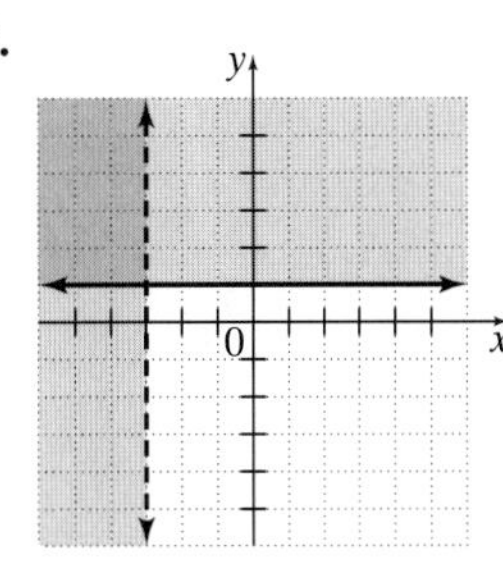

15.

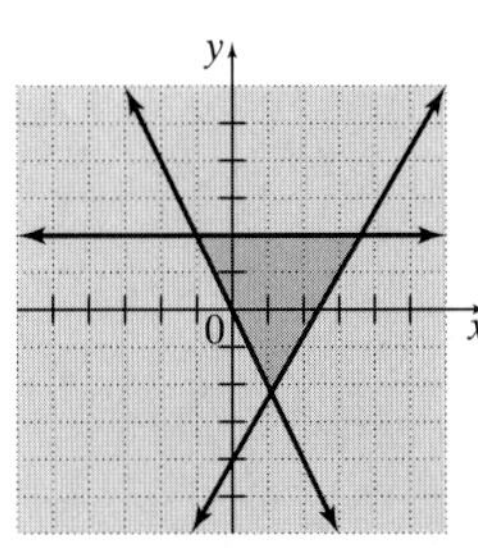

17.

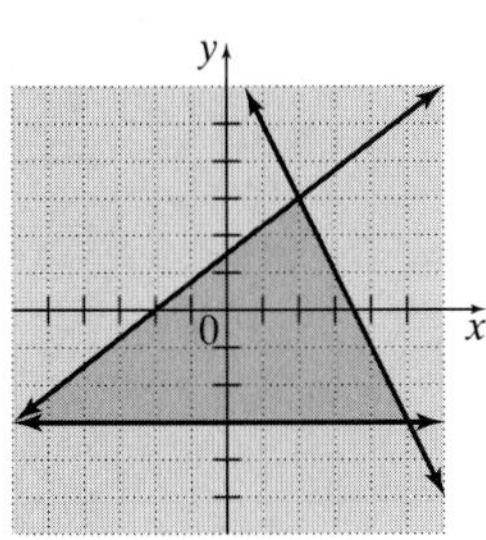

19.

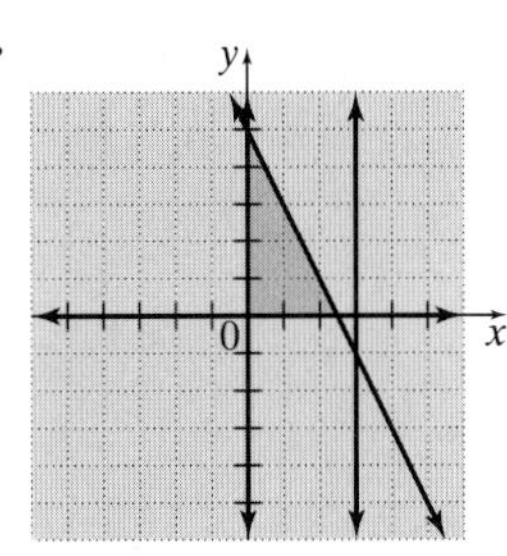

21. C **23.** D **25.** 9 **27.** $\frac{4}{9}$ **29.** 5 **31.** 59

33. a. $\begin{cases} x + y \leq 8 \\ x \quad\quad < 3 \end{cases}$

b.

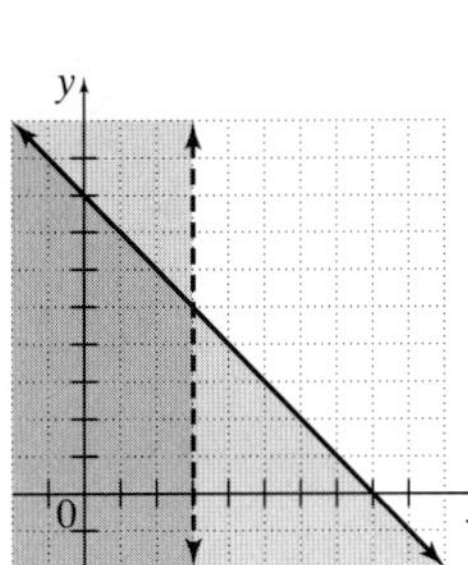

Chapter 4 Review

1. $(-3, 1)$ **2.** $\left(0, \frac{2}{3}\right)$ **3.** $\varnothing$ **4.** $\{(x, y)|3x - 6y = 12\}$ **5.** $\left(3, \frac{8}{3}\right)$ **6.** 1500 backpacks **7.** $(2, 0, 2)$ **8.** $(2, 0, -3)$
9. $\left(-\frac{1}{2}, \frac{3}{4}, 1\right)$ **10.** $(-1, 2, 0)$ **11.** $\varnothing$ **12.** $(5, 3, 0)$ **13.** $(1, 1, -2)$ **14.** $(3, 1, 1)$ **15.** 10, 40, and 48 **16.** 63 and 21
17. 58 mph, 65 mph **18.** width, 37 ft; length, 111 ft **19.** 20 L of 10% solution, 30 L of 60% solution
20. 30 lb of creme-filled; 5 lb of chocolate-covered nuts; 10 lb of chocolate-covered raisins **21.** 17 pennies, 20 nickels, and 16 dimes
22. larger investment, 9.5%; smaller investment, 7.5% **23.** Two sides are 22 cm each; third side is 29 cm. **24.** 120, 115, and 60
25. $(-3, 1)$ **26.** $\{(x, y)|x - 2y = 4\}$ **27.** $\left(-\frac{2}{3}, 3\right)$ **28.** $\left(\frac{1}{3}, \frac{7}{6}\right)$ **29.** $\left(\frac{5}{4}, \frac{5}{8}\right)$ **30.** $(-7, -15)$ **31.** $(1, 3)$ **32.** $(2, 1)$
33. $(1, 2, 3)$ **34.** $(2, 0, -3)$ **35.** $(3, -2, 5)$ **36.** $(-1, 2, 0)$ **37.** $(1, 1, -2)$ **38.** $\varnothing$ **39.** -17 **40.** 17 **41.** 34 **42.** -72
43. $\left(-\frac{2}{3}, 3\right)$ **44.** $\left(\frac{1}{3}, \frac{7}{6}\right)$ **45.** $(-3, 1)$ **46.** $\left(0, \frac{2}{3}\right)$ **47.** $\varnothing$ **48.** $\{(x, y)|x - 2y = 4\}$ **49.** $(1, 2, 3)$ **50.** $(2, 0, -3)$
51. $(2, 1, 0)$ **52.** $\left(\frac{3}{7}, -2, -\frac{1}{7}\right)$ **53.** $\varnothing$ **54.** $(-1, 2, 0)$

55.

56. 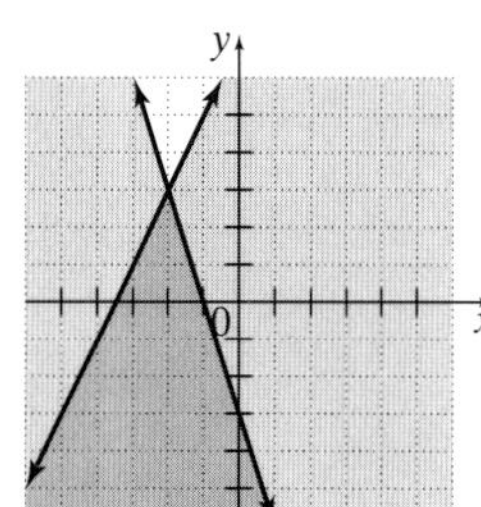

57. 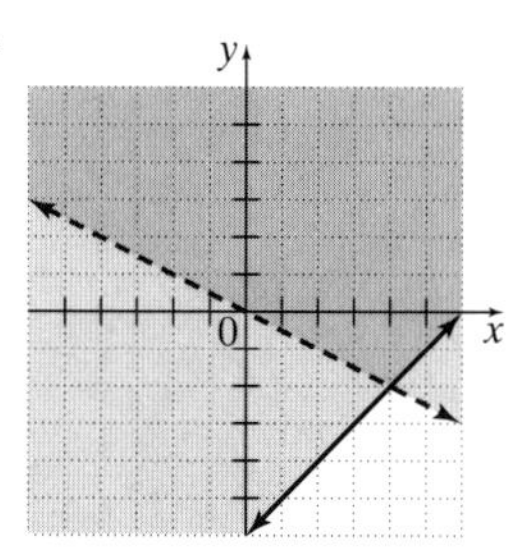

58. 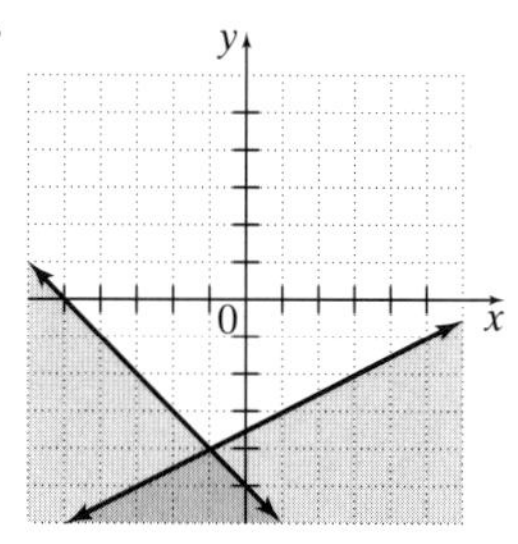

59. 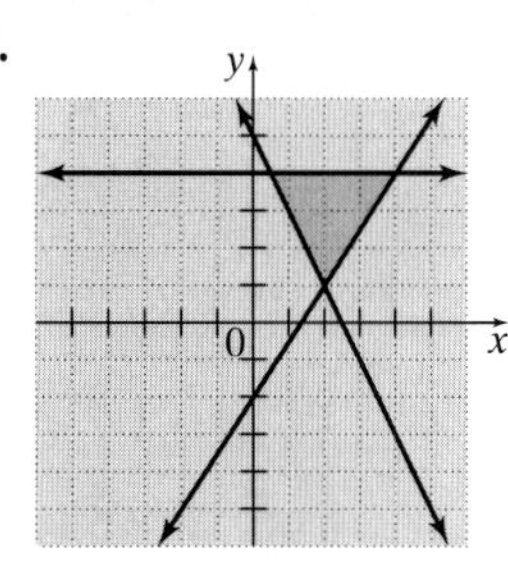

60. 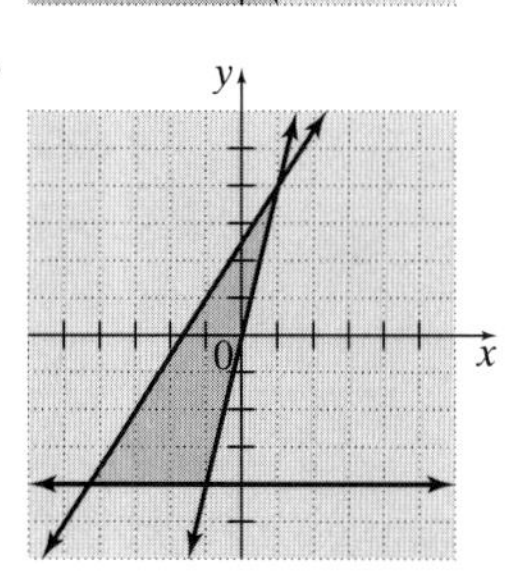

61. 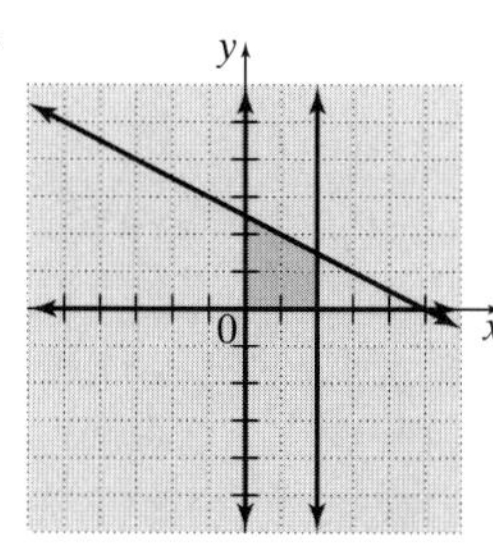

62. 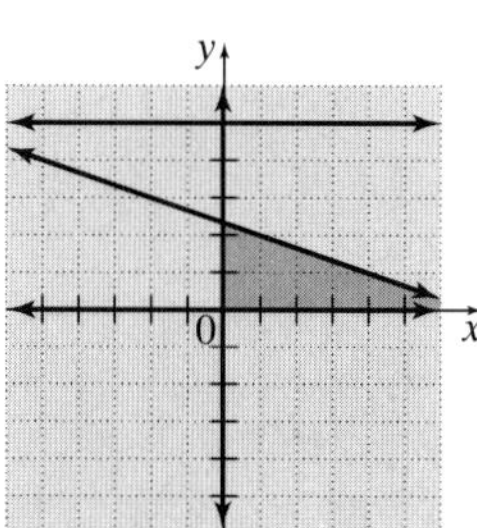

Chapter 4 Test

1. 34 **2.** -6 **3.** $(1, 3)$ **4.** $\varnothing$ **5.** $(2, -3)$ **6.** $\{(x, y)|10x + 4y = 10\}$ **7.** $(-1, -2, 4)$ **8.** $\varnothing$ **9.** $\left(\frac{7}{2}, -10\right)$
10. $(2, -1)$ **11.** $(3, 6)$ **12.** $(3, -1, 2)$ **13.** $(5, 0, -4)$ **14.** $\{(x, y)|x - y = -2\}$ **15.** $(5, -3)$ **16.** $(-1, -1, 0)$ **17.** $\varnothing$
18. 53 double rooms and 27 single rooms **19.** 5 gal of 10%, 15 gal of 20% **20.** 275 frames
21. 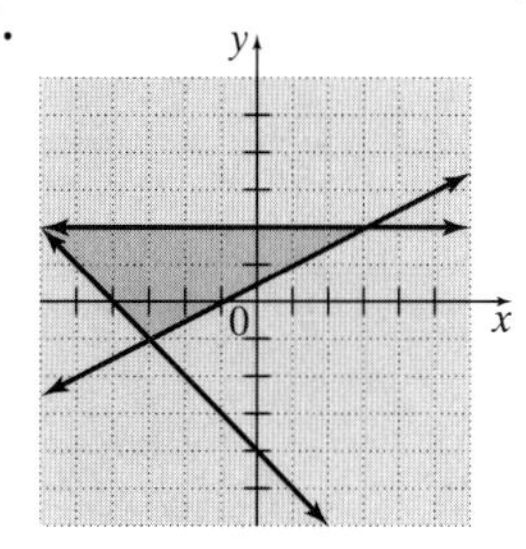

22. 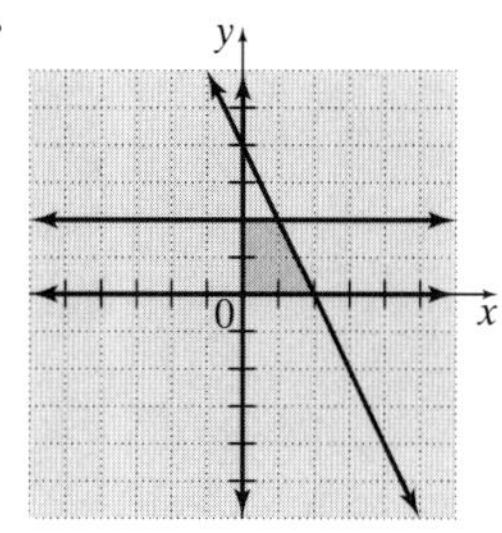

Cumulative Review

1. true; Sec. 1.1, Ex. 7 **2.** false; Sec. 1.1, Ex. 10 **3.** $<$; Sec. 1.2, Ex. 4 **4.** $<$; Sec. 1.2, Ex. 7 **5.** $<$; Sec. 1.2, Ex. 11 **6.** 4; Sec. 1.3; Ex. 3
7. -8; Sec. 1.3, Ex. 5 **8.** -1; Sec. 1.4, Ex. 3 **9.** $\frac{1}{3x}$; Sec. 1.5, Ex. 16 **10.** $\frac{11}{18}$; Sec. 1.5, Ex. 19 **11.** solution; Sec. 2.1, Ex. 1
12. 4; Sec. 2.2, Ex. 2 **13.** $b = \frac{2A - Bh}{h}$; Sec. 2.3, Ex. 3 **14.** $(-2, \infty)$; Sec. 2.4, Ex. 5 **15.** $\{4, 6\}$, Sec. 2.5, Ex. 1
16. $\{0\}$; Sec. 2.6, Ex. 3 **17.** 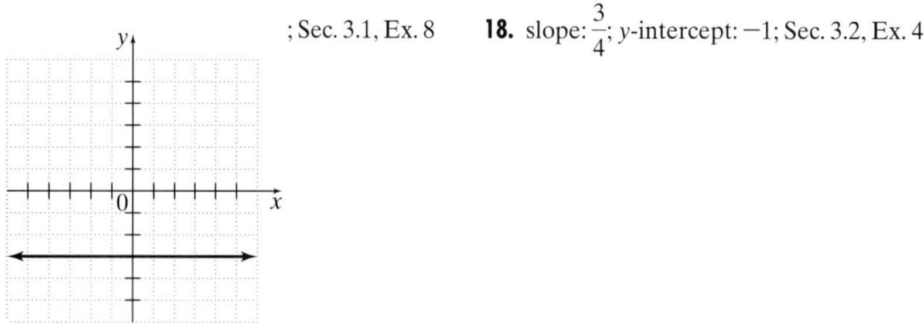 ; Sec. 3.1, Ex. 8 **18.** slope: $\frac{3}{4}$; y-intercept: -1; Sec. 3.2, Ex. 4

19. $y = -3x - 2$; Sec. 3.4, Ex. 1 **20.** function; Sec. 3.5, Ex. 7 **21.** 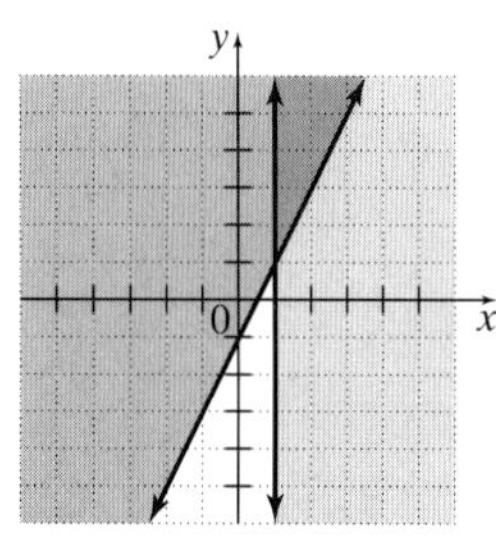 ; Sec. 3.6, Ex. 3

22. $(0, -5)$; Sec. 4.1, Ex. 10 **23.** $\varnothing$; Sec. 4.2, Ex. 2 **24.** 7 and 11; Sec. 4.3, Ex. 1 **25. a.** -2 **b.** -10; Sec. 4.5, Ex. 1

Chapter 5

Chapter 5 Pretest

1. 6; 5.1A **2.** 3; 5.1E **3.** $3x^2 + 4x + 1$; 5.1C **4.** $-12y^2 + 3y + 6$; 5.1D **5.** $6x^2 + 7x - 5$; 5.2B **6.** $64y^2 + 48y + 9$; 5.2C **7.** $4m^2 - 25$; 5.2D **8.** $3t^2 - 2t + \frac{5}{2}$; 5.3A **9.** $x^2 - 4x + 1 - \frac{7}{2x+3}$; 5.3B **10.** $2x^2(3x^2 - 6x + 5)$; 5.4A **11.** $(a + 2b)(3c - 4d)$; 5.4B **12.** $(x + 9)(x - 7)$; 5.5A **13.** $(6x + 1)(x + 3)$; 5.5B **14.** $(x - 9)(x - 6)$; 5.5C **15.** $(2t - 1)(4t^2 + 2t + 1)$; 5.6C **16.** $3x(2x + 3y)(2x - 3y)$; 5.6B **17.** $\left\{-\frac{3}{2}, -5\right\}$; 5.7A **18.** $\{-5, 5, -1\}$; 5.7A **19.** 8 and 15 or -8 and -15; 5.7B **20.**

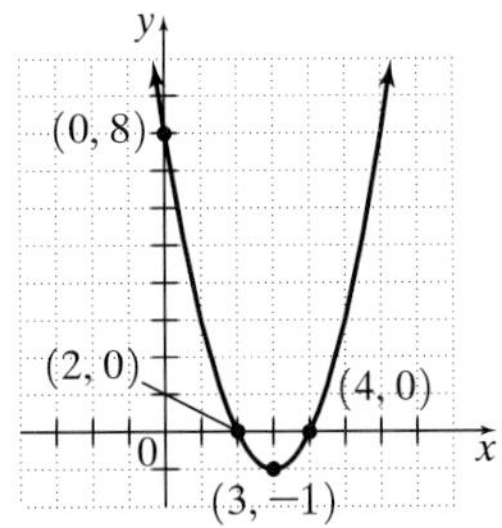

; 5.8B, C

Calculator Explorations

1. $x^3 - 4x^2 + 7x - 8$ **3.** $-2.1x^2 - 3.2x - 1.7$ **5.** $7.69x^2 - 1.26x + 5.3$

Mental Math

1. $10x$ **3.** $5y$ **5.** $-9z$

Exercise Set 5.1

1. 0 **3.** 2 **5.** 3 **7.** binomial of degree 1 **9.** trinomial of degree 2 **11.** monomial of degree 3 **13.** degree 3; none of these **15.** answers may vary **17.** $6y$ **19.** $11x - 3$ **21.** $xy + 2x - 1$ **23.** $18y^2 - 17$ **25.** $3x^2 - 3xy + 6y^2$ **27.** $x^2 - 4x + 8$ **29.** $5x^2 + 22x + 16$ **31.** $-3x^2 + 3$ **33.** $2y^4 - 5y^2 + x^2 + 1$ **35.** $4x - 13$ **37.** $-x^3 + 8a - 12$ **39.** $8xy^2 + 2x^3 + 3x^2 - 3$ **41.** $12x^3y + 8x + 8$ **43.** $4.5x^3 + 0.2x^2 - 3.8x + 9.1$ **45.** $y^2 + 3$ **47.** $-2x^2 + 5x$ **49.** $-2x^2 - 4x + 15$ **51.** $7y^2 - 3$ **53.** $5x^2 - 9x - 3$ **55.** $x^2 + 12$ **57.** $7x^3 + 4x^2 + 8x - 10$ **59.** $-20y^2 + 3yx$ **61.** $15x^2 + 8x - 6$ **63.** $14ab + 10a^2b - 18a^2 + 12b^2$ **65.** $\frac{1}{3}x^2 - x + 1$ **67.** 57 **69.** 499 **71.** 1 **73. a.** 284 ft **b.** 536 ft **c.** 756 ft **d.** 944 ft **e.** answers may vary **f.** 19 seconds **75.** \$80,000 **77.** \$40,000 **79.** $-14z + 42y$ **81.** $-15y^2 - 10y + 35$ **83.** $4x^2 - 3x + 6$ **85.** $2a - 3, -2x - 3, 2x + 2h - 3$ **87.** $12z^{5x} + 13z^{2x} - 2z$

Calculator Explorations

1. $x^2 - 16$ **3.** $9x^2 - 42x + 49$ **5.** $5x^3 - 14x^2 - 13x - 2$

Exercise Set 5.2

1. $-12x^5$ **3.** $12x^2 + 21x$ **5.** $-24x^2y - 6xy^2$ **7.** $-4a^3bx - 4a^3by + 12ab$ **9.** $2x^2 - 2x - 12$ **11.** $2x^4 + 3x^3 - 2x^2 + x + 6$ **13.** $15x^2 - 7x - 2$ **15.** $15m^3 + 16m^2 - m - 2$ **17.** $9x^3 + 30x^2 + 12x - 24$ **19.** $-30a^4b^4 + 36a^3b^2 + 36a^2b^3$ **21.** $10x^5 + 8x^4 + 2x^3 + 25x^2 + 20x + 5$ **23.** $9x^4 + 12x^3 - 2x^2 - 4x + 1$ **25.** $12x^3 - 2x^2 + 13x + 5$ **27.** answers may vary **29.** $x^2 + x - 12$ **31.** $10x^2 + 11xy - 8y^2$ **33.** $3x^2 + 8x - 3$ **35.** $2a^2 - 12a + 16$ **37.** $y^2 - 7y + 12$ **39.** $9x^2 + 18x + 5$ **41.** $16x^2 - \frac{2}{3}x - \frac{1}{6}$ **43.** $5x^4 - 17x^2y^2 + 6y^4$ **45.** $x^2 + 8x + 16$ **47.** $36y^2 - 1$ **49.** $9x^2 - 6xy + y^2$ **51.** $49a^2b^2 - 9c^2$ **53.** $m^2 - 8m + 16$ **55.** $9x^2 + 6x + 1$ **57.** $9b^2 - 36y^2$ **59.** $49x^2 - 9$ **61.** $9x^2 - \frac{1}{4}$ **63.** $36x^2 + 12x + 1$ **65.** $x^4 - 4y^2$ **67.** $16b^2 + 32b + 16$ **69.** $4s^2 - 12s + 8$ **71.** $x^2y^2 - 4xy + 4$ **73.** $2x^3 + 2x^2y + x^2 + xy - x - y$ **75.** $x^4 - 8x^3 + 24x^2 - 32x + 16$ **77.** $x^4 - 625$ **79.** $2x^2$ **81.** $\frac{10a^2b^3}{9}$ **83.** $\frac{2m^3}{3}$ **85.** $\pi(25x^2 - 20x + 4)$ sq. km **87.** $30x^2y^{2n+1} - 10x^2y^n$ **89.** $x^{3a} + 5x^{2a} - 3x^a - 15$ **91.** $6x + 12$; $9x^2 + 36x + 35$; one operation is addition, the other is multiplication **93.** $5x^2 + 25x$ **95.** $a^2 - 3a$ **97.** answers may vary

Exercise Set 5.3

1. $2a + 4$ **3.** $3ab + 4$ **5.** $2y + \frac{3y}{x} - \frac{2y}{x^2}$ **7.** $x + 1$ **9.** $2x - 8$ **11.** $x - \frac{1}{2}$ **13.** $2x^2 - \frac{1}{2}x + 5$ **15.** $2x^2 - 6$ **17.** $2x^2 + 2x + 8 + \frac{28}{x-4}$ **19.** $5x^2 - 6 - \frac{5}{2x-1}$ **21.** $3x^3 + 5x + 4 - \frac{2x}{x^2-2}$ **23.** $2x^3 + \frac{9}{2}x^2 + 10x + 21 + \frac{42}{x-2}$ **25.** $x + 8$ **27.** $x - 1$ **29.** $x^2 - 5x - 23 - \frac{41}{x-2}$ **31.** $4x + 8 + \frac{7}{x-2}$ **33.** $2x^3 - 3x^2 + x - 4$ **35.** $3x^2 + 4x - 8 + \frac{20}{x+1}$ **37.** $3x^2 + 3x - 3$ **39.** $x^2 + x + 1$ **41.** $6x^2 + 23x + 15$ **43.** $(-9, -1)$ **45.** $(-\infty, -8] \cup [1, \infty)$ **47.** $(x^4 + 2x^2 - 6)$ m **49.** $(3x - 7)$ in. **51.** 4 **53.** $x^3 + 2x^2 + 7x + 28$ **55. a.** answers may vary **b.** answers may vary

ANSWERS

MENTAL MATH

1. 6 **3.** 5 **5.** x **7.** $7x$

EXERCISE SET 5.4

1. $6(3x - 2)$ **3.** $4y^2(1 - 4xy)$ **5.** $2x^3(3x^2 - 4x + 1)$ **7.** $4ab(2a^2b^2 - ab + 1 + 4b)$ **9.** $(x + 3)(6 + 5a)$ **11.** $(z + 7)(2x + 1)$ **13.** $(x^2 + 5)(3x - 2)$ **15.** $2\pi r(r + h)$ **17.** $A = P(1 + RT)$ **19.** answers may vary **21.** $(a + 2)(b + 3)$ **23.** $(a - 2)(c + 4)$ **25.** $(x - 2)(2y - 3)$ **27.** $(4x - 1)(3y - 2)$ **29.** $(x^2 + 4)(x + 3)$ **31.** $(x^2 - 2)(x - 1)$ **33.** $(2x + 3y)(x + 2)$ **35.** $(5x - 3)(x + y)$ **37.** $(2x + 3)(3y + 5)$ **39.** $(x + 3)(y - 5)$ **41.** $3b(3ac^2 + 2a^2c - 2a + c)$ **43.** $x^2 - 3x - 10$ **45.** $x^2 + 5x + 6$ **47.** $y^2 - 4y + 3$ **49.** none **51.** $y^n(3 + 3y^n + 5y^{7n})$ **53.** $3x^{2a}(x^{3a} - 2x^a + 3)$ **55. a.** $h(t) = -16(t^2 - 14)$ **b.** 160 ft **c.** answers may vary

MENTAL MATH

1. 5 and 2 **3.** 8 and 3

EXERCISE SET 5.5

1. $(x + 3)(x + 6)$ **3.** $(x - 8)(x - 4)$ **5.** $(x + 12)(x - 2)$ **7.** $(x - 6)(x + 4)$ **9.** $3(x - 2)(x - 4)$ **11.** $4z(x + 2)(x + 5)$ **13.** $2(x + 18)(x - 3)$ **15.** $(x - 27)(x + 3)$ **17.** $(x - 18)(x + 3)$ **19.** $3(x - 1)^2$ **21.** $2(x + 3)(x - 2)$ **23.** $(x + 5y)(x + y)$ **25.** $x(x + 4)(x - 2)$ **27.** $\pm 5, \pm 7$ **29.** $(5x + 1)(x + 3)$ **31.** $(2x - 3)(x - 4)$ **33.** prime polynomial **35.** $(2x - 3)^2$ **37.** $2(3x - 5)(2x + 5)$ **39.** $y^2(3y + 5)(y - 2)$ **41.** $2x(3x^2 + 4x + 12)$ **43.** $(x + 7z)(x + z)$ **45.** $(2x + y)(x - 3y)$ **47.** $(x - 4)(x + 3)$ **49.** $2(7y + 2)(2y + 1)$ **51.** $(2x - 3)(x + 9)$ **53.** $(3x + 1)(x - 2)$ **55.** $(4x - 3)(2x - 5)$ **57.** $3x^2(2x + 1)(3x + 2)$ **59.** $3(a + 2b)^2$ **61.** $x(3x + 1)(2x - 1)$ **63.** $(4a - 3b)(3a - 5b)$ **65.** $(3x + 5)^2$ **67.** $y(3x - 8)(x - 1)$ **69.** $(x^2 + 3)(x^2 - 2)$ **71.** $(5x + 8)(5x + 2)$ **73.** $(x^3 - 4)(x^3 - 3)$ **75.** $(a - 3)(a + 8)$ **77.** $(x + 2)(x - 7)$ **79.** $(2x^3 - 3)(x^3 + 3)$ **81.** $(2x + 13)(x + 3)$ **83.** $(x^2 - 6)(x^2 + 1)$ **85.** $x^2 - 9$ **87.** $4x^2 + 4x + 1$ **89.** $x^3 - 8$ **91.** $(x^n + 2)(x^n + 8)$ **93.** $(x^n - 6)(x^n + 3)$ **95.** $(2x^n + 1)(x^n + 5)$ **97.** $(2x^n - 3)^2$ **99.** $x^2(x + 5)(x + 1)$ **101.** $3x(5x - 1)(2x + 1)$

EXERCISE SET 5.6

1. $(x + 3)^2$ **3.** $(2x - 3)^2$ **5.** $3(x - 4)^2$ **7.** $x^2(3y + 2)^2$ **9.** $(2a + 3)^2$ **11.** $(x + 5)(x - 5)$ **13.** $(3 + 2z)(3 - 2z)$ **15.** $(y + 9)(y - 5)$ **17.** $4(4x + 5)(4x - 5)$ **19.** $2y(3x + 1)(3x - 1)$ **21.** $(3x + 7)(3x - 7)$ **23.** $(x^2 + 9)(x + 3)(x - 3)$ **25.** $(x + 2y + 3)(x + 2y - 3)$ **27.** $(x + 8 + x^2)(x + 8 - x^2)$ **29.** $(x - 5 + y)(x - 5 - y)$ **31.** $(2x + 1 + z)(2x + 1 - z)$ **33.** $(x + 3)(x^2 - 3x + 9)$ **35.** $(z - 1)(z^2 + z + 1)$ **37.** $(m + n)(m^2 - mn + n^2)$ **39.** $y^2(x - 3)(x^2 + 3x + 9)$ **41.** $b(a + 2b)(a^2 - 2ab + 4b^2)$ **43.** $(5y - 2x)(25y^2 + 10xy + 4x^2)$ **45.** $(x^2 - y)(x^4 + x^2y + y^2)$ **47.** $(2x + 3y)(4x^2 - 6xy + 9y^2)$ **49.** $(x - 1)(x^2 + x + 1)$ **51.** $(x + 5)(x^2 - 5x + 25)$ **53.** $3y^2(x^2 + 3)(x^4 - 3x^2 + 9)$ **55.** $\{5\}$ **57.** $\left\{-\frac{1}{3}\right\}$ **59.** $\{0\}$ **61.** $\{5\}$ **63.** $\pi R^2 - \pi r^2 = \pi(R + r)(R - r)$ **65.** $c = 9$ **67.** $c = 49$ **69. a.** $(x + 1)(x^2 - x + 1)(x - 1)(x^2 + x + 1)$ **b.** $(x + 1)(x - 1)(x^4 + x^2 + 1)$ **c.** answers may vary **71.** $(x^n + 6)(x^n - 6)$ **73.** $(5x^n + 9)(5x^n - 9)$ **75.** $(x^{2n} + 25)(x^n + 5)(x^n - 5)$

INTEGRATED REVIEW

1. $2y^2 + 2y - 11$ **2.** $-2z^4 - 6z^2 + 3z$ **3.** $x^2 - 7x + 7$ **4.** $7x^2 - 4x - 5$ **5.** $25x^2 - 30x + 9$ **6.** $x - 3$ **7.** $2x^3 - 4x^2 + 5x - 5 + \frac{8}{x + 2}$ **8.** $4x^3 - 13x^2 - 5x + 2$ **9.** $(x - 4 + y)(x - 4 - y)$ **10.** $2(3x + 2)(2x - 5)$ **11.** $x(x - 1)(x^2 + x + 1)$ **12.** $2x(2x - 1)$ **13.** $2xy(7x - 1)$ **14.** $6ab(4b - 1)$ **15.** $4(x + 2)(x - 2)$ **16.** $9(x + 3)(x - 3)$ **17.** $(3x - 11)(x + 1)$ **18.** $(5x + 3)(x - 1)$ **19.** $4(x + 3)(x - 1)$ **20.** $6(x + 1)(x - 2)$ **21.** $(2x + 9)^2$ **22.** $(5x + 4)^2$ **23.** $(2x + 5y)(4x^2 - 10xy + 25y^2)$ **24.** $(3x - 4y)(9x^2 + 12xy + 16y^2)$ **25.** $8x^2(2y - 1)(4y^2 + 2y + 1)$ **26.** $27x^2y(xy - 2)(x^2y^2 + 2xy + 4)$ **27.** $(x + 5 + y)(x^2 + 10x - xy - 5y + y^2 + 25)$ **28.** $(y - 1 + 3x)(y^2 - 2y + 1 - 3xy + 3x + 9x^2)$ **29.** $(5a - 6)^2$ **30.** $(4r + 5)^2$ **31.** $A = 9 - 4x^2 = (3 + 2x)(3 - 2x)$

MENTAL MATH

1. $\{3, -5\}$ **3.** $\{3, -7\}$ **5.** $\{0, 9\}$

EXERCISE SET 5.7

1. $\left\{-3, \frac{4}{3}\right\}$ **3.** $\left\{-\frac{3}{4}, \frac{5}{2}\right\}$ **5.** $\{-3, -8\}$ **7.** $\left\{\frac{1}{4}, -\frac{2}{3}\right\}$ **9.** $\{1, 9\}$ **11.** $\left\{\frac{3}{5}, -1\right\}$ **13.** $\{0\}$ **15.** $\{6, -3\}$ **17.** $\left\{\frac{2}{5}, -\frac{1}{2}\right\}$ **19.** $\left\{\frac{3}{4}, -\frac{1}{2}\right\}$ **21.** $\left\{-2, 7, \frac{8}{3}\right\}$ **23.** $\{0, 3, -3\}$ **25.** $\{-1, 1, 2\}$ **27.** answers may vary **29.** $\left\{-\frac{7}{2}, 10\right\}$ **31.** $\{0, 5\}$ **33.** $\{-3, 5\}$ **35.** $\left\{-\frac{1}{2}, \frac{1}{3}\right\}$ **37.** $\{-4, 9\}$ **39.** $\left\{\frac{4}{5}\right\}$ **41.** $\{-5, 0, 2\}$ **43.** $\left\{-3, 0, \frac{4}{5}\right\}$ **45.** $\varnothing$ **47.** $\{1\}$ **49.** -11 and -6 or 6 and 11 **51.** 75 ft **53.** 6 ft and 8 ft **55.** 12, 14 or -12, -14 **57.** 10 sec **59.** width, 7ft; length, 13 ft **61.** 2 in. **63.** $(-3, 0), (0, 2)$ **65.** $(-4, 0), (4, 0), (0, 2), (0, -2)$ **67.** $\left\{-3, -\frac{1}{3}, 2, 5\right\}$ **69.** no; answers may vary **71.** answers may vary

CALCULATOR EXPLORATIONS

1. $\{-3.562, 0.562\}$ **3.** $\{-0.874, 2.787\}$ **5.** $\{-0.465, 1.910\}$

Mental Math

1. upward **3.** downward

Exercise Set 5.8

1. a. domain, $(-\infty, \infty)$; range, $(-\infty, 5]$ **b.** x-intercept points, $(-2, 0)$, $(6, 0)$; y-intercept point, $(0, 5)$ **c.** $(0, 5)$ **d.** There is no such point. **3. a.** domain, $(-\infty, \infty)$; range, $[-4, \infty)$ **b.** x-intercept points, $(-3, 0)$, $(1, 0)$; y-intercept point, $(0, -3)$ **c.** There is no such point. **d.** $(-1, -4)$ **5. a.** domain, $(-\infty, \infty)$; range, $(-\infty, \infty)$ **b.** x-intercept points, $(-2, 0)$, $(0, 0)$, $(2, 0)$; y-intercept point, $(0, 0)$ **c.** There is no such point. **d.** There is no such point.

7.

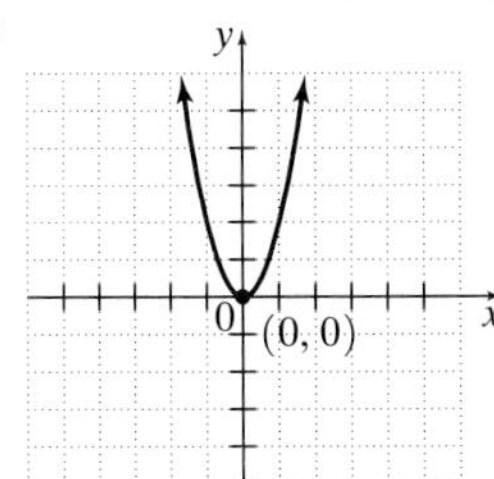

9.

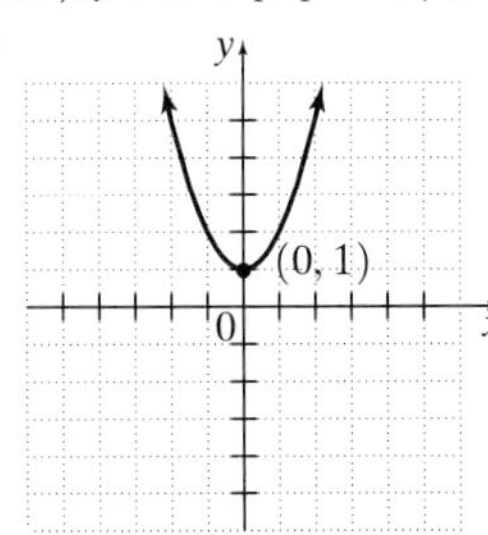

11.

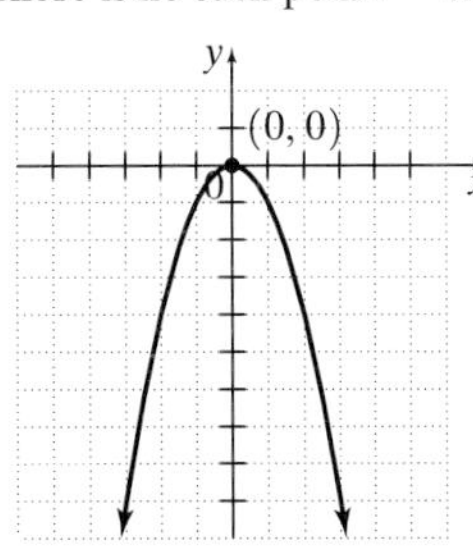

13.

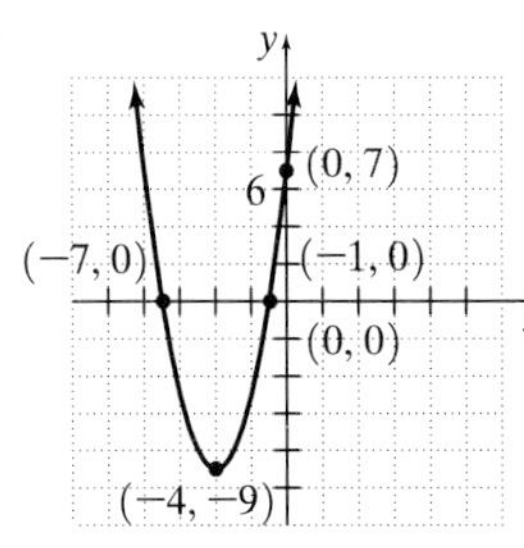

15.

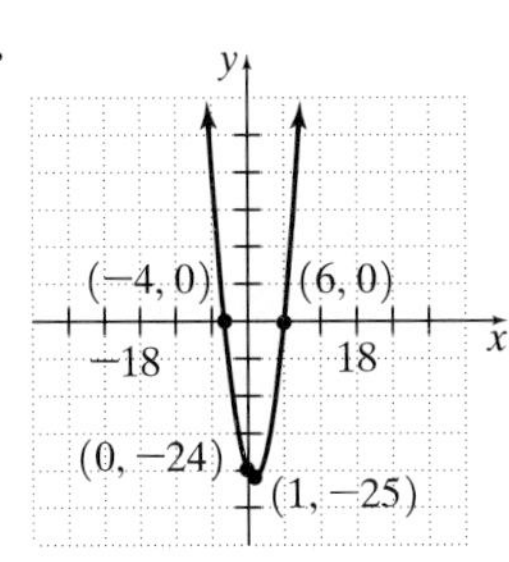

17.

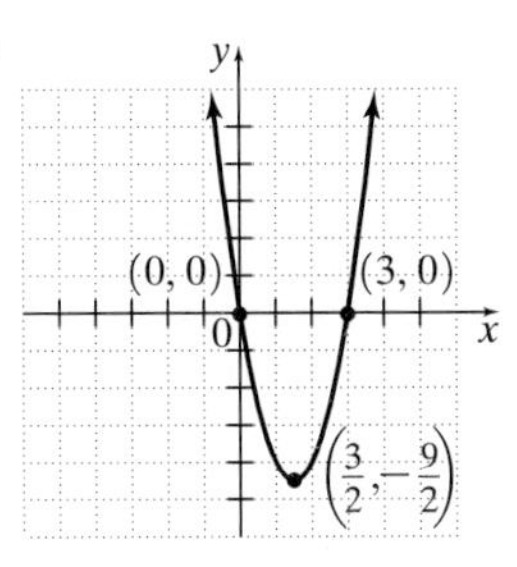

19.

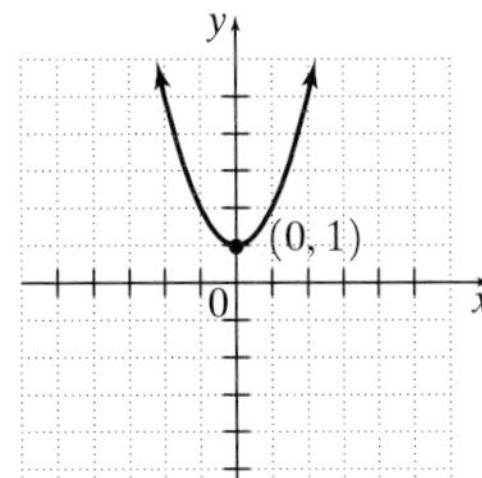
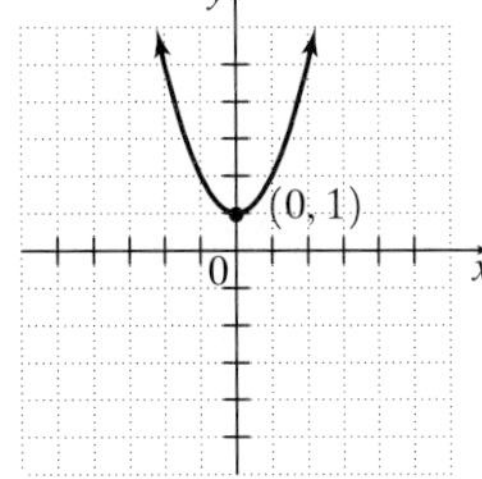

21. 2 **23.** 0

25.

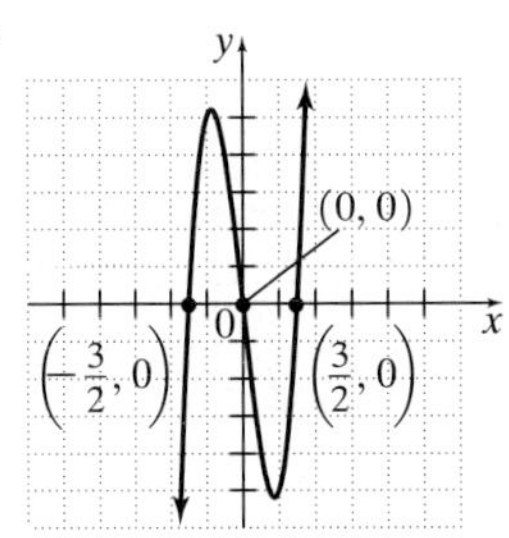

27.

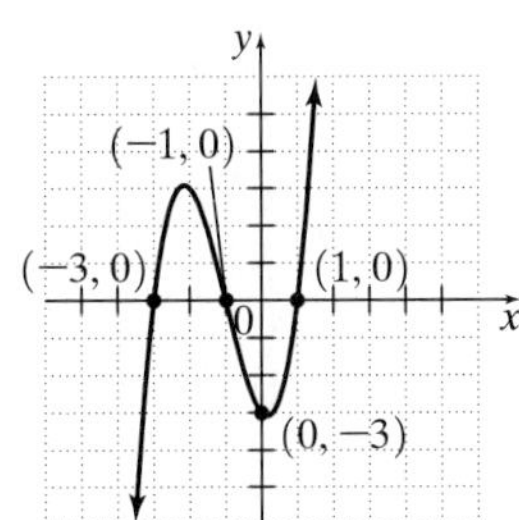

29.

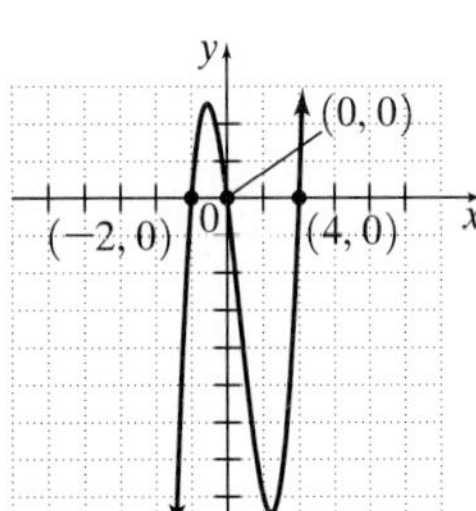

31.

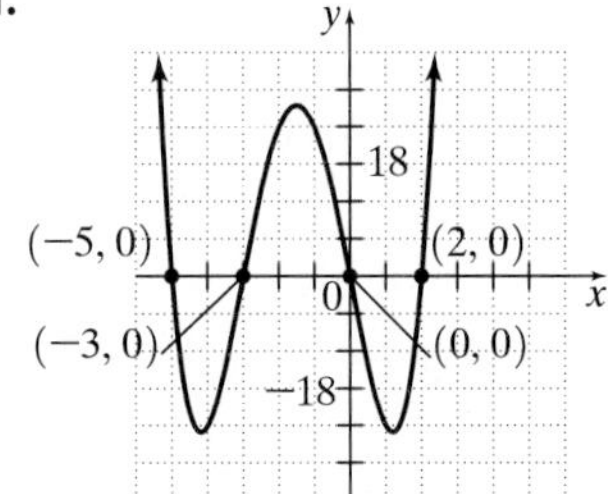

33. no; answers may vary **35.** $-\frac{4}{5}$ **37.** $\frac{x^4}{y^5}$ **39.** $\frac{m^3}{2n^{10}}$ **41.** E **43.** F **45.** B

47.

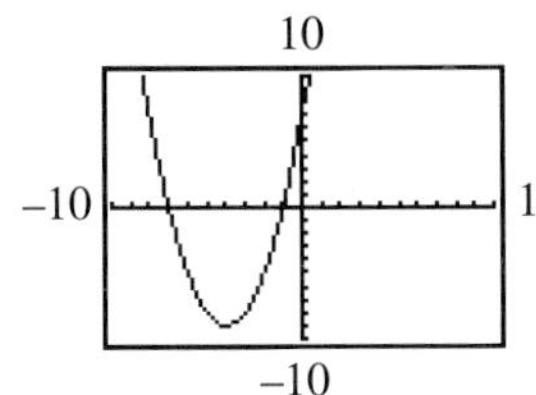

49.

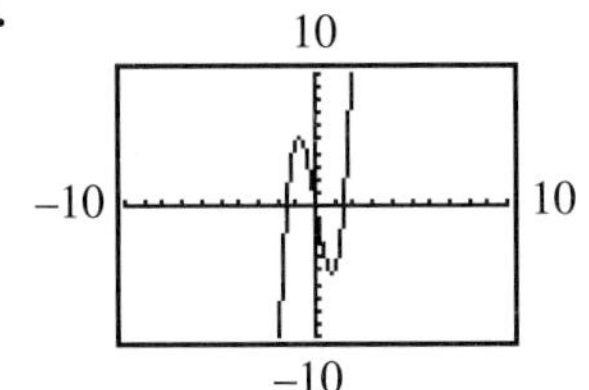

Chapter 5 Review

1. 5 **2.** 1 **3.** $12x - 6x^2 - 6x^2y$ **4.** $-4xy^3 - 3x^3y$ **5.** $4x^2 + 8y + 6$ **6.** $-4x^2 + 10y^2$ **7.** $8x^2 + 2b - 22$ **8.** $-4x^3 + 4x^2 + 16xy - 9x + 18$ **9.** $12x^2y - 7xy + 3$ **10.** $x^2 - 6x + 3$ **11.** $x^3 + x - 2xy^2 - y - 7$ **12.** 290 **13.** 58 **14.** 110 **15.** $x^2 + 4x - 6$ **16.** $-x^2 + 2x + 3$ **17.** $(6x^2y - 12x + 12)$ cm **18.** $-24x^3 + 36x^2 - 6x$ **19.** $-12a^2b^5 - 28a^2b^3 - 4ab^2$ **20.** $2x^2 + x - 36$ **21.** $9x^2a^2 - 24xab + 16b^2$ **22.** $36x^3 - 11x^2 - 8x - 3$ **23.** $15x^2 + 18xy - 81y^2$ **24.** $x^2 + \frac{1}{3}x - \frac{2}{9}$ **25.** $x^4 + 18x^3 + 83x^2 + 18x + 1$ **26.** $2x^3 + 3x^2 - 12x + 5$ **27.** $9x^2 - 6xy + y^2$ **28.** $16x^2 + 72x + 81$ **29.** $x^2 - 9y^2$ **30.** $-9a^2 + 6ab - b^2 + 16$ **31.** $(9y^2 - 49z^2)$ sq. units **32.** $1 + \frac{x}{2y} - \frac{9}{4xy}$ **33.** $\frac{3}{b} + 4b$ **34.** $3x^3 + 9x^2 + 2x + 6 - \frac{2}{x - 3}$ **35.** $2x^3 + 6x^2 + 17x + 56 + \frac{156}{x - 3}$ **36.** $2x^3 + 2x - 2$ **37.** $x^2 + \frac{7}{2}x - \frac{1}{4} + \frac{15}{8\left(x - \frac{1}{2}\right)}$ **38.** $3x^2 + 2x - 1$ **39.** $3x^2 + 6$ **40.** $3x^2 + 6x + 24 + \frac{44}{x - 2}$ **41.** $4x^2 - 4x + 2 - \frac{5}{x + \frac{3}{2}}$ **42.** $x^4 - x^3 + x^2 - x + 1 - \frac{2}{x + 1}$

43. $x^2 + 3x + 9 - \dfrac{54}{x - 3}$ **44.** $3x^3 + 13x^2 + 51x + 204 + \dfrac{814}{x - 4}$ **45.** $3x^3 - 6x^2 + 10x - 20 + \dfrac{50}{x + 2}$ **46.** $8x^2(2x - 3)$
47. $12y(3 - 2y)$ **48.** $2ab(3b + 4 - 2ab)$ **49.** $7ab(2ab - 3b + 1)$ **50.** $(a + 3b)(6a - 5)$ **51.** $(x - 2y)(4x - 5)$
52. $(x - 6)(y + 3)$ **53.** $(a - 8)(b + 4)$ **54.** $(p - 5)(q - 3)$ **55.** $(x^2 - 2)(x - 1)$ **56.** $x(2y - x)$ **57.** $(x - 18)(x + 4)$
58. $(x - 4)(x + 20)$ **59.** $2(x - 2)(x - 7)$ **60.** $3(x + 2)(x + 9)$ **61.** $x(2x - 9)(x + 1)$ **62.** $(3x + 8)(x - 2)$
63. $(6x + 5)(x + 2)$ **64.** $(15x - 1)(x - 6)$ **65.** $2(2x - 3)(x + 2)$ **66.** $3(x - 2)(3x + 2)$ **67.** $(x + 6)^2(y - 3)(y + 1)$
68. $(x + 7)(x + 9)$ **69.** $(x^2 - 8)(x^2 + 2)$ **70.** $(x^2 - 2)(x^2 + 10)$ **71.** $(x + 10)(x - 10)$ **72.** $(x - 9)(x + 9)$
73. $2(x + 4)(x - 4)$ **74.** $6(x - 3)(x + 3)$ **75.** $(9 + x^2)(3 + x)(3 - x)$ **76.** $(4 + y^2)(2 - y)(2 + y)$ **77.** $(y + 7)(y - 3)$
78. $(x - 7)(x + 1)$ **79.** $(x + 6)(x^2 - 6x + 36)$ **80.** $(y + 8)(y^2 - 8y + 64)$ **81.** $(2 - 3y)(4 + 6y + 9y^2)$
82. $(1 - 4y)(1 + 4y + 16y^2)$ **83.** $6xy(x + 2)(x^2 - 2x + 4)$ **84.** $2x^2(x + 2y)(x^2 - 2xy + 4y^2)$ **85.** $(x - 1 + y)(x - 1 - y)$
86. $(x - 3 - 2y)(x - 3 + 2y)$ **87.** $(2x + 3)^2$ **88.** $(4a - 5b)^2$ **89.** $\pi h(R + r)(R - r)$ cu. units **90.** $\left\{\frac{1}{3}, -7\right\}$ **91.** $\left\{-5, \frac{3}{8}\right\}$
92. $\left\{0, 4, \frac{9}{2}\right\}$ **93.** $\left\{-3, -\frac{1}{5}, 4\right\}$ **94.** $\{0, 6\}$ **95.** $\{-3, 0, 3\}$ **96.** $\left\{-\frac{1}{3}, 2\right\}$ **97.** $\{2, 10\}$ **98.** $\{-4, 1\}$ **99.** $\left\{\frac{7}{2}, -5\right\}$
100. $\{0, 6, -3\}$ **101.** $\{-21, 0, 2\}$ **102.** $\{0, -2, 1\}$ **103.** $\left\{-\frac{3}{2}, 0, \frac{1}{4}\right\}$ **104.** $-\frac{15}{2}, 7$ **105.** width, 2 m; length, 8 m **106.** 5 sec
107. domain, $(-\infty, \infty)$; range, $(-\infty, 4]$ **108.** $(-4, 0), (2, 0)$; y-intercept, $(0, 3)$ **109.** $(-1, 4)$ **110.** between $x = -4$ and $x = 2$

111.
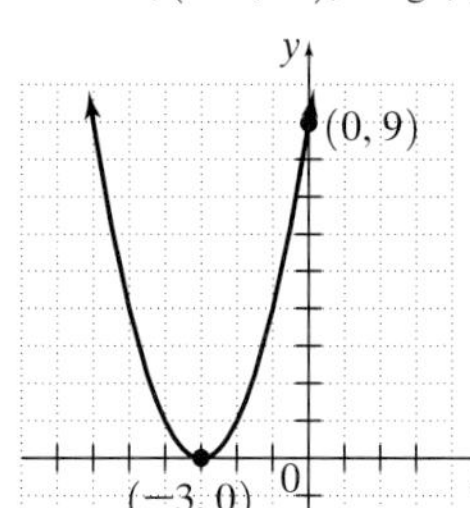

112.
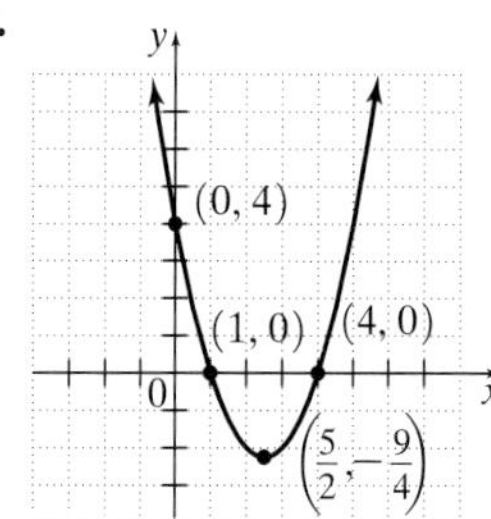

113.
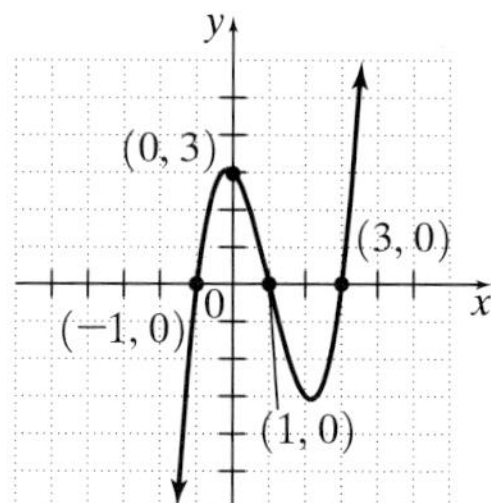

114.
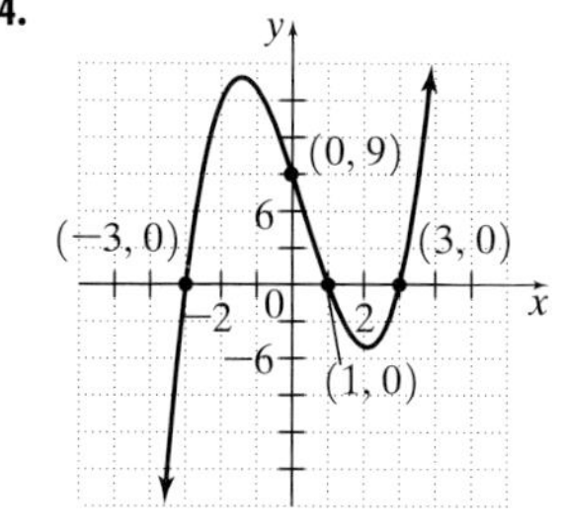

115.
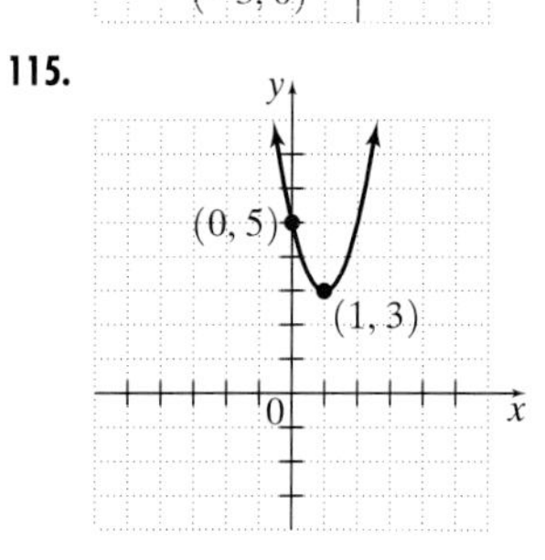

116.
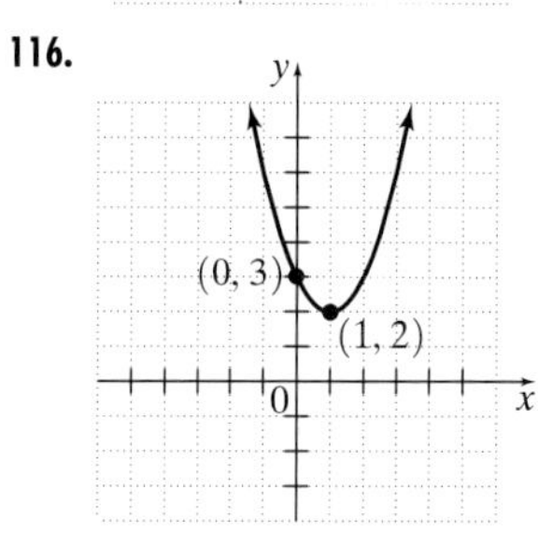

117.
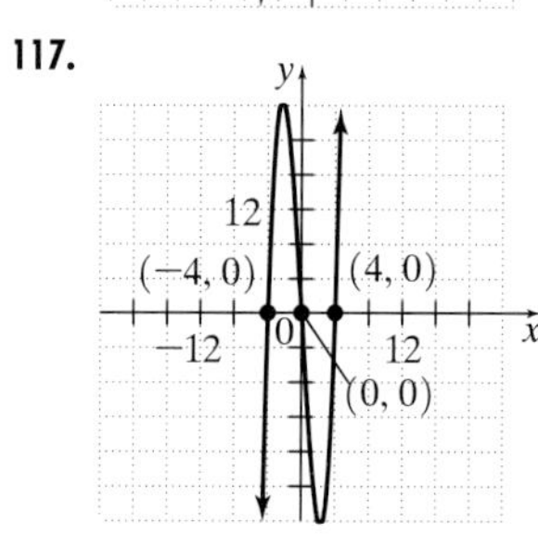

118.
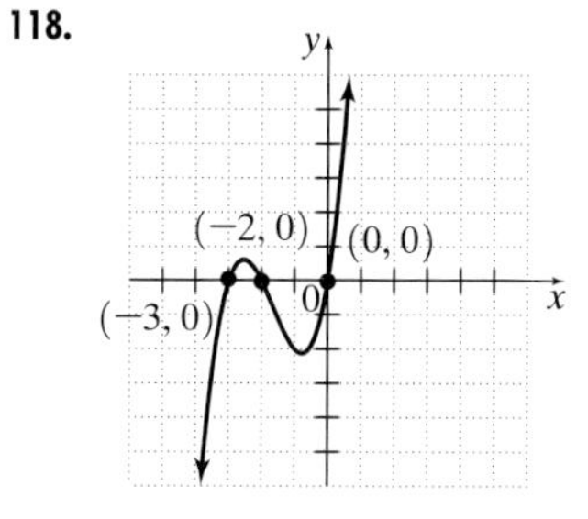

Chapter 5 Test

1. $-5x^3 - 11x - 9$ **2.** $-12x^2y - 3xy^2$ **3.** $12x^2 - 5x - 28$ **4.** $25a^2 - 4b^2$ **5.** $36m^2 + 12mn + n^2$ **6.** $2x^3 - 13x^2 + 14x - 4$
7. $\dfrac{4xy}{3z} + \dfrac{3}{z} + \dfrac{1}{3x}$ **8.** $2x^4 + 2x - 2 + \dfrac{1}{2x - 1}$ **9.** $4x^3 - 15x^2 + 47x - 142 + \dfrac{425}{x + 3}$ **10.** $4x^2y(4x - 3y^3)$ **11.** $(x - 15)(x + 2)$
12. $(2y + 5)^2$ **13.** $3(2x + 1)(x - 3)$ **14.** $(2x + 5)(2x - 5)$ **15.** $(x + 4)(x^2 - 4x + 16)$ **16.** $3y(x + 3y)(x - 3y)$
17. $6(x^2 + 4)$ **18.** $(x + 3)(x - 3)(y - 3)$ **19.** $\left\{4, -\frac{8}{7}\right\}$ **20.** $\{-3, 8\}$ **21.** $\left\{-\frac{5}{2}, -2, 2\right\}$ **22.** $(x + 2y)(x - 2y)$
23. **a.** 960 ft **b.** 953.44 ft **c.** 11 sec **24.**

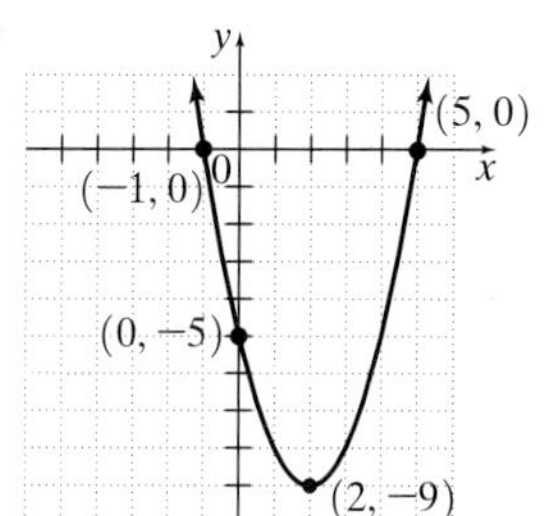

25.
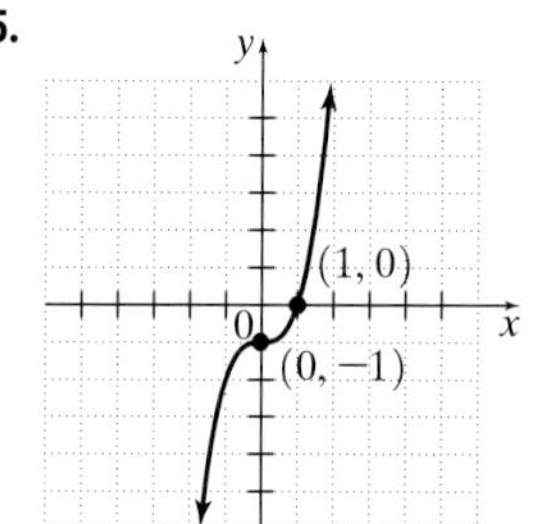

Cumulative Review

1. $-2x - 6$; Sec. 1.4, Ex. 12 **2.** $2x + 23$; Sec. 1.4, Ex. 13 **3.** 6×10^{-5}; Sec. 1.6, Ex. 20 **4.** $\{1\}$; Sec. 2.1, Ex. 6
5. $y = \dfrac{2x + 7}{3}$ or $y = \dfrac{2x}{3} + \dfrac{7}{3}$; Sec. 2.3, Ex. 2 **6.** $[4, \infty)$; Sec. 2.4, Ex. 9 **7.** $\left[-9, -\frac{9}{2}\right]$; Sec. 2.5, Ex. 5 **8.** $\left\{\frac{3}{4}, 5\right\}$; Sec. 2.6, Ex. 7

9. undefined; Sec. 3.2, Ex. 5 **10.** 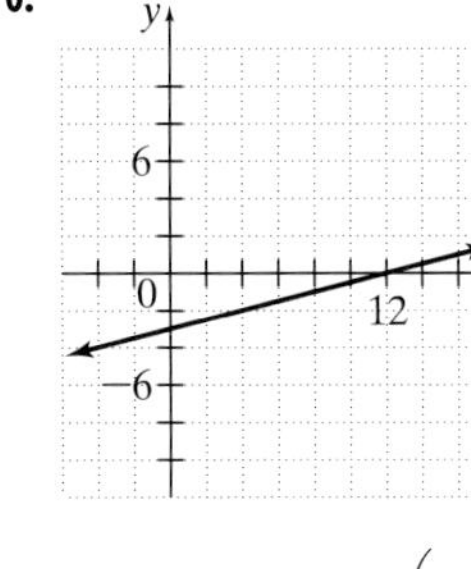

; Sec. 3.3, Ex. 1 **11.** $y = \frac{5}{3}x + \frac{13}{3}$; Sec. 3.4, Ex. 6 **12.** domain: {2, 0, 3}; range: {3, 4, −1}; Sec. 3.5, Ex. 1 **13.** −2; Sec. 3.5, Ex. 19 **14.** 5; Sec. 3.5, Ex. 20

15. 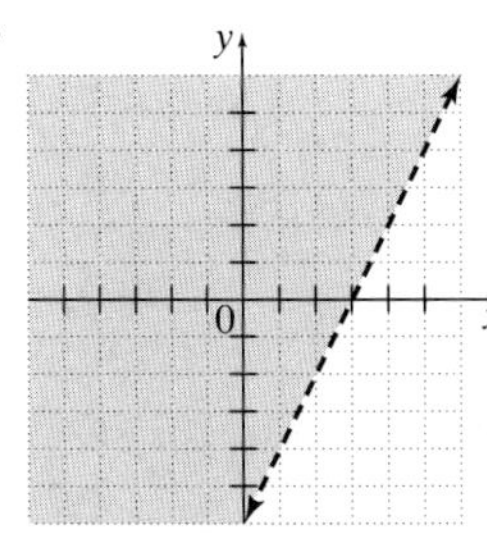

; Sec. 3.6, Ex. 1 **16.** $\left(-4, \frac{1}{2}\right)$; Sec. 4.1, Ex. 6 **17.** (−4, 2, −1); Sec. 4.2, Ex. 1 **18.** (−1, 2); Sec. 4.4, Ex. 1 **19.** (−5, 2); Sec. 4.5, Ex. 2 **20.** {(1, −2, −1)}; Sec. 4.5, Ex. 5

21. 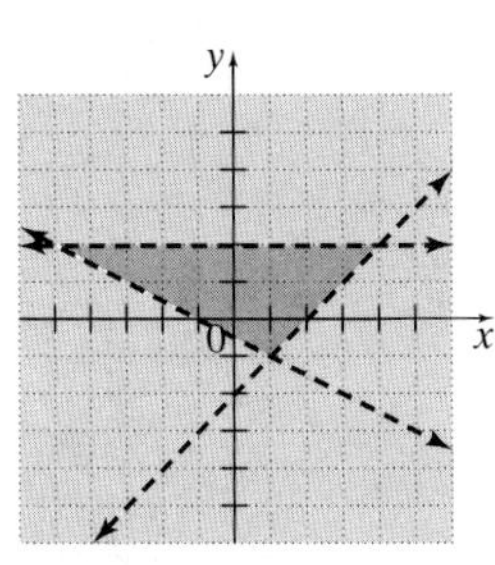

; Sec. 4.6, Ex. 2 **22.** 4; Sec. 5.1, Ex. 9 **23.** $10x^2 - 8x$; Sec. 5.2, Ex. 3 **24.** $-7x^3y^2 - 3x^2y^2 + 11xy$; Sec. 5.2, Ex. 5 **25.** $2x^2 - x + 4$; Sec. 5.3, Ex. 1

Chapter 6

Chapter 6 Pretest

1. $x = 2$ or $x = 3$; (6.1A) **2.** $1 - 2x$; (6.1B) **3.** $\frac{x+2}{x-4}$; (6.1B) **4.** $-\frac{3}{2}$; (6.1C) **5.** $\frac{2x-7}{2x-5}$; (6.1C) **6.** $\frac{2mn^2}{m-5}$; (6.1D) **7.** −1; (6.1D) **8.** $\frac{7+x}{x-7}$; (6.2A) **9.** $\frac{20-6a}{15a^2}$; (6.2C) **10.** $\frac{3(3y+5)}{(y-3)(y+3)}$; (6.2C) **11.** $\frac{x^2+11x+5}{(x+6)(x+1)(2x+1)}$; (6.2C) **12.** {12}; (6.3A) **13.** $\frac{4y+5}{16-25y}$; (6.3B) **14.** $\frac{y}{xy+x^2}$; (6.3C) **15.** {60}; (6.4A) **16.** {−2}; (6.4A) **17.** $\frac{2S - na}{n} = L$; (6.5A) **18.** 2 and 3; (6.5B) **19.** $y = \frac{15}{2}$; (6.6A) **20.** $W = 32$; (6.6B)

Calculator Explorations

1. $\{x|x$ is a real number and $x \neq -2, x \neq 2\}$ **3.** $\left\{x|x \text{ is a real number and } x \neq -4, x \neq \frac{1}{2}\right\}$

Mental Math

1. $\frac{xy}{10}$ **3.** $\frac{2y}{3x}$ **5.** $\frac{m^2}{36}$

Exercise Set 6.1

1. 2 **3.** $-\frac{1}{5}$ **5.** 0 **7.** no real number **9.** 0, −2, 1 **11.** 2, −2 **13.** $\frac{5x^2}{9}$ **15.** $\frac{x^4}{2y^2}$ **17.** $1 - 2x$ **19.** $x + 3$ **21.** $\frac{9}{7}$ **23.** $x - 4$ **25.** −1 **27.** $-(x + 7)$ **29.** $\frac{2x+1}{x-1}$ **31.** $\frac{x^2+5x+25}{2}$ **33.** $\frac{x-2}{2x^2+1}$ **35.** $\frac{1}{3x+5}$ **37.** $\frac{x}{2}$ **39.** $-\frac{4}{5}$ **41.** $-\frac{6a}{2a+1}$ **43.** $\frac{3}{2(x-1)}$ **45.** $\frac{x+2}{x+3}$ **47.** $\frac{3a}{5(a-b)}$ **49.** $\frac{1}{6}$ **51.** $\frac{5}{x-2}$ sq. m **53.** $\frac{x}{2}$ **55.** $\frac{x}{3}$ **57.** $\frac{4a^2}{a-b}$ **59.** $\frac{4}{(x+2)(x+3)}$ **61.** $\frac{1}{2}$ **63.** −1 **65.** $\frac{8(a-2)}{3(a+2)}$ **67.** $\frac{8}{x^2y}$ **69.** $\frac{(y+5)(2x-1)}{(y+2)(5x+1)}$ **71.** $\frac{10}{3}, -8, -\frac{7}{3}$ **73. a.** \$200 million **b.** \$500 million **c.** \$300 million **75.** $\frac{7}{5}$ **77.** $\frac{1}{12}$ **79.** $\frac{11}{16}$ **81.** $\frac{(x+2)(x-1)^2}{x^5}$ ft **83.** −1 **85.** $\frac{1}{x^n - 4}$

ANSWERS

Calculator Explorations

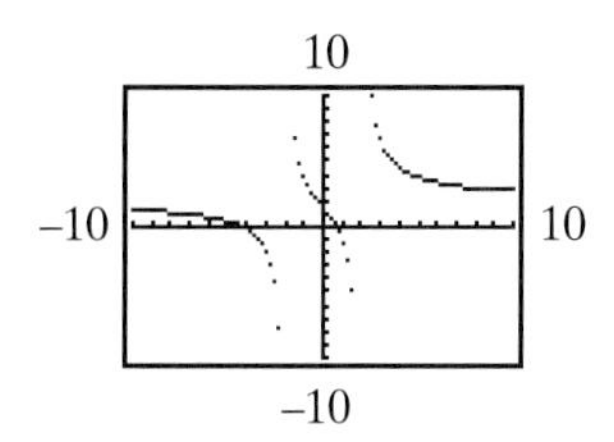

Exercise Set 6.2

1. $-\frac{3}{x}$ **3.** $\frac{x+2}{x-2}$ **5.** $x-2$ **7.** $\frac{-1}{x-2}$ or $\frac{1}{2-x}$ **9.** $-\frac{5}{x}$ **11.** $\frac{4x}{x+5}$ ft **13.** $35x$ **15.** $x(x+1)$ **17.** $(x+7)(x-7)$ **19.** $6(x+2)(x-2)$ **21.** $(a+b)(a-b)^2$ **23.** $-4x(x+3)(x-3)$ **25.** $\frac{17}{6x}$ **27.** $\frac{35-4y}{14y^2}$ **29.** $\frac{-13x+4}{(x+4)(x-4)}$ **31.** $\frac{3}{x+4}$ **33.** $\frac{25}{6(x+5)}$ **35.** $\frac{-2x-1}{x^2(x-3)}$ or $-\frac{2x+1}{x^2(x-3)}$ **37.** 0 **39.** $\frac{1-x}{x-2}$ **41.** $\frac{y^2+2y+10}{(y+4)(y-4)(y-2)}$ **43.** $\frac{x^2+5x+21}{(x-2)(x+1)(x+3)}$ **45.** $\frac{5(x^2+x-4)}{(3x+2)(x+3)(2x-5)}$ **47.** $\frac{5a+1}{(a+1)^2(a-1)}$ **49.** $\frac{3ab-b^2}{2(a+b)(a-b)}$ **51.** $\frac{x+18}{(x+2)^2(x-2)}$ **53.** $\frac{5-2x}{2(x+1)}$ **55.** $\frac{2(x^2+x-21)}{(x+3)^2(x-3)}$ **57.** $\frac{6x}{(x+3)(x-3)^2}$ **59.** $\frac{4}{3}$ **61.** 10 **63.** $4+x^2$ **65.** answers may vary **67.** $\frac{5}{4y}$ **69.** $\frac{3-16x}{48x^2}$

Exercise Set 6.3

1. 4 **3.** $\frac{7}{13}$ **5.** $\frac{4}{x}$ **7.** $\frac{9(x-2)}{9x^2+4}$ **9.** $2x+y$ **11.** $\frac{2(x+1)}{2x-1}$ **13.** $\frac{2x+3}{4-9x}$ **15.** $\frac{1}{x^2-2x+4}$ **17.** $\frac{x}{5x-10}$ **19.** $\frac{x-2}{2x-1}$ **21.** $\frac{xy^2}{x^2+y^2}$ **23.** $\frac{2b^2+3a}{b(b-a)}$ **25.** $\frac{x}{(x+1)(x-1)}$ **27.** $\left\{-\frac{5}{6}\right\}$ **29.** $\{2\}$ **31.** $\{54\}$ **33.** $\frac{770a}{770-s}$ **35.** $2x$ **37.** $-x(x-1)$

Mental Math

1. equation **3.** expression **5.** equation

Exercise Set 6.4

1. $\{72\}$ **3.** $\{3\}$ **5.** $\{-1\}$ **7.** $\{2\}$ **9.** $\{6\}$ **11.** $\{2\}$ **13.** $\left\{-\frac{28}{3}\right\}$ **15.** $\{3\}$ **17.** $\{3\}$ **19.** $\varnothing$ **21.** $\{1\}$ **23.** $\left\{\frac{1}{3}\right\}$ **25.** $\{-5, 5\}$ **27.** $\{3\}$ **29.** $\varnothing$ **31.** $\left\{\frac{4}{3}\right\}$ **33.** $\{-12\}$ **35.** $\left\{1, \frac{11}{4}\right\}$ **37.** $\{-5, -1\}$ **39.** $\left\{-\frac{7}{5}\right\}$ **41.** 5 **43.** length, 15 in.; width, 10 in. **45.** 10% **47.** 25–29 and 30–34 **49.** 5560 **51.** $\left\{\frac{1}{9}, -\frac{1}{4}\right\}$ **53.** $\left\{-\frac{1}{4}, \frac{1}{9}\right\}$ **55.** $\{0.42\}$ **57.** 800 pencil sharpeners

Integrated Review

1. $\left\{\frac{1}{2}\right\}$ **2.** $\{10\}$ **3.** $\frac{1+2x}{8}$ **4.** $\frac{15+x}{10}$ **5.** $\frac{2(x-4)}{(x+2)(x-1)}$ **6.** $-\frac{5(x-8)}{(x-2)(x+4)}$ **7.** $\{4\}$ **8.** $\{8\}$ **9.** $\{-5\}$ **10.** $\left\{-\frac{2}{3}\right\}$ **11.** $\frac{2x+5}{x(x-3)}$ **12.** $\frac{5}{2x}$ **13.** $\{-2\}$ **14.** $-\frac{y}{x}$ **15.** $\frac{(a+3)(a+1)}{a+2}$ **16.** $\frac{-a^2+31a+10}{5(a-6)(a+1)}$ **17.** $\left\{-\frac{1}{5}\right\}$ **18.** $\left\{-\frac{3}{13}\right\}$ **19.** $\frac{4a+1}{(3a+1)(3a-1)}$ **20.** $-\frac{a+8}{4a(a-2)}$ **21.** $\left\{-1, \frac{3}{2}\right\}$ **22.** $\frac{x^2-3x+10}{2(x+3)(x-3)}$ **23.** $\frac{3}{x+1}$ **24.** $\{x \mid x \text{ is a real number and } x \neq 2, x \neq -1\}$ **25.** $\{-1\}$ **26.** $\frac{22z-45}{3z(z-3)}$

Exercise Set 6.5

1. $C = \frac{5}{9}(F-32)$ **3.** $I = A - QL$ **5.** $R = \frac{R_1R_2}{R_1+R_2}$ **7.** $n = \frac{2S}{a+L}$ **9.** $b = \frac{2A-ah}{h}$ **11.** $T_2 = \frac{P_2V_2T_1}{P_1V_1}$ **13.** $f_2 = \frac{f_1f}{f_1-f}$ **15.** $L = \frac{n\lambda}{2}$ **17.** $c = \frac{2Lw}{\theta}$ **19.** 1 and 5 **21.** 5 **23.** 6 ohms **25.** $\frac{1}{R} = \frac{1}{R_1} + \frac{1}{R_2} + \frac{1}{R_3}$; $R = \frac{15}{13}$ ohms **27.** 15.6 hr **29.** 10 min **31.** 200 mph **33.** 15 mph **35.** −8 and −7 **37.** 36 min **39.** 45 mph and 60 mph **41.** 10 mph; 8 mph **43.** 2 hr **45.** 135 mph **47.** 12 mi **49.** $2\frac{2}{9}$ hr **51.** by jet, 3 hr; by car, 4 hr **53.** $1\frac{1}{2}$ min **55.** 63 mph **57.** $\{-5\}$ **59.** $\{2\}$ **61.** 164 lb **63.** higher; F

Exercise Set 6.6

1. $k = \frac{1}{5}; y = \frac{1}{5}x$ **3.** $k = \frac{3}{2}; y = \frac{3}{2}x$ **5.** $k = 14; y = 14x$ **7.** $k = 0.25; y = 0.25x$ **9.** 4.05 lb **11.** $P \approx 566{,}222$ tons **13.** $k = 30; y = \frac{30}{x}$ **15.** $k = 700; y = \frac{700}{x}$ **17.** $k = 2; y = \frac{2}{x}$ **19.** $k = 0.14; y = \frac{0.14}{x}$ **21.** $R = 54$ mph **23.** 72 amps **25.** divided by 4 **27.** $x = kyz$ **29.** $r = kst^3$ **31.** 22.5 tons **33.** 15π cu. in. **35.** 90 hp **37.** $C = 12\pi$ cm; $A = 36\pi$ sq. cm **39.** $C = 14\pi$ m; $C = 49\pi$ sq. m **41.** 0 **43.** -1 **45.** multiplied by 2 **47.** multiplied by 4

Chapter 6 Review

1. none **2.** none **3.** 5 **4.** 4 **5.** $0, -8$ **6.** $-4, 4$ **7.** $\frac{x^2}{3}$ **8.** 1 **9.** $\frac{9m^2p}{5}$ **10.** -1 **11.** $\frac{1}{5}$ **12.** 2 **13.** $\frac{1}{x-1}$ **14.** $\frac{1}{x-7}$ **15.** $\frac{2(x-3)}{x-4}$ **16.** $\frac{y-3}{x+2}$ **17. a.** \$119 **b.** \$77 **c.** decrease **18.** $\frac{1}{3x}$ **19.** $\frac{2x^3}{z^3}$ **20.** $-\frac{3}{2}$ **21.** $\frac{2}{5}$ **22.** $\frac{a-b}{2a}$ **23.** $\frac{1}{6}$ **24.** $\frac{4(x+2)}{5(x-2)}$ **25.** $\frac{3x}{16}$ **26.** $\frac{12}{5}$ **27.** $\frac{3c^2}{14a^2b}$ **28.** $\frac{3}{5x}$ **29.** $\frac{(x+4)(x+5)}{3}$ **30.** $\frac{a-b}{5a}$ **31.** $\frac{7(x-4)}{2(x-2)}$ **32.** $\frac{3(x+1)}{x-7}$ **33.** $-\frac{1}{x}$ **34.** $\frac{5(a-2)}{7}$ **35.** $\frac{8}{9a^2}$ **36.** $-\frac{x+3}{2(x+2)}$ **37.** $\frac{6}{a}$ **38.** 18 **39.** $60x^2y^5$ **40.** $2x(x-2)$ **41.** $5x(x-5)$ **42.** $10x^3(x-4)(x+7)(x-3)$ **43.** $\frac{2}{5}$ **44.** $\frac{4+x}{x-4}$ **45.** $\frac{2}{x^2}$ **46.** $\frac{3}{2(x-2)}$ **47.** $\frac{1}{x-2}$ **48.** $\frac{13}{3x}$ **49.** $\frac{5x^2-3y^2}{15x^4y^3}$ **50.** $\frac{x-2}{x-10}$ **51.** $\frac{-x+5}{(x+1)(x-1)}$ **52.** $\frac{-7x-6}{5(x-3)(x+3)}$ **53.** $\frac{2x^2-5x-4}{x-3}$ **54.** $\frac{5a-1}{(a-1)^2(a+1)}$ **55.** $\frac{3x^2-7x-4}{(3x-4)(9x^2+12x+16)}$ **56.** $\frac{5-2x}{2(x-1)}$ **57.** $\frac{2(7x-20)}{(x+4)^2(x-4)}$ **58.** $\frac{11}{x}$ **59.** $\frac{2}{3}$ **60.** $\frac{4-3x}{8+x}$ **61.** $\frac{2}{15-2x}$ **62.** $\frac{x^3}{3(x+1)}$ **63.** $\frac{y}{2}$ **64.** $\frac{(x+2)(x-2)}{15}$ **65.** $\frac{5(4x-3)}{2(5x^2-2)}$ **66.** $\frac{y}{x-y}$ **67.** $\frac{x(5y+1)}{3y}$ **68.** $\frac{x-3}{3}$ **69.** $\frac{1+x}{1-x}$ **70.** $\frac{2x}{x-2}$ **71.** $\frac{x-1}{3x-1}$ **72.** $\frac{2(x-1)}{x+6}$ **73.** $-\frac{x^2+9}{6x}$ **74.** $\frac{5x}{2}$ **75.** $\{6\}$ **76.** $\{6\}$ **77.** $\{2\}$ **78.** $\{3\}$ **79.** $\left\{\frac{3}{2}\right\}$ **80.** $\{-7, 7\}$ **81.** $\left\{\frac{5}{3}\right\}$ **82.** $\frac{2x+5}{x(x-7)}$ **83.** $\left\{-\frac{1}{3}, 2\right\}$ **84.** $\frac{-5x-30}{2x(x-3)}$ **85.** $a = \frac{2A}{h} - b$ **86.** $R_2 = \frac{RR_1}{R_1 - R}$ **87.** $R = \frac{E}{I} - r$ **88.** $r = \frac{A-P}{Pt}$ **89.** $A = \frac{HL}{k(T_1 - T_2)}$ **90.** $\{1, 2\}$ **91.** 7 **92.** $\frac{23}{25}$ **93.** -10 and -8 **94.** $1\frac{23}{37}$ hr **95.** 12 hr **96.** 10 hr **97.** 490 mph **98.** 6 ohms **99.** 8 mph **100.** 45 mph **101.** 4 mph **102.** $63\frac{2}{3}$ mph and 45 mph **103.** 9 **104.** 4 **105.** 3.125 cu. ft **106.** 64π sq. in.

Chapter 6 Test

1. 1 **2.** $-1, -3$ **3.** $\frac{5x^3}{3}$ **4.** $-\frac{7}{8}$ **5.** $\frac{x}{x+9}$ **6.** $\frac{x+2}{5}$ **7.** $\frac{5}{3x}$ **8.** $\frac{4a^3b^4}{c^6}$ **9.** $\frac{x+2}{2(x+3)}$ **10.** $\frac{-4(2x+9)}{5}$ **11.** $\frac{3}{x^3}$ **12.** -1 **13.** $\frac{5x-2}{(x-3)(x+2)(x-2)}$ **14.** $\frac{-x+30}{6(x-7)}$ **15.** $\frac{3}{2}$ **16.** $\frac{1}{5}$ **17.** $\frac{64}{3}$ **18.** $\frac{(x-3)^2}{x-2}$ **19.** $\{7\}$ **20.** $\{2, -2\}$ **21.** $\{8\}$ **22.** $x = \frac{7a^2+b^2}{4a-b}$ **23.** 5 **24.** $\frac{6}{7}$ hr **25.** 16 **26.** 9 **27.** 256 ft

Cumulative Review

1. 8; Sec. 1.1, Ex. 3 **2.** -1; Sec. 1.5, Ex. 7 **3.** 1; Sec. 1.5, Ex. 8 **4.** $\varnothing$; Sec. 2.5, Ex. 3 **5.** $\{4, -10\}$; Sec. 2.6, Ex. 5 **6.** $(1, 9)$, not a solution; $(0, -12)$, solution; $(2, -6)$, solution; Sec. 3.1, Ex. 3 **7.** parallel; Sec. 3.2, Ex. 7 **8.** $y = \frac{5}{8}x - \frac{5}{2}$; Sec. 3.4, Ex. 2 **9.** domain: $(-\infty, \infty)$; range: $[0, \infty)$; Sec. 3.5, Ex. 15 **10.** domain: $[-4, 4]$; range: $[-2, 2]$; Sec. 3.5, Ex. 16 **11.** $\left(-\frac{21}{10}, \frac{3}{10}\right)$; Sec. 4.1, Ex. 7 **12.** $\left(\frac{1}{2}, 0, \frac{3}{4}\right)$; Sec. 4.2, Ex. 3 **13.** 52 mph and 47 mph; Sec. 4.3, Ex. 2 **14.** $(1, -1, 3)$; Sec. 4.4, Ex. 3 **15.** $(1, 0)$; Sec. 4.5, Ex. 3

16.

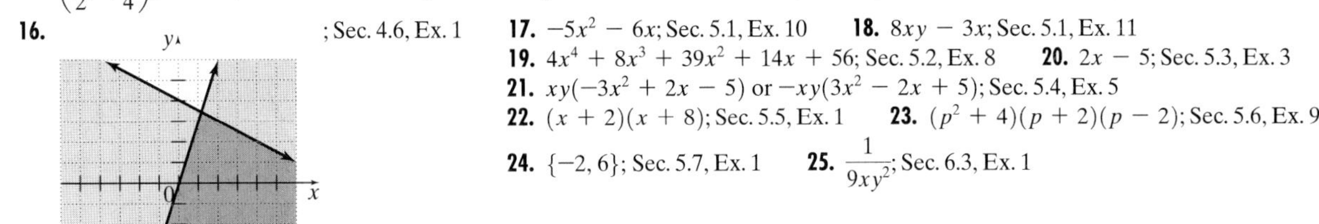

; Sec. 4.6, Ex. 1 **17.** $-5x^2 - 6x$; Sec. 5.1, Ex. 10 **18.** $8xy - 3x$; Sec. 5.1, Ex. 11 **19.** $4x^4 + 8x^3 + 39x^2 + 14x + 56$; Sec. 5.2, Ex. 8 **20.** $2x - 5$; Sec. 5.3, Ex. 3 **21.** $xy(-3x^2 + 2x - 5)$ or $-xy(3x^2 - 2x + 5)$; Sec. 5.4, Ex. 5 **22.** $(x+2)(x+8)$; Sec. 5.5, Ex. 1 **23.** $(p^2+4)(p+2)(p-2)$; Sec. 5.6, Ex. 9 **24.** $\{-2, 6\}$; Sec. 5.7, Ex. 1 **25.** $\frac{1}{9xy^2}$; Sec. 6.3, Ex. 1

Chapter 7

Chapter 7 Pretest

1. $9x$; (7.1A) **2.** $-3y^3$; (7.1B) **3.** x^6; (7.1C) **4.** 4; (7.1E) **5.** 12; (7.2A) **6.** 27; (7.2B) **7.** $\frac{3m^{5/6}}{2}$; (7.2C) **8.** $\frac{27x^6}{y^9}$; (7.2D) **9.** $\sqrt{39x}$; (7.3A) **10.** $\frac{\sqrt{6}}{7}$; (7.3B) **11.** $-21\sqrt{2}$; (7.4A) **12.** $8x - 10\sqrt{x} - 3$; (7.4B) **13.** $5 - 2\sqrt{5y} + y^2$; (7.4B) **14.** $3b^4\sqrt[3]{5a^2}$; (7.3C) **15.** $\frac{\sqrt{3x}}{3x}$; (7.5A) **16.** $-4 + 2\sqrt{11}$; (7.5C) **17.** $\{12\}$; (7.6A) **18.** $4 - 8i$; (7.7B) **19.** $-21 - 20i$; (7.7C) **20.** $\frac{8}{13} + \frac{1}{13}i$; (7.7D)

Exercise Set 7.1

1. 2, −2 **3.** not a real number **5.** 10, −10 **7.** 10 **9.** $\frac{1}{2}$ **11.** 0.01 **13.** −6 **15.** x^5 **17.** $4y^3$ **19.** 2.646 **21.** 6.164 **23.** 14.142 **25.** 4 **27.** $\frac{1}{2}$ **29.** −1 **31.** x^4 **33.** $-3x^3$ **35.** −2 **37.** not a real number **39.** −2 **41.** x^4 **43.** $2x^2$ **45.** $9x^2$ **47.** $4x^2$ **49.** 8 **51.** −8 **53.** $2|x|$ **55.** x **57.** $|x - 2|$ **59.** $|x + 2|$ **61.** $\sqrt{3}$ **63.** −1 **65.** −3 **67.** $\sqrt{7}$ **69.** $-32x^{15}y^{10}$ **71.** $-60x^7y^{10}z^5$ **73.** $\frac{x^9y^5}{2}$ **75.** answers may vary

Exercise Set 7.2

1. 7 **3.** 3 **5.** $\frac{1}{2}$ **7.** 13 **9.** $2\sqrt[3]{m}$ **11.** $3x^2$ **13.** −3 **15.** −2 **17.** 8 **19.** 16 **21.** not a real number **23.** $\sqrt[5]{(2x)^3}$ **25.** $\sqrt[3]{(7x + 2)^2}$ **27.** $\frac{64}{27}$ **29.** $\frac{1}{16}$ **31.** $\frac{1}{16}$ **33.** not a real number **35.** $\frac{1}{x^{1/4}}$ **37.** $a^{2/3}$ **39.** $\frac{5x^{3/4}}{7}$ **41.** answers may vary **43.** $a^{7/3}$ **45.** x **47.** $3^{5/8}$ **49.** $y^{1/6}$ **51.** $8u^3$ **53.** $-b$ **55.** $27x^{2/3}$ **57.** $\sqrt{x}$ **59.** $\sqrt[3]{2}$ **61.** $2\sqrt{x}$ **63.** $\sqrt{x + 3}$ **65.** $\sqrt{xy}$ **67.** $\sqrt[3]{a^2b}$ **69.** $\sqrt[15]{y^{11}}$ **71.** $\sqrt[12]{b^5}$ **73.** $\sqrt[24]{x^{23}}$ **75.** $\sqrt{a}$ **77.** $\sqrt[6]{432}$ **79.** $\sqrt[15]{343y^5}$ **81.** $\sqrt[6]{125r^3s^2}$ **83.** $25 \cdot 3$ **85.** $16 \cdot 3$ or $4 \cdot 12$ **87.** $8 \cdot 2$ **89.** $27 \cdot 2$ **91.** $a^{1/3}$ **93.** $x^{1/5}$ **95.** 1.6818 **97.** $\frac{t^{1/2}}{u^{1/2}}$

Exercise Set 7.3

1. $\sqrt{14}$ **3.** 2 **5.** $\sqrt[3]{36}$ **7.** $\sqrt{6x}$ **9.** $\sqrt{\frac{14}{xy}}$ **11.** $\sqrt[4]{20x^3}$ **13.** $\frac{\sqrt{6}}{7}$ **15.** $\frac{\sqrt{2}}{7}$ **17.** $\frac{\sqrt[4]{x^3}}{2}$ **19.** $\frac{\sqrt[3]{4}}{3}$ **21.** $\frac{\sqrt[4]{8}}{x^2}$ **23.** $\frac{\sqrt[3]{2x}}{3y^4\sqrt[3]{3}}$ **25.** $\frac{x\sqrt{y}}{10}$ **27.** $\frac{\sqrt{5x}}{2y}$ **29.** $-\frac{z^2\sqrt[3]{z}}{3x}$ **31.** $4\sqrt{2}$ **33.** $4\sqrt[3]{3}$ **35.** $25\sqrt{3}$ **37.** $2\sqrt{6}$ **39.** $10x^2\sqrt{x}$ **41.** $2y^2\sqrt[3]{2y}$ **43.** $a^2b\sqrt[4]{b^3}$ **45.** $y^2\sqrt{y}$ **47.** $5ab\sqrt{b}$ **49.** $-2x^2\sqrt[5]{y}$ **51.** $x^4\sqrt[3]{50x^2}$ **53.** $-4a^4b^3\sqrt{2b}$ **55.** $3x^3y^4\sqrt{xy}$ **57.** $5r^3s^4$ **59.** $\sqrt{2}$ **61.** 2 **63.** 10 **65.** x^2y **67.** $24m^2$ **69.** $\frac{15x\sqrt{2x}}{2}$ or $\frac{15x}{2}\sqrt{2x}$ **71.** $2a^2$ **73.** $14x$ **75.** $2x^2 - 7x - 15$ **77.** y^2 **79.** $-3x - 15$ **81.** $x^2 - 8x + 16$ **83. a.** 20π sq. cm **b.** 211.57 sq. ft **85. a.** 3.8 times **b.** 2.9 times **c.** answers may vary

Mental Math

1. $6\sqrt{3}$ **3.** $3\sqrt{x}$ **5.** $12\sqrt[3]{x}$ **7.** 3

Exercise Set 7.4

1. $-2\sqrt{2}$ **3.** $10x\sqrt{2x}$ **5.** $17\sqrt{2} - 15\sqrt{5}$ **7.** $-\sqrt[3]{2x}$ **9.** $5b\sqrt{b}$ **11.** $\frac{31\sqrt{2}}{15}$ **13.** $\frac{\sqrt[3]{11}}{3}$ **15.** $\frac{5\sqrt{5x}}{9}$ **17.** $14 + \sqrt{3}$ **19.** $7 - 3y$ **21.** $6\sqrt{3} - 6\sqrt{2}$ **23.** $-23\sqrt[3]{5}$ **25.** $2b\sqrt{b}$ **27.** $20y\sqrt{2y}$ **29.** $2y\sqrt[3]{2x}$ **31.** $6\sqrt[3]{11} - 4\sqrt{11}$ **33.** $4x\sqrt[4]{x^3}$ **35.** $\frac{2\sqrt{3}}{3}$ **37.** $\frac{5x\sqrt[3]{x}}{7}$ **39.** $\frac{5\sqrt{7}}{2x}$ **41.** $\frac{\sqrt[3]{2}}{6}$ **43.** $\frac{14x\sqrt[3]{2x}}{9}$ **45.** $15\sqrt{3}$ in. **47.** $\sqrt{35} + \sqrt{21}$ **49.** $7 - 2\sqrt{10}$ **51.** $3\sqrt{x} - x\sqrt{3}$ **53.** $6x - 13\sqrt{x} - 5$ **55.** $\sqrt[3]{a^2} + \sqrt[3]{a} - 20$ **57.** $6\sqrt{2} - 12$ **59.** $2 + 2x\sqrt{3}$ **61.** $-16 - \sqrt{35}$ **63.** $x - y^2$ **65.** $3 + 2x\sqrt{3} + x^2$ **67.** $5x - 3\sqrt{10x} - 3\sqrt{15x} + 9\sqrt{6}$ **69.** $2\sqrt[3]{2} - \sqrt[3]{4}$ **71.** $-4\sqrt[6]{x^5} + \sqrt[3]{x^2} + 8\sqrt[3]{x} - 4\sqrt{x} + 7$ **73.** $x + 24 + 10\sqrt{x - 1}$ **75.** $2x + 6 - 2\sqrt{2x + 5}$ **77.** $x - 7$ **79.** $\frac{7}{x + y}$ **81.** $2a - 3$ **83.** $\frac{-2 + \sqrt{3}}{3}$ **85.** $22\sqrt{5}$ ft; 150 sq. ft **87.** $2\sqrt{6} - 2\sqrt{2} - 2\sqrt{3} + 6$

Mental Math

1. $\sqrt{2} - x$ **3.** $5 + \sqrt{a}$ **5.** $7\sqrt{4} - 8\sqrt{x}$

Exercise Set 7.5

1. $\frac{\sqrt{14}}{7}$ **3.** $\frac{\sqrt{5}}{5}$ **5.** $\frac{4\sqrt[3]{9}}{3}$ **7.** $\frac{3\sqrt{2x}}{4x}$ **9.** $\frac{3\sqrt[3]{2x}}{2x}$ **11.** $\frac{3\sqrt{3a}}{a}$ **13.** $\frac{3\sqrt[3]{4}}{2}$ **15.** $\frac{2\sqrt{21}}{7}$ **17.** $\frac{\sqrt{10xy}}{5y}$ **19.** $\frac{2\sqrt[4]{9x}}{3x^2}$

21. $\frac{5a\sqrt[5]{4ab^4}}{2a^2b^3}$ **23.** $\frac{5}{\sqrt{15}}$ **25.** $\frac{6}{\sqrt{10}}$ **27.** $\frac{4x}{7\sqrt{4x}}$ **29.** $\frac{5y}{\sqrt[3]{100xy}}$ **31.** $\frac{2}{\sqrt{10}}$ **33.** $\frac{2x}{11\sqrt{2x}}$ **35.** $\frac{7}{2\sqrt[3]{49}}$ **37.** $\frac{3x^2}{10\sqrt[3]{9x}}$ **39.** $\frac{6x^2y^3}{\sqrt{6z}}$ **41.** answers may vary **43.** $-2(2 + \sqrt{7})$ **45.** $\frac{7(3 + \sqrt{x})}{9 - x}$ **47.** $-5 + 2\sqrt{6}$ **49.** $\frac{2a + 2\sqrt{a} + \sqrt{ab} + \sqrt{b}}{4a - b}$ **51.** $-\frac{8(1 - \sqrt{10})}{9}$ **53.** $\frac{x - \sqrt{xy}}{x - y}$ **55.** $\frac{15 + 9\sqrt{2}}{21}$ **57.** $\frac{-7}{12 + 6\sqrt{11}}$ **59.** $\frac{3}{10 + 5\sqrt{7}}$ **61.** $\frac{x - 9}{x - 3\sqrt{x}}$ **63.** $\frac{1}{3 - 2\sqrt{2}}$ **65.** $\frac{x - 1}{x - 2\sqrt{x} + 1}$ **67.** $\{5\}$ **69.** $\left\{-\frac{1}{2}, 6\right\}$ **71.** $\{2, 6\}$ **73.** $r = \frac{\sqrt{A\pi}}{2\pi}$ **75.** answers may vary

Integrated Review

1. 9 **2.** -2 **3.** $\frac{1}{2}$ **4.** x^3 **5.** y^3 **6.** $2y^5$ **7.** $-2y$ **8.** $3b^3$ **9.** 6 **10.** $\sqrt[4]{3y}$ **11.** $\frac{1}{16}$ **12.** $\sqrt[5]{(x + 1)^3}$ **13.** y **14.** $16x^{1/2}$ **15.** $x^{5/4}$ **16.** $4^{11/15}$ **17.** $2x^2$ **18.** $\sqrt[4]{a^3b^2}$ **19.** $\sqrt[4]{x^3}$ **20.** $\sqrt[6]{500}$ **21.** $2\sqrt{10}$ **22.** $2xy^2\sqrt[4]{x^3y^2}$ **23.** $3x\sqrt[3]{2x}$ **24.** $-2b^2$ **25.** $\sqrt{5x}$ **26.** $4x$ **27.** $7y^2\sqrt{y}$ **28.** $2a^2\sqrt[4]{3}$ **29.** $2\sqrt{5} - 5\sqrt{3} + 5\sqrt{7}$ **30.** $y\sqrt[3]{2y}$ **31.** $\sqrt{15} - \sqrt{6}$ **32.** $10 + 2\sqrt{21}$ **33.** $4x^2 - 5$ **34.** $x + 2 - 2\sqrt{x + 1}$ **35.** $\frac{\sqrt{21}}{3}$ **36.** $\frac{5\sqrt[3]{4x}}{2x}$ **37.** $\frac{13 - 3\sqrt{21}}{5}$ **38.** $\frac{7}{\sqrt{21}}$ **39.** $\frac{3y}{\sqrt[3]{33y^2}}$ **40.** $\frac{x - 4}{x + 2\sqrt{x}}$

Calculator Explorations

1. $\{3.19\}$ **3.** $\varnothing$ **5.** $\{3.23\}$

Exercise Set 7.6

1. $\{8\}$ **3.** $\{7\}$ **5.** $\varnothing$ **7.** $\{7\}$ **9.** $\{6\}$ **11.** $\left\{\frac{-9}{2}\right\}$ **13.** $\{29\}$ **15.** $\{4\}$ **17.** $\{-4\}$ **19.** $\varnothing$ **21.** $\{7\}$ **23.** $\{9\}$ **25.** $\{50\}$ **27.** $\varnothing$ **29.** $\left\{\frac{15}{4}\right\}$ **31.** $\{7\}$ **33.** $\{5\}$ **35.** $\{-12\}$ **37.** $\{9\}$ **39.** $\{-3\}$ **41.** $\{1\}$ **43.** $\{1\}$ **45.** $\left\{\frac{1}{2}\right\}$ **47.** $\{0, 4\}$ **49.** $\left\{\frac{37}{4}\right\}$ **51.** $3\sqrt{5}$ ft **53.** $2\sqrt{10}$ m **55.** $2\sqrt{131}$ m ≈ 22.9 m **57.** $\sqrt{100.84}$ mm ≈ 10.0 mm **59.** 17 ft **61.** 13 ft **63.** 14,657,415 sq. mi **65.** 100 ft **67.** $\frac{x}{4x + 3}$ **69.** $-\frac{4z + 2}{3z}$ **71.** $\{1\}$ **73.** 2743 deliveries **75. a.** answers may vary **b.** answers may vary

Mental Math

1. $9i$ **3.** $i\sqrt{7}$ **5.** -4 **7.** $8i$

Exercise Set 7.7

1. $2i\sqrt{6}$ **3.** $-6i$ **5.** $24i\sqrt{7}$ **7.** $-3\sqrt{6}$ **9.** $-\sqrt{14}$ **11.** $-5\sqrt{2}$ **13.** $4i$ **15.** $i\sqrt{3}$ **17.** $2\sqrt{2}$ **19.** $6 - 4i$ **21.** $2 + 6i$ **23.** $-2 - 4i$ **25.** $2 - i$ **27.** $5 - 10i$ **29.** $8 - i$ **31.** -12 **33.** 63 **35.** -40 **37.** $18 + 12i$ **39.** $27 + 3i$ **41.** $18 + 13i$ **43.** 7 **45.** $12 - 16i$ **47.** 20 **49.** 2 **51.** $17 + 144i$ **53.** $-2i$ **55.** $-4i$ **57.** $\frac{28}{25} - \frac{21}{25}i$ **59.** $-\frac{12}{5} + \frac{6}{5}i$ **61.** $4 + i$ **63.** $-\frac{5}{2} - 2i$ **65.** $-5 + \frac{16}{3}i$ **67.** $\frac{3}{5} - \frac{1}{5}i$ **69.** $\frac{1}{5} - \frac{8}{5}i$ **71.** 1 **73.** i **75.** $-i$ **77.** -1 **79.** -64 **81.** $-243i$ **83.** 5 people **85.** 14 people **87.** 16.7% **89.** $1 - i$ **91.** 0 **93.** $2 + 3i$ **95.** $2 - i\sqrt{2}$ **97.** $\frac{1}{2} - \frac{\sqrt{3}}{2}i$ **99.** answers may vary **101.** $6 - 6i$ **103.** yes

Chapter 7 Review

1. 9 **2.** 3 **3.** -2 **4.** not a real number **5.** $-\frac{1}{7}$ **6.** x^{32} **7.** -6 **8.** 4 **9.** $-a^2b^3$ **10.** $4a^2b^6$ **11.** $2ab^2$ **12.** $-2x^3y^4$ **13.** $\frac{x^6}{6y}$ **14.** $\frac{3y}{z^4}$ **15.** $|x|$ **16.** $|x^2 - 4|$ **17.** -27 **18.** -5 **19.** $-x$ **20.** $2|2y + z|$ **21.** $5|x - y|$ **22.** y **23.** $|x|$ **24.** 3, 6 **25.** 2, $\sqrt[3]{17}$ **26.** $\frac{1}{3}$ **27.** $-\frac{1}{3}$ **28.** $-\frac{1}{3}$ **29.** $-\frac{1}{4}$ **30.** -27 **31.** $\frac{1}{4}$ **32.** not a real number **33.** $\frac{343}{125}$ **34.** $\frac{9}{4}$ **35.** not a real number **36.** $x^{2/3}$ **37.** $5^{1/5}x^{2/5}y^{3/5}$ **38.** $\sqrt[5]{y^4}$ **39.** $5\sqrt[3]{xy^2z^5}$ **40.** $\frac{1}{\sqrt{x + 2y}}$ **41.** $a^{13/6}$ **42.** $\frac{1}{b}$ **43.** $\frac{1}{a^{9/2}}$ **44.** $\frac{y^2}{x}$ **45.** a^4b^6 **46.** $\frac{1}{x^{11/12}}$ **47.** $\frac{b^{5/6}}{49a^{1/4}c^{5/3}}$ **48.** $a - a^2$ **49.** 4.472 **50.** -3.391 **51.** 5.191 **52.** 3.826 **53.** -26.246 **54.** 0.045 **55.** $\sqrt[6]{1372}$ **56.** $\sqrt[12]{81x^3}$ **57.** $2\sqrt{6}$ **58.** $\sqrt[3]{7x^2yz}$ **59.** $2x$ **60.** ab^3 **61.** $2\sqrt{15}$ **62.** $-5\sqrt{3}$ **63.** $3\sqrt[3]{6}$ **64.** $-2\sqrt[3]{4}$ **65.** $6x^3\sqrt{x}$ **66.** $2ab^2\sqrt[3]{3a^2b}$ **67.** $\frac{p^8\sqrt{p}}{11}$ **68.** $\frac{y\sqrt[3]{y^2}}{3x^2}$ **69.** $\frac{y\sqrt[4]{xy^2}}{3}$ **70.** $\frac{x\sqrt{2x}}{7y^2}$ **71. a.** $\frac{5}{\sqrt{\pi}}$ m **b.** 5.75 in. **72.** $-2\sqrt{5}$ **73.** $2x\sqrt{3xy}$ **74.** $9\sqrt[3]{2}$ **75.** $3a\sqrt[4]{2a}$ **76.** $\frac{15 + 2\sqrt{3}}{6}$ **77.** $\frac{3\sqrt{2}}{4x}$ **78.** $17\sqrt{2} - 15\sqrt{5}$ **79.** $-4ab\sqrt[4]{2b}$ **80.** 6 **81.** $x - 6\sqrt{x} + 9$ **82.** $-8\sqrt{5}$ **83.** $4x - 9y$ **84.** $a - 9$

85. $\sqrt[3]{a^2} + 4\sqrt[3]{a} + 4$ **86.** $\sqrt[3]{25x^2} - 81$ **87.** $a + 64$ **88.** $\frac{3\sqrt{7}}{7}$ **89.** $\frac{\sqrt{3x}}{6}$ **90.** $\frac{5\sqrt[3]{2}}{2}$ **91.** $\frac{2x^2\sqrt{2x}}{y}$ **92.** $\frac{x^2y^2\sqrt[3]{15yz}}{z}$

93. $\frac{3\sqrt[4]{2x^2}}{2x^3}$ **94.** $\frac{3\sqrt{y} + 6}{y - 4}$ **95.** $-5 + 2\sqrt{6}$ **96.** $\frac{11}{3\sqrt{11}}$ **97.** $\frac{6}{\sqrt{2y}}$ **98.** $\frac{3}{7\sqrt[3]{3}}$ **99.** $\frac{4x^3}{y\sqrt{2x}}$ **100.** $\frac{xy}{\sqrt[3]{10x^2yz}}$

101. $\frac{x - 25}{-3\sqrt{x} + 15}$ **102.** $\{32\}$ **103.** $\varnothing$ **104.** $\{35\}$ **105.** $\varnothing$ **106.** $\{9\}$ **107.** $\{16\}$ **108.** $3\sqrt{2}$ **109.** $\sqrt{241}$

110. 51.2 ft **111.** 4.24 ft **112.** $2i\sqrt{2}$ **113.** $-i\sqrt{6}$ **114.** $6i$ **115.** $-\sqrt{10}$ **116.** $15 - 4i$ **117.** $-13 - 3i$ **118.** -64

119. $10 + 4i$ **120.** $-12 - 18i$ **121.** $1 + 5i$ **122.** $-5 - 12i$ **123.** 87 **124.** $\frac{3}{2} - i$ **125.** $-\frac{1}{3} + \frac{1}{3}i$

Chapter 7 Test

1. $6\sqrt{6}$ **2.** $-x^{16}$ **3.** $\frac{1}{5}$ **4.** 5 **5.** $\frac{4x^2}{9}$ **6.** $-a^6b^3$ **7.** $\frac{8a^{1/3}c^{2/3}}{b^{5/12}}$ **8.** $a^{7/12} - a^{7/3}$ **9.** $|4xy|$ or $4|xy|$ **10.** -27

11. $\frac{3\sqrt{y}}{y}$ **12.** $\frac{8 - 6\sqrt{x} + x}{8 - 2x}$ **13.** $\frac{2\sqrt[3]{3x^2}}{3x}$ **14.** $\frac{6 - x^2}{8(\sqrt{6} - x)}$ **15.** $-x\sqrt{5x}$ **16.** $4\sqrt{3} - \sqrt{6}$ **17.** $x + 2\sqrt{x} + 1$

18. $\sqrt{6} - 4\sqrt{3} + \sqrt{2} - 4$ **19.** -20 **20.** 23.685 **21.** 0.019 **22.** $\{2, 3\}$ **23.** $\varnothing$ **24.** $\{6\}$ **25.** $i\sqrt{2}$ **26.** $-2i\sqrt{2}$

27. $-3i$ **28.** 40 **29.** $7 + 24i$ **30.** $-\frac{3}{2} + \frac{5}{2}i$ **31.** $x = \frac{5\sqrt{2}}{2}$ **32.** $\sqrt{2}, 5$ **33.** 27 mph **34.** 360 ft

Cumulative Review

1. $4x^b$; Sec. 1.6, Ex. 16 **2.** y^{5a+6}; Sec. 1.6, Ex. 17 **3.** $\{-7, 5\}$; Sec. 2.6, Ex. 12 **4.** $y = -2x + 12$; Sec. 3.4, Ex. 5
5. $(-2, 2)$; Sec. 4.1, Ex. 8 **6.** $\{(x, y, z) | x - 5y - 2z = 6\}$; Sec. 4.2, Ex. 4 **7.** no solution; Sec. 4.4, Ex. 2
8. $13x^3y - xy^3 + 7$; Sec. 5.1, Ex. 14 **9.** -4; Sec. 5.1, Ex. 18 **10.** 11; Sec. 5.1, Ex. 19 **11.** $x^2 + 10x + 25$; Sec. 5.2, Ex. 12
12. $16m^4 - 24m^2n + 9n^2$; Sec. 5.2, Ex. 15 **13.** $2x^2 + 5x + 2 + \frac{7}{x - 3}$; Sec. 5.3, Ex. 7 **14.** $(b - 6)(a + 2)$; Sec. 5.4, Ex. 8
15. $2(n^2 - 19n + 40)$; Sec. 5.5, Ex. 4 **16.** $(4x + 3y)^2$; Sec. 5.5, Ex. 8 **17.** $2(5 + 2y)(5 - 2y)$; Sec. 5.6, Ex. 7
18. $\left\{-\frac{1}{2}, 4\right\}$; Sec. 5.7, Ex. 3 **19.**

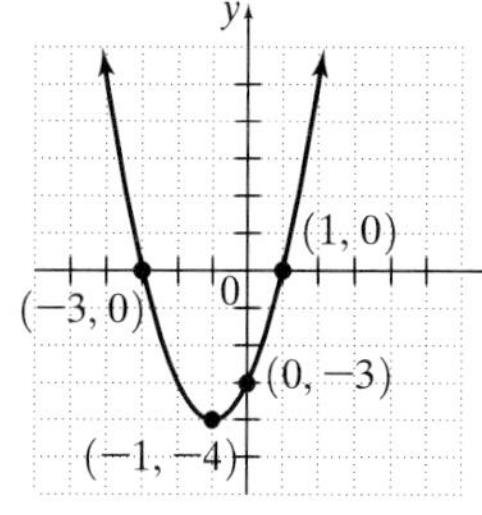

; Sec. 5.8, Ex. 4 **20.** $\frac{3y^4}{x}$; Sec. 6.1, Ex. 4 **21.** $\frac{1}{5x - 1}$; Sec. 6.1, Ex. 5
22. $\frac{5 + x}{7}$; Sec. 6.2, Ex. 1 **23.** $\{-2\}$; Sec. 6.4, Ex. 2
24. $\frac{1}{8}$; Sec. 7.2, Ex. 12 **25.** $\frac{1}{9}$; Sec. 7.2, Ex. 13

Chapter 8

Chapter 8 Pretest

1. $\{-3\sqrt{6}, 3\sqrt{6}\}$; 8.1A **2.** $\left\{\frac{-2 - 2\sqrt{3}}{3}, \frac{-2 + 2\sqrt{3}}{3}\right\}$; 8.1A **3.** $\{-2 - \sqrt{14}, -2 + \sqrt{14}\}$; 8.1C
4. $\left\{\frac{-9 - 2\sqrt{21}}{3}, \frac{-9 + 2\sqrt{21}}{3}\right\}$; 8.1C **5.** $\left\{\frac{1 - i\sqrt{31}}{4}, \frac{1 + i\sqrt{31}}{4}\right\}$; 8.2A **6.** $\left\{\frac{-1 - \sqrt{17}}{12}, \frac{-1 + \sqrt{17}}{12}\right\}$; 8.2A
7. $\{4, -2 - 2i\sqrt{3}, -2 + 2i\sqrt{3}\}$; 8.3A **8.** $\{64, 1\}$; 8.3A **9.** $(-\infty, -5] \cup [6, \infty)$; 8.4A **10.** $(-1, 0)$; 8.4A **11.** $(-\infty, -3)$; 8.4B

12. y, $(-\sqrt{5}, 0)$, $(\sqrt{5}, 0)$, $(0, -5)$
13. y, $(0, 4)$, $(2, 0)$; 8.5A
14. y, $(0, 8)$, $(-4, 0)$, $(-2, 0)$, $(-3, -1)$; 8.5B 8.5C
15.

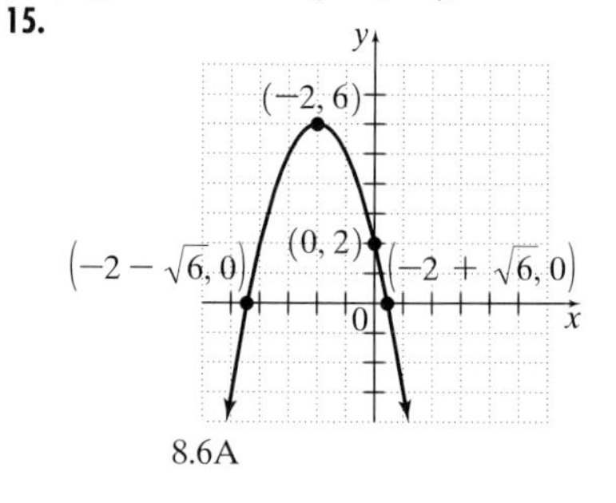

8.6A

16. length: 20 cm; width: 5 cm; 8.2C

Calculator Explorations

1. $\{-1.27, 6.27\}$ **3.** $\{-1.10, 0.90\}$ **5.** $\varnothing$

Exercise Set 8.1

1. $\{-4, 4\}$ **3.** $\{-\sqrt{7}, \sqrt{7}\}$ **5.** $\{-3\sqrt{2}, 3\sqrt{2}\}$ **7.** $\{-\sqrt{10}, \sqrt{10}\}$ **9.** $\{-8, -2\}$ **11.** $\{6 - 3\sqrt{2}, 6 + 3\sqrt{2}\}$ **13.** $\left\{\frac{3 - 2\sqrt{2}}{2}, \frac{3 + 2\sqrt{2}}{2}\right\}$ **15.** $\{-3i, 3i\}$ **17.** $\{-\sqrt{6}, \sqrt{6}\}$ **19.** $\{-2i\sqrt{2}, 2i\sqrt{2}\}$ **21.** $\{1 - 4i, 1 + 4i\}$ **23.** $\{-7 - \sqrt{5}, -7 + \sqrt{5}\}$ **25.** $\{-3 - 2i\sqrt{2}, -3 + 2i\sqrt{2}\}$ **27.** $x^2 + 16x + 64 = (x + 8)^2$ **29.** $z^2 - 12z + 36 = (z - 6)^2$ **31.** $p^2 + 9p + \frac{81}{4} = \left(p + \frac{9}{2}\right)^2$ **33.** $r^2 - 3r + \frac{9}{4} = \left(r - \frac{3}{2}\right)^2$ **35.** $\{-5, -3\}$ **37.** $\{-3 - \sqrt{7}, -3 + \sqrt{7}\}$ **39.** $\left\{\frac{-1 - \sqrt{5}}{2}, \frac{-1 + \sqrt{5}}{2}\right\}$ **41.** $\{-1 - \sqrt{6}, -1 + \sqrt{6}\}$ **43.** $\left\{\frac{6 - \sqrt{30}}{3}, \frac{6 + \sqrt{30}}{3}\right\}$ **45.** $\left\{-4, \frac{1}{2}\right\}$ **47.** $\{-4 - \sqrt{15}, -4 + \sqrt{15}\}$ **49.** $\left\{\frac{-3 - \sqrt{21}}{3}, \frac{-3 + \sqrt{21}}{3}\right\}$ **51.** $\{-1 - i, -1 + i\}$ **53.** $\{3 - \sqrt{6}, 3 + \sqrt{6}\}$ **55.** $\{-2 - i\sqrt{2}, -2 + i\sqrt{2}\}$ **57.** $\left\{\frac{1 - i\sqrt{47}}{4}, \frac{1 + i\sqrt{47}}{4}\right\}$ **59.** $\{-5 - i\sqrt{3}, -5 + i\sqrt{3}\}$ **61.** $\{-4, 1\}$ **63.** $\left\{\frac{2 - i\sqrt{2}}{2}, \frac{2 + i\sqrt{2}}{2}\right\}$ **65.** $\left\{\frac{-3 - \sqrt{69}}{6}, \frac{-3 + \sqrt{69}}{6}\right\}$ **67.** 20% **69.** 11% **71.** answers may vary **73.** simple **75.** $5 - 10\sqrt{3}$ **77.** $\frac{3 - 2\sqrt{7}}{4}$ **79.** $2\sqrt{7}$ **81.** $\sqrt{13}$ **83.** $-6y, 6y$ **85.** 8.11 sec **87.** 6.73 sec

Mental Math

1. $a = 1, b = 3, c = 1$ **3.** $a = 7, b = 0, c = -4$ **5.** $a = 6, b = -1, c = 0$

Exercise Set 8.2

1. $\{-6, 1\}$ **3.** $\left\{-\frac{3}{5}, 1\right\}$ **5.** $\{3\}$ **7.** $\left\{\frac{7 - \sqrt{33}}{2}, \frac{-7 + \sqrt{33}}{2}\right\}$ **9.** $\left\{\frac{1 - \sqrt{57}}{8}, \frac{1 + \sqrt{57}}{8}\right\}$ **11.** $\left\{\frac{7 - \sqrt{85}}{6}, \frac{7 + \sqrt{85}}{6}\right\}$ **13.** $\{1 - \sqrt{3}, 1 + \sqrt{3}\}$ **15.** $\left\{-\frac{3}{2}, 1\right\}$ **17.** $\left\{\frac{3 - \sqrt{11}}{2}, \frac{3 + \sqrt{11}}{2}\right\}$ **19.** $\left\{\frac{-5 - i\sqrt{5}}{10}, \frac{-5 + i\sqrt{5}}{10}\right\}$ **21.** $\left\{\frac{-5 - \sqrt{17}}{2}, \frac{-5 + \sqrt{17}}{2}\right\}$ **23.** $\left\{\frac{5}{2}, 1\right\}$ **25.** $\left\{\frac{3 - \sqrt{29}}{2}, \frac{3 + \sqrt{29}}{2}\right\}$ **27.** $\left\{\frac{-1 - \sqrt{19}}{6}, \frac{-1 + \sqrt{19}}{6}\right\}$ **29.** answers may vary **31.** $\left\{\frac{3 - i\sqrt{87}}{8}, \frac{3 + i\sqrt{87}}{8}\right\}$ **33.** $\{-2 - \sqrt{11}, -2 + \sqrt{11}\}$ **35.** $\{-3 - 2i, -3 + 2i\}$ **37.** $\left\{\frac{-1 - i\sqrt{23}}{4}, \frac{-1 + i\sqrt{23}}{4}\right\}$ **39.** $\{1\}$ **41.** $\{3 + \sqrt{5}, 3 - \sqrt{5}\}$ **43.** 2 real solutions **45.** 1 real solution **47.** 2 real solutions **49.** 2 complex but not real solutions **51.** 14 ft **53.** $2 + 2\sqrt{2}$ cm, $2 + 2\sqrt{2}$ cm, $4 + 2\sqrt{2}$ cm **55.** width, $-5 + 5\sqrt{17}$ ft; length, $5 + 5\sqrt{17}$ ft **57. a.** $50\sqrt{2}$ m **b.** 5000 sq. m **59.** $\frac{1 + \sqrt{5}}{2}$ **61.** 8.9 sec **63.** 2.8 sec **65.** $\left\{\frac{11}{5}\right\}$ **67.** $\{15\}$ **69.** $(x^2 + 5)(x + 2)(x - 2)$ **71.** $(z + 3)(z - 3)(z + 2)(z - 2)$ **73.** $\{0.6, 2.4\}$ **75.** Sunday to Monday **77.** Wednesday **79.** $f(4) = 33$; answers may vary **83. a.** \$3056 million **b.** 2005

Exercise Set 8.3

1. $\{2\}$ **3.** $\{16\}$ **5.** $\{1, 4\}$ **7.** $\{3 - \sqrt{7}, 3 + \sqrt{7}\}$ **9.** $\left\{\frac{3 - \sqrt{57}}{4}, \frac{3 + \sqrt{57}}{4}\right\}$ **11.** $\left\{\frac{1 - \sqrt{29}}{2}, \frac{1 + \sqrt{29}}{2}\right\}$ **13.** $\{-2, 2, -2i, 2i\}$ **15.** $\left\{-\frac{1}{2}, \frac{1}{2}, -i\sqrt{3}, i\sqrt{3}\right\}$ **17.** $\{-3, 3, -2, 2\}$ **19.** $\{125, -8\}$ **21.** $\left\{-\frac{4}{5}, 0\right\}$ **23.** $\left\{-\frac{1}{8}, 27\right\}$ **25.** $\left\{-\frac{2}{3}, \frac{4}{3}\right\}$ **27.** $\left\{-\frac{1}{125}, \frac{1}{8}\right\}$ **29.** $\{-\sqrt{2}, \sqrt{2}, -\sqrt{3}, \sqrt{3}\}$ **31.** $\left\{\frac{-9 - \sqrt{201}}{6}, \frac{-9 + \sqrt{201}}{6}\right\}$ **33.** $\{2, 3\}$ **35.** $\{3\}$ **37.** $\{27, 125\}$ **39.** $\{1, -3i, 3i\}$ **41.** $\left\{\frac{1}{8}, -8\right\}$ **43.** $\left\{-\frac{1}{2}, \frac{1}{3}\right\}$ **45.** $\{4\}$ **47.** $\{-3\}$ **49.** $\{-\sqrt{5}, \sqrt{5}, -2i, 2i\}$ **51.** $\left\{-3, \frac{3 - 3i\sqrt{3}}{2}, \frac{3 + 3i\sqrt{3}}{2}\right\}$ **53.** $\{6, 12\}$ **55.** $\left\{-\frac{1}{3}, \frac{1}{3}, \frac{-i\sqrt{6}}{3}, \frac{i\sqrt{6}}{3}\right\}$ **57.** 5 mph, then 4 mph **59.** inlet pipe, 15.5 hr; hose, 16.5 hr **61.** 55 mph; 66 mph **63.** 8.5 hr **65.** 12 or -8 **67. a.** $x - 6$ **b.** $300 = (x - 6)\cdot(x - 6)\cdot 3$ **c.** 16 cm by 16 cm **69.** $(-\infty, 3]$ **71.** $(-5, \infty)$ **73.** answers may vary **75. a.** 365.36 ft per sec **b.** 358.72 ft per sec **c.** Hearn, 249.11 mph; Moore, 244.58 mph

Integrated Review

1. $\{-\sqrt{10}, \sqrt{10}\}$ **2.** $\{-\sqrt{14}, \sqrt{14}\}$ **3.** $\{1 - 2\sqrt{2}, 1 + 2\sqrt{2}\}$ **4.** $\{-5 - 2\sqrt{3}, -5 + 2\sqrt{3}\}$ **5.** $\{-1 - \sqrt{13}, -1 + \sqrt{13}\}$ **6.** $\{1, 11\}$ **7.** $\left\{\frac{-3 - \sqrt{69}}{6}, \frac{-3 + \sqrt{69}}{6}\right\}$ **8.** $\left\{\frac{-2 - \sqrt{5}}{4}, \frac{-2 + \sqrt{5}}{4}\right\}$ **9.** $\left\{\frac{2 - \sqrt{2}}{2}, \frac{2 + \sqrt{2}}{2}\right\}$ **10.** $\{-3 - \sqrt{5}, -3 + \sqrt{5}\}$ **11.** $\{-2 + i\sqrt{3}, -2 - i\sqrt{3}\}$ **12.** $\left\{\frac{-1 - i\sqrt{11}}{2}, \frac{-1 + i\sqrt{11}}{2}\right\}$ **13.** $\left\{\frac{-3 + i\sqrt{15}}{2}, \frac{-3 - i\sqrt{15}}{2}\right\}$ **14.** $\{3i, -3i\}$ **15.** $\{0, -17\}$ **16.** $\left\{\frac{1 + \sqrt{13}}{4}, \frac{1 - \sqrt{13}}{4}\right\}$ **17.** $\{2 + 3\sqrt{3}, 2 - 3\sqrt{3}\}$ **18.** $\{2 + \sqrt{3}, 2 - \sqrt{3}\}$ **19.** $\left\{-2, \frac{4}{3}\right\}$ **20.** $\left\{\frac{-5 + \sqrt{17}}{4}, \frac{-5 - \sqrt{17}}{4}\right\}$ **21.** $\{1 - \sqrt{6}, 1 + \sqrt{6}\}$ **22.** $10\sqrt{2}$ ft ≈ 14.1 ft **23.** Diane, 9.1 hr; Lucy, 7.1 hr **24.** 5 mph during the first part, then 6 mph

Exercise Set 8.4

1. $(-\infty, -5) \cup (-1, \infty)$ **3.** $[-4, 3]$ **5.** $[2, 5]$ **7.** $\left(-5, -\frac{1}{3}\right)$ **9.** $(2, 4) \cup (6, \infty)$ **11.** $(-\infty, -4] \cup [0, 1]$

13. $(-\infty, -3) \cup (-2, 2) \cup (3, \infty)$ **15.** $(-\infty, -7) \cup (8, \infty)$ **17.** $\left[-\frac{4}{3}, \frac{1}{2}\right]$ **19.** $(-\infty, 0) \cup (1, \infty)$ **21.** $(-\infty, -4] \cup [4, 6]$
23. $(-7, 2)$ **25.** $(-1, \infty)$ **27.** $(-\infty, -1] \cup (4, \infty)$ **29.** $(-\infty, 3)$ **31.** $(0, 10)$ **33.** $(-\infty, -4) \cup [5, \infty)$
35. $(-\infty, -6] \cup (-1, 0] \cup (7, \infty)$ **37.** $(-\infty, 1) \cup (2, \infty)$ **39.** $(-\infty, -8] \cup (-4, \infty)$ **41.** 0, 1, 1, 4, 4 **43.** 0, −1, −1, −4, −4
45. answers may vary **47.** $(-\infty, -1) \cup (0, 1)$ **49.** when x is between 2 and 11 **51.**

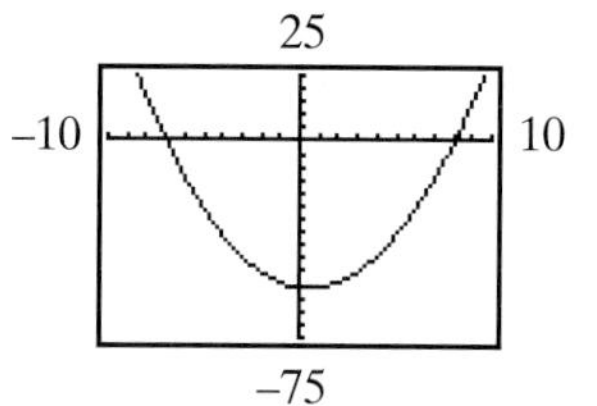

Calculator Explorations

1.

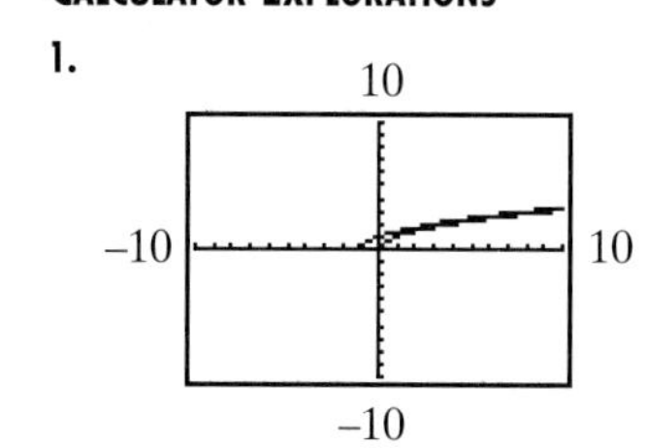

3.

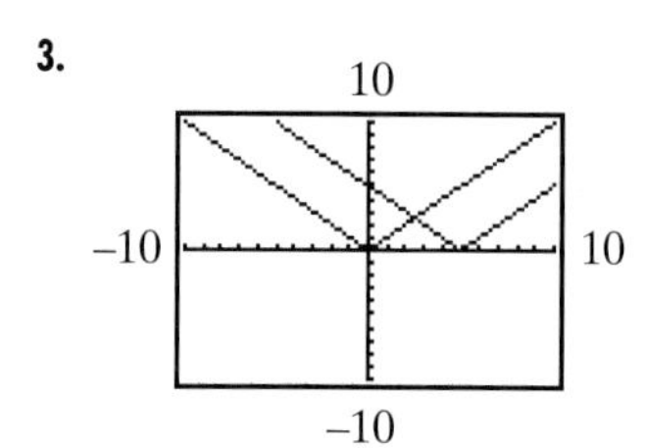

5.

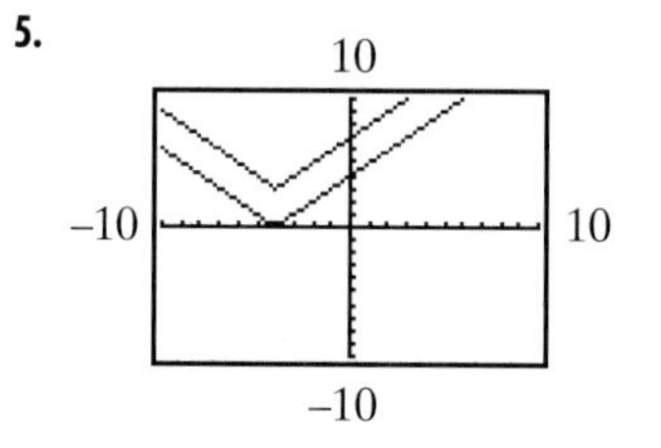

Mental Math

1. $(0, 0)$ **3.** $(2, 0)$ **5.** $(0, 3)$ **7.** $(-1, 5)$

Exercise Set 8.5

1.

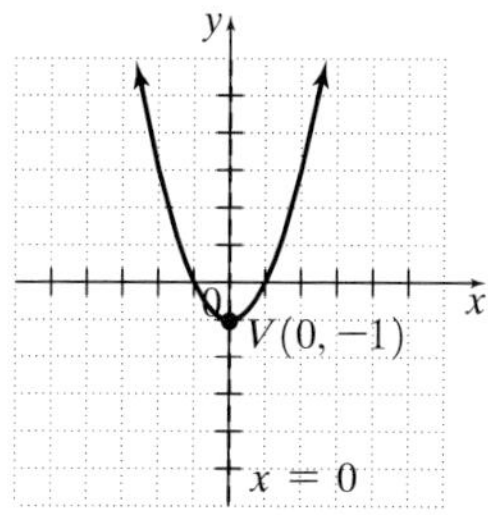

3.

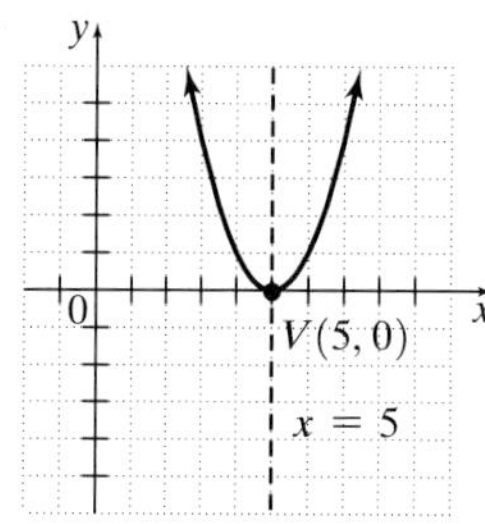

5.

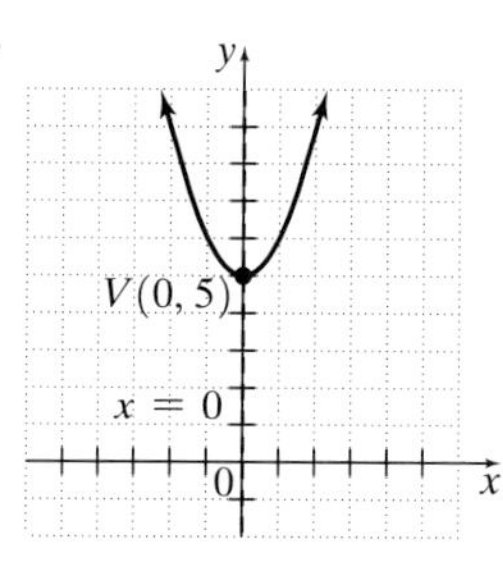

7.

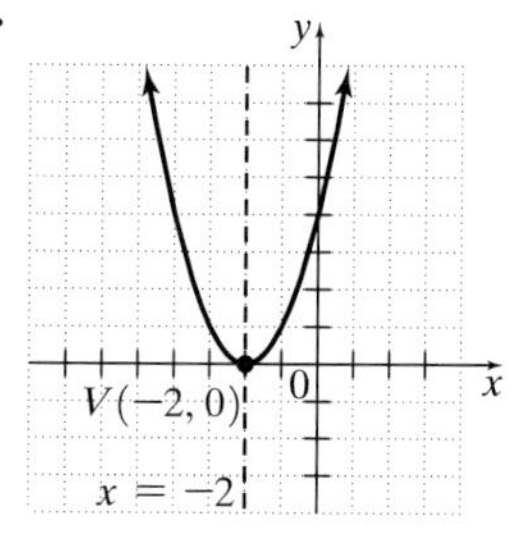

9.

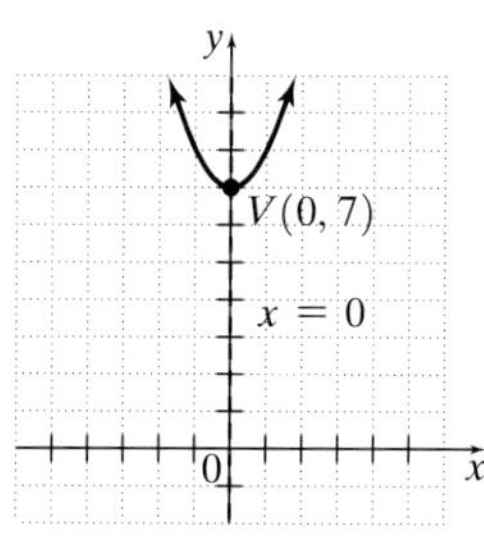

11.

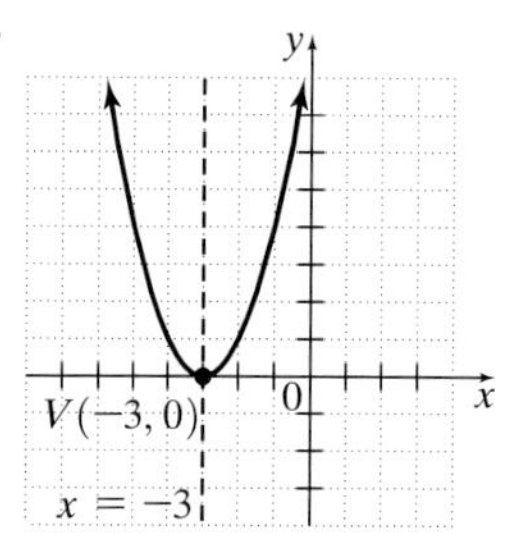

13.

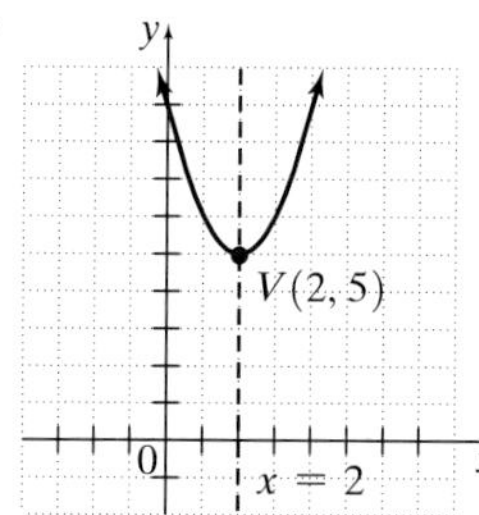

15.

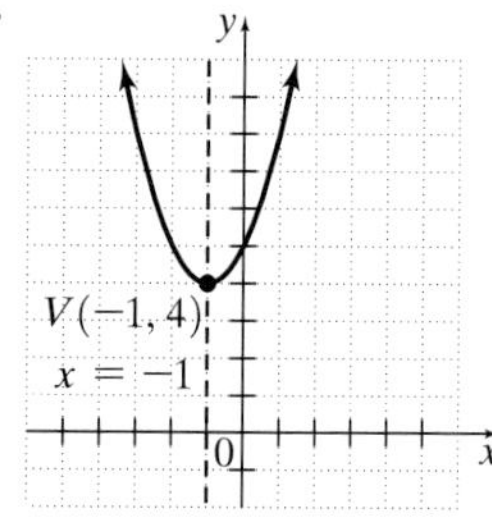

17.

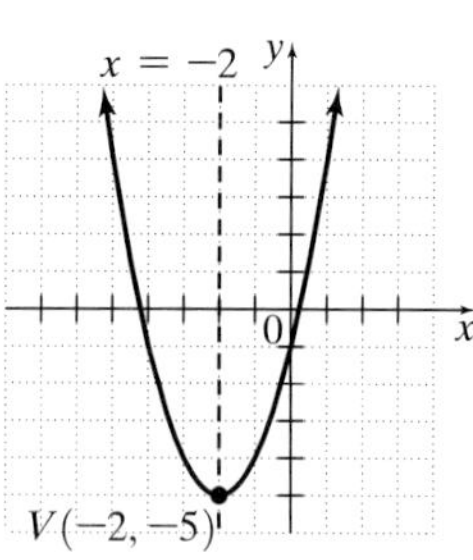

19.

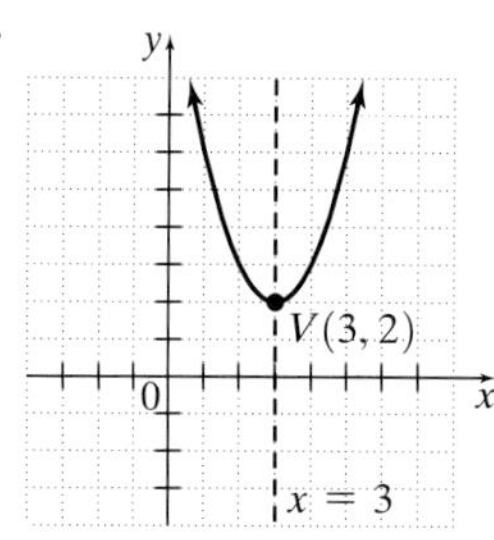

21.

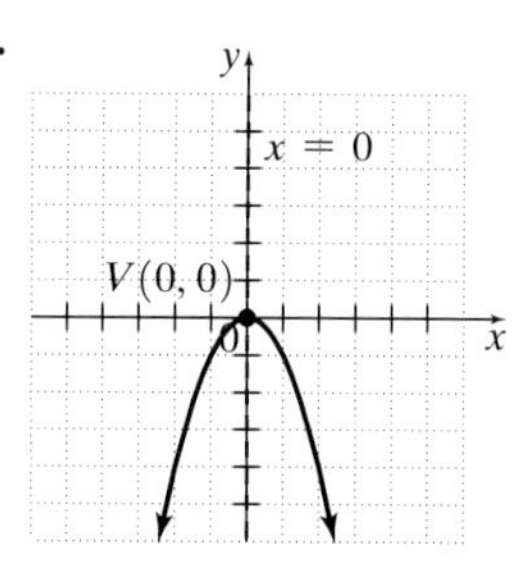

23.

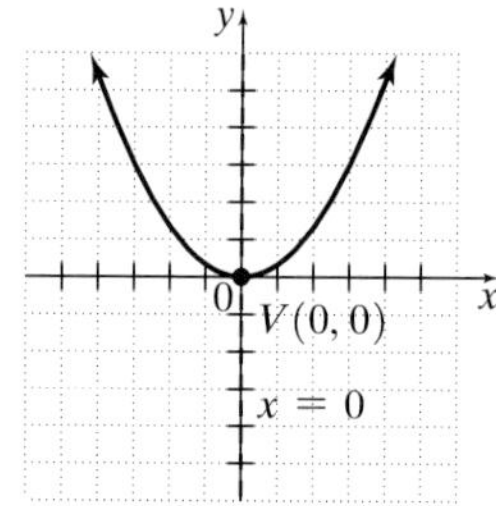

25.

27.

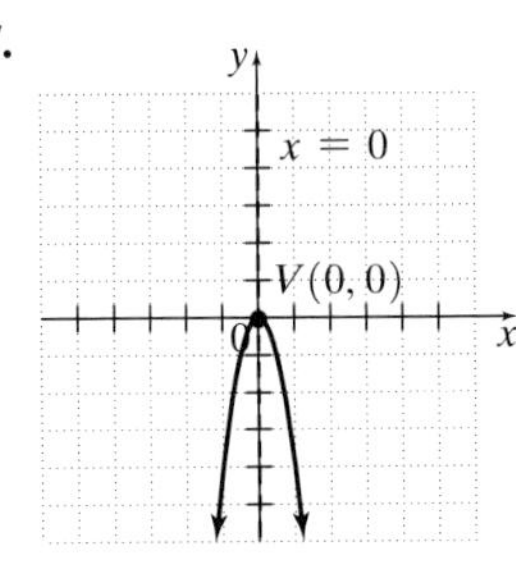

29.

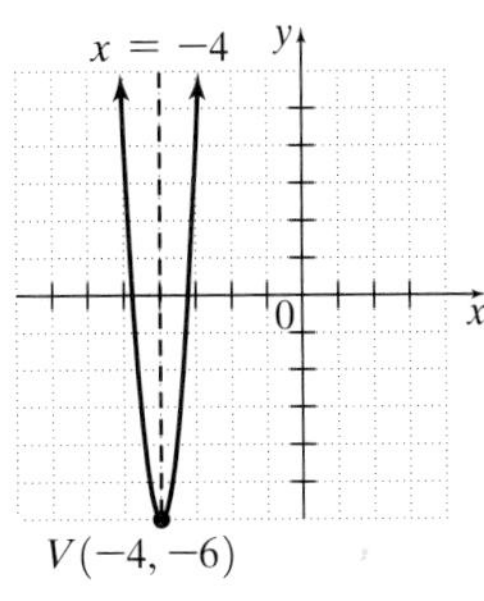

31.

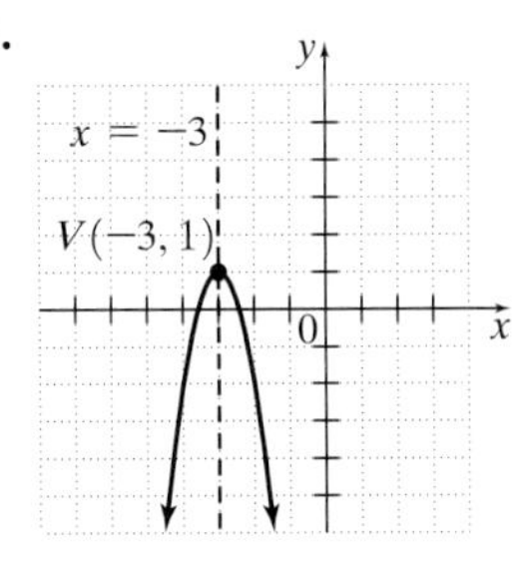

33.

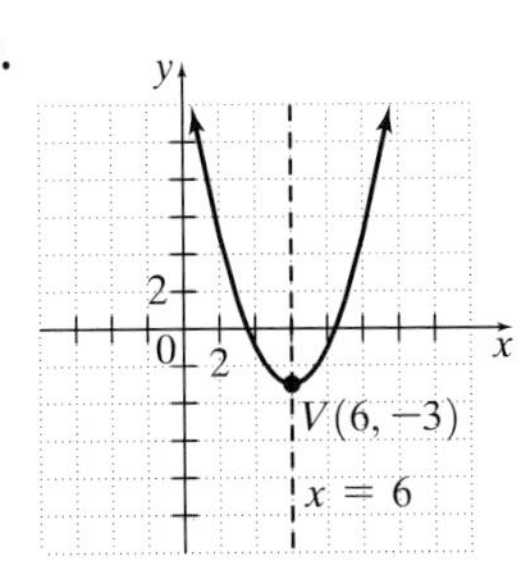

35.

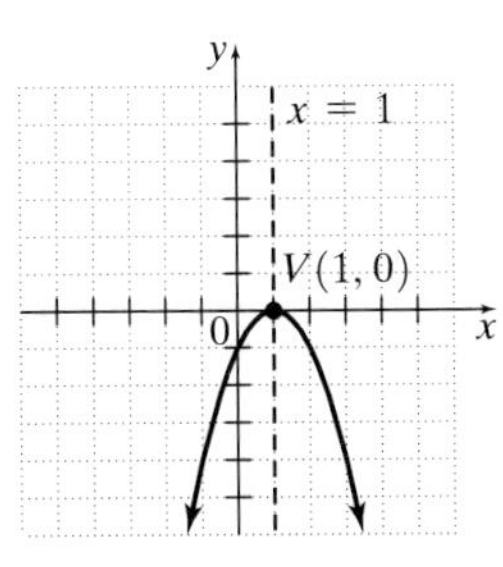

37.

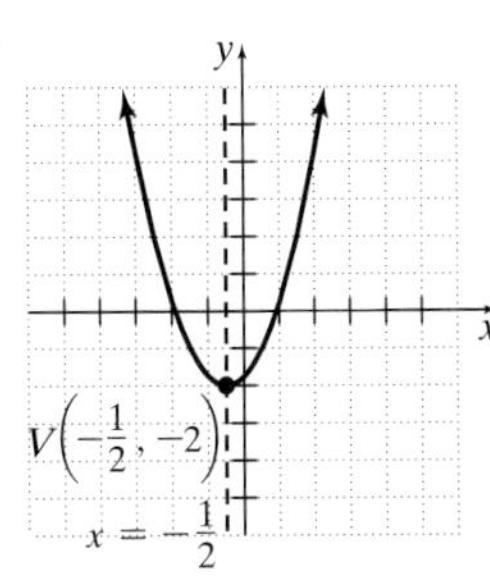

39.

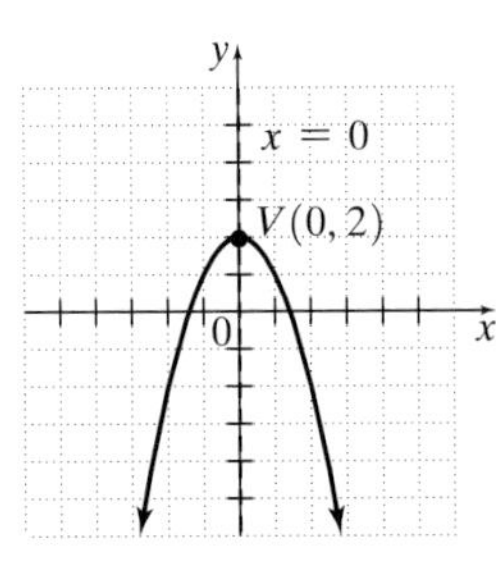

41. $x^2 + 8x + 16$ **43.** $z^2 - 16z + 64$ **45.** $y^2 + y + \frac{1}{4}$ **47.** $5(x - 2)^2 + 3$ **49.** $5(x + 3)^2 + 6$

51.

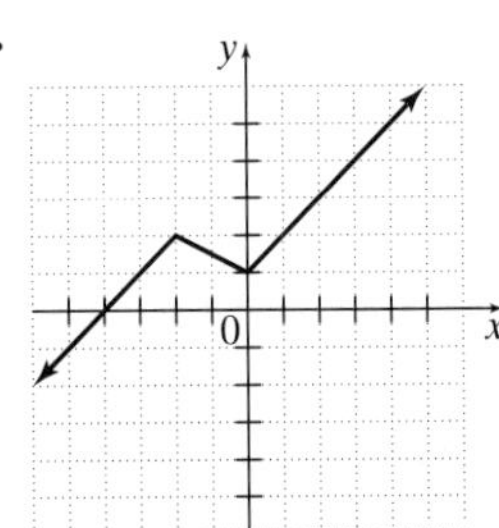

53.

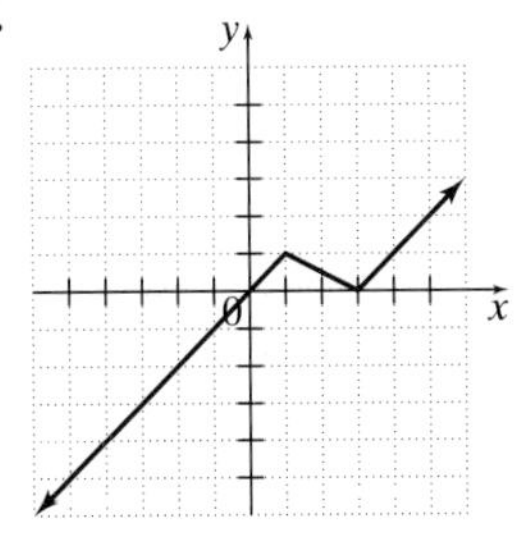

55.

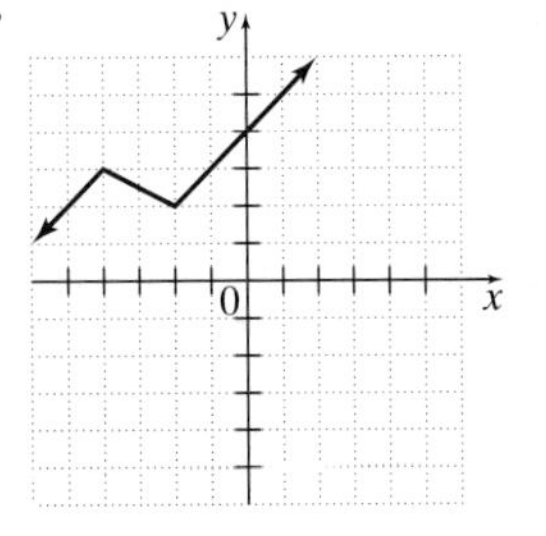

Exercise Set 8.6

1. $(-4, -9)$ **3.** $(5, 30)$ **5.** $(1, -2)$ **7.** $\left(\frac{1}{2}, \frac{5}{4}\right)$ **9.** D **11.** B

13. Vertex: $(-2, -9)$; opens upward; x-intercepts: $-5, 1$; y-intercept: -5

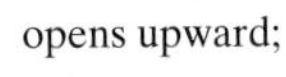

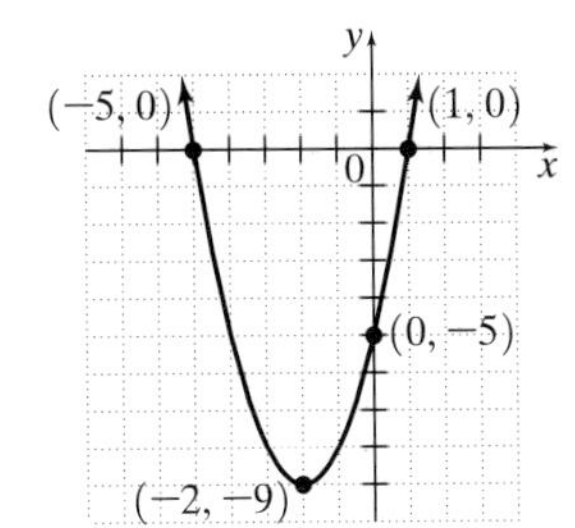

15. Vertex: $(1, 0)$; opens downward; x-intercept: 1; y-intercept: -1

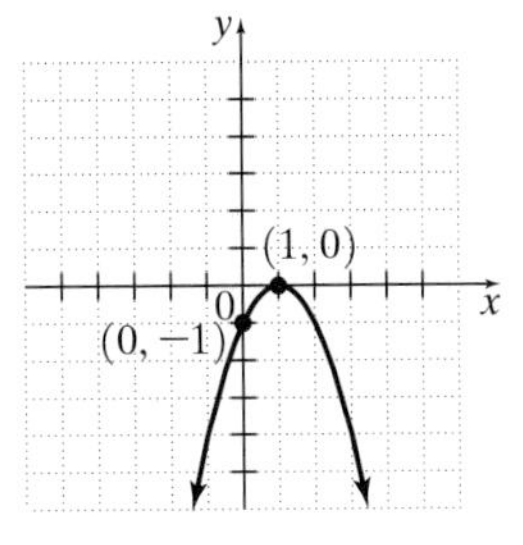

17. Vertex: $(0, -4)$; opens upward; x-intercepts: $-2, 2$; y-intercept: -4

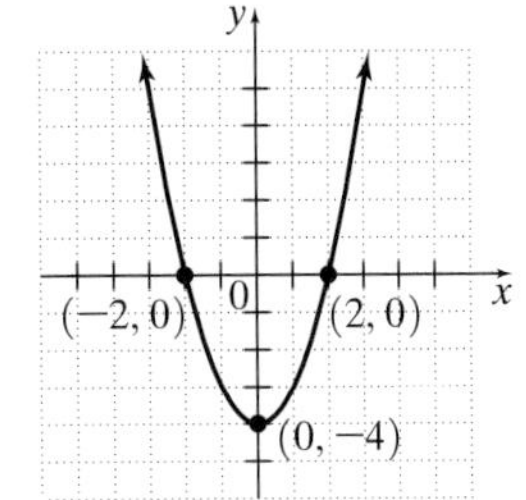

19. Vertex: $\left(-\frac{1}{2}, -4\right)$; opens upward; x-intercepts: $-\frac{3}{2}, \frac{1}{2}$; y-intercept: -3

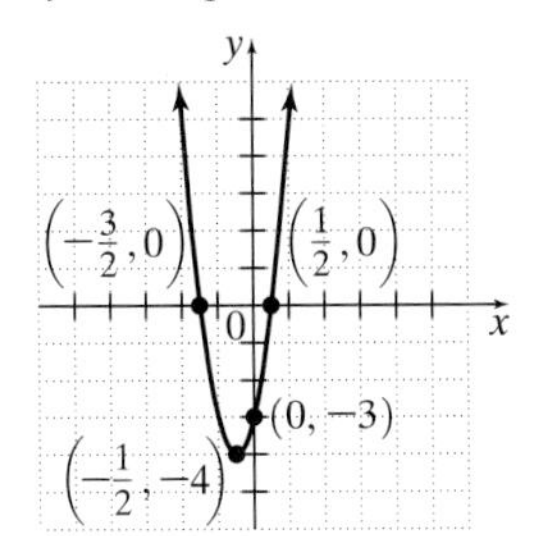

ANSWERS

21. Vertex: $\left(-4, -\frac{1}{2}\right)$; opens upward; x-intercepts: $-5, -3$; y-intercept: $\frac{15}{2}$

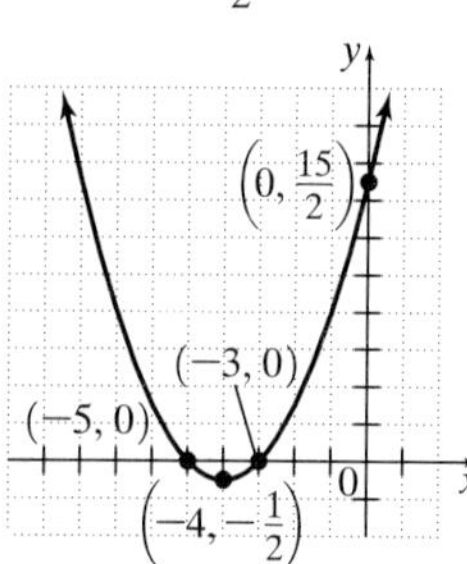

23. Vertex: (2, 1); opens upward; y-intercept: 5

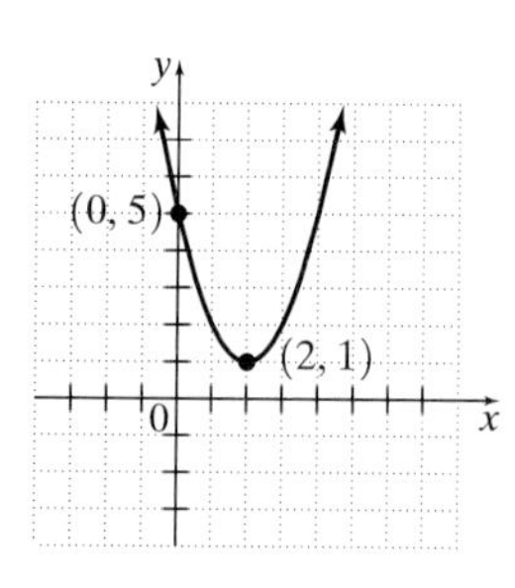

25. Vertex: $(-1, 3)$; opens upward; y-intercept: 5

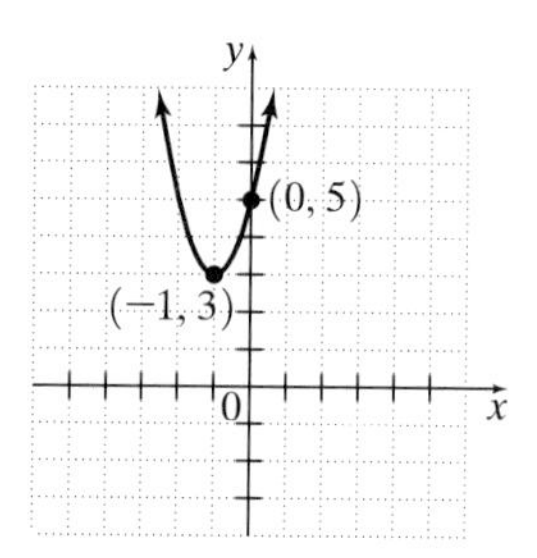

27. Vertex: (3, 18); opens downward; x-intercepts: 0, 6

y-intercept: 0

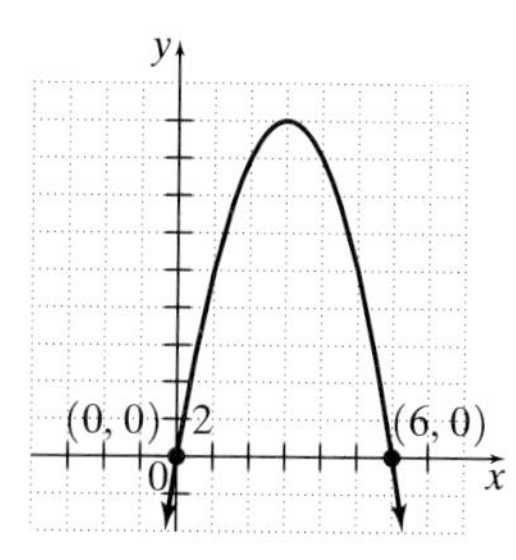

29. 144 ft **31.** 16 ft **33.** 30 and 30 **35.** -5 and 5 **37.** width 20 units, length 20 units **39.** (0, 2) **41.** undefined **43.** $(-5, 2)$ **45.** (4, 1) **47.** Vertex: $(-5, -10)$; opens upward; y-intercept: 15; x-intercepts: $-1.8, -8.2$

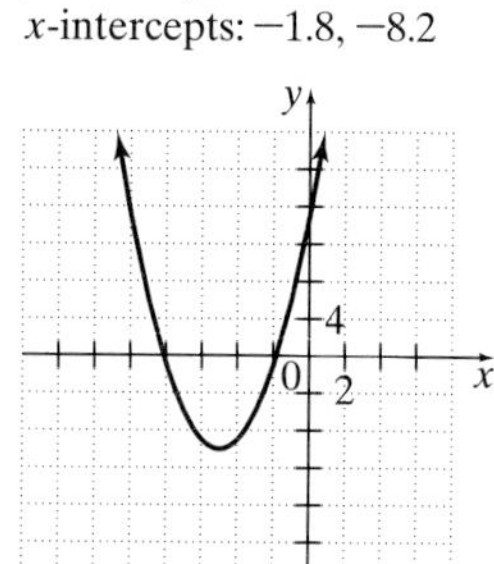

49.

10

−10 10

−10

51. -0.84 **53. a.** maximum **b.** The year 2010. **c.** 2,157,060 inmates

CHAPTER 8 REVIEW

1. {14, 1} **2.** {−5, 6} **3.** $\left\{\frac{4}{5}, -\frac{1}{2}\right\}$ **4.** $\left\{-\frac{6}{7}, 5\right\}$ **5.** $\{-7, 7\}$ **6.** $\{-2, 2\}$ **7.** $\left\{-\frac{4}{9}, \frac{2}{9}\right\}$ **8.** $\left\{\frac{2-\sqrt{2}}{5}, \frac{2+\sqrt{2}}{5}\right\}$ **9.** $\left\{\frac{-3-\sqrt{5}}{2}, \frac{-3+\sqrt{5}}{2}\right\}$ **10.** $\left\{\frac{-1-3i\sqrt{3}}{2}, \frac{-1+3i\sqrt{3}}{2}\right\}$ **11.** $\left\{\frac{-3-i\sqrt{7}}{8}, \frac{-3+i\sqrt{7}}{8}\right\}$ **12.** $\left\{\frac{17-\sqrt{145}}{9}, \frac{17+\sqrt{145}}{9}\right\}$ **13.** 4.25% **14.** $75\sqrt{2}$ mi; 106.1 mi **15.** two complex solutions **16.** two real solutions **17.** two real solutions **18.** one real solution **19.** {8} **20.** {−5, 0} **21.** $\{-i\sqrt{11}, i\sqrt{11}\}$ **22.** $\left\{-\frac{5}{2}, 1\right\}$ **23.** $\left\{\frac{5-i\sqrt{143}}{12}, \frac{5+i\sqrt{143}}{12}\right\}$ **24.** $\left\{\frac{1-i\sqrt{35}}{9}, \frac{1+i\sqrt{35}}{9}\right\}$ **25.** $\left\{\frac{21-\sqrt{41}}{50}, \frac{21+\sqrt{41}}{50}\right\}$ **26.** $\left\{1, \frac{9}{4}\right\}$ **27. a.** 20 ft **b.** $\frac{15+\sqrt{321}}{16}$ sec; 2.1 sec **28.** $(6+6\sqrt{2})$ cm **29.** $\left\{3, \frac{-3+3i\sqrt{3}}{2}, \frac{-3-3i\sqrt{3}}{2}\right\}$ **30.** $\{-4, 2-2i\sqrt{3}, 2+2i\sqrt{3}\}$ **31.** $\left\{\frac{2}{3}, 5\right\}$ **32.** $\left\{\frac{-8\sqrt{7}}{7}, \frac{8\sqrt{7}}{7}\right\}$ **33.** $\{-5, 5, -2i, 2i\}$ **34.** $\left\{-\frac{16}{5}, 1\right\}$ **35.** {1, 125} **36.** {8, 64} **37.** $\{-1, 1, -i, i\}$ **38.** $\left\{-\frac{1}{5}, \frac{1}{4}\right\}$ **39.** Jerome, 10.5 hr; Tim, 9.5 hr **40.** The number is −5. **41.** $[-5, 5]$ **42.** $\left(-\frac{1}{2}, \frac{1}{2}\right)$ **43.** $\left(-\infty, -\frac{5}{4}\right] \cup \left[\frac{3}{2}, \infty\right)$ **44.** $(-\infty, -4) \cup (-1, 1) \cup (4, \infty)$ **45.** (5, 6) **46.** $[-5, 0] \cup \left(\frac{3}{4}, \infty\right)$ **47.** $(-\infty, -6) \cup \left(-\frac{3}{4}, 0\right) \cup (5, \infty)$ **48.** $(-\infty, -5] \cup [-2, 6]$ **49.** $(-5, -3) \cup (5, \infty)$ **50.** $(-\infty, 0)$ **51.** $\left(-\frac{6}{5}, 0\right) \cup \left(\frac{5}{6}, 3\right)$ **52.** $\left(2, \frac{7}{2}\right)$

53.

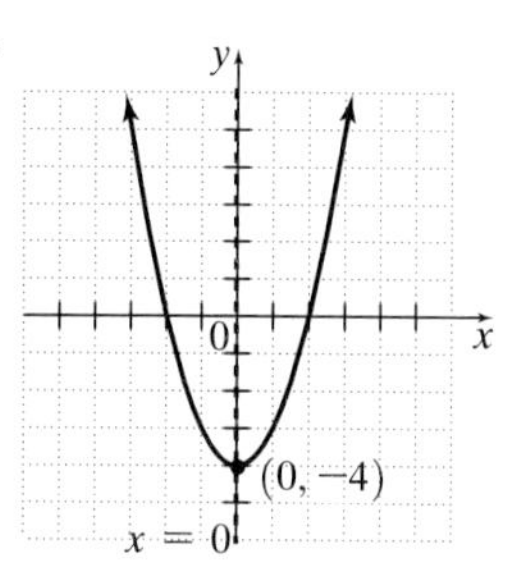

54.

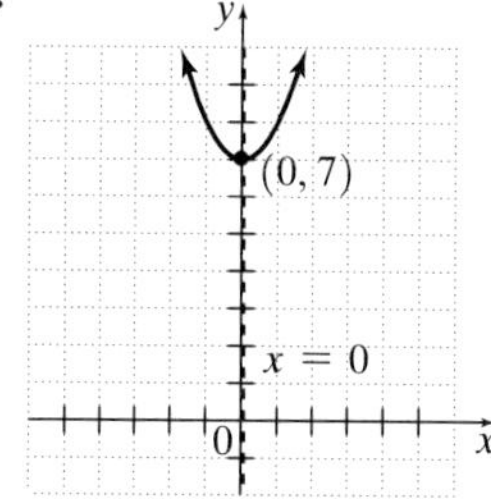

55.

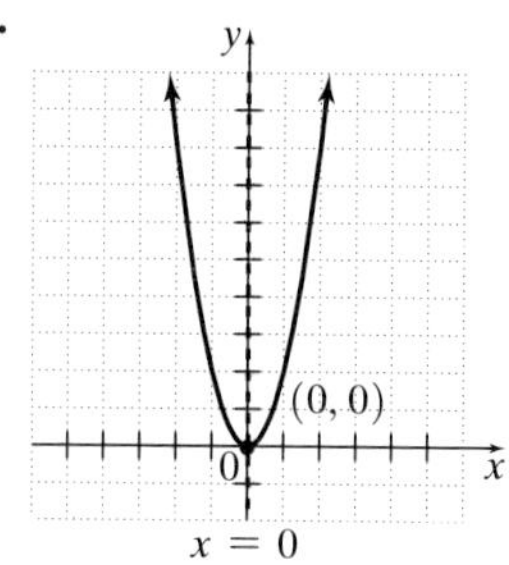

56.

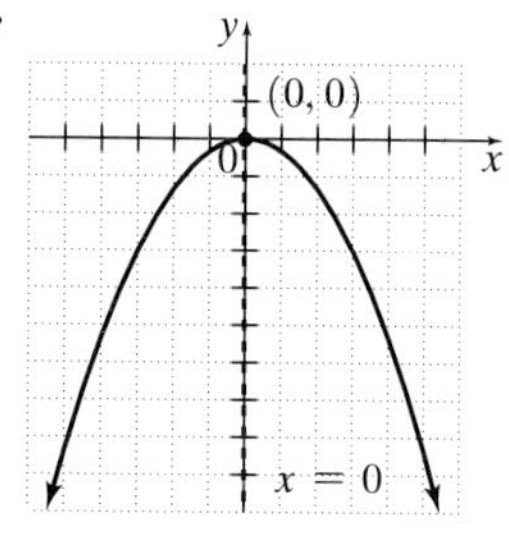

57.

58.

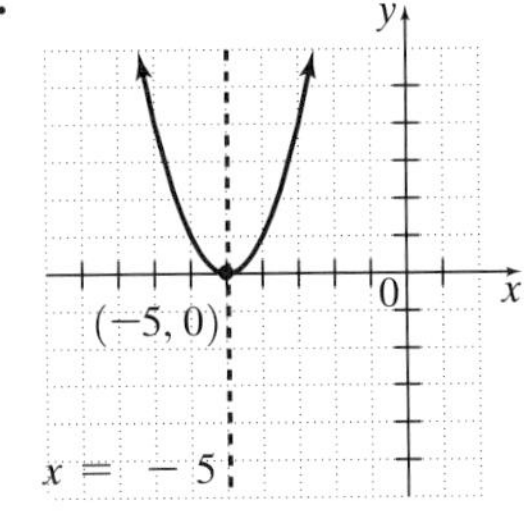

59.

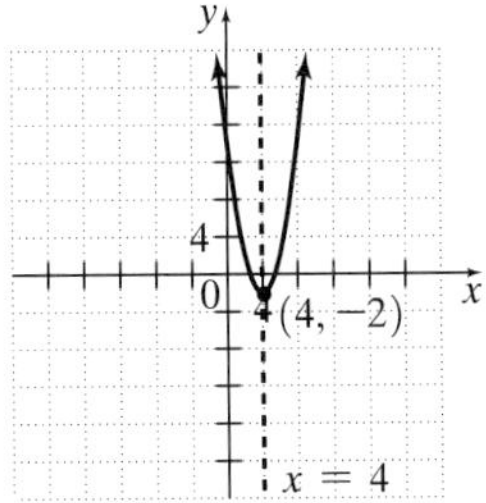

60.

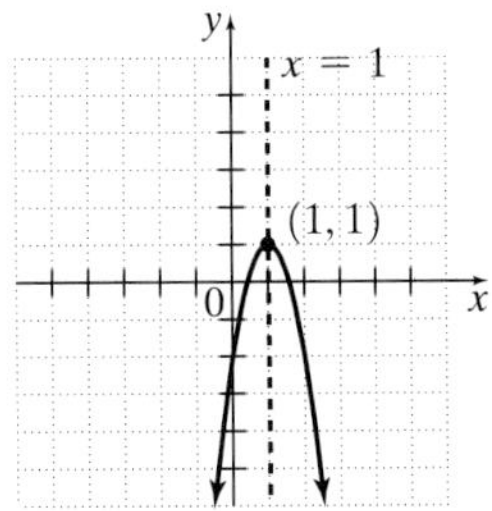

61. Vertex: $(-5, 0)$;

x-intercept: -5; y-intercept: 25

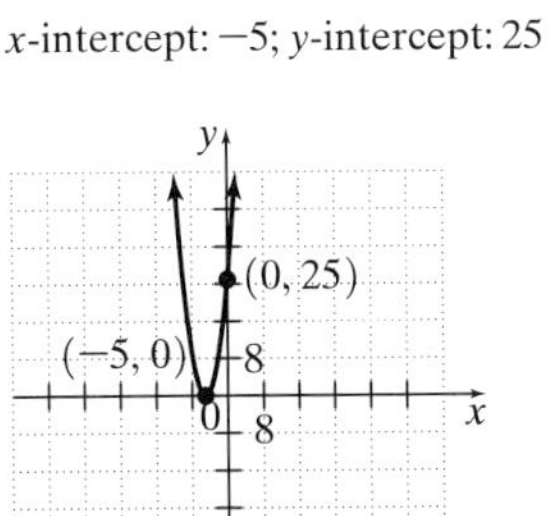

62. Vertex: $(3, 0)$;

x-intercept: 3; y-intercept: -9

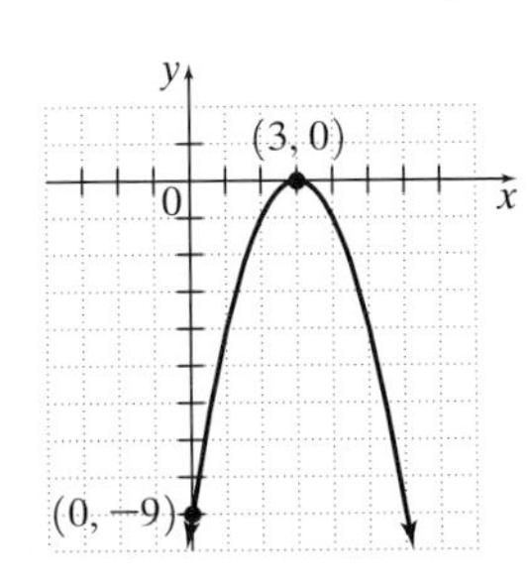

63. Vertex: $(0, -1)$;

x-intercepts: $-\frac{1}{2}, \frac{1}{2}$;

y-intercept: -1

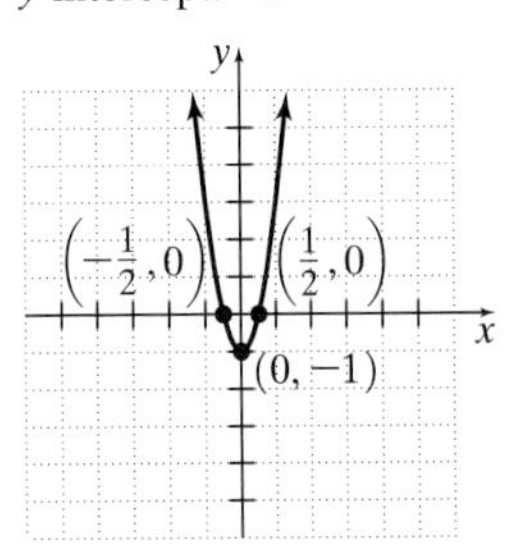

64. Vertex: $(0, 5)$;

x-intercepts: $-1, 1$;

y-intercept: 5

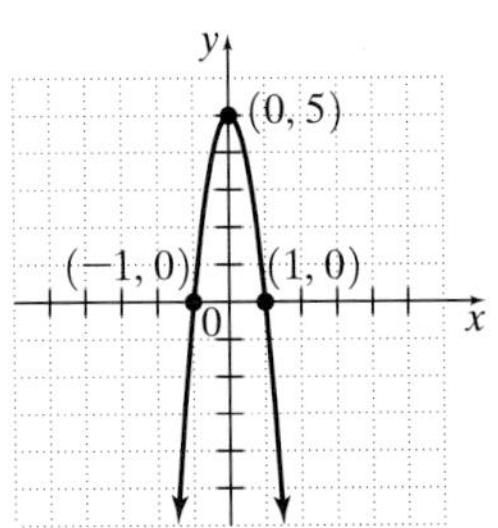

65. Vertex: $\left(-\frac{5}{6}, \frac{73}{12}\right)$;
opens downward;
x-intercepts: $-2.3, 0.6$;
y-intercept: 4

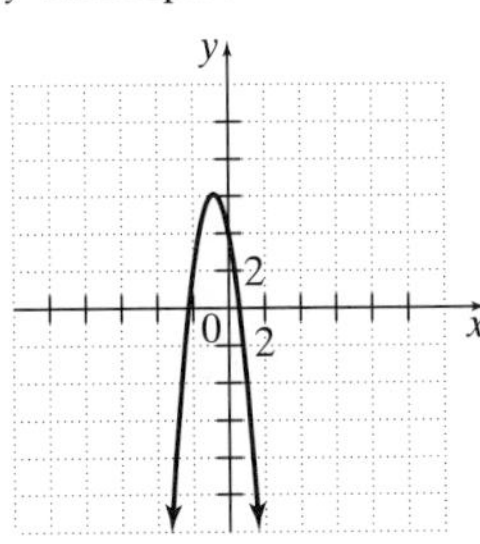
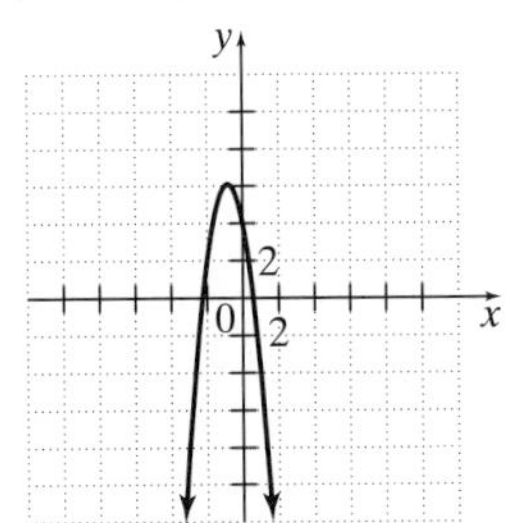

66. a. 0.4 sec and 7.1 sec

b. answers may vary

67. The numbers are both 210.

Chapter 8 Test

1. $\left\{\frac{7}{5}, -1\right\}$ **2.** $\{-1 - \sqrt{10}, -1 + \sqrt{10}\}$ **3.** $\left\{\frac{1 + i\sqrt{31}}{2}, \frac{1 - i\sqrt{31}}{2}\right\}$ **4.** $\{3 - \sqrt{7}, 3 + \sqrt{7}\}$ **5.** $\left\{-\frac{1}{7}, -1\right\}$
6. $\left\{\frac{3 + \sqrt{29}}{2}, \frac{3 - \sqrt{29}}{2}\right\}$ **7.** $\{-2 - \sqrt{11}, -2 + \sqrt{11}\}$ **8.** $\{-3, 3, -i, i\}$ **9.** $\{-1, 1, -i, i\}$ **10.** $\{6, 7\}$
11. $\{3 - \sqrt{7}, 3 + \sqrt{7}\}$ **12.** $\left\{\frac{2 - i\sqrt{6}}{2}, \frac{2 + i\sqrt{6}}{2}\right\}$ **13.** $\left(-\infty, -\frac{3}{2}\right) \cup (5, \infty)$ **14.** $(-\infty, -5) \cup (-4, 4) \cup (5, \infty)$
15. $(-\infty, -3) \cup (2, \infty)$ **16.** $(-\infty, -3) \cup [2, 3)$

17. **18.** **19.** **20.**

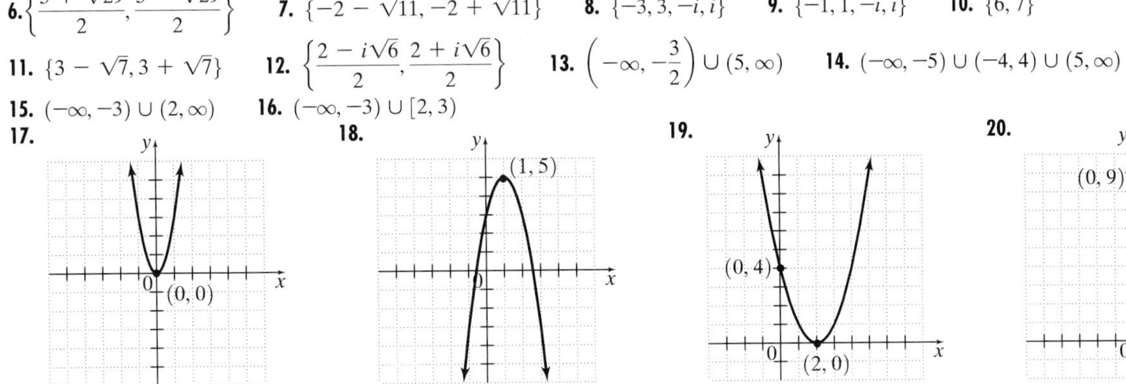

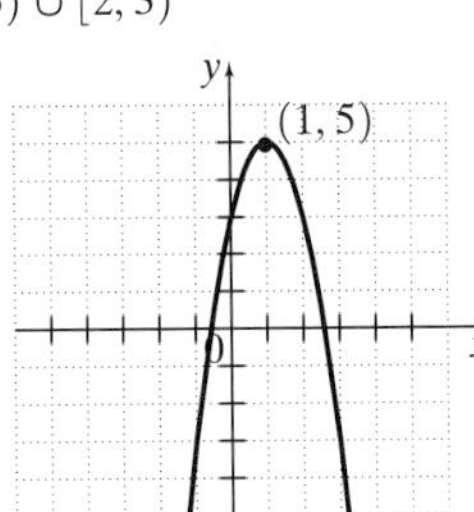

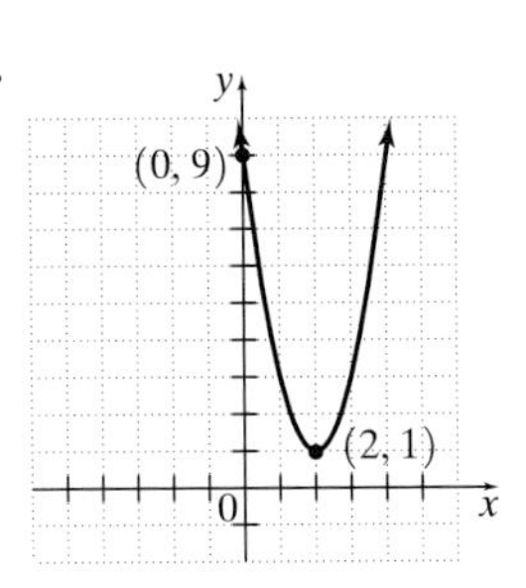

21. $(2 + \sqrt{46})$ ft ≈ 8.8 ft **22.** $(5 + \sqrt{17})$ hr ≈ 9.12 hr **23. a.** 272 ft **b.** 5.12 sec

Cumulative Review

1. 7.3×10^5; Sec. 1.5, Ex. 28 **2.** $\{0, 1\}$; Sec. 2.6, Ex. 6 **3.** not a function; Sec. 3.5, Ex. 8 **4.** $\varnothing$; Sec. 4.1, Ex. 9
5. $6x^3 - 4x^2 - 2$; Sec. 5.1, Ex. 16 **6.** $6x^2 - 29x + 28$; Sec. 5.2, Ex. 10 **7.** $15x^2 - xy - 2y^2$; Sec. 5.2, Ex. 11
8. $3x^2 + 2x + 3 + \frac{-6x + 9}{x^2 - 1}$; Sec. 5.3, Ex. 5 **9.** $3x^2(2 - x)$; Sec. 5.4, Ex. 2 **10.** $(2a - 1)(a + 4)$; Sec. 5.5, Ex. 10

11. $3x(a - 2b)^2$; Sec. 5.6, Ex. 4 **12.** $\left\{-\frac{1}{6}, 3\right\}$; Sec. 5.7, Ex. 5 **13.** $\frac{n-2}{n(n-1)}$; Sec. 6.1, Ex. 11 **14.** $-5x^2$; Sec. 6.1, Ex. 12 **15.** $\frac{6x+5}{3x^3}$; Sec. 6.2, Ex. 6 **16.** $\frac{x^2}{y^2}$; Sec. 6.3, Ex. 5 **17.** $\{-6, -1\}$; Sec. 6.4, Ex. 5 **18.** $x = \frac{yz}{y-z}$; Sec. 6.5, Ex. 1 **19.** $S = krh$; Sec. 6.6, Ex. 5 **20.** -4; Sec. 7.1, Ex. 14 **21.** $\frac{2}{5}$; Sec. 7.1, Ex. 15 **22.** 2; Sec. 7.3, Ex. 3 **23.** $\sqrt{\frac{2b}{3a}}$; Sec. 7.3, Ex. 5 **24.** $-6\sqrt[3]{2}$; Sec. 7.4, Ex. 5 **25.** $\frac{5\sqrt[3]{7x}}{2}$; Sec. 7.4, Ex. 10

Chapter 9

Chapter 9 Pretest

1. $\sqrt{34}$ units; 9.1B **2.** $\left(\frac{19}{2}, \frac{7}{2}\right)$; 9.1B

3.

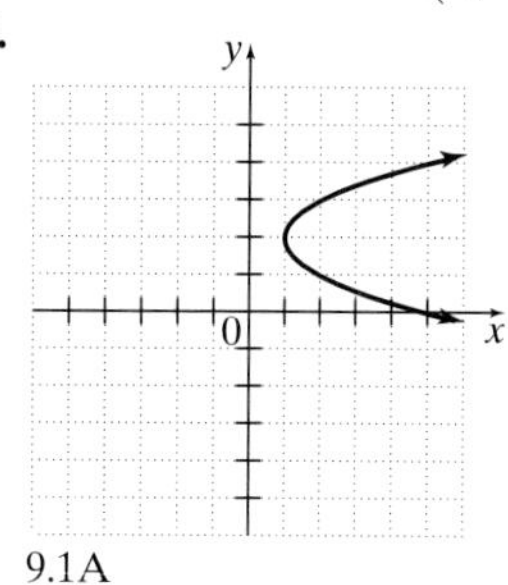

9.1A

4.

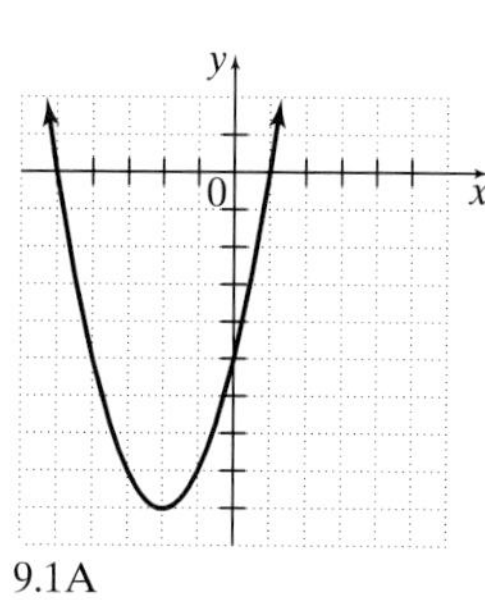

9.1A

5.

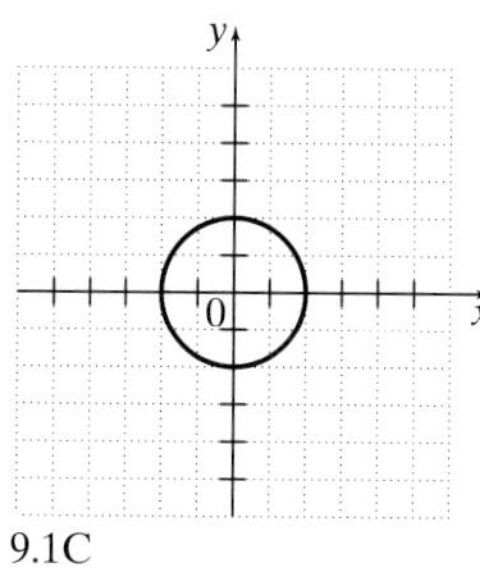

9.1C

6.

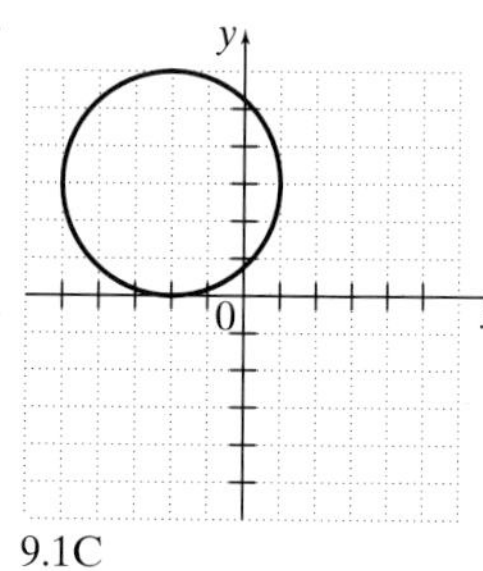

9.1C

7.

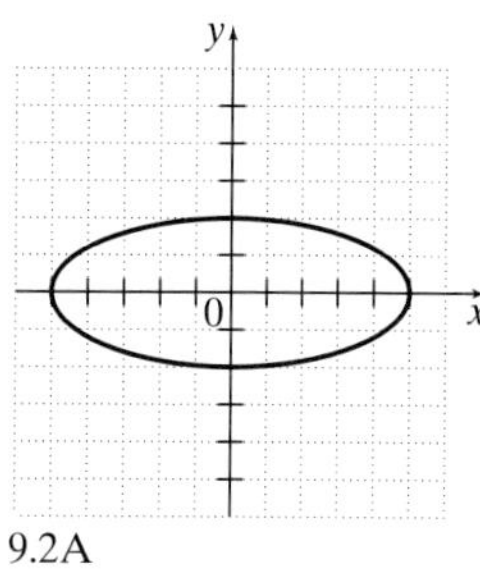

9.2A

8.

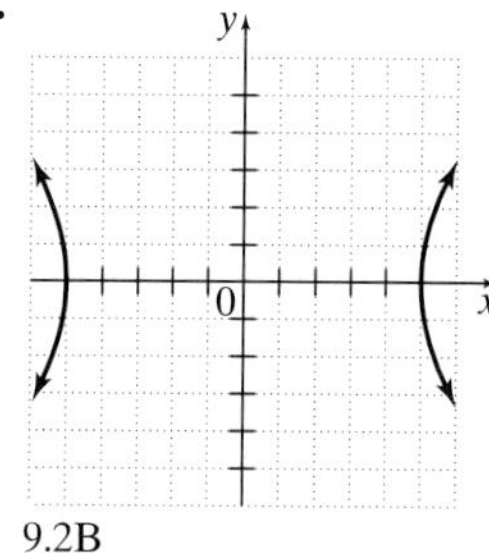

9.2B

9.

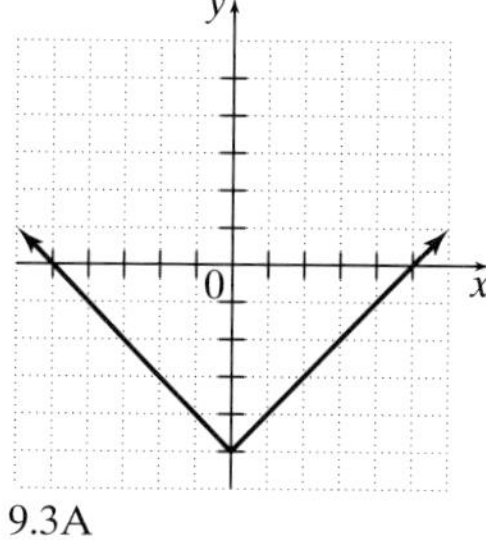

9.3A

10.

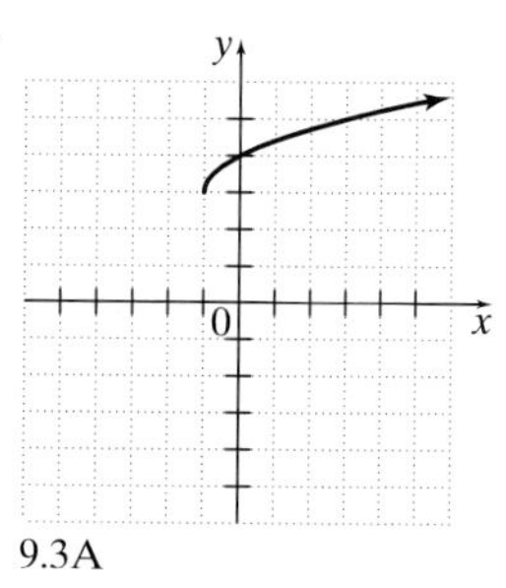

9.3A

11. $\{(0, 3), (-5, 8)\}$; 9.4A, B

12.

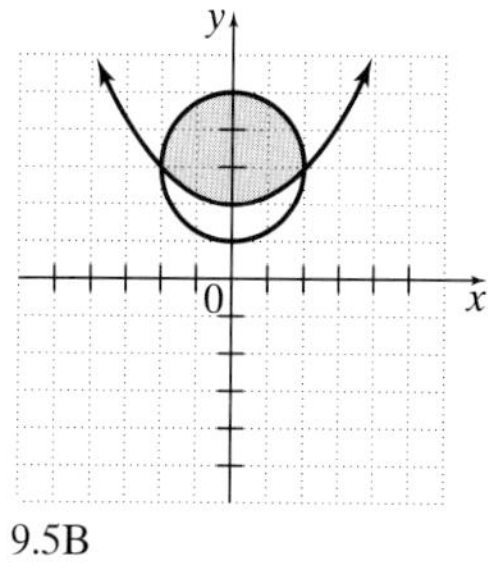

9.5B

13. $(x - 2)^2 + (y - 5)^2 = 64$; 9.1D **14.** center: $(-3, 4)$; radius: 3

Calculator Explorations

1.

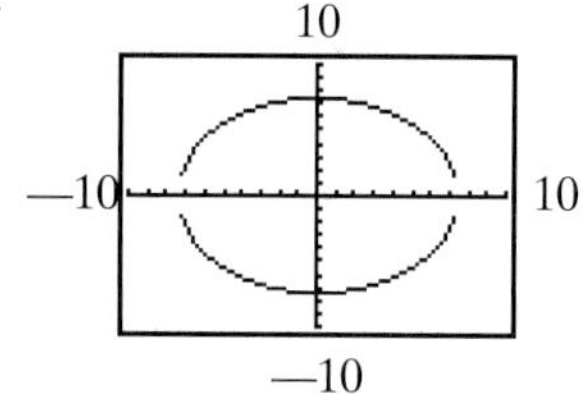

3.

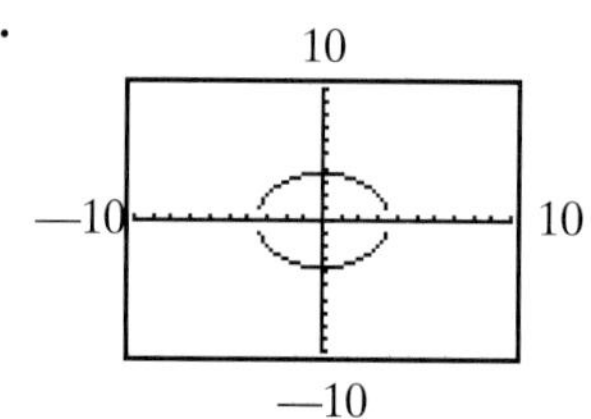

Mental Math

1. upward **3.** to the left **5.** downward

Exercise Set 9.1

1.

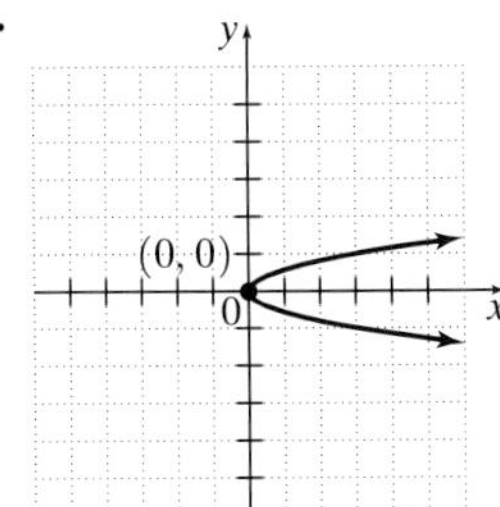

3.

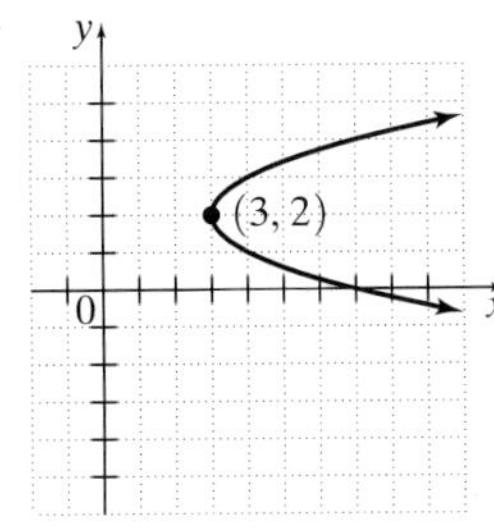

5.

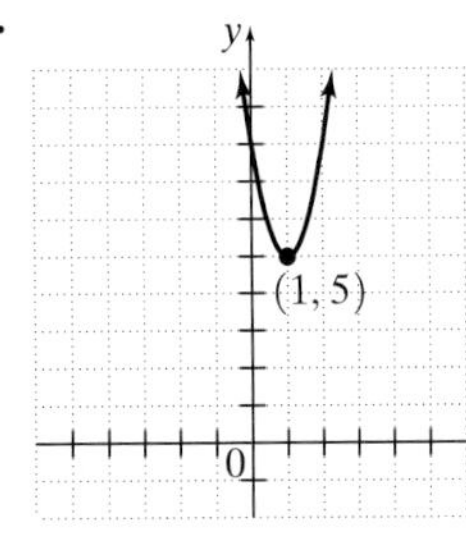

7.

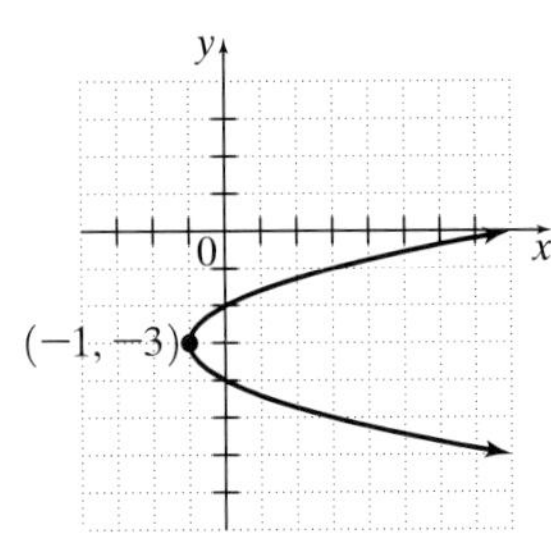

9.

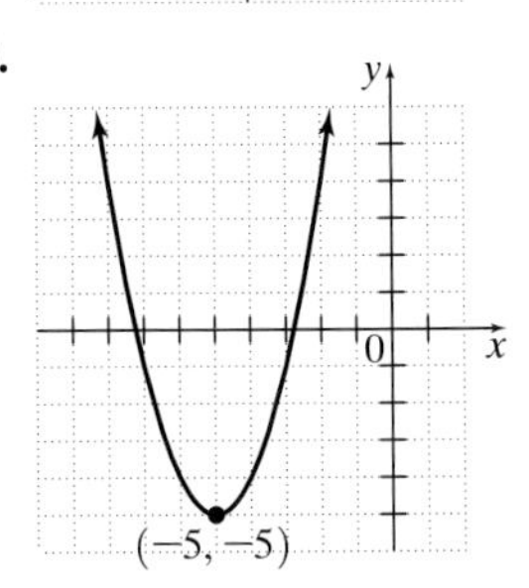

11.

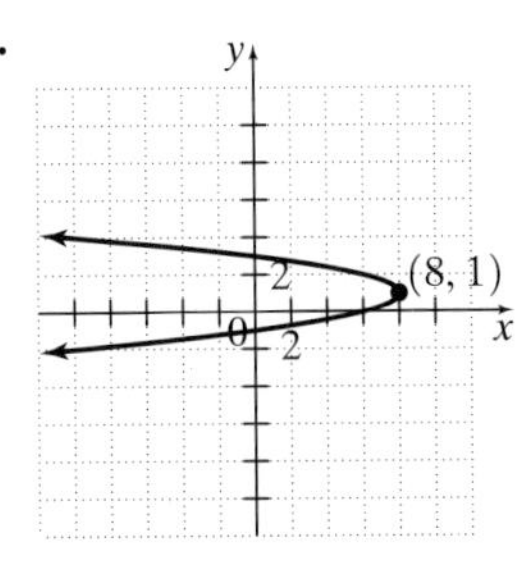

13. 5 units **15.** $\sqrt{41}$ units ≈ 6.403 **17.** $\sqrt{10}$ units ≈ 3.162 **19.** $\sqrt{5}$ units ≈ 2.236 **21.** $\sqrt{192.58}$ units ≈ 13.877 **23.** 9 units **25.** $(4, -2)$ **27.** $\left(-5, \frac{5}{2}\right)$ **29.** $(3, 0)$ **31.** $\left(-\frac{1}{2}, \frac{1}{2}\right)$ **33.** $\left(\sqrt{2}, \frac{\sqrt{5}}{2}\right)$ **35.** $(6.2, -6.65)$

37.

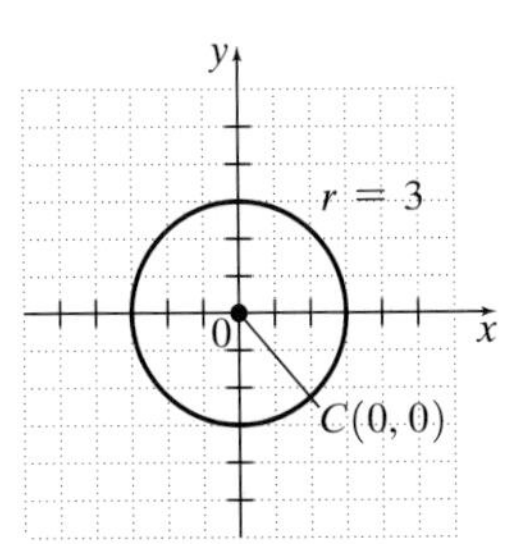

39.

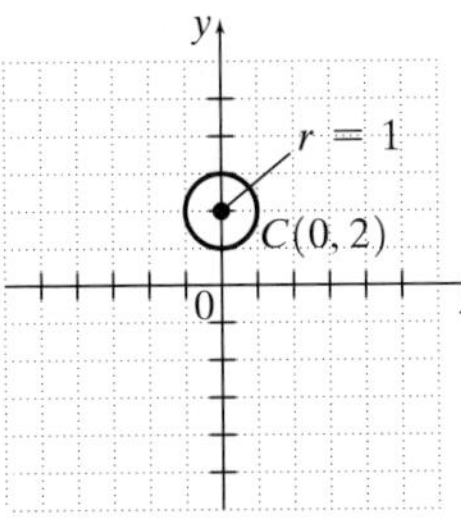

41.

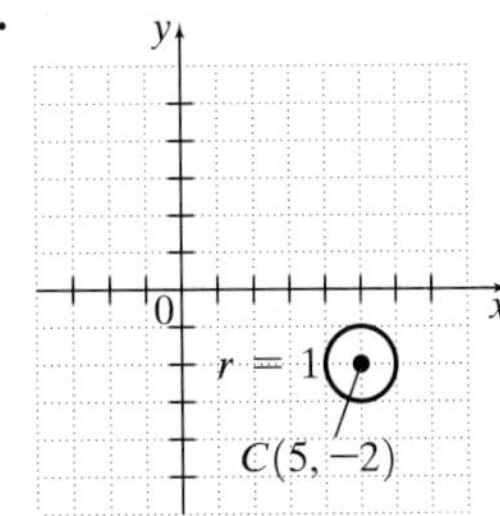

43.

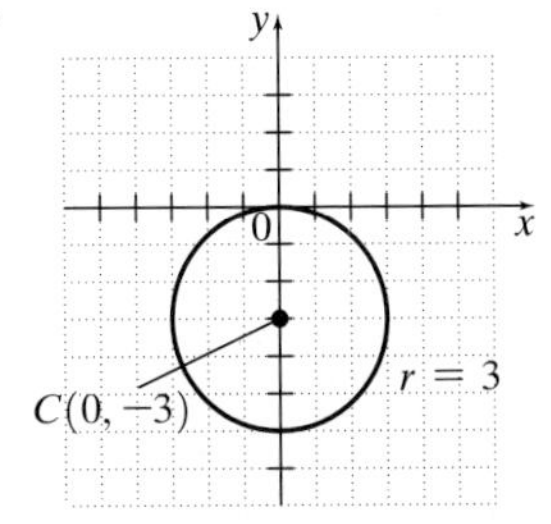

45.

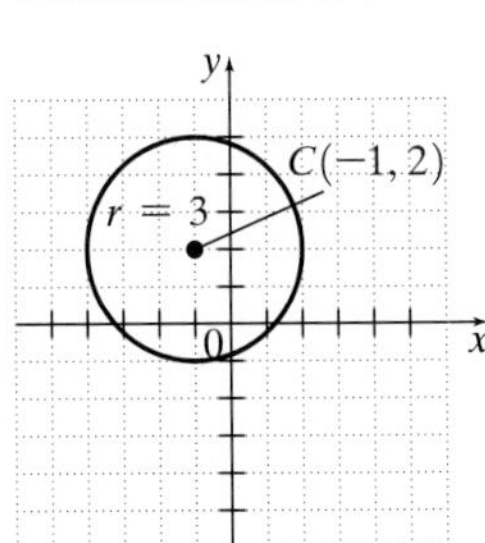

47.

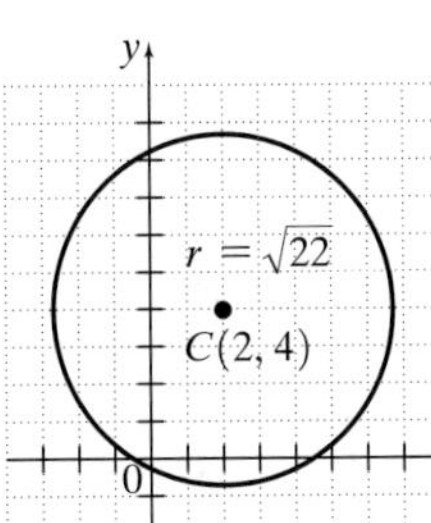

49. $(x - 2)^2 + (y - 3)^2 = 36$ **51.** $x^2 + y^2 = 3$ **53.** $(x + 5)^2 + (y - 4)^2 = 45$

55.

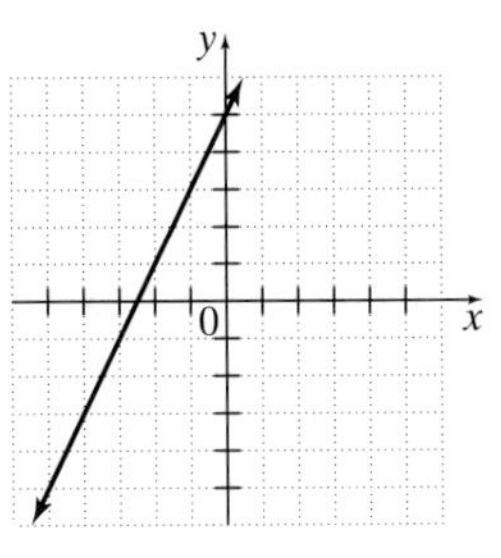

57.

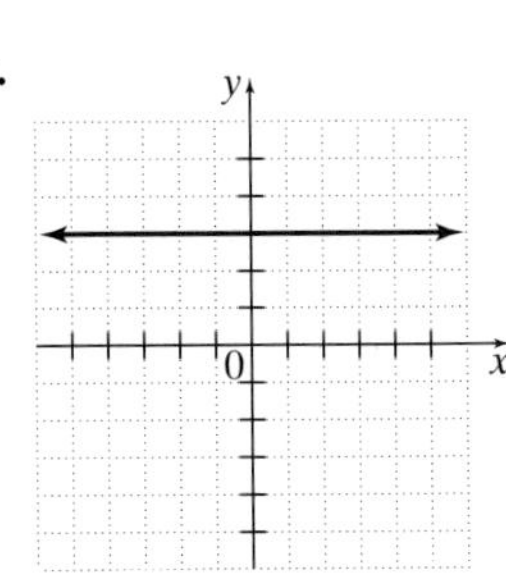

59. $\frac{\sqrt{3}}{3}$ **61.** $\frac{2\sqrt{42}}{3}$ **63.** answers may vary **65.** 20 m **67.** $y = -\frac{2}{125}x^2 + 40$

Calculator Explorations

1.

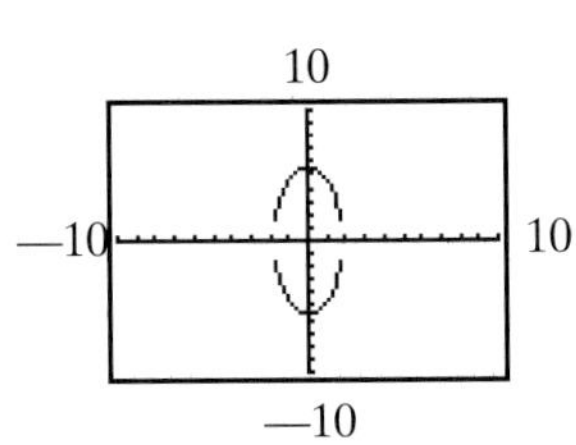

3.

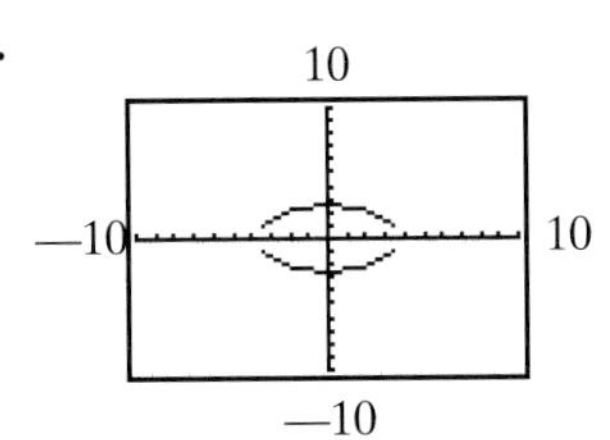

ANSWERS

Mental Math

1. ellipse **3.** hyperbola **5.** hyperbola

Exercise Set 9.2

1.

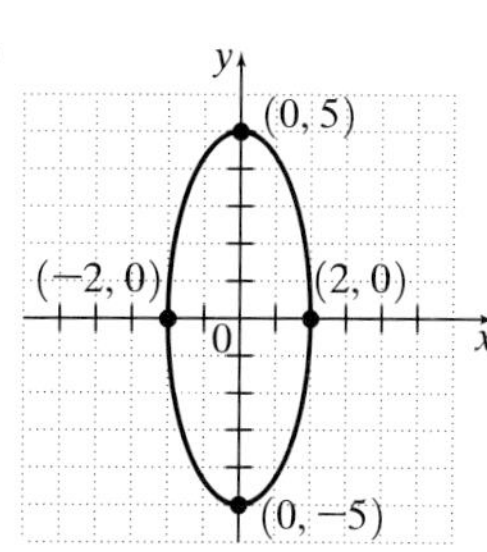

3.

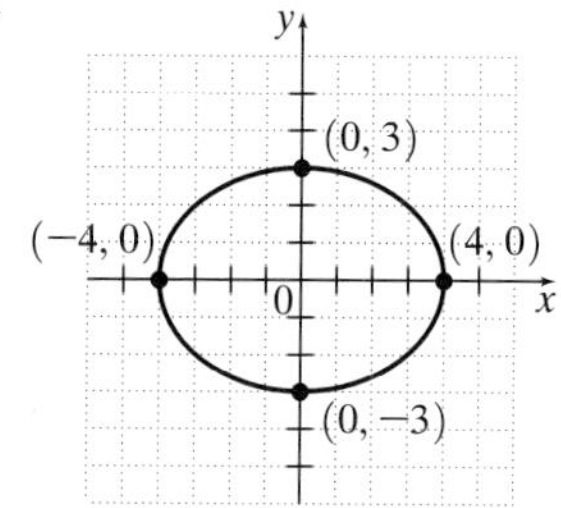

5.

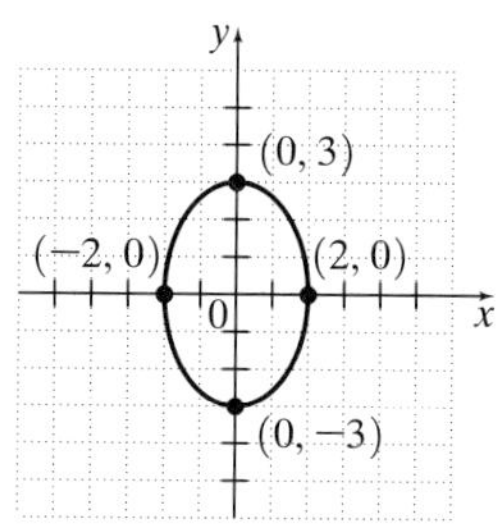

7.

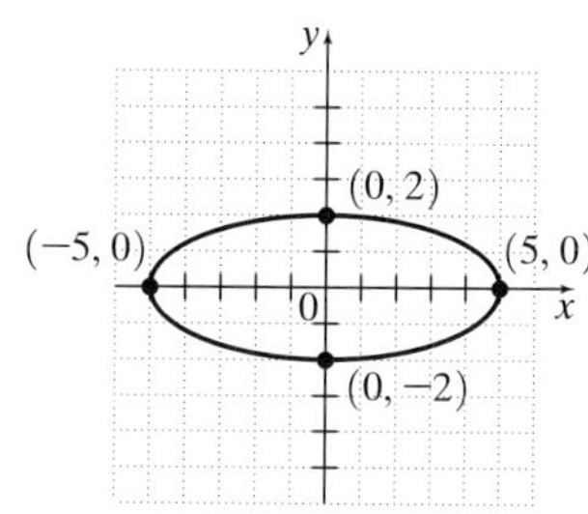

9.

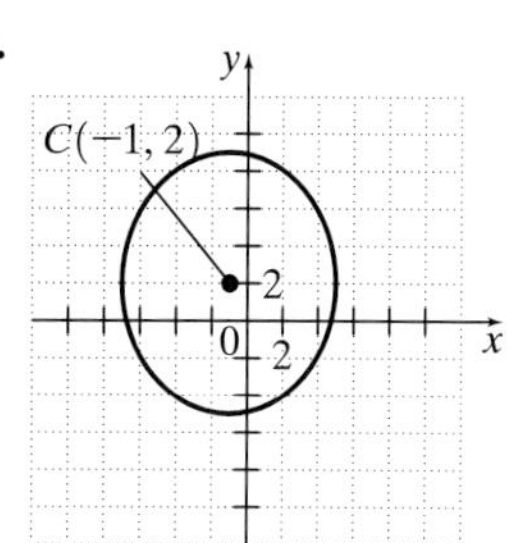

11.

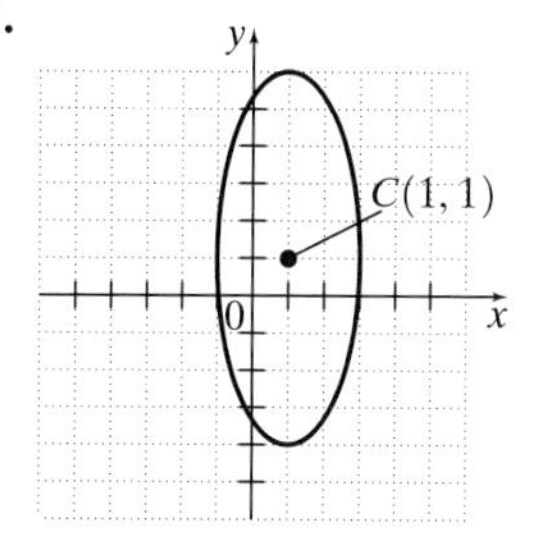

13.

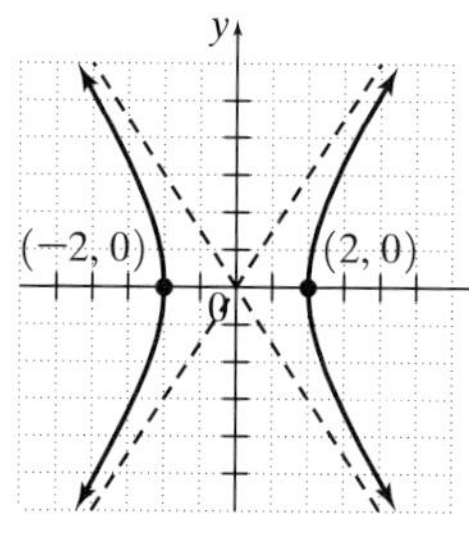

15.

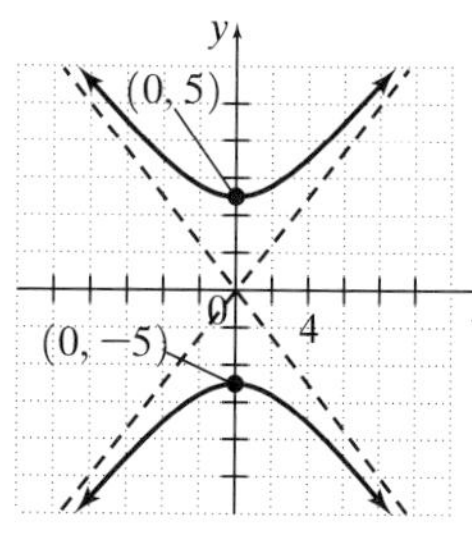

17.

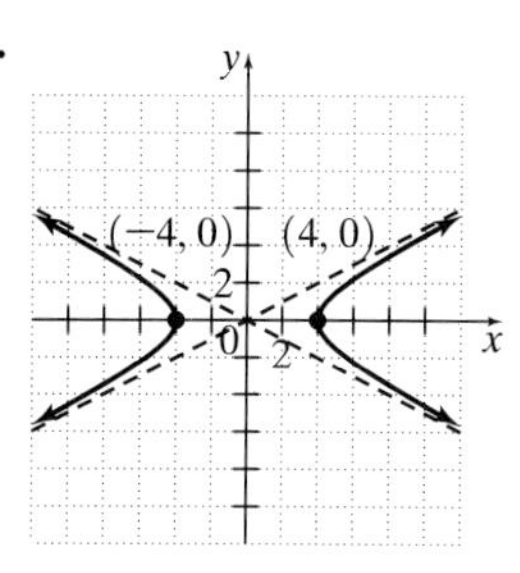

19.

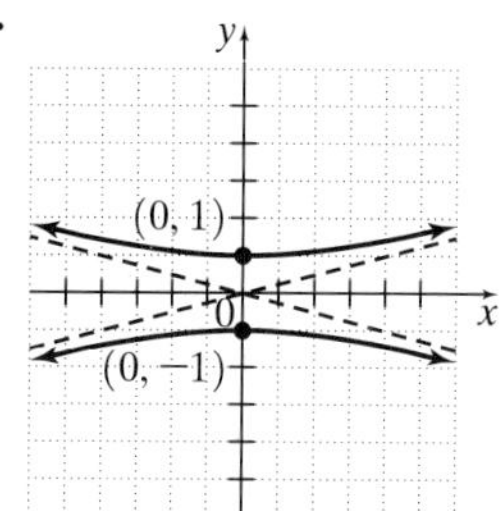

21. $-8x^5$ **23.** $-4x^2$ **25.** $\frac{x^2}{25} + \frac{y^2}{25} = 1$; ellipse; when $a = b$

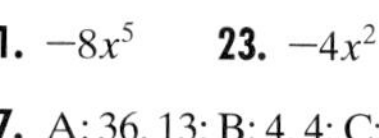

27. A: 36, 13; B: 4, 4; C: 25, 16; D: 39, 25; E: 17, 81; F: 36, 36; G: 16, 65; H: 144, 140 **29.** A: 6; B: 2; C: 5; D: 5; E: 9; F: 6; G: 4; H: 12

31. greater than zero and less than one **33.** greater than one

35. answers may vary

A:

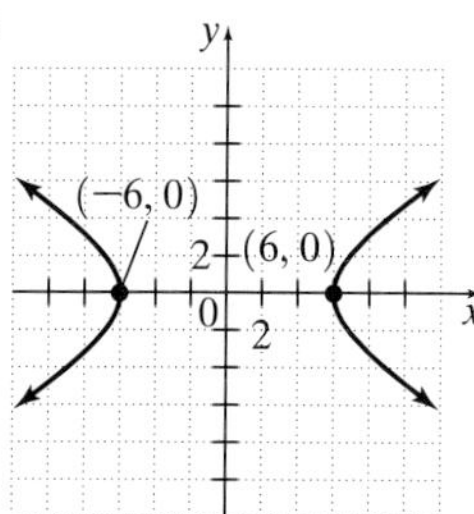

B:

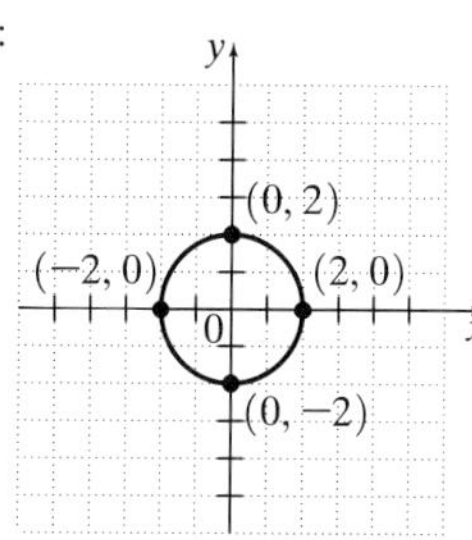

C:

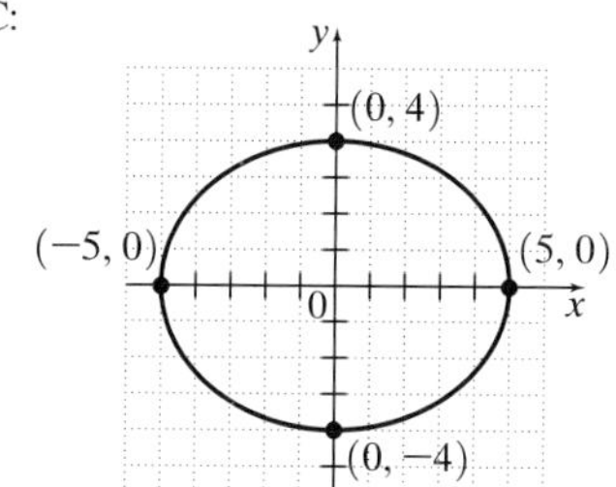

D:

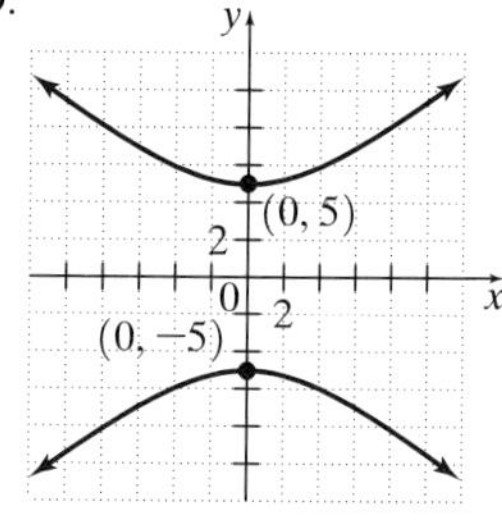

E:

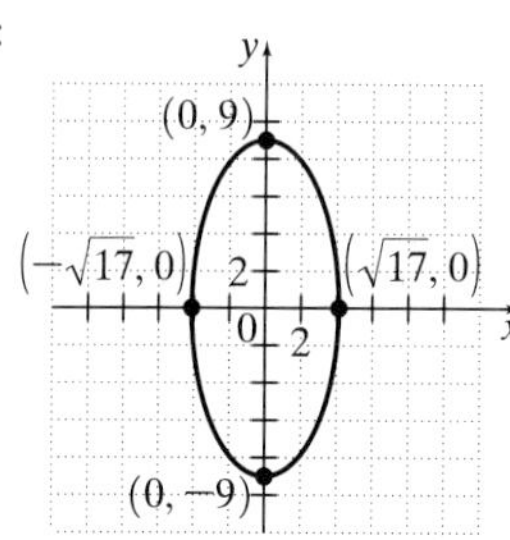

F:

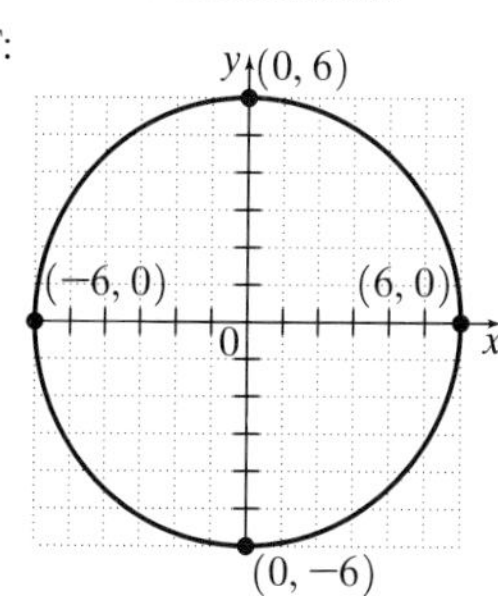

G:

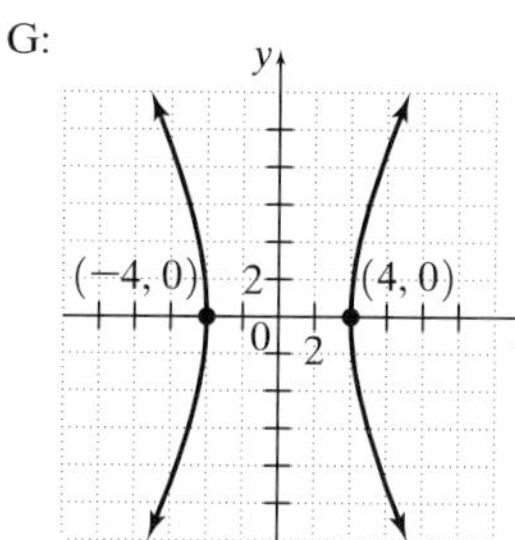

H:

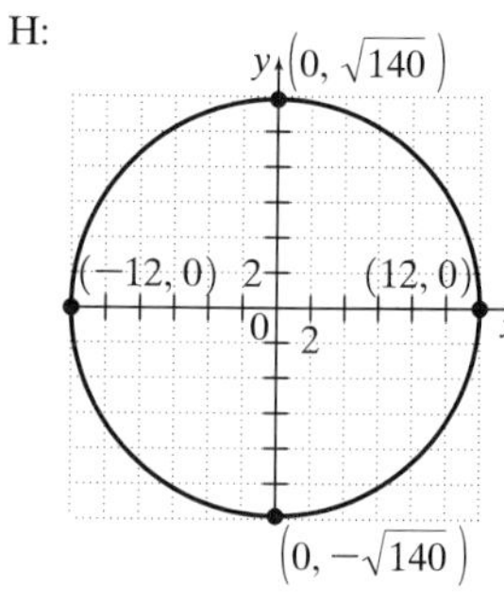

Integrated Review

1. circle;

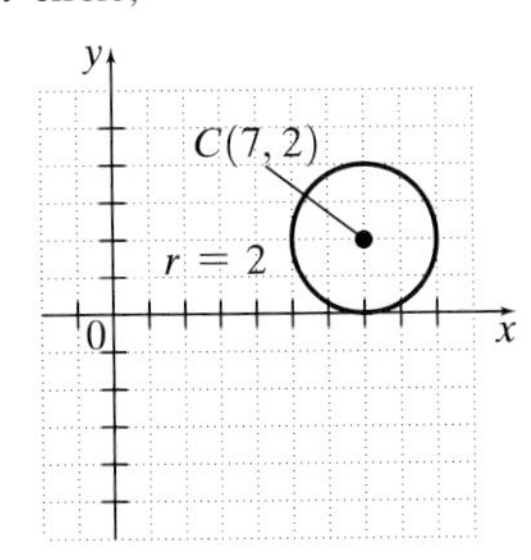

2. parabola;

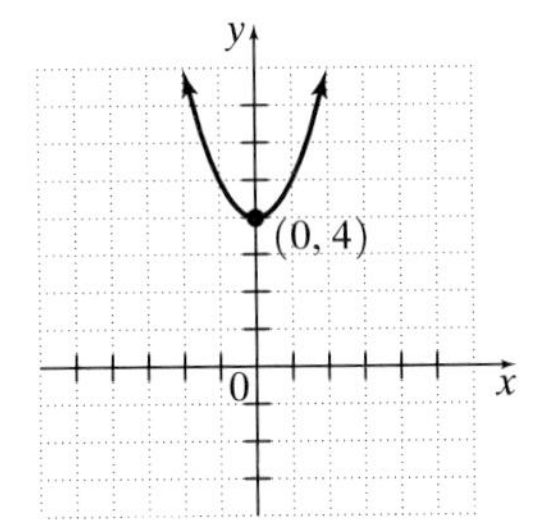

3. parabola;

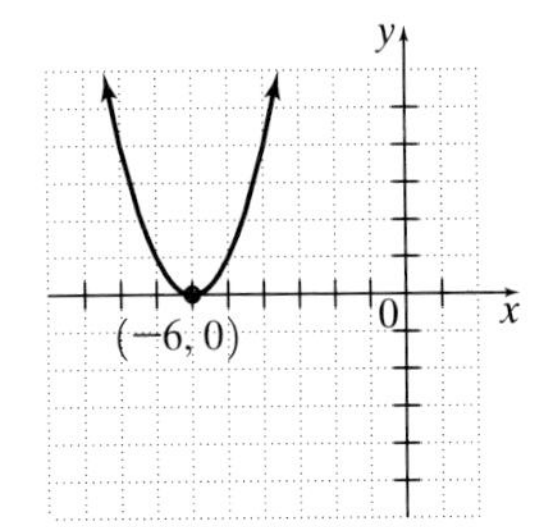

4. ellipse;

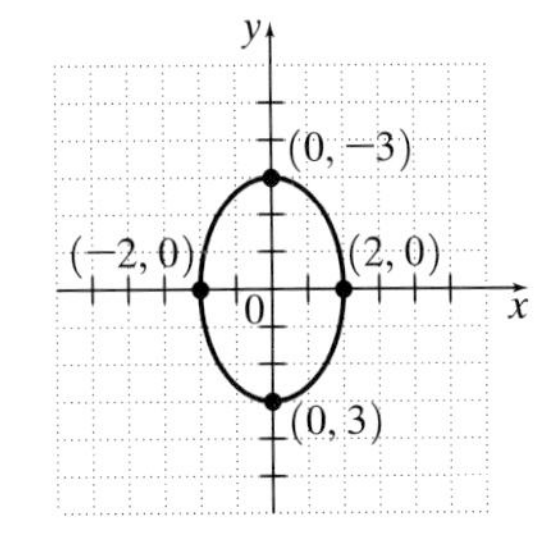

5. hyperbola;

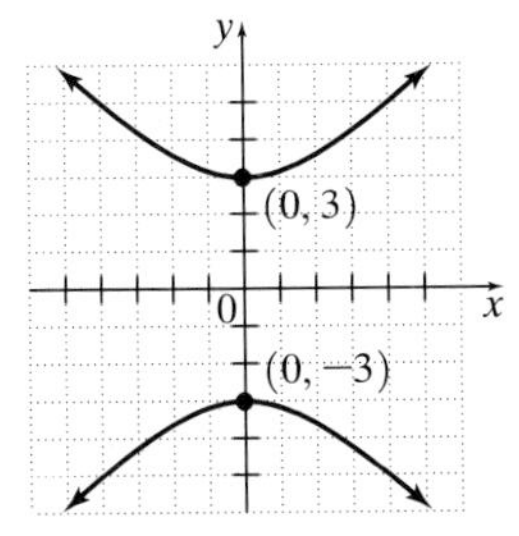

6. hyperbola;

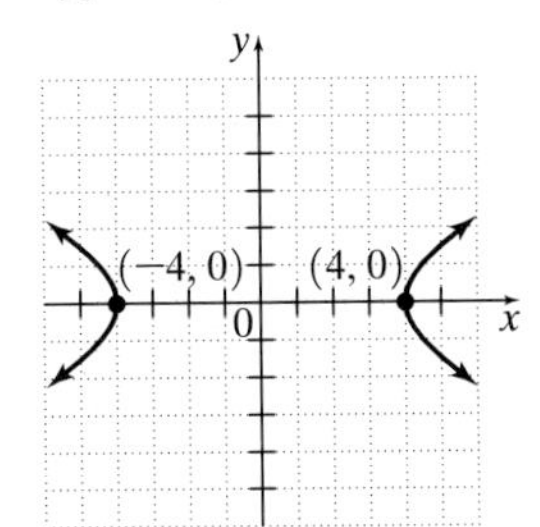

7. ellipse;

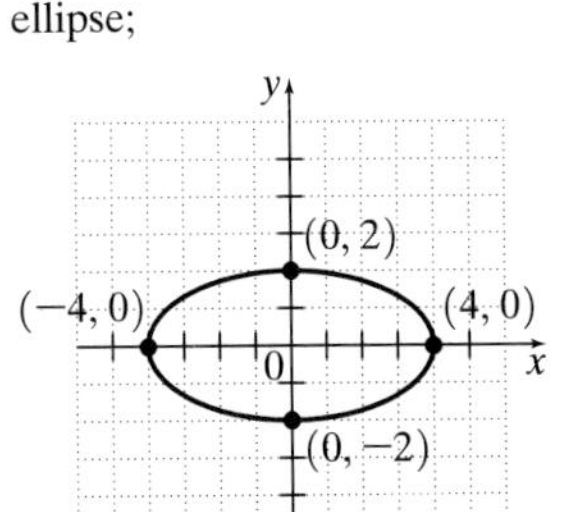

8. circle;

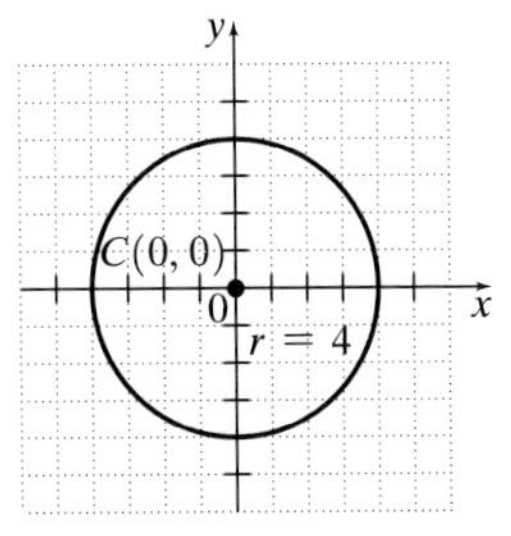

9. parabola;

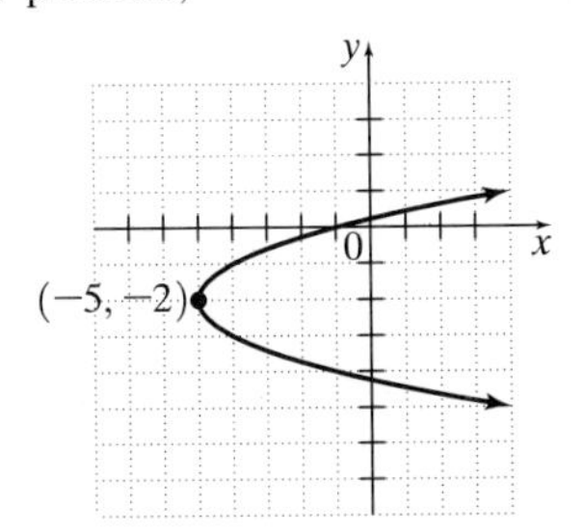

10. parabola;

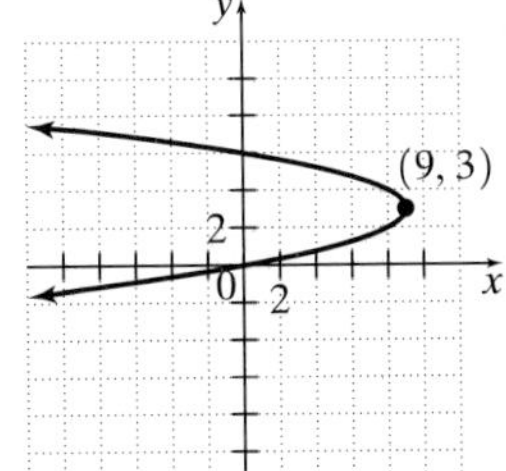

11. hyperbola;

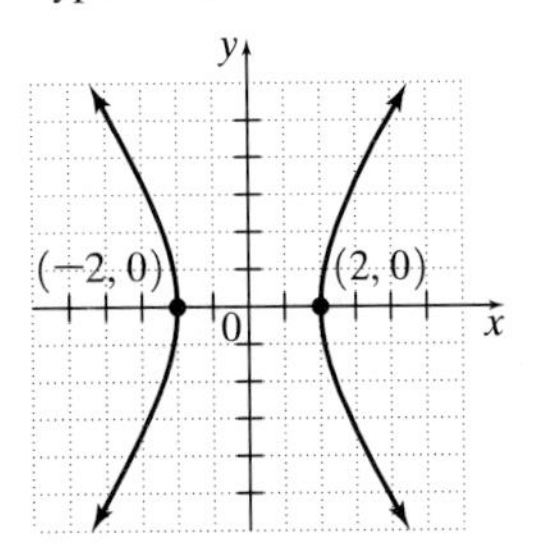

12. ellipse;

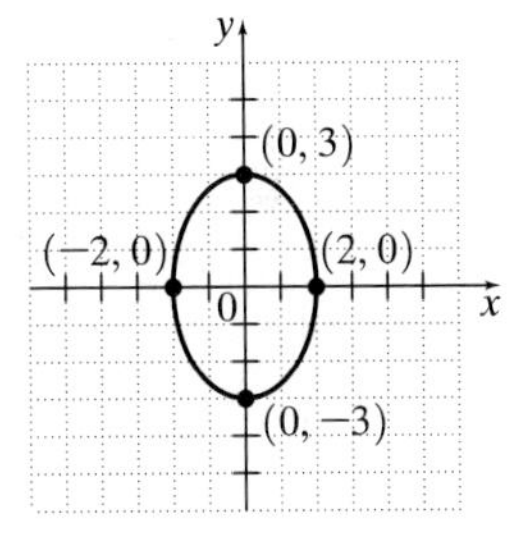

13. ellipse;

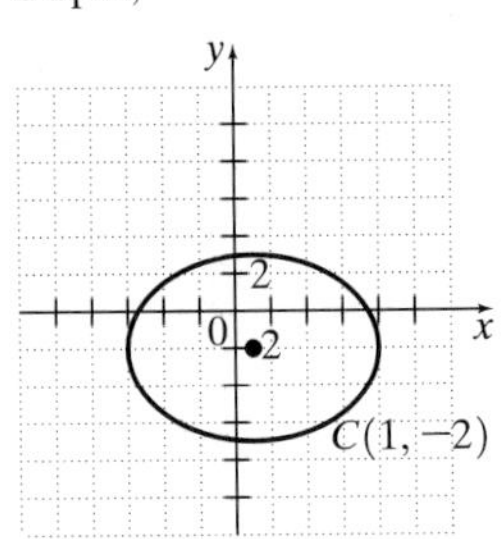

14. hyperbola;

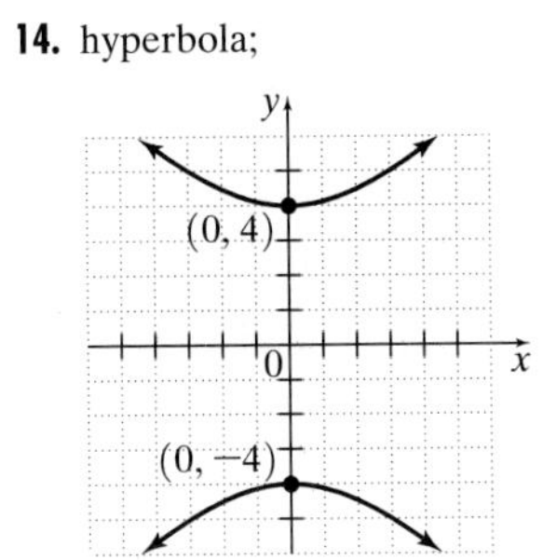

15. circle;

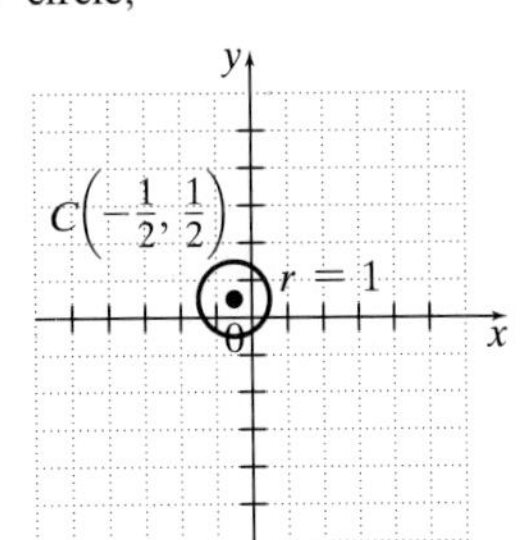

16. parabola;

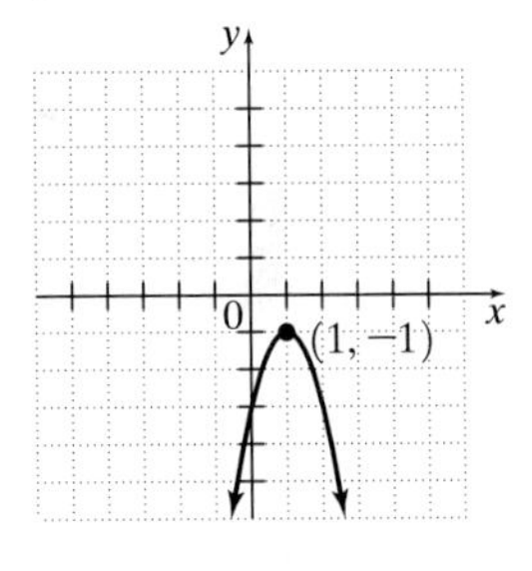

Exercise Set 9.3

1.

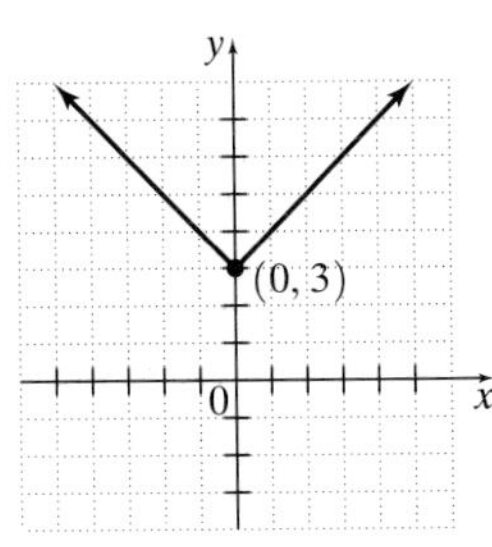

3.

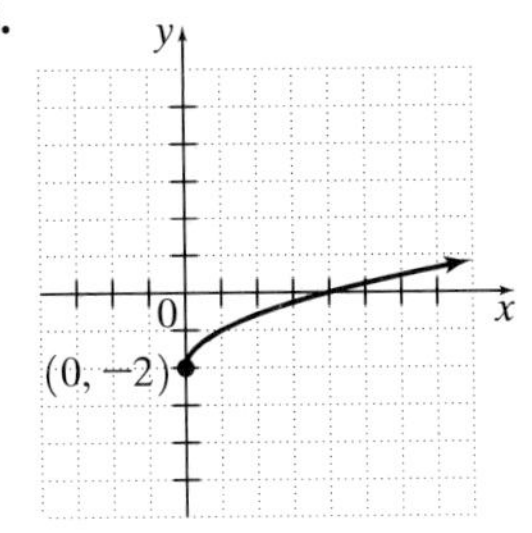

5.

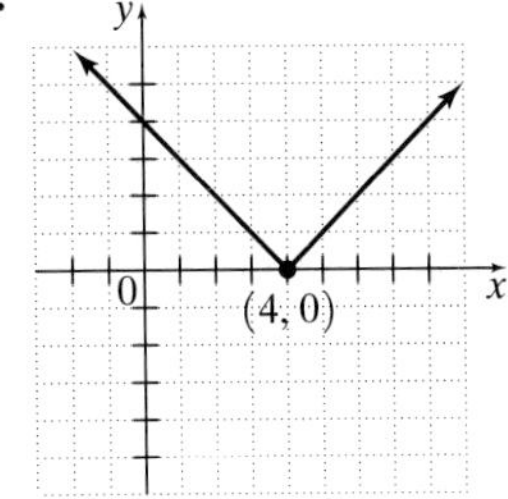

7.

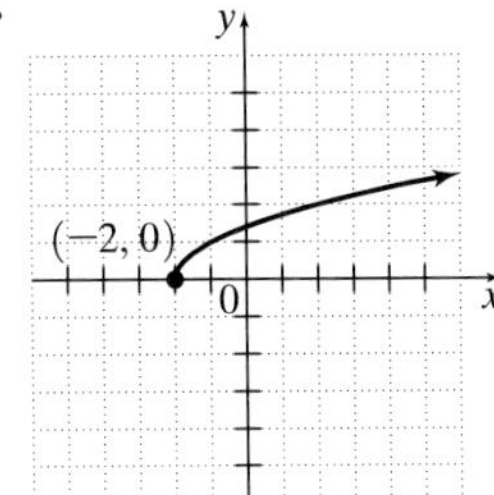

9.

11.

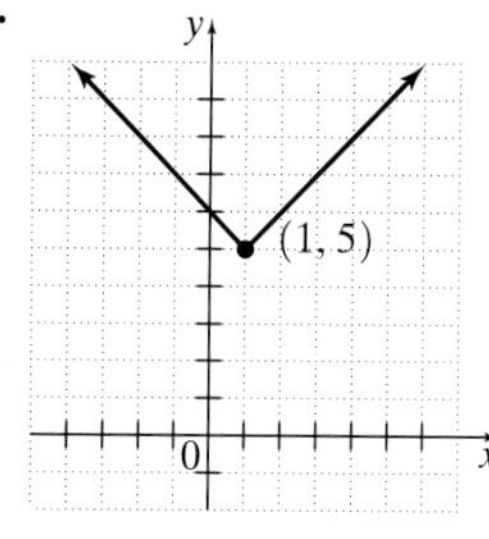

13.

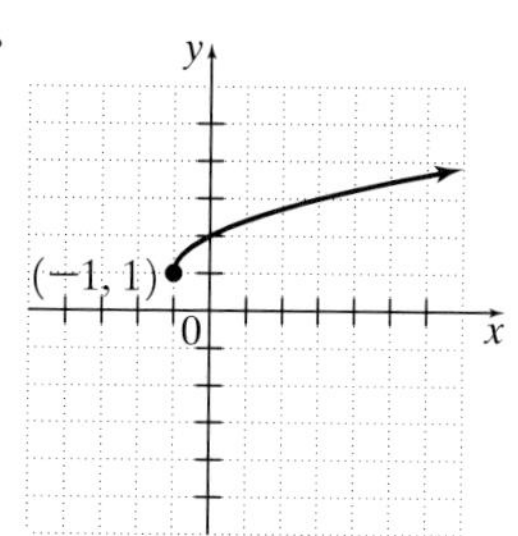

15.

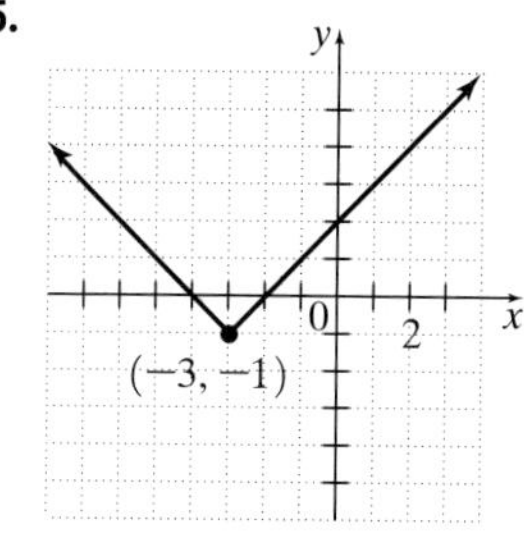

17. (8, −2)

19. (−1, 3)

21.

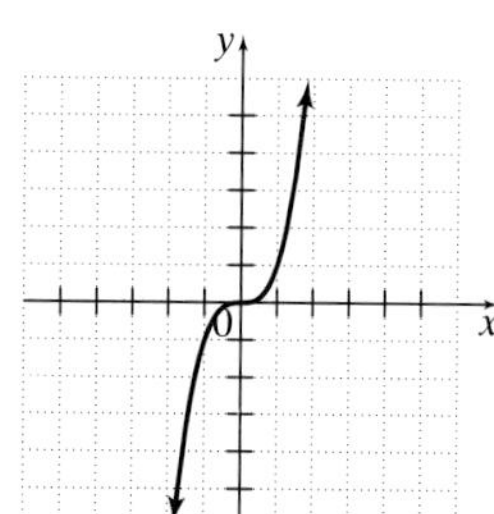

23.

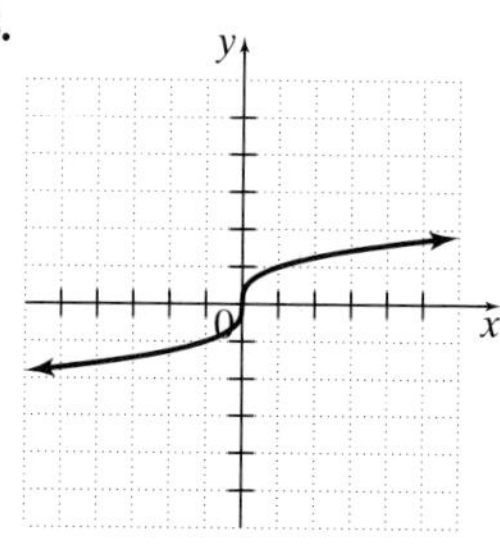

Exercise Set 9.4

1. $\{(3,-4),(-3,4)\}$ **3.** $\{(\sqrt{2},\sqrt{2}),(-\sqrt{2},-\sqrt{2})\}$ **5.** $\{(4,0),(0,-2)\}$ **7.** $\{(-\sqrt{5},-2),(-\sqrt{5},2),(\sqrt{5},-2),(\sqrt{5},2)\}$ **9.** $\emptyset$
11. $\{(1,-2),(3,6)\}$ **13.** $\{(2,4),(-5,25)\}$ **15.** $\emptyset$ **17.** $\{(1,-3)\}$ **19.** $\{(-1,-2),(-1,2),(1,-2),(1,2)\}$ **21.** $\{(0,-1)\}$
23. $\{(-1,3),(1,3)\}$ **25.** $\{(\sqrt{3},0),(-\sqrt{3},0)\}$ **27.** $\emptyset$ **29.** $\{(-6,0),(6,0),(0,-6)\}$ **31.** 0, 1, 2, 3, or 4

33.

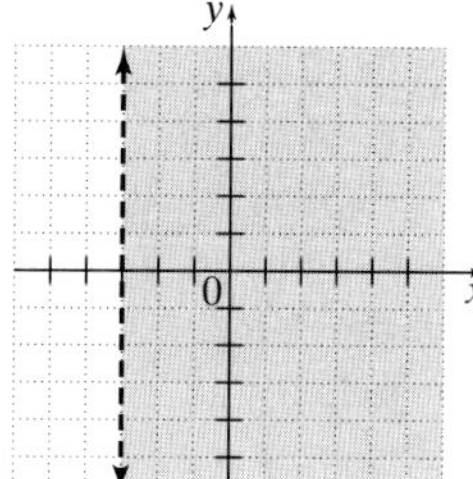

35.

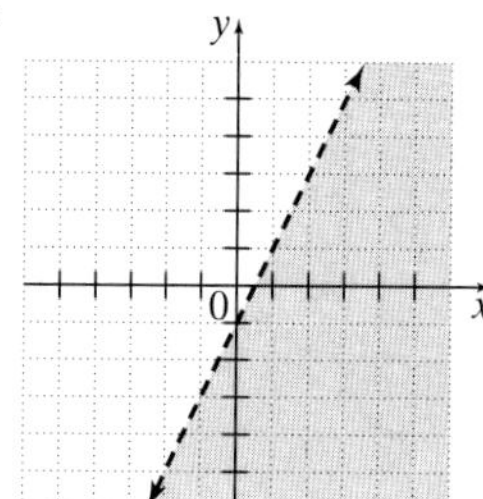

37. 9 and 7; 9 and −7; −9 and 7; −9 and −7
39. 15 cm by 19 cm **41.** 15 thousand compact discs; price, $3.75

Exercise Set 9.5

1.

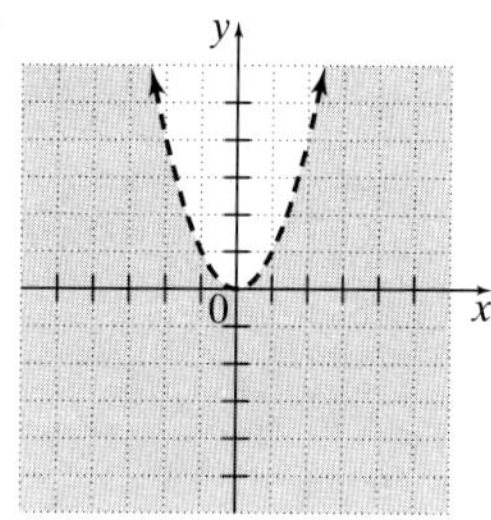

3.

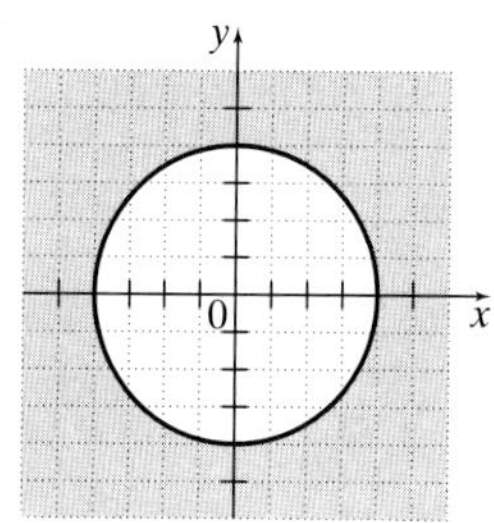

5.

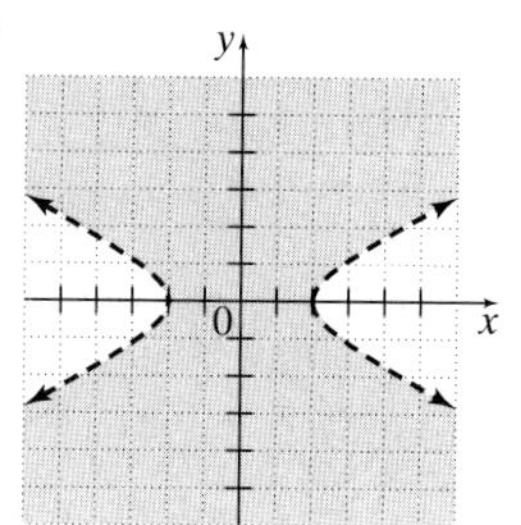

7.

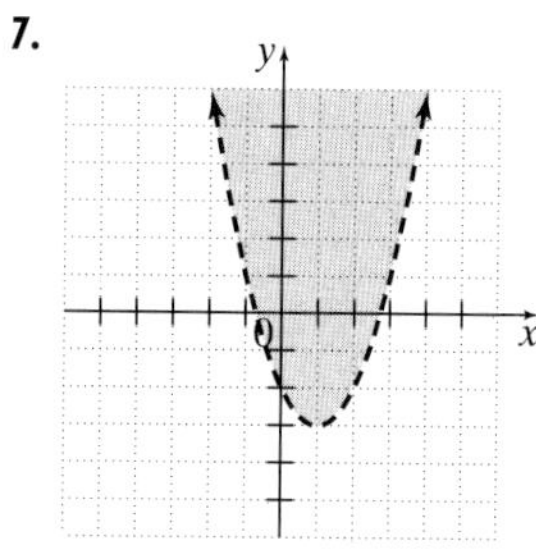

9.

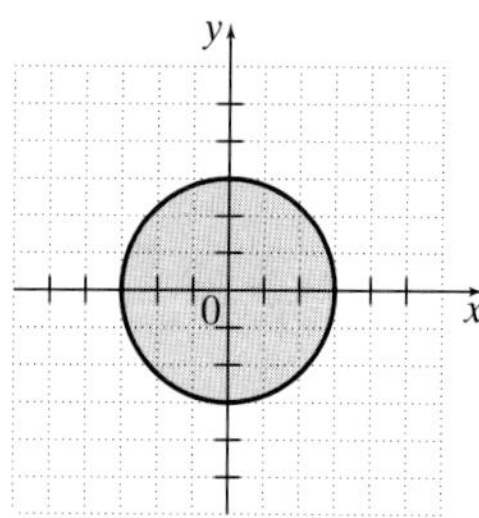

11.

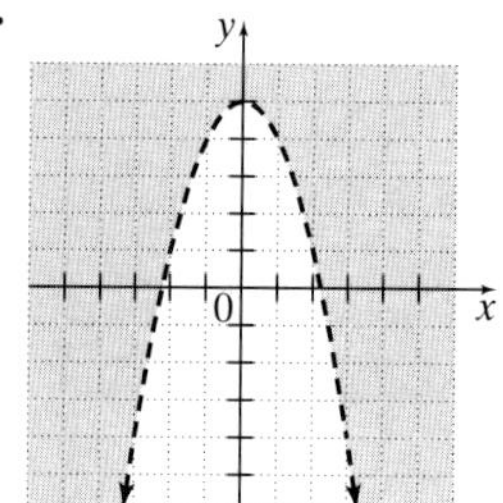

13.

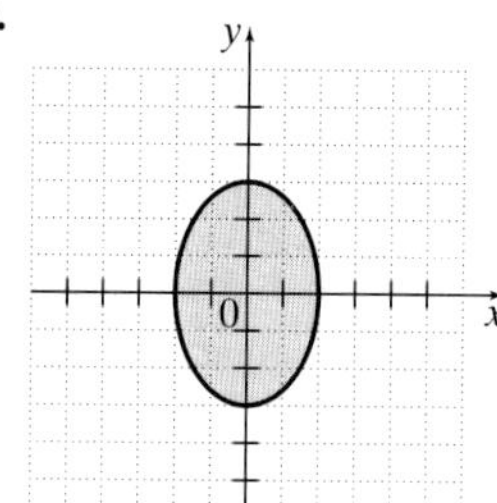

15.

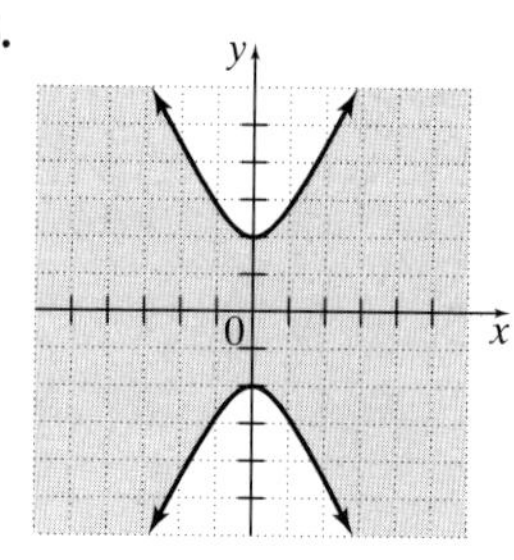

17. 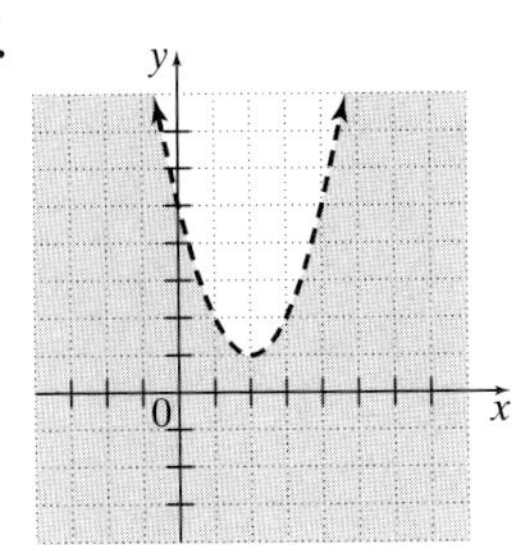

19. 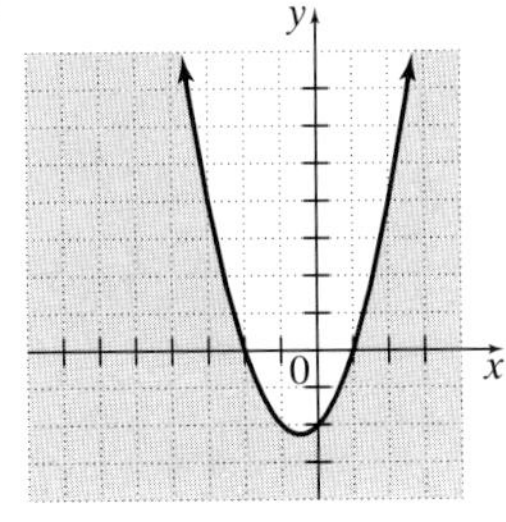

21. 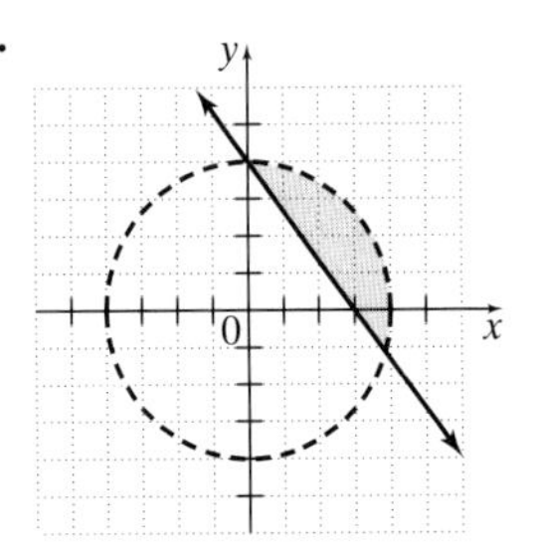

23. 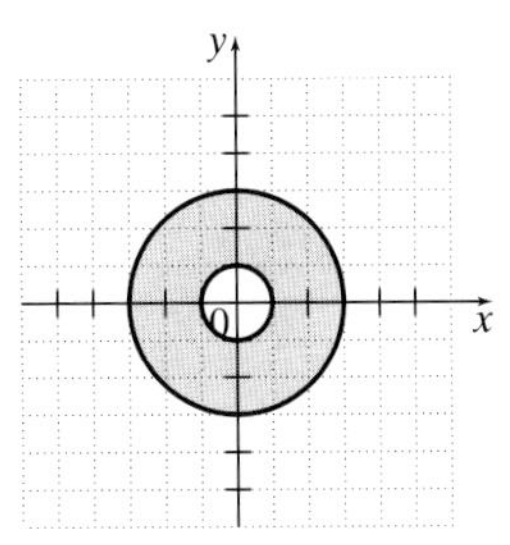

25. 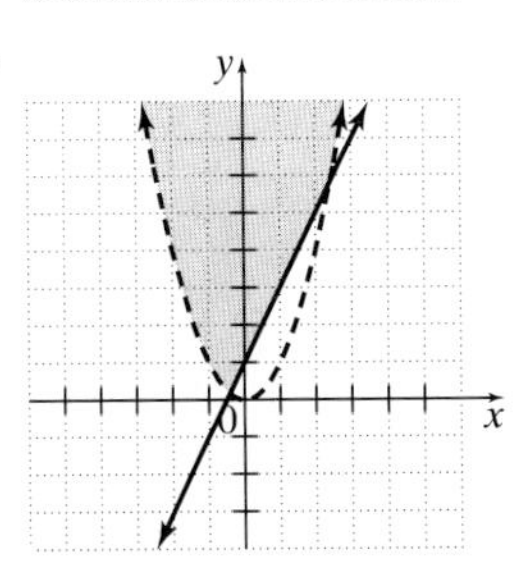

27. 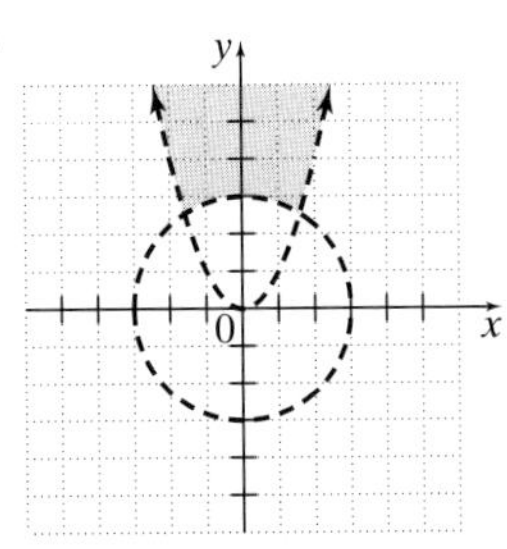

29. 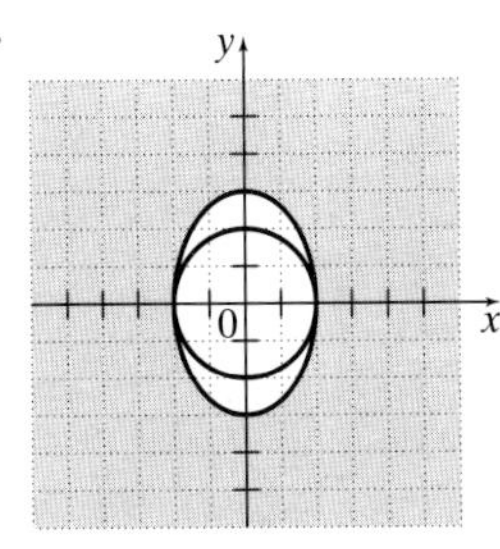

31. 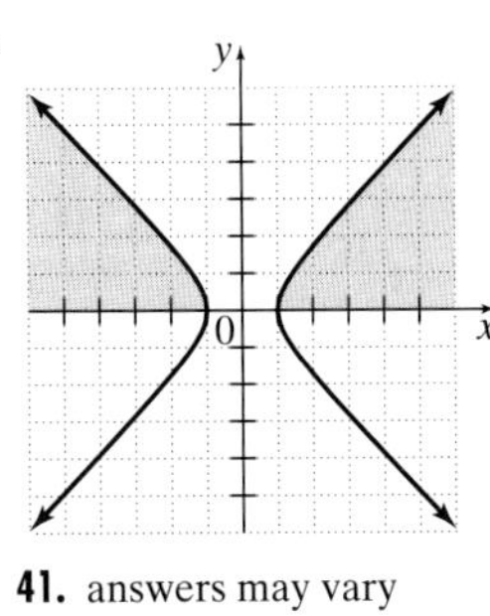

33. 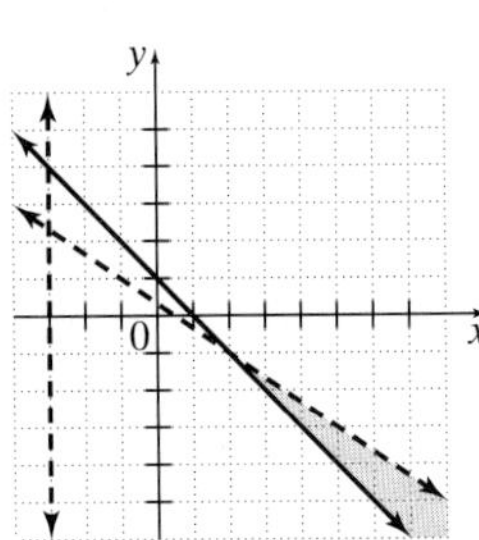

35. 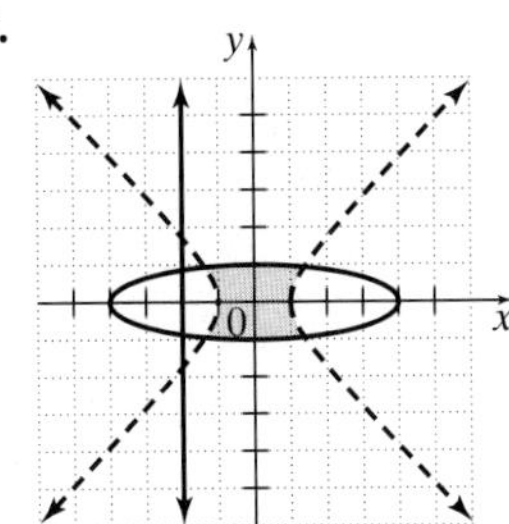

37. not a function **39.** function **41.** answers may vary

43. 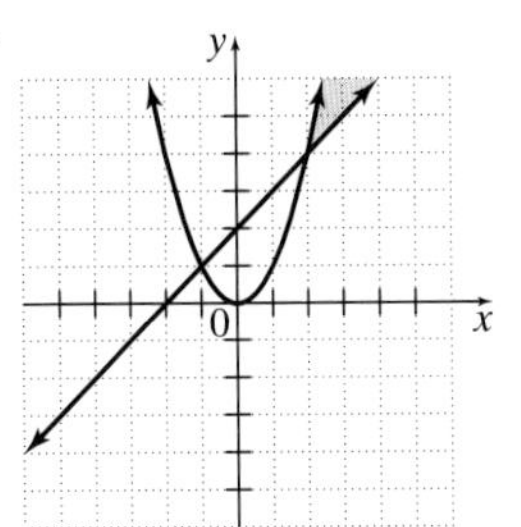

Chapter 9 Review

1. $\sqrt{197}$ units ≈ 14.036 **2.** $\sqrt{41}$ units ≈ 6.403 **3.** $\sqrt{130}$ units ≈ 11.402 **4.** $\sqrt{73}$ units ≈ 8.544
5. $7\sqrt{2}$ units ≈ 9.899 **6.** $2\sqrt{11}$ units ≈ 6.633 **7.** $\sqrt{275.6}$ units ≈ 16.601 **8.** $\sqrt{191.56}$ units ≈ 13.841 **9.** $(-5, 5)$
10. $(4, 16)$ **11.** $\left(-\frac{15}{2}, 1\right)$ **12.** $\left(-\frac{11}{2}, -2\right)$ **13.** $\left(\frac{1}{20}, -\frac{3}{16}\right)$ **14.** $\left(\frac{1}{4}, -\frac{2}{7}\right)$ **15.** $(\sqrt{3}, -3\sqrt{6})$ **16.** $(-4\sqrt{3}, 6\sqrt{7})$
17. $(x + 4)^2 + (y - 4)^2 = 9$ **18.** $(x - 5)^2 + y^2 = 25$ **19.** $(x + 7)^2 + (y + 9)^2 = 11$ **20.** $x^2 + y^2 = \frac{49}{4}$

21. 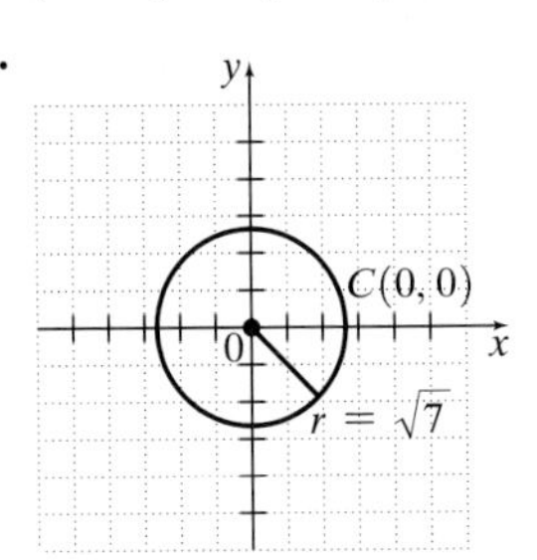

22. 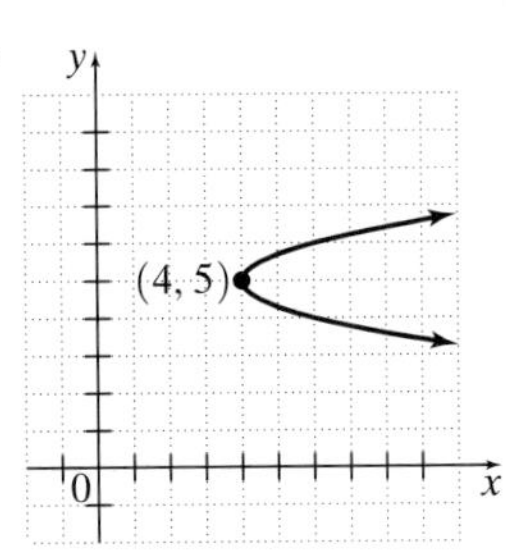

23. 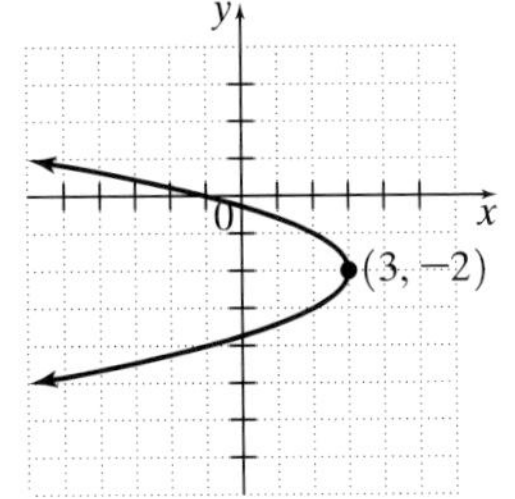

24. 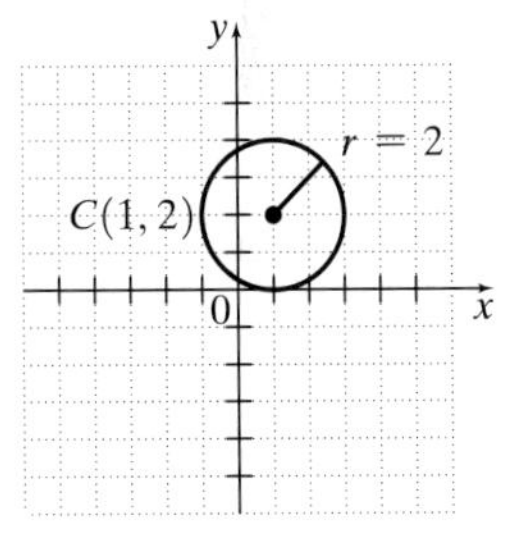

25. 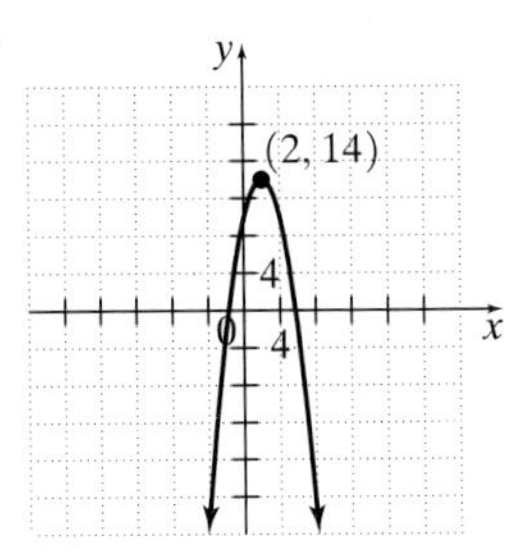

26. 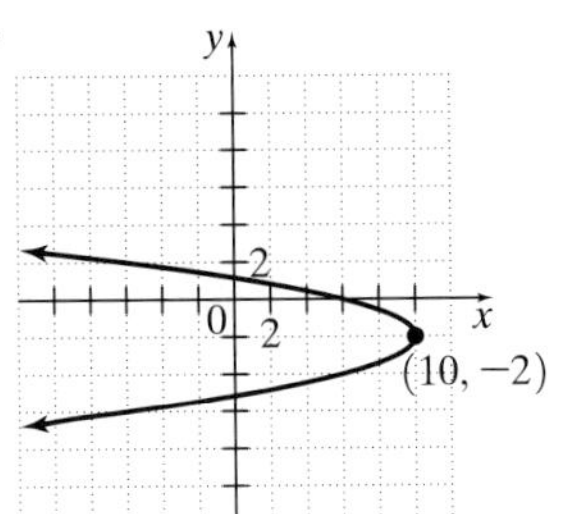

27. 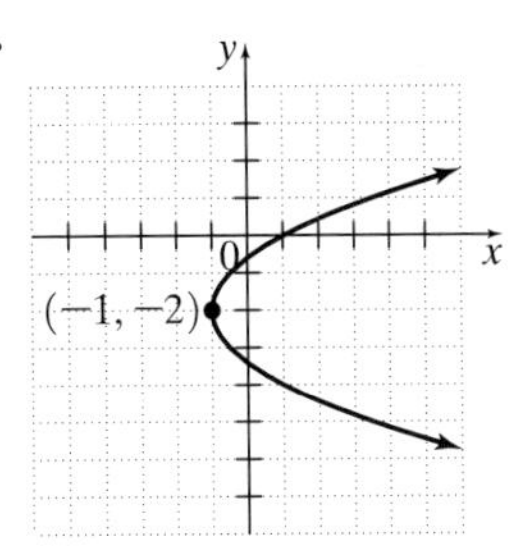

28. 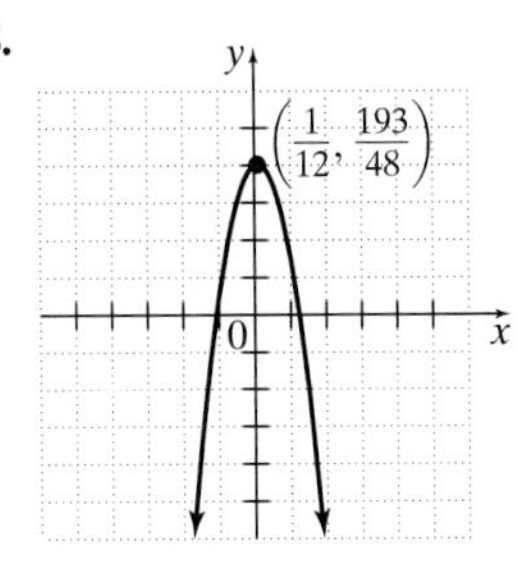

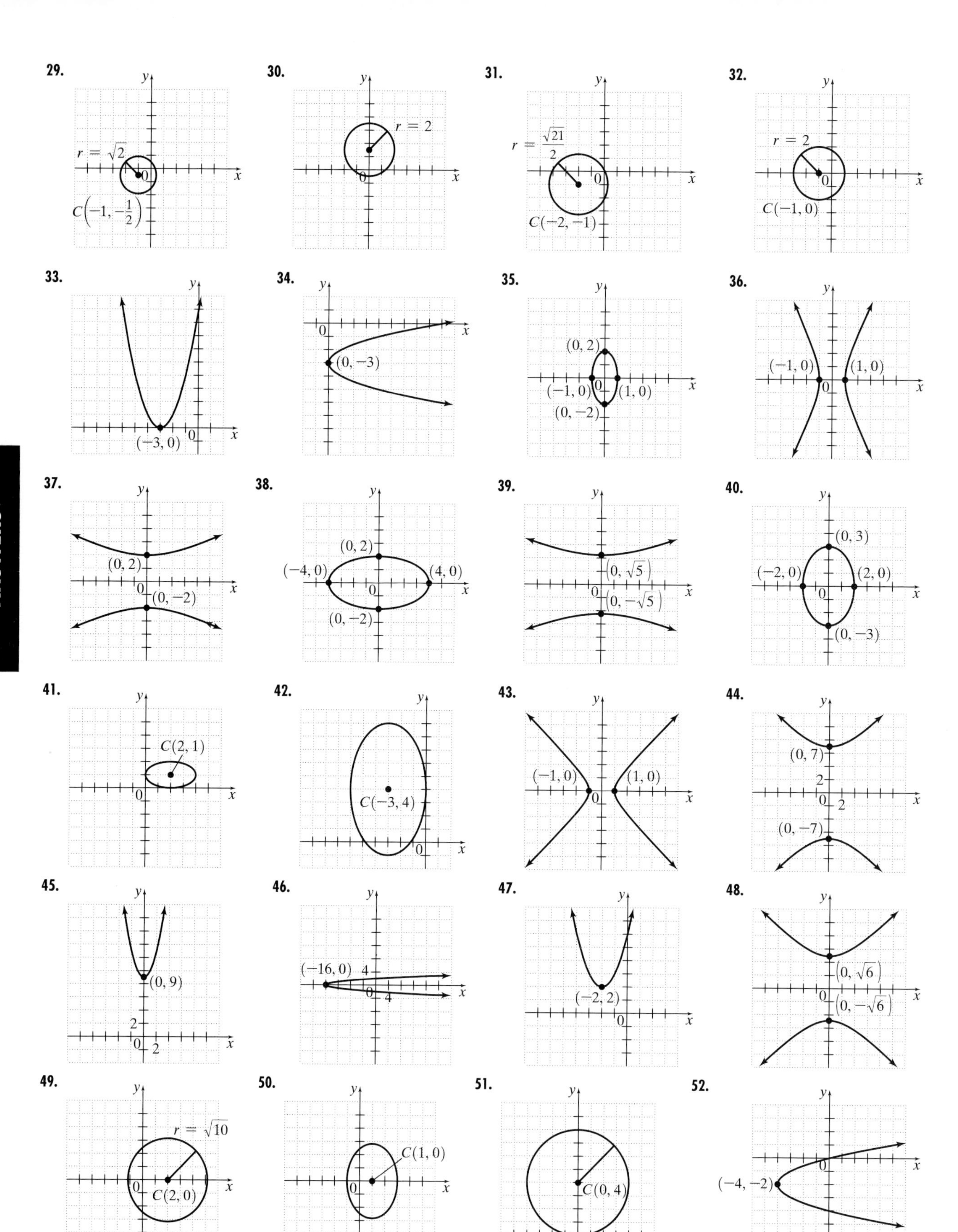
29.
$r = \sqrt{2}$
$C\left(-1, -\frac{1}{2}\right)$
30.
$r = 2$
31.
$r = \frac{\sqrt{21}}{2}$
$C(-2, -1)$
32.
$r = 2$
$C(-1, 0)$
33.
$(-3, 0)$
34.
$(0, -3)$
35.
$(0, 2)$
$(-1, 0)$
$(1, 0)$
$(0, -2)$
36.
$(-1, 0)$
$(1, 0)$
37.
$(0, 2)$
$(0, -2)$
38.
$(0, 2)$
$(-4, 0)$
$(4, 0)$
$(0, -2)$
39.
$(0, \sqrt{5})$
$(0, -\sqrt{5})$
40.
$(0, 3)$
$(-2, 0)$
$(2, 0)$
$(0, -3)$
41.
$C(2, 1)$
42.
$C(-3, 4)$
43.
$(-1, 0)$
$(1, 0)$
44.
$(0, 7)$
$(0, -7)$
45.
$(0, 9)$
46.
$(-16, 0)$
47.
$(-2, 2)$
48.
$(0, \sqrt{6})$
$(0, -\sqrt{6})$
49.
$r = \sqrt{10}$
$C(2, 0)$
50.
$C(1, 0)$
51.
$C(0, 4)$
52.
$(-4, -2)$
ANSWERS

53.

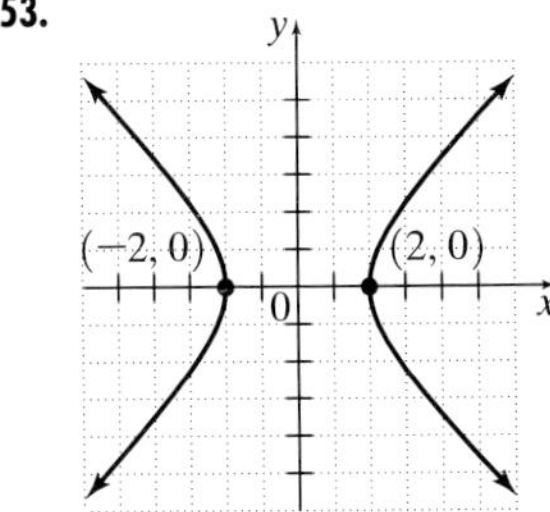

54.

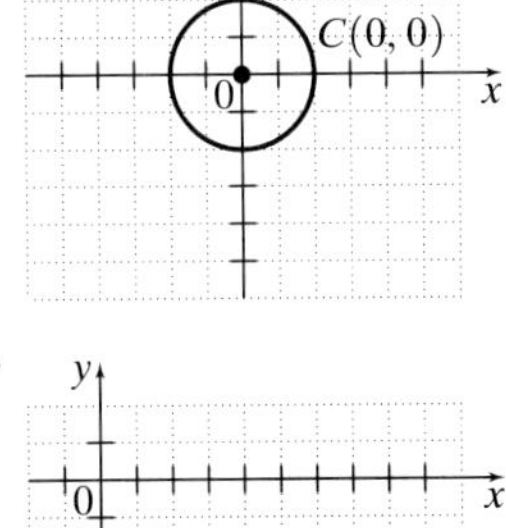

55.

56.

57.

58.

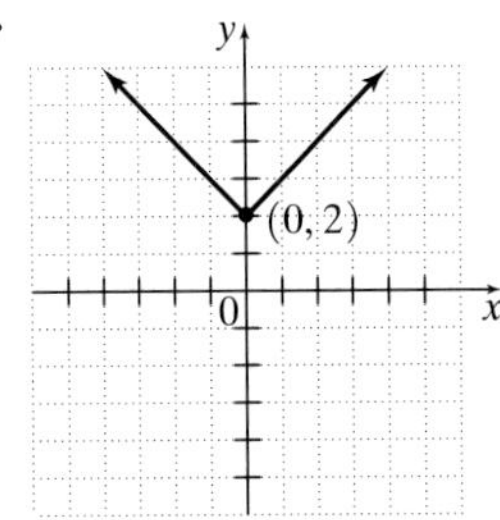

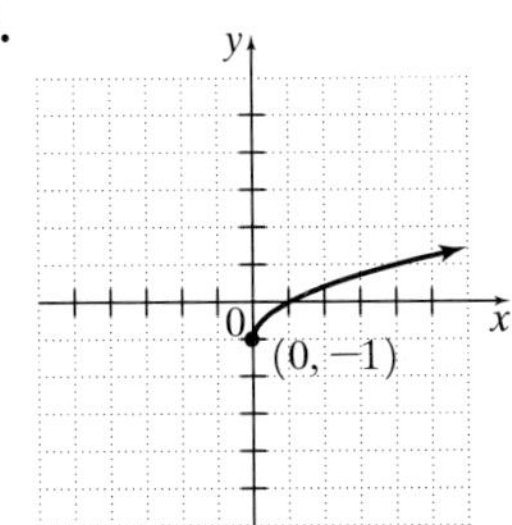

59.

60.

61.

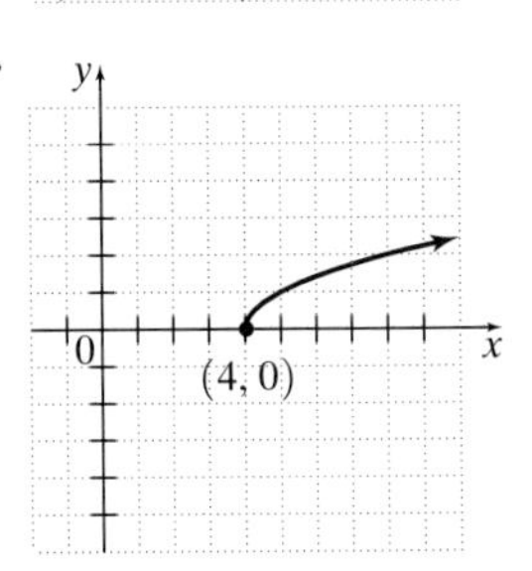

62.

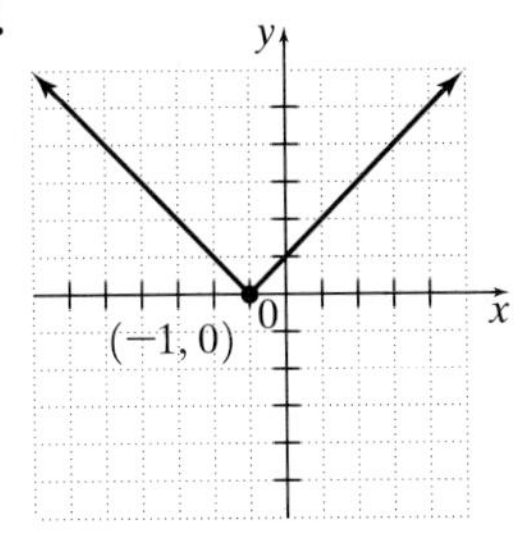

63.

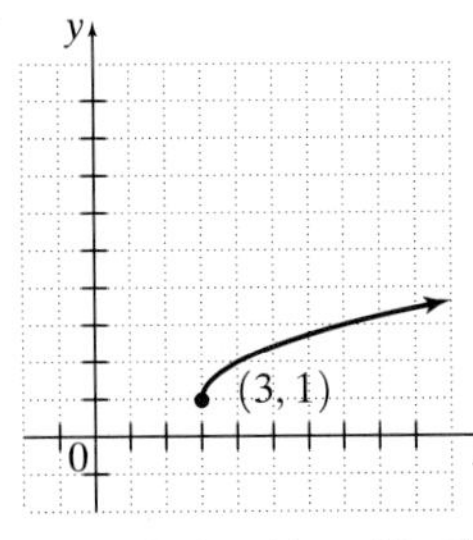

64.

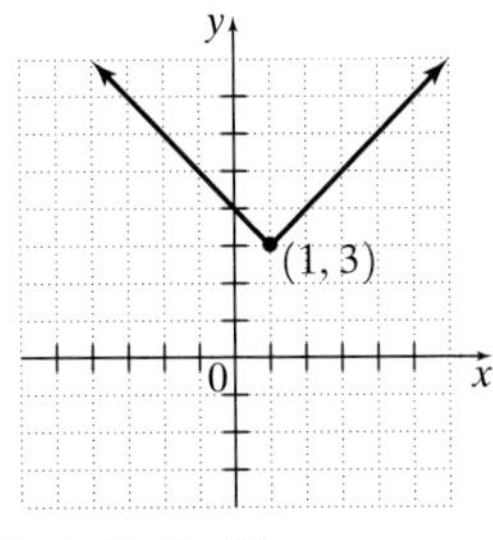

65.

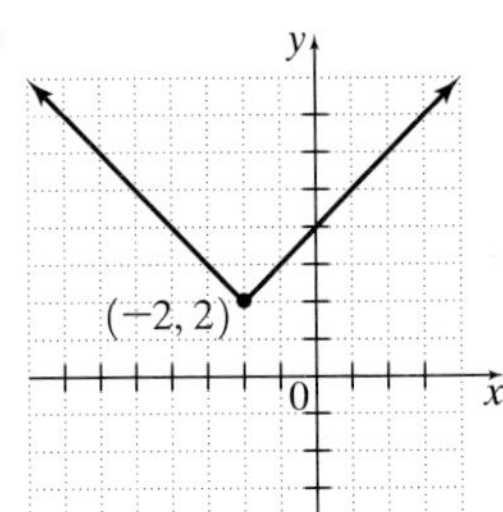

66.

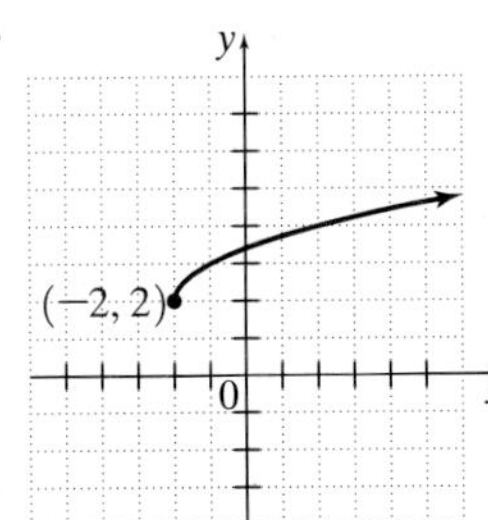

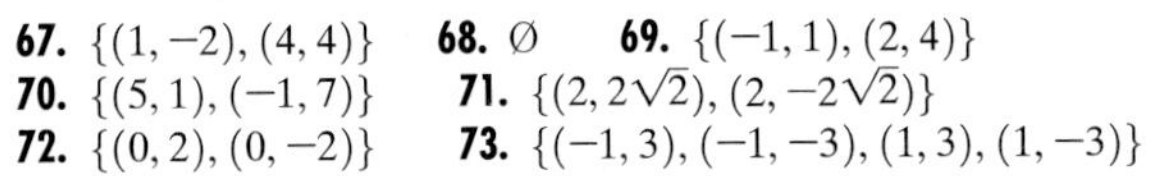

67. $\{(1, -2), (4, 4)\}$ **68.** $\varnothing$ **69.** $\{(-1, 1), (2, 4)\}$
70. $\{(5, 1), (-1, 7)\}$ **71.** $\{(2, 2\sqrt{2}), (2, -2\sqrt{2})\}$
72. $\{(0, 2), (0, -2)\}$ **73.** $\{(-1, 3), (-1, -3), (1, 3), (1, -3)\}$
74. $\left\{\left(2, \frac{5}{2}\right), (-7, -20)\right\}$ **75.** $\{(1, 4)\}$ **76.** $\{(-2, -1),$
$(-2, 1), (2, -1), (2, 1)\}$ **77.** 15 ft by 10 ft **78.** 4

79.

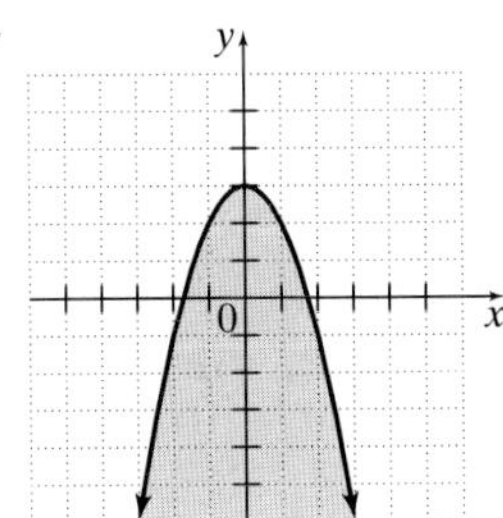

80.

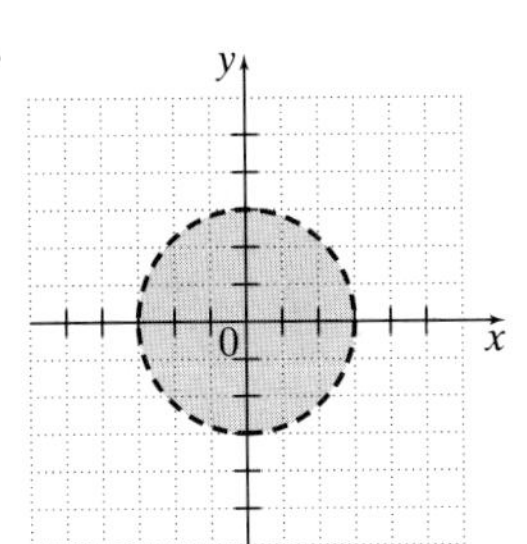

81.

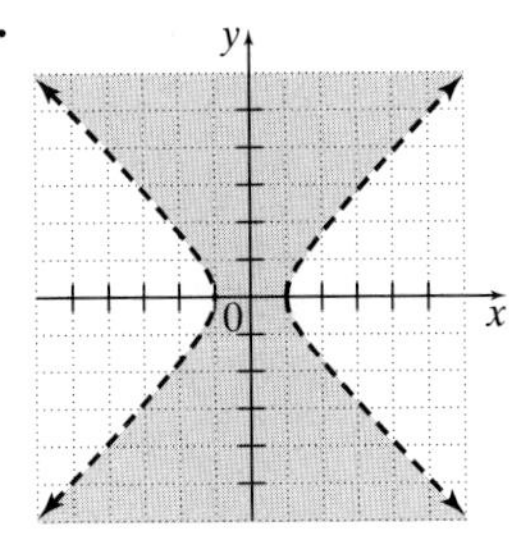

82.

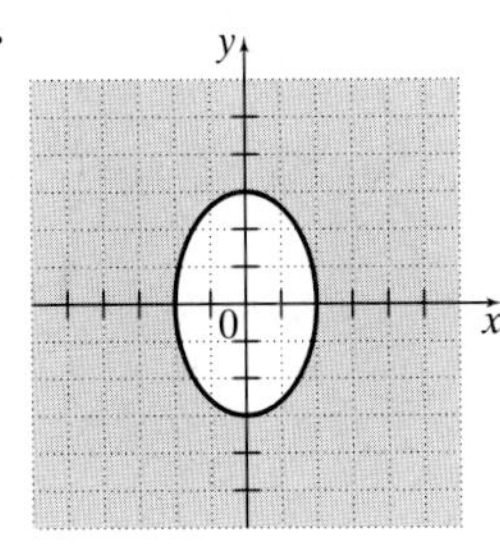

83.

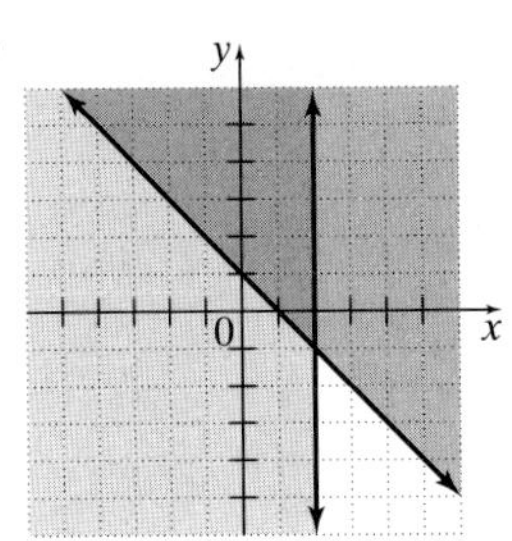

84.

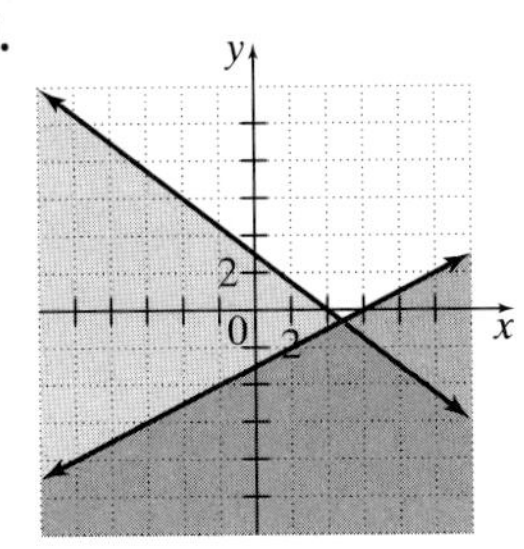

85

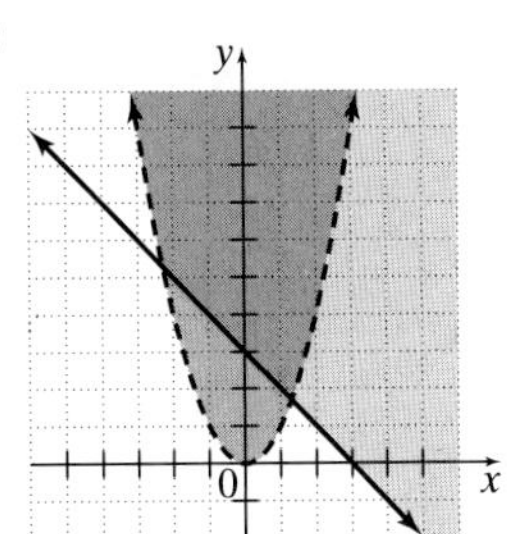

86.

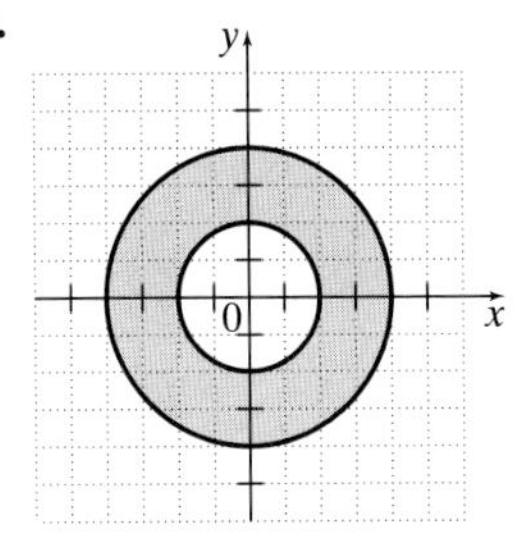

ANSWERS

87.

88.

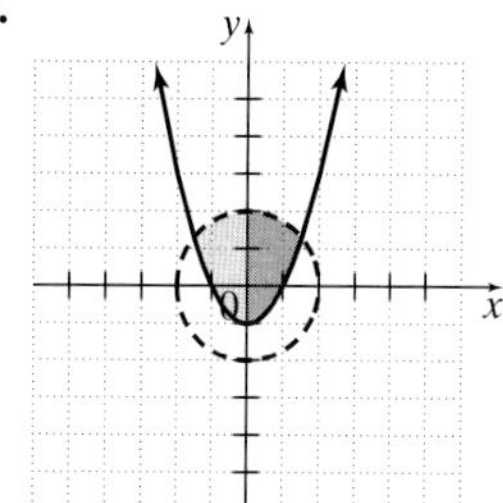

Chapter 9 Test

1. $2\sqrt{26}$ units **2.** $\sqrt{95}$ units **3.** $\left(-4, \frac{7}{2}\right)$ **4.** $\left(-\frac{1}{2}, \frac{3}{10}\right)$

5.

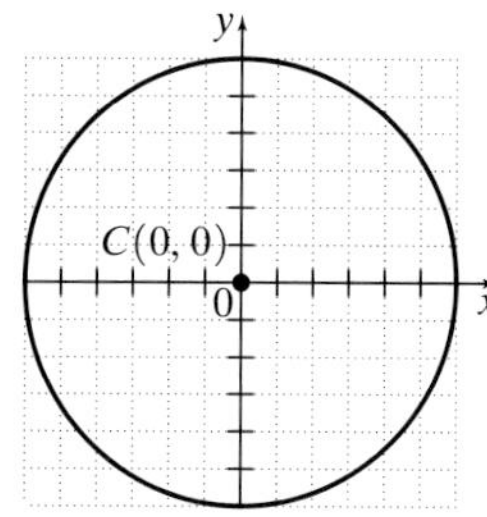

6.

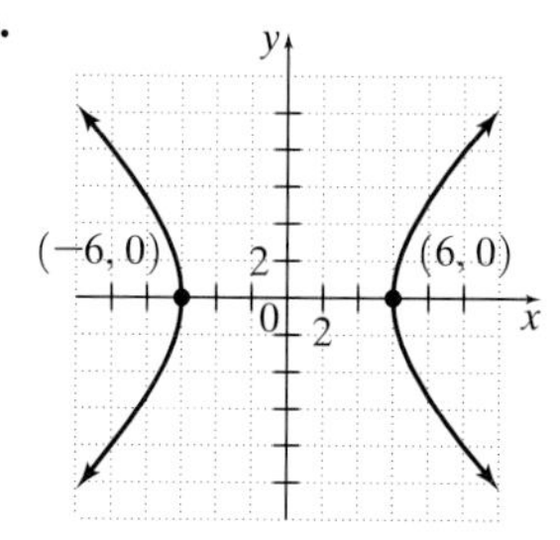

7.

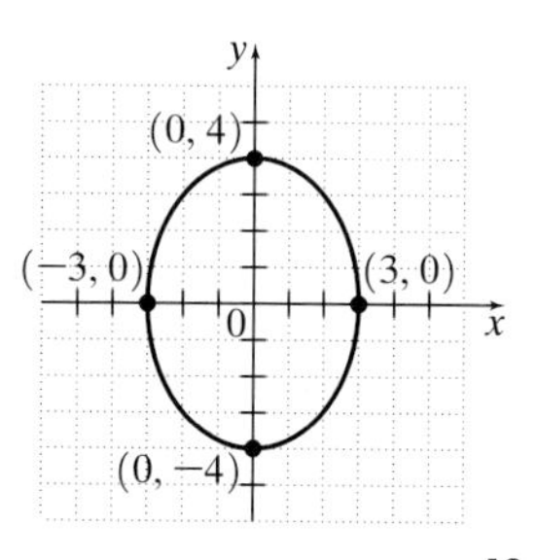

8.

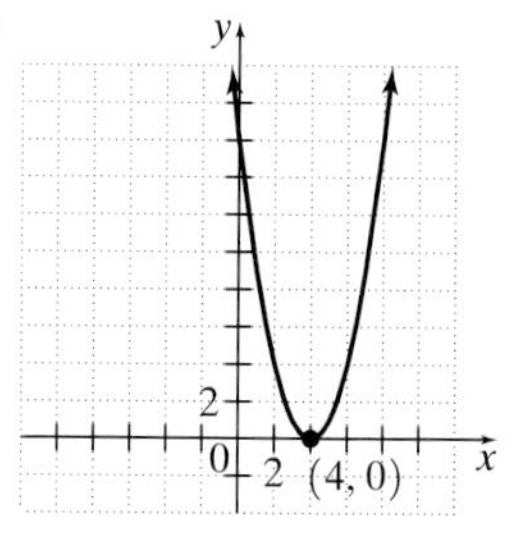

9.

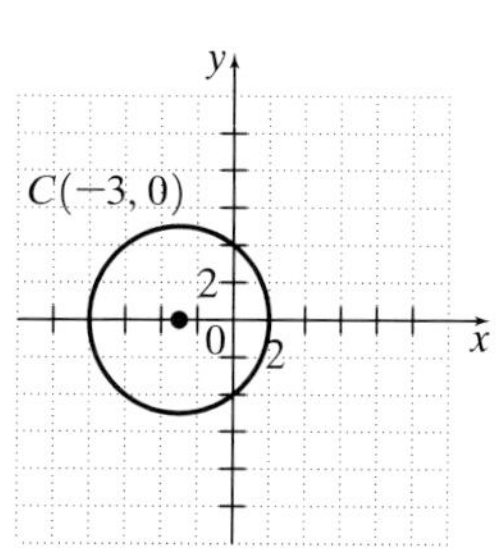

10.

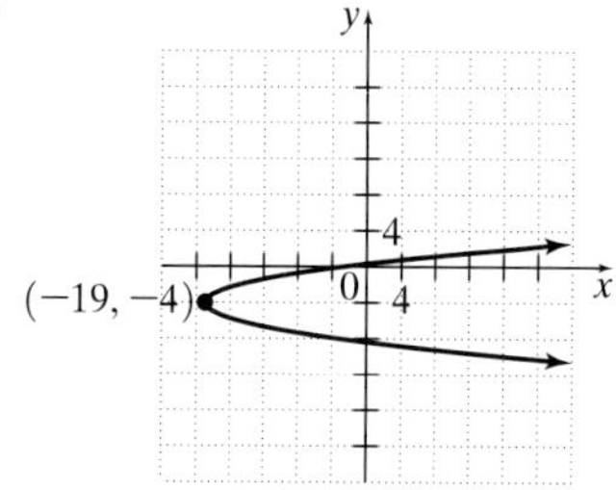

11.

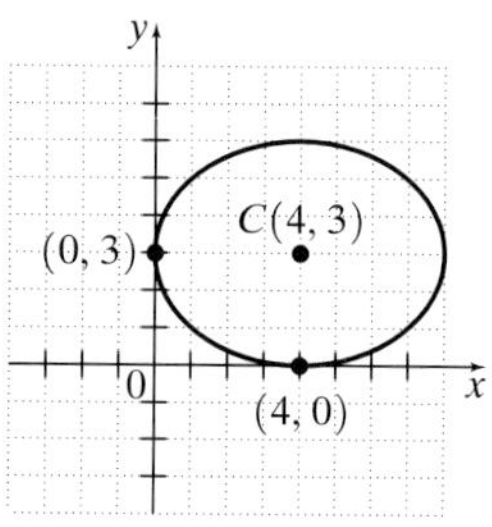

12.

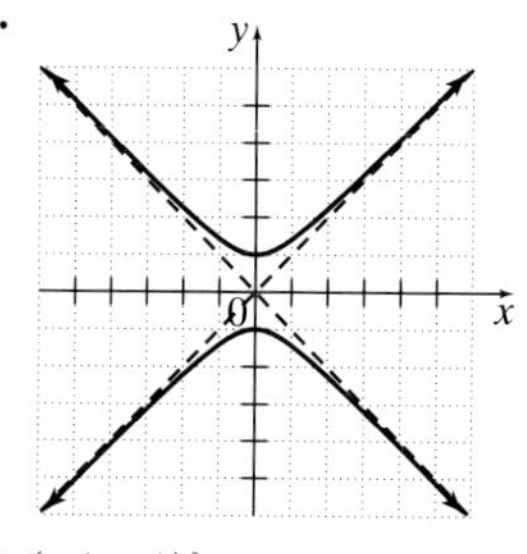

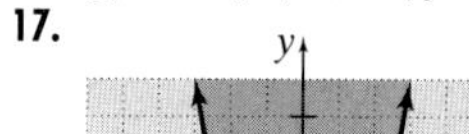

13. $\{(-12, 5), (12, -5)\}$ **14.** $\{(-5, -1), (-5, 1), (5, -1), (5, 1)\}$ **15.** $\{(6, 12), (1, 2)\}$ **16.** $\{(1, 1), (-1, -1)\}$

17.

18.

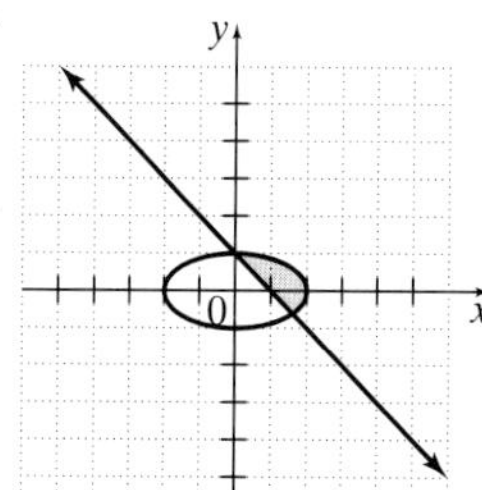

19.

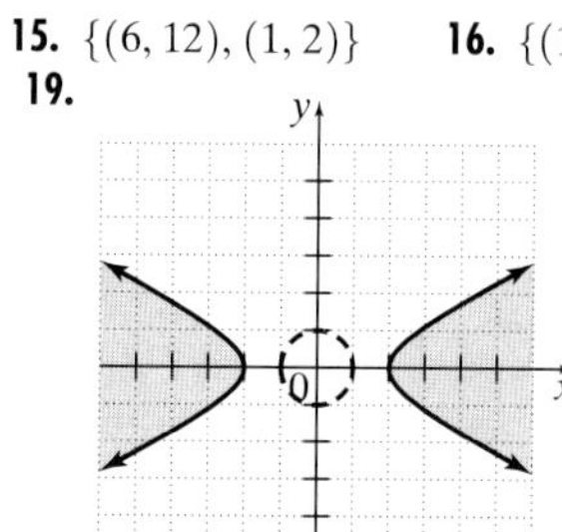

20.

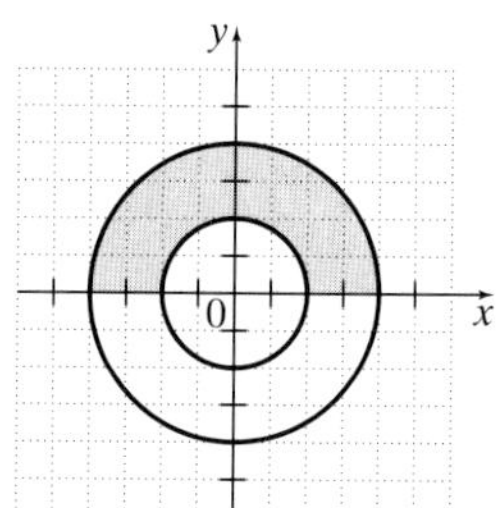

21. B **22.** height, 10 ft; width, 30 ft

23.

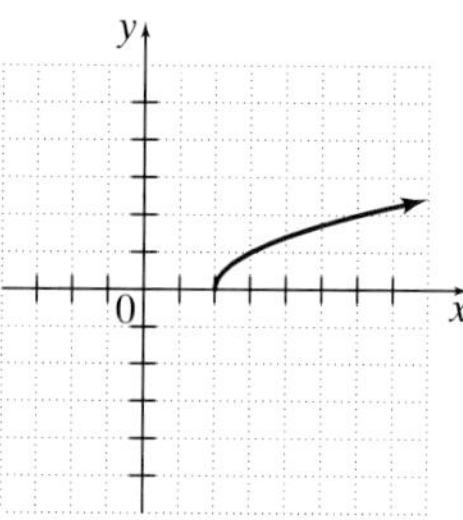

24.

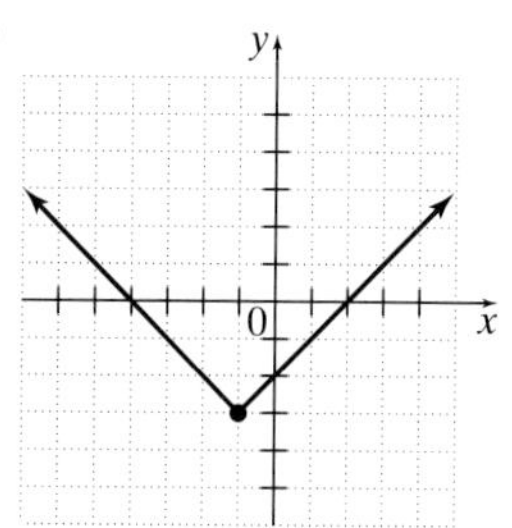

Cumulative Review

1. $12x^3 - 12x^2 - 9x + 2$; Sec. 5.1, Ex. 12 **2.** $2x^2 + 11x + 15$; Sec. 5.2, Ex. 6 **3.** $x^3 - 4x^2 - 3x + 6 + \frac{22}{x + 2}$; Sec. 5.3, Ex. 8

4. $(x - 5)(2 + 3a)$; Sec. 5.4, Ex. 6 **5.** $(x - 5)(x - 7)$; Sec. 5.5, Ex. 2 **6.** $(x + 9)(x - 3)$; Sec. 5.6, Ex. 10 **7.** $\left\{-5, \frac{1}{2}\right\}$; Sec. 5.7, Ex. 2

8. 1; Sec. 6.1, Ex. 6 **9.** −1; Sec. 6.1, Ex. 7 **10.** $x - 7$; Sec. 6.2, Ex. 3 **11.** $-\frac{1}{3y^2}$; Sec. 6.2, Ex. 4 **12.** $\frac{x(x - 2)}{2(x + 2)}$; Sec. 6.3, Ex. 2

13. $\{-6, -1\}$; Sec. 6.4, Ex. 5 **14.** 2; Sec. 6.5, Ex. 2 **15.** 0; Sec. 7.1, Ex. 5 **16.** 0.5; Sec. 7.1, Ex. 7 **17.** $\sqrt{x}$; Sec. 7.2, Ex. 17

18. $\sqrt[3]{rs^2}$; Sec. 7.2, Ex. 19 **19.** $2\sqrt[3]{3}$; Sec. 7.3, Ex. 11 **20.** $2\sqrt[4]{2}$; Sec. 7.3, Ex. 13 **21.** $\frac{2\sqrt{5}}{5}$; Sec. 7.5, Ex. 1

22. $\left\{-\frac{1}{9}, -1\right\}$; Sec. 7.6, Ex. 2 **23.** $13 - 18i$; Sec. 7.7, Ex. 13 **24.** $\{-1 + \sqrt{5}, -1 - \sqrt{5}\}$; Sec. 8.1, Ex. 7

25. $\left\{\frac{2 + \sqrt{10}}{2}, \frac{2 - \sqrt{10}}{2}\right\}$; Sec. 8.2, Ex. 2

Chapter 10

Chapter 10 Pretest

1. $x^2 + 3x + 3$; 10.1A **2.** $25x^2 + 20x + 3$; 10.1B **3.** one-to-one; 10.2A **4.** not one-to-one; 10.2B **5.** $f^{-1}(x) = \frac{x + 12}{7}$; 10.2C

6.

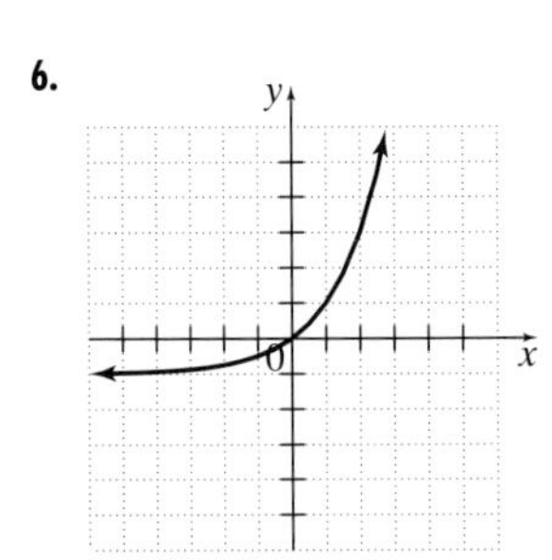

10.3A

7. $\{3\}$; 10.3B **8.** $\left\{-\frac{1}{6}\right\}$; 10.3B **9.** $\{625\}$; 10.4B **10.** $\{-6\}$; 10.4B **11.** $\{10\}$; 10.4B **12.** 5; 10.4B

13.

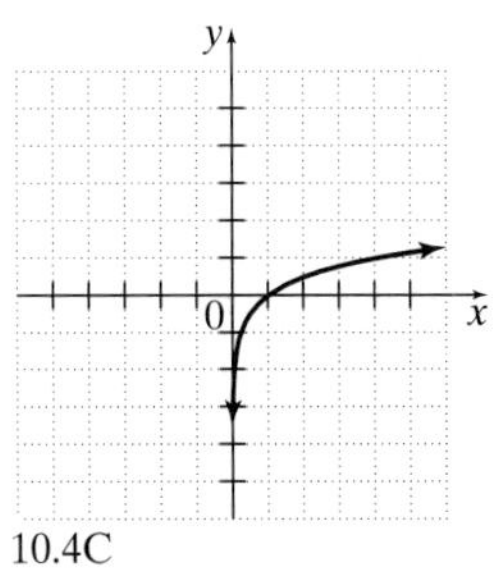

10.4C

14. $\log_5 6$; 10.5A **15.** $\log_6 \frac{a^2}{(a + 1)^7}$; 10.5D **16.** $\log_7 3 + \log_7 y - \log_7 5 - 2\log_7 x$; 10.5D **17.** -2; 10.6B **18.** 8; 10.6B

19. 2.4650 **20.** $\left\{-\frac{21}{5}\right\}$

Exercise Set 10.1

1. a. $3x - 6$ **b.** $-x - 8$ **c.** $2x^2 - 13x - 7$ **d.** $\frac{x - 7}{2x + 1}$ where $x \neq -\frac{1}{2}$ **3. a.** $x^2 + 5x + 1$ **b.** $x^2 - 5x + 1$ **c.** $5x^3 + 5x$ **d.** $\frac{x^2 + 1}{5x}$ where $x \neq 0$ **5. a.** $\sqrt{x} + x + 5$ **b.** $\sqrt{x} - x - 5$ **c.** $x\sqrt{x} + 5\sqrt{x}$ **d.** $\frac{\sqrt{x}}{x + 5}$ where $x \neq -5$ **7. a.** $5x^2 - 3x$ **b.** $-5x^2 - 3x$ **c.** $-15x^3$ **d.** $-\frac{3}{5x}$ where $x \neq 0$ **9.** 42 **11.** -18 **13.** 0 **15.** $(f \circ g)(x) = 25x^2 + 1$; $(g \circ f)(x) = 5x^2 + 5$ **17.** $(f \circ g)(x) = 2x + 11$; $(g \circ f)(x) = 2x + 4$ **19.** $(f \circ g)(x) = -8x^2 - 2x - 2$; $(g \circ f)(x) = -2x^3 - 2x + 4$ **21.** $(f \circ g)(x) = \sqrt{-5x + 2}$; $(g \circ f)(x) = -5\sqrt{x} + 2$ **23.** $H(x) = (g \circ h)(x)$ **25.** $F(x) = (h \circ f)(x)$ **27.** $G(x) = (f \circ g)(x)$ **29.** answers may vary **31.** answers may vary **33.** answers may vary **35.** $y = x - 2$ **37.** $y = \frac{x}{3}$ **39.** $y = -\frac{x + 7}{2}$ **41.** $P(x) = R(x) - C(x)$

Exercise Set 10.2

1. one-to-one; $f^{-1} = \{(-1, -1), (1, 1), (2, 0), (0, 2)\}$ **3.** one-to-one; $h^{-1} = \{(10, 10)\}$ **5.** one-to-one; $f^{-1} = \{(12, 11), (3, 4), (4, 3), (6, 6)\}$ **7.** not one-to-one **9.** one-to-one **11. a.** 3 **b.** 1 **13. a.** 1 **b.** -1 **15.** one-to-one **17.** not one-to-one **19.** one-to-one **21.** not one-to-one

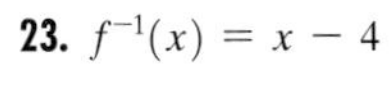

23. $f^{-1}(x) = x - 4$

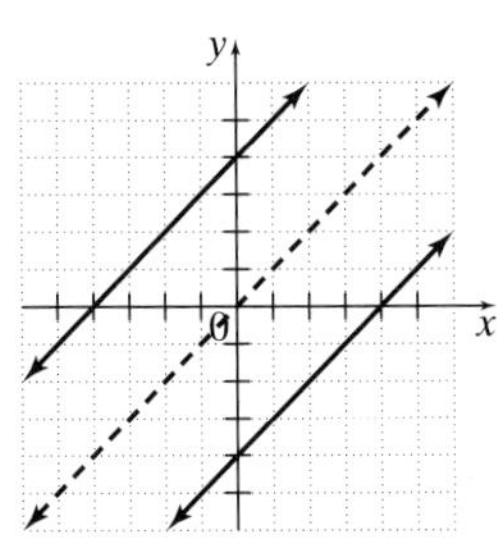

25. $f^{-1}(x) = \frac{x + 3}{2}$

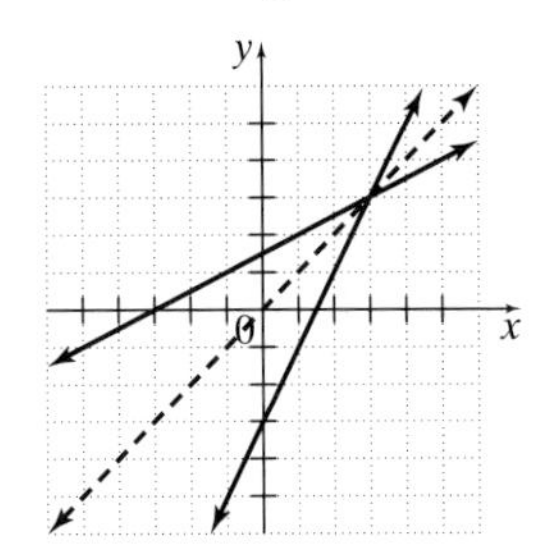

27. $f^{-1}(x) = 2x + 2$

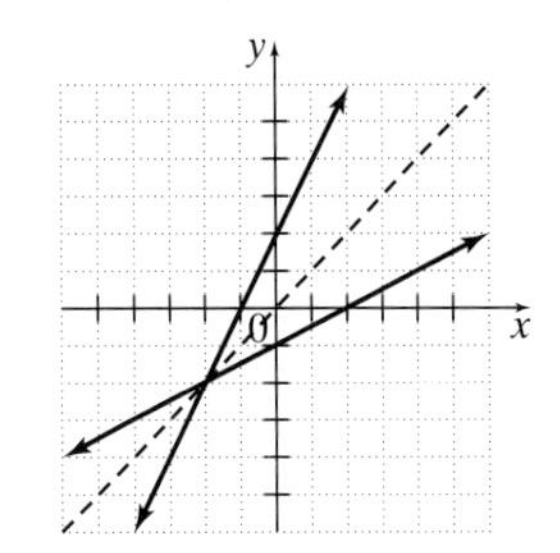

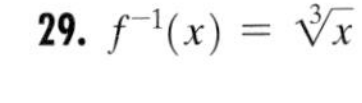

29. $f^{-1}(x) = \sqrt[3]{x}$

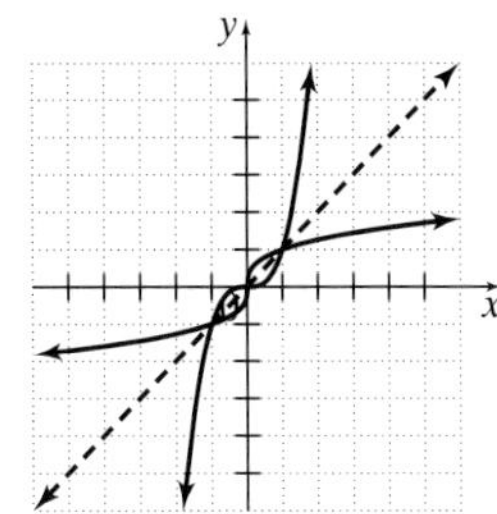

ANSWERS

31. $f^{-1}(x) = 5x + 2$ **33.** $f^{-1}(x) = x^3$ **35.** $f^{-1}(x) = \frac{5}{3x} - \frac{1}{3}$ **37.** $f^{-1}(x) = \sqrt[3]{x} - 2$ **39.**

41.

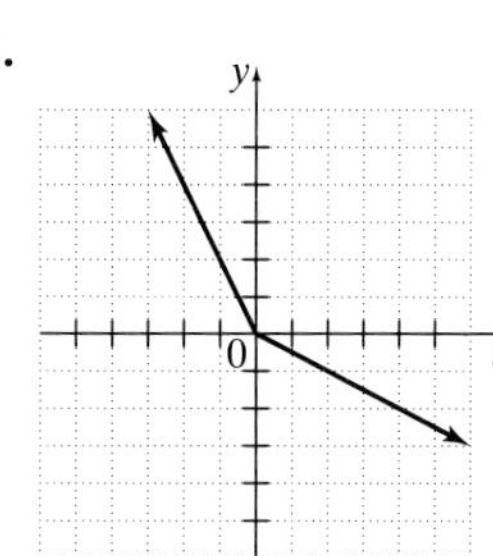

43. 5 **45.** 8 **47.** $\frac{1}{27}$ **49.** 9 **51.** $3^{1/2} \approx 1.73$ **53. a.** $\left(-2, \frac{1}{4}\right), \left(-1, \frac{1}{2}\right), (0, 1), (1, 2), (2, 5)$
b. $\left(\frac{1}{4}, -2\right), \left(\frac{1}{2}, -1\right), (1, 0), (2, 1), (5, 2)$

c.

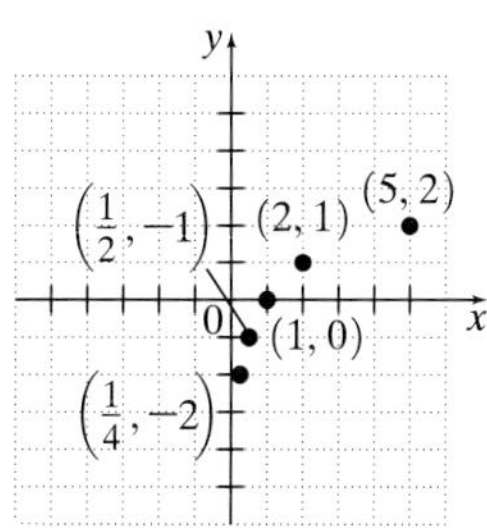

d.

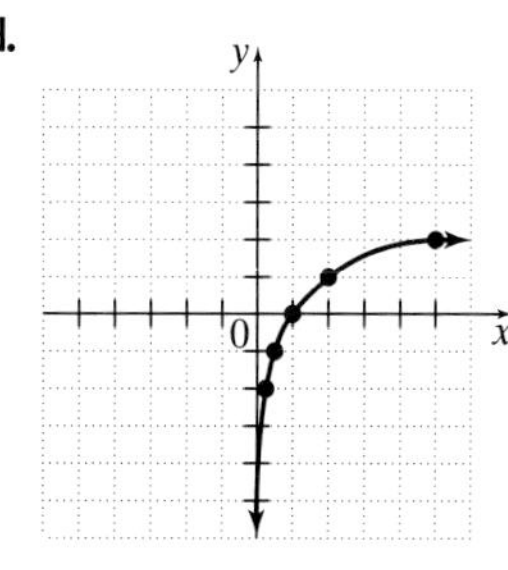

55. $f^{-1}(x) = \frac{x - 1}{3}$

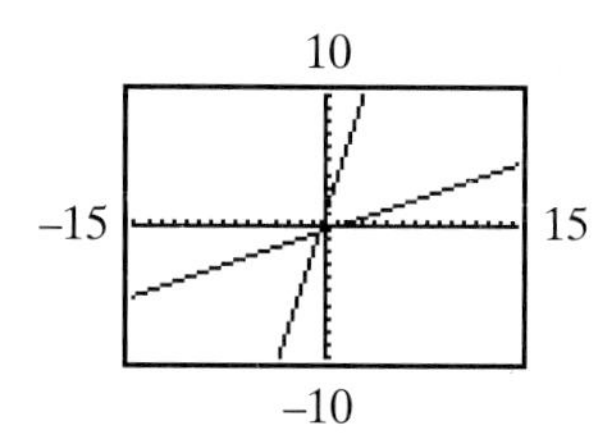

57. $f^{-1}(x) = x^3 - 3$

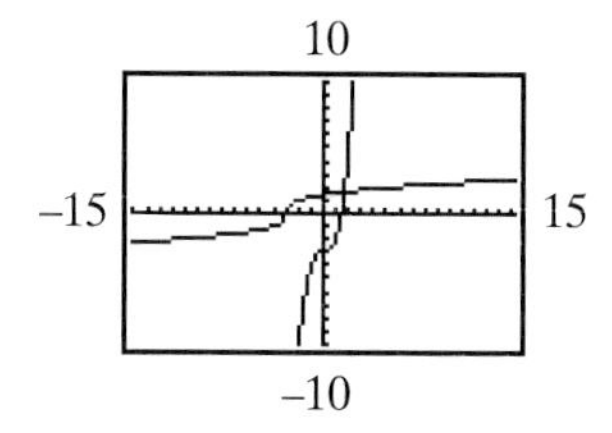

Calculator Explorations

1. 81.98% **2.** 22.54%

Exercise Set 10.3

1.

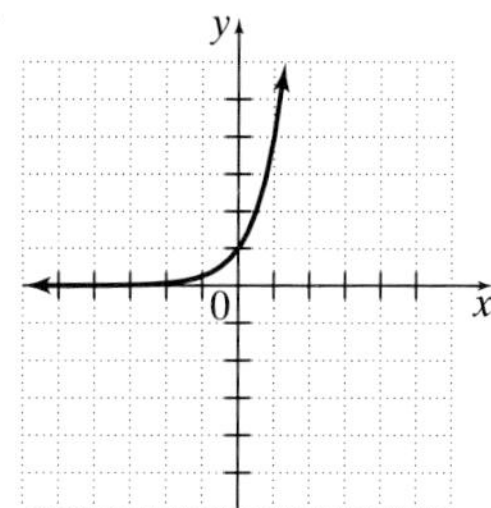

3.

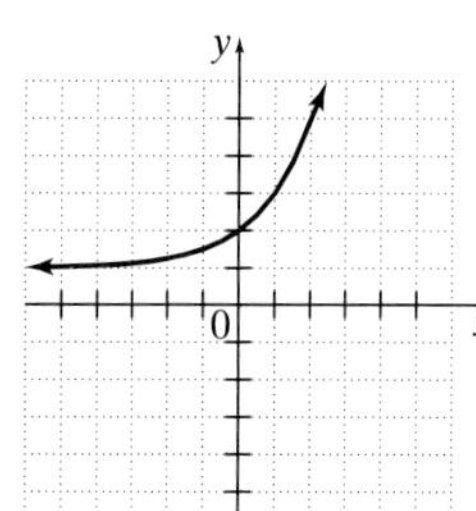

5.

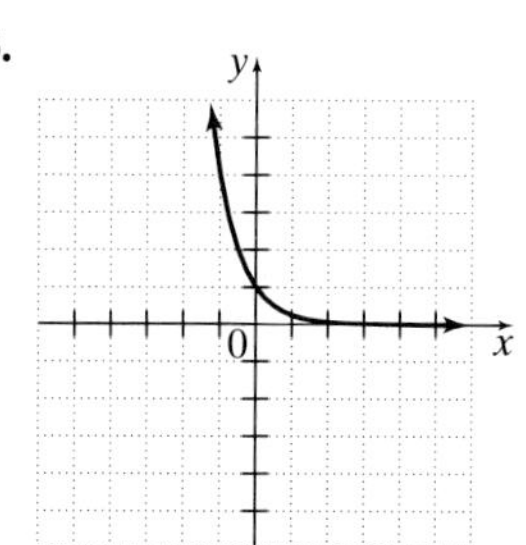

7.

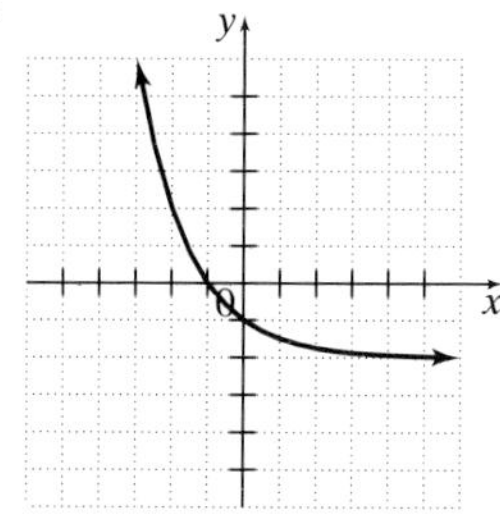

9.

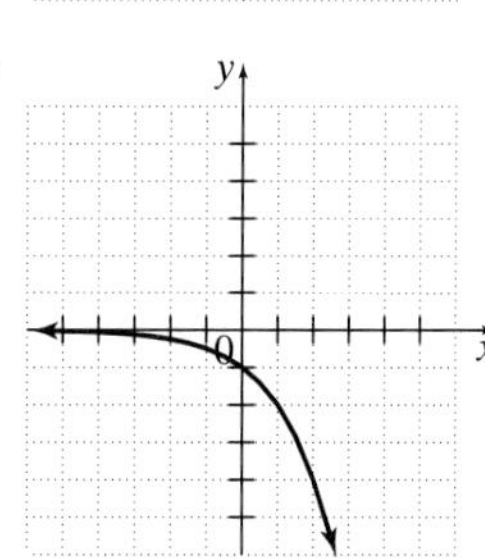

11.

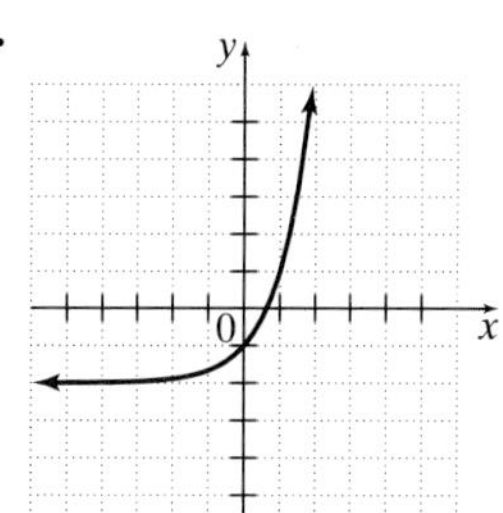

13.

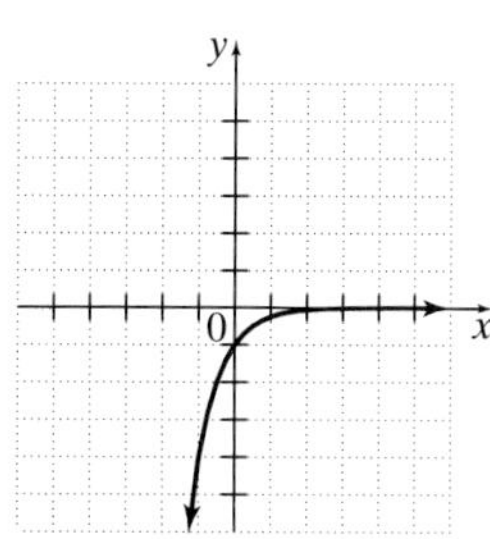

15.

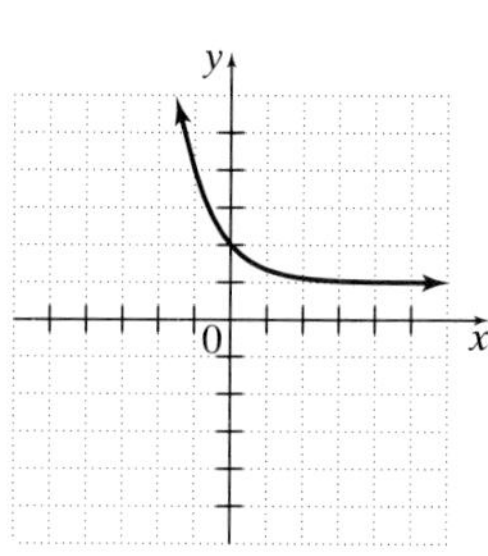

17. $\{3\}$ **19.** $\left\{\frac{3}{4}\right\}$ **21.** $\left\{\frac{8}{5}\right\}$ **23.** $\left\{-\frac{2}{3}\right\}$ **25.** $\left\{\frac{3}{2}\right\}$ **27.** $\left\{-\frac{1}{3}\right\}$ **29.** $\{-2\}$ **31.** 24.6 lb **33.** 333 bison
35. a. \$17.4 billion **b.** \$1490 billion **37.** approximately 1286 million cellular phone users **39.** \$7621.42 **41.** $\{4\}$ **43.** $\varnothing$
45. C **47.** D **49.** answers may vary **53.** 18.62 lb

Exercise Set 10.4

1. $6^2 = 36$ **3.** $3^{-3} = \frac{1}{27}$ **5.** $10^3 = 1000$ **7.** $e^4 = x$ **9.** $e^{-2} = \frac{1}{e^2}$ **11.** $7^{1/2} = \sqrt{7}$ **13.** $\log_2 16 = 4$ **15.** $\log_{10} 100 = 2$

17. $\log_e x = 3$ **19.** $\log_{10} \frac{1}{10} = -1$ **21.** $\log_4 \frac{1}{16} = -2$ **23.** $\log_5 \sqrt{5} = \frac{1}{2}$ **25.** 3 **27.** -2 **29.** $\frac{1}{2}$ **31.** -1 **33.** 0 **35.** 4 **37.** 2 **39.** 5 **41.** 4 **43.** -3 **45.** answers may vary **47.** $\{2\}$ **49.** $\{81\}$ **51.** $\{7\}$ **53.** $\{-3\}$ **55.** $\{-3\}$ **57.** $\{2\}$ **59.** $\{2\}$ **61.** $\left\{\frac{27}{64}\right\}$ **63.** $\{10\}$ **65.** 3 **67.** 3 **69.** 1

71.

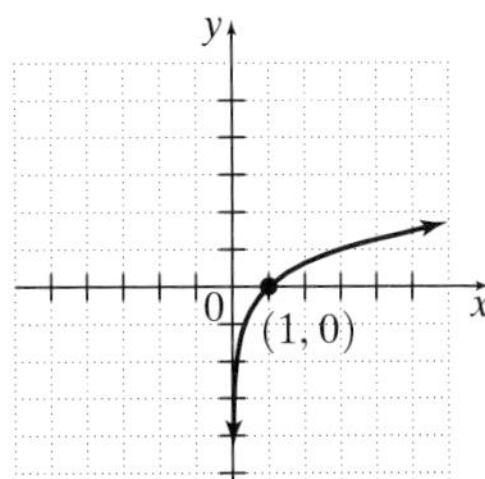

73.

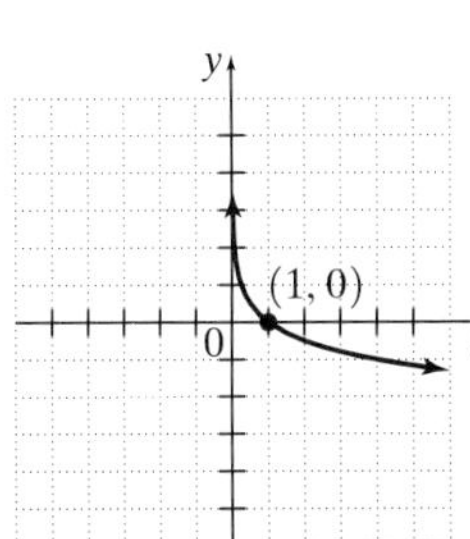

75.

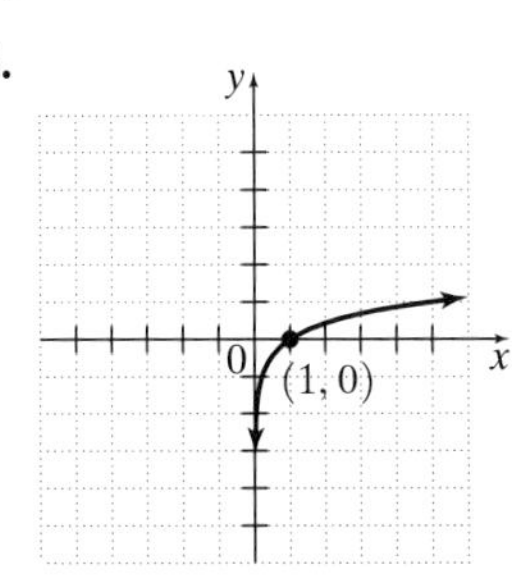

77.

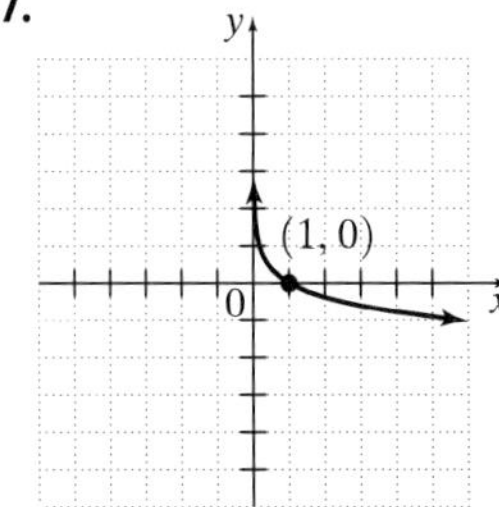

79. 1 **81.** $\frac{x-4}{2}$ **83.**

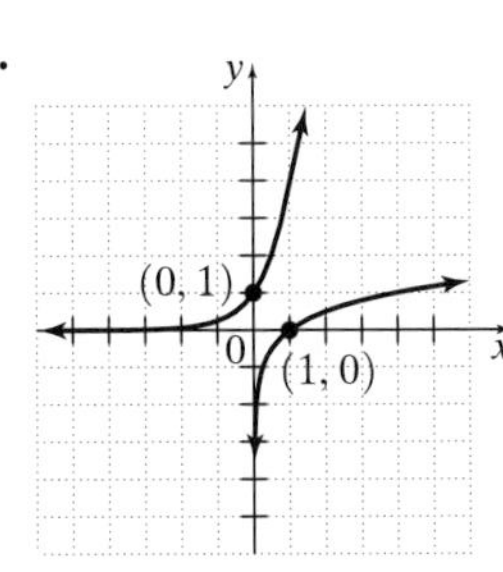

85.

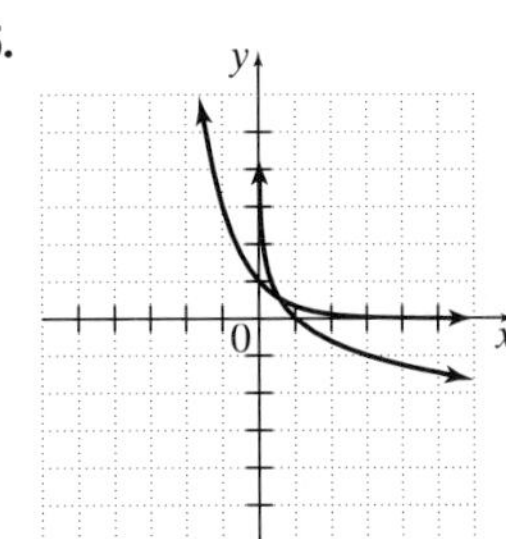

87. 0.0827 **89.** 2 and 3

Exercise Set 10.5

1. $\log_5 14$ **3.** $\log_4 9x$ **5.** $\log_{10}(10x^2 + 20)$ **7.** $\log_5 3$ **9.** $\log_2 \frac{x}{y}$ **11.** $\log_3 4$ **13.** $2\log_3 x$ **15.** $-1\log_4 5$ **17.** $\frac{1}{2}\log_5 y$ **19.** $\log_2 5$ **21.** $\log_5 x^3z^6$ **23.** $\log_{10} \frac{x^3 - 2x}{x + 1}$ **25.** $\log_4 4$, or 1 **27.** $\log_7 \frac{9}{2}$ **29.** $\log_4 48$ **31.** $\log_2 \frac{x^{7/2}}{(x+1)^2}$ **33.** $\log_8 x^{16/3}$ **35.** $\log_3 4 + \log_3 y - \log_3 5$ **37.** $\log_2 x - \log_2 y$ **39.** $\frac{1}{2}\log_b 7 + \frac{1}{2}\log_b x$ **41.** $\log_7 5 + \log_7 x - \log_7 4$ **43.** $3\log_5 x + \log_5(x + 1)$ **45.** $2 \log_6 x - \log_6(x + 3)$ **47.** 0.2 **49.** 1.2 **51.** 0.35 **53.** 1.29 **55.** -0.68 **57.** -0.125 **59.** **61.** $\{-1\}$ **63.** $\left\{\frac{1}{2}\right\}$ **65.** false **67.** true **69.** false

Integrated Review

1. $x^2 + x - 5$ **2.** $-x^2 + x - 7$ **3.** $x^3 - 6x^2 + x - 6$ **4.** $\frac{x-6}{x^2+1}$ **5.** $\sqrt{3x - 1}$ **6.** $3\sqrt{x} - 1$ **7.** one-to-one; $\{(6, -2), (8, 4), (-6, 2), (3, 3)\}$ **8.** not one-to-one **9.** not one-to-one **10.** one-to-one **11.** not one-to-one **12.** $f^{-1}(x) = \frac{x}{3}$ **13.** $f^{-1}(x) = x - 4$ **14.** $f^{-1}(x) = \frac{x+1}{5}$

15.

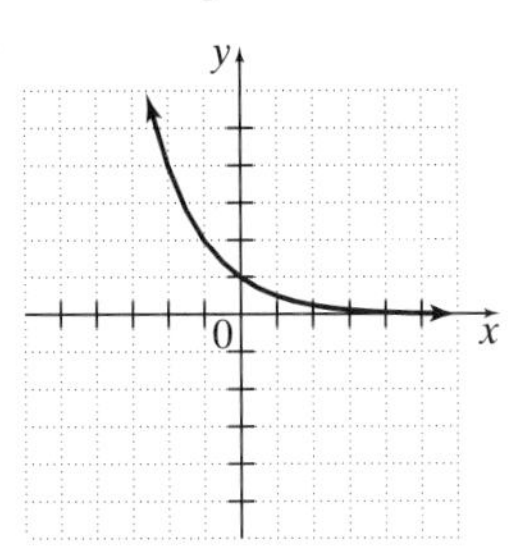

16.

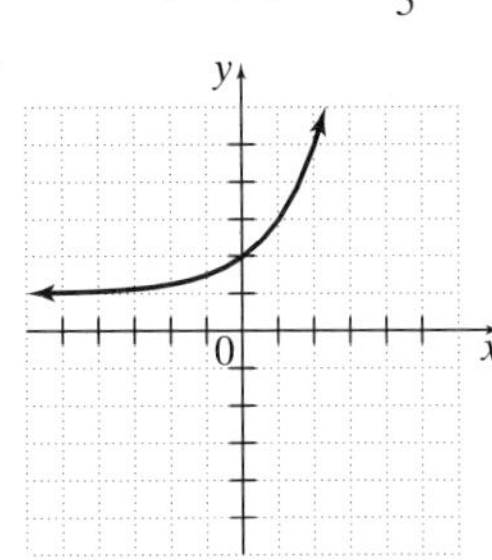

17.

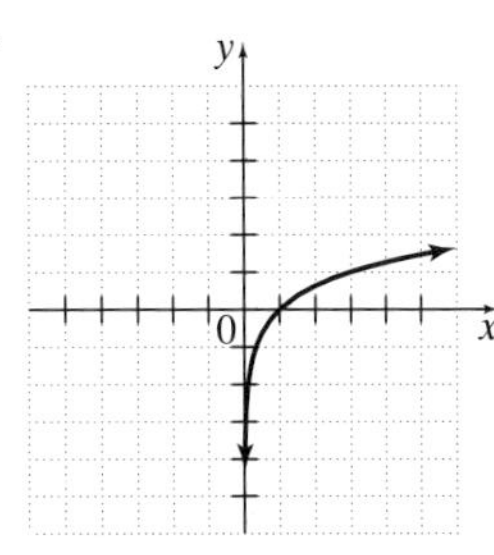

18.

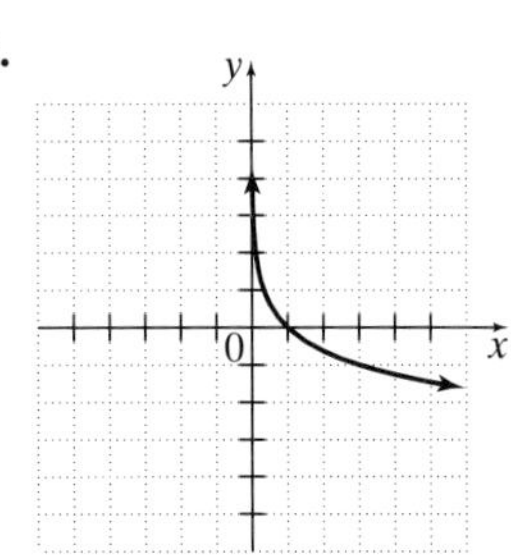

19. $\{3\}$ **20.** $\{-8\}$ **21.** $\{2\}$ **22.** $\left\{\frac{1}{2}\right\}$ **23.** $\{32\}$ **24.** $\{4\}$ **25.** $\{5\}$ **26.** $\left\{\frac{1}{9}\right\}$ **27.** $\log_5 \frac{x^3}{y^5}$ **28.** $\log_2 \frac{x^2 - 3x}{x^2 + 4}$

ANSWERS

Exercise Set 10.6

1. 0.9031 **3.** 0.3636 **5.** 0.6931 **7.** −2.6367 **9.** 1.1004 **11.** 1.6094 **13.** 1.6180 **15.** answers may vary **17.** 2 **19.** −3 **21.** 2 **23.** $\frac{1}{4}$ **25.** 3 **27.** 2 **29.** −4 **31.** $\frac{1}{2}$ **33.** $\{10^{1.3}\}$; {19.9526} **35.** $\left\{\frac{10^{1.1}}{2}\right\}$; {6.2946} **37.** $\{e^{1.4}\}$; {4.0552} **39.** $\left\{\frac{4+e^{2.3}}{3}\right\}$; {4.6581} **41.** $\{10^{2.3}\}$; {199.5262} **43.** $\{e^{-2.3}\}$; {0.1003} **45.** $\left\{\frac{10^{-0.5}-1}{2}\right\}$; {−0.3419} **47.** $\left\{\frac{e^{0.18}}{4}\right\}$; {0.2993} **49.** \$3656.38 **51.** \$2542.50 **53.** 1.5850 **55.** −2.3219 **57.** 1.5850 **59.** −1.6309 **61.** 0.8617 **63.** $\left\{\frac{4}{7}\right\}$ **65.** $x = \frac{3y}{4}$ **67.** {−6, −1} **69.**

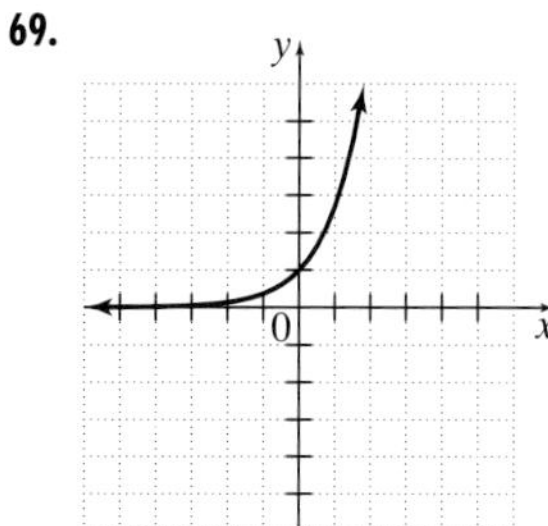

71.

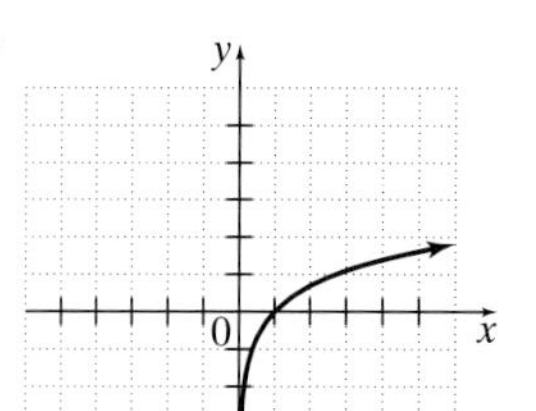

73. answers may vary **75.** 4.2 **77.** 5.3

Calculator Explorations

1. 3.67 yr, or 3 yr and 8 mo **3.** 23.16 yr, or 23 yr and 2 mo

Exercise Set 10.7

1. $\left\{\frac{\log 6}{\log 3}\right\}$; {1.6309} **3.** $\left\{\frac{\log 3.8}{2\log 3}\right\}$; {0.6076} **5.** $\left\{3 + \frac{\log 5}{\log 2}\right\}$; {5.3219} **7.** $\left\{\frac{\log 5}{\log 9}\right\}$; {0.7325} **9.** $\left\{\frac{\log 3}{\log 4} - 7\right\}$; {−6.2075} **11.** $\left\{\frac{1}{3}\left(4 + \frac{\log 11}{\log 7}\right)\right\}$; {1.7441} **13.** $\left\{\frac{\ln 5}{6}\right\}$; {0.2682} **15.** {11} **17.** $\left\{\frac{1}{2}\right\}$ **19.** $\left\{\frac{3}{4}\right\}$ **21.** {2} **23.** $\left\{\frac{1}{8}\right\}$ **25.** {11} **27.** {4, −1} **29.** $\left\{\frac{2}{3}\right\}$ **31.** 103 wolves **33.** 11,094,996 inhabitants **35.** 40.4 yr **37.** 9.9 yr **39.** 1.7 yr **41.** 8.8 yr **43.** 55.7 in. **45.** 11.9 lb/sq. in. **47.** 3.2 mi **49.** 12 weeks **51.** 18 weeks **53.** $-\frac{5}{3}$ **55.** $\frac{17}{4}$ **57.** 4.8% **59.** {6.93}

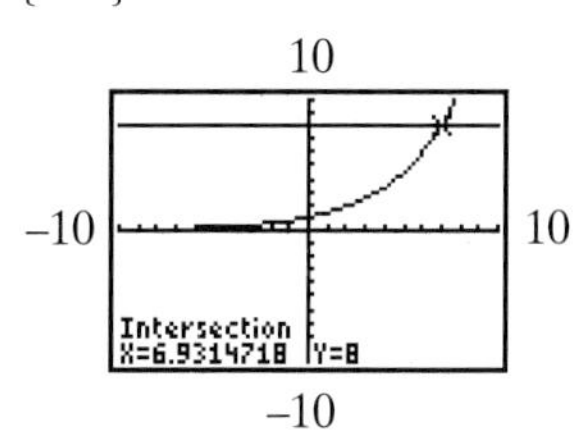

Chapter 10 Review

1. $3x - 4$ **2.** $-x - 6$ **3.** $2x^2 - 9x - 5$ **4.** $\frac{2x+1}{x-5}$ **5.** $x^2 + 2x - 1$ **6.** $x^2 - 1$ **7.** 18 **8.** $x^4 - 4x^2 + 2$ **9.** −2 **10.** 48 **11.** one-to-one; $h^{-1} = \{(14, -9), (8, 6), (12, -11), (15, 15)\}$ **12.** not one-to-one **13.** one-to-one;

Rank in Automobile Thefts (Input)	2	4	1	3
U.S. Region (Output)	West	Midwest	South	Northeast

14. not one-to-one **15.** not one-to-one **16.** not one-to-one **17.** not one-to-one **18.** one-to-one **19.** $f^{-1}(x) = \frac{x-11}{6}$ **20.** $f^{-1}(x) = \frac{x}{12}$ **21.** $f^{-1}(x) = \frac{x+5}{3}$ **22.** $f^{-1}(x) = \frac{x-1}{2}$ **23.**

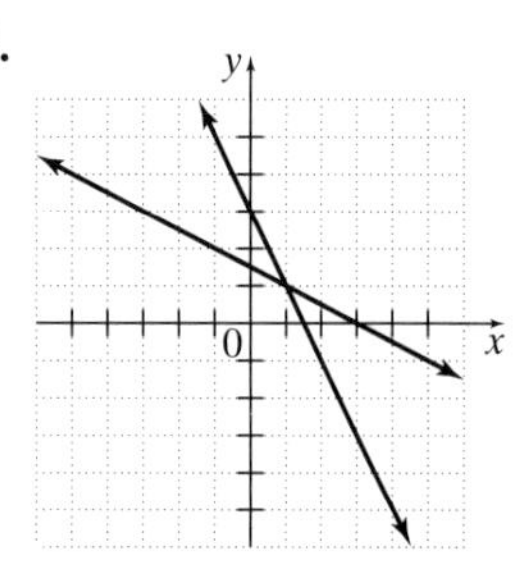

24.

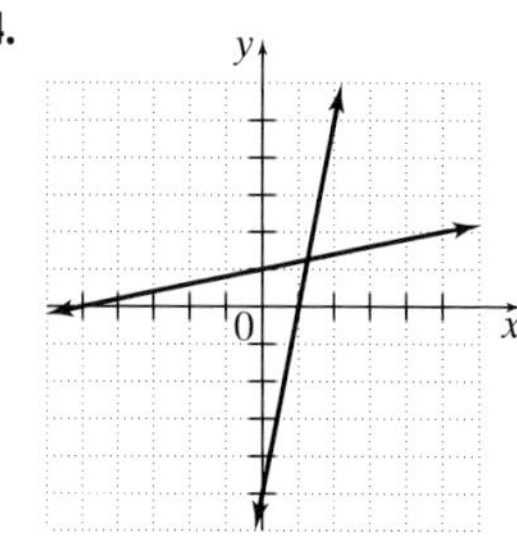

25. {3} **26.** {−2} **27.** $\left\{-\frac{4}{3}\right\}$ **28.** $\left\{\frac{3}{2}\right\}$ **29.** $\left\{\frac{3}{2}\right\}$ **30.** $\left\{\frac{8}{9}\right\}$

31.

32.

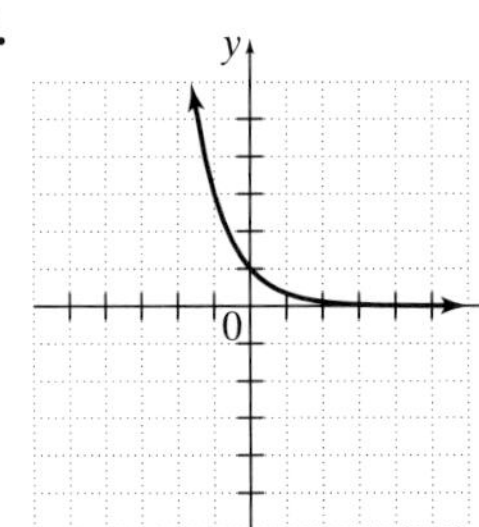

33.

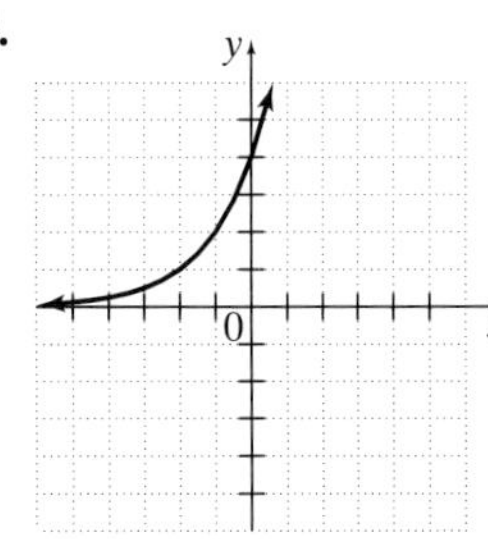

34.

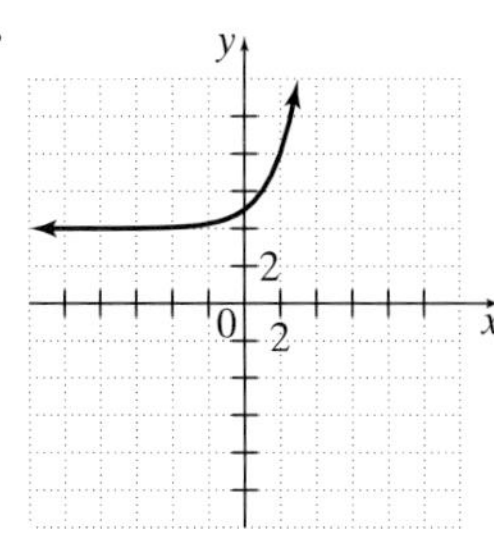

35. \$2963.11 **36.** \$1131.82 **37.** $\log_7 49 = 2$ **38.** $\log_2\left(\frac{1}{16}\right) = -4$ **39.** $\left(\frac{1}{2}\right)^{-4} = 16$ **40.** $0.4^3 = 0.064$ **41.** $\left\{\frac{1}{64}\right\}$

42. $\{9\}$ **43.** $\{0\}$ **44.** $\{3\}$ **45.** $\{8\}$ **46.** $\{3\}$ **47.** $\{5\}$ **48.** $\{-2\}$ **49.** $\{4\}$ **50.** $\{9\}$ **51.** $\left\{\frac{17}{3}\right\}$ **52.** $\{2\}$

53. $\{-1, 4\}$ **54.** $\{-8, 1\}$

55.

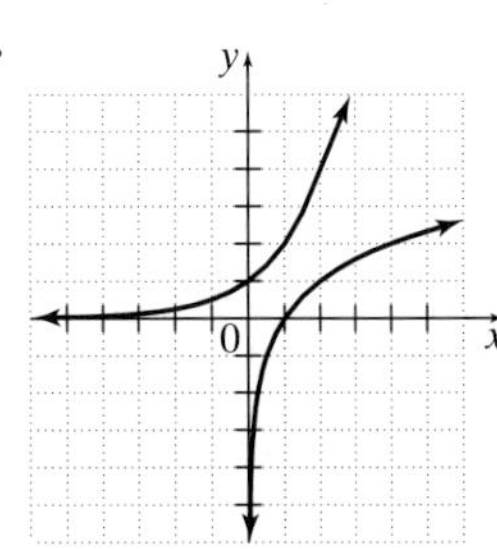

56.

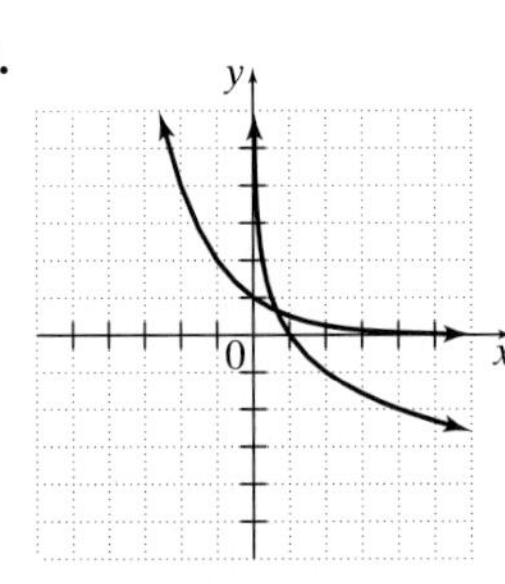

57. $\log_3 32$ **58.** $\log_2 18$ **59.** $\log_7 \frac{3}{4}$ **60.** $\log\left(\frac{3}{2}\right)$

61. $\log_{11} 4$ **62.** $\log_5 2$ **63.** $\log_5 \frac{x^3}{(x + 1)^2}$

64. $\log_3 (x^4 + 2x^3)$ **65.** $3\log_3 x - \log_3(x + 2)$

66. $\log_4(x + 5) - 2\log_4 x$

67. $\log_2 3 + 2\log_2 x + \log_2 y - \log_2 z$

68. $\log_7 y + 3\log_7 z - \log_7 x$ **69.** 2.02 **70.** −0.11

71. 0.5563 **72.** −0.8239 **73.** 0.2231 **74.** 1.5326 **75.** 3

76. −1 **77.** −1 **78.** 4 **79.** $\left\{\frac{e^2}{2}\right\}$ **80.** $\left\{\frac{e^{1.6}}{3}\right\}$ **81.** $\left\{\frac{e^{-1} + 3}{2}\right\}$ **82.** $\left\{\frac{e^2 - 1}{3}\right\}$ **83.** 0.2920 **84.** 1.2619

85. \$1957.30 **86.** \$1307.51 **87.** $\left\{\frac{\log 7}{2\log 3}\right\}$; $\{0.8856\}$ **88.** $\left\{\frac{\log 5}{3\log 6}\right\}$; $\{0.2994\}$ **89.** $\left\{\frac{1}{2}\left(\frac{\log 6}{\log 3} - 1\right)\right\}$; $\{0.3155\}$

90. $\left\{\frac{1}{3}\left(\frac{\log 9}{\log 4} - 2\right)\right\}$; $\{-0.1383\}$ **91.** $\left\{\frac{1}{3}\left(\frac{\log 4}{\log 5} + 5\right)\right\}$; $\{1.9538\}$ **92.** $\left\{\frac{1}{4}\left(\frac{\log 3}{\log 8} + 2\right)\right\}$; $\{0.6321\}$ **93.** $\left\{-\frac{\log 2}{\log 5} + 1\right\}$; $\{0.5693\}$

94. $\left\{\frac{\log \frac{2}{3}}{\log 4} - 5\right\}$; $\{-5.2925\}$ **95.** $\left\{\frac{25}{2}\right\}$ **96.** $\left\{\frac{9}{10}\right\}$ **97.** $\emptyset$ **98.** $\left\{\frac{3e^2}{e^2 - 3}\right\}$ **99.** $\{2\sqrt{2}\}$ **100.** $\emptyset$ **101.** 197,044 ducks

102. 240,091,435 **103.** 53 yr **104.** 63 yr **105.** 36 yr **106.** 8.8 yr **107.** 8.5 yr

Chapter 10 Test

1. $2x^2 - 3x$ **2.** $3 - x$ **3.** 5 **4.** $x - 7$ **5.** $x^2 - 6x - 2$

6.

7. one-to-one **8.** not one-to-one **9.** one-to-one; $f^{-1}(x) = \frac{-x + 6}{2}$

10. one-to-one; $f^{-1} = \{(0, 0), (3, 2), (5, -1)\}$ **11.** not one-to-one **12.** $\log_3 24$

13. $\log_5\left(\frac{x^4}{x + 1}\right)$ **14.** $\log_6 2 + \log_6 x - 3\log_6 y$ **15.** −1.53 **16.** 1.0686 **17.** $\{-1\}$

18. $\left\{\frac{1}{2}\left(\frac{\log 4}{\log 3} - 5\right)\right\}$; $\{-1.8691\}$ **19.** $\left\{\frac{1}{9}\right\}$ **20.** $\left\{\frac{1}{2}\right\}$ **21.** $\{22\}$ **22.** $\left\{\frac{25}{3}\right\}$ **23.** $\left\{\frac{43}{21}\right\}$

24. $\{-1.0979\}$

25.

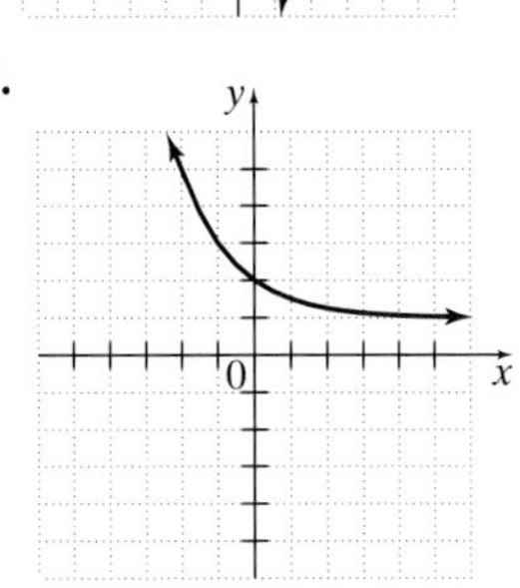

26.

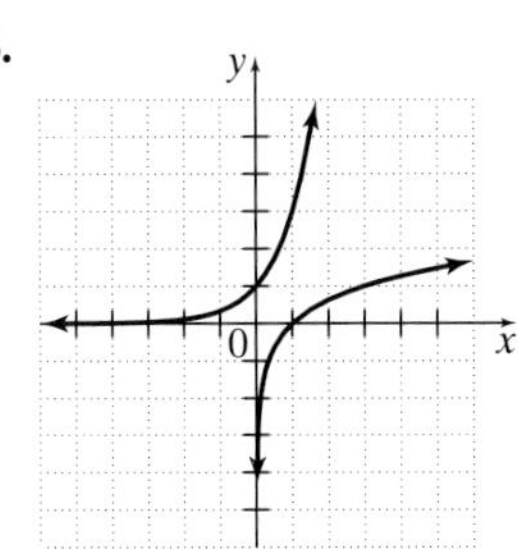

27. \$5234.58 **28.** 6 yr **29.** 64,913 prairie dogs **30.** 15 yr

Cumulative Review

1. 3; Sec. 1.3, Ex. 42 **2.** 2; Sec. 1.3, Ex. 44 **3.** 30°, 40°, and 110°; Sec. 4.3, Ex. 4 **4.** $(x^2 + 5y)(7x - 1)$; Sec. 5.4, Ex. 7

5. $\frac{5k^2 - 7k + 4}{(k + 2)(k - 2)(k - 1)}$; Sec. 6.2, Ex. 9 **6.** 2; Sec. 7.1, Ex. 30 **7.** 1; Sec. 7.1, Ex. 32 **8.** $x^{5/6}$; Sec. 7.2, Ex. 14 **9.** 32; Sec. 7.2, Ex. 16

10. $\frac{\sqrt{x}}{3}$; Sec. 7.3, Ex. 7 **11.** $\frac{\sqrt[4]{3}}{2y}$; Sec. 7.3, Ex. 9 **12.** $2x - 25$; Sec. 7.4, Ex. 13 **13.** $4 - 2\sqrt{3}$; Sec. 7.4, Ex. 14
14. $\frac{7}{3\sqrt{35}}$; Sec. 7.5, Ex. 6 **15.** $\{3\}$; Sec. 7.6, Ex. 4 **16.** $-1 + 5i$; Sec. 7.7, Ex. 8 **17.** $\left\{\frac{6 + i\sqrt{5}}{2}, \frac{6 - i\sqrt{5}}{2}\right\}$; Sec. 8.1, Ex. 9
18. $\{2 + \sqrt{2}, 2 - \sqrt{2}\}$; Sec. 8.2, Ex. 3 **19.** $\{8, 27\}$; Sec. 8.3, Ex. 5 **20.** $\left(-\frac{7}{2}, -1\right)$; Sec. 8.4, Ex. 5

21.

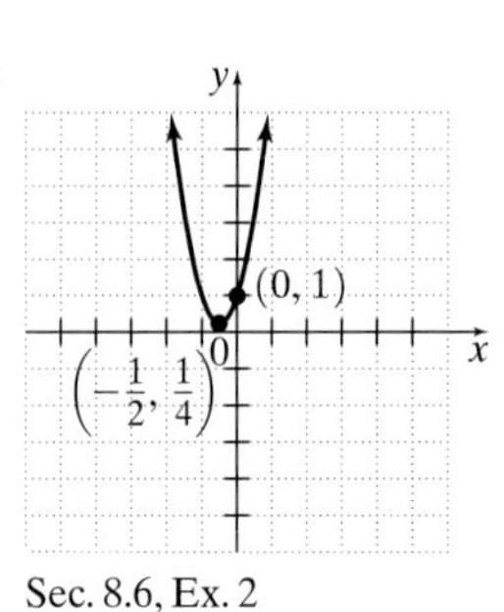

Sec. 8.6, Ex. 2

22. $\left(-1, \frac{3}{2}\right)$; Sec. 9.1, Ex. 6 **23.**

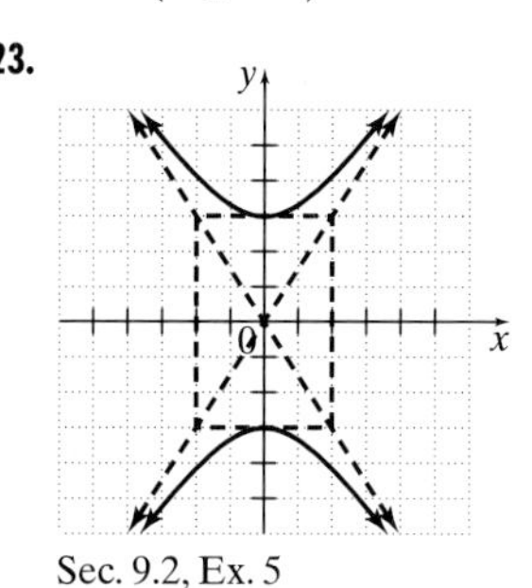

Sec. 9.2, Ex. 5

24. $\{(2, \sqrt{2})\}$; Sec. 9.4, Ex. 2
25. $\left\{\frac{2}{99}\right\}$; Sec. 10.7, Ex. 4

Appendix

Viewing Window and Interpreting Window Settings Exercise Set

1. yes **2.** yes **3.** no **4.** no **5.** answers may vary **6.** answers may vary **7.** answers may vary **8.** answers may vary **9.** answers may vary **10.** answers may vary **11.** Xmin = −12; Xmax = 12; Xscl = 3; Ymin = −12; Ymax = 12; Yscl = 3 **12.** Xmin = −20; Xmax = 20; Xscl = 5; Ymin = −20; Ymax = 20; Yscl = 5 **13.** Xmin = −9; Xmax = 9; Xscl = 1; Ymin = −12; Ymax = 12; Yscl = 2 **14.** Xmin = −27; Xmax = 27; Xscl = 3; Ymin = −6; Ymax = 6; Yscl = 1 **15.** Xmin = −10; Xmax = 10; Xscl = 2; Ymin = −25; Ymax = 25; Yscl = 5 **16.** Xmin = −50; Xmax = 50; Xscl = 10; Ymin = −20; Ymax = 20; Yscl = 4 **17.** Xmin = −10; Xmax = 10; Xscl = 1; Ymin = −30; Ymax = 30; Yscl = 3 **18.** Xmin = −100; Xmax = 100; Xscl = 10; Ymin = −20; Ymax = 20; Yscl = 2 **19.** Xmin = −20; Xmax = 30; Xscl = 5; Ymin = −30; Ymax = 50; Yscl = 10 **20.** Xmin = −500; Xmax = 700; Xscl = 100; Ymin = −800; Ymax = 400; Yscl = 200

Graphing Equations and Square Viewing Window Exercise Set

1. Setting B **2.** Setting B **3.** Setting B **4.** Setting A **5.** Setting B **6.** Setting B

7.

8.

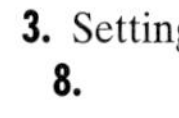

9.

10.

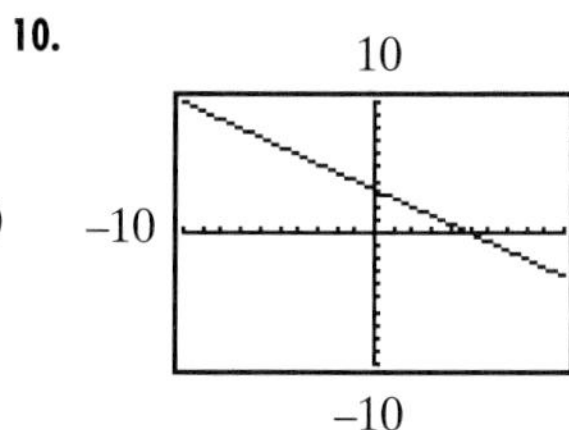

12.

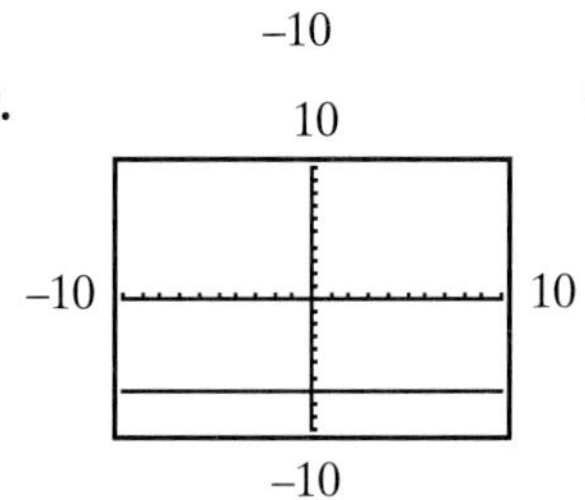

12.

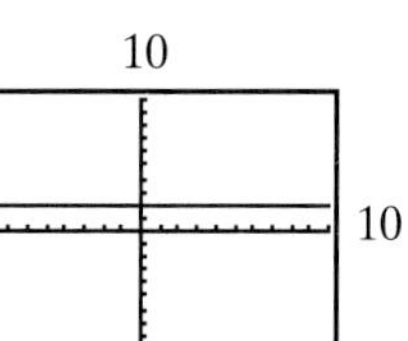

13.

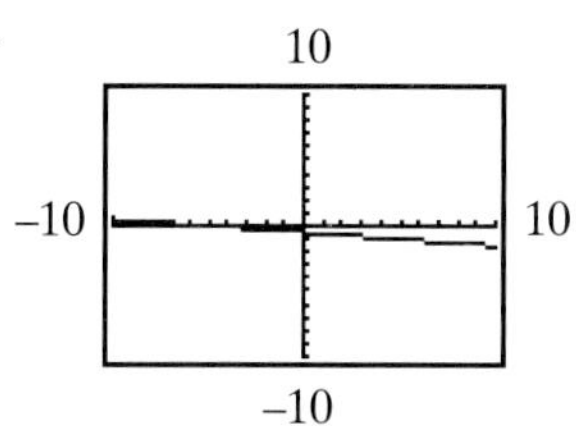

14.

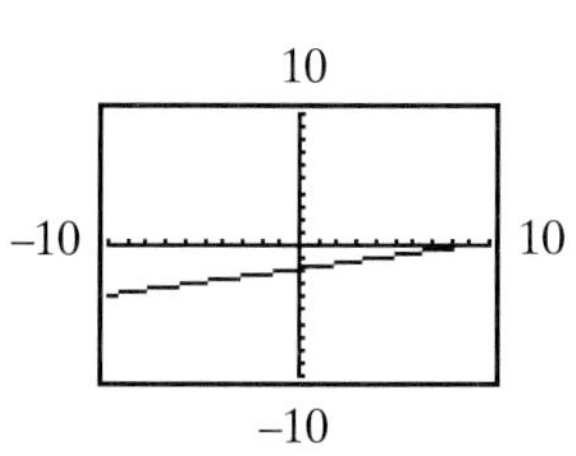

15.

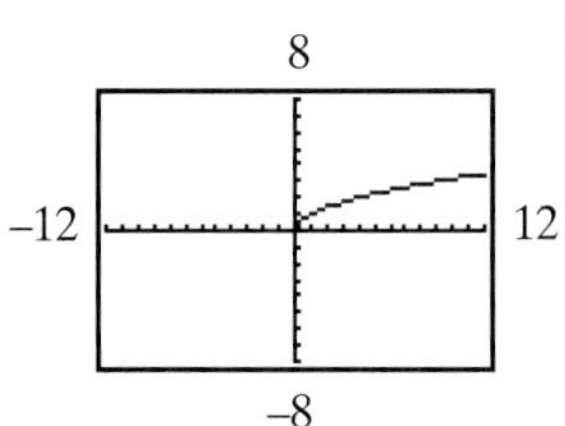

16.

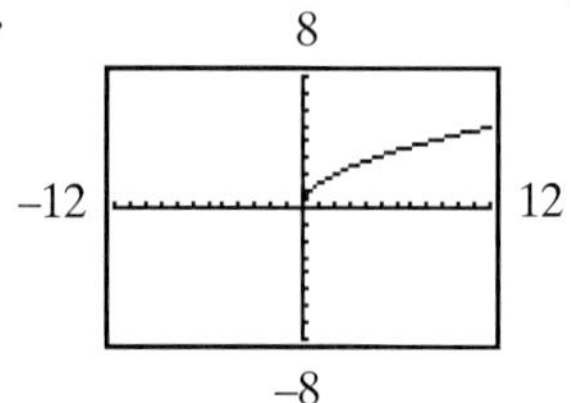

17.

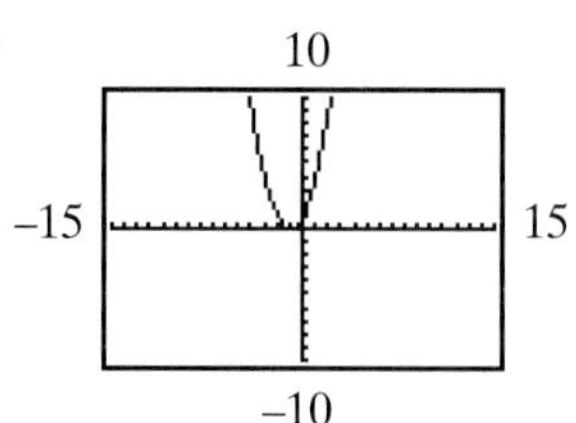

18.

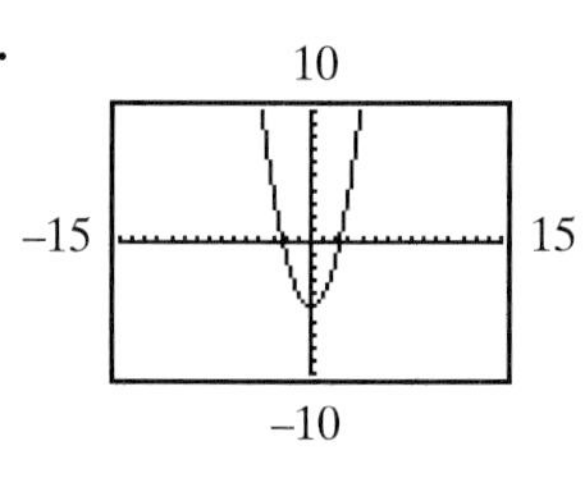

19.

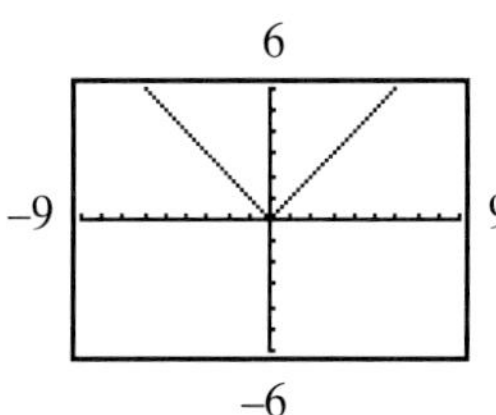

20.

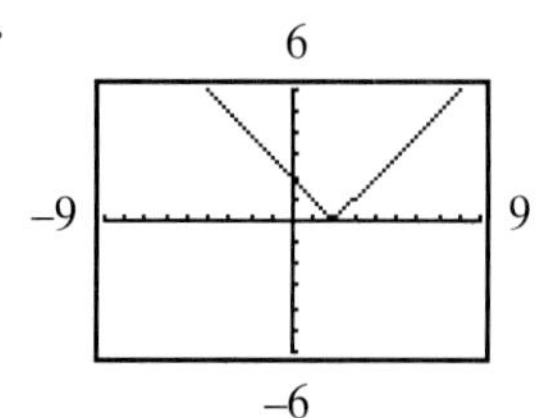

21.

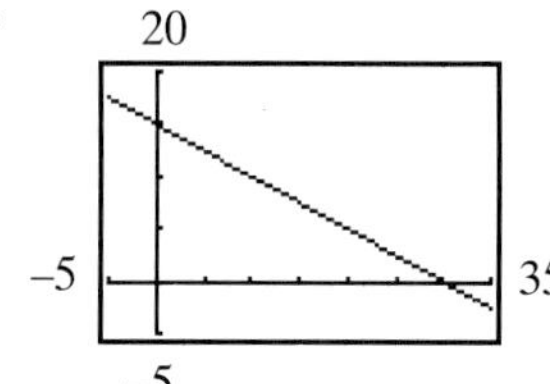

22.

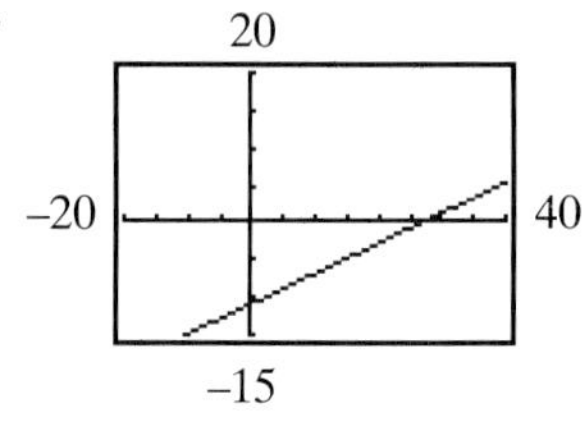

Solutions to Selected Exercises

Chapter 1

Exercise Set 1.1

1. $5x = (5)(7) = 35$

5. $\frac{x}{y} = \frac{18}{3} = 6$

9. $3x + y = (3)(6) + 4 = 18 + 4 = 22$

13. $414t = 414(5) = 2070$
The aircraft flies 2070 miles in 5 hours.

17. $7075t = 7075(5.2) = 36{,}790$
It costs \$36,790 to operate the aircraft for 5.2 hours.

21. $\{11, 12, 13, 14, 15, 16\}$ **25.** $\{0, 2, 4, 6, 8\}$

29. $\{3, \sqrt{36}\}$ **33.** $-11 \in \{x | x \text{ is an integer}\}$

37. $12 \notin \{1, 3, 5, \ldots\}$

41. True. Some real numbers are irrational numbers.

45. $2x$ **49.** $x + 2$

53. $x - 4$ **57.** $10 - x$

61. $x + 9$ **65.** $12 - 3x$

69. $5x - 10$ **73.** answers may vary

77. $8(x - 9)$

Exercise Set 1.2

1. $4c = 7$ **5.** $\frac{n}{5} = 4n$

9. $0 > -2$ **13.** $-7.9 < -7.09$

17. $8.6 > -3.5$

21. $\frac{1}{2} < \frac{5}{8}$ because $\frac{1}{2} = 0.5$ and $\frac{5}{8} = 0.625$ and $0.5 < 0.625$.

25. $-6 \leq 0$ is true since $-6 < 0$ is true.

29. $-14 \geq -1$ is false since neither $-14 > -1$ nor $-14 = -1$ is true.

33. The opposite of -6.2 is $-(-6.2) = 6.2$.

37. The opposite of $-\frac{2}{3}$ is $-\left(-\frac{2}{3}\right) = \frac{2}{3}$.

41. The reciprocal of 5 is $\frac{1}{5}$.

45. The reciprocal of $-\frac{1}{4}$ is $-\frac{4}{1}$ or -4.

49. The reciprocal of $\frac{7}{8}$ is $\frac{8}{7}$.

53. $\frac{2}{3}; -\frac{2}{3}; \frac{3}{2}$

57. $7x + y = y + 7x$

61. $\frac{1}{3} \cdot \frac{x}{5} = \frac{x}{5} \cdot \frac{1}{3}$

65. $(x + 1.2) + y = x + (1.2 + y)$

69. $3(x + 5) = 3 \cdot x + 3 \cdot 5 = 3x + 15$

73. $2(6x + 5y + 2z) = 2 \cdot 6x + 2 \cdot 5y + 2 \cdot 2z = 12x + 10y + 4z$

77. $6x(y - 4) = 6x \cdot y - 6x \cdot 4 = 6xy - 24x$

81. 0 **85.** $(10 \cdot 2)y$

89. $4 + 8y = 8y + 4$ **93.** answers may vary

Exercise Set 1.3

1. $|2| = 2$ **5.** $|0| = 0$

9. $-|-2| = -2$ **13.** $-14 + (-10) = -24$

17. $13 - 17 = -4$

21. $19 - 10 - 11 = 9 - 11 = -2$

25. $-\frac{4}{5} - \left(-\frac{3}{10}\right) = -\frac{8}{10} + \frac{3}{10} = -\frac{5}{10} = -\frac{1}{2}$

29. $-4 + 7 = 3$

33. $-4 - (-19) = -4 + 19 = 15$

37. $16 - 8 - 9 = 8 - 9 = -1$

41. $-5 \cdot 12 = -60$ **45.** $-8(-10) = 80$

49. $-17 \cdot 0 = 0$ **53.** $\frac{-9}{3} = -3$

57. $\frac{-12}{-4} = 3$ **61.** $\frac{0}{-5} = 0$

65. $-4(-2)(-1) = 8(-1) = -8$

69. $-\frac{6}{7} \div 2 = -\frac{6}{7} \cdot \frac{1}{2} = -\frac{3}{7}$

73. $-\frac{1}{6} \div \frac{9}{10} = -\frac{1}{6} \cdot \frac{10}{9} = -\frac{10}{54} = -\frac{5}{27}$

77. $-2(-3.6) = 7.2$

81. $\frac{-5.2}{-1.3} = 4$

85. $\frac{3}{5} \div \left(-\frac{2}{5}\right) = \frac{3}{5} \cdot \left(-\frac{5}{2}\right) = -\frac{15}{10} = -\frac{3}{2}$

89. $-7^2 = -(7)(7) = -49$

93. $(-2)^3 = (-2)(-2)(-2) = 4(-2) = -8$

97. answers may vary

101. $\sqrt{64} = 8$ because $8^2 = 64$.

105. $\sqrt{\frac{1}{16}} = \frac{1}{4}$ because $\left(\frac{1}{4}\right)^2 = \frac{1}{16}$

109. $\sqrt[4]{81} = 3$ because $3^4 = 81$.

113. $1 - \frac{1}{5} - \frac{3}{7} = \frac{35}{35} - \frac{7}{35} - \frac{15}{35} = \frac{28}{35} - \frac{15}{35} = \frac{13}{35}$

117. b

121. Yes. Two players have 6 points each (third player has 0 points), or two players have 5 points each (third has 2 points).

125. $\sqrt{19.6} \approx 4.4272$ **129.** $12.9 - 2.1 = 10.8\%$

Exercise Set 1.4

1. $3(5-7)^4 = 3(-2)^4 = 3(16) = 48$

5. $\dfrac{3-(-12)}{-5} = \dfrac{3+12}{-5} = \dfrac{15}{-5} = -3$

9. $(-3)^2 + 2^3 = 9 + 8 = 17$

13. $-9 \cdot 8 + 5(-6) = -72 + (-30) = -102$

17. $-8\left(-\dfrac{3}{4}\right) - 8 = \dfrac{24}{4} - 8 = 6 - 8 = -2$

21. $5^2 - 3^4 = 25 - 81 = -56$

25. $2 \cdot (7 - 4 \cdot 5) = 2 \cdot (7 - 20) = 2 \cdot (-13) = -26$

29. $\dfrac{(-9+6)(-1^2)}{-2-2} = \dfrac{(-3)(-1)}{-4} = \dfrac{3}{-4} = -\dfrac{3}{4}$

33.
$$\begin{aligned} 12 + \{6 - [5 - 2(-5)]\} &= 12 + \{6 - [5 - (-10)]\} \\ &= 12 + [6 - (5 + 10)] \\ &= 12 + (6 - 15) \\ &= 12 + (-9) \\ &= 3 \end{aligned}$$

37.
$$\begin{aligned} \frac{(3-\sqrt{9})-(-5-1.3)}{-3} &= \frac{(3-3)-(-6.3)}{-3} \\ &= \frac{0+6.3}{-3} \\ &= \frac{6.3}{-3} \\ &= -2.1 \end{aligned}$$

41.
$$\begin{aligned} \frac{3(-2+1)}{5} - \frac{-7(2-4)}{1-(-2)} &= \frac{3(-1)}{5} - \frac{-7(-2)}{1+2} \\ &= \frac{-3}{5} - \frac{14}{3} \\ &= \frac{-9}{15} - \frac{70}{15} \\ &= -\frac{79}{15} \end{aligned}$$

45.
$$\begin{aligned} 3\{-2 + 5[1 - 2(-2+5)]\} &= 3\{-2 + 5[1 - 2(3)]\} \\ &= 3[-2 + 5(1 - 6)] \\ &= 3[-2 + 5(-5)] \\ &= 3[-2 + (-25)] \\ &= 3(-27) \\ &= -81 \end{aligned}$$

49. $\left(\dfrac{5.6 - 8.4}{1.9 - 2.7}\right)^2 = \left(\dfrac{-2.8}{-0.8}\right)^2 = (3.5)^2 = 12.25$

53. a. When $x = 10$, $\dfrac{100x + 5000}{x} = \dfrac{100(10) + 5000}{10} = \dfrac{1000 + 5000}{10} = \dfrac{6000}{10} = 600.$

When $x = 100$, $\dfrac{100x + 5000}{x} = \dfrac{100(100) + 5000}{100} = \dfrac{10{,}000 + 5000}{100} = \dfrac{15{,}000}{100} = 150.$

When $x = 1000$, $\dfrac{100x + 5000}{x} = \dfrac{100(1000) + 5000}{1000} = \dfrac{100{,}000 + 5000}{1000} = \dfrac{105{,}000}{1000} = 105.$

x	10	100	1000
$\dfrac{100x + 5000}{x}$	600	150	105

b. As the number of bookshelves increases, the cost per bookshelf decreases.

57. $9y - 11y = (9 - 11)y = -2y$

61. $19y - y = 19y - 1y = (19 - 1)y = 18y$

65. $9x - 8 - 10x = (9 - 10)x - 8 = -1x - 8 = -x - 8$

69. $-9 + 4x + 18 - 10x = (4 - 10)x + (-9 + 18) = -6x + 9$

73. $x - y + x - y = (1 + 1)x + (-1 - 1)y = 2x - 2y$

77. $\frac{3}{4}b - \frac{1}{2} + \frac{1}{6}b - \frac{2}{3} = \left(\frac{3}{4} + \frac{1}{6}\right)b + \left(-\frac{1}{2} - \frac{2}{3}\right)$
$= \left(\frac{9}{12} + \frac{2}{12}\right)b + \left(-\frac{3}{6} - \frac{4}{6}\right)$
$= \frac{11}{12}b - \frac{7}{6}$

81. $-5(x - 1) = -5(x) - (-5)(1) = -5x + 5$

85. $5k - (3k - 10) = 5k - 3k + 10$
$= (5 - 3)k + 10$
$= 2k + 10$

89. $3(x - 2) + x + 15 = 3x - 6 + x + 15$
$= 3x + x - 6 + 15$
$= (3 + 1)x + 9$
$= 4x + 9$

93. $4(6n - 3) - 3(8n + 4) = 24n - 12 - 24n - 12$
$= 24n - 24n - 12 - 12$
$= (24 - 24)n - 24$
$= 0n - 24$
$= -24$

97. $-1.2(5.7x - 3.6) + 8.75x = -6.84x + 4.32 + 8.75x$
$= -6.84x + 8.75x + 4.32$
$= (-6.84 + 8.75)x + 4.32$
$= 1.91x + 4.32$

101. $(2 + 7) \cdot (1 + 3) = 9 \cdot 4 = 36$

105. The estimated population is 42 million.

109. $\frac{-1.682 - 17.895}{(-7.102)(-4.691)} = -\frac{19.577}{33.315482} \approx -0.5876$

Exercise Set 1.5

1. $4^2 \cdot 4^3 = 4^{2+3} = 4^5$

5. $-7x^3 \cdot 20x^9 = -7 \cdot 20x^{3+9} = -140x^{12}$

9. $(-4x^3p^2)(4y^3x^3) = -16x^{3+3}y^3p^2$
$= -16x^6y^3p^2$

13. $2x^3 \cdot 5x^7 = 2 \cdot 5x^{3+7} = 10x^{10}$

17. $(4x + 5)^0 = 1$

21. $4x^0 + 5 = 4 \cdot 1 + 5 = 9$

25. answers may vary

29. $\frac{x^9y^6}{x^8y^6} = x^{9-8}y^{6-6} = x^1y^0 = x$

33. $\frac{-36a^5b^7c^{10}}{6ab^3c^4} = -6a^{5-1}b^{7-3}c^{10-4} = -6a^4b^4c^6$

37. $4^{-2} = \frac{1}{4^2} = \frac{1}{16}$

41. $5a^{-4} = 5\frac{1}{a^4} = \frac{5}{a^4}$

45. $\frac{8r^4}{2r^{-4}} = 4r^{4-(-4)} = 4r^8$

49. $4^{-1} + 3^{-2} = \frac{1}{4^1} + \frac{1}{3^2}$
$= \frac{1}{4} + \frac{1}{9}$
$= \frac{9}{36} + \frac{4}{36}$
$= \frac{13}{36}$

53. $3x^{-1} = 3 \cdot \frac{1}{x} = \frac{3}{x}$

57. $\frac{x^{-7}y^{-2}}{x^2y^2} = x^{-7-2}y^{-2-2} = x^{-9}y^{-4} = \frac{1}{x^9y^4}$

61. $\frac{(24x^8)x}{20x^{-7}} = \frac{6x^{8+1}}{5x^{-7}} = \frac{6x^{9-(-7)}}{5} = \frac{6x^{16}}{5}$

65. $\frac{x^{3t-1}}{x^t} = x^{3t-1-t} = x^{2t-1}$

69. $\frac{z^{6x}}{z^7} = z^{6x-7}$

73. $31{,}250{,}000 = 3.125 \times 10^7$

77. $67{,}413 = 6.7413 \times 10^4$

81. $0.000053 = 5.3 \times 10^{-5}$

85. $22{,}580{,}000{,}000 = 2.258 \times 10^{10}$

89. $0.001 = 1.0 \times 10^{-3}$

93. $9.3 \times 10^7 = 93{,}000{,}000$

97. $7.35 \times 10^{12} = 7{,}350{,}000{,}000{,}000$

101. answers may vary

Exercise Set 1.6

1. $(3^{-1})^2 = 3^{-1(2)} = 3^{-2} = \frac{1}{3^2} = \frac{1}{9}$

5. $(y)^{-5} = \frac{1}{y^{-5}}$

9. $\left(\frac{2x^5}{y^{-3}}\right)^4 = \frac{2^4(x^5)^4}{(y^{-3})^4} = \frac{16x^{5(4)}}{y^{-3(4)}} = \frac{16x^{20}}{y^{-12}} = 16x^{20}y^{12}$

13. $\left(\frac{x^7y^{-3}}{z^{-4}}\right)^{-5} = \frac{(x^7)^{-5}(y^{-3})^{-5}}{(z^{-4})^{-5}} = \frac{x^{-35}y^{15}}{z^{20}} = \frac{y^{15}}{x^{35}z^{20}}$

17. $(x^7)^{-9} = x^{-63} = \frac{1}{x^{63}}$

21. $(4x^2)^2 = 4^2(x^2)^2 = 16x^4$

25. $\left(\frac{4^{-4}}{y^3x}\right)^{-2} = \frac{(4^{-4})^{-2}}{(y^3)^{-2}x^{-2}} = \frac{4^8}{y^{-6}x^{-2}} = 4^8x^2y^6$

29. $\left(\frac{2a^{-2}b^5}{4a^2b^7}\right)^{-2} = \left(\frac{1}{2a^4b^2}\right)^{-2}$
$= \frac{(1)^{-2}}{(2)^{-2}(a^4)^{-2}(b^2)^{-2}}$
$= \frac{2^2}{1^2a^{-8}b^{-4}}$
$= 4a^8b^4$

33. $\left(\frac{6p^6}{p^{12}}\right)^2 = \frac{6^2p^{12}}{p^{24}} = 6^2p^{12-24} = 36p^{-12} = \frac{36}{p^{12}}$

37. $\left(\frac{x^{-2}y^{-2}}{a^{-3}}\right)^{-7} = \frac{(x^{-2})^{-7}(y^{-2})^{-7}}{(a^{-3})^{-7}} = \frac{x^{14}y^{14}}{a^{21}}$

41. $\left(\frac{1}{4}\right)^{-3} = (4^{-1})^{-3} = 4^3 = 64$

45. $\frac{8p^7}{4p^9} = 2p^{7-9} = 2p^{-2} = \frac{2}{p^2}$

49. $x^6(x^6bc)^{-6} = x^6(x^6)^{-6}b^{-6}c^{-6}$

$= \frac{x^6 \cdot x^{-36}}{b^6c^6}$

$= \frac{x^{-30}}{b^6c^6}$

$= \frac{1}{x^{30}b^6c^6}$

SOLUTIONS

53. $\left(\frac{2x^2}{y^4}\right)^3 \cdot \left(\frac{2x^5}{y}\right)^{-2} = \frac{2^3(x^2)^3 2^{-2}(x^5)^{-2}}{(y^4)^3 y^{-2}}$

$= \frac{8x^6x^{-10}}{2^2y^{12}y^{-2}}$

$= \frac{8x^{6-10}}{4y^{12-2}}$

$= \frac{2x^{-4}}{y^{10}}$

$= \frac{2}{x^4y^{10}}$

57. $\frac{x^{4a}(x^{4a})^3}{x^{4a-2}} = \frac{x^{4a} \cdot x^{12a}}{x^{4a-2}}$

$= \frac{x^{4a+12a}}{x^{4a-2}}$

$= x^{16a-(4a-2)}$

$= x^{12a+2}$

61. $\frac{(y^{2a})^8}{y^{a-3}} = \frac{y^{16a}}{y^{a-3}} = y^{16a-(a-3)} = y^{15a+3}$

65. $(5 \times 10^{11})(2.9 \times 10^{-3}) = 5 \times 2.9 \times 10^{11-3}$

$= 14.5 \times 10^8$

$= 1.45 \times 10^1 \times 10^8$

$= 1.45 \times 10^9$

69. $\frac{3.6 \times 10^{-4}}{9 \times 10^2} = 0.4 \times 10^{-4-2}$

$= 0.4 \times 10^{-6}$

$= 4 \times 10^{-1} \times 10^{-6}$

$= 4 \times 10^{-7}$

73. $\frac{18{,}200 \times 100}{91{,}000} = \frac{1.82 \times 10^4 \times 1 \times 10^2}{9.1 \times 10^4}$

$= \frac{1.82 \times 10^6}{9.1 \times 10^4}$

$= 0.2 \times 10^{6-4}$

$= 2 \times 10^{-1} \times 10^2$

$= 2 \times 10^{2-1}$

$= 2 \times 10^1$

77. $\frac{0.00064 \times 2000}{16{,}000} = \frac{(6.4 \times 10^{-4})(2 \times 10^3)}{1.6 \times 10^4}$

$= \frac{12.8 \times 10^{-4+3}}{1.6 \times 10^4}$

$= 8 \times 10^{-1-4}$

$= 8 \times 10^{-5}$

81. $\dfrac{1.25 \times 10^{15}}{(2.2 \times 10^{-2})(6.4 \times 10^{-5})} = \dfrac{1.25 \times 10^{15}}{14.08 \times 10^{-7}}$

$$\begin{aligned} &\approx 0.08877840909 \times 10^{22} \\ &= 8.877840909 \times 10^{-2} \times 10^{22} \\ &= 8.877840909 \times 10^{20} \end{aligned}$$

85. $1.64 \times 10^{-5} \times 3.8 \times 10^{-6} = 6.232 \times 10^{-5-6}$

$$= 6.232 \times 10^{-11}$$

The size is 6.232×10^{-11} cubic meters.

89. $D = \dfrac{M}{V}$ or $M = DV$

$$\begin{aligned} M &= (3.12 \times 10^{-2})(4.269 \times 10^{14}) \\ &= 13.31928 \times 10^{12} \\ &= 1.331928 \times 10^{13} \end{aligned}$$

The mass is 1.331928×10^{13} tons.

93. $\dfrac{2.680 \times 10^{8}}{3.536 \times 10^{6}} \approx 0.76 \times 10^{8-6}$

$$\begin{aligned} &= 7.6 \times 10^{-1} \times 10^{2} \\ &= 7.6 \times 10^{1} \\ &= 76 \end{aligned}$$

The density is approximately 76 people per square mile.

Chapter 1 Test

1. False, -2.3 lies to the left of 2.33 on the number line.

5. True, natural numbers are positive integers.

9. $(4 - 9)^3 - |-4 - 6|^2 = (-5)^3 - |-10|^2$

$$\begin{aligned} &= -125 - 100 \\ &= -225 \end{aligned}$$

13. $\dfrac{5t - 3q}{3r - 1} = \dfrac{5(1) - 3(4)}{3(-2) - 1}$

$$\begin{aligned} &= \frac{5 - 12}{-6 - 1} \\ &= \frac{-7}{-7} \\ &= 1 \end{aligned}$$

17. $-2 = \dfrac{x}{x + 5}$

21. multiplication property of zero

25. $630{,}000{,}000 = 6.3 \times 10^{8}$

29. $\dfrac{(0.00012)(144{,}000)}{0.0003} = \dfrac{1.2 \times 10^{-4} \times 1.44 \times 10^{5}}{3 \times 10^{-4}}$

$$\begin{aligned} &= \frac{1.728 \times 10^{1}}{3 \times 10^{-4}} \\ &= 0.576 \times 10^{5} \\ &= 5.76 \times 10^{-1} \times 10^{5} \\ &= 5.76 \times 10^{4} \end{aligned}$$

Chapter 2

Exercise Set 2.1

1. $\dfrac{x}{-6} = 4$

$$\frac{-24}{-6} \stackrel{?}{=} 4$$

$4 = 4$ True

-24 is a solution.

SOLUTIONS

5. $5 + 3x = -1$

$5 + 3(-2) \stackrel{?}{=} -1$

$5 - 6 \stackrel{?}{=} -1$

$-1 = -1$ True

-2 is a solution.

9. $4(x - 3) = 12$

$4(5 - 3) \stackrel{?}{=} 12$

$4(2) \stackrel{?}{=} 12$

$8 = 12$ False

5 is not a solution.

13. $-5x = -30$

$\frac{-5x}{-5} = \frac{-30}{-5}$

$x = 6$

Check:

$-5(6) \stackrel{?}{=} -30$

$-30 = -30$ True

The solution set is $\{6\}$.

17. $x + 2.8 = 1.9$

$x + 2.8 - 2.8 = 1.9 - 2.8$

$x = -0.9$

Check:

$-0.9 + 2.8 \stackrel{?}{=} 1.9$

$1.9 = 1.9$ True

The solution set is $\{-0.9\}$.

21. $-4.1 - 7z = 3.6$

$-4.1 - 7z + 4.1 = 3.6 + 4.1$

$-7z = 7.7$

$\frac{-7z}{-7} = \frac{7.7}{-7}$

$z = -1.1$

Check:

$-4.1 - 7(-1.1) \stackrel{?}{=} 3.6$

$-4.1 + 7.7 \stackrel{?}{=} 3.6$

$3.6 = 3.6$ True

The solution set is $\{-1.1\}$.

25. $8x - 5x + 3 = x - 7 + 10$

$3x + 3 = x + 3$

$2x = 0$

$x = 0$

Check:

$8(0) - 5(0) + 3 \stackrel{?}{=} 0 - 7 + 10$

$0 - 0 + 3 \stackrel{?}{=} 3$

$3 = 3$ True

The solution set is $\{0\}$.

29. $3(x - 6) = 5x$

$3x - 18 = 5x$

$-18 = 2x$

$x = -9$

Check:

$3(-9 - 6) \stackrel{?}{=} 5(-9)$

$3(-15) \stackrel{?}{=} -45$

$-45 = -45$ True

The solution set is $\{-9\}$.

33. $-2(5y - 1) - y = -4(y - 3)$

$-10y + 2 - y = -4y + 12$

$-11y + 2 = -4y + 12$

$-7y = 10$

$y = -\frac{10}{7}$

Check:

$-2\left[5\left(-\frac{10}{7}\right) - 1\right] - \left(-\frac{10}{7}\right) \stackrel{?}{=} -4\left(-\frac{10}{7} - 3\right)$

$-2\left(-\frac{50}{7} - 1\right) + \frac{10}{7} \stackrel{?}{=} -4\left(-\frac{31}{7}\right)$

$-2\left(-\frac{57}{7}\right) + \frac{10}{7} \stackrel{?}{=} \frac{124}{7}$

$\frac{114}{7} + \frac{10}{7} \stackrel{?}{=} \frac{124}{7}$

$\frac{124}{7} = \frac{124}{7}$ True

The solution set is $\left\{-\frac{10}{7}\right\}$.

37. a. $4(x + 1) + 1 = 4x + 4 + 1 = 4x + 5$

b. $4(x + 1) + 1 = -7$

$4x + 5 = -7$

$4x = -12$

$x = -3$

The solution set is $\{-3\}$.

c. answers may vary

41. $\frac{3t}{4} - \frac{t}{2} = 1$

$4\left(\frac{3t}{4} - \frac{t}{2}\right) = 4(1)$

$3t - 2t = 4$

$t = 4$

Check:

$\frac{3(4)}{4} - \frac{4}{2} \stackrel{?}{=} 1$

$3 - 2 \stackrel{?}{=} 1$

$1 = 1$ True

The solution set is $\{4\}$.

45. $0.6x - 10 = 1.4x - 14$

$4 = 0.8x$

$x = 5$

Check:

$0.6(5) - 10 \stackrel{?}{=} 1.4(5) - 14$

$3 - 10 \stackrel{?}{=} 7 - 14$

$-7 = -7$ True

The solution set is $\{5\}$.

49. $1.5(4 - x) = 1.3(2 - x)$

$10[1.5(4 - x)] = 10[1.3(2 - x)]$

$15(4 - x) = 13(2 - x)$

$60 - 15x = 26 - 13x$

$-2x = -34$

$x = 17$

Check:

$1.5(4 - 17) \stackrel{?}{=} 1.3(2 - 17)$

$1.5(-13) \stackrel{?}{=} 1.3(-15)$

$-19.5 = -19.5$ True

The solution set is $\{17\}$.

53. $3(x + 1) + 5 = 3x + 2$
$$3x + 3 + 5 = 3x + 2$$
$$3x + 8 = 3x + 2$$
$$3x + 8 - 3x = 3x + 2 - 3x$$
$$8 = 2$$
This is false for any x. Therefore, the solution set is $\emptyset$.

57. $2(x - 8) + x = 3(x - 6) + 2$
$$2x - 16 + x = 3x - 18 + 2$$
$$3x - 16 = 3x - 16$$
This is true for all x.
Therefore, all real numbers are solutions.
The solution set is $\{x|x \text{ is a real number}\}$.

61. $x - 10 = -6x + 4$
$$7x = 14$$
$$x = 2$$
Check:
$$2 - 10 \stackrel{?}{=} -6(2) + 4$$
$$-8 \stackrel{?}{=} -12 + 4$$
$$-8 = -8 \text{ True}$$
The solution set is $\{2\}$.

65. $\frac{1}{4}(a + 2) = \frac{1}{6}(5 - a)$
$$12 \cdot \frac{1}{4}(a + 2) = 12 \cdot \frac{1}{6}(5 - a)$$
$$3(a + 2) = 2(5 - a)$$
$$3a + 6 = 10 - 2a$$
$$5a = 4$$
$$a = \frac{4}{5}$$
Check:
$$\frac{1}{4}\left(\frac{4}{5} + 2\right) \stackrel{?}{=} \frac{1}{6}\left(5 - \frac{4}{5}\right)$$
$$\frac{1}{4}\left(\frac{14}{5}\right) \stackrel{?}{=} \frac{1}{6}\left(\frac{21}{5}\right)$$
$$\frac{7}{10} = \frac{7}{10} \text{ True}$$
The solution set is $\left\{\frac{4}{5}\right\}$.

69. $\frac{m - 4}{3} - \frac{3m - 1}{5} = 1$
$$15\left(\frac{m - 4}{3} - \frac{3m - 1}{5}\right) = 15(1)$$
$$5(m - 4) - 3(3m - 1) = 15$$
$$5m - 20 - 9m + 3 = 15$$
$$-4m - 17 = 15$$
$$-4m = 32$$
$$m = -8$$
Check:
$$\frac{-8 - 4}{3} - \frac{3(-8) - 1}{5} \stackrel{?}{=} 1$$
$$\frac{-12}{3} - \frac{-24 - 1}{5} \stackrel{?}{=} 1$$
$$-4 - \frac{-25}{5} \stackrel{?}{=} 1$$
$$-4 + 5 \stackrel{?}{=} 1$$
$$1 = 1 \text{ True}$$
The solution set is $\{-8\}$.

73. $\frac{8}{x}$

77. $3x + 2$

81. $x(x - 6) + 7 = x(x + 1)$
$$x^2 - 6x + 7 = x^2 + x$$
$$x^2 - 6x + 7 - x^2 = x^2 + x - x^2$$
$$-6x + 7 = x$$
$$-7x = -7$$
$$x = 1$$
Check:
$$1(1 - 6) + 7 \stackrel{?}{=} 1(1 + 1)$$
$$-5 + 7 \stackrel{?}{=} 2$$
$$2 = 2 \text{ True}$$
The solution set is $\{1\}$.

Exercise Set 2.2

1. $4(x - 2) = 2 + 6x$
$$4x - 8 = 2 + 6x$$
$$-10 = 2x$$
$$-5 = x$$

5. $30\% \cdot 260 = 0.30 \cdot 260 = 78$

9. $29\% \cdot 2271 = 0.29 \cdot 2271 = 658.59$
Approximately 658.59 million acres are federally owned.

13. $100 - 12 - 39 - 8 - 21 = 20$
20% of credit union loans are for credit cards and other unsecured loans.

17. Let x = seats in B737-200,
then $x + 104$ = seats in B767-300ER.
$$x + x + 104 = 328$$
$$2x + 104 = 328$$
$$2x = 224$$
$$x = 112$$
$$x + 104 = 216$$
B737-200 has 112 seats.
B767-300ER has 216 seats.

21. Let x = length of a side of the square
$$4x = 3(x + 6)$$
$$4x = 3x + 18$$
$$x = 18$$
$$x + 6 = 24$$
The square's sides are 18 centimeters.
The triangle's sides are 24 centimeters.

25. $20 - 75\% \cdot 20 = 20 - 0.75 \cdot 20 = 20 - 15 = 5$ years

29. Let x = measure of second angle,
then $2x$ = measure of first angle, and
$3x - 12$ = measure of third angle.
$$x + 2x + 3x - 12 = 180$$
$$6x - 12 = 180$$
$$6x = 192$$
$$x = 32$$
$$2x = 64$$
$$3x - 12 = 84$$
The angles measure 64°, 32°, and 84°.

33. Let x = 1st integer,
then $x + 2$ = 2nd integer, and
$x + 4$ = 3rd integer.

$$\begin{aligned} 2x + x + 4 &= 268{,}222 \\ 3x + 4 &= 268{,}222 \\ 3x &= 268{,}218 \\ x &= 89{,}406 \end{aligned}$$

$x + 2 = 89{,}408$, $x + 4 = 89{,}410$

Fallon's zip code is 89406, Fernley's is 89408, and Gardnerville's is 85410.

37. Let x = net income in 1996.

$$\begin{aligned} x + 0.18x &= 4129 \\ 1.18x &= 4129 \\ x &\approx 3499 \end{aligned}$$

Coca-Cola's net income in 1996 was about $3499 million.

41. $$\begin{aligned} 2a + b - c &= 2(5) + (-1) - 3 \\ &= 10 - 1 - 3 \\ &= 6 \end{aligned}$$

45. $n^2 - m^2 = (-3)^2 - (-8)^2 = 9 - 64 = -55$

49. Let x = total trees' worth of newsprint each year.

$$\begin{aligned} 0.27x + 30{,}000{,}000 &= x \\ 30{,}000{,}000 &= 0.73x \\ 41{,}095{,}890 &\approx x \end{aligned}$$

$0.27x \approx 0.27(41{,}095{,}890) \approx 11{,}095{,}900$

Approximately 11 million trees' worth of newsprint is recycled each year.

53. The company makes a profit.

Exercise Set 2.3

1. $$\begin{aligned} d &= rt \\ \frac{d}{r} &= \frac{rt}{r} \\ \frac{d}{r} &= t \end{aligned}$$

5. $$\begin{aligned} P &= a + b + c \\ P - a - b &= a + b + c - a - b \\ P - a - b &= c \end{aligned}$$

9. $$\begin{aligned} P &= 2l + 2w \\ P - 2w &= 2l + 2w - 2w \\ P - 2w &= 2l \\ \frac{P - 2w}{2} &= \frac{2l}{2} \\ \frac{P - 2w}{2} &= l \end{aligned}$$

13. $$\begin{aligned} S &= 2LW + 2LH + 2WH \\ S - 2LW &= 2LW + 2LH + 2WH - 2LW \\ S - 2LW &= 2LH + 2WH \\ S - 2LW &= H(2L + 2W) \\ \frac{S - 2LW}{2L + 2W} &= \frac{H(2L + 2W)}{2L + 2W} \\ \frac{S - 2LW}{2L + 2W} &= H \end{aligned}$$

17. $A = P\left(1 + \frac{r}{n}\right)^{nt} = 6000\left(1 + \frac{0.04}{n}\right)^{5n}$

a. $n = 2$

$A = 6000\left(1 + \frac{0.04}{2}\right)^{5 \cdot 2} \approx 7313.97$

$7313.97

b. $n = 4$

$A = 6000\left(1 + \frac{0.04}{4}\right)^{5 \cdot 4} \approx 7321.14$

$7321.14

c. $n = 12$

$A = 6000\left(1 + \frac{0.04}{12}\right)^{5 \cdot 12} \approx 7325.98$

$7325.98

21. $d = rt$ or $t = \frac{d}{r}$

where $d = 2(90) = 180$ miles and

$r = 50$ miles/hour, so $t = \frac{180}{50} = 3.6$ hours

Thus, she takes 3.6 hours or 3 hours, 36 minutes to make the round trip.

25. $A = \frac{1}{2}bh$ or $h = \frac{2A}{b}$

$h = \frac{2(18)}{4} = 9$

The height is 9 feet.

29. a. $$\begin{aligned} V &= \pi r^2 h \\ &= \pi(4.2)^2(21.2) \\ &\approx 1174.86 \text{ cubic meters} \end{aligned}$$

b. $$\begin{aligned} V &= \frac{4}{3}\pi r^3 \\ &= \frac{4}{3}\pi(4.2)^3 \\ &\approx 310.34 \text{ cubic meters} \end{aligned}$$

c. $1174.86 + 310.34 = 1485.20$ cubic meters

33. $V = \pi r^2 h$ or $r = \sqrt{\frac{V}{\pi h}}$

1 mile = 5280 feet

1.3 miles = 6864 feet

$r = \sqrt{\frac{3800}{\pi(6864)}} \approx 0.42$

The radius is about 0.42 foot.

37. $$\begin{aligned} x + 5 &\le 6 \\ x &\le 1 \end{aligned}$$

$\{-3, -2, -1, 0, 1\}$

SOLUTIONS

41. $AU = \dfrac{\text{miles from sun}}{92{,}900{,}000 \text{ miles}}$

Planet	Miles from the sun	AU from the sun
Mercury	36 million	0.388
Venus	67.2 million	0.723
Earth	92.9 million	1.00
Mars	141.5 million	1.523
Jupiter	483.3 million	5.202
Saturn	886.1 million	9.538
Uranus	1783 million	19.193
Neptune	2793 million	30.065
Pluto	3670 million	39.505

45. answers may vary

Exercise Set 2.4

1. $(-\infty, -3)$

5. $(5, \infty)$

9. $(-1, 5)$

13. $x - 7 \geq -9$
$x \geq -2$
$[-2, \infty)$

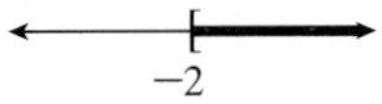

17. $8x - 7 \leq 7x - 5$
$x - 7 \leq -5$
$x \leq 2$
$(-\infty, 2]$

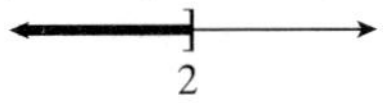

21. $5x < -23.5$
$x < -4.7$
$(-\infty, -4.7)$

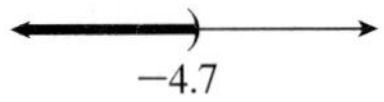

25. $-x < -4$
$x > 4$
$(4, \infty)$

29. $15 + 2x \geq 4x - 7$
$15 \geq 2x - 7$
$22 \geq 2x$
$11 \geq x$
$(-\infty, 11]$

33. $\frac{1}{2} + \frac{2}{3} \geq \frac{x}{6}$
$6\left(\frac{1}{2} + \frac{2}{3}\right) \geq 6\left(\frac{x}{6}\right)$
$3 + 4 \geq x$
$7 \geq x$
$(-\infty, 7]$

37. $\frac{1}{4}(x - 7) \geq x + 2$
$4\left[\frac{1}{4}(x - 7)\right] \geq 4(x + 2)$
$x - 7 \geq 4x + 8$
$-15 \geq 3x$
$-5 \geq x$
$(-\infty, -5]$

41. $4(2x + 1) > 4$
$8x + 4 > 4$
$8x > 0$
$x > 0$
$(0, \infty)$

45. $4(x - 6) + 2x - 4 \geq 3(x - 7) + 10x$
$4x - 24 + 2x - 4 \geq 3x - 21 + 10x$
$6x - 28 \geq 13x - 21$
$-7x \geq 7$
$x \leq -1$
$(-\infty, -1]$

49. Let x = the weight of the luggage and cargo. Then
$6(160) + x \leq 2000$
$960 + x \leq 2000$
$x \leq 1040$
The plane can carry a maximum of 1040 pounds of luggage and cargo.

53. Let n = the number of calls made in a given month. Then solve
$25 < 13 + 0.06n$
$12 < 0.06n$
$200 < n$
$n > 200$
Therefore, Plan 1 is more economical than Plan 2 when more than 200 calls are made.

57. **a.** $35{,}000 \leq 651.2t + 27{,}821$
$7179 \leq 651.2t$
$11 \leq t$
$1989 + 11 = 2000$
Beginning salaries will be greater than $35,000 in the year 2000.

b. answers may vary

61. $t = 2000 - 1990 = 10$
$w = -2.71t + 89.67$
$= -2.71(10) + 89.67$
$= 62.57$
The consumption will be about 62.57 pounds per person per year.

65. answers may vary

69. $x \geq 0$ and $x \leq 7$
The integers are 0, 1, 2, 3, 4, 5, 6, 7 or $\{0, 1, 2, 3, 4, 5, 6, 7\}$.

73. $\{x|-7 < x \le 1\}$
$(-7, 1]$

−7 1

77. $3x + 1 < 3(x - 2)$
$3x + 1 < 3x - 6$
$1 < -6$
This is false for any x. Therefore, the solution set is $\emptyset$.

81. answers may vary

Exercise Set 2.5

1. $A \cap C = \{2, 4\}$

5. $B \cap C = \{3, 5\}$

9. $x \le -3$

−3

$x \ge -2$

−2

$x \le -3$ and $x \ge -2$

13. $x < 5$ and $x > -2$
$-2 < x < 5$
The solution set is $(-2, 5)$.

17. $4x + 2 \le -10$ and $2x < 0$
$4x \le -12$ and $x < 0$
$x \le -3$ and $x < 0$
The solution set is $(-\infty, -3)$.

21. $5 < x - 6 < 11$
$11 < x < 17$
The solution set is $(11, 17)$.

25. $1 \le \frac{2}{3}x + 3 \le 4$
$-2 \le \frac{2}{3}x \le 1$
$-3 \le x \le \frac{3}{2}$
The solution set is $\left[-3, \frac{3}{2}\right]$.

29. $0 \le 2x - 3 \le 9$
$3 \le 2x \le 12$
$\frac{3}{2} \le x \le 6$
The solution set is $\left[\frac{3}{2}, 6\right]$.

33. $A \cup B = \{1, 2, 3, 4, 5, 6, 7, 8\}$

37. $C \cup D = \{2, 4, 6, 8\}$

41. $x \le -4$

−4

$x \ge 1$

1

$x \le -4$ or $x \ge 1$

−4 1

45. $x < -1$ or $x > 0$
The solution set is $(-\infty, -1) \cup (0, \infty)$.

49. $3(x - 1) < 12$ or $x + 7 > 10$
$x - 1 < 4$ or $x > 3$
$x < 5$ or $x > 3$
all real numbers
The solution set is $(-\infty, \infty)$.

53. $3x \ge 5$ or $-x - 6 < 1$
$x \ge \frac{5}{3}$ or $-x < 7$
$x \ge \frac{5}{3}$ or $x > -7$
The solution set is $(-7, \infty)$.

57. $|-7| - |19| = 7 - 19 = -12$

61. $|x| = 7$
$x = -7, 7$

65. The years that the consumption of pork was greater than 48 pounds per person were 1992, 1993, 1994, and 1995.

The years that the consumption of chicken was greater than 48 pounds per person were 1993, 1994, 1995, and 1996.

The years in common are 1993, 1994, and 1995.

69. $x + 3 < 2x + 1 < 4x + 6$
$x + 3 < 2x + 1$ and $2x + 1 < 4x + 6$
$2 < x$ and $-5 < 2x$
$x > 2$ and $-\frac{5}{2} < x$
The solution set is $(2, \infty)$.

Exercise Set 2.6

1. $|x| = 7$
$x = 7$ or $x = -7$
The solution set is $\{7, -7\}$.

5. $|3x| = 12.6$
$3x = 12.6$ or $3x = -12.6$
$x = 4.2$ or $x = -4.2$
The solution set is $\{4.2, -4.2\}$.

9. $|2x - 5| = 9$
$2x - 5 = 9$ or $2x - 5 = -9$
$2x = 14$ or $2x = -4$
$x = 7$ or $x = -2$
The solution set is $\{7, -2\}$.

13. $|z| + 4 = 9$
$|z| = 5$
$z = 5$ or $z = -5$
The solution set is $\{5, -5\}$.

17. $|2x| = 0$
$2x = 0$
$x = 0$
The solution set is $\{0\}$.

21. $|5x - 7| = |3x + 11|$
$$5x - 7 = 3x + 11 \text{ or } 5x - 7 = -(3x + 11)$$
$$2x = 18 \quad \text{or } 5x - 7 = -3x - 11$$
$$x = 9 \quad \text{or} \quad 8x = -4$$
$$x = 9 \quad \text{or} \quad x = -\frac{1}{2}$$
The solution set is $\left\{9, -\frac{1}{2}\right\}$.

25. $|2y - 3| = |9 - 4y|$
$$2y - 3 = 9 - 4y \text{ or } 2y - 3 = -(9 - 4y)$$
$$6y = 12 \quad \text{or } 2y - 3 = -9 + 4y$$
$$y = 2 \quad \text{or} \quad 6 = 2y$$
$$y = 2 \quad \text{or} \quad 3 = y$$
The solution set is $\{2, 3\}$.

29. $|x + 4| = |7 - x|$
$$x + 4 = 7 - x \text{ or } x + 4 = -(7 - x)$$
$$2x = 3 \quad \text{or } x + 4 = -7 + x$$
$$x = \frac{3}{2} \quad \text{or} \quad 4 = -7 \text{ False}$$
The solution set is $\left\{\frac{3}{2}\right\}$.

33. $|x| \le 4$
$$-4 \le x \le 4$$
The solution set is $[-4, 4]$.

−4 4

37. $|x + 3| < 2$
$$-2 < x + 3 < 2$$
$$-5 < x < -1$$
The solution set is $(-5, -1)$.

−5 −1

41. $|x| + 7 \le 12$
$$|x| \le 5$$
$$-5 \le x \le 5$$
The solution set is $[-5, 5]$.

−5 5

45. $|2x + 7| \le 13$
$$-13 \le 2x + 7 \le 13$$
$$-20 \le 2x \le 6$$
$$-10 \le x \le 3$$
The solution set is $[-10, 3]$.

−10 3

49. $|2x - 7| \le 11$
$$-11 \le 2x - 7 \le 11$$
$$-4 \le 2x \le 18$$
$$-2 \le x \le 9$$
The solution set is $[-2, 9]$.

−2 9

53. $6 + |4x - 1| \le 9$
$$|4x - 1| \le 3$$
$$-3 \le 4x - 1 \le 3$$
$$-2 \le 4x \le 4$$
$$-\frac{1}{2} \le x \le 1$$
The solution set is $\left[-\frac{1}{2}, 1\right]$.

$-\frac{1}{2}$ 1

57. $|5x + 3| < -6$
No real numbers.
The solution set is $\emptyset$.

61. $|2x - 3| < 7$
$$-7 < 2x - 3 < 7$$
$$-4 < 2x < 10$$
$$-2 < x < 5$$
The solution set is $(-2, 5)$.

65. $|x - 5| \ge 12$
$$x - 5 \le -12 \text{ or } x - 5 \ge 12$$
$$x \le -7 \quad \text{or} \quad x \ge 17$$
The solution set is $(-\infty, -7] \cup [17, \infty)$.

69. $|2x + 1| + 4 < 7$
$$|2x + 1| < 3$$
$$-3 < 2x + 1 < 3$$
$$-4 < 2x < 2$$
$$-2 < x < 1$$
The solution set is $(-2, 1)$.

73. $|x + 11| = -1$
No real solution.
The solution set is $\emptyset$.

77. $\left|\frac{3x - 5}{6}\right| > 5$
$$\frac{3x - 5}{6} < -5 \text{ or } \frac{3x - 5}{6} > 5$$
$$3x - 5 < -30 \text{ or } 3x - 5 > 30$$
$$3x < -25 \text{ or } \quad 3x > 35$$
$$x < -\frac{25}{3} \text{ or } \quad x > \frac{35}{3}$$
The solution set is $\left(-\infty, -\frac{25}{3}\right) \cup \left(\frac{35}{3}, \infty\right)$.

81. $100 - 24 - 44 = 32$
32% comes from the broadcasting segment.

85.
$$3x - 4y = 12$$
$$3x - 4(-1) = 12$$
$$3x + 4 = 12$$
$$3x = 8$$
$$x = \frac{8}{3}$$

89. $|x| = 2$

93. $|x| > 1$

Chapter 2 Test

1. $8x + 14 = 5x + 44$
$$3x = 30$$
$$x = 10$$
The solution set is $\{10\}$.

5. $\frac{z}{2} + \frac{z}{3} = 10$

$6\left(\frac{z}{2} + \frac{z}{3}\right) = 6(10)$

$3z + 2z = 60$

$5z = 60$

$z = 12$

The solution set is $\{12\}$.

9. $|2x - 3| = |4x + 5|$

$2x - 3 = -(4x + 5)$ or $2x - 3 = 4x + 5$

$2x - 3 = -4x - 5$ or $-2x = 8$

$6x = -2$ or $x = -4$

$x = -\frac{1}{3}$ or $x = -4$

The solution set is $\left\{-4, -\frac{1}{3}\right\}$.

13. $F = \frac{9}{5}C + 32$

$F - 32 = \frac{9}{5}C$

$C = \frac{5}{9}(F - 32)$

17. $|3x + 1| > 5$

$3x + 1 < -5$ or $3x + 1 > 5$

$3x < -6$ or $3x > 4$

$x < -2$ or $x > \frac{4}{3}$

The solution set is $(-\infty, -2) \cup \left(\frac{4}{3}, \infty\right)$.

21. $-x > 1$ and $3x + 3 \geq x - 3$

$x < -1$ and $2x \geq -6$

$x < -1$ and $x \geq -3$

$-3 \leq x < -1$

The solution set is $[-3, -1)$.

25. Recall that $C = 2\pi r$. Here $C = 78.5$.

$78.5 = 2\pi r$

$r = \frac{78.5}{2\pi} = \frac{39.25}{\pi}$

Also, recall that $A = \pi r^2$.

$A = \pi\left(\frac{39.25}{\pi}\right)^2 \approx \frac{39.25^2}{3.14} \approx 490.63$

Dividing this by 60 yields approximately 8.18. Therefore, about 8 hunting dogs could safely be kept in the pen.

Chapter 3

Exercise Set 3.1

1.

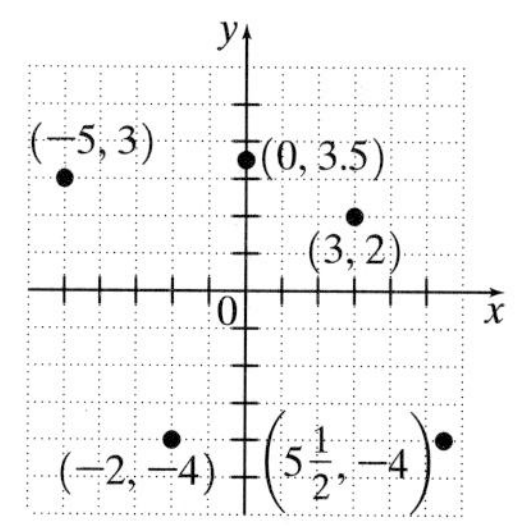

$(3, 2)$ lies in quadrant I.
$(-5, 3)$ lies in quadrant II.
$\left(5\frac{1}{2}, -4\right)$ lies quadrant IV.
$(0, 3.5)$ lies on the y-axis.
$(-2, -4)$ lies in quadrant III.

5. $(x, 0)$ lies on the x-axis.

9.

Year, x	Revenue (billions of dollars), y
1993	64
1994	65
1995	70
1996	75
1997	79

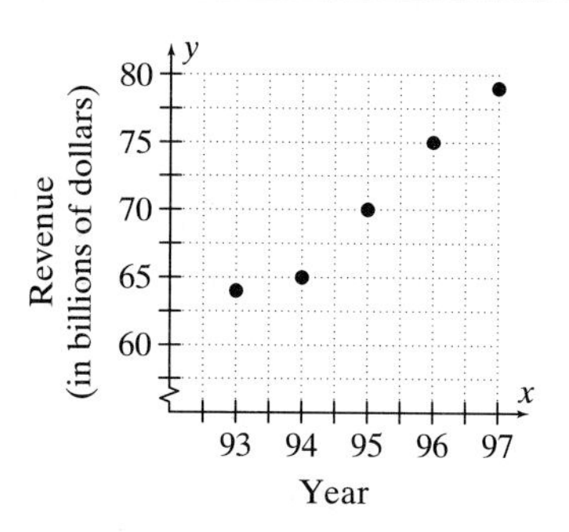

13. Let $x = 1$, $y = 0$.

$-6x + 5y = -6$

$-6(1) + 5(0) \stackrel{?}{=} -6$

$-6 + 0 \stackrel{?}{=} -6$

$-6 \stackrel{?}{=} -6$

True; Yes

Let $x = 2$, $y = \frac{6}{5}$

$-6x + 5y = -6$

$-6(2) + 5\left(\frac{6}{5}\right) \stackrel{?}{=} -6$

$-12 + 6 \stackrel{?}{=} -6$

$-6 \stackrel{?}{=} -6$

True; Yes

17. $x - 2y = 4$
Find three ordered pair solutions.

x	y
4	0
0	−2
−2	−3

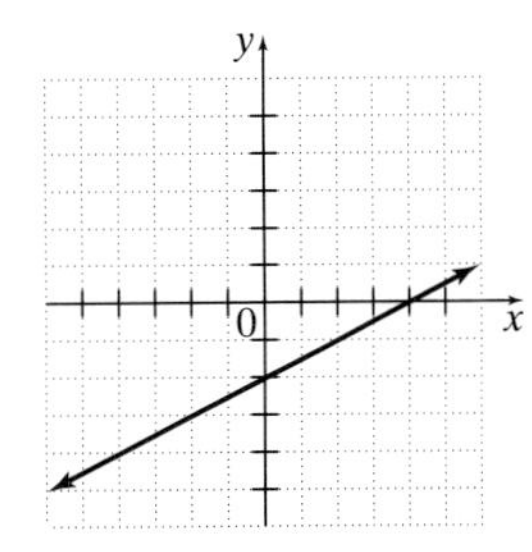

21. $x = 4$
Any ordered pair with an x-coordinate of 4 is a solution to $x = 4$.

x	y
4	−1
4	0
4	1

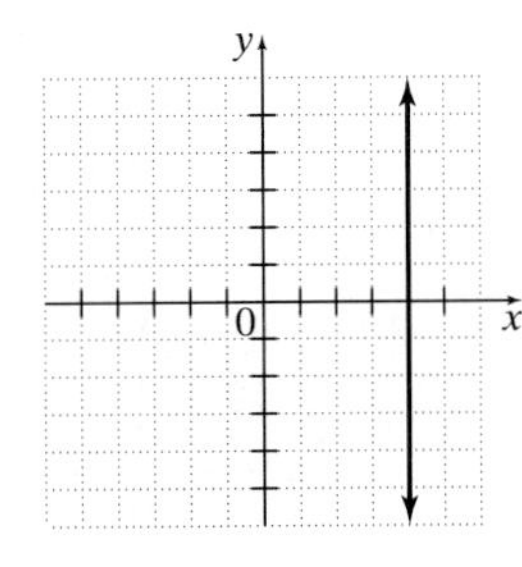

25. $y = 3x$
Find three ordered pair solutions.

x	y
0	0
1	3
−1	−3

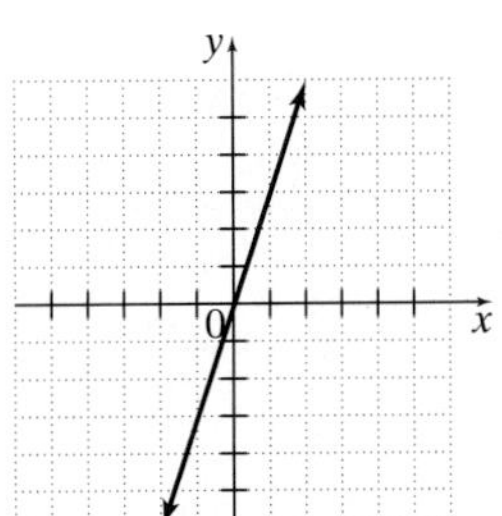

29. $4x + 5y = 15$
Find three ordered pair solutions.

x	y
0	3
5	−1
−5	7

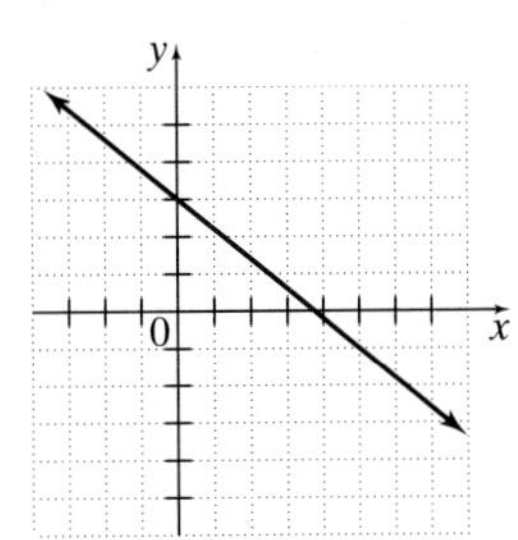

33. $x = \frac{1}{2}$

This is a vertical line with an x-intercept at $\frac{1}{2}$ and no y-intercept.

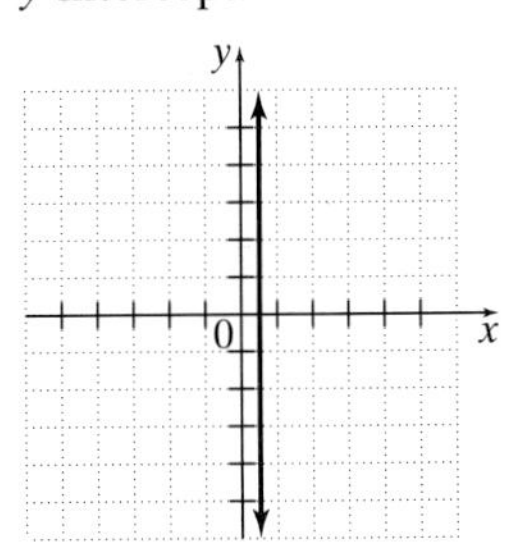

37. $y = -4x + 1$
Find three ordered pair solutions.

x	y
0	1
1	−3
−1	5

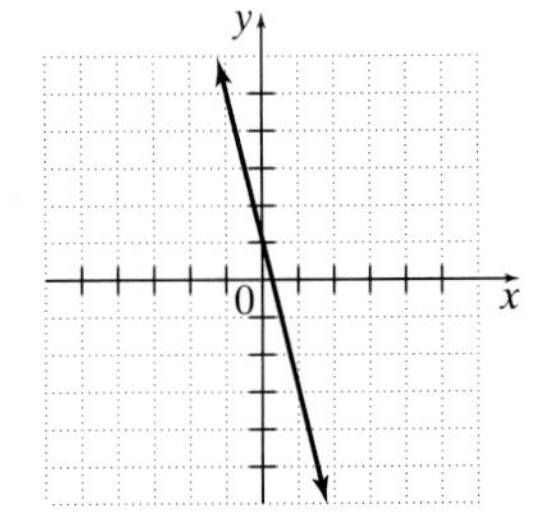

41. $y = 2$ matches graph C.

45. $\frac{-6 - 3}{2 - 8} = \frac{-9}{-6} = \frac{3}{2}$

49. $\frac{0 - 6}{5 - 0} = \frac{-6}{5} = -\frac{6}{5}$

53.

x	$y = 2x$	$y = 2x - 5$	$y = 2x + 5$
0	0	−5	5
1	2	−3	7
−1	−2	−7	3

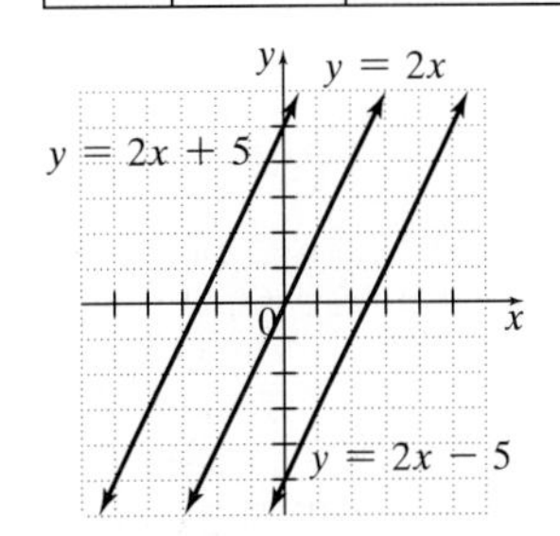

The graphs are parallel.

57. $y = -4x + 1$

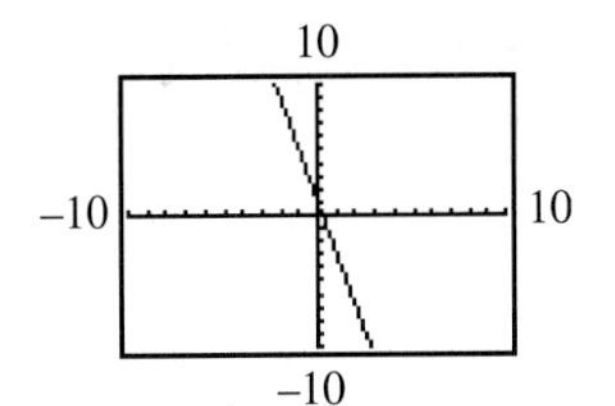

Exercise Set 3.2

1. $m = \frac{11 - 2}{8 - 3} = \frac{9}{5}$

5. $m = \frac{3 - 8}{4 - (-2)} = -\frac{5}{6}$

9. $m = \frac{11 - (-1)}{-12 - (-3)} = \frac{12}{-9} = -\frac{4}{3}$

13. $m = \frac{8}{12} = \frac{2}{3}$

17. $y = 5x - 2$
$m = 5, b = -2$

21. $2x - 3y = 10$
$-3y = -2x + 10$
$y = \frac{2}{3}x - \frac{10}{3}$
$m = \frac{2}{3}, b = -\frac{10}{3}$

25. $3x + 9 = y$
$y = 3x + 9$
$m = 3, b = 9$

29. $y = -2x + 3$
$m = -2, b = 3$
B

33. $y = -x + 5$
$m = -1$

37. $x = 4$
m is undefined

41. $x + 2 = 0$
$x = -2$
m is undefined

45. l_2 has the greater slope because the slope of l_2 is 0 and the slope of l_1 is negative.

49. $y = -3x + 6$ $\quad$ $y = 3x + 5$
$m = -3$ $\quad$ $m = 3$
$b = 6$ $\quad$ $b = 5$
Neither, since their slopes are not equal, nor does their product equal -1.

53. $-2x + 3y = 1$ $\quad$ $3x + 2y = 12$
$3y = 2x + 1$ $\quad$ $2y = -3x + 12$
$y = \frac{2}{3}x + \frac{1}{3}$ $\quad$ $y = -\frac{3}{2}x + 6$
$m = \frac{2}{3}$ $\quad$ $m = -\frac{3}{2}$
$b = \frac{1}{3}$ $\quad$ $b = 6$
Perpendicular, since the product of their slopes is -1.

57. $y = 12x + 6$ $\quad$ $y = 12x - 2$
$m = 12$ $\quad$ $m = 12$
$b = 6$ $\quad$ $b = -2$
Parallel, since they have the same slope but different y-intercepts.

61. $5x - 2y = 6$
$-2y = -5x + 6$
$y = \frac{5}{2}x - 3$
$m = \frac{5}{2}$

The slope of a parallel line is $\frac{5}{2}$.

65. $|2x + 5| > 3$
$2x + 5 < -3$ or $2x + 5 > 3$
$2x < -8$ or $2x > -2$
$x < -4$ or $x > -1$
$(-\infty, -4) \cup (-1, \infty)$

69. m for $l_1 = \frac{-2-4}{2-(-1)} = \frac{-6}{3} = -2$
m for $l_2 = \frac{2-6}{-4-(-8)} = \frac{-4}{4} = -1$
m for $l_3 = \frac{-4-0}{0-(-6)} = \frac{-4}{6} = -\frac{2}{3}$

73. $F(22, 2), G(26, 8)$
$m = \frac{8-2}{26-22} = \frac{6}{4} = \frac{3}{2}$,
or $\frac{3}{2}$ yards per second.

77. a. $y = \frac{1}{2}x + 1$
$y = x + 1$
$y = 2x + 1$

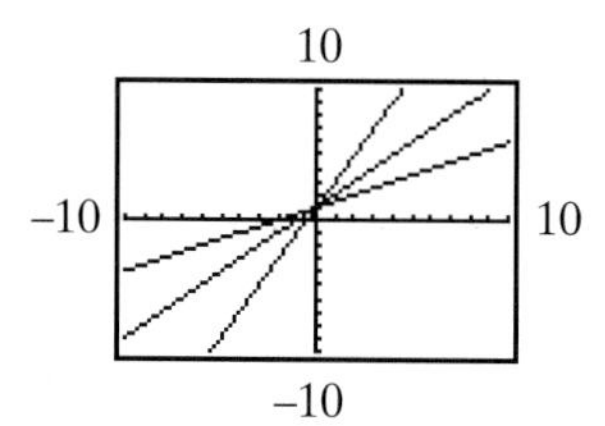

b. $y = -\frac{1}{2}x + 1$
$y = -x + 1$
$y = -2x + 1$

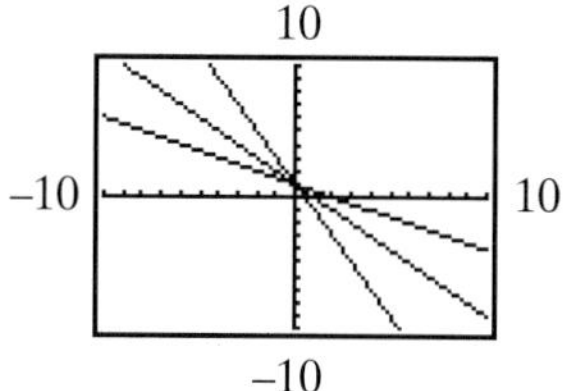

c. True

Exercise Set 3.3

1. Point: $(1, 3)$
Slope: $\frac{3}{2}$

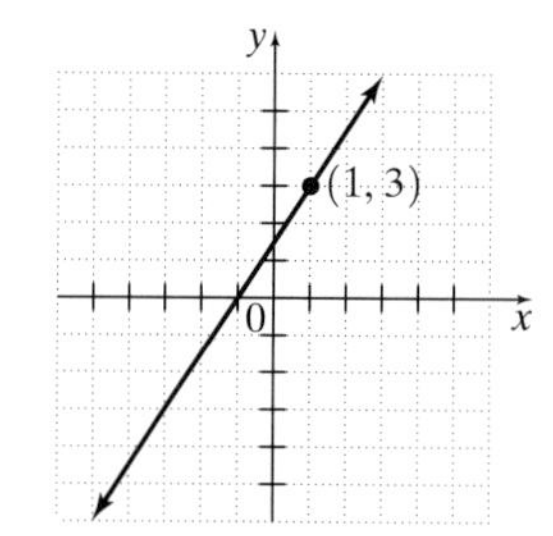

5. Point: (0, 7)
Slope: −1

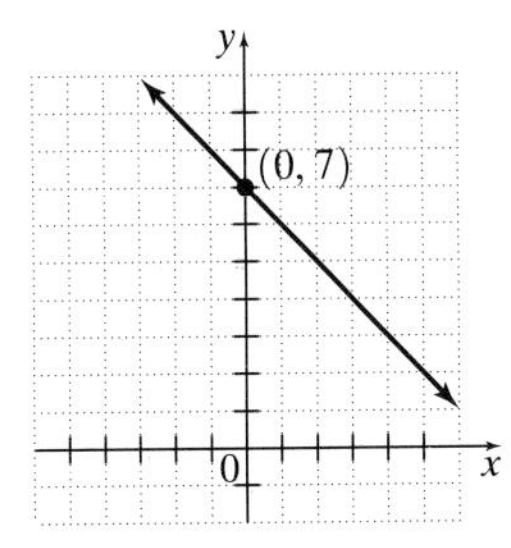

9. $y = -2x + 3$
$m = -2, b = 3$

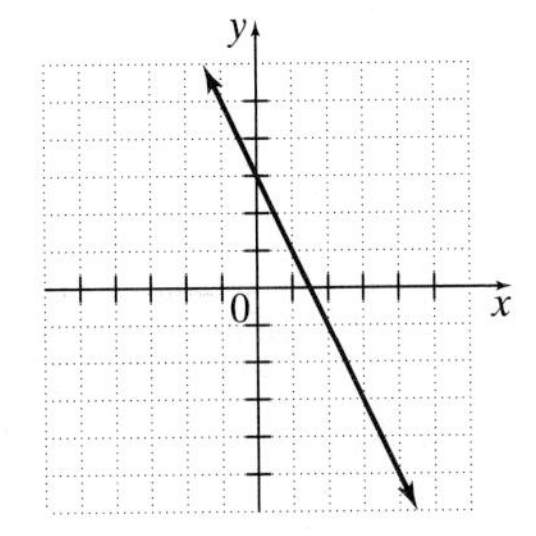

13. $y = \frac{1}{2}x - 4$

$m = \frac{1}{2}, b = -4$

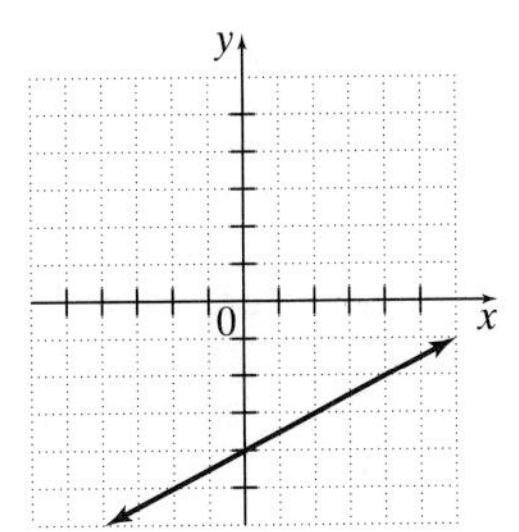

17. $x + 2y = 8$

$$2y = -x + 8$$
$$y = -\frac{1}{2}x + 4$$
$$m = -\frac{1}{2}, b = 4$$

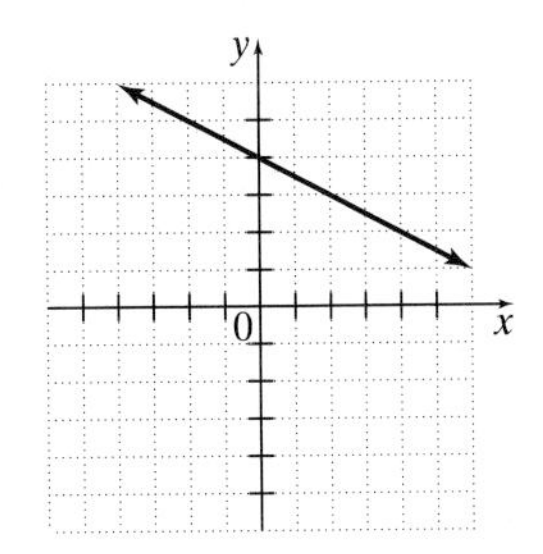

21. $y = 5x + 1$
$m = 5, b = 1$
graph D

25. $y = mx + b$

$y = 2x + \frac{3}{4}$

29. $y = 1431.5x + 31{,}775.2$

a. $x = 1996 - 1992 = 4$
$y = 1431.5(4) + 31{,}775.2$
$y = 37{,}501.2$
The income is \$37,501.20.

b. $m = 1431.5$; The annual average income increases \$1431.50 each year.

c. $b = 31{,}775.2$; At $x = 0$, or 1992, the annual average income was \$31,775.20.

33. $y = 131.4x + 2037.4$

a. $x = 2010 - 1990 = 20$
$y = 131.4(20) + 2037.4 = 4665.4$
The cost is \$4665.40.

b.
$$4000 = 131.4x + 2037.4$$
$$1962.6 = 131.4x$$
$$15 \approx x$$
$1990 + 15 = 2005$
In 2005 the cost will exceed \$4000.

c. answers may vary

37.
$$y - (-1) = 2(x - 0)$$
$$y + 1 = 2x - 0$$
$$y + 1 - 1 = 2x - 1$$
$$y = 2x - 1$$

41. $y = -7x + 500$
where y is the height at time x.

Exercise Set 3.4

1.
$$y - y_1 = m(x - x_1)$$
$$y - 2 = 3(x - 1)$$
$$y - 2 = 3x - 3$$
$$y = 3x - 1$$

5.
$$y - y_1 = m(x - x_1)$$
$$y - 2 = \frac{1}{2}[x - (-6)]$$
$$y - 2 = \frac{1}{2}(x + 6)$$
$$y - 2 = \frac{1}{2}x + 3$$
$$y = \frac{1}{2}x + 5$$

9.
$$y - y_1 = m(x - x_1)$$
$$y - 3 = 2[x - (-2)]$$
$$y - 3 = 2(x + 2)$$
$$y - 3 = 2x + 4$$
$$y = 2x + 7$$

13.
$$m = \frac{6 - 0}{4 - 2} = \frac{6}{2} = 3$$
$$y - 0 = 3(x - 2)$$
$$y = 3x - 6$$

SOLUTIONS

17. $m = \dfrac{-3-(-4)}{-4-(-2)} = \dfrac{1}{-2} = -\dfrac{1}{2}$

$$y-(-4) = -\frac{1}{2}[x-(-2)]$$
$$2(y+4) = -(x+2)$$
$$2y+8 = -x-2$$
$$y = -\frac{1}{2}x - \frac{10}{2}$$
$$y = -\frac{1}{2}x - 5$$

21. $m = \dfrac{-6-(-4)}{0-(-7)} = \dfrac{-2}{7} = -\dfrac{2}{7}$

$$y-(-6) = -\frac{2}{7}(x-0)$$
$$y+6 = -\frac{2}{7}x$$
$$y = -\frac{2}{7}x - 6$$

25. Every vertical line is in the form $x = c$. Since the line passes through the point $(2, 6)$, its equation is $x = 2$.

29. A line with undefined slope is vertical. Every vertical line is in the form $x = c$. Since the line passes through the point $(0, 5)$, its equation is $x = 0$.

33. $y = -2x - 6, m = -2$

so perpendicular slope $= \dfrac{1}{2}$

$$y-(-5) = \frac{1}{2}(x-2)$$
$$y+5 = \frac{1}{2}x - 1$$
$$y = \frac{1}{2}x - 6$$

37. $2x - y = 8$

$y = 2x - 8, m = 2$

so perpendicular slope $= -\dfrac{1}{2}$.

$$y-5 = -\frac{1}{2}(x-3)$$
$$y-5 = -\frac{1}{2}x + \frac{3}{2}$$
$$y = -\frac{1}{2}x + \frac{13}{2}$$

41. $x - 4y = 4$

$$-4y = -x + 4$$
$$y = \frac{1}{4}x - 1, m = \frac{1}{4}$$

so perpendicular slope $= -4$

$$y-5 = -4(x+1)$$
$$y-5 = -4x - 4$$
$$y = -4x + 1$$

45. a. $(1, 30{,}000), (4, 66{,}000)$

$$m = \frac{66{,}000 - 30{,}000}{4-1} = 12{,}000$$
$$y - 30{,}000 = 12{,}000(x-1)$$
$$y = 12{,}000x + 18{,}000$$

b. $y = 12{,}000(7) + 18{,}000 = \$102{,}000$

c. $126{,}000 = 12{,}000x + 18{,}000$

$$x = \frac{126{,}000 - 18{,}000}{12{,}000}$$
$$x = 9 \text{ years}$$

49. a. $(0, 225), (10, 391)$

$$m = \frac{391 - 225}{10 - 0} = 16.6$$
$$y - 225 = 16.6(x-0)$$
$$y = 16.6x + 225$$

b. $x = 2004 - 1996 = 8$

$$y = 16.6(8) + 225$$
$$y = 357.8 \text{ thousand people}$$

53. $y = 4.2x$

$y = 4.2(-2)$

$y = -8.4$

$(-2, -8.4)$

57. True

61. $y = 4x - 2, y = 4x - 4$

10

−10 10

−10

Exercise Set 3.5

1. Domain: $= \{-1, 0, -2, 5\}$
Range $= \{7, 6, 2\}$
The relation is a function.

5. Domain: $= \{1\}$
Range $= \{1, 2, 3, 4\}$
The relation is not a function since 1 is paired with 1, 2, 3 and 4.

9. Domain: $\{-3, 0, 3\}$
Range $= \{-3, 0, 3\}$
The relation is a function.

13. Domain $=$ {Colorado, Alaska, Delaware, Illinois, Connecticut, Texas}
Range $= \{6, 1, 20, 30\}$
The relation is a function.

17. Domain $= \{2, -1, 5, 100\}$
Range $= \{0\}$
The relation is a function.

21. Yes

25. Yes

29. Not a function

33. Domain $= [0, \infty)$
Range $=$ All real numbers
The relation is not a function since it fails the vertical line test (try $x = 1$).

37. Domain $= (-\infty, \infty)$
Range $= (-\infty, -3] \cup [3, \infty)$
The relation is not a function since it fails the vertical line test (try $x = 3$).

41. Domain = $\{-2\}$
Range = $(-\infty, \infty)$
The relation is not a function since it fails the vertical line test.

45. $f(x) = 3x + 3$
$f(4) = 3(4) + 3 = 12 + 3 = 15$

49. $g(x) = 4x^2 - 6x + 3$
$g(2) = 4(2)^2 - 6(2) + 3$
$= 4(4) - 12 + 3$
$= 16 - 12 + 3$
$= 7$

53. $f(x) = \frac{1}{2}x$

a. $f(0) = \frac{1}{2}(0) = 0$

b. $f(2) = \frac{1}{2}(2) = 1$

c. $f(-2) = \frac{1}{2}(-2) = -1$

57. $A(x) = \pi x^2$
$A(5) = \pi(5)^2 = 25\pi$
25π square centimeters

61. $H(f) = 2.59f + 47.24$
$H(46) = 2.59(46) + 47.24$
$= 119.14 + 47.24$
$= 166.38$
166.38 centimeters

65. $C(x) = 1.7x + 88$

a. $C(2) = 1.7(2) + 88 = 91.4$
The per capita consumption of poultry was 91.4 pounds in 1997.

b. $x = 1999 - 1995 = 4$
$C(4) = 1.7(4) + 88 = 94.8$
94.8 pounds

69. $f(x) = -3x$
or
$y = -3x$
where $m = -3$ and $b = 0$

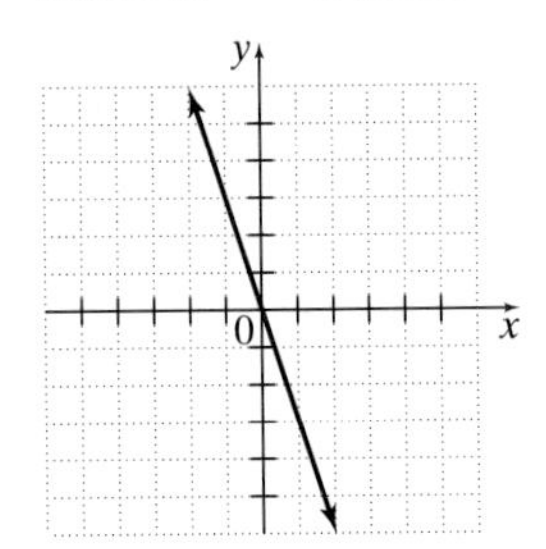

73. $2x - 7 \leq 21$
$2x \leq 28$
$x \leq 14$
$(-\infty, 14]$

77. $\frac{x}{2} + \frac{1}{4} < \frac{1}{8}$
$\frac{x}{2} < \frac{1}{8} - \frac{1}{4}$
$\frac{x}{2} < -\frac{1}{8}$
$x < -\frac{1}{4}$
$\left(-\infty, -\frac{1}{4}\right)$

81. $f(x) = x^2 - 12$

a. $f(12) = (12)^2 - 12 = 144 - 12 = 132$

b. $f(a) = a^2 - 12$

85. answers may vary

Exercise Set 3.6

1. $x < 2$

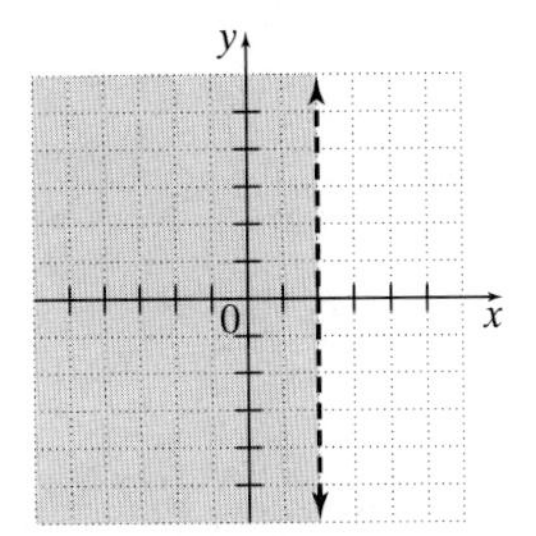

5. $3x + y > 6$
Test (0, 0)
$3(0) + 0 > 6$
$0 > 6$ False
Shade the half-plane that does not contain (0, 0).

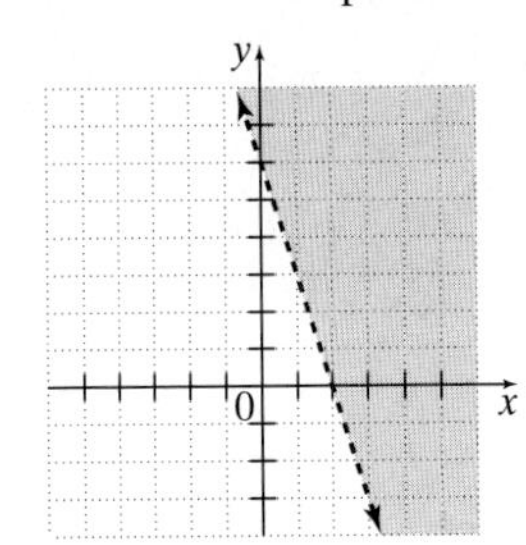

9. $2x + 4y \geq 8$
Test (0, 0)
$2(0) + 4(0) \geq 8$
$0 \geq 8$ False
Shade the half-plane that does not contain the point (0, 0).

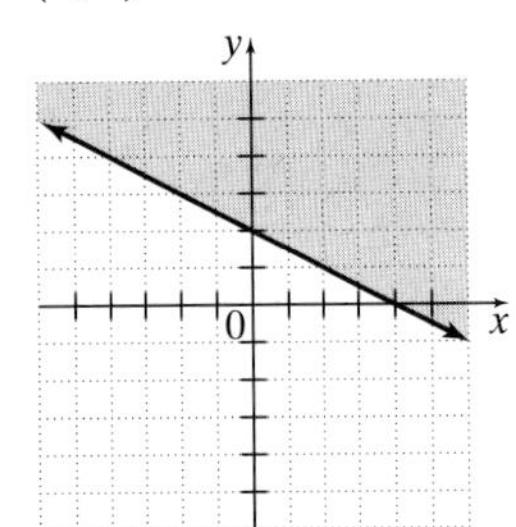

13. A dashed boundary line should be used when the inequality contains a $<$ or $>$.

SOLUTIONS

SOLUTIONS

17. $x \le -2$ or $y \ge 4$

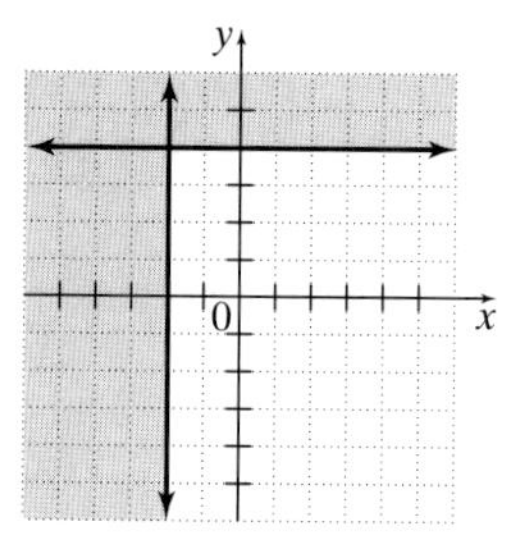

21. $x + y \le 3$ or $x - y \ge 5$

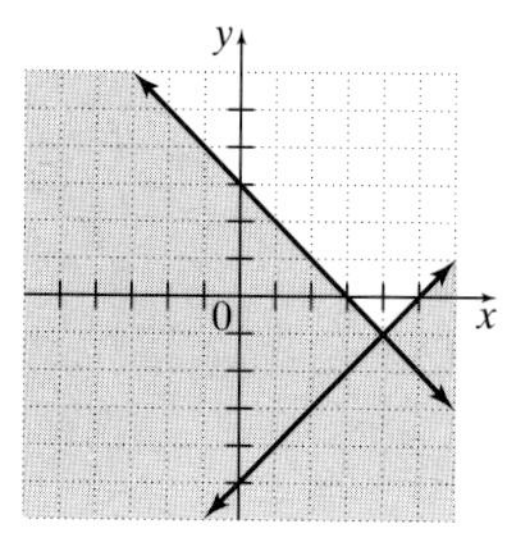

25. $x + y \le 1$ and $y \le -1$

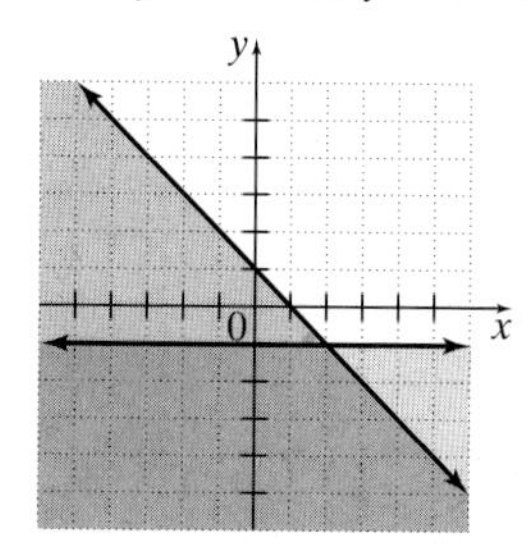

29. $y > 2x + 3$
dashed line
$(0, 0)$ results in a false inequality
graph A

33.
$$\begin{aligned} 3x + 2y &= -12 \\ 3(-4) + 2(0) &\overset{?}{=} -12 \\ -12 &= -12 \text{ True} \end{aligned} \qquad \begin{aligned} x &= 4y \\ -4 &\overset{?}{=} 4(0) \\ -4 &= 0 \text{ False} \end{aligned}$$

No, the ordered pair is not a solution of both equations.

Chapter 3 Test

1.

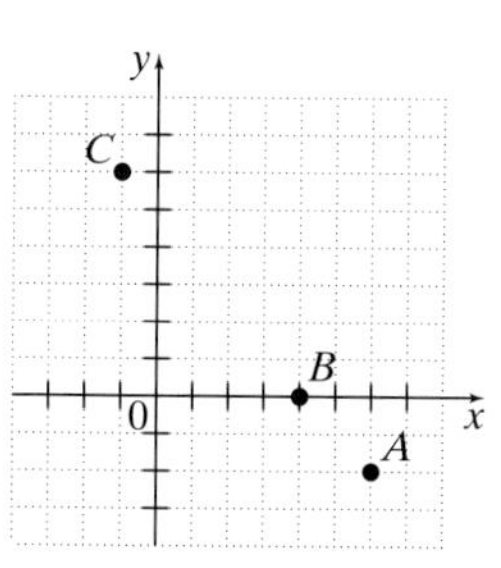

A is in quadrant IV.
B is on the x-axis, no quadrant.
C is in quadrant II.

5.
$$\begin{aligned} 4x + 6y &= 8 \\ 6y &= -4x + 8 \\ y &= -\frac{2}{3}x + \frac{4}{3} \\ m &= -\frac{2}{3}, b = \frac{4}{3} \end{aligned}$$

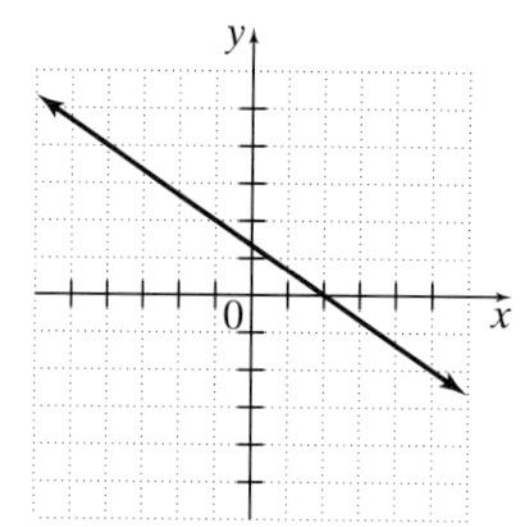

9. $f(x) = 3x + 1$
$m = 3, b = 1$
graph C

13. $y = -8$

17. $y = 5x + (-2)$
$y = 5x - 2$

21.
$$\begin{aligned} 2x - 5y &= 8 \\ 5y &= 2x - 8 \\ y &= \frac{2}{5}x - \frac{8}{5}, \text{ so } m_1 = \frac{2}{5}, \\ m_2 &= \frac{-1-4}{-1-1} = \frac{-5}{-2} = \frac{5}{2} \end{aligned}$$

Therefore, line L_1 and L_2 are neither parallel nor perpendicular.

25. $2x + 4y < 6$ and $y \le -4$

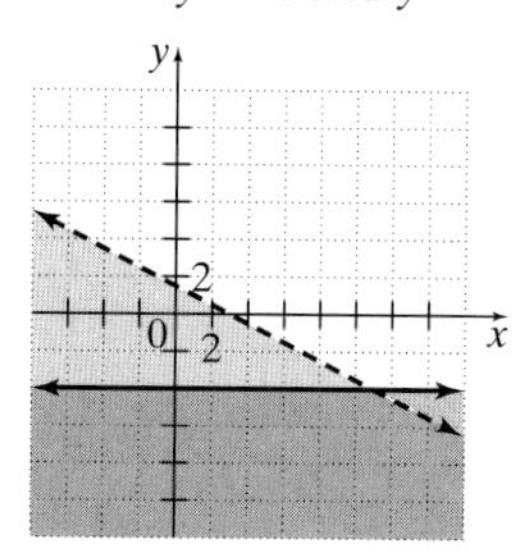

29. Domain: $(-\infty, \infty)$
Range: $(-\infty, \infty)$
Function

Chapter 4

Exercise Set 4.1

1. $\begin{cases} x - y = 3 \\ 2x - 4y = 8 \end{cases}$

$x - y = 3$
$2 - (-1) = 3$
$2 + 1 = 3$
$3 = 3$ True

$2x - 4y = 8$
$2(2) - 4(-1) = 8$
$4 + 4 = 8$
$8 = 8$ True

Yes, $(2, -1)$ is a solution.

5. $\begin{cases} y = -5x \\ x = -2 \end{cases}$

$y = -5x$
$10 = -5(-2)$
$10 = 10$ True

$x = -2$
$-2 = -2$ True

Yes, $(-2, 10)$ is a solution.

9. $\begin{cases} 2y - 4 = 0 \\ x + 2y = 5 \end{cases}$

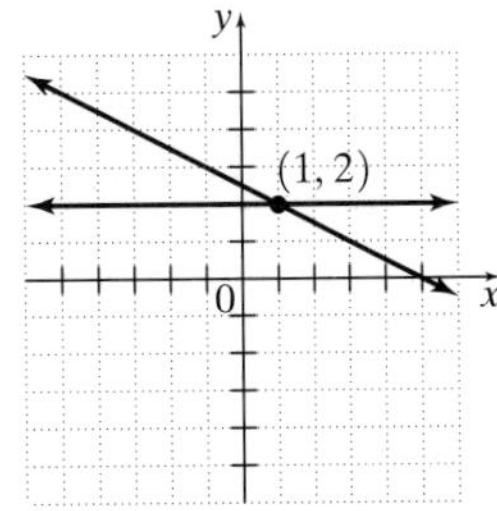

The solution is $(1, 2)$.

13. $\begin{cases} y = -3x \\ 2x - y = -5 \end{cases}$

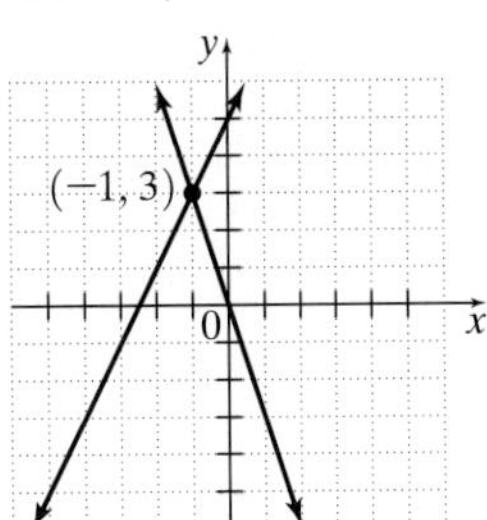

The solution is $(-1, 3)$.

17. $\begin{cases} 4x - y = 9 \\ 2x + 3y = -27 \end{cases}$

Solve the first equation for y.
$4x - y = 9$
$y = 4x - 9$
Replace y with $4x - 9$ in the second equation.
$2x + 3(4x - 9) = -27$
$2x + 12x - 27 = -27$
$14x = 0$
$x = 0$
Replace x with 0 in the first equation.
$4(0) - y = 9$
$y = -9$
The solution is $(0, -9)$.

21. $\begin{cases} \dfrac{x}{3} + y = \dfrac{4}{3} \\ -x + 2y = 11 \end{cases}$

Clear fractions by multiplying the first equation by 3.
$\begin{cases} x + 3y = 4 \\ -x + 2y = 11 \end{cases}$
Solve the second equation for x.
$2y - 11 = x$
$x = 2y - 11$
Replace x with $2y - 11$ in the first equation.
$2y - 11 + 3y = 4$
$5y = 15$
$y = 3$
Replace y with 3 in the equation $x = 2y - 11$.
$x = 2(3) - 11$
$x = -5$
The solution is $(-5, 3)$.

25. $\begin{cases} 2x = 6 \\ y = 5 - x \end{cases}$

Solve the first equation for x.
$x = 3$
Replace x with 3 in the second equation.
$y = 5 - 3$
$y = 2$
The solution is $(3, 2)$.

29. $\begin{cases} 5x + 2y = 1 \\ x - 3y = 7 \end{cases}$

Multiply the second equation by -5.
$\begin{cases} 5x + 2y = 1 \\ -5x + 15y = -35 \end{cases}$
Add the equations.
$5x + 2y = 1$
$-5x + 15y = -35$
$17y = -34$
$y = -2$
Replace y with -2 in the second equation.
$x - 3(-2) = 7$
$x + 6 = 7$
$x = 1$
The solution is $(1, -2)$.

33. $\begin{cases} 3x - 5y = 11 \\ 2x - 6y = 2 \end{cases}$

Multiply the second equation by $\dfrac{1}{2}$.
$\begin{cases} 3x - 5y = 11 \\ x - 3y = 1 \end{cases}$
Multiply the second equation by -3.
$\begin{cases} 3x - 5y = 11 \\ -3x + 9y = -3 \end{cases}$
Add the equations.
$3x - 5y = 11$
$-3x + 9y = -3$
$4y = 8$
$y = 2$
Replace y with 2 in the equation $x - 3y = 1$.
$x - 3(2) = 1$
$x - 6 = 1$
$x = 7$
The solution is $(7, 2)$.

37. $\begin{cases} 3x + y = 1 \\ 2y = 2 - 6x \end{cases}$

$\begin{cases} 3x + y = 1 \\ 6x + 2y = 2 \end{cases}$

Multiply the first equation by −2.

$\begin{cases} -6x - 2y = -2 \\ 6x + 2y = 2 \end{cases}$

Add the equations.

$$\begin{aligned} -6x - 2y &= -2 \\ 6x + 2y &= 2 \\ \hline 0 &= 0 \text{ True} \end{aligned}$$

Dependent system

$\{(x, y) | 3x + y = 1\}$

41. $\begin{cases} 4x + 2y = 5 \\ 2x + y = -1 \end{cases}$

Multiply the second equation by −2.

$\begin{cases} 4x + 2y = 5 \\ -4x - 2y = 2 \end{cases}$

Add the equations.

$$\begin{aligned} 4x + 2y &= 5 \\ -4x - 2y &= 2 \\ \hline 0 &= 7 \text{ False} \end{aligned}$$

Inconsistent system

The solution is $\emptyset$.

45. $\begin{cases} \frac{2}{3}x - \frac{3}{4}y = -1 \\ -\frac{1}{6}x + \frac{3}{8}y = 1 \end{cases}$

Multiply the first equation by 12 and the second equation by 24.

$\begin{cases} 8x - 9y = -12 \\ -4x + 9y = 24 \end{cases}$

Add the equations.

$$\begin{aligned} 8x - 9y &= -12 \\ -4x + 9y &= 24 \\ \hline 4x &= 12 \\ x &= 3 \end{aligned}$$

Replace x with 3 in the equation $-4x + 9y = 24$.

$$\begin{aligned} -4(3) + 9y &= 24 \\ -12 + 9y &= 24 \\ 9y &= 36 \\ y &= 4 \end{aligned}$$

The solution is (3, 4).

49. $\begin{cases} 10y - 2x = 1 \\ 5y = 4 - 6x \end{cases}$

Multiply the second equation by −2.

$\begin{cases} 10y = 2x + 1 \\ -10y = 12x - 8 \end{cases}$

Add the equations.

$$\begin{aligned} 10y &= 2x + 1 \\ -10y &= 12x - 8 \\ \hline 0 &= 14x - 7 \\ 14x &= 7 \\ x &= \frac{1}{2} \end{aligned}$$

Replace with $\frac{1}{2}$ in the first equation.

$$\begin{aligned} 10y - 2\left(\frac{1}{2}\right) &= 1 \\ 10y - 1 &= 1 \\ 10y &= 2 \\ y &= \frac{1}{5} \end{aligned}$$

The solution is $\left(\frac{1}{2}, \frac{1}{5}\right)$.

53.

$$\begin{aligned} 3x - 4y + 2z &= 5 \\ 3(1) - 4(2) + 2(5) &\stackrel{?}{=} 5 \\ 3 - 8 + 10 &\stackrel{?}{=} 5 \\ 5 &= 5 \text{ True} \end{aligned}$$

57.

$$\begin{aligned} 3x + 2y - 5z &= 10 \\ -3x + 4y + z &= 15 \\ \hline 6y - 4z &= 25 \end{aligned}$$

61. no

65. Supply is greater than demand because the supply line is above the demand line and the supply line is increasing linearly while the demand line is decreasing linearly.

Exercise Set 4.2

1. $\begin{cases} x + y = 3 \\ 2y = 10 \\ 3x + 2y - 3z = 1 \end{cases}$

Solve the second equation for y.

$y = 5$

Replace y with 5 in the first equation.

$$\begin{aligned} x + 5 &= 3 \\ x &= -2 \end{aligned}$$

Replace x with −2 and y with 5 in the third equation.

$$\begin{aligned} 3(-2) + 2(5) - 3z &= 1 \\ -6 + 10 - 3z &= 1 \\ 4 - 3z &= 1 \\ -3z &= -3 \\ z &= 1 \end{aligned}$$

The solution is (−2, 5, 1).

5. $\begin{cases} x - 2y + z = -5 & (1) \\ -3x + 6y - 3z = 15 & (2) \\ 2x - 4y + 2z = -10 & (3) \end{cases}$

Multiply equation (2) by $-\frac{1}{3}$ and equation (1) by $\frac{1}{2}$.

$\begin{cases} x - 2y + z = -5 \\ x - 2y + z = -5 \\ x - 2y + z = -5 \end{cases}$

All three equations are identical. There are infinitely many solutions.
The solution is $\{(x, y, z) | x - 2y + z = -5\}$.

9. $\begin{cases} x + 5z = 0 & (1) \\ 5x + y = 0 & (2) \\ y - 3z = 1 & (3) \end{cases}$

Multiply equation (3) by -1 and add to equation (2).

$$\begin{array}{l} -y + 3z = 0 \\ \underline{5x + y \qquad = 0} \\ 5x \qquad + 3z = 0 \quad (4) \end{array}$$

Multiply equation (1) by -5 and add to equation (4).

$$\begin{array}{l} -5x - 25z = 0 \\ \underline{5x + 3z = 0} \\ -22z = 0 \\ z = 0 \end{array}$$

Replace z with 0 in equation (4).

$5x + 3(0) = 0$
$5x = 0$
$x = 0$

Replace x with 0 in equation (2).

$5(0) + y = 0$
$y = 0$

The solution is $(0, 0, 0)$.

13. $\begin{cases} x + y + z = 8 & (1) \\ 2x - y - z = 10 & (2) \\ x - 2y - 3z = 22 & (3) \end{cases}$

Add equations (1) and (2).

$3x = 18$ or $x = 6$

And twice equation 1 to equation 3.

$$\begin{array}{l} 2x + 2y + 2z = 16 \\ \underline{x - 2y - 3z = 22} \\ 3x \qquad - z = 38 \end{array}$$

Replace x with 6 in this equation.

$3(6) - z = 38$
$18 - z = 38$
$-z = 20$
$z = -20$

Replace x with 6 and z with -20 in equation (1).

$6 + y + (-20) = 8$
$y - 14 = 8$
$y = 22$

The solution is $(6, 22, -20)$.

17. $\begin{cases} 2x - 3y + z = 2 & (1) \\ x - 5y + 5z = 3 & (2) \\ 3x + y - 3z = 5 & (3) \end{cases}$

Add -2 times equation (2) to equation (1).

$$\begin{array}{l} 2x - 3y + z = 2 \\ \underline{-2x + 10y - 10z = -6} \\ 7y - 9z = -4 \end{array}$$

Add -3 times equation (2) to equation (3).

$$\begin{array}{l} -3x + 15y - 15z = -9 \\ \underline{3x + y - 3z = 5} \\ 16y - 18z = -4 \end{array}$$

We now have the system:

$\begin{cases} 7y - 9z = -4 & (4) \\ 16y - 18z = -4 & (5) \end{cases}$

Multiply equation (4) by -2 and add to equation (5).

$$\begin{array}{l} -14y + 18z = 8 \\ \underline{16y - 18z = -4} \\ 2y = 4 \\ y = 2 \end{array}$$

Replace y with 2 in equation (4).

$7(2) - 9z = -4$
$-9z = -18$
$z = 2$

Replace y with 2 and z with 2 in equation (1).

$2x - 3(2) + 2 = 2$
$2x = 6$
$x = 3$

The solution is $(3, 2, 2)$.

21. $\begin{cases} 2x + 2y - 3z = 1 & (1) \\ y + 2z = -14 & (2) \\ 3x - 2y = -1 & (3) \end{cases}$

Add equations (1) and (3).

$5x - 3z = 0 \quad (4)$

Multiply equation (2) by 2 and add to equation (3).

$$\begin{array}{l} 2y + 4z = -28 \\ \underline{3x - 2y \qquad = -1} \\ 3x \qquad + 4z = -29 \quad (5) \end{array}$$

Multiply equation (4) by 4, multiply equation (5) by 3, and add.

$$\begin{array}{l} 20x - 12z = 0 \\ \underline{9x + 12z = -87} \\ 29x \qquad = -87 \\ x = -3 \end{array}$$

Replace x with -3 in equation (4).

$5(-3) - 3z = 0$
$3z = -15$
$z = -5$

Replace z with -5 in equation (2).

$y + 2(-5) = -14$
$y - 10 = -14$
$y = -4$

The solution is $(-3, -4, -5)$.

SOLUTIONS

25. $\begin{cases} x + y + z = 1 & (1) \\ 2x - y + z = 0 & (2) \\ -x + 2y + 2z = -1 & (3) \end{cases}$

Multiply equation (3) by 2 and add to equation (2).

$$\begin{array}{r} 2x - y + z = 0 \\ -2x + 4y + 4z = -2 \\ \hline 3y + 5z = -2 \quad (4) \end{array}$$

Multiply equation (1) by -2 and add to equation (2).

$$\begin{array}{r} -2x - 2y - 2z = -2 \\ 2x - y + z = 0 \\ \hline -3y - z = -2 \quad (5) \end{array}$$

Add equations (4) and (5).

$$\begin{array}{r} 3y + 5z = -2 \\ -3y - z = -2 \\ \hline 4z = -4 \\ z = -1 \end{array}$$

Replace z with -1 in equation (4).

$3y + 5(-1) = -2$

$3y = 3$

$y = 1$

Replace y with 1 and z with -1 in equation (1).

$x + 1 + (-1) = 1$

$x = 1$

The solution is $(1, 1, -1)$.

$\frac{1}{24} = \frac{x}{8} + \frac{y}{4} + \frac{z}{3}$

$\frac{1}{24} = \frac{1}{8} + \frac{1}{4} - \frac{1}{3}$

$\frac{1}{24} = \frac{3}{24} + \frac{6}{24} - \frac{8}{24}$

$\frac{1}{24} = \frac{1}{24}$ True

29. $2(x - 1) - 3x = x - 12$

$2x - 2 - 3x = x - 12$

$-2 - x = x - 12$

$10 = 2x$

$5 = x$

The solution is $\{5\}$.

33. answers may vary

37. $\begin{cases} x + y + z + w = 5 & (1) \\ 2x + y + z + w = 6 & (2) \\ x + y + z \quad\quad = 2 & (3) \\ x + y \quad\quad\quad\quad = 0 & (4) \end{cases}$

Multiply equation (1) by -1 and add to equation (2).

$x = 1$

Replace x with 1 in equation (4).

$1 + y = 0$

$y = -1$

Replace x with 1 and y with -1 in equation (3).

$1 + (-1) + z = 2$

$z = 2$

Replace x with 1, y with -1, and z with 2 in equation (1).

$1 + (-1) + 2 + w = 5$

$2 + w = 5$

$w = 3$

The solution is $(1, -1, 2, 3)$.

Exercise Set 4.3

1. Let $m =$ the first number

$n =$ the second number

$\begin{cases} m = n + 2 \\ 2m = 3n - 4 \end{cases}$

Substitute $m = n + 2$ in the second equation.

$2(n + 2) = 3n - 4$

$2n + 4 = 3n - 4$

$n = 8$

Replace n with 8 in the first equation.

$m = 8 + 2 = 10$

The numbers are 10 and 8.

5. Let $x =$ number of quarts of 4% butterfat milk

$y =$ number of quarts of 1% butterfat milk

$\begin{cases} x + y = 60 \\ 0.04x + 0.01y = 0.02(60) \end{cases}$

Multiply the second equation by -100 and add to the first equation.

$$\begin{array}{r} x + y = 60 \\ -4x - y = -120 \\ \hline -3x \quad\quad = -60 \\ x = 20 \end{array}$$

Replace x with 20 in the first equation.

$20 + y = 60$

$y = 40$

20 quarts of 4% butterfat milk and 40 quarts of 1% butterfat milk should be used.

9. Let $m =$ the first number

$n =$ the second number

$\begin{cases} m = n - 2 \\ 2m = 3n + 4 \end{cases}$

Substitute $m = n - 2$ in the second equation.

$2(n - 2) = 3n + 4$

$2n - 4 = 3n + 4$

$n = -8$

Replace n with -8 in the first equation.

$m = -8 - 2 = -10$

The numbers are -10 and -8.

13. Let $p =$ the speed of the plane in still air

$w =$ the speed of the wind

First note:

$\frac{2160 \text{ miles}}{3 \text{ hours}} = 720$ mph and

$\frac{2160 \text{ miles}}{4 \text{ hours}} = 540$ mph

Now,

$\begin{cases} p + w = 720 \\ p - w = 540 \end{cases}$

Add the equations.

$2p = 1260$

$p = 630$

Replace p with 630 in the first equation.

$630 + w = 720$

$w = 90$

The speed of the plane in still air is 630 mph and the speed of the wind is 90 mph.

17. Let $x =$ the first number
$y =$ the second number
$z =$ the third number

$$\begin{cases} x + y + z = 40 \\ x = y + 5 \\ x = 2z \end{cases}$$

$$\begin{cases} x + y + z = 40 \\ y = x - 5 \\ z = \frac{1}{2}x \end{cases}$$

Substitute $y = x - 5$ and $z = \frac{1}{2}x$ in the first equation.

$$x + x - 5 + \frac{1}{2}x = 40$$
$$\frac{5}{2}x - 5 = 40$$
$$\frac{5}{2}x = 45$$
$$x = \frac{2}{5}(45) = 18$$
$$y = 18 - 5 = 13$$
$$z = \frac{1}{2}(18) = 9$$

The three numbers are 18, 13, and 9.

21. Let $x =$ price for template
$y =$ price for pencil
$z =$ price for paper

$$\begin{cases} 3x + y = 6.45 \\ 2z + 4y = 7.50 \\ z = 3y \end{cases}$$

Substitute $z = 3y$ in the second equation.

$$2(3y) + 4y = 7.50$$
$$6y + 4y = 7.50$$
$$10y = 7.50$$
$$y = 0.75$$

Replace y with 0.75 in the third equation.

$$3(0.75) = z$$
$$2.25 = z$$

Replace y with 0.75 in the first equation.

$$3x + 0.75 = 6.45$$
$$3x = 5.70$$
$$x = 1.90$$

\$1.90 for template; \$0.75 for pencil; \$2.25 for paper

25.

Concentration	Amount	Solution
25%	$2x$	$0.25(2x) = 0.5x$
40%	x	$0.4x$
50%	y	$0.5y$
32%	200	$0.32(200) = 64$

$$\begin{cases} 2x + x + y = 200 \\ 0.5x + 0.4x + 0.5y = 64 \end{cases}$$

$$\begin{cases} 3x + y = 200 \\ 0.9x + 0.5y = 64 \end{cases}$$

Solve first equation for y and substitute into second equation.

$$y = 200 - 3x$$
$$0.9x + 0.5(200 - 3x) = 64$$
$$0.9x + 100 - 1.5x = 64$$
$$-0.6x = -36$$
$$x = 60$$

$$2x = 120$$

Replace x with 60 in the equation $y = 200 - 3x$.

$$y = 200 = 3(60) = 20$$

120 liters of 25% solution
60 liters of 40% solution
20 liters of 50% solution

29. $x = 180 - (z + 15)$
$x = 165 - z$

$y = 180 - (z - 13)$
$y = 193 - z$

$360 = x + y + z + 72$
$288 = x + y + z$

$$\begin{cases} x = 165 - z \\ y = 193 - z \\ x + y + z = 288 \end{cases}$$

Substitute $x = 165 - z$ and $y = 193 - z$ in the third equation.

$$165 - z + 193 - z + z = 288$$
$$-z = -70$$
$$z = 70$$

Replace z with 70 in the first and second equations.

$x = 165 - 70 = 95$
$y = 193 - 70 = 123$
$(95°, 123°, 70°)$

33.
$$C(x) = 0.8x + 900$$
$$R(x) = 2x$$
$$0.8x + 900 = 2x$$
$$900 = 1.2x$$
$$750 = x$$

750 units

37. **a.** $R(x) = 31x$

b. $C(x) = 15x + 500$

c. $R(x) = C(x)$
$$31x = 15x + 500$$
$$16x = 500$$
$$x = 31.25$$

32 baskets

SOLUTIONS

41. $\begin{aligned} 2x - 3y + 2z &= 5 \quad (1) \\ x - 9y + z &= -1 \quad (2) \end{aligned}$

$$\begin{array}{r} -6x + 9y - 6z = -15 \\ x - 9y + z = -1 \\ \hline -5x - \qquad 5z = -16 \end{array}$$

Exercise Set 4.4

1. $\begin{cases} x + y = 1 \\ x - 2y = 4 \end{cases}$

$\left[\begin{array}{cc|c} 1 & 1 & 1 \\ 1 & -2 & 4 \end{array}\right]$

Multiply row 1 by -1 and add to row 2.

$\left[\begin{array}{cc|c} 1 & 1 & 1 \\ 0 & -3 & 3 \end{array}\right]$

Divide row 2 by -3.

$\left[\begin{array}{cc|c} 1 & 1 & 1 \\ 0 & 1 & -1 \end{array}\right]$

This corresponds to $\begin{cases} x + y = \ \ 1 \\ \quad\ \ y = -1 \end{cases}$.

$$\begin{aligned} x + (-1) &= 1 \\ x - 1 &= 1 \\ x &= 2 \end{aligned}$$

The solution is $(2, -1)$.

5. $\begin{cases} x - 2y = 4 \\ 2x - 4y = 4 \end{cases}$

$\left[\begin{array}{cc|c} 1 & -2 & 4 \\ 2 & -4 & 4 \end{array}\right]$

Multiply row 1 by -2 and add to row 2.

$\left[\begin{array}{cc|c} 1 & -2 & 4 \\ 0 & 0 & 4 \end{array}\right]$

This is an inconsistent system.
The solution is $\emptyset$.

9. $\begin{cases} x - 4 = 0 \\ x + y = 1 \end{cases}$ or $\begin{cases} \quad\ \ x = 4 \\ x + y = 1 \end{cases}$

$\left[\begin{array}{cc|c} 1 & 0 & 4 \\ 1 & 1 & 1 \end{array}\right]$

Multiply row 1 by -1 and add to row 2.

$\left[\begin{array}{cc|c} 1 & 0 & 4 \\ 0 & 1 & -3 \end{array}\right]$

This corresponds to $\begin{cases} x \quad\ = 4 \\ \quad y = -3 \end{cases}$.

The solution is $(4, -3)$.

13. $\begin{cases} \qquad 2y - z = -7 \\ x + 4y + z = -4 \\ 5x - y + 2z = 13 \end{cases}$

$\left[\begin{array}{ccc|c} 0 & 2 & -1 & -7 \\ 1 & 4 & 1 & -4 \\ 5 & -1 & 2 & 13 \end{array}\right]$

Interchange rows 1 and 2.

$\left[\begin{array}{ccc|c} 1 & 4 & 1 & -4 \\ 0 & 2 & -1 & -7 \\ 5 & -1 & 2 & 13 \end{array}\right]$

Multiply row 1 by -5 and add to row 3.

$\left[\begin{array}{ccc|c} 1 & 4 & 1 & -4 \\ 0 & 2 & -1 & -7 \\ 0 & -21 & -3 & 33 \end{array}\right]$

Divide row 2 by 2.

$\left[\begin{array}{ccc|c} 1 & 4 & 1 & -4 \\ 0 & 1 & -\frac{1}{2} & -\frac{7}{2} \\ 0 & -21 & -3 & 33 \end{array}\right]$

Multiply row 2 by 21 and add to row 3.

$\left[\begin{array}{ccc|c} 1 & 4 & 1 & -4 \\ 0 & 1 & -\frac{1}{2} & -\frac{7}{2} \\ 0 & 0 & -\frac{27}{2} & -\frac{81}{2} \end{array}\right]$

Multiply row 3 by $-\dfrac{2}{27}$.

$\left[\begin{array}{ccc|c} 1 & 4 & 1 & -4 \\ 0 & 1 & -\frac{1}{2} & -\frac{7}{2} \\ 0 & 0 & 1 & 3 \end{array}\right]$

This corresponds to $\begin{cases} x + 4y + z = -4 \\ \quad\ y - \dfrac{1}{2}z = -\dfrac{7}{2} \\ \qquad\qquad z = 3 \end{cases}$.

$$\begin{aligned} y - \frac{1}{2}(3) &= -\frac{7}{2} \\ y - \frac{3}{2} &= -\frac{7}{2} \\ y &= -2 \end{aligned}$$

$$\begin{aligned} x + 4(-2) + 3 &= -4 \\ x - 8 + 3 &= -4 \\ x &= 1 \end{aligned}$$

The solution is $(1, -2, 3)$.

17. $\begin{cases} 4x + y + z = 3 \\ -x + y - 2z = -11 \\ x + 2y + 2z = -1 \end{cases}$

$$\left[\begin{array}{ccc|c} 4 & 1 & 1 & 3 \\ -1 & 1 & -2 & -11 \\ 1 & 2 & 2 & -1 \end{array}\right]$$

Interchange rows 1 and 3.

$$\left[\begin{array}{ccc|c} 1 & 2 & 2 & -1 \\ -1 & 1 & -2 & -11 \\ 4 & 1 & 1 & 3 \end{array}\right]$$

Multiply row 1 by 1 and add to row 2.
Multiply row 1 by −4 and add to row 3.

$$\left[\begin{array}{ccc|c} 1 & 2 & 2 & -1 \\ 0 & 3 & 0 & -12 \\ 0 & -7 & -7 & 7 \end{array}\right]$$

Divide row 2 by 3.

$$\left[\begin{array}{ccc|c} 1 & 2 & 2 & -1 \\ 0 & 1 & 0 & -4 \\ 0 & -7 & -7 & 7 \end{array}\right]$$

Multiply row 2 by 7 and add to row 3.

$$\left[\begin{array}{ccc|c} 1 & 2 & 2 & -1 \\ 0 & 1 & 0 & -4 \\ 0 & 0 & -7 & -21 \end{array}\right]$$

Divide row 3 by −7.

$$\left[\begin{array}{ccc|c} 1 & 2 & 2 & -1 \\ 0 & 1 & 0 & -4 \\ 0 & 0 & 1 & 3 \end{array}\right]$$

This corresponds to $\begin{cases} x + 2y + 2z = -1 \\ \quad\quad y \quad\quad = -4. \\ \quad\quad\quad\quad z = 3 \end{cases}$

$$x + 2(-4) + 2(3) = -1$$
$$x - 8 + 6 = -1$$
$$x = 1$$

The solution is $(1, -4, 3)$.

21. $(4)(-10) - (2)(-2) = -40 + 4 = -36$

25. a. Solve the system $\begin{cases} 2.3x + y = 52 \\ -5.4x + y = 14 \end{cases}$.

$$\left[\begin{array}{cc|c} 2.3 & 1 & 52 \\ -5.4 & 1 & 14 \end{array}\right]$$

Since getting a 1 in the first column would lead to repeating decimals, we multiply row 1 by −1 and add to row 2.

$$\left[\begin{array}{cc|c} 2.3 & 1 & 52 \\ -7.7 & 0 & -38 \end{array}\right]$$

This corresponds to $\begin{cases} 2.3x + y = 52 \\ -7.7x \quad = -38 \end{cases}$.

From the second equation, $x = \frac{-38}{-77} \approx 4.935$.

Thus, the percent of U.S. households owning black-and-white television sets was the same as the percent of U.S. households owning a microwave oven in the end of 1984 (about 4.9 years after 1980).

b. Solving the television equation for y, we get $y = -2.3x + 52$. Thus, for 1980, $y = -2.3x + 52 = 52$, and for 1993, $y = -2.3(13) + 52 = 22.1$.
Solving the microwave oven equation for y, we get $y = 5.4x + 14$. Thus, for 1980, $y = 5.4(0) + 14 = 14$, and for 1993, $y = 5.4(13) + 14 = 84.2$.
In 1980, a greater percent of U.S. households, hence more households, owned black-and-white television sets. In 1993, more households owned a microwave oven. The percent of households owning black-and-white television sets is decreasing and the percent of households owning microwave ovens is increasing. answers may vary

c. The percent will reach 0% when $y = 0$ in the equation $2.3x + y = 52$.

$$2.3x + 0 = 52$$
$$x = \frac{52}{2.3}$$
$$x \approx 22.6$$

According to this model, the percent of U.S. households owning a black-and-white television set will be 0% about 22.6 years after 1980, or sometime in 2002.

Exercise Set 4.5

1. $\begin{vmatrix} 3 & 5 \\ -1 & 7 \end{vmatrix} = 3(7) - 5(-1)$
$= 21 + 5 = 26$

5. $\begin{vmatrix} -2 & 9 \\ 4 & -18 \end{vmatrix} = -2(-18) - 9(4)$
$= 36 - 36 = 0$

9. $\begin{cases} 2y - 4 = 0 \\ x + 2y = 5 \end{cases}$ or $\begin{cases} 2y = 4 \\ x + 2y = 5 \end{cases}$

$D = \begin{vmatrix} 0 & 2 \\ 1 & 2 \end{vmatrix} = 0(2) - 2(1) = 0 - 2 = -2$

$D_x = \begin{vmatrix} 4 & 2 \\ 5 & 2 \end{vmatrix} = 4(2) - 2(5) = 8 - 10 = -2$

$D_y = \begin{vmatrix} 0 & 4 \\ 1 & 5 \end{vmatrix} = 0(5) - 4(1) = 0 - 4 = -4$

$x = \frac{-2}{-2} = 1$ and $y = \frac{-4}{-2} = 2$

The solution is $(1, 2)$.

13. $\begin{cases} 5x - 2y = 27 \\ -3x + 5y = 18 \end{cases}$

$D = \begin{vmatrix} 5 & -2 \\ -3 & 5 \end{vmatrix} = 5(5) - (-2)(-3) = 25 - 6 = 19$

$D_x = \begin{vmatrix} 27 & -2 \\ 18 & 5 \end{vmatrix} = 27(5) - (-2)18 = 135 + 36$
$= 171$

$D_y = \begin{vmatrix} 5 & 27 \\ -3 & 18 \end{vmatrix} = 5(18) - 27(-3) = 90 + 81 = 171$

$x = \frac{171}{19} = 9$ and $y = \frac{171}{19} = 9$

The solution is $(9, 9)$.

17. $\begin{cases} \frac{2}{3}x - \frac{3}{4}y = -1 \\ -\frac{1}{6}x + \frac{3}{4}y = \frac{5}{2} \end{cases}$

$$D = \begin{vmatrix} \frac{2}{3} & -\frac{3}{4} \\ -\frac{1}{6} & \frac{3}{4} \end{vmatrix} = \frac{2}{3}\cdot\frac{3}{4} - \left(-\frac{3}{4}\right)\left(-\frac{1}{6}\right) = \frac{1}{2} - \frac{1}{8} = \frac{3}{8}$$

$$D_x = \begin{vmatrix} -1 & -\frac{3}{4} \\ \frac{5}{2} & \frac{3}{4} \end{vmatrix} = (-1)\frac{3}{4} - \left(-\frac{3}{4}\right)\frac{5}{2} = -\frac{3}{4} + \frac{15}{8} = \frac{9}{8}$$

$$D_y = \begin{vmatrix} \frac{2}{3} & -1 \\ -\frac{1}{6} & \frac{5}{2} \end{vmatrix} = \frac{2}{3}\cdot\frac{5}{2} - (-1)\left(-\frac{1}{6}\right) = \frac{5}{3} - \frac{1}{6} = \frac{3}{2}$$

Thus, $x = \frac{\frac{9}{8}}{\frac{3}{8}} = 3$ and $y = \frac{\frac{3}{2}}{\frac{3}{8}} = 4$

The solution is $(3, 4)$.

21. $\begin{vmatrix} 4 & -6 & 0 \\ -2 & 3 & 0 \\ 4 & -6 & 1 \end{vmatrix}$

$$= 0\begin{vmatrix} -2 & 3 \\ 4 & -6 \end{vmatrix} - 0\begin{vmatrix} 4 & -6 \\ 4 & -6 \end{vmatrix} + 1\begin{vmatrix} 4 & -6 \\ -2 & 3 \end{vmatrix}$$
$$= 0 - 0 + [4(3) - (-6)(-2)]$$
$$= 0$$

25. $\begin{vmatrix} 3 & 6 & -3 \\ -1 & -2 & 3 \\ 4 & -1 & 6 \end{vmatrix}$

$$= 3\begin{vmatrix} -2 & 3 \\ -1 & 6 \end{vmatrix} - 6\begin{vmatrix} -1 & 3 \\ 4 & 6 \end{vmatrix} + (-3)\begin{vmatrix} -1 & -2 \\ 4 & -1 \end{vmatrix}$$
$$= 3[(-2)(6) - 3(-1)] - 6[(-1)(6) - 3(4)] - 3[(-1)(-1) - (-2)4]$$
$$= 3(-9) - 6(-18) - 3(9)$$
$$= -27 + 108 - 27$$
$$= 54$$

29. $\begin{cases} x + y + z = 8 \\ 2x - y - z = 10 \\ x - 2y + 3z = 22 \end{cases}$

$$D = \begin{vmatrix} 1 & 1 & 1 \\ 2 & -1 & -1 \\ 1 & -2 & 3 \end{vmatrix} = 1\begin{vmatrix} -1 & -1 \\ -2 & 3 \end{vmatrix} - 1\begin{vmatrix} 2 & -1 \\ 1 & 3 \end{vmatrix} + 1\begin{vmatrix} 2 & -1 \\ 1 & -2 \end{vmatrix}$$
$$= (-3 - 2) - [6 - (-1)] + [-4 - (-1)] = -5 - 7 - 3 = -15$$

$$D_x = \begin{vmatrix} 8 & 1 & 1 \\ 10 & -1 & -1 \\ 22 & -2 & 3 \end{vmatrix} = 8\begin{vmatrix} -1 & -1 \\ -2 & 3 \end{vmatrix} - 1\begin{vmatrix} 10 & -1 \\ 22 & 3 \end{vmatrix} + 1\begin{vmatrix} 10 & -1 \\ 22 & -2 \end{vmatrix}$$
$$= 8(-3 - 2) - [30 - (-22)] + [-20 - (-22)] = 8(-5) - 52 + 2 = -40 - 52 + 2 = -90$$

$$D_y = \begin{vmatrix} 1 & 8 & 1 \\ 2 & 10 & -1 \\ 1 & 22 & 3 \end{vmatrix} = 1\begin{vmatrix} 10 & -1 \\ 22 & 3 \end{vmatrix} - 8\begin{vmatrix} 2 & -1 \\ 1 & 3 \end{vmatrix} + 1\begin{vmatrix} 2 & 10 \\ 1 & 22 \end{vmatrix}$$
$$= [30 - (-22)] - 8[6 - (-1)] + [44 - 10] = 52 - 8(7) + 34 = 52 - 56 + 34 = 30$$

$$D_z = \begin{vmatrix} 1 & 1 & 8 \\ 2 & -1 & 10 \\ 1 & -2 & 22 \end{vmatrix} = 1\begin{vmatrix} -1 & 10 \\ -2 & 22 \end{vmatrix} - 1\begin{vmatrix} 2 & 10 \\ 1 & 22 \end{vmatrix} + 8\begin{vmatrix} 2 & -1 \\ 1 & -2 \end{vmatrix}$$
$$= [-22 - (-20)] - (44 - 10) + 8[-4 - (-1)] = -2 - 34 + 8(-3) = -36 - 24 = -60$$

$$x = \frac{-90}{-15} = 6,\ y = \frac{30}{-15} = -2,\ z = \frac{-60}{-15} = 4$$

The solution is $(6, -2, 4)$.

SOLUTIONS

33. $\begin{cases} x - 2y + z = -5 \\ 3y + 2z = 4 \\ 3x - y = -2 \end{cases}$

$$D = \begin{vmatrix} 1 & -2 & 1 \\ 0 & 3 & 2 \\ 3 & -1 & 0 \end{vmatrix}$$
$$= 1\begin{vmatrix} 3 & 2 \\ -1 & 0 \end{vmatrix} - 0\begin{vmatrix} -2 & 1 \\ -1 & 0 \end{vmatrix} + 3\begin{vmatrix} -2 & 1 \\ 3 & 2 \end{vmatrix}$$
$$= [0 - (-2)] - 0 + 3[-4 - 3]$$
$$= 2 + 3(-7)$$
$$= 2 - 21$$
$$= -19$$

$$D_x = \begin{vmatrix} -5 & -2 & 1 \\ 4 & 3 & 2 \\ -2 & -1 & 0 \end{vmatrix}$$
$$= 1\begin{vmatrix} 4 & 3 \\ -2 & -1 \end{vmatrix} - 2\begin{vmatrix} -5 & -2 \\ -2 & -1 \end{vmatrix} + 0\begin{vmatrix} -5 & -2 \\ 4 & 3 \end{vmatrix}$$
$$= [-4 - (-6)] - 2[5 - 4] + 0$$
$$= 2 - 2(1)$$
$$= 0$$

$$D_y = \begin{vmatrix} 1 & -5 & 1 \\ 0 & 4 & 2 \\ 3 & -2 & 0 \end{vmatrix}$$
$$= 1\begin{vmatrix} 4 & 2 \\ -2 & 0 \end{vmatrix} - 0\begin{vmatrix} -5 & 1 \\ -2 & 0 \end{vmatrix} + 3\begin{vmatrix} -5 & 1 \\ 4 & 2 \end{vmatrix}$$
$$= [0 - (-4)] - 0 + 3[-10 - 4]$$
$$= 4 + 3(-14)$$
$$= 4 - 42$$
$$= -38$$

$$D_z = \begin{vmatrix} 1 & -2 & -5 \\ 0 & 3 & 4 \\ 3 & -1 & -2 \end{vmatrix}$$
$$= 1\begin{vmatrix} 3 & 4 \\ -1 & -2 \end{vmatrix} - 0\begin{vmatrix} -2 & -5 \\ -1 & -2 \end{vmatrix} + 3\begin{vmatrix} -2 & -5 \\ 3 & 4 \end{vmatrix}$$
$$= [-6 - (-4)] - 0 + 3[-8 - (-15)]$$
$$= -2 + 3(7)$$
$$= -2 + 21$$
$$= 19$$

$$x = \frac{0}{-19} = 0,\ y = \frac{-38}{-19} = 2,$$
$$z = \frac{19}{-19} = -1$$

The solution is $(0, 2, -1)$.

37. $5(x - 6) \le 2(x - 7)$
$$5x - 30 \le 2x - 14$$
$$3x \le 16$$
$$x \le \frac{16}{3}$$
$$x \le 5\frac{1}{3}$$

41. 0; If the elements of a single row (or column) of a determinant are all zero, the value of the determinant will be zero. To see this, consider expanding by the row (or column) containing all zeros.

Exercise Set 4.6

1. $\begin{cases} y \ge x + 1 \\ y \ge 3 - x \end{cases}$

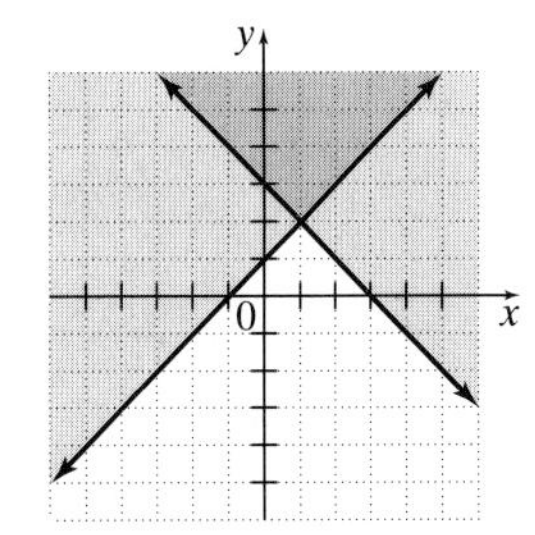

5. $\begin{cases} y \le -2x - 2 \\ y \ge x + 4 \end{cases}$

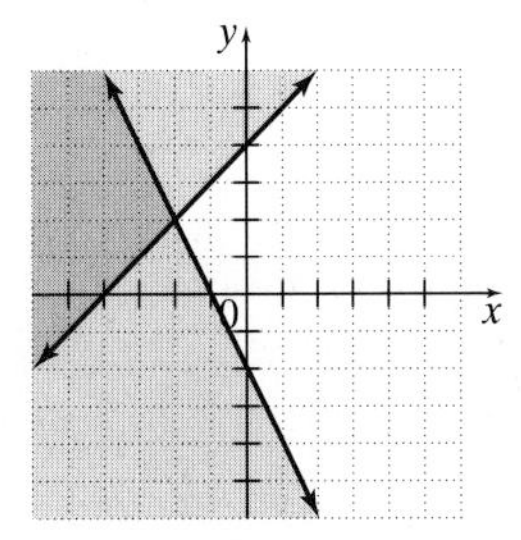

9. $\begin{cases} x \ge 3y \\ x + 3y \le 6 \end{cases}$

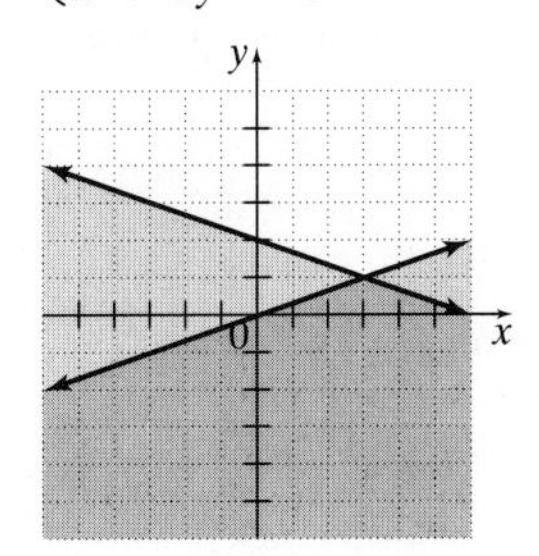

13. $\begin{cases} y \ge 1 \\ x < -3 \end{cases}$

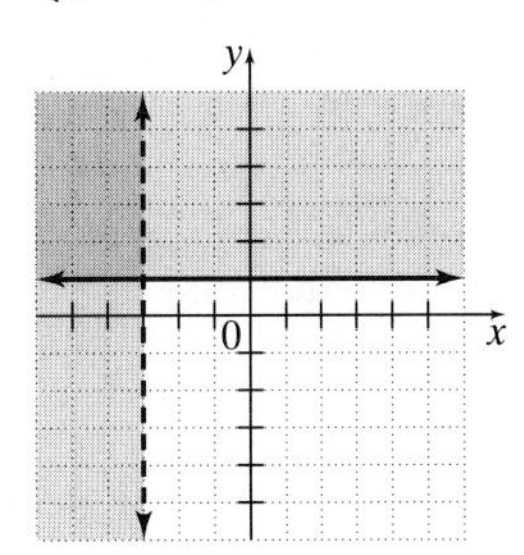

17. $\begin{cases} 3x - 4y \ge -6 \\ 2x + y \le 7 \\ y \ge -3 \end{cases}$

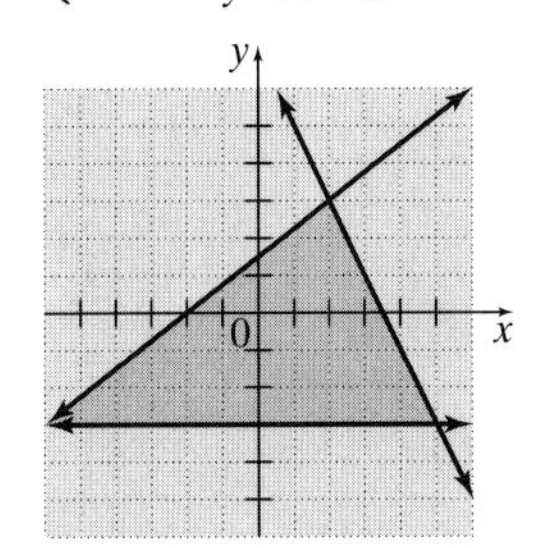

21. $\begin{cases} y < 5 \\ x > 3 \end{cases}$
graph C

25. $(-3)^2 = (-3)(-3) = 9$

29. $(-2)^2 - (-3) + 2(-1) = 4 + 3 - 2 = 5$

33. a. $\begin{cases} x + y \le 8 \\ x < 3 \end{cases}$

b.

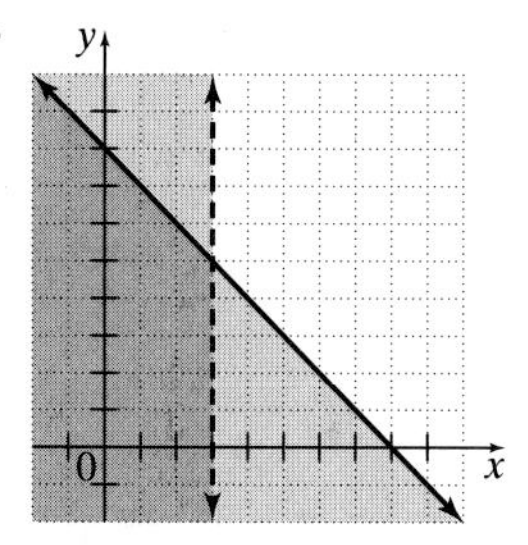

Chapter 4 Test

1. $\begin{vmatrix} 4 & -7 \\ 2 & 5 \end{vmatrix} = 4(5) - (-7)(2)$
$= 20 + 14$
$= 34$

5. $\begin{cases} 4x - 7y = 29 & (1) \\ 2x + 5y = -11 & (2) \end{cases}$

Multiply equation (2) by −2 and add to equation (1).

$$\begin{aligned} 4x - 7y &= 29 \\ -4x - 10y &= 22 \\ \hline -17y &= 51 \\ y &= -3 \end{aligned}$$

Replace y with −3 in equation (1).

$$\begin{aligned} 4x - 7(-3) &= 29 \\ 4x + 21 &= 29 \\ 4x &= 8 \\ x &= 2 \end{aligned}$$

The solution is (2, −3).

9. $\begin{cases} \dfrac{x}{2} + \dfrac{y}{4} = -\dfrac{3}{4} \\ x + \dfrac{3}{4}y = -4 \end{cases}$

Clear fraction by multiplying both equations by 4.

$\begin{cases} 2x + y = -3 & (1) \\ 4x + 3y = -16 & (2) \end{cases}$

Multiply equation (1) by −2 and add to equation (2).

$$\begin{aligned} -4x - 2y &= 6 \\ 4x + 3y &= -16 \\ \hline y &= -10 \end{aligned}$$

Replace y with −10 in equation (1).

$$\begin{aligned} 2x + (-10) &= -3 \\ 2x &= 7 \\ x &= \frac{7}{2} \end{aligned}$$

The solution is $\left(\frac{7}{2}, -10\right)$.

13. $\begin{cases} 3x + 2y + 3z = 3 \\ x \qquad\quad - z = 9 \\ \qquad 4y + z = -4 \end{cases}$

$$\begin{aligned} D &= \begin{vmatrix} 3 & 2 & 3 \\ 1 & 0 & -1 \\ 0 & 4 & 1 \end{vmatrix} \\ &= -1\begin{vmatrix} 2 & 3 \\ 4 & 1 \end{vmatrix} + 0\begin{vmatrix} 3 & 3 \\ 0 & 1 \end{vmatrix} - (-1)\begin{vmatrix} 3 & 2 \\ 0 & 4 \end{vmatrix} \\ &= -(2 - 12) + 0 + (12 - 0) \\ &= -(-10) + 12 \\ &= 10 + 12 \\ &= 22 \end{aligned}$$

$$\begin{aligned} D_x &= \begin{vmatrix} 3 & 2 & 3 \\ 9 & 0 & -1 \\ -4 & 4 & 1 \end{vmatrix} \\ &= -9\begin{vmatrix} 2 & 3 \\ 4 & 1 \end{vmatrix} + 0\begin{vmatrix} 3 & 3 \\ -4 & 1 \end{vmatrix} - (-1)\begin{vmatrix} 3 & 2 \\ -4 & 4 \end{vmatrix} \\ &= -9(2 - 12) + 0 + [12 - (-8)] \\ &= -9(-10) + 20 \\ &= 90 + 20 \\ &= 110 \end{aligned}$$

$$\begin{aligned} D_y &= \begin{vmatrix} 3 & 3 & 3 \\ 1 & 9 & -1 \\ 0 & -4 & 1 \end{vmatrix} \\ &= 3\begin{vmatrix} 9 & -1 \\ -4 & 1 \end{vmatrix} - 1\begin{vmatrix} 3 & 3 \\ -4 & 1 \end{vmatrix} + 0\begin{vmatrix} 3 & 3 \\ 9 & -1 \end{vmatrix} \\ &= 3(9 - 4) - [3 - (-12)] + 0 \\ &= 3(5) - 15 \\ &= 15 - 15 \\ &= 0 \end{aligned}$$

$$\begin{aligned} D_z &= \begin{vmatrix} 3 & 2 & 3 \\ 1 & 0 & 9 \\ 0 & 4 & -4 \end{vmatrix} \\ &= 3\begin{vmatrix} 0 & 9 \\ 4 & -4 \end{vmatrix} - 1\begin{vmatrix} 2 & 3 \\ 4 & -4 \end{vmatrix} + 0\begin{vmatrix} 2 & 3 \\ 0 & 9 \end{vmatrix} \\ &= 3(0 - 36) - (-8 - 12) + 0 \\ &= 3(-36) - (-20) \\ &= -108 + 20 \\ &= -88 \end{aligned}$$

$x = \dfrac{D_x}{D} = \dfrac{110}{22} = 5,\ y = \dfrac{D_y}{D} = \dfrac{0}{22} = 0,$

$z = \dfrac{D_z}{D} = \dfrac{-88}{22} = -4$

The solution is (5, 0, −4).

17. $\begin{cases} 2x - y + 3z = 4 \\ 3x - 3z = -2 \\ -5x + y = 0 \end{cases}$

$$\left[\begin{array}{rrr|r} 2 & -1 & 3 & 4 \\ 3 & 0 & -3 & -2 \\ -5 & 1 & 0 & 0 \end{array}\right]$$

Divide row 1 by 2.

$$\left[\begin{array}{rrr|r} 1 & -\frac{1}{2} & \frac{3}{2} & 2 \\ 3 & 0 & -3 & -2 \\ -5 & 1 & 0 & 0 \end{array}\right]$$

Multiply row 1 by -3 and add to row 2.
Multiply row 1 by 5 and add to row 3.

$$\left[\begin{array}{rrr|r} 1 & -\frac{1}{2} & \frac{3}{2} & 2 \\ 0 & \frac{3}{2} & -\frac{15}{2} & -8 \\ 0 & -\frac{3}{2} & \frac{15}{2} & 10 \end{array}\right]$$

Multiply row 2 by $\frac{2}{3}$.

$$\left[\begin{array}{rrr|r} 1 & -\frac{1}{2} & \frac{3}{2} & 2 \\ 0 & 1 & -5 & -\frac{16}{3} \\ 0 & -\frac{3}{2} & \frac{15}{2} & 10 \end{array}\right]$$

Multiply row 2 by $\frac{3}{2}$ and add to row 3.

$$\left[\begin{array}{rrr|r} 1 & -\frac{1}{2} & \frac{3}{2} & 2 \\ 0 & 1 & -5 & -\frac{16}{3} \\ 0 & 0 & 0 & 2 \end{array}\right]$$

This corresponds to $\begin{cases} x - \frac{1}{2}y + \frac{3}{2}z = 2 \\ y - 5z = -\frac{16}{3}. \\ 0 = 2 \end{cases}$

This is an inconsistent system.
The solution is $\emptyset$.

21. $\begin{cases} 2y - x \geq 1 \\ x + y \geq -4 \\ y \leq 2 \end{cases}$

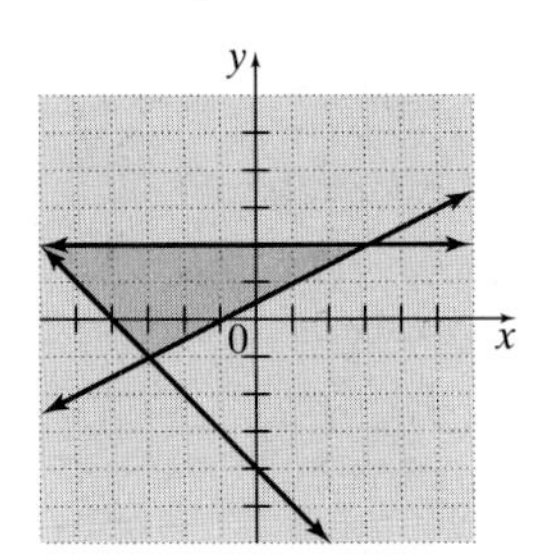

Chapter 5

Exercise Set 5.1

1. 4 has degree 0.

5. $-3xy^2$ has degree $1 + 2 = 3$.

9. $3x^2 - 2x + 5$ has degree 2 and is a trinomial.

13. $x^2y - 4xy^2 + 5x + y$ has degree $2 + 1 = 3$ and is none of these.

17. $5y + y = 6y$

21. $4xy + 2x - 3xy - 1 = xy + 2x - 1$

25.
$$\begin{array}{r} x^2 + xy - y^2 \\ 2x^2 - 4xy + 7y^2 \\ \hline 3x^2 - 3xy + 6y^2 \end{array}$$

29.
$$\begin{array}{r} 3x^2 + 15x + 8 \\ +(2x^2 + 7x + 8) \\ \hline 5x^2 + 22x + 16 \end{array}$$

33. $(5y^4 - 7y^2 + x^2 - 3) + (-3y^4 + 2y^2 + 4)$
$= 5y^4 - 7y^2 + x^2 - 3 - 3y^4 + 2y^2 + 4$
$= 5y^4 - 3y^4 - 7y^2 + 2y^2 + x^2 - 3 + 4$
$= 2y^4 - 5y^2 + x^2 + 1$

37.
$$\begin{array}{r} 3x^3 - b + 2a - 6 \\ -4x^3 + b + 6a - 6 \\ \hline -x^3 + 8a - 12 \end{array}$$

41. $(7x^3y - 4xy + 8) + (5x^3y + 4xy + 8x)$
$= 7x^3y + 5x^3y - 4xy + 4xy + 8x + 8$
$= 12x^3y + 8x + 8$

45. $(9y^2 - 7y + 5) - (8y^2 - 7y + 2)$
$= 9y^2 - 7y + 5 - 8y^2 + 7y - 2$
$= y^2 + 3$

49.
$$\begin{array}{r} 3x^2 - 4x + 8 \\ -\quad (5x^2 - 7) \\ \hline \end{array} \quad \text{or} \quad \begin{array}{r} 3x^2 - 4x + 8 \\ -5x^2 \qquad + 7 \\ \hline -2x^2 - 4x + 15 \end{array}$$

53.
$$\begin{array}{r} 4x^2 - 6x + 2 \\ -(-x^2 + 3x + 5) \\ \hline \end{array} \quad \text{or} \quad \begin{array}{r} 4x^2 - 6x + 2 \\ x^2 - 3x - 5 \\ \hline 5x^2 - 9x - 3 \end{array}$$

57. $(9x^3 - 2x^2 + 4x - 7) - (2x^3 - 6x^2 - 4x + 3)$
$= 9x^3 - 2x^2 + 4x - 7 - 2x^3 + 6x^2 + 4x - 3$
$= 7x^3 + 4x^2 + 8x - 10$

61.
$$\begin{array}{r} 7x^2 + 4x + 9 \\ +(8x^2 + 7x - 8) \\ \hline 15x^2 + 11x + 1 \\ -\qquad (3x + 7) \\ \hline \end{array} \quad \text{or} \quad \begin{array}{r} 15x^2 + 11x + 1 \\ -3x - 7 \\ \hline 15x^2 + 8x - 6 \end{array}$$

65. $\left(\frac{2}{3}x^2 - \frac{1}{6}x + \frac{5}{6}\right) - \left(\frac{1}{3}x^2 + \frac{5}{6}x - \frac{1}{6}\right)$
$= \frac{2}{3}x^2 - \frac{1}{6}x + \frac{5}{6} - \frac{1}{3}x^2 - \frac{5}{6}x + \frac{1}{6}$
$= \frac{1}{3}x^2 - x + 1$

69.
$$\begin{aligned} Q(x) &= 5x^2 - 1 \\ Q(-10) &= 5(-10)^2 - 1 \\ &= 5(100) - 1 \\ &= 500 - 1 \\ &= 499 \end{aligned}$$

SOLUTIONS

73. $P(t) = -16t^2 + 300t$

a. $P(1) = -16(1)^2 + 300(1) = 284$ feet

b. $P(2) = -16(2)^2 + 300(2) = 536$ feet

c. $P(3) = -16(3)^2 + 300(3) = 756$ feet

d. $P(4) = -16(4)^2 + 300(4) = 944$ feet

e. answers may vary

f. When $t = 18$ seconds
$P(18) = -16(18)^2 + 300(18) = 216$ feet
When $t = 19$ seconds
$P(19) = -16(19)^2 + 300(19) = -76$ feet
The object will hit the ground at approximately 19 seconds.

77. $$\begin{aligned} R(x) &= 2x \\ P(20{,}000) &= 2(20{,}000) \\ &= \$40{,}000 \end{aligned}$$

81. $5(-3y^2 - 2y + 7) = -15y^2 - 10y + 35$

85. $$\begin{aligned} P(x) &= 2x - 3 \\ P(a) &= 2a - 3 \\ P(-x) &= 2(-x) - 3 = -2x - 3 \\ P(x + h) &= 2(x + h) - 3 = 2x + 2h - 3 \end{aligned}$$

Exercise Set 5.2

1. $(-4x^3)(3x^2) = -12x^5$

5. $$\begin{aligned} -6xy(4x + y) &= -6xy(4x) - 6xy(y) \\ &= -24x^2y - 6xy^2 \end{aligned}$$

9. $$\begin{aligned} (x - 3)(2x + 4) &= x(2x + 4) - 3(2x + 4) \\ &= 2x^2 + 4x - 6x - 12 \\ &= 2x^2 - 2x - 12 \end{aligned}$$

13. $$\begin{array}{r} 3x - 2 \\ \underline{5x + 1} \\ 3x - 2 \\ \underline{15x^2 - 10x \phantom{{}-2}} \\ 15x^2 - 7x - 2 \end{array}$$

17. $$\begin{array}{r} 3x^2 + 4x - 4 \\ \underline{3x + 6} \\ 18x^2 + 24x - 24 \\ \underline{9x^3 + 12x^2 - 12x \phantom{{}-24}} \\ 9x^3 + 30x^2 + 12x - 24 \end{array}$$

21. $$\begin{aligned} &(2x^3 + 5)(5x^2 + 4x + 1) \\ &= 2x^3(5x^2 + 4x + 1) + 5(5x^2 + 4x + 1) \\ &= 10x^5 + 8x^4 + 2x^3 + 25x^2 + 20x + 5 \end{aligned}$$

25. $$\begin{array}{r} 4x^2 - 2x + 5 \\ \underline{3x + 1} \\ 4x^2 - 2x + 5 \\ \underline{12x^3 - 6x^2 + 15x \phantom{{}+5}} \\ 12x^3 - 2x^2 + 13x + 5 \end{array}$$

29. $$\begin{aligned} (x - 3)(x + 4) &= x^2 + 4x - 3x - 12 \\ &= x^2 + x - 12 \end{aligned}$$

33. $$\begin{aligned} (3x - 1)(x + 3) &= 3x^2 + 9x - x - 3 \\ &= 3x^2 + 8x - 3 \end{aligned}$$

37. $$\begin{aligned} (y - 4)(y - 3) &= y^2 - 3y - 4y + 12 \\ &= y^2 - 7y + 12 \end{aligned}$$

41. $$\begin{aligned} \left(4x + \frac{1}{3}\right)\left(4x - \frac{1}{2}\right) &= 16x^2 - 2x + \frac{4}{3}x - \frac{1}{6} \\ &= 16x^2 - \frac{2}{3}x - \frac{1}{6} \end{aligned}$$

45. $$\begin{aligned} (x + 4)^2 &= x^2 + 2(x)(4) + 4^2 \\ &= x^2 + 8x + 16 \end{aligned}$$

49. $$\begin{aligned} (3x - y)^2 &= (3x)^2 - 2(3x)y + y^2 \\ &= 9x^2 - 6xy + y^2 \end{aligned}$$

53. $$\begin{aligned} (m - 4)^2 &= m^2 - 2(m)(4) + 4^2 \\ &= m^2 - 8m + 16 \end{aligned}$$

57. $(3b - 6y)(3b + 6y) = (3b)^2 - (6y)^2 = 9b^2 - 36y^2$

61. $$\begin{aligned} \left(3x + \frac{1}{2}\right)\left(3x - \frac{1}{2}\right) &= (3x)^2 - \left(\frac{1}{2}\right)^2 \\ &= 9x^2 - \frac{1}{4} \end{aligned}$$

65. $$\begin{aligned} (x^2 + 2y)(x^2 - 2y) &= (x^2)^2 - (2y)^2 \\ &= x^4 - 4y^2 \end{aligned}$$

69. $$\begin{aligned} &[(2s - 3) - 1][(2s - 3) + 1] \\ &= (2s - 3)^2 - 1^2 \\ &= (2s)^2 - 2(2s)(3) + 3^2 - 1 \\ &= 4s^2 - 12s + 9 - 1 \\ &= 4s^2 - 12s + 8 \end{aligned}$$

73. $$\begin{array}{r} x + y \\ 2x - 1 \\ \underline{-x - y} \\ \underline{2x^2 + 2xy \phantom{{}-x-y}} \\ 2x^2 + 2xy - x - y \\ \underline{x + 1} \\ 2x^2 + 2xy - x - y \\ \underline{2x^3 + 2x^2y - x^2 - xy \phantom{{}-x-y}} \\ 2x^3 + 2x^2y + x^2 + xy - x - y \end{array}$$

77. $$\begin{aligned} (x - 5)(x + 5)(x^2 + 25) &= (x^2 - 25)(x^2 + 25) \\ &= (x^2)^2 - (25)^2 \\ &= x^4 - 625 \end{aligned}$$

81. $$\frac{20a^3b^5}{18ab^2} = \frac{2ab^2(10a^2b^3)}{2ab^2(9)} = \frac{10a^2b^3}{9}$$

85. $$\begin{aligned} A &= \pi r^2 \\ A &= \pi(5x - 2)^2 \\ A &= \pi[(5x)^2 - 2(5x)(2) + 2^2] \\ A &= \pi(25x^2 - 20x + 4) \text{ square kilometers} \end{aligned}$$

89. $$\begin{aligned} (x^a + 5)(x^{2a} - 3) &= (x^a)(x^{2a}) - 3x^a + 5x^{2a} - 15 \\ &= x^{3a} - 3x^a + 5x^{2a} - 15 \end{aligned}$$

93. $$\begin{aligned} P(x) \cdot R(x) &= 5x(x + 5) \\ &= 5x^2 + 25x \end{aligned}$$

Exercise Set 5.3

1. $$\frac{4a^2 + 8a}{2a} = \frac{4a^2}{2a} + \frac{8a}{2a} = 2a + 4$$

5. $$\begin{aligned} \frac{4x^2y^2 + 6xy^2 - 4y^2}{2x^2y} &= \frac{4x^2y^2}{2x^2y} + \frac{6xy^2}{2x^2y} - \frac{4y^2}{2x^2y} \\ &= 2y + \frac{3y}{x} - \frac{2y}{x^2} \end{aligned}$$

9.
$$\begin{array}{r} 2x - 8 \\ x + 1\overline{)2x^2 - 6x - 8} \\ \underline{2x^2 + 2x} \\ -8x - 8 \\ \underline{-8x - 8} \\ 0 \end{array}$$

Answer: $2x - 8$

13.
$$\begin{array}{r} 2x^2 - \frac{1}{2}x + 5 \\ 2x + 4\overline{)4x^3 + 7x^2 + 8x + 20} \\ \underline{4x^3 + 8x^2} \\ -x^2 + 8x \\ \underline{-x^2 - 2x} \\ 10x + 20 \\ \underline{10x + 20} \\ 0 \end{array}$$

Answer: $2x^2 - \frac{1}{2}x + 5$

17.
$$\begin{array}{r} 2x^2 + 2x + 8 \\ x - 4\overline{)2x^3 - 6x^2 + 0x - 4} \\ \underline{2x^3 - 8x^2} \\ 2x^2 + 0x - 4 \\ \underline{2x^2 - 8x} \\ 8x - 4 \\ \underline{8x - 32} \\ 28 \end{array}$$

Answer: $2x^2 + 2x + 8 + \dfrac{28}{x - 4}$

21.
$$\begin{array}{r} 3x^3 + 5x + 4 \\ x^2 - 2\overline{)3x^5 + 0x^4 - x^3 + 4x^2 - 12x - 8} \\ \underline{3x^5 - 6x^3} \\ 5x^3 + 4x^2 - 12x \\ \underline{5x^3 - 10x} \\ 4x^2 - 2x - 8 \\ \underline{4x^2 - 8} \\ -2x \end{array}$$

Answer: $3x^3 + 5x + 4 - \dfrac{2x}{x^2 - 2}$

25.
$$\begin{array}{c|ccc} 5 & 1 & 3 & -40 \\ & & 5 & 40 \\ \hline & 1 & 8 & 0 \end{array}$$

$x + 8$

29.
$$\begin{array}{c|cccc} 2 & 1 & -7 & -13 & 5 \\ & & 2 & -10 & -46 \\ \hline & 1 & -5 & -23 & -41 \end{array}$$

$x^2 - 5x - 23 - \dfrac{41}{x - 2}$

33.
$$\begin{array}{c|ccccc} 5 & 2 & -13 & 16 & -9 & 20 \\ & & 10 & -15 & 5 & -20 \\ \hline & 2 & -3 & 1 & -4 & 0 \end{array}$$

$2x^3 - 3x^2 + x - 4$

37.
$$\begin{array}{c|cccc} \frac{1}{3} & 3 & 2 & -4 & 1 \\ & & 1 & 1 & -1 \\ \hline & 3 & 3 & -3 & 0 \end{array}$$

$3x^2 + 3x - 3$

41. $6x(x + 3) + 5(x + 3) = 6x^2 + 18x + 5x + 15$
$= 6x^2 + 23x + 15$

45. $|2x + 7| \geq 9$
$2x + 7 \leq -9$ or $2x + 7 \geq 9$
$2x \leq -16$ or $2x \geq 2$
$x \leq -8$ or $x \geq 1$
$(-\infty, -8] \cup [1, \infty)$

49. Recall that $A = l \cdot w$ so
$$w = \frac{A}{l} = \frac{15x^2 - 29x - 14}{5x + 2}$$
$$\begin{array}{r} 3x - 7 \\ 5x + 2\overline{)15x^2 - 29x - 14} \\ \underline{15x^2 + 6x} \\ -35x - 14 \\ \underline{-35x - 14} \\ 0 \end{array}$$

The width is $(3x - 7)$ in.

53. Multiply $(x^2 - x + 10)$ by $(x + 3)$ and add the remainder, -2.
$(x^2 - x + 10)(x + 3)$
$x^3 - x^2 + 10x + 3x^2 - 3x + 30$
$x^3 + 2x^2 + 7x + 30$
$\underline{ -2}$
$x^3 + 2x^2 + 7x + 28$

Exercise Set 5.4

1. $18x - 12 = 6(3x - 2)$

5. $6x^5 - 8x^4 + 2x^3 = 2x^3(3x^2 - 4x + 1)$

9. $6(x + 3) + 5a(x + 3) = (6 + 5a)(x + 3)$

13. $3x(x^2 + 5) - 2(x^2 + 5) = (3x - 2)(x^2 + 5)$

17. $A = P + PRT$
$A = P(1 + RT)$

21. $ab + 3a + 2b + 6 = a(b + 3) + 2(b + 3)$
$= (a + 2)(b + 3)$

25. $2xy - 3x - 4y + 6 = x(2y - 3) - 2(2y - 3)$
$= (x - 2)(2y - 3)$

29. $x^3 + 3x^2 + 4x + 12 = x^2(x + 3) + 4(x + 3)$
$= (x^2 + 4)(x + 3)$

33. $2x^2 + 3xy + 4x + 6y = x(2x + 3y) + 2(2x + 3y)$
$= (x + 2)(2x + 3y)$

37. $6xy + 10x + 9y + 15 = 2x(3y + 5) + 3(3y + 5)$
$= (2x + 3)(3y + 5)$

41. $9abc^2 + 6a^2bc - 6ab + 3bc$
$= 3b(3ac^2 + 2a^2c - 2a + c)$

45. $(x + 3)(x + 2) = x^2 + 2x + 3x + 6$
$= x^2 + 5x + 6$

SOLUTIONS

49. **a.** $(2 - x)(3 - y) = 6 - 2y - 3x + xy$
$= xy - 3x - 2y + 6$

b. $(-2 + x)(-3 + y) = 6 - 2y - 3x + xy$
$= xy - 3x - 2y + 6$

c. $(x - 2)(y - 3) = xy - 3x - 2y + 6$

d. $(-x + 2)(-y + 3) = xy - 3x - 2y + 6$

The answer is none.

53. $3x^{5a} - 6x^{3a} + 9x^{2a} = 3x^{2a}(x^{3a} - 2x^a + 3)$

Exercise Set 5.5

1. $x^2 + 9x + 18 = (x + 6)(x + 3)$

5. $x^2 + 10x - 24 = (x + 12)(x - 2)$

9. $3x^2 - 18x + 24 = 3(x^2 - 6x + 8)$
$= 3(x - 2)(x - 4)$

13. $2x^2 + 30x - 108 = 2(x^2 + 15x - 54)$
$= 2(x + 18)(x - 3)$

17. $x^2 - 15x - 54 = (x - 18)(x + 3)$

21. $2x^2 + 2x - 12 = 2(x^2 + x - 6)$
$= 2(x + 3)(x - 2)$

25. $x^3 + 2x^2 - 8x = x(x^2 + 2x - 8)$
$= x(x + 4)(x - 2)$

29. $5x^2 + 16x + 3 = (5x + 1)(x + 3)$

33. $2x^2 + 25x - 20$ is prime.

37. $12x^2 + 10x - 50 = 2(6x^2 + 5x - 25)$
$= 2(3x - 5)(2x + 5)$

41. $6x^3 + 8x^2 + 24x = 2x(3x^2 + 4x + 12)$

45. $2x^2 - 5xy - 3y^2 = (2x + y)(x - 3y)$

49. $28y^2 + 22y + 4 = 2(14y^2 + 11y + 2)$
$= 2(7y + 2)(2y + 1)$

53. $3x^2 - 5x - 2 = (3x + 1)(x - 2)$

57. $18x^4 + 21x^3 + 6x^2 = 3x^2(6x^2 + 7x + 2)$
$= 3x^2(3x + 2)(2x + 1)$

61. $6x^3 - x^2 - x = x(6x^2 - x - 1)$
$= x(3x + 1)(2x - 1)$

65. $9x^2 + 30x + 25 = (3x + 5)^2$

69. Let $y = x^2$. Then $y^2 = x^4$.
$x^4 + x^2 - 6 = y^2 + y - 6$
$= (y + 3)(y - 2)$
$= (x^2 + 3)(x^2 - 2)$

73. Let $y = x^3$. Then $y^2 = x^6$.
$x^6 - 7x^3 + 12 = y^2 - 7y + 12$
$= (y - 4)(y - 3)$
$= (x^3 - 4)(x^3 - 3)$

77. Let $y = x - 4$.
$(x - 4)^2 + 3(x - 4) - 18$
$= y^2 + 3y - 18$
$= (y + 6)(y - 3)$
$= [(x - 4) + 6][(x - 4) - 3]$
$= (x + 2)(x - 7)$

81. Let $y = x + 4$.
$2(x + 4)^2 + 3(x + 4) - 5$
$= 2y^2 + 3y - 5$
$= (2y + 5)(y - 1)$
$= [2(x + 4) + 5][(x + 4) - 1]$
$= [2x + 8 + 5][x + 3]$
$= (2x + 13)(x + 3)$

85. $(x - 3)(x + 3) = x^2 - 3^2$
$= x^2 - 9$

89.
$$\begin{array}{r} x^2 + 2x + 4 \\ x - 2 \\ \hline -2x^2 - 4x - 8 \\ x^3 + 2x^2 + 4x \qquad \\ \hline x^3 \qquad\qquad\quad -8 \end{array}$$

93. $x^{2n} - 3x^n - 18 = (x^n - 6)(x^n + 3)$

97. $4x^{2n} - 12x^n + 9 = (2x^n - 3)^2$

101. $30x^3 + 9x^2 - 3x = 3x(10x^2 + 3x - 1)$
$= 3x(5x - 1)(2x + 1)$

Exercise Set 5.6

1. $x^2 + 6x + 9 = x^2 + 2(x)(3) + 3^2 = (x + 3)^2$

5. $3x^2 - 24x + 48 = 3(x^2 - 8x + 16)$
$= 3[x^2 - 2(x)(4) + 4^2]$
$= 3(x - 4)^2$

9. $4a^2 + 12a + 9 = (2a)^2 + 2(2a)(3) + 3^2$
$= (2a + 3)^2$

13. $9 - 4z^2 = (3 + 2z)(3 - 2z)$

17. $64x^2 - 100 = 4(16x^2 - 25)$
$= 4(4x + 5)(4x - 5)$

21. $9x^2 - 49 = (3x + 7)(3x - 7)$

25. $(x + 2y)^2 - 9 = (x + 2y - 3)(x + 2y + 3)$

29. $x^2 - 10x + 25 - y^2 = (x^2 - 10x + 25) - y^2$
$= (x - 5)^2 - y^2$
$= (x - 5 - y)(x - 5 + y)$

33. $x^3 + 27 = x^3 + 3^3$
$= (x + 3)(x^2 - 3x + 9)$

37. $m^3 + n^3 = (m + n)(m^2 - mn + n^2)$

41. $a^3b + 8b^4 = (a^3 + 8b^3)b$
$= (a^3 + (2b)^3)b$
$= (a + 2b)(a^2 - 2ab + 4b^2)b$

45. $x^6 - y^3 = (x^2)^3 - y^3$
$= (x^2 - y)(x^4 + x^2y + y^2)$

49. $x^3 - 1 = x^3 - 1^3$
$= (x - 1)(x^2 + x + 1)$

53. $3x^6y^2 + 81y^2 = 3y^2(x^6 + 27)$
$= 3y^2((x^2)^3 + 3^3)$
$= 3y^2(x^2 + 3)(x^4 - 3x^2 + 9)$

57. $3x + 1 = 0$
$3x = -1$
$x = -\frac{1}{3}$

$\left\{-\frac{1}{3}\right\}$

61. $-5x + 25 = 0$
$-5x = -25$
$x = 5$
$\{5\}$

65. $c = \left(\frac{6}{2}\right)^2 = 3^2 = 9$
$x^2 + 6x + 9 = (x + 3)^2$

69. a. $(x^3)^2 - 1^2$
$= (x^3 + 1)(x^3 - 1)$
$= (x + 1)(x^2 - x + 1)(x - 1)(x^2 + x + 1)$

b. $(x^2)^3 - 1^3 = (x^2 - 1)(x^4 + x^2 + 1)$
$= (x + 1)(x - 1)(x^4 + x^2 + 1)$

c. answers may vary

73. $25x^{2n} - 81 = (5x^n)^2 - 9^2$
$= (5x^n - 9)(5x^n + 9)$

Exercise Set 5.7

1. $(x + 3)(3x - 4) = 0$
$x + 3 = 0$ or $3x - 4 = 0$
$3x = 4$
$x = -3$ or $x = \frac{4}{3}$
$\left\{-3, \frac{4}{3}\right\}$

5. $x^2 + 11x + 24 = 0$
$(x + 3)(x + 8) = 0$
$x + 3 = 0$ or $x + 8 = 0$
$x = -3$ or $x = -8$
$\{-3, -8\}$

9. $z^2 + 9 = 10z$
$z^2 - 10z + 9 = 0$
$(z - 1)(z - 9) = 0$
$z - 1 = 0$ or $z - 9 = 0$
$z = 1$ or $z = 9$
$\{1, 9\}$

13. $x^2 - 6x = x(8 + x)$
$x^2 - 6x = 8x + x^2$
$0 = 14x$
$x = 0$
$\{0\}$

17. $\frac{x^2}{2} + \frac{x}{20} = \frac{1}{10}$
$10x^2 + x = 2$
$10x^2 + x - 2 = 0$
$(5x - 2)(2x + 1) = 0$
$5x - 2 = 0$ or $2x + 1 = 0$
$5x = 2$ or $2x = -1$
$x = \frac{2}{5}$ or $x = -\frac{1}{2}$
$\left\{\frac{2}{5}, -\frac{1}{2}\right\}$

21. $(x + 2)(x - 7)(3x - 8) = 0$
$x + 2 = 0$ or $x - 7 = 0$ or $3x - 8 = 0$
$x = -2$ or $x = 7$ or $x = \frac{8}{3}$
$\left\{-2, 7, \frac{8}{3}\right\}$

25. $x^3 - x = 2x^2 - 2$
$x^3 - 2x^2 - x + 2 = 0$
$x^2(x - 2) - (x - 2) = 0$
$(x^2 - 1)(x - 2) = 0$
$(x + 1)(x - 1)(x - 2) = 0$
$x + 1 = 0$ or $x - 1 = 0$ or $x - 2 = 0$
$x = -1$ or $x = 1$ or $x = 2$
$\{-1, 1, 2\}$

29. $(2x + 7)(x - 10) = 0$
$2x + 7 = 0$ or $x - 10 = 0$
$2x = -7$
$x = -\frac{7}{2}$ or $x = 10$
$\left\{-\frac{7}{2}, 10\right\}$

33. $x^2 - 2x - 15 = 0$
$(x - 5)(x + 3) = 0$
$x - 5 = 0$ or $x + 3 = 0$
$x = 5$ or $x = -3$
$\{5, -3\}$

37. $w^2 - 5w = 36$
$w^2 - 5w - 36 = 0$
$(w - 9)(w + 4) = 0$
$w - 9 = 0$ or $w + 4 = 0$
$w = 9$ or $w = -4$
$\{9, -4\}$

41. $2r^3 + 6r^2 = 20r$
$r^3 + 3r^2 = 10r$
$r^3 + 3r^2 - 10r = 0$
$r(r^2 + 3r - 10) = 0$
$r(r + 5)(r - 2) = 0$
$r = 0$ or $r + 5 = 0$ or $r - 2 = 0$
$r = 0$ or $r = -5$ or $r = 2$
$\{0, -5, 2\}$

45. $2z(z + 6) = 2z^2 + 12z - 8$
$2z^2 + 12z = 2z^2 + 12z - 8$
$0 = -8$ False
No solution exists.
$\emptyset$

49. Let $n =$ one number and $n + 5 =$ the other number.
$n(n + 5) = 66$
$n^2 + 5n - 66 = 0$
$(n + 11)(n - 6) = 0$
$n + 11 = 0$ or $n - 6 = 0$
$n = -11$ or $n = 6$
$n + 5 = -6$ $\quad n + 5 = 11$
There are two solutions: -11 and -6 and 6 and 11.

53. Let $x =$ longest leg
$x - 2 =$ shortest leg
$x^2 + (x - 2)^2 = (10)^2$
$x^2 + x^2 - 4x + 4 = 100$
$2x^2 - 4x - 96 = 0$
$2(x - 8)(x + 6) = 0$
$x = 8$ or $x = -6$ (disregard)
$x - 2 = 6$
The legs are 6 feet and 8 feet.

57. $h = -16t^2 + 1600$
$0 = -16(t^2 - 100)$
$0 = -16(t + 10)(t - 10)$
$t + 10 = 0$ or $t - 10 = 0$
$t = -10$ (disregard) or $t = 10$
They will hit the ground in 10 seconds.

61. Let x = width of border
$(2x + 16)(2x + 12) - 12(16) = 128$
$4x^2 + 24x + 32x + 192 - 192 = 128$
$4x^2 + 56x - 128 = 0$
$4(x^2 + 14x - 32) = 0$
$4(x - 2)(x + 16) = 0$
$x - 2 = 0$ or $x + 16 = 0$
$x = 2$ or $x = -16$ (disregard)
The border is 2 inches.

65. $(-4, 0)$, $(4, 0)$, $(0, 2)$, $(0, -2)$

69. No. Answers may vary.

Exercise Set 5.8

1. a. Domain, $(-\infty, \infty)$; Range, $(-\infty, 5]$
b. x-intercepts, $(-2, 0)$, $(6, 0)$; y-intercept, $(0, 5)$
c. $(0, 5)$
d. There is no such point.

5. a. Domain, $(-\infty, \infty)$; Range, $(-\infty, \infty)$
b. x-intercepts $(-2, 0)$, $(0, 0)$, $(2, 0)$; y-intercept, $(0, 0)$
c. There is no such point.
d. There is no such point.

9. $f(x) = x^2 + 1$

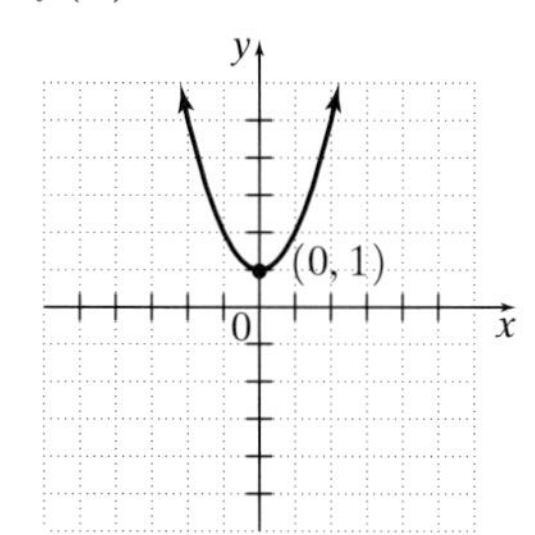

13. $f(x) = x^2 + 8x + 7$
$-\frac{b}{2a} = -\frac{8}{2(1)} = -4$
$f(-4) = (-4)^2 + 8(-4) + 7 = -9$
The vertex is $(-4, -9)$.
$x^2 + 8x + 7 = 0$
$(x + 7)(x + 1) = 0$
$x + 7 = 0$ or $x + 1 = 0$
$x = -7$ or $x = -1$
The x-intercepts are -7 and -1.
If $x = 0$, then $y = f(0) = 7$.
The y-intercept is 7.

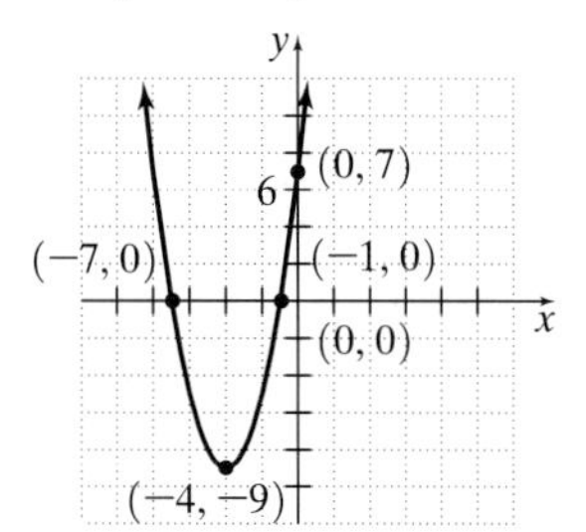

17. $f(x) = 2x^2 - 6x$

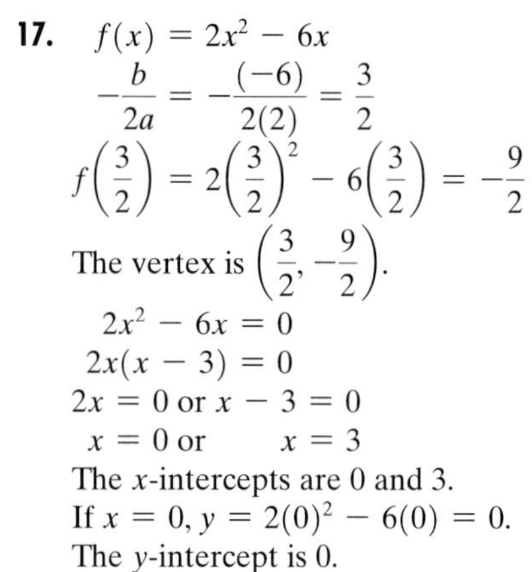

$-\frac{b}{2a} = -\frac{(-6)}{2(2)} = \frac{3}{2}$
$f\left(\frac{3}{2}\right) = 2\left(\frac{3}{2}\right)^2 - 6\left(\frac{3}{2}\right) = -\frac{9}{2}$
The vertex is $\left(\frac{3}{2}, -\frac{9}{2}\right)$.
$2x^2 - 6x = 0$
$2x(x - 3) = 0$
$2x = 0$ or $x - 3 = 0$
$x = 0$ or $x = 3$
The x-intercepts are 0 and 3.
If $x = 0$, $y = 2(0)^2 - 6(0) = 0$.
The y-intercept is 0.

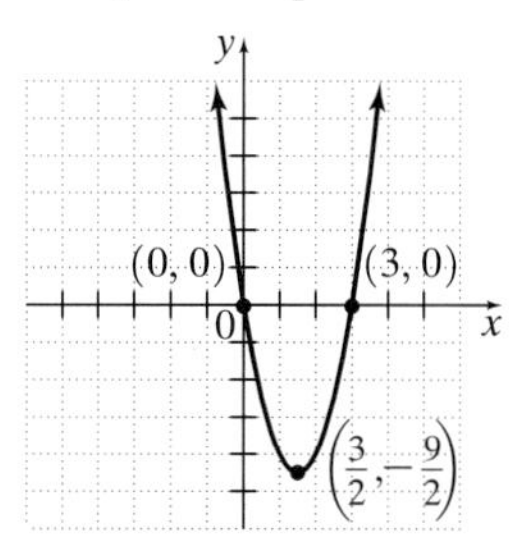

21. Two

25. $f(x) = 4x^3 - 9x$

$$4x^3 - 9x = 0$$
$$x(4x^2 - 9) = 0$$
$$x(2x + 3)(2x - 3) = 0$$
$$x = 0 \text{ or } 2x + 3 = 0 \text{ or } 2x - 3 = 0$$
$$x = 0 \text{ or } \quad x = -\frac{3}{2} \text{ or } \quad x = \frac{3}{2}$$

The x-intercepts are $-\frac{3}{2}, 0, \frac{3}{2}$.

If $x = 0$, $y = 4(0)^3 - 9(0) = 0$.

The y-intercept is 0.

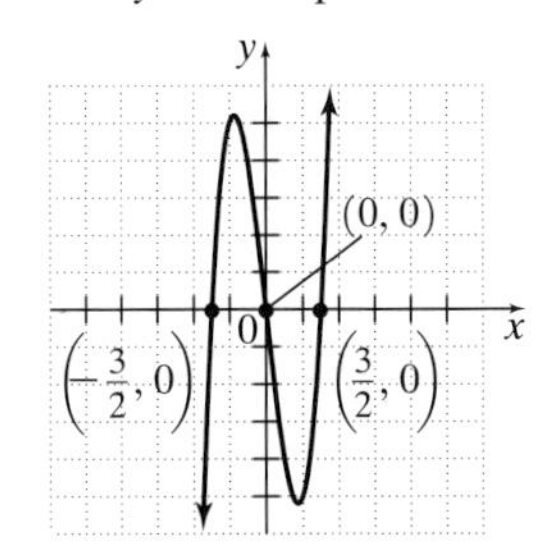

29. $f(x) = x(x - 4)(x + 2)$

If $x(x - 4)(x + 2) = 0$

$x = 0$ or $x - 4 = 0$ or $x + 2 = 0$

$x = 0$ or $\quad x = 4$ or $\quad x = -2$

The x-intercepts are $-2, 0, 4$.

If $x = 0$, $y = 0(0 - 4)(0 + 2) = 0$.

The y-intercept is 0.

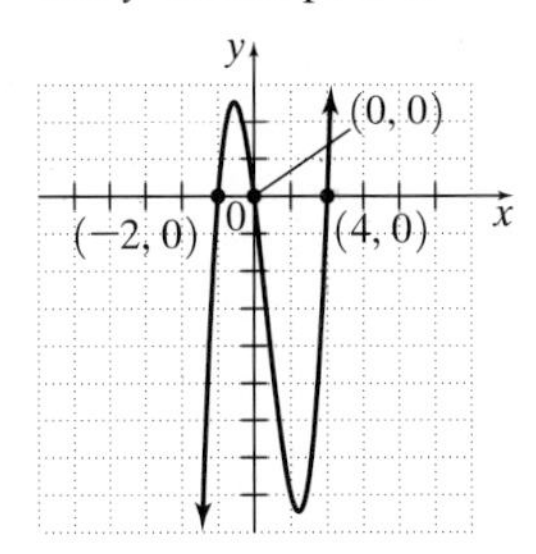

33. No. Answers may vary.

37. $\dfrac{x^7y^{10}}{x^3y^{15}} = \dfrac{x^{7-3}}{y^{15-10}} = \dfrac{x^4}{y^5}$

41. $f(x) = (x - 2)(x + 5)$

Graph E because the x-intercepts are at $(2, 0)$ and $(-5, 0)$

45. $f(x) = 2x^2 + 9x + 4 = (2x + 1)(x + 4)$

Graph B because the x-intercepts are at $\left(-\frac{1}{2}, 0\right)$ and $(-4, 0)$

49. $f(x) = 4x^3 - 9x$

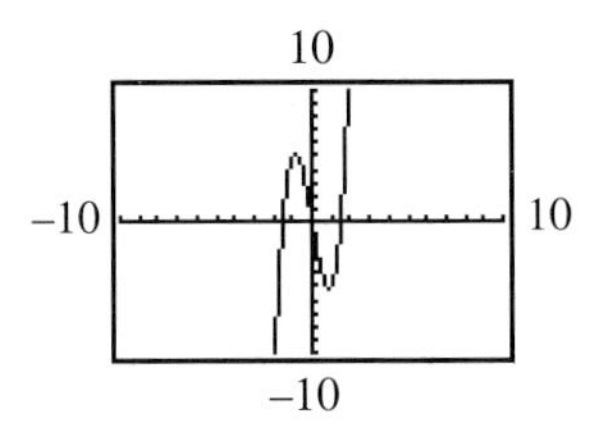

Chapter 5 Test

1. $(4x^3 - 3x - 4) - (9x^3 + 8x + 5)$

$= 4x^3 - 3x - 4 - 9x^3 - 8x - 5$

$= 4x^3 - 9x^3 - 3x - 8x - 4 - 5$

$= -5x^3 - 11x - 9$

5. $(6m + n)^2 = 36m^2 + 12mn + n^2$

9.

−3	4	−3	2	−1	−1
		−12	45	−141	426
	4	−15	47	−142	425

$(4x^4 - 3x^3 + 2x^2 - x - 1) \div (x + 3)$

$= 4x^3 - 15x^2 + 47x - 142 + \dfrac{425}{x + 3}$

13. $6x^2 - 15x - 9 = 3(2x^2 - 5x - 3)$

$= 3(2x + 1)(x - 3)$

17. $6x^2 + 24 = 6(x^2 + 4)$

21.

$$2x^3 - 8x + 5x^2 - 20 = 0$$
$$(2x + 5)(x^2 - 4) = 0$$
$$(2x + 5)(x + 2)(x - 2) = 0$$

$2x + 5 = 0$ or $x + 2 = 0$ or $x - 2 = 0$

$x = -\frac{5}{2}$ or $\quad x = -2$ or $\quad x = 2$

$\left\{-\frac{5}{2}, -2, 2\right\}$

25. $f(x) = x^3 - 1$

$x^3 - 1 = 0$

$(x - 1)(x^2 + x + 1) = 0$

$x - 1 = 0$ or $x^2 + x + 1 = 0$

$x = 1 \quad$ does not factor

The x-intercept is 1.

If $x = 0$, $y = f(0) = 0^3 - 1 = -1$,

so the y-intercept is -1.

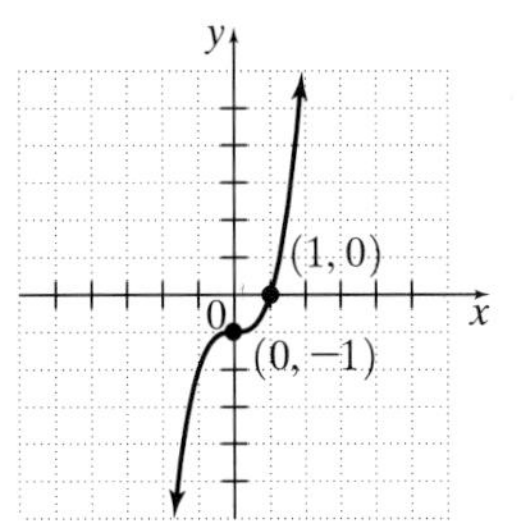

SOLUTIONS

Chapter 6

Exercise Set 6.1

1. $\frac{x+3}{x-2}$ is undefined when $x-2=0$ or $x=2$.

5. $\frac{x^2+1}{3x}$ is undefined when $3x=0$ or $x=0$.

9. $\frac{3x+2}{x^3+x^2-2x}$ is undefined when

$$x^3+x^2-2x=0$$
$$x(x^2+x-2)=0$$
$$x(x+2)(x-1)=0$$
$$x=0 \text{ or } x+2=0 \quad \text{or } x-1=0$$
$$x=0 \text{ or } \quad x=-2 \text{ or } \quad x=1$$

13. $\frac{10x^3}{18x}=\frac{(2x)5x^2}{(2x)9}=\frac{5x^2}{9}$

17. $\frac{8x-16x^2}{8x}=\frac{8x(1-2x)}{8x}=1-2x$

21. $\frac{9y-18}{7y-14}=\frac{9(y-2)}{7(y-2)}=\frac{9}{7}$

25. $\frac{x-9}{9-x}=\frac{-1(9-x)}{9-x}=-1$

29. $\frac{2x^2-7x-4}{x^2-5x+4}=\frac{(2x+1)(x-4)}{(x-1)(x-4)}=\frac{2x+1}{x-1}$

33. $\frac{3x^2-5x-2}{6x^3+2x^2+3x+1}=\frac{(3x+1)(x-2)}{2x^2(3x+1)+1(3x+1)}=\frac{(3x+1)(x-2)}{(3x+1)(2x^2+1)}=\frac{x-2}{2x^2+1}$

37. $\frac{4}{x}\cdot\frac{x^2}{8}=\frac{2\cdot2\cdot x\cdot x}{x\cdot2\cdot2\cdot2}=\frac{x}{2}$

41. $\frac{18a-12a^2}{4a^2+4a+1}\cdot\frac{4a^2+8a+3}{4a^2-9}$
$=\frac{6a(3-2a)(2a+3)(2a+1)}{(2a+1)^2(2a+3)(2a-3)}$
$=\frac{(6a)[-(2a-3)]}{(2a+1)(2a-3)}$
$=\frac{-6a}{2a+1}$

45. $\frac{2x^3-16}{6x^2+6x-36}\cdot\frac{9x+18}{3x^2+6x+12}$
$=\frac{2(x^3-8)\cdot9(x+2)}{6(x^2+x-6)\cdot3(x^2+2x+4)}$
$=\frac{18(x-2)(x^2+2x+4)(x+2)}{18(x+3)(x-2)(x^2+2x+4)}$
$=\frac{x+2}{x+3}$

49. $\frac{x^2-6x-16}{2x^2-128}\cdot\frac{x^2+16x+64}{3x^2+30x+48}$
$=\frac{(x-8)(x+2)(x+8)^2}{2(x^2-64)[3(x^2+10x+16)]}$
$=\frac{(x-8)(x+2)(x+8)^2}{2(x+8)(x-8)3(x+8)(x+2)}$
$=\frac{1}{6}$

53. $\frac{4}{x}\div\frac{8}{x^2}=\frac{4}{x}\cdot\frac{x^2}{8}=\frac{2\cdot2\cdot x\cdot x}{x\cdot2\cdot2\cdot2}=\frac{x}{2}$

57. $\frac{a+b}{ab}\div\frac{a^2-b^2}{4a^3b}=\frac{a+b}{ab}\cdot\frac{4a^3b}{(a+b)(a-b)}$
$=\frac{4a^2}{a-b}$

61. $\frac{x^2-6x-16}{2x^2-128}\div\frac{x^2+10x+16}{x^2+16x+64}$
$=\frac{(x-8)(x+2)}{2(x^2-64)}\cdot\frac{(x+8)^2}{(x+2)(x+8)}$
$=\frac{(x-8)(x+8)}{2(x+8)(x-8)}$
$=\frac{1}{2}$

65. $\frac{8b+24}{3a+6}\div\frac{ab-2b+3a-6}{a^2-4a+4}$
$=\frac{8(b+3)}{3(a+2)}\cdot\frac{(a-2)^2}{(a-2)(b+3)}$
$=\frac{8(a-2)}{3(a+2)}$

SOLUTIONS

69. $\dfrac{3x^2 - 5x - 2}{y^2 + y - 2} \cdot \dfrac{y^2 + 4y - 5}{12x^2 + 7x + 1} \div \dfrac{5x^2 - 9x - 2}{8x^2 - 2x - 1}$

$= \dfrac{3x^2 - 5x - 2}{y^2 + y - 2} \cdot \dfrac{y^2 + 4y - 5}{12x^2 + 7x + 1} \cdot \dfrac{8x^2 - 2x - 1}{5x^2 - 9x - 2}$

$= \dfrac{(3x + 1)(x - 2)(y + 5)(y - 1)(4x + 1)(2x - 1)}{(y + 2)(y - 1)(4x + 1)(3x + 1)(5x + 1)(x - 2)}$

$= \dfrac{(y + 5)(2x - 1)}{(y + 2)(5x + 1)}$

73. $R(x) = \dfrac{1000x^2}{x^2 + 4}$

a. $R(1) = \dfrac{1000(1)^2}{1^2 + 4} = \$200$ million

b. $R(2) = \dfrac{1000(2)^2}{2^2 + 4} = \$500$ million

c. \$500 million $-$ \$200 million $=$ \$300 million

77. $\dfrac{5}{28} - \dfrac{2}{21}$

The LCD is 84.

$\dfrac{5}{28} \cdot \dfrac{3}{3} - \dfrac{2}{21} \cdot \dfrac{4}{4} = \dfrac{15}{84} - \dfrac{8}{84} = \dfrac{7}{84} = \dfrac{7 \cdot 1}{7 \cdot 12} = \dfrac{1}{12}$

81. Since $A = b \cdot h$, $b = \dfrac{A}{h}$. Now,

$b = \dfrac{\frac{x^2 + x - 2}{x^3}}{\frac{x^2}{x - 1}}$

$b = \dfrac{(x + 2)(x - 1)}{x^3} \cdot \dfrac{(x - 1)}{x^2}$

$b = \dfrac{(x + 2)(x - 1)^2}{x^5}$ feet

85. $\dfrac{x^n + 4}{x^{2n} - 16} = \dfrac{x^n + 4}{(x^n - 4)(x^n + 4)} = \dfrac{1}{x^n - 4}$

Exercise Set 6.2

1. $\dfrac{2}{x} - \dfrac{5}{x} = \dfrac{2 - 5}{x} = \dfrac{-3}{x} = -\dfrac{3}{x}$

5. $\dfrac{x^2}{x + 2} - \dfrac{4}{x + 2} = \dfrac{x^2 - 4}{x + 2}$

9. $\dfrac{x - 5}{2x} - \dfrac{x + 5}{2x} = \dfrac{x - 5 - x - 5}{2x} = \dfrac{-10}{2x} = -\dfrac{5}{x}$

13. $7 = 7$
$5x = 5 \cdot x$
$\text{LCD} = 7 \cdot 5 \cdot x = 35x$

17. $x + 7 = x + 7$
$x - 7 = x - 7$
$\text{LCD} = (x + 7)(x - 7)$

21. $a^2 - b^2 = (a + b)(a - b)$
$a^2 - 2ab + b^2 = (a - b)(a - b)$
$\text{LCD} = (a + b)(a - b)(a - b) = (a + b)(a - b)^2$

25. $\dfrac{4}{3x} + \dfrac{3}{2x} = \dfrac{(4)2}{(3x)2} + \dfrac{(3)3}{(2x)3} = \dfrac{8 + 9}{6x} = \dfrac{17}{6x}$

29. $\dfrac{x - 3}{x + 4} - \dfrac{x + 2}{x - 4} = \dfrac{(x - 3)(x - 4)}{(x + 4)(x - 4)} - \dfrac{(x + 2)(x + 4)}{(x - 4)(x + 4)}$

$= \dfrac{-13x + 4}{(x + 4)(x - 4)}$

33. $\dfrac{3}{2x+10}+\dfrac{8}{3x+15}=\dfrac{3}{2(x+5)}+\dfrac{8}{3(x+5)}$
$=\dfrac{9}{6(x+5)}+\dfrac{16}{6(x+5)}$
$=\dfrac{9+16}{6(x+5)}$
$=\dfrac{25}{6(x+5)}$

37. $\dfrac{1}{a-b}+\dfrac{1}{b-a}=\dfrac{1}{a-b}+\dfrac{1}{-(a-b)}$
$=\dfrac{1}{a-b}-\dfrac{1}{a-b}$
$=0$

41. $\dfrac{y+1}{y^2-6y+8}-\dfrac{3}{y^2-16}$
$=\dfrac{y+1}{(y-2)(y-4)}-\dfrac{3}{(y+4)(y-4)}$
$=\dfrac{(y+1)(y+4)}{(y-2)(y-4)(y+4)}$
$-\dfrac{3(y-2)}{(y-2)(y-4)(y+4)}$
$=\dfrac{y^2+5y+4-(3y-6)}{(y-2)(y-4)(y+4)}$
$=\dfrac{y^2+5y+4-3y+6}{(y-2)(y-4)(y+4)}$
$=\dfrac{y^2+2y+10}{(y-2)(y-4)(y+4)}$

45. $\dfrac{x+4}{3x^2+11x+6}+\dfrac{x}{2x^2+x-15}$
$=\dfrac{x+4}{(3x+2)(x+3)}+\dfrac{x}{(2x-5)(x+3)}$
$=\dfrac{(x+4)(2x-5)}{(3x+2)(x+3)(2x-5)}$
$+\dfrac{x(3x+2)}{(2x-5)(x+3)(3x+2)}$
$=\dfrac{2x^2+3x-20+3x^2+2x}{(3x+2)(x+3)(2x-5)}$
$=\dfrac{5x^2+5x-20}{(3x+2)(x+3)(2x-5)}$
$=\dfrac{5(x^2+x-4)}{(3x+2)(x+3)(2x-5)}$

49. $\dfrac{ab}{a^2-b^2}+\dfrac{b}{2a+2b}=\dfrac{ab}{(a+b)(a-b)}+\dfrac{b}{2(a+b)}$
$=\dfrac{ab(2)+b(a-b)}{2(a+b)(a-b)}$
$=\dfrac{2ab+ab-b^2}{2(a+b)(a-b)}$
$=\dfrac{3ab-b^2}{2(a+b)(a-b)}$

53. $\dfrac{2}{x+1}-\dfrac{3x}{3x+3}+\dfrac{1}{2x+2}$
$=\dfrac{2}{x+1}-\dfrac{3x}{3(x+1)}+\dfrac{1}{2(x+1)}$
$=\dfrac{2}{x+1}-\dfrac{x}{x+1}+\dfrac{1}{2(x+1)}$
$=\dfrac{2-x}{x+1}+\dfrac{1}{2(x+1)}$
$=\dfrac{2(2-x)}{2(x+1)}+\dfrac{1}{2(x+1)}$
$=\dfrac{4-2x+1}{2(x+1)}$
$=\dfrac{5-2x}{2(x+1)}$

57. $\dfrac{x}{x^2-9}+\dfrac{3}{x^2-6x+9}-\dfrac{1}{x+3}$
$=\dfrac{x}{(x+3)(x-3)}+\dfrac{3}{(x-3)^2}-\dfrac{1}{(x+3)}$
$=\dfrac{x(x-3)+3(x+3)-(x-3)^2}{(x+3)(x-3)^2}$
$=\dfrac{x^2-3x+3x+9-x^2+6x-9}{(x-3)(x-3)^2}$
$=\dfrac{6x}{(x+3)(x-3)^2}$

61. $12\left(\dfrac{2}{3}+\dfrac{1}{6}\right)=\dfrac{24}{3}+\dfrac{12}{6}=8+2=10$

65. answers may vary

69. $(4x)^{-2}-(3x)^{-1}=\dfrac{1}{(4x)^2}-\dfrac{1}{3x}$
$=\dfrac{1}{16x^2}-\dfrac{1}{3x}$
$=\dfrac{3}{48x^2}-\dfrac{16x}{48x^2}$
$=\dfrac{3-16x}{48x^2}$

Exercise Set 6.3

1. $\dfrac{\frac{10}{3x}}{\frac{5}{6x}}=\dfrac{10}{3x}\cdot\dfrac{6x}{5}=\dfrac{2}{1}\cdot\dfrac{2}{1}=4$

5. $\dfrac{\frac{4}{x-1}}{\frac{x}{x-1}}=\dfrac{4}{x-1}\cdot\dfrac{x-1}{x}=\dfrac{4}{x}$

9. $\dfrac{\frac{4x^2-y^2}{xy}}{\frac{2}{y}-\frac{1}{x}}=\dfrac{\left(\frac{4x^2-y^2}{xy}\right)\cdot xy}{\left(\frac{2}{y}-\frac{1}{x}\right)\cdot xy}$
$=\dfrac{4x^2-y^2}{2x-y}$
$=\dfrac{(2x-y)(2x+y)}{2x-y}$
$=2x+y$

13. $$\frac{\frac{2}{x}+\frac{3}{x^2}}{\frac{4}{x^2}-\frac{9}{x}} = \frac{\left(\frac{2}{x}+\frac{3}{x^2}\right)x^2}{\left(\frac{4}{x^2}-\frac{9}{x}\right)x^2}$$
$$= \frac{2x+3}{4-9x}$$

17. $$\frac{\frac{4}{5-x}+\frac{5}{x-5}}{\frac{2}{x}+\frac{3}{x-5}} = \frac{-\frac{4}{x-5}+\frac{5}{x-5}}{\frac{2(x-5)+3x}{x(x-5)}}$$
$$= \frac{\frac{1}{x-5}}{\frac{2x-10+3x}{x(x-5)}}$$
$$= \frac{1}{x-5}\cdot\frac{x(x-5)}{5x-10}$$
$$= \frac{x}{5x-10}$$

21. $$\frac{x^{-1}}{x^{-2}+y^{-2}} = \frac{\frac{1}{x}}{\frac{1}{x^2}+\frac{1}{y^2}}$$
$$= \frac{\frac{1}{x}(x^2y^2)}{\left(\frac{1}{x^2}+\frac{1}{y^2}\right)(x^2y^2)}$$
$$= \frac{xy^2}{y^2+x^2}$$

25. $$\frac{1}{x-x^{-1}} = \frac{1}{x-\frac{1}{x}}$$
$$= \frac{1\cdot x}{\left(x-\frac{1}{x}\right)\cdot x}$$
$$= \frac{x}{x^2-1}$$
$$= \frac{x}{(x+1)(x-1)}$$

29. $x^2 = 4x - 4$
$x^2 - 4x + 4 = 0$
$(x-2)^2 = 0$
$x - 2 = 0$
$x = 2$
$\{2\}$

33. $$\frac{a}{1-\frac{s}{770}} = \frac{a\cdot 770}{\left(1-\frac{s}{770}\right)\cdot 770}$$
$$= \frac{770a}{770-s}$$

37. $$\frac{x}{1-\frac{1}{1-\frac{1}{x}}} = \frac{x}{1-\frac{1\cdot x}{\left(1-\frac{1}{x}\right)\cdot x}}$$
$$= \frac{x}{1-\frac{x}{x-1}}$$
$$= \frac{x(x-1)}{\left(1-\frac{x}{x-1}\right)(x-1)}$$
$$= \frac{x(x-1)}{(x-1)-x}$$
$$= \frac{x(x-1)}{x-1-x}$$
$$= \frac{x(x-1)}{-1}$$
$$= -x(x-1)$$

Exercise Set 6.4

1. $\frac{x}{2} - \frac{x}{3} = 12$
$3x - 2x = 72$
$x = 72$
$\{72\}$

5. $1 - \frac{4}{a} = 5$
$a - 4 = 5a$
$-4a = 4$
$a = -1$
$\{-1\}$

9. $\frac{2}{x} + \frac{1}{2} = \frac{5}{x}$
$\frac{1}{2} = \frac{3}{x}$
$x = 6$
$\{6\}$

SOLUTIONS

13.

$$\begin{aligned}\frac{5}{x-2}-\frac{2}{x+4}&=-\frac{4}{x^2+2x-8}\\(x-2)(x+4)\left(\frac{5}{x-2}-\frac{2}{x+4}\right)&=(x-2)(x+4)\left[\frac{-4}{(x+4)(x-2)}\right]\\5(x+4)-2(x-2)&=-4\\5x+20-2x+4&=-4\\3x&=-28\\x&=-\frac{28}{3}\end{aligned}$$

$\left\{-\frac{28}{3}\right\}$

17.

$$\begin{aligned}\frac{1}{x-4}-\frac{3x}{x^2-16}&=\frac{2}{x+4}\\(x+4)(x-4)\left[\frac{1}{x-4}-\frac{3x}{(x+4)(x-4)}\right]&=(x+4)(x-4)\left(\frac{2}{x+4}\right)\\x+4-3x&=2x-8\\-4x&=-12\\x&=3\end{aligned}$$

$\{3\}$

21.

$$\begin{aligned}\frac{1}{x-2}-\frac{2}{x^2-2x}&=1\\x(x-2)\left[\frac{1}{x-2}-\frac{2}{x(x-2)}\right]&=x(x-2)\cdot 1\\x-2&=x^2-2x\\0&=x^2-3x+2\\0&=(x-2)(x-1)\end{aligned}$$

$x-2=0$ or $x-1=0$

$x=2$ or $x=1$

We discard z as extraneous.

$\{1\}$

25.

$$\begin{aligned}\frac{1}{x}-\frac{x}{25}&=0\\25x\left(\frac{1}{x}-\frac{x}{25}\right)&=25x\cdot 0\\25-x^2&=0\\x^2-25&=0\\(x+5)(x-5)&=0\end{aligned}$$

$x+5=0$ or $x-5=0$

$x=-5$ or $x=5$

$\{-5, 5\}$

29.

$$\begin{aligned}\frac{x+3}{x+2}&=\frac{1}{x+2}\\(x+2)\left(\frac{x+3}{x+2}\right)&=(x+2)\left(\frac{1}{x+2}\right)\\x+3&=1\\x&=-2\end{aligned}$$

Which we discard as extraneous.

No solution

Ø

33. $$\frac{64}{x^2-16}+1=\frac{2x}{x-4}$$
$$(x+4)(x-4)\left[\frac{64}{(x+4)(x-4)}+1\right]=(x+4)(x-4)\left(\frac{2x}{x-4}\right)$$
$$64+(x+4)(x-4)=2x(x+4)$$
$$64+x^2-16=2x^2+8x$$
$$x^2+8x-48=0$$
$$(x+12)(x-4)=0$$
$x+12=0$ or $x-4=0$
$x=-12$ or $x=4$
We discard 4 as extraneous.
$\{-12\}$

37. $$\frac{28}{x^2-9}+\frac{2x}{x-3}+\frac{6}{x+3}=0$$
$$(x^2-9)\left(\frac{28}{x^2-9}+\frac{2x}{x-3}+\frac{6}{x+3}\right)=(x^2-9)\cdot 0$$
$$28+2x(x+3)+6(x-3)=0$$
$$28+2x^2+6x+6x-18=0$$
$$2x^2+12x+10=0$$
$$2(x+5)(x+1)=0$$
$x+5=0$ or $x+1=0$
$x=-5$ or $x=-1$
$\{-5,-1\}$

41. Let $x=$ the number.
$3x+4=19$
$3x=15$
$x=5$
The number is 5.

45. 10% (reading from the graph)

49. 19% of 29,265
$0.19(29{,}265)\approx 5560$
There were 5560 inmates.

53. $$x^{-2}-5x^{-1}-36=0$$
$$\frac{1}{x^2}-\frac{5}{x}-36=0$$
$$x^2\left(\frac{1}{x^2}-\frac{5}{x}-36\right)=x^2\cdot 0$$
$$1-5x-36x^2=0$$
$$36x^2+5x-1=0$$
$$(9x-1)(4x+1)=0$$
$9x-1=0$ or $4x+1=0$
$9x=1$ or $4x=-1$
$x=\frac{1}{9}$ or $x=-\frac{1}{4}$
$\left\{-\frac{1}{4},\frac{1}{9}\right\}$

57. $$f(x)=20+\frac{4000}{x}$$
$$25=20+\frac{4000}{x}$$
$$25-20=\frac{4000}{x}$$
$$5=\frac{4000}{x}$$
$$5x=4000$$
$$x=\frac{4000}{5}$$
$x=800$ pencil sharpeners

SOLUTIONS

Exercise Set 6.5

1.
$$F = \frac{9}{5}C + 32$$
$$F - 32 = \frac{9}{5}C$$
$$C = \frac{5}{9}(F - 32)$$

5.
$$\frac{1}{R} = \frac{1}{R_1} + \frac{1}{R_2}$$
$$\frac{1}{R} = \frac{R_2 + R_1}{R_1R_2}$$
$$R = \frac{R_1R_2}{R_1 + R_2}$$

9.
$$A = \frac{h(a + b)}{2}$$
$$2A = ah + bh$$
$$2A - ah = bh$$
$$\frac{2A - ah}{h} = b$$

13.
$$f = \frac{f_1f_2}{f_1 + f_2}$$
$$(f_1 + f_2)f = \left(\frac{f_1f_2}{f_1 + f_2}\right)(f_1 + f_2)$$
$$f_1f + f_2f = f_1f_2$$
$$f_1f = f_1f_2 - ff_2$$
$$f_1f = f_2(f_1 - f)$$
$$\frac{f_1f}{f_1 - f} = f_2$$

17.
$$\frac{\theta}{\omega} = \frac{2L}{c}$$
$$c\omega\left(\frac{\theta}{\omega}\right) = \frac{2L}{c}(c\omega)$$
$$c\theta = 2L\omega$$
$$c = \frac{2L\omega}{\theta}$$

21. Let $x =$ the number.
$$\frac{12 + x}{41 + 2x} = \frac{1}{3}$$
$$3(12 + x) = (41 + 2x)\cdot 1$$
$$36 + 3x = 41 + 2x$$
$$x = 5$$
The number is 5.

25. For three resistances R_1, R_2, and R_3 wired in a parallel circuit, the combined resistance R is given by
$$\frac{1}{R} = \frac{1}{R_1} + \frac{1}{R_2} + \frac{1}{R_3}$$
$$\frac{1}{R} = \frac{1}{5} + \frac{1}{6} + \frac{1}{2}$$
$$\frac{1}{R} = \frac{6 + 5 + 15}{30}$$
$$\frac{1}{R} = \frac{26}{30} = \frac{13}{15}$$
$$R = \frac{15}{13}$$

29. Convert each time to a rate.
$$\frac{1 \text{ task}}{20 \text{ minutes}} = \frac{1}{20}\frac{\text{task}}{\text{minute}}$$
$$\frac{1 \text{ task}}{30 \text{ minutes}} = \frac{1}{30}\frac{\text{task}}{\text{minute}}$$
$$\frac{1 \text{ task}}{60 \text{ minutes}} = \frac{1}{60}\frac{\text{task}}{\text{minute}}$$
Let $x =$ amount of time required for all three computers to complete the task.
So $\frac{1 \text{ task}}{x \text{ minute}} = \frac{1}{x}\frac{\text{task}}{\text{minute}}$. Summing we get,
$$\frac{1}{20} + \frac{1}{30} + \frac{1}{60} = \frac{1}{x}$$
$$\frac{3 + 2 + 1}{60} = \frac{1}{x}$$
$$\frac{6}{60} = \frac{1}{x}$$
$$\frac{1}{10} = \frac{1}{x} \text{ so } x = 10 \text{ minutes}$$
It takes 10 minutes for the three computers to complete the task.

33. Let $r =$ the speed of the boat in still water.
Recall that $d = rt$ or $t = \frac{d}{r}$. Using the latter equation we get $\frac{20}{r + 5} = \frac{10}{r - 5}$ where $r + 5$ is the rate of the boat traveling downstream and $r - 5$ is the rate of the boat traveling upstream. Now,
$$\frac{2}{r + 5} = \frac{1}{r - 5}$$
$$2(r - 5) = r + 5$$
$$2r - 10 = r + 5$$
$$r = 15$$
Thus, the speed of the boat in still water is 15 mph.

37. Convert times to rates.
$$\frac{1 \text{ pond}}{45 \text{ minutes}} = \frac{1}{45}\frac{\text{pond}}{\text{minute}}$$
$$\frac{1 \text{ pond}}{20 \text{ minutes}} = \frac{1}{20}\frac{\text{pond}}{\text{minute}}$$
Let $x =$ the number of minutes required for the second hose to fill the pond. Then,
$\frac{1 \text{ pond}}{x \text{ minute}} = \frac{1}{x}\frac{\text{pond}}{\text{minute}}$. Now,
$$\frac{1}{20} = \frac{1}{45} + \frac{1}{x}$$
$$\frac{1}{20} = \frac{x + 45}{45x}$$
$$45x = 20(x + 45)$$
$$45x = 20x + 900$$
$$25x = 900$$
$$x = \frac{900}{25}$$
$$= 36$$
Thus, the second hose alone can fill the pond in 36 minutes.

41.

	distance	= rate	· time
First part	20	r	$\frac{20}{r}$
Cooldown	16	$r - 2$	$\frac{16}{r-2}$

$$\frac{20}{r} = \frac{16}{r-2}$$
$$20(r-2) = 16r$$
$$20r - 40 = 16r$$
$$20r - 16r = 40$$
$$4r = 40$$
$$r = \frac{40}{4}$$
$$= 10$$

The first part is at 10 mph, and the cooldown at 8 mph.

45.

	distance	= rate	· time
with wind	465	$x + 20$	$\frac{465}{x+20}$
against wind	345	$x - 20$	$\frac{345}{x-20}$

$$\frac{465}{x+20} = \frac{345}{x-20}$$
$$465(x-20) = 345(x+20)$$
$$465x - 9300 = 345x + 6900$$
$$120x = 16{,}200$$
$$x = 135 \text{ mph}$$

49.

	Time	In one hour
Experienced	4	$\frac{1}{4}$
Apprentice	5	$\frac{1}{5}$
Together	x	$\frac{1}{x}$

$$\frac{1}{4} + \frac{1}{5} = \frac{1}{x}$$
$$20x\left(\frac{1}{4} + \frac{1}{5}\right) = 20x\left(\frac{1}{x}\right)$$
$$20x\left(\frac{1}{4}\right) + 20x\left(\frac{1}{5}\right) = 20$$
$$5x + 4x = 20$$
$$9x = 20$$
$$x = \frac{20}{9}$$
$$= 2\frac{2}{9} \text{ hours}$$

53.

	Time	In one minute
Belt	2	$\frac{1}{2}$
Smaller	6	$\frac{1}{6}$
Together	x	$\frac{1}{x}$

$$\frac{1}{2} + \frac{1}{6} = \frac{1}{x}$$
$$6x\left(\frac{1}{2} + \frac{1}{6}\right) = 6x\left(\frac{1}{x}\right)$$
$$6x\left(\frac{1}{2}\right) + 6x\left(\frac{1}{6}\right) = 6$$
$$3x + x = 6$$
$$4x = 6$$
$$x = \frac{6}{4}$$
$$= \frac{3}{2}$$
$$= 1\frac{1}{2} \text{ minutes}$$

57.
$$\frac{x}{5} = \frac{x+2}{3}$$
$$3x = 5(x+2)$$
$$3x = 5x + 10$$
$$-10 = 2x$$
$$x = -5$$
$$\{-5\}$$

61.
$$B = \frac{705w}{h^2}$$
$$h = 5'8''$$
$$= 68''$$
$$25 = \frac{705w}{(68)^2}$$
$$25 = \frac{705w}{4624}$$
$$115{,}600 = 705w$$
$$164 \approx w$$

The wight should be 164 pounds.

65. answers may vary

Exercise Set 6.6

1.
$$y = kx$$
$$4 = k(20)$$
$$k = \frac{1}{5}$$
$$y = \frac{1}{5}x$$

5.
$$y = kx$$
$$7 = k\left(\frac{1}{2}\right)$$
$$k = 14$$
$$y = 14x$$

SOLUTIONS

SOLUTIONS

9.
$$\begin{aligned} W &= kr^3 \\ 1.2 &= k \cdot 2^3 \\ k &= \frac{1.2}{8} \\ &= 0.15 \\ W &= 0.15r^3 \\ &= 0.15(3)^3 \\ W &= 0.15(27) \\ &= 4.05 \text{ pounds} \end{aligned}$$

13.
$$\begin{aligned} y &= \frac{k}{x} \\ 6 &= \frac{k}{5} \\ k &= 30 \\ y &= \frac{30}{x} \end{aligned}$$

17.
$$\begin{aligned} y &= \frac{k}{x} \\ \frac{1}{8} &= \frac{k}{16} \\ k &= 2 \\ y &= \frac{2}{x} \end{aligned}$$

21.
$$\begin{aligned} R &= \frac{k}{T} \\ 45 &= \frac{k}{6} \\ k &= 270 \\ R &= \frac{270}{5} \\ R &= 54 \text{ mph} \end{aligned}$$

25. $I_1 = \dfrac{k}{d^2}$

Replace d by $2d$.

$$I_2 = \frac{k}{(2d)^2} = \frac{k}{4d^2} = \frac{1}{4}I_1$$

Thus, the intensity is divided by 4.

29. $r = kst^3$

33.
$$\begin{aligned} V &= kr^2h \\ 32\pi &= k(4)^2(6) \\ k &= \frac{\pi}{3} \\ V &= \frac{\pi}{3}(3)^2(5) \\ V &= (15\pi) \text{ in}^3 \end{aligned}$$

37.
$$\begin{aligned} r &= 6 \text{ cm} \\ C &= 2\pi(6) \\ &= 12\pi \text{ cm} \\ A &= \pi(6^2) \\ &= 36\pi \text{ cm}^2 \end{aligned}$$

41. $(3, 6), (-2, 6)$

$$\begin{aligned} m &= \frac{y_2 - y_1}{x_2 - x_1} \\ m &= \frac{6 - 6}{-2 - 3} \\ &= \frac{0}{-5} \\ &= 0 \end{aligned}$$

45.
$$\begin{aligned} V_1 &= khr^2 \\ V_2 &= k\left(\frac{1}{2}h\right)(2r)^2 \\ V_2 &= 2khr^2 = 2V_1 \end{aligned}$$

It is multiplied by 2.

Chapter 6 Test

1. $\dfrac{5x^2}{1 - x}$

Undefined when

$$\begin{aligned} 1 - x &= 0 \\ 1 &= x \end{aligned}$$

5. $$\frac{x^2 - 4x}{x^2 + 5x - 36} = \frac{x(x - 4)}{(x + 9)(x - 4)} = \frac{x}{x + 9}$$

9.
$$\begin{aligned} &\frac{3x^2 - 12}{x^2 + 2x - 8} \div \frac{6x + 18}{x + 4} \\ &= \frac{3(x^2 - 4)}{(x + 4)(x - 2)} \cdot \frac{x + 4}{6(x + 3)} \\ &= \frac{(x + 2)(x - 2)}{x - 2} \cdot \frac{1}{2(x + 3)} \\ &= \frac{x + 2}{2(x + 3)} \end{aligned}$$

13.
$$\begin{aligned} &\frac{3}{x^2 - x - 6} + \frac{2}{x^2 - 5x + 6} \\ &= \frac{3}{(x - 3)(x + 2)} + \frac{2}{(x - 3)(x - 2)} \\ &= \frac{3(x - 2) + 2(x + 2)}{(x - 3)(x + 2)(x - 2)} \\ &= \frac{3x - 6 + 2x + 4}{(x - 3)(x + 2)(x - 2)} \\ &= \frac{5x - 2}{(x - 3)(x + 2)(x - 2)} \end{aligned}$$

17. $$\frac{\frac{5}{x} - \frac{7}{3x}}{\frac{9}{8x} - \frac{1}{x}} = \frac{\frac{15 - 7}{3x}}{\frac{9 - 8}{8x}} = \frac{8}{3x} \cdot \frac{8x}{1} = \frac{64}{3}$$

21.
$$\begin{aligned} \frac{x}{x - 4} &= 3 - \frac{4}{x - 4} \\ \frac{x}{x - 4} + \frac{4}{x - 4} &= 3 \\ \frac{x + 4}{x - 4} &= 3 \\ x + 4 &= 3(x - 4) \\ x + 4 &= 3x - 12 \\ 16 &= 2x \\ x &= 8 \end{aligned}$$

$\{8\}$

25. $W = \frac{k}{V}$

$20 = \frac{k}{12}$

$k = 240$

so $W = \frac{240}{V}$

$= \frac{240}{15}$

$= 16$

Chapter 7

Exercise Set 7.1

1. Since $2^2 = 4$ and $(-2)^2 = 4$, the square roots of 4 are 2 and -2.

5. Since $10^2 = 100$ and $(-10)^2 = 100$, the square roots of 100 are 10 and -10.

9. $\sqrt{\frac{1}{4}} = \frac{1}{2}$ because $\left(\frac{1}{2}\right)^2 = \frac{1}{4}$.

13. $-\sqrt{36} = -6$ because $(6)^2 = 36$.

17. $\sqrt{16y^6} = \sqrt{16}\sqrt{y^6} = 4y^3$ because $(4y^3)^2 = 16y^6$.

21. $\sqrt{38} \approx 6.164$
Since $36 < 38 < 49$, then $\sqrt{36} < \sqrt{38} < \sqrt{49}$ or $6 < \sqrt{38} < 7$. The approximation is between 6 and 7 and this is reasonable.

25. $\sqrt[3]{64} = 4$ because $(4)^3 = 64$.

29. $\sqrt[3]{-1} = -1$ because $(-1)^3 = -1$.

33. $\sqrt[3]{-27x^9} = -3x^3$ because $(-3x^3)^3 = -27x^9$.

37. $\sqrt[4]{-16}$ is not a real number.

41. $\sqrt[5]{x^{20}} = x^4$ because $(x^4)^5 = x^{20}$.

45. $\sqrt{81x^4} = 9x^2$ because $(9x^2)^2 = 81x^4$.

49. $\sqrt{(-8)^2} = |-8| = 8$ **53.** $\sqrt{4x^2} = 2|x|$

57. $\sqrt[4]{(x-2)^4} = |x-2|$

61. $f(x) = \sqrt{2x+3}$
$f(0) = \sqrt{2 \cdot 0 + 3} = \sqrt{3}$

65. $g(x) = \sqrt[3]{x-8}$
$g(-19) = \sqrt[3]{-19-8} = \sqrt[3]{-27} = -3$

69. $(-2x^3y^2)^5 = (-2)^5x^{3\cdot5}y^{2\cdot5} = -32x^{15}y^{10}$

73. $\frac{7x^{-1}y}{14(x^5y^2)^{-2}} = \frac{7x^{-1}y}{14x^{-10}y^{-4}} = \frac{x^9y^5}{2}$

Exercise Set 7.2

1. $49^{1/2} = \sqrt{49} = 7$

5. $\left(\frac{1}{16}\right)^{1/4} = \sqrt[4]{\frac{1}{16}} = \frac{1}{2}$

9. $2m^{1/3} = 2\sqrt[3]{m}$

13. $(-27)^{1/3} = \sqrt[3]{-27} = -3$

17. $16^{3/4} = (\sqrt[4]{16})^3 = 2^3 = 8$

21. $(-16)^{3/4} = (\sqrt[4]{-16})^3$ is not a real number.

25. $(7x+2)^{2/3} = \sqrt[3]{(7x+2)^2}$ or $(\sqrt[3]{7x+2})^2$

29. $8^{-4/3} = \frac{1}{8^{4/3}} = \frac{1}{(8^{1/3})^4} = \frac{1}{2^4} = \frac{1}{16}$

33. $(-4)^{-3/2} = \frac{1}{(-4)^{3/2}} = \frac{1}{[(-4)^{1/2}]^3}$ is not a real number.

37. $\frac{1}{a^{-2/3}} = a^{2/3}$ **41.** answers may vary

45. $x^{-2/5} \cdot x^{7/5} = x^{-2/5+7/5} = x^{5/5} = x$

49. $\frac{y^{1/3}}{y^{1/6}} = y^{1/3-1/6} = y^{2/6-1/6} = y^{1/6}$

53. $\frac{b^{1/2}b^{3/4}}{-b^{1/4}} = -b^{1/2+3/4-1/4} = -b^{1/2+1/2} = -b^1 = -b$

57. $\sqrt[6]{x^3} = x^{3/6} = x^{1/2} = \sqrt{x}$

61. $\sqrt[4]{16x^2} = 16^{1/4}x^{2/4} = 2x^{1/2} = 2\sqrt{x}$

65. $\sqrt[8]{x^4y^4} = x^{4/8}y^{4/8} = x^{1/2}y^{1/2} = \sqrt{xy}$

69. $\sqrt[3]{y} \cdot \sqrt[5]{y^2} = y^{1/3} \cdot y^{2/5} = y^{5/15} \cdot y^{6/15} = y^{11/15} = \sqrt[15]{y^{11}}$

73. $\sqrt[3]{x} \cdot \sqrt[4]{x} \cdot \sqrt[8]{x^3} = x^{1/3} \cdot x^{1/4} \cdot x^{3/8}$
$= x^{8/24} \cdot x^{6/24} \cdot x^{9/24}$
$= x^{23/24}$
$= \sqrt[24]{x^{23}}$

77. $\sqrt{3} \cdot \sqrt[3]{4} = 3^{1/2} \cdot 4^{1/3}$
$= 3^{3/6} \cdot 4^{2/6}$
$= (3^3 \cdot 4^2)^{1/6}$
$= (432)^{1/6}$
$= \sqrt[6]{432}$

81. $\sqrt{5r} \cdot \sqrt[3]{s} = (5r)^{1/2} \cdot s^{1/3}$
$= (5r)^{3/6} \cdot s^{2/6}$
$= [(5r)^3 \cdot s^2]^{1/6}$
$= (125r^3s^2)^{1/6}$
$= \sqrt[6]{125r^3s^2}$

85. $48 = 16 \cdot 3$ or $4 \cdot 12$ **89.** $54 = 27 \cdot 2$

93. $x^{1/5}$
$\frac{x^{1/5}}{x^{-2/5}} = x^{1/5+2/5} = x^{3/5}$

97. $\frac{\sqrt{t}}{\sqrt{u}} = \frac{t^{1/2}}{u^{1/2}}$

Exercise Set 7.3

1. $\sqrt{7} \cdot \sqrt{2} = \sqrt{7 \cdot 2} = \sqrt{14}$

5. $\sqrt[3]{4} \cdot \sqrt[3]{9} = \sqrt[3]{4 \cdot 9} = \sqrt[3]{36}$

9. $\sqrt{\frac{7}{x}} \cdot \sqrt{\frac{2}{y}} = \sqrt{\frac{7 \cdot 2}{x \cdot y}} = \sqrt{\frac{14}{xy}}$

13. $\sqrt{\frac{6}{49}} = \frac{\sqrt{6}}{\sqrt{49}} = \frac{\sqrt{6}}{7}$ **17.** $\sqrt[4]{\frac{x^3}{16}} = \frac{\sqrt[4]{x^3}}{\sqrt[4]{16}} = \frac{\sqrt[4]{x^3}}{2}$

21. $\sqrt[4]{\frac{8}{x^8}} = \frac{\sqrt[4]{8}}{\sqrt[4]{x^8}} = \frac{\sqrt[4]{8}}{x^2}$

25. $\sqrt{\frac{x^2y}{100}} = \frac{\sqrt{x^2} \cdot \sqrt{y}}{\sqrt{100}} = \frac{x\sqrt{y}}{10}$

29. $-\sqrt[3]{\frac{z^7}{27x^3}} = \frac{-\sqrt[3]{z^7}}{\sqrt[3]{27x^3}}$
$= \frac{-\sqrt[3]{z^6z}}{\sqrt[3]{27} \cdot \sqrt[3]{x^3}}$
$= \frac{-\sqrt[3]{z^6} \cdot \sqrt[3]{z}}{3x}$
$= \frac{-z^2\sqrt[3]{z}}{3x}$

SOLUTIONS

SOLUTIONS

33. $\sqrt[3]{192} = \sqrt[3]{64(3)} = \sqrt[3]{64} \cdot \sqrt[3]{3} = 4\sqrt[3]{3}$

37. $\sqrt{24} = \sqrt{4 \cdot 6} = \sqrt{4} \cdot \sqrt{6} = 2\sqrt{6}$

41. $\sqrt[3]{16y^7} = \sqrt[3]{(8y^6)(2y)}$
$= \sqrt[3]{8} \cdot \sqrt[3]{y^6} \cdot \sqrt[3]{2y}$
$= 2y^2\sqrt[3]{2y}$

45. $\sqrt{y^5} = \sqrt{y^4 y} + \sqrt{y^4} \cdot \sqrt{y} = y^2\sqrt{y}$

49. $\sqrt[5]{-32x^{10}y} = \sqrt[5]{-32} \cdot \sqrt[5]{x^{10}} \cdot \sqrt[5]{y} = -2x^2\sqrt[5]{y}$

53. $-\sqrt{32a^8b^7} = -\sqrt{16a^8b^6(2b)}$
$= -\sqrt{16} \cdot \sqrt{a^8} \cdot \sqrt{b^6} \cdot \sqrt{2b}$
$= -4a^4b^3\sqrt{2b}$

57. $\sqrt[3]{125r^9s^{12}} = 5r^3s^4$

61. $\dfrac{\sqrt[3]{24}}{\sqrt[3]{3}} = \sqrt[3]{\dfrac{24}{3}} = \sqrt[3]{8} = 2$

65. $\dfrac{\sqrt{x^5y^3}}{\sqrt{xy}} = \sqrt{\dfrac{x^5y^3}{xy}}$
$= \sqrt{x^4y^2}$
$= x^2y$

69. $\dfrac{3\sqrt{100x^2}}{2\sqrt{2x^{-1}}} = \dfrac{3}{2}\sqrt{\dfrac{100x^2}{2x^{-1}}}$
$= \dfrac{3}{2}\sqrt{50x^3}$
$= \dfrac{3}{2}\sqrt{25x^2 \cdot 2x}$
$= \dfrac{3}{2} \cdot 5x\sqrt{2x}$
$= \dfrac{15x}{2}\sqrt{2x}$

73. $6x + 8x = 14x$

77. $9y^2 - 8y^2 = y^2$

81. $(x - 4)^2 = x^2 - 2x(4) + 4^2 = x^2 - 8x + 16$

85. $F(x) = 0.6\sqrt{49 - x^2}$

a. $F(3) = 0.6\sqrt{49 - 3^2}$
$= 0.6\sqrt{49 - 9}$
$= 0.6\sqrt{40}$
≈ 3.8

b. $F(5) = 0.6\sqrt{49 - 5^2}$
$= 0.6\sqrt{49 - 25}$
$= 0.6\sqrt{24}$
≈ 2.9

c. answers may vary

Exercise Set 7.4

1. $\sqrt{8} - \sqrt{32} = \sqrt{4(2)} - \sqrt{16(2)}$
$= \sqrt{4}\sqrt{2} - \sqrt{16}\sqrt{2}$
$= 2\sqrt{2} - 4\sqrt{2}$
$= -2\sqrt{2}$

5. $2\sqrt{50} - 3\sqrt{125} + \sqrt{98}$
$= 2\sqrt{25(2)} - 3\sqrt{25(5)} + \sqrt{49(2)}$
$= 2\sqrt{25}\sqrt{2} - 3\sqrt{25}\sqrt{5} + \sqrt{49}\sqrt{2}$
$= 2(5)\sqrt{2} - 3(5)\sqrt{5} + 7\sqrt{2}$
$= 10\sqrt{2} - 15\sqrt{5} + 7\sqrt{2}$
$= 17\sqrt{2} - 15\sqrt{5}$

9. $\sqrt{9b^3} - \sqrt{25b^3} + \sqrt{49b^3}$
$= \sqrt{9b^2(b)} - \sqrt{25b^2(b)} + \sqrt{49b^2(b)}$
$= \sqrt{9b^2}\sqrt{b} - \sqrt{25b^2}\sqrt{b} + \sqrt{49b^2}\sqrt{b}$
$= 3b\sqrt{b} - 5b\sqrt{b} + 7b\sqrt{b}$
$= 5b\sqrt{b}$

13. $\sqrt[3]{\dfrac{11}{8}} - \dfrac{\sqrt[3]{11}}{6} = \dfrac{\sqrt[3]{11}}{\sqrt[3]{8}} - \dfrac{\sqrt[3]{11}}{6}$
$= \dfrac{\sqrt[3]{11}}{2} - \dfrac{\sqrt[3]{11}}{6}$
$= \dfrac{3\sqrt[3]{11} - \sqrt[3]{11}}{6}$
$= \dfrac{2\sqrt[3]{11}}{6}$
$= \dfrac{\sqrt[3]{11}}{3}$

17. $7\sqrt{9} - 7 + \sqrt{3} = 7(3) - 7 + \sqrt{3}$
$= 21 - 7 + \sqrt{3}$
$= 14 + \sqrt{3}$

21. $3\sqrt{108} - 2\sqrt{18} - 3\sqrt{48}$
$= 3\sqrt{36}\sqrt{3} - 2\sqrt{9}\sqrt{2} - 3\sqrt{16}\sqrt{3}$
$= 3(6)\sqrt{3} - 2(3)\sqrt{2} - 3(4)\sqrt{3}$
$= 18\sqrt{3} - 6\sqrt{2} - 12\sqrt{3}$
$= 6\sqrt{3} - 6\sqrt{2}$

25. $\sqrt{9b^3} - \sqrt{25b^3} + \sqrt{16b^3}$
$= \sqrt{9b^2}\sqrt{b} - \sqrt{25b^2}\sqrt{b} + \sqrt{16b^2}\sqrt{b}$
$= 3b\sqrt{b} - 5b\sqrt{b} + 4b\sqrt{b}$
$= (3 - 5 + 4)b\sqrt{b}$
$= 2b\sqrt{b}$

29. $\sqrt[3]{54xy^3} - 5\sqrt[3]{2xy^3} + y\sqrt[3]{128x}$
$= \sqrt[3]{27y^3}\sqrt[3]{2x} - 5\sqrt[3]{y^3}\sqrt[3]{2x} + y\sqrt[3]{64}\sqrt[3]{2x}$
$= 3y\sqrt[3]{2x} - 5y\sqrt[3]{2x} + y(4)\sqrt[3]{2x}$
$= -2y\sqrt[3]{2x} + 4y\sqrt[3]{2x}$
$= 2y\sqrt[3]{2x}$

33. $-2\sqrt[4]{x^7} + 3\sqrt[4]{16x^7} = -2\sqrt[4]{x^4}\sqrt[4]{x^3} + 3\sqrt[4]{16x^4}\sqrt[4]{x^3}$
$= -2x\sqrt[4]{x^3} + 3(2x)\sqrt[4]{x^3}$
$= -2x\sqrt[4]{x^3} + 6x\sqrt[4]{x^3}$
$= 4x\sqrt[4]{x^3}$

37. $\dfrac{\sqrt[3]{8x^4}}{7} + \dfrac{3x\sqrt[3]{x}}{7} = \dfrac{\sqrt[3]{8x^3}\sqrt[3]{x} + 3x\sqrt[3]{x}}{7}$
$= \dfrac{2x\sqrt[3]{x} + 3x\sqrt[3]{x}}{7}$
$= \dfrac{5x\sqrt[3]{x}}{7}$

41. $\sqrt[3]{\dfrac{16}{27}} - \dfrac{\sqrt[3]{54}}{6} = \dfrac{\sqrt[3]{16}}{\sqrt[3]{27}} - \dfrac{\sqrt[3]{27}\sqrt[3]{2}}{6}$
$= \dfrac{\sqrt[3]{8}\sqrt[3]{2}}{3} - \dfrac{3\sqrt[3]{2}}{6}$
$= \dfrac{2(2)\sqrt[3]{2}}{6} - \dfrac{3\sqrt[3]{2}}{6}$
$= \dfrac{4\sqrt[3]{2} - 3\sqrt[3]{2}}{6}$
$= \dfrac{\sqrt[3]{2}}{6}$

45. $P = 2\sqrt{12} + \sqrt{12} + 2\sqrt{27} + 3\sqrt{3}$
$= 2\sqrt{4}\sqrt{3} + \sqrt{4}\sqrt{3} + 2\sqrt{9}\sqrt{3} + 3\sqrt{3}$
$= 2 \cdot 2\sqrt{3} + 2\sqrt{3} + 2 \cdot 3\sqrt{3} + 3\sqrt{3}$
$= (4 + 2 + 6 + 3)\sqrt{3}$
$= 15\sqrt{3}$

The perimeter of the trapezoid is $15\sqrt{3}$ in.

49. $(\sqrt{5}-\sqrt{2})^2 = \sqrt{5}^2 - 2\sqrt{5}\sqrt{2} + \sqrt{2}^2$
$= 5 - 2\sqrt{10} + 2$
$= 7 - 2\sqrt{10}$

53. $(2\sqrt{x}-5)(3\sqrt{x}+1)$
$= (2\sqrt{x})(3\sqrt{x}) + (2\sqrt{x})1 - 5(3\sqrt{x}) - 5\cdot 1$
$= 6x + 2\sqrt{x} - 15\sqrt{x} - 5$
$= 6x - 13\sqrt{x} - 5$

57. $6(\sqrt{2}-2) = 6\sqrt{2} - 6\cdot 2 = 6\sqrt{2} - 12$

61. $(2\sqrt{7}+3\sqrt{5})(\sqrt{7}-2\sqrt{5})$
$= 2\sqrt{7}^2 - (2\sqrt{7})(2\sqrt{5}) + (3\sqrt{5})\sqrt{7} - 3\cdot 2\sqrt{5}^2$
$= 2\cdot 7 - 4\sqrt{35} + 3\sqrt{35} - 6\cdot 5$
$= 14 - \sqrt{35} - 30$
$= -16 - \sqrt{35}$

65. $(\sqrt{3}+x)^2 = \sqrt{3}^2 + 2\sqrt{3}x + x^2$
$= 3 + 2\sqrt{3}x + x^2$

69. $(\sqrt[3]{4}+2)(\sqrt[3]{2}-1)$
$= \sqrt[3]{4}\sqrt[3]{2} - \sqrt[3]{4}\cdot 1 + 2\sqrt[3]{2} - 2\cdot 1$
$= \sqrt[3]{8} - \sqrt[3]{4} + 2\sqrt[3]{2} - 2$
$= 2 - \sqrt[3]{4} + 2\sqrt[3]{2} - 2$
$= -\sqrt[3]{4} + 2\sqrt[3]{2}$

73. $(\sqrt{x-1}+5)^2 = (\sqrt{x-1})^2 + 2\sqrt{x-1}(5) + 5^2$
$= x - 1 + 10\sqrt{x-1} + 25$
$= x + 24 + 10\sqrt{x-1}$

77. $\frac{2x-14}{2} = \frac{2(x-7)}{2} = x - 7$

81. $\frac{6a^2b - 9ab}{3ab} = \frac{3ab(2a-3)}{3ab} = 2a - 3$

85. $P = 2(3\sqrt{20}) + 2\sqrt{125}$
$= 6\sqrt{4}\sqrt{5} + 2\sqrt{25}\sqrt{5}$
$= 6(2)\sqrt{5} + 2(5)\sqrt{5}$
$= 12\sqrt{5} + 10\sqrt{5}$
$= 22\sqrt{5}$ feet

$A = 3\sqrt{20}\cdot\sqrt{125}$
$= 3\cdot 2\sqrt{5}\cdot 5\sqrt{5}$
$= 30(\sqrt{5})^2$
$= 30\cdot 5$
$= 150$ square feet

Exercise Set 7.5

1. $\frac{\sqrt{2}}{\sqrt{7}} = \frac{\sqrt{2}\cdot\sqrt{7}}{\sqrt{7}\cdot\sqrt{7}} = \frac{\sqrt{14}}{7}$

5. $\frac{4}{\sqrt[3]{3}}\cdot\frac{\sqrt[3]{9}}{\sqrt[3]{9}} = \frac{4\sqrt[3]{9}}{\sqrt[3]{27}} = \frac{4\sqrt[3]{9}}{3}$

9. $\frac{3}{\sqrt[3]{4x^2}} = \frac{3}{\sqrt[3]{4x^2}}\cdot\frac{\sqrt[3]{2x}}{\sqrt[3]{2x}} = \frac{3\sqrt[3]{2x}}{\sqrt[3]{8x^3}} = \frac{3\sqrt[3]{2x}}{2x}$

13. $\frac{3}{\sqrt[3]{2}} = \frac{3}{\sqrt[3]{2}}\cdot\frac{\sqrt[3]{4}}{\sqrt[3]{4}} = \frac{3\sqrt[3]{4}}{\sqrt[3]{8}} = \frac{3\sqrt[3]{4}}{2}$

17. $\sqrt{\frac{2x}{5y}} = \frac{\sqrt{2x}}{\sqrt{5y}} = \frac{\sqrt{2x}\cdot\sqrt{5y}}{\sqrt{5y}\cdot\sqrt{5y}} = \frac{\sqrt{10xy}}{5y}$

21. $\frac{5a}{\sqrt[5]{8a^9b^{11}}} = \frac{5a}{ab^2\sqrt[5]{8a^4b}}$
$= \frac{5a\sqrt[5]{4ab^4}}{ab^2\sqrt[5]{8a^4b}\cdot\sqrt[5]{4ab^4}}$
$= \frac{5a\sqrt[5]{4ab^4}}{2a^2b^3}$

25. $\sqrt{\frac{18}{5}} = \frac{\sqrt{18}}{\sqrt{5}}$
$= \frac{\sqrt{9}\sqrt{2}}{\sqrt{5}}$
$= \frac{3\sqrt{2}}{\sqrt{5}}$
$= \frac{3\sqrt{2}}{\sqrt{5}}\cdot\frac{\sqrt{2}}{\sqrt{2}}$
$= \frac{3\sqrt{4}}{\sqrt{10}}$
$= \frac{3(2)}{\sqrt{10}}$
$= \frac{6}{\sqrt{10}}$

29. $\frac{\sqrt[3]{5y^2}}{\sqrt[3]{4x}}\cdot\frac{\sqrt[3]{5^2y}}{\sqrt[3]{5^2y}} = \frac{\sqrt[3]{5^3y^3}}{\sqrt[3]{4(5^2)xy}}$
$= \frac{5y}{\sqrt[3]{100xy}}$

33. $\frac{\sqrt{2x}}{11}\cdot\frac{\sqrt{2x}}{\sqrt{2x}} = \frac{\sqrt{4x^2}}{11\sqrt{2x}} = \frac{2x}{11\sqrt{2x}}$

37. $\frac{\sqrt[3]{3x^5}}{10}\cdot\frac{\sqrt[3]{3^2x}}{\sqrt[3]{3^2x}} = \frac{\sqrt[3]{3^3x^6}}{10\sqrt[3]{3^2x}} = \frac{3x^2}{10\sqrt[3]{9x}}$

41. answers may vary

45. $\frac{-7}{\sqrt{x}-3} = \frac{-7}{\sqrt{x}-3}\cdot\frac{\sqrt{x}+3}{\sqrt{x}+3}$
$= \frac{-7(\sqrt{x}+3)}{\sqrt{x}^2 - 3^2}$
$= \frac{-7(\sqrt{x}+3)}{x-9}$

49. $\frac{\sqrt{a}+1}{2\sqrt{a}-\sqrt{b}}$
$= \frac{\sqrt{a}+1}{2\sqrt{a}-\sqrt{b}}\cdot\frac{2\sqrt{a}+\sqrt{b}}{2\sqrt{a}+\sqrt{b}}$
$= \frac{\sqrt{a}(2\sqrt{a}) + \sqrt{a}\sqrt{b} + 1(2\sqrt{a}) + 1\sqrt{b}}{(2\sqrt{a})^2 - \sqrt{b}^2}$
$= \frac{2a + \sqrt{ab} + 2\sqrt{a} + \sqrt{b}}{4a - b}$

53. $\frac{\sqrt{x}}{\sqrt{x}+\sqrt{y}} = \frac{\sqrt{x}}{\sqrt{x}+\sqrt{y}}\cdot\frac{\sqrt{x}-\sqrt{y}}{\sqrt{x}-\sqrt{y}}$
$= \frac{\sqrt{x}\sqrt{x} - \sqrt{x}\sqrt{y}}{\sqrt{x}^2 - \sqrt{y}^2}$
$= \frac{x - \sqrt{xy}}{x - y}$

57. $\frac{(2-\sqrt{11})}{6}\cdot\frac{(2+\sqrt{11})}{(2+\sqrt{11})} = \frac{(2-\sqrt{11})(2+\sqrt{11})}{6(2+\sqrt{11})}$
$= \frac{4-\sqrt{121}}{12+6\sqrt{11}}$
$= \frac{4-11}{12+6\sqrt{11}}$
$= \frac{-7}{12+6\sqrt{11}}$

61. $\frac{(\sqrt{x}+3)}{\sqrt{x}} \cdot \frac{(\sqrt{x}-3)}{(\sqrt{x}-3)} = \frac{(\sqrt{x}+3)(\sqrt{x}-3)}{\sqrt{x}(\sqrt{x}-3)}$

$= \frac{\sqrt{x^2}-9}{\sqrt{x^2}-3\sqrt{x}}$

$= \frac{x-9}{x-3\sqrt{x}}$

65. $\frac{(\sqrt{x}+1)}{(\sqrt{x}-1)} \cdot \frac{(\sqrt{x}-1)}{(\sqrt{x}-1)} = \frac{(\sqrt{x}+1)(\sqrt{x}-1)}{(\sqrt{x}-1)(\sqrt{x}-1)}$

$= \frac{\sqrt{x^2}-1}{\sqrt{x^2}-2\sqrt{x}+1}$

$= \frac{x-1}{x-2\sqrt{x}+1}$

69. $(x-6)(2x+1) = 0$

$x - 6 = 0$ or $2x + 1 = 0$

$x = 6$ or $\quad x = -\frac{1}{2}$

The solution set is $\left\{-\frac{1}{2}, 6\right\}$.

73. $r = \sqrt{\frac{A}{4\pi}}$

$= \frac{\sqrt{A}}{\sqrt{4\pi}}$

$= \frac{\sqrt{A}}{2\sqrt{\pi}}$

$= \frac{\sqrt{A}\sqrt{\pi}}{2\sqrt{\pi}\sqrt{\pi}}$

$= \frac{\sqrt{A\pi}}{2\pi}$

Exercise Set 7.6

1. $\sqrt{2x} = 4$

$2x = 4^2$

$2x = 16$

$x = 8$

The solution set is $\{8\}$.

5. $\sqrt{2x} = -4$

No solution since a principle square root does not yield a negative number. The solution set is $\emptyset$.

9. $\sqrt{2x-3} - 2 = 1$

$\sqrt{2x-3} = 3$

$2x - 3 = 3^2$

$2x - 3 = 9$

$2x = 12$

$x = 6$

The solution set is $\{6\}$.

13. $\sqrt[3]{x-2} - 3 = 0$

$\sqrt[3]{x-2} = 3$

$x - 2 = 3^3$

$x - 2 = 27$

$x = 29$

The solution set is $\{29\}$.

17. $x - \sqrt{4-3x} = -8$

$x + 8 = \sqrt{4-3x}$

$(x+8)^2 = 4 - 3x$

$x^2 + 16x + 64 = 4 - 3x$

$x^2 + 16x + 64 = 4 - 3x$

$x^2 + 19x + 60 = 0$

$(x+4)(x+15) = 0$

$x + 4 = 0 \quad$ or $x + 15 = 0$

$x = -4 \qquad x = -15$

We discard the -15 as extraneous, leaving $x = -4$ as the only solution. The solution set is $\{-4\}$.

21. $\sqrt{x-3} + \sqrt{x+2} = 5$

$\sqrt{x-3} = 5 - \sqrt{x+2}$

$x - 3 = 25 - 10\sqrt{x+2} + x + 2$

$-3 = 27 - 10\sqrt{x+2}$

$-30 = -10\sqrt{x+2}$

$3 = \sqrt{x+2}$

$9 = x + 2$

$7 = x$

The solution set is $\{7\}$.

25. $-\sqrt{2x} + 4 = -6$

$10 = \sqrt{2x}$

$10^2 = 2x$

$100 = 2x$

$x = 50$

The solution set is $\{50\}$.

29. $\sqrt[4]{4x+1} - 2 = 0$

$\sqrt[4]{4x+1} = 2$

$4x + 1 = 2^4$

$4x + 1 = 16$

$4x = 15$

$x = \frac{15}{4}$

The solution set is $\left\{\frac{15}{4}\right\}$.

33. $\sqrt[3]{6x-3} - 3 = 0$

$\sqrt[3]{6x-3} = 3$

$6x - 3 = 3^3$

$6x - 3 = 27$

$6x = 30$

$x = 5$

The solution set is $\{5\}$.

37. $\sqrt{x+4} = \sqrt{2x-5}$

$x + 4 = 2x - 5$

$9 = x$

$x = 9$

The solution set is $\{9\}$.

41. $\sqrt[3]{-6x-1} = \sqrt[3]{-2x-5}$

$-6x - 1 = -2x - 5$

$4 = 4x$

$x = 1$

The solution set is $\{1\}$.

45. $\sqrt{2x-1} = \sqrt{1-2x}$

$\sqrt{2x-1} = \sqrt{-(2x-1)}$

It follows that $2x - 1 = 0$ (Otherwise one of the radicands would be negative).

So $2x = 1$

$x = \frac{1}{2}$

The solution set is $\left\{\frac{1}{2}\right\}$.

SOLUTIONS

49. $\sqrt{y+3} - \sqrt{y-3} = 1$

$$\sqrt{y+3} = 1 + \sqrt{y-3}$$
$$(\sqrt{y+3})^2 = (1 + \sqrt{y-3})^2$$
$$y + 3 = 1 + 2\sqrt{y-3} + y - 3$$
$$5 = 2\sqrt{y-3}$$
$$25 = 4(y-3)$$
$$\frac{25}{4} = y - 3$$
$$\frac{25}{4} + \frac{12}{4} = y$$
$$\frac{37}{4} = y$$

The solution set is $\left\{\frac{37}{4}\right\}$.

53. Let b = the length of the unknown leg of the right triangle. By the Pythagorean theorem,

$7^2 = 3^2 + b^2$
$49 = 9 + b^2$
$b^2 = 40$
$b = \sqrt{40}$
$= \sqrt{4}\sqrt{10}$
$= 2\sqrt{10}$ meters

57. Let c = the length of the hypotenuse of the right triangle. By the Pythagorean theorem,

$c^2 = 7^2 + (7.2)^2 = 100.84$
so $c = \sqrt{100.84} \approx 10.0$ millimeters

61. $x^2 = (5)^2 + (12)^2$
$x^2 = 25 + 144$
$x^2 = 169$
$x = \sqrt{169}$
$x = 13$
The answer is 13 feet.

65. $v = \sqrt{2gh}$
$80 = \sqrt{2(32)h}$
$(80)^2 = (\sqrt{2(32)h})^2$
$6400 = 2(32) \cdot h$
$100 = h$
The object fell 100 feet.

69. $$\frac{\frac{z}{5} + \frac{1}{10}}{\frac{z}{20} - \frac{z}{5}} = \frac{\frac{2z}{10} + \frac{1}{10}}{\frac{z}{20} - \frac{4z}{20}}$$
$$= \frac{\frac{2z+1}{10}}{\frac{-3z}{20}}$$
$$= \frac{2z+1}{10} \cdot \frac{20}{-3z}$$
$$= \frac{2(2z+1)}{-3z}$$
$$= -\frac{4z+2}{3z}$$

73. We need to solve the equation:

$80\sqrt[3]{x} + 500 = 1620$
$80\sqrt[3]{x} = 1120$
$\sqrt[3]{x} = 14$
$x = 14^3$
$x = 2744$

Thus, the company needs to make fewer than 2744 deliveries per day, or 2743 deliveries.

Exercise Set 7.7

1. $\sqrt{-24} = \sqrt{4}\sqrt{6}\sqrt{-1} = 2i\sqrt{6}$

5. $8\sqrt{-63} = 8\sqrt{9}\sqrt{7}\sqrt{-1} = 8 \cdot 3\sqrt{7}i = 24i\sqrt{7}$

9. $\sqrt{-2}\sqrt{-7} = (\sqrt{2}i)(\sqrt{7}i)$
$= \sqrt{14i^2}$
$= \sqrt{14(-1)}$
$= -\sqrt{14}$

13. $\sqrt{16}\sqrt{-1} = 4i$

17. $\dfrac{\sqrt{-80}}{\sqrt{-10}} = \dfrac{\sqrt{80}i}{\sqrt{10}i} = \sqrt{\dfrac{80}{10}} = \sqrt{8} = \sqrt{4}\sqrt{2} = 2\sqrt{2}$

21. $(6 + 5i) - (8 - i) = (6 - 8) + [5 - (-1)]i$
$= -2 + 6i$

25. $(6 - 3i) - (4 - 2i) = 6 - 3i - 4 + 2i = 2 - i$

29. $(2 + 4i) + (6 - 5i) = (2 + 6) + (4 - 5)i$
$= 8 - i$

33. $(-9i)(7i) = -63i^2 = -63(-1) = 63$

37. $6i(2 - 3i) = 6i(2) - 6i(3i)$
$= 12i = 18i^2$
$= 12i - 18(-1)$
$= 18 + 12i$

41. $(4 + i)(5 + 2i) = 20 + 8i + 5i + 2i^2$
$= 20 + 13i - 2$
$= 18 + 13i$

45. $(4 - 2i)^2 = 16 - 16i + 4i^2$
$= 16 - 16i + 4(-1)$
$= 16 - 4 - 16i$
$= 12 - 16i$

49. $(1 - i)(1 + i) = 1^2 - i^2 = 1^2 + 1^2 = 1 + 1 = 2$

53. $(1 - i)^2 = 1^2 - 2(1)(i) + i^2 = 1 - 2i - 1 = -2i$

57. $$\frac{7}{4+3i} = \frac{7}{4+3i} \cdot \frac{4-3i}{4-3i}$$
$$= \frac{28-21i}{4^2+3^2}$$
$$= \frac{28-21i}{16+9}$$
$$= \frac{28-21i}{25}$$
$$= \frac{28}{25} - \frac{21}{25}i$$

SOLUTIONS

61. $\frac{3+5i}{1+i} = \frac{3+5i}{1+i}\cdot\frac{1-i}{1-i}$
$= \frac{3+5i-3i-5i^2}{1^2+1^2}$
$= \frac{3+2i-5(-1)}{1+1}$
$= \frac{3+5+2i}{2}$
$= \frac{8+2i}{2}$
$= 4+i$

65. $\frac{16+15i}{-3i} = \frac{(16+15i)i}{-3i^2}$
$= \frac{16i+15i^2}{-3(-1)}$
$= \frac{16i+15(-1)}{3}$
$= \frac{-15}{3}+\frac{16}{3}i$
$= -5+\frac{16}{3}i$

69. $\frac{2-3i}{2+i} = \frac{(2-3i)(2-i)}{(2+i)(2-i)}$
$= \frac{4-6i-2i+3i^2}{2^2+1^2}$
$= \frac{4-8i+3(-1)}{4+1}$
$= \frac{4-3-8i}{5}$
$= \frac{1}{5}-\frac{8}{5}i$

73. $i^{21} = i^{20}i = (i^4)^5 i = 1^5 i = 1i = i$

77. $i^{-6} = (i^2)^{-3} = (-1)^{-3} = \frac{1}{(-1)^3} = \frac{1}{-1} = -1$

81. $(-3i)^5 = (-3)^5 i^5$
$= -243\cdot i^4\cdot i$
$= -243(1)(i)$
$= -243i$

85. $5+9=14$

89. $i^3+i^4 = -i+1 = 1-i$

93. $2+\sqrt{-9} = 2+3i$

97. $\frac{5-\sqrt{-75}}{10} = \frac{5-5i\sqrt{3}}{10} = \frac{5}{10}-\frac{5i\sqrt{3}}{10} = \frac{1}{2}-\frac{\sqrt{3}}{2}i$

101. $(8-\sqrt{-4})-(2+\sqrt{-16})$
$= (8-2i)-(2+4i)$
$= (8-2)+(-2i-4i)$
$= 6-6i$

Chapter 7 Test

1. $\sqrt{216} = \sqrt{36\cdot 6} = 6\sqrt{6}$

5. $\left[\frac{8x^3}{27}\right]^{2/3} = \frac{8^{2/3}(x^3)^{2/3}}{27^{2/3}}$
$= \frac{(8^{1/3})^2x^2}{(27^{1/3})^2}$
$= \frac{2^2x^2}{3^2}$
$= \frac{4x^2}{9}$

9. $\sqrt[4]{(4xy)^4} = |4xy|$ or $|4xy|$

13. $\sqrt[3]{\frac{8}{9x}} = \frac{\sqrt[3]{8}}{\sqrt[3]{9x}}$
$= \frac{2\cdot\sqrt[3]{3x^2}}{\sqrt[3]{9x}\cdot\sqrt[3]{3x^2}}$
$= \frac{2\sqrt[3]{3x^2}}{3x}$

17. $(\sqrt{x}+1)^2 = \sqrt{x}^2+2\sqrt{x}+1$
$= x+2\sqrt{x}+1$

21. $386^{-2/3} \approx 0.019$

25. $\sqrt{-2} = i\sqrt{2}$

29. $(4+3i)^2 = 16+24i+9i^2$
$= 16+24i+9(-1)$
$= (16-9)+24i$
$= 7+24i$

33. $V = \sqrt{2.5(300)} \approx 27$ mph

Chapter 8

Exercise Set 8.1

1. $x^2 = 16$
$x = \pm\sqrt{16}$
$x = \pm 4$
The solution set is $\{-4, 4\}$.

5. $x^2 = 18$
$x = \pm\sqrt{18}$
$x = \pm\sqrt{9}\sqrt{2}$
$x = \pm 3\sqrt{2}$
The solution set is $\{-3\sqrt{2}, 3\sqrt{2}\}$.

9. $(x+5)^2 = 9$
$x+5 = \pm\sqrt{9}$
$x+5 = \pm 3$
$x = -5 \pm 3$
$x = -8$ or $x = -2$
The solution set is $\{-8, -2\}$.

13. $(2x-3)^2 = 8$
$2x-3 = \pm\sqrt{8}$
$2x-3 = \pm\sqrt{4}\sqrt{2}$
$2x-3 = \pm 2\sqrt{2}$
$2x = 3 \pm 2\sqrt{2}$
$x = \frac{3\pm 2\sqrt{2}}{2}$
The solution set is $\left\{\frac{3-2\sqrt{2}}{2}, \frac{3+2\sqrt{2}}{2}\right\}$.

17. $x^2 - 6 = 0$

$$x^2 = 6$$
$$x = \pm\sqrt{6}$$

The solution set is $\{-\sqrt{6}, \sqrt{6}\}$.

21. $(x - 1)^2 = -16$

$$x - 1 = \pm\sqrt{-16}$$
$$x - 1 = \pm 4i$$
$$x = 1 \pm 4i$$

The solution set is $\{1 - 4i, 1 + 4i\}$.

25. $(x + 3)^2 = -8$

$$x + 3 = \pm\sqrt{-8}$$
$$x + 3 = \pm\sqrt{4}\sqrt{2}\sqrt{-1}$$
$$x + 3 = \pm 2i\sqrt{2}$$
$$x = -3 \pm 2i\sqrt{2}$$

The solution set is $\{-3 - 2i\sqrt{2}, -3 + 2i\sqrt{2}\}$.

29. $z^2 - 12z + \left(\frac{12}{2}\right)^2 = z^2 - 12z + 36 = (z - 6)^2$

33. $r^2 - 3r + \left(\frac{3}{2}\right)^2 = r^2 - 3r + \frac{9}{4}$

$$= \left(r - \frac{3}{2}\right)^2$$

37.

$$x^2 + 6x + 2 = 0$$
$$x^2 + 6x + \left(\frac{6}{2}\right)^2 = -2 + 9$$
$$(x + 3)^2 = 7$$
$$x + 3 = \pm\sqrt{7}$$
$$x = -3 \pm \sqrt{7}$$

The solution set is $\{-3 - \sqrt{7}, -3 + \sqrt{7}\}$.

41.

$$x^2 + 2x - 5 = 0$$
$$x^2 + 2x + \left(\frac{2}{2}\right)^2 = 5 + 1$$
$$x^2 + 2x + 1 = 6$$
$$(x + 1)^2 = 6$$
$$x + 1 = \pm\sqrt{6}$$
$$x = -1 \pm \sqrt{6}$$

The solution set is $\{-1 - \sqrt{6}, -1 + \sqrt{6}\}$.

45.

$$2x^2 + 7x = 4$$
$$x^2 + \frac{7}{2}x = 2$$
$$x^2 + \frac{7}{2}x + \left(\frac{\frac{7}{2}}{2}\right)^2 = 2 + \frac{49}{16}$$
$$\left(x + \frac{7}{4}\right)^2 = \frac{81}{16}$$
$$x + \frac{7}{4} = \pm\sqrt{\frac{81}{16}}$$
$$x = -\frac{7}{4} \pm \frac{9}{4}$$
$$x = -4 \text{ or } x = \frac{1}{2}$$

The solution set is $\left\{-4, \frac{1}{2}\right\}$.

49.

$$3y^2 + 6y - 4 = 0$$
$$3y^2 + 6y = 4$$
$$y^2 + 2y = \frac{4}{3}$$
$$y^2 + 2y + \left(\frac{2}{2}\right)^2 = \frac{4}{3} + 1$$
$$(y + 1)^2 = \frac{7}{3}$$
$$y + 1 = \pm\sqrt{\frac{7}{3}}$$
$$y + 1 = \pm\frac{\sqrt{7}}{\sqrt{3}} \cdot \frac{\sqrt{3}}{\sqrt{3}}$$
$$y + 1 = \pm\frac{\sqrt{21}}{3}$$
$$y = -1 \pm \frac{\sqrt{21}}{3}$$
$$= \frac{-3 \pm \sqrt{21}}{3}$$

The solution set is $\left\{\frac{-3 - \sqrt{21}}{3}, \frac{-3 + \sqrt{21}}{3}\right\}$.

53.

$$x^2 - 6x + 3 = 0$$
$$x^2 - 6x + \left(\frac{6}{2}\right)^2 = -3 + 9$$
$$(x - 3)^2 = 6$$
$$x - 3 = \pm\sqrt{6}$$
$$x = 3 \pm \sqrt{6}$$

The solution set is $\{3 - \sqrt{6}, 3 + \sqrt{6}\}$.

57.

$$2x^2 - x + 6 = 0$$
$$2x^2 - x = -6$$
$$x^2 - \frac{1}{2}x = -3$$
$$x^2 - \frac{1}{2}x + \left(\frac{\frac{1}{2}}{2}\right)^2 = -3 + \frac{1}{16}$$
$$\left(x - \frac{1}{4}\right)^2 = -\frac{47}{16}$$
$$x - \frac{1}{4} = \pm\sqrt{-\frac{47}{16}}$$
$$x - \frac{1}{4} = \pm\frac{\sqrt{47}\sqrt{-1}}{\sqrt{16}}$$
$$x - \frac{1}{4} = \pm\frac{i\sqrt{47}}{4}$$
$$x = \frac{1 \pm i\sqrt{47}}{4}$$

The solution set is $\left\{\frac{1 + i\sqrt{47}}{4}, \frac{1 - i\sqrt{47}}{4}\right\}$.

61. $z^2 + 3z - 4 = 0$
$$z^2 + 3z = 4$$
$$z^2 + 3z + \left(\frac{3}{2}\right)^2 = 4 + \frac{9}{4}$$
$$\left(z + \frac{3}{2}\right)^2 = \frac{25}{4}$$
$$z + \frac{3}{2} = \pm\sqrt{\frac{25}{4}}$$
$$z + \frac{3}{2} = \pm\frac{5}{2}$$
$$z = -\frac{3}{2} \pm \frac{5}{2}$$
$$z = -4 \text{ or } z = 1$$
The solution set is $\{-4, 1\}$.

65. $3x^2 + 3x = 5$
$$x^2 + x = \frac{5}{3}$$
$$x^2 + x + \left(\frac{1}{2}\right)^2 = \frac{5}{3} + \frac{1}{4}$$
$$\left(x + \frac{1}{2}\right)^2 = \frac{23}{12}$$
$$x + \frac{1}{2} = \pm\sqrt{\frac{23}{12}}$$
$$x + \frac{1}{2} = \pm\frac{\sqrt{23}}{\sqrt{4}\sqrt{3}}$$
$$x + \frac{1}{2} = \pm\frac{\sqrt{23}}{2\sqrt{3}}$$
$$x + \frac{1}{2} = \pm\frac{\sqrt{23}\sqrt{3}}{2\sqrt{3^2}}$$
$$x + \frac{1}{2} = \pm\frac{\sqrt{69}}{2 \cdot 3}$$
$$x + \frac{1}{2} = \pm\frac{\sqrt{69}}{6}$$
$$x = -\frac{1}{2} \pm \frac{\sqrt{69}}{6}$$
$$x = \frac{-3 \pm \sqrt{69}}{6}$$
The solution set is $\left\{\frac{-3 - \sqrt{69}}{6}, \frac{-3 + \sqrt{69}}{6}\right\}$.

69. $A = P(1 + r)^t$
$$1000 = 810(1 + r)^2$$
$$\frac{1000}{810} = (1 + r)^2$$
$$\frac{100}{81} = (1 + r)^2$$
$$\pm\sqrt{\frac{100}{81}} = 1 + r$$
$$\pm\frac{10}{9} = 1 + r$$
$$-1 \pm \frac{10}{9} = r$$
$$r = -1 + \frac{10}{9} \text{ or } r = -1 - \frac{10}{9}$$
$$r = \frac{1}{9} \qquad \text{or } r = -\frac{19}{9}$$
Rate cannot be negative, so $r = \frac{1}{9}$, or about 11%.

73. Simple

77. $\frac{12 - 8\sqrt{7}}{16} = \frac{12}{16} - \frac{8\sqrt{7}}{16}$
$$= \frac{3}{4} - \frac{\sqrt{7}}{2}$$
$$= \frac{3}{4} - \frac{2\sqrt{7}}{4}$$
$$= \frac{3 - 2\sqrt{7}}{4}$$

81. $\sqrt{b^2 - 4ac}$
$a = 1, b = -3, c = -1$
$\sqrt{(-3)^2 - 4(1)(-1)} = \sqrt{9 + 4} = \sqrt{13}$

85. $s(t) = 16t^2$
$1053 = 16t^2$
$$t = \pm\sqrt{\frac{1053}{16}}$$
$t = 8.11$ or -8.11 (disregard)
It would take about 8.11 seconds.

Exercise Set 8.2

1. $m^2 + 5m - 6 = 0$
$a = 1, b = 5, c = -6$
$$m = \frac{-5 \pm \sqrt{5^2 - 4(1)(-6)}}{2(1)}$$
$$m = \frac{-5 \pm \sqrt{25 + 24}}{2} = \frac{-5 \pm \sqrt{49}}{2}$$
$$m = \frac{-5 \pm 7}{2}$$
$m = -6$ or $m = 1$
The solution set is $\{-6, 1\}$.

5. $x^2 - 6x + 9 = 0$
$a = 1, b = -6, c = 9$
$$x = \frac{6 \pm \sqrt{(-6)^2 - 4(1)(9)}}{2(1)}$$
$$x = \frac{6 \pm \sqrt{36 - 36}}{2} = \frac{6 \pm \sqrt{0}}{2} = \frac{6}{2} = 3$$
The solution set is $\{3\}$.

9. $8m^2 - 2m = 7$
$8m^2 - 2m - 7 = 0$
$a = 8, b = -2, c = -7$
$$m = \frac{2 \pm \sqrt{(-2)^2 - 4(8)(-7)}}{2(8)}$$
$$m = \frac{2 \pm \sqrt{4 + 224}}{16} = \frac{2 \pm \sqrt{228}}{16}$$
$$m = \frac{2 \pm \sqrt{4}\sqrt{57}}{16} = \frac{2 \pm 2\sqrt{57}}{16}$$
$$m = \frac{1 \pm \sqrt{57}}{8}$$
The solution set is $\left\{\frac{1 + \sqrt{57}}{8}, \frac{1 - \sqrt{57}}{8}\right\}$.

13. $\frac{1}{2}x^2 - x - 1 = 0$

$x^2 - 2x - 2 = 0$

$a = 1, b = -2, c = -2$

$x = \frac{2 \pm \sqrt{(-2)^2 - 4(1)(-2)}}{2(1)}$

$x = \frac{2 \pm \sqrt{4 + 8}}{2} = \frac{2 \pm \sqrt{12}}{2}$

$x = \frac{2 \pm \sqrt{4}\sqrt{3}}{2} = \frac{2 \pm 2\sqrt{3}}{2}$

$x = 1 \pm \sqrt{3}$

The solution set is $\{1 - \sqrt{3}, 1 + \sqrt{3}\}$.

17. $\frac{1}{3}y^2 - y - \frac{1}{6} = 0$

$2y^2 - 6y - 1 = 0$

$a = 2, b = -6, c = -1$

$y = \frac{6 \pm \sqrt{(-6)^2 - 4(2)(-1)}}{2(2)}$

$y = \frac{6 \pm \sqrt{36 + 8}}{4} = \frac{6 \pm \sqrt{44}}{4}$

$y = \frac{6 \pm \sqrt{4}\sqrt{11}}{4} = \frac{6 \pm 2\sqrt{11}}{4}$

$y = \frac{3 \pm \sqrt{11}}{2}$

The solution set is $\left\{\frac{3 - \sqrt{11}}{2}, \frac{3 + \sqrt{11}}{2}\right\}$.

21. $x^2 + 5x = -2$

$x^2 + 5x + 2 = 0$

$a = 1, b = 5, c = 2$

$x = \frac{-5 \pm \sqrt{5^2 - 4(1)(2)}}{2(1)}$

$x = \frac{-5 \pm \sqrt{25 - 8}}{2} = \frac{-5 \pm \sqrt{17}}{2}$

The solution set is $\left\{\frac{-5 - \sqrt{17}}{2}, \frac{-5 + \sqrt{17}}{2}\right\}$.

25. $\frac{x^2}{3} - x = \frac{5}{3}$

$x^2 - 3x = 5$

$x^2 - 3x - 5 = 0$

$a = 1, b = -3, c = -5$

$x = \frac{3 \pm \sqrt{(-3)^2 - 4(1)(-5)}}{2(1)}$

$x = \frac{3 \pm \sqrt{9 + 20}}{2} = \frac{3 \pm \sqrt{29}}{2}$

The solution set is $\left\{\frac{3 - \sqrt{29}}{2}, \frac{3 + \sqrt{29}}{2}\right\}$.

29. answers may vary

33. $(x + 5)(x - 1) = 2$

$x^2 + 4x - 5 = 2$

$x^2 + 4x - 7 = 0$

$a = 1, b = 4, c = -7$

$x = \frac{-4 \pm \sqrt{4^2 - 4(1)(-7)}}{2(1)}$

$x = \frac{-4 \pm \sqrt{16 + 28}}{2} = \frac{-4 + \sqrt{44}}{2}$

$x = \frac{-4 \pm \sqrt{4}\sqrt{11}}{2} = \frac{-4 \pm 2\sqrt{11}}{2}$

$x = -2 \pm \sqrt{11}$

The solution set is $\{-2 - \sqrt{11}, -2 + \sqrt{11}\}$.

37. $\frac{2}{5}y^2 + \frac{1}{5}y + \frac{3}{5} = 0$

$2y^2 + y + 3 = 0$

$a = 2, b = 1, c = 3$

$y = \frac{-1 \pm \sqrt{1^2 - 4(2)(3)}}{2(2)}$

$y = \frac{-1 \pm \sqrt{1 - 24}}{4} = \frac{-1 \pm \sqrt{-23}}{4}$

$y = \frac{-1 \pm i\sqrt{23}}{4}$

The solution set is $\left\{\frac{-1 - i\sqrt{23}}{4}, \frac{-1 + i\sqrt{23}}{4}\right\}$.

41. $(n - 2)^2 = 2n$

$n^2 - 4n + 4 = 2n$

$n^2 - 6n + 4 = 0$

$a = 1, b = -6, c = 4$

$n = \frac{6 \pm \sqrt{(-6)^2 - 4(1)(4)}}{2(1)}$

$n = \frac{6 \pm \sqrt{36 - 16}}{2} = \frac{6 \pm \sqrt{20}}{2}$

$n = \frac{6 \pm \sqrt{4}\sqrt{5}}{2} = \frac{6 \pm 2\sqrt{5}}{2}$

$n = 3 \pm \sqrt{5}$

The solution set is $\{3 - \sqrt{5}, 3 + \sqrt{5}\}$.

45. $4x^2 + 12x = -9$

$4x^2 + 12x + 9 = 0$

$a = 4, b = 12, c = 9$

$b^2 - 4ac = 12^2 - 4(4)(9)$

$b^2 - 4ac = 144 - 144 = 0$

Therefore, there is 1 real solution.

49. $6 = 4x - 5x^2$

$5x^2 - 4x + 6 = 0$

$a = 5, b = -4, c = 6$

$b^2 - 4ac = (-4)^2 - 4(5)(6)$

$b^2 - 4ac = 16 - 120$

$b^2 - 4ac = -104 < 0$

Therefore, there are 2 complex but not real solutions.

53. Let $x =$ length of leg
$x + 2 =$ length of hypotenuse
$x^2 + x^2 = (x + 2)^2$
$2x^2 = x^2 + 4x + 4$
$x^2 - 4x - 4 = 0$
$a = 1, b = -4, c = -4$

$$x = \frac{4 \pm \sqrt{(-4)^2 - 4(1)(-4)}}{2(1)}$$
$$x = \frac{4 \pm \sqrt{32}}{2}$$
$$x = \frac{4 \pm 4\sqrt{2}}{2}$$

$x = 2 \pm 2\sqrt{2}$ (disregard a negative length)
The sides measure $2 + 2\sqrt{2}$ cm, $2 + 2\sqrt{2}$ cm, and $4 + 2\sqrt{2}$ cm.

57. a. Let $x =$ length
$x^2 + x^2 = 100^2$
$2x^2 - 10{,}000 = 0$
$a = 2, b = 0, c = -10{,}000$

$$x = \frac{0 \pm \sqrt{0^2 - 4(2)(-10{,}000)}}{2(2)}$$
$$x = \frac{\pm\sqrt{80{,}000}}{4}$$
$$x = \frac{\pm 200\sqrt{2}}{4}$$
$$x = \pm 50\sqrt{2}$$

Disregard a negative length. The side measures $50\sqrt{2}$ meters.

b. Area $= s^2$
$= (50\sqrt{2})^2$
$= 50^2(\sqrt{2})^2$
$= 2500 \cdot 2$
$= 5000$
The area is 5000 square meters.

61. $h = -16t^2 + 20t + 1100$
$0 = -16t^2 + 20t + 1100$
$a = -16, b = 20, c = 1100$

$$t = \frac{-20 \pm \sqrt{20^2 - 4(-16)(1100)}}{2(-16)}$$
$$t = \frac{-20 \pm \sqrt{70{,}800}}{-32}$$

$t \approx 8.9$ or $t \approx -7.7$ (disregard)
It will take about 8.9 seconds.

65.
$\sqrt{5x - 2} = 3$
$(\sqrt{5x - 2})^2 = 3^2$
$5x - 2 = 9$
$5x = 11$
$x = \frac{11}{5}$

Check:

$$\sqrt{5\left(\frac{11}{5}\right) - 2} \stackrel{?}{=} 3$$
$$\sqrt{11 - 2} \stackrel{?}{=} 3$$
$$\sqrt{9} \stackrel{?}{=} 3$$
$$3 = 3$$

The solution set is $\left\{\frac{11}{5}\right\}$.

69. $x^4 + x^2 - 20 = (x^2 + 5)(x^2 - 4)$
$= (x^2 + 5)(x + 2)(x - 2)$

73. $2x^2 - 6x + 3 = 0$
$a = 2, b = -6, c = 3$

$$x = \frac{-(-6) \pm \sqrt{(-6)^2 - 4(2)(3)}}{2(2)}$$
$$= \frac{6 \pm \sqrt{36 - 24}}{4}$$
$$= \frac{6 \pm \sqrt{12}}{4}$$
$$= \frac{6 \pm 2\sqrt{3}}{4}$$
$$= \frac{3 \pm \sqrt{3}}{2}$$

The solutions are about $\{0.6, 2.4\}$.

77. Wednesday

Exercise Set 8.3

1.
$2x = \sqrt{10 + 3x}$
$4x^2 = 10 + 3x$
$4x^2 - 3x - 10 = 0$
$(4x + 5)(x - 2) = 0$
$4x + 5 = 0$ or $x - 2 = 0$
$x = -\frac{5}{4}$ or $x = 2$

Discard $x = -\frac{5}{4}$.
The solution set is $\{2\}$.

5. $\sqrt{9x} = x + 2$
$9x = x^2 + 4x + 4$
$0 = x^2 - 5x + 4$
$0 = (x - 4)(x - 1)$
$x - 4 = 0$ or $x - 1 = 0$
$x = 4$ or $x = 1$
The solution set is $\{1, 4\}$.

9.
$$\frac{3}{x} + \frac{4}{x + 2} = 2$$
$$\frac{3(x + 2) + 4x}{x(x + 2)} = 2$$
$$\frac{3x + 6 + 4x}{x^2 + 2x} = 2$$
$7x + 6 = 2(x^2 + 2x)$
$7x + 6 = 2x^2 + 4x$
$2x^2 - 3x - 6 = 0$

$$x = \frac{3 \pm \sqrt{(-3)^2 - 4(2)(-6)}}{2(2)}$$
$$x = \frac{3 \pm \sqrt{57}}{4}$$

The solution set is $\left\{\frac{3 + \sqrt{57}}{4}, \frac{3 - \sqrt{57}}{4}\right\}$.

13.
$p^4 - 16 = 0$
$(p^2 + 4)(p^2 - 4) = 0$
$(p + 2i)(p - 2i)(p + 2)(p - 2) = 0$
$p + 2i = 0$ or $p - 2i = 0$
or $p + 2 = 0$ or $p - 2 = 0$
$p = -2i$ or $p = 2i$
$p = -2$ or $p = 2$
The solution set is $\{-2i, 2i, -2, 2\}$.

17. $z^4 - 13z^2 + 36 = 0$
$(z^2 - 9)(z^2 - 4) = 0$
$(z + 3)(z - 3)(z + 2)(z - 2) = 0$
$z + 3 = 0$ or $z - 3 = 0$ or $z + 2 = 0$ or $z - 2 = 0$
$z = -3$ or $z = 3$ or $z = -2$ or $z = 2$
The solution set is $\{-3, 3, -2, 2\}$.

21. $(5n + 1)^2 + 2(5n + 1) - 3 = 0$
Let $y = 5n + 1$.
$y^2 + 2y - 3 = 0$
$(y + 3)(y - 1) = 0$
$y + 3 = 0$ or $y - 1 = 0$
$y = -3$ or $y = 1$
$5n + 1 = -3$ or $5n + 1 = 1$
$5n = -4$ or $5n = 0$
$n = -\frac{4}{5}$ or $n = 0$
The solution set is $\left\{-\frac{4}{5}, 0\right\}$.

25. $1 + \frac{2}{3t - 2} = \frac{8}{(3t - 2)^2}$
$(3t - 2)^2 + 2(3t - 2) - 8 = 0$
Let $y = 3t - 2$.
$y^2 + 2y - 8 = 0$
$(y + 4)(y - 2) = 0$
$y + 4 = 0$ or $y - 2 = 0$
$y = -4$ or $y = 2$
$3t - 2 = -4$ or $3t - 2 = 2$
$3t = -2$ or $3t = 4$
$t = -\frac{2}{3}$ or $t = \frac{4}{3}$
The solution set is $\left\{-\frac{2}{3}, \frac{4}{3}\right\}$.

29. $a^4 - 5a^2 + 6 = 0$
$(a^2 - 3)(a^2 - 2) = 0$
$a^2 - 3 = 0$ or $a^2 - 2 = 0$
$a^2 = 3$ or $a^2 = 2$
$a = \pm\sqrt{3}$ or $a = \pm\sqrt{2}$
The solution set is $\{\sqrt{3}, -\sqrt{3}, \sqrt{2}, -\sqrt{2}\}$.

33. $(p + 2)^2 = 9(p + 2) - 20$
$(p + 2)^2 - 9(p + 2) + 20 = 0$
Let $x = p + 2$.
$x^2 - 9x + 20 = 0$
$(x - 5)(x - 4) = 0$
$x - 5 = 0$ or $x - 4 = 0$
$x = 5$ or $x = 4$
$p + 2 = 5$ or $p + 2 = 4$
$p = 3$ or $p = 2$
The solution set is $\{2, 3\}$.

37. $x^{2/3} - 8x^{1/3} + 15 = 0$
Let $y = x^{1/3}$.
$y^2 - 8y + 15 = 0$
$(y - 5)(y - 3) = 0$
$y - 5 = 0$ or $y - 3 = 0$
$y = 5$ or $y = 3$
$x^{1/3} = 5$ or $x^{1/3} = 3$
$x = 5^3$ or $x = 3^3$
$x = 125$ or $x = 27$
The solution set is $\{125, 27\}$.

41. $2x^{2/3} + 3x^{1/3} - 2 = 0$
Let $m = x^{1/3}$.
$2m^2 + 3m - 2 = 0$
$(2m - 1)(m + 2) = 0$
$2m - 1 = 0$ or $m + 2 = 0$
$2m = 1$ or $m = -2$
$m = \frac{1}{2}$ or $m = -2$
$x^{1/3} = \frac{1}{2}$ or $x^{1/3} = -2$
$x = \left(\frac{1}{2}\right)^3 = \frac{1}{8}$ or $x = (-2)^3 = -8$
The solution set is $\left\{\frac{1}{8}, -8\right\}$.

45. $x - \sqrt{x} = 2$
$x - 2 = \sqrt{x}$
$x^2 - 4x + 4 = x$
$x^2 - 5x + 4 = 0$
$(x - 4)(x - 1) = 0$
$x = 4$ or $x = 1$ (discard)
The solution set is $\{4\}$.

49. $p^4 - p^2 - 20 = 0$
$(p^2 - 5)(p^2 + 4) = 0$
$p^2 - 5 = 0$ or $p^2 + 4 = 0$
$p^2 = 5$ or $(p + 2i)(p - 2i) = 0$
$p = \pm\sqrt{5}$ or $p + 2i = 0$ or $p - 2i = 0$
$p = \pm\sqrt{5}$ or $p = -2i$ or $p = 2i$
The solution set is $\{\sqrt{5}, -\sqrt{5}, -2i, 2i\}$.

53. $1 = \frac{4}{x - 7} + \frac{5}{(x - 7)^2}$
$(x - 7)^2 = 4(x - 7) + 5$
Let $y = x - 7$.
$y^2 - 4y - 5 = 0$
$(y - 5)(y + 1) = 0$
$y - 5 = 0$ or $y + 1 = 0$
$y = 5$ or $y = -1$
$x - 7 = 5$ or $x - 7 = -1$
$x = 12$ or $x = 6$
The solution set is $\{6, 12\}$.

57. Let x = speed on first part
$x - 1$ = speed on second part
$D = r \cdot t$ or $t = \frac{D}{r}, 1\frac{3}{5} = \frac{8}{5}$
$\frac{3}{x} + \frac{4}{x - 1} = \frac{8}{5}$
$3 \cdot 5(x - 1) + 4 \cdot 5x = 8 \cdot x(x - 1)$
$15x - 15 + 20x = 8x^2 - 8x$
$0 = 8x^2 - 43x + 15$
$0 = (8x - 3)(x - 5)$
$8x - 3 = 0$ or $x - 5 = 20$
$x = \frac{3}{8}$ or $x = 5$
$x - 1 = 4$
Her speeds were 5 mph and then 4 mph.

SOLUTIONS

SOLUTIONS

61. Let x = original speed
$x + 11$ = return speed

$$D = r \cdot t \text{ or } t = \frac{D}{r}$$

$$\frac{330}{x} - \frac{330}{x+11} = 1$$
$$330(x + 11) - 330x = x(x + 11)$$
$$330x + 3630 - 330x = x^2 + 11x$$
$$0 = x^2 + 11x - 3630$$
$$0 = (x - 55)(x + 66)$$
$$x = 55 \text{ or } x = -66 \text{ (discard)}$$
$$x + 11 = 66$$
The speeds are 55 mph and 66 mph.

65. Let x = number
$$x(x - 4) = 96$$
$$x^2 - 4x = 96$$
$$x^2 - 4x - 96 = 0$$
$$(x - 12)(x + 8) = 0$$
$$x - 12 = 0 \quad \text{or } x + 8 = 0$$
$$x = 12 \text{ or } \quad x = -8$$
The number is 12 or −8.

69. $\frac{5x}{3} + 2 \le 7$
$$\frac{5x}{3} \le 5$$
$$\frac{3}{5}\left(\frac{5x}{3}\right) \le \frac{3}{5}(5)$$
$$x \le 3$$
The solution set is $(-\infty, 3]$.

73. answers may vary

Exercise Set 8.4

1. $(x + 1)(x + 5) > 0$
$(x + 1)(x + 5) = 0$
$x + 1 = 0 \quad \text{or } x + 5 = 0$
$x = -1 \text{ or} \quad x = -5$

Region	Interval	Test Point
A	$(-\infty, -5)$	−6
B	$(-5, -1)$	−2
C	$(-1, \infty)$	0

$x = -6$	$(-6 + 1)(-6 + 5) > 0$	True
$x = -2$	$(-2 + 1)(-2 + 5) > 0$	False
$x = 0$	$(0 + 1)(0 + 5) > 0$	True

The solution set is $(-\infty, -5) \cup (-1, \infty)$.

5. $x^2 - 7x + 10 \le 0$
$(x - 5)(x - 2) \le 0$
$(x - 5)(x - 2) = 0$
$x - 5 = 0 \text{ or } x - 2 = 0$
$x = 5 \text{ or} \quad x = 2$

Region	Interval	Test Point
A	$(-\infty, 2)$	0
B	$(2, 5)$	3
C	$(5, \infty)$	6

$x = 0$	$(0 - 5)(0 - 2) \le 0$	False
$x = 3$	$(3 - 5)(3 - 2) \le 0$	True
$x = 6$	$(6 - 5)(6 - 2) \le 0$	False

The solution set is $[2, 5]$.

9. $(x - 6)(x - 4)(x - 2) > 0$
$(x - 6)(x - 4)(x - 2) = 0$
$x - 6 = 0 \text{ or } x - 4 = 0 \text{ or } x - 2 = 0$
$x = 6 \text{ or } x = 4 \text{ or } x = 2$

Region	Interval	Test Point
A	$(-\infty, 2)$	1
B	$(2, 4)$	3
C	$(4, 6)$	5
D	$(6, \infty)$	7

$x = 1$	$(1 - 6)(1 - 4)(1 - 2) > 0$	False
$x = 3$	$(3 - 6)(3 - 4)(3 - 2) > 0$	True
$x = 5$	$(5 - 6)(5 - 4)(5 - 2) > 0$	False
$x = 7$	$(7 - 6)(7 - 4)(7 - 2) > 0$	True

The solution set is $(2, 4) \cup (6, \infty)$.

13. $(x^2 - 9)(x^2 - 4) > 0$
$(x + 3)(x - 3)(x + 2)(x - 2) > 0$
$(x + 3)(x - 3)(x + 2)(x - 2) = 0$
$x + 3 = 0 \text{ or } x - 3 = 0 \text{ or } x + 2 = 0 \text{ or } x - 2 = 0$
$x = -3 \text{ or } x = 3 \text{ or } x = -2 \text{ or } x = 2$

Region	Interval	Test Point
A	$(-\infty, -3)$	−4
B	$(-3, -2)$	$-\frac{5}{2}$
C	$(-2, 2)$	0
D	$(2, 3)$	$\frac{5}{2}$
E	$(3, \infty)$	4

$x = -4$	$[(-4)^2 - 9][(-4)^2 - 4] > 0$	True
$x = -\frac{5}{2}$	$\left[\left(-\frac{5}{2}\right)^2 - 9\right]\left[\left(-\frac{5}{2}\right)^2 - 4\right] > 0$	False
$x = 0$	$(0^2 - 9)(0^2 - 4) > 0$	True
$x = \frac{5}{2}$	$\left[\left(\frac{5}{2}\right)^2 - 9\right]\left[\left(\frac{5}{2}\right)^2 - 4\right] > 0$	False
$x = 4$	$(4^2 - 9)(4^2 - 4) > 0$	True

The solution set is $(-\infty, -3) \cup (-2, 2) \cup (3, \infty)$.

17. $6x^2 + 5x \le 4$
$6x^2 + 5x - 4 \le 0$
$(2x - 1)(3x + 4) = 0$
$2x - 1 = 0 \text{ or } 3x + 4 = 0$
$2x = 1 \text{ or} \quad 3x = -4$
$x = \frac{1}{2} \text{ or} \quad x = -\frac{4}{3}$

Region	Interval	Test Point
A	$\left(-\infty, -\frac{4}{3}\right)$	−2
B	$\left(-\frac{4}{3}, \frac{1}{2}\right)$	1
C	$\left(\frac{1}{2}, \infty\right)$	0

$x = -2$	$6(-2)^2 + 5(-2) \le 4$	False
$x = 0$	$6(0)^2 + 5(0) \le 4$	True
$x = 1$	$6(1)^2 + 5(1) \le 4$	False

The solution set is $\left[-\frac{4}{3}, \frac{1}{2}\right]$.

21. $(2x - 8)(x + 4)(x - 6) \leq 0$

$(2x - 8)(x + 4)(x - 6) = 0$

$2x - 8 = 0$ or $x + 4 = 0$ or $x - 6 = 0$

$2x = 8$ or $x = -4$ or $x = 6$

$x = 4$ or $x = -4$ or $x = 6$

Region	Interval	Test Point
A	$(-\infty, -4)$	-5
B	$(-4, 4)$	0
C	$(4, 6)$	5
D	$(6, \infty)$	7

$x = -5$	$[2(-5) - 8](-5 + 4)(-5 - 6) \leq 0$	True
$x = 0$	$[2(0) - 8](0 + 4)(0 - 6) \leq 0$	False
$x = 5$	$[2(5) - 8](5 + 4)(5 - 6) \leq 0$	True
$x = 7$	$[2(7) - 8)](7 + 4)(7 - 6) \leq 0$	False

The solution set is $(-\infty, -4] \cup [4, 6]$.

25. $\dfrac{5}{x + 1} > 0$

$x + 1 = 0$

$x = -1$

Region	Interval	Test Point
A	$(-\infty, -1)$	-2
B	$(-1, \infty)$	0

$x = -2$	$\dfrac{5}{-2 + 1} > 0$	False
$x = 0$	$\dfrac{5}{0 + 1} > 0$	True

The solution set is $(-1, \infty)$.

29. $\dfrac{x + 2}{x - 3} < 1$

$\dfrac{x + 2}{x - 3} - \dfrac{x - 3}{x - 3} < 0$

$\dfrac{x + 2 - x + 3}{x - 3} < 0$

$\dfrac{5}{x - 3} < 0$

$x = 3$

Region	Interval	Test Point
A	$(-\infty, 3)$	0
B	$(3, \infty)$	4

$x = 0$	$\dfrac{0 + 2}{0 - 3} < 1$	True
$x = 4$	$\dfrac{4 + 2}{4 - 3} < 1$	False

The solution set is $(-\infty, 3)$.

33. $\dfrac{x - 5}{x + 4} \geq 0$

$x - 5 = 0$ or $x + 4 = 0$

$x = 5$ or $x = -4$

Region	Interval	Test Point
A	$(-\infty, -4)$	-5
B	$(-4, 5)$	0
C	$(5, \infty)$	6

$x = -5$	$\dfrac{-5 - 5}{-5 + 4} \geq 0$	True
$x = 0$	$\dfrac{0 - 5}{0 + 4} \geq 0$	False
$x = 6$	$\dfrac{6 - 5}{6 + 4} \geq 0$	True

The solution set is $(-\infty, -4) \cup [5, \infty)$.

37. $\dfrac{-1}{x - 1} > -1$

$\dfrac{1}{x - 1} < 1$

$\dfrac{1}{x - 1} - 1 < 0$

$\dfrac{1 - (x - 1)}{x - 1} < 0$

$\dfrac{1 - x + 1}{x - 1} < 0$

$\dfrac{2 - x}{x - 1} < 0$

$2 - x = 0$ or $x - 1 = 0$

$2 = x$ or $x = 1$

Region	Interval	Test Point
A	$(-\infty, 1)$	0
B	$(1, 2)$	$\dfrac{3}{2}$
C	$(2, \infty)$	3

$x = 0$	$\dfrac{-1}{0 - 1} > -1$	True
$x = \dfrac{3}{2}$	$\dfrac{-1}{\frac{3}{2} - 1} > -1$	False
$x = 3$	$\dfrac{-1}{3 - 1} > -1$	True

The solution set is $(-\infty, 1) \cup (2, \infty)$.

41.

x	x^2	y
0	$0^2 = 0$	0
1	$1^2 = 1$	1
-1	$(-1)^2 = 1$	1
2	$(2)^2 = 4$	4
-2	$(-2)^2 = 4$	4

45. answers may vary

49. $P(x) = -2x^2 + 26x - 44$

$-2x^2 + 26x - 44 > 0$

$-2(x^2 - 13x + 22) > 0$

$-2(x - 11)(x - 2) > 0$

$x - 11 = 0$ or $x - 2 = 0$

$x = 11$ or $x = 2$

Region	Interval	Test Point
A	$(-\infty, 2)$	0
B	$(2, 11)$	3
C	$(11, \infty)$	12

$x = 0$	$-2(0)^2 + 26(0) - 44 > 0$	False
$x = 3$	$-2(3)^2 + 26(3) - 44 > 0$	True
$x = 12$	$-2(12)^2 + 26(12) - 44 > 0$	False

The solution set is $(2, 11)$.

The company makes a profit when x is between 2 and 11.

Exercise Set 8.5

1. $f(x) = x^2 - 1$

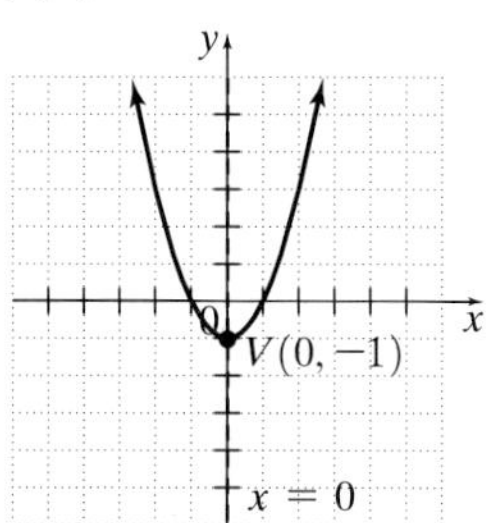

5. $h(x) = x^2 + 5$

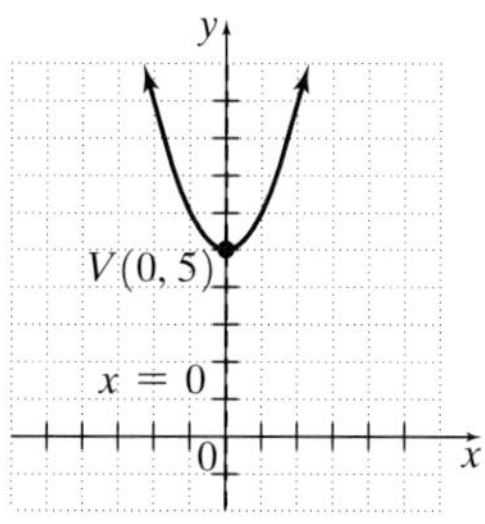

9. $g(x) = x^2 + 7$

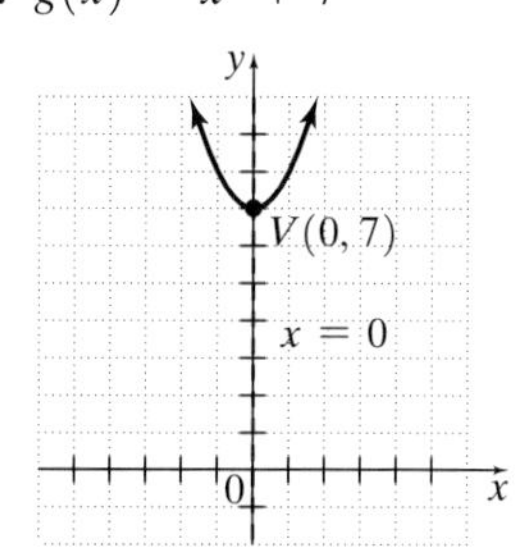

13. $f(x) = (x - 2)^2 + 5$

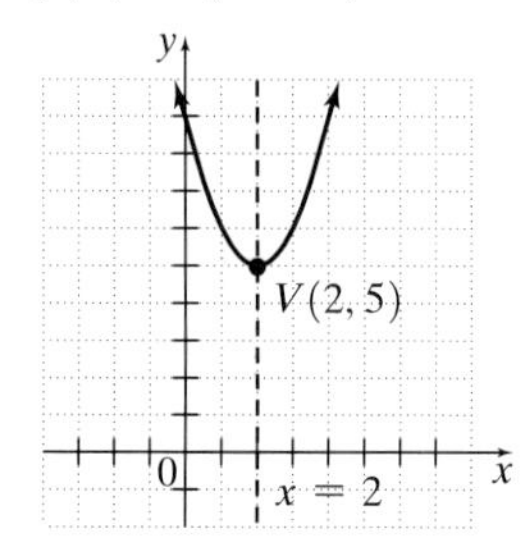

17. $g(x) = (x + 2)^2 - 5$

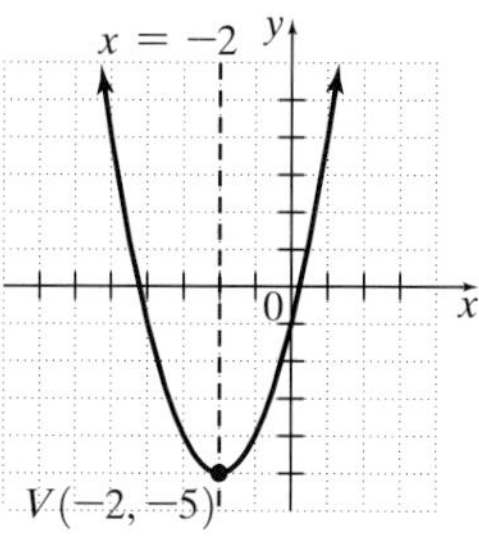

21. $g(x) = -x^2$

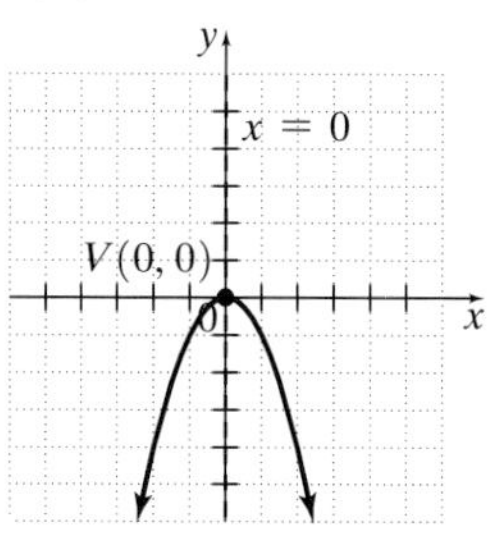

25. $H(x) = 2x^2$

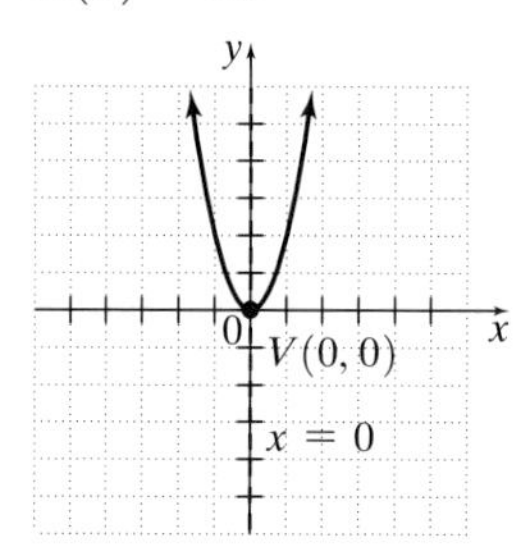

29. $f(x) = 10(x + 4)^2 - 6$

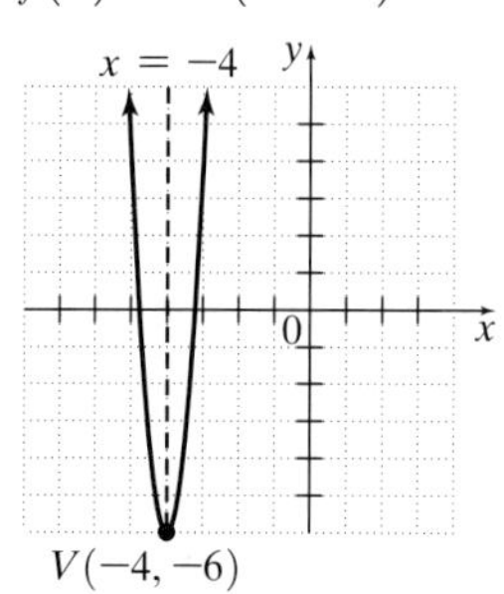

33. $H(x) = \frac{1}{2}(x - 6)^2 - 3$

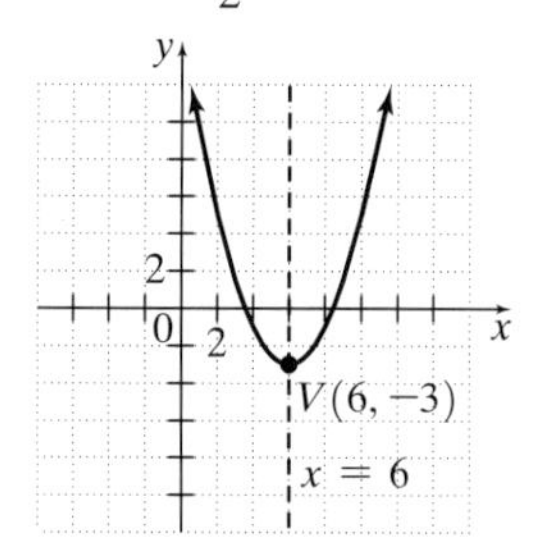

37. $F(x) = \left(x + \frac{1}{2}\right)^2 - 2$

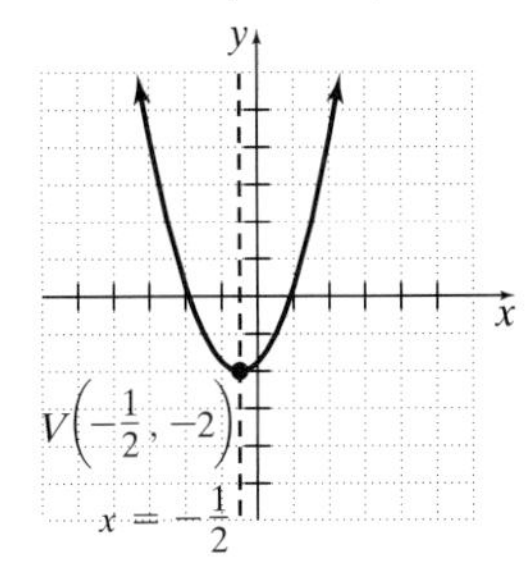

41. $x^2 + 8x$

$\left[\frac{1}{2}(8)\right]^2 = [4]^2 = 16$

$x^2 + 8x + 16$

45. $y^2 + y$

$\left[\frac{1}{2}(1)\right]^2 = \left[\frac{1}{2}\right]^2 = \frac{1}{4}$

$y^2 + y + \frac{1}{4}$

49. $f(x) = 5[(x - (-3))]^2 + 6$

$f(x) = 5(x + 3)^2 + 6$

53. $y = f(x - 3)$

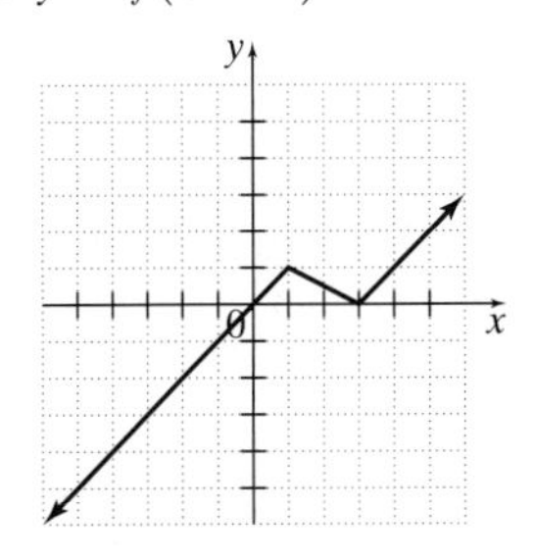

EXERCISE SET 8.6

1. $f(x) = x^2 + 8x + 7$

$\frac{-b}{2a} = \frac{-8}{2(1)} = -4$ and $f(-4) = (-4)^2 + 8(-4) + 7$

$f(-4) = 16 - 32 + 7 = -9$

Thus, $V(-4, -9)$.

5. $f(x) = 5x^2 - 10x + 3$

$\frac{-b}{2a} = \frac{-(-10)}{2(5)} = 1$ and $f(1) = 5(1)^2 - 10(1) + 3$

$f(1) = 5 - 10 + 3 = -2$

Thus, $V(1, -2)$.

9. $f(x) = x^2 - 4x + 3$

$\frac{-b}{2a} = \frac{-(-4)}{2(1)} = 2$

$f(2) = 2^2 - 4(2) + 3 = -1$

$V(2, -1)$

Graph D

13. $f(x) = x^2 + 4x - 5$

$\frac{-b}{2a} = \frac{-4}{2(1)} = -2$ and $f(-2) = (-2)^2 + 4(-2) - 5$

$f(-2) = 4 - 8 - 5 = -9$

Thus, $V(-2, -9)$.

The graph opens upward since $a > 0$.

$x^2 + 4x - 5 = 0$

$(x + 5)(x - 1) = 0$

$x + 5 = 0$ or $x - 1 = 0$

$x = -5$ or $x = 1$

The x-intercepts are -5 and 1.

$f(0) = -5$ is the y-intercept.

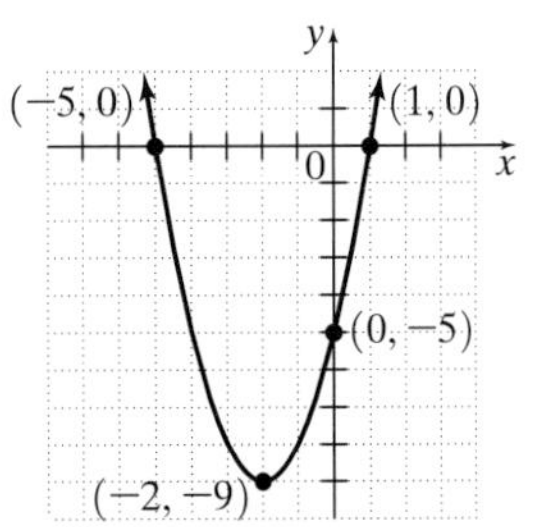

17. $f(x) = x^2 - 4$

$\frac{-b}{2a} = \frac{-0}{2(1)} = 0$ and $f(0) = -4$

Thus, $V(0, -4)$.

The graph opens upward since $a > 0$.

$x^2 - 4 = 0$

$(x + 2)(x - 2) = 0$

$x + 2 = 0$ or $x - 2 = 0$

$x = -2$ or $x = 2$

The x-intercepts are -2 and 2.

$f(0) = -4$ is the y-intercept.

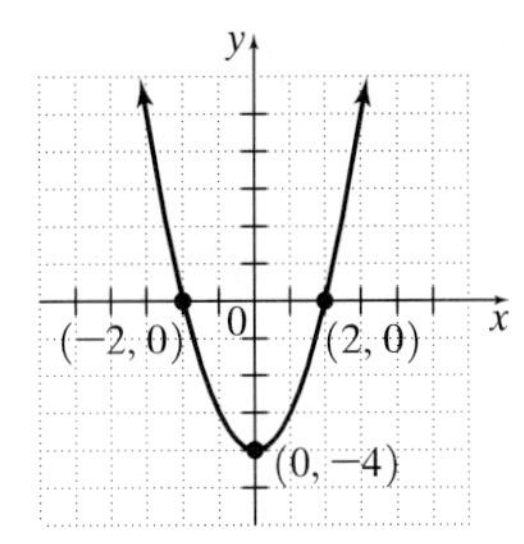

SOLUTIONS

21. $f(x) = \frac{1}{2}x^2 + 4x + \frac{15}{2}$

$\frac{-b}{2a} = \frac{-4}{2\left(\frac{1}{2}\right)} = -4$

$f(-4) = \frac{1}{2}(-4)^2 + 4(-4) + \frac{15}{2} = -\frac{1}{2}$

$V\left(-4, -\frac{1}{2}\right)$

The graph opens upward since $a > 0$.

$\frac{1}{2}x^2 + 4x + \frac{15}{2} = 0$

$x^2 + 8x + 15 = 0$

$(x + 5)(x + 3) = 0$

$x + 5 = 0$ or $x + 3 = 0$

$x = -5$ or $x = -3$

$x = -5, x = -3$

The x-intercepts are -5 and -3.

$f(0) = \frac{15}{2}$ is the y-intercept.

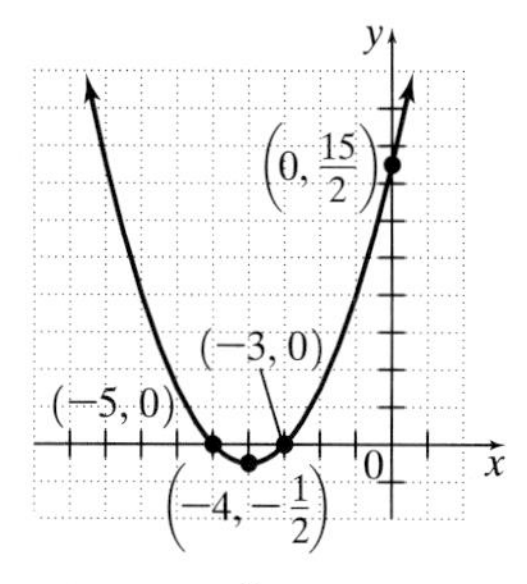

25. $f(x) = 2x^2 + 4x + 5$

$f(x) = 2(x^2 + 2x) + 5$

$f(x) = 2\left[x^2 + 2x + \left(\frac{2}{2}\right)^2\right] + 5 - 2$

$f(x) = 2(x + 1)^2 + 3$

Thus, $V(-1, 3)$.

The graph opens upward since $a > 0$.

$2(x + 1)^2 + 3 = 0$

$2(x + 1)^2 = -3$

Hence, there are no x-intercepts.

$f(0) = 2(0 + 1)^2 + 3 = 5$ is the y-intercept.

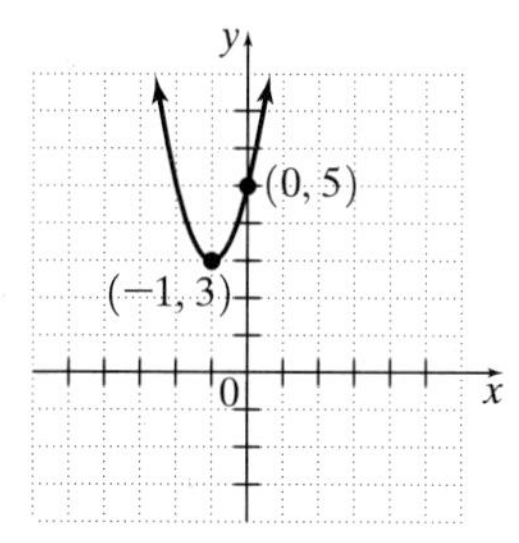

29. $h(t) = -16t^2 + 96t$

$\frac{-b}{2a} = \frac{-96}{2(-16)} = 3$ and $h(3) = -16(3)^2 + 96(3)$

$h(3) = -144 + 288 = 144$

The maximum height of the projectile is 144 ft.

33. Let x = one number.

$60 - x$ = other number

$f(x) = x(60 - x)$

$f(x) = 60x - x^2$

$f(x) = -x^2 + 60x$

$f(x) = -1(x^2 - 60x)$

$f(x) = -1(x^2 - 60x + 900) + 900$

$f(x) = -(x - 30)^2 + 900$

The maximum will occur at the vertex which is (30, 900). The numbers are 30 and 30.

37. Let x = the width

$40 - x$ = the length

$f(x) = x(40 - x)$

$f(x) = 40x - x^2$

$f(x) = -x^2 + 40x$

$f(x) = -1(x^2 - 40x)$

$f(x) = -1(x^2 - 40x + 400) + 400$

$f(x) = -(x - 20)^2 + 400$

The maximum will occur at the vertex which is (20, 400). The width is 20 units and the length is 20 units.

41. $g(x) = x + 2$

The vertex is undefined.

45. $f(x) = 3(x - 4)^2 + 1$

The vertex is (4, 1)

49. $f(x) = \frac{1}{2}x^2 + 4x + \frac{15}{2}$

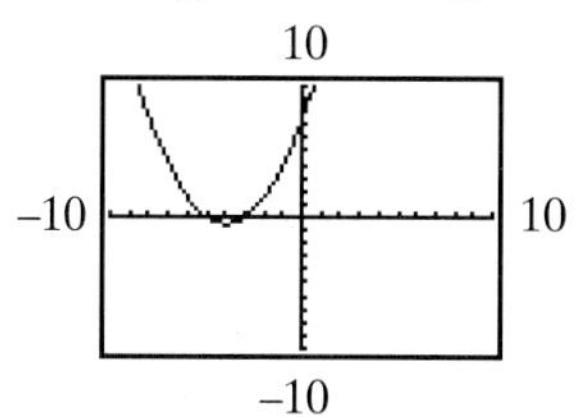

53. a. Because $a < 0$, the parabola will open downward and the function will have a maximum.

b. $\frac{-b}{2a} = \frac{-99{,}699.5}{2(-2464.4)} \approx 20.23$

The number of inmates will be at its maximum about 20 years after 1990 which is the year 2010.

c. $f(20.23)$

$= -2464.4(20.23)^2 + 99{,}699.5(20.23) + 1, 148{,}702$

$\approx 2{,}157{,}060$

The maximum number of inmates predicted for that year is about 2,157,060.

Chapter 8 Test

1. $5x^2 - 2x = 7$

$5x^2 - 2x - 7 = 0$

$(5x - 7)(x + 1) = 0$

$5x - 7 = 0$ or $x + 1 = 0$

$5x = 7$ or $x = -1$

$x = \frac{7}{5}$ or $x = -1$

The solution set is $\left\{\frac{7}{5}, -1\right\}$.

5. $7x^2 + 8x + 1 = 0$
$(7x + 1)(x + 1) = 0$
$7x + 1 = 0 \quad \text{or } x + 1 = 0$
$7x = -1 \quad \text{or} \quad x = -1$
$x = -\frac{1}{7} \quad \text{or} \quad x = -1$
The solution set is $\left\{-\frac{1}{7}, -1\right\}$.

9. $x^6 + 1 = x^4 + x^2$
$x^6 - x^4 - x^2 + 1 = 0$
$x^4(x^2 - 1) - (x^2 - 1) = 0$
$(x^4 - 1)(x^2 - 1) = 0$
$(x^2 + 1)(x^2 - 1)(x^2 - 1) = 0$
$(x^2 + 1)(x^2 - 1)^2 = 0$
$(x^2 + 1)[(x + 1)(x - 1)]^2 = 0$
$(x^2 + 1)(x + 1)^2(x - 1)^2 = 0$
$x^2 + 1 = 0$ or $(x + 1)^2 = 0$ or $(x - 1)^2 = 0$
$x^2 = -1$ or $x + 1 = 0$ or $x - 1 = 0$
$x = \pm\sqrt{-1} = \pm i$ or $x = -1$ or $x = 1$
The solution set is $\{1, -1, i, -i\}$.

13. $2x^2 - 7x > 15$
$2x^2 - 7x - 15 > 0$
$(2x + 3)(x - 5) > 0$
$2x + 3 = 0 \quad \text{or } x - 5 = 0$
$x = -\frac{3}{2} \quad \text{or} \quad x = 5$

Region	Interval	Test Point
A	$\left(-\infty, -\frac{3}{2}\right)$	-2
B	$\left(-\frac{3}{2}, 5\right)$	0
C	$(5, \infty)$	6
$x = -2$	$2(-2)^2 - 7(-2) > 15$	True
$x = 0$	$2(0)^2 - 7(0) > 15$	False
$x = 6$	$2(6)^2 - 7(6) > 15$	True

17. $f(x) = 3x^2$
vertex: $(0, 0)$

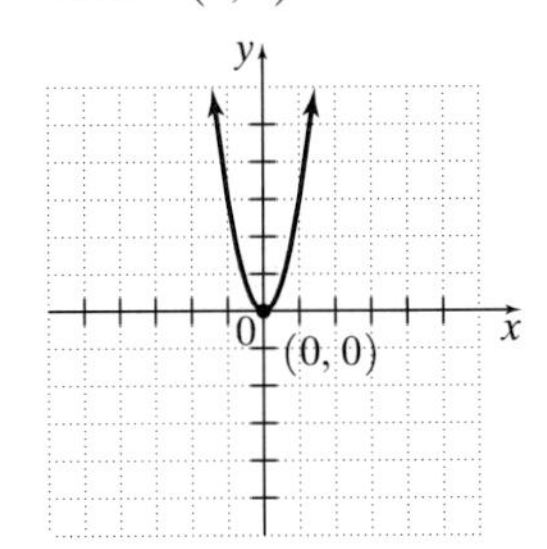

21. $c^2 = a^2 + b^2$
$(10)^2 = x^2 + (x - 4)^2$
$100 = x^2 + x^2 - 8x + 16$
$0 = 2x^2 - 8x - 84$
$0 = x^2 - 4x - 42$
$a = 1, b = -4, c = -42$
$x = \frac{-(-4) \pm \sqrt{(-4)^2 - 4(1)(-42)}}{2(1)}$
$x = \frac{4 \pm \sqrt{16 + 168}}{2}$
$x = \frac{4 \pm \sqrt{184}}{2}$
$x = \frac{4 \pm 2\sqrt{46}}{2}$
$x = 2 \pm \sqrt{46}$
Disregard the negative result.
The top of the ladder is $2 + \sqrt{46} \approx 8.8$ feet from the ground.

Chapter 9

Exercise Set 9.1

1. $x = 3y^2$
$x = 3(y - 0)^2$
$V(0, 0)$

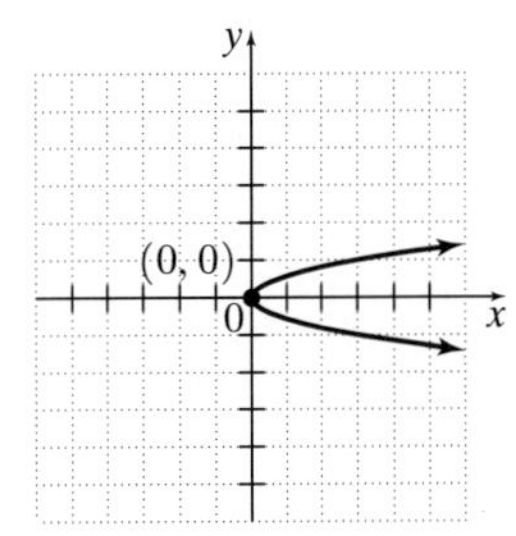

5. $y = 3(x - 1)^2 + 5$
$V(1, 5)$

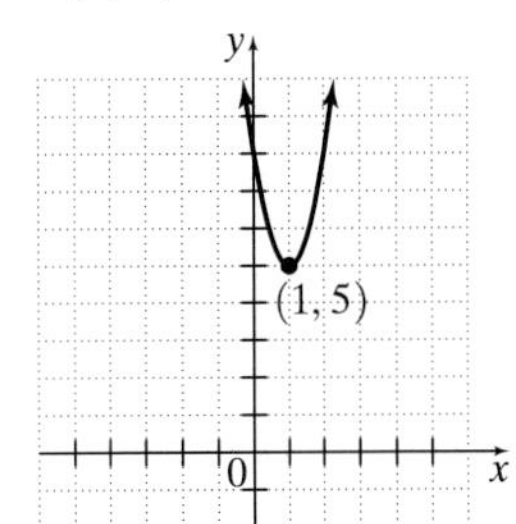

9. $y = x^2 + 10x + 20$
$y = x^2 + 10x + 25 + 20 - 25$
$y = (x + 5)^2 - 5$
$V(-5, -5)$

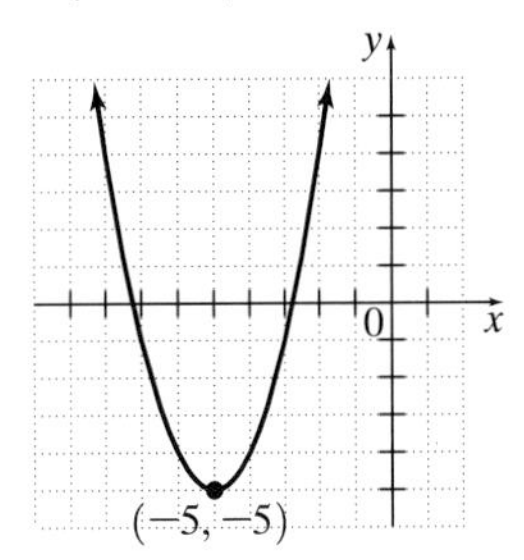

SOLUTIONS

13. $(5, 1), (8, 5)$
$d = \sqrt{(8-5)^2 + (5-1)^2}$
$d = \sqrt{9 + 16}$
$d = \sqrt{25}$
$d = 5$ units

17. $(-9, 4), (-8, 1)$
$d = \sqrt{[-8-(-9)]^2 + (1-4)^2}$
$d = \sqrt{(-8+9)^2 + (-3)^2}$
$d = \sqrt{1+9}$
$d = \sqrt{10} \approx 3.162$ units

21. $(1.7, -3.6), (-8.6, 5.7)$
$d = \sqrt{(-8.6-1.7)^2 + [5.7-(-3.6)]^2}$
$d = \sqrt{(-10.3)^2 + (9.3)^2}$
$d = \sqrt{192.58} \approx 13.877$ units

25. $(6, -8), (2, 4)$
$\left(\frac{6+2}{2}, \frac{-8+4}{2}\right)$
$(4, -2)$

29. $(7, 3), (-1, -3)$
$\left(\frac{7+(-1)}{2}, \frac{3+(-3)}{2}\right)$
$(3, 0)$

33. $(\sqrt{2}, 3\sqrt{5}), (\sqrt{2}, -2\sqrt{5})$
$\left(\frac{\sqrt{2}+\sqrt{2}}{2}, \frac{3\sqrt{5}-2\sqrt{5}}{2}\right)$
$\left(\sqrt{2}, \frac{\sqrt{5}}{2}\right)$

37. $x^2 + y^2 = 9$
$(x-0)^2 + (y-0)^2 = 3^2$
$C(0, 0)$ and $r = 3$

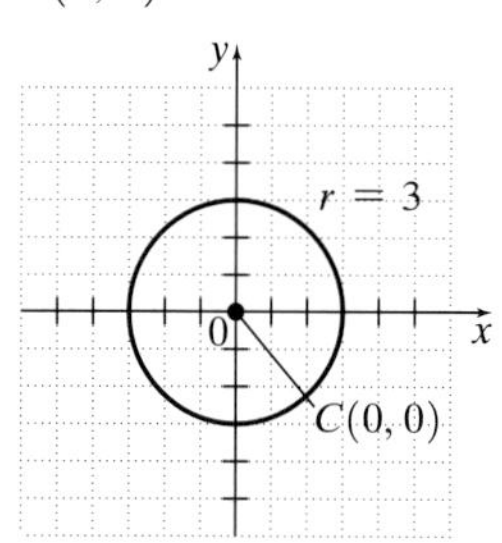

41. $(x-5)^2 + (y+2)^2 = 1$
$(x-5)^2 + (y+2)^2 = 1^2$
$C(5, -2)$ and $r = 1$

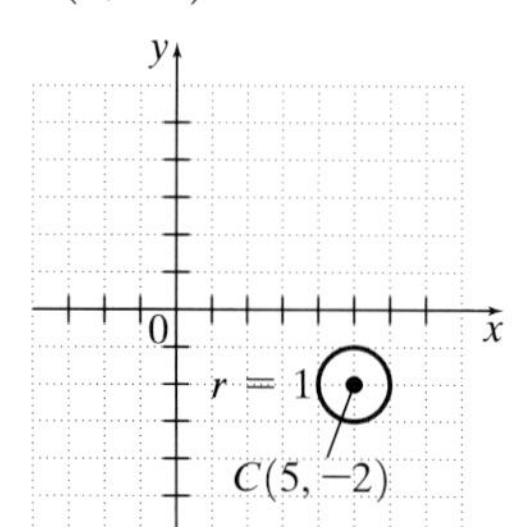

45.
$$x^2 + y^2 + 2x - 4y = 4$$
$$x^2 + 2x + 1 + y^2 - 4y + 4 = 4 + 1 + 4$$
$$(x+1)^2 + (y-2)^2 = 9$$
$$(x+1)^2 + (y-2)^2 = 3^2$$
$C(-1, 2)$ and $r = 3$

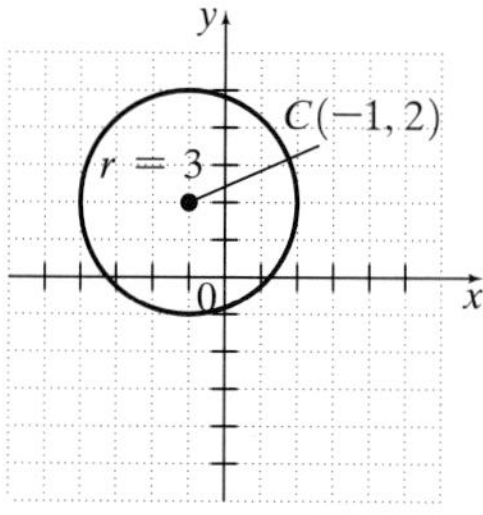

49. $C(2, 3); r = 6$
$(x-2)^2 + (y-3)^2 = 6^2$
$(x-2)^2 + (y-3)^2 = 36$

53. $C(-5, 4); r = 3\sqrt{5}$
$[x-(-5)]^2 + (y-4)^2 = (3\sqrt{5})^2$
$(x+5)^2 + (y-4)^2 = 45$

57. $y = 3$

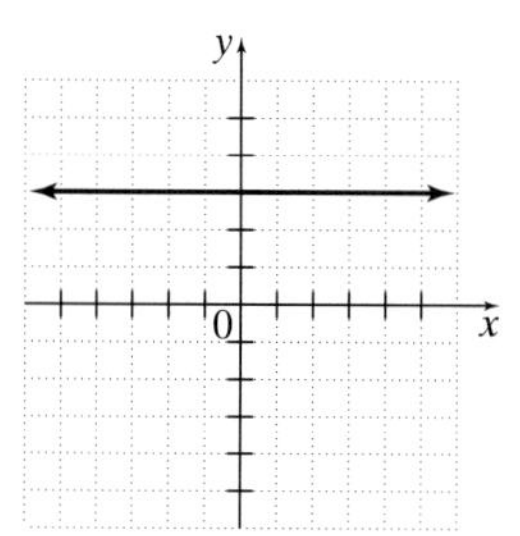

61.
$$\frac{4\sqrt{7}}{\sqrt{6}} = \frac{4\sqrt{7}}{\sqrt{6}} \cdot \frac{\sqrt{6}}{\sqrt{6}} = \frac{4\sqrt{42}}{6} = \frac{2\sqrt{42}}{3}$$

65. $B(3, 1), C(19, 13)$
$d = \sqrt{(19-3)^2 + (13-1)^2}$
$d = \sqrt{256 + 144}$
$d = \sqrt{400} = 20$
The distance is 20 meters.

Exercise Set 9.2

1. $\frac{x^2}{4} + \frac{y^2}{25} = 1$
$\frac{x^2}{2^2} + \frac{y^2}{5^2} = 1$
$C(0, 0)$
x-intercepts: $-2, 2$
y-intercepts: $-5, 5$

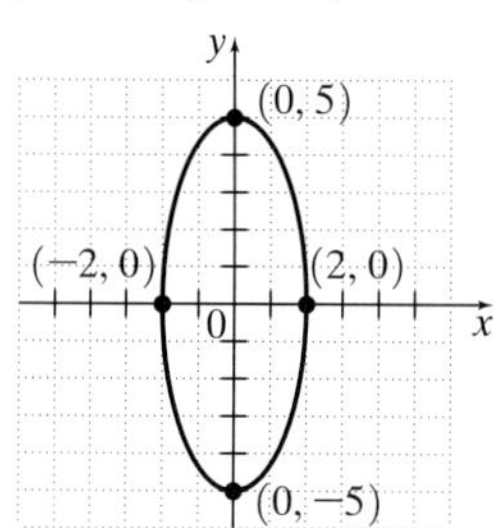

5. $9x^2 + 4y^2 = 36$

$\frac{x^2}{4} + \frac{y^2}{9} = 1$

$\frac{x^2}{2^2} + \frac{y^2}{3^2} = 1$

$C(0, 0)$

x-intercepts: $-2, 2$

y-intercepts: $-3, 3$

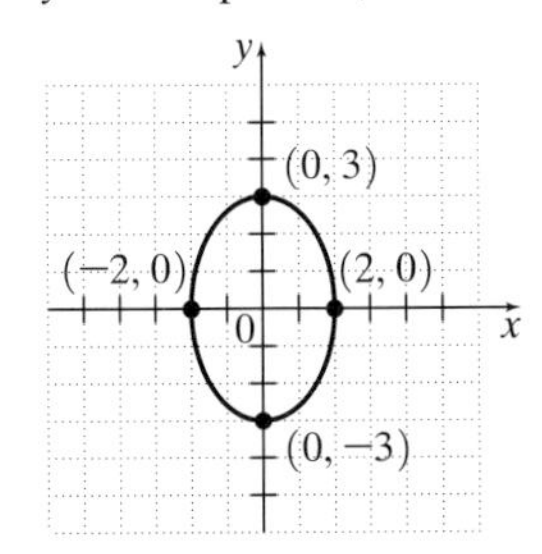

9. $\frac{(x+1)^2}{36} + \frac{(y-2)^2}{49} = 1$

$\frac{(x+1)^2}{6^2} + \frac{(y-2)^2}{7^2} = 1$

$C(-1, 2)$

other points:

$(-1 - 6, 2)$ or $(-7, 2)$

$(-1 + 6, 2)$ or $(5, 2)$

$(-1, 2 - 7)$ or $(-1, -5)$

$(-1, 2 + 7)$ or $(-1, 9)$

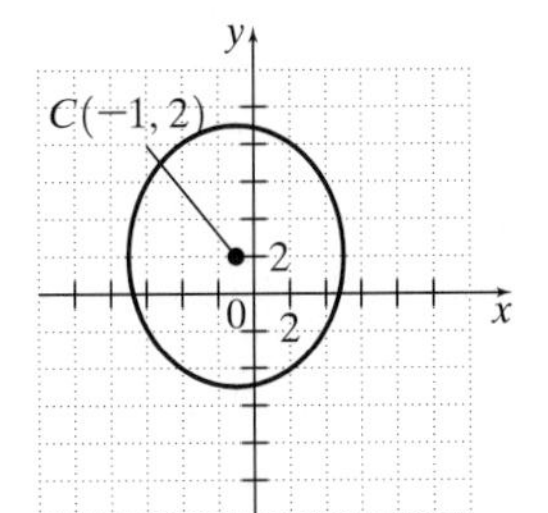

13. $\frac{x^2}{4} - \frac{y^2}{9} = 1$

$\frac{x^2}{2^2} - \frac{y^2}{3^2} = 1$

$a = 2, b = 3$

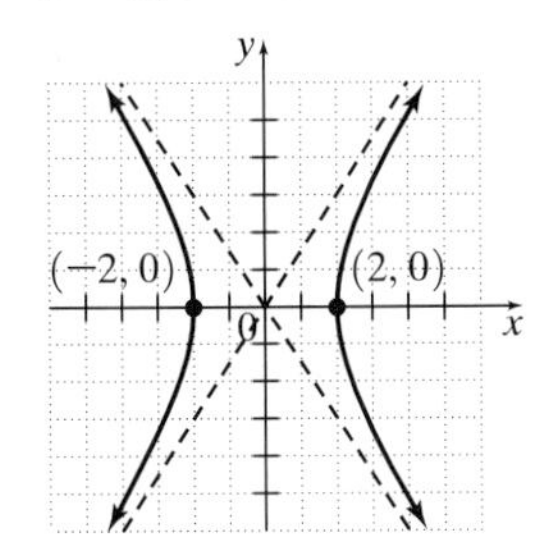

17. $x^2 - 4y^2 = 16$

$\frac{x^2}{16} - \frac{y^2}{4} = 1$

$\frac{x^2}{4^2} - \frac{y^2}{2^2} = 1$

$a = 4, b = 2$

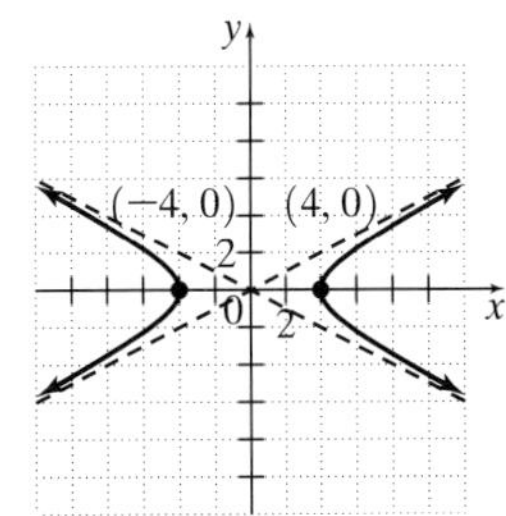

21. $(2x^3)(-4x^2) = 2(-4)x^{3+2}$

$= -8x^5$

25. $x^2 + y^2 = 25$

$\frac{x^2}{25} + \frac{y^2}{25} = \frac{25}{25}$

$\frac{x^2}{25} + \frac{y^2}{25} = 1$

It resembles an ellipse.

An ellipse is a circle when $a = b$.

29. A: $d = 6$

B: $d = 2$

C: $d = 5$

D: $d = 5$

E: $d = 9$

F: $d = 6$

G: $d = 4$

H: $d = 12$

33. They are greater than 1.

Exercise Set 9.3

1. $f(x) = |x| + 3$

Shift $y = |x|$ up 3 units.

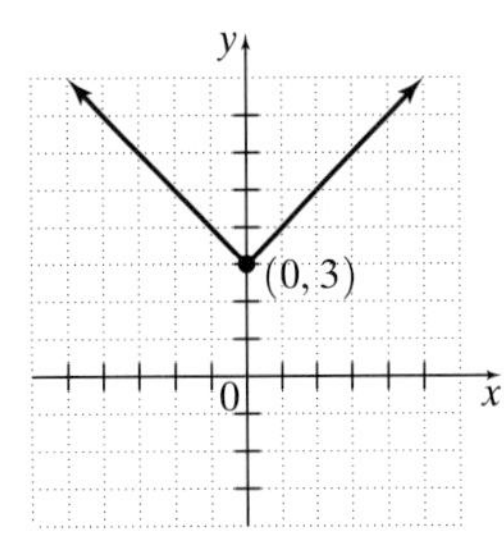

5. $f(x) = |x - 4|$

Shift $y = |x|$ to the right 4 units.

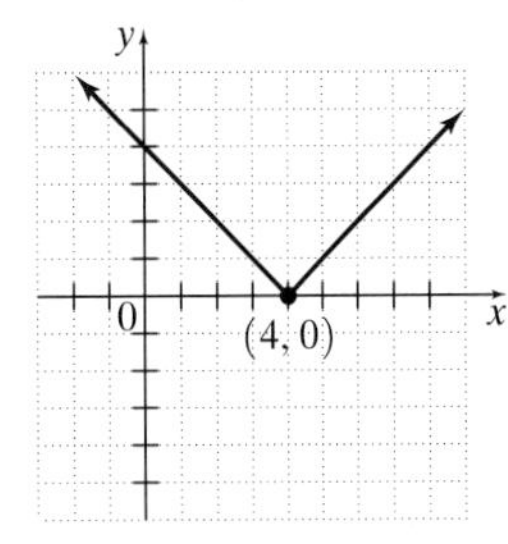

SOLUTIONS

9. $f(x) = \sqrt{x-2} + 3$
Shift $y = \sqrt{x}$ to the right 2 units and up 3 units.

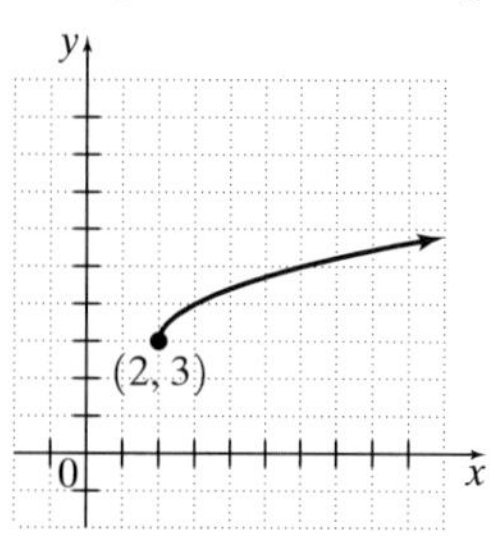

13. $f(x) = \sqrt{x+1} + 1$
Shift $y = \sqrt{x}$ to the left one unit and up one unit.

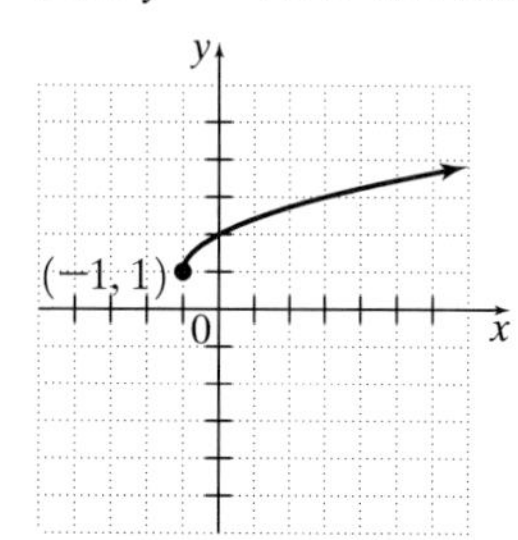

17. $x + y = 6$
$x - y = 10$
$2x = 16$
$x = 8$
Replace x with 8 in the first equation.
$8 + y = 6$
$y = -2$
The solution is $(8, -2)$.

21. $f(x) = x^3$

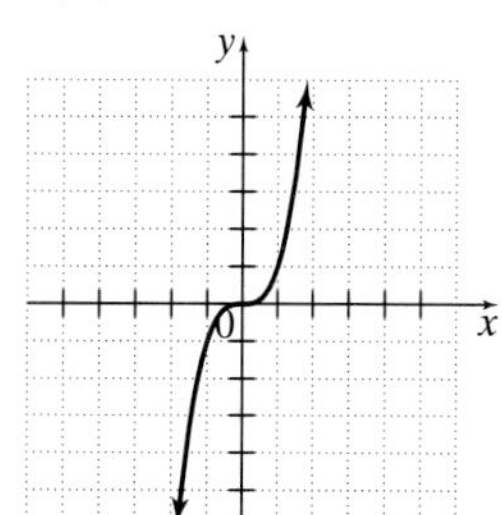

Exercise Set 9.4

1. $\begin{cases} x^2 + y^2 = 25 \\ 4x + 3y = 0 \end{cases}$
Solve equation 2 for y.
$3y = -4x$
$y = \dfrac{-4x}{3}$
Substitute.
$x^2 + \left(-\dfrac{4x}{3}\right)^2 = 25$
$x^2 + \dfrac{16x^2}{9} = 25$
$\dfrac{25}{9}x^2 = 1$
$\dfrac{x^2}{9} = 1$
$x^2 = 9$
$x = \pm\sqrt{9} = \pm 3$
$x = 3{:}\quad y = -\frac{4}{3}(3) = -4$
$x = -3{:}\ y = -\frac{4}{3}(-3) = 4$
The solution set is $\{(3, -4), (-3, 4)\}$.

5. $\begin{cases} y^2 = 4 - x \\ x - 2y = 4 \end{cases}$
$-2y = 4 - x$
Substitute.
$y^2 = -2y$
$y^2 + 2y = 0$
$y(y + 2) = 0$
$y = 0$ or $y + 2 = 0$
$y = -2$
$y = 0{:}\quad x - 2(0) = 4$
$x = 4$
$y = -2{:}\ x - 2(-2) = 4$
$x + 4 = 4$
$x = 0$
The solution set is $\{(4, 0), (0, -2)\}$.

9. $\begin{cases} x^2 + 2y^2 = 2 \\ x - y = 2 \end{cases}$
$x = y + 2$
Substitute.
$(y + 2)^2 + 2y^2 = 2$
$y^2 + 4y + 4 + 2y^2 = 2$
$3y^2 + 4y + 4 = 2$
$3y^2 + 4y + 2 = 0$
$b^2 - 4ac = 4^2 - 4(3)(2)$
$= 16 - 24 = -8 < 0$
Therefore, no real solutions exits.
The solution set is $\emptyset$.

SOLUTIONS

13. $\begin{cases} y = x^2 \\ 3x + y = 10 \end{cases}$

Substitute.

$3x + x^2 = 10$
$x^2 + 3x - 10 = 0$
$(x + 5)(x - 2) = 0$
$x + 5 = 0$ or $x - 2 = 0$
$x = -5$ or $x = 2$
$x = -5$: $y = (-5)^2 = 25$
$x = 2$: $y = 2^2 = 4$

The solution set is $\{(-5, 25), (2, 4)\}$.

17. $\begin{cases} y = x^2 - 4 \\ y = x^2 - 4x \end{cases}$

Substitute.

$x^2 - 4 = x^2 - 4x$
$-4 = -4x$
$x = 1$
$y = 1^2 - 4 = -3$

The solution set is $\{(1, -3)\}$.

21. $\begin{cases} x^2 + y^2 = 1 \\ x^2 + (y + 3)^2 = 4 \end{cases}$

Subtract equation 1 from equation 2.

$(y + 3)^2 - y^2 = 3$
$y^2 + 6y + 9 - y^2 = 3$
$6y + 9 = 3$
$6y = -6$
$y = -1$

Substitute back.

$x^2 + (-1)^2 = 1$
$x^2 + 1 = 1$
$x^2 = 0$
$x = 0$

The solution set is $\{(0, -1)\}$.

25. $\begin{cases} 3x^2 + y^2 = 9 \\ 3x^2 - y^2 = 9 \end{cases}$

Subtract.

$2y^2 = 0$
$y^2 = 0$
$y = 0$

Substitute back.

$3x^2 + 0 = 9$
$3x^2 = 9$
$x^2 = 3$
$x = \pm\sqrt{3}$

The solution set is $\{(\sqrt{3}, 0), (-\sqrt{3}, 0)\}$.

29. $\begin{cases} x^2 + y^2 = 36 \\ y = \frac{1}{6}x^2 - 6 \end{cases}$

$y + 6 = \frac{1}{6}x^2$
$x^2 = 6(y + 6)$

Substitute.

$6(y + 6) + y^2 = 36$
$6y + 36 + y^2 = 36$
$6y + y^2 = 0$
$y(6 + y) = 0$
$y = 0$ or $6 + y = 0$
$y = -6$
$y = 0$: $x^2 + 0^2 = 36$
$x^2 = 36$
$x = \pm 6$
$y = -6$: $x^2 + (-6)^2 = 36$
$x^2 + 36 = 36$
$x^2 = 0$
$x = 0$

The solution set is $\{(6, 0), (-6, 0), (0, -6)\}$.

33. $x > -3$

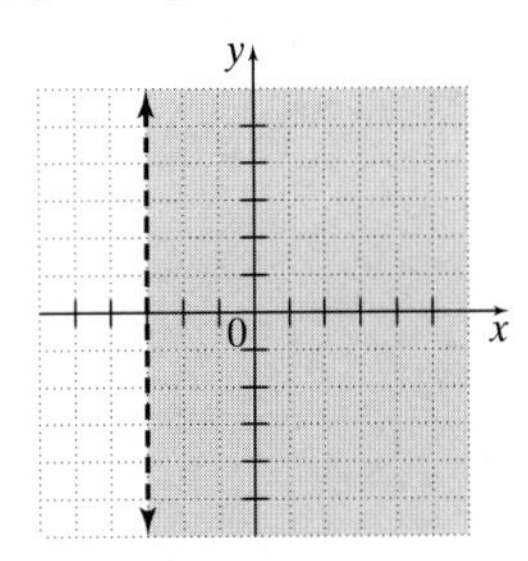

37. $\begin{cases} x^2 + y^2 = 130 \\ x^2 - y^2 = 32 \end{cases}$

Add.

$2x^2 = 162$
$x^2 = 81$
$x = \pm 9$

Substitute back.

$9^2 + y^2 = 130$ $\quad$ $(-9)^2 + y^2 = 130$
$y^2 = 49$ $\quad$ $y^2 = 49$
$y = \pm 7$ $\quad$ $y = \pm 7$

The numbers are 9 and 7, 9 and −7, −9 and 7, or −9 and −7.

41. $p = -0.01x^2 - 0.2x + 9$
$p = 0.01x^2 - 0.1x + 3$
$-0.01x^2 - 0.2x + 9 = 0.01x^2 - 0.01x + 3$
$0 = 0.02x^2 + 0.1x - 6$
$0 = x^2 + 5x - 300$
$0 = (x + 20)(x - 15)$
$x + 20 = 0$ or $x - 15 = 0$
$x = -20$ or $x = 15$

Disregard the negative

$p = -0.01(15)^2 - 0.2(15) + 9$
$p = 3.75$

15 thousand compact discs; price, \$3.75

SOLUTIONS

45. $\begin{cases} y = x^2 + 2 \\ y = -x^2 + 4 \end{cases}$

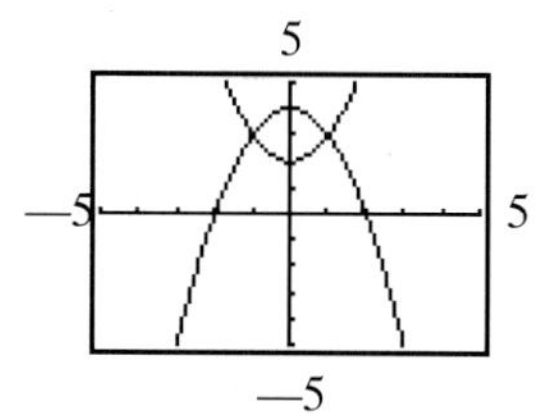

Exercise Set 9.5

1. $y < x^2$

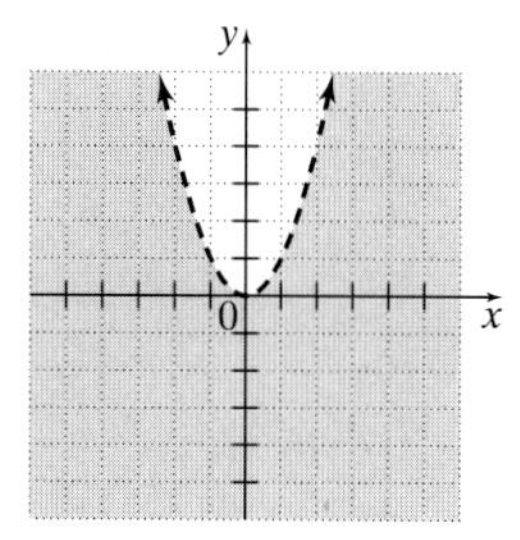

5. $\dfrac{x^2}{4} - y^2 < 1$

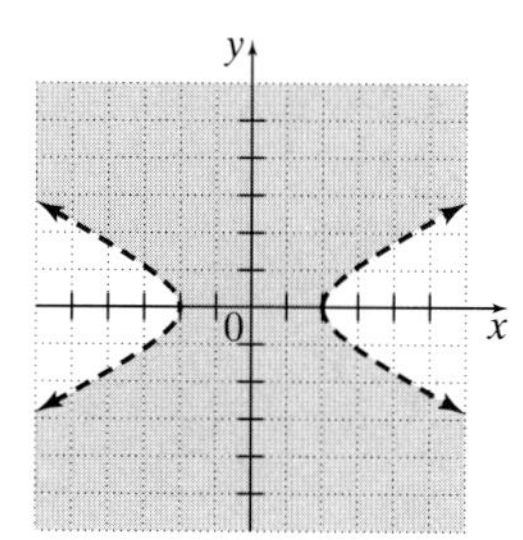

9. $x^2 + y^2 \le 9$

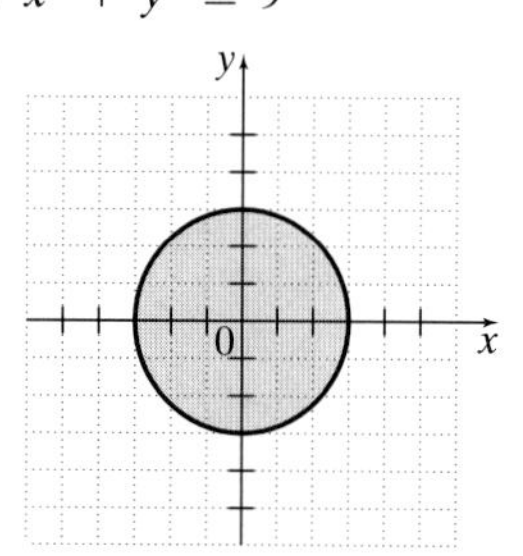

13. $\dfrac{x^2}{4} + \dfrac{y^2}{9} \le 1$

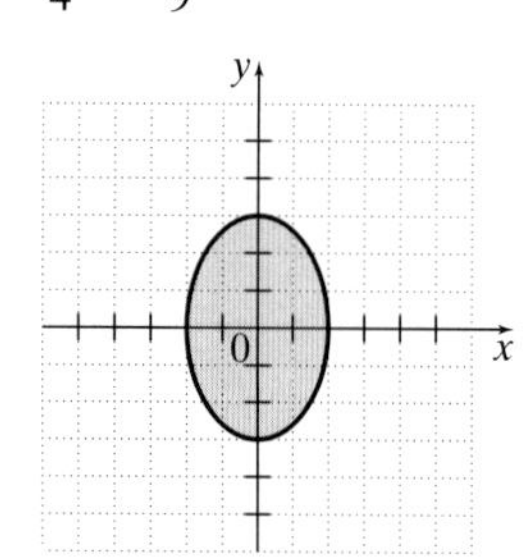

17. $y < (x - 2)^2 + 1$

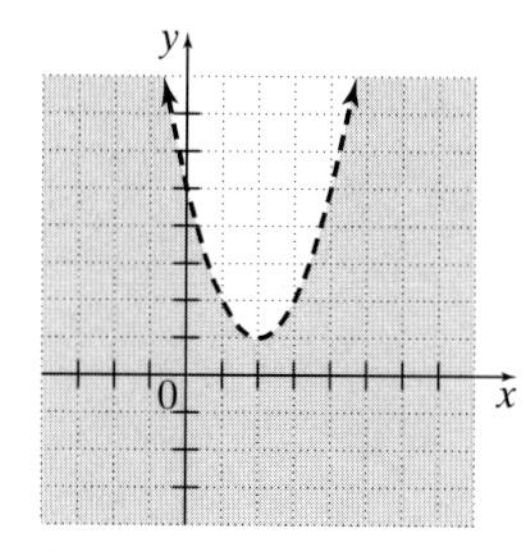

21. $\begin{cases} 4x + 3y \ge 12 \\ x^2 + y^2 < 16 \end{cases}$

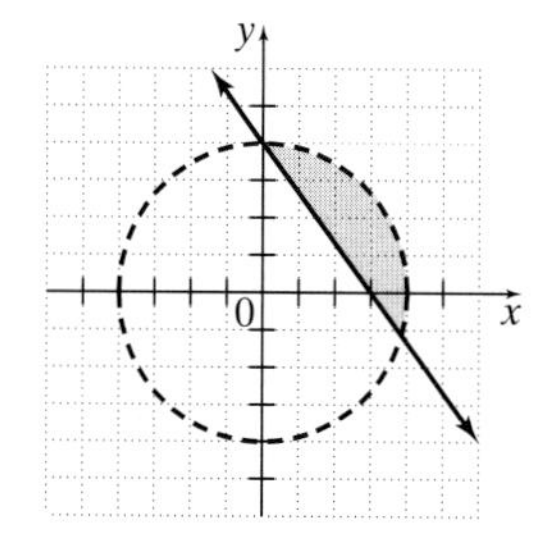

25. $\begin{cases} y > x^2 \\ y \ge 2x + 1 \end{cases}$

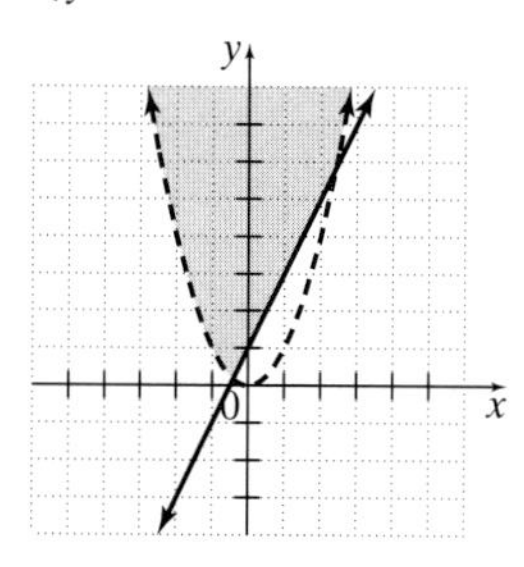

29. $\begin{cases} \dfrac{x^2}{4} + \dfrac{y^2}{9} \ge 1 \\ x^2 + y^2 \ge 4 \end{cases}$

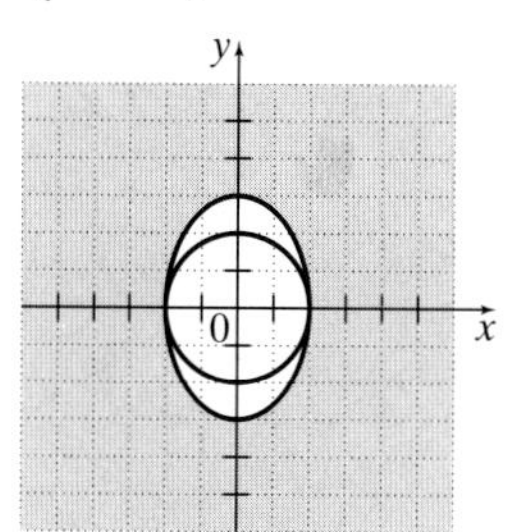

33. $\begin{cases} x + y \ge 1 \\ 2x + 3y < 1 \\ x > -3 \end{cases}$

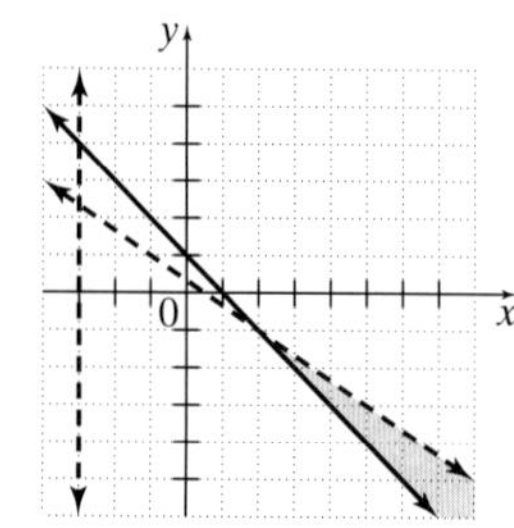

37. This is not a function because a vertical line can cross the graph in two places.

SOLUTIONS

41. answers may vary

Chapter 9 Test

1. $(-6, 3)$ and $(-8, -7)$

$d = \sqrt{(-8 + 6)^2 + (-7 - 3)^2}$
$d = \sqrt{(-2)^2 + (-10)^2}$
$d = \sqrt{4 + 100}$
$d = \sqrt{104}$
$d = 2\sqrt{26}$ units

5. $x^2 + y^2 = 36$ or $(x - 0)^2 + (y - 0)^2 = 6^2$
Circle: $C(0, 0)$, $r = 6$

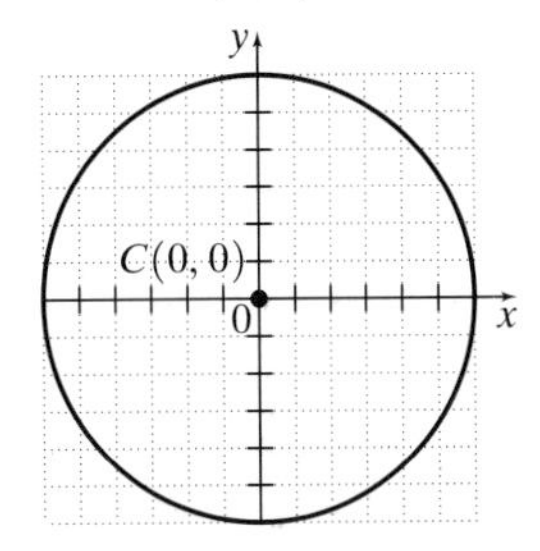

9.
$$x^2 + y^2 + 6x = 16$$
$$x^2 + 6x + y^2 = 16$$
$$(x^2 + 6x + 9) + y^2 = 16 + 9$$
$$(x + 3)^2 + y^2 = 5^2$$
Circle: $C(-3, 0)$, $r = 5$

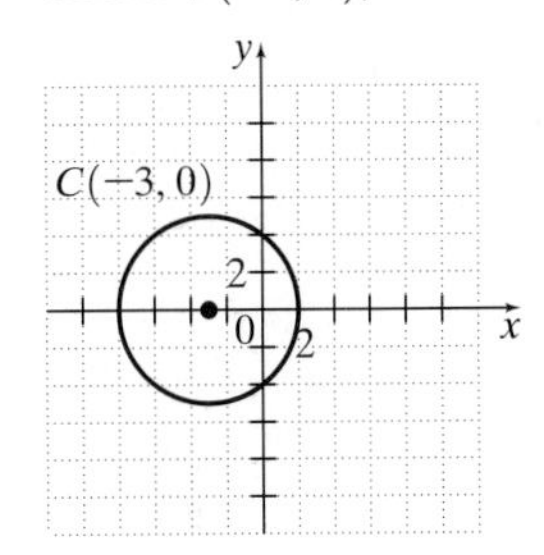

13. $\begin{cases} x^2 + y^2 = 169 \\ 5x + 12y = 0 \end{cases}$

$12y = -5x$
$y = -\dfrac{5x}{12}$

Substitute.

$x^2 + \left(-\dfrac{5x}{12}\right)^2 = 169$
$x^2 + \dfrac{25x^2}{144} = 169$
$\dfrac{169x^2}{144} = 169$

$x^2 = 144$ so $x = \pm 12$

Substitute back.

$x = 12$: $\quad y = -\dfrac{5}{12}(12) = -5$

$x = -12$: $\quad y = -\dfrac{5}{12}(-12) = 5$

$\{(12, -5), (-12, 5)\}$

17. $\begin{cases} 2x + 5y \geq 10 \\ y \geq x^2 + 1 \end{cases}$ First graph.

$\begin{cases} 2x + 5y = 10 \\ y = x^2 + 1 \end{cases}$ or

$\begin{cases} y = -\dfrac{2}{5}x + 2 \\ y = 1 \cdot (x - 0)^2 + 1 \end{cases}$

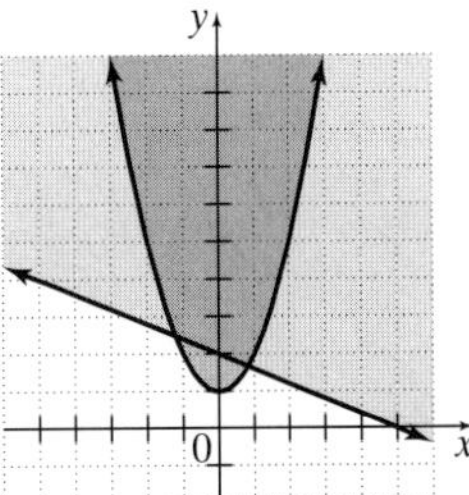

21. Graph B; vertex in third quadrant, opens to the right.

Chapter 10

Exercise Set 10.1

1. a. $(f + g)(x) = x - 7 + 2x + 1$
$= 3x - 6$

b. $(f - g)(x) = x - 7 - (2x + 1)$
$= x - 7 - 2x - 1$
$= -x - 8$

c. $(f \cdot g)(x) = (x - 7)(2x + 1)$
$= 2x^2 - 13x - 7$

d. $\left(\dfrac{f}{g}\right)(x) = \dfrac{x - 7}{2x + 1}$, where $x \neq -\dfrac{1}{2}$

5. a. $(f + g)(x) = \sqrt{x} + x + 5$

b. $(f \circ g)(x) = \sqrt{x} - x - 5$

c. $(f \circ g)(x) = \sqrt{x}(x + 5)$
$= x\sqrt{x} + 5\sqrt{x}$

d. $\left(\dfrac{f}{g}\right)(x) = \dfrac{\sqrt{x}}{x + 5}$, where $x \neq -5$

9. $(f \circ g)(2) = f(g(2))$
$= f(-4)$
$= (-4)^2 - 6(-4) + 2$
$= 16 + 24 + 2$
$= 42$

13. $(g \circ h)(0) = g(h(0))$
$= g(0)$
$= -2(0)$
$= 0$

17. $(f \circ g)(x) = f(g(x))$
$= f(x + 7)$
$= 2(x + 7) - 3$
$= 2x + 14 - 3$
$= 2x + 11$

$(g \circ f)(x) = g(f(x))$
$= g(2x - 3)$
$= (2x - 3) + 7$
$= 2x + 4$

SOLUTIONS

SOLUTIONS

21. $(f \circ g)(x) = f(g(x))$
$= f(-5x + 2)$
$= \sqrt{-5x + 2}$
$(g \circ f)(x) = g(f(x))$
$= g(\sqrt{x})$
$= -5\sqrt{x} + 2$

25. $F(x) = (h \circ f)(x)$
$= h(f(x))$
$= h(3x)$
$= (3x)^2 + 2$
$= 9x^2 + 2$

29. answers may vary

33. answers may vary

37. $x = 3y$
$\frac{x}{3} = \frac{3y}{3}$
$\frac{x}{3} = y$
$y = \frac{x}{3}$

41. $P(x) = R(x) - C(x)$

Exercise Set 10.2

1. $f = \{(-1, -1), (1, 1), (0, 2), (2, 0)\}$ is a one-to-one function.
$f^{-1} = \{(-1, -1), (1, 1), (2, 0), (0, 2)\}$

5. $f = \{(11, 12), (4, 3), (3, 4), (6, 6)\}$ is a one-to-one function.
$f^{-1} = \{(12, 11), (3, 4), (4, 3), (6, 6)\}$

9. This function is one-to-one.

Rank (input)	1	49	12	2	45
State (output)	CA	VT	VA	TX	SD

13. $f(x) = x^3 + 2$
a. $f(-1) = (-1)^3 + 2 = 1$
b. $f^{-1}(1) = -1$

17. The graph does not represent a one-to-one function because it does not pass the horizontal line test.

21. The graph does not represent a one-to-one function because it does not pass the horizontal line test.

25. $f(x) = 2x - 3$
$y = 2x - 3$
$x = 2y - 3$
$2y = x + 3$
$y = \frac{x + 3}{2}$
$f^{-1}(x) = \frac{x + 3}{2}$

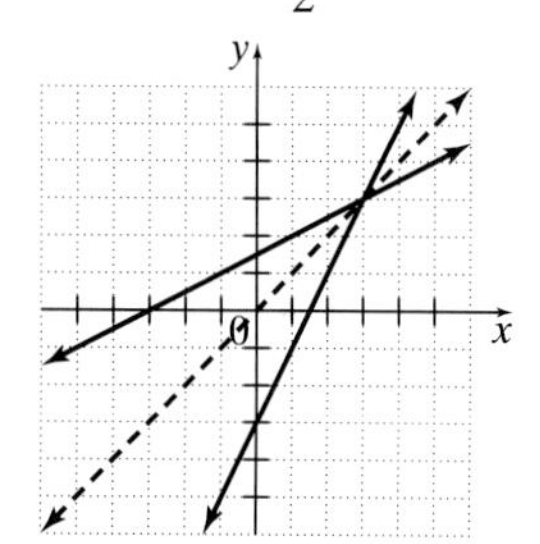

29. $f(x) = x^3$
$y = x^3$
$x = y^3$
$y = \sqrt[3]{x}$
$f^{-1}(x) = \sqrt[3]{x}$

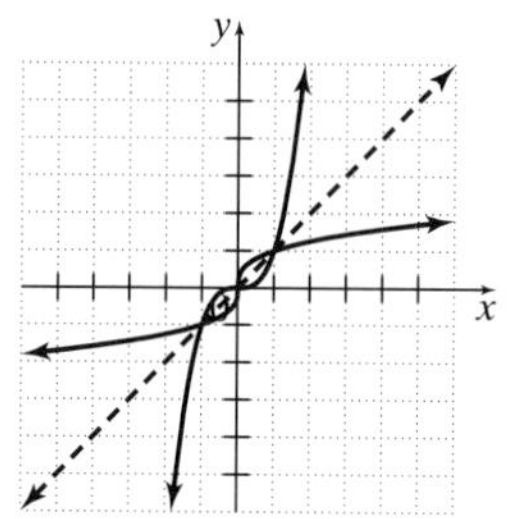

33. $f(x) = \sqrt[3]{x}$
$y = \sqrt[3]{x}$
$x = \sqrt[3]{y}$
$x^3 = y$
$f^{-1}(x) = x^3$

37. $f(x) = (x + 2)^3$
$y = (x + 2)^3$
$x = (y + 2)^3$
$\sqrt[3]{x} = y + 2$
$\sqrt[3]{x} - 2 = y$
$f^{-1}(x) = \sqrt[3]{x} - 2$

41.

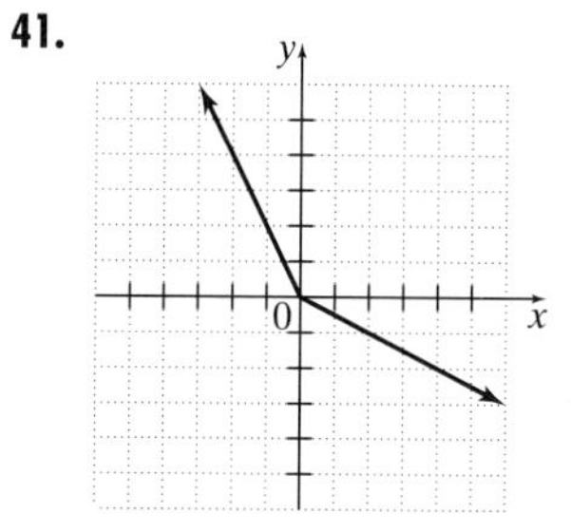

45. $16^{3/4} = (16^3)^{1/4} = (4096)^{1/4} = 8$

49. $f(x) = 3^x$
$f(2) = 3^2$
$= 9$

53. **a.** $\left(-2, \frac{1}{4}\right), \left(-1, \frac{1}{2}\right), (0, 1), (1, 2), (2, 5)$

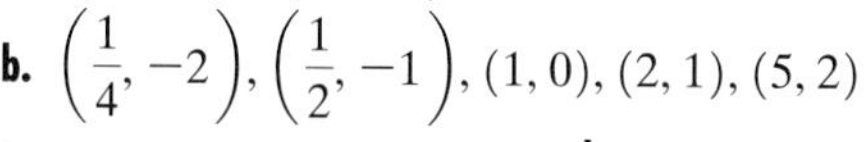

b. $\left(\frac{1}{4}, -2\right), \left(\frac{1}{2}, -1\right), (1, 0), (2, 1), (5, 2)$

c.

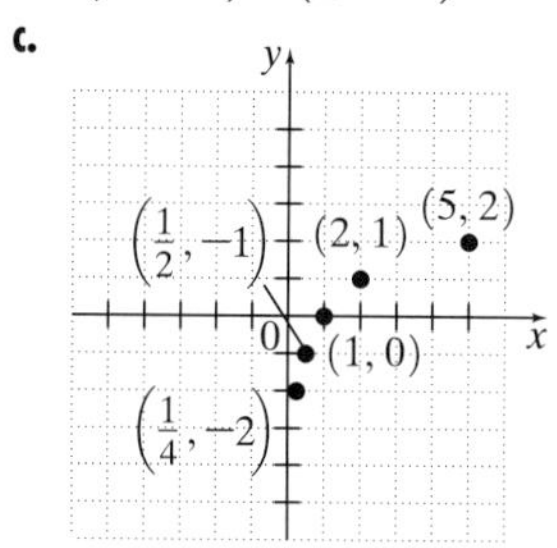

d.

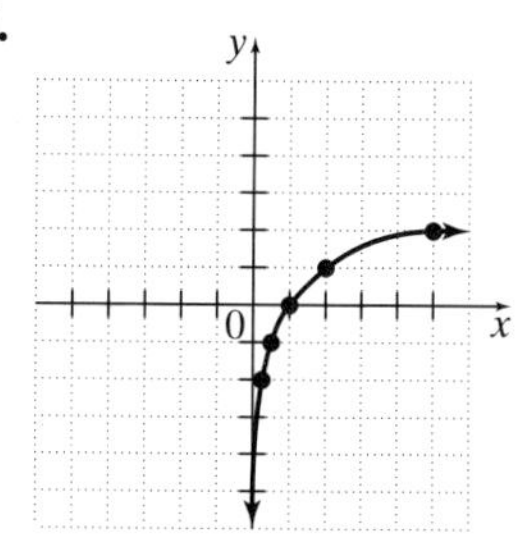

57. $f(x) = \sqrt[3]{x+3}$
$y = \sqrt[3]{x+3}$
$x = \sqrt[3]{y+3}$
$x^3 = y + 3$
$x^3 - 3 = y$
$f^{-1}(x) = x^3 - 3$

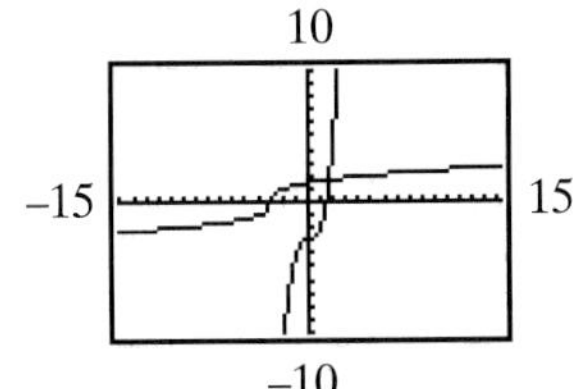

Exercise Set 10.3

1. $y = 4^x$

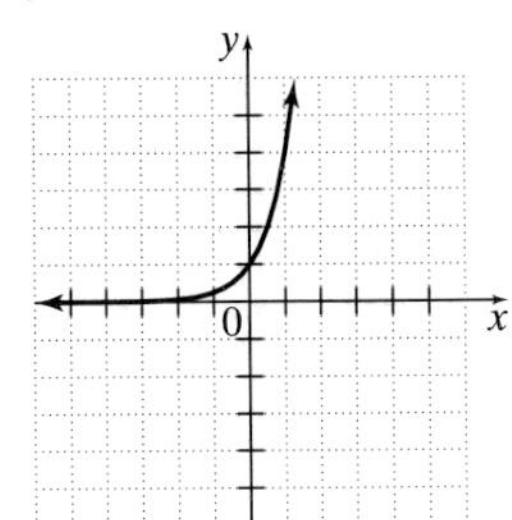

5. $y = \left(\frac{1}{4}\right)^x$

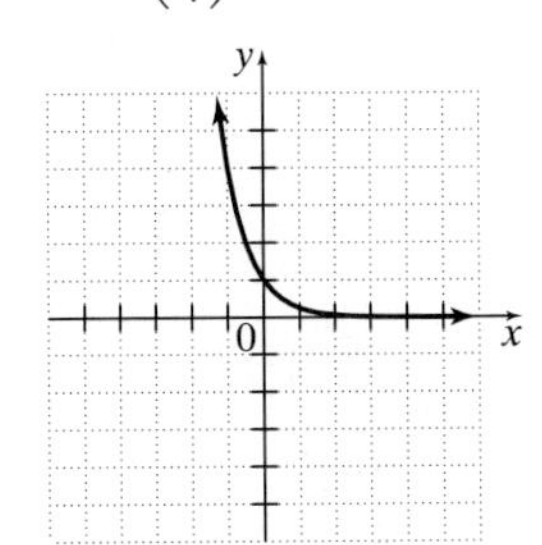

9. $y = -2^x$

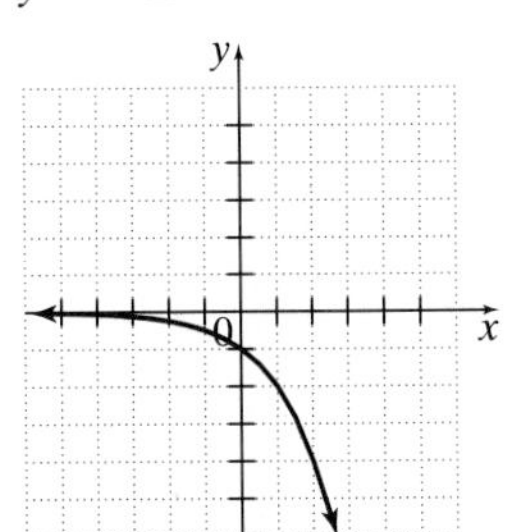

13. $y = -\left(\frac{1}{4}\right)^x$

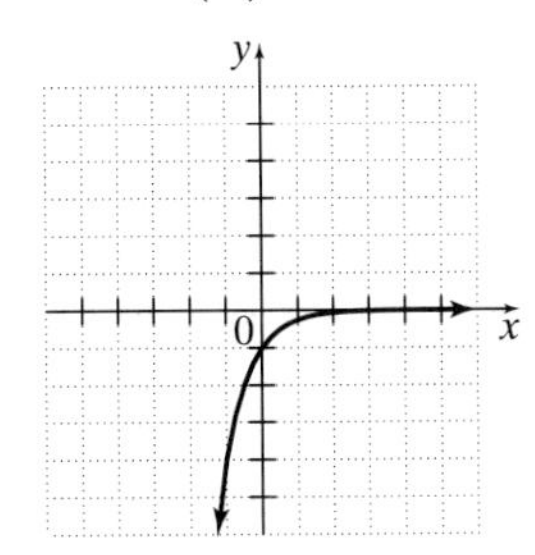

17. $3^x = 27$
$3^x = 3^3$
$x = 3$
The solution set is $\{3\}$.

21. $32^{2x-3} = 2$
$(2^5)^{2x-3} = 2^1$
$10x - 15 = 1$
$10x = 16$
$x = \frac{8}{5}$
The solution set is $\left\{\frac{8}{5}\right\}$.

25. $4^x = 8$
$(2^2)^x = 2^3$
$2^{2x} = 2^3$
$2x = 3$
$x = \frac{3}{2}$
The solution set is $\left\{\frac{3}{2}\right\}$.

29. $81^{x-1} = 27^{2x}$
$(3^4)^{x-1} = (3^3)^{2x}$
$3^{4x-4} = 3^{6x}$
$4x - 4 = 6x$
$-4 = 2x$
$x = -2$
The solution set is $\{-2\}$.

33. $y = 260(2.7)^{0.025t}$, $t = 10$
$y = 260(2.7)^{0.025(10)}$
$y \approx 333$
Approximately 333 bison will remain after 10 years.

37. $y = 3.638(1.443)^x$
$t = 2005 - 1989 = 16$
$y = 3.638(1.443)^{16} \approx 1286$
There will be approximately 1286 million cellular phone users in 2005.

41. $5x - 2 = 18$
$5x = 20$
$x = 4$
The solution set is $\{4\}$.

45. $f(x) = \left(\frac{1}{2}\right)^x$
$b = \frac{1}{2}, 0 < b < 1$
graph C

49. answers may vary

53. At $t = 120$, $y \approx 18.62$.
Approximately 18.62 lb of uranium will be available after 120 days.

Exercise Set 10.4

1. $\log_6 36 = 2$
$6^2 = 36$

5. $\log_{10} 1000 = 3$
$10^3 = 1000$

9. $\log_e \frac{1}{e^2} = -2$
$e^{-2} = \frac{1}{e^2}$

13. $2^4 = 16$
$\log_2 16 = 4$

SOLUTIONS

17. $e^3 = x$
$\log_e x = 3$

21. $4^{-2} = \frac{1}{6}$
$\log_4 \frac{1}{16} = -2$

25. $\log_2 8 = 3$ since $2^3 = 8$

29. $\log_{25} 5 = \frac{1}{2}$ since $25^{1/2} = 5$

33. $\log_6 1 = 0$ since $6^0 = 1$

37. $\log_{10} 100 = 2$ since $10^2 = 100$

41. $\log_3 81 = 4$ since $3^4 = 81$

45. answers may vary

49. $\log_3 x = 4$
$x = 3^4$
$= 81$
The solution set is $\{81\}$.

53. $\log_2 \frac{1}{8} = x$
$2^x = \frac{1}{8}$
$2^x = 2^{-3}$
$x = -3$
The solution set is $\{-3\}$.

57. $\log_8 x = \frac{1}{3}$
$x = 8^{1/3}$
$= 2$
The solution set is $\{2\}$.

61. $\log_{3/4} x = 3$
$\left(\frac{3}{4}\right)^3 = x$
$x = \frac{3^3}{4^3}$
$= \frac{27}{64}$
The solution set is $\left\{\frac{27}{64}\right\}$.

65. $\log_5 5^3 = 3$

69. $\log_9 9 = 1$

73. $f(x) = \log_{1/4} x$ or $y = \log_{1/4} x$
$y = 0$:
$0 = \log_{1/4} x$
$x = \left(\frac{1}{4}\right)^0 = 1$ is the x-intercept.
$x = 0$:
$y = \log_{1/4} 0$ which is not defined.
No y-intercept exists.

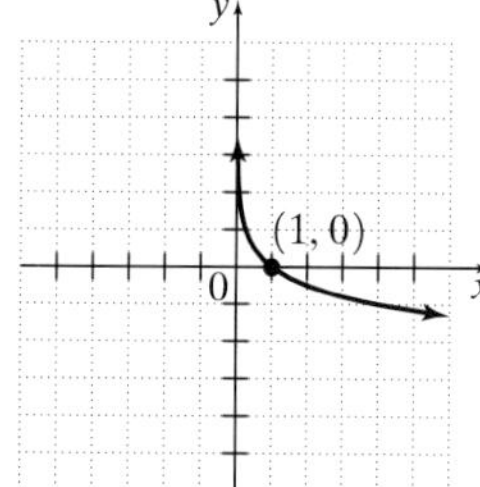

77. $f(x) = \log_{1/6} x$ or $y = \log_{1/6} x$
$x = 0$:
$y = \log_{1/6} 0$ is not defined so there is no y-intercept.
$y = 0$:
$0 = \log_{1/6} x$
$x = \left(\frac{1}{6}\right)^0 = 1$ is the x-intercept.

81. $\frac{x^2 - 8x + 16}{2x - 8} = \frac{(x-4)(x-4)}{2(x-4)} = \frac{x-4}{2}$

85. $y = \left(\frac{1}{3}\right)^x$; $y = \log_{1/3} x$
$x = 0$: $y = \left(\frac{1}{3}\right)^0 = 1$ is the y-intercept of $y = \left(\frac{1}{3}\right)^x$, and hence the x-intercept of $y = \log_{1/3} x$.
$y = 0$: $0 = \left(\frac{1}{3}\right)^x$ has no solution so $y = \left(\frac{1}{3}\right)^x$ has no x-intercept; hence $y = \log_{1/3} x$ has no y-intercept.

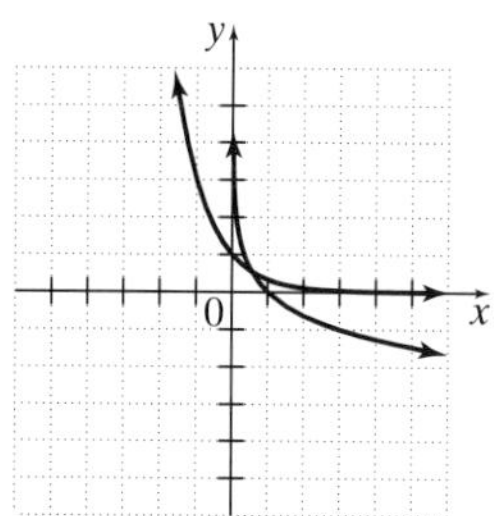

89. $\log_3 10$ is between 2 and 3 because $3^2 = 9$ and $3^3 = 27$

Exercise Set 10.5

1. $\log_5 2 + \log_5 7 = \log_5(2 \cdot 7) = \log_5 14$

5. $\log_{10} 5 + \log_{10} 2 + \log_{10}(x^2 + 2)$
$= \log_{10}[5 \cdot 2(x^2 + 2)]$
$= \log_{10}(10x^2 + 20)$

9. $\log_2 x - \log_2 y = \log_2\left(\frac{x}{y}\right)$

13. $\log_3 x^2 = 2 \log_3 x$

17. $\log_5 \sqrt{y} = \log_5 y^{1/2}$
$= \frac{1}{2} \log_5 y$

21. $3\log_5 x + 6\log_5 z = \log_5 x^3 + \log_5 z^6$
$= \log_5(x^3z^6)$

25. $\log_4 2 + \log_4 10 - \log_4 5 = \log_4 2 \cdot 10 - \log_4 5$
$= \log_4\left(\frac{20}{5}\right)$
$= \log_4 4$
$= 1$

29. $3\log_4 2 + \log_4 6 = \log_4 2^3 + \log_4 6$
$= \log_4 8 + \log_4 6$
$= \log_4(8 \cdot 6)$
$= \log_4 48$

33. $2\log_8 x - \frac{2}{3}\log_8 x + 4\log_8 x = \left(2 - \frac{2}{3} + 4\right)\log_8 x$
$= \frac{16}{3}\log_8 x$
$= \log_8 x^{16/3}$

37. $\log_2\left(\frac{x^3}{y}\right) = \log_2 x^3 - \log_2 y$
$= 3\log_2 x - \log_2 y$

41. $\log_7\left(\frac{5x}{4}\right) = \log_7 5x - \log_7 4$
$= \log_7 5 + \log_7 x - \log_7 4$

45. $\log_6 \frac{x^2}{x + 3} = \log_6 x^2 - \log_6(x + 3)$
$= 2\log_6 x - \log_6(x + 3)$

49. $\log_b 15 = \log_b(5 \cdot 3)$
$= \log_b 5 + \log_b 3 = 0.7 + 0.5 = 1.2$

53. $\log_b 8 = \log_b 2^3 = 3\log_b 2 = 3(0.43) = 1.29$

57. $\log_b \sqrt{\frac{2}{3}} = \log_b\left(\frac{2}{3}\right)^{1/2}$
$= \frac{1}{2}\log_b \frac{2}{3}$
$= \frac{1}{2}(\log_b 2 - \log_b 3)$
$= \frac{1}{2}(0.43 - 0.68)$
$= \frac{1}{2}(-0.25)$
$= -0.125$

61. $\log_{10}\frac{1}{10} = x$
$10^x = \frac{1}{10}$
$10^{-1} = \frac{1}{10}$
$\log_{10}\frac{1}{10} = -1$

65. $\log_3(x + y) = \log_3 x + \log_3 y$
false

69. $(\log_3 6)(\log_3 4) = \log_3 24$
false

Exercise Set 10.6

1. $\log 8 \approx 0.9031$

5. $\ln 2 \approx 0.6931$

9. $\log 12.6 \approx 1.1004$

13. $\log 41.5 \approx 1.6180$

17. $\log 100 = \log 10^2 = 2$

21. $\ln e^2 = 2$

25. $\log 10^3 = 3$

29. $\log 0.0001 = \log 10^{-4}$
$= -4$

33. $\log x = 1.3$
$x = 10^{1.3}$
≈ 19.9526

37. $\ln x = 1.4$
$x = e^{1.4}$
≈ 4.0552

41. $\log x = 2.3$
$x = 10^{2.3}$
≈ 199.5262

45. $\log(2x + 1) = -0.5$
$2x + 1 = 10^{-0.5}$
$2x = 10^{-0.5} - 1$
$x = \frac{10^{-0.5} - 1}{2}$
≈ -0.3419

49. $A = Pe^{rt}, t = 12, P = 1400,$ and $r = 0.08$
$A = 1400e^{(0.08)12} = 1400e^{0.96} \approx 3656.38$
Dana has \$3656.38 after 12 years.

53. $\log_2 3 = \frac{\log 3}{\log 2} \approx 1.5850$

57. $\log_4 9 = \frac{\log 9}{\log 4} \approx 1.5850$

61. $\log_8 6 = \frac{\log 6}{\log 8} \approx 0.8617$

65. $2x + 3y = 6x$
$3y = 4x$
$x = \frac{3y}{4}$

69. $f(x) = e^x$

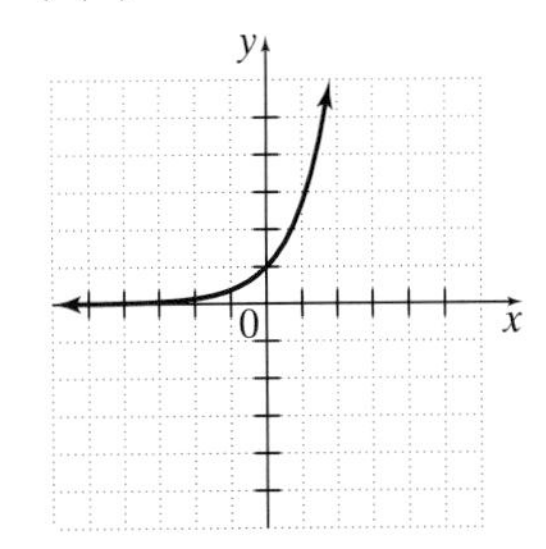

73. answers may vary

77. $R = log\left(\frac{a}{T}\right) + B$
$= \log\left(\frac{400}{2.6}\right) + 3.1$
≈ 5.3

Exercise Set 10.7

1. $3^x = 6$
$x = \log_3 6$
$= \frac{\log 6}{\log 3} \approx 1.6309$

The solution set is $\left\{\frac{\log 6}{\log 3}\right\}$ or approximately $\{1.6309\}$.

5. $2^{x-3} = 5$
$x - 3 = \log_2 5$
$x = 3 + \log_2 5$
$= 3 + \frac{\log 5}{\log 2}$
$= 5.3219$

The solution set is $\left\{3 + \frac{\log 5}{\log 2}\right\}$ or approximately $\{5.3219\}$.

9. $4^{x+7} = 3$
$x + 7 = \log_4 3$
$x = \log_4 3 - 7$
$= \frac{\log 3}{\log 4} - 7$
≈ -6.2075

The solution set is $\left\{\frac{\log 3}{\log 4} - 7\right\}$ or approximately $\{-6.2075\}$.

13. $e^{6x} = 5$
$6x = \ln 5$
$x = \frac{1}{6}\ln 5$
≈ 0.2682

The solution set is $\left\{\frac{1}{6}\ln 5\right\}$ or approximately $\{0.2682\}$.

17. $\log_4 2 + \log_4 x = 0$

$$\log_4(2x) = 0$$
$$2x = 4^0$$
$$2x = 1$$
$$x = \frac{1}{2}$$

The solution set is $\left\{\frac{1}{2}\right\}$.

21. $\log_4 x + \log_4(x + 6) = 2$

$$\log_4 x(x + 6) = 2$$
$$x(x + 6) = 4^2$$
$$x^2 + 6x = 16$$
$$x^2 + 6x - 16 = 0$$
$$(x + 8)(x - 2) = 0$$
$$x + 8 = 0 \quad \text{or } x - 2 = 0$$
$$x = -8 \text{ or} \quad x = 2$$

We discard -8 as extraneous.
The solution set is $\{2\}$.

25. $\log_3(x - 2) = 2$

$$x - 2 = 3^2$$
$$x - 2 = 9$$
$$x = 11$$

The solution set is $\{11\}$.

29. $\log_2 x + \log_2(3x + 1) = 1$

$$\log_2 x(3x + 1) = 1$$
$$x(3x + 1) = 2^1$$
$$3x^2 + x = 2$$
$$3x^2 + x - 2 = 0$$
$$(3x - 2)(x + 1) = 0$$
$$3x - 2 = 0 \text{ or } x + 1 = 0$$
$$3x = 2 \text{ or} \quad x = -1$$
$$x = \frac{2}{3}$$

Discard -1 as extraneous.

The solution set is $\left\{\frac{2}{3}\right\}$.

33. $y = y_0e^{0.033t}$, $y_0 = 9{,}723{,}000$ and $t = 4$

$$y = 9{,}723{,}000e^{0.033(4)} = 9{,}723{,}000e^{0.132}$$
$$= 11{,}094{,}996$$

There will be approximately 11,094,996 inhabitants by 2002.

37. $A = P\left(1 + \frac{r}{n}\right)^{nt}$, $P = 600$,

$A = 2(600) = 1200$, $r = 0.07$, and $n = 12$

$$1200 = 600\left(1 + \frac{0.07}{12}\right)^{12t}$$
$$2 \approx (1.00583)^{12t}$$
$$12t = \log_{1.00583}(2)$$
$$t = \frac{1}{12}\log_{1.00583}(2)$$
$$= \left(\frac{1}{12}\right)\frac{\log 2}{\log(1.00583)} \approx 9.9$$

It would take approximately 9.9 years for the \$600 to double.

41. $A = P\left(1 + \frac{r}{n}\right)^{nt}$, $P = 1000$.

$A = 2(1000) = 2000$, $r = 0.08$, and $n = 2$.

$$2000 = 1000\left(1 + \frac{0.08}{2}\right)^{2t}$$
$$2 = (1.04)^{2t}$$
$$2t = \log_{1.04} 2 = \frac{\log 2}{\log 1.04}$$
$$t = \left(\frac{1}{2}\right)\frac{\log 2}{\log 1.04} \approx 8.8$$

It would take approximately 8.8 years for the \$1000 to double.

45. $P = 14.7e^{-0.21x}$, $x = 1$

$$= 14.7e^{-0.21(1)} = 14.7e^{-0.21} \approx 11.9$$

The average atmospheric pressure of Denver is approximately 11.9 lb/in^2.

49. $t = \frac{1}{c}\ln\left(\frac{A}{A - N}\right)$

$$t = \frac{1}{0.09}\ln\left(\frac{75}{75 - 50}\right)$$
$$t = \frac{1}{0.09}\ln(3)$$
$$t \approx 12.21$$

It will take about 12 weeks.

53. $\frac{x^2 - y + 2z}{3x} = \frac{(-2)^2 - (0) + 2(3)}{3(-2)}$

$$= \frac{4 + 6}{-6}$$
$$= \frac{10}{-6}$$
$$= -\frac{5}{3}$$

57.

$$y = y_0e^{kt}$$
$$1{,}677{,}000 = 1{,}202{,}000e^{k \cdot 7}$$
$$k = \frac{1}{7}\ln\frac{1{,}677{,}000}{1{,}202{,}000}$$
$$k \approx 0.048$$

The growth rate is about 4.8%.

Chapter 10 Test

1. $(f \cdot g)(x) = f(x) \cdot g(x)$

$$= x(2x - 3)$$
$$= 2x^2 - 3x$$

5. $(g \circ h)(x) = g(h(x))$

$$= g(x^2 - 6x + 5)$$
$$= x^2 - 6x + 5 - 7$$
$$= x^2 - 6x - 2$$

9. $y = 6 - 2x$

$f(x) = -2x + 6$ so the function is one-to-one.

$$f^{-1}(x) = \frac{x - 6}{-2} \text{ or } f^{-1}(x) = \frac{-x + 6}{2}$$

13. $\log_5 x + 3\log_5 x - \log_5(x + 1)$

$$= 4\log_5 x - \log_5(x + 1)$$
$$= \log_5 x^4 - \log_5(x + 1)$$
$$= \log_5\frac{x^4}{x + 1}$$

SOLUTIONS

17.

$$8^{x-1} = 8^{-2}$$
$$x - 1 = -2$$
$$x = -1$$

The solution set is $\{-1\}$.

21. $\log_8(3x - 2) = 2$

$$3x - 2 = 8^2$$
$$3x - 2 = 64$$
$$3x = 66$$
$$x = \frac{66}{3}$$
$$= 22$$

The solution set is $\{22\}$.

25. $y = \left(\frac{1}{2}\right)^x + 1$

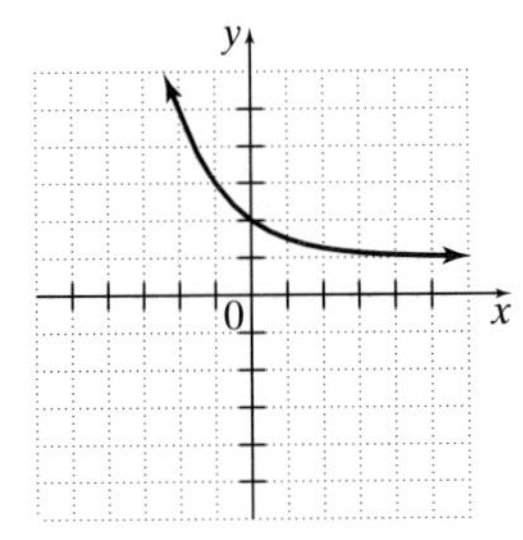

29. $y = y_0e^{kt}$, $y_0 = 57{,}000$, $k = 0.026$, and $t = 5$

$$y = 57{,}000e^{0.026(5)}$$
$$= 57{,}000e^{0.13} \approx 64{,}913$$

There will be approximately 64,913 prairie dogs 5 years from now.

SOLUTIONS

Index

Photo Credits

Chapter 1 CO Alan Schen/The Stock Market, **p. 10** Jeremy Davey/SSC Programme Ltd., **p. 56** Chris Butler/Science Photo Library/Photo Researchers, Inc.

Chapter 2 CO UPI/Corbis-Bettmann, **p. 93** Michael Newman/PhotoEdit, **p. 95** Tomi/PhotoDisc, Inc., **p. 110** Jet Propulsion Laboratory/NASA Headquarters

Chapter 3 CO Kevin Virobik-Adams/Progressive Photo

Chapter 4 CO T.A. Wiewandt/DRK Photo, **p. 294** Doug Densinger/Allsport Photography USA Inc., **p. 294** Kirthmon Dozier/AP/Wide World Photos

Chapter 5 CO Bill Bachmann/Photo Researchers, Inc., **p. 374** The Granger Collection, **p. 407** PhotoDisc, Inc.

Chapter 6 CO Kevork Djansezian/AP/Wide World Photos, **p. 483** John Seralin/Simon & Schuster Corporate Digital Archive, **p. 487** Michael Gadomski/Photo Researchers, Inc., **p. 494** Richard A. Cooke III/Tony Stone Images, **p. 498** Mary Teresa Giancoli

Chapter 7 CO Roger Ressmeyer/Corbis Corporation, **p. 571** Steve Gottlieb/FPG International LLC, **p. 583** Richard Megna/Fundamental Photographs

Chapter 8 CO Bruce Forster/Tony Stone Images, **p. 633** Tim Flach/Tony Stone Images, **p. 633** Arthur S. Aubry Photography/PhotoDisc, Inc.

Chapter 9 CO Bob Daemmrich/The Image Works, **p. 694** Colored French engraving, 1584. The Granger Collection, New York.

Chapter 10 CO John Barr/Liaison Agency, Inc., **p. 773** PhotoDisc, Inc., **p. 801** Tom McHugh/Photo Researchers, Inc., **p. 805** Tony Freeman/PhotoEdit